MODERN MACHINE SHOP's
Handbook
for the
Metalworking Industries

Woodrow Chapman, Editor

PUBLICATIONS

Hanser Gardner Publications

Cincinnati, Ohio

Library of Congress Cataloging-in-Publication Data

Modern machine shop's handbook for the metalworking industries /
Woodrow W. Chapman, editor.

    p. cm.
    ISBN 1-56990-345-X
    1. Machine-shop practice--Handbooks, manuals, etc. I. Chapman,
Woodrow W. (Woodrow Wilson), 1944-
    TJ1165 .M56 2002
    670.42'3--dc21

                       2002005345

A **MODERN MACHINE SHOP** book

Published by

Gardner Publications, Metalworking's Premier Publisher

www.mmsonline.com

Hanser Gardner Publications

6915 Valley Avenue

Cincinnati, OH 45255-3029

www.hansergardner.com

Portions of this book originally appeared in *Handbook of Technical Formulas* by
Karl-Friedrich Fischer (Hanser Gardner Publications, 1999).

        2 3 4 5 6   07 06 05 04 03 02

Since 1928, MODERN MACHINE SHOP magazine has been providing machine shops with the latest information and developments in machining. Today, it is one of the most avidly read magazines in its field, and on the internet mmsonline.com provides a wealth of information and electronic tools for metalworking professionals. With the publication of this handbook, "MODERN" has moved into a new field, offering machinists a one-volume source containing critical up-to-date information as well as traditional reference tables and charts.

Several contributors have written material specifically for this book. Thanks to the following authors (in alphabetical order) for their help. Miles Arnone authored the sections on "Principles of 5-axis Machining," "High-speed Machining," and "Minimizing Door-to-Door Time." Chip Burnham wrote the section on "Waterjet Cutting Technology." James J. Childs provided the sections on "Evaluating CMM Accuracy," and "Geometric Dimensioning and Tolerancing." Christopher DeHut contributed the section on "Computer Numerical Control." Ray Miller contributed the process material in the "Welding, Brazing, and Soldering" section. Nadeem Rizvi wrote "Precision Machining with Lasers." In addition, some material presented in this book originally appeared in MODERN MACHINE SHOP, and in each instance the author has been acknowledged in the text.

Even with the help of the above authors and contributors, this book would not have been possible without the cooperation and contributions of the companies and associations that produce the tools, machines, and raw materials that are used in manufacturing throughout the world. A list of these contributors begins on the following page, but special thanks must be given to Kennametal Inc., not only for providing information, but for being the genesis of this handbook. What was originally conceived as a "brief" book on milling, turning, and drilling based on technical materials supplied by Kennametal, evolved into an expanded reference that covers the full spectrum of machining. Every effort has been made to acknowledge the copyrighted material supplied by each contributor within the text, and any lapses of identification are regretted.

Finally, several individuals who worked tirelessly to make this project a reality are Rhonda Weaver, who single-handedly did the typesetting—a massive job performed with exceptional skill; Sheri Briggs, who has the remarkable talent of being able to turn miserable pencil sketches into precise and admirable illustrations; Susan Hart Anderson, who diligently and expertly proofread the entire manuscript and offered insightful suggestions for improvements; and Lisa Sullivan Chapman who quickly and accurately keyboarded the hundreds of charts and tables.

Compiling a book of this length and scope is an enormous undertaking, and I apologize for any errors that have, in spite of our efforts, found their way into print. It would be a great help if you would share any corrections, suggestions, or criticisms with me by e-mail at wchapman@gardnerweb.com. With your assistance, improvements can be made in future editions.

Woodrow Chapman
Cincinnati, Ohio

AGMA
American Gear Manufacturers' Association
1500 King Street
Arlington, VA 22314-2730
www.agma.org

AISI
American Iron and Steel Institute
1000 16th Street NW
Washington, DC 20006
www.steel.org

ANSI
American National Standards Institute
11 West 42nd Street
New York, NY 10036
www.ansi.org

ASME
American Society of Mechanical Engineers
345 East 47th Street
New York, NY 10017
www.asme.org

AWS
American Welding Society
550 NW Lejeune Road
Miami, FL 33135
www.aws.org

Agie Ltd.
565 Griffith Street
Davidson, NC 28036
www.agieus.com

The Aluminum Association
818 Connecticut Avenue NW
Washington, DC 20006
www.aluminum.org

Alvord-Polk Tool
125 Gearhart Street
Millersburg, PA 17061
www.alvordpolk.com

Amada Cutting Technologies, Inc.
14849 E. Northam Street
La Mirada, CA 90638
www.amadabandsaw.com

American Drill Bushing Co.
2000 Camfield Avenue
Los Angeles, CA 90040
www.americandrillbushing.com

American Fasteners Technologies Corp.
Nine Frontier Drive
Gibsonia, PA 15044-7999
www.americanfastener.com

American Saw and Mfg. Company
301 Chestnut Street
East Longmeadow, MA 01028
www.lenoxsaw.com

Bethlehem Steel Corporation
1170 Eighth Avenue
Bethlehem, PA 18016
www.bethsteel.com

Bostomatic Corporation USA
Granite Park
125 Fortune Blvd.
Milford, MA 01757
www.bostomatic.com

Brown & Sharpe
Precision Park
200 Frenchtown Road
North Kingston, RI 02852
www.brownandsharpe.com

Brubaker Tool Corp.
200 Front Street
Millersburg, PA 17061
www.brubakertool.com

Century Spring Company
222 East 16th Street
Los Angeles, CA 90015
www.centuryspring.com

Cleveland Twist Drill
PO Box 2578
Augusta, GA 30903-2578
www.gfii.com

Copper Development Association
260 Madison Avenue
New York, NY 10016
www.copper.org

Darling Bolt Co.
2941 E. 10 Mile Road
Warren, MI 48090-2035
www.darlingbolt.com

DAPRA Corp.
66 Granby Street
Bloomfield, CT 06002
www.dapra.com

DoAll Co.
254 N. Laurel Avenue
Des Plaines, IL 60016-4398
www.doall.com

Driv-Lok, Inc.
1140 Park Avenue
Sycamore, IL 60178
www.driv-lok.com

Fastbolt Corp.
200 Louis Street
South Hackensack, NJ 07606
www.fastboltcorp.com

Fastcut Tool Corporation
31800 Sherman
Madison Heights, MI 48071-1463
www.fastcut.com

Flow International Corporation
27 Ridgewood Road
Woodbury, CT 06798
www.flowcorp.com

Giddings & Lewis
142 Doty Street
Fond du Lac, WI 54936-0590
www.giddings.com

GISCO Equipment Inc.
100 Laser Court
Hauppauge, NY 11788
www.gisco-equipment.com

Gleason Corporation
1000 University Avenue
Rochester, NY 14692-2970
www.gleason.com

Greenfield Industries, Inc.
PO Box 2587
Augusta, GA 30903-2578
www.gfii.com

Hanson Whitney Company
725 Marshall Phelps Road
Windsor, CT 06095
www.hansonwhitney.com

Hassay Savage Company
10 Industrial Boulevard
Turners Falls, MA 01376-2251
www.hassay-savage.com

Kennametal Inc.
1600 Technology Way
Latrobe, PA 15650
www.kennametal.com

Landis Threading Systems
360 South Church Street
Waynesboro, PA 17268-2659
www.landisthreadingsystems.com

Mahr Federal, Inc.
1144 Eddy Street
Providence, RI 02940
www.fedgage.com

Maryland Metrics
PO Box 261
Owens Mills, MD 21117
www.mdmetric.com

Metric & Multistandard Components Corp.
120 Old Saw Mill River Road
Hawthorne, NY 10532
www.metricmcc.com

Miniature Thread Specialists (MTS)
2005 W. Acoma Blvd., Unit C
Lake Havasu City, AZ 86403
www.minithread.com

M.K. Morse Company
P.O. Box 8677
Canton, OH 44711-8677
www.mkmorse.com

New England Tap Corp.
175 Cocasset Street
Foxboro, MA 02035
www.netap.com

Norton Company
1 New Bond Street
Worcester, MA 01615-0008
www.nortonabrasives.com

OSG Tap & Die, Inc.
676 East Fullerton Avenue
Glendale Heights, IL 60139
www.osg-sossner.com

Pacific Grinding Wheel Co.
13120 State Avenue
Marysville, WA 98270
www.pacificgrindingwheel.com

Pivot Point, Inc.
P.O. Box 488
Hustisford, WI 53034-0488
www.pivotpins.com

Precision Twist Drill
One Precision Plaza
Crystal Lake, IL 60039-9000
www.precisiontwistdrill.com

Quadrant Engineering Plastic Products
2120 Fairmont Avenue
Reading, PA 19612-4235
www.quadrantepp.com

Renishaw Inc.
623 Cooper Court
Schaumburg, IL 60173
www.renishaw.com

SPS Technologies/Unbrako
301 Highland Avenue
Jenkintown, PA 19046
www.spstech.com

SSINA
Specialty Steel Industry of North America
3050 K St. NW
Washington, DC 20007
www.ssina.org

Sodick, Inc.
851 Feehanville Drive
Mount Prospect, IL 60056
www.sodick.co.jp

Schunk
211 Kitty Hawk Drive
Morrisville, NC 27560
www.schunk-usa.com

Smalley Steel Ring Co.
555 Oakwood Road
Lake Zurich, IL 60047
www.smalley.com

L.S. Starrett Company
121 Crescent Street
Athol, MA 01331
www.lsstarrett.com

Stellram USA
One Teledyne Place
La Vergne, TN 37086
www.stellram.com

Surburban Tool
4141 N. Atlantic Blvd.
Auburn Hills, MI 48326
www.subtool.com

Taylor & Jones Limited
17 Southgate Honley
Huddersfield HD7 2NU
England
www.taylorandjones.co.uk

Timken Latrobe Steel
2626 Ligonier Street
Latrobe, PA 15650

Tinnerman Fasteners
Engineered Components LLC
1060 West 130th Street
Brunswick, OH 44212
www.tinnerman.com

Unbrako/SPS Technologies, Inc.
4444 Lee Road
Cleveland, OH 44128
www.spstech.com/unbrako

Vermont Gage Company
6 Brooklyn Street
Swanton, VT 05488
www.vtgage.com

Weldon Tool Company
6030 Carey Drive
Valley View, OH 44125-4218
www.weldonmachinetool.com

West Coast Lockwasher (WCL)
16730 East Johnson Drive
Industry, CA 91744
www.wclco.com

WERKÖ
Werkzeugfabrik GmBh Königsee
Werkstrassa 5, 07426 Königsee/Thuringen
Germany
www.werkoe.de

Wire Rope Industries Ltd.
5501 Trans Canada Highway
Pointe Claire, Quebec
Canada H9R 1B7
www.wirerope.com

The Yankee Corporation
Box 1255 RD#2
Fairfax, VT 05454
www.yankeereamer.com

Review of Basic Mathematics

The widespread availability of hand calculators and desk computers has greatly facilitated solving complicated mathematical problems. This section is intended to be used as a quick review for frequently encountered calculations that are still performed manually. Mathematics signs and symbols that are commonly encountered in shop mathematics are shown in **Table 1.**

Fractions and Decimals

Fractions are used to denote one or more equal parts of a whole. The proper fraction $\frac{3}{16}$, for instance, denotes that three of sixteen possible units are to be taken—i.e., of sixteen equal parts of a whole, three equal parts have been taken. Actually, $\frac{3}{16}$ is simply another way of writing $3 \div 16$, and is another way of expressing the relation of equal parts to the total. If, on the other hand, the fraction contains unknowns such as $\frac{a}{x}$, the unit is divided into x equal parts, and a of these parts are taken.

Addition of fractions

Before fractions can be added, the fractional units must be reduced to a common denominator (in fractions, the *denominator* is the number below the line). For example, $\frac{1}{4}$ and $\frac{1}{3}$ cannot be added until their denominators are the same. The smallest number that 4 and 3 can commonly and evenly be divided into is 12, so the fractions must be converted to 12ths before they can be added. This is done by multiplying both the numerator (the *numerator* is the number above the line) and denominator by the number of times the denominator will go into 12. Since 4 goes into 12 three times, $\frac{1}{4}$ becomes $\frac{3}{12}$. Since 3 goes into 12 four times, $\frac{1}{3}$ becomes $\frac{4}{12}$. The numerators can now be added. For example, to add $\frac{1}{4}$ and $\frac{1}{3}$,

$$\frac{1}{4} + \frac{1}{3} = \frac{3}{12} + \frac{4}{12} = \frac{7}{12}.$$

To add unknowns, the same fundamental rules apply. To add $\frac{d}{2b}$, plus $\frac{d}{10b}$, plus $\frac{d}{3b}$, first find the lowest common denominator. In this instance, 30 is the smallest number into which all three denominators will divide evenly. Since 2 goes into 30 fifteen times, 10 goes into 30 three times, and 3 goes into 30 ten times, therefore,

$$\frac{15d}{30b} + \frac{3d}{30b} + \frac{10d}{30b} = \frac{28d}{30b}.$$

If the problem involves only unknowns, the common denominator must be a combination of the letters representing the unknown values. To add $\frac{a}{x} + \frac{b}{d}$, use the common denominator xd.

$$\frac{a}{x} = \frac{ad}{xd}, \frac{b}{d} = \frac{bx}{xd} \quad \text{Therefore,} \frac{a}{x} + \frac{b}{d} = \frac{ad + bx}{xd}.$$

When adding mixed quantities, such as $15\frac{3}{4}$ and $7\frac{1}{4}$, add the entire quantities and the fractions separately as in: $15 + 7 = 22$, and $\frac{3}{4} + \frac{1}{4} = 1$. These two sums, added together, are $22 + 1 = 23$.

Subtraction of fractions

As in addition, fractions can be subtracted only when they share a common denominator. Once a common denominator is established, the smaller numerator is subtracted from the larger. For example, find the difference between $\frac{5}{6}$ and $\frac{3}{5}$.

$$\frac{5}{6} = \frac{25}{30} \; ; \frac{3}{5} = \frac{18}{30} \; . \text{ Therefore, } \frac{25}{30} - \frac{18}{30} = \frac{7}{30}.$$

Solving for unknowns is similar. For example, find the difference between a and $\frac{b}{d}$. First, a may be written in fractional form with a desired denominator by multiplying a by d, resulting in $a = \frac{ad}{d}$.

$$a = \frac{ad}{d} - \frac{b}{d} = ad - \frac{b}{d}.$$

When subtracting mixed quantities, such as $7\frac{1}{4}$ from $15\frac{3}{4}$, subtract the entire quantities and the fractions separately as in: $15 - 7 = 8$, and $\frac{3}{4} - \frac{1}{4} = \frac{1}{2}$. The two remainders, when added together, give the answer: $8 + \frac{1}{2} = 8\frac{1}{2}$.

Multiplication of fractions

Two or more fractions are multiplied by multiplying the numerators and the denominators individually. Therefore, to multiply $\frac{3}{16}$ times $\frac{5}{8}$,

$$\frac{3}{16} \times \frac{5}{8} = \frac{3 \times 5}{16 \times 8} = \frac{15}{128}.$$

In the event that whole numbers with fractions (commonly called *mixed quantities*) are to be multiplied, the whole numbers must first be converted to improper fractions (in improper fractions, the numerator is larger than the denominator—the reverse is true with proper fractions). For example, $1\frac{3}{8}$ is to be multiplied by $1\frac{5}{8}$,

$$1\frac{3}{8} = \frac{11}{8}, \text{ and } 1\frac{5}{8} = \frac{13}{8}. \text{ Therefore, } \frac{11 \times 13}{8 \times 8} = \frac{143}{64} = 2\frac{15}{64}.$$

Unknowns can be multiplied in the same manner as other fractions. For example, multiply $\frac{a}{4}$ by $\frac{2x}{b}$.

$$\frac{a}{4} \times \frac{2x}{b} = \frac{a \times 2x}{4 \times b} = \frac{2ax}{4b}.$$

Division of fractions

The division of fractions is the reverse of multiplication. In division, the fraction to be divided is the *dividend*, and the whole number or fraction by which we divide is called the *divisor*. The first step in division is to invert the devisor. For example, if the devisor is $\frac{3}{8}$, it is inverted to $\frac{8}{3}$. If it is a whole number, such as 4, it is inverted as $\frac{1}{4}$. The calculation is then handled as in multiplication. To divide $\frac{3}{16}$ by $\frac{5}{8}$,

$$\frac{3}{16} \div \frac{5}{8} = \frac{3}{16} \times \frac{8}{5} = \frac{24}{80} = \frac{12}{40} = \frac{6}{20} = \frac{3}{10}.$$

If $\frac{3}{16}$ is divided by the whole number 5, the procedure is

$$\frac{3}{16} \div \frac{5}{1} = \frac{3}{15} \times \frac{1}{5} = \frac{3}{75} = \frac{1}{25}.$$

Unknowns are divided in exactly the same manner.

$$\frac{b}{a} \div \frac{d}{x} = \frac{b}{a} \times \frac{x}{d} = \frac{bx}{ad}.$$

By using cancellation, some unknowns can be solved. For example, when $\frac{x}{d}$ is divided by x.

$$\frac{x}{d} \div x = \frac{x}{d} \times \frac{1}{x}.$$

In this instance, x appears on both sides of the equation. Since $x = x$, these values cancel themselves and 1 may be substituted for the value.

$$\frac{1}{d} \times \frac{1}{1} = \frac{1}{d}.$$

Cancellation and reduction of fractions

Cancellation can be used in many instances to remove equal values in both numerators and denominators, which can greatly simplify the multiplication of fractions. Often, a fraction may be simplified by *reduction*, which is a form of cancellation. In reduction, the form of a fraction is changed without altering its value. While fractions may be reduced to either higher or lower terms, it is more often useful to reduce the fraction to its most basic form.

To reduce a fraction to higher terms, both the numerator and denominator are multiplied by the same value. For example, to reduce $\frac{3}{5}$ to a higher value, multiply both sides of the fraction by the same value. In this example, 4 is used to increase the value of the fraction times 4.

Table 1. Mathematical Signs.

+	Positive or Plus	lim.	Limit value (of an expression)
-	Negative or Minus	∞	Infinity
± (∓)	Plus or minus (minus or plus)	α	Alpha
×	Multiplied by (multiplication sign)	β	Beta
·	Multiplied by (multiplication sign)	γ	Gamma — commonly used to denote angles
÷	Divided by (division sign)	θ	Theta
:	Divided by (division sign)	φ	Phi
:	Is to (in proportion)	Δ	Delta (difference)
=	Equals	δ	Delta (differential)
::	Equals (in proportion)	μ	Mu (coefficient of friction)
≡	Approximately equals	π	Pi (3.1416)
>	Greater than	Σ	Sigma (sign or summation)
<	Less than	ω	Omega (angles measured in radians)
≥	Greater than or equal to	g	Acceleration due to gravity (32.16 ft. per sec.)
≤	Less than or equal to	sin	Sine
∴	Therefore	cos	Cosine
√	Square root	tan	Tangent
∛	Cube root	(tg)	Tangent
∜	4th root	(tang)	Tangent
ⁿ√	ηth root	cot	Cotangent
$\frac{1}{\eta}$	Reciprocal value of η	(ctg)	Cotangent
		sec	Secant
log	Logarithm	cosec	Cosecant
		versin	Versed sine
hyp. log	Hyperbolic, natural or	covers	Coversed sine
nat. log	Napierian logarithm	∫	Integral (in calculus)
log_e		∫ᵇₐ	Integral (in calculus)
In		!	Factorial, e.g., 5! = 1 × 2 × 3 × 4 × 5

$$\frac{3}{5} \times 4 = \frac{3}{5} \times \frac{4}{4} = \frac{12}{20} = \frac{6}{10} = \frac{3}{5}.$$

As can be seen, even though the value of the fraction is increased by 4, it can be reduced to lower terms by dividing the numerator and denominator by the same number. In the above example, it was multiplied by 4, and then twice divided by 2 to return to the base value. Multiplying fractions is easier when the numerators and denominators are reduced as far as possible. When the dividend and divisor are divided by the same number, the quotient (the *quotient* is the number resulting from the division of one number by another) is unchanged. For example, multiply $\frac{5}{8}$ times $\frac{18}{20}$.

$$\frac{5}{6} \times \frac{18}{20} = \frac{1}{1} \times \frac{3}{4} = \frac{3}{4}.$$

Using cancellation, the numerator 5 and the denominator 20 were both divided by 5, and the denominator 6 and the numerator 18 were both divided by 6, thereby reducing the fractions to lower terms. Consequently, it was possible to quickly and simply reach the final answer without having to multiply 5 by 18 and 6 by 20.

Converting fractions to decimals

To convert a fraction to a decimal, simply divide the numerator by the denominator.

$$\frac{3}{4} = 4 \div 3 = 0.75.$$

The decimal, in turn, can be converted to a percentage by multiplying by 100.

$$0.75 \times 100 = 75\%.$$

All three values— $\frac{3}{4}$, 0.75, and 75% are, of course, different ways of representing the value of three equal parts of a whole composed of four equal parts.

The numbers on either side of a decimal point are assigned names to enable a decimal to be read. These names are

Millions	Hundred-Thousands	Ten-Thousands	Thousands	Hundreds	Tens	Units	Decimal	Tenths	Hundredths	Thousandths	Ten-Thousandths	Hundred-Thousandths	Millionths	Ten-Millionths	Hundred-Millionths
2	4	6	8	5	9	2	.	4	6	3	7	2	8	5	9

The following are examples of how to read numbers to the right of the decimal point. These numbers are known as whole numbers.

8592 = Eight-thousand five-hundred ninety-two, which is usually written as 8,592.

468592 = Four-hundred-sixty-eight thousand five-hundred ninety-two, which is usually written as 468,592.

The following are examples of how to read numbers to the right of the decimal point.

0.4 = Four tenths   0.46 = Forty-six hundredths
0.463 = Four hundred sixty-three thousandths   0.003 = Three thousandths
0.4007 = Four thousand seven ten-thousandths
0.0004 = Four ten-thousandths.

Addition and subtraction of decimals

When adding decimals, the numbers must be aligned so that the decimal points appear directly below each other.

$$
\begin{array}{r}
32.752 \\
+\ 1.25 \\
+\ 45682.1 \\
\hline
45716.102
\end{array}
$$

Decimals are subtracted in exactly the same manner.

$$
\begin{array}{r}
465.1256 \\
-\ 2.05 \\
\hline
463.0756
\end{array}
$$

Multiplication of decimals

Close attention must be given to the decimal point when multiplying decimals. The numbers are first multiplied as with whole numbers. Next, the number of decimal places in the multiplier (the number that is multiplying) and multiplicand (the number being multiplied) are added. The decimal is then placed in the product (the answer) the same number of places from the right.

$$
\begin{array}{r}
56.1875 \\
\times\ 3.14 \\
\hline
2247500 \\
561875 \\
1685625 \\
\hline
176.428750
\end{array}
$$

In the above example, there are four decimal places in the multiplicand and two in the multiplier, for a total of six. Therefore, in the product, the decimal place has been inserted so that there are six figures to the right of the decimal point.

In some instances, there will not be enough places in the product for insertion of the decimal point. In these instances, zeros (known as prefixing ciphers) are added until the needed number of places are available. This often is the case when multiplying only decimals, without whole numbers attached, as in the following example.

$$
\begin{array}{r}
.0055 \\
\times\ .25 \\
\hline
275 \\
110 \\
\hline
.001375
\end{array}
$$

Division of decimals

When dividing decimals, the procedure is exactly the same as with whole numbers. Once the quotient has been obtained, point off in the quotient as many decimal places as those in the dividend exceed those in the divisor. In instances where the divisor has more decimal places than the dividend, ciphers (zeros) must be added to the dividend until each has the same number of decimal places, then divide as with whole numbers. In the first example below, the divisor and dividend have the same number of decimal places, so no decimal places are placed in the quotient.

$$
\begin{array}{r}
37 \\
.005\ )\overline{\ .185} \\
15 \\
\hline
35
\end{array}
$$

In the next example, there are four decimal places in the dividend, but only one in the divisor, so three places are pointed off in the quotient.

$$
\begin{array}{r}
1.026 \\
3.2\ \overline{)\ 3.2832} \\
\underline{32}\ \ \ \ \ \\
08\ \ \ \\
\underline{00}\ \ \ \\
83\ \\
\underline{64}\ \\
192
\end{array}
$$

In the final example, the divisor has more decimal places than the dividend, and ciphers are added to the dividend. In this example, the dividend is 3.25, and the devisor is .0625. Therefore, as shown below, two zeros are added to 3.25, making it 3.2500. Because the divisor and dividend now share the same number of decimal places, no decimal places are added to the quotient.

$$
\begin{array}{r}
52 \\
.0625\ \overline{)\ 3.2500} \\
\underline{3125} \\
1250
\end{array}
$$

Reciprocals, Powers, and Roots of Numbers

Reciprocals

The reciprocal of any number is the quotient of 1 divided by that number. Thus, the reciprocal of 80 can be calculated as $80 \div 1 = 0.0125$. Reciprocals were once widely used to save time in complex long division problems by finding the reciprocal of the divisor, and then multiplying it times the dividend to find the quotient. For example, to find how many times 200 will go into 16,875, find the reciprocal of 200 ($200 \div 1 = 0.005$) and then multiply it by the dividend: $0.005 \times 16875 = 84.375$.

Powers of numbers

The product resulting from multiplying a number by itself one or more times is called the "power" of that number. The second power of a number, also known as the square of the number, is the product of the number being multiplied by itself. The second power, or square, of 4 is 16 ($4 \times 4 = 16$). The third power of a number, or cube of a number, is obtained by multiplying the number by itself twice. The third power, or cube, of 4 is 64 ($4 \times 4 \times 4 = 64$). This process of raising a number to a given power is called *involution*. To indicate that a number is to be taken to a higher power, it is displayed with a superscript number indicating how many times the number is to be multiplied by itself. For example, 4^8 indicates that four is to the eighth power. The small superscript letter is called an *exponent*. Other examples are 4^2, meaning four to the second power, or four squared. 4^3 indicates the third power of four, or four cubed.

Letters as well as numbers may be taken to higher powers, and may be used to represent any value in a formula. In the expression $7x^2$, the exponent applies only to the value of x. However, if expressed as $(7x)^2$, the product of 7 times whatever the value of x is squared.

Powers can be multiplied and divided, but only when like quantities are multiplied or divided. In other words, $5^2 \times 5^3$ can be multiplied, and $a^4 \div a^3$ can be divided, but the following cannot: $5^4 \div 8^3$, or $a^4 \times b^3$. When multiplying powers, the exponents are added together, thus $5^2 \times 5^3 = 5^5$, and $a^4 \times a^3 = a^7$. When dividing powers, the exponent of the divisor is subtracted from that of the dividend. Examples are: $5^5 \div 5^3 = 5^2$, and $a^6 \div a^3 = a^3$.

Powers can also be applied to fractions. The following example shows $\frac{5}{6}$ taken to the third power (cubed).

$$\frac{5}{6} \times \frac{5}{6} \times \frac{5}{6} = \frac{5 \times 5 \times 5}{6 \times 6 \times 6} = \frac{125}{216}.$$

Decimal squares of fractions are given in **Table 2**, and squares, cubes, square roots, and cube roots of whole numbers from 1 to 1,000 are given in **Table 3**.

Roots of numbers

A root of a number is one of the equal factors that, when multiplied together, give that number. Thus, the square root of a given number is a number that, when multiplied by itself, will produce a product equal to the given number. The square root of 25 is 5, because $5 \times 5 = 25$. The cube root of a given number is a number that, when twice multiplied by itself, will produce a product equal to the given number. The cube root of 27 is 3: because $3 \times 3 \times 3 = 27$. The process of finding the root of a number is the reverse of finding the power, and is called extracting the root. Thanks to the availability of calculators, this process is seldom undertaken manually, which is just as well as it can become a lengthy process.

Extracting the square root

The following example demonstrates how to find the square root of 185,761. Normally, this operation is indicated by showing the number within the square root symbol: $\sqrt{185761}$.

1. First, divide the number into groups of two numbers, starting from the right: 18 57 61.
2. Find the largest whole number whose square does not exceed the value of the number in the left-hand group (18). The number 4, which has a square of 16, is the largest number, so 4 becomes the first number of the root.
3. Subtract 16 (the square of 4) from 18 (the number in the first group), which equals 2. Move the 2 to the second group of numbers, which will now contain three figures and becomes 257.
4. Next, multiply the first number of the root (4) by the constant 20 ($4 \times 20 = 80$) and determine how many times the product is divisible into 257. It goes three times ($3 \times 80 = 240$), so 3 becomes the second number of the root.
5. Add the product 80 (from the previous step) to the root just obtained (3), and multiply the sum by the root just obtained ($80 + 3 = 83 \times 3 = 249$). Subtract the answer from 257 ($257 - 249 = 8$). Move the remainder, 8, to the third group of numbers, which will now contain three numbers and becomes 861.
6. Multiply the figures of the root thus far obtained by the constant 20 ($20 \times 43 = 860$) and determine how many times the product is contained in the final three numbers, 861. In this example, it is contained once. Thus, 1 becomes the third figure of the root. Since 1 is also the remainder of 861 minus 860, the number 1 is added to 860, making the new devisor exactly equal to the final three numbers (861). This leaves no remainder, because the divisor is equal to the final group of three figures. Therefore, 431 is the precise square root of 185,761.

In the above example, the answer was exact and did not require a decimal. If, however, there had been a remainder from the last group of figures, it would have been necessary to place a decimal point after the numbers already extracted for the root, add two zeros (ciphers) to the remainder, again multiply the number thus far extracted for the root by 20, and continue until a sufficient number of decimal places have been achieved, or until there is no longer a remainder.

When finding the square root of a number containing a decimal, the initial step is different. Instead of pairing numbers off starting from the right, the numbers are paired off on both sides of the decimal point. For example, the number "24,735.185" would be paired off as "2   47   35.18   5." In this example, the last digit to the right is single, so a cipher (zero) must be added to complete the number and allow the square root to be extracted.

Square roots of common fractions are found by extracting the square root of both the numerator and denominator. This example shows a fraction with perfect squares (squares of whole numbers).

$$\sqrt{\frac{9}{64}} = \frac{\sqrt{9}}{\sqrt{64}} = \frac{3}{8}.$$

When the numerator and/or denominator are not perfect squares, one of two alternatives can be used. The preferred method is to convert the fraction to a decimal and then extract the square of the decimal. The less efficient method is to extract the square of the numerator and/or denominator by usual methods, but since this will result in decimals within the fraction, it will serve very little practical purpose.

Extracting the cube root

Cube roots are extracted in essentially the same manner as square roots, but the entire process is raised a level in complexity. In the following example, the cube root of 80,062,991 will be extracted. Normally, this operation is indicated by showing the number within the cube root symbol: $\sqrt[3]{80062991}$.

1. First, divide the number into groups of three numbers, starting from the right: 80 062 991.
2. Find the largest whole number whose cube does not exceed the value of the number in the left-hand group (80). The number 4, which has a cube of 64, is the largest number, so 4 becomes the first number of the root.
3. Subtract 64 (the cube of 4) from 80 (the number in the first group), which equals 16. Move the 16 to the second group of numbers, which will now contain five figures and becomes 16062.
4. Next, multiply the square of the first number of the root (4) by the constant 300 (16 × 3000 = 4800) and determine how many times the product is divisible into 16062. It goes three times (3 × 4800 = 14400), so 3 becomes the second number of the root.
5. From 16062, subtract the following.
   A. The square of the numbers thus far obtained for the root, with the exception of the last number obtained, multiplied by 300 and by the last number obtained. In this instance, the two numbers so far obtained are 4 and 3, so 4 will be squared and then multiplied by 300 and then by 3. (4 × 4 × 300 × 3 = 14400.)
   B. The numbers thus far obtained for the root, with the exception of the last number obtained, multiplied by 30 and by the square of the last number obtained. (4 × 30 × 3 × 3 = 1080.)
   C. The cube of the last number obtained. (3 × 3 × 3 = 27.)
   The sum of A + B + C is 15507. Subtract this number from 16062 (16062 - 15507 = 555). Add 555 (the remainder from above) to the last three figures, resulting in 555991.
6. Multiply the square of the numbers thus far obtained by the constant 300 (43 × 43

× 300 = 554700) and determine how many times the product is divisible into 555991. The answer is 1, and 1 becomes the third number of the root.

7. Next, repeat steps 5A, 5B, and 5C, using the numbers thus far obtained for the root (431).

A. 43 × 43 × 300 × 1 = 554700.

B. 43 × 30 × 1 × 1 = 1290.

C. 1 × 1 × 1 = 1.

The sum is 555991, which leaves no remainder when subtracted from the sum arrived at in Step 5. Therefore, 431 is the exact cube root of 80,062,991. Had there been a remainder, a decimal would have been placed after the root so far extracted, three ciphers (000) would have been added to the number (80,062,991,000), the number thus far obtained (431) would have been squared and multiplied by 300, and the calculations would have continued as above.

When finding the cube root of a number containing a decimal, the initial step is different. Instead of dividing numbers into groups of three starting from the right, the numbers are divided on both sides of the decimal point. For example, the number "24,735, 254.18562" would be divided as "24 735 254.185 62." In this example, the last digits to the right have only two numbers, so a cipher (zero) must be added to complete the number and allow the cube root to be extracted.

Cubes of fractions are extracted in the same manner as squares. If the numerator and denominator both are not perfect cubes, the fraction should be converted to a decimal and the cube root extracted in the normal manner.

The following Tables provide decimal squares of fractions and squares, cubes, square roots, and cube roots of whole numbers.

Table 2. Squares of Fractions.

Fraction	Square	Fraction	Square	Fraction	Square	Fraction	Square
1/64	0.000244	41/64	0.410400	1 17/64	1.60181	1 57/64	3.57446
1/32	0.000977	21/32	0.430664	1 9/32	1.64160	1 29/32	3.63378
3/64	0.002197	43/64	0.451416	1 19/64	1.68188	1 59/64	3.69360
1/16	0.003906	11/16	0.472656	1 5/16	1.72266	1 15/16	3.75391
5/64	0.006104	45/64	0.494385	1 21/64	1.76392	1 61/64	3.81470
3/32	0.008789	23/32	0.516602	1 11/32	1.80566	1 31/32	3.87589
7/64	0.011963	47/64	0.539307	1 23/64	1.84790	1 63/64	3.93774
1/8	0.015625	3/4	0.562500	1 3/8	1.89063	2	4
9/64	0.019775	49/64	0.586182	1 25/64	1.93384	2 1/64	4.06274
5/32	0.024414	25/32	0.610352	1 13/32	1.97754	2 1/32	4.12598
11/64	0.029541	51/64	0.635010	1 27/64	2.02173	2 3/64	4.18970
3/16	0.035156	13/16	0.660156	1 7/16	2.06641	2 1/16	4.25391
13/64	0.041260	53/64	0.685791	1 29/64	2.11157	2 5/64	4.31860
7/32	0.047852	27/32	0.711914	1 15/32	2.15723	2 3/32	4.38379
15/64	0.054932	55/64	0.738526	1 31/64	2.20337	2 27/64	4.44946
1/4	0.062500	7/8	0.765625	1 1/2	2.25000	2 1/8	4.51563
17/64	0.070557	57/64	0.793213	1 33/64	2.29712	2 9/64	4.58228
9/32	0.079102	29/32	0.821289	1 17/32	2.34473	2 5/32	4.64941
19/64	0.088135	59/64	0.849854	1 35/64	2.39282	2 11/64	4.71704
5/16	0.097656	15/16	0.878906	1 9/16	2.44141	2 3/16	4.78516
21/64	0.107666	61/64	0.908447	1 37/64	2.49048	2 13/64	4.85376
11/32	0.118164	31/32	0.938477	1 19/32	2.54004	2 7/32	4.92285
23/64	0.129151	63/64	0.968994	1 39/64	2.59009	2 15/64	4.99243
3/8	0.140625	1	1	1 5/8	2.64063	2 1/4	5.06250
25/64	0.152588	1 1/64	1.03149	1 41/64	2.69165	2 17/64	5.13306
13/32	0.165039	1 1/32	1.06348	1 21/32	2.74316	2 9/32	5.20410
27/64	0.177979	1 3/64	1.09595	1 43/64	2.79517	2 19/64	5.27563
7/16	0.191406	1 1/16	1.12891	1 11/16	2.84766	2 5/16	5.34766
29/64	0.205322	1 5/64	1.16235	1 45/64	2.90063	2 21/64	5.42017
15/32	0.219726	1 3/32	1.19629	1 23/32	2.95410	2 11/32	5.49316
31/64	0.234619	1 7/64	1.23071	1 47/64	3.00806	2 23/64	5.56665
1/2	0.250000	1 1/8	1.26563	1 3/4	2.06250	2 3/8	5.64063
33/64	0.265869	1 9/64	1.30103	1 49/64	3.11743	2 25/64	5.71509
17/32	0.282227	1 5/32	1.33691	1 25/32	3.17285	2 13/32	5.79004
35/64	0.299072	1 11/64	1.37329	1 51/64	3.22876	2 27/64	5.86548
9/16	0.316406	1 3/16	1.41016	1 13/16	3.28516	2 7/16	5.94141
37/64	0.334229	1 13/64	1.44751	1 53/64	3.34204	2 29/64	6.01782
19/32	0.352539	1 7/32	1.48535	1 27/32	3.39941	2 15/32	6.09473
39/64	0.371338	1 15/64	1.52368	1 55/64	3.45728	2 31/64	6.17212
5/8	0.390625	1 1/4	1.56250	1 7/8	3.51563	2 1/2	6.25000

(Continued)

Table 2. *(Continued)* **Squares of Fractions.**

Fraction	Square	Fraction	Square	Fraction	Square	Fraction	Square
2 33/64	6.32837	3 9/64	9.86353	3 49/64	14.17993	4 25/64	19.27759
2 17/32	6.40723	3 5/32	9.96191	3 25/32	14.29785	4 13/32	19.41504
2 35/64	6.48657	3 11/64	10.06079	3 51/64	14.41626	4 27/64	19.55298
2 9/16	6.56641	3 3/16	10.16016	3 13/16	14.53516	4 7/16	19.69141
2 37/64	6.64673	3 13/64	10.26001	3 53/64	14.65454	4 29/64	19.83032
2 19/32	6.72754	3 7/32	10.36035	3 27/32	14.77441	4 15/32	19.96973
2 39/64	6.80884	3 15/64	10.46118	3 55/64	14.89478	4 31/64	20.10962
2 5/8	6.89063	3 1/4	10.56250	3 7/8	15.01563	4 1/2	20.25000
2 41/64	6.97290	3 17/64	10.66431	3 57/64	15.13696	4 33/64	20.39087
2 21/32	7.05566	3 9/32	10.76660	3 29/32	15.25879	4 17/32	20.53223
2 43/64	7.13892	3 19/64	10.85938	3 59/64	15.38110	4 35/64	20.67407
2 11/16	7.22266	3 5/16	10.97266	3 15/16	15.50319	4 9/16	20.81641
2 45/64	7.30688	3 21/64	11.07642	3 61/64	15.62720	4 37/64	20.95923
2 23/32	7.39160	3 11/32	11.18066	3 31/32	15.75098	4 19/32	21.10254
2 47/64	7.47681	3 23/64	11.28540	3 63/64	15.87524	4 39/64	21.24634
2 3/4	7.56250	3 3/8	11.39063	4	16	4 5/8	21.39063
2 49/64	7.64868	3 25/64	11.49634	4 1/64	16.12524	4 41/64	21.53540
2 25/32	7.73535	3 13/32	11.60254	4 1/32	16.25098	4 21/32	21.68066
2 51/64	7.82251	3 27/64	11.70923	4 3/64	16.37720	4 43/64	21.82642
2 13/16	7.91016	3 7/16	11.81641	4 1/16	16.50391	4 11/16	21.97266
2 53/64	7.99829	3 29/64	11.92407	4 5/64	16.63110	4 45/64	22.11938
2 27/32	8.08691	3 15/32	12.03223	4 3/32	16.75879	4 23/32	22.26660
2 55/64	8.17603	3 31/64	12.14087	4 7/64	16.88696	4 47/64	22.41431
2 7/8	8.26563	3 1/2	12.25000	4 1/8	17.01563	4 3/4	22.56250
2 57/64	8.35571	3 33/64	12.35692	4 9/64	17.14478	4 49/64	22.71118
2 29/32	8.44629	3 17/32	12.46973	4 5/32	17.27441	4 25/32	22.86035
2 59/64	8.53735	3 35/64	12.58032	4 11/64	17.40454	4 51/64	23.01001
2 15/16	8.62891	3 9/16	12.69141	4 3/16	17.53516	4 13/16	23.16016
2 61/64	8.72095	3 37/64	12.80298	4 13/64	17.66626	4 53/64	23.31079
2 31/32	8.81348	3 19/32	12.91504	4 7/32	17.79785	4 27/32	23.46191
2 63/64	8.90649	3 39/64	13.02759	4 15/64	17.92993	4 55/64	23.61353
3	9	3 5/8	13.14063	4 1/4	18.06250	4 7/8	23.76563
3 1/64	9.09399	3 41/64	13.25415	4 17/64	18.19556	4 57/64	23.91827
3 1/32	9.18848	3 21/32	13.36816	4 9/32	18.32910	4 29/32	24.07129
3 3/64	9.28345	3 43/64	13.48267	4 19/64	18.46313	4 59/64	24.22485
3 1/16	9.37891	3 11/16	13.59766	4 5/16	18.59766	4 15/16	24.37891
3 5/64	9.47485	3 45/64	13.71313	4 21/64	18.73267	4 61/64	24.53345
3 3/32	9.57129	3 23/32	13.82910	4 11/32	18.86816	4 31/32	24.68848
3 7/64	9.66821	3 47/64	13.94556	4 23/64	19.00415	4 63/64	24.84399
3 1/8	9.76563	3 3/4	14.06250	4 3/8	19.14063	5	25

Table 3. Squares, Square Roots, Cubes, and Cube Roots.

No.	Square	Square Root	Cube	Cube Root	No.	Square	Square Root	Cube	Cube Root
1	1	1	1	1	51	2,601	7.141428	132,651	3.70843
2	4	1.414214	8	1.25992	52	2,704	7.211103	140,608	3.73251
3	9	1.732051	27	1.44225	53	2,809	7.28011	148,877	3.75629
4	16	2	64	1.58740	54	2,916	7.348469	157,464	3.77976
5	25	2.236068	125	1.70998	55	3,025	7.416198	166,375	3.80295
6	36	2.44949	216	1.81712	56	3,136	7.483315	175,616	3.82586
7	49	2.645751	343	1.91293	57	3,249	7.549834	185,193	3.84850
8	64	2.828427	512	2	58	3,364	7.615773	195,112	3.87088
9	81	3	729	2.08008	59	3,481	7.681146	205,379	3.89300
10	100	3.162278	1,000	2.15442	60	3,600	7.745967	216,000	3.91487
11	121	3.316625	1,331	2.22398	61	3,721	7.81025	226,981	3.93650
12	144	3.464102	1,728	2.28943	62	3,844	7.874008	238,328	3.95789
13	169	3.605551	2,197	2.35133	63	3,969	7.937254	250,047	3.97906
14	196	3.741657	2,744	2.41014	64	4,096	8	262,144	4.00000
15	225	3.872983	3,375	2.46621	65	4,225	8.062258	274,625	4.02073
16	256	4	4,096	2.51984	66	4,356	8.124038	287,496	4.04124
17	289	4.123106	4,913	2.57128	67	4,489	8.185353	300,763	4.06155
18	324	4.242641	5,832	2.62074	68	4,624	8.246211	314,432	4.08166
19	361	4.358899	6,859	2.66840	69	4,761	8.306624	328,509	4.10157
20	400	4.472136	8,000	2.71442	70	4,900	8.3666	343,000	4.12129
21	441	4.582576	9,261	2.75892	71	5,041	8.42615	357,911	4.14082
22	484	4.690416	10,648	2.80204	72	5,184	8.485281	373248	4.16017
23	529	4.795832	12,167	2.84387	73	5,329	8.544004	389,017	4.17934
24	576	4.898979	13,824	2.88450	74	5,476	8.602325	405,224	4.19834
25	625	5	15,625	2.92402	75	5,625	8.660254	421,875	4.21716
26	676	5.09902	17,576	2.96250	76	5,776	8.717798	438,976	4.23582
27	729	5.196152	19,683	3	77	5,929	8.774964	456,533	4.25432
28	784	5.291503	21,952	3.03659	78	6,084	8.831761	474,552	4.27266
29	841	5.385165	24,389	3.07232	79	6,241	8.888194	493,039	4.29084
30	900	5.477226	27,000	3.10723	80	6,400	8.944272	512,000	4.30887
31	961	5.567764	29,791	3.14138	81	6,561	9	531,441	4.32675
32	1,024	5.656854	32,768	3.17480	82	6,724	9.055385	551,368	4.34448
33	1,089	5.744563	35,937	3.20753	83	6,889	9.110434	571,787	4.36207
34	1,156	5.830952	39,304	3.23961	84	7,056	9.165151	592,704	4.37952
35	1,225	5.91608	42,875	3.27107	85	7,225	9.219544	614,125	4.39683
36	1,296	6	46,656	3.30193	86	7,396	9.273618	636,056	4.41400
37	1,369	6.082763	50,653	3.33222	87	7,569	9.327379	658,503	4.43105
38	1,444	6.164414	54,872	3.36198	88	7,744	9.380832	681,472	4.44796
39	1,521	6.244998	59,319	3.39121	89	7,921	9.433981	704,969	4.46475
40	1,600	6.324555	64,000	3.41995	90	8,100	9.486833	729,000	4.48140
41	1,681	6.403124	68,921	3.44822	91	8,281	9.539392	753,571	4.49794
42	1,764	6.480741	74,088	3.47603	92	8,464	9.591663	778,688	4.51436
43	1,849	6.557439	79,507	3.50340	93	8,649	9.643651	804,357	4.53065
44	1,936	6.63325	85,184	3.53035	94	8,836	9.69536	830,584	4.54684
45	2,025	6.708204	91,125	3.55689	95	9,025	9.746794	857,375	4.56290
46	2,116	6.78233	97,336	3.58305	96	9,216	9.797959	884,736	4.57886
47	2,209	6.855655	103,823	3.60883	97	9,409	9.848858	912,673	4.59470
48	2,304	6.928203	110,592	3.63424	98	9,604	9.899495	941,192	4.61044
49	2,401	7	117,649	3.65931	99	9,801	9.949874	970,299	4.62607
50	2,500	7.071068	125,000	3.68403	100	10,000	10	1,000,000	4.64159

(Continued)

Table 3. *(Continued)* Squares, Square Roots, Cubes, and Cube Roots.

No.	Square	Square Root	Cube	Cube Root	No.	Square	Square Root	Cube	Cube Root
101	10,201	10.04988	1,030,301	4.65701	151	22,801	12.28821	3,442,951	5.32507
102	10,404	10.0995	1,061,208	4.67233	152	23,104	12.32883	3,511,808	5.33680
103	10,609	10.14889	1,092,727	4.68755	153	23,409	12.36932	3,581,577	5.34848
104	10,816	10.19804	1,124,864	4.70267	154	23,716	12.40967	3,652,264	5.36011
105	11,025	10.24695	1,157,625	4.71769	155	24,025	12.4499	3,723,875	5.37169
106	11,236	10.29563	1,191,016	4.73262	156	24,336	12.49	3,796,416	5.38321
107	11,449	10.34408	1,225,043	4.74746	157	24,649	12.52996	3,869,893	5.39469
108	11,664	10.3923	1,259,712	4.76220	158	24,964	12.56981	3,944,312	5.40612
109	11,881	10.44031	1,295,029	4.77686	159	25,281	12.60952	4,019,679	5.41750
110	12,100	10.48800	1,331,000	4.79142	160	25,600	12.64911	4,096,000	5.42884
111	12,321	10.53565	1,367,631	4.80590	161	25,921	12.68858	4,173,281	5.44012
112	12,544	10.58301	1,404,928	4.82028	162	26,244	12.72792	4,251,528	5.45136
113	12,769	10.63015	1,442,897	4.83459	163	26,569	12.76715	4,330,747	5.46256
114	12,996	10.67708	1,481,544	4.84881	164	26,896	12.80625	4,410,944	5.47370
115	13,225	10.72381	1,520,875	4.86294	165	27,225	12.84523	4,492,125	5.48481
116	13,456	10.77033	1,560,896	4.87700	166	27,556	12.8841	4,574,296	5.49586
117	13,689	10.81665	1,601,613	4.89097	167	27,889	12.92285	4,657,463	5.50688
118	13,924	10.86278	1,643,032	4.90487	168	28,224	12.96148	4,741,632	5.51785
119	14,161	10.90871	1,685,159	4.91868	169	28,561	13	4,826,809	5.52877
120	14,400	10.95445	1,728,000	4.93242	170	28,900	13.0384	4,913,000	5.53966
121	14,641	11	1,771,561	4.94609	171	29,241	13.0767	5,000,211	5.55050
122	14,884	11.04536	1,815,848	4.95968	172	29,584	13.11488	5,088,448	5.56130
123	15,129	11.09054	1,860,867	4.97319	173	29,929	13.15295	5,177,717	5.57205
124	15,376	11.13553	1,906,624	4.98663	174	30,276	13.19091	5,268,024	5.58277
125	15,625	11.18034	1,953,125	5.00000	175	30,625	13.22876	5,359,375	5.59344
126	15,876	11.22497	2,000,376	5.01330	176	30,976	13.2665	5,451,776	5.60408
127	16,129	11.26943	2,048,383	5.02653	177	31,329	13.30413	5,545,233	5.61467
128	16,384	11.31371	2,097,152	5.03968	178	31,684	13.34166	5,639,752	5.62523
129	16,641	11.35782	2,146,689	5.05277	179	32,041	13.37909	5,735,339	5.63574
130	16,900	11.40175	2,197,000	5.06580	180	32,400	13.41641	5,832,000	5.64622
131	17,161	11.44552	2,248,091	5.07875	181	32,761	13.45362	5,929,741	5.65665
132	17,424	11.48913	2,299,968	5.09164	182	33,124	13.49074	6,028,568	5.66705
133	17,689	11.53256	2,352,637	5.10447	183	33,489	13.52775	6,128,487	5.67741
134	17,956	11.57584	2,406,104	5.11723	184	33,856	13.56466	6,229,504	5.68773
135	18,225	11.61895	2,460,375	5.12993	185	34,225	13.60147	6,331,625	5.69802
136	18,496	11.6619	2,515,456	5.14256	186	34,596	13.63818	6,434,856	5.70827
137	18,769	11.7047	2,571,353	5.15514	187	34,969	13.67479	6,539,203	5.71848
138	19,044	11.74734	2,628,072	5.16765	188	35,344	13.71131	6,644,672	5.72865
139	19,321	11.78983	2,685,619	5.18010	189	35,721	13.74773	6,751,269	5.73879
140	19,600	11.83216	2,744,000	5.19249	190	36,100	13.78405	6,859,000	5.74890
141	19,881	11.87434	2,803,221	5.20483	191	36,481	13.82027	6,967,871	5.75897
142	20,164	11.91638	2,863,288	5.21710	192	36,864	13.85641	7,077,888	5.76900
143	20,449	11.95826	2,924,207	5.22932	193	37,249	13.89244	7,189,057	5.77900
144	20,736	12	2,985,984	5.24148	194	37,636	13.92839	7,301,384	5.78896
145	21,025	12.04159	3,048,625	5.25359	195	38,025	13.96424	7,414,875	5.79889
146	21,316	12.08305	3,112,136	5.26564	196	38,416	14	7,529,536	5.80879
147	21,609	12.12436	3,176,523	5.27763	197	38,809	14.03567	7,645,373	5.81865
148	21,904	12.16553	3,241,792	5.28957	198	39,204	14.07125	7,762,392	5.82848
149	22,201	12.20656	3,307,949	5.30146	199	39,601	14.10674	7,880,599	5.83827
150	22,500	12.24745	3,375,000	5.31329	200	40,000	14.14214	8,000,000	5.84804

(Continued)

Table 3. *(Continued)* Squares, Square Roots, Cubes, and Cube Roots.

No.	Square	Square Root	Cube	Cube Root	No.	Square	Square Root	Cube	Cube Root
201	40,401	14.17745	8,120,601	5.85777	251	63,001	15.84298	15,813,251	6.30799
202	40,804	14.21267	8,242,408	5.86746	252	63,504	15.87451	16,003,008	6.31636
203	41,209	14.24781	8,365,427	5.87713	253	64,009	15.90597	16,194,277	6.32470
204	41,616	14.28286	8,489,664	5.88677	254	64,516	15.93738	16,387,064	6.33303
205	42,025	14.31782	8,615,125	5.89637	255	65,025	15.96872	16,581,375	6.34133
206	42,436	14.3527	8,741,816	5.90594	256	65,536	16	16,777,216	6.34960
207	42,849	14.38749	8,869,743	5.91548	257	66,049	16.03122	16,974,593	6.35786
208	43,264	14.42221	8,998,912	5.92499	258	66,564	16.06238	17,173,512	6.36610
209	43,681	14.45683	9,129,329	5.93447	259	67,081	16.09348	17,373,979	6.37431
210	44,100	14.49138	9,261,000	5.94392	260	67,600	16.12452	17,576,000	6.38250
211	44,521	14.52584	9,393,931	5.95334	261	68,121	16.15549	17,779,581	6.39068
212	44,944	14.56022	9,528,128	5.96273	262	68,644	16.18641	17,984,728	6.39883
213	45,369	14.59452	9,663,597	5.97209	263	69,169	16.21727	18,191,447	6.40696
214	45,796	14.62874	9,800,344	5.98142	264	69,696	16.24808	18,399,744	6.41507
215	46,225	14.66288	9,938,375	5.99073	265	70,225	16.27882	18,609,625	6.42316
216	46,656	14.69694	10,077,696	6.00000	266	70,756	16.30951	18,821,096	6.43123
217	47,089	14.73092	10,218,313	6.00925	267	71,289	16.34013	19,034,163	6.43928
218	47,524	14.76482	10,360,232	6.01846	268	71,824	16.37071	19,248,832	6.44731
219	47,961	14.79865	10,503,459	6.02765	269	72,361	16.40122	19,465,109	6.45531
220	48,400	14.8324	10,648,000	6.03681	270	72,900	16.43168	19,683,000	6.46330
221	48,841	14.86607	10,793,861	6.04594	271	73,441	16.46208	19,902,511	6.47127
222	49,284	14.89966	10,941,048	6.05505	272	73,984	16.49242	20,123,648	6.47922
223	49,729	14.93318	11,089,567	6.06413	273	74,529	16.52271	20,346,417	6.48715
224	50,176	14.96663	11,239,424	6.07318	274	75,076	16.55295	20,570,824	6.49507
225	50,625	15	11,390,625	6.08220	275	75,625	16.58312	20,796,875	6.50296
226	51,076	15.0333	11,543,176	6.09120	276	76,176	16.61325	21,024,576	6.51083
227	51,529	15.06652	11,697,083	6.10017	277	76,729	16.64332	21,253,933	6.51868
228	51,984	15.09967	11,852,352	6.10911	278	77,284	16.67333	21,484,952	6.52652
229	52,441	15.13275	12,008,989	6.11803	279	77,841	16.70329	21,717,639	6.53434
230	52,900	15.16575	12,167,000	6.12693	280	78,400	16.7332	21,952,000	6.54213
231	53,361	15.19868	12,326,391	6.13579	281	78,961	16.76305	22,188,041	6.54991
232	53,824	15.23155	12,487,168	6.14463	282	79,524	16.79286	22,425,768	6.55767
233	54,289	15.26434	12,649,337	6.15345	283	80,089	16.8226	22,665,187	6.56541
234	54,756	15.29706	12,812,904	6.16224	284	80,656	16.8523	22,906,304	6.57314
235	55,225	15.32971	12,977,875	6.17101	285	81,225	16.88194	23,149,125	6.58084
236	55,696	15.36229	13,144,256	6.17975	286	81,796	16.91153	23,393,656	6.58853
237	56,169	15.3948	13,312,053	6.18846	287	82,369	16.94107	23,639,903	6.59620
238	56,644	15.42725	13,481,272	6.19715	288	82,944	16.97056	23,887,872	6.60385
239	57,121	15.45962	13,651,919	6.20582	289	83,521	17	24,137,569	6.61149
240	57,600	15.49193	13,824,000	6.21447	290	84,100	17.02939	24,389,000	6.61911
241	58,081	15.52417	13,997,521	6.22308	291	84,681	17.05872	24,642,171	6.62671
242	58,564	15.55635	14,172,488	6.23168	292	85,264	17.08801	24,897,088	6.63429
243	59,049	15.58846	14,348,907	6.24025	293	85,849	17.11724	25,153,757	6.64185
244	59,536	15.6205	14,526,784	6.24880	294	86,436	17.14643	25,412,184	6.64940
245	60,025	15.65248	14,706,125	6.25732	295	87,025	17.17556	25,672,375	6.65693
246	60,516	15.68439	14,886,936	6.26583	296	87,616	17.20465	25,934,336	6.66444
247	61,009	15.71623	15,069,223	6.27431	297	88,209	17.23369	26,198,073	6.67194
248	61,504	15.74802	15,252,992	6.28276	298	88,804	17.26268	26,463,592	6.67942
249	62,001	15.77973	15,438,249	6.29119	299	89,401	17.29162	26,730,899	6.68688
250	62,500	15.81139	15,625,000	6.29961	300	90,000	17.32051	27,000,000	6.69433

(Continued)

Table 3. *(Continued)* Squares, Square Roots, Cubes, and Cube Roots.

No.	Square	Square Root	Cube	Cube Root	No.	Square	Square Root	Cube	Cube Root
301	90,601	17.34935	27,270,901	6.70176	351	123,201	18.73499	43,243,551	7.05400
302	91,204	17.37815	27,543,608	6.70917	352	123,904	18.76166	43,614,208	7.06070
303	91,809	17.4069	27,818,127	6.71657	353	124,609	18.78829	43,986,977	7.06738
304	92,416	17.4356	28,094,464	6.72395	354	125,316	18.81489	44,361,864	7.07404
305	93,025	17.46425	28,372,625	6.73132	355	126,025	18.84144	44,738,875	7.08070
306	93,636	17.49286	28,652,616	6.73866	356	126,736	18.86796	45,118,016	7.08734
307	94,249	17.52142	28,934,443	6.74600	357	127,449	18.89444	45,499,293	7.09397
308	94,864	17.54993	29,218,112	6.75331	358	128,164	18.92089	45,882,712	7.10059
309	95,481	17.5784	29,503,629	6.76061	359	128,881	18.9473	46,268,279	7.10719
310	96,100	17.60682	29,791,000	6.76790	360	129,600	18.97367	46,656,000	7.11379
311	96,721	17.63519	30,080,231	6.77517	361	130,321	19	47,045,881	7.12037
312	97,344	17.66352	30,371,328	6.78242	362	131,044	19.0263	47,437,928	7.12694
313	97,969	17.69181	30,664,297	6.78966	363	131,769	19.05256	47,832,147	7.13349
314	98,596	17.72005	30,959,144	6.79688	364	132,496	19.07878	48,228,544	7.14004
315	99,225	17.74824	31,255,875	6.80409	365	133,225	19.10497	48,627,125	7.14657
316	99,856	17.77639	31,554,496	6.81128	366	133,956	19.13113	49,027,896	7.15309
317	100,489	17.80449	31,855,013	6.81846	367	134,689	19.15724	49,430,863	7.15960
318	101,124	17.83255	32,157,432	6.82562	368	135,424	19.18333	49,836,032	7.16610
319	101,761	17.86057	32,461,759	6.83277	369	136,161	19.20937	50,243,409	7.17258
320	102,400	17.88854	32,768,000	6.83990	370	136,900	19.23538	50,653,000	7.17905
321	103,041	17.91647	33,076,161	6.84702	371	137,641	19.26136	51,064,811	7.18552
322	103,684	17.94436	33,386,248	6.85412	372	138,384	19.2873	51,478,848	7.19197
323	104,329	17.9722	33,698,267	6.86121	373	139,129	19.31321	51,895,117	7.19840
324	104,976	18	34,012,224	6.86829	374	139,876	19.33908	52,313,624	7.20483
325	105,625	18.02776	34,328,125	6.87534	375	140,625	19.36492	52,734,375	7.21125
326	106,276	18.05547	34,645,976	6.88239	376	141,376	19.39072	53,157,376	7.21765
327	106,929	18.08314	34,965,783	6.88942	377	142,129	19.41649	53,582,633	7.22405
328	107,584	18.11077	35,287,552	6.89643	378	142,884	19.44222	54,010,152	7.23043
329	108,241	18.13836	35,611,289	6.90344	379	143,641	19.46792	54,439,939	7.23680
330	108,900	18.1659	35,937,000	6.91042	380	144,400	19.49359	54,872,000	7.24316
331	109,561	18.19341	36,264,691	6.91740	381	145,161	19.51922	55,306,341	7.24950
332	110,224	18.22087	36,594,368	6.92436	382	145,924	19.54482	55,742,968	7.25584
333	110,889	18.24829	36,926,037	6.93131	383	146,689	19.57039	56,181,887	7.26217
334	111,556	18.27567	37,259,704	6.93823	384	147,456	19.59592	56,623,104	7.26848
335	112,225	18.30301	37,595,375	6.94515	385	148,225	19.62142	57,066,625	7.27479
336	112,896	18.3303	37,933,056	6.95205	386	148,996	19.64688	57,512,456	7.28108
337	113,569	18.35756	38,272,753	6.95894	387	149,769	19.67232	57,960,603	7.28736
338	114,244	18.38478	38,614,472	6.96582	388	150,544	19.69772	58,411,072	7.29363
339	114,921	18.44195	38,958,219	6.97268	389	151,321	19.72308	58,863,869	7.29989
340	115,600	18.43909	39,304,000	6.97953	390	152,100	19.74842	59,319,000	7.30614
341	116,281	18.44619	39,651,821	6.98637	391	152,881	19.77372	59,776,471	7.31238
342	116,964	18.49324	40,001,688	6.99319	392	153,664	19.79899	60,236,288	7.31861
343	117,649	18.52026	40,353,607	7.00000	393	154,449	19.82423	60,698,457	7.32483
344	118,336	18.54724	40,707,584	7.00680	394	155,236	19.84943	61,162,984	7.33104
345	119,025	18.57418	41,063,625	7.01358	395	156,025	19.87461	61,629,875	7.33723
346	119,716	18.60108	41,421,736	7.02035	396	156,816	19.89975	62,099,136	7.34342
347	120,409	18.62794	41,781,923	7.02711	397	157,609	19.92486	62,570,773	7.34960
348	121,104	18.65476	42,144,192	7.03385	398	158,404	19.94994	63,044,792	7.35576
349	121,801	18.68154	42,508,549	7.04058	399	159,201	19.97498	63,521,199	7.36192
350	122,500	18.70829	42,875,000	7.04730	400	160,000	20	64,000,000	7.36806

(Continued)

Table 3. *(Continued)* **Squares, Square Roots, Cubes, and Cube Roots.**

No.	Square	Square Root	Cube	Cube Root	No.	Square	Square Root	Cube	Cube Root
401	160,801	20.02498	64,481,201	7.37420	451	203,401	21.23676	91,733,851	7.66877
402	161,604	20.04994	64,964,808	7.38032	452	204,304	21.26029	92,345,408	7.67443
403	162,409	20.07486	65,450,827	7.38644	453	205,209	21.2838	92,959,677	7.68009
404	163,216	20.09975	65,939,264	7.39254	454	206,116	21.30728	93,576,664	7.68573
405	164,025	20.12461	66,430,125	7.39864	455	207,025	21.33073	94,196,375	7.69137
406	164,836	20.14944	66,923,416	7.40472	456	207,936	21.35416	94,818,816	7.69700
407	165,649	20.17424	67,419,143	7.41080	457	208,849	21.37756	95,443,993	7.70262
408	166,464	20.19901	67,917,312	7.41686	458	209,764	21.40093	96,071,912	7.70824
409	167,281	20.22375	68,417,929	7.42291	459	210,681	21.42429	96,702,579	7.71384
410	168,100	20.24846	68,921,000	7.42896	460	211,600	21.44761	97,336,000	7.71944
411	168,921	20.27313	69,426,531	7.43499	461	212,521	21.47091	97,972,181	7.72503
412	169,744	20.29778	69,934,528	7.44102	462	213,444	21.49419	98,611,128	7.73061
413	170,569	20.3224	70,444,997	7.44703	463	214,369	21.51743	99,252,847	7.73619
414	171,396	20.34699	70,957,944	7.45304	464	215,296	21.54066	99,897,344	7.74175
415	172,225	20.37155	71,473,375	7.45904	465	216,225	21.56386	100,544,625	7.74731
416	173,056	20.39608	71,991,296	7.46502	466	217,156	21.58703	101,194,696	7.75286
417	173,889	20.42058	72,511,713	7.47100	467	218,089	21.61018	101,847,563	7.75840
418	174,724	20.44505	73,034,632	7.47697	468	219,024	21.63331	102,503,232	7.76394
419	175,561	20.46949	73,560,059	7.48292	469	219,961	21.65641	103,161,709	7.76946
420	176,400	20.4939	74,088,000	7.48887	470	220,900	21.67948	103,823,000	7.77498
421	177,241	20.51828	74,618,461	7.49481	471	221,841	21.70253	104,487,111	7.78049
422	178,084	20.54264	75,151,448	7.50074	472	222,784	21.72556	105,154,048	7.78599
423	178,929	20.56696	75,686,967	7.50666	473	223,729	21.74856	105,823,817	7.79149
424	179,776	20.59126	76,225,024	7.51257	474	224,676	21.77154	106,496,424	7.79697
425	180,625	20.61553	76,765,625	7.51847	475	225,625	21.79449	107,171,875	7.80245
426	181,476	20.63977	77,308,776	7.52437	476	226,576	21.81742	107,850,176	7.80793
427	182,329	20.66398	77,854,483	7.53025	477	227,529	21.84033	108,531,333	7.81339
428	183,184	20.68816	78,402,752	7.53612	478	228,484	21.86321	109,215,352	7.81885
429	184,041	20.71232	78,953,589	7.54199	479	229,441	21.88607	109,902,239	7.82429
430	184,900	20.73644	79,507,000	7.54784	480	230,400	21.9089	110,592,000	7.82974
431	185,761	20.76054	80,062,991	7.55369	481	231,361	21.93171	111,284,641	7.83517
432	186,624	20.78461	80,621,568	7.55953	482	232,324	21.9545	111,980,168	7.84059
433	187,489	20.80865	81,182,737	7.56535	483	233,289	21.97726	112,678,587	7.84601
434	188,356	20.83267	81,746,504	7.57117	484	234,256	22	113,379,904	7.85142
435	189,225	20.85665	82,312,875	7.57698	485	235,225	22.02272	114,084,125	7.85683
436	190,096	20.88061	82,881,856	7.58279	486	236,196	22.04541	114,791,256	7.86222
437	190,969	20.90454	83,453,453	7.58858	487	237,169	22.06808	115,501,303	7.86761
438	191,844	20.92845	84,027,672	7.59436	488	238,144	22.09072	116,214,272	7.87299
439	192,721	20.95233	84,604,519	7.60014	489	239,121	22.11334	116,930,169	7.87837
440	193,600	20.97618	85,184,000	7.60590	490	240,100	22.13594	117,649,000	7.88374
441	194,481	21	85,766,121	7.61166	491	241,081	22.15852	118,370,771	7.88909
442	195,364	21.0238	86,650,888	7.61741	492	242,064	22.18107	119,095,488	7.89445
443	196,249	21.04757	86,938,307	7.62315	493	243,049	22.2036	119,823,157	7.89979
444	197,136	21.07131	87,528,384	7.62888	494	244,036	22.22611	120,553,784	7.90513
445	198,025	21.09502	88,121,125	7.63461	495	245,025	22.2486	121,287,375	7.91046
446	198,916	21.11871	88,716,536	7.64032	496	246,016	22.27106	122,023,936	7.91578
447	199,809	21.14237	89,314,623	7.64603	497	247,009	22.2935	122,763,473	7.92110
448	200,704	21.16601	89,915,392	7.65172	498	248,004	22.31591	123,505,992	7.92641
449	201,601	21.18962	90,518,849	7.65741	499	249,001	22.33831	124,251,499	7.93171
450	202,500	21.2132	91,125,000	7.66309	500	250,000	22.36068	125,000,000	7.93701

(Continued)

Table 3. *(Continued)* Squares, Square Roots, Cubes, and Cube Roots.

No.	Square	Square Root	Cube	Cube Root	No.	Square	Square Root	Cube	Cube Root
501	251,001	22.38303	125,751,501	7.	551	303,601	23.47339	167,284,151	8.19818
502	252,004	22.40536	126,506,008	7.94229	552	304,704	23.49468	168,196,608	8.20313
503	253,009	22.42766	127,263,527	7.94757	553	305,809	23.51595	169,112,377	8.20808
504	254,016	22.44994	128,024,064	7.95285	554	306,916	23.5372	170,031,464	8.21303
505	255,025	22.47221	128,787,625	7.95811	555	308,025	23.55844	170,953,875	8.21797
506	256,036	22.49444	129,554,216	7.96337	556	309,136	23.57965	171,879,616	8.22290
507	257,049	22.51666	130,323,843	7.96863	557	310,249	23.60085	172,808,693	8.22783
508	258,064	22.53886	131,096,512	7.97387	558	311,364	23.62202	173,741,112	8.23275
509	259,081	22.56103	131,872,229	7.97911	559	312,481	23.64318	174,676,879	8.23766
510	260,100	22.58318	132,651,000	7.98434	560	313,600	23.66432	175,616,000	8.24257
511	261,121	22.60531	133,432,831	7.98957	561	314,721	23.68544	176,558,481	8.24747
512	262,144	22.62742	134,217,728	8.00000	562	315,844	23.70654	177,504,328	8.25237
513	263,169	22.6495	135,005,697	8.00520	563	316,969	23.72762	178,453,547	8.25726
514	264,196	22.67157	135,796,744	8.01040	564	318,096	23.74868	179,406,144	8.26215
515	265,225	22.69361	136,590,875	8.01559	565	319,225	23.76973	180,362,125	8.26703
516	266,256	22.71563	137,388,096	8.02078	566	320,356	23.79075	181,321,496	8.27190
517	267,289	22.73763	138,188,413	8.02596	567	321,489	23.81176	182,284,263	8.27677
518	268,324	22.75961	138,991,832	8.03113	568	322,624	23.83275	183,250,432	8.28164
519	269,361	22.78157	139,798,359	8.03629	569	323,761	23.85372	184,220,009	8.28649
520	270,400	22.80351	140,608,000	8.04145	570	324,900	23.87467	185,193,000	8.29134
521	271,441	22.82542	141,420,761	8.04660	571	326,041	23.89561	186,169,411	8.29619
522	272,484	22.84732	142,236,648	8.05175	572	327,184	23.91652	187,149,248	8.30103
523	273,529	22.86919	143,055,667	8.05689	573	328,329	23.93742	188,132,517	8.30587
524	274,576	22.89105	143,877,824	8.06202	574	329,476	23.9583	189,119,224	8.31069
525	275,625	22.91288	144,703,125	8.06714	575	330,625	23.97916	190,109,375	8.31552
526	276,676	22.93469	145,531,576	8.07226	576	331,776	24	191,102,976	8.32034
527	277,729	22.95648	146,363,183	8.07737	577	332,929	24.02082	192,100,033	8.32515
528	278,784	22.97825	147,197,952	8.08248	578	334,084	24.04163	193,100,552	8.32995
529	279,841	23	148,035,889	8.08758	579	335,241	24.06242	194,104,539	8.33476
530	280,900	23.02173	148,877,000	8.09267	580	336,400	24.08319	195,112,000	8.33955
531	281,961	23.04344	149,721,291	8.09776	581	337,561	24.10394	196,122,941	8.34434
532	283,024	23.06513	150,568,768	8.10284	582	338,724	24.12468	197,137,368	8.34913
533	284,089	23.08679	151,419,437	8.10791	583	339,889	24.14539	198,155,287	8.35390
534	285,156	23.10844	152,276,304	8.11298	584	341,056	24.16609	199,176,704	8.35868
535	286,225	23.13007	153,130,375	8.11804	585	342,225	24.18677	200,201,625	8.36345
536	287,296	23.15167	153,990,656	8.12310	586	343,396	24.20744	201,230,056	8.36821
537	288,369	23.17326	154,854,153	8.12814	587	344,569	24.22808	202,262,003	8.37297
538	289,444	23.19483	155,720,872	8.13319	588	345,744	24.24871	203,297,472	8.37772
539	290,521	23.21637	156,590,819	8.13822	589	346,921	24.26932	204,336,469	8.38247
540	291,600	23.2379	157,464,000	8.14325	590	348,100	24.28992	205,379,000	8.38721
541	292,681	23.25941	158,340,421	8.14828	591	349,281	24.31049	206,425,071	8.39194
542	293,764	23.28089	159,220,088	8.15329	592	350,464	24.33105	207,474,688	8.39667
543	294,849	23.30236	160,103,007	8.15831	593	351,649	24.35159	208,527,857	8.40140
544	295,936	23.32381	160,989,184	8.16331	594	352,836	24.37212	209,584,584	8.40612
545	297,025	23.34524	161,878,625	8.16831	595	354,025	24.39262	210,644,875	8.41083
546	298,116	23.36664	162,771,336	8.17330	596	355,216	24.41311	211,708,736	8.41554
547	299,209	23.38803	163,667,323	8.17829	597	356,409	24.43358	212,776,173	8.42025
548	300,304	23.4094	164,566,592	8.18327	598	357,604	24.45404	213,847,192	8.42494
549	301,401	23.43075	165,469,149	8.18824	599	358,801	24.47448	214,921,799	8.42964
550	302,500	23.45208	166,375,000	8.19321	600	360,000	24.4949	216,000,000	8.43433

(Continued)

Table 3. *(Continued)* Squares, Square Roots, Cubes, and Cube Roots.

No.	Square	Square Root	Cube	Cube Root	No.	Square	Square Root	Cube	Cube Root
601	361,201	24.5153	217,081,801	8.43901	651	423,801	25.5147	275,894,451	8.66683
602	362,404	24.53569	218,167,208	8.44369	652	425,104	25.53429	277,167,808	8.67127
603	363,609	24.55606	219,256,227	8.44836	653	426,409	25.55386	278,445,077	8.67570
604	364,816	24.57641	220,348,864	8.45303	654	427,716	25.57342	279,726,264	8.68012
605	366,025	24.59675	221,445,125	8.45769	655	429,025	25.59297	281,011,375	8.68455
606	367,236	24.61707	222,545,016	8.46235	656	430,336	25.6125	282,300,416	8.68896
607	368,449	24.63737	223,648,543	8.46700	657	431,649	25.63201	283,593,393	8.69338
608	369,664	24.65766	224,755,712	8.47165	658	432,964	25.65151	284,890,312	8.69778
609	370,881	24.67793	225,866,529	8.47629	659	434,281	25.671	286,191,179	8.70219
610	372,100	24.69818	226,981,000	8.48093	660	435,600	25.69047	287,496,000	8.70659
611	373,321	24.71841	228,099,131	8.48556	661	436,921	25.70992	288,804,781	8.71098
612	374,544	24.73863	229,220,928	8.49018	662	438,244	25.72936	290,117,528	8.71537
613	375,769	24.75884	230,346,397	8.49481	663	439,569	25.74879	291,434,247	8.71976
614	376,996	24.77902	231,475,544	8.49942	664	440,896	25.7682	292,754,944	8.72414
615	378,225	24.79919	232,608,375	8.50403	665	442,225	25.78759	294,079,625	8.72852
616	379,456	24.81935	233,744,896	8.50864	666	443,556	25.80698	295,408,296	8.73289
617	380,689	24.83948	234,885,113	8.51324	667	444,889	25.82634	296,740,963	8.73726
618	381,924	24.85961	236,029,032	8.51784	668	446,224	25.8457	298,077,632	8.74162
619	383,161	24.87971	237,176,659	8.52243	669	447,561	25.86503	299,418,309	8.74598
620	384,400	24.8998	238,328,000	8.52702	670	448,900	25.88436	300,763,000	8.75034
621	385,641	24.91987	239,483,061	8.53160	671	450,241	25.90367	302,111,711	8.75469
622	386,884	24.93993	240,641,848	8.53618	672	451,584	25.92296	303,464,448	8.75904
623	388,129	24.95997	241,804,367	8.54075	673	452,929	25.94224	304,821,217	8.76338
624	389,376	24.97999	242,970,624	8.54532	674	454,276	25.96151	306,182,024	8.76772
625	390,625	25	244,140,625	8.54988	675	455,625	25.98076	307,546,875	8.77205
626	391,876	25.01999	245,314,376	8.55444	676	456,976	26	308,915,776	8.77638
627	393,129	25.03997	246,491,883	8.55899	677	458,329	26.01922	310,288,733	8.78071
628	394,384	25.05993	247,673,152	8.56354	678	459,684	26.03843	311,665,752	8.78503
629	395,641	25.07987	248,858,189	8.56808	679	461,041	26.05763	313,046,839	8.78935
630	396,900	25.0998	250,047,000	8.57262	680	462,400	26.07681	314,432,000	8.79366
631	398,161	25.11971	251,239,591	8.57715	681	463,761	26.09598	315,821,241	8.79797
632	399,424	25.13961	252,435,968	8.58168	682	465,124	26.11513	317,214,568	8.80227
633	400,689	25.15949	253,636,137	8.58620	683	466,489	26.13427	318,611,987	8.80657
634	401,956	25.17936	254,840,104	8.59072	684	467,856	26.15339	320,013,504	8.81087
635	403,225	25.19921	256,047,875	8.59524	685	469,225	26.1725	321,419,125	8.81516
636	404,496	25.21904	257,259,456	8.59975	686	470,596	26.1916	322,828,856	8.81945
637	405,769	25.23886	258,474,853	8.60425	687	471,969	26.21068	324,242,703	8.82373
638	407,044	25.25866	259,694,072	8.60875	688	473,344	26.22975	325,660,672	8.82801
639	408,321	25.27845	260,917,119	8.61325	689	474,721	26.24881	327,082,769	8.83228
640	409,600	25.29822	262,144,000	8.61774	690	476,100	26.26785	328,509,000	8.83656
641	410,881	25.31798	263,374,721	8.62222	691	477,481	26.28688	329,939,371	8.84082
642	412,164	25.33772	264,609,288	8.62671	692	478,864	26.30589	331,373,888	8.84509
643	413,449	25.35744	265,847,707	8.63118	693	480,249	26.32489	332,812,557	8.84934
644	414,736	25.37716	267,089,984	8.63566	694	481,636	26.34388	334,255,384	8.85360
645	416,025	25.39685	268,336,125	8.64012	695	483,025	26.36285	335,702,375	8.85785
646	417,316	25.41653	269,586,136	8.64459	696	484,416	26.38181	337,153,536	8.86210
647	418,609	25.43619	270,840,023	8.64904	697	485,809	26.40076	338,608,873	8.86634
648	419,904	25.45584	272,097,792	8.65350	698	487,204	26.41969	340,068,392	8.87058
649	421,201	25.47548	273,359,449	8.65795	699	488,601	26.43861	341,532,099	8.87481
650	422,500	25.4951	274,625,000	8.66239	700	490,000	26.45751	343,000,000	8.87904

(Continued)

Table 3. *(Continued)* Squares, Square Roots, Cubes, and Cube Roots.

No.	Square	Square Root	Cube	Cube Root	No.	Square	Square Root	Cube	Cube Root
701	491,401	26.4764	344,472,101	8.88327	751	564,001	27.40438	423,564,751	9.08964
702	492,804	26.49528	345,948,408	8.88749	752	565,504	27.42262	425,259,008	9.09367
703	494,209	26.51415	347,428,927	8.89171	753	567,009	27.44085	426,957,777	9.09770
704	495,616	26.533	348,913,664	8.89592	754	568,516	27.45906	428,661,064	9.10173
705	497,025	26.55184	350,402,625	8.90013	755	570,025	27.47726	430,368,875	9.10575
706	498,436	26.57066	351,895,816	8.90434	756	571,536	27.49545	432,081,216	9.10977
707	499,849	26.58947	353,393,243	8.90854	757	573,049	27.51363	433,798,093	9.11378
708	501,264	26.60827	354,894,912	8.91274	758	574,564	27.5318	435,519,512	9.11779
709	502,681	26.62705	356,400,829	8.91693	759	576,081	27.54995	437,245,479	9.12180
710	504,100	26.64583	357,911,000	8.92112	760	577,600	27.5681	438,976,000	9.12581
711	505,521	26.66458	359,425,431	8.92531	761	579,121	27.58623	440,711,081	9.12981
712	506,944	26.68333	360,944,128	8.92949	762	580,644	27.60435	442,450,728	9.13380
713	508,369	26.70206	362,467,097	8.93367	763	582,169	27.62245	444,194,947	9.13780
714	509,796	26.72078	363,994,344	8.93784	764	583,696	27.64055	445,943,744	9.14179
715	511,225	26.73948	365,525,875	8.94201	765	585,225	27.65863	447,697,125	9.14577
716	512,656	26.75818	367,061,696	8.94618	766	586,756	27.67671	449,455,096	9.14976
717	514,089	26.77686	368,601,813	8.95034	767	588,289	27.69476	451,217,663	9.15374
718	515,524	26.79552	370,146,232	8.95450	768	589,824	27.71281	452,984,832	9.15771
719	516,961	26.81418	371,694,959	8.95866	769	591,361	27.73085	454,756,609	9.16169
720	518,400	26.83282	373,248,000	8.96281	770	592,900	27.74887	456,533,000	9.16566
721	519,841	26.85144	374,805,361	8.96696	771	594,441	27.76689	458,314,011	9.16962
722	521,284	26.87006	376,367,048	8.97110	772	595,984	27.78489	460,099,648	9.17359
723	522,729	26.88866	377,933,067	8.97524	773	597,529	27.80288	461,889,917	9.17754
724	524,176	26.90725	379,503,424	8.97938	774	599,076	27.82086	463,684,824	9.18150
725	525,625	26.92582	381,078,125	8.98351	775	600,625	27.83882	465,484,375	9.18545
726	527,076	26.94439	382,657,176	8.98764	776	602,176	27.85678	467,288,576	9.18940
727	528,529	26.96294	384,240,583	8.99176	777	603,729	27.87472	469,097,433	9.19335
728	529,984	26.98148	385,828,352	8.99589	778	605,284	27.89265	470,910,952	9.19729
729	531,441	27	387,420,489	9.00000	779	606,841	27.91057	472,729,139	9.20123
730	532,900	27.01851	389,017,000	9.00411	780	608,400	27.92848	474,552,000	9.20516
731	534,361	27.03701	390,617,891	9.00822	781	609,961	27.94638	476,379,541	9.20910
732	535,824	27.0555	392,223,168	9.01233	782	611,524	27.96426	478,211,768	9.21303
733	537,289	27.07397	393,832,837	9.01643	783	613,089	27.98214	480,048,687	9.21695
734	538,756	27.09243	395,446,904	9.02053	784	614,656	28	481,890,304	9.22087
735	540,225	27.11088	397,065,375	9.02462	785	616,225	28.01785	483,736,625	9.22479
736	541,696	27.12932	398,688,256	9.02871	786	617,796	28.03569	485,587,656	9.22871
737	543,169	27.14774	400,315,553	9.03280	787	619,369	28.05352	487,443,403	9.23262
738	544,644	27.16616	401,947,272	9.03689	788	620,944	28.07134	489,303,872	9.23653
739	546,121	27.18455	403,583,419	9.04097	789	622,521	28.08914	491,169,069	9.24043
740	547,600	27.20294	405,224,000	9.04504	790	624,100	28.10694	493,039,000	9.24434
741	549,081	27.22132	406,869,021	9.04911	791	625,681	28.12472	494,913,671	9.24823
742	550,564	27.23968	408,518,488	9.05318	792	627,264	28.14249	496,793,088	9.25213
743	552,049	27.25803	410,172,407	9.05725	793	628,849	28.16026	498,677,257	9.25602
744	553,536	27.27636	411,830,784	9.06131	794	630,436	28.17801	500,566,184	9.25991
745	555,025	27.29469	413,493,625	9.06537	795	632,025	28.19574	502,459,875	9.26380
746	556,516	27.313	415,160,936	9.06942	796	633,616	28.21347	504,358,336	9.26768
747	558,009	27.3313	416,832,723	9.07347	797	635,209	28.23119	506,261,573	9.27156
748	559,504	27.34959	418,508,992	9.07752	798	636,804	28.24889	508,169,592	9.27544
749	561,001	27.36786	420,189,749	9.08156	799	638,401	28.26659	510,082,399	9.27931
750	562,500	27.38613	421,875,000	9.08560	800	640,000	28.28427	512,000,000	9.28318

(Continued)

Table 3. *(Continued)* Squares, Square Roots, Cubes, and Cube Roots.

No.	Square	Square Root	Cube	Cube Root	No.	Square	Square Root	Cube	Cube Root
801	641,601	28.30194	513,922,401	9.28704	851	724,201	29.1719	616,295,051	9.47640
802	643,204	28.3196	515,849,608	9.29091	852	725,904	29.18904	618,470,208	9.48011
803	644,809	28.33725	517,781,627	9.29477	853	727,609	29.20616	620,650,477	9.48381
804	646,416	28.35489	519,718,464	9.29862	854	729,316	29.22328	622,835,864	9.48752
805	648,025	28.37252	521,660,125	9.30248	855	731,025	29.24038	625,026,375	9.49122
806	649,636	28.39014	523,606,616	9.30633	856	732,736	29.25748	627,222,016	9.49492
807	651,249	28.40775	525,557,943	9.31018	857	734,449	29.27456	629,422,793	9.49861
808	652,864	28.42534	527,514,112	9.31402	858	736,164	29.29164	631,628,712	9.50231
809	654,481	28.44293	529,475,129	9.31786	859	737,881	29.3087	633,839,779	9.50600
810	656,100	28.4605	531,441,000	9.32170	860	739,600	29.32576	636,056,000	9.50969
811	657,721	28.47806	533,411,731	9.32553	861	741,321	29.3428	638,277,381	9.51337
812	659,344	28.49561	535,387,328	9.32936	862	743,044	29.35984	640,503,928	9.51705
813	660,969	28.51315	537,367,797	9.33319	863	744,769	29.37686	642,735,647	9.52073
814	662,596	28.53069	539,353,144	9.33702	864	746,496	29.39388	644,972,544	9.52441
815	664,225	28.5482	541,343,375	9.34084	865	748,225	29.41088	647,214,625	9.52808
816	665,856	28.56571	543,338,496	9.34466	866	749,956	29.42788	649,461,896	9.53175
817	667,489	28.58321	545,338,513	9.34847	867	751,689	29.44486	651,714,363	9.53542
818	669,124	28.6007	547,343,432	9.35229	868	753,424	29.46184	653,972,032	9.53908
819	670,761	28.61818	549,353,259	9.35610	869	755,161	29.47881	656,234,909	9.54274
820	672,400	28.63564	551,368,000	9.35990	870	756,900	29.49576	658,503,000	9.54640
821	674,041	28.6531	553,387,661	9.36370	871	758,641	29.51271	660,776,311	9.55006
822	675,684	28.67054	555,412,248	9.36751	872	760,384	29.52965	663,054,848	9.55371
823	677,329	28.68798	557,441,767	9.37130	873	762,129	29.54657	665,338,617	9.55736
824	678,976	28.7054	559,476,224	9.37510	874	763,876	29.56349	667,627,624	9.56101
825	680,625	28.72281	561,515,625	9.37889	875	765,625	29.5804	669,921,875	9.56466
826	682,276	28.74022	563,559,976	9.38268	876	767,376	29.5973	672,221,376	9.56830
827	683,929	28.75761	565,609,283	9.38646	877	769,129	29.61419	674,526,133	9.57194
828	685,584	28.77499	567,663,552	9.39024	878	770,884	29.63106	676,836,152	9.57557
829	687,241	28.79236	569,722,789	9.39402	879	772,641	29.64793	679,151,439	9.57921
830	688,900	28.80972	571,787,000	9.39780	880	774,400	29.66479	681,472,000	9.58284
831	690,561	28.82707	573,856,191	9.40157	881	776,161	29.68164	683,797,841	9.58647
832	692,224	28.84441	575,930,368	9.40534	882	777,924	29.69848	686,128,968	9.59009
833	693,889	28.86174	578,009,537	9.40911	883	779,689	29.71532	688,465,387	9.59372
834	695,556	28.87906	580,093,704	9.41287	884	781,456	29.73214	690,807,104	9.59734
835	697,225	28.89637	582,182,875	9.41663	885	783,225	29.74895	693,154,125	9.60095
836	698,896	28.91366	584,277,056	9.42039	886	784,996	29.76575	695,506,456	9.60457
837	700,569	28.93095	586,376,253	9.42414	887	786,769	29.78255	697,864,103	9.60818
838	702,244	28.94823	588,480,472	9.42789	888	788,544	29.79933	700,227,072	9.61179
839	703,921	28.9655	590,589,719	9.43164	889	790,321	29.8161	702,595,369	9.61540
840	705,600	28.98275	592,704,000	9.43539	890	792,100	29.83287	704,969,000	9.61900
841	707,281	29	594,823,321	9.43913	891	793,881	29.84962	707,347,971	9.62260
842	708,964	29.01724	596,947,688	9.44287	892	795,664	29.86637	709,732,288	9.62620
843	710,649	29.03446	599,077,107	9.44661	893	797,449	29.88311	712,121,957	9.62980
844	712,336	29.05168	601,211,584	9.45034	894	799,236	29.89983	714,516,984	9.63339
845	714,025	29.06888	603,351,125	9.45407	895	801,025	29.91655	716,917,375	9.63698
846	715,716	29.08608	605,495,736	9.45780	896	802,816	29.93326	719,323,136	9.64057
847	717,409	29.10326	607,645,423	9.46152	897	804,609	29.94996	721,734,273	9.64415
848	719,104	29.12044	609,800,192	9.46525	898	806,404	29.96665	724,150,792	9.64774
849	720,801	29.1376	611,960,049	9.46897	899	808,201	29.98333	726,572,699	9.65132
850	722,500	29.15476	614,125,000	9.47268	900	810,000	30	729,000,000	9.65489

(Continued)

Table 3. *(Continued)* Squares, Square Roots, Cubes, and Cube Roots.

No.	Square	Square Root	Cube	Cube Root	No.	Square	Square Root	Cube	Cube Root
901	811,801	30.01666	731,432,701	9.65847	951	904,401	30.83829	860,085,351	9.83392
902	813,604	30.03331	733,870,808	9.66204	952	906,304	30.8545	862,801,408	9.83737
903	815,409	30.04996	736,314,327	9.66561	953	908,209	30.8707	865,523,177	9.84081
904	817,216	30.06659	738,763,264	9.66918	954	910,116	30.88689	868,250,664	9.84425
905	819,025	30.08322	741,217,625	9.67274	955	912,025	30.90307	870,983,875	9.84769
906	820,836	30.09983	743,677,416	9.67630	956	913,936	30.91925	873,722,816	9.85113
907	822,649	30.11644	746,142,643	9.67986	957	915,849	30.93542	876,467,493	9.85456
908	824,464	30.13304	748,613,312	9.68342	958	917,764	30.95158	879,217,912	9.85799
909	826,281	30.14963	751,089,429	9.68697	959	919,681	30.96773	881,974,079	9.86142
910	828,100	30.16621	753,571,000	9.69052	960	921,600	30.98387	884,736,000	9.86485
911	829,921	30.18278	756,058,031	9.69407	961	923,521	31	887,503,681	9.86827
912	831,744	30.19934	758,550,528	9.69762	962	925,444	31.01612	890,277,128	9.87169
913	833,569	30.21589	761,048,497	9.70116	963	927,369	31.03224	893,056,347	9.87511
914	835,396	30.23243	763,551,944	9.70470	964	929,296	31.04835	895,841,344	9.87853
915	837,225	30.24897	766,060,875	9.70824	965	931,225	31.06445	898,632,125	9.88195
916	839,056	30.26549	768,575,296	9.71177	966	933,156	31.08054	901,428,696	9.88536
917	840,889	30.28201	771,095,213	9.71531	967	935,089	31.09662	904,231,063	9.88877
918	842,724	30.29851	773,620,632	9.71884	968	937,024	31.1127	907,039,232	9.89217
919	844,561	30.31501	776,151,559	9.72236	969	938,961	31.12876	909,853,209	9.89558
920	846,400	30.3315	778,688,000	9.72589	970	940,900	31.14482	912,673,000	9.89898
921	848,241	30.34798	781,229,961	9.72941	971	942,841	31.16087	915,498,611	9.90238
922	850,084	30.36445	783,777,448	9.73293	972	944,784	31.17691	918,330,048	9.90578
923	851,929	30.38092	786,330,467	9.73645	973	946,729	31.19295	921,167,317	9.90918
924	853,776	30.39737	788,889,024	9.73996	974	948,676	31.20897	924,010,424	9.91257
925	855,625	30.41381	791,453,125	9.74348	975	950,625	31.22499	926,859,375	9.91596
926	857,476	30.43025	794,022,776	9.74699	976	952,576	31.241	929,714,176	9.91935
927	859,329	30.44667	796,597,983	9.75049	977	954,529	31.257	932,574,833	9.92274
928	861,184	30.46309	799,178,752	9.75400	978	956,484	31.27299	935,441,352	9.92612
929	863,041	30.4795	801,765,089	9.75750	979	958,441	31.28898	938,313,739	9.92950
930	864,900	30.4959	804,357,000	9.76100	980	960,400	31.30495	941,192,000	9.93288
931	866,761	30.51229	806,954,491	9.76450	981	962,361	31.32092	944,076,141	9.93626
932	868,624	30.52868	809,557,568	9.76799	982	964,324	31.33688	946,966,168	9.93964
933	870,489	30.54505	812,166,237	9.77148	983	966,289	31.35283	949,862,087	9.94301
934	872,356	30.56141	814,780,504	9.77497	984	968,256	31.36877	952,763,904	9.94638
935	874,225	30.57777	817,400,375	9.77846	985	970,225	31.38471	955,671,625	9.94975
936	876,096	30.59412	820,025,856	9.78195	986	972,196	31.40064	958,585,256	9.95311
937	877,969	30.61046	822,656,953	9.78543	987	974,169	31.41656	961,504,803	9.95648
938	879,844	30.62679	825,293,672	9.78891	988	976,144	31.43247	964,430,272	9.95984
939	881,721	30.64311	827,936,019	9.79239	989	978,121	31.44837	967,361,669	9.96320
940	883,600	30.65942	830,584,000	9.79586	990	980,100	31.46427	970,299,000	9.96655
941	885,481	30.67572	833,237,621	9.79933	991	982,081	31.48015	973,242,271	9.96991
942	887,364	30.69202	835,896,888	9.80280	992	984,064	31.49603	976,191,488	9.97326
943	889,249	30.70831	838,561,807	9.80627	993	986,049	31.5119	979,146,657	9.97661
944	891,136	30.72458	841,232,384	9.80974	994	988,036	31.52777	982,107,784	9.97996
945	893,025	30.74085	843,908,625	9.81320	995	990,025	31.54362	985,074,875	9.98331
946	894,916	30.75711	846,590,536	9.81666	996	992,016	31.55947	988,047,936	9.98665
947	896,809	30.77337	849,278,123	9.82012	997	994,009	31.57531	991,026,973	9.98999
948	898,704	30.78961	851,971,392	9.82357	998	996,004	31.59114	994,011,992	9.99333
949	900,601	30.80584	854,670,349	9.82703	999	998,001	31.60696	997,002,999	9.99667
950	902,500	30.82207	857,375,000	9.83048	1000	1,000,000	31.62278	1,000,000,000	10.00000

Negative Numbers

When dealing with most physical objects, we normally count in positive numbers. Three (or one-hundred) jobbers drills can be easily counted and accounted for, but minus three jobbers drills is a difficult concept to imagine. While it is easy to visualize three jobbers drills being discarded from an inventory of one-hundred, it is difficult to conceive of a total inventory of minus three drills. However, negatives do have a role in daily experience as well as mathematics. Negatives are encountered when the quantity falls below a base of zero. In some measurements, negative numbers are commonplace. In temperatures, for example, negative numbers are frequently encountered, especially on the Celsius scale where any number below the freezing temperature of water is a negative number.

Positive numbers, then, are ordinary numbers greater than 0, while negative numbers are less than 0 and noted by a "-" sign before the number. And, while basic mathematical operations with positive numbers are well understood, the rules are different when dealing with negative numbers.

Absolute value is another way to view a number. Both plus 9 and minus 9 have the same absolute value of 9.

Adding negative numbers

When adding negative numbers to positive numbers, subtract the negative from the positive.

$(-8) + 3 = -5$	or	$16 + (-7) = 9$
$(-8) + 7 + (-6) + 2 = -5$	or	$7 + (-2) + 9 + (-5) = 9$

When adding two or more negative numbers, the sum is equal to the value of the numbers and is a negative value.

$$(-4) + (-8) + (-6) = -18$$

Subtracting negative numbers

When subtracting a negative number from a positive number, or vice versa, add the numerical value of the negative number to the positive number.

$18 - (-7) = 25$	or	$12 - (-18) = 30$
$(-18) - 7 = -25$	or	$(-12) - 18 = -30$

When subtracting a negative number from another negative number, the rule is different. The remainder is the difference of the absolute values of the numbers being subtracted, and will have the same sign (minus) as the numbers being subtracted if the *minuend* (the first, or top, number) has a larger value than the *subtrahend* (the second, or bottom, number). If the minuend has a smaller absolute value than the subtrahend, it will have the opposite sign.

$(-16) - (-5) = -11$ $(-5) - (-16) = 11$

Multiplying negative numbers

When multiplying a positive number by a negative number, multiply the numbers in the usual manner, and the product is always negative.

$5 \times (-6) = -30$ or $(-6) \times 5 = 30$

When multiplying a negative number by a negative number, multiply the numbers in the usual manner, and the product will always be positive.

$(-5) \times (-6) = 30$

Dividing negative numbers

When dividing a positive number by a negative number, or vice versa, divide the numbers in the usual manner, and the quotient is always negative.

$$500 \div (-25) = -20 \qquad \text{or} \qquad (-500) \div 25 = -20$$

When dividing a negative number by a negative number, divide the numbers in the usual manner, and the quotient will always be positive.

$$(-500) \div (-25) = 20$$

Zero

Any number multiplied by zero equals zero. When a number is divided by zero, the quotient is infinity, as shown by the infinity sign (∞). This means that zero will go into the number an incalculable, or infinite, number of times.

$$5281 \times 0 = 0 \qquad\qquad 5281 \div 0 = \infty \qquad\qquad 0.0001 \div 0 = \infty$$

Ratio and Proportion

Ratio

Ratio is a means of expressing the relation of one quantity to another quantity of the same kind. The relation is visualized by dividing the first quantity by the second. These two quantities are called the ratio's terms, and each term has a name. The first term is called the *antecedent*, and the second is the *consequent*. The symbol used to express ratio is two dots (:), a short form of the division sign ($\div$). Some examples of ratios are as follows.

The ratio of 500 to 125 = 500 : 125 = $\frac{500}{125}$ = 4.

A ratio is said to be inverse, or reciprocal, if it is inverted. The inverse of the above ratio is

The ratio of 125 to 500 = 125 : 500 = $\frac{125}{500}$ = $\frac{1}{4}$.

Proportion

Proportion is used to express the equality between ratios. Two symbols are used to indicate proportion, either a repeated ratio symbol (::) or an equal sign (=). For example, the proportion 3:12 :: 6:24 can be read as 3 is to 12 as 6 is to 24. By expressing the ratios as fractions, this can be seen more clearly.

$\frac{3}{12} = \frac{6}{24}$, which is equal to $\frac{1}{4} = \frac{1}{4}$.

In proportions, the *terms* contain the antecedents and consequents of two ratios. The first and last terms in a proportion are known as the *extremes*, and the second and third terms are known as the *means*. Together, all four terms are known as the *proportionals*.

In all proportions, the product of the means equals the product of the extremes. In fact, proportions are highly structured, and, by using the following rules that govern them, it is possible to find a fourth term if only three terms of the proportion are known.

1. The product of the extremes, divided by the third term, is equal to the second term.
2. The product of the extremes, divided by the second term, is equal to the third.
3. The product of the means, divided by the fourth term, is equal to the first term.
4. The product of the means, divided by the first term, is equal to the fourth term.

For proportions expressed as fractions, the following rules can be applied to find missing numerators and denominators.

1. The product of the numerator of the first fraction and the denominator of the

second fraction is equal to the product of the numerator of the second fraction and the denominator of the first.

2. The product of the numerator of the second fraction and denominator of the first fraction, divided by the numerator of the first fraction, is equal to the denominator of the second fraction.

3. The product of the numerator of the first fraction and the denominator of the second fraction, divided by the denominator of the first fraction, is equal to the numerator of the second fraction.

Proportions may be either *direct* or *inverse*. Quantities are in direct proportion when their relation is such that when one becomes greater the other is increased at the same rate, or as one decreases, the other becomes less at the same rate. Quantities are in inverse proportion when their relation is such that as one increases, the other becomes smaller at the same rate, or as one decreases, the other grows larger at the same rate.

Equations

Simple equations

The above section on proportions explains how to find unknown quantities if sufficient information is available. A proportion is essentially an equation, an expression of equality between two quantities. Equations contain *members,* which may be letters or numbers, separated by an equals (=) sign. The quantities on the left side of the equals sign are the *first member,* and those to the right of the sign are the *second member.* In an equation, the first member is equal to the second member.

In the equation $x = b + d$, the value, or quantity, of x is equal to the combined values, or quantities, of b and d. In this case, x could stand for 8, b could stand for 2, and d could stand for 6. In numerical terms, it would be stated as $8 = 2 + 6$. Both the left and right member will remain proportional if multiplied or divided by the same quantity, or if raised to the same power. If, for example, the terms of both members are multiplied by 6, the equation will be changed to $48 = 12 + 36$. If the terms of both members are divided by 2, the equation would be $4 = 1 + 3$.

Terms can also be transferred from one member to another, but when this occurs its sign changes. For example, in the equation $x = b + d - c$, let the values be $x = 12$, $b = 10$, $d = 8$, $c = 6$. If c is transferred from the right member, it will change its sign from negative to positive, which will result in: $x + c = b + d$. Substituting the numerical values: $12 + 6 = 10 + 8$. If both x and c are transferred to opposite members, the equation would be $c = b + b - x$. Or, substituting the numerical values, $6 = 10 + 8 - 12$.

Transferring from one member to another is useful for aligning all unknown terms into one member, or when it is desirable to simplify the form of the equation.

$x + 5 = 20 - 2x$ can be transformed to $x + 2x + 5 = 20$, or $3x + 5 = 20$.
In this instance, $x = 5$.

When fractions are involved in equations, each term in the equation must be multiplied by the *least common multiple* of the denominators. For instance, if the fractions in the equation are $\frac{5}{6}$ and $\frac{3}{7}$, the least common multiple is 42 and all terms in the equation must be multiplied by 42. In the following example, the least common multiple principle is used to solve the equation.

$$3x + \frac{1}{2}x = 30 - \frac{1}{4}x.$$

The least common multiple in this equation is 8, so all terms must first be multiplied by 8.

$24\,x + 4\,x = 240 - 2\,x.$

The minus value is now shifted from the second member to the first member and becomes a positive value.

$24\,x + 4\,x + 2\,x = 240.$

Adding the x values, $30\,x = 240$, so $x = \dfrac{240}{30} = 8$.

When more than one unknown value is contained in a simple equation, it is impossible to definitively solve for the unknowns. In the equation $2\,x + 3\,y = 22$, for example, x could equal 2 and y could equal 6, or x could equal 5 and y could equal 4. However, if two equations containing x and y with the same values can be used for comparison, it is possible to definitively solve for both unknowns. Such equations are called *simultaneous simple equations*, and are illustrated in the following example, equations (a) and (b).

(a) $3\,x + 5\,y = 38$
(b) $6\,x - 3\,y = 24.$

Each equation must first be multiplied by some values that will allow for either the x or y value to be cancelled out when the two equations are added together. In this example, multiplying equation (a) by 3, and equation (b) by 5, will accomplish this goal.

$3\,(3\,x + 5\,y = 38) = 9\,x + 15\,y = 114$
$5\,(6\,x - 3\,y = 24) = 30\,x - 15\,y = 120$
$39\,x = 234$, so $x = \dfrac{234}{39} = 6$.

Once x is solved for, y is easily solved for.

$3\,x + 5\,y = 38 = 18 + 5\,y = 38.$
$5\,y = 38 - 18 = 20.$ Therefore, $y = \dfrac{20}{5} = 4.$

The above method for solving for more than one unknown is called eliminating by addition. Another method, which works by transforming the form of the equation, is known as eliminating by substitution. When using this method, a new equation, based on the original, is substituted for one of the unknowns.

Use above equation (b) to solve for x. By transferring values, $x = \dfrac{24 + 3y}{6}$.

Next, substitute the value for x in equation (a). $\dfrac{3(24 + 3y)}{6} + 5\,y = 38$

$72 + 9\,y + 30\,y = 228$
$39\,y = 156.$ Therefore, $y = \dfrac{156}{39} = 4.$

A third method for solving for more than one unknown is called elimination by comparison. This method calls for solving x in terms of y, or y in terms of x. Using equation (a) above,

$x = \dfrac{38 - 5y}{3}$.

And, from equation (b),

$x = \dfrac{24 + 3y}{6}.$

Using both equations, it is now possible to solve for y.

$\dfrac{38 - 5y}{3} = \dfrac{24 + 3y}{6}$

$228 - 30\,y = 72 + 9\,y$
$228 - 72 = 9\,y + 30\,y$
$156 = 39\,y.$ Therefore, $y = \dfrac{156}{39} = 4.$

Factors, Factoring, and Prime Numbers

Factoring

Any number can be classified as an odd number or an even number. A number that , when divided by 2, leaves a remainder (such as 5, 7, 9, etc.) is an *odd number*. An *even number* can be divided by 2 without a remainder. A number is said to be a *prime number* if it cannot be divided evenly by any numbers other than itself and 1. Except for the prime numbers 2 and 5, the right-hand unit or number for *every* prime number is 1, 3, 7, or 9. A listing of prime numbers up to 10,000 is shown in **Table 4**.

A *factor* of a whole number is a number (or numbers) that is contained an exact number of times in the given whole number. 3 is a factor of 9 because it is contained in 9 an exact number of times ($3 \times 3 = 9$). In this example (the factor of 9), 3 is a *prime factor* because it is a factor that is a prime number.

The factors of 18 are 3, 3, and 2, because $3 \times 3 \times 2 = 18$. In this example, both 3 and 2 are prime factors, and 2 is the smallest prime factor. **Table 5** provides smallest prime factors of numbers up to 9,600. Note that the Table does not contain numbers ending in even numbers (2, 4, 6, 8, 0). This is because the smallest prime factor of these numbers is obviously 2. Use the Table to find all prime factors of a number by following the steps in this example.

Find the prime factors of 3,723. First, go to the row for numbers "3700 - 3799" and find 23. If the number had not appeared in this row, it would have been a prime number and could be found in Table 3 for prime numbers.

The listing for 3723 appears as 23/3. The number following the slash is the smallest prime factor for 3723. Divide 3723 by its smallest prime factor ($3723 \div 3 = 1241$).

Next find the prime factor on the table for 1241. Go to the row "1200 - 1299" and find 41, which is followed by "/17." Thus, 17 is the second prime factor of 3723.

Divide 1241 by 17 ($1241 \div 17 = 73$) to determine that 73 is a third prime factor of 3723.

Look for 73 on the row "1 - 100." It does not appear in the row, because it is a prime number as can be confirmed by consulting Table 3. Therefore, 73 joins 3 and 17 to become the third and final prime factor of 3723.

There are several shortcuts that can be taken in factoring a number.

If the right-hand digit of a number is an even number or zero, then 2 is a factor of that number.

If the sum of the digits of a number is divisible by 3, then 3 is a factor of that number. For example, use the number 771: $7 + 7 + 1 = 15$, and 3 divides into 15 exactly 5 times. Therefore 3 is a factor of 771.

If the two right-hand digits of a number are divisible by 4, then 4 is a factor of that number. For example, use the number 9216: $16 \div 4 = 4$. Therefore, 4 is a factor of 9216.

If the right-hand digit of a number is 5 or 0, 5 is a factor of that number.

If an even number can be evenly divided by 3, then 6 is a factor of that number. For example, use the number 150: $150 \div 3 = 50$; and $150 \div 6 = 25$.

If the three right-hand digits of a number are divisible by 8, then 8 is a factor of that number. For example, use the number 174,128: $128 \div 8 = 16$; and $174128 \div 8 = 21766$.

If the sum of all of the digits of a number is divisible by 9, then 9 is a factor of that number. For example, use the number 3,419,361: $3 + 4 + 1 + 9 + 3 + 6 + 1 = 27$; and $27 \div 9 = 3$. Also, $3419361 \div 9 = 379929$.

Table 4. Prime Numbers.

1 - 99	1, 3, 5, 7, 11, 13, 17, 19, 23, 29, 31, 37, 41, 43, 47, 53, 59, 61, 67, 71, 73, 79, 83, 89, 97
100 - 199	101, 103, 107, 109, 113, 127, 131, 137, 139, 149, 151, 157, 163, 167, 173, 179, 181, 191, 193, 197, 199
200 - 299	211, 223, 227, 229, 233, 239, 241, 251, 257, 263, 269, 271, 277, 281, 283, 293
300 - 399	307, 311, 313, 317, 331, 337, 347, 349, 353, 359, 367, 369, 373, 379, 383, 389, 397
400 - 499	401, 409, 419, 421, 431, 433, 439, 443, 449, 457, 461, 463, 467, 479, 487, 491, 499
500 - 599	503, 509, 521, 523, 541, 547, 557, 563, 569, 571, 577, 587, 593, 599
600 - 699	601, 607, 613, 617, 619, 631, 641, 643, 647, 653, 659, 661, 673, 677, 683, 691
700 - 799	701, 709, 719, 727, 733, 739, 743, 751, 557, 561, 769, 773, 787, 797
800 - 899	809, 811, 821, 823, 827, 829, 839, 853, 857, 859, 863, 877, 881, 883, 887
900 - 999	907, 911, 919, 929, 937, 941, 947, 953, 967, 971, 977, 983, 991, 997
1000 - 1099	1009, 1013, 1019, 1021, 1031, 1033, 1039, 1049, 1051, 1061, 1063, 1069, 1087, 1091, 1093, 1097
1100 - 1199	1103, 1109, 1117, 1123, 1129, 1151, 1153, 1163, 1171, 1181, 1187, 1193
1200 - 1299	1201, 1213, 1217, 1223, 1229, 1231, 1237, 1249, 1259, 1277, 1279, 1283, 1289, 1291, 1297
1300 - 1399	1301, 1303, 1307, 1319, 1321, 1327, 1361, 1367, 1373, 1381, 1399
1400 - 1499	1409, 1423, 1427, 1429, 1433, 1439, 1447, 1451, 1453, 1459, 1471, 1481, 1483, 1487, 1489, 1493, 1499
1500 - 1599	1511, 1523, 1531, 1543, 1549, 1553, 1559, 1567, 1571, 1579, 1583, 1597
1600 - 1699	1601, 1607, 1609, 1613, 1619, 1621, 1627, 1637, 1657, 1663, 1667, 1669, 1693, 1697, 169
1700 -1799	1709, 1721, 1723, 1733, 1741, 1747, 1753, 1759, 1777, 1783, 1787, 1789
1800 - 1899	1801, 1811, 1823, 1831, 1847, 1861, 1867, 1871, 1873, 1877, 1879, 1889
1900 - 1999	1901, 1907, 1913, 1931, 1933, 1949, 1951, 1973, 1979, 1987, 1993, 1997, 1999
2000 - 2099	2003, 2011, 2017, 2027, 2029, 2039, 2053, 2063, 2069, 2081, 2083, 2087, 2089, 209
2100 - 2199	2111, 2113, 2129, 2131, 2137, 2141, 2143, 2153, 2161, 2179
2200 - 2299	2203, 2207, 2213, 2221, 2237, 2239, 2243, 2251, 2267, 2269, 2273, 2281, 2287, 2293, 2297
2300 - 2399	2309, 2311, 2333, 2339, 2341, 2347, 2351, 2357, 2371, 2377, 2381, 2383, 2389, 2393, 2399
2400 - 2499	2411, 2417, 2423, 2437, 2441, 2447, 2459, 2467, 2473, 2477
2500 - 2599	2503, 2521, 2531, 2539, 2543, 2549, 2551, 2557, 2579, 2591, 2593
2600 - 2699	2609, 2617, 2621, 2633, 2647, 2657, 2659, 2663, 2671, 2677, 2683, 2687, 2689, 2693, 2699
2700 - 2799	2707, 2711, 2713, 2719, 2729, 2731, 2741, 2749, 2753, 2767, 2777, 2789, 2891, 2797
2800 - 2899	2801, 2803, 2819, 2833, 2837, 2843, 2851, 2857, 2861, 2879, 2887, 2897
2900 - 2999	2903, 2909, 2917, 2927, 2939, 2953, 2957, 2963, 2969, 2971, 2999
3000 - 3099	3001, 3011, 3019, 3023, 3037, 3041, 3049, 3061, 3067, 3079, 3083, 3089
3100 - 3199	3109, 3119, 3121, 3137, 3163, 3167, 3169, 3181, 3187, 3191
3200 - 3299	3203, 3209, 3217, 3221, 3229, 3251, 3253, 3257, 3259, 3271, 3299
3300 - 3399	3301, 3307, 3313, 3319, 3323, 3329, 3331, 3343, 3347, 3359, 3361, 3371, 3373, 3389, 3391
3400 - 3499	3407, 3413, 3433, 3449, 3457, 3461, 3463, 3467, 3469, 3491, 3499
3500 - 3599	3511, 3517, 3527, 3529, 3533, 3539, 3541, 3547, 3557, 3559, 3571, 3581, 3583, 3593
3600 - 3699	3607, 3613, 3617, 3623, 3631, 3637, 3643, 3659, 3671, 3673, 3677, 3691, 3697
3700 - 3799	3701, 3709, 3719, 3727, 3733, 3739, 3761, 3767, 3769, 3779, 3793, 3797

(Continued)

Table 4. *(Continued)* **Prime Numbers.**

3800 - 3899	3803, 3821, 3823, 3833, 3847, 3851, 3853, 3863, 3877, 3881, 3889
3900 - 3999	3907, 3911, 3917, 3919, 3923, 3929, 3931, 3943, 3947, 3967, 3989
4000 - 4099	4001, 4003, 4007, 4013, 4019, 4021, 4027, 4049, 4051, 4057, 4073, 4079, 4091, 4093 4099
4100 - 4199	4111, 4127, 4129, 4133, 4139, 4153, 4157, 4159, 4177
4200 - 4299	4201, 4211, 4217, 4219, 4229, 4231, 4241, 4243, 4253, 4259, 4261, 4271, 4273, 4283, 4289, 4297
4300 - 4399	4327, 4337, 4339, 4349, 4357, 4363, 4373, 4391, 4397
4400 - 4499	4409, 4421, 4423, 4441, 4447, 4451, 4457, 4463, 4481, 4483, 4493
4500 - 4599	4507, 4513, 4517, 4519, 4523, 4547, 4549, 4561, 4567, 4583, 4591, 4597
4600 - 4699	4603, 4621, 4637, 4639, 4643, 4649, 4651, 4657, 4663, 4673, 4679, 4691
4700 - 4799	4703, 4721, 4723, 4729, 4733, 4751, 4759, 4783, 4787, 4789, 4793, 4799
4800 - 4899	4801, 4813, 4817, 4831, 4861, 4871, 4877, 4889
4900 - 4999	4903, 4909, 4919, 4931, 4933, 4937, 4943, 4951, 4957, 4967, 4969, 4973, 4987, 4993, 4999
5000 - 5099	5003, 5009, 5011, 5021, 5023, 5039, 5051, 5059, 5077, 5081, 5087, 5099
5100 - 5199	5101, 5107, 5113, 5119, 5147, 5153, 5167, 5171, 5179, 5189, 5197
5200 - 5299	5209, 5227, 5231, 5233, 5237, 5261, 5273, 5279, 5281, 5297
5300 - 5399	5303, 5309, 5323, 5333, 5347, 5351, 5381, 5393, 5399
5400 - 5499	5407, 5413, 5417, 5419, 5431, 5437, 5441, 5443, 5449, 5471, 5477, 5479, 5483
5500 - 5599	5501, 5503, 5507, 5519, 5521, 5527, 5531, 5557, 5563, 5569, 5573, 5581, 5591
5600 - 5699	5623, 5639, 5641, 5647, 5651, 5653, 5657, 5659, 5669, 5683, 5689, 5693
5700 - 5799	5701, 5711, 5717, 5737, 5741, 5743, 5749, 5779, 5783, 5791
5800 - 5899	5801, 5807, 5813, 5821, 5827, 5839, 5843, 5849, 5851, 5857, 5861, 5867, 5869, 5879, 5881, 5897
5900 - 5999	5903, 5923, 5927, 5939, 5953, 5981, 5987
6000 - 6099	6007, 6011, 6029, 6037, 6043, 6047, 6053, 6067, 6073, 6079, 6089, 6091
6100 - 6199	6101, 6113, 6121, 6131, 6133, 6143, 6151, 6163, 6173, 6197, 6199
6200 - 6299	6203, 6211, 6217, 6221, 6229, 6247, 6257, 6263, 6269, 6271, 6277, 6287, 6299
6300 - 6399	6301, 6311, 6323, 6329, 6337, 6343, 6353, 6359, 6361, 6367, 6373, 6379, 6389, 6397
6400 -6499	6421, 6427, 6449, 6451, 6469, 6473, 6481, 6491
6500 - 6599	6521, 6529, 6547, 6551, 6553, 6563, 6569, 6571, 6577, 6581, 6599
6600 -6699	6607, 6619, 6637, 6653, 6659, 6661, 6673, 6679, 6689, 6691
6700 - 6799	6701, 6703, 6709, 6719, 6733, 6737, 6761, 6763, 6779, 6781, 6791, 6793
6800 - 6899	6803, 6823, 6827, 6829, 6833, 6841, 6857, 6863, 6869, 6871, 6883, 6899
6900 - 6999	6907, 6911, 6917, 6947, 6949, 6959, 6961, 6967, 6971, 6977, 6983, 6991, 6997
7000 - 7099	7001, 7013, 7019, 7027, 7039, 7043, 7057, 7069, 7079
7100 - 7199	7103, 7109, 7121, 7127, 7129, 7151, 7159, 7177, 7187, 7193
7200 - 7299	7207, 7211, 7213, 7219, 7229, 7237, 7243, 7247, 7253, 7283, 7297
7300 - 7399	7307, 7309, 7321, 7331, 7333, 7349, 7351, 7369, 7393
7400 - 7499	7411, 7417, 7433, 7451, 7457, 7459, 7477, 7481, 7487, 7489, 7499
7500 - 7599	7507, 7517, 7523, 7529, 7537, 7541, 7547, 7549, 7559, 7561, 7573, 7577, 7583, 7589, 7591

(Continued)

Table 4. *(Continued)* **Prime Numbers.**

7600 - 7699	7603, 7607, 7621, 7639, 7643, 7649, 7669, 7673, 7681, 7687, 7691, 7699
7700 - 7799	7703, 7717, 7723, 7727, 7741, 7753, 7757, 7759, 7789, 7793
7800 - 7899	7817, 7823, 7829, 7841, 7853, 7867, 7873, 7877, 7879, 7883
7900 - 7999	7901, 7907, 7919, 7927, 7933, 7937, 7949, 7951, 7963, 7993
8000 - 8099	8009, 8011, 8017, 8039, 8053, 8059, 8069, 8081, 8087, 8089, 8093
8100 - 8199	8101, 8111, 8117, 8123, 8147, 8161, 8167, 8171, 8179, 8191
8200 - 8299	8209, 8219, 8221, 8231, 8233, 8237, 8243, 8263, 8269, 8273, 8287, 8291, 8293, 8297
8300 - 8399	8311, 8317, 8329, 8353, 8363, 8369, 8377, 8387, 8389
8400 - 8499	8419, 8423, 8429, 8431, 8443, 8447, 8461, 8467
8500 - 8599	8501, 8513, 8521, 8527, 8537, 8539, 8543, 8563, 8573, 8581, 8597, 8599
8600 - 8699	8609, 8623, 8627, 8629, 8641, 8647, 8663, 8669, 8677, 8681, 8689, 8693, 8699
8700 - 8799	8707, 8713, 8719, 8731, 8737, 8741, 8747, 8753, 8761, 8779, 8783
8800 - 8899	8803, 8807, 8819, 8821, 8831, 8837, 8839, 8849, 8861, 8863, 8867, 8887, 8893
8900 - 8999	8923, 8929, 8933, 8941, 8951, 8963, 8969, 8971, 8999
9000 - 9099	9001, 9007, 9011, 9013, 9029, 9041, 9043, 9049, 9059, 9067, 9091
9100 - 9199	9103, 9109, 9127, 9133, 9137, 9151, 9157, 9161, 9173, 9181, 9187, 9199
9200 - 9299	9203, 9209, 9221, 9227, 9239, 9241, 9257, 9277, 9281, 9283, 9293
9300 - 9399	9311, 9319, 9323, 9337, 9341, 9343, 9349, 9371, 9377, 9391, 9397
9400 - 9499	9403, 9413, 9419, 9421, 9431, 9433, 9437, 9439, 9461, 9463, 9467, 9473, 9479, 9491, 9497
9500 - 9599	9511, 9521, 9533, 9539, 9547, 9551, 9587
9600 - 9699	9601, 9613, 9619, 9623, 9629, 9631, 9643, 9649, 9661, 9677, 9679, 9689, 9697
9700 - 9799	9719, 9721, 9733, 9739, 9743, 9749, 9767, 9769, 9781, 9787, 9791
9800 - 9899	9803, 9811, 9817, 9829, 9833, 9839, 9851, 9857, 9859, 9871, 9883, 9887
9900 - 10,000	9901, 9907, 9923, 9929, 9931, 9941, 9949, 9967, 9973

Table 5. Smallest Prime Factors of Numbers. *(See Notes.)*

1 - 99	9/3, 15/3, 21/3, 25/5, 27/3, 33/3, 35/5, 39/3, 45/3, 49/7, 51/3, 55/5, 57/3, 63/3, 65/5, 69/3, 75/3, 77/7, 81/3, 85/5, 87/3, 91/7, 93/3, 95/5, 99/3
100 - 199	5/3, 11/2, 15/5, 17/3, 19/7, 21/11, 23/3, 25/5, 29/3, 33/7, 35/3, 41/3, 43/11, 45/5, 47/3, 53/3, 55/5, 59/3, 61/7, 65/3, 69/13, 71/3, 75/5, 77/3, 83/3, 85/5, 87/11, 89/3, 95/3
200 - 299	01/3, 03/7, 05/5, 07/3, 09/11, 13/3, 15/5, 17/7, 19/3, 21/13, 25/3, 31/3, 35/5, 37/3, 43/3, 45/5, 47/13, 49/3, 53/11, 55/3, 59/7, 261/3, 65/5, 67/3, 73/3, 75/5, 79/3, 85/3, 87/7, 89/17, 91/3, 95/5, 97/3, 99/13
300 - 399	1/7, 3/3, 5/5, 9/3, 15/3, 19/11, 21/3, 23/17, 25/5, 27/3, 29/7, 33/3, 35/5, 39/3, 41/11, 43/7, 45/3, 51/3, 55/5, 57/3, 61/19, 63/3, 65/5, 69/3, 71/7, 75/3, 77/13, 81/3, 85/5, 87/3, 91/17, 93/3, 95/5, 99/3
400 - 499	3/13, 5/3, 7/11, 11/3, 13/7 15/5, 17/3, 23/3, 25/5, 27/7, 29/3, 35/3, 37/19, 41/3, 45/5, 47/3, 51/11, 53/3, 55/5, 59/3, 65/3, 69/7, 71/3, 73/11, 75/5, 77/3, 81/13, 83/3, 85/5, 89/3, 93/17, 95/3, 97/7
500 - 599	1/3, 5/5, 7/3, 11/7, 13/3, 15/5, 17/11, 19/3, 25/3, 27/17, 29/23, 31/3, 33/13, 35/5, 37/3, 39/7, 43/3, 45/5, 49/3, 51/19, 53/7, 55/3, 59/13, 61/3, 65/5, 67/3, 73/3, 75/5, 79/3, 81/7, 83/11, 85/3, 89/19, 91/3, 95/5, 97/3
600 - 699	3/3, 5/5,9/3, 11/3, 15/3, 21/3, 23/7, 25/5, 27/3, 29/17, 33/3, 35/5, 37/7, 39/3, 45/5, 49/11, 51/3, 55/5, 57/3, 63/3, 65/5, 67/23, 69/3, 71/11, 75/3, 79/7, 81/3, 85/5, 87/3, 89/13, 93/3, 95/5, 97/17, 99/3

(Continued)

Table 5. *(Continued)* **Smallest Prime Factors of Numbers.** *(See Notes.)*

700 - 799	3/19, 5/3, 7/7, 11/3, 13/23, 15/5, 17/3, 21/7, 23/3, 25/5, 29/3, 31/17, 35/3, 37/11, 41/3, 45/5, 47/3, 49/7, 53/3, 55/5, 59/3, 63/7, 65/3, 67/13, 71/3, 75/5, 77/3, 79/19, 81/11, 83/3, 85/5, 89/3, 91/7, 93/13, 95/3, 99/17
800 - 899	1/3, 5/11, 5/5, 7/3, 13/3, 15/5, 17/19, 19/3, 25/3, 31/3, 33/7, 35/5, 37/3, 41/29, 43/3, 45/5, 47/7, 49/3, 51/23, 55/3, 61/3, 65/5, 67/3, 69/11, 71/13, 73/3, 75/5, 79/3, 85/3, 89/7, 91/3, 93/19, 95/5, 97/3, 99/29
900 - 999	1/17, 3/3, 5/5, 9/3, 13/11, 15/3, 17/7, 21/3, 23/13, 25/5, 27/3, 31/7, 33/3, 35/5, 39/3, 43/23, 45/3, 49/13, 51/3, 55/5, 57/3, 59/7, 61/31, 63/3, 65/5, 69/3, 73/7, 75/3, 79/11, 81/3, 85/5, 87/3, 89/23, 93/3, 95/5, 99/3
1000 - 1099	1/7, 3/17, 5/3, 7/19, 11/3, 15/5, 17/3, 23/3, 25/5, 27/13, 29/3, 35/3, 37/17, 41/3, 43/7, 45/5, 47/3, 53/3, 55/5, 57/7, 59/3, 65/3, 67/11, 71/3, 73/29, 75/5, 77/3, 79/13, 81/23, 83/3, 85/5, 89/3, 95/3, 99/7
1100 - 1199	1/3, 5/5, 7/3, 11/11, 13/3, 15/5, 19/3, 21/19, 25/3, 27/7, 31/3, 33/11, 35/5, 37/3, 39/17, 41/7, 43/3, 45/5, 47/31, 49/3, 55/3, 57/13, 59/19, 61/3, 65/5, 67/3, 69/7, 73/3, 75/5, 77/11, 79/3, 83/7, 85/3, 89/29, 91/3, 95/5, 97/3, 99/11
1200 - 1299	3/3, 5/5, 7/17, 9/3, 11/7, 15/3, 19/23, 21/3, 25/5, 27/3, 33/3, 35/5, 39/3, 41/17, 43/11, 45/3, 47/29, 51/3, 53/7, 55/5, 57/3, 61/13, 63/3, 65/5, 67/7, 69/3, 71/31, 73/19, 75/3, 81/3, 85/5, 87/3, 93/3, 95/5, 99/3
1300 - 1399	5/3, 9/7, 11/3, 13/13, 15/5, 17/3, 23/3, 25/5, 29/3, 31/11, 33/31, 35/3, 37/7, 39/13, 41/3, 43/17, 45/5, 47/3, 49/19, 51/7, 53/3, 55/5, 57/23, 59/3, 63/29, 65/3, 69/37, 71/3, 75/5, 77/3, 79/7, 83/3, 85/5, 87/19, 89/3, 91/13, 93/7, 95/3, 97/11
1400 - 1499	1/3, 3/23, 5/5, 7/3, 11/17, 13/3, 15/5, 17/13, 19/3, 21/7, 25/3, 31/3, 35/5, 37/3, 41/11, 43/3, 45/5, 49/3, 55/3, 57/31, 61/3, 63/7, 65/5, 67/3, 69/13, 73/3, 75/5, 77/7, 79/3, 85/3, 91/3, 95/5, 97/3
1500 - 1599	1/19, 3/3, 5/5, 7/11, 9/3, 13/17, 15/3, 17/37, 19/7, 21/3, 25/5, 27/3, 29/11, 33/3, 35/5, 37/29, 39/3, 41/23, 45/3, 47/7, 51/3, 55/5, 57/3, 61/7, 63/3, 65/5, 69/3, 73/11, 75/3, 77/19, 81/3, 85/5, 87/3, 89/7, 91/37, 93/3, 95/5, 99/3
1600 - 1699	3/7, 5/3, 11/3, 15/5, 17/3, 23/3, 25/5, 29/3, 31/7, 33/23, 35/3, 39/11, 41/3, 43/31, 45/5, 47/3, 49/17, 51/13, 53/3, 55/5, 59/3, 61/11, 65/3, 71/3, 73/7, 75/5, 77/3, 79/23, 81/41, 83/3, 85/5, 87/7, 89/3, 91/19, 95/3
1700 -1799	1/3, 3/13, 5/5, 7/3, 11/29, 13/3, 15/5, 17/17, 19/3, 25/3, 27/11, 29/7, 31/3, 35/5, 37/3, 39/37, 43/3, 45/5, 49/3, 51/17, 53/3, 57/7, 61/3, 63/41, 65/5, 67/3, 69/29, 71/7, 73/3, 75/5, 79/3, 81/13, 85/3, 91/3, 93/11, 95/5, 97/3, 99/7
1800 - 1899	3/3, 5/5, 7/13, 9/3, 13/7, 15/3, 17/23, 19/17, 21/3, 25/5, 27/3, 29/31, 33/3, 35/5, 37/11, 39/3, 41/7, 43/19, 45/3, 49/43, 51/3, 53/17, 55/5, 57/3, 59/11, 63/3, 65/5, 69/3, 75/3, 81/3, 83/7, 85/5, 87/3, 91/31, 93/3, 95/5, 97/7, 99/3
1900 - 1999	3/11, 5/3, 9/23, 11/3, 15/5, 17/3, 19/19, 21/17, 23/3, 25/5, 27/41, 29/3, 35/3, 37/13, 39/7, 41/3, 43/29, 45/5, 47/3, 53/3, 55/5, 57/19, 59/3, 61/37, 63/13, 65/3, 67/7, 69/11, 71/3, 75/5, 77/3, 81/7, 83/3, 85/5, 89/3, 91/11, 95/3
2000 - 2099	1/3, 5/5, 7/3, 9/7, 13/3, 15/5, 19/3, 21/43, 23/7, 25/3, 31/3, 33/19, 35/5, 37/3, 41/13, 43/3, 45/5, 47/23, 49/3, 51/7, 55/3, 57/11, 59/29, 61/3, 65/5, 67/3, 71/19, 73/3, 75/5, 77/31, 79/3, 85/3, 91/3, 93/7, 95/5, 97/3
2100 - 2199	1/11, 3/3, 5/5, 7/7, 9/3, 15/3, 17/29, 19/13, 21/3, 23/11, 25/5, 27/3, 33/3, 35/5, 39/3, 45/3, 47/19, 49/7, 51/3, 55/5, 57/3, 59/17, 63/3, 65/5, 67/11, 69/3, 71/13, 73/41, 75/3, 77/7, 81/3, 83/37, 85/5, 87/3, 89/11, 91/7, 93/3, 95/5, 97/13, 99/3
2200 - 2299	1/31, 5/3, 9/47, 11/3, 15/5, 17/3, 19/7, 23/3, 25/5, 27/17, 29/3, 31/23, 33/7, 35/3, 41/3, 45/5, 47/3, 49/13, 53/3, 55/5, 57/37, 59/3, 61/7, 63/31, 65/3, 71/3, 75/5, 77/3, 79/43, 83/3, 85/5, 89/3, 91/29, 95/3, 99/11
2300 - 2399	1/3, 3/7, 5/5, 7/3, 13/3, 15/5, 17/7, 19/3, 21/11, 23/23, 25/3, 27/13, 29/17, 31/3, 35/5, 37/3, 43/3, 45/5, 49/3, 53/13, 55/3, 59/7, 61/3, 63/17, 65/5, 67/3, 69/23, 73/3, 75/5, 79/3, 85/3, 87/7, 91/3, 95/5, 97/3

(Continued)

Table 5. *(Continued)* **Smallest Prime Factors of Numbers.** *(See Notes.)*

2400 - 2499	1/7, 3/3, 5/5, 7/29, 9/3, 13/19, 15/3, 19/41, 21/3, 25/5, 27/3, 29/7, 31/11, 33/3, 35/5, 39/3, 43/7, 45/3, 49/31, 51/3, 53/11, 55/5, 57/3, 61/23, 63/3, 65/5, 69/3, 71/7, 75/3, 79/37, 81/3, 83/13, 85/5, 87/3, 89/19, 91/47, 93/3, 95/5, 97/11, 99/3
2500 - 2599	1/41, 5/3, 7/23, 9/13, 11/3, 13/7, 15/5, 17/3, 19/11, 23/3, 25/5, 27/7, 29/3, 33/17, 35/3, 37/43, 41/3, 45/5, 47/3, 53/3, 55/5, 59/3, 61/13, 63/11, 65/3, 67/17, 69/7, 71/3, 73/31, 75/5, 77/3, 81/29, 83/3, 85/5, 87/13, 89/3, 95/3, 97/7, 99/23
2600 - 2699	1/3, 3/19, 5/5, 7/3, 11/7, 13/3, 15/5, 19/3, 23/43, 25/3, 27/37, 29/11, 31/3, 35/5, 37/3, 39/7, 41/19, 43/3, 45/5, 49/3, 51/11, 53/7, 55/3, 61/3, 65/5, 67/3, 69/17, 73/3, 75/5, 79/3, 81/7, 85/3, 91/3, 95/5, 97/3
2700 - 2799	1/37, 3/3, 5/5, 9/3, 15/3, 17/11, 21/3, 23/7, 25/5, 27/3, 33/3, 35/5, 37/7, 39/3, 43/13, 45/3, 47/41, 51/3, 55/5, 57/3, 59/31, 61/11, 63/3, 65/5, 69/3, 71/17, 73/47, 75/3, 79/7, 81/3, 83/11, 85/5, 87/3, 93/3, 95/5, 99/3
2800 - 2899	5/3, 7/7, 9/53, 11/3, 13/29, 15/5, 17/3, 21/7, 23/3, 25/5, 27/11, 29/3, 31/19, 35/3, 39/17, 41/3, 45/5, 47/3, 49/7, 53/3, 55/5, 59/3, 63/7, 65/3, 67/47, 69/19, 71/3, 73/13, 75/5, 77/3, 81/43, 83/3, 85/5, 89/3, 91/7, 93/11, 95/3, 99/13
2900 - 2999	1/3, 5/5, 7/3, 11/41, 13/3, 15/5, 19/3, 21/23, 23/37, 25/3, 29/29, 31/3, 33/7, 35/5, 37/3, 41/17, 43/3, 45/5, 47/7, 49/3, 51/13, 55/3, 59/11, 61/3, 65/5, 67/3, 73/3, 75/5, 77/13, 79/3, 81/11, 83/19, 85/3, 87/29, 89/7, 91/3, 93/41, 95/5, 97/3
3000 - 3099	3/3, 5/5, 7/31, 9/3, 13/23, 15/3, 17/7, 21/3, 25/5, 27/3, 29/13, 31/7, 33/3, 35/5, 39/3, 43/17, 45/3, 47/11, 51/3, 53/43, 55/5, 57/3, 59/7, 63/3, 65/5, 69/3, 71/37, 73/7, 75/3, 77/17, 81/3, 85/5, 87/3, 91/11, 93/3, 95/5, 97/19, 99/3
3100 - 3199	1/7, 3/29, 5/3, 7/13, 11/3, 13/11, 15/5, 17/3, 23/3, 25/5, 27/53, 29/3, 31/31, 33/13, 35/3, 39/43, 41/3, 43/7, 45/5, 47/3, 49/47, 51/23, 53/3, 55/5, 57/7, 59/3, 61/29, 65/3, 71/3, 73/19, 75/5, 77/3, 79/11, 83/3, 85/5, 89/3, 93/31, 95/3, 97/23, 99/7
3200 - 3299	1/3, 5/5, 7/3, 11/13, 13/3, 15/5, 19/3, 23/11, 25/3, 27/7, 31/3, 33/53, 35/5, 37/3, 39/41, 41/7, 43/3, 45/5, 47/17, 49/3, 55/3, 61/3, 63/13, 65/5, 67/3, 69/7, 73/3, 75/5, 77/29, 79/3, 81/17, 83/7, 85/3, 87/19, 89/11, 91/3, 93/37, 95/5, 97/3
3300 - 3399	3/3, 5/5, 9/3, 11/7, 15/3, 17/31, 21/3, 25/5, 27/3, 33/3, 35/5, 37/47, 39/3, 41/13, 45/3, 49/17, 51/3, 53/7, 55/5, 57/3, 63/3, 65/5, 67/7, 69/3, 75/3, 77/11, 79/31, 81/3, 83/17, 85/5, 87/3, 93/3, 95/5, 97/43, 99/3
3400 - 3499	1/19, 3/41, 5/3, 9/7, 11/3, 15/5, 17/3, 19/13, 21/11, 23/3, 25/5, 27/23, 29/3, 31/47, 35/3, 37/7, 39/19, 41/3, 43/11, 45/5, 47/3, 51/7, 53/3, 55/5, 59/3, 65/3, 71/3, 73/23, 75/5, 77/3, 79/7, 81/59, 83/3, 85/5, 87/11, 89/3, 93/7, 95/3, 97/13
3500 - 3599	1/3, 3/31, 5/5, 7/3, 9/11, 13/3, 15/5, 19/3, 21/7, 23/13, 25/3, 31/3, 35/5, 37/3, 43/3, 45/5, 49/3, 51/53, 53/11, 55/3, 61/3, 63/7, 65/5, 67/3, 69/43, 73/3, 75/5, 77/7, 79/3, 85/3, 87/17, 89/37, 91/3, 95/5, 97/3, 99/59
3600 - 3699	1/13, 3/3, 5/5, 9/3, 11/23, 15/3, 19/7, 21/3, 25/5, 27/3, 29/19, 33/3, 35/5, 39/3, 41/11, 45/3, 47/7, 49/41, 51/3, 53/13, 55/5, 57/3, 61/7, 63/3, 65/5, 67/19, 69/3, 75/3, 79/13, 81/3, 83/29, 85/5, 87/3, 89/7, 93/3, 95/5, 99/3
3700 - 3799	3/7, 5/3, 7/11, 11/3, 13/47, 15/5, 17/3, 21/61, 23/3, 25/5, 29/3, 31/7, 35/3, 37/37, 41/3, 43/19, 45/5, 47/3, 49/23, 51/11, 53/3, 55/5, 57/13, 59/3, 63/53, 65/3, 71/3, 73/7, 75/5, 77/3, 81/19, 83/3, 85/5, 87/7, 89/3, 91/17, 95/3, 99/29
3800 - 3899	1/3, 5/5, 7/3, 9/13, 11/37, 13/3, 15/5, 17/11, 19/3, 25/3, 27/43, 29/7, 31/3, 35/5, 37/3, 39/11, 41/23, 43/3, 45/5, 49/3, 55/3, 57/7, 59/17, 61/3, 65/5, 67/3, 69/53, 71/7, 73/3, 75/5, 79/3, 83/11, 85/3, 87/13, 91/3, 93/17, 95/5, 97/3, 99/7
3900 - 3999	1/47, 3/3, 5/5, 9/3, 13/7, 15/3, 21/3, 25/5, 27/3, 33/3, 35/5, 37/31, 39/3, 41/7, 45/3, 49/11, 51/3, 53/59, 55/5, 57/3, 59/37, 61/17, 63/3, 65/5, 69/3, 71/11, 73/29, 75/3, 77/41, 79/23, 81/3, 83/7, 85/5, 87/3, 91/13, 93/3, 95/5, 97/7, 99/3
4000 - 4099	5/3, 9/19, 11/3, 15/5, 17/3, 23/3, 25/5, 29/3, 31/29, 33/37, 35/3, 37/11, 39/7, 41/3, 43/13, 45/5, 47/3, 53/3, 55/5, 59/3, 61/31, 63/17, 65/3, 67/7, 69/13, 71/3, 75/5, 77/3, 81/7, 83/3, 85/5, 87/61, 89/3, 95/3, 97/17

(Continued)

Table 5. *(Continued)* **Smallest Prime Factors of Numbers.** *(See Notes.)*

Range	Factors
4100 - 4199	1/3, 3/11, 5/5, 7/3, 9/7, 13/3, 15/5, 17/23, 19/3, 21/13, 23/7, 25/3, 31/3, 35/5, 37/3, 41/41, 43/3, 45/5, 47/11, 49/3, 51/7, 53/3, 61/3, 63/23, 65/5, 67/3, 69/11, 71/43, 73/3, 75/5, 79/3, 81/37, 83/47, 85/3, 87/53, 89/59, 91/3, 93/7, 95/5, 97/3, 99/13
4200 - 4299	3/3, 5/5, 7/7, 9/3, 13/11, 15/3, 21/3, 23/41, 25/5, 27/3, 33/3, 35/5, 37/19, 39/3, 45/3, 47/31, 49/7, 51/3, 55/5, 57/3, 63/3, 65/5, 67/17, 69/3, 75/3, 77/7, 79/11, 81/3, 85/5, 87/3, 91/7, 93/3, 95/5, 99/3
4300 - 4399	1/11, 3/13, 5/3, 7/59, 9/31, 11/3, 13/19, 15/5, 17/3, 19/7, 21/29, 23/3, 25/5, 29/3, 31/61, 33/7, 35/3, 41/3, 43/43, 45/5, 47/3, 51/19, 53/3, 55/5, 59/3, 61/7, 65/3, 67/11, 69/17, 71/3, 75/5, 77/3, 79/29, 81/13, 83/3, 85/5, 87/41, 89/3, 93/23, 95/3, 99/53
4400 - 4499	1/3, 3/7, 5/5, 7/3, 11/11, 13/3, 15/5, 17/7, 19/3, 25/3, 27/19, 29/43, 31/3, 33/11, 35/5, 37/3, 39/23, 43/3, 45/5, 49/3, 53/61, 55/3, 59/7, 61/3, 65/5, 67/3, 69/41, 71/17, 73/3, 75/5, 77/11, 79/3, 85/3, 87/7, 89/67, 91/3, 95/5, 97/3, 99/11
4500 - 4599	1/7, 3/3, 5/5, 9/3, 11/13, 15/3, 21/3, 25/5, 27/3, 29/7, 31/23, 33/3, 35/5, 37/13, 39/3, 41/19, 43/7, 45/3, 51/3, 53/29, 55/5, 57/3, 59/47, 63/3, 65/5, 69/3, 71/7, 73/17, 75/3, 77/23, 79/19, 81/3, 85/5, 87/3, 89/13, 93/3, 95/5, 99/3
4600 - 4699	1/43, 5/3, 7/17, 9/11, 11/3, 13/7, 15/5, 17/3, 19/31, 23/3, 25/5, 27/7, 29/3, 31/11, 33/41, 35/3, 41/3, 45/5, 47/3, 53/3, 55/5, 59/3, 61/59, 65/3, 67/13, 69/7, 71/3, 75/5, 77/3, 81/31, 83/3, 85/5, 87/43, 89/3, 93/13, 95/3, 97/7, 99/37
4700 - 4799	1/3, 5/5, 7/3, 9/17, 11/7, 13/3, 15/5, 17/53, 19/3, 25/3, 27/29, 31/3, 35/5, 37/3, 39/7, 41/11, 43/3, 45/5, 47/47, 49/3, 53/7, 55/3, 57/67, 61/3, 63/11, 65/5, 67/3, 69/19, 71/13, 73/3, 75/5, 77/17, 79/3, 81/7, 85/3, 91/3, 95/5, 97/3
4800 - 4899	3/3, 5/5, 7/11, 9/3, 11/17, 15/3, 19/61, 21/3, 23/7, 25/5, 27/3, 29/11, 33/3, 35/5, 37/7, 39/3, 41/47, 43/29, 45/3, 47/37, 49/13, 51/3, 53/23, 55/5, 57/3, 59/43, 63/3, 65/5, 67/31, 69/3, 73/11, 75/3, 79/7, 81/3, 83/19, 85/5, 87/3, 91/67, 93/3, 95/5, 97/59, 99/3
4900 - 4999	1/13, 5/3, 7/7, 11/3, 13/17, 15/5, 17/3, 21/7, 23/3, 25/5, 27/13, 29/3, 35/3, 39/11, 41/3, 45/5, 47/3, 49/7, 53/3, 55/5, 59/3, 61/11, 63/7, 65/3, 71/3, 75/5, 77/3, 79/13, 81/17, 83/3, 85/5, 89/3, 91/7, 95/3, 97/19
5000 - 5099	1/3, 5/5, 7/3, 13/3, 15/5, 17/29, 19/3, 25/3, 27/11, 29/47, 31/3, 33/7, 35/5, 37/3, 41/71, 43/3, 45/5, 47/7, 49/3, 53/31, 55/3, 57/13, 61/3, 63/61, 65/5, 67/3, 69/37, 71/11, 73/3, 75/5, 79/3, 83/13, 85/3, 89/7, 91/3, 93/11, 95/5, 97/3
5100 - 5199	3/3, 5/5, 9/3, 11/19, 15/3, 17/7, 21/3, 23/47, 25/5, 27/3, 29/23, 31/7, 33/3, 35/5, 37/11, 39/3, 41/53, 43/37, 45/3, 49/19, 51/3, 55/5, 57/3, 59/7, 61/13, 63/3, 65/5, 69/3, 73/7, 75/3, 77/31, 81/3, 83/71, 85/5, 87/3, 91/29, 93/3, 95/5, 99/3
5200 - 5299	1/7, 3/11, 5/3, 7/41, 11/3, 13/13, 15/5, 17/3, 19/17, 21/23, 23/3, 25/5, 29/3, 35/3, 39/13, 41/3, 43/7, 45/5, 47/3, 49/29, 51/59, 53/3, 55/5, 57/7, 59/3, 63/19, 65/3, 67/23, 69/11, 71/3, 75/5, 77/3, 83/3, 85/5, 87/17, 89/3, 91/11, 93/67, 95/3, 99/7
5300 - 5399	1/3, 5/5, 7/3, 11/47, 13/3, 15/5, 17/13, 19/3, 21/7, 25/3, 27/7, 29/73, 31/3, 35/5, 37/3, 39/19, 41/7, 43/3, 45/5, 49/3, 53/53, 55/3, 57/11, 59/23, 61/3, 63/31, 65/5, 67/3, 69/7, 71/41, 73/3, 75/5, 77/19, 79/3, 83/7, 85/3, 89/17, 91/3, 95/5, 97/3
5400 - 5499	1/11, 3/3, 5/5, 9/3, 11/7, 15/3, 21/3, 23/11, 25/5, 27/3, 29/61, 33/3, 35/5, 39/3, 45/3, 47/13, 51/3, 53/7, 55/5, 57/3, 59/53, 61/43, 63/3, 65/5, 67/7, 69/3, 73/13, 75/3, 81/3, 85/5, 87/3, 89/11, 91/17, 93/3, 95/5, 97/23, 99/3
5500 - 5599	5/3, 9/7, 11/3, 13/37, 15/5, 17/3, 23/3, 25/5, 29/3, 33/11, 35/3, 37/7, 39/29, 41/3, 43/23, 45/5, 47/3, 49/31, 51/7, 53/3, 55/5, 59/3, 61/67, 65/3, 67/19, 71/3, 75/5, 77/3, 79/7, 83/3, 85/5, 87/37, 89/3, 93/7, 95/3, 97/29, 99/11
5600 - 5699	1/3, 3/13, 5/5, 7/3, 9/71, 11/31, 13/3, 15/5, 17/41, 19/3, 21/7, 25/3, 27/17, 29/13, 31/3, 33/43, 35/5, 37/3, 43/3, 45/5, 49/3, 55/3, 61/3, 63/7, 65/5, 67/3, 71/53, 73/3, 75/5, 77/7, 79/3, 81/13, 85/3, 87/11, 91/3, 95/5, 97/3, 99/41
5700 - 5799	3/3, 5/5, 7/13, 9/3, 13/29, 15/3, 19/7, 21/3, 23/59, 25/5, 27/3, 29/17, 31/11, 33/3, 35/5, 39/3, 45/3, 47/7, 51/3, 53/11, 55/5, 57/3, 59/13, 61/7, 63/3, 65/5, 67/73, 69/3, 71/29, 73/23, 75/3, 77/53, 81/3, 85/5, 87/3, 89/7, 93/3, 95/5, 97/11, 99/3

(Continued)

Table 5. *(Continued)* **Smallest Prime Factors of Numbers.** *(See Notes.)*

5800 - 5899	3/7, 5/3, 9/37, 11/3, 15/5, 17/3, 19/11, 23/3, 25/5, 29/3, 31/7, 33/19, 35/3, 37/13, 41/5, 45/5, 47/3, 53/3, 55/5, 59/3, 63/11, 65/3, 71/3, 73/7, 75/5, 77/3, 83/3, 85/5, 87/7, 89/3, 91/43, 93/71, 95/3, 99/17
5900 - 5999	1/3, 5/5, 7/3, 9/19, 11/23, 13/3, 15/5, 17/61, 19/3, 21/31, 25/3, 29/7, 31/3, 33/17, 35/5, 37/3, 41/13, 43/3, 45/5, 47/19, 49/3, 51/11, 55/3, 57/7, 59/59, 61/3, 63/67, 65/5, 67/3, 69/47, 71/7, 73/3, 75/5, 77/43, 79/3, 83/31, 85/3, 89/53, 91/3, 93/13, 95/5, 97/3, 99/7
6000 - 6099	1/17, 3/3, 5/5, 9/3, 13/7, 15/3, 17/11, 19/13, 21/3, 23/19, 25/5, 27/3, 31/37, 33/3, 35/5, 39/3, 41/7, 45/3, 49/23, 51/3, 55/5, 57/3, 59/73, 61/11, 63/3, 65/5, 69/3, 71/13, 75/3, 77/59, 81/3, 83/7, 85/5, 87/3, 93/3, 95/5, 97/7, 99/3
6100 - 6199	3/17, 5/3, 7/31, 9/41, 11/3, 15/5, 17/3, 19/29, 23/3, 25/5, 27/11, 29/3, 35/3, 37/17, 39/7, 41/3, 45/5, 47/3, 49/11, 53/3, 55/5, 57/47, 59/3, 61/61, 65/3, 67/7, 69/31, 71/3, 75/5, 77/3, 79/37, 81/7, 83/3, 85/5, 87/23, 89/3, 91/41, 93/11, 95/3
6200 - 6299	1/3, 5/5, 7/3, 9/7, 13/3, 15/5, 19/3, 23/7, 25/3, 27/13, 31/3, 33/23, 35/5, 37/3, 39/17, 41/79, 43/3, 45/5, 49/3, 51/7, 53/13, 55/3, 59/11, 61/3, 65/5, 67/3, 73/3, 75/5, 79/3, 81/11, 83/61, 85/3, 89/19, 91/3, 93/7, 95/5, 97/3
6300 - 6399	3/3, 5/5, 7/7, 9/3, 13/59, 15/3, 19/71, 21/3, 25/5, 27/3, 31/13, 33/3, 35/5, 39/3, 41/17, 45/3, 47/11, 49/7, 51/3, 55/5, 57/3, 63/3, 65/5, 69/3, 71/23, 75/3, 77/7, 81/3, 83/13, 85/5, 87/3, 91/7, 93/3, 95/5, 99/3
6400 -6499	1/37, 3/19, 5/3, 7/43, 9/13, 11/3, 13/11, 15/5, 17/3, 19/7, 23/3, 25/5, 29/3, 31/59, 33/7, 35/3, 37/41, 39/47, 41/3, 43/17, 45/5, 47/3, 53/3, 55/5, 57/11, 59/3, 61/7, 63/23, 65/3, 67/29, 71/3, 75/5, 77/3, 79/11, 83/3, 85/5, 87/13, 89/3, 93/43, 95/3, 97/73, 99/67
6500 - 6599	1/3, 3/7, 5/5, 7/3, 9/23, 11/17, 13/3, 15/5, 17/7, 19/3, 23/11, 25/3, 27/61, 31/3, 33/47, 35/5, 37/3, 39/13, 41/31, 43/3, 45/5, 49/3, 55/3, 57/79, 59/7, 61/3, 65/5, 67/3, 73/3, 75/5, 79/3, 83/29, 85/3, 87/7, 89/11, 91/3, 93/19, 95/5, 97/3
6600 -6699	1/7, 3/3, 5/5, 9/3, 11/11, 13/17, 15/3, 17/3, 21/3, 23/37, 25/5, 27/3, 29/7, 31/19, 33/3, 35/5, 39/3, 41/29, 43/7, 45/3, 47/17, 49/61, 51/3, 55/5, 57/3, 63/3, 65/5, 67/59, 69/3, 71/7, 75/3, 77/11, 81/3, 83/41, 85/5, 87/3, 93/3, 95/5, 97/37, 99/3
6700 - 6799	5/3, 7/19, 11/3, 13/7, 15/5, 17/3, 21/11, 23/3, 25/5, 27/7, 29/3, 31/53, 35/3, 39/23, 41/3, 43/11, 45/5, 47/3, 49/17, 51/43, 53/3, 55/5, 57/29, 59/3, 65/3, 67/67, 69/7, 71/3, 73/13, 75/5, 77/3, 83/3, 85/5, 87/11, 89/3, 95/3, 97/7, 99/13
6800 - 6899	1/3, 5/5, 7/3, 9/11, 11/7, 13/3, 15/5, 17/17, 19/3, 21/19, 25/3, 31/3, 35/5, 37/3, 39/7, 43/3, 45/5, 47/41, 49/3, 51/13, 53/7, 55/3, 59/19, 61/3, 65/5, 67/3, 73/3, 75/5, 77/13, 79/3, 81/7, 85/3, 87/71, 89/83, 91/3, 93/61, 95/5, 97/3
6900 - 6999	1/67, 3/3, 5/5, 9/3, 13/31, 15/3, 19/11, 21/3, 23/7, 25/5, 27/3, 29/13, 31/29, 33/3, 35/5, 37/7, 39/3, 41/11, 43/53, 45/3, 51/3, 53/17, 55/5, 57/3, 63/3, 65/5, 69/3, 73/19, 75/3, 79/7, 81/3, 85/5, 87/3, 89/29, 93/3, 95/5, 99/3
7000 - 7099	3/47, 5/3, 7/7, 9/43, 11/3, 15/5, 17/3, 21/7, 23/3, 25/5, 29/3, 31/79, 33/13, 35/3, 37/31, 41/3, 45/5, 47/3, 49/7, 51/11, 53/3, 55/5, 59/3, 61/23, 63/7, 65/3, 67/37, 71/3, 73/11, 75/5, 77/3, 81/73, 83/3, 85/5, 87/19, 89/3, 91/7, 93/41, 95/3, 97/47, 99/31
7100 - 7199	1/3, 5/5, 7/3, 11/13, 13/3, 15/5, 17/11, 19/3, 23/17, 25/3, 31/3, 33/7, 35/5, 37/3, 39/11, 41/37, 43/3, 45/5, 47/7, 49/3, 53/23, 55/3, 57/17, 61/3, 63/13, 65/5, 67/3, 69/67, 71/71, 73/3, 75/5, 79/3, 81/43, 83/11, 85/3, 89/7, 91/3, 95/5, 97/3, 99/23
7200 - 7299	1/19, 3/3, 5/5, 9/3, 15/3, 17/7, 21/3, 23/31, 25/5, 27/3, 31/7, 33/3, 35/5, 39/3, 41/13, 45/3, 49/11, 51/3, 55/5, 57/3, 59/7, 61/53, 63/3, 65/5, 67/13, 69/3, 71/11, 73/7, 75/3, 77/19, 79/29, 81/3, 85/5, 87/3, 89/37, 91/23, 93/3, 95/5, 99/3
7300 - 7399	1/7, 3/67, 5/3, 11/3, 13/71, 15/5, 17/3, 19/13, 23/3, 25/5, 27/17, 29/3, 35/3, 37/11, 39/41, 41/3, 43/7, 45/5, 47/3, 53/3, 55/5, 57/7, 59/3, 61/17, 63/37, 65/3, 67/53, 71/3, 73/73, 75/5, 77/3, 79/47, 81/11, 83/3, 85/5, 87/83, 89/3, 91/19, 95/3, 97/13, 99/7
7400 - 7499	1/3, 3/11, 5/5, 7/3, 9/31, 13/3, 15/5, 19/3, 21/41, 23/13, 25/3, 27/7, 29/17, 31/3, 35/5, 37/3, 39/43, 41/7, 43/3, 45/5, 47/11, 49/3, 53/29, 55/3, 61/3, 63/17, 65/5, 67/3, 69/7, 71/31, 73/3, 75/5, 79/3, 83/7, 85/3, 91/3, 93/59, 95/5, 97/3

(Continued)

Table 5. *(Continued)* **Smallest Prime Factors of Numbers.** *(See Notes.)*

7500 - 7599	1/13, 3/3, 5/5, 9/3, 11/7, 13/11, 15/3, 19/73, 21/3, 25/5, 27/3, 31/17, 33/3, 35/5, 39/3, 43/19, 45/3, 51/3, 53/7, 55/5, 57/3, 63/3, 65/5, 67/7, 69/3, 71/67, 75/3, 79/11, 81/3, 85/5, 87/3, 93/3, 95/5, 97/71, 99/3
7600 - 7699	1/11, 5/3, 9/7, 11/3, 13/23, 15/5, 17/3, 19/19, 23/3, 25/5, 27/29, 29/3, 31/13, 33/17, 35/3, 37/7, 41/3, 45/5, 47/3, 51/7, 53/3, 55/5, 57/13, 59/3, 61/47, 63/79, 65/3, 67/11, 71/3, 75/5, 77/3, 79/7, 83/3, 85/5, 89/3, 93/7, 95/3, 97/43
7700 - 7799	1/3, 5/5, 7/3, 9/13, 11/11, 13/3, 15/5, 19/3, 21/7, 25/3, 29/59, 31/3, 33/11, 35/5, 37/3, 39/71, 43/3, 45/5, 47/61, 49/3, 51/23, 55/3, 61/3, 63/7, 65/5, 67/3, 69/17, 71/19, 73/3, 75/5, 77/7, 79/3, 81/31, 83/43, 85/3, 87/13, 91/3, 95/5, 97/3, 99/11
7800 - 7899	1/29, 3/3, 5/5, 7/37, 9/3, 11/73, 13/13, 15/3, 19/7, 21/3, 25/5, 27/3, 31/41, 33/3, 35/5, 37/17, 39/3, 43/11, 45/3, 47/7, 49/47, 51/3, 55/5, 57/3, 59/29, 61/7, 63/3, 65/5, 69/3, 71/17, 75/3, 81/3, 85/5, 87/3, 89/7, 91/13, 93/3, 95/5, 97/53, 99/3
7900 - 7999	3/7, 5/3, 9/11, 11/3, 13/41, 15/5, 17/3, 21/89, 23/3, 25/5, 29/3, 31/7, 35/3, 39/17, 41/3, 43/13, 45/5, 47/3, 53/3, 55/5, 57/73, 59/3, 61/19, 65/3, 67/31, 69/13, 71/3, 73/7, 75/5, 77/3, 79/79, 81/23, 83/3, 85/5, 87/7, 89/3, 91/61, 95/3, 97/11, 99/1
8000 - 8099	1/3, 3/53, 5/5, 7/3, 13/3, 15/5, 19/3, 21/13, 23/71, 25/3, 27/23, 29/7, 31/3, 33/29, 35/5, 37/3, 41/11, 43/3, 45/5, 47/13, 49/3, 51/83, 55/3, 57/7, 61/3, 63/11, 65/5, 67/3, 71/7, 73/3, 75/5, 77/41, 79/3, 83/59, 85/3, 91/3, 95/5, 97/3, 99/7
8100 - 8199	3/3, 5/5, 7/11, 9/3, 13/7, 15/3, 19/23, 21/3, 25/5, 27/3, 29/11, 31/47, 33/3, 35/5, 37/79, 39/3, 41/7, 43/17, 45/3, 49/29, 51/3, 53/31, 55/5, 57/3, 59/41, 63/3, 65/5, 69/3, 73/11, 75/3, 77/13, 81/3, 83/7, 85/5, 87/3, 89/19, 93/3, 95/5, 97/7, 99/3
8200 - 8299	1/59, 3/13, 5/3, 7/29, 11/3, 13/43, 15/5, 17/3, 23/3, 25/5, 27/19, 29/3, 35/3, 39/7, 41/3, 45/5, 47/3, 49/73, 51/37, 53/3, 55/5, 57/23, 59/3, 61/11, 65/3, 67/7, 71/3, 75/5, 77/3, 79/17, 81/7, 83/3, 85/5, 89/3, 95/3, 99/43
8300 - 8399	1/3, 3/19, 5/5, 7/3, 9/7, 13/3, 15/5, 19/3, 21/53, 23/7, 25/3, 27/11, 31/3, 33/13, 35/5, 37/3, 39/31, 41/19, 43/3, 45/5, 47/17, 49/3, 51/7, 55/3, 57/61, 59/13, 61/3, 65/5, 67/3, 71/11, 73/3, 75/5, 79/3, 81/17, 83/83, 85/3, 91/3, 93/7, 95/5, 97/3, 99/37
8400 - 8499	1/31, 3/3, 5/5, 7/7, 9/3, 11/13, 13/47, 15/3, 17/19, 21/3, 25/5, 27/3, 33/3, 35/5, 37/11, 39/3, 41/23, 45/3, 49/7, 51/3, 53/79, 55/5, 57/3, 59/11, 63/3, 65/5, 69/3, 71/43, 73/37, 75/3, 77/7, 79/61, 81/3, 83/17, 85/5, 87/3, 89/13, 91/7, 93/3, 95/5, 97/29, 99/3
8500 - 8599	3/11, 5/3, 7/47, 9/67, 11/3, 15/5, 17/3, 19/7, 23/3, 25/5, 29/3, 31/19, 33/7, 35/3, 41/3, 45/5, 47/3, 49/83, 51/17, 53/3, 55/5, 57/43, 59/3, 61/7, 65/3, 67/13, 69/11, 71/3, 75/5, 77/3, 79/23, 83/3, 85/3, 89/3, 91/11, 93/13, 95/3
8600 - 8699	1/3, 3/7, 5/5, 7/3, 11/79, 13/3, 15/5, 17/7, 19/3, 21/37, 25/3, 31/3, 33/89, 35/5, 37/3, 39/53, 43/3, 45/5, 49/3, 51/41, 53/17, 55/3, 57/11, 59/7, 61/3, 65/5, 67/3, 71/13, 73/3, 75/5, 79/3, 83/19, 85/3, 87/7, 91/3, 95/5, 97/3
8700 - 8799	1/7, 3/3, 5/5, 9/3, 11/31, 15/3, 17/23, 21/3, 23/11, 25/5, 27/3, 29/7, 33/3, 35/5, 39/3, 43/7, 45/3, 49/13, 51/3, 55/5, 57/3, 59/19, 63/3, 65/5, 67/11, 69/3, 71/7, 73/31, 75/3, 77/67, 81/3, 85/5, 87/3, 89/11, 91/59, 93/3, 95/5, 97/19, 99/3
8800 - 8899	1/13, 5/3, 9/23, 11/3, 13/7, 15/5, 17/3, 23/3, 25/5, 27/7, 29/3, 33/11, 35/3, 41/3, 43/37, 45/5, 47/3, 51/53, 53/3, 55/5, 57/17, 59/3, 65/3, 69/7, 71/3, 73/19, 75/5, 77/3, 79/13, 81/83, 83/3, 85/5, 89/3, 91/17, 95/3, 97/7, 99/11
8900 - 8999	1/3, 3/29, 5/5, 7/3, 9/59, 11/7, 13/3, 15/5, 17/37, 19/3, 21/11, 25/3, 27/79, 31/3, 35/5, 37/3, 39/7, 43/3, 45/5, 47/23, 49/3, 53/7, 55/3, 57/13, 59/17, 61/3, 65/5, 67/3, 73/3, 75/5, 77/47, 79/3, 81/7, 83/13, 85/3, 87/11, 89/89, 91/3, 93/17, 95/5, 97/3
9000 - 9099	3/3, 5/5, 9/3, 15/3, 17/71, 19/29, 21/3, 23/7, 25/5, 27/3, 31/11, 33/3, 35/5, 37/7, 39/3, 45/3, 47/83, 51/3, 53/11, 55/5, 57/3, 61/13, 63/3, 65/5, 69/3, 71/47, 73/43, 75/3, 77/29, 79/7, 81/3, 83/31, 85/5, 87/3, 89/61, 93/3, 95/5, 97/11, 99/3
9100 - 9199	1/19, 5/3, 7/7, 11/3, 13/13, 15/5, 17/3, 19/11, 21/7, 23/3, 25/5, 29/3, 31/23, 35/3, 39/13, 41/3, 43/41, 45/5, 47/3, 49/7, 53/3, 55/5, 59/3, 63/7, 65/3, 67/89, 69/53, 71/3, 75/5, 77/3, 79/67, 83/3, 85/5, 89/3, 91/7, 93/29, 95/3, 97/17

(Continued)

Table 5. *(Continued)* **Smallest Prime Factors of Numbers.** *(See Notes.)*

9200 - 9299	1/3, 5/5, 7/3, 11/61, 13/3, 15/5, 17/13, 19/3, 23/23, 25/3, 29/11, 31/3, 33/7, 35/5, 37/3, 43/3, 45/5, 47/7, 49/3, 51/11, 53/19, 55/3, 59/47, 61/3, 63/59, 65/5, 67/3, 69/13, 71/73, 73/3, 75/5, 79/3, 85/3, 87/37, 89/7, 91/3, 95/5, 97/3, 99/17
9300 - 9399	1/71, 3/3, 5/5, 7/41, 9/3, 13/67, 15/3, 17/7, 21/3, 25/5, 27/3, 29/19, 31/7, 33/3, 35/5, 39/3, 45/3, 47/13, 51/3, 53/47, 55/5, 57/3, 59/7, 61/11, 63/3, 65/5, 67/17, 69/3, 73/7, 75/3, 79/83, 81/3, 83/11, 85/5, 87/3, 89/41, 93/3, 95/5, 99/3
9400 - 9499	1/7, 5/3, 7/23, 9/97, 11/3, 15/5, 17/3, 23/3, 25/5, 27/11, 29/3, 35/3, 41/3, 43/7, 45/5, 47/3, 49/11, 51/13, 53/3, 55/7, 57/3, 59/3, 65/3, 69/17, 71/3, 75/5, 77/3, 81/19, 83/3, 85/7, 87/53, 89/3, 93/11, 95/3, 99/7
9500 - 9599	1/3, 3/13, 5/5, 7/3, 9/37, 13/3, 15/5, 17/31, 19/3, 23/89, 25/3, 27/7, 29/13, 31/3, 35/5, 37/3, 41/7, 43/3, 45/5, 49/3, 53/41, 55/3, 57/19, 59/11, 61/3, 63/73, 65/5, 67/3, 69/7, 71/17, 73/3, 75/5, 77/61, 79/3, 81/11, 83/7, 85/3, 89/43, 91/3, 93/53, 95/5, 97/3, 99/29

Notes: To look up a number on the Table, first locate the number in the number ranges found in the left hand column. For example, the number 9,123 can be found in the range 9,100 - 9,199. In the right column, thousandths and hundredths have not been included for the numbers, so to look up 9,123, look for "23" in the row next to 9,100 - 9,199. The notation for 23 is "23/3," which means that the smallest prime factor of 9,123 is 3. **Numbers ending in even numbers (2, 4, 6, 8, 0), such as 6,758, are not included in the Table because the smallest prime factor of any even number is 2.**

Common Logarithms

Logarithms are a shortcut method devised to save time when multiplying or dividing large numbers, extracting square, cube, and other roots, and obtaining powers of numbers. The invention of logarithms has been attributed to a baron of Scotland by the name of John Napier. The tables of logarithms that are normally called "common" or "Briggs" logarithms were compiled between 1615 and 1628 by an English professor of geometry named Henry Briggs. The compilation of these tables is recognized as one of the great feats of mathematics, and before the invention of the electronic calculator and computer, these tables, like the slide rule, afforded great savings in time and effort in solving mathematical calculations. Today, even though not used as widely as in the past, logarithms can still provide valuable assistance in shortening calculations.

Every logarithm contains two parts, separated by a decimal point. The part to the left side of the decimal is called the *characteristic*, and the part on the right side is the *mantissa*. The mantissa is also called the decimal part of the logarithm. In a typical logarithm such as 23.3504, the two numbers to the left of the decimal (23) are the characteristic, and the numbers to the right of the decimal (3504) are the mantissa.

The characteristic

The characteristic may be either positive or negative, or zero, and it is assigned a value based on the number of figures it contains. The value is the number of figures it contains, minus 1. For example, the characteristic of the logarithm 1.428 is 0, the characteristic of 224.4922 is 2, and of 456249.764 is 5. When all figures of a logarithm are to the right of the decimal, the characteristic is negative. If the first figure to the right of the decimal is any number other than a cipher (0), its characteristic is - 1 (minus 1), regardless of how many numbers it contains. When the first figure to the right of the decimal is a cipher, the characteristic is - 2, and it increases in increments of 1 for each additional cipher to the right of the decimal point. For example, the characteristic of .5243 is - 1, the characteristic of .0524379 is - 2, and of .00000243 is - 6. The minus sign is sometimes written directly above the characteristic, as in 3, because the minus sign applies only to the characteristic. The mantissa is always positive. The complete rule for obtaining the characteristic is:

When the number is 1, or greater than 1, the characteristic is positive and is one less than the number of figures to the left of the decimal point. When the number is less than 1, the characteristic is negative and is one more than the number of ciphers between the decimal point and the first significant figure.

The mantissa

The mantissa is always a decimal number, it is always positive, and it is independent of the decimal point. The logarithm tables in **Table 6** provide the mantissa of the logarithms of numbers from 1 to 10,000. Decimal points in a number are ignored when finding the mantissa for a number. For example, the mantissa for the following numbers is the same: 5214, 521.4, 52.14, 5.214, .5214, 0.5214.

The first line of the logarithms Table is reproduced below. The first column contains numbers whose mantissae are given in the columns to its right. Above the mantissae are additional tenths of the number indicated in the first column. The mantissa for 100.3 is given in the three-tenths column, and is equal to .001301. The mantissa for 100.4 is .001734, etc.

No.	0	1	2	3	4	5	6	7	8	9
100	000000	000434	000868	001301	001734	002166	002598	003029	003461	003891

The characteristic for 100 is 2, which is added to the mantissa to complete the logarithm. Therefore, the log of 100.3 = 2.001301, and the log of 100.4 = 2.001734.

From the following example from the Tables, find the logarithm of 5028.

No.	0	1	2	3	4	5	6	7	8	9
500	698970	699057	699144	699231	699317	699404	699491	699578	699664	699751
501	699838	699924	700011	700098	700184	700271	700358	700444	700531	700617
502	700704	700790	700877	700963	701050	701136	701222	701309	701395	701482
503	701568	701654	701741	701827	701913	701999	702086	702172	702258	702344
504	702431	702517	702603	702689	702775	702861	702947	703033	703119	703205
505	703291	703377	703463	703549	703635	703721	703807	703893	703979	704065

First, find 502 in the first column, and then move to the "8" column to the right. The mantissa in the 502 row, under the "8" column is .701395. The characteristic of 5028 is three, so the log is 3.701395. In this manner, logarithms of numbers larger than those provided in the Number Column can be found. As stated earlier, the mantissa for the numbers 5214, 521.4, 52.14, 5.214, .5214, 0.5214 is exactly the same in each instance, but the characteristic of each is different. Therefore, to look up the mantissa for numbers below 100 (the smallest number on the chart) such as 8 or 80, find the mantissa for 800, which will be the same, but the characteristic of 8 will be zero, and the characteristic of 80 will be 1.

From the above extract from the Table, it is also possible to find a number that corresponds to a given logarithm. For example, find the number corresponding to the logarithm 1.702603. At this point, ignore the characteristic and look up the mantissa, which is .702603. First, find the mantissa section beginning with the prefix "70." By examining the table, it can be seen that 70 prefix mantissae range from the number 5012 to 5128. The mantissa .702603 is in the column for the number 5042. Since the characteristic for the logarithm is 1, the number corresponding to log 1.702603 is 50.42.

Using the above example, but changing the characteristic, illustrates how the characteristic is equal to powers of ten. The following series of logs and their corresponding numbers make this evident.

Logarithm	=	Number	Power
- 3.702603	=	.005042	10^{-3} Thousandths
- 2.702603	=	.05042	10^{-2} Hundredths
- 1.702603	=	.5042	10^{-1} Tenths
.702603	=	5.042	10^0 Units
+ 1.702603	=	50.42	10^1 Tens
+ 2.702603	=	504.2	10^2 Hundreds
+ 3.702603	=	5042	10^3 Thousands
+ 4.702603	=	50420	10^4 Ten-Thousands
+ 5.702603	=	504200	10^5 Hundred-Thousands
+ 6.702603	=	5042000	10^6 Millions

Multiplying with logarithms

To multiply two or more numbers using logarithms, first find the logs of each number, then add the numbers together. For example, multiply 3,424 × 6.92 × 168. First, find the log of each number, then add the three together.

```
log 3,424      = 3.534534
log      6.92   = 0.840106
log    168      = 2.225309
                 6.599949
```

Now find the number that corresponds to the mantissa 599949, which will provide the first four figures of the answer. The following extract from the tables shows that the mantissa falls between two existing mantissae for even numbers, 3980 and 3981. Therefore, the answer falls between these two numbers.

No.	0	1	2	3	4	5	6	7	8	9
398	599883	599992	600101	600210	600319	600428	600537	600646	600755	600864

Next, calculate the difference between the two mantissae.

599992 - 599883 = 109.

Then find the difference between the mantissa found in the answer and the mantissa for 398.

599949 - 599883 = 66.

Divide the larger of these two figures into the smaller.

66 ÷ 109 = 605.

Annex the above quotient to 3980, which will give the figure 3,980,605, which is the answer. (Actually, the exact answer when multiplying 3424 × 6.92 × 168 is 3,980,605.44, but logarithms do not provide decimal point accuracy on large numbers.)

When multiplying with logarithms with negative characteristics, the mantissae are first added, then the total of positive characteristics is added to the sum, and the total of negative characteristics is subtracted from the sum. To begin, list the logarithms of the numbers being multiplied and add their mantissae together.

```
log   37       =   1.568202
log      .46    = - 1.662758
log      .053   = - 2.724276
log   940       =   2.973128
log     7.9     =   0.897627
```

The sum of the mantissae is: .568202 + .662758 + .724276 + .973128 + .897627 = 3.825991.

Next, add the positive characteristics, and subtract the negative characteristics: 1 + 2 + 0 = 3, and (-1) + (-2) = - 3. Subtract the negative from the positive: 3 - 3 = 0. Therefore, as shown below, the characteristic in the answer is 3.

+ 3 characteristics carried over from adding the mantissae
+ 3 positive characteristics in the logs
- 3 negative characteristics in the log
+ 3 positive mantissae.

Dividing with logarithms

To divide one number by another with logarithms, first find the logarithms of each number and subtract the logarithm of the divisor from the logarithm of the dividend. The remainder is the logarithm of the quotient. For example, find the quotient of 102,500 ÷ 625.

$$\begin{array}{rl} \log 102500 &= 5.010724 \\ - \log \quad 625 &= \underline{2.795880} \\ & 2.214844 \end{array}$$

The quotient is the exact log of 164, which is the answer: 102,500 ÷ 625 = 164.

Extracting roots with logarithms

Using logarithms greatly simplifies extracting the root of a number. First, find the logarithm of the number, then divide the logarithm by the index of the root of the number. The quotient will be the logarithm of the answer. For example, find the square root of 961.

log 961 = 2.982723 ÷ 2 = 1.4913615, which is the exact logarithm of 31: $\sqrt[2]{961}$ = 31.

Finding the cube of the same number requires dividing the logarithm of 961 by 3.

log 961 = 2.982723 ÷ 3 = 0.994241. As can be seen from the extract from the table below, the mantissa 994241 lies between the mantissae for 9868 and 9869.

No.	0	1	2	3	4	5	6	7	8	9
986	993877	993921	993965	994009	994053	994097	994141	994185	994229	994273

As explained earlier in the section on multiplying with logarithms, first calculate the difference between the two mantissae.

994273 - 994229 = 44.

Then find the difference between the mantissa found in the answer and the mantissa for 9868.

994241 - 994229 = 12.

Divide the larger of these two figures into the smaller.

12 ÷ 44 = 272.

Annex the above quotient to 9.868, which will give the figure 9.868272, which is the answer.

When extracting the root of numbers with logarithms that have negative characteristics, it is necessary to change the negative characteristic to a positive characteristic that can be evenly divided by the index of the root. Begin by adding 10 to the characteristic—the number added will be subtracted from the logarithm form of the answer. For example, find the cube root of 0.000729.

log 0.000729 = - 4.862728. To change the characteristic to positive, add 10. This is shown as

6.862728 - 10.

However, 10 is not divisible by 3, which is the index of the root, so 2 will be added, and the total added to the characteristic is now 12. The characteristic is now written as

8.862728 - 12.

Divide the logarithm and the negative characteristic by the index of the root. Subtract the characteristics to find the logarithm.

8.862728 ÷ 3 = 2.954243   - 12 ÷ 3 = - 4   2.954243 - (- 4) = - 2.954243.

-2.954243 is the exact log of 0.09, which is the answer.

Obtaining the powers of numbers with logarithms
A number can be raised to any power with logarithms by multiplying the logarithm of the number by the exponent of the number. The product is the logarithm of the value of the power. The process is the reverse of finding the root.

Find the value of 4.28^3. Find the logarithm of 4.28 from the Tables, and multiply it by the exponent, which is 3.

log 4.28 =0.631444 × 3 = 1.894332.
1.894332 is the log of 78.4, which is the answer.

Finding the powers of numbers having logarithms with negative characteristics is more complicated. As discussed above, in the section on extracting roots, the negative characteristics must first be changed to positive characteristics by adding ten or another number to the characteristic. For example, find the value of 0.0785^3.

log 0.0785 = - 2.894870. To change the characteristic to positive, add 10. This is shown as

8.894870 - 10.

Multiply the logarithm and the negative characteristic by the index of the root. Subtract the characteristics to find the logarithm.

8.894870 × 3 = 26.68461.   - 10 × 3 = - 30.   26.68461 - (- 30) = - 4.68461.

- 4.68461 is the logarithm of 0.0004837, which is the answer.

Finding logarithms for numbers higher than those on the Table
The above sections on multiplying and extracting roots with logarithms gave examples of how to derive logarithms for numbers not included in the Table: i.e., any number in excess of 9,999. The process for finding the logarithm of a number with five or more figures is called *interpolation*. As an example, use the following extract from the Table to find the logarithm of the number 14,125.

No.	0	1	2	3	4	5	6	7	8	9
140	146128	146438	146748	147058	147367	147676	147985	148294	148603	148911
141	149219	149527	149835	150142	150449	150756	151063	151370	151676	151982
142	152288	152594	152900	153205	153510	153815	154120	154424	154728	155032

First, find the log for the first four figures of the number.

log 1412 = 3.149835.

Next calculate the difference between the two logarithm for 1412, and the next highest number, 1413.

3.150142 - 3.149835 = 0.000307.

Expressed in tenths, each tenth between 1412 and 1413 on the chart has a value of

0.000307 ÷ 10 = 0.0000307.

14125 is five "tenths" more than 1412, so the value of five times one tenth is added to the value of 1214:

5 × .0000307 = 0.0001535, which is rounded off to 0.000154.

log 14125 = log 1412 + 0.000154 + 1 characteristic.

log 14125 = 3.149835 + 0.000154 + 1 characteristic.

log 14125 = 4.149989.

Note that the characteristic changes with the additional place to the left of the decimal.

To find the logarithm for larger numbers, divisions between the mantissae on the chart are broken down further. Find the logarithm for 141257.

log 14125 = 4.149989 (from above example).

The difference in tenths, also from the above example, has been found to be 0.0000307. The difference in hundredths, therefore, will be 0.00000307, or 0.000003 if rounded to six places. 141257 is 7 "hundredths" larger than 14125.

7 × 0.000003 = 0.000021, which is added to the log of 14125.

log 141257 = log 14125 + 0.000021 + 1 characteristic.

log 141257 = 4.149989 + 0.000021 + 1 characteristic.

log 141257 = 5.150010.

Natural logarithms

Natural logarithms, also known as Napierian (after their creator, John Napier) or hyperbolic logarithms, are used in some advanced areas of mathematics. Natural logarithms use a base of 2.7182818284 +, which is the base of a particular mathematical series. Common, or Briggs, logarithms, of course, use the base number of 10. Natural logarithms are not normally encountered in shop mathematics, and are mentioned here only to clarify the difference between the two types of logarithms.

Table 6. Logarithms of Numbers 100 to 150.9 — Mantissae 000000 to 178689.

No.	0	1	2	3	4	5	6	7	8	9
100	000000	000434	000868	001301	001734	002166	002598	003029	003461	003891
101	004321	004751	005181	005609	006038	006466	006894	007321	007748	008174
102	008600	009026	009451	009876	010300	010724	011147	011570	011993	012415
103	012837	013259	013680	014100	014521	014940	015360	015779	016197	016616
104	017033	017451	017868	018284	018700	019116	019532	019947	020361	020775
105	021189	021603	022016	022428	022841	023252	023664	024075	024486	024896
106	025306	025715	026125	026533	026942	027350	027757	028164	028571	028978
107	029384	029789	030195	030600	031004	031408	031812	032216	032619	033021
108	033424	033826	034227	034628	035029	035430	035830	036230	036629	037028
109	037426	037825	038223	038620	039017	039414	039811	040207	040602	040998
110	041393	041787	042182	042576	042969	043362	043755	044148	044540	044932
111	045323	045714	046105	046495	046885	047275	047664	048053	048442	048830
112	049218	049606	049993	050380	050766	051153	051538	051924	052309	052694
113	053078	053463	053846	054230	054613	054996	055378	055760	056142	056524
114	056905	057286	057666	058046	058426	058805	059185	059563	059942	050320
115	060698	061075	061452	061829	062206	062582	062958	063333	063709	064083
116	064458	064832	065206	065580	065953	066326	066699	067071	067443	067815
117	068186	068557	068928	069298	069668	070038	070407	070776	071145	071514
118	071882	072250	072617	072985	073352	073718	074085	074451	074816	075182
119	075547	075912	076276	076640	077004	077368	077731	078094	078457	078819
120	079181	079543	079904	080266	080626	080987	081347	081707	082067	082426
121	082785	083144	083503	083861	084219	084576	084934	085219	085647	086004
122	086360	086716	087071	087426	087781	088136	088490	088845	089198	089552
123	089905	090258	090611	090963	091315	091667	092018	092370	092721	093071
124	093422	093772	094122	094471	094820	095169	095518	095866	096215	096562
125	096910	097257	097604	097951	098298	098644	098990	099335	099681	100026
126	100371	100715	101059	101403	101747	102091	102434	102777	103119	103462
127	103804	104146	104487	104828	105169	105510	105851	106191	106531	106871
128	107210	107549	107888	108227	108565	108903	109241	109579	109916	110253
129	110590	110926	111263	111599	111934	112270	112605	112940	113275	113609
130	113943	114277	114611	114944	115278	115611	115943	116276	116608	116940
131	117271	117603	117934	118265	118595	118926	119256	119586	119915	120245
132	120574	120903	121231	121560	121888	122216	122544	122871	123198	123525
133	123852	124178	124504	124830	125156	125481	125806	126131	126456	126781
134	127105	127429	127753	128076	128399	128722	129045	129368	129690	130012
135	130334	130655	130977	131298	131619	131939	132260	132580	132900	133219
136	133539	133858	134177	134496	134814	135133	135451	135769	136086	136403
137	136721	137037	137354	137671	137987	138303	138618	138934	139249	139564
138	139879	140194	140508	140822	141136	141450	141763	142076	142389	142702
139	143015	143327	143639	143951	144263	144574	144885	145196	145507	145818
140	146128	146438	146748	147058	147367	147676	147985	148294	148603	148911
141	149219	149527	149835	150142	150449	150756	151063	151370	151676	151982
142	152288	152594	152900	153205	153510	153815	154120	154424	154728	155032
143	155336	155640	155943	156246	156549	156852	157154	157457	157759	158061
144	158362	158664	158965	159266	159567	159868	160168	160469	160769	161068
145	161368	161667	161967	162266	162564	162863	163161	163460	163758	164055
146	164353	164650	164947	165244	165541	165838	166134	166430	166726	167022
147	167317	167613	167908	168203	168497	168792	169086	169380	169674	169968
148	170262	170555	170848	171141	171434	171726	172019	172311	172603	172895
149	173186	173478	173769	174060	174351	174641	174932	175222	175512	175802
150	176091	176381	176670	176959	177248	177536	177825	178113	178401	178689

(Continued)

Table 6. *(Continued)* Logarithms of Numbers 150 to 200.9 — Mantisae 176091 to 302980.

No.	0	1	2	3	4	5	6	7	8	9
150	176091	176381	176670	176959	177248	177536	177825	178113	178401	178689
151	178977	179264	179552	179839	180126	180413	180699	180986	181272	181558
152	181844	182129	182415	182700	182985	183270	183555	183839	184123	184407
153	184691	184975	185259	185542	185825	186108	186391	186674	186956	187239
154	187521	187803	188084	188366	188647	188928	189209	189490	189771	190051
155	190332	190612	190892	191171	191451	191730	192010	192289	192567	192846
156	193125	193403	193681	193959	194237	194514	194792	195069	195346	195623
157	195900	196176	196453	196729	197005	197281	197556	197832	198107	198382
158	198657	198932	199206	199481	199755	200029	200303	200577	200850	201124
159	201397	201670	201943	202216	202488	202761	203033	203305	203577	203848
160	204120	204391	204663	204934	205204	205475	205746	206016	206286	206556
161	206826	207096	207365	207634	207904	208173	208441	208710	208979	209247
162	209515	209783	210051	210319	210586	210853	211121	211388	211654	211921
163	212188	212454	212720	212986	213252	213518	213783	214049	214314	214579
164	214844	215109	215373	215638	215902	216166	216430	216694	216957	217221
165	217484	217747	218010	218273	218536	218798	219060	219323	219585	219846
166	220108	220370	220631	220892	221153	221414	221675	221936	222196	222456
167	222716	222976	223236	223496	223755	224015	224274	224533	224792	225051
168	225309	225568	225826	226084	226342	226600	226858	227115	227372	227630
169	227887	228144	228400	228657	228913	229170	229426	229682	29938	230193
170	230449	230704	230960	231215	231470	231724	231979	232234	232488	232742
171	232996	233250	233504	233757	234011	234264	234517	234770	235023	235276
172	235528	235781	236033	236285	236537	236789	237041	237292	237544	237795
173	238046	238297	238548	238799	239049	239299	239550	239800	240050	240300
174	240549	240799	241048	241297	241546	241795	242044	242293	242541	242790
175	243038	243286	243534	243782	244030	244277	244525	244772	245019	245266
176	245513	245759	246006	246252	246499	246745	246991	247237	247482	247728
177	247973	248219	248464	248709	248954	249198	249443	249687	249932	250176
178	250420	250664	250908	251151	251395	251638	251881	252125	252368	252610
179	252853	253096	253338	253580	253822	254064	254306	254548	254790	255031
180	255273	255514	255755	255996	256237	256477	256718	256958	257198	257439
181	257679	257918	258158	258398	258637	258877	259116	259355	259594	259833
182	260071	260310	260548	260787	261025	261263	261501	261739	261976	262214
183	262451	262688	262925	263162	263399	263636	263873	264109	264346	264582
184	264818	265054	265290	265525	265761	265996	266232	266467	266702	266937
185	267172	267406	267641	267875	268110	268344	268578	268812	269046	269279
186	269513	269746	269980	270213	270446	270679	270912	271144	271377	271609
187	271842	272074	272306	272538	272770	273001	273233	273464	273696	273927
188	274158	274389	274620	274850	275081	275311	275542	275772	276002	276232
189	276462	276692	276921	277151	277380	277609	277838	278067	278296	278525
190	278754	278982	279211	279439	279667	279895	280123	280351	280578	280806
191	281033	281261	281488	281715	281942	282169	282396	282622	282849	283075
192	283301	283527	283753	283979	284205	284431	284656	284882	285107	285332
193	285557	285782	286007	286232	286456	286681	286905	287130	287354	287578
194	287802	288026	288249	288473	288696	288920	289143	289366	289589	289812
195	290035	290257	290480	290702	290925	291147	291369	291591	291813	292034
196	292256	292478	292699	292920	293141	293363	293584	293804	294025	294246
197	294466	294687	294907	295127	295347	295567	295787	296007	296226	296446
198	296665	296884	297104	297323	297542	297761	297979	298198	298416	298635
199	298853	299071	299289	299507	299725	299943	300161	300378	300595	300813
200	301030	301247	301464	301681	301898	302114	302331	302547	302764	302980

(Continued)

Table 6. *(Continued)* Logarithms of Numbers 200 to 250.9 — Mantissae 301030 to 399501.

No.	0	1	2	3	4	5	6	7	8	9
200	301030	301247	301464	301681	301898	302114	302331	302547	302764	302980
201	303196	303412	303628	303844	304059	304275	304491	304706	304921	305136
202	305351	305566	305781	305996	306211	306425	306639	306854	307068	307282
203	307496	307710	307924	308137	308351	308564	308778	308991	309204	309417
204	309630	309843	310056	310268	310481	310693	310906	311118	311330	311542
205	311754	311966	312177	312389	312600	312812	313023	313234	313445	313656
206	313867	314078	314289	314499	314710	314920	315130	315340	315551	315760
207	315970	316180	316390	316599	316809	317018	317227	317436	317646	317854
208	318063	318272	318481	318689	318898	319106	319314	319522	319730	319938
209	320146	320354	320562	320769	320977	321184	321391	321598	321805	322012
210	322219	322426	322633	322839	323046	323252	323458	323665	323871	324077
211	324282	324488	324694	324899	325105	325310	325516	325721	325926	326131
212	326336	326541	326745	326950	327155	327359	327563	327767	327972	328176
213	328380	328583	328787	328991	329194	329398	329601	329805	330008	330211
214	330414	330617	330819	331022	331225	331427	331630	331832	332034	332236
215	332438	332640	332842	333044	333246	333447	333649	333850	334051	334253
216	334454	334655	334856	335057	335257	335458	335658	335859	336059	336260
217	336460	336660	336860	337060	337260	337459	337659	337858	338058	338257
218	338456	338656	338855	339054	339253	339451	339650	339849	340047	340246
219	340444	340642	340841	341039	341237	341435	341632	341830	342028	342225
220	342423	342620	342817	343014	343212	343409	343606	343802	343999	344196
221	344392	344589	344785	344981	345178	345374	345570	345766	345962	346157
222	346353	346549	346744	346939	347135	347330	347525	347720	347915	348110
223	348305	348500	348694	348889	349083	349278	349472	349666	359860	350054
224	350248	350442	350636	350829	351023	351216	351410	351603	351796	351989
225	352183	352375	352568	352761	352954	353147	353339	353532	353724	353916
226	354108	354301	354493	354685	354876	355068	355260	355452	355643	355834
227	356026	356217	356408	356599	356790	356981	357172	357363	357554	357744
228	357935	358125	358316	358506	358696	358886	359076	359266	359456	359646
229	359835	360025	360215	360404	360593	360783	360972	361161	361350	361539
230	361728	361917	362105	362294	362482	362671	362859	363048	363236	363424
231	363612	363800	363988	364176	364363	364551	364739	364926	365113	365301
232	365488	365675	365862	366049	366236	366423	366610	366796	366983	367169
233	367356	367542	367729	367915	368101	368287	368473	368659	368845	369030
234	369216	369401	369587	369772	369958	370143	370328	370513	370698	370883
235	371068	371253	371437	371622	371806	371991	372175	372360	372544	372728
236	372912	373096	373280	373464	373647	373831	374015	374198	374382	374565
237	374748	374932	375115	375298	375481	375664	375846	376029	376212	376394
238	376577	376759	376942	377124	377306	377488	377670	377852	378034	378216
239	378398	378580	378761	378943	379124	379306	379487	379668	379849	380030
240	380211	380392	380573	380754	380934	381115	381296	381476	381656	381837
241	382017	382197	382377	382557	382737	382917	383097	383277	383456	383636
242	383815	383995	384174	384353	384533	384712	384891	385070	385249	385428
243	385606	385785	385964	386142	386321	386499	386677	386856	387034	387212
244	387390	387568	387746	387924	388101	388279	388456	388634	388811	388989
245	389166	389343	389520	389698	389875	390051	390228	390405	390582	390759
246	390935	391112	391288	391464	391641	391817	391993	392169	392345	392521
247	392697	392873	393048	393224	393400	393575	393751	393926	394101	394277
248	394452	394627	394802	394977	395152	395326	395501	395676	395850	396025
249	396199	396374	396548	396722	396896	397071	397245	397419	397592	397766
250	397940	398114	398287	398461	398634	398808	398981	399154	399328	399501

(Continued)

Table 6. *(Continued)* Logarithms of Numbers 250 to 300.9 — Mantissae 397940 to 478422.

No.	0	1	2	3	4	5	6	7	8	9
250	397940	398114	398287	398461	398634	398808	398981	399154	399328	399501
251	399674	399847	400020	400192	400365	400538	400711	400883	401056	401228
252	401401	401573	401745	401917	402089	402261	402433	402605	402777	402949
253	403121	403292	403464	403635	403807	403978	404149	404320	404492	404663
254	404834	405005	405176	405346	405517	405688	405858	406029	406199	406370
255	406540	406710	406881	407051	407221	407391	407561	407731	407901	408070
256	408240	408410	408579	408749	408918	409087	409257	409426	409595	409764
257	409933	410102	410271	410440	410609	410777	410946	411114	411283	411451
258	411620	411788	411956	412124	412293	412461	412629	412796	412964	413132
259	413300	413467	413635	413803	413970	414137	414305	414472	414639	414806
260	414973	415140	415307	415474	415641	415808	415974	416141	416308	416474
261	416641	416807	416973	417139	417306	417472	417638	417804	417970	418135
262	418301	418467	418633	418798	418964	419129	419295	419460	419625	419791
263	419956	420121	420286	420451	420616	420781	420945	421110	421275	421439
264	421604	421768	421933	422097	422261	422426	422590	422754	422918	423082
265	423246	423410	423574	423737	423901	424065	424228	424392	424555	424718
266	424882	425045	425208	425371	425534	425697	425860	426023	426186	426349
267	426511	426674	426836	426999	427161	427324	427486	427648	427811	427973
268	428135	428297	428459	428621	428783	428944	429106	429268	429429	429591
269	429752	429914	430075	430236	430398	430559	430720	430881	431042	431203
270	431364	431525	431685	431846	432007	432167	432328	432488	432649	432809
271	432969	433130	433290	433450	433610	433770	433930	434090	434249	434409
272	434569	434729	434888	435048	435207	435367	435526	435685	435844	436004
273	436163	436322	436481	436640	436799	436957	437116	437275	437433	437592
274	437751	437909	438067	438226	438384	438542	438701	438859	439017	439175
275	439333	439491	439648	439806	439964	440122	440279	440437	440594	440752
276	440909	441066	441224	441381	441538	441695	441852	442009	442166	442323
277	442480	442637	442793	442950	443106	443263	443419	443576	443732	443889
278	444045	444201	444357	444513	444669	444825	444981	445137	445293	445449
279	445604	445760	445915	446071	446226	446382	446537	446692	446848	447003
280	447158	447313	447468	447623	447778	447933	448088	448242	448397	448552
281	448706	448861	449015	449170	449324	449478	449633	449787	449941	450095
282	450249	450403	450557	450711	450865	451018	451172	451326	451479	451633
283	451786	451940	452093	452247	452400	452553	452706	452859	453012	453165
284	453318	453471	453624	453777	453930	454082	454235	454387	454540	454692
285	454845	454997	455150	455302	455454	455606	455758	455910	456062	456214
286	456366	456518	456670	456821	456973	457125	457276	457428	457579	457731
287	457882	458033	458184	458336	458487	458638	458789	458940	459091	4592242
288	459392	459543	459694	459845	459995	460146	460296	460447	460597	460748
289	460898	461048	461198	461348	461499	461649	461799	461948	462098	462248
290	462398	462548	462697	462847	462997	463146	463296	463445	463594	463744
291	463893	464042	464191	464340	464490	464639	464788	464936	465085	465234
292	465383	465532	465680	465829	465977	466126	466274	466423	466571	466719
293	466868	467016	467164	467312	467460	467608	467756	467904	468052	468200
294	468347	468495	468643	468790	468938	469085	469233	469380	469527	469675
295	469822	469969	470116	470263	470410	470557	470704	470851	470998	471145
296	471292	471438	471585	471732	471878	472025	472171	472318	472464	472610
297	472756	472903	473049	473195	473341	473487	473633	473779	473925	474071
298	474216	474362	474508	474653	474799	474944	475090	475235	475381	475526
299	475671	475816	475962	476107	476252	476397	476542	476687	476832	476976
300	477121	477266	477411	477555	477700	477844	477989	478133	478278	478422

(Continued)

Table 6. *(Continued)* **Logarithms of Numbers 300 to 350.9 — Mantissae 477121 to 545183.**

No.	0	1	2	3	4	5	6	7	8	9
300	477121	477266	477411	477555	477700	477844	477989	478133	478278	478422
301	478566	478711	478855	478999	479143	479287	479431	479575	479719	479863
302	480007	480151	480294	480438	480582	480725	480869	481012	481156	481299
303	481443	481586	481729	481872	482016	482159	482302	482445	482588	482731
304	482874	483016	483159	483302	483445	483587	483730	483872	484015	484157
305	484300	484442	484585	484727	484869	485011	485153	485295	485437	485579
306	485721	485863	486005	486147	486289	486430	486572	486714	486855	486997
307	487138	487280	487421	487563	487704	487845	487986	488127	488269	488410
308	488551	488692	488833	488974	489114	489255	489396	489537	489677	489818
309	489958	490099	490239	490380	490520	490661	490801	490941	491081	491222
310	491362	491502	491642	491782	491922	492062	492201	492341	492481	492621
311	492760	492900	493040	493179	493319	493458	493597	493737	493876	494015
312	494155	494294	494433	494572	494711	494850	494989	495128	495267	495406
313	495544	495683	495822	495960	496099	496238	496376	496515	496653	496791
314	496930	497068	497206	497344	497483	497621	497759	497897	498035	498173
315	498311	498448	498586	498724	498862	498999	499137	499275	499412	499550
316	499687	499824	499962	500099	500236	500374	500511	500648	500785	500922
317	501059	501196	501333	501470	501607	501744	501880	502017	502154	502291
318	502427	502564	502700	502837	502973	503109	503246	503382	503518	503655
319	503791	503927	504063	504199	504335	504471	504607	504743	504878	505014
320	505150	505286	505421	505557	505693	505828	505964	506099	506234	506370
321	506505	506640	506776	506911	507046	507181	507316	507451	507586	507721
322	507856	507991	508126	508260	508395	508530	508664	508799	508934	509068
323	509203	509337	509471	509606	509740	509874	510009	510143	510277	510411
324	510545	510679	510813	510947	511081	511215	511349	511482	511616	511750
325	511883	512017	512151	512284	512418	512551	512684	512818	512951	513084
326	513218	513351	513484	513617	513750	513883	514016	514149	514282	514415
327	514548	514681	514813	514946	515079	515211	515344	515476	515609	515741
328	515874	516006	516139	516271	516403	516535	516668	516800	516932	517064
329	517196	517328	517460	517592	517724	517855	517987	518119	518251	518382
330	518514	518646	518777	518909	519040	519171	519303	519434	519566	519697
331	519828	519959	520090	520221	520353	520484	520615	520745	520876	521007
332	521138	521269	521400	521530	521661	521792	521922	522053	522183	522314
333	522444	522575	522705	522835	522966	523096	523226	523356	523486	523616
334	523746	523876	524006	524136	524266	524396	524526	524656	524785	524915
335	525045	525174	525304	525434	525563	525693	525822	525951	526081	526210
336	526339	526469	526598	526727	526856	526985	527114	527243	527372	527501
337	527630	527759	527888	528016	528145	528274	528402	528531	528660	528788
338	528917	529045	529174	529302	529430	529559	529687	529815	529943	530072
339	530200	530328	530456	530584	530712	530840	530968	531096	531223	531351
340	531479	531607	531734	531862	531990	532117	532245	532372	532500	532627
341	532754	532882	533009	533136	533264	533391	533518	533645	533772	533899
342	534026	534153	534280	534407	534534	534661	534787	534914	535041	535167
343	535294	535421	535547	535674	535800	535927	536053	536180	536306	536432
344	536558	536685	536811	536937	537063	537189	537315	537441	537567	537693
345	537819	537945	538071	538197	538322	538448	538574	538699	538825	538951
346	539076	539202	539327	539452	539578	539703	539829	539954	540079	540204
347	540329	540455	540580	540705	540830	540955	541080	541205	541330	541454
348	541579	541704	541829	541953	542078	542203	542327	542452	542576	542701
349	542825	542950	543074	543199	543323	543447	543571	543696	543820	543944
350	544068	544192	544316	544440	544564	544688	544812	544936	545060	545183

(Continued)

Table 6. *(Continued)* Logarithms of Numbers 350 to 400.9 — Mantissae 544068 to 603036.

No.	0	1	2	3	4	5	6	7	8	9
350	544068	544192	544316	544440	544564	544688	544812	544936	545060	545183
351	545307	545431	545555	545678	545802	545925	546049	546172	546296	546419
352	546543	546666	546789	546913	547036	547159	547282	547405	547529	547652
353	547775	547898	548021	548144	548267	548389	548512	548635	548758	548881
354	549003	549126	549249	549371	549494	549616	549739	549861	549984	550106
355	550228	550351	550473	550595	550717	550840	550962	551084	551206	551328
356	551450	551572	551694	551816	551938	552060	552181	552303	552425	552547
357	552668	552790	552911	553033	553155	553276	553398	553519	553640	553762
358	553883	554004	554126	554247	554368	554489	554610	554731	554852	554973
359	555094	555215	555336	555457	555578	555699	555820	555940	556061	556182
360	556303	556423	556544	556664	556785	556905	557026	557146	557267	557387
361	557507	557627	557748	557868	557988	558108	558228	558349	558469	558589
362	558709	558829	558948	559068	559188	559308	559428	559548	559667	559787
363	559907	560026	560146	560265	560385	560504	560624	560743	560863	560982
364	561101	561221	561340	561459	561578	561698	561817	561936	562055	562174
365	562293	562412	562531	562650	562769	562887	563006	563125	563244	563362
366	563481	563600	563718	563837	563955	564074	564192	564311	564429	564548
367	564666	564784	564903	565021	565139	565257	565376	565494	565612	565730
368	565848	565966	566084	566202	566320	566437	566555	566673	566791	566909
369	567026	567144	567262	567379	567497	567614	567732	567849	567967	568084
370	568202	568319	568436	568554	568671	568788	568905	569023	569140	569257
371	569374	569491	569608	569725	569842	569959	570076	570193	570309	570426
372	570543	570660	570776	570893	571010	571126	571243	571359	571476	571592
373	571709	571825	571942	572058	572174	572291	572407	572523	572639	572755
374	572872	572988	573104	573220	573336	573452	573568	573684	573800	573915
375	574031	574147	574263	574379	574494	574610	574726	574841	574957	575072
376	575188	575303	575419	575534	575650	575765	575880	575996	576111	576226
377	576341	576457	576572	576687	576802	576917	577032	577147	577262	577377
378	577492	577607	577722	577836	577951	578066	578181	578295	578410	578525
379	578639	578754	578868	578983	579097	579212	579326	579441	579555	579669
380	579784	579898	580012	580126	580241	580355	580469	580583	580697	580811
381	580925	581039	581153	581267	581381	581495	581608	581722	581836	581950
382	582063	582177	582291	582404	582518	582631	582745	582858	582972	583085
383	583199	583312	583426	583539	583652	583765	583879	583992	584105	584218
384	584331	584444	584557	584670	584783	584896	585009	585122	585235	585348
385	585461	585574	585686	585799	585912	586024	586137	586250	586362	586475
386	586587	586700	586812	586925	587037	587149	587262	587374	587486	587599
387	587711	587823	587935	588047	588160	588272	588384	588496	588608	588720
388	588832	588944	589056	589167	589279	589391	589503	589615	589726	589838
389	589950	590061	590173	590284	590396	590507	590619	590730	590842	590953
390	591065	591176	591287	591399	591510	591621	591732	591843	591955	592066
391	592177	592288	592399	592510	592621	592732	592843	592954	593064	593175
392	593286	593397	593508	593618	593729	593840	593950	594061	594171	594282
393	594393	594503	594614	594724	594834	594945	595055	595165	595276	595386
394	595496	595606	595717	595827	595937	596047	596157	596267	596377	596487
395	596597	596707	596817	596927	597037	597146	597256	597366	597476	597586
396	597695	597805	597914	598024	598134	598243	598353	598462	598572	598681
397	598791	598900	599009	599119	599228	599337	599446	599556	599665	599774
398	599883	599992	600101	600210	600319	600428	600537	600646	600755	600864
399	600973	601082	601191	601299	601408	601517	601625	601734	601843	601951
400	602060	602169	602277	602386	602494	602603	602711	602819	602928	603036

(Continued)

Table 6. *(Continued)* Logarithms of Numbers 400 to 450.9 — Mantissae 602060 to 654080.

No.	0	1	2	3	4	5	6	7	8	9
400	602060	602169	602277	602386	602494	602603	602711	602819	602928	603036
401	603144	603253	603361	603469	603577	603686	603794	603902	604010	604118
402	604226	604334	604442	604550	604658	604766	604874	604982	605089	605197
403	605305	605413	605521	605628	605736	605844	605951	606059	606166	606274
404	606381	606489	606596	606704	606811	606919	607026	607133	607241	607348
405	607455	607562	607669	607777	607884	607991	608098	608205	608312	608419
406	608526	608633	608740	608847	608954	609061	609167	609274	609381	609488
407	609594	609701	609808	609914	610021	610128	610234	610341	610447	610554
408	610660	610767	610873	610979	611086	611192	611298	611405	611511	611617
409	611723	611829	611936	612042	612148	612254	612360	612466	612572	612678
410	612784	612890	612996	613102	613207	613313	613419	613525	613630	613736
411	613842	613947	614053	614159	614264	614370	614475	614581	614686	614792
412	614897	615003	615108	615213	615319	615424	615529	615634	615740	615845
413	615950	616055	616160	616265	616370	616476	616581	616686	616790	616895
414	617000	617105	617210	617315	617420	617525	617629	617734	617839	617943
415	618048	618153	618257	618362	618466	618571	618676	618780	618884	618989
416	619093	619198	619302	619406	619511	619615	619719	619824	619928	620032
417	620136	620240	620344	620448	620552	620656	620760	620864	620968	621072
418	621176	621280	621384	621488	621592	621695	621799	621903	622007	622110
419	622214	622318	622421	622525	622628	622732	622835	622939	623042	623146
420	623249	623353	623456	623559	623663	623766	623869	623973	624076	624179
421	624282	624385	624488	624591	624695	624798	624901	625004	625107	625210
422	625312	625415	625518	625621	625724	625827	625929	626032	626135	626238
423	626340	626443	626546	626648	626751	626853	626956	627058	627161	627263
424	627366	627468	627571	627673	627775	627878	627980	628082	628185	628287
425	628389	628491	628593	628695	628797	628900	629002	629104	629206	629308
426	629410	629512	629613	629715	629817	629919	630021	630123	630224	630326
427	630428	630530	630631	630733	630835	630936	631038	631139	631241	631342
428	631444	631545	631647	631748	631849	631951	632052	632153	632255	632356
429	632457	632559	632660	632761	632862	632963	633064	633165	633266	633367
430	633468	633569	633670	633771	633872	633973	634074	634175	634276	634376
431	634477	634578	634679	634779	634880	634981	635081	635182	635283	635383
432	635484	635584	635685	635785	635886	635986	636087	636187	636287	636388
433	636488	636588	636688	636789	636889	636989	637089	637189	637290	637390
434	637490	637590	637690	637790	637890	637990	638090	638190	638290	638389
435	638489	638589	638689	638789	638888	638988	639088	639188	639287	639387
436	639486	639586	639686	639785	639885	639984	640084	640183	640283	640382
437	640481	640581	640680	640779	640879	640978	641077	641177	641276	641375
438	641474	641573	641672	641771	641871	641970	642069	642168	642267	642366
439	642465	642563	642662	642761	642860	642959	643058	643156	643255	643354
440	643453	643551	643650	643749	643847	643946	644044	644143	644242	644340
441	644439	644537	644636	644734	644832	644931	645029	645127	645226	645324
442	645422	645521	645619	645717	645815	645913	646011	646110	646208	646306
443	646404	646502	646600	646698	646796	646894	646992	647089	647187	647285
444	647383	647481	647579	647676	647774	647872	647969	648067	648165	648262
445	648360	648458	648555	648653	648750	648848	648945	649043	6491140	649237
446	649335	649432	649530	649627	649724	649821	649919	650016	650113	650210
447	650308	650405	650502	650599	650696	650793	650890	650987	651084	651181
448	651278	651375	651472	651569	651666	651762	651859	651956	652053	652150
449	652246	652343	652440	652536	652633	652730	652826	652923	653019	653116
450	653213	653309	653405	653502	653598	653695	653791	653888	653984	654080

(Continued)

Table 6. *(Continued)* **Logarithms of Numbers 450 to 500.9 — Mantissae 653213 to 699751.**

No.	0	1	2	3	4	5	6	7	8	9
450	653213	653309	653405	653502	653598	653695	653791	653888	653984	654080
451	654177	654273	654369	654465	654562	654658	654754	654850	654946	655042
452	655138	655235	655331	655427	655523	655619	655715	655810	655906	656002
453	656098	656194	656290	656386	656482	656577	656673	656769	656864	656960
454	657056	657152	657247	657343	657438	657534	657629	657725	657820	657916
455	658011	658107	658202	658298	658393	658488	658584	658679	658774	658870
456	658965	659060	659155	659250	659346	659441	659536	659631	659726	659821
457	659916	660011	660106	660201	660296	660391	660486	660581	660676	660771
458	660865	660960	661055	661150	661245	661339	661434	661529	661623	661718
459	661813	661907	662002	662096	662191	662286	662380	662475	662569	662663
460	662758	662852	662947	663041	663135	663230	663324	663418	663512	663607
461	663701	663795	663889	663983	664078	664172	664266	664360	664454	664548
462	664642	664736	664830	664924	665018	665112	665206	665299	665393	665487
463	665581	665675	665769	665862	665956	666050	666143	666237	666331	666424
464	666518	666612	666705	666799	666892	666986	667079	667173	667266	667360
465	667453	667546	667640	667733	667826	667920	668013	668106	668199	668293
466	668386	668479	668572	668665	668759	668852	668945	669038	669131	669224
467	669317	669410	669503	669596	669689	669782	669875	669967	670060	670153
468	670246	670339	670431	670524	670617	670710	670802	670895	670988	671080
469	671173	671265	671358	671451	671543	671636	671728	671821	671913	672005
470	672098	672190	672283	672375	672467	672560	672652	672744	672836	672929
471	673021	673113	673205	673297	673390	673482	673574	673666	673758	673850
472	673942	674034	674126	674218	674310	674402	674494	674586	674677	674769
473	674861	674953	675045	675137	675228	675320	675412	675503	675595	675687
474	675778	675870	675962	676053	676145	676236	676328	676419	676511	676602
475	676694	676785	676876	676968	677059	677151	677242	677333	677424	677516
476	677607	677698	677789	677881	677972	678063	678154	678245	678336	678427
477	678518	678609	678700	678791	678882	678973	679064	679155	679246	679337
478	679428	679519	679610	679700	679791	679882	679973	680063	680154	680245
479	680336	680426	680517	680607	680698	680789	680879	680970	681060	681151
480	681241	681332	681422	681513	681603	681693	681784	681874	681964	682055
481	682145	682235	682326	682416	682506	682596	682686	682777	682867	682957
482	683047	683137	683227	683317	683407	683497	683587	683677	683767	683857
483	683947	684037	684127	684217	684307	684396	684486	684576	684666	684756
484	684845	684935	685025	685114	685204	685294	685383	685473	685563	685652
485	685742	685831	685921	686010	686100	686189	686279	686368	686458	686547
486	686636	686726	686815	686904	686994	687083	687172	687261	687351	687440
487	687529	687618	687707	687796	687886	687975	688064	688153	688242	688331
488	688420	688509	688598	688687	688776	688865	688953	689042	689131	689220
489	689309	689398	689486	689575	689664	689753	689841	689930	690019	690107
490	690196	690285	690373	690462	690550	690639	690728	690816	690905	690993
491	691081	691170	691258	691347	691435	691524	691612	691700	691789	691877
492	691965	692053	692142	692230	692318	692406	692494	692583	692671	692759
493	692847	692935	693023	693111	693199	693287	693375	693463	693551	693639
494	693727	693815	693903	693991	694078	694166	694254	694342	694430	694517
495	694605	694693	694781	694868	694956	695044	695131	695219	695307	695394
496	695482	695569	695657	695744	695832	695919	696007	696094	696182	696269
497	696356	696444	696531	696618	696706	696793	696880	696968	697055	697142
498	697229	697317	697404	697491	697578	697665	697752	697839	697926	698014
499	698101	698188	698275	698362	698449	698535	698622	698709	698796	698883
500	698970	699057	699144	699231	699317	699404	699491	699578	699664	699751

(Continued)

Table 6. *(Continued)* Logarithms of Numbers 500 to 550.9 — Mantissae 698970 to 741073.

No.	0	1	2	3	4	5	6	7	8	9
500	698970	699057	699144	699231	699317	699404	699491	699578	699664	699751
501	699838	699924	700011	700098	700184	700271	700358	700444	700531	700617
502	700704	700790	700877	700963	701050	701136	701222	701309	701395	701482
503	701568	701654	701741	701827	701913	701999	702086	702172	702258	702344
504	702431	702517	702603	702689	702775	702861	702947	703033	703119	703205
505	703291	703377	703463	703549	703635	703721	703807	703893	703979	704065
506	704151	704236	704322	704408	704494	704579	704665	704751	704837	704922
507	705008	705094	705179	705265	705350	705436	705522	705607	705693	705778
508	705864	705949	706035	706120	706206	706291	706376	706462	706547	706632
509	706718	706803	706888	706974	707059	707144	707229	707315	707400	707485
510	707570	707655	707740	707826	707911	707996	708081	708166	708251	708336
511	708421	708506	708591	708676	708761	708846	708931	709015	709100	709185
512	709270	709355	709440	709524	709609	709694	709779	709863	709948	710033
513	710117	710202	710287	710371	710456	710540	710625	710710	710794	710879
514	710963	711048	711132	711217	711301	711385	711470	711554	711639	711723
515	711807	711892	711976	712060	712144	712229	712313	712397	712481	712566
516	712650	712734	712818	712902	712986	713070	713154	713238	713323	713407
517	713491	713575	713659	713742	713826	713910	713994	714078	714162	714246
518	714330	714414	714497	714581	714665	714749	714833	714916	715000	715084
519	715167	715251	715335	715418	715502	715586	715669	715753	715836	715920
520	716003	716087	716170	716254	716337	716421	716504	716588	716671	716754
521	716838	716921	717004	717088	717171	717254	717338	717421	717504	717587
522	717671	717754	717837	717920	718003	718086	718169	718253	718336	718419
523	718502	718585	718668	718751	718834	718917	719000	719083	719165	719248
524	719331	719414	719497	719580	719663	719745	719828	719911	719994	720077
525	720159	720242	720325	720407	720490	720573	720655	720738	720821	720903
526	720986	721068	721151	721233	721316	721398	721481	721563	721646	721728
527	721811	721893	721975	722058	722140	722222	722305	722387	722469	722552
528	722634	722716	722798	722881	722963	723045	723127	723209	723291	723374
529	723456	723538	723620	723702	723784	723866	723948	724030	724112	724194
530	724276	724358	724440	724522	724604	724685	724767	724849	724931	725013
531	725095	725176	725258	725340	725422	725503	725585	725667	725748	725830
532	725912	725993	726075	726156	726238	726320	726401	726483	726564	726646
533	726727	726809	726890	726972	727053	727134	727216	727297	727379	727460
534	727541	727623	727704	727785	727866	727948	728029	728110	728191	728273
535	728354	728435	728516	728597	728678	728759	728841	728922	729003	729084
536	729165	729246	729327	729408	729489	729570	729651	729732	729813	729893
537	729974	730055	730136	730217	730298	730378	730459	730540	730621	730702
538	730782	730863	730944	731024	731105	731186	731266	731347	731428	731508
539	731589	731669	731750	731830	731911	731991	732072	732152	732233	732313
540	732394	732474	732555	732635	732715	732796	732876	732956	733037	733117
541	733197	733278	733358	733438	733518	733598	733679	733759	733839	733919
542	733999	734079	734160	734240	734320	734400	734480	734560	734640	734720
543	734800	734880	734960	735040	735120	735200	735279	735359	735439	735519
544	735599	735679	735759	735838	735918	735998	736078	736157	736237	736317
545	736397	736476	736556	736635	736715	736795	736874	736954	737034	737113
546	737193	737272	737352	737431	737511	737590	737670	737749	737829	737908
547	737987	738067	738146	738225	738305	738384	738463	738543	738622	738701
548	738781	738860	738939	739018	739097	739177	739256	739335	739414	739493
549	739572	739651	739731	739810	739889	739968	740047	740126	740205	740284
550	740363	740442	740521	740600	740678	740757	740836	740915	740994	741073

(Continued)

Table 6. *(Continued)* **Logarithms of Numbers 550 to 600.9 — Mantissae 740363 to 778802.**

No.	0	1	2	3	4	5	6	7	8	9
550	740363	740442	740521	740600	740678	740757	740836	740915	740994	741073
551	741152	741230	741309	741388	741467	741546	741624	741703	741782	741860
552	741939	742018	742096	742175	742254	742332	742411	742489	742568	742647
553	742725	742804	742882	742961	743039	743118	743196	743275	743353	743431
554	743510	743588	743667	743745	743823	743902	743980	744058	744136	744215
555	744293	744371	744449	744528	744606	744684	744762	744840	744919	744997
556	745075	745153	745231	745309	745387	745465	745543	745621	745699	745777
557	745855	745933	746011	746089	746167	746245	746323	746401	746479	746556
558	746634	746712	746790	746868	746945	747023	747101	747179	747256	747334
559	747412	747489	747567	747645	747722	747800	747878	747955	748033	748110
560	748188	748266	748343	748421	748498	748576	748653	748731	748808	748885
561	748963	749040	749118	749195	749272	749350	749427	749504	749582	749659
562	749736	749814	749891	749968	750045	750123	750200	750277	750354	750431
563	750508	750586	750663	750740	750817	750894	750971	751048	751125	751202
564	751279	751356	751433	751510	751587	751664	751741	751818	751895	751972
565	752048	752125	752202	752279	752356	752433	752509	752586	752663	752740
566	752816	752893	752970	753047	753123	753200	753277	753353	753430	753506
567	753583	753660	753736	753813	753889	753966	754042	754119	754195	754272
568	754348	754425	754501	754578	754654	754730	754807	754883	754960	755036
569	755112	755189	755265	755341	755417	755494	755570	755646	755722	755799
570	755875	755951	756027	756103	756180	756256	756332	756408	756484	756560
571	756636	756712	756788	756864	756940	757016	757092	757168	757244	757320
572	757396	757472	757548	757624	757700	757775	757851	757927	758003	758079
573	758155	758230	758306	758382	758458	758533	758609	758685	758761	758836
574	758912	758988	759063	759139	759214	759290	759366	759441	759517	759592
575	759668	759743	759819	759894	759970	760045	760121	760196	760272	760347
576	760422	760498	760573	760649	760724	760799	760875	760950	761025	761101
577	761176	761251	761326	761402	761477	761552	761627	761702	761778	761853
578	761928	762003	762078	762153	762228	762303	762378	762453	762529	762604
579	762679	762754	762829	762904	762978	763053	763128	763203	763278	763353
580	763428	763503	763578	763653	763727	763802	763877	763952	764027	764101
581	764176	764251	764326	764400	764475	764550	764624	764699	764774	764848
582	764923	764998	765072	765147	765221	765296	765370	765445	765520	765594
583	765669	765743	765818	765892	765966	766041	766115	766190	766264	766338
584	766413	766487	766562	766636	766710	766785	766859	766933	767007	767082
585	767156	767230	767304	767379	767453	767527	767601	767675	767749	767823
586	767898	767972	768046	768120	768194	768268	768342	768416	768490	768564
587	768638	768712	768786	768860	768934	769008	769082	769156	769230	769303
588	769377	769451	769525	769599	769673	769746	769820	769894	769968	770042
589	770115	770189	770263	770336	770410	770484	770557	770631	770705	770778
590	770852	770926	770999	771073	771146	771220	771293	771367	771440	771514
591	771587	771661	771734	771808	771881	771955	772028	772102	772175	772248
592	772322	772395	772468	772542	772615	772688	772762	772835	772908	772981
593	773055	773128	773201	773274	773348	773421	773494	773567	773640	773713
594	773786	773860	773933	774006	774079	774152	774225	774298	774371	774444
595	774517	774590	774663	774736	774809	774882	774955	775028	775100	775173
596	775246	775319	775392	775465	775538	775610	775683	775756	775829	775902
597	775974	776047	776120	776193	776265	776338	776411	776483	776556	776629
598	776701	776774	776846	776919	776992	777064	777137	777209	777282	777354
599	777427	777499	777572	777644	777717	777789	777862	777934	778006	778079
600	778151	778224	778296	778368	778441	778513	778585	778658	778730	778802

(Continued)

Table 6. *(Continued)* **Logarithms of Numbers 600 to 650.9 — Mantissae 778151 to 813514.**

No.	0	1	2	3	4	5	6	7	8	9
600	778151	778224	778296	778368	778441	778513	778585	778658	778730	778802
601	778874	778947	779019	779091	779163	779236	779308	779380	779452	779524
602	779596	779669	779741	779813	779885	779957	780029	780101	780173	780245
603	780317	780389	780461	780533	780605	780677	780749	780821	780893	780965
604	781037	781109	781181	781253	781324	781396	781468	781540	781612	781684
605	781755	781827	781899	781971	782042	782114	782186	782258	782329	782401
606	782473	782544	782616	782688	782759	782831	782902	782974	783046	783117
607	783189	783260	783332	783403	783475	783546	783618	783689	783761	783832
608	783904	783975	784046	784118	784189	784261	784332	784403	784475	784546
609	784617	784689	784760	784831	784902	784974	785045	785116	785187	785259
610	785330	785401	785472	785543	785615	785686	785757	785828	785899	785970
611	786041	786112	786183	786254	786325	786396	786467	786538	786609	786680
612	786751	786822	786893	786964	787035	787106	787177	787248	787319	787390
613	787460	787531	787602	787673	787744	787815	787885	787956	788027	788098
614	788168	788239	788310	788381	788451	788522	788593	788663	788734	788804
615	788875	788946	789016	789087	789157	789228	789299	789369	789440	789510
616	789581	789651	789722	789792	789863	789933	790004	790074	790144	790215
617	790285	790356	790426	790496	790567	790637	790707	790778	790848	790918
618	790988	791059	791129	791199	791269	791340	791410	791480	791550	791620
619	791691	791761	791831	791901	791971	792041	792111	792181	792252	792322
620	792392	792462	792532	792602	792672	792742	792812	792882	792952	793022
621	793092	793162	793231	793301	793371	793441	793511	793581	793651	793721
622	793790	793860	793930	794000	794070	794139	794209	794279	794349	794418
623	794488	794558	794627	794697	794767	794836	794906	794976	795045	795115
624	795185	795254	795324	795393	795463	795532	795602	795672	795741	795811
625	795880	795949	796019	796088	796158	796227	796297	796366	796436	796505
626	796574	796644	796713	796782	796852	796921	796990	797060	797129	797198
627	797268	797337	797406	797475	797545	797614	797683	797752	797821	797890
628	797960	798029	798098	798167	798236	798305	798374	798443	798513	798582
629	798651	798720	798789	798858	798927	798996	799065	799134	799203	799272
630	799341	799409	799478	799547	799616	799685	799754	799823	799892	799961
631	800029	800098	800167	800236	800305	800373	800442	800511	800580	800648
632	800717	800786	800854	800923	800992	801061	801129	801198	801266	801335
633	801404	801472	801541	801609	801678	801747	801815	801884	801952	802021
634	802089	802158	802226	802295	802363	802432	802500	802568	802637	802705
635	802774	802842	802910	802979	803047	803116	803184	803252	803321	803389
636	803457	803525	803594	803662	803730	803798	803867	803935	804003	804071
637	804139	804208	804276	804344	804412	804480	804548	804616	804685	804753
638	804821	804889	804957	805025	805093	805161	805229	805297	805365	805433
639	805501	805569	805637	805705	805773	805841	805908	805976	806044	806112
640	806180	806248	806316	806384	806451	806519	806587	806655	806723	806790
641	806858	806926	806994	807061	807129	807197	807264	807332	807400	807467
642	807535	807603	807670	807738	807806	807873	807941	808008	808076	808143
643	808211	808279	808346	808414	808481	808549	808616	808684	808751	808818
644	808886	808953	809021	809088	809156	809223	809290	809358	809425	809492
645	809560	809627	809694	809762	809829	809896	809964	810031	810098	810165
646	810233	810300	810367	810434	810501	810569	810636	810703	810770	810837
647	810904	810971	811039	811106	811173	811240	811307	811374	811441	811508
648	811575	811642	811709	811776	811843	811910	811977	812044	812111	812178
649	812245	812312	812379	812445	812512	812579	812646	812713	812780	812847
650	812913	812980	813047	813114	813181	813247	813314	813381	813448	813514

(Continued)

Table 6. *(Continued)* **Logarithms of Numbers 650 to 700.9 — Mantissae 812913 to 845656.**

No.	0	1	2	3	4	5	6	7	8	9
650	812913	812980	813047	813114	813181	813247	813314	813381	813448	813514
651	813581	813648	813714	813781	813848	813914	813981	814048	814114	814181
652	814248	814314	814381	814447	814514	814581	814647	814714	814780	814847
653	814913	814980	815046	815113	815179	815246	815312	815378	815445	815511
654	815578	815644	815711	815777	815843	815910	815976	816042	816109	816175
655	816241	816308	816374	816440	816506	816573	816639	816705	816771	816838
656	816904	816970	817036	817102	817169	817235	817301	817367	817433	817499
657	817565	817631	817698	817764	817830	817896	817962	818028	818094	818160
658	818226	818292	818358	818424	818490	818556	818622	818688	818754	818820
659	818885	818951	819017	819083	819149	819215	819281	819346	819412	819478
660	819544	819610	819676	819741	819807	819873	819939	820004	820070	820136
661	820201	820267	820333	820399	820464	820530	820595	820661	820727	820792
662	820858	820924	820989	821055	821120	821186	821251	821317	821382	821448
663	821514	821579	821645	821710	821775	821841	821906	821972	822037	822103
664	822168	822233	822299	822364	822430	822495	822560	822626	822691	822756
665	822822	822887	822952	823018	823083	823148	823213	823279	823344	823409
666	823474	823539	823605	823670	823735	823800	823865	823930	823996	824061
667	824126	824191	824256	824321	824386	824451	824516	824581	824646	824711
668	824776	824841	824906	824971	825036	825101	825166	825231	825296	825361
669	825426	825491	825556	825621	825686	825751	825815	825880	825945	826010
670	826075	826140	826204	826269	826334	826399	826464	826528	826593	826658
671	826723	826787	826852	826917	826981	827046	827111	827175	827240	827305
672	827369	827434	827499	827563	827628	827692	827757	827821	827886	827951
673	828015	828080	828144	828209	828273	828338	828402	828467	828531	828595
674	828660	828724	828789	828853	828918	828982	829046	829111	829175	829239
675	829304	829368	829432	829497	829561	829625	829690	829754	829818	829882
676	829947	830011	830075	830139	830204	830268	830332	830396	830460	830525
677	830589	830653	830717	830781	830845	830909	830973	831037	831102	831166
678	831230	831294	831358	831422	831486	831550	831614	831678	831742	831806
679	831870	831934	831998	832062	832126	832189	832253	832317	832381	832445
680	832509	832573	832637	832700	832764	832828	832892	832956	833020	833083
681	833147	833211	833275	833338	833402	833466	833530	833593	833657	833721
682	833784	833848	833912	833975	834039	834103	834166	834230	834294	834357
683	834421	834484	834548	834611	834675	834739	834802	834866	834929	834993
684	835056	835120	835183	835247	835310	835373	835437	835500	835564	835627
685	835691	835754	835817	835881	835944	836007	836071	836134	836197	836261
686	836324	836387	836451	836514	836577	836641	836704	836767	836830	836894
687	836957	837020	837083	837146	837210	837273	837336	837399	837462	837525
688	837588	837652	837715	837778	837841	837904	837967	838030	838093	838156
689	838219	838282	838345	838408	838471	838534	838597	838660	838723	838786
690	838849	838912	838975	839038	839101	839164	839227	839289	839352	839415
691	839478	839541	839604	839667	839729	839792	839855	839918	839981	840043
692	840106	840169	840232	840294	840357	840420	840482	840545	840608	840671
693	840733	840796	840859	840921	840984	841046	841109	841172	841234	841297
694	841359	841422	841485	841547	841610	841672	841735	841797	841860	841922
695	841985	842047	842110	842172	842235	842297	842360	842422	842484	842547
696	842609	842672	842734	842796	842859	842921	842983	843046	843108	843170
697	843233	843295	843357	843420	843482	843544	843606	843669	843731	843793
698	843855	843918	843980	844042	844104	844166	844229	844291	844353	844415
699	844477	844539	844601	844664	844726	844788	844850	844912	844974	845036
700	845098	845160	845222	845284	845346	845408	845470	845532	845594	845656

(Continued)

Table 6. *(Continued)* **Logarithms of Numbers 700 to 750.9 — Mantissae 845098 to 875582.**

No.	0	1	2	3	4	5	6	7	8	9
700	845098	845160	845222	845284	845346	845408	845470	845532	845594	845656
701	845718	845780	845842	845904	845966	846028	846090	846151	846213	846275
702	846337	846399	846461	846523	846585	846646	846708	846770	846832	846894
703	846955	847017	847079	847141	847202	847264	847326	847388	847449	847511
704	847573	847634	847696	847758	847819	847881	847943	848004	848066	848128
705	848189	848251	848312	848374	848435	848497	848559	848620	848682	848743
706	848805	848866	848928	848989	849051	849112	849174	849235	849297	849358
707	849419	849481	849542	849604	849665	849726	849788	849849	849911	849972
708	850033	850095	850156	850217	850279	850340	850401	850462	850524	850585
709	850646	850707	850769	850830	850891	850952	851014	851075	851136	851197
710	851258	851320	851381	851442	851503	851564	851625	851686	851747	851809
711	851870	851931	851992	852053	852114	852175	852236	852297	852358	852419
712	852480	852541	852602	852663	852724	852785	852846	852907	852968	853029
713	853090	853150	853211	853272	853333	853394	853455	853516	853577	853637
714	853698	853759	853820	853881	853941	854002	854063	854124	854185	854245
715	854306	854367	854428	854488	854549	854610	854670	854731	854792	854852
716	854913	854974	855034	855095	855156	855216	855277	855337	855398	855459
717	855519	855580	855640	855701	855761	855822	855882	855943	856003	856064
718	856124	856185	856245	856306	856366	856427	856487	856548	856608	856668
719	856729	856789	856850	856910	856970	857031	857091	857152	857212	857272
720	857332	857393	857453	857513	857574	857634	857694	857755	857815	857875
721	857935	857995	858056	858116	858176	858236	858297	858357	858417	858477
722	858537	858597	858657	858718	858778	858838	858898	858958	859018	859078
723	859138	859198	859258	859318	859379	859439	859499	859559	859619	859679
724	859739	859799	859859	859918	859978	860038	860098	860158	860218	860278
725	860338	860398	860458	860518	860578	860637	860697	860757	860817	860877
726	860937	860996	861056	861116	861176	861236	861295	861355	861415	861475
727	861534	861594	861654	861714	861773	861833	861893	861952	862012	862072
728	862131	862191	862251	862310	862370	862430	862489	862549	862608	862668
729	862728	862787	862847	862906	862966	863025	863085	863144	863204	863263
730	863323	863382	863442	863501	863561	863620	863680	863739	863799	863858
731	863917	863977	864036	864096	864155	864214	864274	864333	864392	864452
732	864511	864570	864630	864689	864748	864808	864867	864926	864985	865045
733	865104	865163	865222	865282	865341	865400	865459	865519	865578	865637
734	865696	865755	865814	865874	865933	865992	866051	866110	866169	866228
735	866287	866346	866405	866465	866524	866583	866642	866701	866760	866819
736	866878	866937	866996	867055	867114	867173	867232	867291	867350	867409
737	867467	867526	867585	867644	867703	867762	867821	867880	867939	867998
738	868056	868115	868174	868233	868292	868350	868409	868468	868527	868586
739	868644	868703	868762	868821	868879	868938	868997	869056	869114	869173
740	869232	869290	869349	869408	869466	869525	869584	869642	869701	869760
741	869818	869877	869935	869994	870053	870111	870170	870228	870287	870345
742	870404	870462	870521	870579	870638	870696	870755	870813	870872	870930
743	870989	871047	871106	871164	871223	871281	871339	871398	871456	871515
744	871573	871631	871690	871748	871806	871865	871923	871981	872040	872098
745	872156	872215	872273	872331	872389	872448	872506	872564	872622	872681
746	872739	872797	872855	872913	872972	873030	873088	873146	873204	873262
747	873321	873379	873437	873495	873553	873611	873669	873727	873785	873844
748	873902	873960	874018	874076	874134	874192	874250	874308	874366	874424
749	874482	874540	874598	874656	874714	874772	874830	874888	874945	875003
750	875061	875119	875177	875235	875293	875351	875409	875466	875524	875582

(Continued)

Table 6. *(Continued)* Logarithms of Numbers 750 to 800.9 — Mantissae 875061 to 903578.

No.	0	1	2	3	4	5	6	7	8	9
750	875061	875119	875177	875235	875293	875351	875409	875466	875524	875582
751	875640	875698	875756	875813	875871	875929	875987	876045	876102	876160
752	876218	876276	876333	876391	876449	876507	876564	876622	876680	876737
753	876795	876853	876910	876968	877026	877083	877141	877199	877256	877314
754	877371	877429	877487	877544	877602	877659	877717	877774	877832	877889
755	877947	878004	878062	878119	878177	878234	878292	878349	878407	878464
756	878522	878579	878637	878694	878752	878809	878866	878924	878981	879039
757	879096	879153	879211	879268	879325	879383	879440	879497	879555	879612
758	879669	879726	879784	879841	879898	879956	880013	880070	880127	880185
759	880242	880299	880356	880413	880471	880528	880585	880642	880699	880756
760	880814	880871	880928	880985	881042	881099	881156	881213	881271	881328
761	881385	881442	881499	881556	881613	881670	881727	881784	881841	881898
762	881955	882012	882069	882126	882183	882240	882297	882354	882411	882468
763	882525	882581	882638	882695	882752	882809	882866	882923	882980	883037
764	883093	883150	883207	883264	883321	883377	883434	883491	883548	883605
765	883661	883718	883775	883832	883888	883945	884002	884059	884115	884172
766	884229	884285	884342	884399	884455	884512	884569	884625	884682	884739
767	884795	884852	884909	884965	885022	885078	885135	885192	885248	885305
768	885361	885418	885474	885531	885587	885644	885700	885757	885813	885870
769	885926	885983	886039	886096	886152	886209	886265	886321	886378	886434
770	886491	886547	886604	886660	886716	886773	886829	886885	886942	886998
771	887054	887111	887167	887223	887280	887336	887392	887449	887505	887561
772	887617	887674	887730	887786	887842	887898	887955	888011	888067	888123
773	888179	888236	888292	888348	888404	888460	888516	888573	888629	888685
774	888741	888797	888853	888909	888965	889021	889077	889134	889190	889246
775	889302	889358	889414	889470	889526	889582	889638	889694	889750	889806
776	889862	889918	889974	890030	890086	890141	890197	890253	890309	890365
777	890421	890477	890533	890589	890645	890700	890756	890812	890868	890924
778	890980	891035	891091	891147	891203	891259	891314	891370	891426	891482
779	891537	891593	891649	891705	891760	891816	891872	891928	891983	892039
780	892095	892150	892206	892262	892317	892373	892429	892484	892540	892595
781	892651	892707	892762	892818	892873	892929	892985	893040	893096	893151
782	893207	893262	893318	893373	893429	893484	893540	893595	893651	893706
783	893762	893817	893873	893928	893984	894039	894094	894150	894205	894261
784	894316	894371	894427	894482	894538	894593	894648	894704	894759	894814
785	894870	894925	894980	895036	895091	895146	895201	895257	895312	895367
786	895423	895478	895533	895588	895644	895699	895754	895809	895864	895920
787	895975	896030	896085	896140	896195	896251	896306	896361	896416	896471
788	896526	896581	896636	896692	896747	896802	896857	896912	896967	897022
789	897077	897132	897187	897242	897297	897352	897407	897462	897517	897572
790	897627	897682	897737	897792	897847	897902	897957	898012	898067	898122
791	898176	898231	898286	898341	898396	898451	898506	898561	898615	898670
792	898725	898780	898835	898890	898944	898999	899054	899109	899164	899218
793	899273	899328	899383	899437	899492	899547	899602	899656	899711	899766
794	899821	899875	899930	899985	900039	900094	900149	900203	900258	900312
795	900367	900422	900476	900531	900586	900640	900695	900749	900804	900859
796	900913	900968	901022	901077	901131	901186	901240	901295	901349	901404
797	901458	901513	901567	901622	901676	901731	901785	901840	901894	901948
798	902003	902057	902112	902166	902221	902275	902329	902384	902438	902492
799	902547	902601	902655	902710	902764	902818	902873	902927	902981	903036
800	903090	903144	903199	903253	903307	903361	903416	903470	903524	903578

(Continued)

Table 6. *(Continued)* Logarithms of Numbers 800 to 850.9 — Mantissae 903090 to 929879.

No.	0	1	2	3	4	5	6	7	8	9
800	903090	903144	903199	903253	903307	903361	903416	903470	903524	903578
801	903633	903687	903741	903795	903849	903904	903958	904012	904066	904120
802	904174	904229	904283	904337	904391	904445	904499	904553	904607	904661
803	904716	904770	904824	904878	904932	904986	905040	905094	905148	905202
804	905256	905310	905364	905418	905472	905526	905580	905634	905688	905742
805	905796	905850	905904	905958	906012	906066	906119	906173	906227	906281
806	906335	906389	906443	906497	906551	906604	906658	906712	906766	906820
807	906874	906927	906981	907035	907089	907143	907196	907250	907304	907358
808	907411	907465	907519	907573	907626	907680	907734	907787	907841	907895
809	907949	908002	908056	908110	908163	908217	908270	908324	908378	908431
810	908485	908539	908592	908646	908699	908753	908807	908860	908914	908967
811	909021	909074	909128	909181	909235	909289	909342	909396	909449	909503
812	909556	909610	909663	909716	909770	909823	909877	909930	909984	910037
813	910091	910144	910197	910251	910304	910358	910411	910464	910518	910571
814	910624	910678	910731	910784	910838	910891	910944	910998	911051	911104
815	911158	911211	911264	911317	911371	911424	911477	911530	911584	911637
816	911690	911743	911797	911850	911903	911956	912009	912063	912116	912169
817	912222	912275	912328	912381	912435	912488	912541	912594	912647	912700
818	912753	912806	912859	912913	912966	913019	913072	913125	913178	913231
819	913284	913337	913390	913443	913549	913549	913602	913655	913708	913761
820	913814	913867	913920	913973	914026	914079	914132	914184	914237	914290
821	914343	914396	914449	914502	914555	914608	914660	914713	914766	914819
822	914872	914925	914977	915030	915083	915136	915189	915241	915294	915347
823	915400	915453	915505	915558	915611	915664	915716	915769	915822	915875
824	915927	915980	916033	916085	916138	916191	916243	916296	916349	916401
825	916454	916507	916559	916612	916664	916717	916770	916822	916875	916927
826	916980	917033	917085	917138	917190	917243	917295	917348	917400	917453
827	917506	917558	917611	917663	917716	917768	917820	917873	917925	917978
828	918030	918083	918135	918188	918240	918293	918345	918397	918450	918502
829	918555	918607	918659	918712	918764	918816	918869	918921	918973	919026
830	919078	919130	919183	919235	919287	919340	919392	919444	919496	919549
831	919601	919653	919706	919758	919810	919862	919914	919967	920019	920071
832	920123	920176	920228	920280	920332	920384	920436	920489	920541	920593
833	920645	920697	920749	920801	920853	920906	920958	921010	921062	921114
834	921166	921218	921270	921322	921374	921426	921478	921530	921582	921634
835	921686	921738	921790	921842	921894	921946	921998	922050	922102	922154
836	922206	922258	922310	922362	922414	922466	922518	922570	922622	922674
837	922725	922777	922829	922881	922933	922985	923037	923089	923140	923192
838	923244	923296	923348	923399	923451	923503	923555	923607	923658	923710
839	923762	923814	923865	923917	923969	924021	924072	924124	924176	924228
840	924279	924331	924383	924434	924486	924538	924589	924641	924693	924744
841	924796	924848	924899	924951	925003	925054	925106	925157	925209	925261
842	925312	925364	925415	925467	925518	925570	925621	925673	925725	925776
843	925828	925879	925931	925982	926034	926085	926137	926188	926240	926291
844	926342	926394	926445	926497	926548	926600	926651	926702	926754	926805
845	926857	926908	926959	927011	927062	927114	927165	927216	927268	927319
846	927370	927422	927473	927524	927576	927627	927678	927730	927781	927832
847	927883	927935	927986	928037	928088	928140	928191	928242	928293	928345
848	928396	928447	928498	928549	928601	928652	928703	928754	928805	928857
849	928908	928959	929010	929061	929112	929163	929215	929266	929317	929368
850	929419	929470	929521	929572	929623	929674	929725	929776	929827	929879

(Continued)

Table 6. *(Continued)* Logarithms of Numbers 850 to 900.9 — Mantissae 929419 to 954677.

No.	0	1	2	3	4	5	6	7	8	9
850	929419	929470	929521	929572	929623	929674	929725	929776	929827	929879
851	929930	929981	930032	930083	930134	930185	930236	930287	930338	930389
852	930440	930491	930542	930592	930643	930694	930745	930796	930847	930898
853	930949	931000	931051	931102	931153	931204	931254	931305	931356	931407
854	931458	931509	931560	931610	931661	931712	931763	931814	931865	931915
855	931966	932017	932068	932118	932169	932220	932271	932322	932372	932423
856	932474	932524	932575	932626	932677	932727	932778	932829	932879	932930
857	932981	933031	933082	933133	933183	933234	933285	933335	933386	933437
858	933487	933538	933589	933639	933690	933740	933791	933841	933892	933943
859	933993	934044	934094	934145	934195	934246	934296	934347	934397	934448
860	934498	934549	934599	934650	934700	934751	934801	934852	934902	934953
861	935003	935054	935104	935154	935205	935255	935306	935356	935406	935457
862	935507	935558	935608	935658	935709	935759	935809	935860	935910	935960
863	936011	936061	936111	936162	936212	936262	936313	936363	936413	936463
864	936514	936564	936614	936665	936715	936765	936815	936865	936916	936966
865	937016	937066	937116	937167	937217	937267	937317	937367	937418	937468
866	937518	937568	937618	937668	937718	937769	937819	937869	937919	937969
867	938019	938069	938119	938169	938219	938269	938320	938370	938420	938470
868	938520	938570	938620	938670	938720	938770	938820	938870	938920	938970
869	939020	939070	939120	939170	939220	939270	939320	939369	939419	939469
870	939519	939569	939619	939669	939719	939769	939819	939869	939918	939968
871	940018	940068	940118	940168	940218	940267	940317	940367	940417	940467
872	940516	940566	940616	940666	940716	940765	940815	940865	940915	940964
873	941014	941064	941114	941163	941213	941263	941313	941362	941412	941462
874	941511	941561	941611	941660	941710	941760	941809	941859	941909	941958
875	942008	942058	942107	942157	942207	942256	942306	942355	942405	942455
876	942504	942554	942603	942653	942702	942752	942801	942851	942901	942950
877	943000	943049	943099	943148	943198	943247	943297	943346	943396	943445
878	943495	943544	943593	943643	943692	943742	943791	943841	943890	943939
879	943989	944038	944088	944137	944186	944236	944285	944335	944384	944433
880	944483	944532	944581	944631	944680	944729	944779	944828	944877	944927
881	944976	945025	945074	945124	945173	945222	945272	945321	945370	945419
882	945469	945518	945567	945616	945665	945715	945764	945813	945862	945912
883	945961	946010	946059	946108	946157	946207	946256	946305	946354	946403
884	946452	946501	946551	946600	946649	946698	946747	946796	946845	946894
885	946943	946992	947041	947090	947139	947189	947238	947287	947336	947385
886	947434	947483	947532	947581	947630	947679	947728	947777	947826	947875
887	947924	947973	948022	948070	948119	948168	948217	948266	948315	948364
888	948413	948462	948511	948560	948609	948657	948706	948755	948804	948853
889	948902	948951	948999	949048	949097	949146	949195	949244	949292	949341
890	949390	949439	949488	949536	949585	949634	949683	949731	949780	949829
891	949878	949926	949975	950024	950073	950121	950170	950219	950267	950316
892	950365	950414	950462	950511	950560	950608	950657	950706	950754	950803
893	950851	950900	950949	950997	951046	951095	951143	951192	951240	951289
894	951338	951386	951435	951483	951532	951580	951629	951677	951726	951775
895	951823	951872	951920	951969	952017	952066	952114	952163	952211	952260
896	952308	952356	952405	952453	952502	952550	952599	952647	952696	952744
897	952792	952841	952889	952938	952986	953034	953083	953131	953180	953228
898	953276	953325	953373	953421	953470	953518	953566	953615	953663	953711
899	953760	953808	953856	953905	953953	954001	954049	954098	954146	954194
900	954243	954291	954339	954387	954435	954484	954532	954580	954628	954677

(Continued)

Table 6. *(Continued)* **Logarithms of Numbers 900 to 950.9 — Mantissae 954243 to 978135.**

No.	0	1	2	3	4	5	6	7	8	9
900	954243	954291	954339	954387	954435	954484	954532	954580	954628	954677
901	954725	954773	954821	954869	954918	954966	955014	955062	955110	955158
902	955207	955255	955303	955351	955399	955447	955495	955543	955592	955640
903	955688	955736	955784	955832	955880	955928	955976	956024	956072	956120
904	956168	956216	956265	956313	956361	956409	956457	956505	956553	956601
905	956649	956994	956745	956793	956840	956888	956936	956984	957032	957080
906	957128	957176	957224	957272	957320	957368	957416	957464	957512	957559
907	957607	957655	957703	957751	957799	957847	957894	957942	957990	958038
908	958086	958134	958181	958229	958277	958325	958373	958421	958468	958516
909	958564	958612	958659	958707	958755	958803	958850	958898	958946	958994
910	959041	959089	959137	959185	959232	959280	959328	959375	959423	959471
911	959518	959566	959614	959661	959709	959757	959804	959852	959900	959947
912	959995	960042	960090	960138	960185	960233	960280	960328	960376	960423
913	960471	960518	960566	960613	960661	960709	960756	960804	960851	960899
914	960946	960994	961041	961089	961136	961184	961231	961279	961326	961374
915	961421	961469	961516	961563	961611	961658	961706	961753	961801	961848
916	961895	961943	961990	962038	962085	962132	962180	962227	962275	962322
917	962369	962417	962464	962511	962559	962606	962653	962701	962748	962795
918	962843	962890	962937	962985	963032	963079	963126	963174	963221	963268
919	963316	963363	963410	963457	963504	963552	963599	963646	963693	963741
920	963788	963835	963882	963929	963977	964024	964071	964118	964165	964212
921	964260	964307	964354	964401	964448	964495	964542	964590	964637	964684
922	964731	964778	964825	964872	964919	964966	965013	965061	965108	965155
923	965202	965249	965296	965343	965390	965437	965484	965531	965578	965625
924	965672	965719	965766	965813	965860	965907	965954	966001	966048	966095
925	966142	966189	966236	966283	966329	966376	966423	966470	966517	966564
926	966611	966658	966705	966752	966799	966845	966892	966939	966986	967033
927	967080	967127	967173	967220	967267	967314	967361	967408	967454	967501
928	967548	967595	967642	967688	967735	967782	967829	967875	967922	967969
929	968016	968062	968109	968156	968203	968249	968296	968343	968390	968436
930	968483	968530	968576	968623	968670	968716	968763	968810	968856	968903
931	968950	968996	969043	969090	969136	969183	969229	969276	969323	969369
932	969416	969463	969509	969556	969602	969649	969695	969742	969789	969835
933	969882	969928	969975	970021	970068	970114	970161	970207	970254	970300
934	970347	970393	970440	970486	970533	970579	970626	970672	970719	970765
935	970812	970858	970904	970951	970997	971044	971090	971137	971183	971229
936	971276	971322	971369	971415	971461	971508	971554	971601	971647	971693
937	971740	971786	971832	971879	971925	971971	972018	972064	972110	972157
938	972203	972249	972295	972342	972388	972434	972481	972527	972573	972619
939	972666	972712	972758	972804	972851	972897	972943	972989	973035	973082
940	973128	973174	973220	973266	973313	973359	973405	973451	973497	973543
941	973590	973636	973682	973728	973774	973820	973866	973913	973959	974005
942	974051	974097	974143	974189	974235	974281	974327	974374	974420	974466
943	974512	974558	974604	974650	974696	974742	974788	974834	974880	974926
944	974972	975018	975064	975110	975156	975202	975248	975294	975340	975386
945	975432	975478	975524	975570	975616	975662	975707	975753	975799	975845
946	975891	975937	975983	976029	976075	976121	976167	976212	976258	976304
947	976350	976396	976442	976488	976533	976579	976625	976671	976717	976763
948	976808	976854	976900	976946	976992	977037	977083	977129	977175	977220
949	977266	977312	977358	977403	977449	977495	977541	977586	977632	977678
950	977724	977769	977815	977861	977906	977952	977998	978043	978089	978135

(Continued)

Table 6. *(Continued)* Logarithms of Numbers 950 to 1000 — Mantissae 977724 to 000000.

No.	0	1	2	3	4	5	6	7	8	9
950	977724	977769	977815	977861	977906	977952	977998	978043	978089	978135
951	978181	978226	978272	978317	978363	978409	978454	978500	978546	978591
952	978637	978683	978728	978774	978819	978865	978911	978956	979002	979047
953	979093	979138	979184	979230	979275	979321	979366	979412	979457	979503
954	979548	979594	979639	979685	979730	979776	979821	979867	979912	979958
955	980003	980049	980094	980140	980185	980231	980276	980322	980367	980412
956	980458	980503	980549	980594	980640	980685	980730	980776	980821	980867
957	980912	980957	981003	981048	981093	981139	981184	981229	981275	981320
958	981366	981411	981456	981501	981547	981592	981637	981683	981728	981773
959	981819	981864	981909	981954	982000	982045	982090	982135	982181	982226
960	982271	982316	982362	982407	982452	982497	982543	982588	982633	982678
961	982723	982769	982814	982859	982904	982949	982994	983040	983085	983130
962	983175	983220	983265	983310	983356	983401	983446	983491	983536	983581
963	983626	983671	983716	983762	983807	983852	983897	983942	983987	984032
964	984077	984122	984167	984212	984257	984302	984347	984392	984437	984482
965	984527	984572	984617	984662	984707	984752	984797	984842	984887	984932
966	984977	985022	985067	985112	985157	985202	985247	985292	985337	985382
967	985426	985471	985516	985561	985606	985651	985696	985741	985786	985830
968	985875	985920	985965	986010	986055	986100	986144	986189	986234	986279
969	986324	986369	986413	986458	986503	986548	986593	986637	986682	986727
970	986772	986817	986861	986906	986951	986996	987040	987085	987130	987175
971	987219	987264	987309	987353	987398	987443	987488	987532	987577	987622
972	987666	987711	987756	987800	987845	987890	987934	987979	988024	988068
973	988113	988157	988202	988247	988291	988336	988381	988425	988470	988514
974	988559	988604	988648	988693	988737	988782	988826	988871	988916	988960
975	989005	989049	989094	989138	989183	989227	989272	989316	989361	989405
976	989450	989494	989539	989583	989628	989672	989717	989761	989806	989850
977	989895	989939	989983	990028	990072	990117	990161	990206	990250	990294
978	990339	990383	990428	990472	990516	990561	990605	990650	990694	990738
979	990783	990827	990871	990916	990960	991004	991049	991093	991137	991182
980	991226	991270	991315	991359	991403	991448	991492	991536	991580	991625
981	991669	991713	991758	991802	991846	991890	991935	991979	992023	992067
982	992111	992156	992200	992244	992288	992333	992377	992421	992465	992509
983	992554	992598	992642	992686	992730	992774	992819	992863	992907	992951
984	992995	993039	993083	993127	993172	993216	993260	993304	993348	993392
985	993436	993480	993524	993568	993613	993657	993701	993745	993789	993833
986	993877	993921	993965	994009	994053	994097	994141	994185	994229	994273
987	994317	994361	994405	994449	994493	994537	994581	994625	994669	994713
988	994757	994801	994845	994889	994933	994977	995021	995065	995108	995152
989	995196	995240	995284	995328	995372	995416	995460	995504	995547	995591
990	995635	995679	995723	995767	995811	995854	995898	995942	995986	996030
991	996074	996117	996161	996205	996249	996293	996337	996380	996424	996468
992	996512	996555	996599	996643	996687	996731	996774	996818	996862	996906
993	996949	996993	997037	997080	997124	997168	997212	997255	997299	997343
994	997386	997430	997474	997517	997561	997605	997648	997692	997736	997779
995	997823	997867	997910	997954	997998	998041	998085	998129	998172	998216
996	998259	998303	998347	998390	998434	998477	998521	998564	998608	998652
997	998695	998739	998782	998826	998869	998913	998956	999000	999043	999087
998	999131	999174	999218	999261	999305	999348	999392	999435	999479	999522
999	999565	999609	999652	999696	999739	999783	999826	999870	999913	999957
1000	00000									

Basic Geometry

Geometry is the branch of mathematics that deals with the measurement of properties and relationships of points, lines, angles, surfaces, and solids. Some definitions are helpful to understanding these terms.

Lines and angles

Lines are measured in linear units. Their only dimension is length. There are several different types of lines.

A **straight line** has the same direction throughout its length. It is the shortest distance between two points.

A **curved line** is one that is continually changing direction. Commonly called a **curve**.

A **broken line** is made up of several connected straight lines having different directions.

Parallel lines are lines that lie in the same plane and are equally distant from each other at all points.

A **point** has position only. It lacks length, breadth, or thickness.

Angles are formed when two straight lines meet at a point. The angle is the measure of the difference in direction of the meeting lines, which make up the **sides** of the angle. The point where the lines meet is called the **vertex** of the angle.

When one straight line meets another so that the opposite adjacent angles that are formed are equal, those lines are said to be **perpendicular** and the angles formed are **right angles**. A right angle contains 90 degrees (90°).

An **acute angle** is any angle less than a right angle.

An **obtuse angle** is any angle greater than a right angle, but less than two right angles.

When the sum of two angles is equal to a right angle, these angles are said to be **complementary**.

When the sum of two angles is equal to two right angles, these angles are said to be **supplementary**.

The sum of all angles about a point on one side of a straight line is equal to two right angles.

The sum of all angles about any point is equal to four right angles (360°).

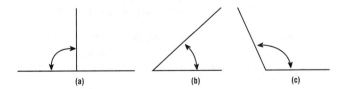

Figure 1. (a) Right angle. (b) Acute angle. (c) Obtuse angle.

Surfaces and plane figures

A **surface** has two dimensions: length and width (breadth). Therefore, it extends in two directions. It is said to be a **plane surface** if any two points within the surface can be connected by a straight line that is contained entirely within the surface.

A **plane figure** is a plane surface bounded by either straight or curved lines. The distance around a plane figure is called its **perimeter**, and the **area** of a plane figure is measured as the number of square units it contains.

A **polygon** is a plane figure bounded by straight lines. The lines are its sides, and the sum of the sides is its perimeter. If all sides are equal in length, it is said to be **equilateral**. If all of the angles within the polygon are equal, it is said to be **equiangular**. If all sides and all angles are equal, it is said to be a **regular polygon**. See **Table 7** for geometrical formulas for plane polygons.

A **triangle** is a polygon with three sides.

A **quadrilateral** is a polygon with four sides.

A **pentagon** is a polygon with five sides.

A **hexagon** is a polygon with six sides.

A **heptagon** is a polygon with seven sides.

An **octagon** is a polygon with eight sides.

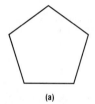   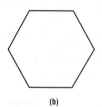   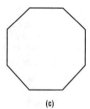

(a)   (b)   (c)

Figure 2. (a) Pentagon. (b) Hexagon. (c) Octagon.

Triangles. In any triangle, the sum of the angles is equal to two right angles (180°). Therefore, if any two angles are known, the third can be found by subtracting the sum of the two known angles from 180°.

A **right-angled triangle** is a three-sided polygon that includes a right angle. The longest side, opposite the right angle, is called the **hypotenuse**.

Oblique triangle is a term used to describe any triangle that is not a right-angled triangle.

In an **acute-angled triangle**, each of the angles is smaller than a right angle.

An **obtuse-angled triangle** is one that includes an obtuse angle.

In an **equilateral triangle**, all three sides are equal in length, and all three angles are equal (60° each).

In an **isosceles triangle**, two of the sides are equal in length. The angles opposite the equal sides are also equal.

In a **scalene triangle**, no two sides are equal.

Although any side of a triangle can be considered its **base**, the base is normally considered to be its lowest side. In an isosceles triangle, the side that is not equal is normally considered the base.

The **altitude** of a triangle is a perpendicular line drawn from the base to the vertex of the angle opposite the base.

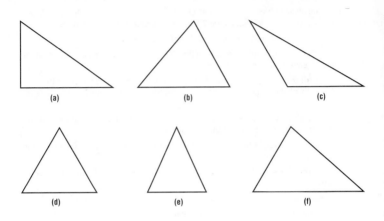

*Figure 3. (a) Right-angled triangle. (b) Acute-angled triangle. (c) Obtuse-angled triangle.
(d) Equilateral triangle. (e) Isosceles triangle. (f) Scalene triangle.*

Quadrilaterals are polygons bounded by four straight lines. The sum of the interior angles of a quadrilateral is always 360°.

A **trapezoid** is a quadrilateral with two parallel sides. Either parallel side may be called the base, and the perpendicular distance between the bases is the altitude.

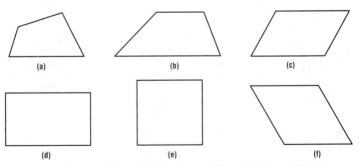

Figure 4. (a) Trapezium. (b) Trapezoid. (c) Parallelogram. (d) Rectangle. (e) Square. (f) Rhombus.

A **parallelogram** is a quadrilateral in which both sets of opposite sides are parallel.

A **square** is a parallelogram with four equal length sides and four right angles.

A **rectangle** is a parallelogram with four right angles.

A **rhombus** is a parallelogram with four equal sides, but no right angles.

Circles are plane figures bounded by a curved line, every point of which is equidistant from a point called the center. See **Table 7** for geometrical formulas for plane circles.

The **circumference** is the boundary line of a circle.

The **diameter** of a circle is a straight line drawn through the center and terminating at both ends at the circumference.

The **radius** of a circle is a straight line joining the center and the circumference. All radii of the same circle are equal, and the radius is equal to one-half the diameter.

An **arc** is any part of the circumference of a circle. If it is equal to one-half the circumference, it is a **semi-circumference.**

A **chord** is a straight line connecting the extremities of an arc, or any two points on the circumference of a circle.

A **secant** is a straight line that passes through the circumference of a circle at two points.

A **segment** is the area of a circle included between an arc and a chord.

A **sector** is the area of a circle included between an arc and two radii drawn to the extremities of the arc.

A **tangent** is a straight line that touches the circumference at a single point. This point is called the **point of tangency.**

A **quadrant** is an arc equal to one-fourth the circumference. When two diameters are drawn at right angles to each other, four quadrants are formed.

Circles are said to be **concentric** when they share the same central point. When the circumferences of two circles touch at a single point, they are said to be **tangent** to each other.

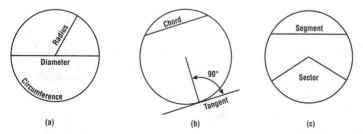

Figure 5. (a) Circumference, diameter, and radius. (b) Chord and tangent. (c) Segment and sector.

Solid Figures

A **solid figure** has three dimensions: length, width, and thickness (height). Therefore, it extends in three directions. A solid may be in any shape, but the most commonly encountered solids in engineering are cones, cylinders, prisms, pyramids, and spheres. **Table 8** provides geometrical formulas for solid figures.

A **cone** is bounded by a conical surface and a plane that cuts the conical surface. The conical surface is called the **lateral area**, and the plane is the **base**. The _____ ___ ꞏnd of the conical surface is the **vertex**.

The **altitude** of a cone is the perpendicular distance from the vertex to the base.

A straight line extended from the vertex to the perimeter of the base is the **element** of the cone. This line represents the **slant height** of the cone.

If a cone has a circle as its base, it is a **circular cone**. The axis of a **right circular cone**, or **cone of revolution**, is perpendicular to its base.

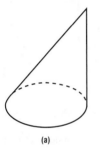 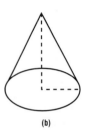

(a)   (b)

Figure 6. (a) Circular cone. (b) Cone of revolution.

A **cylinder** consists of two equal parallel plane surfaces, the **bases**, and a continuous lateral face generated by a straight line connecting the bases and moving along their circumferences. Generally, the bases are circles, and cylinders with circular bases are known as **circular cylinders**. If the side of a cylinder is perpendicular to its base, the cylinder is called a **right cylinder**.

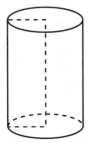

Figure 7. Right cylinder.

Prisms have two equal and parallel opposite faces, called **bases**, and other faces, called **lateral faces**, that are parallelograms. A prism takes its name from the shape of its bases, and may be triangular, rectangular, hexagonal, octagonal, etc.

A prism with lateral faces perpendicular to the base is called a **right prism**.

A right prism that has regular polygons for bases is a **regular prism**.

A **parallelepiped** is a prism that has parallelogram bases. If all the edges are perpendicular to the bases, it is called a right parallelepiped.

A right parallelepiped whose bases and lateral faces are rectangles is a **rectangular parallelepiped.** If all the bases and lateral faces are squares, it is called a **cube**.

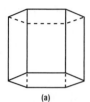 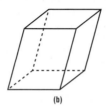 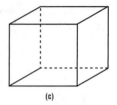

(a)   (b)   (c)

Figure 8. (a) Right prism. (b) Parallelepiped. (c) Rectangular parallelepiped (cube).

A **pyramid** has a polygon shaped base and rectangular sides. They are named for the type of polygon forming the base: i.e., hexagonal pyramid, octagonal pyramid, etc.

The **vertex** of a pyramid is the point where the vertices of the triangles meet.

The **altitude** of a pyramid is the perpendicular distance from the base to the vertex.

A pyramid whose base is a regular polygon, and whose vertex lies perpendicular to the center of the base, is a **regular pyramid**.

The **slant height** of a regular pyramid is equal to a line drawn from the vertex perpendicular to a side of the base. The slant height is equal to the altitude of a triangle forming one of its sides.

 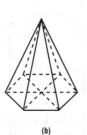 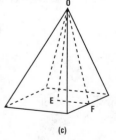

(a)   (b)   (c)

Figure 9. (a) Regular pyramid. (b) Hexagonal pyramid. (c) The altitude, line E,O, and the slant height, line F,O, of a pyramid.

A **sphere** is bounded by a curved surface with every point on the surface an equal distance from an internal point called the center. A ball is a sphere.

The **diameter** of a sphere is a straight line drawn through the center and terminating at both ends at the curved surface.

A **plane is tangent to a sphere** when it touches the sphere at only one point.

Table 7. Areas and Dimensions of Plane Polygons and Circles. *A* = Area.

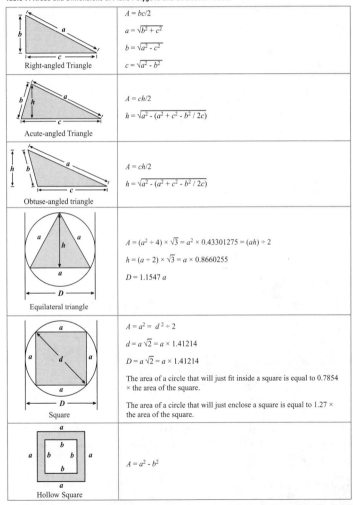

Right-angled Triangle	$A = bc/2$ $a = \sqrt{b^2 + c^2}$ $b = \sqrt{a^2 - c^2}$ $c = \sqrt{a^2 - b^2}$
Acute-angled Triangle	$A = ch/2$ $h = \sqrt{a^2 - (a^2 + c^2 - b^2 / 2c)}$
Obtuse-angled triangle	$A = ch/2$ $h = \sqrt{a^2 - (a^2 + c^2 - b^2 / 2c)}$
Equilateral triangle	$A = (a^2 \div 4) \times \sqrt{3} = a^2 \times 0.43301275 = (ah) \div 2$ $h = (a \div 2) \times \sqrt{3} = a \times 0.8660255$ $D = 1.1547\,a$
Square	$A = a^2 = d^2 \div 2$ $d = a\sqrt{2} = a \times 1.41214$ $D = a\sqrt{2} = a \times 1.41214$ The area of a circle that will just fit inside a square is equal to 0.7854 × the area of the square. The area of a circle that will just enclose a square is equal to 1.27 × the area of the square.
Hollow Square	$A = a^2 - b^2$

(Continued)

Table 7. *(Continued)* **Areas and Dimensions of Plane Polygons and Circles.** *A* = Area.

 Rectangle	$A = ab$ $d = \sqrt{a^2 + b^2}$
 Parallelogram	$A = ah$ $h = b \times \sin \theta$ $b = h \times \sin \theta$ $d_1 = \sqrt{(a + h \times \sin \theta)^2 + h^2}$ $d_2 = \sqrt{(a - h \times \sin \theta)^2 + h^2}$
 Rhombus	$A = ah$ $A = (d_1 \times d_2) \div 2$ $h = a (\sin \theta)$ $d_1 = (a\sqrt{2}) \times \sqrt{(1 + \cos \theta)^2}$ $d_1 = (a\sqrt{2}) \times \sqrt{(1 - \cos \theta)^2}$
 Trapezoid	$A = (h \times [a + b]) \div 2$ $h = c (\sin \theta) = d (\sin \beta)$
 Trapezium	$A = A_1 + A_2$ $A_1 = ch/2$ $h = \sqrt{a^2 - (a^2 + c^2 - b^2 / 2c)}$ $A_2 = (H \times [C + c]) \div 2$ $H = d (\sin \beta) = e (\sin \theta)$

Table 7. *(Continued)* **Areas and Dimensions of Plane Polygons and Circles. *A* = Area.**

 Regular Pentagon	$A = 0.625\ r^2 \times \sqrt{10 + (2 \times 2.236068)}$ $a = 0.5\ r^2 \times \sqrt{10 - (2 \times 2.236068)}$
 Regular Hexagon	$A = 2.5981\ a^2 = 0.866\ h^2$ $a = R = 1.1547\ r$ $r = 0.866\ a = 0.866\ R$ $h = 2\ a = 1.1547\ f$ (width across corners) $f = 1.732\ a$ (width across flats) $\theta = 120°$
 Regular Octagon	$A = 4.828\ a^2 = 2\ af$ $a = 0.414\ f = 0.765\ R = 0.828\ r$ $f = 2.414\ a = 0.924\ h$ (width across flats) $h = 1.083\ f$ (width across corners) $R = 1.307\ a = 1.082\ r$ $r = 1.207\ a = 0.924\ R$ $\theta = 135°$
 Circle	$A = (\pi/4) \times d^2 = \pi\ r^2$ $\pi = 3.1416$ Circumference $= C = \pi\ d = 2\ \pi\ r$ $r = C \div 6.2832 = 0.564\ \sqrt{A} = {}^1/_2\ d$ $d = C \div 3.1416 = 1.128\ \sqrt{A}$
 Annulus (Hollow Circle)	$A = \pi\ (R^2 - r^2) = (\pi \div 4) \times (D^2 - d^2)$ $b = R - r = (D - d) \div 2$

(Continued)

Table 7. *(Continued)* **Areas and Dimensions of Plane Polygons and Circles.** *A* **= Area.**

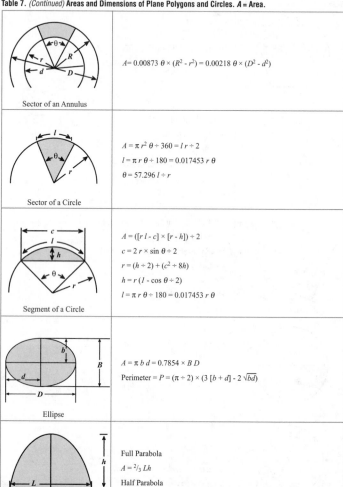

Sector of an Annulus	$A = 0.00873\ \theta \times (R^2 - r^2) = 0.00218\ \theta \times (D^2 - d^2)$
Sector of a Circle	$A = \pi r^2 \theta \div 360 = l r \div 2$ $l = \pi r \theta \div 180 = 0.017453\ r\ \theta$ $\theta = 57.296\ l \div r$
Segment of a Circle	$A = ([r\ l - c] \times [r - h]) \div 2$ $c = 2\ r \times \sin \theta \div 2$ $r = (h \div 2) + (c^2 \div 8h)$ $h = r\ (1 - \cos \theta \div 2)$ $l = \pi r \theta \div 180 = 0.017453\ r\ \theta$
Ellipse	$A = \pi b d = 0.7854 \times B\ D$ Perimeter $= P = (\pi \div 2) \times (3\ [b + d] - 2\ \sqrt{bd})$
Parabola	Full Parabola $A = {}^2\!/_3\ Lh$ Half Parabola $A = {}^2\!/_3\ lh$

Table 8. Areas and Dimensions of Solid Figures. V = Volume, S = Total Surface Area, S_L = Lateral Surface Area.

 Circular Cone	$V = (\pi \div 3) \times r^2 h$ $S = \pi r (r + a)$ $S_L = \pi r a$ $a = \sqrt{h^2 + r^2}$
 Frustum of Cone/Truncated Cone	$V = (\pi \div 12) h (D^2 + Dd + d^2)$ $S = \pi (R^2 + r^2 + a (R + r))$ $S_L = \pi a (R + r)$ $a = \sqrt{h^2 (R - r)^2}$
 Cylinder	$V = {}^1/_4 \pi d^2 h$ $S = 2 \pi r h$ $S_L = 2 \pi r (r + h)$
 Hollow Cylinder	$V = {}^1/_4 \pi h (D^2 - d^2)$
 Portion of Cylinder	$V = {}^1/_4 \pi d^2 h$ $S = \pi r (h_1 + h_2 + r + \sqrt{[r^2 + \{h_1 - h_2\}^2]} \div 4$ $S_L = \pi d h$

(Continued)

Table 8. *(Continued)* **Areas and Dimensions of Solid Figures.** V = Volume, S = Total Surface Area, S_L = Lateral Surface Area.

 Cube	$V = a^3$ $S = 6a^3$ $D = a\sqrt{3}$ $d = a\sqrt{2}$
 Cuboid/Square Prism	$V = abc$ $S = 2(ab + ac + bc)$ $d = \sqrt{(a^2 + b^2 + c^2)}$
 Pyramid	$V = \frac{1}{3} hA$ (A = area of base)
 Frustum of Pyramid	A_1 = area of top, A_2 = area of base $V = \frac{1}{3} h(A_1 + A_2 + \sqrt{A_1 A_2})$
 Sphere	$V = 4\pi r^3 \div 3$ $S = 4\pi r^2 = \pi d^2$

(Continued)

Table 8. *(Continued)* **Areas and Dimensions of Solid Figures.** V = Volume, S = Total Surface Area, S_L = Lateral Surface Area.

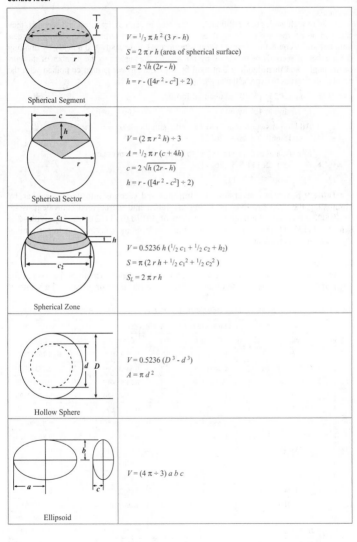

Spherical Segment	$V = \frac{1}{3}\pi h^2 (3r - h)$ $S = 2\pi r h$ (area of spherical surface) $c = 2\sqrt{h(2r - h)}$ $h = r - ([4r^2 - c^2] \div 2)$
Spherical Sector	$V = (2\pi r^2 h) \div 3$ $A = \frac{1}{2}\pi r (c + 4h)$ $c = 2\sqrt{h(2r - h)}$ $h = r - ([4r^2 - c^2] \div 2)$
Spherical Zone	$V = 0.5236\, h\, (\frac{1}{2} c_1 + \frac{1}{2} c_2 + h_2)$ $S = \pi\, (2r h + \frac{1}{2} c_1^2 + \frac{1}{2} c_2^2)$ $S_L = 2\pi r h$
Hollow Sphere	$V = 0.5236\, (D^3 - d^3)$ $A = \pi d^2$
Ellipsoid	$V = (4\pi \div 3)\, a\, b\, c$

Expressing Angles in Decimals, Percentages, and Radians

Decimals
    Even though angles are traditionally expressed in degrees, minutes, and seconds, they can also be expressed as a decimal, as a percentage, or in radians. Converting degrees, minutes, and seconds to a decimal form is performed quite easily and quickly by dividing the number of seconds of the angle by 60, adding the answer to the number of minutes of the angle, and then dividing that total by 60. That answer is then combined with the number of degrees to express the angle as a decimal.

Example. Convert 35° 15' 40" to decimal format.

First, divide 40" by 60 $40 \div 60 = .66667$.

Add the answer (.66667) to 15' and again divide by 60
    $15.66667 \div 60 = .26111$.

Finally, add the answer to 35° to complete the conversion
    $35 + .26111 = 35.26111$.

35° 15' 40" = 35.26111.

Table 9 provides conversions for minutes and seconds into decimals, thereby eliminating the need for division. Using the Table, find the decimal equivalent to 15' (0.250002), then add the decimal equivalent of 40" (0.0111112), which provides the sum 0.2611132. This is slightly more accurate and less time consuming than performing the necessary mathematics.

Percentages
    When measuring angles in percentages, a ratio between the length of the base of the angle and the amount of the rise of the angle is established. If, for instance, the length of

Table 9. Minutes and Seconds Converted to Decimals of a Degree. 1 min. = 0.01666680 Degree.

Minutes to Decimals of a Degree							
Min.	Decimal	Min.	Decimal	Min.	Decimal	Min.	Decimal
1	0.01666680	16	0.26666880	31	0.51667080	46	0.76667280
2	0.03333360	17	0.28333560	32	0.53333760	47	0.78333960
3	0.05000040	18	0.30000240	33	0.55000440	48	0.80000640
4	0.06666720	19	0.31666920	34	0.56667120	49	0.81667320
5	0.08333400	20	0.33333600	35	0.58333800	50	0.83334000
6	0.10000080	21	0.35000280	36	0.60000480	51	0.85000680
7	0.11666760	22	0.36666960	37	0.61667160	52	0.86667360
8	0.13333440	23	0.38333640	38	0.63333840	53	0.88334040
9	0.15000120	24	0.40000320	39	0.65000520	54	0.90000720
10	0.16666800	25	0.41667000	40	0.66667200	55	0.91667400
11	0.18333480	26	0.43333680	41	0.68333880	56	0.93334080
12	0.20000160	27	0.45000360	42	0.70000560	57	0.95000760
13	0.21666840	28	0.46667040	43	0.71667240	58	0.96667440
14	0.23333520	29	0.48333720	44	0.73333920	59	0.98334120
15	0.25000200	30	0.50000400	45	0.75000600	60	1.00000000

(Continued)

Table 9. *(Continued)* **Minutes and Seconds Converted to Decimals of a Degree.** 1 min. = 0.01666680 Degree.

			Seconds to Decimals of a Degree				
Sec.	Decimal	Sec.	Decimal	Sec.	Decimal	Sec.	Decimal
1	0.00027778	16	0.00444448	31	0.00861118	46	0.01277788
2	0.00055556	17	0.00472226	32	0.00888896	47	0.01305566
3	0.00083334	18	0.00500004	33	0.00916674	48	0.01333344
4	0.00111112	19	0.00527782	34	0.00944452	49	0.01361122
5	0.00138890	20	0.00555560	35	0.00972230	50	0.01388900
6	0.00166668	21	0.00583338	36	0.01000008	51	0.01416678
7	0.00194446	22	0.00611116	37	0.01027786	52	0.01444456
8	0.00222224	23	0.00638894	38	0.01055564	53	0.01472234
9	0.00250002	24	0.00666672	39	0.01083342	54	0.01500012
10	0.00277780	25	0.00694450	40	0.01111120	55	0.01527790
11	0.00305558	26	0.00722228	41	0.01138898	56	0.01555568
12	0.00333336	27	0.00750006	42	0.01166676	57	0.01583346
13	0.00361114	28	0.00777784	43	0.01194454	58	0.01611124
14	0.00388892	29	0.00805562	44	0.01222232	59	0.01638902
15	0.00416670	30	0.00833340	45	0.01250010	60	0.01666680

the base of the angle is 20 inches, and the rise over that distance is four inches, the angle, expressed as a percentage, would be 20%. This is solved by dividing the amount of rise by the base length, and multiplying the answer by 100 as follows:

$$4 \div 20 \times 100 = 20\%$$

Another way of expressing this angle would be to call it a 1 in 5 taper. In this case, the name refers to the observation that for every five inches of base, the rise increases by one inch. Tapers are covered in detail in another section.

Radians

The expression of angles as radians has been popular with mathematicians, and the concept is based on the circumference of a circle, which is twice the radius times π ($C = 2 \pi r$). Therefore, a full circle can be seen as corresponding to an angle of 2π radians ($360° = 2 \pi$ and $180° = \pi$). One degree can be expressed as $2 \pi \div 360 = 0.01745329252$ radians. One minute will therefore be equal to $0.01745329252 \div 60 = 0.0002908882$ radians, and one second will be equal to $0.0002928882 \div 60 = 0.0000048481$ radians.

A radian actually represents the proportionate length of arc resulting from an angle within a circle. For example, $90°$ is equal to 1.5707963268 radians, which is exactly one-quarter of 2π. Because they are directly proportional to the angle, radians offer a shortcut for computing the area of a sector of a circle, which is usually solved by the formula $A = \pi r^2 \alpha \div 360$. The formula requires that the angle of the sector be multiplied by the radius squared and then π, before being divided by 360. However, when using radians, the same formula can be solved with the equation $A = \frac{1}{2} \alpha r^2$.

For example, find the area of a sector when the angle, α, equals $18°$, and the radius is 2.5 inches.

$A = \frac{1}{2} \alpha r^2$, and $\frac{1}{2} \alpha = 9°$
α, in radians = 0.1570796327 (from **Table 10**)
$A = 0.1570796327 \times (2.5 \times 2.5) = 0.981747704375$ in.^{2}.

(Text continued on p. 77)

Table 10. Degrees Converted to Radians. 2π Radians = 360°

Degrees	Radians	Degrees	Radians	Degrees	Radians
0.01	0.0001745329	33	0.5759586532	74	1.2915436465
0.02	0.0003490659	34	0.5934119457	75	1.3089969390
0.03	0.0005235988	35	0.6108652382	76	1.3264502315
0.04	0.0006981317	36	0.6283185307	77	1.3439035240
0.05	0.0008726646	37	0.6457718232	78	1.3613568166
0.06	0.0010471976	38	0.6632251158	79	1.3788101091
0.07	0.0012217305	39	0.6806784083	80	1.3962634016
0.08	0.0013962634	40	0.6981317008	81	1.4137166941
0.09	0.0015707963	41	0.7155849933	82	1.4311699866
1	0.0174532925	42	0.7330382858	83	1.4486232792
2	0.0349065850	43	0.7504915784	84	1.4660765717
3	0.0523598776	44	0.7679448709	85	1.4835298642
4	0.0698131701	45	0.7853981634	86	1.5009831567
5	0.0872664626	46	0.8028514559	87	1.5184364492
6	0.1047197551	47	0.8203047484	88	1.5358897418
7	0.1221730476	48	0.8377580410	89	1.5533430343
8	0.1396263402	49	0.8552113335	90	1.5707963268
9	0.1570796327	50	0.8726646260	91	1.5882496193
10	0.1745329252	51	0.8901179185	92	1.6057029118
11	0.1919862177	52	0.9075712110	93	1.6231562044
12	0.2094395102	53	0.9250245036	94	1.6406094969
13	0.2268928028	54	0.9424777961	95	1.6580627894
14	0.2443460953	55	0.9599310886	96	1.6755160819
15	0.2617993878	56	0.9773843811	97	1.6929693744
16	0.2792526803	57	0.9948376736	98	1.7104226670
17	0.2967059728	58	1.0122909662	99	1.7278759595
18	0.3141592654	59	1.0297442587	100	1.7453292520
19	0.3316125579	60	1.0471975512	101	1.7627825445
20	0.3490658504	61	1.0646508437	102	1.7802358370
21	0.3665191429	62	1.0821041362	103	1.7976891296
22	0.3839724354	63	1.0995574288	104	1.8151424221
23	0.4014257280	64	1.1170107213	105	1.8325957146
24	0.4188790205	65	1.1344640138	106	1.8500490071
25	0.4363323130	66	1.1519173063	107	1.8675022996
26	0.4537856055	67	1.1693705988	108	1.8849555922
27	0.4712388980	68	1.1868238914	109	1.9024088847
28	0.4886921906	69	1.2042771839	110	1.9198621772
29	0.5061454831	70	1.2217304764	111	1.9373154697
30	0.5235987756	71	1.2391837689	112	1.9547687622
31	0.5410520681	72	1.2566370614	113	1.9722220548
32	0.5585053606	73	1.2740903540	114	1.9896753473

(Continued)

Table 10. *(Continued)* **Degrees Converted to Radians.** 2π Radians = 360°

Degrees	Radians	Degrees	Radians	Degrees	Radians
115	2.0071286398	156	2.7227136300	197	3.4382986225
116	2.0245819323	157	2.7401669225	198	3.4557519150
117	2.0420352248	158	2.7576202150	199	3.4732052075
118	2.0594885174	159	2.7576202150	200	3.4906585000
119	2.0769418099	160	2.7925268000	201	3.5081117925
120	2.0943951024	161	2.8099800925	202	3.5255650850
121	2.1118483949	162	2.8274333850	203	3.5430183775
122	2.1293016874	163	2.8448866775	204	3.5604716700
123	2.1467549800	164	2.8623399700	205	3.5779249625
124	2.1642082725	165	2.8797932625	206	3.5953782550
125	2.1816615650	166	2.8972465550	207	3.6128315475
126	2.1991148575	167	2.9146998475	208	3.6302848400
127	2.2165681500	168	2.9321531400	209	3.6477381325
128	2.2340214426	169	2.9496064325	210	3.6651914250
129	2.2514747351	170	2.9670597250	211	3.6826447175
130	2.2689280276	171	2.9845130175	212	3.7000980100
131	2.2863813201	172	3.0019663100	213	3.7175513025
132	2.3038346126	173	3.0194196025	214	3.7350045950
133	2.3212879052	174	3.0368728950	215	3.7524578875
134	2.3387411977	175	3.0543261875	216	3.7699111800
135	2.3561944902	176	3.0717794800	217	3.7873644725
136	2.3736477800	177	3.0892327725	218	3.8048177650
137	2.3911010725	178	3.1066860650	219	3.8222710575
138	2.4085543650	179	3.1241393575	220	3.8397243500
139	2.4260076575	180	3.1415926500	221	3.8571776425
140	2.4434609500	181	3.1590459425	222	3.8746309350
141	2.4609142425	182	3.1764992350	223	3.8920842275
142	2.4783675350	183	3.1939525275	224	3.9095375200
143	2.4958208275	184	3.2114058200	225	3.9269908125
144	2.5132741200	185	3.2288591125	226	3.9444441050
145	2.5307274125	186	3.2463124050	227	3.9618973975
146	2.5481807050	187	3.2637656975	228	3.9793506900
147	2.5656339975	188	3.2812189900	229	3.9968039825
148	2.5830872900	189	3.2986722825	230	4.0142572750
149	2.6005405825	190	3.3161255750	231	4.0317105675
150	2.6179938750	191	3.3335788675	232	4.0491638600
151	2.6354471675	192	3.3510321600	233	4.0666171525
152	2.6529004600	193	3.3684854525	234	4.0840704450
153	2.6703537525	194	3.3859387450	235	4.1015237375
154	2.6878070450	195	3.4033920375	236	4.1189770300
155	2.7052603375	196	3.4208453300	237	4.1364303225

(Continued)

Table 10. *(Continued)* **Degrees Converted to Radians.** 2π Radians = 360°

Degrees	Radians	Degrees	Radians	Degrees	Radians
238	4.1538836150	279	4.8694686075	320	5.5850536000
239	4.1713369075	280	4.8869219000	321	5.6025068925
240	4.1887902000	281	4.9043751925	322	5.6199601850
241	4.2062434925	282	4.9218284850	323	5.6374134775
242	4.2236967850	283	4.9392817775	324	5.6548667700
243	4.2411500775	284	4.9567350700	325	5.6723200625
244	4.2586033700	285	4.9741883625	326	5.6897733550
245	4.2760566625	286	4.9916416550	327	5.7072266475
246	4.2935099550	287	5.0090949475	328	5.7246799400
247	4.3109632475	288	5.0265482400	329	5.7421332325
248	4.3284165400	289	5.0440015325	330	5.7595865250
249	4.3458698325	290	5.0614548250	331	5.7770398175
250	4.3633231250	291	5.0789081175	332	5.7944931100
251	4.3807764175	292	5.0963614100	333	5.8119464025
252	4.3807764175	293	5.1138147025	334	5.8293996950
253	4.4156830025	294	5.1312679950	335	5.8468529875
254	4.4331362950	295	5.1487212875	336	5.8643062800
255	4.4505895875	296	5.1661745800	337	5.8817595725
256	4.4680428800	297	5.1836278725	338	5.8992128650
257	4.4854961725	298	5.2010811650	339	5.9166661575
258	4.5029494650	299	5.2185344575	340	5.9340094500
259	4.5204027575	300	5.2359877500	341	5.9515727425
260	4.5378560500	301	5.2534410425	342	5.9690260350
261	4.5553093425	302	5.2708943350	343	5.9864793275
262	4.5727626350	303	5.2883476275	344	6.0039326200
263	4.5902159275	304	5.3058009200	345	6.0213859125
264	4.6076692200	305	5.3232542125	346	6.0388392050
265	4.6251225125	306	5.3407075050	347	6.0562924975
266	4.6425758050	307	5.3581607975	348	6.0737457900
267	4.6600290975	308	5.3756140900	349	6.0911990825
268	4.6774823900	309	5.3930673825	350	6.1086523750
269	4.6949356825	310	5.4105206750	351	6.1261056675
270	4.7123889750	311	5.4279739675	352	6.1435589600
271	4.7298422675	312	5.4454272600	353	6.1610122525
272	4.7472955600	313	5.4628805525	354	6.1784655450
273	4.7647488525	314	5.4803338450	355	6.1959188375
274	4.7822021450	315	5.4977871375	356	6.2133721300
275	4.7996554375	316	5.5152404300	357	6.2308254225
276	4.8171087300	317	5.5326937225	358	6.2482787150
277	4.8345620225	318	5.5501470150	359	6.2657320075
278	4.8520153150	319	5.5676003075	360	6.2831853000

Using the standard formula would require the following steps.

$A = \pi\, r^2\, \alpha \div 360$
$A = (3.1416 \times [2.5 \times 2.5] \times 18) \div 360$
$A = 353.45 \div 360 = 0.98175$ in.^{2}.

As can be seen, the result is also more accurate using the radian measure.

Table 10 provides degrees converted to radians, **Table 11** shows minutes and seconds of degrees converted to radians, **Table 12** converts radians to decimal degrees, and **Table 13** gives radians converted to degrees, minutes, and seconds.

Table 11. Minutes and Seconds Converted to Radians. 2π Radians = 360°

Minutes to Radians							
Min.	Radians	Min.	Radians	Min.	Radians	Min.	Radians
1	0.0002908882	16	0.0046542112	31	0.0090175342	46	0.0133808572
2	0.0005817764	17	0.0049450994	32	0.0093084224	47	0.0136717454
3	0.0008726646	18	0.0052359876	33	0.0095993106	48	0.0139626336
4	0.0011635528	19	0.0055268758	34	0.0098901988	49	0.0142535218
5	0.0014544410	20	0.0058177640	35	0.0101810870	50	0.0145444100
6	0.0017453292	21	0.0061086522	36	0.0104719752	51	0.0148352982
7	0.0020362174	22	0.0063995404	37	0.0107628634	52	0.0151261864
8	0.0023271056	23	0.0066904286	38	0.0110537516	53	0.0154170746
9	0.0026179938	24	0.0069813168	39	0.0113446398	54	0.0157079628
10	0.0029088820	25	0.0072722050	40	0.0116355280	55	0.0159988510
11	0.0031997702	26	0.0075630932	41	0.0119264162	56	0.0162897392
12	0.0034906584	27	0.0078539814	42	0.0122173044	57	0.0165806274
13	0.0037815466	28	0.0081448696	43	0.0125081926	58	0.0168715156
14	0.0040724348	29	0.0084357578	44	0.0127990808	59	0.0171624038
15	0.0043633230	30	0.0087266460	45	0.0130899690	60	0.0174532920
Seconds to Radians							
Sec.	Radians	Sec.	Radians	Sec.	Radians	Sec.	Radians
1	0.0000048481	16	0.0000775702	31	0.0001018109	46	0.0002230143
2	0.0000096963	17	0.0000824183	32	0.0001551404	47	0.0002278624
3	0.0000145444	18	0.0000872665	33	0.0001599885	48	0.0002327106
4	0.0000193925	19	0.0000921146	34	0.0001648366	49	0.0002375587
5	0.0000242407	20	0.0000969627	35	0.0001696848	50	0.0002424068
6	0.0000290888	21	0.0001018109	36	0.0001745329	51	0.0002472550
7	0.0000339370	22	0.0001066590	37	0.0001793811	52	0.0002521031
8	0.0000387851	23	0.0001115071	38	0.0001842292	53	0.0002569512
9	0.0000436332	24	0.0001163553	39	0.0001890773	54	0.0002617994
10	0.0000484814	25	0.0001212034	40	0.0001939255	55	0.0002666475
11	0.0000533295	26	0.0001260516	41	0.0001987736	56	0.0002714957
12	0.0000581776	27	0.0001308997	42	0.0002036217	57	0.0002763438
13	0.0000630258	28	0.0001357478	43	0.0002084699	58	0.0002811919
14	0.0000678739	29	0.0001405960	44	0.0002133180	59	0.0002860401
15	0.0000727221	30	0.0001454441	45	0.0002181662	60	0.0002908882

Table 12. Radians Converted to Decimal Degrees. 2π Radians = 360°

Radians	Degrees	Radians	Degrees	Radians	Degrees
0.0001	0.0057295780	0.5	28.6478897566	26	1489.6902673406
0.0002	0.0114591559	0.6	34.3774677079	27	1546.9860468537
0.0003	0.0171887339	0.7	40.1070456592	28	1604.2818263668
0.0004	0.0229183118	0.8	45.8366236105	29	1661.5776058799
0.0005	0.0286478898	0.9	51.5662015618	30	1718.8733853930
0.0006	0.0343774677	1	57.2957795131	31	1776.1691649061
0.0007	0.0401070457	2	114.5915590262	32	1833.4649444192
0.0008	0.0458366236	3	171.8873385393	33	1890.7607239323
0.0009	0.0515662016	4	229.1831180524	34	1948.0565034454
0.001	0.0572957795	5	286.4788975655	35	2005.3522829585
0.002	0.1145915590	6	343.7746770786	36	2062.6480624716
0.003	0.1718873385	7	401.0704565917	37	2119.9438419847
0.004	0.2291831181	8	458.3662361048	38	2177.2396214978
0.005	0.2864788976	9	515.6620156179	39	2234.5354010109
0.006	0.3437746771	10	572.9577951310	40	2291.8311805240
0.007	0.4010704566	11	630.2535746441	41	2349.1269600371
0.008	0.4583662361	12	687.5493541572	42	2406.4227395502
0.009	0.5156620156	13	744.8451336703	43	2463.7185190633
0.01	0.5729577951	14	802.1409131834	44	2521.0142985764
0.02	1.1459155903	15	859.4366926965	45	2578.3100780895
0.03	1.7188733854	16	916.7324722096	46	2635.6058576026
0.04	2.2918311805	17	974.0282517227	47	2692.9016371157
0.05	2.8647889757	18	1031.3240312358	48	2750.1974166288
0.06	3.4377467708	19	1088.6198107489	49	2807.4931961419
0.07	4.0107045659	20	1145.9155902620	50	2864.7889756550
0.08	4.5836623610	21	1203.2113697751	60	3437.7467707860
0.09	5.1566201562	22	1260.5071492882	70	4010.7045659170
0.1	5.7295779513	23	1317.8029288013	80	4583.6623610480
0.2	11.4591559026	24	1375.0987083144	90	5156.6201561790
0.3	17.1887338539	25	1432.3944878275	100	5729.5779513100
0.4	22.9183118052				

Table 13. Radians Converted to Angles (Degrees, Minutes, Seconds). 2π Radians = 360°

Radians	Angle	Radians	Angle	Radians	Angle
0.0001	0° 0' 21"	0.5	28° 38' 52"	27	1546° 59' 10"
0.0002	0° 0' 41"	0.6	34° 22' 39"	28	1604° 16' 54"
0.0003	0° 1' 2"	0.7	40° 6' 25"	29	1661° 34' 39"
0.0004	0° 1' 23"	0.8	45° 50' 12"	30	1718° 52' 24"
0.0005	0° 1' 43"	0.9	51° 33' 58"	31	1776° 10' 9"
0.0006	0° 2' 4"	1	57° 17' 45"	32	1833° 27' 54"
0.0007	0° 2' 24"	2	114° 35' 30"	33	1890° 45' 39"
0.0008	0° 2' 45"	3	171° 53' 14"	34	1948° 3' 23"
0.0009	0° 3' 6"	4	229° 10' 59"	35	2005° 21' 8"
0.001	0° 3' 26"	5	286° 28' 44"	36	2062° 38' 53"
0.002	0° 6' 53"	6	343° 46' 29"	37	2119° 56' 38"
0.003	0° 10' 19"	7	401° 4' 14"	38	2177° 14' 23"
0.004	0° 13' 45"	8	458° 21' 58"	39	2234° 32' 7"
0.005	0° 17' 11"	9	515° 39' 43"	40	2291° 49' 52"
0.006	0° 20' 38"	10	572° 57' 28"	41	2349° 7' 37"
0.007	0° 24' 4"	11	630° 15' 13"	42	2406° 25' 22"
0.008	0° 27' 30"	12	687° 32' 58"	43	2463° 43' 7"
0.009	0° 30' 56"	13	744° 50' 42"	44	2521° 0' 51"
0.01	0° 34' 23"	14	802° 8' 27"	45	2578° 18' 36"
0.02	1° 8' 45"	15	859° 26' 12"	46	2635° 36' 21"
0.03	1° 43' 8"	16	916° 43' 57"	47	2692° 54' 6"
0.04	1° 17' 31"	17	974° 1' 42"	48	2750° 11' 51"
0.05	2° 51' 53"	18	1031° 19' 26"	49	2807° 29' 36"
0.06	3° 26' 16"	19	1088° 37' 11"	50	2864° 47' 20"
0.07	4° 0' 39"	20	1145° 54' 56"	60	3437° 44' 48"
0.08	4° 35' 2"	21	1203° 12' 41"	70	4010° 42' 17"
0.09	5° 9' 24"	22	1260° 30' 26"	80	4583° 39' 45"
0.1	5° 43' 46"	23	1317° 48' 10"	90	5156° 37' 13"
0.2	11° 27' 33"	24	1375° 5' 55"	100	5729° 34' 41"
0.3	17° 11' 19"	25	1432° 23' 40"		
0.4	22° 55' 6"	26	1489° 41' 25"		

Segments and Sectors of Circles

Table 14 provides values for segments and sectors of a circle, given the central angle (α) in degrees (column 1) and the radius (r), which is equal to 1 (for all units of measurement – inch, centimeter, etc.). From these variables, the arc length (b), height of the circle segment (h), the chord length (s), the area of the segment (A), the area of the sector (A'), and the centers of gravity of both the segment (cg) and sector (cg') can be derived. If the radius and central angle (α) are unknown, but the height and chord length can be measured, divide the height by the chord length and consult the final column of the Table (h/s) to find which two central angles the value falls between. By subtracting the h/s value of the lower angle from the higher angle, and dividing by 60, the figure can be broken approximately into minute increments.

Example. If $h = 0.061$ and $s = 0.685$, then $h \div s = 0.08905$. This value falls between the h/s ratio values on the chart for angles of 40° ($h/s = 0.088167$) and 41° ($h/s = 0.090417$). Next subtract the lower from the higher and divide by 60 to derive the value of 1 minute.

$$0.090417 - 0.088167 = 0.00225 \div 60 = .0000375.$$

Therefore, each increase of 0.0000375 over the base h/s value for 40° equals 1 minute. Our example, 0.08905 contains 23.5 increases of this value, which means the central angle, α, is 40° 23' 30" (30" = 0.5 minute).

$$0.08905 - 0.088167 = .000883 \div .0000375 = 23.5.$$

Then find the radius, r, with the formula

$$r = (h \div 2) + (s^2 \div 8h)$$

With a, r, h, and s known, all other values can be calculated.

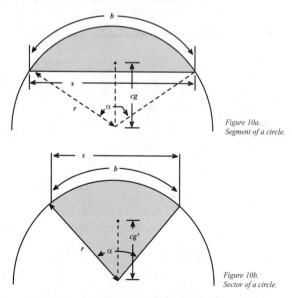

Figure 10a.
Segment of a circle.

Figure 10b.
Sector of a circle.

NOTE: Although $r = 1$ on the Table, the values for s (chord length), h (segment height), and b (arc length) can be multiplied by a given radius to solve for larger circles. The center of gravity (cg), which is the distance from the point of center of gravity to the center of the circle, can also be multiplied by the radius to find the center of gravity for segments and sectors in larger circles. The areas must be multiplied by r^2 to solve for other size circles.

The formulas given below were used to prepare **Table 14**.

Height of Segment	$h = r (1 - \cos [\alpha \div 2])$
Length of Chord	$s = 2r \times \sin (\alpha \div 2)$
Length of Arc	$b = (\pi \times r \times \alpha) \div 180$
Area of Segment	$A = (r^2 \div 2) \times (\alpha \text{ [expressed in radians]} - \sin \alpha)$
Center of Gravity of Segment	$cg = s^3 \div (12 \times A)$
Area of Sector	$A' = (l \times r) \div 2$
Center of Gravity of Sector	$cg' = (2 \times r s) \div (3 \times b)$

A comparison between **Table 14** and **Table 10** gives a clear indication of how radians and angles interrelate. By examining both Tables, it can be seen that the Length of Arc (b) in **Table 14** is actually expressed in radians.

Table 14. Chord Length, Arc Height, Arc Length, Area, and Center of Gravity for Segments and Sectors of a Circle. Radius = 1.

Degrees α	Chord Length s	Arc Height h	Arc Length b	Area of Segment A	Segment Center of Gravity cg	Area of Sector A'	Sector Center of Gravity cg'	h/s ratio
1	0.0174530	0.00004	0.0174533	0.0000006463	Area too small to accurately calculate center of gravity	0.008727	0.666655	0.002292
2	0.0349048	0.00015	0.0349066	0.0000089250		0.017453	0.666633	0.004297
3	0.0523538	0.00034	0.0523599	0.0000119388		0.026180	0.666589	0.006494
4	0.0697990	0.00061	0.0698132	0.0000285850		0.034907	0.666531	0.008739
5	0.0872386	0.00095	0.0872665	0.0000552313		0.043633	0.666454	0.010890
6	0.1046800	0.00137	0.1047198	0.0000958776		0.052360	0.666414	0.013088
7	0.1221000	0.00187	0.1221730	0.0001520238		0.061087	0.666268	0.015315
8	0.1395200	0.00244	0.1396263	0.0002266701	0.998468	0.069813	0.666159	0.017489
9	0.1569180	0.00308	0.1570796	0.0003228163	0.997428	0.078540	0.665981	0.019628
10	0.1743000	0.00381	0.1745329	0.0004424626	0.997320	0.087266	0.665777	0.021859
11	0.1916800	0.00460	0.1919862	0.0005886089	0.997062	0.095993	0.665603	0.023998
12	0.2090600	0.00548	0.2094395	0.0007637551	0.996959	0.104720	0.665459	0.026213
13	0.2264000	0.00643	0.2268928	0.0009709014	0.996031	0.113446	0.665219	0.028401
14	0.2437400	0.00745	0.2443461	0.0012120476	0.995588	0.122173	0.665013	0.030565
15	0.2610600	0.00856	0.2617994	0.0014901939	0.994940	0.130900	0.664784	0.032789
16	0.2783400	0.00973	0.2792527	0.0018078402	0.993998	0.139626	0.664488	0.034957
17	0.2956200	0.01099	0.2967060	0.0021669864	0.993491	0.148353	0.664227	0.037176
18	0.3128600	0.01231	0.3141593	0.0025711327	0.992532	0.157080	0.663910	0.039347
19	0.3301000	0.01372	0.3316126	0.0030222789	0.991792	0.165806	0.663626	0.041563
20	0.3473000	0.01519	0.3490659	0.0035229252	0.990900	0.174533	0.663294	0.043737
21	0.3646600	0.01675	0.3665191	0.0040755715	0.989874	0.183260	0.662921	0.045933
22	0.3816200	0.01837	0.3839724	0.0046827177	0.989041	0.191986	0.662582	0.048137

(Continued)

Table 14. *(Continued)* Chord Length, Arc Height, Arc Length, Area, and Center of Gravity for Segments and Sectors of a Circle. Radius = 1.

Degrees α	Chord Length s	Arc Height h	Arc Length b	Area of Segment A	Segment Center of Gravity cg	Area of Sector A'	Sector Center of Gravity cg'	h/s ratio
23	0.3987400	0.02008	0.4014257	0.0053473640	0.987981	0.200713	0.662206	0.050359
24	0.4158200	0.02185	0.4188790	0.0060710102	0.986902	0.209440	0.661798	0.052547
25	0.4328800	0.02370	0.4363323	0.0068571565	0.985774	0.218166	0.661392	0.054750
26	0.4499000	0.02563	0.4537856	0.0077073028	0.984610	0.226893	0.660958	0.056968
27	0.4668800	0.02763	0.4712389	0.0086244490	0.983339	0.235619	0.660500	0.059180
28	0.4838400	0.02971	0.4886922	0.0096100953	0.982192	0.244346	0.660047	0.061405
29	0.5007600	0.03185	0.5061455	0.0106677415	0.980924	0.253073	0.659573	0.063603
30	0.5176400	0.03407	0.5235988	0.0117993878	0.979586	0.261799	0.659080	0.065818
31	0.5344800	0.03637	0.5410521	0.0130070341	0.978216	0.270526	0.658569	0.068047
32	0.5512800	0.03874	0.5585054	0.0142931803	0.976802	0.279253	0.658042	0.070273
33	0.5680200	0.04118	0.5759587	0.0156598266	0.975265	0.287979	0.657478	0.072497
34	0.5847400	0.04370	0.5934119	0.0171094728	0.973802	0.296706	0.656924	0.074734
35	0.6014000	0.04628	0.6108652	0.0186446191	0.972200	0.305433	0.656337	0.076954
36	0.6180400	0.04840	0.6283185	0.0202692654	0.970578	0.314159	0.655761	0.078312
37	0.6346000	0.05168	0.6457718	0.0219784116	0.968997	0.322886	0.655133	0.081437
38	0.6511400	0.05448	0.6632251	0.0237820579	0.967370	0.331613	0.654519	0.083669
39	0.6676200	0.05736	0.6806784	0.0256792041	0.965662	0.340339	0.653877	0.085917
40	0.6840400	0.06031	0.6981317	0.0276718504	0.963885	0.349066	0.653210	0.088167
41	0.7004200	0.06333	0.7155850	0.0297629967	0.962094	0.357792	0.652538	0.090417
42	0.7167400	0.06642	0.7330383	0.0319536429	0.960248	0.366519	0.651844	0.092670
43	0.7330000	0.06958	0.7504916	0.0342467892	0.958321	0.375246	0.651129	0.094925
44	0.7492200	0.07282	0.7679449	0.0366434354	0.956424	0.383972	0.650411	0.097194
45	0.7653600	0.07612	0.7853982	0.0391455817	0.954406	0.392699	0.649658	0.099456
46	0.7814600	0.07950	0.8028515	0.0417557280	0.952408	0.401426	0.648904	0.101733
47	0.7975000	0.08294	0.8203047	0.0444773742	0.950324	0.410152	0.648133	0.104000
48	0.8134800	0.08646	0.8377580	0.0473065205	0.948284	0.418879	0.647347	0.106284
49	0.8293800	0.09004	0.8552113	0.0502506667	0.946101	0.427606	0.646530	0.108563
50	0.8452400	0.09369	0.8726646	0.0533103130	0.943947	0.436332	0.645716	0.110844
51	0.8610200	0.09742	0.8901179	0.0564859593	0.941712	0.445059	0.644873	0.113145
52	0.8767400	0.10121	0.9075712	0.0597851055	0.939373	0.453786	0.644019	0.115439
53	0.8924000	0.10507	0.9250245	0.0631942518	0.937173	0.462512	0.643154	0.117739
54	0.9079800	0.10899	0.9424778	0.0667303980	0.934811	0.471239	0.642264	0.120036
55	0.9235000	0.11299	0.9599311	0.0703895443	0.932441	0.479966	0.641365	0.122350
56	0.9389400	0.11705	0.9773844	0.0741731906	0.930005	0.488692	0.640444	0.124662
57	0.9543200	0.12118	0.9948377	0.0780833368	0.927561	0.497419	0.639515	0.126980
58	0.9696200	0.12538	1.0122910	0.0821214831	0.925053	0.506145	0.638565	0.129308
59	0.9848400	0.12965	1.0297443	0.0862886293	0.922491	0.514872	0.637595	0.131646

(Continued)

Table 14. *(Continued)* Chord Length, Arc Height, Arc Length, Area, and Center of Gravity for Segments and Sectors of a Circle. Radius = 1.

Degrees α	Chord Length s	Arc Height h	Arc Length b	Area of Segment A	Segment Center of Gravity cg	Area of Sector A'	Sector Center of Gravity cg'	h/s ratio
60	1.0000000	0.13397	1.0471976	0.0905862756	0.919933	0.523599	0.636620	0.133970
61	1.0150800	0.13837	1.0646508	0.0950154219	0.917330	0.532325	0.635626	0.136314
62	1.0300800	0.14283	1.0821041	0.0995780681	0.914677	0.541052	0.634615	0.138659
63	1.0450000	0.14736	1.0995574	0.1042752144	0.911983	0.549779	0.633588	0.141014
64	1.0598400	0.15195	1.1170107	0.1091083606	0.909247	0.558505	0.632545	0.143371
65	1.0746000	0.15661	1.1344640	0.1140780069	0.906478	0.567232	0.631488	0.145738
66	1.0892800	0.16133	1.1519173	0.1191861532	0.903673	0.575959	0.630416	0.148107
67	1.1038800	0.16612	1.1693706	0.1244327994	0.900844	0.584685	0.629330	0.150487
68	1.1183800	0.17096	1.1868239	0.1298199457	0.897936	0.593412	0.628220	0.152864
69	1.1328200	0.17587	1.2042772	0.1353485919	0.895051	0.602139	0.627109	0.155250
70	1.1471600	0.18085	1.2217305	0.1410187382	0.892101	0.610865	0.625975	0.157650
71	1.1614000	0.18589	1.2391838	0.1468323845	0.889083	0.619592	0.624820	0.160057
72	1.1755600	0.19098	1.2566371	0.1527900307	0.886050	0.628319	0.623654	0.162459
73	1.1896400	0.19614	1.2740904	0.1588926770	0.883002	0.637045	0.622478	0.164873
74	1.2036200	0.20127	1.2915436	0.1651408232	0.879898	0.645772	0.621282	0.167221
75	1.2175200	0.20665	1.3089969	0.1715354695	0.876785	0.654498	0.620078	0.169730
76	1.2313200	0.21199	1.3264502	0.1780771158	0.873622	0.663225	0.618855	0.172165
77	1.2450200	0.21739	1.3439035	0.1847667620	0.870410	0.671952	0.617614	0.174608
78	1.2586400	0.22285	1.3613568	0.1916044083	0.867197	0.680678	0.616365	0.177056
79	1.2721600	0.22838	1.3788101	0.1985915545	0.863939	0.689405	0.615100	0.179521
80	1.2855800	0.23396	1.3962634	0.2057277008	0.860644	0.698132	0.613819	0.181988
81	1.2989000	0.23959	1.4137167	0.2130143471	0.857308	0.706858	0.612523	0.184456
82	1.3121200	0.24529	1.4311700	0.2204509933	0.853940	0.715585	0.611211	0.186942
83	1.3252400	0.25105	1.4486233	0.2280386396	0.850538	0.724312	0.609885	0.189437
84	1.3382600	0.25686	1.4660766	0.2357772858	0.847107	0.733038	0.608545	0.191936
85	1.3511800	0.26272	1.4835299	0.2436674321	0.843647	0.741765	0.607191	0.194437
86	1.3640000	0.26865	1.5009832	0.2517095784	0.840160	0.750492	0.605825	0.196957
87	1.3767000	0.27463	1.5184364	0.2599032246	0.836614	0.759218	0.604438	0.199484
88	1.3893200	0.28066	1.5358897	0.2682493709	0.833080	0.767945	0.603047	0.202012
89	1.4018200	0.28675	1.5533430	0.2767475171	0.829491	0.776672	0.601636	0.204556
90	1.4142200	0.29289	1.5707963	0.2853981634	0.825883	0.785398	0.600214	0.207104
91	1.4265000	0.29909	1.5882496	0.2942008097	0.822224	0.794125	0.598772	0.209667
92	1.4386800	0.30534	1.6057029	0.3031559559	0.818550	0.802851	0.597321	0.212236
93	1.4507400	0.31165	1.6231562	0.3122631022	0.814830	0.811578	0.595851	0.214821
94	1.4627000	0.31800	1.6406095	0.3215227484	0.811097	0.820305	0.594373	0.217406
95	1.4745600	0.32441	1.6580628	0.3309338947	0.807356	0.829031	0.592885	0.220005
96	1.4862800	0.33087	1.6755161	0.3404970410	0.803540	0.837758	0.591372	0.222616

(Continued)

Table 14. *(Continued)* Chord Length, Arc Height, Arc Length, Area, and Center of Gravity for Segments and Sectors of a Circle. Radius = 1.

Degrees α	Chord Length s	Arc Height h	Arc Length b	Area of Segment A	Segment Center of Gravity cg	Area of Sector A'	Sector Center of Gravity cg'	h/s ratio
97	1.4979000	0.33738	1.6929694	0.3502116872	0.799717	0.846485	0.589851	0.225235
98	1.5094200	0.34394	1.7104227	0.3600773335	0.795890	0.855211	0.588322	0.227862
99	1.5208200	0.35055	1.7278760	0.3700939797	0.792027	0.863938	0.586778	0.230501
100	1.5320800	0.35721	1.7453293	0.3802606260	0.788101	0.872665	0.585211	0.233154
101	1.5432400	0.36392	1.7627825	0.3905777723	0.784173	0.881391	0.583638	0.235816
102	1.5543000	0.37068	1.7802358	0.4010439185	0.780246	0.890118	0.582058	0.238487
103	1.5652200	0.37749	1.7976891	0.4116595648	0.776259	0.898845	0.580456	0.241174
104	1.5760200	0.38434	1.8151424	0.4224232110	0.772247	0.907571	0.578842	0.243867
105	1.5867000	0.39124	1.8325957	0.4333348573	0.768209	0.916298	0.577214	0.246575
106	1.5972600	0.39819	1.8500490	0.4443935036	0.764149	0.925025	0.575574	0.249296
107	1.6077200	0.40518	1.8675023	0.4555986498	0.760094	0.933751	0.573929	0.252021
108	1.6180400	0.41222	1.8849556	0.4669492961	0.755991	0.942478	0.572265	0.254765
109	1.6282200	0.41930	1.9024089	0.4784449423	0.751841	0.951204	0.570582	0.257520
110	1.6383000	0.42642	1.9198622	0.4900845886	0.747701	0.959931	0.568895	0.260282
111	1.6482600	0.43359	1.9373155	0.5018677349	0.743544	0.968658	0.567197	0.263059
112	1.6580800	0.44081	1.9547688	0.5137923811	0.739346	0.977384	0.565482	0.265856
113	1.6677600	0.44806	1.9722221	0.5258585274	0.735107	0.986111	0.563750	0.268660
114	1.6773400	0.45536	1.9896753	0.5380651736	0.730882	0.994838	0.562015	0.271477
115	1.6867800	0.46270	2.0071286	0.5504103199	0.726620	1.003564	0.560263	0.274310
116	1.6961000	0.47008	2.0245819	0.5628939662	0.722348	1.012291	0.558502	0.277153
117	1.7052800	0.47750	2.0420352	0.5755141124	0.718042	1.021018	0.556726	0.280013
118	1.7143400	0.48496	2.0594885	0.5882702587	0.713728	1.029744	0.554940	0.282884
119	1.7232600	0.49246	2.0769418	0.6011609049	0.709382	1.038471	0.553140	0.285772
120	1.7320600	0.50000	2.0943951	0.6141850512	0.705031	1.047198	0.551332	0.288674
121	1.7407000	0.50758	2.1118484	0.6273406975	0.700627	1.055924	0.549503	0.291595
122	1.7492400	0.51519	2.1293017	0.6406268437	0.696245	1.064651	0.547673	0.294522
123	1.7564400	0.52284	2.1467550	0.6540419900	0.690420	1.073377	0.545456	0.297670
124	1.7659000	0.53053	2.1642083	0.6675851362	0.687401	1.082104	0.543971	0.300430
125	1.7740200	0.53825	2.1816616	0.6812547825	0.682943	1.090831	0.542101	0.303407
126	1.7820200	0.54601	2.1991149	0.6950489288	0.678486	1.099557	0.540223	0.306399
127	1.7898600	0.55380	2.2165682	0.7089660750	0.673985	1.108284	0.538328	0.309410
128	1.7975880	0.56163	2.2340214	0.7230052213	0.669496	1.117011	0.536428	0.312435
129	1.8051700	0.56949	2.2514747	0.7371643675	0.664980	1.125737	0.534515	0.315477
130	1.8126160	0.57738	2.2689280	0.7514420138	0.660451	1.134464	0.532591	0.318535
131	1.8199200	0.58531	2.2863813	0.7658356601	0.655904	1.143191	0.530655	0.321613
132	1.8270800	0.59326	2.3038346	0.7803448063	0.651336	1.151917	0.528707	0.324704
133	1.8341200	0.60125	2.3212879	0.7949669526	0.646775	1.160644	0.526754	0.327814

(Continued)

Table 14. *(Continued)* Chord Length, Arc Height, Arc Length, Area, and Center of Gravity for Segments and Sectors of a Circle. Radius = 1.

Degrees α	Chord Length s	Arc Height h	Arc Length b	Area of Segment A	Segment Center of Gravity cg	Area of Sector A'	Sector Center of Gravity cg'	h/s ratio
134	1.8410000	0.60927	2.3387412	0.8097005988	0.642178	1.169371	0.524784	0.330945
135	1.8477600	0.61732	2.3561945	0.8245437451	0.637590	1.178097	0.522809	0.334091
136	1.8543600	0.62539	2.3736478	0.8394948914	0.632970	1.186824	0.520819	0.337254
137	1.8608400	0.63350	2.3911011	0.8545515376	0.628359	1.195551	0.518824	0.340438
138	1.8671600	0.64163	2.4085544	0.8697116839	0.623718	1.204277	0.516813	0.343640
139	1.8733400	0.64979	2.4260077	0.8849743301	0.619067	1.213004	0.514794	0.346862
140	1.8793800	0.65798	2.4434610	0.9003364764	0.614409	1.221730	0.512764	0.350105
141	1.8852800	0.66619	2.4609142	0.9157971227	0.609743	1.230457	0.510726	0.353364
142	1.8910400	0.67423	2.4783675	0.9313532689	0.605071	1.239184	0.508679	0.356539
143	1.8966400	0.68270	2.4958208	0.9470029152	0.600374	1.247910	0.506618	0.359952
144	1.9021200	0.69098	2.5132741	0.9627445614	0.595692	1.256637	0.504553	0.363268
145	1.9074400	0.69930	2.5307274	0.9785757077	0.590986	1.265364	0.502475	0.366617
146	1.9126000	0.70763	2.5481807	0.9944938540	0.586258	1.274090	0.500383	0.369983
147	1.9176400	0.71599	2.5656340	1.0104975002	0.581547	1.282817	0.498289	0.373370
148	1.9225200	0.72436	2.5830873	1.0265841465	0.576815	1.291544	0.496181	0.376776
149	1.9272600	0.73276	2.6005406	1.0427512927	0.572083	1.300270	0.494067	0.380208
150	1.9318600	0.74118	2.6179939	1.0589969390	0.567350	1.308997	0.491944	0.383661
151	1.9363000	0.74962	2.6354472	1.0753185853	0.562600	1.317724	0.489809	0.387140
152	1.9405800	0.75808	2.6529005	1.0917142315	0.557833	1.326450	0.487662	0.390646
153	1.9447400	0.76656	2.6703538	1.1081818778	0.553086	1.335177	0.485514	0.394171
154	1.9487400	0.77505	2.6878070	1.1247180240	0.548323	1.343904	0.483353	0.397719
155	1.9526000	0.78356	2.7052603	1.1413211703	0.543564	1.352630	0.481186	0.401291
156	1.9563000	0.79209	2.7227136	1.1579883166	0.538792	1.361357	0.479007	0.404892
157	1.9598400	0.80063	2.7401669	1.1747179628	0.534007	1.370083	0.476818	0.408518
158	1.9632600	0.80919	2.7576202	1.1915066091	0.529244	1.378810	0.474627	0.412166
159	1.9665000	0.81777	2.7750735	1.2083527553	0.524453	1.387537	0.472420	0.415850
160	1.9696200	0.82635	2.7925268	1.2252534016	0.519685	1.396263	0.470212	0.419548
161	1.9725600	0.83495	2.8099801	1.2422060479	0.514892	1.404990	0.467989	0.423282
162	1.9753800	0.84357	2.8274334	1.2591631941	0.510139	1.413717	0.465765	0.427042
163	1.9780200	0.85219	2.8448867	1.2762573404	0.505327	1.422443	0.463526	0.430830
164	1.9805400	0.86083	2.8623400	1.2933514866	0.500556	1.431170	0.461287	0.434644
165	1.9828800	0.86947	2.8797933	1.3104871329	0.495764	1.439897	0.459033	0.438488
166	1.9851000	0.87813	2.8972466	1.3276622792	0.490996	1.448623	0.456779	0.442361
167	1.9871400	0.88680	2.9146999	1.3448744254	0.486208	1.457350	0.454510	0.446270
168	1.9890400	0.89547	2.9321531	1.3621205717	0.481431	1.466077	0.452236	0.450202
169	1.9908000	0.90416	2.9496064	1.3793987179	0.476663	1.474803	0.449958	0.454169
170	1.9923800	0.91285	2.9670597	1.3967058642	0.471879	1.483530	0.447667	0.458171

(Continued)

Table 14. *(Continued)* Chord Length, Arc Height, Arc Length, Area, and Center of Gravity for Segments and Sectors of a Circle. Radius = 1.

Degrees α	Chord Length s	Arc Height h	Arc Length b	Area of Segment A	Segment Center of Gravity cg	Area of Sector A'	Sector Center of Gravity cg'	h/s ratio
171	1.9938400	0.92154	2.9845130	1.4140395105	0.467120	1.492257	0.445375	0.462194
172	1.9951200	0.93024	3.0019663	1.4313966567	0.462345	1.500983	0.443070	0.466258
173	1.9962600	0.93895	3.0194196	1.4487753030	0.457582	1.509710	0.440760	0.470355
174	1.9972600	0.94766	3.0368729	1.4661724492	0.452832	1.518436	0.438447	0.474480
175	1.9981000	0.95638	3.0543262	1.4835850955	0.448082	1.527163	0.436125	0.478645
176	1.9987800	0.96510	3.0717795	1.5010117418	0.443333	1.535890	0.433794	0.482845
177	1.9993200	0.97382	3.0892328	1.5184483880	0.438597	1.544616	0.431460	0.487076
178	1.9997000	0.98254	3.1066861	1.5358935343	0.433863	1.553343	0.429117	0.491344
179	1.9999200	0.99127	3.1241394	1.5533436805	0.429130	1.562070	0.426767	0.495655
180	2.0000000	1.00000	3.1415927	1.5707963268	0.424413	1.570796	0.424413	0.500000

Using Chordal Lengths to Divide Circles

Although a rotary indexing table can be used to precisely divide the circumference of a circle, the procedure can also be done mathematically if exact measurements are taken. Some years ago, the makers of Krueger multiple spindle heads provided a simple chart and explanation for dividing a circle into a number of equally spaced points around its radius. The chart, shown in **Table 15**, has often been duplicated and remains useful for laying out a circle. The version shown here has been updated to give decimal location to six places. The system relies on three formulas that provide the lengths of the longest (chord A), second longest (chord B), and shortest (chord C) chords used to space the layout for the holes. Once three locations are determined, additional holes can be spaced using the gaps between the existing points to locate additional hole centers around the circle.

The chord lengths on the chart are for a circle with a diameter of 1 in any measurement unit (inch, centimeter, etc.). Chord size for larger circles can be calculated by multiplying the given chord lengths by the given diameter.

The formulas used to calculate the chord lengths are given below. They can be calculated for any diameter circle.

Odd number of holes

A = cos 90° ÷ N × diameter

B = cos 270° ÷ N × diameter

C = sin 180° ÷ N × diameter

Even number of holes

A = 1

B = cos 180° ÷ N × diameter

C = sin 180° ÷ N × diameter

Other Dimensions of Circles and Spheres

Area and circumference of circles, and the volume of spheres, ranging in diameter from 1 to 10, are given in **Table 16**. As in the previous two tables, the diameter can be in any measurement unit, and the values may be easily converted to larger circles and spheres.

To convert the **circumference** of a circle on the chart, multiply it by the number of times the size is being increased. For example, to double it, multiply the circumference by 2. To quadruple it, multiply by 4, etc.

Table 15. Circle Division Using Chordal Dimensions.

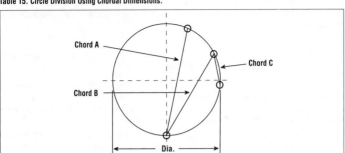

No. of Divisions	Chords, in inches			No. of Divisions	Chords, in inches		
	A	B	C		A	B	C
1	0.000000	0.000000	0.000000	31	0.998716	0.988468	0.101168
2	1.000000	1.000000	1.000000	32	1.000000	0.995185	0.098017
3	0.866025	0.866025	0.866025	33	0.998867	0.989821	0.95056
4	1.000000	0.707107	0.707107	34	1.000000	0.995734	0.092268
5	0.951056	0.587785	0.587785	35	0.998993	0.990950	0.089639
6	1.000000	0.866025	0.500000	36	1.000000	0.996195	0.087156
7	0.974928	0.781830	0.433884	37	0.999099	0.991900	0.084806
8	1.000000	0.923880	0.382683	38	1.000000	0.996584	0.082579
9	0.984808	0.866025	0.342020	39	0.999189	0.992089	0.080466
10	1.000000	0.951056	0.309017	40	1.000000	0.996917	0.078459
11	0.989821	0.909632	0.281732	41	0.999266	0.993402	0.076549
12	1.000000	0.965926	0.258819	42	1.000000	0.997204	0.074730
13	0.992709	0.935016	0.239316	43	0.999333	0.994001	0.072995
14	1.000000	0.974928	0.222521	44	1.000000	0.997452	0.071339
15	0.994522	0.951056	0.207912	45	0.999391	0.994522	0.069756
16	1.000000	0.980785	0.195090	46	1.000000	0.997669	0.068242
17	0.995734	0.961826	0.183750	47	0.999442	0.994978	0.066792
18	1.000000	0.984808	0.173648	48	1.000000	0.997859	0.065403
19	0.996584	0.969400	0.164594	49	0.999486	0.995379	0.064702
20	1.000000	0.987688	0.156434	50	1.000000	0.998027	0.062791
21	0.997204	0.974928	0.149042	51	0.999526	0.995734	0.061561
22	1.000000	0.989821	0.142315	52	1.000000	0.998176	0.060378
23	0.997669	0.979084	0.136167	53	0.999561	0.996050	0.059241
24	1.000000	0.991445	0.130526	54	1.000000	0.998308	0.058145
25	0.998027	0.982287	0.125333	55	0.999592	0.996332	0.057089
26	1.000000	0.992709	0.120537	56	1.000000	0.998427	0.056070
27	0.998308	0.984808	0.116093	57	0.999620	0.996584	0.055088
28	1.000000	0.993712	0.111964	58	1.000000	0.998533	0.054139
29	0.998533	0.986826	0.108119	59	0.999646	0.996812	0.053222
30	1.000000	0.994522	0.104528	60	1.000000	0.998630	0.052336

To find the **circumference** of a 40 inch diameter circle, multiply the circumference for a 4 inch diameter circle by 10. From the chart, a 4 inch diameter has a circumference of 12.5664.

$$10 \times 12.5664 = 125.664 \text{ inches.}$$

This can be easily checked by multiplying the value of a 10 inch diameter circle by 4 to find the circumference of a 40 inch circle.

$$4 \times 31.4160 = 125.644 \text{ inches.}$$

To find the **area** of a larger circle than given in the Table, multiply the area by the square of the multiplier. For example, to double it, multiply the area by 4 (because 4 is the square of 2). To triple it, multiply by 9 (because 9 is the square of 3), etc.

To find the area of a 25.4 centimeter diameter circle, multiply the area of a 2.54 cm diameter circle by 100 (the size is being increased ten times, so the multiplier is the square of 10). From the chart, the area of a 2.54 cm circle is 5.0671 cm^2.

$$100 \times 5.0671 = 506.71 \text{ cm}^2.$$

Since a 25.4 cm diameter circle is exactly the same size as a 10 inch diameter circle, the answer can be checked by converting the area to in.^{2} and finding the value on the table.

$$506.71 \text{ cm}^2 \times 0.1550003 = 78.540 \text{ in.}^2.$$

Since the table shows that the area of a 10 inch diameter circle is 78.5398 in.^{2}, the answer is correct to 0.0002 in.^{2}.

To find the **volume** of a larger sphere than given in the Table, multiply the area by the cube of the multiplier. For example, to double it, multiply the area by 8 (because 8 is the cube of 2). To triple it, multiply by 27 (because 27 is the cube of 3), etc.

Find the volume of a 25 inch diameter sphere, multiply the volume of a 5 inch diameter sphere by 125 (the size is being increased five times, so the multiplier is the cube of 5). From the chart, the volume of a 5 inch sphere is 65.45 in.^{3}.

$$125 \times 65.45 = 8181.25 \text{ in.}^3.$$

This can be verified by multiplying the volume of a 6.25 inch sphere by 64 (the size is being increased four times, so the multiplier is the cube of 4).

$$64 \times 127.832 = 8181.248 \text{ in.}^3.$$

Area and circumferences for fractional size circles are given in **Table 17**.

Dimensions of Polygons

Regular polygons, from triangles to multiple sided figures, can be inscribed within circles so that every corner of the polygon touches the circumference. In addition, a circle can be inscribed within the polygon, and the circumference of that circle will touch the center point of every side of the polygon. **Table 18** contains an illustration of such a hexagon, together with values for solving the area and side lengths of the polygons, and the radius of both circles. In regular polygons with an even number of sides, the distance across the points (corners) is equal to the diameter of the circumscribing circle, and the distance across the flats is equal to the diameter of the inscribed circle.

For example, using the Table, the distance across the corners and flats of any size regular hexagon can be easily found.

Find the distance across corners and distance across flats of a regular octagon. First, find the distance across corners. Using the Table, find the number of sides of the polygon

(Text continued on p. 102)

Table 16. Areas and Circumferences of Circles, and Volumes of Spheres 1 to 1.6 in Diameter.

Diameter		See text for units of measure and how to find dimensions of circles larger than 10.									
		0	.01	.02	.03	.04	.05	.06	.07	.08	.09
1.0	Area	0.7854	0.8012	0.8171	0.8332	0.8495	0.8659	0.8825	0.8992	0.9161	0.9331
	Volume	0.5236	0.5395	0.5556	0.5722	0.5890	0.6061	0.6236	0.6414	0.6596	0.6781
	Circumference	3.1416	3.1730	3.2044	3.2358	3.2673	3.2987	3.3301	3.3615	3.3929	3.4243
1.1	Area	0.9503	0.9677	0.9852	1.0029	1.0207	1.0387	1.0568	1.0751	1.0936	1.1122
	Volume	0.6969	0.7161	0.7356	0.7555	0.7757	0.7963	0.8173	0.8386	0.8603	0.8823
	Circumference	3.4558	3.4872	3.5186	3.5500	3.5814	3.6128	3.6443	3.6757	3.7071	3.7385
1.2	Area	1.1310	1.1499	1.1690	1.1882	1.2076	1.2272	1.2469	1.2668	1.2868	1.3070
	Volume	0.9048	0.9276	0.9508	0.9743	0.9983	1.0227	1.0474	1.0725	1.0981	1.1240
	Circumference	3.7699	3.8013	3.8328	3.8642	3.8956	3.9270	3.9584	3.9898	4.0212	4.0527
1.3	Area	1.3273	1.3478	1.3685	1.3893	1.4103	1.4314	1.4527	1.4741	1.4957	1.5175
	Volume	1.1503	1.1771	1.2043	1.2318	1.2598	1.2883	1.3171	1.3464	1.3761	1.4062
	Circumference	4.0841	4.1155	4.1469	4.1783	4.2097	4.2412	4.2726	4.3040	4.3354	4.3668
1.4	Area	1.5394	1.5614	1.5837	1.6061	1.6286	1.6513	1.6742	1.6972	1.7203	1.7437
	Volume	1.4368	1.4678	1.4992	1.5311	1.5635	1.5963	1.6295	1.6632	1.6974	1.7320
	Circumference	4.3982	4.4297	4.4611	4.4925	4.5239	4.5553	4.5867	4.6182	4.6496	4.6810
1.5	Area	1.7671	1.7908	1.8146	1.8385	1.8626	1.8869	1.9113	1.9359	1.9607	1.9856
	Volume	1.7672	1.8027	1.8388	1.8753	1.9123	1.9498	1.9878	2.0263	2.0652	2.1047
	Circumference	4.7124	4.7438	4.7752	4.8066	4.8381	4.8695	4.9009	4.9323	4.9637	4.9951
1.6	Area	2.0106	2.0358	2.0612	2.0867	2.1124	2.1382	2.1642	2.1904	2.2167	2.2432
	Volume	2.1447	2.1851	2.2261	2.2676	2.3096	2.3521	2.3951	2.4386	2.4827	2.5273
	Circumference	5.0266	5.0580	5.0894	5.1208	5.1522	5.1836	5.2151	5.2465	5.2779	5.3093

(Continued)

Table 16. (Continued) Areas and Circumferences of Circles, and Volumes of Spheres 1.7 to 2.3 in Diameter.

See text for units of measure and how to find dimensions of circles larger than 10.

Diameter		0	.01	.02	.03	.04	.05	.06	.07	.08	.09
1.7	Area	2.2698	2.2966	2.3235	2.3506	2.3779	2.4053	2.4328	2.4606	2.4885	2.5165
	Volume	2.5724	2.6181	2.6643	2.7111	2.7583	2.8062	2.8545	2.9035	2.9530	3.0030
	Circumference	5.3407	5.3721	5.4036	5.4350	5.4664	5.4978	5.5292	5.5606	5.5920	5.6235
1.8	Area	2.5447	2.5730	2.6016	2.6302	2.6590	2.6880	2.7172	2.7465	2.7759	2.8055
	Volume	3.0536	3.1048	3.1566	3.2089	3.2618	3.3152	3.3693	3.4239	3.4792	3.5350
	Circumference	5.6549	5.6863	5.7177	5.7491	5.7805	5.8120	5.8434	5.8748	5.9062	5.9376
1.9	Area	2.8353	2.8652	2.8953	2.9255	2.9559	2.9865	3.0172	3.0481	3.0791	3.1103
	Volume	3.5914	3.6484	3.7060	3.7642	3.8230	3.8824	3.9425	4.0031	4.0644	4.1263
	Circumference	5.9690	6.0005	6.0319	6.0633	6.0947	6.1261	6.1575	6.1890	6.2204	6.2518
2.0	Area	3.1416	3.1731	3.2047	3.2365	3.2685	3.3006	3.3329	3.3654	3.3979	3.4307
	Volume	4.1888	4.2519	4.3157	4.3801	4.4452	4.5109	4.5772	4.6442	4.7118	4.7801
	Circumference	6.2832	6.3146	6.3460	6.3774	6.4089	6.4403	6.4717	6.5031	6.5345	6.5659
2.1	Area	3.4636	3.4967	3.5299	3.5633	3.5968	3.6305	3.6644	3.6984	3.7325	3.7668
	Volume	4.8491	4.9187	4.9889	5.0599	5.1315	5.2037	5.2767	5.3503	5.4246	5.4996
	Circumference	6.5974	6.6288	6.6602	6.6916	6.7230	6.7544	6.7859	6.8173	6.8487	6.8801
2.2	Area	3.8013	3.8360	3.8708	3.9057	3.9408	3.9761	4.0115	4.0471	4.0828	4.1187
	Volume	5.5753	5.6517	5.7287	5.8065	5.8850	5.9641	6.0440	6.1246	6.2059	6.2879
	Circumference	6.9115	6.9429	6.9744	7.0058	7.0372	7.0686	7.1000	7.1314	7.1628	7.1943
2.3	Area	4.1548	4.1910	4.2273	4.2638	4.3005	4.3374	4.3744	4.4115	4.4488	4.4863
	Volume	6.3706	6.4541	6.5383	6.6232	6.7088	6.7952	6.8823	6.9702	7.0588	7.1481
	Circumference	7.2257	7.2571	7.2885	7.3199	7.3513	7.3828	7.4142	7.4456	7.4770	7.5084

(Continued)

Table 16. (Continued) Areas and Circumferences of Circles, and Volumes of Spheres 2.4 to 3.0 in Diameter.

See text for units of measure and how to find dimensions of circles larger than 10.

Diameter		0	.01	.02	.03	.04	.05	.06	.07	.08	.09
2.4	Area	4.5239	4.5617	4.5996	4.6377	4.6759	4.7144	4.7529	4.7916	4.8305	4.8695
	Volume	7.2382	7.3291	7.4207	7.5131	7.6062	7.7001	7.7948	7.8902	7.9865	8.0835
	Circumference	7.5398	7.5713	7.6027	7.6341	7.6655	7.6969	7.7283	7.7598	7.7912	7.8226
2.5	Area	4.9087	4.9481	4.9876	5.0273	5.0671	5.1071	5.1472	5.1875	5.2279	5.2685
	Volume	8.1813	8.2798	8.3792	8.4793	8.5803	8.6820	8.7846	8.8879	8.9921	9.0970
	Circumference	7.8540	7.8854	7.9168	7.9482	7.9797	8.0111	8.0425	8.0739	8.1053	8.1367
2.6	Area	5.3093	5.3502	5.3913	5.4325	5.4739	5.5155	5.5572	5.5990	5.6410	5.6832
	Volume	9.2028	9.3094	9.4168	9.5250	9.6341	9.7440	9.8547	9.9663	10.0787	10.1919
	Circumference	8.1682	8.1996	8.2310	8.2624	8.2938	8.3252	8.3567	8.3881	8.4195	8.4509
2.7	Area	5.7256	5.7680	5.8107	5.8535	5.8965	5.9396	5.9828	6.0263	6.0699	6.1136
	Volume	10.3060	10.4210	10.5367	10.6534	10.7709	10.8892	11.0085	11.1286	11.2495	11.3714
	Circumference	8.4823	8.5137	8.5452	8.5766	8.6080	8.6394	8.6708	8.7022	8.7336	8.7651
2.8	Area	6.1575	6.2016	6.2458	6.2902	6.3347	6.3794	6.4242	6.4692	6.5144	6.5597
	Volume	11.4941	11.6177	11.7421	11.8675	11.9937	12.1209	12.2489	12.3779	12.5077	12.6384
	Circumference	8.7965	8.8279	8.8593	8.8907	8.9221	8.9536	8.9850	9.0164	9.0478	9.0792
2.9	Area	6.6052	6.6508	6.6966	6.7426	6.7887	6.8349	6.8813	6.9279	6.9746	7.0215
	Volume	12.7701	12.9026	13.0361	13.1705	13.3058	13.4421	13.5792	13.7173	13.8563	13.9963
	Circumference	9.1106	9.1421	9.1735	9.2049	9.2363	9.2677	9.2991	9.3306	9.3620	9.3934
3.0	Area	7.0686	7.1158	7.1631	7.2107	7.2583	7.3062	7.3542	7.4023	7.4506	7.4991
	Volume	14.1372	14.2790	14.4218	14.5656	14.7103	14.8559	15.0025	15.1501	15.2986	15.4481
	Circumference	9.4248	9.4562	9.4876	9.5190	9.5505	9.5819	9.6133	9.6447	9.6761	9.7075

(Continued)

Table 16. *(Continued)* Areas and Circumferences of Circles, and Volumes of Spheres 3.1 to 3.7 in Diameter.

Diameter		See text for units of measure and how to find dimensions of circles larger than 10.								
		0	.01	.02	.03	.04	.05	.06	.07	.08	.09
3.1	Area	7.5477	7.5964	7.6454	7.6945	7.7437	7.7931	7.8427	7.8924	7.9423	7.9923
	Volume	15.5986	15.7500	15.9024	16.0558	16.2102	16.3656	16.5219	16.6793	16.8376	16.9970
	Circumference	9.7390	9.7704	9.8018	9.8332	9.8646	9.8960	9.9275	9.9589	9.9903	10.217
3.2	Area	8.0425	8.0928	8.1433	8.1940	8.2448	8.2958	8.3469	8.3982	8.4496	8.5012
	Volume	17.1573	17.3187	17.4810	17.6444	17.8088	17.9742	18.1406	18.3081	18.4766	18.6461
	Circumference	10.0531	10.0845	10.1160	10.1474	10.1788	10.2102	10.2416	10.2730	10.3044	10.3359
3.3	Area	8.5530	8.6049	8.6570	8.7092	8.7616	8.8141	8.8668	8.9197	8.9727	9.0259
	Volume	18.8166	18.9882	19.1608	19.3345	19.5092	19.6849	19.8617	20.0396	20.2185	20.3985
	Circumference	10.3673	10.3987	10.4301	10.4615	10.4929	10.5244	10.5558	10.5872	10.6186	10.6500
3.4	Area	9.0792	9.1327	9.1863	9.2401	9.2941	9.3482	9.4025	9.4569	9.5115	9.5662
	Volume	20.5796	20.7617	20.9449	21.1291	21.3145	21.5009	21.6884	21.8770	22.0667	22.2575
	Circumference	10.6814	10.7129	10.7443	10.7757	10.8071	10.8385	10.8699	10.9014	10.9328	10.9642
3.5	Area	9.6211	9.6762	9.7314	9.7868	9.8423	9.8980	9.9538	10.0098	10.0660	10.1223
	Volume	22.4494	22.6423	22.8364	23.0316	23.2279	23.4253	23.6238	23.8234	24.0242	24.2261
	Circumference	10.9956	11.0270	11.0584	11.0898	11.1213	11.1527	11.1841	11.2155	11.2469	11.2783
3.6	Area	10.1788	10.2354	10.2922	10.3491	10.4062	10.4635	10.5209	10.5784	10.6362	10.6941
	Volume	24.4291	24.6332	24.8385	25.0449	25.2525	25.4612	25.6710	25.8820	26.0941	26.3074
	Circumference	11.3098	11.3412	11.3726	11.4040	11.4354	11.4668	11.4983	11.5297	11.5611	11.5925
3.7	Area	10.7521	10.8103	10.8687	10.9272	10.9858	11.0447	11.1036	11.1628	11.2221	11.2815
	Volume	26.5219	26.7375	26.9543	27.1723	27.3914	27.6117	27.8332	28.0559	28.2797	28.5048
	Circumference	11.6239	11.6553	11.6868	11.7182	11.7496	11.7810	11.8124	11.8438	11.8752	11.9067

(Continued)

Table 16. *(Continued)* Areas and Circumferences of Circles, and Volumes of Spheres 3.8 to 4.4 in Diameter.

Diameter		*See text for units of measure and how to find dimensions of circles larger than 10.*									
		0	.01	.02	.03	.04	.05	.06	.07	.08	.09
3.8	Area	11.3411	11.4009	11.4608	11.5209	11.5812	11.6416	11.7021	11.7628	11.8237	11.8847
	Volume	28.7310	28.9584	29.1870	29.4168	29.6479	29.8801	30.1135	30.3482	30.5840	30.8211
	Circumference	11.9381	11.9695	12.0009	12.0323	12.0637	12.0952	12.1266	12.1580	12.1894	12.2208
3.9	Area	11.9459	12.0072	12.0687	12.1304	12.1922	12.2542	12.3163	12.3786	12.4410	12.5036
	Volume	31.0594	31.2990	31.5397	31.7817	32.0249	32.2694	32.5151	32.7621	33.0103	33.2597
	Circumference	12.2522	12.2837	12.3151	12.3465	12.3779	12.4093	12.4407	12.4722	12.5036	12.5350
4.0	Area	12.5664	12.6293	12.6923	12.7556	12.8190	12.8825	12.9462	13.0100	13.0740	13.1382
	Volume	33.5104	33.7624	34.0156	34.2701	34.5258	34.7828	35.0411	35.3007	35.5615	35.8236
	Circumference	12.5664	12.5978	12.6292	12.6606	12.6921	12.7235	12.7549	12.7863	12.8177	12.8491
4.1	Area	13.2025	13.2670	13.3317	13.3965	13.4614	13.5265	13.5918	13.6572	13.7228	13.7885
	Volume	36.0870	36.3517	36.6177	36.8850	37.1536	37.4235	37.6946	37.9671	38.2409	38.5160
	Circumference	12.8806	12.9120	12.9434	12.9748	13.0062	13.0376	13.0691	13.1005	13.1319	13.1633
4.2	Area	13.8544	13.9205	13.9867	14.0530	14.1196	14.1863	14.2531	14.3201	14.3872	14.4545
	Volume	38.7925	39.0702	39.3493	39.6297	39.9114	40.1945	40.4789	40.7646	41.0517	41.3401
	Circumference	13.1947	13.2261	13.2576	13.2890	13.3204	13.3518	13.3832	13.4146	13.4460	13.4775
4.3	Area	14.5220	14.5896	14.6574	14.7253	14.7934	14.8617	14.9301	14.9987	15.0674	15.1363
	Volume	41.6299	41.9210	42.2135	42.5073	42.8025	43.0990	43.3969	43.6962	43.9969	44.2989
	Circumference	13.5089	13.5403	13.5717	13.6031	13.6345	13.6660	13.6974	13.7288	13.7602	13.7916
4.4	Area	15.2053	15.2745	15.3438	15.4134	15.4830	15.5528	15.6228	15.6930	15.7633	15.8337
	Volume	44.6023	44.9071	45.2133	45.5209	45.8299	46.1402	46.4520	46.7651	47.0797	47.3957
	Circumference	13.8230	13.8545	13.8859	13.9173	13.9487	13.9801	14.0115	14.0430	14.0744	14.1058

(Continued)

Table 16. (*Continued*) Areas and Circumferences of Circles, and Volumes of Spheres 4.5 to 5.1 in Diameter.

See text for units of measure and how to find dimensions of circles larger than 10.

Diameter		0	.01	.02	.03	.04	.05	.06	.07	.08	.09
4.5	Area	15.9043	15.9751	16.0460	16.1171	16.1883	16.2597	16.3313	16.4030	16.4748	16.5468
	Volume	47.7131	48.0318	48.3521	48.6737	48.9967	49.3212	49.6471	49.9745	50.3033	50.6335
	Circumference	14.1372	14.1686	14.2000	14.2314	14.2629	14.2943	14.3257	14.3571	14.3885	14.4199
4.6	Area	16.6190	16.6914	16.7638	16.8365	16.9093	16.9823	17.0554	17.1287	17.2021	17.2757
	Volume	50.9651	51.2982	51.6328	51.9688	52.3062	52.6452	52.9855	53.3274	53.6707	54.0155
	Circumference	14.4514	14.4828	14.5142	14.5456	14.5770	14.6084	14.6399	14.6713	14.7027	14.7341
4.7	Area	17.3494	17.4233	17.4974	17.5716	17.6460	17.7205	17.7952	17.8701	17.9451	18.0203
	Volume	54.3617	54.7095	55.0587	55.4094	55.7615	56.1152	56.4704	56.8270	57.1852	57.5448
	Circumference	14.7655	14.7969	14.8284	14.8598	14.8912	14.9226	14.9540	14.9854	15.0168	15.0483
4.8	Area	18.0956	18.1710	18.2467	18.3225	18.3984	18.4745	18.5508	18.6272	18.7038	18.7805
	Volume	57.9060	58.2686	58.6328	58.9985	59.3657	59.7344	60.1047	60.4765	60.8498	61.2246
	Circumference	15.0797	15.1111	15.1425	15.1739	15.2053	15.2368	15.2682	15.2996	15.3310	15.3624
4.9	Area	18.8574	18.9345	19.0117	19.0890	19.1665	19.2442	19.3220	19.4000	19.4782	19.5565
	Volume	61.6010	61.9789	62.3584	62.7394	63.1220	63.5061	63.8917	64.2790	64.6677	65.0581
	Circumference	15.3938	15.4253	15.4567	15.4881	15.5195	15.5509	15.5823	15.6138	15.6452	15.6766
5.0	Area	19.6350	19.7136	19.7923	19.8713	19.9504	20.0296	20.1090	20.1886	20.2683	20.3482
	Volume	65.4500	65.8435	66.2385	66.6352	67.0334	67.4332	67.8346	68.2376	68.6421	69.0483
	Circumference	15.7080	15.7394	15.7708	15.8022	15.8336	15.8650	15.8965	15.9279	15.9593	15.9907
5.1	Area	20.4282	20.5084	20.5887	20.6692	20.7499	20.8307	20.9117	20.9928	21.0741	21.1556
	Volume	69.4561	69.8654	70.2764	70.6890	71.1032	71.5190	71.9364	72.3555	72.7761	73.1984
	Circumference	16.0222	16.0536	16.0850	16.1164	16.1478	16.1792	16.2107	16.2421	16.2735	16.3049

(Continued)

Table 16. (Continued) Areas and Circumferences of Circles, and Volumes of Spheres 5.2 to 5.8 in Diameter.

See text for units of measure and how to find dimensions of circles larger than 10.

Diameter		0	.01	.02	.03	.04	.05	.06	.07	.08	.09
5.2	Area	21.2372	21.3189	21.4008	21.4829	21.5651	21.6475	21.7301	21.8128	21.8956	21.9787
	Volume	73.6223	74.0479	74.4751	74.9039	75.3344	75.7666	76.2003	76.6358	77.0728	77.5116
	Circumference	16.3363	16.3677	16.3992	16.4306	16.4620	16.4934	16.5248	16.5562	16.5876	16.6191
5.3	Area	22.0618	22.1452	22.2286	22.3123	22.3961	22.4801	22.5642	22.6484	22.7329	22.8175
	Volume	77.9520	78.3941	78.8378	79.2832	79.7303	80.1791	80.6295	81.0816	81.5354	81.9910
	Circumference	16.6506	16.6819	16.7133	16.7447	16.7761	16.8076	16.8390	16.8704	16.9018	16.9332
5.4	Area	22.9022	22.9871	23.0722	23.1574	23.2428	23.3283	23.4140	23.4998	23.5858	23.6720
	Volume	82.4482	82.9070	83.3676	83.8299	84.2939	84.7596	85.2271	85.6962	86.1671	86.6396
	Circumference	16.9646	16.9961	17.0275	17.0589	17.0903	17.1217	17.1531	17.1846	17.2160	17.2474
5.5	Area	23.7583	23.8448	23.9314	24.0182	24.1051	24.1922	24.2795	24.3669	24.4545	24.5422
	Volume	87.1140	87.5900	88.0677	88.5472	89.0285	89.5114	89.9962	90.4826	90.9708	91.4608
	Circumference	17.2788	17.3102	17.3416	17.3730	17.4045	17.4359	17.4673	17.4987	17.5301	17.5615
5.6	Area	24.6301	24.7181	24.8063	24.8947	24.9832	25.0719	25.1607	25.2497	25.3388	25.4281
	Volume	91.9525	92.4460	92.9413	93.4383	93.9371	94.4376	94.9399	95.4440	95.9499	96.4576
	Circumference	17.5930	17.6244	17.6558	17.6872	17.7186	17.7500	17.7815	17.8129	17.8443	17.8757
5.7	Area	25.5176	25.6072	25.6970	25.7869	25.8770	25.9672	26.0576	26.1482	26.2389	26.3298
	Volume	96.9671	97.4783	97.9913	98.5062	99.0228	99.5413	100.0615	100.5836	101.1074	101.6331
	Circumference	17.9071	17.9385	17.9700	18.0014	18.0328	18.0642	18.0956	18.1270	18.1584	18.1899
5.8	Area	26.4208	26.5120	26.6033	26.6948	26.7865	26.8783	26.9703	27.0624	27.1547	27.2471
	Volume	102.1606	102.6900	103.2211	103.7541	104.2889	104.8256	105.3641	105.9044	106.4466	106.9906
	Circumference	18.2213	18.2527	18.2841	18.3155	18.3469	18.3784	18.4098	18.4412	18.4726	18.5040

(Continued)

96   CIRCLES

Table 16. *(Continued)* Areas and Circumferences of Circles, and Volumes of Spheres 5.9 to 6.5 in Diameter.

Diameter		See text for units of measure and how to find dimensions of circles larger than 10.								
		0	.01	.02	.03	.04	.05	.06	.07	.08	.09
5.9	Area	27.3397	27.4325	27.5254	27.6184	27.7117	27.8051	27.8986	27.9923	28.0861	28.1802
	Volume	107.5364	108.0842	108.6337	109.1852	109.7385	110.2937	110.8507	111.4096	111.9704	112.5331
	Circumference	18.5354	18.5669	18.5983	18.6297	18.6611	18.6925	18.7239	18.7554	18.7868	18.8182
6.0	Area	28.2743	28.3687	28.4631	28.5578	28.6526	28.7475	28.8426	28.9379	29.0333	29.1289
	Volume	113.0976	113.6640	114.2324	114.8026	115.3747	115.9487	116.5246	117.1024	117.6821	118.2637
	Circumference	18.8496	18.8810	18.9124	18.9438	18.9753	19.0067	19.0381	19.0695	19.1009	19.1323
6.1	Area	29.2247	29.3206	29.4166	29.5128	29.6092	29.7057	29.8024	29.8992	29.9962	30.0934
	Volume	118.8473	119.4327	120.0201	120.6094	121.2006	121.7937	122.3888	122.9858	123.5848	124.1857
	Circumference	19.1638	19.1952	19.2266	19.2580	19.2894	19.3208	19.3523	19.3837	19.4151	19.4465
6.2	Area	30.1907	30.2882	30.3858	30.4836	30.5815	30.6796	30.7779	30.8763	30.9748	31.0736
	Volume	124.7885	125.3933	126.0001	126.6088	127.2194	127.8320	128.4466	129.0631	129.6817	130.3021
	Circumference	19.4779	19.5093	19.5408	19.5722	19.6036	19.6350	19.6664	19.6978	19.7292	19.7607
6.3	Area	31.1724	31.2715	31.3707	31.4700	31.5695	31.6692	31.7690	31.8690	31.9692	32.0694
	Volume	130.9246	131.5490	132.1755	132.8039	133.4343	134.0667	134.7011	135.3374	135.9758	136.6162
	Circumference	19.7921	19.8235	19.8549	19.8863	19.9177	19.9492	19.9806	20.0120	20.0434	20.0748
6.4	Area	32.1699	32.2705	32.3713	32.4722	32.5733	32.6745	32.7759	32.8775	32.9792	33.0810
	Volume	137.2586	137.9030	138.5494	139.1979	139.8483	140.5008	141.1553	141.8118	142.4704	143.1310
	Circumference	20.1062	20.1377	20.1691	20.2005	20.2319	20.2633	20.2947	20.3262	20.3576	20.3890
6.5	Area	33.1831	33.2852	33.3876	33.4901	33.5927	33.6955	33.7985	33.9016	34.0049	34.1083
	Volume	143.7937	144.4583	145.1251	145.7938	146.4647	147.1376	147.8125	148.4895	149.1686	149.8497
	Circumference	20.4204	20.4518	20.4832	20.5146	20.5461	20.5775	20.6089	20.6403	20.6717	20.7031

(Continued)

Table 16. (Continued) Areas and Circumferences of Circles, and Volumes of Spheres 6.6 to 7.2 in Diameter.

See text for units of measure and how to find dimensions of circles larger than 10.

Diameter		0	.01	.02	.03	.04	.05	.06	.07	.08	.09
6.6	Area	34.2119	34.3157	34.4196	34.5237	34.6279	34.7323	34.8368	34.9415	35.0463	35.1514
	Volume	150.5329	151.2182	151.9055	152.5950	153.2865	153.9801	154.6758	155.3736	156.0734	156.7754
	Circumference	20.7346	20.7660	20.7974	20.8288	20.8602	20.8916	20.9231	20.9545	20.9859	21.0173
6.7	Area	35.2565	35.3618	35.4673	35.5730	35.6787	35.7847	35.8908	35.9971	36.1035	36.2101
	Volume	157.4795	158.1857	158.8940	159.6044	160.3169	161.0315	161.7483	162.4672	163.1882	163.9113
	Circumference	21.0487	21.0801	21.1116	21.1430	21.1744	21.2058	21.2372	21.2686	21.3000	21.3315
6.8	Area	36.3168	36.4237	36.5307	36.6380	36.7453	36.8528	36.9605	37.0684	37.1763	37.2845
	Volume	164.6366	165.3640	166.0935	166.8252	167.5591	168.2951	169.0332	169.7735	170.5159	171.2605
	Circumference	21.3629	21.3943	21.4257	21.4571	21.4885	21.5200	21.5514	21.5828	21.6142	21.6456
6.9	Area	37.3928	37.5013	37.6099	37.7187	37.8276	37.9367	38.0459	38.1553	38.2649	38.3746
	Volume	172.0073	172.7563	173.5074	174.2607	175.0161	175.7738	176.5336	177.2956	178.0598	178.8262
	Circumference	21.6770	21.7085	21.7399	21.7713	21.8027	21.8341	21.8655	21.8970	21.9284	21.9598
7.0	Area	38.4845	38.5945	38.7047	38.8151	38.9256	39.0362	39.1471	39.2580	39.3692	39.4805
	Volume	179.5948	180.3656	181.1386	181.9138	182.6912	183.4708	184.2526	185.0367	185.8230	186.6115
	Circumference	21.9912	22.0226	22.0540	22.0854	22.1169	22.1483	22.1797	22.2111	22.2425	22.2739
7.1	Area	39.5919	39.7035	39.8153	39.9272	40.0393	40.1515	40.2639	40.3764	40.4892	40.6020
	Volume	187.4022	188.1952	188.9900	189.7878	190.5874	191.3893	192.1935	192.9999	193.8086	194.6195
	Circumference	22.3054	22.3368	22.3682	22.3996	22.4310	22.4624	22.4939	22.5253	22.5567	22.5881
7.2	Area	40.7150	40.8282	40.9415	41.0550	41.1687	41.2825	41.3964	41.5106	41.6248	41.7393
	Volume	195.4327	196.2481	197.0658	197.8858	198.7080	199.5325	200.3593	201.1884	202.0197	202.8534
	Circumference	22.6195	22.6509	22.6824	22.7138	22.7452	22.7766	22.8080	22.8394	22.8708	22.9023

(Continued)

Table 16. *(Continued)* Areas and Circumferences of Circles, and Volumes of Spheres 7.3 to 7.9 in Diameter.

See text for units of measure and how to find dimensions of circles larger than 10.

Diameter		0	.01	.02	.03	.04	.05	.06	.07	.08	.09
7.3	Area	41.85539	41.9686	42.0835	42.1986	42.3138	42.4292	42.5447	42.6604	42.7762	42.8922
	Volume	203.6893	204.5275	205.3681	206.2109	207.0560	207.9034	208.7532	209.6052	210.4596	211.3163
	Circumference	22.9337	22.9651	22.9965	23.0279	23.0593	23.0908	23.1222	23.1536	23.1850	23.2164
7.4	Area	43.0084	43.1247	43.2412	43.3578	43.4746	43.5916	43.7087	43.8259	43.9433	44.0609
	Volume	212.1753	213.0366	213.9003	214.7663	215.6346	216.5053	217.3783	218.2536	219.1313	220.0114
	Circumference	23.2478	23.2793	23.3107	23.3421	23.3735	23.4049	23.4363	23.4678	23.4992	23.5306
7.5	Area	44.1786	44.2965	44.4146	44.5328	44.6511	44.7696	44.8883	45.0072	45.1261	45.2453
	Volume	220.8938	221.7785	222.6656	223.5551	224.4469	225.3411	226.2377	227.1367	228.0380	228.9417
	Circumference	23.5620	23.5934	23.6248	23.6562	23.6877	23.7191	23.7505	23.7819	23.8133	23.8447
7.6	Area	45.3646	45.4840	45.6037	45.7234	45.8434	45.9635	46.0837	46.2041	46.3247	46.4454
	Volume	229.8478	230.7563	231.6672	232.5805	233.4961	234.4142	235.3347	236.2576	237.1829	238.1106
	Circumference	23.8762	23.9076	23.9390	23.9704	24.0018	24.0332	24.0647	24.0961	24.1275	24.1589
7.7	Area	46.5662	46.6873	46.8085	46.9298	47.0513	47.1730	47.2948	47.4168	47.5389	47.6612
	Volume	239.0407	239.9732	240.9082	241.8456	242.7854	243.7276	244.6723	245.6194	246.5690	247.5210
	Circumference	24.1903	24.2217	24.2532	24.2846	24.3160	24.3474	24.3788	24.4102	24.4416	24.4731
7.8	Area	47.7836	47.9062	48.0290	48.1519	48.2750	48.3982	48.5216	48.6451	48.7688	48.8927
	Volume	248.4754	249.4323	250.3917	251.3535	252.3178	253.2845	254.2537	255.2254	256.1995	257.1761
	Circumference	24.5045	24.5359	24.5673	24.5987	24.6301	24.6616	24.6930	24.7244	24.7558	24.7872
7.9	Area	49.0167	49.1409	49.2652	49.3897	49.5143	49.6391	49.7641	49.8892	50.0145	50.1399
	Volume	258.1552	259.1368	260.1209	261.1074	262.0965	263.0880	264.0820	265.0786	266.0776	267.0791
	Circumference	24.8186	24.8501	24.8815	24.9129	24.9443	24.9757	25.0071	25.0386	25.0700	25.1014

(Continued)

Table 16. (Continued) Areas and Circumferences of Circles, and Volumes of Spheres 8.0 to 8.6 in Diameter.

See text for units of measure and how to find dimensions of circles larger than 10.

Diameter		0	.01	.02	.03	.04	.05	.06	.07	.08	.09
8.0	Area	50.2655	50.3912	50.5171	50.6432	50.7694	50.8958	51.0223	51.1490	51.2758	51.4028
	Volume	268.0832	269.0898	270.0989	271.1105	272.1246	273.1412	274.1604	275.1821	276.2064	277.2332
	Circumference	25.1328	25.1642	25.1956	25.2270	25.2585	25.2899	25.3213	25.3527	25.3841	25.4155
8.1	Area	51.5300	51.6573	51.7847	51.9124	52.0402	52.1681	52.2962	52.4245	52.5529	52.6814
	Volume	278.2625	279.2944	280.3288	281.3658	282.4053	283.4474	284.4920	285.5392	286.5890	287.6414
	Circumference	25.4470	25.4784	25.5098	25.5412	25.5726	25.6040	25.6355	25.6669	25.6983	25.7297
8.2	Area	52.8102	52.9390	53.0681	53.1973	53.3266	53.4562	53.5858	53.7156	53.8456	53.9758
	Volume	288.6963	289.7538	290.8139	291.8765	292.9418	294.0096	295.0800	296.1530	297.2286	298.3069
	Circumference	25.7611	25.7925	25.8240	25.8554	25.8868	25.9182	25.9496	25.9810	26.0124	26.0439
8.3	Area	54.1061	54.2365	54.3671	54.4979	54.6288	54.7599	54.8912	55.0225	55.1541	55.2858
	Volume	299.3877	300.4711	301.5571	302.6458	303.7371	304.8310	305.9275	307.0266	308.1284	309.2328
	Circumference	26.0753	26.1067	26.1381	26.1695	26.2009	26.2324	26.2638	26.2952	26.3266	26.3580
8.4	Area	55.4177	55.5497	55.6819	55.8142	55.9467	56.0794	56.2122	56.3452	56.4783	56.6116
	Volume	310.3398	311.4495	312.5618	313.6768	314.7944	315.9146	317.0376	318.1631	319.2914	320.4223
	Circumference	26.3894	26.4209	26.4523	26.4837	26.5151	26.5465	26.5779	26.6093	26.6408	26.6722
8.5	Area	56.7450	56.8786	57.0124	57.1463	57.2803	57.4146	57.5489	57.6835	57.8182	57.9530
	Volume	321.5559	322.6921	323.8310	324.9726	326.1169	327.2638	328.4134	329.5658	330.7208	331.8785
	Circumference	26.7036	26.7350	26.7664	26.7978	26.8293	26.8607	26.8921	26.9235	26.9549	26.9863
8.6	Area	58.0880	58.2232	58.3585	58.4940	58.6296	58.7654	58.9014	59.0375	59.1738	59.3102
	Volume	333.0389	334.2020	335.3679	336.5364	337.7076	338.8816	340.0582	341.2376	342.4198	343.6046
	Circumference	27.0178	27.0492	27.0806	27.1120	27.1434	27.1748	27.2063	27.2377	27.2691	27.3005

(Continued)

Table 16. *(Continued)* Areas and Circumferences of Circles, and Volumes of Spheres 8.7 to 9.3 in Diameter.

See text for units of measure and how to find dimensions of circles larger than 10.

Diameter		0	.01	.02	.03	.04	.05	.06	.07	.08	.09
8.7	Area	59.4468	59.5835	59.7204	59.8575	59.9947	60.1320	60.2696	60.4072	60.5451	60.6831
	Volume	344.7922	345.9825	347.1755	348.3713	349.5698	350.7711	351.9751	353.1819	354.3914	355.6037
	Circumference	27.3319	27.3633	27.3948	27.4262	27.4576	27.4890	27.5204	27.5518	27.5832	27.6147
8.8	Area	60.8212	60.9595	61.0980	61.2366	61.3754	61.5143	61.6534	61.7927	61.9321	62.0717
	Volume	356.8187	358.0365	359.2571	360.4805	361.7066	362.9355	364.1672	365.4016	366.6389	367.8789
	Circumference	27.6461	27.6775	27.7089	27.7403	27.7717	27.8032	27.8346	27.8660	27.8974	27.9288
8.9	Area	62.2114	62.3513	62.4913	62.6315	62.7718	62.9123	63.0530	63.1938	63.3348	63.4759
	Volume	369.1218	370.3674	371.6158	372.8671	374.1211	375.3779	376.6376	377.9001	379.1654	380.4335
	Circumference	27.9602	27.9917	28.0231	28.0545	28.0859	28.1173	28.1487	28.1802	28.2116	28.2430
9.0	Area	63.6172	63.7587	63.9003	64.0421	64.1840	64.3261	64.4683	64.6107	64.7532	64.8959
	Volume	381.7044	382.9782	384.2548	385.5342	386.8164	388.1015	389.3895	390.6803	391.9739	393.2704
	Circumference	28.744	28.3058	28.3372	28.3686	28.4001	28.4315	28.4629	28.4943	28.5257	28.5571
9.1	Area	65.0388	65.1818	65.3250	65.4683	65.6118	65.7555	65.8993	66.0433	66.1874	66.3317
	Volume	394.5698	395.8720	397.1771	398.4850	399.7958	401.1095	402.4260	403.7455	405.0678	406.3930
	Circumference	28.5886	28.6200	28.6514	28.6828	28.7142	28.7456	28.7771	28.8085	28.8399	28.8713
9.2	Area	66.4761	66.6207	66.7654	66.9103	67.0554	67.2006	67.3460	67.4915	67.6372	67.7831
	Volume	407.7210	409.0520	410.3859	411.7226	413.0623	414.4049	415.7503	417.0987	418.4500	419.8042
	Circumference	28.9027	28.9341	28.9656	28.9970	29.0284	29.0598	29.0912	29.1226	29.1540	29.1855
9.3	Area	67.9291	68.0752	68.2216	68.3680	68.5147	68.6615	68.8084	68.9555	69.1028	69.2502
	Volume	421.1613	422.5214	423.8843	425.2502	426.6191	427.9908	429.3655	430.7432	432.1238	433.5073
	Circumference	29.2169	29.2483	29.2797	29.3111	29.3425	29.3740	29.4054	29.4368	29.4682	29.4996

(Continued)

Table 16. (Continued) Areas and Circumferences of Circles, and Volumes of Spheres 9.4 to 10 in Diameter.

See text for units of measure and how to find dimensions of circles larger than 10.

Diameter		0	.01	.02	.03	.04	.05	.06	.07	.08	.09
9.4	Area	69.3978	69.5455	69.6934	69.8414	69.9896	70.1380	70.2865	70.4352	70.5840	70.7330
	Volume	434.8938	436.2832	437.6756	439.0710	440.4693	441.8706	443.2748	444.6820	446.0922	447.5054
	Circumference	29.5310	29.5625	29.5939	29.6253	29.6567	29.6881	29.7195	29.7510	29.7824	29.8138
9.5	Area	70.8822	71.0315	71.1809	71.3306	71.4803	71.6303	71.7804	71.9306	72.0810	72.2316
	Volume	448.9216	450.3407	451.7628	453.1879	454.6160	456.0472	457.4813	458.9184	460.3585	461.8016
	Circumference	29.8452	29.8766	29.9080	29.9394	29.9709	30.0023	30.0337	30.0651	30.0965	30.1279
9.6	Area	72.3823	72.5332	72.6842	72.8354	72.9867	73.1382	73.2899	73.4417	73.5937	73.7458
	Volume	463.2478	464.6969	466.1491	467.6043	469.0625	470.5238	471.9881	473.4554	474.9257	476.3991
	Circumference	30.1594	30.1908	30.2222	30.2536	30.2850	30.3164	30.3479	30.3793	30.4107	30.4421
9.7	Area	73.8981	74.0505	74.2031	74.3559	74.5088	74.6619	74.8151	74.9685	75.1221	75.2758
	Volume	477.8756	479.3551	480.8376	482.3232	483.8119	485.3036	486.7983	488.2962	489.7971	491.3011
	Circumference	30.4735	30.5049	30.5364	30.5678	30.5992	30.6306	30.6620	30.6934	30.7248	30.7563
9.8	Area	75.4296	75.5836	75.7378	75.8921	76.0466	76.2013	76.3561	76.5110	76.6662	76.8214
	Volume	492.8081	494.3183	495.8315	497.3478	498.8672	500.3897	501.9152	503.4439	504.9757	506.5106
	Circumference	30.7877	30.8191	30.8505	30.8819	30.9133	30.9448	30.9762	31.0076	31.0390	31.0704
9.9	Area	76.9769	77.1324	77.2882	77.4441	77.6001	77.7564	77.9127	78.0693	78.2260	78.3828
	Volume	580.0486	509.5897	511.1339	512.6812	514.2316	515.7852	517.3419	518.9017	520.4647	522.0308
	Circumference	31.1018	31.1333	31.1647	31.1961	31.2275	31.2589	31.2903	31.3218	31.3532	31.3846
10	Area	78.5398	78.6970	78.8543	79.0117	79.1694	79.3272	79.4851	79.6432	79.8015	79.9599
	Volume	523.6000	525.1724	526.7479	528.3266	529.9084	531.4933	533.0815	534.6727	536.2672	537.8648
	Circumference	31.4160	31.4474	31.4788	31.5102	31.5417	31.5731	31.6045	31.6359	31.6673	31.6987

Table 17. Circumferences and Areas of Fractional Circles.

Diameter	Circumference	Area	Diameter	Circumference	Area	Diameter	Circumference	Area
1/64	0.04909	0.00019	11/32	1.080	0.09281	43/64	2.111	0.3545
1/32	0.09817	0.00077	23/64	1.129	0.1014	11/16	2.160	0.3712
3/64	0.1473	0.00173	3/8	1.178	0.1104	45/64	2.209	0.3883
1/16	0.1963	0.00307	25/64	1.227	0.1198	23/32	2.258	0.4057
5/64	0.2454	0.00479	13/32	1.276	0.1296	47/64	2.307	0.4236
3/32	0.2945	0.00690	27/64	1.325	0.1398	3/4	2.356	0.4418
7/64	0.3436	0.00940	7/16	1.374	0.1503	49/64	2.405	0.4604
1/8	0.3927	0.01227	29/64	1.424	0.1613	25/32	2.454	0.4794
9/64	0.4418	0.01553	15/32	1.473	0.1726	51/64	2.503	0.4987
5/32	0.4909	0.01917	31/64	1.522	0.1843	13/16	2.553	0.5185
11/64	0.5400	0.02320	1/2	1.571	0.1963	53/64	2.602	0.5386
3/16	0.5890	0.02761	33/64	1.620	0.2088	27/32	2.651	0.5591
13/64	0.6381	0.03241	17/32	1.669	0.2217	55/64	2700	0.5800
7/32	0.6872	0.03758	35/64	1.718	0.2349	7/8	2.749	0.6013
15/64	0.7363	0.04314	9/16	1.767	0.2485	57/64	2.798	0.6230
1/4	0.7854	0.04909	37/64	1.816	0.2625	29/32	2.847	0.6450
17/64	0.8345	0.05542	19/32	1.865	0.2769	59/64	2.896	0.6675
9/32	0.8836	0.06213	39/64	1.914	0.2916	15/16	2.945	0.6903
19/64	0.9327	0.06922	5/8	1.963	0.3068	61/64	2.994	0.7135
5/16	0.9817	0.07670	41/64	2.013	0.3223	31/32	3.043	0.7371
21/64	1.031	0.08456	21/32	2.062	0.3382	63/64	3.093	0.7610

(8) in the left column, and follow the row to the two columns headed by "$R =$." These columns are used to determine the radius of the circumscribing circle, and the answer can be determined if either the length of the sides (s = six-inches) or the radius of the inscribed circle (r) are known. Since the length of the sides is given, it can be seen that $R = 1.307\ s$. Therefore, $R = 1.307 \times 6 = 7.842$. Distance across corners is equal to $2R$, so the distance across the corners of a regular octagon with six-inch sides is 15.684 inches.

The distance across flats is equal to twice the radius of the inscribed circle, so follow the same row to the two columns headed "$r =$." These columns are used to determine the radius of the inscribed circle, and the answer can be determined if either the length of the sides (s = six-inches) or the radius of the circumscribing circle (R) are known. Since the length of the sides is given, it can be seen that $r = 1.207$. Therefore, $r = 1.207 \times 6 = 7.242$. Distance across flats is equal to $2R$, so the distance across the flats of a regular octagon with six-inch sides is 14.484 inches.

The table can also be used to find the area of regular polygons with 3 to 64 sides.

Other Useful Calculations

The derivation of certain other dimensions of lines, circles, and curves is shown in **Table 19**. Some of these formulas require more advanced mathematics, but their use will simplify many layout problems that are commonly encountered.

Table 18. Formulas for Determining Dimensions of Regular Polygons.

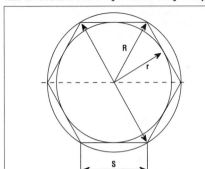

n = number of sides
$\alpha = 360/n$
s = length of side
R = radius of circumscribed circle
r = radius of inscribed circle
A = area

n	α	$A =$			$R =$		$s =$		$r =$	
3	120°	$0.4330\,s^2$	$1.299\,r^2$	$5.196\,r^2$	$0.5774\,s$	$2.000\,r$	$1.732\,R$	$3.464\,r$	$0.5000\,R$	$0.2887\,r$
4	90°	$1.000\,s^2$	$2.000\,R^2$	$4.000\,r^2$	$0.7071\,s$	$1.414\,r$	$1.414\,R$	$2.000\,r$	$0.7071\,R$	$0.5000\,r$
5	72°	$1.721\,s^2$	$2.378\,R^2$	$3.633\,r^2$	$0.8507\,s$	$1.236\,r$	$1.176\,R$	$1.453\,r$	$0.8090\,R$	$0.6882\,r$
6	60°	$2.598\,s^2$	$2.598\,R^2$	$3.464\,r^2$	$1.0000\,s$	$1.155\,r$	$1.000\,R$	$1.155\,r$	$0.8660\,R$	$0.8660\,r$
7	51°.43	$3.634\,s^2$	$2.736\,R^2$	$3.371\,r^2$	$1.152\,s$	$1.110\,r$	$0.8678\,R$	$0.9631\,r$	$0.9010\,R$	$1.038\,r$
8	45°	$4.828\,s^2$	$2.828\,R^2$	$3.314\,r^2$	$1.307\,s$	$1.082\,r$	$0.7654\,R$	$0.8284\,r$	$0.9239\,R$	$1.207\,r$
9	40°	$6.182\,s^2$	$2.893\,R^2$	$3.276\,r^2$	$1.462\,s$	$1.064\,r$	$0.6840\,R$	$0.7279\,r$	$0.9397\,R$	$1.374\,r$
10	36°	$7.694\,s^2$	$2.939\,R^2$	$3.249\,r^2$	$1.618\,s$	$1.052\,r$	$0.6180\,R$	$0.6498\,r$	$0.9511\,R$	$1.539\,r$
12	30°	$11.20\,s^2$	$3.000\,R^2$	$3.215\,r^2$	$1.932\,s$	$1.035\,r$	$0.5176\,R$	$0.5359\,r$	$0.9659\,R$	$1.866\,r$
15	24°	$17.64\,s^2$	$3.051\,R^2$	$3.188\,r^2$	$2.405\,s$	$1.022\,r$	$0.4158\,R$	$0.4251\,r$	$0.9781\,R$	$2.352\,r$
16	22°.50	$20.11\,s^2$	$3.062\,R^2$	$3.183\,r^2$	$2.563\,s$	$1.020\,r$	$0.3902\,R$	$0.3978\,r$	$0.9808\,R$	$2.514\,r$
20	18°	$31.57\,s^2$	$3.090\,R^2$	$3.168\,r^2$	$3.196\,s$	$1.013\,r$	$0.3129\,R$	$0.3168\,r$	$0.9877\,R$	$3.157\,r$
24	15°	$45.58\,s^2$	$3.106\,R^2$	$3.160\,r^2$	$3.831\,s$	$1.009\,r$	$0.2611\,R$	$0.2633\,r$	$0.9914\,R$	$3.798\,r$
32	11°.25	$81.23\,s^2$	$3.121\,R^2$	$3.152\,r^2$	$5.101\,s$	$1.005\,r$	$0.1960\,R$	$0.1970\,r$	$0.9952\,R$	$5.077\,r$
48	7°.50	$183.1\,s^2$	$3.133\,R^2$	$3.146\,r^2$	$7.645\,s$	$1.002\,r$	$0.1308\,R$	$0.1311\,r$	$0.9979\,R$	$7.629\,r$
64	5°.625	$325.7\,s^2$	$3.137\,R^2$	$3.144\,r^2$	$10.19\,s$	$1.001\,r$	$0.0981\,R$	$0.0983\,r$	$0.9988\,R$	$10.18\,r$

Table 19. Analytic Geometry in the Plane.

Distance Between Two Points

$P_1(x_1, y_1)$, $P_2(x_2, y_2)$:

(length of a line segment)

$$|\overline{P_1 P_2}| = \sqrt{(x_2 - x_1)^2 + (y_2 - y_1)^2}$$

Straight Line

General equation

$$Ax + By + C = 0$$

Normal form

$$\frac{Ax + By + C}{\sqrt{A^2 + B^2}} = 0$$

(Continued)

Table 19. *(Continued)* **Analytic Geometry in the Plane.**

Slope-intercept form

$y = mx + b$ where

m slope of straight line,

$$m = \tan \alpha = \frac{y_2 - y_1}{x_2 - x_1}$$

Two-point form

$$\frac{y - y_1}{x - x_1} = \frac{y_2 - y_1}{x_2 - x_1}$$

Point-slope form

$$\frac{y - y_1}{x - x_1} = m$$

Intercept form

$$\frac{x}{a} + \frac{y}{b} = 1, \ a \neq 0, \ b \neq 0,$$

The coordinate axes are intersected at the points $P(a,0)$, $P(0,b)$.

Angle of intersection between two lines
g_1, g_2:

$g_1: y = m_1 x + b_1, \ m_1 = \tan \alpha_1$

$g_2: y = m_2 x + b_2, \ m_2 = \tan \alpha_2$

Angle of intersection β such that

$$\tan \beta = \frac{m_2 - m_1}{1 + m_1 m_2}$$

$$g_1 \perp g_2: m_2 = -\frac{1}{m_1}$$

$$g_1 \| g_2: m_1 = m_2$$

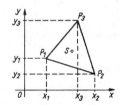

Triangle

If the points $P(x_1, y_1)$, $P(x_2, y_2)$, $P(x_3, y_3)$ do not all lie on a single line, then they form the corners of a triangle.

Area

$$A = \frac{1}{2} \begin{vmatrix} x_1 & y_1 & 1 \\ x_2 & y_2 & 1 \\ x_3 & y_3 & 1 \end{vmatrix}$$

Center of area $S(x_S, y_S)$

$$x_S = \frac{1}{3}(x_1 + x_2 + x_3), \ y_S = \frac{1}{3}(y_1 + y_2 + y_3)$$

(Continued)

Table 19. *(Continued)* **Analytic Geometry in the Plane.**

Circle

Center $M(x_m, y_m)$, radius r

Center-radius equation

$$(x - x_m)^2 + (y - y_m)^2 = r^2$$

Special $M(0, 0)$: $x^2 + y^2 = r^2$

Line tangent to the circle at the point $P(x_0, y_0)$:

$$(x - x_m)(x_0 - x_m) + (y - y_m)(y_0 - y_m) = r^2$$

Special $M(0, 0)$: $xx_0 + yy_0 = r^2$

If $P(x_0, y_0)$ lies outside the circle, the formulas represent the equation of the chord connecting the points of tangency.

Parabola

Vertex equation for vertex at $S(x_S, y_S)$:

$$(y - y_s)^2 = 2p(x - x_s)$$

Special $S(0, 0)$: $y^2 = 2px$

Assumes that parabola opens in direction of x; other common directions:

opening in direction of y	$x^2 = 2py$
opening in direction of $(-x)$	$y^2 = -2px$
opening in direction of $(-y)$	$x^2 = -2py$

Line tangent to the parabola
at point $P(x_0, y_0)$: $\qquad\qquad\qquad\qquad yy_0 = 2p(x - x_0)$

Focus F $(p/2, 0)$, semilatus rectum p

Focal radius $\qquad\qquad\qquad\qquad r = |\overline{FP_0}| = x_0 + \dfrac{p}{2}$

Circle of curvature at the radius: $\qquad\qquad$ center $P(0, p)$, radius p

Table 19. *(Continued)* **Analytic Geometry in the Plane.**

Ellipse

Center $M(x_m, y_m)$, semiaxes a, b

Center-radius equation

$$\frac{(x - x_m)^2}{a^2} + \frac{(y - y_m)^2}{b^2} = 1$$

Special $M(0, 0)$: $\dfrac{x^2}{a^2} + \dfrac{y^2}{b^2} = 1$

Equation of the line tangent to
the ellipse at point $P(x_0, y_0)$

$$\frac{(x - x_m)(x - x_0)}{a^2} + \frac{(y - y_m)(y - y_0)}{b^2} = 1$$

Vertex equation for vertex at $A_1(0, 0)$:	$y^2 = 2px - \dfrac{p}{a}x^2$
Semilatus rectum	$p = \dfrac{b^2}{a}$
Linear eccentricity	$e^2 = a^2 - b^2$
Numerical eccentricity	$\varepsilon = \dfrac{e}{a} < 1$
Focal radii	$r_1 = a + \varepsilon x_0, \ r_2 = a - \varepsilon x_0$
Defining property	$r_1 + r_2 = 2a$
Radiuses of the circles of curvature in A_1 or A_2:	$\varrho_A = \dfrac{b^2}{a}$
$\qquad B_1$ or B_2:	$\varrho_B = \dfrac{a^2}{b}$

Hyperbola

Center $M(x_m, y_m)$, semiaxes a, b

Center-radius equation

$$\frac{(x - x_m)^2}{a^2} - \frac{(y - y_m)^2}{b^2} = 1$$

Special $M(0, 0)$: $\dfrac{x^2}{a^2} - \dfrac{y^2}{b^2} = 1$

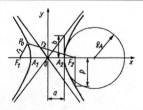

Table 19. *(Continued)* **Analytic Geometry in the Plane.**

Equation of the line tangent to the hyperbola at point $P(x_0, y_0)$

$$\frac{(x - x_m)(x - x_0)}{a^2} - \frac{(y - y_m)(y - y_0)}{b^2} = 1$$

Vertex equation for vertex at $A_2(0, 0)$:	$y^2 = 2px + \dfrac{p}{a}x^2$
Semilatus rectum	$p = \dfrac{b^2}{a}$
Linear eccentricity	$e^2 = a^2 + b^2$
Numerical eccentricity	$\varepsilon = \dfrac{e}{a} > 1$
Focal radii	$r_1 = a + \varepsilon x_0,\ r_2 = -a + \varepsilon x_0$
Defining property	$r_1 - r_2 = 2a$
Radius of the circle of curvature in A_1 or A_2:	$\varrho_A = \dfrac{b^2}{a}$
Asymptotes	$y = \dfrac{b}{a}x,\ y = -\dfrac{b}{a}x$

Second-Degree Curves (Quadric Curves)

General representation of a second-degree curve (translated and rotated with respect to the ordinary position)

$$a_{11}x^2 + 2a_{12}xy + a_{22}y^2 + 2a_{13}x + 2a_{23}y + a_{33} = 0$$

The specially represented curve can be determined from the coefficients by analyzing the quantities:

$$\Delta = \begin{vmatrix} a_{11} & a_{12} & a_{13} \\ a_{12} & a_{22} & a_{23} \\ a_{13} & a_{23} & a_{33} \end{vmatrix}, \quad \delta = \begin{vmatrix} a_{11} & a_{12} \\ a_{21} & a_{22} \end{vmatrix}, \quad S = a_{11} + a_{22}$$

$\Delta \neq 0,$	$\delta > 0,$	$\Delta S < 0$: real ellipse
		$\Delta S > 0$: imaginary ellipse
$\Delta \neq 0,$	$\delta < 0$:	hyperbola
$\Delta \neq 0,$	$\delta = 0$:	parabola
$\Delta = 0,$	$\delta > 0$:	pair of intersecting imaginary straight lines with real point of intersection
	$\delta < 0$:	pair of intersecting real straight lines
	$\delta = 0$:	pair of parallel straight lines

Solving Triangles

Right-angled triangles

To understand triangles and trigonometric functions, it is important to know how the sides and angles of all triangles relate to each other. These relationships are best understood with a right-angled triangle, as shown in *Figure 11*. The sides of the triangle are identified by the letters a, b, and c, and the three angles are named A, B, and C. Angle A is the right angle, and the side opposite the right angle (side a) is known as the **hypotenuse**. Side b, which is opposite angle B, is the *side adjacent* to angle C. Side c is the *side opposite* angle C, but is the side adjacent to angle B.

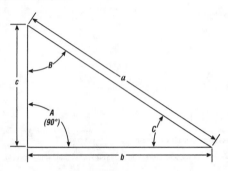

Figure 11. A right-angled triangle.

The following definitions explain the relationships of angles with the sides of any right-angled triangle. The sides and angles are identified in *Figure 11*. Note that when solving for angles, a solution for angle A is not given because A always equals 90°.

The **sine** (usually abbreviated as *sin*) of an angle is equal to its opposite side divided by the hypotenuse.

$$\sin = \frac{\text{side opposite}}{\text{hypotenuse}}$$

Therefore, $\sin B = b \div a$, and $\sin C = c \div a$. In every triangle, the sines of its angles are proportional to the opposite sides.

$$\frac{\sin A}{a} = \frac{\sin B}{b} = \frac{\sin C}{c}$$

$$\frac{a}{b} = \frac{\sin A}{\sin B} \qquad \frac{b}{c} = \frac{\sin B}{\sin C} \qquad \frac{a}{c} = \frac{\sin A}{\sin C}$$

$$a = \frac{b \sin A}{\sin B} = \frac{c \sin A}{\sin C}$$

$$b = \frac{a \sin B}{\sin A} = \frac{c \sin B}{\sin C}$$

$$c = \frac{b \sin C}{\sin A} = \frac{c \sin C}{\sin B}.$$

These proportional relationships are known as the *Law of Sines*, which is used to solve for an unknown side when two sides are known, or an unknown angle when two angles are known. This law, and the law of cosines below, applies to any triangle.

The **cosine** (*cos*) of an angle is equal to its adjacent side divided by the hypotenuse.

$$\cos = \frac{\text{side adjacent}}{\text{hypotenuse}}$$

Therefore, $\cos B = c \div a$, and $\cos C = b \div a$. In every triangle, the square of any side equals the sum of the squares of the other sides, minus twice the product of the given sides times the cosine of the angle between the two given sides (the *included angle*).

$a^2 = b^2 + c^2 - 2bc(\cos A)$	or	$\cos A = (a^2 + b^2 + c^2) \div 2bc$
$b^2 = a^2 + c^2 - 2ac(\cos B)$	or	$\cos B = (a^2 + b^2 + c^2) \div 2ac$
$c^2 = a^2 + b^2 - 2ab(\cos C)$	or	$\cos C = (a^2 + b^2 + c^2) \div 2ab.$

This is known as the *Law of Cosines*, and can be used to find an unknown side when any two sides and the included angle between them are known. This law, and the law of sines above, applies to any triangle.

The **tangent** (*tan*) of an angle is equal to the opposite side divided by the adjacent side. Therefore, $\tan B = b \div c$, and $\tan C = c \div b$.

$$\tan = \frac{\text{side opposite}}{\text{side adjacent}}$$

The **cotangent** (*cot*) of an angle is equal to the adjacent side divided by the opposite side. Therefore, $\cot B = c \div b$, and $\cot C = b \div c$.

$$\cot = \frac{\text{side adjacent}}{\text{side opposite}}$$

The **secant** (*sec*) of an angle is equal to the hypotenuse divided by the adjacent side. Therefore, $\sec B = a \div c$, and $\sec C = a \div b$.

$$\sec = \frac{\text{hypotenuse}}{\text{side adjacent}}$$

The **cosecant** (*cosec*) of an angle is equal to the hypotenuse divided by the opposite side. Therefore, $\csc B = a \div b$, and $\csc C = a \div c$.

$$\csc = \frac{\text{hypotenuse}}{\text{side opposite}}$$

The following are some useful rules for determining the length of the hypotenuse, the side adjacent, and the side opposite when an angle and side are known.

Length of the hypotenuse is equal to
Side opposite × cosecant	Side opposite ÷ sine
Side adjacent × secant	Side adjacent ÷ cosine.

Length of the side opposite is equal to
Hypotenuse × sine	Hypotenuse ÷ cosecant
Side adjacent × tangent	Side adjacent ÷ cotangent.

Length of the side adjacent is equal to
Hypotenuse × cosine	Hypotenuse ÷ secant
Side opposite × cotangent	Side opposite ÷ tangent.

These and similar formulas can be used to find unknown sides and angles in right-angled triangles.

Given: the length of two sides. Find the length of the third side.

$a = \sqrt{b_2 + c_2}$ $b = \sqrt{a_2 - c_2}$ $c = \sqrt{a_2 - b_2}$

Given: The hypotenuse (a), and one acute angle B. Find b, c, and C.

$b = a \times \sin B$ $c = a \times \cos B$ $C = 90° - B$

Given: The hypotenuse (a), and one angle C. Find b, c, and B.

$b = a \times \cos C$ $c = a \times \sin C$ $B = 90° - C$

Given: Angle B and its adjacent side c. Find a, b, and C.

$a = c \div \cos B$ $b = $ c $\times \tan B$ $C = 90° - B$

Given: Angle C and its adjacent side b. Find a, c, and B.

$a = b \div \cos C$ $c = b \times \tan C$ $B = 90° - C$

Given: Angle B and the opposite side b. Find a, c, and C.

$a = b \div \sin B$ $c = b \times \cot B$ $C = 90° - B$

Given: Angle C and the opposite side c. Find a, b, and B.

$a = c \div \sin C$ $b = c \times \cot C$ $B = 90° - C$

Given B: $C = 90° - B$

Given C: $B = 90° - C$

To find the **area** of a right-angled triangle.

area $= b \times c \div 2$.

Oblique-angled triangles

Oblique angles may be either acute or obtuse. If acute, all angles within the triangle are less than 90°. If obtuse, one of the angles within the triangle is more than 90°. In most instances, the Law of Sines and the Law of Cosines can be used to find unknown sides and angles. Several of the solutions below are derived from these laws, plus the fact that the sum of all angles in a triangle always equals 180°.

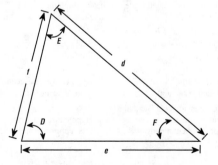

Figure 12. An oblique-angled triangle.

In the following examples, two unknown sides are found when one side and two angles are known. The unknown side of an oblique-angled triangle is equal to the known side times the sine of the angle opposite the unknown side, divided by the sine of the angle opposite the known side.

Given: One side d and two angles, D and E. Find e, f, and F.

$$e = (d \times \sin E) \div \sin D \qquad f = (d \times \sin F) \div \sin D \qquad F = 180° - (D + E).$$

Given: One side e and two angles, E and F. Find d, f, and D.

$$d = (e \times \sin D) \div \sin E \qquad f = (e \times \sin F) \div \sin E \qquad D = 180° - (E + F).$$

Given: One side f and two angles, D and F. Find d, e, and E.

$$d = (f \times \sin D) \div \sin F \qquad e = (f \times \sin E) \div \sin F \qquad E = 180° - (D + F).$$

In the next examples, two sides and one opposite angle are given.

Given: Sides d and e, and opposite angle D. Find E, then find F and f.

$$\sin E = (e \times \sin D) \div d \qquad F = 180° - (D + E) \qquad f = (d \times \sin F) \div \sin D.$$

Given: Sides e and f, and opposite angle E. Find F, then find D and d.

$$\sin F = (f \times \sin E) \div e \qquad D = 180° - (E + F) \qquad d = (e \times \sin D) \div \sin E.$$

Given: Sides d and f, and opposite angle F. Find D, then find E and e.

$$\sin D = (d \times \sin F) \div f \qquad E = 180° - (D + F) \qquad e = (f \times \sin E) \div \sin F.$$

These examples have two given sides, and the included angle is also known.

Given: Sides d and e, and included angle F. Find D, then find E and f.

$$\tan D = (d \times \sin F) \div (e - [d \times \cos F]) \quad E = 180° - (D + E) \ f = (d \times \sin F) \div \sin D.$$

Given: Sides e and f, and included angle D. Find E, then find F and d.

$$\tan E = (e \times \sin D) \div (f - [e \times \cos D]) \quad F = 180° - (D + E) \ d = (e \times \sin D) \div \sin E.$$

Given: Sides d and f, and included angle E. Find F, then find D and e.

$$\tan F = (f \times \sin E) \div (d - [f \times \cos E]) \quad D = 180° - (E + F) \ e = (f \times \sin E) \div \sin F.$$

In the following examples, all sides are known, but no angles. Find the angles.

$$\cos D = (e^2 + f^2 - d^2) \div 2\,ef \qquad \cos E = (d^2 + f^2 - e^2) \div 2\,df$$
$$\cos F = (d^2 + e^2 - f^2) \div 2\,d\,e.$$

The area of an oblique triangle is equal to the product of two sides times the sine of the included angle.

$$\text{Area} = (e\,f \times \sin D) \div 2 = (d\,f \times \sin E) \div 2 = (d\,e \times \sin F) \div 2.$$

Trigonometric functions

Definitions of the basic trigonometric functions of angles were given in the section on right-angled triangles, and values for the functions for all angles are given in the unnumbered Table that follows ("Natural Trigonometric Functions").

Not all trigonometric functions have positive values, and should be stated with a minus sign if appropriate. Functions are either positive or negative, in segments of 90°. For example, the sine of 234° is - 0.80902, while the sine of 54° is + 0.80902.

0° to 90°   all values positive

Above 90° to 180°   sine and cosecant positive values, all other functions negative

Above 180° to 270° tangent and cotangent positive values, all other functions negative

Above 270° to 360° cosine and secant positive, all other values negative.

The following trigonometric functions apply to all triangles.

$$\sin \alpha = \frac{\cos \alpha}{\cot \alpha} = \cos \alpha \times \tan \alpha = \frac{1}{\csc \alpha}$$

$$\sin^2 \alpha + \cos 2\alpha = \sin \alpha \times \csc \alpha = \cos \alpha \times \sec \alpha = \tan \times \cot \alpha = 1$$

$$\sin^2 \alpha = 1 - \cos \alpha = \frac{\tan^2 \alpha}{1 + \tan^2 \alpha} = \frac{1}{1 + cor^2 \alpha}$$

$$\sin 2\alpha = 2 \left(\sin \alpha \times \cot \alpha \right)$$

$$\cos \alpha = \frac{\sin \alpha}{\tan \alpha} = \sin \alpha \times \cot \alpha = \frac{1}{\sec \alpha}$$

$$\cos^2 \alpha = 1 - \sin \alpha = \frac{\cot^2 \alpha}{1 + \cot^2 \alpha} = \frac{1}{1 + \tan^2 \alpha}$$

$$\cos 2\alpha = \cos^2 \alpha - \sin^2 \alpha$$

$$\tan \alpha = \frac{\sin \alpha}{\cos \alpha} = \sin \alpha \times \sec \alpha = \frac{1}{\cot \alpha}$$

$$\tan^2 \alpha = \frac{1 - \cos^2 \alpha}{\cos^2 \alpha} = \frac{\sin^2 \alpha}{1 - \sin^2 \alpha} = \frac{1}{\cot^2 \alpha}$$

$$\tan 2\alpha = \frac{2 \tan \alpha}{1 - \tan \alpha}$$

$$\cot \alpha = \frac{\cos \alpha}{\sin \alpha} = \cos \alpha \times \csc \alpha = \frac{1}{\tan \alpha}$$

$$\cot^2 \alpha = \frac{1 - \sin^2 \alpha}{\sin^2 \alpha} = \frac{\cos^2 \alpha}{1 - \cos^2 \alpha} = \frac{1}{\tan^2 \alpha}$$

$$\cot 2\alpha = \frac{\cot^2 \alpha - 1}{2 \cot \alpha}$$

$$\sec \alpha = \frac{\tan \alpha}{\sin \alpha} = \frac{1}{\cos \alpha}$$

$$\csc \alpha = \frac{\cot \alpha}{\cos \alpha} = \frac{1}{\sin \alpha}$$

$$\sin x + \sin y = 2 \sin \frac{x+y}{2} \times \cos \frac{x-y}{2}$$

$$\sin x - \sin y = 2 \cos \frac{x+y}{2} \times \sin \frac{x-y}{2}$$

$$\cos x + \cos y = 2 \cos \frac{x+y}{2} \times \cos \frac{x-y}{2}$$

$$\cos x - \cos y = -2 \sin \frac{x+y}{2} \times \sin \frac{x-y}{2}$$

$$\tan x + \tan y = \frac{\sin (x+y)}{\cos x \times \cos y}$$

$$\tan x - \tan y = \frac{\sin (x-y)}{\cos x \times \cos y}$$

$$\cot x + \cot y = \frac{\sin (x+y)}{\sin x \times \sin y}$$

$$\tan x - \tan y = \frac{\sin (x-y)}{\sin x \times \sin y}$$

Using the "Natural Trigonometric Functions" table

The Table is structured in a format designed to use the least amount of space possible. Therefore, depending on the angle desired, it may be necessary to enter the Table from any of the four corners. Use the following guide for looking up angles and minutes.

Angles from	To Find Function	To Determine Minutes
0° to 45°	Enter at top left	Read from left side
45° to 90°	Enter from bottom right	Read from right side
90° to 135°	Enter from bottom left	Read from left side
135° to 180°	Enter from top right	Read from right side
180° to 225°	Enter from top left	Read from left side
225° to 270°	Enter from bottom right	Read from right side
270° to 315°	Enter from bottom left	Read from left side
315° to 360°	Enter from top right	Read from right side

M	Sine	Cosine	Tan.	Cotan.	Secant	Cosec.	M
0	0.00000	1.0000	0.00000	Infinite	1.0000	Infinite	60
1	0.00029	1.0000	0.00029	3437.7	1.0000	3437.7	59
2	0.00058	1.0000	0.00058	1718.9	1.0000	1718.9	58
3	0.00087	1.0000	0.00087	1145.9	1.0000	1145.9	57
4	0.00116	1.0000	0.00116	859.44	1.0000	859.44	56
5	0.00145	1.0000	0.00145	687.55	1.0000	687.55	55
6	0.00174	1.0000	0.00174	572.96	1.0000	572.96	54
7	0.00204	1.0000	0.00204	491.11	1.0000	491.11	53
8	0.00233	1.0000	0.00233	429.72	1.0000	429.72	52
9	0.00262	1.0000	0.00262	381.97	1.0000	381.97	51
10	0.00291	0.99999	0.00291	343.77	1.0000	343.77	50
11	0.00320	0.99999	0.00320	312.52	1.0000	312.52	49
12	0.00349	0.99999	0.00349	286.48	1.0000	286.48	48
13	0.00378	0.99999	0.00378	264.44	1.0000	264.44	47
14	0.00407	0.99999	0.00407	245.55	1.0000	245.55	46
15	0.00436	0.99999	0.00436	229.18	1.0000	229.18	45
16	0.00465	0.99999	0.00465	214.86	1.0000	214.86	44
17	0.00494	0.99999	0.00494	202.22	1.0000	202.22	43
18	0.00524	0.99999	0.00524	190.98	1.0000	190.98	42
19	0.00553	0.99998	0.00553	180.93	1.0000	180.93	41
20	0.00582	0.99998	0.00582	171.88	1.0000	171.88	40
21	0.00611	0.99998	0.00611	163.70	1.0000	163.70	39
22	0.00640	0.99998	0.00640	156.26	1.0000	156.26	38
23	0.00669	0.99998	0.00669	149.46	1.0000	149.47	37
24	0.00698	0.99997	0.00698	143.24	1.0000	143.24	36
25	0.00727	0.99997	0.00727	137.51	1.0000	137.51	35
26	0.00756	0.99997	0.00756	132.22	1.0000	132.22	34
27	0.00785	0.99997	0.00785	127.32	1.0000	127.32	33
28	0.00814	0.99997	0.00814	122.77	1.0000	122.77	32
29	0.00843	0.99996	0.00844	118.54	1.0000	118.54	31
30	0.00873	0.99996	0.00873	114.59	1.0000	114.59	30
31	0.00902	0.99996	0.00902	110.89	1.0000	110.90	29
32	0.00931	0.99996	0.00931	107.43	1.0000	107.43	28
33	0.00960	0.99995	0.00960	104.17	1.0000	104.17	27
34	0.00989	0.99995	0.00989	101.11	1.0000	101.11	26
35	0.01018	0.99995	0.01018	98.218	1.0000	98.223	25
36	0.01047	0.99994	0.01047	95.489	1.0000	95.495	24
37	0.01076	0.99994	0.01076	92.908	1.0000	92.914	23
38	0.01105	0.99994	0.01105	90.463	1.0001	90.469	22
39	0.01134	0.99993	0.01134	88.143	1.0001	88.149	21
40	0.01163	0.99993	0.01164	85.940	1.0001	85.946	20
41	0.01193	0.99993	0.01193	83.843	1.0001	83.849	19
42	0.01222	0.99992	0.01222	81.847	1.0001	81.853	18
43	0.01251	0.99992	0.01251	79.943	1.0001	79.950	17
44	0.01280	0.99992	0.01280	78.126	1.0001	78.133	16
45	0.01309	0.99991	0.01309	76.390	1.0001	76.396	15
46	0.01338	0.99991	0.01338	74.729	1.0001	74.736	14
47	0.01367	0.99991	0.01367	73.139	1.0001	73.146	13
48	0.01396	0.99990	0.01396	71.615	1.0001	71.622	12
49	0.01425	0.99990	0.01425	70.153	1.0001	70.160	11
50	0.01454	0.99989	0.01455	68.750	1.0001	68.757	10
51	0.01483	0.99989	0.01484	67.402	1.0001	67.409	9
52	0.01512	0.99988	0.01513	66.105	1.0001	66.113	8
53	0.01542	0.99988	0.01542	64.858	1.0001	64.866	7
54	0.01571	0.99988	0.01571	63.657	1.0001	63.664	6
55	0.01600	0.99987	0.01600	62.499	1.0001	62.507	5
56	0.01629	0.99987	0.01629	61.383	1.0001	61.391	4
57	0.01658	0.99987	0.01658	60.306	1.0001	60.314	3
58	0.01687	0.99986	0.01687	59.266	1.0001	59.274	2
59	0.01716	0.99985	0.01716	58.261	1.0001	58.270	1
60	0.01745	0.99985	0.01746	57.290	1.0001	57.299	0
M	**Cosine**	**Sine**	**Cotan.**	**Tan.**	**Cosec.**	**Secant**	**M**

1° or 181° Natural Trigonometric Functions **178° or 358°**

M	Sine	Cosine	Tan.	Cotan.	Secant	Cosec.	M
0	0.01745	0.99985	0.01746	57.290	1.0001	57.299	60
1	0.01774	0.99984	0.01775	56.350	1.0001	56.359	59
2	0.01803	0.99984	0.01804	55.441	1.0001	55.450	58
3	0.01832	0.99983	0.01833	54.561	1.0002	54.570	57
4	0.01861	0.99983	0.01862	53.708	1.0002	53.718	56
5	0.01891	0.99982	0.01891	52.882	1.0002	52.891	55
6	0.01920	0.99981	0.01920	52.081	1.0002	52.090	54
7	0.01949	0.99981	0.01949	51.303	1.0002	51.313	53
8	0.01978	0.99980	0.01978	50.548	1.0002	50.558	52
9	0.02007	0.99980	0.02007	49.816	1.0002	49.826	51
10	0.02036	0.99979	0.02036	49.104	1.0002	49.114	50
11	0.02065	0.99979	0.02066	48.412	1.0002	48.422	49
12	0.02094	0.99978	0.02095	47.739	1.0002	47.750	48
13	0.02123	0.99977	0.02124	47.085	1.0002	47.096	47
14	0.02152	0.99977	0.02153	46.449	1.0002	46.460	46
15	0.02181	0.99976	0.02182	45.829	1.0002	45.840	45
16	0.02210	0.99975	0.02211	45.226	1.0002	45.237	44
17	0.02240	0.99975	0.02240	44.638	1.0002	44.650	43
18	0.02269	0.99974	0.02269	44.066	1.0002	44.077	42
19	0.02298	0.99974	0.02298	43.508	1.0003	43.520	41
20	0.02326	0.99973	0.02327	42.964	1.0003	42.976	40
21	0.02356	0.99972	0.02357	42.433	1.0003	42.445	39
22	0.02385	0.99971	0.02386	41.916	1.0003	41.928	38
23	0.02414	0.99971	0.02415	41.140	1.0003	41.423	37
24	0.02443	0.99970	0.02444	40.917	1.0003	40.930	36
25	0.02472	0.99969	0.02473	40.436	1.0003	40.448	35
26	0.02501	0.99969	0.02502	39.965	1.0003	39.978	34
27	0.02530	0.99968	0.02531	39.506	1.0003	39.518	33
28	0.02559	0.99967	0.02560	39.057	1.0003	39.069	32
29	0.02589	0.99966	0.02589	38.618	1.0003	38.631	31
30	0.02618	0.99966	0.02618	38.188	1.0003	38.201	30
31	0.02647	0.99965	0.02648	37.769	1.0003	37.782	29
32	0.02676	0.99964	0.02677	37.358	1.0003	37.371	28
33	0.02705	0.99963	0.02706	36.956	1.0004	36.969	27
34	0.02734	0.99963	0.02735	36.563	1.0004	36.576	26
35	0.02763	0.99962	0.02764	36.177	1.0004	36.191	25
36	0.02792	0.99961	0.02793	35.800	1.0004	35.814	24
37	0.02821	0.99960	0.02822	35.431	1.0004	35.445	23
38	0.02850	0.99959	0.02851	35.069	1.0004	35.084	22
39	0.02879	0.99958	0.02881	34.715	1.0004	34.729	21
40	0.02908	0.99958	0.02910	34.368	1.0004	34.382	20
41	0.02937	0.99957	0.02939	34.027	1.0004	34.042	19
42	0.02967	0.99956	0.02968	33.693	1.0004	33.708	18
43	0.02996	0.99955	0.02997	33.366	1.0004	33.381	17
44	0.03025	0.99954	0.03026	33.045	1.0004	33.060	16
45	0.03054	0.99953	0.03055	32.730	1.0005	32.745	15
46	0.03083	0.99952	0.03084	32.421	1.0005	32.437	14
47	0.03112	0.99951	0.03113	32.118	1.0005	32.134	13
48	0.03141	0.99951	0.03143	31.820	1.0005	31.836	12
49	0.03170	0.99950	0.03172	31.528	1.0005	31.544	11
50	0.03199	0.99949	0.03201	31.241	1.0005	31.257	10
51	0.03228	0.99948	0.03230	30.960	1.0005	30.976	9
52	0.03257	0.99947	0.03259	30.683	1.0005	30.699	8
53	0.03286	0.99946	0.03288	30.411	1.0005	30.428	7
54	0.03315	0.99945	0.03317	30.145	1.0005	30.161	6
55	0.03344	0.99944	0.03346	29.882	1.0005	29.899	5
56	0.03374	0.99943	0.03375	29.624	1.0006	29.641	4
57	0.03403	0.99942	0.03405	29.371	1.0006	29.388	3
58	0.03432	0.99941	0.03434	29.122	1.0006	29.139	2
59	0.03461	0.99940	0.03463	28.877	1.0006	28.894	1
60	0.03490	0.99939	0.03492	28.636	1.0006	28.654	0
M	Cosine	Sine	Cotan.	Tan.	Cosec.	Secant	M

91° or 271° **88° or 268°**

    Natural Trigonometric Functions

M	Sine	Cosine	Tan.	Cotan.	Secant	Cosec.	M
0	0.03490	0.99939	0.03492	28.636	1.0006	28.654	60
1	0.03519	0.99938	0.03521	28.399	1.0006	28.417	59
2	0.03548	0.99937	0.03550	28.166	1.0006	28.184	58
3	0.03577	0.99936	0.03579	27.937	1.0006	27.955	57
4	0.03606	0.99935	0.03608	27.712	1.0006	27.730	56
5	0.03635	0.99934	0.03638	27.490	1.0007	27.508	55
6	0.03664	0.99933	0.03667	27.271	1.0007	27.290	54
7	0.03693	0.99932	0.03696	27.056	1.0007	27.075	53
8	0.03722	0.99931	0.03725	26.845	1.0007	26.864	52
9	0.03751	0.99930	0.03754	26.637	1.0007	26.655	51
10	0.03781	0.99928	0.03783	26.432	1.0007	26.450	50
11	0.03810	0.99927	0.03812	26.230	1.0007	26.249	49
12	0.03839	0.99926	0.03842	26.031	1.0007	26.050	48
13	0.03868	0.99925	0.03871	25.835	1.0007	25.854	47
14	0.03897	0.99924	0.03900	25.642	1.0008	25.661	46
15	0.03926	0.99923	0.03929	25.452	1.0008	25.471	45
16	0.03955	0.99922	0.03958	25.264	1.0008	25.284	44
17	0.03984	0.99921	0.03987	25.080	1.0008	25.100	43
18	0.04013	0.99919	0.04016	24.898	1.0008	24.918	42
19	0.04042	0.99918	0.04045	24.718	1.0008	24.739	41
20	0.04071	0.99917	0.04075	24.542	1.0008	24.562	40
21	0.04100	0.99916	0.04104	24.367	1.0008	24.388	39
22	0.04129	0.99915	0.04133	24.196	1.0008	24.216	38
23	0.04158	0.99913	0.04162	24.026	1.0009	24.047	37
24	0.04187	0.99912	0.04191	23.859	1.0009	23.880	36
25	0.04217	0.99911	0.04220	23.694	1.0009	23.716	35
26	0.04246	0.99910	0.04249	23.532	1.0009	23.553	34
27	0.04275	0.99908	0.04279	23.372	1.0009	23.393	33
28	0.04304	0.99907	0.04308	23.214	1.0009	23.235	32
29	0.04333	0.99906	0.04337	23.058	1.0009	23.079	31
30	0.04362	0.99905	0.04366	22.904	1.0009	22.925	30
31	0.04391	0.99903	0.04395	22.752	1.0010	22.774	29
32	0.04420	0.99902	0.04424	22.602	1.0010	22.624	28
33	0.04449	0.99901	0.04453	22.454	1.0010	22.476	27
34	0.04478	0.99900	0.04483	22.308	1.0010	22.330	26
35	0.04507	0.99898	0.04512	22.164	1.0010	22.186	25
36	0.04536	0.99897	0.04541	22.022	1.0010	22.044	24
37	0.04565	0.99896	0.04570	21.881	1.0010	21.904	23
38	0.04594	0.99894	0.04599	21.742	1.0010	21.765	22
39	0.04623	0.99893	0.04628	21.606	1.0011	21.629	21
40	0.04652	0.99892	0.04657	21.470	1.0011	21.494	20
41	0.04681	0.99890	0.04687	21.337	1.0011	21.360	19
42	0.04711	0.99889	0.04716	21.205	1.0011	21.228	18
43	0.04740	0.99888	0.04745	21.075	1.0011	21.098	17
44	0.04769	0.99886	0.04774	20.946	1.0011	20.970	16
45	0.04798	0.99885	0.04803	20.819	1.0011	20.843	15
46	0.04827	0.99883	0.04832	20.693	1.0012	20.717	14
47	0.04856	0.99882	0.04862	20.569	1.0012	20.593	13
48	0.04885	0.99881	0.04891	20.446	1.0012	20.471	12
49	0.04914	0.99879	0.04920	20.325	1.0012	20.350	11
50	0.04943	0.99878	0.04949	20.205	1.0012	20.230	10
51	0.04972	0.99876	0.04978	20.087	1.0012	20.112	9
52	0.05001	0.99875	0.05007	19.970	1.0012	19.995	8
53	0.05030	0.99873	0.05037	19.854	1.0013	19.880	7
54	0.05059	0.99872	0.05066	19.740	1.0013	19.766	6
55	0.05088	0.99870	0.05095	19.627	1.0013	19.653	5
56	0.05117	0.99869	0.05124	19.515	1.0013	19.541	4
57	0.05146	0.99867	0.05153	19.405	1.0013	19.431	3
58	0.05175	0.99866	0.05182	19.296	1.0013	19.322	2
59	0.05204	0.99864	0.05212	19.188	1.0013	19.214	1
60	0.05234	0.99863	0.05241	19.081	1.0014	19.107	0
M	Cosine	Sine	Cotan.	Tan.	Cosec.	Secant	M

3° or 183° **Natural Trigonometric Functions** **176° or 356°**

M	Sine	Cosine	Tan.	Cotan.	Secant	Cosec.	M
0	0.05234	0.99863	0.05241	19.081	1.0014	19.107	60
1	0.05263	0.99861	0.05270	18.975	1.0014	19.002	59
2	0.05292	0.99860	0.05299	18.871	1.0014	18.897	58
3	0.05321	0.99858	0.05328	18.763	1.0014	18.794	57
4	0.05350	0.99857	0.05357	18.665	1.0014	18.692	56
5	0.05379	0.99855	0.05387	18.564	1.0014	18.591	55
6	0.05408	0.99854	0.05416	18.464	1.0015	18.491	54
7	0.05437	0.99852	0.05445	18.365	1.0015	18.393	53
8	0.05466	0.99850	0.05474	18.268	1.0015	18.295	52
9	0.05495	0.99849	0.05503	18.171	1.0015	18.198	51
10	0.05524	0.99847	0.05532	18.075	1.0015	18.103	50
11	0.05553	0.99846	0.05562	17.980	1.0015	18.008	49
12	0.05582	0.99844	0.05591	17.886	1.0016	17.914	48
13	0.05611	0.99842	0.05620	17.793	1.0016	17.821	47
14	0.05640	0.99841	0.05649	17.701	1.0016	17.730	46
15	0.05669	0.99839	0.05678	17.610	1.0016	17.639	45
16	0.05698	0.99837	0.05707	17.520	1.0016	17.549	44
17	0.05727	0.99836	0.05737	17.431	1.0016	17.460	43
18	0.05756	0.99834	0.05766	17.343	1.0017	17.372	42
19	0.05785	0.99832	0.05795	17.256	1.0017	17.285	41
20	0.05814	0.99831	0.05824	17.169	1.0017	17.198	40
21	0.05843	0.99829	0.05853	17.084	1.0017	17.113	39
22	0.05872	0.99827	0.05883	16.999	1.0017	17.028	38
23	0.05902	0.99826	0.05912	16.915	1.0017	16.944	37
24	0.05931	0.99824	0.05941	16.832	1.0018	16.861	36
25	0.05960	0.99822	0.05970	16.750	1.0018	16.779	35
26	0.05989	0.99820	0.05999	16.668	1.0018	16.698	34
27	0.06018	0.99819	0.06029	16.587	1.0018	16.617	33
28	0.06047	0.99817	0.06058	16.507	1.0018	16.538	32
29	0.06076	0.99815	0.06087	16.428	1.0018	16.459	31
30	0.06105	0.99813	0.06116	16.350	1.0019	16.380	30
31	0.06134	0.99812	0.06145	16.272	1.0019	16.303	29
32	0.06163	0.99810	0.06175	16.195	1.0019	16.226	28
33	0.06192	0.99808	0.06204	16.119	1.0019	16.150	27
34	0.06221	0.99806	0.06233	16.043	1.0019	16.075	26
35	0.06250	0.99804	0.06262	15.969	1.0019	16.000	25
36	0.06279	0.99803	0.06291	15.894	1.0020	15.926	24
37	0.06308	0.99801	0.06321	15.821	1.0020	15.853	23
38	0.06337	0.99799	0.06350	15.748	1.0020	15.780	22
39	0.06366	0.99797	0.06379	15.676	1.0020	15.708	21
40	0.06395	0.99795	0.06408	15.605	1.0020	15.637	20
41	0.06424	0.99793	0.06437	15.534	1.0021	15.566	19
42	0.06453	0.99791	0.06467	15.464	1.0021	15.496	18
43	0.06482	0.99790	0.06496	15.394	1.0021	15.427	17
44	0.06511	0.99788	0.06525	15.325	1.0021	15.358	16
45	0.06540	0.99786	0.06554	15.257	1.0021	15.290	15
46	0.06569	0.99784	0.06583	15.189	1.0022	15.222	14
47	0.06598	0.99782	0.06613	15.122	1.0022	15.155	13
48	0.06627	0.99780	0.06642	15.056	1.0022	15.089	12
49	0.06656	0.99778	0.06671	14.990	1.0022	15.023	11
50	0.06685	0.99776	0.06700	14.924	1.0022	14.958	10
51	0.06714	0.99774	0.06730	14.860	1.0023	14.893	9
52	0.06743	0.99772	0.06759	14.795	1.0023	14.829	8
53	0.06772	0.99770	0.06788	14.732	1.0023	14.765	7
54	0.06801	0.99768	0.06817	14.668	1.0023	14.702	6
55	0.06830	0.99766	0.06846	14.606	1.0023	14.640	5
56	0.06859	0.99764	0.06876	14.544	1.0024	14.578	4
57	0.06888	0.99762	0.06905	14.482	1.0024	14.517	3
58	0.06918	0.99760	0.06934	14.421	1.0024	14.456	2
59	0.06947	0.99758	0.06963	14.361	1.0024	14.395	1
60	0.06976	0.99756	0.06993	14.301	1.0024	14.335	0
M	Cosine	Sine	Cotan.	Tan.	Cosec.	Secant	M

93° or 273° **86° or 266°**

Natural Trigonometric Functions

M	Sine	Cosine	Tan.	Cotan.	Secant	Cosec.	M
0	0.06976	0.99756	0.06993	14.301	1.0024	14.335	60
1	0.07005	0.99754	0.07022	14.241	1.0025	14.276	59
2	0.07034	0.99752	0.07051	14.182	1.0025	14.217	58
3	0.07063	0.99750	0.07080	14.123	1.0025	14.159	57
4	0.07092	0.99748	0.07110	14.065	1.0025	14.101	56
5	0.07121	0.99746	0.07139	14.008	1.0025	14.043	55
6	0.07150	0.99744	0.07168	13.951	1.0026	13.986	54
7	0.07179	0.99742	0.07197	13.894	1.0026	13.930	53
8	0.07208	0.99740	0.07226	13.838	1.0026	13.874	52
9	0.07237	0.99738	0.07256	13.782	1.0026	13.818	51
10	0.07266	0.99736	0.07285	13.727	1.0026	13.763	50
11	0.07295	0.99733	0.07314	13.672	1.0027	13.708	49
12	0.07324	0.99731	0.07343	13.617	1.0027	13.654	48
13	0.07353	0.99729	0.07373	13.563	1.0027	13.600	47
14	0.07382	0.99727	0.07402	13.510	1.0027	13.547	46
15	0.07411	0.99725	0.07431	13.457	1.0027	13.494	45
16	0.07440	0.99723	0.07460	13.404	1.0028	13.441	44
17	0.07469	0.99721	0.07490	13.351	1.0028	13.389	43
18	0.07498	0.99718	0.07519	13.299	1.0028	13.337	42
19	0.07527	0.99716	0.07548	13.248	1.0028	13.286	41
20	0.07556	0.99714	0.07577	13.197	1.0029	13.235	40
21	0.07585	0.99712	0.07607	13.146	1.0029	13.184	39
22	0.07614	0.99710	0.07636	13.096	1.0029	13.134	38
23	0.07643	0.99707	0.07665	13.046	1.0029	13.084	37
24	0.07672	0.99705	0.07694	12.996	1.0029	13.034	36
25	0.07701	0.99703	0.07724	12.947	1.0030	12.985	35
26	0.07730	0.99701	0.07753	12.898	1.0030	12.937	34
27	0.07759	0.99698	0.07782	12.849	1.0030	12.888	33
28	0.07788	0.99696	0.07812	12.801	1.0030	12.840	32
29	0.07817	0.99694	0.07841	12.754	1.0031	12.793	31
30	0.07846	0.99692	0.07870	12.706	1.0031	12.745	30
31	0.07875	0.99689	0.07899	12.659	1.0031	12.698	29
32	0.07904	0.99687	0.07929	12.612	1.0031	12.652	28
33	0.07933	0.99685	0.07958	12.566	1.0032	12.606	27
34	0.07962	0.99682	0.07987	12.520	1.0032	12.560	26
35	0.07991	0.99680	0.08016	12.474	1.0032	12.514	25
36	0.08020	0.99678	0.08046	12.429	1.0032	12.469	24
37	0.08049	0.99675	0.08075	12.384	1.0032	12.424	23
38	0.08078	0.99673	0.08104	12.339	1.0033	12.379	22
39	0.08107	0.99671	0.08134	12.295	1.0033	12.335	21
40	0.08136	0.99668	0.08163	12.250	1.0033	12.291	20
41	0.08165	0.99666	0.08192	12.207	1.0033	12.248	19
42	0.08194	0.99664	0.08221	12.163	1.0034	12.204	18
43	0.08223	0.99661	0.08251	12.120	1.0034	12.161	17
44	0.08252	0.99659	0.08280	12.077	1.0034	12.118	16
45	0.08281	0.99656	0.08309	12.035	1.0034	12.076	15
46	0.08310	0.99654	0.08339	11.992	1.0035	12.034	14
47	0.08339	0.99652	0.08368	11.950	1.0035	11.992	13
48	0.08368	0.99649	0.08397	11.909	1.0035	11.950	12
49	0.08397	0.99647	0.08426	11.867	1.0035	11.909	11
50	0.08426	0.99644	0.08456	11.826	1.0036	11.868	10
51	0.08455	0.99642	0.08485	11.785	1.0036	11.828	9
52	0.08484	0.99639	0.08514	11.745	1.0036	11.787	8
53	0.08513	0.99637	0.08544	11.704	1.0036	11.747	7
54	0.08542	0.99634	0.08573	11.664	1.0037	11.707	6
55	0.08571	0.99632	0.08602	11.625	1.0037	11.668	5
56	0.08600	0.99629	0.08632	11.585	1.0037	11.628	4
57	0.08629	0.99627	0.08661	11.546	1.0037	11.589	3
58	0.08658	0.99624	0.08690	11.507	1.0038	11.550	2
59	0.08687	0.99622	0.08719	11.468	1.0038	11.512	1
60	0.08715	0.99619	0.08749	11.430	1.0038	11.474	0
M	Cosine	Sine	Cotan.	Tan.	Cosec.	Secant	M

5° or 185° **Natural Trigonometric Functions** **174° or 354°**

M	Sine	Cosine	Tan.	Cotan.	Secant	Cosec.	M
0	0.08715	0.99619	0.08749	11.430	1.0038	11.474	60
1	0.08744	0.99617	0.08773	11.392	1.0038	11.436	59
2	0.08773	0.99614	0.08807	11.354	1.0039	11.398	58
3	0.08802	0.99612	0.08837	11.316	1.0039	11.360	57
4	0.08831	0.99609	0.08866	11.279	1.0039	11.323	56
5	0.08860	0.99607	0.08895	11.242	1.0039	11.286	55
6	0.08889	0.99604	0.08925	11.205	1.0040	11.249	54
7	0.08918	0.99601	0.08954	11.168	1.0040	11.213	53
8	0.08947	0.99599	0.08983	11.132	1.0040	11.176	52
9	0.08976	0.99596	0.09013	11.095	1.0040	11.140	51
10	0.09005	0.99594	0.09042	11.059	1.0041	11.104	50
11	0.09034	0.99591	0.09071	11.024	1.0041	11.069	49
12	0.09063	0.99588	0.09101	10.988	1.0041	11.033	48
13	0.09092	0.99586	0.09130	10.953	1.0041	10.998	47
14	0.09121	0.99583	0.09159	10.918	1.0042	10.963	46
15	0.09150	0.99580	0.09189	10.883	1.0042	10.929	45
16	0.09179	0.99578	0.09218	10.848	1.0042	10.894	44
17	0.09208	0.99575	0.09247	10.814	1.0043	10.860	43
18	0.09237	0.99572	0.09277	10.780	1.0043	10.826	42
19	0.09266	0.99570	0.09306	10.746	1.0043	10.792	41
20	0.09295	0.99567	0.09335	10.712	1.0043	10.758	40
21	0.09324	0.99564	0.09365	10.678	1.0044	10.725	39
22	0.09353	0.99562	0.09394	10.645	1.0044	10.692	38
23	0.09382	0.99559	0.09423	10.612	1.0044	10.659	37
24	0.09411	0.99556	0.09453	10.579	1.0044	10.626	36
25	0.09440	0.99553	0.09482	10.546	1.0045	10.593	35
26	0.09469	0.99551	0.09511	10.514	1.0045	10.561	34
27	0.09498	0.99548	0.09541	10.481	1.0045	10.529	33
28	0.09527	0.99545	0.09570	10.449	1.0046	10.497	32
29	0.09556	0.99542	0.09599	10.417	1.0046	10.465	31
30	0.09584	0.99540	0.09629	10.385	1.0046	10.433	30
31	0.09613	0.99537	0.09658	10.354	1.0046	10.402	29
32	0.09642	0.99534	0.09688	10.322	1.0047	10.371	28
33	0.09671	0.99531	0.09717	10.291	1.0047	10.340	27
34	0.09700	0.99528	0.09746	10.260	1.0047	10.309	26
35	0.09729	0.99525	0.09776	10.229	1.0048	10.278	25
36	0.09758	0.99523	0.09805	10.199	1.0048	10.248	24
37	0.09787	0.99520	0.09834	10.168	1.0048	10.217	23
38	0.09816	0.99517	0.09864	10.138	1.0048	10.187	22
39	0.09845	0.99514	0.09893	10.108	1.0049	10.157	21
40	0.09874	0.99511	0.09922	10.078	1.0049	10.127	20
41	0.09903	0.99508	0.09952	10.048	1.0049	10.098	19
42	0.09932	0.99505	0.09981	10.019	1.0050	10.068	18
43	0.09961	0.99503	0.10011	9.9893	1.0050	10.039	17
44	0.09990	0.99500	0.10040	9.9601	1.0050	10.010	16
45	0.10019	0.99497	0.10069	9.9310	1.0050	9.9812	15
46	0.10048	0.99494	0.10099	9.9021	1.0051	9.9525	14
47	0.10077	0.99491	0.10128	9.8734	1.0051	9.9239	13
48	0.10106	0.99488	0.10158	9.8448	1.0051	9.8955	12
49	0.10134	0.99485	0.10187	9.8164	1.0052	9.8672	11
50	0.10163	0.99482	0.10216	9.7882	1.0052	9.8391	10
51	0.10192	0.99479	0.10246	9.7601	1.0052	9.8112	9
52	0.10221	0.99476	0.10275	9.7322	1.0053	9.7834	8
53	0.10250	0.99473	0.10305	9.7044	1.0053	9.7558	7
54	0.10279	0.99470	0.10334	9.6768	1.0053	9.7283	6
55	0.10308	0.99467	0.10363	9.6493	1.0053	9.7010	5
56	0.10337	0.99464	0.10393	9.6220	1.0054	9.6739	4
57	0.10366	0.99461	0.10422	9.5949	1.0054	9.6469	3
58	0.10395	0.99458	0.10452	9.5679	1.0054	9.6200	2
59	0.10424	0.99455	0.10481	9.5411	1.0055	9.5933	1
60	0.10453	0.99452	0.10510	9.5144	1.0055	9.5668	0
M	Cosine	Sine	Cotan.	Tan.	Cosec.	Secant	M

95° or 275° **84° or 264°**

6° or 186°   **Natural Trigonometric Functions**   **173° or 353°**

M	Sine	Cosine	Tan.	Cotan.	Secant	Cosec.	M
0	0.10453	0.99452	0.10510	9.5144	1.0055	9.5668	60
1	0.10482	0.99449	0.10540	9.4878	1.0055	9.5404	59
2	0.10511	0.99446	0.10569	9.4614	1.0056	9.5141	58
3	0.10540	0.99443	0.10599	9.4351	1.0056	9.4880	57
4	0.10568	0.99440	0.10628	9.4090	1.0056	9.4620	56
5	0.10597	0.99437	0.10657	9.3831	1.0057	9.4362	55
6	0.10626	0.99434	0.10687	9.3572	1.0057	9.4105	54
7	0.10655	0.99431	0.10716	9.3315	1.0057	9.3850	53
8	0.10684	0.99428	0.10746	9.3060	1.0057	9.3596	52
9	0.10713	0.99424	0.10775	9.2806	1.0058	9.3343	51
10	0.10742	0.99421	0.10805	9.2553	1.0058	9.3092	50
11	0.10771	0.99418	0.10834	9.2302	1.0058	9.2842	49
12	0.10800	0.99415	0.10863	9.2051	1.0059	9.2593	48
13	0.10829	0.99412	0.10893	9.1803	1.0059	9.2346	47
14	0.10858	0.99409	0.10922	9.1555	1.0059	9.2100	46
15	0.10887	0.99406	0.10952	9.1309	1.0060	9.1855	45
16	0.10916	0.99402	0.10981	9.1064	1.0060	9.1612	44
17	0.10944	0.99399	0.11011	9.0821	1.0060	9.1370	43
18	0.10973	0.99396	0.11040	9.0579	1.0061	9.1129	42
19	0.11002	0.99393	0.11069	9.0338	1.0061	9.0890	41
20	0.11031	0.99390	0.11099	9.0098	1.0061	9.0651	40
21	0.11060	0.99386	0.11128	8.9860	1.0062	9.0414	39
22	0.11089	0.99383	0.11158	8.9623	1.0062	9.0179	38
23	0.11118	0.99380	0.11187	8.9387	1.0062	8.9944	37
24	0.11147	0.99377	0.11217	8.9152	1.0063	8.9711	36
25	0.11176	0.99373	0.11246	8.8918	1.0063	8.9479	35
26	0.11205	0.99370	0.11276	8.8686	1.0063	8.9248	34
27	0.11234	0.99367	0.11305	8.8455	1.0064	8.9018	33
28	0.11262	0.99364	0.11335	8.8225	1.0064	8.8790	32
29	0.11291	0.99360	0.11364	8.7996	1.0064	8.8563	31
30	0.11320	0.99357	0.11393	8.7769	1.0065	8.8337	30
31	0.11349	0.99354	0.11423	8.7542	1.0065	8.8112	29
32	0.11378	0.99350	0.11452	8.7317	1.0065	8.7888	28
33	0.11407	0.99347	0.11482	8.7093	1.0066	8.7665	27
34	0.11436	0.99344	0.11511	8.6870	1.0066	8.7444	26
35	0.11465	0.99341	0.11541	8.6648	1.0066	8.7223	25
36	0.11494	0.99337	0.11570	8.6427	1.0067	8.7004	24
37	0.11523	0.99334	0.11600	8.6208	1.0067	8.6786	23
38	0.11551	0.99330	0.11629	8.5989	1.0067	8.6569	22
39	0.11580	0.99327	0.11659	8.5772	1.0068	8.6353	21
40	0.11609	0.99324	0.11688	8.5555	1.0068	8.6138	20
41	0.11638	0.99320	0.11718	8.5340	1.0068	8.5924	19
42	0.11667	0.99317	0.11747	8.5126	1.0069	8.5711	18
43	0.11696	0.99314	0.11777	8.4913	1.0069	8.5499	17
44	0.11725	0.99310	0.11806	8.4701	1.0069	8.5289	16
45	0.11754	0.99307	0.11836	8.4489	1.0070	8.5079	15
46	0.11783	0.99303	0.11865	8.4279	1.0070	8.4871	14
47	0.11811	0.99300	0.11895	8.4070	1.0070	8.4663	13
48	0.11840	0.99296	0.11924	8.3862	1.0071	8.4457	12
49	0.11869	0.99293	0.11954	8.3655	1.0071	8.4251	11
50	0.11898	0.99290	0.11983	8.3449	1.0071	8.4046	10
51	0.11927	0.99286	0.12013	8.3244	1.0072	8.3843	9
52	0.11956	0.99283	0.12042	8.3040	1.0072	8.3640	8
53	0.11985	0.99279	0.12072	8.2837	1.0073	8.3439	7
54	0.12014	0.99276	0.12101	8.2635	1.0073	8.3238	6
55	0.12042	0.99272	0.12131	8.2434	1.0073	8.3039	5
56	0.12071	0.99269	0.12160	8.2234	1.0074	8.2840	4
57	0.12100	0.99265	0.12190	8.2035	1.0074	8.2642	3
58	0.12129	0.99262	0.12219	8.1837	1.0074	8.2446	2
59	0.12158	0.99258	0.12249	8.1640	1.0075	8.2250	1
60	0.12187	0.99255	0.12278	8.1443	1.0075	8.2055	0
M	Cosine	Sine	Cotan.	Tan.	Cosec.	Secant	M

96° or 276°   **83° or 263°**

7° or 187°   **Natural Trigonometric Functions**   **172° or 352°**

M	Sine	Cosine	Tan.	Cotan.	Secant	Cosec.	M
0	0.12187	0.99255	0.12278	8.1443	1.0075	8.2055	60
1	0.12216	0.99251	0.12308	8.1248	1.0075	8.1861	59
2	0.12245	0.99247	0.12337	8.1053	1.0076	8.1668	58
3	0.12273	0.99244	0.12367	8.0860	1.0076	8.1476	57
4	0.12302	0.99240	0.12396	8.0667	1.0076	8.1285	56
5	0.12331	0.99237	0.12426	8.0476	1.0077	8.1094	55
6	0.12360	0.99233	0.12456	8.0285	1.0077	8.0905	54
7	0.12389	0.99229	0.12485	8.0095	1.0078	8.0717	53
8	0.12418	0.99226	0.12515	7.9906	1.0078	8.0529	52
9	0.12447	0.99222	0.12544	7.9717	1.0078	8.0342	51
10	0.12476	0.99219	0.12574	7.9530	1.0079	8.0156	50
11	0.12504	0.99215	0.12603	7.9344	1.0079	7.9971	49
12	0.12533	0.99211	0.12633	7.9158	1.0079	7.9787	48
13	0.12562	0.99208	0.12662	7.8973	1.0080	7.9604	47
14	0.12591	0.99204	0.12692	7.8789	1.0080	7.9421	46
15	0.12620	0.99200	0.12722	7.8606	1.0080	7.9240	45
16	0.12649	0.99197	0.12751	7.8424	1.0081	7.9059	44
17	0.12678	0.99193	0.12781	7.8243	1.0081	7.8879	43
18	0.12706	0.99189	0.12810	7.8062	1.0082	7.8700	42
19	0.12735	0.99186	0.12840	7.7882	1.0082	7.8522	41
20	0.12764	0.99182	0.12869	7.7703	1.0082	7.8344	40
21	0.12793	0.99178	0.12899	7.7525	1.0083	7.8168	39
22	0.12822	0.99174	0.12928	7.7348	1.0083	7.7992	38
23	0.12851	0.99171	0.12958	7.7171	1.0084	7.7817	37
24	0.12879	0.99167	0.12988	7.6996	1.0084	7.7642	36
25	0.12908	0.99163	0.13017	7.6821	1.0084	7.7469	35
26	0.12937	0.99160	0.13047	7.6646	1.0085	7.7296	34
27	0.12966	0.99156	0.13076	7.6473	1.0085	7.7124	33
28	0.12995	0.99152	0.13106	7.6300	1.0085	7.6953	32
29	0.13024	0.99148	0.13136	7.6129	1.0086	7.6783	31
30	0.13053	0.99144	0.13165	7.5957	1.0086	7.6613	30
31	0.13081	0.99141	0.13195	7.5787	1.0087	7.6444	29
32	0.13110	0.99137	0.13224	7.5617	1.0087	7.6276	28
33	0.13139	0.99133	0.13254	7.5449	1.0087	7.6108	27
34	0.13168	0.99129	0.13284	7.5280	1.0088	7.5942	26
35	0.13197	0.99125	0.13313	7.5113	1.0088	7.5776	25
36	0.13226	0.99121	0.13343	7.4946	1.0089	7.5611	24
37	0.13254	0.99118	0.13372	7.4780	1.0089	7.5446	23
38	0.13283	0.99114	0.13402	7.4615	1.0089	7.5282	22
39	0.13312	0.99110	0.13432	7.4451	1.0090	7.5119	21
40	0.13341	0.99106	0.13461	7.4287	1.0090	7.4957	20
41	0.13370	0.99102	0.13491	7.4124	1.0090	7.4795	19
42	0.13399	0.99098	0.13520	7.3961	1.0091	7.4634	18
43	0.13427	0.99094	0.13550	7.3800	1.0091	7.4474	17
44	0.13456	0.99090	0.13580	7.3639	1.0092	7.4315	16
45	0.13485	0.99086	0.13609	7.3479	1.0092	7.4156	15
46	0.13514	0.99083	0.13639	7.3319	1.0092	7.3998	14
47	0.13543	0.99079	0.13669	7.3160	1.0093	7.3840	13
48	0.13571	0.99075	0.13698	7.3002	1.0093	7.3683	12
49	0.13600	0.99071	0.13728	7.2844	1.0094	7.3527	11
50	0.13629	0.99067	0.13757	7.2687	1.0094	7.3372	10
51	0.13658	0.99063	0.13787	7.2531	1.0094	7.3217	9
52	0.13687	0.99059	0.13817	7.2375	1.0095	7.3063	8
53	0.13716	0.99055	0.13846	7.2220	1.0095	7.2909	7
54	0.13744	0.99051	0.13876	7.2066	1.0096	7.2757	6
55	0.13773	0.99047	0.13906	7.1912	1.0096	7.2604	5
56	0.13802	0.99043	0.13935	7.1759	1.0097	7.2453	4
57	0.13831	0.99039	0.13965	7.1607	1.0097	7.2302	3
58	0.13860	0.99035	0.13995	7.1455	1.0097	7.2152	2
59	0.13888	0.99031	0.14024	7.1304	1.0098	7.2002	1
60	0.13917	0.99027	0.14054	7.1154	1.0098	7.1853	0
M	Cosine	Sine	Cotan.	Tan.	Cosec.	Secant	M

97° or 277°   **82° or 262°**

Natural Trigonometric Functions

M	Sine	Cosine	Tan.	Cotan.	Secant	Cosec.	M
0	0.13917	0.99027	0.14054	7.1154	1.0098	7.1853	60
1	0.13946	0.99023	0.14084	7.1004	1.0099	7.1704	59
2	0.13975	0.99019	0.14113	7.0854	1.0099	7.1557	58
3	0.14004	0.99015	0.14143	7.0706	1.0099	7.1409	57
4	0.14032	0.99010	0.14173	7.0558	1.0100	7.1263	56
5	0.14061	0.99006	0.14202	7.0410	1.0100	7.1117	55
6	0.14090	0.99002	0.14232	7.0264	1.0101	7.0972	54
7	0.14119	0.98998	0.14262	7.0117	1.0101	7.0827	53
8	0.14148	0.98994	0.14291	6.9972	1.0102	7.0683	52
9	0.14176	0.98990	0.14321	6.9827	1.0102	7.0539	51
10	0.14205	0.98986	0.14351	6.9682	1.0102	7.0396	50
11	0.14234	0.98982	0.14380	6.9538	1.0103	7.0254	49
12	0.14263	0.98978	0.14410	6.9395	1.0103	7.0112	48
13	0.14292	0.98973	0.14440	6.9252	1.0104	7.9971	47
14	0.14320	0.98969	0.14470	6.9110	1.0104	7.9830	46
15	0.14349	0.98965	0.14499	6.8969	1.0104	7.9690	45
16	0.14378	0.98961	0.14529	6.8828	1.0105	7.9550	44
17	0.14407	0.98957	0.14559	6.8687	1.0105	7.9411	43
18	0.14436	0.98952	0.14588	6.8547	1.0106	7.9273	42
19	0.14464	0.98948	0.14618	6.8408	1.0106	7.9135	41
20	0.14493	0.98944	0.14648	6.8269	1.0107	6.8998	40
21	0.14522	0.98940	0.14677	6.8131	1.0107	6.8861	39
22	0.14551	0.98936	0.14707	6.7993	1.0107	6.8725	38
23	0.14579	0.98931	0.14737	6.7856	1.0108	6.8589	37
24	0.14608	0.98927	0.14767	6.7720	1.0108	6.8454	36
25	0.14637	0.98923	0.14796	6.7584	1.0109	6.8320	35
26	0.14666	0.98919	0.14826	6.7448	1.0109	6.8185	34
27	0.14695	0.98914	0.14856	6.7313	1.0110	6.8052	33
28	0.14723	0.98910	0.14886	6.7179	1.0110	6.7919	32
29	0.14752	0.98906	0.14915	6.7045	1.0111	6.7787	31
30	0.14781	0.98901	0.14945	6.6911	1.0111	6.7655	30
31	0.14810	0.98897	0.14975	6.6779	1.0111	6.7523	29
32	0.14838	0.98893	0.15004	6.6646	1.0112	6.7392	28
33	0.14867	0.98889	0.15034	6.6514	1.0112	6.7262	27
34	0.14896	0.98884	0.15064	6.6383	1.0113	6.7132	26
35	0.14925	0.98880	0.15094	6.6252	1.0113	6.7003	25
36	0.14953	0.98876	0.15123	6.6122	1.0114	6.6874	24
37	0.14982	0.98871	0.15153	6.5992	1.0114	6.6745	23
38	0.15011	0.98867	0.15183	6.5863	1.0115	6.6617	22
39	0.15040	0.98862	0.15213	6.5734	1.0115	6.6490	21
40	0.15068	0.98858	0.15243	6.5605	1.0115	6.6363	20
41	0.15097	0.98854	0.15272	6.5478	1.0116	6.6237	19
42	0.15126	0.98849	0.15302	6.5350	1.0116	6.6111	18
43	0.15155	0.98845	0.15332	6.5223	1.0117	6.5985	17
44	0.15183	0.98840	0.15362	6.5097	1.0117	6.5860	16
45	0.15212	0.98836	0.15391	6.4971	1.0118	6.5736	15
46	0.15241	0.98832	0.15421	6.4845	1.0118	6.5612	14
47	0.15270	0.98827	0.15451	6.4720	1.0119	6.5488	13
48	0.15298	0.98823	0.15481	6.4596	1.0119	6.5365	12
49	0.15328	0.98818	0.15511	6.4472	1.0119	6.5243	11
50	0.15356	0.98814	0.15540	6.4348	1.0120	6.5121	10
51	0.15385	0.98809	0.15570	6.4225	1.0120	6.4999	9
52	0.15413	0.98805	0.15600	6.4103	1.0121	6.4878	8
53	0.15442	0.98800	0.15630	6.3980	1.0121	6.4757	7
54	0.15471	0.98796	0.15659	6.3859	1.0122	6.4637	6
55	0.15500	0.98791	0.15689	6.3737	1.0122	6.4517	5
56	0.15528	0.98787	0.15719	6.3616	1.0123	6.4398	4
57	0.15557	0.98782	0.15749	6.3496	1.0123	6.4279	3
58	0.15586	0.98778	0.15779	6.3376	1.0124	6.4160	2
59	0.15615	0.98773	0.15809	6.3257	1.0124	6.4042	1
60	0.15643	0.98769	0.15838	6.3137	1.0125	6.3924	0
M	Cosine	Sine	Cotan.	Tan.	Cosec.	Secant	M

9° or 189° **Natural Trigonometric Functions** **170° or 350°**

M	Sine	Cosine	Tan.	Cotan.	Secant	Cosec.	M
0	0.15643	0.98769	0.15838	6.3137	1.0125	6.3924	60
1	0.15672	0.98764	0.15868	6.3019	1.0125	6.3807	59
2	0.15701	0.98760	0.15898	6.2901	1.0125	6.3690	58
3	0.15730	0.98755	0.15928	6.2783	1.0126	6.3574	57
4	0.15758	0.98750	0.15958	6.2665	1.0126	6.3458	56
5	0.15787	0.98746	0.15987	6.2548	1.0127	6.3343	55
6	0.15816	0.98741	0.16017	6.2432	1.0127	6.3228	54
7	0.15844	0.98737	0.16047	6.2316	1.0128	6.3113	53
8	0.15873	0.98732	0.16077	6.2200	1.0128	6.2999	52
9	0.15902	0.98727	0.16107	6.2085	1.0129	6.2885	51
10	0.15931	0.98723	0.16137	6.1970	1.0129	6.2772	50
11	0.15959	0.98718	0.16167	6.1856	1.0130	6.2659	49
12	0.15988	0.98714	0.16196	6.1742	1.0130	6.2546	48
13	0.16017	0.98709	0.16226	6.1628	1.0131	6.2434	47
14	0.16045	0.98704	0.16256	6.1515	1.0131	6.2322	46
15	0.16074	0.98700	0.16286	6.1402	1.0132	6.2211	45
16	0.16103	0.98695	0.16316	6.1290	1.0132	6.2100	44
17	0.16132	0.98690	0.16346	6.1178	1.0133	6.1990	43
18	0.16160	0.98685	0.16376	6.1066	1.0133	6.1880	42
19	0.16189	0.98681	0.16405	6.0955	1.0134	6.1770	41
20	0.16218	0.98676	0.16435	6.0844	1.0134	6.1661	40
21	0.16246	0.98671	0.16465	6.0734	1.0135	6.1552	39
22	0.16275	0.98667	0.16495	6.0624	1.0135	6.1443	38
23	0.16304	0.98662	0.16525	6.0514	1.0136	6.1335	37
24	0.16333	0.98657	0.16555	6.0405	1.0136	6.1227	36
25	0.16361	0.98652	0.16585	6.0296	1.0136	6.1120	35
26	0.16390	0.98648	0.16615	6.0188	1.0137	6.1013	34
27	0.16419	0.98643	0.16644	6.0080	1.0137	6.0906	33
28	0.16447	0.98638	0.16674	5.9972	1.0138	6.0800	32
29	0.16476	0.98633	0.16704	5.9865	1.0138	6.0694	31
30	0.16505	0.98628	0.16734	5.9758	1.0139	6.0588	30
31	0.16533	0.98624	0.16764	5.9651	1.0139	6.0483	29
32	0.16562	0.98619	0.16794	5.9545	1.0140	6.0379	28
33	0.16591	0.98614	0.16824	5.9439	1.0140	6.0274	27
34	0.16619	0.98609	0.16854	5.9333	1.0141	6.0170	26
35	0.16648	0.98604	0.16884	5.9228	1.0141	6.0066	25
36	0.16677	0.98600	0.16914	5.9123	1.0142	5.9963	24
37	0.16705	0.98595	0.16944	5.9019	1.0142	5.9860	23
38	0.16734	0.98590	0.16973	5.8915	1.0143	5.9758	22
39	0.16763	0.98585	0.17003	5.8811	1.0143	5.9655	21
40	0.16791	0.98580	0.17033	5.8708	1.0144	5.9554	20
41	0.16820	0.98575	0.17063	5.8605	1.0144	5.9452	19
42	0.16849	0.98570	0.17093	5.8502	1.0145	5.9351	18
43	0.16878	0.98565	0.17123	5.8400	1.0145	5.9250	17
44	0.16906	0.98560	0.17153	5.8298	1.0146	5.9150	16
45	0.16935	0.98556	0.17183	5.8196	1.0146	5.9049	15
46	0.16964	0.98551	0.17213	5.8095	1.0147	5.8950	14
47	0.16992	0.98546	0.17243	5.7994	1.0147	5.8850	13
48	0.17021	0.98541	0.17273	5.7894	1.0148	5.8751	12
49	0.17050	0.98536	0.17303	5.7794	1.0148	5.8652	11
50	0.17078	0.98531	0.17333	5.7694	1.0149	5.8554	10
51	0.17107	0.98526	0.17363	5.7594	1.0150	5.8456	9
52	0.17136	0.98521	0.17393	5.7495	1.0150	5.8358	8
53	0.17164	0.98516	0.17423	5.7396	1.0151	5.8261	7
54	0.17193	0.98511	0.17453	5.7297	1.0151	5.8163	6
55	0.17221	0.98506	0.17483	5.7199	1.0152	5.8067	5
56	0.17250	0.98501	0.17513	5.7101	1.0152	5.7970	4
57	0.17279	0.98496	0.17543	5.7004	1.0153	5.7874	3
58	0.17307	0.98491	0.17573	5.6906	1.0153	5.7778	2
59	0.17336	0.98486	0.17603	5.6809	1.0154	5.7683	1
60	0.17365	0.98481	0.17633	5.6713	1.0154	5.7588	0
M	Cosine	Sine	Cotan.	Tan.	Cosec.	Secant	M

99° or 279° **80° or 260°**

Natural Trigonometric Functions

M	Sine	Cosine	Tan.	Cotan.	Secant	Cosec.	M
0	0.17365	0.98481	0.17633	5.6713	1.0154	5.7588	60
1	0.17393	0.98476	0.17663	5.6616	1.0155	5.7493	59
2	0.17422	0.98471	0.17693	5.6520	1.0155	5.7398	58
3	0.17451	0.98465	0.17723	5.6425	1.0156	5.7304	57
4	0.17479	0.98460	0.17753	5.6329	1.0156	5.7210	56
5	0.17508	0.98455	0.17783	5.6234	1.0157	5.7117	55
6	0.17537	0.98450	0.17813	5.6140	1.0157	5.7023	54
7	0.17565	0.98445	0.17843	5.6045	1.0158	5.6930	53
8	0.17594	0.98440	0.17873	5.5951	1.0158	5.6838	52
9	0.17622	0.98435	0.17903	5.5857	1.0159	5.6745	51
10	0.17651	0.98430	0.17933	5.5764	1.0159	5.6653	50
11	0.17680	0.98425	0.17963	5.5670	1.0160	5.6561	49
12	0.17708	0.98419	0.17993	5.5578	1.0160	5.6470	48
13	0.17737	0.98414	0.18023	5.5485	1.0161	5.6379	47
14	0.17766	0.98409	0.18053	5.5393	1.0162	5.6288	46
15	0.17794	0.98404	0.18083	5.5301	1.0162	5.6197	45
16	0.17823	0.98399	0.18113	5.5209	1.0163	5.6107	44
17	0.17852	0.98394	0.18143	5.5117	1.0163	5.6017	43
18	0.17880	0.98388	0.18173	5.5026	1.0164	5.5928	42
19	0.17909	0.98383	0.18203	5.4936	1.0164	5.5838	41
20	0.17937	0.98378	0.18233	5.4845	1.0165	5.5749	40
21	0.17966	0.98373	0.18263	5.4755	1.0165	5.5660	39
22	0.17995	0.98368	0.18293	5.4665	1.0166	5.5572	38
23	0.18023	0.98362	0.18323	5.4575	1.0166	5.5484	37
24	0.18052	0.98357	0.18353	5.4486	1.0167	5.5396	36
25	0.18080	0.98352	0.18383	5.4396	1.0167	5.5308	35
26	0.18109	0.98347	0.18413	5.4308	1.0168	5.5221	34
27	0.18138	0.98341	0.18444	5.4219	1.0169	5.5134	33
28	0.18166	0.98336	0.18474	5.4131	1.0169	5.5047	32
29	0.18195	0.98331	0.18504	5.4043	1.0170	5.4960	31
30	0.18223	0.98325	0.18534	5.3955	1.0170	5.4874	30
31	0.18252	0.98320	0.18564	5.3868	1.0171	5.4788	29
32	0.18281	0.98315	0.18594	5.3780	1.0171	5.4702	28
33	0.18309	0.98309	0.18624	5.3694	1.0172	5.4617	27
34	0.18338	0.98304	0.18654	5.3607	1.0172	5.4532	26
35	0.18366	0.98299	0.18684	5.3521	1.0173	5.4447	25
36	0.18395	0.98293	0.18714	5.3434	1.0174	5.4362	24
37	0.18424	0.98288	0.18745	5.3349	1.0174	5.4278	23
38	0.18452	0.98283	0.18775	5.3263	1.0175	5.4194	22
39	0.18481	0.98277	0.18805	5.3178	1.0175	5.4110	21
40	0.18509	0.98272	0.18835	5.3093	1.0176	5.4026	20
41	0.18538	0.98267	0.18865	5.3008	1.0176	5.3943	19
42	0.18567	0.98261	0.18895	5.2923	1.0177	5.3860	18
43	0.18595	0.98256	0.18925	5.2839	1.0177	5.3777	17
44	0.18624	0.98250	0.18955	5.2755	1.0178	5.3695	16
45	0.18652	0.98245	0.18985	5.2671	1.0179	5.3612	15
46	0.18681	0.98240	0.19016	5.2588	1.0179	5.3530	14
47	0.18709	0.98234	0.19046	5.2505	1.0180	5.3449	13
48	0.18738	0.98229	0.19076	5.2422	1.0180	5.3367	12
49	0.18767	0.98223	0.19106	5.2339	1.0181	5.3286	11
50	0.18795	0.98218	0.19136	5.2257	1.0181	5.3205	10
51	0.18824	0.98212	0.19166	5.2174	1.0182	5.3124	9
52	0.18852	0.98207	0.19197	5.2092	1.0182	5.3044	8
53	0.18881	0.98201	0.19227	5.2011	1.0183	5.2963	7
54	0.18909	0.98196	0.19257	5.1929	1.0184	5.2883	6
55	0.18938	0.98190	0.19287	5.1848	1.0184	5.2803	5
56	0.18967	0.98185	0.19317	5.1767	1.0185	5.2724	4
57	0.18995	0.98179	0.19347	5.1686	1.0185	5.2645	3
58	0.19024	0.98174	0.19378	5.1606	1.0186	5.2566	2
59	0.19052	0.98168	0.19408	5.1525	1.0186	5.2487	1
60	0.19081	0.98163	0.19438	5.1445	1.0187	5.2408	0
M	Cosine	Sine	Cotan.	Tan.	Cosec.	Secant	M

11° or 191°   **Natural Trigonometric Functions**   **168° or 348°**

M	Sine	Cosine	Tan.	Cotan.	Secant	Cosec.	M
0	0.19081	0.98163	0.19438	5.1445	1.0187	5.2408	60
1	0.19109	0.98157	0.19468	5.1366	1.0188	5.2330	59
2	0.19138	0.98152	0.19498	5.1286	1.0188	5.2252	58
3	0.19166	0.98146	0.19529	5.1207	1.0189	5.2174	57
4	0.19195	0.98140	0.19559	5.1128	1.0189	5.2097	56
5	0.19224	0.98135	0.19589	5.1049	1.0190	5.2019	55
6	0.19252	0.98129	0.19619	5.0970	1.0191	5.1942	54
7	0.19281	0.98124	0.19649	5.0892	1.0191	5.1865	53
8	0.19309	0.98118	0.19680	5.0814	1.0192	5.1788	52
9	0.19338	0.98112	0.19710	5.0736	1.0192	5.1712	51
10	0.19366	0.98107	0.19740	5.0658	1.0193	5.1636	50
11	0.19395	0.98101	0.19770	5.0581	1.0193	5.1560	49
12	0.19423	0.98095	0.19800	5.0504	1.0194	5.1484	48
13	0.19452	0.98090	0.19831	5.0427	1.0195	5.1409	47
14	0.19480	0.98084	0.19861	5.0350	1.0195	5.1333	46
15	0.19509	0.98078	0.19891	5.0273	1.0196	5.1258	45
16	0.19537	0.98073	0.19921	5.0197	1.0196	5.1183	44
17	0.19566	0.98067	0.19952	5.0121	1.0197	5.1109	43
18	0.19595	0.98061	0.19982	5.0045	1.0198	5.1034	42
19	0.19623	0.98056	0.20012	4.9969	1.0198	5.0960	41
20	0.19652	0.98050	0.20042	4.9894	1.0199	5.0886	40
21	0.19680	0.98044	0.20073	4.9819	1.0199	5.0812	39
22	0.19709	0.98039	0.20103	4.9744	1.0200	5.0739	38
23	0.19737	0.98033	0.20133	4.9669	1.0201	5.0666	37
24	0.19766	0.98027	0.20163	4.9594	1.0201	5.0593	36
25	0.19794	0.98021	0.20194	4.9520	1.0202	5.0520	35
26	0.19823	0.98016	0.20224	4.9446	1.0202	5.0447	34
27	0.19851	0.98010	0.20254	4.9372	1.0203	5.0375	33
28	0.19880	0.98004	0.20285	4.9298	1.0204	5.0302	32
29	0.19908	0.97998	0.20315	4.9225	1.0204	5.0230	31
30	0.19937	0.97992	0.20345	4.9151	1.0205	5.0158	30
31	0.19965	0.97987	0.20375	4.9078	1.0205	5.0087	29
32	0.19994	0.97981	0.20406	4.9006	1.0206	5.0015	28
33	0.20022	0.97975	0.20436	4.8933	1.0207	4.9944	27
34	0.20051	0.97969	0.20466	4.8860	1.0207	4.9873	26
35	0.20079	0.97963	0.20497	4.8788	1.0208	4.9802	25
36	0.20108	0.97957	0.20527	4.8716	1.0208	4.9732	24
37	0.20136	0.97952	0.20557	4.8644	1.0209	4.9661	23
38	0.20165	0.97946	0.20588	4.8573	1.0210	4.9591	22
39	0.20193	0.97940	0.20618	4.8501	1.0210	4.9521	21
40	0.20222	0.97934	0.20648	4.8430	1.0211	4.9452	20
41	0.20250	0.97928	0.20679	4.8359	1.0211	4.9382	19
42	0.20279	0.97922	0.20709	4.8288	1.0212	4.9313	18
43	0.20307	0.97916	0.20739	4.8217	1.0213	4.9243	17
44	0.20336	0.97910	0.20770	4.8147	1.0213	4.9175	16
45	0.20364	0.97904	0.20800	4.8077	1.0214	4.9106	15
46	0.20393	0.97899	0.20830	4.8007	1.0215	4.9037	14
47	0.20421	0.97893	0.20861	4.7937	1.0215	4.8969	13
48	0.20450	0.97887	0.20891	4.7867	1.0216	4.8901	12
49	0.20478	0.97881	0.20921	4.7798	1.0216	4.8833	11
50	0.20506	0.97875	0.20952	4.7728	1.0217	4.8765	10
51	0.20535	0.97869	0.20982	4.7659	1.0218	4.8697	9
52	0.20563	0.97863	0.21012	4.7591	1.0218	4.8630	8
53	0.20592	0.97857	0.21043	4.7522	1.0219	4.8563	7
54	0.20620	0.97851	0.21073	4.7453	1.0220	4.8496	6
55	0.20649	0.97845	0.21104	4.7385	1.0220	4.8429	5
56	0.20677	0.97839	0.21134	4.7317	1.0221	4.8362	4
57	0.20706	0.97833	0.21164	4.7249	1.0221	4.8296	3
58	0.20734	0.97827	0.21195	4.7181	1.0222	4.8229	2
59	0.20763	0.97821	0.21225	4.7114	1.0223	4.8163	1
60	0.20791	0.97815	0.21256	4.7046	1.0223	4.8097	0
M	Cosine	Sine	Cotan.	Tan.	Cosec.	Secant	M

    Natural Trigonometric Functions

M	Sine	Cosine	Tan.	Cotan.	Secant	Cosec.	M
0	0.20791	0.97815	0.21256	4.7046	1.0223	4.8097	60
1	0.20820	0.97809	0.21286	4.6979	1.0224	4.8032	59
2	0.20848	0.97803	0.21316	4.6912	1.0225	4.7966	58
3	0.20876	0.97797	0.21347	4.6845	1.0225	4.7901	57
4	0.20905	0.97790	0.21377	4.6778	1.0226	4.7835	56
5	0.20933	0.97784	0.21408	4.6712	1.0226	4.7770	55
6	0.20962	0.97778	0.21438	4.6646	1.0227	4.7706	54
7	0.20990	0.97772	0.21468	4.6580	1.0228	4.7641	53
8	0.21019	0.97766	0.21499	4.6514	1.0228	4.7576	52
9	0.21047	0.97760	0.21529	4.6448	1.0229	4.7512	51
10	0.21076	0.97754	0.21560	4.6382	1.0230	4.7448	50
11	0.21104	0.97748	0.21590	4.6317	1.0230	4.7384	49
12	0.21132	0.97741	0.21621	4.6252	1.0231	4.7320	48
13	0.21161	0.97735	0.21651	4.6187	1.0232	4.7257	47
14	0.21189	0.97729	0.21682	4.6122	1.0232	4.7193	46
15	0.21218	0.97723	0.21712	4.6057	1.0233	4.7130	45
16	0.21246	0.97717	0.21742	4.5993	1.0234	4.7067	44
17	0.21275	0.97711	0.21773	4.5928	1.0234	4.7004	43
18	0.21303	0.97704	0.21803	4.5864	1.0235	4.6942	42
19	0.21331	0.97698	0.21834	4.5800	1.0235	4.6879	41
20	0.21360	0.97692	0.21864	4.5736	1.0236	4.6817	40
21	0.21388	0.97686	0.21895	4.5673	1.0237	4.6754	39
22	0.21417	0.97680	0.21925	4.5609	1.0237	4.6692	38
23	0.21445	0.97673	0.21956	4.5546	1.0238	4.6631	37
24	0.21473	0.97667	0.21986	4.5483	1.0239	4.6569	36
25	0.21502	0.97661	0.22017	4.5420	1.0239	4.6507	35
26	0.21530	0.97655	0.22047	4.5357	1.0240	4.6446	34
27	0.21559	0.97648	0.22078	4.5294	1.0241	4.6385	33
28	0.21587	0.97642	0.22108	4.5232	1.0241	4.6324	32
29	0.21615	0.97636	0.22139	4.5169	1.0242	4.6263	31
30	0.21644	0.97630	0.22169	4.5107	1.0243	4.6201	30
31	0.21672	0.97623	0.22200	4.5045	1.0243	4.6142	29
32	0.21701	0.97617	0.22230	4.4983	1.0244	4.6081	28
33	0.21729	0.97611	0.22261	4.4921	1.0245	4.6021	27
34	0.21757	0.97604	0.22291	4.4860	1.0245	4.5961	26
35	0.21786	0.97598	0.22322	4.4799	1.0246	4.5901	25
36	0.21814	0.97592	0.22353	4.4737	1.0247	4.5841	24
37	0.21843	0.97585	0.22383	4.4676	1.0247	4.5782	23
38	0.21871	0.97579	0.22414	4.4615	1.0248	4.5722	22
39	0.21899	0.97573	0.22444	4.4555	1.0249	4.5663	21
40	0.21928	0.97566	0.22475	4.4494	1.0249	4.5604	20
41	0.21956	0.97560	0.22505	4.4434	1.0250	4.5545	19
42	0.21985	0.97553	0.22536	4.4373	1.0251	4.5486	18
43	0.22013	0.97547	0.22566	4.4313	1.0251	4.5428	17
44	0.22041	0.97541	0.22597	4.4253	1.0252	4.5369	16
45	0.22070	0.97534	0.22628	4.4194	1.0253	4.5311	15
46	0.22098	0.97528	0.22658	4.4134	1.0253	4.5253	14
47	0.22126	0.97521	0.22689	4.4074	1.0254	4.5195	13
48	0.22155	0.97515	0.22719	4.4015	1.0255	4.5137	12
49	0.22183	0.97508	0.22750	4.3956	1.0255	4.5079	11
50	0.22211	0.97502	0.22781	4.3897	1.0256	4.5021	10
51	0.22240	0.97495	0.22811	4.3838	1.0257	4.4964	9
52	0.22268	0.97489	0.22842	4.3779	1.0257	4.4907	8
53	0.22297	0.97483	0.22872	4.3721	1.0258	4.4850	7
54	0.22325	0.97476	0.22903	4.3662	1.0259	4.4793	6
55	0.22353	0.97470	0.22934	4.3604	1.0260	4.4736	5
56	0.22382	0.97463	0.22964	4.3546	1.0260	4.4679	4
57	0.22410	0.97457	0.22995	4.3488	1.0261	4.4623	3
58	0.22438	0.97450	0.23025	4.3430	1.0262	4.4566	2
59	0.22467	0.97443	0.23056	4.3372	1.0262	4.4510	1
60	0.22495	0.97437	0.23087	4.3315	1.0263	4.4454	0
M	Cosine	Sine	Cotan.	Tan.	Cosec.	Secant	M

13° or 193°   **Natural Trigonometric Functions**   **166° or 346°**

M	Sine	Cosine	Tan.	Cotan.	Secant	Cosec.	M
0	0.22495	0.97437	0.23087	4.3315	1.0263	4.4454	60
1	0.22523	0.97430	0.23117	4.3257	1.0264	4.4398	59
2	0.22552	0.97424	0.23148	4.3200	1.0264	4.4342	58
3	0.22580	0.97417	0.23179	4.3143	1.0265	4.4287	57
4	0.22608	0.97411	0.23209	4.3086	1.0266	4.4231	56
5	0.22637	0.97404	0.23240	4.3029	1.0266	4.4176	55
6	0.22665	0.97398	0.23270	4.2972	1.0267	4.4121	54
7	0.22693	0.97391	0.23301	4.2916	1.0268	4.4065	53
8	0.22722	0.97384	0.23332	4.2859	1.0268	4.4011	52
9	0.22750	0.97378	0.23363	4.2803	1.0269	4.3956	51
10	0.22778	0.97371	0.23393	4.2747	1.0270	4.3901	50
11	0.22807	0.97364	0.23424	4.2691	1.0271	4.3847	49
12	0.22835	0.97358	0.23455	4.2635	1.0271	4.3792	48
13	0.22863	0.97351	0.23485	4.2579	1.0272	4.3738	47
14	0.22892	0.97344	0.23516	4.2524	1.0273	4.3684	46
15	0.22920	0.97338	0.23547	4.2468	1.0273	4.3630	45
16	0.22948	0.97331	0.23577	4.2413	1.0274	4.3576	44
17	0.22977	0.97324	0.23608	4.2358	1.0275	4.3522	43
18	0.23005	0.97318	0.23639	4.2303	1.0276	4.3469	42
19	0.23033	0.97311	0.23670	4.2248	1.0276	4.3415	41
20	0.23061	0.97304	0.23700	4.2193	1.0277	4.3362	40
21	0.23090	0.97298	0.23731	4.2139	1.0278	4.3309	39
22	0.23118	0.97291	0.23762	4.2084	1.0278	4.3256	38
23	0.23146	0.97284	0.23793	4.2030	1.0279	4.3203	37
24	0.23175	0.97277	0.23823	4.1976	1.0280	4.3150	36
25	0.23203	0.97271	0.23854	4.1921	1.0280	4.3098	35
26	0.23231	0.97264	0.23885	4.1867	1.0281	4.3045	34
27	0.23260	0.97257	0.23916	4.1814	1.0282	4.2993	33
28	0.23288	0.97250	0.23946	4.1760	1.0283	4.2941	32
29	0.23316	0.97244	0.23977	4.1706	1.0283	4.2888	31
30	0.23344	0.97237	0.24008	4.1653	1.0284	4.2836	30
31	0.23373	0.97230	0.24039	4.1600	1.0285	4.2785	29
32	0.23401	0.97223	0.24069	4.1546	1.0285	4.2733	28
33	0.23429	0.97216	0.24100	4.1493	1.0286	4.2681	27
34	0.23458	0.97210	0.24131	4.1440	1.0287	4.2630	26
35	0.23486	0.97203	0.24162	4.1388	1.0288	4.2579	25
36	0.23514	0.97196	0.24192	4.1335	1.0288	4.2527	24
37	0.23542	0.97189	0.24223	4.1282	1.0289	4.2476	23
38	0.23571	0.97182	0.24254	4.1230	1.0290	4.2425	22
39	0.23599	0.97175	0.24285	4.1178	1.0291	4.2375	21
40	0.23627	0.97169	0.24316	4.1126	1.0291	4.2324	20
41	0.23655	0.97162	0.24346	4.1073	1.0292	4.2273	19
42	0.23684	0.97155	0.24377	4.1022	1.0293	4.2223	18
43	0.23712	0.97148	0.24408	4.0970	1.0293	4.2173	17
44	0.23740	0.97141	0.24439	4.0918	1.0294	4.2122	16
45	0.23768	0.97134	0.24470	4.0867	1.0295	4.2072	15
46	0.23797	0.97127	0.24501	4.0815	1.0296	4.2022	14
47	0.23825	0.97120	0.24531	4.0764	1.0296	4.1972	13
48	0.23853	0.97113	0.24562	4.0713	1.0297	4.1923	12
49	0.23881	0.97106	0.24593	4.0662	1.0298	4.1873	11
50	0.23910	0.97099	0.24624	4.0611	1.0299	4.1824	10
51	0.23938	0.97092	0.24655	4.0560	1.0299	4.1774	9
52	0.23966	0.97086	0.24686	4.0509	1.0300	4.1725	8
53	0.23994	0.97079	0.24717	4.0458	1.0301	4.1676	7
54	0.24023	0.97072	0.24747	4.0408	1.0302	4.1627	6
55	0.24051	0.97065	0.24778	4.0358	1.0302	4.1578	5
56	0.24079	0.97058	0.24809	4.0307	1.0303	4.1529	4
57	0.24107	0.97051	0.24840	4.0257	1.0304	4.1481	3
58	0.24136	0.97044	0.24871	4.0207	1.0305	4.1432	2
59	0.24164	0.97037	0.24902	4.0157	1.0305	4.1384	1
60	0.24192	0.97029	0.24933	4.0108	1.0306	4.1336	0
M	Cosine	Sine	Cotan.	Tan.	Cosec.	Secant	M

103° or 283°   **76° or 256°**

Natural Trigonometric Functions

M	Sine	Cosine	Tan.	Cotan.	Secant	Cosec.	M
0	0.24192	0.97029	0.24933	4.0108	1.0306	4.1336	60
1	0.24220	0.97022	0.24964	4.0058	1.0307	4.1287	59
2	0.24249	0.97015	0.24995	4.0009	1.0308	4.1239	58
3	0.24277	0.97008	0.25025	3.9959	1.0308	4.1191	57
4	0.24305	0.97001	0.25056	3.9910	1.0309	4.1144	56
5	0.24333	0.96994	0.25087	3.9861	1.0310	4.1096	55
6	0.24361	0.96987	0.25118	3.9812	1.0311	4.1048	54
7	0.24390	0.96980	0.25149	3.9763	1.0311	4.1001	53
8	0.24418	0.96973	0.25180	3.9714	1.0312	4.0953	52
9	0.24446	0.96966	0.25211	3.9665	1.0313	4.0906	51
10	0.24474	0.96959	0.25242	3.9616	1.0314	4.0859	50
11	0.24502	0.96952	0.25273	3.9568	1.0314	4.0812	49
12	0.24531	0.96944	0.25304	3.9520	1.0315	4.0765	48
13	0.24559	0.96937	0.25335	3.9471	1.0316	4.0718	47
14	0.24587	0.96930	0.25366	3.9423	1.0317	4.0672	46
15	0.24615	0.96923	0.25397	3.9375	1.0317	4.0625	45
16	0.24643	0.96916	0.25428	3.9327	1.0318	4.0579	44
17	0.24672	0.96909	0.25459	3.9279	1.0319	4.0532	43
18	0.24700	0.96901	0.25490	3.9231	1.0320	4.0486	42
19	0.24728	0.96894	0.25521	3.9184	1.0320	4.0440	41
20	0.24756	0.96887	0.25552	3.9136	1.0321	4.0394	40
21	0.24784	0.96880	0.25583	3.9089	1.0322	4.0348	39
22	0.24813	0.96873	0.25614	3.9042	1.0323	4.0302	38
23	0.24841	0.96865	0.25645	3.8994	1.0323	4.0256	37
24	0.24869	0.96858	0.25676	3.8947	1.0324	4.0211	36
25	0.24897	0.96851	0.25707	3.8900	1.0325	4.0165	35
26	0.24925	0.96844	0.25738	3.8853	1.0326	4.0120	34
27	0.24953	0.96836	0.25769	3.8807	1.0327	4.0074	33
28	0.24982	0.96829	0.25800	3.8760	1.0327	4.0029	32
29	0.25010	0.96822	0.25831	3.8713	1.0328	3.9984	31
30	0.25038	0.96815	0.25862	3.8667	1.0329	3.9939	30
31	0.25066	0.96807	0.25893	3.8621	1.0330	3.9894	29
32	0.25094	0.96800	0.25924	3.8574	1.0330	3.9850	28
33	0.25122	0.96793	0.25955	3.8528	1.0331	3.9805	27
34	0.25151	0.96785	0.25986	3.8482	1.0332	3.9760	26
35	0.25179	0.96778	0.26017	3.8436	1.0333	3.9716	25
36	0.25207	0.96771	0.26048	3.8390	1.0334	3.9672	24
37	0.25235	0.96763	0.26079	3.8345	1.0334	3.9627	23
38	0.25263	0.96756	0.26110	3.8299	1.0335	3.9583	22
39	0.25291	0.96749	0.26141	3.8254	1.0336	3.9539	21
40	0.25319	0.96741	0.26172	3.8208	1.0337	3.9495	20
41	0.25348	0.96734	0.26203	3.8163	1.0338	3.9451	19
42	0.25376	0.96727	0.26234	3.8118	1.0338	3.9408	18
43	0.25404	0.96719	0.26266	3.8073	1.0339	3.9364	17
44	0.25432	0.96712	0.26297	3.8027	1.0340	3.9320	16
45	0.25460	0.96704	0.26328	3.7983	1.0341	3.9277	15
46	0.25488	0.96697	0.26359	3.7938	1.0341	3.9234	14
47	0.25516	0.96690	0.26390	3.7893	1.0342	3.9190	13
48	0.25544	0.96682	0.26421	3.7848	1.0343	3.9147	12
49	0.25573	0.96675	0.26452	3.7804	1.0344	3.9104	11
50	0.25601	0.96667	0.26483	3.7759	1.0345	3.9061	10
51	0.25629	0.96660	0.26514	3.7715	1.0345	3.9018	9
52	0.25657	0.96652	0.26546	3.7671	1.0346	3.8976	8
53	0.25685	0.96645	0.26577	3.7627	1.0347	3.8933	7
54	0.25713	0.96638	0.26608	3.7583	1.0348	3.8890	6
55	0.25741	0.96630	0.26639	3.7539	1.0349	3.8848	5
56	0.25769	0.96623	0.26670	3.7495	1.0349	3.8805	4
57	0.25798	0.96615	0.26701	3.7451	1.0350	3.8763	3
58	0.25826	0.96608	0.26732	3.7407	1.0351	3.8721	2
59	0.25854	0.96600	0.26764	3.7364	1.0352	3.8679	1
60	0.25882	0.96593	0.26795	3.7320	1.0353	3.8637	0
M	Cosine	Sine	Cotan.	Tan.	Cosec.	Secant	M

15° or 195°   **Natural Trigonometric Functions**   **164° or 344°**

M	Sine	Cosine	Tan.	Cotan.	Secant	Cosec.	M
0	0.25882	0.96593	0.26795	3.7320	1.0353	3.8637	60
1	0.25910	0.96585	0.26826	3.7277	1.0353	3.8595	59
2	0.25938	0.96577	0.26857	3.7234	1.0354	3.8553	58
3	0.25966	0.96570	0.26888	3.7191	1.0355	3.8512	57
4	0.25994	0.96562	0.26920	3.7147	1.0356	3.8470	56
5	0.26022	0.96555	0.26951	3.7104	1.0357	3.8428	55
6	0.26050	0.96547	0.26982	3.7062	1.0358	3.8387	54
7	0.26078	0.96540	0.27013	3.7019	1.0358	3.8346	53
8	0.26107	0.96532	0.27044	3.6976	1.0359	3.8304	52
9	0.26135	0.96524	0.27076	3.6933	1.0360	3.8263	51
10	0.26163	0.96517	0.27107	3.6891	1.0361	3.8222	50
11	0.26191	0.96509	0.27138	3.6848	1.0362	3.8181	49
12	0.26219	0.96502	0.27169	3.6806	1.0362	3.8140	48
13	0.26247	0.96494	0.27201	3.6764	1.0363	3.8100	47
14	0.26275	0.96486	0.27232	3.6722	1.0364	3.8059	46
15	0.26303	0.96479	0.27263	3.6679	1.0365	3.8018	45
16	0.26331	0.96471	0.27294	3.6637	1.0366	3.7978	44
17	0.26359	0.96463	0.27326	3.6596	1.0367	3.7937	43
18	0.26387	0.96456	0.27357	3.6554	1.0367	3.7897	42
19	0.26415	0.96448	0.27388	3.6512	1.0368	3.7857	41
20	0.26443	0.96440	0.27419	3.6470	1.0369	3.7816	40
21	0.26471	0.96433	0.27451	3.6429	1.0370	3.7776	39
22	0.26499	0.96425	0.27482	3.6387	1.0371	3.7736	38
23	0.26527	0.96417	0.27513	3.6346	1.0371	3.7697	37
24	0.26556	0.96409	0.27544	3.6305	1.0372	3.7657	36
25	0.26584	0.96402	0.27576	3.6263	1.0373	3.7617	35
26	0.26612	0.96394	0.27607	3.6222	1.0374	3.7577	34
27	0.26640	0.96386	0.27638	3.6181	1.0375	3.7538	33
28	0.26668	0.96378	0.27670	3.6140	1.0376	3.7498	32
29	0.26696	0.96371	0.27701	3.6100	1.0376	3.7450	31
30	0.26724	0.96363	0.27732	3.6059	1.0377	3.7420	30
31	0.26752	0.96355	0.27764	3.6018	1.0378	3.7380	29
32	0.26780	0.96347	0.27795	3.5977	1.0379	3.7341	28
33	0.26808	0.96340	0.27826	3.5937	1.0380	3.7302	27
34	0.26836	0.96332	0.27858	3.5896	1.0381	3.7263	26
35	0.26864	0.96324	0.27889	3.5856	1.0382	3.7224	25
36	0.26892	0.96316	0.27920	3.5816	1.0382	3.7186	24
37	0.26920	0.96308	0.27952	3.5776	1.0383	3.7147	23
38	0.26948	0.96301	0.27983	3.5736	1.0384	3.7108	22
39	0.26976	0.96293	0.28014	3.5696	1.0385	3.7070	21
40	0.27004	0.96285	0.28046	3.5656	1.0386	3.7031	20
41	0.27032	0.96277	0.28077	3.5616	1.0387	3.6993	19
42	0.27060	0.96269	0.28109	3.5576	1.0387	3.6955	18
43	0.27088	0.96261	0.28140	3.5536	1.0388	3.6917	17
44	0.27116	0.96253	0.28171	3.5497	1.0389	3.6878	16
45	0.27144	0.96245	0.28203	3.5457	1.0390	3.6840	15
46	0.27172	0.96238	0.28234	3.5418	1.0391	3.6802	14
47	0.27200	0.96230	0.28266	3.5378	1.0392	3.6765	13
48	0.27228	0.96222	0.28297	3.5339	1.0393	3.6727	12
49	0.27256	0.96214	0.28328	3.5300	1.0393	3.6689	11
50	0.27284	0.96206	0.28360	3.5261	1.0394	3.6651	10
51	0.27312	0.96198	0.28391	3.5222	1.0395	3.6614	9
52	0.27340	0.96190	0.28423	3.5183	1.0396	3.6576	8
53	0.27368	0.96182	0.28454	3.5144	1.0397	3.6539	7
54	0.27396	0.96174	0.28486	3.5105	1.0398	3.6502	6
55	0.27424	0.96166	0.28517	3.5066	1.0399	3.6464	5
56	0.27452	0.96158	0.28549	3.5028	1.0399	3.6427	4
57	0.27480	0.96150	0.28580	3.4989	1.0400	3.6390	3
58	0.27508	0.96142	0.28611	3.4951	1.0401	3.6353	2
59	0.27536	0.96134	0.28643	3.4912	1.0402	3.6316	1
60	0.27564	0.96126	0.28674	3.4874	1.0403	3.6279	0
M	Cosine	Sine	Cotan.	Tan.	Cosec.	Secant	M

105° or 285°   **74° or 254°**

Natural Trigonometric Functions

M	Sine	Cosine	Tan.	Cotan.	Secant	Cosec.	M
0	0.27564	0.96126	0.28674	3.4874	1.0403	3.6279	60
1	0.27592	0.96118	0.28706	3.4836	1.0404	3.6243	59
2	0.27620	0.96110	0.28737	3.4798	1.0405	3.6206	58
3	0.27648	0.96102	0.28769	3.4760	1.0406	3.6169	57
4	0.27675	0.96094	0.28800	3.4722	1.0406	3.6133	56
5	0.27703	0.96086	0.28832	3.4684	1.0407	3.6096	55
6	0.27731	0.96078	0.28863	3.4646	1.0408	3.6060	54
7	0.27759	0.96070	0.28895	3.4608	1.0409	3.6024	53
8	0.27787	0.96062	0.28926	3.4570	1.0410	3.5987	52
9	0.27815	0.96054	0.28958	3.4533	1.0411	3.5951	51
10	0.27843	0.96045	0.28990	3.4495	1.0412	3.5915	50
11	0.27871	0.96037	0.29021	3.4458	1.0413	3.5879	49
12	0.27899	0.96029	0.29053	3.4420	1.0413	3.5843	48
13	0.27927	0.96021	0.29084	3.4383	1.0414	3.5807	47
14	0.27955	0.96013	0.29116	3.4346	1.0415	3.5772	46
15	0.27983	0.96005	0.29147	3.4308	1.0416	3.5736	45
16	0.28011	0.95997	0.29179	3.4271	1.0417	3.5700	44
17	0.28039	0.95989	0.29210	3.4234	1.0418	3.5665	43
18	0.28067	0.95980	0.29242	3.4197	1.0419	3.5629	42
19	0.28094	0.95972	0.29274	3.4160	1.0420	3.5594	41
20	0.28122	0.95964	0.29305	3.4124	1.0420	3.5559	40
21	0.28150	0.95956	0.29337	3.4087	1.0421	3.5523	39
22	0.28178	0.95948	0.29368	3.4050	1.0422	3.5488	38
23	0.28206	0.95940	0.29400	3.4014	1.0423	3.5453	37
24	0.28234	0.95931	0.29432	3.3977	1.0424	3.5418	36
25	0.28262	0.95923	0.29463	3.3941	1.0425	3.5383	35
26	0.28290	0.95915	0.29495	3.3904	1.0426	3.5348	34
27	0.28318	0.95907	0.29526	3.3868	1.0427	3.5313	33
28	0.28346	0.95898	0.29558	3.3832	1.0428	3.5279	32
29	0.28374	0.95890	0.29590	3.3795	1.0428	3.5244	31
30	0.28401	0.95882	0.29621	3.3759	1.0429	3.5209	30
31	0.28429	0.95874	0.29653	3.3723	1.0430	3.5175	29
32	0.28457	0.95865	0.29685	3.3687	1.0431	3.5140	28
33	0.28485	0.95857	0.29716	3.3651	1.0432	3.5106	27
34	0.28513	0.95849	0.29748	3.3616	1.0433	3.5072	26
35	0.28541	0.95840	0.29780	3.3580	1.0434	3.5037	25
36	0.28569	0.95832	0.29811	3.3544	1.0435	3.5003	24
37	0.28597	0.95824	0.29843	3.3509	1.0436	3.4969	23
38	0.28624	0.95816	0.29875	3.3473	1.0437	3.4935	22
39	0.28652	0.95807	0.29906	3.3438	1.0438	3.4901	21
40	0.28680	0.95799	0.29938	3.3402	1.0438	3.4867	20
41	0.28708	0.95791	0.29970	3.3367	1.0439	3.4833	19
42	0.28736	0.95782	0.30001	3.3332	1.0440	3.4799	18
43	0.28764	0.95774	0.30033	3.3296	1.0441	3.4766	17
44	0.28792	0.95765	0.30065	3.3261	1.0442	3.4732	16
45	0.28820	0.95757	0.30096	3.3226	1.0443	3.4698	15
46	0.28847	0.95749	0.30128	3.3191	1.0444	3.4665	14
47	0.28875	0.95740	0.30160	3.3156	1.0445	3.4632	13
48	0.28903	0.95732	0.30192	3.3121	1.0446	3.4598	12
49	0.28931	0.95724	0.30223	3.3087	1.0447	3.4565	11
50	0.28959	0.95715	0.30255	3.3052	1.0448	3.4532	10
51	0.28987	0.95707	0.30287	3.3017	1.0448	3.4498	9
52	0.29014	0.95698	0.30319	3.2983	1.0449	3.4465	8
53	0.29042	0.95690	0.30350	3.2948	1.0450	3.4432	7
54	0.29070	0.95681	0.30382	3.2914	1.0451	3.4399	6
55	0.29098	0.95673	0.30414	3.2879	1.0452	3.4366	5
56	0.29126	0.95664	0.30446	3.2845	1.0453	3.4334	4
57	0.29154	0.95656	0.30478	3.2811	1.0454	3.4301	3
58	0.29181	0.95647	0.30509	3.2777	1.0455	3.4268	2
59	0.29209	0.95639	0.30541	3.2742	1.0456	3.4236	1
60	0.29237	0.95630	0.30573	3.2708	1.0457	3.4203	0
M	Cosine	Sine	Cotan.	Tan.	Cosec.	Secant	M

17° or 197°   Natural Trigonometric Functions   **162° or 342°**

M	Sine	Cosine	Tan.	Cotan.	Secant	Cosec.	M
0	0.29237	0.95630	0.30573	3.2708	1.0457	3.4203	60
1	0.29265	0.95622	0.30605	3.2674	1.0458	3.4170	59
2	0.29293	0.95613	0.30637	3.2640	1.0459	3.4138	58
3	0.29321	0.95605	0.30668	3.2607	1.0460	3.4106	57
4	0.29348	0.95596	0.30700	3.2573	1.0461	3.4073	56
5	0.29376	0.95588	0.30732	3.2539	1.0461	3.4041	55
6	0.29404	0.95579	0.30764	3.2505	1.0462	3.4009	54
7	0.29432	0.95571	0.30796	3.2472	1.0463	3.3977	53
8	0.29460	0.95562	0.30828	3.2438	1.0464	3.3945	52
9	0.29487	0.95554	0.30859	3.2405	1.0465	3.3913	51
10	0.29515	0.95545	0.30891	3.2371	1.0466	3.3881	50
11	0.29543	0.95536	0.30923	3.2338	1.0467	3.3849	49
12	0.29571	0.95528	0.30955	3.2305	1.0468	3.3817	48
13	0.29598	0.95519	0.30987	3.2271	1.0469	3.3785	47
14	0.29626	0.95511	0.31019	3.2238	1.0470	3.3754	46
15	0.29654	0.95502	0.31051	3.2205	1.0471	3.3722	45
16	0.29682	0.95493	0.31083	3.2172	1.0472	3.3690	44
17	0.29710	0.95485	0.31115	3.2139	1.0473	3.3659	43
18	0.29737	0.95476	0.31146	3.2106	1.0474	3.3627	42
19	0.29765	0.95467	0.31178	3.2073	1.0475	3.3596	41
20	0.29793	0.95459	0.31210	3.2041	1.0476	3.3565	40
21	0.29821	0.95450	0.31242	3.2008	1.0477	3.3534	39
22	0.29848	0.95441	0.31274	3.1975	1.0478	3.3502	38
23	0.29876	0.95433	0.31306	3.1942	1.0478	3.3471	37
24	0.29904	0.95424	0.31338	3.1910	1.0479	3.3440	36
25	0.29932	0.95415	0.31370	3.1877	1.0480	3.3409	35
26	0.29959	0.95407	0.31402	3.1845	1.0481	3.3378	34
27	0.29987	0.95398	0.31434	3.1813	1.0482	3.3347	33
28	0.30015	0.95389	0.31466	3.1780	1.0483	3.3316	32
29	0.30043	0.95380	0.31498	3.1748	1.0484	3.3286	31
30	0.30070	0.95372	0.31530	3.1716	1.0485	3.3255	30
31	0.30098	0.95363	0.31562	3.1684	1.0486	3.3224	29
32	0.30126	0.95354	0.31594	3.1652	1.0487	3.3194	28
33	0.30154	0.95345	0.31626	3.1620	1.0488	3.3163	27
34	0.30181	0.95337	0.31658	3.1588	1.0489	3.3133	26
35	0.30209	0.95328	0.31690	3.1556	1.0490	3.3102	25
36	0.30237	0.95319	0.31722	3.1524	1.0491	3.3072	24
37	0.30265	0.95310	0.31754	3.1492	1.0492	3.3042	23
38	0.30292	0.95301	0.31786	3.1460	1.0493	3.3011	22
39	0.30320	0.95293	0.31818	3.1429	1.0494	3.2981	21
40	0.30348	0.95284	0.31850	3.1397	1.0495	3.2951	20
41	0.30375	0.95275	0.31882	3.1366	1.0496	3.2921	19
42	0.30403	0.95266	0.31914	3.1334	1.0497	3.2891	18
43	0.30431	0.95257	0.31946	3.1303	1.0498	3.2861	17
44	0.30459	0.95248	0.31978	3.1271	1.0499	3.2831	16
45	0.30486	0.95239	0.32010	3.1240	1.0500	3.2801	15
46	0.30514	0.95231	0.32042	3.1209	1.0501	3.2772	14
47	0.30542	0.95222	0.32074	3.1177	1.0502	3.2742	13
48	0.30569	0.95213	0.32106	3.1146	1.0503	3.2712	12
49	0.30597	0.95204	0.32138	3.1115	1.0504	3.2683	11
50	0.30625	0.95195	0.32171	3.1084	1.0505	3.2653	10
51	0.30653	0.95186	0.32203	3.1053	1.0506	3.2624	9
52	0.30680	0.95177	0.32235	3.1022	1.0507	3.2594	8
53	0.30708	0.95168	0.32267	3.0991	1.0508	3.2565	7
54	0.30736	0.95159	0.32299	3.0960	1.0509	3.2535	6
55	0.30763	0.95150	0.32331	3.0930	1.0510	3.2506	5
56	0.30791	0.95141	0.32363	3.0899	1.0511	3.2477	4
57	0.30819	0.95132	0.32395	3.0868	1.0512	3.2448	3
58	0.30846	0.95124	0.32428	3.0838	1.0513	3.2419	2
59	0.30874	0.95115	0.32460	3.0807	1.0514	3.2390	1
60	0.30902	0.95106	0.32492	3.0777	1.0515	3.2361	0
M	Cosine	Sine	Cotan.	Tan.	Cosec.	Secant	M

    Natural Trigonometric Functions

M	Sine	Cosine	Tan.	Cotan.	Secant	Cosec.	M
0	0.30902	0.95160	0.32492	3.0777	1.0515	3.2361	60
1	0.30929	0.95097	0.32524	3.0746	1.0516	3.2332	59
2	0.30957	0.95088	0.32556	3.0716	1.0517	3.2303	58
3	0.30985	0.95079	0.32588	3.0686	1.0518	3.2274	57
4	0.31012	0.95070	0.32621	3.0655	1.0519	3.2245	56
5	0.31040	0.95061	0.32653	3.0625	1.0520	3.2216	55
6	0.31068	0.95051	0.32685	3.0595	1.0521	3.2188	54
7	0.31095	0.95042	0.32717	3.0565	1.0522	3.2159	53
8	0.31123	0.95033	0.32749	3.0535	1.0523	3.2131	52
9	0.31150	0.95024	0.32782	3.0505	1.0524	3.2102	51
10	0.31178	0.95015	0.32814	3.0475	1.0525	3.2074	50
11	0.31260	0.95006	0.32846	3.0445	1.0526	3.2045	49
12	0.31233	0.94997	0.32878	3.0415	1.0527	3.2017	48
13	0.31261	0.94988	0.32910	3.0385	1.0528	3.1989	47
14	0.31289	0.94979	0.32943	3.0356	1.0529	3.1960	46
15	0.31316	0.94970	0.32975	3.0326	1.0530	3.1932	45
16	0.31344	0.94961	0.33007	3.0296	1.0531	3.1904	44
17	0.31372	0.94952	0.33039	3.0267	1.0532	3.1876	43
18	0.31399	0.94942	0.33072	3.0237	1.0533	3.1848	42
19	0.31427	0.94933	0.33104	3.0208	1.0534	3.1820	41
20	0.31454	0.94924	0.33136	3.0178	1.0535	3.1792	40
21	0.31482	0.94915	0.33169	3.0149	1.0536	3.1764	39
22	0.31510	0.94906	0.33201	3.0120	1.0537	3.1736	38
23	0.31537	0.94897	0.33233	3.0090	1.0538	3.1708	37
24	0.31565	0.94888	0.33265	3.0061	1.0539	3.1681	36
25	0.31592	0.94878	0.33298	3.0032	1.0540	3.1653	35
26	0.31620	0.94869	0.33330	3.0003	1.0541	3.1625	34
27	0.31648	0.94860	0.33362	2.9974	1.0542	3.1598	33
28	0.31675	0.94851	0.33395	2.9945	1.0543	3.1570	32
29	0.31703	0.94841	0.33427	2.9916	1.0544	3.1543	31
30	0.31730	0.94832	0.33459	2.9887	1.0545	3.1515	30
31	0.31758	0.94823	0.33492	2.9858	1.0546	3.1488	29
32	0.31786	0.94814	0.33524	2.9829	1.0547	3.1461	28
33	0.31813	0.94805	0.33557	2.9800	1.0548	3.1433	27
34	0.31841	0.94795	0.33589	2.9772	1.0549	3.1406	26
35	0.31868	0.94786	0.33621	2.9743	1.0550	3.1379	25
36	0.31896	0.94777	0.33654	2.9714	1.0551	3.1352	24
37	0.31923	0.94767	0.33686	2.9686	1.0552	3.1325	23
38	0.31951	0.94758	0.33718	2.9657	1.0553	3.1298	22
39	0.31978	0.94749	0.33751	2.9629	1.0554	3.1271	21
40	0.32006	0.94740	0.33783	2.9600	1.0555	3.1244	20
41	0.32034	0.94730	0.33816	2.9572	1.0556	3.1217	19
42	0.32061	0.94721	0.33848	2.9544	1.0557	3.1190	18
43	0.32089	0.94712	0.33880	2.9515	1.0558	3.1163	17
44	0.32116	0.94702	0.33913	2.9487	1.0559	3.1137	16
45	0.32144	0.94693	0.33945	2.9459	1.0560	3.1110	15
46	0.32171	0.94684	0.33978	2.9431	1.0561	3.1083	14
47	0.32199	0.94674	0.34010	2.9403	1.0562	3.1057	13
48	0.32226	0.94665	0.34043	2.9375	1.0563	3.1030	12
49	0.32254	0.94655	0.34075	2.9347	1.0565	3.1004	11
50	0.32282	0.94646	0.34108	2.9319	1.0566	3.0977	10
51	0.32309	0.94637	0.34140	2.9291	1.0567	3.0951	9
52	0.32337	0.94627	0.34173	2.9263	1.0568	3.0925	8
53	0.32364	0.94618	0.34205	2.9235	1.0569	3.0898	7
54	0.32392	0.94608	0.34238	2.9208	1.0570	3.0872	6
55	0.32419	0.94599	0.34270	2.9180	1.0571	3.0846	5
56	0.32447	0.94590	0.34303	2.9152	1.0572	3.0820	4
57	0.32474	0.94580	0.34335	2.9125	1.0573	3.0793	3
58	0.32502	0.94571	0.34368	2.9097	1.0574	3.0767	2
59	0.32529	0.94561	0.34400	2.9069	1.0575	3.0741	1
60	0.32557	0.94552	0.34433	2.9042	1.0576	3.0715	0
M	Cosine	Sine	Cotan.	Tan.	Cosec.	Secant	M

19° or 199°   **Natural Trigonometric Functions**   **160° or 340°**

M	Sine	Cosine	Tan.	Cotan.	Secant	Cosec.	M
0	0.32557	0.94552	0.34433	2.9042	1.0576	3.0715	60
1	0.32584	0.94542	0.34465	2.9015	1.0577	3.0690	59
2	0.32612	0.94533	0.34498	2.8987	1.0578	3.0664	58
3	0.32639	0.94523	0.34530	2.8960	1.0579	3.0638	57
4	0.32667	0.94514	0.34563	2.8933	1.0580	3.0612	56
5	0.32694	0.94504	0.34595	2.8905	1.0581	3.0586	55
6	0.32722	0.94495	0.34628	2.8878	1.0582	3.0561	54
7	0.32749	0.94485	0.34661	2.8851	1.0584	3.0535	53
8	0.32777	0.94476	0.34693	2.8824	1.0585	3.0509	52
9	0.32804	0.94466	0.34726	2.8797	1.0586	3.0484	51
10	0.32832	0.94457	0.34758	2.8770	1.0587	3.0458	50
11	0.32859	0.94447	0.34791	2.8743	1.0588	3.0433	49
12	0.32887	0.94438	0.34824	2.8716	1.0589	3.0407	48
13	0.32914	0.94428	0.34856	2.8689	1.0590	3.0382	47
14	0.32942	0.94418	0.34889	2.8662	1.0591	3.0357	46
15	0.32969	0.94409	0.34921	2.8636	1.0592	3.0331	45
16	0.32996	0.94399	0.34954	2.8609	1.0593	3.0306	44
17	0.33024	0.94390	0.34987	2.8582	1.0594	3.0281	43
18	0.33051	0.94380	0.35019	2.8555	1.0595	3.0256	42
19	0.33079	0.94370	0.35052	2.8529	1.0596	3.0231	41
20	0.33106	0.94361	0.35085	2.8502	1.0598	3.0206	40
21	0.33134	0.94351	0.35117	2.8476	1.0599	3.0181	39
22	0.33161	0.94341	0.35150	2.8449	1.0600	3.0156	38
23	0.33189	0.94332	0.35183	2.8423	1.0601	3.0131	37
24	0.33216	0.94322	0.35215	2.8396	1.0602	3.0106	36
25	0.33243	0.94313	0.35248	2.8370	1.0603	3.0081	35
26	0.33271	0.94303	0.35281	2.8344	1.0604	3.0056	34
27	0.33298	0.94293	0.35314	2.8318	1.0605	3.0031	33
28	0.33326	0.94283	0.35346	2.8291	1.0606	3.0007	32
29	0.33353	0.94274	0.35379	2.8265	1.0607	2.9982	31
30	0.33381	0.94264	0.35412	2.8239	1.0608	2.9957	30
31	0.33408	0.94254	0.35445	2.8213	1.0609	2.9933	29
32	0.33435	0.94245	0.35477	2.8187	1.0611	2.9908	28
33	0.33463	0.94235	0.35510	2.8161	1.0612	2.9884	27
34	0.33490	0.94225	0.35543	2.8135	1.0613	2.9859	26
35	0.33518	0.94215	0.35576	2.8109	1.0614	2.9835	25
36	0.33545	0.94206	0.35608	2.8083	1.0615	2.9810	24
37	0.33572	0.94196	0.35641	2.8057	1.0616	2.9786	23
38	0.33600	0.94186	0.35674	2.8032	1.0617	2.9762	22
39	0.33627	0.94176	0.35707	2.8006	1.0618	2.9738	21
40	0.33655	0.94167	0.35739	2.7980	1.0619	2.9713	20
41	0.33682	0.94157	0.35772	2.7954	1.0620	2.9689	19
42	0.33709	0.94147	0.35805	2.7929	1.0622	2.9665	18
43	0.33737	0.94137	0.35838	2.7903	1.0623	2.9641	17
44	0.33764	0.94127	0.35871	2.7878	1.0624	2.9617	16
45	0.33792	0.94118	0.35904	2.7852	1.0625	2.9593	15
46	0.33819	0.94108	0.35936	2.7827	1.0626	2.9569	14
47	0.33846	0.94098	0.35969	2.7801	1.0627	2.9545	13
48	0.33874	0.94088	0.36002	2.7776	1.0628	2.9521	12
49	0.33901	0.94078	0.36035	2.7751	1.0629	2.9497	11
50	0.33928	0.94068	0.36068	2.7725	1.0630	2.9474	10
51	0.33956	0.94058	0.36101	2.7700	1.0632	2.9450	9
52	0.33983	0.94049	0.36134	2.7675	1.0633	2.9426	8
53	0.34011	0.94039	0.36167	2.7650	1.0634	2.9402	7
54	0.34038	0.94029	0.36199	2.7625	1.0635	2.9379	6
55	0.34065	0.94019	0.36232	2.7600	1.0636	2.9355	5
56	0.34093	0.94009	0.36265	2.7575	1.0637	2.9332	4
57	0.34120	0.93999	0.36298	2.7550	1.0638	2.9308	3
58	0.34147	0.93989	0.36331	2.7525	1.0639	2.9285	2
59	0.34175	0.93979	0.36364	2.7500	1.0641	2.9261	1
60	0.34202	0.93969	0.36397	2.7475	1.0642	2.9238	0
M	Cosine	Sine	Cotan.	Tan.	Cosec.	Secant	M

109° or 289°   **70° or 250°**

　　Natural Trigonometric Functions　　

M	Sine	Cosine	Tan.	Cotan.	Secant	Cosec.	M
0	0.34202	0.93969	0.36397	2.7475	1.0642	2.9238	60
1	0.34229	0.93959	0.36430	2.7450	1.0643	2.9215	59
2	0.34257	0.93949	0.36463	2.7425	1.0644	2.9191	58
3	0.34284	0.93939	0.36496	2.7400	1.0645	2.9168	57
4	0.34311	0.93929	0.36529	2.7376	1.0646	2.9145	56
5	0.34339	0.93919	0.36562	2.7351	1.0647	2.9122	55
6	0.34366	0.93909	0.36595	2.7326	1.0648	2.9098	54
7	0.34393	0.93899	0.36628	2.7302	1.0650	2.9075	53
8	0.34421	0.93889	0.36661	2.7277	1.0651	2.9052	52
9	0.34448	0.93879	0.36694	2.7252	1.0652	2.9029	51
10	0.34475	0.93869	0.36727	2.7228	1.0653	2.9006	50
11	0.34502	0.93859	0.36760	2.7204	1.0654	2.8983	49
12	0.34530	0.93849	0.36793	2.7179	1.0655	2.8960	48
13	0.34557	0.93839	0.36826	2.7155	1.0656	2.8937	47
14	0.34584	0.93829	0.36859	2.7130	1.0658	2.8915	46
15	0.34612	0.93819	0.36892	2.7106	1.0659	2.8892	45
16	0.34639	0.93809	0.36925	2.7082	1.0660	2.8869	44
17	0.34666	0.93799	0.36958	2.7058	1.0661	2.8846	43
18	0.34693	0.93789	0.36991	2.7033	1.0662	2.8824	42
19	0.34712	0.93779	0.37024	2.7009	1.0663	2.8801	41
20	0.34748	0.93769	0.37057	2.6985	1.0664	2.8778	40
21	0.34775	0.93758	0.37090	2.6961	1.0666	2.8756	39
22	0.34803	0.93748	0.37123	2.6937	1.0667	2.8733	38
23	0.34830	0.93738	0.37156	2.6913	1.0668	2.8711	37
24	0.34857	0.93728	0.37190	2.6889	1.0669	2.8688	36
25	0.34884	0.93718	0.37223	2.6865	1.0670	2.8666	35
26	0.34912	0.93708	0.37256	2.6841	1.0671	2.8644	34
27	0.34939	0.93698	0.37289	2.6817	1.0673	2.8621	33
28	0.34966	0.93687	0.37322	2.6794	1.0674	2.8599	32
29	0.34993	0.93677	0.37355	2.6770	1.0675	2.8577	31
30	0.35021	0.93667	0.37388	2.6746	1.0676	2.8554	30
31	0.35048	0.93657	0.37422	2.6722	1.0677	2.8532	29
32	0.35075	0.93647	0.37455	2.6699	1.0678	2.8510	28
33	0.35102	0.93637	0.37488	2.6675	1.0679	2.8488	27
34	0.35130	0.93626	0.37521	2.6652	1.0681	2.8466	26
35	0.35157	0.93616	0.37554	2.6628	1.0682	2.8444	25
36	0.35184	0.93606	0.37587	2.6604	1.0683	2.8422	24
37	0.35211	0.93596	0.37621	2.6581	1.0684	2.8400	23
38	0.35239	0.93585	0.37654	2.6558	1.0685	2.8378	22
39	0.35266	0.93575	0.37687	2.6534	1.0686	2.8356	21
40	0.35293	0.93565	0.37720	2.6511	1.0688	2.8334	20
41	0.35320	0.93555	0.37754	2.6487	1.0689	2.8312	19
42	0.35347	0.93544	0.37787	2.6464	1.0690	2.8290	18
43	0.35375	0.93534	0.37820	2.6441	1.0691	2.8269	17
44	0.35402	0.93524	0.37853	2.6418	1.0692	2.8247	16
45	0.35429	0.93513	0.37887	2.6394	1.0694	2.8225	15
46	0.35456	0.93503	0.37920	2.6371	1.0695	2.8204	14
47	0.35483	0.93493	0.37953	2.6348	1.0696	2.8182	13
48	0.35511	0.93482	0.37986	2.6325	1.0697	2.8160	12
49	0.35538	0.93472	0.38020	2.6302	1.0698	2.8139	11
50	0.35565	0.93462	0.38053	2.6279	1.0699	2.8117	10
51	0.35592	0.93451	0.38086	2.6256	1.0701	2.8096	9
52	0.35619	0.93441	0.38120	2.6233	1.0702	2.8074	8
53	0.35647	0.93431	0.38153	2.6210	1.0703	2.8053	7
54	0.35674	0.93420	0.38186	2.6187	1.0704	2.8032	6
55	0.35701	0.93410	0.38220	2.6164	1.0705	2.8010	5
56	0.35728	0.93400	0.38253	2.6142	1.0707	2.7989	4
57	0.35755	0.93389	0.38286	2.6119	1.0708	2.7968	3
58	0.35782	0.93379	0.38320	2.6096	1.0709	2.7947	2
59	0.35810	0.93368	0.38353	2.6073	1.0710	2.7925	1
60	0.35837	0.93358	0.38386	2.6051	1.0711	2.7904	0
M	Cosine	Sine	Cotan.	Tan.	Cosec.	Secant	M

21° or 201°   **Natural Trigonometric Functions**   **158° or 338°**

M	Sine	Cosine	Tan.	Cotan.	Secant	Cosec.	M
0	0.35837	0.93358	0.38386	2.6051	1.0711	2.7904	60
1	0.35864	0.93348	0.38420	2.6028	1.0713	2.7883	59
2	0.35891	0.93337	0.38453	2.6006	1.0714	2.7862	58
3	0.35918	0.93327	0.38486	2.5983	1.0715	2.7841	57
4	0.35945	0.93316	0.38520	2.5960	1.0716	2.7820	56
5	0.35972	0.93306	0.38553	2.5938	1.0717	2.7799	55
6	0.36000	0.93295	0.38587	2.5916	1.0719	2.7778	54
7	0.36027	0.93285	0.38620	2.5893	1.0720	2.7757	53
8	0.36054	0.93274	0.38654	2.5871	1.0721	2.7736	52
9	0.36081	0.93264	0.38687	2.5848	1.0722	2.7715	51
10	0.36108	0.93253	0.38720	2.5826	1.0723	2.7694	50
11	0.36135	0.93243	0.38754	2.5804	1.0725	2.7674	49
12	0.36162	0.93232	0.38787	2.5781	1.0726	2.7653	48
13	0.36189	0.93222	0.38821	2.5759	1.0727	2.7632	47
14	0.36217	0.93211	0.38854	2.5737	1.0728	2.7611	46
15	0.36244	0.93201	0.38888	2.5715	1.0729	2.7591	45
16	0.36271	0.93190	0.38921	2.5693	1.0731	2.7570	44
17	0.36298	0.93180	0.38955	2.5671	1.0732	2.7550	43
18	0.36325	0.93169	0.38988	2.5649	1.0733	2.7529	42
19	0.36352	0.93158	0.39022	2.5627	1.0734	2.7509	41
20	0.36379	0.93148	0.39055	2.5605	1.0736	2.7488	40
21	0.36406	0.93137	0.39089	2.5583	1.0737	2.7468	39
22	0.36433	0.93127	0.39122	2.5561	1.0738	2.7447	38
23	0.36460	0.93116	0.39156	2.5539	1.0739	2.7427	37
24	0.36488	0.93105	0.39189	2.5517	1.0740	2.7406	36
25	0.36515	0.93095	0.39223	2.5495	1.0742	2.7386	35
26	0.36542	0.93084	0.39257	2.5473	1.0743	2.7366	34
27	0.36569	0.93074	0.39290	2.5451	1.0744	2.7346	33
28	0.36596	0.93063	0.39324	2.5430	1.0745	2.7325	32
29	0.36623	0.93052	0.39357	2.5408	1.0747	2.7305	31
30	0.36650	0.93042	0.39391	2.5386	1.0748	2.7285	30
31	0.36677	0.93031	0.39425	2.5365	1.0749	2.7265	29
32	0.36704	0.93020	0.39458	2.5343	1.0750	2.7245	28
33	0.36731	0.93010	0.39492	2.5322	1.0751	2.7225	27
34	0.36758	0.92999	0.39525	2.5300	1.0753	2.7205	26
35	0.36785	0.92988	0.39559	2.5278	1.0754	2.7185	25
36	0.36812	0.92978	0.39593	2.5257	1.0755	2.7165	24
37	0.36839	0.92967	0.39626	2.5236	1.0756	2.7145	23
38	0.36866	0.92956	0.39660	2.5214	1.0758	2.7125	22
39	0.36893	0.92945	0.39694	2.5193	1.0759	2.7105	21
40	0.36921	0.92935	0.39727	2.5171	1.0760	2.7085	20
41	0.36948	0.92924	0.39761	2.5150	1.0761	2.7065	19
42	0.36975	0.92913	0.39795	2.5129	1.0763	2.7045	18
43	0.37002	0.92902	0.39828	2.5108	1.0764	2.7026	17
44	0.37029	0.92892	0.39862	2.5086	1.0765	2.7006	16
45	0.37056	0.92881	0.39896	2.5065	1.0766	2.6986	15
46	0.37083	0.92870	0.39930	2.5044	1.0768	2.6967	14
47	0.37110	0.92859	0.39963	2.5023	1.0769	2.6947	13
48	0.37137	0.92848	0.39997	2.5002	1.0770	2.6927	12
49	0.37164	0.92838	0.40031	2.4981	1.0771	2.6908	11
50	0.37191	0.92827	0.40065	2.4960	1.0773	2.6888	10
51	0.37218	0.92816	0.40098	2.4939	1.0774	2.6869	9
52	0.37245	0.92805	0.40132	2.4918	1.0775	2.6849	8
53	0.37272	0.92794	0.40166	2.4897	1.0776	2.6830	7
54	0.37299	0.92784	0.40200	2.4876	1.0778	2.6810	6
55	0.37326	0.92773	0.40233	2.4855	1.0779	2.6791	5
56	0.37353	0.92762	0.40267	2.4834	1.0780	2.6772	4
57	0.37380	0.92751	0.40301	2.4813	1.0781	2.6752	3
58	0.37407	0.92740	0.40335	2.4792	1.0783	2.6733	2
59	0.37434	0.92729	0.40369	2.4772	1.0784	2.6714	1
60	0.37461	0.92718	0.40403	2.4751	1.0785	2.6695	0
M	Cosine	Sine	Cotan.	Tan.	Cosec.	Secant	M

111° or 291°   **68° or 248°**

Natural Trigonometric Functions

M	Sine	Cosine	Tan.	Cotan.	Secant	Cosec.	M
0	0.37461	0.92718	0.40403	2.4751	1.0785	2.6695	60
1	0.37488	0.92707	0.40436	2.4730	1.0787	2.6675	59
2	0.37514	0.92696	0.40470	2.4709	1.0788	2.6656	58
3	0.37541	0.92686	0.40504	2.4689	1.0789	2.6637	57
4	0.37568	0.92675	0.40538	2.4668	1.0790	2.6618	56
5	0.37595	0.92664	0.40572	2.4647	1.0792	2.6599	55
6	0.37622	0.92653	0.40606	2.4627	1.0793	2.6580	54
7	0.37649	0.92642	0.40640	2.4606	1.0794	2.6561	53
8	0.37676	0.92631	0.40673	2.4586	1.0795	2.6542	52
9	0.37703	0.92620	0.40707	2.4565	1.0797	2.6523	51
10	0.37730	0.92609	0.40741	2.4545	1.0798	2.6504	50
11	0.37757	0.92598	0.40775	2.4525	1.0799	2.6485	49
12	0.37784	0.92587	0.40809	2.4504	1.0801	2.6466	48
13	0.37811	0.92576	0.40843	2.4484	1.0802	2.6447	47
14	0.37838	0.92565	0.40877	2.4463	1.0803	2.6428	46
15	0.37865	0.92554	0.40911	2.4443	1.0804	2.6410	45
16	0.37892	0.92543	0.40945	2.4423	1.0806	2.6391	44
17	0.37919	0.92532	0.40979	2.4403	1.0807	2.6372	43
18	0.37946	0.92521	0.41013	2.4382	1.0808	2.6353	42
19	0.37972	0.92510	0.41047	2.4362	1.0810	2.6335	41
20	0.37999	0.92499	0.41081	2.4342	1.0811	2.6316	40
21	0.38026	0.92488	0.41115	2.4322	1.0812	2.6297	39
22	0.38053	0.92477	0.41149	2.4302	1.0813	2.6279	38
23	0.38080	0.92466	0.41183	2.4282	1.0815	2.6260	37
24	0.38107	0.92455	0.41217	2.4262	1.0816	2.6242	36
25	0.38134	0.92443	0.41251	2.4242	1.0817	2.6223	35
26	0.38161	0.92432	0.41285	2.4222	1.0819	2.6205	34
27	0.38188	0.92421	0.41319	2.4202	1.0820	2.6186	33
28	0.38214	0.92410	0.41353	2.4182	1.0821	2.6168	32
29	0.38241	0.92399	0.41387	2.4162	1.0823	2.6150	31
30	0.38268	0.92388	0.41421	2.4142	1.0824	2.6131	30
31	0.38295	0.92377	0.41455	2.4122	1.0825	2.6113	29
32	0.38322	0.92366	0.41489	2.4102	1.0826	2.6095	28
33	0.38349	0.92354	0.41524	2.4083	1.0828	2.6076	27
34	0.38376	0.92343	0.41558	2.4063	1.0829	2.6058	26
35	0.38403	0.92332	0.41592	2.4043	1.0830	2.6040	25
36	0.38429	0.92321	0.41626	2.4023	1.0832	2.6022	24
37	0.38456	0.92310	0.41660	2.4004	1.0833	2.6003	23
38	0.38483	0.92299	0.41694	2.3984	1.0834	2.5985	22
39	0.38510	0.92287	0.41728	2.3964	1.0836	2.5967	21
40	0.38537	0.92276	0.41762	2.3945	1.0837	2.5949	20
41	0.38564	0.92265	0.41797	2.3925	1.0838	2.5931	19
42	0.38591	0.92254	0.41831	2.3906	1.0840	2.5913	18
43	0.38617	0.92242	0.41865	2.3886	1.0841	2.5895	17
44	0.38644	0.92231	0.41899	2.3867	1.0842	2.5877	16
45	0.38671	0.92220	0.41933	2.3847	1.0844	2.5859	15
46	0.38698	0.92209	0.41968	2.3828	1.0845	2.5841	14
47	0.38725	0.92197	0.42002	2.3808	1.0846	2.5823	13
48	0.38751	0.92186	0.42036	2.3789	1.0847	2.5805	12
49	0.38778	0.92175	0.42070	2.3770	1.0849	2.5787	11
50	0.38805	0.92164	0.42105	2.3750	1.0850	2.5770	10
51	0.38832	0.92152	0.42139	2.3731	1.0851	2.5752	9
52	0.38859	0.92141	0.42173	2.3712	1.0853	2.5734	8
53	0.38886	0.92130	0.42207	2.3692	1.0854	2.5716	7
54	0.38912	0.92118	0.42242	2.3673	1.0855	2.5699	6
55	0.38939	0.92107	0.42276	2.3654	1.0857	2.5681	5
56	0.38966	0.92096	0.42310	2.3635	1.0858	2.5663	4
57	0.38993	0.92084	0.42344	2.3616	1.0859	2.5646	3
58	0.39019	0.92073	0.42379	2.3597	1.0861	2.5628	2
59	0.39046	0.92062	0.42413	2.3577	1.0862	2.5610	1
60	0.39073	0.92050	0.42447	2.3558	1.0864	2.5593	0
M	Cosine	Sine	Cotan.	Tan.	Cosec.	Secant	M

23° or 203°   **Natural Trigonometric Functions**   **156° or 336°**

M	Sine	Cosine	Tan.	Cotan.	Secant	Cosec.	M
0	0.39073	0.92050	0.42447	2.3558	1.0864	2.5593	60
1	0.33910	0.92039	0.42482	2.3539	1.0865	2.5575	59
2	0.39126	0.92028	0.42516	2.3520	1.0866	2.5558	58
3	0.39153	0.92016	0.42550	2.3501	1.0868	2.5540	57
4	0.39180	0.92005	0.42585	2.3482	1.0869	2.5526	56
5	0.39207	0.91993	0.42619	2.3463	1.0870	2.5506	55
6	0.39234	0.91982	0.42654	2.3445	1.0872	2.5488	54
7	0.39260	0.91971	0.42688	2.3426	1.0873	2.5471	53
8	0.39287	0.91959	0.42722	2.3407	1.0874	2.5453	52
9	0.39314	0.91948	0.42757	2.3388	1.0876	2.5436	51
10	0.39341	0.91936	0.42791	2.3369	1.0877	2.5419	50
11	0.39367	0.91925	0.42826	2.3350	1.0878	2.5402	49
12	0.39394	0.91913	0.42860	2.3332	1.0880	2.5384	48
13	0.39421	0.91902	0.42894	2.3313	1.0881	2.5367	47
14	0.39448	0.91891	0.42929	2.3294	1.0882	2.5350	46
15	0.39474	0.91879	0.42963	2.3276	1.0884	2.5333	45
16	0.39501	0.91868	0.42998	2.3257	1.0885	2.5316	44
17	0.39528	0.91856	0.43032	2.3238	1.0886	2.5299	43
18	0.39554	0.91845	0.43067	2.3220	1.0888	2.5281	42
19	0.39581	0.91833	0.43101	2.3201	1.0889	2.5264	41
20	0.39608	0.91822	0.43136	2.3183	1.0891	2.5247	40
21	0.39635	0.91810	0.43170	2.3164	1.0892	2.5230	39
22	0.39661	0.91798	0.43205	2.3145	1.0893	2.5213	38
23	0.39688	0.91787	0.43239	2.3127	1.0895	2.5196	37
24	0.39715	0.91775	0.43274	2.3109	1.0896	2.5179	36
25	0.39741	0.91764	0.43308	2.3090	1.0897	2.5163	35
26	0.39768	0.91752	0.43343	2.3072	1.0899	2.5146	34
27	0.39795	0.91741	0.43377	2.3053	1.0900	2.5129	33
28	0.39821	0.91729	0.43412	2.3035	1.0902	2.5112	32
29	0.39848	0.91718	0.43447	2.3017	1.0903	2.5095	31
30	0.39875	0.91706	0.43481	2.2998	1.0904	2.5078	30
31	0.39901	0.91694	0.43516	2.2980	1.0906	2.5062	29
32	0.39928	0.91683	0.43550	2.2962	1.0907	2.5045	28
33	0.39955	0.91671	0.43585	2.2944	1.0908	2.5028	27
34	0.39981	0.91659	0.43620	2.2925	1.0910	2.5011	26
35	0.40008	0.91648	0.43654	2.2907	1.0911	2.4995	25
36	0.40035	0.91636	0.43689	2.2889	1.0913	2.4978	24
37	0.40061	0.91625	0.43723	2.2871	1.0914	2.4961	23
38	0.40088	0.91613	0.43758	2.2853	1.0915	2.4945	22
39	0.40115	0.91601	0.43793	2.2835	1.0917	2.4928	21
40	0.40141	0.91590	0.43827	2.2817	1.0918	2.4912	20
41	0.40168	0.91578	0.43862	2.2799	1.0920	2.4895	19
42	0.40195	0.91566	0.43897	2.2781	1.0921	2.4879	18
43	0.40221	0.91554	0.43932	2.2763	1.0922	2.4862	17
44	0.40248	0.91543	0.43966	2.2745	1.0924	2.4846	16
45	0.40275	0.91531	0.44001	2.2727	1.0925	2.4829	15
46	0.40301	0.91519	0.44036	2.2709	1.0927	2.4813	14
47	0.40328	0.91508	0.44070	2.2691	1.0928	2.4797	13
48	0.40354	0.91496	0.44105	2.2673	1.0929	2.4780	12
49	0.40381	0.91484	0.44140	2.2655	1.0931	2.4764	11
50	0.40408	0.91472	0.44175	2.2637	1.0932	2.4748	10
51	0.40434	0.91461	0.44209	2.2619	1.0934	2.4731	9
52	0.40461	0.91449	0.44244	2.2602	1.0935	2.4715	8
53	0.40487	0.91437	0.44279	2.2584	1.0936	2.4699	7
54	0.40514	0.91425	0.44314	2.2566	1.0938	2.4683	6
55	0.40541	0.91414	0.44349	2.2548	1.0939	2.4666	5
56	0.40567	0.91402	0.44383	2.2531	1.0941	2.4650	4
57	0.40594	0.91390	0.44418	2.2513	1.0924	2.4634	3
58	0.40620	0.91378	0.44453	2.2495	1.0943	2.4618	2
59	0.40647	0.91366	0.44488	2.2478	1.0945	2.4602	1
60	0.40674	0.91354	0.44523	2.2460	1.0946	2.4586	0
M	Cosine	Sine	Cotan.	Tan.	Cosec.	Secant	M

113° or 293°   **66° or 246°**

Natural Trigonometric Functions

M	Sine	Cosine	Tan.	Cotan.	Secant	Cosec.	M
0	0.40674	0.91354	0.44523	2.2460	1.0946	2.4586	60
1	0.40700	0.91343	0.44558	2.2443	1.0948	2.4570	59
2	0.40727	0.91331	0.44593	2.2425	1.0949	2.4554	58
3	0.40753	0.91319	0.44627	2.2408	1.0951	2.4538	57
4	0.40780	0.91307	0.44662	2.2390	1.0952	2.4522	56
5	0.40806	0.91295	0.44697	2.2373	1.0953	2.4506	55
6	0.40833	0.91283	0.44732	2.2355	1.0955	2.4490	54
7	0.40860	0.91271	0.44767	2.2338	1.0956	2.4474	53
8	0.40886	0.91260	0.44802	2.2320	1.0958	2.4458	52
9	0.40913	0.91248	0.44837	2.2303	1.0959	2.4442	51
10	0.40939	0.91236	0.44872	2.2286	1.0961	2.4426	50
11	0.40966	0.91224	0.44907	2.2268	1.0962	2.4411	49
12	0.40992	0.91212	0.44942	2.2251	1.0963	2.4395	48
13	0.41019	0.91200	0.44977	2.2234	1.0965	2.4379	47
14	0.41045	0.91188	0.45012	2.2216	1.0966	2.4363	46
15	0.41072	0.91176	0.45047	2.2199	1.0968	2.4347	45
16	0.41098	0.91164	0.45082	2.2182	1.0969	2.4332	44
17	0.41125	0.91152	0.45117	2.2165	1.0971	2.4316	43
18	0.41151	0.91140	0.45152	2.2147	1.0972	2.4300	42
19	0.41178	0.91128	0.45187	2.2130	1.0973	2.4285	41
20	0.41204	0.91116	0.45222	2.2113	1.0975	2.4269	40
21	0.41231	0.91104	0.45257	2.2096	1.0976	2.4254	39
22	0.41257	0.91092	0.45292	2.2079	1.0978	2.4238	38
23	0.41284	0.91080	0.45327	2.2062	1.0979	2.4222	37
24	0.41310	0.91068	0.45362	2.2045	1.0981	2.4207	36
25	0.41337	0.91056	0.45397	2.2028	1.0982	2.4191	35
26	0.41363	0.91044	0.45432	2.2011	1.0984	2.4176	34
27	0.41390	0.91032	0.45467	2.1994	1.0985	2.4160	33
28	0.41416	0.91020	0.45502	2.1977	1.0986	2.4145	32
29	0.41443	0.91008	0.45537	2.1960	1.0988	2.4130	31
30	0.41469	0.90996	0.45573	2.1943	1.0989	2.4114	30
31	0.41496	0.90984	0.45608	2.1926	1.0991	2.4099	29
32	0.41522	0.90972	0.45643	2.1909	1.0992	2.4083	28
33	0.41549	0.90960	0.45678	2.1892	1.0994	2.4068	27
34	0.41575	0.90948	0.45713	2.1875	1.0995	2.4053	26
35	0.41602	0.90936	0.45748	2.1859	1.0997	2.4037	25
36	0.41628	0.90924	0.45783	2.1842	1.0998	2.4022	24
37	0.41654	0.90911	0.45819	2.1825	1.1000	2.4007	23
38	0.41681	0.90899	0.45854	2.1808	1.1001	2.3992	22
39	0.41707	0.90887	0.45889	2.1792	1.1003	2.3976	21
40	0.41734	0.90875	0.45924	2.1775	1.1004	2.3961	20
41	0.41760	0.90863	0.45960	2.1758	1.1005	2.3946	19
42	0.41787	0.90851	0.45995	2.1741	1.1007	2.3931	18
43	0.41813	0.90839	0.46030	2.1725	1.1008	2.3916	17
44	0.41839	0.90826	0.46065	2.1708	1.1010	2.3901	16
45	0.41866	0.90814	0.46101	2.1692	1.1011	2.3886	15
46	0.41892	0.90802	0.46136	2.1675	1.1013	2.3871	14
47	0.41919	0.90790	0.46171	2.1658	1.1014	2.3856	13
48	0.41945	0.90778	0.46206	2.1642	1.1016	2.3841	12
49	0.41972	0.90765	0.46242	2.1625	1.1017	2.3826	11
50	0.41998	0.90753	0.46277	2.1609	1.1019	2.3811	10
51	0.42024	0.90741	0.46312	2.1592	1.1020	2.3796	9
52	0.42051	0.90729	0.46348	2.1576	1.1022	2.3781	8
53	0.42077	0.90717	0.46383	2.1559	1.1023	2.3766	7
54	0.42103	0.90704	0.46418	2.1543	1.1025	2.3751	6
55	0.42130	0.90692	0.46454	2.1527	1.1026	2.3736	5
56	0.42156	0.90680	0.46489	2.1510	1.1028	2.3721	4
57	0.42183	0.90668	0.46524	2.1494	1.1029	2.3706	3
58	0.42209	0.90655	0.46560	2.1478	1.1031	2.3691	2
59	0.42235	0.90643	0.46595	2.1461	1.1032	2.3677	1
60	0.42262	0.90631	0.46631	2.1445	1.1034	2.3662	0
M	Cosine	Sine	Cotan.	Tan.	Cosec.	Secant	M

25° or 205°   **Natural Trigonometric Functions**   **154° or 334°**

M	Sine	Cosine	Tan.	Cotan.	Secant	Cosec.	M
0	0.42262	0.90631	0.46631	2.1445	1.1034	2.3662	60
1	0.42288	0.90618	0.46666	2.1429	1.1035	2.3647	59
2	0.42314	0.90606	0.46702	2.1412	1.1037	2.3632	58
3	0.42341	0.90594	0.46737	2.1396	1.1038	2.3618	57
4	0.42367	0.90581	0.46772	2.1380	1.1040	2.3603	56
5	0.42394	0.90569	0.46808	2.1364	1.1041	2.3588	55
6	0.42420	0.90557	0.46843	2.1348	1.1043	2.3574	54
7	0.42446	0.90544	0.46879	2.1331	1.1044	2.3559	53
8	0.42473	0.90532	0.46914	2.1315	1.1046	2.3544	52
9	0.42499	0.90520	0.46950	2.1299	1.1047	2.3530	51
10	0.42525	0.90507	0.46985	2.1283	1.1049	2.3515	50
11	0.42552	0.90495	0.47021	2.1267	1.1050	2.3501	49
12	0.42578	0.90483	0.47056	2.1251	1.1052	2.3486	48
13	0.42604	0.90470	0.47092	2.1235	1.1053	2.3472	47
14	0.42630	0.90458	0.47127	2.1219	1.1055	2.3457	46
15	0.42657	0.90445	0.47163	2.1203	1.1056	2.3443	45
16	0.42683	0.90433	0.47199	2.1187	1.1058	2.3428	44
17	0.42709	0.90421	0.47234	2.1171	1.1059	2.3414	43
18	0.42736	0.90408	0.47270	2.1155	1.1061	2.3399	42
19	0.42762	0.90396	0.47305	2.1139	1.1062	2.3385	41
20	0.42788	0.90383	0.47341	2.1123	1.1064	2.3371	40
21	0.42815	0.90371	0.47376	2.1107	1.1065	2.3356	39
22	0.42841	0.90358	0.47412	2.1092	1.1067	2.3342	38
23	0.42867	0.90346	0.47448	2.1076	1.1068	2.3328	37
24	0.42893	0.90333	0.47483	2.1060	1.1070	2.3313	36
25	0.42920	0.90321	0.47519	2.1044	1.1072	2.3299	35
26	0.42946	0.90308	0.47555	2.1028	1.1073	2.3285	34
27	0.42972	0.90296	0.47590	2.1013	1.1075	2.3271	33
28	0.42998	0.90283	0.47626	2.0997	1.1076	2.3256	32
29	0.43025	0.90271	0.47662	2.0981	1.1078	2.3242	31
30	0.43051	0.90258	0.47697	2.0965	1.1079	2.3228	30
31	0.43077	0.90246	0.47733	2.0950	1.1081	2.3214	29
32	0.43104	0.90233	0.47769	2.0934	1.1082	2.3200	28
33	0.43130	0.90221	0.47805	2.0918	1.1084	2.3186	27
34	0.43156	0.90208	0.47840	2.0903	1.1085	2.3172	26
35	0.43182	0.90196	0.47876	2.0887	1.1087	2.3158	25
36	0.43208	0.90183	0.47912	2.0872	1.1088	2.3143	24
37	0.43235	0.90171	0.47948	2.0856	1.1090	2.3129	23
38	0.43261	0.90158	0.47983	2.0840	1.1092	2.3115	22
39	0.43287	0.90145	0.48019	2.0825	1.1093	2.3101	21
40	0.43313	0.90133	0.48055	2.0809	1.1095	2.3087	20
41	0.43340	0.90120	0.48091	2.0794	1.1096	2.3073	19
42	0.43366	0.90108	0.48127	2.0778	1.1098	2.3059	18
43	0.43392	0.90095	0.48162	2.0763	1.1099	2.3046	17
44	0.43418	0.90082	0.48198	2.0747	1.1101	2.3032	16
45	0.43444	0.90070	0.48234	2.0732	1.1102	2.3018	15
46	0.43471	0.90057	0.48270	2.0717	1.1104	2.3004	14
47	0.43497	0.90044	0.48306	2.0701	1.1106	2.2990	13
48	0.43523	0.90032	0.48342	2.0686	1.1107	2.2976	12
49	0.43549	0.90019	0.48378	2.0671	1.1109	2.2962	11
50	0.43575	0.90006	0.48414	2.0655	1.1110	2.2949	10
51	0.43602	0.89994	0.48449	2.0640	1.1112	2.2935	9
52	0.43628	0.89981	0.48485	2.0625	1.1113	2.2921	8
53	0.43654	0.89968	0.48521	2.0609	1.1115	2.2907	7
54	0.43680	0.89956	0.48557	2.0594	1.1116	2.2894	6
55	0.43706	0.89943	0.48593	2.0579	1.1118	2.2880	5
56	0.43732	0.89930	0.48629	2.0564	1.1120	2.2866	4
57	0.43759	0.89918	0.48665	2.0548	1.1121	2.2853	3
58	0.43785	0.89905	0.48701	2.0533	1.1123	2.2839	2
59	0.43811	0.89892	0.48737	2.0518	1.1124	2.2825	1
60	0.43837	0.89879	0.48773	2.0503	1.1126	2.2812	0
M	Cosine	Sine	Cotan.	Tan.	Cosec.	Secant	M

115° or 295°   **64° or 244°**

Natural Trigonometric Functions

M	Sine	Cosine	Tan.	Cotan.	Secant	Cosec.	M
0	0.43837	0.89879	0.48773	2.0503	1.1126	2.2812	60
1	0.43863	0.89867	0.48809	2.0488	1.1127	2.2798	59
2	0.43889	0.89854	0.48845	2.0473	1.1129	2.2784	58
3	0.43915	0.89841	0.48881	2.0458	1.1131	2.2771	57
4	0.43942	0.89828	0.48917	2.0443	1.1132	2.2757	56
5	0.43968	0.89815	0.48953	2.0427	1.1134	2.2744	55
6	0.43994	0.89803	0.48989	2.0412	1.1135	2.2730	54
7	0.44020	0.89790	0.49025	2.0397	1.1137	2.2717	53
8	0.44046	0.89777	0.49062	2.0382	1.1139	2.2703	52
9	0.44072	0.89764	0.49098	2.0367	1.1140	2.2690	51
10	0.44098	0.89751	0.49134	2.0352	1.1142	2.2676	50
11	0.44124	0.89739	0.49170	2.0338	1.1143	2.2663	49
12	0.44150	0.89726	0.49206	2.0323	1.1145	2.2650	48
13	0.44177	0.89713	0.49242	2.0308	1.1147	2.2636	47
14	0.44203	0.89700	0.49278	2.0293	1.1148	2.2623	46
15	0.44229	0.89687	0.49314	2.0278	1.1150	2.2610	45
16	0.44255	0.89674	0.49351	2.0263	1.1151	2.2596	44
17	0.44281	0.89661	0.49387	2.0248	1.1153	2.2583	43
18	0.44307	0.89649	0.49423	2.0233	1.1155	2.2570	42
19	0.44333	0.89636	0.49459	2.0219	1.1156	2.2556	41
20	0.44359	0.89623	0.49495	2.0204	1.1158	2.2543	40
21	0.44385	0.89610	0.49532	2.0189	1.1159	2.2530	39
22	0.44411	0.89597	0.49568	2.0174	1.1161	2.2517	38
23	0.44437	0.89584	0.49604	2.0159	1.1163	2.2503	37
24	0.44463	0.89571	0.49640	2.0145	1.1164	2.2490	36
25	0.44489	0.89558	0.49677	2.0130	1.1166	2.2477	35
26	0.44516	0.89545	0.49713	2.0115	1.1167	2.2464	34
27	0.44542	0.89532	0.49749	2.0101	1.1169	2.2451	33
28	0.44568	0.89519	0.49785	2.0086	1.1171	2.2438	32
29	0.44594	0.89506	0.49822	2.0071	1.1172	2.2425	31
30	0.44620	0.89493	0.49858	2.0057	1.1174	2.2411	30
31	0.44646	0.89480	0.49894	2.0042	1.1176	2.2398	29
32	0.44672	0.89467	0.49931	2.0028	1.1177	2.2385	28
33	0.44698	0.89454	0.49967	2.0013	1.1179	2.2372	27
34	0.44724	0.89441	0.50003	1.9998	1.1180	2.2359	26
35	0.44750	0.89428	0.50040	1.9984	1.1182	2.2346	25
36	0.44776	0.89415	0.50076	1.9969	1.1184	2.2333	24
37	0.44802	0.89402	0.50113	1.9955	1.1185	2.2320	23
38	0.44828	0.89398	0.50149	1.9940	1.1187	2.2307	22
39	0.44854	0.89376	0.50185	1.9926	1.1189	2.2294	21
40	0.44880	0.89363	0.50222	1.9912	1.1190	2.2282	20
41	0.44906	0.89350	0.50258	1.9897	1.1192	2.2269	19
42	0.44932	0.89337	0.50295	1.9883	1.1193	2.2256	18
43	0.44958	0.89324	0.50331	1.9868	1.1195	2.2243	17
44	0.44984	0.89311	0.50368	1.9854	1.1197	2.2230	16
45	0.45010	0.89298	0.50404	1.9840	1.1198	2.2217	15
46	0.45036	0.89285	0.50441	1.9825	1.1200	2.2204	14
47	0.45062	0.89272	0.50477	1.9811	1.1202	2.2192	13
48	0.45088	0.89258	0.50514	1.9797	1.1203	2.2179	12
49	0.45114	0.89245	0.50550	1.9782	1.1205	2.2166	11
50	0.45140	0.89232	0.50587	1.9768	1.1207	2.2153	10
51	0.45166	0.89219	0.50623	1.9754	1.1208	2.2141	9
52	0.45191	0.89206	0.50660	1.9739	1.1210	2.2128	8
53	0.45217	0.89193	0.50696	1.9725	1.1212	2.2115	7
54	0.45243	0.89180	0.50733	1.9711	1.1213	2.2103	6
55	0.45269	0.89166	0.50769	1.9697	1.1215	2.2090	5
56	0.45295	0.89153	0.50806	1.9683	1.1217	2.2077	4
57	0.45321	0.89140	0.50843	1.9668	1.1218	2.2065	3
58	0.45347	0.89127	0.50879	1.9654	1.1220	2.2052	2
59	0.45373	0.89114	0.50916	1.9640	1.1222	2.2039	1
60	0.45399	0.89101	0.50952	1.9626	1.1223	2.2027	0
M	Cosine	Sine	Cotan.	Tan.	Cosec.	Secant	M

27° or 207°   **Natural Trigonometric Functions**   **152° or 332°**

M	Sine	Cosine	Tan.	Cotan.	Secant	Cosec.	M
0	0.45399	0.89101	0.50952	1.9626	1.1223	2.2027	60
1	0.45425	0.89087	0.50989	1.9612	1.1225	2.2014	59
2	0.45451	0.89074	0.51026	1.9598	1.1226	2.2002	58
3	0.45477	0.89061	0.51062	1.9584	1.1228	2.1989	57
4	0.45503	0.89048	0.51099	1.9570	1.1230	2.1977	56
5	0.45528	0.89034	0.51136	1.9556	1.1231	2.1964	55
6	0.45554	0.89021	0.51172	1.9542	1.1233	2.1952	54
7	0.45580	0.89008	0.51209	1.9528	1.1235	2.1939	53
8	0.45606	0.88995	0.51246	1.9514	1.1237	2.1927	52
9	0.45632	0.88981	0.51283	1.9500	1.1238	2.1914	51
10	0.45658	0.88968	0.51319	1.9486	1.1240	2.1902	50
11	0.45684	0.88955	0.51356	1.9472	1.1242	2.1889	49
12	0.45710	0.88942	0.51393	1.9458	1.1243	2.1877	48
13	0.45736	0.88928	0.51430	1.9444	1.1245	2.1865	47
14	0.45761	0.88915	0.51466	1.9430	1.1247	2.1852	46
15	0.45787	0.88902	0.51503	1.9416	1.1248	2.1840	45
16	0.45813	0.88888	0.51540	1.9402	1.1250	2.1828	44
17	0.45839	0.88875	0.51577	1.9388	1.1252	2.1815	43
18	0.45865	0.88862	0.51614	1.9375	1.1253	2.1803	42
19	0.45891	0.88848	0.51651	1.9361	1.1255	2.1791	41
20	0.45917	0.88835	0.51687	1.9347	1.1257	2.1778	40
21	0.45942	0.88822	0.51724	1.9333	1.1258	2.1766	39
22	0.45968	0.88808	0.51761	1.9319	1.1260	2.1754	38
23	0.45994	0.88795	0.51798	1.9306	1.1262	2.1742	37
24	0.46020	0.88781	0.51835	1.9292	1.1264	2.1730	36
25	0.46046	0.88768	0.51872	1.9278	1.1265	2.1717	35
26	0.46072	0.88755	0.51909	1.9264	1.1267	2.1705	34
27	0.46079	0.88741	0.51946	1.9251	1.1269	2.1693	33
28	0.46123	0.88728	0.51983	1.9237	1.1270	2.1681	32
29	0.46149	0.88714	0.52020	1.9223	1.1272	2.1669	31
30	0.46175	0.88701	0.52057	1.9210	1.1274	2.1657	30
31	0.46201	0.88688	0.52094	1.9196	1.1275	2.1645	29
32	0.46226	0.88674	0.52131	1.9182	1.1277	2.1633	28
33	0.46252	0.88661	0.52168	1.9169	1.1279	2.1620	27
34	0.46278	0.88647	0.52205	1.9155	1.1281	2.1608	26
35	0.46304	0.88634	0.52242	1.9142	1.1282	2.1596	25
36	0.46330	0.88620	0.52279	1.9128	1.1284	2.1584	24
37	0.46355	0.88607	0.52316	1.9115	1.1286	2.1572	23
38	0.46381	0.88593	0.52353	1.9101	1.1287	2.1560	22
39	0.46407	0.88580	0.52390	1.9088	1.1289	2.1548	21
40	0.46433	0.88566	0.52427	1.9074	1.1291	2.1536	20
41	0.46458	0.88553	0.52464	1.9061	1.1293	2.1525	19
42	0.46484	0.88539	0.52501	1.9047	1.1294	2.1513	18
43	0.46510	0.88526	0.52538	1.9034	1.1296	2.1501	17
44	0.46536	0.88512	0.52575	1.9020	1.1298	2.1489	16
45	0.46561	0.88499	0.52612	1.9007	1.1299	2.1477	15
46	0.46587	0.88485	0.52650	1.8993	1.1301	2.1465	14
47	0.46613	0.88472	0.52687	1.8980	1.1303	2.1453	13
48	0.46639	0.88458	0.52724	1.8967	1.1305	2.1441	12
49	0.46664	0.88444	0.52761	1.8953	1.1306	2.1430	11
50	0.46690	0.88431	0.52798	1.8940	1.1308	2.1418	10
51	0.46716	0.88417	0.52836	1.8927	1.1310	2.1406	9
52	0.46741	0.88404	0.52873	1.8913	1.1312	2.1394	8
53	0.46767	0.88390	0.52910	1.8900	1.1313	2.1382	7
54	0.46793	0.88376	0.52947	1.8887	1.1315	2.1371	6
55	0.46819	0.88363	0.52984	1.8873	1.1317	2.1359	5
56	0.46844	0.88349	0.53022	1.8860	1.1319	2.1347	4
57	0.46870	0.88336	0.53059	1.8847	1.1320	2.1335	3
58	0.46896	0.88322	0.53096	1.8834	1.1322	2.1324	2
59	0.46921	0.88308	0.53134	1.8820	1.1324	2.1312	1
60	0.46947	0.88295	0.53171	1.8807	1.1326	2.1300	0
M	Cosine	Sine	Cotan.	Tan.	Cosec.	Secant	M

117° or 297°   **62° or 242°**

　　Natural Trigonometric Functions　　

M	Sine	Cosine	Tan.	Cotan.	Secant	Cosec.	M
0	0.46947	0.88295	0.53171	1.8807	1.1326	2.1300	60
1	0.46973	0.88281	0.53208	1.8794	1.1327	2.1289	59
2	0.46998	0.88267	0.53245	1.8781	1.1329	2.1277	58
3	0.47024	0.88254	0.53283	1.8768	1.1331	2.1266	57
4	0.47050	0.88240	0.53320	1.8754	1.1333	2.1254	56
5	0.47075	0.88226	0.53358	1.8741	1.1334	2.1242	55
6	0.47101	0.88213	0.53395	1.8728	1.1336	2.1231	54
7	0.47127	0.88199	0.53432	1.8715	1.1338	2.1219	53
8	0.47152	0.88185	0.53470	1.8702	1.1340	2.1208	52
9	0.47178	0.88171	0.53507	1.8689	1.1341	2.1196	51
10	0.47204	0.88158	0.53545	1.8676	1.1343	2.1185	50
11	0.47229	0.88144	0.53582	1.8663	1.1345	2.1173	49
12	0.47255	0.88130	0.53619	1.8650	1.1347	2.1162	48
13	0.47281	0.88117	0.53657	1.8637	1.1349	2.1150	47
14	0.47306	0.88103	0.53694	1.8624	1.1350	2.1139	46
15	0.47332	0.88089	0.53732	1.8611	1.1352	2.1127	45
16	0.47357	0.88075	0.53769	1.8598	1.1354	2.1116	44
17	0.47383	0.88061	0.53807	1.8585	1.1356	2.1104	43
18	0.47409	0.88048	0.53844	1.8572	1.1357	2.1093	42
19	0.47434	0.88034	0.53882	1.8559	1.1359	2.1082	41
20	0.47460	0.88020	0.53919	1.8546	1.1361	2.1070	40
21	0.47486	0.88006	0.53957	1.8533	1.1363	2.1059	39
22	0.47511	0.87992	0.53995	1.8520	1.1365	2.1048	38
23	0.47537	0.87979	0.54032	1.8507	1.1366	2.1036	37
24	0.47562	0.87965	0.54070	1.8495	1.1368	2.1025	36
25	0.47588	0.87951	0.54107	1.8482	1.1370	2.1014	35
26	0.47613	0.87937	0.54145	1.8469	1.1372	2.1002	34
27	0.47639	0.87923	0.54183	1.8456	1.1373	2.0991	33
28	0.47665	0.87909	0.54220	1.8443	1.1375	2.0980	32
29	0.47690	0.87895	0.54258	1.8430	1.1377	2.0969	31
30	0.47716	0.87882	0.54295	1.8418	1.1379	2.0957	30
31	0.47741	0.87868	0.54333	1.8405	1.1381	2.0946	29
32	0.47767	0.87854	0.54371	1.8392	1.1382	2.0935	28
33	0.47792	0.87840	0.54409	1.8379	1.1384	2.0924	27
34	0.47818	0.87826	0.54446	1.8367	1.1386	2.0912	26
35	0.47844	0.87812	0.54484	1.8354	1.1388	2.0901	25
36	0.47869	0.87798	0.54522	1.8341	1.1390	2.0890	24
37	0.47895	0.87784	0.54559	1.8329	1.1391	2.0879	23
38	0.47920	0.87770	0.54597	1.8316	1.1393	2.0868	22
39	0.47946	0.87756	0.54635	1.8303	1.1395	2.0857	21
40	0.47971	0.87742	0.54673	1.8291	1.1397	2.0846	20
41	0.47997	0.87728	0.54711	1.8278	1.1399	2.0835	19
42	0.48022	0.87715	0.54748	1.8265	1.1401	2.0824	18
43	0.48048	0.87701	0.54786	1.8253	1.1402	2.0812	17
44	0.48073	0.87687	0.54824	1.8240	1.1404	2.0801	16
45	0.48099	0.87673	0.54862	1.8227	1.1406	2.0790	15
46	0.48124	0.87659	0.54900	1.8215	1.1408	2.0779	14
47	0.48150	0.87645	0.54937	1.8202	1.1410	2.0768	13
48	0.48175	0.87631	0.54975	1.8190	1.1411	2.0757	12
49	0.48201	0.87617	0.55013	1.8177	1.1413	2.0746	11
50	0.48226	0.87603	0.55051	1.8165	1.1415	2.0735	10
51	0.48252	0.87588	0.55089	1.8152	1.1417	2.0725	9
52	0.48277	0.87574	0.55127	1.8140	1.1419	2.0714	8
53	0.48303	0.87560	0.55165	1.8127	1.1421	2.0703	7
54	0.48328	0.87546	0.55203	1.8115	1.1422	2.0692	6
55	0.48354	0.87532	0.55241	1.8102	1.1424	2.0681	5
56	0.48379	0.87518	0.55279	1.8090	1.1426	2.0670	4
57	0.48405	0.87504	0.55317	1.8078	1.1428	2.0659	3
58	0.48430	0.87490	0.55355	1.8065	1.1430	2.0648	2
59	0.48455	0.87476	0.55393	1.8053	1.1432	2.0637	1
60	0.48481	0.87462	0.55431	1.8040	1.1433	2.0627	0
M	Cosine	Sine	Cotan.	Tan.	Cosec.	Secant	M

29° or 209° **Natural Trigonometric Functions** **150° or 330°**

M	Sine	Cosine	Tan.	Cotan.	Secant	Cosec.	M
0	0.48481	0.87462	0.55431	1.8040	1.1433	2.0627	60
1	0.48506	0.87448	0.55469	1.8028	1.1435	2.0616	59
2	0.48532	0.87434	0.55507	1.8016	1.1437	2.0605	58
3	0.48557	0.87420	0.55545	1.8003	1.1439	2.0594	57
4	0.48583	0.87405	0.55583	1.7991	1.1441	2.0583	56
5	0.48608	0.87391	0.55621	1.7979	1.1443	2.0573	55
6	0.48633	0.87377	0.55659	1.7966	1.1445	2.0562	54
7	0.48659	0.87363	0.55697	1.7954	1.1446	2.0551	53
8	0.48684	0.87349	0.55735	1.7942	1.1448	2.0540	52
9	0.48710	0.87335	0.55774	1.7930	1.1450	2.0530	51
10	0.48735	0.87320	0.55812	1.7917	1.1452	2.0519	50
11	0.48760	0.87306	0.55850	1.7905	1.1454	2.0508	49
12	0.48786	0.87292	0.55888	1.7893	1.1456	2.0498	48
13	0.48811	0.87278	0.55926	1.7881	1.1458	2.0487	47
14	0.48837	0.87264	0.55964	1.7868	1.1459	2.0476	46
15	0.48862	0.87250	0.56003	1.7856	1.1461	2.0466	45
16	0.48887	0.87235	0.56041	1.7844	1.1463	2.0455	44
17	0.48913	0.87221	0.56079	1.7832	1.1465	2.0444	43
18	0.48938	0.87207	0.56117	1.7820	1.1467	2.0434	42
19	0.48964	0.87193	0.56156	1.7808	1.1469	2.0423	41
20	0.48989	0.87178	0.56194	1.7795	1.1471	2.0413	40
21	0.49014	0.87164	0.56232	1.7783	1.1473	2.0402	39
22	0.49040	0.87150	0.56270	1.7771	1.1474	2.0392	38
23	0.49065	0.87136	0.56309	1.7759	1.1476	2.0381	37
24	0.49090	0.87121	0.56347	1.7747	1.1478	2.0370	36
25	0.49116	0.87107	0.56385	1.7735	1.1480	2.0360	35
26	0.49141	0.87093	0.56424	1.7723	1.1482	2.0349	34
27	0.49166	0.87078	0.56462	1.7711	1.1484	2.0339	33
28	0.49192	0.87064	0.56500	1.7699	1.1486	2.0329	32
29	0.49217	0.87050	0.56539	1.7687	1.1488	2.0318	31
30	0.49242	0.87035	0.56577	1.7675	1.1489	2.0308	30
31	0.49268	0.87021	0.56616	1.7663	1.1491	2.0297	29
32	0.49293	0.87007	0.56654	1.7651	1.1493	2.0287	28
33	0.49318	0.86992	0.56692	1.7639	1.1495	2.0276	27
34	0.49343	0.86978	0.56731	1.7627	1.1497	2.0266	26
35	0.49369	0.86964	0.56769	1.7615	1.1499	2.0256	25
36	0.49394	0.86949	0.56808	1.7603	1.1501	2.0245	24
37	0.49419	0.86935	0.56846	1.7591	1.1503	2.0235	23
38	0.49445	0.86921	0.56885	1.7579	1.1505	2.0224	22
39	0.49470	0.86906	0.56923	1.7567	1.1507	2.0214	21
40	0.49495	0.86892	0.56962	1.7555	1.1508	2.0204	20
41	0.49521	0.86877	0.57000	1.7544	1.1510	2.0194	19
42	0.49546	0.86863	0.57039	1.7532	1.1512	2.0183	18
43	0.49571	0.86849	0.57077	1.7520	1.1514	2.0173	17
44	0.49596	0.86834	0.57116	1.7508	1.1516	2.0163	16
45	0.49622	0.86820	0.57155	1.7496	1.1518	2.0152	15
46	0.49647	0.86805	0.57193	1.7484	1.1520	2.0142	14
47	0.49672	0.86791	0.57232	1.7473	1.1522	2.0132	13
48	0.49697	0.86776	0.57270	1.7461	1.1524	2.0122	12
49	0.49723	0.86762	0.57309	1.7449	1.1526	2.0111	11
50	0.49748	0.86748	0.57348	1.7437	1.1528	2.0101	10
51	0.49773	0.86733	0.57368	1.7426	1.1530	2.0091	9
52	0.49798	0.86719	0.57425	1.7414	1.1531	2.0081	8
53	0.49823	0.86704	0.57464	1.7402	1.1533	2.0071	7
54	0.49849	0.86690	0.57503	1.7390	1.1535	2.0061	6
55	0.49874	0.86675	0.57541	1.7379	1.1537	2.0050	5
56	0.49899	0.86661	0.57580	1.7367	1.1539	2.0040	4
57	0.49924	0.86646	0.57619	1.7355	1.1541	2.0030	3
58	0.49950	0.86632	0.57657	1.7344	1.1543	2.0020	2
59	0.49975	0.86617	0.57696	1.7332	1.1545	2.0010	1
60	0.50000	0.86603	0.577355	1.7320	1.1547	2.0000	0
M	Cosine	Sine	Cotan.	Tan.	Cosec.	Secant	M

119° or 299° **60° or 240°**

30° or 210°   **Natural Trigonometric Functions**   **149° or 329°**

M	Sine	Cosine	Tan.	Cotan.	Secant	Cosec.	M
0	0.50000	0.86603	0.57735	1.7320	1.1547	2.0000	60
1	0.50025	0.86588	0.57774	1.7309	1.1549	1.9990	59
2	0.50050	0.86573	0.57813	1.7297	1.1551	1.9980	58
3	0.50075	0.86559	0.57851	1.7286	1.1553	1.9970	57
4	0.50101	0.86544	0.57890	1.7274	1.1555	1.9960	56
5	0.50126	0.86530	0.57929	1.7262	1.1557	1.9950	55
6	0.50151	0.86515	0.57968	1.7251	1.1559	1.9940	54
7	0.50176	0.86500	0.58007	1.7239	1.1561	1.9930	53
8	0.50201	0.86486	0.58046	1.7228	1.1562	1.9920	52
9	0.50226	0.86471	0.58085	1.7216	1.1564	1.9910	51
10	0.50252	0.86457	0.58123	1.7205	1.1566	1.9900	50
11	0.50277	0.86442	0.58162	1.7193	1.1568	1.9890	49
12	0.50302	0.86427	0.58201	1.7182	1.1570	1.9880	48
13	0.50327	0.86413	0.58240	1.7170	1.1572	1.9870	47
14	0.50352	0.86398	0.58279	1.7159	1.1574	1.9860	46
15	0.50377	0.86383	0.58318	1.7147	1.1576	1.9850	45
16	0.50402	0.86369	0.58357	1.7136	1.1578	1.9840	44
17	0.50428	0.86354	0.58396	1.7124	1.1580	1.9830	43
18	0.50453	0.86339	0.58435	1.7113	1.1582	1.9820	42
19	0.50478	0.86325	0.58474	1.7101	1.1584	1.9811	41
20	0.50503	0.86310	0.58513	1.7090	1.1586	1.9801	40
21	0.50528	0.86295	0.58552	1.7079	1.1588	1.9791	39
22	0.50553	0.86281	0.58591	1.7067	1.1590	1.9781	38
23	0.50578	0.86266	0.58630	1.7056	1.1592	1.9771	37
24	0.50603	0.86251	0.58670	1.7044	1.1594	1.9761	36
25	0.50628	0.86237	0.58709	1.7033	1.1596	1.9752	35
26	0.50653	0.86222	0.58748	1.7022	1.1598	1.9742	34
27	0.50679	0.86207	0.58787	1.7010	1.1600	1.9732	33
28	0.50704	0.86192	0.58826	1.6999	1.1602	1.9722	32
29	0.50729	0.86178	0.58865	1.6988	1.1604	1.9713	31
30	0.50754	0.86163	0.58904	1.6977	1.1606	1.9703	30
31	0.50779	0.86148	0.58944	1.6965	1.1608	1.9693	29
32	0.50804	0.86133	0.58983	1.6954	1.1610	1.9683	28
33	0.50829	0.86118	0.59022	1.6943	1.1612	1.9674	27
34	0.50854	0.86104	0.59061	1.6931	1.1614	1.9664	26
35	0.50879	0.86089	0.59100	1.6920	1.1616	1.9654	25
36	0.50904	0.86074	0.59140	1.6909	1.1618	1.9645	24
37	0.50929	0.86059	0.59179	1.6898	1.1620	1.9635	23
38	0.50954	0.86044	0.59218	1.6887	1.1622	1.9625	22
39	0.50979	0.86030	0.59258	1.6875	1.1624	1.9616	21
40	0.51004	0.86015	0.59297	1.6864	1.1626	1.9606	20
41	0.51029	0.86000	0.59336	1.6853	1.1628	1.9596	19
42	0.51054	0.85985	0.59376	1.6842	1.1630	1.9587	18
43	0.51079	0.85970	0.59415	1.6831	1.1632	1.9577	17
44	0.51104	0.85955	0.59454	1.6820	1.1634	1.9568	16
45	0.51129	0.85941	0.59494	1.6808	1.1636	1.9558	15
46	0.51154	0.85926	0.59533	1.6797	1.1638	1.9549	14
47	0.51179	0.85911	0.59572	1.6786	1.1640	1.9539	13
48	0.51204	0.85896	0.59612	1.6775	1.1642	1.9530	12
49	0.51229	0.85881	0.59651	1.6764	1.1644	1.9520	11
50	0.51254	0.85866	0.59691	1.6753	1.1646	1.9510	10
51	0.51279	0.85851	0.59730	1.6742	1.1648	1.9501	9
52	0.51304	0.85836	0.59770	1.6731	1.1650	1.9491	8
53	0.51329	0.85821	0.59809	1.6720	1.1652	1.9482	7
54	0.51354	0.85806	0.59849	1.6709	1.1654	1.9473	6
55	0.51379	0.85791	0.59888	1.6698	1.1656	1.9463	5
56	0.51404	0.85777	0.59928	1.6687	1.1658	1.9454	4
57	0.51429	0.85762	0.59967	1.6676	1.1660	1.9444	3
58	0.51454	0.85747	0.60007	1.6665	1.1662	1.9435	2
59	0.51479	0.85732	0.60046	1.6654	1.1664	1.9425	1
60	0.51504	0.85717	0.600865	1.6643	1.1666	1.9416	0
M	Cosine	Sine	Cotan.	Tan.	Cosec.	Secant	M

31° or 211°   **Natural Trigonometric Functions**   **148° or 328°**

M	Sine	Cosine	Tan.	Cotan.	Secant	Cosec.	M
0	0.51504	0.85717	0.60086	1.6643	1.1666	1.9416	60
1	0.51529	0.85702	0.60126	1.6632	1.1668	1.9407	59
2	0.51554	0.85687	0.60165	1.6621	1.1670	1.9397	58
3	0.51578	0.85672	0.60205	1.6610	1.1672	1.9388	57
4	0.51603	0.85657	0.60244	1.6599	1.1674	1.9378	56
5	0.51628	0.85642	0.60284	1.6588	1.1676	1.9369	55
6	0.51653	0.85627	0.60324	1.6577	1.1678	1.9360	54
7	0.51678	0.85612	0.60363	1.6566	1.1681	1.9350	53
8	0.51703	0.85597	0.60403	1.6555	1.1683	1.9341	52
9	0.51728	0.85582	0.60443	1.6544	1.1685	1.9332	51
10	0.51753	0.85566	0.60483	1.6534	1.1687	1.9322	50
11	0.51778	0.85551	0.60522	1.6523	1.1689	1.9313	49
12	0.51803	0.85536	0.60562	1.6512	1.1691	1.9304	48
13	0.51827	0.85521	0.60602	1.6501	1.1693	1.9295	47
14	0.51852	0.85506	0.60642	1.6490	1.1695	1.9285	46
15	0.51877	0.85491	0.60681	1.6479	1.1697	1.9276	45
16	0.51902	0.85476	0.60721	1.6469	1.1699	1.9267	44
17	0.51927	0.85461	0.60761	1.6458	1.1701	1.9258	43
18	0.51952	0.85446	0.60801	1.6447	1.1703	1.9248	42
19	0.51977	0.85431	0.60841	1.6436	1.1705	1.9239	41
20	0.52002	0.85416	0.60881	1.6425	1.1707	1.9230	40
21	0.52026	0.85400	0.60920	1.6415	1.1709	1.9221	39
22	0.52051	0.85385	0.60960	1.6404	1.1712	1.9212	38
23	0.52076	0.85370	0.61000	1.6393	1.1714	1.9203	37
24	0.52101	0.85355	0.61040	1.6383	1.1716	1.9193	36
25	0.52126	0.85340	0.61080	1.6372	1.1718	1.9184	35
26	0.52151	0.85325	0.61120	1.6361	1.1720	1.9175	34
27	0.52175	0.85309	0.61160	1.6350	1.1722	1.9166	33
28	0.52200	0.85294	0.61200	1.6340	1.1724	1.9157	32
29	0.52225	0.85279	0.61240	1.6329	1.1726	1.9148	31
30	0.52250	0.85264	0.61280	1.6318	1.1728	1.9139	30
31	0.52275	0.85249	0.61320	1.6308	1.1730	1.9130	29
32	0.52299	0.85234	0.61360	1.6297	1.1732	1.9121	28
33	0.52324	0.85218	0.61400	1.6286	1.1734	1.9112	27
34	0.52349	0.85203	0.61440	1.6276	1.1737	1.9102	26
35	0.52374	0.85188	0.61480	1.6265	1.1739	1.9093	25
36	0.52398	0.85173	0.61520	1.6255	1.1741	1.9084	24
37	0.52423	0.85157	0.61560	1.6244	1.1743	1.9075	23
38	0.52448	0.85142	0.61601	1.6233	1.1745	1.9066	22
39	0.52473	0.85127	0.61641	1.6223	1.1747	1.9057	21
40	0.52498	0.85112	0.61681	1.6212	1.1749	1.9048	20
41	0.52522	0.85096	0.61721	1.6202	1.1751	1.9039	19
42	0.52547	0.85081	0.61761	1.6191	1.1753	1.9030	18
43	0.52572	0.85066	0.61801	1.6181	1.1756	1.9021	17
44	0.52597	0.85050	0.61842	1.6170	1.1758	1.9013	16
45	0.52621	0.85035	0.61882	1.6160	1.1760	1.9004	15
46	0.52646	0.85020	0.61922	1.6149	1.1762	1.8995	14
47	0.52671	0.85004	0.61962	1.6139	1.1764	1.8986	13
48	0.52695	0.84989	0.62003	1.6128	1.1766	1.8977	12
49	0.52720	0.84974	0.62043	1.6118	1.1768	1.8968	11
50	0.52745	0.84959	0.62083	1.6107	1.1770	1.8959	10
51	0.52770	0.84943	0.62123	1.6097	1.1772	1.8950	9
52	0.52794	0.84928	0.62164	1.6086	1.1775	1.8941	8
53	0.52819	0.84912	0.62204	1.6076	1.1777	1.8932	7
54	0.52844	0.84897	0.62244	1.6066	1.1779	1.8924	6
55	0.52868	0.84882	0.62285	1.6055	1.1781	1.8915	5
56	0.52893	0.84866	0.62325	1.6045	1.1783	1.8906	4
57	0.52918	0.84851	0.62366	1.6034	1.1785	1.8897	3
58	0.52942	0.84836	0.62406	1.6024	1.1787	1.8888	2
59	0.52967	0.84820	0.62446	1.6014	1.1790	1.8879	1
60	0.52992	0.84805	0.62487	1.6003	1.1792	1.8871	0
M	Cosine	Sine	Cotan.	Tan.	Cosec.	Secant	M

121° or 301°   **58° or 238°**

    Natural Trigonometric Functions

M	Sine	Cosine	Tan.	Cotan.	Secant	Cosec.	M
0	0.52992	0.84805	0.62487	1.6003	1.1792	1.8871	60
1	0.53016	0.84789	0.62527	1.5993	1.1794	1.8862	59
2	0.53041	0.84774	0.62568	1.5983	1.1796	1.8853	58
3	0.53066	0.84758	0.62608	1.5972	1.1798	1.8844	57
4	0.53090	0.84743	0.62649	1.5962	1.1800	1.8836	56
5	0.53115	0.84728	0.62689	1.5952	1.1802	1.8827	55
6	0.53140	0.84712	0.62730	1.5941	1.1805	1.8818	54
7	0.53164	0.84697	0.62770	1.5931	1.1807	1.8809	53
8	0.53189	0.84681	0.62811	1.5921	1.1809	1.8801	52
9	0.53214	0.84666	0.62851	1.5910	1.1811	1.8792	51
10	0.53238	0.84650	0.62892	1.5900	1.1813	1.8783	50
11	0.53263	0.84635	0.62933	1.5890	1.1815	1.8775	49
12	0.53288	0.84619	0.62973	1.5880	1.1818	1.8766	48
13	0.53312	0.84604	0.63014	1.5869	1.1820	1.8757	47
14	0.53337	0.84588	0.63055	1.5859	1.1822	1.8749	46
15	0.53361	0.84573	0.63095	1.5849	1.1824	1.8740	45
16	0.53386	0.84557	0.63136	1.5839	1.1826	1.8731	44
17	0.53411	0.84542	0.63177	1.5829	1.1828	1.8723	43
18	0.53435	0.84526	0.63217	1.5818	1.1831	1.8714	42
19	0.53460	0.84511	0.63258	1.5808	1.1833	1.8706	41
20	0.53484	0.84495	0.63299	1.5798	1.1835	1.8697	40
21	0.53509	0.84479	0.63339	1.5788	1.1837	1.8688	39
22	0.53533	0.84464	0.63380	1.5778	1.1839	1.8680	38
23	0.53558	0.84448	0.63421	1.5768	1.1841	1.8671	37
24	0.53583	0.84433	0.63462	1.5757	1.1844	1.8663	36
25	0.53607	0.84417	0.63503	1.5747	1.1846	1.8654	35
26	0.53632	0.84402	0.63543	1.5737	1.1848	1.8646	34
27	0.53656	0.84386	0.63584	1.5727	1.1850	1.8637	33
28	0.53681	0.84370	0.63625	1.5717	1.1852	1.8629	32
29	0.53705	0.84355	0.63666	1.5707	1.1855	1.8620	31
30	0.53730	0.84339	0.63707	1.5697	1.1857	1.8611	30
31	0.53754	0.84323	0.63748	1.5687	1.1859	1.8603	29
32	0.53779	0.84308	0.63789	1.5677	1.1861	1.8595	28
33	0.53803	0.84292	0.63830	1.5667	1.1863	1.8586	27
34	0.53828	0.84276	0.63871	1.5657	1.1866	1.8578	26
35	0.53852	0.84261	0.63912	1.5646	1.1868	1.8569	25
36	0.53877	0.84245	0.63953	1.5636	1.1870	1.8561	24
37	0.53901	0.84229	0.63994	1.5626	1.1872	1.8552	23
38	0.53926	0.84214	0.64035	1.5616	1.1874	1.8544	22
39	0.53950	0.84198	0.64076	1.5606	1.1877	1.8535	21
40	0.53975	0.84182	0.64117	1.5596	1.1879	1.8527	20
41	0.53999	0.84167	0.64158	1.5586	1.1881	1.8519	19
42	0.54024	0.84151	0.64199	1.5577	1.1883	1.8510	18
43	0.54048	0.84135	0.64240	1.5567	1.1886	1.8502	17
44	0.54073	0.84120	0.64281	1.5557	1.1888	1.8493	16
45	0.54097	0.84104	0.64322	1.5547	1.1890	1.8485	15
46	0.54122	0.84088	0.64363	1.5537	1.1892	1.8477	14
47	0.54146	0.84072	0.64404	1.5527	1.1894	1.8468	13
48	0.54171	0.84057	0.64446	1.5517	1.1897	1.8460	12
49	0.54195	0.84041	0.64487	1.5507	1.1899	1.8452	11
50	0.54220	0.84025	0.64528	1.5497	1.1901	1.8443	10
51	0.54244	0.84009	0.64569	1.5487	1.1903	1.8435	9
52	0.54268	0.83993	0.64610	1.5477	1.1906	1.8427	8
53	0.54293	0.83978	0.64652	1.5467	1.1908	1.8418	7
54	0.54317	0.83962	0.64693	1.5458	1.1910	1.8410	6
55	0.54342	0.83946	0.64734	1.5448	1.1912	1.8402	5
56	0.54366	0.83930	0.64775	1.5438	1.1915	1.8394	4
57	0.54391	0.83914	0.64817	1.5428	1.1917	1.8385	3
58	0.54415	0.83899	0.64858	1.5418	1.1919	1.8377	2
59	0.54439	0.83883	0.64899	1.5408	1.1921	1.8369	1
60	0.54464	0.83867	0.64941	1.5399	1.1924	1.8361	0
M	Cosine	Sine	Cotan.	Tan.	Cosec.	Secant	M

33° or 213°   **Natural Trigonometric Functions**   **146° or 326°**

M	Sine	Cosine	Tan.	Cotan.	Secant	Cosec.	M
0	0.54464	0.83867	0.64941	1.5399	1.1924	1.8361	60
1	0.54488	0.83851	0.64982	1.5389	1.1926	1.8352	59
2	0.54513	0.83835	0.65023	1.5379	1.1928	1.8344	58
3	0.54537	0.83819	0.65065	1.5369	1.1930	1.8336	57
4	0.54561	0.83804	0.65106	1.5359	1.1933	1.8328	56
5	0.54586	0.83788	0.65148	1.5350	1.1935	1.8320	55
6	0.54610	0.83772	0.65189	1.5340	1.1937	1.8311	54
7	0.54634	0.83756	0.65231	1.5330	1.1939	1.8303	53
8	0.54659	0.83740	0.65272	1.5320	1.1942	1.8295	52
9	0.54683	0.83724	0.65314	1.5311	1.1944	1.8287	51
10	0.54708	0.83708	0.65355	1.5301	1.1946	1.8279	50
11	0.54732	0.83692	0.65397	1.5291	1.1948	1.8271	49
12	0.54756	0.83676	0.65438	1.5282	1.1951	1.8263	48
13	0.54781	0.83660	0.65480	1.5272	1.1953	1.8255	47
14	0.54805	0.83644	0.65521	1.5262	1.1955	1.8246	46
15	0.54829	0.83629	0.65563	1.5252	1.1958	1.8238	45
16	0.54854	0.83613	0.65604	1.5243	1.1960	1.8230	44
17	0.54878	0.83597	0.65646	1.5233	1.1962	1.8222	43
18	0.54902	0.83581	0.65688	1.5223	1.1964	1.8214	42
19	0.54926	0.83565	0.65729	1.5214	1.1967	1.8206	41
20	0.54951	0.83549	0.65771	1.5204	1.1969	1.8198	40
21	0.54975	0.83533	0.65813	1.5195	1.1971	1.8190	39
22	0.54999	0.83517	0.65854	1.5185	1.1974	1.8182	38
23	0.55024	0.83501	0.65896	1.5175	1.1976	1.8174	37
24	0.55048	0.83485	0.65938	1.5166	1.1978	1.8166	36
25	0.55072	0.83469	0.65980	1.5156	1.1980	1.8158	35
26	0.55097	0.83453	0.66021	1.5147	1.1983	1.8150	34
27	0.55121	0.83437	0.66063	1.5137	1.1985	1.8142	33
28	0.55145	0.83421	0.66105	1.5127	1.1987	1.8134	32
29	0.55169	0.83405	0.66147	1.5118	1.1990	1.8126	31
30	0.55194	0.83388	0.66188	1.5108	1.1992	1.8118	30
31	0.55218	0.83372	0.66230	1.5099	1.1994	1.8110	29
32	0.55242	0.83356	0.66272	1.5089	1.1997	1.8102	28
33	0.55266	0.83340	0.66314	1.5080	1.1999	1.8094	27
34	0.55291	0.83324	0.66356	1.5070	1.2001	1.8086	26
35	0.55315	0.83308	0.66398	1.5061	1.2004	1.8078	25
36	0.55339	0.83292	0.66440	1.5051	1.2006	1.8070	24
37	0.55363	0.83276	0.66482	1.5042	1.2008	1.8062	23
38	0.55388	0.83260	0.66524	1.5032	1.2010	1.8054	22
39	0.55412	0.83244	0.66566	1.5023	1.2013	1.8047	21
40	0.55436	0.83228	0.66608	1.5013	1.2015	1.8039	20
41	0.55460	0.83211	0.66650	1.5004	1.2017	1.8031	19
42	0.55484	0.83195	0.66692	1.4994	1.2020	1.8023	18
43	0.55509	0.83179	0.66734	1.4985	1.2022	1.8015	17
44	0.55533	0.83163	0.66776	1.4975	1.2024	1.8007	16
45	0.55557	0.83147	0.66818	1.4966	1.2027	1.7999	15
46	0.55581	0.83131	0.66860	1.4957	1.2029	1.7992	14
47	0.55605	0.83115	0.66902	1.4947	1.2031	1.7984	13
48	0.55629	0.83098	0.66944	1.4938	1.2034	1.7976	12
49	0.55654	0.83082	0.66986	1.4928	1.2036	1.7968	11
50	0.55678	0.83066	0.67028	1.4919	1.2039	1.7960	10
51	0.55702	0.83050	0.67071	1.4910	1.2041	1.7953	9
52	0.55726	0.83034	0.67113	1.4900	1.2043	1.7945	8
53	0.55750	0.83017	0.67155	1.4891	1.2046	1.7937	7
54	0.55774	0.83001	0.67197	1.4881	1.2048	1.7929	6
55	0.55799	0.82985	0.67239	1.4872	1.2050	1.7921	5
56	0.55823	0.82969	0.67282	1.4863	1.2053	1.7914	4
57	0.55847	0.82952	0.67324	1.4853	1.2055	1.7906	3
58	0.55871	0.82936	0.67366	1.4844	1.2057	1.7898	2
59	0.55895	0.82920	0.67408	1.4835	1.2060	1.7891	1
60	0.55919	0.82904	0.67451	1.4826	1.2062	1.7883	0
M	Cosine	Sine	Cotan.	Tan.	Cosec.	Secant	M

123° or 303°   **56° or 236°**

M	Sine	Cosine	Tan.	Cotan.	Secant	Cosec.	M
0	0.55919	0.82904	0.67451	1.4826	1.2062	1.7883	60
1	0.55943	0.82887	0.67493	1.4816	1.2064	1.7875	59
2	0.55967	0.82871	0.67535	1.4807	1.2067	1.7867	58
3	0.55992	0.82855	0.67578	1.4798	1.2069	1.7860	57
4	0.56016	0.82839	0.67620	1.4788	1.2072	1.7852	56
5	0.56040	0.82822	0.67663	1.4779	1.2074	1.7844	55
6	0.56064	0.82806	0.67705	1.4770	1.2076	1.7837	54
7	0.56088	0.82790	0.67747	1.4761	1.2079	1.7829	53
8	0.56112	0.82773	0.67790	1.4751	1.2081	1.7821	52
9	0.56136	0.82757	0.67832	1.4742	1.2083	1.7814	51
10	0.56160	0.82741	0.67875	1.4733	1.2086	1.7806	50
11	0.56184	0.82724	0.67917	1.4724	1.2088	1.7798	49
12	0.56208	0.82708	0.67960	1.4714	1.2091	1.7791	48
13	0.56232	0.82692	0.68002	1.4705	1.2093	1.7783	47
14	0.56256	0.82675	0.68045	1.4696	1.2095	1.7776	46
15	0.56280	0.82659	0.68087	1.4687	1.2098	1.7768	45
16	0.56304	0.82643	0.68130	1.4678	1.2100	1.7760	44
17	0.56328	0.82626	0.68173	1.4669	1.2103	1.7753	43
18	0.56353	0.82610	0.68215	1.4659	1.2105	1.7745	42
19	0.56377	0.82593	0.68258	1.4650	1.2107	1.7738	41
20	0.56401	0.82577	0.68301	1.4641	1.2110	1.7730	40
21	0.56425	0.82561	0.68343	1.4632	1.2112	1.7723	39
22	0.56449	0.82544	0.68386	1.4623	1.2115	1.7715	38
23	0.56473	0.82528	0.68429	1.4614	1.2117	1.7708	37
24	0.56497	0.82511	0.68471	1.4605	1.2119	1.7700	36
25	0.56521	0.82495	0.68514	1.4595	1.2122	1.7693	35
26	0.56545	0.82478	0.68557	1.4586	1.2124	1.7685	34
27	0.56569	0.82462	0.68600	1.4577	1.2127	1.7678	33
28	0.56593	0.82445	0.68642	1.4568	1.2129	1.7670	32
29	0.56617	0.82429	0.68685	1.4559	1.2132	1.7663	31
30	0.56641	0.82413	0.68728	1.4550	1.2134	1.7655	30
31	0.56664	0.82396	0.68771	1.4541	1.2136	1.7648	29
32	0.56688	0.82380	0.68814	1.4532	1.2139	1.7640	28
33	0.56712	0.82363	0.68857	1.4523	1.2141	1.7633	27
34	0.56736	0.82347	0.68899	1.4514	1.2144	1.7625	26
35	0.56760	0.82330	0.68942	1.4505	1.2146	1.7618	25
36	0.56784	0.82314	0.68985	1.4496	1.2149	1.7610	24
37	0.56808	0.82297	0.69028	1.4487	1.2151	1.7603	23
38	0.56832	0.82280	0.69071	1.4478	1.2153	1.7596	22
39	0.56856	0.82264	0.69114	1.4469	1.2156	1.7588	21
40	0.56880	0.82247	0.69157	1.4460	1.2158	1.7581	20
41	0.56904	0.82231	0.69200	1.4451	1.2161	1.7573	19
42	0.56928	0.82214	0.69243	1.4442	1.2163	1.7566	18
43	0.56952	0.82198	0.69286	1.4433	1.2166	1.7559	17
44	0.56976	0.82181	0.69329	1.4424	1.2168	1.7551	16
45	0.57000	0.82165	0.69372	1.4415	1.2171	1.7544	15
46	0.57023	0.82148	0.69415	1.4406	1.2173	1.7537	14
47	0.57047	0.82131	0.69459	1.4397	1.2175	1.7529	13
48	0.57071	0.82115	0.69502	1.4388	1.2178	1.7522	12
49	0.57095	0.82098	0.69545	1.4379	1.2180	1.7514	11
50	0.57119	0.82082	0.69588	1.4370	1.2183	1.7507	10
51	0.57143	0.82065	0.69631	1.4361	1.2185	1.7500	9
52	0.57167	0.82048	0.69674	1.4352	1.2188	1.7493	8
53	0.57191	0.82032	0.69718	1.4343	1.2190	1.7485	7
54	0.57214	0.82015	0.69761	1.4335	1.2193	1.7478	6
55	0.57238	0.81998	0.69804	1.4326	1.2195	1.7471	5
56	0.57262	0.81982	0.69847	1.4317	1.2198	1.7463	4
57	0.57286	0.81965	0.69891	1.4308	1.2200	1.7456	3
58	0.57310	0.81948	0.69934	1.4299	1.2203	1.7449	2
59	0.57334	0.81932	0.69977	1.4290	1.2205	1.7442	1
60	0.57358	0.81915	0.70021	1.4281	1.2208	1.7434	0
M	Cosine	Sine	Cotan.	Tan.	Cosec.	Secant	M

35° or 215°   **Natural Trigonometric Functions**   **144° or 324°**

M	Sine	Cosine	Tan.	Cotan.	Secant	Cosec.	M
0	0.57358	0.81915	0.70021	1.4281	1.2208	1.7434	60
1	0.57381	0.81898	0.70064	1.4273	1.2210	1.7427	59
2	0.57405	0.81882	0.70107	1.4264	1.2213	1.7420	58
3	0.57429	0.81865	0.70151	1.4255	1.2215	1.7413	57
4	0.57453	0.81848	0.70194	1.4246	1.2218	1.7405	56
5	0.57477	0.81832	0.70238	1.4237	1.2220	1.7398	55
6	0.57500	0.81815	0.70281	1.4228	1.2223	1.7391	54
7	0.57524	0.81798	0.70325	1.4220	1.2225	1.7384	53
8	0.57548	0.81781	0.70368	1.4211	1.2228	1.7377	52
9	0.57572	0.81765	0.70412	1.4202	1.2230	1.7369	51
10	0.57596	0.81748	0.70455	1.4193	1.2233	1.7362	50
11	0.57619	0.81731	0.70499	1.4185	1.2235	1.7355	49
12	0.57643	0.81714	0.70542	1.4176	1.2238	1.7348	48
13	0.57667	0.81698	0.70586	1.4167	1.2240	1.7341	47
14	0.57691	0.81681	0.70629	1.4158	1.2243	1.7334	46
15	0.57714	0.81664	0.70673	1.4150	1.2245	1.7327	45
16	0.57738	0.81647	0.70717	1.4141	1.2248	1.7319	44
17	0.57762	0.81630	0.70760	1.4132	1.2250	1.7312	43
18	0.57786	0.81614	0.70804	1.4123	1.2253	1.7305	42
19	0.57809	0.81597	0.70848	1.4115	1.2255	1.7298	41
20	0.57833	0.81580	0.70891	1.4106	1.2258	1.7291	40
21	0.57857	0.81563	0.70935	1.4097	1.2260	1.7284	39
22	0.57881	0.81546	0.70979	1.4089	1.2263	1.7277	38
23	0.57904	0.81530	0.71022	1.4080	1.2265	1.7270	37
24	0.57928	0.81513	0.71066	1.4071	1.2268	1.7263	36
25	0.57952	0.81496	0.71110	1.4063	1.2270	1.7256	35
26	0.57975	0.81479	0.71154	1.4054	1.2273	1.7249	34
27	0.57999	0.81462	0.71198	1.4045	1.2276	1.7242	33
28	0.58023	0.81445	0.71241	1.4037	1.2278	1.7234	32
29	0.58047	0.81428	0.71285	1.4028	1.2281	1.7227	31
30	0.58070	0.81411	0.71329	1.4019	1.2283	1.7220	30
31	0.58094	0.81395	0.71373	1.4011	1.2286	1.7213	29
32	0.58118	0.81378	0.71417	1.4002	1.2288	1.7206	28
33	0.58141	0.81361	0.71461	1.3994	1.2291	1.7199	27
34	0.58165	0.81344	0.71505	1.3985	1.2293	1.7192	26
35	0.58189	0.81327	0.71549	1.3976	1.2296	1.7185	25
36	0.58212	0.81310	0.71593	1.3968	1.2298	1.7178	24
37	0.58236	0.81293	0.71637	1.3959	1.2301	1.7171	23
38	0.58259	0.81276	0.71681	1.3951	1.2304	1.7164	22
39	0.58283	0.81259	0.71725	1.3942	1.2306	1.7157	21
40	0.58307	0.81242	0.71769	1.3933	1.2309	1.7151	20
41	0.58330	0.81225	0.71813	1.3925	1.2311	1.7144	19
42	0.58354	0.81208	0.71857	1.3916	1.2314	1.7137	18
43	0.58378	0.81191	0.71901	1.3908	1.2316	1.7130	17
44	0.58401	0.81174	0.71945	1.3899	1.2319	1.7123	16
45	0.58425	0.81157	0.71990	1.3891	1.2322	1.7116	15
46	0.58448	0.81140	0.72034	1.3882	1.2324	1.7109	14
47	0.58472	0.81123	0.72078	1.3874	1.2327	1.7102	13
48	0.58496	0.81106	0.72122	1.3865	1.2329	1.7095	12
49	0.58519	0.81089	0.72166	1.3857	1.2332	1.7088	11
50	0.58543	0.81072	0.72211	1.3848	1.2335	1.7081	10
51	0.58566	0.81055	0.72255	1.3840	1.2337	1.7075	9
52	0.58590	0.81038	0.72299	1.3831	1.2340	1.7068	8
53	0.58614	0.81021	0.72344	1.3823	1.2342	1.7061	7
54	0.58637	0.81004	0.72388	1.3814	1.2345	1.7054	6
55	0.58661	0.80987	0.72432	1.3806	1.2348	1.7047	5
56	0.58684	0.80970	0.72477	1.3797	1.2350	1.7040	4
57	0.58708	0.80953	0.72521	1.3789	1.2353	1.7033	3
58	0.58731	0.80936	0.72565	1.3781	1.2355	1.7027	2
59	0.58755	0.80919	0.72610	1.3772	1.2358	1.7020	1
60	0.58778	0.80902	0.72654	1.3764	1.2361	1.7013	0
M	Cosine	Sine	Cotan.	Tan.	Cosec.	Secant	M

36° or 216°   **Natural Trigonometric Functions**   **143° or 323°**

M	Sine	Cosine	Tan.	Cotan.	Secant	Cosec.	M
0	0.58778	0.80902	0.72654	1.3764	1.2361	1.7013	60
1	0.58802	0.80885	0.72699	1.3755	1.2363	1.7006	59
2	0.58825	0.80867	0.72743	1.3747	1.2366	1.6999	58
3	0.58849	0.80850	0.72788	1.3738	1.2368	1.6993	57
4	0.58873	0.80833	0.72832	1.3730	1.2371	1.6986	56
5	0.58896	0.80816	0.72877	1.3722	1.2374	1.6979	55
6	0.58920	0.80799	0.72921	1.3713	1.2376	1.6972	54
7	0.58943	0.80782	0.72966	1.3705	1.2379	1.6965	53
8	0.58967	0.80765	0.73010	1.3697	1.2282	1.6959	52
9	0.58990	0.80747	0.73055	1.3688	1.2384	1.6952	51
10	0.59014	0.80730	0.73100	1.3680	1.2387	1.6945	50
11	0.59037	0.80713	0.73144	1.3672	1.2389	1.6938	49
12	0.59060	0.80696	0.73189	1.3663	1.2392	1.6932	48
13	0.59084	0.80679	0.73234	1.3655	1.2395	1.6925	47
14	0.59107	0.80662	0.73278	1.3647	1.2397	1.6918	46
15	0.59131	0.80644	0.73323	1.3638	1.2400	1.6912	45
16	0.59154	0.80627	0.73368	1.3630	1.2403	1.6905	44
17	0.59178	0.80610	0.73412	1.3622	1.2405	1.6898	43
18	0.59201	0.80593	0.73457	1.3613	1.2408	1.6891	42
19	0.59225	0.80576	0.73502	1.3605	1.2411	1.6885	41
20	0.59248	0.80558	0.73547	1.3597	1.2413	1.6878	40
21	0.59272	0.80541	0.73592	1.4097	1.2416	1.6871	39
22	0.59295	0.80524	0.73637	1.4089	1.2419	1.6865	38
23	0.59318	0.80507	0.73681	1.4080	1.2421	1.6858	37
24	0.59342	0.80489	0.73726	1.4071	1.2424	1.6851	36
25	0.55965	0.80472	0.73771	1.4063	1.2427	1.6845	35
26	0.59389	0.80455	0.73816	1.4054	1.2429	1.6838	34
27	0.59412	0.80437	0.73861	1.4045	1.2432	1.6831	33
28	0.59435	0.80420	0.73906	1.4037	1.2435	1.6825	32
29	0.59459	0.80403	0.73951	1.4028	1.2437	1.6818	31
30	0.59482	0.80386	0.73996	1.4019	1.2440	1.6812	30
31	0.59506	0.80368	0.74041	1.4011	1.2443	1.6805	29
32	0.59529	0.80351	0.74086	1.4002	1.2445	1.6798	28
33	0.59552	0.80334	0.74131	1.3994	1.2448	1.6792	27
34	0.59576	0.80316	0.74176	1.3985	1.2451	1.6785	26
35	0.59599	0.80299	0.74221	1.3976	1.2453	1.6779	25
36	0.59622	0.80282	0.74266	1.3968	1.2456	1.6772	24
37	0.59646	0.80264	0.74312	1.3959	1.2459	1.6766	23
38	0.59669	0.80247	0.74357	1.3951	1.2461	1.6759	22
39	0.59692	0.80230	0.74402	1.3942	1.2464	1.6752	21
40	0.59716	0.80212	0.74447	1.3933	1.2467	1.6746	20
41	0.59739	0.80195	0.74492	1.3925	1.2470	1.6739	19
42	0.59762	0.80177	0.74538	1.3916	1.2472	1.6733	18
43	0.59786	0.80160	0.74583	1.3908	1.2475	1.6726	17
44	0.59809	0.80143	0.74628	1.3899	1.2478	1.6720	16
45	0.59832	0.80125	0.74673	1.3891	1.2480	1.6713	15
46	0.59856	0.80108	0.74719	1.3882	1.2483	1.6707	14
47	0.59879	0.80090	0.74764	1.3874	1.2486	1.6700	13
48	0.59902	0.80073	0.74809	1.3865	1.2488	1.6694	12
49	0.59926	0.80056	0.74855	1.3857	1.2491	1.6687	11
50	0.59949	0.80038	0.74900	1.3848	1.2494	1.6681	10
51	0.59972	0.80021	0.74946	1.3840	1.2497	1.6674	9
52	0.59995	0.80003	0.74991	1.3831	1.2499	1.6668	8
53	0.60019	0.79986	0.75037	1.3823	1.2502	1.6661	7
54	0.60042	0.79968	0.75082	1.3814	1.2505	1.6655	6
55	0.60065	0.79951	0.75128	1.3806	1.2508	1.6648	5
56	0.60088	0.79933	0.75173	1.3797	1.2510	1.6642	4
57	0.60112	0.79916	0.75219	1.3789	1.2513	1.6636	3
58	0.60135	0.79898	0.75264	1.3781	1.2516	1.6629	2
59	0.60158	0.79881	0.75310	1.3772	1.2519	1.6623	1
60	0.60181	0.79863	0.75355	1.3764	1.2521	1.6616	0
M	Cosine	Sine	Cotan.	Tan.	Cosec.	Secant	M

37° or 217°   **Natural Trigonometric Functions**   **142° or 322°**

M	Sine	Cosine	Tan.	Cotan.	Secant	Cosec.	M
0	0.60181	0.79863	0.75355	1.3270	1.2521	1.6616	60
1	0.60205	0.79846	0.75401	1.3262	1.2524	1.6610	59
2	0.60228	0.79828	0.75447	1.3254	1.2527	1.6603	58
3	0.60251	0.79811	0.75492	1.3246	1.2530	1.6597	57
4	0.60274	0.79793	0.75538	1.3238	1.2532	1.6591	56
5	0.60298	0.79776	0.75584	1.3230	1.2535	1.6584	55
6	0.60320	0.79758	0.75629	1.3222	1.2538	1.6578	54
7	0.60344	0.79741	0.75675	1.3214	1.2541	1.6572	53
8	0.60367	0.79723	0.75721	1.3206	1.2543	1.6565	52
9	0.60390	0.79706	0.75767	1.3198	1.2546	1.6559	51
10	0.60413	0.79688	0.75812	1.3190	1.2549	1.6552	50
11	0.60437	0.79670	0.75858	1.3182	1.2552	1.6546	49
12	0.60460	0.79653	0.75904	1.3174	1.2554	1.6540	48
13	0.60483	0.79635	0.75950	1.3166	1.2557	1.6533	47
14	0.60506	0.79618	0.75996	1.3159	1.2560	1.6527	46
15	0.60529	0.79600	0.76042	1.3151	1.2563	1.6521	45
16	0.60552	0.79582	0.76088	1.3143	1.2565	1.6514	44
17	0.60576	0.79565	0.76134	1.3135	1.2568	1.6508	43
18	0.60599	0.79547	0.76179	1.3127	1.2571	1.6502	42
19	0.60622	0.79530	0.76225	1.3119	1.2574	1.6496	41
20	0.60645	0.79512	0.76271	1.3111	1.2577	1.6489	40
21	0.60668	0.79494	0.76317	1.3103	1.2579	1.6483	39
22	0.60691	0.79477	0.76364	1.3095	1.2582	1.6477	38
23	0.60714	0.79459	0.76410	1.3087	1.2585	1.6470	37
24	0.60737	0.79441	0.76456	1.3079	1.2588	1.6464	36
25	0.60761	0.79424	0.76502	1.3071	1.2591	1.6458	35
26	0.60784	0.79406	0.76548	1.3064	1.2593	1.6452	34
27	0.60807	0.79388	0.76594	1.3056	1.2596	1.6445	33
28	0.60830	0.79371	0.76640	1.3048	1.2599	1.6439	32
29	0.60853	0.79353	0.76686	1.3040	1.2602	1.6433	31
30	0.60876	0.79335	0.76733	1.3032	1.2605	1.6427	30
31	0.60899	0.79318	0.76779	1.3024	1.2607	1.6420	29
32	0.60922	0.79300	0.76825	1.3016	1.2610	1.6414	28
33	0.60945	0.79282	0.76871	1.3009	1.2613	1.6408	27
34	0.60968	0.79264	0.76918	1.3001	1.2616	1.6402	26
35	0.60991	0.79247	0.76964	1.2993	1.2619	1.6396	25
36	0.61014	0.79229	0.77010	1.2985	1.2622	1.6389	24
37	0.61037	0.79211	0.77057	1.2977	1.2624	1.6383	23
38	0.61061	0.79193	0.77103	1.2970	1.2627	1.6377	22
39	0.61084	0.79176	0.77149	1.2962	1.2630	1.6371	21
40	0.61107	0.79158	0.77196	1.2954	1.2633	1.6365	20
41	0.61130	0.79140	0.77242	1.2946	1.2636	1.6359	19
42	0.61153	0.79122	0.77289	1.2938	1.2639	1.6352	18
43	0.61176	0.79104	0.77335	1.2931	1.2641	1.6346	17
44	0.61199	0.79087	0.77382	1.2923	1.2644	1.6340	16
45	0.61222	0.79069	0.77428	1.2915	1.2647	1.6334	15
46	0.61245	0.79051	0.77475	1.2907	1.2650	1.6328	14
47	0.61268	0.79033	0.77521	1.2900	1.2653	1.6322	13
48	0.61290	0.79015	0.77568	1.2892	1.2656	1.6316	12
49	0.61314	0.78998	0.77614	1.2884	1.2659	1.6309	11
50	0.61337	0.78980	0.77661	1.2876	1.2661	1.6303	10
51	0.61360	0.78962	0.77708	1.2869	1.2664	1.6297	9
52	0.61383	0.78944	0.77754	1.2861	1.2667	1.6291	8
53	0.61405	0.78926	0.77801	1.2853	1.2670	1.6285	7
54	0.61428	0.78908	0.77848	1.2845	1.2673	1.6279	6
55	0.61451	0.78890	0.77895	1.2838	1.2676	1.6273	5
56	0.61474	0.78873	0.77941	1.2830	1.2679	1.6267	4
57	0.61497	0.78855	0.77988	1.2822	1.2681	1.6261	3
58	0.61520	0.78837	0.78035	1.2815	1.2684	1.6255	2
59	0.61543	0.78819	0.78082	1.2807	1.2687	1.6249	1
60	0.61566	0.78801	0.78128	1.2799	1.2690	1.6243	0
M	Cosine	Sine	Cotan.	Tan.	Cosec.	Secant	M

127° or 307°   **52° or 232°**

38° or 218°   **Natural Trigonometric Functions**   **141° or 321°**

M	Sine	Cosine	Tan.	Cotan.	Secant	Cosec.	M
0	0.61566	0.78801	0.78128	1.2799	1.2690	1.6243	60
1	0.61589	0.78783	0.78175	1.2792	1.2693	1.6237	59
2	0.61612	0.78765	0.78222	1.2784	1.2696	1.6231	58
3	0.61635	0.78747	0.78269	1.2776	1.2699	1.6224	57
4	0.61658	0.78729	0.78316	1.2769	1.2702	1.6218	56
5	0.61681	0.78711	0.78363	1.2761	1.2705	1.6212	55
6	0.61703	0.78693	0.78410	1.2753	1.2707	1.6206	54
7	0.61726	0.78675	0.78457	1.2746	1.2710	1.6200	53
8	0.61749	0.78657	0.78504	1.2738	1.2713	1.6194	52
9	0.61772	0.78640	0.78551	1.2730	1.2716	1.6188	51
10	0.61795	0.78622	0.78598	1.2723	1.2719	1.6182	50
11	0.61818	0.78604	0.78645	1.2715	1.2722	1.6176	49
12	0.61841	0.78586	0.78692	1.2708	1.2725	1.6170	48
13	0.61864	0.78563	0.78739	1.2700	1.2728	1.6164	47
14	0.61886	0.78550	0.78786	1.2692	1.2731	1.6159	46
15	0.61909	0.78532	0.78834	1.2685	1.2734	1.6153	45
16	0.61932	0.78514	0.78881	1.2677	1.2737	1.6147	44
17	0.61955	0.78496	0.78928	1.2670	1.2739	1.6141	43
18	0.61978	0.78478	0.78975	1.2662	1.2742	1.6135	42
19	0.62001	0.78460	0.79022	1.2655	1.2745	1.6129	41
20	0.62023	0.78441	0.79070	1.2647	1.2748	1.6123	40
21	0.62046	0.78423	0.79117	1.2639	1.2751	1.6117	39
22	0.62069	0.78405	0.79164	1.2632	1.2754	1.6111	38
23	0.62092	0.78387	0.79212	1.2624	1.2757	1.6105	37
24	0.62115	0.78369	0.79259	1.2617	1.2760	1.6099	36
25	0.62137	0.78351	0.79306	1.2609	1.2763	1.6093	35
26	0.62160	0.78333	0.79354	1.2602	1.2766	1.6087	34
27	0.62183	0.78315	0.79401	1.2594	1.2769	1.6081	33
28	0.62206	0.78297	0.79449	1.2587	1.2772	1.6077	32
29	0.62229	0.78279	0.79496	1.2579	1.2775	1.6070	31
30	0.62251	0.78261	0.79543	1.2572	1.2778	1.6064	30
31	0.62274	0.78243	0.79591	1.2564	1.2781	1.6058	29
32	0.62297	0.78224	0.79639	1.2557	1.2784	1.6052	28
33	0.62320	0.78206	0.79686	1.2549	1.2787	1.6046	27
34	0.62342	0.78188	0.79734	1.2542	1.2790	1.6040	26
35	0.62365	0.78170	0.79781	1.2534	1.2793	1.6034	25
36	0.62388	0.78152	0.79829	1.2527	1.2795	1.6029	24
37	0.62411	0.78134	0.79876	1.2519	1.2798	1.6023	23
38	0.62433	0.78116	0.79924	1.2512	1.2801	1.6017	22
39	0.62456	0.78097	0.79972	1.2504	1.2804	1.6011	21
40	0.62479	0.78079	0.80020	1.2497	1.2807	1.6005	20
41	0.62501	0.78061	0.80067	1.2489	1.2810	1.6000	19
42	0.62524	0.78043	0.80115	1.2482	1.2813	1.5994	18
43	0.62547	0.78025	0.80163	1.2475	1.2816	1.5988	17
44	0.62570	0.78007	0.80211	1.2467	1.2819	1.5982	16
45	0.62592	0.77988	0.80258	1.2460	1.2822	1.5976	15
46	0.62615	0.77970	0.80306	1.2452	1.2825	1.5971	14
47	0.62638	0.77952	0.80354	1.2445	1.2828	1.5965	13
48	0.62660	0.77934	0.80402	1.2437	1.2831	1.5959	12
49	0.62683	0.77915	0.80450	1.2430	1.2834	1.5953	11
50	0.62706	0.77897	0.80498	1.2423	1.2837	1.5947	10
51	0.62728	0.77879	0.80546	1.2415	1.2840	1.5942	9
52	0.62751	0.77861	0.80594	1.2408	1.2843	1.5936	8
53	0.62774	0.77842	0.80642	1.2400	1.2846	1.5930	7
54	0.62796	0.77824	0.80690	1.2393	1.2849	1.5924	6
55	0.62819	0.77806	0.80738	1.2386	1.2852	1.5919	5
56	0.62841	0.77788	0.80786	1.2378	1.2855	1.5913	4
57	0.62864	0.77769	0.80834	1.2371	1.2858	1.5907	3
58	0.62887	0.77751	0.80882	1.2364	1.2861	1.5901	2
59	0.62909	0.77733	0.80930	1.2356	1.2864	1.5896	1
60	0.62932	0.77715	0.80978	1.2349	1.2867	1.5890	0
M	Cosine	Sine	Cotan.	Tan.	Cosec.	Secant	M

128° or 308°   **51° or 231°**

39° or 219° **Natural Trigonometric Functions** **140° or 320°**

M	Sine	Cosine	Tan.	Cotan.	Secant	Cosec.	M
0	0.62932	0.77715	0.80978	1.2349	1.2867	1.5890	60
1	0.62955	0.77696	0.81026	1.2342	1.2871	1.5884	59
2	0.62977	0.77678	0.81075	1.2334	1.2874	1.5879	58
3	0.63000	0.77660	0.81123	1.2327	1.2877	1.5873	57
4	0.63022	0.77641	0.81171	1.2320	1.2880	1.5867	56
5	0.63045	0.77623	0.81219	1.2312	1.2883	1.5862	55
6	0.63067	0.77605	0.81268	1.2305	1.2886	1.5856	54
7	0.63090	0.77586	0.81316	1.2297	1.2889	1.5850	53
8	0.63113	0.77568	0.81364	1.2290	1.2892	1.5845	52
9	0.63135	0.77549	0.81413	1.2283	1.2895	1.5839	51
10	0.63158	0.77531	0.81461	1.2276	1.2898	1.5833	50
11	0.63180	0.77513	0.81509	1.2268	1.2901	1.5828	49
12	0.63203	0.77494	0.81558	1.2261	1.2904	1.5822	48
13	0.63225	0.77476	0.81606	1.2254	1.2907	1.5816	47
14	0.63248	0.77458	0.81655	1.2247	1.2910	1.5811	46
15	0.63270	0.77439	0.81703	1.2239	1.2913	1.5805	45
16	0.63293	0.77421	0.81752	1.2232	1.2916	1.5799	44
17	0.63315	0.77402	0.81800	1.2225	1.2919	1.5794	43
18	0.63338	0.77384	0.81849	1.2218	1.2922	1.5788	42
19	0.63360	0.77365	0.81898	1.2210	1.2926	1.5783	41
20	0.63383	0.77347	0.81946	1.2203	1.2929	1.5777	40
21	0.63405	0.77329	0.81995	1.2196	1.2932	1.5771	39
22	0.63428	0.77310	0.82043	1.2189	1.2935	1.5766	38
23	0.63450	0.77292	0.82092	1.2181	1.2938	1.5760	37
24	0.63473	0.77273	0.82141	1.2174	1.2941	1.5755	36
25	0.63495	0.77255	0.82190	1.2167	1.2944	1.5749	35
26	0.63518	0.77236	0.82238	1.2160	1.2947	1.5743	34
27	0.63540	0.77218	0.82287	1.2152	1.2950	1.5738	33
28	0.63563	0.77199	0.82336	1.2145	1.2953	1.5732	32
29	0.63585	0.77181	0.82385	1.2138	1.2956	1.5727	31
30	0.63608	0.77162	0.82434	1.2131	1.2960	1.5721	30
31	0.63630	0.77144	0.82482	1.2124	1.2963	1.5716	29
32	0.63653	0.77125	0.82531	1.2117	1.2966	1.5710	28
33	0.63675	0.77107	0.82580	1.2109	1.2969	1.5705	27
34	0.63697	0.77088	0.82629	1.2102	1.2972	1.5699	26
35	0.63720	0.77070	0.82678	1.2095	1.2975	1.5694	25
36	0.63742	0.77051	0.82727	1.2088	1.2978	1.5688	24
37	0.63765	0.77033	0.82776	1.2081	1.2981	1.5683	23
38	0.63787	0.77014	0.82825	1.2074	1.2985	1.5677	22
39	0.63810	0.76996	0.82874	1.2066	1.2988	1.5672	21
40	0.63832	0.76977	0.82923	1.2059	1.2991	1.5666	20
41	0.63854	0.76958	0.82972	1.2052	1.2994	1.5661	19
42	0.63877	0.76940	0.83022	1.2045	1.2997	1.5655	18
43	0.63899	0.76921	0.83071	1.2038	1.3000	1.5650	17
44	0.63921	0.76903	0.83120	1.2031	1.3003	1.5644	16
45	0.63944	0.76884	0.83169	1.2024	1.3006	1.5639	15
46	0.63966	0.76865	0.83218	1.2016	1.3010	1.5633	14
47	0.63989	0.76847	0.83267	1.2009	1.3013	1.5628	13
48	0.64011	0.76828	0.83317	1.2002	1.3016	1.5622	12
49	0.64033	0.76810	0.83366	1.1995	1.3019	1.5617	11
50	0.64056	0.76791	0.83415	1.1988	1.3022	1.5611	10
51	0.64078	0.76772	0.83465	1.1981	1.3025	1.5606	9
52	0.64100	0.76754	0.83514	1.1974	1.3029	1.5600	8
53	0.64123	0.76735	0.83563	1.1967	1.3032	1.5595	7
54	0.64145	0.76716	0.83613	1.1960	1.3035	1.5590	6
55	0.64167	0.76698	0.83662	1.1953	1.3038	1.5584	5
56	0.64189	0.76679	0.83712	1.1946	1.3041	1.5579	4
57	0.64212	0.76660	0.83761	1.1939	1.3044	1.5573	3
58	0.64234	0.76642	0.83811	1.1932	1.3048	1.5568	2
59	0.64256	0.76623	0.83860	1.1924	1.3051	1.5563	1
60	0.64279	0.76604	0.83910	1.1917	1.3054	1.5557	0
M	Cosine	Sine	Cotan.	Tan.	Cosec.	Secant	M

    Natural Trigonometric Functions

M	Sine	Cosine	Tan.	Cotan.	Secant	Cosec.	M
0	0.64279	0.76604	0.83910	1.1917	1.3054	1.5557	60
1	0.64301	0.76586	0.83959	1.1910	1.3057	1.5552	59
2	0.64323	0.76567	0.84009	1.1903	1.3060	1.5546	58
3	0.64345	0.76548	0.84059	1.1896	1.3064	1.5541	57
4	0.64368	0.76530	0.84108	1.1889	1.3067	1.5536	56
5	0.64390	0.76511	0.84158	1.1882	1.3070	1.5530	55
6	0.64412	0.76492	0.84208	1.1875	1.3073	1.5525	54
7	0.64435	0.76473	0.84257	1.1868	1.3076	1.5520	53
8	0.64457	0.76455	0.84307	1.1861	1.3080	1.5514	52
9	0.64479	0.76436	0.84357	1.1854	1.3083	1.5509	51
10	0.64501	0.76417	0.84407	1.1847	1.3086	1.5503	50
11	0.64523	0.76398	0.84457	1.1840	1.3089	1.5498	49
12	0.64546	0.76380	0.84506	1.1833	1.3092	1.5493	48
13	0.64568	0.76361	0.84556	1.1826	1.3096	1.5487	47
14	0.64590	0.76342	0.84606	1.1819	1.3099	1.5482	46
15	0.64612	0.76323	0.84656	1.1812	1.3102	1.5477	45
16	0.64635	0.76304	0.84706	1.1805	1.3105	1.5471	44
17	0.64657	0.76286	0.84756	1.1798	1.3109	1.5466	43
18	0.64679	0.76267	0.84806	1.1791	1.3112	1.5461	42
19	0.64701	0.76248	0.84856	1.1785	1.3115	1.5456	41
20	0.64723	0.76229	0.84906	1.1778	1.3118	1.5450	40
21	0.64745	0.76210	0.84956	1.1771	1.3121	1.5445	39
22	0.64768	0.76191	0.85006	1.1764	1.3125	1.5440	38
23	0.64790	0.76173	0.85056	1.1757	1.3128	1.5434	37
24	0.64812	0.76154	0.85107	1.1750	1.3131	1.5429	36
25	0.64834	0.76135	0.85157	1.1743	1.3134	1.5424	35
26	0.64856	0.76116	0.85207	1.1736	1.3138	1.5419	34
27	0.64878	0.76097	0.85257	1.1729	1.3141	1.5413	33
28	0.64900	0.76078	0.85307	1.1722	1.3144	1.5408	32
29	0.64923	0.76059	0.85358	1.1715	1.3148	1.5403	31
30	0.64945	0.76041	0.85408	1.1708	1.3151	1.5398	30
31	0.64967	0.76022	0.85458	1.1702	1.3154	1.5392	29
32	0.64989	0.76003	0.85509	1.1695	1.3157	1.5387	28
33	0.65011	0.75984	0.85559	1.1688	1.3161	1.5382	27
34	0.65033	0.75965	0.85609	1.1681	1.3164	1.5377	26
35	0.65055	0.75946	0.85660	1.1674	1.3167	1.5371	25
36	0.65077	0.75927	0.85710	1.1667	1.3170	1.5366	24
37	0.65100	0.75908	0.85761	1.1660	1.3174	1.5361	23
38	0.65121	0.75889	0.85811	1.1653	1.3177	1.5356	22
39	0.65144	0.75870	0.85862	1.1647	1.3180	1.5351	21
40	0.65166	0.75851	0.85912	1.1640	1.3184	1.5345	20
41	0.65188	0.75832	0.85963	1.1633	1.3187	1.5340	19
42	0.65210	0.75813	0.86013	1.1626	1.3190	1.5335	18
43	0.65232	0.75794	0.86064	1.1619	1.3193	1.5330	17
44	0.65254	0.75775	0.86115	1.1612	1.3197	1.5325	16
45	0.65276	0.75756	0.86165	1.1605	1.3200	1.5319	15
46	0.65298	0.75737	0.86216	1.1599	1.3203	1.5314	14
47	0.65320	0.75718	0.86267	1.1592	1.3207	1.5309	13
48	0.65342	0.75700	0.86318	1.1585	1.3210	1.5304	12
49	0.65364	0.75680	0.86368	1.1578	1.3213	1.5299	11
50	0.65386	0.75661	0.86419	1.1571	1.3217	1.5294	10
51	0.65408	0.75642	0.86470	1.1565	1.3220	1.5289	9
52	0.65430	0.75623	0.86521	1.1558	1.3223	1.5283	8
53	0.65452	0.75604	0.86572	1.1551	1.3227	1.5278	7
54	0.65474	0.75585	0.86623	1.1544	1.3230	1.5273	6
55	0.65496	0.75566	0.86674	1.1537	1.3233	1.5268	5
56	0.65518	0.75547	0.86725	1.1531	1.3237	1.5263	4
57	0.65540	0.75528	0.86775	1.1524	1.3240	1.5258	3
58	0.65562	0.75509	0.86826	1.1517	1.3243	1.5253	2
59	0.65584	0.75490	0.86878	1.1510	1.3247	1.5248	1
60	0.65606	0.75471	0.86929	1.1504	1.3250	1.5242	0
M	Cosine	Sine	Cotan.	Tan.	Cosec.	Secant	M

41° or 221° **Natural Trigonometric Functions** **138° or 318°**

M	Sine	Cosine	Tan.	Cotan.	Secant	Cosec.	M
0	0.65606	0.75471	0.86929	1.1504	1.3250	1.5242	60
1	0.65628	0.75452	0.86980	1.1497	1.3253	1.5237	59
2	0.65650	0.75433	0.87031	1.1490	1.3257	1.5232	58
3	0.65672	0.75414	0.87082	1.1483	1.3260	1.5227	57
4	0.65694	0.75394	0.87133	1.1477	1.3263	1.5222	56
5	0.65716	0.75375	0.87184	1.1470	1.3267	1.5217	55
6	0.65737	0.75356	0.87235	1.1463	1.3270	1.5212	54
7	0.65759	0.75337	0.87287	1.1456	1.3274	1.5207	53
8	0.65781	0.75318	0.87338	1.1450	1.3277	1.5202	52
9	0.65803	0.75299	0.87389	1.1443	1.3280	1.5197	51
10	0.65825	0.75280	0.87441	1.1436	1.3284	1.5192	50
11	0.65847	0.75261	0.87492	1.1430	1.3287	1.5187	49
12	0.65869	0.75241	0.87543	1.1423	1.3290	1.5182	48
13	0.65891	0.75222	0.87595	1.1416	1.3294	1.5177	47
14	0.65913	0.75203	0.87646	1.1409	1.3297	1.5171	46
15	0.65934	0.75184	0.87698	1.1403	1.3301	1.5166	45
16	0.65956	0.75165	0.87749	1.1396	1.3304	1.5161	44
17	0.65978	0.75146	0.87801	1.1389	1.3307	1.5156	43
18	0.66000	0.75126	0.87852	1.1383	1.3311	1.5151	42
19	0.66022	0.75107	0.87904	1.1376	1.3314	1.5146	41
20	0.66044	0.75088	0.87955	1.1369	1.3318	1.5141	40
21	0.66066	0.75069	0.88007	1.1363	1.3321	1.5136	39
22	0.66087	0.75049	0.88058	1.1356	1.3324	1.5131	38
23	0.66109	0.75030	0.88110	1.1349	1.3328	1.5126	37
24	0.66131	0.75011	0.88162	1.1343	1.3331	1.5121	36
25	0.66153	0.74992	0.88213	1.1336	1.3335	1.5116	35
26	0.66175	0.74973	0.88265	1.1329	1.3338	1.5111	34
27	0.66197	0.74953	0.88317	1.1323	1.3342	1.5106	33
28	0.66218	0.74934	0.88369	1.1316	1.3345	1.5101	32
29	0.66240	0.74915	0.88421	1.1309	1.3348	1.5096	31
30	0.66262	0.74895	0.88472	1.1303	1.3352	1.5092	30
31	0.66284	0.74876	0.88524	1.1296	1.3355	1.5087	29
32	0.66305	0.74857	0.88576	1.1290	1.3359	1.5082	28
33	0.66327	0.74838	0.88628	1.1283	1.3362	1.5077	27
34	0.66349	0.74818	0.88680	1.1276	1.3366	1.5072	26
35	0.66371	0.74799	0.88732	1.1270	1.3369	1.5067	25
36	0.66393	0.74780	0.88784	1.1263	1.3372	1.5062	24
37	0.66414	0.74760	0.88836	1.1257	1.3376	1.5057	23
38	0.66436	0.74741	0.88888	1.1250	1.3379	1.5052	22
39	0.66458	0.74722	0.88940	1.1243	1.3383	1.5047	21
40	0.66479	0.74702	0.88992	1.1237	1.3386	1.5042	20
41	0.66501	0.74683	0.89044	1.1230	1.3390	1.5037	19
42	0.66523	0.74664	0.89097	1.1224	1.3393	1.5032	18
43	0.66545	0.74644	0.89149	1.1217	1.3397	1.5027	17
44	0.66566	0.74625	0.89201	1.1211	1.3400	1.5022	16
45	0.66588	0.74606	0.89253	1.1204	1.3404	1.5018	15
46	0.66610	0.74586	0.89306	1.1197	1.3407	1.5013	14
47	0.66631	0.74567	0.89358	1.1191	1.3411	1.5008	13
48	0.66653	0.74548	0.89410	1.1184	1.3414	1.5003	12
49	0.66675	0.74528	0.89463	1.1178	1.3418	1.4998	11
50	0.66697	0.74509	0.89515	1.1171	1.3421	1.4993	10
51	0.66718	0.74489	0.89567	1.1165	1.3425	1.4988	9
52	0.66740	0.74470	0.89620	1.1158	1.3428	1.4983	8
53	0.66762	0.74450	0.89672	1.1152	1.3432	1.4979	7
54	0.66783	0.74431	0.89725	1.1145	1.3435	1.4974	6
55	0.66805	0.74412	0.89777	1.1139	1.3439	1.4969	5
56	0.66826	0.74392	0.89830	1.1132	1.3442	1.4964	4
57	0.66848	0.74373	0.89882	1.1126	1.3446	1.4959	3
58	0.66870	0.74353	0.89935	1.1119	1.3449	1.4954	2
59	0.66891	0.74334	0.89988	1.1113	1.3453	1.4949	1
60	0.66913	0.74314	0.90040	1.1106	1.3456	1.4945	0
M	Cosine	Sine	Cotan.	Tan.	Cosec.	Secant	M

131° or 311° **48° or 228°**

       Natural Trigonometric Functions

M	Sine	Cosine	Tan.	Cotan.	Secant	Cosec.	M
0	0.66913	0.74314	0.90040	1.1106	1.3456	1.4945	60
1	0.66935	0.74295	0.90093	1.1100	1.3460	1.4940	59
2	0.66956	0.74275	0.90146	1.1093	1.3463	1.4935	58
3	0.66978	0.74256	0.90198	1.1086	1.3467	1.4930	57
4	0.66999	0.74236	0.90251	1.1080	1.3470	1.4925	56
5	0.67021	0.74217	0.90304	1.1074	1.3474	1.4921	55
6	0.67043	0.74197	0.90357	1.1067	1.3477	1.4916	54
7	0.67064	0.74178	0.90410	1.1061	1.3481	1.4911	53
8	0.67086	0.74158	0.90463	1.1054	1.3485	1.4906	52
9	0.67107	0.74139	0.90515	1.1048	1.3488	1.4901	51
10	0.67129	0.74119	0.90568	1.1041	1.3492	1.4897	50
11	0.67150	0.74100	0.90621	1.1035	1.3495	1.4892	49
12	0.67172	0.74080	0.90674	1.1028	1.3499	1.4887	48
13	0.67194	0.74061	0.90727	1.1022	1.3502	1.4882	47
14	0.67215	0.74041	0.90780	1.1015	1.3506	1.4877	46
15	0.67237	0.74022	0.90834	1.1009	1.3509	1.4873	45
16	0.67258	0.74002	0.90887	1.1003	1.3513	1.4868	44
17	0.67280	0.73983	0.90940	1.0996	1.3517	1.4863	43
18	0.67301	0.73963	0.90993	1.0990	1.3520	1.4858	42
19	0.67323	0.73944	0.91046	1.0983	1.3524	1.4854	41
20	0.67344	0.73924	0.91099	1.0977	1.3527	1.4849	40
21	0.67366	0.73904	0.91153	1.0971	1.3531	1.4844	39
22	0.67387	0.73885	0.91206	1.0964	1.3534	1.4839	38
23	0.67409	0.73865	0.91250	1.0958	1.3538	1.4835	37
24	0.67430	0.73845	0.91312	1.0951	1.3542	1.4830	36
25	0.67452	0.73826	0.91366	1.0945	1.3545	1.4825	35
26	0.67473	0.73806	0.91419	1.0939	1.3549	1.4821	34
27	0.67495	0.73787	0.91473	1.0932	1.3552	1.4816	33
28	0.67516	0.73767	0.91526	1.0926	1.3556	1.4811	32
29	0.67537	0.73747	0.91580	1.0919	1.3560	1.4806	31
30	0.67559	0.73728	0.91633	1.0913	1.3563	1.4802	30
31	0.67580	0.73708	0.91687	1.0907	1.3567	1.4797	29
32	0.67602	0.73688	0.91740	1.0900	1.3571	1.4792	28
33	0.67623	0.73669	0.91794	1.0894	1.3574	1.4788	27
34	0.67645	0.73649	0.91847	1.0888	1.3578	1.4783	26
35	0.67666	0.73629	0.91901	1.0881	1.3581	1.4778	25
36	0.67688	0.73610	0.91955	1.0875	1.3585	1.4774	24
37	0.67709	0.73590	0.92008	1.0868	1.3589	1.4769	23
38	0.67730	0.73570	0.92062	1.0862	1.3592	1.4764	22
39	0.67752	0.73551	0.92116	1.0856	1.3596	1.4760	21
40	0.67773	0.73531	0.92170	1.0849	1.3600	1.4755	20
41	0.67794	0.73511	0.92223	1.0843	1.3603	1.4750	19
42	0.67816	0.73491	0.92277	1.0837	1.3607	1.4746	18
43	0.67837	0.73472	0.92331	1.0830	1.3611	1.4741	17
44	0.67859	0.73452	0.92385	1.0824	1.3614	1.4736	16
45	0.67880	0.73432	0.92439	1.0818	1.3618	1.4732	15
46	0.67901	0.73412	0.92493	1.0812	1.3622	1.4727	14
47	0.67923	0.73393	0.92547	1.0805	1.3625	1.4723	13
48	0.67944	0.73373	0.93601	1.0799	1.3629	1.4718	12
49	0.67965	0.73353	0.92655	1.0793	1.3633	1.4713	11
50	0.67987	0.73333	0.92709	1.0786	1.3636	1.4709	10
51	0.68008	0.73314	0.92763	1.0780	1.3640	1.4704	9
52	0.68029	0.73294	0.92817	1.0774	1.3644	1.4699	8
53	0.68051	0.73274	0.92871	1.0767	1.3647	1.4695	7
54	0.68072	0.73254	0.92926	1.0761	1.3651	1.4690	6
55	0.68093	0.73234	0.92980	1.0755	1.3655	1.4686	5
56	0.68115	0.73215	0.93034	1.0749	1.3658	1.4681	4
57	0.68136	0.73195	0.93088	1.0742	1.3662	1.4676	3
58	0.68157	0.73175	0.93143	1.0736	1.3666	1.4672	2
59	0.68178	0.73155	0.93197	1.0730	1.3669	1.4667	1
60	0.68200	0.73135	0.93251	1.0724	1.3673	1.4663	0
M	Cosine	Sine	Cotan.	Tan.	Cosec.	Secant	M

43° or 223°   **Natural Trigonometric Functions**   **136° or 316°**

M	Sine	Cosine	Tan.	Cotan.	Secant	Cosec.	M
0	0.68200	0.73135	0.93251	1.0724	1.3673	1.4663	60
1	0.68221	0.73115	0.93306	1.0717	1.3677	1.4658	59
2	0.68242	0.73096	0.93360	1.0711	1.3681	1.4654	58
3	0.68264	0.73076	0.93415	1.0705	1.3684	1.4649	57
4	0.68285	0.73056	0.93469	1.0699	1.3688	1.4644	56
5	0.68306	0.73036	0.93524	1.0692	1.3692	1.4640	55
6	0.68327	0.73016	0.93578	1.0686	1.3695	1.4635	54
7	0.68349	0.72996	0.93633	1.0680	1.3699	1.4631	53
8	0.68370	0.72976	0.93687	1.0674	1.3703	1.4626	52
9	0.68391	0.72956	0.93742	1.0667	1.3707	1.4622	51
10	0.68412	0.72937	0.93797	1.0661	1.3710	1.4617	50
11	0.68433	0.72917	0.93851	1.0655	1.3714	1.4613	49
12	0.68455	0.72897	0.93906	1.0649	1.3718	1.4608	48
13	0.68476	0.72877	0.93961	1.0643	1.3722	1.4604	47
14	0.68497	0.72857	0.94016	1.0636	1.3725	1.4599	46
15	0.68518	0.72837	0.94071	1.0630	1.3729	1.4595	45
16	0.68539	0.72817	0.94125	1.0624	1.3733	1.4590	44
17	0.68561	0.72797	0.94180	1.0618	1.3737	1.4586	43
18	0.68582	0.72777	0.94235	1.0612	1.3740	1.4581	42
19	0.68603	0.72757	0.94290	1.0605	1.3744	1.4577	41
20	0.68624	0.72737	0.94345	1.0599	1.3748	1.4572	40
21	0.68645	0.72717	0.94400	1.0593	1.3752	1.4568	39
22	0.68666	0.72697	0.94455	1.0587	1.3756	1.4563	38
23	0.68688	0.72677	0.94510	1.0581	1.3759	1.4559	37
24	0.68709	0.72657	0.94565	1.0575	1.3763	1.4554	36
25	0.68730	0.72637	0.94620	1.0568	1.3767	1.4550	35
26	0.68751	0.72617	0.94675	1.0562	1.3771	1.4545	34
27	0.68772	0.72597	0.94731	1.0556	1.3774	1.4541	33
28	0.68793	0.72577	0.94786	1.0550	1.3778	1.4536	32
29	0.68814	0.72557	0.94841	1.0544	1.3782	1.4532	31
30	0.68835	0.72537	0.94896	1.0538	1.3786	1.4527	30
31	0.68856	0.72517	0.94952	1.0532	1.3790	1.4523	29
32	0.68878	0.72497	0.95007	1.0525	1.3794	1.4518	28
33	0.68899	0.72477	0.95062	1.0519	1.3797	1.4514	27
34	0.68920	0.72457	0.95118	1.0513	1.3801	1.4510	26
35	0.68941	0.72437	0.95173	1.0507	1.3805	1.4505	25
36	0.68962	0.72417	0.95229	1.0501	1.3809	1.4501	24
37	0.68983	0.72397	0.95284	1.0495	1.3813	1.4496	23
38	0.69004	0.72377	0.95340	1.0489	1.3816	1.4492	22
39	0.69025	0.72357	0.95395	1.0483	1.3820	1.4487	21
40	0.66946	0.72337	0.95451	1.0476	1.3824	1.4483	20
41	0.69067	0.72317	0.95506	1.0470	1.3828	1.4479	19
42	0.69088	0.72297	0.95562	1.0464	1.3832	1.4474	18
43	0.69109	0.72277	0.95618	1.0458	1.3836	1.4470	17
44	0.69130	0.72256	0.95673	1.0452	1.3839	1.4465	16
45	0.69151	0.72236	0.95729	1.0446	1.3843	1.4461	15
46	0.69172	0.72216	0.95785	1.0440	1.3847	1.4457	14
47	0.69193	0.72196	0.95841	1.0434	1.3851	1.4452	13
48	0.69214	0.72176	0.95896	1.0428	1.3855	1.4448	12
49	0.69235	0.72156	0.95952	1.0422	1.3859	1.4443	11
50	0.69256	0.72136	0.96008	1.0416	1.3863	1.4439	10
51	0.69277	0.72115	0.96064	1.0410	1.3867	1.4435	9
52	0.69298	0.72095	0.96120	1.0404	1.3870	1.4430	8
53	0.69319	0.72075	0.96176	1.0397	1.3874	1.4426	7
54	0.69340	0.72055	0.96232	1.0391	1.3878	1.4422	6
55	0.69361	0.72035	0.96288	1.0385	1.3882	1.4417	5
56	0.69382	0.72015	0.96344	1.0379	1.3886	1.4413	4
57	0.69403	0.71994	0.96400	1.0373	1.3890	1.4408	3
58	0.69424	0.71974	0.96456	1.0367	1.3894	1.4404	2
59	0.69445	0.71954	0.96513	1.0361	1.3898	1.4400	1
60	0.69466	0.71934	0.96569	1.0355	1.3902	1.4395	0
M	Cosine	Sine	Cotan.	Tan.	Cosec.	Secant	M

44° or 224°   **Natural Trigonometric Functions**   **135° or 315°**

M	Sine	Cosine	Tan.	Cotan.	Secant	Cosec.	M
0	0.69466	0.71934	0.96569	1.0355	1.3902	1.4395	60
1	0.69487	0.71914	0.96625	1.0349	1.3905	1.4391	59
2	0.69508	0.71893	0.96681	1.0343	1.3909	1.4387	58
3	0.69528	0.71873	0.96738	1.0337	1.3913	1.4382	57
4	0.69549	0.71853	0.96794	1.0331	1.3917	1.4378	56
5	0.69570	0.71833	0.96850	1.0325	1.3921	1.4374	55
6	0.69591	0.71813	0.96907	1.0319	1.3925	1.4370	54
7	0.69612	0.71792	0.96963	1.0313	1.3929	1.4365	53
8	0.69633	0.71772	0.97020	1.0307	1.3933	1.4361	52
9	0.69654	0.71752	0.97076	1.0301	1.3937	1.4357	51
10	0.69675	0.71732	0.97133	1.0295	1.3941	1.4352	50
11	0.69696	0.71711	0.97189	1.0289	1.3945	1.4348	49
12	0.69716	0.71691	0.97246	1.0283	1.3949	1.4344	48
13	0.69737	0.71671	0.97302	1.0277	1.3953	1.4339	47
14	0.69758	0.71650	0.97359	1.0271	1.3957	1.4335	46
15	0.69779	0.71630	0.97416	1.0265	1.3960	1.4331	45
16	0.69800	0.71610	0.97472	1.0259	1.3964	1.4327	44
17	0.69821	0.71589	0.97529	1.0253	1.3968	1.4322	43
18	0.69841	0.71569	0.97586	1.0247	1.3972	1.4318	42
19	0.69862	0.71549	0.97643	1.0241	1.3976	1.4314	41
20	0.69883	0.71529	0.97700	1.0235	1.3980	1.4310	40
21	0.69904	0.71508	0.97756	1.0229	1.3984	1.4305	39
22	0.69925	0.71488	0.97813	1.0223	1.3988	1.4301	38
23	0.69945	0.71468	0.97870	1.0218	1.3992	1.4297	37
24	0.69966	0.71447	0.97927	1.0212	1.3996	1.4292	36
25	0.69987	0.71427	0.97984	1.0206	1.4000	1.4288	35
26	0.70008	0.71406	0.98041	1.0200	1.4004	1.4284	34
27	0.70029	0.71386	0.98098	1.0194	1.4008	1.4280	33
28	0.70049	0.71366	0.98155	1.0188	1.4012	1.4276	32
29	0.70070	0.71345	0.98212	1.0182	1.4016	1.4271	31
30	0.70091	0.71325	0.98270	1.0176	1.4020	1.4267	30
31	0.70112	0.71305	0.98327	1.0170	1.4024	1.4263	29
32	0.70132	0.71284	0.98384	1.0164	1.4028	1.4259	28
33	0.70153	0.71264	0.98441	1.0158	1.4032	1.4254	27
34	0.70174	0.71243	0.98499	1.0152	1.4036	1.4250	26
35	0.70194	0.71223	0.98556	1.0146	1.4040	1.4246	25
36	0.70215	0.71203	0.98613	1.0141	1.4044	1.4242	24
37	0.70236	0.71182	0.98671	1.0135	1.4048	1.4238	23
38	0.70257	0.71162	0.98728	1.0129	1.4052	1.4233	22
39	0.70277	0.71141	0.98786	1.0123	1.4056	1.4229	21
40	0.70298	0.71121	0.98843	1.0117	1.4060	1.4225	20
41	0.70319	0.71100	0.98901	1.0111	1.4065	1.4221	19
42	0.70339	0.71080	0.98958	1.0105	1.4069	1.4217	18
43	0.70360	0.71059	0.99016	1.0099	1.4073	1.4212	17
44	0.70381	0.71039	0.99073	1.0093	1.4077	1.4208	16
45	0.70401	0.71018	0.99131	1.0088	1.4081	1.4204	15
46	0.70422	0.70998	0.99189	1.0082	1.4085	1.4200	14
47	0.70443	0.70977	0.99246	1.0076	1.4089	1.4196	13
48	0.70463	0.70957	0.99304	1.0070	1.4093	1.4192	12
49	0.70484	0.70936	0.99362	1.0064	1.4097	1.4188	11
50	0.70505	0.70916	0.99420	1.0058	1.4101	1.4183	10
51	0.70525	0.70895	0.99478	1.0052	1.4105	1.4179	9
52	0.70546	0.70875	0.99536	1.0047	1.4109	1.4175	8
53	0.70566	0.70854	0.99593	1.0041	1.4113	1.4171	7
54	0.70587	0.70834	0.99651	1.0035	1.4117	1.4167	6
55	0.70608	0.70813	0.99709	1.0029	1.4122	1.4163	5
56	0.70628	0.70793	0.99767	1.0023	1.4126	1.4159	4
57	0.70649	0.70772	0.99826	1.0017	1.4130	1.4154	3
58	0.70669	0.70752	0.99884	1.0012	1.4134	1.4150	2
59	0.70690	0.70731	0.99942	1.0006	1.4138	1.4146	1
60	0.70711	0.70711	1.00000	1.0000	1.4142	1.4142	0
M	Cosine	Sine	Cotan.	Tan.	Cosec.	Secant	M

134° or 314°   **45° or 225°**

Measuring Angles with a Sine Bar

Precision gage blocks

When used with a sine bar, precision gage blocks play a vital role in assuring dimensional quality control in the manufacture of parts. The four primary characteristics of a precision gage block are accuracy, surface finish, wear resistance, and dimensional stability. Other factors are corrosion resistance, hardness, thermal conductivity, and coefficient of expansion. While gage blocks are traditionally square or rectangular, angle blocks are also available and will be discussed later.

Gage blocks are manufactured from different materials, which greatly impacts their cost and utility. Starrett-Webber precision gage blocks, for example, are made from four different base materials. Traditional high-grade steel gage blocks are generally used in shop floor environments; hardness is approximately R_C 64-65. Tungsten carbide gage blocks are harder and longer wearing than high-grade steel and provide increased wear resistance. Ceramic gage blocks have certain advantages over steel: they outwear steel 10 to 1, and they do not corrode. However, because they are a relatively new material for gage blocks, their stability has not stood the test of time. Chromium carbide gage blocks are considered the best available. Not only do they outwear steel 30 to 1, but they will not corrode and they have exceptional stability and accuracy. Their hardness is approximately R_C 71-73, and their fine, hard grain structure provides exceptional resistance to wear and abrasion.

Thermal conductivity and coefficient of expansion are important considerations when choosing a set of gage blocks. Ideally, all setups will be performed with the room and blocks at a stable temperature of 68° F (20° C). On the shop floor, where measurements are rarely finer than 0.0002" or 0.005 mm, the coefficient of resistance of steel, chromium carbide, and ceramics is so close as to be negligible. But for more critical measurements, these characteristics can dictate the choice of the most desirable material. Typical coefficients of thermal expansion for these materials, as stated by Starrett, follow.

SAE 52100 Steel	6.4×10^{-6} in./°F/in.	11.5×10^{-6} mm/°C/mm
Ceramic	5.5×10^{-6} in./°F/in.	9.9×10^{-6} mm/°C/mm
Chromium Carbide	4.7×10^{-6} in./°F/in.	8.5×10^{-6} mm/°C/mm.

Accuracy ratings

Gage blocks are manufactured to stringent standards: U.S. Federal Standard GGG-G-15C; British Standard 4311 (1993); and German Standard DIN 861 (1980). Accuracy standards are further divided into four grades.

Federal Grade 0.5, which exceeds German Grade 0, and DIN, ISO, and BS Grades K.

Federal Grade 1 (also called Class AA), which equals German Grade 00.

Federal Grade 2 (also called Class A), which exceeds DIN, ISO, and BS Grades K.

Federal Grade 3 (also called Class B), which exceeds DIN, ISO, and BS Grades 0.

Table 20 provides standards for accuracy for the four Federal Grades. It should be noted that Grades 0.5 and 1 are both considered Laboratory Master grades that are primarily used in temperature controlled environments to check the accuracy of other gages and measuring tools. Grade 2 is commonly used for inspection purposes. Sets of Grade 3 blocks are often known as "working sets" as they are normally used for tool setup and layout work in the shop.

Gage block sets

While English unit gage block sets may contain as many as 115 or as few as eight blocks (in sets designed to check and calibrate micrometers), typical sets contain 81 blocks in the following sizes.

9 blocks sized 0.1001 through 0.1009 in increments of 0.001 inch.

49 blocks sized 0.101 through 0.149 in increments of 0.001 inch.

19 blocks sized 0.050 through 0.950 in increments of 0.050 inch.

4 blocks sized 1.000 through 4.000 in increments of 1 inch.

This set has a measuring range of 0.100 - 12 inches in steps of 0.001 inch, or 0.200 to 12 inches in steps of 0.0001 inch.

Metric sets are usually available with either a one or two millimeter base. A 112 block set of one millimeter base blocks will typically contain the following sizes.

1 block sized 5.0 mm.

1 block sized 1.0005 mm

9 blocks sized 1.001 through 0.009 in increments of 0.001 mm.

49 blocks sized 1.01 through 1.49 in increments of 0.01 mm.

48 blocks sized 1 through 24.5 in increments of 0.5 mm.

4 blocks sized 25 through 100 in increments of 25 mm.

This set has a measuring range of 3.0 through 250 mm in steps of 0.0005 mm; 2.0 through 250 mm in steps of 0.001 mm; 1.0 through 250 mm in steps of 0.01 mm; or 1.0 through 250 mm in steps of 0.1 mm.

Angle gage blocks

Angle gage blocks are an alternative to measuring with a sine bar. A typical set contains up to 16 angle blocks made up of the following sizes.

5 blocks sized 1, 3, 5, 20, and 30 seconds.

5 blocks sized 1, 3, 5, 20, and 30 minutes.

6 blocks sized 1, 3, 5, 15, 30, and 45 degrees.

From the above set, it is possible to construct angles from 0° to 99° in increments of 1 second.

Table 20. Gage Block Accuracy by Federal Grade. *(Source, The L.S. Starrett Company.)*

Length of Block	Federal Grade 0.5		Federal Grade 1		Federal Grade 2		Federal Grade 3	
	Tolerance (Length)	Flatness & Parallelism	Tolerance (Length)	Flatness & Parallelism	Tolerance (Length)	Flatness & Parallelism	Tolerance (Length)	Flatness & Parallelism
English System: Tolerances expressed in microinches (0.000001") - 1 millionth of an inch								
Through 1"	±1	1	±2	2	+4 - 2	4	+8 - 4	5
2"	±2	1	±4	2	+8 - 4	4	+16 - 8	5
3"	±3	1	±5	3	+10 - 5	4	+20 - 10	5
4"	±4	1	±6	3	+12 - 6	4	+24 - 12	5
5"			±7	3	+14 - 7	4	+28 - 14	5
6"			±8	3	+16 - 8	4	+32 - 16	5
7"			±9	3	+18 - 9	4	+36 - 18	5
8"			±10	3	+20 - 10	4	+40 - 20	5
10"			±12	4	+24 - 12	5	+48 - 24	6

(Continued)

Table 20. *(Continued)* **Gage Block Accuracy by Federal Grade.** *(Source, The L.S. Starrett Company.)*

Length of Block	Federal Grade 0.5		Federal Grade 1		Federal Grade 2		Federal Grade 3	
	Tolerance (Length)	Flatness & Parallelism	Tolerance (Length)	Flatness & Parallelism	Tolerance (Length)	Flatness & Parallelism	Tolerance (Length)	Flatness & Parallelism
English System: Tolerances expressed in microinches (0.000001") - 1 millionth of an inch								
12"			±14	4	+28 - 14	5	+56 - 28	6
16"			±18	4	+36 - 18	5	+72 - 36	6
20"			±20	4	+40 - 20	5	+80 - 40	6
	Surface Finish Equal to Quartz Flat		Surface Finish Croblox® .1 - .3 Steel .3 - .5		Surface Finish Croblox® .3 Steel .6 - .8		Surface Finish .8 Arithmetical Average	
Metric System: Tolerances expressed in microns (0.001 mm)								
Through 10 mm	±.03	.03	±.05	.05	+.10 - .05	.10	+.20 - .10	.12
25 mm	±.04	.03	±.08	.05	+.15 - .08	.10	+.30 - .15	.12
50 mm	±.05	.03	±.10	.05	+.20 - .10	.10	+.40 - .20	.12
75 mm	±.08	.03	±.13	.07	+.25 - .12	.10	+.45 - .22	.12
100 mm	±.10	.03	±.15	.07	+.30 - .15	.10	+.60 - .30	.12
125 mm			±.18	.07	+.35 - .18	.10	+.70 - .35	.12
150 mm			±.20	.07	+.40 - .20	.10	+.80 - .40	.12
175 mm			±.23	.07	+.45 - .22	.10	+.90 - .45	.12
200 mm			±.25	.07	+.50 - .25	.10	+1.0 - .50	.12
250 mm			±.30	.10	+.60 - .30	.12	+1.2 - .60	.15
300 mm			±.35	.10	+.70 - .35	.12	+1.4 - .70	.15
400 mm			±.45	.10	+.90 - .45	.12	+1.8 - .90	.15
500 mm			±.50	.10	+1.0 - .50	.12	+2.0 - 1.0	.15
	Surface Finish Equal to Quartz Flat		Surface Finish Croblox® .003 - .008; Steel .008-.012		Surface Finish Croblox® .008 Steel .015 - .020		Surface Finish .020 Arithmetical Average	

Sine bars

Sine bars are usually made from tool steel that has been stabilized, hardened, ground, and finally lapped to high accuracy. The top and bottom are parallel to 0.0002 inch. The bar is mounted on two cylinders spaced at a known distance. The distance between the centers of the two cylinders is accurate to within 0.0003 inch. The distance between the centers of the cylinders on inch measurement sine bar cylinders is normally set at 2.75 inches, 5 inches, or 10 inches. On Metric sine bars, the distances are usually 100 mm, 125 mm, or 250 mm. A typical sine bar is illustrated in *Figure 13.*

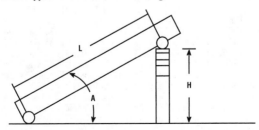

Figure 13. A sine bar. "A" is the angle being set, "L" is the center distance (the hypotenuse), and "H" is the height of the gage blocks.

When used in measuring, the sine bar becomes the hypotenuse of a right-angled triangle. Gage blocks are stacked to form the side opposite, and the surface plate upon which the sine bar sits represents the side adjacent. Therefore, by calculating the height of the side opposite (the gage blocks) for any angle between 0° and 90°, it is possible to create that angle between the sine bar and the surface plate.

For example, determine the gage block height necessary to establish an angle of 32° 15'. The sine bar length is 5 inches.

In this problem, the hypotenuse equals 5 inches, and one angle is given. First, find the sine of 32° 15', which is 0.53361. The unknown height is equal to the hypotenuse times the sine of the angle, so the answer is 5 × 0.53361 = 2.66805 inches.

There is an easier and more accurate method of solving the problem, and that is to use sine bar constants. Since most trigonometric function tables only give the answer to five decimal places, measurements can become slightly distorted when measuring angles in degrees and minutes. On the other hand, **Table 21** provides sine bar constants that are accurate to six decimal places. By finding 30° across the top row of the table, and then locating 15 minutes in the left-hand column, it can be seen that the suggested sine bar height is 2.668073 inches. The chart can be used for 10 inch sine bars as well, simply by multiplying the suggested height by 2. **Table 22** provides heights for 100 mm sine bars.   *(Text continued on p. 175)*

Table 21. Five Inch Sine Bar Constants. All Dimensions in Inches. *(Source, Suburban Tool Co.)*

Min.	0°	1°	2°	3°	4°	5°	6°	7°	8°
0	0.000000	0.087262	0.174497	0.261680	0.348782	0.435779	0.522642	0.609347	0.695866
1	0.001454	0.088716	0.175951	0.263132	0.350233	0.437228	0.524089	0.610790	0.697306
2	0.002909	0.090170	0.177405	0.264585	0.351684	0.438676	0.525535	0.612234	0.698746
3	0.004363	0.091625	0.178858	0.266037	0.353135	0.440125	0.526982	0.613677	0.700186
4	0.005818	0.093079	0.180312	0.267489	0.354586	0.441574	0.528428	0.615121	0.701626
5	0.007272	0.094533	0.181765	0.268942	0.356036	0.443023	0.529874	0.616564	0.703066
6	0.008727	0.095987	0.183219	0.270394	0.357487	0.444471	0.531320	0.618007	0.704506
7	0.010181	0.097441	0.184672	0.271846	0.358938	0.445920	0.532767	0.619451	0.705946
8	0.011636	0.098896	0.186125	0.273299	0.360389	0.447369	0.534213	0.620894	0.707386
9	0.013090	0.100350	0.187579	0.274751	0.361839	0.448817	0.535659	0.622337	0.708826
10	0.014544	0.101804	0.189032	0.276203	0.363290	0.450266	0.537105	0.623780	0.710265
11	0.015999	0.103258	0.190486	0.277655	0.364740	0.451714	0.538551	0.625223	0.711705
12	0.017453	0.104712	0.191939	0.279108	0.366191	0.453163	0.539997	0.626666	0.713145
13	0.018908	0.106166	0.193392	0.280560	0.367642	0.454611	0.541443	0.628109	0.714584
14	0.020362	0.107620	0.194846	0.282012	0.369092	0.456060	0.542889	0.629552	0.716024
15	0.021817	0.109074	0.196299	0.283464	0.370542	0.457508	0.544334	0.630995	0.717463
16	0.023271	0.110529	0.197752	0.284916	0.371993	0.458956	0.545780	0.632438	0.718902
17	0.024725	0.111983	0.199206	0.286368	0.373443	0.460405	0.547226	0.633880	0.720342
18	0.026180	0.113437	0.200659	0.287820	0.374894	0.461853	0.548672	0.635323	0.721781
19	0.027634	0.114891	0.202112	0.289272	0.376344	0.463301	0.550117	0.636766	0.723220
20	0.029089	0.116345	0.203565	0.290724	0.377794	0.464749	0.551563	0.638208	0.724659
21	0.030543	0.117799	0.205019	0.292176	0.379245	0.466197	0.553008	0.639651	0.726098
22	0.031997	0.119253	0.206472	0.293628	0.380695	0.467646	0.554454	0.641093	0.727537
23	0.033452	0.120707	0.207925	0.295080	0.382145	0.469094	0.555899	0.642536	0.728976
24	0.034906	0.122161	0.209378	0.296532	0.383595	0.470542	0.557345	0.643978	0.730415
25	0.036361	0.123615	0.210831	0.297984	0.385045	0.471990	0.558790	0.645420	0.731854
26	0.037815	0.125069	0.212285	0.299436	0.386495	0.473437	0.560235	0.646863	0.733293

(Continued)

Table 21. *(Continued)* **Five Inch Sine Bar Constants.** All Dimensions in Inches. *(Source, Suburban Tool Co.)*

Min.	0°	1°	2°	3°	4°	5°	6°	7°	8°
27	0.039270	0.126523	0.213738	0.300887	0.387945	0.474885	0.561681	0.648305	0.734731
28	0.040724	0.127977	0.215191	0.302339	0.389395	0.476333	0.563126	0.649747	0.736170
29	0.042178	0.129431	0.216644	0.303791	0.390846	0.477781	0.564571	0.651189	0.737609
30	0.043633	0.1308855	0.218097	0.305243	0.392295	0.479229	0.566016	0.652631	0.739047
31	0.045087	0.132339	0.219550	0.306694	0.393745	0.480676	0.567461	0.654073	0.740485
32	0.046541	0.133793	0.221003	0.308146	0.395195	0.482124	0.568906	0.655515	0.741924
33	0.047996	0.135247	0.222456	0.309598	0.396645	0.483572	0.570351	0.656957	0.743362
34	0.049450	0.136700	0.223909	0.311049	0.398095	0.485019	0.571796	0.658399	0.744800
35	0.050905	0.138154	0.225362	0.312501	0.399545	0.486467	0.573241	0.659840	0.746239
36	0.052359	0.139608	0.226815	0.313953	0.400995	0.487914	0.574686	0.661282	0.747677
37	0.053813	0.141062	0.228268	0.315404	0.402444	0.489362	0.576131	0.662724	0.749115
38	0.055268	0.142516	0.229721	0.316856	0.403894	0.490809	0.577575	0.664165	0.750553
39	0.056722	0.143970	0.231174	0.318307	0.405344	0.492257	0.579020	0.665607	0.751991
40	0.058176	0.145424	0.232627	0.319759	0.406793	0.493704	0.580465	0.667048	0.753429
41	0.059631	0.146877	0.234079	0.321210	0.408243	0.495151	0.581909	0.668490	0.754866
42	0.061085	0.148331	0.235532	0.322662	0.409693	0.496599	0.583354	0.669931	0.756304
43	0.062539	0.149785	0.236985	0.324113	0.411142	0.498046	0.584798	0.671372	0.757742
44	0.063994	0.151239	0.238438	0.325564	0.412592	0.499493	0.586243	0.672813	0.759179
45	0.065448	0.152693	0.239891	0.327016	0.414041	0.500940	0.587687	0.674255	0.760617
46	0.066902	0.154146	0.241343	0.328467	0.415490	0.502387	0.589131	0.675696	0.762054
47	0.068357	0.155600	0.242796	0.329918	0.416940	0.503834	0.590576	0.677137	0.763492
48	0.069811	0.157054	0.244249	0.331370	0.418389	0.505281	0.592020	0.678578	0.764929
49	0.071265	0.158508	0.245702	0.332821	0.419839	0.506728	0.593464	0.680019	0.766366
50	0.072719	0.159961	0.247154	0.334272	0.421288	0.508175	0.594908	0.681460	0.767804
51	0.074174	0.161415	0.248607	0.335723	0.422737	0.509622	0.596352	0.682901	0.769241
52	0.075628	0.162869	0.250060	0.337174	0.424186	0.511069	0.597796	0.684341	0.770678
53	0.077082	0.164322	0.251512	0.338625	0.425635	0.512516	0.599240	0.685782	0.772115
54	0.078537	0.165776	0.252965	0.340076	0.427085	0.513963	0.600684	0.687223	0.773552
55	0.079991	0.167230	0.254417	0.341528	0.428534	0.515409	0.602128	0.688663	0.774989
56	0.081445	0.168683	0.255870	0.342979	0.429983	0.516856	0.603572	0.690104	0.776426
57	0.082899	0.170137	0.257322	0.344430	0.431432	0.518303	0.605016	0.691544	0.777862
58	0.084354	0.171590	0.258775	0.345881	0.432881	0.519749	0.606459	0.692985	0.779299
59	0.085808	0.173044	0.260227	0.347331	0.434330	0.521196	0.607903	0.694425	0.780736
Min.	**9°**	**10°**	**11°**	**12°**	**13°**	**14°**	**15°**	**16°**	**17°**
0	0.782172	0.868241	0.954045	1.039558	1.124755	1.209609	1.294095	1.378187	1.461859
1	0.783609	0.869673	0.955473	1.040981	1.126172	1.211021	1.295500	1.379585	1.463249
2	0.785045	0.871105	0.956900	1.042404	1.127589	1.212432	1.296905	1.380983	1.464640
3	0.786482	0.872538	0.958328	1.043826	1.129006	1.213843	1.298309	1.382381	1.466031
4	0.787918	0.873970	0.959755	1.045248	1.130423	1.215254	1.299714	1.383778	1.467421
5	0.789354	0.875402	0.961183	1.046671	1.131840	1.216664	1.301118	1.385176	1.468811
6	0.790790	0.876834	0.962610	1.048093	1.133257	1.218075	1.302523	1.386573	1.470202
7	0.792226	0.878265	0.964037	1.049515	1.134673	1.219486	1.303927	1.387971	1.471592
8	0.793662	0.879697	0.965464	1.050937	1.136090	1.220896	1.305331	1.389368	1.472982
9	0.795098	0.881129	0.966891	1.052359	1.137506	1.222306	1.306735	1.390765	1.474371
10	0.796534	0.882561	0.968318	1.053781	1.138922	1.223717	1.308139	1.392162	1.475761
11	0.797970	0.883992	0.969745	1.055202	1.140338	1.225127	1.309542	1.393559	1.477151
12	0.799406	0.885424	0.971172	1.056624	1.141754	1.226537	1.310946	1.394956	1.478540
13	0.800842	0.886855	0.972598	1.058046	1.143170	1.227947	1.312349	1.396352	1.479930
14	0.802277	0.888286	0.974025	1.059467	1.144586	1.229357	1.313753	1.397749	1.481319

(Continued)

Table 21. *(Continued)* **Five Inch Sine Bar Constants.** All Dimensions in Inches. *(Source, Suburban Tool Co.)*

Min.	9°	10°	11°	12°	13°	14°	15°	16°	17°
15	0.8033713	0.889718	0.975452	1.060888	1.146002	1.230766	1.315156	1.399145	1.482708
16	0.805148	0.891149	0.976878	1.062310	1.147418	1.232176	1.316559	1.400541	1.484097
17	0.806584	0.892580	0.978304	1.063731	1.148833	1.233586	1.317962	1.401938	1.485486
18	0.808019	0.894011	0.979731	1.065152	1.150249	1.234995	1.319365	1.403334	1.486874
19	0.809454	0.895442	0.981157	1.066573	1.151664	1.236404	1.320768	1.404729	1.488263
20	0.810890	0.896873	0.982583	1.067994	1.153079	1.237814	1.322171	1.406125	1.489651
21	0.812325	0.898304	0.984009	1.069415	1.154495	1.239223	1.323573	1.407521	1.491040
22	0.813760	0.899734	0.985435	1.070835	1.155910	1.240632	1.324976	1.408917	1.492428
23	0.815195	0.901165	0.986861	1.072256	1.157325	1.242041	1.326378	1.410312	1.493816
24	0.816630	0.902596	0.988287	1.073677	1.158740	1.243449	1.327781	1.411707	1.495204
25	0.818065	0.904026	0.989712	1.075097	1.160154	1.244858	1.329183	1.413102	1.496592
26	0.819499	0.905457	0.991138	1.076517	1.161569	1.246267	1.330585	1.414498	1.497979
27	0.820934	0.906887	0.992564	1.077938	1.162984	1.247675	1.331987	1.415893	1.499367
28	0.822369	0.908317	0.993989	1.079358	1.164398	1.249084	1.333389	1.417287	1.500754
29	0.823803	0.909748	0.995414	1.080778	1.165813	1.250492	1.334790	1.418682	1.502142
30	0.825238	0.911178	0.996840	1.082198	1.167227	1.251900	1.336192	1.420077	1.503529
31	0.826672	0.912608	0.998265	1.083618	1.168641	1.253308	1.337593	1.421471	1.504916
32	0.828107	0.914038	0.999690	1.085038	1.170055	1.254716	1.338995	1.422866	1.506303
33	0.829541	0.915468	1.001115	1.086458	1.171469	1.256124	1.340396	1.424260	1.507690
34	0.830975	0.916897	1.002540	1.087877	1.172883	1.257532	1.341797	1.425654	1.509076
35	0.832410	0.918327	1.003965	1.089297	1.174297	1.258939	1.343198	1.427048	1.510463
36	0.833844	0.919757	1.005390	1.090716	1.175711	1.260347	1.344599	1.428442	1.511849
37	0.835278	0.921186	1.006814	1.092136	1.177124	1.261754	1.346000	1.429836	1.513236
38	0.836712	0.922616	1.008239	1.093555	1.178538	1.263162	1.347401	1.431229	1.514622
39	0.838146	0.924045	1.009663	1.094974	1.179951	1.264569	1.348801	1.432623	1.516008
40	0.839579	0.925475	1.011088	1.096393	1.181364	1.265976	1.350202	1.434016	1.517394
41	0.841013	0.926904	1.012512	1.097812	1.182778	1.267383	1.351602	1.435409	1.518780
42	0.842447	0.928333	1.013936	1.099231	1.184191	1.268790	1.353002	1.436803	1.520165
43	0.843881	0.929762	1.015361	1.100650	1.185604	1.270197	1.354402	1.438196	1.521551
44	0.845314	0.931191	1.016785	1.102069	1.187017	1.271603	1.355802	1.439589	1.522936
45	0.846748	0.932620	1.018209	1.103487	1.188429	1.273010	1.357202	1.440981	1.524321
46	0.848181	0.934049	1.019633	1.104906	1.189842	1.274416	1.358602	1.442374	1.525707
47	0.849614	0.935478	1.021057	1.106324	1.191255	1.275823	1.360002	1.443767	1.527092
48	0.851047	0.936907	1.022480	1.107742	1.192667	1.277229	1.361401	1.445159	1.528477
49	0.852481	0.938335	1.023904	1.109161	1.194080	1.278635	1.362801	1.446551	1.529861
50	0.853914	0.939764	1.025327	1.110579	1.195492	1.280041	1.364200	1.447943	1.531246
51	0.855347	0.941192	1.026751	1.111997	1.196904	1.281447	1.365599	1.449336	1.532630
52	0.856780	0.942621	1.028174	1.113415	1.198316	1.282853	1.366998	1.450727	1.534015
53	0.858213	0.944049	1.029598	1.114833	1.199728	1.284258	1.368397	1.452119	1.535399
54	0.859646	0.945477	1.031021	1.116251	1.201140	1.285664	1.369796	1.453511	1.536783
55	0.861078	0.946905	1.032444	1.117668	1.202552	1.287069	1.371195	1.454903	1.538167
56	0.862511	0.948333	1.033867	1.119086	1.203964	1.288475	1.372593	1.456294	1.539551
57	0.863944	0.949761	1.035290	1.120503	1.205375	1.289880	1.373992	1.457685	1.540935
58	0.865376	0.951189	1.036713	1.121921	1.206787	1.291285	1.375390	1.459076	1.542318
59	0.866809	0.952617	1.038136	1.123338	1.208198	1.292690	1.376789	1.460468	1.543702
Min.	18°	19°	20°	21°	22°	23°	24°	25°	26°
0	1.545085	1.627841	1.710101	1.791840	1.873033	1.953656	2.033683	2.113091	2.191856
1	1.546468	1.629216	1.711467	1.793198	1.874381	1.954994	2.035012	2.114409	2.193163
2	1.547851	1.630591	1.712834	1.794555	1.875730	1.956333	2.036340	2.115727	2.194470

(Continued)

Table 21. *(Continued)* **Five Inch Sine Bar Constants.** All Dimensions in Inches. *(Source, Suburban Tool Co.)*

Min.	18°	19°	20°	21°	22°	23°	24°	25°	26°
3	1.549234	1.631966	1.714200	1.795913	1.877078	1.957671	2.037669	2.117045	2.195777
4	1.550617	1.633340	1.715566	1.797270	1.878426	1.959010	2.038997	2.118363	2.197083
5	1.552000	1.634715	1.716933	1.798627	1.879774	1.960348	2.040325	2.119680	2.198390
6	1.553382	1.636089	1.718298	1.799984	1.881121	1.961686	2.041652	2.120997	2.199696
7	1.554765	1.637464	1.719664	1.801341	1.882469	1.963023	2.042980	2.122314	2.201002
8	1.556147	1.638838	1.721030	1.802698	1.883816	1.964361	2.044307	2.123631	2.202308
9	1.557529	1.640212	1.722395	1.804054	1.885163	1.965698	2.045635	2.124948	2.203613
10	1.558911	1.641586	1.723761	1.805411	1.886510	1.967036	2.046962	2.126264	2.204919
11	1.560293	1.642960	1.725126	1.806767	1.887857	1.968373	2.048288	2.127580	2.206224
12	1.561675	1.644333	1.726491	1.808123	1.889204	1.969710	2.049615	2.128896	2.207529
13	1.563056	1.645707	1.727856	1.809479	1.890550	1.971046	2.050942	2.130212	2.208834
14	1.564438	1.647080	1.729221	1.810835	1.891897	1.972383	2.052268	2.131528	2.210139
15	1.565819	1.648453	1.730585	1.812190	1.893243	1.973719	2.053594	2.132844	2.211443
16	1.567200	1.649826	1.731950	1.813546	1.894589	1.975056	2.054920	2.134159	2.212748
17	1.568581	1.651199	1.733314	1.814901	1.895935	1.976392	2.056246	2.135474	2.214052
18	1.569962	1.652572	1.734678	1.816256	1.897281	1.977728	2.057572	2.136789	2.215356
19	1.571343	1.653945	1.736042	1.817611	1.898626	1.979063	2.058897	2.138104	2.216660
20	1.572724	1.655317	1.737406	1.818966	1.899972	1.980399	2.060223	2.139419	2.217963
21	1.574104	1.656689	1.738770	1.820321	1.901317	1.981734	2.061548	2.140733	2.219267
22	1.575485	1.658062	1.740133	1.821675	1.902662	1.983069	2.062873	2.142048	2.220570
23	1.576865	1.659434	1.741497	1.823030	1.904007	1.984405	2.064198	2.143362	2.221873
24	1.578245	1.660806	1.742860	1.824384	1.905352	1.985739	2.065522	2.144676	2.223176
25	1.579625	1.662177	1.744223	1.825738	1.906696	1.987074	2.066847	2.145989	2.224479
26	1.581005	1.663549	1.745586	1.827092	1.908041	1.988409	2.068171	2.147303	2.225781
27	1.582385	1.664921	1.746949	1.828446	1.909385	1.989743	2.069495	2.148616	2.227083
28	1.583764	1.666292	1.748312	1.829799	1.910729	1.991077	2.070819	2.149930	2.228385
29	1.585144	1.667663	1.749674	1.831153	1.912073	1.992411	2.072143	2.151243	2.229687
30	1.586523	1.669034	1.751037	1.832506	1.913417	1.993745	2.073466	2.152555	2.230989
31	1.587902	1.670405	1.752399	1.833859	1.914761	1.995079	2.074790	2.153868	2.232291
32	1.589282	1.671776	1.753761	1.835212	1.916104	1.996413	2.076113	2.155181	2.233592
33	1.590661	1.673147	1.755123	1.836565	1.917448	1.997746	2.077436	2.156493	2.234893
34	1.592039	1.674517	1.756485	1.837918	1.918791	1.999079	2.078759	2.157805	2.236194
35	1.593418	1.675888	1.757847	1.839270	1.920134	2.000412	2.080081	2.159117	2.237495
36	1.594797	1.677258	1.759208	1.840623	1.921477	2.001745	2.081404	2.160429	2.238795
37	1.596175	1.678628	1.760570	1.841975	1.922819	2.003078	2.082726	2.161740	2.240096
38	1.597553	1.679998	1.761931	1.843327	1.924162	2.004410	2.084048	2.163052	2.241396
39	1.598931	1.681368	1.763292	1.844679	1.925504	2.005743	2.085370	2.164363	2.242696
40	1.600309	1.682737	1.764653	1.846031	1.926846	2.007075	2.086692	2.165674	2.243996
41	1.601687	1.684107	1.766014	1.847382	1.928188	2.008407	2.088014	2.166985	2.245296
42	1.603065	1.685476	1.767374	1.848734	1.929530	2.009739	2.089335	2.168295	2.246595
43	1.604443	1.686846	1.768735	1.850085	1.930872	2.011071	2.090657	2.169606	2.247894
44	1.605820	1.688215	1.770095	1.851436	1.932213	2.012402	2.091978	2.170916	2.249193
45	1.607197	1.689584	1.771455	1.852787	1.933555	2.013733	2.093299	2.172226	2.250492
46	1.608575	1.690952	1.772815	1.854138	1.934896	2.015065	2.094619	2.173536	2.251791
47	1.609952	1.692321	1.774175	1.855489	1.936237	2.016396	2.095940	2.174846	2.253089
48	1.611328	1.693690	1.775535	1.856839	1.937578	2.017726	2.097260	2.176155	2.254388
49	1.612705	1.695058	1.776894	1.858190	1.938919	2.019057	2.098581	2.177465	2.255686
50	1.614082	1.696426	1.778254	1.859540	1.940259	2.020388	2.099901	2.178774	2.256984
51	1.615458	1.697794	1.779613	1.860890	1.941600	2.021718	2.101221	2.180083	2.258281

(Continued)

Table 21. *(Continued)* **Five Inch Sine Bar Constants.** All Dimensions in Inches. *(Source, Suburban Tool Co.)*

Min.	18°	19°	20°	21°	22°	23°	24°	25°	26°
52	1.616835	1.699162	1.780972	1.862240	1.942940	2.023048	2.102540	2.181392	2.259579
53	1.618211	1.700530	1.782331	1.863589	1.944280	2.024378	2.103860	2.182700	2.260876
54	1.619587	1.701898	1.783690	1.864939	1.945620	2.025708	2.105179	2.184009	2.262174
55	1.620963	1.703265	1.785049	1.866288	1.946959	2.027038	2.106498	2.185317	2.263471
56	1.622339	1.704633	1.786407	1.867638	1.948299	2.028367	2.107817	2.186625	2.264767
57	1.623715	1.706000	1.787766	1.868987	1.949638	2.029696	2.109136	2.187933	2.266064
58	1.625090	1.707367	1.789124	1.870336	1.950978	2.031025	2.110455	2.189241	2.267360
59	1.626466	1.708734	1.790482	1.871684	1.952317	2.032354	2.111773	2.190548	2.268656

Min.	27°	28°	29°	30°	31°	32°	33°	34°	35°
0	2.269952	2.347358	2.424048	2.500000	2.575190	2.649596	2.723195	2.795965	2.867882
1	2.271248	2.348642	2.425320	2.501259	2.576437	2.650830	2.724415	2.797170	2.869073
2	2.272544	2.349926	2.426592	2.502519	2.577683	2.652063	2.725634	2.798376	2.870265
3	2.273839	2.351210	2.427863	2.503778	2.578929	2.653296	2.726854	2.799581	2.871455
4	2.275135	2.352493	2.429135	2.505037	2.580175	2.654528	2.728073	2.800786	2.872646
5	2.276430	2.353776	2.430406	2.506295	2.581421	2.655761	2.729291	2.801990	2.873836
6	2.277725	2.355059	2.431677	2.507554	2.582667	2.656993	2.730510	2.803195	2.875026
7	2.279019	2.356342	2.432948	2.508812	2.583912	2.658225	2.731728	2.804399	2.876216
8	2.280314	2.357625	2.434218	2.510070	2.585157	2.659457	2.732946	2.805603	2.877406
9	2.281608	2.358908	2.435489	2.511328	2.586402	2.660688	2.734164	2.806807	2.878595
10	2.282902	2.360190	2.436759	2.512585	2.587646	2.661919	2.735382	2.808011	2.879784
11	2.284196	2.361472	2.438029	2.513843	2.588891	2.663151	2.736599	2.809214	2.880973
12	2.285490	2.362754	2.439298	2.515100	2.590135	2.664381	2.737816	2.810417	2.882162
13	2.286783	2.364036	2.440568	2.516357	2.591379	2.665612	2.739033	2.811620	2.883350
14	2.288076	2.365317	2.441837	2.517613	2.592623	2.666842	2.740250	2.812822	2.884538
15	2.289370	2.366598	2.443106	2.518870	2.593866	2.668073	2.741466	2.814025	2.885726
16	2.290662	2.367879	2.444375	2.520126	2.595110	2.669303	2.742682	2.815227	2.886914
17	2.291955	2.369160	2.445644	2.521382	2.596353	2.670532	2.743898	2.816429	2.888101
18	2.293248	2.370441	2.446912	2.522638	2.597596	2.671762	2.745114	2.817630	2.889288
19	2.294540	2.371722	2.448181	2.523894	2.598838	2.672991	2.746330	2.818832	2.890475
20	2.295832	2.373002	2.449449	2.525149	2.600081	2.674220	2.747545	2.820033	2.891662
21	2.297124	2.374282	2.450716	2.526404	2.601323	2.675449	2.748760	2.821234	2.892848
22	2.298416	2.375562	2.451984	2.527659	2.602565	2.676677	2.749975	2.822434	2.894034
23	2.299708	2.376842	2.453252	2.528914	2.603807	2.677906	2.751189	2.823635	2.895220
24	2.300999	2.378121	2.454519	2.530169	2.605048	2.679134	2.752404	2.824835	2.896406
25	2.302290	2.379400	2.455786	2.531423	2.606289	2.680362	2.753618	2.826035	2.897591
26	2.303581	2.380679	2.457053	2.532677	2.607531	2.681590	2.754832	2.827235	2.898776
27	2.304872	2.381958	2.458319	2.533931	2.608771	2.682817	2.756045	2.828434	2.899961
28	2.306162	2.383237	2.459586	2.535185	2.610012	2.684044	2.757259	2.829633	2.901146
29	2.307453	2.384516	2.460852	2.536439	2.611253	2.685271	2.758472	2.830832	2.902331
30	2.308743	2.385794	2.462118	2.537692	2.612493	2.686498	2.759685	2.832031	2.903515
31	2.310033	2.387072	2.463384	2.538945	2.613733	2.687725	2.760898	2.833230	2.904699
32	2.311323	2.388350	2.464649	2.540198	2.614973	2.688951	2.762110	2.834428	2.905882
33	2.312612	2.389627	2.465915	2.541450	2.616212	2.690177	2.763322	2.835626	2.907066
34	2.313902	2.390905	2.467180	2.542703	2.617452	2.691403	2.764534	2.836824	2.908249
35	2.315191	2.392182	2.468445	2.543955	2.618691	2.692629	2.765746	2.838021	2.909432
36	2.316480	2.393459	2.469709	2.545207	2.619930	2.693854	2.766958	2.839219	2.910615
37	2.317769	2.394736	2.470974	2.546459	2.621168	2.695079	2.768169	2.840416	2.911797
38	2.319058	2.396013	2.472238	2.547710	2.622407	2.696304	2.769380	2.841613	2.912980
39	2.320346	2.397289	2.473502	2.548962	2.623645	2.697529	2.770591	2.842800	2.914162

(Continued)

Table 21. *(Continued)* **Five Inch Sine Bar Constants.** All Dimensions in Inches. *(Source, Suburban Tool Co.)*

Min.	27°	28°	29°	30°	31°	32°	33°	34°	35°
40	2.321634	2.398566	2.474766	2.550213	2.624883	2.698753	2.771802	2.844006	2.915343
41	2.322922	2.399842	2.476030	2.551464	2.626121	2.699978	2.773012	2.845202	2.916525
42	2.324210	2.401117	2.477293	2.552715	2.627358	2.701202	2.774222	2.846398	2.917706
43	2.325498	2.402393	2.478557	2.553965	2.628596	2.702425	2.775432	2.847593	2.918887
44	2.326785	2.403669	2.479820	2.555215	2.629833	2.703649	2.776642	2.848789	2.920068
45	2.328073	2.404944	2.481083	2.556465	2.631070	2.704872	2.777851	2.849984	2.921248
46	2.329360	2.406219	2.482345	2.557715	2.632306	2.706095	2.779060	2.851179	2.922429
47	2.330647	2.407490	2.483608	2.558965	2.633543	2.707318	2.780269	2.852373	2.923609
48	2.331933	2.408768	2.484870	2.560214	2.634779	2.708541	2.781478	2.853568	2.924788
49	2.333220	2.410043	2.486132	2.561464	2.636015	2.709763	2.782687	2.854762	2.925968
50	2.334506	2.411317	2.487394	2.562713	2.637251	2.710986	2.783895	2.855956	2.927147
51	2.335792	2.412591	2.488655	2.563961	2.638486	2.712208	2.785103	2.857150	2.928326
52	2.337078	2.413865	2.489917	2.565210	2.639722	2.713429	2.786311	2.858343	2.929505
53	2.338364	2.415139	2.491178	2.566458	2.640957	2.714651	2.787518	2.859536	2.930683
54	2.339649	2.416412	2.492439	2.567706	2.642192	2.715872	2.788726	2.860729	2.931862
55	2.340934	2.417685	2.493699	2.568954	2.643426	2.717093	2.789933	2.861922	2.933040
56	2.342219	2.418958	2.494960	2.570202	2.644661	2.718314	2.791139	2.863115	2.934218
57	2.343504	2.420231	2.496220	2.571449	2.645895	2.719535	2.792346	2.864307	2.935395
58	2.344789	2.421504	2.497480	2.572697	2.647129	2.720755	2.793552	2.865499	2.936572
59	2.346074	2.422776	2.498740	2.573944	2.648363	2.721975	2.794759	2.866691	2.937749
Min.	36°	37°	38°	39°	40°	41°	42°	43°	44°
0	2.938926	3.009705	3.078307	3.146602	3.213938	3.280295	3.345653	3.409992	3.473292
1	2.940103	3.010237	3.079453	3.147732	3.215053	3.281393	3.346734	3.411055	3.474338
2	2.941279	3.011398	3.080599	3.148862	3.216166	3.282490	3.347814	3.412119	3.475384
3	2.942455	3.012559	3.081745	3.149992	3.217279	3.253587	3.348894	3.413182	3.476429
4	2.943631	3.013719	3.082890	3.151121	3.218393	3.284684	3.349974	3.414244	3.477474
5	2.944806	3.014880	3.084035	3.152250	3.219505	3.285780	3.351054	3.415307	3.478519
6	2.945980	3.016040	3.085179	3.153379	3.220618	3.286876	3.352133	3.416360	3.479564
7	2.947157	3.017200	3.086324	3.154508	3.221731	3.287972	3.353212	3.417431	3.480608
8	2.948332	3.018360	3.087468	3.155636	3.222843	3.289068	3.354291	3.418492	3.481652
9	2.949506	3.019519	3.088612	3.156764	3.223955	3.290163	3.355369	3.419553	3.482696
10	2.950680	3.020678	3.089755	3.157892	3.225066	3.291258	3.356447	3.420614	3.483740
11	2.951855	3.021837	3.090899	3.159017	3.226177	3.292353	3.357525	3.421675	3.484783
12	2.953028	3.022996	3.092042	3.160147	3.227288	3.293447	3.358603	3.422736	3.485826
13	2.954202	3.024154	3.093184	3.161273	3.228399	3.294542	3.359680	3.423796	3.486868
14	2.955375	3.025312	3.094327	3.162400	3.229510	3.295635	3.360757	3.424855	3.487910
15	2.956548	3.026470	3.095470	3.163527	3.230620	3.296729	3.361834	3.425915	3.488952
16	2.757720	3.027628	3.096612	3.164653	3.231730	3.297822	3.362910	3.426974	3.489994
17	2.958894	3.028785	3.097754	3.165779	3.232840	3.298916	3.363987	3.428033	3.491035
18	2.960066	3.029942	3.098895	3.166904	3.233949	3.300008	3.365063	3.429092	3.492076
19	2.961238	3.031099	3.100036	3.168030	3.235058	3.301101	3.366138	3.430150	3.493117
20	2.962410	3.032255	3.101177	3.169155	3.226167	3.302193	3.367213	3.431208	3.494158
21	2.943581	3.033412	3.102318	3.170280	3.237275	3.303285	3.368289	3.432266	3.495198
22	2.964753	3.034568	3.103459	3.171401	3.238384	3.304377	3.369383	3.433323	3.496238
23	2.965924	3.035724	3.104599	3.172529	3.239492	3.305468	3.370438	3.434381	3.497277
24	2.967094	3.036879	3.105739	3.173653	3.240600	3.306559	3.371512	3.435438	3.498317
25	2.968265	3.038034	3.106879	3.174776	3.241707	3.307650	3.372586	3.436494	3.499356
26	2.969435	3.039190	3.108018	3.175900	3.242814	3.308741	3.373659	3.437550	3.500394
27	2.970605	3.040344	3.109157	3.177023	3.243921	3.309831	3.374733	3.438607	3.501433

(Continued)

Table 21. *(Continued)* **Five Inch Sine Bar Constants.** All Dimensions in Inches. *(Source, Suburban Tool Co.)*

Min.	36°	37°	38°	39°	40°	41°	42°	43°	44°
28	2.971775	3.041199	3.110296	3.178146	3.245028	3.310921	3.375806	3.439662	3.502471
29	2.972945	3.042653	3.111435	3.179269	3.246134	3.312011	3.376879	3.440718	3.503509
30	2.974114	3.043807	3.112573	3.180391	3.247240	3.313100	3.377951	3.441773	3.504546
31	2.975283	3.044961	3.113711	3.181513	3.248346	3.314189	3.379023	3.442828	3.505584
32	2.976452	3.046114	3.114849	3.182635	3.249452	3.315278	3.380095	3.443882	3.506620
33	2.977620	3.047628	3.115987	3.183757	3.250557	3.316367	3.381167	3.444937	3.507657
34	2.978789	3.048421	3.117124	3.184878	3.251662	3.317455	3.382238	3.445991	3.508693
35	2.979957	3.049573	3.118261	3.185999	3.252767	3.318543	3.383309	3.447044	3.509730
36	2.981124	3.050726	3.119398	3.187120	3.253871	3.319631	3.384380	3.448098	3.510765
37	2.982292	3.051878	3.120536	3.188240	3.254975	3.320719	3.385450	3.449151	3.511801
38	2.983459	3.053030	3.121671	3.189361	3.256079	3.321806	3.386520	3.450204	3.512836
39	2.984626	3.054182	3.122807	3.190481	3.257183	3.322893	3.387950	3.451256	3.513871
40	2.985793	3.055333	3.123943	3.191600	3.258286	3.323979	3.388860	3.452308	3.514905
41	2.986959	3.056484	3.125078	3.192720	3.259389	3.325066	3.389729	3.453360	3.515940
42	2.988126	3.056736	3.126213	3.193839	3.260492	3.326152	3.390798	3.454412	3.516974
43	2.989292	3.058786	3.127348	3.194959	3.261595	3.327238	3.391867	3.455463	3.518007
44	2.990457	3.059936	3.128483	3.196077	3.262697	3.328323	3.392936	3.456514	3.519041
45	2.991623	3.061086	3.129617	3.197195	3.263799	3.329408	3.394004	3.457565	3.520074
46	2.992788	3.062236	3.130752	3.198313	3.264900	3.330493	3.395072	3.458616	3.521106
47	2.993953	3.063386	3.131885	3.199431	3.266002	3.331578	3.396139	3.459666	3.522139
48	2.995118	3.064535	3.133019	3.200548	3.267103	3.332662	3.397207	3.460716	3.523171
49	2.996282	3.065684	3.134152	3.201666	3.268204	3.333746	3.398274	3.461765	3.524203
50	2.997447	3.066833	3.135286	3.202783	3.269304	3.334830	3.399340	3.462815	3.525235
51	2.998611	3.067982	3.136418	3.203900	3.270405	3.335914	3.400407	3.463864	3.526266
52	2.999774	3.069130	3.137551	3.205016	3.271505	3.336997	3.401473	3.464913	3.527297
53	3.000938	3.070278	3.138683	3.206132	3.272605	3.338080	3.402539	3.465961	3.528327
54	3.002100	3.071426	3.139815	3.207248	3.273704	3.339163	3.403604	3.467009	3.529358
55	3.003264	3.072574	3.140947	3.208364	3.274803	3.340245	3.404670	3.468057	3.530388
56	3.004427	3.073721	3.142079	3.209479	3.275902	3.341327	3.405735	3.469105	3.531418
57	3.005589	3.074868	3.143210	3.210594	3.277001	3.342409	3.406799	3.470152	3.532447
58	3.006751	3.076015	3.144341	3.211709	3.278099	3.343491	3.407864	3.471199	3.533476
59	3.007913	3.077161	3.145472	3.212824	3.279197	3.344572	3.408928	3.472245	3.534505

Table 22. 100 Millimeter Sine Bar Constants. All Dimensions in Millimeters. *(Source, Suburban Tool Co.)*

Min.	0°	1°	2°	3°	4°	5°	6°	7°	8°
0	0.00000	1.74524	3.48995	5.23360	6.97565	8.71557	10.45285	12.18693	13.91731
1	0.02909	1.77432	3.51902	5.26264	7.00467	8.74455	10.48178	12.21581	13.94612
2	0.05818	1.80341	3.54809	5.29169	7.03368	8.77353	10.51070	12.24468	13.97492
3	0.08727	1.83249	3.57716	5.32074	7.06270	8.80251	10.53963	12.27355	14.00372
4	0.11636	1.86158	3.60623	5.34979	7.09171	8.83148	10.56856	12.30241	14.03252
5	0.14544	1.89066	3.63530	5.37883	7.12073	8.86046	10.59748	12.33128	14.06132
6	0.17453	1.91974	3.66437	5.40788	7.14974	8.88943	10.62641	12.36015	14.09012
7	0.20362	1.94883	3.69344	5.43693	7.17876	8.91840	10.65533	12.38901	14.11892
8	0.23271	1.97791	3.72251	5.46597	7.20777	8.94738	10.68425	12.41788	14.14772
9	0.26180	2.00699	3.75158	5.49502	7.23678	8.97635	10.71318	12.44674	14.17651
10	0.29089	2.03608	3.78065	5.52406	7.26580	9.00532	10.74210	12.47560	14.20531
11	0.31998	2.06516	3.80971	5.55311	7.29481	9.03429	10.77102	12.50446	14.23410
12	0.34907	2.09424	3.83878	5.58215	7.32382	9.06326	10.79994	12.53332	14.26289
13	0.37815	2.12332	3.86785	5.61119	7.35283	9.09223	10.82885	12.56218	14.29268

(Continued)

Table 22. *(Continued)* **100 Millimeter Sine Bar Constants.** All Dimensions in Millimeters.
(Source, Suburban Tool Co.)

Min.	0°	1°	2°	3°	4°	5°	6°	7°	8°
14	0.40724	2.15241	3.89692	5.64024	7.38184	9.12119	10.85777	12.59104	14.32047
15	0.43633	2.18149	3.92598	5.66928	7.41085	9.15016	10.88669	12.61990	14.34926
16	0.46542	2.21057	3.95505	5.69832	7.43986	9.17913	10.91560	12.64875	14.37805
17	0.49451	2.23965	3.98411	5.72736	7.46887	9.20809	10.94452	12.67761	14.40684
18	0.52360	2.26873	4.01318	5.75640	7.49787	9.23706	10.97343	12.70646	14.43562
19	0.55268	2.29781	4.04224	5.78544	7.52688	9.26602	11.00234	12.73531	14.46440
20	0.58177	2.32690	4.07131	5.81448	7.55589	9.29499	11.03126	12.76416	14.49319
21	0.61086	2.35598	4.10037	5.84352	7.58489	9.32395	11.06017	12.79302	14.52197
22	0.63995	2.38506	4.12944	5.87256	7.61390	9.35291	11.08908	12.82186	14.55075
23	0.66904	2.41414	4.15850	5.90160	7.64290	9.38187	11.11799	12.85071	14.57953
24	0.69813	2.44322	4.18757	5.93064	7.67190	9.41083	11.14689	12.87965	14.60830
25	0.72721	2.47230	4.21663	5.95967	7.70091	9.43979	11.17580	12.90841	14.63708
26	0.75630	2.50138	4.24569	5.98871	7.72991	9.46875	11.20471	12.93725	14.66585
27	0.78539	2.53046	4.27475	6.01775	7.75891	9.49771	11.23361	12.96609	14.69463
28	0.81448	2.55954	4.30382	6.04678	7.78791	9.52666	11.26252	12.99494	14.72340
29	0.84357	2.58862	4.33288	6.07582	7.81691	9.55562	11.29142	13.02378	14.75217
30	0.87265	2.61769	4.36194	6.10485	7.84591	9.58458	11.32032	13.05262	14.78094
31	0.90174	2.64677	4.39100	6.13389	7.87491	9.61353	11.34922	13.08146	14.80971
32	0.93083	2.67585	4.42006	6.16292	7.90391	9.64248	11.37812	13.11030	14.83848
33	0.95992	2.70493	4.44912	6.19196	7.93290	9.67144	11.40702	13.13913	14.86724
34	0.98900	2.73401	4.47818	6.22099	7.96190	9.70039	11.43592	13.16797	14.89601
35	1.01809	2.76309	4.50724	6.25002	7.99090	9.72934	11.46482	13.19681	14.92477
36	1.04718	2.79216	4.53630	6.27905	8.01989	9.75829	11.49372	13.22564	14.95353
37	1.07627	2.82124	4.56536	6.30808	8.04889	9.78724	11.52261	13.25447	14.98230
38	1.10535	2.85032	4.59442	6.33711	8.07788	9.81619	11.55151	13.28330	15.01106
39	1.13444	2.87940	4.62347	6.36614	8.10687	9.84514	11.58040	13.31213	15.03981
40	1.16353	2.90847	4.65253	6.39517	8.13587	9.87408	11.60929	13.34096	15.06857
41	1.19261	2.93755	4.68159	6.42420	8.16486	9.90303	11.63818	13.36979	15.09733
42	1.22170	2.96662	4.71065	6.45323	8.19385	9.93197	11.66707	13.39862	15.12608
43	1.25079	2.99570	4.73970	6.48226	8.22284	9.96092	11.69596	13.42744	15.15484
44	1.27987	3.02478	4.76876	6.51129	8.25183	9.98986	11.72485	13.45627	5.18359
45	1.30896	3.05385	4.79781	6.54031	8.28082	10.01881	11.75374	13.48509	15.21234
46	1.33805	3.08293	4.82687	6.56934	8.30981	10.04775	11.78263	13.51392	15.24109
47	1.36713	3.11200	4.85592	6.59836	8.33880	10.07669	11.81151	13.54274	15.26984
48	1.39622	3.14108	4.88498	6.62739	8.36778	10.10563	11.84040	13.57156	15.29858
49	1.42530	3.17015	4.91403	6.65641	8.39677	10.13457	11.86928	13.60038	15.32733
50	1.45439	3.19922	4.94308	6.68544	8.42576	10.16351	11.89816	13.62919	15.35607
51	1.48348	3.22830	4.97214	6.71446	8.45474	10.19245	11.92704	13.65801	15.38482
52	1.51256	3.25737	5.00119	6.74348	8.48373	10.22138	11.95593	13.68683	15.41356
53	1.54165	3.28644	5.03024	6.77251	8.51271	10.25032	11.98481	13.71564	15.44230
54	1.57073	3.31552	5.05929	6.80153	8.54169	10.27925	12.01368	13.74445	15.47104
55	1.59982	3.34459	5.08835	6.83055	8.57067	10.30819	12.04256	13.77327	15.49978
56	1.62890	3.37366	5.11740	6.85957	8.59966	10.33712	12.07144	13.80208	15.52851
57	1.65799	3.40274	5.14645	6.88859	8.62864	10.36605	12.10031	13.83089	15.55725
58	1.68707	3.43181	5.17550	6.91761	8.65762	10.39499	12.12919	13.85970	15.58598
59	1.71616	3.46088	5.20455	6.94663	8.68660	10.42392	12.15806	13.88850	15.61472

Min.	9°	10°	11°	12°	13°	14°	15°	16°	17°
0	15.64345	17.36482	19.08090	20.79117	22.49511	24.19219	25.88190	27.56374	29.23717
1	15.67218	17.39346	19.10945	20.81962	22.52345	24.22041	25.91000	27.59170	29.26499

(Continued)

Table 22. *(Continued)* **100 Millimeter Sine Bar Constants.** All Dimensions in Millimeters.
(Source, Suburban Tool Co.)

Min.	9°	10°	11°	12°	13°	14°	15°	16°	17°
2	15.70091	17.42211	19.13801	20.84807	22.55179	24.24863	25.93810	27.61965	29.29280
3	15.72963	17.45075	19.16656	20.87652	22.58013	24.27685	25.96619	27.64761	29.32061
4	15.75836	17.47939	19.19510	20.90497	22.60846	24.30507	25.99428	27.67556	29.34842
5	15.78708	17.50803	19.22365	20.93341	22.63680	24.33329	26.02237	27.70352	29.37623
6	15.81581	17.53667	19.25220	20.96186	22.66513	24.36150	26.05045	27.73147	29.40403
7	15.84453	17.56531	19.28074	20.99030	22.69346	24.38971	26.07853	27.75941	29.43183
8	15.87325	17.59395	19.30928	21.01874	22.72179	24.41792	26.10662	27.78736	29.45963
9	15.90197	17.62258	19.33782	21.04718	22.75012	24.44613	26.13469	27.81530	29.48743
10	15.93069	17.65121	19.36636	21.07561	22.77844	24.47433	26.16277	27.84324	29.51522
11	15.95940	17.67984	19.39490	21.10405	22.80677	24.50254	26.19085	27.87118	29.54302
12	15.98812	17.70847	19.42344	21.13248	22.83509	24.53074	26.21892	27.89911	29.57080
13	16.01683	17.73710	19.45197	21.16091	22.86341	24.55894	26.24699	27.92704	29.59859
14	16.04555	17.76573	19.48050	21.18934	22.89172	24.58713	26.27506	27.95497	29.62638
15	16.07426	17.79435	19.50903	21.21777	22.92004	24.61533	26.30312	27.98290	29.65416
16	16.10297	17.82298	19.53756	21.24619	22.94835	24.64352	26.33118	28.01083	29.68194
17	16.13167	17.85160	19.56609	21.27462	22.97666	24.67171	26.35925	28.03875	29.70971
18	16.16038	17.88022	19.59461	21.30304	23.00497	24.69990	26.38730	28.06667	29.73749
19	16.18909	17.90884	19.62314	21.33146	23.03328	24.72809	26.41536	28.09459	29.76526
20	16.21779	17.93746	19.65166	21.35988	23.06159	24.75627	26.44342	28.12251	29.79303
21	16.24650	17.96607	19.68018	21.38829	23.08989	24.78445	26.47147	28.15042	29.82079
22	16.27520	17.99469	19.70870	21.41671	23.11819	24.81263	26.49952	28.17833	29.84856
23	16.30390	18.02330	19.73722	21.44512	23.14649	24.84081	26.52757	28.20624	29.87632
24	16.33260	18.05191	19.76573	21.47353	23.17479	24.86899	26.55561	28.23415	29.90408
25	16.36129	18.08052	19.79425	21.50194	23.20309	24.89716	26.58365	28.26205	29.93184
26	16.38999	18.10913	19.82276	21.53035	23.23138	24.92533	26.61170	28.28995	29.95959
27	16.41868	18.13774	19.85127	21.55876	23.25967	24.95350	26.63973	28.31785	29.98734
28	16.44738	18.16635	19.87978	21.58716	23.28796	24.98167	26.66777	28.34575	30.01509
29	16.47607	18.19495	19.90829	21.61556	23.31625	25.00984	26.69581	28.37364	30.04284
30	16.50476	18.22355	19.93679	21.64396	23.34454	25.03800	26.72384	28.40153	30.07058
31	16.53345	18.25215	19.96530	21.67236	23.37282	25.06616	26.75187	28.42942	30.09832
32	16.56214	18.28075	19.99380	21.70076	23.40110	25.09432	26.77989	28.45731	30.12606
33	16.59082	18.30935	20.02230	21.72915	23.42938	25.12248	26.80792	28.48520	30.15380
34	16.61951	18.33795	20.05080	21.75754	23.45766	25.15063	26.83594	28.51308	30.18153
35	16.64819	18.36654	20.07930	21.78593	23.48594	25.17879	26.86396	28.54096	30.20926
36	16.67687	18.39514	20.10779	21.81432	23.51421	25.20694	26.89198	28.56884	30.23699
37	16.70556	18.42373	20.13629	21.84271	23.54248	25.23508	26.92000	28.59671	30.26471
38	16.73423	18.45232	20.16478	21.87110	23.57075	25.26323	26.94801	28.62458	30.29244
39	16.76291	18.48091	20.19327	21.89948	23.59902	25.29137	26.97602	28.65246	30.32016
40	16.79159	18.50949	20.22176	21.92786	23.62729	25.31952	27.00403	28.68032	30.34788
41	16.82026	18.53808	20.25024	21.95624	23.65555	25.34766	27.03204	28.70819	30.37559
42	16.84894	18.56666	20.27873	21.98462	23.68381	25.37579	27.06004	28.73605	30.40331
43	16.87761	18.59524	20.30721	22.01300	23.71207	25.40393	27.08805	28.76391	30.43102
44	16.90628	18.62382	20.33569	22.04137	23.74033	25.43206	27.11605	28.79177	30.45872
45	16.93495	18.65240	20.36418	22.06974	23.76859	25.46019	27.14404	28.81963	30.48643
46	16.96362	18.68098	20.39265	22.09811	23.79684	25.48832	27.17204	28.84748	30.51413
47	16.99228	18.70956	20.42113	22.12648	23.82510	25.51645	27.20003	28.87533	30.54183
48	17.02095	18.73813	20.44961	22.15485	23.85335	25.54458	27.22802	28.90318	30.56953
49	17.04961	18.76670	20.47808	22.18321	23.88159	25.57270	27.25601	28.93103	30.59723
50	17.07828	18.79528	20.50655	22.21158	23.90984	25.60082	27.28400	28.95887	30.62492

(Continued)

Table 22. *(Continued)* **100 Millimeter Sine Bar Constants.** All Dimensions in Millimeters.
(Source, Suburban Tool Co.)

Min.	9°	10°	11°	12°	13°	14°	15°	16°	17°
51	17.10694	18.82385	20.53502	22.23994	23.93808	25.62894	27.31198	28.98671	30.65261
52	17.13560	18.85241	20.56349	22.26830	23.96633	25.65705	27.33997	29.01455	30.68030
53	17.16425	18.88098	20.59195	22.29666	23.99457	25.68517	27.36794	29.04239	30.70798
54	17.19291	18.90954	20.62042	22.32501	24.02280	25.71328	27.39592	29.07022	30.73566
55	17.22156	18.93811	20.64888	22.35337	24.05104	25.74139	27.42390	29.09805	30.76334
56	17.25022	18.96667	20.67734	22.38172	24.07927	25.76950	27.45187	29.12588	30.79102
57	17.27887	18.99523	20.70580	22.41007	24.10751	25.79760	27.47984	29.15371	30.81869
58	17.30752	19.02379	20.73426	22.43842	24.13574	25.82570	27.50781	29.18153	30.84636
59	17.33617	19.05234	20.76272	22.46676	24.16396	25.85381	27.53577	29.20935	30.87403

Min.	18°	19°	20°	21°	22°	23°	24°	25°	26°
0	30.90170	32.55682	34.20201	35.83679	37.46066	39.07311	40.67366	42.26183	43.83711
1	30.92936	32.58432	34.22935	35.86395	37.48763	39.09989	40.70024	42.28819	43.86326
2	30.95702	32.61182	34.25668	35.89110	37.51459	39.12666	40.72681	42.31455	43.88940
3	30.98468	32.63932	34.28400	35.91825	37.54156	39.15343	40.75337	42.34090	43.91553
4	31.01234	32.66681	34.31133	35.94540	37.56852	39.18019	40.77993	42.36725	43.94166
5	31.03999	32.69430	34.33865	35.97254	37.59547	39.20695	40.80649	42.39360	43.96779
6	31.06764	32.72179	34.36597	35.99968	37.62243	39.23371	40.83305	42.41994	43.99392
7	31.09529	32.74928	34.39329	36.02682	37.64938	39.26047	40.85960	42.44628	44.02004
8	31.12294	32.77676	34.42060	36.05395	37.67632	39.28722	40.88615	42.47262	44.04615
9	31.15058	32.80424	34.44791	36.08108	37.70327	39.31397	40.91269	42.49895	44.07227
10	31.17822	32.83172	34.47521	36.10821	37.73021	39.34071	40.93923	42.52528	44.09838
11	31.20586	32.85919	34.50252	36.13534	37.75714	39.36745	40.96577	42.55161	44.12448
12	31.23349	32.88666	34.52982	36.16246	37.78408	39.39419	40.99230	42.57793	44.15059
13	31.26112	32.91413	34.55712	36.18958	37.81101	39.42093	41.01883	42.60425	44.17668
14	31.28875	32.94160	34.58441	36.21669	37.83794	39.44766	41.04536	42.63056	44.20278
15	31.31638	32.96906	34.61171	36.24380	37.86486	39.47439	41.07189	42.65687	44.22887
16	31.34400	32.99653	34.63900	36.27091	37.89178	39.50111	41.09841	42.68318	44.25496
17	31.37163	33.02398	34.66628	36.29802	37.91870	39.52783	41.12492	42.70949	44.28104
18	31.39925	33.05144	34.69357	36.32512	37.94562	39.55455	41.15144	42.73579	44.30712
19	31.42686	33.07889	34.72085	36.35222	37.97253	39.58127	41.17795	42.76208	44.33319
20	31.45448	33.10634	34.74812	36.37932	37.99944	39.60798	41.20445	42.78838	44.35927
21	31.48209	33.13379	34.77540	36.40641	38.02634	39.63468	41.23096	42.81467	44.38534
22	31.50969	33.16123	34.80267	36.43351	38.05324	39.66139	41.25745	42.84095	44.41140
23	31.53730	33.18867	34.82994	36.46059	38.08014	39.68809	41.28395	42.86723	44.43746
24	31.56490	33.21611	34.85720	36.48768	38.10704	39.71479	41.31044	42.89351	44.46352
25	31.59250	33.24355	34.88447	36.51476	38.13393	39.74148	41.33693	42.91979	44.48957
26	31.62010	33.27098	34.91173	36.54184	38.16082	39.76818	41.36342	42.94606	44.51562
27	31.64770	33.29841	34.93898	36.56891	38.18770	39.79486	41.38990	42.97233	44.54167
28	31.67529	33.32584	34.96624	36.59599	38.21459	39.82155	41.41638	42.99859	44.56771
29	31.70288	33.35326	34.99349	36.62306	38.24147	39.84823	41.44285	43.02485	44.59375
30	31.73047	33.38069	35.02074	36.65012	38.26834	39.87491	41.46932	43.05111	44.61978
31	31.75805	33.40810	35.04798	36.67719	38.29522	39.90158	41.49579	43.07736	44.64581
32	31.78563	33.43552	35.07523	36.70425	38.32209	39.92825	41.52226	43.10361	44.67184
33	31.81321	33.46293	35.10246	36.73130	38.34895	39.95492	41.54872	43.12986	44.69786
34	31.84079	33.49034	35.12970	36.75836	38.37582	39.98158	41.57517	43.15610	44.72388
35	31.86836	33.51775	35.15693	36.78541	38.40268	40.00825	41.60163	43.18234	44.74990
36	31.89593	33.54516	35.18416	36.81246	38.42953	40.03490	41.62808	43.20857	44.77591
37	31.92350	33.57256	35.21139	36.83950	38.45639	40.06156	41.65453	43.23481	44.80192
38	31.95106	33.59996	35.23862	36.86654	38.48324	40.08821	41.68097	43.26103	44.82792

(Continued)

Table 22. *(Continued)* **100 Millimeter Sine Bar Constants.** All Dimensions in Millimeters. *(Source, Suburban Tool Co.)*

Min.	18°	19°	20°	21°	22°	23°	24°	25°	26°
39	31.97863	33.62735	35.26584	36.89358	38.51008	40.11486	41.70741	43.28726	44.85392
40	32.00619	33.65475	35.29306	36.92061	38.53693	40.14150	41.73385	43.31348	44.87992
41	32.03374	33.68214	35.32027	36.94765	38.56377	40.16814	41.76028	43.33970	44.90591
42	32.06130	33.70953	35.34748	36.97468	38.59060	40.19478	41.78671	43.36591	44.93190
43	32.08885	33.73691	35.37469	37.00170	38.61744	40.22141	41.81313	43.39212	44.95789
44	32.11640	33.76429	35.40190	37.02872	38.64427	40.24804	41.83956	43.41832	44.98387
45	32.14395	33.79167	35.42910	37.05574	38.67110	40.27467	41.86597	43.44453	45.00984
46	32.17149	33.81905	35.45630	37.08276	38.69792	40.30129	41.89239	43.47072	45.03582
47	32.19903	33.84642	35.48350	37.10977	38.72474	40.32791	41.91880	43.49692	45.06179
48	32.22657	33.87379	35.51070	37.13678	38.75156	40.35453	41.94521	43.52311	45.08775
49	32.25411	33.90116	35.53789	37.16379	38.77837	40.38114	41.97161	43.54930	45.11372
50	32.28164	33.92852	35.56508	37.19079	38.80518	40.40775	41.99801	43.57548	45.13967
51	32.30917	33.95589	35.59226	37.21780	38.83199	40.43436	42.02441	43.60166	45.16563
52	32.33670	33.98325	35.61944	37.24479	38.85880	40.46096	42.05080	43.62784	45.19158
53	32.36422	34.01060	35.64662	37.27179	38.88560	40.48756	42.07719	43.65401	45.21753
54	32.39174	34.03795	35.67380	37.29878	38.91239	40.51416	42.10358	43.68018	45.24347
55	32.41926	34.06531	35.70097	37.32577	38.93919	40.54075	42.12996	43.70634	45.26941
56	32.44678	34.09265	35.72814	37.35275	38.96598	40.56734	42.15634	43.73251	45.29535
57	32.47429	34.12000	35.75531	37.37973	38.99277	40.59393	42.18272	43.75866	45.32128
58	32.50180	34.14734	35.78248	37.40671	39.01955	40.62051	42.20909	43.78482	45.34721
59	32.52931	34.17468	35.80964	37.43369	39.04633	40.64709	42.23546	43.81097	45.37313

Min.	27°	28°	29°	30°	31°	32°	33°	34°	35°
0	45.39905	46.94716	48.48096	50.00000	51.50381	52.99193	54.46390	55.91929	57.35764
1	45.42497	46.97284	48.50640	50.02519	51.52874	53.01659	54.48830	55.94340	57.38147
2	45.45088	46.99852	48.53184	50.05037	51.55367	53.04125	5451269	55.96751	57.40529
3	45.47679	47.02419	48.55727	50.07556	51.57859	53.06591	54.53707	55.99162	57.42911
4	45.50269	47.04986	48.58270	50.10073	51.60351	53.09057	54.56145	56.01572	57.45292
5	45.52859	47.07553	48.60812	50.12591	51.62842	53.11521	54.58583	56.03981	57.47672
6	45.55449	47.10119	48.63354	50.15107	51.65333	53.13986	54.61020	56.06390	57.50053
7	45.58038	47.12685	48.65895	50.17624	51.67824	53.16450	54.63456	56.08798	57.52432
8	45.60627	47.15250	48.68436	50.20140	51.70314	53.18913	54.65892	56.11206	57.54811
9	45.63216	47.17815	48.70977	50.22655	51.72804	53.21376	54.68328	56.13614	57.57190
10	45.65804	47.20380	48.73517	50.25170	51.75293	53.23839	54.70763	56.16021	57.59568
11	45.68392	47.22944	48.76057	50.27685	51.77782	53.26301	54.73198	56.18428	57.61946
12	45.70979	47.25508	48.78597	50.30199	51.80270	53.28763	54.75632	56.20834	57.64323
13	45.73566	47.28071	48.81136	50.32713	51.82758	53.31224	54.78066	56.23239	57.66700
14	45.76153	47.30634	48.83674	50.35227	51.85246	53.33685	54.80499	56.25645	57.69076
15	45.78739	47.33197	48.86212	50.37740	51.87733	53.36145	54.82932	56.28049	57.71452
16	45.81325	47.35759	48.88750	50.40252	51.90219	53.38605	54.85365	56.30453	57.73827
17	45.83910	47.38321	48.91288	50.42765	51.92705	53.41064	54.87797	56.32857	57.76202
18	45.86496	47.40882	48.93825	50.45276	51.95191	53.43523	54.90228	56.35260	57.78576
19	45.89080	47.43443	48.96361	50.47788	51.97676	53.45982	54.92659	56.37663	57.80950
20	45.91665	47.46004	48.98897	50.50298	52.00161	53.48440	54.95090	56.40066	57.83323
21	45.94248	47.48564	49.01433	50.52809	52.02646	53.50898	54.97520	56.42467	57.85696
22	45.96832	47.51124	49.03968	50.55319	52.05130	53.53355	54.99950	56.44869	57.88069
23	45.99415	47.53683	49.06503	50.57828	52.07613	53.55812	55.02379	56.47270	57.90440
24	46.01998	47.56242	49.09038	50.60338	52.10096	53.58268	55.04807	56.49670	57.92812
25	46.04580	47.58801	49.11572	50.62846	52.12579	53.60724	55.07236	56.52070	57.95183
26	46.07162	47.61359	49.14105	50.65355	52.15061	53.63179	55.09663	56.54469	57.97553

(Continued)

Table 22. *(Continued)* **100 Millimeter Sine Bar Constants.** All Dimensions in Millimeters.
(Source, Suburban Tool Co.)

Min.	27°	28°	29°	30°	31°	32°	33°	34°	35°
27	46.09744	47.63917	49.16638	50.67863	52.17543	53.65634	55.12091	56.56868	57.99923
28	46.12325	47.66474	49.19171	50.70370	52.20024	53.68089	55.14518	56.59267	58.02292
29	46.14906	47.69031	49.21704	50.72877	52.22505	53.70543	55.16944	56.61665	58.04661
30	46.17486	47.71588	49.24236	50.75384	52.24986	53.72996	55.19370	56.64062	58.07030
31	46.20066	47.74144	49.26767	50.77890	52.27466	53.75449	55.21795	56.66459	58.09397
32	46.22646	47.76700	49.29298	50.80396	52.29945	53.77902	55.24220	56.68856	58.11765
33	46.25225	47.79255	49.31829	50.82901	52.32424	53.80354	55.26645	56.71252	58.14132
34	46.27804	47.81810	49.34359	50.85406	52.34903	53.82806	55.29069	56.73648	58.16498
35	46.30382	47.84364	49.36889	50.87910	52.37381	53.85257	55.31492	56.76043	58.18864
36	46.32960	47.86919	49.39419	50.90414	52.39859	53.87708	55.33915	56.78437	58.21230
37	46.35538	47.89472	49.41948	50.92918	52.42336	53.90158	55.36338	56.80832	58.23595
38	46.38115	47.92026	49.44476	50.95421	52.44813	53.92608	55.38760	56.83225	58.25959
39	46.40692	47.94579	49.47005	50.97924	52.47290	53.95058	55.41182	56.85619	58.28323
40	46.43269	47.97131	49.49532	51.00426	52.49766	53.97507	55.43603	56.88011	58.30687
41	46.45845	47.99683	49.52060	51.02928	52.52241	53.99955	55.46024	56.90403	58.33050
42	46.48420	48.02235	49.54587	51.05429	52.54717	54.02403	55.48444	56.92795	58.35412
43	46.50996	48.04786	49.57003	51.07930	52.57191	54.04851	55.50864	56.95187	58.37774
44	46.53571	48.07337	49.59639	51.10431	52.59665	54.07298	55.53283	56.97577	58.40136
45	46.56145	48.09888	49.62165	51.12931	52.62139	54.09745	55.55702	56.99968	58.42497
46	46.58719	48.12438	49.64690	51.15431	52.64613	54.12191	55.58121	57.02357	58.44857
47	46.61293	48.14987	49.67215	51.17930	52.67085	54.14637	55.60539	57.04747	58.47217
48	46.63866	48.17537	49.69740	51.20429	52.69558	54.17082	55.62956	57.07136	58.49577
49	46.66439	48.20086	49.72264	51.22927	52.72030	54.19527	55.65373	57.09524	58.51936
50	46.69012	48.22634	49.74787	51.25425	52.74502	54.21971	55.67790	57.11912	58.54294
51	46.71584	48.25182	49.77310	51.27923	52.76973	54.24415	55.70206	57.14299	58.56652
52	46.74156	48.27730	49.79833	51.30420	52.79443	54.26859	55.72621	57.16686	58.59010
53	46.76727	48.30277	49.82355	51.32916	52.81914	54.29302	55.75036	57.19073	58.61367
54	46.79298	48.32824	49.84877	51.35413	52.84383	54.31744	55.77451	57.21459	58.63724
55	46.81869	48.35370	49.87399	51.37908	52.86853	54.34187	55.79865	57.23844	58.66080
56	46.84439	48.37916	49.89920	51.40404	52.89322	54.36628	55.82279	57.26229	58.68435
57	46.87009	48.40462	49.92441	51.42899	52.91790	54.39069	55.84692	57.28614	58.70790
58	46.89578	48.43007	49.94961	51.45393	52.94258	54.41510	55.87105	57.30998	58.73145
59	46.92147	48.45552	49.97481	51.47887	52.96726	54.43951	55.89517	57.33381	58.75499
Min.	36°	37°	38°	39°	40°	41°	42°	43°	44°
0	58.77853	60.18150	61.56615	62.93204	64.27876	65.60590	66.91306	68.19984	69.46584
1	58.80206	60.20473	61.58907	62.95464	64.30104	65.62785	66.93467	68.22111	69.48676
2	58.82558	60.22795	61.61198	62.97724	64.32332	6564980	66.95628	68.24237	69.50767
3	58.84910	60.25117	61.63489	62.99983	64.34559	65.67174	66.97789	68.26363	69.52858
4	58.87262	60.27439	61.65779	63.02242	64.36785	65.69367	66.99948	68.28489	69.54949
5	58.89613	60.29760	61.68069	63.04500	64.39011	65.71560	67.02108	68.30613	69.57039
6	58.91964	60.32080	61.70359	63.06758	64.41236	65.73752	67.04266	68.32738	69.59128
7	58.94314	60.34400	61.72648	63.09015	64.43461	6575944	67.06424	68.34861	69.61217
8	58.96663	60.36719	61.74936	63.11272	64.45685	65.78135	67.08582	6836984	69.63305
9	58.99012	60.39038	61.77224	63.13528	64.47909	65.80326	67.10739	68.39107	69.65392
10	59.01361	60.41356	61.79511	63.15784	64.50132	65.82516	67.12895	68.41229	69.67479
11	59.03709	60.43674	61.81798	63.18039	64.52355	65.84706	67.15051	68.43350	69.69565
12	59.06057	60.45991	61.84084	63.20293	64.54577	6586895	67.17206	68.45471	69.71651
13	59.08404	60.48308	61.86370	63.22547	64.56798	65.89083	67.19361	68.47591	69.73736
14	59.10750	60.50624	61.88655	63.24800	64.59019	65.91271	67.21515	68.49711	69.75821

(Continued)

Table 22. *(Continued)* **100 Millimeter Sine Bar Constants.** All Dimensions in Millimeters. *(Source, Suburban Tool Co.)*

Min.	36°	37°	38°	39°	40°	41°	42°	43°	44°
15	59.13096	60.52940	61.90939	63.27053	64.61240	65.93458	67.23668	68.51830	69.77905
16	59.15442	60.55255	61.93224	63.29306	64.63460	65.95645	67.25821	68.53948	69.79988
17	59.17787	60.57570	61.95507	63.31557	64.65679	65.97831	67.27973	6856066	69.82071
18	59.20132	60.59884	61.97790	63.33809	64.67898	66.00017	67.30125	68.58184	69.84153
19	59.22476	60.62198	62.00073	63.36059	64.70116	66.02202	67.32276	68.60300	69.86234
20	59.24819	60.64511	62.02355	63.38310	64.72334	66.04386	67.34427	68.62416	69.88315
21	59.27163	60.66824	62.04636	63.40559	64.74551	66.06570	67.36577	68.64532	69.90396
22	59.29505	60.69136	62.06917	63.42808	64.76767	66.08754	67.38727	68.66647	69.92476
23	59.31847	60.71447	62.09198	63.45057	64.78984	66.10936	67.40876	68.68761	69.94555
24	59.34189	60.73758	62.11478	63.47305	64.81199	66.13119	67.43024	68.70875	69.96633
25	59.36530	60.76069	62.13757	63.49553	64.83414	66.15300	67.45172	68.72988	69.98711
26	59.38871	60.78379	62.16036	63.51800	64.85628	66.17481	67.47319	68.75101	70.00789
27	59.41211	60.80689	62.18314	63.54046	64.87842	66.19662	67.49466	68.77213	70.02866
28	59.43550	60.82998	62.20592	63.56292	64.90056	66.21842	67.51612	68.79325	70.04942
29	59.45889	60.85306	62.22870	63.58537	64.92268	66.24022	67.53757	68.81435	70.07018
30	59.48228	60.87614	62.25146	63.60782	64.94480	66.26200	67.55902	68.83546	70.09093
31	59.50566	60.89922	62.27423	63.63026	64.96692	66.28379	67.58046	68.85655	70.11167
32	59.52904	60.92229	62.29698	63.65270	64.98903	66.30557	67.60190	68.87765	70.13241
33	59.55241	60.94535	62.31974	63.67513	65.01114	66.32734	67.62333	68.89873	70.15314
34	59.57577	60.96841	62.34248	63.69756	65.03324	66.34910	67.64476	68.91981	70.17387
35	59.59913	60.99147	62.36522	63.71998	65.05533	66.37087	67.66618	68.94089	70.19459
36	59.62249	61.01452	62.38796	63.74240	65.07742	66.39262	67.68760	68.96195	70.21531
37	59.64584	61.03756	62.41069	63.76481	6509951	66.41437	67.70901	68.98302	70.23601
38	59.66918	61.06060	62.43342	63.78721	65.12158	66.43612	67.73041	69.00407	70.25672
39	59.69252	61.08363	62.45614	63.80961	65.14366	66.45785	67.75181	69.02512	70.27741
40	59.71586	61.10666	62.47885	63.83201	65.16572	66.47959	67.77320	69.04617	70.29811
41	59.73919	61.12969	62.50156	63.85440	65.18778	66.50131	67.79459	69.06721	70.31879
42	59.76251	61.15270	62.52427	63.87678	6520984	66.52304	67.81597	69.08824	70.33947
43	59.78583	61.17572	62.54696	63.89916	65.23189	66.54475	67.83734	69.10927	70.36014
44	59.80915	61.19873	62.56966	63.92153	65.25394	66.56646	67.85871	69.13029	70.38081
45	59.83246	61.22173	62.59235	63.94390	65.27598	66.58817	67.88007	69.15131	70.40147
46	59.85577	61.24473	62.61503	63.96626	65.29801	66.60987	67.90143	69.17232	70.42213
47	59.87906	61.26772	62.63771	63.98862	6532004	66.63156	67.92278	69.19332	70.44278
48	59.90236	61.29071	62.66038	64.01097	65.34206	66.65325	67.94413	69.21432	70.46342
49	59.92565	61.31369	62.68305	64.03332	65.36408	66.67493	67.96547	69.23531	70.48406
50	59.94893	61.336666	62.70571	64.05566	65.38609	66.69661	67.98681	69.25630	70.50469
51	59.97221	61.25964	62.72837	64.07799	65.40810	66.71828	68.00813	69.27728	70.52532
52	59.99549	61.38260	62.75102	64.10032	6543010	66.73994	68.02946	69.29825	70.54594
53	60.01876	61.40556	62.77366	64.12264	65.45209	66.76160	68.05078	69.31922	70.56655
54	60.04202	61.42852	62.79631	64.14496	65.47408	66.78326	68.07209	69.34018	70.58716
55	60.06528	61.45147	62.81894	64.16728	65.49607	66.80490	68.09339	69.36114	70.60776
56	60.08854	61.47442	62.84157	64.18958	65.51804	66.82655	68.11469	69.38209	70.62835
57	60.11179	61.49736	62.86420	64.21189	65.54002	66.84818	68.13599	69.40304	70.64894
58	60.13503	61.52029	62.88682	64.23418	65.56198	66.86981	68.15728	69.42398	70.66953
59	60.15827	61.54322	62.90943	64.25647	65.58395	66.89144	68.17856	69.44491	70.69011

Sine bars can also be used to measure the angle on a finished part by finding the angle corresponding to a given measurement rather than the measurement for setting the bar to a given angle. This is done by setting the bar to the exact angle of the existing part, and then finding the sum of the gage blocks. That value can then be found on the sine bar constant Table to determine the angle. For example, if the height is 53.36 mm, and a 100 mm sine bar is being used to measure the angle, find 53.36 mm on the 100 mm sine bar constant Table to determine that the angle is 32° 15'.

By using the compound sine plate, compound angles (angles in two dimensions) can be set. With compound angles, it is necessary to compensate for the first angle set before setting the second angle. The compensation for the part shown in *Figure 14 (b)* can be determined with the following procedure.

> A = first desired angle on part
>
> B = second desired angle on part
>
> C = true angle setting required to obtain B.

First, set angle A on the top section of the compound sine plate.

Next, calculate the tangent of angle C with the formula

> $\tan C = \tan B \times \cos A.$

Then, determine the identity of angle C by looking up its tangent in the "Natural Trigonometric Functions" Table.

The lower section can now be set to compensate for angle A. The compound sine plate is now set to obtain the desired angles (A and B) in their respective planes. See *Figure 14 (a)*.

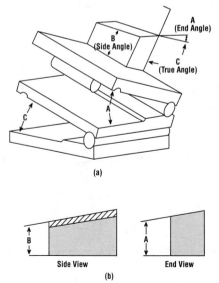

Figure 14. (a) Compound sine plate. (b) Profiles of part. (Source, Suburban Tool.)

Calculating Tapers

Tapers are usually expressed as a ratio. A taper of 1:15, for example, would indicate a taper of 1 inch in 15 inches, or perhaps 1 mm in 15 mm. In English units, tapers are normally calculated in inches per foot, so a ratio of 3.5:12 would be a common way of expressing a taper. It is important to note that the parts shown in *Figures 15* and *16* are conical in shape, and therefore the top and bottom taper an equal amount over the length of the part. The equations given below for solving unknown values in these illustrations are based on dividing the part down its center and solving for the taper on only one side of the part.

Tapers can be calculated with a sine bar, but first it is necessary to determine the angle of the taper to find the height of the gage blocks necessary to duplicate the angle. A simple explanation of this procedure can be shown with the part illustrated in *Figure 15*. First, determine the length of the part, which in this example will be 12 inches. Next determine the height of the part at each end and subtract the smaller from the larger. Then divide the difference by two to find the taper. In the example part, the height of the large end is four inches, and the height of the small end is two inches, so the part tapers two inches over a distance of twelve inches. This means that the taper is 1 inch per foot on the top, and one inch per foot on the bottom. To solve the problem, the taper on one side is used, which means the taper per foot (*tpf*) is 1. The angle at which to set the sine bar can now be found with the following formula.

$$\tan \alpha \div 2 = tpf \div 24$$

$$\tan \alpha = 1 \div 24 = .041666 = 2° 23' 10" \times 2$$

$$\alpha = 4° 46' 20".$$

Using the tapered part shown in *Figure 15*, it is also possible to find the length of the part if the *tpf* and the end diameters are known.

$$\tan \alpha \div 2 = tpf \div 24$$

$$1 \div 24 = .041666 = 2° 23' 10" \times 2 = 4° 46' 20"$$

Next find the sine of 4° 46' 20", and divide it into the difference between the large end and small end of the part.

$$\sin 4° 46' 20" = .08320$$

$$2 \div (2 \times .08320) = 12"$$

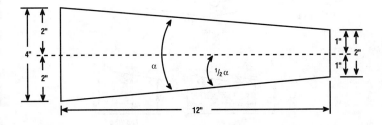

Figure 15. The angle formed by the taper of one inch per foot.

When the taper per foot is not known, the angle with the centerline ($^1/_2\ \alpha$) can be calculated. The included angle is represented in *Figure 16* by angle α, and the angle with the centerline equals $^1/_2\ \alpha$.

$$\tan \frac{\alpha}{2} = \frac{\dfrac{E \text{-} C}{2}}{L}$$

If, for example, $E = 3$ inches, and C equals 1 inch, and L equals 10 inches

$$\tan \frac{\alpha}{2} = \frac{\dfrac{3 \text{-} 1}{2}}{10} = 0.1 = 5° \ 42' \ 40'' \text{ (which is the angle with the centerline)} \times 2$$

$\alpha = 11° \ 25' \ 40''$.

The length in this example is 10 inches rather than a foot, but determining the taper per foot is not difficult. Subtract the diameter of the smaller end from the diameter of the larger end, divide by the length of the taper, and multiply by 12. In the above example, the *tpf* would be

$$\frac{3 \text{-} 1}{10} \times 12 = 2.4.$$

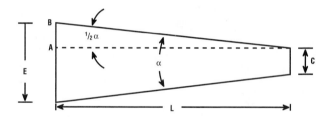

Figure 16. Included and centerline angles within a tapered part.

Two tables are included which greatly simplify finding tapers. **Table 23** provides centerline angles and included angles expressed in both degree and radians, for taper per foot ranging from $^1/_{64}$ inch to 10 inches. Use this table when the taper per foot is known. **Table 24** gives taper per inch for angles and minutes (in increments of 10 minutes) for included angles from 0° to 90°. Any number on this table can be converted to taper per foot by multiplying the value by 12. For example, to find the taper per foot for an included angle of 30° 20', find 30° in the left column, then find 20' in the top row and follow it down until it intersects with the 30° row. The taper per inch is 0.54214, so the taper per foot will be 12 × .54214 = 6.50568 inches. The taper of the angle with centerline would be one-half the taper of the included angle, or 3.25284 inches. The answer can be approximately checked by finding the taper per foot for 6.5 inches in **Table 23**, which is 30° 18' 28". The taper per inch values in **Table 24** can be multiplied by 25.4 to convert the values to "tapers per millimeter."

Calculating Dovetails

Dimensioning dovetails and gibs (gibs are wedge shaped pieces of material that hold the slide in place, or are used to adjust a bearing, etc.—they are commonly used to adjust

(Text continued on p. 183)

Table 23. Taper Per Foot Expressed in Angles and Radians.

Taper per foot	Angle with Centerline	Angle with Centerline in Radians	Included Angle	Included Angle in Radians
1/64	0° 2' 15"	0.0006535283	0° 4' 30"	0.0013070566
1/32	0° 4' 30"	0.0013070566	0° 8' 59"	0.0026141131
3/64	0° 6' 44"	0.0019605848	0° 13 29"	0.0039221415
1/16	0° 8' 57"	0.0026040291	0° 17 54"	0.0052068968
5/64	0° 11' 12"	0.0032555234	0° 22' 23"	0.0065110467
3/32	0° 13' 26"	0.0039090516	0° 26' 53"	0.0078181033
7/64	0° 15' 41"	0.0045625799	0° 31' 22"	0.0091241924
1/8	0° 17' 56"	0.0052161082	0° .35' 52"	0.0104331882
9/64	0° 20' 8"	0.0058575184	0° 40' 16"	0.0117150368
5/32	0° 22' 23"	0.0065110467	0° 44' 46"	0.0130220934
11//64	0° 24' 38"	0.0071645750	0° 49' 16"	0.0143310914
3/16	0° 26' 53"	0.0078181033	0° 53' 45"	0.0156352391
13/64	0° 29' 5"	0.0084595135	0° 58' 10"	0.0169190270
7/32	0° 31' 20"	0.0091130418	1° 2' 39"	0.0182260841
15/64	0° 33' 35"	0.0097665701	1° 7' 9"	0.0195331428
1/4	0° 35' 49"	0.0104200983	1° 11' 39"	0.0208421386
17/64	0° 38' 4"	0.0110716411	1° 16' 7"	0.0221432827
9/32	0° 40' 16"	0.0117150368	1° 20' 33"	0.0234300742
19/64	0° 42' 31"	0.0123675955	1° 25' 2"	0.0247351937
5/16	0° 44' 46"	0.0130220934	1° 29' 32"	0.0260441895
21/64	0° 47' 0"	0.0136731998	1° 34' 1"	0.0273464002
11/32	0° 49' 16"	0.0143306066	1° 38' 32"	0.0286612137
23/64	0° 51' 28"	0.0149705602	1° 42' 56"	0.0299411209
3/8	0° 53' 43"	0.0156240885	1° 47' 25"	0.0312462404
25/64	0° 55' 43"	0.0162058649	1° 51' 25"	0.0324097932
13/32	0° 58' 12"	0.0169291595	1° 56' 24"	0.0338583195
27/64	1° 0' 25"	0.0175725558	2° 0' 49"	0.0351451116
7/16	1° 2' 39"	0.0182261325	2° 4' 79"	0.0364522651
29/64	1° 4' 54"	0.0188786427	2° 9' 48"	0.0377572876
15/32	1° 7' 9"	0.0195311551	2° 14' 17"	0.0390623102
31/64	1° 9' 23"	0.0201846834	2° 18' 47"	0.0403693668
1/2	1° 11' 36"	0.0208281276	2° 23' 12"	0.0416551908
33/64	1° 13' 51"	0.0214816559	2° 27' 42"	0.0429641866
17/32	1° 16' 5"	0.0221331502	2° 32' 11"	0.0442663003
35/64	1° 18' 20"	0.0227866784	2° 36' 40"	0.0455733569
9/16	1° 20' 35"	0.0234402067	2° 41' 10"	0.0468814822
37/64	1° 22' 48"	0.0240836510	2° 45' 35"	0.0481662375
19/32	1° 25' 2"	0.0247351452	2° 50' 4"	0.0494702905
39/64	1° 27' 17"	0.0253886735	2° 54' 34"	0.0507773470
5/8	1° 29' 32"	0.0260422018	2° 59' 3"	0.0520835331
41/64	1° 31' 44"	0.0266856460	3° 3' 29"	0.0533731371
21/32	1° 33' 59"	0.0273391743	3° 7' 58"	0.0546772848
43/64	1° 36' 13"	0.0279906686	3° 12' 27"	0.0559813377
11/16	1° 38' 28"	0.0286441969	3° 16' 57"	0.0572883943

(Continued)

Table 23. *(Continued)* **Taper Per Foot Expressed in Angles and Radians.**

Taper per foot	Angle with Centerline	Angle with Centerline in Radians	Included Angle	Included Angle in Radians
45/64	1° 40' 41"	0.0292876411	3° 21' 22"	0.0585751880
23/32	1° 42' 56"	0.0299411694	3° 25' 52"	0.0598841838
47/64	1° 45' 10"	0.0305926637	3° 30' 20"	0.0611853278
3/4	1° 47' 25"	0.0312462404	3° 34' 50"	0.0624924814
49/64	1° 49' 38"	0.0318896362	3° 39' 15"	0.0637772389
25/32	1° 51' 53"	0.0325431645	3° 43' 45"	0.0650862347
51/64	1° 54' 7"	0.0331947072	3° 48' 14"	0.0663894149
13/16	1° 56' 22"	0.0338482655	3° 42' 43"	0.0647875895
53/64	1° 58' 34"	0.0344916312	3° 47' 9"	0.0660752559
27/32	2° 0' 49"	0.0351451600	4° 1' 38"	0.0702882861
55/64	2° 3' 4"	0.0357967028	4° 6' 7"	0.0715934056
7/8	2° 5' 16"	0.0364400986	4° 10' 33"	0.0728801971
57/64	2° 7' 31"	0.0370936753	4° 15' 2"	0.0741861893
29/32	2° 9' 46"	0.0377472036	4° 19' 32"	0.0754951851
59/64	2° 12' 2"	0.0384069396	4° 23' 4"	0.0765229911
15/16	2° 14' 13"	0.0390421421	4° 28' 26"	0.0780842842
61/64	2° 16' 28"	0.0396956704	4° 32' 56"	0.0793913408
31/32	2° 18' 43"	0.0403491987	4° 37' 25"	0.0806972360
63/64	2° 20' 55"	0.0409925944	4° 41' 51"	0.0819868394
1	2° 23' 10"	0.0416441372	4° 46' 19"	0.0832882743
1 1/16	2° 32' 6"	0.0442461322	5° 4' 13"	0.0884922650
1 1/8	2° 41' 1"	0.0468380433	5° 22' 2"	0.0936760870
1 3/16	2° 49' 56"	0.0494319398	5° 39' 52"	0.0988632036
1 1/4	2° 58' 55"	0.0520440674	5° 57' 50"	0.1040894950
1 5/16	3° 7' 49"	0.0546359305	6° 16' 39"	0.1095630422
1 3/8	3° 16' 46"	0.0572379740	6° 33' 32"	0.1144742049
1 7/16	3° 25'41"	0.0598298366	6° 51' 32"	0.1197101925
1 1/2	3° 34' 35"	0.0624217476	7° 9' 11"	0.1248443705
1 9/16	3° 43' 30"	0.0650136586	7° 27' 0"	0.1300270290
1 5/8	3° 52' 25"	0.0676055696	7° 44' 49"	0.1352111398
1 11/16	4° 1' 19"	0.0701974812	8° 2' 39"	0.1403949623
1 3/4	4° 10' 14"	0.0727893437	8° 20' 28"	0.1455786874
1 13/16	4° 19' 9"	0.0753812547	8° 38' 17"	0.1507625095
1 7/8	4° 28' 1"	0.0779630817	8° 56' 2"	0.1559261634
1 15/16	4° 36' 56"	0.0805569782	9° 73' 52"	0.1785665725
2	4° 45' 51"	0.0831488893	9° 31' 41"	0.1662959390
2 1/16	4° 54' 43"	0.0857306678	9° 49' 26"	0.1714592051
2 1/8	5° 3' 36"	0.0883119135	10° 7' 11"	0.1766224717
2 3/16	5° 12' 28"	0.0908942737	10° 24' 57"	0.1817885475
2 1/4	5° 21' 21"	0.0934761007	10° 42' 42"	0.1869522014
2 5/16	5° 30' 13"	0.0960578792	11° 0' 27"	0.1921157590
2 3/8	5° 39' 6"	0.0986397062	11° 18' 12"	0.1972794129
2 7/16	5° 47' 59"	0.1012235187	11° 35' 58"	0.2024484945
2 1/2	5° 56' 49"	0.1037952131	11° 53' 39"	0.2075923682

(Continued)

Table 23. *(Continued)* **Taper Per Foot Expressed in Angles and Radians.**

Taper per foot	Angle with Centerline	Angle with Centerline in Radians	Included Angle	Included Angle in Radians
2 9/16	6° 5' 42"	0.1063770406	12° 11' 24"	0.2127556348
2 5/8	6° 14' 32"	0.1089487351	12° 29' 5"	0.2178995085
2 11/16	6° 23' 23"	0.1115204295	12° 46' 45"	0.2230408590
2 3/4	6° 32' 13"	0.1140920270	13° 4' 26"	0.2281840545
2 13/16	6° 41' 2"	0.1166537344	13° 32' 3"	0.2362163512
2 7/8	6° 49' 52"	0.1192274628	13° 39' 45"	0.2384556071
2 15/16	6° 58' 43"	0.1217991572	13° 57' 26"	0.2435994808
3	7° 7' 31"	0.1243607682	14° 15' 3"	0.2487239626
3 1/8	7° 25' 8"	0.1294839405	14° 50' 16"	0.2589678810
3 1/4	7° 42' 45"	0.1346091468	15° 25' 30"	0.2692170358
3 3/8	8° 0' 18"	0.1397121515	16° 0' 35"	0.2794243031
3 1/2	8° 17' 51"	0.1448171897	16° 35' 41"	0.2896325394
3 5/8	8° 35' 21"	0.1499101099	17° 10' 42"	0.2998202204
3 3/4	8° 52' 52"	0.1550050156	17° 45' 44"	0.3100092582
3 7/8	9° 10' 18"	0.1600777197	18° 20' 37"	0.3201554394
4	9° 27' 47"	0.1651625414	18° 55' 34"	0.3303229518
4 1/8	9° 45' 9"	0.1702132346	19° 30' 18"	0.3404264697
4 1/4	10° 2' 31"	0.1752633443	20° 5' 1"	0.3505251395
4 3/8	10° 19' 52"	0.1803135021	20° 39' 45"	0.3606286547
4 1/2	10° 37' 10"	0.1853442696	21° 14' 20"	0.3706885397
4 5/8	10° 54' 28"	0.1903749864	21° 48' 55"	0.3807499734
4 3/4	11° 11' 43"	0.1953960561	22° 23' 27"	0.3907937627
4 7/8	11° 28' 55"	0.2003977328	22° 57' 50"	0.4007954678
5	11° 46' 7"	0.2053993633	23° 32' 13"	0.4107987271
5 1/8	12° 3' 14"	0.2103815027	24° 6' 29"	0.4207630053
5 1/4	12° 20' 22"	0.2153639324	24° 40' 44"	0.4307278649
5 3/8	12° 37' 26"	0.2203264850	25° 14' 51"	0.4406529705
5 1/2	12° 54' 27"	0.2252798262	25° 48' 55"	0.4505596529
5 5/8	13° 11' 27"	0.2302234717	26° 22' 54"	0.4604469433
5 3/4	13° 28' 25"	0.2351569356	26° 56' 49"	0.4703138712
5 7/8	13° 45' 18"	0.2400714920	27° 30' 37"	0.4801429845
6	14° 2' 10"	0.2449744134	28° 4' 19"	0.4899488269
6 1/8	14° 19' 0"	0.2498729711	28° 38' 0"	0.4997459422
6 1/4	14° 35' 48"	0.2547589215	29° 31' 36"	0.5153375489
6 3/8	14° 52' 32"	0.2596269361	29° 45' 04"	0.5192548445
6 1/2	15° 9' 14"	0.2644862246	30° 18' 28"	0.5289724492
6 5/8	15° 25' 55"	0.2693362021	30° 51' 49"	0.5386716307
6 3/4	15° 42' 31"	0.2741669833	31° 25' 2"	0.5483339693
6 7/8	15° 59' 6"	0.2789884562	31° 58' 11"	0.5579769128
7	16° 15' 38"	0.2838007644	32° 31' 16"	0.5676004644
7 1/8	16° 32' 5"	0.2885843736	33° 4' 10"	0.5771687477
7 1/4	16° 48' 32"	0.2933687563	33° 37' 3"	0.5867360609
7 3/8	17° 4' 57"	0.2981435403	34° 9' 53"	0.5962868888
7 1/2	17° 21' 15"	0.3028873465	34° 42' 30"	0.6057746931

(Continued)

Table 23. *(Continued)* **Taper Per Foot Expressed in Angles and Radians.**

Taper per foot	Angle with Centerline	Angle with Centerline in Radians	Included Angle	Included Angle in Radians
7 5/8	17° 37' 32"	0.3076219392	35° 15' 4"	0.6152479536
7 3/4	17° 53' 46"	0.3123473690	35° 47' 33"	0.6246969709
7 7/8	18° 9' 57"	0.3170546674	36° 19' 54"	0.6341072039
8	18° 26' 6"	0.3217526592	36° 52' 13"	0.6435053184
8 1/8	18° 42' 12"	0.3264326138	37° 24' 23"	0.6528652281
8 1/4	18° 58' 13"	0.3310944365	37° 56' 26"	0.6621888735
8 3/8	19° 14' 13"	0.3357471453	38° 28' 26"	0.6714942906
8 1/2	19° 30' 9"	0.3403816248	39° 0' 18"	0.6807632500
8 5/8	19° 46' 3"	0.3450070382	39° 32' 6"	0.6900140770
8 3/4	20° 1' 53"	0.3496144151	40° 3' 46"	0.6992273780
8 7/8	20° 17' 40"	0.3542034679	40° 35' 19"	0.7084049017
9	20° 33' 23"	0.3587742917	41° 6' 45"	0.7175485840
9 1/8	20° 49' 2"	0.3633269352	41° 38' 3"	0.7266538710
9 1/4	21° 4' 39"	0.3678736624	42° 9' 19"	0.7357483935
9 3/8	21° 20' 14"	0.3724029865	42° 40' 27"	0.7448059730
9 1/2	21° 35' 44"	0.3769147099	43° 11' 28"	0.7538270954
9 5/8	21° 51' 11"	0.3814087883	43° 42' 22"	0.7628175772
9 3/4	22° 6' 35"	0.3858852180	44° 13' 9"	0.7717700504
9 7/8	22° 21' 56"	0.3903528721	44° 43' 52"	0.7807051647
10	22° 37' 12"	0.3947940578	45° 14' 24"	0.7895881161

Table 24. Taper Per Inch for a Given Angle. Taper/in. = tan $\alpha/2 \times 2$.

Deg.	Minutes						
	0'	10'	20'	30'	40'	50'	60'
0	.00000	.00290	.00582	.00872	.01164	.01454	.01746
1	.01746	.02036	.02326	.02618	.02910	.03200	.03492
2	.03492	0.3782	.04072	.04364	.04656	.04946	.05238
3	.5238	.05528	.05820	.06110	.06402	.06692	.06984
4	.06984	.07276	.07566	.07858	.08150	.08440	.08732
5	.08732	.09024	.09316	.09606	.09898	.10190	.10482
6	.10482	.10774	.11066	.11356	.11648	.11940	.12232
7	.12232	.12524	.12816	.13108	.13400	.13694	.13986
8	.13986	.14278	.14570	.14862	.15156	.15448	.15740
9	.15740	.16034	.16326	.16618	.16912	.17204	.17498
10	.17498	.17790	.18084	.18378	.18670	.18964	.19258
11	.19258	.19552	.19846	.20138	.20432	.20726	.21020
12	.21020	.21314	.21610	.21904	.22198	.22492	.22788
13	.22788	.23082	.23376	.23672	.23966	.24262	.24556
14	.24556	.24852	.25148	.25444	.25738	.26034	.26330
15	.26330	.26626	.26922	.27218	.27516	.27812	.28108
16	.28108	.28404	.28702	.28998	.29296	.29592	.29890
17	.29890	.30188	.30486	.30782	.31080	.31378	.31676
18	.31676	.31976	.32274	.32572	.32870	.33170	.33468
19	.33468	.33768	.34066	.34366	.34666	.34966	.35266

(Continued)

Table 24. *(Continued)* **Taper Per Inch for a Given Angle.** Taper/in. = tan α/2 × 2.

Deg.	Minutes						
	0'	10'	20'	30'	40'	50'	60'
20	.35266	.35566	.35866	.36166	.36466	.36768	.37068
21	.37068	.37368	.37670	.37972	.38272	.38574	.38876
22	.38876	.39178	.39480	.39782	.40084	.40388	.40690
23	.40690	.40994	.41296	.41600	.41904	.42208	.42512
24	.42512	.42816	.43120	.43424	.43728	.44034	.44338
25	.44338	.44644	.44950	.45256	.45562	.45868	.46174
26	.46174	.46480	.46786	.47094	.47400	.47708	.48016
27	.48016	.48324	.48632	.48940	.49248	.49556	.49866
28	.49866	.50174	.50484	.50794	.51004	.51414	.51724
29	.51724	.52034	.52344	.52656	.52966	.53278	.53590
30	.53590	.53902	.54214	.54526	.54838	.55152	.55464
31	.55464	.55778	.56092	.56406	.56720	.57034	.57350
32	.57350	.57664	.57980	.58294	.58610	.58926	.59242
33	.59242	.59560	.59876	.60194	.60510	.60828	.61146
34	.61146	.61464	.61782	.62102	.62420	.62740	.63060
35	.63060	.63380	.63700	.64020	.64342	.64662	.64984
36	.64984	.65306	.65628	.65950	.66272	.66596	.66920
37	.66920	.67242	.67566	.67890	.68216	.68540	.68866
38	.68866	.69192	.69516	.69844	.70170	.70496	.70824
39	.70824	.71152	.71480	.71808	.72136	.72464	.72794
40	.72794	.73124	.73454	.73784	.74114	.74446	.74776
41	.74776	.75108	.75440	.75774	.76106	.76440	.76772
42	.76772	.77106	.77442	.77776	.78110	.78446	.78782
43	.78782	.79118	.79454	.79792	.80130	.80468	.80806
44	.80806	.81144	.81482	.81822	.82162	.82502	.82842
45	.82842	.83184	.83526	.83866	.84210	.84552	.84894
46	.84894	.85238	.85582	.85926	.86272	.86616	.86962
47	.86962	.87308	.87656	.88002	.88350	.88698	.89046
48	.89046	.89394	.89744	.90094	.90444	.90794	.91146
49	.91146	.91496	.91848	.92200	.92554	.92908	.93262
50	.93262	.93616	.93970	.94326	.94682	.95038	.95396
51	.95396	.95752	.96110	.96468	.96828	.97186	.97546
52	.97546	.97906	.98268	.98630	.98990	.99354	.99716
53	.99716	1.00080	1.00444	1.00808	1.01174	1.01538	1.01906
54	1.01906	1.02272	1.02638	1.03006	1.03376	1.03744	1.04114
55	1.04114	1.04484	1.04854	1.05226	1.05596	1.05970	1.06342
56	1.06342	1.06716	1.07090	1.07464	1.07840	1.08214	1.08592
57	1.08592	1.08968	1.09346	1.09724	1.10102	1.10482	1.10862
58	1.10862	1.11242	1.11624	1.12006	1.12388	1.12770	1.13154
59	1.13154	1.13538	1.13924	1.14310	1.14696	1.15082	1.15470
60	1.15470	1.15858	1.16248	1.16636	1.17026	1.17418	1.17810
61	1.17810	1.18202	1.18594	1.18988	1.19382	1.19776	1.20172
62	1.20172	1.20568	1.20966	1.21362	1.21762	1.22160	1.22560
63	1.22560	1.22960	1.23362	1.23764	1.24166	1.24570	1.24974

(Continued)

Table 24. *(Continued)* **Taper Per Inch for a Given Angle.** Taper/in. = tan α/2 × 2.

Deg.	Minutes						
	0'	10'	20'	30'	40'	50'	60'
64	1.24974	1.25378	1.25784	1.26190	1.26598	1.27006	1.27414
65	1.27414	1.27824	1.28234	1.28644	1.29056	1.29468	1.29882
66	1.29882	1.30296	1.30710	1.31120	1.31542	1.31960	1.32378
67	1.32378	1.32796	1.33216	1.33636	1.34056	1.34478	1.34902
68	1.34902	1.35326	1.35750	1.36176	1.36602	1.37028	1.37456
69	1.37456	1.37984	1.38314	1.38744	1.39176	1.39608	1.40042
70	1.40042	1.40476	1.40910	1.41346	1.41782	1.42220	1.42658
71	1.42658	1.43098	1.43538	1.43980	1.44422	1.44864	1.45308
72	1.45308	1.45754	1.46200	1.46646	1.47094	1.47542	1.47992
73	1.47992	1.48442	1.48894	1.49348	1.49800	1.50256	1.50710
74	1.50710	1.51168	1.51624	1.52084	1.52544	1.53004	1.53466
75	1.53466	1.53928	1.54392	1.54856	1.55322	1.55790	1.56258
76	1.56258	1.56726	1.57196	1.57668	1.58140	1.58612	1.59088
77	1.59088	1.59562	1.60040	1.60516	1.60996	1.61476	1.61966
78	1.61956	1.62440	1.62922	1.63406	1.63892	1.64380	1.64868
79	1.64868	1.65356	1.65846	1.66338	1.66830	1.67324	1.67820
80	1.67820	1.68316	1.68814	1.69312	1.69812	1.70314	1.70816
81	1.70816	1.71320	1.71824	1.72332	1.72836	1.73348	1.73858
82	1.73858	1.74368	1.74882	1.75396	1.75910	1.76428	1.76946
83	1.76946	1.77464	1.77984	1.78506	1.79030	1.79554	1.80080
84	1.80080	1.80608	1.81138	1.81668	1.82198	1.82732	1.83266
85	1.83266	1.83802	1.84340	1.84878	1.85418	1.85960	1.86504
86	1.86504	1.87048	1.87594	1.88142	1.88690	1.89240	1.89792
87	1.89792	1.90346	1.90902	1.91458	1.92016	1.92576	1.93138
88	1.93138	1.93700	1.94266	1.94832	1.95400	1.95968	1.96540
89	1.96540	1.97112	1.97686	1.98262	1.98840	1.99420	2.00000
90	2.00000	-	-	-	-	-	-

dovetailed parts.). In *Figure 17*, a dovetailed slide with accommodation for a $1/4$" gib is shown. The dimensions for machining the dovetail must be determined beforehand, using the formulas found below. In the drawing, the angle is given as 60°, the height of the slide as $5/8$ inch, and the width of the slide as 2 inches. Determine what the dimensions A and B must be to accommodate a $1/4$ inch gib.

First, split the opening into two 1 inch sections. Dimension A is equal to 1 inch minus the amount of relief that must be cut to allow for the $5/8$ inch high 60° angle. Drawing a line perpendicular to the base creates a right-angled triangle (as shown by the dotted line that forms the base of the triangle). One angle is given (60°) and the length of the base is $5/8$ inch (.625 inch). The depth of the dovetail is

side ÷ tan of angle opposite = .625 ÷ 1.7321 = .360 inch

Dimension *A*, therefore, is 1" - .360" = .640 inch.

Dimension *B* must accommodate the gib, which is $1/4$" thick, plus the amount of relief necessary to accommodate the 60° dovetail. The diameter of the base of the $1/4$" gib is found with the formula

$$D = (h \div \sin \alpha)$$

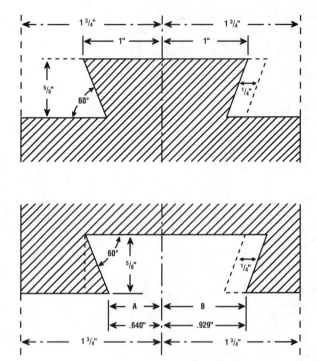

Figure 17. Dimensions of external and internal sections of a dovetail slide.

$D = (.250 \div .86603) = .289$

Therefore, dimension *B* must equal 1 inch, less the amount required for relief of the dovetail, plus the amount required to fit the gib.

$1 - .360 + .289 = .929$ inch.

Table 25 provides dimensions for the amount of relief necessary for cutting slides of different heights and angles. Dimensions for a 45° angle are not included in the Table, but for 45° the amount of relief required is exactly the same as the height of the dovetail.

A cylindrical pin or wire is often used to measure an existing dovetailed slide. By inserting the pin or wire of a known diameter within an internal dovetail, or on the outside of an external dovetail, measurements can be taken to confirm the dimensions of the dovetail, as shown in *Figure 18*. In the illustration, the dimensions of *x* and *y* can be calculated with the following equations.

$x = d \left(1 + \cot \tfrac{1}{2} \, \alpha\right) + A$

$y = B - d \left(1 + \cot \tfrac{1}{2} \, \alpha\right).$

In the equations for solving for x and y, "$d \times (1 + \cot \frac{1}{2}\alpha)$" is a constant relating to the angle and pin size. Therefore, A can be solved with the equation

$A = x - d(1 + \cot \frac{1}{2}\alpha)$,

and B can be solved with the equation

$B = y + d(1 + \cot \frac{1}{2}\alpha)$.

Constants for standard size pins are given in **Table 26**. To use the Table, determine the pin diameter and the angle, and then find the value for the constant. For example, if the pin size is $\frac{1}{8}$ inch, and the angle is $60°$, the value from the Table is 0.34151. A and B can therefore be found as follows.

$A = x - 0.34151$, and $B = y + 0.34151$.

Finally, both A and B can be solved for if the height (h) and angle of the dovetail is known, with the following equations.

$A = B - (h \times 2 \tan \alpha)$, and $B = A + (h \times 2 \tan \alpha)$.

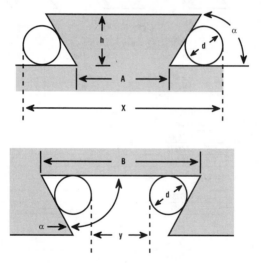

Figure 18. Using pins to measure dovetailed slides.

Table 25. Dimensions of Dovetailed Slides and Gibs.

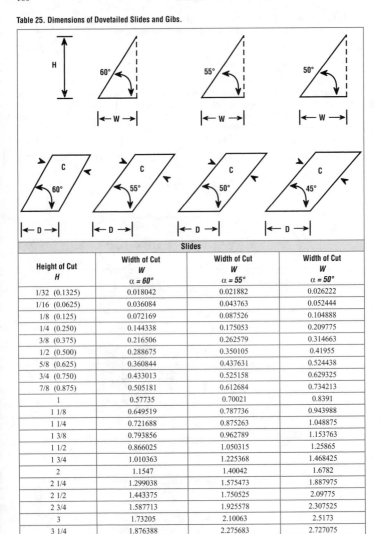

Slides			
Height of Cut **H**	**Width of Cut** **W** $\alpha = 60°$	**Width of Cut** **W** $\alpha = 55°$	**Width of Cut** **W** $\alpha = 50°$
1/32 (0.1325)	0.018042	0.021882	0.026222
1/16 (0.0625)	0.036084	0.043763	0.052444
1/8 (0.125)	0.072169	0.087526	0.104888
1/4 (0.250)	0.144338	0.175053	0.209775
3/8 (0.375)	0.216506	0.262579	0.314663
1/2 (0.500)	0.288675	0.350105	0.41955
5/8 (0.625)	0.360844	0.437631	0.524438
3/4 (0.750)	0.433013	0.525158	0.629325
7/8 (0.875)	0.505181	0.612684	0.734213
1	0.57735	0.70021	0.8391
1 1/8	0.649519	0.787736	0.943988
1 1/4	0.721688	0.875263	1.048875
1 3/8	0.793856	0.962789	1.153763
1 1/2	0.866025	1.050315	1.25865
1 3/4	1.010363	1.225368	1.468425
2	1.1547	1.40042	1.6782
2 1/4	1.299038	1.575473	1.887975
2 1/2	1.443375	1.750525	2.09775
2 3/4	1.587713	1.925578	2.307525
3	1.73205	2.10063	2.5173
3 1/4	1.876388	2.275683	2.727075
3 1/2	2.020725	2.450735	2.93685
3 3/4	2.165063	2.625788	3.146625
4	2.3094	2.80084	3.3564
4 1/4	2.453738	2.975893	3.566175
4 1/2	2.598075	3.150945	3.77595

(Continued)

Table 25. *(Continued)* **Dimensions of Dovetailed Slides and Gibs.**

Gibs				
Width of Gibs C	Width of Cut D α = 60°	Width of Cut D α = 55°	Width of Cut D α = 50°	Width of Cut D α = 45°
1/8 (0.125)	0.144337	0.152597	0.163177	0.176776
3/16 (0.1875)	0.216505	0.228896	0.244765	0.265164
1/4 (0.250)	0.288674	0.305194	0.326354	0.353552
5/16 (0.3125)	0.360842	0.381493	0.407942	0.44194
3/8 (0.375)	0.43301	0.457792	0.489531	0.530328
1/2 (0.500)	0.577347	0.610389	0.652707	0.707104
5/8 (0.625)	0.721684	0.762986	0.815884	0.883879
3/4 (0.750)	0.866021	0.915583	0.979061	1.060655
7/8 (0.875)	1.010358	1.06818	1.142238	1.237431
1	1.154694	1.220778	1.305415	1.414207

All dimensions in inches. See text for formulas used to construct this Table. For 45° angle dovetail, $W = H$.

Table 26. Constants for Calculating Dovetail Dimensions.

Width or Pin Dia.	60° Angle	55° Angle	50° Angle	45° Angle	40° Angle	35° Angle
1/16 (0.1325)	0.17076	0.18256	0.19653	0.21339	0.23422	0.26073
1/8 (0.125)	0.34151	0.36513	0.39306	0.42678	0.46844	0.52145
3/16 (0.1875)	0.51227	0.54769	0.58959	0.64016	0.70266	0.78218
1/4 (0.250)	0.68303	0.73025	0.78613	0.85355	0.93688	1.04290
5/16 (0.3125)	0.85378	0.91281	0.98266	1.06694	1.17109	1.30363
3/8 (0.375)	1.02454	1.09538	1.17919	1.28033	1.40531	1.56435
1/2 (0.500)	1.36605	1.46050	1.57225	1.70710	1.87375	2.08580
5/8 (0.625)	1.70756	1.82563	1.96531	2.13388	2.34219	2.60725
3/4 (0.750)	2.04908	2.19075	2.35838	2.56065	2.81063	3.12870
7/8 (0.875)	2.39059	2.55588	2.75144	2.98743	3.27906	3.65015
1	2.73210	2.92100	3.14450	3.41420	3.74750	4.17160
1 1/4	3.41513	3.65125	3.93063	4.26775	4.68438	5.21450
1 1/2	4.09815	4.38150	4.71675	5.12130	5.62125	6.25740

All dimensions in inches.

Statics

Central Two-Dimensional Force Systems

A central two-dimensional force system exists when the action lines of all forces lie in a common plane and intersect in a single point.

Resolution of a force $\vec{F}$ into two perpendicular components in the cartesian coordinate system x, y:

$$\vec{F} = F_x \vec{e}_x + F_y \vec{e}_y$$

where
$F_x = F \cos \alpha, \quad F_y = F \sin \alpha$
$F_x = F \sin \beta, \quad F_y = F \cos \beta$

F_x, F_y perpendicular components of force $\vec{F}$

α angle between the abscissa (x axis) and the force vector in the mathematically positive sense of rotation

$\beta = 90° - \alpha$ complementary angle to α

$\vec{e}_x, \vec{e}_y$ unit vectors in the direction of the coordinate axes x, y

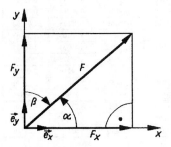

Magnitude of a force, dependent on its two perpendicular components:

$$F = |\vec{F}| = \sqrt{F_x^2 + F_y^2}$$

Addition of forces — determination of the resultant

The resultant $\vec{F}_R$ of n single forces $\vec{F}_i$ is obtained by vector addition. The single forces F_i are resolved into their respective perpendicular components $F_{ix}, F_{i\,y}$ Combining these in the x, y direction yields the perpendicular components of the resultant F_{Rx}, F_{Ry}:

Resultant

$$\vec{F}_R = F_{Rx}\vec{e}_x + F_{Ry}\vec{e}_y$$

Components

$$F_{Rx} = \sum_{i=1}^{n} F_{ix} = \sum_{i=1}^{n} F\cos\alpha_i$$

$$F_{Ry} = \sum_{i=1}^{n} F_{iy} = \sum_{i=1}^{n} F\sin\alpha_i$$

$$\tan\alpha_R = \frac{F_{Ry}}{F_{Rx}} \quad \text{direction}$$

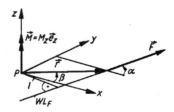

α_R angle between positive abscissa
and force vector for $\vec{F}_R$

Magnitude of resultant

$$F_R = \sqrt{F_{Rx}^2 + F_{Ry}^2}$$

Equilibrium — equilibrium conditions

A central two-dimensional force system is in equilibrium when the resultant becomes zero.

Equilibrium condition:

$$\vec{F}_R = \vec{0} \quad \text{(null vector)}$$

Component notation:

$$\sum_{i=1}^{n} F_{ix} = 0, \quad \sum_{i=1}^{n} F_{iy} = 0$$

General two-dimensional force system

A general two-dimensional force system exists when the action lines of all forces lie in a common plane but do not intersect in a common point.

Moment of a Force

Moment vector (free vector)

$$\vec{M} = \vec{r} \times \vec{F} \quad \text{or}$$

$$\vec{M} = \begin{vmatrix} \vec{e}_x & \vec{e}_y & \vec{e}_z \\ r_x & r_y & 0 \\ F_x & F_y & 0 \end{vmatrix}$$

$$= (r_x F_y - r_y F_x)\,\vec{e}_z = M_z\vec{e}_z$$

$$\vec{F} = F_x \vec{e}_x + F_y \vec{e}_y$$

force vector

$$\vec{r} = r_x \vec{e}_x + r_y \vec{e}_y$$

The position vector describes the location of the point of application of the force in relation to the selected reference point. (In this case, the reference point is the origin P.)

$$M_z = r_x F_y - r_y F_x$$

M_z is the component of the moment of the force F with respect to P ($M_x = M_y = 0$); the sign of $M = M_z$ is positive when the moment with respect to P causes a mathematically positive rotation $\circlearrowleft$ in the x, y plane.

Magnitude of vector product

$$M = |\vec{M}| = |\vec{r}||\vec{F}| \sin(\alpha - \beta) = Fl$$

Symbols for representing moment: double arrow for moment vector
$\twoheadrightarrow$, curl for sense of rotation of the moment $\circlearrowleft$

Right-hand rule:
When the curled fingers of the right hand are pointing in the direction of rotation brought about by the moment (curl), the outstretched thumb points in the direction of the moment vector (double arrow).

Resolution of a force into a force and a couple

A force can be shifted parallel to its action line by a distance of l_V, provided that the change in moment action is offset by a couple moment M_V:

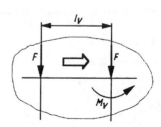

$$M_V = Fl_V$$

Resultant Moment from n Single Forces

$$\vec{M}_R = \sum_{i=1}^{n} \vec{M}_i = \sum_{i=1}^{n} \vec{r}_i \times \vec{F}_i = \sum_{i=1}^{n} M_{iz} \vec{e}_z = \sum_{i=1}^{n} \left(r_{ix} F_{iy} - r_{iy} F_{ix} \right) \vec{e}_z$$

$$\vec{M}_{\mathrm{R}} = M_{\mathrm{R}z}\vec{e}_z$$

resultant moment, dependent on reference point P

$$\vec{F}_i = F_{ix}\vec{e}_x + F_{iy}\vec{e}_y$$

force vectors

$$\vec{r}_i = r_{ix}\vec{e}_x + r_{iy}\vec{e}_y$$

position vectors for common reference point P

Determination of the resultant $\vec{F}_{\mathrm{R}}$

The resultant must possess the same actions of force and moment as all the single forces combined (static equivalence); resolution of all forces into forces and couples with respect to the moment reference point P, in consideration of the couple moments; determination of the magnitude, direction, and location of the resultant:

— For magnitude and direction equations, see the earlier section "Central Two-Dimensional Force System."

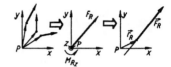

— The location can be obtained by the principle of static moments ($F_{\mathrm{R}} > 0$):

The resultant moment is equal to the vector product of the position vector of the resultant force $\vec{r}_{\mathrm{R}}$ and $\vec{F}_{\mathrm{R}}$:

$$\vec{M}_{\mathrm{R}} = \vec{r}_{\mathrm{R}} \times \vec{F}_{\mathrm{R}}$$

$$\vec{F}_{\mathrm{R}} = F_{\mathrm{R}x}\vec{e}_x + F_{\mathrm{R}y}\vec{e}_y$$

resultant

$$\vec{r}_{\mathrm{R}} = r_{\mathrm{R}x}\vec{e}_x + r_{\mathrm{R}y}\vec{e}_y$$

position vector of the resultant in relation to P

Straight-line equation of the action line of $\vec{F}_{\mathrm{R}}$ for $F_{\mathrm{R}} > 0$:

$$y = \frac{F_{\mathrm{R}y}}{F_{\mathrm{R}x}}x - \frac{M_{\mathrm{R}z}}{F_{\mathrm{R}x}}$$

Special case: $F_{\mathrm{R}} = 0$, moment due to couple

Equilibrium — equilibrium conditions

A general two-dimensional force system is in equilibrium when the resultant force and the resultant moment equal zero.

Equilibrium conditions $\qquad\qquad \vec{F}_{\mathrm{R}} = \vec{0},\ \vec{M}_{\mathrm{R}} = \vec{0}$

Component notation

$$\sum_{i=1}^{n} F_{ix} = 0, \quad \sum_{i=1}^{n} F_{iy} = 0$$

$$\sum_{i=1}^{n} M_i = 0, \quad M_i = M_{iz}$$

Symbols for equilibrium conditions
P arbitrary moment reference point

$\rightarrow : 0 =, \quad \uparrow : 0 =, \quad \circlearrowleft P : 0 =$

Three-Dimensional Force Systems

Central three-dimensional force system

Resultant force of n forces F_i

$$F_{\mathrm{R}} = F_{\mathrm{R}x}\vec{e}_x + F_{\mathrm{R}y}\vec{e}_y + F_{\mathrm{R}z}\vec{e}_z$$

$$F_{\mathrm{R}x} = \sum_{i=1}^{n} F_i \cos \alpha_i,$$

$$F_{\mathrm{R}y} = \sum_{i=1}^{n} F_i \cos \beta_i,$$

$$F_{\mathrm{R}z} = \sum_{i=1}^{n} F_i \cos \gamma_i$$

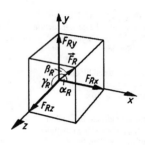

Magnitude

$$F_{\mathrm{R}} = |\vec{F}_{\mathrm{R}}| = \sqrt{F_{\mathrm{R}x}^2 + F_{\mathrm{R}y}^2 + F_{\mathrm{R}z}^2}$$

$\alpha_i, \beta_i, \gamma_i$ angles between the
action line of F_i and the axes

The direction of each force is determined by two direction cosines of its line of action
with respect to the axes of the cartesian coordinate system (example $\vec{F}_{\mathrm{R}}$):

$$\cos \alpha_{\mathrm{R}} = \frac{F_{\mathrm{R}x}}{F_{\mathrm{R}}}, \quad \cos \beta_{\mathrm{R}} = \frac{F_{\mathrm{R}y}}{F_{\mathrm{R}}}$$

$$\cos \gamma_{\mathrm{R}} = \frac{F_{\mathrm{R}z}}{F_{\mathrm{R}}}$$

Equilibrium, if the equilibrium condition $\vec{F}_{\mathrm{R}} = \vec{0}$ is fulfilled

Component notation:

$$\sum_i F_{ix} = 0, \quad \sum_i F_{iy} = 0$$

$$\sum_i F_{iz} = 0$$

General three-dimensional force system

Moment of a force

$$\vec{M} = \vec{r} \times \vec{F}$$
$$= \begin{vmatrix} \vec{e}_x & \vec{e}_y & \vec{e}_z \\ r_x & r_y & r_z \\ F_x & F_y & F_z \end{vmatrix}$$

$$\vec{M} = M_x \vec{e}_x + M_y \vec{e}_y + M_z \vec{e}_z$$

Components:

$$M_x = r_y F_z - r_z F_y$$
$$M_y = r_z F_x - r_x F_z$$
$$M_z = r_x F_y - r_y F_x$$

Resultant moment of n single forces

$$\vec{M}_R = \sum_{i=1}^{n} \vec{M}_i = \sum_{i=1}^{n} \vec{r}_i \times \vec{F}_i$$

$$\vec{F}_i = F_{ix} \vec{e}_x + F_{iy} \vec{e}_y + F_{iz} \vec{e}_z$$ force vectors
$$\vec{r}_i = r_{ix} \vec{e}_x + r_{iy} \vec{e}_y + r_{iz} \vec{e}_z$$ position vectors for common
reference point P

Determination of resultant:

The magnitude and direction of $\vec{F}_R$ are determined as for the central three-dimensional force system. In contrast to the general two-dimensional force system, the vectors $\vec{F}_R$ and $\vec{M}_R$ are not perpendicular to one another, as a rule, so the principle of static moments for the two-dimensional force system is not applicable here. The location of $\vec{F}_R$ is determined by introducing a moment which acts parallel to the action line of $\vec{F}_R$.

Equilibrium, if the equilibrium conditions $\vec{F}_R = \vec{0}$, $\vec{M}_R = \vec{0}$ are fulfilled.

Component notation:

$$\sum_i F_{ix} = 0, \quad \sum_i F_{iy} = 0, \quad \sum_i F_{iz} = 0$$

$$\sum_i M_{ix} = 0, \quad \sum_i M_{iy} = 0, \quad \sum_i M_{iz} = 0$$

Models of Rigid Bodies

Model	Sketch	Features
Uniaxial Model **Bar**	Elongation in one dimension (axial direction) is much greater than the maximum elongation in the other two dimensions.	
Frame member		transmits forces only in the direction of the longitudinal axis
Beam/girder		transmits any desired forces and moments
Biaxial Models	Thickness h is much less than the elongation in other dimensions.	
Vertical plate		loads in the midplane
Horizontal plate		loads perpendicular to the midplane
Shell		generally features a doubly curved midplane, ϱ_1, ϱ_2 radii of the circles of principal curvature of the middle surface
Triaxial (Three-Dimensional) Models	Examples include the fullspace, halfspace, rectangular solid, cylinder, and sphere.	

Models of Various Supports and Connections

Model	Symbol / Sketch	Free-Body Diagram
Models of two-dimensional supports		
Supports with one unknown		
Shear leg		
Roller support		
Supports with two unknowns		
Pin support		
Sliding support		
Planar linkage (frictionless)		
Supports with three unknowns		
Fixed support		

Model	Symbol / Sketch	Free-Body Diagram
Models of three-dimensional supports		
Roller support (one unknown)		
Pin support (three unknowns)		
Fixed support (six unknowns)		

Load models

Categorized according to the elongation or geometry of the region of mechanical interaction:

Single force   (point load)

Uniformly distributed load   (line load, e.g., static weight of members over length)

A uniformly distributed load is represented in terms of the intensity $q(s)$ along the length l. It can be replaced by a statically equivalent single force, as follows:

Magnitude

$$F_q = \int_0^l q(s)\,\mathrm{d}s$$

Location of action line

$$s_q = \frac{1}{F_q} \int_0^l q(s)\,s\,\mathrm{d}s$$

Note: The determination of the magnitude and location of F_q is expressed in words as follows: F_q equals the area of the uniformly distributed load across the length l; the action line of F_q passes through the centroid of this area. (The direction of F_q is determined by the direction of the intensity.)

Special cases:

Uniform load across length *l*,

$q = $ const

$$F_q = ql, \quad s_q = \frac{1}{2}l$$

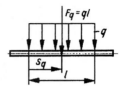

Triangular load across length *l*

$$q(s) = q_0\frac{s}{l}, \quad F_q = \frac{1}{2}q_0l, \quad s_q = \frac{2}{3}l$$

$q(s)$ intensity function

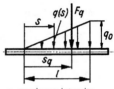

q_0 maximum intensity

Surface load (surface force)

(snow loads on planar trusses/roofs, pressure distribution on dams, containers under internal and external pressure, flow-obstructing biaxial and triaxial structures, e.g., automobile bodies)

Body load (body force)

(inertial forces, e.g., centrifugal forces in rotating structures/rotors as a result of centripetal acceleration)

Planar Trusses

Designations

Simple truss   Modeling from one rigid body and supports

Compound truss   Modeling from multiple rigid bodies,
                                           connections, and supports. For example,
                                           a frame is a system of frame members,
                                           which are connected to one another in a
                                           frictionless manner at joints, or so-called
                                           "nodes"; a triangular truss is a system
                                           of members connected to one another.

Static determinacy

All support reactions, connection reactions (or joint reactions), and internal forces of
the truss (see section 1.8), which may be composed of multiple rigid bodies, can
be determined from the equilibrium conditions. The necessary condition for static
determinacy is $n = 0$, where:

Triangular trusses: $n = 3k - a - v$ k number of rigid bodies

                                          a number of support reactions

                                          v number of connection reactions

Frames: $n = 2g - s - t$ g number of joints/nodes in
                                                the frame

                                          s number of frame members

                                          t number of supporting
                                                members/support reactions

$n > 0$ mechanism/power transmission
                                          with n degrees of freedom

$n < 0$ The truss is statically indeterminate
                                          by a degree of $|n|$.

Calculation of support reactions and connection reactions

Solution principle: a truss is in equilibrium when each rigid body (subsystem) is in
equilibrium itself.

Example of solution:

Composite truss
(three-hinged arch)

Given: q, a
Determine: support and
joint reactions

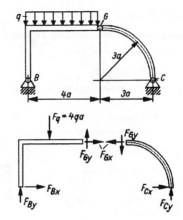

Solution:

— $k = 2, a = 4, v = 2 \rightarrow n = 0$

  Necessary condition for static
  determinacy is fulfilled.

— Draw free-body diagram.

— Equilibrium conditions

  Left body

  $\rightarrow : 0 = F_{Bx} + F_{Gx}$

  $\uparrow : \; 0 = F_{By} + F_{Gy} - F_q$

  $\circlearrowleft B: 0 = F_{Gy} \cdot 4a - F_{Gx} \cdot 3a - F_q \cdot 2a$

  Right body

  $\rightarrow : 0 = -F_{Gx} + F_{Cx}$

  $\uparrow : \; 0 = -F_{Gy} + F_{Cy}$

  $\circlearrowleft C: 0 = (F_{Gx} + F_{Gy}) \cdot 3a$

— Solution of system of equations

  $F_{Bx} = -F_{Cx} = F_{Cy} = F_{Gy} = -F_{Gx} = \dfrac{8}{7}qa, \quad F_{By} = \dfrac{20}{7}qa$

Tips on determining the member forces / axial forces in frames

— Apply the method of joints, free-body diagram, and equilibrium conditions at each
  joint/node as a central force system.

— Apply Ritter's method of sections, choosing the sections such that general force
  systems are obtained; e.g., section through three members whose lines of action
  do not intersect in a common point.

— For statically indeterminate frames, it is preferable to apply energetic methods
  (e.g., Castigliano's theorem).

Tips on space trusses

The solution principle and procedure are similar to those given for planar trusses.
For the sake of a clearer solution, it is often advisable to treat the three-dimensional
model as three perpendicular planes and then solve these as separate two-dimensional
problems.

Internal Forces

Internal forces in a uniformly loaded member

F_L; (N) axial or normal forces

F_Q; (Q) shearing force

M; (M_b) internal moment (bending moment)

Definition of internal forces

Directions of internal forces at edge of cut z:

z coordinate of the cut location

v coordinate of the deflection
    (see section 2.5)

F_L, F_Q, and M are functions of the
coordinate z, as a rule.

Relationships between the intensity of the uniformly distributed load, shearing force, and internal moment

$$\frac{dF_Q}{dz} = -q(z), \quad \frac{dM}{dz} = F_Q$$

$$\frac{d^2M}{dz^2} = -q(z)$$

$q(z)$ intensity of the uniformly
distributed load (perpendicular
to the z axis)

Conclusions:

— The internal forces are constant in regions without uniform loading.

— Points of zero shear are peaks of the moment curve, i.e., points of
  maximum positive or maximum negative moment.

Internal forces in a member subjected to three-dimensional loading

z coordinate of the cut location

Internal forces

F_L; (N) axial forces
F_{Qx}, F_{Qy}; (Q_x, Q_y) shearing forces

Internal moments

$M_x = M_{bx}, M_y = M_{by}$ bending moments
$M_t = M_z$ torque

Relationships between shearing forces and bending moments

$$F_{Qy} = \frac{dM_x}{dz}, \quad F_{Qx} = -\frac{dM_y}{dz}$$

Shearing Force Curves and Moment Curves for Selected Trusses

Truss	Internal Force Curves
	$F_A = F\dfrac{b}{l}$, $F_B = F\dfrac{a}{l}$ $M_1 = F_A z_1$, $M_2 = F_B z_2$ $M_{max} = F\dfrac{ab}{l}$
	$F_A = F\left(1 + \dfrac{l_l}{l}\right)$, $F_B = F\dfrac{l_l}{l}$ $M_1 = -F z_1$, $M_2 = F_B z_2$ $|M|_{max} = F l_l$
	$F_A = F_B = \dfrac{1}{2}ql$ $F_Q = ql\left(\dfrac{1}{2} - \dfrac{z}{l}\right)$ $M = \dfrac{1}{2}qz(l - z)$ $M_{max} = \dfrac{1}{8}ql^2$
	$F_A = ql_l\left(1 + \dfrac{l_l}{2l}\right)$ $F_B = ql_l\dfrac{l_l}{2l}$ $|M|_{max} = \dfrac{1}{2}ql_l^2$ $F_{Q1} = -qz_1$, $M_1 = -\dfrac{1}{2}qz_1^2$, $F_{Q2} = F_B$, $M_2 = -F_B z_2$
	$F_A = F$, $M_A = Fl$, $F_Q = -F$, $M = -Fz$ $|M|_{max} = Fl$
	$F_A = ql$, $M_A = \dfrac{1}{2}ql^2$ $F_Q = -qz$, $M = -\dfrac{1}{2}qz^2$ $|M|_{max} = \dfrac{1}{2}ql^2$

Tips:
— As a rule, internal forces are constant only within certain regions of the member.

— Region boundaries are determined by geometry (ends of members, bends, branches, discontinuous changes in curvature) and loading (points of application of forces and moments, beginning and end of uniformly distributed loads of constant intensity).

Friction

Bodies in contact can only be moved in relation to one another by overcoming the resisting forces (frictional forces).

Static friction or adhesion: resistance in effect until motion begins

Law of adhesion (empirical)

$F_R \leqq \mu_0 F_N$

F_R frictional force
This reaction force is tangent to the plane of contact and opposes the possible relative motion of the other body.

F_N normal force
This reaction force is perpendicular to the plane of contact (and thus to F_R) and acts upon the other body.

μ_0 coefficient of static friction

Sliding friction: resistance during relative motion between two bodies in contact

Coulomb's law of sliding friction

$F_R = \mu F_N$

μ coefficient of sliding friction, such that $\mu < \mu_0$

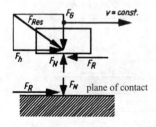

F_R Friction force: this impressed force opposes the relative motion of the other body. It is applicable up to a certain maximum value of F_N and at a low constant velocity v.

F_R and F_N are determined by the methods of statics (free-body diagram, equilibrium conditions).

Relationship between maximum inclination of F_{Res} and the coefficients of friction:

friction cone (defined by the bounding line of action of F_{Res})

$\mu_0 = \tan \varrho_0$, $\mu = \tan \varrho$, $\varrho < \varrho_0$

ϱ_0, ϱ friction angle (half the static or sliding friction cone angle)

plane of contact

If the action line of F_{Res} lies within the cone ϱ_0, the system is self-locking.

Selected coefficients of static and sliding friction

Material Combination	Coefficient of Static Friction μ_0		Coefficient of Sliding Friction μ	
	Dry	Lubricated	Dry	Lubricated
Steel/steel	0.15 - 0.4	0.1	0.1 - 0.2	0.05 - 0.1
Steel/gray cast iron	0.18 - 0.24	0.1	0.17 - 0.2	0.02 - 0.1
Wood/metal	0.5 - 0.65	0.1	0.2 - 0.5	0.02 - 0.1
Wood/wood	0.4 - 0.65	0.16 - 0.2	0.2 - 0.4	0.04 - 0.16
Rubber tire/ road surface	0.7		0.3 - 0.5	0.15 - 0.2 (with water)
Steel-ice	0.027		0.014	

Friction in guideways

$F_A = \mu' F_Q$

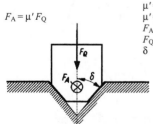

$\mu' = \mu$ flat way guidance
$\mu' = \mu/\sin \delta$ V-way (vee way) guidance
F_A feed force
F_Q operating load on the guideway
δ angle of inclination between the side of the V and the vertical

Friction in screw threads

This is a relationship between the moment required for tightening or loosening a threaded bolt and the axial force of the connection between the bolt and nut.

Drive screw thread (square thread):

$M_A = F_Q r_m \tan(\varrho + \alpha)$

$M_L = F_Q r_m \tan(\varrho - \alpha)$

Fastening screw thread:
ϱ is replaced by ϱ',

$$\tan \varrho' = \frac{\mu}{\cos \beta}$$

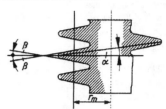

F_Q axial force

M_A moment required to raise the load; tightening moment

M_L moment required to lower the load; loosening moment

r_m mean radius of thread

α pitch angle

β half the angle of the flanks

Cable friction

Cable friction law of adhesion

$F_{S2} = F_{S1} e^{m0a}, F_{S2} > F_{S1}$

F_{S1}, F_{S2} cable forces

α arc of contact

   $[\alpha] = \text{rad}$

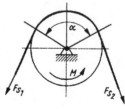

Rolling friction

Law of rolling friction

$$F_R = \frac{f}{R} F_N$$

Distinction between sliding and rolling:

$\mu_0 < f / R$ for sliding,
$\mu_0 > f / R$ for rolling

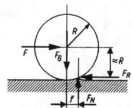

F_R rolling resistance
f lever arm of rolling friction
R radius of wheel

Stress Analysis

Calculation of Centroid Location

$S(\overline{x}_S, \overline{y}_S, \overline{z}_S)$ centroid coordinates

$\overline{x}, \overline{y}, \overline{z}$ reference coordinate system

Center of mass / center of volume
(of homogeneous bodies)

$$\overline{x}_S = \frac{1}{V} \int\limits_{(V)} \overline{x}\, dV, \ \overline{y}_S = \frac{1}{V} \int\limits_{(V)} \overline{y}\, dV$$

$$\overline{z}_S = \frac{1}{V} \int\limits_{(V)} \overline{z}\, dV$$

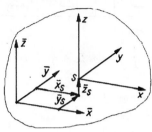

Centroid (Center of Area)

$$\overline{x}_S = \frac{1}{A} \int\limits_{(A)} \overline{x}\, dA, \ \overline{y}_S = \frac{1}{A} \int\limits_{(A)} \overline{y}\, dA$$

Calculation of centroid for composite areas

$$\overline{x}_S = \frac{1}{A} \sum_{i=1}^{n} \overline{x}_{Si} A_i$$

$$\overline{y}_S = \frac{1}{A} \sum_{i=1}^{n} \overline{y}_{Si} A_i$$

$$A = \sum_{i=1}^{n} A_i$$

n number of composite parts

A_i area of the ith composite part

$\overline{x}_{Si}, \overline{y}_{Si}$ centroid coordinates for the ith composite part

Practical calculation:

— determination of the reference coordinate system

— separation into composite parts, each with a known centroid location

   (figure shows example with two composite parts)

— calculation with the following table

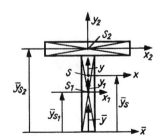

Table for determining the coordinates of the centroid (center of area)

i	A_i	$\overline{x}_{Si}$	$\overline{y}_{Si}$	$A_i\overline{x}_{Si}$	$A_i\overline{y}_{Si}$
1	A_1	$\overline{x}_1$	$\overline{y}_1$	$A_1\overline{x}_{S1}$	$A_1\overline{y}_{S1}$
n	A_n	$\overline{x}_n$	$\overline{y}_n$	$A_n\overline{x}_{Sn}$	$A_n\overline{y}_{Sn}$
$\displaystyle\sum_i$	$\displaystyle\sum_i A_i$			$\displaystyle\sum_i A_i\overline{x}_{Si}$	$\displaystyle\sum_i A_i\overline{y}_{Si}$

Line centroid

$$\overline{x}_S = \frac{1}{L}\int\limits_{(L)} \overline{x}\,\mathrm{d}s, \quad \overline{y}_S = \frac{1}{L}\int\limits_{(L)} \overline{y}\,\mathrm{d}s$$

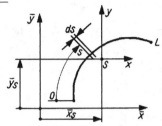

s coordinate along the line

$\mathrm{d}s$ line element

L length of line

Area Moment of Inertia

Preconditions and Definitions

x, y origin at centroid S

$\overline{x}, \overline{y}$ arbitrary

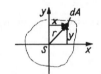

n^{th} moment of an area:

$$\int\limits_{(A)} (x^s y^t)\,\mathrm{d}A, \quad n = s + t, \ n, s, t \in \mathbb{N}$$

0th moment: area

$$A = \int\limits_{(A)} \mathrm{d}A$$

1st moment: static moments

$$S_x := \int\limits_{(A)} y\,\mathrm{d}A, \quad S_y := \int\limits_{(A)} x\,\mathrm{d}A, \quad S_x = S_y = 0$$

2nd moment: area moments of inertia

$$I_{xx} := \int_{(A)} y^2 \, dA, \quad I_{yy} := \int_{(A)} x^2 \, dA$$

I_{xx}, I_{yy} axial area moments of inertia

$$I_{xy} := -\int_{(A)} xy \, dA$$

I_{xy} product of inertia or centrifugal moment

$$I_p := I_{xx} + I_{yy} = \int_{(A)} r^2 \, dA$$

I_p polar moment of inertia

Transformation of area moments of inertia between parallel coordinate systems

(Steiner's parallel axis theorem)

$$I_{\overline{xx}} = I_{xx} + \overline{y}_S^2 A$$
$$I_{\overline{yy}} = I_{yy} + \overline{x}_S^2 A$$
$$I_{\overline{xy}} = I_{xy} - \overline{x}_S \overline{y}_S A$$

Method for determining area moments of inertia for composite areas:

— Area moments of inertia for different areas may be added (or subtracted), provided that they exist in relation to the same coordinate system.

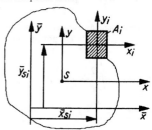

— Determination of the area moments of inertia of n composite parts with regard to their respective centroid coordinate systems; designation: $I_{x_i x_i}, I_{y_i y_i}, I_{x_i, y_i}$

— Transformation of these area moments of inertia to the x, y coordinate system by means of Steiner's theorem:

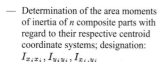
$$I_{xx_i} = I_{x_i x_i} + (\overline{y}_S - \overline{y}_{Si})^2 A_i$$
$$I_{yy_i} = I_{y_i y_i} + (\overline{x}_S - \overline{x}_{Si})^2 A_i$$
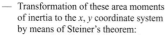
$$I_{xy_i} = I_{x_i y_i} - (\overline{x}_S - \overline{x}_{Si})(\overline{y}_S - \overline{y}_{Si}) A_i$$

— Summation of the area moments of inertia for the composite parts (sign is reversed for shaded areas)

Area moments of inertia in rotation of the cartesian coordinate system

Rotation transformation of the cartesian coordinates

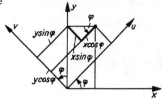

$u = x \cos \varphi + y \sin \varphi$
$v = -x \sin \varphi + y \cos \varphi$

$$I_{uu} = \frac{I_{xx} + I_{yy}}{2} + \frac{I_{xx} - I_{yy}}{2} \cos 2\varphi + I_{xy} \sin 2\varphi$$

$$I_{vv} = \frac{I_{xx} + I_{yy}}{2} - \frac{I_{xx} - I_{yy}}{2} \cos 2\varphi - I_{xy} \sin 2\varphi$$

$$I_{uv} = \qquad\quad -\frac{I_{xx} - I_{yy}}{2} \sin 2\varphi + I_{xy} \cos 2\varphi$$

Principal moments and principal axes

Extreme values of the axial area moments of inertia are called principal moments I_1, I_2. I_1 assumes a maximum value, while I_2 assumes a minimum value:

$$I_{1,2} = \frac{I_{xx} + I_{yy}}{2} \pm \sqrt{\left(\frac{I_{xx} - I_{yy}}{2}\right)^2 + I_{xy}^2}$$

The corresponding axes are called principal axes. The product of inertia for these axes vanishes, $I_{uv}(\varphi_0) = 0$. The principal axes are obtained at the following angles of rotation φ_0:

$$\tan 2\varphi_0 = \frac{2I_{xy}}{I_{xx} - I_{yy}}$$

$$\tan \varphi_0 = \frac{I_{xy}}{I_{xx} - I_2}$$

Tips:

— Axes of symmetry are always principal axes.

— If $I_{xx} = I_{yy}$ and $I_{xy} = 0$ for a particular area, then all centroidal axes are principal axes, or the area moments of inertia are invariant with respect to rotation of the coordinate system.

— Area moments of inertia are components of the inertia tensor.

Centroid Locations and Area Moments of Inertia for Selected Areas

Cross Section	Centroid Location	Area Moments of Inertia	Section Moduli
Square box section		$I_{xx} = I_{yy} = \dfrac{1}{12}\left(H^4 - h^4\right)$	$W_x = W_y =$ $\dfrac{H^3}{6}\left[1 - \left(\dfrac{h}{H}\right)^4\right]$
Rectangle	$\overline{x}_S = \dfrac{B}{2}$ $\overline{y}_S = \dfrac{H}{2}$	$I_{xx} = \dfrac{BH^3}{12},\ I_{yy} = \dfrac{HB^3}{12}$ $I_{\overline{xx}} = \dfrac{BH^3}{3},\ I_{\overline{yy}} = \dfrac{HB^3}{3}$ $I_{\overline{xy}} = -\dfrac{B^2 H^2}{4}$	$W_x = \dfrac{BH^2}{6}$ $W_y = \dfrac{HB^2}{6}$
Arbitrary triangle	$\overline{x}_S = \dfrac{1}{3}(b_2 - b_1)$ $\overline{y}_S = \dfrac{1}{3}H$ $B = b_1 + b_2$	$I_{xx} = \dfrac{BH^3}{36},\ I_{yy} = \dfrac{HB}{36}(B^2 - b_1 b_2)$ $I_{xy} = \dfrac{H^2}{72}(b_2^2 - b_1^2)$ $I_{\overline{xx}} = \dfrac{BH^3}{12},\ I_{\overline{yy}} = \dfrac{H}{12}(b_1^3 + b_2^3)$ $I_{\overline{xy}} = -\dfrac{H^2}{72}(b_1^2 - b_2^2)$	

(Continued)

Centroid Locations and Area Moments of Inertia for Selected Areas

Cross Section	Centroid Location	Area Moments of Inertia	Section Moduli
Special case: right triangle	$\overline{x}_S = \dfrac{B}{3}$ $\overline{y}_S = \dfrac{H}{3}$	$I_{xx} = \dfrac{BH^3}{36}, \; I_{yy} = \dfrac{HB^3}{36}, \; I_{xy} = \dfrac{B^2 H^2}{72}$ $I_{\overline{x}\overline{x}} = \dfrac{BH^3}{12}, \; I_{\overline{y}\overline{y}} = \dfrac{HB^3}{12}, \; I_{\overline{x}\overline{y}} = -\dfrac{B^2 H^2}{24}$	
Sector of an annulus		$I_{\overline{x}\overline{x}} = \dfrac{\varkappa}{8}\left[(\alpha_2 - \alpha_1) - \dfrac{1}{2}(\sin 2\alpha_2 - \sin 2\alpha_1)\right]$ $I_{\overline{y}\overline{y}} = \dfrac{\varkappa}{8}\left[(\alpha_2 - \alpha_1) + \dfrac{1}{2}(\sin 2\alpha_2 - \sin 2\alpha_1)\right]$ $I_{\overline{x}\overline{y}} = \dfrac{\varkappa}{16}(\cos \alpha_2 - \cos \alpha_1)$ $\varkappa = R_a^4 - R_i^4, \; \alpha_1, \alpha_2 \text{ in rad}$	
Special cases: circle / annulus		circle $\quad I_{xx} = I_{yy} = \dfrac{\pi}{4} R^4 = \dfrac{\pi}{64} D^4$ annulus $\quad I_{xx} = I_{yy} = \dfrac{\pi}{4} R_a^4 (1 - \lambda^4)$ $\qquad I_{xx} = I_{yy} = \dfrac{\pi}{64} D_a^4 (1 - \lambda^4)$ $\qquad \lambda = \dfrac{R_i}{R_a} = \dfrac{D_i}{D_a}$	$W_x = W_y$ $\quad = \dfrac{\pi R^3}{4} = \dfrac{\pi D^3}{32}$ $W_x = \dfrac{\pi R_a^3}{4}(1 - \lambda^4)$ $W_y = W_x$

(Continued)

Centroid Locations and Area Moments of Inertia for Selected Areas

Cross Section	Centroid Location	Area Moments of Inertia	Section Moduli
Semicircle	$\bar{y}_S = \dfrac{4R}{3\pi}$	$I_{xx} = R^4 \left(\dfrac{\pi}{8} - \dfrac{8}{9\pi} \right) \approx 0,11 R^4$ $I_{yy} = \dfrac{\pi}{8} R^4$ $I_{\bar{x}\bar{x}} = I_{\bar{y}\bar{y}} = \dfrac{\pi}{8} R^4$	$W_x = \dfrac{R^3}{24} \left(\dfrac{9\pi^2 - 64}{3\pi - 4} \right)$ $W_x \approx 0,191 R^3$ $W_y = \dfrac{\pi}{8} R^3$
Quadrant	$\bar{x}_S = \bar{y}_S = \dfrac{4R}{3\pi}$	$I_{xx} = I_{yy} = R^4 \left(\dfrac{\pi}{16} - \dfrac{4}{9\pi} \right) \approx 0,0549 R^4$ $I_{xy} = R^4 \left(\dfrac{4}{9\pi} - \dfrac{1}{8} \right) \approx 0,0165 R^4$ $I_{\bar{x}\bar{x}} = I_{\bar{y}\bar{y}} = \dfrac{\pi}{16} R^4, \ I_{\bar{x}\bar{y}} = -\dfrac{R^4}{8}$	
Thin-walled annulus		$I_{xx} = I_{yy} \approx \pi R^3 \delta, \ \delta \ll R$	$W_x = W_y \approx \pi R^2 \delta$

(Continued)

Centroid Locations and Area Moments of Inertia for Selected Areas

Cross Section	Centroid Location	Area Moments of Inertia	Section Moduli
Ellipse		$I_{xx} = \frac{\pi}{4}ab^3$, $I_{yy} = \frac{\pi}{4}ba^3$	$W_x = \frac{\pi}{4}ab^2$ $W_y = \frac{\pi}{4}ba^2$
Composite cross sections I profile		$I_{xx} = \frac{1}{12}\left[BH^3 - (B-b)h^3\right]$ $I_{yy} = \frac{1}{12}\left[B^3(H-h) + b^3h\right]$	$W_x = 2\frac{I_{xx}}{H}$ $W_y = 2\frac{I_{yy}}{B}$
⊔ profile	$\bar{y}_S = \dfrac{BH^2 - bh^2}{2A}$ $A = BH - bh$	$I_{xx} = I_{\bar{x}\bar{x}} - A\bar{y}_S^2$, $I_{yy} = \frac{1}{12}\left(B^3H - b^3h\right)$ $I_{\bar{x}\bar{x}} = \frac{1}{3}(BH^3 - bh^3)$, $I_{\overline{yy}} = I_{yy}$	$W_x = \dfrac{I_{xx}}{\bar{y}_S}$ $W_y = 2\dfrac{I_{yy}}{B}$

(Continued)

Centroid Locations and Area Moments of Inertia for Selected Areas

Cross Section	Centroid Location	Area Moments of Inertia	Section Moduli
T profile	$\overline{y}_S = \dfrac{BH^2 - (B-b)(H-h)^2}{2A}$ $A = Bh + b(H-h)$	$I_{xx} = I_{\overline{x}\overline{x}} - A\overline{y}_S^2,$ $I_{yy} = \dfrac{1}{12}\left[B^3 h + b^3(H-h)\right]$ $I_{\overline{x}\overline{x}} = \dfrac{1}{3}\left[BH^3 - (B-b)(H-h)^3\right]$ $I_{\overline{y}\overline{y}} = I_{yy}$	$W_x = \dfrac{I_{xx}}{\overline{y}_S}$ $W_y = 2\dfrac{I_{yy}}{B}$

Tips:

— For coordinate systems in which at least one axis is an axis of symmetry (principal axis), the product of inertia or centrifugal moment assumes a value of zero.

— The section moduli are based on a bending type of stress. In this table, they are specified only for the principal axes depicted here.

Fundamental Concepts

Assumptions

Deformable solid body

Linear theory	Applicability of principle of superposition (resultant action is sum of individual actions); linear equations; small deformations in relation to dimensions of structural component.
Classical field theory / classical field model	Materials properties are unaffected by separating body into parts; structural influences are disregarded; laws describe manifestations (phenomena) and not causes.
Isotropy, homogeneity	Materials properties are independent of orientation and location.

Concept of stress — state of stress (Two-dimensional analysis)

Mechanical stress as a measure of loading

Stress vector

$$\vec{S} := \frac{\mathrm{d}\vec{F}}{\mathrm{d}A} = \sigma \vec{e}_n + \tau \vec{e}_t$$

$$\sigma := \frac{\mathrm{d}F_n}{\mathrm{d}A}, \quad \tau := \frac{\mathrm{d}F_t}{\mathrm{d}A}$$

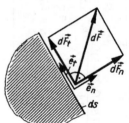

The component of the stress vector σ
in the normal direction is called
"normal stress." The component τ in
the tangent direction is called "shearing stress."

Stress components are components of the so-called "stress tensor." These components collectively define the state of stress.

Uniaxial stress

Stress components along an
oblique section of a member:

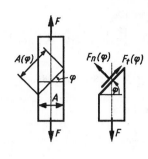

$$\sigma(\varphi) = \frac{F_n(\varphi)}{A(\varphi)}$$

$$= \frac{1}{2}\sigma_0(1 + \cos 2\varphi)$$

$$\tau(\varphi) = \frac{F_t(\varphi)}{A(\varphi)} = \frac{1}{2}\sigma_0 \sin 2\varphi$$

$$\sigma_0 := \frac{F}{A}$$

or

$$\left[\sigma(\varphi) - \frac{1}{2}\sigma_0\right]^2 + \tau^2(\varphi) = \left(\frac{1}{2}\sigma_0\right)^2$$

Circle equation in the σ, τ
plane, radius $\sigma_0/2$,
center $M(\sigma_0/2; 0)$,
Mohr's circle

σ_0 nominal stress, uniform/
homogeneous normal stress
distribution in a smooth
member with an unperturbed
cross section

Law of equivalence of conjugate shearing stresses:

$$\tau\left(\varphi + \frac{\pi}{2}\right) = -\tau(\varphi)$$

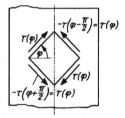

Shearing stresses on perpendicular sectional
surfaces possess the same magnitude but
different directions with respect to the
surface tangent.

Biaxial stress

Superposition of uniaxial tension
and plane shear with σ_0, τ_0

$$\sigma(\varphi) = \frac{1}{2}\sigma_0 + \frac{1}{2}\sigma_0 \cos 2\varphi - \tau_0 \sin 2\varphi$$

$$\tau(\varphi) = \frac{1}{2}\sigma_0 \sin 2\varphi + \tau_0 \cos 2\varphi$$

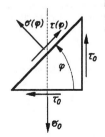

Mohr's Circle

$$\left[\sigma(\varphi) - \frac{1}{2}\sigma_0\right]^2 + \tau^2(\varphi) = \left(\frac{1}{2}\sigma_0\right)^2 + \tau_0^2$$

Principal Normal Stresses — Principal Stresses

$\sigma_1 > \sigma_2$: extremal normal stresses along perpendicular sectional surfaces that are free of shearing stress.

$$\sigma_{1,2} = \frac{1}{2}\sigma_0 \pm \sqrt{\left(\frac{1}{2}\sigma_0\right)^2 + \tau_0^2}$$

$$\tan(2\varphi_0) = -\frac{2\tau_0}{\sigma_0}$$

φ_0 angle of inclination of the shear-free sectional surfaces (the state of stress is described in terms of the number of non-zero principal normal stresses)

Maximum shearing stress

$$\tau_{max} = \sqrt{\left(\frac{1}{2}\sigma_0\right)^2 + \tau_0^2}$$

$$\tan 2\varphi_1 = \frac{\sigma_0}{2\tau_0}$$

φ_1 angle of inclination of the section surfaces with τ_{max}

Notch effect

Geometric causes:
abrupt changes in cross section,
such as notches, drill holes,
recesses, indentations, punctures, etc.

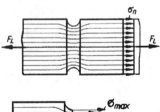

Loading is increased because:
— the local maximum stress (σ_{max})
  in certain regions is substantially
  higher than the nominal stress
  σ_n; α_K form factor or stress
  concentration factor:
  $\sigma_{max} = \alpha_K \sigma_n$

— the state of stress is multiaxial
  in this region

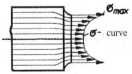

Form factors can be found in standard tables; if $\alpha_K > 10$ (cracks), stress is no longer meaningful as a measure of loading. The methods of fracture mechanics should then be applied.

Stresses and strains

Displacement vector
(two-dimensional analysis)

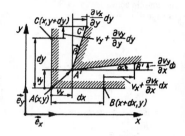

$$\vec{v}(x,y) = v_x(x,y)\vec{e}_x + v_y(x,y)\vec{e}_y$$

Strains (elongation, slip,
or shear strain)

Elongation

$$\varepsilon_x = \frac{\overline{A'B'} - \overline{AB}}{\overline{AB}} = \frac{\partial v_x}{\partial x}$$

$$\varepsilon_y = \frac{\overline{A'C'} - \overline{AC}}{\overline{AC}} = \frac{\partial v_y}{\partial y}$$

Slip

$$\gamma_{xy} = \angle BAC - \angle B'A'C' = \alpha + \beta \approx \frac{\partial v_y}{\partial x} + \frac{\partial v_x}{\partial u}$$

Strains are components of the strain tensor. They collectively define the state of strain.

Rotation of the coordinate system by φ

$$\varepsilon(\varphi) = \frac{\varepsilon_x + \varepsilon_y}{2} - \frac{\varepsilon_x - \varepsilon_y}{2}\cos 2\varphi + \frac{1}{2}\gamma_{xy}\sin 2\varphi$$

$$\frac{1}{2}\gamma(\varphi) = -\frac{\varepsilon_x - \varepsilon_y}{2}\sin 2\varphi - \frac{1}{2}\gamma_{xy}\cos 2\varphi$$

Principal strains for directions of vanishing slip j_0

$$\varepsilon_{1,2} = \frac{\varepsilon_x + \varepsilon_y}{2} \pm \sqrt{\left(\frac{\varepsilon_x - \varepsilon_y}{2}\right)^2 + \left(\frac{\gamma_{xy}}{2}\right)^2}, \quad \tan 2\varphi_0 = \frac{\gamma_{xy}}{\varepsilon_x - \varepsilon_y}$$

Constitutive law

Empirically determined relationship between components of the stress tensor and strain tensor.

Linear elastic material properties: linear

relationship between stress and strain

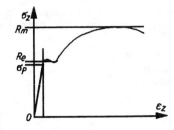

Hooke's Law

$$\sigma_z = E\varepsilon_z, \ 0 \leqq \sigma_z \leqq \sigma_P$$
$$\sigma_z = \frac{F_L}{A_0}, \ \varepsilon_z = \frac{\Delta l}{l_0}$$

E modulus of elasticity
    (material characteristic)
F_L axial force in the sample bar
Δl change in length
l_0 initial length
A_0 initial cross section
σ_P proportional limit
R_e yield strength
R_m tensile strength

Characteristic values in the tensile test (continued)

$$\varepsilon_q = -\nu\varepsilon_z, \ \varepsilon_q = \frac{\Delta d}{d_0} < 0$$

shear strain, shear strain in round bar
d_0 initial diameter
Δd change in diameter

$$\nu := \left| \frac{\varepsilon_q}{\varepsilon_z} \right|, \ 0 < \nu < 0,5$$

Poisson ratio
(material characteristic)

$$e := \frac{\Delta V}{V_0} = \varepsilon_x + \varepsilon_y + \varepsilon_z$$
$$= \varepsilon_z(1 - 2\nu)$$

volumetric strain
(incompressible material
$e = 0 \Rightarrow \nu = 0.5$)

Hooke's Law for Shear

$$\tau = G\gamma, \ G := \frac{E}{2(1 + \nu)}$$

G modulus of elasticity in shear
(material characteristic)

Thermal expansion (when temperature changes)

$\varepsilon_{th} = \alpha_{th}\Delta T$, $\Delta T = T_{final} - T_{initial}$

α_{th} linear coefficient of thermal expansion (material characteristic)

$T_{initial}$, T_{final} temperatures of the initial and final states, respectively

Total expansion

Example: tension member

$\varepsilon_{total} = \varepsilon_{el} + \varepsilon_{th}$

$\varepsilon_{total} = \dfrac{\sigma_z}{E} + \alpha_{th}\Delta T$

Thermal stress when expansion is completely constrained

$\sigma_z = \sigma_{th} = -E\alpha_{th}\Delta T$

Material characteristics E, G, α_{th} for selected materials			
Material	**E/GPa**	**G/GPa**	**$\alpha_{th}/10^{-6}$ K^{-1}**
steel, cast steel	210	81	13
gray cast iron	90	40	10
aluminum	72	28	23
copper	125	48	16
bronze	115	44	18
wood	11	5.5	8
aluminum oxide ceramics	380	145	8

Fundamentals of structural component evaluation

$\sigma_{ultimate} \leq \sigma_{allowable}$, $\tau_{ultimate} \leq \tau_{allowable}$

The maximum ultimate stress ($\sigma_{ultimate}$, $\tau_{ultimate}$) may not exceed the allowable stress ($\sigma_{allowable}$, $\tau_{allowable}$).

Allowable stresses are obtained from applicable material characteristics, using safety factors. For normal stresses under static loading:

$\sigma_{allowable} = \dfrac{R_m}{S_m}$, $S_m > 1$

or

$\sigma_{allowable} = \dfrac{R_e}{S_e}$, $S_e > 1$

tensile strength R_m for brittle materials, yield strength R_e for tough (ductile) materials, S_m, S_e safety factors.

Basic engineering tasks

Given	Determine	Task
loading, material characteristics	dimensions	design calculations and/or dimensioning
dimensions, material characteristics	loading	calculation of load limits
loading, dimensions, material characteristics	stresses deformations	stress analysis deformation analysis

Solution of statically indeterminate problems:

In addition to the equilibrium conditions of statics, constraints are formulated as statements concerning deformation at supports and connections, for example. (Static indeterminacy by a degree of $|n| \Rightarrow |n|$ constraints.)

Stress/Strain Loads

Internal force: axial force F_L ($F_L > 0$ tensile loading, $F_L < 0$ compressive loading)

Assumptions: Straight member or slightly curved member with cross section that is either constant or displays a constant rate of change $A(z)$, longitudinal coordinate z

Stress Distribution (homogeneous, i.e., constant along $A(z)$)

$$\sigma_z(z) = \frac{F_L(z)}{A(z)}.$$

Tips:

— Notch effects must be considered separately.
— Members under compressive loading are in danger of buckling.

Length at rupture

$$l_{rupture} = \frac{R_m}{\varrho g}$$

R_m tensile strength

g gravitational acceleration

ϱ density

(length at which rupture occurs when the specimen is hung vertically, as a result of the static weight of the specimen)

Deformations

Case: Member of constant cross section and constant axial force

$$\Delta l = \left[\frac{F_L}{EA} + \alpha_{th} \Delta T \right] l_0$$

change in length of member
l_0 initial length of member
       segment
EA ensile stiffness
α_{th} linear coefficient of thermal
       expansion
ΔT temperature change

$$\Delta l_i = \frac{F_{Si} l_i}{(EA)_i}$$

isothermal elongation of ith
frame member

$$\varepsilon_q = -\nu \varepsilon_z$$
$$= -\nu \left[\frac{F_L}{EA} + \alpha_{th} \Delta T \right]$$

shear strain

$$\Delta d = -\nu \frac{F_L d_0}{EA}$$

isothermal change in diameter
(for circular cross section)

Case: member of variable axial force $F_L(z)$ and cross section with a constant rate of change $A(z)$

$$v_z(z) - v_z(0) =$$
$$= \int_0^z \left[\frac{F_L(\bar{z})}{EA(\bar{z})} + \alpha_{th} \Delta T \right] d\bar{z}$$

change in length of the member
segment between points 0 and z

Special case: prismatic member
under static weight

$$v_z(z) = \frac{\varrho g l^2}{E} \left[\frac{z}{l} - \frac{1}{2} \left(\frac{z}{l} \right)^2 \right]$$
$$\Delta l = v_z(l) = \frac{F_G l}{2EA}, \quad F_G = mg$$

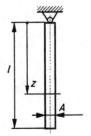

g gravitational acceleration
ϱ density
m mass of member

Surface pressure

Compressive stress distribution at the contact surface between two bodies being pressed together.

$$p = \frac{F_N}{A}$$

approximation for surface pressure on plane contact surfaces
F_N normal force
A contact surface area

Average surface pressure on curved contact surfaces:

— journal bearing, loading by surface pressure p

— bolt / shank / rivet hole, loading by internal surface contact $\sigma_L = p$

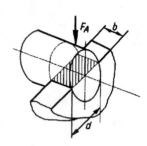

$$p = \frac{F_A}{A_{\text{proj}}}$$

$A_{\text{proj}} = bd$ contact surface projected onto plane (cylinder jacket)
b width of bearing, length of shank
d diameter of bearing or bolt

The theory of Hertz pressure makes it possible to describe the stress distribution in greater detail in contact-related problems of this nature.

Approximation solutions for maximum surface pressure at a contact force of F:

Maximum Hertz surface pressure

Body Combination	p_{max}	Remarks	
point contact: sphere against plane sphere 1 against sphere 2	$0.388 \sqrt[3]{\dfrac{FE^2}{R^2}}$	$R = \dfrac{R_1 R_2}{R_1 + R_2}$	sphere: radius R, E_1 plane: E_2 sphere 1: R_1, E_1 sphere 2: R_2, E_2
line contact: cylinder against plane cylinder 1 against cylinder 2 cylinder lengths l	$0.418 \sqrt{\dfrac{FE}{lR}}$	$E = \dfrac{2E_1 E_2}{E_1 + E_2}$	cylinder 1: R_1, E_1 plane: E_2 cylinder 1: R_1, E_1 cylinder 2: R_2, E_2

Shearing

The member is loaded predominantly by shearing force (a force couple, with both components on parallel lines of action, perpendicular or transverse to the longitudinal axis of the member or structural component).

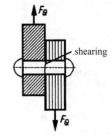

Example: shearing of bolts, parting or stamping of sheet metal

Shearing stress (average)

$$\tau_a = \frac{F_Q}{A}$$

A surface area across which F_Q is transmitted (for cylindrical bolt: circular cross section; blank to be removed: $A = Uh$, U circumference of blank h thickness of sheet metal

Approximation for shear strength:

$\tau_{aB} \approx 0.8\, R_m$

Bending

Internal force

$$M_b = \sqrt{M_x^2 + M_y^2}$$

resultant bending moment of the member M_b (pure bending)

If $F_Q = dM_b$ and $dz \neq 0$, the term "transverse bending" is used.

Assumptions:

The member axis is straight or only slightly curved. The cross sections of the member remain plane under bending loads (Bernoulli hypothesis); this condition is ordinarily fulfilled only in pure bending, but can be fulfilled by transverse bending, given sufficient precision. Strain and stress $\sigma_b = \sigma_z$ (normal stress in the z direction) are thus linearly distributed along the cross section. There is a strain-free "neutral surface" within the member. The intersection of this surface with the cross section of the member is called the "neutral axis" or "line of zero stress."

Stress distribution (general, unsymmetric, or biaxial bending; x, y arbitrary centroid coordinate system):

$$\sigma_b(x,y) = \frac{1}{I_{xx}I_{yy} - I_{xy}^2}\left[(M_x I_{yy} - M_y I_{xy})y + (M_x I_{xy} - M_y I_{xx})x\right]$$

Remarks: stress of greatest magnitude at the point (or points) of the cross-sectional contour at the greatest distance from the line of zero stress, point $P_0(x_0, y_0)$.

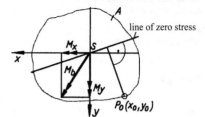

Equation of the line of zero stress:

$$y = \frac{-M_x I_{xy} + M_y I_{xx}}{M_x I_{yy} - M_y I_{xy}}x$$

In practical calculations, the line of zero stress is entered into the cross section, and the point P_0 is determined, at which:

$$|\sigma_b|_{max} = |\sigma_b(x_0, y_0)|$$

Special cases:

1. Bending around two principal axes ($M_x \neq 0$, $M_y \neq 0$, x and y principal axes, $I_{xy} = 0$)

$$\sigma_b(x,y) = \frac{M_x}{I_{xx}}y - \frac{M_y}{I_{yy}}x$$ bending stress

$$y = \frac{M_y I_{xx}}{M_x I_{yy}}x$$ line of zero stress (superposition of stress distributions for simultaneous simple bending around the x and y axes)

Tip: If the positions of the principal axes and principal moments are known, a case of unsymmetric bending can be treated as a case of bending around two principal axes.

2. Simple bending (occurs when the resultant bending moment vector coincides with a principal axis of the cross section)

Bending around principal axis	x	y
Bending moment	$M_x \neq 0,\ M_y = 0$	$M_y \neq 0,\ M_x = 0$
Stress distribution	$\sigma_b(y) = \dfrac{M_x}{I_{xx}} y$	$\sigma_b(x) = -\dfrac{M_y}{I_{yy}} x$
Line of zero stress	$y = 0$ (x axis)	$x = 0$ (y axis)

Maximum bending stress in simple bending around the x axis

$$|\sigma_b|_{max} = \frac{|M_b|_{max}}{I_{xx}} |y|_{max}$$

$$= \frac{|M_b|_{max}}{W_b}$$

$$W_b = W_x := \frac{I_{xx}}{|y|_{max}}$$

W_b resisting moment (values for semifinished products can be found in standard tables, similar to tables of area moment of inertia)

Calculation of Deformation

Designations:

$v = v(z)$ Displacement perpendicular to the z axis (deflection).

$v' := \dfrac{dv}{dz}$ inclination of member axis

1. Differential equation of the elastic curve:

$$EIv'' = -M_b,\ [EIv'']'' = q(z)$$

formula for simple bending around the x axis, where $I := I_{xx}$, $M_b = M_x$; functions EI (flexural rigidity) and M_b are continuous within the region

Ordinary, linear, inhomogeneous differential equation, solution by use of indefinite integrals for each region $\Rightarrow$ general solution for $EI = $ const

$$EIv(z) = -\int \left[\int M_b(z)\, dz \right] + C_1 z + C_2$$

example of second-order differential equation

Integration constants must be determined by means of the boundary conditions of the deformations (kinematic boundary conditions) and the internal forces (dynamic boundary conditions) $\Rightarrow$ special solution; see table above. *(Text continued on page 228)*

Equations of the Elastic Curve for Selected Trusses

Truss	Equation of the Elastic Curve
1	$$v = \frac{Fab}{6EIl}\begin{cases} bz_1\left(1+\dfrac{l}{b}\right) & ; \ -l_1 \leqq z_1 \leqq 0 \\[1ex] bz_1\left(1+\dfrac{l}{b}-\dfrac{z_1^2}{ab}\right) & ; \ 0 \leqq z_1 \leqq a \\[1ex] az_2\left(1+\dfrac{l}{a}-\dfrac{z_2^2}{ab}\right) & ; \ 0 \leqq z_2 \leqq b \\[1ex] az_2\left(1+\dfrac{l}{a}\right) & ; \ -l_r \leqq z_2 \leqq 0 \end{cases}$$
2	$$v = \frac{-Fzl_1^2}{6EI}\begin{cases} 2\dfrac{l}{l_1}-3\dfrac{z}{l_1}-\left(\dfrac{z}{l_1}\right)^2 & ; \ -l_1 \leqq z \leqq 0 \\[1ex] 2\dfrac{l}{l_1}-3\dfrac{z}{l_1}+\dfrac{z^2}{ll_1} & ; \ 0 \leqq z \leqq l \\[1ex] \dfrac{l_1^2}{l}\dfrac{l}{l_1}\left(1-\dfrac{z}{l}\right) & ; \ l \leqq z \leqq (l+l_r) \end{cases}$$
3	$$v = \frac{ql^4}{24EI}\begin{cases} \lambda & ; \ -l_1 \leqq z \leqq 0 \\ \lambda - 2\lambda^3 + \lambda^4 & ; \ 0 \leqq z \leqq l \\ 1-\lambda & ; \ l \leqq z \leqq (l+l_r) \end{cases}$$ $$\lambda := \frac{z}{l}$$

(Continued)

Equations of the Elastic Curve for Selected Trusses

Truss	Equation of the Elastic Curve
4 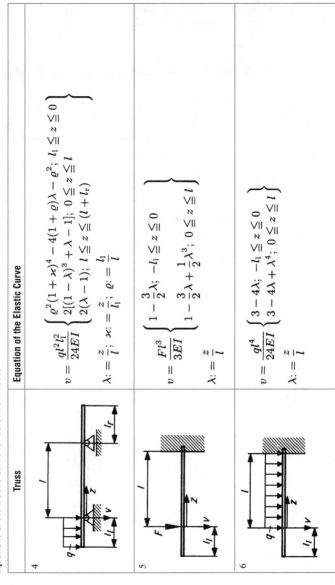	$v = \dfrac{q l^2 l_1^2}{24 E I} \left\{ \begin{array}{l} \varrho^2 (1 + \varkappa)^4 - 4(1 + \varrho)\lambda - \varrho^2; \ l_1 \leqq z \leqq 0 \\ 2[(1 - \lambda)^3 + \lambda - 1]; \ 0 \leqq z \leqq l \\ 2(\lambda - 1); \ l \leqq z \leqq (l + l_r) \end{array} \right\}$ $\lambda := \dfrac{z}{l}; \ \varkappa := \dfrac{z}{l_1}; \ \varrho := \dfrac{l_1}{l}$
5	$v = \dfrac{F l^3}{3 E I} \left\{ \begin{array}{l} 1 - \dfrac{3}{2}\lambda; \ -l_1 \leqq z \leqq 0 \\ 1 - \dfrac{3}{2}\lambda + \dfrac{1}{2}\lambda^3; \ 0 \leqq z \leqq l \end{array} \right\}$ $\lambda := \dfrac{z}{l}$
6	$v = \dfrac{q l^4}{24 E I} \left\{ \begin{array}{l} 3 - 4\lambda; \ -l_1 \leqq z \leqq 0 \\ 3 - 4\lambda + \lambda^4; \ 0 \leqq z \leqq l \end{array} \right\}$ $\lambda := \dfrac{z}{l}$

Deformation with unsymmetric bending: superposition of two bending components along the principal axis, taking the sign into account

2. Method of superposition

The deformation at one point on the truss as the result of a combined load can be calculated by adding the deformations resulting from the component loads (principle of superposition).

$$v_j = \sum_{i=1}^{m} v_{ji}, \ \ v_j' = \sum_{i=1}^{m} v_{ji}'$$

j point at which deformation is being calculated

i designation of the component load case

m number of component load cases

3. Other methods $\Rightarrow$ Castigliano's Theorem

Members of nonhomogeneous material composition, laminated beams

When a member of nonhomogeneous material composition is subjected to simple bending, the line of zero stress and the centroidal axis generally do not coincide.

$$M_b = M_x, \ \ F_L = 0, \ \ E = E(\overline{y}) \qquad \text{simple bending around the } x \text{ axis}$$

Stress distribution

$$\sigma_b(\overline{y}) = \frac{M_b}{I(\overline{y})}\overline{y} \qquad\qquad I(\overline{y}) = \frac{1}{E(\overline{y})} \int\limits_{(A)} E(\overline{y})\overline{y}^2 \, \mathrm{d}A$$

Two-ply laminated beam with rectangular cross section

Position of line of zero stress (coordinate a)

$$a = \frac{E_2 h_2^2 - E_1 h_1^2}{2(E_1 h_1 + E_2 h_2)}$$

Stress distribution across the ply thicknesses (with abrupt change in stress at the ply interface)

($i = 1$: ply 1

$i = 2$: ply 2):

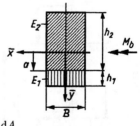

$$\sigma_{bi} = \frac{M_b}{I_i(\overline{y})}\overline{y}, \quad I_i(\overline{y}) = \frac{1}{E_i} \int\limits_{(A)} E(\overline{y})\overline{y}^2 \, \mathrm{d}A$$

$$\int\limits_{(A)} E(\bar{y})\bar{y}^2 \, \mathrm{d}A = \frac{B}{3} \left[E_2(h_2^3 + 3ah_2^2 + 3a^2 h_2) \right.$$

$$\left. + E_1(h_1^3 + 3h_1^2 a + 3h_1 a^2) \right]$$

Ply 2: $-(h_2 - a) \leqq \bar{y} \leqq a$ \qquad Ply 1: $a \leqq \bar{y} \leqq a + h_1$

Curved member, girder

Assumptions: $h \lesssim R$, radius of the centroidal surface R, depth of cross section h, cross-sectional area of the curved member A, centroid coordinates x, y are principal axes, internal forces F_L, $M_b = M_x$

Stress distribution

$$\sigma_z(y) = \frac{F_L}{A} + \frac{M_b}{AR} + \frac{M_b R}{Z} \frac{y}{R+y}$$

(The line of zero stress and the centroidal axis x do not coincide when $F_L = 0$; the curve is nonlinear.)

$$Z := \int\limits_{(A)} \frac{y^2}{1 + \dfrac{y}{R}} \, \mathrm{d}A$$

$$y_N = \frac{F_L R^2 Z + M_b Z R}{F_L R Z + M_b Z + M_b R^2 A}$$

coordinate of the line of zero stress

Values for Z integral

$$Z = \begin{cases} -AR^2 + BR^3 \ln \dfrac{1 + \varkappa}{1 - \varkappa}, & \varkappa := \dfrac{H}{2R}, & \text{rectangle} \\[2ex] AR^2 \left[2\lambda(\lambda - \sqrt{\lambda^2 - 1}) - 1 \right], & \lambda := \dfrac{R}{r_0}, & \text{circle} \end{cases}$$

Rectangle: width B, height H, circle: radius r_0

Torsion

Internal force: torque M_t of the member. For shafts, determination can be made as follows:

$P = M_t \omega = M_t(2\pi n_D)$ \qquad

P \quad power
ω \quad angular velocity
n_D \quad rotational speed

Assumptions: The member theoretically consists of infinitesimally small plates that maintain their initial cross-sectional shape (Saint Venant hypothesis). Circular cross sections, annular cross sections, and thin-walled closed cross sections in the shape of tangent polygons also remain plane or undistorted. Warping (local displacement in the direction of the longitudinal axis of the member) are not restrained.

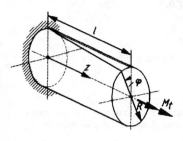

φ angle of torsion of the cross section

$\vartheta := d\varphi//dz$ twist, relative or specific angle of torsion

τ_t, τ_{tmax} torsional shearing stress / maximum value

I_p polar moment of inertia

I_t torsional moment of inertia, torsional resistance, $I_t < I_p$ (circular or annular cross section $I_t = I_p$)

GI_t torsional rigidity

W_t torsional section modulus

Tips:

— linear stress distribution in circular or annular cross section:

$$\tau_t(r) = \frac{M_t}{I_p} r$$

Thin-walled cross sections:
— The wall thickness $\delta(s)$ is negligibly small in comparison with the length of the center line of the wall section.
— The position on the contour is defined by the coordinate s.
— shear flow (a loading characteristic):

$$t(s): = \tau_t(s)\delta(s)$$

(constant across δ in closed cross sections, variable across δ in open cross sections)

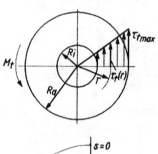

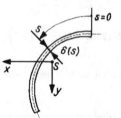

Maximum torsional shearing stress

$$\tau_{max} = \frac{M_t}{W_t}$$

Torsional Moment of Inertia and Torsional Section Modulus for Selected Cross Sections

Cross Section	Torsional Moment of Inertia I_t	Torsional Section Modulus W_t	Remarks
circular or annular cross section	$\dfrac{\pi D_a^4}{32}(1-\lambda^4) = \dfrac{\pi R_a^4}{2}(1-\lambda^4)$, $\lambda := \dfrac{D_i}{D_a} = \dfrac{R_i}{R_a}$	$\dfrac{\pi D_a^3}{16}(1-\lambda^4)$ $= \dfrac{\pi R_a^3}{2}(1-\lambda^4)$	Cross sections remain undistorted, D_a, D_i, R_a, R_i outside and inside diameters and radii, respectively $\lambda = 0 \Rightarrow$ circular cross section
thin-walled, closed cross section (one cell)	$\dfrac{4A_m^2}{\oint \dfrac{ds}{\delta(s)}}$ (Bredt's second formula) n wall elements of constant wall thickness δ_i: $\oint \dfrac{ds}{\delta(s)} = \displaystyle\sum_{i=1}^{n} \dfrac{L_i}{\delta_i}$	$2A_m\delta_{min}$ (Bredt's first formula)	shear flow $t = \dfrac{M_t}{2A_m} = $ const A_m surface area bounded by center line of wall section, δ_{min} minimum wall thickness, L_i lengths of wall elements with thickness δ_i

(Continued)

Torsional Moment of Inertia and Torsional Section Modulus for Selected Cross Sections *(Continued)*

Cross Section	Torsional Moment of Inertia I_t	Torsional Section Modulus W_t	Remarks
thin-walled, open cross section	n wall elements of constant thickness δ_i: $$\frac{1}{3}\,\varkappa\sum_{i=1}^{n}L_i\delta_i^3$$ $\varkappa = \begin{cases} 1 \text{ rectangle; } 1,3 \text{ } \text{I-cross section} \\ 1,12 \text{ T-/}\sqcup\text{- cross section} \end{cases}$	$\dfrac{I_t}{\delta_{max}}$	L_i lengths of wall elements with thickness δ_i; geometrically determined correction factor $\varkappa$ takes torsion restraint into account. For comparable cross sections: $I_{open} \ll I_{closed}$
noncircular solid sections	ellipse: $$\frac{a^3b^3}{\pi\,a^2+b^2}$$ rectangle: $\approx c_1 n b^4$ $$c_1 = \frac{1}{3}\left(1 - \frac{0.63}{n} + \frac{0.052}{n^5}\right)$$	$$\frac{\pi}{2}ab^2$$ $\approx \dfrac{c_1}{c_2}nb^3$ $$c_2 = 1 - \frac{0.65}{1+n^3}$$	a, b major and minor semiaxes of ellipse B, H width, height $n := H/B \geq 1$ t_{max} occurs in the centers of the "longer edges" of the ellipse and rectangle, point A

Twist, angle of torsion of a region of a member

$$\vartheta = \frac{M_t}{GI_t}, \quad \varphi = \int_{(z)} \frac{M_t(z)}{GI_t(z)} \, dz$$

(For specifications of I_t, W_t, see above table.)

$$\varphi_{ges} = \sum_{i=1}^{n} \frac{M_{ti} l_i}{(GI_t)_i}$$

total angle of torsion in the shafting with n regions of length l_i and constant values of $(GI_t)_i$ and M_{ti}

Shear

Internal force:

The following specifications apply to the shearing force $F_Q = F_{Qy}$ as the causal internal force. The method for $F_Q = F_{Qx}$ is similar.

Assumptions: For geometry, see the earlier section on "Bending"; linear deformation transverse bending around the x axis such $M_b = M_x \neq 0$; x, y that principal axes; cross section becomes warped due to slip (designation:$\gamma_s := \gamma_{zy} = \gamma_{yz}$); this leads to shearing stress (designation:$\tau_s := \tau_{zy} = \tau_{yz}$), which occurs both between the regions and tangentially to the cross section. The upper and lower edges of the cross section are free of shearing stress.

Simply connected solid sections

Shearing stress distribution along the cross section

$$\tau_s(y_R) = \frac{F_Q S_{x\,Rest}}{I_{xx} b(y_R)}$$

y_R coordinate of the region in which τ_s is effective

$b(y_R)$ thickness of region

y_{SRest} y coordinate of the centroid of the residual area

$$S_{x\,Rest} = \int_{(A_{Rest})} y \, dA = \int_{y_R}^{e} b(y) y \, dy$$

$S_{x\,Rest}$ static moment of the residual area A_{rest} or x axis

$$S_{x\,Rest} = y_{SRest} A_{Rest}$$

Special case: rectangle with $A = HB$

Static moment of the residual area

$$S_{x\,\text{Rest}}(y_R) = B \int\limits_{y_R}^{\frac{1}{2}H} y \, dy$$

$$= \frac{1}{2} B \left[\left(\frac{H}{2} \right)^2 - y_R^2 \right]$$

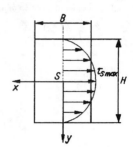

Shearing stress distribution

$$\tau_s(y_R) = \frac{3F_Q}{2A} \left[1 - \left(\frac{2y_R}{H} \right)^2 \right]$$

$$-\frac{1}{2}H \leq y_R \leq \frac{1}{2}H$$

$$\tau_{s\max} = \frac{3}{2}\tau_{sm}, \quad \tau_{sm} = \frac{F_Q}{A}$$

Parabolic curve (shearing stress in the y direction; sketch shows collapsed view); for the neutral surface, the shearing stress assumes a maximum value.

Deformation

$$\gamma_S(y_R) = \frac{F_Q S_{x\,\text{Rest}}}{G I_{xx} b(y_R)}$$

Slip in the region y_R

The influence on the deflection of members is usually negligible.

Thin-walled cross sections

Shear flow as loading characteristic (see earlier section on "Torsion")

$$t(s) := \tau_s \delta(s) \qquad\qquad \delta(s) \quad \text{wall thickness}$$

$$\qquad\qquad\qquad\qquad s \quad \text{coordinate along the center line of the wall section}$$

Thin-walled, closed cross section
Shear flow distribution
(one-cell cross sections)

$$t(s) = -\frac{F_Q}{I_{xx}} \left[S_x(s) - \frac{\oint \dfrac{S_x(s)}{\delta(s)}\,ds}{\oint \dfrac{ds}{\delta(s)}} \right]$$

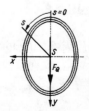

$$S_x(s) := \int_{(s)} \delta(s)y\,\mathrm{d}s$$

S_x (s) static moment of the residual
         area or x axis

Thin-walled, open cross section

Shear flow distribution

$$t(s) = -\frac{F_Q}{I_{xx}}S_x(s)$$

(approximation for narrow web of

height h: $t \approx \dfrac{F_Q}{h}$)

Shear center (center of twist)

If the action line of the shearing force passes through the so-called shear center (center of twist) $T(x_T, y_T)$, then torque does not occur in transverse bending. The location of T is determined by the geometry of the cross section. T lies on the axis or axes of symmetry of the cross section.

This is especially relevant for cross sections
of low torsional rigidity (e.g., thin-walled,
open cross sections), such that the action
line of the external load passes through T.
The coordinates of T for thin-walled,
open cross sections:

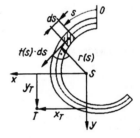

$$x_T = -\frac{1}{I_{xx}}\int_0^L S_x(s)r(s)\,\mathrm{d}s$$

$$y_T = \frac{1}{I_{yy}}\int_0^L S_y(s)r(s)\,\mathrm{d}s$$

L length of center line of
    wall section

S_x (s), S_y (s) static moments of the
            residual area with respect
            to x and y axes

Thin-walled ⊔ cross section
(wall thickness h = const)

$$x_T = \frac{4B^2(3B + H)}{(2B + H)(6B + H)}$$

The figure shows the position of
T and the qualitative shear flow
curve at a shearing force of $F_Q = F_{Qy}$.

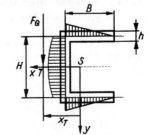

Compound Stresses

Structural components are usually subjected to stresses composed of multiple simultaneous internal forces, i.e., compound stresses.

Technical requirement for evaluation of structural components:

The compound stress must be less than an allowable uniaxial reference stress. The reference stress is usually an allowable stress $\sigma_{\text{allowable}}$.

Explanation of the method applied to a beam:

Superposition of tension/compression and bending around two principal axes

Stress distribution

$$\sigma_{\text{compound}}(x, y) = \frac{F_{\text{L}}}{A} + \left[\frac{M_x}{I_{xx}} y - \frac{M_y}{I_{yy}} x \right]$$

Line of zero stress

$$y = \left(\frac{M_y I_{xx}}{M_x I_{yy}} \right) x - \frac{F_{\text{L}} I_{xx}}{M_x A}$$

The line of zero stress is generally inclined in relation to the principal axes and does not pass through the center of mass.

Superposition of torsion and shear

Combination of τ_t and τ_s in the cross section, accounting for the applicable shearing stress distribution.

Example: circular cross section
The following compound shearing stresses result at the four points indicated on the edge:

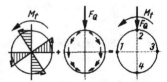

$\tau_{\text{compound1}} = \tau_t + \tau_s, \ \tau_{\text{compound2}} = \tau_t$

$\tau_{\text{compound3}} = \tau_t - \tau_s, \ \tau_{\text{compound4}} = \tau_t$

Compound normal stress and tangential stress

Loading and damaging of structural components occurs in a multiaxial state of stress (see also section 2.11).

The uniaxial equivalent stress σ_v is determined by means of suitable hypotheses of strength, rupture, or failure.

Stress components in the beam cross section

$\sigma = \sigma_z$ stress resulting from axial force
                                            and/or bending

$\tau = \tau_{yz}$ stress resulting from torsion
                                            and/or shearing force

Selected strength hypotheses and reference stresses

Principal normal stress hypothesis

$$\sigma_{v1} := \sigma_1 = \frac{1}{2}\sigma + \sqrt{\left(\frac{\sigma}{2}\right)^2 + \tau^2}$$

(Failure is determined by the principal normal stress σ_1. This hypothesis is applicable to structural components that are prone to brittle fracture.)

Principal strain hypothesis

$$\sigma_{v2} := E\varepsilon_1$$
$$= (1-\nu)\frac{\sigma}{2} + (1+\nu)\sqrt{\left(\frac{\sigma}{2}\right)^2 + \tau^2}$$

(Failure is determined by the maximum principal strain ε_1.)

Maximum shearing stress hypothesis

$$\sigma_{v3} := 2\tau_{max} = \sqrt{\sigma^2 + 4\tau^2}$$

(Failure is determined by τ_{max}. This hypothesis is applicable to structural components that are prone to ductile fracture.)

Shear strain energy hypothesis (according to Huber, von Mises, Hencky)

$$\sigma_{v4} = \sqrt{\sigma^2 + 3\tau^2}$$

(Failure is determined by the shear strain energy as a proportion of the deformation energy. This hypothesis is widely accepted and preferred in mechanical engineering.)

The state of stress known as "pure shear" is expressed by introducing the strain ratio (Bach's correction factor) α_0.

$$\alpha_0 := \frac{\sigma_{allowable}}{\sqrt{3}\tau_{allowable}}$$
$$\sigma_{v4} = \sqrt{\sigma^2 + 3(\alpha_0\tau)^2}$$

Graphic interpretation as a breaking load (ultimate load) curve

$$\left(\frac{\sigma}{\sigma_{\text{allowable}}}\right)^2 + \left(\frac{\tau}{\tau_{\text{allowable}}}\right)^2 = 1$$

Evaluation of shafts with circular or annular cross sections

Internal forces:

Bending moment
$$M_{\text{b}} = \sqrt{M_{\text{b}x}^2 + M_{\text{b}y}^2}$$
and torque M_{t}

(maximum stress on the edge of the cross section, at the most heavily loaded point in the z direction; may be necessary to distinguish between different cases).

Reference stress

$$\sigma_{\text{v4}} = \frac{1}{W_{\text{b}}}\sqrt{M_{\text{b}}^2 + \frac{3}{4}(\alpha_0 M_{\text{t}})^2}$$

$$W_{\text{b}} = \frac{\pi}{32}D_{\text{a}}^3(1 - \lambda^4), \quad \lambda := \frac{D_{\text{i}}}{D_{\text{a}}}$$

(The expression

$$M_{\text{v}} := \sqrt{M_{\text{b}}^2 + \frac{3}{4}(\alpha_0 M_{\text{t}})^2}$$

is also called the "reference moment."

$$D_{\text{aerf}} \geq \sqrt[3]{\frac{32\sqrt{M_{\text{b}}^2 + \frac{3}{4}(\alpha_0 M_{\text{t}})^2}}{(1 - \lambda^4)\pi\sigma_{\text{allowable}}}}$$

for dimensioning applications

Deformation Energies

Assumptions: linear elastic material properties, l length of region

Deformation energy density
$$W_{\text{F}}^* = \int_{(\varepsilon)} \sigma(\varepsilon)\,\mathrm{d}\varepsilon = \frac{\sigma^2}{2E}$$

(for each normal stress component and shearing stress component)

$$W_{\text{F}}^* = \int_{(\gamma)} \tau(\gamma)\,\mathrm{d}\gamma = \frac{\tau^2}{2G}$$

Deformation energy

$$W_{\mathrm{F}} = \int\limits_{(V)} W_{\mathrm{F}}^* \, \mathrm{d}V$$

Deformation energy for primary stresses in the beam

Primary Stress	Deformation Energy Density	Deformation Energy
tension/ compression	$W_{F_{z/d}}^* = \dfrac{\sigma_{z/d}^2}{2E}$ $= \dfrac{F_{\mathrm{L}}^2}{2A(z)EA(z)}$	$W_{F_{z/d}} = \int\limits_{(l)} \dfrac{F_{\mathrm{L}}^2}{2EA(z)} \, \mathrm{d}z$ For $A = \text{const}$, $F_{\mathrm{L}} = \text{const}$: $W_{F_{z/d}} = \dfrac{F_{\mathrm{L}}^2 l}{2EA}$
bending (pure)	symmetric bending: $W_{F_{\mathrm{b}}}^* = \dfrac{\sigma_{\mathrm{b}}^2}{2E}$ $= \dfrac{M_{\mathrm{b}}^2}{2I_{xx}EI_{xx}} y^2$	symmetric bending: $W_{F_{\mathrm{b}}} = \int\limits_{(l)} \dfrac{M_{\mathrm{b}}^2}{2EI_{xx}} \, \mathrm{d}z$ bending around two principal axes: $W_{F_{\mathrm{b}}} = \int\limits_{(l)} \left(\dfrac{M_x^2}{2EI_{xx}} + \dfrac{M_y^2}{2EI_{yy}} \right) \mathrm{d}z$
torsion	$W_{F_t}^* = \dfrac{\tau_t^2}{2G}$	$W_{F_t} = \int\limits_{(l)} \dfrac{M_t^2}{2GI_t} \, \mathrm{d}z$
shear	$F_{\mathrm{Q}} = F_{\mathrm{Q}y}$ $W_{F_{\mathrm{s}}}^* = \dfrac{\tau_{\mathrm{s}}^2}{2G}$	$W_{F_{\mathrm{s}}} = \varkappa \int\limits_{(l)} \dfrac{F_{\mathrm{Q}}^2}{2GA} \, \mathrm{d}z$ $\varkappa := \dfrac{A}{I_{xx}^2} \int\limits_{(A)} \left(\dfrac{S_{x\,\mathrm{Rest}}}{b(y_{\mathrm{R}})} \right)^2 \mathrm{d}A$ Rest = residual Shear distribution $\chi = 6/5$ (rectangle) and $10/9$ (circle)

Tip: In frame members, it is most important to consider bending energy and torsional energy. In trusses and cables, it is most important to consider the energy of axial forces.

Castigliano's Theorem

$$\frac{\partial W_a}{\partial F_j} = v_j, \qquad \frac{\partial W_a}{\partial M_j} = v_j' = \varphi_j$$

Partial derivation of the external work W_a with respect to a force component yields the displacement of the beam at the point of application of, and in the direction of, this force. Partial derivation of the external work with respect to a moment yields the inclination of the beam at the point of application and in the direction of the moment.

Practical application by means of equivalence of $W_a = W_F$.

Determination of displacement and inclination at point j with x, y principal axes, isothermal loading:

$$v_j = \frac{\partial W_F}{\partial F_j} = \sum_{i=1}^{m} \int_{(l_i)} \left[\frac{F_{Li}}{(EA)_i} \frac{\partial F_{Li}}{\partial F_j} + \frac{M_{xi}}{(EI_{xx})_i} \frac{\partial M_{xi}}{\partial F_j} \right.$$

$$\left. + \frac{M_{yi}}{(EI_{yy})_i} \frac{\partial M_{yi}}{\partial F_j} + \frac{M_{ti}}{(GI_t)_i} \frac{\partial M_{ti}}{\partial F_j} \right] dz_i$$

$$v_j' = \frac{\partial W_F}{\partial M_j} = \sum_{i=1}^{m} \int_{(l_i)} \left[\frac{F_{Li}}{(EA)_i} \frac{\partial F_{Li}}{\partial M_j} + \frac{M_{xi}}{(EI_{xx})_i} \frac{\partial M_{xi}}{\partial M_j} \right.$$

$$\left. + \frac{M_{yi}}{(EI_{yy})_i} \frac{\partial M_{yi}}{\partial M_j} + \frac{M_{ti}}{(GI_t)_i} \frac{\partial M_{ti}}{\partial M_j} \right] dz_i$$

m number of regions

l_i lengths of regions

The deformation energy can be found in the above table. It is also necessary to take the energy of the shearing force into account.

If it is desirable to calculate deformations at points where no force component or no moment is applied, then zero forces ($F_H = 0$) or zero moments ($M_H = 0$) should be introduced at these points. Zero loads must be included in support reactions and internal forces for the purpose of determining the partial derivatives.

Calculation of statically indeterminate trusses

(Menabrea's Theorem)

$$\frac{\partial W_a}{\partial X_p} = \frac{\partial W_F}{\partial X_p} = 0$$

$|n|$ degree of static indeterminacy

$$1 \leq p \leq |n|$$

X_p statically indeterminate reactions; unknown support reactions, restraint reactions, or internal forces that can not be determined by application of equilibrium conditions alone

Partial derivation of the external work or the deformation energy with respect to the statically indeterminate reactions yields a result of zero.

Multiaxial States of Stress

Components of the stress tensor

$$\begin{pmatrix} \sigma_x & \tau_{xy} & \tau_{xz} \\ \tau_{yx} & \sigma_y & \tau_{yz} \\ \tau_{zx} & \tau_{zy} & \sigma_z \end{pmatrix}$$

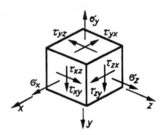

Law of equivalence of conjugate shearing stresses (symmetry of the stress tensor)

$\tau_{ij} = \tau_{ji,}$ $(i, j = x, y, z),$ where $i \neq j$

Reference stress according to the shear strain energy hypothesis

$$\sigma_{v4} = \sqrt{\frac{1}{2}\left[(\sigma_x - \sigma_y)^2 + (\sigma_x - \sigma_z)^2 + (\sigma_y - \sigma_z)^2\right] + 3(\tau_{xy}^2 + \tau_{xz}^2 + \tau_{yz}^2)}$$

$$\sigma_{v4} = \sqrt{\frac{1}{2}\left[(\sigma_1 - \sigma_2)^2 + (\sigma_1 - \sigma_3)^2 + (\sigma_2 - \sigma_3)^2\right]}$$

$\sigma_1 > \sigma_2 > \sigma_3$ principal normal stresses

Components of the (symmetrical) strain tensor

$$
\begin{pmatrix}
\varepsilon_x & \dfrac{1}{2}\gamma_{xy} & \dfrac{1}{2}\gamma_{xz} \\[2mm]
\dfrac{1}{2}\gamma_{yx} & \varepsilon_y & \dfrac{1}{2}\gamma_{yz} \\[2mm]
\dfrac{1}{2}\gamma_{zx} & \dfrac{1}{2}\gamma_{zy} & \varepsilon_z
\end{pmatrix}
$$

Strain-stress relationships

$$
\varepsilon_i = \frac{\partial v_i}{\partial i}, \quad \gamma_{ij} = \frac{\partial v_i}{\partial j} + \frac{\partial v_j}{\partial i} \qquad (i, j = x, y, z)
$$

Generalized Hooke's Law $(i, j, k = x, y, z)$

$$
\varepsilon_i = \frac{1}{E}\left[\sigma_i - \nu(\sigma_j + \sigma_k)\right] + \alpha_{\mathrm{th}}\Delta T \qquad \gamma_{ij} = \frac{1}{G}\tau_{ij}
$$

$$
\sigma_i = \frac{E}{(1+\nu)}\left(\varepsilon_i + \frac{\nu}{(1-2\nu)}e\right) - \frac{E}{(1-2\nu)}\alpha_{\mathrm{th}}\Delta T
$$

$$
\varepsilon = \varepsilon_x + \varepsilon_y + \varepsilon_z \qquad\qquad\qquad \text{volumetric strain}
$$

Elastostatics problem:

Approximation methods are generally the only means for solving the 15 field equations (3 equilibrium conditions, 6 strain-stress relationships, 6 Hooke's laws) for the 15 field quantities (3 displacements, 6 stresses, 6 strains), taking the boundary conditions into account $\Rightarrow$ numerical methods.

Simplified modeling:

— plane stress, plate stress	Stress components perpendicular to the plane of the plate are negligible.
— plane strain	Strain components perpendicular to parallel planes are negligible.
— rotationally symmetrical state	Geometry, boundary conditions, and loading are rotationally symmetrical with respect to one axis and independent of the polar angle.

Thin-walled containers under internal pressure (Membrane Theory)

Assumptions: Closed contains under isothermal internal pressure; wall thickness is small in relation to radii of curvature; unit vector in the normal direction of the middle surface of the shell $\vec{e}_n$ is always directed toward the outside; radial stress is negligible.

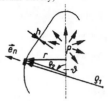

Pressure vessel formula

$$\frac{\sigma_\varphi}{\varrho_2} + \frac{\sigma_\vartheta}{\varrho_1} = \frac{p}{h}$$

p internal pressure

h wall thickness

ϱ_1, ϱ_2 radii of the circles of principal curvature

Case: container closed along the axis of rotation

$$\sigma_\vartheta = \frac{pr}{2h\sin\vartheta} = \frac{p\varrho_2}{2h}$$

$$\sigma_\varphi = \frac{p\varrho_2}{2h}\left(2 - \frac{\varrho_2}{\varrho_1}\right)$$

σ_ϑ meridianal stress

σ_φ circumferential stress

Membrane stresses in selected closed containers

Designation	ϱ_1	ϱ_2	σ_ϑ	σ_φ	Remarks
spherical pressure vessel (radius R)	R	R	$\dfrac{pR}{2h}$	$\dfrac{pR}{2h}$	$\sigma_\vartheta = \sigma_\varphi$
shell-type boiler (radius R)	∞	R	$\sigma_l = \dfrac{pR}{2h}$	$\dfrac{pR}{h}$	$\sigma_\varphi = 2\sigma_l$ σ_l longitudinal stress
one vessel $\varkappa = \dfrac{\tan\alpha}{\cos\alpha}$ α half the cone angle, z coordinate of apex	∞	$\varkappa z$	$\dfrac{p\varkappa z}{2h}$	$\dfrac{p\varkappa z}{h}$	$\sigma_\varphi = 2\sigma_\vartheta$

Case: container open along the axis of rotation (toroidal vessel)

$$\sigma_\vartheta = \frac{p(r^2 - R_0^2)}{2rh\sin\vartheta} = \frac{p\varrho_2}{2h}\left[1 - \left(\frac{R_0}{r}\right)^2\right]$$

$$\sigma_\varphi = \frac{p\varrho_2}{2h}\left\{2 - \frac{\varrho_2}{\varrho_1}\left[1 - \left(\frac{R_0}{r}\right)^2\right]\right\}$$

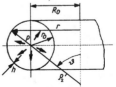

R_0 radius of the generatrix

Rotationally symmetrical problems

Use of cylindrical coordinates r, φ, z

Designation of the partial derivative: $(\ )_{,i} := \dfrac{\partial}{\partial i}(\ )$

Field quantities

Displacements $v_r, \ v_\varphi, \ v_z$

Stresses $\sigma_r, \sigma_\varphi, \sigma_z$
(principal normal stresses)

Strains (elongations) $\varepsilon_r, \varepsilon_\varphi, \varepsilon_z$

Reference stress

$$\sigma_{v4} = \sqrt{\sigma_r^2 + \sigma_\varphi^2 + \sigma_z^2 - (\sigma_r\sigma_\varphi + \sigma_r\sigma_z + \sigma_\varphi\sigma_z)}$$

Strain-stress relationships

$$\varepsilon_r = v_{r,r}, \quad \varepsilon_\varphi = \frac{v_r}{r}, \quad \varepsilon_z = v_{z,z}$$

Hooke's Law

$$\left\{\begin{matrix} \varepsilon_r \\ \varepsilon_\varphi \\ \varepsilon_z \end{matrix}\right\} = \frac{1}{E} \left\{\begin{matrix} \sigma_r - \nu(\sigma_\varphi + \sigma_z) + E\alpha_{th}\Delta T \\ \sigma_\varphi - \nu(\sigma_r + \sigma_z) + E\alpha_{th}\Delta T \\ \sigma_z - \nu(\sigma_r + \sigma_\varphi) + E\alpha_{th}\Delta T \end{matrix}\right\} \text{ or}$$

$$\left\{\begin{matrix} \sigma_r \\ \sigma_\varphi \\ \sigma_z \end{matrix}\right\} =$$

$$= \frac{E}{(1+\nu)(1-2\nu)} \left\{\begin{matrix} \varepsilon_r(1-\nu) + \nu(\varepsilon_\varphi + \varepsilon_z) - (1+\nu)\alpha_{th}\Delta T \\ \varepsilon_\varphi(1-\nu) + \nu(\varepsilon_z + \varepsilon_r) - (1+\nu)\alpha_{th}\Delta T \\ \varepsilon_z(1-\nu) + \nu(\varepsilon_r + \varepsilon_\varphi) - (1+\nu)\alpha_{th}\Delta T \end{matrix}\right\}$$

Rotating disk (plane stress such that $\sigma_z = 0$)

ϱ density

ω angular velocity

$h(r) = h_0 r^n$ profile function of
disk thickness

h_0 disk thickness on inside edge

General solution for radial displacement

$$v_r = A_1 r^{\lambda_1} + A_2 r^{\lambda_2} \qquad\qquad \lambda_{1,2} = -\frac{n}{2} \pm \sqrt{\left(\frac{n}{2}\right)^2 - \nu n + 1}$$
$$- \frac{\varrho\omega^2(1-\nu^2)}{E(8+3n+\nu n)} r^3$$

Example: disk of constant thickness ($n = 0$, $h = h_0$):

$$v_r = A_1 r + \frac{A_2}{r} - \frac{\varrho\omega^2(1-\nu^2)}{8E} r^3$$

Stress distribution

$$\left\{ \begin{matrix} \sigma_r \\ \sigma_\varphi \end{matrix} \right\} = \frac{\varrho\omega^2}{8} \left\{ \begin{matrix} B_1 - \dfrac{B_2}{r^2} - (3+\nu)r^2 \\[2mm] B_1 + \dfrac{B_2}{r^2} - (1+3\nu)r^2 \end{matrix} \right\}$$

$$B_1 = \frac{8E}{\varrho\omega^2(1-\nu)} A_1$$

$$B_2 = \frac{8E}{\varrho\omega^2(1+\nu)} A_2$$

Integration constants A_1, A_2, and B_1, B_2 are determined by boundary conditions.

Special case: rotating solid disk (radius R)

$$\left\{ \begin{matrix} \sigma_r \\ \sigma_\varphi \end{matrix} \right\} = \frac{\varrho\omega^2(3+\nu)}{8} R^2 \left\{ \begin{matrix} 1 - \left(\dfrac{r}{R}\right)^2 \\[2mm] 1 - \dfrac{1+3\nu}{3+\nu}\left(\dfrac{r}{R}\right)^2 \end{matrix} \right\}$$

(Graph based on $\overline{\sigma}_{r,\varphi} = \sigma_{r,\varphi}\left[\dfrac{1}{8}R^2\varrho\omega^2(3+\nu)\right]^{-1}$, $\nu = \dfrac{1}{3}$)

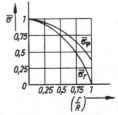

Rotating disk of uniform strength (Laval disk)

$$h(r) = h_0\, e^{-\frac{\varrho\omega^2}{2\sigma_0}(r^2 - R_i^2)}$$

Profile function is such that
$\sigma_r(r) = \sigma_\varphi(r) = \sigma_0 = $ const,
$R_i \leqq r \leqq R_a$

Thick-walled long tube

Loads:
p_i internal pressure
p_a external pressure

Dimensions:
R_i, R_a inside radius, outside radius

$$\lambda = \frac{R_i}{R_a}$$

Radial displacement for plane strain such that $\varepsilon_z = 0$

$$v_r = \frac{(1+\nu)R_a\lambda}{E(1-\lambda^2)} \left\{ p_i\lambda \left[\left(\frac{R_a}{r}\right) + (1-2\nu)\left(\frac{r}{R_a}\right) \right] \right.$$
$$\left. -p_a \left[\left(\frac{R_i}{r}\right) + (1-2\nu)\left(\frac{r}{R_i}\right) \right] \right\}$$

Stress distribution

$$\sigma_r = \frac{\lambda^2}{1-\lambda^2} \left\{ p_i \left[1 - \left(\frac{R_a}{r}\right)^2 \right] - \frac{p_a}{\lambda^2} \left[1 - \left(\frac{R_i}{r}\right)^2 \right] \right\}$$

$$\sigma_\varphi = \frac{\lambda^2}{1-\lambda^2} \left\{ p_i \left[1 + \left(\frac{R_a}{r}\right)^2 \right] - \frac{p_a}{\lambda^2} \left[1 + \left(\frac{R_i}{r}\right)^2 \right] \right\}$$

The support affects only the axial or longitudinal stress σ_z:

$$\sigma_z = \nu(\sigma_r + \sigma_\varphi)$$
$$= \frac{2\nu}{1-\lambda^2}(\lambda^2 p_i - p_a)$$

case of a tube fixed at both ends (plane strain such that $\varepsilon_z = 0$)

Special case:

Tube under internal pressure such that

$R_i = R, R_a = 2R, \lambda = 1/2$

$$\sigma_r = \frac{p_i}{3} \left[1 - \left(\frac{2R}{r}\right)^2 \right]$$

$$\sigma_\varphi = \frac{p_i}{3} \left[1 + \left(\frac{2R}{r}\right)^2 \right]$$

$$v_r = \frac{p_i R(1+\nu)}{3E} \left[\left(\frac{4R}{r}\right) + (1-2\nu)\left(\frac{r}{R}\right) \right]$$

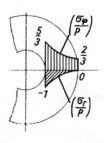

Press fitting or shrink fitting of two tubes with the same modulus of elasticity and Poisson ratio

Dimensions: tube 1 (inside) R_{i1}, R_{a1}
                                                tube 2 (outside) R_{i2}, R_{a2}

Interference (before assembly)

$\Delta R = R_{a1} - R_{i2}$, $R_{a1} > R_{i2}$

Shrinkage

$\epsilon = \dfrac{\Delta R}{R}$, $R \approx R_{a1} \approx R_{i2}$

Relationship between shrinkage and fitting pressure p

$$\epsilon = \frac{p}{E}\left(\frac{R_{a2}^2 + R^2}{R_{a2}^2 - R^2} + \frac{R^2 + R_{i1}^2}{R^2 - R_{i1}^2} \right)$$

Stability Problems

Important stability problems in engineering
(failure due to unstable equilibrium)

— buckling of columns (see below),
   loading: axial compressive force
   (buckling force)

— torsional buckling of thin-walled
   columns, loading: moment around
   the longitudinal axis of the column

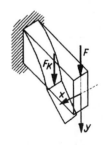

— lateral-torsional buckling of
   girders, loading: buckling moment,
   e.g., due to loading in the plane
   described by the longitudinal axis
   of the girder and the principal axis s_2,
   such that $I_{xx} = I_1 > I_{yy} = I_2$; see figure

— compression buckling of plates
   (compressive loading in the midplane
   of the plate) and shells (e.g., axial
   compressive loading of cylindrical shells)

Buckling of columns

(prismatic, straight, or slightly curved)

Elastic buckling

Buckling force F_K (cases of buckling
according to Euler):

$$F_K = \frac{E I_{\min} \pi^2}{l_K^2}$$

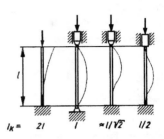

$I_{\min} = I_2$ minimum principal moment
l_K effective length
A cross-sectional area of column

Applicability of $\sigma_K < \sigma_P$, see diagram,
such that:

Buckling stress

$$\sigma_K := \frac{F_K}{A} = \frac{E \pi^2}{\lambda^2}$$

Proportional limit σ_P

Slenderness ratio

$$\lambda := l_K \sqrt{\frac{A}{I_{\min}}}$$

Slenderness ratio limit

$$\lambda_P := \sqrt{\frac{E}{\sigma_P}}$$

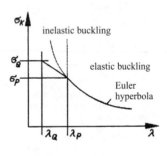

Inelastic buckling (according to Tetmajer)

Buckling force F_K from σ_K for $\sigma_Q < \sigma_K < \sigma_P$:

$$\sigma_K = a\lambda^2 + b\lambda + c,$$
a, b, c material characteristics
(see table)

$$\lambda_Q < \lambda < \lambda_P$$
σ_Q compressive yield strength

$$\lambda_Q := \sqrt{\frac{E}{\sigma_Q}}$$

Characteristic values for inelastic buckling

Material	a/MPa	b/MPa	c/MPa	λ_P
St 37	0	-1.14	310	104
St 50	0	-0.62	335	88
GG	0.053	-12	776	80
lumber	0	-0.194	29.3	100

The slenderness ratio of the column is used to distinguish between elastic and inelastic buckling. If $\lambda < \lambda_Q$, no buckling occurs.

Kinematics

Particle Kinematics

Motion in cartesian coordinates

Position vector $\vec{r}(t)$ as a function of time t (particle path / displacement)

$$\vec{r}(t) = x(t)\vec{e}_x + y(t)\vec{e}_y + z(t)\vec{e}_z$$

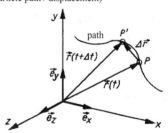

Three coordinates x, y, z are used to describe the three possible direction components or degrees of freedom of the freely moving particle.

If a particle or body possesses n degrees of freedom, then n independent coordinates are needed to describe its motion.

Velocity as a change in the position vector over time

$$\vec{v}(t) := \lim_{\Delta t \to 0} \frac{\vec{r}(t + \Delta t) - \vec{r}(t)}{\Delta t} = \lim_{\Delta t \to 0} \frac{\Delta \vec{r}}{\Delta t} = \frac{d\vec{r}}{dt} = \dot{\vec{r}}$$

$$\vec{v} = \dot{\vec{r}} = v_x \vec{e}_x + v_y \vec{e}_y + v_z \vec{e}_z \qquad \text{components}$$

$$= \dot{x}(t)\vec{e}_x + \dot{y}(t)\vec{e}_y + \dot{z}(t)\vec{e}_z$$

$$v = \sqrt{v_x^2 + v_y^2 + v_z^2} \qquad \text{magnitude}$$

$$= \sqrt{\dot{x}^2 + \dot{y}^2 + \dot{z}^2}$$

Acceleration as a change in the velocity vector over time

$$\vec{a}(t) := \lim_{\Delta t \to 0} \frac{\vec{v}(t + \Delta t) - \vec{v}(t)}{\Delta t} = \lim_{\Delta t \to 0} \frac{\Delta \vec{v}}{\Delta t} = \frac{d\vec{v}}{dt} = \dot{\vec{v}} = \ddot{\vec{r}}$$

$$\vec{a} = \dot{\vec{v}} = a_x \vec{e}_x + a_y \vec{e}_y + a_z \vec{e}_z$$
$$= \ddot{x}(t)\vec{e}_x + \ddot{y}(t)\vec{e}_y + \ddot{z}(t)\vec{e}_z$$

components

$$a = \sqrt{a_x^2 + a_y^2 + a_z^2}$$
$$= \sqrt{\ddot{x}^2 + \ddot{y}^2 + \ddot{z}^2}$$

magnitude

Planar motion in polar coordinates r, φ

Position vector

$$\vec{r}(t) = r(t)\vec{e}_r(t)$$

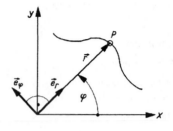

Velocity

$$\vec{v}(t) = \dot{\vec{r}}(t) = \dot{r}\vec{e}_r + r\dot{\varphi}\vec{e}_\varphi$$

$v_r = \dot{r}$ radial component

$v_\varphi = r\dot{\varphi} = r\omega$ circular component

Angular velocity

$$\omega := \dot{\varphi}$$

Acceleration

$$\vec{a}(t) = (\ddot{r} - r\dot{\varphi}^2)\vec{e}_r + (2\dot{r}\dot{\varphi} + r\ddot{\varphi})\vec{e}_\varphi$$

components:
$a_r = \ddot{r} - r\dot{\varphi}^2$ radial
component
$a_\varphi = 2\dot{r}\dot{\varphi} + r\ddot{\varphi}$ circular
component

Angular acceleration

$$\ddot{\varphi} = \dot{\omega}$$

Centripetal acceleration

$$-r\dot{\varphi}^2 = -r\omega^2$$

Coriolis acceleration

$$2\dot{r}\dot{\varphi} = 2\dot{r}\omega$$

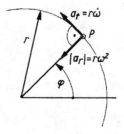

Special case: circular motion
($r(t) = r = $ const)

Position vector, velocity, acceleration

$$\vec{r} = r\vec{e}_r, \quad \vec{v} = r\dot{\varphi}\vec{e}_\varphi = r\omega\vec{e}_\varphi$$
$$\vec{a} = -r\omega^2\vec{e}_r + r\dot{\omega}\vec{e}_\varphi$$

Velocity component

$v_\varphi = v_t = r\omega$ $\qquad\qquad$ tangential velocity

Acceleration components

$a_r = -r\omega^2$ $\qquad\qquad$ centripetal acceleration

$a_\varphi = a_t = r\dot\omega$ $\qquad\qquad$ tangential acceleration

Description of motion in other coordinates

Particle motion in cylindrical coordinates
r, φ, z

$\vec r = r\vec e_r + z\vec e_z$

$\vec v = \dot r\vec e_r + r\dot\varphi\vec e_\varphi + \dot z\vec e_z$

$\vec a = (\ddot r - \dot r\dot\varphi^2)\vec e_r + (2\dot r\dot\varphi + r\ddot\varphi)\vec e_\varphi + \ddot z\vec e_z$

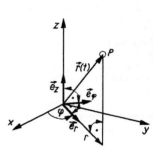

Particle motion in natural coordinates

The coordinate system moves with the particle along the path.

Velocity (in the direction tangent to the path)

$\vec v = v(t)\vec e_t$

Acceleration

$\vec a = \dot v\vec e_t + v\dot\varphi\vec e_n$

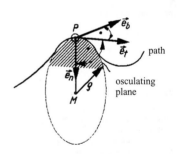

path

osculating
plane

Angular velocity in relation to the
instantaneous center of rotation M

$\dot\varphi = \omega = \dfrac{v}{\varrho}$

Tangential and normal components
of acceleration, respectively

ϱ instantaneous radius of
curvature of the path

$a_t = \dot v, \quad a_n = \dfrac{v^2}{\varrho}$

Rigid-body kinematics

Types of motion

Translation: all points on the body move in congruent paths and thus share the same velocity and acceleration at a given moment in time → described in terms of particle kinematics.

Rotation (around one axis): all points on the body move in concentric circular paths around the axis and thus share the same angular velocity and angular acceleration at a given moment in time.

Planar motion: the paths of all points on the body lie in parallel planes.

General motion: superposition of translation and rotation (e.g., translation of the center of mass of the body, and rotation around an axis through the center of mass).

Kinematics of rotation around a fixed axis

Velocity and acceleration of a point P on a body

$$\vec{v}_P = \vec{\omega} \times \vec{r}_P$$
$$\vec{a}_P = \dot{\vec{\omega}} \times \vec{r}_P + \vec{\omega} \times (\vec{\omega} \times \vec{r}_P)$$

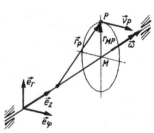

Representation in cylindrical coordinates

$$\vec{v}_P = \omega r_{MP}\vec{e}_\varphi$$
$$\vec{a}_P = \dot{\omega} r_{MP}\vec{e}_\varphi - \omega^2 r_{MP}\vec{e}_r$$

Kinematics of general motion

Translation at $\vec{v}_A$, $\vec{a}_A$ and rotation around the instantaneous axis of rotation at $\vec{\omega}$ (basic formulas of rigid-body dynamics, Eulerian equations), such that

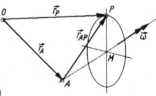

$$\vec{r}_P = \vec{r}_A + \vec{r}_{AP}$$
$$\vec{v}_P = \vec{v}_A + \vec{\omega} \times \vec{r}_{AP}$$
$$\vec{a}_P = \vec{a}_A + \dot{\vec{\omega}} \times \vec{r}_{AP} + \vec{\omega} \times (\vec{\omega} \times \vec{r}_{AP})$$

Special case: Planar motion

Rotation transformation of the unit vectors

$$\vec{e}_r = \cos\varphi\,\vec{e}_x + \sin\varphi\,\vec{e}_y$$
$$\vec{e}_\varphi = -\sin\varphi\,\vec{e}_x + \cos\varphi\,\vec{e}_y$$

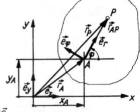

Components of the position vector
to the body-fixed point P in relation to the
space-fixed coordinate system

$$\vec{r}_P = (x_A + r_{AP}\cos\varphi)\vec{e}_x + (y_A + r_{AP}\sin\dot{\varphi})\vec{e}_y$$

Total velocity and total acceleration at point P

$$\vec{v}_P = (\dot{x}_A - r_{AP}\omega\sin\varphi)\vec{e}_x + (\dot{y}_A + r_{AP}\omega\cos\varphi)\vec{e}_y$$
$$\vec{a}_P = (\ddot{x}_A - r_{AP}\omega^2\cos\varphi - r_{AP}\dot{\omega}\sin\varphi)\vec{e}_x$$
$$\quad + (\ddot{y}_A - r_{AP}\omega^2\sin\varphi + r_{AP}\dot{\omega}\cos\varphi)\vec{e}_y$$

Relative motion of a point P

Space-fixed coordinate system $\qquad\qquad x, y, z$

Body-fixed coordinate system $\qquad\qquad x', y', z'$

Description of absolute path (in relation to x, y, z)

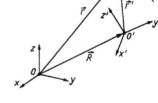

Position vector
$$\vec{r} = \vec{R} + \vec{r}'$$

Velocity of origin displacement
$$\vec{v}_F = \dot{\vec{R}} + \vec{\omega} \times \vec{r}'$$

Relative velocity
$$\vec{v}_{rel} = \frac{d'\vec{r}}{dt}$$

Absolute velocity
(velocity in relation to x, y, z)

$$\vec{v} = \vec{v}_F + \vec{v}_{rel} = \dot{\vec{R}} + \omega \times \vec{r}' + \vec{v}_{rel}$$

$\vec{r}'$ relative path (in relation to x', y', z')

$\vec{R}$ origin displacement

$\dfrac{d'}{dt}$ time derivative in relation to the displaced coordinate system

Absolute acceleration
$$\vec{a} = \vec{a}_F + \vec{a}_{rel} + \vec{a}_c$$

Acceleration of origin
$$\vec{a}_F = \ddot{\vec{R}} + \dot{\vec{\omega}} \times \vec{r}' + \vec{\omega} \times (\vec{\omega} \times \vec{r}')$$

Relative acceleration

$$\vec{a}_{\text{rel}} = \frac{d'\vec{v}_{\text{rel}}}{dt} = \frac{d'^2\vec{r}'}{dt^2}$$

Coriolis acceleration

$$\vec{a}_{\text{c}} = 2(\vec{\omega} \times \vec{v}_{\text{rel}})$$

Kinetics

Kinetics of the Center-of-Mass

Fundamental law of kinetics

Momentum

$$\vec{p} = m\vec{v}$$ m mass of the mass point

Law of inertia (Newton's first law)

$$\vec{p} = m\vec{v} = \text{const}$$

Law of motion (Newton's second law)

$$\vec{F} = \frac{d(m\vec{v})}{dt} = \frac{d\vec{p}}{dt}$$

Mass m = const (fundamental law of kinetics)

$$\vec{F} = m\frac{d\vec{v}}{dt} = m\vec{a}$$

Kinetostatic method ("conversion" of the dynamic problem to a static problem)

$$\vec{F} + \vec{F}_{\text{T}} = \vec{0}$$ d'Alembert's inertial force, such that $F_{\text{T}} = -ma$

Procedure:

— introduction of suitable motion coordinates

— creation of free-body diagram ($\rightarrow$ Statics) for the mass-point (body) under consideration and application of all forces (external forces and d'Alembert's inertial forces, the latter opposed to the coordinate directions)

— statement of equilibrium conditions / motion equations

— solution of motion equations for the desired quantities

Work-kinetic energy theorem, first version:

$$W = \int\limits_{\vec{r}_1}^{\vec{r}_2} \vec{F} \cdot \mathrm{d}\vec{r} = E_{\mathrm{kin2}} - E_{\mathrm{kin1}}$$

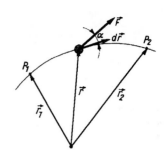

The work performed by the impressed forces in displacing the mass point is equal to the change in kinetic energy.

Kinetic energy $E_{\mathrm{kin}} := \dfrac{m}{2} v^2$

Power

$$P = \frac{\mathrm{d}W}{\mathrm{d}t}$$

If force remains constant:
$$P = \vec{F} \cdot \vec{v}$$

Potential

If the work performed by the impressed forces is independent of the displacement, then these forces possess a potential E_{pot}.

$$F_x = -\frac{\partial E_{\mathrm{pot}}}{\partial x}, \quad F_y = -\frac{\partial E_{\mathrm{pot}}}{\partial y}$$

$$F_z = -\frac{\partial E_{\mathrm{pot}}}{\partial z}$$

condition for forces that possess potential (potential forces, conservative forces)

Work-kinetic energy theorem, second version:

$$W_{\mathrm{oP}} = \int\limits_{\vec{r}_1}^{\vec{r}_2} \vec{F}_{\mathrm{oP}} \cdot \mathrm{d}\vec{r}$$

$$= (E_{\mathrm{kin2}} + E_{\mathrm{pot2}})$$
$$- (E_{\mathrm{kin1}} + E_{\mathrm{pot1}})$$

W_{oP} work performed by forces without potential

$\vec{F}_{\mathrm{oP}}$ (e.g., frictional forces)

$E_{\mathrm{pot1}}, E_{\mathrm{pot2}}$ potential energies

The work performed by nonconservative forces in displacing a mass point is equal to the change in mechanical energy. The mechanical energy is composed of the potential energy and the kinetic energy.

Law of conservation of energy

$\vec{F}_{oP} = \vec{0}$, conservative force field:

$$E_{kin2} + E_{pot2} = E_{kin1} + E_{pot1} = const$$

In the motion of a mass point in a conservative force field, the mechanical energy (the sum of the kinetic energy and the potential energy) remains constant.

Potential energies

Potential energy of gravitational force

$$E_{pot} = mgh$$

g acceleration of gravity

h height difference

Potential energy of a spring

$$E_{pot} = \frac{c}{2}x^2$$

c elastic constant / stiffness constant

x elongation or compression of the spring

Theory of vector momentum, impulse for time interval $t_2 - t_1$

$$\int_{t_1}^{t_2} \vec{F}\, dt = m\vec{v}_2 - m\vec{v}_1 = \vec{p}_2 - \vec{p}_1$$

principle of conservation of momentum for vanishing impulse
$$\vec{p} = m\vec{v} = const$$

Angular momentum (with respect to an arbitrary point 0)

$$\vec{L}_0 = \vec{r}_0 \times \vec{p} = \vec{r}_0 \times (m\vec{v})$$

Angular momentum theorem

$$\vec{M}_0 = \frac{d}{dt}(\vec{r}_0 \times m\vec{v}) = \frac{d}{dt}(\vec{r}_0 \times \vec{p})$$
$$= \frac{d}{dt}\vec{L}_0 = \dot{\vec{L}}_0$$

The moment of the resultant force acting on the mass point with respect to the space-fixed point O is equal to the change in total angular momentum over time with respect to the same point O.

Principle of conservation of angular momentum for $\vec{M}_0 = 0$

$$\vec{L}_0 = \vec{r}_0 \times \vec{p} = const$$

Kinetics of the Center-of-Mass Point System

Assumption: system of N mass points
$$m = \sum_{i=1}^{N} m_i$$

Center of mass

$$\vec{r}_S := \frac{1}{m} \sum_{i=1}^{N} m_i \vec{r}_i$$
position vector of center of mass

Center-of-mass law (motion equation of the mass point system)

$$\sum_{i=1}^{N} \vec{F}_{ia} = m \frac{d^2 \vec{r}_S}{dt^2} = m\ddot{\vec{r}}_S$$
$\vec{F}_{ia}$ total external force at m_i

Work-kinetic energy theorem

$$W_{oP} = \sum_{i=1}^{N} W_{oPi}$$
$$= (E_{kin2} + E_{pot2})$$
$$- (E_{kin1} + E_{pot1})$$

The sum of the work performed by all external and internal forces (without potential) acting on the mass points is equal to the change in the total mechanical energy of the system. The mechanical energy is composed of kinetic and potential energy.

Law of conservation of energy for $W_{oP} = 0$

$$E_{kin2} + E_{pot2} = E_{kin1} + E_{pot1}$$
$$= \text{const}$$

The mechanical energy, as the sum of the kinetic and potential energy, remains constant during motion in a conservative mass point system.

Total momentum

$$\vec{p} = \frac{d}{dt} \sum_{i=1}^{N} m_i \vec{r}_i = \frac{d}{dt}(m\vec{r}_S) = m \frac{d\vec{r}_S}{dt} = m\vec{v}_S$$

Theory of Vector Momentum

$$\int_{t_1}^{t_2} \vec{F}_{\mathrm{a}}\, \mathrm{d}t = m\vec{v}_{\mathrm{S}2} - m\vec{v}_{\mathrm{S}1} = \vec{p}_2 - \vec{p}_1 \qquad \text{where}$$

$$\sum_{i=1}^{N} \vec{F}_{i\mathrm{a}} = \vec{F}_{\mathrm{a}} = m\dot{\vec{v}}_{\mathrm{S}} = \frac{\mathrm{d}\vec{p}}{\mathrm{d}t}$$

Principle of conservation of momentum ($\vec{F}_{\mathrm{a}} = 0$)

$$\vec{p} = m\vec{v}_{\mathrm{S}} = \text{const}$$

The center of mass moves uniformly and in a straight line.

Angular momentum with respect to point 0

$$\vec{L}_0 = \sum_{i=1}^{N} \vec{L}_{i0} = \sum_{i=1}^{n} (\vec{r}_{i0} \times m_i \vec{v}_i)$$

Angular momentum theorem

$$\vec{M}_0 = \sum_{i=1}^{N} \vec{M}_{i0} = \sum_{i=1}^{N} (\vec{r}_{i0} \times \vec{F}_{i\mathrm{a}}), \quad \vec{M}_0 = \frac{\mathrm{d}\vec{L}}{\mathrm{d}t} = \dot{\vec{L}}_0$$

Principle of conservation of angular momentum

$$\vec{M}_0 = \vec{0} \quad \text{or} \quad \vec{L}_0 = \text{const}$$

The angular momentum remains constant.

Rotation of a Rigid Body About a Fixed Axis

Fundamental law of kinetics

$$M_z = J_z \dot{\omega} = J_z \ddot{\varphi}$$

M_z external moment around the fixed z axis

$$J_z := \int_{(m)} r^2\, \mathrm{d}m = \int_{(m)} (x^2 + y^2)\, \mathrm{d}m$$

axial mass moment of inertia with respect to the axis of rotation z; see "Mass Moments of Inertia"

Kinetostatic method

$$M_z + M_{\mathrm{T}} = 0, \text{ where } M_{\mathrm{T}} = -J_z\,\ddot{\varphi}$$

M_{T} moment of inertia (d'Alembert's moment of inertia)

Work-kinetic energy theorem

$$\int_{\varphi_1}^{\varphi_2} M_z \, \mathrm{d}\varphi = E_{\mathrm{kin}2} - E_{\mathrm{kin}1}$$

When a rigid body rotates around
a fixed axis, the work performed
by the resultant moment causes a
change in the kinetic energy between
the initial position and final position.

Kinetic energy in rotation

$$E_{\mathrm{kin}} := \frac{1}{2} J_z \omega^2$$

Power for $M_z = const$

$$P = \frac{\mathrm{d}W}{\mathrm{d}t} = M_z \omega, \ M_z = \frac{P}{2\pi n_\mathrm{D}}$$

n_D speed of rotation

Angular momentum

$$\vec{L}_0 = \omega J_{xz}\vec{e}_x + \omega J_{yz}\vec{e}_y + \omega J_z\vec{e}_z$$

for body-fixed cartesian
coordinate system with origin 0

Angular momentum theorem

$$\vec{M}_0 = \frac{\mathrm{d}\vec{L}_0}{\mathrm{d}t} = \frac{\mathrm{d}'\vec{L}_0}{\mathrm{d}t} + \vec{\omega} \times \vec{L}_0$$

with respect to the body-fixed
(rotating) frame of reference

Components

$$M_{0x} = J_{xz}\dot{\omega} - J_{yz}\omega^2$$
$$M_{0y} = J_{yz}\dot{\omega} + J_{xz}\omega^2$$
$$M_{0z} = J_z\dot{\omega}$$

(Euler equations for the rotation
of a rigid body around the fixed
z axis; also see "Mass Moments
of Inertia")

M_{0x}, M_{0y} gyroscopic moments
(perpendicular to axis of
rotation)

Case:

$$\vec{M}_z = J_z\dot{\omega}\vec{e}_z$$
$$\vec{L}_z = J_z\omega\vec{e}_z = J_z\vec{\omega}$$

The axis of rotation z
coincides with the principal axis.

Mass Moments of Inertia

Assumptions: r radius coordinate perpendicular to the reference axis; x, y, z center-of-mass coordinate system of the body; ϱ density

Axial Mass Moments of Inertia

$$J_x := \int\limits_{(m)} (y^2 + z^2)\, \mathrm{d}m$$

$$J_y := \int\limits_{(m)} (x^2 + z^2)\, \mathrm{d}m$$

$$J_z := \int\limits_{(m)} (x^2 + y^2)\, \mathrm{d}m$$

mass moment of inertia for
reference axes x, y, z
m mass of body

(For mass moments of inertia for simple bodies, see the table below.)

Products of inertia

$$J_{xy} := -\int\limits_{(m)} xy\, \mathrm{d}m, \quad J_{xz} := -\int\limits_{(m)} xz\, \mathrm{d}m, \quad J_{yz} := -\int\limits_{(m)} yz\, \mathrm{d}m$$

Transformation between parallel coordinate systems

(Steiner's parallel axis theorem, see "Area Moments of Inertia", section 2.2)

$$J_{\bar{x}} = J_x + (\bar{y}_S^2 + \bar{z}_S^2)m$$
$$J_{\bar{y}} = J_y + (\bar{x}_S^2 + \bar{z}_S^2)m$$
$$J_{\bar{z}} = J_z + (\bar{x}_S^2 + \bar{y}_S^2)m$$
$$J_{\overline{xy}} = J_{xy} - \bar{x}_S\bar{y}_S m$$
$$J_{\overline{yz}} = J_{yz} - \bar{y}_S\bar{z}_S m$$
$$J_{\overline{zx}} = J_{zx} - \bar{z}_S\bar{y}_S m$$

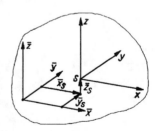

reference coordinate system
(arbitrary): $\bar{x}$, $\bar{y}$, $\bar{z}$
center-of-mass coordinates:
$\bar{x}_S$, $\bar{y}_S$, $\bar{z}_S$

Mass Moments of Inertia for Selected Bodies

Body	Principal Moments
bar 	$J_x = J_y = \dfrac{1}{12}ml^2, \quad J_z = 0$ $J_{\overline{x}} = J_{\overline{y}} = \dfrac{1}{3}ml^2$
rectangular solid 	$J_x = \dfrac{1}{12}m(b^2 + c^2)$ $J_y = \dfrac{1}{12}m(a^2 + c^2)$ $J_z = \dfrac{1}{12}m(a^2 + b^2)$ $m = \varrho abc$
prismatic body 	principal moments around the z axis: $J_z = \dfrac{m}{A}(I_{xx} + I_{yy})$ $J_z = \varrho H(I_{xx} + I_{yy})$ $m = \varrho AH$ A cross-sectional area of the body
straight circular cone 	$J_x = J_y = \dfrac{3}{5}m\left(\dfrac{1}{4}R^2 + H^2\right)$ $J_z = \dfrac{3}{10}mR^2$ $m = \varrho\dfrac{1}{3}\pi R^2 H, \quad \overline{z}_S = \dfrac{1}{4}H$

(Continued)

Mass Moments of Inertia for Selected Bodies (continued)

Body	Principal Moments
cylinder / disk 	$J_x = J_y = \frac{1}{4}m\left(R^2 + \frac{1}{3}H^2\right)$ $J_z = \frac{1}{2}mR^2$ $m = \varrho\pi R^2 H$
tube / hollow cylinder 	$J_x = J_y = \frac{1}{4}mR_a^2\left[(1+\lambda^2) + \frac{1}{3}\left(\frac{H}{R_a}\right)^2\right]$ $J_z = \frac{1}{2}mR_a^2(1+\lambda^2)$ $m = \varrho\pi R_a^2 H(1-\lambda^2), \quad \lambda = \frac{R_i}{R_a}$
thin tire 	$J_x = J_y = \frac{1}{2}mR^2$ $J_z = mR^2$ $m = \varrho 2\pi RA$ A cross-sectional area of the tire
sphere 	$J_x = J_y = J_z = \frac{2}{5}mR^2$ $m = \varrho\frac{4}{3}\pi R^3$

Inertia tensor

$$J = \begin{pmatrix} J_x & J_{xy} & J_{xz} \\ J_{xy} & J_y & J_{yz} \\ J_{xz} & J_{yz} & J_z \end{pmatrix}$$

Axial mass moments of inertia and products of inertia form the components of the symmetrical inertia tensor (second-rank tensor; see "Area Moments of Inertia").

When the coordinate system is rotated, the components change according to the same laws that govern the area moments of inertia.

For each reference point (origin), there are three mutually perpendicular axes, for which all products of inertia vanish. These axes are called "principal axes." The corresponding axial mass moments of inertia assume extreme values and are described as "principal moments" ($J_1 > J_2 > J_3$).

For homogeneous bodies, axes of symmetry are always principal axes!

Mechanical Properties in Translation and Rotation

Rectilinear motion (translation, in the x direction here) and rotation of a rigid body about a fixed z axis are motions with one degree of freedom ($n = 1$) $\rightarrow$ analogy between important properties and laws pertaining to both types of motion:

Translation in the x Direction	Rotation About the Fixed Axis z
displacement x	angle φ
velocity $v = \dot{x}$	angular velocity $\omega = \dot{\varphi}$
acceleration $a = \dot{v} = \ddot{x}$	angular acceleration $\dot{\omega} = \ddot{\varphi}$
force F_x	moment M_z
mass m	mass moment of inertia J_z
fundamental law of kinetics	
$F_x = m\ddot{x}$	$M_z = J_z\ddot{\varphi}$
kinetostatic method	
$F_x + F_T = 0$	$M_z = + M_T = 0$
d'Alembert's inertial force	d'Alembert's moment of inertia
$|F_T| = m\ddot{x}$	$|M_T| = J_z\ddot{\varphi}$

(Continued)

Translation in the x Direction	Rotation About the Fixed Axis z
kinetic energy $$E_{\text{kin}} = \frac{1}{2}mv^2$$	$$E_{\text{kin}} = \frac{1}{2}J_z\omega^2$$
work-kinetic energy theorem $$W = \int_{x_1}^{x_2} F_x \, dx = E_{\text{kin2}} - E_{\text{kin1}}$$	$$W = \int_{\varphi_1}^{\varphi_2} M_z \, d\varphi = E_{\text{kin2}} - E_{\text{kin1}}$$
power $\;\; P = F_x v$	$P = M_z\omega$
momentum $\;\; p = mv$ theory of vector momentum $$F_x = \frac{dp}{dt}$$	angular momentum $L_z = J_z\omega$ angular momentum theorem $$M_z = \frac{dL_z}{dt}$$

Planar Motion of a Rigid Body

Assumptions: motion of a rigid body (mass m, axial mass moment of inertia J_S) as the superposition of translation of the center of mass S (velocity v_S) and rotation around the gravitational axis (angular velocity φ)

components of the resultant external force: $F_{\text{Res}x}$, $F_{\text{Res}y}$, resultant external moment around the z axis: $M_{\text{Res}z}$

Kinetostatic method (kinetic equilibrium conditions)

$\rightarrow : 0 = F_{\text{Res}x} - m\ddot{x}_S$

$\uparrow \;\; : 0 = F_{\text{Res}y} - m\ddot{y}_S$

$\circlearrowleft S : 0 = M_{\text{Res}z} - J_S\ddot{\varphi}$

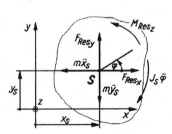

Kinetic energy

$$E_{\text{kin}} = \frac{1}{2}mv_S^2 + \frac{1}{2}J_S\omega^2$$

Work

$$W = \int_{\vec{r}_{S1}}^{\vec{r}_{S2}} \vec{F}_{\text{Res}} \, d\vec{r} + \int_{\varphi_1}^{\varphi_2} M_{\text{Res}z} \, d\varphi$$

Motion from initial position $1(\vec{r}_{S1}, \varphi_1)$ to final position $2(\vec{r}_{S2}, \varphi_2)$

Work-kinetic energy theorem

First version:

$W = E_{kin2} - E_{kin1}$

Second version: W_{oP} work performed by forces and
                                                            moments without potential

$W_{oP} = \quad (E_{kin2} + E_{pot2})$
$\qquad\quad -(E_{kin1} + E_{pot1})$

Law of conservation of energy ($W_{oP} = 0$):

$(E_{kin2} + E_{pot2}) = (E_{kin1} + E_{pot1}) = E_{kin} + E_{pot} = const$

Theory of vector momentum and angle momentum theorem

For the time interval $t_2 - t_1$, such that: $\quad \dot{x}_{Si} = \dot{x}_S(t_i)$, $\dot{y}_{Si} = \dot{y}_S(t_i)$,
$\dot{\varphi}_i = \dot{\varphi}(t_i)$

$$\int_{t_1}^{t_2} F_{Resx}(t)\, dt = m\dot{x}_{S2} - m\dot{x}_{S1} \qquad \int_{t_1}^{t_2} F_{Resy}(t)\, dt = m\dot{y}_{S2} - m\dot{y}_{S1}$$

$$\int_{t_1}^{t_2} M_{Resz}(t)\, dt = J_S\dot{\varphi}_2 - J_S\dot{\varphi}_1$$

Time integrals of the force
components or moment identify
the impulses or rotational impulse.

Planar Motion of a Rigid-Body System

Assumptions: system / gearing system with N rigid bodies / elements (masses m_i, axial
mass moments of inertia around the centroidal axis J_{Si})

Motion coordinates of the ith body: $x_{Si}, y_{Si}, \varphi_i$

Constraints

Constrained motion of the system restricts its freedom of motion. Constraints are
determined by the geometry of the system.

The number of constraints is calculated by subtracting the number of degrees of freedom
n from the number of selected motion coordinates.

Example of constraints:

$x_S = \varphi R$ rolling condition for a wheel of
                                                  radius R

$$\varphi_1 R_1 = \varphi_2 R_2$$

turning of two meshed gears of radii R_1, R_2 without slippage

$$x_{S1} = x_{S2}$$

rigid connection between two bodies in rectilinear motion

Kinetostatic method

Kinetic equilibrium conditions are formulated for each rigid body in the system.

Kinetic energy

$$E_{kin} = \sum_{i=1}^{N} \left(\frac{1}{2} m_i v_{Si}^2 + \frac{1}{2} J_{Si} \omega_i^2 \right)$$

Work-kinetic energy theorem

First version:

$$W = \sum_{i=1}^{N} \left(\int_{\vec{r}_{Si1}}^{\vec{r}_{Si2}} \vec{F}_{Resi}\, d\vec{r}_i + \int_{\varphi_{i1}}^{\varphi_{i2}} \vec{M}_{Resi}\, d\vec{\varphi}_i \right) = E_{kin2} - E_{kin1}$$

The work performed by the impressed forces and moments in moving a many-body system from an initial position 1 to a final position 2 is equal to the change in the kinetic energy of the system.

Second version:

$$W_{oP} = \begin{array}{l} (E_{kin2} + E_{pot2}) \\ -(E_{kin1} + E_{pot1}) \end{array}$$

W_{oP} work performed by forces without potential

Collision Problems

Collision

Sudden impact between two (or more) bodies. Contact occurs in two phases: compression (during which the deformation increases) and restitution (during which the original form is partially or completely restored).

Direct collision (collision in translational motion of two point masses along a straight line)

Velocities after collision

$$v_1^* = v_1 - \frac{m_2}{m_1+m_2}(k+1)(v_1-v_2)$$

$$v_2^* = v_2 + \frac{m_1}{m_1+m_2}(k+1)(v_1-v_2)$$

m_1, m_2 masses

v_1, v_2 velocities before collision

k coefficient of restitution (ratio of momentum losses during restitution and compression, so-called "Newton's law of restitution," where $k = 1$ represents perfectly elastic collision and $k = 0$ represents perfectly plastic collision)

Difference in velocity after collision
$$v_2^* - v_1^* = k(v_1 - v_2)$$

$$k = \frac{v_2^* - v_1^*}{v_1 - v_2}$$

Energy loss

$$\Delta E = \frac{1-k^2}{2}\frac{m_1 m_2}{m_1 + m_2}(v_1-v_2)^2$$

dissipation by plastic deformation accompanied by warming

Special cases:

Perfectly elastic collision ($k = 1$, no permanent deformation, compression equals restitution), velocities after collision:

$$v_1^* = v_1 - \frac{2m_2}{m_1 + m_2}(v_1 - v_2)$$

$$v_2^* = v_2 + \frac{2m_1}{m_1 + m_2}(v_1 - v_2)$$

The magnitudes of the velocity differences of the bodies before and after the collision are equal:
$$v_2^* - v_1^* = v_1 - v_2$$

Case $m_1 = m_2$:

$$v_1^* = v_2, \; v_2^* = v_1$$

The velocities are "exchanged."

Ideal plastic collision ($k = 0$, assuming a rigid plastic material $\rightarrow$ no restitution). Both bodies move at the same velocity v^* after the collision:

$$v_1^* = v_2^* = v^* = \frac{1}{(m_1 + m_2)}(m_1 v_1 + m_2 v_2)$$

Rotational collision, coupling impact (coupling collision)
where k = 0

Common angular velocity
after rotational collision

$$\omega^* = \frac{J_1 \omega_1 + J_2 \omega_2}{J_1 + J_2}$$

Energy loss

$$\Delta E = \frac{1}{2} \frac{J_1 J_2}{J_1 + J_2} (\omega_1 - \omega_2)^2$$

Mechanical Vibrations

Periodic vibration
Magnitude of vibration

$q(t + T) = q(t)$

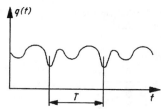

Frequency of vibration

$$f = \frac{1}{T}$$

T time of vibration, period

Harmonic vibration
Magnitude of vibration

$q(t) = \hat{q} \sin(\omega t + \alpha)$ or

$q(t) = A_1 \cos(\omega t) + A_2 \sin(\omega t)$

such that

$$\hat{q} = \sqrt{A_1^2 + A_2^2}, \quad \tan \alpha = \frac{A_1}{A_2}$$

$\hat{q}$ amplitude

ω angular frequency

$\omega t + \alpha$ phase angle

α zero phase angle

Relationship between angular frequency and frequency

$$\omega = \frac{2\pi}{T} = 2\pi f$$

Free undamped vibration with one degree of freedom

General motion equation, differential equation of vibration

$$\ddot{q} + \omega_0^2 q = 0$$

q magnitude of vibration; equals displacement in case of translating oscillators or linear oscillators, equals angle in case of rotational, torsional or pendulum oscillators

$\ddot{q}$ second time derivative of the magnitude of vibration (acceleration)

ω_0 natural angular frequency of the free undamped vibration

Natural angular frequencies of undamped oscillators with one degree of freedom

Linear oscillator:

$$\omega_0 = \sqrt{\frac{c}{m}}$$

c elastic constant

m mass

Torsional oscillator:

$$\omega_0 = \sqrt{\frac{c_t}{J}}$$

c_t torsional spring constant (torsional stiffness constant)

J mass moment of inertia of the rotational mass with respect to the axis of rotation

Simple pendulum:

$$\omega_0 = \sqrt{\frac{g}{l}}$$

l length of pendulum

Compound pendulum:

$$\omega_0 = \sqrt{\frac{mgl}{J_A}}$$

J_A mass moment of inertia with respect to axis of rotation A

l distance between axis of rotation and center of mass

General solution of the differential equation of vibration

$q(t) = \widehat{q}\sin(\omega_0 t + \alpha)$ or
$q(t) = A_1 \cos(\omega_0 t) + A_2 \sin(\omega_0 t)$

integration constants $\widehat{q}$, α, or A_1, A_2 from initial conditions

Equivalent stiffness for oscillators with multiple springs:
(also applicable to torsional spring constants)

$$c = \sum_k c_k$$

parallel connection (equal elongation or compression of springs)

$$\frac{1}{c} = \sum_k \frac{1}{c_k} \qquad \text{series connection (equal elasticity)}$$

Free damped vibration with one degree of freedom

Type of vibration: damping force F_W is proportional to the velocity

$$F_W = kv \qquad\qquad k \quad \text{damping constant (resistance)}$$

General equation of motion

$$\ddot{q} + 2\delta\dot{q} + \omega_0^2 q = 0 \qquad\qquad \delta \quad \text{damping constant (resistance)}$$

ω_0 natural angular frequency of the undamped vibration

Special case: linear oscillator

$$\delta = \frac{k}{2m}, \quad \omega_0 = \sqrt{\frac{c}{m}}$$

Solution cases determined by (Lehr's) attenuation constant (attenuation equivalent)

$$D = \frac{\delta}{\omega_0}$$

1. $D > 1$, strong damping:

$$q(t) = e^{-\delta t}(B_1 e^{\varkappa t} + B_2 e^{-\varkappa t}) \qquad \text{no vibration (aperiodic phenomenon)}$$

where

$$\varkappa = \sqrt{\delta^2 - \omega_0^2} = \omega_0\sqrt{D^2 - 1}$$

2. $D = 1$, critically damped case:

$$q(t) = e^{-\delta t}(B_1 + B_2 t) \qquad \text{exponentially decaying motion (aperiodic critically damped case)}$$

3. $D < 1$, weak damping:

$$q(t) = e^{-\delta t}(A_1 \cos \omega t + A_2 \sin \omega t)$$

$$q(t) = e^{-\delta t} C \sin (\omega t + \alpha)$$

Vibration with exponentially decreasing amplitude. Graphs of the functions $\pm C\, e^{-dt}$ envelop the decay curve (envelope).

Period of vibration T_D

$$T_D = \frac{2\pi}{\omega} = \frac{2\pi}{\omega_0 \sqrt{1 - D^2}}, \quad T_D > T$$

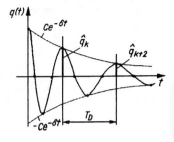

Logarithmic decrement

$$\vartheta = \ln \frac{\widehat{q}_k}{\widehat{q}_{k+2}} = \delta T_D = \frac{2\pi D}{\sqrt{1 - D^2}}$$

Attenuation constant

$$D = \frac{\vartheta}{\sqrt{4\pi^2 + \vartheta^2}}$$

Forced damped vibration with one degree of freedom

General equation of motion with harmonic excitation

$$\ddot{q} + 2\delta\dot{q} + \omega_0^2 q = q_0 \sin \Omega t$$

Coefficients for the following types of exciters with a linear oscillator:

Type of Exciter	Force	Spring Origin	Unbalance
δ	$\dfrac{k}{2m}$	$\dfrac{k}{2m}$	$\dfrac{k}{2(m + m_1)}$
ω_0	$\sqrt{\dfrac{c}{m}}$	$\sqrt{\dfrac{c}{m}}$	$\sqrt{\dfrac{c}{m + m_1}}$
q_0	$\dfrac{\widehat{F}}{m}$	$\dfrac{c\widehat{u}}{m}$	$\dfrac{m_1}{m + m_1} r_0 \Omega^2$

m oscillator mass, c elastic constant, k damping constant, $\widehat{F}$ amplitude of force excitation, Ω angular frequency of the exciter, $\widehat{u}$ amplitude of the displacement excitation at the spring origin, m_1 unbalance mass, r_0 radius of unbalance

Solution of the motion equation for the types of exciters listed above:

Superposition of general solution q_h of the homogeneous differential equation (free damped vibration) and the particular solution q_p of the inhomogeneous differential equation (steady-state vibration):

$q = q_h + q_p$

Particular solution $q_p = \widehat{q}_p \sin(\Omega t - \psi)$

$$\hat{q}_p = \frac{q_0}{\sqrt{(\omega_0^2 - \Omega^2)^2 + (2\delta\Omega)^2}}$$ $\hat{q}_p$ amplitude

$$= \frac{q_0}{\omega_0^2} \frac{1}{\sqrt{(1 - \eta^2)^2 + (2D\eta)^2}}$$

$$= \frac{q_0}{\omega_0^2} V_1$$

$$\tan\psi = \frac{2\delta\Omega}{\omega_0^2 - \Omega^2} = \frac{2D\eta}{1 - \eta^2}$$ φ phase angle

$$\eta = \frac{\Omega}{\omega_0}$$ frequency ratio, tuning ratio

$$V_1 := \frac{1}{\sqrt{(1 - \eta^2)^2 + (2D\eta)^2}}$$ gain function

In force excitation and spring origin excitation, the gain function V_1 characterizes the ratio of the output variable (amplitude of vibration) to the input variable (excitation amplitude).

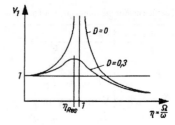

Resonant amplitude for $0 < D < 0.5$:

$$V_{1max} = V_1(\eta = \eta_{res})$$

$$= \frac{1}{2D\sqrt{1 - D^2}}$$

$$\eta_{res} = \sqrt{1 - 2D^2}$$

Metrology

A wide range of topics are covered in this section, from the very basics to the advanced. Measurement tools play an important role in design, layout, machining, and inspection throughout industry.

Using a micrometer

Micrometers are measuring devices used to obtain precise measurements in many situations. Although today's electronic micrometers offer accuracy of ±0.0001 inch or ±0.002 mm, with the dimension displayed on a high-contract LCD readout, manual micrometers remain in use throughout industry. The most common type of micrometer is the "outside measurement" version, which will be discussed in more detail below, but inside micrometers and depth micrometers are also in wide use. The following instructional information on how to read an outside micrometer, and the following sections on slide calipers and vernier protractors, have been adapted from information published by The L.S. Starrett Company.

Micrometers graduated in inches

The pitch of the screw thread on the spindle of most English measurement unit micrometers is 40 threads per inch. One full revolution of the thimble therefore advances the spindle face toward or away from the anvil face precisely $1/40$ inch (0.025"). The reading line on the sleeve (sometimes called the barrel or barrel sleeve) is divided into 40 equal parts indicated by vertical lines. Therefore, each vertical line designates $1/40$ inch (0.025"). The lines on the sleeve vary in length, and every fourth line is longer than the others, and designates hundred thousandths of an inch. For example, the line marker "1" represents 0.100", the line marked "2" represents 0.200", etc.

The beveled edge of the thimble is divided into 25 equal parts with each line representing 0.001", and every line numbered consecutively. Rotating the thimble from

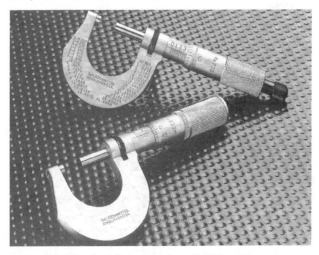

Figure 1. Typical manual micrometers. (Source, The L.S. Starrett Company.)

one of these lines to the next moves the spindle longitudinally 1/25 of 0.025", or 0.001". Rotating two divisions represents 0.002", etc. Twenty-five divisions indicate a complete revolution of the thimble of $^{1}/_{40}$ inch (0.025").

To read the micrometer in thousandths, multiply the number of vertical divisions visible on the sleeve by 0.025", and to this add the number of thousandths indicated by the line on the thimble that coincides with the reading on the sleeve.

Example. In _Figure 2_, the "1" line on the sleeve is visible, representing 0.100". There

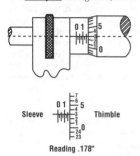

are three additional lines visible, each representing 0.025", so all three represent 0.075". Line "3" on the thimble coincides with the reading line on the sleeve, and, since each line represents 0.001", the value is 0.003". Therefore, the total micrometer reading is 0.100" + 0.075" + 0.003" = 0.178".

Reading to ten-thousandths of an inch. The above example provided a reading in thousandths of an inch, but more accurate readings are possible with micrometers graduated to ten-thousandths (0.0001) of an inch. The additional reading in ten-thousandths is obtained from a vernier scale (named after Peter Vernier who invented the scale in 1631) on the sleeve. The vernier consists of ten divisions on the sleeve, which occupy the same space as the nine divisions on the thimble. Therefore, the difference between the width of one of the ten spaces on the vernier and

Reading .178"

Figure 2. Reading a micrometer graduated in thousandths of an inch.
(Source, The L.S. Starrett Company.)

one of the nine spaces on the thimble is one-tenth of a division on the thimble, or one ten-thousandth inch (0.0001").

To read a ten-thousandths micrometer, first obtain the thousandths reading, then see which of the lines on the vernier coincides with a line on the thimble. If it is the line marked "1" on the sleeve, add one ten-thousandth, if it is the line marker "2," add two ten-thousandths, etc.

Example. In _Figure 3_, the "2" on the sleeve is visible, representing 0.200". There are two additional lines visible, and since each represents 0.025", the value is 0.050". The reading on the sleeve lies between the "0" and the "1" on the thimble, indicating that a vernier reading must be added. The "7" is the only line on the vernier that coincides with a line on the thimble, representing 7 × 0.0001" = 0.0007". Therefore, the total micrometer reading is 0.200" + 0.050" + 0.0007" = 0.2507".

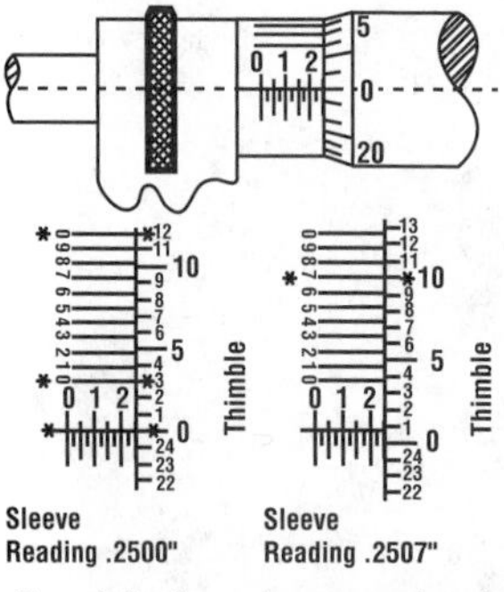

Micrometers graduated in millimeters

The basic Metric units micrometer is graduated in hundredths of a millimeter (0.01 mm). The pitch of the screw thread is one-half millimeter (0.5 mm), so one full revolution of the spindle advances the spindle face forward or away from the anvil face precisely 0.5 mm.

The reading line on the sleeve is graduated above the line in millimeters (1.0 mm) with every fifth millimeter being numbered from 0 to 25. Each millimeter is also divided in half (0.5 mm)

Sleeve
Reading .2500"

Sleeve
Reading .2507"

Figure 3. Reading a micrometer graduated in ten-thousandths of an inch.
(Source, The L.S. Starrett Company.)

below the reading line. Two revolutions of the thimble are required to advance the spindle 1.0 mm.

The beveled edge of the thimble is divided into 50 equal parts, with each line representing 0.01 mm and every fifth line being numbered from 0 to 50. Rotating the thimble from one of these lines to the next moves the thimble longitudinally 0.01 mm, rotating two divisions represents 0.02 mm, etc.

To read the micrometer, add the number of millimeters and half-millimeters visible on the sleeve to the number of hundredths of a millimeter indicated by the thimble graduation which coincides with the reading line on the sleeve.

Example. In *Figure 4*, the 5 mm sleeve graduation is visible and indicates 5.00 mm. One additional 0.5 mm line is visible on the sleeve. Line 28 on the thimble coincides with the reading line on the sleeve, so 28 × 0.01 = 0.28 mm. Therefore, the total micrometer reading is 5.00 + 0.50 + 0.28 = 5.78 mm.

Reading to two-thousandths of a millimeter. Metric vernier micrometers graduated in 0.002 mm are used like those graduated in hundredths of a millimeter except that an additional reading in two-thousandths of a millimeter (0.002 mm) is obtained from a vernier scale on the sleeve. The vernier consists of five divisions on the sleeve, which occupy the same space as nine divisions on the thimble. Therefore, the difference between the width of one of the five spaces on the vernier

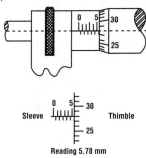

Figure 4. Reading a micrometer graduated
in hundredths of a millimeter.
(Source, The L.S. Starrett Company.)

and one of the nine spaces on the thimble is one-fifth or two-tenths of a division on the thimble, or two-thousandths of a millimeter (0.002 mm).

To read a 0.002 mm micrometer, first obtain the hundredth of a millimeter (0.01 mm) reading, then see which of the lines on the vernier coincides with a line on the thimble. If it is the line marked "2," add 0.002 mm, if it is the line marked "4," add 0.004 mm, etc.

Example. In *Figure 5*, the 5 mm sleeve is visible and indicates 5.00 mm. No additional lines on the sleeve are visible. The reading line on the sleeve lies between zero and the first line on the thimble, indicating that a vernier reading must be added. Line "8" on the vernier is the only line that coincides with a line on the thimble, representing 0.008 mm. Therefore, the total micrometer reading is 5.00 + 0.008 = 5.008 mm.

Reading to one-thousandth of a millimeter. Reading a 0.001 mm micrometer is exactly like reading a 0.002 mm micrometer except that there are ten divisions on the vernier occupying the same space as nine divisions on the thimble. Therefore, the difference between the width of one of the spaces on the vernier and one of the nine spaces on the thimble is one-tenth of a division on the thimble, or one-thousandth of a millimeter (0.001 mm).

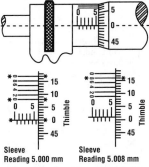

Figure 5. Reading a micrometer graduated
in two-thousandths of a millimeter.
(Source, The L.S. Starrett Company.)

To read a 0.001 mm micrometer, first obtain the hundredth of a millimeter reading. Next, see which of the lines on the vernier coincide with a line on the thimble. If it is the first line, add 0.001 mm to the reading, if it is the second line, add 0.002 mm, etc. Only every second vernier line is numbered on a 0.001 mm reading tool due to space congestion.

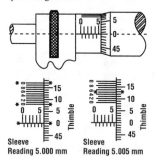

Figure 6. Reading a micrometer graduated in one-thousandth of a millimeter.
(Source, The L.S. Starrett Company.)

Example. In *Figure 6*, the 5 mm sleeve is visible and indicates 5.00 mm. No additional lines on the sleeve are visible. The reading line on the sleeve lies between zero and the first line on the thimble, indicating that a vernier reading must be added. Line "5" on the vernier is the only line that coincides with a line on the thimble, representing 0.005 mm. Therefore, the total micrometer reading is 5.00 + 0.005 = 5.005 mm.

Vernier, dial, and electronic slide calipers

The vernier caliper was originally designed and graduated similar to a micrometer in that it incorporated a scale equal in value to 0.25 (as in the single revolution of the micrometer thimble). Later, the graduations were increased to 50 to make the caliper easier to read. In order to read a caliper with 50 graduations, an etched line on the tool's vernier plate must be aligned with an equally spaced line running the length of the tool's master bar. Because of this complication in using a vernier caliper, dial calipers have steadily grown in popularity over the last thirty years. Dial calipers allow for internal and external dimensions to be read quickly and accurately, and they can also be used for depth measurement.

The measurement reporting dial on a dial caliper is fixed to the moveable jaw and rides along the tool's bar (or slide), meshing with a toothed rack. The rack is typically a ground bar so that each tooth is equal to 0.025, or one-quarter of the dial resolution, and four teeth equal 0.100. The bar is graduated in 0.100 increments instead of the more complex 0.025 or 0.050 of the vernier style caliper, and the dial is graduated from 0.001 up to one-hundred-thousandths. To obtain a measurement, the number on the bar is read first, and then the dial reading is added to get the dimension to the nearest 0.001".

As with the vernier caliper, the dial caliper does have some disadvantages. The most obvious is that the toothed rack necessary to carry the dial pinion gear is subject to contamination and must be kept clean since fine metal chips or similar materials can become lodged in the gullets of the rack. If the pinion gear runs over chips or similar contamination, the pinion may "jump" to the next gullet, thereby providing an inaccurate reading.

Electronic calipers have been available since the late 1970s and their construction is similar to dial calipers. New models not only perform outside, inside, and depth readings, but they can also convert English to Metric readings, provide high and low limit alerts, and they have the ability to set zero position anywhere along the bar. Since the measurement sensing function of the electronic caliper has no moving parts, foreign matter and other contaminants are less of a problem. Water, on the other hand, should not be present around the equipment unless the caliper was specially designed to work in harsh conditions.

While slide calipers do not have the precision of a micrometer, they are versatile and have more range than a micrometer. The best digital and dual slide calipers, regardless of resolution, are accurate to within 0.001" in 6 inches, or within 0.03 mm in 150 mm. The best vernier calipers are accurate to 0.0005" per foot, or 0.013 mm in 300 mm. Height gages are a form of slide caliper and are usually read in the same manner. Sometimes it is necessary to decide whether to use a micrometer or a caliper for a measurement, in which case the "ten times rule" should be applied: the tool to be used must be accurate to at least ten times the tolerance involved with the construction. If the part is toleranced to plus or minus 0.001", calipers are not used because they are only accurate to plus or minus one-thousandths of an inch. Instead, a micrometer graduated to 0.0001" is required.

Reading a vernier caliper gage

In *Figure 7*, a vernier caliper gage with both inch and metric readings is pictured.

Inch reading. Refer to the lower master bar graduations in the illustration. Inches are numbered in sequence over the full range of the master bar. Every second graduation between the inch lines is numbered and equals 0.100". Each bar graduation is 0.050".

The inch vernier plate is divided into 50 parts, each representing 0.001". Every fifth line is numbered 5, 10, 15, 20, etc., for easy counting.

To read the gage, first count how many inches and how many 0.050" lines lie between the zero line on the master bar and the zero line on the inch vernier plate, and add them. Then count the number of graduations on the inch vernier plate from its zero line to the line that coincides with a line on the master bar. Multiply the number of vernier plate graduations that were counted by 0.001" and add this figure to the number of inches and 0.050" lines that were counted on the bar. This is the total reading.

Example. In *Figure 7*, the vernier plate zero line is one inch (1.000") plus 0.100" beyond the zero line on the bar, or 1.100". The ninth graduation on the vernier plate coincides with a line on the bar (as indicated by stars in the illustration), 9, so 9 × 0.001 (0.009") is added to the 1.100" reading, giving a total reading of 1.109".

Metric reading. Refer to the upper master bar graduations and the millimeter vernier plate in the illustration. Each master bar graduation is 1.00 millimeter, and every tenth graduation is numbered in sequence, 10, 20, 30, 40, etc., over the full range of the bar. This provides for direct reading in millimeters.

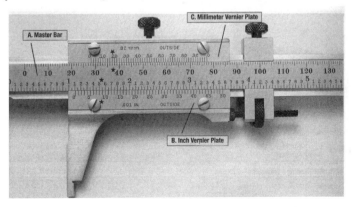

Figure 7. Reading a vernier caliper gage. (Source, The L.S. Starrett Company.)

The millimeter vernier plate is divided into 50 parts, each representing 0.02 mm. Every fifth line is numbered in sequence, 0.10 mm, 0.20 mm, 0.30 mm, etc., to provide for direct reading in hundredths of a millimeter.

To read the gage, first count how many millimeters lie between the zero line on the master bar and the zero line on the millimeter vernier plate. Then find the graduation on the vernier plate that coincides with a line on the bar and note its value in hundredths of a millimeter. Add the vernier plate reading in hundredths of a millimeter to the number of millimeters that were counted on the bar. This is the total reading.

Example. In *Figure 7*, the vernier plate zero line is 28 millimeters beyond the zero line on the bar, and the 0.18 mm graduation on the vernier plate coincides with a line on the bar (as indicated by stars in the illustration). 0.18 mm is therefore added to the 28 mm bar reading, giving a total reading of 28.18 mm.

Vernier protractors

Universal bevel protractors that incorporate a vernier scale are designed for precision measurement of angles, and to aid in laying out angles. They are capable of reading to five minutes of a degree and widely used by toolmakers, inspectors, and machinists.

Figure 8 shows the protractor dial scale and vernier scale of a vernier protractor. The dial is graduated both to the right and left of zero up to 90°. The vernier is graduated to the right and left of zero up to 60 minutes (60'). Each of the 12 vernier graduations each side of 0 represents 5'. Any angle can be measured, with the vernier being read in the same direction from zero as the protractor dial scale, either left or right.

Since twelve graduations of the vernier scale occupy the same space as 23 graduations, or 23°, on the protractor dial, each vernier graduation is $1/12$ degree or 5' shorter than two graduations on the protractor dial. Therefore, if the zero graduation on the vernier scale coincides with a graduation on the protractor, the reading is in exact degrees. However, if some other graduation on the vernier scale coincides with a protractor graduation, the number of vernier graduations multiplied by 5' must be added to the number of degrees read between the zeros on the protractor dial and vernier scale.

Example. In *Figure 8*, the zero on the vernier scale lies between the "50" and "51" on the protractor dial to the left of the zero, indicating 50 whole degrees. Also reading to

Figure 8. Vernier scale on a universal bevel protractor. (Source, The L.S. Starrett Company.)

the left, the fourth line on the vernier coincides with a graduation on the protractor dial (as indicated by stars in the illustration). Therefore, 4 × 5', or 20', are to be added to the number of degrees. The reading is 50° 20'.

Thread measurement with the three-wire method

Several gages are available to measure the pitch diameter of threads, including the thread micrometer, which uses a pointed anvil and spindle to measure the minor diameter of an external thread. Another method provides for locating segments of wire of a known diameter at three places on a thread and using a micrometer to measure the diameter over the three points of the wire, as shown in *Figure 9*. The key to an exact reading is in selecting the proper size wire for the measurement. The best size wires are those that touch the thread at the pitch diameter, and the size will obviously vary with the angle of the thread. The best size wire can be identified by using the following equation to identify a constant for the angle, and then dividing the constant by the pitch (number of threads per inch).

Constant = sec $^1/_2$ angle of the thread × 0.5.

For 60° threads, this means that the best wire size is 0.57735 ÷ pitch; and for 55° threads, the best wire size is 0.56369 ÷ pitch.

Therefore, the best wire size for measuring a $^7/_{16}$ × 24 thread would be found by dividing the pitch by the constant, or 0.57735 ÷ 24 = 0.02405625. Best wire sizes and constants for Unified threads are given in **Table 1**, and for Metric threads in **Table 2**. Metric constants are derived from a different formula based on pitch and the wire diameter rather than the number of threads per inch. Wires used in thread measurement should be hardened and lapped steel wires with an accuracy of 0.0002 inch.

After three wires of equal diameter have been selected, they are positioned in the thread grooves as shown in *Figure 9*. The anvil and spindle of an ordinary micrometer are then placed against the three wires and a measurement is taken. To determine what the reading of the micrometer *should* be if the thread is the correct finish size, use one of the following formulas.

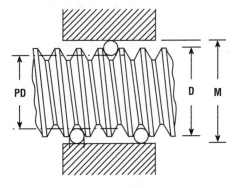

Figure 9. Measuring threads with the three-wire method.

For Unified threads $M = (D + 3W) - (1.5155 \div P)$

where M = Measurement over the best wire size, in inches
             D = Outside diameter of the thread, in inches
             W = Diameter of the best wire size
             P = Pitch (number of threads per inch).

For Metric threads $M = P + C$

where M = Measurement over the best wire size, in millimeters
             P = Pitch, in millimeters
             C = Constant (from **Table 2**).

Note that **Table 2** has pitch, best wire size, and the constant stated in millimeters and converted to decimal inches, so M can be determined using the inch conversions in the Table. When the conversions are used, the value of M will be in inches, rather than millimeters.

Table 1. Three-Wire Measurement for Unified Threads.

Pitch	Best Wire Size	Pitch	Best Wire Size
4	0.144338	20	0.028868
4.5	0.128300	22	0.026243
5	0.115470	24	0.024056
5.5	0.104973	26	0.022205
6	0.096225	27	0.021383
7	0.082478	28	0.020619
7.5	0.076980	30	0.019245
8	0.072169	32	0.018042
9	0.064150	36	0.016037
10	0.057735	40	0.014434
11	0.052538	44	0.013121
11.5	0.050204	48	0.012028
12	0.048112	50	0.011547
13	0.044411	56	0.010310
14	0.041239	64	0.090211
16	0.036084	72	0.008018
18	0.032075	80	0.007217

All dimensions in inches.

Table 2. Three-Wire Thread Measurement for Metric Threads.

Pitch		Best Wire Size		Constant	
Millimeters	Inches	Millimeters	Inches	Millimeters	Inches
0.35	0.01378	0.2021	0.00796	0.3031	0.01193
0.40	0.01576	0.2309	0.00909	0.3464	0.01364
0.45	0.01772	0.2598	0.01023	0.3897	0.01534
0.50	0.01989	0.2887	0.01137	0.4330	0.01705
0.60	0.02362	0.3464	0.01364	0.5195	0.02046

(Continued)

Table 2. *(Continued)* Three-Wire Thread Measurement for Metric Threads.

Pitch		Best Wire Size		Constant	
Millimeters	Inches	Millimeters	Inches	Millimeters	Inches
0.70	0.02756	0.4041	0.01591	0.6062	0.02387
0.80	0.03150	0.4619	0.01818	0.6928	0.02728
1.00	0.03937	0.5774	0.02273	0.8660	0.03410
1.25	0.04921	0.7217	0.02841	1.0815	0.04262
1.50	0.05906	0.8660	0.03410	1.2990	0.05114
1.75	0.06890	1.0104	0.03978	1.5155	0.05967
2.00	0.07874	1.1547	0.04646	1.7321	0.06819
2.50	0.09843	1.4434	0.05688	2.1651	0.08524
3.00	0.11811	1.7321	0.06819	2.5981	0.10229
3.50	0.13780	2.0207	0.07956	3.0311	0.11933
4.00	0.15748	2.3094	0.09092	3.4641	0.13638
4.50	0.17717	2.5981	0.10229	3.8971	0.15343
5.00	0.19685	2.8868	0.11365	4.3301	0.17048
5.50	0.21654	3.1754	0.12602	4.7631	0.18753
6.00	0.23622	3.4641	0.13638	5.1962	0.20457

Machine Tool Probes

This introduction to probes and probing was written by Dave Bozich of Renishaw, Inc., and appeared in *Modern Machine Shop* magazine.

The machine tool probe is a relatively inexpensive and easy-to-use machine tool accessory that can deliver significant reductions in production time and cost. Long used mainly for setup, on-machine probing is gaining widespread application for process improvement—automating and speeding part processing, even eliminating parts of the process. The growth is being driven by shifts to flexible machining, shorter lead times, tighter accuracy specifications and automated processing—all areas where on-machine probing yields time, quality, and productivity improvements.

A traditional objection to on-machine probing—that it diverts machine time away from making chips—can be overcome by measuring productivity in terms of total in-process time, rather than machining cycle time. The view that probing steals machining time focuses on the 10 seconds or so that checking a part feature can require. However, this focus overlooks the fact that checking the part off-line, a step that probing may make unnecessary, can impose the need for additional part handling and another setup, adding to in-process time (not to mention introducing the potential of fixturing error).

Considered in the overall context of part processing time, rather than simply machining time, probing can produce vital gains in processing efficiency and manufacturing productivity. In flexible machining operations with small batch sizes, the time that probing saves up front on setup, and at the end on inspection, more than offsets total probing time.

Machine tool probing can contribute to nearly every stage of the process.

- Setup. Used to locate the part automatically and establish a work coordinate system, probing cuts setup time, increases spindle availability, lowers fixture costs, and eliminates nonproductive machining passes. On complex parts, 45 minutes of fixture alignment can be replaced by 45 seconds of probing—performed automatically by the CNC.

- Fail-safe operation. Data such as work coordinate updates, tool geometry changes, part measurement, etc., can be automatically determined by the CNC after the successful completion of a probing cycle. This eliminates costly errors resulting from manually miskeying or miscalculating the information.
- Part identification. Probing can determine that the correct part has been loaded during automated part processing. This part ID procedure lets the CNC call up the correct program, while delivering an added level of protection for the machine tool.
- Toolsetting. In addition to the spindle-mounted probe, another type is the table-mounted toolsetting probe. A toolsetting probe is an economical solution for on-machine verification of tool geometry and condition. The toolsetting probe can automatically set length and diameter, and identify broken tools. Once tool condition is known, a "sister" tool may be called or an operator message may be issued.
- In-process control. This uses the spindle-mounted probe to monitor size and position of machined features during the cutting process. A probe can be programmed to monitor the process and automatically apply cutter compensation or adjust work coordinates—all helping to eliminate scrap, reprogramming, or potential rework. What results is a closed-loop process that requires no on-the-fly operator intervention.
- First-off part inspection. Probing can make first-part inspection seamless and automatic. Probing eliminates the delays and lost spindle time for manual inspection, especially when the operator needs to tear down the setup to inspect the part, then reinstall it for any compensating cuts.
- Final inspection. Used to inspect parts after machining, probing reduces the need for offline inspection and in some cases can eliminate it altogether. Through the use of a traceable artifact, probing cycles can be structured to compare final dimensions of a machined feature to a known dimension for the machine-resident article. When making this comparison, the CNC can determine if the specified machining tolerances were actually achieved. Based on these results, an intelligent decision can be made: proceed with machining, adjust tool geometry, modify work coordinate system, and so on.

Using on-machine probing to "buy off" a part while still fixtured can greatly shorten in-process time, eliminating the need to remove clamps, transport the part to a CMM, refixture the part, account for thermal effects, and reestablish datum points before measuring. Inspecting on the machine is particularly beneficial with large, expensive workpieces such as molds or large aerospace parts that can be difficult and time-consuming to move.

The most expensive, nonvalue-added process in most shops is piece-part inspection. When the machine tool is well-maintained, automated probing can assure that the process stays in control and the parts coming off the machine stay on spec. That's why keeping the machine well-maintained is also essential if piece-part inspection is to be reduced.

Establish machine capabilities

Today's standard machine tools deliver accuracy and repeatability approaching levels formerly available only on CMMs. In addition, technology advances are making these machines easier to maintain.

Shops investing in the health of their machines can do many more things with their capital equipment. When the ability of the machine tool to move a probe around accurately and repeatably can be trusted, the shop can inspect the part immediately after cutting it without extra handling and refixturing.

In order to move toward 100 percent good parts, it is essential to understand all process variables. This is a requirement of ISO 9000 and QS 9000. A shop must be able to document the process capability and the accuracy of its machines. One way to do this is to inspect them to a nationally recognized and accepted standard, such as ISO 230 or ASME B5.54. Both call for a ballbar and laser to be used with a recommended procedure for checking machine tool accuracy.

The purpose of these standards is not to specify an accuracy the machine must meet, but to find out its process capability or what accuracy level it can meet. The part print dictates the accuracy that a machine must have to make good parts—that is, where to set the accuracy bar. Testing reveals how well a machine can perform. As long as the machine can top the bar, it will have process capability.

The industry trend is to calibrate the machine on need, not time. There is no reason for the maintenance department to pull a perfectly good machine out of production for calibration. Instead, let the ballbar and the accuracy of the parts determine when something has gone awry.

More accurate, versatile, and affordable probes

The growth in on-machine probing is also being driven by technology advances that make machine tool probes more accurate, more productive, easier to use, and easier to afford. Four relatively new probe developments illustrate this trend.

- *Strain-gage probe*. This probe has a strain-gage bridge in place of a mechanical switch to minimize pretravel variation. The effect on probing precision is to provide levels of performance never before possible on a machine tool probe—that is, unidirectional repeatability of 0.5 micron with a 50 mm stylus. Strain-gage sensors deliver low trigger force and a uniform 3D triggering pattern. This allows faster probing on complex parts while eliminating the programmer's requirement to compensate for lobing characteristics that can affect probing precision.

  A multichannel digital filter recognizes and ignores unintended triggers resulting from machine vibration and high acceleration/deceleration forces. This makes the strain-gage probe particularly suited to high-speed machining centers.

- *High-power optical system*. This new method of transmission, available for strain gage or conventional switching probes, extends transmission range to 12 meters, while omnidirectional signal coverage permits transmission from any spindle orientation. The new system uses a high-power infrared transmitter, modified optical receiver, and machine interface with high speed digital switch. It is designed for applications requiring extended-range transmission, environments that are not conducive to the use of RF transmission, or tilt-spindle applications that position the probe at complex vectors that traditionally result in loss of an optical signal.

  The system is particularly suited to very large five-axis machine tools, such as gantry profilers in the aerospace industry. It provides the high positioning accuracy needed to meet aerospace requirements that are driving the industry to automated "shimless" assembly.

- *Radio probe system*. This new probe takes radio technology out of the crystal age with 40 user-selectable channels. It extends the advantages of automated on-machine probing to applications not suitable for optical transmission. Unlike single channel systems that require crystal replacement to change a channel, this probe permits channel selection using a simple optical transmission link similar to the setup procedure for a garage door opener.

- *Noncontact toolsetting*. A new laser-based toolsetting system takes the complexity (and cost premium) out of noncontact toolsetting. A solution for high-speed, high-precision tool setting and broken tool detection, the laser-based system rapidly measures tool length and diameter at normal speeds to minimize impact on cycle times. Laser checking at working spindle speeds identifies errors caused by clamping inconsistencies and radial run-out of the spindle, tool, and toolholders—not feasible with static toolsetting systems. The toolsetting system also performs broken tool detection "on the fly," even at maximum rapid traverse, to minimize out-of-cut time.

As the tool moves through the laser beam, system electronics detect when the beam is broken and issue an output signal to the controller. The system accurately measures tools as small as 0.2 mm diameter anywhere in the beam. The system triggers when the laser beam is broken beyond a 50 percent threshold by the tool being checked. The noncontact toolsetting system uses a visible-red diode laser proven reliable in machining conditions. The visible-red laser operates at 670 nm wavelength at low-power output of less than 1 mW.

Advanced electronics permitting simplified design for the first time make noncontact toolsetting an affordable alternative to contact systems. The design avoids the mechanical elements and potentially complex installation of other toolsetting systems—the brackets and actuators with contact-based systems, and the problems with shutters, switches and solenoids on other noncontact designs. A protection system features positive internal air pressure that supplies a continuous stream of air through the laser beam apertures to keep out contaminants. Probing represents a fundamental change in the way inspection is performed. Rather than a postprocess check performed off-line on parts after machining, probing on the machine makes inspection part of the process, as well as a powerful process improvement tool for machining parts to spec in the shortest possible time.

Alternative uses for spindle-mounted probes

The following article on the use of spindle-mounted probes, which was written by Pete Zelinski with additional information supplied by Andy Haggerty, originally appeared in *Modern Machine Shop* magazine.

The spindle-mounted probe as a tool for in-process error compensation may be difficult to imagine. Probing is generally used as a productivity tool, to let the CNC locate a feature automatically, instead of having an operator locate it by hand. However, when the position of the feature probed is known already, probing becomes an accuracy tool. It can let a machining process beat its own inherent error, to hold tolerances tighter than its accuracy margin would otherwise allow.

Figure 10. A spindle-mounted probe.
(Source, Renishaw, Inc.)

The technique, called reference comparison, involves employing a reference object close to the workpiece as an objective measure of positioning error.

Just prior to making a critical machining pass, the CNC probes the reference. Comparing its own "sense" of position to the known position of this surface—which is precisely determined prior to machining—lets the CNC measure its own error. It programs a temporary offset equal to the discrepancy. By adding this offset to the critical machining move, the CNC edits the measured error out of the process. Reference comparison assumes that the error in the probing move and the critical pass is nearly the same—which it should be, if reference and workpiece are close to one another and physically similar. If true, then the critical pass will come in on target, despite whatever error accumulated in the moves to load the tool and return to the part.

Cincinnati Machine, Cincinnati, Ohio, has demonstrated the effectiveness of this technique. Using its own patented version of reference comparison, Cincinnati Machine routinely holds ±0.0005-inch tolerances in its own production operation, using a process with a total accuracy margin (the sum of machine accuracy, plus error from effects like changing ambient and coolant temperatures) measured at 0.0028 inch. Implementing reference comparison not only reduced scrap and rework, but also eliminated the need for a high accuracy boring mill for one of the company's critical parts.

Andy Haggerty, Cincinnati Machine emeritus R&D engineer, has studied probing for nearly 20 years. He was instrumental in developing Cincinnati Machine's reference comparison approach. This article summarizes his insights into the technique, and his advice for other shops hoping to apply it to their own high precision parts.

Where the laser leaves off

Compensating for error instead of correcting it has been popular almost since the dawn of NC. No one gets dirty hands that way. But a more important reason is that many machining center errors are either untraceable or impractical to fix for any number of other reasons.

Early error compensation took the form of "funny tapes"—programs adjusted to individual machine tools, employing lopsided positioning commands to make up for errors consistently observed on a given machine. Though this approach often did improve the process, it is employed less and less in an age when distributed numerical control and flexible cells require programs to be accessible to multiple CNCs and multiple machines.

A more effective technique is linear error compensation using a laser measurement system. The laser system—often directly interfaced to the CNC—takes precise linear position measurements at regular points along one or more lines of the machine's travel. These measurements are compared to the CNC's programmed position at each point, with discrepancies entered as correction offsets in a CNC axis compensation table.

This method of laser calibration remains an effective technique for improving the accuracy of a machining center across its work zone. Reference comparison can't replace it, but can complement it. While a common misconception says that laser calibration "corrects" a machine tool, it actually leaves many sources of error untouched.

For example, errors related to squareness and angular effects of the machine axes remain after the laser is done. And off-line calibration is powerless to address machine errors that will change during the process, due to the changing temperatures. In fact, machine positioning is truly "correct" only for moves at slide positions and machine component temperatures matching those of the laser compensation move, using a tool length equal to the distance from the spindle nose to the laser optic.

As the process departs from any of these conditions, positioning error increases, leaving room to improve the process through reference comparison.

The basics

A machining program using reference comparison doesn't trust the program coordinates to locate the tool point accurately enough for critical machining. Instead, it references each critical machining pass to a reference surface with a known location.

This reference surface can be a premachined feature of the part itself, but can also be a feature of a separate reference object. In the latter case, the object is positioned on the fixture prior to machining, using precise measuring equipment to locate it with respect to the position of the part. Often, programs employ multiple reference surfaces to locate a machined feature precisely in multiple axes.

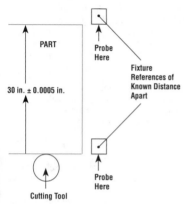

Reference comparison recognizes that errors accumulate over long machine moves. Consider the part in *Figure 11*, which requires opposing machined surfaces to locate 30 inches apart in Y, to an accuracy of ±0.0005 inch. By the time the CNC moves the slides 30 inches from one surface to the other, machine or process errors might cause the axes to stray from this position by more than ±0.0005 inch. Therefore, instead of relying on a constant set of program coordinates, a reference comparison machining program checks positioning against two reference surfaces, one near each machined surface.

The distance between these surfaces need only be approximately 30 inches, but its true distance is precisely measured before machining begins—ideally on a CMM in a temperature-controlled room. The reference comparison program treats this distance as a dimensional master, using it to update its own sense of positioning through the use of program offsets.

Figure 11. Normal machine and process errors may make it impossible to hold the 30-inch dimension to the given tolerance, using just a straight line move between the two surfaces. A more accurate method is reference comparison, where CNC probes references of known position to update its position prior to a critical move.

The procedure is simple. Assume that the machining center used for the operation in *Figure 11* is a horizontal, and that the references are indeed 30.0000 inches apart. The program would then be written to include offsets that assign coordinates to the two surfaces that are separated by the same distance: say, Y = 0.0000 and Y = 30.0000.

The program begins by probing the lower reference surface. Because of the influence of machine or process error, the program coordinates may show the surface to lie at Y = -0.0003, Y = 0.0001 or some other slightly inaccurate coordinate. However, the reference has already been positioned at Y = 0.0000, so the program is justified in adjusting the coordinates to match this value. It does this by subtracting the coordinate measured through probing from the known coordinate and storing the difference as an offset.

This offset is activated when the machine loads the milling cutter. With the offset in place, a machining pass programmed to bring the surface in line with the level Y = 0.0000 will meet that mark more accurately, because the machine already made its "mistakes" when it probed the reference surface.

Accomplishing that, the CNC probes the upper reference surface with the offset deactivated. Again, the difference between the known coordinate (30.0000) and the coordinate measured through probing is entered as an offset. The program activates this offset for the machining operation, allowing the second surface to be machined to the same coordinate.

The result will be a part with surfaces 30 inches apart, to approximately the same accuracy as the original off-line measurement of the distance separating the two reference surfaces, regardless of machine or process error.

Maximizing accuracy

Reference comparison can improve on the machining accuracy achieved through laser compensation for one reason: it lets the process avoid the errors that still remain after laser compensation is complete. These can include tool tip, axis squareness, and thermal errors.

Maximizing the accuracy gains from reference comparison means minimizing the chance for errors such as these to creep in between the probing and machining moves. This is a three-step process.

1. *Match probe length and tool length to overcome tool tip error.*

   Several geometric errors—including X-yaw, Y-pitch, X- and Y-roll, and XY-axis out-of-squareness—can place the tip of the tool in a position different from what the X- and Y-scale coordinates indicate. When such an error does exist, its magnitude is a function of tool length. Laser compensation "corrects" tool positioning only when tool length is the same as the distance from the laser optic to the spindle nose. *Figure 12* shows why.

   The horizontal machine in the *Figure 12* has a not-uncommon condition: a "bowed" Y-axis (exaggerated in the illustration), causing the Y-axis pitch to increase as the spindle travels up. The result is a tool tip that rises somewhat faster in Y than does the Y-axis slide.

   Even limiting the process to tools that match the optic-to-spindle length fails to solve the problem. Pitch error changes with changing temperature (the concern of

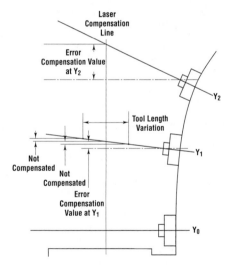

Figure 12. When the Y-axis is bowed so that pitch error increases as the spindle rises, laser compensation "corrects" the pitch error only when tool length matches the length from laser optic to the nose of the spindle.

point 3, below), so not only tool length, but also the machine element temperatures would have to be tightly controlled.

A better solution is just to avoid the error, by matching the length of the probe to the tool length of the critical machining pass. The tool tip error is then identical for both, making this error just one more component of the total error compensated by the reference comparison offset.

The magnitude of tool tip errors is generally small enough that the match between tool and probe lengths doesn't have to be precise. Agreement of ±1 inch is sufficient—a match that can probably be achieved solely through the choice of the probe's stylus from within a standard kit.

2. *Match axis coordinates to overcome squareness error.*
   Another source of inherent machine error is an out-of-square condition between the X- and Y-axes. The magnitude of this error grows as the machine travels away from the line of travel along which it was laser compensated. The error can be the result of out-of-square machine axes, or a part or fixture mounted out-of-square with respect to the machine.

   To address this problem using reference comparison, the probing move and the machining move should have similar coordinates in as many axes as possible. The separation should be no more than 5 inches, but identical coordinates are even better. If the machine is spared from moving in a given axis, it's also spared from accumulating squareness error.

   Figure 12 illustrates how the lack of X-Y squareness affects accuracy. The part shown requires three holes to be bored along the line X = 20.0000, at Y = 5.0000, 15.0000, and 25.0000. The right face of the part corresponds to the line X = 16.0000. Though the machine has been laser compensated at the level Y = 15.0000, a small squareness error translates to 0.00004 inch of deviation in X for every 1 inch of movement away from this value in the Y axis.

   Given the squareness error, a cycle that bored all three holes using the same set of program coordinates could accurately position only the hole at Y = 15.0000. The upper and lower holes would position incorrectly in X, by -0.0004 inch and +0.0004 inch, respectively.

   However, because the distance of each bore from the right face of the part is precisely defined—4.0000 inches—both the upper and lower bore can be positioned accurately, in spite of the squareness error, by using this reference.

   The reference comparison portion of the cycle would probe this face three times immediately before boring—once at the Y-axis position of each cut. (Even the middle hole is probed. In an actual process, setup or process related errors could cause the surface to locate incorrectly at this level, too, and the extra probing pass would add little to cycle time in this instance.)

   For example, to position the lower hole, the CNC would probe the right face along the level Y = 5.0000. However, because of the squareness error, it would meet the surface 0.0004 inch sooner than expected. To correct for this, the program would enter an offset value to be added to the X-axis coordinate of the machining pass; in this case, -0.0004. The same steps would be repeated at Y = recommendation to match tool length to probe length (above) will automatically make slide and tool tip positions the same.15.0000 and Y = 25.0000 to position the other two bores. A similar degree of inaccuracy may result from an angular error along the Z-axis. To overcome this error, the probing pass would be made at a slide position in Z approximating that of the machining pass: for example, at the midpoint of the boring move. Note that this is slide position, not Z-axis tool-tip position.

However, following the recommendation to match tool length to probe length (above) will automatically make slide and tool-tip positions the same.

3. *Match materials and machine soon to overcome thermal error.*
   Machine tool straightness, squareness, and angular errors are generally dynamic. They change over time, as thermal effects distort machine elements, and change their positions relative to one another. Even on a "perfect" machine, thermal effects would still affect the process, as changing temperature caused the part and setup to change position relative to the machine.

   Thermal effects are the most important reason to resort to reference comparison. If error never changed, then permanent offsets could adjust for them and in- process compensation would be unnecessary. However, precision machinists know that a real process is always in a state of thermal and geometric flux.

   Flood coolant can improve thermal stability, but only partially. Unless a coolant temperature control system is used, applying flood coolant just introduces changing coolant temperature as another thermal effect. And coolant temperature can at times vary significantly throughout the machining run.

   Avoiding thermal errors through reference comparison takes a two-part strategy: 1) choose the reference object to have thermal expansion characteristics similar to the part, and 2) perform the critical machining as soon as possible after the reference probing move, leaving little time for temperature to change.

   The reference object material should be the same alloy as the workpiece, in the same metallurgical condition. The object geometry should also be chosen to produce a thermal expansion response closely matching that of the workpiece. This means similar orientation, as well as similar length and cross-section—or more precisely, a similar ratio of volume to surface area.

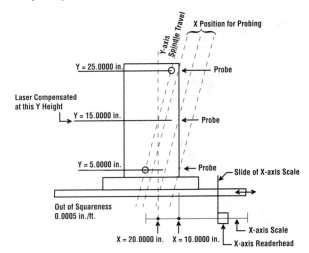

Figure 13. Despite a squareness error of 0.00004 inch in X for every one inch in Y away from the line of laser compensation, the machine in this illustration can position the three bores correctly through reference comparison. The reference is the surface at X = 16, probed at the level of each bore immediately before machining.

But even if the dimensions of the workpiece and reference object are changing identically, they are still changing. In most processes, this means data measured through probing become increasingly less accurate with each minute that passes. For this reason, the critical machining pass should be performed immediately after the probing pass used to locate it. No other machining operations should separate the two.

Return on investment

The costs of reference comparison are obvious. The technique adds time and complexity not only to the machining cycle, but also to setup. It's not for every part—not even for every precision part. However, the technique has proven itself a profitable tool to Cincinnati Machine, and several other high precision manufacturers.

When reference comparison tightens the tolerance band that a machine can reliably hold, it can improve the process by reducing scrap, and eliminating the extra transport and setup costs associated with high precision operations on separate machines. The "cost" in setup and cycle time then becomes an investment that pays its dividends downstream.

Example 1: Precision Link: The cast iron link shown in *Figure 14*, which is used in toggle mechanisms of Cincinnati Machine injection molding machines, features two 2-inch diameter bores whose center-to-center distance of 26.380 inches must be accurate to within ±0.0005 inch. Though holding this tolerance used to require a manual boring mill, reference comparison now allows Cincinnati Machine to bore the holes accurately on a CNC machining center as part of a process with an accuracy margin measured at 0.0028 inch.

Because the part is cast, no surface of the part itself can be trusted as a reference for the bores. Therefore, Cincinnati Machine uses a cast iron mounting plate set adjacent and

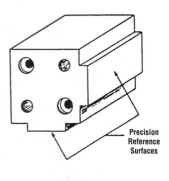

Precision
Reference
Surfaces

Figure 14. The precision link part (left) and the references (right) used to locate two bores. Although there are two small reference blocks, they are bolted to a long connecting bar to give the entire assembly a set of thermal expansion characteristics similar to that of the part itself.

parallel to the part, supporting two reference blocks positioned so that their Y reference surfaces are 27.500 inches apart.

Each block provides ground and hardened reference surfaces that the CNC probes in X and Y to counter the effects of any axis positioning errors immediately before the bores are machined. The blocks are positioned repeatably on the mounting plate using two locator pin holes. Both blocks are probed after the holes have been rough-bored in order to locate the holes for finish machining.

Example 2: Opposing Bores: Reference comparison can also be used to correct errors associated with workpiece rotation, whether manual or through B-axis indexing.

The part shown in *Figure 15* includes opposing bores that must share a common centerline. Though the ideal way to ensure this concentricity would be to bore straight through in a single operation, the part does not leave enough room within the machine's work zone to accommodate such a long boring bar. The part must instead be rotated.

However, the spindle in this example has a drift error of +0.0020 inch with respect to the pallet's center of rotation. Even with perfect concentricity between the coordinate axes and pallet rotation, this error would still cause the bore centerlines to differ by 0.0040 inch, which is the combined effect of the displacement of the near hole by +0.0020 inch, and the far hole by -0.0020 inch.

The solution is to place reference surfaces along the bores' shared centerline, on each side, and to probe these surfaces to locate each bore immediately before machining. This not only eliminates the influence of the spindle drift, but can also compensate for the part being positioned off-center with respect to the axis of rotation of a B-axis move.

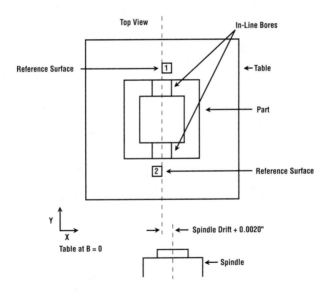

Figure 15. To overcome spindle drift error in opposing bores machined through B-axis indexing, reference surfaces are placed in front of the bores along the common centerline.

Review of basics

1. Use the same approach direction for probe and tool. When the coordinates of a machining pass are based on data gathered through probing, have the machine move in the same direction to position the tool that it did to position the probe.

2. On a horizontal, position tool and probe up in Y. That is, against gravity. This also reduces lost motion error, by tending to compress the axis ballscrew.

3. Recalibrate daily. Recalibrate also after adjustment or alteration of the probe, particularly a stylus change.

4. Clean surfaces first. Even a speck of dirt added to a probing measurement will result in out-of-tolerance parts in precision machining operations. Cleaning can be automatic, via a programmed coolant wash. If this doesn't remove all chips and debris, consider a dwell cycle in the program, to give the operator time to clean the part by hand.

5. Average three hits to reduce repeatability error. Consider replacing a standard "touch and go" probe move with a routine for probing the same spot three times and taking the average. This reduces random repeatability error.

Coordinate Measuring Machines (CMM)

CMM systems

CMMs consist of the machine itself, a probing system, a computer system, and computer software. L. S. Starrett Co. recommends that the following factors should be evaluated in the selection of a machine.

Frame and base. All precision measuring must be secure from vibration and thermally induced variations, and CMMs rely on their frame construction and bearings to minimize irregularities. While both aluminum and granite frames are common, granite bases, in spite of their high weight, provided they are thermally stable (aluminum expands at 0.000012" per inch per °F, while granite expands at just 0.000003" per °F), exhibit a very low coefficient of expansion, and have excellent vibration dampening characteristics. The working surfaces of granite surface plates are precision lapped, making them excellent for use with the air bearings often employed with CMMs.

Bearings. Air bearings have, to a large degree, replaced roller-bearing systems on CMMs. Their attributes include zero wear because there is no contact between the bearing and the bearing surface during use, and no moving parts to produce noise or vibration. Some systems use porous graphite as bearing material (instead of aluminum). Porous graphite allows compressed air to pass directly through the material, resulting in an evenly distributed cushion of air across the bearing surface. This layer of air is very thin (approximately 0.0002"), especially when contrasted with conventional orificed aluminum bearings, which require a larger air gap (0.001 to 0.003"). A small gap reduces a machine's tendency to bounce on the bearing's air cushion (as on a spring) and provides a more rigid and repeatable machine, resulting in more accurate measurements.

Scales. On each axis of a CMM, a device is mounted that identifies the position of each axis within the measuring range of the machine. These are typically glass or reflective scales. When the reader head, which is attached to the moving member of the axis, moves across the surface of the scale, a pulse is generated that is instantly converted into a digital reading. The finer the markings on the scale, the finer the resolution of the reading. It should be possible to receive accuracy data on any CMM. In addition, the individual machine should have been checked for accuracy before leaving its factory. Accuracy testing should be done in accordance with ANSI B-89.1.1, and all testing equipment should be to levels required by the National Institute of Standards and Technology (NIST).

Manual versus Distributed Numerical Control (DNC) machines. Manual CMMs are ideal for use when the machine is used primarily for first-article inspection or reverse engineering. However, even though DNC machines are more expensive to purchase, they are preferred for use in a production oriented manufacturing environment. When evaluating a DNC machine, it is important that no hysteresis (backlash) is present in the drive system. Friction drives incorporate a direct shaft drive to a precision drive band to eliminate backlash. Because friction drives eliminate the reciprocating shaft found in screw drives (which can introduce vibration), they induce minimal vibration and, in spite of their name, operate on very low amounts of friction.

Probe heads. The probe head mounts into the ram (or arm) of the CMM. It may have a trigger mechanism built into the head with the stylus mounted directly to the head, or a stylus extension may be used to extend its reach. Extensions are commonly used on smaller machines to increase the reach of the stylus and thereby improve the machine's maximum measuring volume. Nonarticulating probe heads measure in only one direction—straight down—and are used extensively for measuring machined or stamped impressions in flat sheet metal or similar applications. They can be used on both manual and DNC machines. Articulating probe heads allow the probe to be pointed in many different directions. Some manufacturers, such as Renishaw, use two angles to control the probe. The first angle determines the number of degrees the probe deviates from straight down, while the second angle determines, in increments of either 7.5° or 15°, the direction that the probe points in the horizontal plane. An articulated probe allows for measurements on the back side of a part. They are available in manual and DNC configurations.

Probes can be triggered by various mechanisms. Common examples are touch trigger probes, analogue probes, and piezo electric probes that make contact with the part, or optical or laser probes that are used when physical contact is impractical.

Evaluating accuracy

The following article by James J. Childs investigates what a manufacturer's rating of CMM accuracy really means.

What does it mean when a manufacturer lists the accuracy of one of its CMMs as "2.3 + L/600"? This expression follows the popular international standard ISO 10360-2. The "L" is expressed in millimeters (mm) of travel and the accuracy is noted in micrometers (μm). In the rating 2.3 + L/600, 2.3 is a constant, as is the 600 figure, and both are determined by the builder and are peculiar to the accuracy of a CMM model. What do these constants mean, and how are they determined?

The accuracy of a CMM is expressed by the straight-line formula

$$E = A + L \div K, \qquad \text{or} \qquad E = A + L \,(1 \div K)$$

where E = Accuracy as read on the ordinate (Y-axis) of a graph
  A = A constant that would be equal to E if the length of travel (L) were "0" (zero)
  L = Length of travel of the measuring probe
  $1 \div K$ = A constant and the slope of the straight-line.

The constants are determined in the following manner. *Seven* configurations (lines of different lengths and orientations) are established within the X, Y, and Z travel volume of the CMM. (See *Figure 16.*) Each of these lines has *five* target points.

The CMM is programmed to move to each target point *three* times, for a total of 105 readings (7 × 5 × 3). The absolute differences between the target points and the actual CMM readings are then plotted on a graph where the X-axis is the length of travel between the points, and the Y-axis is the "error of indication," or accuracy. Such a graph is shown in *Figure 17.*

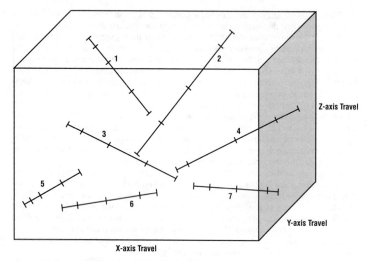

Figure 16. Establishing X-, Y-, and Z-axis references.

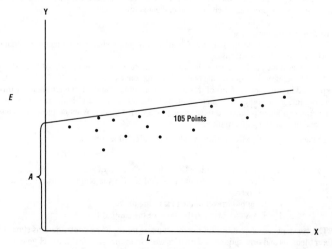

Figure 17. Graph of target points.

A, in the straight-line formula, would be where the line crosses the Y-axis (intercept) in *Figure 17*, or the error of indication *E* when there is no travel, or when *L* = 0. Therefore,

$$E = A + L \, (1 \div K)$$
$$E = A + 0 \, (1 \div K)$$
$$E = A.$$

As the value of *L* increases, so does the total error for the CMM that is dependent on the slope of the line, which is $1 \div K$. The machine's builder determines the value of *K* by fitting a line over all the 105 points, then finding the slope, and then calculating *K*. Manufacturers will normally note the accuracy of their CMM with a formula in which the constants *A* and *K* are inserted. In the example cited in the beginning of this discussion on accuracy, the value "2.3" would be *A*, and the value "600" would be *K*.

CMMs on the shop-floor

In many instances, CMMs have moved from an inspection tool to an essential part of the manufacturing environment. The following article, which originally appeared in *Modern Machine Shop* magazine, was written by David Genest of Brown & Sharpe.

The practical need to accurately measure part dimensions has been the catalyst in the development of a host of precision measuring instruments, including calipers and micrometers, height gages, and other graduated devices. Of all the tools used to measure part dimensions, the coordinate measuring machine (CMM) is the most capable. Coordinate measuring machines collect detailed dimensional data by moving a sensing probe along workpiece surfaces. Most CMMs acquire data using a touch trigger probe that contacts individual points on the workpiece. This single-point measurement technique can generally collect data at a maximum rate of about 50 to 60 points per minute, considerably faster than manual measuring instruments.

CMMs as process control tools

What makes coordinate metrology valuable as a process control tool is that it can be used to accurately measure objects of widely varying sizes and geometric configurations, and discern the relationship between features of a workpiece. This flexibility, and the ability of the coordinate measuring operation to be performed quickly when compared with surface plate techniques or fixed gages, means that measurement results can be used to cost effectively refine manufacturing process applications and analyze process trends.

The difficulty with coordinate metrology has been that, as valuable as it is, measurement routines often had to be performed off-line by specially trained technicians who were removed from manufacturing operations. Off-line inspection of this type developed into more of a quality control and quality assurance tool to verify setups or discover nonconforming parts before they were shipped to customers than a true means of process control. With the introduction of statistical process control techniques, however, quality technicians could use the dimensional data gathered by CMMs to study trends in nonconformance and offer that analysis to manufacturing personnel. Manufacturing personnel, in turn, used the information to correct process variations that caused out-of-tolerance conditions. But those process corrections still often occurred after parts had been scrapped or reworked to meet specifications.

Moving CMMs to the shop-floor

During the last two decades, there has been a trend toward integrating the speed, flexibility, and accuracy of the CMM into shop-floor operations. One of the factors driving this trend is the increasing demand for close tolerance production where even the slightest process aberrations can quickly produce useless parts.

Increasingly, measurement is being integrated with the machining process itself. In a transfer line, for example, a measurement station is physically attached to the line. In other cases, measurement integration may be facilitated by some type of transfer mechanism such as a rail guided vehicle, cart, lift-and-carry device, or overhead conveyor that transfers workpieces from the machine tool to the CMM for inspection.

The advantage of integrated shop-floor gaging is that, by moving dimensional inspection closer to manufacturing operations, much greater control can be exerted over the process than if the data is gathered in a remote location. Increased control reduces the probability of producing out-of-tolerance parts, meaning less scrap and rework. There are other benefits as well. Integrated shop-floor gaging is most often performed by machine tool operators, which reduces the need for special inspectors. Moreover, shop-floor gaging can be used to compile a historical database of how machines, cutting tools, parts, pallets, and fixtures behave during the actual machining process.

Coordinate measuring machines used in these applications have the advantage of being flexible—that is, with a simple software program change, they can be used to inspect workpieces with different dimensions, as well as fixtures, pallets, and tooling.

Making shop-floor measurement work

The integration of CMMs on the shop floor has been a technological challenge for CMM manufacturers. The challenge is how to guarantee high accuracy performance at speeds that meet throughput objectives.

The biggest obstacle is temperature variation and the effect it has on the measuring machine and the workpiece. An early solution to this problem was to enclose a traditional CMM to protect it from thermal gradients and to condition the air inside the enclosure to hold the temperature constant. In effect, this approach enclosed the CMM in a laboratory environment. While this can be a good solution in some circumstances, it does have a drawback in that throughput is sacrificed for accuracy. The enclosure creates some difficulties in workpiece loading and unloading, plus the workpiece has to "soak" at the enclosure temperature before accurate measurements can be taken.

A new approach is to compensate for thermal expansion and thermal distortion errors with a mix of hardware and software solutions that include a web of sensors placed at critical points in the machine structure. The sensors read the temperature on the structure of the machine and a powerful algorithm extrapolates expansion and distortion values from the data. With this data the software is able to compensate for the current thermal state of the machine so that the influence of temperature variations is virtually canceled over a wide range. The result is that on the shop floor, dimensional inspection can be performed with accuracy comparable to that of a lab.

Software makes integration easier

While hardware is important to the functioning of the coordinate measuring machine on the shop floor, software has an equally important role.

CMM software has been refined so that computer programming knowledge or experience is seldom required to run even the most sophisticated programs. Virtually all new CMM software consists of off-the-shelf, menu-driven programs with a comprehensive array of help screens. The result is that software for any particular routine is very easy to use and can be customized to fit individual applications using plain English rather than a specific programming language tied to a complicated operating system. A wide range of software is available. For example, there are programs for statistical process control, sheet metal applications, and turbine blade and gear programs along with a host of others.

Bridging design and manufacturing

Next-generation software is being developed to provide the information conduit that genuinely links the design function with manufacturing through the common language of metrology. This new software establishes a bridge between design and manufacturing by means of a common implementation of design tolerances, and between manufacturing and design by providing online and on-demand process information to design engineers.

For example, new software is designed to interface seamlessly with CAD/CAM and off-line inspection systems so that CAD models do not have to be translated when they are downloaded to the inspection system. This gives both the design and manufacturing groups an accurate and consistent analysis of the tolerances that the manufacturing process must maintain for complete and unambiguous assessment of design intent.

Most measurement and inspection software available today is proprietary to the vendor and does not easily interact with other systems. Next-generation software, however, will fit everywhere in the manufacturing operation. The operator interface of this new software will be designed to be sensitive to the type of measuring device it is running. It will know if the operator is using a manual measuring instrument, such as a caliper, or a full-featured coordinate measuring machine. The software will present only those interactions required for whatever application is being used. This capability will allow users to select the measurement tool most appropriate to the task, maximizing the use of measurement equipment. This approach to software development will enable users to add value to all of their measuring systems by making them more easily integrated into the manufacturing process, rather than isolating them in inspection laboratories.

New software, even today, often features an open architecture, object-oriented approach to development. This means that users will be able to add their own, in-house developed inspection routines and analysis packages to the operating system. Third party developers can also create metrology packages that can be incorporated into the basic software system. The potential for the development of new and unique approaches to measurement and inspection applications is virtually limitless.

Scanning adds new dimensions to shop-floor measurement

As manufactured components become smaller and tolerances become tighter, more data points must be collected and analyzed to help determine the viability of the manufacturing process. CMMs capable of scanning provide the means of gathering these data points. Scanning is simply a way of automatically collecting data points to accurately define the shape of a workpiece.

Scanning capability was once considered the province of only high technology manufacturers because sophisticated CMMs capable of scanning and the software to run them were expensive and not readily available. Today, however, advances in software and sensor technology have made scanning CMMs considerably more affordable.

New sensors will combine the elements of optics, video, and laser technologies into devices that can rapidly scan complex shapes and surfaces, and accurately gather dimensional data. Some of these new sensors can gather up to 20,000 data points per second with extreme accuracy. Combined with these sensors will be powerful mathematical engines that will quickly analyze the billions of bits of dimensional data these systems can generate.

Is the integration of coordinate metrology on the shop floor a concept that is here to stay, or just a passing fancy? It is safe to say that the closer dimensional inspection can be placed to the process, the more valuable the process control capabilities become to the user. The key issues are accuracy and throughput. CMMs coming on the market today are faster, more accurate, and less costly than their predecessors of only a few years ago, plus, most are being designed with shop-floor integration in mind. The ultimate value of shop-floor metrology is not only its ability to be integrated into the process, however, but also the link it forges between design intent and manufacturing capability. That is the quantum leap that will make continuous quality improvement an integral part of design, engineering, and manufacturing.

Measuring Surface Finish

Additional information on surface finish can be found in the "Insert Selection," "Abrasive Machining," and "Turning" sections of this book. The following article, by Alex Tabenkin of Mahr Federal, Inc., originally appeared in *Modern Machine Shop* magazine.

Surface finish, or *texture*, can be viewed from two very different perspectives. From the machinist's point of view, texture is a result of the manufacturing process. By altering the process, the texture can be changed. From the part designer's point of view, surface finish is a condition that affects the functionality of the part to which it applies. By changing the surface finish specification, the part's functionality can be altered—and hopefully, improved.

Bridging the gap between these two perspectives is the manufacturing engineer, who must determine how the machinist is to produce the surface finish specified by the design engineer. The methods one chooses to measure surface finish, therefore, depend upon perspective, and upon what one hopes to achieve.

Defining surface texture?

Turning, milling, grinding, and all other machining processes impose characteristic irregularities on a part's surface. Additional factors such as cutting tool selection, machine tool condition, speeds, feeds, vibration, and other environmental conditions further influence these irregularities.

Texture consists of the peaks and valleys that make up a surface and their direction on the surface. On analysis, texture can be broken down into three components: roughness, waviness, and form.

Roughness is essentially synonymous with tool marks. Every pass of a cutting tool leaves a groove of some width and depth. In the case of grinding, the individual abrasive granules on the wheel constitute millions of tiny cutting tools, each of which leaves a mark on the surface.

Waviness is the result of small fluctuations in the distance between the cutting tool and the workpiece during machining. These changes are caused by cutting tool instability and by vibration, several sources of which affect the stability of every machine tool. Some of these sources are external and sporadic—for example, a passing forklift, and the operation of other machines on the shop floor. Other vibration sources are internal, such as worn spindle bearings, power motor vibration, and so on.

Assuming that the part is nominally straight and/or flat, errors of *form* are due to a lack of straightness or flatness in the machine tool's ways. This is a highly repeatable type of irregularity, as the machine will always follow the same out-of-straight path.

All three surface finish components exist simultaneously, superimposed over one another. In many cases it is desirable to examine each condition independently. We approach this problem by making the assumption—a correct one, in most cases—that roughness has a shorter wavelength than waviness, which in turn has a shorter wavelength than does form.

Gages separate surface finish components using discrete units of length, called cutoffs.

The length of the cutoff selected and implemented by various electrical filtering techniques permits the measurement of roughness by itself, waviness by itself, or "total profile," which combines roughness, waviness, and form.

For parts produced by modern machine tools at typical speeds and feeds, roughness may be defined, for example, as any irregularity with a wavelength shorter than 0.030 inch; waviness as between 0.030 inch and 0.300 inch, and form errors as having wavelengths greater than 0.300 inch. These figures are quite flexible, and standards exist for roughness measurements with wavelengths from below 0.003 inch and up to 1 inch.

Surface finish tends to be a stable condition; it should not change from part to part, unless process conditions change. Manufacturing engineers, in fact, can generally predict the surface finish that a process will generate, given a known material, machine tool, cutting tool, coolant, speed, feed rate, and depth of cut. For this reason, surface finish measurements have historically been used primarily as a means of monitoring the stability of manufacturing processes. By taking an occasional measurement, a machinist can establish that the entire process is running as it should.

If the measurement changes, it is a signal that some element of the process has changed significantly—perhaps the cutting tool has reached its wear limits, the coolant needs changing, or a new source of vibration has arisen. By examining roughness, waviness, or total profile separately, machinists can narrow the search for sources of error, and take effective action to reduce or eliminate them.

Parameters

Parameters are the quantitative methods used to describe and compare surface characteristics. These are defined by the algorithms that are used to turn raw measurement data into a numerical value. Although more than 100 parameters exist, machinists have traditionally relied upon just one or two parameters.

Currently, R_a, or average roughness, is the parameter most widely specified and measured. The algorithm for R_a calculates the average height of the *entire* surface, within the sampling length, from the mean line. It serves as an effective means of monitoring process stability, which explains why it is the predominant parameter in use today (see *Figure 18*).

Of more than a dozen roughness parameters specified by ASME in Standard B46.1-1995, two others that are widely used on the shop floor are R_{max} and R_z. R_{max} measures the vertical distance from the highest peak to the lowest valley within five sampling lengths, and selects the largest of the five values. It is, therefore, very sensitive to anomalies such as scratches and burrs on the part's surface, and specifically useful for inspecting for these conditions. But because a single scratch or burr is often *not* the result of a symptomatic problem in the manufacturing process, this parameter is

Figure 18. The R_a parameter measures average roughness. It is not sensitive to occasional spikes and gouges. On the other hand, the R_{max} parameter is designed to detect these anomalies. The two surfaces shown here have nearly identical R_a values but very different R_{max} values.

not so useful for monitoring process stability. On the other hand, R_a, as an averaging function, is fairly insensitive to occasional anomalies, and is therefore not useful to detect the presence of these features.

R_z is widely used in Germany and elsewhere in Europe, in preference to R_a. Like R_{max}, R_z is based on the evaluation of five sampling lengths. But instead of selecting the largest peak-to-valley distance of the five, it averages the five values.

Design and engineering influences

Thus, even within the single component of roughness, the specification of surface finish goes far beyond the notion of a simple smooth/rough continuum. Through parameters, surfaces can be defined and described in great detail, and engineers have made correlations between parameters and part performance under various conditions.

In some applications, a single scratch may render a part unacceptable from a design-engineering point of view, regardless of how fine its average roughness value. These same considerations apply to the multiple parameters for waviness and total profile.

When a design engineer specifies a surface finish parameter and value, therefore, he must do so with an understanding of how they will affect the part's performance. Selecting the ideal parameter(s) for a given application is somewhat complicated by the great number of parameters in existence, but most of these have limited applications. The majority of applications can be successfully specified using a few well-known parameters.

Smoother, of course, is not always better. There are obvious economic benefits to machining parts as quickly as possible, and to minimizing the amount of secondary work performed. Additionally, there are applications in which a certain degree of roughness enhances functionality, and specifications may specify minimum as well as maximum roughness values. Having some definite roughness to the surface, for example, often enhances adhesion of paint or other coatings.

Some parts that perform multiple functions require complex surfaces to perform optimally. Engine cylinder walls, for example, must be smooth enough to provide a good sealing surface for the piston rings, to promote compression, and prevent blow-by. At the same time, they must have "pockets" of sufficient size, number, and distribution, to hold lubricating oil. The R_k family of parameters was developed to describe such complex, multifunctional surfaces. This is an example of the parameters that were developed as design, rather than inspection, tools.

Once a surface has been defined and specified, the manufacturing engineer must determine how to produce it reliably and cost-effectively. In the case of surfaces specified only by the R_a parameter, this is usually quite easy, because the actual shape of the surface can vary considerably and still meet a given R_a value. Finer R_a values can be achieved by many alternate approaches, including slowing the speeds or feeds, making shallower cuts, or following the primary cutting process with a secondary process such as fine-grinding, honing, lapping, and so on. If R_a is the only parameter specified, the manufacturing engineer can choose the most economical and efficient approach.

But where R_a is strictly a quantitative parameter, the R_k parameters are both quantitative and qualitative, in that they define the shape of the surface. The manufacturing engineer is faced with a more complex task. In the case of the cylinder wall described above, the desired surface requires two-step processing, at minimum. The first step, which may be boring, grinding, or rough honing, produces a relatively rough surface, with many prominent peaks and valleys. The second step, plateau honing, knocks the tops off the peaks, but does not extend to the bottoms of the valleys, leaving a mainly smooth surface with a number of oil pockets (see *Figure 19*).

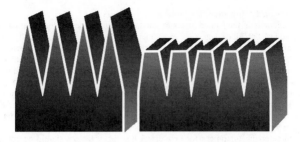

Figure 19. The R_k family of parameters describes multifunctional surfaces, such as cylinder walls. The part section on the left shows the surface after rough cutting. Plateau honing eliminates the tops of the peaks (right). In operation, this surface has sufficient area to promote piston compression, and retains enough pockets to retain lubricating oil.

Inspection

After the part has been designed and manufactured, it must, of course, be inspected. For surfaces specified only by a roughness parameter, this is a simple matter. Pocket-sized, battery-powered gages that offer a small number of roughness parameters are available at low cost (below $2,000), are extremely easy to use, and can be very flexible in application.

More complex parameters require full-featured instruments that are run by computers, and these devices may cost over $10,000. Some of these surface analysis systems are hardened for shop-floor use, and recent advances in software have made even complex measurements relatively easy to perform.

Existing standards are written around the use of instruments that measure part texture by moving a stylus in a straight line across the surface, and by monitoring the vertical movement of the stylus. Generally, the less expensive stylus-type gages that measure only roughness use the surface of the part itself as a reference. These are called skidded gages. In contrast, the full-featured instruments incorporate a precision internal reference surface, which enables them to measure waviness and total profile in addition to roughness. These are called skidless gages.

Traditional stylus-type inspection is not feasible in all instances, however. Gold-plated surfaces, for example, may be scratched by the stylus. Some high-speed or continuous manufacturing processes require faster throughput than stylus instruments allow. And some design applications require analysis of the surface over an area rather than in a straight line. For such applications, instruments using optical and other noncontact methods, or special area-measuring stylus methods, are available. These are generally quite expensive, however, and standards regulating their performance are still under development, so their use is generally restricted to applications in which traditional stylus methods are impractical.

Inexpensive, compact "roughness" gages, however, often retain their traditional utility, even where more complex parameters are specified. A shop may maintain one skidless gage for manufacturing engineering and quality assurance purposes, while making several of the more economical skidded gages readily available to machinists. Once the process is established and confirmed on the skidless gage, machinists use the skidded gages to measure parts for R_a or another roughness parameter, strictly as a means of ensuring process stability. This often represents a practical approach to meeting surface finish specifications.

Tolerances and Allowances for Hole and Shaft Fits

First attempts in the United States to standardize limits and fits for metal parts were undertaken in 1920. Since then, these recommended dimensions were frequently reviewed to recognize the increasing accuracy of machine tools. During World War II, revisions and reviews became the responsibility of a joint committee of members from the United States, Great Britain, and Canada (or ABC conference, as these meetings came to be known) in an effort to standardize production for wartime. In 1947, in recognition of the efforts of the ABC committee, the American Standards Association (the ASA, and later the United States of America Standards Institute, the USAS) issued standard B4.1-1947 "Limits and Fits for Engineering and Manufacturing," and, with periodic updates, this standard has remained the source for selecting standard tolerances for design.

When selecting tolerances for cylindrical fits, many factors should be considered including length of engagement, bearing load, lubrication, operating temperature and humidity (the reference temperature for the standard is 68° F or 20 °C: see the discussion below for length differences induced by temperature change), surface texture, and material. Therefore, the recommendations are intended for guidance and for use where they might serve to improve and simplify products, practices, and facilities.

The following definitions are useful for interpreting limits and fits.

Tolerance. The total permissible variation of a size. It is the difference between the limits of size.

Tolerance Limit. The variation, either positive of negative, by which an actual size is permitted to depart from the design size.

Actual Size. The measured size.

Design Size. The basic size with allowance applied.

Basic Size. The size from which the limits of size are derived by the application of allowances and tolerances.

Allowance. The prescribed distance between the maximum material limits of mating parts.

Maximum Material Limit. The limit of size that provides the maximum amount of material for a part. Normally the maximum limit of size for external dimensions, and the minimum limit of size for internal dimensions.

Minimum Material Limit. The limit of size that provides the minimum amount of material for a part. Normally the minimum limit of size for external dimensions, and the maximum limit of size for internal dimensions.

Fit. The range of tightness or looseness resulting from the allowances and tolerances in mating parts.

Clearance Fit. When a clearance always exists between mating assembled mating parts.

Transition Fit. When either a clearance or an interference may result when mating parts are assembled.

Interference Fit. When an interference always exists between assembled mating parts.

Unilateral Tolerance. Where the tolerance variation is permitted only on direction from the design size.

Bilateral Tolerance. Where the tolerance variation is permitted on both directions from the design size (usually expressed as a plus or minus [±]).

Basic Hole System. When the design size of the hole is the basic size, and any allowance is applied to the shaft.

Basic Shaft System. When the design size of the shaft is the basic size, and any allowance is applied to the hole.

Preferred basic sizes, standard tolerances, and standard fits

For aid in designing parts, preferred sizes are specified. When using inch units, preferred sizes are expressed in either fractional or decimal sizes, even though clearance limits in the inch standard are given for specific ranges of sizes. Therefore, the design size must refer to the appropriate decimal range of sizes on the limits chart for the desired fit. Limits charts for metric units are expressed in exact sizes, which are the preferred basic sizes for designing mating parts for assembly. Preferred basic sizes for inch units are given in **Table 1**.

Standard tolerances, arranged by grade, have been specified so that comparable machining operations can be referenced in order to select a suitable process for a particular grade. In **Table 2**, the grades 4 through 13 indicate machining processes that can normally

Table 1. Preferred Basic Sizes for Mating Parts.

Fractional Inch Sizes			
1/64, 1/32, 1/16, 3/32, 1/8, 5/32, 3/16, 1/4, 5/16, 3/8, 7/16, 1/2, 9/16, 5/8, 11/16, 3/4, 7/8		6, 6 1/2	7, 7 1/2
		8, 8 1/2	9, 9 1/2
1, 1 1/4, 1 1/2, 1 3/4,		10, 10 1/2	11, 11 1/2
2, 2 1/4, 2 1/2, 2 3/4,		12, 12 1/2	13, 13 1/2
3, 3 1/4, 3 1/2, 3 3/4,		14, 14 1/2	15, 15 1/2
4, 4 1/4, 1 4/2, 4 3/4,		16, 16 1/2	17, 17 1/2
5, 1 5/4, 5 1/2, 5 3/4		18, 18 1/2	19, 19 1/2	20

Decimal Inch Sizes				
0.010	0.30	3.00	6.00	13.50
0.012	0.40	3.20	6.50	14.00
0.016	0.50	3.40	7.00	14.50
0.020	0.06	3.60	7.50	15.00
0.025	0.08	3.80	8.00	15.50
0.032	1.00	4.00	8.50	16.00
0.040	1.20	4.20	9.00	16.50
0.05	1.40	4.40	9.50	17.00
0.06	1.60	4.60	10.00	17.50
0.08	1.80	4.80	10.50	18.00
0.10	2.0	5.00	11.00	18.50
0.12	2.20	5.20	11.50	19.00
0.16	2.40	5.40	12.00	19.50
0.20	2.60	5.60	12.50	20.00
0.24	2.80	5.80	13.00	–

be expected to produce work within the tolerances indicated by the grade, as follows: *Grade 4*, lapping and honing; *Grade 5*, lapping and honing, cylindrical grinding, surface grinding, diamond turning, diamond boring, broaching; *Grade 6*, same processes as Grade 5, plus reaming, *Grade 7*, same as Grade 6, plus turning; *Grade 8*, surface grinding, broaching, reaming, turning, boring; *Grade 9*, reaming, turning, boring; *Grade 10*, reaming, turning, boring, milling, planing, shaping, drilling; *Grades 11–13*, turning, boring, milling, planing, shaping, drilling.

Five classes of standard fits have been designated for shaft and hole clearances.

Table 2. Standard Tolerances. *(Source USAS B4.1.)*

Nominal Size Over to	Grade 4	Grade 5	Grade 6	Grade 7	Grade 8	Grade 9	Grade 10	Grade 11	Grade 12	Grade 13
0 - 0.12	0.12	0.15	0.25	0.4	0.6	1.0	1.6	2.5	4	6
0.12 - 0.24	0.15	0.20	0.3	0.5	0.7	1.2	1.8	3.0	5	7
0.24 - 0.40	0.15	0.25	0.4	0.6	0.9	1.4	2.2	3.5	6	9
0.40 - 0.71	0.2	0.3	0.4	0.7	1.0	1.6	2.8	4.0	7	10
0.71 - 1.19	0.25	0.4	0.5	0.8	1.2	2.0	3.5	5.0	8	12
1.19 - 1.97	0.3	0.4	0.6	1.0	1.6	2.5	4.0	6	10	16
1.97 - 3.15	0.3	0.5	0.7	1.2	1.8	3.0	4.5	7	12	18
3.15 - 4.73	0.4	0.6	0.9	1.4	2.2	3.5	5	9	14	22
4.73 - 7.09	0.5	0.7	1.0	1.6	2.5	4.0	6	10	16	25
7.09 - 9.85	0.6	0.8	1.2	1.8	2.8	4.5	7	12	18	28
9.85 - 12.41	0.6	0.9	1.2	2.0	3.0	5.0	8	12	20	30
12.41 - 15.75	0.7	1.0	1.4	2.2	3.5	6	9	14	22	35
15.75 - 19.69	0.8	1.0	1.6	2.5	4	6	10	16	25	40
19.69 - 30.09	0.9	1.2	2.0	3	5	8	12	20	30	50
30.09 - 41.49	1.0	1.6	2.5	4	6	10	16	25	40	60
41.49 - 56.19	1.2	2.0	3	5	8	12	20	30	50	80
56.19 - 76.39	1.6	2.5	4	6	10	16	25	40	60	100
76.39 - 100.9	2.0	3	5	8	12	20	30	50	80	125
100.9 - 131.9	2.5	4	6	10	16	25	40	60	100	160
131.9 - 171.9	3	5	8	12	20	30	50	80	125	200
171.9 - 200	4	6	10	16	25	40	60	100	160	250

Nominal sizes in inches. Tolerance values are in thousandths of an inch.

Running and Sliding Fits

This designation is intended to provide a similar running performance, with suitable lubrication allowance, throughout the range of sizes. The first two sizes, RC1 and RC2, which are designed to be used primarily for slide fits, increase more slowly with diameter than the other classes. Clearance limits for Running and Sliding fits are given in **Table 3**.

RC1 – Close Sliding Fit. Intended for use on parts that are to be assembled without noticeable play.

RC2 – Sliding Fit. Intended to accurately locate the parts, but with greater maximum clearance than RC1. RC2 parts should move and turn easily, but not freely. Larger sizes may be prone to seizure with small changes in temperature.

RC3 – Precision Running Fit. Intended for precision work at slow speeds and light journal pressures, this is the closest fit that can be expected to run freely. Not suitable for conditions where wide variations in temperature may be encountered.

RC4 – Close Running Fit. Intended primarily for running fits on machinery with moderate surface speeds and journal pressures. Provides locational accuracy with minimal play.

RC5 and RC6 – Medium Running Fit. For use where higher running speeds and/or heavy journal pressure is encountered.

RC7 – Free Running Fit. For use when accuracy is not a primary consideration, and/or large temperature variations are likely.

RC 8 and RC9 – Loose Running Fit. For use when wide commercial tolerances and an allowance on the external member are necessary.

Locational Fits

This designation is intended to determine only the *location* of the mating parts. Part mating may be rigid or accurate (as with interference fits), or provide freedom of location (as with clearance fits).

LC – Locational Clearance Fit. Intended for parts that, although normally stationary, will be disassembled on occasion. **Table 4** provides clearance for class LC.

LT – Locational Transition Fit. This is a compromise between clearance and interference fits. **Table 5** provides clearance limits for class LT.

LN – Locational Interference Fit. Intended for use when accuracy and precision are overriding considerations, and bore pressure is not specified. **Table 5** provides clearance limits for class LN.

Force or Shrink Fits

This designation is used where constant bore pressure must be maintained. The applied pressure is constant throughout the range. Clearance limits for Force fits are given in **Table 6**.

FN1 – Light Drive Fit. Intended for parts that will likely be permanent assemblies. Requires light assembly pressure, and is suitable for thin sections, long fits, or in cast iron external members.

FN2 – Medium Drive Fit. Useful for ordinary steel parts, or for shrink fits on light sections. The tightest class of fit that can be used on cast iron external members.

FN3 – Heavy Drive Fit. Intended for heavier steel parts or for shrink fits in medium sections.

FN4 and FN5 – Force Fit. Intended for parts that can be highly stressed, or for shrink fits where the heavy pressing forces required are impractical.

In **Tables 3** through **6**, a hole and shaft designation is given for each class of fit. For

(Text continued on p. 321)

Table 3. Clearance Limits for Running Sliding (RC) Fits for Cylindrical Parts. (Sizes 0 through 1.19 inches.) (Limits are in thousandths of an inch.)

Class	Hole/Shaft Desig.	0 to 0.12 Limits Low	High	Clear. Limits	0.121 to 0.24 Limits Low	High	Clear. Limits	0.241 to 0.40 Limits Low	High	Clear. Limits	0.401 to 0.71 Limits Low	High	Clear. Limits	0.711 to 1.19 Limits Low	High	Clear. Limits
RC 1	Hole H5	0	+0.2	0.1 to 0.45	0	+0.2	0.15 to 0.5	0	+0.25	0.2 to 0.6	0	+0.3	0.25 to 0.75	0	+0.4	0.3 to 0.95
	Shaft g4	-0.1	-0.25		-0.15	-0.3		-0.2	-0.35		-0.25	-0.45		-0.3	-0.55	
RC 2	Hole H6	0	+0.25	0.1 to 0.55	0	+0.3	0.15 to 0.65	0	+0.4	0.2 to 0.85	0	+0.4	0.25 to 0.95	0	+0.5	0.3 to 1.2
	Shaft g5	-0.1	-0.3		-0.15	-0.35		-0.2	-0.45		-0.25	-0.55		-0.3	-0.7	
RC 3	Hole H7	0	+0.4	0.3 to 0.95	0	+0.5	0.4 to 1.12	0	+0.6	0.5 to 1.5	0	+0.7	0.6 to 1.7	0	+0.8	0.8 to 2.1
	Shaft f6	-0.3	-0.55		-0.4	-0.7		-0.5	-0.9		-0.6	-1.0		-0.8	-1.3	
RC 4	Hole H8	0	+0.6	0.3 to 1.3	0	+0.7	0.4 to 1.6	0	+0.9	0.5 to 2.0	0	+1.0	0.6 to 2.3	0	+1.2	0.8 to 2.8
	Shaft f7	-0.3	-0.7		-0.4	-0.9		-0.5	-1.1		-0.6	-1.3		-0.8	-1.6	
RC 5	Hole H8	-0	+0.6	0.6 to 1.6	-0	+0.7	0.8 to 2.0	-0	+0.9	1.0 to 2.5	-0	+1.0	1.2 to 2.9	-0	+1.2	1.6 to 3.6
	Shaft e7	-0.6	-1.0		-0.8	-1.3		-1.0	-1.6		-1.2	-1.9		-1.6	-2.4	
RC 6	Hole H9	-0	+1.0	0.6 to 2.2	-0	+1.2	0.8 to 2.7	0	+1.4	1.0 to 3.3	-0	+1.6	1.2 to 3.8	-0	+2.0	1.6 to 4.8
	Shaft e8	-0.6	-1.2		-0.8	-1.5		-1.0	-1.9		-1.2	-2.2		-1.6	-2.8	
RC 7	Hole H9	0	+1.0	1.0 to 2.6	0	+1.2	1.2 to 3.1	0	+1.4	1.6 to 3.9	0	+1.6	2.0 to 4.6	0	+2.0	2.5 to 5.7
	Shaft d8	-1.0	-1.6		-1.2	-1.9		-1.6	-2.5		-2.0	-3.0		-2.5	-3.7	
RC 8	Hole H10	0	+1.6	2.5 to 5.1	0	+1.8	2.8 to 5.8	0	+2.2	3.0 to 6.6	0	+2.8	3.5 to 7.9	0	+3.5	4.5 to 10.0
	Shaft c9	-2.5	-3.5		-2.8	-4.0		-3.0	-4.4		-3.5	-5.1		-4.5	-6.5	
RC 9	Hole H11	0	+2.5	4.0 to 8.1	0	+3.0	4.5 to 9.0	0	+3.5	5.0 to 10.7	0	+4.0	6.0 to 12.8	0	+5.0	7.0 to 15.5
	Shaft	-4.0	-5.6		-4.5	-6.0		-5.0	-7.2		-6.0	-8.8		-7.0	-10.5	

(Continued)

All dimensions in inches. Limit dimensions are in thousandths of an inch. Limits are in agreement with USAS B4.1. (Originally issued as ASA B4.1-1947. Currently issued as ANSI B4.1.)

Table 3. *(Continued)* Clearance Limits for Running Sliding (RC) Fits for Cylindrical Parts. (Sizes 1.191 through 9.85 inches.) (Limits are in thousandths of an inch.)

Class	Hole/Shaft Desig.	1.191 to 1.97 Limits Low	High	Clear. Limits	1.971 to 3.15 Limits Low	High	Clear. Limits	3.151 to 4.73 Limits Low	High	Clear. Limits	4.731 to 7.09 Limits Low	High	Clear. Limits	7.091 to 9.85 Limits Low	High	Clear. Limits
RC 1	Hole H5	0	+0.4	0.4 to 1.1	0	+0.5	0.4 to 1.2	0	+0.6	0.5 to 1.5	0	+0.7	0.6 to 1.8	0	+0.8	0.6 to 2.0
	Shaft g4	-0.4	-0.7		-0.4	-0.7		-0.5	-0.9		-0.6	-1.1		-0.6	-1.2	
RC 2	Hole H6	0	+0.6	0.4 to 1.4	0	+0.7	0.4 to 1.6	0	+0.9	0.5 to 2.0	-	+1.0	0.6 to 2.3	0	+1.2	0.6 to 2.6
	Shaft g5	-0.4	-0.8		-0.4	-0.9		-0.5	-1.1		-0.6	-1.3		-0.6	-1.4	
RC 3	Hole H7	0	+1.0	1.0 to 2.6	0	+1.2	1.2 to 3.1	0	+1.4	1.4 to 3.7	0	+1.6	1.6 to 4.2	0	+1.8	2.0 to 5.0
	Shaft f6	-1.0	-1.6		-1.2	-1.9		-1.4	-2.3		-1.6	-2.6		-2.0	-3.2	
RC 4	Hole H8	0	+1.6	1.0 to 3.6	0	+1.8	1.2 to 4.2	0	+2.2	1.4 to 5.0	0	+2.5	1.6 to 5.7	0	+2.8	2.0 to 6.6
	Shaft f7	-1.0	-2.0		-1.2	-2.4		-1.4	-2.8		-1.6	-3.2		-2.0	-3.8	
RC 5	Hole H8	-0	+1.6	2.0 to 4.6	-0	+1.8	2.5 to 5.5	-0	+2.2	3.0 to 6.6	-0	+2.5	3.5 to 7.6	-0	+2.8	4.0 to 8.6
	Shaft e7	-2.0	-3.0		-2.5	-3.7		-3.0	-4.4		-3.5	-5.1		-4.0	-5.8	
RC 6	Hole H9	-0	+2.5	2.0 to 6.1	-0	+3.0	2.5 to 7.3	-0	+3.5	3.0 to 8.7	-0	+4.0	3.5 to 10.0	0	+4.5	4.0 to 11.3
	Shaft e8	-2.0	-3.6		-2.5	-4.3		-3.0	-5.2		-3.5	-6.0		-4.0	-6.8	
RC 7	Hole H9	0	+2.5	3.0 to 7.1	0	+3.0	4.0 to 8.8	0	+3.5	5.0 to 10.7	-	+4.0	6.0 to 12.5	0	+4.5	7.0 to 14.3
	Shaft d8	-3.0	-4.6		-4.0	-5.8		-5.0	-7.2		-6.0	-8.5		-7.0	-9.8	
RC 8	Hole H10	0	+4.0	5.0 to 11.5	0	+4.5	6.0 to 13.5	0	+5.0	7.0 to 15.5	0	+6.0	8.0 to 18.0	0	+7.0	10.0 to 21.5
	Shaft c9	-5.0	-7.5		-6.0	-9.0		-7.0	-10.5		-8.0	-12.0		-10.0	-14.5	
RC 9	Hole H11	0	+6.0	8.0 to 18.0	0	+7.0	9.0 to 20.5	0	+9.0	10.0 to 24.0	0	+10.0	12.0 to 28.0	0	+12.0	15.0 to 34.0
	Shaft	-8.0	-12.0		-9.0	-13.5		-10.0	-15.0		-12.0	-18.0		-15.0	-22.0	

(Continued)

All dimensions in inches. Limit dimensions are in thousandths of an inch. Limits are in agreement with USAS B4.1. (Originally issued as ASA B4.1-1947. Currently issued as ANSI B4.1.)

Table 3. *(Continued)* **Clearance Limits for Running Sliding (RC) Fits for Cylindrical Parts.** (Sizes 9.851 through 41.49 inches.) (Limits are in thousandths of an inch.)

Hole or Shaft Size

Class	Hole/Shaft Desig.	9.851 to 12.41			12.411 to 15.75			15.751 to 19.69			19.691 to 30.09			30.091 to 41.49		
		Low	High	Clear. Limits	Low	High	Clear. Limits	Low	High	Clear. Limits	Low	High	Clear. Limits	Low	High	Clear. Limits
RC 1	Hole H5	0	+0.9	0.8 to 2.3	0	+1.0	1.0 to 2.7	0	+1.0	1.2 to 3.0	0	+1.2	1.6 to 3.7	0	+1.6	2.0 to 4.6
	Shaft g4	-0.8	-1.4		-1.0	-1.7		-1.2	-2.0		-1.6	-2.5		-2.0	-3.0	
RC 2	Hole H6	0	+1.2	0.8 to 2.9	0	+1.4	1.0 to 3.4	0	+1.6	1.2 to 3.8	0	+2.0	1.6 to 4.8	0	+2.5	2.0 to 6.1
	Shaft g5	-0.8	-1.7		-1.0	-2.0		-1.2	-2.2		-1.6	-2.8		-2.0	-3.6	
RC 3	Hole H7	0	+2.0	2.5 to 5.7	0	+2.2	3.0 to 6.6	0	+2.5	4.0 to 8.1	0	+3.0	5.0 to 10.0	0	+4.0	6.0 to 12.5
	Shaft f6	-2.5	-3.7		-3.0	-4.4		-4.0	-5.6		-5.0	-7.0		-6.0	-8.5	
RC 4	Hole H8	0	+3.0	2.5 to 7.5	0	+3.5	3.0 to 8.7	0	+4.0	4.0 to 10.5	0	+5.0	5.0 to 13.0	0	+6.0	6.0 to 16.0
	Shaft f7	-2.5	-4.5		-3.0	-5.2		-4.0	-6.5		-5.0	-8.0		-6.0	-10.0	
RC 5	Hole H8	0	+3.0	5.0 to 10.0	0	+3.5	6.0 to 11.7	0	+4.0	8.0 to 14.5	0	+5.0	10.0 to 18.0	0	+6.0	12.0 to 22.0
	Shaft e7	-5.0	-7.0		-6.0	-8.2		-8.0	-10.5		-10.0	-13.0		-12.0	-16.0	
RC 6	Hole H9	0	+5.0	5.0 to 13.0	0	+6.0	6.0 to 15.5	0	+6.0	8.0 to 18.0	0	+8.0	10.0 to 23.0	0	+10.0	12.0 to 28.0
	Shaft e8	-5.0	-8.0		-6.0	-9.5		-8.0	-12.0		-10.0	-15.0		-12.0	-18.0	
RC 7	Hole H9	0	+5.0	8.0 to 16.0	0	+6.0	10.0 to 19.5	0	+6.0	12.0 to 22.0	0	+8.0	16.0 to 29.0	0	+10.0	20.0 to 36.0
	Shaft d8	-8.0	-11.0		-10.0	-13.5		-12.0	-16.0		-16.0	-21.0		-20.0	-26.0	
RC 8	Hole H10	0	+8.0	12.0 to 25.0	0	+9.0	14.0 to 29.0	0	+10.0	16.0 to 32.0	0	+12.0	20.0 to 40.0	0	+16.0	25.0 to 51.0
	Shaft c9	-12.0	-17.0		-14.0	-20.0		-16.0	-22.0		-20.0	-28.0		-25.0	-35.0	
RC 9	Hole H11	0	+12.0	18.0 to 38.0	0	+14.0	22.0 to 45.0	0	+16.0	25.0 to 51.0	0	+20.0	30.0 to 62.0	0	+25.0	40.0 to 81.0
	Shaft	-18.0	-26.0		-22.0	-31.0		-25.0	-35.0		-30.0	-42.0		-40.0	-56.0	

All dimensions in inches. Limit dimensions are in thousandths of an inch. Limits are in agreement with USAS B4.1. (Originally issued as ASA B4.1-1947. Currently issued as ANSI B4.1.)

(Continued)

Table 3. *(Continued)* **Clearance Limits for Running Sliding (RC) Fits for Cylindrical Parts.** (Sizes 41.491 through 171.9 inches.) (Limits are in thousandths of an inch.)

Class	Hole/Shaft Desig.	41.491 to 56.19			56.191 to 76.39			76.391 to 100.9			100.91 to 131.9			131.91 to 171.9		
		Low	High	Clear. Limits	Low	High	Clear. Limits	Low	High	Clear. Limits	Low	High	Clear. Limits	Low	High	Clear. Limits
RC 1	Hole H5	0	+2.0	2.5 to 5.7	0	+2.5	3.0 to 7.1	0	+3.0	4.0 to 9.0	0	+4.0	5.0 to 11.5	0	+5.0	6.0 to 14.0
	Shaft g4	-2.5	-3.7		-3.0	-4.6		-4.0	-6.0		-5.0	-7.5		-6.0	-9.0	
RC 2	Hole H6	0	+3.0	2.5 to 7.5	0	+4.0	3.0 to 9.5	0	+5.0	4.0 to 12.0	0	+6.0	5.0 to 15.0	0	+8.0	6.0 to 19.0
	Shaft g5	-2.5	-4.5		-3.0	-5.5		-4.0	-7.0		-5.0	-9.0		-6.0	-11.0	
RC 3	Hole H7	0	+5.0	8.0 to 16.0	0	+6.0	10.0 to 20.0	0	+8.0	12.0 to 25.0	0	+10.0	16.0 to 32.0	0	+12.0	18.0 to 38.0
	Shaft f6	-8.0	-11.0		-10.0	-14.0		-12.0	-17.0		-16.0	-22.0		-18.0	-26.0	
RC 4	Hole H8	0	+8.0	8.0 to 21.0	0	+10.0	10.0 to 26.0	0	+12.0	12.0 to 32.0	0	+16.0	16.0 to 42.0	0	+20.0	18.0 to 50.0
	Shaft f7	-8.0	-13.0		-10.0	-16.0		-12.0	-20.0		-16.0	-26.0		-18.0	-30.0	
RC 5	Hole H8	0	+8.0	16.0 to 29.0	0	+10.0	20.0 to 36.0	0	+12.0	25.0 to 45.0	0	+16.0	30.0 to 56.0	0	+20.0	35.0 to 67.0
	Shaft e7	-16.0	-21.0		-20.0	-26.0		-25.0	-33.0		-30.0	-40.0		-35.0	-47.0	
RC 6	Hole H9	0	+12.0	16.0 to 36.0	0	+16.0	20.0 to 46.0	0	+20.0	25.0 to 57.0	0	+25.0	30.0 to 71.0	0	+30.0	35.0 to 85.0
	Shaft e8	-16.0	-24.0		-20.0	-30.0		-25.0	-37.0		-30.0	-46.0		-35.0	-55.0	
RC 7	Hole H9	0	+12.0	25.0 to 45.0	0	+16.0	30.0 to 56.0	0	+20.0	40.0 to 72.0	0	+25.0	50.0 to 91.0	0	+30.0	60.0 to 110.0
	Shaft d8	-25.0	-33.0		-30.0	-40.0		-40.0	-52.0		-50.0	-66.0		-60.0	-80.0	
RC 8	Hole H10	0	+20.0	30.0 to 62.0	0	+25.0	40.0 to 81.0	0	+30.0	50.0 to 100	0	+40.0	60.0 to 125	0	+50.0	80.0 to 160
	Shaft c9	-30.0	-42.0		-40.0	-56.0		-50.0	-70.0		-60.0	-85.0		-80.0	-110	
RC 9	Hole H11	0	+30.0	50.0 to 100	0	+40.0	60.0 to 125	0	+50.0	80.0 to 160	0	+60.0	100 to 200	0	+80.0	130 to 260
	Shaft	-50.0	-70.0		-60.0	-85.0		-80.0	-110		-100	-140		-130	-180	

All dimensions in inches. Limit dimensions are in thousandths of an inch. Limits are in agreement with USAS B4.1. (Originally issued as ASA B4.1-1947. Currently issued as ANSI B4.1.)

Table 4. Clearance Limits for Locational Clearance (LC) Fits for Cylindrical Parts. (Sizes 0 through 1.19 inches.) (Limits are in thousandths of an inch.)

Hole or Shaft Size

Class	Hole/Shaft Desig.	0 to 0.12			0.121 to 0.24			0.241 to 0.40			0.401 to 0.71			0.711 to 1.19		
		Low	High	Clear. Limits	Low	High	Clear. Limits	Low	High	Clear. Limits	Low	High	Clear. Limits	Low	High	Clear. Limits
LC 1	Hole H6	-0	+0.25	0 to 0.45	-0	+0.3	0 to 0.5	-0	+0.4	0 to 0.65	-0	+0.4	0 to 0.7	-0	+0.5	0 to 0.9
	Shaft h5	+0	-0.2		+0	-0.2		+0	-0.25		+0	-0.3		+0	-0.4	
LC 2	Hole H7	-0	+0.4	0 to 0.65	-0	+0.5	0 to 0.8	-0	+0.6	0 to 1.0	-0	+0.7	0 to 1.1	-0	+0.8	0 to 1.3
	Shaft h6	+0	-0.25		+0	-0.3		+0	-0.4		+0	-0.4		+0	-0.5	
LC 3	Hole H8	-0	+0.6	0 to 1	-0	+0.7	0 to 1.2	-0	+0.9	0 to 1.5	-0	+1.0	0 to 1.7	-0	+1.2	0 to 2
	Shaft h7	+0	-0.4		+0	-0.5		+0	-0.6		+0	-0.7		+0	-0.8	
LC 4	Hole H10	-0	+1.6	0 to 2.6	-0	+1.8	0 to 3.0	-0	+2.2	0 to 3.6	-0	+2.8	0 to 4.4	-0	+3.5	0 to 5.5
	Shaft h9	+0	-1.0		+0	-1.2		+0	-1.4		+0	-1.6		+0	-2.0	
LC 5	Hole H7	-0	+0.4	0.1 to 0.75	-0	+0.5	0.15 to 0.95	-0	+0.6	0.2 to 1.2	-0	+0.7	0.25 to 1.35	-0	+0.8	0.3 to 1.6
	Shaft g6	-0.1	-0.35		-0.15	-0.45		-0.2	-0.6		-0.25	-0.65		-0.3	-0.8	
LC 6	Hole H9	-0	+1.0	0.3 to 1.9	-0	+1.2	0.4 to 2.3	-0	+1.4	0.5 to 2.8	-0	+1.6	0.6 to 3.2	-0	+2.0	0.8 to 4.0
	Shaft f8	-0.3	-0.9		-0.4	-1.1		-0.5	-1.4		-0.6	-1.6		-0.8	-2.0	
LC 7	Hole H10	-0	+1.6	0.6 to 3.2	-0	+1.8	0.8 to 3.8	-0	+2.2	1.0 to 4.6	-0	+2.8	1.2 to 5.6	-0	+3.5	1.6 to 7.1
	Shaft e9	-0.6	-1.6		-0.8	-2.0		-1.0	-2.4		-1.2	-2.8		-1.6	-3.6	
LC 8	Hole H10	-0	+1.6	1.0 to 3.6	-0	+1.8	1.2 to 4.2	-0	+2.2	1.6 to 5.2	-0	+2.8	2.0 to 6.4	-0	+3.5	2.5 to 8.0
	Shaft d9	-1.0	-2.0		-1.2	-2.4		-1.6	-3.0		-2.0	-3.6		-2.5	-4.5	
LC 9	Hole H11	-0	+2.5	2.5 to 6.6	-0	+3.0	2.8 to 7.6	-0	+3.5	3.0 to 8.7	-0	+4.0	3.5 to 10.3	-0	+5.0	4.5 to 13.0
	Shaft c10	-2.5	-4.1		-2.8	-4.6		-3.0	-5.2		-3.5	-6.3		-4.5	-8.0	
LC 10	Hole H12	-0	+4	4 to 12	-0	+5	4.5 to 14.5	-0	+6	5 to 17	-0	+7	6 to 20	-0	+8	7 to 23
	Shaft	-4	-8		-4.5	-9.5		-5	-11		-6	-13		-7	-15	
LC 11	Hole H13	-0	+6	5 to 17	-0	+7	6 to 20	-0	+9	7 to 25	-0	+10	8 to 28	-0	+12	10 to 34
	Shaft	-5	-11		-6	-13		-7	-16		-8	-18		-10	-22	

(Continued)

All dimensions in inches. Limits are in agreement with USAS B4.1. (Originally issued as ASA B4.1-1947. Currently issued as ANSI B4.1.)

Table 4. *(Continued)* Clearance Limits for Locational Clearance (LC) Fits for Cylindrical Parts. (Sizes 1.191 through 9.85 inches.) (Limits are in thousandths of an inch.)

Class	Hole/Shaft Desig.	1.191 to 1.97 Low	High	Clear. Limits	1.971 to 3.15 Low	High	Clear. Limits	3.151 to 4.73 Low	High	Clear. Limits	4.731 to 7.09 Low	High	Clear. Limits	7.091 to 9.85 Low	High	Clear. Limits
LC 1	Hole H6	-0	+0.6	0 to 1.0	-0	+0.7	0 to 1.2	-0	+0.9	0 to 1.5	-0	+1.0	0 to 1.7	-0	+1.2	0 to 2.0
	Shaft h5	+0	-0.4		+0	-0.5		+0	-0.6		+0	-0.7		+0	-0.8	
LC 2	Hole H7	-0	+1.0	0 to 1.6	-0	+1.2	0 to 1.9	-0	+1.4	0 to 2.3	-0	+1.6	0 to 2.6	-0	+1.8	0 to 3.0
	Shaft h6	+0	-0.6		+0	-0.7		+0	-0.9		+0	-1.0		+0	-1.2	
LC 3	Hole H8	-0	+1.6	0 to 2.6	-0	+1.8	0 to 3	-0	+2.2	0 to 3.6	-0	+2.5	0 to 4.1	-0	+2.8	0 to 4.6
	Shaft h7	+0	-1		+0	-1.2		+0	-1.4		+0	-1.6		+0	-1.8	
LC 4	Hole H10	-0	+4.0	0 to 6.5	-0	+4.5	0 to 7.5	-0	+5.0	0 to 8.5	-0	+6.0	0 to 10	-0	+7.0	0 to 11.5
	Shaft h9	+0	-2.5		+0	-3		+0	-3.5		+0	-4		+0	-4.5	
LC 5	Hole H7	-0	+1.0	0.4 to 2.0	-0	+1.2	0.4 to 2.3	-0	+1.4	0.5 to 2.8	-0	+1.6	0.6 to 3.2	-0	+1.8	0.6 to 3.6
	Shaft g6	-0.4	-1.0		-0.4	-1.1		-0.5	-1.4		-0.6	-1.6		-0.6	-1.8	
LC 6	Hole H9	0	+2.5	1.0 to 5.1	0	+3.0	1.2 to 6.0	0	+3.5	1.4 to 7.1	0	+4.0	1.6 to 8.1	0	+4.5	2.0 to 9.3
	Shaft f8	-1.0	-2.6		-1.2	-3.0		-1.4	-3.6		-1.6	-4.1		-2.0	-4.8	
LC 7	Hole H10	0	+4.0	2.0 to 8.5	0	+4.5	2.5 to 10.0	0	+5.0	3.0 to 11.5	0	+6.0	3.5 to 13.5	0	+7.0	4.0 to 15.5
	Shaft e9	-2.0	-4.5		-2.5	-5.5		-3.0	-6.5		-3.5	-7.5		-4.0	-8.5	
LC 8	Hole H10	-0	+4.0	3.0 to 9.5	-0	+4.5	4.0 to 11.5	-0	+5.0	5.0 to 13.5	-0	+6	6 to 16	-0	+7	7 to 18.5
	Shaft d9	-3.0	-5.5		-4.0	-7.0		-5.0	-8.5		-6	-10		-7	-11.5	
LC 9	Hole H11	-0	+6	5 to 15	-0	+7	6 to 17.5	-0	+9	7 to 21	-0	+10	8 to 24	-0	+12	10 to 29
	Shaft c10	-5	-9		-6	-10.5		-7	-12		-8	-14		-10	-17	
LC 10	Hole H12	-0	+10	8 to 28	-0	+12	10 to 34	-0	+14	11 to 39	-0	+16	12 to 44	-0	+18	16 to 52
	Shaft	-8	-18		-10	-22		-11	-25		-12	-28		-16	-34	
LC 11	Hole H13	-0	+16	12 to 44	-0	+18	14 to 50	-0	+22	16 to 60	-0	+25	18 to 68	-0	+28	22 to 78
	Shaft	-12	-28		-14	-32		-16	-38		-18	-43		-22	-50	

All dimensions in inches. Limits are in agreement with USAS B4.1. (Originally issued as ASA B4.1-1947. Currently issued as ANSI B4.1.)

(Continued)

Table 4. *(Continued)* Clearance Limits for Locational Clearance (LC) Fits for Cylindrical Parts. (Sizes 9.851 through 41.49 inches.) (Limits are in thousandths of an inch.)

Class	Hole/Shaft Desig.	9.851 to 12.41 Limits Low	High	Clear. Limits	12.411 to 15.75 Limits Low	High	Clear. Limits	15.751 to 19.69 Limits Low	High	Clear. Limits	19.691 to 30.09 Limits Low	High	Clear. Limits	30.091 to 41.49 Limits Low	High	Clear. Limits
LC 1	Hole H6	-0	+1.2	0 to 2.1	-0	+1.4	0 to 2.4	-0	+1.6	0 to 2.6	-0	+2.0	0 to 3.2	-0	+2.5	0 to 4.1
	Shaft h5	+0	-0.9		+0	-1.0		+0	-1.0		+0	-1.2		+0	-1.6	
LC 2	Hole H7	-0	+2.0	0 to 3.2	-0	+2.2	0 to 3.6	-0	+2.5	0 to 4.1	-0	+3	0 to 5.0	-0	+4	0 to 6.5
	Shaft h6	+0	-1.2		+0	-1.4		+0	-1.6		+0	-2		+0	-2.5	
LC 3	Hole H8	-0	+3.0	0 to 5	-0	+3.5	0 to 5.7	-0	+4	0 to 6.5	-0	+5	0 to 8	-0	+6	0 to 10
	Shaft h7	+0	-2.0		+0	-2.2		+0	-2.5		+0	-3		+0	-4	
LC 4	Hole H10	-0	+8.0	0 to 13	-0	+9.0	0 to 15	-0	+10.0	0 to 16	-0	+12.0	0 to 20	-0	+16.0	0 to 26
	Shaft h9	+0	-5		+0	-6		+0	-6		+0	-8		+0	-10	
LC 5	Hole H7	-0	+2.0	0.7 to 3.9	-0	+2.2	0.7 to 4.3	-0	+2.5	0.8 to 4.9	-0	+3.0	0.9 to 5.9	-0	+4.0	1.0 to 7.5
	Shaft g6	-0.7	-1.9		-0.7	-2.1		-0.8	-2.4		-0.9	-2.9		-1.0	-3.5	
LC 6	Hole H9	0	+5.0	2.2 to 10.2	0	+6.0	2.5 to 12.0	0	+6.0	2.8 to 12.8	0	+8.0	3.0 to 16.0	0	+10.0	3.5 to 19.5
	Shaft f8	-2.2	-5.2		-2.5	-6.0		-2.8	-6.8		-3.0	-8.0		-3.5	-9.5	
LC 7	Hole H10	0	+8.0	4.5 to 17.5	0	+9.0	5.0 to 20.0	0	+10.0	5.0 to 21.0	0	+12.0	6.0 to 26.0	0	+16	7.0 to 33.0
	Shaft e9	-4.5	-9.5		-5	-11		-5	-11		-6	-14		-7	-17	
LC 8	Hole H10	-0	+8	7 to 20	-0	+9	8 to 23	-0	+10	9 to 25	-0	+12	10 to 30	-0	+16	12 to 38
	Shaft d9	-7	-12		-8	-14		-9	-15		-10	-18		-12	-22	
LC 9	Hole H11	-0	+12	12 to 32	-0	+14	14 to 37	-0	+16	16 to 42	-0	+20	18 to 50	-0	+25	20 to 61
	Shaft c10	-12	-20		-14	-23		-16	-26		-18	-30		-20	-36	
LC 10	Hole H12	-0	+20	20 to 60	-0	+22	22 to 66	-0	+25	25 to 75	-0	+30	28 to 88	-0	+40	30 to 110
	Shaft	-20	-40		-22	-44		-25	-50		-28	-58		-30	-70	
LC 11	Hole H13	-0	+30	28 to 88	-0	+35	30 to 100	-0	+40	35 to 115	-0	+50	40 to 140	-0	+60	45 to 165
	Shaft	-28	-58		-30	-65		-35	-75		-40	-90		-45	-105	

(Continued)

All dimensions in inches. Limits are in agreement with USAS B4.1. (Originally issued as ASA B4.1-1947. Currently issued as ANSI B4.1.)

Table 4. (Continued) Clearance Limits for Locational Clearance (LC) Fits for Cylindrical Parts. (Sizes 41.491 through 171.9 inches.) (Limits are in thousandths of an inch.)

Class	Hole/Shaft Desig.	41.491 to 56.19			56.191 to 76.39			76.391 to 100.9			100.91 to 131.9			131.91 to 171.9		
		Low	High	Clear. Limits	Low	High	Clear. Limits	Low	High	Clear. Limits	Low	High	Clear. Limits	Low	High	Clear. Limits
LC 1	Hole H6	-0	+3.0	0 to 5.0	-0	+4.0	0 to 6.5	-0	+5.0	0 to 8.0	-0	+6.0	0 to 10.0	-0	+8.0	0 to 13.0
	Shaft h5	+0	-2.0		+0	-2.5		+0	-3.0		+0	-4.0		+0	-5.0	
LC 2	Hole H7	-0	+5	0 to 8.0	-0	+6	0 to 10	-0	+8	0 to 13	-0	+10	0 to 16	-0	+12	0 to 20
	Shaft h6	+0	-3		+0	-4		+0	-5		+0	-6		+0	-8	
LC 3	Hole H8	-0	+8	0 to 13	-0	+10	0 to 16	-0	+12	0 to 20	-0	+16	0 to 26	-0	+20	0 to 32
	Shaft h7	+0	-5		+0	-6		+0	-8		+0	-10		+0	-12	
LC 4	Hole H10	-0	+20	0 to 32	-0	+25.0	0 to 41	-0	+30.0	0 to 50	-0	+40.0	0 to 65	-0	+50.0	0 to 80
	Shaft h9	+0	-12		+0	-16		+0	-20		+0	-25		+0	-30	
LC 5	Hole H7	-0	+5.0	1.2 to 9.2	-0	+6.0	1.2 to 11.2	-0	+8.0	1.4 to 14.4	-0	+10.0	1.6 to 17.6	-0	+12.0	1.8 to 21.8
	Shaft g6	-1.2	-4.2		-1.2	-5.2		-1.4	-6.4		-1.6	-7.6		-1.8	-9.8	
LC 6	Hole H9	0	+12.0	4.0 to 24.0	0	+16.0	4.5 to 30.5	0	+20.0	5.0 to 37.0	0	+25.0	6.0 to 47.0	0	+30.0	7.0 to 57.0
	Shaft f8	-4.0	-12.0		-4.5	-14.5		-5	-17		-6	-22		-7	-27	
LC 7	Hole H10	-0	+20.0	8.0 to 40.0	-0	+25.0	9.0 to 50.0	-0	+30.0	10.0 to 60.0	-0	+40.0	12.0 to 67.0	-0	+50.0	14.0 to 94.0
	Shaft e9	-8	-20		-9	-25		-10	-30		-12	-27		-14	-44	
LC 8	Hole H10	-0	+20	14.0 to 46.0	-0	+25	16.0 to 57.0	-0	+30	18.0 to 68.0	-0	+40	20.0 to 85.0	-0	+50	25.0 to 105.0
	Shaft d9	-14	-26		-16	-32		-18	-38		-20	-45		-25	-55	
LC 9	Hole H11	-0	+30	25.0 to 75.0	-0	+40	30.0 to 95.0	-0	+50	35.0 to 115.0	-0	+60	40.0 to 140.0	-0	+80	50.0 to 180.0
	Shaft c10	-25	-45		-30	-55		-35	-65		-40	-80		-50	-100	
LC 10	Hole H12	-0	+50	40.0 to 140.0	-0	+60	50.0 to 170.0	-0	+80	50.0 to 210.0	-0	+100	60.0 to 260.0	-0	+125	80.0 to 330.0
	Shaft	-40	-90		-50	-110		-50	-130		-60	-160		-80	-205	
LC 11	Hole H13	-0	+80	60.0 to 220.0	-0	+100	70.0 to 270.0	-0	+125	80.0 to 330.0	-0	+160	90.0 to 410.0	-0	+200	100 to 500
	Shaft	-60	-140		-70	-170		-80	-205		-90	-250		-100	-300	

All dimensions in inches. Limits are in agreement with USAS B4.1. (Originally issued as ASA B4.1 1947. Currently issued as ANSI B4.1.)

Table 5. Interference Limits for Locational Transitional (LT) Fits and Locational Interference (LN) Fits for Cylindrical Parts. (Sizes 0 through 1.19 inches.) (Limits are in thousandths of an inch.)

Hole or Shaft Size

Class	Hole/Shaft Desig.	Low (0 to 0.12)	High	Inter. Limits	Low (0.12 to 0.24)	High	Inter. Limits	Low (0.24 to 0.40)	High	Inter. Limits	Low (0.40 to 0.71)	High	Inter. Limits	Low (0.71 to 1.19)	High	Inter. Limits
LT 1	Hole H7	-0	+0.4		-0	+0.5		-0	+0.6		-0	+0.7		-0	+0.8	
	Shaft js6	-0.10	+0.10	-0.10 to +0.50	-0.15	+0.15	-0.15 to +0.65	-0.2	+0.2	-0.2 to +0.8	-0.2	+0.2	-0.2 to +0.9	-0.25	+0.25	-0.25 to +1.05
LT 2	Hole H8	-0	+0.6		-0	+0.7		-0	+0.9		-0	+1.0		-0	+1.2	
	Shaft js7	-0.2	+0.2	-0.2 to +0.8	-0.25	+0.25	-0.25 to +0.95	-0.3	+0.3	-0.3 to +1.2	-0.35	+0.35	-0.35 to +1.35	-0.4	+0.4	-0.4 to +1.6
LT 3	Hole H7							-0	+0.6		-0	+0.7		-0	+0.8	
	Shaft k6							+0.1	+0.5	-0.5 to +0.5	+0.1	+0.6	-0.6 to +0.6	+0.1	+0.7	-0.6 to +0.7
LT 4	Hole H8							-0	+0.9		-0	+1.0		-0	+1.2	
	Shaft k7							+0.1	+0.7	-0.7 to +0.8	+0.1	+0.8	-0.8 to +0.9	+0.1	+0.9	-0.9 to +1.1
LT 5	Hole H7	-0	+0.4		-0	+0.5		-0	+0.6		-0	+0.7		-0	+0.8	
	Shaft n6	+0.25	+0.5	-0.5 to +0.15	+0.3	+0.6	-0.6 to +0.2	+0.4	+0.8	-0.8 to +0.2	+0.5	+0.9	-0.9 to +0.2	+0.6	+0.8	-1.1 to +0.2
LT 6	Hole H7	-0	+0.4		-0	+0.5		-0	+0.6		-0	+0.7		-0	+0.8	
	Shaft n7	+0.25	+0.65	-0.65 to +0.15	+0.3	+0.8	-0.8 to +0.2	+0.4	+1.0	-1.0 to +0.2	+0.5	+1.1	-1.2 to +0.2	+0.6	+1.4	-1.4 to +0.2
LN 1	Hole H6	0	+0.25		0	+0.3		0	+0.4		0	+0.4		0	+0.5	
	Shaft n5	+0.25	+0.45	0 to 0.45	+0.3	+0.5	0 to 0.5	+0.4	+0.65	0 to 0.65	+0.4	+0.8	0 to 0.8	+0.5	+1	0 to 1.0
LN 2	Hole H7	-0	+0.4		-0	+0.5		-0	+0.6		-0	+0.7		-0	+0.8	
	Shaft p6	+0.4	+0.65	0 to 0.65	+0.5	+0.8	0 to 0.8	+0.6	+1.0	0 to 1.0	+0.7	+1.1	0 to 1.1	+0.8	+1.3	0 to 1.3
LN 3	Hole H7	-0	+0.5		-0	+0.6		-0	+0.8		-0	+1.0		-0	+1.2	
	Shaft r6	+0.4	+0.75	0.1 to 0.75	+0.5	+0.9	0.1 to 0.9	+0.6	+1.2	0.2 to 1.2	+0.7	+1.4	0.3 to 1.4	+0.8	+1.7	0.4 to 1.7

All dimensions in inches. Limit dimensions are in thousandths of an inch. Limits are in agreement with USAS B4.1 (Originally issued as ASA B4.1-1947. Currently issued as ANSI B4.1.)

(Continued)

Table 5. *(Continued)* Interference Limits for Locational Transitional (LT) Fits and Locational Interference (LN) Fits for Cylindrical Parts. *(Sizes 1.191 through 7.09 inches.)* (Limits are in thousandths of an inch.)

Class	Hole/Shaft Desig.	1.19 to 1.97			1.97 to 3.15			3.15 to 4.73			4.73 to 7.09		
		Limits		Inter. Limits	Limits		Inter. Limits	Limits		Inter. Limits	Limits		Inter. Limits
		Low	High		Low	High		Low	High		Low	High	
LT 1	Hole H7	-0	1.00	-0.3 to +1.3	-0	+1.2	-0.3 to +1.5	-0	+1.4	-0.4 to +1.8	-0	+1.6	-0.5 to +2.1
	Shaft js6	-0.3	+0.3		-0.3	+0.3		-0.4	+0.4		-0.5	+0.5	
LT 2	Hole H8	-0	+1.6	-0.5 to +2.1	-0	+1.8	-0.6 to +2.4	-0	+2.2	-0.7 to +2.9	-0	+2.5	-0.8 to +3.3
	Shaft js7	-0.5	+0.5		-0.6	+0.6		-0.7	+0.7		-0.8	+0.8	
LT 3	Hole H7	-0	+1.0	-0.7 to +0.9	-0	+1.2	-0.8 to +1.1	-0	+1.4	-1.0 to +1.3	-0	+1.6	-1.1 to +1.5
	Shaft k6	+0.7	+0.1		+0.1	+0.8		+0.1	+1.0		+0.1	+1.1	
LT 4	Hole H8	-0	+1.6	-1.1 to +1.5	-0	+1.8	-1.3 to +1.7	-0	+2.2	-1.5 to +2.1	-0	+2.5	-1.7 to +2.4
	Shaft k7	+0.1	+1.1		+0.1	+1.3		+0.1	+1.5		+0.1	+1.7	
LT 5	Hole H7	-0	+1.0	-1.3 to +0.3	-0	+1.2	-1.5 to +0.4	-0	+1.4	-1.9 to +0.4	-0	+1.6	-2.2 to +0.4
	Shaft n6	+0.7	+1.3		+0.8	+1.5		+1.0	+1.9		+1.2	+2.2	
LT 6	Hole H7	-0	+1.0	-1.7 to +0.3	-0	+1.2	-2.0 to +0.4	-0	+1.4	-2.4 to +0.4	-0	+1.6	-2.8 to +0.4
	Shaft n7	+0.7	+1.7		+0.8	+2.0		+1.0	+2.4		+1.2	+2.8	
LN 1	Hole H6	-0	+0.6	0 to 1.1	-0	+0.7	0.1 to 1.3	-0	+1.0	0.1 to 1.6	-0	+1.2	0.2 to 1.9
	Shaft n5	+0.6	+1.1		+0.7	+1.3		+0.9	+1.6		+1.0	+1.9	
LN 2	Hole H7	-0	+1.0	0 to 1.6	-0	+1.4	0.2 to 2.1	-0	+1.6	0.2 to 2.5	-0	+1.8	0.2 to 2.8
	Shaft p6	+1.0	+1.6		+1.2	+2.1		+1.4	+2.5		+1.6	+2.8	
LN 3	Hole H7	-0	+1.4	0.4 to 2.0	-0	+1.6	0.4 to 2.3	-0	+2.0	0.6 to 2.9	-0	+2.5	0.9 to 3.5
	Shaft r6	+1.0	+2.0		+1.2	+2.3		+1.4	+2.9		+1.6	+3.5	

All dimensions in inches. Limit dimensions are in thousandths of an inch. Limits are in agreement with USAS B4.1 (Originally issued as ASA B4.1-1947. Currently issued as ANSI B4.1.)

(Continued)

Table 5. *(Continued)* **Interference Limits for Locational Transitional (LT) Fits and Locational Interference (LN) Fits for Cylindrical Parts.** (Sizes 7.09 through 19.69 inches.) (Limits are in thousandths of an inch.)

Class	Hole/Shaft Desig.	7.09 to 9.85 Limits Low	High	Inter. Limits	9.85 to 12.41 Limits Low	High	Inter. Limits	12.41 to 15.75 Limits Low	High	Inter. Limits	15.75 to 19.69 Limits Low	High	Inter. Limits
LT 1	Hole H7	-0	+1.8	-0.6 to +2.4	-0	+2.0	-0.6 to +2.6	-0	+2.2	-0.7 to +2.9	-0	+2.5	-0.8 to +3.3
	Shaft js6	-0.6	+0.6		-0.6	+0.6		-0.7	+0.7		-0.8	+0.8	
LT 2	Hole H8	-0	+2.8	-0.9 to +3.7	-0	+3.0	-1.0 to +4.0	-0	+3.5	-1.0 to +4.5	-0	+4.0	-1.2 to +5.2
	Shaft js7	-0.9	+0.9		-1.0	+1.0		-1.0	+1.0		-1.2	+1.2	
LT 3	Hole H7	-0	+1.8	-1.4 to +1.6	-0	+2.0	-1.4 to +1.8	-0	+2.2	-1.6 to +2.0	-0	+2.5	-1.8 to +2.3
	Shaft k6	+0.2	+1.4		+0.2	+1.4		+0.2	+1.6		+0.2	+1.8	
LT 4	Hole H8	-0	+2.8	-2.0 to +2.6	-0	+3.0	-2.2 to +2.8	-0	+3.5	-2.4 to +3.3	-0	+4.0	-2.7 to +3.8
	Shaft k7	+0.2	+2.0		+0.2	+2.2		+0.2	+2.4		+0.2	+2.7	
LT 5	Hole H7	-0	+1.8	-2.6 to +0.4	-0	+2.0	-2.6 to +0.6	-0	+2.2	-3.0 to +0.6	-0	+2.5	-3.4 to +0.7
	Shaft n6	+1.4	+2.6		+1.4	+2.6		+1.6	+3.0		+1.8	+3.4	
LT 6	Hole H7	-0	+1.8	-3.2 to +0.4	-0	+2.0	-3.4 to +0.6	-0	+2.2	-3.8 to +0.6	-0	+2.5	-4.3 to +0.7
	Shaft n7	+1.4	+3.2		+1.4	+3.4		+1.6	+3.8		+1.8	+4.3	
LN 1	Hole H6	-0	+1.4	0.2 to 2.2	-0	+1.4	0.2 to 2.3	-0	+1.6	0.2 to 2.6	-0	+1.8	0.2 to 2.8
	Shaft n5	+1.2	+2.2		+1.2	+2.3		+1.4	+2.6		+1.6	+2.8	
LN 2	Hole H7	-0	+2.0	0.2 to 3.2	-0	+2.2	0.2 to 3.4	-0	+2.5	0.3 to 3.9	-0	+2.8	0.3 to 4.4
	Shaft p6	+1.8	+3.2		+2.0	+3.4		+2.2	+3.9		+2.5	+4.4	
LN 3	Hole H7	-0	+3.0	2.0 to 4.2	-0	+3.5	2.0 to 4.7	-0	+4.5	2.2 to 5.9	-0	+5.0	2.5 to 6.6
	Shaft r6	+1.8	+4.2		+2.0	+4.7		+2.2	+5.9		+2.5	+6.6	

All dimensions in inches. Limit dimensions are in thousandths of an inch. Limits are in agreement with USAS B4.1 (Originally issued as ASA B4.1-1947. Currently issued as ANSI B4.1.)

Table 6. Interference Limits for Force and Shrink (FN) Fits for Cylindrical Parts. (Sizes 0 through 1.97 inches.) (Limits are in thousandths of an inch.)

Class	Hole/Shaft Desig.	Hole or Shaft Size								
		0 to 0.12			0.121 to 0.24			0.241 to 0.40		
		Limits		Inter. Limits	Limits		Inter. Limits	Limits		Inter. Limits
		Low	High		Low	High		Low	High	
FN 1	Hole H6	0	+0.25	0.05 to 0.5	0	+0.3	0.1 to 0.6	0	+0.4	0.1 to .75
	Shaft	+0.3	+0.5		+0.4	+0.6		+0.5	+0.75	
FN 2	Hole H7	0	+0.4	0.2 to 0.85	0	+0.5	0.2 to 1.0	0	+0.6	0.4 to 1.4
	Shaft s6	+0.6	+0.85		+0.7	+1.0		+1.0	+1.4	
FN 3	Hole H7	-	-		-	-		-	-	
	Shaft t6	-	-		-	-		-	-	
FN 4	Hole H7	0	+0.4	0.3 to 0.95	0	+0.5	0.4 to 1.2	0	+0.6	0.6 to 1.6
	Shaft u6	+0.7	+0.95		+0.9	+1.2		+1.2	+1.6	
FN 5	Hole H8	0	+0.6	0.3 to 1.3	0	+0.7	0.5 to 1.7	0	+0.9	0.5 to 2.0
	Shaft x7	+0.9	+1.3		+1.2	+1.7		+1.4	+2.0	

Class	Hole/Shaft Desig.	Hole or Shaft Size								
		0.401 to 0.56			0.561 to 0.71			0.711 to 0.95		
		Limits		Inter. Limits	Limits		Inter. Limits	Limits		Inter. Limits
		Low	High		Low	High		Low	High	
FN 1	Hole H6	0	-0.4	0.1 to 1.8	0	+0.4	0.2 to 0.9	0	+0.7	0.2 to 1.1
	Shaft	+0.5	+0.8		+0.6	+0.9		+0.5	+1.1	
FN 2	Hole H7	0	+0.7	0.5 to 1.6	0	+0.7	0.5 to 1.6	0	+0.8	0.6 to 1.9
	Shaft s6	+1.2	+1.6		+1.2	+1.6		+1.4	+1.9	
FN 3	Hole H7	-	-		-	-		-	-	
	Shaft t6	-	-		-	-		-	-	
FN 4	Hole H7	0	+0.7	0.7 to 1.8	0	+0.7	0.7 to 1.8	0	+0.8	0.8 to 2.1
	Shaft u6	+1.4	+1.8		+1.4	+1.8		+1.6	+2.1	
FN 5	Hole H8	0	+1.0	0.6 to 2.3	0	+1.0	0.8 to 2.5	0	+1.2	1.0 to 3.0
	Shaft x7	+1.6	+2.3		+1.8	+2.5		+2.2	+3.0	

Class	Hole/Shaft Desig.	Hole or Shaft Size								
		0.951 to 1.19			1.191 to 1.58			1.581 to 1.97		
		Limits		Inter. Limits	Limits		Inter. Limits	Limits		Inter. Limits
		Low	High		Low	High		Low	High	
FN 1	Hole H6	0	+0.5	0.3 to 1.2	0	+0.6	0.3 to 1.3	0	+0.6	0.4 to 1.4
	Shaft	+0.8	+1.2		+0.9	+1.3		+1.0	+1.4	
FN 2	Hole H7	0	+0.8	0.6 to 1.9	0	+1.0	0.8 to 2.4	0	+1.0	0.8 to 2.4
	Shaft s6	+1.4	+1.9		+1.8	+2.4		+1.8	+2.4	
FN 3	Hole H7	0	+0.8	0.8 to 2.1	0	+1.0	1.0 to 2.6	0	+1.0	1.2 to 2.8
	Shaft t6	+1.6	+2.1		+2.0	+2.6		+2.2	+2.8	
FN 4	Hole H7	0	+1.8	1.0 to 2.3	0	+1.0	1.5 to 3.1	0	+1.0	1.8 to 3.4
	Shaft u6	+0.8	+2.3		+2.5	+3.1		+2.8	+3.4	
FN 5	Hole H8	0	+1.2	1.3 to 3.3	0	+1.6	1.4 to 4.0	0	+1.6	2.4 to 5.0
	Shaft x7	+2.5	+3.3		+3.0	+4.0		+4.0	+5.0	

All dimensions in inches. Limits are in agreement with USAS B4.1 (Originally issued as ASA B4.1-1947. Currently issued as ANSI B4.1.)

(Continued)

Table 6. *(Continued)* **Interference Limits for Force and Shrink (FN) Fits for Cylindrical Parts.** (Sizes 1.971 through 8.86 inches.) (Limits are in thousandths of an inch.)

All dimensions in inches. Limits are in agreement with USAS B4.1 (Originally issued as ASA B4.1-1947. Currently issued as ANSI B4.1.)

Class	Hole/Shaft Desig.	1.971 to 2.56			2.561 to 3.15			3.151 to 3.94		
		Limits Low	High	Inter. Limits	Limits Low	High	Inter. Limits	Limits Low	High	Inter. Limits
FN 1	Hole H6	0	+0.7	0.6 to 1.8	0	+0.7	0.7 to 1.9	0	+0.9	0.9 to 2.4
	Shaft	+1.3	+1.8		+1.4	+1.9		+1.8	+2.4	
FN 2	Hole H7	0	+1.2	0.8 to 2.7	0	+1.2	1.0 to 2.9	0	+1.4	1.4 to 3.7
	Shaft s6	+2.0	+2.7		+2.2	+2.9		+2.8	+3.7	
FN 3	Hole H7	0	+1.2	1.3 to 3.2	0	+1.2	1.8 to 3.7	0	+1.4	2.1 to 4.4
	Shaft t6	+2.5	+3.2		+3.0	+3.7		+3.5	+4.4	
FN 4	Hole H7	0	+1.2	2.3 to 4.2	0	+1.2	2.8 to 4.7	0	+1.4	3.6 to 5.9
	Shaft u6	+3.5	+4.2		+4.0	+4.7		+5.0	+5.9	
FN 5	Hole H8	0	+1.8	3.2 to 6.2	0	+1.8	4.2 to 7.2	0	+2.2	4.8 to 8.4
	Shaft x7	+5.0	+6.2		+6.0	+7.2		+7.0	+8.4	

Class	Hole/Shaft Desig.	3.941 to 4.73			4.731 to 5.52			5.521 to 6.30		
		Limits Low	High	Inter. Limits	Limits Low	High	Inter. Limits	Limits Low	High	Inter. Limits
FN 1	Hole H6	0	+0.9	1.1 to 2.6	0	+1.0	1.2 to 2.9	0	+1.0	1.5 to 3.2
	Shaft	+2.0	+2.6		+2.2	+2.9		+2.5	+3.2	
FN 2	Hole H7	0	+1.4	1.6 to 3.9	0	+1.6	1.9 to 4.5	0	+1.6	2.4 to 5.0
	Shaft s6	+3.0	+3.9		+3.5	+4.5		+4.0	+5.0	
FN 3	Hole H7	0	+1.4	2.6 to 4.9	0	+1.6	3.4 to 6.0	0	+1.6	3.4 to 6.0
	Shaft t6	+4.0	+4.9		+5.0	+6.0		+5.0	+6.0	
FN 4	Hole H7	0	+1.4	4.6 to 6.9	0	+1.6	5.4 to 8.0	0	+1.6	5.4 to 8.0
	Shaft u6	+6.0	+6.9		+7.0	+8.0		+7.0	+8.0	
FN 5	Hole H8	0	+2.2	5.8 to 9.4	0	+2.5	7.5 to 11.6	0	+2.5	9.5 to 13.6
	Shaft x7	+8.0	+9.4		+10.0	+11.6		+12.0	+13.6	

Class	Hole/Shaft Desig.	6.31 to 7.09			7.091 to 7.88			7.881 to 8.86		
		Limits Low	High	Inter. Limits	Limits Low	High	Inter. Limits	Limits Low	High	Inter. Limits
FN 1	Hole H6	0	+1.0	1.8 to 3.5	0	+1.2	1.8 to 3.8	0	+1.2	2.3 to 4.3
	Shaft	+2.8	+3.5		+3.0	+3.8		+3.5	+4.3	
FN 2	Hole H7	0	+1.6	2.9 to 5.5	0	+1.8	3.2 to 6.2	0	+1.8	3.2 to 6.2
	Shaft s6	+4.5	+5.5		+5.0	+6.2		+5.0	+6.2	
FN 3	Hole H7	0	+1.6	4.4 to 7.0	0	+1.8	5.2 to 8.2	0	+1.8	5.2 to 8.2
	Shaft t6	+6.0	+7.0		+7.0	+8.2		+7.0	+8.2	
FN 4	Hole H7	0	+1.6	6.4 to 9.0	0	+1.8	7.2 to 10.2	0	+1.8	8.2 to 11.2
	Shaft u6	+8.0	+9.0		+9.0	+10.2		+10.0	+11.2	
FN 5	Hole H8	0	+2.5	9.5 to 13.6	0	+2.8	11.2 to 15.8	0	+2.8	13.2 to 17.8
	Shaft x7	+12.0	+13.6		+14.0	+15.8		+16.0	+17.8	

(Continued)

Table 6. *(Continued)* **Interference Limits for Force and Shrink (FN) Fits for Cylindrical Parts.** (Sizes 8.861 through 30.09 inches.) (Limits are in thousandths of an inch.)

Class	Hole/Shaft Desig.	Hole or Shaft Size								
		8.861 to 9.85			9.851 to 11.03			11.031 to 12.41		
		Limits		Inter. Limits	Limits		Inter. Limits	Limits		Inter. Limits
		Low	High		Low	High		Low	High	
FN 1	Hole H6	0	+1.2	2.3 to 4.3	0	+1.2	2.8 to 4.9	0	+1.2	2.8 to 4.9
	Shaft	+3.5	+4.3		+4.0	+4.9		+4.0	+4.9	
FN 2	Hole H7	0	+1.8	4.2 to 7.2	0	+2.0	4.0 to 7.2	0	+2.0	5.0 to 8.2
	Shaft s6	+6.0	+7.2		+6.0	+7.2		+7.0	+8.2	
FN 3	Hole H7	0	+1.8	6.2 to 9.2	0	+2.0	7.0 to 10.2	0	+2.0	7.0 to 10.2
	Shaft t6	+8.0	+9.2		+9.0	+10.2		+9.0	10.2	
FN 4	Hole H7	0	+1.8	10.2 to 13.2	0	+2.0	10.0 to 13.2	0	+2.0	12.2 to 15.2
	Shaft u6	+12.0	+13.2		+12.0	+13.2		+14.0	+15.2	
FN 5	Hole H8	0	+2.8	13.2 to 17.8	0	+3.0	15.0 to 20.0	0	+3.0	17.0 to 22.0
	Shaft x7	+16.0	+17.7		+18.0	+20.0		+20.0	+22.0	

Class	Hole/Shaft Desig.	Hole or Shaft Size								
		12.411 to 13.98			13.981 to 15.75			15.751 to 17.72		
		Limits		Inter. Limits	Limits		Inter. Limits	Limits		Inter. Limits
		Low	High		Low	High		Low	High	
FN 1	Hole H6	0	+1.4	3.1 to 5.5	0	+1.4	3.6 to 6.1	0	+1.6	4.4 to 7.0
	Shaft	+4.5	+5.5		+5.0	+6.1		+6.0	+7.0	
FN 2	Hole H6	0	+2.2	5.8 to 9.4	0	+2.2	5.8 to 9.4	0	+2.5	6.5 to 10.6
	Shaft	+8.0	+9.4		+8.0	+9.4		+9.0	+10.6	
FN 3	Hole H7	0	+2.2	7.8 to 11.4	0	+2.2	9.8 to 13.4	0	+2.5	9.5 to 13.6
	Shaft s6	+10.0	+11.4		+12.0	+13.4		+12.0	+13.6	
FN 4	Hole H7	0	+2.2	13.8 to 17.4	0	+2.2	15.8 to 19.4	0	+2.5	17.5 to 21.6
	Shaft t6	+16.0	+17.4		+18.0	+19.4		+20.0	+21.6	
FN 5	Hole H7	0	+3.5	18.5 to 24.2	0	+3.5	21.5 to 27.2	0	+4.0	24.0 to 30.5
	Shaft u6	+22.0	+24.2		+25.0	+27.2		+28.0	+30.5	

Class	Hole/Shaft Desig.	Hole or Shaft Size								
		17.721 to 19.69			19.691 to 24.34			24.341 to 30.09		
		Limits		Inter. Limits	Limits		Inter. Limits	Limits		Inter. Limits
		Low	High		Low	High		Low	High	
FN 1	Hole H6	0	+1.6	4.4 to 7.0	0	+2.0	6.0 to 9.2	0	+2.0	7.0 to 10.2
	Shaft	+6.0	+7.0		+8.0	+9.2		+9.0	+10.2	
FN 2	Hole H7	0	+2.5	7.5 to 11.6	0	+3.0	9.0 to 14.0	0	+3.0	11.0 to 16.0
	Shaft s6	+10.0	+11.6		+12.0	+14.0		+14.0	+16.0	
FN 3	Hole H7	0	+2.5	11.5 to 15.6	0	+3.0	15.0 to 20.0	0	+3.0	17.0 to 22.0
	Shaft t6	+14.0	+15.6		+18.0	+20.0		+20.0	+22.0	
FN 4	Hole H7	0	+2.5	19.5 to 23.6	0	+3.0	22.0 to 27.0	0	+3.0	27.0 to 32.0
	Shaft u6	+22.0	+23.6		+25.0	+27.0		+30.0	+32.0	
FN 5	Hole H8	0	+4.0	26.0 to 32.5	0	+5.0	30.0 to 38.0	0	+5.0	35.0 to 43.0
	Shaft x7	+30.0	+32.5		+35.0	+38.0		+40.0	+43.0	

All dimensions in inches. Limits are in agreement with USAS B4.1 (Originally issued as ASA B4.1-1947. Currently issued as ANSI B4.1.)

(Continued)

Table 6. *(Continued)* **Interference Limits for Force and Shrink (FN) Fits for Cylindrical Parts.** (Sizes 30.091 through 115.3 inches.) (Limits are in thousandths of an inch.)

Class	Hole/Shaft Desig.	Hole or Shaft Size								
		30.091 to 35.47			35.471 to 41.49			41.491 to 48.28		
		Limits		Inter. Limits	Limits		Inter. Limits	Limits		Inter. Limits
		Low	High		Low	High		Low	High	
FN 1	Hole H6	0	+2.5	7.5 to 11.6	0	+2.5	9.5 to 13.6	0	+3.0	11.0 to 16.0
	Shaft	+10.0	+11.6		+12.0	+13.6		+14.0	+16.0	
FN 2	Hole H7	0	+4.0	14.0 to 20.5	0	+4.0	16.0 to 22.5	0	+5.0	17.0 to 25.0
	Shaft s6	+18.0	+20.5		+20.0	+22.5		+22.0	+25.0	
FN 3	Hole H7	0	+4.0	21.0 to 27.5	0	+4.0	24.0 to 30.5	0	+5.0	30.0 to 38.0
	Shaft t6	+25.0	+27.5		+28.0	+30.5		+35.0	+38.0	
FN 4	Hole H7	0	+4.0	31.0 to 37.5	0	+4.0	36.0 to 43.5	0	+5.0	45.0 to 53.0
	Shaft u6	+35.0	+37.5		+40.0	+43.5		+50.0	+53.0	
FN 5	Hole H8	0	+6.0	44.0 to 54.0	0	+6.0	54.0 to 64.0	0	+8.0	62.0 to 75.0
	Shaft x7	+50.0	+54.0		+60.0	+64.0		+70.0	+75.0	

Class	Hole/Shaft Desig.	Hole or Shaft Size								
		48.281 to 56.19			56.191 to 65.54			65.541 to 76.39		
		Limits		Inter. Limits	Limits		Inter. Limits	Limits		Inter. Limits
		Low	High		Low	High		Low	High	
FN 1	Hole H6	0	+3.0	13.0 to 18.0	0	+4.0	14.0 to 20.5	0	+4.0	18.0 to 24.5
	Shaft	+16.0	+18.0		+18.0	+20.5		+22.0	+24.5	
FN 2	Hole H7	0	+5.0	20.0 to 28.0	0	+6.0	24.0 to 34.0	0	+6.0	29.0 to 39.0
	Shaft s6	+25.0	+28.0		+30.0	+34.0		+35.0	+39.0	
FN 3	Hole H7	0	+5.0	35.0 to 43.0	0	+6.0	29.0 to 49.0	0	+6.0	44.0 to 54.0
	Shaft t6	+40.0	+43.0		+45.0	+49.0		+50.0	+54.0	
FN 4	Hole H7	0	+5.0	55.00 to 63.0	0	+6.0	64.0 to 74.0	0	+6.0	74.0 to 84.0
	Shaft u6	+60.0	+63.0		+70.0	+74.0		+80.0	+84.0	
FN 5	Hole H8	0	+8.0	72.0 to 85.0	0	+10.0	90.0 to 106	0	+10.0	110 to 126
	Shaft x7	+80.0	+85.0		+100	+106		+120	+126	

Class	Hole/Shaft Desig.	Hole or Shaft Size								
		76.391 to 87.79			87.791 to 100.9			100.91 to 115.3		
		Limits		Inter. Limits	Limits		Inter. Limits	Limits		Inter. Limits
		Low	High		Low	High		Low	High	
FN 1	Hole H6	0	+5.0	20.0 to 28.0	0	+5.0	23.0 to 31.0	0	+6.0	24.0 to 34.0
	Shaft	+25.0	+28.0		+28.0	+31.0		+30.0	+34.0	
FN 2	Hole H7	0	+8.0	32.0 to 45.0	0	+8.0	37.0 to 50.0	0	+10.0	40.0 to 56.0
	Shaft s6	+40.0	+45.0		+45.0	+50.0		+50.0	+56.0	
FN 3	Hole H7	0	+8.0	52.0 to 65.0	0	+8.0	62.0 to 75.0	0	+10.0	70.0 to 86.0
	Shaft t6	+60.0	+65.0		+70.0	+75.0		+80.0	+86.0	
FN 4	Hole H7	0	+8.0	82.0 to 95.0	0	+8.0	92.0 to 105	0	+10.0	110 to 126
	Shaft u6	+90.0	+95.0		+100	+105		+120	+126	
FN 5	Hole H8	0	+12.0	128 to 148	0	+12.0	148 to 168	0	+16.0	164 to 190
	Shaft x7	+140	+148		+160	+168		+180	+190	

instance, in **Table 4**, the first listing, for class LC1, specifies that the hole designation is "H6" and the shaft designation is "h5." These hole and shaft designations are provided in full in the standard, but have not been reproduced here because the inclusion of clearance limit charts for all classes of fit have been provided in the tables. The number in the designation refers to the grades discussed above and reflect the standard of machining required.

Special/Modified Classes of Fit

For bilateral hole or shaft system fits, the limits of size are calculated for holes and shafts and will differ from the specifications given in the Tables. Commonly, holes designated as bilateral are created with drills or reamers. In order to increase tool life, they are designated bilateral by changing the plus limit to one grade finer (from **Table 1**) than the one shown on the clearance limit Tables. The minus limit equals the amount by which the plus limit was lowered, and the shaft limits are each lowered by the same amount as the lower limit of size of the hole. The code used for a bilateral fit is followed by a "B." For example, LC4B would be a locational clearance fit, class 4, bilateral hole tolerance.

Basic shaft fits use similar logic. The maximum size of the shaft is basic, and the allowance is applied to the hole. The symbol is "S" (i.e., LC4S). When using basic shaft fits, the specified limits for both hole and shaft are increased for clearance fits, and decreased for transition or interference fits, by the value of the upper shaft limit, which is the amount required to change the maximum shaft to the basic size.

Metric classes of fit

Rather than using a series of codes, the metric clearance limits charts in this section use a descriptive term to designate the ten recognized classes, and sizes are given for two series of fits: hole size preferred, and shaft size preferred. Normally, the hole size preferred is used, but when a single shaft is to be mated to more than one hole, the shaft size preferred is to be used. Grades have been specified, and are identified by the number following the tolerance position letter in the International Tolerance (IT) system, so that comparable machining operations can be referenced in order to select a suitable process for a particular grade. The grade indicates machining processes that can normally be expected to produce work within the tolerances indicated by the tolerance position and grade, as follows: *Grade 4*, lapping, honing; *Grade 5*, lapping, honing, cylindrical grinding, surface grinding, diamond turning, diamond boring, broaching; *Grade 6*, cylindrical grinding, surface grinding, diamond turning, diamond boring, broaching, powder metal, reaming, turning; *Grade 7*, cylindrical grinding, surface grinding, diamond turning, diamond boring, broaching, powder metal, reaming, turning, powder metal-sintered, boring; *Grade 8*, surface grinding, powder metal, reaming, turning, powder metal-sintered, boring; *Grade 9*, reaming, turning, powder metal-sintered, boring; *Grade 10*, reaming, turning, powder metal-sintered, boring, milling, planning, shaping, drilling, punching; *Grade 11*, turning, boring, milling, planning, shaping, drilling, punching, die casting.

Even though the classes of fits in the metric clearance limits charts are not segmented into larger groups, they can roughly be assigned as follows.

Clearance Fits. Loose Running, Free Running, Close Running, Sliding, and Locational Clearance fits.

Transition Fits. Locational Transition—two different ranges that share the same name are designated.

Interference Fits. Locational Interference, Medium Drive, and Force fits.

See **Tables 7** and **8** for full specifications for metric hole and shaft classes of fit.

Table 7. Clearance, Transitional, and Interference Limits for Metric Cylindrical Parts, Hole Size Preferred. (Sizes 1 through 2 millimeters.)

Class	Hole/Shaft Desig.	1mm Limits Min.	1mm Limits Max.	1mm Fit	1.2mm Limits Min.	1.2mm Limits Max.	1.2mm Fit	1.6mm Limits Min.	1.6mm Limits Max.	1.6mm Fit	2mm Limits Min.	2mm Limits Max.	2mm Fit
Loose Running	Hole H11	0	0.06	.06 to .18	0	0.06	.06 to .18	0	0.06	.06 to .18	0	0.06	.06 to .18
	Shaft c11	-0.12	-0.06		-0.12	-0.06		-0.12	-0.06		-0.12	-0.06	
Free Running	Hole H9	0	0.025	.02 to .07	0	0.025	.02 to .07	0	0.025	.02 to .07	0	0.025	.02 to .07
	Shaft d9	-0.045	-0.02		-0.045	-0.02		-0.045	-0.02		-0.045	-0.02	
Close Running	Hole H8	0	0.014	.006 to .03	0	0.014	.006 to .03	0	0.014	.006 to .03	0	0.014	.006 to .03
	Shaft f7	-0.016	-0.006		-0.016	-0.006		-0.016	-0.006		-0.016	-0.006	
Sliding	Hole H7	0	0.01	.002 to .018	0	0.01	.002 to .018	0	0.01	.002 to .018	0	0.01	.002 to .018
	Shaft g6	-0.008	-0.002		-0.008	-0.002		-0.008	-0.002		-0.008	-0.002	
Locational Clearance	Hole H7	0	0.01	0 to .016	0	0.01	0 to .016	0	0.01	0 to .016	0	0.01	0 to .016
	Shaft h6	-0.006	0		-0.006	0		-0.006	0		-0.006	0	
Locational Transition	Hole H7	0	0.01	-.006 to .01	0	0.01	-.006 to .01	0	0.01	-.006 to .01	0	0.01	-.006 to .01
	Shaft k6	0	0.006		0	0.006		0	0.006		0	0.006	
Locational Transition	Hole H7	0	0.01	-.01 to .006	0	0.01	-.01 to .006	0	0.01	-.01 to .006	0	0.01	-.01 to .006
	Shaft n6	0.004	0.01		0.004	0.01		0.004	0.01		0.004	0.01	
Locational Interference	Hole H7	0	0.01	-.012 to .004	0	0.01	-.012 to .004	0	0.01	-.012 to .004	0	0.01	-.012 to .004
	Shaft p6	0.006	0.012		0.006	0.012		0.006	0.012		0.006	0.012	
Medium Drive	Hole H7	0	0.01	-.02 to .004	0	0.01	-.02 to .004	0	0.01	-.02 to .004	0	0.01	-.02 to .004
	Shaft s6	0.014	0.02		0.014	0.02		0.014	0.02		0.014	0.02	
Force	Hole H7	0	0.01	-.024 to -.008	0	0.01	-.024 to -.008	0	0.01	-.024 to -.008	0	0.01	-.024 to -.008
	Shaft u6	0.018	0.024		0.018	0.024		0.018	0.024		0.018	0.024	

Hole or Shaft Size

(Continued)

Table 7. *(Continued)* **Clearance, Transitional, and Interference Limits for Metric Cylindrical Parts, Hole Size Preferred.** (Sizes 2.5 through 5 millimeters.)

Class	Hole/Shaft Desig.	2.5mm Limits Min.	Max.	Fit	3mm Limits Min.	Max.	Fit	4mm Limits Min.	Max.	Fit	5mm Limits Min.	Max.	Fit
Loose Running	Hole H11	0	0.06	.06 to .18	0	0.06	.06 to .18	0	0.075	.07 to .22	0	0.075	.07 to .22
	Shaft c11	-0.12	-0.06		-0.12	-0.06		-0.145	-0.07		-0.145	-0.07	
Free Running	Hole H9	0	0.025	.02 to .07	0	0.025	.02 to .07	0	0.03	.03 to .09	0	0.03	.03 to .09
	Shaft d9	-0.045	-0.02		-0.045	-0.02		-0.06	-0.03		-0.06	-0.03	
Close Running	Hole H8	0	0.014	.006 to .03	0	0.014	.006 to .03	0	0.018	.01 to .04	0	0.018	.01 to .04
	Shaft f7	-0.016	-0.006		-0.016	-0.006		-0.022	-0.01		-0.022	-0.01	
Sliding	Hole H7	0	0.01	.002 to .018	0	0.01	.002 to .018	0	0.012	.004 to .024	0	0.012	.004 to .024
	Shaft g6	-0.008	-0.002		-0.008	-0.002		-0.012	-0.004		-0.012	-0.004	
Locational Clearance	Hole H7	0	0.01	0 to .016	0	0.01	0 to .016	0	0.012	0 to .02	0	0.012	0 to .02
	Shaft h6	-0.006	0		-0.006	0		-0.008	0		-0.008	0	
Locational Transition	Hole H7	0	0.01	-.006 to .01	0	0.01	-.006 to .01	0	0.012	-.009 to .011	0	0.012	-.009 to .011
	Shaft k6	0	0.006		0	0.006		0.001	0.009		0.001	0.009	
Locational Transition	Hole H7	0	0.01	-.01 to .006	0	0.01	-.01 to .006	0	0.012	-.016 to .004	0	0.012	-.016 to .004
	Shaft n6	0.004	0.01		0.004	0.01		0.008	0.016		0.008	0.016	
Locational Interference	Hole H7	0	0.01	-.012 to .004	0	0.01	-.012 to .004	0	0.012	-.02 to 0	0	0.012	-.02 to 0
	Shaft p6	0.006	0.012		0.006	0.012		0.012	0.02		0.012	0.02	
Medium Drive	Hole H7	0	0.01	-.02 to .004	0	0.01	-.02 to .004	0	0.012	-.027 to -.007	0	0.012	-.027 to -.007
	Shaft s6	0.014	0.02		0.014	0.02		0.019	0.027		0.019	0.027	
Force	Hole H7	0	0.01	-.024 to -.008	0	0.01	-.024 to -.008	0	0.012	-.031 to -.011	0	0.012	-.031 to -.011
	Shaft u6	0.018	0.024		0.018	0.024		0.023	0.031		0.023	0.031	

(Continued)

Table 7. *(Continued)* Clearance, Transitional, and Interference Limits for Metric Cylindrical Parts, Hole Size Preferred. (Sizes 6 through 12 millimeters.)

Class	Hole/Shaft Desig.	6mm Limits Min.	6mm Limits Max.	6mm Fit	8mm Limits Min.	8mm Limits Max.	8mm Fit	10mm Limits Min.	10mm Limits Max.	10mm Fit	12mm Limits Min.	12mm Limits Max.	12mm Fit
Loose Running	Hole H11	0	0.075	.07 to .22	0	0.09	.08 to .26	0	0.09	.08 to .26	0	0.11	.095 to .315
	Shaft c11	-0.145	-0.07		-0.17	-0.08		-0.17	-0.08		-0.205	-0.095	
Free Running	Hole H9	0	0.03	.03 to .09	0	0.036	.04 to .112	0	0.036	.04 to .112	0	0.043	.05 to .136
	Shaft d9	-0.06	-0.03		-0.076	-0.04		-0.076	-0.04		-0.093	-0.044	
Close Running	Hole H8	0	0.018	.01 to .04	0	0.022	.013 to .05	0	0.022	.013 to .05	0	0.027	.016 to .061
	Shaft f7	-0.022	-0.01		-0.028	-0.013		-0.028	-0.013		-0.034	-0.016	
Sliding	Hole H7	0	0.012	.004 to .024	0	0.015	.005 to .029	0	0.015	.005 to .029	0	0.018	.006 to .035
	Shaft g6	-0.012	-0.004		-0.014	-0.005		-0.014	-0.005		-0.017	-0.006	
Locational Clearance	Hole H7	0	0.012	0 to .02	0	0.015	0 to .024	0	0.015	0 to .024	0	0.018	0 to .029
	Shaft h6	-0.008	0		-0.009	0		-0.009	0		-0.011	0	
Locational Transition	Hole H7	0	0.012	-.009 to .011	0	0.015	-.01 to .014	0	0.015	-.01 to .014	0	0.018	-.012 to .017
	Shaft k6	0.001	0.009		0.001	0.01		0.001	0.01		0.001	0.012	
Locational Transition	Hole H7	0	0.012	-.016 to .004	0	0.015	-.019 to .005	0	0.015	-.019 to .005	0	0.018	-.023 to .006
	Shaft n6	0.008	0.016		0.01	0.019		0.01	0.019		0.012	0.023	
Locational Interference	Hole H7	0	0.012	-.02 to 0	0	0.015	-.024 to 0	0	0.015	-.024 to 0	0	0.018	-.029 to 0
	Shaft p6	0.012	0.02		0.015	0.024		0.015	0.024		0.018	0.029	
Medium Drive	Hole H7	0	0.012	-.027 to -.007	0	0.015	-.032 to -.008	0	0.015	-.032 to -.008	0	0.018	-.039 to -.01
	Shaft s6	0.019	0.027		0.023	0.032		0.023	0.032		0.028	0.039	
Force	Hole H7	0	0.012	-.031 to -.011	0	0.015	-.037 to -.013	0	0.015	-.037 to -.013	0	0.018	-.044 to -.015
	Shaft u6	0.023	0.031		0.028	0.037		0.028	0.037		0.033	0.044	

(Continued)

Table 7. *(Continued)* **Clearance, Transitional, and Interference Limits for Metric Cylindrical Parts, Hole Size Preferred.** *(Sizes 16 through 30 millimeters.)*

		Hole or Shaft Size											
		16mm			20mm			25mm			30mm		
		Limits			Limits			Limits			Limits		
Class	Hole/Shaft Desig.	Min.	Max.	Fit	Min.	Max.	Fit	Min.	Max.	Fit	Min.	Max.	Fit
Loose Running	Hole H11	0	0.11	.095 to .315	0	0.13	.11 to .37	0	0.13	.11 to .37	0	0.13	.11 to .37
	Shaft c11	-0.205	-0.095		-0.24	-0.11		-0.24	-0.11		-0.24	-0.11	
Free Running	Hole H9	0	0.043	.05 to .136	0	0.052	.065 to .169	0	0.052	.065 to .169	0	0.052	.065 to .169
	Shaft d9	-0.093	-0.044		-0.117	-0.065		-0.117	-0.065		-0.117	-0.065	
Close Running	Hole H8	0	0.027	.016 to .061	0	0.033	.02 to .074	0	0.033	.02 to .074	0	0.033	.02 to .074
	Shaft f7	-0.034	-0.016		-0.041	-0.02		-0.041	-0.02		-0.041	-0.02	
Sliding	Hole H7	0	0.018	.006 to .035	0	0.021	.007 to .041	0	0.021	.007 to .041	0	0.021	.007 to .041
	Shaft g6	-0.017	-0.006		-0.02	-0.007		-0.02	-0.007		-0.02	-0.007	
Locational Clearance	Hole H7	0	0.018	0 to .029	0	0.021	0 to .034	0	0.021	0 to .034	0	0.021	0 to .034
	Shaft h6	-0.011	0		-0.013	0		-0.013	0		-0.013	0	
Locational Transition	Hole H7	0	0.018	-.012 to .017	0	0.021	-.015 to .019	0	0.021	-.015 to .019	0	0.021	-.015 to .019
	Shaft k6	0.001	0.012		0.002	0.015		0.002	0.015		0.002	0.015	
Locational Transition	Hole H7	0	0.018	-.023 to .006	0	0.021	-.028 to .006	0	0.021	-.028 to .006	0	0.021	-.028 to .006
	Shaft n6	0.012	0.023		0.015	0.028		0.015	0.028		0.015	0.028	
Locational Interference	Hole H7	0	0.018	-.029 to 0	0	0.021	-.035 to -.001	0	0.021	-.035 to -.001	0	0.021	-.035 to -.001
	Shaft p6	0.018	0.029		0.022	0.035		0.022	0.035		0.022	0.035	
Medium Drive	Hole H7	0.018	0	-.039 to -.01	0	0.021	-.048 to -.014	0	0.021	-.048 to -.014	0	0.021	-.048 to -.014
	Shaft s6	0.028	0.039		0.035	0.048		0.035	0.048		0.035	0.048	
Force	Hole H7	0.018	0	-.044 to -.015	0	0.021	-.054 to -.02	0	0.021	-.061 to -.027	0	0.021	-.061 to -.027
	Shaft u6	0.033	0.044		0.041	0.054		0.048	0.061		0.048	0.061	

(Continued)

Table 7. *(Continued)* Clearance, Transitional, and Interference Limits for Metric Cylindrical Parts, Hole Size Preferred. (Sizes 40 through 80 millimeters.)

Class	Hole/Shaft Desig.	40mm Limits		40mm	50mm Limits		50mm	60mm Limits		60mm	80mm Limits		80mm
		Min.	Max.	Fit	Min.	Max.	Fit	Min.	Max.	Fit	Min.	Max.	Fit
Loose Running	Hole H11	0	0.16	.12 to .44	0	0.16	.13 to .45	0	0.019	.14 to .52	0	0.19	.15 to .53
	Shaft c11	-0.28	-0.12		-0.29	-0.13		-0.33	-0.14		-0.34	-0.15	
Free Running	Hole H9	0	0.062	.08 to .204	0	0.062	.08 to.204	0	0.074	.1 to .248	0	0.074	.1 to .248
	Shaft d9	-0.142	-0.08		-0.142	-0.08		-0.174	-0.1		-0.174	-0.1	
Close Running	Hole H8	0	0.039	.025 to .089	0	0.039	.025 to .089	0	0.046	.03 to .106	0	0.046	.03 to .106
	Shaft f7	-0.05	-0.025		-0.05	-0.025		-0.06	-0.03		-0.06	-0.03	
Sliding	Hole H7	0	0.025	.009 to .05	0	0.025	.009 to .05	0	0.03	.01 to .059	0	0.03	.01 to .059
	Shaft g6	-0.025	-0.009		-0.025	-0.009		-0.029	-0.01		-0.029	-0.01	
Locational Clearance	Hole H7	0	0.025	0 to .041	0	0.025	0 to .041	0	0.03	0 to .049	0	0.03	0 to .049
	Shaft h6	-0.016	0		-0.016	0		-0.019	0		-0.019	0	
Locational Transition	Hole H7	0	0.025	-.018 to .023	0	0.025	-.018 to .023	0	0.03	-.021 to .028	0	0.03	-.021 to .028
	Shaft k6	0.002	0.018		0.002	0.018		0.002	0.021		0.002	0.021	
Locational Transition	Hole H7	0	0.025	-.033 to .008	0	0.025	-.033 to .008	0	0.03	-.039 to .01	0	0.03	-.039 to .01
	Shaft n6	0.017	0.033		0.017	0.033		0.02	0.039		0.02	0.039	
Locational Interference	Hole H7	0	0.025	-.042 to -.001	0	0.025	-.042 to -.001	0	0.03	-.051 to -.002	0	0.03	-.051 to -.002
	Shaft p6	0.026	0.042		0.026	0.042		0.032	0.051		0.032	0.051	
Medium Drive	Hole H7	0	0.025	-.059 to -.018	0	0.025	-.059 to -.018	0	0.03	-.072 to -.023	0	0.03	-.078 to -.029
	Shaft s6	0.043	0.059		0.043	0.059		0.053	0.072		0.059	0.078	
Force	Hole H7	0	0.025	-.076 to -.035	0	0.025	-.086 to -.045	0	0.03	-.106 to -.057	0	0.03	-.121 to -.072
	Shaft u6	0.06	0.076		0.07	0.086		0.087	0.106		0.102	0.121	

(Continued)

Table 7. *(Continued)* Clearance, Transitional, and Interference Limits for Metric Cylindrical Parts, Hole Size Preferred. (Sizes 100 through 200 millimeters.)

Class	Hole/Shaft Desig.	100mm			120mm			160mm			200mm		
		Limits		Fit	Limits		Fit	Limits		Fit	Limits		Fit
		Min.	Max.		Min.	Max.		Min.	Max.		Min.	Max.	
Loose Running	Hole H11	0	0.22	.17 to .61	0	0.22	.18 to .62	0	0.25	.21 to .71	0	0.29	.24 to .82
	Shaft c11	-0.39	-0.17		-0.4	-0.18		-0.46	-0.21		-0.53	-0.24	
Free Running	Hole H9	0	0.087	.12 to .294	0	0.087	.12 to .294	0	0.1	.145 to .345	0	0.115	.17 to .40
	Shaft d9	-0.207	-0.12		-0.207	-0.12		-0.245	-0.145		-0.285	-0.17	
Close Running	Hole H8	0	0.054	.036 to .125	0	0.054	.036 to .125	0	0.063	.043 to .146	0	0.072	.05 to .168
	Shaft f7	-0.071	-0.036		-0.071	-0.036		-0.083	-0.043		-0.096	-0.05	
Sliding	Hole H7	0	0.035	.012 to .069	0	0.035	.012 to .069	0	0.04	.014 to .079	0	0.046	.015 to .09
	Shaft g6	-0.034	-0.012		-0.034	-0.012		-0.039	-0.014		-0.044	-0.015	
Locational Clearance	Hole H7	0	0.035	0 to .057	0	0.035	0 to .057	0	0.04	0 to .065	0	0.046	0 to .075
	Shaft h6	-0.022	0		-0.022	0		-0.025	0		-0.029	0	
Locational Transition	Hole H7	0	0.035	-.025 to .032	0	0.035	-.025 to .032	0	0.04	-.028 to .037	0	0.046	-.033 to .042
	Shaft k6	0.003	0.025		0.003	0.025		0.003	0.028		0.004	0.033	
Locational Transition	Hole H7	0	0.035	-.045 to .012	0	0.035	-.045 to .012	0	0.04	-.052 to .013	0	0.046	-.06 to .015
	Shaft n6	0.023	0.045		0.023	0.045		0.027	0.052		0.031	0.06	
Locational Interference	Hole H7	0	0.035	-.059 to -.002	0	0.035	-.059 to -.002	0	0.04	-.068 to -.003	0	0.046	-.079 to -.004
	Shaft p6	0.037	0.059		0.037	0.059		0.043	0.068		0.05	0.079	
Medium Drive	Hole H7	0	0.035	-.093 to -.036	0	0.035	-.101 to -.044	0	0.04	-.125 to -.06	0	0.046	-.151 to -.076
	Shaft s6	0.071	0.093		0.079	0.101		0.1	0.125		0.122	0.151	
Force	Hole H7	0	0.035	-.146 to -.089	0	0.035	-.166 to -.109	0	0.04	-.215 to -.15	0	0.046	-.265 to -.19
	Shaft u6	0.124	0.146		0.144	0.166		0.19	0.215		0.236	0.265	

(Continued)

Table 7. *(Continued)* **Clearance, Transitional, and Interference Limits for Metric Cylindrical Parts, Hole Size Preferred.** (Sizes 250 through 500 millimeters.)

Class	Hole/Shaft Desig.	250mm Min	250mm Max	250mm Fit	300mm Min	300mm Max	300mm Fit	400mm Min	400mm Max	400mm Fit	500mm Min	500mm Max	500mm Fit
Loose Running	Hole H11	0	0.029	.28 to .86	0	0.32	.33 to .97	0	0.36	.4 to 1.12	0	0.4	.48 to 1.28
	Shaft c11	-0.57	-0.28		-0.65	-0.33		-0.76	-0.4		-0.88	-0.48	
Free Running	Hole H9	0	0.115	.17 to .4	0	0.13	.19 to .45	0	0.14	.21 to .49	0	0.155	.23 to .54
	Shaft d9	-0.285	-0.17		-0.32	-0.19		-0.35	-0.21		-0.385	-0.23	
Close Running	Hole H8	0	0.072	.05 to .168	0	0.081	.056 to .189	0	0.089	.062 to .208	0	0.097	.068 to .228
	Shaft f7	-0.096	-0.05		-0.108	-0.056		-0.119	-0.062		-0.131	-0.068	
Sliding	Hole H7	0	0.046	.015 to .09	0	0.052	.017 to .101	0	0.057	.018 to .111	0	0.063	.02 to .123
	Shaft g6	-0.044	-0.015		-0.049	-0.017		-0.054	-0.018		-0.06	-0.02	
Locational Clearance	Hole H7	0	0.046	0 to .075	0	0.052	0 to .084	0	0.057	0 to .093	0	0.063	0 to .103
	Shaft h6	-0.029	0		-0.032	0		-0.036	0		-0.04	0	
Locational Transition	Hole H7	0	0.046	-.033 to .042	0	0.052	-.036 to .048	0	0.057	-.04 to .053	0	0.063	-.045 to .058
	Shaft k6	0.004	0.033		0.004	0.036		0.004	0.04		0.005	0.045	
Locational Transition	Hole H7	0	0.046	-.06 to .015	0	0.052	-.066 to .018	0	0.057	-.073 to .02	0	0.063	-.08 to .023
	Shaft n6	0.031	0.06		0.034	0.066		0.037	0.073		0.04	0.08	
Locational Interference	Hole H7	0	0.046	-.079 to -.004	0	0.052	-.088 to -.004	0	0.057	-.098 to -.005	0	0.063	-.108 to -.005
	Shaft p6	0.05	0.079		0.056	0.088		0.062	0.098		0.068	0.108	
Medium Drive	Hole H7	0	0.046	-.169 to -.094	0	0.052	-.202 to -.118	0	0.057	-.244 to -.151	0	0.063	-.292 to -.189
	Shaft s6	0.14	0.169		0.17	0.202		0.208	0.244		0.252	0.292	
Force	Hole H7	0	0.046	-.313 to -.238	0	0.052	-.382 to -.298	0	0.057	-.471 to -.378	0	0.063	-.58 to -.477
	Shaft u6	0.284	0.313		0.35	0.382		0.435	.471		0.54	0.58	

Table 8. Clearance, Transitional, and Interference Limits for Metric Cylindrical Parts, Shaft Size Preferred. (Sizes 1 through 2 millimeters.)

Class	Hole/Shaft Desig.	1mm Limits Min.	1mm Limits Max.	1mm Fit	1.2mm Limits Min.	1.2mm Limits Max.	1.2mm Fit	1.6mm Limits Min.	1.6mm Limits Max.	1.6mm Fit	2mm Limits Min.	2mm Limits Max.	2mm Fit
Loose Running	Hole C11	0.06	0.12	.06 to .18	0.06	0.12	.06 to .18	0.06	0.12	.06 to .18	0.06	0.12	.06 to .18
	Shaft h11	-0.06	0		-0.06	0		-0.06	0		-0.06	0	
Free Running	Hole D9	0.02	0.045	.02 to .07	0.02	0.045	.02 to .07	0.02	0.045	.02 to .07	0.02	0.045	.02 to .07
	Shaft h9	-0.025	0		-0.025	0		-0.025	0		-0.025	0	
Close Running	Hole F8	0.006	0.02	.006 to .03	0.006	0.02	.006 to .03	0.006	0.02	.006 to .03	0.006	0.02	.006 to .03
	Shaft h7	-0.1	0		-0.1	0		-0.1	0		-0.1	0	
Sliding	Hole G7	0.002	0.012	.002 to .018	0.002	0.012	.002 to .018	0.002	0.012	.002 to .018	0.002	0.012	.002 to .018
	Shaft h6	-0.006	0		-0.006	0		-0.006	0		-0.006	0	
Locational Clearance	Hole H7	0	0.01	0 to .16	0	0.01	0 to .16	0	0.01	0 to .16	0	0.01	0 to .16
	Shaft h6	-0.006	0		-0.006	0		-0.006	0		-0.006	0	
Locational Transition	Hole K7	-0.01	0	-.01 to .006	-0.01	0	-.01 to .006	-0.01	0	-.01 to .006	-0.01	0	-.01 to .006
	Shaft H6	-0.006	0		-0.006	0		-0.006	0		-0.006	0	
Locational Transition	Hole N7	-0.014	-0.004	-.014 to .002	-0.014	-0.004	-.014 to .002	-0.014	-0.004	-.014 to .002	-0.014	-0.004	-.014 to .002
	Shaft h6	-0.046	0		-0.046	0		-0.046	0		-0.046	0	
Locational Interference	Hole P7	-0.016	-0.006	-.016 to 0	-0.016	-0.006	-.016 to 0	-0.016	-0.006	-.016 to 0	-0.016	-0.006	-.016 to 0
	Shaft h6	-0.006	0		-0.006	0		-0.006	0		-0.006	0	
Medium Drive	Hole S7	-0.024	-0.014	-.024 to -.008	-0.024	-0.014	-.024 to -.008	-0.024	-0.014	-.024 to -.008	-0.024	-0.014	-.024 to -.008
	Shaft h6	-0.006	0		-0.006	0		-0.006	0		-0.006	0	
Force	Hole U7	-0.028	-0.018	-.028 to -.012	-0.028	-0.018	-.028 to -.012	-0.028	-0.018	-.028 to -.012	-0.028	-0.018	-.028 to -.012
	Shaft h6	-0.006	0		-0.006	0		-0.006	0		-0.006	0	

(Continued)

Table 8. (Continued) Clearance, Transitional, and Interference Limits for Metric Cylindrical Parts, Shaft Size Preferred. (Sizes 2.5 through 5 millimeters.)

Class	Hole/Shaft Desig.	Hole or Shaft Size											
		2.5mm			3mm			4mm			5mm		
		Limits		Fit	Limits		Fit	Limits		Fit	Limits		Fit
		Min.	Max.		Min.	Max.		Min.	Max.		Min.	Max.	
Loose Running	Hole C11	0.06	0.12	.06 to .18	0.06	0.12	.06 to .18	0.07	0.145	.07 to .22	0.07	0.145	.07 to .22
	Shaft h11	-0.06	0		-0.06	0		-0.075	0		-0.075	0	
Free Running	Hole D9	0.02	0.045	.02 to .07	0.02	0.045	.02 to .07	0.03	0.06	.03 to .09	0.03	0.06	.03 to .09
	Shaft h9	-0.025	0		-0.025	0		-0.03	0		-0.03	0	
Close Running	Hole F8	0.006	0.02	.006 to .03	0.006	0.02	.006 to .03	0.01	0.028	.01 to .04	0.01	0.028	.01 to .04
	Shaft h7	-0.1	0		-0.1	0		-0.012	0		-0.012	0	
Sliding	Hole G7	0.002	0.012	.002 to .018	0.002	0.012	.002 to .018	0.004	0.016	.004 to .024	0.004	0.016	.004 to .024
	Shaft h6	-0.006	0		-0.006	0		-0.008	0		-0.008	0	
Locational Clearance	Hole H7	0	0.01	0 to .16	0	0.01	0 to .16	0	0.012	0 to .02	0	0.012	0 to .02
	Shaft h6	-0.006	0		-0.006	0		-0.008	0		-0.008	0	
Locational Transition	Hole K7	-0.01	0	-.01 to .006	-0.01	0	-.01 to .006	-0.009	0.003	-.009 to .011	-0.009	0.003	-.009 to .011
	Shaft H6	-0.006	0		-0.006	0		-0.008	0		-0.008	0	
Locational Transition	Hole N7	-0.014	-0.004	-.014 to .002	-0.014	-0.004	-.014 to .002	-0.016	-0.004	-.016 to .004	-0.016	-0.004	-.016 to .004
	Shaft h6	-0.046	0		-0.046	0		-0.008	0		-0.008	0	
Locational Interference	Hole P7	-0.016	-0.006	-.016 to 0	-0.016	-0.006	-.016 to 0	-0.02	-0.008	-.02 to 0	-0.02	-0.008	-.02 to 0
	Shaft h6	-0.006	0		-0.006	0		-0.008	0		-0.008	0	
Medium Drive	Hole S7	-0.024	-0.014	-.024 to -.008	-0.024	-0.014	-.024 to -.008	-0.027	-0.015	-.027 to -.007	-0.027	-0.015	-.027 to -.007
	Shaft h6	-0.006	0		-0.006	0		-0.008	0		-0.008	0	
Force	Hole U7	-0.028	-0.018	-.028 to -.012	-0.028	-0.018	-.028 to -.012	-0.031	-0.019	-.031 to -.011	-0.031	-0.019	-.031 to -.011
	Shaft h6	-0.006	0		-0.006	0		-0.008	0		-0.008	0	

(Continued)

Table 8. *(Continued)* Clearance, Transitional, and Interference Limits for Metric Cylindrical Parts, Shaft Size Preferred. (Sizes 6 through 12 millimeters.)

Class	Hole/Shaft Desig.	6mm Limits Min.	6mm Limits Max.	6mm Fit	8mm Limits Min.	8mm Limits Max.	8mm Fit	10mm Limits Min.	10mm Limits Max.	10mm Fit	12mm Limits Min.	12mm Limits Max.	12mm Fit
Loose Running	Hole C11	0.07	0.145	.07 to .22	0.08	0.17	.08 to .26	0.08	0.17	.08 to .26	0.095	0.205	.095 to .315
	Shaft h11	-0.075	0		-0.09	0		-0.09	0		-0.11	0	
Free Running	Hole D9	0.03	0.06	.03 to .09	0.04	0.076	.04 to .112	0.04	0.076	.04 to .112	0.05	0.093	.05 to .136
	Shaft h9	-0.03	0		-0.036	0		-0.036	0		-0.043	0	
Close Running	Hole F8	0.01	0.028	.01 to .04	0.013	0.035	.013 to .05	0.013	0.035	.013 to .05	0.016	0.043	.016 to .061
	Shaft h7	-0.012	0		-0.015	0		-0.015	0		-0.018	0	
Sliding	Hole G7	0.004	0.016	.004 to .024	0.005	0.02	.005 to .029	0.005	0.02	.005 to .029	0.006	0.024	.006 to .035
	Shaft h6	-0.008	0		-0.009	0		-0.009	0		-0.011	0	
Locational Clearance	Hole H7	0	0.012	0 to .02	0	0.015	0 to .024	0	0.015	0 to .024	0	0.018	0 to .029
	Shaft h6	-0.008	0		-0.009	0		-0.009	0		-0.011	0	
Locational Transition	Hole K7	-0.009	0.003	-.009 to .011	-0.01	0.005	-.01 to .014	-0.01	0.005	-.01 to .014	-0.012	0.006	-.012 to .017
	Shaft h6	-0.008	0		-0.009	0		-0.009	0		-0.011	0	
Locational Transition	Hole N7	-0.016	-0.004	-.016 to .004	-0.019	-0.004	-.019 to .005	-0.019	-0.004	-.019 to .005	-0.023	-0.005	-.023 to .006
	Shaft h6	-0.008	0		-0.009	0		-0.009	0		-0.011	0	
Locational Interference	Hole P7	-0.02	-0.008	-.02 to 0	-0.024	-0.009	-.024 to 0	-0.024	-0.009	-.024 to 0	-0.029	-0.011	-.029 to 0
	Shaft h6	-0.008	0		-0.009	0		-0.009	0		-0.011	0	
Medium Drive	Hole S7	-0.027	-0.015	-.027 to -.007	-0.032	-0.017	-.032 to -.008	-0.032	-0.017	-.032 to -.008	-0.039	-0.021	-.039 to -.01
	Shaft h6	-0.008	0		-0.009	0		-0.009	0		-0.011	0	
Force	Hole U7	-0.031	-0.019	-.031 to -.011	-0.037	-0.022	-.037 to -.013	-0.037	-0.022	-.037 to -.013	-0.044	-0.026	-.044 to -.015
	Shaft h6	-0.008	0		-0.009	0		-0.009	0		-0.011	0	

(Continued)

Table 8. *(Continued)* **Clearance, Transitional, and Interference Limits for Metric Cylindrical Parts, Shaft Size Preferred.** (Sizes 16 through 30 millimeters.)

Class	Hole/Shaft Desig.	16mm Min.	16mm Max.	16mm Fit	20mm Min.	20mm Max.	20mm Fit	25mm Min.	25mm Max.	25mm Fit	30mm Min.	30mm Max.	30mm Fit
Loose Running	Hole C11	0.095	0.205	.095 to .315	0.11	0.24	.11 to .37	0.11	0.24	.11 to .37	0.11	0.24	.11 to .37
	Shaft h11	-0.11	0		-0.13	0		-0.13	0		-0.13	0	
Free Running	Hole D9	0.05	0.093	.05 to .136	0.065	0.117	.065 to .169	0.065	0.117	.065 to .169	0.065	0.117	.065 to .169
	Shaft h9	-0.043	0		-0.052	0		-0.052	0		-0.052	0	
Close Running	Hole F8	0.016	0.043	.016 to .061	0.02	0.053	.02 to .074	0.02	0.053	.02 to .074	0.02	0.053	.02 to .074
	Shaft h7	-0.018	0		-0.021	0		-0.021	0		-0.021	0	
Sliding	Hole G7	0.006	0.024	.006 to .035	0.007	0.028	.007 to .041	0.007	0.028	.007 to .041	0.007	0.028	.007 to .041
	Shaft h6	-0.011	0		-0.013	0		-0.013	0		-0.013	0	
Locational Clearance	Hole H7	0	0.018	0 to .029	0	0.021	0 to .034	0	0.021	0 to .034	0	0.021	0 to .034
	Shaft h6	-0.011	0		-0.013	0		-0.013	0		-0.013	0	
Locational Transition	Hole K7	-0.012	0.006	-.012 to .017	-0.015	0.006	-.015 to .019	-0.015	0.006	-.015 to .019	-0.015	0.006	-.015 to .019
	Shaft H6	-0.011	0		-0.013	0		-0.013	0		-0.013	0	
Locational Transition	Hole N7	-0.023	-0.005	-.023 to .006	-0.028	-0.007	-.028 to .006	-0.028	-0.007	-.028 to .006	-0.028	-0.007	-.028 to .006
	Shaft h6	-0.011	0		-0.013	0		-0.013	0		-0.013	0	
Locational Interference	Hole P7	-0.029	-0.011	-.029 to 0	-0.035	-0.014	-.035 to -.001	-0.035	-0.014	-.035 to -.001	-0.035	-0.014	-.035 to -.001
	Shaft h6	-0.011	0		-0.013	0		-0.013	0		-0.013	0	
Medium Drive	Hole S7	-0.039	-0.021	-.039 to -.01	-0.048	-0.027	-.048 to -.014	-0.048	-0.027	-.048 to -.014	-0.048	-0.027	-.048 to -.014
	Shaft h6	-0.011	0		-0.013	0		-0.013	0		-0.013	0	
Force	Hole U7	-0.044	-0.026	-.044 to -.015	-0.054	-0.033	-.054 to -.02	-0.061	-0.04	-.061 to -.027	-0.061	-0.04	-.061 to -.027
	Shaft h6	-0.011	0		-0.013	0		-0.013	0		-0.013	0	

(Continued)

Table 8. *(Continued)* Clearance, Transitional, and Interference Limits for Metric Cylindrical Parts, Shaft Size Preferred. (Sizes 40 through 80 millimeters.)

Class	Hole/Shaft Desig.	40mm Min.	40mm Max.	40mm Fit	50mm Min.	50mm Max.	50mm Fit	60mm Min.	60mm Max.	60mm Fit	80mm Min.	80mm Max.	80mm Fit
Loose Running	Hole C11	0.12	0.28	.12 to .44	0.013	0.29	.13 to .45	0.14	0.33	.14 to .52	0.15	0.34	.15 to .53
	Shaft h11	-0.16	0		-0.16	0		-0.19	0		-0.19	0	
Free Running	Hole D9	0.08	0.142	.08 to .204	0.08	0.142	.08 to .204	0.1	0.174	.1 to .248	0.1	0.174	.1 to .248
	Shaft h9	-0.062	0		-0.062	0		-0.074	0		-0.074	0	
Close Running	Hole F8	0.025	0.064	.025 to .089	0.025	0.064	.025 to .089	0.03	0.076	.03 to .106	0.03	0.076	.03 to .106
	Shaft h7	-0.025	0		-0.025	0		-0.03	0		-0.03	0	
Sliding	Hole G7	0.009	0.034	.009 to .054	0.009	0.034	.009 to .05	0.01	0.04	.01 to .059	0.01	0.04	.01 to .059
	Shaft h6	-0.016	0		-0.016	0		-0.019	0		-0.019	0	
Locational Clearance	Hole H7	0	0.025	0 to .041	0	0.025	0 to .041	0	0.03	0 to .049	0	0.03	0 to .049
	Shaft h6	-0.016	0		-0.016	0		-0.019	0		-0.019	0	
Locational Transition	Hole K7	-0.018	0.007	-.018 to .023	-0.018	0.007	-.018 to 0.23	-0.021	0.009	-.021 to .028	-0.021	0	-.021 to .028
	Shaft H6	-0.016	0		-0.016	0		-0.019	0		-0.019	0	
Locational Transition	Hole N7	-0.033	-0.008	-.033 to .008	-0.033	-0.008	-.033 to .008	-0.039	-0.009	-.039 to .01	-0.039	-0.009	-.039 to .01
	Shaft h6	-0.016	0		-0.016	0		-0.019	0		-0.019	0	
Locational Interference	Hole P7	-0.042	-0.017	-.042 to -.001	-0.042	-0.017	-.042 to -.001	-0.051	-0.021	-.051 to -.002	-0.051	-0.021	-.051 to -.002
	Shaft h6	-0.016	0		-0.016	0		-0.019	0		-0.019	0	
Medium Drive	Hole S7	-0.059	-0.034	-.059 to -.018	-0.059	-0.034	-.059 to -.018	-0.072	-0.042	-.072 to -.023	-0.078	-0.048	-.078 to -.029
	Shaft h6	-0.016	0		-0.016	0		-0.019	0		-0.019	0	
Force	Hole U7	-0.076	-0.051	-.076 to -.035	-0.086	-0.061	-.086 to -.045	-0.106	-0.076	-.106 to -.087	-0.121	-0.091	-.121 to -.072
	Shaft h6	-0.016	0		-0.016	0		-0.019	0		-0.019	0	

(Continued)

Table 8. *(Continued)* **Clearance, Transitional, and Interference Limits for Metric Cylindrical Parts, Shaft Size Preferred.** (Sizes 100 through 200 millimeters.)

Class	Hole/Shaft Desig.	100mm			120mm			160mm			200mm		
		Limits		Fit	Limits		Fit	Limits		Fit	Limits		Fit
		Min.	Max.		Min.	Max.		Min.	Max.		Min.	Max.	
Loose Running	Hole C11	0.17	0.39	.17 to .61	0.18	0.4	.18 to .62	0.21	0.46	.21 to .71	0.24	0.53	.24 to .82
	Shaft h11	-0.22	0		-0.22	0		-0.25	0		-0.29	0	
Free Running	Hole D9	0.12	0.207	.12 to .294	0.12	0.207	.12 to .294	0.145	0.245	.145 to .345	0.17	0.285	.17 to .4
	Shaft h9	-0.087	0		-0.087	0		-0.1	0		-0.115	0	
Close Running	Hole F8	0.036	0.09	.036 to .125	0.036	0.09	.036 to .125	0.043	0.106	.043 to .146	0.05	0.122	.05 to .168
	Shaft h7	-0.035	0		-0.035	0		-0.04	0		-0.046	0	
Sliding	Hole G7	0.012	0.047	.012 to .069	0.012	0.047	.012 to .069	0.014	0.054	.014 to .079	0.015	0.061	.015 to .09
	Shaft h6	-0.022	0		-0.022	0		-0.025	0		-0.029	0	
Locational Clearance	Hole H7	0	0.035	0 to .057	0	0.035	0 to .057	0	0.04	0 to .065	0	0.046	0 to .075
	Shaft h6	-0.022	0		-0.022	0		-0.025	0		-0.029	0	
Locational Transition	Hole K7	-0.025	0.01	-.025 to .032	-0.025	0.01	-.025 to .032	-0.028	0.012	-.028 to .037	-0.033	0.013	-.033 to .042
	Shaft H6	-0.022	0		-0.022	0		-0.025	0		-0.029	0	
Locational Transition	Hole N7	-0.045	-0.01	-.045 to .012	-0.045	-0.01	-.045 to .012	-0.052	-0.012	-.052 to .013	-0.06	-0.014	-.06 to .015
	Shaft h6	-0.022	0		-0.022	0		-0.025	0		-0.029	0	
Locational Interference	Hole P7	-0.059	-0.024	-.059 to -.002	-0.059	-0.024	-.059 to -.002	-0.068	-0.028	-.068 to -.003	-0.079	-0.033	-.079 to -.004
	Shaft h6	-0.022	0		-0.022	0		-0.025	0		-0.029	0	
Medium Drive	Hole S7	-0.093	-0.058	-.093 to -.036	-0.101	-0.066	-.101 to -.044	-0.125	-0.085	-.125 to -.06	-0.151	-0.105	.151 to -.076
	Shaft h6	-0.022	0		-0.022	0		-0.025	0		-0.029	0	
Force	Hole U7	-0.146	-0.111	-.146 to -.089	-0.166	-0.131	-.166 to -.109	-0.215	-0.175	-.215 to -.15	-0.265	-0.219	-.265 to -.19
	Shaft h6	-0.022	0		-0.022	0		-0.025	0		-0.029	0	

(Continued)

Table 8. (Continued) Clearance, Transitional, and Interference Limits for Metric Cylindrical Parts, Shaft Size Preferred. (Sizes 250 through 500 millimeters.)

Class	Hole/Shaft Desig.	250mm Limits Min	250mm Limits Max	250mm Fit	300mm Limits Min	300mm Limits Max	300mm Fit	400mm Limits Min	400mm Limits Max	400mm Fit	500mm Limits Min	500mm Limits Max	500mm Fit
Loose Running	Hole C11	0.28	0.57	.28 to .86	0.33	0.65	.33 to .97	0.4	0.76	.4 to 1.12	0.48	0.88	.48 to 1.28
	Shaft h11	-0.29	0		-0.32	0		-0.36	0		-0.4	0	
Free Running	Hole D9	0.17	0.285	.17 to .4	0.19	0.32	.19 to .45	0.21	0.35	.21 to .49	0.23	0.385	.23 to .54
	Shaft h9	-0.115	0		-0.13	0		-0.14	0		-0.155	0	
Close Running	Hole F8	0.05	0.122	.05 to .168	0.056	0.137	.056 to .189	0.062	0.151	.062 to .208	0.068	0.165	.068 to .228
	Shaft h7	-0.046	0		-0.052	0		-0.057	0		-0.063	0	
Sliding	Hole G7	0.015	0.061	.015 to .09	0.017	0.069	.017 to .101	0.018	0.075	.018 to .111	0.02	0.083	.02 to .123
	Shaft h6	-0.029	0		-0.032	0		-0.036	0		-0.04	0	
Locational Clearance	Hole H7	0	0.046	0 to .075	0	0.052	0 to .084	0	0.057	0 to .093	0	0.063	0 to .103
	Shaft h6	-0.029	0		-0.032	0		-0.036	0		-0.04	0	
Locational Transition	Hole K7	-0.033	0.013	-.033 to .042	-0.036	0.016	-.036 to .048	-0.04	0.017	-.04 to .053	-0.045	0.018	-.045 to .058
	Shaft H6	-0.029	0		-0.032	0		-0.036	0		-0.04	0	
Locational Transition	Hole N7	-0.06	-0.014	-.06 to .015	-0.066	-0.014	-.066 to .018	-0.073	-0.016	-.073 to .02	-0.08	-0.017	-.08 to .023
	Shaft h6	-0.029	0		-0.032	0		-0.036	0		-0.04	0	
Locational Interference	Hole P7	-0.079	-0.033	-.079 to -.004	-0.088	-0.036	-.088 to -.004	-0.098	-0.041	-.098 to -.005	-0.108	-0.045	-.108 to -.005
	Shaft h6	-0.029	0		-0.032	0		-0.036	0		-0.04	0	
Medium Drive	Hole S7	-0.169	-0.123	-.169 to -.094	-0.202	-0.15	-.202 to -.118	-0.244	-0.187	-.244 to -.151	-0.292	-0.229	-.292 to -.189
	Shaft h6	-0.029	0		-0.032	0		-0.036	0		-0.04	0	
Force	Hole U7	-0.313	-0.267	-.313 to -.238	-0.382	-0.33	-.382 to -.298	-0.471	-0.414	-.471 to -.378	-0.58	-0.517	-.58 to -.477
	Shaft h6	-0.029	0		-0.032	0		-0.036	0		-0.04	0	

Press, shrink, and expansion fits

Press Fit

Press fits, also called force fits, are used when the mating parts are to be considered permanently assembled. Their assembly requires force. Normally, the shaft is slightly bigger than the hole, and the aim is to avoid overstressing the material around the hole. Selecting the best pressure, requires specifying the allowance and calculating the dimensions of the fit. A good rule of thumb for calculating the allowance is to use 0.001 inch (0.0254 millimeters) per inch (25.4 millimeters) of diameter, although this allowance sometimes runs as high as 0.0025 inches (0.0635 millimeters) for some materials. A pressure factor, which is given for several diameters in **Table 9**, is used in the calculation. It should be noted that pressure factors for diameters not given can be calculated as follows: to find the pressure factor for a diameter that is twice the size of a given diameter, multiply the given diameter's factor by 48%. For example, the pressure factor for a 10" diameter is 43.8. To find the pressure factor for a 20" diameter, multiply 43.8 × .48 = 21.

Table 9. Pressure Factors for Press Fits.

Diameter	Factor	Diameter	Factor	Diameter	Factor	Diameter	Factor
1	500	3.25	143	6.50	68.6	11	39.4
1.25	395	3.50	132.5	7	63.6	11.50	37.4
1.50	325	3.75	123	7.50	59	12	35.9
1.75	276	4	115.2	8	55.3	12.50	34.6
2	240	4.25	108	8.50	51.8	13	32.9
2.25	212	4.50	101.8	9	48.8	13.50	31.7
2.50	190	5	91.2	9.50	46.1	14	30.5
2.75	171	5.50	82	10	43.8	14.50	29.3
3	156	6	74.9	10.50	41.3	15	28.3

All diameters in inches.

To find the pressure required, in tons, for force fitting cylindrical parts, use the following formula.

$$P = (A \times a \times F) \div 2$$

where P = Pressure, in tons
  A = Area of contact of the parts, in in.^{2}
  a = Total allowance in size
  F = Pressure factor, from Table.

Example. Find the pressure to force a 4 inch bar into a hole 3 inches long with an allowance of 0.001" per inch of diameter. To find A, the area, first find the circumference of the part: $4 \times \pi = 12.5664$. Next, find the area of the contacting surfaces: circumference × length = $12.5664 \times 3 = 37.699$. Find the pressure factor from the Table for a 4" diameter part = 115.2. The allowance, a, is 0.001" per inch of diameter = 0.004".

$$P = (37.699 \times .004 \times 115.2) \div 2 = 8.685 \text{ tons of pressure.}$$

A variation of the same formula can be used to approximate the allowance necessary to require a given pressure.

$$a = 2P \div AF.$$

Using the parameters from the above formula, it can be seen that the allowance necessary to require 8.685 tons of pressure can be found.

$$a = (2 \times 8.685) \div (37.699 \times 115.2) = 0.003999".$$

Shrink Fit

A shrink fit is obtained by heating the outer member of two mating parts until the heat expands it sufficiently to insert the inner member, which is slightly larger than the hole in its ambient temperature state. As the outer piece cools, it shrinks to form a tight fit that is meant to be a permanent and immovable mating of the two parts. Normally, the allowance for shrink fits is slightly smaller than for force fits. Given the same dimensions and allowances, shrink fits are over three and a half times more resistant to slippage than force fits, and slightly more than three times less likely to rotate than shrink fits. Unfortunately, most of the suggested allowances and dimensions for shrink fits date back to railroad data from the early part of the last century, when pages of tables for heating large diameter driving wheels to mate them with wheel centers (axles) were published. In all cases, steel wheels were mated to cast iron hubs, with the goal being not to exceed the elasticity of the cast iron, which could result in breakage of the hub. Today shrink fits are normally used when the forces required for pressing are impractical. It is recommended that class FN2, FN3, FN4, or FN5 limits for force fits be used to determine interference limits and part dimensions (see **Table 6**).

Expansion Fit

An expansion fit is created by supercooling the inner member of two mating parts in dry ice, liquid nitrogen, or another medium, until the cold makes it shrink sufficiently to allow its insertion into the outer mating member. As the inner piece warms, it expands and makes an extremely tight fit with its mate. This method is most commonly used in situations where it is more practical to cool the inner member than heat the outer member. Temperatures of minus 110° F (−79° C) can be achieved with dry ice or solid CO_2, and minus 310° F (−190° C) can be reached with pure liquid nitrogen. Using the coefficient of expansion for steel (in the range of 7) and the formula from the "Temperature affect on length" section above, it can be seen that steel, when lowered 350° beneath the standard temperature of 68° F, will contract 0.0025 in./in., from its room temperature length. A so-called "compound fit" can be achieved by heating the outer mating piece and cooling the inner mating piece, but controlling the expansion/contraction, plus handling both parts at extreme temperatures, can make precise assembly difficult.

Temperature and Its Effect on Dimensions

The "standard" temperature of 68° F or 20° C was set in the first half of the last century. Initially, it was intended as the ambient temperature for calibrating gages, but it logically spread to be the temperature of record for measurements of all kinds. At 68° F/20° C, part size should be precisely that size given in the tables for specified limits and fits. However, as temperature rises or falls, shaft size changes. **Table 10** for inch sizes and **Table 11** for metric sizes provide length differences that will be encountered at different temperatures. The numbers 1 to 10 across the top of each Table represent coefficients of thermal expansion. If, for example, the coefficient of expansion of a material is 5, the changes experienced by that material at a given temperature, in μin./in. or μm/m, can be found in the column headed 5. If the coefficient needed is above 10, add or multiply values to arrive at the amount of expansion or contraction. If, for example, the coefficient is 13, add the value for 10 to the values for 2 and 1. For reference, the coefficient of thermal expansion is quantified as the coefficient $\times 10^{-6}$. Therefore, the length/degree change for Monel, with a coefficient of 8.7, would be $8.7 \times 10^{-6} = 0.0000087$ in./in. At 78° F, which is ten degrees more than the ambient of 68°, the expansion will be $10 \times .0000087 = .0000870$ in./in., or 87 μin./in. This can be confirmed on the Table by adding the values for 5 (50) and 3 (30) and 0.7 (.7 × 10 [the value for 1] = 7. The total is 87 μin./in. (50 + 30 +7).

Table 10. Length Differences per Inch from Standard for Temperatures 38° to 98° F. *(See Notes.)*

Temp. °F	Coefficient of Thermal Expansion of Material per Degree F × 10⁻⁶					
	1	2	3	4	5	10
	Total Change in Length from Standard, Microinches per Inch of Length (μin./in.)					
38	-30	-60	-90	-120	-150	-300
39	-29	-58	-87	-116	-150	-290
40	-28	-56	-84	-112	-145	-280
41	-27	-54	-81	-108	-140	-270
42	-26	-52	-78	-104	-135	-260
43	-25	-50	-75	-100	-130	-250
44	-24	-48	-72	-96	-125	-240
45	-23	-46	-69	-92	-120	-230
46	-22	-44	-66	-88	-115	-220
47	-21	-42	-63	-84	-110	-210
48	-20	-40	-60	-80	-105	-200
49	-19	-38	-57	-76	-100	-190
50	-18	-36	-54	-72	-95	-180
51	-17	-34	-51	-68	-90	-170
52	-16	-32	-48	-64	-85	-160
53	-15	-30	-45	-60	-80	-150
54	-14	-28	-42	-56	-75	-140
55	-13	-26	-39	-52	-70	-130
56	-12	-24	-36	-48	-65	-120
57	-11	-22	-33	-44	-60	-110
58	-10	-20	-30	-40	-55	-100
59	-9	-18	-27	-36	-50	-90
60	-8	-16	-24	-32	-45	-80
61	-7	-14	-21	-28	-40	-70
62	-6	-12	-18	-24	-35	-60
63	-5	-10	-15	-20	-30	-50
64	-4	-8	-12	-16	-25	-40
65	-3	-6	-9	-12	-20	-30
66	-2	-4	-6	-8	-15	-20
67	-1	-2	-3	-4	-5	-10
68	0	0	0	0	0	0
69	1	2	3	4	5	10
70	2	4	6	8	10	20
71	3	6	9	12	15	30
72	4	8	12	16	20	40
73	5	10	15	20	25	50
74	6	12	18	24	30	60
75	7	14	21	28	35	70

(Continued)

Table 10. *(Continued)* **Length Differences per Inch from Standard for Temperatures 38° to 98° F.** *(See Notes.)*

Temp. °F	Coefficient of Thermal Expansion of Material per Degree F × 10⁻⁶					
	1	2	3	4	5	10
	Total Change in Length from Standard, Microinches per Inch of Length (μin./in.)					
76	8	16	24	32	40	80
77	9	18	27	36	45	90
78	10	20	30	40	50	100
79	11	22	33	44	55	110
80	12	24	36	48	60	120
81	13	26	39	52	65	130
82	14	28	42	56	70	140
83	15	30	45	60	75	150
84	16	32	48	64	80	160
85	17	34	51	68	85	170
86	18	36	54	72	90	180
87	19	38	57	76	95	190
88	20	40	60	80	100	200
89	21	42	63	84	105	210
90	22	44	66	88	110	220
91	23	46	69	92	115	230
92	24	48	72	96	120	240
93	25	50	75	100	125	250
94	26	52	78	104	130	260
95	27	54	81	108	135	270
96	28	56	84	112	140	280
97	29	58	87	116	145	290
98	30	60	90	120	150	300

Notes. To calculate for coefficients of thermal expansion not given, combine values given to give the sum needed. For example, for a coefficient of expansion of 25, double the "10" column and add the "5" column. For 7, add the values for "5" and "2." Fractional interpolation may be similarly calculated. For examples of how to use this Table, see text. Source, USAS B4.1. (Originally issued as ASA B4.1-1947. Currently issued as ANSI B4.1.)

Table 11. Length Differences per Centimeter from Standard for Temperatures 0° to 40° C. *(See Notes.)*

Temp. °C	Coefficient of Thermal Expansion of Material per Degree C × 10⁻⁶					
	1	2	3	4	5	10
	Total Change in Length from Standard, Micron per Meter of Length (μm/m)					
0	-20	-40	-60	-80	-100	-200
1	-19	-38	-57	-76	-95	-190
2	-18	-36	-54	-72	-90	-180
3	-17	-34	-51	-68	-85	-170
4	-16	-32	-48	-64	-80	-160
5	-15	-30	-45	-60	-75	-150

(Continued)

Table 11. *(Continued)* **Length Differences per Centimeter from Standard for Temperatures 0° to 40° C.** *(See Notes.)*

Temp. °C	Coefficient of Thermal Expansion of Material per Degree C × 10⁻⁶					
	1	2	3	4	5	10
	Total Change in Length from Standard, Micron per Meter of Length (μm/m)					
6	-14	-28	-42	-56	-70	-140
7	-13	-26	-39	-52	-65	-130
8	-12	-24	-36	-48	-60	-120
9	-11	-22	-33	-44	-55	-110
10	-10	-20	-30	-40	-50	-100
11	-9	-18	-27	-36	-45	-90
12	-8	-16	-24	-32	-40	-80
13	-7	-14	-21	-28	-35	-70
14	-6	-12	-18	-24	-30	-60
15	-5	-10	-15	-20	-25	-50
16	-4	-8	-12	-16	-20	-40
17	-3	-6	-9	-12	-15	-30
18	-2	-4	-6	-8	-10	-20
19	-1	-2	-3	-4	-5	-10
20	0	0	0	0	0	0
21	1	2	3	4	5	10
22	2	4	6	8	10	20
23	3	6	9	12	15	30
24	4	8	12	16	20	40
25	5	10	15	20	25	50
26	6	12	18	24	30	60
27	7	14	21	28	35	70
28	8	16	24	32	40	80
29	9	18	27	36	45	90
30	10	20	30	40	50	100
31	11	22	33	44	55	110
32	12	24	36	48	60	120
33	13	26	39	52	65	130
34	14	28	42	56	70	140
35	15	30	45	60	75	150
36	16	32	48	64	80	160
37	17	34	51	68	85	170
38	18	36	54	72	90	180
39	19	38	57	76	95	190
40	20	40	60	80	100	200

Notes. To calculate for coefficients of thermal expansion not given, combine values given to give the sum needed. For example, for a coefficient of expansion of 25, double the "10" column and add the "5" column. For 7, add the values for "5" and "2." Fractional interpolation may be similarly calculated. For examples of how to use this Table, see text. Source, USAS B4.1. (Originally issued as ASA B4.1-1947. Currently issued as ANSI B4.1.)

Compensation for thermal effects

The ability to meet higher part tolerance is crucial in today's metal cutting business. Manufacturing engineers expect the high accuracy performance of a machining center to yield high accuracy parts. But this is not necessarily the case. Thermal errors introduced during part processing can have a significant impact on accuracy. The following article by Satish Shivaswamy, R&D Technology Analysis Engineer at Cincinnati Machine, investigates thermal errors. It originally appeared in *Modern Machine Shop on line*.

Effects on part accuracy

Part accuracy is affected by various errors, which may be broadly classified into geometric and processing errors. Geometric errors include positioning accuracy, roll, pitch, yaw, straightness, and squareness. Processing errors include tooling, fixturing, NC programming, and thermal errors. Thermal errors play a considerable role in part accuracy. The effects of part temperature and the coefficient of thermal expansion (CTE) of the part material are described in this article, and a simple method to overcome thermal errors is discussed. Keep in mind that inspection errors must also be considered, but are not specifically discussed in this article.

Effects of thermal expansion

Many manufacturing plants require machining materials as varied as titanium, cast iron, steel, brass, aluminum, and magnesium on a single machine. However, the coefficients of thermal expansion (CTE) of these materials are all different. The standards organizations throughout the world have adopted 68° F or 20° C as the standard temperature for the measurement of length. This means that all materials at the standard temperature of 68° F or 20° C will have the same length. At any other temperature, the length will vary according to the CTE of that material. Because the CTEs of various materials are different, the length of similar parts of different materials will be unique unless measured at 68° F or 20° C. **Table 12** shows CTEs for some of the commonly used materials in machining.

Table 12. Coefficient of Thermal Expansion for Commonly Machined Materials.

Material	CTE (α) μin./°F	CTE (α) μm/°C
Glass Scale	4.4 – 5.6	8 – 10
Titanium	4.7	8.4
Cast Iron	5.7	10.5
Stainless Steel	5.5 – 9.6	9.9 – 17.3
Brass	11.3	20.3
Aluminum	13.2	23.8
Magnesium	14.1	25.2

The thermal expansion or contraction in length as a result of a change in temperature of the part may be computed when the CTE of the part is known, and the equation is:

$$\Delta L = L \times \alpha \times \Delta T$$

where: ΔL = Change in length (meters or inches) due to change in temperature
L = Nominal distance (meters or inches)
α = Coefficient of thermal expansion in μm/° C or μin /° F
ΔT = (T - 20) when L is in meters and α is in μm/°C or
ΔT = (T - 68) when L is in inches and α is in μm/° F.

Current compensation methods

Modern machine tools are equipped with a feature called axis compensation. The concept of axis compensation is similar to the pitch error compensation on older NC machines. In axis compensation, the machine movement to a commanded position will be adjusted based on the values entered in a look up table to correct for positioning errors.

Often the machine thermal behavior tracks closely to that of steel or cast iron. Steel ballscrews, cast iron and steel structures, and glass scale systems all typically have CTEs close to that of cast iron or steel. However, wet machining, and machining parts of materials other than steel or cast iron will lead to inaccurate parts. By combining the knowledge of the part temperature and its CTE with the feature of axis compensation, the part accuracy can be significantly improved.

Example process

Consider, for example, a machining center that is equipped with a coolant chiller and a glass scale for the machine's feedback system. Let us consider a hypothetical situation where two bored holes spaced 400 mm apart are machined both in steel and aluminum parts. Based on how the machine's axes are compensated, different results are obtained for distance between the bored holes in the steel and aluminum parts.

Conditions:
  Aluminum CTE = 23.8 μm/°C (23.8 % 10^{-6}/°C)
  Steel CTE = 11.7 μm /°C
  Glass Scale CTE = 8 μm /°C
  Air/Glass Scale/Pallet Temperature = 24 °C
  Coolant Temperature = 26.1°C
  Distance Between the Bored Holes = 400 mm.

Machine compensated using scale temperature (or pallet temperature) and CTE of glass scale:

A. Compensation value generated at the time of machine calibration
    0.4 m × 8 × 10^{-6} per °C × (24 − 20)°C ≈ 13 microns

B. Aluminum part soaked in coolant is machined
    0.4 m × 23.8 × 10^{-6} per °C × (26.1 − 20)°C ≈ 58 microns

    Error in aluminum part when measured at 20° C = 13 − 58 = - 45 microns (*SHORT*)

C. Steel part soaked in coolant is machined
    0.4 m × 11.7 × 10^{-6} per °C × (26.1 − 20)°C ≈ 29 microns

    Error in steel part when measured at 20° C = 13 − 29 = - 16 microns (*SHORT*)

If the axes of the machine were compensated using the CTE of steel and the coolant temperature, the steel part when measured at 20°C would have no error. Inspection of the aluminum part at 20°C will reveal that the distance between the bored holes would be spaced 15 microns (0.0006") short of 400 mm. If the CTE of aluminum were used instead, the steel part would show that the distance between the bored holes is farther apart by 29 microns (0.00114"), while the aluminum part would have no error when measured at 20°C.

Temperatures must remain stable after the compensations are established. Continued temperature changes will cause an error.

Calculations demonstrate that the part accuracy may be enhanced using the CTE of the part and its temperature. To counter the thermal growth of the part, the axis compensation tables in the machine control may be suitably changed to match the dimensional change of the part using the part CTE value and the coolant temperature to which the part temperature will be soaked.

Multiple compensation tables

If the coolant temperature is controlled, it is possible to generate a set of axis compensation tables for various CTE values. This can be done while generating axis compensation data using the laser, and modifying these values for the various CTEs. The operator can then choose and load the appropriate axis compensation file into the control based on the particular material being processed. This process of first order thermal compensation may be used on any type of controller.

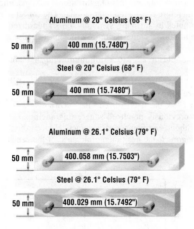

Figure 1. The coefficient of thermal expansion (CTE) of the part material and the coolant temperature play an important role in the final accuracy of the part. The coolant temperature may be controlled to within ±1° C using optional coolant chiller units. The compression values should be generated using the CTE of the part material and the coolant temperature for enhancing part accuracy.

Geometric Dimensioning and Tolerancing (GD&T)[1]

Note: All dimensions are shown in inches. Also, dimensions are not necessarily shown proportional in order to better demonstrate the examples shown in the figures. In addition, some notes have been added to the sketches for explanation purposes, and are normally not shown on a print. These are shown with a line (no arrow) pointing to the section of the part being explained.

What is geometric dimensioning and tolerancing?

Geometric Dimensioning and Tolerancing, also referred to as GD&T, is an accepted and standardized way of denoting dimensions and tolerances on an engineering drawing. The current standard in the U.S. is ANSI/ASME Y14.5M-1994, and is being utilized by design engineering, manufacturing, and inspection. Most of the notations and practices noted in the ISO standards (International Organization for Standardization) conform to this standard and are therefore considered international. The concept replaces voluminous engineering drawing notes by utilizing symbols and standardized nomenclature. The symbol concept, in addition to being far more efficient, can also be international and nonrestrictive because it is not language dependent. This is not to say that notes are prohibited, but they should be kept to a minimum, and the standard notations used wherever possible.

Round versus square tolerancing

First appearing in 1956, the standardized rules incorporate a very beneficial concept of round rather than previously practiced square tolerancing. This allows for greater positioning leeway without sacrificing part accuracy. If, according to the old tolerance version as shown in *Figure 1*, the positioning tolerance of the center of a hole from two

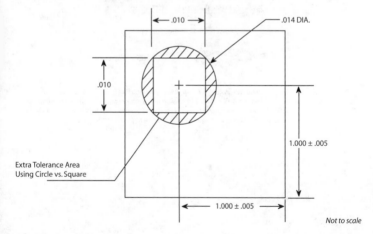

Figure 1. By using a radius tolerance, as proposed in GD&T, instead of a square area tolerance, the total tolerance area is increased.

---

[1]This section on GD&T was contributed by James J. Childs.

given fixed positions, such as the sides of a part, were ± .005, the allowance would be a .010 square, and the center of the hole could fall anywhere within the square—even near the corners which would mean that the center of the hole could be off from the target by as much as a diagonal ±.007. Therefore, the center of the hole could fall anywhere within a radius of .007 of the true center, or within a circle of .014 diameter, and still be within tolerance. The circle would therefore allow considerably more leeway, as shown in the shaded area, than the square and the center would still be within tolerance.

Part example with some common symbols and nomenclature

(Covers: features, datum feature, datum, feature control frame, basic dimension, true position, maximum material condition, and least material condition.)

The sketch of the flat plate shown in *Figure 2a* has a number of *features,* which refer to a physical portion of a part, such as a hole or surface. If the feature is used as a reference, it is called a *datum feature.* In this example, datum features, which can also be called *datums*, are the three surfaces noted as A (bottom), B (side), and C (side) in the small boxes shown in the rectangular *feature control frame* in the upper right corner of *Figure 2a*. Several layers in a feature control frame may also be shown in order to refine the feature. This generally concerns patterns: e.g., holes, where the top line applies to the location of the pattern, and the second line—which usually notes a smaller position tolerance—applies to the holes or other features within the pattern. In all cases, feature control frames are read from left to right, and the frames themselves are read from top to bottom. The letters need not be in alphabetical order.

Referring again to *Figure 2a*, the center of the hole is located 2.000 from datum C, and 1.000 from datum B. These dimensions are called *basic* dimensions and describe the *true position* of a feature, in this case the center of the hole from the datums. The circle, or hole, is 1.000 in diameter, noted by the symbol ∅, and the size can vary ±.005, or can be as small as .995 and as large as 1.005. The diameter symbol ∅ shown in the *feature control*

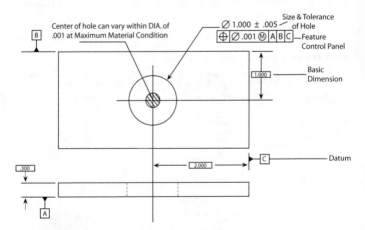

Figure 2a. The sketch above shows a 1.000 diameter hole whose center must lie within a diameter of .001 when the hole is the smallest, or at maximum material condition, noted by the capital letter M within a circle, in the feature control frame.

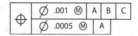

Figure 2b. The second line of the feature control frame, in the above case, refines the first line.

frame notes that the position of the center of the hole can vary within a diameter of .001 when the material is at its *maximum material condition*, as noted by the capital M within a circle in the feature control frame. This is when the hole is geometrically the smallest, or measures .995 diameter. If the letter L were used, it would indicate a *least material condition* and the diameter of the hole would be at its allowable largest, or 1.005. Also note that the identification of the datums A, B, and C is shown in boxes with indicator lines that have reverse solid arrowheads on the extension lines. Previous to this 1994 standard the datums were identified by a letter, with dashes on both sides and located on a rectangular box.

Greater tolerance when maximum material condition is noted

(Covers: maximum material condition, MMC; least material condition, LMC; and regardless of feature size, RFS.)

As the hole shown in *Figure 2a* becomes larger, so does the diameter tolerance for the center of the hole by the same increment as does the size of the hole. For example, if the diameter of the hole were 1.000 instead of .995, the center of the hole could vary within a diameter of .001 + .005, or .006, and still be within tolerance. This would continue until a situation known as *least material condition* was met and the diameter of the hole was 1.005. The tolerance diameter for the center of the hole would then be .011, i.e.: the tolerance diameter of the center of the hole would increase the same amount as the increase in the diameter of the hole. See *Figure 3a*. This is shown, on an incremental basis, in **Table 1**.

The concept of *maximum material condition* would be opposite for an external feature, such as a pin. In this latter case, the *maximum material condition* would indicate the largest

Table 1. Relationship of Increase in Hole Size to Tolerance.

As the hole size increases and goes from a Maximum Material Condition (MMC) to a Least Material Condition (LMC), the tolerance for the center of the hole increases by the same amount.		
Size of Hole	**Tolerance for Center or Hole**	**Condition**
.995	.001	MMC
.996	.002	
.997	.003	
.998	.004	
.999	.005	
1.000	.006	
1.001	.007	
1.002	.008	
1.003	.009	
1.004	.010	
1.005	.011	LMC

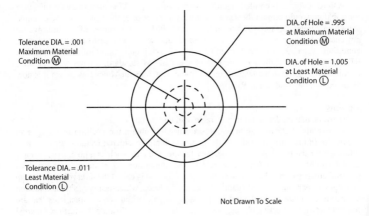

Tolerance DIA. = .001
Maximum Material
Condition Ⓜ

DIA. of Hole = .995
at Maximum Material
Condition Ⓜ

DIA. of Hole = 1.005
at Least Material
Condition Ⓛ

Tolerance DIA. = .011
Least Material
Condition Ⓛ

Not Drawn To Scale

Figure 3a. With maximum material condition, MMC, the diametric tolerance for the center of the hole increases by the same amount as the diameter of the hole, until a least material condition, LMC, is met.

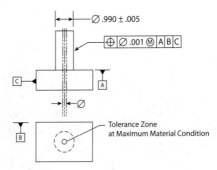

Ø .990 ± .005

⊕ Ø .001 Ⓜ A B C

C

A

Ø

B

Tolerance Zone
at Maximum Material Condition

Figure 3b. The feature control frame points to a cylinder that has a radius of .990 ± .005, and has a diametric tolerance of .001 at maximum material condition, for the center of the cylinder. As the diameter of the cylinder decreases, and approaches a least material condition, the tolerance allowance for the axis of the cylinder increases.

diameter of the pin, or .995. See *Figure 3b*. If the diameter of the pin were to get smaller and approach a *least material condition,* the diameter tolerance for the center of the pin would get larger. If this seems backwards, consider that the pin has to fit into a hole, and the smaller the diameter, the more room the pin has to "play" with.

A *least material condition* is the opposite of a *maximum material condition*, and is noted by a capital letter L, with a circle around it, in a feature control frame. The least material condition situation is not too common and used far less than the *maximum material condition.*

A third condition is entitled *regardless of feature size*, or *RFS*. In this case the tolerance for a feature would remain the same regardless of the size of the feature. In the example shown in *Figure 2a*, the diameter tolerance of .001 for the center of the hole would not change as the size of the feature changed, and went from an M condition to an L condition. Again, as with the L condition, the M condition is far more common than an RFS condition. Until the 1994 version of the Y14.5M standard, a letter S for an RFS condition was noted in the feature control frame. If neither an M nor L is shown it is now assumed that the condition is RFS.

Datums

(Covers: datum, datum reference frame, and datum axis.)

As explained above, datum features are used to obtain the position of the hole in *Figure 2a*. In this case, the sides of the part, which are considered features, are used as a reference for the position of the hole. Any feature of a part, such as a hole, axis, surface, or slot, can be used as a *datum* from which other dimensions, or characteristics—such as perpendicularity or parallelism—can be taken. *Figure 4* is called the *data reference frame* and shows a part that is fitted against three datum planes, which is a very similar situation as shown in *Figure 2a*. Cylindrical type parts are referenced about *datum axes* that are formed by two intersecting planes, as shown in *Figure 4*. To illustrate this further, *Figure 5* shows a cylindrical part with the primary datum plane shown at the bottom of the cylinder and a datum axis running up through the center of the cylinder.

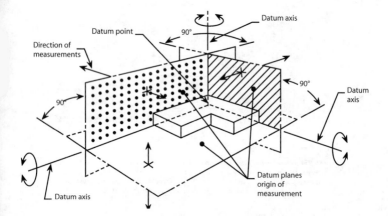

Figure 4. A datum reference frame is shown above with a rectangular type part fitted against the three major datum planes. Cylindrical type parts can also be described using datum axes, as shown. Points, lines, or areas can also be described by using datum targets. The "datum point" indicated in Figure 4 would be a datum target. (Courtesy of American Society of Mechanical Engineers, ASME Y-14.5M-1994.)

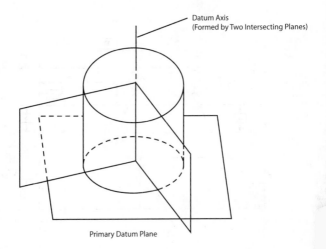

Datum Axis
(Formed by Two Intersecting Planes)

Primary Datum Plane

Figure 5. A cylinder can be shown with a primary datum plane and a datum axis formed by two intersecting planes.

Datum targets (⊖)

(Covers: datum target points, datum target symbols, datum target lines, datum target surfaces, and datum axis.)

Datum targets, also called *datum target symbols*, call out points, lines, and specific areas when it is not possible to establish a normal datum feature (e.g., a plane). This may be the situation if a part has a rough surface, such as with a casting or forging. Although suitable for workholding fixtures, datum targets are particularly helpful for inspection fixtures. Referring again to *Figure 2a*, surface A would require three points, not in a straight line, to establish datum plane A. Surface B would require two points, and surface C would require one point. The symbol for a datum target is a circle with a horizontal line through the center, e.g., ⊖. The bottom half of the circle would note the *set* of the datum target followed by the number of the datum target within the set. The top half of the circle would note the diameter of a datum target area, e.g., ∅2. The datum target is then directed, via a thin leader line, to an "X" spot noted on the part drawing. See *Figure 6a*. The dimensions that note the location of the datum target are generally *basic*, and do not consider a tolerance.

An example of datum targets on a cylindrical part is shown in *Figure 6b*. In this case, the datum targets are located on the circumference of the cylinder and establish a datum axis.

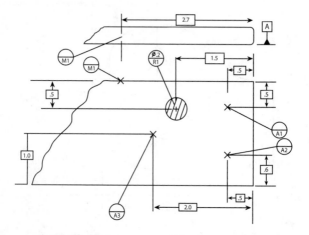

Figure 6a. Three datum target points ⊕ ⊕ ⊕ *are shown on one surface with their datum target symbols. The three points, providing they are all not in a straight line, can establish a datum plane from which other dimensions can be taken. Also shown is a datum line* ⊕ *and a datum target area* ⊕ *with a diameter of .2 in.*

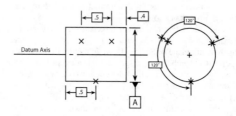

Figure 6b. A primary datum axis ⟂A *is established for the RFS (Regardless of Feature Size) cylinder shown above using three target points.*

Tolerance accumulation

Two of the ways of describing dimensional tolerancing are *chain dimensioning* and *base line dimensioning.* The results are quite different (see *Figures 7a* and *7b*).

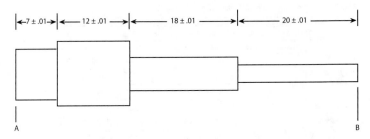

Figure 7a. Chain dimensioning results in the greatest tolerance accumulation which, in the case above and looking on the high side, results in .04 inches between surfaces A and B.

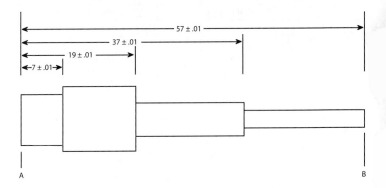

Figure 7b. Base line dimensioning is limited to a tolerance allowance of .01 inches between surfaces A and B.

Geometric Symbols

Geometric symbols may be shown almost anywhere on an engineering drawing but are shown mostly in feature control frames. They are an extremely efficient shorthand way of describing the characteristics of a part, and as importantly are, for the most part, internationally accepted. The geometric feature symbols are divided into five "types of tolerances," viz. *form, profile, orientation, location,* and *runout.* These are shown in **Table 2** and are described with illustrated examples below. The order of the descriptions noted below follows the order in **Table 2**, and references to figure numbers in the right-hand column are those shown in this text.

Straightness (symbol ——)

Straightness can apply to cylindrical or flat type parts. For cylindrical parts, each of the elements (in this case, a line on the surface of the part), or an axis, has to be within two parallel lines prescribed by the tolerance. See *Figure 8a,* in which the tolerance is noted as .002. Also, since this is normally an RFS (Regardless of Feature Size) condition, the tolerance does not change as the feature size changes, and would remain at .002.

Table 2. Geometric Dimensioning and Tolerancing Feature Symbols.

This Table illustrates the various feature symbols used in ASME Y-14.5M-1994. The column on the right of the Table refers to the figure numbers in the text.				
	Type of Tolerance	Characteristic	Symbol	Figure Numbers
For Individual Features	Form	Straightness	—	8a; 8b; 9a; 9b; 9c; 9d
		Flatness	▱	10a; 10b; 10c; 10d
		Circularity (Roundness)	○	11a; 11b; 11c; 11d
		Cylindricity	⌭	12a; 12b
For Individual or Related Features	Profile	Profile of a Line	⌒	13a; 13b; 13c; 13d; 13e; 13f
		Profile of a Surface	⌓	14a; 14b
For Related Features	Orientation	Angularity	∠	15a; 15b; 15c; 15d
		Perpendicularity	⊥	16a; 16b; 16c; 16d
		Parallelism	//	17a; 17b; 17c; 17d
	Location	Position	⊕	2a; 2b
		Concentricity	◎	18a; 18b; 18c
	Runout	Circular Runout	↗	19a; 19b; 19c; 19d
		Total Runout	↗↗	19c; 19d

Figure 8b illustrates the tolerance zone that would be acceptable were the cylinder to be bowed as shown.

Straightness on a flat surface can be applied using line elements lying on that surface, in either one direction or in two. In either case, the elements must lie within the tolerance zone; the distance between is equal to the tolerance noted in the feature control frame. See *Figure 9a*. With the feature control frame shown in *Figure 9b*, the line elements would fall across the broader part of the surface and, with the feature control frame in *Figure 9c*—across the narrower part. The three dimensional view in *Figure 9d* illustrates two of the line elements.

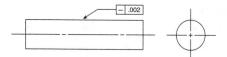

Figure 8a. Tolerance zone noted in a feature control frame.

Figure 8b. Tolerance allowance shown with a bowed part.

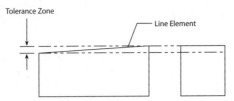

Figure 9a. Tolerance zone shown against a line element on the surface of a part.

Figures 9b, 9c. The tolerance for line elements need not be the same in all directions.
The feature control frame in 9b applies to lines running longitudinally; and in 9c, the lines run across the part and, in this case, at right angles.

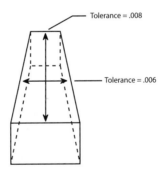

Figure 9d. A three-dimensional view of two of the line elements on a part.

Flatness (symbol ⌭)

Flatness is a condition in which all points on a surface fall between two parallel planes. The planes need not be horizontal if the surface is not. This is shown on a drawing via the feature control frame and an extension line running to the surface—see *Figure 10a*. *Figure 10b* shows the tolerance zone. Flatness can also apply to different zones on the same surface. *Figure 10c* shows a 10 × 10 patch having a flatness tolerance of .01 while the surrounding area has a surface tolerance of .02. See *Figure 10d* for the relevant feature control frame. As noted earlier, in instances of two feature control frames, one above another, the lower one is a subdefinition of the upper one. In this case, the patch is part of the whole upper surface, but calling for a finer tolerance. As with any feature control frame, the reading is from left to right and the complete frames are read from top to bottom.

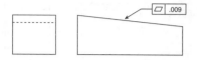

Figure 10a. Feature control frame pointing to the flatness zone.

Figure 10b. Two imaginary parallel planes control the tolerance zone for flatness, as shown above.

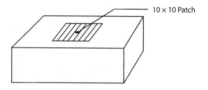

Figure 10c. Two different flatness tolerances; one with a 10 × 10 patch.
For clarity, the dimensioned location of the patch, which would be required, has not been shown.

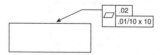

Figure10d. Feature control frame for two different surface flatnesses.

Circularity (symbol ◯)

Circularity applies to cylinders principally, and also to spheres. If we consider the cylinder in *Figure 11a*, the feature control frame may be attached to either the side or end view. In this case the tolerance range is .010, and this means that if two concentric circles are drawn on one plane that is perpendicular to the axis and any place along it, their radii would be different by .010. See *Figure 11b*. As with straightness and flatness, the tolerance has to be less than the feature allowance which, in this case, is the allowance for the diameter. Circularity is not confined to cylinders but can also apply to cones, since the main concern is not with the diameter but rather the tolerance established by any cross section that is perpendicular to the axis. See *Figure 11c.*

A sphere can also have a circularity tolerance, and this is accomplished by having two concentric circles, lying on the same plane that passes through the center of the sphere, and whose radii differ by the allowable tolerance. See *Figure 11d.*

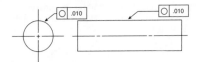

Figure 11a. Cylinder with feature control frame showing circularity tolerance.

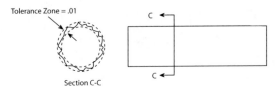

Figure 11b. Cross-section showing tolerance zone.

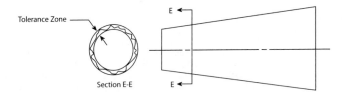

Figure 11c. Circular cross-section of cone.

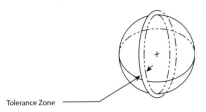

Figure 11d. Cross-section of sphere showing tolerance zone.

Cylindricity (symbol ⌭)

A cylindricity tolerance zone is established by two concentric cylinders between which a cylinder must lie. This differs from circularity in that the entire surface of the cylinder must lie within the zone; whereas with circularity, only individual cross sections of a cylinder that are perpendicular to the axis are considered. The feature control frame and cylinder are noted in *Figure 12a*. *Figure 12b* shows a cylinder that is narrower at the middle, yet does fit between two concentric imaginary cylinders that define the tolerance. As in the cases for straightness, flatness, and circularity, this tolerance must be less than the size tolerance.

Figure 12a. Feature control frame directed at a cylinder indicating the cylindricity tolerance.

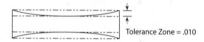

Figure 12b. The tolerance is represented by outer and inner cylinders.
The actual cylinder must lie completely within these two imaginary cylinders.

Profile – Line (symbol ⌒)

The profile of a line is developed from a cross section of a plane that intersects a surface. The element, or line, can be a continuous curve, can reverse itself, or even be a straight line. *Figure 13a* shows the profile of a line that falls between two points, C and D, and has a slight reversal. The line, in most cases, requires relationship to a datum: in this case datum A. The tolerance is bilateral, meaning that the tolerance zone is split; one-half outside (above), and one-half inside (below) the profile line. See *Figure 13b*. If a tolerance line were shown outside (above) the profile line, i.e., on the original drawing, all of the tolerance (i.e., .01) would apply outside the profile line as in *Figure 13c*. If a tolerance line were shown below (inside) the profile line, i.e., all of the tolerance (.01) would apply below, or inside, the profile line as shown in *Figure 13d*. In a case where the tolerance is to be all around the profile line, the symbol ─○ is used with the feature control frame. See *Figure 13e*. The tolerance lines are shown in *Figure 13f*. In airframe structures, this line is often referred to as the BCL, or *Basic Contour Line*.

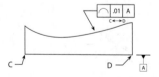

Figure 13a. The feature control frame notes a tolerance zone of .01 with the profile line going from C to D. Since there is no other notation, the tolerance is bilateral, with one-half outside (above) the profile line and one-half inside (see Figure 13b).

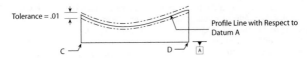

Figure 13b. Tolerance when neither above nor below the profile line is shown.

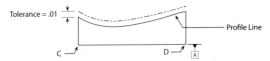

Figure 13c. Unilateral tolerance in which all of the tolerance is outside (above) the profile line.
Generally, only a short section of the tolerance line, above the profile line, is shown on the original print.

Figure 13d. Unilateral tolerance in which all of the tolerance is inside (below) the profile line. Generally, only a short section of the tolerance line, below the profile line, is shown on the original print.

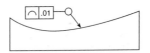

Figure 13e. Feature control frame for a profile line that extends around the whole part; noted by the all-around symbol.

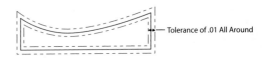

Figure 13f. Effect of the all-around symbol for a profile line.

Profile - Surface (symbol ⌒)

The principle of a profile tolerance for a surface does not differ greatly from that of a line, in that the profile plane has to lie within two tolerance planes instead of two tolerance lines. *Figure 14a* shows a three-dimensional sketch of a profile plane. One tolerance plane would lie .005 above the profile plane and one would lie .005 below, thus creating a tolerance zone of 0.010. The feature control frame is shown in *Figure 14b*.

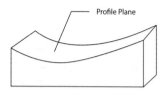

Figure 14a. Three-dimensional drawing showing the profile plane. The tolerance planes would fall above and below this plane.

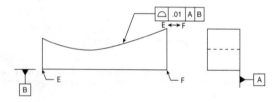

Figure 14b. The feature control frame is directed at the profile plane.

Angularity (symbol ∠)

Angularity is the angle that a surface or axis makes with a datum plane or axis. The angularity tolerance specifies the tolerance zone between two parallel planes or a cylinder around an axis. *Figure 15a* shows a feature control frame that would apply to a sloped surface: in this case 20°. Also, since there is no modifier such as M or L after the letter "A," an RFS (regardless of feature size) applies. The two imaginary planes are .02 apart, which is the allowance shown in the feature control frame. See *Figure 15b*. The same consideration of two planes being .01 apart applies in *Figures 15c and 15d*. In addition, the axis must be within positional tolerance (i.e., within an imaginary cylinder).

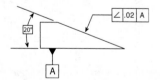

Figure 15a. Plane at 20° angle with datum A. The tolerance of .02 is shown in the feature control frame.

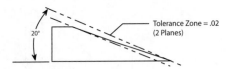

Figure 15b. The tolerance zone consists of two geometric planes that are .02 apart.

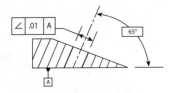

Figure 15c. The tolerance for the cylinder axis is noted in the feature control frame for the cross-section view.

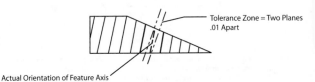

Figure 15d. The tolerance zone consists of two geometric planes that are .01 apart. This applies only to the view that is shown.

Perpendicularity (symbol ⊥)

Perpendicularity is a condition in which a plane or axis is at a right angle to a datum plane or datum axis. There are instances when a perpendicularity tolerance is not shown, and in these cases, if the features appear perpendicular, it can be assumed that they are. As with angularity, the tolerance zone consists of two geometric planes that are spaced in accordance with the tolerance allowed in the feature control frame. *Figure 16a* shows a plane that is declared perpendicular to a datum plane (A) within an allowance of .002. *Figure 16b* illustrates this allowance. Perpendicularity also applies to cylinders, or pins, and holes. However, in these cases the tolerance is generally a geometric cylinder, the diameter of which is noted in the feature control frame. See *Figure 16c*. The tolerance range is illustrated in *Figure 16d*, where the axis cylinder has been tilted, but is still within the tolerance range.

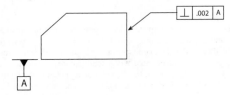

Figure 16a. Side of part (plane) noted as being perpendicular to the datum.

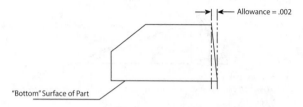

Figure 16b. Illustration of the .002 allowance.

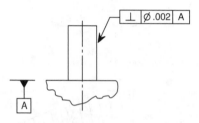

Figure 16c. Perpendicularity allowance for the cylinder shown above is set by a small geometric cylinder around the axis, as shown in the feature control frame. Since no modifier (i.e., M of L) is noted, it is assumed that an RFS condition exists, and the geometric tolerance cylinder around the axis will not change as the diameter of the feature (cylinder) changes.

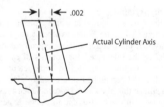

Figure 16d. The cylinder may slant to a degree that does not exceed the tolerance zone set by the geometric cylinder that has a diameter of .002, and is perpendicular to the datum A.

Parallelism (symbol //)

The concept is similar to perpendicularity, only in this case a plane surface must lie between two parallel planes that are parallel to a datum. See *Figures 17a* and *17b*. For a cylinder or hole, the parallel plane concept can also hold, providing the cross-section view is specified. In cases for a cylinder or hole, it is generally more practicable to note the tolerance as an imaginary geometric cylinder, the diameter being noted in the feature control frame. See *Figure 17c*. The tolerance situation in this case is shown in *Figure 17d*.

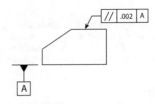

Figure 17a. The surface indicated by the feature control frame is to be parallel to datum A.

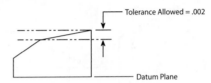

Figure 17b. The upper surface may vary between the allowable tolerance planes that are parallel to the

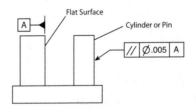

Figure 17c. In a case such as a cylinder, whether external such as a pin, or internal such as a hole, it is the axis that must be parallel to the datum, as shown above. Note the diameter symbol in the feature control frame.

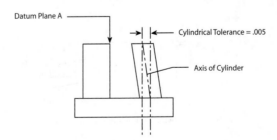

Figure 17d. The slant of a cylinder is shown whose axis is within the allowable tolerance, which is a geometric cylinder with diameter .005.

Position (symbol ⊕)

The position symbol is likely the most common symbol, and what follows it in the feature control frame are the relationships of the various features with respect to their positions to one another. *Figure 2a*, for example, shows the relationship of the center, or axis, of the hole with respect to the sides (features) of the part which are shown as datums.

Concentricity (symbol ◎)

Concentricity applies when two or more cylinders are required to relate to each other on an RFS basis, and within the tolerance as specified for their respective axes. The axes may be parallel as shown in *Figure 18a*, or off at an angle as shown in *Figure 18c*.

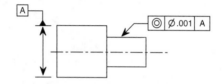

Figure 18a. The concentricity symbol is shown in the feature control frame and the smaller cylinder refers to the larger one that is noted as datum A. Since there is no modifier (M or L) after A, the condition is RFS, meaning Regardless of Feature Size.

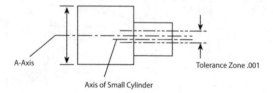

Figure 18b. Shown above is the amount that the axis of the smaller cylinder can be off, but parallel with respect to the axis of datum A. The tolerance zone for concentricity in this case is .001.

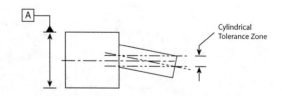

Figure 18c. Concentricity can also be off at an angle, as shown above.

Runout – circular (symbol ⟋)

There are two kinds of runout with two kinds of symbols, viz. circular and total. There are also two kinds of circular runout, viz. radial and axial. Radial runout applies to the surface of a single circular element about an axis that must lie within a given tolerance with respect to that axis. In *Figure 19a*, the axis is noted as datum A. Axial runout applies to a surface that lies perpendicular to the axis and checks the "wobble" of the surface which is due to the tolerance runout for the axis. Sometimes this is called "camming." The extent of the dial movement for a full rotation is called FIM, for *Full Indicator Movement*. See *Figure 19b*.

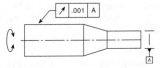

Figure 19a. The feature control frame points to the surface that is to be checked for circular runout. This can be done by checking several places on the indicated surface.

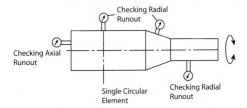

Figure 19b. Gages are shown along two surfaces that are parallel to the axes and one that is at an angle to the axis. Radial runout would be checked along these surfaces. Axial runout, as shown perpendicular to the axis, would check the wobble of the axis.

Runout – Total (symbol ⟋⟋)

Total runout affects the entire surface that is indicated by the feature control frame and is applied to surfaces constructed around a datum axis. See *Figure 19c*. The tolerance allowance for the surface noted is shown in *Figure 19d*.

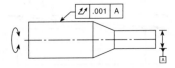

Figure 19c. The feature control frame points to the surface that is affected by the runout tolerance.

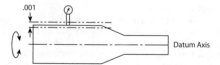

Figure 19d. The runout tolerance of .001 is shown for the surface indicated in the feature control frame.

Other Symbols

Other symbols include modifying symbols that describe geometry or datum targets, such as a counterbore or countersink symbol. Descriptions and illustrations of these symbols are provided below.

Projected tolerance zone (symbol Ⓟ)

This symbol refers generally to the extension of the tolerance of a threaded or pin hole where the tolerance is extended, or projected, so that the bolt or fitted pin does not interfere with a mating part. The amount of the projected tolerance zone is noted in the feature control frame following the encircled letter P. See *Figure 20*.

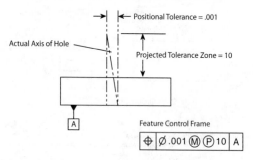

Figure 20. An example of a projected tolerance zone is shown above. The extent of this projected tolerance zone is shown in the feature control frame after the encircled letter P.

Tangent plane (symbol Ⓣ)

A tangent plane is one that touches the high points of a surface and is still within the tolerance required. See *Figure 21*.

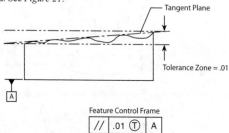

Figure 21. The feature control frame shows a requirement in which two parallel planes establish a tolerance zone of .01 and incorporates a tangent plane, noted by the encircled letter T that connects the high points.

Free state (symbol Ⓕ)

This symbol applies to the tolerance of a part after it has been removed from a fixture in which it has been restrained, and the part has a tendency to distort. This is very typical of thin airframe parts that are machined, and then released from a holding fixture; especially with forged parts for which the "skin" has been broken by the machining process. A feature control frame showing a surface profile, in a free state tolerance, would appear as follows:

Reference dimension (symbol (25))

A reference dimension is used as an aid, and for information purposes, i.e., the part is not to be produced to this dimension. It is shown in parentheses.

Dimension origin (symbol ⊕▶)

Indicates that a dimension originates from one of two features.

Counterbore (symbol ⌴), and Countersink (symbol ⌵)

A counterbore is shown in *Figure 22a*. A countersink is shown in *Figure 22b*. The diameter dimension for either is shown immediately after the symbol. It would also normally show the depth of the counterbore (see symbol below) and the diameter of the hole. A countersink would normally show the diameter of the countersink, the angle of the countersink, and the diameter of the hole.

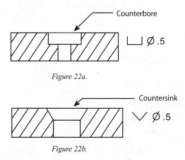

Figure 22a.

Figure 22b.

Depth (symbol ↧)

The depth symbol directly precedes the amount of the depth. See *Figure 23*.

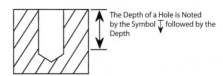

Figure 23.

Arc length (symbol ⌒)

    A short arc with a number, describing the length of the arc, under it. See *Figure 24*.

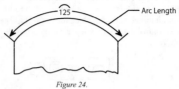

Figure 24.

Number of times (nx)

    A quantity number shown after a symbol.

Square (symbol □)

    Distinguished from a cylinder by a square symbol followed by the dimension. See *Figure 25*.

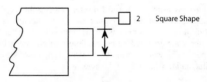

Figure 25.

Radius (symbol R)

Spherical radius (symbol SR)

Spherical diameter (symbol SØ)

Controlled radius (symbol CR)

    Radius is tangent to adjacent surfaces.

Between (symbol A⎯⎯B)

    Means the symbol applies to a limited section of the surface. See *Figure 26*. Also shown in *Figures 13a* and *14b*.

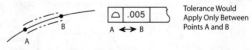

Figure 26.

Properties of Materials

Selection of a material for a specific application requires a full understanding of its mechanical properties including tensile properties, elastic and plastic behavior, brittle fracture strength, and hardness. In addition, physical properties such as corrosion resistance, thermal characteristics, and density must be considered. Choosing the correct material for use in a specific application may be a simple task or a complex one, depending on the requirements of the application. In most cases, a material is selected for use in a particular application for either or both of the following reasons. 1) The part or structure made from the material will satisfy, as completely as possible, all of the essential requirements of the application. 2) The part or structure made from the material can be produced and maintained for a lower total cost than is possible with any other material.

The raw material is an important economic factor in any selection criteria, as initial purchase costs, manufacturing costs, inspection costs, maintenance costs, etc., will in some ways be a direct result of the chosen material. The effect of these costs should be analyzed before final selection, as raw material costs alone very often do not reflect the total life cycle expenses of any given material. However, in some instances, requirements are so stringent or unique that it is necessary to conduct a test program to evaluate various materials that may satisfy the design profile.

As with any engineering problem, a logical, organized approach is beneficial in solving selection problems. This section examines many of the characteristics of materials and explains the standard tests that are used to evaluate materials. By testing and using the comparative data, it is possible to assure that the end product will meet the essential functional requirements of the application.

Mechanical Properties

Elasticity

Tensile Stress. The rod illustrated in *Figure 1* is subjected to tensile loading, due to applied loads in opposite directions on each end. The intensity of this load creates stress in the rod, and the magnitude of the stress can be calculated with the following equation.

$$f = P \div A_0$$

where f = tensile stress

P = applied load (force)

A_0 = original cross-sectional area of the rod.

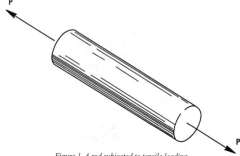

Figure 1. A rod subjected to tensile loading.

Shear Stress. Tensile stress, as shown on the rod in the figure, would make it longer. Compressive stress would be directed in the opposite directions, and would make it shorter. If a force acts normal to a given cross section, the resulting stress is normal stress. If, however, the force acts parallel to a cross section, as shown in *Figure 2*, the resultant stress is shear stress, and its magnitude can be calculated with the following equation.

$$f_S = P \div A$$

where　f_S = shear stress
　　　　P = applied force acting parallel to the cross section
　　　　A = cross-sectional area.

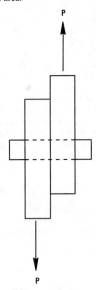

Figure 2. Shear loading on a pin.

Axial Strain. Every stress is accompanied by a corresponding deflection, or change in dimension, in the stressed member. For the tensile loading shown in *Figure 3*, the rod will increase in length an amount δ, provided that the magnitude of the load remains constant. When the deflection (δ) is divided by the original length, a value defined as the normal conventional strain can be calculated with the following equation.

$$e = \delta \div L_0$$

where　e = axial strain
　　　　δ = length of deflection
　　　　L_0 = original length.

Shear Strain. If an element of uniform thickness is subjected to pure shear stress, there will be a displacement on each side of the element relative to the opposite side (shear strain). The shear strain is obtained by dividing this displacement by the distance between the sides of the element. It should be noted that shear strain is obtained by dividing a

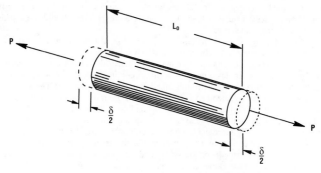

Figure 3. Deflection in a rod subjected to tensile load.

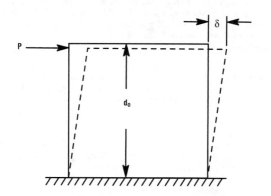

Figure 4. Shear deflection.

displacement by a distance at right angles to the displacement, whereas axial strain is obtained by dividing the deformation by a length measured in the same direction as the deformation. *Figure 4* illustrates shear strain (deflection), which can be calculated by the following equation.

$$e_S = \delta \div d_0$$

where e_S = shear strain

δ = length of deflection

d_0 = cross-sectional length.

Normal and Principal Strains; Poisson's Ratio. Normal strain is the strain associated with a normal stress, and it results in a change in length. Increases in length are denoted as positive strains and decreases in length are called negative strains. Positive strains are identified with the "+" sign and negative with the "−" sign. If one projects three mutually perpendicular planes through a point, there is always some orientation of this system such that only normal strains exist, and all shear strains are zero. The normal strains are called principal strains. Their direction coincides with the direction of the principal *stresses* only for materials that have uniform properties in all directions, which are known as isotropic materials.

In uniaxial loading, strain in the direction of the applied stress varies with that stress. The ratio of stress to strain has a constant value (E) within the elastic range of the material, but it decreases when plastic strain is encountered (in elastic deformation, any deformation caused by stress disappears upon removal of that stress; in plastic deformation, the stress strains the material beyond its elastic limit, and deformation is permanent). The axial strain is always accompanied by lateral strains of opposite sign in the two directions mutually perpendicular to the axial strain. Under uniaxial conditions, the absolute value of the ratio of either of the lateral strains to the axial strain is called *Poisson's ratio*. Simply stated, Poisson's ratio (normally represented by μ) describes unit lateral deflection and is the relationship of the lateral force to the axial force. For stresses within the elastic range, this ratio is approximately constant. For stresses beyond the elastic limit, this ratio is a function of the axial strain and is sometimes called the lateral contraction ratio.

Poisson's ratio can be viewed as a tensile stress acting along the x-axis of the beam, as shown in *Figure 5*. This force will produce a strain, e_X. The force will also produce transverse strains e_y and e_z. According to Hooke's Law (see below), $e_X = f \div E$. Forces e_y and e_z will not be as large as e_X and will be negative as the bar will contract in these two directions. The ratio of e_y and e_z to e_X is Poisson's ratio, which is a nondimensional term.

$$\mu = e_y \div e_X = e_z \div e_X \quad \text{or} \quad e_y = e_z = \mu \, e_X.$$

In multiaxial loading, the strains resulting from the application of each of the stresses are additive: thus, the strains in each of the principal directions must be calculated, taking into account each of the principal stresses and Poisson's ratio.

Modulus of Elasticity. Also known as *Young's Modulus*, the modulus of elasticity is a measurement of the rigidity or stiffness of a material, and it is calculated as the ratio of stress below the proportional limit (which is discussed in the section on Stress-Strain Diagrams) to the corresponding strain. This constant value (E) of the ratio of stress to strain is determined by dividing the amount of tensile stress that an element is subjected to by the elongation of the element, stated as a percentage, at that stress. If, for example, a tensile stress of 4 ksi (27.6 MPa) produces an elongation of 2% in a material, E for that material will be 200 ksi (1380 MPa).

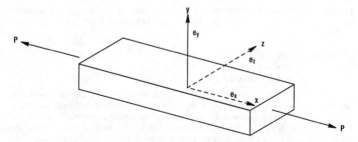

Figure 5. Three-dimensional strain (Poisson's ratio).

Modulus of Rigidity. This is sometimes referred to as the modulus of elasticity in shear, or simply the shear modulus, and it is the ratio of the shear stress to the shear strain at shear stresses below the proportional limit of the material. It can be calculated with the following equation.

$$G = E \div (2 [1 + \mu])$$

where G = modulus of rigidity
E = modulus of elasticity
μ = Poisson's ratio.

Hooke's Law and Stress-Strain Diagrams. Hooke's Law states that stress is directly proportional to strain, and it is expressed as:

$$f \div e = E.$$

This law is applicable to all solid materials below the proportional limit where materials lose their elasticity and become plastic. This can be demonstrated by subjecting a specimen to an increasing tensile load until failure of that specimen occurs. Depending on the instrumentation of the test, a load deflection or load strain curve is usually obtained. The curves shown in *Figure 6* are typical, and it will be noted that the first part of the diagram is typically a straight line ascending vertically. This indicates a constant ratio between stress and strain over that range. The numerical value or the ratio is the modulus of elasticity for the material, and, up to the point where the line begins to curve, the material is in its elastic region. In the elastic region, stress and strain are related by $f = Ee$, or stress is equal to the elasticity modulus times the strain.

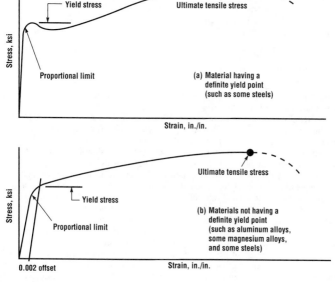

Figure 6. Typical stress-strain diagrams.

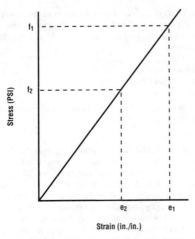

Figure 7. The elastic segment of a stress-strain diagram.

If a material is loaded to a stress f_1 in the elastic region, a corresponding strain e_1 will occur, as shown in *Figure 7*. If the load is removed, the total elastic strain will be recovered. If, rather than removed, the load is reduced to f_2, then the total existing strain will be e_2, and the amount of strain recovered will be $e = e_1 - e_2$. This strain can be calculated by the equation $f = Ee$.

$$f_1 = e_1E \quad \text{or} \quad e_1 = f_1 \div E$$
$$f_2 = e_2E \quad \text{or} \quad e_2 = f_2 \div E$$
$$\text{therefore,} \quad e_1 - e_2 = (f_1 \div E) - (f_2 \div E)$$
$$\Delta e = (f_1 - f_2) \div E.$$

No permanent deflection or strain will occur if a material is stressed in the elastic region and the load is later released.

A material will behave elastically only until a characteristic stress is reached. At this stress, the straight line portion of the stress-strain curve ends. This point is known as the proportional limit, or the tensile proportional limit, and it is the maximum stress in which strain remains directly proportional to stress. The value that shows how much stress a material can withstand before plastic deformation occurs is its yield strength. However, except in the softer ferrous metals this point can be very difficult to pinpoint, and it is almost impossible to precisely determine this point on a stress-strain diagram. Therefore, it is customary to assign a small value of permanent strain and identify the corresponding stress value (at the intersection of the stress-strain curve) as the proportional limit. The selected permanent strain offset value should be stated when using the proportional limit. Ultimate tensile stress ratings, on the other hand, recognize the stress at the maximum load reached in a test. They are not necessarily an indication of the proportional limit of the material. Ultimate strength of a material is the maximum stress a material can withstand without fracturing. After the proportional limit has been exceeded, stress and strain are no longer related by Hooke's Law. In this plastic region, the total strain is composed of elastic strain plus plastic strain. Plastic strain is permanent strain that cannot be recovered after the load has been removed. Even though plastic strain is occurring, the amount of elastic strain can still be measured by Hooke's Law. As shown in *Figure 8*, a material loaded to point f_1 will follow the stress-strain curve (O-A-B in the figure) and the total strain will be e_2. When unloaded, it will unload along the line B-C, which is parallel to the original elastic line O-A. The remaining plastic strain at $f = 0$ will be e_1. The recovered elastic strain will be $e_2 - e_1$.

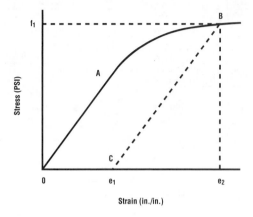

Figure 8. Material behavior in the plastic zone.

Mechanical Properties–Plastic Strain Characteristics

When a metal is deformed beyond its proportional limit, the magnitude of the stress required for further deformation increases. For example, during a tensile test, as a tensile specimen elongates, the load will increase due to strain hardening. Strain hardening (or work hardening) is the point at which a metal, which has become harder and harder under progressive deformation, cannot deform further without fracturing. Tensile strength and hardness are at their maximum and ductility is at its minimum. During strain hardening, the cross-sectional area of the specimen will begin to decrease. During the first part of the test, strain hardening predominates and the load increases. During the later stages of the test, strain hardening becomes less pronounced and the load reaches a maximum. Then it progressively decreases until fracture occurs. Strain will be uniform along the entire test section until the maximum is reached. At this load, some cross section of the bar that is infinitesimally weaker than the rest of the specimen will stretch, as plastic flow takes over, under constant load, while the other portions of the specimen will not. Because of this, localized strain will occur and the specimen will get thinner or "neck" at this particular location. The load will then begin to decrease, the neck will continue to grow thinner, and fracture will eventually occur at the neck.

Testing Mechanical Properties

As shown in **Table 1**, several tests are used to determine the mechanical properties of materials. Most of the properties discussed up to this point are measured with one or more of several common tension tests.

Table 1. Standard Testing Procedures to Determine Mechanical Properties of Materials.
(Source, Bethlehem Steel.)

Mechanical Property	Test Name	Units: English (Metric)
Cold Bending	Cold-bend	Angular Degrees (radians)
Compressive Strength	Compression	psi or ksi (MPa)
Corrosion-fatigue Limit	Corrosion-fatigue	psi or ksi (MPa)
Creep Strength	Creep	psi or ksi (MPa) per time and temperature
Elastic Limit	Tension; Compression	psi or ksi (MPa)
Elongation	Tension	Percent of a specific specimen gage length
Endurance Limit	Fatigue	psi or ksi (MPa)
Hardness	Static: Brinell; Rockwell; Vickers Dynamic: Shore (Scleroscope)	Empirical Numbers Empirical Numbers
Impact, Bending	Bend	ft-lb (Joule)
Impact, Torsional	Torsion-impact	ft-lb (Joule)
Modulus of Rupture	Bend	psi or ksi (MPa)
Proof Stress	Tension; Compression	psi or ksi (MPa)
Proportional Limit	Tension; Compression	psi or ksi (MPa)
Reduction of Area	Tension	percent
Shear Strength	Shear	psi or ksi (MPa)
Tensile Strength	Tension	psi or ksi (MPa)
Torsional Strength	Torsion	psi or ksi (MPa)
Yield Point	Tension	psi or ksi (MPa)
Yield Strength	Tension	psi or ksi (MPa)

Tension Tests

The requirements of this test are set down in ASTM specification E8. The basic methodology of the test is as follows.

There are two standard tensile test specimens: one with a rectangular cross section, and the other round. The grips can be of any configuration that will allow tensile loads to be applied axially through the specimen. Typical tensile test specimens are shown in *Figure 9*. General requirements for a specimen are as follows.

The gage section should be of uniform cross section along its length. It may taper slightly so that at the center of the gage length the width of the flat specimen may be reduced to a width that is 0.010 inch (0.254 mm) less than the width of the specimen at the ends of the gage length. For round specimens, the permissible reduction is limited to 1% of the diameter of the specimen at the center of the gage length.

The gage section should be free from burrs, scratches, pits, or other surface defects.

The ratio of the gage section length to specimen width or diameter should be at least 4:1.

The grip length should be long enough to prevent slippage or fracture in the grips.

The properties most often obtained from a tension test are yield strength, ultimate strength, percent elongation, and reduction of area. Other properties, less frequently measured, are the modulus of elasticity, Poisson's ratio, and the strain hardening exponent. The typical sequence of events in a tension test is as follows.

The cross-sectional area of the gage section is measured.

The specimen is marked with standard gage lengths for subsequent elongation

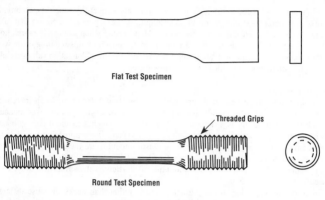

Figure 9. Flat and round tensile test specimens.

measurements, or fitted with an extensometer that will precisely measure changes in length.

The specimen is placed into a tensile machine and a suitable strain measuring device is attached.

The specimen is loaded to fracture.

For those alloys having a distinct yield point, the stress at which the yield point occurs is the yield strength. For those alloys without a distinct yield point, the yield strength is commonly designated as the stress at which 0.2% plastic strain has occurred. The ultimate strength is obtained by dividing the maximum load that the specimen carried by the original cross-sectional area. This load can normally be read from a dial on the tensile machine.

Percent elongation is the ratio of the increase in length of the gage section of the specimen to its original length expressed as a percent. It can be measured by indexing a gage length—usually 1 or 2 inches long—on the specimen gage section and then measuring this length after fracture. The percent elongation can be determined with the following equation.

% elongation = (change in length ÷ original length) × 100.

Elongation measurements made in this manner are only an approximation of the true uniform strain that an alloy can undergo before necking. The measurement is made across the fracture, introducing an error, and localized strain occurring in the necked portion introduces additional error.

Reduction in area is the ratio of the decrease in area at the fracture cross section (the neck) to the original area. It can be measured with the following equation.

% reduction in area = (original cross-sectional area – final cross-sectional area)
÷ original cross-sectional area × 100.

The application of tensile properties to design is straightforward. The yield and ultimate strengths designate how much load a structure can carry under ideal conditions. However, in many applications another criteria, such as fatigue characteristics or sheer stress, may determine the actual working stress of a structure.

Elongation and reduction in area are measurements of ductility (ductility, commonly expressed as percentage elongation in gage length, is the degree of extension that takes place in a material in tension before fracture) and have primary applications in fabrications and forming work. Elongation values are the main consideration in stretching or drawing operations, while the reduction rate is important in roll forming. Elongation values are also a qualitative indication of brittleness of an alloy or temper.

Compression Tests

Compression properties are determined by subjecting a specimen to an increasing compressive load until general yielding has occurred. Compression tests measure malleability (malleability is a measure of the extent to which a material can withstand deformation in compression before failure occurs), and in compressive testing only compressive yield strength and compressive elastic modulus are measured. This is done in a manner similar to the methods of the tensile test. Theoretically, these values should be the same as the tensile yield and modulus of elasticity values but, in reality, there is usually a small difference.

In compression testing, the specimen "barrels" rather than necks. Because different factors are at work in barreling, malleable specimens flow in response to the load and actually compress rather than fracture. As the material yields, it swells out (barrels), so that its increasing area continues to support the increasing load. Therefore, there is no measurement point for pinpointing the ultimate tensile strength of the material. Brittle specimens, on the other hand, will fracture into two or more pieces, making it possible to establish an ultimate tensile strength for these materials.

Compression tests are perhaps most useful for predicting how materials will flow in forging, extruding, rolling, and other operations. Generally, no significant data are obtained from a compression test that cannot be obtained from a tension test. Since a material fails by crack formation and propagation, a crack cannot form in malleable materials under compressive loading conditions. Failures under compressive loading can normally be attributed to buckling instability and in some cases shear or bending stresses.

Torsion Tests

Shear properties are usually determined through the torsion test. In this test, shear modulus, shear yield strength, shear ultimate strength (modulus of rupture), and the shear modulus are measured. Although the testing criteria for torsion testing have not been standardized, and it is almost never specified, the data is normally presented as a torque-twist curve on which the degree of twist in a specimen is plotted against the applied torque.

Torsion tests are made by restraining a cylindrical specimen at one end and subjecting it to a twisting movement at the other. The following equation can be used to predict the maximum shear stress of a thin walled tube. When using this and the following equation, a torque angle twist curve is plotted and the shear yield strength is determined at the proportional limit or at a specified permanent twist such as 0.001 radians per inch. It should be noted that this equation is slightly in error, but provides a useful approximation.

$$f_S = T \div (2r^2 \times t)$$

where T = torque
      r = outer radius of tube
      t = tube wall thickness

For a solid cylinder, the equation is:

    $F_S = 16T \div (\pi \times d^3)$, where d is the diameter of the cylinder. Like the above equation, this can only be applied when strain is proportional to stress. The modulus of rigidity (or modulus of elasticity in shear) can also be determined once the maximum shear stress is known.

$$G = (f_S \times 1) \div (r \times \theta)$$

where 1 = length of test specimen
r = radius of the cylinder
θ = angle of twist (in radians) in length (l).

Additional information on torsion and other methods of materials testing appears in the Statics and Stress Analysis technical units that preceded this section.

Brittle Fracture Tests

The catastrophic failures of welded ships and tankers during World War II brought to the attention of engineers that structural steels can fail at very low stress levels in some environments. These failures occurred at very low ambient temperatures, and there was generally a notch, crack, or other defect present at the failure origin. This type of failure has appropriately been named brittle fracture, because the failure is preceded by little or no plastic strain. However, it is not the brittleness of the fracture that is important. The important consideration is that the failure stress may be considerably less than the yield strength of the material. The engineering methods used to prevent such low strength failures have taken both quantitative and qualitative approaches.

Quantitative Methods

Quantitative methods allow for establishing limiting parameters, such as minimum or maximum operating temperatures; maximum permissible defect size, or maximum safe operating stress for a particular alloy in a given application. These data initiating these parameters were all easily converted to design and operational reference values. In contrast, when using qualitative methods, data are not readily converted to design parameters.

Transition temperature method. The drop weight test described in ASTM E208-63T is a test method that was developed to determine the nil-ductility transition temperature of ferritic steels. The test relies on the concept that ferritic steels in the notched condition are markedly affected by temperature so that there is a characteristic temperature below which a given steel will fail in a brittle manner, but above which brittle fracture will not occur. The nil-ductility transition temperature as determined by the drop weight test is defined as the temperature at which, in a series of tests conducted under controlled conditions, specimens break, while at a temperature 10° F (5.5° C) higher, under duplicate test conditions, no break performance is obtained from near identical specimens.

The drop weight test is simple and inexpensive, and its significant feature of all sizes of specimens is the weld bead deposited on the tension side of the specimen along its longitudinal centerline, as shown in *Figure 10*. The test is conducted by positioning the specimen in a fixture, where it is struck by 60 or 100 pound weights dropped from a predetermined height. The weight is sufficient to deflect the specimen until it impinges on the anvil (about 5°) as shown in *Figure 11*. A cleavage crack forms in the weld bead as soon as incipient yield occurs, at about 3° deflection. A series of specimens are tested over a range of temperatures, and from these tests the nil-ductility transition temperature (which is the temperature at which the steel, in the presence of a cleavage crack, will not deform plastically before fracturing, but will fracture at the moment of yielding) is determined. A specimen is considered broken if it fractures to one or both edges of the tensile surfaces. Complete separation at the compression side is not required If a specimen develops a crack that does not extend to either edge of the tensile surface, it is regarded as a no-break performance.

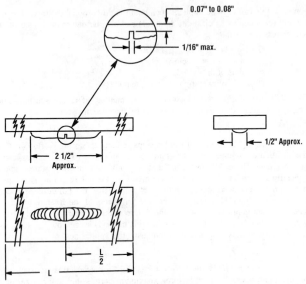

Figure 10. A drop weight test specimen.

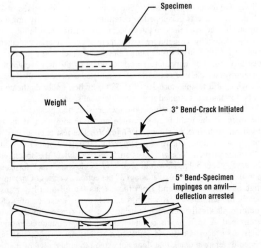

Figure 11. The drop weight test.

After the nil-ductility transition temperature of an alloy has been established, brittle fracture can be avoided by specifying the alloy for use only where it will be in service above the transition temperature. Or, if the operating temperature is fixed, an alloy can be selected that has a transition temperature below the lowest operating temperature. There are, however, two major disadvantages of the transition temperature approach to avoiding brittle fracture. First, it is sometimes necessary to use an alloy at a temperature below its transition temperature, and in these instances it would be desirable to know what level of stress can be considered safe. Secondly, the higher strength steel alloys do not have definite transition temperatures. To satisfy these two problems, the fracture mechanics method can be used.

Fracture mechanics method. This approach to brittle fracture analysis determines the fracture strength of an alloy in the presence of a defect of known geometry. It was developed from Griffith's elastic analysis of the fracture of brittle materials. A. A. Griffith was an English physicist who, in 1924, developed a method of putting crack prediction on an analytical basis. It is based on the concept that small elliptical cracks in a material act as local stress risers that cause stresses to exceed the strength of the material, even though the nominal stress across the section may be quite low. Furthermore, a crack will begin to propagate when the elastic energy released by propagation is equal to, or greater than, the energy of formation of the two new surfaces. Since the elastic energy increases with increasing stress, it is apparent that, at some value of stress, the strain energy released by the crack will exceed the energy of formation of the surfaces, and the crack will become self-propagating under its own stress concentration. Griffith's theory was developed for completely brittle materials that do not deform plastically, and it does not strictly apply to metals, which do undergo plastic deformation.

To account for plastic deformation, a modification of Griffith's theory allows the fracture stress of an alloy containing a crack of known size to be accurately predicted. The fracture stress is characterized by the strain energy release rate (G), an experimentally determined parameter. G increases with crack length, and the stress at which a crack of known size becomes self-propagating is called the critical strain energy rate (denoted by the subscript C, as in G_C or K_C). G_C denotes the critical G determined under plane stress conditions, and G_{IC} is the critical G determined under plane strain conditions. G can also be expressed in terms of the stress intensity factor, K.

$$K_C = \sqrt{G_C \times E}$$
and, $$K_{IC} = \sqrt{(G_{IC} \times E)} \div (1 - \mu^2)$$
where E = modulus of elasticity
μ = Poisson's ratio.

G and K are both referred to as fracture toughness values. Since these values vary considerably from each other, it is important to specify which is being discussed.

By definition, a plane stress condition is one in which the stress in at least one direction is zero, as can be illustrated by a thin-walled pressure vessel or a thin sheet loaded in tension. In each instance, the stress through the thickness is zero. As applied to fracture mechanics, plane stress actually describes the stress state or restraint at the leading crack edge. Therefore, a through crack in thin material is in plane stress conditions because the stress in the thickness direction at the crack tip is zero or very small. For thicker material, the restraint at the crack front increases until full restraint exists and the stress in the thickness direction is quite high. The fully restrained condition is the plane strain condition. The thickness at which full restraint is reached differs for different materials, but is in the neighborhood of $1/4$ to $1/2$ inch.

The significance of the plane stress to plane strain transition is that a much lower fracture stress is required for a given defect size when plane strain conditions exist. This is apparent from the difference in magnitude between K_C and K_{IC}. As the thickness increases, the fracture toughness in the plane stress region (K_C) decreases to a constant value of K_{IC}, corresponding to plane strain or full restraint.

There are three basic types of cracks. 1) Through cracks which extend completely through the thickness of the material. 2) Surface cracks, which can be viewed on the outside of the material but do not pass through. 3) Embedded cracks which cannot be seen from the outside but are in the interior of the material. A fracture initiated by a through crack may occur under either plane stress or plane strain conditions, depending on the material thickness. The initial propagation of both the surface and embedded cracks is always under plane strain conditions. However, once a crack travels through the thickness it will be identical to a through crack and may be in either the plane stress or plane strain stress state depending, again, on the thickness.

The stress at which a tension member fails in the presence of a crack is dependent on the fracture toughness of the material, and the defect size. This relationship, sometimes called the inverse square root law, is the basis of fracture mechanics and is expressed by the following equation.

$$S_f = A \div \sqrt{c_l}$$

where S_f = fracture stress
  A = constant expressing the fracture toughness.
  c_l = crack length.

For example, using this expression, if a stress of 100,000 PSI will cause failure in a material in which a $^1/_4$ inch long crack is present, then it can be determined that a stress of 50,000 PSI would cause failure if the crack were 1 inch long.

$$S_{f1} = A \div \sqrt{c_{lA}} = 100,000 = A \div \sqrt{0.25}$$

therefore, $S_{f2} = (A \div \sqrt{0.25}) \div \sqrt{1} = 100,000 \times 0.5,$
  $S_{f2} = 50,000$ PSI.

For the case of a crack in an infinitely wide solid, such as a small crack in a pressure vessel, this equation takes the forms:

$$S_f = K_C \div \sqrt{\pi \times 0.5 \, c}$$

where K_C = fracture toughness in the plane stress region,

  or $S_f = K_{IC} \div \sqrt{(\pi \times 0.5 \, c)} \times (1 - \mu^2)$

where K_{IC} = fracture toughness in the plane strain region.

The above equations are standard ways of presenting fracture strength as a function of defect size. As discussed previously, the K_C values for thin material are considerably higher than K_{IC} values. Plotting the lower K_{IC} values in place of the K_C values would show that it takes a considerably longer crack to cause failure under plane stress conditions than it does under plane strain conditions at a given stress level.

Fracture toughness testing. The fracture toughness of a material is determined by loading a fatigue cracked specimen in tension and recording the load at which the crack begins to propagate, and the load at which it fails. Fracture toughness testing methods are sometimes very complex, but typical specimens used in the tests are shown in *Figure 12*. Specimen widths usually range from 1 to 4 inches, with the thickness normally being equal to that of the part that is to be made from the material. The major restriction on testing is that fracture toughness data may only be obtainable in the elastic region of the tensile curve. If yielding of the specimen occurs, then the equations that are used to determine K are no longer valid, since they are based on the elastic theory of stress analysis.

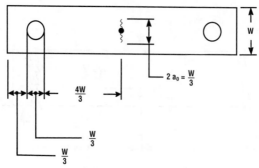

a. Center-cracked specimen.

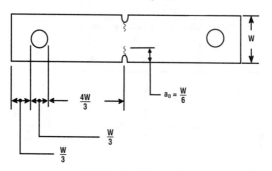

b. Double-edge-notched specimen.

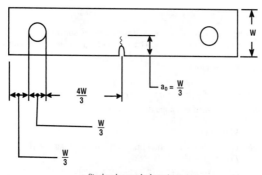

c. Single-edge-notched specimen.

Figure 12a-c. Common fracture toughness test specimens.

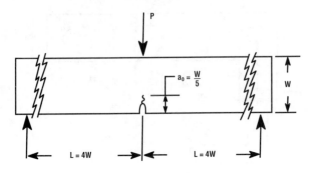

d. Three point bend specimen.

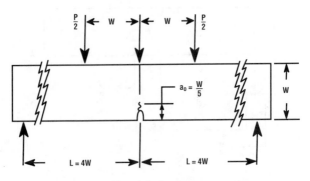

e. Four point bend specimen.

Figure 12d-e. Common fracture toughness test specimens.

The following equations accompany the specimens shown in *Figure 12*, and are used to calculate the K and G values. The K_{IC} and G_{IC} values are obtained by using the load at which the crack begins to propagate, and the K_C and G_C are determined by using the load at fast fracture.

Figure 12a: $G = (P^2 \div EWt^2) \times \tan(\pi a \div W)$
$EG_C = (1 - \mu^2) \times K_{IC}^2 = EG_{IC}$

where   a = one-half crack length at fracture
            t = thickness
            W = width

Figure 12b: $G = (P^2 \div EWt^2) \times (\tan[\pi a \div W] + 0.1 \sin[2\pi a \div W])$
$EG_C = (1 - \mu^2) \times K_{IC}^2 = EG_{IC}.$

Figure 12c: $G_{IC} = (P^2 \div EWt^2) \times (7.59[a \div W] - 32[a \div W]^2 + 117[a \div W]^3)$
$EG_C = (1 - \mu^2) \times K_{IC}^2 = EG_{IC}.$

Figure 12d: $G_{IC} = (P^2 L^2 \div Et^2W^3) \times (31.7 [a \div W] - 64.8 [a \div W]^2 + 211 [a \div W]^3)$
$EG_C = (1 - \mu^2) \times K_{IC}^2 = EG_{IC}.$

Figure 12e: $G_{IC} = (P^2 L^2 \div Et^2W^3) \times (34.7 [a \div W] - 55.2 [a \div W]^2 + 196 [a \div W]^3)$
$EG_C = (1 - \mu^2) \times K_{IC}^2 = EG_{IC}.$

The load at which the crack first begins to propagate (the "pop-in" point) is measured by recording the strain occurring across the crack, in much the same way as in the tensile stress. A load-strain curve (as shown in *Figure 13*) can be obtained from the data collected, and the deviation from linearity is the pop-in point. Two types of behavior are demonstrated. In the first, pop-in occurs and then the load increases until failure under plane stress condition. In the second case, failure occurs immediately as pop-in initiates, indicating that at one-half inch thickness full restraint exists and the plane strain conditions prevail.

Designing against low strength failures. There are three basic fracture mechanics philosophies to designing against low strength failures: proof testing, leak before failure, and stress analysis method.

Proof testing postulates that if a structure contains a crack and it is loaded to a particular stress and the crack does not propagate, then it may be safely used at a slightly lower static operating stress. This is the basis of implementing fracture toughness through proof testing. For example, if a material has a characteristic fracture stress flaw curve such as the one shown in *Figure 14*, and is proof tested at point Sp, then the largest flaw that can possibly be present is slightly smaller than c_1. The static operating stress in this case is So, and the curve shows that for a failure to occur at So, a crack of size c_2 must be present. However, it has already been shown that the largest possible flaw that can exist is c_1, and since c_2 is larger than c_1, the structure can safely operate at So.

Leak before failure is primarily used in pressure vessel design. The success of this method is dependent on the strain-plane stress transition. A surface or embedded flaw in a pressure vessel may propagate at the proof, or operating, pressure of the vessel. The material to be used is selected on the basis that it will be able to tolerate a crack having a length of at least twice the wall thickness of the vessel at the required stress level, under plane strain conditions. Then, if a surface or embedded crack exists, it may propagate at the operating stress. In this case, it will pop through the thickness when its length is approximately twice the wall thickness. Provided the wall is thin enough, the crack

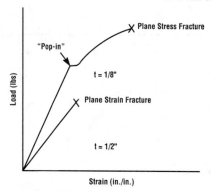

Figure 13. Fracture toughness load curves.

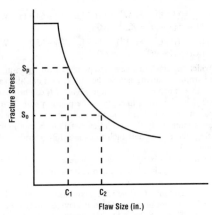

Figure 14. Fracture stress flaw size curve.

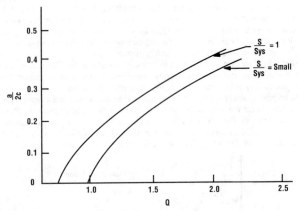

Figure 15. Flaw shape parameter curve.

is under plane stress conditions and additional load or stress is required for further crack propagation. Therefore, the vessel will leak and the defect can be easily detected before brittle failure can occur.

Stress analysis uses fracture toughness of the material and the defect size that can be detected by nondestructive inspection to limit the working stress of a part. The defect shape and size are both considered, and are expressed in terms of the defect size parameter, a/Q. The parameter is the ratio of a, the crack depth, to Q, the flaw shape parameter. Q, in turn, is a function of the ratio of crack depth to crack length: Q is equal to tensile stress times the crack depth, divided by the crack length. This function is plotted in *Figure 15*. To determine the fracture strength of a part having a known defect size, the value of Q is determined from the curve on the figure, and fracture stress is then determined with the following equation.

$$S = K_{IC} \div \sqrt{1.21} \times (a \div Q)$$

where S = applied stress

K_{IC} = plain strain fracture toughness.

Qualitative Methods

Charpy impact tests, the Izod test, and notched tensile tests are classified as qualitative brittle fracture tests. Essentially, the Charpy is a simple beam test, and the Izod is a cantilever beam test. They measure fracture toughness, and are useful for evaluating the transition from brittle to ductile structure in steels. Both are easy to conduct. The Charpy has gained wide recognition and acceptance because it can be done quickly, easily, and with reliable results.

In the Charpy test, a swinging pendulum is released from a known height to strike a specially prepared notched specimen (as shown in *Figure 16*) that is positioned in the anvil of an impact machine. The pendulum's knife-edged striker contacts the specimen at the bottom of the swing, at which point the kinetic energy of the pendulum reaches a maximum value of 25 to 240 ft lbs (35 to 325 J) on impact. After breaking the specimen, the pendulum continues in its swing to a height that is measured. Since the pendulum is released from a known height, its kinetic energy at impact is a known quantity. The energy expended in breaking the specimen can be expressed as a function of the difference in the release height of the pendulum and the height it attains after the specimen is broken. Therefore,

$$E_A = W (h_o - h_f)$$

where E_A = energy absorbed by the specimen in foot pounds or Joules

W = weight of the pendulum

h_o = release height

h_f = final height.

Normally, swinging pendulum impact machines are equipped with scales from which the energy absorbed by the specimen can be read directly in foot pounds or Joules.

The Charpy test is a convenient, though very definitely destructive, test method for determining the change in the fracture mode of a steel as a function of temperature. A series of specimens can be tested over a range of temperatures and the data thus obtained can be plotted to develop a brittle to ductile transition curve. The transition from a brittle to a ductile material usually occurs over a range of temperatures. Likewise, the fracture of a specimen changes from 100% cleavage (a bright, faceted appearance), to 100% shear (a silky, fibrous appearance). The temperature at which specimens show a fracture of 50% cleavage and 50% shear is frequently defined as the transition temperature.

The Charpy test is also a convenient method for comparing the notch toughness characteristics of various materials.

Methods and procedures for conducting Charpy tests are specified in ASTM E23. It should also be noted that the NATO nations use the Charpy test as the standard impact test for steel for guns.

The Izod test supports the specimen on one end only, leaving usually up to 75% of its area projecting from a fixture. The notched side of the specimen faces the anvil of the pendulum, and the force required to fracture the specimen is recorded, much as in a Charpy test. The singular drawback of the Izod test is that setup can be lengthy, making it impractical for testing materials at different temperatures.

Notched tensile tests are performed on cylindrical specimens having an annular notch,

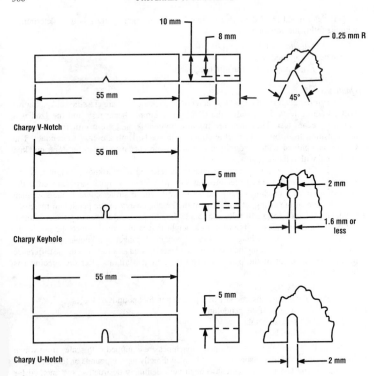

Figure 16. Charpy impact test specimens.

and on flat specimens having double edge notches. The notches act as stress concentrators and affect the load carrying capabilities of the specimens. The stress concentration, due to a notch or an abrupt change of section, is described in terms of K_t, the theoretical stress concentration factor. K_t is the ratio of the maximum stress, due to the notch, to the nominal stress across the area beneath the notch. It should not be confused with K_C or K_{IC}, the fracture toughness values. Values for K_t have been established for various geometries, the most common of which are shown in *Figure 17*. K_t is a function of both the specimen size and the notch radius, and is often approximated with the equation $K_t = \sqrt{a/r}$.

In the most commonly performed notch tests, a round specimen with a 50% notch depth is used. The restraint caused by the notch sets up a state of triaxial tension near the notch root. This stress state increases the flow resistance of the metal and thus decreases the ductility at failure, but it may also increase the fracture strength. The notch strength ratio (NSR—the ratio of notched strength to unnotched strength) has been shown to increase with increasing notch sharpness, at a constant notch depth, to a maximum value at which

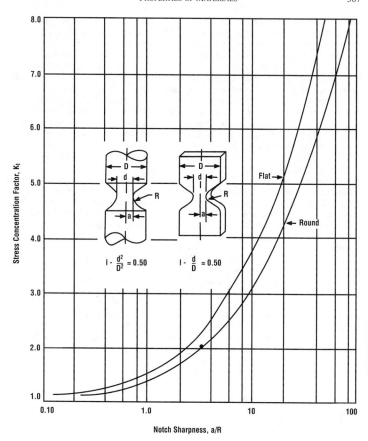

Figure 17. Effect of notch sharpness on the stress concentration factor.

the triaxiality becomes constant. Increasing the sharpness beyond this point causes a reduction in the NSR for notch-sensitive materials. Increasing the notch depth will increase the triaxiality, and will increase the notch strength of the specimen. The primary use of notched tensile tests is to screen materials.

Fatigue Tests

Fatigue has been defined as the "process of progressive localized permanent structural change occurring in material subjected to conditions that produce fluctuating stresses and strains at some point or points, and which may culminate in cracks or complete fracture after a sufficient number of fluctuations. Traditionally, tests have been performed on specimens having simple geometries in attempts to characterize the failure properties of particular materials. Fatigue tests have also been conducted for many reasons, including fatigue life information for design purposes, to evaluate the differences between materials, or to investigate the influences of heat treatment or mechanical working.

A material that is repeatedly subjected to tensile loading may fail even though the stresses imposed on the specimen are below yield strength. Failure may occur after a few load applications or after years of repeated applications. This phenomenon is known as fatigue, and fatigue life is dependent on many variables such as maximum stress, average stress, alternating stress, the ratio of alternating to average stress, yield strength, surface condition, and environment. Since the first indication of fatigue in a material is cracking, fatigue tests generally fall into two categories: those designed to initiate cracking, and those that propagate cracks to determine their growth rate and failure point. Most fatigue tests are performed by repeatedly rotating a simple beam until cracking is determined or it breaks, or by subjecting the specimen to alternating axial tension and strain loads. In rotating tests, the specimen is subjected 1800 RPM or more. The pressures on the specimen provide axial-loads and axial-strains, subjecting the specimen to repeated cyclic loading until failure occurs, or until the testers and machine admit defeat and discontinue the test. The loading cycles are classed as tension-compression, tension-tension, or zero stress-tension. The maximum stress, $S\,max$, is the highest tension stress in the cycle, and the minimum stress, $S\,min$, is the largest compressive stress for compression-tension cycling, or the minimum tensile stress for tension-tension cycling. The mean (or average) of $S\,max$ and $S\,min$ is Sm, and Sa is the alternating stress, calculated as $S\,max$ minus Sm.

The basic method of presenting fatigue data is on an S-N curve that plots maximum stress as a function of the number of cycles to failure. When reading an S/N curve, it is necessary to identify the "A" ratio (alternating stress, Sa, divided by the mean stress, Sm) or the stress ratio, "R" ($S\,min$ divided by $S\,max$), because the fatigue life of a material is dependent on the maximum stress and the mean stress of the applied cyclic load. At a specified maximum stress, the fatigue life of a material is the lowest under reversed bending, when Sm equals zero and Sa = T. For a given maximum stress, the fatigue life increases as Sm increases and Sa decreases. On the example S-A chart shown in *Figure 18*, the curved horizontal lines labeled "endurance limit" are stresses below which a material will not fail by fatigue, under the indicated conditions.

A Soderberg diagram is another, and more useful, method of presenting fatigue data.

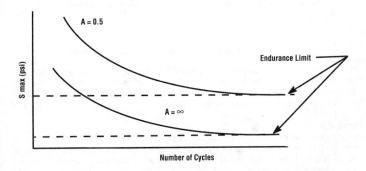

Figure 18. An S-N curve for predicting fatigue life.

Also called a "modified Goodman diagram," it plots the alternating stress (Sa) against the mean stress (Sm). *Figure 19* illustrates a sample Soderberg diagram, which shows the dependence of fatigue life on the alternating and mean stress, as well as on the A ratio. If the fatigue life is to be determined under the conditions of $Sa = Sm$ (A = 1) at a given maximum stress, and if $Sa = S\ max - Sm = Sa_1$, then the stress Sa_1 is read off the vertical axis and the fatigue life at the proper A ratio is read from the diagram—in this case 10^4 cycles. It can also be noted that at an A ratio of 5, the alternating stress that could be endured would be Sm_2 for 10^4 cycles. Ideally, the Soderberg diagram should be constructed by determining the S-N curve for several A ratios and then plotting these data. The diagram can, however, be determined by testing with only one A ratio. To do this, the S-N curve is established and the points are plotted on the appropriate A ratio line. As an example, data might be obtained from an A ratio of 1. These data are plotted on the 45° (A = 1) line as shown in *Figure 20*, and then a straight line can be drawn connecting these points and the material ultimate strength value plotted on the Sm axis. This will provide a Soderberg diagram that is conservative, but accurate enough for many engineering applications. This method offers considerable utility because most of the fatigue data found in the literature is in the form of S-N curves for reversed bending (A ratio = T). Using this method, it is possible to use such data for applications in which the A ratio is other than T. The dashed line in *Figures 19* and *20* that connects the yield strengths plotted on each axis limits the range over which the Soderberg diagram may be used. If values beyond this line are selected, the yield strength of the material is exceeded. Since yielding is usually a criteria for failure, fatigue tests in which the maximum stress is greater than yield are seldom needed.

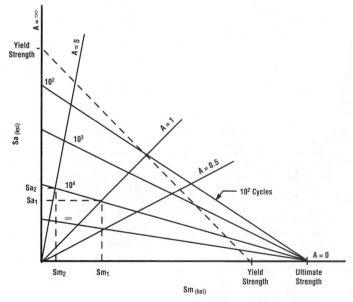

Figure 19. A typical Soderberg diagram.

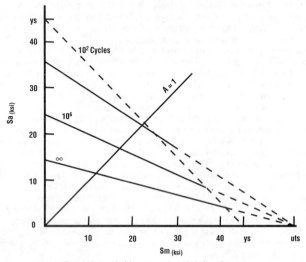

Figure 20. A method for constructing a Soderberg diagram.

There are many factors that affect fatigue life, and the presence of a notch or any other defect that causes a stress concentration can seriously reduce the fatigue life of a metal. The amount of the reduction in fatigue strength is dependent on the stress concentration factor, K_t. For a notched specimen with a K_t of 2, the fatigue strength would be one-half of the unnotched fatigue strength for a given number of cycles. The actual amount that the fatigue strength is reduced is expressed by the fatigue notch factor, K_f, which is the ratio of the unnotched strength to the notched strength for a given number of cycles. Therefore,

$$K_f = Su \div Sn,$$

where Su is the unnotched fatigue strength and Sn is the notched fatigue strength. For small values, K_f is equal to K_t, but as K_t increases, the ratio of the two values increases. K_f is determined by obtaining S-N curves for notched and unnotched specimens of the same material. Then, for a given number of cycles, the ratio of $Su \div Sn$ is determined from these curves. K_f is also dependent on the notch radius and sharpness, and may have different values for the same K_t, when K_t is developed from nonidentical geometries.

For loadings where the cyclic loading does not vary (i.e., where Sa and Sm remain constant), the fatigue life can be read directly from the S-N curve or the Soderberg diagram. When the loading cycle varies, a method such as the linear damage principle is used. The linear damage principle can be stated as "the total fatigue life is the sum of the number of cycles experienced at a given stress, divided by the fatigue life at that stress." In equation form, it is:

$$(M_1 \div N_1) + (M_2 \div N_2) + (M_3 \div N_3) + (Mi \div Ni) = 1,$$

where   M = cycles at a given stress
   N = fatigue life at that stress.

For example, if a material having the fatigue characteristics shown below had been subjected to 20,000 cycles at 20,000 PSI, how many additional cycles could it withstand at 50,000 PSI before it would fail?

Sample characteristics

$S\ max$ (PSI)	Cycles to Failure
10,000	∞
20,000	70,000,000
30,000	20,000,000
50,000	6,000,000

In this example, $M_1 = 20,000,000$ for $S = 20,000$ PSI and $N_1 = 70,000,000$. M_2 is unknown for $S = 50,000$ PSI. $N_2 = 6,000,000$.

$$M_1 \div N_1 + M_2 \div N_2 = 1,$$
$$20.000,000 \div 70,000,000 + M_2 \div 6,000,000 = 1, \text{ and}$$
$$M_2 = 6,000,000 \times 1 - (20,000,000 \div 70,000,000),$$
$$1 = 7/7, \text{ and } 20,000,000 \div 70,000,000 = 2/7$$

therefore, $M_2 = 6,000,000 \times 5/7 = 4,280,000$.

As can be determined by the calculation, the part could be expected to withstand an additional 4,280,000 cycles at 50,000 PSI.

It is good design practice to eliminate stress concentrations. This applies not only to stress concentrations designed into a part, but also to those resulting from surface roughness and defects in the material. The effect of surface finish on fatigue life, as reported by G. Dieter Jr. in his book *Mechanical Metallurgy*, can be seen from the following readings obtained on 3130 steel specimens tested under completely reversed stress at 95,000 PSI.

Condition	Surface Roughness	Average Fatigue Life (Cycles)
As machined by lathe	105	24,000
Light hand polishing	6	91,000
Hand-polished	5	137,000
Superfinished	7	212,000
Ground	7	217,000
Ground and polished	2	234,000

Other surface properties can be adjusted to improve fatigue life. Carburizing, nitriding, and cold working will also increase fatigue life. Conversely, decarburization on the surface is particularly damaging to fatigue properties. Shot-peening introduces compressive residual stresses into the surface, as well as increasing strength by cold work. Both of these factors increase fatigue life, but shot-peening also increases surface roughness which of course decreases fatigue life.

Corrosion in metal

Most briefly stated, corrosion is the deterioration of a metal due to chemical or electrochemical action. Because it is so prevalent, corrosion losses are one of the most expensive problems faced in the application of metals.

The galvanic cell, in its simplest form, consists of two electrodes immersed in a conducting solution (the electrolyte). The electrodes may be two dissimilar metals, a metal and a conducting nonmetal (e.g., carbon), or a metal and an oxide. In each case, an electrical potential, which may produce current when the electrodes are joined by a suitable conductor, is induced between the electrodes. When current is flowing, reactions take place at each electrode as seen in *Figure 21*.

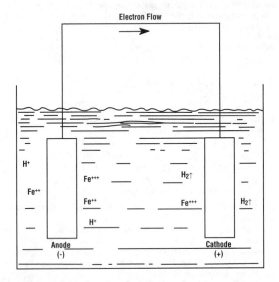

Anode Reaction: Fe → Fe++ + 2e⁻ → Fe+++ + 3e⁻
Cathode Reaction: 2H+ + 2e⁻ → H₂ (gas)

Figure 21. A galvanic cell.

In a cell, dissolution of the metal occurs at the negative electrode (the anode), while hydrogen is evolved at the positive electrode (the cathode). The anodic metal is corroded and the cathode is protected. The rate of corrosion of the anodic metal depends on the degree of separation of the two metals concerned in the practical galvanic series of metals and alloys (see **Table 2**). The corrosion rate is also dependent on the conductivity and composition of the electrolyte and the relative area of the two metals. Galvanic corrosion in service applications occurs through the formation of composition cells or concentration cells.

Composition cells may occur between dissimilar metals or in a single metal that has areas of differing electrode potential. An example of the dissimilar metal cell is galvanized (Zn coated) steel. The zinc coating is anodic to the steel base metal. Therefore, when an electrolyte, such as water, is present between the Fe and Zn, a galvanic cell is produced. Since Zn is anodic to steel, it is preferentially attacked, thereby protecting the steel. A composition cell may be set up in a single metal because of the differences in electrode potential between different phases, or between the grain boundaries and the matrix. An example of this is intergranular corrosion. This type of attack takes place when the grain boundaries are anodic to the matrix. The occurrence of intergranular corrosion is strongly dependent on the thermal and mechanical treatment given the metal. Generally, the stronger the alloy is made through heat treatment or cold work, the less corrosion resistant it becomes.

Table 2. Galvanic Series of Metals and Alloys.

Anodic End	Magnesium
	Magnesium Alloys
	Zinc
	Alclad
	Aluminum 6053
	Cadmium
	Aluminum 2024
	Cast Iron
	Wrought Iron
	Mild Steel
	13% Chromium Steel Type 410 (Active)
	18-8-3 Chromium Nickel Stainless Steel Type 316
	18-8 Chromium Nickel Stainless Steel Type 304 (Active)
	Tin
	Lead
	Lead Tin Solders
	Naval Brass
	Manganese Bronze
	Muntz Metal
	76 Ni-16 Cr-7 Fe Alloy (Active)
	Nickel (Active)
	Silicon Bronze
	Copper
	Red Brass
	Aluminum Brass
	Admiralty Brass
	Yellow Brass
	76 Ni-16 Cr-7 Fe Alloy (Passive)
	Nickel (Passive)
	Silver Solder
	70-30 Cupro-Nickel
	Monel
	Titanium
	13% Chromium Steel Type 410 (Passive)
	18-8-3 Chromium Nickel Stainless Steel Type 316 (Passive)
	18-8 Chromium Nickel Stainless Steel Type 304 (Passive)
	Silver
	Graphite
	Gold
Cathodic End	Platinum

Concentration cells are formed when there is a difference in the concentration of the electrolyte between the areas in contact. The area with the weakest concentration of electrolyte is attacked. The same type of cell, but of considerably more importance, is the oxygen concentration cell, which is characterized by areas that have different levels of oxygen concentration. The area having the lowest oxygen concentration is anodic and is attacked. Corrosion caused by an oxygen concentration cell can usually be found under surface dirt, mill scale, or other areas that may be oxygen deficient.

Stress corrosion may occur in a susceptible material when it is subjected to residual or applied surface tensile stresses while exposed to a corrosive environment. Stress corrosion cracks initiate and propagate transverse to the loading direction, and a low strength failure results. The time required for stress corrosion to occur is dependent on several variables.

<u>Strength level.</u> As with general corrosion, the higher the strength level obtained through heat treatment or cold working of a given alloy, the lower its resistance to stress corrosion.

<u>Grain orientation.</u> The direction of the applied stress with reference to the grain orientation is of prime importance. The stress corrosion resistance is lowest in the short transverse grain direction and highest in the longitudinal direction, with the long transverse being intermediate. The stress corrosion resistance of a material in the short transverse may be only 10% as good as that in the longitudinal. The resistance in the long transverse direction is dependent on the grain geometry. A material with narrow elongated grains would have a long transverse susceptibility to stress corrosion nearly as poor as that in the short transverse direction. For equiaxed grains (a grain structure with approximately the same dimensions in all directions), such as found in cross rolled material, there is no difference between the longitudinal and the long transverse direction.

<u>Stress level.</u> At a high stress level, a susceptible alloy, when exposed to a corrosive medium, may fail in a matter of minutes. As the stress is decreased, the time to failure increases. This is the basis of the threshold stress concept, which postulates that for stress corrosion susceptible materials, there is a stress value (called threshold stress) below which stress corrosion cracking will not occur, while at higher levels stress corrosion cracking should be expected to occur.

Corrosion protection is afforded by two basic methods: surface protection and galvanic protection. Surface protection involves keeping corrosive environments away from the metal and is commonly achieved by plating, cladding, or painting. Galvanic protection is achieved by electrically connecting the metal with a sacrificial anode, such as zinc coating steels or burying magnesium bars in contact with underground pipelines. Stress corrosion is not usually a problem with carbon steel and wrought iron, but the higher strength alloy steels may be susceptible. It is prevented primarily with surface protection, proper grain direction exposure, and by control of the magnitude of applied stresses.

Physical properties

Thermal conductivity, designated by the letter K, is the measure of the ability of a material to carry or conduct heat. It is analogous to electrical conductivity, and is used in the analysis of heat transfer calculations.

Thermal expansion measures the change in dimension in a solid when it is subjected to a change in temperature. It can be expressed as:

$$\alpha = (\Delta L \div L_0) \times (1 \div T)$$

where α = thermal expansion, expressed as in./in./°F *or* mm/mm/°C
ΔL = change in dimension with temperature
L_0 = original dimension
T = temperature, in °F or °C.

As can be seen from the equation, thermal expansion is actually the strain in a solid due to a temperature change of 1° F or 1° C. The coefficient of thermal expansion depends on the temperature at which the measurement is taken, so the range of temperatures over which α is valid is always specified. The coefficient of thermal expansion is used to calculate volume changes and to determine stresses in restrained members that are subjected to temperature changes. As an example, if a restrained rod is heated from 68 to 268° F (20 to 131° C), which is a temperature change of 200° F, or 111° C, the amount it would increase in length is:

$$\Delta L = L_0 \times \alpha \times \Delta T,$$

or, $\quad \Delta L \div L_0 = \alpha \times \Delta T = e$

where $\quad e = \text{strain}$

$\quad \Delta T = \text{temperature change}.$

Since the rod is restrained, it cannot increase in length and a compressive stress (S) is induced into the rod. By definition,

$$S = E \times e.$$

The two equations above, when combined, show compressive stress as:

$$S = E \times \alpha \times \Delta T.$$

Using the above example and equations, determine the compressive strain on a restrained rod made of steel with a coefficient of thermal expansion of 9.9 in./in./°F $\times$ 10^{-6} (17.82 m/m/°K $\times 10^{-6}$) and a modulus of elasticity of 28,000 ksi (190 GPa). The change in temperature is 200° F (111° C).

$$S = 28,000 \times (9.9 \times 10^{-6}) \times 200 = 55.4 \text{ ksi,}$$

or, $\quad S = 190 \times (17.82 \times 10^{-6}) \times 111 = 0.3578 \text{ GPa (375,823,800 Pascals)}.$

Therefore, a compressive force of 55.4 ksi (0.3578 GPa) would be induced in the rod by the indicated temperature change.

Density is the mass per unit volume of a solid, but it is commonly expressed as weight per unit volume, as pounds per cubic foot, pounds per cubic inch, kilograms per cubic meter, and grams per cubic centimeter. **Table 3** provides weights for selected materials. (It should be mentioned that mass density is measured in slugs/ft^3 in U.S. Customary Units and in kg/m^3 in the metric system.) Density becomes a design consideration when the weight of a structure must be controlled. The selection of an alloy is then made on the basis of a high yield strength to density ratio (also called the strength to weight ratio).

Hardness tests

In simple terms, hardness is a material's ability to resist surface abrasion. The relative hardness of materials can be seen on Mohs' hardness scale, which is based on fifteen materials, each of which is capable of scratching the one below it. Any material can be positioned on Mohs' scale according to which of the other materials it is able to scratch.

Mohs' Hardness Scale

Mineral	Hardness index
Diamond	15
Boron Carbide	14
Silicon Carbide	13
Fused Alumina	12
Fused Zirconia	11
Garnet	10
Topaz	9
Quartz or Stellite	8
Vitreous Silica	7

(Continued)

<u>Mohs' Hardness Scale</u> *(Continued)*

Mineral	Hardness index
Orthoclase (feldspar)	6
Apatite	5
Fluorite	4
Calcite	3
Gypsum	2
Talc	1

While this scale can be useful in determining the approximate relative hardness of a material, it is obviously not capable of recognizing the comparative hardness of steels and other metals. In material selection, hardness plays a large part in determining the machinability and strength, so several tests have been devised for accurately cataloging hardness values.

Table 3. Weights of Materials.

Material	Weight, Expressed in			
	Pounds per Cubic Foot	Pounds per Cubic Inch	Kilograms per Cubic Meter	Grams per Cubic Centimeter
Aluminum	168.5	.0975	2699.11	2.6988
Brass, 80% C, 20% Z	536.6	.3105	8595.51	8.5946
Brass, 70% C, 30% Z	526.7	.3048	8436.92	8.4368
Brass, 60% C, 40% Z	521.7	.3019	8356.83	8.3538
Brass, 50% C, 50% Z	511.7	.2961	8196.65	8.1960
Brick, common	112	.0648	1794.07	1.7937
Brick, fire	143	.0827	2290.64	2.2891
Brick, pressed	137	.0793	2194.53	2.1950
Brick, hard	125	.0723	2002.31	2.0013
Bronze, 90% C, 10% T	547.9	.3171	8776.51	8.7773
Cement, portland, loose	90	.0521	1441.66	1.4421
Cement, portland, set	183	.1059	2931.38	2.9313
Chromium	432.4	.2502	6926.38	6.9255
Clay, loose	63	.0365	1009.16	1.0103
Coal broken, loose, anthracite	54	.0313	864.99	0.8664
Coal broken, loose, bituminous	49	.0294	784.90	0.7861
Concrete	137	.0793	2194.53	2.1950
Copper	554.7	.3210	8885.44	8.8852
Earth, common, loam	75	.0434	1201.38	1.2013
Earth, packed	100	.0579	1601.85	1.6027
Glass	162	.0938	2594.99	2.5964
Gravel, dry, loose	90 to 106	-	1441.66 to 1697.96	-
Gravel, well shaken	99 to 117	-	1585.83 to 1874.16	-
Gold	1204.3	.6969	19291.03	19.2901
Ice	56	.0324	897.03	0.8968
Iron, cast	450	.2604	708.31	7.2078

(Continued)

Table 3. Weights of Materials. *(Continued)*

Material	Weight, Expressed in			
	Pounds per Cubic Foot	Pounds per Cubic Inch	Kilograms per Cubic Meter	Grams per Cubic Centimeter
Iron, wrought	486.7	.2817	7796.18	7.7974
Lead	707.7	.4095	11336.26	11.3349
Lime	53	.0307	848.98	0.8498
Magnesium	108.6	.0628	1739.60	1.7383
Masonry	150	.0868	2402.77	2.4026
Masonry, dry rubble	138	.0799	2210.55	2.2116
Molybdenum	636.5	.3683	10195.75	10.1945
Mortar, set	103	.0590	1649.90	1.6497
Nickel	549	.3177	8794.13	8.7939
Petroleum, benzene	46	.0266	736.85	0.7363
Petroleum, gasoline	42	.0243	672.78	0.6726
Plaster of Paris	112	.0648	1794.07	1.7937
Platinum	1333.5	.7717	21360.62	21.3606
Quartz	162	.0938	2594.99	2.5964
Salt, common	48	.0278	768.88	0.7695
Sand, dry, loose	90 to 106	-	1441.66 to 1697.96	-
Sand, well shaken	99 to 106	-	1585.83 to 1874.16	-
Silver	657	.3802	10524.13	10.5239
Snow, freshly fallen	5 to 12	-	80.09 to 192.22	-
Snow, wet and compacted	15 to 50	-	240.28 to 800.92	-
Steel	490	.2836	7849.05	7.8500
Stone, gneiss	168	.0972	2691.10	2.6905
Stone, granite	168	.0972	2691.10	2.6905
Stone, limestone	162	.0938	2594.99	2.5964
Stone, marble	168	.0972	2691.10	2.6905
Stone, sandstone	143	.0828	2290.64	2.2919
Stone, shale	162	.0938	2594.99	2.5964
Stone, slate	175	.1013	21803.23	2.8040
Tar	75	.0434	1201.38	1.2013
Tin	455	.2632	7288.40	7.2881
Titanium	280.1	.1621	4486.77	4.4869
Tungsten	1192	.6898	19094.00	19.0936
Water, fresh	62.5	.0362	1001.15	1.0020
Water, sea water	64	.0370	1025.18	1.0242
Zinc	439.3	.2542	7036.91	7.0362
Wood, dry				
Ash, black	28	.0162	448.52	0.4484
Ash, white	41	.0237	656.76	0.6560

(Continued)

Table 3. Weights of Materials. *(Continued)*

Material	Weight, Expressed in			
	Pounds per Cubic Foot	Pounds per Cubic Inch	Kilograms per Cubic Meter	Grams per Cubic Centimeter
Wood, dry				
Beech	45	.0260	720.83	0.7197
Birch	44	.0255	704.81	0.7058
Birch, paper	38	.0220	608.70	0.6090
Cedar, Alaska	31	.0179	496.57	0.4955
Cedar, eastern red	33	.0191	528.61	0.5287
Cedar, southern white	23	.0133	368.42	0.3681
Cedar, western red	23	.0133	368.42	0.3681
Cherry	42	.0203	672.78	0.6726
Cherry, black	35	.0203	560.65	0.5619
Chestnut	41	.0237	656.76	0.6560
Cypress	30	.0174	480.55	0.4816
Elm	45	.0260	720.83	0.7197
Hemlock	29	.0168	464.54	0.4650
Hickory	49	.0284	784.90	0.7861
Locust	46	.0266	736.85	0.7363
Mahogany	53	.0307	848.98	0.8498
Maple, hard	43	.0249	688.79	0.6892
Maple, white	33	.0191	528.61	0.5287
Oak, chestnut	54	.0313	864.99	0.8664
Oak, live	59	.0341	954.09	0.9439
Oak, red, black	41	.0237	656.76	0.6560
Oak, white	46	.0266	738.85	0.7363
Pine, white	26	.0150	416.48	0.4152
Pine, yellow, longleaf	44	.0255	656.76	0.7058
Pine, yellow, shortleaf	36	.0208	576.66	0.5757
Poplar	28	.0162	448.52	0.4484
Redwood, California	26	.0150	416.48	0.4152
Spruce, white, black	27	.0156	432.50	0.4318
Sycamore	37	.0214	592.68	0.5923
Walnut, black	38	.0220	608.70	0.6089
Walnut, white	26	.0150	416.48	0.4152

There are three basic types of hardness tests: indention hardness, scratch hardness, and rebound hardness. As the name implies, indention hardness is measured by indenting the metal with a suitable load and indenting mechanism. The hardness value obtained is actually a measure of the material's resistance to plastic deformation, but through empirical calculations, hardness gives the designer an indication of the strength of the alloy or metal being tested. The advantage of indention hardness testing, which is the hardness test commonly applied to metals, is that it is fast, inexpensive, and virtually nondestructive. In the majority of cases, a small indentation less than $1/16$ inch in diameter is produced. The three common indention tests are Brinell, Rockwell, and Vickers. Conversion charts for relative hardnesses on these three scales, plus Scleroscope hardness, are given in **Table 4**, which provides Rockwell C conversions, and **Table 5**, which gives Rockwell B scale conversions.

Table 4. Hardness Conversion Scale for Brinell, Rockwell C, Vickers, Scleroscope, and Knoop Hardness Scales.

Brinell Hardness	Rockwell C Hardness	Vickers Hardness	Scleroscope Hardness	Knoop Hardness	Tensile Strength ksi	Tensile Strength MPa
3000 kg	150 kg					
767	66.5	882	93	870	-	-
745	65.3	842	91	584	-	-
722	64	800	88	822	-	-
705	63	772	85	799	-	-
688	62	746	84	776	-	-
670	61	720	83	754	-	-
654	60	697	81	732	-	-
634	59	674	80	710	351	2420
615	58	653	79	690	338	2330
595	57	633	77	670	325	2241
577	56	613	75	650	313	2158
560	55	595	74	630	301	2075
543	54	577	72	612	292	2013
525	53	560	71	594	283	1951
512	52	544	70	576	273	1882
496	51	528	68	558	264	1820
481	50	513	67	542	255	1758
469	49	498	66	526	246	1696
455	48	484	65	510	237	1634
443	47	471	63	495	229	1579
432	46	458	62	480	221	1524
421	45	446	60	466	215	1482
409	44	434	59	452	208	1434
400	43	423	58	438	201	1386
390	42	412	56	426	195	1344
381	41	402	55	414	188	1296
371	40	392	54	402	182	1255
362	39	382	52	391	177	1220
353	38	372	51	380	171	1179
344	37	363	50	370	166	1145
336	36	354	49	360	161	1110

(Continued)

Table 4. *(Continued)* **Hardness Conversion Scale for Brinell, Rockwell C, Vickers, Scleroscope, and Knoop Hardness Scales.**

Brinell Hardness 3000 kg	Rockwell C Hardness 150 kg	Vickers Hardness	Scleroscope Hardness	Knoop Hardness	Tensile Strength ksi	Tensile Strength MPa
327	35	345	48	351	155	1069
319	34	336	47	342	152	1048
311	33	327	46	334	149	1027
301	32	318	45	326	146	1007
294	31	310	44	318	141	972
286	30	302	42	311	138	951
279	29	294	41	304	135	931
271	28	286	40	297	131	903
264	27	279	39	290	128	883
258	26	272	38	284	126	869
253	25	266	-	278	123	848
247	24	260	37	272	119	820
243	23	254	36	265	117	807
237	22	248	35	261	115	793
231	21	243	34	256	112	772
226	20	238	34	251	110	758

Brinell ratings 500 and above obtained with tungsten carbide ball. All Brinell with 3,000 kg load.
Rockwell C ratings obtained with diamond indenter and 150 kg load.

Table 5. Hardness Conversion Scale for Brinell, Rockwell B, Vickers, Scleroscope, and Knoop Hardness Scales.

Brinell Hardness 3000 kg	Brinell Hardness 500 kg	Rockwell B Hardness 100 kg	Vickers Hardness	Scleroscope Hardness	Knoop Hardness	Tensile Strength ksi	Tensile Strength MPa
240	201	100	240	36	261	116	800
234	195	99	234	35	258	114	786
228	189	98	228	34	253	109	752
222	184	97	222	-	247	105	724
216	179	96	216	33	240	102	703
210	175	95	210	32	233	100	689
206	171	94	206	31	225	98	676
200	167	93	200	30	218	94	648
195	163	92	195	29	211	92	634
190	160	91	190	-	206	90	621
185	157	90	185	28	201	89	614
180	154	89	180	27	196	88	607
176	151	88	176	-	192	86	593

(Continued)

Table 5. *(Continued)* **Hardness Conversion Scale for Brinell, Rockwell B, Vickers, Scleroscope, and Knoop Hardness Scales.**

Brinell Hardness 3000 kg	Brinell Hardness 500 kg	Rockwell B Hardness 100 kg	Vickers Hardness	Scleroscope Hardness	Knoop Hardness	Tensile Strength ksi	Tensile Strength MPa
172	148	87	172	26	188	84	579
169	145	86	169	-	184	83	572
165	142	85	165	25	180	82	565
162	140	84	162	-	176	81	558
159	137	83	159	24	173	80	552
156	135	82	156	-	170	76	524
153	133	81	153	23	167	73	503
150	130	80	150	-	164	72	496
147	128	79	147	22	161	70	483
144	126	78	144	-	158	69	476
141	124	77	141	21	155	68	469
139	122	76	139	-	152	67	462
137	120	75	137	-	150	66	455
135	118	74	135	-	147	65	448
132	116	73	132	-	145	64	441
130	114	72	130	20	143	63	434
127	112	71	127	-	141	62	427
125	110	70	125	19	139	61	421
123	109	69	123	-	137	60	414
121	108	68	121	18	135	59	407
119	106	67	119	-	133	58	400
117	104	66	117	17	131	57	393
116	102	65	116	16	129	56	386
114	100	64	114	15	127	-	-
112	99	63	112	-	125	-	-
110	98	62	110	-	124	-	-
108	96	61	108	-	122	-	-
107	95	60	107	-	120	-	-
106	94	59	106	-	118	-	-
104	92	58	104	-	117	-	-
103	91	57	103	-	115	-	-
101	90	56	101	-	114	-	-
100	89	55	100	-	112	-	-
-	87	54	-	-	111	-	-
-	86	53	-	-	110	-	-
-	85	52	-	-	109	-	-
-	84	51	-	-	108	-	-

(Continued)

Table 5. *(Continued)* **Hardness Conversion Scale for Brinell, Rockwell B, Vickers, Scleroscope, and Knoop Hardness Scales.**

Brinell Hardness 3000 kg	Brinell Hardness 500 kg	Rockwell B Hardness 100 kg	Vickers Hardness	Scleroscope Hardness	Knoop Hardness	Tensile Strength ksi	Tensile Strength MPa
-	83	50	-	-	107	-	-
-	82	49	-	-	106	-	-
-	81	48	-	-	105	-	-
-	80	47	-	-	104	-	-
-	80	46	-	-	103	-	-
-	79	45	-	-	102	-	-
-	78	44	-	-	101	-	-
-	77	43	-	-	100	-	-
-	76	42	-	-	99	-	-
-	75	41	-	-	98	-	-
-	75	40	-	-	97	-	-

Brinell readings are given for both 500 kg and 3,000 kg load with 10 mm ball.
Rockwell B ratings obtained with 100 kg weight and $^1/_{16}$ inch ball.

Brinell Hardness Testing

A Swede, Johann Brinell, invented the test named for him in 1900, and it remains in use as the most often cited method of hardness testing, especially on castings and forgings. Brinell hardness is measured by indenting the surface of the metal with a hardened steel ball (10 mm in diameter) under a constant load of 500 to 3,000 kg (1100 to 6614 lbs). The lightest load is normally used for testing nonferrous alloys, and the highest load is used on harder materials such as steels and cast irons. The applied load is usually held for thirty seconds on nonferrous materials and for one-third to one-half that time on harder materials. The diameter of the indentation is then measured with a microscope having an ocular scale, and the Brinell Hardness (BHN, or HB) is calculated with the following equation.

$$BHN = P \div \text{surface area of the impression,}$$
or, $$BHN = P \div ([\pi D \div 2] \times [D - \sqrt{D^2 - d^2}])$$
where P = applied load in kilograms
      D = diameter of the ball
      d = diameter of the indentation, in millimeters.

For materials with BHN of 444 to 627 (which is the upper limit of BHN ratings), the steel ball must be substituted with a ball made of tungsten carbide.

When conducting Brinell tests, there are several standards that should be followed. Perhaps of primary importance is to be sure that the test is performed on a flat surface, not one that is curved or has irregularities. Indentations should not be made near an edge of the material, as there may be insufficient supporting material on the edge side of the indentation. Since indentations cause localized cold working of the material, indentations should not be made within three diameters of each other. The material's thickness at the location of the test should be a minimum of ten times the depth of the indentation.

Brinell Hardness can be converted to Rockwell C scale and Rockwell B scale hardness numbers with the following conversion formulas, courtesy of Kennametal Inc. It should be noted that even though these conversion formulas provide a reasonable degree of accuracy, the validity of converting from one measurement system to another cannot always be assured. The conversion charts on the Tables in this section may also be used for converting from one scale to another.

Rockwell C Hardness (HRC)		Equation for Converting HRC to Brinell Hardness
from	to	
21	30	BHN = 5.97 × HRC + 104.7
31	40	BHN = 8.57 × HRC + 27.6
41	50	BHN = 11.158 × HRC + 79.6
51	60	BHN = 17.515 × HRC – 401
Rockwell B Hardness (HRB)		Equation for Converting HRB to Brinell Hardness
from	to	
55	69	BHN = 1.646 × HRB + 8.7
70	79	BHN = 2.394 × HRB – 42.7
80	89	BHN = 3.279 × HRB – 114
90	100	BHN = 5.582 × HRB – 319

Rockwell Hardness Testing

The most widely used hardness test in the U.S. is the Rockwell test. The indenter in this test is specified by ASTM specifications, and it has a diamond cone point (called a Brale indenter) with a radius of 200 ± 10 μm (0.0079 in.). Unfortunately, for years the indenter size was specified as 192 μm, and older tests are not compatible with newer tests, especially in the HRC (or R_C) 63 range. In substitution for the diamond indenter, Rockwell tests may make use of hardened steel and cemented carbide balls. Standard ball diameters are $^1/_{16}$, $^1/_8$, $^1/_4$, and $^1/_2$ inch, and they are usually employed to test softer materials including soft steels and aluminum and copper alloys.

There are several different Rockwell "scales," the two most common being the B and C. The scale is determined by the indenter and load used in the test. HRB (R_B) uses a $^1/_{16}$" ball and 100 kg load and is used to measure soft steels, malleable irons, and copper and aluminum alloys as well as other softer materials. HRC uses the Brale indenter and 150 kg load and is used to measure materials harder than HRB 100 including steels, hard cast irons, pearlitic malleable iron, titanium, and deep case-hardened steel. Most of the remainder of the alphabet is used up with the other scales, which are: HRA–Brale indenter and 60 kg load, for testing shallow, case-hardened steel, cemented carbide; HRD–Brale indenter and 100 kg load for measuring medium case-hardened steel, some pearlitic malleable cast irons; HRE–uses a $^1/_8$" ball and 100 kg load to test cast iron, magnesium and some aluminum alloys, bearing metals; HRF–uses a $^1/_{16}$" ball indenter and 60 kg load to measure annealed copper alloys, thin soft sheet metals; HRG–uses a $^1/_{16}$" ball indenter and 150 kg load to test phosphorus bronze, beryllium copper, malleable irons; the upper limit for the scale is 92 HRG; HRH–Uses $^1/_8$" ball and 60 kg load for testing aluminum, zinc, lead. The following suffixes are used for bearing metals and very soft or thin materials: K ($^1/_8$" ball, 150 kg load), L ($^1/_4$" ball, 60 kg load), M ($^1/_4$" ball, 100 kg load), P ($^1/_4$" ball, 150 kg load), R ($^1/_2$" ball, 60 kg load), S ($^1/_2$" ball, 100 kg load), V ($^1/_2$" ball, 150 kg load); and there are designations for the T, W, and X suffixes which are also used for very soft materials. See **Tables 4, 5**, and **6** for Rockwell hardness charts.

In the actual test, a minor load of 10 kg is first applied to stabilize the indenter and specimen. This is followed by the major load, which may be 60, 100, or 150 kg, and the

Table 6. Rockwell Hardness and Rockwell Superficial Hardness Comparative Scales. (See Notes.)

| | Rockwell Hardness Scales | | | | | Rockwell Superficial Hardness Scales | | | | | |
A Scale	B Scale	C Scale	D Scale	E Scale	F Scale	15-N Scale	30-N Scale	45-N Scale	15-T Scale	30-T Scale	45-T Scale
85.0	–	66.0	76.0	–	–	93.0	83.0	73.0	–	–	–
84.0	–	64.0	74.0	–	–	92.0	81.0	71.0	–	–	–
83.0	–	62.0	73.0	–	–	91.0	79.0	69.0	–	–	–
81.0	–	60.0	71.0	–	–	90.0	78.0	67.0	–	–	–
80.0	–	58.0	69.0	–	–	89.0	76.0	64.0	–	–	–
79.0	–	56.0	68.0	–	–	88.0	74.0	62.0	–	–	–
78.0	–	54.0	66.0	–	–	87.0	72.0	60.0	–	–	–
77.0	–	52.0	65.0	–	–	86.0	70.0	57.0	–	–	–
75.5	–	50.0	63.0	–	–	85.5	68.0	54.5	–	–	–
74.5	–	48.0	61.5	–	–	84.5	66.5	52.5	–	–	–
73.5	–	46.0	60.0	–	–	83.5	64.5	50.0	–	–	–
72.5	–	44.0	58.5	–	–	82.5	63.0	47.5	–	–	–
71.5	–	42.0	57.0	–	–	81.5	61.0	45.5	–	–	–
70.5	–	40.0	55.5	–	–	80.5	59.5	43.0	–	–	–
69.5	–	38.0	54.0	–	–	79.5	58.0	41.0	–	–	–
68.5	–	36.0	52.5	–	–	78.5	56.0	38.5	–	–	–
67.5	–	34.0	50.5	–	–	77.5	54.5	36.0	–	–	–
66.5	106	32.0	49.5	–	116.5	76.5	52.5	34.0	94.5	85.5	77.0
64.5	104	28.5	46.5	–	115.5	75.0	49.5	30.0	94.0	84.5	75.0
63.0	102	25.5	44.5	–	114.5	73.5	47.0	26.5	93.0	83.0	73.0
61.5	100	22.5	42.0	–	113.0	72.0	44.5	23.0	92.5	81.5	71.0
60.5	98	20.0	40.0	–	112.0	70.5	42.0	20.0	92.0	80.5	69.0
59.0	96	17.0	38.0	–	111.0	69.0	39.5	17.0	91.0	79.0	67.0
57.5	94	14.5	36.0	–	110.0	68.0	37.5	14.0	90.5	77.5	65.0
56.5	92	12.0	34.0	–	108.5	66.5	35.5	11.0	89.5	76.0	63.0
55.0	90	9.0	32.0	108.5	107.5	65.0	32.5	7.5	89.0	75.0	61.0
53.5	88	6.5	30.0	107.0	106.5	64.0	30.5	5.0	88.0	73.5	59.5
52.5	86	4.0	28.0	106.0	105.0	62.5	28.5	2.0	87.5	72.0	57.5
51.5	84	2.0	26.5	104.5	104.0	61.5	26.5	0.5	87.0	70.5	55.5
50.0	80	–	24.5	103.0	103.0	–	–	–	86.0	69.5	53.5
49.0	78	–	22.5	102.0	101.5	–	–	–	85.5	68.0	51.5
47.5	76	–	21.0	100.5	100.5	–	–	–	84.5	66.5	49.5

(Continued)

Table 6. *(Continued)* Rockwell Hardness and Rockwell Hardness Superficial Hardness Comparative Scales. *(See Notes.)*

	Rockwell Hardness Scales						Rockwell Superficial Hardness Scales					
A Scale	B Scale	C Scale	D Scale	E Scale	F Scale	15-N Scale	30-N Scale	45-N Scale	15-T Scale	30-T Scale	45-T Scale
46.5	76	-	19.0	99.5	99.5	-	-	-	84.0	65.5	47.5
45.5	74	-	17.5	98.0	98.5	-	-	-	83.0	64.0	45.5
44.0	72	-	16.0	97.0	97.0	-	-	-	82.5	62.5	43.5
43.0	70	-	14.5	95.5	96.0	-	-	-	82.0	61.0	41.5
42.0	68	-	13.0	94.5	95.0	-	-	-	81.0	60.0	39.5
41.0	66	-	11.5	93.0	93.5	-	-	-	80.5	58.5	37.5
4.0.	64	-	10.0	91.5	92.5	-	-	-	79.5	57.0	35.5
39.0	62	-	8.0	90.5	91.5	-	-	-	79.0	56.0	33.5
-	60	-	-	89.0	90.0	-	-	-	78.5	54.5	31.5
-	58	-	-	88.0	89.0	-	-	-	77.5	53.0	29.5
-	56	-	-	86.5	88.0	-	-	-	77.0	51.5	27.5
-	54	-	-	85.5	87.0	-	-	-	76.0	50.5	25.5
-	52	-	-	84.0	85.5	-	-	-	75.5	49.0	23.5
-	50	-	-	83.0	84.5	-	-	-	74.5	47.5	21.5
-	48	-	-	81.5	83.5	-	-	-	74.0	46.5	19.5
-	46	-	-	80.5	82.0	-	-	-	73.5	45.0	17.0

Notes: **Hardness Scales.** A scale readings are obtained with a Brale indenter and 60 kg load. B scale readings are obtained with a $1/16$ inch ball and 100 kg load. C scale readings are obtained with a Brale indenter and 150 kg load. D scale readings are obtained with a Brale indenter and 100 kg load. E scale readings are obtained with a $1/8$ inch ball and 100 kg load. F scale readings are obtained with a $1/16$ inch ball and 60 kg load. **Superficial Hardness Scales.** 15-N scale readings are obtained with an N Brale indenter and 15 kg load. 30-N scale readings are obtained with an N Brale indenter and 30 kg load. 45-N scale readings are obtained with an N Brale indenter and 45 kg load. 15-T scale readings are obtained with a $1/16$ inch ball and 15 kg load. 30-T scale readings are obtained with a $1/16$ inch ball and 30 kg load. 45-T scale readings are obtained with a $1/16$ inch ball and 45 kg load.

penetration is read on a dial gage. The hardness number read from the gage varies directly with the actual material hardness, and inversely with the depth of penetration. A hard material will not allow as deep a penetration of the indenter as will a soft material. The dial gage used to measure the indentation contains 100 divisions, each of which corresponds to a penetration of 0.00008 inch.

Another version of the Rockwell test is the Superficial Rockwell Hardness Test. The test is conducted as with the standard test, but very light loads are used. The minor (initial load is 3 kg, and the major load may be 15, 30, or 45 kg. Superficial Rockwell values always give the hardness number followed by the letters N (for Brale tester), T ($^1/_{16}$" ball), W ($^1/_8$" ball), X ($^1/_4$" ball), or Y ($^1/_2$" ball) that indicate the type of tester used. A typical Superficial Rockwell designation would be 70HR45W, indicating that the material has a superficial hardness of 70HR and it was tested with a 45 kg load using a $^1/_8$" ball.

Formulas for converting HRB and HRC to BHN hardness are provided in the section immediately above on Brinell hardness.

Vickers Hardness Testing

Also known as the Diamond Pyramid Hardness Test, this test was developed in the U.K. in 1925 and uses a square base diamond indenter. The main attraction of this test is that it uses an identical indenter for all tests, thereby making comparative analysis of different materials easy because they can all be placed on the same scale. The load is applied smoothly in a Vickers test, and may vary from 1 to 120 kg. After being applied for 10 to 15 seconds, it is removed and the impression is measured. A diamond pyramid hardness (DPH) or Vickers hardness (VHN or HV) number obtained in this way is defined as the load divided by area of indentation, and is calculated with the following equation.

$$VHN = (2 P \times \sin \theta/2) \div d^2$$
or, $$VHN = 1.8544 P \div d^2$$
where P = load, in kg
       θ = the angle (136°) between opposite faces in the diamond
       d = average length of the diagonals in the indentation, in millimeters.

A metallurgically polished surface is required for Vickers testing.

Other Hardness Tests

Scleroscope hardness testing (also known as Shore's Scleroscope Test) uses a diamond tipped hammer dropped from a fixed and standardized height. The height of the rebound of the hammer is measured and used as an indicator of the hardness of the material. It offers the ability and flexibility to perform tests rapidly, in an almost production line setting. Also, it usually leaves the specimen completely unmarked, and, like the Vickers test, all Scleroscope hardness ratings are made on a single, easily comparable scale. On the other (negative) hand, the readings are very sensitive to the surface condition of the material, and measurements of the rebound of the hammer are sometimes not as precise as a microscopic measurement of an indented material. See **Tables 4** and **5**.

Knoop indention testing is used for microhardness testing, when it is necessary to perform tests on smaller samples than required by other methods of testing. For example, to test the variation of the hardness through the thickness of a carburized case, Knoop tests could be used. A microhardness test is analogous to ordinary indention testing, the only difference being a smaller indenter and the use of much lighter loads. It should be noted that although the Knoop indenter, in conjunction with the Tukon tester, is widely used for microhardness testing in the U.S., the Vickers indenter test is prevalent in Europe. See **Tables 4** and **5**.

The Knoop indenter is a pyramidal shaped diamond, and its impression, viewed normal to the specimen surface, is rhombic in shape with the long diagonal perpendicular to, and seven times the length of, the shorter diagonal. The loads applied by the Tukon tester are from 1 to 1000 g. The length of the long diagonal of the impression is measured precisely with a microscope, and the Knoop hardness (KHN or HK) is calculated as follows.

$$KHN = P \div (L^2 \times C)$$

where P = load, in kg

L = length of longest diagonal, in millimeters

C = a constant of the indenter, relating projected area of the indentation to the square of the length of the long diagonal.

A metallurgically polished surface is required for Knoop testing.

IHRD (International Rubber Hardness Degrees) tests are carried out on rubber parts, especially O-rings. It uses a light initial force (8.46 grams for micro scales and 295.74 grams for regular scales) applied through an indenter to zero the testing apparatus scale, followed by a test force (ranging from 15.7 grams for micro scales to 597 grams for regular scales). The degree of indentation is measured to determine the hardness. Readings indicate the durometer (testing apparatus) type (either A or D), a test "value," and duration of the test. For instance, a D/50/15 reading indicated that a D type durometer was used, the test value is 50, and the test duration was 15 seconds.

Machinability ratings of materials

The machinability of a workpiece material can be defined as the comparative ease with which a material removal operation can be performed upon the material relative to the same operation on a benchmark material. A material's rating can help in the selection and adjustment of speeds and feeds for unfamiliar materials. For example, SAE-AISI resulfurized carbon steel 1212 has a machinability rating of 100%, while SAE-AISI 1030 has a machinability rating of 70%. This means that if 3 in.^{3} (49.16 cm^3) of SAE-AISI 1112 can be removed per minute with a specific tooling setup, that same setup should remove 70% of that amount, or 2.1 in.^{3} (34.41 cm^3) of 1030 carbon steel. See **Table 7** for comparative machinability ratings.

Table 7. Machinability Rating Guide. *(See text for explanation.)*

SAE-AISI Designation	Rating %	SAE-AISI Designation	Rating %	SAE-AISI Designation	Rating %	SAE-AISI Designation	Rating %	SAE-AISI Designation	Rating %
Carbon Steels									
1005	45	1019	78	1034	70	1046	57	1069A	49
1006	50	1020	72	1035	70	1049	54	1070A	49
1008	55	1021	78	1037	70	1050	54	1074A	45
1010	55	1022	78	1038	64	1050A	70	1075A	45
1011	53	1023	76	1039	64	1053	54	1078A	45
1012	55	1025	72	1040	64	1054	54	1080A	42
1013	53	1026	78	1042	64	1055A	51	1084A	45
1015	72	1029	70	1043	57	1059A	51	1085A	42
1016	78	1030	70	1044	57	1060A	51	1086A	42
1017	72	1031	70	1045	57	1064A	49	1090A	42
1018	78	1033	70	1045A	72	1065A	49	1095A	42
Resulfurized Carbon Steel									
1106	79	1117	91	1132	76	1141A	76	1151	70
1108	80	1118	91	1137	72	1144	72	1151A	81
1109	81	1119	100	1138	76	1144A	76	1211	94
1110	81	1120	81	1139	76	1145	76	1212	100
1115	81	1125	81	1140	72	1145A	72	1213	136
1116	94	1126	81	1141	70	1146	70	1215	136
High Manganese Carbon Steel									
1524	66	1536	64	1548	55	1552	49	1566A	49
1527	66	1541	57	1551	54	1561A	49	1572A	49

(Continued)

(Continued)

Table 7. (Continued) Machinability Rating Guide. (See text for explanation.)

SAE-AISI Designation	Rating %	SAE-AISI Designation	Rating %	SAE-AISI Designation	Rating %	SAE-AISI Designation	Rating %	SAE-AISI Designation	Rating %
Carbon Steel (Leaded)									
10L18	92	10L45	66	10L45	84	10L50	60	10L50A	78
Free Machining Steel (Leaded)									
11L17	104	11L41	79	11L44	87	12L13	170	12L15	170
11L37	84	11L41A	94	11L44A	98	12L14	170	-	-
Alloy Steel (Leaded)									
41L40A	77	41L50A	70	86L20A	77	-	-	-	-
Manganese Alloy Steel									
1320A	57	1330A	60	1335A	60	1340A	57	1345A	57
Molybdenum Steel									
4012	78	4037A	72	4142A	66	4340A	57	4640	66
4017	78	4047A	66	4145A	64	4419	78	4718	60
4023	78	4118	78	4147A	64	4615	66	4720	60
4024	78	4130A	72	4150A	60	4620	66	4815A	51
4027	66	4137A	70	4161A	60	4621	66	4817A	49
4028	72	4140A	66	4320A	60	4626	60	4820	49
Nickel Chromium Molybdenum Steel									
8615	70	8625	64	8637A	70	8650A	60	8720	66
8617	66	8627	64	8640A	66	8653A	56	8822	64
8620	66	8630A	72	8645A	64	8655A	57	9255A	54
8622	66	8635A	70	8647A	60	8660A	54	9260A	51

"A" indicates annealed. Carbon Steel 1212 (100% rating) is the comparison material for ratings. The information on this Table was supplied by DoAll Company and Texaco.

Table 7. (Continued) Machinability Rating Guide. (See text for explanation.)

SAE-AISI Designation	Rating %	SAE-AISI Designation	Rating %	SAE-AISI Designation	Rating %	SAE-AISI Designation	Rating %	SAE-AISI Designation	Rating %
Chromium Steel									
5015	78	5130	57	5140A	70	5150A	64	5160A	60
5060A	60	5132A	72	5145A	66	5152A	64	E51100A	40
5120	76	5135A	72	5147A	66	5155A	60	E52100A	40
Chromium Vanadium Steel									
6102	57	6118	66	6145	66	6150A	60	6152A	60
Alloy Steel - Boron									
50B44A	70	50B50A	70	51B60A	60	94B17	66	94B30A	72
50B46A	70	50B60A	64	81B45A	66	-	-	-	-
Stainless Steel									
301	55	308	27	317	35	403	55	418	40
302	50	309	28	321	36	405	60	420	45
303	65	310	30	330	30	410	55	430F	65
304	40	314	32	347	40	416	90	440	50
Tool Steel									
A2, A3, A4	16	D5, D7	11	H24, H25	15	O1, O2, O7	16	S1, S2, S5	20
A6, A8, A9	16	H10, H11	20	H26, H42	15	O6	38	T1	14
A7	11	H13, H14	20	M2	14	P2, P3, P4	25	T4	11
A10	27	H19	20	M3	11	P5, P6	25	Ta5	8
D2, D3, D4	11	H21, H22	15	M15	8	P20, P21	22	W (All)	30

"A" indicates annealed. Carbon Steel 1212 (100% rating) is the comparison material for ratings. The information on this Table was supplied by DoAll Company and Texaco.

(Continued)

Table 7. *(Continued)* **Machinability Rating Guide**. *(See text for explanation.)*

			Other Materials (Non-Steel)			
Material	**Type**	**Rating %**	**Material**	**Type**	**Rating %**	
Cast and Malleable Iron	Soft Cast	81	Cast Steel	BHN 120	85	
	Medium Cast	64		BHN 220	50	
	Hard Cast	47		BHN 245	44	
	Malleable Iron	80–106	Bronze	Al Bronze	60	
Brass	Yellow	80		Mn Bronze	60	
	Red	60		Ph Bronze	40	
	Leaded	280		Ph, Leaded	140	
	Red, Leaded	180		Si Bronze	60	
Aluminum	2S	300–1500	Magnesium	Dow H	500–2000	
	11S-T3	500–2000		Dow J	500–2000	
	17S-T	300–1500		Dow R	1150	
Nickel and Nickel Alloy	BHN 135	26	Copper and Copper Alloy	Cast	70	
	Monel, Cast	45		Rolled	60	
	Monel, Rolled	55		Everdur	60	
	Monel "K"	35		Everdur, Leaded	120	
	Inconel (B)	35		Gun Metal, Cast	60	

"A" indicates annealed. Carbon Steel 1212 (100% rating) is the comparison material for ratings. The information on this Table was supplied by DoAll Company and Texaco.

Overview of Materials Classifications

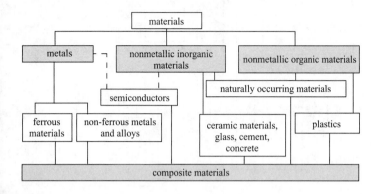

Solid-body structures as a basis of materials properties

Solid-body properties are determined by the structure:

Type of structural components	atoms, ions, molecules
Arrangement of structural components	amorphous (disordered) crystalline (ordered)
Chemical bonding	metallic bonding, valence bonding, ionic bonding, bonding by van der Waals attraction

Characterization of crystalline structure

Unit cell	smallest spatial unit of the crystal lattice
Characteristic values for the unit cell	a) lattice constants: a, b, c b) angle of crystal lattice: α, β, γ c) occupation number: number of atoms per unit cell

    d)   coordination number:
        number of nearest neighbors
    e)   packing index:
        ratio of volume of atoms per
        unit cell to overall volume of
        unit cell

Crystal systems

cubic, tetragonal, hexagonal, monoclinic, orthorhombic (rhombic), trigonal (rhombohedral), triclinic

Structure and Properties of Metals

Properties of metals

These include crystalline structure; reflectivity and opacity; conductivity of heat and electricity; permanent plastic deformation with solidification; varying properties with respect to strength, hardness, modulus of elasticity, magnetism, and melting point; formation of salts in acids.

Ideal structure

Characterization of selected crystal lattices

Lattice type	cubic body-centered	cubic face-centered	hexagonal close-packed
Unit cell			
Intercepts	$a = b = c$	$a = b = c$	$a = b \neq c$
Angle	$\alpha = \beta = \gamma = 90°$	$\alpha = \beta = \gamma = 90°$	$\alpha = \beta = 90°, \gamma = 120°$
Occupation number	2	4	6
Packing index	0.68	0.74	0.74
Coordination number	8	12	12
Examples	α-Fe, δ-Fe, Cr, Mo, V, Ta	γ-Fe, Al, Cu, Ag, Au, Pb	Mg, Zn, Cd, α-Ti

Real crystal structures (lattice defects)

Zero-dimensional lattice defects
(point defects)

vacancy, interstitial, foreign atom
(included, substituted)

One-dimensional lattice defects
(line defects)

dislocations (basis of plastic
deformation)

edge dislocation

screw dislocation

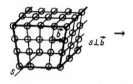

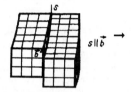

s, dislocation line
b Burgers vector

Two-dimensional lattice defects
(surface defects)

grain boundaries (large-angle
and small-angle grain
boundaries), twin boundaries,
phase boundaries, stacking faults

Structure and Properties of Plastics

Synthesis reactions

Creation of macromolecules by synthesis reactions

Synthesis Reaction	Description / Example
polymerization	connection of monomers containing double bonds, producing chain-like macromolecules without separation of by-products; *example:* polyethylene
polycondensation	connection of monomers (with functional groups), producing a macromolecule while separating a by-product of low molecular weight (H_2O, NH_3); *example:* reaction of phenol with formaldehyde, producing phenolic resin while separating water
polyaddition	combination of different types of monomers (with functional groups), producing a macro molecule without separation of by-products; *example:* polyurethane

Polymer material classes

Structure of polymers

Material	Structure / Examples
plastomers	threadlike macromolecules without cross-linking (amorphous, semicrystalline); *examples:* PVC, PS, PC, PE, PP, PA
thermosets	densely cross-linked macromolecules; *examples:* PF, UF, MF, UP, EP
elastomers	loosely cross-linked macromolecules; *examples:* BR, SBR, NBR, PUR, SI

Influence of temperature on properties (schematic)

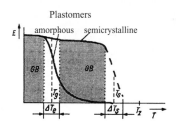

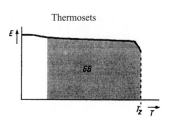

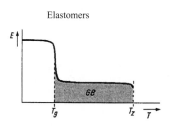

ΔT_e	glass transition temperature range or softening temperature range
T_g	glass transition temperature
T_S	melting temperature
ΔT_S	melting temperature range
T_z	decomposition temperature
E	property (e.g., strength)
GB	use temperature range

Structure and Properties of Ceramics and Glass

Ceramics

Produced by sintering of powders; consists of stable compounds of metals and nonmetals: oxides, carbides, nitrides, borides, (e.g., SiO_2, SiC, Si_3N_4, Al_2O_3, BN, WC)

Glass materials

Glasses are constructed from irregular arrangements of tetrahedral SiO_4 molecules and are therefore amorphous. Glasses are a mixture of metal oxides and other chemical elements and compounds.

Tetrahedral SiO_4 molecules:

Basic structural component of glass materials and many ceramic materials, produced by polycondensation of orthosilicic acid

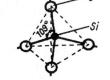

The connection of tetrahedral SiO_4 molecules produces chains, bands, layers, and spatial structures.

Alloying of Metals (Crystalline Structure)

Mixed crystals

Components A and B miscible, form solid atomic solution, homogeneous structure

Types of mixed crystals:

Inclusion crystal	condition: $r_B/r_A \leq 0.58$; $r_{A,B}$ atomic radii of A, B
Exchange or replacement crystal	conditions: — components of same lattice type — $0.85 \leq r_A/r_B \leq 1.15$ — low degree of chemical affinity

Crystal mixtures

Structural components (A and B, mixed crystal, mixed crystal and intermetallic phase) coexist as different types of crystals. The properties vary in proportion to the composition. The structure is heterogeneous.

Intermetallic phases

Components form a new phase with a different, relatively complex lattice. Intermetallic phases are hard and brittle. For example, Fe_3C (6.67% C) is an important intermetallic phase in the Fe-C system.

Ferrous Metals

Ferrous metals—cast iron, steel, and various alloy steels—contain iron as their base metal. By adding alloying elements, they may be altered to produce specific physical and chemical properties. Cast irons are hard, brittle iron alloys containing approximately 2 to 4% carbon and 1 to 3% silicon that possess excellent wear resistance characteristics. In the form of white cast iron, which is essentially free of graphite, it is very hard and is used where abrasion and wear resistance are required. Unalloyed white cast iron typically contains the following elements: carbon, 1.8 to 3.6%; silicon, 0.5 to 1.9%; manganese, 0.25 to 0.8%; phosphorus, 0.06 to 0.2%; and sodium, 0.06 to 0.2%.

White iron can be made into malleable iron by bringing it to an elevated temperature (960° F [1760° C]) for an extended period of time (around 60 hours) and then allowing it to cool very slowly to provide prolonged annealing. Malleable iron is stronger and tougher than white cast iron, but much more expensive to produce. Its ductility and toughness falls between that of ductile cast iron and gray cast iron. Unalloyed malleable cast iron typically contains the following elements: carbon, 2.2 to 2.9%; silicon, 0.9 to 1.9%; manganese, 0.15 to 1.2%; phosphorus, 0.02 to 0.2%; and sodium, 0.02 to 0.2%.

Another form of cast iron is gray iron, which is widely used for castings because of its excellent flow characteristics. It can be machined relatively easily, and it is widely used as a base material to dampen vibration or reduce thermal shock. Unalloyed gray cast iron typically contains the following elements: carbon, 2.5 to 4%; silicon, 1.0 to 3%; manganese, 0.2 to 1%; phosphorus, 0.002 to 1%; and sodium, 0.02 to 0.25%.

Ductile cast iron, which is also called modular iron or spheroidal graphite cast iron, is derived from the same basic materials as gray iron, but the purity requirements are more strict. Its castability is approximately equal to that of gray iron, but its ductility and strength characteristics are appreciably better. Unalloyed ductile cast iron typically contains the following elements: carbon, 3 to 4%; silicon, 1.8 to 2.8%; manganese, 0.1 to 1%; phosphorus, 0.01 to 0.1%; and sodium, 0.01 to 0.03%.

Compacted graphite cast iron is used extensively in industry as the material widely used for the heads of diesel engines and for disk brake rotors. Its properties can be changed radically by the inclusion of magnesium and titanium. Unalloyed compacted graphite cast iron typically contains the following elements: carbon, 2.5 to 4%; silicon, 1 to 3%; manganese, 0.2 to 1%; phosphorus, 0.01 to 0.1%; and sodium, 0.01 to 0.03%.

In addition to the basic forms of cast iron, the metal is commonly alloyed to increase characteristics desirable for a specific application. These alloyed varieties are commonly separated into five defining categories. The following contents relate to the individual range, and not to a particular iron alloy. 1) Abrasion resistant white irons. These alloys gain abrasion resistance from chromium contents of up to 28% (high chromium iron), as much as 7% nickel (martensitic nickel high chromium iron), and up to 3.5% molybdenum (martensitic chromium-molybdenum iron). 2) Corrosion resistant irons. Alloying elements to improve corrosion resistance include up to 4.5% manganese (nickel-chromium ductile iron), as much as 17% silicon (high silicon iron), and nickel contents as high as 36% (nickel-chromium gray iron and nickel-chromium ductile iron). 3) High resistant gray irons. Alloying elements include as much as 7% silicon (medium silicon iron), 43% nickel nickel-chromium-silicon iron), 25% aluminum (high aluminum iron), and 10% copper (nickel-chromium-silicon iron). 4) Heat resistant ductile irons. These alloys include increased silicon (up to 6%) in medium silicon ductile iron), and nickel (up to 36%) in nickel-chromium ductile iron. 5) Heat resistant white irons. These alloys contain up to 15% nickel (austentitic grade), and 35% chromium (ferritic grade).

MATERIALS

Steel making process

Iron and steel production begins with mining iron rich mineral deposits known as iron ores. Iron is present in the ores in the form of chemical compounds, not as free iron, because the element iron is so active chemically that it does not exist in nature in the free state. The important iron compounds, as related to iron and steel production, are the oxides of iron. Important ores are hematite, magnetite, and limonite, which contain ferric oxide (Fe_2O_3), ferrosoferric oxide (Fe_3O_4), and hydrated ferric oxide ($2Fe_2O_3 \cdot 3H_2O$), respectively. Today in the U.S., the most available iron ore is taconite, which contains less iron ore and must be upgraded in the iron making process.

Iron is extracted from the ore in a blast furnace which is essentially a tall steel shell erected on a base, often made of concrete. The blast furnace process of extracting iron from ore is a reduction smelting process. Ore and other solids (chiefly coke and limestone) are fed into the top of the furnace, and iron and waste materials are drawn off at the bottom. During the process, preheated air is blown into the furnace through tuyeres (nozzles), and within the furnace this oxygen reacts with the carbon in the coke to form carbon dioxide which, at the prevailing elevated temperature, reacts with the excess carbon present to form carbon monoxide which reacts with the iron oxides in the ore to convert them to free iron, carbon dioxide, and carbon monoxide. Reaction of the oxides with carbon monoxide accounts for a majority of the iron reduced in a blast furnace, although 20% is produced by direct reduction of the oxides reacting with carbon.

When the molten iron is drawn off, it is used either as a molten charge in steel making, or is taken to a casting machine and cast into small blocks called "pigs," each weighing approximately 100 pounds (hence the term "pig iron"). Pig iron is not pure iron. It contains considerable amounts of carbon, manganese, phosphorus, sulfur, and silicon. It must be transformed to steel using one of four different processes: the open-hearth process, the acid Bessemer process, the basic oxygen process, or the electric-arc process. Between them, these processes account for nearly all the steel tonnage produced in the U.S. Until 1969, the open-hearth furnace was the nation's major source of steel, but today the basic oxygen furnace has assumed that position. In each process, refining the raw materials is an essential operation. The raw materials used to make the steel contain various metallic and nonmetallic elements as impurities. To make steel, many of these impurities must be removed or reduced to an acceptable level. In all four major steel making processes, refining is accomplished by promoting chemical reactions—principally but not exclusively oxidation reactions. Carbon, manganese, iron, silicon, and other elements are oxidized during the refining period when many of the oxides evolve as gases and others enter the slag. Many desirable elements such as carbon and manganese are removed along with the impurities. Consequently, each process allows for the addition of selected materials to the molten metal so that the amounts of carbon, manganese, and other desirable alloying elements in the steel can be adjusted.

Steel making processes are classed as either basic or acid processes, depending on whether the furnace lining used is an acid or basic refractory material. The slag that forms during the process must be compatible with the lining: that is, a basic slag for a basic lining and an acid slag for an acid lining. Silica (SiO_2) is a common acid furnace lining, while mangensite ($MgCO_3$) and dolomite ($MgCO_3 + CaCO_3$) are basic linings. The open-hearth, Bessemer, and electric-arc processes may be operated as basic or as acid processes. The basic oxygen process is basic only. All processes conclude with the refined steel being poured into a large ladle and the slag containing impurities overflows into a slag pot. The steel is now ready to be poured into ingots, which are produced in a variety of sizes and shapes depending on the wrought product to be manufactured.

Steel nomenclature

Annealing. A heat treat process of heating a metal to, and holding it at, a particular temperature, and then cooling it at a suitable rate to achieve a desired result such as reducing hardness, grain refinement, developing a particular microstructure, stress relieving, etc. Also see full annealing.

Austempering. A heat treat process that consists of quenching a ferrous alloy from a temperature above its transformation range in a medium having a sufficiently high rate of heat abstraction to prevent the formation of high temperature transformation products. The alloy is held at a temperature below that of pearlite formation and above that of martensite formation until transformation is complete.

Austenite. The solid solution of one or more elements in a gamma-iron (face-centered cubic iron). The solution is assumed to be carbon, unless otherwise designated; e.g., nickel austenite.

Austenitizing. The process of forming austenite in a ferrous alloy by heating it to a temperature within the transformation range (partial austenitizing), or to a temperature above the critical range (complete austenitizing).

Bainite. A transformation product of austenite. Bainite is formed by isothermal transformation of austenite at temperatures lower than that for fine pearlite and above the M_S (see the list of "transformation temperatures" in the section on Heat Treatment of Steel) temperature. Bainite is an aggregate of ferrite and cementite—when formed at temperatures in the upper portion of the transformation range it has a feathery appearance (upper bainite), while that formed at lower temperatures has an acicular or needlelike structure (lower bainite).

Basic-oxygen process. A process of steel making in which molten pig iron and scrap are refined in a converter by blowing high-purity oxygen onto the surface of the charge.

Bessemer process. A process of steel making in which air is blown through the charge of molten pig iron contained in a Bessemer converter.

Billet. A semifinished hot-rolled, forged, or extruded product, generally with a round or square cross section. For ferrous materials, the minimum diameter or thickness for a billet is a 1.5 inches, and the cross-sectional area may range from 2.25 to 36 square inches.

Blast furnace. A shaft furnace in which solid fuel is burned with an air blast to smelt ore in a continuous operation. The molten metal and molten slag are collected at the bottom of the furnace and are drawn off periodically.

Bloom. A semifinished hot-rolled product, usually with a square or rectangular cross section. If rectangular, the width is no greater than twice the thickness. The cross-sectional area of a bloom usually exceeds 36 square inches.

Blue brittleness. Brittleness that occurs in some steels after being heated to within the temperature range of 400 to 700° F (204 to 371° C), or more especially after being worked within this range. Killed steels are virtually free of blue brittleness.

Carburizing. A process in which an austenitized ferrous material is brought into contact with a carbonaceous atmosphere or medium of sufficient carbon potential as to cause absorption of carbon at the surface and, by diffusion, create a concentration gradient. Hardening by quenching follows.

Case hardening. A term used to describe one or more processes of hardening steel in which the outer portion (or case) is made substantially harder than the inner portion (or core). Most of the processes involve either enriching the surface layer with carbon and/or nitrogen, usually followed by quenching or tempering, or the selective hardening of the surface layer by means of flame or induction hardening.

Cementite. Iron carbide: a hard and brittle compound of iron and carbon (Fe_3C).

Cold working. Plastic deformation of a metal at a temperature below its recrystallization temperature. Cold working produces two effects: the surface of the steel is improved by comparison to hot worked material, and the ultimate and yield strengths of the steel are increased while ductility and toughness decrease.

Creep. A time dependent deformation of steel occurring under conditions of elevated temperature accompanied by stress intensities well within the apparent elastic limit for the temperature involved.

Critical cooling rate. The slowest rate at which steel can be cooled from above the upper critical temperature to prevent the transformation of austenite at any temperature above the M_S (see the list of "transformation temperatures" in the section on Heat Treatment of Steel) temperature.

Critical point. In an equilibrium diagram, the specific combination of composition, temperature, and pressure at which the phases of a heterogeneous system are in equilibrium.

Critical range. In ferrous alloys, the temperature ranges within which austenite forms on heating, and transforms on cooling. The heating and cooling range may overlap but they never coincide.

Critical temperature. Synonymous with critical point when pressure is constant.

Decarburization. The loss of carbon from the surface of a ferrous alloy as the result of heating in a medium that reacts with the carbon at the surface of the material.

Deoxidizer. A substance that can be added to molten metal to react with free or combined oxygen to facilitate its removal.

Ductility. The ability of a material to deform plastically without fracturing, usually measured by elongation or reduction of area in a tension test, or, for flat products such as sheet, by height of cupping in an Erichsen test.

Elastic deformation. The change in dimensions accompanying stress in the elastic region. The original dimensions are restored upon complete release of the stress.

Elastic limit. The greatest stress that a material can withstand without permanent strain remaining upon complete release of the stress.

Element. A substance that cannot be decomposed by ordinary chemical reactions: the fundamental substance from which other substances are made.

Elongation. The increase in gage length, measured after fracture of the tensile specimen within the gage length, expressed as a percentage of the original gage length. It is usually measured in two inches for rectangular tensile specimens and in 4D for round (not welded) specimens. It is a measurement of ductility.

Endothermic. A reaction attended by the absorption of heat.

Equilibrium. A dynamic condition of balance between atomic movements where the resultant change is zero and the condition appears to be static rather than dynamic.

Equilibrium diagram. A graph of the composition, temperature, and pressure limits of the phase fields in an alloy system under equilibrium conditions. In metal systems pressure is usually considered to be constant.

Etch test. An inspection procedure in which a sample is deep etched with acid and visually examined for the purpose of evaluating its structural homogeneity. Also known as a macroetch.

Exothermic. A reaction attended by the liberation of heat.

Fatigue. The phenomenon which results in the fracture of a material under repeated cyclic stresses having a maximum value lower than the tensile yield strength of the material.

Fatigue limit. The maximum or limit stress value below which a material can presumably endure an infinite number of stress cycles. Also referred to as endurance limit, and measured by the rotating beam fatigue test.

Fatigue strength. The maximum stress that can be sustained by a material for a specified number of cycles without failure. Completely reversed loading is implied unless otherwise qualified.

Ferrite. The solid solution of one or more elements (usually carbon) in body-centered cubic iron.

Flakes. In steel, internal fissures can occur in wrought products during cooling from hot forging or rolling. Their occurrence may be minimized by effective control of hydrogen, either in melting or in cooling from hot work.

Ferroalloy. An alloy of iron that contains a sufficient amount of some other element(s) to be useful as an agent for introducing the other element(s) into molten metal.

Full annealing. A thermal treatment for steel intended to decrease hardness. It is accompanied by heating above the transformation range, holding for the specified time interval, and controlled slow cooling to below that range. Subsequent cooling to ambient temperature may be accompanied either in air or in the furnace.

Gamma iron. A face-centered cubic form of pure iron, stable from 1670 to 2535° F (910 – 1390° C).

Grain. An individual crystal in a polycrystalline metal or alloy. A grain size number designation is an arbitrary value calculated from the average number of individual crystals, or grains, that appear on the etched surface of a specimen at 100 diameters magnification.

Hardenability. The property of a ferrous alloy that determines the depth and distribution of hardness that may be introduced by quenching.

Heat treatment. The operation or series of operations of heating and cooling a metal or alloy in the solid state to develop specific desired properties or characteristics.

Hooke's Law. Stress is proportional to strain. This law is valid only to the elastic point.

Hot shortness. Brittleness in metal when hot.

Hot working. The plastic deformation of a metal at a temperature above the recrystallization temperature of the metal.

Induction hardening. A quench hardening process in which heat is generated by electrical induction.

Ingot. A special kind of casting intended for rolling or forging to wrought iron.

Jominy test. A method for determining the hardenability of steel by water-quenching one end of an austentized cylindrical test specimen and measuring the resulting hardness at specified distances from the quenched end. Also called an end-quench hardenability test.

Killed steel. Steel that has been deoxidized with a strong deoxidizer such as aluminum so that little or no reaction occurs between carbon and oxygen during solidification. Generally, killed steels are used when a sound, homogeneous structure is required. Alloy steels, carburizing steels, and steels for forgings are typical applications for killed steels.

Martempering. A heat treat process that consists of quenching a ferrous alloy from an austenitizing temperature to a temperature slightly above or within the martensitic transformation range, holding the material in the quench medium until the temperature throughout the mass is essentially uniform, after which the material is removed from the quench medium and cooled slowly in air.

Martensite. A transformation product of austenite that forms below the M_S (see *Transformation temperature* below for list of abbreviations) temperature. In a metastable-phase, as formed, alpha-martensite is a supersaturated interstitial solid solution of iron carbide in ferrite having a body-centered tetragonal crystal structure characterized by an acicular or needlelike microstructural appearance. Aging or tempering alpha-martensite converts the tetragonal crystal structure to the body-centered cubic structure in which form it is called beta-martensite.

Modulus of elasticity. Within the proportional limit, the ratio of stress to corresponding strain. A measure of the rigidity of a metal. Also known as Young's modulus.

Nitriding. A surface hardening process in which certain steels are heated to, and held at, a temperature below the transformation range in contact with gaseous ammonia or other source of nascent nitrogen in order to affect a transfer of nitrogen to the surface layer of the steel. The nitrogen combines with certain alloying elements, resulting in a thin case of very high hardness.

Normalizing. A heat treat process for ferrous materials that consists of cooling the material in air from a temperature slightly above the transformation range to room temperature.

Open-hearth process. A steel making process in which pig iron and scrap are refined in a reverbatory furnace having a shallow hearth and low roof. The charge is heated by a long sweeping flame that passes over it.

Pearlite. The lamellar aggregate of ferrite and iron-carbide that results from the transformation of austenite at the lower critical point on cooling.

Phase change. The transformation of a material from one physical state to another. For example, changing from a liquid to a solid.

Poisson's ratio. The ratio of the traverse strain to the corresponding axial strain in a body subjected to uniaxial stress. It is usually applied to elastic conditions.

Proportional limit. The maximum stress at which strain remains directly proportional to stress.

Quench hardening. Hardening a ferrous alloy by austenitizing and then cooling rapidly enough so that all or some of the austenite transforms to martensite. Austenitizing temperatures for the hypoeutectoid steels are usually above the A_3 (see *Transformation temperature* below for list of abbreviations) temperature. For the hypereutectoid steels austenitizing temperatures are usually between the A_1 and A_{cm} temperatures.

Recrystallization temperature. The point at which deformed grains in heated steel break up and reform into new grains. The new grain size depends on the temperature of the metal; grain size will increase with temperature.

Reduction in area. The difference, expressed as a percentage of the original area, between the original cross section area of the tensile stress specimen and the minimum cross-section area measured after failure of the specimen. It is a measure of ductility.

Rimmed steel. A low carbon steel containing sufficient iron oxide to promote a continuous evolution of carbon monoxide during solidification of the ingot. Rimmed steel has a good surface finish and a high degree of ductility, making it useful for thin sheet, wire, tinplate, and similar products.

Semikilled steel. Steel that is incompletely deoxidized so that sufficient oxygen remains to react with carbon to form carbon monoxide to offset solidification shrinkage. Semikilled steel is used extensively for heavy structural shapes and plate.

Slab. A semifinished hot rolled product of rectangular cross section with a width greater than twice the thickness. The minimum thickness of a slab is 2 inches, and the minimal cross-sectional area is 16 square inches.

Spheroidizing. Heating and cooling to produce a spheroizoidal or globular form of carbide in steel. Also called spheroidize annealing.

Strain. The unit of change in size or shape of a body due to outside forces.

Stress. Internal forces that resist a change in the volume or shape of a material. Normal stress forces are tension, compression, and shear.

Stress relieving. A heat treatment that consists of heating a metal to a suitable temperature (normally 1000 to 1200° F [538 to 649° C]) and holding it at the temperature long enough to reduce residual stresses, and then cooling it slowly to minimize the possibility of developing new residual stresses.

Tempering. Reheating a quench hardened or normalized ferrous alloy to a temperature below the transformation range and then cooling at any suitable rate.

Tensile strength. The maximum tensile stress a material is capable of sustaining, as measured in a tension test.

Tension test. A test in which a machined or full section specimen is subjected to a measured axial load sufficient to cause fracture. The usual information derived from the test includes the elastic properties, ultimate tensile strength, elongation, and reduction in area.

Toughness. A material's ability to absorb energy and deform plastically before fracturing.

Transformation temperature. The temperature at which a phase change occurs. Often used to denote the limiting temperature of a transformation range. See the list of "transformation temperatures" in the section on Heat Treatment of Steel.

Yield point. The minimum stress at which a marked increase in strain occurs without an increase in stress.

Yield strength. The stress at which a material exhibits a specified deviation from the proportionality of stress to strain. The deviation is expressed in terms of strain, and in the offset method, a strain of 0.2% is usually specified.

Carbon and alloy steel

Carbon Steel. (AISI definition) Carbon steel is classed as such when no minimum content is specified or guaranteed for aluminum, chromium, columbium, molybdenum, nickel, titanium, tungsten, vanadium, or zirconium; when the minimum for copper does not exceed 0.40%; or when the maximum content specified or guaranteed for any of the following elements does not exceed the percentage noted: manganese 1.65%; silicon 0.60%; copper 0.60%.

The American Iron and Steel Institute (AISI) and the Society of Automotive Engineers (SAE) have developed similar designation systems by which steels are identified by chemical composition. In addition, ASTM and SAE have jointly devised a Unified Numbering System for metals and alloys in an effort to reduce confusion. **Table 1** provides numerical designations for steels by composition. In this Table, each of the carbon and alloy steels is identified by an assigned number. It will be noted that the AISI and SAE designations are essentially the same, and the UNS number is usually similar to the SAE designation in instances where AISI and SAE are different.

There are four distinct grades of carbon steels. The first two nonresulfurized grades are both referred to as "plain carbon steels." *Plain carbon steels* with 1% manganese maximum are represented by the 10*xx* designated grades. The second *plain carbon steel* group has a manganese range of 1 to 1.65% and is comprised of the 15*xx* designated grades. The chemical compositions of both of the plain carbon grades are given in **Table 2**. Plain carbon steels with low carbon contents tend to be tough and gummy in machining operations. Increases in carbon and manganese increase the hardness and provide better surface finish and chip character. For carbon contents up to 0.20/0.25%, increases in carbon result in improved machinability for both hot-rolled and cold-drawn steels. As carbon increases above this level, hardness increases to the point that tool life is affected, leading to a decrease in the machinability rating. Most carbon steels are machined in the as-rolled, or as-rolled, cold-drawn condition. Higher carbon content steels are frequently annealed

(Text continued on p. 427)

Table 1. Numerical Designations of Steels by Composition.

Classification	Groupings	Series Designation		
		AISI	SAE	UNS
Carbon and Alloy Steels				
Plain Carbon Steels (Nonresulfurized)	Plain Carbon (Mn 1.0 max.)	10xx	10xx	G10xxx
	Plain Carbon (Mn 1.0 - 1.65)	15xx	15xx	G15xxx
Carbon Steels (Free Machining)	Resulphurized	11xx	11xx	G11xxx
	Resulphurized and Rephosphorized	12xx	12xx	G12xxx
	Acid Bessemer	B11xx	11xx	-
Alloy Steels	Manganese Steel (Mn 1.75)	13xx	13xx	G13xxx
	Boron Steels	14xx	14xx	-
	Nickel Steels (Ni 3.50) (Ni 5.0)	23xx 25xx	23xx 25xx	- -
	Nickel-Chromium Steels (Ni 1.25, Cr 0.65 and 0.80) (Ni 1.75, Cr 1.07) (Ni 3.50, Cr 1.50 and 1.57) (Ni 3.00, Cr 0.77)	31xx 32xx 33xx 34xx	31xx 32xx 33xx 34xx	- - - -
	Molybdenum Steels (Mo 0.20 and 0.25) (Mo 0.40 and 0.52)	40xx 44xx	40xx 44xx	G40xxx G44xxx
	Chromium-Molybdenum Steels (Cr 0.50, 0.80, and 0.95, Mo 0.12, 0.20, 0.25, and 0.30)	41xx	41xx	G41xxx
	Nickel-Chromium-Molybdenum Steels (Ni 1.82, Cr 0.50 and 0.80, Mo .25) (Ni 1.82, Cr 0.50, Mo 0.12 and 0.25, V 0.03 min.) (Ni 1.05, Cr 0.45, Mo 0.20 and 0.35) (Ni 0.30, Cr 0.40, Mo 0.12) (Ni 0.55, Cr 0.50, Mo 0.20) (Ni 0.55, Cr 0.50, Mo 0.25) (Ni 0.55, Cr 0.50, Mo 0.35) (Ni 3.25, Cr 1.20, Mo 0.12) (Ni 0.45, Cr 0.40, Mo 0.12) (Ni 0.55, Cr 0.20, Mo 0.20) (Ni 1.00, Cr 0.80, Mo 0.25)	43xx 43BVxx 47xx 81xx 86xx 87xx 88xx 93xx 94xx 97xx 98xx	43xx 43BVxx 47xx 81xx 86xx 87xx 88xx 93xx 94xx 97xx 98xx	G43xxx - G47xxx G81xxx G86xxx G87xxx G88xxx G93xxx G94xxx G97xxx G98xxx
	Nickel-Molybdenum Steels (Ni 0.85 and1.82, Mo 0.20 and 0.25) (Ni 3.50, Mo 0.25)	46xx 48xx	46xx 48xx	G46xxx G48xx
	Chromium Steels (Cr 0.27, 0.40, 0.50, and 0 .65) (Cr 0.80, 0.87, 0.92, 0.95, 1.00, and 1.05) (Cr 0.50, C 1.00 min.) (bearing steel) (Cr 1.02, C 1.00 min.) (bearing steel) (Cr 1.45, C 1.00 min.) (bearing steel)	50xx 51xx 50xxx 51xxx 52xxx	50xx 51xx 501xx 511xx 521xx	G50xxx G51xxx - - -
	Chromium-Vanadium Steels	61xx	61xx	G61xxx
	Tungsten-Chromium Steels	72xx	72xx	G72xxx
	Silicon-Manganese Steels	92xx	92xx	G92xxx

(Continued)

Table 2. *(Continued)* **Composition of Nonresulfurized Plain Carbon Steel.** *(Source, Bethlehem Steel.)*

AISI/SAE Number	UNS Number	Carbon %	Manganese %	Phosphorus % Max.	Sulphur %
Plain Carbon Steels (Mn 1-1.65%)					
1526	G15260	.22/.29	1.10/1.40	.040	.050
1527	G15270	.22/.29	1.20/1.50	.040	.050
1536	G15360	.30/.37	1.20/1.50	.040	.050
1541	G15410	.36/.44	1.35/1.65	.040	.050
1547	G15470	.45/.51	1.35/1.65	.040	.050
1548	G15480	.44/.52	1.10/1.40	.040	.050
1551	G15510	.45/.56	.85/1.15	.040	.050
1552	G15520	.47/.55	1.20/1.50	.040	.050
1561	G15610	.55/.65	.75/1.05	.040	.050
1566	G15660	.60/.71	.85/1.15	.040	.050
1572	G15720	.65/.76	1.00/1.30	.040	.050

*Standard grades for wire rods and wire only.   † SAE only.

Notes for Carbon Steels:
In the case of certain qualities, the foregoing standard steels are ordinarily furnished to lower phosphorus and lower sulfur maxima.

BARS AND SEMI-FINISHED:
Silicon. When silicon ranges or limits are required, the values shown in the table for Ladle Chemical Ranges and Limits apply.

RODS:
Silicon. When silicon is required, the following ranges and limits are commonly used for nonresulfurized carbon steels:

0.10 per cent maximum	0.10 to 0.20 percent	0.20 to 0.40 percent
0.07 to 0.15 percent	0.15 to 0.30 percent	0.30 to 0.60 percent

ALL PRODUCTS:
Boron. Standard killed carbon steels may be produced with a boron addition to improve hardenability. Such steels can be expected to contain 0.0005 percent minimum boron. These steels are identified by inserting the letter "B" between the second and third numerals of the AISI number, e.g., 10B46.

Lead. Standard carbon steels can be produced to a lead range of 0.15 to 0.35 percent to improve machinability. Such steels are identified by inserting the letter "L" between the second and third numerals of the AISI number, e.g., 10L45.

Copper. When copper is required, 0.20 percent minimum is generally used.

to improve machinability, particularly when they are to be cold-drawn prior to machining. The other two carbon steel groups are referred to as the free machining (or free cutting) carbon steels. Increased percentages of sulfur and/or phosphorus in these grades greatly enhances machinability. The third class of carbon steels are *resulphurized steels*, which are designated 11*xx*. The composition of these steels appears in **Table 3**. The final carbon steels group, the *resulphurized and rephosphorized steels*, is designated 12*xx* and their composition is shown in **Table 4**.

Two subcategories of the carbon steel group are "H" and "M" carbon steels. The M (for merchant) prefix steels are 10*xx* series metal with marginally altered contents of carbon and/or sulfur. Their composition is shown in **Table 5**. The H (for hardenability) suffix steels are derived from both the 10*xx* and 15*xx* series and they possess hardenability requirements in addition to composition limits. Their composition is shown in **Table 6**. Mechanical properties of selected carbon and alloy steels are given in **Table 7**.

Alloy Steel. (AISI definition) By common custom, steel is considered to be alloy steel when the maximum range given for content of alloying elements exceeds one or more of the following limits: manganese 1.65%; silicon 0.60%; copper 0.60%; or in which a definite range or a definite minimum quantity of any of the following elements is specified or required within the limits of the recognized field of constructional alloy steels; aluminum, boron, chromium up to 3.99%, cobalt, columbium, molybdenum, nickel, titanium, tungsten, vanadium, zirconium, or any other alloying element added to obtain a desired alloying effect. Small quantities of certain elements are present in alloy steels that are not specified or required. These elements are considered as incidental and may be present in the following maximum amounts: copper 0.35%; nickel 0.25%; chromium 0.20%; molybdenum 0.06%.

Low alloy steels are generally considered to obtain a total alloy content of 5% or less. All alloy steels have one or more elements, in addition to carbon, added to the iron specifically to affect a change in the properties that cannot be obtained with plain carbon steel. In alloy steels, the first two digits of the designation indicate the primary alloying agent(s) and the last two or three digits indicate the approximate middle of the carbon percentage range of the steel, in hundredths of a percent. In 4161 steel, for example, the "41" indicates that the principal alloying agent is molybdenum, and the "61" indicates that the carbon content is approximately 61% (in fact, 4160 has a carbon content range of 0.56 to 0.64%). **Table 8** provides composition information on common alloy steels. Alloy steels with specific hardenability requirements are given an 'H' suffix. Composition limits for H alloy steels are given in **Tables 9** and **10**. Mechanical properties of selected carbon and alloy steels are given in **Table 7**. Elements added to alloy steels, and their intended effect, are described below in the following section.

Carbon and alloy steels are commonly purchased as round or square bars (solid) or tubes (hollow). Approximate weights and areas for square and round bars and tubes, ranging in size or diameter from $^1/_{16}$ inch through 7 $^{15}/_{16}$ inches, are given in **Table 11a & b**.

Effects of alloying elements

Early civilizations grouped everything in their universe into four essential "elements" or essences identified by Aristotle: air, water, fire, and earth. A later Greek philosopher/scientist named Democritus was the first to propose that all materials are comprised of small particles that he called *atomos*, after the Greek word for invisible. It wasn't until 1868, however, that the Russian chemist Dimitri Mendeléef revealed his classification of the known elements, which became the basis for the Periodic Table of the Elements. As defined in the nomenclature section above, an element cannot be further broken down or subdivided into other substances. An element, in other words, be it gas, liquid, or solid, is a fundamental substance composed of atoms of only one kind, and they represent the simplest form of matter. There are just over 100 cataloged elements, and their selected properties are given in **Table 12**.

The essential elements of steel are iron (Fe) and carbon (C). By combining other elements, as described below, the characteristics of steel can be modified to suit an almost infinite variety of requirements.

Carbon (chemical symbol C) is the principal hardening element in steel. As carbon content increases above 0.85%, additional increases in strength and hardness through adding additional carbon are proportionally less than when below this percentage. Upon quenching, the maximum attainable hardness also increases with higher percentages of carbon content, but above a content of 0.60%, the rate of increase is very small. Ductility, toughness, and weldability generally decrease as carbon content increases.

Manganese (chemical symbol Mn) is an active deoxidizer and, in addition, combines

(Text continued on p. 448)

Table 3. Composition of Free Machining (Resulfurized) Carbon Steel. *(Source, Bethlehem Steel.)*

AISI/SAE Number	UNS Number	Carbon %	Manganese %	Phosphorus % Max.	Sulphur %
1108	G11080	.08/.13	.50/.80	.040	.08/.13
1109	-	.08/.13	.60/.90	.040	.08/.13
1110	G11100	.08/.13	.30/.60	.040	.08/.13
1116	-	.14/.20	1.10/1.40	.040	.16/.23
1117	G11170	.14/.20	1.00/1.30	.040	.08/.13
1118	G11180	.14/.20	1.30/1.60	.040	.08/.13
1119	-	.14/.20	1.00/1.30	.040	.24/.33
1132	-	.27/.34	1.35/1.65	.040	.08/.13
1137	G11370	.32/.39	1.35/1.65	.040	.08/.13
1139	G11390	.35/.43	1.35/1.65	.040	.13/.20
1140	G11400	.37/.44	.70/1.00	.040	.08/.13
1141	G11410	.37/.45	1.35/1.65	.040	.08/.13
1144	G11440	.40/.48	1.35/1.65	.040	.24/.33
1145	-	.42/.49	.70/1.00	.040	.04/.07
1146	G11460	.42/.49	.70/1.00	.040	.08/.13
1151	G11510	.48/.55	.70/1.00	.040	.08/.13

BARS AND SEMI-FINISHED:

Silicon. When silicon ranges and limits are required, the values shown in the table for Ladle Chemical Ranges and Limits apply.

RODS:

Silicon. When silicon is required, the following ranges and limits are commonly used:

Standard Steel Designations	Silicon Ranges or Limits, percent
Up to 1110 incl.	0.10 max.
1116 and over	0.10 max.; or
	0.10 to 0.20; or
	0.15 to 0.30

Table 4. Composition of Free Machining (Rephosphorized and Resulfurized) Carbon Steel.
(Source, Bethlehem Steel.)

AISI/SAE Number	UNS Number	Carbon %	Manganese %	Phosphorus %	Sulphur %	Lead %
1211	G12110	.13 max	.60/.90	.07/.12	.10/.15	-
1212	G12120	.13 max	.70/1.00	.07/.12	.16/.23	-
1213	G12130	.13 max	.70/1.00	.07/.12	.24/.33	-
12L14	G12144	.15 max	.85/1.15	.04/.09	.26/.35	.15/.35
1215	G12150	.09 max	.75/1.05	.04/.09	.26/.35	-

Silicon. It is not common practice to produce these steels to specified limits for silicon because of its adverse effect on machinability.

Nitrogen. These grades are normally nitrogen treated unless otherwise specified.

Table 5. Composition of AISI Merchant (M) Series Carbon Steel. *(Source, Bethlehem Steel.)*

AISI/SAE Number	Carbon %	Manganese %	Phosphorus % Max.	Sulphur %
M1008	.10 max.	.25/.60	.04	0.5
M1010	.07/.14	.25/.60	.04	0.5
M1012	.09/.16	.25/.60	.04	0.5
M1015	.12/.19	.25/.60	.04	0.5
M1017	.14/.21	.25/.60	.04	0.5
M1020	.17/.24	.25/.60	.04	0.5
M1023	.19/.27	.25/.60	.04	0.5
M1025	.20/.30	.25/.60	.04	0.5
M1031	.26/.36	.25/.60	.04	0.5
M1044	.40/.50	.25/.60	.04	0.5

Table 6. Composition of Carbon and Carbon-Boron "H" Series Steel. *(Source, Bethlehem Steel.)*

AISI/SAE Number	UNS Number	Carbon %	Manganese %	Phosphorus % Max.	Sulphur % Max.	Silicon %
1038 H	H10380	.34/.43	.500/1.00	.040	.050	.15/.30
1045 H	H10450	.42/.51	.500/1.00	.040	.050	.15/.30
1522 H	H15220	.17/.25	1.00/1.50	.040	.050	.15/.30
1524 H	H15240	.18/.26	1.25/1.75*	.040	.050	.15/.30
1526 H	H15260	.21/.30	1.00/1.50	.040	.050	.15/.30
1541 H	H15410	.35/.45	1.25/1.75*	.040	.050	.15/.30
15B21 H	H15211	.17/.24	.70/1.20	.040	.050	.15/.30
15B35 H	H15351	.31/.39	.70/1.20	.040	.050	.15/.30
15B37 H	H15371	.30/.39	1.00/1.50	.040	.050	.15/.30
15B41 H	H15411	.35/.45	1.25/1.75*	.040	.050	.15/.30
15B48 H	H15481	.43/.53	1.00/1.50	.040	.050	.15/.30
15B62 H	H15621	.54/.67	1.00/1.50	.040	.050	.40/.60

* Standard H-Steels with 1.75 percent maximum manganese are classified as carbon steels.

Table 7. Mechanical Properties of Selected Carbon and Alloy Steels. See Notes. *(Source, Bethlehem Steel.)*

AISI No.	Description	Tensile Strength PSI	Yield Point PSI	Elongation %	Area Reduction %	Hardness BHN
Carbon Steel Carburizing Grades						
1015	Annealed 1600°F	56,000	41,500	37.0	69.7	111
	Normalized 1700°F	61,500	47,000	37.0	69.6	121
	M-C 1675°F, T 350°F	75,500	44,000	30.0	69.0	156
1020	As Rolled	68,500	55,750	32.0	66.5	137
	Annealed 1600°F	57,250	42,750	36.5	66.0	111
	Normalized 1700°F	64,000	50,250	35.8	67.9	131
	M-C 1675°F, T 350°F	87,000	54,000	23.0	64.2	179

(Continued)

Table 7. *(Continued)* **Mechanical Properties of Selected Carbon and Alloy Steels.** See Notes.
(Source, Bethlehem Steel.)

AISI No.	Description	Tensile Strength PSI	Yield Point PSI	Elongation %	Area Reduction %	Hardness BHN
	Carbon Steel Carburizing Grades					
1022	As Rolled	70,250	52,250	33.0	65.2	137
	Annealed 1600°F	65,250	46,000	35.0	63.6	137
	Normalized 1700°F	70,000	52,000	34.0	67.5	143
	M-C 1675°F, 350°F	89,000	55,000	25.5	57.3	179
1117	As Rolled	69,750	49,500	33.5	61.1	149
	Annealed 1575°F	62,250	40,500	32.8	58.0	121
	Normalized 1650°F	67,750	44,000	33.5	63.8	137
	M-C 1700°F, T 350°F	89,500	50,500	22.3	48.8	183
1118	As Rolled	70,500	51,500	32.3	63.0	143
	Annealed 1450°F	65,250	41,250	34.5	66.8	131
	Normalized 1700°F	69,250	46,250	33.5	65.9	143
	M-C 1700°F, T 350°F	102,500	59,250	19.0	48.9	207
	Carbon Steel Water and Oil Hardening Grades—Water or Oil Quenched					
1030	Annealed 1550°F	67,250	49,500	31.2	57.9	126
	Normalized 1700°F	75,500	50,000	32.0	60.8	149
	W Q 1600°F, T 1000°F	88,000	68,500	28.0	68.6	179
	W Q 1600°F, Temp. 1100°F	85,250	63,000	29.0	70.8	170
	W Q 1600°F, Temp. 1200°F	84,500	61,500	28.5	71.4	170
1040	Annealed 1450°F	75,250	51,250	30.2	57.2	149
	Normalized 1650°F	85,500	54,250	28.0	54.9	170
	O Q 1575°F, T 1000°F	96,250	68,000	26.5	61.1	197
	O Q 1575°F, T 1100°F	91,500	64,250	28.2	63.5	187
	O Q 1575°F, T 1200°F	85,250	60,250	30.0	67.4	170
	W Q 1550°F, T 1000°F	107,750	78,500	23.2	62.6	217
	W Q 1550°F, T 1100°F	100,000	69,500	26.0	65.0	207
	W Q 1550°F, T 1200°F	93,500	68,000	27.0	67.9	197
1050	Annealed 1450°F	92,250	53,000	23.7	39.9	187
	Normalized 1650°F	108,500	62,000	20.0	39.4	217
	O Q 1550°F, T 1000°F	123,500	76,000	20.2	53.3	248
	O Q 1550°F, T 1100°F	114,000	70,500	23.5	57.6	223
	O Q 1550°F, T 1200°F	106,000	64,250	24.7	60.5	217
	W Q 1525°F, T 1000°F	131,250	92,250	20.0	55.2	262
	W Q 1525°F, T 1100°F	118,000	80,000	22.5	59.9	241
	W Q 1525°F, T 1200°F	109,000	76,500	23.7	61.2	229
1060	Annealed 1450°F	90,750	54,000	22.5	38.2	179
	Normalized 1650°F	112,500	61,000	18.0	37.2	229
	O Q 1550°F, T 900°F	145,500	93,000	16.2	44.0	293
	O Q 1550°F, T 1000°F	136,500	85,750	17.7	48.0	269
	O Q 1550°F, T 1100°F	127,750	79,000	20.0	51.7	255
1080	Annealed 1450°F	89,250	54,500	24.7	45.0	174
	Normalized 1650°F	146,500	76,000	11.0	20.6	293
	O Q 1500°F, T 900°F	181,500	112,500	13.0	35.8	352
	O Q 1500°F, T 1000°F	166,000	103,500	15.0	37.6	331
	O Q 1500°F, T 1100°F	150,000	97,000	16.5	40.3	302
1095	Annealed 1450°F	95,250	55,000	13.0	20.6	192
	Normalized 1650°F	147,000	72,500	9.5	13.5	293
	O Q 1475°°F, T 900°F	175,750	102,250	10.0	23.4	352
	O Q 1475°F, T 1000°F	159,750	95,250	13.2	32.4	321
	O Q 1475°F, T 1100°F	139,750	79,000	17.2	38.8	277
	W Q 1450°F, T 900°F	182,000	121,000	13.0	37.3	363
	W Q 1450°F, T 1000°F	165,000	102,500	16.0	41.4	311
	W Q 1450°F, T 1100°F	143,000	96,500	16.7	43.7	293

(Continued)

Table 7. *(Continued)* **Mechanical Properties of Selected Carbon and Alloy Steels.** See Notes.
(Source, Bethlehem Steel.)

AISI No.	Description	Tensile Strength PSI	Yield Point PSI	Elongation %	Area Reduction %	Hardness BHN
	**Carbon Steel Water and Oil Hardening Grades—Water or Oil Quenched**					
1137	Annealed 1450°F	84,750	50,000	26.8	53.9	174
	Normalized 1650°F	97,000	57,500	22.5	48.5	197
	O Q 1575°F, T 1000°F	108,000	75,750	21.3	56.0	223
	O Q 1575°F, T 1100°F	100,750	68,750	23.5	60.1	207
	O Q 1575°F, T 1200°F	97,750	68,750	23.5	60.8	201
	W Q 1550°F, T 1000°F	122,000	98,000	16.9	51.2	248
	W Q 1550°F, T 1100°F	107,750	87,750	21.0	59.2	223
	W Q 1550°F, T 1200°F	102,500	81,750	22.3	58.8	217
1141	Annealed 1500°F	86,800	51,200	25.5	49.3	163
	Normalized 1650°F	102,500	58,750	22.7	55.5	201
	O Q 1500°F, T 1000°F	110,200	75,300	23.5	58.7	229
	O Q 1500°F, T 1100°F	103,000	69,800	23.8	62.2	207
	O Q 1500°F, T 1200°F	96,300	69,600	24.8	64.1	197
1144	Annealed 1450°F	84,750	50,250	24.8	41.3	167
	Normalized 1650°F	96,750	58,000	21.0	40.4	197
	O Q 1550°F, T 1000°F	108,500	72,750	19.3	46.0	223
	O Q 1550°F, T 1100°F	102,750	68,000	21.5	51.4	212
	O Q 1550°F, T 1200°F	97,000	68,000	23.0	52.4	201
	**Alloy Steel Carburizing Grades**					
4118	Annealed 1600°F	75,000	53,000	33.0	63.7	137
	Normalized 1670°F	84,500	56,000	32.0	71.0	156
	M-C 1700°F, T 300°F	119,000	64,500	21.0	37.5	241
	M-C 1700°F, T 450°F	115,000	64,000	22.0	49.0	235
4320	Annealed 1560°F	84,000	61,625	29.0	58.4	163
	Normalized 1640°F	115,000	67,250	20.8	50.7	235
	M-C 1700°F, T 300°F	152,500	107,250	17.0	51.0	302
	M-C 1700°F, T 450°F	148,750	105,000	17.8	55.2	285
4419	Annealed 1675°F	64,750	48,000	31.2	62.8	121
	Normalized 1750°F	75,250	51,000	32.5	69.4	143
	M-C 1700°F, T 300°F	97,250	62,750	24.2	66.4	201
	M-C 1700°F, T 450°F	94,250	58,750	25.0	68.6	197
4620	Annealed 1575°F	74,250	54,000	31.3	60.3	149
	Normalized 1650°F	83,250	53,125	29.0	66.7	174
	M-C 1700°F, T 300°F	98,000	67,000	25.8	70.0	197
	M-C 1700°F, T 450°F	98,000	66,250	27.5	68.9	192
4820	Annealed 1500°F	98,750	67,250	22.3	58.8	197
	Normalized 1580°F	109,500	70,250	24.0	59.2	229
	M-C 1700°F, T 300°F	169,500	126,500	15.0	51.0	352
	M-C 1700°F, T 450°F	163,250	120,500	15.5	53.1	331
8620	Annealed 1600°F	77,750	55,875	31.3	62.1	149
	Normalized 1675°F	91,750	51,750	26.3	59.7	183
	M-C 1700°F, T 300°F	126,750	83,750	20.8	52.7	255
	M-C 1700°F, T 450°F	124,250	80,750	19.5	54.2	248
E9310	Annealed 1550°F	119,000	63,750	17.3	42.1	241
	Normalized 1630°F	131,500	82,750	18.8	58.1	269
	M-C 1700°F, T 300°F	159,000	122,750	15.5	57.5	321
	M-C 1700°F, T 450°F	157,500	123,000	16.0	61.7	321

(Continued)

Table 7. *(Continued)* **Mechanical Properties of Selected Carbon and Alloy Steels.** See Notes.
(Source, Bethlehem Steel.)

AISI No.	Description	Tensile Strength PSI	Yield Point PSI	Elongation %	Area Reduction %	Hardness BHN
Alloy Steel Water Hardening Grades						
4027	Annealed 1585°F	75,000	47,250	30.0	52.9	143
	Normalized 1660°F	93,250	61,250	25.8	60.2	179
	W Q 1585°F, T 900°F	150,000	133,000	16.0	57.8	311
	W Q 1585°F, T 1000°F	139,250	122,250	18.8	60.1	285
	W Q 1585°F, T 1100°F	114,250	93,250	23.0	67.6	229
4130	Annealed 1585°F	81,250	52,250	28.2	55.6	156
	Normalized 1600°F	97,000	63,250	25.5	59.5	197
	W Q 1575°F, T 900°F	161,000	137,500	14.7	54.4	321
	W Q 1575°F, T 1000°F	144,500	129,500	18.5	61.8	293
	W Q 1575°F, T 1100°F	128,000	113,250	21.2	67.5	262
8630	Annealed 1550°F	81,750	54,000	29.0	58.9	156
	Normalized 1600°F	94,250	62,250	23.5	53.5	187
	W Q 1550°F, T 900°F	146,750	131,750	16.2	56.5	293
	W Q 1550°F, T 1000°F	134,750	123,000	18.7	59.6	269
	W Q 1550°F, T 1100°F	118,000	101,250	18.7	58.2	241
Alloy Steel Oil Hardening Grades						
1340	Annealed 1475°F	102,000	63,250	25.5	57.3	207
	Normalized 1600°F	121,250	81,000	22.0	62.9	248
	O Q 1525°F, T 1000°F	137,750	121,000	19.2	57.4	285
	O Q 1525°F, T 1100°F	118,000	98,250	21.7	60.1	241
	O Q 1525°F, T 1200°F	112,000	96,000	23.2	62.4	229
4140	Annealed 1500°F	95,000	60,500	25.7	56.9	197
	Normalized 1600°F	148,000	95,000	17.7	46.8	302
	O Q 1550°F, T 1000°F	156,000	143,250	15.5	56.9	311
	O Q 1550°F, T 1100°F	140,250	135,000	19.5	62.3	285
	O Q 1550°F, T 1200°F	132,750	122,500	21.0	65.0	269
4150	Annealed 1525°F	105,750	55,000	20.2	40.2	197
	Normalized 1600°F	167,500	106,500	11.7	30.8	321
	O Q 1525°F, T 1000°F	175,250	159,500	14.0	46.5	352
	O Q 1525°F, T 1100°F	165,500	150,000	15.7	51.1	331
	O Q 1525°F, T 1200°F	141,000	127,500	18.7	55.7	285
4340	Annealed 1490°F	108,000	68,500	22.0	49.9	217
	Normalized 1600°F	185,500	125,000	12.2	36.3	363
	O Q 1475°F, T 1000°F	175,000	166,000	14.2	45.9	352
	O Q 1475°F, T 1100°F	164,750	159,000	16.5	54.1	331
	O Q 1475°F, T 1200°F	139,000	128,000	20.0	59.7	277
5140	Annealed 1255°F	83,000	42,500	28.6	57.3	167
	Normalized 1600°F	115,000	68,500	22.7	59.2	229
	O Q 1550°F, T 1000°F	141,000	121,500	18.5	58.9	293
	O Q 1550°F, T 1100°F	127,250	105,000	20.5	61.7	262
	O Q 1550°F, T 1200°F	117,000	94,500	22.5	63.5	235
5150	Annealed 1520°F	98,000	51,750	22.0	43.7	197
	Normalized 1600°F	126,250	76,750	20.7	58.7	255
	O Q 1525°F, T 1000°F	153,000	131,750	17.0	54.1	302
	O Q 1525°F, T 1100°F	137,000	115,250	20.2	59.5	277
	O Q 1525°F, T 1200°F	128,000	108,000	21.2	61.9	255

(Continued)

Table 7. *(Continued)* **Mechanical Properties of Selected Carbon and Alloy Steels.** See Notes.
(Source, Bethlehem Steel.)

AISI No.	Description	Tensile Strength PSI	Yield Point PSI	Elongation %	Area Reduction %	Hardness BHN
Alloy Steel Oil Hardening Grades						
5160	Annealed 1495°F	104,750	40,000	17.2	30.6	197
	Normalized 1575°F	138,750	77,000	17.5	44.8	269
	O Q 1525°F, T 1000°F	165,500	145,500	14.5	45.7	341
	O Q 1525°F, T 1100°F	145,250	126,000	18.0	53.6	302
	O Q 1525°F, T 1200°F	128,750	110,750	20.7	55.6	262
6150	Annealed 1500°F	96,750	59,750	23.0	48.4	197
	Normalized 1600°F	136,250	89,250	21.8	61.0	269
	O Q 1550°F, T 1000°F	173,500	167,750	14.5	48.2	352
	O Q 1550°F, T 1100°F	158,250	150,500	16.0	53.2	311
	O Q 1550°F, T 1200°F	141,250	129,500	18.7	56.3	293
8650	Annealed 1465°F	103,750	56,000	22.5	46.4	212
	Normalized 1600°F	148,500	99,750	14.0	40.4	302
	O Q 1475°F, T 1000°F	172,500	159,750	14.5	49.1	352
	O Q 1475°F, T 1100°F	153,500	142,750	17.7	57.3	311
	O Q 1475°F, T 1200°F	141,000	132,000	19.5	59.8	285
8740	Annealed 1500°F	100,750	60,250	22.2	46.4	201
	Normalized 1600°F	134,750	88,000	16.0	47.9	269
	O Q 1525°F, T 1000°F	178,500	164,250	16.0	53.0	352
	O Q 1525°F, T 1100°F	149,250	134,500	18.2	59.9	302
	O Q 1525°F, T 1200°F	138,000	123,000	20.0	60.7	285
9255	Annealed 1550°F	112,750	70,500	21.7	41.1	229
	Normalized 1650°F	135,250	84,000	19.7	43.4	269
	O Q 1625°F, T 1000°F	164,250	133,750	16.7	38.3	321
	O Q 1625°F, T 1100°F	150,000	118,000	19.2	44.8	293
	O Q 1625°F, T 1200°F	138,000	106,500	21.2	48.2	277

Note: The following abbreviations are used in this Table. M-C = Mock-Carburization Temperature; T = Tempering Temperature; W Q = Water Quenching Temperature; O T = Oil Quenching Temperature. All values are for a one inch round bar, treated as indicated.

Table 8. **Composition of Alloy Steels** (Billets, Blooms, Slabs, Hot Rolled Bars, Cold Rolled Bars).
(Source, Bethlehem Steel.)

AISI/SAE Number	UNS Number	Carbon %	Manganese %	Nickel %	Chromium %	Molybdenum %	Other Elements
1330	G13300	.28/.33	1.60/1.90	-	-	-	-
1335	G13350	.33/.38	1.60/1.90	-	-	-	-
1340	G13400	.38/.43	1.60/1.90	-	-	-	-
1345	G13450	.43/.48	1.60/1.90	-	-	-	-
4012††	-	.09/.14	.75/1.00	-	-	.15/.25	-
4023	G40230	.20/.25	.70/.90	-	-	.20/.30	-
4024	G40240	.20/.25	.70/.90	-	-	.20/.30	-
4027	G40270	.25/.30	.70/.90	-	-	.20/.30	S - .035/.050
4028	G40280	.25/.30	.70/.90	-	-	.20/.30	-
4032††	G40320	.30/.35	.70/.90	-	-	.20/.30	S - .035/.050
4037	G40370	.35/.40	.70/.90	-	-	.20/.30	-

(Continued)

Table 8. *(Continued)* **Composition of Alloy Steels** (Billets, Blooms, Slabs, Hot Rolled Bars, Cold Rolled Bars). *(Source, Bethlehem Steel.)*

AISI/SAE Number	UNS Number	Carbon %	Manganese %	Nickel %	Chromium %	Molybdenum %	Other Elements
4042††	G40420	.40/.45	.70/.90	-	-	.20/.30	-
4047	G40470	.45/.50	.70/.90	-	-	.20/.30	-
4118	G41180	.18/.23	.70/.90	-	.40/.60	.08/.15	-
4130	G41300	.28/.33	.40/.60	-	.80/1.10	.15/.25	-
4135††	G41350	.33/.38	.70/.90	-	.80/1.10	.15/.25	-
4137	G41370	.35/.40	.70/.90	-	.80/1.10	.15/.25	-
4140	G41400	.38/.43	.75/1.00	-	.80/1.10	.15/.25	-
4142	G41420	.40/.45	.75/1.00	-	.80/1.10	.15/.25	-
4145	G41450	.43/.48	.75/1.00	-	.80/1.10	.15/.25	-
4147	G41470	.45/.50	.75/1.00	-	.80/1.10	.15/.25	-
4150	G41500	.48/.53	.75/1.00	-	.80/1.10	.15/.25	-
4161	G41610	.56/.64	.75/1.00	-	.70/.90	.25/.35	-
4320	G43200	.17/.22	.45/.65	1.65/2.00	.40/.60	.20/.30	-
4340	G43400	.38/.43	.60/.80	1.65/2.00	.70/.90	.20/.30	-
E4340	G43406	.38/.43	.65/.85	1.65/2.00	.70/.90	.20/.30	-
4419††	-	.18/.23	.45/.65	-	-	.45/.60	-
4422††	G44220	.20/.25	.70/.90	-	-	.35/.45	-
4427††	G44270	.24/.29	.70/.90	-	-	.35/.45	-
4615	G46150	.13/.18	.45/.65	1.65/2.00	-	.20/.30	-
4617††	G47170	.15/.20	.45/.65	1.65/2.00	-	.20/.30	-
4620	G46200	.17/.22	.45/.65	1.65/2.00	-	.20/.30	-
4621††	-	.18/.23	.70/.90	1.65/2.00	-	.20/.30	-
4626	G46260	.24/.29	.45/.65	.70/1.00	-	.15/.25	-
4718††	G47180	.16/.21	.70/.90	.90/1.20	.35/.55	.30/.40	-
4720	G47200	.17/.22	.50/.70	.90/1.20	.35/.55	.15/.25	-
4815	G48150	.13/.18	.40/.60	3.25/3.75	-	.20/.30	-
4817	G48170	.15/.20	.40/.60	3.25/3.75	-	.20/.30	-
4820	G48200	.18/.23	.50/.70	3.25/3.75	-	.20/.30	-
5015††	-	.12/.17	.30/.50	-	.30/.50	-	-
5046††	G50460	.43/.48	.75/1.00	-	.20/.35	-	-
5060††	G50600	.56/.64	.75/1.00	-	.40/.60	-	-
5115††	G51150	.13/.18	.70/.90	-	.70/.90	-	-
5120	G51200	.17/.22	.70/.90	-	.70/.90	-	-
5130	G51300	.28/.33	.70/.90	-	.80/1.10	-	-
5132	G51320	.30/.35	.60/.80	-	.75/1.00	-	-
5135	G51350	.33/.38	.60/.80	-	.80/1.05	-	-
5140	G51400	.38/.43	.70/.90	-	.70/.90	-	-
5145††	-	.43/.48	.70/.90	-	.70/.90	-	-

(Continued)

Table 8. *(Continued)* **Composition of Alloy Steels** (Billets, Blooms, Slabs, Hot Rolled Bars, Cold Rolled Bars). *(Source, Bethlehem Steel.)*

AISI/SAE Number	UNS Number	Carbon %	Manganese %	Nickel %	Chromium %	Molybdenum %	Other Elements
5147††	G51470	.46/.51	.70/.95	-	.85/1.15	-	-
5150	G51500	.48/.53	.70/.90	-	.70/.90	-	-
5155	G51550	.51/.59	.70/.90	-	.70/.90	-	-
5160	G51600	.56/.64	.75/1.00	-	.70/.90	-	-
50100††	G50986	.98/1.10	.25/.45	-	.40/.60	-	-
E51100	G51986	.98/1.10	.25/.45	-	.90/1.15	-	-
E52100	G52986	.98/1.10	.25/.45	-	1.30/1.60	-	-
6118	G61180	.16/.21	.50/.70	-	.50/.70	-	V - .10/.15
6150	G61500	.48/.53	.70/.90	-	.80/1.10	-	V - .15 min
8115††	-G81150	.13/.18	.70/.90	.20/.40	.30/.50	.08/.15	-
8615	G86150	.13/.18	.70/.90	.40/.70	.40/.60	.15/.25	-
8617	G86170	.15/.20	.70/.90	.40/.70	.40/.60	.15/.25	-
8620	G86200	.18/.23	.70/.90	.40/.70	.40/.60	.15/.25	-
8622	G86220	.20/.25	.70/.90	.40/.70	.40/.60	.15/.25	-
8625	G86250	.23/.28	.70/.90	.40/.70	.40/.60	.15/.25	-
8627	G86270	.25/.30	.70/.90	.40/.70	.40/.60	.15/.25	-
8630	G86300	.28/.33	.70/.90	.40/.70	.40/.60	.15/.25	-
8637	G86370	.35/.40	.75/1.00	.40/.70	.40/.60	.15/.25	-
8640	G86400	.38/.43	.75/1.00	.40/.70	.40/.60	.15/.25	-
8642	G86420	.40/.45	.75/1.00	.40/.70	.40/.60	.15/.25	-
8645	G86450	.43/.48	.75/1.00	.40/.70	.40/.60	.15/.25	-
8650††	G86500	.48/.53	.75/1.00	.40/.70	.40/.60	.15/.25	-
8655	G86550	.51/.59	.75/1.00	.40/.70	.40/.60	.15/.25	-
8660††	G86600	.56/.64	.75/1.00	.40/.70	.40/.60	.15/.25	-
8720	G87200	.18/.23	.70/.90	.40/.70	.40/.60	.20/.30	-
8740	G87400	.38/.43	.75/1.00	.40/.70	.40/.60	.20/.30	-
8822	G88220	.20/.25	.75/1.00	.40/.70	.40/.60	.30/.40	-
9254††	G92540	.51/.59	.60/.80	-	.60/.80	-	Si - 1.20/1.60
9255††	-	.51/.59	.70/.95	-	-	-	Si - 1.80/2.20
9260	G92600	.56/.64	.75/1.00	-	-	-	Si - 1.80/2.20
9310††	-	.08/.13	.45/.65	3.00/3.50	1.00/1.40	.08/.15	-

†† SAE only   (See Alloy Steel Notes following Table 10)

Table 9. Composition of Alloy "H" Series Steels. *(Source, Bethlehem Steel.)*

AISI/SAE Number	UNS Number	Carbon %	Manganese %	Nickel %	Chromium %	Molybdenum %	Other Elements
1330 H	H13300	.27/.33	1.45/2.05	-	-	-	-
1335 H	H13350	.32/.38	1.45/2.05	-	-	-	-
1340 H	H13400	.37/.44	1.45/2.05	-	-	-	-
1345 H	H13450	.42/.49	1.45/2.05	-	-	-	-
4027 H	H40270	.24/.30	.60/1.00	-	-	.20/.30	-
4028 H	H40280	.24/.30	.60/1.00	-	-	.20/.30	S - .035/.050
4032 H ††	H40320	.29/.35	.60/1.00	-	-	.20/.30	-
4037 H	H40370	.34/.41	.60/1.00	-	-	.20/.30	-
4042 H ††	H40420	.39/.46	.60/1.00	-	-	.20/.30	-
4047 H	H40470	.44/51	.60/1.00	-	-	.20/.30	-
4118 H	H41180	.17/.23	.60/1.00	-	.30/.70	.08/.15	-
4130 H	H41300	.27/.33	.30/.70	-	.75/1.20	.15/.25	-
4135 H ††	H41350	.32/.38	.60/1.00	-	.75/1.20	.15/.25	-
4137 H	H41370	.34/.41	.60/1.00	-	.75/1.20	.15/.25	-
4140 H	H41400	.37/.44	.60/1.10	-	.75/1.20	.15/.25	-
4142 H	H41420	.39/.46	.60/1.10	-	.75/1.20	.15/.25	-
4145 H	H41450	.42/.49	.60/1.10	-	.75/1.20	.15/.25	-
4147 H	H41470	.44/.51	.60/1.10	-	.75/1.20	.15/.25	-
4150 H	H41500	.47/.54	.60/1.10	-	.75/1.20	.15/.25	-
4161 H	H41610	.55/.65	.60/1.10	-	.60/.95	.25/.35	-
4320 H	H43200	.17/.23	.40/.70	1.55/2.00	.35/.65	.20/.30	-
4340 H	H43400	.37/.44	.55/.90	1.55/2.00	.65/.95	.20/.30	-
E4340 H	H43406	.37/.44	.60/.95	1.55/2.00	.65/.95	.20/.30	-
4419 H ††	-	.17/.23	.35/.75	-	-	.45/.60	-
4620 H	H46200	.17/.23	.35/.75	1.55/2.00	-	.20/.30	-
4621 H ††	-	.17/.23	.60/1.00	1.55/2.00	-	.20/.30	-
4626 H †	-	.23/29	.40/.70	.65/1.05	-	.15/.25	-
4718 H ††	H47180	.15/.21	.60/.95	.85/1.25	.30/.60	.30/.40	-
4720 H	H47200	.17/.23	.45/.75	.85/1.25	.30/.60	.15/.25	-
4815 H	H48150	.12/.18	.30/.70	3.20/3.80	-	.20/.30	-
4817 H	H48170	.14/.20	.30/.70	3.20/3.80	-	.20/.30	-
4820 H	H48200	.17/.23	.40/.80	3.20/3.80	-	.20/.30	-
5046 H ††	H50450	.43/.50	.65/1.10	-	.13/.43	-	-
5120 H	H51200	.17/.23	.60/1.00	-	.60/1.00	-	-
5130 H	H51300	.27/.33	.60/1.00	-	.75/1.20	-	-
5132 H	H51320	.29/.35	.50/.90	-	.65/1.10	-	-
5135 H	H51350	.32/.38	.50/.90	-	.70/1.15	-	-
5140 H	H51400	.37/.44	.60/1.00	-	.60/1.00	-	-
5145 H ††	-	.42/49	.60/1.00	-	.60/1.00	-	-

(Continued)

Table 9. (Continued) Composition of Alloy "H" Series Steels. (Source, Bethlehem Steel.)

AISI/SAE Number	UNS Number	Carbon %	Manganese %	Nickel %	Chromium %	Molybdenum %	Other Elements
5147 H ††	H51470	.45/.52	.60/1.05	-	.80/1.25	-	-
5150 H	H51500	.47/.54	.60/1.00	-	.60/1.00	-	-
5155 H	H51550	.50/.60	.60/1.00	-	.60/1.00.	-	-
5160 H	H51600	.55/.65	.60/1.10	-	60/1.00	-	-
6118 H	H61180	.15/.21	.40/.80	-	.40/.80	-	V - .10/.15
6150 H	H61500	.47/.54	.60/1.00	-	.75/1.20	-	V - .15 min
8617 H	H86170	.14/.20	.60/.95	.35/.75	.35/.65	.15/.25	-
8620 H	H86200	.17/.23	.60/.95	.35/.75	.35/.65	.15/.25	-
8622 H	H86220	.19/.25	.60/.95	.35/.75	.35/.65	.15/.25	-
8625 H	H86250	.22/.28	.60/.95	.35/.75	.35/.65	.15/.25	-
8627 H	H86270	.24/.30	.60/.95	.35/.75	.35/.65	.15/.25	-
8630 H	H86300	.27/.33	.60/.95	.35/.75	.35/.65	.15/.25	-
8637 H	H86370	.34/.41	.70/.105	.35/.75	.35/.65	.15/.25	-
8640 H	H86400	.37/.44	.70/.105	.35/.75	.35/.65	.15/.25	-
8642 H	H86420	.39/.46	.70/.105	.35/.75	.35/.65	.15/.25	-
8645 H	H86450	.42/.49	.70/.105	.35/.75	.35/.65	.15/.25	-
8650 ††	H86500	.47/.54	.70/.105	.35/.75	.35/.65	.15/.25	-
8655 H	H86550	.50/.60	.70/.105	.35/.75	.35/.65	.15/.25	-
8660 H ††	H86600	.55/.65	.70/.105	.35/.75	.35/.65	.15/.25	-
8720 H	H87200	.17/.23	.60/.95	.35/.75	.35/.65	.20/.30	-
8740 H	H87400	.37/.44	.70/1.05	.35/.75	.35/.65	.20/.30	-
8822 H	H88220	.19/.25	.70/1.05	.35/.75	.35/.65	.30/.40	-
9260 H	H92600	.55/.65	.65/1.10	-	-	-	Si - 1.70/2.20
9310 H	H93100	.07/.13	.40/.70	2.95/3.55	1.00/1.45	.08/.15	-

† AISI only   †† SAE only   (See Alloy Steel Notes following Table 10)

Table 10. Composition of Alloy Boron and Boron "H" Series Steels. (Source, Bethlehem Steel.)

AISI/SAE Number	UNS Number	Carbon %	Manganese %	Nickel %	Chromium %	Molybdenum %
50B40 ††	-	.38/.43	.75/1.00	-	.40/.60	-
50B44	G50441	.43/.48	.75/1.00	-	.40/.60	-
50B46	G50461	.44/.49	.75/1.00	-	.20/.35	-
50B50	G50501	.48/.53	.75/1.00	-	.40/.60	-
50B60	G50601	.56/.64	.75/1.00	-	.40/.60	-
51B60	G51601	.56/.64	.75/1.00	-	.70/.90	-
81B45	G81451	.43/.48	.75/1.00	.20/.40	.35/.55	.08/.15
86B45 ††	-	.43/.48	.75/1.00	.40/.70	.40/.60	.15/.25
94B15 ††	-	.13/.18	.75/1.00	.30/.60	.30/.50	.08/.15
94B17	G94171	.15/.20	.75/1.00	.30/.60	.30/.50	.08/.15

(Continued)

Table 10. *(Continued)* **Composition of Alloy Boron and Boron "H" Series Steels.** *(Source, Bethlehem Steel.)*

AISI/SAE Number	UNS Number	Carbon %	Manganese %	Nickel %	Chromium %	Molybdenum %
94B30	G94301	.28/.33	.75/1.00	.30/.60	.30/.50	.08/.15
50B40 H ††	H50401	.37/.44	.65/1.10	-	.30/.70	-
50B44 H	H50441	.42/.49	.65/1.10	-	.30/.70	-
50B46 H	H50461	.43/.50	.65/1.10	-	.13/.43	-
50B50 H	H50501	.47/.54	.65/1.10	-	.30/.70	-
50B60 H	H50601	.55/.65	.65/1.10	-	.30/.70	-
51B60 H	H51601	.55/.65	.65/1.10	-	.60/1.00	-
81B45 H	H81451	.42/.49	.70/1.05	.15/.45	.30/.60	.08/.15
86B30 H	H86301	.27/.33	.60/.95	.35/.75	.35/.65	.15/.25
86B45 H ††	-	.42/.49	.70/1.05	.35/.75	.35/.65	.15/.25
94B15 H ††	H94151	.12/.18	.70/1.05	.25/.65	.25/.55	.08/.15
94B17 H	H94171	.14/.20	.70/1.05	.25/.65	.25/.55	.08/.15
94B30 H	H94301	.27/.33	.70/1.05	.25/.65	.25/.55	.08/.15

†† SAE only

Notes on Alloy Steel Tables:

1. Grades shown with prefix letter E are made only by the basic electric furnace process.
   All others are normally manufactured by the basic open hearth or basic oxygen processes, but may be manufactured by the basic electric furnace process with adjustments in phosphorus and sulfur.

2. The phosphorus and sulfur limitations for each process are as follows:

	Maximum (percent)	
	P	**S**
Basic electric	0.025	0.025
Basic open hearth or basic oxygen	0.035	0.040
Acid electric or acid open hearth	0.050	0.050

3. Minimum silicon limit for acid open hearth or acid electric furnace alloy steel is .15%.

4. Small quantities of certain elements are present in alloy steels, but are not specified or required. These elements are considered as incidental and may be present in the following maximum percentages: copper, .35; nickel, .25; chromium, .20; molybdenum, .06.

5. The listing of minimum and maximum sulfur content indicates a resulfurized steel.

6. Standard alloy steels can be produced to a lead range of .15/.35% to improve machinability.

7. Silicon range for all standard alloy steels except where noted is .15/.30%.

Table 11a. Weights and Areas of Square and Round Steel Bars ¹/₁₆ Through 2 ¹/₁₆ Inches. *(Source, Bethlehem Steel.)*

Size or Dia. inches	Weight (lb. per ft.) Square ■	Weight (lb. per ft.) Round ●	Area (in)² Square □	Area (in)² Round ○	Size or Dia. inches	Weight (lb. per ft.) Square ■	Weight (lb. per ft.) Round ●	Area (in)² Square □	Area (in)² Round ○
1/16	.013	.010	.0039	.0031	51/64	2.159	1.696	.6350	.4987
5/64	.021	.016	.0061	.0048	13/16	2.245	1.763	.6602	.5185
3/32	.030	.023	.0088	.0069	53/64	2.332	1.831	.6858	.5386
7/64	.041	.032	.0120	.0094	27/32	2.420	1.901	.7119	.5591
1/8	.053	.042	.0156	.0123	55/64	2.511	1.972	.7385	.5800
9/64	.067	.053	.0198	.0155	7/8	2.603	2.044	.7656	.6013
5/32	.083	.065	.0244	.0192	57/64	2.697	2.118	.7932	.6230
11/64	.100	.079	.0295	.0232	29/32	2.792	2.193	.8213	.6450
3/16	.120	.094	.0352	.0276	59/64	2.889	2.270	.8498	.6675
13/64	.140	.110	.0413	.0324	15/16	2.988	2.347	.8789	.6903
7/32	.163	.128	.0479	.0376	61/64	3.089	2.426	.9084	.7135
15/64	.187	.147	.0549	.0431	31/32	3.191	2.506	.9385	.7371
1/4	.212	.167	.0625	.0491	63/64	3.294	2.587	.9689	.7610
17/64	.240	.188	.0706	.0554	1	3.400	2.670	1.0000	.7854
9/32	.269	.211	.0791	.0621	1-1/32	3.616	2.840	1.0635	.8353
19/64	.300	.235	.0881	.0692	1-1/16	3.838	3.014	1.1289	.8866
5/16	.332	.261	.0977	.0767	1-3/32	4.067	3.194	1.1963	.9396
21/64	.366	.288	.1077	.0846	1-1/8	4.303	3.379	1.2656	.9940
11/32	.402	.316	.1182	.0928	1-5/32	4.545	3.570	1.3369	1.0500
23/64	.439	.345	.1292	.1014	1-3/16	4.795	3.766	1.4102	1.1075
3/8	.478	.376	.1406	.1104	1-7/32	5.050	3.966	1.4853	1.1666
25/64	.519	.407	.1526	.1198	1-1/4	5.312	4.173	1.5625	1.2272

(Continued)

Table 11a. (Continued) Weights and Areas of Square and Round Steel Bars $1/16$ Through $2 1/16$ Inches. (Source, Bethlehem Steel.)

Size or Dia. inches	Weight (lb. per ft.) Square ■	Weight (lb. per ft.) Round ●	Area (in)² Square □	Area (in)² Round ○	Size or Dia. inches	Weight (lb. per ft.) Square ■	Weight (lb. per ft.) Round ●	Area (in)² Square □	Area (in)² Round ○
13/32	.561	.441	.1650	.1296	1-9/32	5.581	4.384	1.6416	1.2893
27/64	.605	.475	.1780	.1398	1-5/16	5.857	4.600	1.7227	1.3530
7/16	.651	.511	.1914	.1503	1-11/32	6.139	4.822	1.8056	1.4182
29/64	.698	.548	.2053	.1613	1-3/8	6.428	5.049	1.8906	1.4849
15/32	.747	.587	.2197	.1726	1-13/32	6.724	5.281	1.9775	1.5532
31/64	.798	.627	.2346	.1843	1-7/16	7.026	5.518	2.0664	1.6230
1/2	.850	.668	.2500	.1963	1-15/32	7.334	5.761	2.1572	1.6943
33/64	.904	.710	.2659	.2088	1-1/2	7.650	6.008	2.2500	1.7671
17/32	.960	.754	.2822	.2217	1-17/32	7.972	6.261	2.3447	1.8415
35/64	1.017	.799	.2991	.2349	1-9/16	8.301	6.520	2.4414	1.9175
9/16	1.076	.845	.3164	.2485	1-19/32	8.636	6.783	2.5400	1.9949
37/64	1.136	.893	.3342	.2625	1-5/8	8.978	7.051	2.6406	2.0739
19/32	1.199	.941	.3525	.2769	1-21/32	9.327	7.325	2.7431	2.1545
39/64	1.263	.992	.3713	.2916	1-11/16	9.682	7.604	2.8477	2.2365
5/8	1.328	1.043	.3906	.3068	1-23/32	10.044	7.889	2.9541	2.3202
41/64	1.395	1.096	.4104	.3223	1-3/4	10.413	8.178	3.0625	2.4053
21/32	1.464	1.150	.4307	.3382	1-25/32	10.788	8.473	3.1728	2.4920
43/64	1.535	1.205	.4514	.3545	1-13/16	11.170	8.773	3.2852	2.5802
11/16	1.607	1.262	.4727	.3712	1-27/32	11.558	9.078	3.3994	2.6699
45/64	1.681	1.320	.4944	.3883	1-7/8	11.953	9.388	3.5156	2.7612
23/32	1.756	1.379	.5166	.4057	1-29/32	12.355	9.704	3.6337	2.8540
47/64	1.834	1.440	.5393	.4236	1-15/16	12.763	10.024	3.7539	2.9483

(Continued)

Table 11a. *(Continued)* Weights and Areas of Square and Round Steel Bars $^1/_{16}$ Through 2 $^1/_{16}$ Inches. *(Source, Bethlehem Steel.)*

Size or Dia. inches	Weight (lb. per ft.) Square ■	Weight (lb. per ft.) Round ●	Area (in)² Square □	Area (in)² Round ○	Size or Dia. inches	Weight (lb. per ft.) Square ■	Weight (lb. per ft.) Round ●	Area (in)² Square □	Area (in)² Round ○
3/4	1.913	1.502	.5625	.4418	1-31/32	13.178	10.350	3.8760	3.0442
49/64	1.993	1.565	.5862	.4604	2	13.600	10.681	4.0000	3.1416
25/32	2.075	1.630	.6103	.4794	2-1/16	14.463	11.359	4.2539	3.3410

Table 11b. Weights and Areas of Square and Round Steel Bars 2 $^1/_8$ Through 7 $^5/_{16}$ Inches. *(Source, Bethlehem Steel.)*

Size or Dia. inches	Weight (lb. per ft.) Square ■	Weight (lb. per ft.) Round ●	Area (in)² Square □	Area (in)² Round ○	Size or Dia. inches	Weight (lb. per ft.) Square ■	Weight (lb. per ft.) Round ●	Area (in)² Square □	Area (in)² Round ○
2-1/8	15.353	12.058	4.5156	3.5466	5-1/16	87.138	68.438	25.629	20.129
2-3/16	16.270	12.778	4.7852	3.7583	5-1/8	89.303	70.139	26.266	20.629
2-1/4	17.213	13.519	5.0625	3.9761	5-3/16	91.495	71.860	26.910	21.135
2-5/16	18.182	14.280	5.3477	4.2000	5-1/4	93.713	73.602	27.563	21.648
2-3/8	19.178	15.062	5.6406	4.4301	5-5/16	95.957	75.364	28.223	22.166
2-7/16	20.201	15.866	5.9414	4.6664	5-3/8	98.228	77.148	28.891	22.691
2-1/2	21.250	16.690	6.2500	4.9087	5-7/16	100.53	78.953	29.566	23.221
2-9/16	22.326	17.535	6.5664	5.1572	5-1/2	102.85	80.778	30.250	23.758
2-5/8	23.428	18.400	6.8906	5.4119	5-9/16	105.20	82.624	30.941	24.301
2-11/16	24.557	19.287	7.2227	5.6727	5-5/8	107.58	84.492	31.641	24.850
2-3/4	25.713	20.195	7.5625	5.9396	5-11/16	109.98	86.380	32.348	25.406
2-13/16	26.895	21.123	7.9102	6.2126	5-3/4	112.41	88.289	33.063	25.967
2-7/8	28.103	22.072	8.2656	6.4918	5-13/16	114.87	90.218	33.785	26.535
2-15/16	29.338	23.042	8.6289	6.7771	5-7/8	117.35	92.169	34.516	27.109

(Continued)

Table 11b. *(Continued)* Weights and Areas of Square and Round Steel Bars 2 $\frac{1}{8}$ Through 7 $\frac{5}{16}$ Inches. *(Source, Bethlehem Steel.)*

Size or Dia. inches	Weight (lb. per ft.) Square ■	Round ●	Area (in)² Square □	Round ○	Size or Dia. inches	Weight (lb. per ft.) Square ■	Round ●	Area (in)² Square □	Round ○
3	30.600	24.033	9.0000	7.0686	5-15/16	119.86	94.140	35.254	27.688
3-1/16	31.888	25.045	9.3789	7.3662	6	122.40	96.133	36.000	28.274
3-1/8	33.203	26.078	9.7656	7.6699	6-1/16	124.96	98.146	36.754	28.866
3-3/16	34.545	27.131	10.160	7.9798	6-1/8	127.55	100.18	37.516	29.465
3-1/4	35.913	28.206	10.563	8.2958	6-3/16	130.17	102.23	38.285	30.069
3-5/16	37.307	29.301	10.973	8.6179	6-1/4	132.81	104.31	39.063	30.680
3-3/8	38.728	30.417	11.391	8.9462	6-5/16	135.48	106.41	39.848	31.296
3-7/16	40.176	31.554	11.816	9.2806	6-3/8	138.18	108.52	40.641	31.919
3-1/2	41.650	32.712	12.250	9.6211	6-7/16	140.90	110.66	41.441	32.548
3-9/16	43.151	33.891	12.691	9.9678	6-1/2	143.65	112.82	42.250	33.183
3-5/8	44.678	35.090	13.141	10.321	6-9/16	146.43	115.00	43.066	33.824
3-11/16	46.232	36.311	13.598	10.680	6-5/8	149.23	117.20	43.891	34.472
3-3/4	47.813	37.552	14.063	11.045	6-11/16	152.06	119.43	44.723	35.125
3-13/16	49.420	38.814	14.535	11.416	6-3/4	154.91	121.67	45.563	35.785
3-7/8	51.053	40.097	15.016	11.793	6-13/16	157.79	123.93	46.410	36.450
3-15/16	52.713	41.401	15.504	12.177	6-7/8	160.70	126.22	47.266	37.122
4	54.400	42.726	16.000	12.566	6-15/16	163.64	128.52	48.129	37.800
4-1/16	56.113	44.071	16.504	12.962	7	166.60	130.85	49.000	38.485
4-1/8	57.853	45.438	17.016	13.364	7-1/16	169.59	133.19	49.879	39.175
4-3/16	59.620	46.825	17.535	13.772	7-1/8	172.60	135.56	50.766	39.871
4-1/4	61.413	48.233	18.063	14.186	7-3/16	175.64	137.95	51.660	40.574
4-5/16	63.232	49.662	18.598	14.607	7-1/4	178.71	140.36	52.563	41.282

(Continued)

Table 11b. *(Continued)* **Weights and Areas of Square and Round Steel Bars 2 1/8 Through 7 5/16 Inches.** *(Source, Bethlehem Steel.)*

Size or Dia. inches	Weight (lb. per ft.) Square ■	Weight (lb. per ft.) Round ●	Area (in)² Square □	Area (in)² Round ○	Size or Dia. inches	Weight (lb. per ft.) Square ■	Weight (lb. per ft.) Round ●	Area (in)² Square □	Area (in)² Round ○
4-3/8	65.078	51.112	19.141	15.033	7-5/16	181.81	142.79	53.473	41.997
4-7/16	66.951	52.583	19.691	15.466	7-3/8	184.93	145.24	54.391	42.718
4-1/2	68.850	54.075	20.250	15.904	7-7/16	188.08	147.71	55.316	43.445
4-9/16	70.776	55.587	20.816	16.349	7-1/2	191.25	150.21	56.250	44.179
4-5/8	72.728	57.121	21.391	16.800	7-9/16	194.45	152.72	57.191	44.918
4-11/16	74.707	58.675	21.973	17.257	7-5/8	197.68	155.26	58.141	45.664
4-3/4	76.713	60.250	22.563	17.721	7-11/16	200.93	157.81	59.098	46.415
4-13/16	78.745	61.846	23.160	18.190	7-3/4	204.21	160.39	60.063	47.173
4-7/8	80.803	63.463	23.766	18.665	7-13/16	207.52	162.99	61.035	47.937
4-15/16	82.888	65.100	24.379	19.147	7-7/8	210.85	165.60	62.016	48.707
5	85.000	66.759	25.000	19.635	7-15/16	214.21	168.24	63.004	49.483

Table 12. Selected Properties of the Elements.

Element/Symbol	Atomic Number	Relative Atomic Mass	Density		Melting Point		Boiling Point		Modulus of Elasticity	
			lb/in.³	g/cm.³	°F	°C	°F	°C	10⁶ PSI	GPa
Aluminum/ Al	13	26.9815	0.09751	2.699	1,220.4	660	4,520	2,480	10	69
Antimony/ Sb	51	121.75	0.239	6.62	1,166.9	630.5	2,620	1610	11.3	77
Arsenic/As	33	74.9216	0.207	5.73	1,497	(817) (36 at.)	1,130	616	11	75.8
Barium/Ba	56	137.34	0.13	3.5	1,300	710	2,980	(1700)	1.8	12.4
Beryllium/Be	4	9.01218	0.0658	1.85	2,340	1284	5,020	(2400)	37	255
Bismuth/Bi	83	208.9806	0.654	9.80	520.3	271	2,590	1680	4.6	31.7
Boron/B	5	10.811	0.083	2.3	3,812	(2300)	4,620	(2550)	-	-
Cadmium/Cd	48	112.40	0.313	8.65	609.3	321	1,409	765	8	55
Calcium/Ca	20	40.88	0.056	1.55	1,560	843	2,625	2075	3	20.6
Carbon/C	6	12.01115	0.0802	2.25 (graphite) 3.51 (diamond)	6,700	(5000) Gr.	8,730	(5000)	0.7	4.8
Cerium/Ce	58	140.12	0.25	6.77	1,460	804	4,380	3500	-	-
Chromium/Cr	24	51.996	0.260	7.19	3,350	1900	4,500	2690	36	248
Cobalt/Co	27	58.9332	0.32	8.90	2,723	1492	6,420	(2900)	30	206.8
Columbium (Niobium)/Cb	41	92.9064	0.310	8.57	4,380	2468	5,970	4735	15	103.4
Copper/Cu	29	63.546	0.324	8.96	1,981.4	1083	4,700	2575	16	110.3
Gadolinium/Gd	64	157.25	0.287	7.87	-	1350	-	(3000)	-	-
Gallium/Ga	31	69.72	0.216	5.91	85.5	29.8	3,600	2420	1	6.9
Germanium/Ge	32	72.59	0.192	5.36	1,756	937	4,890	2700	11.4	79
Gold/Au	79	196.9665	0.698	19.32	1,945.4	1063	5,380	(2950)	12	82.7

(Continued)

Table 12. (Continued) Selected Properties of the Elements.

Element/Symbol	Atomic Number	Relative Atomic Mass	Density		Melting Point		Boiling Point		Modulus of Elasticity	
			lb/in.³	g/cm.³	°F	°C	°F	°C	10⁶ PSI	GPa
Hafnium/Hf	72	178.49	0.473	11.4	3,865	2220	9,700	4450	20	137.9
Hydrogen/H	1	1.0080	3.026 × 10⁻⁶	0.0899‡	-434.6	-259.2	-422.9	-252.5	-	-
Indium/In	49	114.82	0.264	7.31	313.5	156.4	3,630	2075	1.57	11
Iridium/Ir	77	192.22	0.813	22.5	4,449	2443	9,600	4400	75	517
Iron/Fe	26	55.847	0.284	7.87	2,802	1536	4,960	(3070)	28.5	199.4
Lanthanum/La	57	138.9055	0.223	6.16	1,535	920	8,000	(3420)	5	34.4
Lead/Pb	82	207.2	0.4097	11.34	621.3	327.4	3,160	1740	2.6	17.9
Lithium/Li	3	6.941	0.019	0.534	367	180.5	2,500	1329	1.7	11.7
Magnesium/Mg	12	24.305	0.0628	1.74	1,202	650	2,030	1103	6.5	44.8
Manganese/Mn	25	54.9380	0.268	7.43	2,273	1244	3,900	2020	23	158.5
Mercury/Hg	80	200.59	0.4896	13.55	-37.97	-38.9	675	357	-	-
Molybdenum/Mo	42	95.94	0.369	10.2	4,750	2620	8,670	4650	50	344.7
Nickel/Ni	28	58.71	0.322	8.90	2,651	1453	4,950	(3000)	30	206.8
Niobium (see Columbium)/Nb	41	92.9064	0.310	8.57	4,380	2468	5,970	4735	-	-
Nitrogen/N	7	14.0067	0.042 × 10⁻³	1.2506‡	-346	-210.0	-320.4	-195.8	-	-
Osmium/Os	76	190.2	0.813	22.5	4,900	3045	9,900	>5300	80	551.5
Oxygen/O	8	15.9994	0.048 × 10⁻³	1.429‡	-361.8	-218.8	-297.4		-	-
Palladium/Pd	46	106.4	0.434	12.0	2,829	1552	7,200		17	117.2
Phosphorus/P	15	30.9738	0.0658	1.82	111.4	44.1	536		-	-
Platinum/Pt	78	195.09	0.7750	21.45	3,224.3	1769	7,970		21	114.7
Plutonium/Pu	94	(244)	0.686	(19.0)	1,229	640	-		-	-
Potassium/K	19	39.102	0.031	0.86	145	63.6	1,420		0.5	3.4

(Continued)

Table 12. *(Continued)* Selected Properties of the Elements.

Element/Symbol	Atomic Number	Relative Atomic Mass	Density		Melting Point		Boiling Point		Modulus of Elasticity	
			lb/in.³	g/cm.³	°F	°C	°F	°C	10⁶ PSI	GPa
Radium/Ra	88	226.0254	0.18	5.0	1,300	960	-	-	-	-
Rhenium/Re	75	186.2	0.765	20.0	5,733	3150	10,700		75	517
Rhodium/Rh	45	102.9055	0.4495	12.44	3,571	1960	8,100		54	372.3
Selenium/Se	34	78.96	0.174	4.81	428	220.5	1,260		8.4	57.9
Silicon/Si	14	28.086	0.084	2.4	2,605	1412	4,200		16	110.3
Silver/Ag	47	107.868	0.379	10.49	1,760.9	960.8	4,010		11	75.8
Sodium/Na	11	22.9898	0.035	0.97	207.9	97.8	1,638		1.3	8.9
Sulfer/S	16	32.06	0.0748	2.07	246.2	112.8	832.3		-	-
Tantalum/Ta	73	180.9479	0.600	16.6	5,420	3010	9,570		27	186
Tellurium/Te	52	127.60	0.225	6.24	840	449.8	2,530		6	41.3
Thallium/Tl	81	204.37	0.428	11.85	577	303	2,655		1.2	8.2
Thorium/Th	90	232.0381	0.422	11.5	3,348	1750	8,100		11.4	78
Tin/Sn	50	118.69	0.264	7.298	449.4	231.9	4,120		6	41.3
Titanium/Ti	22	47.90	0.164	4.54	3,074	1667	6,395		16.8	116
Tungsten/W	74	183.85	0.697	19.3	6,150	3380	10,700		50	344.7
Uranium/U	92	238.029	0.687	18.7	2,065	1130	7,100		29.7	205.8
Vanadium/V	23	50.9414	0.217	6.0	3,452	1920	5,430		18.4	127
Zinc/Zn	30	65.37	0.258	7.133	787	419.5	1,663		12	82.7
Zirconium/Zr	40	91.22	0.23	6.5	3,326	1860	9,030		11	75.8

‡ Density of gases are given g/dm³ (grams/decimeter3).

with sulfur to form manganese sulfide. This enhances the machining characteristics of the free machining carbon steels. Manganese also enhances the hardenability of steel because it decreases the critical (minimum) cooling rate necessary for hardening a steel. It is present (0.3 to 0.8%) in all commercial steels. Its effectiveness depends largely upon, and is directly proportional to, the carbon content of the steel.

Phosphorus (chemical symbol P) is normally regarded as an impurity except where it is added to increase machinability and atmospheric corrosion. While it increases strength and hardness to about the same degree as carbon, it also tends to decrease ductility and toughness, particularly in steel in the quenched and tempered condition. The maximum content of most steels is up to 0.04%, but in free-machining steels the phosphorus content, which is boosted by adding phosphorus at the ladle (known as rephosphorizing), can be as high as 0.12%.

Sulfur (chemical symbol S) is not a desirable element except when machinability is an important consideration. Increasing sulfur increases machinability at all carbon levels in both alloy and plain carbon grades. Increases in sulfur up to 0.06% markedly increase machinability, but above this level machinability improves at a lower rate. Sulfur also impairs weldability, and has an adverse effect on surface quality.

Silicon (chemical symbol Si) acts as a deoxidizer in carbon and alloy steels. In the company of manganese, it produces extremely high strength steel that is combined with good ductility and shock resistance in the quenched and tempered condition. In larger quantities it adversely affects machinability and increases susceptibility to decarburization and graphitization.

Nickel (chemical symbol Ni) increases impact resistance and toughness, especially at low temperature. It also lowers the critical temperature of steel for heat treatment and enhances hardenability and it does not form carbides or other compounds that might be difficult to dissolve during heating for austenitizing. Nickel's insensitivity to variations in quenching conditions provides insurance against costly failures to attain the desired properties, particularly when the furnace is not equipped for precision control.

Chromium (chemical symbol Cr) enhances hardenability, improves wear and abrasion resistance, and promotes carburization. Of the common alloying elements, chromium is surpassed only by manganese and molybdenum in its effect on hardenability.

Molybdenum (chemical symbol Mo) enhances hardness and widens the temperature range for heat treatment. It also increases steel's high temperature tensile and creep strength and reduces susceptibility to temper brittleness.

Vanadium (chemical symbol V) is a strong deoxidizer and inhibits grain growth over a fairly large quenching range. With vanadium additions of 0.04 to 0.05%, hardenability of medium carbon steels is increased with minimum affect on grain size. For higher vanadium content, hardenability can be increased with higher austenitizing temperatures.

Copper (chemical symbol Cu) is used to improve steel's resistance to corrosion. In amounts of 0.20 to 0.50%, it does not significantly increase the mechanical properties of the steel.

Boron (chemical symbol B) increases hardenability, even when present in amounts as small as 0.0005%. Because boron is ineffective when allowed to combine with oxygen or nitrogen, its use is limited to aluminum-killed steels.

Lead (chemical symbol Pb) does not alloy with steel, but it is retained in its elemental state as a fine dispersion within the steel's structure. It is most widely used in the free-machining carbon grades where, in amounts of 0.15 to 0.35%, it effectively lubricates the cutting edge of the tool and permits an increase in cutting speed and feed and an improvement in surface finish quality without an attendant decrease in tool life.

Nitrogen (chemical symbol N) is present in almost all steels in small amounts that do not affect the characteristics of the steel. When nitrogen content reaches 0.004% or more, it combines with other elements to precipitate as a nitride, thereby increasing the steel's hardness and tensile and yield strengths while reducing its ductility and toughness.

Aluminum (chemical symbol Al) is a strong deoxidizer and inhibitor of grain growth, and it improves the nitriding characteristics of steel when present in amounts of approximately 1%.

Multiple alloying elements. A combination of two or more of the above alloying elements usually imparts some of the characteristic properties of each. Constructional chromium-nickel alloy steels, for example, develop good hardening properties, with excellent ductility, while chromium-molybdenum combinations develop excellent hardenability with satisfactory ductility and a certain amount of heat resistance. The combined effect of the two or more alloying elements on the hardenability of a steel is considerably greater than the sum of the same alloying elements if used separately. The general effectiveness of the nickel-chromium-molybdenum steels, with and without boron, can be accounted for in this way. Various combinations of alloying elements are employed by the different producers of high strength low alloy steels to achieve the mechanical properties, resistance to atmospheric corrosion, and other properties that characterize those steels. Carbon is generally maintained at a level that will insure freedom from excessive hardening after welding and will retain ductility. Manganese is used primarily as a strengthening element. Phosphorus is sometimes employed as a strengthening element and to enhance resistance to atmospheric corrosion. Copper is used to enhance resistance to atmospheric corrosion and as a strengthening element. Silicon, nickel, chromium, molybdenum, vanadium, aluminum, titanium, zirconium, and other elements are sometimes used, singly or in combination, for their beneficial effects on strength, toughness, corrosion resistance, and other desirable properties.

Manufacturing considerations—carbon and low alloy steels

Machining. The low carbon grades (0.30% carbon and less) are soft and gummy in the annealed condition and are preferably machined in the cold-worked or the normalized condition. Medium carbon grades (0.30 to 0.50% carbon) are best machined in the annealed condition, and high carbon grades (0.50 to 0.90% carbon) in the spheroidized condition. Finish machining must often be done in the fully heat treated condition for dimensional accuracy. The resulfurized grades are well known for good machinability. Nearly all carbon steels are available with 0.15 to 0.35% lead, added to improve machinability. However, resulfurized and leaded steels are not recommended for highly stressed parts. Alloy steels are generally harder than unalloyed steels of the same carbon content. Consequently, the low carbon alloy steels are somewhat easier to finish machine than their counterparts in the carbon steels. It is usually desirable to finish machine the carburizing and through-hardening grades in the final heat treated stage for better dimensional accuracy. This often requires two steps: rough machining in the annealed or hot-finished condition, then finish machining after heat treating. The latter operation, because of the relative high hardness of the material, necessitates the use of sharp, well designed, high speed steel cutting tools, proper feeds, speeds, and a generous supply of coolant. Medium and high carbon alloy grades are usually spheroidized for optimum machinability and, after heat treatment, may be finished by grinding. Many of the alloy steels are available with added sulfur or lead for improved machinability. These resulfurized and leaded alloy steels are not recommended for highly stressed applications, such as in aircraft parts.

Cold forming. The very low carbon grades have excellent cold-forming characteristics when in the annealed or normalized conditions. Medium carbon grades show progressively poorer formality with higher carbon content, and more frequent annealing is required.

The high carbon grades require special softening treatments for cold forming. Many carbon steels are embrittled by warm working or prolonged exposure in the temperature range of 300 to 700° F (150 to 370° C). The alloy steels are normally cold formed in annealed condition. Their formability depends on their carbon content. Little cold forming is done on alloy steels in the heat treated condition because of their high strength and limited ductility.

Forging. All the carbon steels exhibit excellent forgeability in the austenitic state provided the proper forging temperatures are used. As the carbon content is increased, the maximum forging temperature is decreased. At high temperatures, these steels are soft and ductile and exhibit little or no tendency to work harden. The resulfurized grades (free-machining steels) exhibit a tendency to rupture when deformed in certain high temperature ranges. Close control of forging temperatures is required. The alloy steels are only slightly more difficult to forge than carbon steels. However, maximum recommended forging temperatures are generally about 50% lower than for carbon steels of the same carbon content. Slower heating rates, shorter soaking period, and slower cooling rates are also required for alloy steels.

Welding. The low carbon grades are readily welded or brazed by all techniques. The medium carbon grades are also readily weldable but may require postwelding heat treatment. The high carbon grades are difficult to weld—preheating and postwelding treatment are usually mandatory, and special care must be taken to avoid overheating. Furnace brazing has been done successfully with all grades. Alloy steels with low carbon content are readily welded or brazed by all techniques. Alloy welding rods, comparable in strength to the base metal, are used, and moderate preheating (200–600° F *or* 98–315° C) is usually necessary. At higher carbon levels, alloy steels require higher preheating temperatures and often postwelding stress relieving. Certain alloy steels can be welded without loss of strength in the heat-affected zone provided that the welding heat is carefully controlled. If the composition and strength of the steel are such that the welded joint is reduced, the strength of the joint may be restored by heat treatment after welding. See the Welding section of this book for more detailed information.

Heat treatment. Due to the poor oxidation resistance of carbon steels, protective atmospheres must be employed during heat treatment if scaling on the surface cannot be tolerated. Also, these steels are subject to decarburization at elevated temperatures and, where surface carbon content is critical, should be heated in reducing atmospheres. For low alloy steels, various heat treatment procedures may be used to achieve any number of specific mechanical properties. In general, the annealed condition is achieved by heating to a specific condition and holding for a specified period of time. Annealing generally softens the material, producing the lowest mechanical properties. The normalized condition is achieved by holding to a slightly higher temperature than annealing, but for a shorter period of time. The purpose of normalizing varies depending on the desired properties—it can be used to increase or decrease mechanical properties. The quenched and tempered condition provides the highest mechanical properties with relative high toughness. Maximum hardness is obtained in the as-quenched condition, but alloy steels may be embrittled in this condition.

Stainless Steels

Stainless Steel. (Specialty Steel Industry of North America definition.) Stainless steels are iron-based alloys containing 10.5% or more of chromium, which is the alloying element that imparts to stainless steels their corrosion-resistant qualities. It does this by combining with oxygen to form a thin, transparent chromium-oxide protective film on the metal surface.

Austenitic stainless steels are those containing chromium, nickel, and manganese (AISI 200 series), or just chromium and nickel (AISI 300 series) as the principal alloying elements. These steels can be hardened only by cold working, heat treatment serves only to soften them. In the annealed condition, they are not attracted to a magnet, although some may become magnetic after cold working. They have excellent corrosion resistance, good formability, and the ability to develop excellent strength characteristics by cold working. When annealed, they possess maximum corrosion resistance, good yield and tensile strength, and freedom from notch effect. Some austenitic stainless steels contain titanium and niobium that are added to form more stable carbides than can be obtained with chromium. These stabilizing elements reduce susceptibility to intergranular corrosion, and prevent sensitization (the precipitation of chromium carbides, usually along grain boundaries, at temperatures ranging from 1,000 to 1,550° F [538 to 843° C]—thereby leaving the grain boundaries depleted of carbon) during welding and heat treatment. The most commonly used stabilized grades are 321 and 327.

Ferritic stainless steels (AISI 400 series) are straight chromium steels with a low carbon to chromium ratio that eliminates the effects of thermal transformation—they are not hardenable by heat treatment but can be marginally hardened by cold working. These steels are strong, attracted to a magnet, have good ductility, and are resistant to corrosion and oxidation. Their corrosion resistance can be improved by increasing the chromium or molybdenum content, while ductility can be improved by reducing carbon and nitrogen content. The ferritic steels, like the austenitic steels, are also subject to intergranular corrosion, but at temperatures in the region of 1,700° F (927° C).

Martensitic stainless steels (Also AISI 400 series) are straight chromium types having a higher carbon to chromium ratio than the ferritic group. Consequently, when cooled rapidly, they do harden (in some cases to tensile strengths of 200,000 PSI). These steels resist corrosion in mild environments, have fairly good ductility, and are always strongly magnetic. Some of the martensitic types, such as 416, 420F, and 44F, have been modified to improve machinability.

Precipitation hardened grades (AISI and UNS types S13800, S15500, S17400, and S17700) can achieve tensile and yield strengths as high as 300,000 PSI through the use of low temperature (about 900° F) aging in combination with cold working. With these steels, fabrication can be completed in an annealed condition, and then the component can be uniformly hardened without cold working or the high temperature thermal treatments that can cause distortion and heavy scaling. Prior to hardening, the machinability of these steels is approximately equal to AISI type 304 in the annealed condition.

Duplex stainless steels are dual phase materials with austenite and ferrite in close to 50-50 balance. They have excellent strength, corrosion resistance, and fabrication properties (especially those grades that contain nitrogen). Newer duplex materials with nitrogen are readily weldable and are often used in oil and gas production and chemical processing.

Basic chemical, mechanical, and physical properties of selected stainless steels are given in **Tables 13**, **14**, and **15**.

Effects of machining and fabrication on stainless steel

In all cutting operations on stainless steels, the following guidelines should be followed for maintaining corrosion resistance.

1) No contamination by ferrous (iron or steel) materials or particles should take place.

2) Mechanically cut edges will naturally form the corrosion resistant passive film that is a characteristic of stainless steel. The formation of this film on cut edges will be enhanced

(Text continued on p. 456)

Table 13. Chemical Composition of Wrought Stainless Steels.

AISI Type	UNS No.	C	Mn	P	S	Si	Cr	Ni	Mo	Other
					Austenitic Grades					
201	S20100	0.15	5.50-7.50	0.060	0.030	0.75	16.00-18.00	3.50-5.50	-	0.25N
202	S20200	0.15	7.50-10.00	0.060	0.030	0.75	17.00-19.00	4.00-6.00	-	0.25N
203	S20300	0.08	5.00-6.50	0.040	0.18-0.35	1.00	16.00-19.00	5.00-6.50	0.50	1.75/2.25 Cu
205	S20500	0.12-0.25	14.00-15.50	0.060	0.030	0.75	16.50-18.00	1.00-1.75	-	0.32-0.40N
301	S30100	0.15	2.00	0.045	0.030	0.75	16.00-18.00	6.00-8.00	-	-
302	S30200	0.15	2.00	0.045	0.030	0.75	17.00-19.00	8.00-10.00	-	0.10N
303	S30300	0.15	2.00	0.20	0.15 (min.)	1.00	17.00-19.00	8.00-10.00	-	-
303Se	S30323	0.15	2.00	0.20	0.060	1.00	17.00-19.00	8.00-10.00	-	0.15(min.)Se
304	S30400	0.08	2.00	0.045	0.030	0.75	18.00-20.00	8.00-10.50	-	0.10N
305	S30500	0.12	2.00	0.045	0.030	0.75	17.00-19.00	10.50-13.00	-	-
308	S30800	0.08	2.00	0.045	0.030	1.00	19.00-21.00	10.00-12.00	-	-
309	S30900	0.20	2.00	0.045	0.030	1.00	22.00-24.00	12.00-15.00	-	-
310	S31000	0.25	2.00	0.045	0.030	1.50	24.00-26.00	19.00-22.00	-	-
314	S31400	0.25	2.00	0.045	0.030	1.50-3.00	23.00-26.00	19.00-22.00	-	-
316	S31600	0.08	2.00	0.045	0.030	0.75	16.00-18.00	10.00-14.00	2.00-3.00	0.10N
316H	S31609	0.04-0.10	2.00	0.045	0.030	0.75	16.00-18.00	10.00-14.00	2.00-3.00	-
316F	S31620	0.08	2.00	0.20	0.10 (min.)	1.00	16.00-18.00	10.00-14.00	1.75-2.50	-
317	S31700	0.08	2.0	0.045	0.030	0.75	18.00-20.00	11.00-15.00	3.00-4.00	0.10(max)N
329	S32900	0.08	2.00	0.040	0.030	0.75	23.00-28.00	2.50-5.00	1.00-2.00	-
330	S33000	0.08	2.00	0.040	0.030	0.75-1.50	17.00-20.00	34.00-37.00	-	-
384	S38400	0.08	2.00	0.045	0.030	1.00	15.00-17.00	17.00-19.00	-	-

(Continued)

Table 13. *(Continued)* Chemical Composition of Wrought Stainless Steels.

AISI Type	UNS No.	C	Mn	P	S	Si	Cr	Ni	Mo	Other
Ferritic Grades										
429	S42900	0.12	1.00	0.040	0.030	1.00	14.00-16.00	0.75	-	-
430	S43000	0.12	1.00	0.040	0.030	1.00	16.00-18.00	0.75	-	-
430F	S43020	0.12	1.25	0.060	0.15 (min.)	1.00	16.00-18.00	-	-	-
430FSe	S43023	0.12	1.25	0.060	0.060	1.00	16.00-18.00	-	-	0.15(min)Se
434	S43400	0.12	1.00	0.040	0.030	1.00	16.00-18.00	-	0.75-1.25	-
442	S44200	0.20	1.00	0.40	0.300	1.00	18.00-23.00	-	-	-
446	S44600	0.20	1.50	0.040	0.030	1.00	23.00-27.00	-	-	0.25N
Martensitic Grades										
403	S40300	0.15	1.00	0.040	0.030	0.50	11.50-13.00	0.60	-	-
410	S41000	0.15	1.00	0.040	0.30	1.00	11.50-13.50	0.75	-	-
414	S41400	0.15	1.00	0.040	0.030	1.00	11.50-13.50	1.25-2.50	-	-
416	S41600	0.15	1.25	0.060	0.15 (min.)	1.00	12.00-14.00	-	-	-
416Se	S41623	0.15	1.25	0.060	0.060	1.00	12.00-14.00	-	-	-
420	S42000	0.15 (min)	1.00	0.040	0.030	1.00	12.00-14.00	-	-	-
420F	S42020	Over 0.15	1.25	0.060	0.15	1.00	12.00-14.00	-	-	-
422	S42200	0.20-0.25	0.50-1.00	0.25	0.25	0.50	11.00-12.50	0.50-1.00	0.90-1.25	0.20-0.30V 0.90-1.25W
431	S43000	0.20	1.00	0.040	0.030	1.00	15.00-17.00	1.25-2.50	-	-
440A	S44002	0.60-0.75	1.00	0.040	0.030	1.00	16.00-18.00	-	0.75	-
440B	S44003	0.75-0.95	1.00	0.040	0.030	1.00	16.00-18.00	-	0.75	-
440C	S44004	0.95-1.20	1.25	0.40	0.030 (min.)	1.00	16.00/18.00	-	-	-
440F	S44020	0.95 (min.)	1.25	0.040	0.10 (min.)	1.00	16.00/18.00	-	-	-

Notes: Data are for information only and should not be used for design purposes. For design and specification, refer to appropriate ASTM specifications. Data were obtained from various sources, including AISI Steel Products Manuals, ASTM specification, and SSINA.

454 STAINLESS STEELS

Table 14. Typical Mechanical Properties of Stainless Steels (Annealed Bar). (Source, SSINA.)

	AISI Type	UNS No.	Tensile Strength		Yield Strength (0.2% offset)		Elongation In 2" %	Hardness HBN (max)
			ksi (min.)	MPa	ksi (min.)	MPa		
Austenitic	203	S20300	75	517	30	207	40	262
	303	S30300	75	517	30	207	35	262
	303Se	S30323	75	517	30	207	35	262
	304	S30400	75 (125)*	517 (861)	30 (100)*	207 (690)	40 (10)*	-
	316	S31600	75 (125)*	517 (861)	30 (100)*	207 (690)	40 (10)*	-
	316H	S31609	75	517	30	207	40	-
	316F	S31620	75	517	30	207	40	262
Ferritic	430	S43000	70	483	40	276	20	-
	430F	S43020	70	483	40	276	20	262
Martensitic	416	S41600	75	517	40	276	22	262
	420	S42000	75	517	40	276	25	241
	420F	-	75	517	40	276	22	262
	440C	S44004	110	758	65	448	14	262
	440F	S44020	110	758	65	448	14	262

Notes: Data are for information only and should not be used for design purposes. For design and specification, refer to appropriate ASTM specifications. Data were obtained from various sources, including AISI Steel Products Manuals, ASTM specification, and individual company literature.

* Cold finished, for sizes up to ³/₄ inch.

Table 15. Physical Properties of Selected Stainless Steels. *(Source, SSINA.)*

Stainless Steel Type	Density lb./in.³	Modulus of Elasticity PSI % 10⁶	Specific Electrical Resistance at 68° F (microhm-cm)	Specific Heat Capacity Btu/lb. °F	Thermal Conductivity Btu/ft. hr. °F
Austenitic Grades					
203	0.29	28.0	74	-	9.5
303	0.29	28.0	72	0.12	9.4
303Se	0.29	28.0	72	0.12	9.4
304	0.29	28.0	72	0.12	9.4
316	0.29	28.0	74	0.12	9.4
316H	0.29	28.0	74	0.12	9.4
316F	0.29	29.0	74	0.116	8.3
Ferritic Grades					
430	0.28	29.0	60	0.11	13.8
430F	0.28	29.0	60	0.11	15.1
Martensitic Grades					
416	0.28	29.0	57	0.11	14.4
420	0.28	29.0	55	0.11	14.4
420F	0.28	29.0	55	0.11	14.5
440C	0.28	29.0	60	0.11	14.0
440F	0.28	29.0	60	0.11	14.0

Stainless Steel Type	Mean Coefficient of Thermal Expansion in./in./°F %10⁻⁶					Magnetic Permeability (max)	Annealing Temperature °F
	32-212°F	32-600°F	32-1000°F	32-1200°F	32-1600°F		
Austenitic Grades							
203	9.4	-	-	-	11.6	1.02	1850-2050 [A]
303	9.6	9.9	10.2	10.4	-	[1.02]	1850-2050 [A]
303Se	9.6	9.9	10.2	10.4	-	[1.02]	1850-2050 [A]
304	9.6	9.9	10.2	10.4	11.0	1.02	1850-2050 [A]
316	8.9	9.0	9.7	10.3	11.1 [D]	1.02	1850-2050 [A]
316H	8.9	9.0	9.7	10.3	11.1 [D]	1.02	1850-2050 [A]
316F	9.2	9.7	10.1	-	-	-	2000 [A]
Ferritic Grades							
430	5.8	6.1	6.3	6.6	6.9 [D]	-	1250-1400
430F	5.8	6.1	6.3	6.6	6.9	-	1250-1400 [C]
Martensitic Grades							
416	5.5	6.1	6.4	6.5	-	-	1500-1650 [B] 1200-1400 [C]
420	5.7	6.0	6.5	-	-	-	1550-1650 [B] 1350-1450 [C]
420F	5.7	-	-	-	-	-	1550-1650 [B] 1350-1450 [C]
440C	5.6	-	-	-	-	-	1550-1650 [B]
440F	5.6	-	-	-	-	-	1350-1450 [C]

Notes: Data are for information only and should not be used for design purposes. For design and specification, refer to appropriate ASTM specifications. Data were obtained from various sources, including AISI Steel Products Manuals, ASTM specification, and individual company literature.

[A] = Cool rapidly from these annealing temperatures.
[B] = Full annealing-cool slowly.
[C] = Process annealing.
[D] = Thermal range is 32-1500° F.

by a chemical (acid) passivation (see below) treatment with nitric acid.

3) Thermally cut edges will be affected in terms of chemical composition and metallurgical structure. Removal of surface layers by dressing is necessary so that impaired areas of mechanical and corrosion resistant properties are minimized.

Passivation. The thin transparent film on the surface of stainless steel is uniform, stable, and passive. This film will form spontaneously, or repair itself if damaged in air (due to the presence of oxygen) and when immersed in solutions that contain oxygen or oxidizing agents; passivity appears to be enhanced by the immersion method. Nitric acid is normally used as the oxidizing agent. It does not corrode stainless steel, does not alter critical dimensions, and will not remove heat tint, embedded iron, or other embedded surface contamination. The standard nitric acid passivation solution is made up of 10–15% (by volume) of nitric acid in water. Best results are obtained if the solution is 150° F (65° C) for the austenitic (300 series) stainless steels and 120° F (50° C) for the ferritic and martensitic (400 series) plain chromium stainless steels. The suggested immersion time is thirty minutes, followed by thorough water washing. If the component cannot be thoroughly immersed, cold swabbing with cold acid solution is normally used. Under these conditions, the following guidelines are recommended. For austentitic (300 series), use a solution of 15% nitric acid (by volume) and apply at a temperature between 65–80° F (20–25° C) for thirty to ninety minutes. For ferritics (400 series), use a solution of 12% nitric acid (by volume) and apply at 65–80° F (20–25° C) for thirty to forty five minutes. The solution should be continually swabbed on with sponges, paint brushes, or nylon pads. Thorough water rinsing must follow passivitating treatments.

Descaling. When stainless steel is heated to elevated temperatures, such as during annealing and welding, an oxide scale will form on the surface unless the material is completely surrounded by a protective atmosphere. These oxides should be removed to restore the stainless steel to its optimum corrosion resistance condition. Pickling is one of the most common methods used to remove this oxide scale. To remove scale caused by annealing austenitic stainless steel, pickling the component in a solution of 10–15% nitric acid plus 1 to 3% hydrofluoric acid (by volume), warmed to 120–140° F (50–60° C) effectively removes oxides and loosely imbedded iron and chromium depleted layers, and it leaves the surface in a clean, passivated condition. For ferritic and martensitic stainless steels, a solution of 8–12% sulfuric acid and 2% rock salt (by volume), warmed to 150–170° F (65–71° C) should be used. Pickling times are normally five to ten minutes and, upon removal, the component should be scrubbed and rinsed to remove sludge. Pickling can be done at room temperature, but exposure times must be increased.

Scale can also be removed by grit blasting, sand blasting, or glass bead blasting. Glass bead media is preferred as it is less likely to damage or contaminate the surface.

Machinability of stainless steels

Machinability of stainless steels differs substantially from carbon or low alloy steels. Most standard stainless steels are more difficult to machine, but the free-machining types are routinely machined on high production equipment. The easiest to machine are the 400 series, and the 200 and 300 series are often characterized as being the most difficult, primarily because of their gumminess and their propensity to work harden at a very rapid rate. *Figures 1* and *2* show the comparative machinability of stainless steel and other common metals.

Free-machining Stainless Steels. Certain alloying elements in stainless steels, such as sulfur, selenium, lead, copper, aluminum, calcium, or phosphorus can be added or adjusted during melting to alter the machining characteristics. These elements serve to reduce the friction created by the tool, thereby minimizing the tendency of the chip to weld to the

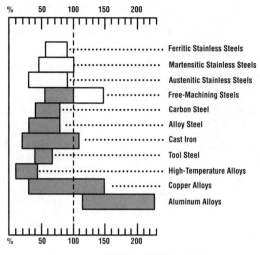

Figure 1. Comparative machinability of common metals. (Source, SSINA.)

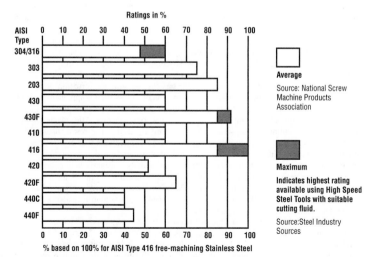

Figure 2. Comparative machinability of frequently used stainless steels and their free-machining counterparts. (Source, SSINA.)

tool. Sulfur can also be added to induce inclusions that reduce the friction forces and transverse ductility of the chips, causing them to break more readily. The improvement in machinability in the free-machining stainless steels—namely types 303, 303 Se, 203, 430F, 416, and 420F—compared to other frequently used stainless steels, can be seen in *Figure 3*. The free-machining grades can be competitive with the tougher grades and, where ease of production is an issue, should be used whenever possible. If, for example, a design demands excellent corrosion and strength of type 304, but a high machining rate is required, free-machining 303 might be a suitable substitution. While each has similar chromium, nickel, and sulfur content, type 303 can be machined at speeds 25 to 30% faster than type 304. Type 303 Se is a free-machining variation of type 304 that contains selenium rather than sulfur. Therefore, when surface finish is a primary consideration, or when cold-working may be involved (staking, swaging, spinning, or severe thread rolling in addition to machining), selecting 303 Se instead of type 304 can be advantageous. Type 203 is an austenitic free-machining grade that is a modification of the 200 series. Although, like the 200 series, it contains significant additions of manganese, it also contains nickel and copper. In the 400 series, if end-use conditions call for type 430, type 430F might be considered. The composition of type 430F is adjusted to enhance its machining characteristics while preserving as many of type 430's qualities as possible. The free-machining variation of type 410 is 416, and for type 420 the free-machining variation is 420F. The alloying elements used to improve free-machining characteristics can adversely affect corrosion resistance, transverse ductility, and other qualities including weldability. Free-machining grades are normally not used when welding, cold forming, or hot forging is specified.

Type 203. This is an austenitic grade containing increased levels of manganese in order to obtain maximum machining speeds. It is especially well suited to high production, high volume automatic screw machine work. This material is similar to type 303 but requires slightly higher machining speeds: typical turning speeds are 125–155 sfm (38–47 smm) for high-speed steel tools. Higher speeds are possible with carbide tooling. It resists a variety of organic and inorganic compounds and is comparable in corrosion resistance to type 303.

Type 303 and 303 Se. These grades are particularly well suited to screw machining operations. They provide longer tool life and higher cutting speeds than type 304. Type 303 is widely used for the following applications: shafting, valve bodies, valve trim, and valve fittings. It has desirable nongalling properties that make disassembly of parts easy. Type 303 Se is used for similar applications, and it has slightly superior corrosion properties. It also has better formability characteristics for hot- or cold-working applications. Both types machine easily with a brittle chip. In turning operations, they can be used at speeds up to 100–130 sfm (31–40 smm). Moderate cold working increases the machinability. Polishing and grinding can be satisfactorily performed. Both types resist rusting from normal atmospheric source, and they can be used in applications that expose them to sterilizing solutions, most organic chemicals and dyestuffs, and a wide variety of inorganic chemicals. They are also very nitric acid resistant, moderately resistant to sulfuric acids, and they have poor resistance to halogen acids. If, during fabrication, components are heated and cooled in the range of 800–1650° F (425–900° C), a corrective thermal treatment consisting of heating to 1900° F (1037° C) followed by quenching in water should be applied to reduce the risk of intergranular corrosion.

Type 416. This is the most readily machinable of all stainless steels, and it is particularly well suited for automatic screw machine operations (where speeds are about 165 sfm *or* 50 smm) because its use is beneficial to tool life. It is widely used for fittings, gears, housings, lead screws, shafts, valve bodies, valve stems, and valve trim, or for any part requiring considerable machining work. While not as corrosion resistant as the 300 series grades, type 416 resists atmospheric environments, fresh water, mine water, steam, carbonic acid, gasoline, crude oil, blood, alcohol, ammonia, soap, sugar solutions, etc. Maximum corrosion resistance is achieved in the heat treated condition. It resists scaling at elevated temperatures and can be used for continuous service to about 1200° F (356° C).

Type 420F. This grade is easy to machine, grind, and polish, and it has certain anti-galling and nonseizing properties. It is often used for parts made on automatic screw machines such as valve trim, pump shafts, needle valves, ball check valves, gears, cams, and pivots. It provides high hardness and good corrosion resistance. It is comparable to AISI 2315 and 2340 for automatic screw machine operations. In turning operations involving heavy-duty equipment, speeds up to 90–110 sfm (27–34 smm) and feeds up to 0.0008–0.0020 inches (0.02 to 0.05 mm) are suggested. 420F should always be used in the hardened condition for maximum corrosion resistance, and the surface should be free of scale which can be removed by grinding and polishing, or pickling (acid pickling is immersion in an aqueous solution containing about 10% nitric acid and 2% hydrofluoric acid at a maximum temperature of 140°F [60° C]). If pickled after hardening, the parts should be baked at 250–300° F (121–148° C) for at least one hour to remove acid brittleness. In the hardened condition, type 420F will resist corrosion from atmosphere, fresh water, mine water, steam, carbonic acid, crude oil, gasoline, blood, alcohol, ammonia, mercury, sterilizing solutions, soap, etc. Passivation after machining is recommended.

Type 430F. When a 16–18% straight chromium stainless steel is specified, this grade is recommended for faster cutting and reduced costs. It does not harden heat treatment. 430F is commonly used for parts requiring good corrosion resistance, such as solenoid valves, aircraft parts, and gears. It machines at approximately the same speeds (125–155 sfm *or* 38–47 smm) as AISI 1120 and 1130 steels. 430F offers superior corrosion resistance from atmospheric, fresh water, nitric acid, and dairy products, but parts must be free of scale and foreign particles. As a final treatment, after scale removal and machining, all parts should be passivated.

Fabrication of stainless steel

Stainless steels can be fabricated by methods similar to those used for carbon steels, but some modifications must be made for their yield strength and their rate of work hardening. All have work hardening rates higher than common carbon steels, but the austenitics are characterized by large increases in strength and hardness with cold work. With the exception of the resulfurized free-machining grades (303 is the common type), all stainless steels are suitable for crimping or flattening operations. The free-machining grades will withstand mild longitudinal deformation but may exhibit some tendency to splitting. In spite of their higher hardness, most martensitic and all of the ferritic grades can be successfully fabricated. **Table 16** shows the relative fabrication characteristics of three groups of stainless steels.

Tool Steels

Tool steels are used to make tools for cutting or shaping other materials. Therefore, they are usually subjected to extreme repetitive loads and their survival depends on their characteristics. Defining characteristics of tool steels are hot hardness (resistance to softening at elevated temperature), resistance to abrasive wear, resistance to breakage, resistance to deformation, resistance to adhesive wear, and toughness. There are six general classes of tool steels, plus two special purpose steels. Much of the information that follows on specific tool steels and their characteristics was extracted from *Heat Treatment, Selection, and Application of Tool Steels* by Bill Bryson (Hanser Gardner Publications)

Table 16. Relative Fabrication Characteristics of Stainless Steels. (Source, SSINA.)

Group	Austenitic					Ferritic		Martensitic		
Type Number	201, 202, 301, 302, 304, 304L, 305	303*	309S, 310S	316, 316L, 317, 317LMN	321, 347	430, 439	405, 442, 446	403, 410	420	440A, 440C
Characteristic										
Air Hardening	No	No	No	No	No	No	No	Yes	Yes	Yes
Blanking	F	F	F	F	F	E	E	E	E	G
Brazing, Silver	G	-	G	G	G	G	G	G	G	G
Buffing	G	-	G	G	G	G	G	G	G	G
Drawing, Deep	E	-	G	E	G	E	F	E	NR	NR
Forming, Hot	G	F	G	G	G	G	G	G	G	G
Forming, Cold	G	F	G	G	G	G	F	G	F	NR
Grinding, Ease of	F	G	F	F	F	F	F	G	E	E
Grinding, (magnetic)	No	No	No	No	No	Yes	Yes	Yes	Yes	Yes
Hardenable by Heat Treatment	No	No	No	No	No	No	No	Yes	Yes	Yes
Punching (perforating)	F	-	F	F	F	G	G	G	G	G
Polishing	G	G	G	G	G	G	G	G	G	G
Riveting, Hot	G	F	G	G	G	G	G	G	NR	NR
Riveting, Cold	G	F	G	G	G	G	G	G	NR	NR
Shearing, Cold	F	F	F	F	F	G	G	G	G	F
Soldering	G	G	G	G	G	G	G	G	G	G
Brazing	G	G	G	G	G	G	G	G	G	G

(Continued)

Table 16. (Continued) Relative Fabrication Characteristics of Stainless Steels. (Source, SS/NA.)

Group	Austenitic					Ferritic			Martensitic	
Type Number	201, 202, 301, 302, 304, 304L, 305	303*	309S, 310S	316, 316L, 317, 317LMN	321, 347	430, 439	405, 442, 446	403, 410	420	440A, 440C
	Characteristic									
Spinning	G	-	G	G	G	G	F	F	NR	NR
Welding	E	NR	E	E	E	F	F	F	F	F
Machining	F	E	F	F	F	G	F	G	F	NR

* Chemistry designed for improved machining (as are other grades, i.e., 416, 420F, 430F, 440F)

Note: Code: E = Excellent; G = Good; F = Fair; NR = Not Generally Recommended (Poor).

which contains detailed analysis of heat treating and selecting many of the popular tool steels. The machinability ratings are taken from the Materials Selection section of this book (Table 7) and are based on AISI having a rating of 100%. Each class and subcategory is explained below, and the influencing elements in the composition of tool steels are provided in **Table 17**.

Water-hardened tool steels (W series) contain few alloying agents other than carbon. These steels are generally described as shallow hardening, and the hardened zone is thin. Their low hardenability allows for high surface hardness, and the prescribed heat treatment for these steels is relatively simple and inexpensive. Their name is derived from the fact that they are normally water quenched from austenizing temperature, but it should be remembered that water-hardening steels are the steels most subject to cracking and deformation due to the severe shock of water quenching. Also, decarb and scale are most prominent in the water hardening grades. Because of their high strength, excellent wear resistance, and resistance to splitting and galling, W series tool steels are commonly used to produce arbors, stone chisels, collets, burnishing dies, coining dies, cold extrusion dies, threading dies, dowels, gun drills, files, plug gages, taps, and most striking and woodworking tools. In fact, these were the first modern heat treatable steels and were probably first used for swords and axes. Their carbon content is in the area of 1%, and the achievable hardness range is typically between 63 and 65 R_C on the case (0.050 inch [1.27 mm] depth), making it possible to develop the sharpest edge of any steel in the marketplace. Interestingly, W1 tool steel has the same basic chemistry as AISI 1095, but the two differ in manufacturing method. Tool steels are generally manufactured under very tight rules concerning percentage of scrap versus raw material, and the grain size on the annealed material is controlled with greater accuracy. Both W1 and W2 should be preheated to 1,200° F (649° C) and hardened at 1,475° F (802° C). The optimum tempering temperature for W1 and W2 tool steels is 300° F (149° C), which will yield the following approximate results: hardness of 64 R_C, Charpy value of 80 ft lbs (108 Joules) for W1 and 77 ft lbs (104 Joules) for W2, and a machinability rating of 30 for both W1 and W2.

Cold-work tool steels are comprised of the O, A, and D series steels.

O series cold-work tool steels are oil hardening, and therefore they can be quenched with less distortion than is often experienced with the W series steels. This series is comparatively inexpensive to purchase and features high resistance to wear at normal temperatures even though softening and reduced wear resistance occur at higher temperatures. O1 is considered the work horse of general purpose tool steel, but it suffers from decarb formation during heat treatment and is subject to cracking from thermal shock when quenching. Except for machinability, O1 and O6 have very similar properties. Preheat of both steels should be done at 1,200° F (649° C), and both should be hardened at 1,475° F (802° C). When tempered at 300° F (140° C), they have a hardness of 63 R_C, a Charpy value of 33 ft lbs (45 Joules) and machinability ratings of 16 for O1 and 38 for O6. These steels are widely used for cold drawing dies, drilling bushings (O1); blanking and forming dies, threading dies, bending dies (O2); cold stamping dies, punches, and thread gages (O1 and O2). Type O7 is often used for tools that require sharp and wear resistant cutting edges.

A series cold-work tool steels are air hardening. They are primarily used for cold working, as they relinquish hardness as temperatures rise. The A7 grade, with its high percentage of carbon and chromium, has the highest wear resistance in this series. Currently, A2 is probably the most popular grade of tool steel in the U.S., with excellent wear resistance resulting from increased carbon and chrome (its wear resistance is 20 to 25% better than A6's). Its preheat temperature is 1,200°F (649° C) and its hardening

(Text continued on p. 468)

Table 17. Chemical Composition and Typical Hardness of Tool Steels. *(See Note.) (Source, Bethlehem Steel.)*

AISI Type	UNS No.	Identifying Elements %								Hardness R$_c$
		C	Mn	Cr	Si	W	Mo	Other		
WATER-HARDENING TOOL STEELS										
W1	T72301	0.60/1.40*	-	-	-	-	-	-		64
W2	T72302	0.60/1.40*	-	-	-	-	-	0.25V		64
W5	T72305	1.10	-	0.50	-	-	-	-		64
		*Other carbon contents may be available.								
COLD-WORK TOOL STEELS										
Oil-Hardening Types										
O1	T31501	0.90	1.00	0.50	-	0.50	-	-		62
O2	T31502	0.90	1.60	-	-	-	-	-		62
O6	T31506	1.45	0.80	-	1.00	-	0.25	-		63
O7	T31507	1.20	-	0.75	-	1.75	-	-		64
Medium Alloy Air-Hardening Types										
A2	T30102	1.00	-	5.00	-	-	1.00	-		62
A3	T30103	1.25	-	5.00	-	-	1.00	1.00V		65
A4	T30104	1.00	2.00	1.00	-	-	1.00	-		62
A6	T30106	0.70	2.00	1.00	-	-	1.25	-		60
A7	T30107	2.25	-	5.25	-	1.00†	1.00	4.75		67
A8	T30108	0.55	-	5.00	-	1.25	1.25	-		60
A9	T30109	0.50	-	5.00	-	-	1.40	1.00V 1.50Ni		56
A10	T30110	1.35	1.80	-	1.25	-	1.50	1.80Ni		62

† Optional

(Continued)

Table 17. *(Continued)* Chemical Composition and Typical Hardness of Tool Steels. *(See Note.)* *(Source, Bethlehem Steel.)*

AISI Type	UNS No.	C	Mn	Cr	Si	W	Mo	V	Other	Hardness R_c
					Identifying Elements %					
				COLD-WORK TOOL STEELS						
				High Carbon–High Chromium Types						
D2	T30402	1.50	-	12.00	-	-	1.00	1.00V		61
D3	T30403	2.25	-	12.00	-	-	-	-		61
D4	T30404	2.25	-	12.00	-	-	1.00	-		61
D5	T30405	1.50	-	12.00	-	-	1.00	3.00Co		61
D7	T30407	2.35	-	12.00	-	-	1.00	4.00V		65
				SHOCK-RESISTING TOOL STEELS						
S1	T41901	0.50	-	1.50	-	2.50	-	-		58
S2	T41902	0.50	-	-	1.00	-	0.50	-		60
S5	T41905	0.55	0.80	-	2.00	-	0.40	-		60
S6	T41906	0.45	1.40	1.50	2.25	-	0.40	-		56
S7	T41907	0.50	-	3.25	-	-	1.40	-		57

AISI Type	C	Mn	Cr	Si	W	Mo	V	Co	Hardness R_c
				Identifying Elements %					
				HOT-WORK TOOL STEELS					
				Chromium Types					
H10	0.40	-	3.25	-	-	2.50	0.40	-	56
H11	0.35	-	5.00	-	-	1.50	0.40	-	54
H12	0.35	-	5.00	-	1.50	1.50	0.40	-	55
H13	0.35	-	5.00	-	-	1.50	1.00	-	53
H14	0.40	-	5.00	-	5.00	-	-	-	47
H19	0.40	-	4.25	-	4.25	-	2.00	4.25	57

(Continued)

Table 17. *(Continued)* **Chemical Composition and Typical Hardness of Tool Steels.** *(See Note.)* *(Source, Bethlehem Steel.)*

| AISI | | | Identifying Elements % | | | | | | | | Hardness |
Type		C	Mn	Cr	Si	W	Mo	V	Co		R_c
				HOT-WORK TOOL STEELS							
						Tungsten Types					
H21	T20821	0.35	-	3.50	-	9.00	-	-	-		54
H22	T20822	0.35	-	2.00	-	11.00	-	-	-		52
H23	T20823	0.30	-	12.00	-	12.00	-	-	-		47
H24	T20824	0.45	-	3.00	-	15.00	-	-	-		55
H25	T20825	0.25	-	4.00	-	15.00	-	-	-		44
H26	T20826	0.50	-	4.00	-	18.00	-	1.00	-		58
						Molybdenum Types					
H42	T20842	0.55	-	4.00	-	-	8.00	2.00	-		60
				HIGH-SPEED TOOL STEELS							
						Tungsten Types					
T1	T12001	0.75*	-	4.00	-	18.00	-	1.00	-		65
T2	T12002	0.80	-	4.00	-	18.00	-	2.00	-		66
T4	T12004	0.75	-	4.00	-	18.00	-	1.00	5.00		66
T5	T12005	0.80	-	4.00	-	18.00	-	2.00	8.00		65
T6	T12006	0.80	-	4.50	-	20.00	-	1.50	12.00		65
T8	T12008	0.75	-	4.00	-	14.00	-	2.00	5.00		65
T15	T12015	1.50	-	4.00	-	12.00	-	5.00	5.00		68
						Molybdenum Types					
M1	T11301	0.85*	-	4.00	-	1.50	8.50	1.00	-		65
M2	T11302	0.85/1.00*	-	4.00	-	6.00	5.00	2.00	-		65
M3-1	T11313	1.05	-	4.00	-	6.00	5.00	2.40	-		66

(Continued)

Table 17. (Continued) Chemical Composition and Typical Hardness of Tool Steels. (See Note.) (Source, Bethlehem Steel.)

AISI Type		C	Mn	Cr	Si	W	Mo	V	Co	Hardness Rc
					Identifying Elements %					
				HIGH-SPEED TOOL STEELS						
				Molybdenum Types						
M3-2	T11323	1.20	-	4.00	-	6.00	5.00	3.00	-	66
M4	T11304	1.30	-	4.00	-	5.50	4.50	4.00	-	66
M6	T11306	0.80	-	4.00	-	4.00	5.00	1.50	12.00	66
M7	T11307	1.00	-	4.00	-	1.75	8.75	2.00	-	66
M10	T11310	0.85/1.00*	-	4.00	-	-	8.00	2.00	-	65
M30	T11330	0.80	-	4.00	-	2.00	8.00	1.25	5.00	65
M33	T11333	0.90	-	4.00	-	1.50	9.50	1.15	8.00	65
M34	T11334	0.90	-	4.00	-	2.00	8.00	2.00	8.00	65
M36	T11336	0.80	-	4.00	-	6.00	5.00	2.00	8.00	65
M41	T11341	1.10	-	4.25	-	6.75	3.75	2.00	5.00	70
M42	T11342	1.10	-	3.75	-	1.50	9.50	1.15	8.00	70
M43	T11343	1.20	-	3.75	-	2.75	8.00	1.60	8.25	70
M44	T11344	1.15	-	4.25	-	5.25	6.25	2.00	12.00	70
M46	T11346	1.25	-	4.00	-	2.00	8.25	3.20	8.25	69
M47	T11347	1.10	-	3.75	-	1.50	9.50	1.25	5.00	70
M62	T11362	1.25	-	3.75	-	6.50	10.50	2.00	-	70

* Other carbon contents may be available.

AISI Type		C	Mn	Cr	Si	W	Mo	Ni	Other	Hardness Rc
					Identifying Elements %					
				PLASTIC-MOLD STEELS						
P1	T51601	0.10	0.20	-	-	-	-	-	0.10V	64‡

(Continued)

Table 17. *(Continued)* Chemical Composition and Typical Hardness of Tool Steels. *(See Note.)* *(Source ,Bethlehem Steel.)*

AISI Type		C	Mn	Cr	Si	W	Mo	Ni	Other	Hardness R_c
PLASTIC-MOLD STEELS										
P2	T51602	0.07	0.50	2.00	-	-	0.20	0.50	-	64 ‡
P3	T51603	0.10	0.50	0.60	-	-	-	1.25	-	64 ‡
P4	T51604	0.07	0.50	5.00	-	-	0.75	-	-	64 ‡
P5	T51605	0.10	0.30	2.25	-	-	-	-	-	64 ‡
P6	T51606	0.10	0.40	1.50	-	-	-	3.50	-	61 ‡
P20	T51620	0.35	0.75	1.70	-	-	0.40	-	-	37
P21	T51621	0.20	0.30	-	-	-	-	4.00	1.20Al	40
‡ Hardness of carburized case										
SPECIAL-PURPOSE TOOL STEELS										
Low Alloy Types										
L2	T61202	0.50/1.10*	-	1.00	-	-	-	-	0.20V	63
L6	T61206	0.70	-	0.75	-	-	0.25†	1.50	-	62
* Other carbon contents may be available. † Optional										

Note: The percentages of elements in this Table are given for identification purposes only and are not to be considered as the means of the composition ranges for the elements.

temperature is 1,775° F (968° C). When tempered at 400° F (204° C) it has a hardness of 60 R$_C$, a Charpy value of 17 ft lbs (23 Joules), and a machinability rating of 16. A6 has a low chrome content and is an ideal tool steel for roll forming applications in situations where galling can be a problem. It should be preheated at 1,200° F (649° C) and hardened at 1,575° F (857° C). When tempered at 400° F (204° C), it has a hardness of 59 R$_C$, a Charpy value of 28 ft lbs (38 Joules), and a machinability rating of 16. Because they exhibit excellent dimensional stability properties, these steels are often used for precision measuring tools, as well as cold blanking, cold coining, cold stamping, cold extrusion dies, tool shanks, engraver's tools (A2); embossing dies, powder metal dies, tool shanks, and thread roller dies (A6); and thread roller dies (A7).

D series cold-work tool steels contain high carbon and very high (12%) chromium contents. Except for D3, they are all air hardened. D3 is usually oil quenched, resulting in increased likelihood of distortion and cracking in tools made of this grade. Like the A series, they are primarily used in cold working applications, but D7, in particular has good resistance to softening at elevated temperatures, primarily due to its high vanadium content. Dimensional stability is comparable to the A series. Due to their high chrome content, these steels are difficult to work or grind. The preheat temperature for D2 is 1,200° F (649° C), and its hardening temperature is 1,850° F (1,010° C). Recommended tempering temperature is 960° F (515° C), which produces a hardness rating of 58/60 R$_C$, a Charpy value of 8 ft lbs (11 Joules), and a machinability rating of 11. D3 should be preheated to 1,200° F (649° C) and hardened at 1,725° C (940° C). When tempered at 400° F (204° C), its hardness is 62 R$_C$, its Charpy value is 6 ft lbs (8 Joules), and its machinability rating is 11. Since D3 is only available from a limited number of suppliers, substituting it with M2 should be considered. The vanadium in D7 controls and promotes a very fine grain structure in the hardened steel, and its molybdenum content (1%) provides a very slight heat resistance. D7 should be preheated to 1,400° F (760° C) and hardened at 1,975° F (1,079° C). Tempering at 400° F (204° C) results in a hardness of 62 R$_C$, a Charpy value of 6 ft lbs (8 Joules), and a machinability rating of 11. These steels are often used for long run dies and for precision gages, as well as for broaches, cold burnishing dies, cold cutting dies, cold drawing dies, cold extrusion dies, cold stamping dies, thread roller dies, wire drawing dies, cold mandrels, and cutters for glass and wire (D2 and D3).

Shock-resisting tool steels are the S series steels. These steels have high strength, moderate wear resistance, and a high level of toughness, making them ideal for applications where high impact loads are encountered (chisels, punches, hammers, etc.), but they are also employed in certain structural applications. They are quenched using different media: S1, S5, and S6 are usually oil quenched, S2 is water quenched, and S7, in large sections, is oil quenched while small parts are air quenched. Specific applications include screw driver bits, wood chisels, cold chisels, bolt clippers, shear blades, pipe cutters, wire cutters (S1 and S2); embossing dies (S4); swaging dies, air hammers, shear blades (S5); hot forging dies, and hot shear blades (S7). While S1 is best known for its use in chisels, it is no longer widely available. It has better compression strength in conjunction with shock resistance than S7. The preheat temperature for S1 is 1,200° F (649° C), and the hardening temperature is 1,725° F (940° C). When tempered at 400° F (204° C), its hardness is 56 R$_C$, its Charpy value is 185 ft lbs (250 Joules), and its machinability rating is 20. S7 has the highest shock resistance of all tool steels, and it should be preheated at 1,200° F (649° C) and hardened at 1,725° F (940° C). When tempered at 450° F (232° C), it achieves a hardness of 58 R$_C$, a Charpy value of 240 ft lbs, and a machinability rating of 20.

Hot-work tool steels are designated H series steels, and they are segmented into three distinct groups: chromium types, tungsten types, and molybdenum types.

Chromium hot-work tool steels, grades H10 to H19, have chromium contents ranging from 3.25 to 5%, with relatively low carbon content (0.35 to 0.40%). In thicknesses of up to 1.25 inches (30 mm), they can be air hardened. In the past, H11, H12, and H13 were the most widely used of all the hot-work tool steels (in tonnage), as they are well suited for hot die work due to their ability to endure continued exposure to high temperatures. H11, for example, provides tensile strengths of up to 250 to 300 ksi (1723 to 2068 MPa) at working temperatures of 1,000° F (540° C). Typical uses of these steels include hot extrusion dies, hot mandrels, dowel pins (H11); die casting dies (H12 and H13); and hot header dies (H13 and H19). H11 and H13 are also widely used in the construction of light metal die-casting molds. In fact, H13 is considered an excellent steel for plastic molding or die cast molds, as it can withstand temperatures up to 1,300° F (704° C) and is readily available on the market. H13 should be preheated to 1,500° F (815° C) and hardened at 1,875° F (1024° C). When tempered at 1,000° F (538° C), it has a hardness of 54 R_C, a Charpy rating of 68 ft lbs, and a machinability rating of 15. H19 has superior resistance to heat check in die casting applications, and it provides excellent wear resistance in high temperature applications. It should be preheated at 1,500° F (815° C) and hardened at 1,850° F (1,010° C). When tempered at 1,100° F (593° C), it has a hardness of 52 R_C, a Charpy value of 43 ft lbs, and a machinability rating of 15.

Tungsten hot-work tool steels (H21 to H26) contain tungsten in the range of 9 to 18%. The high alloy content makes this series slightly more brittle than other tool steels, but also significantly boosts their resistance to high temperature softening. Even though these steels can be air hardened, scaling can be minimized by quenching in oil or molten salt. Common applications for this group include forging dies for brass, hot coining dies, hot drawing dies, and hot punches (H21).

Molybdenum hot-work tool steels (H42 and H43) have similar properties to the tungsten hot-work tool steels. They do possess slightly better resistance to heat checking (parallel surface cracks caused by rapid heating and cooling), and are slightly lower priced than the tungsten grades.

High-speed tool steels are made up of two different types: the M (molybdenum) type and the T (tungsten) type. The M steels account for 95% of all high speed steel made in the U.S. The overwhelming popularity of the M series is due to the fact that they are 40% less expensive to purchase than the T series steels. As the name of this group of tool steels suggests, these steels are designed for machining at high speeds, and they have been a mainstay of the cutting tool industry since they were introduced, thanks to the addition of vanadium to tool steels in 1904. Cobalt was introduced in 1912, and the final crucial ingredient, molybdenum, was added in 1930.

Molybdenum high-speed tool steels (M1 to M62) have marginally better toughness than the tungsten steels of the same hardness, but in other respects, the M series and T series virtually share mechanical properties. Within the range, physical properties can be manipulated by choosing a steel with increased percentages of carbon and vanadium, which will result in better wear resistance, but the additional carbon will reduce the ductility; or selecting one with more cobalt, which will enhance red hardness but lower impact strength. While all M series steels, except M10, contain tungsten in amounts ranging from 3.75 to 11%, the T series steels contain only minimal amounts of molybdenum (from 0 to 1.25% maximum). In both series, achievable hardness increases proportionally to the amount of molybdenum and/or tungsten in the composition of the steel. M2 is an excellent, and extremely popular, general purpose high speed steel that has good toughness and very acceptable heat resistance. It also has the most forgiving hardening range of all of the high speed steels. M2 should be preheated at 1,500° F (816° C) and hardened at 2,200° F (1,204° C). When tempered at 1,000° F (528° C), it has a hardness of 66 R_C, a Charpy value of 10 ft lbs (13.5 Joules), and a machinability rating of 14. The M3 grade is available in two versions. Both have excellent wear and heat resistance, plus very good edge strength. M3-2 has slightly higher carbon content than M3-1, allowing it to develop a better cutting edge while losing a small amount of compressive strength. M42 can achieve a maximum hardness of 70 R_C which, along with a few other M series steels, is the highest hardness recorded for tool steels. It also holds an edge without losing hardness or significant wear resistance. M42 should be preheated at 1,500° F (816° C) and hardened at 2,175° F (1,190° C). When tempered at 950° F (510° C), it reaches a hardness of 68 R_C, achieves a Charpy value of 10 ft lbs (13.5 Joules), and has a machinability rating of 13.

Tungsten high-speed tool steels (T1 to T15) have very high red hardness and exceptional wear resistance. These steels are used primarily for making drills, form cutters, hacksaws, circular saws, slotting saws, and many other varieties of cutting tools, dies, and punches. T5 has better wear resistance, but slightly inferior shock resistance, than M42. T5 should be preheated at 1,500° F (816° C) and hardened at 2,350° F (1,288° C). Tempering at 1,000° F (528° C) will produce a hardness of 64 R_C, a Charpy value of 5 ft lbs (7 Joules), and a machinability rating of 11. Thanks to the combination of tungsten, cobalt, high carbon, and vanadium, T15 provides maximum wear resistance. Its preheat temperature is 1,500° F (816° C), and its hardening temperature is 2,250° F (1,232° C). When tempered at 1,000° F (528° C), it has a hardness of 67 R_C, a Charpy value of 3-4 ft lbs (4-5 Joules), and a machinability rating of 8.

Plastic mold steels, P2 to P21, are favored for plastic molds where heavy pressures and high temperatures (350 to 390° F [176 to 199° C]) are commonly encountered. There are several additional factors that make these steels well suited for plastic molds including good wear resistance, corrosion resistance, good thermal conductivity, and excellent toughness. The mold steel must also have certain manufacturing properties that are satisfactorily met by the P series tool steels, including acceptable machinability, dimensional stability not only during heat treatment, but also during molding operations, good polishing properties, and, in some cases, good hubbing characteristics (hubbing, which is discussed later in this section, is also called hobbing, and it is forcing a steel master "hub" or "hob"—which is an exact replica of the part to be formed—into a softer die bank to create a mold).

These properties are essential because the cost and time required to prepare a mold make it imperative that it remains in service for as long as possible. Wear resistance is important because, unless the steel has a hard, abrasion resistant surface, the repeated flow of material into the mold will rapidly erode the surface of the mold. Because many plastics are highly corrosive, and rust or other surface deterioration will destroy the surface of the mold, the steel must have good corrosion resistance. Toughness is essential because the opening and closing of the die or mold, as well as the high velocity injection of materials, expose the mold steel to repeated shocks. To avoid excessive machining, die materials should retain their size, with as little deformation as possible, during heat treating, and they must not distort from the heat generated by normal operations. The surface texture of the mold is important because the surface of the finished product will show any imperfections in the mold's surface, therefore the steel must possess structural uniformity and high surface hardness to allow it to be polished to a smooth finish. A final important characteristic of P series tool steels is that they resist losing hardness on tempering. This becomes very important to molds that are repeatedly exposed to elevated temperatures during operation, which can cause some steels to lose their hardness.

Although the P series steels remain the most popular choice for plastic molds, maraging (a combination of the words *mar*tensite and *aging*, because they are age, or precipitation, hardened at 840 to 930° F [449 to 409° C] for around three hours) steels, containing high nickel content, are also used, especially in plastic injection molding. Although these are not classified as "tool steels," their use in molds, but not in cutting operations, is responsible for their inclusion in this section. The hardness of maraging steels is between 50 and 58 R_C, and, after aging, they can obtain tensile strength ratings of 29 ksi (200 MPa). Their primary application is for use in intricate mold designs, because they shrink uniformly during heat treatment and can therefore be made to very close tolerances. Also, since carbon is not an alloying agent in these steels, decarburization does not have to be considered. The maraging steels are much more expensive than the P series steels, and they are not commonly available in large sizes, so their use remains limited.

Another non-tool steel used for molds, especially injection or compression molds with large cavities, is AISI 6F7, which has good toughness qualities and hardness between 52 and 54 R_C. GF7 is a through- (rather than case-) hardened steel. Other tool steels used for mold making include H11, H13, O1, O2, D2, and M2 and, even though these were briefly discussed, their mold steel characteristics will be briefly examined here. These through-hardened steels have an obvious advantage over surface hardened steels: modifications can be made to the cavity without requiring a new heat treatment. They also have one major disadvantage: they do not have a tough core, and therefore are more likely to experience cracking. H11 is commonly used for molds for highly abrasive plastics, usually after nitride treatment to increase surface hardness. O2, because of its high hardness and carbon content, has good wear resistance, but is not as tough as other mold steels. It is used for shallow cavities only, and especially in those with high parting line stresses. D2 is very wear resistant, but the difficulty in machining (even when annealed) and polishing D2 is widely known. Properties of the P series tool steels, plus other steels used for molds, are shown in **Table 18**.

Of the P series steels, P2 through P6, as will be discussed later, are normally used for hubbed and/or carburized cavities, while P20 and P21 are used for machined cavities. Distortion and changes in size of the latter two steels are avoided because they are supplied heat treated to hardnesses ranging from R_C 30 to R_C 60, and therefore do not require additional heat treatment after machining. In spite of this, when being used for plastics molding applications, P20 is commonly carburized to bring its surface hardness up to as high as R_C 65, but carburizing temperatures in excess of 1,600° F (871° C) can be detrimental to the polishing properties of the material. P20+Ni (3.5% Ni) can be carburized without concern about degrading its polishability. **Table 19** provides heat treating temperatures and resulting hardness values for P20 tool steel.

P21 contains nickel and aluminum as its principal alloying agents, and the aluminum content allows it to age harden in the same manner as many aluminum alloys. To obtain full hardness, it may be nitrated during aging to provide the steel with high surface hardness (to a depth of 0.006 to 0.008 inch (0.1524 to 0.2032 mm) by holding it in a gas nitriding furnace for 24 hours at 950 to 975° F (510 to 524° C). With this treatment, surface hardnesses of approximately 70 R_C can be obtained. P21 polishes to a high mirror finish, making it the preferred material for molds with critical finishing requirements.

Hubbing can, in some instances, replace machining or EDM as the preferred method for creating mold cavities. It is especially economical for making multiple cavity molds for injection molding, and it can save time and money while providing high accuracy, repeatability, and surface quality. In the process, a hubbing punch, or master hob, is slowly pressed into annealed steel under very high pressure, thereby creating the shape of the hob's face in a cavity in the annealed steel. Master hobs (or hubs) are commonly made from shock resistant S4 or S1, or cold worked die steels such as A2 or D2, and, generally, the steels most often hobbed for plastics mold cavities are P1 (the most easily hobbed mold tool steel, mainly due to its vanadium content that allows for it to have a fine grained structure after carburizing, but not widely in use), P2, P4, and P6. P5 is also hobbed without serious difficulty, but P3 is more troublesome and requires more frequent annealing. All

Table 18. Properties of Steels Commonly Used for Molds.

Type	Tensile Strength[1]		Hardness[1]	Coefficient of Heat Expansion[2]				Heat Conductivity[3]		
	ksi	MPa	BHN	68-212°F	68-752°F	20-100°C	20-400°C	660°F	350°C	
P2	103	710	210	6.8	7.7	12.2	13.9	21.5	36.6	
P4	52	360	120	6.6	7.3	12.0	13.2	21.4	36.5	
P20	160	1100	325	6.2	7.7	11.1	13.8	19.6	33.4	
P20+Ni	160	1100	325	6.2	7.7	11.1	13.8	19.7	33.5	
P20+S	195	1350	400	6.2	7.7	11.1	13.9	19.6	33.4	
P21	123	850	250	5.7	7.1	10.4	12.7	18.9	32.2	
D2	123	850	250	5.8	6.8	10.5	12.2	12.0	20.5	
H11	117	810	240	5.5	6.2	10.0	11.1	17.9	30.5	
GF7	123	850	250	6.8	7.7	12.3	13.8	17.5	29.8	
Maraging	157	1080	320	5.7	6.4	10.3	11.5	10.9	18.5	

Notes: [1] Maximum tensile strength and hardness values are given. [2] Coefficient of heat expansion expressed in 10^{-6} in./in. °F, and 10^{-6} m/(mK) for degrees Celsius.
[3] Heat conductivity expressed in Btu/ft h °F, and W/(mK) for degrees Celsius.

Table 19. Heat Treatment Recommendations for P20 Mold Steel, Oil Quenched and Carburized.

Preheat Temperature - 1200° F (650° C) *		
Hardening Temperature - 1525° F (830° C)		
Chemistry	**Tempering Temperature**	**Hardness (R$_c$)**
Carbon 0.30%	As quenched	51
Manganese 0.75% Silicon 0.50%	400° F/205° C	49
Chromium 1.65%	600° F/315° C	47
Molybdenum 0.40%	800° F/425° C	44
	1000° F/450° C	39
	1100° F/595° C	33
	1200° F/650° C	26
	1250° F/675° C	21
* = if used on larger mass parts		
Gas Carburize Temperature - 1600°F (870° C)		
Furnace Cool To - 1475° F (800° C)		
Tempering Temperature	**Case Hardness (R$_c$)**	**Core Hardness (R$_c$)**
600° F/315° C	58	47
650° F/354° C	57	46
700° F/370° C	56	45
750° F/400° C	55	44
800° F/425° C	54	43
900° F/480° C	52	40
Carburizing Time (Hours)	**Average Case Depth**	
4	0.030"/0.762mm	
8	0.042"/1.0668mm	
12	0.055"/1.397mm	
16	0.062"/1.5748mm	

Reproduced from *Heat Treatment, Selection, and Application of Tool Steels* by Bill Bryson (Hanser Gardner Publications).

steels used for hubbing should be at their lowest annealed hardness, and the annealed hardnesses for these tool steels range from 110 HBN for P4, to 180 for P6. In no instance should hubbing be carried out on tool steels with hardnesses in excess of 200 HBN. Other considerations in hubbing are the diameter of the hubbing blank, which should ideally be four times the diameter of the hob, and the thickness of the blank which should be 2.5 times the depth of hubbing. Because of the extreme amounts of force required in hubbing, specialized hubbing presses are used. Smaller presses generate up to 36,000 pounds of force (1,600 kN), while the largest presses are capable of creating over 5.5 million pounds of force (25,000 kN).

Low alloy AISI tool steels, also known as the L series, were once a much larger group of tool steels, but only two remain. These steels are most commonly used for machine parts (collets, cams, forming dies, punches, etc.) as they possess good strength and toughness. Though they have similar characteristics to the W series tool steels, these steels have greater wear resistance, and L6 has nickel, which increases toughness and improves hardenability, and this steel can achieve through hardness in thicknesses up to three inches (76 mm).

L6 should be preheated to 1,500° F (815° C) and hardened at 1,525° F (830° C). When tempered at 300° F (149° C) it will obtain a hardness of 60/61 R_C, its Charpy value will be 72 ft lbs (98 Joules), and its machinability rating will be 20.

Heat Treatment of Steel - Theory

Equilibrium diagrams are graphs or maps of the temperature and composition boundaries of the phase fields that exist in an alloy system under conditions of complete equilibrium. A phase is a microscopically homogeneous body of matter. As related to alloys, a phase may be considered as a structurally homogeneous and physically distinct portion of the alloy system. For example, when an alloy solidifies, or freezes, the phases formed may be a chemical compound and a solid solution, or two solid solutions. Before an alloy solidifies completely, a solid solution may exist in equilibrium with a liquid. In each of these examples, two phases of the alloy are involved. The temperatures that appear on an equilibrium diagram are called critical temperatures or critical points. The loci of these critical temperatures are the Ae_1, Ae_3, Ae_4, and A_{cm} lines or curves. The letter A derives from the word "arrest," since these temperatures designate the point at which an arrest in the heating or cooling curve occurs. The temperature interval between the Ae_1 and Ae_3 critical points for a particular alloy is designated as the *critical range*. Ae_1 is the lower critical temperature, and Ae_3 is the upper critical temperature. Common A designations used to indicate transformation temperatures are as follows.

Ac_{cm}	For hypereutectoid steel, the temperature at which the solution of the cementite in austenite is completed on heating.
Ac_1	The temperature at which austenite begins to form on heating.
Ac_3	The temperature at which the transformation of ferrite to austenite is completed on heating.
Ac_4	The temperature at which austenite transforms to delta ferrite on heating.
Ar_{cm}	For hypereutectoid steel, the temperature at which cementite begins to precipitate from austenite on cooling.
Ar_1	The temperature at which the transformation of austenite to ferrite, or to ferrite plus cementite is completed on cooling.
Ar_3	The temperature at which austenite begins to transform to ferrite on cooling.
Ar_4	The temperature at which delta ferrite transforms to austenite on cooling.
Ae_1, Ae_3, Ae_4, Ae_{cm}	Transformation temperatures under equilibrium conditions.
M_s	(Martensite start) The temperature at which austenite starts to transform to martensite.
$M_f Ar$	(Martensite finish) The temperature at which the formation of martensite is completed. The temperature at which austenite begins to transform to pearlite on cooling.

Critical temperatures for selected carbon and alloy steels are shown in **Table 20**.

The rate at which an iron-carbon alloy is heated or cools affects the critical temperature of the alloy. In practice, during manufacture, strict equilibrium conditions are not maintained during heating and cooling because equilibrium could be achieved only if the heating or cooling rate was extremely slow. In commercial enterprises, strictly adhering to equilibrium rates is not feasible, and generally, the greater the rate of heating or cooling, the greater the deviation from the equilibrium critical temperatures. As can be seen from the display above, the critical points on heating are identified by the symbol Ac followed

Table 20. Critical Temperatures for Selected Carbon and Alloy Steels. See Notes. *(Source, Bethlehem Steel.)*

AISI No.	A$_{c1}$ Temperature		A$_{c3}$ Temperature		A$_{r1}$ Temperature		A$_{r3}$ Temperature	
	°F	°C	°F	°C	°F	°C	°F	°C
1015	1390	754	1560	849	1390	754	1510	821
1020	1350	732	1540	838	1340	727	1470	799
1022	1360	738	1530	832	1300	704	1440	782
1117	1345	729	1540	838	1340	727	1450	788
1118	1330	721	1515	824	1175	635	1385	752
1030	1350	732	1485	807	1250	677	1395	757
1040	1340	727	1445	785	1250	677	1350	732
1050	1340	727	1420	771	1250	677	1320	716
1060	1355	735	1400	760	1250	677	1300	704
1080	1350	732	1370	743	1250	677	1280	693
1095	1350	732	1365	741	1265	685	1320	716
1337	1330	721	1450	788	1180	638	1310	710
1141	1330	721	1435	779	1190	643	1230	666
1144	1335	724	1400	760	1200	649	1285	696
4118	1380	749	1520	827	1260	682	1430	777
4320	1350	732	1485	807	840	449	1330	721
4419	1380	749	1600	871	1420	771	1510	821
4620	1300	704	1490	810	1220	660	1335	724
4820	1310	710	1440	782	780	416	1215	657
8620	1380	749	1520	827	1200	649	1400	760
E9310	1350	732	1480	804	810	432	1210	654
4027	1370	743	1510	821	1320	716	1410	766
4130	1400	760	1510	821	1305	707	1400	760
8630	1365	741	1465	796	1205	652	1335	724
1340	1340	727	1420	771	1160	627	1195	646
4140	1395	757	1450	788	1280	693	1330	721
4150	1390	754	1450	788	1245	674	1290	699
4340	1350	732	1415	768	720	382	890	477
5140	1370	743	1440	782	1260	682	1320	716
5150	1345	729	1445	785	1240	671	1310	710
5160	1380	749	1420	771	1280	693	1310	710
6150	1395	757	1445	785	1290	699	1315	713
8650	1325	718	1390	754	910	488	1230	666
8740	1370	743	1435	779	1160	627	1265	685
9255	1410	766	1480	804	1270	688	1330	721

A$_{c1}$ is the temperature at which austenite begins to form on heating.
A$_{c3}$ is the temperature at which the transformation of ferrite to austenite is completed on heating.
A$_{r1}$ is the temperature at which the transformation of austenite to ferrite, or to ferrite plus cementite, is completed on cooling.
A$_{r3}$ is the temperature at which austenite begins to transform to ferrite on cooling.

by a number subscript. The letter "c" is from the French word for heating, "chauffage." The symbol Ar, plus a number subscript, indicates critical points on cooling, and the letter "r" is taken from "refroidissement," the French word for cooling. When it is necessary to identify critical points for equilibrium, the symbol Ae is normally used used, with the "e" inserted to designate equilibrium. However, equilibrium temperatures are sometimes given as A_1, etc., without the "e."

A typical iron-carbon equilibrium diagram is shown in *Figure 3*. Point "C" on the diagram indicates the eutectic point where a liquid solution is converted into two or more intimately mixed solids on cooling; the number of solids formed is the same as the number of components in the solution. A eutectic is defined as an intimate mechanical mixture of two or more phases having a definite melting point and a definite combustion. Point "S" is the eutectoid point where an isothermal reversible reaction in which a solid solution is converted into two or more intimately mixed solids on cooling. The number of solids formed is the same as the number of components in the solution. Therefore, a eutectoid is a mechanical mixture of two or more phases having a definite composition and a definite temperature of transformation within the solid state.

The heat treatment of steels is essentially a process of controlled departure from equilibrium heating and cooling. When a steel is cooled, under equilibrium conditions, from above the Ae_3 critical temperature, the austenite remaining when the Ae_1 temperature

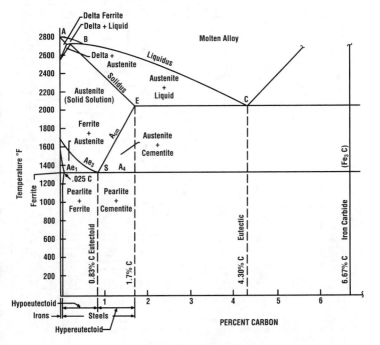

Figure 3. An iron-carbon equilibrium diagram.

is reached transforms into pearlite. With more rapid cooling, the faster the cooling rate, the further the transformation temperature Ar_1 is depressed below the Ae_1 temperature. The eutectoid, pearlite, that forms when austenite is cooled below the Ar_1 temperature, is a lamellar product made up of alternate plates of ferrite and cementite. The pearlite formed under equilibrium cooling is a coarse lamellar product. The process of transformation of pearlite from austenite involves diffusion of carbon, and since the diffusion rate is a function of temperature, faster cooling rates with attendant lower transformation temperatures produce pearlite with a finer lamellar structure. A finer dispersion of phases tends to produce greater strength and hardness and reduces the ductility of the steel. With very rapid cooling, the austenite transformation occurs at a low temperature and the resulting structure is not pearlite, but a structure known as martensite. Therefore, by controlling the cooling rate, the temperature of the austenite transformation and the structure and properties of the steel are controlled.

Fine grain size is usually specified for applications involving hardening by thermal treatment. As carbon and alloy steels are heated to a temperature just above their upper critical temperature, they transform to austenite of uniformly fine grain size. If heated to higher temperatures, coarsening of the grains will eventually occur. The temperature at which this transformation occurs is to some extent dependent on the composition of the steel, but the primary influence is the type and degree of deoxidation used in the steel making process. Time at temperature also influences the degree of coarsening. Deoxidizers such as aluminum, and alloying elements such as vanadium, titanium, and columbium, inhibit grain growth and thereby increase the temperature at which coarsening of the austenite grains occurs. Aluminum is most commonly used to control grain size because of its low cost and dependability. For steels used in the quenched and tempered condition, a fine grain size at the quenching temperature is almost always preferred, because fine austenitic grain size is conducive to good ductility and toughness. Coarse grain size enhances hardenability, but also increases the tendency of the steel to crack during thermal treatment. Grain size is observed by examining a polished and etched specimen under 100 diameters magnification, and comparing it with a standard. It is impossible to produce steel with a single grain size, so a range of numbers is usually recorded. For identification purposes, a steel is considered to be fine grained if it is "5 to 8 inclusive," and coarse grained if it is "1 to 5 inclusive." These requirements are usually considered fulfilled if 70% of the grains observed fall within these ranges.

The time-temperature relationships for austenite transformations are graphically presented with isothermal transformation diagrams, commonly known as Time-Temperature-Transformation (T-T-T) diagrams. These diagrams, which are a plot of the times of the beginning and end of austenite transformations as a function of temperature, have been developed for various plain carbon and alloy steels. The data required for plotting a T-T-T diagram are developed through a series of isothermal transformation studies that determine the austenite transformation(s) that occur in a steel held at a constant temperature for increasing periods of time. These studies consist of heating small, thin pieces of specimen steel above the upper critical temperature, and holding it at the temperature long enough for homogeneous austenite to form. Each specimen is then removed and quenched in a lead or salt bath that is held at the desired study temperature. Thin specimens are used so that cooling to the bath temperature can be assumed to be instantaneous, and each specimen is held in the isothermal bath for a different period of time, ranging from seconds to hours. Each is then quenched in brine or water to transform any remaining austenite to martensite. The structure of each specimen is then examined to determine which transformation product(s), if any, formed during the isothermal cycle, and what percentage of the total structure transformed into each product. The collected data are

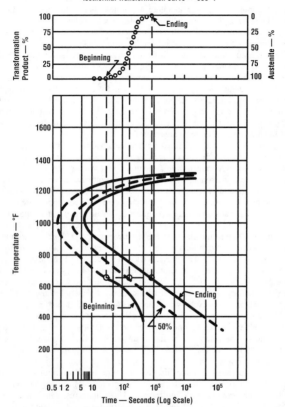

Figure 4. The use of transformational curves to develop T-T-T diagrams.

used to construct T-T-T diagrams in which time, from the beginning to the end of the transformation process, is plotted as a function of temperature. The basic methodology for creating a T-T-T diagram is shown in *Figure 4*.

Depending on the transformation temperatures, and the composition of the steel, austenite will transform into one or more of the final constituents: proeutectoid ferrite, proeutectoid cementite, pearlite, upper bainite, lower bainite, or martensite. Generally, for carbon and low alloy steels, the temperature ranges over which austenite transforms into the various constituents, under isothermal cooling, are as indicated on the transformation diagram shown in *Figure 5*. As can be seen, pearlitic microstructures are formed from about 1,300 to 1,000° F (705 to 538° C). Pearlite, as we have seen, is a constituent with a lamellar structure of alternating plates of ferrite and cementite. When the transformation occurs at

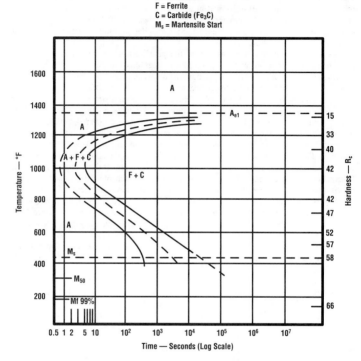

A = Austenite
F = Ferrite
C = Carbide (Fe₃C)
Mₛ = Martensite Start

Figure 5. The isothermal transformation diagram of a eutectoid carbon steel.

about 1,000° F, the lamellar structure is very fine and difficult to resolve with an optical microscope. The transformation of austenite to pearlite is a process of nucleation and growth. As the nucleus grows into the grain, it absorbs carbon atoms from the surrounding austenite. When the carbon concentration of the surrounding austenite has been sufficiently reduced, ferrite nucleates and grows along the surface of the cementite plate. Because ferrite can dissolve only about 0.02% carbon, carbon atoms are rejected by the ferrite as it forms, resulting in a buildup of carbon at the ferrite-austenite interface that continues to increase until the concentration level is sufficiently high, and at that point a new cementite nucleus forms. The sideways nucleation is repeated while, simultaneously, growth occurs at the edges of the ferrite and cementite plates. The alternating lamellae of cementite and ferrite originate from a single cementite nucleus called a pearlite colony.

The rate of nucleation and the rate of growth of pearlite colonies increase as the temperature decreases from Ae₁ down to about 1,100° F (593° C). As a result, at lower transformation temperatures the spacing between the pearlite lamellae becomes smaller and the metal becomes harder. Pearlite formed just below the Ae₁ temperature (about

1,300° F) is coarse pearlite with a lamellar spacing in the order of 10^{-3} mm, and a hardness of about R_C 15. The pearlite formed at about 1,100° F, however, is fine pearlite with spacing in the order of 10^{-4} mm and a hardness of about R_C 40. The fine pearlite formed at lower transformation temperatures is harder, tougher, and more ductile than pearlite formed at the higher temperatures.

As temperatures near the low end of the pearlite transformation range, another constituent forms from austenite. In plain carbon steel, this constituent, known as bainite, is formed only by isothermal transformation treatments and does not form when the steel is cooled continuously from above the critical temperature. Bainite forms from about 1,050° F (566° C) down to about 400° F (205° C). Steels that are transformed in the range where bainite and pearlite transformation temperatures overlap have structures containing both pearlite and bainite but, as the transformation temperatures are lowered, bainite becomes the predominant constituent and pearlite disappears. Like pearlite, bainite is a mixture of ferrite and cementite, but the two are not arranged in lamellar form as in a pearlite structure. The transformation of austenite to bainite is also thought to involve a process of nucleation and growth accompanied by carbon diffusion. Bainite apparently grows from a ferrite nucleus into a plate like structure with each plate composed of ferrite matrix in which carbide particles are embedded. When viewed in section, bainite has a characteristic acicular (needle like) appearance. Bainite formed at the higher temperatures in the transformation range has a feathery appearance and is commonly referred to as upper-bainite, while that formed at lower temperatures (lower-bainite) has a pronounced acicular structure. As with pearlite, the hardness and toughness of bainite both increase as the transformation temperature is lowered. Pearlite is usually tougher than upper-bainite of similar hardness, while lower-bainite compares favorably with tempered martensite on the basis of toughness.

The transformations of austenite to pearlite or bainite are time and temperature dependent, while the transformation of austenite to martensite is an athermal change, meaning that the transformation is dependent primarily on temperature and is essentially independent of time. Therefore, the martensite transformation process is considerably different from pearlite and bainite transformations. As can be seen in *Figure 5*, neither pearlite nor bainite forms immediately upon reaching an isothermal reaction temperature. An incubation period is required, and the steel must be held at temperature for a sufficient period of time for the reaction to be completed. By contrast, when austenite reaches the M_s (martensite start) temperature, the austenite immediately transforms and, if the steel is held at that temperature, the small amount of martensite that has formed is all that will form. Unless the steel is cooled to a still lower temperature, the transformation is arrested.

At some temperature below the M_s temperature, the transformation of austenite to martensite will essentially be complete. This temperature is designated M_f (martensite finish). At temperatures between M_s and M_f, fractional transformation will occur. This makes it possible to quench a steel to the temperature at which 50% of the austenite will transform to martensite, and then isothermally transform the remaining austenite to lower-bainite.

The transformation of austenite to martensite does not involve the diffusion of carbon. The instantaneous transformation of a small volume of austenite to form a martensite needle involves a shear displacement of the iron atoms in the austenite space lattice. (A space lattice is a geometric construction formed by a regular array of points—called lattice points—that represent the three-dimensional arrangement of atoms in space for a specific crystal structure.) The martensite formed in this way has a body-centered tetragonal (having four angles and four sides) crystal structure. In this form, it is called alpha-martensite and consists of ferrite and carbon, or finely divided iron-carbide (Fe_3C) in a metastable (unstable) structure that is considered to be in transition between the face-centered cubic structure of austenite, and the body-centered cubic structure of ferrite. The instantaneous athermal transformation of austenite to martensite at relatively low temperatures does not allow the carbon atoms to diffuse out of the lattice, so they remain in solution in the highly stressed transition lattice. The tetragonal crystal structure of alpha martensite transforms to the body-centered cubic structure known as beta-martensite upon slight heating or long standing. Essentially, the trapped carbon escapes or is thrown out of the crystal lattice, allowing the stressed lattice to shrink down to the body-centered cubic structure.

Alpha-martensite has an acicular structure similar to that of lower-bainite. It is the hardest and most brittle of the microstructures that can be obtained in a given steel. Its hardness is a function of carbon content, ranging from a theoretical maximum of R_C 65 in eutectoid alloys to R_C 40 or less in low carbon steels. Alpha-martensite is usually tempered to increase its ductility and toughness, although these changes are usually at the expense of hardness and strength. In tempering, the steel is heated to some temperature below the critical temperatures and cooled at a suitable rate. The microstructure and mechanical properties of tempered steel depend on the temperature and duration of the tempering cycle. As temper time and temperature increase, the carbide particles agglomerate and become progressively larger. Therefore, an increase in temperature and/or time at temperature usually results in lowering the hardness and strength of a steel while increasing its ductility and toughness.

To develop a martensitic structure, a steel must be austenitized at a rate sufficient to prevent the formation of ferrite, pearlite, or bainite. Most steels must be cooled very rapidly if pearlite formation is to be avoided. This is evident in *Figure 4*, which shows that the transformation of austenite to pearlite begins in one second or less between the temperatures of 1,100 and 950° F (593 and 510° C). In this range, often referred to as the pearlite nose of the transformation curve, the transformation is complete in less than ten seconds. An isothermal transformation diagram shows the changes in microstructure that occur when a steel is cooled instantly to some reaction temperature, and held at that temperature long enough for the reaction to go to completion. The diagram for a given steel shows what structure or structures are formed by isothermal transformation at any selected temperature, the time required for the steel to be at temperature before a reaction will begin, and the time needed to complete the reaction.

Fine grained austenite transforms to pearlite more rapidly than does coarser grained austenite. This can be explained by the fact that, as grain size decreases, the proportion of grain boundary material to total mass increases. Because pearlite nucleates at the grain boundaries in homogeneous austenite, transformation to pearlite begins more quickly and progresses faster in fine grained austenite than coarse grained austenite.

Figures 5 and *6* demonstrate the effect of carbon on the isothermal transformation reactions in plain carbon steels. For the steel with the higher carbon content (*Figure 5*), the transformations are shown to start later and to progress more slowly than do comparable reactions in the lower carbon steel (*Figure 6*). In effect, higher carbon contents shift the isothermal transformation curves to the right. In other words, transformations start later and proceed more slowly as carbon content is increased. Most of the common metallic alloying elements also tend to retard the start of isothermal transformations and increase the length of time required to complete them.

Compared to isothermal transformations, transformations under continuous cooling take longer to start and begin at lower temperatures. In effect, the isothermal transformation curve is shifted downward and to the right. This can be seen by the reactions of a plain carbon steel of eutectoid composition. In actual practice, the steel can be fully hardened

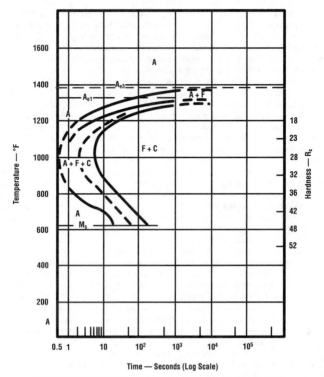

Figure 6. The isothermal transformation diagram of a hypoeutectoid carbon steel.

by cooling at a rate of 250° F per second (140° C/s). However, the isothermal diagram for such steel, as seen in *Figure 7*, indicates that a cooling rate of at least 400° F per second (220° C/s) is required to prevent the cooling rate curve from intersecting the transformation start curve at the pearlite knee. Theoretically, when the two curves intersect, transformation to pearlite should begin and should continue until the cooling curve passes out of the pearlite field.

If a steel is to be hardened to a martensitic structure, it must be austenitized and then cooled at a rate that is fast enough to prevent the formation of any of the other transformation products such as pearlite, ferrite, or bainite. The slowest rate that avoids all other transformation is called the critical cooling rate for martensite. As would be expected, different steels have different critical cooling rates, and the variations can be significant. Steels with relatively low critical cooling rates are high hardenability steels, and steels that must be cooled rapidly in order to obtain a martensitic structure are low hardenability steels. Hardenability is essentially a measure of the ability of a steel to transform to martensite—that is, to harden. It is also a measure of the depth to which a

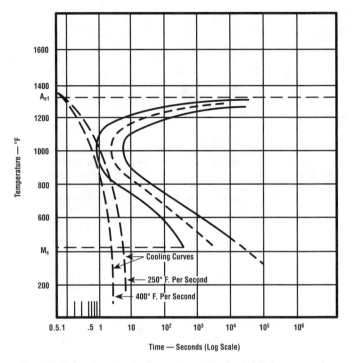

Figure 7. The isothermal transformation diagram of a eutectoid carbon steel. Cooling curves are shown.

steel will transform to martensite under given cooling conditions. Low hardenability steels respond to only a limited depth, which classes them as shallow hardening steels. With high hardenability steels, it is possible to harden thicker sections to greater depths.

The most common method of cooling a steel to obtain a martensitic structure is by quenching, which is rapid cooling of the steel from above the critical temperature by immersion in a medium that is capable of cooling it at the required rate. The cooling rate is determined by the hardenability of the steel and the size of the piece being quenched. Brine, water, and oils, listed in order of decreasing severity of quench, are the most common quenching media. In a piece of steel of any appreciable size, the cooling rates at the surface and at the center are different, and the difference increases with the severity of the quench.

Jominy end-quench test. The most widely used hardenability test is the Jominy end-quench test. Its popularity is attributable to its convenience, as only a single specimen is required, and in one operation the specimen is exposed to a wide range of cooling rates varying from a rapid water quench on one end of the specimen to a slow air quench on the other end. The test specimen is a bar four inches long and one inch in diameter that is heated to an austenitizing temperature and held there long enough to develop a uniform austenite structure, and then placed in a fixture and quenched. Quenching is accomplished by a gentle stream of water that is directed at, and only allowed to contact, one flat end of the specimen. The test bar is thereby subjected to a full range of cooling rates that vary from a rapid water quench on one end to a slow air cool on the other. After quenching, two flat surfaces are ground on the specimen. These surfaces are positioned 180° apart, are ground to a depth of at least 0.015 inch (0.381 mm), and run the full length of the specimen. Starting at the quenched end, Rockwell C hardness measurements are performed at $1/16$ inch (1.6 mm) apart along the length of the bar over a span of at least 2.5 inches (63.5 mm). The results are then plotted to show hardness versus distance from the quenched end.

Heat Treatment of Steel – Procedures

Most steel is used in the as-rolled condition, but thermal treatment is capable of broadening the properties attainable. There are two general categories of thermal treatment: 1) those that increase the strength, hardness, and toughness of steel by cooling rapidly from above the transformation range; and 2) those that decrease hardness and promote uniformity either by cooling slowly from above the transformation range, or by being subjected to prolonged heating within the transformation range, and then cooled slowly. The first category usually involves through hardening that is induced by quenching and tempering, or one of many methods that are capable of enhancing the surface of the steel to a predetermined depth. The second category encompasses normalizing together with annealing in order to improve machinability, toughness or cold forming characteristics, or to relieve stresses and restore ductility after some form of cold working.

Furnaces of various descriptions (and ages) are in use for heat treating. The essential quality of a heat treating furnace is that it must maintain indicated uniform temperatures within a relatively narrow thermal range, and the sensing gage must be able to accurately report the temperature within the furnace. Most furnaces in use are nonatmospheric, meaning that they do not have the ability to control the precise mixture of gases present during treatment. While this is not normally a problem, precautions must be taken—especially at temperatures in excess of 1,200° F (649° C)—to protect the surfaces of steel parts during heating, otherwise scaling, carburization, or decarburization may result. The most modern furnaces are capable of creating a vacuum environment, which allows customizing the pressure so that it is compatible with the steel being treated.

As reviewed earlier, the predictable change to the microstructure that can be affected by heating and cooling solid steel at selected rates under controlled rates is the basis for the heat treatment of steel. The practical aspects of steel heat treatment are considered in the following discussion of annealing, normalizing, quench and tempering, martempering, and austempering. Consult **Tables 21** through **26** for detailed information on temperatures required for thermally treating carbon and low alloy steels, and see **Tables 26** through **29** for details for thermally treating stainless steels.

Annealing

Steel is annealed to reduce hardness, relieve stresses, to develop a particular microstructure with specific physical and mechanical properties, to improve machinability and formability. Basically, it consists of a heating cycle, a holding period, and a controlled cooling cycle. Various annealing procedures are in use, each of which is identified by a descriptive term.

Full annealing is a relatively straightforward heat treatment in which the steel is heated to a temperature above the A_3 critical temperature and held at the temperature long enough to allow the solution of carbon and other alloying elements in the austenite. The

(Text continued on p. 490)

Table 21. Thermal Treatments for Carbon Steels-Water and Oil Hardening Grades. *(Source, Bethlehem Steel.)*

Type[1]	Normalizing Temp.		Annealing Temp.		Hardening Temp.		Quenching Medium
	°F	°C	°F	°C	°F	°C	
1030	-	-	-	-	1575-1600	857-871	Water or Caustic
1035	-	-	-	-	1550-1600	829-871	Water or Caustic
1037	-	-	-	-	1525-1575	802-857	Water or Caustic
1038 [2]	-	-	-	-	1525-1575	802-857	Water or Caustic
1039 [2]	-	-	-	-	1525-1575	802-857	Water or Caustic
1040 [2]	-	-	-	-	1525-1575	802-857	Water or Caustic
1042	-	-	-	-	1500-1550	816-829	Water or Caustic
1043 [2]	-	-	-	-	1500-1550	816-829	Water or Caustic
1045 [2]	-	-	-	-	1500-1550	816-829	Water or Caustic
1046 [2]	-	-	-	-	1500-1550	816-829	Water or Caustic
1050 [2]	1600-1700	871-927	-	-	1500-1550	816-829	Water or Caustic
1053	1600-1700	871-927	-	-	1500-1550	816-829	Water or Caustic
1060	1600-1700	871-927	1400-1500	760-816	1575-1625	857-885	Oil
1074 †	1550-1650	829-899	1400-1500	760-816	1575-1625	857-885	Oil
1080	1550-1650	829-899	1400-1500 [3]	760-816 [3]	1575-1625	857-885	Oil [4]
1084	1550-1650	829-899	1400-1500 [3]	760-816 [3]	1575-1625	857-885	Oil [4]
1085 †	1550-1650	829-899	1400-1500 [3]	760-816 [3]	1575-1625	857-885	Oil [4]
1090	1550-1650	829-899	1400-1500 [3]	760-816 [3]	1575-1625	857-885	Oil [4]
1095	1550-1650	829-899	1400-1500 [3]	760-816 [3]	1575-1625	857-885	Water and Oil
1137	-	-	-	-	1550-1600	829-871	Oil
1141	1600-1700	871-927	1400-1500	760-816	1500-1550	816-829	Oil
1144	1600-1700	871-927	-	-	1500-1550	816-829	Oil
1145	-	-	-	-	1475-1500	802-816	Water or Oil
1146	-	-	-	-	1475-1500	802-816	Water or Oil
1151	1600-1700	871-927	-	-	1475-1500	802-816	Water or Oil
1536	1600-1700	871-927	-	-	1500-1550	816-829	Water or Oil
1541	1600-1700	871-927	1400-1500	760-816	1500-1550	816-829	Oil
1548	1600-1700	871-927	-	-	1500-1550	816-829	Oil
1552	1600-1700	871-927	-	-	1500-1550	816-829	Oil
1566	1600-1700	871-927	-	-	1575-1625	857-885	Oil

Notes: When tempering is required, temperature should be selected to effect desired hardness.

† Indicates SAE designation only. No AISI number assigned.

[1] AISI/SAE designation numbers.

[2] These grades are commonly used for parts where induction hardening is employed, although all steels from 1030 up can be induction hardened.

[3] Spheroidal structures are often required for machining purposes and should be cooled very slowly or be isothermally transformed to produce the desired structure.

[4] May be water or brine quenched by special techniques such as partial immersion, or time quenched; otherwise subject to quench cracking.

Table 22. Thermal Treatments for Carbon Steels-Carburizing Grades. (Source, Bethlehem Steel.)

Type[1]	Carburizing Temp.		Cooling Medium	Reheat Temp.		Cooling Medium	Carbonitriding Temp.		Cooling Medium
	°F	°C		°F	°C		°F	°C	
1010	-	-	-	-	-	-	1450-1650	788-899	Oil
1015	-	-	-	-	-	-	1450-1650	788-899	Oil
1016	1650-1700	899-927	Water or Caustic [2]	-	-	-	1450-1650	788-899	Oil
1018	1650-1700	899-927	Water or Caustic [2]	1450	788	Water or Caustic [2]	1450-1650	788-899	Oil
1019	1650-1700	899-927	Water or Caustic [2]	1450	788	Water or Caustic [2]	1450-1650	788-899	Oil
1020	1650-1700	899-927	Water or Caustic [2]	1450	788	Water or Caustic [2]	1450-1650	788-899	Oil
1022	1650-1700	899-927	Water or Caustic [2]	1450	788	Water or Caustic [2]	1450-1650	788-899	Oil
1026	1650-1700	899-927	Water or Caustic [2]	1450	788	Water or Caustic [2]	1450-1650	788-899	Oil
1030	1650-1700	899-927	Water or Caustic [2]	1450	788	Water or Caustic [2]	1450-1650	788-899	Oil
1109	1650-1700	899-927	Water or Oil	1400-1450	760-788	Water or Caustic [2]	-	-	-
1117	1650-1700	899-927	Water or Oil	1450-1600	788-871	Water or Caustic [2]	1450-1650	788-899	Oil
1118	1650-1700	899-927	Oil	1450-1600	788-871	Water or Caustic [2]	-[3]	-	-
1513	1650-1700	899-927	Oil	1450	788	Oil	-[3]	-	-
1518	1650-1700	899-927	Oil	1450	788	Oil	-[3]	-	-
1522	1650-1700	899-927	Oil	1450	788	Oil	-[3]	-	-
1524	1650-1700	899-927	Oil	1450	788	Oil	-[3]	-	-
1525	1650-1700	899-927	Oil	1450	788	Oil	-[3]	-	-
1526	1650-1700	899-927	Oil	1450	788	Oil	-[3]	-	-
1527	1650-1700	899-927	Oil	1450	788	Oil	-[3]	-	-

Notes: Normalizing is generally unnecessary for fulfilling either dimensional or machinability requirements of parts made from the above grades. Where dimensioning is of vital importance, normalizing temperatures of at least 50° F (28° C) above the carburizing temperatures are sometimes required to minimize distortion. Tempering temperatures are usually 250-400° F (121-204° C), but higher temperatures may be used when permitted by the hardness specification for the finished parts.

[1] AISI/SAE designation numbers.
[2] 3% sodium hydroxide.
[3] Higher manganese steels, such as 1118 and the 1500 series are not usually carbonitrided. If carbonitriding is performed, care must be taken to limit the nitrogen content because high nitrogen will increase their tendency to retain austenite.

Table 23. Representative Heat Treatments for Selected Low-Carbon and Low-Alloy Steels.
(Source, MIL-HDBK-723.)

Stress Relief Temperatures			
Material	**°F**	**°C**	**Soak Time at Temp.**
Low-Alloy Steels (after heat treat at 150 to 180,000 psi)	700) 25	371) 14	1 hour per inch (25.4 mm) of cross section

Annealing Cycle						
Material	**Number**	**Annealing Temp.**		**Furnace Cooling Cycle***		**Soak Time (hr.)**	**Heatup Time**
		°F	**°C**	**°F**	**°C**		
Low-Carbon	1018	1575 to 1650	857 to 899	From 1575 to 1300	857 to 704	1 hr. for sections to 1 in. thick. Add 1/2 hr. for each additional 1 in. (25.4 mm) of thickness.	1 hr. per in. (25.4 mm) of material thickness.
	1020	1575 to 1650	857 to 899	From 1575 to 1290	857 to 699		
	1025	1575 to 1650	857 to 899	From 1575 to 1290	857 to 699		
	1030	1550 to 1625	829 to 885	From 1550 to 1200	829 to 649		
	1035	1550 to 1625	829 to 885	From 1550 to 1200	829 to 649		
Low-Alloy	4130	1450 to 1550	788 to 829	From 1450 to 900	788 to 482		
	4140	1450 to 1550	788 to 829	From 1450 to 900	788 to 482		
	4340	1450 to 1550	788 to 829	From 1450 to 900	788 to 482		
	5150	1500 to 1600	816 to 871	From 1500 to 900	816 to 482		
	6150	1550 to 1650	829 to 899	From 1550 to 1000	829 to 538		

* Cooling Rate is 50° F (28° C) per hour. After reaching lower temperature, rate of cooling is unimportant.

Normalizing Cycle					
Material	**Number**	**Normalizing Temp.**		**Soak Time (hr.)**	**Heatup Time**
		°F	**°C**		
Low-Carbon	1015	1650 to 1700	899 to 927	1 hr. for sections to 1 in. thick. Add 1/2 hr. for each additional 1 in. of thickness	1 hr. per in. of material thickness.
	1020	1650 to 1700	899 to 927		
	1025	1625 to 1675	885 to 913		
	1030	1625 to 1675	885 to 913		
	1035	1600 to 1650	871 to 899		
Low-Alloy	4130	1600 to 1750	871 to 954		
	4140	1550 to 1700	829 to 927		
	4340	1550 to 1700	829 to 927		
	5150	1550 to 1700	829 to 927		
	6150	1600 to 1750	871 to 954		

Austenizing Cycle							
Material	**Number**	**Austenizing Temp.**		**Soak Time**			**Heatup Time**
		°F	**°C**	**Thickness inch**	**Thickness mm**	**Time (Hours)**	
Low-Carbon	1025	1575 to 1650	857 to 899	1/2 or less	12.7 or less	1/4	1 hr. per in. of material thickness.
	1030	1550 to 1600	829 to 871	1	25.4	1/3	
	1035	1525 to 1575	802 to 857	2	50.8	1/2	
Low-Alloy	4130	1500 to 1600	816 to 871	3	76.2	3/4	
	4140	1550 to 1600	829 to 871	4	101.6	1-1/4	

(Continued)

Table 23. *(Continued)* **Representative Heat Treatments for Selected Low-Carbon and Low-Alloy Steels.** *(Source, MIL-HDBK-723.)*

Austenizing Cycle							
Material	Number	Austenizing Temp.		Soak Time			Heatup Time
		°F	°C	Thickness inch	Thickness mm	Time (Hours)	
Low-Alloy	4340	1500 to 1550	816 to 829	5	127	1-1/2	1 hr. per in. of material thickness.
	5150	1475 to 1550	802 to 829	-	-	-	
	6150	1550 to 1625	829 to 885	-	-	-	

Tempering Cycle for Low-Alloy Steels							
Type	Tempering Temperature for Indicated Tensile Strength						Soak Time	Heatup Time
	125 ksi - °F	860 MPa - °C	150 ksi - °F	1034 MPa - °C	180 ksi - °F	1240 MPa - °C		
4130	950 to 1150	510 to 621	800 to 1000	427 to 538	700 to 900	371 to 482	1 hr. per in. of material thickness.	1 hr. per in. of material thickness.
4140	1050 to 1250	566 to 677	950 to 1150	510 to 621	800 to 1000	527 to 538		
4340	1075 to 1225	579 to 663	975 to 1075	524 to 579	850 to 975	454 to 524		

The information presented in this table is intended for general cases. Temperatures and times should be adjusted to compensate for differences in equipment, chemical composition, size and shape of parts being treated, and other variables.

Table 24. Thermal Treatments for Alloy Steels-Directly Hardenable Grades. *(Source, Bethlehem Steel.)*

Type[1]	Normalizing Temp.		Annealing Temp.[4]		Hardening Temp.[5]		Quenching Medium
	°F	°C	°F	°C	°F	°C	
1330	1600-1700 [2]	871-927 [2]	1550-1650	843-899	1525-1575	802-857	Water or Oil
1335	1600-1700 [2]	871-927 [2]	1550-1650	843-899	1500-1550	829-829	Oil
1340	1600-1700 [2]	871-927 [2]	1550-1650	843-899	1500-1550	829-829	Oil
1345	1600-1700 [2]	871-927 [2]	1550-1650	843-899	1500-1550	829-829	Oil
4037	-	-	1550-1575	843-857	1525-1575	802-857	Oil
4042 †	-	-	1550-1575	843-857	1525-1575	802-857	Oil
4047	-	-	1450-1550	788-843	1500-1575	829-857	Oil
4130	1600-1700 [2]	871-927 [2]	1450-1550	788-843	1500-1600	829-871	Water or Oil
4135 †	-	-	1450-1550	788-843	1500-1600	829-871	Oil
4137	-	-	1450-1550	788-843	1500-1600	829-871	Oil
4140	-	-	1450-1550	788-843	1500-1600	829-871	Oil
4142	-	-	1450-1550	788-843	1500-1600	829-871	Oil
4145	-	-	1450-1550	788-843	1500-1550	829-829	Oil
4147	-	-	1450-1550	788-843	1500-1550	829-829	Oil
4150	-	-	1450-1550	788-843	1500-1550	829-829	Oil
4161	-	-	1450-1550	788-843	1500-1550	829-829	Oil [6]
4340	1600-1700 [2]	871-927 [2]	1450-1550	788-843	1500-1550	829-829	Oil [3]
50B40 †	1600-1700 [2]	871-927 [2]	1500-1600	829-871	1500-1550	829-829	Oil
50B44	1600-1700 2	871-927 2	1500-1600	829-871	1500-1550	829-829	Oil
5046 †	1600-1700 2	871-927 2	1500-1600	829-871	1500-1550	829-829	Oil
50B46	1600-1700 2	871-927 2	1500-1600	829-871	1500-1550	829-829	Oil

(Continued)

Table 24. *(Continued)* **Thermal Treatments for Alloy Steels-Directly Hardenable Grades.**
(Source, Bethlehem Steel.)

Type[1]	Normalizing Temp.		Annealing Temp.[4]		Hardening Temp.[5]		Quenching Medium
	°F	°C	°F	°C	°F	°C	
50B50	1600-1700 [2]	871-927 [2]	1500-1600	829-871	1475-1550	802-829	Oil
5060 †	1600-1700 [2]	871-927 [2]	1500-1600	829-871	1475-1550	802-829	Oil
50B60	1600-1700 [2]	871-927 [2]	1500-1600	829-871	1475-1550	802-829	Oil
5130	1600-1700 [2]	871-927 [2]	1450-1550	788-843	1525-1575	802-857	Water, Caustic, or Oil
5132	1600-1700 [2]	871-927 [2]	1450-1550	788-843	1525-1575	802-857	Oil
5135	1600-1700 [2]	871-927 [2]	1500-1600	829-871	1500-1550	829-829	Oil
5140	1600-1700 [2]	871-927 [2]	1500-1600	829-871	1500-1550	829-829	Oil
5145 †	1600-1700 [2]	871-927 [2]	1500-1600	829-871	1500-1550	829-829	Oil
5147 †	1600-1700 [2]	871-927 [2]	1500-1600	829-871	1475-1550	802-829	Oil
5150	1600-1700 [2]	871-927 [2]	1500-1600	829-871	1475-1550	802-829	Oil
5155	1600-1700 [2]	871-927 [2]	1500-1600	829-871	1475-1550	802-829	Oil
5160	1600-1700 [2]	871-927 [2]	1500-1600	829-871	1475-1550	802-829	Oil
51B60	1600-1700 [2]	871-927 [2]	1500-1600	829-871	1475-1550	802-829	Oil
50100 †	-	-	1350-1450	732-788	1425-1475	774-802	Water
					1500-1600	829-871	Oil
51100	-	-	1350-1450	732-788	1425-1475	774-802	Water
					1500-1600	829-871	Oil
52100	-	-	1350-1450	732-788	1425-1475	774-802	Water
					1500-1600	829-871	Oil
6150	-	-	1550-1650	843-899	1500-1625	829-885	Oil
81B45	1600-1700 [2]	871-927 [2]	1550-1650	843-899	1500-1575	829-857	Oil
8630	1600-1700 [2]	871-927 [2]	1450-1550	788-843	1525-1600	802-871	Water or Oil
8637	-	-	1500-1600	829-871	1525-1575	802-857	Oil
8640	-	-	1500-1600	829-871	1525-1575	802-857	Oil
8642	-	-	1500-1600	829-871	1500-1575	829-857	Oil
8645	-	-	1500-1600	829-871	1500-1575	829-857	Oil
86B45	-	-	1500-1600	829-871	1500-1575	829-857	Oil
8650 †	-	-	1500-1600	829-871	1500-1575	829-857	Oil
8655	-	-	1500-1600	829-871	1475-1550	802-829	Oil
8660 †	-	-	1500-1600	829-871	1475-1550	802-829	Oil
8740	-	-	1500-1600	829-871	1525-1575	802-857	Oil
8254 †	-	-	-	-	1500-1650	829-899	Oil
9255 †	-	-	-	-	1500-1650	829-899	Oil
9260	-	-	-	-	1500-1650	829-899	Oil
94B30	1600-1700 [2]	871-927 [2]	1450-1550	788-843	1500-1625	829-885	Oil

Note: When tempering is required, temperature should be selected to effect desired hardness (see footnotes 3 and 6).
† Indicates SAE designation only. No AISI number assigned. *(Continued)*

Table 24. *(Continued)* **Thermal Treatments for Alloy Steels-Directly Hardenable Grades.**
(Source, Bethlehem Steel.)

[1]   All steels are fine grain.

[2]   These steels should be either normalized or annealed for optimum machinability.

[3]   Temper at 1,100-1,225° F (593-663° C).

[4]   The specific annealing cycle is dependent on the alloy content of the steel, the type of subsequent machining, and the desired surface finish.

[5]   With the exception of 4340, 50100, 51100, and 52100, these steels are frequently hardened and tempered to a final machinable hardness without preliminary thermal treatment.

[6]   Temper above 700° F (371° C).

steel is cooled from the annealing temperature at a slow rate so that the transformation is completed in the high temperature region of the pearlite range. Full annealing produces a structure of relatively soft coarse pearlite and, depending on the composition of the steel, ferrite or carbide may also be present. This process is also known as *Solution Annealing*. It softens the steel, but it is primarily used to improve the machinability of medium carbon steels. While it is simple, it is also a very slow process, as the steel must be cooled very slowly from the annealing (austenitizing) temperature to a temperature below that at which the transformation is completed.

Isothermal annealing is effective in obtaining either a lamellar or spheroidized structure. If a lamellar pearlitic structure is desired, the steel is austenitized above the upper transformation temperature and then cooled to and held at the proper temperature for austenitization to occur. To obtain a spheroidized structure, a lower austenitizing temperature is used so that some carbide remains undissolved. Cooling and transformation as for the pearlitic structure above will result in a spheroidized structure. An advantage of isothermal annealing is that it accelerates the cooling and transformation temperature, as well as the cooling subsequent to transformation, allowing for appreciable time savings compared with some other processes.

Spheroidize annealing is used to achieve a spheroidal (globular) carbide in a ferrite matrix , primarily to obtain optimum cold forming characteristics. One method consists of holding the steel at a temperature just below Ae_1 (holding between Ac_1 and Ac_3 for at least part of the time is normally involved). In order to achieve full spheroidization of the carbides by this method, the steel usually must be held at the temperature for long periods of time. Heating and cooling the steel to temperatures slightly above and slightly below the Ae_1 temperature is another method that will produce a spheroidized structure. Also, if the carbide is not completely dissolved during austenitizing, and the steel is slowly cooled in a manner similar to full annealing, a spheroidized structure can be developed. A spheroidized structure is sometimes desirable to develop minimum hardness and maximum ductility to facilitate cold forming, or to improve the machinability of high carbon steels.

Process annealing is also known as *stress relief annealing* and it is carried out at subcritical temperatures (below Ac_1) on cold worked, normalized, welded, or straightened steel to restore its ductility and reduce its hardness through recrystallization. It consists of heating steel to some temperature below, and usually near, Ac_1 and holding it at that temperature for an appropriate period, then air cooling it to ambient temperature.

Normalizing

Normalizing is performed by reheating steel to about 150 to 200° F (83 to 111° C) above its critical temperature, and then cooling it in air. The resulting fine grained pearlitic structure enhances the uniformity of mechanical properties and, for some grades, improves machinability. Notch toughness, in particular, is much better than in the as-rolled condition. For large sections, and when freedom from residual stresses or lower hardness are

(Text continued on p. 496)

Table 25. Thermal Treatments for Alloy Steels-Carburizing Grades. *(Source, Bethlehem Steel.)*

Type[1]	Pretreatments	Carburizing Temp.[5] °F	°C	Cooling Method	Reheat temp. °F	°C	Quenching Medium	Reheat temp. °F	°C
4012 †	Normalize [2]	1650-1700	899-927	Quench in Oil [7]	-	-	-	250-350	121-177
4023	Normalize [2]	1650-1700	899-927	Quench in Oil [7]	-	-	-	250-350	121-177
4024	Normalize [2]	1650-1700	899-927	Quench in Oil [7]	-	-	-	250-350	121-177
4027	Normalize [2]	1650-1700	899-927	Quench in Oil [7]	-	-	-	250-350	121-177
4028	Normalize [2]	1650-1700	899-927	Quench in Oil [7]	-	-	-	250-350	121-177
4032	Normalize [2]	1650-1700	899-927	Quench in Oil [7]	-	-	-	250-350	121-177
4118	Normalize [2]	1650-1700	899-927	Quench in Oil [7]	-	-	-	250-350	121-177
4320	Normalize [2] and Cycle Anneal [4]	1650-1700	899-927	Quench in Oil [7]	-	-	-	250-350	121-177
				Cool Slowly	1525-1550 [9]	829-843 [9]	Oil	250-350	121-177
4419 †	Normalize [2] and Cycle Anneal [4]	1650-1700	899-927	Quench in Oil [7]	-	-	-	250-350	121-177
4422 †	Normalize [2] and Cycle Anneal [4]	1650-1700	899-927	Quench in Oil [7]	-	-	-	250-350	121-177
4427 †	Normalize [2] and Cycle Anneal [4]	1650-1700	899-927	Quench in Oil [7]	-	-	-	250-350	121-177
4615	Normalize [2] and Cycle Anneal [4]	1650-1700	899-927	Quench in Oil [7]	-	-	-	250-300	121-149
				Cool Slowly	1525-1550 [9]	829-843 [9]	Oil		
				Quench in Oil	1525-1550 [8]	829-843 [9]	Oil		
4617 †	Normalize [2] and Cycle Anneal [4]	1650-1700	899-927	Quench in Oil [7]	-	-	-	250-300	121-149
				Cool Slowly	1525-1550 [9]	829-843 [9]	Oil		
				Quench in Oil	1525-1550 [8]	829-843 9	Oil		
4620	Normalize [2] and Cycle Anneal [4]	1650-1700	899-927	Quench in Oil [7]	-	-	-	250-300	121-149
				Cool Slowly	1525-1550 [9]	829-843 [9]	Oil		
				Quench in Oil	1525-1550 [8]	829-843 [9]	Oil		

(Continued)

Table 25. *(Continued)* **Thermal Treatments for Alloy Steels-Carburizing Grades.** *(Source, Bethlehem Steel.)*

Type[1]	Pretreatments	Carburizing Temp.[5] °F	°C	Cooling Method	Reheat temp. °F	°C	Quenching Medium	Reheat temp. °F	°C
4621 †	Normalize[2] and Cycle Anneal[4]	1650-1700	899-927	Quench in Oil[7] Cool Slowly Quench in Oil	- 1525-1550[9] 1525-1550[8]	- 829-843[9] 829-843[9]	- Oil Oil	250-300	121-149
4626	Normalize[2] and Cycle Anneal[4]	1650-1700	899-927	Quench in Oil[7] Cool Slowly Quench in Oil	- 1525-1550[9] 1525-1550[8]	- 829-843[9] 829-843[9]	- Oil Oil	250-300	121-149
4718 †	Normalize[2] and Cycle Anneal[4]	1650-1700	899-927	Quench in Oil[7] Cool Slowly Quench in Oil	- 1525-1550[9] 1525-1550[8]	- 829-843[9] 829-843[9]	- Oil Oil	250-300	121-149
4720	Normalize[2] and Cycle Anneal[4]	1650-1700	899-927	Quench in Oil[7]	1525-1550[8]	829-843[9]	Oil	250-350	121-177
4815	Normalize,[2] Temper,[3] and Cycle Anneal[4]	1650-1700	899-927	Cool Slowly Quench in Oil	1475-1525[9] 1475-1525[8]	802-829 802-829	Oil Oil	250-325	121-1
4817	Normalize,[2] Temper,[3] and Cycle Anneal[4]	1650-1700	899-927	Quench in Oil[7] Cool Slowly Quench in Oil	- 1475-1525[9] 1475-1525[8]	- 802-829 802-829	- Oil Oil	250-325	121-163
4820	Normalize,[2] Temper,[3] and Cycle Anneal[4]	1650-1700	899-927	Quench in Oil[7] Cool Slowly Quench in Oil	- 1475-1525[9] 1475-1525[8]	- 802-829 802-829	- Oil Oil	250-325	121-163
5015 †	Normalize[2]	1650-1700	899-927	Quench in Oil[7]			-	250-325	121-163
5115 †	Normalize[2]	1650-1700	899-927	Quench in Oil[7]			-	250-325	121-163
5120	Normalize[2]	1650-1700	899-927	Quench in Oil[7]			-	250-325	121-163

(Continued)

Table 25. *(Continued)* Thermal Treatments for Alloy Steels-Carburizing Grades. *(Source, Bethlehem Steel.)*

Type[1]	Pretreatments	Carburizing Temp.[5] °F	Carburizing Temp.[5] °C	Cooling Method	Reheat temp. °F	Reheat temp. °C	Quenching Medium	Reheat temp. °F	Reheat temp. °C
6118	Normalize[2]	1650	899	Quench in Oil[7]	-	-	-	325	163
8115 †	Normalize[2] and Cycle Anneal[4]	1650-1700	899-927	Quench in Oil[7]	-	-	-	250-350	121-177
				Cool Slowly	1550-1600[9]	843-871	Oil		
				Quench in Oil	1550-1600[8]	843-871	Oil		
8615	Normalize[2] and Cycle Anneal[4]	1650-1700	899-927	Quench in Oil[7]	-	-	-	250-350	121-177
				Cool Slowly	1550-1600[9]	843-871	Oil		
				Quench in Oil	1550-1600[8]	843-871	Oil		
8617	Normalize[2] and Cycle Anneal[4]	1650-1700	899-927	Quench in Oil[7]	-	-	-	250-350	121-177
				Cool Slowly	1550-1600[9]	843-871	Oil		
				Quench in Oil	1550-1600[8]	843-871	Oil		
8620	Normalize[2]	1650-1700	899-927	Quench in Oil[7]	-	-	-	250-350	121-177
				Cool Slowly	1550-1600[9]	843-871	Oil		
				Quench in Oil	1550-1600[8]	843-871	Oil		
8622	Normalize[2]	1650-1700	899-927	Quench in Oil[7]	-	-	-	250-350	121-177
				Cool Slowly	1550-1600[9]	843-871	Oil		
				Quench in Oil	1550-1600[8]	843-871	Oil		
8625	Normalize[2]	1650-1700	899-927	Quench in Oil[7]	-	-	-	250-350	121-177
				Cool Slowly	1550-1600[9]	843-871	Oil		
				Quench in Oil	1550-1600[8]	843-871	Oil		
8627	Normalize[2]	1650-1700	899-927	Quench in Oil[7]	-	-	-	250-350	121-177
				Cool Slowly	1550-1600[9]	843-871	Oil		
				Quench in Oil	1550-1600[8]	843-871	Oil		

(Continued)

Table 25. *(Continued)* **Thermal Treatments for Alloy Steels-Carburizing Grades.** *(Source, Bethlehem Steel.)*

Type[1]	Pretreatments	Carburizing Temp.[5] °F	°C	Cooling Method	Reheat temp. °F	°C	Quenching Medium	Reheat temp. °F	°C
8720	Normalize[2]	1650-1700	899-927	Quench in Oil[7]	-	-	-	250-350	121-177
				Cool Slowly	1550-1600[9]	843-871	Oil		
				Quench in Oil	1550-1600[8]	843-871	Oil		
8822	Normalize[2]	1650-1700	899-927	Quench in Oil[7]	-	-	-	250-350	121-177
				Cool Slowly	1550-1600[9]	843-871	Oil		
				Quench in Oil	1550-1600[8]	843-871	Oil		
9310	Normalize[2] and Temper[3]	1600-1700	871-927	Quench in Oil	1450-1525[8]	788-829	Oil	250-325	121-177
				Cool Slowly	1450-1525[9]	788-829	Oil		
94B15	Normalize[2]	1650-1700	899-927	Quench in Oil[7]	-	-	-	250-350	121-177
94B17	Normalize[2]	1650-1700	899-927	Quench in Oil[7]	-	-	-	250-350	121-177

[†] Indicates SAE designation only. No AISI number assigned.

[1] All steels are fine grain. Indicates heat treatments may not be appropriate for coarse grain steels.

[2] Normalizing temperatures should be at least equal to carburizing temperature, followed by air cooling.

[3] After normalizing, reheat to temperature of 1,100-1,200° F (593-649° C) and hold at temperature for approximately one hour per inch (25.4 mm) of maximum section, or for four hours minimum.

[4] Where cycle annealing is desired, heat to at least as high as the carburizing temperature, hold for uniformity, cool rapidly to 1,000-1,250° F (538-677° C), hold one to three hours, then air cool or furnace cool to obtain a structure suitable for machining and finishing.

[5] It is general practice to reduce carburizing temperatures to approximately 1,550° F (843° C) before quenching to minimize distortion and retained austenite. For 4800 series steels, the carburizing temperature is reduced to approximately 1,500° F (816° C) before quenching.

[6] Temperatures higher than those shown are used in some instances where application applies.

[7] This treatment is most commonly used and generally produces a minimum of distortion.

[8] This treatment is used where maximum grain refinement is required and/or where parts are subsequently ground on critical dimensions. A combination of good case and core properties is secured with somewhat greater distortion than is obtained by a single quench from the carburizing treatment.

[9] In this treatment, the parts are slowly cooled—preferably under a protective atmosphere. They are then reheated and oil quenched. A tempering operation follows as required. This treatment is used when machining must be done between carburizing and hardening, or when facilities for quenching from the carburizing cycle are not available. Distortion is at least equal to that obtained by a single quench from the carburizing cycle, as described in note 5.

Table 26. Recommended Heating and Holding Time for Annealing, Normalizing, Austenitizing, and Stress Relieving Carbon Steel, Low Alloy Steel, and Martensitic Stainless Steel. *(Source, MIL-H-6875G.)*

Thickness[5]		Heat-up Time (Minutes)[4]		Minimum Holding Time [2,3]
Inches	Millimeters	Furnace[1]	Salt Bath	Minutes
0.250 and under	6.35 and under	20	10	15
0.251-0.500	6.3754 to 12.7	30	10	25
0.501-1.000	12.7254 to 25.4	45	10	30
1.001-1.500	25.4254 to 38.1	60	15	30
1.501-2.000	38.1254 to 50.8	75	20	30
2.001-2.500	50.8245 to 63.5	90	25	40
2.501-3.000	63.5254 to 76.2	105	30	45
3.001-3.500	76.2254 to 88.9	120	35	55
3.501-4.000	88.9254 to 101.6	135	40	60
4.001-5.000	101.6254 to 127	165	50	75
5.001-6.000	127.0254 to 152.4	195	60	90
6.001-7.000	152.4254 to 177.8	225	75	105
7.001-8.000	177.8254 to 203.2	255	90	120

[1] For unplated parts only. Copper plated parts require at least fifty percent longer heat-up time and the heat-treating facility should: (a) determine the appropriate heat-up time as a function of maximum part thickness and (b) establish suitable process controls for ensuring that the parts reach the required heat-treat temperature prior to start of holding time.

[2] Maximum holding time should not exceed twice the recommended minimum time. In all cases, holding time shall not start until parts or material have reached specified heat-treat temperature.

[3] Minimum stress relieving time shall be one hour for stress relieving temperatures up to 850° (F 454° C), inclusive, and 2 hours for higher stress relieving temperatures.

[4] Heat-up time starts when all temperature indicators rise to within 10° F (5.5° C) of set temperature. These times are suitable for simple solid shapes heated from all surfaces. Longer times are necessary for complex shapes and/or parts not uniformly heated.

[5] Thickness is minimum dimension of heaviest section.

Table 27. Recommended Holding Time for Annealing Austenitic Stainless Steel. *(Source, MIL-H-6875G.)*

Thickness[1]		Minimum Holding Time for Full Annealing (Minutes)
Inches	Millimeters	Atmosphere Furnace[2]
Up to 0.100	Up to 2.54	20
0.101 to 0.250	2.5654 to 6.35	25
0.251 to 0.500	6.3754 to 12.7	45
0.501 to 1.00	12.7254 to 25.4	60
1.01 to 1.50	25.5654 to 38.1	75
1.51 to 2.00	38.354 to 50.8	90
2.01 to 2.50	51.054 to 63.5	105
2.51 to 3.00	63.754 to 76.2	120

[1] Thickness is the minimum dimension of the heaviest section of a part or the minimum dimension of the heaviest section of a multi-layer load.

[2] Holding time starts when all temperature indicators rise to within 10° F (5.5° C) of set temperature. For continuous and repetitive batch heat treatment, the holding time may be lowered provided the solution of carbides is assured per ASTM A 262.

Table 28. Annealing Procedures for Austenitic Stainless Steels. *(Source, MIL-H-6875H.)*

Type	Annealing Temperature		Quenching Medium[1]
	°F	°C	
201, 202, 301, 302, 303, 304, 304L, and 308 [2]	1850-2050	1010-1121	Water
309, 310, 316, 316L [2]	1900-2050	1038-1121	Water
321 [3]	1750-2050	954-1121	Air or Water
347, 348 [3]	1800-2050	982-1121	Air or Water

[1]   Other means of cooling are permitted provided that the cooling rate is rapid enough to prevent carbide precipitation.

[2]   Do not stress relieve the unstabilized grades, except 304L and 316L, between 875° ± 25° F and 1,500° F (468° ± 14° C and 815° C). Stress relieving of stabilized grains should be at 1,650° F (899° C) for one hour.

[3]   When stress relieving after welding is specified, hold for one-half hour minimum at specified temperature or hold for two hours at 1,650° ± 25° F (899° ± 14° C).

desired, it can be followed by a stress relief treatment. As a preliminary treatment to quenching and tempering, it is used to develop a more uniform microstructure and facilitate the solution of carbides and alloying elements. When necessary, normalizing can be followed by tempering, usually at 1,000 to 1,300° F (538 to 704° C), to reduce hardness and improve toughness.

Quenching and Tempering

Hardening by quenching and tempering is the heat treatment commonly used to develop martensitic structures with the desired combination of toughness and strength. The process is divided into separate operations: heating, quenching, and tempering.

Austenitizing is performed prior to hardening by heating steel to a temperature above its critical range. The steel should be held at the temperature long enough for the carbon and other alloying elements to dissolve, but not long enough for excessive grain growth to occur.

Quenching is the rapid cooling of steel from the austenitizing temperature to a temperature below M_s (martensite start temperature). The cooling rate must be rapid enough to prevent the formation of other transformation products such as pearlite, bainite, ferrite, or cementite. The choice of cooling medium is dependent on the desired cooling rate, which is determined by the composition (hardenability) of the steel, and the size and shape of the section being quenched. Quenching sets up high thermal and transformation stresses that can cause distortion or cracking of the part, therefore it is usually desirable to keep those stresses at a minimum by cooling at a rate that is just slightly faster than the critical cooling rate as determined by the hardenability of the steel and the size and shape of the part being quenched. The most common quenching media are water, mineral oil, and, of course, forced air.

Quenching water is normally held at about 65° F (18° C). As the water temperature increases during the quenching process, the treated steel tends to develop a surrounding envelope of steam that inhibits the water's cooling capacity. The tendency to form a steam envelope can be reduced by adding salt to the water to create brine (5 to 10% sodium chloride). Sodium hydroxide is even more efficient, and both solutions are generally used on very shallow hardening steels to attain high surface hardness while retaining a ductile core. Brine solutions are commonly used for carbon and low alloy steel as well as martensitic stainless steel. In both brine and water quenching, agitation of the quenching medium is especially important because it produces more uniform cooling as well as accelerates the rate of cooling.

Table 29. Thermal Treatment Procedures for Martensitic Stainless Steels. (Source, MIL-H-6875H.)

Type	Annealing °F		Annealing °C		Transformation Temp.		Quenching Medium	Subcritical Anneal Temp. Air Cool	
	Temp.	Cool Rate[1]	Temp.	Cool Rate[1]	°F	°C		°F	°C
403	1500-1600	25 to 50°/hr to 1100°	829-871	14 to 28°/hr to 593°	1750-1850	954-1010	Oil or Air	1200-1450	649-788
410	1500-1600	25 to 50°/hr to 1100°	829-871	14 to 28°/hr to 593°	1750-1850	954-1010	Oil or Air	1200-1450	649-788
416	1500-1650	25 to 50°/hr to 1100°	829-899	14 to 28°/hr to 593°	1750-1850	954-1010	Oil or Air	1200-1450	649-788
420	1550-1650 for 6 hours	25 to 50°/hr to 1100° then water quench	829-899 for 6 hours	14 to 28°/hr to 593° then water quench	1750-1850	954-1010	Oil or Air	1350-1450	732-788
440C	1550-1600 for 6 hours + 1300 for 4 hours	25 to 50°/hr to 1100°	829-899 for 6 hours + 704 for 4 hours	14 to 28°/hr to 593°	1900-1950	1037-1065	Oil or Air[2]	1250-1350	677-732

Tempering Temperature to Achieve Indicated Tensile Strength

Type	100 ksi 690 MPa		120 ksi 830 MPa		Avoid Tempering or Holding Within Range[3]		180 ksi 1240 MPa		200 ksi 1380 MPa	
	°F	°C	°F	°C	°F	°C	°F	°C	°F	°C
403	1300	704	1100	593	700-1100	371-593	500	260	-	-
410	1300	704	1100	593	700-1100	371-593	500	260	-	-
416	1300	704	1075	579	700-1075	371-579	500	260	-	-
420	1300 4	704 4	1075	579	700-1075	371-579	-	-	600	315
440C	Temper at 325° F (163° C) for R_c 58 minimum. Temper at 375° F (190° C) for R_c 57 minimum. Temper at 450° F (232° C) for R_c 55 minimum.									

1 Cooling rate performed in furnace.

2 Quench for 440C shall be followed by refrigeration to -100° F (-73° C) or lower for two hours. Double temper to remove retained austenite.

3 Tempering these alloys in the range listed results in decreased impact strength and reduced corrosion resistance. However, tempering in this range is sometimes necessary to obtain the strength and ductility required.

4 Temper 420 steel 300° F (149° C) for RC 52 minimum; 400° F (204° C) for RC 50 minimum; 600° F (315° C) for RC 48 minimum.

Quenching oils are mineral oils that usually have a viscosity in the range of 100 to 110 SUS at 40° C. They can be used to provide a wide variety of cooling rates dependent on viscosity and their mineral base. Normally, quenching oils have a temperature of between 60 and 160° F (15 and 71° C) and are generally maintained at temperatures below 200° F (93° C) during the quench. Due to the quenching properties of oils, which are less effective than water, they are not usually used for low-hardenability steels.

Polymer solutions are also used for quenching, but only three have achieved wide acceptance in the U.S.: PAG (polyalkylene glycol), PVP (polyvinyl pyrrolidone), and PVA (polyvinyl alcohol). Proponents of polymer quenching fluids attest that they are superior to either water or oil.

Tempering is the process of heating quench hardened or normalized steel to a temperature below the transformation range, holding it at that temperature for a suitable time, and cooling it at an appropriate rate. The martensite formed during quenching is very hard, brittle, and highly stressed, and tempering is used to remove these stresses and to improve the ductility of the steel, usually to the detriment of strength and hardness. The stress relief and the recovery of ductility are achieved by the precipitation of carbide from the supersaturated unstable alpha martensite and through diffusion and coalescence of the carbide as tempering proceeds.

Tempering is usually carried out at a temperature range between 350 and 1,300° F (177 to 704° C) and the usual time at temperature ranges from thirty minutes to four hours. For many carbon and low alloy steels, ductility and toughness increase when tempered at temperatures up to 400° F (204° C). In the approximate temperature range of 450 to 700° F (232 to 371° C), notch toughness decreases as tempering temperatures are increased—consequently, quench hardened steels are rarely hardened in this range. Tempering in the range of 200 to 400° F (93 to 204° C) is used when it is important to retain hardness and strength and effect a modest improvement in toughness. In the highest temperature range of 700 to 1,250° F (371 to 677° C), tempering causes an appreciable increase in toughness and ductility of the steel as well as a simultaneous decrease in hardness and strength.

Martempering is a process used to reduce high stresses, that can lead to distortion and cracking, caused by the transformation of austenite to martensite during rapid cooling of a steel through the martensite transformation range. It is an interrupted quenching process in which steel is quenched from the austenitizing temperature into hot oil or brine at a temperature near, but slightly above, the M_s temperature of the alloy. The steel is held in the quenching medium long enough for its temperature to stabilize, after which time it is removed and allowed to cool slowly in air. While the steel cools slowly, the transformation of austenite to martensite takes place. Due to the slow cooling rate, a relative uniform temperature is maintained throughout the mass of the steel, and the severe thermal gradients that are characteristic of conventional quenching are thereby avoided. The martensite forms at a uniform rate throughout the piece, and the stresses developed during transformation are much lower than for conventional quenching. The lower stresses induced by martempering, in turn, lessen the distortion of the treated part.

Martempering is normally used on steels of medium hardenability, such as those that are conventionally hardened by oil quenching. A modified martempering process consists of quenching from the austenitizing temperature to a temperature slightly below M_s. The higher cooling rates obtained by the lower temperature quench, in effect, allows the treatment of steels of lower hardenability that can be treated by the standard tempering process, and alloy steels are generally more suitable than carbon steels for martempering. An isothermal transformation diagram illustrating the martempering cycle is shown in *Figure 8*.

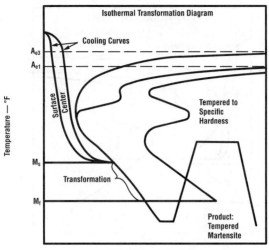

Figure 8. Schematic representation of the martempering cycle.

Austempering is used to isothermally transform austenite to lower bainite. Lower bainite compares favorably with tempered martensite with respect to strength and hardness and, at comparable hardness, it exhibits superior ductility. Austempering is, therefore, an alternate method of heat treating steels to develop high strength and hardness in combination with good ductility and toughness. The process consists of the following steps.

Heating the steel to a temperature within its austenitizing range, which usually occurs between 1,450 and 1,600° F (788 and 871° C).

Quenching the steel in a constant temperature bath at the desired transformation temperature in the lower bainite region, normally in the range of 500 to 750° F (260 to 399° C).

Holding the steel in the bath for a sufficient period of time to allow for the austenite to transform isothermally to bainite.

Cooling the steel to room temperature in still air.

Since austempering is an isothermal transformation process carried out at relatively high temperatures, transformation stresses induced by normal quenching are reduced, and distortion is minimized. It is usually substituted for conventional quenching and tempering in order to achieve higher ductility or notch toughness at a given hardness, and/or to decrease the distortion and cracking associated with normal quenching. *Figure 9* shows an isothermal transformation diagram for the austempering cycle.

Surface Hardening Steel

In many industrial applications it is necessary to develop a high surface hardness on a steel part so that it can resist wear and abrasion. This can be achieved by increasing the hardness of a high carbon steel, but high hardness is then accompanied by low ductility and toughness. In many applications, the poor ductility and toughness cannot be tolerated throughout the entire part, and another solution to the problem must be found.

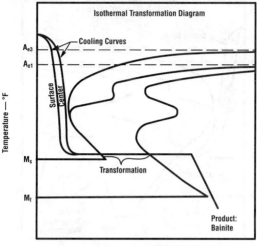

Figure 9. Schematic representation of the austempering cycle.

Low carbon steels can be treated to develop a hard surface, or "case," while the interior of the steel, or core, is unaffected and retains its normal ductility and toughness. Surface hardening (or case hardening) processes may be divided into two classifications, 1) those in which the composition of the surface materials must be changed, and 2) those in which the composition of the surface materials is not changed. Carburizing, nitriding, carbonitriding, and cyaniding processes are used to satisfy the first classification, and flame hardening and induction hardening can be used to create the characteristics of the second classification.

Carburizing is usually applied to plain carbon or low alloy steels with less than 0.20% carbon. Low alloy steels are heated in contact with a carbonaceous material to develop a surface layer on the steel that has a high carbon content. Upon quenching, the high carbon core remains relatively soft, resulting in a steel with a wear resistant exterior surface and a soft, tough core. There are three basic methods of carburizing steel: pack carburizing, gas carburizing, and liquid carburizing.

Pack carburizing is a process in which steel parts are placed in containers and carbonaceous solids (usually charcoal, or coke mixed with a suitable energizing catalyst such as barium carbonate, potassium carbonate carbonate, or sodium carbonate) are packed around them. The carbon monoxide gas that forms is actually the carburizing medium. The chemical reaction is $3Fe + 2CO \rightarrow Fe_3C + CO_2$.

The carbon atoms diffuse into the steel, which, for carburizing, is heated above the critical range, and goes into solid solution in the austenite. The temperature range for pack carburizing is from 1,500 to 1,750° F (815 to 954° C), with the temperature selected being about 100° F (56° C) above the A_{c3} point of the alloy being treated. The depth of core and the carbon concentration gradient of carburized parts are both governed by the carburizing temperature, time, and the original composition of the steel, but depths are usually at least 0.04 inch (1.0160 mm), and up to 0.060 inch

(1.5240 mm) or more can be obtained.

Gas carburizing exposes the steel parts to carburizing gases such as methane, propane, butane, or from vaporized hydrocarbon liquids. The process can be performed in either batch or continuous furnaces at temperatures from 1,550 to 1,750° F (843 to 954° C), with 1,700° F (927° C) being a popular choice as it allows for rapid carburizing with minimum wear to furnace components. Case depths of between 0.10 and 0.40 inch (0.254 and 1.0160 mm) are common, with depth being dependent on time and temperature. Gas carburizing is generally more versatile and efficient than pack carburizing.

Liquid carburizing immerses the steel in a molten salt bath containing barium cyanide or sodium cyanide at temperatures ranging from 1,550 to 1,700° F (843 to 927° C) for depths up to 0.030 inch (0.7620 mm), but higher temperature baths can penetrate depths up to 0.250 inch (6.35 mm) in some circumstances. The primary application of liquid carburizing is to rapidly induce case depths of between 0.040 and 0.080 inch (1.0160 and 2.0320 mm). The case usually absorbs some nitrogen in addition to carbon, which enhances surface hardness.

Vacuum carburizing can provide good uniformity and reliability, but it requires a vacuum furnace. In vacuum carburizing, austenitization takes place in a vacuum, carburization is induced by introducing hydrocarbon gas under partial pressure, and quenching is done in either oil or gas. It is performed as follows. The first step is to bring the steel to carburization temperature, typically 1,550 to 1900° F (843 to 1038° C), and hold it until it has reached a uniform temperature throughout. This is performed in a vacuum of 0.30 to 0.50 torr (40 to 67 Pa). Next, hydrocarbon gas (usually methane or propane, or a mixture) is added to the atmosphere under partial pressure (usually between 10 and 50 torr (1.3 and 6.6 kPa) for graphite furnaces, and 100 to 200 torr (13 to 25 kPa) in ceramic furnaces for a period of approximately one-half hour. A partial vacuum of 0.50 to 1.0 torr (67 to 135 Pa) is then reintroduced, and the steel is held at carburiding temperature for another half-hour or longer while the carbon diffuses inward from the surface. Finally, the steel is quenched in oil, which normally takes place under a partial pressure of nitrogen.

Plasma or *Ion carburizing* requires an oxygen-free vacuum and an electric flow between the part (the cathode) and a counter electrode (the anode). It normally takes place in a methane gas atmosphere of 1 to 20 torr (0.20 and 2.7 kPa), and at temperatures 1,650 to 1,830° F (899 to 1000° C). The process is essentially insensitive to the composition of the steel, provides uniform treatment results, and, because the glow discharge from the anode accelerates ions bearing carbon directly toward the steel surface, it is fast.

Many carbon and low alloy steels are carburized, although steels containing between 0.15 and 0.20% carbon are usually selected. The heat treatment selected to carburize the steel will depend on the carburizing temperature, composition of the core and case, and the properties that the steel must obtain. In some cases, the steel may be quenched directly from the carburizing temperature and then retempered, but in some instances it is desirable to slowly cool the steel from the carburizing temperature and then reheat it to a temperature slightly above Ae$_4$, and then quench and retemper. Double reheat and quench operations are also employed for some steels. In some circumstances it is necessary to develop a case locally, and not over the entire surface of a part. In such cases, carburization can be preserved in local surface areas by protecting those areas with copper plating, or covering them with a copper bearing commercial paste, thereby allowing the carbon to penetrate only the exposed areas. Another effective method is to machine the part after carburization to remove the case from those areas where a soft surface is required.

Gas nitriding is a surface hardening process that employs nitrogen to induce the formation of nitrides on the case. For successful nitriding, it is necessary to use alloy steels that contain aluminum, chromium, molybdenum, tungsten, and/or vanadium, which are elements that combine with nitrogen to form hard nitrides. The nitriding medium is ammonia gas, and the process temperature is 900 to 1,100° F (482 to 593° C). Case depth is dependent on time and temperature, but is usually in the range of 0.015 to 0.020 inch (0.3810 to 0.5080 mm). A significant advantage is that parts may be hardened, tempered, *and* machined prior to nitriding, because the process does not require post tempering or quenching that can impair dimensional instability or cause distortion. Nitrided parts exhibit exceptional wear resistance qualities and have very little tendency to gall and seize, making them especially well suited for applications involving metal-to-metal wear. Additional positive effects are high fatigue resistance and improved corrosion resistance.

Liquid nitriding uses a molten salt bath containing either cyanides or cyanates, heated to 900 to 1,100° F (same range as for gas nitriding). The process is very similar to gas nitriding, and treated parts exhibit nearly identical qualities. Parts to be treated must have the elements necessary for nitriding.

Plasma or *Ion nitriding* uses the same principles as plasma carburizing, which is discussed above. It is performed at relatively low temperatures (700° F [371° C]) because the glow discharge of the anode to the grounded workpiece introduces nitrogen directly to its surface.

Carbonitriding exposes carbon and alloy steels to a gaseous atmosphere from which they simultaneously absorb carbon and nitrogen. The process is a modified gas carburizing process in which ammonia is introduced into the gas carburizing atmosphere. Operating temperatures range from 1,425 to 1,650° F (774 to 899° C) for parts that are to be quenched, to 1,200 to 1,450° F (649 to 788° C) when a liquid quench is not required. Because nitrogen increases the hardenability of the case, full hardness can be achieved by less severe quenching, and distortion is minimized. Case depths of 0.003 to 0.0.0 inch (0.0762 to 0.7620 mm) are developed.

Cyaniding, sometimes called *liquid carbonitriding*, is similar to liquid carburizing except that the molten salt bath contains higher percentages of sodium cyanide, ranging from 30 to 97%. The steel absorbs both carbon and nitrogen from the molten bath and, by controlling the bath composition and operating temperature, it is possible to regulate, within limits, the relative amount of carbon and nitrogen in the case. Operating temperatures range from 1,400 to 1,600° F (760 to 871° C), just above Ac_1 for the steel being treated. The case is hard and somewhat brittle (due to the presence of nitrides), and rather thin, usually less than 0.010 inch (0.2540 mm) deep. Cyanided parts are extremely wear resistant.

Flame hardening consists of rapidly heating the surface area to be hardened with a flame to a temperature above the upper critical temperature of the alloy. With oxy-gas fuels hardness can be induced to a depth of less than 0.125 inch (3.18 mm). The hardened surface is then quenched in a suitable manner. Heating and cooling cycles are precisely controlled to attain the desired depth of hardening consistency. Steels suitable for flame hardening are usually in the range of 0.30 to 0.60% carbon, with hardenability appropriate for the depth to be hardened and the quenchant used. Various quenching media can be used, and are usually sprayed directly on the surface a short distance behind the flame. Immediate tempering is required to avoid cracking caused by residual stresses, and may be done by conventional furnace tempering or flame tempering.

Induction hardening uses an induced electrical current to heat steel above the upper critical temperature. When the high frequency alternating current is sent through a coil (or inductor) surrounding a steel part (or conductor) a magnetic field is created that heats the part by induced energy. Heating results primarily from the resistance of the part to the flow of currents created by the induced voltage, and from hysteresis losses caused by the rapidly alternating magnetic field if the part is magnetic. Therefore, most plain carbon and alloy steels heat most rapidly below the Curie point (approximately the upper critical temperature–the Curie point is the temperature that marks the transition between ferromagnetism and paramagnetism, or between the ferroelectric phase and the paraelectric phase), where they are ferromagnetic, and less rapidly above that point.

With conventional induction generators, heat is developed primarily on the surface of the part, and the total depth of heating is dependent on the frequency of the alternating current passing through the coil, the rate at which heat is conducted from the surface to the interior, and the length of the heating cycle. For surface hardening, frequencies of 10,000 to 50,000 cycles per second, using high power and short heating cycles, are used, while lower frequencies and long heating cycles are preferred for through hardening by induction. Quenching may be done with a water spray, but oil quenching is also possible by immersing the part in an oil bath after reaching hardening temperature.

The hardness in induction heating is, as in conventional heating, a function of the carbon content of the steel, and higher hardness values for a given carbon content can be obtained for induction surface hardened parts. The increment of added hardness may be as much as 5 R_C points for steels containing 0.30% carbon, and less for those with lower carbon content. Conventional hardening temperatures can generally be used when induction heating plain carbon grades, and alloy grades containing carbide forming elements such as chromium, molybdenum, and vanadium. The hardening temperature must be increased, however, if the normal influence of the alloying element is desired. Increased hardening temperatures do not increase the austenitic grain size, because grain growth is inhibited by the undissolved carbides. Generally, steels heated to conventional hardening temperatures by induction heating show a similar or somewhat finer grain size than steels heated in the furnace for hardening.

Nonferrous Metals

Metals that contain no iron are known as nonferrous. Seven of these nonferrous elements are produced as pure metals or alloys in significant quantities: aluminum, copper, lead, magnesium, nickel, tin, and zinc.

Aluminum

Aluminum alloys have found wide acceptance in engineering design primarily because they are relative lightweight, have a high strength to weight ratio, have superior corrosion resistance, and they are comparatively inexpensive. For some applications they are favored because of their high thermal and electrical conductivity, ease of fabrication, and ready availability. In fact, aluminum is the fourth most widely distributed of the elements, following oxygen, nitrogen, and silicon. Aluminum alloys weigh approximately 0.1 pound per cubic inch, which is one-third the weight of iron (0.28 lb/in.^{3}), and copper (0.32 lb/in.^{3}). It is just slightly heavier than magnesium (0.066 lb/in.^{3}) and somewhat lighter than titanium (0.163 lb/in.^{3}). In its commercially pure state, aluminum is a relatively weak metal, having a tensile strength of approximately 12,000 PSI. However, with the addition of such alloying elements as manganese, silicon, copper, magnesium, and/or zinc, and with proper heat treatment or cold working, the tensile strength of aluminum can approach 100,000 PSI. Physical property ranges for aluminum are given in **Table 1**.

Generally, the strengths of aluminum alloys decrease and toughness increases with increases in temperature, and with time at temperature above room temperature. The effect is usually greatest over the temperature range of 212 to 400° F (100 to 204° C). Exceptions to the general trends are tempers developed by solution heat treating without subsequent aging, for which the initial elevated temperature exposure results in some age hardening and reduction in toughness. Further time at temperature beyond that required to achieve peak hardness results in the aforementioned decrease in strength and increase in toughness.

Corrosion resistance of aluminum is attributable to its self-healing nature; a thin, invisible skin of aluminum oxide forms when the metal is exposed to the atmosphere. Pure aluminum will form a continuous protective oxide film, while high strength alloyed forms will sometimes become pitted as a result of localized galvanic corrosion at sites of alloying-constituent concentration. As a conductor of electricity, aluminum compares favorably with copper. Although the conductivity of the electroconductor grade of aluminum is only 62% that of the International Annealed Copper Standard (IACS), on a pound for pound basis the power loss for aluminum is less than half that of copper—an advantage where weight and space are primary requirements. As a heat conductor, aluminum ranks high among the metals. It is especially useful in heat exchangers and other applications requiring rapid heat dissipation. As a reflector of radiant energy, aluminum is excellent throughout the entire range of wavelengths—from the ultraviolet end of the spectrum through the visible and infrared bands, to the electromagnetic wave frequencies of radio and radar.

Alclad sheet (sometimes called clad sheet) is a product consisting of an aluminum alloy sheet having on one or both surfaces a layer of aluminum or aluminum alloy integrally bonded to the surface of the base metal. In general, alclad sheets have mechanical properties slightly lower than those of the bare alloy sheets of the same thickness. However, the corrosion resistant qualities of the aluminum alloy sheet are improved by the cladding.

Aluminum is easily fabricated. It can be cast by any method, rolled to any reasonable thickness, stamped, hammered, forged, or extruded. It is readily turned, milled, bored, or machined. It can be joined by several welding processes (see the Welding section

Table 1. Physical Property Ranges for Wrought and Cast Aluminum Alloys. *(Source, MIL-HDBK-694A.)*

Property	Range		Notes	
	Wrought Alloys	**Cast Alloys**		
Specific Gravity	2.70 to 2.82	2.57 to 2.95	About one-third that of steel.	
Weight (lb/cubic in.)	0.095 to 0.102	0.093 to 0.107	Approximately 173 lb./cubic ft	
Electrical Conductivity (Int'l. Annealed Copper Std.)	30% to 60%	21% to 47%	About 59% for 99.9% Al	Values for electrical and thermal conductivity depend on the composition and condition of the alloys. Both are increased by annealing, and decreased by adding alloying elements to pure aluminum. Both are also decreased by heat treatment, cold work, and aging.
Thermal Conductivity (cgs units at 77° F)	0.29 to 0.56	0.21 to 0.40	About 0.53 for 99% Al	
Coefficient of Thermal Expansion	10.8 to 13.2	11.0 to 14.0	Roughly double that of ordinary steels and cast irons; substantially greater than copper-alloy materials. Alloying elements other than silicon have little effect on the expansion of aluminum. Considerable amounts of silicon (12%) appreciably decrease the dimensional changes induced by varying temperatures. Where a low coefficient of thermal expansion is desirable (as in engine pistons) an aluminum alloy containing a relatively high percentage of silicon may be specified.	
Reflectivity	See Notes		Greater than for any other metal. Suitably treated, aluminum sheet of high purity may yield a reflectivity for light greater than 80%.	
Modulus of Elasticity	In the range of 10.0 to 10.6 PSI ×10⁶ (69 to 73 GPa) for wrought alloys. In the range of 10.0 to 11.9 PSI × 10⁶ (69 to 82 GPa) for casting alloys.			
Modulus of Rigidity	3.9×10^6 (average)		Average, all alloys.	
Poisson's Ratio	0.33		Average, all alloys.	
Ultimate Torsional Strength (% of Ultimate Tensile Strength)	65		Average all alloys.	

of this book for guidelines on welding, brazing, and soldering aluminum). Aluminum can also be coated with a wide variety of surface finishes for decorative and protective purposes.

Types of Aluminum

The Aluminum Association Alloy and Temper Designation System is used to identify wrought and casting alloys. With few exceptions, aluminum alloys are designed either for casting or for use in wrought products, but not for both—some general purpose alloys are available but, on the whole, compositions are formulated to satisfy specific requirements. The first digit in the alloy designation system identifies the principal alloying constituent of the metal. For the wrought alloys, the second digit notes variations of the initial alloy. For the casting alloys, the second digit, together with the third, are used to set the designation off from others—even though they have no reference to a specific alloy, these numbers make the designation unique. For wrought alloys, the third and fourth digits serve the same purpose as the second and third for the casting alloys—they are chosen to make the designation unique. The fourth digit on the casting alloys designation indicates casting (0) or ingot (1.2). Variations of similar casting alloys are preceded by A, B, or C. Tempers use a letter plus number designation system. The four-digit designation system, plus the letters of the temper designation system, is detailed in **Table 2**. The letter-number combination designations used for tempering are detailed later in this section. See **Table 3** for chemical compositions of wrought alloys, **Table 4** for their mechanical properties, and **Table 5** for their physical properties. Chemical compositions for casting alloys are given in **Table 6**. Mechanical properties of sand casting alloys are provided in **Table 7**, and the same information for permanent mold casting alloys appears in **Table 8**. Physical properties of casting alloys are shown in **Table 9**.

Table 2. Aluminum Association Alloy and Temper Designation System.

Wrought Alloy System	Cast Alloy System	Temper System
1xxx – Pure Al (99.00% or greater)	1xx.x – Pure Al (99.00% or greater)	F – As fabricated
2xxx – Al-Cu Alloys	2xx.x – Al-Cu Alloys	O – Annealed
3xxx – Al-Mn Alloys	3xx.x – Al-Si + Cu and/or Mg	H – Strain hardened (wrought alloys only)
4xxx – Al-Si Alloys	4xx.x – Al-Si	W – Solution heat treated
5xxx – Al-Mg Alloys	5xx.x – Al-Mg	T – Thermally treated to produce tempers other than F, O, H (usually solution heat treated, quenched, and precipitation hardened)
6xxx – Al-Mg-Si Alloys	7xx.x – Al-Zn	
7xxx – Al-Zn Alloys	8xx.x – Al-Sn	
8xxx– Al + Other Elements	9xx.x – Al + Other Elements	
9xxx – Unused	6xx.x – Unused	

Characteristics of Wrought Alloys by Series Designations

1xxx Series – Pure Aluminum. This series is made up of commercially pure aluminum, ranging from the baseline 1100 (99.00% minimum Al) to relatively purer 1050/1350 (99.50% min.) and 1175 (99.75% min.). Some alloys in this series, such as 1350, which is used especially for electrical applications, have relatively tight controls on those impurities that might impair performance of the alloy for its intended use. The 1xxx series are strain hardenable, but would not be used where strength is a prime consideration. Instead, the emphasis would be on those applications where extremely high corrosive resistance, high formability, and/or good electrical conductivity are required. Examples are foil and strip for packaging, chemical equipment, tank, car, or truck bodies, spun hollowware, and elaborate sheet metal work.

2xxx Series – Al-Cu Alloys. The alloys in this series are heat treatable and possess, in individual alloys, good combinations of high strength (especially at elevated temperatures), toughness, and, in specific cases, weldability. They are not resistant to atmospheric corrosion, so painting or cladding is suggested to resist environmental exposure. The higher strength 2xxx series are used primarily for aircraft (2024 is an example) and truck bodies (2014), usually in bolted or riveted construction. Specific members of this series (2219 and 2048) are readily welded and are often used in aerospace applications where welding is the preferred joining method. Several of these alloys have been developed for the aircraft industry and have high toughness (e.g., 2124, 2324, 2419), and higher control on any impurities that may diminish their resistance to unstable fracture. Alloys 2011, 2017, and 2117 are widely used for fasteners and screw machine stock. A recent addition to the 2xxx series is 2195, a "weldalite" alloy that, with 2219 and 2419, has been used for fuel tanks and booster rockets on the space shuttle. As a group, these alloys are noteworthy for their excellent strengths at elevated and cryogenic temperatures, and creep resistance at elevated temperatures. Typically, 2xxx alloys have a tensile strength range of 27,000 to 62,000 PSI (186 to 427 MPa).

(Text continued on p. 526)

Table 3. Chemical Composition of Wrought Aluminum and Aluminum Alloys. *(Source, Aluminum Association, Inc.)*

Alloy	Si	Fe	Cu	Mn	Mg	Cr	Zn	Ti	V	Other (Specified)	Other (Unspecified) Each	Total	Al
1050	0.25	0.40	0.05	0.05	0.05	-	0.05	0.03	0.05	-	0.03	-	99.50 min.
1060	0.25	0.35	0.05	0.03	0.03	-	0.05	0.03	0.05	-	0.03	-	99.60 min.
1100	0.95 Si + Fe	Si + Fe	0.05 - 0.20	0.05	-	-	0.10	-	-	-	0.05	0.15	99.0 min.
1200	1.00 Si + Fe	Si + Fe	0.05	0.05	-	-	0.10	0.05	-	-	0.05	0.15	99.0 min.
1230	0.70 Si + Fe	Si + Fe	0.10	0.05	0.05	-	0.10	0.03	0.05	-	0.03	-	99.30 min
1235	0.65 Si + Fe	Si + Fe	0.05	0.05	0.05	-	0.10	0.06	0.05	-	0.03	-	99.35 min.
1145	0.55 Si + Fe	Si + Fe	0.05	0.05	0.05	-	0.05	0.03	0.05	-	0.03	-	99.45 min.
1345	0.30	0.40	0.10	0.05	0.05	-	0.05	0.03	0.05	-	0.03	-	99.45 min.
1350	0.10	0.40	0.05	0.01	-	0.01	0.05	0.02 V + Ti	-	0.03 Ga	0.03	0.10	99.50 min.
1175	0.15 Si + Fe	Si + Fe	0.10	0.02	0.02	-	0.04	0.02	0.05	0.03 Ga	0.02	-	99.75 min.
2011	0.40	0.7	5.0-6.0	-	-	-	0.30	-	-	-	0.05	0.15	Balance
2014	0.50-1.2	0.7	3.9-5.0	0.40-1.2	0.20-0.8	0.10	0.25	0.15	-	-	0.05	0.15	Balance
2017	0.20-0.8	0.7	3.5-4.5	0.40-1.0	0.40-0.8	0.10	0.25	0.15	-	-	0.05	0.15	Balance
2117	0.8	0.7	2.2-3.0	0.20	0.20-0.50	0.10	0.25	-	-	-	0.05	0.15	Balance
2018	0.9	1.0	3.5-4.5	0.20	0.45-0.9	0.10	0.25	-	-	1.7-2.3 Ni	0.05	0.15	Balance
2218	0.9	1.0	3.5-4.5	0.20	1.2-1.8	0.10	0.25	-	-	1.7-2.3 Ni	0.05	0.15	Balance
2219	0.20	0.30	5.8-6.8	0.20-0.40	0.02	-	0.10	0.02-0.10	0.05-0.15	-	0.05	0.15	Balance
2419	0.15	0.18	5.8-6.8	0.20-0.40	0.02	-	0.10	0.02-0.10	0.05-0.15	-	0.05	0.15	Balance
2024	0.50	0.50	3.8-4.9	0.30-0.9	1.2-1.8	0.10	0.25	0.15	-	-	0.05	0.15	Balance
2124	0.20	0.30	3.8-4.9	0.30-0.9	1.2-1.8	0.10	0.25	0.15	-	-	0.05	0.15	Balance
2025	0.50-1.2	1.0	3.9-5.0	0.40-1.2	0.05	0.10	0.25	0.15	-	-	0.05	0.15	Balance
2036	.050	0.50	2.2-3.0	0.10-0.4	0.30-0.6	0.10	0.25	0.15	-	-	0.05	0.15	Balance
2048	0.15	0.20	2.8-3.8	0.20-0.6	1.2-1.8	-	0.25	0.10	-	-	0.05	0.15	Balance

(Continued)

Table 3. (Continued) Chemical Composition of Wrought Aluminum and Aluminum Alloys. (Source, Aluminum Association, Inc.)

Alloy	Si	Fe	Cu	Mn	Mg	Cr	Zn	Ti	V	Other (Specified)	Other (Unspecified) Each	Total	Al
2195	0.12	0.15	3.7-4.3	0.25	0.25-0.8	-	0.25	0.10	-	-	0.05	0.15	Balance
3003	0.6	0.7	0.05-0.20	1.0-1.5	-	-	0.10	-	-	-	0.05	0.15	Balance
3004	0.30	0.7	0.25	1.0-1.5	0.8-1.3	-	0.25	-	-	-	0.05	0.15	Balance
3104	0.6	0.8	0.05-0.25	0.8-1.4	0.8-1.3	-	0.25	0.10	0.05	0.05 Ga	0.05	0.15	Balance
3005	0.6	0.7	0.30	1.0-1.5	0.20-0.6	0.10	0.25	0.10	-	-	0.05	0.15	Balance
3105	0.6	0.7	0.30	0.30-0.8	0.20-0.8	0.20	0.40	0.10	-	-	0.05	0.15	Balance
4032	11.0-13.5	1.0	0.50-1.3	-	0.8-1.3	0.10	0.25	-	-	0.50-1.3 Ni	0.05	0.15	Balance
4043	4.5-6.0	0.8	0.30	0.05	0.05	-	0.10	0.20	-	-	0.05	0.15	Balance
4045	9.0-11.0	0.8	0.30	0.05	0.05	-	0.10	0.20	-	-	0.05	0.15	Balance
4047	11.0-13.0	0.8	0.30	0.15	0.10	-	0.20	-	-	-	0.05	0.15	Balance
5005	0.30	0.7	0.20	0.20	0.50-1.1	0.10	0.25	-	-	-	0.05	0.15	Balance
5050	0.40	0.7	0.20	0.10	1.1-1.8	0.10	0.25	-	-	-	0.05	0.15	Balance
5052	0.25	0.40	0.10	0.10	2.2-2.8	0.15-0.35	0.10	-	-	-	0.05	0.15	Balance
5252	0.08	0.10	0.10	0.10	2.2-2.8	-	0.05	-	0.05	-	0.03	0.10	Balance
5154	0.25	0.40	0.10	0.10	3.1-3.9	0.15-0.35	0.20	0.20	-	-	0.05	0.15	Balance
5254	0.45 Si + Fe	Si + Fe	0.05	0.01	3.1-3.9	0.15-0.35	0.20	0.05	-	-	0.05	0.15	Balance
5454	0.25	0.40	0.10	0.50-1.0	2.4-3.0	0.05-0.20	0.25	0.20	-	-	0.05	0.15	Balance
5754	0.40	0.40	0.10	0.50	2.6-3.6	0.30	0.20	0.15	-	-	0.05	0.15	Balance
5056	0.30	0.40	0.10	0.05-0.20	4.5-5.6	0.05-0.20	0.10	0.06-0.20	-	-	0.05	0.15	Balance
5356	0.25	0.40	0.10	0.05-0.20	4.5-5.5	0.05-0.20	0.10	0.06-0.20	-	-	0.05	0.15	Balance
5456	0.25	0.40	0.10	0.50-1.0	4.7-5.5	0.05-0.20	0.25	0.20	-	-	0.05	0.15	Balance
5556	0.25	0.40	0.10	0.50-1.0	4.7-5.5	0.05-0.20	0.25	0.05-0.20	-	-	0.05	0.15	Balance
5457	0.08	0.10	0.20	0.15-0.45	0.8-1.2	-	0.05	-	0.05	-	0.03	0.10	Balance
5657	0.08	0.10	0.10	0.03	0.6-1.0	-	0.05	-	0.05	0.03 Ga	0.02	0.05	Balance

(Continued)

Table 3. *(Continued)* Chemical Composition of Wrought Aluminum and Aluminum Alloys. *(Source, Aluminum Association, Inc.)*

Alloy	Si	Fe	Cu	Mn	Mg	Cr	Zn	Ti	V	Other (Specified)	Other (Unspecified) Each	Other (Unspecified) Total	Al
5182	0.20	0.35	0.15	0.20-0.50	4.0-5.0	0.10	0.25	0.10	-	-	0.05	0.15	Balance
5083	0.40	0.40	0.10	0.40-1.0	4.0-4.9	0.05-0.25	0.25	0.15	-	-	0.05	0.15	Balance
5086	0.40	0.50	0.10	0.20-0.7	3.5-4.5	0.05-0.25	0.25	0.15	-	-	0.05	0.15	Balance
6101	0.30-0.7	0.50	0.10	0.03	0.35-0.8	0.03	0.10	-	-	-	0.03	0.10	Balance
6201	0.50-0.9	0.50	0.10	0.03	0.6-0.9	0.03	0.10	-	-	-	0.03	0.10	Balance
6111	0.6-1.1	0.40	0.50-0.9	0.10-0.45	0.50-1.0	0.10	0.15	0.10	-	-	0.10	0.05	Balance
6351	0.7-1.3	0.50	0.10	0.40-0.8	0.40-0.8	-	0.20	0.20	-	-	0.20	0.05	Balance
6061	0.40-0.8	0.7	0.15-0.40	0.15	0.8-1.2	0.04-0.35	0.25	0.15	-	-	0.05	0.15	Balance
6262	0.40-0.8	0.7	0.15-0.40	0.15	0.8-1.2	0.04-0.14	0.25	0.15	-	-	0.05	0.15	Balance
6063	0.20-0.6	0.35	0.10	0.10	0.45-0.9	0.10	0.10	0.10	-	-	0.05	0.15	Balance
6463	0.20-0.6	0.15	0.20	0.05	0.45-0.9	-	0.05	-	-	-	0.05	0.15	Balance
6066	0.9-1.8	0.50	0.7-1.2	0.6-1.1	0.8-1.4	0.40	0.25	0.20	-	-	0.05	0.15	Balance
6070	1.0-1.7	0.50	0.15-0.40	0.40-1.0	0.50-1.2	0.10	0.25	0.15	-	-	0.05	0.15	Balance
7049	0.25	0.35	1.2-1.9	0.20	2.0-2.9	0.10-0.22	7.2-8.2	0.10	-	-	0.05	0.15	Balance
7050	0.12	0.15	2.0-2.6	0.10	1.9-2.6	0.04	5.7-6.7	0.06	-	-	0.05	0.15	Balance
7150	0.12	0.15	1.9-2.5	0.10	2.0-2.7	0.04	5.9-6.9	0.06	-	-	0.05	0.15	Balance
7075	0.40	0.50	1.2-2.0	0.30	2.1-2.9	0.18-0.28	5.1-6.1	0.20	-	-	0.05	0.15	Balance
7175	0.15	0.20	1.2-2.0	0.10	2.1-2.9	0.18-0.28	5.1-6.1	0.10	-	-	0.05	0.15	Balance
7475	0.10	0.12	1.2-1.9	0.06	1.9-2.6	0.18-0.25	5.2-6.2	0.06	-	-	0.05	0.15	Balance
7178	0.40	0.50	1.6-2.4	0.30	2.4-3.1	0.18-0.28	6.3-7.3	0.20	-	-	0.05	0.15	Balance
8017	0.10	0.55-0.8	0.10-0.20	-	0.01-0.05	-	0.05	-	-	-	0.03	0.10	Balance
8030	0.10	0.30-0.8	0.15-0.30	-	0.05	-	0.05	-	-	-	0.03	0.10	Balance
8176	0.03-0.15	0.40-1.0	-	-	-	-	0.10	-	-	0.03 Ga	0.05	0.15	Balance
8177	0.10	0.25-0.45	0.04	-	0.04-0.12	-	0.05	-	-	-	0.03	0.10	Balance

Table 4. Mechanical Properties of Wrought Aluminum Alloys. (Source, Aluminum Association, Inc.)

Alloy	Temper	Ultimate Tensile Strength		Tensile Yield Strength		Ultimate Shearing Strength		Fatigue Endurance Limit		Elongation %[1]		Brinell Hardness
		ksi	MPa	ksi	MPa	ksi	MPa	ksi	MPa	0.0625" Thick	0.50" Dia.	BHN
1060	O	10	69	4	28	7	48	3	21	43	-	19
	H18	19	131	18	124	11	76	6.5	45	6	-	35
1100	O	13	90	5	34	9	62	5	34	35	45	23
	H18	24	165	22	152	13	90	9	62	5	15	44
1350	O	12	83	4	28	8	55	-	-	-	(d)	-
2011	T3	55	379	43	296	32	221	18	124	-	15	95
2014	O	27	186	14	97	18	124	13	90	-	18	45
	T4, T451	62	427	42	290	38	262	20	138	-	20	105
	T6, T651	70	483	60	414	42	290	18	124	-	13	135
2017	O	26	179	10	69	18	124	13	90	-	22	45
	T4, T451	61	421	37	255	37	255	-	-	22	-	-
2018	T61	61	421	46	317	39	269	17	117	-	12	120
2024	O	27	186	11	76	18	124	13	90	20	22	47
	T4, T351	68	469	47	324	41	283	20	138	20	19	120
2025	T6	58	400	37	255	35	241	18	124	-	19	110
2117	T4	43	296	24	165	28	193	14	97	-	27	70
2218	T72	48	331	37	255	30	207	-	-	-	11	95
2219	O	25	172	11	76	-	-	-	-	18	-	-
	T31, T351	52	359	36	248	-	-	-	-	17	-	-
	T37	57	393	46	317	-	-	-	-	11	-	-
	T62	60	414	42	290	-	-	15	103	10	-	-

(Continued)

Table 4. *(Continued)* **Mechanical Properties of Wrought Aluminum Alloys.** *(Source, Aluminum Association, Inc.)*

Alloy	Temper	Ultimate Tensile Strength		Tensile Yield Strength		Ultimate Shearing Strength		Fatigue Endurance Limit		Elongation %[1]		Brinell Hardness
		ksi	MPa	ksi	MPa	ksi	MPa	ksi	MPa	0.0625" Thick	0.50" Dia.	BHN
2219	T81, T851	66	455	51	352	-	-	15	103	10	-	-
	T87	69	476	57	393	-	-	15	103	10	-	-
3003	O	16	110	6	41	11	76	7	48	30	40	28
	H12	19	131	18	124	12	83	8	55	10	20	35
	H18	29	200	27	186	16	110	19	131	4	10	55
3004	O	26	179	10	69	16	110	14	97	20	25	45
	H32	31	214	25	172	17	117	15	103	10	17	52
	H34	35	241	29	200	18	124	15	103	9	12	63
	H36	38	262	33	228	20	138	16	110	5	9	70
	H38	41	283	36	248	21	145	16	110	5	6	77
4032	T6	55	379	46	317	38	262	16	110	-	9	120
5005	O	18	124	6	41	11	76	-	-	25	-	28
	H12	20	138	19	131	14	97	-	-	10	-	-
	H14	23	159	22	152	14	97	-	-	6	-	-
	H16	26	179	25	172	15	103	-	-	5	-	-
	H18	29	200	28	193	16	110	-	-	4	-	-
	H32	20	138	17	117	14	97	-	-	11	-	36
	H34	23	159	20	138	14	97	-	-	8	-	41
	H36	26	179	24	165	15	103	-	-	6	-	46
	H38	29	200	27	186	16	110	-	-	5	-	51
5050	O	21	145	8	55	15	103	12	83	24	-	36

(Continued)

Table 4. (Continued) Mechanical Properties of Wrought Aluminum Alloys. (Source, Aluminum Association, Inc.)

Alloy	Temper	Ultimate Tensile Strength		Tensile Yield Strength		Ultimate Shearing Strength		Fatigue Endurance Limit		Elongation %[1]		Brinell Hardness
		ksi	MPa	ksi	MPa	ksi	MPa	ksi	MPa	0.0625" Thick	0.50" Dia.	BHN
5050	H32	25	172	21	145	17	117	13	90	9	-	46
	H34	28	193	24	165	18	124	13	90	8	-	53
	H36	30	207	26	179	19	131	14	97	7	-	58
	H38	32	221	29	200	20	138	14	97	6	-	63
5052	O	28	193	13	90	18	124	16	110	25	30	47
	H32	33	228	28	193	20	138	17	117	12	18	60
	H34	38	262	31	214	21	145	18	124	10	14	68
	H36	40	276	35	241	23	159	19	131	8	10	73
	H38	42	290	37	255	24	165	20	138	7	8	77
5056	O	42	290	22	152	26	179	20	138	-	35	65
	H18	63	434	59	407	34	234	22	152	-	10	105
5083	O	42	290	21	145	25	172	-	-	-	22	-
5086	O	38	262	17	117	23	159	-	-	22	-	-
	H32, H116	42	290	12	83	-	-	-	-	12	-	-
	H34	47	324	37	255	27	186	-	-	10	-	-
	H112	39	269	19	131	-	-	-	-	14	-	-
5154	O	35	241	17	117	22	152	17	117	27	-	58
	H32	39	269	30	207	22	152	18	124	15	-	67
	H34	42	290	33	228	24	165	19	131	13	-	73
	H36	45	310	36	248	26	179	20	138	12	-	78
	H38	48	331	39	269	28	193	21	145	10	-	80

(Continued)

Table 4. (Continued) Mechanical Properties of Wrought Aluminum Alloys. (Source, Aluminum Association, Inc.)

Alloy	Temper	Ultimate Tensile Strength		Tensile Yield Strength		Ultimate Shearing Strength		Fatigue Endurance Limit		Elongation %[1]		Brinell Hardness
		ksi	MPa	ksi	MPa	ksi	MPa	ksi	MPa	0.0625" Thick	0.50" Dia.	BHN
5154	H112	35	241	17	117	-	-	17	117	25	-	63
5252	H25	34	234	25	172	21	145	-	-	11	-	68
	H38, H28	41	283	35	241	23	159	-	-	5	-	75
5254	O	35	241	17	117	22	152	17	117	27	-	58
	H32	39	269	30	207	22	152	18	124	15	-	67
	H34	42	290	33	228	24	165	19	131	13	-	73
	H36	45	310	36	248	26	179	20	138	12	-	78
	H38	48	331	39	269	28	193	21	145	10	-	80
	H112	35	241	17	117	-	-	17	117	25	-	63
5454	O	36	248	17	117	23	159	-	-	22	-	62
	H34	44	303	35	241	26	179	-	-	10	-	81
5456	O	45	310	23	159	-	-	-	-	-	24	-
5457	O	19	131	7	48	12	83	-	-	22	-	32
	H25	26	179	23	159	16	110	-	-	12	-	48
	H38, H28	30	207	27	186	18	124	-	-	6	-	55
5657	H25	23	159	20	138	14	97	-	-	12	-	40
	H38, H28	28	193	24	165	15	103	-	-	7	-	50
6061	O	18	124	8	55	12	83	9	62	25	30	30
	T4, T451	35	241	21	145	24	165	14	97	22	25	65
6063	O	13	90	7	48	10	69	8	55	-	-	25
	T4	25	172	13	90	-	-	-	-	22	-	-

(Continued)

Table 4. *(Continued)* **Mechanical Properties of Wrought Aluminum Alloys.** *(Source, Aluminum Association, Inc.)*

Alloy	Temper	Ultimate Tensile Strength		Tensile Yield Strength		Ultimate Shearing Strength		Fatigue Endurance Limit		Elongation %[1]		Brinell Hardness
		ksi	MPa	ksi	MPa	ksi	MPa	ksi	MPa	0.0625" Thick	0.50" Dia.	BHN
6063	T6	35	241	31	214	22	152	10	69	12	-	73
6066	O	22	152	12	83	14	97	-	-	-	18	43
	T6, T651	57	393	52	359	34	234	16	110	-	12	120
6101	T6	32	221	28	193	20	138	-	-	15	-	71
7049	T73	75	517	65	448	44	303	-	-	-	12	135
7075	T6, T651	83	572	73	503	48	331	23	159	11	11	150
7178	T6, T651	88	607	78	538	-	-	-	-	10	11	-

[1] Elongation is the percent elongation in two inch (50.8 mm) specimens—one $^1/_{16}$" thick (1.6 mm), the other $^1/_2$" thick (12.7 mm).

Table 5. Physical Properties of Wrought Aluminum Alloys. (Source, MIL-HDBK-694A.)

Alloy	Temper	Density lb/cu in.	Thermal Conductivity CGS[1]	Thermal Conductivity Eng.[2]	Coef. Thermal Expan. μin °F	Coef. Thermal Expan. μm °C	Electrical Conductivity[3]	Melting Point °F/°C Solidus	Melting Point °F/°C Liquidus[4]
1060	O	0.098	0.56	1625	13.1	23.6	62	1195/645	1215/655
	H18	0.098	0.53	1540			61	1195/645	1215/655
1100	O	0.098	0.53	1540	13.1	23.6	59	1190/643	1215/655
	H18	0.098	0.52	1510			57	1190/643	1215/655
2011	T3	0.102	0.34	990	12.7	22.9	39	1005/540	1190(d)/643(d)
2014	O	0.101	0.46	1340	12.8	23.0	50	945/507	1180(e)/638(e)
	T4	0.101	0.29	840			34	945/507	1180(e)/638(e)
	T6	0.101	0.37	1070			40	945/507	1180(e)/638(e)
2017	O	0.099	0.41	1190	13.2	23.6	50	955/513	1185(e)/640(e)
	T4	0.099	0.29	840			34	955/513	1185(e)/640(e)
2018	T61	0.101	0.37	1070	12.4	22.3	40	945/507	1180(d)/638(d)
2024	O	0.100	0.45	1310	12.9	23.2	50	935/500	1180(e)/638(e)
	T4	0.100	0.29	840			30	935/500	1180(e)/638(e)
2025	T6	0.101	0.37	1070	12.6	22.7	40	970/520	1185(e)/640(e)
2117	T4	0.099	0.37	1070	13.2	23.75	40	1030/555	1200(d)/650(d)
2218	T72	0.102	0.37	1070	12.4	22.3	40	940/505	1175(e)/635(e)
2219	O	0.103	0.41	1190	12.4	22.3	44	1010/543	1190(e)/643(e)
	T31, T37	0.103	0.27	780			28	1010/543	1190(e)/643(e)
	T62, T81, T87	0.103	0.30	870			30	1010/543	1190(e)/643(e)
3003	O	0.099	0.46	1340	12.9	23.2	50	1190/643	1210/655
	H12	0.099	0.39	1130			42	1190/643	1210/655
	H18	0.099	0.37	1070			41	1190/643	1210/655
3004	A11	0.098	0.39	1130	13.3	23.9	42	1165/630	1210/655

(Continued)

Table 5. *(Continued)* Physical Properties of Wrought Aluminum Alloys. *(Source, MIL-HDBK-694A.)*

Alloy	Temper	Density lb/cu in.	Thermal Conductivity CGS[1]	Thermal Conductivity Eng.[2]	Coef. Thermal Expan. μin °F	Coef. Thermal Expan. μm °C	Electrical Conductivity[3]	Melting Point °F/°C Solidus	Melting Point °F/°C Liquidus[4]
4032	T6	0.097	0.33	960	10.8	19.4	40	990/532	1060(e)/570(te)
5005	A11	0.098	0.48	1390	13.2	23.75	52	1170/632	1210/655
5050	A11	0.097	0.46	1340	13.2	23.75	50	1155/625	1205/650
5052	A11	0.097	0.33	960	13.2	23.75	35	1125/607	1200/650
5056	O	0.095	0.28	810	13.4	24.1	29	1055/568	1180/638
	H38	0.095	0.26	750			27	1055/568	1180/638
5083	O	0.096	0.28	810	13.2	23.75	29	1095/590	1180/638
5086	A11	0.096	0.30	870	13.2	23.75	31	1085/585	1185/640
5154	A11	0.096	0.30	870	13.3	23.9	32	1100/593	1190/643
5252	A11	0.097	0.33	960	13.2	23.75	35	1125/607	1200/650
5254	A11	0.096	0.30	870	13.3	23.9	32	1100/593	1190/643
5357	A11	0.098	0.40	1160	13.2	-	-	-	-
5454	O	0.097	0.32	930	13.1	23.6	34	1115/600	1195/645
	H38	0.097	0.32	930			34	1115/600	1195/645
5456	O	0.096	0.28	810	13.3	23.9	29	1055/568	1180/638
5457	A11	0.098	0.45	1310	13.1	23.75	46	1165/630	1210/655
5657	A11	0.098	0.33	960	13.1	23.75	54	1180/638	1215/657
6061	O	0.098	0.41	1190	13.1	23.6	47	1080/580	1205(d)/650(d)
	T4	0.098	0.37	1070			40	1080/580	1205(d)/650(d)
6063	T6	0.098	0.48	1390	13.0	23.4	58	1140/615	1210/655
	T42	0.098	0.46	1340			50	1140/615	1210/655
6066	O	0.098	0.37	1070	12.9	23.2	40	1045/565	1195(e)/645(e)
	T6	0.098	0.35	1010			37	1045/565	1195(e)/645(e)

(Continued)

Table 5. *(Continued)* Physical Properties of Wrought Aluminum Alloys. *(Source, MIL-HDBK-694A.)*

Alloy	Temper	Density lb/cu in.	Thermal Conductivity		Coef. Thermal Expan.		Electrical Conductivity[3]	Melting Point °F/°C	
			CGS[1]	Eng.[2]	μin °F	μm °C		Solidus	Liquidus[4]
6101	T6	0.098	0.52	1510	13.0	23.4	57	1150/620	1210/655
	T61	0.098	0.53	1540			59	1150/620	1210/655
	T62	0.098	0.52	1510			58	1150/620	1210/655
	T64	0.098	0.54	1570			60	1150/620	1210/655
7001	T6	0.102	0.29	840	13.0	-	-	-	1215/655
7072	O	0.098	0.53	1540	13.1	23.6	59	1185/640	-
7075	T6	0.101	0.29	840	13.1	23.6	3	890/475	1175(g)/635(g)
7178	T6	0.102	0.30	870	13.0	23.4	31	890/475	1165(g)/630(g)
7079	T6	0.099	0.29	840	13.1	-	-	-	-

Notes: [1] CGS = cal/cm2/°C/sec at 25° C (77° F).

   [2] Eng. = English Units = btu/in./ft2/°F/hour at 77° F.

   [3] At 68° F (20° C). International Annealed Copper Standard (IACS). Coefficient of Thermal Expansion is based on change in length per °F or °C, from 68 to 212° F (20 to 100° C).

   [4] Letter symbols represent points on phase equilibrium diagram for two component systems.

Table 6. Chemical Composition Limits of Aluminum Casting Alloys. (Source, Aluminum Association, Inc.)

Alloy	Use*	Si	Cu	Zn	Fe	Mg	Ni	Mn	Ti	Other (specified)	Others Each	Others Total
201.0	S	0.10	4.0-5.2	-	0.15	0.15-0.55	-	0.20-0.50	0.15-0.35	-	0.05	0.10
204.0	S&P	0.20	4.2-5.0	0.10	0.35	0.15-0.35	0.05	0.10	0.15-0.30	-	0.05	0.15
208.0	S&P	2.5-3.5	3.5-4.5	1.0	1.2	0.10	0.35	0.50	0.25	-	-	0.50
222.0	S&P	2.0	9.2-10.7	0.8	1.5	0.15-0.35	0.50	0.50	0.25	-	-	0.35
242.0	S	0.7	3.5-4.5	0.35	1.0	1.2-1.8	1.7-2.3	0.35	0.25	0.25 Cr	0.05	0.15
295.0	S	0.7-1.5	4.0-5.0	0.35	1.0	0.03	-	0.35	0.25	-	0.05	0.15
296.0	P	2.0-3.0	4.0-5.0	0.50	1.2	0.05	0.35	0.35	0.25	-	-	0.35
308.0	P	5.0-6.0	4.0-5.0	1.0	1.0	0.10	-	0.50	0.25	-	-	0.50
319.0	S&P	5.5-6.5	3.0-4.0	1.0	1.0	0.10	0.35	0.50	0.25	-	-	0.50
328.0	S	7.5-8.5	1.0-2.0	1.5	1.0	0.20-0.6	0.25	0.20-0.6	0.25	0.35 Cr	-	0.50
332.0	P	8.5-10.5	2.0-4.0	1.0	1.2	0.50-1.5	0.50	0.50	0.25	-	-	0.50
333.0	P	8.0-10.0	3.0-4.0	1.0	1.0	0.05-0.50	0.50	0.50	0.25	-	-	0.50
336.0	P	11.0-13.0	0.50-1.5	0.35	1.2	0.7-1.3	2.0-3.0	0.35	0.25	-	0.05	-
354.0	S&P	8.6-9.4	1.6-2.0	0.10	0.20	0.40-0.6	-	0.10	0.20	-	0.05	0.15
355.0	S&P	4.5-5.5	1.0-1.5	0.35	0.6	0.40-0.6	-	0.50	0.25	0.25 Cr	0.05	0.15
C355.0	S&P	4.5-5.5	1.0-1.5	0.10	0.20	0.40-0.6	-	0.10	0.20	-	0.05	0.15
356.0	S&P	6.5-7.5	0.25	0.35	0.6	0.20-0.45	-	0.35	0.25	-	0.05	0.15
A356.0	S&P	6.5-7.5	0.20	0.10	0.20	0.25-0.45	-	0.10	0.20	-	0.05	0.15
357.0	S&P	6.5-7.5	0.05	0.05	0.15	0.45-0.6	-	0.03	0.20	-	0.05	0.15
A357.0	S&P	6.5-7.5	0.20	0.10	0.20	0.40-0.7	-	0.10	0.04-0.20	-	0.05	0.15
359.0	S&P	8.5-9.5	0.20	0.10	0.20	0.50-0.7	-	0.10	0.20	-	0.05	0.15
360.0	D	9.0-10.0	0.6	0.50	2.0	0.40-0.6	0.50	0.35	-	0.15 Sn	-	0.25
A360.0	D	9.0-10.0	0.6	0.50	1.3	0.40-0.6	0.50	0.35	-	0.15 Sn	-	0.25

(Continued)

Table 6. (Continued) Chemical Composition Limits of Aluminum Casting Alloys. (Source, Aluminum Association, Inc.)

Alloy	Use*	Si	Cu	Zn	Fe	Mg	Ni	Mn	Ti	Other (specified)	Others Each	Others Total
380.0	D	7.5-9.5	3.0-4.0	3.0	2.0	0.10	0.50	0.50	-	0.35 Sn	-	0.50
A380.0	D	7.5-9.5	3.0-4.0	3.0	1.3	0.10	0.50	0.50	-	0.35 Sn	-	0.50
383.0	D	9.5-11.5	2.0-3.0	3.0	1.3	0.10	0.30	0.50	-	0.15 Sn	-	0.50
384.0	D	10.5-12.0	3.0-4.5	3.0	1.3	0.10	0.50	0.50	-	0.35 Sn	-	0.50
390.0	D	16.0-18.0	4.0-5.0	0.10	1.3	0.45-0.65	-	0.10	0.20	-	-	0.20
B390.0	D	16.0-18.0	4.0-5.0	1.5	1.3	0.45-0.65	0.10	0.50	0.10	-	-	0.20
392.0	D	18.0-20.0	0.40-0.80	0.50	1.5	0.80-1.20	0.50	0.20-0.60	0.20	0.30 Sn	-	0.50
413.0	D	11.0-13.0	1.0	0.50	2.0	0.10	0.50	0.35	-	0.15 Sn	-	0.25
A413.0	D	11.0-13.0	1.0	0.50	1.3	0.10	0.50	0.35	-	0.15 Sn	-	0.25
C433.0	D	4.5-6.0	0.6	0.50	2.0	0.10	0.50	0.35	-	0.15 Sn	-	0.25
443.0	S&P	4.5-6.0	0.6	0.50	0.8	0.05	-	0.50	0.25	0.25 Cr	-	0.35
B443.0	S&P	4.5-6.0	0.15	0.35	0.8	0.05	-	0.35	0.25	-	0.05	0.15
A444.0	P	6.5-7.5	0.10	0.10	0.20	0.05	-	0.10	0.20	-	0.05	0.15
512.0	S	1.4-2.2	0.35	0.35	0.6	3.5-4.5	-	0.8	0.25	0.25 Cr	0.05	0.15
513.0	P	0.30	0.10	1.4-2.2	0.40	3.5-4.5	-	0.30	0.20	-	0.05	0.15
514.0	S	0.35	0.15	0.15	0.50	3.5-4.5	-	0.35	0.25	-	0.05	0.15
518.0	D	0.35	0.25	0.15	1.8	7.5-8.5	0.15	0.35	-	0.15 Sn	-	0.25
520.0	S	0.25	0.25	0.15	0.30	9.5-10.6	-	0.15	0.25	-	0.05	0.15
535.0	S&P	0.15	0.05	-	0.15	6.2-7.5	-	0.10-0.25	0.10-0.25	-	0.05	0.15
705.0	S&P	0.20	0.20	2.7-3.3	0.8	1.4-1.8	-	0.40-0.6	0.25	0.20-0.40 Cr	0.05	0.15
707.0	S&P	0.20	0.20	4.0-4.5	0.8	1.8-2.4	-	0.40-0.6	0.25	0.20-0.40 Cr	0.05	0.15
710.0	S	0.15	0.35-0.65	6.0-7.0	0.50	0.6-0.8	-	0.05	0.25	-	0.05	0.15
711.0	P	0.30	0.35-0.65	6.0-7.0	0.7-1.4	0.25-0.45	-	0.05	0.20	-	0.05	0.15

(Continued)

Table 6. *(Continued)* **Chemical Composition Limits of Aluminum Casting Alloys.** *(Source, Aluminum Association, Inc.)*

Alloy	Use*	Si	Cu	Zn	Fe	Mg	Ni	Mn	Ti	Other (specified)	Others Each	Others Total
712.0	S	0.30	0.25	5.0-6.5	0.50	0.50-0.65	-	0.10	0.15-0.25	0.40-0.6 Cr	0.05	0.20
713.0	S&P	0.25	0.40-1.0	7.0-8.0	1.1	0.20-0.50	0.15	0.6	0.25	0.35 Cr	0.10	0.25
771.0	S	0.15	0.10	6.5-7.5	0.15	0.8-1.0	-	0.10	0.10-0.20	0.06-0.20 Cr	0.05	0.15
850.0	S&P	0.7	0.7-1.3	-	0.7	0.10	0.7-1.3	0.10	0.20	-	-	0.30
851.0	S&P	2.0-3.0	0.7-1.3	-	0.7	0.10	0.30-0.7	0.10	0.20	-	-	0.30
852.0	S&P	0.40	1.7-2.3	-	0.7	0.6-0.9	0.9-1.5	0.10	0.20	-	-	0.30

* S = sand casting, P = permanent mold casting, D = die casting.

Table 7. Mechanical Properties for Aluminum Sand Casting Alloys.
(Source, Aluminum Association, Inc. and QQ-A-601F.)

Alloy	Temper	Ultimate Tensile Strength		Yield Tensile Strength		% Elongation in Two Inches, or 4 × Dia.	Hardness BHN
		ksi	MPa	ksi	MPa		
201.0	T7	60	414	50	345	3.0	110-140
204.0	T4	45	310	28	193	6.0	-
208.0	F	19	131	12	83	1.5	40-70
208.0	T55	21	145	-	-	-	75
222.0	0	23	159	20	138	-	65-95
222.0	T61	30	207	40	276	-	100-130
242.0	0	23	159	18	124	-	55-85
242.0	T571	29	200	30	207	-	70-100
242.0	T61	32	221	20	138	-	90-120
242.0	T77	24	165	13	90	1.0	60-90
295.0	T4	29	200	13	90	6.0	45-75
295.0	T6	32	221	20	138	3.0	60-90
295.0	T62	36	248	28	193	-	80-110
295.0	T7	29	200	16	110	3.0	55-85
319.0	F	23	159	13	90	1.5	55-85
319.0	T5	25	172	26	179-	-	65-95
319.0	T6	31	214	20	138	1.5	65-95
328.0	F	25	172	14	97	1.0	45-75
328.0	T6	34	234	21	145	1.0	65-95
355.0	T51	25	172	18	124	-	50-80
355.0	T6	32	221	20	138	2.0	70-105
355.0	T7	35	241	36	248	-	70-100
355.0	T71	30	207	22	152	-	60-95
C355.0	T6	36	248	25	172	2.5	75-105
356.0	F	19	131	-	-	2.0	40-70
356.0	T51	23	159	16	110	-	45-75
356.0	T6	30	207	20	138	3.0	55-90
356.0	T7	31	214	29	200	-	60-90
356.0	T71	25	172	18	124	3.0	45-75
A356.0	T6	34	234	24	165	3.5	70-105
443.0	F	17	117	7	49	3.0	25-55
B433.0	F	17	117	6	41	3.0	25-55
512.0	F	17	117	10	69	-	35-65
514.0	F	22	152	9	62	6.0	35-65
520.0	T4	42	290	22	152	12.0	60-90
535.0	0	35	241	18	124	9.0	-
535.0	F or T5	35	241	18	124	9.0	60-90
705.0	F or T5	30	207	17	117	5.0	50-80

(Continued)

Table 7. *(Continued)* **Mechanical Properties for Aluminum Sand Casting Alloys.**
(Source, Aluminum Association, Inc. and QQ-A-601F.)

Alloy	Temper	Ultimate Tensile Strength		Yield Tensile Strength		% Elongation in Two Inches, or 4 × Dia.	Hardness BHN
		ksi	MPa	ksi	MPa		
707.0	T5	33	228	22	152	2.0	70-100
707.0	T7	37	255	30	207	1.0	65-95
710.0	F or T5	32	221	20	138	2.0	60-90
712.0	F or T5	34	234	25	172	4.0	60-90
713.0	F or T5	32	221	22	152	3.0	60-90
771.0	0	36	248	27	186	1.5	-
771.0	T5	42	290	38	262	1.5	85-115
771.0	T51	32	221	27	186	3.0	70-100
771.0	T52	36	248	30	207	1.5	70-100
771.0	T53	36	248	27	186	1.5	-
771.0	T6	42	290	35	241	5.0	75-105
771.0	T71	48	331	45	310	2.0	105-135
850.0	T5	16	110	11	76	5.0	30-60
851.0	T5	17	117	11	76	3.0	30-60
852.0	T5	24	165	18	124	-	45-75

Table 8. Mechanical Properties for Aluminum Permanent Mold Casting Alloys.
(Source, Aluminum Association, Inc. and QQ-A-596E.)

Alloy	Temper	Ultimate Tensile Strength		Yield Tensile Strength		% Elongation in Two Inches, or 4 × Dia.	Hardness BHN
		ksi	MPa	ksi	MPa		
204.0	T4	48	331	29	200	8.0	-
208.0	T4	33	228	15	103	4.5	60-90
208.0	T6	35	241	22	152	2.0	75-105
208.0	T7	33	228	16	110	3.0	65-95
213.0	F	23	159	-	-	-	-
222.0	T551	30	207	-	-	-	100-130
222.0	T65	40	276	-	-	-	125-155
242.0	T571	34	234	-	-	-	90-120
242.0	T61	40	276	-	-	-	95-125
296.0	T4	33	228	-	-	4.5	-
296.0	T6	35	241	-	-	2.0	75-105
296.0	T7	33	228	-	-	3.0	-
308.0	F	24	165	-	-	-	55-85
319.0	F	28	193	14	97	1.5	70-100
319.0	T6	34	234	-	-	2.0	75-105
332.0	T5	31	214	-	-	-	90-120
333.0	F	28	193	-	-	-	65-100
333.0	T5	30	207	-	-	-	70-105

(Continued)

Table 8. *(Continued)* **Mechanical Properties for Aluminum Permanent Mold Casting Alloys.**
(Source, Aluminum Association, Inc. and QQ-A-596E.)

Alloy	Temper	Ultimate Tensile Strength		Yield Tensile Strength		% Elongation in Two Inches, or 4 × Dia.	Hardness BHN
		ksi	MPa	ksi	MPa		
333.0	T6	35	241	-	-	-	85-115
333.0	T7	31	214	-	-	-	75-105
336.0	T551	31	214	-	-	-	90-120
336.0	T65	40	276	-	-	-	110-140
354.0	T61	48	331	37	255	3.0	-
354.0	T62	52	359	42	290	2.0	-
355.0	T51	27	186	-	-	-	60-90
355.0	T6	37	255	-	-	1.5	75-105
355.0	T62	42	290	-	-	-	90-120
355.0	T7	36	248	-	-	-	70-100
355.0	T71	34	234	27	186	-	65-95
C355.0	T61	40	276	30	207	3.0	75-105
356.0	F	21	145	-	-	3.0	40-70
356.0	T51	25	172	-	-	-	55-85
356.0	T6	33	228	22	152	3.0	65-95
356.0	T7	25	172	-	-	3.0	60-90
356.0	T71	25	172	-	-	3.0	60-90
A356.0	T61	37	255	26	179	5.0	70-100
357.0	T6	45	310	-	-	3.0	75-105
A357.0	T61	45	310	36	248	3.0	85-115
359.0	T61	45.0	310	34	234	4.0	75-105
359.0	T62	47	324	38	262	3.0	85-115
443.0	F	21	145	7	49	2.0	30-60
B443.0	F	21	145	6	41	2.5	30-60
A444.0	T4	20	138	-	-	20.0	-
513.0	F	22	152	12	83	2.5	45-75
535.0	F	35	241	18	124	8.0	60-90
705.0	T5	37	255	17	117	10.0	55-85
707.0	T5	42	290	-	-	4.0	-
707.0	T7	45	310	35	241	3.0	80-110
711.0	T1	28	193	18	124	7.0	55-85
713.0	T5	32	221	22	152	4.0	60-90
850.0	T5	18	124	-	-	8.0	30-60
851.0	T5	17	117	-	-	3.0	30-60
851.0	T6	18	124	-	-	8.0	-
852.0	T5	27	186	-	-	3.0	55-85

Note: The values on this Table represent properties obtained from separately cast test bars. Average properties of specimens cut from castings shall be not less than 75% of tensile and yield strength values and shall be not less than 25% of elongation values given above.

Table 9. Physical Properties of Aluminum Casting Alloys.

Alloy	Temper	Use[5]	Density lb/cu in.	Thermal Conductivity CGS[1]	Thermal Conductivity Eng.[2]	Coef. Thermal Expan.[4] μin °F	Coef. Thermal Expan.[4] μm °C	Electrical Conductivity[3]	Melting Range (Approx.) °F	Melting Range (Approx.) °C
201.0	T7	P	0.101	0.29	840	19.3	34.7	32-34	1060-1200	570-650
208.0	F	S	0.101	0.29	840	12.2	22.0	31	970-1170	520-630
222.0	O	S	0.107	0.38	1100	-	-	41	970-1160	520-625
	T61	S	0.107	0.31	900	12.3	22.1	33	970-1160	520-625
242.0	O	S	0.102	0.40	1160	-	-	44	990-1180	530-635
	T77	S	0.102	0.36	1040	12.3	22.1	38	980-1180	525-635
	T571	P	0.102	0.32	930	12.5	22.5	34	980-1180	525-635
	T61	P	0.102	0.32	930	12.5	22.5	33	980-1180	525-635
295.0	T4	S	0.102	0.33	960	12.7	22.9	35	970-1190	520-645
	T62	S	0.102	0.34	990	12.7	22.9	35	970-1190	520-645
296.0	T4	P	0.101	0.32	930	12.2	22.0	33	970-1170	520-630
	T6	P	0.101	0.32	930	12.2	22.0	33	970-1170	520-630
308.0	F	P	0.101	0.34	990	11.9	21.4	37	970-1140	520-615
319.0	F	S	0.101	0.27	780	12.0	21.6	27	970-1120	520-605
	F	P	0.101	0.28	810	12.0	21.6	28	970-1120	520-605
332.0	T5	P	0.100	0.25	720	11.5	20.7	26	970-1080	520-580
333.0	F	P	0.100	0.25	720	11.5	20.7	26	970-1090	520-585
	T5	P	0.100	0.29	840	11.5	20.7	29	970-1090	520-585
	T6	P	0.100	0.28	810	11.5	20.7	29	970-1090	520-585
	T7	P	0.100	0.34	990	11.5	20.7	35	970-1090	520-585
355.0	T51	S	0.098	0.40	1160	12.4	22.3	43	1020-1150	550-620
	T6	S	0.098	0.34	990	12.4	22.3	36	1020-1150	550-620
	T7	S	0.098	0.39	1130	12.4	22.3	42	1020-1150	550-620
	T6	P	0.098	0.36	1040	12.4	22.3	39	1020-1150	550-620

(Continued)

Table 9. *(Continued)* **Physical Properties of Aluminum Casting Alloys.**

Alloy	Temper	Use[5]	Density lb/cu in.	Thermal Conductivity		Coef. Thermal Expan.[4]		Electrical Conductivity[3]	Melting Range (Approx.)	
				CGS[1]	Eng.[2]	μin °F	μm °C		°F	°C
C355.0	T61	S	0.098	0.35	1010	12.4	22.3	39	1020-1150	550-620
356.0	T51	S	0.097	0.40	1160	11.9	21.4	43	1040-1140	560-615
	T6	S	0.097	0.36	1040	11.9	21.4	39	1040-1140	560-615
	T7	S	0.097	0.37	1070	11.9	21.4	40	1040-1140	560-615
A356.0	T6	S	0.098	0.36	1040	11.9	21.4	40	1040-1130	560-610
357.0	T6	S	0.098	0.36	1040	11.9	21.4	39	1040-1140	560-615
443.0	F	S	0.097	0.35	1010	12.3	22.1	37	1070-1170	575-630
512.0	F	S	0.096	0.35	1010	12.7	22.9	38	1090-1170	590-630
513.0	F	P	0.097	0.32	930	13.3	23.9	34	1080-1180	580-640
514.0	F	S	0.096	0.33	960	13.3	23.9	35	1110-1180	600-640
520.0	T4	S	0.093	0.21	600	14.0	25.2	21	840-1110	450-600
535.0	F	S	0.091	0.24	690	13.1	23.6	23	1020-1170	550-630
705.0	F	S	0.100	0.25	720	13.1	23.6	25	1110-1180	600-640
710.0	F	S	0.102	0.33	960	13.4	24.1	35	1110-1200	600-650
712.0	F	S	0.102	0.38	1100	13.1	23.6	40	1110-1180	600-640
713.0	F	S	0.104	0.37	1070	13.3	23.9	37	1100-1170	595-630
850.0	T5	S	0.103	0.44	1280	-	-	47	440-1200	225-650
851.0	T5	S	0.102	0.40	1160	12.6	22.7	43	450-1170	230-630
852.0	T5	S	0.104	0.42	1210	12.9	23.2	45	410-1180	210-635

Notes: [1] CGS = cal/cm/cm²/°C/sec at 25° C (77° F).
[2] Eng. = English Units = btu/in./ft²/°F/hour at 77° F.
[3] At 68° F (20° C). International Annealed Copper Standard (IACS).
[4] Coefficient of Thermal Expansion is based on change in length per °F or °C, from 68 to 212° F (20 to 100° C).
[5] S = sand casting, P = permanent mold casting, D = die casting.

3xxx Series – Al-Mn Alloys. These medium strength alloys are strain hardenable, have excellent corrosion resistance, and are readily welded, brazed, or soldered. Because of its superiority in handling many foods and chemicals, alloy 3003 is widely used in cooking utensils as well as in builder's hardware. Alloy 3105 is widely used for roofing and siding, and variations of the 3xxx series are used in sheet and tubular form for heat exchangers in vehicles and power plants. Because of its use in the bodies of beverage cans, alloy 3004 (and its modification 3104) is among the most widely used of all aluminum alloys. Typically, 3xxx alloys have a tensile strength range of 16,000 to 41,000 PSI (110 to 283 MPa).

4xxx Series – Al-Si Alloys. These are medium high strength heat treatable alloys that are widely used in complex shaped forgings. Alloy 4032 is used principally for forgings such as aircraft pistons. 4043, on the other hand, is one of the most widely used filler alloys for gas metal arc and gas tungsten arc welding of 6xxx series alloys for use in structural and automotive applications. These alloys have good flow characteristics as a result of their high silicon content which, in the case of forgings, ensures the filling of complex dies, and in welding allows complete filling of crevices and seams in welded joints. Typically, 4xxx alloys have a tensile strength range of 25,000 to 55,000 PSI (172 to 379 MPa).

5xxx Series – Al-Mg Alloys. These strain hardenable alloys have moderately high strength, excellent corrosion resistance (even in salt water), and very high toughness even at cryogenic temperatures to near absolute zero. They can be welded with a variety of techniques, even at thicknesses up to 7 $7/8$ inches (20 cm), which has contributed to their popularity in bridge construction and other building applications, marine use, and for use in storage tanks and pressure vessels where temperatures as low as –270° F (–168° C) may be encountered. Alloys 5052, 5086, and 5083 are usually chosen for structural use. Specialty alloys include 5182, used for beverage can ends, 5754 which is used for automotive panels and frames, and 5252, 5454, and 5657 which are all used for bright trim applications. In this series, strength increases with higher magnesium content, but care must be taken to avoid the use of 5xxx alloys with more than 3% Mg content (such as 5454 and 5754) in applications where they receive continuous exposure to temperatures above 212° F (100° C) where they become sensitized and susceptible to stress corrosion cracking. Typically, 5xxx alloys have a tensile strength of 18,000 to 51,000 PSI (124 to 352 MPa).

6xxx Series – Al-Mg-Si Alloys. The alloys in this series are heat treatable, have moderately high strength, excellent extrudability properties, and excellent corrosion resistance. Because they are easily welded and extruded, they are the first choice for architectural and structural members where particular strength or stiffness criticality is important. Alloy 6061 is often used in welded structures such as vehicle frames, railroad cars, and pipelines. 6063 is widely used in extruded form for bridge construction and automobile frames. Specialty alloys in the series include 6066-T6 (used for high strength forgings), 6111 (used for automotive body panels), and 6101 and 6201, which are used for high strength electrical bus and electrical conductor wire, respectively. Typically, 6xxx alloys have a tensile strength range of 18,000 to 58,000 PSI (124 to 400 MPa).

7xxx Series – Al-Zn Alloys. These are, among the Al-Zn-Mg-Cu aluminum alloy versions, the highest strength of all aluminum alloys. They are heat treatable. Alloys 7150 and 7475, especially, are noted for their combination of strength and fracture toughness. Historically, the widest application of the 7xxx alloys has been in the aircraft industry where fracture critical design concepts have led to the development of high-toughness alloys. Although this series is not weldable by routine commercial processes, it is regularly used in riveted construction. Because their atmospheric corrosion resistance is not as high as the 5xxx or 6xxx series, 7xxx series alloys are usually coated or, for sheet and plate, used in an alclad version. Special tempers such as the T73 type are required whenever stress corrosion cracking may be a problem. Typically, 7xxx alloys have a tensile strength range of 32,000 to 88,000 PSI (220 to 607 MPa).

8xxx Series – Alloys with Al + other elements not covered by other series. These alloys contain less frequently used alloying elements such as lead, nickel, and lithium. They are heat treatable and have high conductivity, strength, and hardness. Applications include electrical conductors, bearings, and the aerospace industry. Typically, 8xxx alloys have a tensile strength in the range of 17,000 to 35,000 PSI (117 to 241 MPa).

Characteristics of Casting Alloys by Series Designation

The casting aluminum alloys contain larger percentages of alloying agents such as silicon and copper than are found in the wrought alloys. Their elongation and strength (especially fatigue) properties are relatively lower than the wrought products, mainly because current casting practice is unable to reliably prevent casting defects, even though recent innovations such as squeeze casting and thixocasting have resulted in considerable improvements.

2xx.x Series – Al-Cu Alloys. Both sand and permanent mold castings are produced from this series, which is heat treatable. These alloys possess high strength at room and elevated temperatures, and many of the metals in this group are high toughness alloys. Heat treated alloy 201.0 is the strongest of the casting alloys. However, since its castability is somewhat limited by a tendency to microporosity and hot tearing, it is best suited to investment casting. 201.0 is commonly used in machine tool construction and aircraft construction. Typically, 2xx.x alloys have a tensile strength range of 19,000 to 65,000 PSI (131 to 448 MPa).

3xx.x Series – Al-Si+Cu or Mg Alloys. These are among the most widely used casting alloys. Their high silicon content contributes to fluidity, and their heat treatability provides high strength options. They may be cast by sand and permanent mold techniques (319.0 and 356.0/A356.0), die casting (360.0, 380.0/A380.0, and 390.0) and other methods such as squeeze and forge casing (357.0/A357.0). Typically, 3xx.x alloys have a tensile strength range of 19,000 to 40,000 PSI (131 to 276 MPa).

4xx.x Series – Al-Si Alloys. Although not heat treatable, this series features very good castability and excellent weldability. They have a low melting point (570° F [299° C]), moderate strength, good corrosion resistance, and high elongation before rupture. This series is widely used when intricate, thin walled, leak-proof, fatigue resistant castings are required. 4xx.x alloys have excellent fluidity and can be used in sand, permanent mold, and die casting operations. Typically, 4xx.x alloys have a tensile strength in the range of 17,000 to 25,000 PSI (117 to 172 MPa).

5xx.x Series – Al-Mg alloys. This series is relatively tough to cast, but offers good finishing characteristics and machinability, and excellent corrosion resistance—they are suitable for salt water and other similar corrosive environments. The 512.0 and 514.0 is commonly used for door and window frames where they can be anodized or color coated. Other alloys in this series are used for cooking utensils and aircraft fittings. These alloys are not heat treatable and are sand, permanent mold, and die castable. Typically, 5xx.x alloys have a tensile strength range of 17,000 to 25,000 PSI (117 to 172 MPa).

7xx.x Series – Al-Zn Alloys. Because these alloys are more difficult to cast, they tend to be used only when excellent finishing and machinability characteristics are important (such as in furniture, office machinery, etc.). They are heat treatable, and may be cast with both sand and permanent mold techniques. Typically, 7xx.x alloys have a tensile strength range of 10,000 to 55,000 PSI (69 to 379 MPa).

8xx.x Series – Al-Sn Alloys. These alloys are also relatively difficult to cast, but they are uniquely suited for bushings and have excellent machinability characteristics. They are heat treatable, and can cast with sand or permanent molds. Bearings and bushings are the most common applications for these alloys. Typically, 8xx.x alloys have a tensile strength in the range of 15,000 to 30,000 PSI (103 to 207 MPa).

Temper designation system for aluminum alloys

Temper designations indicate mechanical or thermal treatment of the alloy. It follows the alloy number and is always preceded by a dash, as in 2014–T6. Basic designations consist of letters. Subdivisions of the basic designations, where required, are indicated by one or more digits following the letter. These designate specific sequences of basic treatments, but only operations recognized as significantly influencing the characteristics of the product are indicated. Should some other variation of the same sequence of basic operations be applied to the same alloy, resulting in different characteristics, additional digits are added to the designation.

The basic temper designations and subdivisions are as follows.

–F   *As Fabricated.* Applies to products that acquire some temper from shaping processes not having special control over the amount of strain hardening or thermal treatment. For wrought products, there are no mechanical property limits.

–O   *Annealed.* Applies to wrought products that are annealed to obtain the lowest strength temper, and to cast products that are annealed to improve ductility and dimensional stability. The O may be followed by a digit other than zero, indicating a product in the annealed condition that has specific characteristics. It should be noted that variations of the –O temper shall not apply to products that are strain hardened after annealing and in which the effect of strain hardening is recognized in the mechanical properties or other characteristics. The following temper designation has been assigned for wrought products that are high temperature annealed to accentuate ultrasonic response and provide dimensional stability.

–O1   *Thermally Treated at Approximately the Same Time and Temperature Required for Solution Heat Treatment and Slow Cooled to Room Temperature.* Applicable to products that are to be machined prior to solution heat treatment by the user. Mechanical property limits are not applicable.

–H   *Strain Hardened.* Wrought products only. Applies to products that have their strength increased by strain hardening with or without supplemental thermal treatments to produce partial softening. The –H is always followed by *two* or more digits. The first digit indicates the specific combination of basic operations as follows.

–H1   *Strain Hardened Only.* Applies to products that are strain hardened to obtain the desired mechanical properties without supplementary thermal treatment. The number that follows this designation indicates the degree of strain hardening (see below).

–H2   *Strain Hardened and then Partially Annealed.* Applies to products that are strain hardened more than the desired final amount and then reduced in strength to the desired level by partial annealing. For alloys that age-soften at room temperature, the –H2 tempers have approximately the same ultimate strength as the corresponding –H3 tempers. For other alloys, the –H2 tempers have approximately the same ultimate strength and slightly longer elongations as the corresponding –H1 tempers. The number following this designation indicates the degree of strain hardening remaining after the product has been partially annealed (see below).

-H3   *Strain Hardened and then Stabilized.* Applies to products that are strain hardened and then stabilized by low temperature heating in order to slightly lower their strength and increase ductility. This designation applies only to the alloys that contain magnesium: unless stabilized gradually, they will age soften at room temperature. The number following this designation indicates the degree of strain hardening remaining after the product has been strain hardened a specified amount and then stabilized (see below).

-H4   *Strain Hardened and Lacquered or Painted.* Applied to products that are strain hardened and subsequently subjected to a thermal operation during lacquering or painting. The number following this designation indicates the degree of strain hardening remaining after the product has been thermally treated as part of the lacquering or painting cure operation (see below).

The digit following the –H1, –H2, –H3, and –H4 designation indicates the final degree of strain hardening. The hardest commercially practical temper is designated by the numeral 8 (full hard). Tempers between –O (annealed) and 8 (full hard) are designated by numerals 1 through 7. Materials having an ultimate strength about midway between that of the –O temper and that of the 8 temper are designated by the numeral 4 (half hard); between –O and 4 by the numeral 2 (quarter hard); between 4 and 8 by the numeral 6 (three-quarter hard); etc. Numeral 9 indicates extra hard tempers with tensile strength exceeding the –Hx8 temper by 2 ksi (13.79 MPa) or more.

A third digit, when used, indicates that the degree of control of temper, or the mechanical properties, are different from, but within the range of, those for the two digit –H temper designation to which it is added. Numerals 1 through 9 may be arbitrarily assigned and registered with the Aluminum Association for an alloy and product to indicate a specific degree of control of temper or specific mechanical property limits. Zero has been assigned to indicate degrees of control of temper, or mechanical property limits negotiated between the manufacturer and purchaser that are not used widely enough to justify registration.

The following three-digit –H temper designations have been assigned for wrought products in all alloys.

-Hx11   Applies to products that are strain hardened less than the amount required for a controlled –Hx1 temper.

-H112   Applies to products that acquire some temper from shaping processes not having special control over the amount of strain hardening or thermal treatment, but for which there are mechanical property limits or mechanical property testing is required.

-H311   Applies to products that are strain hardened less than the amount required for a controlled –H31 temper.

The following three-digit –H temper designations have been assigned.

Patterned or Embossed Sheet	*Fabricated From*	*Patterned or Embossed Sheet*	*Fabricated From*
–H114	–O	–H264	–H25
–H124	–H11	–H364	–H35
–H224	–H21	–H174	–H16
–H324	–H31	–H274	–H26
–H134	–H12	–H374	–H36
–H234	–H22	–H184	–H17

(Continued)

Patterned or Embossed Sheet	Fabricated From	Patterned or Embossed Sheet	Fabricated From
–H334	–H32	–H284	–H27
–H144	–H13	–H384	–H37
–H244	–H23	–H194	–H18
–H344	–H33	–H 294	–H28
–H154	–H14	–H394	–H38
–H254	–H24	–H195	–H19
–H354	–H34	–H295	–H29
–H164	–H15	–H395	–H39

–W   *Solution Heat Treated.* Applies to products whose strength naturally changes at room temperature after solution heat treatment. The change might take place over weeks, or years, and the designation is specific only when the period of natural aging is determined: for example,W 2.5 hours.

–T   *Thermally Treated to Produce Tempers Other than – F, – O, or – H.* Applies to products that are thermally treated with or without supplementary strain hardening to produce stable tempers. The –T is always followed by one or more digits. Numerals 1 to 10 have been assigned to indicate specific sequences of basic treatment, as follows.

   –T1   *Cooled from an Elevated Temperature Shaping Process and then Naturally Aged to a Substantially Stable Condition.* Applies to castings, extrusions, etc., that are not cold worked after an elevated temperature shaping process. Also applies to products that are cold worked by flattened or straightened after cooling, but any effects incurred by these processes are not accounted for in the property limits.

   –T2   *Cooled from an Elevated Temperature Shaping Process and then Cold Worked and Naturally Aged to a Substantially Stable Condition.* Applies to products that are cold worked to improve strength after cooling from a hot working process. Also applies to products that are cold worked by flattening or straightening after cooling, and any effects incurred in these processes are accounted for in the property limits.

   –T3   *Solution Heat Treated and then Cold Worked and Naturally Aged to a Substantially Stable Condition.* Applies to products that are cold worked to improve strength, or in which the effects of cold work in flattening or straightening is recognized in applicable specifications.

   –T4   *Solution Heat Treated and Naturally Aged to a Substantially Stable Condition.* Applies to products that are not cold worked after solution heat treatment, but in which the effect of cold work in flattening or straightening may be recognized in applicable specifications.

   –T5   *Cooled from an Elevated Temperature Shaping Process and then Artificially Aged.* Applies to products that are artificially aged after an elevated temperature rapid cool fabrication process, such as casting or extrusion, to improve mechanical properties and/or dimensional stability.

   –T6   *Solution Heat Treated and then Artificially Aged.* Applies to products that are not cold worked after solution heat treatment, but in which the effect of cold work in flattening or straightening may be recognized in applicable specifications.

   –T7   *Solution Heat Treated and then Stabilized or Overaged.* Applies to products that are stabilized to carry them beyond the point of maximum hardness, providing control of growth and/or residual stress. Cast products that are artificially aged after solution treatment to provide dimensional and strength stability are covered by this designation.

–T8 *Solution Heat Treated, Cold Worked, and then Artificially Aged.* Applies to products that are cold worked to improve strength, or in which the effect of cold work in flattening or straightening is recognized in applicable specifications.

–T9 *Solution Heat Treated, Artificially Aged, and then Cold Worked.* Applies to products that are cold worked specifically for the purpose of gaining strength.

–T10 *Cooled from an Elevated Temperature Shaping Process, Cold Worked, and then Artificially Aged.* Applies to products that are artificially aged after an elevated temperature rapid cool fabrication process, such as casting or extrusion, and then cold worked to improve strength.

A period of natural aging at room temperature may occur between or after the operations listed for tempers –T3 through –T10. Control of this period is exercised when it is metallurgically important. Additional digits may be added to designations –T1 through –T10 to indicate a variation in treatment that significantly alters the characteristics of the product. These may be assigned for an alloy and product to indicate a specific treatment or specific mechanical property limits. The following additional digits have been assigned for wrought products in all alloys.

–T*x*51 *Stress Relieving by Stretching.* Applies to products that are stress relieved by stretching the following amounts after solution heat treatment or after cooling from an elevated temperature shaping process. The products receive no further straightening after stretching.

Plate: 1.5 to 3% permanent set.

Rolled or cold finished rod, and bar: 1 to 3% permanent set.

Die or ring forgings and rolled rings: 1 to 5% permanent set.

Applies directly to plate and rolled or cold finished rod and bar. These products receive no further straightening after stretching. Applies to extruded rod, bar, and shapes when designated as follows.

–T*x*510 Applies to extruded rod, bar, shapes, and tube, and to drawn tube, when stretched the indicated amounts amounts after solution heat treatment or after cooling from an elevated temperature shaping process. These products receive no further straightening after stretching.

Extruded rod, bar, shapes, and tube: 1 to 3% permanent set.

Drawn tube: 0.50 to 3% permanent set.

–T*x*511 Applies to extruded rod, bar, shapes, and tube, and tube to be drawn, that receive minor straightening after stretching to comply with standard tolerances.

–T*x*52 *Stress Relieved by Compressing.* Applies to products that are stress relieved by compression by the following amounts after solution heat treatment.

Extrusions: 1 to 5% permanent set.

Forgings: 1 to 5% permanent set.

–T*x*53 *Stress Relieved by Thermal Treatment.*

–T*x*54 *Stress Relieved by Combined Stretching and Compression.* Applies to die forgings that are stress relieved by restriking cold in the finish die.

The following two-numeral temper designations have been assigned for wrought products in all alloys.

–T42 Applies to products solution heat treated and naturally aged by the user that attain mechanical properties different from the –T4 temper.

–T62   Applies to products solution heat treated and artificially aged by the user that attain mechanical properties different from the –T6 temper.

Unregistered Tempers. The letter P has been assigned to denote –H, –T, and –O temper variations that are negotiated between manufacturer and purchaser. The letter P immediately follows the temper designation that most nearly pertains.

Heat treating aluminum alloys

The heat treatment processes commonly used to improve the properties of aluminum alloys are solution heat treatment, precipitation (age) hardening, and annealing. Dimensional changes in aluminum alloys during thermal treatment are generally negligible, but in some cases the changes may have to be considered in manufacturing. Because of the many variables involved, there are no tabular values for these dimensional changes, but some guidelines can be followed. For instance, in the artificial aging of alloy 2219 from the "–T42," "–T351," and "–T37" tempers to the "–T62," "–T851," and "–T87" tempers, respectively, a net dimensional growth of 0.00010 to 0.0015 inch per inch (0.00254 to 0.0381 mm per 25.4 mm) may be anticipated. Additional growth of as much as 0.0010 in./in. (0.0254 mm/25.4 mm) may occur during subsequent service of a year or more at 300° F (149° C), or equivalent shorter exposure at higher temperatures. The dimensional changes that occur during the artificial aging of other wrought heat treatable alloys are less than one-half that for alloy 2219 under the same conditions.

Solution heat treatment. Solution heat treatment (–W temper) is used to redistribute the alloying constituents that segregate from the aluminum during cooling from the molten state. It consists of 1) heating the alloy to a temperature at which the soluble constituents will form a homogeneous mass by solid diffusion, 2) holding the mass at that temperature until diffusion takes place, and then 3) quenching the alloy rapidly to retain the homogeneous condition. **Table 10** shows suitable temperatures for solution heat treating wrought alloys and casting alloys. If the specified maximum temperature is exceeded, there is the danger of localized melting and lowering the mechanical properties of the alloy. Excessive overheating will cause severe blistering. If the temperature is below the minimum specified, solution will not be complete, resulting in underdevelopment of the product's physical properties and lowered corrosion resistance. After solution heat treating, the product is soaked the required time to bring about the necessary degree of solid solution. Soaking times vary with thickness, and **Tables 11** and **12** provide minimal soaking times for wrought alloys and casting alloys. Alclad products should be soaked at the minimal temperature necessary to develop the required mechanical properties. Longer soaking may allow the alloying constituents of the base metal to diffuse through the alclad coating. When this occurs, corrosion resistance is adversely affected.

Precipitation (or "age") hardening. In the quenched condition, heat treated alloys are supersaturated solid solutions that are comparatively soft and workable, but unstable, depending on composition. At room temperature, the alloying constituents of some alloys (–W temper) tend to precipitate from the solution spontaneously, causing the metal to harden in about four days. This is called "natural aging." It can be retarded or even arrested to facilitate fabrication by holding the alloy at subzero temperatures until ready for forming. Other alloys age more slowly at room temperature, and may take years to reach maximum strength and hardness. These alloys can be aged artificially (called "artificial aging" or "elevated precipitation heat treatment"), to stabilize them and improve their qualities, by heating them to moderately elevated temperatures for specified lengths of time. A small amount of cold working, after solution heat treatment, produces a substantial increase in yield strength, some increase in tensile strength, and some loss of ductility. The effect on the properties developed will vary with different compositions.

(Text continued on p. 538)

Table 10. Recommended Temperatures for Solution Heat Treating of Aluminum Alloys. *(Source, MIL-H-6088G.)*

Alloy	Products/ Limitations[1]	Metal Temperature[5]		Temper Designation		
		°F	°C	Immediately After Quenching[2]	After Natural Aging[3]	After Stress Relief[4]
Wrought Products (Excluding Forgings)						
2011	Wire, rod, bar	945-995	507-535	-W	-T3[6], -T4	-T451
2014	Flat sheet	925-945	496-507	-W	-T3[6], -T42	-
	Coiled sheet	925-945	496-507	-W	-T4, -T42	-
	Plate	925-945	496-507	-W	-T4, -T42	-T451
	Wire, rod, bar	925-945	496-507	-W	-T4	-T451
	Extrusions	925-945	496-507	-W	-T4, -T42	-T4510, -T4511
	Drawn tube	925-945	496-507	-W	-T4	-
2017	Wire, rod, bar	925-950	496-510	-W	-T4	-T451
	Rivets	925-950	496-510	-W	-T4	-
2024	Flat sheet	910-930	488-499	-W	-T3[6], -T361[6], -T42	-
	Coiled sheet	910-930	488-499	-W	-T4, -T42, -T3[6]	-
	Rivets	910-930	488-499	-W	-T4	-
	Plate	910-930	488-499	-W	-T4, -T42, -T361[6]	-T351
	Wire, rod, bar	910-930[7]	488-499	-W	-T4, -T36[6], -T42	-T351
	Extrusions	910-930	488-499	-W	-T3[6], -T42	-T3510, T3511
	Drawn tube	910-930	488-499	-W	-T3[6], -T42	-
2048	Sheet, plate	910-930	488-499	-W	-T4, -T42	-T351
2117	Wire, rod, bar	925-950	496-510	-W	-T4	-
	Rivets	890-950	477-510	-W	-T4	-
2124	Plate	910-930	488-499	-W	-T4[6], -T42	-T351
2219	Sheet	985-1005	529-540	-W	-T31[6], -T37[6], -T42	-
	Plate	985-1005	529-540	-W	-T31[6], -T37[6], -T42	-T351
	Rivets	985-1005	529-540	-W	-T4	-
	Wire, rod, bar	985-1005	529-540	-W	-T31[6], -T42	-T351
	Extrusions	985-1005	529-540	-W	-T31[6], -T42	-T3510, -T3511
6010	Sheet	1045-1065	563-574	-W	-T4	-
6013	Sheet	1045-1065	563-574	-W	-T4	-
6061	Sheet	960-1075[8]	516-579	-W	-T4, -T42	-
	Plate	960-1075	516-579	-W	-T4, -T42	-T451
	Wire, rod, bar	960-1075	516-579	-W	-T4, -T42	-T451
	Extrusions	960-1075	516-579	-W	-T4, -T42	-T4510, -T4511
	Drawn tube	960-1075	516-579	-W	-T4, -T42	-
6063	Extrusions	960-985	516-529	-W	-T4, -T42	-T4510, -T4511
	Drawn tube	960-980	516-527	-W	-T4, -T42	N/A
6066	Extrusions	960-1010	516-543	-W	-T4, -T42	-T4510, -T4511
	Drawn tube	960-1010	516-543	-W	-T4, -T42	-

(Continued)

Table 10. *(Continued)* Recommended Temperatures for Solution Heat Treating of Aluminum Alloys. *(Source, MIL-H-6088G.)*

Alloy	Products/Limitations[1]	Metal Temperature[5]		Temper Designation		
		°F	°C	Immediately After Quenching[2]	After Natural Aging[3]	After Stress Relief[4]
Wrought Products (Excluding Forgings)						
6262	Wire, rod, bar	960-1050	516-543	-W	-T4	-T451
	Extrusions	960-1050	516-543	-W	-T4	-T4510, -T4511
	Drawn tube	960-1050	516-543	-W	-T4	-
6951	Sheet	975-995	524-535	-W	-T4, -T42	-
7001	Extrusions	860-880	460-471	-W	-	-W510[2], -W511[2]
7010	Plate	880-900	471-482	-W	-	-W51[2]
7039	Sheet	840-860[9]	449-460	-W	-	-
	Plate	840-860[9]	449-460	-W	-	-W51[2]
7049/ 7149	Extrusions	860-885	460-474	-W	-	-W510[2] -W511[2]
7050	Sheet	880-900	471-482	-W	-	-
	Plate	880-900	471-482	-W	-	-W51[2]
	Extrusions	880-900	471-482	-W	-	-W510[2] -W511[2]
	Wire, rod, rivets	880-900	471-482	-W	-	-
7075	Sheet	860-930[10]	460-499	-W	-	-
	Plate[11]	860-930	460-499	-W	-	-W51[2]
	Wire, rod, bar[11]	860-930	460-930	-W	-	-W51[2]
	Extrusions	860-880	460-471	-W	-	-W510[2] -W511[2]
	Drawn tube	860-880	460-471	-W	-	-
7150	Extrusions	880-900	471-482	-W	-	-W510[2] -W511[2]
	Plate	880-895	471-479	-W	-	-W51[2]
7178	Sheet[13]	860-930	460-499	-W	-	-
	Plate[13]	860-910	460-488	-W	-	-W51[2]
	Extrusions	860-880	460-471	-W	-	-W510[2] -W511[2]
7475	Sheet	880-970	471-521	-W	-	-
	Plate	880-970	471-521	-W	-	-
7475 Alclad	Sheet	880-945	471-507	-W	-	-
Forgings[14]						
2014	Die forgings	925-945	496-507	-W	-T4, -T41	-
	Hand forgings	925-945	496-507	-W	-T4, -T41	-T452
2018	Die forgings	940-970	504-521	-W	-T4, -T41	-

(Continued)

Table 10. *(Continued)* **Recommended Temperatures for Solution Heat Treating of Aluminum Alloys.** *(Source, MIL-H-6088G.)*

Alloy	Products/ Limitations[1]	Metal Temperature[5]		Temper Designation		
		°F	°C	Immediately After Quenching[2]	After Natural Aging[3]	After Stress Relief[4]
Forgings[14]						
2024	Die & hand forgings	910-930	488-499	-W	-T4	-T352
2025	Die forgings	950-970	510-521	-W	-T4	-
2218	Die forgings	940-960	504-516	-W	-T4, -T41	-
2219	Die & hand forgings	985-1005	529-540	-W	-T4	-T352
2618	Die & hand forgings	975-995	524-535	-W	-T4, -T41	-
4032	Die forgings	940-970	504-521	-W	-T4	-
6053	Die forgings	960-980	516-527	-W	-T4	-
6061	Die & hand forgings	960-1075	516-579	-W	-T4, -T41	-T452
	Rolled rings	960-1025	516-552	-W	-T4, -T41	-T452
6066	Die forgings	960-1010	516-543	-W	-T4	-
6151	Die forgings	950-980	510-527	-W	-T4	-
	Rolled rings	950-980	510-527	-W	-T4	-T452
7049/ 7149	Die & hand forgings	860-885	460-474	-W	-	-W52[2]
7050	Die & hand forgings	880-900	471-482	-W	-	-W52[2]
7075	Die & hand forgings	860-890[9]	460-477	-W	-	-W52[2]
	Rolled rings	860-890[9]	460-477	-W	-	-W52[2]
7076	Die & hand forgings	850-910	454-488	-W	-	-
7175	Die forgings	See Note 15		-W	-	-
	Hand forgings	See Note 15		-W	-	-
Castings (all mold practices)[16]						
A201.0[18]	-	945-965 followed by 970-995	496-518 521-535	-	-T4	-
A206.0 (206)[18]	-	945-965 followed by 970-995	496-518 521-535	-	-T4	-
222.0 (122)	-	930-960	499-516	-	-T4	-
242.0 (142)	-	950-980	510-527	-	-T4, -T41	-
295.0 (195)	-	940-970	504-521	-	-T4	-

(Continued)

Table 10. *(Continued)* **Recommended Temperatures for Solution Heat Treating of Aluminum Alloys.**
(Source, MIL-H-6088G.)

Alloy	Products/ Limitations[1]	Metal Temperature[5]		Temper Designation		
		°F	°C	Immediately After Quenching[2]	After Natural Aging[3]	After Stress Relief[4]
Castings (all mold practices)[16]						
296.0 (B295.0)	-	935-965	502-518	-	-T4	-
319.0 (319)	-	920-950	493-510	-	-T4	-
328.0 (Red X-8)	-	950-970	510-521	-	-T4	-
333.0 (333)	-	930-950	499-510	-	-T4	-
336.0 (A332.0)	-	950-970	510-521	-	-T45	-
A336.0 (A332.0)	-	940-970	504-521	-	-T45	-
354.0 (354)	-	980-995	527-535	-	-T4	-
355.0 (355), C355.0	-	960-995	516-535	-	-T4	-
356.0 (356), A356.0 (A356)	-	980-1025[12]	527-552	- -	-T4 -T4	- -
357.0 (357), A357.0 (A357)	-	980-1025[12]	527-552	- -	-T4 -T4	- -
359.0 (359)	-	980-1010	527-543	-	-T4	-
520.0 (220)	-	800-820	427-438	-	-T4	-
705.0[17]	-			-	T1 T5	-
707.0[17]	-	-	-		T1	-
712.0[17]	-	990 -	532	-	T4 T1	-
713.0[17]	-	-	-	-	T1	-
850.0[17]	-	-	-	-	T1	-
851.0[17]	-	-	-	-	T1	-
852.0[17]	-	-	-	-	T1	-

[1] The term "wire, rod, and bar" as used herein refers to rolled or cold finished wire, rod, and bar. The term "extrusions" refers to extruded wire, rod, bar, shapes, and tube.

[2] This temper is unstable and generally not available.

[3] Applied only to those alloys that will naturally age to a substantially stable condition.

(Continued)

Table 10. *(Continued)* **Recommended Temperatures for Solution Heat Treating of Aluminum Alloys.**
(Source, MIL-H-6088G.)

⁴ For rolled or extruded products, metal is stress relieved by stretching after quenching; and for forgings, metal is stress relieved by stretching or compression after quenching.

⁵ When a difference between the maximum and minimum temperatures of a range listed herein exceeds 20° F (11° C), any 20° F temperature range (or 30° F [17° C] range for 6061) within the entire range may be utilized, provided that no exclusions or qualifying criteria are cited herein or in the applicable material specification.

⁶ Cold working subsequent to solution heat treatment and prior to any precipitation heat treatment is necessary.

⁷ Temperatures as low as 900° F may be used, provided that every heat treat lot is tested to show that the requirements of the applicable material specification are met, and analysis of test data to show statistic conformance to the specification limits is available for review.

⁸ Maximum temperature for alclad 6061 sheet should not exceed 1000° F (538° C).

⁹ Other temperatures may be necessary for certain sections, conditions, and requirements.

¹⁰ It must be recognized that under some conditions melting can occur when heating 7075 alloy above 900° F (482° C) and that caution should be exercised to avoid this problem. In order to minimize diffusion between the cladding and the core, alclad 7075 sheet in thicknesses of 0.020 inch (0.508 mm) or less may be solution heat-treated at 850° F to 930° F (454 to 499° C).

¹¹ For plate thicknesses over 4 inches and for rod diameters or bar thicknesses over 4 inches (101.6 mm), a maximum temperature of 910° F (488° C) F is recommended to avoid melting.

¹² Heat treatment above 1010° F (593° C) may require an intermediate solution heat treatment of one hour at 1000-1010° F (538-543° C) to prevent eutectic melting of magnesium rich phases.

¹³ Under some conditions melting can occur when heating this alloy above 900° F (482° C).

¹⁴ Unless otherwise indicated, hand forgings include rolled rings, and die forgings include rolled rings, and die forgings include impacts.

¹⁵ Heat-treating procedures are at present proprietary among producers. At least one such procedure is patented (U.S. Patent Number 3,791,876).

¹⁶ Former commercial designation is shown in parentheses.

¹⁷ Unless otherwise specified, solution heat treatment is not required. Castings should be quickly cooled after shake-out or stripping from molds, so as to obtain a fine tin distribution.

¹⁸ In general, product should be soaked for two hours in the range 910-930° F (488-499° C) prior to heating into the solution heat-treating range. Other presolution heat-treating temperature ranges may be necessary for some configurations and sizes.

Table 11. Recommended Soaking Time for Solution Heat Treatment of Wrought Products.
(Source, MIL-6088G.)

| Thickness² | | Soaking Time (Minutes)¹ | | | |
| | | Salt Bath³ | | Air Furnace⁴ | |
Inches	Millimeters	Min.	Max. (Alclad)⁵	Min.	Max. (Alclad)⁵
0.016 and under	0.4064 and under	10	15	20	25
0.017 to 0.020 incl.	0.4381 to 0.5080 incl.	10	20	20	30
0.021 to 0.032 incl.	0.5334 to 0.8128 incl.	15	25	25	35
0.033 to 0.063 incl.	0.8382 to 1.6002 incl.	20	30	30	40
0.064 to 0.090 incl.	1.6256 to 2.2860 incl.	25	35	35	45
0.091 to 0.124 incl.	2.3114 to 3.1496 incl.	30	40	40	50
0.125 to 0.250 incl.	3.1750 to 6.3500 incl.	35	45	50	60
0.251 to 0.500 incl.	6.3754 to 12.700 incl.	45	55	60	70
0.501 to 1.000 incl.	12.7254 to 25.400 incl.	60	70	90	100
1.001 to 1.500 incl.	25.425 to 38.100 incl.	90	100	120	130
1.501 to 2.000 incl.	38.125 to 50.800 incl.	105	115	150	160
2.001 to 2.500 incl.	50.825 to 63.500 incl.	120	130	180	190

(Continued)

Table 11. *(Continued)* **Recommended Soaking Time for Solution Heat Treatment of Wrought Products.** *(Source, MIL-H-6088G.)*

Thickness[2]		Soaking Time (Minutes)[1]			
		Salt Bath[3]		Air Furnance[4]	
Inches	Millimeters	Min.	Max. (Alclad)[5]	Min.	Max. (Alclad)[5]
2.501 to 3.000 incl.	63.525 to 76.200 incl.	135	160	210	220
3.001 to 3.500 incl.	76.225 to 88.900 incl.	150	175	240	250
3.501 to 4.000 incl.	88.925 to 101.600 incl.	165	190	270	280

[1] Longer soaking times may be necessary for specific forgings. Shorter soaking times are satisfactory when the soak time is accurately determined by thermocouples attached to the load or when other metal temperature-measuring devices are used.

[2] The thickness is the minimum dimension of the heaviest section.

[3] Soaking time in salt-bath furnaces begins at time of immersion, except when, owing to a heavy charge, the temperature of the bath drops below the specified minimum; in such cases, soaking time begins when the bath reaches the specified minimum.

[4] Soaking time in air furnaces begins when all furnace control instruments indicate recovery to the minimum of the process range.

[5] For alclad metals, the maximum recovery time (time between charging furnace and recovery of furnace instruments) should not exceed 30 minutes for thicknesses up to 0.050 inch (12.700 mm), 60 minutes for 0.050 or greater but less than 0.102 inch (2.591 mm), and 120 minutes for 0.102 or greater.

Table 12. Recommended Soaking Time for Solution Heat Treatment of Cast Alloys. *(Source, MIL-H-6088G.)*

Alloy	Soaking Time (Hours)
A201.0 and A206.0	2 at 910-930° F (488-409° C) followed by 2-8 at 945-965° F (340-518° C) followed by 8-24 at 970-995° F (521-535° C)
222.0	6 to 18 incl.
242.0	2 to 10 incl.
295.0	6 to 18 incl.
296.0	4 to 12 incl.
319.0	6 to 18 incl.
328.0	12
336.0, A336.0	8 hr. then water quench to 150-212° F (66-100° C)
354.0	10 to 12 incl.
355.0 and C355.0	6 to 24 incl.
356.0 and A356.0	6 to 24 incl.
357.0 and A357.0	8 to 24 incl.
359.0	10 to 14 incl.
520.0	18

Table 13 shows recommended precipitation hardening treatments for wrought products, forgings, and castings.

Annealing. Annealing is used to effect recrystallization, essentially complete precipitation, or to remove internal stresses (annealing for the purpose of obliterating the hardening effects of cold working will remove the effects of heat treating). For most alloys annealing consists of heating to approximately 650° F (343° C) at a controlled rate. The rate is dependent on such factors as thickness, type of anneal desired, and method employed. Cooling rate is unimportant, but drastic quenching is not recommended because of the strains produced. **Table 14** lists recommended annealing conditions for work hardened wrought aluminum alloys. In order to avoid excessive oxidation and grain growth, annealing temperatures should not exceed 775° F (412° C). Soaking castings for two hours at 650 to 750° F (343 to 393° C) and then cooling them for two hours will relieve them of residual stress and provide dimensional stability.

Salt baths versus air chamber furnaces

Salt baths. The time required to bring the product to temperature is shortened, and uniform temperature is more easily maintained in salt baths than in air chamber furnaces. Also, when solution heat treating in molten salt, the danger of generating porosity is greatly diminished. After prolonged use, however, there is some decomposition of the sodium nitrate to form compounds that, when dissolved in the quenching water, attack the aluminum alloys. The addition to the salt bath of about 0.50 ounce of sodium or potassium dichromate per 100 pounds of nitrate tends to inhibit this attack. It should be noted that nitrate salt baths can present an explosive hazard when heat treating 5xx.x series casting alloys.

Air chamber furnaces. Air chamber furnaces are more flexible and economical for handling large volumes of work. When solution heat treating certain aluminum alloys, it is necessary to control the atmosphere in order to avoid the generation of porosity. Such porosity lowers the mechanical properties of aluminum alloys and may be manifested as large numbers of minute blisters over the surface of the product. In severe cases, the product may crack when it is quenched. Furnace products of combustion contain water vapor and may contain gaseous compounds of sulfur, both of which tend to cause porosity during solution heat treatment. For this reason, furnaces that permit their products of combustion to come into contact with the load are not recommended for the solution heat treatment of alloys that may become porous during such treatment. Anodic oxide films or the metal coating on alclad products protect the underlying material from this effect. To some degree, certain fluoroborates will also protect against or minimize porosity.

Effects of quenching

Quenching is the sudden chilling of the metal in oil or water. It increases the strength and corrosion resistance of the alloy and "freezes" the structure and distribution of the alloying constituents that existed at the temperature just prior to cooling. The properties of the alloy are governed by its composition and characteristics, its thickness of cross section, and the rate at which it is cooled. This rate is controlled by proper choice of both type and temperature of cooling medium. Rapid quenching, as in cold water, will provide maximum corrosion resistance and is used for items made from sheet, tube, extrusions, and small forgings. Rapid quenching is preferred over a less dramatic quench that would increase the mechanical properties. A slower quench, done in boiling or hot water, is used for heavy sections and large forgings. It tends to minimize distortion and cracking that can result from uneven cooling. The corrosion resistance of forging alloys is not affected by the temperature of the quench water. Also, corrosion resistance of thicker sections is generally less critical than that of thinner sections. **Table 15** provides maximum time delays, from furnace to immersion, for various product thicknesses.

Formability and machinability of aluminum and aluminum alloys

Aluminum alloys are readily formed hot or cold by common fabrication processes. Generally, pure aluminum is more easily worked than the alloys, and annealed tempers

(Text continued on p. 550)

Table 13. Recommended Precipitation Hardening Heat Treating Condition. *(Source, MIL-H-6088G.)*

Alloy	Temper Before Aging	Limitations	Age Hardening Heat Treatment[1]		Aging Time [2, 13]	Temper Designation After Indicated Treatment
			Metal Temperature[4]			
			°F	°C		
Wrought Products (Excluding Forgings)						
2011	-W	-	Room Temp.	Room Temp.	96 (minimum)	-T4, -T42
	-T3	-	310-330	154-166	14	-T8
	-T4	-	-	-	-	-
	-T451	-	-	-	-	-
2014	-W	-	Room Temp.	Room Temp.	96 (minimum)	-T4, -T42
	-T3	Flat sheet	310-330	154-166	18	-T6
	-T4, -T42 [3]	-	340-360	171-182	10	-T6, -T62
	-T451 [3]	-	340-360	171-182	10	-T651
	-T4510	Extrusions	340-360	171-182	10	-T6510
	-T4511	Extrusions	340-360	171-182	10	-T6511
2017	-W	-	Room Temp.	Room Temp.	96 (minimum)	-T4
	-T4	-	-	-	-	-
	-T451	-	-	-	-	-
2024	-W	-	Room Temp.	Room Temp.	96 (minimum)	-T4, -T42
	-T3	Sheet and drawn tube	365-385	185-196	12	-T81
	-T4	Wire, rod, bar	365-385	185-196	12	-T6
	-T3	Extrusions	365-385	185-196	12	-T81
	-T36	Wire	365-385	185-196	8	-T86
	-T42	Sheet and plate	365-385	185-196	9	-T62
	-T42	Sheet only	365-385	185-196	16	-T72
	-T42	Other than sheet and plate	365-385	185-196	16	-T62
	-T351	Sheet and plate	365-385	185-196	12	-T851
	-T361		365-385	185-196	8	-T861
	-T3510	Extrusions	365-385	185-196	12	-T8510
	-T3511		365-385	185-196	12	-T8511
2048	-W	-	Room Temp.	Room Temp.	96 (minimum)	-T4, -T42
	-T42	Sheet and plate	365-385	185-196	9	-T62
	-T351	-	365-385	185-196	12	-T851
2117	-W	Wire, rod, bar and rivets	Room Temp.	Room Temp.	96 (minimum)	-T4
2124	-W	Plate	Room Temp.	Room Temp.	96 (minimum)	-T4, -T42
	-T4		365-385	185-196	9	-T6
	-T42		365-385	185-196	9	-T62
	-T351		365-385	185-196	12	-T851
2219	-W	-	Room Temp.	Room Temp.	96 (minimum)	-T4, -T42
	-T31	Sheet	340-360	171-182	18	-T81

(Continued)

Table 13. *(Continued)* **Recommended Precipitation Hardening Heat Treating Condition.** *(Source, MIL-H-6088G.)*

Alloy	Temper Before Aging	Limitations	Age Hardening Heat Treatment[1]		Aging Time [2, 13]	Temper Designation After Indicated Treatment
			Metal Temperature[4]			
			°F	°C		
Wrought Products (Excluding Forgings)						
2219	-T31	Extrusions	365-385	185-196	18	-T81
	-T31	Rivets	340-360	171-182	18	-T81
	-T37	Sheet	315-335	157-168	24	-T87
	-T37	Plate	340-360	171-182	18	-T87
	-T42	-	365-385	185-196	36	-T62
	-T351	-	340-360	171-182	18	-T851
	-T351	Rod and bar	365-385	185-196	18	-T851
	-T3510	Extrusions	365-385	185-196	18	-T8510
	-T3511		365-385	185-196	18	-T8511
6010	-W	Sheet	340-360	171-182	8	-T6
6013	-W	Sheet	Room Temp.	Room Temp.	336	-T4
	-T4 [22]	-	365-385	185-196	4	-T6
6061	-W	-	Room Temp.	Room Temp.	96 (minimum)	-T4, -T42
	-T1	Rods, bar, shapes, and tube, extruded	340-360	171-182	8	-T5
	-T4 [14]	Except extrusions	310-330	154-166	18	-T6
	-T451		310-330	154-166	18	-T651
	-T42		310-330	154-166	18	-T62
	-T4	Extrusions	340-360	171-182	8	-T6
	-T42		340-360	171-182	8	-T62
	-T4510		340-360	171-182	8	-T6510
	-T4511		340-360	171-182	8	-T6511
6063	-W	Extrusions	Room Temp.	Room Temp.	96 (minimum)	-T4, -T42
	-T1	-	350-370	177-188	3	-T5, -T52
	-T1	-	415-435	213-224	1-2	-T5, -T52
	-T4	-	340-360	171-182	8	-T6
	-T4	-	350-370	166-166	6	-T6
	-T42	-	340-360	171-182	8	-T62
	-T42	-	350-370	154-166	6	-T62
	-T4510	-	340-360	171-182	8	-T6510
	-T4511	-	340-360	171-182	8	-T6511
6066	-W	Extrusions	Room Temp.	Room Temp.	96 (minimum)	-T4, -T42
	-T4	-	340-360	171-182	8	-T6
	-T42	-	340-360	171-182	8	-T62
	-T4510	-	340-360	171-182	8	-T6510
	-T4511	-	340-360	171-182	8	-T6511
6262	-W	-	Room Temp.	Room Temp.	96 (minimum)	-T4

(Continued)

Table 13. *(Continued)* **Recommended Precipitation Hardening Heat Treating Condition.** *(Source, MIL-H-6088G.)*

Alloy	Temper Before Aging	Limitations	Age Hardening Heat Treatment[1]		Aging Time [2, 13]	Temper Designation After Indicated Treatment
			Metal Temperature[4]			
			°F	°C		
Wrought Products (Excluding Forgings)						
6262	-T4	Wire, rod, bar and drawn tube	330-350	166-177	8	-T6
	-T451	-	330-350	166-177	8	-T651
	-T4	Extrusions	340-360	171-182	12	-T6
	-T4510		340-360	171-182	12	-T6510
	-T4511		340-360	171-182	12	-T6511
6951	-W	-	Room Temp.	Room Temp.	96 (minimum)	-T4, -T42
	-T4	Sheet	310-330	154-166	18	-T6
	-T42	-	310-330	154-166	18	-T62
7001	-W	Extrusions	240-260	116-127	24	-T6
	-T510		240-260	116-127	24	-T6510
	-T511		240-260	116-127	24	-T6511
7010	-W51 [21]	Plate	240-260 plus 330-350	116-127 plus 166-177	6-24 6-15	- -T7651
			240-260 plus 330-350	116-127 plus 166-177	6-24 9-18	-T7451 [17]
			240-260 plus 330-350	116-127 plus 166-177	6-24 15-24	T7351
7039	-W [15]	Sheet	165-185 plus 310-330	74-85 plus 154-166	16 14	-T61
	-W51 [15]	Plate	165-185 plus 310-330	74-85 plus 154-166	16 14	-T64
7049, 7149	-W511	Extrusions	Room Temp. followed by 240-260 followed by 320-330	Room Temp. followed by 116-127 followed by 160-166	48 24 12-14	-T76510, -T76511
			Room Temp. followed by 240-260 followed by 325-335	Room Temp. followed by 116-127 followed by 163-168	48 24-25 12-21	-T73510, -T73511 [19]
7050	-W51 [8]	Plate	240-260 plus 315-335	116-127 plus 157-168	3-6 12-15	-T7651
			240-260 plus 315-335	116-127 plus 157-168	3-6 24-30	-T7451 [17]

(Continued)

Table 13. *(Continued)* **Recommended Precipitation Hardening Heat Treating Condition.** *(Source, MIL-H-6088G.)*

Alloy	Temper Before Aging	Limitations	Age Hardening Heat Treatment[1]			Temper Designation After Indicated Treatment
			Metal Temperature[4]		Aging Time [2],[13]	
			°F	°C		
Wrought Products (Excluding Forgings)						
7050	-W510 [8]	Extrusions	240-260 plus 315-335	116-127 plus 157-168	3-8 15-18	-T76510
	-W511 [8]		240-260 plus 315-335	116-127 plus 157-168	3-8 15-18	-T76511
	-W [8]	Wire, rod, rivets	245-255 plus 350-360	118-124 plus 177-182	4 min. 8 min.	-T73
7075	-W [7]	-	240-260	116-127	24	-T6, -T62
	-W [5, 8, 11]	Sheet and plate	215-235 plus 315-335	101-113 plus 157-168	6-8 24-30	-T73
	-W [8, 11]		240-260 plus 315-335	116-127 plus 157-168	3-5 15-18	-T76
	-W [6, 8, 11]	Wire, rod, bar	215-235 plus 340-360	101-113 plus 171-182	6-8 8-10	-T73
	-W [5, 8, 11]	Extrusions	215-235 plus 340-360	101-113 plus 171-182	6-8 6-8	-T73
	-W [8, 11]		240-260 plus 310-330	116-127 plus 154-166	3-5 18-21	-T76
	-W51 [5, 8, 11]	Plate	215-235 plus 315-335	101-235 plus 157-168	6-8 24-30	-T7351
	-W51 [8, 11]		240-260 plus 315-335	116-127 plus 157-168	3-5 15-18	-T7651
	-W51 [10]	-	240-260	116-127	24	-T651
	-W51 [6, 8, 11]	Wire, rod, bar	215-235 plus 340-360	101-113 plus 171-182	6-8 8-10	-T7351
	-W510 [7]	Extrusions	240-260	116-127	24	-T6510
	-W511 [7]		240-260	116-127	24	-T6511
	-W50 [5, 8, 11]		215-235 plus 340-360	101-113 plus 171-182	6-8 6-8	-T73510
	-W511 [5, 8, 11]		215-235 plus 340-360	101-113 plus 171-182	6-8 6-8	-T73511

(Continued)

Table 13. *(Continued)* **Recommended Precipitation Hardening Heat Treating Condition.** *(Source, MIL-H-6088G.)*

Alloy	Temper Before Aging	Limitations	Age Hardening Heat Treatment[1]			Temper Designation After Indicated Treatment
			Metal Temperature[4]		Aging Time [2, 13]	
			°F	°C		
Wrought Products (Excluding Forgings)						
7075	-W510 [5, 8, 11]	Extrusions	240-260 plus 310-330	116-127 plus 154-166	3-5 18-21	-T76510
	-W511 [8, 11]		240-260 plus 310-330	116-127 plus 154-166	3-5 18-21	-T76511
	-T6 [8]	Sheet	315-335	157-168	24-30	-T73
	-T6 [8]	Wire, rod, bar	340-360	171-182	8-10	-T73
	-T6 [8]	Extrusions	340-360 310-330	171-182 154-166	6-8 18-21	-T73 -T76
	-T651 [8]	Plate	315-335 315-335	157-168 157-168	24-30 15-18	-T7351 -T7651
	-T651 [8]	Wire, rod, bar	340-360	171-182	8-10	-T7351
	-T6510 [8]	Extrusions	340-360 310-330	171-182 154-166	6-8 18-21	-T73510 -T76510
	-T6511 [8]		340-360 310-330	171-182 154-166	6-8 18-21	-T73511 -T76511
7150	-W510, -W511	Extrusions	240-260 plus 310-330	116-127 plus 154-166	8 4-6 [20]	-T6510, -T6511
	-W51	Plate	240-260 plus 300-320	116-127 plus 149-160	24 12	-T651
7178	-W	-	240-260	116-127	24	-T6, -T62
	-W [8, 11]	Sheet	240-260 plus 315-335	116-127 plus 157-168	3-5 15-18	-T76
	-W [8, 11]	Extrusions	240-260 plus 310-330	116-127 plus 154-166	3-5 18-21	-T76
	-W51	Plate	240-260	116-127	24	-T651
	-W51 [8, 11]		240-260 plus 315-335	116-127 plus 157-168	3-5 15-18	-T7651
	-W510	Extrusions	240-260	116-127	24	-T6510
	-W510 [8, 11]	-	240-260 plus 310-330	116-127 plus 154-166	3-5 18-21	-T76510
	-W511	Extrusions	240-260	116-127	24	-T6511
	-W511 [8, 11]		240-260 followed by 310-330	116-127 followed by 154-166	3-5 18-21	-T76511

(Continued)

Table 13. *(Continued)* **Recommended Precipitation Hardening Heat Treating Condition.** *(Source, MIL-H-6088G.)*

Alloy	Temper Before Aging	Limitations	Age Hardening Heat Treatment[1]		Aging Time [2, 13]	Temper Designation After Indicated Treatment
			Metal Temperature[4]			
			°F	°C		
Wrought Products (Excluding Forgings)						
7475	-W	Sheet	240-260 followed by 315-325	116-127 followed by 157-163	3 8-10	-T761
	-W51	Plate	240-260	116-127	24	-T651
7475 Alclad	-W	Sheet	250-315	121-157	3	-T61
Forgings						
2014	-W	-	Room Temp.	Room Temp.	96 (minimum)	-T4
	-T4	-	330-350	166-177	10	-T6
	-T41	-	340-360	171-182	5-14	-T61
	-T452	Hand forgings	330-350	166-177	10	-T652
2018	-W	Die forgings	Room Temp.	Room Temp.	96 (minimum)	-T4
	-T41	Die forgings	330-350	166-177	10	-T61
2024	-W	Die and hand forgings	Room Temp.	Room Temp.	96 (minimum)	-T4
	-W52	Hand forgings	Room Temp.	Room Temp.	96 (minimum)	-T352
	-T4	Die and hand forgings	365-385	171-182	12	-T6
	-T352	Hand forgings	365-385	171-182	12	-T852
2025	-W	Die forgings	Room Temp.	Room Temp.	96 (minimum)	-T4
	-T4	Die forgings	330-350	166-177	10	-T6
2218	-W	Die forgings	Room Temp.	Room Temp.	96 (minimum)	-T4, -T41
	-T4	Die forgings	330-350	166-177	10	-T61
	-T41	Die forgings	450-470	232-243	6	-T72
2219	-W	-	Room Temp.	Room Temp.	96 (minimum)	-T4
	-T4	-	365-385	171-182	26	-T6
	-T352	Hand forgings	340-360	171-182	18	-T852
2618	-W	-	Room Temp.	Room Temp.	96 (minimum)	-T4
	-T41	Die forgings	380-400	193-204	20	-T61
4032	-W	Die forgings	Room Temp.	Room Temp.	96 (minimum)	-T4
	-T4	Die forgings	330-350	166-177	10	-T6
6053	-W	Die forgings	Room Temp.	Room Temp.	96 (minimum)	-T4
	-T4	Die forgings	330-350	166-177	10	-T6
6061	-W	Die and hand forgings	Room Temp.	Room Temp.	96 (minimum)	-T4
	-T41	Die and hand forgings	340-360	171-182	8	-T61
	-T452	Rolled rings and hand forgings	340-360	171-182	8	-T652
6066	-W	Die forgings	Room Temp.	Room Temp.	96 (minimum)	-T4
	-T4	Die forgings	340-360	171-182	8	-T6

(Continued)

Table 13. *(Continued)* **Recommended Precipitation Hardening Heat Treating Condition.** *(Source, MIL-H-6088G.)*

Alloy	Temper Before Aging	Limitations	Age Hardening Heat Treatment[1]		Aging Time [2, 13]	Temper Designation After Indicated Treatment
			Metal Temperature[4]			
			°F	°C		
Forgings						
6151	-W	Die forgings	Room Temp.	Room Temp.	96 (minimum)	-T4
	-T4	Die forgings	330-350	166-177	10	-T6
	-T452	Rolled Rings	330-350	166-177	10	-T652
7049	-W -W52	Die and hand forgings	Room Temp. followed by 240-260 followed by 320-330	Room Temp. followed by 116-127 followed by 160-166	48 24 10-16	-T73 -T7352
7050			240-260 plus 340-360	116-127 plus 171-182	3-6 6-12	
	-W	Die forgings				-T74 [16]
	-W52	Hand forgings	240-260 plus 340-360	116-127 plus 171-182	3-6 6-8	-T7452 [18]
7075	-W	-	240-260	116-127	24	-T6
	-W [8, 11]	- - -	215-235 plus 340-360	101-113 plus 171-182	6-8 8-10	-T73
	-W52	Hand forgings	240-260	116-127	24	-T652
	-W52 [8, 11]	- -	215-235 plus 340-360	101-113 plus 171-182	6-8 6-8	-T7352
	-W51	Rolled Rings	215-235 plus 340-360	101-113 plus 171-182	6-8 6-8	-T7351
	-W	Die and hand forgings	215-235 plus 340-360	101-113 plus 171-182	6-8 6-8	-T74 [16]
7076	-W	Die and hand forgings	265-285	129-140	14	-T6
7149	-W	Die and hand forgings	Room Temp.	Room Temp.	48	
	-W52	-	240-260 plus 320-340	116-127 plus 160-171	24 10-16	-T73, -T7352
7175	-W52	Hand forgings	240-260	116-127		
	-W	Die and hand forgings	215-235 plus 340-360	101-113 plus 171-182	6-8 6-8	-T74 [16]
Castings (all mold practices)						
201.0	-T4	-	300-320	149-160	10-24	-T6
A201.0	-T4	-	360-380	177-193	5 (minimum)	-T7
A206.0	-T4	-	380-400	193-204	5 (minimum)	-T7
222.0	-F	-	330-350	166-177	16-22	-T551
	-T4	-	380-400	193-204	10-12	-T61
	-T4	-	330-350	166-177	7-9	-T65

(Continued)

Table 13. *(Continued)* Recommended Precipitation Hardening Heat Treating Condition. *(Source, MIL-H-6088G.)*

Alloy	Temper Before Aging	Limitations	Age Hardening Heat Treatment[1]		Aging Time [2, 13]	Temper Designation After Indicated Treatment
			Metal Temperature[4]			
			°F	°C		
Castings (all mold practices)						
242.0	-F	-	320-350	160-177	22-26	-T571
	-T41	-	400-450	204-232	1-3	-T61
295.0	-T4	-	300-320	149-160	12-20	-T62
296.0	-T4	-	300-320	149-160	1-8	-T6
	-T4	-	490-510	254-266	4-6	-T7
319.0	-T4	-	300-320	149-160	1-6	-T6
328.0	-T4	-	300-320	149-160	2-5	-T6
333.0	-F	-	390-410	199-210	7-9	-T5
	-T4	-	300-320	149-160	2-5	-T6
	-T4	-	490-510	254-266	4-6	-T7
336.0	-T45	-	300-350	149-177	14-18	-T65
354.0	-T41	-	300-320	149-160	10-12	-T61
	-T41	-	330-350	166-177	6-10	-T62
355.0 and C355.0	-F	-	430-450	221-232	7-9	-T51
	-T4	-	300-320	149-160	1-6	-T6
	-T4	-	300-320	149-160	10-12	-T61
	-T4	-	330-350	166-177	14-18	-T62
	-T4	-	430-450	221-232	3-5	-T7
	-T4	-	465-485	240-252	4-6	-T71
356.0 and A356.0	-T4	-	-	-	-	-
	-F	-	430-450	221-232	6-12	-T51
	-T4	-	300-320	149-160	1-6	-T6
	-T4	-	300-320	149-160	6-10	-T61
357.0 and A357.0	-T4		300-340	149-171	2-12	-T6
359.0	-T4	-	300-320	149-160	8-12	-T61
	-T41	-	330-350	166-177	6-10	-T62
520.0	-T4	-	300-320	149-160	20-12	-T61
	-T41	-	330-350	166-177	6-10	-T62
705.0	-W	-	200-220 or Room Temp.	93-104 or Room Temp.	10 21 days	-T5
707.0	-F	-	300-320 or Room Temp.	149-160 or Room Temp.	3-5 21 days	-T5
712.0	-F	-	300-320 or Room Temp.	149-160 or Room Temp.	9-11 21 days	-T5
	-F	-	Room Temp.	Room Temp.	96 (minimum)	-T1
713.0	-F	-	240-260 or Room Temp.	116-127 or Room Temp.	16 21 days	-T5

(Continued)

Table 13. *(Continued)* **Recommended Precipitation Hardening Heat Treating Condition.** *(Source, MIL-H-6088G.)*

Alloy	Temper Before Aging	Limitations	Age Hardening Heat Treatment[1]		Aging Time [2, 13]	Temper Designation After Indicated Treatment
			Metal Temperature[4]			
			°F	°C		
Castings (all mold practices)						
850.0	-F	-	420-440	216-226	7-9	-T5
851.0	-F	-	420-440	216-226	7-9	-T5
852.0	-F	-	420-440	216-226	7-9	-T5

[1] To produce the stress-relieved tempers, metal that has been solution heat-treated in accordance with -W temper, material must be stretched or compressed as required before aging. In instances where a multiple stage aging treatment is used, the metal may be, but need not be, removed from the furnace and cooled between aging steps.

[2] The time at temperature will depend on time required for load to reach temperature. The times shown are based on rapid heating with soaking time measured from the time the load reached the minimum temperature shown.

[3] Alternate treatment of 18 hours at 305°-330° F (151.6°-165.6° C) may be used for sheet and plate.

[4] When the interval of the specified temperature range exceeds 20° F (11.1- C), any 20° temperature range (or 30° [16.6° C] range for 6061) within the entire range may be utilized provided that no exclusions or qualifying criteria are cited herein or in the applicable material specification.

[5] Alternate treatment of 6 to 8 hours at 215° to 235° F (101.6° to 112.7° C) followed by a second stage of 14 to 18 hours at 325° to 345° F (162.7° to 173.8° C) may be used providing a heating-up rate of 25° F (13.8° C) per hour is used.

[6] Alternate treatment of 10 to 14 hours at 340° to 360° F (171.1° to 182.2° C) may be used providing a heating-up rate of 25° F (13.8° C) per hour is used.

[7] For extrusions an alternate three-stage treatment comprised of 5 hours at 200° to 220° F (93.3° to 104.4° C) followed by 4 hours at 240° to 260° F (115.6° to 126.7° C) followed by 4 hours at 290° to 310° F (143.3° to 154.4° C) may be used.

[8] The aging of aluminum alloys 7049, 7050, 7075 and 7178 from any temper to the -T7 type tempers requires closer control on aging practice variables such as time, temperature, heating-up rates, etc., for any given item. In addition to the above, when re-aging material in the -T6 temper series to the -T7 type temper series, the specific condition of the T6 temper material (such as its property level and other effects of processing variables) is extremely important and will affect the capability of the re-aged material to conform to the requirements specified for the applicable -T7 type tempers.

[9] Old or former commercial designation is shown in parentheses.

[10] For plate, an alternate treatment of 4 hours at 195°-215° F (90.6°-101.6° C) followed by a second stage of 8 hours at 305°-325° F (151.6°-162.7° C) may be used.

[11] With respect to -T73, -7351, -T73510, -T73511, -T7352, -T76, -T76510 and -T76511 tempers, a license has been granted to the public under U.S. Patent 3,198,676 and these times and temperatures are those generally recommended by the patent holder. Counterpart patents exist in several countries other than the United States. Licenses to operate under these counterpart patents should be obtained from the patent holder.

[12] A heating-up rate of 50°-75° F (27.7°-47.6° C) per hour is recommended.

[13] The 96 hour minimum aging time required for each alloy listed with temper designation -W is not necessary if artificial aging is to be employed to obtain tempers other than that derived from room temperature aging. (For example, natural aging—96 hours—to achieve the -T4 or -T42 temper for 2014 alloy is not necessary prior to artificial aging to obtain a -T6 or -T62 temper.)

[14] An alternate treatment comprised of 8 hours at 350° F (176.7° C) also may be used.

[15] A heating-up rate of 35° F (19.4° C) per hour from 135° F (57.2° C) is recommended.

[16] Formerly designated as -T736 temper.

[17] Formerly designated as -T73651 temper.

[18] Formerly designated as -T73652 temper.

[19] Longer times are to be used with section thicknesses less than 2 inches (50.5 mm).

[20] Soak time of 4 hours for extrusions with leg thickness less than 0.8 inch (20.3 mm) and 6 hours for extrusions having thicker legs.

[21] An alternative treatment is to omit the first stage and heat at a rate no greater than 36° F (20° C) per hour.

[22] Does not require the 14-day room temperature age.

Table 14. Recommended Annealing Conditions for Wrought Aluminum and Aluminum Alloys.[1]
(Source, MIL-H-6088G.)

-O Temper is Obtained After These Annealing Conditions						
Alloy	Metal Temperature[3]		Time at Temperature	Alloy	Metal Temperature		Time at Temperature
	°F	°C			°F	°C	
1060	650	349	See Note 2	5086	650	349	See Note 2
1100	650	349	See Note 2	5154	650	349	See Note 2
1350	650	349	See Note 2	5254	650	349	See Note 2
2014	760	404	See Note 4, 2-3 hours	5454	650	349	See Note 2
2017	760	404	See Note 4, 2-3 hours	5456	650	349	See Note 2
2024	760	404	See Note 4, 2-3 hours	5457	650	349	See Note 2
2036	725	385	See Note 4, 2-3 hours	5652	650	349	See Note 2
2117	760	404	See Note 4, 2-3 hours	6005	760	404	See Note 4, 2-3 hours
2219	760	404	See Note 4, 2-3 hours	6013	775	413	See Note 4, 2-3 hours
3003	775	413	See Note 2	6053	760	404	See Note 4, 2-3 hours
3004	650	349	See Note 2	6061	760	404	See Note 4, 2-3 hours
3105	650	349	See Note 2	6063	760	404	See Note 4, 2-3 hours
5005	650	349	See Note 2	6066	760	404	See Note 4, 2-3 hours
5050	650	349	See Note 2	7001	760	404	See Note 5, 2-3 hours
5052	650	349	See Note 2	7075	760	404	See Note 5, 2-3 hours
5056	650	349	See Note 2	7175	760	404	See Note 5, 2-3 hours
5083	650	349	See Note 2	7178	760	404	See Note 5, 2-3 hours

1 This table should be used for information and guidance purposes only. It is desirable to test specimen before establishing final temperatures.

2 Time in furnace should be no longer than necessary to get center of load to the desired temperature, taking into consideration the thickness or diameter of metal. Rate of cooling is unimportant.

3 Metal temperature variation in the annealing furnace should not be greater than +10° F, -15° F (+5.5° C, -8.3° C).

4 This annealing removes the effects of the solution heat treatment. Cooling rate must be 50° F (28° C) per hour from annealing temperature to 500° F (260° C). The rate of subsequent cooling is unimportant.

5 This annealing removes the effects of the solution heat treatment by cooling at an uncontrolled rate in the air to 400° F (204° C) or less followed by a reheating to 450° F (232° C) for 4 hours and cooling at room atmosphere conditions.

Table 15. Maximum Quench Delay for Immersion Quenching. (See Note.) *(Source, MIL-H-6088G.)*

Nominal Thickness		Maximum Time (Seconds)[1]
Up to 0.016 inch incl.	Up to 0.4064 mm incl.	5
0.017 to 0.031 inch incl.	0.4318 to 0.7874 mm incl.	7
0.032 to 0.090 inch incl.	0.8128 to 2.2860 mm incl.	10
0.091 inch and over	2.3114 mm and over	15

Note: Quench delay time begins when the furnace door starts to open or when the first corner of the load emerges from a salt bath, and ends when the last corner of the load is immersed in the quenchant. With the exception of alloy 2219, the maximum quench delay times may be exceeded (for example, with extremely large loads or long lengths) if performance tests indicate that all portions of the load will be above 775° F (413° C) when quenched. For alloy 2219, the maximum quench delay times may be exceeded if performance tests indicate all parts will be above 900° F when quenched.

[1] Shorter times than shown may be necessary to ensure that the minimum temperature of 7178 alloy is above 775° F (412° C) when quenched.

are more easily worked than the hard tempers. Also, the naturally aged tempers afford better formability than the artificially aged tempers. For example, the 99% metal (alloy 1100), in the annealed temper "–O," has the best forming characteristics, and alloy 7075, in the full heat treated temper "–T6," is among the most difficult to form because of its hardness.

In the process of forming, the metal hardens and strengthens due to the working effect. In cold drawing, the changes in tensile strength and other properties can become quite large depending on the amount of work and the alloy's composition. In bending (a form of cold working), the bend radius and the thickness of the metal must also be considered (see **Table 16** for appropriate bend radii for specified thicknesses). Most forming is done when cold, and choosing the proper temper usually permits the completion of the fabrication without the necessity of intermediate anneal. In some difficult drawing operations, however, intermediate annealing may be required between successive draws. Hot forming is usually done at temperatures of 300–400° F (149–204° C), which allows the metal to be easily worked without appreciably lowering its strength provided that the heating periods do not exceed thirty minutes. In general, a combination of the shortest possible time with the lowest temperature that will provide the desired results is recommended. Forming is also done in the as-quenched condition on those alloys that age spontaneously at room temperature after solution heat treatment (–W temper). In these instances, the quenched material should be refrigerated to retard hardening until forming is complete.

Selecting the proper temper is important when specifying aluminum for forming operations. When nonheat treatable alloys are to be formed, the temper chosen should be sufficiently soft to permit the desired bend radius or draw depth. In more difficult forming operations, material in the annealed temper "–O" should be used. For less severe forming requirements, material in one of the harder tempers, such as "–H14," may be satisfactorily formed. When heat treatable alloys are to be used for forming, the shape should govern the selection of the alloy and its temper. Maximum formability of the heat treatable alloys is obtained in the annealed temper. However, limited formability can be effected in the fully heat treated temper, provided the bend radii are large enough.

Clues to the formability of an alloy are its percent of elongation, and the difference between the yield and ultimate tensile strengths. As a rule, the higher the elongation value and/or the wider the range between the yield and tensile strengths, the better the forming characteristics.

Machinability—the ease with which a material can be finished by cutting—is characterized by fast cutting speeds, small chip size, smoothness of the machined surface,

Table 16. Approximate Radii for 90° Cold Bend of Wrought Aluminum Alloys. (Source, MIL-HDBK-694A.)

| Designation | | Radius Required (in terms of sheet thickness) | | | | | | | |
Alloy	Temper	1/64 inch .397 mm	1/32 inch .794 mm	1/16 inch 1.588 mm	1/8 inch 3.175 mm	3/16 inch 4.762 mm	1/4 inch 6.35 mm	3/8 inch 9.525 mm	1/2 inch 12.7 mm
1100	-O	0	0	0	0	0	0	0	1-2
	-H14	0	0	0	0	0-1	0-1	0-1	2-3
	-H18	0-1	0.5-1.5	1-2	1.5-3	2-4	2-4	3-5	3-6
3003	-O	0	0	0	0	0	0	0	1-2
	-H14	0	0	0	0-1	0-1	1-1.5	1-2.5	1.5-3
	-H18	0.5-1.5	1-2	1.5-3	2-4	3-5	4-6	4-7	5-8
5052	-O	0	0	0	0	0-1	0-1	0.5-1.5	1-2
	-H34	0.5-1.5	0	0-1	0.5-1.5	1-2	1.5-3	2-3	2.5-3.5
	-H38	0.5-1.5	1-2	1.5-3	2-4	3-5	4-6	4-7	5-8
5083	-O	-	-	0-0.5	0-1	0-1	0.5-1.5	1.5-2	1.5-2.5
2014 Clad	-O	0	0	0	0	0-1	0-1	1.5-3	3-5
	-T3	1-2	1.5-3	2-4	3-5	4-6	4-6	5-7	5.5-8
	-T4	1-2	1.5-3	2-4	3-5	4-6	4-6	5-7	5.5-8
	-T6	2-4	3-5	3-5	4-6	5-7	6-10	7-10	8-11
2024	-O	0.5-1.5	0	0	0	0-1	0-1	1.5-3	3-5
	-T3	1.5-3	2-4	3-5	4-6	4-6	5-7	6-8	6-9
	-T4	1.5-3	2-4	3-5	4-6	4-6	5-7	6-8	6-9
	-T81	3.5-5	4.5-6	5-7	6.5-8	7-9	8-10	9-11	9-12
5456	-O	-	-	-	0-1	0.5-1	0.5-1	0.5-1.5	0.5-2
	-H321	-	-	-	2-3	2-3	3-4	3-4	3-4
6061	-O	0	0	0	0	0	0	0.5-2	1-2.5
	-T4	0-1	0-1	0.5-1.5	1-2	1.5-3	2-4	2.5-4	3-5
	-T6	0-1	0.5-1.5	1-2	1.5-3	2-4	3-4	3.5-5.5	4-6
7075	-O	0	0	0-1	0.5-1.5	1-2	1.5-3	2.5-4	3-5
	-T6	2-4	3-5	4-6	5-7	5-7	6-10	7-11	7-12

and good tool life. While some aluminum alloys are excellent for machining, others are very troublesome. The troublesome alloys are soft and gummy, produce long and stringy chips, and must be machined at slow cutting rates. The harder alloys and tempers afford better machinability. In general, alloys containing copper, zinc, or magnesium as the principal alloying constituents are the most readily machined. Other compositions, containing bismuth and lead, are also usually machinable as they were specially designed for high speed screw machine work. Compositions containing more than 10% silicon are usually the most difficult to machine, and even content of 5% silicon alloys exhibit a gray surface rather than machining to a bright finish.

Wrought alloys that have been heat treated have fair to good machining characteristics. These alloys are easier to machine to a good finish when they are in the full hard temper condition, rather than annealed. Wrought alloys that are not heat treated, regardless of temper, tend to gumminess. Finally, wrought compositions having copper as the principal alloying element are more easily machined than those that have been hardened mainly by magnesium silicide.

Corrosion Resistance and Protective Finishes

Aluminum and aluminum alloys owe their corrosion resistance to the oxide film that forms on the surface upon exposure to oxygen. This coating prevents further oxidation, but, in some environments, supplementary protection is required. The degree of inherent corrosion resistance of the alloy depends on the composition and the thermal history of the metal. Magnesium, silicon, or magnesium silicide enhance the corrosion resistance properties of aluminum alloys.

Cladding, chemical treatment, electrolytic oxide finishing, electroplating, and applications of organic or inorganic coatings are used for supplementary protection of aluminum alloys. Cladding is perhaps the most effective means of protection, and the process consists of applying layers (approximately 2 to 15% of the total thickness) of pure aluminum or a corrosion resistant aluminum alloy to the surface of the ingot, and then hot working the ingot to cause the cladding metal to weld to the core. In subsequent hot working and fabricating, the cladding becomes alloyed with the core and is reduced in thickness proportionally. The cladding not only serves to protect the core metal, it also affords protection by electrolytic action, even when the surface is scratched, because the cladding is anodic to the base metal and therefore corrodes sacrificially.

Some chemical treatments result in the formation of oxide films, while others etch the metal and lower the corrosion resistance by removing the oxide film. Chemical finishes, though widely used as paint bases because they are slightly porous, are not as successful as electrolytic finishes, and electrolytic oxide finishing is perhaps the most widely used method for protecting aluminum. It consists of treating the metal in an electrolyte capable of giving off oxygen, using the metal as an anode. The film formed by the process is an aluminum oxide that is thin, hard, inert, and minutely porous, and can be left as is, painted, or dyed. Electroplating plating aluminum alloys requires great care in preparing the surface of the metal. It must be buffed to remove any scratches or defects, cleaned thoroughly to remove all grease, dirt, or other foreign material, and it must be given a coating of pure zinc (by immersion in a zincate solution). After plating, the surface is buffed and finished like other metals.

Organic and inorganic coatings range from paints and lacquers to vitreous enamels. When providing a protective (rather than decorative) painted coating, the surface should first be etched with a cleaner containing phosphoritic acid (or similar) to remove contaminates and deposit a thin phosphate film. A prime coat such as zinc chromate can then be applied, followed by paint, varnish, or lacquer.

Stress-Corrosion Cracking

The high strength heat treatable wrought alloys in certain tempers are susceptible to stress-corrosion cracking, depending on product, section size, and the direction and magnitude of stress. These alloys include 2014, 2025, 2618, 7075, 7150, 7175, and 7475 in the "–T6" temper, and 2014, 2024, 2124, and 2219 in "–T3" and "–T4" tempers. Other alloy-temper combinations—notably 2024, 2124, 2219, and 2519 in the "–T6" or "–T8" type tempers, and 7010, 7049, 7050, 7075, 7149, 7175, and 7475 in the "–T3" type temper—are decidedly more resistant, and sustained tensile stresses of 50 to 75% of the minimum yield strength may be permitted without concern about stress corrosion cracking. The "–T74" and "–T76" tempers of 7010, 7075, 7475, 7049, 7149, and 7050 provide an intermediate degree of resistance to stress-corrosion cracking (superior to the "–T6" temper but not as good as the "–T73" temper of 7075). To assist in the selection of materials, letter ratings indicating the relative resistance to stress-corrosion cracking of various mill products are presented in **Table 17**. This table is based on ASTM G64, which contains more detailed information regarding the rating system and the procedures used to determine the ratings.

Where short periods at elevated temperatures of 150 to 500° F (66 to 260° C) may be encountered, the precipitation heat treated tempers of 2024 and 2219 are recommended over the naturally aged tempers. Alloys 5083, 5086, and 5456 should not be used under high constant applied stress for continuous service at temperatures exceeding 150° F (66° C), because of the hazard of developing susceptibility to stress-corrosion cracking. In general, the "–H34" through "–H38" tempers of 5086, and the "–H32" through "–H38" tempers of 5083 and 5456 are not recommended, because these tempers can become susceptible to stress-corrosion cracking.

To avoid stress-corrosion cracking, practices such as the use of press or shrink fits, taper pins, clevis joints (in which tightening of the bolt imposes a bending load on female lugs), and straightening or assembly operations that result in sustained surface stresses (especially when acting in the short transverse grain orientation) should be avoided in these high strength alloys: 2014–T451, –T4, –T6, –T651, –T652; 2024–T3, –T351, –T4; 7075–T6, –T651, –T652; 7150–T6151; and 7475–T6, –T651. When straightening or forming is necessary, it should be performed when the material is in the freshly quenched condition or at an elevated temperature to minimize the residual stresses induced. Where elevated temperature forming is performed on 2014–T4 and –T451, or 2024–T3 and –T351, a subsequent precipitation heat treatment for the "–T6" or "–T651," or "–T81" or "–T851," temper is recommended.

It is good engineering practice to control short transverse tensile stress at the surface of structural parts at the lowest practicable level. Careful attention should be given in all stages of manufacturing, beginning with design of the part configuration, to choose practices in the heat treatment, fabrication, and assembly that avoid unfavorable combinations of end-grain microstructure and sustained tensile stress. The greatest danger arises when residual, assembly, and service stress combine to produce high sustained tensile stress at the metal surface. It is imperative that, for materials with low resistance to stress-corrosion cracking in the short transverse grain orientation, every effort be taken to keep the level of sustained tensile stress close to zero.

Cryogenic effects

In general, the strengths (including fatigue strength) of aluminum alloys increase with decreases in temperature below room temperature. The increase is greatest over the range of minus 100 to minus 423° F (minus 73 to minus 253° C), with the upper range being the temperature of liquid hydrogen. The strengths at minus 452° F (minus 269°

(Text continued on p. 557)

Table 17. Resistance to Stress-Corrosion Ratings for High Strength Aluminum Alloy Products. See Note. *(Source, MIL-HDBK-5H.)*

Alloy and Temper[1]	Test Direction[2]	Rolled Plate	Rod and Bar[3]	Extruded Shapes	Forging
2014-T6	L	A	A	A	B
	LT	B[4]	D	B[4]	B[4]
	ST	D	D	D	D
2024-T3, T4	L	A	A	A	5
	LT	B[4]	D	B[4]	5
	ST	D	D	D	5
2024-T6	L	5	A	5	A
	LT	5	B	5	A[4]
	ST	5	B	5	D
2024-T8	L	A	A	A	A
	LT	A	A	A	A
	ST	B	A	B	C
2124-T8	L	A	5	5	5
	LT	A	5	5	5
	ST	B	5	5	5
2219-T351X, T37	L	A	5	A	5
	LT	B	5	B	5
	ST	D	5	D	5
2219-T6	L	A	A	A	A
	LT	A	A	A	A
	ST	A	A	A	A
2219-T85XX, T87	L	A	5	A	A
	LT	A	5	A	A
	ST	A	5	A	A
6061-T6	L	A	A	A	A
	LT	A	A	A	A
	ST	A	A	A	A
7049-T73	L	A	5	A	A
	LT	A	5	A	A
	ST	A	5	B	A
7049-T76	L	5	5	A	5
	LT	5	5	A	5
	ST	5	5	C	5
7050-T74	L	A	5	A	A
	LT	A	5	A	A
	ST	B	5	B	B
7050-T76	L	A	A	A	5
	LT	A	B	A	5
	ST	C	B	C	5
7075-T76	L	A	A	A	A
	LT	B[4]	D	B[4]	B[4]
	ST	D	D	D	D
7075-T73	L	A	A	A	A
	LT	A	A	A	A
	ST	A	A	A	A
7075-T74	L	5	5	5	A
	LT	5	5	5	A
	ST	5	5	5	B

(Continued)

Table 17. *(Continued)* **Resistance to Stress-Corrosion Ratings for High Strength Aluminum Alloy Products.**
See Note. *(Source, MIL-HDBK-5H.)*

Alloy and Temper[1]	Test Direction[2]	Rolled Plate	Rod and Bar[3]	Extruded Shapes	Forging
7075-T76	L	A	[5]	A	[5]
	LT	A	[5]	A	[5]
	ST	C	[5]	C	[5]
7149-T73	L	[5]	[5]	A	A
	LT	[5]	[5]	A	A
	ST	[5]	[5]	B	A
7175-T74	L	[5]	[5]	[5]	A
	LT	[5]	[5]	[5]	A
	ST	[5]	[5]	[5]	B
7475-T6	L	A	[5]	[5]	[5]
	LT	B[4]	[5]	[5]	[5]
	ST	D	[5]	[5]	[5]
7475-T73	L	A	[5]	[5]	[5]
	LT	A	[5]	[5]	[5]
	ST	A	[5]	[5]	[5]
7475-T76	L	A	[5]	[5]	[5]
	LT	A	[5]	[5]	[5]
	ST	C	[5]	[5]	[5]

Note: Ratings were determined from stress corrosion tests performed on at least ten random lots for which test results showed 95% conformance at the 95% confidence level when tested at the stresses indicated below. A practical interpretation of these ratings follows the rating definition.

A - Equal or greater than 75% of the specified minimum yield strength. Very high. No record of service problems and SCC not anticipated in general applications.

B - Equal or greater than 50% of the specified minimum yield strength. High. No record of service problems and SCC not anticipated at stresses of the magnitude caused by solution heat treatment. Precautions must be taken to avoid high sustained tensile stress exceeding 50% of the minimum specified yield strength produced by any combination of sources including heat treatment, straightening, forming, fit-up, and sustained service loads.

C - Equal or greater than 25% of the specified minimum yield strength. Intermediate. SCC not anticipated if the total sustained tensile strength is less than 25% of the minimum specified yield strength. This rating is designated for the short transverse direction in improved products used primarily for high resistance to exfoliation corrosion in relatively thin structures where applicable short transverse stresses are unlikely.

D - Fails to meet the criterion for the rating C. Low. SCC failures have occurred in service or would be anticipated if there is any sustained tensile stress in the designated test direction. This rating currently is designated only for the short transverse direction in certain materials.

The above stress levels are not to be interpreted as "threshold" stresses, and are not recommended for design. Other documents, such as MIL-STD-1568, NAS SD-24, and MSFC-SPEC-522A, should be consulted for design recommendations.

[1] The ratings apply to standard mill products in the types of tempers indicated, including stress-relieved tempers, and could be invalidated in some cases by application of nonstandard thermal treatments or mechanical deformation at room temperature by the user.

[2] Test direction refers to orientation of the stressing direction relative to the directional grain structure typical of wrought materials, which in the case of extrusions and forgings may not be predictable from the geometrical cross section of the product.

L-Longitudinal: parallel to the direction of principal metal extension during manufacture of the product.

LT-Long Transverse: perpendicular to direction of principal metal extension. In products whose grain structure clearly shows directionality (width to thickness ratio greater than two) it is that perpendicular direction parallel to the major grain dimension.

ST-Short Transverse: perpendicular to direction of principal metal extension and parallel to minor dimension of grains in products with significant grain directionality.

[3] Sections with width-to-thickness ratio equal to or less than two for which there is no distinction between LT and ST.

[4] Rating is one class lower for thicker sections: extrusion, 1 inch and over; plate and forgings, 1.5 inches and over.

[5] Ratings not established because the product is not offered commercially.

This table is based upon ASTM G 64.

Table 18. Typical Fatigue Strengths of Wrought Aluminum Alloys. *(Source, MIL-HDBK-694.)*

Alloy	Temper	Repeated Flexure Fatigue Strength				
		0.1 ksi/0.7 MPa	1.0 ksi/6.9 MPa	10 ksi/69 MPa	100 ksi/689 MPa	500 ksi/3450 MPa
		Million Cycles to Failure				
1100	-O	-	6.5	5.5	5	5
	-H16	14	11.5	10	9	8
3003	-O	10.5	9	8	7.5	7
	-H14	17	12	10	9	9
	-H18	19	14	11.5	10.5	10
5052	-O	23.5	19.5	17.5	16.5	16
	-H34	26	20.5	19	18	18
	-H38	29.5	24	22.5	21	20
2011	-T3	35	26.5	22.5	19.5	18
2014	-T6	39	30	24	19	18
2017	-T4	42	34	27	22	20
2018	-T61	42	29	23	19.5	17
2024	-T4	43	31	24	21	20
4032	-T6	37	30	23.5	18	16
6061	-T6	31	23	17	14.5	13.5
6063	-T42	19.5	16	13.5	11	9.5
	-T5	20.5	15.5	12	10.5	9.5
	-T6	23.5	16.5	13.5	11	9.5
6151	-T6	30	22	17	13	12
7075	-T6	40	29	24	22	22

Alloy	Temper	75°F/24°C	300°F/149°C	400°F/204°C	500°F/260°C
		Fatigue Strength ksi/MPa			
3003	-H18	10	7.5	5	3.5
2014	-T6	18	12	8	5
2024	-T4	20	14	9	6
5052	-H36	18.5	12.5	9.5	6
6061	-T6	16	11	7.5	4.5
7075	-T6	22	12	8.5	7

Table 19. Effect of Temperature on Thermal Coefficient of Linear Expansion for Wrought Aluminum Alloys. *(Source, MIL-HDBL-694.)*

Alloy	Average Coefficient, 10^{-6} in./in./°F			
	-58 to +68° F -49 to 20° C	68 to 212° F 20 to 100° C	68 to 392° F 20 to 200° C	68 to 572° F 20 to 300° C
1100	12.2	13.1	13.7	14.2
2011	11.9	12.8	13.4	13.9
2014	12.0	12.3	13.1	13.6
2017	12.1	12.7	13.3	13.9
2018	11.7	12.4	12.9	13.4

(Continued)

Table 19. *(Continued)* **Effect of Temperature on Thermal Coefficient of Linear Expansion for Wrought Aluminum Alloys.** *(Source, MIL-HDBL-694.)*

Alloy	Average Coefficient, 10^{-6} in./in./°F			
	-58 to +68° F -49 to 20° C	68 to 212° F 20 to 100° C	68 to 392° F 20 to 200° C	68 to 572° F 20 to 300° C
2024	11.9	12.6	13.2	13.7
2025	12.1	12.6	13.1	13.6
2117	12.1	13.0	13.6	14.0
2218	11.7	12.4	13.0	13.5
3003	12.0	12.9	13.5	13.9
4032	10.3	10.8	11.3	11.7
5052	12.3	13.2	13.8	14.3
5056	12.5	13.4	14.0	14.5
6053	12.1	12.8	13.4	14.0
6061	12.1	13.0	13.5	14.1
6063	12.1	13.0	13.6	14.2
6151	12.1	12.8	13.4	13.9
7075	12.1	12.9	13.5	14.4

C)—the temperature of liquid helium—are nearly the same as at minus 423° F. For most alloys, elongation and various indices of toughness remain nearly constant or increase with decrease in temperature, but the 7000 series experiences modest reductions. None of the alloys exhibits a marked transition in fracture resistance over a narrow range of temperature indicative of embrittlement. The tensile and shear moduli of aluminum alloys also increase with decreasing temperature, so that at minus100, minus 320, and minus 423° F (minus 73, minus 196, and minus 253° C), they are approximately 5, 12, and 16% (respectively) above the room temperature values.

Selecting aluminum alloys

As in selecting any other material used in engineering design, several factors must be considered when choosing an aluminum alloy for maximum value and optimum performance. The factors include service conditions, production run, and the relative costs of suitable machining and fabricating processes. Within limits, the selection of a specific composition for a particular application may be simplified by determining the requirements for mechanical or physical properties, determining which alloys meet the selected criteria, and from the chosen alloys, pick the one best suited for the fabrication and finish machining requirements. One critical characteristic when considering wrought alloys is fatigue strength, and typical fatigue strengths of selected alloys are given in **Table 18**. The effects of temperature on aluminum's characteristics are shown in **Table 19** (coefficient of linear expansion), **Table 20** (ultimate tensile strength), **Table 21** (yield strength), and **Table 22** (elongation).

The choice of an alloy for casting is, to a great extent, governed by the type of mold that is to be used. The type of mold sets guidelines for such factors as intricacy of design, size, cross section, tolerance, surface finish, and number of castings to be produced. Sand molds are particularly suited to large castings, large tolerances, and small runs. They are not acceptable for the production of thin sections (less than $^3/_{16}$ inch [4.76 mm]), or smooth finishes. Permanent molds, which are particularly well suited to cast iron, provide better

(Text continued on p. 561)

Table 20. Typical Effect of Temperature on Ultimate Tensile Strength for Wrought Aluminum Alloys. (Source, MIL-HDBK-694.)

Alloy	Temper	Percent of Ultimate Strength at 75°F/24°C								
		-320°F -196°C	-110°F -79°C	-20°F -29°C	+212°F +100°C	300°F 149°C	400°F 204°C	500°F 260°C	600°F 316°C	700°F 371°C
1100	-O	189	115	104	77	65	46	31	19	16
	-H18	144	109	104	92	75	25	17	10	8
3003	-O	206	122	107	81	69	53	38	25	19
	-H14	164	109	103	95	82	64	34	18	14
	-H18	143	110	104	90	79	48	26	14	10
5052	-O	158	106	101	100	86	64	43	27	18
	-H34	144	106	101	100	82	60	32	20	13
	-H38	141	105	101	98	81	55	29	18	12
5083	-O	141	105	101	100	68	52	36	25	14
2011	-T3	-	-	-	85	51	29	12	6	4
2014	-T6	120	104	103	89	57	23	13	9	6
2017	-T4	128	104	103	90	64	36	19	10	7
2018	-T61	118	104	103	92	74	31	16	9	7
2024	-T3	-	-	-	94	79	41	17	11	8
	-T4	127	106	104	94	66	40	18	12	8
	-T81	-	-	-	94	79	41	17	11	8
2117	-T4	-	-	-	84	70	37	17	10	7
2218	-T61	-	-	-	95	70	37	17	9	7
4032	-T6	119	105	102	91	67	24	14	9	6
6053	-T6	-	-	-	86	68	35	15	11	8
6061	-T6	133	110	105	93	76	42	17	10	7
6063	-T42	153	120	108	100	95	41	20	14	11
	-T5	138	108	105	89	74	33	17	11	9
	-T6	135	109	103	89	60	26	13	9	7
6151	-T6	125	107	104	88	56	25	14	10	8
7075	-T6	123	105	103	80	30	17	13	10	8
7079	-T6	-	-	-	86	44	20	14	10	6

Table 21. Typical Effect of Temperature on Yield Strength for Wrought Aluminum Alloys. *(Source, MIL-STD-694.)*

Alloy	Temper	Percent of Yield Strength at 75°F/24°C								
		-320°F -196°C	-110°F -79°C	-20°F -29°C	+212°F +100°C	300°F 149°C	400°F 204°C	500°F 260°C	600°F 316°C	700°F 371°C
1100	-O	123	107	103	100	90	70	40	30	20
	-H18	-	-	-	82	64	18	9	7	4
3003	-O	145	116	105	92	83	75	58	42	33
	-H14	120	104	101	90	76	43	19	12	10
	-H18	122	107	103	78	59	33	15	9	7
5052	-O	121	100	100	100	100	85	62	38	23
	-H34	118	103	100	97	87	48	26	16	10
	-H38	116	101	100	100	78	40	22	14	8
5083	-O	-	-	-	100	82	77	50	34	20
2011	-T3	-	-	-	79	44	26	9	5	4
2014	-T6	113	103	102	93	58	22	12	8	6
2017	-T4	132	104	101	92	75	42	24	12	9
2018	-T61	110	103	101	94	87	28	14	6	5
2024	-T3	133	106	101	96	92	44	18	12	8
	-T4	-	-	-	96	77	43	19	13	8
	-T81	-	-	-	95	78	34	14	9	6
2117	-T4	-	-	-	88	71	50	23	15	8
2218	-T61	-	-	-	95	80	36	14	7	6
4032	-T6	103	100	100	96	72	20	12	6	4
6053	-T6	-	-	-	88	75	38	12	8	6
6061	-T6	116	105	103	95	78	38	12	6	5
6063	-T42	126	119	110	108	115	50	26	19	16
	-T5	116	105	104	95	86	31	17	12	10
	-T6	116	106	102	90	64	21	11	8	6
6151	-T6	115	106	104	91	58	22	13	10	8
7075	-T6	124	105	102	85	29	16	12	9	6
7079	-T6	-	-	-	88	44	19	12	9	6

Table 22. Typical Effect of Temperature on Elongation for Wrought Aluminum Alloys. (Source, MIL-HDBK-694.)

Alloy	Temper	Percent Elongation									
		-320°F -196°C	-110°F -79°C	-20°F -29°C	+75°F +24°C	+212°F +100°C	300°F 149°C	400°F 204°C	500°F 260°C	600°F 316°C	700°F 371°C
1100	-O	-	-	-	45	45	55	65	75	80	85
	-H18	-	-	-	15	15	20	65	75	80	85
3003	-O	49	45	44	43	40	47	60	65	70	70
	-H14	32	18	16	16	16	16	20	60	70	70
	-H18	-	-	-	10	10	11	18	60	70	70
5052	-O	-	-	-	30	35	45	65	80	100	120
	-H32	30	21	18	14	16	25	40	80	100	120
	-H38	-	-	-	8	9	20	40	80	100	120
5083	-O	-	-	-	25	35	45	60	70	95	120
2011	-T3	-	-	-	15	16	25	35	45	90	125
2014	-T6	-	-	-	13	14	15	35	45	65	70
2017	-T4	-	-	-	22	18	16	28	45	65	100
2018	-T61	-	-	-	12	12	12	25	40	60	100
2024	-T3	-	-	-	17	16	11	23	55	75	100
	-T4	-	-	-	19	19	17	27	55	75	100
	-T81	-	-	-	7	8	11	23	55	75	100
2117	-T4	-	-	-	27	16	20	35	55	80	110
2218	-T61	-	-	-	13	14	17	30	70	85	100
4032	-T6	-	-	-	9	9	9	30	50	70	90
6053	-T6	-	-	-	13	13	13	25	70	80	90
6061	-T6	25	20	19	17	18	20	28	60	85	90
6063	-T42	-	-	-	33	18	20	40	75	80	105
	-T5	-	-	-	22	18	20	40	75	80	105
	-T6	-	-	-	18	15	20	40	75	80	105
6151	-T6	-	-	-	17	19	22	40	50	50	50
7075	-T6	-	-	-	11	15	30	60	65	80	65
7079	-T6	-	-	-	13	18	37	60	100	175	175

surface finishes and closer tolerances than sand molds, but the minimum thicknesses that can be produced are about the same. Permanent molds are better suited for longer runs because they do not require the pattern equipment or molding operations needed for sand castings. Dies are especially well suited for long production runs and, even though they are relatively expensive, their initial cost can be justified by savings in finishing costs, plus their high production rate. Other advantages are the ability to produce thinner cross sections, closer tolerances, smoother surfaces, and more intricate designs. **Table 23** gives casting properties of several alloys, plus their principal use.

Choosing a wrought alloy is influenced by the proposed method of fabrication and the design requirements of the finished part. Although a variety of compositions and tempers will generally produce the desired mechanical and physical properties, the number of compositions and tempers amenable to the various fabrication techniques is, in some instances, limited. The fabrication technique that will provide the greatest economy is

Table 23. Casting Properties and Principal Uses of Selected Aluminum Casting Alloys. *(Source, QQ-A-610F.)*

Alloy	Specific Gravity*	Casting Properties	Response to Heat Treatment	Principal Use	Machinability
208.0	2.79	Good	Yes	Miscellaneous castings. Composition is such as to make it more available during periods of shortages of primary aluminum.	Good
213.0	2.87	Good	No	General	Excellent
222.0	2.91	Good	Yes	Same as alloy 242.0. Automotive pistons.	Excellent
242.0	2.78	Fair	Yes	Castings requiring strength and hardness at elevated temperatures, such as air-cooled cylinder heads and pistons. Aircraft and diesel engine pistons.	Good
295.0	2.81	Good	Yes	General, where high strength ductility and resistance to shock are required.	Very good
296.0	2.78	Good	Yes	General, where high strength and high ductility are required.	Good
308.0	2.77	Excellent	No	General	Good
319.0	2.79	Excellent	Yes	High strength and general purpose alloy for intricate castings.	Good
328.0	2.73	Excellent	Yes	General, where high strength is required.	Good
332.0	2.76	Fair	Yes	Automotive and diesel pistons.	Fair
333.0	2.77	Good	Yes	General, high strength castings.	Good
336.0	2.70	Fair	Yes	Automotive and diesel pistons.	Fair
355.0	2.69	Good to Excellent	Yes	General, where high strength and corrosion resistance are required. Also retains strength at elevated temperatures. General, where pressure tightness such as in pump bodies, liquid cooled cylinder heads, is required.	Good
C355.0	2.69	Good	Yes	Same as alloy 355.0.	Good
356.0	2.66	Excellent	Yes	General, where high strength and corrosion resistance are required.	Good
A356.0	2.66	Excellent	Yes	Same as alloy 356.0.	Good
357.0	2.66	Excellent	Yes	Same	Good

(Continued)

Table 23. *(Continued)* **Casting Properties and Principal Uses of Selected Aluminum Casting Alloys.** *(Source, QQ-A-596A.)*

Alloy	Specific Gravity*	Casting Properties	Response to Heat Treatment	Principal Use	Machinability
B443.0	2.64	Excellent	No	General, with maximum corrosion resistance and for leak proof castings of intricate design.	Fair
512.0	2.65	Good	No	Same as alloy 514.0.	Good
513.0	2.68	Fair	No	General	Excellent
514.0	2.65	Fair	No	Castings requiring superior resistance to corrosion.	Very good
520.0	2.57	Fair	Yes	Castings requiring maximum strength, elongation and resistance to shock, require special founding practice.	Excellent
535.0	2.62	Good	Yes	Applications requiring excellent shock resistance and corrosion resistance.	Excellent
705.0	2.76	Good	Note [1]	Same as alloy 712.0. General, where high strength and ductility are required, particularly without heat treatment, in addition to excellent corrosion resistance.	Excellent
707.0	2.77	Good	Yes, but not for normal application [1]	Same as alloy 712.0.	Excellent
710.0	2.76	Fair	Note [1]	Same as alloy 712.0.	Excellent
712.0	2.81	Good	Note [1]	General, where high strength and ductility are required, particularly without heat treatment, in addition to excellent corrosion resistance.	Excellent
713.0	2.81	Good	Note [1]	Same as alloy 712.0.	Excellent
771.0	2.81	Good	Yes	Applications requiring excellent dimensional stability or shock resistance combined with very high yield strength.	Excellent
850.0	2.88	Fair	Yes	Bearings	Excellent
851.0	2.83	Fair	Yes	Bearings	Excellent
852.0	2.88	Fair	Yes	Bearings	Excellent

* Specific gravity value is approximate.

[1] These alloys, when properly cast, develop their highest strength after aging at room temperature for several weeks, or after short artificial aging at slightly elevated temperatures.

Table 24. Principal Characteristics and Uses of Wrought Aluminum Alloys. *(Source, MIL-HDBK-694.)*

Alloy	Outstanding Characteristics	Recommended Uses
Nonheat Treatable Alloys		
1100	Very good formability, weldability, and resistance to corrosion. Relatively low strength but high ductility.	General purpose material for drawing and stamping, and for a miscellany of parts where high strength is not required.
3003	Good formability and weldability, very good resistance to corrosion. Appreciably higher strength than 1100.	General purpose material for drawing and stamping. Miscellaneous parts where higher strength is needed than that provided by 1100.

(Continued)

Table 24. *(Continued)* **Principal Characteristics and Uses of Wrought Aluminum Alloys.**
(Source, MIL-HDBK-694.)

Alloy	Outstanding Characteristics	Recommended Uses
Nonheat Treatable Alloys		
5052	Moderate mechanical properties, stronger and harder than 1100 and 3003. Fairly good formability. Readily weldable. Excellent resistance to corrosion by salt water.	General purpose alloy where fairly high strength is required. For marine and outside applications, fuel and hydraulic lines, and tanks.
Heat Treatable Alloys		
2011	Excellent free-machining qualities. Fairly high mechanical properties.	Stock for screw-machine products. Bolts, nuts, screws, and a great diversity of parts made on automatic screw machines.
2014	High mechanical properties including yield and tensile strength, fatigue, and hardness. Fair formability and forging qualities. Readily machinable.	Most commonly used alloy where high strength is required. General structural applications, heavy duty forgings, and strong fittings.
Clad 2014	A sheet product that combines the high mechanical properties of 2014 with the good corrosion resistance of 6053.	For structures requiring high unit strength together with good resistance to various corrosive environments.
2017	A bar, rod, and wire alloy having relatively high strength, and good machining qualities.	Screw-machine products, fittings, and structural applications where relatively high strength is required. Now largely superseded by newer alloys.
2018 2218	Both retain strength well at elevated temperatures.	Forged pistons and cylinder head for internal-combustion engines. Suitable for various types of high temperature services. Forged cylinder heads and pistons.
2024	A high strength alloy with mechanical properties intermediate between 2014 and 6061.	General purpose material for various structural applications where good strength is required. Fittings and screw-machine products.
Clad 2024	A sheet product that combines the mechanical properties of 2024 with the corrosion resistance of 1230 aluminum alloy.	For structural applications requiring good strength together with resistance to corrosion.
2025	Fairly high mechanical properties. Good forging qualities.	Specialty forging alloy. Applications mostly confined to propellers for superchargers and engines.
4032	Retains strength well at elevated temperatures.	Forged pistons for internal-combustion engines.
6151	Fairly good mechanical properties. Excellent forging qualities. Good resistance to corrosion.	General purpose material for ordinary forgings. Small press forgings and intricate pieces which are difficult to forge in the harder alloys.
6061	Good mechanical properties. Superior brazing and welding qualities. Good forging characteristics, workability, and resistance to corrosion.	General structural purposes. Marine and outside work. Transportation equipment. Many small forged parts. Various extrusion applications.
7075	Affords maximum strength and endurance limit. Not readily formed. Poorest forging qualities.	Structural applications requiring maximum yield and tensile strength. Section thickness limited to 3 inches.
Clad 7075	A sheet product that combines the mechanical properties of 7075 with improved corrosion resistance.	Structural applications where the highest strength together with maximum corrosion resistance is necessary.

to some degree governed by the quantity to be manufactured, making it necessary to take into account the processes and tooling that must be used for each method. Although aluminum can be formed by any of the conventional methods, it is especially well suited to extrusion, drawing, and forging. Principal characteristics of selected wrought alloys are given in **Table 24**. When choosing an aluminum alloy for any wrought product, it should be remembered that for corresponding tempers, the ease of fabricating *decreases* as the strength increases. Also, as strength increases, the cost increases, so strength is related to cost.

Aluminum extrusions have numerous applications, and are especially useful for producing architectural shapes. Extrusion is less expensive than roll forming, but it cannot produce sections as thin, and the die design requirements for extrusion require careful thought so that the metal flows uniformly in both thick and thin sections. Extrusion alloys are specifically designed for their intended use. Alloy 7075–T6 is often chosen when high strength is required. Alloy 2014–T6 has similar qualities, but is not as strong. Alloy 2024 –T6 is useful for thinner sections, and 6061 has superior forming qualities, resistance to corrosion, and high yield strength. Alloy 6063, either as extruded (–T42), or artificially aged (–T5), provides adequate strength for many applications and does not discolor from anodic oxide finishing. When high resistance to corrosion is required, extruded shapes of alloy 1100 and 3003 are commonly used.

For aluminum alloys, drawing is much the same as with other metals. While more expensive than extrusion, it yields products with much closer tolerances. In drawing aluminum, tool radii are important, and a tool thickness of four to eight times metal thickness is usually satisfactory. Too small a radius can cause tensile fracture, and too large a radius may induce wrinkling. Alloys of the nonhardenable variety, such as 1100, 3003, 5050, and 5052, are often used as they may be deformed to a greater extent before rupturing.

Forgings are used when higher strength is required, or where the forging process is especially suited for manufacturing the part. Aluminum may be either press forged or drop forged, using special forging stock produced in the form of an extruded bar or shape. Press forging, while slower than drop forging, affords greater flexibility in design, higher accuracy, and lower die cost. Ratings of alloys for forging are shown in **Table 25**.

Table 25. Relative Ratings of Aluminum Alloys for Forging. *(Source, MIL-HDBK-694R.)*

Alloy	Strength	Cold Weldability	Corrosion Resistance	Machinability	Electrical Conductance	Hardness	Forgability
1100	4 - 3	1 - 3	1	4 - 3	2	4 - 3	1
2011	2	3 - 4	3 - 4	1	3	2	-
2014	1	3 - 4	3 - 4	2	3	1 - 2	3
Alclad 2014	1	3 - 4	1	2	3	-	-
2017	1	3	3 - 4	2	4	2	-
2018	1	-	3	2	3	2	3
2024	1	3 - 4	3 - 4	2	4	1	-
Alclad 2024	1	4	1	2	4	-	-
2117	3	2	3	3	-	-	-
2218	2	-	3	2	3	2	4
3003	4 - 3	1 - 3	1	4 - 3	3	4 - 3	1

Table 25. *(Continued)* **Relative Ratings of Aluminum Alloys for Forging.** *(Source, MIL-HDBK-694R.)*

Alloy	Strength	Cold Weldability	Corrosion Resistance	Machinability	Electrical Conductance	Hardness	Forgability
4032	2	-	3	3	4	2	3
5052	3	1 - 3	1	4 - 3	4	3 - 2	-
5056	2	1 - 3	1 - 3	4 - 3	4	-	-
5083	2	3	1	4 - 3	4	2	-
5456	2	3	1 - 2	4 - 3	4	2	-
6061	3 - 2	3 - 4	1	3	3	3 - 2	-
6063	3 - 2	2 - 3	1	3	2	3 - 2	-
6151	2	-	2	3	3	2	1
7075	1	4	3	2	4	1	4
Alclad 7075	1	4	1	2	4	-	-
7079	1	4	3	2	4	1	-

Note: Relative ratings are in decreasing order of merit. The first number in numbered pairs is a rating of the softest temper, and the second number is a rating of the hardest temper.

Copper

Properties of copper and copper alloys

While copper ranks third behind iron/steel and aluminum in production, almost 50% of the total is consumed in the manufacture of copper wire. Commercially available copper and its alloys are identified by Unified Numbering System Designations that use the former Copper Alloy Number as the base of the new number. The old Copper Alloy Number 377, for example, now has the UNS designation C37700. New copper alloy designations are assigned by the Copper Development Association, and are awarded to new alloys on the following basis: 1) the full chemical composition must be disclosed; 2) the alloy must be in commercial use, or intended for commercial use; and 3) the composition must not fall within the limits of any designated alloy already available. In the designation system, numbers from C10000 through C79999 denote wrought alloys, and numbers from C80000 through C99999 denote cast alloys.

In Europe, CEN and ISO designations are used for copper alloys. The CEN system uses a six-character system, with the first character being C for copper, the second character being W (wrought), B (ingots), C (castings), or M (master alloys). The last character is a letter used to designate the material group. The ISO system is based on the element symbols of the material, in descending order of magnitude. **Table 1** explains the CEN system, and **Table 2** shows selected ISO/CEN conversions.

Description of wrought copper alloys

Characteristics and popular uses of wrought copper rod and bar alloys are given below. Composition of wrought copper alloys are given in **Tables 3** through **8**; **Table 9** provides mechanical properties; and **Table 10** provides physical properties.

Coppers. C10100 – C15815 are wrought coppers, with a minimum copper content of 99.3%.

C10100, C10200, C10300, C10800: Conductivity 101% IACS (International Annealed Copper Standard), the highest among coppers. Used for electric and electronic conductors, wave guides, cavity resonators, superconductor matrixes, vacuum tube and solid-state devices, glass-to-metal seals.

C10400, C10500, C10700: Conductivity 100% IACS. Very ductile, good hot strength and annealing resistance. Used for electrical conductors operating at elevated temperatures where pure coppers would soften too readily.

C11000, C11300, C11400, C11500, C11600: Conductivity 10% IACS. Ductile, anneal resistant. C11000 is primarily used for plumbing fittings and some electrical components not requiring extensive machining. Silver bearing grades have good annealing resistance and are used for electrical components and commutator components.

C12000, C12100, C12200, C12900: Deoxidizing provides improved embrittlement resistance during welding. Conductivity 95% IACS (C12000), and 85% IACS (12200). Used for electromechanical hardware conductors, bus bars, etc., where welding is required.

C14500, C14520, C14700: Conductivity 90–96% IACS. Free machining coppers. Used for welding and cutting torch tips, products requiring high conductivity and high machinability. C14700 has less directionality in ductility than C14500, and is better for crimped conductors, cold-formed parts, but is somewhat more costly than C14500.

C15000: High conductivity combined with good elevated temperature strength and deformation resistance. Used primarily for resistance welding caps RWMA Class I, and sometimes used for Class II applications.

C15715, C15725, C15760: Good resistance to softening after high temperature

(Text continued on p. 601)

Table 1. CEN European Numbering System for Copper and Copper Alloys.

Material Groups	Number Ranges Available for Positions 3, 4, and 5	Final Letter Designating Material Group	Number Range Allocated to Materials
Copper	001-999 001-999	A B	001-049A 050-099B
Miscellaneous Copper Alloys	001-999 001-999	C D	100-149C 150-199D
Miscellaneous Copper Alloys	001-999 001-999	E F	200-249E 250-299F
Copper-aluminum Alloys	001-999	G	300-349G
Copper-nickel Alloys	001-999	H	350-399H
Copper-nickel-zinc Alloys	001-999	J	400-449J
Copper-tin Alloys	001-999	K	459-499K
Copper-zinc Alloys, binary	001-999	L M	500-549L 550-599M
Copper-zinc-lead Alloys	001-999	N P	600-649P 650-699P
Copper-zinc Alloys, complex	001-999	R	700-749R 750-799S
Copper material not standardized by CEN/TC 133	800-999	A-S*	800-999*
* Letter as appropriate for the Material Group			

The first character in CEN designations is C (copper), followed by one of four second letters: W (wrought), B (ingot), C (casting), or M (master alloy).

Table 2. Equivalent ISO/CEN Copper and Copper Alloy Materials Designations.

Materials	Material Designations	
	ISO Symbols	CEN Numbers
Coppers	Cu-ETP Cu-OF	CW 004A CW 008A
Wrought Brasses	CuZn37 CuZn39Pb3 CuZn20Al2As CuZn40Mn1Pb1AlFeSn	CW 508L CW 614N CW 702R CW 721R
Other Wrought Alloys	CuNi2Si CuAl10Fe1 CuNi30Mn1Fe	CW 111C CW 305G CW 354H
Cast Alloys	CuZn33Pb2-GB CuZn33Pb2-GS CuSn12-GB CuSn12-GS	CB 750S CC 750S CB 483K CC 483K
Master Alloys	CuAl50 (A)-M CuCr10-M CuS20-M	CM 344G CM 204E CM 220E

Table 3. Chemical Composition of Wrought Coppers. *(Source, Copper Development Association.)*

Copper Number	Designation	Designation	Cu + Ag % (min.)	Ag (min.) %	Ag (min.) Troy Oz.	As	Sb	P	Te	Other (Named)
C10100 [1]	OFE	Oxygen Free Electronic	99.99 [2]	-	-	0.0005	0.0004	0.0003	0.0002	note [3]
C10200 [1]	OF	Oxygen Free	99.95 [2]	-	-	-	-	-	-	0.0010 Oxygen
C10300	OFXLP	-	99.95 [4]	-	-	-	-	0.001–0.005	-	-
C10400 [1]	OFS	Oxygen Free with Ag	99.95 [2]	0.027	8	-	-	-	-	0.0010 Oxygen
C10500 [1]	OFS	Oxygen Free with Ag	99.95 [2]	0.034	10	-	-	-	-	0.0010 Oxygen
C10700 [1]	OFS	Oxygen Free with Ag	99.95 [2]	0.085	25	-	-	-	-	-
C10800 [1]	OFLP	-	99.95 [4]	-	-	-	-	0.005–0.012	-	0.005 Oxygen
C10910 [1]	-	-	99.95 [2]	-	-	-	-	-	-	0.02 Oxygen
C10920	-	-	99.90	-	-	-	-	-	-	0.02 Oxygen
C10930	-	-	99.90	0.044	13	-	-	-	-	0.02 Oxygen
C10940	-	-	99.90	0.085	25	-	-	-	-	note [5]
C11000 [1]	ETP	Electrolytic Tough Pitch	99.90	-	-	-	-	-	-	note [5]
C11010 [1]	RHC	Remelted High Conductivity	99.90	-	-	-	-	-	-	note [5]

(Continued)

Table 3. (Continued) Chemical Composition of Wrought Coppers. (Source, Copper Development Association.)

Copper Number	Designation	Designation	Cu + Ag % (min.)	Ag (min.)		As	Sb	P	Te	Other (Named)
				%	Troy Oz.					
C11020[1]	FRHC	Fire-Refined High Conductivity	99.90	-	-	-	-	-	-	note[5]
C11030[1]	CRTP	Chemically Refined Tough Pitch	99.90	-	-	-	-	-	-	note[5]
C11040[1]	-	-	99.90	-	-	0.0005	0.0004	-	0.0002	note[6]
C11100[1]	-	Electrolytic Tough Pitch, Anneal Resistant	99.90	-	-	-	-	-	-	note[7]
C11300[1]	STP	Tough Pitch With Ag	99.90	0.027	8	-	-	-	-	note[5]
C11400[1]	STP	Tough Pitch With Ag	99.90	0.034	10	-	-	-	-	note[5]
C11500[1]	STP	Tough Pitch With Ag	99.90	0.054	16	-	-	-	-	note[5]
C11600[1]	STP	Tough Pitch With Ag	99.90	0.085	25	-	-	-	-	note[5]
C11700	-	-	99.90[8]	-	-	-	-	0.04	-	0.004-0.02 B
C12000	DLP	Phosphorus-Deoxidized, Low Residual Phosphorus	99.90	-	-	-	-	0.004-0.012	-	-
C12100	-	-	99.90	0.014	4	-	-	0.005-0.012	-	-
C12200[9]	DHP	Phosphorus-Deoxidized, High Residual Phosphorus	99.90	-	-	-	-	0.015-0.040	-	-

(Continued)

Table 3. *(Continued)* **Chemical Composition of Wrought Coppers.** *(Source, Copper Development Association.)*

Copper Number	Designation	Designation	Cu + Ag % (min.)	Ag (min.) %	Ag (min.) Troy Oz.	As	Sb	P	Te	Other (Named)
C12210	-	-	99.90	-	-	-	-	0.015–0.025	-	-
C12220	-	-	99.90	-	-	-	-	0.040–0.065	-	-
C12300	-	-	99.90	-	-	-	-	0.015–0.040	-	-
C12500	FRTP	-	99.88	-	-	0.012	0.003	-	-	0.025 Te + Se 0.003 Bi 0.004 Pb 0.050 Ni
C12510	-	-	99.90	-	-	-	0.003	0.03	-	0.025 Te + Se 0.005 Bi 0.020 Pb 0.050 Ni 0.05 Fe 0.05 Sn 0.080 Zn
C12900	FRSTP	Fire-Refined Tough Pitch with Ag	99.88	0.054	16	0.012	0.003	-	0.025 [10]	0.050 Ni 0.003 Bi 0.004 Pb
C14180	-	-	99.90	-	-	-	-	0.075	-	0.02 Pb 0.01 Al
C14181	-	-	99.90	-	-	-	-	0.002	-	0.002 Cd 0.005 C 0.002 Pb 0.002 Zn
C14200	DPA	Phosphorus-Deoxidized, Arsenical	99.40	-	-	0.15–0.50	-	0.015–0.040	-	-
C14300	-	Cadmium Copper, Deoxidized	99.90 [11]	-	-	-	-	-	-	0.05–0.15 Cd

(Continued)

Table 3. (Continued) Chemical Composition of Wrought Coppers. (Source, Copper Development Association.)

Copper Number	Designation	Designation	Cu + Ag % (min.)	Ag (min.)		As	Sb	P	Te	Other (Named)
				%	Troy Oz.					
C14410	-	-	99.90[12]	-	-	-	-	0.005-0.020	-	0.05 Fe 0.05 Pb 0.10-0.20 Sn
C14415	-	-	99.96[12]	-	-	-	-	-	-	0.10-0.15 Sn
C14420	-	-	99.90[13]	-	-	-	-	-	0.005-0.05	0.04-0.15 Sn
C14500[14]	-	Tellurium-Bearing	99.90[62]	-	-	-	-	0.004-0.012	0.40-0.7	-
C14510	-	Tellurium-Bearing	99.85[62]	-	-	-	-	0.010-0.030	0.30-0.7	0.05 Pb
C14520	DPTE	Phosphorus-Deoxidized, Tellurium-Bearing	99.90[62]	-	-	-	-	0.004-0.020	0.40-0.7	-
C14530	-	-	99.90[10,12]	-	-	-	-	0.001-0.010	0.003-0.023[10]	0.0030.023 Sn
C14700	-	Sulfur-Bearing	99.90[14,65]	-	-	-	-	0.002-0.005	-	0.20-0.50 S[65]
C15000	-	Zirconium Copper	99.80[19]	-	-	-	-	-	-	0.10-0.20 Zr
C15100	-	-	99.85[19]	-	-	-	-	-	-	0.05-0.15 Zr
C15150	-	-	99.9 min.	-	-	-	-	-	-	0.15-0.030 Zr
C15500	-	-	99.75	0.027-0.10	8-30	-	-	0.040-0.080	-	0.08-0.13 Mg

	Cu + Ag % (min.)		Al[15]	Fe	Pb	Oxygen			Boron
C15715	99.62		0.13-0.17	0.01	0.01	0.12-0.19			-
C15720	99.52		0.18-.22	0.01	0.01	0.16-0.24			-
C15725	99.43		0.23-0.27	0.01	0.01	0.20-0.28			-
C15760	98.77		0.58-0.62	0.01	0.01	0.52-0.59			-
C15815	97.82		0.13-0.17	0.01	0.01	0.19			1.2-1.8

(Continued)

Table 3. *(Continued)* **Chemical Composition of Wrought Coppers.** *(Source, Copper Development Association.)*

Notes:
1. These are high conductivity coppers which have in the annealed condition a minimum conductivity of 100% IACS, except for Alloy C10100 which has a minimum conductivity of 101% IACS.
2. Copper is determined by the difference between the impurity total and 100%.
3. The following additional maximum limits shall apply: Bi, 1 ppm (0.0001%); Cd, 1.0 ppm (0.0001%); Fe, 10 ppm (0.0010%); Pb, 5 ppm (0.0005%); Mn, 0.5 ppm (0.00005%); Ni, 10 ppm (0.0010%); Oxygen, 5 ppm (0.0005%); Se, 3 ppm (0.0003%); Ag, 25 ppm (0.0025%); S, 15 ppm (0.0015%); Sn, 2 ppm (0.0002%); Zn, 1 ppm (0.0001%).
4. Includes P.
5. Oxygen and trace elements may vary depending on the process.
6. The following additional maximum limits shall apply: Se, 2 ppm (0.0002%); Bi, 1.0 ppm (0.00010%); Group Total, Te + Se + Bi, 3 ppm (0.0003%). Sn, 5 ppm (0.0005%); Pb, 5 ppm (0.0005%); Fe, 10 ppm (0.0010%); Ni, 10 ppm (0.0010%); S, 15 ppm (0.0015%); Ag, 25 ppm (0.0025%); Oxygen, 100-650 ppm (0.010-0.065%). The total maximum allowable of 65 ppm (0.0065%) does not include oxygen.
7. Small amounts of Cd or other elements may be added by agreement to improve the resistance to softening at elevated temperatures.
8. Includes B + P.
9. This includes oxygen-free copper which contains P in an amount agreed upon.
10. Includes Te + Se.
11. Includes Cd. Deoxidized with lithium or other suitable elements as agreed upon.
12. Includes Cu + Ag + Sn.
13. Includes Te + Sn.
14. Includes oxygen-free or deoxidized grades with deoxidizers (such as phosphorus, boron, lithium or others) in an amount agreed upon.
15. All aluminum present as Al₂O₃; 0.04% oxygen present as Cu₂O with a negligible amount in solid solution with copper.
19. Cu + Sum of Named Elements, 99.9% min.
62. Includes Te.
65. Includes Cu + Ag + S + P.

Table 4. Chemical Composition of Wrought High Copper Alloys. (Source, Copper Development Association.)

Copper Alloy Number	Cu + Ag	Fe	Sn	Ni	Co	Cr	Si	Be	Other (named)
C16200	Rem.[16]	0.02	-	-	-	-	-	-	0.7-1.2 Cd
C16500	Rem.[16]	0.02	0.50-0.7	-	-	-	-	-	0.6-1.0 Cd
C17000	Rem.[16]	[17]	-	[17]	[17]	-	0.20	1.60-1.79	0.20 Al
C17200	Rem.[16]	[17]	-	[17]	[17]	-	0.20	1.80-2.00	0.20 Al
C17300	Rem.[16]	[17]	-	[17]	[17]	-	0.20	1.80-2.00	0.20 Al 0.20-0.60 Pb
C17410	Rem.[16]	0.20	::	-	0.35-0.6	-	0.20	0.15-0.50	0.20 Al
C17450	Rem.[16]	0.20	0.25	0.50-1.0	-	-	0.20	0.15-0.50	0.20 Al 0.50 Zr
C17455	Rem.[16]	0.20	0.25	0.50-1.0	-	-	0.20	0.15-0.50	0.20 Al 0.50 Zr 0.20-0.60 Pb
C17460	Rem.[16]	0.20	0.25	1.0-1.4	-	-	0.20	0.15-0.50	0.20 Al 0.50 Zr
C17465	Rem.[16]	0.20	0.25	1.0-1.4	-	-	0.20	0.15-0.50	0.20 Al 0.50 Zr 0.20-0.60 Pb
C17500	Rem.[16]	0.10	-	-	2.4-2.7	-	0.20	0.40-0.7	0.20 Al
C17510	Rem.[16]	0.10	-	1.4-2.2	0.30	-	0.20	0.20-0.6	0.20 Al
C17530	Rem.[16]	0.20	-	1.8-2.5[18]	-	-	0.20	0.20-0.40	0.6 Al
C18000	Rem.[16]	0.15	-	1.8-3.0[18]	-	0.10-0.8	0.40-0.8	-	-
C18030	Rem.[19]	-	0.08-0.12	-	-	0.10-0.20	-	-	0.005-0.015 P
C18040	Rem.[14]	-	0.20-0.30	-	-	0.25-0.35	-	-	0.005-0.015 P 0.05-0.15 Zn
C18045	99.1 min.[19]	-	0.20-0.30	-	-	0.20-0.35	0.05	-	0.15-0.30 Zn

(Continued)

Table 4. (Continued) Chemical Composition of Wrought High Copper Alloys. (Source, Copper Development Association.)

Copper Alloy Number	Cu + Ag	Fe	Sn	Ni	Co	Cr	Si	Be	Other (named)
C18050	Rem. [20]	-	-	-	-	0.05-0.15	-	-	0.005-0.015 Te
C18070	Rem. [20]	-	-	-	-	0.15-0.40	0.02-0.07	-	0.01-0.40 Ti
C18080	Rem. [20]	0.02-0.20	-	-	-	0.20-0.7	0.01-0.10	-	0.01-0.30 Ag 0.01-0.15 Ti
C18100	98.7 min. [16]	-	-	-	-	0.40-1.2	-	-	0.03-0.06 Mg 0.08-0.20 Zr
C18135	Rem. [16]	-	-	-	-	0.20-0.6	-	-	0.20-0.6 Cd
C18140	Rem. [16]	-	-	-	-	0.15-0.45	0.005-0.05	-	0.05-0.25 Zr
C18145	Rem. [16]	-	-	-	-	0.10-0.30	-	-	0.10-0.30 Zn 0.05-0.15 Zr
C18150	Rem. [21]	-	-	-	-	0.50-1.5	-	-	0.05-0.25 Zr
C18200	Rem. [16]	0.10	-	-	-	0.6-1.2	0.10	-	0.05 Pb
C18400	Rem. [16]	0.15	-	-	-	0.40-1.2	0.10	-	0.005 As 0.005 Ca 0.05 Li 0.05 P 0.7 Zn
C18600	Rem. [16]	0.25-0.8	-	0.25	0.10	0.10-1.0	-	-	0.05-0.50 Ti 0.05-0.40 Zr
C18610	Rem. [16]	0.10	-	0.25	0.25-0.8	0.10-1.0	-	-	0.05-0.50 Ti 0.05-0.45 Zr
C18665	99.0 min.	-	-	-	-	-	-	-	0.40-0.9 Mg 0.002-0.04 P
C18700	99.5 min. [63]	-	-	-	-	-	-	-	0.80-1.50 Pb
C18835	99.0 min. [16]	0.10	0.15-0.55	-	-	-	-	-	0.01 P, 0.05 Pb 0.02 0.30 Zn

(Continued)

Table 4. (Continued) Chemical Composition of Wrought High Copper Alloys. (Source, Copper Development Association.)

Copper Alloy Number	Cu + Ag	Fe	Sn	Ni	Co	Cr	Si	Be	Other (named)
C18900	Rem. [16]	-	0.6-0.9	-	-	-	0.15-0.40	-	0.05 P, 0.02 Pb 0.10-0.30 Mn 0.10 Zn 0.10 Al
C18980	Rem. [16]	-	1.0	-	-	-	0.50	-	0.50 Mn 0.15 P, 0.02 Pb
C18990	Rem. [19]	-	1.8-2.2	-	-	0.10-0.20	-	-	0.005-0.015 P
C19000	Rem. [16]	0.10	-	0.9-1.3	-	-	-	-	0.8 Zn, 0.05 Pb 0.15-0.35 P
C19010	Rem. [16]	-	-	0.8-1.8	-	-	0.15-0.35	-	0.01-0.05 P
C19015	Rem. [20]	-	-	0.50-2.4	-	-	0.10-0.40	-	0.02-0.20 P 0.02-0.15 Mg
C19020	Rem. [20]	-	0.30-0.9	0.50-3.0	-	-	-	-	0.01-0.20 P
C19025	Rem. [21]	0.10	0.7-1.1	0.8-1.2	-	-	-	-	0.03-0.07 P 0.20 Zn
C19030	Rem. [21]	0.10	1.0-1.5	1.5-2.0	-	-	-	-	0.01-0.03 P 0.02 Pb
C19100	Rem. [16]	0.20	-	0.9-1.3	-	-	-	-	0.50 Zn, 0.10 Pb 0.35-0.6 Te 0.15-0.35 P
C19140	Rem. [16]	0.05	0.05	0.8-1.2	-	-	-	-	0.50 Zn 0.40-0.80 Pb 0.15-0.35 P
C19150	Rem. [16]	0.05	0.05	0.8-1.2	-	-	-	-	0.50 Zn 0.50-1.0 Pb 0.15-0.35 P
C19160	Rem. [16]	0.05	0.05	0.8-1.2	-	-	-	-	0.50 Zn 0.80 to 1.2 Pb 0.15-0.35 P

(Continued)

Table 4. (Continued) Chemical Composition of Wrought High Copper Alloys. (Source, Copper Development Association.)

	Cu	Fe	Sn	Zn	Al	Pb	P	Other (named)
C19200	98.5 min.[20]	0.8-1.2	-	0.20	-	-	0.01-0.04	-
C19210	Rem.[20]	0.05-0.15	-	-	-	-	0.025-0.040	-
C19215	Rem.[20]	0.05-0.20	-	1.1-3.5	-	0.02	0.025-0.050	-
C19220	Rem.[20]	0.10-0.30	0.05-0.10	-	-	-	0.03-0.07	0.005-0.015 B 0.10-0.25 Ni
C19260	98.5 min.[19]	0.40-0.8	-	-	-	-	-	0.20-0.40 Ti 0.02-0.15 Mg
C19280	Rem.[20]	0.50-1.5	0.30-0.7	0.30-0.7	-	-	0.005-0.015	-
C19400	97.0 min.	2.1-2.6	-	0.05-0.20	-	0.03	0.015-0.15	-
C19410	Rem.[20]	1.8-2.3	0.6-0.9	0.10-0.20	-	-	0.015-0.050	-
C19450	Rem.[20]	1.5-3.0	0.8-2.5	-	-	-	0.005-0.05	-
C19500	96.0 min.[20]	1.0-2.0	0.10-1.0	0.20	0.02	0.02	0.01-0.35	0.30-1.3 Co
C19520	96.6 min.[20]	0.50-1.5	-	-	-	0.01-3.5	-	-
C19700	Rem.[20]	0.30-1.2	0.20	0.20	-	0.05	0.10-0.40	0.01-0.20 Mg 0.05 Ni 0.05 Co 0.05 Mn
C19710	Rem.[16]	0.05-0.40	0.20	0.20	-	0.05	0.07-0.15	0.10 Ni + Co 0.05 Mn 0.03-0.06 Mg
C19720	Rem.[16]	0.05-0.50	0.20	0.20	-	0.05	0.05-0.15	0.06-0.20 Mg 0.05 Mn 0.10 Ni + Co
C19750	Rem.[20]	0.35-1.2	0.05-0.40	0.20	-	0.05	0.10-0.40	0.01-0.20 Mg 0.05 Ni 0.05 Co 0.05 Mn

(Continued)

Table 4. *(Continued)* **Chemical Composition of Wrought High Copper Alloys.** *(Source, Copper Development Association.)*

-	Cu	Fe	Sn	Zn	Al	Pb	P	Other (named)
C19800	Rem. [20]	0.02-0.50	0.10-1.0	0.30-1.5	-	-	0.01-0.10	0.10-1.0 Mg
C19810	Rem. [20]	1.5-3.0	-	1.0-5.0	-	-	0.10	0.10 Cr 0.10 Mg 0.10 Ti 0.10 Zr
C19900	Rem. [16]	-	-	-	-	-	-	2.9-3.4 Ti

Notes: [14] Includes oxygen-free or deoxidized grades with deoxidizers (such as phosphorus, boron, lithium or others) in an amount agreed upon.
[16] Cu + Sum of Named Elements, 99.5% min.
[17] Ni + Co, 0.20% min.; Ni + Fe + Co, 0.6% max.
[18] Ni values include Co.
[19] Cu + Sum of Named Elements, 99.9% min.
[20] Cu + Sum of Named Elements, 99.8% min.
[21] Cu + Sum of Named Elements, 99.7% min.
[63] Includes Pb.

Table 5. Chemical Composition of Wrought Brasses and Tin Brasses.
(Source, Copper Development Association.)

Copper Alloy No.	Cu	Pb	Fe	Zn	Other (named)
Copper-Zinc Alloys *(Brasses)*					
C21000	94.0-96.0 [20]	0.03 [40]	0.05	Rem.	-
C22000	89.0-91.0 [20]	0.05	0.05	Rem.	-
C22600	86.0-89.0 [20]	0.05	0.05	Rem.	-
C23000	84.0-86.0 [20]	0.05	0.05	Rem.	-
C23030	83.5-85.5 [20]	0.05	0.05	Rem.	0.20-0.40 Si
C23400	81.0-84.0 [20]	0.05	0.05	Rem.	-
C24000	78.5-81.5 [20]	0.05	0.05	Rem.	-
C24080	78.0-82.0 [20]	0.20	-	Rem.	0.10 Al
C25600	71.0-73.0 [21]	0.05	0.05	Rem.	-
C26000	68.5-71.5 [21]	0.07	0.05	Rem.	-
C26130	68.5-71.5 [21]	0.05	0.05	Rem.	0.02-0.08 As
C26200	67.0-70.0 [21]	0.07	0.05	Rem.	-
C26800	64.0-68.5 [21]	0.15	0.05	Rem.	-
C27000	63.0-68.5 [21]	0.10	0.07	Rem.	-
C27200	62.0-65.0 [21]	0.07	0.07	Rem.	-
C27400	61.0-64.0 [21]	0.10	0.05	Rem.	-
C28000	59.0-63.0 [21]	0.30	0.07	Rem.	-
Copper-Zinc-Lead Alloys *(Brasses)*					
C31200	87.5-90.5 [22]	0.7-1.2	0.10	Rem.	0.25 Ni
C31400	87.5-90.5 [22]	1.3-2.5	0.10	Rem.	0.7 Ni
C31600	87.5-90.5 [22]	1.3-2.5	0.10	Rem.	0.7-1.2 Ni 0.04-0.10 P
C32000	83.5-86.5 [22]	1.5-2.2	0.10	Rem.	0.25 Ni
C33000	65.0-68.0 [22]	.25-0.7	0.07	Rem.	-
C33200	65.0-68.0 [22]	1.5-2.5	0.07	Rem.	-
C33500	62.0-65.0 [22]	0.25-0.7	0.15 [24]	Rem.	-
C34000	62.0-65.0 [22]	0.8-1.5	0.15 [24]	Rem.	-
C34200	62.0-65.0 [22]	1.5-2.5	0.15 [24]	Rem.	-
C34500	62.0-65.0 [22]	1.5-2.5	0.15	Rem.	-
C35000	60.0-63.0 [22, 25]	0.8-2.0	0.15 [24]	Rem.	-
C35300	60.0-63.0 [16, 25]	1.5-2.5	0.15 [24]	Rem.	-
C35330	59.5-64.0 [16]	1.5-3.5 [26]	-	Rem.	0.02-0.25 As
C35600	60.0-63.0 [16]	2.0-3.0	0.15 [24]	Rem.	-
C36000	60.0-63.0 [16]	2.5-3.7	0.35	Rem.	-
C36500	58.0-61.0 [22]	0.25-0.7	0.15	Rem.	0.25 Sn
C37000	59.0-62.0 [22]	0.8-1.5	0.15	Rem.	-
C37100	58.0-62.0 [22]	0.6-1.2	0.15	Rem.	-
C37700	58.0-61.0 [16]	1.5-2.5	0.30	Rem.	-

(Continued)

Table 5. *(Continued)* Chemical Composition of Wrought Brasses and Tin Brasses.
(Source, Copper Development Association.)

Copper Alloy No.	Cu	Pb	Fe	Zn	Other (named)
Copper-Zinc-Lead Alloys *(Brasses)*					
C37710	56.5-60.0 [16]	1.0-3.0	0.30	Rem.	-
C38000	55.0-60.0 [16]	1.5-2.5	0.35	Rem.	0.50 Al 0.30 Sn
C38500	55.0-59.0 [16]	2.5-3.5	0.35	Rem.	-

Alloy No.	Cu	Pb	Fe	Sn	Zn	P	Other (named)
Copper-Zinc-Tin Alloys *(Tin Brasses)*							
C40400	Rem. 21	-	-	0.35-0.7	2.0-3.0	-	-
C40500	94.0-96.0 [21]	0.05	0.05	0.7-1.3	Rem.	-	-
C40810	94.0-96.0 [21]	0.05	0.08-0.12	1.8-2.2	Rem.	0.028-.04	0.11-0.20 Ni
C40820	94.0 min. [16]	0.02	-	1.0-2.5	.20-2.5	0.05	0.10-0.50 Ni
C40850	94.5-96.5 [21]	0.05	0.05-0.20	2.6-4.0	Rem.	0.02-0.04	0.05-0.20 Ni
C40860	94.0-96.0 [21]	0.05	0.01-0.05	1.7-2.3	Rem.	0.02-0.04	0.05-0.20 Ni
C41000	91.0-93.0 [21]	0.05	0.05	2.0-2.8	Rem.	-	-
C41100	89.0-92.0 [21]	0.10	0.05	0.30-0.7	Rem.	-	-
C41120	89.0-92.0 [21]	0.05	0.05-0.20	0.30-0.7	Rem.	0.02-0.05	0.05-0.20 Ni
C41300	89.0-93.0 [21]	0.10	0.05	0.7-1.3	Rem.	-	-
C41500	89.0-93.0 [21]	0.10	0.05	1.5-2.2	Rem.	-	-
C42000	88.0-91.0 [21]	-	-	1.5-2.0	Rem.	0.25	-
C42200	86.0-89.0 [21]	0.05	0.05	0.8-1.4	Rem.	0.35	-
C42220	88.0-91.0 [21]	0.05	0.05-0.20	0.7-1.4	Rem.	0.02-0.05	0.20 Ni
C42500	87.0-90.0 [21]	0.05	0.05	1.5-3.0	Rem.	0.35	-
C42520	88.0-91.0 [21]	0.05	0.05-0.20	1.5-3.0	Rem.	0.02-0.04	0.05-0.20 Ni
C42600	87.0-90.0 [21]	0.05	0.05-0.20	2.5-4.0	Rem.	0.02-0.05	0.05-0.20 Ni (incl. Co)
C43000	84.0-87.0 [21]	0.10	0.05	1.7-2.7	Rem.	-	-
C43400	84.0-87.0 [21]	0.05	0.05	0.40-1.0	Rem.	-	-
C43500	79.0-83.0 [21]	0.10	0.05	0.6-1.2	Rem.	-	-
C43600	80.0-83.0 [21]	0.05	0.05	0.20-0.50	Rem.	-	-
C44300	70.0-73.0 [22]	0.07	0.06	0.8-1.2 [27]	Rem.	-	0.02-0.06 As
C44400	70.0-73.0 [22]	0.07	0.06	0.8-1.2 [27]	Rem.	-	0.02-0.10 Sb
C44500	70.0-73.0 [22]	0.07	0.06	0.8-1.2 [27]	Rem.	0.02-0.10	-
C46200	62.0-65.0 [22]	0.20	0.10	0.50-1.0	Rem.	-	-
C46400	59.0-62.0 [22]	0.20	0.10	0.50-1.0	Rem.	-	-
C46500	59.0-62.0 [22]	0.20	0.10	0.50-1.0	Rem.	-	0.02-0.06 As
C47000	57.0-61.0 [22]	0.05	-	0.25-1.0	Rem.	-	0.01 Al
C47940	63.0-66.0 [22]	1.0-2.0	0.10-1.0	1.2-2.0	Rem.	-	0.10-0.50 Ni (incl. Co)
C48200	59.0-62.0 [22]	0.40-1.0	0.10	0.50-1.0	Rem.	-	-

(Continued)

Table 5. *(Continued)* **Chemical Composition of Wrought Brasses and Tin Brasses.**
(Source, Copper Development Association.)

Alloy No.	Cu	Pb	Fe	Sn	Zn	P	Other (named)
Copper-Zinc-Tin Alloys *(Tin Brasses)*							
C48500	59.0-62.0 [22]	1.3-2.2	0.10	0.50-1.0	Rem.	-	-
C48600	59.0-62.0 [22]	1.0-2.5	-	0.30-1.5	Rem.	-	0.2-0.25 As

Notes: [16] Cu + Sum of Named Elements, 99.5% min.
[20] Cu + Sum of Named Elements, 99.8% min.
[21] Cu + Sum of Named Elements, 99.7% min.
[22] Cu + Sum of Named Elements, 99.6% min.
[24] For flat products, the iron shall be .10% max.
[25] Cu, 61.0% min. for rod.
[26] Pb may be reduced to 1.0% by agreement.
[27] For tubular products, the minimum Sn content may be .9%.
[40] 0.05% Pb, max., for rod wire and tube.

Table 6. Chemical Composition of Wrought Bonzes and Copper Zinc Alloys.
(Source, Copper Development Association.)

Copper Alloy No.	Cu[16]	Pb	Fe	Sn	Zn	P	Other (named)
Copper-Tin-Phosphorus Alloys *(Phosphor Bronzes)*							
C50100	Rem.	0.05	0.05	0.50-0.8	-	0.01-0.05	-
C50200	Rem.	0.05	10	1.0-1.5	-	0.04	-
C50500	Rem.	0.05	0.10	1.0-1.7	0.30	0.03-0.35	-
C50510	Rem. [21]	-	-	1.0-1.5	0.10-0.25	0.02-0.07	0.15-0.40 Ni
C50580	Rem.	0.05	0.05-0.20	1.0-1.7	0.30	0.02-0.10	0.05-0.20 Ni
C50590	97.0 min.	0.02	0.05-0.40	0.50-1.5	0.50	0.02-0.15	-
C50700	Rem.	0.05	0.10	1.5-2.0	-	0.30	-
C50705	96.5 min.	0.02	0.10-0.40	1.5-2.0	0.50	0.04-0.15	-
C50710	Rem.	-	-	1.7-2.3	-	0.15	0.10-0.40 Ni
C50715	Rem. [28]	0.02	0.05-0.15	1.7-2.3	-	0.025-0.04	-
C50725	94.0 min.	0.02	0.05-0.20	1.5-2.5	1.5-3.0	0.02-0.06	-
C50780	Rem.	0.05	0.05-0.20	1.7-2.3	0.30	0.02-0.10	0.05-0.20 Ni
C50900	Rem.	0.05	0.10	2.5-3.8	0.30	0.03-0.30	-
C51000	Rem.	0.05	0.10	4.2-5.8	0.30	0.03-0.35	-
C51080	Rem.	0.05	0.05-0.20	4.8-5.8	0.30	0.02-0.10	0.05-0.20 Ni
C51100	Rem.	0.05	0.10	3.5-4.9	0.30	0.03-0.35	-
C51180	Rem.	0.05	0.05-0.20	3.5-4.9	0.30	0.02-0.10	0.11-0.20 Ni
C51190	Rem.	0.02	0.05-0.15	3.0-6.5	-	0.025-0.045	0.15 Co
C51800	Rem.	0.02	-	4.0-6.0	-	0.10-0.35	0.01 Al
C51900	Rem.	0.05	0.10	5.0-7.0	0.30	0.03-0.35	-
C51980	Rem.	0.05	0.05-0.20	5.5-7.0	0.30	0.02-0.10	0.05-0.20 Ni
C52100	Rem.	0.05	0.10	7.0-9.0	0.20	0.03-0.35	-
C52180	Rem.	0.05	0.05-0.20	7.0-9.0	0.30	0.02-0.10	0.05-0.20 Ni

(Continued)

Table 6. *(Continued)* Chemical Composition of Wrought Bonzes and Copper Zinc Alloys.
(Source, Copper Development Association.)

Copper Alloy No.	Cu[16]	Pb	Fe	Sn	Zn	P	Other (named)
Copper-Tin-Phosphorus Alloys *(Phosphor Bronzes)*							
C52400	Rem.	0.05	0.10	9.0-11.0	0.20	0.03-0.35	-
C52480	Rem.	0.05	0.05-0.20	9.0-11.0	0.30	0.02-0.10	0.05-0.20 Ni

Alloy No.		Cu[16]	Pb	Fe	Sn	Zn	P
Copper-Tin-Lead-Phosphorus Alloys *(Leaded Phosphor Bronzes)*							
C53400		Rem.	0.8-1.2	0.10	3.5-5.8	0.30	0.03-0.35
C54400		Rem.	3.5-4.5	0.10	3.5-4.5	1.5-4.5	0.01-0.50

Alloy No.	Cu[29]	Ag	P
Copper-Phosphorus and Copper-Silver-Phophorus Alloys *(Brazing Alloys)*			
C55180	Rem.	-	4.8-5.2
C55181	Rem.	-	7.0-7.5
C55280	Rem.	1.8-2.2	6.8-7.2
C55281	Rem.	4.8-5.2	5.8-6.2
C55282	Rem.	4.8-5.2	6.5-7.0
C55283	Rem.	5.8-6.2	7.0-7.5
C55284	Rem.	14.5-15.5	4.8-5.2

Alloy No.	Cu	Ag	Zn
Copper-Silver-Zinc Alloys			
C56000	Rem. [16]	29.0-31.0	30.0-34.0

(Continued)

Table 6. *(Continued)* Chemical Composition of Wrought Bonzes and Copper Zinc Alloys. *(Source, Copper Development Association.)*

Alloy No.	Cu[16] incl. Ag	Pb	Fe	Sn	Zn	Al	Mn	Si	Ni	Other (named)
			Copper-Aluminum Alloys *(Aluminum Bronzes)*							
C60800	Rem.	0.10	0.10	-	-	-	5.0-6.5	-	-	-	0.02-0.35 As
C61000	Rem.	0.02	0.50	-	0.20	6.0-8.5	-	0.10	-	-
C61300	Rem.[20]	0.01	2.0-3.0	0.20-0.50	0.10[30]	6.0-7.5	0.20	0.10	0.15	0.015 P[30]
C61400	Rem.	0.01	1.5-3.5	-	0.20	6.0-8.0	1.0	-	-	0.015 P
C61500	Rem.	0.015	-	-	-	7.7-8.3	-	-	1.8-2.2	-
C61550	Rem.	0.05	0.20	0.05	0.8	5.5-6.5	1.0	-	1.5-2.5	-
C61800	Rem.	0.02	0.50-1.5	-	0.02	8.5-11.0	-	0.10	-	-
C61900	Rem.	0.02	3.0-4.5	0.6	0.8	8.5-10.0	-	-	-	-
C62200	Rem.	0.02	3.0-4.2	-	0.02	11.0-12.0	-	0.10	-	-
C62300	Rem.	-	2.0-4.0	0.6	-	8.5-10.0	0.50	0.25	1.0	-
C62400	Rem.	-	2.0-4.5	0.20	-	10.0-11.5	0.30	0.25	-	-
C62500	Rem.	-	3.5-5.5	-	-	12.5-13.5	2.0	-	-	-
C62580	Rem.	0.02	3.0-5.0	-	0.02	12.0-13.0	-	0.04	-	-
C62581	Rem.	0.02	3.0-5.0	-	0.02	13.0-14.0	-	0.04	-	-
C62582	Rem.	0.02	3.0-5.0	-	0.02	14.0-15.0	-	0.04	-	-
C63000	Rem.	-	2.0-4.0	0.20	0.30	9.0-11.0	1.5	0.25	4.0-5.5	-
C63010	78.0 min.[20]	-	2.0-3.5	0.20	0.30	9.7-10.9	1.5	-	4.5-5.5	-
C63020	74.5 min.	0.03	4.0-5.5	0.25	0.30	10.0-11.0	1.5	-	4.2-6.0	0.20 Co 0.05 Cr
C63200	Rem.	0.02	3.5-4.3[31]	-	-	8.7-9.5	1.2-2.0	0.10	4.0-4.8[31]	-
C63280	Rem.	0.02	3.0-5.0	-	-	8.5-9.5	0.6-3.5	-	4.0-5.5	-
C63380	Rem.	0.02	2.0-4.0	-	0.15	7.0-8.5	11.0-14.0	0.10	1.5-3.0	-

(Continued)

Table 6. (Continued) Chemical Composition of Wrought Bonzes and Copper Zinc Alloys. (Source, Copper Development Association.)

Alloy No.	Cu[16] incl. Ag	Pb	Fe	Sn	Zn	Al	Mn	Si	Ni	Other (named)
					Copper-Aluminum Alloys (Aluminum Bronzes)					
C63400	Rem.	0.05	0.15	0.20	0.50	2.6-3.2	-	0.25-0.45	0.15	0.15 As
C63600	Rem.	0.05	0.15	0.20	0.50	3.0-4.0	-	0.7-1.3	0.15	0.15 As
C63800	Rem.	0.05	0.20	-	0.8	2.5-3.1	0.10	1.5-2.1	0.20[32]	0.25-0.55 Co
C64200	Rem.	0.05	0.30	0.20	0.50	6.3-7.6	0.10	1.5-2.2	0.25	0.15 As
C64210	Rem.	0.05	0.30	0.20	0.50	6.3-7.0	0.10	1.5-2.0	0.25	0.15 As

Alloy No.	Cu[16] incl. Ag	Pb	Fe	Sn	Zn	Al	Mn	Ni	Si	Other (named)
					Copper-Silicon Alloys (Silicon Bronzes)					
C64700	Rem.	0.10	0.10	-	0.50	-	0.40-8	0.15	1.6-2.2	-
C64710	95.0 min.	-	-	-	0.20-0.50	0.10	0.50-9		2.9-3.5	0.01 Ca 0.20 Mg 0.20 Cr
C64725	95.0 min.	0.01	0.25	0.20-0.8	0.50-1.5	-	0.20-8		1.3-2.7	
C64730	93.5 min.	-	-	1.0-1.5	0.20-0.50	0.10	0.50-9		2.9-3.5	-
C64740	95.0 min.	0.10 max.	-	1.5-2.5	0.20-1.0	-	0.05-0.50		1.0-2.0	0.01 Ca 0.05 Mg
C64750	Rem.	-	1.0	0.05-0.8	1.0	-	0.10-0.7		0.50-3.0	0.10 Mg 0.10 P 0.10 Zr
C64760	93.5 min.	0.02	-	0.30	0.20-2.5	-	0.05-0.6		0.40-2.5	0.05 Mg
C64780	90.0	0.02	-	0.10-2.0	0.20-2.5	0.01-1.0	0.20-0.9		1.0-3.5	0.01 Cr 0.01 Mg 0.01 Ti 0.01 Zr

(Continued)

Table 6. (Continued) Chemical Composition of Wrought Bonzes and Copper Zinc Alloys. (Source, Copper Development Association.)

Alloy No.	Cu[16] incl. Ag	Pb	Fe	Sn	Zn	Mn	Si	Ni	Other (named)
				Copper-Silicon Alloys (Silicon Bronzes)					
C64900	Rem.	0.05	0.10	1.2-1.6	1.20	-	0.8-1.2	0.10	0.10 Al
C65100	Rem.	0.05	0.8	-	1.5	0.7	0.8-2.0	-	-
C65400	Rem.	0.05	-	1.2-1.9	0.50	-	2.7-3.4	-	0.01-0.12 Cr
C65500	Rem.	0.05	0.8	-	1.5	0.50-1.3	2.8-3.8	0.6	-
C65600	Rem.	0.02	0.50	1.5	1.5	1.5	2.8-4.0	-	0.01 Al
C66100	Rem.	0.20-8	0.25	-	1.5	1.5	2.8-3.5	-	-
Alloy No.	**Cu[16] incl. Ag**	**Pb**	**Fe**	**Sn**	**Zn**	**Ni incl. Co**	**Al**	**Si**	**Other (named)**
				Other Copper-Zinc Alloys					
C66200	86.6-91.0	0.05	0.05	0.20-0.7	Rem.	0.30-1.0	-	-	0.05-0.20 P
C66300	84.5-87.5	0.05	1.4-2.4[61]	1.5-3.0	Rem.	-	-	-	0.35-0.20 Co
C66400	Rem.	0.015	1.3-1.7[33]	0.05	11.0-12.0	-	-	-	0.30-0.70 Co[33]
C66410	Rem.	0.015	1.8-2.3	0.05	11.0-12.0	-	-	-	-
C66420	Rem.	-	0.50-1.5	-	12.0-17.0	-	-	-	-
C66430	Rem.	0.05	0.6-0.9	0.6-0.9	13.0-15.0	-	-	-	-
C66700	68.5-71.5	0.07	0.10	-	Rem.	-	-	-	0.10 P
C66800	60.0-63.0	0.50	0.35	0.30	Rem.	-	0.25	0.50-1.5	0.8-1.5 Mn
C66900	62.5-64.5[20]	0.05	0.25	-	Rem.	0.25	-	-	2.0-3.5 Mn
C66950	Rem.	0.01	0.50	-	14.0-15.0	-	1.0-1.5	-	11.5-12.5 Mn
C67000	63.0-68.0	0.20	2.0-4.0	0.50	Rem.	-	3.0-6.0	-	2.5-5.0 Mn
C67300	58.0-63.0	0.40-3.0	0.50	0.30	Rem.	0.25	0.25	0.50-1.5	2.0-3.5 Mn
C67400	57.0-60.0	0.50	0.35	0.30	Rem.	0.25	0.50-2.0	0.50-1.5	2.0-3.5 Mn

(Continued)

Table 6. (Continued) Chemical Composition of Wrought Bronzes and Copper Zinc Alloys. *(Source, Copper Development Association.)*

Alloy No.	Cu[16] incl. Ag	Pb	Fe	Sn	Zn	Ni incl. Co	Al	Si	Other (named)
					Other Copper-Zinc Alloys				
C67420	57.0-58.5	0.25-8	0.15-55	0.35	Rem.	0.25	1.0-2.0	0.25-7	1.5-2.5 Mn
C67500	57.0-60.0	0.20	0.8-2.0	0.50-1.5	Rem.	-	0.25	0.50-1.5	0.05-0.50 Mn
C67600	57.0-60.0	0.50-1.0	0.40-1.3	0.50-1.5	Rem.	-	-	0.50-1.5	0.05-0.50 Mn
C68000	56.0-60.0	0.05	0.25-1.25	0.75-1.10	Rem.	0.20-8	0.01	0.04-0.15	0.01-0.50 Mn
C68100	56.0-60.0	0.05	0.25-1.25	0.75-1.10	Rem.	-	0.01	0.04-0.15	0.01-0.50 Mn
C68700	76.0-79.0	0.07	0.06	-	Rem.	-	1.8-2.5	-	0.02-0.06 As
C68800	Rem.	0.05	0.20	-	21.3-24.1 [34]	-	3.0-3.8 [34]	-	0.25-0.55 Co
C69050	70.0-75.0	-	-	-	Rem.	0.50-1.5	3.0-4.0	0.10-0.6	0.01-0.20 Zr
C69100	81.0-84.0	0.05	0.25	0.10	Rem.	0.8-1.4	0.7-1.2	0.8-1.3	0.10 min. Mn
C69400	80.0-83.0	0.30	0.20	-	Rem.	-	-	3.5-4.5	-
C69430	80.0-83.0	0.30	0.20	-	Rem.	-	-	3.5-4.5	0.03-0.06 As
C69700	75.0-80.0	0.50-1.5	0.20	-	Rem.	-	-	2.5-3.5	0.40 Mn
C69710	75.0-80.0	0.50-1.5	0.20	-	Rem.	-	-	2.5-3.5	0.03-0.06 As, 0.040 Mn

Notes: [16] Cu + Sum of Named Elements, 99.5% min.
[20] Cu + Sum of Named Elements, 99.8% min.
[21] Cu + Sum of Named Elements, 99.7% min.
[28] Cu + Sn + Fe + P, 99.5% min.
[29] Cu + Sum of Named Elements, 99.85% min.
[30] When the products is for subsequent welding applications and is so specified by the purchaser, Cr, Cd, Zr and Zn shall each be 0.05% max.
[31] Fe content shall not exceed Ni content.
[32] Not including Co.
[33] Fe + Co, 1.8-2.3%.
[34] Al + Zn, 25.1-27.1%.
[61] Fe + Co, 1.4-2.4%.

Table 7. Chemical Composition of Wrought Copper-Nickel Alloys. *(Source, Copper Development Association.)*

Copper Alloy No.	Cu incl. Ag	Pb	Fe	Zn	Ni	Sn	Mn	Other (named)
C70100	Rem.[16]	-	.05	.25	3.0-4.0	-	.50	-
C70200	Rem.[16]	.05	.10	-	2.0-3.0	-	.40	-
C70230	Rem.[16]	-	-	.50-2.0	2.2-3.2	.10-.50	-	.40-.80 Si .10 Ag+B
C70250	Rem.[16]	.05	.20	1.0	2.2-4.2	-	.10	.05-.30 Mg .25-1.2 Si
C70260	Rem.[16]	-	-	-	1.0-3.0	-	-	.20-.70 Si .005 P
C70270	Rem.[16]	.05	.28-1.0	1.0	1.0-3.0	.10-1.0	.15	.20-1.0 Si
C70280	Rem.[16]	.02	.015	.30	1.3-1.7	1.0-1.5	-	.02-.04 P .22-.30 Si
C70290	Rem.[16]	.02	.015	.30	1.3-1.7	2.1-2.7	-	.02-.04 P .22-.30 Si
C70400	Rem.[16]	.05	1.3-1.7	1.0	4.8-6.2	-	.30-.80	-
C70500	Rem.[16]	.05	.10	.20	5.8-7.8	-	.15	-
C70600	Rem.[16]	.05	1.0-1.8	1.0	9.0-11.0	-	1.0	-
C70610	Rem.[16]	.01	1.0-2.0	-	10.0-11.0	-	.50-1.0	.05 S .05 C
C70620	86.5 min.[16]	.02	1.0-1.8	.50	9.0-11.0	-	1.0	.05 C .02 P .02 S
C70690	Rem.[16]	.001	.005	.001	9.0-11.0	-	.001	[36]
C70700	Rem.[16]	-	.05	-	9.5-10.5	-	.50	-
C70800	Rem.[16]	.05	.10	.20	10.5-12.5	-	.15	-
C71000	Rem.[16]	.05	1.0	1.0	19.0-23.0	-	1.0	-
C71100	Rem.[16]	.05	.10	.20	22.0-24.0	-	.15	-
C71300	Rem.[16]	.05	.20	1.0	23.5-26.5	-	1.0	-

(Continued)

Table 7. (Continued) Chemical Composition of Wrought Copper-Nickel Alloys. (Source, Copper Development Association.)

Copper Alloy No.	Cu incl. Ag	Pb	Fe	Zn	Ni	Sn	Mn	Other (named)
C71500	Rem.[16]	.05	.40-1.0	1.0	29.0-33.0	-	1.0	-
C71520	65.0 min.[16]	.02	.40-1.0	.50	29.0-33.0	-	1.0	.05 C / .02 P / .02 S
C71580	Rem.[16]	.05	.50	.50	29.0-33.0	-	.30	[37]
C71581	Rem.[16]	.02	.40-.7	-	29.0-32.0	-	1.0	[38]
C71590	Rem.	.001	.15	.001	29.0-31.0	.001	.50	[36]
C71640	Rem.[16]	.01	1.7-2.3	-	29.0-32.0	-	1.5-2.5	.03 S / .06 C
C71700	Rem.[16]	-	.40-1.0	-	29.0-33.0	-	-	.30-.7 Be
C71900	Rem.[16]	.015	.50	.05	28.0-33.0	-	.20-1.0	2.2-3.0 Cr / .02-.35 Zr / .01-20 Ti / .04C / .25 Si / .015 S / .02 P
C72150	Rem.[16]	.05	.10	.20	43.0-46.0	-	.05	.10 C / .50 Si
C72200	Rem.[20]	.05 [35]	.50-1.0	1.0 [35]	15.0-18.0	-	1.0	.30-.70 Cr / .03 Si / .03Ti [35]
C72420	Rem.[21]	.02	.7-1.2	.20	13.5-16.5	.10	3.5-5.5	1.0-2.0 Al / .50 Cr / .15 Si / .05 Mg / .15 S / .01 P / .05 C

(Continued)

Table 7. *(Continued)* Chemical Composition of Wrought Copper-Nickel Alloys. *(Source, Copper Development Association.)*

Copper Alloy No.	Cu incl. Ag	Pb	Fe	Zn	Ni	Sn	Mn	Other (named)
C72500	Rem.[20]	.05	.6	.50	8.5-10.5	1.8-2.8	.20	-
C72650	Rem.[21]	.01	.10	.10	7.0-8.0	4.5-5.5	.10	-
C72700	Rem.[21]	.02[39]	.50	.50	8.5-9.5	5.5-6.5	.05-.30	.10 Nb .15 Mg
C72800	Rem.[21]	.005	.50	1.0	9.5-10.5	7.5-8.5	.05-.30	.10 Al .001 B .001 Bi .10-.30 Nb .005-.15 Mg .005 P .0025 S .02 Sb .05 Si .01 Ti
C72900	Rem.[21]	.02[39]	.50	.50	14.5-15.5	7.5-8.5	.30	.10 Nb .15 Mg
C72950	Rem.[21]	.05	.6	-	20.0-22.0	4.5-5.7	.6	-

Notes: [16] Cu + Sum of Named Elements, 99.5% min.
[20] Cu + Sum of Named Elements, 99.8% min.
[21] Cu + Sum of Named Elements, 99.7% min.
[35] The following additional maximum limits shall apply: When the product is for subsequent welding applications and is so specified by the purchaser, .50% Zn, .02% P, .02% Pb, .02% S (.008% S for C71110) and .05% C.
[36] The following additional maximum limits shall apply: .02% C, .015% Si, .003% S, .002% Al, .001% P, .0005% Hg, .001% Ti, .001% Sb, .001% As, .001% Bi, .05% Co, .10% Mg, and .005% Oxygen. For C70690, Co shall be .02% max.
[37] The following additional maximum limits shall apply: .07% C, .15% Si, .024% S, .05% Al and .03% P.
[38] .02% P, max.; .25% Si, max.; .01% S, max.; .20-.50% Ti.
[39] .005% Pb, max., for hot rolling.

Table 8. Chemical Composition of Wrought Nickel-Silvers. *(Source, Copper Development Association.)*

Copper Alloy No.	Cu incl. Ag	Pb	Fe	Zn	Ni	Mn	Other (named)
C73500	70.5-73.5 [16]	.10	.25	Rem.	16.5-19.5	.50	-
C74000	69.0-73.5 [16]	.10	.25	Rem.	9.0-11.0	.50	-
C74300	63.0-66.0 [16]	.10	.25	Rem.	7.0-9.0	.50	-
C74400	62.0-66.0 [21]	.05	.05	Rem.	2.0-4.0	-	-
C74500	63.5-66.5 [16]	.10 [40]	.25	Rem.	9.0-11.0	.50	-
C75200	63.5-66.5 [16]	.05	.25	Rem.	16.5-19.5	.50	-
C75400	63.5-66.5 [16]	.10	.25	Rem.	14.0-16.0	.50	-
C75700	63.5-66.5 [16]	.05	.25	Rem.	11.0-13.0	.50	-
C76000	60.0-63.0 [16]	.10	.25	Rem.	7.0-9.0	.50	-
C76200	57.0-61.0 [16]	.10	.25	Rem.	11.0-13.5	.50	-
C76400	58.5-61.5 [16]	.05	.25	Rem.	16.5-19.5	.50	-
C76700	55.0-58.0 [16]	-	-	Rem.	14.0-16.0	.50	-
C77000	53.5-56.5 [16]	.05	.25	Rem.	16.5-19.5	.50	-
C77300	46.0-50.0 [16]	.05	-	Rem.	9.0-11.0	-	.01Al .25P .04- .25Si
C77400	43.0-47.0 [16]	.20	-	Rem.	9.0-11.0	-	-
C78200	63.0-67.0 [16]	1.5-2.5	.35	Rem.	7.0-9.0	.50	-
C79000	63.0-67.0 [16]	1.5-2.2	.35	Rem.	11.0-13.0	.50	-
C79200	59.0-66.5 [16]	.8-1.4	.25	Rem.	11.0-13.0	.50	-
C79800	45.5-48.5 [16]	1.5-2.5	.25	Rem.	9.0-11.0	1.5-2.5	-
C79830	45.5-47.0 [16]	1.0-2.5	.45	Rem.	9.0-10.5	.15- .55	-

Notes: [16] Cu + Sum of Named Elements, 99.5% min.

[21] Cu + Sum of Named Elements, 99.7% min.

[40] .05% Pb, max., for rod wire and tube.

COPPER

Table 9. Selected Mechanical Properties of Copper Alloy Rod [1] for Machined Products. *(Source, Copper Development Association)*

Copper Alloy No.	Temper	Tensile Strength		Yield Strength [2]		% Elongation [3]	Hardness	Shear Strength		Fatigue Strength [4]	
		KSI	MPa	KSI	MPa			KSI	MPa	KSI	MPa
C10100 C10200	H04 [5]	48	331	44	303	16	HRF87 HRB47	27	186	17	117
C10400 C10500 C10700 C11000 C11300 C11400 C11500 C11600 C12000 C12100	M20	32	221	10	69	55	HRF40	22	152	-	-
C10300 C10800	H04 [5]	48	331	44	303	16	HRF87 HRB47	27	186	-	-
C12200	H04	45	310	40	276	20	HRF85 HRB45	26	179	-	-
C12900	H04 [5]	48	331	44	303	16	HRF87 HRB47	27	186	-	-
	M20	32	221	10	69	55	HRF40	22	152	-	-
C14500 C14520	H04 [5]	48	331	44	303	20	HRB48	27	186	-	-
C14700	H04 [6]	46	317	43	296	11	HRB46	27	186	-	-
C15000	Note [7]	62	427	60	414	15	-	-	-	-	-
C15715	M30	57	393	-	-	27	HRB57	-	-	-	-
C15725	M30	60	414	-	-	19	HRB68	-	-	-	-
C15760	M30 [9]	80	552	-	-	22	HRB80	-	-	-	-
C16200	H04 [10]	73	503	69	474	9	HRB73	56	386	30	207

(Continued)

Table 9. (Continued) Selected Mechanical Properties of Copper Alloy Rod [1] for Machined Products. (Source, Copper Development Association)

Copper Alloy No.	Temper	Tensile Strength		Yield Strength[2]		% Elongation[3]	Hardness	Shear Strength		Fatigue Strength[4]	
		KSI	MPa	KSI	MPa			KSI	MPa	KSI	MPa
C16500	H04 [5,10]	65	448	55	379	15	HRB75	-	-	-	-
C17200 C17300	TH04 [11]	205	1,415	-	-	3	HRC42	120	825	-	-
C17410	TH04 [12]	120	825	-	-	8	HRB102	70	483	-	-
C17500	TB00	45	310	-	-	28	HRB40	25	172	-	-
C17510	TF00 [13]	120	827	-	-	18	HRB95	70	483	40	276
C18000	TH04	100	690	-	-	13	-	-	-	-	-
C18135	TH01 [10,14]	62	430	53	368	30	HRB71	-	-	25	172
C18150	TH04	78	538	-	-	13	-	-	-	-	-
C18200 C18400	TF00	72	496	65	448	18	HRB80	-	-	-	-
C18700	H04 [5]	48	331	42	290	15	HRB48	27	186	-	-
C19100	TH04 [15]	78	538	68	469	27	HRB84	41	283	33	228
C19150	TH04 [16]	90	621	-	-	-	HRB80	50	345	-	-
C21000	H04 [17]	55	379	35	241	10	HRB60	37	255	-	-
C22000	H00 [10]	45	310	35	241	25	HRB42	33	228	-	-
C22600	H04 [17]	67	462	50	345	10	HRB65	40	276	-	-
C23000	H04 [17]	70	483	52	359	10	HRB65	43	296	-	-
C24000	H04 [17]	80	552	63	434	12	HRB74	46	317	-	-
C26000 C26130 C26200	H02 [18]	70	483	52	359	30	HRB80	42	290	22 [19]	152 [19]
C26800 C27000	H00	55	379	40	276	48	HRB55	36	248	-	-

(Continued)

Table 9. *(Continued)* Selected Mechanical Properties of Copper Alloy Rod [1] for Machined Products. *(Source, Copper Development Association)*

Copper Alloy No.	Temper	Tensile Strength		Yield Strength [2]		% Elongation [3]	Hardness	Shear Strength		Fatigue Strength [4]	
		KSI	MPa	KSI	MPa			KSI	MPa	KSI	MPa
C28000	H01	72	496	50	345	25	HRB78	45	310	-	-
	M30	52	359	20	138	52	HRF78	39	269	-	-
	060	54	372	21	145	50	HRF80	40	276	-	-
C31400	H02 [18]	52	359	45	310	18	HRB58	30	207	-	-
C31600	H04 [5]	65	448	57	393	15	HRB70	39	269	-	-
C32000	H04	70	483	60	414	10	-	-	-	-	-
C34000	H01	55	379	42	290	40	HRB60	36	248	-	-
C34200 C35300 C35330	H02 [18]	58	400	45	310	25	HRB75	34	234	-	-
C34500	H02	70	483	58	400	22	HRB80	42	290	-	-
C35000	H02 [10, 18]	70	483	52	359	22	HRB80	42	290	-	-
C35600	H02 [18]	58	400	45	310	25	HRB78	34	234	-	-
C36000	060	49	338	18	124	53	HRF68	30	207	-	-
C37700	M30	52	359	20	138	45	HRF78	30	207	-	-
C46200	H02	70	483	50	345	2	HRB75	44	303	-	-
C46400	H02 [18]	75	517	53	365	20	HRB82	44	303	-	-
	H01 [20]	69	476	46	317	27	HRB78	43	296	-	-
	050	63	434	30	207	40	HRB60	42	290	-	-
	060	57	393	25	172	47	HRB55	40	276	-	-
C48200	060	57	393	25.0	172	40	HRB55	38	262	-	-
	050	63	434	30.0	207	35	HRB60	39	269	-	-
	H01 [20]	69	476	46	317	20	HRB78	40	276	-	-
	H02 [18]	75	517	53	365	15	HRB82	41	283	-	-

(Continued)

Table 9. (Continued) Selected Mechanical Properties of Copper Alloy Rod [1] for Machined Products. (Source, Copper Development Association)

Copper Alloy No.	Temper	Tensile Strength		Yield Strength[2]		% Elongation[3]	Hardness	Shear Strength		Fatigue Strength[4]	
		KSI	MPa	KSI	MPa			KSI	MPa	KSI	MPa
C48500	H02[18]	75	517	53	365	15	HRB82	40	276	-	-
	H01[20]	69	476	46	317	20	HRB78	39	269	-	-
	060	57	393	25	172	40	HRB55	36	248	-	-
C51000	H02[18]	70	483	58	400	25	HRB78	50	345	-	-
C52100	H02[10,18]	80	552	65	448	33	HRB85	54	372	-	-
C54400	H04[5,10]	75	517	63	434	15	HRB83	52	359	-	-
	H04[22]	68	469	57	393	20	HRB80	54	372	-	-
C61000	H04[21]	70	483	-	-	-	HRB80	-	-	-	-
C61300	H04[22]	82	565	55	379	35	HRB90	45	310	-	-
C61400	H04	82	565	40	276	35	HRB90	45	310	-	-
C61800	H02[23]	85	586	42	293	23	HRB89	47	324	28	193
C62300	H02[23]	95	655	50	345	25	HRB88	50	345	30	207
	M30[24]	75	517	35	241	35	HRB80	35	241	25	172
C62400	H02[25]	105	724	52	359	14	HRB92	65	448	36	248
	M30[24]	90	621	40	276	18	HRB87	58	407	32	221
C62500	M30	100	690	55	379	1	HRC29	60	414	67	462
C63000	H02[25]	118	814	75	517	15	HRB98	70	483	38	262
	M30[24]	100	690	60	414	15	HRB96	62	427	36	248
C63020	TQ30	145	1,000	-	-	8	HRC29	78	538	51	352
C63200	050	105	724	53	365	22	HRB96	-	-	-	-
C64200	H04[21,23]	102	703	68	469	22	HRB94	59	407	50	345
	M30[21]	75	517	35	241	32	HRB77	-	-	-	-
	050[21]	90	621	55	379	28	HRB89	-	-	-	-

(Continued)

Table 9. *(Continued)* Selected Mechanical Properties of Copper Alloy Rod [1] for Machined Products. *(Source, Copper Development Association)*

Copper Alloy No.	Temper	Tensile Strength		Yield Strength[2]		% Elongation[3]	Hardness	Shear Strength		Fatigue Strength[4]	
		KSI	MPa	KSI	MPa			KSI	MPa	KSI	MPa
C65100	H04 [5]	70	483	55	379	15	HRB80	45	310	-	-
C65500	H04 [5]	92	634	55	379	22	HRB90	58	400	-	-
	H02 [18]	78	538	45	310	35	HRB85	52	359	-	-
C65600	H04	92	634	55	379	20	HRB90	58	400	-	-
C66100	H04	90	621	55	379	15	HRB90	58	400	-	-
C67400	H02 [21]	92	634	55	379	20	HRB87	48	331	29	200
	060 [21]	70	483	34	234	28	HRB78	-	-	-	-
C67500	H02 [18]	84	579	60	414	19	HRB90	48	331	-	-
	H01 [25]	77	531	45	310	23	HRB83	47	324	-	-
	060	65	448	30	207	33	HRB65	42	290	-	-
C67600	H02	75	517	45	310	18	HRB85	-	-	-	-
C69400	060	85	586	43	296	25	HRB85	-	-	-	-
C69430	H04	85	586	50	345	20	HRB90	-	-	-	-
C69710	H04	70	483	40	276	22	HRB90	-	-	-	-
C70600	061 [27]	47	324	34	234	-	-	-	-	-	-
C71500	H02 [18]	75	517	70	483	15	HRB80	42	290	-	-
C75200	H02 [10, 18]	70	483	60	414	20	HRB78	-	-	-	-
C75700	H04	80	550	53	365	23	-	-	-	-	-
C79200	H04 [21]	72	496	62	427	15	HRB78	-	-	-	-

(Continued)

Table 9. *(Continued)* **Selected Mechanical Properties of Copper Alloy** [1] **for Machined Products.** *(Source, Copper Development Association)*

Notes: All properties as measured at room temperature (68° F / 20° C).

1 Section size diameter one-inch (25.4 mm) unless otherwise specified.
2 Yield strength at 0.5% extension under load unless otherwise specified.
3 Elongation measured in two-inches (50.8 mm).
4 Fatigue strength for 100×10^6 cycles unless otherwise specified.
5 Specimen 35% cold worked.
6 Specimen 29% cold worked.
7 Specimen condition solution heat treated, cold worked 48%, aged and cold worked 47%.
8 Yield strength for this alloy measured at 0.2% offset.
9 Section size diameter 0.54 inch (13.7 mm).
10 Section size diameter 0.5 inch (12.7 mm).
11 Specimen condition hard and precipitation heat treated.
12 Section size diameter <0.375 inch (<9.5 mm).
13 Specimen precipitation hardened.
14 Specimen heat treated, cold worked 40%, and aged.
15 Specimen 35% cold worked, and heat treated.
16 Section size diameter 0.375–0.500 inch (9.5–12.7 mm).
17 Section size diameter 0.312 inch (7.9 mm).
18 Specimen 20% cold worked.
19 Fatigue strength recorded at 50×10^6 cycles.
20 Specimen 8% cold worked.
21 Section size diameter 0.75 inch (19.0 mm).
22 Specimen 25% cold worked.
23 Specimen 15% cold worked.
24 Section size diameter 4 inches (101.6 mm).
25 Specimen 10% cold worked.
26 Fatigue strength recorded at 300×10^6 cycles.
27 Section size diameter <2.5 inches (<63.5 mm).

Table 10. Selected Physical Properties of Wrought Copper Alloys for Machined Products. *(Source, Copper Development Association.)*

Copper Alloy No.	Coef. Thermal Exp.[1]		Melting Point-Liquidus		Melting Point-Solidus		Elastic Modulus		Shear Modulus	
	10⁻⁶/°F	10⁻⁶/°C	°F	°C	°F	°C	KSI	MPa	KSI	MPa
C10100	9.6	17.4	1,981	1,083	1,981	1,083	17,000	117,000	6,400	44,100
C10200	9.6	17.4	1,981	1,083	1,981	1,083	17,000	117,000	6,400	44,100
C10300	9.6	17.4	1,981	1,083	1,981	1,083	17,000	117,000	6,400	44,100
C10400	9.6	17.4	1,981	1,083	1,981	1,083	17,000	117,000	6,400	44,100
C10500	9.6	17.4	1,981	1,083	1,981	1,083	17,000	117,000	6,400	44,100
C10700	9.6	17.4	1,981	1,083	1,981	1,083	17,000	117,000	6,400	44,100
C10800	9.6	17.4	1,981	1,083	1,981	1,083	17,000	117,000	6,400	44,100
C11000	9.6	17.4	1,981	1,083	1,949	1,065	17,000	117,000	6,400	44,100
C11300	-	-	1,980	1,082	-	-	17,000	117,000	6,400	44,100
C11400	-	-	1,980	1,082	-	-	17,000	117,000	6,400	44,100
C11500	-	-	1,980	1,082	-	-	17,000	117,000	6,400	44,100
C11600	-	-	1,980	1,082	-	-	17,000	117,000	6,400	44,100
C12000	9.6	17.4	1,981	1,083	1,981	1,083	17,000	117,000	6,400	44,100
C12100	9.6	17.4	1,981	1,083	1,981	1,083	17,000	117,000	6,400	44,100
C12200	9.5	17.2	1,981	1,083	-	-	17,000	117,000	6,400	44,100
C12900	9.6	17.4	1,981	1,083	-	-	17,000	117,000	6,400	44,100
C14500	9.7	17.5	1,976	1,080	1,924	1,051	17,000	117,000	6,400	44,100
C14520	9.7	17.5	1,967	1,075	1,924	1,051	17,000	117,000	6,400	44,100
C14700	9.6	17.4	1,969	1,076	1,953	1,067	17,000	117,000	6,400	44,100
C15000	-	-	1,976	1,080	1,796	980	18,700	129,000	-	-
C15715	9.7	17.5	1,981	1,083	1,981	1,083	19,000	130,000	-	-
C15725	9.9	17.9	1,981	1,083	-	-	19,000	130,000	-	-
C15760	10.0	18.0	1,981	1,083	1,981	1,083	19,000	130,000	-	-

(Continued)

Table 10. *(Continued)* Selected Physical Properties of Wrought Copper Alloys for Machined Products. *(Source, Copper Development Association.)*

Copper Alloy No.	Coef. Thermal Exp.¹		Melting Point-Liquidus		Melting Point-Solidus		Elastic Modulus		Shear Modulus	
	10^{-6}/°F	10^{-6}/°C	°F	°C	°F	°C	KSI	MPa	KSI	MPa
C16200	9.6	17.4	1,969	1,076	1,886	1,030	17,000	117,000	6,400	44,100
C16500	9.8	17.7	1,958	1,070	-	-	17,000	117,000	-	-
C17000	9.9	17.9	1,800	982	1,590	866	18,500	128,000	7,300	50,300
C17200	9.9	17.9	1,800	982	1,590	866	18,500	128,000	7,300	50,300
C17300	9.9	17.9	1,800	982	1,590	866	18,500	128,000	7,300	50,300
C17410	-	-	1,950	1,066	1,875	1,024	19,000	131,000	-	-
C17500	9.8	17.7	1,955	1,068	1,885	1,029	19,000	131,000	7,500	51,700
C17510	9.8	17.7	1,955	1,068	1,885	1,029	19,000	131,000	7,500	51,700
C18000	-	-	-	-	-	-	-	-	-	-
C18100	10.2	18.4	1,967	1,075	-	-	18,200	125,000	6,800	46,900
C18135	9.8	17.7	1,976	1,080	-	-	20,000	141,180	7,500	51,700
C18150	-	-	-	-	-	-	-	-	-	-
C18200	-	-	1,967	1,075	1,958	1,070	17,000	117,000	7,200	50,630
C18400	-	-	1,967	1,075	1,958	1,070	17,000	117,000	7,200	50,630
C18700	9.8	17.7	1,976	1,080	1,747	953	17,000	117,000	6,400	44,100
C19100	9.8	17.7	1,980	1,082	1,900	1,038	16,000	110,000	6,000	41,400
C19150	-	-	1,980	1,082	-	-	-	-	-	-
C21000	-	-	1,949	1,065	1,920	1,049	17,000	117,000	6,400	44,100
C22000	10.2	18.4	1,910	1,043	1,870	1,021	17,000	117,000	6,400	44,100
C22600	10.2	18.4	1,895	1,035	1,840	1,004	17,000	117,000	6,400	44,100
C23000	-	-	1,877	1,027	1,810	988	17,000	117,000	6,400	44,100
C24000	-	-	1,832	1,000	1,770	966	16,000	110,000	6,000	41,400
C26000	-	-	1,750	954	1,680	916	16,000	110,000	6,000	41,400

(Continued)

Table 10. (Continued) Selected Physical Properties of Wrought Copper Alloys for Machined Products. (Source, Copper Development Association.)

Copper Alloy No.	Coef. Thermal Exp.[1]		Melting Point-Liquidus		Melting Point-Solidus		Elastic Modulus		Shear Modulus	
	10⁻⁶/°F	10⁻⁶/°C	°F	°C	°F	°C	KSI	MPa	KSI	MPa
C26130	-	-	1,750	954	1,680	916	16,000	110,000	6,000	41,400
C26200	-	-	1,750	954	1,680	916	16,000	110,000	6,000	41,400
C26800	-	-	1,710	932	1,660	904	15,000	103,000	5,600	38,600
C27000	-	-	1,710	932	1,660	904	15,000	103,000	5,600	38,600
C28000	-	-	1,660	904	1,650	899	15,000	103,000	5,600	38,600
C31400	-	-	1,900	1,038	1,850	1,010	17,000	117,000	6,400	44,100
C31600	-	-	1,900	1,038	1,850	1,010	17,000	117,000	-	-
C32000	-	-	1,875	1,024	1,820	993	17,000	117,000	6,400	44,100
C33500	-	-	1,700	927	1,650	899	15,000	103,000	5,600	38,600
C34000	-	-	1,700	927	1,630	888	15,000	103,000	5,600	38,600
C34200	-	-	1,670	910	1,630	888	15,000	103,000	5,600	38,600
C34500	-	-	1,688	920	-	-	-	-	-	-
C35000	-	-	1,680	916	1,640	893	15,000	103,000	5,600	38,600
C35300	-	-	1,670	910	1,630	888	15,000	103,000	5,600	38,600
C35330	-	-	1,650	899	1,630	888	14,000	96,500	5,300	36,500
C35600	-	-	1,650	899	1,630	888	14,000	96,500	5,300	36,500
C36000	-	-	1,650	899	1,630	888	14,000	96,500	5,300	36,500
C37000	-	-	1,650	899	1,630	888	15,000	103,000	5,600	38,600
C37700	-	-	1,640	893	1,620	882	15,000	103,000	5,600	38,600
C38500	-	-	1,630	888	1,610	877	14,000	96,500	5,300	36,500
C46200	-	-	1,670	910	-	-	-	-	-	-
C46400	-	-	1,650	899	1,630	888	15,000	103,000	5,600	38,600
C48200	-	-	1,650	899	1,630	888	15,000	103,000	5,600	38,600

(Continued)

Table 10. (Continued) Selected Physical Properties of Wrought Copper Alloys for Machined Products. (Source, Copper Development Association.)

Copper Alloy No.	Coef. Thermal Exp.[1]		Melting Point-Liquidus		Melting Point-Solidus		Elastic Modulus		Shear Modulus	
	10⁻⁶/°F	10⁻⁶/°C	°F	°C	°F	°C	KSI	MPa	KSI	MPa
C48500	-	-	1,650	899	1,630	888	15,000	103,000	5,600	38,600
C50700	-	-	1,958	1,070	-	-	-	-	-	-
C51000	-	-	1,920	1,049	1,750	954	16,000	110,000	6,000	41,400
C52100	-	-	1,880	1,027	1,620	882	16,000	110,000	6,000	41,400
C54400	-	-	1,830	999	1,700	927	15,000	103,000	5,600	38,600
C61000	-	-	1,904	1,040	-	-	17,000	117,000	6,400	44,100
C61300	-	-	1,915	1,046	1,905	1,041	17,000	117,000	-	-
C61400	-	-	1,915	1,046	1,905	1,041	17,000	117,000	6,400	44,100
C61800	-	-	1,913	1,045	1,904	1,040	17,000	117,000	6,400	44,100
C62300	-	-	1,915	1,046	1,905	1,041	17,000	117,000	6,400	44,100
C62400	-	-	1,900	1,038	1,880	1,027	17,000	117,000	6,400	44,100
C62500	-	-	1,925	1,052	1,917	1,047	16,000	110,000	-	-
C63000	-	-	1,930	1,054	1,895	1,035	17,500	121,000	6,400	44,100
C63020	-	-	-	-	-	-	18,000	124,000	-	-
C63200	-	-	1,940	1,060	1,905	1,041	17,000	117,000	6,400	44,100
C64200	-	-	1,840	1,004	1,800	982	16,000	110,000	6,000	41,400
C64700	-	-	1,994	1,090	1,940	1,060	-	-	-	-
C65100	-	-	1,940	1,060	1,890	1,032	17,000	117,000	6,400	44,100
C65500	-	-	1,880	1,027	1,780	971	15,000	103,000	5,600	38,600
C65600	-	-	1,866	1,019	-	-	-	-	-	-
C66100	-	-	1,866	1,019	-	-	-	-	-	-
C67000	-	-	1,652	900	-	-	-	-	-	-
C67300	-	-	-	-	-	-	-	-	-	-

(Continued)

Table 10. (Continued) **Selected Physical Properties of Wrought Copper Alloys for Machined Products.** (Source, Copper Development Association.)

Copper Alloy No.	Coef. Thermal Exp.[1]		Melting Point-Liquidus		Melting Point-Solidus		Elastic Modulus		Shear Modulus	
	10⁻⁶/°F	10⁻⁶/°C	°F	°C	°F	°C	KSI	MPa	KSI	MPa
C67400	-	-	1,625	885	1,590	866	16,000	110,000	6,000	41,400
C67500	-	-	1,630	888	1,590	866	15,000	103,000	5,600	38,600
C67600	-	-	1,652	900	-	-	-	-	-	-
C69400	-	-	1,685	918	1,510	821	16,000	110,000	-	-
C69430	-	-	1,685	918	1,510	821	16,000	110,000	-	-
C69710	-	-	1,706	930	-	-	-	-	-	-
C70600	9.3	16.7	2,093	1,145	-	-	18,000	124,000	6,800	46,900
C71500	9.0	16.3	2,260	1,238	2,140	1,171	22,000	15,200	8,300	57,200
C74500	9.1	16.5	1,870	1,021	-	-	17,500	121,000	6,600	45,500
C75200	9.0	16.3	2,030	1,110	1,960	1,071	18,000	124,000	6,800	46,900
C75400	9.0	16.3	1,970	1,077	1,900	1,038	18,000	124,000	6,800	46,900
C75700	9.0	16.3	1,900	1,038	-	-	18,000	124,000	6,800	46,900
C79200	9.0	16.2	-	-	-	-	-	-	-	-

Note: [1] Coefficient of thermal expansion for temperature range 68 to 392° F (20 to 200° C).

exposure (1650° F / 900° C). Physical properties similar to pure copper. Excellent elevated temperature mechanical properties. C15760 rod used primarily for RWMA Class II and Class III resistance welding applications. C15715 used for electrical conductors and springs.

High Copper Alloys. C16200 – C19900 are high copper alloys with designated copper content of less than 99.3%, but more than 96%, that do not fall into any other copper alloy group.

C16200, C16500: Moderate strength, good wear resistance. Mostly used for trolley wire, occasionally specified for machined products. C16500 is slightly stronger than C16200. C16200 used for RWMA Class I applications.

C17000, C17200, C17300, C17410, C17500, C17510: Both C17200 and C17300 can develop more than 200 ksi (1380 MPa) tensile strength. C17410, C17500, and C17510 have conductivities of about 50% IACS and moderate strength. These beryllium copper alloys are available in ductile heat treatable (hardenable) tempers as well as in mill hardened tempers. Alloy C17300, available in either rod or wire, is leaded for increased machinability. These alloys are used in a wide range of applications requiring very high strength and stiffness with good conductivity. Typical uses include electrical/electronic connectors, current carrying springs, precision screw machined parts, welding electrodes, bearings, plastic molds, and corrosion resistant components.

C18000, C18100, C18135, C18150: Good wear resistance and high strength at elevated temperatures. Used for high stress, high temperature applications such as welding electrodes, electrical components, contacts, and studs.

C18200, C18400: Moderately high elevated temperature strength. Conductivity 80% IACS. Used for resistance welding equipment components such as tips, clamps, and wheels, and for other electrical and mechanical power transmission devices, circuit breaker parts, and high strength fasteners for elevated temperature service.

C18700: Free machining copper with 96% IACS conductivity and good corrosion resistance. Used for electrical/electronic connectors, welding torch tips, and products requiring a significant amount of machining.

C19100, C19150: Moderately high strength alloys with good conductivity (55% IACS) and corrosion resistance. Highly machinable: C19100 uses tellurium (Te) to achieve free machining properties, and C19150 uses lead (Pb). Used for cylindrical connector contacts, bolts, bushings, electrical and mechanical components, gears, and marine hardware.

Copper-Zinc Alloys (Brasses). C21000 – C28000 are brasses with Cu and Zn as the principal alloying agents.

C21000: Good formability and corrosion resistance. More often cold formed than machined. Used for ammunition components, coinage, and medallions.

C22000: Similar to C21000, but stronger. Used for corrosion resistant products, decorative items, and fasteners. Better suited to cold forming than machining.

C22600, C23000: Similar to, but stronger than, C22000. Used for cold formed electrical and hardware components, jewelry, decorative chain. Sometimes specified for dezincification-resistant fasteners and fittings.

C24000: Similar to, but stronger than, C23000.

C26000, C26130, C26200, C26800, C27700, C28000: C26000 has the highest ductility in the yellow brass series, and is also known as Jewelry brass. It is easily machined, but more often cold formed. C27000 hot forms more readily and is stronger than C26000. Used for both hot and cold formed products.

Copper-Zinc-Lead Alloys (Leaded Brasses). C31200 – C38500 are leaded brasses with Cu, Zn, and Pb as the principal alloying agents.

C31400, C31600: Free machining versions of the 20000 series yellow brasses. Moderate corrosion resistance. Used for architectural hardware, pole line hardware, and fasteners. C31600 is somewhat stronger than C31400.

C32000, C33500, C34000, C34200, C34500, C35000, C35300, C35330, C35600: Free machining brasses with good formability. C34500 has the best combination of machining and formability. C35300 is dezincification resistant, and better for acid waters. These brasses are used for screw machine products requiring some cold formability, as for knurling, peening, crimping, thread rolling. Typical uses include hose fittings; watch, clock, and lock parts; bicycle spoke nipples; and plumbing valve components.

C36000: Highest machinability of all copper alloys. Used for screw machine products in a wide range of applications. Commonly called free cutting brass.

C37000: Normally plate and tube alloy, but also produced as rods, bars, and shapes.

C37700: Known as forging brass, it has excellent forgability, and is free cutting. Used for forgings requiring extensive machining, architectural hardware, fittings, mechanical components, and specialty fasteners.

C38500: Free machining, used as-extruded or as-forged, but best applied where products require machining as the secondary operation. Used mostly for architectural hardware and trim. Most easily hot worked. Light pink in color. Often called Architectural bronze.

Copper-Zinc-Tin Alloys (Tin Brasses). C40400 – C48600 are tin brasses, with Cu, Zn, Sn, and Pb as the principal alloying agents.

C46200, C46400, C48200, C48500: These are commonly known as Naval brass. Excellent hot forgability, good corrosion resistance. Used for fasteners and hardware for corrosion resistant service, including marine applications. C46200 has higher dezincification resistance than C46400. These brasses are used for high strength cold headed products, including fasteners.

Copper-Tin-Phosphorus Alloys (Phosphor Bronzes). C50100 – C52480 are phosphor bronzes with Cu, Sn, and P as the principal alloying agents.

C50700, C51000, C52100: Good combination of strength and ductility. Excellent spring qualities.

Copper-Tin-Phosphorus-Lead Alloys (Leaded Phosphor Bronze). C53400 – C54400 are leaded phosphor bronzes with Cu, Sn, Pb, and P as the principal alloying agents.

C55180 – C55299 are Copper Phosphorus and Copper Silver Phosphorus *Brazing Alloys* with Cu, P, and Ag as the principal alloying agents.

C55300 – C60799 are Copper Silver Zinc Bronzes with Cu, Ag, and Zn as the principal alloying agents.

C54400: Free machining, with good cold working properties. Very good wear resistance. Often used for bearings, cam followers, and similar products.

Copper-Aluminum Alloys (Aluminum Bronzes). C60800 – C64210 are aluminum bronzes, and Cu, Al, Ni, Fe, Si, and Sn are the alloying agents.

C61000, C61300, C61400, C61800, C62300, C62400, C62500, C63000, C63020, C63200, C64200: Aluminum bronzes and their alloyed modifications are best known for moderate to high strength, and excellent corrosion resistance. C61300 has good formability; C61800 has good corrosion resistance; C62300 has good acid resistance; C62400 is heat treatable to high strength; C63000 and C63020 have typical tensile strengths of 118 and 145 ksi (814 and 1000 MPa), respectively; C63400 has moderate strength and corrosion resistance; and C63400 is free machining and has good antigalling and antiseizing properties. These alloys are used for valve and pump components for industrial process streams, and for marine equipment, high strength fasteners, and pole line hardware.

Copper-Silicon Alloys (Silicon Bronzes). C64700 – C67000 are silicon bronzes with Cu, Si, and Sn as the principal alloying agents.

C64700, C65100, C65500, C65600, C66100, C67000: Moderate strength, good corrosion resistance, similar to C64200 in properties and applications. Used for valve guides, valve stems, fasteners, pole line hardwire, marine fittings. Moderate to high strength, and good corrosion resistance. Choice of alloy depends on corrosive environment and manufacturing process. C66100 is free cutting and adaptable to high speed screw machines. Applications similar to aluminum bronzes: high strength fasteners, marine and pole hardware.

Miscellaneous Copper-Zinc Alloys. C67200 – C69710 is reserved for "other copper alloys," and no alloys are specified.

C67300, C67400, C67500, C67600, C69400, C69430, C69710: High strength, hot forgable alloys. Some with free machining qualities that allow them to run on high speed machining centers. Used in heavy duty mechanical components including bearings. Arsenical alloys C69430 and C69710 resist dezincification and are better for use in acid waters. Used principally for valve stems and pump components.

Copper-Nickel Alloys. C70100 – C72950 are copper nickel alloys with Cu, Ni, and Fe as the principal alloying agents.

C70600, C71500: Excellent corrosion resistance, especially in marine environments. Moderately high strength, good creep resistance at elevated temperatures. Properties generally increase with nickel content. Copper-nickels are used mainly for seawater service as forged and machined valve and pump components, fittings, and hardware. Relatively high in cost compared to copper-aluminum and other alloys with similar mechanical properties. Often specified for use where high corrosion resistance is required and where concern over chloride stress-corrosion cracking prevents use of stainless steels.

Copper-Nickel-Zinc Alloys (Nickel Silvers). C73500 – C79830 are nickel silvers with Cu, Ni, and Zn as the principal alloying agents.

C74500, C75200, C75400, C75700, C79200: Good formability, good corrosion- and tarnish resistance. Alloys have pleasing silver-like color. C75400 and C76390 are similar to, but stronger than, C75200. Uses include decorative items, jewelry, musical instrument valves and components, optical instrument components, fittings for food and dairy equipment, screws, rivets, and slide fasteners.

Description of cast copper alloys

Characteristics and popular uses of cast copper alloys are given below. Composition of cast copper alloys are given in **Tables 11** through **17**; **Table 18** provides mechanical properties; and **Table 19** provides physical properties.

Coppers. C80100 – C81200 are cast coppers with a minimum copper content of 99.3%.

These coppers are used almost exclusively for their unsurpassed electrical and thermal conductivities in products such as terminals, connectors, and (water cooled) hot metal handling equipment. They also have high corrosion resistance, but this is usually a secondary consideration.

High Copper Alloys. C81400 – C82800 are high copper alloys with designated copper content in excess of 94%, to which Ag may be added for special properties.

These alloys are used primarily for their combination of high strength and good conductivity. Chromium coppers (C81400 and C81500) with tensile strength of 45 ksi (310 MPa) and conductivity of 82% IACS (as heat treated) are used in electrical contacts, clamps, welding gear, and similar electromechanical hardware. At more than 160 ksi (1,100 MPa), the beryllium coppers (C82800) have the highest tensile strength of all the copper

(Text continued on p. 619)

Table 11. Chemical Composition of Cast Coppers and Cast High Copper Alloys. (Source, Copper Development Association.)

Coppers

Copper Number	Cu (incl. Ag) % (min.)	P	Other (named) Zn
C80100	99.95	-	-
C80410	99.9	-	.10
C81100	99.70	-	-
C81200	99.9	.045-.065	-

High Copper Alloys

Copper Alloy No.	Cu[16]	Be	Co	Si	Ni	Fe	Al	Sn	Pb	Zn	Cr
C81400	Rem.	.02-.10	-	-	-	-	-	-	-	-	.6-1.0
C81500	Rem.	-	-	.15	-	.10	.10	.10	.02	.10	.40-1.5
C81540	95.1 min.[41]	-	-	.40-.8	2.0-3.0[42]	.15	.10	.10	.02	.10	.10-.60
C82000	Rem.	.45-.80	2.40-2.70[42]	.15	.20	.10	.10	.10	.02	.10	.10
C82200	Rem.	.35-.80	.30	-	1.0-2.0	-	-	-	-	-	-
C82400	Rem.	1.60-1.85	.20-.65	-	.20	.20	.15	.10	.02	.10	.10
C82500	Rem.	1.90-2.25	.35-.70[42]	.20-.35	.20	.25	.15	.10	.02	.10	.10
C82510	Rem.	1.90-2.15	1.0-1.2	.20-.35	.20	.25	.15	.10	.02	.10	.10
C82600	Rem.	2.25-2.55	.35-.65	.20-.35	.20	.25	.15	.10	.02	.10	.10
C82700	Rem.	2.35-2.55	-	.15	1.0-1.5	.25	.15	.10	.02	.10	.10
C82800	Rem.	2.50-2.85	.35-.70[42]	.20-.35	.20	.25	.15	.10	.02	.10	.10

Notes: [16] Cu + Sum of Named Elements, 99.5% min.
[41] Includes Ag.
[42] Ni + Co.

Table 12. Chemical Composition of Cast Brasses. (Source, Copper Development Association.)

Copper Alloy No.	Cu 43, 44	Sn	Pb	Zn	Ni (incl. Co)	Other (named)
	Copper-Tin-Zinc and Copper-Tin-Zinc-Lead Alloys (Red and Leaded Red Brasses)					
C83300	92.0-94.0	1.0-2.0	1.0-2.0	2.0-6.0	-	-
C83400	88.0-92.0	0.20	0.50	8.0-12.0	1.0	0.25 Fe, 0.25 Sb, 0.08 S, 0.03 P [45], 0.005 Al, 0.005 Si
C83450	87.0-89.0	2.0-3.5	1.5-3.5	5.5-7.5	0.8-2.0	0.30 Fe, 0.25 Sb, 0.08 S, 0.03 P [45], 0.005 Al, 0.005 Si
C83500	86.0-88.0	5.5-6.5	3.5-5.5	1.0-2.5	0.50-1.0	0.25 Fe, 0.25 Sb, 0.08 S, 0.03 P [45], 0.005 Al, 0.005 Si
C83600	84.0-86.0	4.0-6.0	4.0-6.0	4.0-6.0	1.0	0.30 Fe, 0.25 Sb, 0.08 S, 0.05 P [45], 0.005 Al, 0.005 Si
C83800	82.0-83.8	3.3-4.2	5.0-7.0	5.0-8.0	1.0	0.30 Fe, 0.25 Sb, 0.08 S, 0.03 P [45], 0.005 Al, 0.005 Si
C83810	Rem.	2.0-3.5	4.0-6.0	7.5-9.5	2.0	0.50 Fe [46], Sb [46], As [46], 0.005 Al, .10 Si
	Copper-Tin-Zinc and Copper-Tin-Zinc-Lead Alloys (Semi-Red and Leaded Semi-Red Brasses)					
C84200	78.0-82.0	4.0-6.0	2.0-3.0	10.0-16.0	0.8	0.40 Fe, 0.25 Sb, 0.08 S, 0.05 P [45], 0.005 Al, 0.005 Si
C84400	78.0-82.0	2.3-3.5	6.0-8.0	7.0-10.0	1.0	0.40 Fe, 0.25 Sb, 0.08 S, 0.02 P [45], 0.005 Al, 0.005 Si
C84410	Rem.	3.0-4.5	7.0-9.0	7.0-11.0	1.0	Fe [48], Sb [48], .01 Al, .20 Si, .05 Bi
C84500	77.0-79.0	2.0-4.0	6.0-7.5	10.0-14.0	1.0	0.40 Fe, 0.25 Sb, 0.08 S, 0.02 P [45], 0.005 Al, 0.005 Si
C84800	75.0-77.0	2.0-3.0	5.5-7.0	13.0-17.0	1.0	0.40 Fe, 0.25 Sb, 0.08 S, 0.02 P [45], 0.005 Al, 0.005 Si
Alloy No.	Cu [43]	Sn	Pb	Zn	Ni + Co	Other (named)
	Copper-Tin-Zinc and Copper-Tin-Zinc-Lead Alloys (Yellow and Leaded Yellow Brasses)					
C85200	70.0-74.0 [49]	0.7-2.0	1.5-3.8	20.0-27.0	1.0	0.6 Fe, 0.20 Sb, 0.05 S, 0.02 P, 0.005 Al, 0.05 Si
C85400	65.0-70.0 [52]	0.50-1.5	1.5-3.8	24.0-32.0	1.0	0.7 Fe, 0.35 Al, 0.05 Si
C85500	59.0-63.0 [49]	0.20	0.20	Rem.	0.20	0.20 Fe, 0.20 Mn
C85700	58.0-64.0 [53]	0.50-1.5	0.8-1.5	32.0-40.0	1.0	0.7 Fe, 0.8 Al, 0.05 Si
C85800	57.0 min. [53]	1.5	1.5	31.0-41.0	0.50	0.50 Fe, 0.05 Sb, 0.25 Mn, 0.05 As, 0.05 S, 0.01 P, 0.55 Al, 0.25 Si

(Continued)

Table 12. (Continued) Chemical Composition of Cast Brasses. (Source, Copper Development Association.)

Manganese Bronze and Leaded Manganese Bronze Alloys (High Strength and Leaded High Strength Yellow Brasses)

Alloy No.	Cu [43,50]	Sn	Pb	Zn	Fe	Ni + Co	Al	Mn	Si [45]
C86100	66.0-68.0	0.20	0.20	Rem.	2.0-4.0	-	4.5-5.5	2.5-5.0	-
C86200	60.0-66.0	0.20	0.20	22.0-28.0	2.0-4.0	1.0	3.0-4.9	2.5-5.0	-
C86300	60.0-66.0	0.20	0.20	22.0-28.0	2.0-4.0	1.0	5.0-7.5	2.5-5.0	-
C86400	56.0-62.0	0.50-1.5	0.50-1.5	34.0-42.0	0.40-2.0	1.0	0.50-1.5	0.10-1.5	-
C86500	55.0-60.0	1.0	0.40	36.0-42.0	0.40-2.0	1.0	0.50-1.5	0.10-1.5	-
C86550	57.0 min.	1.0	0.50	Rem.	0.7-2.0	1.0	0.50-2.5	0.10-3.0	0.10
C86700	55.0-60.0	1.5	0.50-1.5	30.0-38.0	1.0-3.0	1.0	1.0-3.0	0.10-3.5	-
C86800	53.5-57.0	1.0	0.20	Rem.	1.0-2.5	2.5-4.0	2.0	2.5-4.0	-

Copper Silicon Alloys (Silicon Bronzes and Silicon Brasses)

Alloy No.	Cu [16]	Pb	Zn	Fe	Si	Al + Co	Bi	Ni + Co	Other (named)
C87300	94.0 min.	0.20	0.25	0.20	3.5-4.5	-	-	-	0.8-1.5 Mn
C87400	79.0 min. [47]	1.0	12.0-16.0	-	2.5-4.0	0.8	-	-	-
C87500	79.0 min.	0.50	12.0-16.0	-	3.0-5.0	0.50	-	-	-
C87600	88.0 min.	0.50	4.0-7.0	0.20	3.5-5.5	-	-	-	0.25 Mn
C87610	90.0 min.	0.20	3.0-5.0	0.20	3.0-5.0	-	-	-	0.25 Mn
C87800	80.0 min.	0.15	12.0-16.0	0.15	3.8-4.2	0.15	-	-	0.25 Sn, 0.15 Mn, 0.01 Mg, 0.20 Ni + Ag, 0.05 S, 0.01 P, 0.05 As, 0.05 Sb

Copper-Bismuth and Copper-Bismuth-Selenium Alloys (High Strength and Leaded High Strength Yellow Brasses)

Alloy No.	Cu	Sn	Zn	Ni + Co	Bi	Other (named)
C89320	87.0-91.0 [16]	5.0-7.0	1.0	1.0	4.0-6.0	0.09 Pb, 0.20 Fe, 0.35 Sb, 0.08 S, 0.30 P, 0.005 Al, 0.005 Si
C89325	84.0-88.0 [50]	9.0-11.0	1.0	1.0	2.7-3.7	0.10 Pb, 0.15 Fe, 0.50 Sb, 0.08 S, 0.10 P, 0.005 Al, 0.005 Si, [54]
C89510	86.0-88.0 [16]	4.0-6.0	1.0	1.0	0.50-1.5 [66]	0.25 Pb, 0.20 Fe, 0.25 Sb, 0.08 S, 0.005 P, 0.005 Al, 0.005 Si, 0.35-0.7 Se [66]

(Continued)

Table 12. (Continued) Chemical Composition of Cast Brasses. (Source, Copper Development Association.)

Alloy No.	Cu	Sn	Zn	Ni + Co	Bi	Other (named)
		Copper-Bismuth and Copper-Bismuth-Selenium Alloys (High Strength and Leaded High Strength Yellow Brasses)				
C89520	85.0-87.0 [16]	5.0-6.0	4.0-6.0	1.0	1.6-2.2	0.25 Pb, 0.20 Fe, 0.25 Sb, 0.08 S, 0.05 P, 0.005 Al, 0.005 Si, 0.8-1.1 Se, [64]
C89550	58.0-64.0 [16]	0.00-1.2	32.0-38.0	1.0	0.6-1.2	0.10 Pb, 0.50 Fe, 0.05 Sb, 0.05 S, 0.01 P, 0.10-0.6 Al, 0.25 Si, 0.01-0.10 Se
C89831	87.0-91.0 [50]	2.7-3.7	2.0-4.0	1.0	2.7-3.7	0.10 Pb, 0.30 Fe, 0.25 Sb, 0.08 S, 0.050 P, 0.005 Al, 0.005 Si, [54]
C89833	87.0-91.0 [50]	4.0-6.0	2.0-4.0	1.0	1.7-2.7	0.10 Pb, 0.30 Fe, 0.25 Sb, 0.08 S, 0.050 P, 0.005 Al, 0.005 Si, [54]
C89835	85.0-89.0 [50]	6.0-7.5	2.0-4.0	1.0	1.7-2.7	0.10 Pb, 0.20 Fe, 0.35 Sb, 0.08 S, 0.10 P, 0.005 Al, 0.005 Si, [54]
C89837	84.0-88.0 [50]	3.0-4.0	6.0-10.0	1.0	0.7-1.2	0.10 Pb, 0.30 Fe, 0.25 Sb, 0.08 S, 0.050 P, 0.005 Al, 0.005 Si, [54]
C89844	83.0-86.0 [44]	3.0-5.0	7.0-10.0	1.0	2.0-4.0	0.20 Pb, 0.30 Fe, 0.25 Sb, 0.08 S, 0.05 P, 0.005 Al, 0.005 Si
C89940	64.0-68.0 [16]	3.0-5.0	3.0-5.0	20.0-23.0	4.0-5.5	0.01 Pb, 0.7-2.0 Fe, 0.10 Sb, 0.05 S, 0.10-0.15 P, 0.005 Al, 0.15 Si, 0.20 Mn

Notes: [16] Cu +Sum of Named Elements, 99.5% min.

[43] In determining copper min., copper may be calculated as Cu + Ni.

[44] Cu +Sum of Named Elements, 99.3% min.

[45] For continuous castings, P shall be 1.5%, max.

[46] Fe + Sb + as shall be .50% max.

[47] Cu +Sum of Named Elements, 99.2% min.

[48] Fe + Sb + as shall be .8% max.

[49] Cu +Sum of Named Elements, 99.1% min.

[50] Cu +Sum of Named Elements, 99.0% min.

[52] Cu +Sum of Named Elements, 98.9% min.

[53] Cu +Sum of Named Elements, 98.7% min.

[54] .01-2.0% as any single or combination of Ce, La or other rare earth* elements, as agreed upon. *ASM International definition: one of the group of chemically similar metals with atomic numbers 57 through 71, commonly referred to as lanthanides.

[64] Bi : Se ³ 2:1.

[66] Experience favors Bi : Se ³ 2:1.

Table 13. Chemical Composition of Cast Bronzes. *(Source, Copper Development Association.)*

Copper Alloy No.	Cu[43, 51]	Sn	Zn	Ni (incl. Co)	P[45]	Other (named)
Copper-Tin Alloys *(Tin Bronzes)*						
C90200	91.0-94.0	6.0-8.0	0.50	0.50	0.05	0.30 Pb, 0.20 Fe, 0.20 Sb, 0.05 S, 0.005 Al, 0.005 Si
C90300	86.0-89.0	7.5-9.0	3.0-5.0	1.0	0.05	0.30 Pb, 0.20 Fe, 0.20 Sb, 0.05 S, 0.005 Al, 0.005 Si
C90500	86.0-89.0 [21]	9.0-11.0	1.0-3.0	1.0	0.05	0.30 Pb, 0.20 Fe, 0.20 Sb, 0.05 S, 0.005 Al, 0.005 Si
C90700	88.0-90.0	10.0-12.0	0.50	0.50	0.30	0.50 Pb, 0.15 Fe, 0.20 Sb, 0.05 S, 0.005 Al, 0.005 Si
C90710	Rem.	10.0-12.0	0.05	0.10	0.05-1.2	0.25 Pb, 0.10 Fe, 0.20 Sb, 0.05 S, 0.005 Al, 0.005 Si
C90800	85.0-89.0	11.0-13.0	0.25	0.50	0.30	0.25 Pb, 0.15 Fe, 0.20 Sb, 0.05 S, 0.005 Al, 0.005 Si
C90810	Rem.	11.0-13.0	0.30	0.50	0.15-0.8	0.25 Pb, 0.15 Fe, 0.20 Sb, 0.05 S, 0.005 Al, 0.005 Si
C90900	86.0-89.0	12.0-14.0	0.25	0.50	0.05	0.25 Pb, 0.15 Fe, 0.20 Sb, 0.05 S, 0.005 Al, 0.005 Si
C91000	84.0-86.0	14.0-16.0	1.5	0.8	0.05	0.20 Pb, 0.10 Fe, 0.20 Sb, 0.05 S, 0.005 Al, 0.005 Si
C91100	82.0-85.0	15.0-17.0	0.25	0.50	1.0	0.25 Pb, 0.25 Fe, 0.20 Sb, 0.05 S, 0.005 Al, 0.005 Si
C91300	79.0-82.0	18.0-20.0	0.25	0.50	1.0	0.25 Pb, 0.25 Fe, 0.20 Sb, 0.05 S, 0.005 Al, 0.005 Si
C91600	86.0-89.0	9.7-10.8	0.25	1.2-2.0	0.30	0.25 Pb, 0.20 Fe, 0.20 Sb, 0.05 S, 0.005 Al, 0.005 Si
C91700	84.0-87.0	11.3-12.5	0.25	1.2-2.0	0.30	0.25 Pb, 0.20 Fe, 0.20 Sb, 0.05 S, 0.005 Al, 0.005 Si
Alloy No.	**Cu[43, 44]**	**Sn**	**Pb**	**Zn**	**Ni + Co**	**Other (named)**
Copper-Tin Lead Alloys *(Leaded Tin Bronzes)*						
C92200	86.0-90.0	5.5-6.5	1.0-2.0	3.0-5.0	1.0	0.25 Fe, 0.25 Sb, 0.05 S, 0.05 P [45], 0.005 Al, 0.005 Si
C92210	86.0-89.0	4.5-5.5	1.7-2.5	3.0-4.5	0.7-1.0	0.25 Fe, 0.20 Sb, 0.05 S, 0.03 P [45], 0.005 Al, 0.005 Si
C92220	86.0-88.0 [43]	5.0-6.0	1.5-2.5	3.0-5.5	0.50-1.0	0.25 Fe, 0.05 P [45]
C92300	85.0-89.0	7.5-9.0	0.30-1.0	2.5-5.0	1.0	0.25 Fe, 0.25 Sb, 0.05 S, 0.05 P [45], 0.005 Al, 0.005 Si
C92310	Rem.	7.5-8.5	0.30-1.5	3.5-4.5	1.0	0.005 Al, 0.005 Si, 0.03 Mn
C92400	86.0-89.0	9.0-11.0	1.0-2.5	1.0-3.0	1.0	0.25 Fe, 0.25 Sb, 0.05 S, 0.05 P [45], 0.005 Al, 0.005 Si
C92410	Rem.	6.0-8.0	2.5-3.5	1.5-3.0	0.20	0.20 Fe, 0.25 Sb, 0.005 Al, 0.005 Si, 0.05 Mn
C92500	85.0-88.0	10.0-12.0	1.0-1.5	0.50	0.8-1.5	0.30 Fe, 0.25 Sb, 0.05 S, 0.30 P [45], 0.005 Al, 0.005 Si
C92600	86.0-88.5	9.3-10.5	0.8-1.5	1.3-2.5	0.7	0.20 Fe, 0.25 Sb, 0.05 S, 0.03 P [45], 0.005 Al, 0.005 Si

(Continued)

Table 13. *(Continued)* **Chemical Composition of Cast Bronzes.** *(Source, Copper Development Association.)*

Alloy No.	Cu[43, 44]	Sn	Pb	Zn	Ni + Co	Other (named)
Copper-Tin Lead Alloys *(Leaded Tin Bronzes)*						
C92610	Rem.	9.5-10.5	0.30-1.5	1.7-2.8	1.0	0.15 Fe, 0.005 Al, 0.005 Si, 0.03 Mn
C92700	86.0-89.0	9.0-11.0	1.0-2.5	0.7	1.0	0.20 Fe, 0.25 Sb, 0.05 S, 0.25 P [45], 0.005 Al, 0.005 Si
C92710	Rem.	9.0-11.0	4.0-6.0	1.0	2.0	0.20 Fe, 0.25 Sb, 0.05 S, 0.10 P [45], 0.005 Al, 0.005 Si
C92800	78.0-82.0	15.0-17.0	4.0-6.0	0.8	0.8	0.20 Fe, 0.25 Sb, 0.05 S, 0.05 P [45], 0.005 Al, 0.005 Si
C92810	78.0-82.0	12.0-14.0	4.0-6.0	0.50	0.8-1.2	0.50 Fe, 0.25 Sb, 0.05 S, 0.05 P [45], 0.005 Al, 0.005 Si
C92900	82.0-86.0	9.0-11.0	2.0-3.2	0.25	2.8-4.0	0.20 Fe, 0.25 Sb, 0.05 S, 0.05 P [45], 0.005 Al, 0.005 Si
Copper-Tin-Lead Alloys *(Hi-Leaded Tin Bronzes)*						
C93100	Rem. [43, 50]	6.5-8.5	2.0-5.0	2.0	1.0	0.25 Fe, 0.25 Sb, 0.05 S, 0.30 P [45], 0.005 Al, 0.005 Si
C93200	81.0-85.0 [43, 50]	6.3-7.5	6.0-8.0	1.0-4.0	1.0	0.20 Fe, 0.35 Sb, 0.08 S, 0.15 P [45], 0.005 Al, 0.005 Si
C93400	82.0-85.0 [43, 50]	7.0-9.0	7.0-9.0	0.8	1.0	0.20 Fe, 0.50 Sb, 0.08 S, 0.50 P [45], 0.005 Al, 0.005 Si
C93500	83.0-86.0 [43, 50]	4.3-6.0	8.0-10.0	2.0	1.0	0.20 Fe, 0.30 Sb, 0.08 S, 0.05 P [45], 0.005 Al, 0.005 Si
C93600	79.0-83.0 [44]	6.0-8.0	11.0-13.0	1.0	1.0	0.20 Fe, 0.55 Sb, 0.08 S, 0.15 P [45], 0.005 Al, 0.005 Si
C93700	78.0-82.0 [50]	9.0-11.0	8.0-11.0	0.8	0.50	0.7 [55] Fe, 0.50 Sb, 0.08 S, 0.10 P [45], 0.005 Al, 0.005 Si
C93720	83.0 min. [50]	3.5-4.5	7.0-9.0	4.0	0.50	0.7 Fe, 0.50 Sb, 0.10 P [45]
C93800	75.0-79.0 [50]	6.3-7.5	13.0-16.0	0.8	1.0	0.15 Fe, 0.8 Sb, 0.08 S, 0.05 P [45], 0.005 Al, 0.005 Si
C93900	76.5-79.5 [52]	5.0-7.0	14.0-18.0	1.5	0.8	0.40 Fe, 0.50 Sb, 0.08 S, 1.5 P [45], 0.005 Al, 0.005 Si
C94000	69.0-72.0 [53]	12.0-14.0	14.0-16.0	0.50	0.50-1.0	0.25 Fe, 0.50 Sb, 0.08 [56] S, 0.05 P [45], 0.005 Al, 0.005 Si
C94100	72.0-79.0 [53]	4.5-6.5	18.0-22.0	1.0	1.0	0.25 Fe, 0.8 Sb, 0.08 [56] S, 0.05 P [45], 0.005 Al, 0.005 Si
C94300	67.0-72.0 [50]	4.5-6.0	23.0-27.0	0.8	1.0	0.15 Fe, 0.8 Sb, 0.08 [56] S, 0.08 P [45], 0.005 Al, 0.005 Si
C94310	Rem. [50]	1.5-3.0	27.0-34.0	0.50	0.25-1.0	0.50 Fe, 0.50 Sb, 0.05 P [45]
C94320	Rem. [50]	4.0-7.0	24.0-32.0	-	-	0.35 Fe
C94330	68.5-75.5 [50]	3.0-4.0	21.0-25.0	3.0	0.50	0.7 Fe, 0.50 Sb, 0.10 P [45]
C94400	Rem. [50]	7.0-9.0	9.0-12.0	0.8	1.0	0.15 Fe, 0.8 Sb, 0.08 S, 0.50 P [45], 0.005 Al, 0.005 Si
C94500	Rem. [50]	6.0-8.0	16.0-22.0	1.2	1.0	0.15 Fe, 0.8 Sb, 0.08 S, 0.05 P [45], 0.005 Al, 0.005 Si

(Continued)

Table 13. *(Continued)* **Chemical Composition of Cast Bronzes.** *(Source, Copper Development Association.)*

Alloy No.	Cu (incl. Ag)	Su	Pb	Zn	Ni + Co	Other (named)
Copper-Tin-Nickel Alloys *(Tin-Nickel Bronzes)*						
C94700	85.0-90.0 [53]	4.5-6.0	0.10 [57]	1.0-2.5	4.5-6.0	0.25 Fe, 0.15 Sb, 0.20 Mn, 0.05 S, 0.05 P, 0.005 Al, 0.005 Si
C94800	84.0-89.0 [53]	4.5-6.0	0.30-1.0	1.0-2.5	4.5-6.0	0.25 Fe, 0.15 Sb, 0.20 Mn, 0.05 S, 0.05 P, 0.005 Al, 0.005 Si
C94900	79.0-81.0 [51]	4.0-6.0	4.0-6.0	4.0-6.0	4.0-6.0	0.30 Fe, 0.25 Sb, 0.10 Mn, 0.08 S, 0.05 P, 0.005 Al, 0.005 Si

Alloy No.	Cu	Fe	Ni + Co	Al	Mn	Other (named)
Copper-Aluminum-Iron and Copper-Aluminum-Iron-Nickel Alloys *(Aluminum Bronzes)*						
C95200	86.0 min. [50]	2.5-4.0	-	8.5-9.5	-	-
C95210	86.0 min. [50]	2.5-4.0	1.0	8.5-9.5	1.0	0.05 Pb, 0.05 Mg, 0.25 Si, 0.50 Zn, 0.10 Sn
C95220	Rem. [16]	2.5-4.0	2.5	9.5-10.5	0.50	-
C95300	86.0 min. [50]	0.8-1.5	-	9.0-11.0	-	-
C95400	83.0 min. [16]	3.0-5.0	1.5	10.0-11.5	0.50	-
C95410	83.0 min. [16]	3.0-5.0	1.5-2.5	10.0-11.5	0.50	-
C95420	83.5 min. [16]	3.0-4.3	0.50	10.5-12.0	0.50	-
C95500	78.0 min. [16]	3.0-5.0	3.0-5.5	10.0-11.5	3.5	-
C95510	78.0 min. [20]	2.0-3.5	4.5-5.5	9.7-10.9	1.5	0.30 Zn, 0.20 Sn
C95520	74.5 min. [16]	4.0-5.5	4.2-6.0	10.5-11.5	1.5	0.03 Pb, 0.15 Si, 0.30 Zn, 0.25 Sn, 0.20 Co, 0.05 Cr
C95600	88.0 min. [50]	-	0.25	6.0-8.0	-	1.8-3.2 Si
C95700	71.0 min. [16]	2.0-4.0	1.5-3.0	7.0-8.5	11.0-14.0	0.10 Si
C95710	71.0 min. [16]	2.0-4.0	1.5-3.0	7.0-8.5	11.0-14.0	0.05 Pb, 0.15 Si, 0.50 Zn, 1.0 Sn, 0.05 P
C95720	73.0 min. [16]	1.5-3.5	3.0-6.0	6.0-8.0	12.0-15.0	0.03 Pb, 0.10 Si, 0.10 Zn, 0.10 Sn, 0.20 Cr
C95800	79.0 min. [16]	3.5-4.5 [31]	4.0-5.0 [31]	8.5-9.5	0.8-1.5	0.03 Pb, 0.10 Si
C95810	79.0 min. [16]	3.5-4.5 [31]	4.0-5.0 [31]	8.5-9.5	0.8-1.5	0.10 Pb, 0.05 Mg, 0.10 Si, 0.50 Zn
C95820	77.5 min. [47]	4.0-5.0	4.5-5.8	9.0-10.0	1.5	0.02 Pb, 0.10 Si, 0.20 Zn, 0.20 Sn
C95900	Rem. [16]	3.0-5.0	0.50	12.0-13.5	1.5	-

Notes: [16] Cu + Sum of Named Elements, 99.5% min.
[20] Cu + Sum of Named Elements, 99.8% min.
[21] Cu + Sum of Named Elements, 99.7% min.
[31] Fe content shall not exceed Ni content.
[43] In determining copper min., copper may be calculated as Cu + Ni.
[44] Cu + Sum of Named Elements, 99.3% min.
[45] For continuous castings, P shall be 1.5%, max.
[47] Cu + Sum of Named Elements, 99.2% min.
[50] Cu + Sum of Named Elements, 99.0% min.
[51] Cu + Sum of Named Elements, 99.4% min.

(Continued)

Table 13. *(Continued)* **Chemical Composition of Cast Bronzes.** *(Source, Copper Development Association.)*

Notes: [52] Cu + Sum of Named Elements, 98.9% min.
[53] Cu + Sum of Named Elements, 98.7% min.
[55] Fe shall be 0.35% max., when used for steel-backed bearings.
[56] For continuous castings, S shall be 0.25%, max.
[57] The mechanical properties of C94700 (heat treated) may not be attainable if the lead content exceeds 0.01%.

Table 14. Chemical Composition of Cast Copper-Nickels. *(Source, Copper Development Association.)*

Copper Alloy No.	Cu[16]	Fe	Ni (incl. Co)	Mn	Nb	Other (named)
Copper-Nickel-Zinc Alloys (Copper Nickels)						
C96200	Rem.	1.0-1.8	9.0-11.0	1.5	1.0 [58]	.01 Pb, .50 Si, .10 C, .02 S, .02 P
C96300	Rem.	.50-1.5	18.0-22.0	.25-1.5	.50-1.5	.01 Pb, .50 Si, .15 C, .02 S, .02 P
C96400	Rem.	.25-1.5	28.0-32.0	1.5	.50-1.5	.01 Pb, .50 Si, .15 C, .02 S, .02 P
C96600	Rem.	.8-1.1	29.0-33.0	1.0	-	.01 Pb, .15 Si, .40-.7 Be
C96700	Rem.	.40-1.0	29.0-33.0	.40-1.0	-	.01 Pb, .15 Si, 1.1-1.2 Be, .15-.35 Zr, .15-.35 Ti
C96800	Rem.	.50	9.5-10.5	.05-.30	.10-.30	.005 Pb, .05 Si, [59]
C96900	Rem.	.50	14.5-15.5	.05-.30	.10	.02 Pb, .15 Mg, 7.5-8.5 Sn, .50 Zn
C96950	Rem.	.50	11.0-15.5	.05-.40	.10	.02 Pb, .30 Si, 5.8-8.5 Sn, .15 Mg

Notes: [16] Cu + Sum of Named Elements, 99.5% min.
[58] When the product or casting is intended for subsequent welding applications, and so specified by the purchaser, the Nb content shall be 0.40% max.
[59] The following additional maximum impurity limits shall apply: 0 .10% Al, 0.001% B, 0 .001% Bi, 0.005-.15% Mg, 0.005% P, 0.0025% S, 0.02% Sb, 7.5-8.5% Sn, 0.01% Ti, 1.0% Zn.

Table 15. Chemical Composition of Cast Nickel Silvers. *(Source, Copper Development Association.)*

Copper Alloy No.	Cu	Sn	Pb	Zn (incl. Co)	No (incl. Co)	Other (named)
Copper-Nickel-Zinc Alloys (Nickel Silvers)						
C97300	53.0-58.0 [50]	1.5-3.0	8.0-11.0	17.0-25.0	11.0-14.0	1.5 Fe, .35 Sb, .08 S, .05 P, .005 Al, .50 Mn, .15 Si
C97400	58.0-61.0 [50]	2.5-3.5	4.5-5.5	Rem.	15.5-17.0	1.5 Fe, .50 Mn
C97600	63.0-67.0 [21]	3.5-4.5	3.0-5.0	3.0-9.0	19.0-21.5	1.5 Fe, .25 Sb, .08 S, .05 P, .005 Al, 1.0 Mn, .15 Si
C97800	64.0-67.0 [22]	4.0-5.5	1.0-2.5	1.0-4.0	24.0-27.0	1.5 Fe, .20 Sb, .08 S, .05 P, .005 Al, 1.0 Mn, .15 Si

Notes: [21] Cu + Sum of Named Elements, 99.7% min.
[22] Cu + Sum of Named Elements, 99.6% min.
[50] Cu + Sum of Named Elements, 99.0% min.

Table 16. Chemical Composition of Cast Leaded Coppers. *(Source, Copper Development Association.)*

Copper Alloy No.	Cu	Sn	Pb	P (incl. Co)	Fe	Other (named)
Copper-Lead Alloys						
C98200	Rem. [16]	.60-2.0	21.0-27.0	.10	.70	.50 Zn, .50 Ni, .50 Sb
C98400	Rem. [16]	.50	26.0-33.0	.10	.70	1.5 Ag, .50 Zn, .50 Ni, .50 Sb
C98600	60.0-70.0	.50	30.0-40.0	-	.35	1.5 Ag
C98800	56.5-62.5 [41]	.25	37.5-42.5 [60]	.02	.35	5.5 [60] Ag, .10 Zn
C98820	Rem.	1.0-5.0	40.0-44.0	-	.35	-
C98840	Rem.	1.0-5.0	44.0-58.0	-	.35	-

Notes: [16] Cu + Sum of Named Elements, 99.5% min.
 [41] Includes Ag.
 [60] Pb and Ag may be adjusted to modify the alloy hardness.

Table 17. Chemical Composition of Cast Special Copper Alloys. *(Source, Copper Development Association.)*

Copper Alloy No.	Cu[21]	Ni	Fe	Al	Mn	Other (named)
C99300	Rem.	13.5-16.5	4.0-1.0	10.7-11.5	-	.05 Sn, .02 Pb, 1.0-2.0 Co, .02 Si
C99350	Rem.	14.5-16.0 [18]	1.0	9.5-10.5	.25	.15 Pb, 7.5-9.5 Zn
C99400	Rem.	1.0-3.5	1.0-3.0	.50-2.0	.50	.25 Pb, .50-2.0 Si, .50-5.0 Zn
C99500	Rem.	3.5-5.5	3.0-5.0	.50-2.0	.50	.25 Pb, .50-2.0 Si, .50-5.0 Zn
C99600	Rem.	.20	.20	1.0-2.8	39.0-45.0	.10 Sn, .02 Pb, .20 Co, .10 Si, .20 Zn, .05 C
C99700	54.0 min.	4.0-6.0	1.0	.50-3.0	11.0-15.0	1.0 Sn, 2.0 Pb, 19.0-25.0 Zn
C99750	55.0-61.0	5.0	1.0	.25-3.0	17.0-23.0	.50-2.5 Sn, 17.0-23.0 Zn

Notes: [18] Includes Co.
 [21] Cu + Sum of named elements, 99.7% min.

Table 18. Mechanical Properties of Selected Copper Casting Alloys. *(Source, Copper Development Association.)*

UNS Number	App. [1]	Temper [2]	Tensile Strength [3]		Yield Strength [4]		% Elong. [5]	Hardness
			ksi	MPa	ksi	MPa		
C80100	S	M01	25	172	9	62	40	-
C81100	S	M01	25	172	9	62	40	-
C81400	S	M01	30	207	12 [6]	83 [6]	35	R_B 62
C81400	S	TF00	53	365	36 [6]	248 [6]	11	R_B 69
C81500	S	TF00	51	352	40	276	17	-
C82000	S	M01	50	345	20 [6]	138 [6]	20	R_B 55
C82000	S	O11	65	448	37 [6]	255 [6]	12	-
C82000	S	TB00	47	324	15 [6]	103 [6]	25	R_B 40
C82000	S	TF00	96	662	75 [6]	517 [6]	6	R_B 96
C82200	S	M01	50	345	25 [6]	172 [6]	20	R_B 55

(Continued)

Table 18. *(Continued)* **Mechanical Properties of Selected Copper Casting Alloys.**
(Source, Copper Development Association.)

UNS Number	App. [1]	Temper [2]	Tensile Strength [3] ksi	Tensile Strength [3] MPa	Yield Strength [4] ksi	Yield Strength [4] MPa	% Elong. [5]	Hardness
C82200	S	O11	65	448	40 [6]	276 [6]	15	R_B 75
C82200	S	TB00	45	310	12 [6]	83 [6]	30	R_B 30
C82200	S	TF00	95	655	75 [6]	517 [6]	7	R_B 96
C82400	S	O11	100	690	80 [6]	551 [6]	3	R_C 21
C82400	S	TB00	60	414	20 [6]	138 [6]	40	R_B 59
C82400	S	TF00	155	1,068	145 [6]	1,000 [6]	1	R_C 38
C82500	S	M01	75	517	40 [6]	276 [6]	15	R_B 81
C82500	S	O11	120	827	105 [6]	724 [6]	2	R_C 30
C82500	S	TB00	60	414	25 [6]	172 [6]	35	R_B 63
C82500	S	TF00	160	1,103	150 [6]	1,034 [6]	1	R_C 43
C82600	S	M01	80	552	50 [6]	345 [6]	10	R_B 68
C82600	S	O11	120	827	105 [6]	724 [6]	2	R_C 31
C82600	S	TB00	70	483	30 [6]	207 [6]	12	R_B 75
C82600	S	TF00	165	1,138	155 [6]	1,069 [6]	1	R_C 45
C82700	S	TF00	155 [7]	1,069 [7]	130 [8]	896 [8]	2	R_C 39
C82800	S	M01	80	552	50 [6]	345 [6]	10	R_B 88
C82800	S	O11	125	862	110 [6]	758 [6]	2	R_C 31
C82800	S	TB00	80	552	35 [6]	241 [6]	10	R_B 58
C82800	S	TF00	165	1,138	155 [6]	1,069 [6]	1	R_C 46
C83300	S	M01	32	221	10	69	35	R_B 35
C83400	S	M01	35	241	10	69	30	R_F 50
C83600	S, CL	M01, M02	37	255	17	117	30	-
C83600	C	M07	36 [7]	248 [7]	19 [9]	131 [9]	15 [10]	-
C83600	C	M07	50 [7]	345 [7]	25 [9]	170 [9]	12 [10]	-
C83800	S, CL	M01, M02	35	241	16	110	25	-
C83800	C	M07	30 [7]	207 [7]	15 [9]	103 [9]	16 [10]	-
C84200	S	M01	35	241	14	97	27	-
C84400	S	M01	34	234	15	103	26	-
C84500	S	M01	35	241	14	97	28	-
C84800	S	M01	37	255	14	97	35	-
C85200	S, CL	M01, M02	38	262	13	90	35	-
C85400	S, CL	M01, M02	34	234	12	83	35	-
C85500	S	M01	60	414	23	159	40	R_B 55
C85700	S, CL	M01, M02	50	345	18	124	40	-
C85800	D	M04	55	379	30 [6]	207 [6]	15	R_B 55
C86100	S	M01	95	655	50 [6]	345 [6]	20	-
C86200	S, CL, C	M01, M02, M07	95	655	48 [6]	331 [6]	20	-

(Continued)

Table 18. *(Continued)* **Mechanical Properties of Selected Copper Casting Alloys.**
(Source, Copper Development Association.)

UNS Number	App. [1]	Temper [2]	Tensile Strength [3]		Yield Strength [4]		% Elong. [5]	Hardness
			ksi	MPa	ksi	MPa		
C86300	S	M01	119	821	67 [6]	462 [6]	18	-
C86300	S, CL	M01, M02	110 [7]	758 [7]	60 [8]	414 [8]	12 [10]	-
C86300	C	M07	110 [7]	758 [7]	62 [8]	427 [8]	14 [10]	-
C86400	S	M01	65	448	25 [6]	172 [6]	20	-
C86500	S, CL	M01, M02	71	490	29	200	30	-
C86500	C	M07	70 [7]	483 [7]	25 [8]	172 [8]	25 [10]	-
C86700	S	M01	85	586	42	290	20	R_B 80
C86800	S	M01	82	565	38	262	22	-
C87300	S, CL	M01, M02	55	379	25	172	30	-
C87400	S, CL	M01, M02	55	379	24	165	30	-
C87500	S, CL	M01, M02	67	462	30	207	21	-
C87600	S	M01	66	455	32	221	20	R_B 76
C87800	D	M04	85	586	50 [6]	345 [6]	25	R_B 85
C90200	S	M01	38	124	16	110	30	-
C90300	S, CL	M01, M02	45	310	21	145	30	-
C90300	C	M07	44 [7]	303 [7]	22 [9]	152 [9]	18 [10]	-
C90500	S, CL	M01, M02	45	310	22	152	25	-
C90500	C	M07	44 [7]	303 [7]	25 [9]	172 [9]	10 [10]	-
C90700	S	M01	44	303	22	152	20	-
C90700	CL, PM	M02, M05	55	379	30	207	16	-
C90700	C	M07	40 [7]	276 [7]	25	172	10	-
C90900	S	M01	40	276	20	138	15	-
C91000	S	M01	32	221	25	172	2	-
C91100	S	M01	35	241	25	172	2	-
C91300	S	M01	35	241	30	207	0.5	-
C91600	S	M01	44	303	22	152	16	-
C91600	CL, PM	M02, M05	60	414	32	221	16	-
C91700	S	M01	44	303	22	152	16	-
C91700	CL, PM	M02, M05	60	414	32	221	16	-
C92200	S, CL	M01, M02	40	276	20	138	30	-
C92200	C	M07	38 [7]	262 [7]	19 [9]	131 [9]	18 [10]	-
C92300	S, CL	M01, M02	40	276	20	138	25	-
C92300	C	M07	40 [7]	276 [7]	19 [9]	131 [9]	16 [10]	-
C92500	S	M01	44	303	20	138	20	-
C92500	C	M07	40 [7]	276 [7]	24 [9]	166 [9]	10 [10]	-
C92600	S	M01	44	303	20	138	30	R_F 78
C92700	S	M01	42	290	21	145	20	-
C92700	C	M07	38 [7]	262 [7]	20 [9]	138 [9]	8	-

(Continued)

Table 18. *(Continued)* **Mechanical Properties of Selected Copper Casting Alloys.**
(Source, Copper Development Association.)

UNS Number	App. [1]	Temper [2]	Tensile Strength [3]		Yield Strength [4]		% Elong. [5]	Hardness
			ksi	MPa	ksi	MPa		
C92800	S	M01	40	276	30	207	1	R_B 80
C92900	S, PM, C	M01, M05, M07	47	324	26	179	20	-
C93200	S, CL	M01, M02	35	241	18	124	20	-
C93200	C	M07	35 [7]	241 [7]	20 [9]	138 [9]	10 [10]	-
C93400	S	M01	32	221	16	110	20	-
C93500	S, CL	M01, M02	32	221	16	110	20	-
C93500	C	M07	30 [7]	207 [7]	16 [9]	110 [9]	12 [10]	-
C93700	S, CL	M01, M02	35	241	18	124	20	-
C93700	C	M07	35 [7]	241 [7]	20 [9]	138 [9]	6 [10]	-
C93700	C	M07	40 [7]	276 [7]	25 [9]	172 [9]	6 [10]	-
C93800	S, CL	M01, M02	30	207	16	110	18	-
C93800	CL	M02	33	228	20	138	12	-
C93800	C	M07	25 [7]	172 [7]	-	-	5 [10]	-
C93900	C	M07	32	221	22	152	7	-
C94300	S	M01	27	186	13	90	15	-
C94300	S, CL	M01, M02	21 [7]	145 [7]	-	-	10 [10]	-
C94300	C	M07	21 [7]	145 [7]	15 [9]	103 [9]	7 [10]	-
C94400	S	M01	32	221	16	110	18	-
C94500	S	M01	25	172	12	83	12	-
C94700	S, C	M01, M07	50	345	23	159	35	-
C94700	S, C	TX00	85	586	60	414	10	-
C94800	S, C	M01, M07	45	310	23	159	35	-
C94800	S	TX00	60	414	30	207	8	-
C95200	S, CL	M01, M02	80	552	27	186	35	R_B 64
C95200	C	M07	68 [7]	469 [7]	26 [9]	179 [9]	20 [10]	-
C95300	S, CL	M01, M02	75	517	27	186	25	R_B 67
C95300	C	M07	70 [7]	483 [7]	26 [9]	179 [9]	25 [10]	-
C95300	S, CL, C	TQ50	85	586	42	290	15	R_B 81
C95400	S, CL	M01, M02	85	586	35	241	18	-
C95400	C	M07	85 [7]	586 [7]	32 [9]	221 [9]	12 [10]	-
C95400	S, CL	TQ50	105	724	54	372	8	-
C95400	C	TQ50	95 [7]	655 [7]	45 [9]	310 [9]	10 [10]	-
C95410	S	M01	85	586	35	241	18	-
C95410	S	TQ50	105	724	54	372	8	-
C95500	S, CL	M01, M02	100	690	44	303	12	R_B 87
C95500	C	M07	95 [7]	665 [7]	42 [9]	290 [9]	10 [10]	-

(Continued)

Table 18. *(Continued)* **Mechanical Properties of Selected Copper Casting Alloys.**
(Source, Copper Development Association.)

UNS Number	App. [1]	Temper [2]	Tensile Strength [3]		Yield Strength [4]		% Elong. [5]	Hardness
			ksi	MPa	ksi	MPa		
C95500	S, CL	TQ50	120	827	68	469	10	R_B 96
C95600	S	M01	75	517	34	234	18	-
C95700	S	M01	95	655	45	310	26	-
C95800	S, CL	M01, M02	95	655	38	262	25	-
C95800	C	M07	90 [7]	621 [7]	38 [9]	262 [9]	18 [10]	-
C96200	S	M01	45 [7]	310 [7]	25 [9]	172 [9]	20 [10]	-
C96300	S	M01	75 [7]	517 [7]	55 [9]	379 [9]	10 [10]	-
C96400	S	M01	68	469	37	255	28	-
C96600	S	TB00	75	517	38	262	12	R_B 74
C96600	S	TF00	120	827	75	517	12	R_C 24
C97300	S	M01	35	241	17	117	20	-
C97400	S	M01	38	262	17	117	20	-
C97600	S	M01	45	310	24	165	20	-
C97800	S	M01	55	379	30	207	15	-
C99300	S	M01	95	655	55	379	2	-
C99400	S	M01	66	455	34	234	25	-
C99400	S	TF00	79	545	54	372	-	-
C99500	S	M01	70 [7]	483 [7]	40 [9]	276 [9]	12 [10]	-
C99500	S	TF00	86	593	62	427	8	-
C99700	S	M01	55	379	25	172	25	-
C99700	D	M04	65	448	27	186	15	-
C99750	S	M01	65	448	32	221	30	R_B 77
C99750	S	TQ50	75	517	40	276	20	R_B 82

Notes:
[1] Application abbreviations are: S = Sand. C = Continuous. CL = Centrifugal. D = Die. I = Investment. P = Plaster. PM = Permanent Mold.
[2] Temper designations given in text.
[3] Typical tensile strength given unless otherwise indicated.
[4] Typical yield strength at 0.5% extension unless otherwise indicated.
[5] Typical percent elongation in 2 inches (50.8 mm) given unless otherwise indicated.
[6] Typical yield strength at 0.2% offset.
[7] Minimum tensile strength.
[8] Minimum yield strength at 2% offset.
[9] Minimum yield strength at 0.5% extension.
[10] Minimum percent elongation in 2 inches (50.8 mm).

Table 19. Physical Properties of Selected Copper Casting Alloys. *(Source, Copper Development Association)*

UNS Number	Liquidus Point		Density lb/cu in.	Coef. Thermal Expan.[1]		Thermal Conduct.[2]	Electrical Conduct.[3]	Elastic Modulus	
	°F	°C		μin. °F	μm °C			ksi	MPa
C80100	1,981	1,083	0.323	9.4	16.9	226	100	17,000	117,000
C81100	1,981	1,083	0.323	9.4	16.9	200	92	17,000	117,000
C81400	2,000	1,093	0.318	10.0	18.0	150	60	16,000	110,000
C81500	1,985	1,085	0.319	9.5	17.1	182	82	16,000	114,000
C82000	1,990	1,088	0.311	9.9	17.8	150	45	17,000	117,000
C82200	2,040	1,116	0.316	9.0 [4]	16.2 [4]	106	45	16,000	114,000
C82400	1,825	996	0.304	9.4 [4]	16.9 [4]	76.9	25	18,000	128,000
C82500	1,800	982	0.302	9.4 [4]	16.9 [4]	74.9	20	18,000	128,000
C82600	1,750	954	0.302	9.4 [4]	16.9 [4]	73.0	19	19,000	131,000
C82700	1,750	954	0.292	9.4 [4]	16.9 [4]	74.9	20	19,000	132,000
C82800	1,710	932	0.294	9.4 [4]	16.9 [4]	70.8	18	19,000	133,000
C83300	1,940	1,060	0.318	-	-	-	32	15,000	103,000
C83400	1,910	1,043	0.318	10.0	18.0	109	44	15,000	103,000
C83600	1,850	1,010	0.318	10.0 [4]	18.0 [4]	41.6	15	13,000	93,000
C83800	1,840	1,004	0.312	10.0 [4]	18.0 [4]	41.8	15	13,000	91,000
C84200	1,820	993	0.311	10.0 [4]	18.0 [4]	41.8	16	14,000	96,000
C84400	1,840	1,004	0.314	10.0	18.0	41.8	16	13,000	89,000
C84500	1,790	977	0.312	10.0	18.0	41.6	16	14,000	96,000
C84800	1,750	954	0.310	10.0 [4]	18.0 [4]	41.6	16	15,000	103,000
C85200	1,725	941	0.307	11.5 [5]	20.8 [5]	48.5	18	11,000	75,000
C85400	1,725	941	0.305	11.1 [5]	20.0 [5]	50.8	20	12,000	82,000
C85500	1,652	900	0.304	11.8 [5]	21.3 [5]	67.0	26	15,000	103,000
C85700	1,725	941	0.304	12.0	21.6	48.5	22	14,000	96,000
C85800	1,650	899	0.305	-	-	48.5	20	15,000	103,000
C86100	1,725	941	0.288	12.0	21.6	20.5	8	15,000	103,000
C86200	1,725	941	0.288	12.0	21.6	20.5	8	15,000	103,000
C86300	1,693	923	0.283	12.0	21.6	20.5	8	14,000	97,000
C86400	1,616	880	0.301	11.0 [4]	19.8 [4]	51.0	19	14,000	96,000
C86500	1,616	880	0.301	11.3 [5]	20.4 [5]	49.6	22	15,000	103,000
C86700	1,616	880	0.301	11.0 [4]	19.8 [4]	-	17	15,000	103,000
C86800	1,652	900	0.290	-	-	-	9	15,000	103,000
C87300	1,780	971	0.302	10.9	19.6	16.4	6	15,000	103,000
C87400	1,680	916	0.300	10.9	19.6	16.0	7	15,000	106,000
C87500	1,680	916	0.299	10.9	19.6	16.0	7	15,000	106,000
C87600	1,780	971	0.300	-	-	16.4	6	17,000	117,000
C87800	1,680	916	0.300	10.9	19.6	16.0	7	20,000	138,000
C90200	1,915	1,046	0.318	10.1	18.2	36.0	13	16,000	110,000
C90300	1,832	1,000	0.318	10.0 [4]	18.0 [4]	43.2	12	14,000	96,000

(Continued)

Table 19. *(Continued)* Physical Properties of Selected Copper Casting Alloys.
(Source, Copper Development Association.)

UNS Number	Liquidus Point		Density lb/cu in.	Coef. Thermal Expan.[1]		Thermal Conduct.[2]	Electrical Conduct.[3]	Elastic Modulus	
	°F	°C		μin. °F	μm °C			ksi	MPa
C90500	1,830	999	0.315	11.0	19.8	43.2	11	15,000	103,000
C90700	1,830	999	0.317	10.2 [4]	18.4 [4]	40.8	10	15,000	103,000
C90900	1,792	978	-	-	-	-	-	16,000	110,000
C91000	1,760	960	-	-	-	-	9	16,000	110,000
C91100	1,742	950	-	-	-	-	8	15,000	103,000
C91300	1,632	889	-	-	-	-	7	16,000	110,000
C91600	1,887	1,031	0.320	9.0 [4]	16.2 [4]	40.8	10	16,000	110,000
C91700	1,859	1,015	0.316	9.0 [4]	16.2 [4]	40.8	10	15,000	103,000
C92200	1,810	988	0.312	10.0	18.0	40.2	14	14,000	96,000
C92300	1,830	999	0.317	10.0 [4]	18.0 [4]	43.2	12	14,000	96,000
C92500	-	-	0.317	-	-	-	-	16,000	110,000
C92600	1,800	982	0.315	10.0 [4]	18.0 [4]	-	9	15,000	103,000
C92700	1,800	982	0.317	10.0 [4]	18.0 [4]	27.2	11	16,000	110,000
C92800	1,751	955	-	-	-	-	-	16,000	110,000
C92900	1,887	1,031	0.320	9.5 [4]	17.1 [4]	33.6	9	14,000	96,000
C93200	1,790	977	0.322	10.0 [5]	18.0 [5]	33.6	12	14,000	100,000
C93400	-	-	0.320	10.0 [4]	18.0 [4]	33.6	12	11,000	75,000
C93500	1,830	999	0.320	9.9 [4]	17.8 [4]	40.7	15	14,000	100,000
C93700	1,705	929	0.320	10.3 [4]	18.5 [4]	27.1	10	11,000	75,000
C93800	1,730	943	0.334	10.3 [4]	18.5 [4]	30.2	11	10,000	72,000
C93900	1,730	943	0.334	10.3 [4]	18.5 [4]	30.2	11	11,000	75,000
C94300	-	-	0.336	-	-	36.2	9	10,000	72,000
C94400	1,725	941	0.320	10.3 [4]	18.5 [4]	30.2	10	11,000	75,000
C94500	1,475	802	0.340	10.3 [4]	18.5 [4]	30.2	10	10,000	72,000
C94700	1,660	904	0.320	10.9 [4]	19.6 [4]	31.2	-	15,000	103,000
C94800	1,660	904	0.320	10.9	19.6	22.3	12	15,000	103,000
C95200	1,913	1,045	0.276	9.0	16.2	29.1	11	15,000	103,000
C95300	1,913	1,045	0.272	9.0	16.2	36.3	13	16,000	110,000
C95400	1,900	1,038	0.269	9.0	16.2	33.9	13	15,000	107,000
C95410	1,900	1,038	0.269	9.0	16.2	33.9	13	15,000	107,000
C95500	1,930	1,054	0.272	9.0	16.2	24.2	8	16,000	110,000
C95600	1,840	1,004	0.278	9.2	16.6	22.3	8	15,000	103,000
C95700	1,814	990	0.272	9.8	17.6	7.0	3	18,000	124,000
C95800	1,940	1,060	0.276	9.0	16.2	20.8	7	16,000	114,000
C96200	2,100	1,149	0.323	9.5	17.1	26.1	11	18,000	124,000
C96300	2,190	1,199	0.323	9.1	16.4	21.3	6	20,000	138,000
C96400	2,260	1,238	0.323	9.0	16.2	16.4	5	21,000	145,000
C96600	2,160	1,182	0.318	9.0	16.2	17.4	4	22,000	152,000

(Continued)

Table 19. *(Continued)* **Physical Properties of Selected Copper Casting Alloys.**
(Source, Copper Development Association.)

UNS Number	Liquidus Point		Density lb/cu in.	Coef. Thermal Expan.[1]		Thermal Conduct.[2]	Electrical Conduct.[3]	Elastic Modulus	
	°F	°C		μin. °F	μm °C			ksi	MPa
C97300	1,904	1,040	0.321	9.0	16.2	16.5	6	16,000	110,000
C97400	2,012	1,100	0.320	9.2	16.6	15.8	6	16,000	110,000
C97600	2,089	1,143	0.321	9.3	16.7	13.0	5	19,000	131,000
C97800	2,156	1,180	0.320	9.7	17.5	14.7	4	19,000	131,000
C99300	1,970	1,077	0.275	9.2	16.6	25.4	9	18,000	124,000
C99400	-	-	0.300	-	-	-	12	19,000	133,000
C99500	-	-	0.300	8.3	14.9	-	10	19,000	131,000
C99700	1,655	902	0.296	-	-	-	3	16,000	114,000
C99750	1,550	843	0.290	13.5 [4]	24.3 [4]	-	2	17,000	117,000

Notes. [1] At 68 to 572° F (20 to 300° C) unless otherwise indicated.
[2] Btu/ft²/ft/h/°F.
[3] % IACS At 68° F (20° C).
[4] Coefficient of thermal expansion measured at 68 to 392° F (20 to 200° C).
[5] Coefficient of thermal expansion measured at 68 to 212° F (20 to 100° C).

alloys—they are used in heavy duty mechanical and electromechanical equipment requiring ultrahigh strength and good electrical and/or thermal conductivity, but require reliable lubrication and well-aligned shafts when used as bearing material. Corrosion resistance for high copper alloys is as good or better than that of pure copper.

Brasses. C83300 – C83810 are red and leaded red brasses with Cu, Zn, Sn, and Pb as the principal alloying agents.

C84200 – C84800 are semi-red and leaded semi-red brasses with Cu, Zn, Sn, and Pb as the principal alloying agents.

C85200 – C85800 are yellow and leaded yellow brasses with Cu, Zn, Sn, and Pb as the principal alloying agents.

C86100 – C86800 are high strength and leaded high strength yellow brasses with Cu, Zn, Mn, Fe, and Pb as the principal alloying agents.

C87300 – C87900 are silicon bronzes and silicon brasses with Cu, Zn, and Si as the principal alloying agents.

C89320 – C89940 are high strength and leaded high strength yellow brasses with Cu, Sn, and Bi as the principal alloying agents.

The most important brasses in terms of tonnage poured are the leaded red brass C83600, and the leaded semi-red brasses C84400, C84500, and C84800. All of these alloys are widely used in water valves, pumps, pipe fittings, and plumbing hardware. Leaded yellow brasses such as C85400, C85700, and C85800 are relatively low in cost and have excellent castability, high machinability, and favorable machining characteristics. These materials are commonly used for mechanical products such as gears and machine components. The high strength and leaded high strength yellow brasses are the strongest, as cast, of all the copper alloys. They are used primarily for heavy duty mechanical products requiring moderately good corrosion resistance at a reasonable cost. The silicon bronzes/brasses have moderate strength and good corrosion resistance, and are well suited for die, permanent mold, and investment casting methods. Applications range from bearings and gears to plumbing goods and intrically shaped pump and valve components.

Tin Bronzes. C90200 – C91700 are tin bronzes with Cu and Sn as the principal alloying agents.

C92200 – C92900 are leaded tin bronzes with Cu, Sn, Zn, and Pb as the principal alloying agents.

C93100 – C94500 are high leaded tin bronzes with Cu, Sn, and Pb as the principal alloying agents.

Nickel bronzes offer excellent corrosion resistance, reasonably high strength, and good wear resistance. Unleaded tin bronze C90300 is used for bearings, pump impellers, piston rings, valve fittings, and other mechanical products. C92300, which is leaded, has similar uses and is specified when better machinability and/or pressure tightness are needed. C90500 is hard and strong, and is especially resistant to seawater corrosion. C93200 is the best known bronze bearing alloy and has unsurpassed wear performance against steel journals. C93500 combines favorable antifriction properties with good load carrying capabilities, and it conforms well to slight shaft misalignments. Lead weakens all these bearing alloys, but imparts the ability to tolerate interrupted lubrication. It also allows dirt particles to become harmlessly imbedded in the bearing's surface, thereby protecting the journal. The "premier" bearing alloys, C93800 and C94300, also wear well with steel and are best known for their ability to conform to slightly misaligned shafts.

Nickel Tin Bronzes. C94700 – C94900 are nickel tin bronzes with Cu, Sn, Zn, and Ni as the principal alloying agents.

These alloys have moderate strength and very good corrosion resistance, especially in aqueous media. One member of the group, C94700, can be age hardened to tensile strengths as high as 75 ksi (517 MPa). Nickel tin bronzes are used for bearings, but are more frequently used as valve and pump components, gears, shifter forks, and circuit breaker parts.

Aluminum Bronzes. C95200 – C95500 are aluminum bronzes with Cu, Al, Fe, and Ni as the principal alloying agents.

Aluminum strengthens copper and imparts oxidation resistance. This group is best known for high corrosion and oxidation resistance combined with exceptionally good mechanical properties. The alloys are readily fabricated and welded and are used in cast structures.

Copper Nickels. C96200 – C96950 are copper nickels with Cu, Ni, and Fe as the principal alloying agents.

These alloys offer excellent resistance to seawater corrosion, high strength, and good fabricability. They are widely used in marine equipment as pumps, impellers, valves, tailshaft sleeves, and centrifugally cast pipe.

Nickel Silvers. C97300 – C97800 are nickel silvers with Cu, Ni, Zn, Pb, and Sn as the principal alloying agents.

These alloys have excellent corrosion resistance and very good machinability. Valves, fittings, and hardware cast in nickel silvers are used in food and beverage handling equipment, and as seals and labyrinth rings in steam turbines.

Leaded Coppers. C98200 – C98840 are leaded coppers with Cu and Pb as the primary alloying agents.

These alloys provide the high corrosion resistance of copper along with the favorable lubricity and low friction characteristics of high leaded bronzes. They operate well under intermittent, unreliable, or dirty lubrication, and can operate underwater with water lubrication. Used for light to moderate loads and high speeds, as in rod bushings and main bearings for refrigeration compressors, or hydraulic pump bushings.

Special Alloys. C99300 – C99750 are special alloys whose compositions do not fall into any of the above categories.

Copper product forms

Specific terminology has been assigned to the physical forms of copper materials, and should be specified when ordering. *Rod* products may have round, hexagonal, or octagonal shapes, and they are supplied in straight lengths (not coiled). *Bar* products are square or rectangular, and supplied in straight lengths. *Shapes* are also supplied in straight lengths, but may have oval, half-round, geometric, or custom ordered cross sections. *Wire* products may have any cross section, but they are always supplied in coils or on spools.

Wrought copper alloy selection based on machinability

All copper alloys are machinable in the sense that they can be cut with standard machine tooling. In fact, high speed steel tools work well with all but the hardest alloys. Since chip appearance is a good indicator of a material's machinability, copper alloys are divided into three groups based on chip characteristics. Small, fragmented chips are produced by the leaded free cutting alloys, which are designated "Type I Free Cutting" materials. Curled but brittle chips are produced by those alloys designated "Type II Short-Chip" materials. Long, stringy chips are associated with metals with homogeneous, or single phase, structures, and these alloys are designated "Type III Long-Chip" materials. Comparative machinability of wrought copper alloys is shown in **Table 20**.

Type I Free Cutting Alloys

The lead bearing copper alloys are more like composite structures than true alloys because lead is insoluble in copper and appears as a dispersion of microscopic globules. The stress raising effect of these lead particles causes chips to break up into tiny flakes as the metal passes over the tool face. Chips from leaded alloys remain in contact with the tool face for a very short time before the energy of fracture propels them away from the cutting tool. The short contact time reduces friction, which in turn minimizes tool wear and energy consumption. It is also thought that the lead globules may actually act as an internal lubricant as they pass over the tool face. The beneficial effect of lead on free cutting behavior increases with lead content, but the rate of improvement decreases as lead content rises. Significantly improved machinability can be detected and measured in alloys containing less than 0.5% Pb, but maximum free cutting behavior occurs at concentrations between 0.5% and 3.5%.

Although lead does not have a significant effect on strength, leaded alloys can be difficult to cold work extensively. The effect becomes more pronounced as lead content rises. Therefore, alloys such as C34500 and C35300, which have lower lead content than C36000 (Pb = 3.1%), may be better choices for products that require both high speed machining and extensive cold deformation. There are no hard and fast rules, unfortunately, because free cutting brass's nominal lead content is usually low enough to permit even deep knurling and coarse rolled threads. Other detrimental effects of lead include an impairment in welding and brazing properties (for further information about welding copper alloys, see the Welding section of this book), as well as related problems in welding electrodes and cutting torch tips that operate at high temperatures.

Lead does not significantly alter an alloy's corrosion behavior, but trace quantities can be dissolved form machined surfaces by appropriately aggressive media, including some potable waters. If lead exposure is a problem, wetted surfaces can be protected with electroplated or organic coatings. Surface lead may also be removed by relatively simple etching processes that leave only "pure" brass exposed to the environment.

Additions of tellurium, sulfur, and bismuth also promote free cutting behavior in copper alloys. These elements can actually be more efficient than lead in terms of the amounts needed to produce optimum improvements, but they are not without shortcomings.

Table 20. Wrought Copper Alloys Ranked by Machinability. (Source, Copper Development Association.)

Copper Alloy No.	Traditional Name *Descriptive Name*	Machinability Index Rating	Copper Alloy No.	Traditional Name *Descriptive Name*	Machinability Index Rating
Type I: Free Cutting Copper Alloys for Screw Machine Production (Suitable for automatic machining at the highest available cutting speeds.)					
C36000	Free-Cutting Brass	100	C37700	Forging Brass	80
C35600	Extra High Leaded Brass	100	C54400	Phosphor Bronze, B-2	80
C32000	Leaded Red Brass	90	C19100	*Nickel Copper with Tellurium*	75
C34200	High Leaded Brass, 64 1/2%	90	C19150	*Leaded Nickel Copper*	75
C35300	High Leaded Brass, 62%	90	C34000	Medium Leaded Brass, 64 1/2%	70
C35330	*Arsenical Free- Cutting Brass*	90	C35000	Medium Leaded Brass, 62%	70
C38500	Architectural Bronze	90	C37000	Free-Cutting Muntz Metal	70
C14500	Tellurium-Bearing Copper	85	C48500	Naval Brass, High Leaded	70
C14520	*Phosphorus Deoxidized, Tellurium-Bearing Copper*	85	C67300	Leaded Silicon-Manganese-Bronze	70
			C69710	*Leaded Arsenical Silicon Red Brass*	70
C14700	Sulfur Bearing Copper	85	C17300	*Leaded Beryllium Copper*	60
C18700	*Leaded Copper*	85	C33500	Low Leaded Brass	60
C31400	Leaded Commercial Bronze	80	C67600	Leaded Manganese Bronze	60
C31600	Leaded Commercial Bronze (Nickel-Bearing)	80	C48200	Naval Brass, Medium Leaded	50
			C66100	*Leaded Silicon Bronze*	50
C34500	*Leaded Brass*	80	C79200	*Leaded Nickel Silver, 12%*	50
Type II: Short Chip Copper Alloys (Usually multiphase alloys. Short, curled, or serrated chips. Screw machine production depends on type of cutting operation.)					
C64200	*Arsenical Silicon-Aluminum Bronze*	60	C67400	*Silicon-Manganese-Aluminum-Brass*	30
C62300	*Aluminum Bronze, 9%*	50	C67500	Manganese Bronze A	30
C62400	*Aluminum Bronze, 10 1/2%*	50	C69400	Silicon Red Brass	30
C17410	Beryllium Copper	40	C69430	*Arsenical Silicon Red Brass*	30

(Continued)

Table 20. *(Continued)* **Wrought Copper Alloys Ranked by Machinability.** *(Source, Copper Development Association.)*

Copper Alloy No.	Traditional Name / *Descriptive Name*	Machinability Index Rating	Copper Alloy No.	Traditional Name / *Descriptive Name*	Machinability Index Rating
	Type II: Short Chip Copper Alloys				
	(Usually multiphase alloys. Short, curled, or serrated chips. Screw machine production depends on type of cutting operation.)				
C17500	Beryllium Copper	40	C15715	*Aluminum Oxide Dispersion-Strengthened Copper*	20
C17510	Beryllium Copper	40			
C28000	Muntz Metal, 60%	40	C15725	*Aluminum Oxide Dispersion-Strengthened Copper*	20
C61800	*Aluminum Bronze, 10%*	40			
C63000	*Nickel-Aluminum Bronze, 10%*	40	C15760	*Aluminum Oxide Dispersion-Strengthened Copper*	20
C63020	*Nickel-Aluminum Bronze, 11%*	40			
C63200	*Nickel-Aluminum Bronze*	40	C17000	Beryllium Copper	20
C46200	Naval Brass, 63 1/2%	30	C17200	Beryllium Copper	20
C46400	Naval Brass, Uninhibited	30	C62500	*Aluminum Bronze, 13%*	20
	Type III: Long Chip Copper Alloys				
	(Usually single phase alloys. Stringy or tangled chips; somewhat gummy behavior. Not suitable for screw machine work.)				
C63200	*Nickel-Aluminum Bronze, 9%*	40	C12100	*Phosphorus-Deoxidized, Low Residual Phosphorus Copper*	20
C22600	Jewelry Bronze, 87 1/2%	30			
C23000	Red Brass, 85%	30	C12200	*Phosphorus-Deoxidized, High Residual Phosphorus Copper*	20
C24000	Low Brass, 80%	30			
C26000	Cartridge Brass, 70%	30	C12900	*Fire-Refined Tough Pitch Copper with Silver*	20
C26130	*Arsenical Cartridge Brass, 70%*	30			
C26800	Yellow Brass, 66%	30	C15000	Zirconium Copper	20
C27000	Yellow Brass, 65%	30	C16200	Cadmium Copper	20
C61000	*Aluminum Bronze, 7%*	30	C16500	*Cadmium-Tin Copper*	20
C61300	*Aluminum Bronze, 7%*	30	C18000	*Nickel-Chromium-Silicon Copper*	20

(Continued)

Table 20. *(Continued)* Wrought Copper Alloys Ranked by Machinability. *(Source, Copper Development Association.)*

Copper Alloy No.	Traditional Name *Descriptive Name*	Machinability Index Rating	Copper Alloy No.	Traditional Name *Descriptive Name*	Machinability Index Rating
	Type III: Long Chip Copper Alloys				
	(Usually single phase alloys. Stringy or tangled chips; somewhat gummy behavior. Not suitable for screw machine work.)				
C61400	*Aluminum Bronze, 7%*	30	C18100	*Chromium-Zirconium-Magnesium Copper*	20
C65100	Low Silicon Bronze B	30	C18135	*Chromium-Cadmium-Copper*	20
C65500	High Silicon Bronze A	30	C18150	*Chromium-Zirconium-Copper*	20
C65600	*Silicon Bronze*	30	C18200	Chromium Coppers	20
C67000	Manganese Bronze B	30	C18400	Chromium Coppers	20
C10100	Oxygen-Free Electronic Copper	20	C21000	Gilding, 95%	20
C10200	Oxygen-Free Copper	20	C22000	Commercial Bronze, 90%	20
C10300	Oxygen-Free Extra Low Phosphorus Copper	20	C50700	*Tin (Signal) Bronze*	
C10400	Oxygen-Free Coppers with Silver	20	C51000	Phosphor Bronze, 5% A	20
C10500	Oxygen-Free Coppers with Silver	20	C52100	Phosphor Bronze, 8% B-2	20
C10700	Oxygen-Free Coppers with Silver	20	C61000	*Aluminum Bronze, 7%*	20
C10800	Oxygen-Free Low Phosphorus Copper	20	C64700	*Silicon-Bronze*	20
C11000	Electrolytic Tough Pitch Copper	20	C70600	Copper-Nickel, 10%	20
C11300	Tough Pitch Coppers with Silver	20	C71500	Copper-Nickel, 30%	20
C11400	Tough Pitch Coppers with Silver	20	C74500	Nickel-Silver, 65-10	20
C11500	Tough Pitch Coppers with Silver	20	C75200	Nickel-Silver, 65-18	20
C11600	Tough Pitch Coppers with Silver	20	C75400	Nickel-Silver, 65-15	20
C12000	Phosphorus-Deoxidized, Low Residual Phosphorus Copper	20	C75700	Nickel-Silver, 65-12	20

Bismuth and tellurium can cause serious embrittlement and/or directionally sensitive ductility when present in uncontrolled concentrations, and sulfur cannot be used in many alloy systems due to its reactivity. Generally, unleaded free cutting alloys are considerably more expensive than leaded versions, and therefore should be considered only when lead must be avoided entirely.

Type II Short-Chip Alloys

Two or more phases may appear in the microstructure when alloy concentration is sufficiently high. The beta phase in heterogeneous (two or more phases) Type II alloys makes cold work more difficult, but considerably improves the capacity for hot deformation. Type II alloys tend to be stronger than single-phase materials, but ductility is correspondingly reduced.

As Type II alloys pass over the cutting tool, the metal tends to shear laterally in a series of closely spaced steps that produce ridges on the short, helical chips. This intermittent shear process raises the potential for tool chatter and poor surface quality, but these problems can be avoided by adjusting machining parameters appropriately. Chip breakers can be used to reduce the length of the coiled chips.

The machinability of Type II alloys is dependent on the complex relationships between alloy microstructure and mechanical properties. Power consumption varies with mechanical properties and work hardening rate, while the shape of the chips depends on the metal's ductility. In part because of their relatively manageable chips, multiphase alloys are considered to have better machinability than ductile Type III metals. In fact, the higher power consumption for the harder Type II alloys is generally taken as being less important than tool wear or chip management—both of which are less favorable in Type III alloys. Type II alloys are often processed on automatic screw machines, although production rates are considerably lower than those attainable with free cutting alloys.

Type III Long-Chip Alloys

The copper alloys with essentially uniform microstructures are the simplest. In pure copper, the structure contains only one phase (or crystal form), commonly designated "alpha." This single-phase structure is retained within fairly broad limits when alloying elements are added, but the alloy content at which additional phases begin to appear differs with the individual alloying element and with processing conditions. For example, up to 39% zinc can be added to copper to form a single-phase alpha brass. The alpha structure can also accommodate up to 9% aluminum and remain homogeneous. The equilibrium room temperature solubility limit for tin in copper is quite low (near 1%, by weight), but copper-tin alloys (tin bronzes) that are heated during processing remain structurally homogeneous up to almost 15% tin because the transformation that produces the second phase is comparatively sluggish. Copper-nickel alloys have homogeneous alpha structures no matter how much nickel is present.

Type III alloys are soft and ductile in the annealed state. Their mechanical properties are governed by alloying and the degree of cold work. Pure copper, for example, can be strengthened only by cold working, while the strength of single-phase brasses, bronzes, aluminum bronzes, and copper-nickels is derived from the combined effects of cold work and alloying. Even highly alloyed single-phase alloys retain a considerable degree of ductility, and while their chips are long and stringy, they have a smooth and uniform surface that reflects the uninterrupted passage of the cutting tool. The chip is thicker than the feed rate because the copper is upset as it passes over the face of the tool, while the accompanying cold work makes the chip hard and springy while consuming energy that, when converted to heat in the chip and cutting tool, increases tool wear.

Initial hardness, work hardening rate, and chip appearance can be related to the machinability of Type III alloys.

Initial hardness, either from prior cold work or from alloy content. Soft materials generally consume less energy than harder alloys, and usually produce less tool wear. However, softer materials tend to deflect more under the pressure of the cutting tool, which can reduce dimensional accuracy. When improperly ground tools are employed on softer materials, chatter and poor surface can result from the tool's tendency to dig into these metals.

Work hardening rate, as a function of deformation. High work hardening rates result in high energy consumption, hard chips, and high tool wear. Severely work hardened chips will tear away from the underlying soft matrix, causing smearing and poor surface finish.

Chip appearance. Although Type III metals are machinable, their tendency to form long, stringy chips makes them an unlikely choice for high speed production on automatic screw machines where clearing chip tangles can cause difficulties.

Other factors influence the machinability of all three types of copper alloys.

Finer grain sizes are generally beneficial to material strength and ductility. The effect is most pronounced in leaded multiphase alloys, although some improvements in machinability will be experienced in fine-grained Type III materials as well.

Texture. The grain structure in wrought, rolled, or extruded metals will reflect the direction of deformation—the metal's grains, grain boundaries, and second phases, if any, tend to become elongated in the direction of hot or cold work, leading to the fibrous texture of heavily wrought metals. Since any nonuniformity in a metal's structure can influence chip behavior, machinability is somewhat enhanced by cutting in a direction that crosses the direction of deformation. For example, rods and bars machine better circumferentially than longitudinally, and plates machine better in the transverse or through-thickness direction than parallel to the rolling direction.

Temper. Generally, the harder the metal, the higher the power needed, but the economic effect of temper on attainable cutting speeds and surface finish is usually more important than power consumption. Temper is discussed in more detail later in this section.

Heat treatment. Beryllium coppers are rough machined best in the solution annealed state, then heat treated to full hardness before finish machining, or, if necessary, grinding. This will also avoid the possibility of distortion caused by the slight (0.5%) volume change that occurs during heat treatment.

Tempers for wrought copper alloys

Copper alloys are said to have a harder temper if they have been cold worked and/or heat treated, and a softer temper when they are in an as-hot-formed state or when the effects of cold work and/or heat treatment have been removed by annealing. As can be expected, higher strength and hardness (harder tempers) are gained at the cost of reduced ductility. As applied to heat treated copper alloys, temper carries exactly the opposite meaning than for heat treated steels where tempering generally implies softening.

Temper designations for wrought copper alloys are defined in ASTM B 601, and the basic designations are given in **Table 21**. The designations imply specific mechanical properties only when used in association with a particular alloy, product form and size. In order to specify a specific copper alloy, it is necessary to provide the UNS Number, the product form and size (for example, $^1/_4$ inch [6.3 mm] round rod), and the desired temper. Familiarity with the temper system is important because properties vary considerably for different forms and tempers of the same alloy. For example, a one-inch rod of the free cutting brass alloy C3600 in the Half Hard (H02) temper has a yield strength of 45 ksi (310

Table 21. Temper Designations for Wrought Copper Alloys.

Annealed - O	Temper Names	Cold Worked with Added Treatments - HR	Temper Name
O25	Hot Rolled & Annealed	HR50	Drawn & Stress Relieved
O30	Hot Extruded & Annealed	**As Manufactured - M**	**Temper Names**
O50	Light Anneal	M10	As Hot Forged- Air Cooled
O60	Soft Anneal	M11	As Hot Forged- Quenched
O61	Annealed (Also Mill Annealed)	M20	As Hot Rolled
O70	Dead Soft Anneal	M30	As Extruded
Annealed with Grain Size Prescribed - OS	**Nominal Average Grain Size, mm**	**Solution Heat Treated - TB**	**Temper Name**
OS005	0.005	TB00	Solution Heat Treated (A)
OS010	0.010	**Solution Treated and Cold Worked- TD**	**Temper Names**
OS015	0.015		
OS020	0.020	TD00	Solution Treated & Cold Worked: 1/8 Hard
OS025	0.025		
OS035	0.035	TD01	Solution Treated & Cold Worked: 1/4 Hard
OS050	0.050		
OS060	0.060	TD02	Solution Treated & Cold Worked: 1/2 Hard
OS070	0.070		
OS100	0.100	TD03	Solution Treated & Cold Worked: 3/4 Hard
OS120	0.120		
OS150	0.150	TD04	Solution Treated & Cold Worked: Hard (H)
OS200	0.200		
Cold Worked. Based on Cold Rolling or Drawing - H	**Temper Names**	**Solution Heat Treated and Precipitation Heat Treated - TF**	**Temper Name**
H00	1/8 Hard	TF00	Solution Heat Treated & Aged (AT)
H01	1/4 Hard	**Quench Hardened - TQ**	**Temper Names**
H02	1/2 Hard	TQ00	Quenched Hardened
H03	3/4 Hard	TQ30	Quenched Hardened and Tempered
H04	Hard		
H06	Extra Hard	**Solution Heat Treated, Cold Worked, & Precipitation Heat Treated - TH**	**Temper Names**
H08	Spring		
H10	Extra Spring		
H12	Special Spring	TH01	1/4 Hard and Precipitation Heat Treated (1/4 HT)
H13	Ultra Spring		
H14	Super Spring		

(Continued)

Table 21. *(Continued)* **Temper Designations for Wrought Copper Alloys.**

Cold Worked. Based on Particular Products (Wire) - H	Temper Names	Solution Heat Treated, Cold Worked, & Precipitation Heat Treated - TH	Temper Names
H60	Cold Heading, Forming	TH02	1/2 Hard and Precipitation Heat Treated (1/2 HT)
H63	Rivet		
H64	Screw	TH03	3/4 Hard and Precipitation Heat Treated (3/4 HT)
H66	Bolt		
H70	Bending	TH04	Hard and Precipitation Heat Treated (HT)
H80	Hard Drawn		
Cold Worked and Stress Relieved - HR	**Temper Names**	**Mill Hardened Tempers - TM**	**Manufacturing Designation**
HR01	1/4 Hard & Stress Relieved	TM00	AM
HR02	1/2 Hard & Stress Relieved	TM01	1/4 HM
HR04	Hard & Stress Relieved	TM02	1/2 HM
HR08	Spring & Stress Relieved	TM04	HM
HR010	Extra Spring & Stress Relieved	TM06	XHM
		TM08	XHMS
		Temper Designations Based on ASTM B 601:Temper Designations for Copper and Wrought Alloys - Wrought and Cast.	

MPa), while the same alloy produced as a 1 in. × 6-in. (25.4 mm × 152.4 mm) bar in the Soft Anneal (O60) temper has a yield strength of only 20 ksi (138 MPa).

The choice of temper depends on the properties required and the type of processing to be done. The Half Hard (H02) cold worked temper is the most frequently specified for screw machine products because it combines the best levels of strength and ductility to suit both machinability and functional requirements. Annealed tempers such as O50 and O60 refer to soft, formable structures that are usually specified for cold formed rather than machined products. Annealed alloys may have inferior machinability, with poor surface finishes, because of a tendency for chips to tear away from the work during cutting. Lightly cold worked tempers such as Eighth Hard (H00) and Quarter Hard (H01) improve machinability yet retain sufficient ductility for forming operations. Hard (H04), Extra Hard (H06), and Spring (H08) tempers produce maximum strength at the expense of ductility.

Electrical and thermal conductivity vary with the degree of temper, but the nature and extent of the effect depend strongly on the type of alloy and its metallurgical condition. Chemical properties such as corrosion resistance and plateability are not strongly affected by temper, but residual cold work-tensile stresses render some copper alloys more susceptible to stress corrosion cracking than would be the case in the annealed state.

Tempers for copper casting alloys are given in **Table 22**.

Cold forming and heat treatment of wrought copper alloys

Cold forming. As previously discussed, the highest machinability ratings for copper belong to the leaded copper alloys, particularly to the free cutting brasses. It was also pointed out that lead lowers ductility and can make alloys difficult to cold form. For this

Table 22. Standard Temper Designations for Copper Casting Alloys.

Annealed - O	Temper Names	Heat Treated - TQ	Temper Names
O10	Cast & Annealed (Homogenized)	TQ00	Quench Hardened
O11	As Cast and Precipitation Heat Treated	TQ30	Quench Hardened & Tempered
As Manufactured - M	**Temper Names**	TQ50	Quench Hardened & Temper Annealed
M01	As Sand Cast	**Solution Heat Treated & Spinodal Heat Treated - TX**	**Temper Name**
M02	As Centrifugal Cast		
M03	As Plaster Cast		
M04	As Pressure Die Cast	TX00	Spinodal Hardened
M05	As Permanent Mold Cast	**Solution Heat Treated - TB**	**Temper Name**
M06	As Investment Cast		
M07	As Continuous Cast	TB00	Solution Heat Treatment
Temper Designations Based on ASTM B 601		**Solution Heat Treated & Precipitation Heat Treated - TF**	**Temper Name**
		TF00	Precipitation Hardened

reason, parts requiring deep knurls or high pitch rolled threads should be specified in alloys with reduced lead content. Alloys C34000 (0.80–1.5% Pb), C34500, and C35300 (both 1.5–2.5% Pb) are good choices. Similar conditions apply to parts such as hose fittings and automobile sensor housings, since both frequently incorporate spun flanges to crimp the parts in place. Alloy C36000 is ductile enough to permit such bends in most instances, but severe deformation may require a lower lead free cutting alloy. The Half Hard temper normally used for screw machine products offers adequate formability for routine work, but a softer temper may be required when extensive cold work must be performed.

Electrical crimp connectors made from free cutting coppers frequently demand a high degree of cold formability. In this case, it is important to take the directionality of the metal's ductility into account. Tellurium bearing coppers and alloys can be significantly less ductile in the radial or transverse direction (with regard to extrusion and drawing direction) than along the long axis of the tool.

Heat treatment. The mechanical properties of several of the machining rod copper alloys can be improved through heat treatment. Heat treatments are sometimes combined with mechanical working to enhance the strengthening effect. Zirconium copper, C15000, can be strengthened by a solution heat treatment at 1,650–1,740° F (900–950° C), followed by quenching, and aging at 930–1,020– F (500–550° C). Cold working following the solution anneal increases the final strength. Cold work also benefits the heat treatable beryllium coppers. Low-beryllium copper alloys (<1% Be, known as the "high conductivity" alloys), such as C17500 and its derivatives, are heat treated by solution annealing at 1,700° F (925° C), rapid quenching, and aging at 900° F (480° C) for two to three hours. Intermediate cold working, up to about 40%, results in an additional strength increase of about 5% above that attainable by heat treatment alone. So-called high strength beryllium coppers are solution annealed at 1,450° F (788° C), quenched, and precipitation hardened at 600° F (315° C).

Chromium-zirconium-magnesium copper (C18100) is solution annealed at 1,650–1,790° F (900–977° C), quenched and hardened at 750–930° F (400–500° C). Chromium-cadmium copper (C18135) is solution heat treated at 1,700–1,760° F (925–960° C), and aged at 850–1,050° F (455–565° C). The same temperatures apply to solution annealed and cold worked material. Chromium coppers (C18200 and C18400) are solution treated at 1,800–1,850° F (980–1010° C), and aged at 800–930° F (425–500° C). A minimum of 50% cold work can be applied after solution annealing to increase the alloy's final strength. Nickel-tellurium copper (C19100) can be age hardened at 800–900° F (425–480° C) after solution annealing at 1,300–1,450° F (705–790° C), or hot working at slightly higher temperatures.

Copper casting alloy selection based on fabricability

Castings usually require further processing after shake-out and cleaning. Machining is the most common secondary operation, and welding is often required to repair minor defects or to join several castings into a larger assembly. Factors to be considered before selecting cast copper alloys for a process or application are shown in **Table 23**.

Machinability

As a class, cast copper alloys can be described as being relatively easy to machine, when compared to steel, and far easier to machine when compared to stainless steels, nickel base alloys, and titanium. The easiest casting alloys to machine are those that contain more than 2% lead. These alloys are free cutting, tool wear is minimal, and surface finishes are generally excellent. High speed steel is the accepted tooling material for these alloys, but carbides are often used for the stronger leaded compositions. Cutting fluids help in the reduction of airborne lead-bearing particles, but they are not otherwise needed when cutting the highly leaded brasses and bronzes. Leaded cast copper alloys have machinability ratings greater than 70 (with wrought free cutting brass C36000 being 100).

Next in order of machinability are moderate to high strength alloys that contain sufficient alloying elements to form second phases in their microstructure (often called duplex or multiphase alloys). Examples include unleaded yellow brasses, manganese bronzes, and silicon brasses and bronzes. These alloys have short, brittle, tightly curled chips that tend to break into manageable segments, and using chipbreaker geometries facilitates this action. Surface finishes can be quite good for the duplex alloys, but cutting speeds will be lower, and tool wear higher, than with the free cutting grades.

The unleaded, single phase alpha alloys (which include high conductivity coppers, high copper alloys such as chromium coppers, beryllium coppers, tin bronzes, red brasses, aluminum bronzes, and copper nickels), as a group, have a general tendency to form long, stringy chips that interfere with machining operations. In addition, pure copper and high nickel alloys tend to weld to the tool face, which leads to impaired surface finishes. Cutting tools used with these alloys should be highly polished and ground with generous rake angles to help ease the flow of chips away from the workpiece. Adequate relief angles will help avoid trapping particles between the tool and workpiece, where they might scratch the freshly machined surface, and cutting fluids should be used to provide adequate lubrication.

Information on welding casting copper alloys can be found in the Welding section of this book.

Table 23. Technical Factors in the Choice of Casting Method for Copper Alloys. *(Source, Copper Development Association.)*

Casting Method	Copper Alloys	Size Range	General Tolerances	Surface Finish	Min. Section Thickness	Ordering Quantities	Relative Cost[1]
Sand	All	All sizes, depends on foundry capability.	±1/32 in up to 3 in.; ±3/64 in 3-6 in.; add ±0.003 in./in above 6 in.; add ±0.020 to ±0.060 in across parting line.	150-500 μin. rms	1/8-1/4 in.	All	1-3
No-Bake	All	All sizes, but usually > 10 lbs.	Same as sand casting.	Same as sand casting	Same as sand casting	All	1-3
Shell	All	Typical maximum mold area = 550 in.² typical maximum thickness = 6 in.	±0.005-0.010 in up to 3 in.; add ±0.002 in/in. above 3 in.; add ±0.005 to 0.010 in. across parting line.	125-200 μin. rms	3/32 in.	≥100	2-3
Permanent Mold	Coppers, high copper alloys, yellow brasses, high strength brasses, silicon bronze, high zinc silicon brass, most tin bronzes, aluminum bronzes, some nickel silvers.	Depends on foundry capability; best = 50 lbs. Best max. thickness = 2 in.	Usually ±0.010 in.; optimum ±0.005 in., ±0.002 in part-to-part.	150-200 μin. rms, best = 70 μin. rms	1/8-1/4 in.	100-1,000, depending on size	2-3
Die	Limited to C85800, C86200, C86500, C87800, C87900, C99700, C99750 & some proprietary alloys.	Best for small, thin parts; max. area ≤ 3 ft.².	±0.002 in/in; not less than 0.002 in. on any one dimension; add ±0.010 in. on dimensions affected by parting line.	32-90 μin. rms	0.05-0.125 in.	≥1000	1
Plaster	Coppers, high copper alloys, silicon bronze, manganese bronze, aluminum bronze, yellow brass.	Up to 800 in.², but can be larger.	One side of parting line, ±0.015 in up to 3 in.; add ±0.002 in./in above 3 in.; add 0.010 in across parting line, and allow for parting line shift of 0.015 in.	63-125 μin. rms, best = 32 μin. rms	0.060 in.	All	4

(Continued)

Table 23. *(Continued)* **Technical Factors in the Choice of Casting Method for Copper Alloys.** *(Source, Copper Development Association.)*

Casting Method	Copper Alloys	Size Range	General Tolerances	Surface Finish	Min. Section Thickness	Ordering Quantities	Relative Cost[1]
Investment	Almost all	Fraction of an ounce to 150 lbs., up to 48 in.	±0.003 in less than 1/4; ±0.004 in between 1/4 to 1/2 in.; ±0.005 in./in between 1/2-3 in.; add ±0.003 in./in above 3 in.	63-125 μin. rms	0.030 in.	>100	5
Centrifugal	Almost all	Ounce to 25,000 lbs. Depends on foundry capacity.	Castings are usually rough machined by foundry.	Not applicable	1/4 in.	All	1-3

[1] 1 = low cost, 5 = high cost.

Magnesium Alloys

Magnesium is usually one of the first materials considered when light weight is a designer's priority—it is the lightest mass produced metal that remains stable under normal operating conditions. It is also the eighth most abundant element in the earth's crust, and the third most abundant element in seawater. As with other metals, magnesium alloys increase in tensile strength, yield strength, and hardness in low temperature environments, but ductility suffers. Tensile properties reverse at elevated temperatures, and most suffer a reduction of 40 to 75% when operating temperatures increase from room temperature (68° F, 20° C) to 600° F (315° C). The density of the magnesium alloys ranges from a low of 1.74 gr/cm^3 (alloy K1A, which is often used for its dampening characteristics) to highs of 1.86 gr/cm^3 (alloy ZH62A), but most are in the range of 1.80 to 1.83. Electrical conductivity of magnesium is 38.6 IACS. Both wrought and cast alloys are available. The wrought is available in extrusions, forgings, sheet, plate, and bar, and the cast for sand, permanent mold, investment, and die castings.

Pure magnesium ignites easily. The alloys are less susceptible to ignition, but care must be taken during machining, and fine chips should be regarded as fire hazards. The larger chips produced by roughing cuts and medium finishing cuts present very little danger of igniting, and even fine chip fires are unlikely at cutting speeds below 700 sfm (213 smm). Feed rates should never be in excess of 0.001 ipr (0.02 mm/rev). Low viscosity cutting fluids minimize risk and reduce the possibility of workpiece distortion due to overheating. Cutting solutions containing water should never be used as water increases the risk of scrap igniting during shipping and storage, and also reduces the value of the scrap.

Composition and properties of magnesium alloys

Although UNS designations have been assigned to magnesium alloys, they are more commonly identified by a four-part ASTM number that reflects the composition of the alloy in the first three parts, and the temper designation in the fourth. For example, the alloy AZ31B-F has aluminum and zinc (AZ) as its principal alloying agents, in amounts of 3% aluminum and 1% zinc (stated as 31 in the alloy number). The B indicates that this is the second alloy standardized with 6% Al and 3% Zn, and the F indicates that the alloy is as fabricated. For a description of the ASTM designation system, see **Table 1**. A more detailed explanation of the temper system can be found below.

Tables 2 and **3** provide chemical compositions for casting and wrought magnesium alloys, and **Tables 4** and **5** show their mechanical and thermal properties. Characteristics of some popular alloys follow. References below to AMS stand for Aerospace Materials Specifications, and those to ASTM stand for American Society for Testing and Materials.

Wrought Alloys

AZ31B. A wrought magnesium base alloy containing aluminum and zinc. It is available as sheet and plate (AMS 4375 and AMS 4377), plate (AMS 4376), extrusion (ASTM B 107), and forging (ASTM B 91). It has good room temperature strength and ductility and is used primarily for applications where the temperature does not exceed 300° F (149° C). Increased strength is obtained in the sheet and plate form by strain hardening with a subsequent partial anneal (H24 and H26 tempers). No treatments are available for increasing the strength of this alloy after fabrication, but AZ31B has moderate strength and is widely used. Forming this alloy must be done at elevated temperatures if small radii or deep draws are required. If the temperatures used are too high or the time is too great, H24 and H26 temper will be softened. It is readily welded but must be stress relieved after welding to prevent stress corrosion cracking.

AZ61A. A wrought magnesium base alloy containing aluminum and zinc. It is

Table 1. Standard ASTM Designation System for Magnesium Alloys and Tempers.

Part	Purpose	Form	Symbols Used
First Part	Indicates the two principal alloying elements.	Two code letters represent the two main alloying elements (if only one alloying element is present, only one letter is used), with the element with the highest percentage listed first. If percentage content of the two primary alloying agents are equal, they are listed alphabetically.	A, Aluminum M, Manganese B, Bismuth N, Nickel C, Copper P, Lead D, Cadmium Q, Silver E, Rare Earth R, Chromium F, Iron S, Silicon G, Magnesium T, Tin H, Thorium W, Yttrium K, Zirconium Y, Antimony L, Lithium Z, Zinc
Second Part	Indicates the percentage content of the two principal alloying elements.	Two numbers correspond to rounded off percentages of the two main alloying elements. They are arranged in the same order as the alloy designations in the first part.	Whole numbers are used.
Third Part	Distinguishes between different alloys with the same percentages of the two principal alloying elements.	One letter is used.	A, First composition, registered with ASTM B, Second composition, registered C, Third composition, registered D, High purity, registered E, High corrosion resistance, registered X, Not registered with ASTM
Fourth Part	Indicates Temper. See text for full and expanded description of tempers.	One letter and, as required, one number, separated from the third part by a hyphen.	F, As fabricated O, Annealed H10, H11, Slightly strain hardened H23, H24, H26, Strain hardened and partially annealed T4, Solution heat treated T5, Artificially aged only T6, Solution heat treated and artificially aged T8, Solution heat treated, cold worked, and artificially aged

Table 2. Chemical Composition of Magnesium Casting Alloys.

ASTM No.	UNS No.	Nominal Composition — Expressed as Percent by Weight
		Sand and Permanent Mold Castings
AM100A	M10100	Al 9.3-10.7; Cu 0.1 max.; Mg 90; Mn 0.1 min.; Ni 0.01 max.; Si 0.3 max.; Zn 0.3 max.; Other 0.3 max.
AZ63A	M11630	Al 5.3-6.7; Cu 0.25 max.; Mg 91; Mn 0.15 min.; Ni 0.01 max.; Si 0.3 max.; Zn 2.5-3.5.; Other 0.3 max.
AZ81A	M11810	Al 7-8.1; Cu 0.1 max.; Mg 92; Mn 0.13 min.; Ni 0.01 max.; Si 0.3 max.; Zn 0.4-1.; Other 0.3 max.
AZ91C	M11914	Al 8.1-9.3; Cu 0.1 max.; Mg 90; Mn 0.13 min.; Ni 0.01 max.; Si 0.3 max.; Zn 0.4-1.; Other 0.3 max.
AZ91E	M11921	Al 8.1-9.3; Cu 0.015 max.; Fe 0.005 max.; Mg 90; Mn 0.17-0.35; Ni 0.001 max.; Si 0.2 max.; Zn 0.35-1.; Other 0.3 max.

(Continued)

Table 2. *(Continued)* **Chemical Composition of Magnesium Casting Alloys.**

ASTM No.	UNS No.	Nominal Composition — Expressed as Percent by Weight
		Sand and Permanent Mold Castings
AZ92A	M11920	Al 8.3-9.7; Cu 0.25 max.; Mg 89; Mn 0.1 min.; Ni 0.01 max.; Si 0.3 max.; Zn 1.6-2.4.; Other 0.3 max.
EZ33A	M12330	Cu 0.1 max.; Mg 93; Nd 2.5-4; Ni 0.01 max.; Zn 2-3.1; Zr 0.5-1; Other 0.3 max.
HK31A	M13310	Cu 0.1 max.; Mg 96; Ni 0.01 max.; Th 2.5-4; Zn 0.3 max.; Zr 0.4-1; Other 0.3 max.
HZ32A	M13320	Cu 0.1 max.; Mg 94; Ni 0.01 max.; Rare Earths 0.1 max.; Th 2.5-4; Zn 1.7-2.5; Zr 0.5-1; Other 0.3 max.
K1A	M18010	Mg 99; Zr 0.4-1; Other 0.3 max.
QE22A	M18220	Ag 2-3; Cu 0.1 max.; Mg 95; Nd 1.7-2.5; Ni 0.01 max.; Zr 0.4-1; Other 0.03 max.
QH21A	-	Ag 2-3; Cu 0.1 max.; Mg 95; Nd 0.6-1.5; Ni 0.01 max.; Th 0.6-1.6; Zn 0.2 max.; Zr 0.4-1; Other 0.03 max.
ZE41A	M16410	Cu 0.1 max.; Mg 94; Mn 0.15 max.; Nd 0.75-1.75; Ni 0.01 max.; Zn 3.5-5; Zr 0.4-1; Other 0.3 max.
ZE63A	M16630	Cu 0.1 max.; Mg 91; Nd 2.1-3; Ni 0.01 max.; Zn 5.5-6; Zr 0.4-1; Other 0.3 max.
ZH62A	M16620	Cu 0.1 max.; Mg 92; Ni 0.01 max.; Th 1.4-2.2; Zn 5.2-6.2; Zr 0.5-1; Other 0.3 max.
ZK61A	M16610	Cu 0.1 max.; Mg 93; Ni 0.01 max.; Zn 5.5-6.5; Zr 0.6-1; Other 0.3 max.
		Die Casting
AM60A	M10600	Al 5.5-6.5; Cu 0.35 max.; Mg 94; Mn 0.13 min.; Ni 0.03 max.; Si 0.5 max.; Zn 0.22 max.
AM60B	M10603	Al 5.5-6.5; Cu 0.01 max.; Fe 0.005 max.; Mg 94; Mn 0.25 min.; Ni 0.002 max.; Si 0.1 max.; Zn 0.22 max.; Other 0.003 max.
AS21X1	-	Al 1.7; Mg 97; Mn 0.4 min.; Si 1.1
AS41XA	M10410	Al 3.5-5; Cu 0.06 max.; Mg 94; Mn 0.2-0.5; Ni 0.03 max.; Si 0.5-1.5; Zn 0.12 max.; Other 0.3 max.
AZ91A	M11910	Al 8.3-9.7; Cu 0.1 max.; Mg 90; Mn 0.13 min.; Ni 0.03 max.; Si 0.5 max.; Zn 0.35-1 max.; Other 0.3 max.
AZ91B	M11912	Al 8.3-9.7; Cu 0.35 max.; Mg 90; Mn 0.13 min.; Ni 0.03 max.; Si 0.5 max.; Zn 0.35-1; Other 0.3 max.

Table 3. Chemical Composition of Magnesium Extruded, Forging, and Sheet and Plate Alloys.

ASTM No.	UNS No.	Nominal Composition — Expressed as Percent by Weight
		Extruded
AZ10A	M11100	Al 1-1.5; Ca 0.04 max.; Cu 0.1 max.; Fe 0.005 max.; Mg 98; Mn 0.2 min.; Ni 0.005 max.; Si 0.1 max.; Zn 0.2-0.6
AZ31B	M11311	Al 2.5-3.5; Ca 0.04 max.; Cu 0.05 max.; Fe 0.005 max.; Mg 97; Mn 0.2 min.; Ni 0.005 max.; Si 0.1 max.; Zn 0.6-1.4; Other 0.3 max.
AZ61A	M11610	Al 5.8-7.2; Cu 0.05 max.; Fe 0.005 max.; Mg 92; Mn 0.15 min.; Ni 0.005 max.; Si 0.1 max.; Zn 0.4-1.5; Other 0.3 max.
AZ80A	M11800	Al 7.8-9.2; Cu 0.05 max.; Fe 0.005 max.; Mg 91; Mn 0.12 min.; Ni 0.005 max.; Si 0.1 max.; Zn 0.2-0.8; Other 0.3 max.
EA55RS	-	Al 5; Mg 85; Nd 4.9; Zn 5
HM31A	M13312	Mg 96; Mn 1.2 min.; Th 2.5-3.5; Other 0.3 max.

(Continued)

Table 3. *(Continued)* **Chemical Composition of Magnesium Extruded, Forging, and Sheet and Plate Alloys.**

ASTM No.	UNS No.	Nominal Composition — Expressed as Percent by Weight
Extruded		
M1A	M15100	Ca 0.3 max.; Cu 0.05 max.; Mg 99; Mn 1.2 min.; Ni 0.01 max.; Si 0.1 max.; Other 0.3 max.
ZK40A	M16400	Mg 95; Zn 3.5-4.5; Zr 0.45 min.; Other 0.3 max.
ZK60A	M16600	Mg 94; Zn 4.8-6.2; Zr 0.45 min.; Other 0.3 max.
Forgings		
AZ31B	M11311	Al 2.5-3.5; Ca 0.04 max.; Cu 0.05 max.; Fe 0.005 max.; 97 min 0.2.; Max. 0.005; Si 0.1 max.; Zn 0.6-1.4; Other 0.3 max.
AZ61A	M11610	Al 5.8-7.2; Cu 0.05 max.; Fe 0.005 max.; 92 min.0.15; 0.005 max,; Si 0.1 max.; Zn 0.4-1.5; Other 0.3 max.
AZ80A	M11800	Al 7.8-9.2; Cu 0.05 max.; Fe 0.005 max.; 91 min.0.12; 0.005 max,; Si 0.1 max.; Zn 0.2-0.8; Other 0.3 max.
HM21A	M13210	Mg 97; Mn 0.45-1.1; Th.1.5-2.5; Other 0.3 max.
M1A	M15100	Ca 0.3 max.; Cu 0.05 max.; Mg 99; Mn 1.2 min.; Ni 0.01 max.; Si 0.1 max.; Other 0.3 max.
ZK60A	M16600	Mg 94; Zn 4.8-6.2; Zr 0.45 min.; Other 0.3 max.
Sheet and Sheet and Plate		
AZ31B	M11311	Al 2.5-3.5; Ca 0.04 max.; Cu 0.05 max.; Fe 0.005 max.; 97 min.0.2; 0.005 max.; Si 0.1 max.; Zn 0.6-1.4; Other 0.3 max.
HK31A	M13310	Cu 0.1 max.; Mg 96; Ni 0.01 max.; Th 2.5-4; Zn 0.3 max.; Zr 0.4-1; Other 0.3 max.
HM21A	M13210	Mg 97; Mn 0.45-1.1; Th 1.5-2.5; Other 0.3 max.
M1A	M15100	Ca 0.3 max.; Cu 0.05 max.; Mg 99; Mn 1.2 min.; Ni 0.01 max.; Si 0.1 max.; Other 0.3 max.

available as extrusion (ASM 4350) and forging (ASTM B91). Much like AZ31B in general characteristics, but the increased aluminum content slightly increases the strength while decreasing the ductility. Artificial aging will also increase strength but reduce ductility. Severe forming must be done at elevated temperatures. It is readily welded but must be stress relieved after welding to prevent stress corrosion cracking.

ZK60A. A wrought magnesium base alloy containing zinc and zirconium. It is available as extrusion (ASTM B107 and AMS 4352), and die and hand forgings (AMS 4362). Increased strength is obtained by artificial aging (T5) from the as fabricated (F) temper. It has the best combination of high room temperature strength and ductility of the wrought magnesium base alloys, and is used primarily at temperatures below 300° F (149° C). ZK60A has good ductility as compared with other high strength magnesium alloys, and can be formed or bent cold into shapes not possible with those alloys having less ductility. It is not considered weldable.

Casting Alloys

AM100A. A magnesium base casting alloy containing aluminum and a small amount of manganese. It is available as investment casting (AMS 4455), permanent mold casting (AMS 4483), and as a casting alloy as specified by MIL-M-46062. Primarily used for permanent mold castings, AM100A has about the same characteristics as AZ92A. It has less tendency to microshrinkage and hot shortness than the Mg-Al-Zn alloys, plus good weldability and fair pressure tightness.

(Text continued on p. 643)

Table 4. Mechanical Properties of Magnesium Alloys.

ASTM No.	Ultimate Tensile Strength		Yield Strength		Ultimate Bearing Strength		Bearing Yield Strength		Modulus of Elasticity		Elong. %¹	HB²
	PSI	MPa	PSI	MPa	PSI	MPa	PSI	MPa	ksi	GPa		
Sand and Permanent Mold Castings												
AM100A-T6	39,885	275	15,954	110	-	-	-	-	6,527	45	4	69
AZ63A-T6	39,885	275	18,855	130	-	-	52,214	360	6,527	45	5	73
AZ81A-T4	39,885	275	12,038	85	58,015	400	34,809	240	6,527	45	15	55
AZ91C-T6	39,885	275	21,031	145	74,695	515	52,214	360	6,498	44.8	6	70
AZ91E-T6	39,885	275	21,031	145	74,695	515	52,214	360	6,498	44.8	6	70
AZ92A-T6	39,885	275	21,756	150	79,771	550	65,267	450	6,527	45	3	84
EZ33A-T5	23,206	160	15,954	110	57,290	395	39,885	275	6,527	45	2	50
HK31A-T6	31,908	220	15,229	105	60,916	420	39,885	275	6,527	45	8	55
HZ32A-T5	26,832	185	13,053	90	59,466	410	36,985	255	6,527	45	4	55
K1A-F	26,107	180	7,977	55	45,977	317	18,130	125	5,511	38	19	-
QE22A-T6	37,710	260	28,282	195	-	-	-	-	6,527	45	3	80
QH21A-T6	39,885	275	29,733	205	-	-	-	-	6,527	45	4	-
ZE41A-T5	29,733	205	20,305	140	70,343	485	50,763	350	6,527	45	3.5	62
ZE63A-T6	43,511	300	27,557	190	-	-	-	-	6,527	45	10	70
ZH62A-T5	34,809	240	24,656	170	71,794	495	49,313	340	6,527	45	4	70
ZK61A-T5	44,962	310	26,832	185	-	-	-	-	6,527	45	10	68
ZK61A-T5	44,962	310	28,282	195	-	-	-	-	6,527	45	10	70
Die Castings												
AM60A-F	29,733	205	16,679	115	-	-	-	-	6,527	45	6	-
AM60B-F	29,733	205	16,679	115	-	-	-	-	6,527	45	6	-
AS21X1	-	-	18,855	130	-	-	-	-	6,527	45	9	-

Notes: ¹ Elongation in 50 mm (2 inches). ² Hardness Brinell with 500 kg load and 10 mm ball.

(Continued)

Table 4. *(Continued)* **Mechanical Properties of Magnesium Alloys.**

ASTM No.	Ultimate Tensile Strength		Yield Strength		Ultimate Bearing Strength		Bearing Yield Strength		Modulus of Elasticity		Elong. %[1]	HB[2]
	PSI	MPa	PSI	MPa	PSI	MPa	PSI	MPa	ksi	GPa		
Die Castings												
AS41XA-F	31,908	220	21,756	150	-	-	-	-	6,527	45	4	-
AZ91A-F	33,359	230	21,756	150	-	-	-	-	6,498	44.8	3	63
AZ91B-F	33,359	230	21,756	150	-	-	-	-	6,498	44.8	3	63
K1A-F	23,931	165	12,038	83	-	-	-	-	5,511	38	8	-
Extruded Bars, Rods and Shapes												
AZ10A-F	34,809	240	22,481	155	-	-	-	-	6,527	45	10	-
AZ31B-F	37,710	260	29,008	200	55,840	385	33,359	230	6,527	45	15	49
AZ61A-F	44,962	310	33,359	230	68,168	470	41,336	285	6,527	45	16	60
AZ80A-F	49,313	340	36,260	250	79,771	550	50,763	350	6,527	45	7	67
AZ80A-T5	55,114	380	39,885	275	-	-	-	-	6,527	45	7	82
EA55RS-T4	61,641	425	53,664	370	-	-	-	-	6,527	45	14	-
HM31A-T5	43,511	300	39,160	270	68,168	470	47,863	330	6,527	45	10	-
M1A-F	36,985	255	26,107	180	50,763	350	28,282	195	6,527	45	12	44
ZK40A-T5	40,030	276	36,985	255	-	-	-	-	6,527	45	4	-
ZK60A-F	49,313	340	37,710	260	79,771	550	55,114	380	6,527	45	11	75
ZK60A-T5	52,939	365	44,237	305	84,847	585	58,740	405	6,527	45	11	88
Forgings												
AZ31B	37,710	260	24,656	170	-	-	-	-	6,527	45	15	-
AZ61A-F	42,786	295	26,107	180	72,519	500	41,336	285	6,527	45	12	55
AZ80A-F	47,863	330	33,359	230	-	-	-	-	6,527	45	11	69
AZ80A-T5	50,038	345	36,260	250	-	-	-	-	6,527	45	11	72

Notes: [1] Elongation in 50 mm (2 inches), [2] Hardness Brinell with 500 kg load and 10 mm ball.

(Continued)

Table 4. *(Continued)* Mechanical Properties of Magnesium Alloys.

ASTM No.	Ultimate Tensile Strength		Yield Strength		Ultimate Bearing Strength		Bearing Yield Strength		Modulus of Elasticity		Elong. %[1]	HB[2]
	PSI	MPa	PSI	MPa	PSI	MPa	PSI	MPa	ksi	GPa		
Forgings												
AZ80A-T6	49,313	340	36,260	250	-	-	-	-	6,527	45	5	72
HM21A-F	33,359	230	20,305	140	-	-	-	-	6,527	45	15	-
HM21A-T5	33,359	230	21,756	150	36,260	250	23,206	160	6,527	45	9	50
M1A	36,260	250	23,206	160	-	-	-	-	6,527	45	7	47
ZK60A-T5	44,237	305	31,183	215	60,916	420	41,336	285	6,527	45	16	65
ZK60A-T6	47,137	325	39,160	270	65,267	450	46,412	320	6,527	45	11	75
Sheet and Sheet and Plate												
AZ31B-O	36,985	255	21,756	150	70,343	485	42,061	290	6,527	45	21	56
AZ31B-H24	42,061	290	31,908	220	71,794	495	47,137	325	6,527	45	15	73
AZ31B-H26	39,885	275	27,557	190	71,794	495	39,885	275	6,527	45	10	-
HK31A-O	29,008	200	18,130	125	-	-	-	-	6,527	45	30	-
HK31A-H24	36,260	250	27,557	190	60,916	420	41,336	285	6,527	45	15	58
HM21A-T8	34,084	235	24,656	170	60,191	415	39,160	270	6,527	45	11	-
HM21A-T81	36,985	255	30,458	210	65,992	455	46,412	320	6,527	45	6	-
M1A-F	33,359	230	18,130	125	50,763	350	29,008	200	6,527	45	17	48

Notes: [1] Elongation in 50 mm (2 inches). [2] Hardness Brinell with 500 kg load and 10 mm ball.

Table 5. Thermal Properties of Magnesium Alloys.

ASTM No.	Process Temperature[1] °C	CTE[2] µin./in.	CTE[2] µm/m	CTE[3] µin./in.	CTE[3] µm/m	Thermal Conductivity[4] Engl.	Thermal Conductivity[4] metric	Solidus °F	Solidus °C	Liquidus °F	Liquidus °C
					Sand and Permanent Mold Castings						
AM100-T6	735-845	14	25	16	28	507	73	865	463	1,103	595
AZ63A-T6	705-845	15	26.1	16	28	534	77	851	455	1,130	610
AZ81A-T4	705-845	14	25	16	28	355	51.1	914	490	1,130	610
AZ91C-T6	625-700	14	26	16	28	505	72.7	878	470	1,103	595
AZ91E-T6	625-700	14	26	16	28	505	72.7	878	470	1,103	595
AZ92A-T6	700-845	14	26	16	28	500	72	833	445	1,103	595
EZ33A-5	750-820	15	26.4	16	28	694	100	1,013	545	1,193	645
HK31A-6	-	15	26.8	16	28	638	92	1,094	590	1,202	650
HZ32A-5	750-820	15	26.7	16	28	763	110	1,022	550	1,202	650
K1A-F	750-820	15	27	16	28	847	122	1,202	650	1,202	650
QE22A-T6	750-820	15	26.7	16	28	784	113	1,022	550	1,193	645
QH21A-T6	750-820	15	26.7	16	28	784	113	995	535	1,184	640
ZE41A-T5	750-820	14	26	16	28	784	113	977	525	1,193	645
ZE63A-T6	750-820	15	26.5	16	28	756	109	950	510	1,175	635
ZH62A-T5	750-820	15	27.1	16	28	763	110	968	520	1,166	630
ZK61A-T5	705-815	15	27	16	28	833	120	986	530	1,175	635
ZK61A-T6	705-815	15	27	16	28	833	120	986	530	1,175	635

Notes: [1] Process temperature is the casting temperature for sand, permanent mold, and die castings, and the hot-working temperature for extruded materials, forgings, sheet, and plate. [2] Coefficient of thermal expansion, from 32 to 212° F (0 to 100° C) measured in µin./in.-°F, and µm/m-°C. [3] Coefficient of thermal expansion, from 32 to 480° F (0 to 250° C) measured in µin./in.-°F, and µm/m-°C. [4] Metric measurement in W/m-K. English measurement in Btu-in./hr-ft²-°F.

(Continued)

Table 5. (Continued) Thermal Properties of Magnesium Alloys.

ASTM No.	Process Temperature[1] °C	CTE[2] μin./in.	CTE[2] μm/m	CTE[3] μin./in.	CTE[3] μm/m	Thermal Conductivity[4] Engl.	Thermal Conductivity[4] metric	Solidus °F	Solidus °C	Liquidus °F	Liquidus °C
					Die Castings						
AM60A-F	650–695	14	25.6	16	28	430	62	1,004	540	1,139	615
AM60B-F	650–695	14	25.6	16	28	430	62	1,004	540	1,139	615
AS21X1	-	14	26	16	28	-	-	-	-	-	-
AS41XA	650–695	15	26.1	16	28	472	68	1,049	565	1,148	620
AZ91A-F	625–700	14	26	16	28	505	72.7	878	470	1,103	595
AZ91B-F	625–700	14	26	16	28	505	72.7	878	470	1,103	595
K1A-F	750–820	15	27	16	28	847	122	1,202	650	1,202	650
				Extruded Bars, Rods and Shapes							
AZ10A-F	-	15	26.6	16	28	763	110	1,166	630	1,193	645
AZ31B-F	230–425	14	26	16	28	666	96	1,121	605	1,166	630
AZ61A-F	230–400	14	26	16	28	486	70	977	525	1,148	620
AZ80A-F	320–400	14	26	16	28	527	76	914	490	1,130	610
AZ80A-T5	320–400	14	26	16	28	527	76	914	490	1,130	610
HM31A-T5	370–540	14	26	16	28	722	104	1,121	605	1,202	650
M1A-F	-	14	26	16	28	-	-	1,198	648	1,200	649
ZK40A-T5	-	14	26	16	28	763	110	-	-	-	-
ZK60A-F	-	14	26	16	28	833	120	968	520	1,175	635
ZK60A-T5	-	14	26	16	28	833	120	968	520	1,175	635

Notes: [1] Process temperature is the casting temperature for sand, permanent mold, and die castings, the hot-working temperature for extruded materials, forgings, sheet, and plate. [2] Coefficient of thermal expansion, from 32 to 212° F (0 to 100° C) measured in μin./in.-°F, and μm/m-°C. [3] Coefficient of thermal expansion, from 32 to 480° F (0 to 250° C) measured in μin./in.-°F, and μm/m-°C. [4] Metric measurement in W/m-K. English measurement in Btu-in./hr-ft2-°F.

(Continued)

Table 5. *(Continued)* Thermal Properties of Magnesium Alloys.

ASTM No.	Process Temperature[1] °C	CTE[2] µin./in.	CTE[2] µm/m	CTE[3] µin./in.	CTE[3] µm/m	Thermal Conductivity[4] Engl.	Thermal Conductivity[4] metric	Solidus °F	Solidus °C	Liquidus °F	Liquidus °C
					Forgings						
AZ31B	230–425	14	26	16	28	666	96	1,121	605	1,166	630
AZ61A-F	230–400	14	26	16	28	486	70	977	525	1,148	620
AZ80A-F	320–400	14	26	16	28	527	76	914	490	1,130	610
AZ80A-T5	320–400	14	26	16	28	527	76	914	490	1,130	610
AZ80A-T6	320–400	14	26	16	28	527	76	914	490	1,130	610
HM21A-F	455–595	15	26.8	16	28	937	135	1,121	605	1,202	650
HM21A-T5	455–595	15	26.8	16	28	937	135	1,121	605	1,202	650
M1A	-	14	26	16	28	-	-	1,198	648	1,200	649
ZK60A-T5	-	14	26	16	28	833	120	968	520	1,175	635
ZK60A-T6	-	15	27	16	28	833	120	968	520	1,175	635
					Sheet and Sheet and Plate						
AZ31B-O	230–425	14	26	16	28	666	96	1,121	605	1,166	630
AZ31B-H24	230–425	14	26	16	28	666	96	1,121	605	1,166	630
AZ31B-H26	230–425	14	26	16	28	666	96	1,121	605	1,166	630
HK31A-O	-	15	26.8	16	28	638	92	1,094	590	1,202	650
HK31A-H24	-	15	26.8	16	28	638	92	1,094	590	1,202	650
HK21A-T8	455–595	15	26.8	16	28	937	135	1,121	605	1,202	650
HM21A-T81	455–595	15	26.8	16	28	937	135	1,121	605	1,202	650
M1A-F	-	14	26	16	28	-	-	1,198	648	1,200	649

Notes: [1] Process temperature is the casting temperature for sand, permanent mold, and die castings, and the hot-working temperature for extruded materials, forgings, sheet, and plate. [2] Coefficient of thermal expansion, from 32 to 212° F (0 to 100° C) measured in µin./in.-°F, and µm/m-°C. [3] Coefficient of thermal expansion, from 32 to 480° F (0 to 250° C) measured in µin./in.-°F, and µm/m-°C. [4] Metric measurement in W/m·K. English measurement in Btu-in./hr-ft²-°F.

(Continued)

AZ91C/AZ91D/AZ91E. These alloys are available as sand casting (AMS 4437 and AMS 4446), investment casting (AMS 4452), and as a casting alloy as specified by MIL-M-46062. AZ91C is a magnesium base casting alloy containing aluminum and zinc. A slightly purer version, AZ91D, has excellent castability and saltwater resistance and has overtaken the once dominant AZ91B as the workhorse for die castings. AZ91E is a version that contains a significantly lower level of impurities resulting in improved corrosion resistance. These alloys have good castability with a good combination of ductility and strength. AZ91C and AZ91E are the most commonly used sand castings for temperatures under 300° F (149° C). Both have fair weldability and pressure tightness.

AZ92A. A magnesium base casting alloy containing aluminum and zinc. It is available as sand casting (AMS 4434), permanent mold casting (AMS 4484), and as a casting alloy as specified by MIL-M-46062. It is slightly stronger and less ductile than AZ91C, but they are much alike in other characteristics. It has fair weldability and pressure tightness.

EZ33A. A magnesium base casting alloy containing rare earths, zinc, and zirconium. It is available as sand casting (AMS 4442) in the artificially aged (T5) temper. EZ33A has lower strength than the Mg-Al-Zn alloys at room temperature but is less affected by increasing temperatures. It is generally used for applications in the temperature range of 300 to 500° F (149 to 260° C). EZ33A castings are very sound and are sometimes specified for pressure tightness. It has good stability in the T5 temper and excellent weldability. It is sometimes used in applications requiring good dampening ability.

QE22A. A magnesium base casting alloy containing silver, rare earths in the form of didymium, and zirconium. It is available as sand casting (AMS 4418) and as a casting alloy as specified by MIL-M-46062 for sand and permanent type mold castings. It is used in the solution heat treated and artificially aged (T6) condition where a high yield strength is needed at temperatures up to 600° F (316° C). QE22A has good weldability and fair pressure tightness.

ZE41A. A magnesium base casting alloy containing zinc, zirconium, and rare earth elements. It is available as sand (AMS 4439) or permanent mold castings in the artificially aged temper (T5). ZE41A has a higher yield strength than the Mg-Al-Zn alloys at room temperature, and is more stable at elevated temperatures. It is useful for applications at temperatures up to 320° F (160° C). ZE41A castings possess good weldability and are pressure tight.

Temper designation system for magnesium

The following temper designation system is used on all forms of wrought and cast magnesium and magnesium alloy products except ingots. Only operations recognized as significantly influencing the characteristics of the product are indicated.

- –F   *As Fabricated.* Applies to the products of shaping in which no special control over thermal conditions or strain hardening is employed.
- –O   *Annealed Recrystallized (Wrought Products Only).* Applies to wrought products that are annealed to obtain the lowest strength temper.
- –H   *Strain Hardened (Wrought Products Only).* Applies to products that have their strength increased by strain hardening, with or without supplementary thermal treatments to produce some reduction in strength. The H is always followed by two or more digits.
  - –H1   *Strain Hardened Only.* Applies to products that are strain hardened to obtain the desired strength without supplementary thermal treatment. The number following this designation indicates the degree of strain hardening.

-H2 *Strain Hardened and Partially Annealed.* Applies to products that are strain hardened more than the desired final amount and then reduced in strength to the desired level by partial annealing. The number following this designation indicated the degree of strain hardening remaining after the product has been partially annealed.

-H3 *Strain Hardened and Stabilized.* Applies to products that are strain hardened and whose mechanical properties are stabilized by a low temperature thermal treatment to slightly lower strength and increase ductility. The number following this designation indicates the degree of strain hardening remaining after the stabilization treatment.

The digit following the designations –H1, –H2, and –H3 indicates the final degree of strain hardening. Tempers between 0 (annealed) and 8 (full hard) are designated by numerals 1 through 7. Material having an ultimate tensile strength about midway between that of the 0 temper and that of the 8 temper is designated by the numeral 4; about midway between the 0 and 4 tempers by the numeral 2; and about midway between the 4 and 8 tempers by the numeral 6, etc. Number 9 designates tempers whose minimum ultimate tensile strength exceeds that of the 8 temper.

The third digit, when used, indicates a variation of a two-digit temper. It is used when the degree of control of temper or the mechanical properties of both differ from, but are close to, that (or those) for the two digit H temper designation to which it is added. Numerals 1 through 9 may be arbitrarily assigned as the third digit for an alloy and product to indicate a specific degree of control of temper or special mechanical property limits.

-W *Solution Heat Treated.* An unstable temper applicable only to alloys that spontaneously age at room temperature after solution heat treatment. This designation is specific only when the period of natural aging is indicated. For example, W ½ hour.

-T *Thermally Treated to Product Stable Temperatures Other Than –F, –O, or –H.* Applies to products that are thermally treated, with or without supplementary strain hardening, to product stable tempers. The T is always followed by one or more digits.

Numerals 1 through 10 following the –T indicate specific sequences of basic treatments, as follows:

-T1 *Cooled from an Elevated Temperature Shaping Process and Naturally Aged to a Substantially Stable Condition.* Applies to products for which the rate of cooling from an elevated temperature shaping process, such as casting or extrusion, is such that their strength is increased by room temperature aging.

-T2 *Annealed (Castings Only).* Applies to a type of annealing treatment used to improve ductility and increase stability.

-T3 *Solution Heat Treated and Cold Worked.* Applies to products that are cold worked to improve strength after solution heat treatment, or in which the effect of cold work in flattening or straightening is recognized in mechanical property limits.

-T4 *Solution Heat Treated and Naturally Aged to Substantially Stable Condition.* Applies to products that are not cold worked after solution heat treatment, or in which the effect of cold work in flattening or straightening may not be recognized in mechanical property limits.

-T5 *Cooled from an Elevated Temperature Shaping Process and Artificially Aged.* Applies to products that are cooled from an elevated temperature shaping process, such as casting or extrusion, and artificially aged to improve mechanical properties or dimensional stability, or both.

−T6 *Solution Heat Treated and Artificially Aged.* Applies to products that are not cold worked after solution heat treatment, or in which the effect of cold work in flattening or straightening may not be recognized in mechanical property limits.

−T7 *Solution Heat Treated and Stabilized.* Applies to products that are stabilized after solution heat treatment to carry them beyond a point of maximum strength to provide control of some special characteristic.

−T8 *Solution Heat Treated, Cold Worked, and Artificially Aged.* Applies to products that are cold worked to improve strength, or in which the effect of cold work in flattening or straightening is recognized in mechanical property limits.

−T9 *Solution Heat Treated, Artificially Aged, and Cold Worked.* Applies to products that are cold worked to improve strength.

−T10 *Cooled from an Elevated Temperature Shaping Process, Artificially Aged, and Cold Worked.* Applies to products that are artificially aged after cooling from an elevated temperature shaping process, such as extrusion, and cold worked to further improve strength.

Additional digits, the first of which shall not be zero, may be added to designations −T1 through −T10 to indicate a variation in treatment that significantly alters the product characteristics (other than mechanical properties) that are or would be obtained using the basic treatment.

Heat treatment procedures for magnesium alloys

A potential fire hazard exists in the heat treatment of magnesium alloys. If, through oversight or failure of the temperature control equipment, the temperature of the furnace appreciably exceeds the maximum solution heat treating temperature of the alloy, the alloy may ignite and burn. A suitable sulfur dioxide (0.5 – 1.0%) or carbon dioxide (3.0 – 3.5%) atmosphere prevents ignition until the upper allowable limits have been exceeded by a considerable amount. If a fire does ignite, the sulfur dioxide or carbon dioxide supply to the furnace should be shut off, because the burning magnesium unites with the oxygen in these materials. An effective method of extinguishing a magnesium fire in a gas tight furnace is to introduce boron trifluoride gas (BF_3) through a small opening into the closed furnace.

The temperatures for solution treatment shown in **Table 6** are the maximum temperatures to which the alloys may be heated without danger of high temperature deterioration or fusion of the eutectic. The alloys may be heat treated at lower temperatures, but in such cases a longer time at temperature than shown in this table will be necessary in order to develop satisfactory properties.

AZ63A, AZ81A, AZ91C, AM100A, and ZK61A castings will be irreversibly harmed if not brought slowly to the solution heat treating temperature. Certain eutectic constituents in these alloys melt at a temperature lower than that required for the solution heat treatment. Consequently, time should be allowed for the constituents to dissolve before their melting point is reached.

The aging treatments recommended in **Table 7** for "as cast" materials are used to improve mechanical properties, to provide stress relief, and to stabilize the alloys in order to prevent dimensional change later (especially during machining). Both yield strength and hardness are increased somewhat by this treatment at the expense of a slight amount of ductility. This treatment is often recommended for those applications where "as cast" mechanical properties suffice, and dimensional stability is essential. See notes following the table.

Table 6. Temperatures for Solution Heat Treating Magnesium Alloys. *(Source, MIL-M-6857D.)*

Alloy	Temp. Range °F (°C)	Time at Temp. (Hours)[1]	Max. Permissible Temp. °F (°C)
AM100A	790 - 810 (421 - 432)	16 - 24	810 (432)
AZ63A	720 - 735 (382 - 390)	10 - 14	735 (390)
AZ81A	770 - 785 (410 - 418)	16 - 24	785 (418)
AZ91C and E	770 - 785 (410 - 418)	16 - 24	785 (418)
AZ92A	760 - 775 (404 - 413)	16 - 24	775 (413)
ZE63A	900 - 910 (482 - 488)	10 - 72	915 (488)
ZK61A	920 - 935 (493 - 502)	2	935 (502)
HK31A	1,040 - 1,060 (560 - 571)	2	1,060 (571)
QH21A	980 - 1,000 (527 - 538)	4 - 8	1,000 (538)
QE22A [2]	975 - 995 (524 - 535)	4 - 8	1,000 (538)
EQ21A [2]	960 - 980 (516 - 527)	4 - 8	990 (532)
WE54A [2]	970 - 990 (521 - 532)	6 - 8	1,000 (538)

Notes: [1] Heavy section castings, one inch (25.4 mm) thick or over, may require longer times than indicated.
[2] Must be quenched in water held at 140 - 196°F (60 - 91° C) or other suitable media.

Table 7. Aging Treatments for Magnesium Alloys. *(Source, MIL-M-6857D.)*

Alloy and Temper	Time and Temperature Required to Produce Temper °F (°C)
AM100A-T5 [1]	5 hr. at 450° F (232° C)
AM100A-T6 [2]	5 hr. at 450° F (232° C) or 24 hr. at 400° F (204° C)
AZ63A-T5 [1]	4 hr. at 500° F (260° C) or 5 hr. at 450° F (232° C)
AZ63A-T6 [2]	5 hr. at 435° F (224° C) or 5 hr. at 450° F (232° C)
AZ91C-T5 [1]	4 hr. at 425° F (219° C) or 16 hr. at 335° F (169° C)
AZ91C-T6 [2]	4 hr. at 425° F (219° C) or 16 hr. at 335° F (169° C)
AZ91E-T6 [2]	4 hr. at 425° F (219° C) or 16 hr. at 335° F (169° C)
AZ92A-T5 [1]	5 hr. at 450° F (232° C)
AZ92A-T6 [2]	5 hr. at 425° F (219° C)
ZK51A-T5 [1]	8 hr. at 425° F (219° C) or 12 hr. at 350° F (177° C)
ZK61A-T5 [1]	48 hr. at 300° F (149° C)
ZK61A-T6 [2]	48 hr. at 265° F (130° C)
ZE41A-T5 [1]	24 hr. at 480° F (249° C) or 1-6 hr. at 620-680° F (327-360° C) + air cool
ZE63A-T6 [2]	48 hr. at 285° F (141° C)
EZ33A-T5 [1]	5 hr. at 425° F (219° C) or 2 hr. at 650° F (343° C) + 5 hr. at 425° F (219° C)
HK31A-T6 [2, 3]	16 hr. at 400° F (204° C)
HZ32A-T5 [1]	16 hr. at 600° F (316° C)
QH21A-T6 [2]	8 hr. at 400° F (204° C)
QE22A-T6 [2]	8 hr. at 400° F (204° C)
EQ21-T6	8 to 16 hr. at 400° F (204° C)
WE54A-T6 [2]	10 to 20 hr. at 480° F (249° C)

Notes: [1] The T5 temper is obtained by artificially aging from the as-cast (F) temper. [2] The T6 temper is obtained by artificially aging from the solution heat treated (T4) temper. [3] HK318-T4 should be brought to aging temperature as rapidly as possible to minimize grain growth.

Titanium

Material properties

The material properties of titanium and its alloys are mainly determined by their alloy content and heat treatment, both of which are influential in determining the form in which the material is bound. Titanium is one of the few allotropic metals, which means that it can exist in two different crystallographic forms. Under equilibrium conditions, pure titanium has an "alpha" (the "α" symbol is sometimes used) structure up to 1,620° F (900° C), above which it transforms to a "beta" (or "β") structure. The inherent properties of these two structures are quite different. The alpha structure is closely packed and hexagonally shaped, and the beta form is a body centered cubic structure. Through alloying and heat treatment, one or the other or a combination of these two structures can be made to exist at service temperatures, and the properties of the material vary accordingly.

Titanium and titanium alloys of the alpha and alpha-beta type exhibit crystallographic textures in sheet form in which certain crystallographic planes or directions are closely aligned with the direction of prior working. The presence of textures in these materials leads to anisotrophy (exhibiting different values of a property in different directions with respect to a fixed reference system in the material) with respect to mechanical and physical properties—especially Poisson's ratio and Young's modulus. Wide variations experienced in these properties both within and between sheets of titanium alloys have been qualitatively related to variations of texture. In general, the degree of texturing, and hence the variations of Poisson's ratio and Young's modulus, that is developed for alpha-beta alloys tends to be less than that developed in all alpha titanium alloys. Rolling temperature has a pronounced effect on the texturing of titanium alloys which may not in general be affected by subsequent thermal treatments. At present, these variations in mechanical properties have been documented on sheet only, and may not be present in other products. **Tables 1**, **2**, and **3** give composition, mechanical properties, and thermal properties of selected titanium alloys.

Beta transus temperature. The descriptions of individual unalloyed and alloyed titanium materials in this section often contain references to their beta transus temperature. This temperature is determined by metallographically examining quenched specimens after solution heat treating. Magnification of 500X is used to determine the amount of primary phase still present. The temperature at which this phase is no longer present is deemed the beta transus temperature. Beta transus temperatures are given in **Table 3**, but it should be remembered that the composition of the alloy will impact the temperature, and that beta transus temperatures may, and often do, vary from lot to lot.

An important engineering characteristics of titanium is that it is about 40% lighter than steel and 60% heavier than aluminum, but its high strength and moderate weight give it the highest strength-to-weight ratio of any structural metal. Another plus is that titanium is naturally corrosion resistant, due to a tough surface film of oxide.

Titanium alloy product designations reflect the alloy's constituent elements, and the percentage of each element contained in the alloy. Since these designations can become rather complex (as in Ti-6Al-2Sn-4Zr-6Mo) they are often identified by abbreviated designations (such as Ti-6246 for the example just given) or by their UNS number (R56260 for this example). The source of much of the information in this section was MIL-HDBK-5H.

Environmental considerations

Below 300° F (149° C), and above 700° F (371° C), creep deformation of titanium alloys can be expected at stresses below the yield strength. Room temperature creep

Table 1. Chemical Composition of Titanium and Titanium Alloys.

Designation	UNS No.	Nominal Composition — Expressed as Percent by Weight
Unalloyed Grades		
Grade 1	R50250	C 0.1 max.; Fe 0.2 max.; H 0.015 max.; N 0.03 max.; O 0.18 max.; Ti 99.5
Grade 2	R50400	C 0.1 max.; Fe 0.3 max.; H 0.015 max.; N 0.03 max.; O 0.25 max.; Ti 99.2
Grade 3	R50550	C 0.1 max.; Fe 0.3 max.; H 0.015 max.; N 0.05 max.; O 0.35 max.; Ti 99.1
Grade 4	R50700	C 0.1 max.; Fe 0.5 max.; H 0.015 max.; N 0.05 max.; O 0.4 max.; Ti 99
Grade 7	R52400	C 0.1 max.; Fe 0.3 max.; H 0.015 max.; N 0.03 max.; O 0.25 max.; Pd 0.2; Ti 99
Grade 11	R52250	Pd 0.2; Ti 99
Grade 12	R53400	Mo 0.3; Ni 0.8; Ti 99
Alpha and Near-Alpha Alloys		
Ti-3Al-2.5V	R56320	Al 3; Ti 95; V 2.5
Ti-5Al-2.5Sn	R54520	Al 5; Fe 0.5 max.; O 0.2 max.; Sn 2.5; Ti 92.5
Ti-5Al-2.5Sn, ELI	R54521	Al 5; Fe 0.25 max.; O 0.12 max.; Sn 2.5; Ti 92.5
Ti-6Al-2Nb-1Ta-0.8Mo	R56210	Al 6; Mo 0.8; Nb 2; Sn 2.5; Ta 1; Ti 90
Ti-6Al-2Sn-4Zr-2Mo	R54620	Al 6; Mo 2; Ti 88; Zr 4
Ti-8Al-1Mo-1V	R54810	Al 8; Mo 1; Ti 90; V 1
Alpha-Beta Alloys		
Ti-4Al-4Mo-2Sn-0.5Si	-	Al 4; Mo 4; Si 0.5; Sn 2; Ti 89
Ti-6Al-4V	R65400	Al 6; Fe 0.25 max.; O 0.2 max.; Ti 90; V 4
Ti-6Al-4V, ELI	R65400	Al 6; Fe 0.14 max.; O 0.13 max.; Ti 90; V 4
Ti-6Al-6V-2Sn	R56620	Al 6; Sn 2; Ti 86; V 6
Ti-6Al-2Sn-2Zr-2Mo-2Cr-.25Si	-	Al 6; Cr 2; Mo 2; Si 0.25; Sn 2; Ti 86; Zr 2
Beta Alloys		
Ti-8Mo-8V-2Fe-3Al	-	Al 3; Fe 2; Mo 8; Ti 79; V 8
Ti-15Mo-5Zr-3Al	-	Al 3; Mo 15; Ti 77; Zr 5
Ti-10V-2Fe-3Al	-	Al 3; Fe 2; Ti 85; V 10
Ti-13V-2.7Al-7Sn-2Zr	-	Al 2.7; Sn 7; Ti 75; V 13; Zr 2
Beta III	R58030	Mo 11.5; Sn 4.5; Ti 78; Zr 6

of unalloyed titanium may exceed 0.2 percent creep-strain in 1,000 hours at stresses that exceed 50% of the design tensile yield stress. While the characteristics of the alloys are generally superior, creep should be a consideration when specifying titanium or titanium alloys.

The use of titanium and its alloys in contact with either liquid oxygen or gaseous oxygen at cryogenic temperatures should be avoided, since either the presentation of a fresh surface (such as produced by tensile rupture) or impact may initiate a violent reaction. Impact of the surface in contact with liquid oxygen will result in a reaction even at very low energy levels. In gaseous oxygen, a partial pressure of 50 PSI is sufficient to ignite a fresh titanium surface over the temperature range of –250° F (–157° C) to room temperature

(Text continued on p. 654)

Table 2. Mechanical Properties of Titanium Alloys.

Designation	Ultimate Tensile Strength		Yield Strength		Compressive Yield Strength		Modulus of Elasticity		Shear Modulus		Elong. %[1]	Hardness
	PSI	MPa	PSI	MPa	PSI	MPa	ksi	MPa	ksi	GPa		
Unalloyed Grades												
Grade 1	34,809	240	24,656	170	-	-	15,229	105	6,525	45	24	RA 70
Grade 1, Annealed	47,863	330	34,809	240	49,313	340	14,504	100	5,510	38	30	HB 120
Grade 2	49,893	344	39,885	275	-	-	15,229	105	6,525	45	20	RB 80
Grade 3	63,817	440	55,114	380	65,267	450	15,229	105	6,525	45	18	RC 16
Grade 4	79,771	550	69,618	480	-	-	15,229	105	5,800	40	15	RC 23
Grade 7	49,893	344	39,885	275	-	-	15,229	105	6,525	45	20	RB 75
Grade 11	34,809	240	24,656	170	-	-	15,229	105	6,525	45	24	-
Grade 12	65,267	450	55,114	380	-	-	15,229	105	6,525	45	12	HB 200
Alpha and Near-Alpha Alloys												
Ti-3Al-2.5V [2]	89,924	620	72,519	500	100,076	690	14,504	100	6,380	44	15	RC 24
Ti-5Al-2.5Sn	124,878	861	119,946	827	120,382	830	16,679	115	6,960	48	15	RC 36
Ti-5Al-2.5Sn, ELI [3]	112,404	775	104,427	720	-	-	15,954	110	6,960	48	12	RC 33
Ti-6Al-2Nb-1Ta-0.8Mo	120,382	830	110,229	760	118,931	820	16,969	117	7,395	51	10	RC 30
Ti-6Al-2Sn-4Zr-2Mo [4]	146,488	1,010	143,588	990	156,641	1,080	17,405	120	7,540	52	3	RC 34
Ti-8Al-1Mo-1V	135,901	937	131,985	910	-	-	17,405	120	6,670	46	18	RC 36
Alpha-Beta Alloys												
Ti-4Al-4Mo-2Sn-0.5Si [5]	159,542	1,100	136,336	940	-	-	16,679	115	7,105	49	7	360

Notes: [1] Elongation percentage at break. [2] Alpha annealed. [3] Annealed. [4] Sheet. [5] Solution treated and aged (STA). [6] Alpha-beta processed. [7] Solution treated. [8] Annealed and aged.

(Continued)

Table 2. *(Continued)* Mechanical Properties of Titanium Alloys.

Designation	Ultimate Tensile Strength		Yield Strength		Compressive Yield Strength		Modulus of Elasticity		Shear Modulus		Elong. %[1]	Hardness
	PSI	MPa	PSI	MPa	PSI	MPa	ksi	MPa	ksi	GPa		
Alpha-Beta Alloys												
Ti-5Al-2Sn-2Zr-4Mo-4Cr[6]	171,870	1,185	165,343	1,140	-	-	16,679	115	7,105	49	8	RC 40
Ti-6Al-4V[7]	137,786	950	127,633	880	140,687	970	16,505	113.8	6,380	44	14	RC 36
Ti-6Al-4V, ELI[3]	124,733	860	114,580	790	124,733	860	16,505	113.8	6,380	44	15	RC 35
Ti-6Al-6V-2Sn[3]	152,290	1,050	142,137	980	-	-	15,998	110.3	6,525	45	14	RC 38
Ti-6Al-2Sn-4Zr-6Mo[5]	229,160	1,580	204,504	1,410	-	-	16,534	114	7,105	49	4	RC 39
Ti-6Al-2Sn-2Zr-2Mo-2Cr-.25Si[3]	149,389	1,030	140,687	970	-	-	17,695	122	6,670	46	-	-
Beta Alloys												
Ti-8Mo-8V-2Fe-3Al[5]	188,549	1,300	-	-	-	-	15,954	110	6,815	47	3	410
Ti-15Mo-5Zr-3Al[5]	213,931	1,475	-	-	-	-	14,504	100	6,235	43	14	412
Ti-10V-2Fe-3Al[7]	145,038	1,000	134,885	930	-	-	15,954	110	5,945	41	17	-
Ti-13V-2.7Al-7Sn[5]	166,794	1,150	159,542	1,100	-	-	15,229	105	6,525	45	6	RC 39
Ti-13V-11Cr-3Al[7]	143,588	990	120,382	830	-	-	14,359	99	6,235	43	23	RC 30
Ti-13V-11Cr-3Al[8]	175,496	1,210	165,343	1,140	155,191	1,070	15,954	110	6,235	43	7	RC 43

Notes: [1] Elongation percentage at break. [2] Alpha annealed. [3] Annealed. [4] Sheet. [5] Solution treated and aged (STA). [6] Alpha-beta processed. [7] Solution treated. [8] Annealed and aged.

(Continued)

TITANIUM ALLOYS 651

Table 2. *(Continued)* **Mechanical Properties of Titanium Alloys.**

Designation	Ultimate Tensile Strength		Yield Strength		Compressive Yield Strength		Modulus of Elasticity		Shear Modulus		Elong. % [1]	Hardness
	PSI	MPa	PSI	MPa	PSI	MPa	ksi	MPa	ksi	GPa		
					Beta Alloys							
Ti-15V-3Cr-3Al-3Sn [7]	114,580	790	111,679	770	-	-	11,893	82	5,075	35	22	RB 95
Beta III [5]	182,023	1,255	171,145	1,180	162,443	1,120	15,664	108	5,945	41	5	325

Notes: [1] Elongation percentage at break. [2] Alpha annealed. [3] Annealed. [4] Sheet. [5] Solution treated and aged (STA). [6] Alpha-beta processed. [7] Solution treated. [8] Annealed and aged.

Table 3. Thermal Properties of Titanium Alloys.

Designation	Beta Transus		CTE		CTE		Thermal Conductivity[1]		Solidus		Liquidus	
	°F	°C	µin/in	µm/m	µin/in	µm/m	Engl.	metric	°F	°C	°F	°C
							Unalloyed Grades					
Grade 1	1630	888	4.8 [2]	8.6 [2]	5.4 [3]	9.7 [3]	111	16	-	-	3,038	1,670
Grade 1, Annealed	1630	888	4.8 [4]	8.6 [4]	5.1 [5]	9.2 [5]	111	16	-	-	3,038	1,670
Grade 2	1675	913	4.8 [2]	8.6 [2]	5.4 [5]	9.7 [5]	114	16.4	-	-	3,029	1,665
Grade 3	1688	920	4.8 [2]	8.6 [2]	5.1 [5]	9.2 [5]	138	19.9	-	-	3,020	1,660
Grade 4	1742	950	4.8 [2]	8.6 [2]	5.1 [5]	9.2 [5]	119	17.2	-	-	3,020	1,660
Grade 7	1675	913	4.8 [2]	8.6 [2]	5.1 [5]	9.2 [5]	114	16.4	-	-	3,029	1,665
Grade 11	1630	888	4.8 [2]	8.6 [2]	5.1 [5]	9.2 [5]	118	17	-	-	3,038	1,670
Grade 12	1634	890	4.8 [2]	8.6 [2]	5.3 [5]	9.5 [5]	132	19	-	-	3,020	1,660

Notes: **CTE** is Coefficient of Thermal Expansion measured in µin/in.-°F, and µm/m-°C. [1] Metric measurement for thermal conductivity in W/m-K; English measurement in Btu-in./hr-ft2-°F. [2] CTE from 32 to 212° F (0 to 100° C). [3] CTE from 32 to 1,000° F (0 to 540° C). [4] CTE from -7 to 200° F (20 to 93° C). [5] CTE from 32 to 600° F (0 to 315° C). [6] Alpha annealed. [7] CTE from -7 to 600° F (20 to 315° C). [8] Annealed. [9] CTE from -7 to 1,200° F (20 to 650° C). [10] Sheet. CTE from 32 to 480° F (0 to 250° C). [12] CTE from 32 to 930° F (0 to 500° C). [13] Solution treated and aged (STA). [14] Alpha-beta processed. [15] CTE from -7 to 750° F (20 to 400° C). [16] Solution treated. [17] Annealed and aged.

(Continued)

Table 3. (*Continued*) Thermal Properties of Titanium Alloys.

Designation	Beta Transus °F	Beta Transus °C	CTE μin/in	CTE μm/m	CTE μin/in	CTE μm/m	Thermal Conductivity[1] Engl.	Thermal Conductivity[1] metric	Solidus °F	Solidus °C	Liquidus °F	Liquidus °C
					Alpha and Near-Alpha Alloys							
Ti-3Al-2.5V [6]	1715	935	5.3 [4]	9.61 [4]	5.5 [7]	9.86 [7]	58	8.3	-	-	1,700	3,092
Ti-5Al-2.5Sn	1904-1994	1040-1090	5.2 [2]	9.4 [2]	5.3 [7]	9.5 [7]	54	7.8	-	-	2,894	1,590
Ti-5Al-2.5Sn, ELI [8]	1904-1994	1040-1090	5.2 [2]	9.4 [2]	5.3 [7]	9.5 [7]	54	7.8	-	-	2,894	1,590
Ti-6Al-2Nb-1Ta-0.8Mo	2030	1110	5.1 [4]	9.2 [4]	5 [9]	9 [9]	44	6.4	-	-	3,002	1,650
Ti-6Al-2Sn-4Zr-2Mo [10]	1814	990	4.3 [4]	7.7 [4]	4.5 [7]	8.1 [7]	49	7.1	-	-	3,092	1,700
Ti-8Al-1Mo-1V	1904	1040	5.1 [11]	9.2 [11]	5.6 [12]	10.2 [12]	42	6	-	-	2,804	1,540
					Alpha-Beta Alloys							
Ti-4Al-4Mo-2Sn-0.5Si [13]	1787	975	4.9 [4]	8.8 [4]	5.1 [7]	9.2 [7]	52	7.5	-	-	3,002	1,650
Ti-5Al-2Sn-2Zr-4Mo-4Cr [14]	1634	890	4.7 [4]	8.5 [4]	5.4 [15]	9.7 [15]	54	7.8	-	-	-	-
Ti-6Al-4V [16]	1796	980	4.8 [4]	8.6 [4]	5.1 [7]	9.2 [7]	46	6.7	2,919	1,604	3,020	1,660
Ti-6Al-4V, ELI [8]	1796	980	5.1 [7]	9.2 [7]	5.4 [9]	9.7 [9]	46	6.7	2,919	1,604	3,020	1,660
Ti-6Al-6V-2Sn [8]	1733	945	5 [4]	9 [4]	5.2 [7]	9.4 [7]	46	6.6	2,961	1,627	3,000	1,649
Ti-6Al-2Sn-4Zr-6Mo [13]	1715	935	5.5 [11]	9.9 [11]	5.8 [3]	10.4 [3]	53	7.7	2,903	1,595	3,047	1,675
Ti-6Al-2Sn-2Zr-2Mo-2Cr-0.25Si [8]	1760	960	5.2 [4]	9.4 [4]	5.1 [7]	9.2 [7]	49	7	-	-	3,002	1,650

Notes: **CTE** is Coefficient of Thermal Expansion measured in μin./in.-°F, and μm/m-°C. [1] Metric measurement for thermal conductivity in W/m-K; English measurement in Btu-in./hr-ft2-°F. [2] CTE from 32 to 212° F (0 to 100° C). [3] CTE from 32 to 1,000° F (0 to 540° C). [4] CTE from -7 to 200° F (20 to 93° C). [5] CTE from 32 to 600° F (0 to 315° C). [6] Alpha annealed. [7] CTE from -7 to 600° F (20 to 315° C). [8] Annealed. [9] CTE from -7 to 1,200° F (20 to 650° C) Sheet. [11] CTE from 32 to 480° F (0 to 250° C). [12] CTE from 32 to 930° F (0 to 500° C). [13] Solution treated and aged (STA). [14] Alpha-beta processed. [15] CTE from -7 to 750° F (20 to 400° C). [16] Solution treated. [17] Annealed and aged.

(Continued)

Table 3. *(Continued)* Thermal Properties of Titanium Alloys.

Designation	Beta Transus		CTE		CTE		Thermal Conductivity[1]		Solidus		Liquidus	
	°F	°C	µin/in	µm/m	µin/in	µm/m	Engl.	metric	°F	°C	°F	°C
Beta Alloys												
Ti-8Mo-8V-2Fe-3Al [13]	1427	775	4.7 [4]	8.5 [4]	-	-	52	7.5	-	-	-	-
Ti-15Mo-5Zr-3Al [13]	1445	785	4.7 [4]	8.5 [4]	-	-	52	7.5	-	-	-	-
Ti 10V-2Fe-3Al [16]	1400	760	5.4 [4]	9.7 [4]	-	-	54	7.8	-	-	-	-
Ti-13V-2.7Al-7Sn [13]	1382	750	4.7 [4]	8.5 [4]	-	-	54	7.8	-	-	-	-
Ti-13V-11Cr-3Al [16]	1292	700	5.4 [4]	9.67 [4]	5.6 [7]	9.99 [7]	48	6.9	-	-	-	-
Ti-13V-11Cr-3Al [17]	1292	700	5.4 [4]	9.67 [4]	-	-	48	6.9	-	-	-	-
Ti-15V-3Cr-3Al-3Sn [16]	1400	760	4.7 [4]	8.5 [4]	5.1 [11]	9.1 [11]	56	8.08	-	-	-	-
Beta III [13]	1400	760	4.2 [4]	7.6 [4]	4.5 [7]	8.1 [7]	44	6.28	2,863	1,572	3,074	1,690

Notes: **CTE** is Coefficient of Thermal Expansion measured in µin./in.-°F, and µm/m-°C. [1] Metric measurement for thermal conductivity in W/m-K; English measurement in Btu-in./hr-ft2-°F. [2] CTE from 32 to 212° F (0 to 100° C). [3] CTE from 32 to 1,000° F (0 to 540° C). [4] CTE from -7 to 200° F (20 to 93° C). [5] CTE from 32 to 600° F (0 to 315° C). [6] Alpha annealed. [7] CTE from -7 to 600° F (20 to 315° C). [8] Annealed. [9] CTE from -7 to 1,200° F (20 to 650° C) [10] Sheet. [11] CTE from 32 to 480° F (0 to 250° C). [12] CTE from 32 to 930° F (0 to 500° C). [13] Solution treated and aged (STA). [14] Alpha-beta processed. [15] CTE from -7 to 750° F (20 to 400° C). [16] Solution treated. [17] Annealed and aged.

or higher. Also, under certain conditions, titanium, when in contact with cadmium, silver, mercury, or certain of their compounds, may become embrittled.

Titanium is susceptible to stress corrosion cracking in certain anhydrous chemicals including methyl alcohol and nitrogen tetroxide. Traces of water tend to inhibit the reaction in either environment. Red fuming nitric acid with less than 1.5% water and 10 to 20% NO_2 can crack the metal and result in a pyrophoric reaction. Titanium alloys are also susceptible to stress corrosion by dry sodium chloride at elevated temperatures. Cleaning with a nonchlorinated solvent to remove salt deposits (including fingerprints) is recommended on parts used above 450° F (232° C).

Unalloyed titanium

Unalloyed (commercially pure) titanium is available in all familiar product forms and is noted for its excellent formability, but it is used primarily where strength is not the primary requirement. It has excellent corrosion resistance properties. These materials are supplied in the annealed condition, which permits extensive forming at room temperature. Severe forming operations can also be accomplished at elevated temperatures (300 to 900° F [149 to 482° C]). Property degradation can be experienced after severe forming if as-received material properties are not restored by reannealing. They can be readily welded by several methods. Atmospheric shielding is preferable although spot or seam welding may be accomplished without shielding. Brazing requires protection from the atmosphere, which may be obtained by fluxing as well as by inert gas or vacuum shielding.

Titanium has an unusually high affinity for oxygen, nitrogen, and hydrogen temperature above 1,050° F (566° C). As this results in embrittlement of the material, usage should be limited to temperatures below that indicated. Additional chemical reactivity between titanium and selected environments such as methyl alcohol, chloride salt solutions, hydrogen, and liquid metal can take place at lower temperatures.

Commercially pure titanium is fully annealed by heating to 1,000 to 1,300° F (538 to 704° C) for 10 to 30 minutes. These materials cannot be hardened by heat treatment.

Alpha and near-alpha titanium alloys

The alpha titanium alloys contain essentially a single phase at room temperature, similar to that of unalloyed titanium. Alloys identified as near-alpha titanium have principally an all-alpha structure but contain small quantities of a beta phase because the composition contains some beta stabilizing elements. In both alloy types, alpha phase is stabilized by aluminum, tin, and zirconium. These elements, especially aluminum, contribute greatly to strength. The beta stabilizing additions (molybdenum and vanadium) improve fabricability and metallurgical stability of highly alpha alloyed materials.

These alloys have toughness at extremely low temperatures as well as long-term strength at elevated temperatures—they are well suited to cryogenic applications and uses requiring good elevated temperature creep strength. The characteristics of near-alpha alloys are predictably between those of all-alpha and alpha-beta alloys in regard to fabricability, weldability, and elevated temperature strength. The hot workability of both alpha and near-alpha alloys is inferior to that of the alpha-beta or beta alloys, and the cold workability is very limited at the high strength level of these grades. However, considerable forming is possible if correct forming temperatures and procedures are used.

Selected Alloys

Ti-5Al-2.5Sn. This all-alpha alloy is available in many product forms and at two purity levels. The high purity grade is used principally for cryogenic applications and may be characterized as having lower strength but higher ductility and toughness than the standard grade. The normal purity grade may also be used at low temperatures but is primarily suited for room-to-elevated-temperature applications (up to 900° F [482° C] or to 1,100° F [593° C] for short periods), where weldability is an important consideration. *Manufacturing Considerations*: Ti-5Al-2.5Sn is not so readily formed into complex shapes as other alloys with similar room temperature properties, but it far surpasses them in weldability—inert gas, vacuum shielded arc, or spot and seam accomplished without atmospheric shielding. Brazing requires protection from the atmospheres. Except for some forging operations, fabrication is conducted at temperatures where the structure remains all-alpha. Severe forming operations may be accomplished at temperatures up to 1,200° F (649° C). Moderately severe forming can be done at 300 to 600° F (149 to 316° C0, and simple forming can be done at room temperature. Most forming and welding operations are followed by an annealing treatment to relieve residual stress. Standard grade has been used at moderately low cryogenic temperatures, but the ELI (extra low interstitial) grade has higher toughness and has been used in temperatures down to –423° F (–253° C). *Heat Treatment*: Anneal by heating to 1,400° F (760° C) for 60 minutes and 1,600° F (871° C) for 10 minutes and cooling in air. Stress relieving requires 1 to 2 hours at 1,000 to 1,200° F (538 to 649° C). This alloy cannot be hardened by heat treatment.

Ti-8Al-1Mo-1V. This near-alpha composition was developed for improved creep resistance and thermal stability up to about 850° F (454° C). It is available a billet, bar, plate, sheet, strip, extrusion, and forgings. *Manufacturing Considerations*: Room temperature forming of sheet is somewhat more difficult than with Ti-6Al-4V, and hot forming is required for severe operations. Ti-8Al-1Mo-1V can be readily fusion welded with inert gas protection or spot welding without atmosphere protection. Weld strengths are comparable to those of the parent metal although ductility is somewhat lower in the weldment. Oxidation resistance and thermal stability up to 850° F (454° C) are good, but extended exposure to temperatures in excess of 600° F (316° C) adversely affects room temperature spot weld tension strength. Not recommended for structural applications at liquid hydrogen temperatures (–423° F [–253° C]), and the alloy is susceptible to chloride stress corrosion attack in either elevated temperature or ambient temperature chloride environments. *Heat Treatment*: Three treatments are used with this alloy. 1) Single anneal: 1,450° F (788° C) for 8 hours, surface cool. 2) Duplex anneal: 1,450° F for 8 hours, furnace cool, then 1,450° F for 15 to 20 minutes, air cool. 3) Solution treated and stabilized. 1,825° F (996° C) for 1 hour, air cool, then 1,075°F (579° C) for 8 hours, air cool. The single anneal is used to obtain the highest room temperature mechanical properties. The duplex anneal is used to obtain the highest fracture toughness. Both the single anneal and the duplex anneal are compatible with hot forming operations. The solution treated and stabilized condition is used for forgings.

Ti-6Al-2Sn-4Zr-2Mo(+Si). This near-alpha titanium composition was developed for improved elevated temperature performance. Initially, it was used in advanced performance gas turbine engine applications, and in the Si version alloy contains approximately 0.08% silicon to improve creep strength without affecting thermal stability. With silicon, it is creep resistant and relatively stable to about 1,050°F (565° C). It is available in bar, billet, plate, sheet, strip, and extrusion. This is the most commonly used titanium alloy for elevated temperature operations. *Manufacturing Considerations*. Elevated temperatures may be used for severe sheet forming operations, while room temperature forming may be used for mild contouring. The material can be welded using TIG or MIG fusion processes to achieve 100% joint efficiencies but with limited weld zone ductility. As in welding any titanium alloy, shielding from atmospheric contamination is required except for spot and seam welding. Forging at temperatures below the beta transus temperature is recommended—for optimum creep properties, beta forging or a modification of the process is recommended with some loss of ductility likely. *Heat Treatment*: Several treatments are available, as follows. For <u>sheet</u> and <u>strip</u>: 1) Duplex anneal: 1,650° F (899° C) for $^1/_2$ hour, air cool, then 1,450° F (788° C) for $^1/_4$ hour, air cool. 2) Triplex anneal: 1,650° F for $^1/_2$ hour, air cool, then 1,450° F for $^1/_4$ hour, air cool, then 1,100° F (593° C) for 2 hours, air cool. For <u>plate</u>: 1) Duplex anneal: 1,650° F for 1 hour, air cool, then 1,100° F for 8 hours, air cool. 2) Triplex anneal: 1,650° F for $^1/_2$ hour, air cool, then 1,450° F for 2 hours, air cool. For <u>bars</u> and <u>forgings</u>: 1) Duplex anneal: Solution anneal 25 to 50° (14 to 28° C) below beta transus temperature for 1 hour, air cool (or faster), then 1,100° F for 8 hours, air cool.

Alpha-beta titanium alloys

These alloys contain both alpha and beta phases at room temperature. The alpha phase is similar to that of unalloyed titanium but is strengthened by alpha stabilizing additions (aluminum). The beta phase is the high temperature phase of titanium, but is stabilized to room temperature by sufficient quantities of beta stabilizing elements such as vanadium, molybdenum, iron, or chromium. In addition to strengthening of titanium by alloying additions, alpha-beta alloys may be further strengthened by heat treatment. Alpha-beta alloys have good strength at room temperature and for short times at elevated temperatures, but they are not noted for long-time creep strength. With the exception of annealed Ti-6Al-4V, these alloys are not recommended for cryogenic applications. Because of their two-phase microstructure, the weldability of most alloys in this group is relatively poor.

Selected Alloys

Ti-6Al-4V. This alloy is available in all mill product forms as well as castings and powder metallurgy forms. It can be used in either the annealed or solution treated plus aged (STA) conditions. For maximum toughness, it should be used in the annealed or duplex annealed conditions. For maximum strength, the STA condition should be used. The useful temperature range for this alloy is −320 to +750° F (−196 to 399° C). This alloy is responsible for about 45% of all industrial applications of titanium alloys. *Manufacturing Considerations*: Forging should be done above the beta transus temperature using procedures to promote a high toughness material. It is routinely finished below the beta transus temperature for good combinations of fabricability, strength, ductility, and toughness. Elevated temperatures are used to form flat rolled products, but extensive forming may be accomplished at room temperature. This alloy can be spot welded and, in certain applications, fusion welded. Stress relief annealing after welding is recommended. It is resistant to hot-salt stress corrosion to about its maximum use temperature, but it is marginally susceptible to aqueous chloride solution stress corrosion. *Heat Treatment*: This alloy is commonly specified in either the annealed condition or in the fully heat treated condition. Annealing requires 1 hour at 1,300° F (704° C) followed by furnace cooling if maximum ductility is required. The specified fully heat treated, or solution treated and aged (STA), condition for 1) <u>sheet</u> is as follows. Solution treat at 1,700° F (927° C) for 5 to 25 minutes, quench in water. Age at 975° F (524° C) for 4 to 6 hours, air cool. For 2) <u>bars</u> and <u>forgings</u>, the procedure is: Solution treat at 1,700° F for one hour, quench in water. Age at 1,000° F (538° C) for 3 hours, air cool.

Ti-6Al-6V-2Sn. Although similar to Ti-6Al-4V in many respects, this alloy has higher strength and deeper hardenability. Available forms include billet, bar, plate, sheet, strip, and extrusions, and these may be used in either the annealed or the STA condition. Maximum strength is developed in SAT condition in sections up to about two inches in thickness. *Manufacturing Considerations*. To ensure optimum properties for forgings, at least 50% reduction should be done at temperatures below the beta transus temperature (<1,730°

F [<945° C]). This alloy is readily formable in the annealed condition. In sheet or plate forms, the alloy is generally used in the annealed condition, although it is capable of heat treatment to higher strength levels with some loss of toughness. When sheet or plate is hot formed at any temperature over 1,000° F (538° C) and air cooled, it should be restabilized by reheating to 1,000° F followed by air cooling. Welding is not recommended although limited weld joining operations are possible if the assembly can be thermally treated after welding to restore ductility to the weld and heat affected zones. Creep strength above 650° F (343° C) and long-term stability at temperatures above 800° F (427° C) are not good. Oxidation resistance is satisfactory in short-term exposures to 1,000° F, and this alloy is nearly equivalent to Ti-6Al-4V in terms of hot salt and aqueous chloride solution stress corrosion resistance. *Heat Treatment*: This alloy is commonly specified in either the annealed condition or the STA condition. The STA procedure is as follows: Solution treat at 1,625° F (885° C) for $^1/_2$ to 1 hour, quench in water. Age at 975 to 1,025° F (524 to 552° C) for 4 to 8 hours, air cool.

Beta, near-beta, and metastable-beta titanium alloys

Although there is no clear-cut definition for beta titanium alloys, conventional terminology usually refers to near-beta alloys and metastable-beta alloys as classes of beta titanium alloys. A near-beta alloy is generally one that has appreciably higher beta stabilizer content than a conventional alpha-beta alloy, but is not quite sufficiently stabilized to readily retain an all beta structure with an air cool of thin sections. Instead, a water quench is required, even for thin sections. Metastable alloys are even more heavily alloyed with beta stabilizers than near-beta alloys and readily retain an all beta structure upon air cooling of thin sections. Due to the added stability of these alloys, it is not necessary to heat treat below the beta transus to enrich the beta phase. Therefore, these alloys do not normally contain primary alpha since they are usually solution treated above the beta transus. They are termed "metastable" because the resultant beta phase is not really stable—it can be aged to precipitate alpha for strengthening purposes.

Stable beta alloys are so heavily alloyed with beta stabilizers that the beta phase will not decompose to alpha plus beta upon subsequent aging. An example would be Ti-30Mo, but there are presently no such alloys being produced commercially.

Selected Alloys

Ti-13V-11Cr-3Al. This is a heat treatable alloy possessing good workability and toughness in the annealed condition and high strength in the heat treated condition. It is noted for its exceptional ability to harden in heavy sections (up to 6 inch [15.24 cm] or greater) to tensile strengths of 170 ksi (1172 MPa). *Manufacturing Considerations*: It has good formability at room temperature, and stretch forming is conducted at 500° F (260° C). It is readily spot welded, and arc welding will provide very ductile joints that have low strength. Stability is good up to 550° F (288° C) in the annealed condition, and up to 600° F (316° C) in the SAT condition. Prolonged exposures above these temperatures may result in loss of ductility. Hot salt stress corrosion is possible at temperatures as low as 500° F (260° C) in highly stressed applications such as rivet heads. *Heat Treatment*: This alloy is commonly specified in either the annealed condition or in the fully heat treated condition. The full heat treatment procedure is: Solution heat at 1,450° F (788° C) for 15 to 60 minutes, air cool (water quench if material is over 2 inches (50.8 mm) thick). Age at 900° F (482° C) for 2 to 60 hours depending on strength level.

Ti-15V-3Cr-3Sn-3Al (Ti-15-3). This is a metastable beta-titanium alloy. It was developed primarily to lower the cost of titanium sheet metal parts by reducing materials and processing costs. Unlike conventional alpha-beta alloys, this alloy is strip producible and has excellent room temperature formability characteristics. It can be aged to a wide range of strength levels to meet a variety of application needs. Although originally developed as a sheet alloy, it has expanded into other areas such as fasteners, foil, tubing, castings, and forgings. *Manufacturing Considerations*: Usually supplied in the solution annealed form, this alloy is easily cold formed due to its single phase (beta) structure. After cold forming, it can be resolution treated in the 1,450 to 1,500° F (788 to 816° C) range and subsequently aged in the 900 to 1,100° F (482 to 593° C) range, depending upon desired strength. Care should be taken to ensure that no surface contamination results from the solution treatment. The alloy can also be hot formed, but heating time prior to forming should be minimized to prevent aging prior to forming. It is readily welded by standard titanium welding techniques. This alloy should not be used in the solution treated condition. Long time exposure of solution treated and cold worked material to service temperatures of approximately 300° F (149° C), or solution treated material to service temperatures above approximately 400° F (204° C), can result in low ductility. *Heat Treatment*: This alloy should be solution treated for 10 to 30 minutes in the 1,450 to 1,550° F (788 to 816° C) range, cooled at a rate approximating an air cool of 0.125 inch (3.175 mm) sheet and subsequently aged. Aging is generally conducted in the 900 to 1,100° F (482 to 593° C) range, followed by an air cool.

Ti-10V-2Fe-3Al (Ti-10-2-3). This is a solute lean beta (near-beta) titanium alloy that was developed primarily as a high strength forging alloy. Its forging characteristics are excellent, possessing flow properties at 1,500° F (816° C) similar to Ti-6Al-4V at 1,700° F (927° C), allowing for the possibility of lower die costs and better die fill capacity. This alloy provides the best combination of strength and toughness of any commercially available titanium alloy, and it is considered to be deep hardenable, capable of generating high strengths in section thicknesses up to approximately 5 inches (12.5 cm). *Manufacturing Considerations*: Ti-10-2-3 is usually supplied as bar or billet that has been finish forged or rolled in the alpha-beta field. In order to optimize the microstructure for the high strength condition, the forging is usually given a pre-form forge above the beta transus, followed by a 15 to 25% reduction below the beta transus. Ideally, the beta forging operation is finished through the beta transus, followed by a quench. The intent of the two-step forging process is to develop a structure without grain boundary alpha, but with elongated primary alpha needles in an aged beta matrix. The alloy is readily weldable by conventional titanium welding techniques. In the solution treated plus aged condition, the material exhibits excellent resistance to stress corrosion cracking. In the solution treated condition, it should not be subjected to long-term exposure in the 500 to 800° F (260 to 427° C) range, since such exposure could result in high strength, low ductility conditions. *Beta Flecks*: This is a segregation-prone alloy that can exhibit a microstructural phenomenon known as beta flecking. Certain areas may possess a lower beta transus than the matrix (due primarily to beta stabilizer enrichment) and, as such, can fully transform during heat treatment just below the matrix transus. In severe cases, this can lead to lower ductility and a reduction in fatigue strength due to grain boundary alpha formation in the flecked region. *Heat Treatment*: For the high strength condition, the alloy is generally solution treated approximately 65° F (36° C) below the beta transus (which is typically 1,400 to 1,480° F [760 to 804° C]), followed by a water quench and an 8 hour age at 900 to 950° F (482 to 510° C). Overaging in the 950 to 1,150° F (510 to 621° C) range may also be used to obtain lower strength levels.

Heat treating titanium alloys

The Tables in this section give temperature ranges and schedules for solution heat treating, aging, stress relieving, and annealing various alloys. These ranges encompass the more exact temperatures given above in descriptions of specific materials. For complete accuracy, the more specific temperatures and times should be followed, and the values in the Tables should be used as reliable ranges. When heat treating titanium, maximum quench delays can be critical and should not be exceeded, even though the beta-stabilized alloys are more tolerant to delay times than the alpha materials. Quench delay time begins when the furnace door starts to open, and ends when the last corner of the load is immersed in the quenchant. Schedules are as follows.

Nominal Thickness	**Maximum Delay Time**
Up to 0.25 inch (6.4 mm)	6 seconds
0.26 to 0.99 inch (6.5 to 25.3 mm)	8 seconds
1.00 inch and over (25.4 mm and over)	10 seconds

Table 4 gives solution heat treating schedules, **Table 5** provides aging schedules, **Table 6** gives the stress relief schedule, and **Table 7** shows annealing temperatures and times.

Table 4. Solution Heat Treating Schedules for Titanium Alloy Raw Materials and Semi-finished Parts.
(Source, MIL-H-81200B.)

Designation	Solution Heat Treating Temperature				Soaking Time-in Minutes[1]		Cooling Method[8]
	Sheet, Strip, Plate		Bars, Forgings, Castings		Sheet, Strip, Plate	Bars, Forgings, Castings[2]	
	°F	°C	°F	°C			
Alpha Alloys							
Ti-8Al-1Mo-1V [6]	-	-	1650-1850	900-1010	-	20-90	Note 3
Alpha-Beta Alloys							
Ti-6Al-4V	1650-1775	900-970	1650-1775	900-970	2-90	20-120	Water quench
Ti-6Al-6V-2Sn [5]	1550-1700	870-925	1550-1700	870-925	2-60	20-90	Water quench
Ti-6Al-2Sn-4Zr-2Mo	1500-1675	815-915	1650-1800	900-980	2-90	20-120	Air cooled
Ti-6Al-2Sn-4Zr-6Mo	1500-1675	815-915	1500-1675	815-915	2-90	20-120	Note 5
Ti-11Sn-5Zr-2Al-1Mo [6]	-	-	1625-1675	885-915	-	20-120	Air cooled
Ti-6Al-2Sn-2Zr-2Mo-2Cr-0.25Si	1600-1700	870-925	1600-1700	870-925	2-60	20-120	Water quench
Ti-5Al-2Sn-2Zr-4Mo-4Cr [6]	-	-	1450-1500	790-815	-	20-120	Water quench
Beta Alloys							
Ti-13V-11Cr-3Al [4]	1400-1500	760-815	1400-1500	760-815	2-60	2-60	Note 9
Ti-15V-3Al-3Cr-3Sn [7]	1400-1500	760-815	-	-	2-30	20-90	Note 3
Ti-10V-2Fe-3Al [6]	-	-	1300-1425	705-775	-	60-120	Water quench

Notes: [1] Soaking time shall be considered to begin as soon as the lowest reading control thermocouple is at the lower limit of the specified solution treating temperature range. [2] Longer soaking times may be necessary for specific forgings. Shorter soaking times are satisfactory when soak time is accurately determined by thermocouples attached to the load. Soaking time should be measured from the time the furnace charge reaches the soaking temperature.

(Continued)

Table 4. *(Continued)* Solution Heat Treating Schedules for Titanium Alloy Raw Materials and Semi-finished Parts. *(Source, MIL-H-81200B.)*

Notes: [3] Air cool or faster. [4] For material less than 0.100 inch (2.54 mm). For thickness greater than 0.100 inch, duplex solution treatment is applicable as follows: 1300°F to 1375°F (705 to 745°C) for 50 to 80 minutes, air-cooled, then 1400°F to 1450°F (760 to 790°C), 20 to 60 minutes. [5] Air cooling may be applied in relatively thin sections. Water quench is required for thick sections. [6] No solution heat treating cycle is specified for sheet, strip and plate. [7] No solution heat treating cycle is specified for bars, forgings and castings. [8] When vacuum furnace equipment is used, inert gas cooling may be applied in lieu of air cooling. [9] An air cool may be applied to sections up to 0.50 inch thick (12.7 mm). A water quench shall be applied to sections greater than 0.50 inch thick.

Table 5. Aging Schedule for Titanium Alloys. *(Source, MIL-H-81200B.)*

Designation	Aging Temperature		Soaking Times Hours[1]
	°F	°C	
Alpha Alloys			
Ti-8Al-1Mo-1V	1000-1150	540-620	8-24 [2]
Ti-11Sn-5Zr-2Al-1Mo	900-1000	480-540	20-30 [2, 3]
Alpha-Beta Alloys			
Ti-6A1-4V	900-1275	480-690	2-8 [2, 3]
Ti-6A1-6V-2Sn	875-1150	470-620	2-10
Ti-6A1-2Sn-4Zr-2Mo	1050-1150	565-620	2-8 [2]
Ti-6A1-2Sn-4Zr-6Mo	1050-1250	480-675	4-8
Ti-6A1-2Sn-2Zr-2Mo-2Cr-0.25Si	900-1250	480-675	2-10
Ti-5A1-2Sn-2Zr-4Mo-4Cr	1100-1250	480-675	4-8 [3]
Beta Alloys			
Ti-13V-11Cr-3A1 [4]	825-975	440-525	2-60 [3]
Ti-15V-3A1-3Cr-3Sn	900-1250	480-675	2-24 [3]
Ti-10V-2Fe-3A1	900-1150	480-620	8-10 [3]

Notes: [1] Soaking time shall be considered to begin as soon as the lowest reading control thermocouple is at the lower limit of the specified solution treating temperature range. [2] See Table V for duplex annealing. An 8-hour stabilizing treatment at 1050-1100°F (565-595°C) can be considered an aging treatment. [3] Aging time and temperature depends on strength level desired. [4] Springs may be aged at 800°F (425°C).

Table 6. Stress Relief Schedule for Titanium and Titanium Alloys. *(Source, MIL-H-81200B.)*

Designation	Stress Relief Temperature		Soaking Times Hours
	°F	°C	
Unalloyed Titanium			
Commercially Pure, All Grades	900-1100	480-595	15-240
Alpha Alloys			
Ti-5Al-2.5Sn	1000-1200	540-650	15-360
Ti-5Al-2.5Sn ELI	1000-1200	540-650	15-360
Ti-6A1-2Cb-1Ta-0.8Mo	1000-1200	540-650	15-60
Ti-8A1-1Mo-1V	1100-1400	595-760	10-75
Ti-11Sn-5Zr-2A1-1Mo	900-1000	480-540	120-480

(Continued)

Table 6. *(Continued)* **Stress Relief Schedule for Titanium and Titanium Alloys.** *(Source, MIL-H-81200B.)*

Designation	Stress Relief Temperature		Soaking Times Hours
	°F	°C	
Alpha-Beta Alloys			
Ti-3Al-2.5V	700-1100	370-595	15-240
Ti-6Al-4V [1]	900-1200	480-650	60-240
Ti-6Al-4V ELI [1]	900-1200	480-650	60-240
Ti-6Al-6V-2Sn	900-1200	480-650	60-240
Ti-6Al-2Sn-4Zr-2Mo	900-1200	480-650	60-240
Ti-5Al-2Sn-2Zr-4Mo-4Cr	900-1200	480-650	60-240
Ti-6Al-2Sn-2Zr-2Mo-2Cr-0.25Si	900-1200	480-650	60-240
Alpha-Beta Alloys			
Ti-13V-11Cr-3Al	1300-1350	480-730	30-60
Ti-15V-3Al-3Cr-3Sn	1450-1500	790-815	30-60
Ti-10V-2Fe-3Al	1250-1300	675-705	30-60

Note: [1] Stress relief may be accomplished simultaneously with hot forming at temperatures of 1400°F (760°C) to 1450°F (790°C). Caution should be exercised in stress relieving titanium alloys that have been strengthened by solution treating and aging. The stress relieving temperature should not exceed the aging temperature used in heat treatment.

Table 7. Annealing Schedule for Titanium and Titanium Alloys. *(Source, MIL-H-81200B.)*

Designation	Annealing Temperature				Soak Time[11, 12]	
	Sheet, Strip, Plate		Bars, Forgings		Sheet, Strip, Plate	Bars, Forgings
	°F	°C	°F	°C		
Unalloyed Titanium						
Commercially Pure, All Grades	1200-1500 [5]	650-815 [5]	1200-1450 [8]	650-815 [8]	15-120 min.	1-2 hrs.
Alpha Alloys						
Ti-5Al-2.5Sn	1300-1550 [4]	705-845 [4]	1300-1550 [4]	705-845 [4]	10-120 min.	1-4 hrs.
Ti-5Al-2.5Sn ELI	1300-1650 [4]	700-900 [4]	1300-1550 [4]	705-845 [4]	10-120 min.	1-4 hrs.
Ti-6Al-2Cb-1Ta-0.8Mo	1450-1650 [5]	790-900 [5]	1450-1650 [5]	790-900 [5]	30-120 min.	1-2 hrs.
Ti-8Al-1Mo-1V	1400-1500 [1]	760-815 [1]	1650-1850 [2, 3]	900-1000 [2, 3]	1-8 hrs.	1-2 hrs.
Alpha-Beta Alloys						
Ti-3Al-2.5V	1200-1450 [5]	650-790 [5]	1200-1450 [5]	650-790 [5]	30-120 min.	1-3 hrs.
Ti-6Al-4V [10]	1300-1600 [5, 6]	705-870 [4]	1300-1450 [4]	705-790 [4]	15-60 min.	1-2 hrs.
Ti-6Al-4V ELI [10]	1300-1600 [5, 6]	705-870 [4]	1300-1450 [4]	705-790 [4]	15-60 min.	1-2 hrs.
Ti-6Al-6V-2Sn	1300-1500 [5]	705-815 [5]	1300-1450 [5]	705-790 [5]	10-120 min.	1-2 hrs.
Ti-6Al-2Sn-4Zr-2Mo	1600-1700 [7, 13] 1600-1700 [8, 14]	870-925 [7] 870-925 [8]	1764-1781 [4, 8] -	962-976 [4, 8] -	10-60 min. [13] 30-120 min. [14]	1-2 hrs. -
Ti-6Al-2Sn-4Zr-6Mo	-	-	1500-1675 [4, 9]	815-915 [4, 9]	-	1-2 hrs.

Table 7. *(Continued)* **Annealing Schedule for Titanium and Titanium Alloys.** *(Source, MIL-H-81200B.)*

Designation	Annealing Temperature				Soak Time[11, 12]	
	Sheet, Strip, Plate		Bars, Forgings		Sheet, Strip, Plate	Bars, Forgings
	°F	°C	°F	°C		
Alpha-Beta Alloys						
Ti-11Sn-5Zr-2A-l-1Mo	-	-	1625-1675 [2]	885-915 [2]	-	1-2 hrs.
Ti-6Al-2Sn-2Zr-2Cr-2Mo	1275-1600 [4]	690-870 [4]	1275-1600 [4]	690-870 [4]	15-360 min.	15-360 min.
Beta Alloys						
Ti-13V-11Cr-3Al	1400-1500 [2, 9]	760-815 [2, 9]	1400-1500 [2, 9]	760-815 [2, 9]	10-60 min.	30-120 min.
Ti-15V-3Al-3Cr-3Sn	1400-1500 [2, 9]	760-815 [2, 9]	-	-	3-30 min.	-
Ti-10V-2Fe-3Al	-	-	1300-1340 [2, 9]	705-727 [2, 9]	-	30-120 min.

Notes: BT denotes that the annealing temperature is the respective beta transus temperature minus the temperature range given at the right. [1] Furnace cool to below 900°F (480°C) - duplex anneal requires second anneal at 1450°F (790°C) for 15 min. followed by air cool. [2] Air cool or faster. [3] Followed by stabilization at 1100°F (595°C) for 8 hrs. and air cool. [4] Air cool. [5] Air cool or slower. [6] When duplex anneal (solution treat and anneal) is specified for 6A1-4V, the annealing treatment is as follows: Heat to beta transus temperature minus 50 to 75°F (14 to 32°C), hold for 1 to 2 hours for bars/forgings/plate, air cool or faster, reheat 1300-1400°F (705-760°C) for 1 or 2 hrs. and air cool. [7] Air cool followed by 1450°F (790°C) for 15 min. and air cool. [8] Air cool followed by 1100°F (595°C) for 8 hrs. and air cool. [9] Followed by aging at a temperature to develop required properties. [10] When recrystallization anneal is specified to optimize fracture toughness, it is generally as follows: Heat to beta transus temperature minus 50 to 75°F (14 to 32°C), hold for 1 to 4 hrs., air cool or slower, reheat between 1300 to 1400°F (705-760°C) for 1 or 2 hrs. and air cool. [11] Soaking time shall be considered to begin as soon as the furnace recovers to the specified soaking temperature. The table below offers guidance in selecting soaking times:

Thickness, inches (mm)	Soaking time, minutes ± 5
0.125 (3.19) and under	20
0.126-0.250 (3.20-6.35)	30
0.251-0.500 (6.36-12.70)	40
0.501-0.750 (12.71-19.07)	50
0.751-1.000 (19.08-25.40)	60
Over 1.000 (25.40)	1 hour minimum plus add fifteen minutes for each quarter inch of thickness over one inch.

[12] Annealing times for flat-rolled product processed as a continuous coil can be shortened to 2 minutes minimum. [13] For sheet only. [14] For plate only.

Heat-resistant Superalloys and Other Exotic Alloys

Heat-resistant, or "super," alloys

Heat-resistant, or "Super," nickel type alloys (superalloys) are arbitrarily defined as iron alloys richer in alloy content than the 18% chromium, 8% nickel types, or as alloys with a base element other than iron, and which are intended for elevated temperature service. Superalloys may be nickel, iron-chromium-nickel, or cobalt based, and they are usually specified for applications where operating temperatures are 1,000° F (540° C) or higher. Oxidation resistance at elevated temperatures is a requirement, and they must be able to withstand elevated temperatures without special surface protection.

Mechanical properties

Most of these alloys are available from the mill in a wide variety of heat treatments, and their mechanical properties are normally affected by relatively minor variations in chemistry (composition is given in **Table 1**), processing, and heat treatment. Consequently, the mechanical properties, given in **Table 2** or in the descriptions of individual alloys, apply only to the form (shape), size (thickness), and heat treatment indicated.

Strength properties of the heat-resistant alloys generally decrease with increasing temperatures or increasing time at temperature. There are exceptions, particularly in the case of age hardening alloys that may actually show an increase in strength with temperature or time, within a limited range, as a result of further aging. In most cases, however, this increase in strength is temporary and cannot usually be taken advantage of in service. At cryogenic temperatures, the strength properties of the heat-resistant alloys are generally higher than at room temperature, provided that some ductility is retained at the low temperatures.

Ductility varies with temperature and is somewhat erratic. Generally, ductility decreases with increasing temperature from room temperature up to about 1,200 to 1,400° F (649 to 760° C), where it reaches a minimum value, then increases with higher temperatures. Prior creep exposure may also adversely affect ductility.

Physical properties are shown in **Table 3**.

Iron-chromium-nickel base heat-resistant alloys

The alloys in this group, in terms of cost and in maximum service temperature, generally fall between the austenitic stainless steels and the nickel-cobalt based alloys. They are specified for operating conditions in the temperature range of 1,000 to 1,200° F (538 to 649° C), in applications where the stainless steels are inadequate but service conditions do not justify the more expensive nickel or cobalt alloys.

The complex-base alloys comprising this group range from those in which iron is considered the base element, to those which border on the nickel-base alloys. All contain sufficient alloying elements to place them in the superalloy category, yet contain enough iron to reduce their cost considerably. Chromium, in amounts ranging from 10 to 20% or higher, primarily increases oxidation resistance and contributes to strengthening these alloys. Nickel and cobalt strengthen these materials. Molybdenum, tungsten, and columbium contribute to hardness and strength, particularly at higher temperatures. Titanium and aluminum are added to provide age hardening. Heat treatment is performed with conventional equipment and fixtures such as would be used for austenitic stainless steels. Since these alloys are susceptible to carburization during heat treatment, it is good practice to remove all grease, oil, or cutting lubricant from the surface before treating. A low-sulfur and natural or slightly oxidizing furnace atmosphere is recommended for heating.

Table 1. Nominal Composition of Selected Wrought "Superalloys."

Material	UNS No.	Composition %
A-286	S66286	Ni 26, Cr 15, Ti 2, Mo 1.3, Al 0.2, B 0.015, Fe bal.
Hastelloy G	N06007	Cr 22, Fe 19.5, Mo 6.5, Co 2.5 max., Cb+Ta 2, Cu 2, Mn 1.5, Si 1 max., W 1 max., C 0.05 max., Ni bal.
Hastelloy C-276	N10276	Mo 16, Cr 15.5, Fe 5, W 4, Co 2.5 max., Mn 1 max., V 0.35 max., Si 0.08 max., C 0.02 max., Ni bal.
Hastelloy X	N06002	Cr 22, Fe 18.5, Mo 9, Co 1.5, Mn 1 max., Si 1 max., W 0.6, C 0.10, Ni bal.
214	N07214	Ni 75, Cr 16, Al 4.5, Fe 3, Mn 0.2, Si 0.1, C 0.05, Other Y.01
230	N06230	Ni 57, Cr 22, W 14, Co 5 max., Mo 2, Mn 0.5, Si 0.4, Al 0.3, Fe 0.3 max., C 0.10, La 0.02
Haynes 188	R30188	Cr 22, Ni 22, W 14, Fe 3 max., Mn 1.25 max., Si 0.35, La 0.04, C 0.10, Co bal.
IN-102	N06102	Cr 15, Fe 7, Mo 3, W 3, Cb 3, Ti 0.6, Al 0.4, Zr 0.03, B 0.005, Ni bal.
Incoloy 800	N08800	Fe 45.5, Ni 32.5, Cr 21, Al 0.4, Ti 0.4, C 0.05
800HT	N08811	Fe 46, Ni 33, Cr 21, Mn 0.8, Si 0.5, Al 0.4, Ti 0.4, C 0.08
Incoloy 825	N08825	Ni 38-46, Cr 19.5-23.5, Fe 22 min, Mo 2.5-3.5, Cu 1.5-3, Ti 0.6-1.2, Mn 1 max, Si 0.5 max
907	N19907	Fe 42, Ni 38, Co 13, Nb 4.7, Ti 1.5, Si 0.15, Al 0.03
909	N19909	Fe 42, Ni 38, Co 13, Nb 4.7, Ti 1.5, Si 0.4, C 0.01, B 0.001
Inconel 600	N06600	Ni 76, Cr 15, Fe 8, Mn 0.5, Cu 0.2, Si 0.2, C 0.08
Inconel 625	N06625	Ni 61, Cr 21.5, Mo 9, Fe 4, Cb 3.6, Mn 0.2, Si 0.2, C 0.05
Inconel 690	N06690	Ni 61, Cr 29, Fe 9, Cu 0.2, Mn 0.2, Si 0.2, C 0.02
Inconel 718	N07718	Ni 52.5, Cr 19, Fe 18.5, Cb 5.1, Mo 3.0, Ti 0.9, Al 0.5, Mn 0.2, Si 0.2, C 0.04
Inconel X750	N07750	Ni 73, Cr 15.5, Fe 7, Ti 2.5, Cb 1, Ti 0.9, Al 0.7, Mn 0.5, Si 0.2, C 0.04
L-605	R30605	Cr 20, W 15, Ni 10, Fe 3 max., Mn 1.5, Si 1 max., C 0.10, Co bal.
M-252	N07252	Cr 20, Co 10, Mo 10, Ti 2.6, Al 1, B 0.005, Ni bal.
N-155	R30155	Cr 21, Co 20, Ni 20, Mo 3, W 2.5, Mn 1.5, Cb+Ta 1, Si 1 max., N 0.15, C 0.10, Fe bal.
Nimonic 80A	N07080	Ni 74.5, Cr 20.5, Ti 2.4, Al 1.25, Fe 0.5, C 0.05
901	N09901	Ni 43, Fe 35, Cr 12, Mo 6, Ti 3, Co 1 max., Mn 0.5 max., Si 0.4 max., Al 0.2, C 0.05
Rene 41	N07041	Cr 19, Co 11, Mo 10, Ti 3.1, Fe 2, Al 1.5, B 0.005, Ni bal.
Udimet 500	N07500	Co 18.5, Cr 18, Mo 4, Al 2.9, Ti 2.9, Zr 0.05, B 0.006, Ni bal.
Ultimet	R31233	Co 54, Cr 26, Ni 9, Mo 5, Fe 3, W 2, Mn 0.8, Si 0.3, C 0.05, B 0.015
Waspaloy	N07001	Cr 19.5, Co 13.5, Mo 4.3, Ti 3, Fe 2, Al 1.3, Zr 0.04, B 0.006, Ni bal.

Note: Y denotes Y_2O_3

Table 2. Mechanical Properties of Selected Wrought "Superalloys."

Material	Form and Condition	Yield Strength[1]		Tensile Strength[2]		% Elong.[3]	Hardness
		ksi	MPa	ksi	MPa		
A-286	Bar 1800° F OQ + 1325° F AC	105	724	146	1007	25	R_c 28
Hastelloy G	Sheet Solution Heat Treated	46	317	102	703	62	R_B 84
Hastelloy C-276	Sheet Solution Heat Treated	52	359	115	793	61	R_B 90

(Continued)

Table 2. *(Continued)* Mechanical Properties of Selected Wrought "Superalloys."

Material	Form and Condition	Yield Strength[1]		Tensile Strength[2]		% Elong.[3]	Hardness
		ksi	MPa	ksi	MPa		
Hastelloy X	Sheet Solution Heat Treated	52	359	114	786	43	R_B 90
214	Typical as Supplied Condition	83	573	135	930	42	-
230	Typical as Supplied Condition	57	393	125	861	49	-
Haynes 188	Sheet 2150° F WQ	70	483	139	958	56	R_B 96
IN-102	Bar 1800° F AC	73	503	139	958	47	HB 215
Incoloy 800	Rod Hot Rolled Annealed	42	290	87	600	44	HB 150
800HT	Typical as Supplied Condition	35	241	83	572	49	-
Incoloy 825	Typical as Supplied Condition	45	310	100	690	45	
907	Typical as Supplied Condition	162	1116	195	1344	15	-
909	Typical as Supplied Condition	148	1020	190	1310	16	-
Inconel 600	Rod Hot Rolled Annealed	41	283	96	662	45	HB 150
Inconel 625	Rod Hot Rolled Annealed	70	483	140	965	50	HB 180
Inconel 690	Rod Hot Rolled Annealed	51	352	104	717	47	R_B 90
Inconel 718	Rod Hot Rolled Annealed Aged	180	1,241	208	1,434	20	R_C 45
Inconel X750	Rod Hot Rolled,1625° F + Aged, Sheet Annealed + Aged	126 / 122	869 / 841	184 / 177	1,269 / 1,220	25 / 27	R_C 36 / R_C 36
L-605	Sheet Solution Heat Treated	67	462	146	1,007	59	R_C 24
M-252	Bar 1900° F AC + 1400° F AC	122	841	180	1,241	16	R_C 39
N-155	Sheet Solution Heat Treated	58	400	118	814	49	R_B 92
Nimonic 80A	1975° F, Air Cool + 1300° F	90	621	154	1,062	39	R_C 34
901	Typical as Supplied Condition	130	896	175	1207	15	-
Rene 41	Bar 1950° F AC + 1400° F AC	154	1,062	206	1,420	14	-
Udimet 500	Bar 1975° F AC + 1550° F AC + 1400° F AC	110	758	176	1,213	16	-
Ultimet	Typical as Supplied Condition	80	551	145	1000	35	-
Waspaloy	Bar 1975° F AC + 1550° F AC + 1400° F AC	115	793	185	1,276	25	-

Notes: Codes used for condition are WQ = Water Quench; AC = Air Cool; OQ = Oil Quench. [1] Yield strength at 0.2% offset, measured at room temperature. [2] Ultimate tensile strength at room temperature. [3] Elongation in 2 inches (50.4 mm).

Table 3. Physical Properties of Selected Wrought "Superalloys."

Material[1]	Density		CTE[2]		Modulus of Elasticity		Melting Point	
	lbs/cu in.	gr/cu cm	Eng.	Metric	ksi	GPa	°F	°C
A-286	0.286	7.92	9.17 [3]	16.49 [3]	29.1	201	2550	1399
Hastelloy G	0.300	8.31	7.5	13.49	27.8	191.5	2450	1343
Hastelloy C-276	0.321	8.89	6.2	11.15	29.8	205.3	2500	1371
Hastelloy X	0.297	8.22	7.70	13.84	28.5	196.4	2470	1354

(Continued)

Table 3. *(Continued)* **Physical Properties of Selected Wrought "Superalloys."**

Material[1]	Density		CTE[2]		Modulus of Elasticity		Melting Point	
	lbs/cu in.	gr/cu cm	Eng.	Metric	ksi	GPa	°F	°C
214	0.291	8.06	-	-	31.6	217.7	2471	1355
230	0.324	8.97	7.0	12.59	30.6	210.8	2462	1350
Haynes 188	0.324	8.97	6.6	11.87	33.6	231.5	2425	1329
IN-102	0.309	8.56	7.32	13.16	29.7	204.6	2530	1388
Incoloy 800	0.287	7.95	7.9	14.20	28	192.9	2525	1385
800 HT	0.287	7.95	7.9	14.20	28.5	196.4	2498	1370
Incoloy 825	0.294	8.14	7.7	14.00	-	-	2525	1385
907	0.299	8.28	4.4	7.91	23.9	164.7	2462	1350
909	0.296	8.20	4.4	7.91	23.0	158.5	2570	1410
Inconel 600	0.304	8.42	7.4	13.31	30	207	2575	1413
Inconel 625	0.305	8.45	7.1	12.77	30.1	207.4	2460	1349
Inconel 690	0.296	8.20	7.8	14.02	30.6	210.8	2510	1377
Inconel 718	0.296	8.20	7.2	12.95	29	199.8	2437	1336
Inconel X750	0.298	8.25	7.0	12.59	31	213.6	2600	1427
L-605	0.330	9.14	6.8	12.23	32.6	224.6	2570	1410
M-252	0.298	8.25	5.9	10.61	29.8	205.3	2500	1371
N-155	0.296	8.20	7.9	14.20	29.3	201.9	2470	1354
Nimonic 80A	0.297	8.22	6.2	11.15	32.1	221.2	2500	1371
901	0.294	8.14	7.5	13.49	30	207	2409	1321
Rene 41	0.298	8.25	6.6	11.87	31.9	219.8	2500	1371
Udimet 500	0.290	8.03	6.8	12.14	32.1	221.2	2540	1393
Ultimet	0.305	8.45	7.1	12.77	-	-	2534	1390
Waspaloy	0.296	8.20	6.8	12.23	30.9	212.9	2475	1357

Notes: [1] For material condition, see "Form and Condition" column of the mechanical properties Table. [2] Thermal Coefficient of Expansion (CTE) measured from 32 to 212° F (0 to 100° C) and expressed in μin./°F for English units, and μm/°C for Metric units. [3] CTE measured between 70 and 200° F (21 and 93° C).

These alloys closely resemble the austenitic stainless steels insofar as forging, cold forming, machining, welding, and brazing are concerned. Their higher strength may require the use of heavier forging or forming equipment, and machining is somewhat more difficult than for the stainless steels.

Selected Alloys

Alloy A-286. This precipitation hardening iron-base alloy was developed for parts requiring high strength up to 1,300° F (704° C) and oxidation resistance to 1,500°F (816° C). It is often specified for jet engines and gas turbines for parts such as turbine buckets, bolts, and discs, and for sheet metal assemblies. It is available in the usual mill forms. *Manufacturing Considerations:* A-286 is somewhat harder to hot or cold work than the austenitic stainless steels. Its forging range is 1,800 to 2,150° F (982 to 1,176° C). When finishing below 1,800° F, light reductions (under 15%) must be avoided to prevent grain coarsening during subsequent heat treatment. A-286 is readily machined in the partially or fully aged condition, but is soft and gummy in the solution treated condition. It should be welded in the solution treated condition, but fusion welding is difficult with large sections sizes and moderately difficult for small cross sections and sheet. Cracking may be encountered in the welding of heavy sections or parts under high restraint. A dimensional contraction of 0.0008 inch (0.02 mm) per inch (per 25.4 mm) is experienced during aging. Oxidation resistance is equivalent to Type 310 stainless steel up to 1,800° F (982° C). *Heat Treatment:* Sheet, strip and plate are normally solution treated (1,800° F/982° C). Bar, forging, tubing, and ring are either solution heat treated (1,800° F), or solution heat treated (1,800° F) and aged. Bar, forging, and tubing is either solution treated (1,650° F/899° C), or solution treated (1,650° F) and aged.

Alloy N-155. Also known as Multimet, N-155 is a solid solution alloy developed to withstand high stresses up to 1,500° F (816° C). It has good oxidation properties and good ductility, and can be readily fabricated by conventional means. It has been used in many aircraft applications including afterburner parts, combustion chambers, exhaust assemblies, turbine parts, and bolting. *Manufacturing Considerations:* N-155 is forged readily between 1,650 and 2,200° F (899 and 1,204° C). It is easily formed by conventional methods, but intermediate anneals may be required to restore ductility. This alloy is machinable in all conditions, using low cutting speeds and ample flow of coolant. Weldability is comparable to austenitic stainless steels, and oxidation resistance of N-155 sheet is good up to 1,500° F (816° C). *Heat Treatment:* Sheet, tubing, bar, and forging materials are normally solution treated, but bar and forging can also be solution treated and aged.

Nickel base heat-resistant alloys

Nickel is the base element for most of the higher temperature heat-resistant alloys. While it is more expensive than iron, nickel provides an austenitic structure that has greater toughness and workability than ferritic structures of the same strength level.

The common alloying agents for nickel are cobalt, iron, chromium, molybdenum, titanium, and aluminum. Cobalt, when substituted for a portion of the nickel in the matrix, improves high temperature strength. Small additions of iron tend to strengthen the nickel matrix and reduce the cost. Chromium is added to increase strength and oxidation resistance at very high temperatures. Molybdenum contributes to solid solution strengthening. Titanium and aluminum are added to most nickel base heat-resistant alloys to permit age hardening by the formation of Ni_3 (Ti, Al) precipitates, and aluminum also contributes to oxidation resistance. The nature of the alloying elements in the age hardening nickel base alloys makes vacuum melting of those alloys advisable. However, the additional cost pf vacuum melting is more than compensated for in elevated temperature properties.

Manufacturing Considerations

Forging: The wrought alloys can, to some degree, be forged. The matrix-strengthened alloys can be forged with proper consideration for cooling rates, atmosphere, etc. Most of the precipitation hardening grades can be forged, but heavier equipment is required and a smaller range of reductions can be safely attained. *Cold Forming:* Almost all wrought nickel base alloys in sheet form are cold formable. The lower strength alloys offer few problems, but the higher strength alloys require forming pressures and more frequent anneals. *Machining:* These alloys are readily machinable, provided the optimum conditions of heat treatment, type of tool, speed, feed, depth of cut, etc., are achieved. *Welding:* The matrix strengthening type alloys offer no serious problems in welding. All of the common resistance and fusion welding processes (except submerged arc) have been successfully employed. For the age hardening alloys, it is necessary to observe the following precautions. 1) Welding should be confined to annealed material where design permits. In full aged hardened materials, the hazard of cracking in the weld and/or parent material is great. 2) If design permits, join some portions only after age hardening. The parts to be joined should be "safe ended" with a matrix strengthened type alloy and then age hardened. Welding should then be carried out on the safe ends. 3) Parts that are severely worked or deformed should be annealed before welding. 4) After welding, the weldment will often require stress relieving before aging. 5) Material must be heated rapidly to the stress relieving temperature. 6) In many age hardening alloys, fusion welds fusion welds may exhibit only 70 to 80% of the rupture strength of the parent material. *Brazing:* Solid solution type chromium containing alloys respond well to brazing using techniques and brazing alloys applicable to the austenitic stainless steels. The aluminum-titanium age hardened nickel base alloys are difficult to braze unless some method of fluxing, solid or gaseous, is used, Most of the high-temperature alloys of the nickel base type are brazed with Ni-Cr-Si-B and Ni-Cr-Si types of brazing alloy.

Heat Treatment

The nickel base alloys are heat treated with conventional equipment and fixtures such as would be used with austenitic stainless steels. Since nickel base alloys are more susceptible to sulfur embrittlement than are the iron base alloys, it is essential that sulfur bearing materials such as grease, oil, cutting lubricants, marking paints, etc., be removed before heat treatment. Mechanical cleaning, such as wire brushing, is not adequate and if used should be followed by washing with a suitable solvent or by vapor degreasing. A low-sulfur content furnace atmosphere should be used. Good furnace control with respect to time and temperature is desirable since overheating some of the alloys by as little as 35° F (19.5° C) impairs strength and corrosion resistance. When it is necessary to anneal the age hardening alloys, a protective atmosphere (such as argon) lessens the possibility of surface contamination or depletion of the precipitation hardening elements. This precaution is less critical in heavier sections because the oxidized surface layer is a smaller percentage of the cross section. After solution annealing, the alloys are generally quenched in water. Heavy sections may require air cooling to avoid surface cracking from thermal stress. In stress relief annealing of a structure or assembly composed of an aluminum-titanium hardened alloy, it is vitally important to heat the structure rapidly through the age hardening temperature range of 1,200 to 1,400° F (649 to 760° C), which is also the low ductility range, so that stress relief can be achieved before any aging takes place. Parts that are likely to be used in the fully heat treated condition should be solution treated, air cooled, and subsequently aged. Little difficulty can be expected with distortion under rapid heating conditions, and distortion of weldments of substantial size is less than can be expected with slow heating methods.

Selected Alloys

Hastelloy X. (Hastelloy is a registered trademark of Haynes International, Inc.) This nickel base alloy is used for combustor-liner parts, turbine exhaust weldments, afterburner parts, and other parts requiring oxidation resistance and moderately high strength above 1,450° F (788° C). It is not hardenable except by cold working, and is used in the solution treated (annealed) condition. It is available in all common mill forms. *Machining Considerations:* Hastelloy X is somewhat difficult to forge. Forging should be started at 2,150 to 2,200° F (1,117 to 1,204° C) and continued as long as the material flows freely. It should be in the annealed condition for optimum cold forming, and severely formed detail parts should be solution treated at 2,150° F for seven to ten minutes and cooled rapidly after forming. Machinability is similar to that of austenitic stainless steel. The alloy is tough and requires low cutting speeds and ample cutting fluids. It can be resistance or fusion welded or brazed, but large or complex fusion weldments require stress relief at 1,600° F (871° C) for one hour. Hastelloy X has good oxidation resistance up to 2,100° F (1149° C). It age hardens somewhat during long exposure between 1,200 and 1,800° F (649 and 892° C).

Inconel 625. (Inconel is a registered trademark of INCO International, Inc.) This is a solid solution, matrix strengthened nickel base alloy primarily used for applications requiring good corrosion and oxidation resistance at temperatures up to approximately 1,800° F (892° C) and also where such parts may require welding. The strength of the alloy is derived from the strengthening effects of molybdenum and columbium. Therefore, precipitation hardening is not required and the alloy is used in the annealed condition. Its strength is greatly affected by the amount of cold work prior to annealing, and by the annealing temperature. The material is normally annealed at 1,700 to 1,900° F (927 to 1,038° C) for a period of time commensurate with thickness. *Machining Considerations:* Because the alloy was developed to retain high strength at elevated temperatures, it resists deformation at hot working temperatures but can be readily fabricated with adequate equipment. The combination of strength, corrosion resistance, and ability to be fabricated, including welding by common industrial practices, are the alloy's outstanding features.

Inconel 706. (Inconel is a registered trademark of INCO International, Inc.) This nickel base vacuum melted precipitation hardened alloy has characteristics similar to Inconel 718 (description follows) except that Inconel 706 has greatly improved machinability. It has good formability and weldability, and excellent resistance to postweld strain age cracking. Depending on choice of heat treatment, this alloy can be used for applications requiring either 1) high resistance to creep and stress rupture up to 1,300° F (704° C), or 2) high tensile strength at cryogenic temperatures or elevated temperatures for short periods. The creep resistant heat treatment is characterized by an intermediate stabilizing treatment before precipitation hardening. Inconel 706 also has good resistance to oxidation and corrosion over a broad range of temperatures and environments. Creep resistant solution treatment temperature is 1,750° F (954° C). Tensile strength solution treatment temperature is 1,800° F (982° C).

Inconel 718. (Inconel is a registered trademark of INCO International, Inc.) Inconel 718 is a vacuum melted, precipitation hardened nickel base alloy that can be easily welded and excels in its resistance to strain age cracking. It is also readily formable. Depending on choice of heat treatments, this alloy finds applications requiring either 1) high resistance to creep and stress rupture to 1,300° F (704° C), or 2) high strength at cryogenic temperatures. It also has good oxidation resistance up to 1,800° F (982° C). It is available in all wrought forms and investment castings.

Inconel X-750. (Inconel is a registered trademark of INCO International, Inc.) This high strength oxidation resistant nickel base alloy is used for parts requiring high strength up to 1,000° F (538° C), or high creep strength up to 1,500°F (816° C), and for low stressed parts operating at temperatures up to 1,900° F (1,038° C). It is hardenable by various combinations of solution treatment and aging, depending on its form and application. Inconel X-750 is available in all the usual wrought mill forms. *Machining Considerations:* This alloy can be readily forged between 1,900 and 2,225° F (1,038 and 1,218° C). "Hot-cold" working between 1,200 and 1,600° F (649 and 871° C) is harmful and should be avoided. Inconel X-750 is readily formed but should be solution treated at 1,925° F (1,052° C) for seven to ten minutes following severe forming operations. It is somewhat more difficult to machine than austenitic stainless steels. Rough machining is easier in the solution treated condition. Finish machining is easier in the partially or fully aged condition. Fusion welding is difficult for large section sizes and moderately difficult for small cross sections and sheet. It must be welded in the annealed or solution treated condition, and weldments should be stress relieved at 1,650° F (899° C) for two hours before aging. Nickel brazing, followed by precipitation heat treatment of the brazed assembly, results in strength nearly equal to the fully heat treated material. Oxidation resistance of Inconel X-750 is good to 1,900° F (1,038° C), but the beneficial effects of aging are lost above 1,500° F (816° C). This alloy is subject to attack in sulfur containing atmospheres. *Heat Treatment:* Annealed and aged for strip and plate; mill annealed plus 1,300° F (704° C) for 20 hours, then air cool. Equalized and aged for bar and forging; 1,625° F (885° C) for four hours, air cool, then 1,300° F (704° C) for 24 hours.

Rene 41. (Rene is a registered trademark of the General Electric Corp.) This vacuum precipitation hardening nickel base alloy was developed for highly stressed parts operating between 1,200 and 1,800°F (649 and 982° C). Its applications include afterburner parts, turbine castings, and high temperature bolts and fasteners. It is available in the form of sheet, bars, and forgings. *Machining Considerations:* Rene 41 is forged between 1,900 and 2,150° F (1,038 and 1,177° C). Small reductions must be made when breaking up an as-cast structure as cracking may be encountered in finishing below 1,850° F (1,010° C). This alloy work hardens rapidly and frequent anneals are required. To anneal, heat rapidly to 1,950° F (1,066° C) for 30 minutes and quench. Machining is difficult. In the soft solution annealed condition it is gummy. Therefore, it should be in the fully aged condition for optimum machinability, and tungsten carbide tools should be used. It can be welded satisfactorily in the solution treated condition, and after welding the parts should be solution treated for stress relief. Rene 41 should not be exposed to temperatures above 2,050° F (1,121° C) during latter stages of hot working or during subsequent operations or severe intergranular cracking may be encountered. Oxidation resistance is good to 1,800° F (982° C). Lengthy exposure above the aging temperature of 1,400 to 1,650° F (760 to 899° C) results in loss of strength and room temperature ductility.

Waspaloy. (Waspaloy is a registered trademark of Pratt and Witney Aircraft Div., United Technologies.) The precipitation of titanium and aluminum compounds, plus the solid solution effects of chromium, molybdenum, and cobalt, strengthen this vacuum melted precipitation hardened nickel base alloy. It was designed for highly stressed parts operating at temperatures up to 1,550° F (843° C) such as aircraft gas turbine blades and discs, and rocket engine parts. Waspaloy is available in all the usual mill forms. *Machining Considerations:* The optimum range for forging is 1,900 to 2,050° F (1,038 to 1,121° C). The alloy should not be worked below 1,900° F due to danger of cracking and also decreasing the stress rupture life. Sufficient soaking time between heating is necessary to ensure complete recrystallization, but excessive long-term soaking at the high forging temperature should be avoided. Furnace atmospheres should be either neutral or slightly oxidizing to prevent carburization and to minimize scaling. This alloy is rather difficult to machine. Drilling, turning, etc., can be best accomplished in the solution treated and partially aged condition. Generally, carbide tools are preferred, and slow feeds are required to avoid work hardening. For finish machining, grinding is preferable. Waspaloy is susceptible to hot cracking or "hot shortness" above 2,150° F (1,177° C), so extreme care should be exercised in the design of weldments so that restraint can be minimized. Welding should be done in the annealed condition, with minimum heat input, and cooling should be accelerated by chill bars and gas backup. The alloy has good resistance to oxidation at temperatures up to 1,750° F (954° C), and to combustion products encountered in gas turbines. *Heat Treatment:* Two heat treatments are used for this material. 1) For optimum tensile strength, solution treat at 1,825 to 1,900° F (996 to 1,038° C), stabilize at 1,550° F (843° C), air cool for 24 hours, age for 16 hours at 1,400° F (760° C), air cool. 2) For optimum stress rupture properties, solution treat at

1,975° F (1,079° C), stabilize at 1,550° F (843° C), air cool for 24 hours, age at 1,400° F (760° C) for 16 hours, air cool.

Cobalt base heat-resistant alloys

Very few of the heat-resistant alloys can be considered to have a cobalt base, as cobalt is seldom the predominant element. Common alloying elements for cobalt are chromium, nickel, carbon, molybdenum, and tungsten. Chromium is added to increase strength and oxidation resistance at very high temperatures; nickel to increase toughness; carbon to increase both hardness and strength (especially when combined with chromium and the other carbide formers, molybdenum and tungsten); molybdenum and tungsten also contribute to solid solution strengthening. Vacuum melting is not required for these alloys, allowing the cobalt based alloys to be competitively priced with vacuum melted nickel base alloys, but higher in price than the nickel alloys.

Manufacturing Considerations

Forging: Because these alloys are designed to have very high strength at temperatures near the forging range, they require the use of heavy forging equipment. However, the forgability of these alloys is good over a fairly wide range of temperatures. Hot cold working is neither required nor recommended for these alloys. *Cold Forming:* These alloys, when the solution treated condition, have excellent ductility and are readily cold formed. Because of their capacity for work hardening, they require higher forming pressures and frequent anneals. *Machining:* These alloys are tough and work harden rapidly. Consequently, heavy-duty vibration-free machine tools, sharp cutting tools (high speed steel or carbide tipped), and low cutting speeds are required. *Welding:* The weldability of the cobalt base is comparable to that of the austenitic stainless steels. Welding may be accomplished with all commonly used welding processes, but large or complex weldments require stress relief.

Heat Treatment

The cobalt base alloys are heat treated with conventional equipment and fixtures such as those used with austenitic stainless steels. The use of good heat treating practices is recommended, although not as critical as in the case of the nickel base alloys.

Special Precaution

If the cobalt base alloys have not been exposed to neutron radiation, no special safety precautions in handling are required. However, neutron radiation creates a very dangerous radioactive isotope, cobalt 60, which has a half-life of 5.2 years. Special precautions must be employed to protect personnel from the radioactive material.

Selected Alloys

L-605. (This alloy is also known as Haynes Alloy 25.) L-605 is a corrosion and heat-resistant cobalt base alloy used for moderately stressed parts operating between 1,000 and 1,900° F (538 and 1,038° C). Its applications include gas turbine blades and rotors, combustion chambers, and afterburner parts. It is hardenable only by cold working, and is usually used in the annealed condition. All usual mill forms are available. *Machining Considerations:* L-606 forges moderately well between 1,900 and 2,250° F (1,038 and 1,232° C). In the annealed condition it has excellent formability at room temperature, but severely formed parts should be annealed at 2,225° F (1,218° C) for seven to ten minutes. This alloy is difficult to machine. Its toughness and capacity for work hardening necessitates the use of sharp tools and low cutting speeds. High speed steel or carbide cutting tools are recommended. L-605 can be fusion or resistance welded or brazed, but large or complex fusion weldments should be stress relieved at 1,300° F (704° C) for two hours. Oxidation resistance is excellent to 1,900° F (1,038° C).

HS 188. (This alloy is also known as Haynes 188.) This corrosion and heat-resistant cobalt base alloy is used for moderately stressed parts up to 2,100° F (1,149° C). It has outstanding oxidation resistance up to 2,100° F resulting from the addition of minute amounts of lanthanum (chemical symbol La) to the alloy system. HS 188 exhibits excellent post-aged ductility after prolonged heating of 1,000 hours at temperatures up to 1,600° F (871° C) inclusive. Gas turbine applications include transition ducts, combustion cans, spray bars, flame holders, and liners. It is not hardenable except by cold working, and is used in the solution treated condition. *Machining Considerations:* The alloy can be forged and welded. Welding can be accomplished by both manual and automatic welding methods including electron beam, gas tungsten air, and resistance welding. Like other cobalt base alloys, machining is difficult, necessitating the use of sharp tools and low cutting speeds. High speed steel or carbide cutting tools are recommended.

Other exotic alloys and materials

Additional nickel alloys, plus beryllium, will be discussed in this section.

Nickel Alloys

Monel contains nickel as the base metal, comprising at least 63% of the material content. The crystal structure is face centered cubic, and its high strength and excellent corrosion resistance in freshwater, seawater, chlorinated solvents, many acids (including sulfuric and hydrofluoric), and alkalis makes it a popular choice for pump and marine applications. Monel is a registered trademark of INCO International, Inc.

Monel 400. UNS N04400. Contents: 63 – 66% nickel; 28 – 31.5% copper; 2.5% iron (max.); 2% manganese (max); 0.5% silicon (max); 0.3% carbon (max); and 0.024 sulfur (max). Nickel content includes cobalt. Primarily used for marine engineering, chemical and hydrocarbon processing equipment, valves, pump shafts, fittings, fasteners, and heat exchangers. Standard product forms are round, hexagon, flats, forging stock, pipe, tube, plate, sheet, strip, and wire. This alloy is hardenable only by cold working. In the annealed state, tensile strength at room temperature is 79.8 ksi (550 MPa), and at 800° F (425° C) it is 65.3 ksi (450 MPa). Yield strength is 34.8 ksi (240 MPa) at room temperature, and 24.6 ksi (170 MPa). Density is 0.319 lb/cu in. (8.83 Mb/cu m).

Monel R-405. UNS N04405. Contents: 63 – 66% nickel; 28 – 31.5% copper; 2.5% iron (max.); 2% manganese (max); 0.5% silicon (max); 0.3% carbon (max); and 0.025 – 0.06% sulfur. Nickel content contains cobalt. This is a free machining version of Monel 400. A controlled amount of sulfur is added to the alloy to provide sulfide inclusions that act as chip breakers. Used for meter and valve parts, fasteners, and screw machine products. Standard product forms are round, hexagon, flats, and wire. This alloy is hardenable only by cold working. Mechanical properties are the same as Monel 404.

Monel K-500. UNS N05500. Contents: 63 – 65.5% nickel; 27 – 33% copper; 2.3 – 3.15% aluminum; 2% iron (max.); 1.5% manganese (max); 0.5% silicon (max); 0.35 – 0.85% titanium; 0.25% carbon (max); and 0.01% sulfur. Nickel content contains cobalt. This precipitation hardenable alloy combines the corrosion resistance of Monel 400 with greater strength and hardness. It has low permeability and is nonmagnetic to under –150° F (–101° C). In the age hardened condition, Monel K-500 is more prone to stress corrosion than Monel 400. Used for pump shafts, oil well tools and instruments, springs, valve trim, fasteners, and marine propeller shafts. Product forms are round, hexagon, flats, forging stock, pipe, tube, plate, sheet, strip, and wire. In the precipitation hardened condition, tensile strength at room temperature is 159.5 ksi (1,100 MPa), and at 800° F (425° C) it is 124.7 ksi (860 MPa). Yield strength is 114.5 ksi (790 MPa) at room temperature, and 91.3 ksi (630 MPa). Density is 0.305 lb/cu in. (8.47 Mb/cu m).

MP35N. UNS R30035. (Also known as Carpenter MP35N. MP35N is a registered trademark of SPS Technologies, Inc.) Contents: 33 – 37% nickel; 33% cobalt; 19 – 21% chromium; 9 – 10.5% molybdenum; 1% iron (max.); 1% titanium (max.); 0.15% manganese (max.); 0.15% silicon (max.); 0.025% carbon (max.); 0.015% phosphorus (max.); 0.01% boron; 0.01% sulfur (max.). This alloy is more properly designated a nickel-cobalt-chromium-molybdenum alloy due to the high percentage of each of these elements in its composition. It has ultrahigh tensile strength, good ductility and toughness, and excellent corrosion resistance. It also has excellent resistance to sulfidation, high temperature oxidation, and hydrogen embrittlement. Its properties are developed through work hardening, phase transformation, and aging. In the fully work hardened condition, service temperatures up to 750° F (399° C) are suggested. It is used for such diverse applications as fasteners, springs, nonmagnetic electrical components, and instrument components in medical, seawater, oil well, and chemical and food processing environments. This material is heat treated by aging at 1,000 to 1,200° F (538 to 649° C) for four hours, and air cooling. In its normal state, with 0% cold reduction, tensile strength is 135 ksi (931 MPa), and yield strength is 60 ksi (414 MPa). Cold drawn 35%, and aged at 1,050° F (565° C) for four hours and air cooled, tensile strength is 227 psi (1,565 MPa), and yield strength is 217 ksi (1,496 MPa).

Beryllium

Particles of beryllium and its components, such as beryllium oxide, are toxic. Special precautions must be taken to prevent inhalation. It is the second lightest metal, second only to magnesium, with a density of 0.0067 lb/cu in. (1.855 gr/cu cm). Pure beryllium has a tensile strength of 53.7 ksi (370 MPa), yield strength of 34.8 ksi (240 MPa), and a modulus of elasticity of 43,950 ksi (303 GPa). Its hardness is in the range of R_B 75 – 85. Standard grade beryllium bars, rods, tubing, and machined shapes are produced from vacuum pressed powder with a 1.50% maximum beryllium oxide content. Sheet and plate are fabricated from vacuum hot pressed powder with 2% maximum beryllium oxide content. Beryllium hot pressed block can be forged and rolled, but requires temperatures of 700° F (371° C) and higher because of brittleness. A temperature range of 1,000 to 1,400° F (538 to 760° C) is recommended. Beryllium sheet should be formed at 1,300 to 1,350° F (704 to 732° C), holding at temperature no more than 1.5 hours for maximum springback. Forming above 1,400 C° F (760° C) will result in a reduction in strength.

Carbide tools are most often used when machining beryllium. Mechanical metal removal techniques cause microcracks and metallographic twins. Finishing cuts are usually 0.002 – 0.005 inch (0.05 – 0.127 mm) in depth to minimize surface damage. Finish machining should be followed by chemical etching at least 0.002 inch (0.05 mm) from the surface to remove machining damage. A combination of 1.350° F 732° C) stress relief followed by a 0.0005 inch (0.013 mm) etch may be necessary for close tolerance parts. Damage free machining techniques include chemical milling and electrochemical machining.

Fusion welding is not recommended. Parts may be joined by brazing, soldering, braze welding, adhesive bonding, diffusion bonding, squeeze riveting, and bolting, threading, or press fitting techniques specifically designed to avoid damage.

Beryllium is alloyed in beryllium coppers (covered in the Copper section), and beryllium nickel (approximately 2% Be, 0.5 Ti, and the remainder Ni), and aluminum. Beralcast 191 (trademark of the Starmet Corp.) contains 61 – 68% beryllium; 27.5 – 34.5%

aluminum; 1.65 – 2.35 silver; 1.65 – 2.5% silicon; and 0.2% iron (max.). It is lighter than aluminum and titanium, but has greater ductility than pure beryllium. It is several times stiffer than aluminum, magnesium, aluminum, or aluminum base metal matrix composites. Its uses include satellite components, avionics packaging, aerospace systems, and golf chubs. The density of Beralcast 191 is 2.16 gr/cu cm, tensile strength is 28.5 ksi (196.5 MPa), yield strength is 20 ksi (137.9 MPa), and modulus of elasticity is 29,300 ksi (202 GPa).

Nonmetallic Engineering Materials

Plastics

When applied to materials, "plastics" usually refers to a class of synthetic organic materials which, though solid in the finished form, were at some stage in their processing fluid enough to be shaped. These materials may be divided into two general categories: thermoplastics which, like paraffin, can be softened and resoftened repeatedly without undergoing a change in chemical composition; and thermosetting resins which undergo a chemical change with the application of heat and pressure, and cannot be resoftened. Plastics in finished form are made up of long chain molecules (polymers) that are built by combining single molecules (monomers) under heat and pressure. Cross-linking is a permanent connection between two polymer molecules binding them together through a system involving primary chemical bonds.

Plastics have many desirable characteristics including light weight, high strength-to-weight ratio, and ease and economy of fabrication. Their specific gravity roughly ranges from 0.92 to 2.3, and their weight from 0.033 to 0.079 pounds per cubic inch (by comparison, the specific gravity for aluminum ranges from 0.091 to 0.108, and for steel it is 0.283). In addition, most plastics have excellent electrical resistance, high heat insulation properties, and good resistance to corrosion and chemical action. Many plastics have self-lubricating characteristics, and many can be made transparent, translucent, opaque, and colored. Substantial increases in temperature will cause a decline in the physical properties of most plastics—although some plastics soften at 180° F (82° C), most can resist temperatures up to 300° F (149° C)—but some can be used successfully in temperatures reaching 500 to 600° F (260 to 316° C). While many plastics should not be used at temperatures below –40° F (–40° C), a few can be used at cryogenic temperature. Plastics are subject to higher thermal expansion, creep, cold flow, low temperature embrittlement, and deformation under load than metals, some varieties change dimensions through solvent and moisture absorption, while others are degraded by ultraviolet and nuclear radiation.

The cost per pound for plastics exceeds that of steel and some other materials. However, they may be among the cheapest to use. Despite their low operating temperatures, plastics may be superior to metals as high temperature heat shields for short exposures (thermal conductivity values for plastics are in the range of 0.002; for aluminum the value is 0.500). Some reinforced plastics (glass reinforced epoxies, polyesters, and phenolics) are now nearly as strong and rigid (particularly in relation to weight) as most steels, and they may be more dimensionally stable. Some oriented films and sheets (oriented polyesters) may have greater strength-to-weight ratios than cold rolled steels.

Definitions of plastics terms

Acetal resins. Copolymers containing the acetal linkage ($-CH_2-O-$), e.g., polyoxymethylene.

Acrylic resin. A synthetic resin prepared from acrylic acid or from a derivative of acrylic acid.

Alkyd resins. Polyesters made from dicarboxylic acids and diols, primarily used as coatings, modified with vegetable oil, fatty acids, etc.

Arc resistance. Time required for a given applied electrical voltage to render the surface of a material conductive because of carbonization by the arc discharge.

Autoclave. A closed vessel producing an environment of fluid pressure, with or without heat, to an enclosed object that is undergoing a chemical reaction or other operation.

Autoclave molding. A process similar to the pressure bag technique. The lay-up is covered by a pressure bag and the entire assembly is placed in an autoclave capable of providing heat and pressure for curing the part. The pressure bag is normally vented to the outside.

Bag molding. A method of molding or laminating that involves the application of a fluid pressure to a flexible material which transmits the pressure to the material being molded or bonded. Fluid pressure is usually applied by means of air, steam, water, or vacuum.

Breakdown voltage. The voltage required, under specific conditions, to cause the failure of an insulating material. See also dielectric strength in the next section on standardized plastics tests.

Butadiene. A diene monomer with the structure $CH_2=CH-CH=CH_2$. May be copolymerized with styrene and with acrylonitrile. Its homopolymer is used as a synthetic rubber.

Cellulose acetate. An acetic acid ester of cellulose obtained by the action, under rigidly controlled conditions, of acetic acid and acetic anhydride on purified cellulose usually obtained from cotton liners. All three available hydroxyl groups in each glucose unit of the cellulose can be acetylated, but in the preparation of cellulose acetate it is usual to acetylate fully and then to lower the acetyl value (expressed as acetic acid) to 52 to 56% by partial hydrolysis. When compounded with suitable plasticizers, the result is a tough thermoplastic material.

Cellulose acetate butyrate. An ester of cellulose made by the action of a mixture of acetic and butyric acids and their anhydrides on a purified cellulose. It is used in the manufacture of plastics that are similar in their general properties to cellulose acetate but are tougher and have better moisture resistance and dimensional stability.

Clarity. The characteristic of a transparent body that allows distinct high-contrast images or high-contrast objects to be observable through the body.

Crazing. Fine cracks that may extend in a network on or under the surface or through a layer of a plastic material. Usually occurs in the presence of an organic liquid or vapor, with or without the application of mechanical stress.

Creep. The dimensional change with time of a material under load that follows the initial instantaneous elastic deformation. Creep at room temperature is called cold flow.

Cross-linking. The formation of primary valence bonds between polymer molecules. When extensive, as in thermosetting resins, cross-linking makes one infusible, insoluble supermolecule of all the chains.

Crystallinity. A state of molecular structure in some resins that denotes stereo-regularity and compactness of the molecular chains forming the polymer. Normally can be attributed to the formation of solid crystals having a definite geometric form.

Denier. A direct numbering system for expressing linear density, equal to the mass in grams per 9,000 meters of yarn, filament, fiber, or other textile strand.

Dimensional stability. The ability of a plastic part to retain its original dimensions during its service life.

Elastomer. A material that at room temperature stretches under low stress to at least twice its length and snaps back to the original length upon release of the stress.

Encapsulating. Encasing an article (usually an electronic component) in a closed envelope of plastic by immersing the object in a casting resin and allowing the resin to polymerize or, if hot, to cool.

Epoxy resins. Based on ethylene oxide, its derivatives, or homologs, epoxy resins form straight-chain thermoplastics and thermosetting resins, e.g., by the condensation of bisphenol and epichlorohydrin to yield a thermoplastic that is converted to a thermoset by active hydrogen-containing compounds, e.g., polyamines and dianhydrides.

Ethylene-vinyl acetate. Copolymer of ethylene and vinyl acetate having many of the properties of polyethylene but of considerably increased flexibility for its density; elongation and impact resistance are also increased.

Fiber. This term usually refers to relatively short lengths of very small cross sections of various materials. Fibers can be made by chopping filaments (converting). Staple fibers may be 0.5 in. (13 mm) to several inches or centimeters in length and usually from one to five denier.

Fiberglass. A widely used reinforcement for plastics that consists of fibers made from borosilicate and other formulations of glass. The reinforcements are in the form of roving (continuous or chopped), yarns, mat, milled, or woven fabric.

Filament winding. Roving or single strands of glass, metal, or other reinforcement are wound in a predetermined pattern onto a suitable mandrel. The pattern is designed to give maximum strength in the directions required. The strands can be run from a creel through a resin bath before winding, or preimpregnated materials can be used. When the right number of layers has been applied, the wound mandrel is cured at room temperature or in an oven.

Filler. An inexpensive, inert substance added to a plastic to make it less costly. Fillers may also improve physical properties, particularly hardness, stiffness, and impact strength. The particles are usually small in contrast to those of reinforcement, but there is some overlap between the functions of the two types of material.

Film. An optional term for sheeting having a nominal thickness not greater than 0.01 in. (0.254 mm).

Flame-retardant resin. A resin compounded with certain chemicals to reduce or eliminate its tendency to burn. For polyethylene and similar resins, chemicals such as antimony trioxide and chlorinated paraffins are useful.

Furan resins. Dark-colored, thermosetting resins that are primarily liquids ranging from low-viscosity polymers to thick, heavy syrups. Based on furfuryl or furfuryl alcohol.

Heat distortion point. The temperature at which a standard test bar deflects 0.01 in. (0.254 mm) under a stated pressure of either 66 or 264 PSI (0.455 or 1.82 MPa).

Homopolymer. A polymer consisting of only one monomeric species.

Melamine formaldehyde resin. A synthetic resin derived from the reaction of melamine (2,4,6-triamino-1,3,5-triazine) with formaldehyde.

Monomer. A relatively simple compound that can react to form a polymer. See also Polymer.

Nylon. The generic name for all synthetic fiber-forming polyamides. They can be formed into monofilaments and yarns characterized by great toughness, strength and elasticity, high melting point, and good resistance to water and chemicals. The material is widely used for bristles in industrial and domestic brushes and for many textile applications; it is also used in injection molding gears, bearings, combs, etc.

Olefins. A group of unsaturated hydrocarbons of the general formula $C_x H_{2y}$ and named after the corresponding paraffins by the addition of "ene" or "ylene" to the stem. Examples are ethylene and pentene-1.

Organic. A material or compound composed of hydrocarbons or their derivatives, or those materials found naturally or derived from plant or animal origin.

Permeability. (1) The passage or diffusion of vapor, liquid, or solid through a barrier without physically or chemically affecting the barrier. (2) The rate of such passage.

Phenolic resin. A synthetic resin produced by the condensation of phenol with formaldehyde. Phenolic resins form the basis of a family of thermosetting molding materials, laminated sheet and oven-drying varnishes. They are also used as impregnating agents and as components of paints, varnishes, lacquers, and adhesives.

Plasticizer. Chemical agents added to plastic compositions to improve flow and processability and to reduce brittleness. These improvements are achieved by lowering the glass transition temperature.

Polyamide. A polymer in which structural units are linked by amide grouping. Many polyamides are fiber formers.

Polycarbonate resins. Polymers derived from the direct reaction between aromatic and aliphatic dihydroxy compounds with phosgene or by the ester exchange reaction with appropriate phosgene-derived precursors. Structural units are linked by carbonate groups.

Polyester. A resin formed by the reaction between a dibasic acid and a dihydroxyalcohol–both organic–or by the polymerization of a hydroxy carboxylic acid. Modification with multifunctional acids and/or alcohols and some unsaturated reactants permit cross-linking to thermosetting resins.

Polyethylene. A thermoplastic material composed solely of ethylene. It is normally a translucent, tough, waxy solid that is unaffected by water and by a large range of chemicals.

Polyimide resins. Aromatic polyimides made by reacting pyrometallic dianhydride with aromatic diamines. Characterized by high resistance to thermal stress. Applications include components for internal combustion engines.

Polymer. A high-molecular-weight compound, natural or synthetic, whose structure can usually be represented by a repeated small unit, the -mer, e.g., polyethylene, rubber, and cellulose. Synthetic polymers are formed by addition or condensation polymerization of monomers. Some polymers are elastomers, some are plastics, and some are fibers.

Polymerization. A chemical reaction in which the high-molecular-weight molecules are formed from the original substances. When two or more monomers are involved, the process is called copolymerization or heteropolymerization.

Polyurethane resins. A family of resins produced by reacting diisocyanates in excess with glycols to form polymers having free isocyanate groups. Under the influence of heat or certain catalysts, these groups will react with each other or with water, glycols, etc., to form a thermoset.

Prepreg. Ready to mold or cure material in sheet form which may be tow, tape, cloth, or mat impregnated with resin. It may be stored before use.

Reinforcement. A strong inert material put into a plastic to improve its strength, stiffness, and impact resistance. Reinforcements are usually long fibers of glass, sisal, cotton, etc., in woven or nonwoven form. To be effective, the reinforcing material must form a strong adhesive bond with the resin.

Resin. Any of a class of solid or semisolid organic products of natural or synthetic origin, generally of high molecular weight, and with no definite melting point. Most resins are polymers. The term "resin" is often used synonymously with "plastic" or "polymer," but it more accurately refers to any thermoplastic or thermosetting type plastic existing in either the solid or liquid state before processing.

Resistivity. The ability of a material to resist passage of electrical current either through its bulk or on a surface. The unit of volume resistivity is the ohm-cm, and of surface resistivity it is the ohm.

Roving. A number of strands, tows, or ends collected into a parallel bundle with little or no twist. In spun yarn production, an intermediate state between silver and yarn.

Self-extinguishing. A somewhat loosely used term describing the ability of a material to cease burning once the source of flame has been removed.

Sheet (thermoplastic). A flat section of a thermoplastic resin with the length considerably greater than the width and 0.01 in. (0.254 mm) or greater in thickness.

Stabilizer. An ingredient used in the formulation of some polymers to assist in maintaining the physical and chemical properties of the compounded materials at their initial values throughout the processing and service life of the material, e.g., heat and ultraviolet stabilizers.

Tack. Stickiness of the prepreg.

Thermal expansion (coefficient of). The fractional change in length (sometimes volume specified) of a material for a unit change in temperature. Values for plastics range from 10^{-5} to 2×10^{-4} mm/mm °C (1.8×10^{-5} to 36×10^{-5} in./in. °F).

Twist. The number of turns about its axis per unit of length in a yarn or other textile strand. It may be expressed as turns per inch (tpi) or turns per centimeter (tpcm).

Vacuum bag molding. A process in which the lay-up is cured under pressure generated by drawing a vacuum in the space between the lay-up and a flexible sheet placed over it and sealed at the edges.

Volume resistivity (specific insulation resistance). The electrical resistance between opposite faces of a 1-centimeter (0.39-inch) cube of insulating material. It is measured under prescribed conditions using a direct current potential after a specified time of electrification. It is commonly expressed in ohm-centimeters. The recommended test is ASTM D 257.

Weathering. The process of degradation and decompostion that results from exposure to the atmosphere, to chemical action, and to the action of other natural environmental factors.

Standard tests for rating plastic materials

Testing the mechanical, thermal, and electrical properties of plastics is carried out under a series of tests defined primarily by ASTM. The following descriptions of these tests and rating criteria were provided by Quadrant Engineered Plastics Products.

Coefficient of Friction (ASTM D 3702). Coefficient of Friction (COF) is the measure of resistance to the sliding of one surface over another. Testing can be conducted in a variety of ways although thrust washers testing is most common. The results do not have a unit of measure associated with them since the COF is the ratio of sliding force to normal force acting on two mating surfaces. COF values are useful to compare the relative "slickness" of various materials, usually run unlubricated over or against polished steel. Since the value reflects sliding resistance, the lower the value, the "slicker" the bearing material. Two values are usually given for COF. "Static" COF refers to the resistance at initial movement from a bearing "at rest." "Dynamic" COF refers to the resistance once the bearing or mating surface is in motion at a given speed.

Coefficient of Linear Thermal Expansion (E 831 TMA). The coefficient of linear thermal expansion (CLTE) is the ratio of the change in a linear dimension to the original dimensions of the material for a unit change of temperature. It is usually measured in units of in./in./°F. CLTE is a very important consideration if dissimilar materials are to be assembled in applications involving large temperature changes. A thermoplastic's CLTE can be decreased (making it more dimensionally stable) by reinforcing it with glass fibers or other additives. The CLTE of plastics varies widely. The most stable plastics approach the CLTE of aluminum but exceed that of steel by up to ten times.

Compressive Strength (ASTD D 695). Compressive strength measures a material's ability to support a compressive force.

Continuous Service Temperature. This value is most commonly defined as the maximum ambient service temperature (in air) that a material can withstand and retain at least 50% of its initial physical properties after long-term service (approximately 10 years). Most thermoplastics can withstand short-term exposure to higher temperatures without significant deterioration. When selecting materials for high temperature service, both heat deflection temperature and continuous service temperature need to be considered.

Dielectric Constant (ASTM D 150$_{(2)}$). The Dielectric Constant, or permittivity, is a measure of the ability of a material to store electrical energy. Polar molecules and induced dipoles in a plastic will align themselves with an applied electric field. It takes energy to make this alignment occur. Some of the energy is converted to heat in the process. This loss of electrical energy in the form of heat is called dielectric loss, and is related to the dissipation factor. The rest of the electrical energy required to align the electric dipoles is stored in the material. It can be released at a later time to do work.

The higher the dielectric constant, the more electrical energy can be stored. A low dielectric constant is desirable in an insulator, whereas someone wanting to build a capacitor will look for materials with high dielectric constants. Dielectric constants are dependent on frequency, temperature, moisture, chemical contamination, and other factors.

Dielectric Strength (ASTM D 149). When an insulator is subjected to increasingly high voltages, it eventually breaks down and allows a current to pass. The voltage reached before breakdown divided by the sample thickness is the dielectric strength of the material, measured in volts/mil. It is generally measured by putting electrodes on either side of a test specimen and increasing the voltage at a controlled rate. Factors that affect dielectric strength in applications include: temperature, sample thickness, conditioning of the sample, rate of increase in voltage, and duration of test. Contamination or internal voids in the sample also affect dielectric strength.

Dissipation Factor (ASTM D 150). The dissipation factor, or dielectric loss tangent, indicates the ease with which molecular ordering occurs under the applied voltage. It is most commonly used in conjunction with dielectric constant to predict power loss in an insulator.

Elongation (ASTM D 638). Elongation, which is always associated with tensile strength, is the increase in length at fracture.

Flammability. In electrical applications (or any applications where plastic constitutes a significant percentage of an enclosed space), the consequences of exposure to an actual flame must be considered (i.e., plastic panels used in the interior of an aircraft cabin). Flammability tests measure combustibility, smoke generation, and ignition temperatures of materials.

UL 94 Flammability Class (HB, V-2, V-1, V-0, 5V). In this test, specimens are subjected to a specified flame exposure. The relative ability to continue burning after the flame is removed is the basis for classification. In general, the more favorable ratings are given to materials that extinguish themselves rapidly and do not drip flaming particles. Each rating is based on a specific material thickness (i.e., UL 94–V1 @ $^1/_8$" thick). The UL rating scale from highest burn rate (HB) to most flame retardant is HB, V-2, V-1, V-0, 5V.

Flexural Strength (ASTM D 790). Flexural properties measure a material's resistance to bending under load. For plastics, the data is usually calculated at 5% deformation/strain, which is the loading necessary to stretch the outer surface 5%.

Glass Transition (Tg) Temperature (ASTM D 3418). The glass transition temperature, Tg, is the temperature above which an amorphous polymer becomes soft and rubbery. Except when thermoforming, it is important to ensure that an amorphous polymer is used below its Tg if reasonable mechanical performance is expected.

Heat Deflection Temperature (ASTM D 648). The heat deflection temperature is the temperature at which a $^1/_2$" thick test bar, loaded to a specified bending stress, deflects by 0.010 in. It is sometimes called the "heat distortion temperature" (HDT). This value is used as a relative measure of the ability of various materials to perform at elevated temperatures short term, while supporting loads.

Melting Point (ASTM D 3418). The temperature at which a crystalline thermoplastic changes from a solid to a liquid.

Specific Gravity (ASTM D 792). Specific gravity is the ratio of the mass of a given volume of material compared to the mass of the same volume of water, both measured at 73°F (23°C). (Density of a material divided by the density of water.) Since it is a dimensionless quantity, it is commonly used to compare materials. Specific gravity is used extensively to determine part cost and weight. Materials with specific gravities less than 1.0 (such as polyethylene and polypropylene) float in water. This can help with identification of an unknown plastic.

Surface Resistivity (ASTM D 257). This test measures the ability of current to flow over the surface of a material. Unlike the volume resistivity test, the test electrodes are both placed on the same side of the test specimen. However, like volume resistivity, surface resistivity is affected by environmental changes such as moisture absorption. Surface resistivity is used to evaluate and select materials for testing when static charge dissipation or other surface characteristics are critical.

Tensile Strength (ASTM D 638). Ultimate tensile strength is the force per unit area required to break a material under tension. It is expressed in PSI or Pascals. The force required to pull apart one square inch of plastic may range from 1,000 to 50,000 PSI (6.8 to 345 MPa).

Volume Resistivity (ASTM D 257). The volume resistivity of a material is its ability to resist the flow of electricity, expressed in ohms-cm. The more readily the current flows, the lower the volume resistivity. Volume resistivity can be used to predict the current flow from an applied voltage as demonstrated by Ohm's Law.

$$V = IR$$

where: V = Applied voltage (volts)
   I = Electrical current (amperes)
   R = Resistance of the wire (ohms).

As the Resistivity Continuum in *Figure 1* indicates:
- insulators exhibit resistivities of 10^{12} and higher
- antistatic/partially conductive products exhibit resistivities of 10^5 to 10^{12}
- conductive products exhibit resistivities of 10^6 to 10^5.

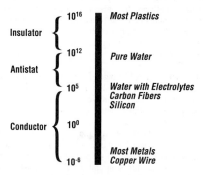

Figure 1. The resistivity continuum. (Source, Quadrant Engineered Plastic Products.)

Water Absorption (ASTM D 570). Water absorption is the percentage increase in weight of a material due to absorption of water. Standard test specimens are first dried then weighed before and after immersion in 73°F (23°C) water. Weight gain is recorded after 24 hours, and again when saturation is reached. Both percentages are important since they reflect absorption rate. Mechanical and electrical properties, and dimensional stability are affected by moisture absorption.

Thermoplastics

Thermoplastics (TP) are a relatively large group of synthetic organic materials that may be divided into three broad categories: 1) rigid load bearing materials, 2) environmental resistant materials, and 3) general purpose materials. Thermoplastics can be repeatedly softened and hardened by heating and cooling, and are susceptible to the effects of temperatures, time, environment, loading rate, and processing. These effects must be charted for each material in order to evaluate engineering usage in relation to performance requirements. **Table 1** gives abbreviations (which are routinely substituted for the full name) for popular thermoplastics and thermoset plastics, and **Table 2** provides selected properties for plastics commonly used in manufacturing.

Rigid Load Bearing Materials

These load bearing plastic materials approach Hooken behavior (Hooke's law relates to elasticity and essentially states that stress is proportional to strain up to the proportional limit of the material). Therefore, reasonable predictability of their behavior may be based on the established equations of state that are used for metals.

ABS (acrylonitrile-butadiene-styrene) resins. The three basic monomers that are used in ABS plastic are combined in varying proportions to form the commercially available resins. The acrylonitrile provides chemical resistance, the butadiene gives greater toughness and impact resistance, and the styrene-acrylonitrile copolymer provides good strength and rigidity. The main variations are as follows. 1) Medium impact, which is a hard, rigid, tough material used for appearance parts requiring high strength, good fatigue resistance, surface hardness, and gloss. 2) High impact, which is used where additional impact strength is required at the expense of hardness and rigidity. 3) Extra high impact, which has the highest impact strength, but further decreases in strength, rigidity, and hardness. 4) Low temperature impact, which is designed for high impact strength at temperatures as low as –40° F (–40° C). Strength, rigidity, and heat resistance are lowered. 5) High strength and heat resistant, which provides maximum heat resistance (heat distortion point at 264 PSI [1820 kPa] is about 229° F [110° C]). Its impact strength is comparable to the high impact type, but it has higher tensile and flexural strength, stiffness, and hardness. ABS resins are available as compounds for injection molding, blow molding, extrusion, and calendaring as well as sheet for thermoforming.

The ultimate tensile strength of ABS ranges from 3,000–9,000 PSI (20.7–62 MPa), and it has excellent resistance to many corrosive materials. It is used for a wide variety of products including instrument panels and interior automotive trim, keyboard housings, portable appliance housings, pipes, fittings, helmets, electrical connectors, luggage shells, and deflectors for hot air systems.

Acetals. There are two basic acetal resins—a homopolymer and a copolymer. In general, the copolymer has better stability and resistance to heat aging, and the homopolymer offers slightly better "as molded" mechanical properties. These are among the strongest and stiffest thermoplastics, and compression and tensile strengths show linearity with temperatures ranging from –45 to 250° F (–43 to 121° C). They have the highest fatigue endurance limits of all thermoplastics, and are only lightly affected by moisture. Good

Table 1. Abbreviations Used for Selected Plastics, Polymers, Elastomers, and Chemicals.

ABA	acrylonitrile-butadiene-acrylate	ABS	acrylonitrile-butadiene-styrene copolymer
ACM	acrylic acid ester rubber	ACS	acrylonitrile-chlorinated PE-styrene
AES	acrylonitrile-ethylene-propylene-styrene	AMMA	acrylonitrile-methyl methacrylate
AN	acrylonitrile	AO	antioxidant
APET	amorphous polyethylene terephthalate	APP	atactic polypropylene
ASA	acrylic-styrene-acrylonitrile	ATH	aluminum trihydrate
AZ(O)	azodicarbonamide	BMI	bismaleimide
BO	biaxially-oriented (film)	BOPP	biaxially-oriented polypropylene
BR	butadiene rubber	BS	butadiene styrene rubber
CA	cellulose acetate	CAB	cellulose acetate butyrate
CAP	cellulose acetate propionate	CAP	controlled atmosphere packaging
CF	cresol formaldehyde	CFC	chlorofluorocarbons
CHDM	cyclohexanedimethanol	CN	cellulose nitrate
COP	copolyester	COPA	copolyamide
COPE	copolyester	CP	cellulose propionate
CPE	chlorinated polyethylene	CPET	crystalline polyethylene terephthalate
CPP	cast polypropylene	CPVC	chlorinated polyvinyl chloride
CR	chloroprene rubber	CS	casein
CTA	cellulose triacetate	DABCO	diazobicyclooctane
DAM	diallyl maleate	DAP	diallyl phthalate
DCPD	dicyclopentadiene	DE	diotamaceous earth
DEA	dielectric analysis	DETDA	diethyltoluenediamine
DMT	dimethyl ester of terephthalate	EAA	ethylene acrylic acid
EB	electron beam	EBA	ethylene butyl acrylate
EC	ethyl cellulose	ECTFE	ethylene-chlorotrifluoroethylene copolymer
EEA	ethylene-ethyl acrylate	EG	ethylene glycol
EMA	ethylene-methyl acrylate	EMAA	ethylene-methacrylic acid
EMAC	ethylene-methyl acrylate copolymer	EMPP	elastomer modified polypropylene
EnBA	ethylene normal butyl acrylate	EP	epoxy resin, also ethylene-propylene
EPDM	ethylene-propylene terpolymer rubber	EPM	ethylene-propylene rubber
EPS	expandable polystyrene	ESI	ethylene-styrene copolymers
ETFE	ethylene-tetrafluoroethylene copolymer	EVA(C)	polyethylene-vinyl acetate
EVOH	polyethylene-vinyl alcohol copolymers	FEP	fluorinated ethylene propylene copolymer
FPVC	flexible polyvinyl chloride	GPC	gel permeation chromotography
GPPS	general purpose polystyrene	GTP	group transfer polymerization
HALS	hindered amine light stabilizer	HAS	hindered amine stabilizers
HCFC	hydrochlorofluorocarbons	HCR	heat-cured rubber
HDI	hexamethylene diisocyanate	HDPE	high-density polyethylene
HDT	heat deflection temperature	HFC	hydrofluorocarbons
HIPS	high-impact polystyrene	HMDI	diisocyanato dicyclohexylmethane
HNP	high nitrile polymer	IPI	isophorone diisocyanate

(Continued)

Table 1. *(Continued)* **Abbreviations Used for Selected Plastics, Polymers, Elastomers, and Chemicals.**

IV	intrinsic viscosity	LCP	liquid crystal polymers
LDPE	low-density polyethylene	LLDPE	linear low-density polyethylene
LP	low-profile resin	MbOCA	3,3'-dichloro-4,4-diamino-diphenylmethane
MBS	methacrylate-butadiene-styrene	MC	methyl cellulose
MDI	methylene diphenylene diisocyanate	MEKP	methyl ethyl ketone peroxide
MF	melamine formaldehyde	MMA	methyl methacrylate
MPE	metallocene polyethylenes	MPF	melamine-phenol-formaldehyde
NBR	nitrile rubber	NDI	naphthalene diisocyanate
NR	natural rubber	ODP	ozone depleting potential
OFS	organofunctional silanes	OPET	oriented polyethylene terephthalate
OPP	oriented polypropylene	O-TPV	olefinic thermoplastic vulcanizate
OEM	original equipment manufacturer	OSA	olefin-modified styrene-acrylonitrile
PA	polyamide	PAEK	polyaryletherketone
PAI	polyamide imide	PAN	polyacrylonitrile
PB	polybutylene	PBA	physical blowing agent
PBAN	polybutadiene-acrylonitrile	PBI	polybenzimidazole
PBN	polybutylene naphthalate	PBS	polybutadiene styrene
PBT	polybutylene terephthalate	PC	polycarbonate
PCC	precipitated calcium carbonate	PCD	polycarboiimide
PCR	post-consumer recyclate	PCT	polycyclohexylenedimethylene terephthalate
PCTA	copolyester of CHDM and PTA	PCTFE	polychlorotrifluoroethylene
PCTG	glycol-modified PCT copolymer	PE	polyethylene
PEBA	polyether block polymide	PEC	chlorinated polyethylene
PEDT	3,4 polyethylene dioxithiophene	PEEK	polyetheretherketone
PEI	polyether imide	PEK	polyetherketone
PEKEKK	polyetherketoneetherketoneketone	PEN	polyethylene naphthalate
PES	polyether sulfone	PET	polyethylene terephthalate
PETG	PET modified with CHDM	PF	phenol formaldehyde
PFA	perfluoroalkoxy resin	PI	polyimide
PID	proportional, integral, derivative	PIBI	butyl rubber
PMDI	polymeric methylene diphenylene	PMMA	polymethyl methacrylate
PMP	polymethylpentene	PO	polyolefins
POM	polyacetal	PP	polypropylene
PPA	polyphthalamide	PPC	chlorinated polypropylene
PPE	polypropylene ether, modified	PPO	polyphenylene oxide
PPS	polyphenylene sulfide	PPSU	polyphenylene sulfone
PS	polystyrene	PSU	polysulfone
PTA	purified terephthalic acid	PTFE	polytetrafluoroethylene
PU	polyurethane	PUR	polyurethane
PVC	polyvinyl chloride	PVCA	polyvinyl chloride acetate
PVDA	polyvinylidene acetate	PVDC	polyvinylidene chloride

(Continued)

Table 1. *(Continued)* **Abbreviations Used for Selected Plastics, Polymers, Elastomers, and Chemicals.**

PVDF	polyvinylidene fluoride	PVF	polyvinyl fluoride
PVOH	polyvinyl alcohol	RHDPE	recycled high density polyethylene
RPET	recycled polyethylene terephthalate	SI	silicone plastic
SAN	styrene acrylonitrile copolymer	SB	styrene butadiene copolymer
SBC	styrene block copolymer	SBR	styrene butadiene rubber
SMA	styrene maleic anhydride	SMC	sheet molding compound
SMC-C	SMC-continuous fibers	SMC-D	SMC-directionally oriented
SMC-R	SMC-randomly oriented	TA	terephthalic acid
TDI	toluene diisocyanate	TEO	thermoplastic elastomeric olefin
TLCP	thermoplastic liquid crystal polymer	T/N	terephthalate/naphthalate
TPA	terephthalic acid	TP	thermoplastic
TPE	thermoplastic elastomer	TPO	thermoplastic olefins
TPU	thermoplastic polyurethane	TPV	thermoplastic vulcanizate
TS	thermoset	UF	urea formaldehyde
ULDPE	ultralow-density polyethylene	UP	unsaturated polyester resin
UR	urethane	VA(C)	vinyl acetate
VC	vinyl chloride	VDC	vinylidene chloride
VLDPE	very low-density polyethylene	ZNC	Ziegler-Natta catalyst

Table 2. Properties of Plastics Used in Manufacturing. *(Source, Plastics Technology Handbook.)*

Application (See Notes)	Melt Flow Rate	Density	Tensile Strength	Elong. at Yield	Flexibility Modulus	Notched Izod Impact[2]	Deflection Temp. °F	UL 94 Rating[3]
	gr/10 min.	gr/cu. cm.	ksi	%	10 E+5	Ft/lb in.	66 psi/264 psi	1/8 in.
ABS (Acrylonitrile-Butadiene-Styrene)								
BM	0.80	1.02-1.05	5.0-6.67	-	2.4-2.95	8.0-8.15	212/192	HB
EX	0.4-2.5	1.02-1.07	4.4-6.7	2-40	2.4-3.5	3.5-8.4	195/187	HB
IM	8.0-5.8	1.03-1.06	5.6-7.7	4-36	3.2-4.0	2.0-6.0	209/185	V-0, HB
ABS/Polycarbonate Alloy								
IM	2.8-10	1.2-1.8	7.7-8.5	3-5	2.7-4.2	3.0-9.0	250/235	V-0, HB
Acetal								
IM	6.5-9.0	1.42-1.54	8.4-9.6	10-30	3.3-6.5	1.2-1.3	325/250	HB
Acrylics								
EX/IM	3.3-7.5	1.17-1.19	7.0-10.0	3-5	3.6-4.7	0.04-0.06	185/170	HB
Cellulose Acetate Butyrate								
IM	-	1.17-1.21	3.5-6.2	-	1.6-2.5	3.0-6.5	195/175	-
PTFE (Polytetrafluoroethlene)								
CM	-	2.1-2.19	2.5-4.2	200-400	0.7-0.82	3.5	-	V-0
Ionomers								
EBF	1.1-1.7	0.94	-	350-400	-	No Break	-	-

(Continued)

Table 2. *(Continued)* **Properties of Plastics Used in Manufacturing.** *(Source, Plastics Technology Handbook.)*

Application (See Notes)	Melt Flow Rate	Density	Tensile Strength	Elong. at Yield	Flexibility Modulus	Notched Izod Impact[2]	Deflection Temp. °F	UL 94 Rating[3]
	gr/10 min.	gr/cu. cm.	ksi	%	10 E+5	Ft/lb in.	66 psi/264 psi	1/8 in.
Liquid Crystal Polymers								
IM	-	1.3-1.65	24.0-26.0	-	16.0-26.0	1.7-2.3	470/445	V-0
Nylon Type 6								
EX	2.6-4.5	1.12-1.13	3.8-10.5	5-200	3.5-4.6	0.9-2.3	250/150	V-2, HB
IM	-	1.1-1.3	9.0-20.0	3-15	3.4-10.0	1.0-3.5	390/175	V-0, HB
Nylon Type 12								
IM	3.0-9.0	1.02-1.25	5.0-13.5	6-20	1.6-7.0	1.5-4.7	275/135	HB
Nylon Type 66								
IM	-	1.15-1.39	9.5-25	-	4.1-12.0	0.9-4.5	475/450	V-0, HB
Polycarbonate (PC)								
EX	5.0-10.0	1.2-1.17	8.5-9.5	6-7	3.3-3.5	6.0-15.0	290/270	V-0, HB
IM	7.5-16.5	1.2-1.25	8.7-9.5	6-8	3.1-8.5	4.0-15.0	275-265	V-0, HB
Thermoplastic Polyesters (PBT)								
IM	8.0-12.0	1.41-1.65	6.0-17.4	2.5-7	5.1-14.0	1.1-10.0	400/350	V-0, HB
Thermoplastic Polyesters (PET)								
IM	-	1.55-1.70	17.2-22.4	1.5-3.0	10.2-16.8	1.2-1.9	460/440	V-0, HB
Polyethersulfone (PES)								
IM	-	1.4-1.65	10.0-21.5	1.3-2.9	3.8-12.0	1.3-1.6	410/400	V-0
Polyethylene (LDPE-unfilled, 0.910-0.940 gm/cu cm)								
EX-Film	0.25-6.5	0.92-0.924	1.2-1.7	100-600	-	-	-	-
IM	5-65	0.92-0.924	1.25-1.7	190-400	0.18-0.45	-	-	-
Polyethylene (HDPE-unfilled, 0.940 gm/cu cm and higher)								
BM	0.4-9.5	0.95-0.96	3.5-4.5	10-800	1.1-2.2	1.9-8.8	175 @ 66psi	-
IM	9.0-80.0	0.95-0.96	3.5-4.2	10-800	1.2-2.2	0.35-1.2	175 @ 66psi	-
Polyphenylene Ether/Oxide (PPO/PPE) Based Resins								
IM	-	1.07-1.27	6.0-12.5	-	3.3-10.0	2.0-6.5	270/260	V-0, HB
Polyphenylene Sulfide (PPS)								
IM	-	1.5-1.75	4.5-23.0	1.1-3.4	5.5-20.0	0.9-1.5	480/470	V-0
Polypropylene-Unfilled								
EX-Film	3.5-12.0	0.89-0.91	3.1-4.8	6-14	1.5-2.3	-	220 @ 66	-
IM	2.0-65.0	0.9-1.04	3.7-6.2	9-14	2.6-2.4	0.5-2.0	210 @ 66 psi	-
Polypropylene-Filled								
IM	5.0-18.0	1.05-1.2	3.5-10.8	-	2.2-6.5	0.6-1.4	260/170	HB
Polystyrene (PS general purpose)								
IM	4.0-10.0	1.04-1.2	4.2-6.8	-	3.7-4.5	0.3-1.0	200/175	V-0, HB
Polystyrene-Impact								
IM	5.0-9.5	1.03-1.05	3.1-5.7	-	2.7-3.5	1.1-3.2	190/175	V-0, HB

(Continued)

Table 2. *(Continued)* **Properties of Plastics Used in Manufacturing.** *(Source, Plastics Technology Handbook.)*

Application (See Notes)	Melt Flow Rate	Density	Tensile Strength	Elong. at Yield	Flexibility Modulus	Notched Izod Impact[2]	Deflection Temp. °F	UL 94 Rating[3]
	gr/10 min.	gr/cu. cm.	ksi	%	10 E+5	Ft/lb in.	66 psi/264 psi	1/8 in.
Styreneacrylonitrile Copolymers (SAN)								
IM	2.0-8.5	1.07-1.35	9.3-14.0	1.2-1.5	1.3-5.5	0.3-0.7	210/200	HB
Polyvinyl Chloride (PVC-rigid)								
IM	-	1.3-1.4	6.5-9.0	-	3.5-4.3	1.5-15.0	168/156	V-0 / 5V
Polyvinyl Chloride (PVC-flexible)								
EX	-	1.25-1.55	1.7-2.6	250-350	-	-	-	-
IM	-	1.22-1.36	1.4-2.7	275-400	-	-	-	-
Thermoplastic Elastomers-Olefinic Type								
IM	7.0-10.0	0.89-0.91	1.8-2.5	500-700	-	-	200/135	-
Thermoplastic Elastomers-Polyurethane Type								
IM	10.0-30.0	1.13-1.4	5.5-6.9	420-550	-	-	-	-
Thermoplastic Elastomers-Styrene Type								
IM	2.0-11.3	0.9-1.5	-	500-900	-	-	-	-

Notes.

Application abbreviations:

BM = Blow Molding   EX = Extrusion   IM = Injection Molding   CM = Compression Molding

EBF = Extrusion Blown Film

[1] Values are medians (not minimums and maximums) expressed in ranges based on properties provided by the manufacturer of the particular plastic and are comparative purposes only. Consult manufacturer for exact specifications for a specific material.

[2] Izod impact test conducted with $1/8$ in. indenter.

[3] Underwriters' Laboratories Rating for fire retardance. Best achievable rating is 5V.

electrical properties, which are relatively unaffected by aging, make them suitable for electrical parts. Acetals are available as compounds for extrusion, injection, and blow molding. Fiberglass reinforced acetals are also available.

Acetels are used as replacements for metals where the higher ultimate strength of metal is not required, and costly finishing and assembly operations can be eliminated. Typical parts are pump impellers, molded and machined rollers, gears, appliance cases, automobile dashes, pipes, fittings, bearings, and zippers.

Polyallomers are highly crystalline polymers prepared from two or more monomers (namely propylene and ethylene) by a polymerization process. Formulations are available for high stiffness, medium impact strength and moderately high stiffness, high impact strength, and extra high impact strength. They may be injection molded, extruded, and thermoformed and are used in wire and cable jacketing, bottles, pipe fittings, and utility boxes and containers. Properties are very similar to HDPE and PP.

Polyamides (nylons). Polyamide (PA), or nylon, was the first of the thermoplastic engineering plastics. Some of the nylons are identified by the number of carbon atoms in the parent diamine and dibasic acid (i.e., Nylon 6, Nylon 12, etc.). Nylons have high strength and elongation, high modulus of flexure, good impact strength, and good resistance to nonpolar solvents. They are softened by polar materials such as alcohol and glycols, and are attacked by strong aids and oxidizing agents.

General purpose nylon molding materials are available for extrusion, injection molding, blow molding, and casting. Sheet and film are also available. Molybdenum disulfide, a solid lubricant, is used as a filler for type 6, 6/6, 6/10, and 6/12 nylons to improve wear and abrasion resistance, frictional characteristics, flexure strength, stiffness, and heat resistance. Glass fiber reinforced nylons show substantial improvement in tensile strength, heat distortion temperatures, and impact strength (in some cases). Nylon 11 and Nylon 12 have exceptional dimensional stability because of their low moisture rates and they offer complete resistance to zinc chloride.

Nylons are used extensively in the automobile industry, and for gears, living hinges, gaskets, wire insulation, high pressure flexible tubing, and chemical containers.

Polycarbonates. Polycarbonate (PC) resins are derived from aromatic and aliphatic dihydroxy compounds. They have properties that include excellent rigidity, toughness, and impact strength coupled with dimensional stability over a wide temperature range. Their heat and flame resistance is excellent: heat distortion temperatures range from 270 to 280° F (132 to 138° C) when stressed at 264 PSI (1820 kPa). Other features are unusually good electrical properties, but they should not be used where high arcing is involved. They are stain resistant and have been approved by the U.S. F.D.A. for safe use with all types of foodstuffs. Fatigue resistance is low, and PCs should not be stressed above 10% of their tensile or compressive strength in long term loading.

Polycarbonate molding compounds are available for extrusions, injection molding, and blow molding. Glass reinforced PCs have twice the tensile strength and tensile modulus, and four times the flexure modulus of nonreinforced resins. Typical applications include safety shields, lenses, electrical relay covers, pump impellers, fuse caps, electrical switch components, and appliance housings.

Polyethylenes (high-density). In volume, polyethylene (PE) is the most widely used plastic. The terms low-, medium-, and high-density refer to ASTM designations, and the low- and medium- grades are discussed under "Environmental Resistant Materials," which follows this section. As density increases, hardness, heat distortion point thickness, ultimate strength, and impermeability to liquids and gases increases. As density decreases, impact strength, mold shrinkage, and stress crack resistance increases. High-density PE has a density range of 0.941 to 0.965 gr/cu cm; stiffness in flexure of 81,000 to 150,000 PSI (558 to 1034 MPa), and a maximum service temperature of 180° F (82° C). Most high-density materials are classified as homopolymers (0.96 densities) and copolymers (0.95 densities). The copolymers are usually butene or hexane copolymers. The homopolymers are stiffer and better suited for thin wall containers where stress is not likely, while the copolymers are more stress crack resistant and more suitable for other applications (molded containers and blown bottles).

Polyethylenes have excellent chemical resistance and electrical properties, but their load bearing characteristics and weatherability are not good. It is flammable, and heat distortion temperatures are fairly low. Although it rates well in processing ability, mold shrinkage is above average. PE compounds can be extruded, injection molded, blow molded, and centrifugally cast. Film and sheet are available. Cellular PE, rigid foam, and fine powder are also available. Applications include containers, electrical insulation, pipes and tubing, grocery sacks, and fuel tanks for automobiles.

Polypropylenes. Polypropylene (PP) offers a good balance of properties. It has a good balance of strength, rigidity, and hardness, making it a good choice for parts requiring stiffness in thick sections, and flexibility in thin areas, and it retains strength up to 184° F (140° C), at which point it quickly looses stiffness. Chemical and biological resistance are outstanding, it has good resistance to environmental stress cracking, does not absorb moisture, and its electrical properties are excellent at high frequencies, heat, and humidity. As the lightest of plastics, it offers a high strength to weight ratio, high surface hardness, and excellent dimensional stability, but its load bearing ability is significantly affected by time. Rigidity, strength, dimensional stability, and heat resistance characteristics can be significantly increased with glass fiber reinforcements.

Polypropylenes are available as molding compounds for extrusion, injection molding, thermoforming, and blow molding, as well as in film and sheet. PPs are used in houseware containers, "living hinges" for a variety of applications, unbreakable medical and hygienic equipment that can be sterilized, appliance parts such as washer tubs, appliance cabinets, pipes and fittings, automotive interior parts, electrical connectors, extrusion coating, and packaging film.

Polystyrenes (modified). Polystyrene (PS) is available in many grades, and modified versions feature altered properties that are influenced by the addition of modifying agents such as rubber (for impact strength), methyl or alpha methyl styrene (for heat resistance), methyl methacrylate (for light stability), and acrylonitrile (for chemical resistance). The high heat resistant materials are produced as copolymers, while the high impact materials are produced by blending.

Polystyrenes are relatively inexpensive, and their clear, colorless properties permit an almost infinite range of color possibilities. They feature low moisture absorption, excellent dimensional stability, high hardness, good fatigue life. Their mechanical properties are affected by temperature, time, and environment, and, though flammable, PSs burn at a slow rate. They are available for extrusion, injection molding, compression, rotational and blow molding, and as polystyrene film, sheet, and foam. Uses include pipes and fittings, automobile parts, large appliance panels, battery housings, packaging materials, and housewares.

Polyvinyl chloride. Polyvinyl chloride (PVC) represents a major portion of the plastic consumed worldwide. PVC and chlorinated polyvinyl chloride (CPVC) retain good strength properties over their temperature range. CPVC has higher tensile strength and modulus, plus a high heat resistance that allows it to resist heat deflection under load to a temperature range of 180 to 220° F (82 to 102° C), versus 158° F (70° C) for PVC. At 212° F (100° C) CPVC retains a strength of 2,100 PSI (14 MPa). These rigid vinyls withstand strong acids and alkalies, and a wide variety of other corrosive liquids. They can be extruded, cast, coated, injection molded, blow molded, rotationally molded, and calendered, and are available as both film and sheet as well as rigid and flexible foams. Primary uses are pipe and fittings, insulation, electrical wire, chemical storage tanks, house siding, and window sash.

Environmental Resistant Materials

These materials do not possess outstanding load bearing characteristics and are non-Hookean (stress is *not* proportional to strain). They are, however, extremely durable in one or more properties such as wear and abrasion resistance, hardness, chemical resistance, heat resistance, and weathering.

Acrylics. These resins feature crystal clarity, outstanding weatherability, and excellent dimensional stability and stain resistance. Impact strength is good to –40° F (–40° C), and resistance to cracking after thermal shock is also good, as are electrical properties. Acrylics are also immune to attack by fungal organisms. Negatives are that they have a tendency to cold flow, a low softening point (160 to 200° F [66 to 93° C]), low scratch resistance, and they are affected by oxidizing acids and other chemicals.

Acrylics are available as compounds for extrusion, injection molding, blow molding, and casting, and in sheet or film forms. They are used for transparent enclosures in aircraft, lenses, containers, optical systems, architectural panels and facia, and outdoor signs and displays.

Cellulosics (cellulose acetate butyrate and ethyl cellulose). Cellulose plastics are not synthesized by the usual polymerization of a monomer—they are prepared by the chemical modification of the natural polymer cellulose. In general, these materials are soluble in ketones and esters, will slightly soften in alcohol, and will decompose in strong acids and bases. They do, however, have good resistance to oil and grease and they have very good molding characteristics. Cellulose acetate butyrate (CAB) can withstand sharp blows without shattering, has exceptional clarity, and can be colored in a wide range of transparent to translucent colors. It is used in warning lights, light filters, instrument panels, protective cases, tool handles, tubing, and developing tanks. Ethyl cellulose (EC) has good low temperature impact resistance, performs well at a wide range of ambient temperatures, has good gloss, colorability, clarity, heat resistance, and excellent dimensional stability. It is used in helmets, cases, gears, refrigerators, slides, tubing, and tool handles. Cellulosics are available for extrusion, injection molding, and blow molding, and sheet and film are available.

Chlorinated polyethers. Chlorinated polyether (CP) is a corrosion resistant thermoplastic that is widely used for production equipment in the chemical and processing industries. They provide excellent chemical and corrosion resistance up to 275° F (135° C), and maintain their strength and mechanical properties to 280° F (138° C). Their thermal conductivity is lower than most thermoplastics, and they have excellent creep resistance and unusually low water absorption properties. They are, however, susceptible to attack by fuming nitric and sulfuric acids.

Chlorinated polyethers are available for injection molding, extrusion, and fluidizing bed coatings. Typical applications are pumps, linings, pipes and fittings, valves, tanks, and meters used in the production of chemicals.

Flexible vinyls are included in this category. These may be plasticized polyvinyl chloride (PVC) or polyvinyl plastic (PVB). The properties of PVC were covered in the previous section.

Fluorocarbons. Also called fluoroplastics or fluoropolymers, these materials are based on polymers made with monomers composed of fluorine and carbon. There are several varieties.

Polytetrafluoroethylene (PTFE or TFE) and *fluorinated ethylene propylene* (FEP) resins are relatively impermeable to many chemicals. Only molten alkali metals, gaseous fluoride at high temperature and pressure, and a few halogenated compounds show any effect on these florocarbons. They are strong and tough down to –450° F (–268° C), and are rated for continuous service to 550° F (288° C) for TFE, and 400° F (204° C) for FEP. Other features are excellent weather resistance and electrical characteristics. However, due to its high viscosity, TFE cannot be processed with conventional melt extrusion and injection molding techniques—ram extrusion and paste extrusion are feasible. TFE is available as granular powder for compression molding or ram extrusion, as powders for lubricated extrusion, and as dispersions for dip coating and impregnating. TFE is used in nonlubricated bearings, chemical resistant pipe and pump parts, high temperature electrical parts, wire, and cable. FEP is supplied in pellets for melt extrusion and molding, and as dispersions. FEP is used for wire insulation and jackets, high frequency connectors, microwave components, electrical terminals, and terminal insulators.

Polychlorotrifluoroethylene (PCTFE) resists most corrosive chemicals and has a temperature range of –400 to + 400° F (–240 to + 204° C) with a low rate of thermal expansion/contraction (it also remains flexible and tough at cryogenic temperatures). However, its hardness falls off rapidly after 300° F (149° C). Its permeability to water and other fluids is among the very lowest for any plastic, and ultraviolet and weather resistance is excellent. PCTFE is available for compression injection, transfer molding, extrusion, and dispersion coatings. It is used for corrosion resistant liners, oxygen and liquid oxygen seals, pump stators, valve linings, beakers, test tubes, syringes, printed circuits, self-locking bolts, O-rings, jacketed cables, and many other applications.

Polyvinylidene fluoride (PVDF) is higher in tensile strength, has a lower specific gravity, and lower service temperature (300° F [149° C]) than TFE. Its useful temperature range is –420 to + 300° F (–251 to + 149° C), and it is flame and weather resistant. It is attacked by sulfuric acid. PVDF resins are available for compression, injection, transfer, and blow molding, and for extrusion and dispersion coatings. It is used in processing tanks, valves, tubes, pump liners, and pump impellers. Films and coatings are used in the packaging industry.

Ethylene-tetrafluoroethylene (ETFE) has a molding rate faster than the other fluoropolymers, high flex life, and exceptional impact strength, even at very low temperatures. Its thermal properties are good, with a long term continuous service temperature of 300° F (149°C), and intermittent service ceiling of 400° F (204° C). Electrical properties are outstanding, and flammability rating is very good. ETFE resins are available for extrusion and injection molding, and are available as fiber and film. It is used in labware, valve liners, and as electrical insulation on wire for most applications.

Ethylene-chlorotrifluoroethylene (ECTFE) has high impact resistance and tensile strength. Its useful properties are available from cryogenic temperatures to 325° F (163° C) and it is highly resistant to corrosives even at elevated temperatures. Electrical properties are good, as are weather and abrasion resistance. ECTFE is available in pellet and powder forms and can be extruded, injection, transfer, compression molded, as well as dispersion coated, rotationally molded, and powder coated. It is used for wire and cable coatings, and for chemically resistant linings in mixing tanks.

Polyethylenes (low- and medium-density). General polyethylene (PE) characteristics were covered earlier under "Polyethylenes (high-density). Low-density PE has a density range of 0.910–0.925 gr/cu cm, and a stiffness modulus of 13,000–30,000 PSI (90–207 MPa). These materials have high impact strength but relatively low heat resistance—the maximum recommended continuous service temperature is 140° F (60° C). Usually available in cube and pellet form, it is now available in powders. Medium-density PE has a density range of 0.926–0.940 gr/cu cm, and a stiffness modulus of 31,000–80,000 PSI (214–552 MPa). It has a maximum service temperature of 160, and a stiffness modulus of 31,000–80,000 PSI (214–552 MPa). It has a maximum service temperature of 160° F (71° C). Film and sheeting is the largest single use for low- and medium-density PEs.

General Purpose Materials

These materials are not outstanding in load bearing or environmental resistance, but they feature ease of processing and good appearance factors.

Cellulosics (cellulose acetate and cellulose propionate). Cellulose acetate (CA) has exceptional clarity and is used in photo and x-ray film, safety glasses and shields, optical frames, knobs, and handles. Cellulose propionate (CP) maintains a hard glossy surface, and is used in telephone headsets, pens, typing keys on keyboards, toothbrush handles, and face shields. See the section above on Cellulosics for more information on cellulose plastics.

Polystyrenes (general). General polystyrenes are clear, and can be colored to an infinite range of hues. They are commonly used for indoor lighting. See the section above on "Polystyrenes (modified)" for more information.

Thermoset plastics

Once thermoset (TS) plastics are cross-linked, they will not soften under heat. With the application of heat, they undergo a series of changes that are irreversible. The chemical change that takes place is known as polymerization. Thermosetting molding and casting materials permit a high degree of freedom for enhancing the existing properties of the basic polymers, and the use of reinforcements and fillers are the rule rather than the exception because the tightly cross-linked, high molecular weight thermosetting polymers are inherently weak and easily fractured. They may not be used alone in structural applications and must be filled or reinforced with strengthening materials. The reinforcements and fillers may be organic or inorganic, and may take the form of powders, discrete fibers, or woven or macerated fabric. The particle geometry and the effects of processing play an important role in the final properties of the compound. **Table 1** gives abbreviations (which are routinely substituted for the full name) for popular thermosets and thermoplastics, and **Table 2** provides selected properties for plastics commonly used in manufacturing.

Alkyds. Structurally, alkyd resins are modified polyesters, but the term "polyester" is used almost exclusively for linear polyesters derived from dihydric alcohols and dibasic acids, whereas alkyds (a word derived from alcohol+acid) are formed by the condensation of a dibasic acid or anhydride with a polyhydric alcohol. The formulations for alkyd resins can be extensively varied by the introduction of different polyhydric alcohols and anhydrides. Due to their excellent weatherability, toughness, adhesion, flexibility, and ease of application, alkyd resins have become the major synthetic resin for use in surface coatings. Alkyd molding compounds use an unsaturated polymerizable alkyd as their base resin, and these compounds cure at high speed and require low molding pressure.

Alkyds have excellent dimensional stability, excellent fungus resistance, and are suitable for use at 400° F (204° C) without degradation. They possess excellent dielectric strength, arc, and dry insulation resistance. Their most notable disadvantages are cost, very low impact strength, and difficulty in molding to close tolerances. Fillers can be used to improve these characteristics. Alkyds are available as flakes, ropes, slugs, and sheets, granular powder, or putty, and they are used in electrical insulation, auto and boat bodies, transformer components, and automobile ignition systems.

Allylics. These are among the most versatile, yet unappreciated, of the thermoset plastics. Low pressure is generally adequate for their molding and curing as no water or other volatiles are formed during their polymerization. Diallyl phthalate (DAP) is the most widely used allylic, and it offers good shelf life, excellent laminating qualities, and good surface hardness. Allylics have the best moisture resistance of the thermosets, exhibit very low mold and post mold shrinkage, and the resin is completely inert in the presence of metals. Heat resistance in continuous service (up to 350° F [177° C]) for DAP is outstanding, and it remains stable in storage at ambient temperatures while cure at temperatures above 200° F (93° C) is normal and rapid. Common fillers include Dacron for improved shock and moisture resistance; Orlon for the same properties but lower heat resistance; and Nylon to increase shock and wear resistance at slight sacrifice to moisture absorption properties.

Compression transfer molding compounds are available in a range of colors. Typical applications include encapsulating shells, insulators, decorative laminates, heat resistant handles, protective insulating coatings, and air ducts.

Amino resins. Amino plastics are obtained by a condensation reaction between formaldehyde and such compounds as urea, melamine, dicyandiamide, ethylene urea, and sulfonamide. The resins with the lowest formaldehyde are used in molding compound production. The urea and melamine amino compounds are the most widely used, and the following is restricted to a review of these two materials.

These materials can be produced in an unlimited range of stable colors, but the melamine products undergo color changes at temperatures above 210° F (99° C) and the ureas will change color at 170° F (77° C). Pronounced color change and blistering occur after one-half hour at 300° F (149° C). These plastics have excellent insulation and arc resistance properties and are classified as self-extinguishing. Physical properties of melamine based products are relatively unaffected between –70 and + 210° F (–57 and + 99° C), while maximum extended temperature range for the ureas is –70 to + 170° F (–57 to + 77° C), and below –70° F urea moldings are subject to embrittlement. Both materials are subject to initial mold shrinkage as well as after mold shrinkage. They are among the hardest plastics available, and their high compression strength provides high resistance to deformation under load. The molding compounds are available in a large range of plasticities, and the rate of cure can be controlled during manufacture. Amino plastics are used in closures for glass and metal containers, buttons, handles, toaster bases, food service trays, ignition parts, and electrical housings and sockets.

Epoxy resins. Epoxy thermoset resins (EPs) are generally derived by reacting bisphenol-A and epichlorohydrin. In fact, approximately 85% of the world's epoxy resins are formed in this way. Other popular forms are epoxidized novolacs, cycloaliphatic resins, and phenoxy resins. The entire family offers an extremely broad capability for blending specific properties through the use of resin systems, fillers, and additives. Epoxies have high dimensional stability over long periods of time and humidity, excellent mechanical, chemical, and thermal shock resistance, excellent tensile, flexural, and compressive strengths, and have a service range of –85 to + 275° F (–65 to + 135° C). Applications are wide and include pressure bottles, oil storage tanks, printed circuit boards, boat bodies, encapsulation material for electrical components, body sealers, and surface coatings.

Phenolics. These are the oldest and least expensive of the thermosetting plastics. They have excellent electrical properties, good impact strength and high flexural modulus, and have high resistance to deformation under load. Most phenolics can operate to temperatures up to 300° F (149° C), and thermal conductivity is low. Colors are restricted to black, dark browns, and tans. Applications include camera cases, electrical sockets, soldering gun handles, small motor housings, distributor caps and rotors, insulation for cables, and pump impellers.

Polyesters. There are a many polyester plastics, and three classifications will be covered here: unsaturated polyester, high temperature aromatic polyester, and thermoplastic aromatic polyester.

Unsaturated polyester. These plastics are produced from an unsaturated dibasic acid and a glycol, and are usually cross-linked with styrene, vinyl toluene, methyl methacrylate, or diallyl phthalate. They can be formed to a wide range of physical properties and may be brittle and hard, tough and resilient, or soft and flexible. They have good weathering and chemical resistance, and fire retardance is achievable. They can be used in open mold casting, vacuum bag molding, die molding, and injection molding. Commonly used in furniture, bowling balls, boats, and vehicle bodies.

High temperature aromatic polyester. These materials have excellent stability to air above 600° F (316° C). It has good thermal conductivity and insulating properties. Machinability of molded shapes is excellent, and it has self-lubricating characteristics. It can be used for self-lubricating bearings and seals, as well as other parts of processing pumps where wear resistance and corrosion resistance are critical.

Thermoplastic aromatic polyester. These polyesters are produced in pellet form for injection molding, general purpose extrusion, film extrusion, and extrusion coating. Powders are available for electrostatic spray coatings. They are tough, even at low temperatures, and have outstanding surface hardness, lubricity, and abrasion resistance. Dimensional stability is excellent. Commonly used in pump components, gears, spray nozzles, sunglass lenses, alternator components, and control knobs.

Polyurethanes. Polyurethane plastic (PUR) products are available as flexible foams, rigid foams, and as an elastomer. Generally, flexible urethane foams are prepared commercially from polyether or polyester resins, diisocyanates, and water in the presence of catalysts. The reaction between the water and the isocyantes liberates carbon dioxide which functions as a blowing agent to create an open cellular structure. By juggling reactants, catalysts, and emulsifiers, either rigid or flexible foams with a wide range of properties can be created. *Flexible foams* are effective sound deadeners and energy absorbers. The polyether type is slightly more flammable than the polyester version, but both can be self-extinguishing with the addition of retardants. *Rigid foams* can also be made self-extinguishing, and have excellent low and high temperature insulating characteristics. Their dimensional stability is good, but they can be attacked by strong acids and bases. Foams are used in cushions, upholstery, padding, clothing, gaskets, insulation in cabin ventilating systems, nonsinkable boats and pontoons, and refrigerator insulation. PUR RIM (reaction injection molding) foams have been widely used in automobile bumpers. *Elastomers* are used in tires, seals and gaskets, bushings, soles and heels, and rollers used in printing. Polyurethane elastomers have the highest abrasion resistance of all elastomers, and higher load bearing capacity than other elastomers of comparable hardness. Their low temperature properties are excellent and continuous service temperature range is -30 to $+180°$ F (-34 to $+82°$ C).

Silicones. Silicones come in a very wide variety of forms, including rigid thermosetting, laminating, and molding formulas that may be available in fluids, elastomers, gels, or resins. They have an expansive temperature ranging from -70 to $+500°$ F (-57 to $260°$ C), but some varieties can withstand even more extreme temperatures. Although these materials have superior temperature ratings and resistance to chemicals, their tensile strength, and abrasion resistance are poor, and they are expensive.

Process technology

A major factor contributing to the growth of plastics use is the innovations made in processing technology. For example, injection molding of thermosetting materials and the use of structural foams formed from existing injection molding equipment have become commonplace. Each process method is virtually unique in its operation and tooling, and secondary processes may be required. The most common primary processes are outlined in **Table 3** and discussed in more detail in this section. Fabrication and finishing are covered in the next section.

Blow Molding. Blow molding rapidly produces thin-walled, hollow thermoplastic parts. There are two different processes in common use: extrusion blow molding (EBM), and injection blow molding (IBM).

Extrusion blow molding is the most common of the processes, and constitutes approximately 75% of total production. It is a hybridization of basic extrusion and molding techniques and rapidly produces thin-walled, hollow thermoplastic parts in the following sequence. 1) The extruder produces a thin, hollow cylinder of molten plastic (called a "parison"). 2) The parison is then captured between the chilled halves of a split mold. 3) Compressed air is introduced and forces the molten parison to expand against the walls of the mold. 4) The formed part is cooled and released from the mold. Flash forms when molten material is forced between the mold halves, and it can be reclaimed and reused. Systems are comprised of a hopper where the material is prepared, an extruder, an extruder head incorporating a die, a clampable mold, and an air-injector capable of pressure of 100 PSI or greater. The process is excellent for producing thin-walled parts such as bottles and other containers, automobile heater ducts, and packaging units. Materials suitable for the process include polyethylene (LDPE and HDPE) polyvinyl chloride (PVC), polypropylene (PP), polyethylene terephthalate (PET), polycarbonate (PC), polyolefins (PO), and polyamide (PA).

Injection blow molding incorporates the principles of injection molding and blow molding. It uses a three-stage process on a rotary indexing table. At the first stage, the parison is preformed. At the second, the preform is blown into the finishing container, and the finished part is removed in stage three. Much higher rates of production can be achieved with EBM, and tooling costs are lower. However, with IBM there is no flash to trim (because the molds are designed to extremely tight tolerances), and wall thickness control is better, which results in less scrap. IBM also provides better surface finishes and superior distribution of material, resulting is more reliable formation of critical neck areas. Materials that can be used include LDPE, HDPE, PP, PC, PET, polystyrene (PS), styrene acrylonitrile copolymer (SAN), polyethylene-vinyl acetate (EVA), polyurethane (PU), polyacrylonitrile (PAN), and polyethylene naphthalate (PEN).

A recent and growing method is *sequential blow molding* (SBM). The perceived advantage is that, unlike EBM, it will be able to shape sharply three-dimensionally shaped parts, while simultaneously reducing flash and pinch lines. This is achieved partially through the use of programmable manipulators or six-axis robots that are used to manipulate the extruded parison. Improved wall thickness control is also claimed through the use of a radial wall thickness system developed by Krupp Kautex. These machines are currently in limited use.

Casting. Casting differs from other processing methods in that external pressure is not required. Gravity and heat settle and harden the mass. Both thermoplastic and thermoset resin liquids or melts can be poured into casting molds to set to a solid by curing with catalysts, at room or higher temperatures (thermoset resins), or by chilling (thermoplastic resins). Casting of film is performed on a continuously moving turntable or belt, or by precipitation in an aqueous chemical bath. In the turntable or belt process, the liquid resin is spread to the desired thickness and, as it dries, the film is stripped off. The most extensively used casting materials are the epoxies, phenolics, polyesters, stryenes, acrylics, and silicones. Mostly, the process is used for rods, tubes, cylinders, sheets, and slabs for further fabrication into parts, and for encapsulation of electrical components. Some advantages of casting are: better optical and strength properties can be achieved with casting than with extrusion or molding; molds may be made from low cost materials such as wood, plaster of paris, or sand-resin formulations; materials can be compounded with reinforced fillers prior to casting. Disadvantages are: high cost due to the amount of hand labor required. Incomplete removal of air bubbles can lead to high scrap rates; relatively few liquid plastics are available for casting.

Calendering. This process is restricted to thermoplastic materials—primarily PVC, but cellulosics, and styrenes, ABS, and EVA can also be calendered—and is used to produce continuous sheets and film. It is also used for applying plastic coatings to textiles, paper, or other supporting material. In the process, the plastic material is softened by heat, plasticizer, and possibly some solvent, and is then fed between a series of several pairs of large heated revolving rollers that squeeze the material between them into the desired

Table 3. Primary Methods of Plastics Processing. *(Source, MIL-HDBK-755.)*

Process Method	Description	Remarks
Blow Molding	Shaping a thermoplastic material into a hollow form by forcing material into a closed mold by internal air pressure. Normally initiated with a form made by extrusion; it is now possible to injection mold the initial form.	High production rate process for manufacturer of hollow items such as bottles, tanks, and drums; uses relatively inexpensive molds.
Casting	Forming a solid part by pouring a liquid resin into a mold and removing the part following curing or solidification.	Practical for small production where inexpensive tooling is employed; primarily used for encapsulation of components or for parts having thick sections.
Compression Molding	Principally used for thermoset parts formed by placing material into an open mold and curing the part by use of heat and pressure after the material is confined with a plunger.	Principally for thermosetting materials of simple shapes having heavy cross sections, high impact fillers, or large deep-draw areas. Cycles are relatively slow, and finishing of parts is required.
Extrusion	A process for making continuous forms by forcing a thermoplastic material in the plastic state through an orifice or die. Thermosets are now somewhat adaptable to extrusion.	Limited to the continuous, low-cost production of rods, tubes, sheets, or other profile shapes; capable of extruding solid, foamed shapes.
Injection Molding	A basic molding process wherein a heat-softened thermoplastic material is forced under pressure into a closed mold. Upon cooling and solidification, the part is ejected. Now widely used as a process method for thermosets and for thermoplastic forms.	A major, high production rate process used for manufacture of intricate shapes. Process maintains good dimensional accuracy. Mold costs are relatively high.
Reaction Injection Molding	A process that involves the mixing of two liquid components, injecting the liquid stream into a closed mold at relatively low pressure, and removing the part following cure.	An injection-molding-type process used for the manufacture of large, solid, or cellular parts. Low pressure and ambient temperature materials eliminate the need for expensive machinery.
Rotational Molding	The forming of a hollow part from thermoplastic resin within a closed mold by heating, rotating, and cooling the material for subsequent removal.	Useful for manufacture of hollow forms with practically no limit to size and shape. Either rigid or flexible parts can be made with inexpensive tooling.
Thermoforming	The forming of thermoplastic sheet material into a three-dimensional shape by heat-softening the sheet and forcing it to conform to the shape of the mold by pressure or vacuum; followed by cooling.	Limited to simple, three-dimensional shapes for signs, trays, cups, domes, and packaging. Tooling is relatively inexpensive.
Transfer Molding	A basic molding process for thermosets; part is formed by transferring the molding compound, which has been heated in a loading well, under pressure into a closed mold where curing takes place.	A process used exclusively for thermoset materials in the manufacture of intricate shapes and parts with fragile inserts. Process maintains good dimensional accuracy.

thickness. The space between the last series of revolving rollers controls the thickness of the end item, film, or sheet. Thicknesses range from 0.00315 to 0.315 in. (0.08 to 0.80 mm) as rolled from the calender, but can be reduced to as thin as 0.00118 in. (0.03 mm) after stretching. Widths can be as wide as 118 inches (3,000 mm) with existing machinery. Tolerance control is very good with this process, and patterned or textured effects are obtainable at low cost.

Compression Molding. In compression molding, a plastic material, usually partially preformed, is placed in a heated mold cavity. The mold is then closed and heat and high pressure are applied to the plastic, thereby forcing the plastic to fill the mold cavity while simultaneously undergoing a chemical reaction that cross-links the polymer chains and hardens the plastic into a finished form. Low cost, quality parts are produced with compression molding. Normally, thermosetting resins in the form of sheet molding compounds (SMC) or bulk loading compounds (BMS), diallyl phthalate (DAP), melamines, phenolics, ureas, silicones, or epoxies are shaped with this process, but glass mat thermoplastics (GMT) such as Azdel, and long fiber thermoplastics (LFT) can also be compression molded. While some materials can be compression molded at room temperature, the cycle is lengthy, so the process is usually performed in the temperature range of 250 to 400° F (121 to 204° C). Compression pressures in the range of 5,000 PSI (34.5 MPa) are normal, but both temperature and pressure will vary with material. The process is usually used to form large parts such as housings, furniture, vehicle body panels, appliance cabinets, and washing machine agitators. There is little waste and finishing costs are low, but intricate parts with undercuts or delicate details are not practical. Tolerances closer than ± 0.005 in. (0.127 mm) are difficult to achieve.

Extrusion. In extrusion, plastic granules or powder are fed through a hopper to a heated plasticizing cylinder where they are melted, then driven by rotating screws through a shaped die that determines the shape of the extruded part. The process is used for the fabrication of pipe, flexible tubing, rods, sheets, tubes, bars, house siding, and other shapes. Single screw extruders are most common. They normally use pellets, screw rotation rates range from 20 to 200 rpm, and screw shapes are varied depending on material application. Twin screw units are sometimes counter-rotating and usually use powders. They incorporate screw geometries that can produce back pressure, knead the material, and affect the volume of the extrusion. Materials commonly used in extrusion are PVC, ABS, polystyrene, thermoplastic elastomers, nylons, and polyolefins.

Extrusion, Blown Film. This process is used in the production of plastic film. Polyethylene (low- to high-density) are the polymers most often used, but other resins such as polypropylene and polycarbonate are also used. In the process, plastic particles are fed into a hopper, melted, and then metered at a controlled rate through an extruder to a ring shaped blown film die that releases the melt in a bubble form. An air stream is fed into the bubble as it is pulled (usually vertically, though some machines are horizontally inclined) by nip or pinch rollers. Before the bubble passes through the rollers, it is flattened by a collapsing frame. The film then is "pinched" to size by the rollers and carried to a spool where it is accumulated on a large roll. Blown films are used for packaging, including food coverings, and in agricultural and building materials. A somewhat similar system, known as *Flat Web Extrusion*, is a continuous system for forming plastics into belt shaper sheet or film. It employs a calender takeoff system for producing sheet and film in thicknesses ranging from 0.007 to over 0.500 in. (0.178 to over 12.7 mm).

Injection Molding. Injection molding is capable of producing complex three-dimensional parts to high quality and tolerance standards. The process is characterized by the fact that the molding mix is preheated to a temperature high enough for it to become a quasi-liquid. When the material reached the desired temperature, a reciprocating screw forces the melt through a nozzle into a controlled temperature closed mold. Basic machines use the screw to force the melt into the mold, but machines intended for high production rates use a plunger device to hasten the injection. Pressures of up to 30,000 PSI (207 MPa) are used to insure that the mold is fully filled. It is common to use a multicavity mold to form several parts simultaneously. The cavities are joined by sprue and runners that fill with scrap which can usually be returned to the process. Thermoplastics are most commonly used in injection molding, but some thermosets (epoxy, melamine, DAP, and others) containing high volumes of fillers to reduce cost and shrinkage are also used. Another development in the field is *Powder Injection Molding* (PIM) for the manufacture of metallic or ceramic moldings. With the exception of aluminum, almost any metal that can be powderized can be used in PIM. Polyolefin wax mixtures and polyacetals are normally used as binders. Typical applications for plastic injection molded parts include cases for appliances, handles, knobs, gears, plumbing and hardware, steering wheels, vehicle grills, fan blades, and fasteners. **Table 4** provides process guidelines for injection molding.

Reaction Injection Molding. Reaction injection molding (RIM) requires the high pressure impingement mixing and polymerization of two or more reactive liquid

Table 4. Guidelines for Injection Molding of Selected Polymers.

Polymer	Processing Temperature		Pre-Dry Time		Mold Temperature		Molding Shrinkage
	°F	°C	°F/hours	°C/hours	°F	°C	
ABS	392-500	200-260	158-176/2	70-80/2	122-176	50 to 80	0.4-0.7
ABS foamed	392-500	200-260	158-176/2	70-80/2	50-104	10 to 40	0.4-0.7
ASA	428-500	220-260	158-176/2-4	70-80/2-4	122-185	50 to 85	0.4-0.7
CA	356-428	180-220	176/2-4	80/2-4	104-176	40 to 80	0.4-0.7
CAB	356-428	180-220	176/2-4	80/2-4	104-176	40 to 80	0.4-0.7
CP	374-446	190-230	176/2-4	80/2-4	104-176	40 to 80	0.4-0.7
EVA	266-464	130-240	-	-	50-122	10 to 50	0.8-2.2
PB	392-554	200-290	-	-	50-140	10 to 60	1.5-2.6
PBT	446-536	230-280	230/4	120/4	104-176	40 to 80	1.0-2.2
PC	518-716	270-380	230-248/4	110-120/4	176-248	80 to 120	0.6-0.7
PE-HD	392-572	200-300	-	-	50-140	10 to 60	1.5-3.0
PE-HD foamed	392-500	200-260	-	-	50-68	10 to 20	1.5-3.0
PE-LD	320-518	160-270	-	-	68-140	20 to 60	1.0-3.0
PES	608-734	320-390	320/5	160/5	212-320	100 to 160	0.6
PET	500-572	260-300	230/4	120/4	266-302	130 to 150	1.6-2.0
PMMA	374-554	190-290	158-212/2-6	70-100/2-6	104-194	40 to 90	0.3-0.8
POM	356-446	180-230	230/2	110/2	140-248	60 to 120	1.5-2.5
PP	392-572	200-300	-	-	68-194	20 to 90	1.3-2.5
PP foamed	392-554	200-290	-	-	50-68	10 to 20	1.5-2.5
PPE/PS	500-590	260-310	212/2	100/2	104-230	40 to 110	1.5-2.5
PS	338-536	170-280	-	-	50-140	10 to 60	0.4-0.7
PSU	644-734	340-390	230/5	120/5	212-320	100 to 160	0.4-0.7
PVC-P	320-374	160-190	-	-	68-140	20 to 60	0.7-3.0
PVC-U	338-410	170-210	-	-	68-140	20 to 60	0.4-0.8
SAN	392-500	200-260	185/2-4	85/2-4	122-176	50 to 80	0.4-0.6
SB	374-536	190-280	-	-	50-176	10 to 80	0.4-0.7

Notes: U = unplasticized   P = plasticized

components, the injection of the mixture into a closed mold to form a solid or foamed part. Polyurethanes are normally used in the process, but other potential materials are nylon, epoxy, and polyester. Three RIM processes are usually identified. 1) Standard RIM does not use fillers or reinforcements. 2) Reinforced RIM (known as RRIM) uses fillers such as glass fiber or mica. 3) Structural RIM (SRIM) injects the material onto and around a glass fiber preform encased in a mold to produce a high strength and durable part. RIM is used to produce vehicle body and instrument panel parts, wheel covers, and seat frames. Parts are typically light in weight, have good impact strength, and can be produced in a wide variety of shapes and sizes.

Rotational Molding. Rotational molding is a relatively simple process used to manufacture hollow objects from thermoplastics and, to a lesser extent, thermoset materials. The solid or liquid polymer is placed in a mold which is first heated, then cooled while being rotating around two perpendicular axes simultaneously. In the initial phase of the heating process, a porous skin is formed on the mold surface which gradually melts to form a homogeneous layer of uniform thickness. When molding a liquid material, the liquid tends to flow and coat the mold surface until the gel temperature of the resin is reached. The mold is then cooled with forced air, water spray, or both. The mold is then opened, the finished part removed, and the mold is recharged for the following cycle. About 80% of the materials used in rotational molding are low- to high-density and cross-linkable polyethylenes (LDPE, LLDPE, HDPE). The other 20% is a mix of polyvinyl chloride, polycarbonate, polypropylene, and polyesters. Early emphasis on rotational molding was in the toy industry, and almost every plastic hobbyhorse and squeeze toy is made with the process. It is also used to produce refuse containers, boat and auto parts, and chemical storage tanks.

Thermoforming. There are many techniques for thermoforming plastics, and seven of them will be discussed here. Each of the techniques involves the heating of plastic sheet or film material until it becomes limp, and then causing the material to slump over a mold or form. Vacuum, air, or mechanical pressure is used to create close conformity. Sheet thermoforming is used to produce such items as luggage cases, hot tubs, refrigerator door liners, and vehicle interior panels. Typical materials are ABS, PVC, PE, PP, PC, polysulfone (PSU), and styrene acrylonitrile copolymer (SAN).

Straight vacuum forming. The sheet is clamped in a stationary frame, then heated, and the frame is clamped to the mold cavity. A vacuum is then applied through holes in the mold cavity to draw the plastic sheet into contact with the inner mold surfaces. It is then chilled and the part removed.

Drape vacuum forming. In this process, the heated sheet is draped over the male form, and a vacuum is used to pull it down onto the form. Relief maps and embossed and texture mats are made with this process.

Male form forced above sheet. The heated sheet is secured in a frame and a male plug is descended from above by means of hydraulic pressure. Partial forming to the shape of the male plug takes place, and then vacuum applied through holes in the plug draws the sheet around the plug. The formed sheet is then chilled.

Vacuum snap-back forming. The heated plastic sheet is positioned over a cavity and partially pulled into the cavity by the application of a vacuum. A male plug is then moved to a predetermined position and the vacuum is released, which allows the stretched sheet to snap back against the male plug. Vacuum is then applied through holes in the plug. The formed sheet is then chilled.

Plug and ring forming. The heated sheet is placed over a ring and clamped into position. A male plug, mounted above, is forced into the sheet, stretching it to conform to the shape of the plug. The formed sheet is then chilled.

Air pressure forming. Many variations are used to make use of air pressure (higher than that available from a vacuum) for forming. In many of these operations, this is accomplished by placing the mold in an autoclave where it can be heated by placing vacuum on one side of the heated sheet, and high pressure on the other. In one instance, the heated sheet is first clamped into position above the cavity of the mold. A cored plug then pushes the heated sheet into the cavity and tightly seals the mold. Air pressure is then introduced through holes in the plug, pushing the sheets against the sides of the female mold. Holes in the latter allow the air to escape from the underside of the sheet. In another case, the sheet is clamped over the female cavity mold, heated to the softening point, and then air is applied on the underside to blow the sheet into a hemispherical shape of the desired size. Vacuum is then applied within the mold cavity, inverting the blister and sucking it into contact with the surfaces of the female mold where it is hardened.

Matched metal mold forming. This process is somewhat similar to compression molding. A heated sheet of plastic resin impregnated fabric or compound thermoset resin is placed between matched male and female dies. Hydraulic pressure is then applied on the two parts of the mold to compress the sheet within the opening between the two halves. The sheet is then cooled by water coils embedded in the metal molds.

Transfer molding. Transfer molding is a combination of injection and compression molding. The plastic (thermosetting resins) is first melted in a heated cylinder, then fed by plunger through sprues and runners to a closed mold where it is compressed into the desired shape and polymerization is completed. Transfer molding differs from compression molding as follows: in compression molding the molding material is charged directly to the mold, but in transfer molding it is first heated to melt temperature in a separate chamber. The essential difference between transfer and injection molding is that in transfer molding, only enough material is heated at each cycle to fill the mold cavities, whereas in injection molding a reservoir of molten plastic is maintained in the heat cylinder. In fact, transfer molding is usually chosen when the designated resin cannot be injection molded. Production rates are high, and dimensional accuracy is good. Molds are usually elaborate, and often encapsulate sensitive electronic components, but waste is comparatively high because runners and sprues cannot be reused. Epoxies and phenolics are common materials, but some elastomeric materials can also be processed. The process is popular for making semiconductors and resistors.

Machining and fabrication of plastics

Practically all plastics, except for the very softest, can be machined with conventional machinery. However, allowances must be made for the greater heat sensitivity of plastics (compared to metals), and for the greater thermal expansion of plastics. Because plastics are poor heat conductors, they sustain high heat build-up during machining, which, in the case of thermoplastics, can cause softening and heat distortion. The high cutting speeds used with metals should be avoided. Instead, a light cut and slow feed are recommended. Coolants or cutting fluids can be used: plain or soapy water, cool air, and less frequently mixtures of water and oil are permissible. Tools should be designed to provide adequate clearance so as to clear away plastic chips as fast as they are formed.

Because thermosetting plastics do not suffer to the same degree from heat distortion, their machining is more straightforward. Coolants may be required to keep surfaces from burning, but chip removal can be done with a vacuum hose as chips are reduced to powder form by machining.

Plastics are routinely machined by grinding, turning, sawing, milling, routing, drilling, reaming, and tapping. When turning plastics, positive geometry inserts with ground peripheries are recommended to help reduce material build-up on the cutting tool. Rake angles of 0 to 15°, and clearance angles of 5 to 15° are appropriate. Cutting speeds average 655 to 1,640 sfm (200 to 500 smm) with a rate of advance of 0.0040 to 0.020 in./rev (0.1 to 0.5 mm/rev).

For drilling, low helix drills having two flutes and a point angle in the range of 70° to 110° are well suited for drilling holes up to one-inch in diameter in plastics, and peck drilling is preferred as it will allow for the removal of swarf. For larger holes, a low helix drill with a 110° point angle and 9° to 16° lip clearance is recommended, and a pilot hole one-half inch in diameter should be drilled before it is enlarged to full size. Cutting speeds for drilling range from 160 to 325 sfm (50 to 100 smm), and rate of advance should be approximately 0.0040 to 0.014 in./rev (0.1 to 0.3 mm/rev).

Sawing (circular and band) speeds of approximately 10,000 sfm (3,000 smm) with a feed rate of 0.0040 to 0.014 in/tooth can be used for most materials. The tooth rake angle should be 0 to 15°, and the clearance angle 10 to 15° for circular saws. For band saws, the rake angle should be 0 to 5°, and the clearance angle 30 to 40°. Plastics are also easily cut by lasers. A 250 W CO_2 laser can typically cut 5/32 inch (4 mm) thick thermoplastics at speeds ranging from 40 to 85 inches (1,000 to 2,100 mm) per minute.

Plastic actually dulls the edges of cutting tools more rapidly than metal, so frequent resharpening is usually required. Tool steel grade tools are normally satisfactory for plastics, but glass reinforced materials may often require tougher tools.

Many plastics can be welded with hot gas (usually nitrogen) equipment. The temperatures should be 600 to 640° F (316 to 338° C) for high-density polyethylene and polystyrene, 640 to 660° F (338 to 349° F) for polypropylene, and 660 to 700° F (349 to 371° C) for acetal. Filler rods should be of the same, or similar, material as the weldment, and range from $^1/_{16}$ to $^3/_{16}$ inch (1.5875 to 4.7625 mm) in diameter. Round rods are preferable, and the diameter should be constant through the length of the rod. Good plastic welds resemble those produced on metal by electric arc welding. No preparation is required for filet welds. For butt welds, the two pieces should be welded as in arc welding, but at a smaller angle (60 to 80°) and with no sharp edges. A sealing run on the reverse side will assure higher tensile strength in butt welds. In plastics welding, a simple bonding process takes place between filler rod and parent material, since only the actual mating surfaces melt. Other parts, such as the center of the rod, remain relatively unaffected and rigid. This allows for the exertion of slight pressure on the rod to force it into the weld and thereby combining the melted surfaces into one homogeneous mass. To assure uniform coalescence, the surfaces and the filler rod should be preheated before insertion into the weld. Overheating must be avoided as it will degrade the material. Tensile strength for butt welds may be as much as 90% of the parent tensile strength. Filet weld strength is somewhat lower, but can be strengthened with reinforcements.

Lasers, which are commonly used to cut plastics, can also be used for welding. CO_2 and Nd:YAG lasers are both used to produce good welds, especially with robot systems. Microwave welding is another recent development. It involves placing a conducting polymer between the sections to be welded and then placing the assembly in a microwave field, and then applying pressure as the mated pieces cool. This system can be employed on three-dimensional joints and assemblies and offers exciting potential in the vehicle and appliance industries.

Several other methods for joining plastics with heat are used for special applications. Most call for heating the material with metal plates, or electric current, or friction to sufficiently melt thermoplastics for joining. With the exception of spin (friction) welding, which is restricted to circular sections, these methods cannot approach the tensile strength of joining with the methods described above. They can, however, be useful for joining panels tightly enough to provide leak free joints.

After fabrication or molding, most plastic parts require some form of finishing, which may consist of removing tool marks, mold flashing, or polishing. Care in machining will minimize the amount of finishing required. Thermoplastics such as the acetates, butyrates, and acrylics that have been dulled by buffing, sanding, or polishing, can often achieve a high gloss finish by dipping them in acetone, butyl acetate, ethylene dichloride, or other suitable solvent. Wiping them with these solvents can achieve the same result.

Compression mold design and construction

Modern, intricate mold design is often performed with computer aided design technology, as there are many variables to be considered. The first consideration is whether the mold will have manual, semiautomatic, or automatic operation. Automatic molds are operator-free, but require sophisticated controls. Operator involvement varies with semiautomatic molds, but manual versions require the operator to handle all phases of operation. Injection and transfer molds are almost universally designed for semi-automatic or automatic operation, but molds used in compression molding vary from the simplest hand-operated types to fully automatic molds.

Well-designed, durable molds are essential to any successful molding operation. The length of the run, the rate of production, and the accuracy of the detail and dimension of the part are some of the major factors to be considered. Additional consideration must be given to the following variables. 1) The type of plastics material being used. This is essential for predetermining the bulk factor and mold shrinkage. 2) The size of the press, especially the size of the platens and the maximum space between them. 3) Molding pressure to be used. 4) Determining the dividing point between the mold halves, and where the parting line will appear on the molded part. 5) Size and shape of the part, which will determine the number of cavities per mold. 6) Inclusion of any inserts (usually metal) that will require provisions for side-core pins and unscrewing threaded pins. 7) Heat requirements for proper spacing of heating coils or wires needed to rapidly and uniformly heat the mold. 8) Ejector assembly for removing molded parts.

Molds for compression molding consist of a cavity, which forms the outside shape of the part, and a plunger which informs the inside shape. Usually, the lower half is secured to the stationary platen, and the upper half (normally the plunger) is secured to a moving platen that is led by alignment guide pins when it is lowered. The material is loaded in the cavity and is compressed and shaped by the plunger upon closing. The material is confined within the area between the force and cavity while it hardens under heat and pressure for the time required to permanently form the part. Provisions for heating the mold can be made directly in the mold, or by use of heated platens mounted on the press. Ejection pins or forced air are usually used to remove the part from the mold.

Generally, there are five basic designs used in compression molding.

Positive type. The plunger descends into the filled cavity. The pressure is almost entirely directed at the material, but tight clearances of 0.0015 to 0.005 in. (0.04 to 0.13 mm) between the cavity and the plunger allow very little material to escape. The material that does escape becomes flash that must be removed from the part. Flash can be controlled by careful measurement of the amount of mold material placed in the cavity. An additional disadvantage of this type of mold is that the vertical flash excreted from the mold leads to eventual degradation of the cavity sides, making part ejection difficult.

Flash type. The procedure is similar to the positive type, except that the flash is forced out of the lands between the cavity and plunger at the top of the cavity. Flash is horizontal, and the flash does not inflict damage to the mold. This method is normally used on parts with shallow thickness such as dinnerware plates.

Semi-Positive type. As the two halves of the mold begin to close, flash is allowed to escape. Then, as full pressure is reached, clearance between the plunger and cavity is minimized, resulting in very thin and relatively easily removed vertical flash at the parting line. This mold combines the advantage of free material flow (flash mold) with the ability to produce dense parts (positive mold).

Land Plunger type. Very similar in appearance to semi-positive molds, but the land does not contact the mold. It is stopped by external landing bars approximately 0.001 in. (0.025 mm) before contact. Flash escapes through the gap, leaving a horizontal flash at the parting line.

Split Wedge type. These molds are more complicated and are used when the part has projections or undercuts. The cavity is in two or more parts, with the inside surface of the part on one part and the outside on the other. As the mold closes, the outside impression presses toward the inside impression through clamping action. This is accomplished with spring loaded pins or other mechanical sliding devices.

Fiber Reinforced Resins

Carbon fibers are produced by the pyrolysis (a chemical change induced by thermal action) of organic precursor fibers such as rayon, polyacrylonitrile (PAN), and pitch in an inert atmosphere. The term carbon fiber is often used interchangeably with "graphite," but carbon fibers and graphite fibers differ in the temperature at which the fibers are made and heat treated, and in the amount of carbon produced. Carbon fibers are typically carbonized at 3,400 to 5,440° F (1,900 to 2,000° C), and assay at more than 99% elemental carbon. Important carbon and other reinforcement fibers, as well as the resin materials used in composite materials, are discussed in this section. Much of the material in this section, and most of the property values, are from MIL-HDBK-17-3E.

Reinforcement fiber materials

Aramid. Kevlar [TM] aramid fiber, an organic fiber with high tensile modulus and strength, was introduced by the Du Pont Company in the 1970s. It was the first organic fiber to be used as reinforcement in advanced composites. Today, it is used in various structural parts including reinforced plastics, ballistics, tires, high performance vehicle components, ropes, cables, asbestos replacement, coated fabrics, and protective apparel. Aramid fiber is manufactured by extruding a polymer solution through a spinneret. It is available in many forms, including continuous filament yarns, rovings, chopped fiber, spun-laced sheet, and thermoformable composite sheets.

Important generic properties of aramid fibers include low density, high tensile strength, high tensile stiffness, low compressive properties (nonlinear), and exceptional toughness characteristics. Its density is 0.052 lb/in.^{3} (1.44 gr/cm^3), which is about 40% lower than glass and about 20% lower than commonly used carbons. Aramids do not melt—they decompose at 900° F (500° C). Tensile strength of various aramid yarns varies from 500 to 600 ksi (3.4–4.1 GPa). The nominal coefficient of thermal expansion is 3×10^{-6} in./in./°F (-5×10^6 m/m/°C) in the axial direction. Aramid fibers, being aromatic polyamide polymers, have high thermal stability and dielectric and chemical properties. Their excellent ballistic performance and general damage tolerance is derived from fiber toughness, both in fabric and composite form.

Composite systems reinforced with aramid have excellent vibration dampening characteristics, and they resist shattering upon impact. Continuous service temperatures range from −33 to + 390° F (−36 to + 200° C). These composites are ductile under compression or flexure, but ultimate strength in this condition is lower than glass or carbon composites. Composite systems reinforced with aramid are resistant to fatigue and stress rupture, and epoxy reinforced aramid (60% fiber volume) specimens under tension/tension fatigue testing survive 3,000,000 cycles at 50% of their ultimate stress. Thermoplastic composites reinforced with aramid have recently been developed. These systems have exhibited mechanical properties comparable to those of thermoset systems, plus offer lower processing costs and the potential for bonding and repair. Properties are shown below.

Nominal Composite Properties Reinforced with Aramid Fiber (60%)					
		Thermoset (epoxy)		Thermoplastic	
Tensile Property	Units	Unidirectional	Fabric [1]	Unidirectional	Fabric [1]
Modulus	Msi (GPa)	11 (68.5)	6 (41)	10.5–11.5 (73–79)	5.1–5.8 (35–40)
Strength	ksi (GPa)	200 (1.4)	82 (0.56)	180–200 (1.2–1.4)	77–83 (0.53–0.57)

Note: [1] Fabric aramid volume 40%.

Aramid fibers are available in many forms with different fiber modulus. Kevlar [TM] 29 has the lowest modulus and highest toughness, and is used primarily in ballistics, protective apparel, ropes, tires, etc., and in composites where maximum impact and damage tolerance is critical and stiffness is less important. Kevlar [TM] 49 is predominately used in reinforced plastics—in both thermosetting and thermoplastic resin systems. It is also used as core in fiber optic cable, high pressure rubber hoses, conveyor belts, etc. Recently, Kevlar [TM] 149 with an ultra-high modulus has become available. Nominal properties for these fibers are given below.

		Type of Kevlar [TM]		
Tensile Property	Units	29	49	149
Modulus	Msi (GPa)	12 (83)	18 (124)	25 (173)
Strength	ksi (GPa)	525 (3.6)	525–600 (3.6–4.1)	500 (3.4)

Aramid fiber is available in various weights, weave patterns, and constructions: from very thin (0.0002 in. [0.005 mm]), lightweight (275 gr/m^2), to thick (0.026 in. [0.66 mm]), heavy (2.8 gr/m^2 roving). Woven fiber prepreg is the most common form used in thermoset composites. Chopped aramid fiber is available in lengths of 6 mm to 100 mm, with the shorter lengths being used primarily to reinforce thermoset, thermoplastic, and elastomeric resins in automotive brake and clutch linings, gaskets, and electrical parts. Aramid short fibers can be processed into spun-laced and wet-laid papers that are useful for surfacing veil, thin-printed wiring boards, and gasket material. Uniform dispersion of aramid short fiber in resin formulations is achieved through special mixing methods and equipment.

Glass. Although the rapid evolution of carbon and aramid fibers has resulted in stronger and lighter materials, glass composite products have prevailed in certain applications. Cost per weight or volume, chemical or galvanic corrosion resistance, electrical properties, and availability of many product forms are among their advantages. Coefficient of thermal expansion and modulus properties may be considered disadvantages when compared to carbon composites. Compared to aramid composites, the tensile properties of glass are not as good, but ultimate compression, shear properties, and moisture absorption properties are superior.

Typical glass compositions used for reinforcement are electrical/Grade "E" glass, a calcium aluminoborosil mica composition with an alkali content of less than 2%; and chemical resistant "C" glass made from soda-lime-borosilicates and high strength S-2 glass (a low alkali magnesi-silicate composition). Glass roving products (untwisted) type yarns are most often directly finished with the final coupling agents during the filament manufacturing step. Other finishes include Volan finishes, with the most recognized variant, Volan A, providing good wet and dry strength properties when used with polyester, epoxy, and phenolic resins. Saline finishes are also often used on epoxy. Most of these finishes are formulated to enhance laminate wet-out, and some also provide high laminate clarity or good composite properties in aqueous environments. Although other finishes are used in combination with matrix materials other than epoxy, these finishes may have proprietary formulations or varied designations relative to the particular glass manufacturer or weaver.

There are many forms of glass products. The continuous filament forms are continuous rovings, yarn for fabrics or braiding, mats, and chopped strand. They are available with a variety of physical surface treatments and finishes, but most structural applications utilize fabric, roving, or rovings converted to unidirectional tapes. Perhaps the most versatile fiber type used to produce glass fiber forms is "E" glass, identified for electrical applications. "E" glass is available in eight or more standard diameters, ranging from 1.4 to 5.1 mils (3.5 to 13 micrometers), which facilitates very thin product forms. The "S" glasses are identified as such to signify high strength. S-2 glasses are available in only one filament diameter, but S-2 rovings are available in yields of 250, 750, and 1,250 yards per pound (500, 1,500, and 2,500 m/kg).

Property	Units	Type of Glass	
		E	S-2
Density	lb/in³ (gr/cm³)	0.094 (2.59)	0.089 (2.46)
Tensile Strength	ksi (MPa)	500 (34,450)	665 (45,820)
Modulus of Elasticity	Msi (GPa)	10.5 (72.35)	12.6 (86.81)
Percent Elongation	–	4.8	5.4
Coeff. Thermal Expan.	10^6 in./in./°F (10^6 m/m/°C)	2.8 (5.1)	1.3 (2.6)
Softening Point	°F (°C)	1,530 (832)	1,810 (988)
Annealing Point	°F (°C)	1,210 (654)	1,510 (821)

Boron. Boron fiber is unmatched for its combination of strength, stiffness, and density. The tensile modulus and strength of boron fiber are 60 % 10^6 PSI and 0.52 % 10^6 PSI (40 GPa and 3,600 MPa), respectively. Thermal conductivity and thermal expansion are both low—the coefficient of thermal expansion is 2.5–3.0 % 10^{-6}/°F (4.5–5.4 % 10^{-6}/°C). Boron is available as a cylindrical fiber in two nominal diameters, 4 and 5.6 mils (10 and 14 micrometers) which have a density of 0.0929 and 0.0900 lb/in.³ (2.57 and 2.49 gr/cm³), respectively. Available in filament or epoxy matrix prepreg form, boron fiber has been used for aerospace applications requiring high strength and/or stiffness, and as reinforcement in selected sporting goods.

Typical End-use Properties of Unidirectional Boron/Epoxy Laminate (60% Volume)			
	Property	Units	Value
Moduli	Tensile, Longitudinal	ksi (MPa)	30 (207)
	Tensile, Transverse		2.7 (19)
Strength	Tensile, Longitudinal		192 (1,323
	Tensile, Transverse		10.4 (72)
	Compressive, Longitudinal		353 (2,432)

Alumina. Continuous polycrystalline alumina fiber is ideally suited for the reinforcement of a variety of materials including plastics, metals, and ceramics. It is prepared as continuous yarn containing a nominal 200 filaments, and supplied in bobbins containing continuous filament yarn, and aluminum/aluminum and aluminum/magnesium plates. Alumina staple is also available for short fiber reinforcement. Its high modulus of 55 Msi (380 GPa) is comparable to that of boron and carbon. Average filament strength is 200 ksi (1,379 MPa) minimum. Since alumina is a good insulator, it can be used in applications where conductive fibers cannot. Alumina, in continuous form, offers many advantages for composite fabrication including ease of handling, the ability to align fibers in desired directions, and filament winding capability. The fact that alumina is an electrical insulator, combined with its high modulus and compressive strength, make it of interest for polymer matrix composite applications. For example, alumina/epoxy and aramid/epoxy hybrid composites reinforced with alumina and aramid fibers have been fabricated and are of interest for radar transparent structures, circuit boards, and antenna supports.

Nominal Properties of Alumina			
Property	*Value*	*Property*	*Value*
Composition	>99% α–Al$_2$O$_3$	Filaments/yarn	200, nominal
Melting Point	3,713 °F (2,045 °C)	Tensile Modulus	55 Msi (385 GPa)
Filament Diameter	0.8×10^{-3} in. (20 μm)	Tensile Strength	200 ksi (1,379 MPa) min.
Length/Weight	~ 4.7 m/gr	Density	0.14 lb/in.^{3} (3.9 (gr/cc)

Silicon Carbide. Silicon fibers are produced with a nominal 0.0055 in. (140 micrometer) filament diameter and are characteristically found to have high strength, modulus, and density. Practically all silicon carbide monofilament fibers are currently produced for metal composite reinforcement. Alloys employing aluminum, titanium, and molybdenum have been produced. General processing for epoxy, bisimide, and polyimide resin can be either a solvated or solventless film impregnation process, with cure cycles equivalent to those for carbon or glass reinforced products. Organic matrix silicon carbide impregnated products may be press, autoclave, or vacuum bag oven cured. Lay-up on tooling proceeds as with carbon or glass composite products with all bleeding, damming, and venting as required for part fabrication. General temperature and pressure ranges for the cure of the selected matrix resins used in silicon carbide products will not adversely affect the fiber morphology. Silicon carbide ceramic composites engineered to provide high service temperatures (in excess of 2,640° F [1,450° C]) are unique in several thermal properties.

Nominal Properties of Alumina Composite (50–55% Volume)					
Property		*Value*		*Property*	*Value*
Moduli	Tensile, Axial	30–32 Msi (210–220 GPa)	Strength	Tensile, Axial	80 ksi (600 MPa)
	Tensile, Transverse	20–22 Msi (140–150 GPa)		Tensile, Transverse	26–30 ksi (130–210 MPa)
	Shear	7 Msi (50 GPa)		Shear	12–17 ksi (85–120 GPa)
Fatigue–Axial Endurance Limit		10^7 cycles at 75% static ultimate	Thermal Conductivity *		22–29 Btu/lbm-°F (38–50 J/m-s-°C)
Average	Axial	4.0 μin./in./°F (7.2 μm/m/°C)	Specific Heat *		0.19–0.12 Btu/lbm-°F (0.8–0.5 J/gr-°C)
Thermal Exp.	Transverse	11 μin./in./°F (20 μm/m/°C)	Density		0.12 lbm/in.^{3} (3.3 gr/cc)

* Thermal conductivity and specific heat at 68–750 ° F (20–400° C).

Quartz. Quartz fiber is very pure (99.95%) fused silica glass fiber. It is produced as continuous strands consisting of 120 to 240 individual filaments of 9 micron nominal diameter. The single strands are twisted and plied into heavier yarns. Quartz fibers are generally coated with an organic binder containing a silane coupling agent that is compatible with many resin systems. Quartz rovings are continuous reinforcements formed by combining a number of 300 2.0 zero twist strands. End counts of 8, 12, and 20 are available with yields ranging from 750 to 1,875 yards/pound. Quartz fibers are also available in the form of chopped fiber in cut lengths of $1/8$ inch to 2 inches (3 to 50 mm). Quartz fiber nomenclature is the same as for "E" or "S" glass except that the glass composition is designated by the letter Q.

Quartz fibers with a filament tensile strength of 850 ksi (5,900 MPa) have the highest strength to weight ratio, exceeding virtually all other high temperature materials. Therefore, quartz fibers can be used at temperatures much higher than "E" or "S" glass, with service temperatures up to 1,920° F (1,050° C) possible. Quartz fibers do not melt or vaporize until the temperature exceeds 3,000° F (1,650° C), and the fibers retain virtually all characteristics and properties of solid quartz up to failure. The fibers are chemically stable, but should not be used in environments where strong concentrations of alkalies are present.

Properties of Quartz Fiber			
Property	*Value*	*Property*	*Value*
Specific Gravity	2.20	Coef. Thermal Exp.	0.3 10⁻⁶ in./in./°F (0.54 10⁻⁶ cm/cm/°C)
Density	0.0795 lb/in.³ (2.20 gr/cc)	Thermal Conductivity	0.80 Btu/hr/ft/°F (0.0033 Cal/sec/cm/°C)
Tensile Strength	870 ksi (5,900 GPa)	Specific Heat	1.80 Btu/lb/°F (7,500 J/kg/°C)
Modulus	10 ksi (72 MPa)	Dielectric Constant	10 GHz, 75°F (24°C)

Typical Properties for Quartz Epoxy (laminate resin content 32–33.5% by weight)		
Property	*Room Temperature*	*1/2 Hour at 350° F (180° C)*
Tensile Strength	74.9–104 ksi (516–717 MPa)	65.4–92.9 ksi (451–636 MPa)
Tensile Modulus	3.14–4.09 Msi (21.7–28.2 GPa)	2.83–3.67 Msi (19.5–25.3 (GPa)
Flexure Strength	95.5–98.9 ksi (658–682 MPa)	53.9–75.9 ksi (372–523 MPa)
Flexure Modulus	3.27–3.46 Msi (22.5–23.8 GPa)	2.78–3.08 Msi (19.2–21.2 GPa)
Compressive Strength	66.4–72.4 ksi (459–499 MPa)	42.6–49.9 ksi (294–344 MPa)
Compressive Modulus	3.43–3.75 Msi (23.6–25.9 GPa)	3.10–3.40 Msi (21.4–23.4 GPa)
Specific Gravity	1.73–1.77	

Typical Properties for Quartz Polyimide (laminate resin content 36.2% by weight)		
Property	*Room Temperature*	*1/2 Hour at 350° F (180° C)*
Tensile Strength	79.1–105 ksi (545–724 MPa)	–
Tensile Modulus	3.9 Msi (27 GPa)	–
Flexure Strength	93.7–102 ksi (646–703 MPa)	62.4–68.3 ksi (430–471 MPa)
Flexure Modulus	3.2 Msi (22 GPa)	2.6–2.8 Msi (18–19 GPa)
Compressive Strength	67–67.4 ksi (462–465 MPa)	38.6–45.2 ksi (266–312 MPa)
Compressive Modulus	3.5–3.7 Msi (24–26 GPa)	2.8 Msi (19 GPa)

Reinforcement resin materials—thermosets

Resin is a generic term used to designate the polymer, polymer precursor material, and/or mixture or formulation thereof with various additives or chemically reactive compounds. The resin, its chemical composition, and its physical properties fundamentally affect the processing, fabrication, and ultimate properties of composite materials.

Epoxy. The term epoxy is a general description of a family of polymers that are based on molecules that contain epoxide groups. Epoxies are widely used in resins for prepregs and structural adhesives. Their advantages are high strength and modulus, low levels of volatiles, excellent adhesion, low shrinkage, good chemical resistance, and ease of processing. Their major disadvantages are brittleness and the reduction of properties in the presence of moisture. The processing or curing of epoxies is slower than for polyester resins, and the cost of the resin is also higher than for the polyesters. Processing techniques include autoclave molding, filament winding, press molding, vacuum bag molding, resin transfer molding, and pultrusion. Curing temperatures vary from room temperature to approximately 350° F (180° C). Higher temperature cures generally yield greater temperature resistance in the cured product.

Polyester (thermosetting). Generally, for a fiber reinforced resin system, the advantage of a polyester is its low cost and its ability to be processed quickly. Common processing methods include matched metal molding, wet lay-up, press (vacuum bag) molding, injection molding, filament winding, pultrusion, and autoclaving. Polyesters can be formulated to cure more rapidly than phenolics during the thermoset molding process. While phenolic processing, for example, is dependent on a time/temperature relationship, polyester processing is primarily dependent on temperature. Depending on the formulation, polyesters can be processed from room temperature to 350° F (180° C). If the proper temperature is applied, a quick cure will occur. Compared to epoxies, polyesters process more easily and are much tougher, whereas phenolics are more difficult to process and are brittle, but they do have higher service temperatures.

Phenolics. Phenolics, in general, cure by condensation through the off-gassing of water. The resulting matrix is characterized by both chemicals and thermal resistance, as well as hardness, plus low smoke and toxic degradation products. Phenolics are often called either resole or novolac resins. The basic difference between the two is that the novolacs have no methylol groups, and therefore they require an extension agent of paraformaldehyde, hexamethyleneetetraamine, or additional formaldehyde as a curative. Since the additives have higher molecular weights and viscosities than either parent material, they allow either press or autoclave cure and allow relatively high temperature free-standing postcures.

Bismaleimide. This type of thermosetting resin only recently became available commercially in prepreg tapes, fabrics, rovings, and sheet molding compound. The physical form of bismaleimide (BMI) resin depends on the requirements of the final application. At room temperature, their form can vary from a solid to a pourable liquid. For aerospace prepregs, stick resins are required. The advantages of BMI resins are best discussed in comparison to epoxy resins. Emerging data suggests that BMIs are versatile resins with many applications in the electronic and aerospace industries. Their primary advantage over epoxy resins is their high glass transition temperature, which is in the range of 500–600° F (260–320° C). Glass transition temperatures for high temperature epoxies are generally less than 500° F. Another BMI advantage is high elongation with the corresponding high service temperature capabilities. BMI resins are processed with essentially the same methods used for epoxies—they are suited for standard autoclave processing, injection molding, resin transfer molding, and sheet molding compounds, among others. The processing time is similar to epoxies, but to prepare for high service temperatures a free-standing postcure is required. At present, there are no room temperature curing BMIs. Due to their recent introduction, these resins are not as widely available as others, and costs are generally higher.

Polyimides. Polyimide matrix composites excel in high temperature environments where their thermal resistance, oxidative stability, low coefficient of thermal expansion, and solvent resistance benefit the design. Their primary uses are circuit boards and hot engine and aerospace structures. A polyimide may be either a thermoset resin or a thermoplastic. The thermoplastic variety will be discussed later, but, because partially cured thermoset polyimides containing residual plasticizing solvents can exhibit thermoplastic behavior, it is difficult to state with certainty that a particular polyimide is indeed a thermoset or a thermoplastic. Polymides, therefore, represent a transition between these two polymer classifications.

Most polyimide resin monomers are powders (some bismaleimides are an exception), and solvents are added to the resin to enable impregnation of unidirectional fiber and woven fabrics. Commonly, a 50/50 by weight mixture is used for fabrics, and a 90/10 by weight high solids mixture is used to produce a film for unidirectional fiber and low areal weight fabric prepregs. Solvents are further used to control prepreg handling qualities, such as tack and drape. Most of the solvents are removed in a drying process during impregnation, but total prepreg volatiles contents range between 2 and 8% by weight. Polyimides require high cure temperatures, usually in excess of 550° F (288° C). Consequently, normal epoxy composite consumable materials are not usable, and steel tooling becomes a necessity. Polyimide bagging and release films, such as Kapton and Upilex, replace the lower cost nylon bagging and polytetrafluoroethylene (PTFE) release films common to epoxy composite processing. Fiberglass fabrics must be used for bleeder and breather materials rather than polyester mat materials.

Reinforcement resin materials—thermoplastics

Semi-crystallines. These materials are so named because a percentage of their volume consists of a crystalline morphology. The remaining volume has a random molecular orientation (called amorphous). A partial list of semi-crystalline thermoplastics includes polyethylene, polypropylene, polyamides, polyphenylene sulfide, polyetheretherketone, and polyarylketone. The inherent speed of processing, ability to produce complicated, detailed parts, excellent thermal stability, and corrosion resistance have enabled them to become established in the automotive, electronic, and chemical processing industries. Processing speed is, in fact, the primary advantage of thermoplastic materials. However, thermoplastic impregs are typically boardy and do not exhibit the tack and drape of thermosets although drapeable forms are available that have commingled (interlaced together) thermoplastic and reinforcing fibers. High performance thermoplastics are slightly more expensive than equivalent performance epoxies, and tooling costs may be higher as well. However, final part cost may be less due to decreased processing time, and the ability to reprocess molder thermoplastics.

Some semi-crystalline thermoplastics possess properties of inherent flame resistance, superior toughness, good mechanical properties at elevated temperatures and after impact, and low moisture absorption—factors that have led to their wide use in the aerospace and automotive industries. Their primary disadvantage is the lack of a design database.

Amorphous polymers. The majority of thermoplastic polymers are composed of a random molecular orientation and are termed amorphous. This group includes polysulfone, polyamide-imide, polyphenylsulfone, polyphenylene sulfine sulfone, polyether sulfone, polystyrene, polyetherimide, and polyarylate. Amorphous thermoplastics are available in many forms, including films, filaments, and powders. Combined with reinforcing fibers, they are also available in injection molding compounds, compression moldable random sheets, unidirectional tapes, woven prepregs, etc. The fibers used are primarily carbon, aramid, and glass.

The primary advantages of amorphous thermoplastics in continuous fiber reinforced composites are potential low cost process at high production rates, high temperature capability, good mechanical properties before and after impact, and chemical stability. A service temperature of 350° F (177° C), and toughness two or three times that of conventional thermoset polymers are typical, and cycle times in production are less than for thermosets since no chemical reaction occurs during the forming process. Disadvantages are similar to those of semi-crystalline thermoplastics, including a lack of an extensive design database and reduced compression properties compared to 350° F (177° C) cure thermosets. Solvent resistance, which is good for semi-crystalline thermoplastics, is a concern for most amorphous ones. Processing methods include stamp molding, thermoforming, autoclave molding, roll forming, filament winding, and pultrusion. The high melting temperatures require process temperatures ranging from 500 to 700° F (260 to 370° C). Thermal expansion differences between the tool and the thermoplastic material should be addressed due to the high processing temperatures. Forming pressures range from 100 PSI (0.7 MPa) for thermoforming, to 5,000 PSI (34.4 MPa) for stamp molding.

Amorphous thermoplastics are used in many applications including the medical, communication, transportation, chemical processing, electronic, and aerospace industries. The majority of applications use unfilled and short-fiber forms—some uses for the unfilled polymers include cookware, power tools, business machines, corrosion resistant piping, medical instruments, and aircraft canopies. Uses for short-fiber reinforced forms include printed circuit boards, transmission parts, under-hood automotive applications, electrical connectors, and jet engine components.

Cure and consolidation processes

Vacuum bag molding. In vacuum bag molding, the lay-up is cured under pressure generated by drawing a vacuum in the space between the lay-up and a flexible sheet placed over it and sealed at the edges. The reinforcement is generally placed in the mold by hand lay-up using prepreg or wet resin. High flow resins are preferred for vacuum bag molding. The following steps are followed. 1) Place composite material for part into mold. 2) Install bleeder and breather material. 3) Place vacuum bag over part. 4) Seal bag and check for leaks. 5) Place tool and part in oven and cure as required at elevated temperature. 6) Remove part from mold. Parts fabricated using vacuum bag oven cure have lower fiber volumes and higher void contents. This is a low cost method of fabrication and uses low cost tooling for short production runs.

Autoclave cure. This process uses a pressurized vessel to apply both pressure and heat to parts that have been sealed in a vacuum bag. Generally, autoclaves operate at 10 to 300 PSI (10 to 30 kPa) and up to 800° F (420° C). Heat transfer and pressure application to the part is achieved by circulation (convection) of pressurized gas (usually air, nitrogen, or carbon dioxide) with the autoclave. Composite materials that are typically processed in autoclaves include adhesives, reinforced matrix (epoxy, bismaleimide, etc.) laminates and reinforced matrix laminates. In the case of the thermoset resin systems, the cure cycle is developed to induce specific chemical reactions within the polymer matrix by exposing the material to elevated temperatures while simultaneously applying vacuum and pressure to consolidate individual plies and compress voids. The cure cycle and vacuum bagging procedure affect such cured product characteristics as degree of cure, glass transformation temperature, void content percentage, cured resin content/fiber volume, residual stress, dimensional tolerances, and mechanical properties.

Pultrusion. Pultrusion is an automated process for the continuous manufacture of composites with a constant cross sectional area. It can be dry, employing prepreg thermosets or thermoplastics; or wet, where the continuous fiber bundle is resin-impregnated in a resin bath. In pultrusion, the material is drawn through a heated die that is specially designed for the shape being made. The key elements in the process consist of a reinforcement delivery platform, resin bath (for wet pultrusion), preform dies, a heated curing die, a pulling system, and a cutoff station. In general, the following process is used. 1) Reinforcements are threaded through the reinforcement delivery station. 2) The fiber bundle is pulled through the resin bath (wet pultrusion) and die preforms. 3) A strap is used to initiate the process by pulling the resin impregnated bundle through the preheated die. 4) As the impregnated fiber bundle is pulled through the heated die, the die temperature and pulling rate are controlled so that the cure to the product (or thermosets) is completed prior to exiting the heated die. 5) The composite parts are cut off by sawing when they reach the desired length.

Resin transfer molding and *thermoforming* are also used for fiber reinforced resins. These processes were discussed in the earlier section on "Process Technology."

Wood Products

Commercially important woods

Native species of trees are divided into two classes: hardwoods which have broad leaves; and softwoods which have scale-like leaves (such as cedars), or needle-like leaves (such as pines). Native softwoods, with the exception of cypress, tamarack, and western larch, are evergreen. Softwoods are also called "conifers," because all native species bear cones of one kind or another. The terms hardwood and softwood are not descriptively exact. Some softwoods, such as southern yellow pine and Douglas fir, are harder than some hardwoods, such as basswood and cottonwood.

Measuring wood products

The standard unit of measurement in the U.S.A. for wood products is the board foot. Lumber less than one inch thick is based on surface measure, and some types of finish lumber, such as moldings and trim, are sold by the linear foot. Common commercial quantity units are "per 1,000 feet board measure" or "per MBF," or "per 100 lineal feet," or "per CLF," and prices are generally quoted in terms of these quantity units. A board foot represents the quantity of lumber contained in a board that is one inch thick, twelve inches wide, and one foot long, or its cubic equivalent. For example, a board one inch thick, six inches wide, and two feet long is also one board foot. It is important to note that, in practice, the board foot calculations for lumber are based on its nominal thickness, nominal width, and actual length. Nominal sizes are not always actual sizes.

Table 1 provides the actual board foot content of various sizes of boards and dimension lumber. The cross-sectional sizes given are "nominal," meaning the sizes commonly used for rough lumber. This Table can be used to compute board footage if the number of pieces needed is known. Thus, 1,000 2 × 4s, each twelve feet long, would contain 8,000 board feet of lumber.

Board measure can be computed mathematically. Multiply the number of pieces by the nominal thickness in inches by the nominal width in inches by the length in feet, then divide the result by twelve.

Pieces × thickness (inches) × width (inches) × length (feet) ÷ 12 = Board Feet.

To calculate surface measure, in square feet, the thickness is not required.

Pieces × width (inches) × length (feet) ÷ 12 = Square Feet.

For items sold by the lineal foot, calculate the total lineal feet on the basis of a board foot estimate. When the thickness and width of the product is such that they must be produced from lumber nominally one inch thick and wide, or more, use the following formula.

(Board Foot measure × 12) ÷ (thickness (inches) × width (inches) = Lineal Feet.

Mechanical properties of wood

Wood can be considered an orthotropic material, which means that it has unique and independent mechanical properties in three mutually perpendicular axes: longitudinal, radial, and tangential. As shown in *Figure 1*, the longitudinal axis, L, is parallel to the grain; the radial axis, R, is normal to the growth rings (perpendicular to the grain in the radial direction); and the tangential axis, T, is perpendicular to the grain but tangent to the growth rings.

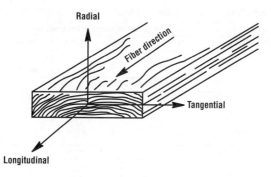

Figure 1. Three principal axes of wood.

Elastic Properties

Twelve constants (nine of which are independent) are needed to describe the elastic behavior of wood. These are three moduli of elasticity, E; three moduli of rigidity, G; and six Poisson's ratios, μ. The moduli of elasticity and the Poisson's ratios are related as follows.

$$\mu_{ij} \div E_i = \mu_{ji} \div E_j, \quad i \neq j \; i,j = L, R, T.$$

Elasticity implies that deformations produced by low stress are completely recoverable after loads are removed. When loaded to higher stress levels, plastic deformation or failure occurs. Wood's three moduli of elasticity, denoted by E_L, E_R, and E_T, are, respectively, the elastic moduli along the longitudinal, radial, and tangential axes. Average values of E_R, and E_T for samples from selected species (along with modulus of rigidity—see below) are shown in **Table 2** as ratios with E_L. It should be remembered that elastic ratios, as well as the elastic constants themselves, vary within and between species and with moisture content and specific gravity. The test results in this section were obtained either from "green" specimens, or from specimens with approximately 12% moisture content.

When a member is loaded axially, the deformation perpendicular to the direction of the load is proportional to the deformation parallel to the direction of the load, and the ratio of the transverse to axial strain in Poisson's ratio. The various ratios are denoted by μ_{LR}, μ_{RL}, μ_{LT}, μ_{TL}, μ_{RT}, and μ_{TR}. The initial letter of the subscript indicates the direction of applied stress, and the second letter the direction of lateral deformation. For example, μ_{LR} indicates deformation along the radial axis caused by stress along the longitudinal axis. Average values of Poisson's ratio for selected species are shown in **Table 3**. Poisson's ratio varies between species and is affected by moisture content and specific gravity.

The modulus of rigidity (shear modulus) for wood indicates its resistance to deflection caused by shear stresses. The three moduli of rigidity, denoted by G_{LR}, G_{LT}, and G_{RT}, are the elastic constants in the LR, LT, and RT planes, respectively. For example, G_{LR} is the modulus of rigidity based on shear strain in the LR plane, and shear stresses in the Lt and RT planes. Average values of shear moduli for samples of selected species, expressed as ratios with E_L, are given in **Table 2**. As with the modulus of elasticity, the modulus of rigidity will vary between species and with moisture content and specific gravity.

Table 1. Lumber Footage by Lengths, Widths, and Thicknesses. *(Source, MIL-HDBK-7B.)*

Nominal Cross Section (inches)	Footage Tally (in Board Feet) for Length of—					
	8 Feet	10 Feet	12 Feet	14 Feet	16 Feet	18 Feet
1 × 3	2	2 1/2	3	3 1/2	4	4 1/2
1 × 4	2 2/3	3 1/3	4	4 2/3	5 1/3	6
1 × 6	4	5	6	7	8	9
1 × 8	5 1/3	6 2/3	8	9 1/3	10 2/3	12
1 × 10	6 2/3	8 1/3	10	11 2/3	13 1/3	15
1 × 12	8	10	12	14	16	18
1 1/2 × 4	4	5	6	7	8	9
1 1/2 × 6	6	7 ½	9	10 1/2	12	13 1/2
1 1/2 × 8	8	10	12	14	16	18
1 1/2 × 10	10	12 1/2	15	17 1/2	20	22 1/2
1 1/2 × 12	12	15	18	21	24	27
2 × 4	5 1/3	6 2/3	8	9 1/3	10 2/3	12
2 × 6	8	10	12	14	16	18
2 × 8	10 2/3	13 1/3	16	18 2/3	21 1/3	24
2 × 10	13 1/3	16 2/3	20	23 1/3	26 2/3	30
2 × 12	16	20	24	28	32	36
3 × 4	8	10	12	14	16	18
3 × 6	12	15	18	21	24	27
3 × 8	16	20	24	28	32	36
3 × 10	20	25	30	35	40	45
3 × 12	24	30	36	42	48	54
4 × 4	10 2/3	13 1/3	16	18 2/3	21 1/3	24
4 × 6	16	20	24	28	32	36
4 × 8	21 1/3	26 2/3	32	37 1/3	42 2/3	48
4 × 10	26 2/3	33 1/3	40	46 2/3	53 1/3	60
4 × 12	32	40	48	56	64	72
6 × 6	24	30	36	42	48	54
6 × 7	28	35	42	49	56	63
6 × 8	32	40	48	56	64	72
6 × 10	40	50	60	70	80	90
6 × 12	48	60	72	84	96	108
7 × 9	42	52 1/2	63	73 1/2	84	94 1/2
8 × 8	42 2/3	43 1/3	64	74 2/3	85 1/3	96
8 × 10	53 1/3	66 2/3	80	93 1/3	106 2/3	120
8 × 12	64	80	96	112	128	144

Table 2. Elastic Ratios for Selected Species (Approximately 12% Moisture Content). See Text for Explanation. *(Source, Wood Handbook, U.S.D.A.)*

Species	E_T/E_L	E_R/E_L	G_{LR}/E_L	G_{LT}/E_L	G_{RT}/E_L
Hardwoods					
Ash, White	0.080	0.125	0.109	0.077	-
Balsa	0.015	0.046	0.054	0.037	0.005
Basswood	0.027	0.066	0.056	0.046	-
Birch, yellow	0.050	0.078	0.074	0.068	0.017
Cherry, black	0.086	0.197	0.147	0.097	-
Cottonwood, eastern	0.047	0.083	0.076	0.052	-
Mahogany, African	0.050	0.111	0.088	0.059	0.021
Mahogany, Honduras	0.064	0.107	0.066	0.086	0.028
Maple, sugar	0.065	0.132	0.111	0.063	-
Maple, red	0.067	0.140	0.133	0.074	-
Oak, red	0.082	0.154	0.089	0.081	-
Oak, white	0.072	0.163	0.086	-	-
Sweet gum	0.050	0.115	0.089	0.061	0.021
Walnut, black	0.056	0.106	0.085	0.062	0.021
Yellow-poplar	0.043	0.092	0.075	0.069	0.011
Softwoods					
Baldcypress	0.039	0.084	0.063	0.054	0.007
Cedar, northern white	0.081	0.183	0.210	0.187	0.015
Cedar, western red	0.055	0.081	0.087	0.086	0.005
Douglas-fir	0.050	0.068	0.064	0.078	0.007
Fir, subalpine	0.039	0.102	0.070	0.058	0.006
Hemlock, western	0.031	0.058	0.038	0.032	0.003
Larch, western	0.065	0.079	0.063	0.069	0.007

Pine
  Loblolly
  Lodgepole
  Longleaf
  Pond
  Ponderosa
  Red
  Slash
  Sugar
  Western white
	
	0.078
	0.068
	0.055
	0.041
	0.083
	0.044
	0.045
	0.087
	0.038
		
		0.113
		0.102
		0.102
		0.071
		0.122
		0.088
		0.074
		0.131
		.078
			
			0.082
			0.049
			0.071
			0.050
			0.138
			0.096
			0.055
			0.124
			0.052
				
				0.081
				0.046
				0.060
				0.045
				0.115
				0.081
				0.053
				0.113
				0.048
					
					0.013
					0.005
					0.012
					0.009
					0.017
					0.011
					0.010
					0.019
					0.005

Redwood	0.089	0.087	0.066	0.077	0.011
Spruce, Sitka	0.043	0.078	0.064	0.061	0.003
Spruce, Engelmann	0.059	0.128	0.124	0.120	0.010

Table 3. Poisson's Ratios for Selected Species (Approximately 12% Moisture Content). See Text for Explanation. *(Source, Wood Handbook, U.S.D.A.)*

Species	μ_{LR}	μ_{LT}	μ_{RT}	μ_{TR}	μ_{RL}	μ_{TL}
Hardwoods						
Ash, White	0.371	0.440	0.684	0.360	0.059	0.051
Aspen, Quaking	0.489	0.374	-	0.496	0.054	0.022
Balsa	0.229	0.488	0.665	0.231	0.018	0.009
Basswood	0.364	0.406	0.912	0.346	0.034	0.022
Birch, yellow	0.426	0.451	0.697	0.426	0.043	0.024
Cherry, black	0.392	0.428	0.695	0.282	0.086	0.048
Cottonwood, eastern	0.344	0.420	0.875	0.292	0.043	0.018
Mahogany, African	0.297	0.641	0.604	0.264	0.033	0.032
Mahogany, Honduras	0.314	0.533	0.600	0.326	0.033	0.034
Maple, sugar	0.424	0.476	0.774	0.349	0.065	0.037
Maple, red	0.434	0.509	0.762	0.354	0.063	0.044
Oak, red	0.350	0.448	0.560	0.292	0.064	0.033
Oak, white	0.369	0.428	0.618	0.300	0.074	0.036
Sweet gum	0.325	0.403	0.682	0.309	0.044	0.023
Walnut, black	0.495	0.632	0.718	0.378	0.052	0.035
Yellow-poplar	0.318	0.392	0.703	0.329	0.030	0.019
Softwoods						
Baldcypress	0.338	0.326	0.411	0.356	-	-
Cedar, northern white	0.337	0.340	0.458	0.345	-	-
Cedar, western red	0.378	0.296	0.484	0.403	-	-
Douglas-fir	0.292	0.449	0.390	0.374	0.036	0.029
Fir, subalpine	0.341	0.332	0.437	0.336	-	-
Hemlock, western	0.485	0.423	0.442	0.382	-	-
Larch, western	0.355	0.276	0.389	0.352	-	-

Pine
  Loblolly
  Lodgepole
  Longleaf
  Pond
  Ponderosa
  Red
  Slash
  Sugar
  Western White
	0.328
	0.316
	0.332
	0.280
	0.337
	0.347
	0.392
	0.356
	0.329
		0.292
		0.347
		0.365
		0.364
		0.400
		0.315
		0.444
		0.349
		0.344
			0.382
			0.469
			0.384
			0.389
			0.426
			0.408
			0.447
			0.428
			0.410
				0.362
				0.381
				0.342
				0.320
				0.359
				0.308
				0.387
				0.358
				0.334
					-
					-
					-
					-
					-
					-
					-
					-
					-
						-
						-
						-
						-
						-
						-
						-
						-
						-

Redwood	0.360	0.346	0.373	0.400	-	-
Spruce, Sitka	0.372	0.467	0.435	0.245	0.040	0.025
Spruce, Engelmann	0.422	0.462	0.530	0.255	0.083	0.058

Strength Properties

The mechanical properties that are most commonly measured and represented as strength properties for design include modulus of rupture in bending, maximum stress in compression perpendicular to grain, and shear strength parallel to grain. Additional measurements are often made to evaluate work in bending, impact bending strength, tensile strength perpendicular to grain, and hardness. These properties are shown in **Table 4** (English units) and **Table 5** (Metric units), and are defined below.

Modulus of rupture reflects the maximum load carrying capacity of a member in bending, and is proportional to maximum moment borne by the specimen. Modulus of rupture is an accepted criterion of strength, but it is not a true stress because the formula by which it is computed is valid only to the elastic limit.

Work to maximum load in bending demonstrates the material's ability to absorb shock with some permanent deformation and injury to the specimen. Work to maximum load is a measurement of the combined strength and toughness of wood under bending stresses.

Impact bending is measured in the impact bending test in which a hammer of given weight is dropped upon a beam from successively increased heights until rupture occurs, or the beam deflects 6 inches (152 mm) or more. The height of the maximum drop, or the drop that causes failure, is a comparative value that represents the ability of wood to absorb shocks that cause stresses beyond the proportional limit.

Compressive strength parallel to grain. This is the maximum stress sustained by a compression parallel to grain specimen having a ratio of length to least dimension of less than 11.

Compressive stress perpendicular to grain. Reported as stress at proportional limit, there is no clearly defined ultimate stress for this property.

Shear strength parallel to grain is the ability to resist internal slippage of one part upon another along the grain. Values presented on the Tables are average strength in radial and tangential shear planes.

Tensile strength perpendicular to grain is a measurement of the resistance of wood to forces acting across the grain that tend to split a member. Values presented are the average of radial and tangential observations.

Hardness is defined as the resistance to indentation using a modified Janka hardness test, measured by the load required to embed a 0.444 inch (11.28 mm) ball to one-half its diameter. Values presented are the average of radial and tangential penetrations.

Tensile strength parallel to grain is the maximum tensile stress sustained in the direction parallel to grain, and is not represented on the Table because little information is available on the tensile strength of various species of clear wood parallel to grain. The modulus of rupture is considered to be a conservative estimate of tensile strength for clear specimens (this is not true for lumber).

Working qualities of wood

The ease of working wood with hand tools generally varies directly with the specific gravity of the wood (specific gravities are given in **Tables 4** and **5**). The lower the specific gravity, the easier it is to cut the wood with a sharp tool. A wood species that is easy to cut does not necessarily develop a smooth surface when it is machined. Consequently, tests have been conducted on many domestic hardwoods to evaluate their machining qualities (see **Table 6**).

Machining evaluation tests are not available for many imported woods, but three major

(Text continued on p. 746)

Table 4. Strength Properties of Selected Commercial Woods Grown in the United States[a] (English Units). *(Source, Wood Handbook, U.S.D.A.)*

| Common Species Names | Moisture Content | Specific Gravity[b] | Static Bending | | | Impact Bending (in.) | Compression Parallel to Grain (lbf/in²) | Compression Perpendicular to Grain (lbf/in²) | Shear Parallel to Grain (lbf/in²) | Tension Perpendicular to Grain (lbf/in²) | Side Hardness (lbf) |
			Modulus of Rupture (lbf/in²)	Modulus of Elasticity[c] (×10⁶ lbf/in²)	Work to Maximum Load (in-lbf/in³)						
Hardwoods											
Alder, Red	Green	0.37	6,500	1.17	8.0	22	2,960	250	770	390	440
	12%	0.41	9,800	1.38	8.4	20	5,820	450	1,080	420	590
Ash											
Black	Green	0.45	6,000	1.04	12.1	33	2,300	350	860	490	520
	12%	0.49	12,600	1.60	14.9	35	5,970	760	1,570	700	850
Blue	Green	0.53	9,600	1.24	14.7	–	4,180	810	1,540	–	–
	12%	0.58	13,800	1.40	14.4	–	6,980	1,420	2,030	–	–
Green	Green	0.53	9,500	1.40	11.8	35	4,200	730	1,260	590	870
	12%	0.56	14,100	1.66	13.4	32	7,080	1,310	1,910	700	1,200
Oregon	Green	0.50	7,600	1.13	12.2	39	3,510	530	1,190	590	790
	12%	0.55	12,700	1.36	14.4	33	6,040	1,250	1,790	720	1,160
White	Green	0.55	9,500	1.44	15.7	38	3,990	670	1,350	590	960
	12%	0.60	15,000	1.74	16.6	43	7,410	1,160	1,910	940	1,320
Aspen											
Bigtooth	Green	0.36	5,400	1.12	5.7	–	2,500	210	730	–	–
	12%	0.39	9,100	1.43	7.7	–	5,300	450	1,080	–	–
Quaking	Green	0.35	5,100	0.86	6.4	22	2,140	180	660	230	300
	12%	0.38	8,400	1.18	7.6	21	4,250	370	850	260	350
Basswood, American	Green	0.32	5,000	1.04	5.3	16	2,220	170	600	280	250
	12%	0.37	8,700	1.46	7.2	16	4,730	370	990	350	410
Beech, American	Green	0.56	8,600	1.38	11.9	43	3,550	540	1,290	720	850
	12%	0.64	14,900	1.72	15.1	41	7,300	1,010	2,010	1,010	1,300

(Continued)

Table 4. *(Continued)* Strength Properties of Selected Commercial Woods Grown in the United States[a] (English Units). *(Source, Wood Handbook, U.S.D.A.)*

Common Species Names	Moisture Content	Specific Gravity[b]	Modulus of Rupture (lbf/in²)	Modulus of Elasticity[c] (× 10⁶ lbf/in²)	Work to Maximum Load (in-lbf/in³)	Impact Bending (in.)	Compression Parallel to Grain (lbf/in²)	Compression Perpendicular to Grain (lbf/in²)	Shear Parallel to Grain (lbf/in²)	Tension Perpendicular to Grain (lbf/in²)	Side Hardness (lbf)
				Static Bending							
						Hardwoods					
Birch											
Paper	Green	0.48	6,400	1.17	16.2	49	2,360	270	840	380	560
	12%	0.55	12,300	1.59	16.0	34	5,690	600	1,210	–	910
Sweet	Green	0.60	9,400	1.65	15.7	48	3,740	470	1,240	430	970
	12%	0.65	16,900	2.17	18.0	47	8,540	1,080	2,240	950	1,470
Yellow	Green	0.55	8,300	1.50	16.1	48	3,380	430	1,110	430	780
	12%	0.62	16,600	2.01	20.8	55	8,170	970	1,880	920	1,260
Butternut	Green	0.36	5,400	0.97	8.2	24	2,420	220	760	430	390
	12%	0.38	8,100	1.18	8.2	24	5,110	460	1,170	440	490
Cherry, Black	Green	0.47	8,000	1.31	12.8	33	3,540	360	1,130	570	660
	12%	0.50	12,300	1.49	11.4	29	7,110	690	1,700	560	950
Chestnut, American	Green	0.40	5,600	0.93	7.0	24	2,470	310	800	440	420
	12%	0.43	8,600	1.23	6.5	19	5,320	620	1,080	460	540
Cottonwood											
Balsam, Poplar	Green	0.31	3,900	0.75	4.2	–	1,690	140	500	–	–
	12%	0.34	6,800	1.10	5.0	–	4,020	300	790	–	–
Black	Green	0.31	4,900	1.08	5.0	20	2,200	160	610	270	250
	12%	0.35	8,500	1.27	6.7	22	4,500	300	1,040	330	350
Eastern	Green	0.37	5,300	1.01	7.3	21	2,280	200	680	410	340
	12%	0.40	8,500	1.37	7.4	20	4,910	380	930	580	430

(Continued)

Table 4. (Continued) Strength Properties of Selected Commercial Woods Grown in the United States[a] (English Units). (Source, Wood Handbook, U.S.D.A.)

| Common Species Names | Moisture Content | Specific Gravity[b] | Static Bending | | | Impact Bending (in.) | Compression Parallel to Grain (lbf/in²) | Compression Perpendicular to Grain (lbf/in²) | Shear Parallel to Grain (lbf/in²) | Tension Perpendicular to Grain (lbf/in²) | Side Hardness (lbf) |
			Modulus of Rupture (lbf/in²)	Modulus of Elasticity[c] (× 10⁶ lbf/in²)	Work to Maximum Load (in-lbf/in³)						
Hardwoods											
Elm											
American	Green	0.46	7,200	1.11	11.8	38	2,910	360	1,000	590	620
	12%	0.50	11,800	1.34	13.0	39	5,520	690	1,510	660	830
Rock	Green	0.57	9,500	1.19	19.8	54	3,780	610	1,270	–	940
	12%	0.63	14,800	1.54	19.2	56	7,050	1,230	1,920	640	1,320
Slippery	Green	0.48	8,000	1.23	15.4	47	3,320	420	1,110	530	660
	12%	0.53	13,000	1.49	16.9	45	6,360	820	1,630		860
Hackberry	Green	0.49	6,500	0.95	14.5	48	2,650	400	1,070	630	700
	12%	0.53	11,000	1.19	12.8	43	5,440	890	1,590	580	880
Hickory, Pecan											
Bitternut	Green	0.60	10,300	1.40	20.0	66	4,570	800	1,240	–	–
	12%	0.66	17,100	1.79	18.2	66	9,040	1,680	–	–	–
Nutmeg	Green	0.56	9,100	1.29	22.8	54	3,980	760	1,030	–	–
	12%	0.60	16,600	1.70	25.1	–	6,910	1,570	–	–	–
Pecan	Green	0.60	9,800	1.37	14.6	53	3,990	780	1,480	680	1,310
	12%	0.66	13,700	1.73	13.8	44	7,850	1,720	2,080	–	1,820
Water	Green	0.61	10,700	1.56	18.8	56	4,660	880	1,440	–	–
	12%	0.62	17,800	2.02	19.3	53	8,600	1,550	–	–	–

(Continued)

Table 4. *(Continued)* **Strength Properties of Selected Commercial Woods Grown in the United States[a] (English Units).** *(Source, Wood Handbook, U.S.D.A.)*

| Common Species Names | Moisture Content | Specific Gravity[b] | Static Bending | | | | Impact Bending (in.) | Compression Parallel to Grain (lbf/in²) | Compression Perpendicular to Grain (lbf/in²) | Shear Parallel to Grain (lbf/in²) | Tension Perpendicular to Grain (lbf/in²) | Side Hardness (lbf) |
			Modulus of Rupture (lbf/in²)	Modulus of Elasticity[c] (× 10⁶ lbf/in²)	Work to Maximum Load (in-lbf/in³)							
Hardwoods												
Hickory, True												
Mockernut	Green	0.64	11,100	1.57	26.1	88	4,480	810	1,280	–	–
	12%	0.72	19,200	2.22	22.6	77	8,940	1,730	1,740	–	–
Pignut	Green	0.66	11,700	1.65	31.7	89	4,810	920	1,370	–	–
	12%	0.75	20,100	2.26	30.4	74	9,190	1,980	2,150	–	–
Shagbark	Green	0.64	11,000	1.57	23.7	74	4,580	840	1,520	–	–
	12%	0.72	20,200	2.16	25.8	67	9,210	1,760	2,430	–	–
Shellbark	Green	0.62	10,500	1.34	29.9	104	3,920	810	1,190	–	–
	12%	0.69	18,100	1.89	23.6	88	8,000	1,800	2,110	–	–
Honeylocust	Green	0.60	10,200	1.29	12.6	47	4,420	1,150	1,660	930	1,390
	12%	-	14,700	1.63	13.3	47	7,500	1,840	2,250	900	1,580
Locust, Black	Green	0.66	13,800	1.85	15.4	44	6,800	1,160	1,760	770	1,570
	12%	0.69	19,400	2.05	18.4	57	10,180	1,830	2,480	640	1,700
Magnolia												
Cucumbertree	Green	0.44	7,400	1.56	10.0	30	3,140	330	990	440	520
	12%	0.48	12,300	1.82	12.2	35	6,310	570	1,340	660	700
Southern	Green	0.46	6,800	1.11	15.4	54	2,700	460	1,040	610	740
	12%	0.50	11,200	1.40	12.8	29	5,460	860	1,530	740	1,020

(Continued)

Table 4. *(Continued)* Strength Properties of Selected Commercial Woods Grown in the United States[a] (English Units). *(Source, Wood Handbook, U.S.D.A.)*

Common Species Names	Moisture Content	Static Bending				Impact Bending (in.)	Compression Parallel to Grain (lbf/in²)	Compression Perpendicular to Grain (lbf/in²)	Shear Parallel to Grain (lbf/on²)	Tension Perpendicular to Grain (lbf/in²)	Side Hardness (lbf)
		Specific Gravity[b]	Modulus of Rupture (lbf/in²)	Modulus of Elasticity[c] (× 10⁶ lbf/in²)	Work to Maximum Load (in-lbf/in³)						
Hardwoods											
Maple											
Bigleaf	Green	0.44	7,400	1.10	8.7	23	3,240	450	1,110	600	620
	12%	0.48	10,700	1.45	7.8	28	5,950	750	1,730	540	850
Black	Green	0.52	7,900	1.33	12.8	48	3,270	600	1,130	720	840
	12%	0.57	13,300	1.62	12.5	40	6,680	1,020	1,820	670	1,180
Red	Green	0.49	7,700	1.39	11.4	32	3,280	400	1,150	-	700
	12%	0.54	13,400	1.64	12.5	32	6,540	1,000	1,850	-	950
Silver	Green	0.44	5,800	0.94	11.0	29	2,490	370	1,050	560	590
	12%	0.47	8,900	1.14	8.3	25	5,220	740	1,480	500	700
Sugar	Green	0.56	9,400	1.55	13.3	40	4,020	640	1,460	-	970
	12%	0.63	15,800	1.83	16.5	39	7,830	1,470	2,330	-	1,450
Oak, Red											
Black	Green	0.56	8,200	1.18	12.2	40	3,470	710	1,220	-	1,060
	12%	0.61	13,900	1.64	13.7	41	6,520	930	1,910	-	1,210
Cherrybark	Green	0.61	10,800	1.79	14.7	54	4,620	760	1,320	800	1,240
	12%	0.68	18,100	2.28	18.3	49	8,740	1,250	2,000	840	1,480
Laurel	Green	0.56	7,900	1.39	11.2	39	3,170	570	1,180	770	1,000
	12%	0.63	12,600	1.69	11.8	39	6,980	1,060	1,830	790	1,210
Northern Red	Green	0.56	8,300	1.35	13.2	44	3,440	610	1,210	750	1,000
	12%	0.63	14,300	1.82	14.5	43	6,760	1,010	1,780	800	1,290
Pin	Green	0.58	8,300	1.32	14.0	48	3,680	720	1,290	800	1,070
	12%	0.63	14,000	1.73	14.8	45	6,820	1,020	2,080	1,050	1,510
Scarlet	Green	0.60	10,400	1.48	15.0	54	4,090	830	1,410	700	1,200
	12%	0.67	17,400	1.91	20.5	53	8,330	1,120	1,890	870	1,400
Southern Red	Green	0.52	6,900	1.14	8.0	29	3,030	550	930	480	860
	12%	0.59	10,900	1.49	9.4	26	6,090	870	1,390	510	1,060

(Continued)

Table 4. *(Continued)* Strength Properties of Selected Commercial Woods Grown in the United States[a] (English Units). *(Source, Wood Handbook, U.S.D.A.)*

| Common Species Names | Moisture Content | Specific Gravity[b] | Static Bending | | | | Impact Bending (in.) | Compression Parallel to Grain (lbf/in²) | Compression Perpendicular to Grain (lbf/in²) | Shear Parallel to Grain (lbf/in²) | Tension Perpendicular to Grain (lbf/in²) | Side Hardness (lbf) |
			Modulus of Rupture (lbf/in²)	Modulus of Elasticity[c] (×10⁶ lbf/in²)	Work to Maximum Load (in-lbf/in³)							
							Hardwoods					
Oak, Red												
Water	Green	0.56	8,900	1.55	11.1	39	3,740	620	1,240	820	1,010
	12%	0.63	15,400	2.02	21.5	44	6,770	1,020	2,020	920	1,190
Willow	Green	0.56	7,400	1.29	8.8	35	3,000	610	1,180	760	980
	12%	0.69	14,500	1.90	14.6	42	7,040	1,130	1,650	-	1,460
Oak, White												
Bur	Green	0.58	7,200	0.88	10.7	44	3,290	680	1,350	800	1,110
	12%	0.64	10,300	1.03	9.8	29	6,060	1,200	1,820	680	1,370
Chestnut	Green	0.57	8,000	1.37	9.4	35	3,520	530	1,210	690	890
	12%	0.66	13,300	1.59	11.0	40	6,830	840	1,490	-	1,130
Live	Green	0.80	11,900	1.58	12.3	-	5,430	2,040	2,210	-	-
	12%	0.88	18,400	1.98	18.9	-	8,900	2,840	2,660	-	-
Overcup	Green	0.57	8,000	1.15	12.6	44	3,370	540	1,320	730	960
	12%	0.63	12,600	1.42	15.7	38	6,200	810	2,000	940	1,190
Post	Green	0.60	8,100	1.09	11.0	44	3,480	860	1,280	790	1,130
	12%	0.67	13,200	1.51	13.2	46	6,600	1,430	1,840	780	1,360
Swamp Chestnut	Green	0.60	8,500	1.35	12.8	45	3,540	570	1,260	670	1,110
	12%	0.67	13,900	1.77	12.0	41	7,270	1,110	1,990	690	1,240
Swamp White	Green	0.64	9,900	1.59	14.5	50	4,360	760	1,300	860	1,160
	12%	0.72	17,700	2.05	19.2	49	8,600	1,190	2,000	830	1,620
White	Green	0.60	8,300	1.25	11.6	42	3,560	670	1,250	770	1,060
	12%	0.68	15,200	1.78	14.8	37	7,440	1,070	2,000	800	1,360

(Continued)

Table 4. *(Continued)* Strength Properties of Selected Commercial Woods Grown in the United States[a] (English Units). *(Source, Wood Handbook, U.S.D.A.)*

| Common Species Names | Moisture Content | Specific Gravity[b] | Static Bending | | | Impact Bending (in.) | Compression Parallel to Grain (lbf/in²) | Compression Perpendicular to Grain (lbf/in²) | Shear Parallel to Grain (lbf/in²) | Tension Perpendicular to Grain (lbf/in²) | Side Hardness (lbf) |
			Modulus of Rupture (lbf/in²)	Modulus of Elasticity[c] ($\times 10^6$ lbf/in²)	Work to Maximum Load (in-lbf/in³)						
					Hardwoods						
Sassafras	Green	0.42	6,000	0.91	7.1	-	2,730	370	950	-	-
	12%	0.46	9,000	1.12	8.7	-	4,760	850	1,240	-	-
Sweetgum	Green	0.46	7,100	1.20	10.1	36	3,040	370	990	540	600
	12%	0.52	12,500	1.64	11.9	32	6,320	620	1,600	760	850
Sycamore, American	Green	0.46	6,500	1.06	7.5	26	2,920	360	1,000	630	610
	12%	0.49	10,000	1.42	8.5	26	5,380	700	1,470	720	770
Tanoak	Green	0.58	10,500	1.55	13.4	-	4,650	-	-	-	-
	12%	-	-	-	-	-	-	-	-	-	-
Tupelo Black	Green	0.46	7,000	1.03	8.0	30	3,040	480	1,100	570	640
	12%	0.50	9,600	1.20	6.2	22	5,520	930	1,340	500	810
Water	Green	0.46	7,300	1.05	8.3	30	3,370	480	1,190	600	710
	12%	0.50	9,600	1.26	6.9	23	5,920	870	1,590	700	880
Walnut, Black	Green	0.51	9,500	1.42	14.6	37	4,300	490	1,220	570	900
	12%	0.55	14,600	1.68	10.7	34	7,580	1,010	1,370	690	1,010
Willow, Black	Green	0.36	4,800	0.79	11.0	-	2,040	180	680	-	-
	12%	0.39	7,800	1.01	8.8	-	4,100	430	1,250	-	-
Yellow-Poplar	Green	0.40	6,000	1.22	7.5	26	2,660	270	790	510	440
	12%	0.42	10,100	1.58	8.8	24	5,540	500	1,190	540	540

(Continued)

Table 4. *(Continued)* Strength Properties of Selected Commercial Woods Grown in the United States[a] (English Units). *(Source, Wood Handbook, U.S.D.A.)*

| Common Species Names | Moisture Content | Specific Gravity[b] | Static Bending | | | Impact Bending (in.) | Compression Parallel to Grain (lbf/in²) | Compression Perpendicular to Grain (lbf/in²) | Shear Parallel to Grain (lbf/in²) | Tension Perpendicular to Grain (lbf/in²) | Side Hardness (lbf) |
			Modulus of Rupture (lbf/in²)	Modulus of Elasticity[c] ($\times 10^6$ lbf/in²)	Work to Maximum Load (in-lbf/in³)						
						Softwoods					
Baldcypress	Green	0.42	6,600	1.18	6.6	25	3,580	400	810	300	390
	12%	0.46	10,600	1.44	8.2	24	6,360	730	1,000	270	510
Cedar											
Atlantic White	Green	0.31	4,700	0.75	5.9	18	2,390	240	690	180	290
	12%	0.32	6,800	0.93	4.1	13	4,700	410	800	220	350
East. Redcedar	Green	0.44	7,000	0.65	15.0	35	3,570	700	1,010	330	650
	12%	0.47	8,800	0.88	8.3	22	6,020	920	-	-	-
Incense	Green	0.35	6,200	0.84	6.4	17	3,150	370	830	280	390
	12%	0.37	8,000	1.04	5.4	17	5,200	590	880	270	470
North. White	Green	0.29	4,200	0.64	5.7	15	1,990	230	620	240	230
	12%	0.31	6,500	0.80	4.8	12	3,960	310	850	240	320
Port-Orford	Green	0.39	6,600	1.30	7.4	21	3,140	300	840	180	380
	12%	0.43	12,700	1.70	9.1	28	6,250	720	1,370	400	630
West. Redcedar	Green	0.31	5,200	0.94	5.0	17	2,770	240	770	230	260
	12%	0.32	7,500	1.11	5.8	17	4,560	460	990	220	350
Yellow	Green	0.42	6,400	1.14	9.2	27	3,050	350	840	330	440
	12%	0.44	11,100	1.42	10.4	29	6,310	620	1,130	360	580
Douglas-Fir[d]											
Coast	Green	0.45	7,700	1.56	7.6	26	3,780	380	900	300	500
	12%	0.48	12,400	1.95	9.9	31	7,230	800	1,130	340	710
Interior West	Green	0.46	7,700	1.51	7.2	26	3,870	420	940	290	510
	12%	0.50	12,600	1.83	10.6	32	7,430	760	1,290	350	660
Interior North	Green	0.45	7,400	1.41	8.1	22	3,470	360	950	340	420
	12%	0.48	13,100	1.79	10.5	26	6,900	770	1,400	390	600
Interior South	Green	0.43	6,800	1.16	8.0	15	3,110	340	950	250	360
	12%	0.46	11,900	1.49	9.0	20	6,230	740	1,510	330	510

(Continued)

Table 4. (Continued) Strength Properties of Selected Commercial Woods Grown in the United States[a] (English Units). (Source, Wood Handbook, U.S.D.A.)

| Common Species Names | Moisture Content | Specific Gravity[b] | Static Bending | | | Impact Bending (in.) | Compression Parallel to Grain (lbf/in²) | Compression Perpendicular to Grain (lbf/in²) | Shear Parallel to Grain (lbf/in²) | Tension Perpendicular to Grain (lbf/in²) | Side Hardness (lbf) |
			Modulus of Rupture (lbf/in²)	Modulus of Elasticity[c] (×10⁶ lbf/in²)	Work to Maximum Load (in-lbf/in³)						
						Softwoods					
Fir											
Balsam	Green	0.33	5,500	1.25	4.7	16	2,630	190	662	180	290
	12%	0.35	9,200	1.45	5.1	20	5,280	404	944	180	400
Calif. Red	Green	0.36	5,800	1.17	6.4	21	2,760	330	770	380	360
	12%	0.38	10,500	1.50	8.9	24	5,460	610	1,040	390	500
Grand	Green	0.35	5,800	1.25	5.6	22	2,940	270	740	240	360
	12%	0.37	8,900	1.57	7.5	28	5,290	500	900	240	490
Noble	Green	0.37	6,200	1.38	6.0	19	3,010	270	800	230	290
	12%	0.39	10,700	1.72	8.8	23	6,100	520	1,050	220	410
Pacific Silver	Green	0.40	6,400	1.42	6.0	21	3,140	220	750	240	310
	12%	0.43	11,000	1.76	9.3	24	6,410	450	1,220	-	430
Subalpine	Green	0.31	4,900	1.05	-	-	2,300	190	700	-	260
	12%	0.32	8,600	1.29	5.6	-	4,860	390	1,070	-	350
White	Green	0.37	5,900	1.16	5.6	22	2,900	280	760	300	340
	12%	0.39	9,800	1.50	7.2	20	5,800	530	1,100	300	480
Hemlock											
Eastern	Green	0.38	6,400	1.07	6.7	21	3,080	360	850	230	400
	12%	0.40	8,900	1.20	6.8	21	5,410	650	1,060	-	500
Mountain	Green	0.42	6,300	1.04	11.0	32	2,880	370	930	330	470
	12%	0.45	11,500	1.33	10.4	32	6,440	860	1,540	-	680
Western	Green	0.42	6,600	1.31	6.9	22	3,360	280	860	290	410
	12%	0.45	11,300	1.63	8.3	23	7,200	550	1,290	340	540
Larch, Western	Green	0.48	7,700	1.46	10.3	29	3,760	400	870	330	510
	12%	0.52	13,000	1.87	12.6	35	7,620	930	1,360	430	830

(Continued)

Table 4. (Continued) Strength Properties of Selected Commercial Woods Grown in the United States[a] (English Units). (Source, Wood Handbook, U.S.D.A.)

Common Species Names	Moisture Content	Specific Gravity[b]	Static Bending			Impact Bending (in.)	Compression Parallel to Grain (lbf/in²)	Compression Perpendicular to Grain (lbf/in²)	Shear Parallel to Grain (lbf/in²)	Tension Perpendicular to Grain (lbf/in²)	Side Hardness (lbf)
			Modulus of Rupture (lbf/in²)	Modulus of Elasticity[c] (×10⁶ lbf/in²)	Work to Maximum Load (in-lbf/in³)						
Pine						Softwoods					
East. White	Green	0.34	4,900	0.99	5.2	17	2,440	220	680	250	290
	12%	0.35	8,600	1.24	6.8	18	4,800	440	900	310	380
Jack	Green	0.40	6,000	1.07	7.2	26	2,950	300	750	360	400
	12%	0.43	9,900	1.35	8.3	27	5,660	580	1,170	420	570
Loblolly	Green	0.47	7,300	1.40	8.2	30	3,510	390	860	260	450
	12%	0.51	12,800	1.79	10.4	30	7,130	790	1,390	470	690
Lodgepole	Green	0.38	5,500	1.08	5.6	20	2,610	250	680	220	330
	12%	0.41	9,400	1.34	6.8	20	5,370	610	880	290	480
Longleaf	Green	0.54	8,500	1.59	8.9	35	4,320	480	1,040	330	590
	12%	0.59	14,500	1.98	11.8	34	8,470	960	1,510	470	870
Pitch	Green	0.47	6,800	1.20	9.2	-	2,950	360	860	-	-
	12%	0.52	10,800	1.43	9.2	-	5,940	820	1,360	-	-
Pond	Green	0.51	7,400	1.28	7.5	-	3,660	440	940	-	-
	12%	0.56	11,600	1.75	8.6	-	7,540	910	1,380	-	-
Ponderosa	Green	0.38	5,100	1.00	5.2	21	2,450	280	700	310	320
	12%	0.40	9,400	1.29	7.1	19	5,320	580	1,130	420	460
Red	Green	0.41	5,800	1.28	6.1	26	2,730	260	690	300	340
	12%	0.46	11,000	1.63	9.9	26	6,070	600	1,210	460	560
Sand	Green	0.46	7,500	1.02	9.6	-	3,440	450	1,140	-	-
	12%	0.48	11,600	1.41	9.6	-	6,920	836	-	-	-
Shortleaf	Green	0.47	7,400	1.39	8.2	30	3,530	350	910	320	440
	12%	0.51	13,100	1.75	11.0	33	7,270	820	1,390	470	690
Slash	Green	0.54	8,700	1.53	9.6	-	3,820	530	960	-	-
	12%	0.59	16,300	1.98	13.2	-	8,140	1,020	1,680	-	-

(Continued)

Table 4. *(Continued)* Strength Properties of Selected Commercial Woods Grown in the United States[a] (English Units). *(Source, Wood Handbook, U.S.D.A.)*

Common Species Names	Moisture Content	Specific Gravity[b]	Static Bending		Work to Maximum Load (in-lbf/in³)	Impact Bending (in.)	Compression Parallel to Grain (lbf/in²)	Compression Perpendicular to Grain (lbf/in²)	Shear Parallel to Grain (lbf/in²)	Tension Perpendicular to Grain (lbf/in²)	Side Hardness (lbf)
			Modulus of Rupture (lbf/in²)	Modulus of Elasticity[c] ($\times 10^6$ lbf/in²)							
Softwoods											
Pine											
Spruce	Green	0.41	6,000	1.00	–	–	2,840	280	900	–	450
	12%	0.44	10,400	1.23	–	–	5,650	730	1,490	–	660
Sugar	Green	0.34	4,900	1.03	5.4	17	2,460	210	720	270	270
	12%	0.36	8,200	1.19	5.5	18	4,460	500	1,130	350	380
Virginia	Green	0.45	7,300	1.22	10.9	34	3,420	390	890	400	540
	12%	0.48	13,000	1.52	13.7	32	6,710	910	1,350	380	740
West. White	Green	0.35	4,700	1.19	5.0	19	2,430	190	680	260	260
	12%	0.38	9,700	1.46	8.8	23	5,040	470	1,040	–	420
Redwood											
Old-growth	Green	0.38	7,500	1.18	7.4	21	4,200	420	800	260	410
	12%	0.40	10,000	1.34	6.9	19	6,150	700	940	240	480
Young-growth	Green	0.34	5,900	0.96	5.7	16	3,110	270	890	300	350
	12%	0.35	7,900	1.10	5.2	15	5,220	520	1,110	250	420
Spruce											
Black	Green	0.38	6,100	1.38	7.4	24	2,840	240	739	100	370
	12%	0.42	10,800	1.61	10.5	23	5,960	550	1,230	–	520
Engelmann	Green	0.33	4,700	1.03	5.1	16	2,180	200	640	240	260
	12%	0.35	9,300	1.30	6.4	18	4,480	410	1,200	350	390
Red	Green	0.37	6,000	1.33	6.9	18	2,720	260	750	220	350
	12%	0.40	10,800	1.61	8.4	25	5,540	550	1,290	350	490
Sitka	Green	0.37	5,700	1.23	6.3	24	2,670	280	760	250	350
	12%	0.40	10,200	1.57	9.4	25	5,610	580	1,150	370	510
White	Green	0.33	5,000	1.14	6.0	22	2,350	210	640	220	320
	12%	0.36	9,400	1.43	7.7	20	5,180	430	970	360	480

(Continued)

Table 4. *(Continued)* **Strength Properties of Selected Commercial Woods Grown in the United States[a] (English Units).** *(Source, Wood Handbook, U.S.D.A.)*

| Common Species Names | Moisture Content | Specific Gravity[b] | Static Bending | | | Impact Bending (in.) | Compression Parallel to Grain (lbf/in²) | Compression Perpendicular to Grain (lbf/in²) | Shear Parallel to Grain (lbf/in²) | Tension Perpendicular to Grain (lbf/in²) | Side Hardness (lbf) |
			Modulus of Rupture (lbf/in²)	Modulus of Elasticity[c] (× 10⁶ lbf/in²)	Work to Maximum Load (in-lbf/in³)						
						Softwoods					
Tamarack	Green	0.49	7,200	1.24	7.2	28	3,480	390	860	260	380
	12%	0.53	11,600	1.64	7.1	23	7,160	800	1,280	400	590

Notes:

[a] Results of tests on small clear specimens in the green and air-dried conditions. Definition of properties: impact bending is height of drop that causes complete failure, using 0.71-kg (50-lb) hammer; compression parallel to grain is also called maximum crushing strength; compression perpendicular to grain is fiber stress at proportional limit; shear is maximum shearing strength; tension is maximum tensile strength; and side hardness is hardness measured when load is perpendicular to grain.

[b] Specified gravity is based on weight when ovendry and volume when green or at 12% moisture content.

[c] Modulus of elasticity measured from a simply supported, center-loaded beam, on a span depth ratio of 14/1. To correct for shear deflection, the modulus can be increased by 10%.

[d] Coast Douglas-fir is defined as Douglas-fir growing in Oregon and Washington State west of the Cascade Mountains summit. Interior West includes California and all counties in Oregon and Washington east of, but adjacent to, the Cascade summit; Interior North, the remainder of Oregon and Washington plus Idaho, Montana, and Wyoming; and Interior South, Utah, Colorado, Arizona, and New Mexico.

Table 5. Strength Properties of Selected Commercial Woods Grown in the United States[a] (Metric Units). (Source, Wood Handbook, U.S.D.A.)

| Common Species Names | Moisture Content | Specific Gravity[b] | Static Bending | | | Impact Bending (mm) | Compression Parallel to Grain (kPa) | Compression Perpendicular to Grain (kPa) | Shear Parallel to Grain (kPa) | Tension Perpendicular to Grain (kPa) | Side Hardness (N) |
			Modulus of Rupture (kPa)	Modulus of Elasticity[c] (MPa)	Work to Maximum Load (kJ/m³)						
						Hardwoods					
Alder, Red	Green	0.37	45,000	8,100	55	560	20,400	1,700	5,300	2,700	2,000
	12%	0.41	68,000	9,500	58	510	40,100	3,000	7,400	2,900	2,600
Ash											
Black	Green	0.45	41,000	7,200	83	840	15,900	2,400	5,900	3,400	2,300
	12%	0.49	87,000	11,000	103	890	41,200	5,200	10,800	4,800	3,800
Blue	Green	0.53	66,000	8,500	101	-	24,800	5,600	10,600	-	-
	12%	0.58	95,000	9,700	99	-	48,100	9,800	14,000	-	-
Green	Green	0.53	66,000	9,700	81	890	29,000	5,000	8,700	4,100	3,900
	12%	0.56	97,000	11,400	92	810	48,800	9,000	13,200	4,800	5,300
Oregon	Green	0.50	52,000	7,800	84	990	24,200	3,700	8,200	4,100	3,500
	12%	0.55	88,000	9,400	99	840	41,600	8,600	12,300	5,000	5,200
White	Green	0.55	66,000	9,900	108	970	27,500	4,600	9,300	4,100	4,300
	12%	0.60	103,000	12,000	115	1,090	51,100	8,000	13,200	6,500	5,900
Aspen											
Bigtooth	Green	0.36	37,000	7,700	39	-	17,200	1,400	5,000	-	-
	12%	0.39	63,000	9,900	53	-	36,500	3,100	7,400	-	-
Quaking	Green	0.35	35,000	5,900	44	560	14,800	1,200	4,600	1,600	1,300
	12%	0.38	58,000	8,100	52	530	29,300	2,600	5,900	1,800	1,600
Basswood, American	Green	0.32	34,000	7,200	37	410	15,300	1,200	4,100	1,900	1,100
	12%	0.37	60,000	10,100	50	410	32,600	2,600	6,800	2,400	1,800
Beech, American	Green	0.56	59,000	9,500	82	1,090	24,500	3,700	8,900	5,000	3,800
	12%	0.64	103,000	11,900	104	1,040	50,300	7,000	13,900	7,000	5,800

(Continued)

Table 5. (Continued) Strength Properties of Selected Commercial Woods Grown in the United States[a] (Metric Units). (Source, Wood Handbook, U.S.D.A.)

| Common Species Names | Moisture Content | Specific Gravity[b] | Static Bending | | | Impact Bending (mm) | Compression Parallel to Grain (kPa) | Compression Perpendicular to Grain (kPa) | Shear Parallel to Grain (kPa) | Tension Perpendicular to Grain (kPa) | Side Hardness (N) |
			Modulus of Rupture (kPa)	Modulus of Elasticity[c] (MPa)	Work to Maximum Load (kJ/m³)						
Hardwoods											
Birch											
Paper	Green	0.48	44,000	8,100	112	1,240	16,300	1,900	5,800	2,600	2,500
	12%	0.55	85,000	11,000	110	860	39,200	4,100	8,300	–	4,000
Sweet	Green	0.60	65,000	11,400	108	1,220	25,800	3,200	8,500	3,000	4,300
	12%	0.65	117,000	15,000	124	1,190	58,900	7,400	15,400	6,600	6,500
Yellow	Green	0.55	57,000	10,300	111	1,220	23,300	3,000	7,700	3,000	3,600
	12%	0.62	114,000	13,900	143	1,400	56,300	6,700	13,000	6,300	5,600
Butternut	Green	0.36	37,000	6,700	57	610	16,700	1,500	5,200	3,000	1,700
	12%	0.38	56,000	8,100	57	610	36,200	3,200	8,100	3,000	2,200
Cherry, Black	Green	0.47	55,000	9,000	88	840	24,400	2,500	7,800	3,900	2,900
	12%	0.50	85,000	10,300	79	740	49,000	4,800	11,700	3,900	4,200
Chestnut, American	Green	0.40	39,000	6,400	48	610	17,000	2,100	5,500	3,000	1,900
	12%	0.43	59,000	8,500	45	480	36,700	4,300	7,400	3,200	2,400
Cottonwood											
Balsam Poplar	Green	0.31	27,000	5,200	29	–	11,700	1,000	3,400	–	–
	12%	0.34	47,000	7,600	34	–	27,700	2,100	5,400	–	–
Black	Green	0.31	34,000	7,400	34	510	15,200	1,100	4,200	1,900	1,100
	12%	0.35	59,000	8,800	46	560	31,000	2,100	7,200	2,300	1,600
Eastern	Green	0.37	37,000	7,000	50	530	15,700	1,400	4,700	2,800	1,500
	12%	0.40	59,000	9,400	51	510	33,900	2,600	6,400	4,000	1,900

(Continued)

Table 5. (Continued) Strength Properties of Selected Commercial Woods Grown in the United States[a] (Metric Units). (Source, Wood Handbook, U.S.D.A.)

Common Species Names	Moisture Content	Specific Gravity[b]	Static Bending			Impact Bending (mm)	Compression Parallel to Grain (kPa)	Compression Perpendicular to Grain (kPa)	Shear Parallel to Grain (kPa)	Tension Perpendicular to Grain (kPa)	Side Hardness (N)
			Modulus of Rupture (kPa)	Modulus of Elasticity[c] (MPa)	Work to Maximum Load (kJ/m³)						
Hardwoods											
Elm											
American	Green	0.46	50,000	7,700	81	970	20,100	2,500	6,900	4,100	2,800
	12%	0.50	81,000	9,200	90	990	38,100	4,800	10,400	4,600	3,700
Rock	Green	0.57	66,000	8,200	137	1,370	26,100	4,200	8,800	-	-
	12%	0.63	102,000	10,600	132	1,420	48,600	8,500	13,200	-	-
Slippery	Green	0.48	55,000	8,500	106	1,190	22,900	2,900	7,700	4,400	2,900
	12%	0.53	90,000	10,300	117	1,140	43,900	5,700	11,200	3,700	3,800
Hackberry	Green	0.49	45,000	6,600	100	1,220	18,300	2,800	7,400	4,300	3,100
	12%	0.53	76,000	8,200	88	1,090	37,500	6,100	11,000	4,000	3,900
Hickory, Pecan											
Bitternut	Green	0.60	71,000	9,700	138	1,680	31,500	5,500	8,500	-	-
	12%	0.66	118,000	12,300	125	1,680	62,300	11,600	-	-	-
Nutmeg	Green	0.56	63,000	8,900	157	1,370	27,400	5,200	7,100	-	-
	12%	0.60	114,000	11,700	173	-	47,600	10,800	-	-	-
Pecan	Green	0.60	68,000	9,400	101	1,350	27,500	5,400	10,200	4,700	5,800
	12%	0.66	94,000	11,900	95	1,120	54,100	11,900	14,300	-	8,100
Water	Green	0.61	74,000	10,800	130	1,420	32,100	6,100	9,900	-	-
	12%	0.62	123,000	13,900	133	1,350	59,300	10,700	-	-	-

(Continued)

Table 5. *(Continued)* Strength Properties of Selected Commercial Woods Grown in the United States[a] (Metric Units). *(Source, Wood Handbook, U.S.D.A.)*

Common Species Names	Moisture Content	Specific Gravity[b]	Modulus of Rupture (kPa)	Modulus of Elasticity[c] (MPa)	Work to Maximum Load (kJ/m³)	Impact Bending (mm)	Compression Parallel to Grain (kPa)	Compression Perpendicular to Grain (kPa)	Shear Parallel to Grain (kPa)	Tension Perpendicular to Grain (kPa)	Side Hardness (N)
					Hardwoods						
Hickory, True											
Mockernut	Green	0.64	77,000	10,800	180	2,240	30,900	5,600	8,800	–	–
	12%	0.72	132,000	15,300	156	1,960	61,000	11,900	12,000	–	–
Pignut	Green	0.66	81,000	11,400	219	2,260	33,200	6,300	9,400	–	–
	12%	0.75	139,000	15,600	210	1,880	63,400	13,700	14,800	–	–
Shagbark	Green	0.64	76,000	10,800	163	1,880	31,600	5,800	10,500	–	–
	12%	0.72	139,000	14,900	178	1,700	63,500	12,100	16,800	–	–
Shellbark	Green	0.62	72,000	9,200	206	2,640	27,000	5,600	8,200	–	–
	12%	0.69	125,000	13,000	163	2,240	55,200	12,400	14,500	–	–
Honeylocust	Green	0.60	70,000	8,900	87	1,190	30,500	7,900	11,400	6,400	6,200
	12%	–	101,000	11,200	92	1,190	51,700	12,700	15,500	6,200	7,000
Locust, Black	Green	0.66	95,000	12,800	106	1,120	46,900	8,000	12,100	5,300	7,000
	12%	0.69	134,000	14,100	127	1,450	70,200	12,600	17,100	4,400	7,600
Magnolia											
Cucumber Tree	Green	0.44	51,000	10,800	69	760	21,600	2,300	6,800	3,000	2,300
	12%	0.48	85,000	12,500	84	890	43,500	3,900	9,200	4,600	3,100
Southern	Green	0.46	47,000	7,700	106	1,370	18,600	3,200	7,200	4,200	3,300
	12%	0.50	77,000	9,700	88	740	37,600	5,900	10,500	5,100	4,500

(Continued)

Table 5. (Continued) Strength Properties of Selected Commercial Woods Grown in the United States[a] (Metric Units). (Source, Wood Handbook, U.S.D.A.)

| Common Species Names | Moisture Content | Specific Gravity[b] | Static Bending | | | Impact Bending (mm) | Compression Parallel to Grain (kPa) | Compression Perpendicular to Grain (kPa) | Shear Parallel to Grain (kPa) | Tension Perpendicular to Grain (kPa) | Side Hardness (N) |
			Modulus of Rupture (kPa)	Modulus of Elasticity[c] (MPa)	Work to Maximum Load (kJ/m³)						
Hardwoods											
Maple											
Bigleaf	Green	0.44	51,000	7,600	60	580	22,300	3,100	7,700	4,100	2,800
	12%	0.48	74,000	10,000	54	710	41,000	5,200	11,900	3,700	3,800
Black	Green	0.52	54,000	9,200	88	1,220	22,500	4,100	7,800	5,000	3,700
	12%	0.57	92,000	11,200	86	1,020	46,100	7,000	12,500	4,600	5,200
Red	Green	0.49	53,000	9,600	79	810	22,600	2,800	7,900	-	3,100
	12%	0.54	92,000	11,300	86	810	45,100	6,900	12,800	-	4,200
Silver	Green	0.44	40,000	6,500	76	740	17,200	2,600	7,200	3,900	2,600
	12%	0.47	61,000	7,900	57	640	36,000	5,100	10,200	3,400	3,100
Sugar	Green	0.56	65,000	10,700	92	1,020	27,700	4,400	10,100	-	4,300
	12%	0.63	109,000	12,600	114	990	54,000	10,100	16,100	-	6,400
Oak, Red											
Black	Green	0.56	57,000	8,100	84	1,020	23,900	4,900	8,400	-	4,700
	12%	0.61	96,000	11,300	94	1,040	45,000	6,400	13,200	5,500	5,400
Cherrybark	Green	0.61	74,000	12,300	101	1,370	31,900	5,200	9,100	5,800	5,500
	12%	0.68	125,000	15,700	126	1,240	60,300	8,600	13,800	5,300	6,600
Laurel	Green	0.56	54,000	9,600	77	990	21,900	3,900	8,100	5,400	4,400
	12%	0.63	87,000	11,700	81	990	48,100	7,300	12,600	5,200	5,400
Northern Red	Green	0.56	57,000	9,300	91	1,120	23,700	4,200	8,300	5,500	4,400
	12%	0.63	99,000	12,500	100	1,090	46,600	7,000	12,300	5,500	5,700
Pin	Green	0.58	57,000	9,100	97	1,220	25,400	5,000	8,900	7,200	4,800
	12%	0.63	97,000	11,900	102	1,140	47,000	7,000	14,300	4,800	6,700
Scarlet	Green	0.60	72,000	10,200	103	1,370	28,200	5,700	9,700	6,000	5,300
	12%	0.67	120,000	13,200	141	1,350	57,400	7,700	13,000	-	6,200

(Continued)

Table 5. *(Continued)* **Strength Properties of Selected Commercial Woods Grown in the United States[a] (Metric Units).** *(Source, Wood Handbook, U.S.D.A.)*

| Common Species Names | Moisture Content | Specific Gravity[b] | Static Bending | | Work to Maximum Load (kJ/m³) | Impact Bending (mm) | Compression Parallel to Grain (kPa) | Compression Perpendicular to Grain (kPa) | Shear Parallel to Grain (kPa) | Tension Perpendicular to Grain (kPa) | Side Hardness (N) |
			Modulus of Rupture (kPa)	Modulus of Elasticity[c] (MPa)							
Hardwoods											
Oak, Red											
Southern Red	Green	0.52	48,000	7,900	55	740	20,900	3,800	6,400	3,300	3,800
	12%	0.59	75,000	10,300	65	660	42,000	6,000	9,600	3,500	4,700
Water	Green	0.56	61,000	10,700	77	990	25,800	4,300	8,500	5,700	4,500
	12%	0.63	106,000	13,900	148	1,120	46,700	7,000	13,900	6,300	5,300
Willow	Green	0.56	51,000	8,900	61	890	20,700	4,200	8,100	5,200	4,400
	12%	0.69	100,000	13,100	101	1,070	48,500	7,800	11,400	-	6,500
Oak, White											
Bur	Green	0.58	50,000	6,100	74	1,120	22,700	4,700	9,300	5,500	4,900
	12%	0.64	71,000	7,100	68	740	41,800	8,300	12,500	4,700	6,100
Chestnut	Green	0.57	55,000	9,400	65	890	24,300	3,700	8,300	4,800	4,000
	12%	0.66	92,000	11,000	76	1,020	47,100	5,800	10,300	-	5,000
Live	Green	0.80	82,000	10,900	85	-	37,400	14,100	15,200	-	-
	12%	0.88	127,000	13,700	130	-	61,400	19,600	18,300	-	-
Overcup	Green	0.57	55,000	7,900	87	1,120	23,200	3,700	9,100	5,000	4,300
	12%	0.63	87,000	9,800	108	970	42,700	5,600	13,800	6,500	5,300
Post	Green	0.60	56,000	7,500	76	1,120	24,000	5,900	8,800	5,400	5,000
	12%	0.67	91,000	10,400	91	1,170	45,300	9,900	12,700	5,400	6,000
Swamp Chestnut	Green	0.60	59,000	9,300	88	1,140	24,400	3,900	8,700	4,600	4,900
	12%	0.67	96,000	12,200	83	1,040	50,100	7,700	13,700	4,800	5,500
Swamp White	Green	0.64	68,000	11,000	100	1,270	30,100	5,200	9,000	5,900	5,200
	12%	0.72	122,000	14,100	132	1,240	59,300	8,200	13,800	5,700	7,200
White	Green	0.60	57,000	8,600	80	1,070	24,500	4,600	8,600	5,300	4,700
	12%	0.68	105,000	12,300	102	940	51,300	7,400	13,800	5,500	6,000

(Continued)

Table 5. (Continued) Strength Properties of Selected Commercial Woods Grown in the United States^a (Metric Units). (Source, Wood Handbook, U.S.D.A.)

Common Species Names	Moisture Content	Specific Gravity^b	Static Bending		Work to Maximum Load (kJ/m³)	Impact Bending (mm)	Compression Parallel to Grain (kPa)	Compression Perpendicular to Grain (kPa)	Shear Parallel to Grain (kPa)	Tension Perpendicular to Grain (kPa)	Side Hardness (N)
			Modulus of Rupture (kPa)	Modulus of Elasticity^c (MPa)							
Hardwoods											
Sassafras	Green	0.42	41,000	6,300	49	–	18,800	2,600	6,600	–	–
	12%	0.46	62,000	7,700	60	–	32,800	5,900	8,500	–	–
Sweetgum	Green	0.46	49,000	8,300	70	910	21,000	2,600	6,800	3,700	2,700
	12%	0.52	86,000	11,300	82	810	43,600	4,300	11,000	5,200	3,800
Sycamore, American	Green	0.46	45,000	7,300	52	660	20,100	2,500	6,900	4,300	2,700
	12%	0.49	69,000	9,800	59	660	37,100	4,800	10,100	5,000	3,400
Tanoak	Green	0.58	72,000	10,700	92	–	32,100	–	–	–	–
	12%	–	–	–	–	–	–	–	–	–	–
Tupelo											
Black	Green	0.46	48,000	7,100	55	760	21,000	3,300	7,600	3,900	2,800
	12%	0.50	66,000	8,300	43	560	38,100	6,400	9,200	3,400	3,600
Water	Green	0.46	50,000	7,200	57	760	23,200	3,300	8,200	4,100	3,200
	12%	0.50	66,000	8,700	48	580	40,800	6,000	11,000	4,800	3,900
Walnut, Black	Green	0.51	66,000	9,800	101	940	29,600	3,400	8,400	3,900	4,000
	12%	0.55	101,000	11,600	74	860	52,300	7,000	9,400	4,800	4,500
Willow, Black	Green	0.36	33,000	5,400	76	–	14,100	1,200	4,700	–	–
	12%	0.39	54,000	7,000	61	–	28,300	3,000	8,600	–	–
Yellow-poplar	Green	0.40	41,000	8,400	52	660	18,300	1,900	5,400	3,500	2,000
	12%	0.42	70,000	10,900	61	610	38,200	3,400	8,200	3,700	2,400

(Continued)

Table 5. *(Continued)* Strength Properties of Selected Commercial Woods Grown in the United States[a] (Metric Units). *(Source, Wood Handbook, U.S.D.A.)*

Common Species Names	Moisture Content	Specific Gravity[b]	Static Bending			Impact Bending (mm)	Compression Parallel to Grain (kPa)	Compression Perpendicular to Grain (kPa)	Shear Parallel to Grain (kPa)	Tension Perpendicular to Grain (kPa)	Side Hardness (N)
			Modulus of Rupture (kPa)	Modulus of Elasticity[c] (MPa)	Work to Maximum Load (kJ/m³)						
			Softwoods								
Baldcypress	Green	0.42	46,000	8,100	46	640	24,700	2,800	5,600	2,100	1,700
	12%	0.46	73,000	9,900	57	610	43,900	5,000	6,900	1,900	2,300
Cedar											
Atlantic White	Green	0.31	32,000	5,200	41	460	16,500	1,700	4,800	1,200	1,300
	12%	0.32	47,000	6,400	28	330	32,400	2,800	5,500	1,500	1,600
Eastern Redcedar	Green	0.44	48,000	4,500	103	890	24,600	4,800	7,000	2,300	2,900
	12%	0.47	61,000	6,100	57	560	41,500	6,300	–	–	4,000
Incense	Green	0.35	43,000	5,800	44	430	21,700	2,600	5,700	1,900	1,700
	12%	0.37	55,000	7,200	37	430	35,900	4,100	6,100	1,900	2,100
Northern White	Green	0.29	29,000	4,400	39	380	13,700	1,600	4,300	1,700	1,000
	12%	0.31	45,000	5,500	33	300	27,300	2,100	5,900	1,700	1,400
Port-Orford	Green	0.39	45,000	9,000	51	530	21,600	2,100	5,800	1,200	1,700
	12%	0.43	88,000	11,700	63	710	43,100	5,000	9,400	2,800	2,800
Western Redcedar	Green	0.31	35,900	6,500	34	430	19,100	1,700	5,300	1,600	1,200
	12%	0.32	51,700	7,700	40	430	31,400	3,200	6,800	1,500	1,600
Yellow	Green	0.42	44,000	7,900	63	690	21,000	2,400	5,800	2,300	2,000
	12%	0.44	77,000	9,800	72	740	43,500	4,300	7,800	2,500	2,600
Douglas-Fir [d]											
Coast	Green	0.45	53,000	10,800	52	660	26,100	2,600	6,200	2,100	2,200
	12%	0.48	85,000	13,400	68	790	49,900	5,500	7,800	2,300	3,200
Interior West	Green	0.46	53,000	10,400	50	660	26,700	2,900	6,500	2,000	2,300
	12%	0.50	87,000	12,600	73	810	51,200	5,200	8,900	2,400	2,900

(Continued)

Table 5. *(Continued)* Strength Properties of Selected Commercial Woods Grown in the United States[a] (Metric Units). *(Source, Wood Handbook, U.S.D.A.)*

Common Species Names	Moisture Content	Specific Gravity[b]	Static Bending			Impact Bending (mm)	Compression Parallel to Grain (kPa)	Compression Perpendicular to Grain (kPa)	Shear Parallel to Grain (kPa)	Tension Perpendicular to Grain (kPa)	Side Hardness (N)
			Modulus of Rupture (kPa)	Modulus of Elasticity[c] (MPa)	Work to Maximum Load (kJ/m³)						
Softwoods											
Douglas-Fir[d]											
Interior North	Green	0.45	51,000	9,700	56	560	23,900	2,500	6,600	2,300	1,900
	12%	0.48	90,000	12,300	72	660	47,600	5,300	9,700	2,700	2,700
Interior South	Green	0.43	47,000	8,000	55	380	21,400	2,300	6,600	1,700	1,600
	12%	0.46	82,000	10,300	62	510	43,000	5,100	10,400	2,300	2,300
Fir											
Balsam	Green	0.33	38,000	8,600	32	410	18,100	1,300	4,600	1,200	1,300
	12%	0.35	63,000	10,000	35	510	36,400	2,800	6,500	1,200	1,800
California Red	Green	0.36	40,000	8,100	44	530	19,000	2,300	5,300	2,600	1,600
	12%	0.38	72,400	10,300	61	610	37,600	4,200	7,200	2,700	2,200
Grand	Green	0.35	40,000	8,600	39	560	20,300	1,900	5,100	1,700	1,600
	12%	0.37	61,400	10,800	52	710	36,500	3,400	6,200	1,700	2,200
Noble	Green	0.37	43,000	9,500	41	480	20,800	1,900	5,500	1,600	1,300
	12%	0.39	74,000	11,900	61	580	42,100	3,600	7,200	1,500	1,800
Pacific Silver	Green	0.40	44,000	9,800	41	530	21,600	1,500	5,200	1,700	1,400
	12%	0.43	75,800	12,100	64	610	44,200	3,100	8,400	1,700	1,900
Subalpine	Green	0.31	34,000	7,200	-	-	15,900	1,300	4,800	-	1,200
	12%	0.32	59,000	8,900	-	-	33,500	2,700	7,400	-	1,600
White	Green	0.37	41,000	8,000	39	560	20,000	1,900	5,200	2,100	1,500
	12%	0.39	68,000	10,300	50	510	40,000	3,700	7,600	2,100	2,100

(Continued)

Table 5. *(Continued)* Strength Properties of Selected Commercial Woods Grown in the United States[a] (Metric Units). *(Source, Wood Handbook, U.S.D.A.)*

Common Species Names	Moisture Content	Specific Gravity[b]	Static Bending			Impact Bending (mm)	Compression Parallel to Grain (kPa)	Compression Perpendicular to Grain (kPa)	Shear Parallel to Grain (kPa)	Tension Perpendicular to Grain (kPa)	Side Hardness (N)
			Modulus of Rupture (kPa)	Modulus of Elasticity[c] (MPa)	Work to Maximum Load (kJ/m³)						
Softwoods											
Hemlock											
Eastern	Green	0.38	44,000	7,400	46	530	21,200	2,500	5,900	1,600	1,800
	12%	0.40	61,000	8,300	47	530	37,300	4,500	7,300	-	2,200
Mountain	Green	0.42	43,000	7,200	76	810	19,900	2,600	6,400	2,300	2,100
	12%	0.45	79,000	9,200	72	810	44,400	5,900	10,600	-	3,000
Western	Green	0.42	46,000	9,000	48	560	23,200	1,900	5,900	2,000	1,800
	12%	0.45	78,000	11,300	57	580	49,000	3,800	8,600	2,300	2,400
Larch, Western	Green	0.48	53,000	10,100	71	740	25,900	2,800	6,000	2,300	2,300
	12%	0.52	90,000	12,900	87	890	52,500	6,400	9,400	3,000	3,700
Pine											
Eastern White	Green	0.34	34,000	6,800	36	430	16,800	1,500	4,700	1,700	1,300
	12%	0.35	59,000	8,500	47	460	33,100	3,000	6,200	2,100	1,700
Jack	Green	0.40	41,000	7,400	50	660	20,300	2,100	5,200	2,500	1,800
	12%	0.43	68,000	9,300	57	690	39,000	4,000	8,100	2,900	2,500
Loblolly	Green	0.47	50,000	9,700	57	760	24,200	2,700	5,900	1,800	2,000
	12%	0.51	88,000	12,300	72	760	49,200	5,400	9,600	3,200	3,100
Lodgepole	Green	0.38	38,000	7,400	39	510	18,000	1,700	4,700	1,500	1,500
	12%	0.41	65,000	9,200	47	510	37,000	4,200	6,100	2,000	2,100
Longleaf	Green	0.54	59,000	11,000	61	890	29,800	3,300	7,200	2,300	2,600
	12%	0.59	100,000	13,700	81	860	58,400	6,600	10,400	3,200	3,900
Pitch	Green	0.47	47,000	8,300	63	-	20,300	2,500	5,900	-	-
	12%	0.52	74,000	9,900	63	-	41,000	5,600	9,400	-	-
Pond	Green	0.51	51,000	8,800	52	-	25,200	3,000	6,500	-	-
	12%	0.56	80,000	12,100	59	-	52,000	6,300	9,500	-	-
Ponderosa	Green	0.38	35,000	6,900	36	530	16,900	1,900	4,800	2,100	1,400
	12%	0.40	65,000	8,900	49	480	36,700	4,000	7,800	2,900	2,000

(Continued)

Table 5. *(Continued)* Strength Properties of Selected Commercial Woods Grown in the United States[a] (Metric Units). *(Source, Wood Handbook, U.S.D.A.)*

Common Species Names	Moisture Content	Specific Gravity[b]	Static Bending			Impact Bending (mm)	Compression Parallel to Grain (kPa)	Compression Perpendicular to Grain (kPa)	Shear Parallel to Grain (kPa)	Tension Perpendicular to Grain (kPa)	Side Hardness (N)
			Modulus of Rupture (kPa)	Modulus of Elasticity[c] (MPa)	Work to Maximum Load (kJ/m³)						
					Softwoods						
Hemlock											
Eastern	Green	0.38	44,000	7,400	46	530	21,200	2,500	5,900	1,600	1,800
	12%	0.40	61,000	8,300	47	530	37,300	4,500	7,300	-	2,200
Mountain	Green	0.42	43,000	7,200	76	810	19,900	2,600	6,400	2,300	2,100
	12%	0.45	79,000	9,200	72	810	44,400	5,900	10,600	-	3,000
Western	Green	0.42	46,000	9,000	48	560	23,200	1,900	5,900	2,000	1,800
	12%	0.45	78,000	11,300	57	580	49,000	3,800	8,600	2,300	2,400
Larch, Western	Green	0.48	53,000	10,100	71	740	25,900	2,800	6,000	2,300	2,300
	12%	0.52	90,000	12,900	87	890	52,500	6,400	9,400	3,000	3,700
Pine											
Eastern White	Green	0.34	34,000	6,800	36	430	16,800	1,500	4,700	1,700	1,300
	12%	0.35	59,000	8,500	47	460	33,100	3,000	6,200	2,100	1,700
Jack	Green	0.40	41,000	7,400	50	660	20,300	2,100	5,200	2,500	1,800
	12%	0.43	68,000	9,300	57	690	39,000	4,000	8,100	2,900	2,500
Loblolly	Green	0.47	50,000	9,700	57	760	24,200	2,700	5,900	1,800	2,000
	12%	0.51	88,000	12,300	72	760	49,200	5,400	9,600	3,200	3,100
Lodgepole	Green	0.38	38,000	7,400	39	510	18,000	1,700	4,700	1,500	1,500
	12%	0.41	65,000	9,200	47	510	37,000	4,200	6,100	2,000	2,100
Longleaf	Green	0.54	59,000	11,000	61	890	29,800	3,300	7,200	2,300	2,600
	12%	0.59	100,000	13,700	81	860	58,400	6,600	10,400	3,200	3,900
Pitch	Green	0.47	47,000	8,300	63	-	20,300	2,500	5,900	-	-
	12%	0.52	74,000	9,900	63	-	41,000	5,600	9,400	-	-
Pond	Green	0.51	51,000	8,800	52	-	25,200	3,000	6,500	-	-
	12%	0.56	80,000	12,100	59	-	52,000	6,300	9,500	-	-
Ponderosa	Green	0.38	35,000	6,900	36	530	16,900	1,900	4,800	2,100	1,400
	12%	0.40	65,000	8,900	49	480	36,700	4,000	7,800	2,900	2,000

(Continued)

Table 5. (Continued) Strength Properties of Selected Commercial Woods Grown in the United States[a] (Metric Units). (Source, Wood Handbook, U.S.D.A.)

Common Species Names	Moisture Content	Specific Gravity[b]	Static Bending			Impact Bending (mm)	Compression Parallel to Grain (kPa)	Compression Perpendicular to Grain (kPa)	Shear Parallel to Grain (kPa)	Tension Perpendicular to Grain (kPa)	Side Hardness (N)
			Modulus of Rupture (kPa)	Modulus of Elasticity[c] (MPa)	Work to Maximum Load (kJ/m³)						
Softwoods											
Pine											
Red	Green	0.41	40,000	8,800	42	660	18,800	1,800	4,800	2,100	1,500
	12%	0.46	76,000	11,200	68	660	41,900	4,100	8,400	3,200	2,500
Sand	Green	0.46	52,000	7,000	66	-	23,700	3,100	7,900	-	-
	12%	0.48	80,000	9,700	66	-	47,700	5,800	-	-	-
Shortleaf	Green	0.47	51,000	9,600	57	760	24,300	2,400	6,300	2,200	2,000
	12%	0.51	90,000	12,100	76	840	50,100	5,700	9,600	3,200	3,100
Slash	Green	0.54	60,000	10,500	66	-	26,300	3,700	6,600	-	-
	12%	0.59	112,000	13,700	91	-	56,100	7,000	11,600	-	-
Spruce	Green	0.41	41,000	6,900	-	-	19,600	1,900	6,200	-	2,000
	12%	0.44	72,000	8,500	-	-	39,000	5,000	10,300	-	2,900
Sugar	Green	0.34	34,000	7,100	37	430	17,000	1,400	5,000	1,900	1,200
	12%	0.36	57,000	8,200	38	460	30,800	3,400	7,800	2,400	1,700
Virginia	Green	0.45	50,000	8,400	75	860	23,600	2,700	6,100	2,800	2,400
	12%	0.48	90,000	10,500	94	810	46,300	6,300	9,300	2,600	3,300
Western White	Green	0.36	32,000	8,200	34	480	16,800	1,300	4,700	1,800	1,200
	12%	0.38	67,000	10,100	61	580	34,700	3,200	7,200	-	1,900
Redwood											
Old-growth	Green	0.38	52,000	8,100	51	530	29,000	2,900	5,500	1,800	1,800
	12%	0.40	69,000	9,200	48	480	42,400	4,800	6,500	1,700	2,100
Young-growth	Green	0.34	41,000	6,600	39	410	21,400	1,900	6,100	2,100	1,600
	12%	0.35	54,000	7,600	36	380	36,000	3,600	7,600	1,700	1,900

(Continued)

Table 5. *(Continued)* **Strength Properties of Selected Commercial Woods Grown in the United States[a] (Metric Units).** *(Source, Wood Handbook, U.S.D.A.)*

| Common Species Names | Moisture Content | Specific Gravity[b] | Static Bending | | | Impact Bending (mm) | Compression Parallel to Grain (kPa) | Compression Perpendicular to Grain (kPa) | Shear Parallel to Grain (kPa) | Tension Perpendicular to Grain (kPa) | Side Hardness (N) |
			Modulus of Rupture (kPa)	Modulus of Elasticity[c] (MPa)	Work to Maximum Load (kJ/m³)						
Softwoods											
Spruce											
Black	Green	0.38	42,000	9,500	51	610	19,600	1,700	5,100	700	1,600
	12%	0.46	74,000	11,100	72	580	41,100	3,800	8,500	-	2,300
Engelmann	Green	0.33	32,000	7,100	35	410	15,000	1,400	4,400	1,700	1,150
	12%	0.35	64,000	8,900	44	460	30,900	2,800	8,300	2,400	1,750
Red	Green	0.37	41,000	9,200	48	460	18,800	1,800	5,200	1,500	1,600
	12%	0.40	74,000	11,100	58	640	38,200	3,800	8,900	2,400	2,200
Sitka	Green	0.33	34,000	7,900	43	610	16,200	1,400	4,400	1,700	1,600
	12%	0.36	65,000	9,900	65	640	35,700	3,000	6,700	2,600	2,300
White	Green	0.37	39,000	7,400	41	560	17,700	1,700	4,800	1,500	1,400
	12%	0.40	68,000	9,200	53	510	37,700	3,200	7,400	2,500	2,100
Tamarack	Green	0.49	50,000	8,500	50	710	24,000	2,700	5,900	1,800	1,700
	12%	0.53	80,000	11,300	49	580	49,400	5,500	8,800	2,800	2,600

Notes:

[a] Results of tests on small clear specimens in the green and air-dried conditions, converted to metric units directly from Table 4. Definition of properties: impact bending is height of drop that causes complete failure, using 0.71-kg (50-lb) hammer; compression parallel to grain is also called maximum crushing strength; compression perpendicular to grain is fiber stress at proportional limit; shear is maximum shearing strength; tension is maximum tensile strength; and side hardness is hardness measured when load is perpendicular to grain.

[b] Specific gravity is based on weight when ovendry and volume when green or at 12% moisture content.

[c] Modulus of elasticity measured from a simply supported, center-loaded beam, on a span depth ratio of 14/1. To correct for shear deflection, the modulus can be increased by 10%.

[d] Coast Douglas-fir is defined as Douglas-fir growing in Oregon and Washington State west of the Cascade Mountains summit. Interior West includes California and all counties in Oregon and Washington east of, but adjacent to, the Cascade summit; Interior North, the remainder of Oregon and Washington plus Idaho, Montano, and Wyoming; and Interior South, Utah, Colorado, Arizona, and New Mexico.

Table 6. Machining and Related Properties of Selected Domestic Hardwoods. *(Source, Wood Handbook, U.S.D.A.)*

Type of Wood[a]	Planing: Perfect Pieces (%)	Shaping: Good to Excellent Pieces (%)	Turning: Good to Excellent Pieces (%)	Boring: Good to Excellent Pieces (%)	Mortising: Fair to Excellent Pieces (%)	Sanding: Good to Excellent Pieces (%)	Steam Bending: Unbroken Pieces (%)	Nail Splitting: Pieces Free From Complete Splits (%)	Screw Splitting: Pieces Free From Complete Splits (%)
Alder, Red	61	20	88	64	52	-	-	-	-
Ash	75	55	79	94	58	75	67	65	71
Aspen	26	7	65	78	60	-	-	-	-
Basswood	64	10	68	76	51	17	2	79	68
Beech	83	24	90	99	92	49	75	42	58
Birch	63	57	80	97	97	34	72	32	48
Birch, paper	47	22	-	-	-	-	-	-	-
Cherry, black	80	80	88	100	100	-	-	-	-
Chestnut	74	28	87	91	70	64	56	66	60
Cottonwood[b]	21	3	70	70	52	19	44	82	78
Elm, soft[b]	33	13	65	94	75	66	74	80	74
Hackberry	74	10	77	99	72	-	94	63	63
Hickory	76	20	84	100	98	80	76	35	63
Magnolia	65	27	79	71	32	37	85	73	76
Maple, bigleaf	52	56	80	100	80	-	-	-	-
Maple, hard	54	72	82	99	95	38	57	27	52
Maple, soft	41	25	76	80	34	37	59	58	61
Oak, red	91	28	84	99	95	81	86	66	78
Oak, white	87	35	85	95	99	83	91	69	74
Pecan	88	40	89	100	98	-	78	47	69

[a] Commercial lumber nomenclature.
[b] Interlocked grain present.

(Continued)

Table 6. *(Continued)* **Machining and Related Properties of Selected Domestic Hardwoods.** *(Source, Wood Handbook, U.S.D.A.)*

Type of Wood[a]	Planing: Perfect Pieces (%)	Shaping: Good to Excellent Pieces (%)	Turning: Good to Excellent Pieces (%)	Boring: Good to Excellent Pieces (%)	Mortising: Fair to Excellent Pieces (%)	Sanding: Good to Excellent Pieces (%)	Steam Bending: Unbroken Pieces (%)	Nail Splitting: Pieces Free From Complete Splits (%)	Screw Splitting: Pieces Free From Complete Splits (%)
Sweetgum [b]	51	28	86	92	58	23	67	69	69
Sycamore [b]	22	12	85	98	96	21	29	79	74
Tanoak	80	39	81	100	100	-	-	-	-
Tupelo, water [b]	55	52	79	62	33	34	46	64	63
Tupelo, black [b]	48	32	75	82	24	21	42	65	63
Walnut, black [b]	62	34	91	100	98	-	78	50	59
Willow	52	5	58	71	24	24	73	89	62
Yellow-poplar	70	13	81	87	63	19	58	77	67

[a] Commercial lumber nomenclature.
[b] Interlocked grain present.

Table 7. Dimensions of Commonly Available Nails.

Size	Gauge	Length		Diameter	
		inch	mm	inch	mm
Bright Common Wire Nails (Penny Nails)					
6d	11-1/2	2	50.8	0.113	2.87
8d	10-1/4	2-1/2	63.5	0.131	3.33
10d	9	3	76.2	0.148	3.76
12d	9	3-1/4	82.6	0.148	3.76
16d	8	3-1/2	88.9	0.162	4.11
20d	6	4	101.6	0.192	4.88
30d	5	4-1/2	114.3	0.207	5.26
40d	4	5	127.0	0.225	5.72
50d	3	5-1/2	139.7	0.244	6.20
60d	2	6	152.4	0.262	6.65
Smooth Box Nails					
3d	14-1/2	1-1/4	31.8	0.076	1.93
4d	14	1-1/2	38.1	0.080	2.03
5d	14	1-3/4	44.5	0.080	2.03
6d	12-1/2	2	50.8	0.098	2.49
7d	12-1/2	2-1/4	57.2	0.098	2.49
8d	11-1/2	2-1/2	63.5	0.113	2.87
10d	10-1/2	3	76.2	0.128	3.25
16d	10	3-1/2	88.9	0.135	3.43
20d	9	4	101.6	0.148	3.76

Size	Length		Diameter	
	inch	mm	inch	mm
Helical and Annularly Threaded Nails				
6d	2	50.8	0.120	3.05
8d	2-1/2	63.5	0.120	3.05
10d	3	76.2	0.135	3.43
12d	3-1/4	82.6	0.135	3.43
16d	3-1/2	88.9	0.148	3.76
20d	4	101.6	0.177	4.50
30d	4-1/2	114.3	0.177	4.50
40d	5	127.0	0.177	4.50
50d	5-1/2	139.7	0.177	4.50
60d	6	152.4	0.177	4.50
70d	7	177.8	0.207	5.26
80d	8	203.2	0.207	5.26
90d	9	228.6	0.207	5.26

factors other than density can affect production of smooth surfaces during wood machining: interlocked and variable grain; hard mineral deposits; and reaction wood—particularly tension wood in hardwoods. Interlocking grain is characteristic of a few domestic species and many tropical species, and it presents difficulty in planning quartersawn boards unless attention is paid to feed rate, cutting angles, and sharpness of cutters. Hard deposits in cells, such as calcium carbonite and silica, can have a pronounced effect on all cutting edges. This dulling effect becomes more pronounced as the wood is dried to normal inservice requirements. Tension wood can cause fibrous and fuzzy surfaces, and can be very troublesome on species of lower density. Reaction wood can also be responsible for the pinching effect on saws as a result of stress relief. The pinching can result in burning and dulling of saw teeth.

Fasteners for wood

Nails

Nails, of many types, forms, and shapes, are the most common fasteners for wood construction. Nails in use must resist withdrawal and/or lateral loads, and their resistance is affected by the species of wood and the type of nail. Sizes of readily available nails are given in **Table 7**. Because international nail producers do not always adhere to the dimensions in the Table, it is advisable to specify length and diameters when purchasing nails.

The resistance of a nail to withdraw from a piece of wood is dependent on the density of the wood, the diameter of the nail, and the depth of penetration. For bright common wire nails driven into the side grain of seasoned wood, or unseasoned wood that remains wet, the maximum withdrawal load can be obtained with the following equation.

For withdrawal load in pounds $\qquad p = 7,850 \times G^{5/2} \times D \times L$

For withdrawal load in Newtons $\qquad p = 54.12 \times G^{5/2} \times D \times L$

> where $\quad p$ = maximum load in pounds or Newtons
> $\quad G$ = specific gravity of the wood at 12% moisture content
> (see **Tables 4** and **5**)
> $\quad D$ = diameter of nail in inches or millimeters
> $\quad L$ = depth of penetration of the nail in inches or millimeters.

The resistance of nails to withdrawal is generally greatest when they are driven perpendicular to the grain of the wood. When the nail is driven parallel to the wood fibers (into the end of the piece), withdrawal resistance in the softer woods drops by 25 to 50% of the resistance obtained when the nail is driven perpendicular to the grain. The difference between side- and end-grain withdrawal loads is less for dense woods than for softer woods. Withdrawal resistance is also affected by other factors including the type of nail point, type of shank, time the nail remains in the wood, surface coatings of the nail, and moisture content changes in the wood. Another factor to consider when nailing wood is that the less dense species do not split as readily as the denser ones, thus offering the opportunity to increase the diameter, length, and number of nails to compensate for the wood's lower resistance to nail withdrawal. Withdrawal resistance in plywood is approximately 15 to 30% less than solid wood of the same thickness, primarily because the fiber distortion is less uniform in plywood.

Wood Screws

The common types of wood screws have flat, oval, or round heads, as shown in *Figure 2*. For flush surfaces, flat heads are essential, but round head screws are required when countersinking is objectionable. The root diameter for most screws averages about

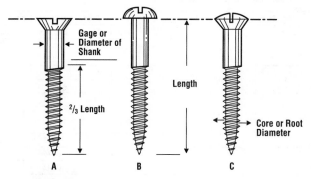

Figure 2. Common wood screws: A) flathead, B) roundhead, C) ovalhead.

two-thirds the shank diameter. Resistance of wood screw shanks to withdrawal varies directly with the square of the specific gravity of the wood. Within limits, the withdrawal load varies directly with the depth of penetration of the threaded portion and the diameter of the screw, provided that the screw does not fail in tension. Failure in tension results when the screw's strength is exceeded by the withdrawal strength from the wood, and the limiting length to cause a tension failure decreases as the density of the wood increases because the withdrawal strength of the wood increases with density. The longer lengths of standard screws are therefore superfluous in dense hardwoods.

Ultimate test values for withdrawal loads of wood screws inserted into the side grain of seasoned wood may be obtained with the following equation.

For withdrawal load in pounds $\quad p = 15{,}700 \times G^2 \times D \times L$

For withdrawal load in Newtons $\quad p = 108.25 \times G^2 \times D \times L$

where p = maximum load in pounds or Newtons
G = specific gravity of the wood at 12% moisture content
(see **Tables 4** and **5**)
D = shank diameter of screw in inches or millimeters
L = depth of penetration of the threaded part of the screw in inches or millimeters.

This equation is applicable when the prebored screw lead hole has a diameter of about 70% of the root diameter of the threads in softwoods, and about 90% in hardwoods. The equation values are applicable to the screw sizes listed in **Table 8**, and when lengths and gages are outside the indicated limits, the actual values are likely to be less than the equation values. The withdrawal loads of screws inserted in the end grain of wood are somewhat erratic, but when splitting is avoided, they should average 75% of the load sustained by screws inserted in the side grain.

Withdrawal resistance of tapping screws (screws that have threads the full length of the shank–commonly known as sheet metal screws) is in general about 10% greater than that for wood screws of comparative diameter and length of threaded portion. The ratio between the withdrawal resistance of tapping screws and wood screws varies from 1.16 in denser woods such as oak, to 1.05 in lighter woods such as redwood.

Table 8. Screw Sizes and Gage Diameters.

Screw Length		Gage Limits	Gage Conversion to Diameter		
inch	mm		Gage Number	Dia. inch	Dia. mm
1/2	12.7	4 to 6	4	0.112	2.84
			5	0.125	3.18
3/4	19.0	4 to 11	6	0.138	3.51
			7	0.151	3.84
1	25.4	4 to 12	8	0.164	4.17
			9	0.177	4.50
1 1/2	38.1	5 to 14	10	0.190	4.83
			11	0.203	5.16
2	50.8	7 to 16	12	0.216	5.49
			14	0.242	6.15
2 1/2	63.5	9 to 18	16	0.268	6.81
			18	0.294	7.47
3	76.2	12 to 20	20	0.320	8.13
			24	0.372	9.45

Lag Screws

Lag screws are normally used where it would be difficult to fasten a bolt, or in instances where a nut on the surface would be objectionable. Commonly available lag screws range from about 0.2 to one inch (5.1 to 25.4 mm) in diameter, and from 1 to 16 inches (25.4 to 406 mm) in length. The thread length varies with screw length and ranges from $3/4$ inch (19 mm) in lengths up to $1 \ 1/4$ inch (31.8 mm), to one-half the overall screw length in lengths 10 inches (254 mm) or more. The head on lag screws is hexagonal shaped to provide for tightening by wrench. The equation for withdrawal resistance of lag screws, which follows, is based on the screw material having a tensile yield strength of 45 ksi (310 MPa) and an average ultimate tensile strength of 77 ksi (531 MPa).

For withdrawal load in pounds $\quad p = 125.4 \times G^{3/2} \times D^{3/4} \times L$

For withdrawal load in Newtons $\quad p = 108.25 \times G^{3/2} \times D^{3/4} \times L$

where p = maximum load in pounds or Newtons
G = specific gravity of the wood at 12% moisture content
(see **Tables 4** and **5**)
D = shank diameter of screw in inches or millimeters
L = depth of penetration of the threaded part of the screw in inches or millimeters.

Lag screws, like wood screws, require prebored lead holes, and the diameter of the lead hole for the threaded section varies with the density of the wood. For low density softwoods, such as cedars and white pines, the lead hole should be 40 to 70% of the shank diameter; for Douglas fir and Southern pine, 60 to 75%; and for dense hardwoods, 65 to 85%. The smaller percentage in each range applies to lag screws of the smaller diameters, and the larger percentage to lag screws with larger diameters. A lead hole should also be prebored to the depth of penetration of the shank, equal to the diameter of the shank.

Machining Centers

Typically, machining centers are numerically controlled machines with multipurpose (usually milling, drilling, boring, tapping, and reaming) capabilities. This flexibility not only allows for productivity improvements, it also permits one machine to replace several single purpose machines. Machining centers provide the ability to perform several operations on a workpiece, from roughing to finishing, with a single setup, resulting in a significant increase in productivity. Standard configurations are horizontal centers and vertical centers, which refers to the inclination of the spindle on the machine. Costs vary greatly depending on the machine complexity and capabilities. Basic machines can actually be very economical when compared to the total costs of the machines they can replace on the shop floor. Since both three- and five-axis machining centers are manufactured by machine tool builders throughout the world, they are available in an almost bewildering number of configurations. In addition, customized machines can be built for specialized applications. The descriptions provided in this section in no way represent the full range of available products, and the purchase of new machining centers should be the result of careful study and planning.

Machining center movement takes place on at least three planes—the X-, Y-, and Z-axes ("axis" refers to any rotational or linear machine move that can be controlled by a CNC command, but it does not include the spindle rotation which is never considered an axis). In a vertical three-axis machining center, the X-axis moves right to left, parallel to the workholding surface. The Y-axis moves front to back, perpendicular to both X and Z. The Z-axis moves up and down. The movement of the machine is accomplished in a variety of ways, depending on the manufacturer. The head containing the spindle may be fixed and the machine table will move in all three axes, or the table may be stationary and the spindle will move in all three axes. *Figure 1* illustrates a three-axis milling station, and *Figure 2* shows a five-axis station.

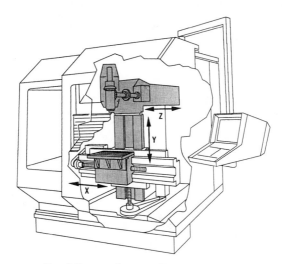

Figure 1. Three-axis milling station. (Source, Kennametal, Inc.)

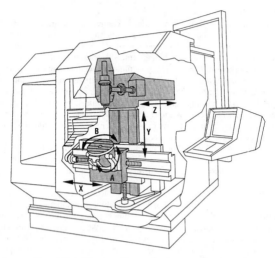

Figure 2. Typical five-axis milling station. (Source, Kennametal, Inc.)

In a five-axis machining center, multifunctional features provide two additional axes. Fitting a three-axis machine with a table or spindle head that both rotates and pivots produces five functional axes. The X-, Y-, and Z-movements of the spindle provide the first three axes, while the tilting and rotary movements of the machine account for the additional axes (A-, B-, or C-axis). A rotary tilting table offers the advantage of mechanical and thermal stability, plus improved access for swarf removal. The pivoting head allows for machining larger pallet loaded components. See *Figure 3* for a graphic of axis layouts, and the section beginning on p. 779 titled *Principles of 5-axis Machining* for an in-depth discussion of this technology.

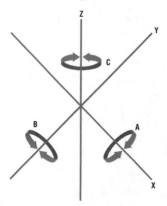

Figure 3. Linear axes in machining.

Spindle adaptation

Face mills, end mills, and slotting mills are rotating cutting tools that require a high degree of mounting accuracy and rigidity for optimal performance in terms of tool life and productivity. Therefore, the adaptation of the tool to the machine spindle face is critical and can be accomplished by either mounting the tool directly to the spindle face (located with a centering plug), or by using an intermediate tapered shank adapter. See *Figure 4*. Adapters are available in many varieties, including end mill adapters, shell mill adapters, stub arbors, and collet chucks. Most adapter systems in use are based on some type of tapered shank that is held in the machine spindle by either a threaded drawbar or a drawbar that grips a retention knob at the back of the adapter. The adapter is then driven by keys mounted on the spindle face. Adapter size is a good indicator of machine rigidity. The larger the adapter diameter, the more rigid the connection. Retention knobs may be unique to each machine tool manufacturer and each spindle size, and caution must be taken to assure that the correct knob is employed.

The majority of machine tools employ a standardized taper with a ratio of 7 to 24 (3.500" per foot, *or* 291.663 mm per meter = 8° 17' 50") that is commonly known as a "steep" taper, American Standard taper, or ISO taper to mate the adapter to the spindle. By definition, a "steep taper" is a taper "having an angle sufficiently large to assure the easy or self-releasing feature." Even though several national standards have evolved to mate adapters to spindles, their tapers are all compatible with the 7 to 24 taper, with other dimensions defined by ANSI Standard B5.18-1972 for milling machines (see **Table 1**). The continued popularity of the standard 7/24-taper is due to several inherent features of the design. First, since the only machined dimension that must be highly accurate is the taper, the adapters are comparatively simple and inexpensive to manufacture. In addition, the adapter is positioned and held deep within the spindle, which reduces the amount of tool overhang from the face of the spindle. Finally, tapered holders do not lock into place or become jammed in the spindle, which enhances the ease of removal. The disadvantage of the steep taper system is that the tapered surface must locate the adapter relative to the spindle, and clamp it rigidly into place. Since continuous face and taper contact would require extremely high machining accuracy in excess of the requirements of the

(Text continued on p. 756)

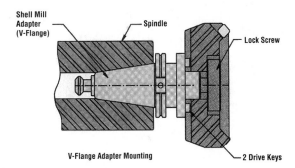

Figure 4. Adapter mounting accommodates cutters up to six inches in diameter. The cutter is located by the pilot and drive keys, and is held in place with a lock screw or socket-head cap screw. (Source, Kennametal, Inc.)

Table 1. Milling Machine Spindle Nose Dimensions.

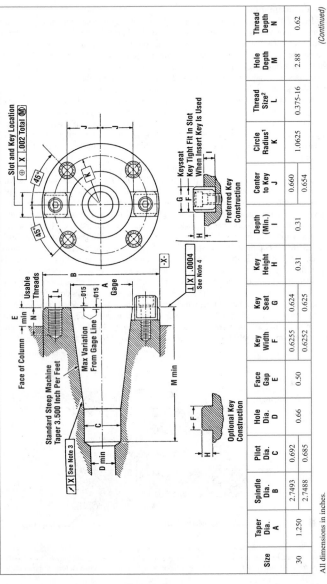

Size	Taper Dia. A	Spindle Dia. B	Pilot Dia. C	Hole Dia. D	Face Gap E	Key Width F	Key Seat G	Key Height H	Depth (Min.) I	Center to Key J	Circle Radius[1] K	Thread Size[2] L	Hole Depth M	Thread Depth N
30	1.250	2.7493 2.7488	0.692 0.685	0.66	0.50	0.6255 0.6252	0.624 0.625	0.31	0.31	0.660 0.654	1.0625	0.375-16	2.88	0.62

(Continued)

All dimensions in inches.

Table 1. *(Continued)* Milling Machine Spindle Nose Dimensions.

Size	Taper Dia. A	Spindle Dia. B	Pilot Dia. C	Hole Dia. D	Face Gap E	Key Width F	Key Seat G	Key Height H	Depth (Min.) I	Center to Key J	Circle Radius¹ K	Thread Size² L	Hole Depth M	Thread Depth N
40	1.750	3.4993	1.005	0.66	0.62	0.6255	0.625	0.31	0.31	0.910	1.3125	0.500-13	3.88	0.81
		3.4988	0.997			0.6252	0.624			0.904				
45	2.250	3.9993	1.286	0.78	0.62	0.7505	0.749	0.38	0.38	1.160	1.500	0.500-13	4.75	0.81
		3.9988	1.278			0.7502	0.750			1.154				
50	2.750	5.0618	1.568	1.06	0.75	1.0006	0.999	0.50	0.50	1.410	2.000	0.625-11	5.50	1.00
		5.0613	1.559			1.0002	1.000			1.404				
60	4.250	8.7180	2.381	1.38	1.50	1.0006	0.999	0.50	0.50	2.420	3.500	0.750-10	8.62	1.25
		8.7175	2.371			1.0002	1.000			2.414				

All dimensions in inches.

Tolerances:

Tolerance on rate of taper is 0.001 in./ft applied only in the direction that decreases rate of taper.

Keyway or solid key is central with axis of taper within 0.002 in. at maximum material condition (0.002" total indicator variation).

Notes: [1] For 30, 40, and 45 taper spindle noses, holes are to be spaced as shown and located within 0.006 in. diameter of true position. For 50 and 60 sizes, holes are to be spaced as shown and located within 0.010 in. diameter of true position.

[2] Threads are UNC-2B.

[3] Maximum runout on test plug not to exceed 0.0004 in. at one inch projection from gage line, and 0.001 in. at 12 inch projection from gage line.

[4] Squareness of mounting face to be measured near mounting bolt hole circle. Source: Extracted from ANSI B5.18-1872, published by the American Society of Mechanical Engineers.

Table 2. Toolholders (Shanks) for Milling Machines.

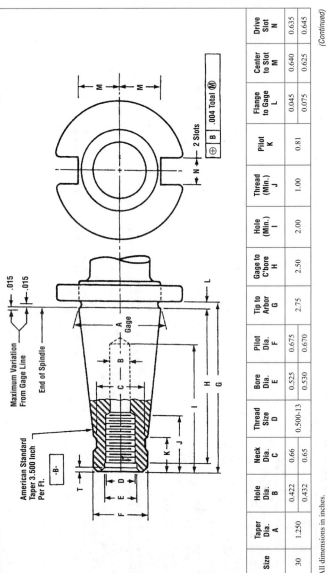

Size	Taper Dia. A	Hole Dia. B	Neck Dia. C	Thread Size D	Bore Dia. E	Pilot Dia. F	Tip to Arbor G	Gage to C'bore H	Hole (Min.) I	Thread (Min.) J	Pilot K	Flange to Gage L	Center to Slot M	Drive Slot N
30	1.250	0.422	0.66	0.500-13	0.525	0.675	2.75	2.50	2.00	1.00	0.81	0.045	0.640	0.635
		0.432	0.65		0.530	0.670						0.075	0.625	0.645

All dimensions in inches.

(Continued)

Table 2. *(Continued)* Toolholders (Shanks) for Milling Machines.

Size	Taper Dia. A	Hole Dia. B	Neck Dia. C	Thread Size D	Bore Dia. E	Pilot Dia. F	Tip to Arbor G	Gage to C'bore H	Hole (Min.) I	Thread (Min.) J	Pilot K	Flange to Gage L	Center to Slot M	Drive Slot N
40	1.750	0.531	0.94	0.625-11	0.650	0.987	3.75	3.50	2.25	1.12	1.00	0.045	0.890	0.635
		0.541	0.93		0.655	0.980						0.075	0.875	0.645
45	2.250	0.656	1.19	0.750-10	0.775	1.268	4.38	4.06	2.75	1.50	1.00	0.105	1.140	0.760
		0.666	1.18		0.780	1.260						0.135	1.125	0.770
50	2.750	0.875	1.50	1.000-8	1.025	1.550	5.12	4.75	3.50	1.75	1.00	0.105	1.390	1.010
		0.885	1.49		1.030	1.540						0.135	1.375	1.020
60	4.250	1.109	2.28	1.250-7	1.307	2.360	8.25	7.81	4.25	2.25	1.75	0.105	2.400	1.010
		1.119	2.27		1.312	2.350						0.135	2.385	1.020

All dimensions in inches.

Notes: The depth of 60° Center (T) is 0.05/0.07 on 30, 40, and 45 Tapers, and 0.05/0.12 on 50 and 60 Tapers. Tolerance on rate of taper is 0.001 in./ft applied only in the direction that decreases rate of taper. Centrality of drive slot with axis of taper shank 0.004 inch at maximum material condition (0.004 inch total indicator variation). Source: Extracted from ANSI B5.18-1972, published by the American Society of Mechanical Engineers.

Table 3. Draw-in Bolt End Dimensions.

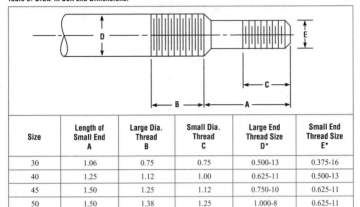

Size	Length of Small End A	Large Dia. Thread B	Small Dia. Thread C	Large End Thread Size D*	Small End Thread Size E*
30	1.06	0.75	0.75	0.500-13	0.375-16
40	1.25	1.12	1.00	0.625-11	0.500-13
45	1.50	1.25	1.12	0.750-10	0.625-11
50	1.50	1.38	1.25	1.000-8	0.625-11
60	1.75	2.00	1.37	1.250-7	1.000-8

* UNC-2A Thread.

All dimensions in inches. Source: ANSI B5.18-1972, published by the American Society of Mechanical Engineers.

standard (0.001" per foot on both the spindle and the adapter, and 0.004" centrality of the drive slot with the axis of the taper for the adapter), tapered adapters are potentially subject to unbalance and radial location problems that are especially pronounced at higher revolutions. ANSI B.510-1981 provides dimensions for twelve steep machining tapers, from number 5 to number 60, in increments of five. Numbers 10, 20, 30, 40, 50, and 60 are identified as the "preferred series," and all other sizes are classified as "intermediate series." See **Table 2** for tapered shank dimensions and **Table 3** for draw-in bolt dimensions. CAT V and other systems that employ ISO steep tapers are described in the following section, "Adapter systems."

Various other tapers are used for dedicated machining processes. In addition to the popular steep taper adapters system just described, Morse tapers, Jarno tapers, and Brown and Sharpe tapers are commonly used. While *Morse tapers* are sometimes found on lathes, their primary application is drilling, and specifications for these tapers can be found on page 1100 in the Drilling section of this book. Morse tapers have a nominal taper per foot of $5/8$ in., and American Standard Tapers 1, 2, 3, 4, 4 $1/2$, 5, 6, and 7, are based on these tapers. *Jarno tapers* were originated by Oscar J. Beale of Brown & Sharpe (see **Table 4**), and have a taper of 0.600 in. per foot (0.050 in. per in.). They are usually found on die-sinking and profiling machines, plus lathes and grinding machine centers. *Brown & Sharpe tapers* are popular for use on milling machines, grinding machines, and dividing head spindles (see **Table 5**). They have a taper per foot ranging from 0.5000 in. per foot to 0.51612 in. per foot (0.041666 in. per in. to 0.043010 in. per in.). Morse, American Standard, Jarno, and Brown & Sharpe tapers belong to a group referred to as "self-holding tapers," meaning that the taper is of an angle "small enough to hold a shank in place ordinarily by friction without other holding means (sometimes referred to as a slow taper) and one that assures the rotation of the tool within the socket."

Table 4. Jarno Taper Shank Dimensions. Taper per foot is 0.600″ for all sizes.

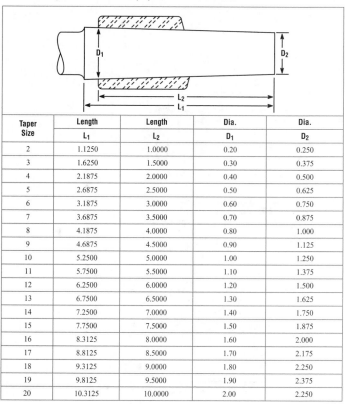

Taper Size	Length L₁	Length L₂	Dia. D₁	Dia. D₂
2	1.1250	1.0000	0.20	0.250
3	1.6250	1.5000	0.30	0.375
4	2.1875	2.0000	0.40	0.500
5	2.6875	2.5000	0.50	0.625
6	3.1875	3.0000	0.60	0.750
7	3.6875	3.5000	0.70	0.875
8	4.1875	4.0000	0.80	1.000
9	4.6875	4.5000	0.90	1.125
10	5.2500	5.0000	1.00	1.250
11	5.7500	5.5000	1.10	1.375
12	6.2500	6.0000	1.20	1.500
13	6.7500	6.5000	1.30	1.625
14	7.2500	7.0000	1.40	1.750
15	7.7500	7.5000	1.50	1.875
16	8.3125	8.0000	1.60	2.000
17	8.8125	8.5000	1.70	2.175
18	9.3125	9.0000	1.80	2.250
19	9.8125	9.5000	1.90	2.375
20	10.3125	10.0000	2.00	2.250

All dimensions in inches.

Adapter systems

CAT V system

   Caterpillar tapered shank adapters (CAT V system) mount directly into the spindle and are secured by a drawbar that grasps the retention knob at the back of the adapter. They employ the $7/24$ (3.5 in. per foot) steep taper. The CAT V system tapers are universally available in 30, 40, 45, 50, or 60 sizes, with the spindle's taper size normally selected by the machine builder based on the available horsepower of the machine or the length of the tools to be used on the machine. The flange (V), which is used by the machine tool change device to pick a tool from the tool changer or magazine and place it in the tapered spindle hole, is unique to the CAT V system. The rear of the adapter is threaded so a variety of

Table 5. Brown and Sharpe Taper Dimensions.

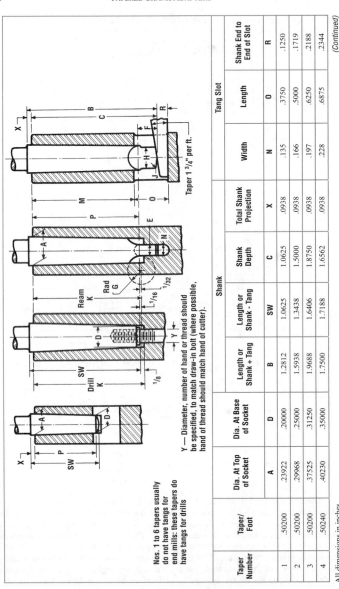

Nos. 1 to 6 tapers usually do not have tangs for end mills; these tapers do have tangs for drills

Y — Diameter, number of hand or thread should be specified, to match draw-in bolt (where possible, hand of thread should match hand of cutter).

Taper 1 3/4" per ft.

Taper Number	Taper/ Foot	Dia. At Top of Socket	Dia. At Base of Socket	Length or Shank + Tang	Length or Shank - Tang	Shank Depth	Total Shank Projection	Width	Length	Shank End to End of Slot
		A	D	B	SW	C	X	N	O	R
1	.50200	.23922	.20000	1.2812	1.0625	1.0625	.0938	.135	.3750	.1250
2	.50200	.29968	.25000	1.5938	1.3438	1.5000	.0938	.166	.5000	.1719
3	.50200	.37525	.31250	1.9688	1.6406	1.8750	.0938	.197	.6250	.2188
4	.50240	.40230	.35000	1.7500	1.7188	1.6562	.0938	.228	.6875	.2344

(Shank columns: Length or Shank + Tang = B, Length or Shank − Tang = SW, Shank Depth = C, Total Shank Projection = X. Tang Slot columns: Width = N, Length = O, Shank End to End of Slot = R.)

All dimensions in inches.

(Continued)

Table 5. (Continued) Brown and Sharpe Taper Dimensions.

Taper Number	Taper/ Foot	Dia. At Top of Socket A	Dia. At Base of Socket D	Shank: Length or Shank + Tang B	Length or Shank - Tang SW	Shank Depth C	Total Shank Projection X	Tang Slot: Width N	Length O	Shank End to End of Slot R
5	.50160	.52320	.45000	2.2812	1.9062	2.1875	.0938	.260	.7500	.2500
6	.50329	.59960	.50000	2.9688	2.5312	2.8750	.0938	.291	.8750	.2969
7	.50147	.72540	.60000	3.6250	3.0938	3 17/32	.0938	.322	.9375	.3125
8	.50100	.89870	.75000	4.2500	3.6875	4.1250	.1250	.353	1.0000	.3281
9	.50085	1.0671	.90010	4.7500	4.1250	4.6250	.1250	.385	1.1250	.3750
10	.51612	1.2893	1.04465	6.5312	5.8125	6.4062	.1250	.447	1.3125	.4375
11	.50100	1.5318	1.24995	7.5938	6.8750	7.4688	.1250	.447	1.3125	.4375
12	.49973	1.7968	1.50010	8.0625	7.2500	7.9375	.1250	.510	1.5000	.5000
13	.50020	2.0731	1.75005	8.6875	7.8750	8.5625	.1250	.510	1.5000	.5000
14	.50000	2.3438	2.00000	9.2812	8.3750	9.1562	.1250	.572	1.6875	.5625
15	.50000	2.6146	2.25000	9.7812	8.8750	9.6562	.1250	.572	1.6875	.5625
16	.50000	2.8854	2.50000	10.3750	9.3750	10.2500	.1250	.635	1.8750	.8750
17	.50000	3.1563	2.75000	-	9.8750	-	.1250	-	-	-
18	.50000	3.4271	3.00000	-	10.3750	-	.1250	-	-	-

Taper Number	Taper/ Inch	Socket End to Tang Slot K	Min. Depth of Tapered Hole K: Drill	Ream	Plug Depth P	Thickness D	Tang: Length F	Radius of Mill G	Diameter O	Radius R
1	.041833	.9375	1.0625	1.0000	.9375	.125	.1875	3/16	.170	.030

All dimensions in inches.

(Continued)

Table 5. *(Continued)* Brown and Sharpe Taper Dimensions.

Taper Number	Taper/ Inch	Socket			Plug Depth P	Tang				
		Socket End to Tang Slot K	Min. Depth of Tapered Hole K			Thickness D	Length F	Radius of Mill G	Diameter O	Radius R
			Drill	Ream						
2	.041833	1.1719	1.3125	1.2500	1.1875	.156	.2500	1/4	.220	.030
3	.041833	1.4688	1.6250	1.5625	1.5000	.188	.3125	5/16	.282	.040
4	.041867	1.2031	1.3750	1.3125	1.2500	.218	.3438	11/32	.320	.050
5	.041800	1.6875	1.8750	1.8125	1.7500	.250	.3750	3/8	.420	.060
6	.041941	2.2969	2.5000	2.4375	2.3750	.281	.4375	7/16	.460	.060
7	.041789	2.4062	3.1250	3.0625	3.0000	.312	.4688	15/32	.560	.070
8	.041750	3.4531	3.6875	3.6250	3.5625	.343	.5000	1/2	.710	.080
9	.041738	3.8750	4.1250	4.0625	4.0000	.375	.5625	9/16	.860	.100
10	.043010	5.5312	5.8125	5.7500	5.6875	.437	.6562	21/32	1.010	.110
11	.041750	6.5938	6.8750	6.8750	6.7500	.437	.6562	21/32	1.210	.130
12	.041644	6.9375	7.2500	7.1875	7.1250	.500	.7500	3/4	1.460	.150
13	.041683	7.5625	7.8750	7.8125	7.7500	.500	.7500	3/4	1.710	.170
14	.041666	8.0312	8.3750	8.3125	8.2500	.562	.8438	27/32	1.960	.190
15	.041666	8.5312	8.8750	8.8125	8.7500	.562	.8438	27/32	2.210	.210
16	.041666	9.0000	9.3750	9.3125	9.2500	.625	.9375	15/16	2.450	.230
17	.041666	-	9.8750	9.8125	9.7500	-	-	-	-	-
18	.041666	-	10.3750	10.3125	10.2500	-	-	-	-	-

All dimensions in inches. Source: ASA B.510-1963, published by the American Society of Mechanical Engineers.

retention knobs can be used in a single holder to accommodate different machine grippers (see *Figure 5*). The CAT V system is the American standard adapter. The BT tapered shank adapter and the ISO 7388 (DIN 69871) adapter share the same taper configuration as the CAT V, but their tool change gripper grooves are unique. These systems also utilize threaded retention knobs, but since they have metric threads, the knobs are not interchangeable with the CAT V system. The BT system is the Japanese standard adapter, and the DIN 69871 is the German standard adapter.

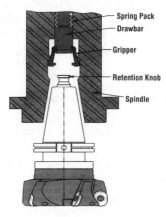

Figure 5. Adapter pullback system.

National Machine Tool Builder Standard (NMTB) tapered shank adapters have two distinct differences from the adapters described above. First, the machine drawbar threads directly into the back of the adapter, rather than attaching onto a retention knob, to secure the tool axially in the machine. The second major difference is that there is no tool change groove, making the NMTBA system impractical for machines with automatic tool changers.

HSK system

HSK (Hollow Shank Kegel) adapters, ("Kegel" is the German word for taper or cone shape) are widely used for high-speed operations. The shank, which is very short and compact when compared to steep taper shanks, is made as a hollow taper with a ratio of 1 to 10 (1.2" per foot *or* 100 mm per meter = taper angle 2° 51' 53"). *Figure 6* demonstrates the size difference between the two systems. There are six different shank configurations, designated by the letters A–F. In addition, each model is identified by the diameter of the shank's flange (in millimeters). Styles A–D were designed for lower machine speeds, and styles E and F were designed for high-speed cutting (these shanks are completely symmetrical, as are the E and F receivers). Most machine tool manufacturers offer machining centers with HSK spindles. *Figure 7* illustrates the six different HSK adapters. German Standards DIN 69063 and DIN 69893 define the specifications of the spindle and shank for HSK tooling, and the essential differences between shank styles are as follows.

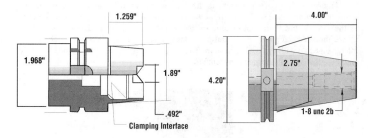

Figure 6. Equivalent sizes of HSK and V-taper adapters. (Source, DAPRA Corp.)

HSK-A – Equipped with automatic tool change provision.

HSK-B – Same features as A, but taper is one size smaller than an A-style shank with a flange of the same specification. The larger flange (shoulder) leaves adequate space on the flange face to locate drive or gripper slots, or to route coolant through the face of the shank.

HSK-C – Same dimensions as A, but exclusively for manual use.

HSK-D – Same dimensions as B, but exclusively for manual use.

HSK-E – Symmetrical, without drive slots. Torque is transmitted from spindle to shank exclusively by friction between the mating surfaces. For high-speed use. Equipped with automatic tool change provision.

HSK-F – Same features as E, but taper is one size smaller than an E-style shank with a flange of the same specification.

The HSK connection relies on a combination of axial clamping forces and taper-shank interference. These adapters simultaneously locate on the socket taper and a flat shoulder face perpendicular to the spindle axis, providing a two-plane fit between holder and machine spindle. At high speeds, the hollow body slightly expands as the rotational speed increases, allowing the adapter to exert pressure on the inner wall of the spindle and assuring dynamic stiffness. Another advantage of the HSK interface is that its mass and weight are reduced by the hollow design, allowing for reductions in centrifugal force at high speeds. HSK adapters are available in both automatic and manual tool change versions. HSK adapters can be retrofitted to existing American Standard spindle noses with intermediate adapters that can be aligned so that runout is close to 0", but, in most cases, these toolholders must be changed manually. See page 806 for additional information on HSK tooling.

Hydraulic and shrink fit adapters

Hydraulic and shrink fit toolholders use hydraulic pressure or thermal contraction to precisely, and automatically, center the cutting tool and minimize runout. They are available in both ISO steep taper and HSK versions. In both applications, pressure is applied evenly all the way around the tool surface, acting as a very accurate collet that essentially automates the tool centering process. Inside the hydraulic toolholder, there is an expanding steel sleeve filled with oil. Turning an actuating screw in the side of the toolholder increases the pressure on the fluid to as much as fifteen tons per square inch, causing the walls of the sleeve to bulge and solidly clamp the tool shank. The metal of the

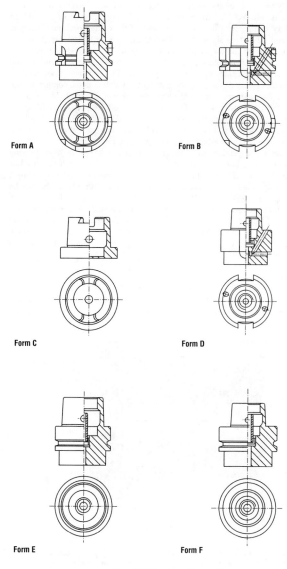

Form A

Form B

Form C

Form D

Form E

Form F

Figure 7. HSK adapters. (Source, Schunk, Inc.)

sleeve expands only within its elastic limits, so it relaxes back to its original dimensions when the pressure is relieved. The diameter of the tool gripping shaft of a shrink fit toolholder is slightly smaller than the shank diameter of the cutting tool. For example, a one-half inch shank fits into a shrink fit toolholder diameter of 0.495". An induction heating system is used to expand the adapter, allowing the tool shank to slip easily inside the adapter and then, as the adapter cools, the contraction exerts a uniform pressure around the surface of the shank. The heating cycle takes approximately 30 seconds. Upon cooling, the shrinking diameter creates a clamping force of up to 10,000 pounds, compared with about 6,000 pounds for a typical spindle drawbar. For removal, the toolholder is again heated and the tool can be easily removed. These toolholders offer especially high rigidity and gripping power and are commonly used for high-speed and high-precision applications. Shrink fit toolholders are less expensive than hydraulic ones, but the cost of the required heating unit makes the purchase of either system a significant investment.

In some instances, milling cutters can be mounted directly to the spindle face, allowing the drive keys and threaded bolt mounting holes on the spindle to match the keyslots and align with the bolt hole pattern on the cutter as shown in *Figure 8*. After inserting a centering plug in the spindle taper, the cutter can then be mounted directly to the spindle face. This mounting system is called "flat back mounting," and it provides maximum rigidity since there is no extension or other overhang from the spindle face. If the flat back drive of a face mill is different than the spindle design, or to allow for quick cutter changes or different types of tooling, an adapter is used to mount the cutter. *Figure 9* pictures the tooling tree for machining centers, and *Figure 10* shows the wide variety of adapters that are available and add versatility to tooling systems.

Balanced tooling

Standard adapters and tooling are normally satisfactory for spindle speeds up to 8,000 rpm. Above that speed, balanced or balanceable tooling is recommended and can be critical for machining within design tolerances and to specified surface finishes. Unbalance in a

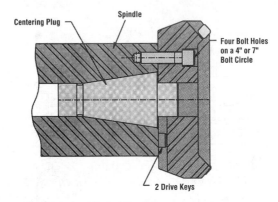

Flat Back Mounting

Figure 8. For cutters eight inches and larger, a centering plug is used to locate the cutter, which is held in place by four bolts on a four- or seven-inch bolt circle. (Source, Kennametal, Inc.)

tool is equal to the tool's mass times its eccentricity. Eccentricity is the distance from the tool's center of rotation to its true center of mass or, more simply, the extent to which the tool's weight is off center (see *Figure 11*). If eccentricity is measured in microns, and tool mass in kilograms, unbalance will be measured in gram-millimeters. Any two sets of mass and eccentricity that yield the same unbalance value will have the same effect on the tool, as long as the unbalance is in the same plane perpendicular to the rotation axis. An inherent unbalance of 6 g-mm would, therefore, be equivalent to attaching a 1-gram weight to the circumference of an otherwise balanced tool, at a distance of 6 mm from the center of mass of this weight to the axis of rotation as shown in *Figure 11*. A tool with the resultant unbalance of 6 g-mm could be put in proper balance by removing 1.2 grams, as long as the center of mass of the removed material was 5 mm from the center of rotation (because $1.2 \times 5 = 6$).

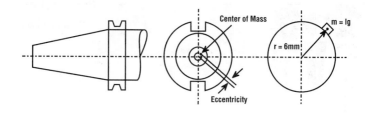

Figure 11. Eccentricity on an adapter shank.

ISO 1940-1, the standard often used to set targets for tool and toolholder balance, uses G-class values intended to establish the suitability of tooling operating at various rpm ranges. The common target value is G2.5, which is discussed below. Equation 1 can be used to determine the G value.

$$G = (u \div m) \times (9549 \div rpm) \qquad \{\text{Eq. 1}\}$$
where G = G-class value, in mm/sec
       u = unbalance, in gram-millimeters
       m = tool mass, in kg
       rpm = spindle speed.

The following equations define the maximum unbalance allowed by a given G-class (Equation 2), and the centrifugal force corresponding to this unbalance (Equation 3).

$$u = m \times ([9549 \times G] \div rpm) \qquad \{\text{Eq. 2}\}$$
$$F = u \times (rpm \div 9549)^2 \qquad \{\text{Eq. 3}\}$$
where F = force from unbalance, in Newtons.

For practical applications, the commonly accepted target G-value of 2.5 establishes a balance requirement that is too aggressive for many processes. To prove this point, Peter Zelinski, in an article in *Modern Machine Shop*, takes the case of a tool/toolholder assembly weighing 1 kg total, to be used at 12,000 rpm. Equation 2 shows that meeting G2.5 would mean balancing to within 2 g-mm. Using Equation 3, it can then be determined that the force corresponding to this maximum allowable unbalance is 3 N. By comparison,

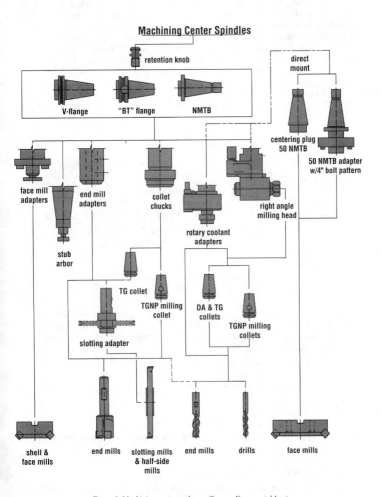

Figure 9. Machining center tool tree. (Source, Kennametal Inc.)

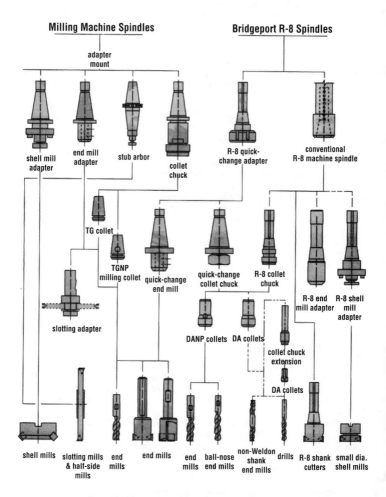

Figure 10. Milling machine adapters. (Source, Kennametal Inc.)

the same tool is likely to experience 100 N of cutting force, even in a relatively shallow cut. Given the total force exerted, the maximum allowable force of 3 N is arguably too strict. Partial blame for these overly strict values can be placed on the origin of ISO 1940-1, which was originally written to apply not to machine tools, but for rigid rotors, particularly those found in power plant turbines. (ANSI standard ANSI S.19-1975, "Balance Quality of Rigid Rotating Bodies," also suggests that toolholders should be compatible with the G2.5 value.) Another weakness of the standard is that it does not accurately predict the way tools behave. Equation 2, for example, states that for a given G class, at constant rpm, increasing the mass of the tool also increases the amount of allowable unbalance. In other words, the standard dictates that balance requirements become less critical as tool weight increases. This is paradoxical, especially to high-speed machining, because balance requirements actually become *more* critical as tool/toolholder weight increases. For example, using Equation 3 it can be calculated that negligible unbalance forces at 1,000 rpm become 100 times greater at 10,000 rpm, and 400 times greater at 20,000 rpm! In the same article, Mark Stover of Sandvik Coromant offers his belief that it is a mistake to set a single numerical target for tool unbalance, and apply it broadly across many unrelated processes. A balance requirement can be considered too strict when it aims to hold the centrifugal force from unbalance to a value significantly lower than the force of the cut, or when it fails to solve the unbalance problem. Balance requirements are best evaluated one process at a time, so long as the process can be repeated enough to make some experimentation worthwhile. Step one is to evaluate whether balancing is warranted. Can the cutting force be determined? If so, compare that force to the one likely to result from an unbalanced tool. To estimate this force, several tool/toolholder assemblies should be measured in search of a maximum realistic unbalance, then the force corresponding to this unbalance should be calculated using Equation 3. So long as the total unbalance force is less than the cutting force, further balancing is unlikely to improve the process. A possible exception to this guideline is an operation with very smooth surface finish requirements.

Taper fit between a tool body and machine spindle can also be a cause of unbalance and runout problems. By specifying toolholders and collet chucks made to the ISO 1947 AT3 standard, and spindles made to ISO 1947 AT2, these problems can be minimized. The ISO standards have twice the accuracy of the American standard related to taper fit, ANSI/ASME B5.10. Commercially available balanceable tooling with precision ground AT-3 tolerance tapers employ several versatile mechanical balancing devices including balance rings with removable counterweights. These rings contain removable counterweights to allow for fine adjustment and to compensate for cutter unbalance. In addition, balanceable locknuts allow for the entire assembly, not just the adapter, to be balanced.

To find the right unbalance target, repeated trial runs of an operation should be performed using tools balanced to a variety of different values, starting with an unbalance of 10 g-mm. After each run, progress to a more balanced tool and repeat until the point is reached where further improvements fail to improve the accuracy or surface finish of the workpiece, or to the point at which the process can easily hold the specified tolerances.

Accuracy and repeatability in machining centers

Two of the key specifications closely scrutinized by potential machine tool buyers are accuracy and repeatability. However, care must be taken because comparing individual manufacturer's specifications can be very misleading. The following information was extracted from a *Modern Machine Shop* article written by Steven Klabunde and Ronald Schmidt of Giddings & Lewis.

There are several positioning standards that provide methods by which an accuracy or repeatability specification can be determined. The same machine tool will give different accuracy and repeatability numbers depending on the standard used. This makes comparison shopping between various machine tools difficult, since various standards and guidelines are in use. Before comparing the standards, it is necessary to define some critical terminology.

Accuracy. Ideally, a machine tool will be at a precise predetermined location every time it is set in motion. Each standard defines how accuracy will be measured. Machine tool builders talk about three kinds of accuracy: unidirectional forward, unidirectional reverse, and bidirectional. These can be compared to a contestant in a shooting contest who first has to walk toward the target while shooting, and then reverse and continue shooting while walking away. The accuracy of the shots made while walking toward the target would be unidirectional forward accuracy, and those made while walking away would be unidirectional reverse accuracy. The overall accuracy would consider the arithmetic average (or mean) of all shots and this would be bidirectional accuracy because both directions are taken into account.

Repeatability. In *Figure 12*, four targets from the shooting contest are shown. In the first (a), the distance between each hit indicates repeatability. Even though all of the shots have landed within the bull's-eye, chances are that the person firing the gun wavered slightly, because the holes are in a pattern. If each bullet had landed on top of the others, making just one hole rather than a pattern of holes, the repeatability would have been perfect. The pattern, however, represents repeatability. Like the holes in target "a," machine tool slides also vary about the target point. On target "b," there is also a tight pattern, which means that repeatability was very good. Even so, the shooter must have pulled slightly to the left with every shot, or the rifle was slightly off to the left, resulting in poor accuracy (no shots landed in the bull's-eye). The shots recorded on target "c" were made while the contestant was walking toward the target, and those on "d" were made while walking away. So "c" indicates forward repeatability, while "d" represents reverse repeatability.

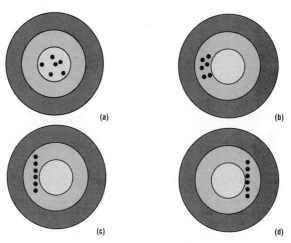

(a)

(b)

(c)

(d)

Figure 12. Repeatability examples.

Repeatability is tested over the total range, so if the bottom and top bullet holes on target "c" are 0.5 inch apart, then this distance represents the repeatability. There is also bidirectional repeatability which takes into account both target "c" and "d," and as a value would be the distance from the bottom bullet hole on target "c" to the top bullet hole on target "d."

Repeatability on a machine tool may be described as forward repeatability, reverse repeatability, bidirectional repeatability, and scatter, and each is calculated differently.

Lost Motion. Lost motion means motion is initiated, but no results can be seen until the lost motion is accounted for—like taking up slack on a block and tackle. The marksman who fired at targets "c" and "d" had lost motion. On "c," all holes are to the left of center, and on "d," all the holes are right of center. The difference between them is lost motion. Other terms for lost motion are reversal error and mean reversal error.

Data Collection. Assume that a machine tool has six locational targets, and positions to each target from an arbitrary starting point designated "zero." If the machine positions to all six targets, reverses and repositions to the same six targets on its way back to zero, and all stops are measured for error (the distance from the stopping point to the target), twelve data references will be collected from each round trip over the targets. If the process is repeated seven times, 84 data references can be collected. From this data (each known as a "sample," and collectively called a "universe" or "population"), accuracy, repeatability, and lost motion specifications can be determined using normal curve calculations. The first step in doing this is to find the mean (arithmetic average) of the 84 samples by adding them all and dividing by 84. The mean will be the central backbone of a bell shaped curve and is represented by the symbol "$\bar{x}$". Next find the standard deviation (sigma, or σ), which is the average distance from the mean. As can be seen in *Figure 13*, which illustrates the normal bell curve, a distance of one standard deviation (one- σ) comprises 68.27% of the curve's area, a distance of two- σ takes 95.45% of the curve, and three- σ occupies 99.73% of the curve. Basically, this indicates that 68.27% of the samples should fall within one (plus and minus) σ, and 99.73% of all of the samples should fall within three (plus and minus) σ. The following formula is used to find the standard deviation of a small size sample such as 84.

$$\Sigma = \sqrt{\Sigma\,(X-\bar{x})^2 \div (n-1)}$$

where Σ indicates "the sum of"
         X is an individual measurement
         $\bar{x}$ is the mean
         n is the total number of measurements. With a large sample, divide by n instead of $n-1$.

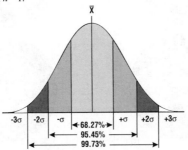

Figure 13. Normal bell shaped curve.

Interpreting the Standards. With the exception of JIS B 6336, each of the standards briefly reviewed below output total values for the values, even though many builders advertise ± values regardless of which standard is used. This may render the appearance of a lower number. NMTBA is the only standard that statistically calculates using bidirectional data. The other statistical standards generate their bidirectional data from the individual forward and reverse information. While this is not necessarily better or worse than NMTBA, it definitely does result in smaller figures that can unfairly give the impression of superior values. NMTBA is, in fact, the most stringent. In summary, any of the standards described below can be used to evaluate a machine tool's linear positioning. However, it should be kept in mind that one standard may allow the machine to appear to be more accurate than it will be when evaluated using another standard.

Table 6 shows comparative data for all six standards based on tests on the same machine. The data is in the inch system of measurement, and all measurements are in μin. (microinches, or millionths of an inch).

NMTBA (U.S.A.) is the standard that has been the reigning monarch of American machining centers for several decades, and in this discussion it will be considered the standard of comparison for the others.

ISO 230-2-1988 (Europe) calculates forward, reverse, and bidirectional repeatability. Its bidirectional values differ from those of NMTBA, because ISO uses the largest of six-sigma (σ) forward and reverse, or the sum of three σ forward plus three σ reverse plus the absolute value of lost motion, whereas NMTBA statistically calculates bidirectional repeatability. ISO also uses the average error for lost motion instead of the point with the largest lost motion (NMTBA). This results in about 40% of the values of NMTBA. Accuracy in ISO is calculated by finding the largest signed number in the $\bar{x} + 3\sigma$ forward and reverse data, then subtracting the smallest signed number from the forward/reverse $\bar{x} + 3\sigma$ data. NMTBA calculates accuracy forward, reverse, and bidirectionally—all statistically. ISO uses only forward and reverse data.

BSI BS 4656 Part 16 (Great Britain) is identical to ISO except for an additional term in BSI called "mean bidirectional position deviation" (D). The D value is the largest average of the forward and reverse averages, less the smallest of the average forward + reverse averages. There is no equivalent to D in NMTBA.

VDI/DGQ 3441 (Germany) defines forward and reverse accuracy and forward and reverse repeatability the same way that NMTBA does. Positioning uncertainty (P) is related to bidirectional accuracy but is calculated using averages and averages of averages. This procedure reduces the actual numbers to about 40% of the NMTBA value (1966 versus 851) in the example shown in **Table 6**. Positional deviation (P_a) is mean-to-mean accuracy: the average of the difference between the average of the forward runs and the average of the reverse runs. Again, this results in a very small number without correlation to any NMTBA value. Positional scatter (P_s) is related to bidirectional repeatability. VDI averages σ from forward and reverse, gets the absolute value, then multiplies the value by six and keeps the largest value. NMTBA statistically calculates the bidirectional σ value, multiplies it by three, and uses this 3 σ value as bidirectional repeatability. Reversal error (U) compares to NMTBA "lost motion" and has the same value. It should be noted that VDI/DGQ 3441 is a guide (recommendation) —not a standard.

JIS B 6336-1986 (Japan) treats accuracy and repeatability differently from most other standards. Accuracy is measured by positioning to equal increments covering almost the entire travel of the slide, once in each direction. The accuracy is the largest error range of *either* the plus or minus direction. For repeatability, the standard requires positioning to a point at one end of travel seven times. This procedure is then repeated at the center of travel and again at the opposite end of travel. The largest error at any of the three points is

Table 6. Numerical Comparisons of Positioning Standards.

Standard	NMTBA	ISO	VDI	BSI	JIS	B5.54
Accuracy						
Forward Accuracy	982	NA	982	NA	NA	NA
Reverse Accuracy	1030	NA	1030	NA	NA	NA
Bidirectional Accuracy	1966	NA	NA	NA	NA	NA
Accuracy	NA	A 1109	NA	A 1109	NA	A 937
Positional Deviation	NA	NA	Pa 449	NA	NA	NA
Positional Uncertainty	NA	NA	P 851	NA	NA	NA
Mean Position Deviation	NA	NA	NA	D 799	NA	NA
Forward Absolute Accuracy	NA	NA	NA	NA	600	NA
Reverse Absolute Accuracy	NA	NA	NA	NA	480	NA
Bidirectional Absolute Accuracy	NA	NA	NA	NA	NA	NA
Repeatability						
Forward Repeatability	182	R ↑ 182	182	RU+182	NA	NA
Reverse Repeatability	280	R ↓ 280	280	RU-280	NA	NA
Bidirectional Repeatability	1966	R 820	NA	RB 820	NA	NA
Positional Scatter	NA	NA	Ps 222	NA	NA	NA
Forward Absolute Repeatability	NA	NA	NA	NA	420	NA
Reverse Absolute Repeatability	NA	NA	NA	NA	NA	NA
Bidirectional Absolute Repeatability	NA	NA	NA	NA	NA	NA
Lost Motion						
Lost Motion	629	NA	NA	NA	180	NA
Mean Reversal Error	NA	B 203	NA	B 203	NA	NA
Reversal Error	NA	NA	U 629	NA	NA	R_{max} 629

saved. The process is then repeated in the opposite direction and the largest error in either direction is divided by two to give a ± value. Lost motion is calculated by positioning to a reference point, continuing in the same direction by a known increment, then reversing by the same increment. The procedure is repeated at both ends and the center of travel. At each location, data are collected seven times, averaged, and the largest of the three averages is called lost motion.

ASME B5.54-92 (U.S.A.) determines accuracy by the range of the average deviations. From the forward $\bar{x}$ the largest and smallest values are used, regardless if in forward or reverse. The standard specifies three round trips. Since the data used to compile **Table 6** was collected on seven round trips, the numbers on the Table reflect a wider range than normal. An additional linear performance attribute can also be obtained from the averages. This figure is the "maximum reversal error" (R_{max}) and is defined as the maximum difference between the forward $\bar{x}$ and the reverse $\bar{x}$, which is the same criterion used by many of the statistical standards. ASME B5.54-92 also calls for bidirectional repeatability to be tested in a fashion similar to JIS 6336. However, since most manufacturers and most laser software calculate the repeatability directly from the accuracy runs, which provides more data that is spread over all target points (compared to the ten bidirectional hits at two locations as called for in the standard), a request for a change in this section of the standard is under review.

Tool measurement and setting

Exact knowledge of a cutter's length, diameter, and profile, and the ability to measure and monitor these dimensions over time, can reduce variability and help optimize the process. Most tool setting is performed off-line in a tool setting gage that verifies the tool's dimensions. The cutter is then loaded into the machine and its dimensional data are read into the control via a bar code or manually. When the cutter is brought within proximity of the workpiece, a gage block is used to establish the workpiece tool offset. This is a proven system, based on a series of static measurements, but it does not translate how the toolholder, drawbar, and spindle are all going to align from tool to tool. Therefore, perfect repeatability between a cutting tool and spindle is rarely achievable.

With laser measurement systems, dynamic measurement of each tool as it rotates in the spindle, within the workzone, can be achieved. As shown in this article by Chris Koepfer, lasers offer two distinct advantages: the beam is basically a "one size fits all" gage, and it is a noncontact system that is virtually impervious to contamination. The laser acts like a high-precision photoelectric beam that is activated when it is broken. When a cutter interrupts the scanner beam, an output is sent to the machine's CNC or PLC, and resident software compares and coordinates the point of beam contact with known values in the control's tool data tables. The scanner can be mounted directly in the cutting zone, on the worktable, or any other convenient location. Its beam, generated by a diode laser, is a visible red light, and power used to run the system is typically less than one Watt, and special operator eye protection is not normally required.

Use of the laser as a measuring device provides a flexible tool for determining cutter dimensions. It can measure concentric tools, such as drills, or nonconcentric tools, such as end mills and face mills. It can measure single cutting edge radii as found on ball nose end mills and periodically recheck them for wear.

The laser scanner can accommodate very small diameters. Cutters with a diameter down to 0.012 inch (0.3 mm) can be measured with accuracy within a few microns.

Larger tools as well can be accurately measured without any changeover of hardware or software. Basically one scanner fits all cutters.

Before the laser can be used to measure, it must be calibrated. This process is necessary to determine the precise trigger point position for the laser light beam and must be established for each axis. Calibrating the laser scanner is a relatively simple procedure. A cylindrical reference-pin with two known diameters and a known radius is used for the procedure. The known dimensions of length, height, and radius are entered into the calibration table in the machine control, as shown in *Figure 14*.

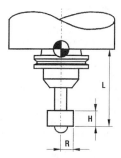

Figure 14. Calibrating the laser system.

Passing the spindle-mounted reference-pin through the beam establishes the measurements in space. Deviations from the known measurement are established as axis offsets. With these values in the tool table, any cutter can be accurately measured. For example, center-cutting tools such as drills are quickly measured for length with two passes through the beam in the Z-axis. First, at rapid traverse with the spindle at operational rotation, the drill breaks the beam. This triggers an axis retract which is followed by a second Z-axis fine feed to a point where the cutter tip breaks the beam. This establishes the actual length of the cutter in the spindle. It all takes a few seconds. The measurement result is written under the tool number into the tool-offset table in the CNC. The measurement routine is an automatic macro built into the measurement software.

Cutting tools such as end mills, face mills, milling cutters, and others with unknown length and diameters, can be measured at spindle speed using the laser scanning gage. Tool length is measured first by moving the cutter periphery into the laser beam with a Z-axis move. Next, the diameter is calculated by moving one side of the cutter through the beam. Diameter and length measurements can be made on a variety of cutter configurations. *Figure 15* shows two points being measured to determine the length and diameter of the tools. The endmill on the right is rotating as the measurement cycle is run.

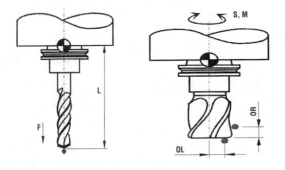

Figure 15. Measuring tool features.

Because the tool is rotating and the laser is triggered by the first cutting edge that breaks the beam, any misalignment of inserts or high edges can be identified. With this data, routines in the software can calculate spindle runout where each cutting edge is monitored for being within a programmed runout tolerance. Measuring accuracy of ±1 micron can be achieved using these routines and is accomplished within a few seconds. When measuring cutter diameter, the laser scanner simply detects the first edge to break the beam. With the cutter rotating, this is the maximum or longest cutting edge.

In order to monitor radial runout, which can be caused by errors in tool clamping and flight circle deviations, or to monitor the radial tolerance between the maximum and minimum cutting edges, the cutter runs into the beam to a depth equal to the tolerance. Once in the beam, the cutter is moved through the beam taking what is in effect a scanning stroke. If the beam remains broken during this routine, the tool is in radial runout tolerance.

Cutters with rounded edges or forms such as ball nose end mills or taper tools can be measured with a laser scanner. This is accomplished by establishing the largest permissible outside cutter edge, the cutting edge outline, and the starting and aiming point of the contour. These are established using parameters in the measurement software.

Tool breakage detection is an important asset of a tool measurement and monitoring system. The fine measurement resolution of the laser makes it possible to determine the condition of individual cutting edges. Wear, chipping, or fracture of each cutting tooth can be determined quickly and checked as often as necessary. As shown in *Figure 16*, the optimum contour on the outside of a ball nose end mill's radii, as well as the minimum deviation on the inside, can be checked to determine that it remains within tolerance.

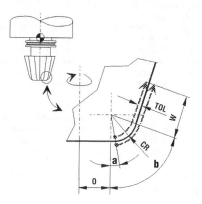

Figure 16. Monitoring the cutting edge of a tool.

All machine tools have temperature drift to some extent. Part of the machine tool designer's job is to understand and anticipate this fact when putting together the components, which comprise the machine. In most machining operations, the dimensional effects of thermal growth are gradual and, over time, predictable. The laser system for measuring tools can be used as a thermal sensing gage for the machine tool. To accomplish this, a shop uses the laser calibration tool to check each machine axis for growth. The tool has known dimensions in X, Y, and Z axes. Comparing laser measurements taken over time can help a shop detect axis drift and compensate for it.

Rotary and linear encoders

Machine tools use encoders to measure their own movements to keep cutting tools on target. The following article by Rick Korte of Heidenhain Corporation appeared in *Modern Machine Shop* magazine.

Machines that move need a means of measuring movement. Since the machine tools, inspection machines, material handling equipment, and the like have themselves evolved from basic rudimentary manual machines to highly sophisticated automated pieces, so have the internal measuring mechanisms. The most common type of measurement component today is the encoder.

Encoders can be generally categorized into optical (photoelectric), magnetic, and mechanical contact types. Photoelectric encoders in particular—due to their high accuracy, high reliability, and relatively low cost—play a significant role in machine tool technology.

There are two basic types of encoders: rotary and linear. While the technical principles behind them are similar, their specific applications most often are not. And while the basic principle of operation developed many years ago is still the basis of today's encoders, a revision of even that technology, highlighted for especially high accuracy needs, is now available.

The basic principle

Most of today's linear and rotary encoders operate on the principle of the photo-electrical scanning of very fine gratings.

The so-called scanning unit in an encoder consists of a light source, a condenser lens for collimating the light beam, the scanning reticle with the index gratings, and silicon photovoltaic cells. When the scale is moved relative to the scanning unit, the lines of the scale coincide alternately with the lines or spaces in the index grating. The periodic fluctuation of light intensity is converted by photovoltaic cells into electrical signals. These signals result from the averaging of a large number of lines. The output signals are two sinusoidal signals that are then interpolated or digitized as necessary.

Rotary encoders

In various sectors of machine technology, angular positions and angular motions need to be transduced into electrical signals, either for display, automation, or numerical control. Rotary encoders are used for this purpose of measuring rotational movement drives. They are also often used in measuring linear movements, for example, when used with spindles and especially with recirculating ballscrews.

The most significant characteristics of rotary encoders are summarized in **Table 7**.

Table 7. Categories of Rotary Encoders.

Incremental Rotary Encoders		Absolute Rotary Encoders	
Rotary encoders for measurement of angles	Rotary encoders for measurement of linear position (used in conjunction with measuring ballscrews or rack and pinion)	Absolute rotary encoders for one rotation (single-turn rotary encoders for measurement of angles)	Absolute rotary encoders for several rotations (multi-turn rotary encoders, mainly used with measuring ballscrews)

Incremental rotary encoders. The output signals of *incremental* rotary encoders are evaluated by an electronic counter in which the measured value is determined by counting "increments." These encoders form the majority of all rotary encoders being used today.

Incremental rotary encoders for *length measurement* are usually found on slides of machines and fixtures that are generally driven by leadscrews. The encoders are normally positioned and coupled to the rear of the servomotor or to the screw and opposite to the drive via a backlash-free precision coupling.

Standard rotary encoders for general length measurement applications—and in particular the measurement of slide movements using a recirculating ballscrew as the scale—are represented by a shaft encoder that incorporates digitizing electronics. The number of square-wave cycles of the output signals per shaft rotation is identical to the number of lines on the graduation disc.

Incremental rotary encoders with *integral couplings* used for length measurement are also available. This design features some favorable characteristics, especially the version in which the coupling is not mounted on the rotor side, that is, between the driving spindle or motor shaft and the encoder shaft, but is permanently fixed to the stator. In these cases, the spindle or motor shaft is directly connected to the shaft of the rotary encoder. The scanning unit is connected to the encoder shaft by ball bearings, however, without a rigid connection to the housing. Instead, a coupling is located between these components and compensates for alignment errors between both shafts.

All of these rotary encoder versions are, in principle, angular measuring systems and are, providing accuracy requirements are fulfilled, used in many cases for this purpose. The resolution of such encoders can be increased by means of electronic interpolation. There are, of course, the precision rotary encoders specifically designed for *angle measurement.*

If finer resolution is required, standard rotary encoders often utilize electronic signal interpolation. Rotary encoders for applications in dividing heads and rotary tables, with very small measuring steps (down to 0.36 arc second), have in principle the same basic design features as standard rotary encoders, but incorporate some overall varying construction.

Absolute rotary encoders. Absolute rotary encoders provide an angular position value which is derived from the pattern of the coded disc. The code signal is processed within a computer or in a numerical control. After system switch-on, such as following a power interruption, the position value is immediately available. Since these encoder types require more sophisticated optics and electronics than incremental versions, a higher price is normally to be expected.

The most commonly used coder is the Gray coder, a unit-distance coder where only one single bit changes with the transition from one measuring step to the next. Natural binary coder is also frequently used for very high resolutions. With this code, more than one signal may change when going from one measuring step to the next. Precautions must therefore be taken to avoid ambiguities.

Apart from these two codes, a range of other codes have been employed, though they are losing their significance since modern computer programs usually are based on the binary system for reasons of high speed.

There are many versions of absolute encoders available today, such as single-turn or multi-stage versions to name only two, and each must be evaluated based on its intended application.

Linear encoders

As the present trend of machine tools evolves toward increasingly higher accuracy and resolution, increased reliability and speeds, and more efficient working ranges, so too must feedback systems. Currently, *linear* feedback systems are available that will achieve resolutions in the submicron range.

Submicron resolutions, for example, are required in the semiconductor industry and in ultra-precision machining. Achieving these resolutions is possible with the use of linear scales that transmit displacement information directly to a digital readout, NC controller, or other peripheral device for display or evaluation.

As in rotary, linear scales operate on the same photoelectric scanning principle, but the linear scales are comprised in an overall straight construction, and their output signals are interpolated or digitized differently in a direct manner. One of these signals is always used by the accompanying digital readout or numerical control to determine and establish home position on the linear machine axis in case of a power interruption or for workpiece referencing.

Overall, there are two physical versions of a linear scale: exposed or enclosed. The versions themselves dictate, fairly accurately, the type of application.

With an *enclosed* or "sealed" scale, the scanning unit is mounted on a small carriage guided by ball bearings along the glass scale; the carriage is connected to the machine slide by a backlash-free coupling that compensates for alignment errors between the scale and the machine tool guideways. A set of sealing lips protects the scale from contamination. The typical applications for the enclosed linear encoders are primarily machine tools and cutting type machines, or they may be any type of machine located in harsh environments.

Exposed linear encoders also consist of a glass scale and scanning unit, but the two components are physically separated. The typical advantages of the noncontact system are easier mounting and higher traversing speeds since no contact or friction between the scanning unit and scale exists. Exposed linear scales can be found in coordinate measuring machines, translation stages, and material handling equipment.

Another version of the scale and scanning unit arrangement is one that uses a metal base rather than glass for the scale. With a metal scale, the line grating is a deposit of highly reflective material such as gold that reflects light back to the scanning unit onto the photovoltaic cells. The advantage of this type of scale is that it can be manufactured in extremely great lengths, up to 30 meters, for larger machines. Glass scales are limited in length, typically three meters.

There are several mechanical considerations that need to be understood when discussing linear encoders. It is not a simple matter to select an encoder based just on length or dimensional profile and install the encoder onto a machine. These characteristic considerations include permissible traversing speeds, accuracy and resolution requirements, thermal behavior and mounting guidelines.

Interferential scanning principle

The newest linear encoders operate on a principle of a unique diffraction of light waves to obtain very fine measuring steps, and in turn offer machinists an even rarer combination of high accuracy, extremely fine resolution, and excellent repeatability of the measuring values not found in any other optical system before.

The semiconductor light source together with a condenser produces a plane light wave, which is represented by a beam normal to the wave front. While striking the scanning grating A, the plane wave is essentially diffracted into three directions. At the phase grating of the scale M, the light is reflected and diffracted once again. After another diffraction series, they come to interference. Three interfering unidirectional light beams are collected by the lens and projected onto three solar cells that convert the light intensities into electrical signals. Since the interferential scanning signals are inherently free of harmonic distortions, the resulting signal periods can undergo a high subdivision.

Measuring systems that operate on the interferential scanning principle combine the advantages of a very fine grating (high resolution and accuracy) with a wide tolerance for the gap between scanner and scale. Due to its high accuracy, extremely fine resolution, and, above all, the excellent repeatability of the measuring values, this device can be considered an alternative to the laser interferometer for shorter distances (less than or equal to one meter). Moreover, a measuring system having a steel scale such as that of an interferential scanning setup is better matched to the thermal expansion of a steel workpiece. Therefore, in many cases, it proved superior to a laser interferometer.

Principles of 5-axis Machining

This section, and the following one on high-speed machining, were prepared by Miles Arnone, author of *High Performance Machining* (Hanser Gardner Publications).

Five-axis machining has long been applied within specialized machining operations. The overwhelming majority of 5-axis machining centers are producing components for the aerospace and other turbo-machinery industries (i.e., steam turbines, air separation, and high pressure or flow pumps). To a lesser extent, 5-axis machining technology has been employed for the production of specialized medical components (joint implants, heart valves), cams, tire molds, and the porting of inlet and exhaust ports on certain racing engine heads. As the technology associated with 5-axis machining—machines, CAM, tooling, process methodology—becomes increasingly well developed and standardized, 5-axis machining is entering the mainstream of process choices for a broad range of industries.

Over the next ten years, 5-axis machining (5M) processes will most rapidly penetrate the mold and die, and the general machining industries. This article explores the reasons for the accelerating popularity of 5-axis machining, and provides a road map for shops considering its adoption. Specifically, the following topics will be addressed.

- Introduction to 5-axis machining concepts
- Motivation for the adoption of 5-axis machining in traditionally 3-axis applications
- Benefits of 5M technology
- Drawbacks of 5M technology
- Machine selection for 5M technology
- CAM selection for 5M technology

Introduction to 5-axis machining concepts

In 3-axis machine tools, the relative *orientation* of the cutting tool and the workpiece is fixed; only the *position* of the cutter relative to the part changes. In a vertical 3-axis machining center, the cutter is always perpendicular to the surface of the machine table. On a horizontal 3-axis machining center, the cutter axis is always parallel to the surface of the machine table. The cutter is translated in X, Y, and Z on such machines.

A 5-axis machine also translates the cutting tool along the X-, Y-, and Z-axes. But it also controls the orientation of the cutter relative to the workpiece with respect to two of these axes. In other words, a 5-axis machine supports both translation and rotation. These rotations can occur around any two of the three axes ({X,Y} {Y,Z} {X, Z}), depending upon the design of the 5-axis machine. Whether the cutter or workpiece is rotated about two axes, or each is rotated about a single axis, depends upon the design of the machine in question and the requirements of the application. The possible rotations undertaken by a 5-axis machine are labeled A, B, or C. "A" corresponds to rotation about the X-axis, "B" corresponds to rotation about the Y-axis, and "C" corresponds to rotation about the Z-axis. Only two of these three rotations will be undertaken by a given 5-axis machine configuration.

5-axis machines provide either 5-axis positioning capability or 5-axis contouring capability. The distinction lies in the capabilities of the rotational axes. In 5-axis positioning, the two rotational axes are capable only of indexing. Their motions cannot be coordinated with the linear {X, Y, Z} axes. When using such a machine, the operational sequence is as follows. 1) Stop all machine motion. 2) Unlock rotary axes' brakes. 3) Rotate rotary axes to new angular positions (may be tool, workpiece, or both). 4) Lock rotary axes' brakes. 5) Resume translational {X, Y, Z} motion. *Figure 1* illustrates the resulting tool path for such a process. Note that while the tool is in motion over the part surface, the orientation

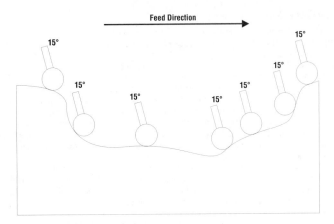

Figure 1. When the orientation of the cutter is fixed relative to the part (here at 15°), the contact angle between the part and the tool varies as the cutter moves across the surface.

of the cutter to the part is fixed. Essentially, 5-axis positioning machines are only capable of providing simultaneous coordinated motion with three linear axes. As a result, 5-axis positioning machines can only machine 3-axis features. The principal benefit of such machines is that they can access multiple sides of a workpiece in one setup, produce undercuts, and improve access to difficult-to-reach areas.

Figure 2 illustrates a part that should be produced on a 5-axis machine with at least positioning capability. The part consists of 8 panels around the circumference of the part, each panel having a curved surface and several hundred holes. For this part, we require three linear axes to move the tool with respect to the part. We must also rotate the part around its major axis to present each panel to the cutting tool (axis number four). Then, we must rotate the cutting tool so that each hole can be produced normal to the panel

surface (axis number five). If the holes did not need to be normal to the part surface, we could produce this part with a 4-axis machine. Note that if we were willing to refixture the part after machining each panel, we could produce the curved panel surfaces (less the holes) using a ball end mill on a 3-axis machine. This part could also be produced on a machine with 5-axis contouring capability, but it is not required.

A machine capable of 5-axis simultaneous contouring is required when the orientation of the cutter must change at the <u>same time</u> as the position of the cutter changes.[1] For example, consider repairs to a titanium impeller. A weld bead will be built up on the edge of each blade edge as part of a repair process. To complete the repair, the weld and blades must be flank milled to reestablish the original part geometry.

Figure 2. Production of this graphite part requires a 5–axis machine, because each of the holes is positioned normal to the surface of the part.

---

[1] From here forward, changes will be described in orientation of the cutter. This implies "changes in orientation between the cutter and workpiece." This does not exclude workpiece rotation.

To access the weld along the entire edge of each blade—while maintaining the desired contact angle between the tool and workpiece—the part *and* tool must be rotated as the tool is translated in three axes. This part could not be produced using only 5-axis positioning.

The geometry of parts that require 5-axis simultaneous contouring falls into two broad categories: ruled surfaces and arbitrary surfaces. Ruled surfaces can be described by sweeping a line through space. For example, if one were to hold a pencil and move it through the air, the surface swept out by the pencil would be a ruled surface. At any given point on the surface, once could extend a line from one edge of the surface to the other. Ruled surfaces are easier to machine because they allow a broader range of machining strategies. In addition to point milling, where only the tip of a ball-, bull- or tapered ball end mill is used, flank milling can be employed. Flank milling, which allows the side of a tapered or cylindrical end mill to be used, can significantly reduce overall cycle times, although the process is often less robust than a point milling process. Many 5-axis parts contain both ruled and arbitrary surface. For example, the blades on an impeller may be ruled surfaces (allowing flank milling), but the hub and the radii connecting the blades to the hub are often arbitrary surfaces that require point milling.

5-axis machining is mandatory for a wide range of parts, and specifically those for which simultaneous 2-axis angular rotation and 3-axis translation of the tool is required. While there are many such parts, there are many more parts for which 5-axis machining is "optional" but highly beneficial. These parts include those with geometry that can be produced using only three translational axes, but that would require many refixturings to properly present each surface to the cutter. Parts with 3-D contoured surfaces that do not explicitly require 5-axis machining (such as mold cavities, electrodes, etc.) are also prime candidates for 5-axis machining, because of the benefits that 5-axis processes provide through improved access and cutter orientation control. These benefits are described in more detail below.

Motivation for the adoption of 5M in traditionally 3-axis applications

In the past ten to fifteen years, there has been a dramatic increase in the complexity of machined parts. Specifically, the amount of curvature—ruled, but increasingly arbitrary—has grown rapidly. This growth has been driven by the advent of improved CAD technology, the explosive growth in cost-effective computer processing speed and data storage, and the increasingly competitive global market. Ten years ago, 3-D solid modeling and analysis was the province of larger firms with large engineering budgets and Ph.D. level staffs. Now, 3-D design and manufacturing tools are widely available. Any organization that wants to design 3-D surfaces into their parts can do so as easily as it was to use 2.5-D ten years ago. The widespread availability of this technology has allowed engineers to radically improve their designs, both aesthetically and functionally. Engineers in all industries can now design structures and surfaces to optimize stress fields, weight and inertia properties, thermal characteristics, etc. At the same time, designers can bring ever more complex and "futuristic" shapes into the mainstream of consumer and industrial products. Not convinced? Compare the dashboard of a 1985 pickup truck to a 2000 model. While the advent of more complex curved surfaces is good for design engineers and the consumer, it makes life much more difficult for the manufacturing staff! 5-axis machine tools allow manufacturing organizations to keep pace with the changes taking place in the engineering side of the house.

While the advancement of technology in computers and CAD has made life more difficult for manufacturing (through the creation of more complex parts), the world of CAM and manufacturing tools has similarly advanced. The capabilities of today's CNC and CAM systems are such that it is much easier to adopt 5-axis machining technology than it was in the late 1980s or early 1990s. While CAM advancements will forever lag behind those in CAD—almost by definition CAM must play catch-up to the design side of the house—manufacturing technology has been by no means standing still. Whereas many firms using 5-axis technology ten years ago had to develop their own CAM tools in-house, now there are a wide range of commercial solutions. It is true that commercially available 5-axis CAM packages are of widely varying quality, and that 5-axis tool path generation is still much more complex than 3-axis tool path generation. Nonetheless, the state of the art has improved dramatically.

Another motivation for adopting 5-axis machine tools is the increased need for automated production processes. Talented machinists and process engineers are harder and harder to find as other sectors of the economy, notably computer hardware & software and service industries, attract a greater proportion of young people entering the job market. In many countries this effect is compounded by severe restrictions on the length of the workweek. In many European countries, for example, the workweek is not allowed to extend beyond 35 hours! While the efficacy of such laws, which are intended to boost overall employment, can certainly be debated, the reality for the machine shop owner is clear. Workers, let alone good workers, are hard to find! Further, because the capital equipment required to produce today's complex parts is increasingly expensive, shop owners must achieve higher utilization of their machines. This often means there is a need to run machine tools 24-hours per day, 7-days per week, even in industries such as mold and die that did not traditionally require such an approach (as compared to mass production facilities). The desire to run 24/7 is accentuated by the need to achieve shorter lead times. Highly automated machining processes are required to reduce labor content, shorten cycle times, improve quality, and to cut costs. As part of a comprehensive approach to automation, 5-axis machine tools enable a high degree of unattended operation. 5-axis machines eliminate setups and enable the use of more robust machining processes than can be achieved with conventional 3-axis machine tools.

One of the greatest opportunities for machining operations to reduce costs, regardless of the level of automation pursued, is to eliminate nonvalue added work and queue time. Five-axis machine tools can significantly reduce both setup time and the number of different machines required to produce a complete part. The latter directly drives the percentage of time a part will spend in queues as opposed to being worked on. This in turn affects total product lead time.

The increasing popularity of difficult-to-machine materials further strengthens the case for adopting 5M technology. For example, in mold and die, it is often desirable to machine mold cavities and inserts directly into hardened steel (Rc 50-62), rather than using EDM. Hard steel machining processes are clearly less robust than graphite machining for electrodes, and so the improved process control afforded by 5-axis can be extremely beneficial. The same can be said for firms whose parts are made of titanium or high nickel content alloys, for example.

Benefits of 5-axis machining technology

This section will review the benefits of 5-axis machining in cases where 5-axis machining is not a mandatory requirement to produce the part.

One of the most readily apparent benefits of 5-axis machining technology is the ability to access five sides of a part in a single setup. As illustrated by the part in *Figure 2*, 5-axis positioning can eliminate a large number setups. In this example, seven setups were eliminated for production of the panels, plus countless more for the hole drilling operations. A prototype for a shoe sole can also demonstrate how 5-axis machining can

Figure 3. Prototype shoe sole pattern produced on a 5–axis machining center. Both the bottom of the sole and the bead on the side were machined with the same setup.

eliminate part setups (see *Figure 3*). Previously, the sole pattern may have been machined in a single setup using a Z-level cutting approach with a 3-axis machine. The semicircular "bead" running around the side wall of the shoe would have to be produced by hand. If it had been produced using a 3-axis machine with a ball end mill, the width of the ensuing groove would increase as the Z-axis moved up or down. With a 5-axis machine, the sole profile and the bead can be produced in the same setup. Each of the plateaus on the sole has tapered walls between the top of the plateau and the floor of the sole. In the previous 3-axis machining process, these tapered walls had to be produced by Z-level cutting around each plateau. With a 5-axis approach, these ruled surfaces were produced in a single pass. Total 5-axis machining time for this pattern was 1.3 hours, significantly less than when using a 3-axis approach or an alternative process such as stereolithography. The 5-axis process provided similar benefits when producing the female version of the sole for an aluminum mold.

5-axis machine tools can also greatly improve access to the work. 5-axis machines improve access to deep cavities while reducing the length and aspect ratio of tool assemblies. *Figure 4* illustrates this effect for an aluminum cavity used for molding plastic cake and pastry covers as might be seen in a supermarket. When using a 3-axis approach, a 6" long cutter is necessary. Because the walls of the cavity are so steep, the cutter body must be narrow so that the corner radius between the floor and walls can be produced. The lack of stiffness in this assembly increases cycle time and results in a lower quality finish. With a 5-axis machine, the overall extension of the cutter from the holder is reduced to 1", and the overall length of the tool assembly from gage line is reduced from 6" to 4". Further, the cross section of the tool assembly is such that significantly greater stiffness is achieved. The result is a better part in significantly less time.

The greatest potential benefit of 5-axis machining, particularly as it relates to 3-D contoured surfaces as typical of mold and die components, is tool orientation control. This benefit is of greatest value when working with difficult-to-machine materials, where tool breakage or chatter are significant problems. As previously discussed, the orientation of the cutting tool relative to the workpiece is fixed when working with a 3-axis machine. *Figure 5* illustrates that if the orientation of the tool is fixed, the point of contact between

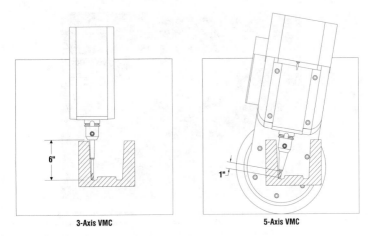

3-Axis VMC   **5-Axis VMC**

Figure 4. The use of 5–axis machine tools improves access to concave shapes such as this mold cavity while employing shorter, stiffer tooling.

the tool and workpiece changes as a function of the local geometry of the part. This is a problem because the surface speed of a ball mill cutter depends upon the contact angle between the tool and the surface. When the bottom of the ball mill is in contact with the part, the surface speed is zero. The surface speed is at maximum when the tool contacts the part at its equator. So, with a 3-axis machine, as the cutter moves across the surface the cutting conditions vary widely. This variation in surface speed causes: 1) variations in surface finish, 2) premature and unpredictable tool wear (particularly if the bottom of the tool is in constant contact with the workpiece), and 3) lower surface speeds and feed rates than could be achieved if the angle between the cutter and local workpiece geometry were actively controlled.

Cutter orientation can be controlled statically or actively. Static orientation involves setting the tool angle before bringing the tool and workpiece into contact. Essentially, static orientation is a variant of 5-axis positioning. Once the tool is positioned, it is held

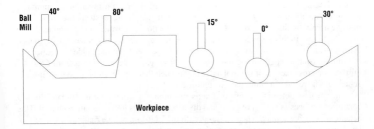

Figure 5. When using a 3–axis machining center, the contact angle (θ) between a ball end mill and the workpiece depends on the local part geometry. As θ varies between 0° and 90°, surface speed increases from zero to the maximum value.

Figure 6. Machining of this steel insert for a plastic lens mold benefits from orienting the part at an angle relative to the spindle.

at a fixed angle for the duration of machining. (*Figure 1* illustrated this effect). Static orientation provides the greatest benefit when the surface to be machined is fairly regular. If the surface normal of the part changes significantly, the angle between the tool (which is at a fixed angle) and the part will vary excessively, and the benefit will be minimal. *Figure 6* illustrates a part for which static tool orientation control would be effective. This part, an insert for a headlamp mold, has a relatively slight amount of curvature, and this curvature is fairly close to the horizontal plane. Therefore, if we only used a 3-axis machine, the bottom of the ball end mill would be in contact with much of this part. If, on the other hand, the tool were cocked at a 45° angle relative to the part, for example, the surface speed, and therefore the feed rate, could be much higher. A better surface finish and lower cycle time would result. If the angle was selected properly, the bottom of the tool would never make contact with the part surface.

Active tool orientation involves altering the angle of the tool relative to the machine axes in real time. Put another way, active tool orientation maintains a constant angular relationship between the cutter and the workpiece surface normal at all times. As a result, the effective surface speed of a ball end mill (for example) can be held constant, allowing higher programmed feed rates and better surface finishes in less time. *Figure 7* illustrates the process.

Tool orientation control can also be used with flat bottom or bull-nose end mills. Typically, the orientation of these mills will be controlled so as to maintain a 5° or 10° angle between the cutter axis and the plane that is normal both to the surface tangent plane and the direction of motion plane (*Figure 8*). By angling the tool in this manner, the effective radius of the tool, as seen by the part, becomes much larger than a ball end mill of the same diameter. This reduces the scallop height significantly, improving surface finish, and reducing the number of tool passes required. Tool orientation control can also optimize machining processes by enabling a constant depth-of-cut to be achieved, and adjusting the machining process to account for variations in the stock-on condition of the part to maintain constant load on the cutter.

As previously discussed, 5-axis machining eliminates the need for multiple discrete setups, either on the same machine or across multiple machines. Besides eliminating queue times, this allows tighter tolerances to be achieved between features on a single part. Significant inaccuracies are introduced each time a part must be refixtured, even if this refixturing takes place on the same machine. If a part can be fixtured once, and machined complete, tolerances can be improved by several ten-thousandths of an inch. The machine shown in *Figure 9* illustrates the epitome of the "single fixturing" approach enabled by

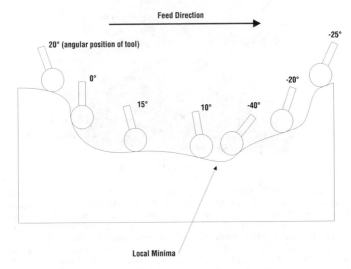

Figure 7. Actively controlling the angle of the ball mill allows it to make contact with the workpiece at a larger effective radius, increasing the surface speed.

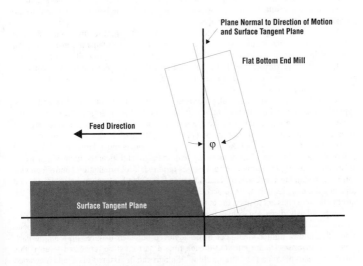

Figure 8. Active control of tool orientation relative to the direction of motion and the surface provides better surface finishes.

Figure 9. Bostomatic 405G machining center outfitted for producing mold components in steel, copper, and graphite.

5-axis machine tools. This 5-axis machine has been outfitted for the production of copper or graphite electrodes as well as hard steel components. A robot automatically loads parts into a high precision receiver mounted to an A-axis. This machine, which utilizes a B-axis (rotation about Y), can be configured with its second rotary axis as either an A-axis (pictured: rotation about X) or C-axis (rotation about Z).

Throughout this section we have addressed 5-axis machine tools independent of high speed machining. 5-axis machine tool technology is inherently compatible with high speed machining principles in that it supports the goal of reducing the door-to-door time (i.e., the time between when an order for a part is received and when it is finally shipped) for machined parts. It further supports high speed machining processes by enabling better parts to be produced in less time. That said, the benefits accrued through 5-axis machining apply equally well to so-called conventional (i.e., not high speed) machining processes.

Finally, besides improving part quality and throughput, 5-axis technology can help a machining operation to differentiate itself from competitors. A shop that develops proficiency with 5-axis machine tools will be able to offer new capabilities and services to its customers by 1) increasing the design space available to engineers, 2) reducing lead times, and 3) improving part quality (as measured by surface finish and tolerance).

Drawbacks of 5-axis machining technology

As with any technology, 5-axis technology is not without weaknesses. To begin, the adoption of 5-axis machining is far from "plug and play." As a result, it is not as widely accessible nor as widely adopted as it should be. Fortunately, 5-axis machining is increasingly better understood and is much easier to adopt today than it was ten years ago. The greatest difficulties lie with development of 5-axis machining processes, and the ability of CAM systems to support 5-axis machining. Because 5-axis machining strategy must be tailored to the specific part family of interest, it is difficult to find a CAM system that excels across a wide spectrum of 5-axis applications. Whereas one system may be ideal for tire molds, that same system may prove wholly inadequate for head porting or mold cavity machining. The CAM manufacturer's task is made more difficult by the wide range of 5-axis machine geometries. Unlike 3-axis machines, each 5-axis machine geometry requires a tool path that is matched to its construction. This is, unfortunately, not simply a matter of syntax, as it is when posting a tool path for one 3-axis machine's CNC or another.

5-axis technology is not cheap. Beyond the more sophisticated CAM system, the machines themselves cost more. To begin, there are two additional degrees of freedom (axes) that must be integrated into the machine both mechanically and electrically. A more capable CNC is required to coordinate all five axes as well, particularly when simultaneous 5-axis machining is undertaken. Even a relatively inaccurate 5-axis solution, consisting of a double-tilt rotary table mounted on a conventional 3-axis machining center, can add $60,000 to the cost of a machine tool. A high performance 5-axis machining center, aimed at producing small to mid-sized parts to tolerances of approximately 0.0005", can cost over $120,000 more than a 3-axis machine capable of the same tolerances.

Another drawback of 5-axis machine tools is that they are inherently less accurate than a 3-axis machine built using comparable construction methods and components. The additional two axes each contain a measure of inaccuracy and compliance. As a result, the net accuracy of the machine must be lower than a 3-axis machine built in the same way. For example, state-of-the-art 5-axis airfoil machining processes result in part tolerances typically no better than 0.0008". Compare this to what the best mold makers achieve on complex electrodes or cavities when working with high performance 3-axis VMCs (around 0.0004" total deviation over the entire surface).

The addition of two more axes means that 5-axis machines are less stiff than comparably constructed 3-axis machines. For example, consider a small 3-axis and 5-axis VMC built around the same basic design. The 5-axis design adds a rotary A-axis mounted on the table, and a B-axis built into the Z-axis slide. While the Z-axis stiffness of the 3-axis machine is 66,300 lb×ft/in., the Z-axis stiffness of the 5-axis machine is only 24,500 lb×ft/in. or 37% that of the 3-axis machine.

5M is a higher risk, higher yield technology than so-called "high speed" or "conventional" 3-axis VMCs. If 3-axis machine tools are bonds, then 5-axis machines are stocks. To successfully adopt 5M, a shop should have a strong infrastructure of machining and CAM expertise in-house. The shop should also expect a steep learning curve associated with adopting 5M. If a shop believes it can purchase a 5-axis machine tool and begin utilizing it to its full capabilities immediately, problems will ensue. Purchase a 5-axis machine with the understanding that initially it will be used in a manner very similar to 3-axis machines. As the shop staff becomes more comfortable with the machine itself and the development of 5-axis processes, significant benefits will accrue. This learning process is not only unavoidable, it is essential to building a competitive advantage relative to one's competition.

Machine selection for 5-axis machining technology

In 5-axis machine tools, a number of different geometries have evolved, including compound rotary tables, compound spindle heads, pendulum head with rotary table, and tilting spindle with rotary table.

Compound Rotary Tables. Compound tables are the most straightforward and economical method of achieving high performance 5-axis machining. While these tables can be installed upon 3-axis machine tools, they are less stiff and possess higher angular errors than more integrated approaches. Because the rotary axes are stacked on top of one another, angular errors are magnified. Also, the large stack-up between workpiece and rotary axes when using a double tilt rotary means that high torque loads are placed upon the gear sets, reducing life and increasing deflection. The nature of the compound rotary also requires large axis travels to maintain contact between the workpiece and tool tip. This further increases errors and limits effective material removal rates. Finally, a large machine tool is required relative to the part size when using a compound rotary table.

Compound rotary tables are best suited to 5-axis positioning (v. contouring) applications, and 5-axis contouring of less demanding small components.

2-Axis Spindle Heads. 2-axis (sometimes called "nutating") spindle heads (*Figure 10*) are typically employed on gantry-type machining centers, and possess the same weaknesses as double-tilt rotary tables, including lower accuracy and stiffness than other designs. Further, the rotational range of these heads is often limited. In a typical VMC arrangement, the C-axis will typically have unlimited range of motion, while the carried ("A") axis might be limited to ±90° of motion from the vertical position. The principal strength of this design is that it provides good access to large and deep cavities, and is therefore used extensively in the production of very large mold and die components, particularly in the automotive sector. 2-axis heads are not well suited to relatively long slender parts, such as blades and propellers, where 360° access to the part is required. 2-axis spindle heads are exceptionally well suited to parts where machining is required in a wide range of orientations, such as mold cavities, and large dies.

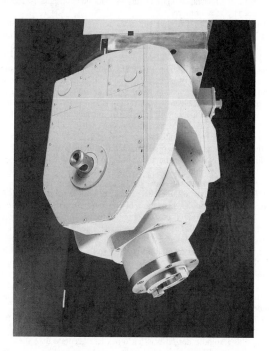

Figure 10. A 2–axis spindle head manufactured by Dayton Machine Tool Company.

Single-Axis Spindle Heads (Pendulum Type). Single-axis spindle heads, when combined with a rotary table, can provide better accuracy than double-tilt tables or multi-axis spindle heads because the rotary axes are not stacked atop one another. When utilized with a VMC, this means that the axes are typically distributed in a three-axis set (X, Y, and A or C), and a two axis pair (Z, and B or A). The principal weakness of this design is that the center of rotation of the spindle head (as with 2-axis heads) is located relatively far from the tool tip (typically 8" or more), causing angular errors and decreasing overall system stiffness. The range of angular motion is often limited, limiting the range of applications to which this design can be successfully applied. Nonetheless, this geometry has advantages over previously discussed designs with respect to high performance machining a range of components such as impellers, blades, and other parts, where the part itself can be rotated to present different faces to the spindle.

Tilting Heads. The tilting spindle combined with a single-axis rotary table is the ideal choice for 5-axis machining of many small- to mid-sized components, and is applicable to both prismatic and long, slender parts. With this system, the B-axis is designed so that the spindle rotates around the tool tip. This minimizes angular errors, and keeps torsional loads on the B-axis at a minimum. Even when the tool tip is not exactly aligned with the center of B-axis rotation, the distance between the tool tip and center of rotation is usually less than 2 or 3 inches, compared with 8–12" for other designs. As a result, angular errors are reduced by 75% or more.

The tilting spindle design further enhances accuracy by ensuring that linear axis motions are kept at a minimum when large B-axis angular motions are required. Other benefits of the tilting spindle design include the separation of the two rotary axes to minimize stack-up, and the ability of the spindle head to rotate ±100° from the vertical. This gives the machine a higher degree of flexibility than tilting heads or double-tilt tables, which cannot easily present the sides of parts to the spindle, making it difficult or impossible to machine undercuts into parts. This is particularly important when machining components such as tire patterns, complex electrodes, or turbo-machinery components.

The tilting spindle head is less well suited to the machining of cavities than the 2-axis spindle head, because it is unable to access very deep cavities in large diameter parts. This restriction exists because of the limited throat distance between the spindle centerline and the face of the B-axis gear set. As a result, the tilting spindle head design is best suited to small- to mid-sized concave parts, and a wider range of convex parts.

Effect of 5-axis design upon machine performance

Figures 11 and *12* illustrate the relative merits of typical 5-axis designs. Each figure illustrates the distance through which the tool-part interface will move to accommodate a 5° angular motion of one rotary axis. Rotation of the second axis would add additional movements to the part or tool. On the tilting spindle machine pictured in *Figure 11*, the tool extends 0.5" below the center of B-axis rotation. *Figure 12* illustrates a pendulum type spindle, where the center of rotation is 12.2" from gage line. The tool length from gage line is the same for *Figures 11* and *12*. Two axis heads and part carrying compound rotaries demonstrate similar performance to the pendulum type in this regard. **Table 1** compares the linear moves required for each 5° rotation. In each case, the distance moved also relates to the amount of axis velocity needed to maintain tool-part contact, and the multiplication of angular errors.

Table 1 indicates that angular errors are multiplied by a factor of over 200 for both the pendulum and double-tilt rotary, compared to the tilting spindle design. Even if the B-axis of the tilting spindle design was ten times less accurate than the other models, it would make significantly more accurate parts, would require lower axis feed rates to

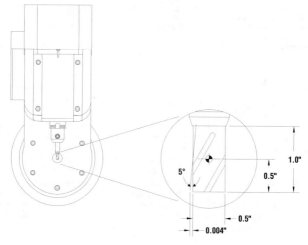

Figure 11. Tool movement along the X-axis is .004" for a 5° rotation of the B-axis. The tool measures 5" from gage line.

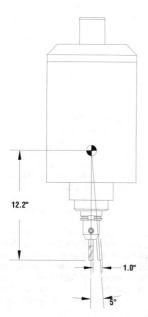

Figure 12. The pendulum spindle head requires 1" of X-axis motion to maintain part–tool contact for a 5° rotation. The tool measures 5" from gage line.

Table 1. Displacement of Tool-part Interface for a 5° Rotation of the Axis Furthest from the Part in Various 5-axis Machine Geometries.
Required axis velocities and the multiplication of angular errors grow with increasing displacement.

5-axis Design	Displacement due to 5° move of one rotary axis	Relative velocities and multiplication of angular errors
Tilting Spindle	0.004"	1
Single Axis Spindle Head	1.0"	250

achieve the same surface speed over the part surface, and would make the same parts with smaller axis travels.

The importance of minimizing the distance between axes of rotation and the tool-part interface cannot be overemphasized. Consider a double-tilt rotary on which the tilt axis (rotation about the X-axis) is accurate to ±5 arc seconds, and the distance between the face of the rotary and the tilt axis is four inches. If a four inch high part is placed on top of this rotary axis, the displacement error along the Y-axis will be ±0.0002". At this stage, only one source of positioning error has been considered. If the rotary axis was accurate to only ±10 arc seconds, as is the case with conventional rotary axes, the error would amount to ±0.0004". Add errors due to positioning of the remaining four axes, and it is likely that the error will grow to ±0.001" before considering dynamic errors, thermal effects, tool run-out and deflection, etc.

When machining 5-axis components, a fast CNC is required. The amount of data that must be processed increases substantially relative to 3-axis machining because of the additional axes that must be controlled. In the case of 5-axis machining, the amount of data to be processed can also increase because many linear axes moves are required to maintain tool-work contact. This is particularly true for those designs where the centers of rotation are located far from the tool-work interface (e.g., double-tilt rotaries, 2-axis spindle heads). When evaluating the fitness of a CNC for a 5-axis machine tool, it is important to determine if the advertised block processing rate applies only to 3-axis machining, as is often the case. Block processing rates can drop by 66% when a CNC must control a 5-axis machine. A well-designed high performance CNC will still exceed 1,000 blocks-per-second when performing 5-axis simultaneous contouring.

CNC and motion performance

A CNC for 5-axis machining should also contain features like 5-axis tool length compensation and part compensation. These features are important because they can prevent the tool path from having to be reposted if the actual tool length, or part size, differs from the nominal value. Ethernet, and a large capacity hard disk drive, should also be part of the system to accommodate the transfer and storage of the large part programs needed to create 5-axis parts. Further, because 5-axis parts can often take many days to machine (some impellers can take weeks, for example), and are quite expensive, it is important that the CNC have well-designed error recovery routines. The ability to retract the tool along its major axis is absolutely critical to prevent part damage in the event of a power failure or simply a broken cutter.

When selecting a 5-axis machine tool, the natural tendency is to evaluate the machine on the basis of its axis feed rates the same way one would judge a 3-axis machine. Unfortunately, axis feed rates cannot be evaluated in the same manner. To begin, the machine geometry and application will dictate whether the rotary axes or the linear axes are going to limit overall performance. Based on the information in **Table 2**, most buyers would expect machine "B" to produce parts faster because the maximum feed rate of the linear axes and B-axis is higher. Not so! Because of the construction of the spindle head the B-axis limits the effective feed rate that can be achieved. Machine "A", on the other hand, can achieve higher cutting feed rates due to the pendulum head design employed. Unfortunately, this increased speed comes at a high price—significantly reduced accuracy (approximately 250× worse) and lower stiffness.

Table 2. Comparison of Achievable Cutting Feed Rates for Two Different 5-axis Geometries.

	Construction	Linear Feed Rate	B-axis Speed (rpm)	Achievable Cutting Feed Rate with 6" Tool Assembly
A	Pendulum	200 in./min	10	200 (limit is linear feed rate)
B	Tilting	300 in./min	30	95 (limit is B-axis rotation)

Spindle selection

Spindle selection for 5-axis machine tools closely resembles that for 3-axis machines in most regards. Top end spindle speed must be balanced against the need for low-end torque, a tradeoff that exists for all machines, and particularly in the context of high performance machining. That said, 5-axis machines present a number of unique challenges for spindles. First, a 5-axis spindle often needs to operate in a wide range of orientations, including "nose up." As a result, the quality of the front-bearing sealing arrangement is particularly critical to ensure long spindle life. Second, the tool-spindle interface must be considered more carefully than in 3-axis applications. 5-axis machines are more likely to be working in difficult-to-machine materials (e.g., titanium and hardened tool steel). Also, 5-axis machines are expected to provide access into deep cavities and complex geometries. As a result of these factors, it is highly recommended that HSK interfaces be used in 5-axis spindles, particularly when working with 0.5" ∅ tools and larger. Compared to CAT/ISO 30, 40, or 50 taper geometries, the HSK design is significantly more rigid and accurate. For example, an HSK63A design can be built into a spindle with the same nose diameter as a CAT/ISO 40 taper. But the HSK63A design provides the rigidity of a CAT/ISO 50 taper system! HSK spindles have proven to be significantly more resistant to long-term fretting, as occurs when flank milling hard materials over extended periods of time or when working with long tools to gain access to complex geometry.

Thermal performance

5-axis machines, like their 3-axis counterparts, are susceptible to thermal growth. For designs that use double-tilt rotaries or pendulum type spindles, these problems are exacerbated. Some 5-axis machine tool builders now incorporate CNC-based thermal growth control into their machines. These systems can reduce the effects of spindle thermal growth by as much as 85% and are highly recommended. Likewise, the use of direct feedback (linear scales on X, Y, Z, and high resolution output shaft encoders on rotary axes) is highly recommended to minimize the effects of thermal growth in 5-axis machine tools. Without these countermeasures, thermal growth can negatively affect accuracy by 0.002" or more over several hours.

CAM selection for 5-axis machining technology

When evaluating CAM packages for 5-axis machining processes, have a clear understanding of what you intend to use 5-axis machining for. Do you want to employ 5M for a specific task or a range of applications? Compared to 3-axis CAM, the market for 5-axis CAM is not such a level playing field. In other words, the functionality, usability, and performance of 5-axis CAM varies much more than it does for 3-axis CAM. Today there are a large number of good 3-axis CAM systems. The market is almost commodity-like in nature. This cannot be said of the 5-axis CAM market, in which there are wide disparities in the cost and effectiveness of the various offerings.

To generate productive 5-axis tool path, the CAM system requires much more knowledge about the machine's geometry and machine's limitations (feeds, accelerations, and servo gains). And, as noted previously, the machining strategies employed with 5M vary widely based upon the application in question. For example, the strategies used to machine titanium airfoils have little in common with those used to produce tire segment molds or patterns. These, in turn, have little to do with the 5-axis machining of indexing cams or racing head ports and so on. Therefore, it is best to select a 5-axis CAM package in the context of the specific application in which you are interested. Check that the CAM vendor's staff has some experience in your application and will be available to help you refine your own machining approach. Be wary of piloting a CAM package's new "5-axis features." Without an excellent CAM system, the entire 5M investment in machinery, personnel, and software can be easily wasted.

When a CAM provider says, "Yes, we have 5-axis capability," push for further clarification. Often this statement means only that the CAM system provides static tool orientation control (i.e., no simultaneous 5-axis machining capabilities). Explore how the CAM system controls orientation of the tool, and how the tool path is programmed. Seek power and flexibility over a simpler, but less powerful approach. One can grow into more capability, even if it takes longer to become comfortable with the system at the onset. Also query the CAM provider as to how they develop their 5-axis features. Do they have a 5-axis machine on-site to test their software and develop new features? Amazingly, most CAM manufacturers do not have such capability in-house. This is a severe drawback, as what makes sense to a CAM system developer working at his desk may not translate well into the reality at the machine tool.

Finally, collision detection and control is a critically important part of 5-axis tool path development. If the CAM package you are considering does not include this functionality, be sure to purchase a secondary package to evaluate tool paths before they are run on a machine. In 3-axis machining, collision detection is generally a minor consideration. But as soon as the part and/or tool begins to rotate as well as translate, the probability that the toolholder or spindle will collide with the part during machining increases dramatically.

Conclusion

5-axis machine tool technology has been used within highly specialized industries for many years. Now, as 5-axis CAM and machine technology matures, 5-axis technology is poised to become a mainstream tool for machining operations. Adopting 5-axis machining processes can increase a shop's profitability and competitiveness. Successful integration of 5-axis technology into a machining operation requires careful consideration of the technology's applicability, and thoughtful selection of machines and CAM. Most important, a shop must possess the patience and human resources to develop the appropriate 5-axis machining process, so as to derive improved product quality and throughput.

High-speed Machining

What is high-speed machining, and what tooling is required to operate with reliability and accuracy at high speeds? Commercial considerations require shops to strive for higher rates of productivity, and this growing technology offers significant opportunities for increasing throughput. However, providing a definition for high-speed machining is difficult, because machining speed is very application specific. Labeling a process "high-speed" is actually a comparison between current and previous performance capabilities. At one end of the spectrum, high-speed might mean changing from an HSS tool to a carbide one, allowing for significant increases in machine speeds and feeds. The other extreme is defined by the rotational speed of the spindle, with speeds of 15,000 to more than 20,000 rpm being achievable for typical machining operations. A consensus definition is "any machining operation where the feed rates and speeds used are at least 50% higher than those conventionally employed."

Several interesting phenomena occur in high-speed machining. Since cutter speed is the major influence in creating heat at the cutting edge of the tool, maintaining a high chip load or feed helps dissipate heat. Correct feed, combined with proper cutter rake angle for the material being machined, produces a chip of sufficient density to carry heat away from the cutting zone, so work hardening can be avoided. For the same reason, coolant may be less necessary in high-speed machining. At elevated feed rates and speeds, chips are cut and evacuated so rapidly that very little of the heat is transferred to the workpiece. This is a significant benefit because at higher rpms, tool rotation throws coolant away from the cutting zone, making traditional coolant application methods impractical. Directing air blasts at the cut can be an effective means of evacuating chips.

As high-speed machining alters the nature of individual machining operations, it also changes the requirements for machine tool design, construction, and application. This section will explore the performance requirements of high-speed machine tools and the machining strategies related to the process.

Implementing high-speed machining

As the world of manufacturing becomes ever more competitive, parts need to be produced at higher and higher rates of speed. Production and design lead times are growing shorter and shorter. As a result, there is increased pressure on machining operations to produce components, whether they are prototypes or part of large quantity production runs, in time frames that would have been considered nearly impossible just a few years ago. The motivation for these changes is clear. A greater portion of the world has become industrialized. Developing countries are able to adopt new technologies and the latest manufacturing equipment at a rate that greatly exceeds the rate at which their wage or infrastructure costs increase. This puts pressure on the more well-established industrial economies (i.e., North America, Western Europe, Japan) to either produce products in less time, or to be satisfied with the loss of manufacturing superiority (and subsequently the loss of manufacturing jobs).

The drive towards shorter manufacturing cycle times is not the only development shaping the machining industry. To improve the performance and reliability of mechanical systems—from combustion engines to pumps to medical devices—the tolerances to which parts must be machined have tightened so that many parts must be machined to tolerances of less than ±0.001" (±0.0254 mm), and a rapidly growing number of components must be precision machined to tolerances of ±0.0005" (±0.127 mm) or better. In addition, the increased sophistication of CAD-CAM systems means that parts are also becoming more complex. Three-dimensional contoured parts are becoming the norm, as engineers seek optimal mechanical properties and aesthetics for their designs.

These changes—tighter part tolerances, increased part complexity, the need for shorter manufacturing cycle times—pose a unique set of challenges and opportunities for the metal cutting industry. While there will always be parts manufactured to wide open tolerances of ± 0.005" (±0.127 mm), machining such parts is becoming a commodity business, with intense price competition. Therefore, while it is critical that shops adopt high-speed machining technologies, they must do so in the context of tighter tolerances and increased part complexity.

Today, many machine shops and machine tool builders claim that they possess high-speed machining capabilities. It is commonplace for machine tool builders to speak of their ability to machine three-dimensional contoured parts, like mold cavities, to tolerances of ±0.0002" (±0.0508 mm). Further investigation usually finds these claims to be false, or at least misleading. The machine tool and process requirements necessary to achieve *high-speed machining* in the context of 3-D components with tolerances of better than ±0.001" will be considered in this section. High-speed machining in the context of vertical machining centers, with the understanding that similar principles but different tolerance and surface finish expectations would apply to other machining processes, will also be explored.

High-speed machining requires high performance in a large number of disciplines. These disciplines must be mastered by the machine tool builder, user, or both. This section surveys these seven areas as they relate to high-speed machining.

1. Machine geometry and construction [machine tool builder].
2. Motion control [machine tool builder].
3. Thermal growth and environmental control [builder/user].
4. Spindle technology [machine tool builder].
5. Tooling selection and application [user].
6. Machining strategy [user].
7. Real-time performance monitoring and correction [builder/user].

Machine geometry and construction

High-quality machine geometry and construction forms the basis for high-speed machining, and without it, performance in the other six disciplines is meaningless. The following characteristics are necessary in a high-speed machining center.

Stiffness. The machine structure must be stiff enough to withstand the dynamic loads created during cutting. Today, with the advent of high-speed spindles, the cutting loads are typically less than in low rpm/high torque machining operations, but are still high enough to cause significant accuracy errors or poor surface finishes. Machine stiffness is largely a function of the frame material, frame geometry, and the compliance of the way systems and drive train.

Of the many frame materials used today, the most common are cast iron, steel weldments, and polymer concrete. Countless variations on these three themes also exist, typically providing a combination of properties. Cast iron structures provide a high degree of stiffness, assuming a high tensile strength iron is used. Most castings are made from 40,000 psi tensile strength (class 40) cast iron, while a select number of builders use 50,000 psi tensile strength iron. The latter is more expensive to pour and machine, but does provide more stiffness than class 40 iron. Polymer concrete castings provide up to ten times more damping than cast iron, but are only about one-third as stiff as class 40 iron. Further, the thermal conductivity is worse, which promotes the creation of "hot spots" and, therefore, asymmetric thermal growth in the casting. **Table 1** identifies some of the key properties of ITW Philadelphia Resin's polymer concrete as compared to Class 40 iron.

Table 1. Properties of ITW Philadelphia's Polymer Concrete versus Class 40 Cast Iron.[1]

Material Property	ITW Phil. Polymer Concrete	Class 40 Cast Iron
Compressive Strength (psi)	20,000	130,000
Tensile Strength (psi)	2,000	40,000
Compressive Modulus (psi)	4.2E6	15.0E6
Coefficient of Thermal Expansion (in/in°F)	6.8E-6	6.7E-6
Thermal conductivity (BTU/ft-hr-°F)	91.2	1,300

[1] ITW Philadelphia Resins PolyCAST product bulletin.

Of equal importance to stiffness is the geometry of the machine structure. A compact machine structure is always preferred to minimize machine compliance, so the smallest possible machine should always be used to make a given part. Large cross-section castings should be employed to maximize resistance to bending and torsion. Joints between components should have bearing surfaces as large as possible, achieved through hand scraping or grinding rather than simply machining the mating surfaces. Overhung and unsupported structural members should be avoided or minimized. In C-frame VMCs, the overhang of the headstock can become a major source of deflection during cutting. Stiffness of the headstock decreases as it is extended forward of the Z-ways, and also decreases as the Z-ways are brought closer together. Similar arguments apply to the stiffness of the X- and Y-axis assemblies, as well as the Z-axis assembly on bridge-type machine tools.

Way system compliance can greatly undermine the stiffness of a machine tool, and is often the weakest link in the structure. There are three basic types of linear bearing suitable for machine tools. They are hydrodynamic bearings, hydrostatic bearings, and rolling element bearings. Hydrodynamic bearings, sometimes known as "hardened ground ways," ride on a film of oil that is only stable when the elements of the bearing are in motion. When the bearing stops, the oil film decreases in thickness and accuracy is lost. Further, slides riding on hydrodynamic way systems tend to rise and fall with the pulsing of oil through the system, causing errors of up to 0.0003" in each axis over the period of oil dispensing. The oil film is also dependent on temperature, and therefore steps must be taken to stabilize the temperature of the machine. Hydrodynamic bearings are therefore suitable for machines that do not stop, start, or change direction frequently. They are often used on large reciprocating grinding machines where the end of the motion can be ignored.

Hydrostatic bearings also rely on a film of fluid, usually oil, but the film is supported by pressure. Thus, hydrostatic bearings can be used for machines that may stop or change direction frequently. The main drawbacks of hydrostatic bearings are that they are very expensive to install and are not generally suitable for machines that may have a large variation in the weight of the part, since the bearing must be matched to the applied load. Hydrostatic bearings are suitable for small grinding and milling machines where the table weight is large compared to the part. Both hydrodynamic and hydrostatic bearings generate high frictional forces due to the shearing of the fluid film. They also require additional pumps and filters, which add to the maintenance burden.

Rolling element bearings have low friction and are not subject to variations in fluid viscosity. They are usually modular and replacement is therefore straightforward. Rolling element bearings may use rollers or balls. Roller bearings have higher stiffness and load capacity, but are more subject to bearing misalignment and structural deflection. Ball based systems are more suitable for light-duty machine tools that are not subjected to very high loads such as milling machines used for graphite and polymers.

Roller bearings using cylindrical rollers on a flat surface have the highest stiffness values, but require stiffer machine structures to avoid deflection, which will cause the rollers to skew and scuff. For general purpose machine tools, where loads can be in many directions, it is important to utilize bearings that are not subject to that restriction. The roller products manufactured by Thomson Industries (USA) and Schneeberger (Switzerland) have features that reduce these effects.

Table 2 outlines the relative merits of popular way systems, as well as hydrostatic way systems, which are expected to become popular over the next 10 years.

Table 2. Comparison of Properties of Popular Way System Types. The comparison applies to high performance components. Lower grade components have correspondingly lower performance characteristics.

Characteristic	Hydrodynamic	Ball Linear Way	Roller Linear Way	Hydrostatic
Stiffness[†]	Low	Medium	High	High
Geometric Accuracy	Medium	High	High	Very High
Friction	High	Low	Low	Almost Zero
Damping	High	Low	Low	Very High
Cost	Medium	Low	Medium	High
Maintenance Requirements	Medium	Low	Low	Medium -High
Longevity	High	Medium	Medium	High

[†]Stiffness measure includes the effects of intrinsic play.

Damping. Damping is critical to high-speed machining. It is derived from the materials used in the structural members, the manner in which they are assembled, and the characteristics of the way system and drive train. Even where specific structural members are highly damped (e.g., polymer concrete base castings), it is important to evaluate the damping of the machine tool as a whole. A poorly damped machine structure will lead to severe restrictions on the speed and acceleration that can be achieved during machining processes.

Unfortunately, increased damping is often achieved at the expense of reduced stiffness in the machine structure and, therefore, a careful balance must be struck. The preferred approach, which is difficult to achieve in practice, is to decouple the damping and stiffness functions of the machine through innovative geometries and methods, such as the use of shear tube dampers, which are currently employed in high precision machine tools, such as Weldon's Gold Line grinders.

Geometric Accuracy. Perhaps the most critical element of machine geometry and construction as it relates to high-speed machining is the quality of the geometry itself. Slight errors in squareness, parallelism, flatness, or straightness can wreak havoc on the accuracy of machined parts. **Table 3** details the minimal geometric accuracy standards for a precision high-speed VMC. Table rise-and-fall and spindle tram, as measured on the table, are particularly good indicators of overall geometric integrity, as is the use of the ball-bar test, which in addition to testing servo system performance will highlight squareness errors in the machine tool. Poor geometric accuracy, combined with overhung structures, measuring devices, and drive trains mounted off-center, and large stack up heights between axes, are particularly detrimental to high-speed machining. For example, consider the X-Y stacked axes depicted in *Figure 1.* If the Y-axis has a pitch error of only 0.0001" over 12" of Y-travel, a positioning error at the workpiece of 0.00053" will result.

Angular errors are of great concern, because their effect is multiplied with increasing distance from the source of the angular error. The same effect comes into play when using

Table 3. Minimum Geometric Requirements for High-speed Machining Centers Producing Tight Tolerance Components.

Geometric Accuracy Specification	Minimum Requirements for High Performance Machining (small to mid-sized components)
Squareness (X-Y, X-Z, Y-Z)	0.0002" / 12", 0.0004" over full travel
Straightness (all axes, pitch, yaw, roll)	0.0002"/12", 0.0004" over full travel
Table Flatness	0.0004"/12", 0.001" over complete surface
Table Rise & Fall	0.0004"/12", 0.001" over complete surface
Spindle Tram to Table	0.0005" TIR on a 12" sweep
Spindle Parallel to X-Z, Y-Z planes	±0.0002" over 6"
Ball Bar (12"⌀, 20 in./min, all three planes)	less than 0.0003" TIR

rotary 4th and 5th axes. The greater the distance between the tool tip and the center of rotation of each axis, the larger the positioning error due to angular errors. It is for this reason that compound rotary axes are inherently less accurate than individually mounted rotaries used to achieve 5-axis positioning. On a compound rotary table with ±10 arc seconds accuracy per rotational axis, the faceplate of the rotary axis is 4 to 6 inches from the centerline of the tilting axis. Add a 4" part on top of this, and the positioning error due to the tilting axis amounts to over ±0.0004". Once the positioning error of the rotary axis and linear axes is included, total positioning error can easily exceed ±0.001".

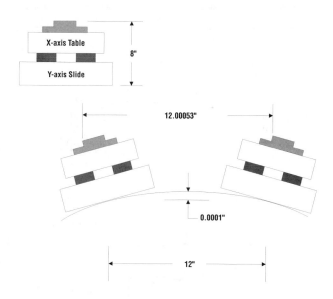

Figure 1. The impact of angular errors is magnified with distance. As a result, it is important to minimize the distance between the axes of motion in a machine tool–part interface.

Motion control

The ability of a machine's servo system to provide highly accurate motion, at high feed rates, is a critical element of high-speed machining, and one that is often overlooked. Many machine tool buyers incorrectly assume that the positioning accuracy specifications provided with machine tool literature describe the ability of the machine to provide accurate motion. For high-speed machining, we are most concerned with the ability of the machine to provide accurate positioning while cutting at high rates of speed!

Linear and angular positioning specifications are static measurements (whether they be ISO or JIS specifications) which are independently collected, and corrected, for each axis. These values represent best-case values for machine performance, and are not reflective of the dynamic performance of the machine while performing two- or three-dimensional contouring.

During actual machining conditions, the machine tool's various axes are not simply positioning point-to-point in "exact-stop" mode, as is the case during positioning accuracy tests. Rather, each axis is continuously accelerating and decelerating through a continuous series of points. These motions must be tightly coordinated between all axes. The performance of the machine's servo system, as embodied within its CNC, servo motors, amplifiers, and feedback devices, determines the quality of the motion generated by a machine tool and, therefore, the accuracy and speed which can be achieved during machining.

CNC machine tools use proportional servo systems, in which the velocity of each axis is proportional to the distance (or error) between the actual position and the command position during each sample period. This error signal is used by the system to determine acceleration and deceleration, as well as steady state velocities. The distance between actual and commanded position is commonly referred to as "servo lag." The effect of servo lag can be illustrated by considering the milling of two straight lines joining a semi-circle (*Figure 2*). The mill cuts at a constant feed rate from point A until the command dimension reaches point B. At that point, due to servo lag, the slide will only have reached point B_L. Since the command is moving ahead at a constant rate, it begins to generate commands

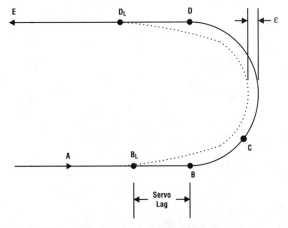

Figure 2. Servo lag causes the tool path to deviate from the desired tool path by an amount ε, which is a function of CNC performance and feed rate.

toward point C. This will cause the slide to move away from the desired path at B_L. The actual path will thus deviate from the ideal by an amount ε. The higher the feed rate, the larger the error, ε, for a given servo system. Similar effects can be seen when the tool approaches the intersection of two line segments. The true geometry of the corner will never be achieved without coming to a complete stop at the intersection of the two segments that form the corner.

Ideally, the error ε would be reduced to zero. But, because the servo system depends upon servo lag to drive corrections to errors in motion, that ideal can never be realized. The more responsive the servo system, the lower the lag and the less contouring error.

Servo gain is the measure of the servo's responsiveness and can be likened to the frequency response of the system. Servo gain is limited by the interaction of the servo system with the natural frequencies of the machine tool. To achieve high gains and, therefore, tight cornering performance, the CNC must be able to issue high frequency motion commands. When the frequency of these commands "bumps" up against a machine resonance, the motion control system can become unstable. To avoid this situation, servo tuning is used to limit performance of the motion control system to a frequency spectrum that the machine tool can handle. Servo lag can be computed in terms of servo gain as follows:

Servo Lag [in.] = Feed rate [in./min] ÷ Gain [in./min/.001"].

Table 4 shows the effects of various gains and feed rates upon the radius error ε in *Figure 2*.

Table 4. Contouring Error as a Function of Servo Gain and Feed Rate.

Feed Rate (in./min)	Radius Gain of 2	Radius Error for Servo Gain of 4	Radius Error for Servo
10	1.0	.000004"	.000001"
10	0.1	.000041"	.00001"
100	1.0	.00042"	.00010"
100	0.1	.00416"	.00103"
200	1.0	.00167"	.00042"
200	0.1	.0146"	.00403"

Note that even a machine with a moderate gain (e.g., 4) will experience a contour deviation of over 0.0004" when feeding at 200 in./min around a 1" radius. Conventional machining centers, using servo gains of 2 or less, are significantly less accurate. The only way to improve accuracy is to slow down, or to specify a machine with a higher servo gain. A high-speed machining center should provide servo gains of 6 or 8 to minimize servo lag during high-speed contouring operations.

Contour optimization and look ahead software systems within CNCs address the problem of servo lag by modifying feed rates in real time to hold a specified tolerance, or by preprocessing part programs and modifying the position commands to achieve the desired geometry. Such systems are a necessity to achieve high precision with high throughput. Real-time systems are recommended over off-line systems to ensure that all system inputs, including modifications to the feed rate override knob, are reflected in modifications to the cutter path, and to ensure that calculations made within the CNC to translate g-code into servo motion commands do not lead to rounding errors or other inaccuracies in the final cutter trajectory. In either case, contour optimization systems augment high gain servo systems, and do not replace the need for highly capable motion control.

High-speed machining requires an in-depth understanding of the limits of the machine tool's motion control system. Exceptionally high feed rates, while useful for high-speed roughing operations, are often impractical for machining precision components to tolerances better than ±0.0005", as **Table 4** illustrates. High acceleration and deceleration capabilities are typically more important than top end feed rates when machining contoured parts. The more severe the changes in curvature over the part, the more frequently the tool will need to change direction. In such cases, the top feed rate cannot be achieved. High-speed machining of contoured surfaces can be likened to a race through city streets, as compared to a drag race. In the drag race, high top-end speeds are important, but maneuverability is not. In the city rally, the ability of the car to hold corners, and to rapidly brake and accelerate, is much more important than top end speed. The same can be said for machining precision contoured surfaces as opposed to machining simpler, less accurate, parts. This fact is illustrated in **Table 5**. This Table illustrates that the distance required to reach full feed rate is quite long, both for machines with very low accelerations (i.e., 0.25g), as well as those seeking to reach very high feed rates with high accelerations. For example, a machine capable of 1g accelerations and programmed to move at 1200 in./min would require a distance of over $^{1}/_{2}$" to reach that feed rate. On a contoured surface the tool is likely to change direction much more often than every $^{1}/_{2}$". Therefore, providing a maximum feed rate of 1200 in./min is probably not practical because the 1200 in./min feed rate will never be achieved in practice. Such a high feed rate, even with a high acceleration of 1g, is practical only for simpler machining operations such as slotting or pocketing of simple shapes.

Table 5. Distance in Inches for a Machine to Reach the Specified Feed Rate as a Function of Available Machine Acceleration.

Acceleration	Distance (inch) to reach full speed		
	400 in./min	800 in./min	1200 in./min
0.25g	0.2300	0.9200	2.0704
0.5g	0.1150	0.4600	1.0352
0.75g	0.0767	0.3067	0.6900
1g	0.0575	0.2300	0.5176

Note also that high feed rates are achieved at the expense of positioning accuracy; there is always a tradeoff. This point is made clear in **Table 6.** Note that the two machines with the highest feed rates have the worst accuracies. This is in part a function of the ball screw lead, which is quite high. To achieve high feed rates, the builders of these machines have sacrificed accuracy. Note that the use of glass scales does not fully correct for the inaccuracy of these high lead screws. Further, system stiffness is significantly reduced when using high lead screws. Machine 4 in **Table 6**, on the other hand, provides a good mix of feed rate, acceleration, and accuracy. It is better suited to high-speed machining of precision components than the other machines.

Another critical aspect of motion control is the circle cutting capability of the machine tool. This capability is best measured through actual cutting tests, or the use of the ball bar test, which measures both the capability of the servo system and the quality of machine geometry. A precision VMC should provide circularity of better than 0.0003" at 20 in./min in all planes during a 6" diameter ball bar test. High-speed machining centers achieve these results through high servo gains, and the use of specialized friction compensation to address the zero velocity condition at each 90 degree "cross-over" point. In addition to checking circularity at low feed rates, it is instructive when evaluating machine tools

Table 6. Comparison of Machine Characteristics in the Context of Overall Price and Accuracy. Accuracy figures are for a single axis as per ISO 230-2.

Characteristic	Machine #1	Machine #2	Machine #3	Machine #4
Machine Construction	Fixed Rail Bridge	C-frame	Fixed Rail Bridge	Fixed Rail Bridge
Travels	28" × 22" × 18"	40" × 22" × 22"	30" × 22" × 18"	15" × 14" × 10"
Maximum Feed Rate	787 in./min (1575 R)	400 in./min (787R)	1200 in./min	600 in./min
Maximum Acceleration	1g	0.75g	1g	1g
Ball Screw Support	Fixed-Fixed	Fixed-Free	Fixed-Fixed	Fixed-Supported
Ball Screw Lead	20 mm	8 mm	20 mm	5 mm
Feedback	0.1m glass scales	0.5m glass scales	0.1m glass scales	0.5m glass scales
CNC Type	Brand H	Brand B	Brand R	Brand B
Machine Price	$300,000	$190,000	$270,000	$150,000
Accuracy (ISO 230-2)	0.00048"	0.0002"	0.0004"	0.00015"

to evaluate the circularity achieved at progressively higher feed rates. This will provide a meaningful appraisal of the machine's true motion control capabilities, and will indicate the practical limits for accurate contouring. If, for example, a ball bar test with 6" diameter at 150 in./min indicates a total circularity of 0.0008", one should expect actual machining results to be worse than this figure when contouring at 150 in./min, due to the presence of errors in the tooling, fixturing, spindle, and thermal effects, in addition to errors contributed by motion control.

Thermal growth and environmental control

Control of thermal growth effects is critical to high-speed machining. The impact of thermal growth upon the dimensional and geometric accuracy of a machine tool can be quite large, and the problem is exacerbated by high-speed motion and high-speed spindles.

The sources of heat, and therefore thermal growth in machine tools, are numerous. Flood coolant pumps, spindle motors, spindles, cutting processes, ball screws, way systems, servo motors, ambient temperature changes, and transformers can all contribute to thermal growth. Of these, the most overlooked detriment to precision is change in ambient temperature. A popular misconception is that as long as the entire machine tool is at the same temperature, the effect of system-wide temperature changes is not important. In reality, changes in ambient temperature cause the entire machine and workpiece to expand or contract, altering the dimensions of machined parts. And, if the air temperature in a machine shop is stratified, the geometry of the machine tool can be adversely affected. Machine shops that are not temperature controlled can experience very large swings in temperature over the course of one day—as much as 20°F in some climates. Such a change would lead to a dimensional increase of 0.003" over a 12" long aluminum section, and a 0.0015" change over a 12" steel piece. If the entire machine structure, and all tooling, workpiece materials, and gauging, had the same coefficient of thermal expansion, changes in ambient temperature would not be a problem. But, when the coefficient of thermal expansion of the machine tool and the workpiece are different (as they almost always are), the impact of changes in ambient temperature can wreak havoc on workpiece accuracy, and on the ability to machine the same part repeatedly throughout the day.

Thermal growth in ball screws is another significant error source in machine tools, particularly when working at high feed rates. Preloaded ball screws can generate a considerable amount of heat, and because motor encoders are used to derive machine position based upon a fixed ball screw pitch, changes in the length of the ball screw adversely affect positioning accuracy. As an example, a C-frame VMC without scales, machining 3" × 3" electrodes continuously over 10 hours at 60 in./min, experienced a shift in part-zero of 0.0005" in X and 0.0008" in Y over that period, even though ambient temperature was tightly controlled.

Several approaches to combating the effects of thermal growth in ball screws are in use. Some machine builders attempt to control the temperature of the ball screws by running liquid through the center of the screws. This method can be very effective (although it can be quite costly to build and maintain), if the heat-transfer fluid is controlled to an absolute temperature. Some systems mistakenly attempt to maintain the temperature of each axis ball screw to within ±1° or ±2° of ambient temperature. This approach is flawed for two reasons. First, the system is still sensitive to changes in room temperature. Only by controlling the screws to a fixed temperature can repeatable parts be manufactured over a wide range of conditions. Second, such systems cannot compensate effectively for localized heating of the ball screw, as experienced when machining small components at high feed rates with a large number of reversals (e.g., electrodes, small molds and dies, medical parts, microwave components, jewelry, etc.).

The most effective way to combat thermal growth in ball screws is to use linear scales. Mounted as close to the ball screw as practical, scales provide an absolute measure of axis position, irrespective of thermal growth in the screw. Whereas thermal control systems for ball screws possess time constants and cannot fully eliminate local "hot spots" caused by highly localized motion, linear scales will provide accurate part positioning data regardless of the thermal state of the screw, and are highly recommended for precision machining applications. The same C-frame VMC used to run the above test, but now outfitted with linear scales, reduced the shift in part-zero to 0.0002" in X and 0.00025" in Y over a ten hour span with no warm-up period.

Spindle growth is another significant source of error in a machine tool. A typical 15,000 rpm belted, chilled spindle can undergo over 0.003" of Z-axis thermal growth between its cold state and full speed operation. The time to steady state for this growth can be up to 1.5 hours, making it essential to warm-up the machine prior to use. X- and Y-axis thermal growth due to spindle heat generation can also be extremely detrimental to part-to-part repeatability, particularly on multi-spindle machines, in which case changes in spindle centerline distances cannot be compensated for via shifts in work offsets.

An alternate approach preheats the spindle and surrounding castings to their expected running temperature, and maintains that fixed temperature (about 100 °F) during spindle operation. This reduces the overall level of Z-axis thermal growth to approximately 0.002" over a wide range of operating conditions, and also minimizes Y-axis movement of the spindle center line. Again, it is important when heating or chilling a spindle that an absolute temperature be maintained, rather than controlling the spindle to the ambient temperature, which invariably changes over the course of a day.

Further reductions in spindle thermal growth can be achieved through the use of CNC-based thermal growth compensation. Such systems, which employ a model of the thermal behavior of the spindle as a function of rpm and time, further reduce the magnitude of thermal growth, and can also be employed with high frequency spindles, which typically must be refrigerated. *Figure 3* illustrates typical results achieved using this thermal growth control system in conjunction with spindle preheat on a 15,000 rpm #40 spindle over a range of operating speeds.

Spindle technology

Once an accurate and stable machine geometry has been established, and capable

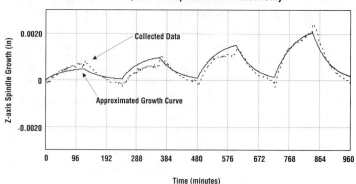

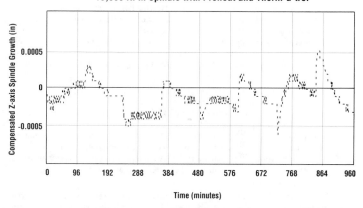

*Figure 3. Thermal growth in a 15,000 rpm belted spindle with preheat is compared to the same system with preheat and CNC-based thermal growth compensation.
The use of CNC-based compensation reduces growth by 50%.*

motion control has been achieved, the spindle becomes the most critical part of a high-speed machine tool. To achieve high material removal rates and quality surface finishes, the spindle must be accurate, stiff, and thermally stable.

A wide range of motorized, belted, and geared spindles are available from machine tool builders today. To achieve high-speed machining, belted or motorized units are recommended to ensure that vibration from a gear train is not transmitted into the workpiece during machining, degrading surface finish. For speeds at or below 15,000 rpm, a belted unit is recommended because it provides a practical combination of good low-end torque, reliability, and economy. Once speeds over 15,000 rpm are sought, a motorized spindle must be used. An integral motorized spindle is the only way that balance and reliability can be achieved at speeds of twenty-, forty-, sixty-thousand rpm, etc.

Spindle run-out, both axial and radial, is critical to achieving tight tolerances in machining. Radial run-out should not exceed 0.0004", measured 6" from the spindle nose, and axial runout should not exceed 0.0001" on a #40 or HSK63A spindle. Accuracy of the spindle will also be determined in large part by the taper type employed. While #40 and #50 tapers dominate the U.S. machine tool market today, HSK tooling is quickly becoming the preferred standard for high-speed machining applications. While it is not within the scope of this article to fully describe the features and benefits of HSK tooling, the following observations are offered.

> HSK spindles, while more expensive to make, provide a more highly repeatable spindle-tool holder interface, due to the face contact between spindle and holder. Improved retention at high rpms is also provided due to the "hollow cup" design of HSK tooling, which allows centrifugal forces to improve gripper performance.

> HSK 63 spindles provide significantly more stiffness than #40 taper, and appear to be comparable in stiffness to #50 taper units, while providing the additional benefits of higher repeatability.

> HSK tooling, while more expensive than standard tapered tooling, is rapidly decreasing in price. Many shops resist switching to HSK spindles because they own a large stock of #40 (or other) tapered tooling. This resistance is often misguided because high rpm spindles require balanced tooling. The use of well-worn tooling in high-speed (≥ 12k rpm) spindles can lead to spindle damage quite quickly. So, if tooling must be purchased to support new high-speed spindles, HSK should be considered.

> Small HSK tooling (e.g., HSK40 and HSK32), such as that used to replace ISO20 or ISO25 tapers, does not appear to provide the same marginal benefit relative to ISO tooling as does HSK50 or HSK63 tooling. This is thought to be a function of the "hollow cup" design of the HSK tooling, which requires a greater overall length from gage line to accommodate collets, leading to a greater extension from gage line to the end of the cutter compared to the comparable tapered toolholder. This effect, while still present on HSK63A tooling, diminishes in importance with increasing holder size because of the increased diameter of the tooling, which provides additional stiffness.

Spindle stiffness is in large measure a function of bearing selection. Larger bearing diameters will provide increased stiffness, but will also bring higher ball velocities, in turn raising operating temperatures, reducing grease life and increasing bearing fatigue. The use of multiple bearing sets, and placement of bearings as close to the nose as possible, will also increase spindle stiffness, as will the use of hybrid ceramic bearings as opposed to steel ball bearings. In addition to high stiffness, hybrid ceramic bearings provide the additional benefits of improved grease life, and higher crash resistance as compared to steel ball bearings. High drawbar pressures (2,000 psi is desirable) will also increase spindle stiffness by increasing the contact between tool and spindle. A high quality taper, with 80% to 90% bearing, is also a necessity.

Thermal stability of the spindle is critical to achieving tight tolerances over extended running periods. As previously discussed, thermal growth of spindles can easily exceed 0.003". Interestingly, the Z-axis thermal growth of most spindles is not a simple exponential curve. When a spindle at rest is initially powered up, convective effects can cause a brief period of cooling to occur (accompanied by contraction of the spindle nose) before the expected thermal growth pattern is observed. Likewise, when a running spindle is brought to zero rpm, a growth spurt can be observed due to the elimination of this convective cooling effect. Second and third order effects such as these make CNC-based spindle growth compensation a valuable tool in controlling spindle thermal growth.

High-speed/lower torque spindles are preferred over units with higher torque but lower maximum speeds. High-speed machining approaches are built around the concept of higher feed rates but lower depths of cut. As a result, less spindle torque but higher rotational speeds, are necessary. Further, higher speed spindles are better able to provide the rotation rates required by smaller cutting tools. And, a higher speed spindle can be used for both roughing and finishing. True, roughing operations will take longer with a high-speed/low-torque spindle than when using a low-rpm/high-torque spindle, but, in the final analysis, finishing operations take longer than roughing and are more closely linked to final part quality. A high torque/low-rpm spindle will rough faster, but will not be able to support the smaller cutters necessary to obtain high quality finished parts, both in terms of finish and tolerance. If in doubt, err towards higher speeds and lower torque.

Another factor to consider when evaluating a spindle is that high power/high-speed spindles typically are less reliable than high-speed spindles with low or moderate power because of the higher levels of heat generated in their bearing systems. For processes requiring both heavy stock removal and fine, high accuracy, finishing, a heavy duty main spindle, plus high-speed auxiliary spindle (*Figure 4*) is recommended. Such machines can improve part accuracy and reduce cycle times by eliminating multiple setups, and by allowing the cutting process to be more finely tuned to the requirements at hand. Such configurations are also extremely cost effective for smaller shops, enabling, for example, the machining of steel molds and graphite electrodes on the same machine without compromise.

When considering a spindle for high-speed machining, seek to obtain a unit with a maximum speed in excess of current requirements by 15 to 20%. This will provide room for growth as new tool coatings and materials appear on the market. High-speed machines are expensive, so it is very important to guard against obsolescence.

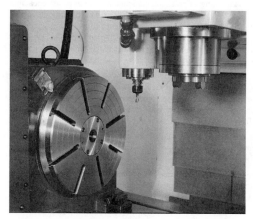

Figure 4. A main + auxiliary spindle combination. The motorized auxiliary spindle provides 2.5 HP and speeds to 40,000 rpm and supports tools up to $^3/_8$" Ø. The main spindle provides 12,000 rpm and 15 HP.

Tooling selection and application

Without discussing the particulars of selecting tooling for specific machining applications, the following should be noted regarding the selection of tooling and tool holders for high-speed machining applications.

As previously discussed, HSK tooling is highly recommended for high-speed machining applications. Even so, wide variation in the axial and radial run-out of such tooling can exist between vendors. For the utmost in precision, hydraulic, heat-shrink, or rear drawn in toolholders are recommended. These holders are capable of minimizing run-out of the tools they are clamping, and are extremely repeatable. When possible, "thin-neck" toolholders (such as those manufactured by Nikken) should be employed, thereby increasing the supported extension of the tool from spindle gage line. Conventional, "wide neck," toolholders greatly reduce the ability to access deep cavities without using long unsupported tool lengths. To achieve machining accuracies of ±0.0005", toolholders with run-out of less than 0.0002" should be used. If run-out is not suitably controlled, damage to high-speed spindles running at 20,000+ rpm can easily occur.

The achievement of minimal run-out in the spindle+toolholder+tool assembly is difficult, but not impossible. Boston Scientific (Watertown, MA) manufactures balloon molds for medical applications to tolerances of 0.0002". To achieve these tolerances, they use Technara Deta-one toolholders, specially ordered end mills, and a well-refined setup process to achieve a net run-out of 0.0001" in the spindle on each of the more than 20 tools used to manufacture a set of molds. Fortunately, the run-out of the toolholder and tool does not individually need to be less than the total desired run-out. Through careful adjustment, some of the run-out in each element can be canceled out as a function of the position of the tool with respect to the toolholder.

Machining strategy

The low cycle times afforded by high-speed machine tools and tooling can be easily defeated by a poorly laid out machining strategy. The way in which a part to be precision machined at high-speed is fixtured and machined will dictate how much effort will be necessary to repeatably attain tight tolerances. Again, the subject of machining strategy is too broad to fully address here, but the following observations are offered.

Fixturing for high-speed machining of precision components should be as repeatable as possible. For smaller parts, the use of EROWA or System 3R workholding devices is an excellent start. These systems are highly repeatable, and easy to use. If less accurate fixturing methods are to be used, such as conventional pallet shuttles (which are typically repeatable to no better than 0.002"), the use of probing routines to reestablish work offsets, and to rotate part programs to account for errors in pallet orientation after each pallet change, should be considered. Other fixturing guidelines follow.

> Minimize stack-up. Keep the part and fixturing as close to the machine's work table as possible. Increased stack-ups will lead to increased displacement errors due to angular errors (pitch, yaw, roll) in the machine's way systems. By the same logic, tool lengths from gage line should be minimized.

> Make the fixturing process as repeatable as possible. If vises are used, employ torque limiting handles. Always use positive stops to locate parts. Regularly check all fixtures for alignment.

> Avoid multiple stations. Minimize the number of times that each part needs to be handled and refixtured. Refixturing immediately contributes positioning errors of at least 0.0001" (on a very, very good day), and more typically 0.0005". The use of rotary tables to enable four-sided machining, or 5-axis machines with independently

mounted rotaries to achieve five-sided machining, can greatly improve overall part accuracy by reducing part handling. The same principle applies to the use of mill-turn methodology to reduce the handling of cylindrical parts.

In terms of the machining process itself, the key is to make it as repeatable and machine friendly as possible. A few suggestions follow.

Minimize on-part reversals. Most CAM systems accommodate this requirement by machining parts using a series of X- or Y-axis sweeps, thereby taking the X- or Y-axis reversals off-part. The notable exception is Z-level cutting, in which a series of "loops" or spirals are made around the part. This machining strategy is superior to uni- or bi-directional sweeps because it reduces machining time by eliminating off-part reversals. At the same time it requires a higher degree of circle cutting performance than conventional CAM cutter paths. As a result, a better machine tool is required to effectively utilize Z-level cutting to achieve a given tolerance. Flow machining, a more sophisticated strategy than Z-level cutting that follows part topology so as to minimize rapid changes in the cutter's motion vector, can also reduce overall cycle time while improving tolerance—so long as the machine tool is capable enough to undertake the strategy without leaving witness lines on the part at reversals, etc.

Exercise the machine. Use of an exercise program that approximates the machining process for at least one hour prior to machining is critical to achieving the tightest possible tolerances. Many machines can initiate such sequences automatically prior to the start of the workday. Machine warm-up can shave up to 0.001" or more off the total tolerance band for a given production run. High-speed machines generate a lot of heat. If the machine is exercised beforehand, it can be brought to a steady state condition and this effect will be minimized.

Minimize loads applied to the part. Substitute higher spindle speeds and feed rates for low rpms and larger depths of cut when possible, keeping in mind contouring accuracy limitations at high feed rates. Such an approach will reduce cutter deflection, and the amount of heat dumped into the part, both of which detract from overall part accuracy. The downside of high-speed machining is that cutter wear can increase relative to conventional machining approaches. Tools with higher wear resistance, achieved through specialized coatings, are a necessity in high-speed machining applications.

Real-time performance monitoring and correction

Everything we have discussed so far provides an open-loop approach to high-speed machining. To "close the loop," and keep the machining process under control, requires that we measure the performance of the machine tool and machining process in real time or near-real time.

Real-time and near-real-time performance monitoring and correction typically encompasses the following methodologies.

- Tool management
- Thermal growth control
- Fixture offset recalibration
- Part inspection and process modification

The effectiveness of these methodologies depends upon 1) the frequency with which measurements are made relative to the frequency of disturbances or errors, and 2) the ability of the system (man-machine) to initiate corrective action based upon the measurements made.

Tool management systems are very useful in high-speed machining to ensure that worn or broken tools are quickly replaced before scrap parts are manufactured. In its simplest form, tool management consists of a machine mounted measurement probe, to which each tool is brought before each tool change. If the tool's dimensions are outside specified boundaries, a redundant tool is exchanged for the rejected tool. Such systems are very useful for lights-out operations, but the frequency of measurement is not high enough to respond to sudden tool failures.

More sophisticated tool management systems make use of the CNC and spindle drives to monitor cutting conditions in real time. The system is first trained by successfully machining a part while monitoring spindle torque and speed levels. Boundaries are then set around the recorded levels to establish acceptable operating conditions. During the machining of successive parts, should the spindle torque or speed levels exceed these bounds, the tool is automatically retracted from the work. The operator is then notified and/or a redundant tool replaces the current tool. Some of the more sophisticated tool management systems can differentiate between gradual tool wear and catastrophic tool failure, and are capable of acting appropriately in each situation. It is the ability of a tool management system to react to gradual changes in tool wear that is most valuable for precision machining.

Thermal growth control, while previously discussed, deserves brief mention here. For high-speed machining, it is highly recommended that the machine tool be capable of managing the temperature of the tool coolant, and be capable of reacting to changes in the temperature of the spindle, bolster casting, and column casting. These areas are considered of greatest importance because they directly affect the position of the spindle nose relative to the work, and cannot be corrected for through the use of linear scales.

The use of touch probes to recalibrate fixture offsets is a necessity for the semi-automated production of precision parts in quantities larger than five to ten pieces. Stored in the tool changer, an infrared probe can be automatically loaded into the spindle, and under CNC control it can measure the location and orientation of parts and fixtures. Some CNCs are capable of integrating this information directly into part programs through the use of variables to automatically rotate part programs and shift work offset values to reflect changes in fixture position.

This same concept can be taken further when used to measure parts after they have been machined. Boston Scientific uses a Renishaw MP700 probe on a Bostomatic 32 VMC to pursue its policy of achieving "Quality at the Source" by measuring aluminum balloon molds on the VMC after they have been machined. Their manufacturing manager explains that they are able to use their VMC as a CMM. "We have qualified measurements on the machine to be repeatable to within 50 millionths and accurate to within 0.0003" anywhere within the work cube (32" × 16" × 20")," in accordance to ANSI standard B89.1.12M-1990, Methods for Performance Evaluation of Coordinate Measuring Machines. By qualifying that the parts have been machined to tolerance on the mill, Boston Scientific can, "certify that the parts are good, download inspection data from the CNC to a PC, print out an inspection report, and speed downstream steps in the manufacturing process." The on-machine inspection process also enables Boston Scientific to catch process errors almost immediately, as there is a minimal time delay between when the parts are machined and when they are inspected.

Computer Numerical Control

NC (Numerical Control) or CNC (Computer Numerical Control) has come of age as we enter the 21st century. NC machines arose out of the need to machine complex tooling fixtures used in the manufacture of helicopter components and tooling fixtures. The first NC machine was a milling machine converted over to NC. This machine tool system was developed by John Parsons, along with others at the Massachusetts Institute of Technology, between the years 1949 and 1952.

Conceptually, the effort to create a numerically controlled machine tool was borne of the desire to be able to position a cutting tool with great accuracy without the need to make hard tooling, such as cams or other mechanical positioning devices. By positioning a cutting tool with numbers, a machine tool would at once become more flexible and therefore more adaptable to machining a variety of parts. The changeover time from one part to another would also be greatly reduced as the need for customized hard tooling would be minimized or eliminated.

In 1972, mass production of NC controls that utilized minicomputers began. Within a few years, the microprocessor replaced the minicomputer in NC machine tool controls. At that time, the terminology transitioned from Numerical Control into Computerized Numerical Control. However, using the term NC is still correct and accurate, as the CNC machine tools of today are still numerically controlled.

The real benefits of NC/CNC would not be realized until machine tools were designed from the ground up as NC machines. Prior to 1965, most NC machine tools were manual machines retrofitted with NC controls. Although this allowed for automation of the cutting, changing tools and parts was still a manual operation, thus limiting the benefits of automation through NC. By 1970, automatic tool changers and automatic pallet changers improved the application of NC through enhanced automation. Today, nearly all CNC machines are designed from the ground up as a CNC machine and not a retrofit.

CNC machine tools come in many varieties today. A partial list of CNC machines that are found in common use include:

- Machining centers
- Turning centers
- Multiple spindle machines (6 or more spindles)
- Cylindrical grinders
- Surface grinders
- Tool and cutter grinders
- Pipe or tube bending machines
- Turret presses
- Gear hobbing machines
- Die sinking EDM machines
- Wire cut EDM machines
- Waterjet cutting machines
- Plasma arc cutting machines
- Flame cutting machines
- Injection molding machines
- Laser cutting machines
- Laser marking machines
- Dot matrix marking machines
- Stereolithography machines
- Coordinate measuring machines
- Wood routers
- Press brakes

The two most common types of CNC machine tools are machining centers and turning centers. Within each type of machine tool, there are also many subcategories.

Machining Centers
        Bed frame vertical
        Knee type vertical
        Bridge type vertical
        Horizontal

Turning Centers
        Flat bed
        Slant bed
        Horizontal
        Vertical
        Single spindle
        Twin spindle
        Multiple spindle (6 or 8 spindles)
        Subspindle (part transfers from main spindle to subspindle)
        2 axes
        3 axes
        4 axes
        Turning centers with milling capabilities

This is only a partial list of the varieties of machining centers and turning centers. The CNC industry is very diverse and in a constant state of improvement.

Throughout the last 50 years, machine tools changed substantially with the advent of NC/CNC. Some of the evolutionary changes are listed here.

Lead screws $\rightarrow$ Ball screws $\rightarrow$ Elimination of ball screws

Manual tool changers $\rightarrow$ Automatic tool changers $\rightarrow$ High capacity tool changers

Open loop positioning feedback systems $\rightarrow$ Closed loop positioning feedback systems

Manual parts loading $\rightarrow$ Automatic pallet changers for machining centers

Rapid traverse speeds 100 IPM $\rightarrow$ 343 IPM $\rightarrow$ 473 IPM $\rightarrow$ 760 IPM $\rightarrow$ 1000 IPM $\rightarrow$ 4000 IPM

Manual parts loading $\rightarrow$ Automatic bar feeds for turning centers

Dove tail ways $\rightarrow$ Box ways $\rightarrow$ Linear bearing ways

Tool change time 17 seconds $\rightarrow$ 12 seconds $\rightarrow$ 6 seconds $\rightarrow$ 1 second

Analog servomotors $\rightarrow$ Digital servomotors $\rightarrow$ Linear servomotors (also eliminated ball screws)

LED display $\rightarrow$ Monochrome CRT (Cathode Ray Tube) displays $\rightarrow$ Color CRTs $\rightarrow$ Color LCD displays

Punched cards $\rightarrow$ Paper tape $\rightarrow$ RS232 serial interface $\rightarrow$ Open architecture systems using a LAN

Terminology

Absolute Coordinates. When programming with absolute coordinates, there is one point on the part that is called program zero. All positions that the tool is programmed to move to are relative to the program zero point. The command: G00 X1.25 Y.1 tells the tool to move to the position 1.25 inches to the right of program zero and .1 inch above the program zero on Y axis.

ATC. Machining centers use an ATC (Automatic Tool Changer) to move tools into and out of the spindle of the machine. There are many styles of ATC devices, but the two that are most common are single arm and double arm. With a single arm system, the tool is first removed from the spindle and placed into the magazine. The magazine rotates to the next tool and the arm grabs the tool and places it into the spindle. Alternatively, a double arm ATC can remove the tool from the spindle and the magazine at the same time, rotate 180 degrees, and insert the tools into the spindle and the magazine at the same time. A double arm ATC is much faster than a single arm system.

Older machines use hydraulic motors and cylinders to move and rotate the ATC arm. On current machine tools, an electric motor coupled to a specialized cam moves the ATC arm through its motions. The hydraulic systems typically take 10 or more seconds to change a tool, whereas electromechanical systems can change tools in under 2 seconds.

Automated Cells. Many machining centers can be configured with large pallet pools. Sometimes these pools contain 20 or more pallets with each pallet holding a different part. In cases like this, a host computer usually runs the cell by directing pallets into and out of the machine. When a pallet is delivered into the machine, the host computer sends the CNC program to the CNC control and instructs the machine tool to run the cycle. These cells often consist of more than one CNC machine tool.

Automatic Mode. The automatic mode is used to run the program. The operator presses the cycle start button, and the machine begins to execute the instructions stored within the CNC program.

Ball Screw. Ball screws have replaced conventional lead screws in modern machine tools. The ball screw is a precision ground screw that is attached to a servomotor and to one of the axes via a special nut. A ball screw uses a recirculating ball bearing nut that is fastened to the axes. This type of nut offers minimal resistance and friction, resulting in a longer life span with greater accuracy due to the low wear factor.

Circular Interpolation. Circular interpolation is a type of motion in which a programmed command specifies the end point, radius of the circle, and the rate of feed. The control then takes the command and moves the tool from the current position to the end point in a circular motion with a given radius at a constant velocity (feed rate). Circular interpolation can be performed on two axes.

Control. The control is the computer that runs the machine. Years ago, the controls were very large and required many separate cabinets to hold all of the electronics. Today, CNC controls are small and they easily fit inside a cabinet attached to the machine. The control must provide a means to both operate the machine manually during setup and to run the machine during automatic cycle.

A control is also a term used to describe a computer when it is used in industrial applications. Many modern CNC machine tools are being built using standard desktop computer components. This is often referred to as an open architecture system. The main benefit of this configuration is that the CNC machine tool can be easily connected to the company's existing LAN. Other benefits include inexpensive upgrades of computer components, and the ability to run other software applications on the CNC control.

CRT. CRT stands for Cathode Ray Tube, which is a computer or television screen. Some of the newer machines use what is called a gas plasma display or an LCD (liquid crystal display). A gas plasma or LCD is generally much smaller in size and mass (they can be as thin as 1 inch).

Cycle Start. The cycle start button is used to start the cycle of the machine. In order to run a program, the machine must be in the automatic mode. When using MDI mode (see MDI Mode) to enter a CNC command, use cycle start to execute that command as well.

DNC. DNC has two meanings—Direct Numerical Control and Distributed Numerical Control.

Direct Numerical Control is the ability of the machine to be run by a host computer. In this scenario, the host computer sends the CNC program to the machine, while the machine executes the program code. In some configurations, the host computer will also issue the cycle start command to the machine. There are many different levels of direct numerical control, but the two most common uses are as follows.

1) When running programs that are too large to fit inside the control's memory, as

when machining large molds or complex parts, the CNC program can be very large in size, sometimes in excess of 1 million characters. Memory for many models of controls is either very expensive, or expansion possibilities may be limited. To handle programs that are very large, the control has a DNC mode which is similar to the automatic mode, but instead of reading the program commands from memory, the commands are read through an electronic interface, usually RS232C.

2) When a host computer sends CNC commands to the CNC control, which stores a small portion of the program in a buffer until the commands are completed. During the machining, the control requests data from the host computer as needed. When the buffer in the CNC control is full, the CNC control sends a stop transmission command to the host computer. This type of operation is sometimes called "drip feeding" the program.

Distributed Numerical Control refers the distribution of NC programs to CNC machines. This form of DNC came into being as computer interfaces were being standardized (RS232), and paper tape was being eliminated as the basic media for storing CNC programs. With distributed numerical control, the CNC programs are loaded into the memory of the CNC control and the CNC control runs the program from memory. This form of DNC cannot start the machine into the automatic cycle, which direct numerical control can do. Many distributed numerical control systems allow the machine operator to save and retrieve programs from the CNC control without having to interact with the host computer directly. In a limited fashion, some of the distributed numerical control systems can also perform direct numerical control operations.

DOC. This usually refers to the depth of cut for cutting conditions. For a turning operation, a DOC (or d.o.c.) of 0.100" would reduce the diameter of the part by .200".

Dry Run. Depending on the machine and how dry run is set up via the parameters, dry run is used to slow down the machine during setup. When dry run is on, the jog rate override controls how fast the axes will move when the program is run. On most machines, the dry run switch overrides the feed rate and the rapid moves, giving the setup person greater control over how fast the axes move. Dry run also allows the setup person to run the cycle without having the spindle in the "on" position. Dry run should never be active when cutting the actual workpiece.

Edit Mode. The edit mode is used to load and alter the CNC program. When in the edit mode, the user can make changes to the CNC program. Several basic features are built into the control for editing, such as: searching for a specific code, altering codes, inserting codes, and deletion of codes.

Programs can also be loaded into the machine from an outside storage device, and personal computers are commonly used for the storage of CNC programs. While in the edit mode, a program can be loaded into memory from the personal computer, or saved to the personal computer. The information is transmitted electronically through the RS232 (serial) port of the machine via a cable that is connected to the personal computer.

Emergency Stop. The emergency stop button is used as a "panic" button when something is not right with the machine or its operation. When the machine is stopped by emergency stop, everything stops including: axes motion, spindle, coolant, and hydraulics. To reset the emergency stop condition, the emergency stop button will either have to be rotated $^1/_4$ turn so that it can pop back out, or the switch may have to be pulled out. Then, the hydraulic on button can be pressed as during a normal power up condition. On most machines, the axes will have to return to home position after reset.

Feed Hold. Feed hold is used to temporarily stop the axes. It is important to understand that the spindle will continue to rotate and the coolant will continue to flow. To continue with the program execution, the operator would press cycle start.

Home Position. Each axis of a CNC machine needs a reference position. This position establishes reference between the machine tool and its slides. Although not standardized, the home position of the machine is usually the farthest position that the axes can move in the plus direction for each axis. The machine tool provides a mode of operation called "zero return" that allows the operator to manually move the axes to the home position. Within the program the machine can be instructed to move to the home position with the G28 command.

Incremental Coordinates. Programming done with incremental coordinates tells the tool how far to move and in which direction. To program the machine using incremental coordinates, use the G91 command to set the machine into incremental mode. Then, each coordinate that is commanded will specify a distance and direction for the tool to move. On turning centers, substitute the letter U for X and W for Z, instead of commanding G91 for the incremental mode.

IPM. IPM (also in./min) stands for Inches Per Minute, which relates to the rate of feed for the tool. 15.0 IPM means that the tool will travel 15 inches in one minute of time. Machining centers use this type of feed rate designation. (Also see MMPM.)

IPR. IPR (also in./rev) stands for Inches Per Revolution, which relates to the rate of feed for the tool. .015 IPR means that the tool will travel .015 inches per revolution of the spindle. Turning centers use this type of feed rate designation for turning operations. (Also see MMPR.)

Jog Mode. The machine can be moved manually during setup by selecting the jog mode on the mode selector switch, and then pressing the jog button located on the operator panel. Other features on the operator panel allow the operator to specify which axes to move and in which direction. Further control is provided via an override control that allows the operator to select the speed at which the slide moves.

Linear Interpolation. Linear interpolation is a type of motion in which a programmed command specifies the end point and the rate of feed. The control then takes the command and moves the tool from its current position to the end point in a straight line at a constant velocity (feed rate). Linear interpolation can be performed on one or more axes.

Machine Lock. Machine lock allows the machine to cycle through a program without any movement taking place on the axes. Originally, this was used to determine if there were syntax errors within the program. By running the cycle through without the tool moving, a quick check could be made to make sure there were no programming errors. When using machine lock, a complete cycle must be run from beginning to end, otherwise the coordinate system of the machine may be shifted, possibly resulting in a crash.

Magazine. On a machining center, the magazine stores all of the tools. Magazines can be disk style, which are large round disks, or chain drive systems, which can be any shape. Tool magazines can hold anywhere from 20 to 200 tools. The capacity of the magazine will dictate how complex the parts are that can be machined in one operation. On horizontal machining centers, many different parts may need to be run when there is a large pallet pool. In systems like this, a large capacity tool magazine is usually required.

Main Program. A main program is the highest order of the entire program. Many programs exist only as a main program. However, a main program can have many smaller parts, known as subprograms, that are called into execution by the main program.

MDI Mode. The Manual Data Input mode allows the operator to enter and execute an NC command. The command is retained by the control until it is executed by pressing the cycle start button, then it is forgotten. The MDI command is often used to start the spindle or to call up a specific tool during setup or manual intervention.

MMPM. MMPM (also mm/min) stands for Millimeters Per Minute, which relates to the rate of feed for the tool. 150.00 MMPM means that the tool will travel 150 millimeters in one minute of time. Machining centers use this type of feed rate designation. (Also see IPM.)

MMPR. MMPR (also mm/rev) stands for Millimeters Per Revolution, which relates to the rate of feed for the tool. .15 MMPR means that the tool will travel .15 millimeters per revolution of the spindle. Turning centers use this type of feed rate designation for turning operations. (Also see IPR.)

Offsets. An offset is used to shift (offset) the tool path to compensate for something that may vary from setup to setup. On CNC machines there are several different types of offsets. On machining centers, tool length offsets compensate for the length of the tool. Tool radius compensation compensates for the radius of the tool. The work shift offset is used to shift (offset) the reference point of each axis, which ultimately shifts all of the programmed positions.

Turning centers use tool geometry offsets to establish a reference point between the tool and the workpiece. Wear offsets are used to compensate for the wear of the tool during use. A work shift offset is used to shift all of the coordinates in a program to line up with the origin point on the workpiece.

Parameter. On CNC machines, parameters refer to special settings within the control that affect how the machine operates. The parameter settings are also used to tailor or customize the control to a specific machine tool. Parameters are used to control such things as: default values for canned cycles, serial communications speeds and settings, rapid traverse speed, etc.

NEVER CHANGE THE PARAMETERS OF THE MACHINE WITHOUT ASSISTANCE FROM THE CONTROL OR MACHINE TOOL BUILDER!

Program. A program is a set of instructions that tells a computer what functions to perform, and in what order. A CNC program provides all of the instructions needed to perform a sequence of actions.

Program Zero. Program zero is a point on the workpiece to which all of the programmed coordinates are referenced. Program zero is not used with the incremental coordinate system. It is used with the absolute coordinate system.

On machining centers, the finished top face of the part is usually considered Z zero. Generally speaking, X and Y zero can be anywhere on the part. However, it is a good practice to place the X and Y zero location where the stops locate the part on the fixture.

On a turning center, program zero for the X axis is at the centerline of the part (zero diameter). For the Z axis, the finished face, which is away from the spindle, is usually Z zero.

RPM. RPM stands for Revolutions Per Minute, which relates to the rotational speed of the spindle. Specifying the spindle speed with RPM applies to both turning centers and machining centers. The spindle speed of a machining center is always specified with RPM. On a turning center, SFM, SMM, or RPM can be specified.

Servomotor. A servomotor is a very high torque motor that is used to rotate the ball screws of the machine, which causes the axes to move. Each axis will have its own servomotor.

The servomotor moves a certain distance specified by a program command, such as G01 Z-1.375. During the motion, the servomotor provides feedback to the control as to the current position. If for some reason the axes can't get to the specified position, the machine will be sent into an alarm state alerting the operator of a problem.

There are many different types of servomotors in use today. Electric servomotors come in three basic varieties: rotary analog, rotary digital, and linear. Hydraulic servomotors have also been used, and are still finding new applications today.

With a rotary servo, rotational power is applied to a screw, which is affixed to a slide with a nut, thereby converting rotary motion to linear motion. This configuration has several drawbacks, including limitations in velocity and acceleration. Linear servomotors, which represent the most current technology, consist of only two parts—a moving coil (attached to the slide) and a stationary magnet track. There is no physical contact between the coil and magnetic track, which reduces wear to an absolute minimum. Maximum velocity with rotary servos and ball screws is approximately 1400 IPM, whereas a linear servo can efficiently operate at a velocity of 7000 IPM and, in special instances, up to 18,000 IPM.

SFM. SFM stands for Surface Feet per Minute. This is a measure of the cutting speed as measured at the cutting diameter and is relative to the circumference at that diameter. The circumference of the part at the cutting diameter is considered the surface. Increases in the amount of surface that passes the tool in one minute generally results in additional friction (heat buildup). For proper machining efficiency, this friction needs to be maintained at a certain level in order for the material to be cut efficiently. With carbide tooling, higher speeds (and heat) are generally achieveable when compared to high-speed steel tooling. (Also see SMM.)

Single Block. Single block is a conditional switch that causes the machine to execute one program line at a time, and then stop. By pressing the cycle start button, the machine will continue to the next line. This is used during setup so that the setup operator has full control of the machine.

SMM. SMM stands for Surface Meters per Minute. This is a measure of the cutting speed as measured at the cutting diameter and is relative to the circumference at that diameter. The circumference of the part at the cutting diameter is considered the surface. Increases in the amount of surface that passes the tool in one minute generally results in additional friction (heat build up). For proper machining efficiency, this friction needs to be maintained at a certain level in order for the material to be cut efficiently. With carbide tooling, higher speeds (and heat) are generally achieveable when compared to high-speed steel tooling. (Also see SFM.)

Soft Key. A soft key is a software driven key located at the bottom of the CRT. The function of the key changes depending on which screen is being observed on the CRT.

The reason for soft keys is that many different control buttons are needed to enable all of the features of a modern machine. Instead of having independent buttons for each feature of the machine, which would require a large number of buttons, control builders install a series of buttons at the bottom or side of the CRT. The buttons can then be used for multiple purposes, as their function shifts dependent on what operational mode the machine is performing. Hard keys are buttons whose functions remain constant regardless of the operational mode of the machine.

Software. Software generally refers to a computer program. A CNC program is software. The term came about to distinguish between hardware elements (the physical components of the computer) and the software (instructions that drive the hardware).

A CNC controller contains software (an operating system) that performs all of the operations of the machine tool. Most, if not all, CNC controls conceal this software from the end user and do not allow modification of this software. This resident software is sometimes referred to as the executive program.

Step Mode. Some machines do not have a hand wheel (manual pulse generator) for accurately jogging the machine. On these machines there is usually a step mode. When the machine is in the step mode, press the jog button and the machine will move a precise amount. The distance the axis would move is dependent on the step setting which typically offers these selections: .0001", .0010", .0100", or .1000".

Subprogram. A subprogram is a part of the whole program. Although a subprogram is a separate program, it is called into execution to be the main program. Generally, a sub program only contains certain functions that are recalled repeatedly by the main program, as in the case of hole locations or a contour that is machined with several tools.

Syntax. Computer programming has a basic set of rules as to how a program should be written. CNC programming language is fairly straightforward and simple when compared to other programming languages. For example, one of the syntax rules specifies that each command (sometimes referred to as an address) begins with a single letter and is followed by a number (G01 X1.2887).

Turret. On a turning center, a turret refers to the device where the tools are mounted on the machine. On most machines, the turret can hold 8 or 12 tools. To index from tool to tool, the turret rotates around a center shaft and couples to the turret body with a curvic coupling.

Zero Return. Zero return refers to the process of sending the axes to the home position. The process can be performed manually or through program control by the G28 command. This command must be performed when the operator turns on the power of the machine. During the program execution, it is not necessary to send the axes to home position between tools if geometry offsets are used. However, when using G50 commands (G92 on a machining center) to define the tool's position relative to program zero, it is necessary to zero return the axes between each tool change.

Zero Return Mode. The zero return mode allows the user to jog the axes to the home position of the machine. This is usually only needed when the machine is turned on or after manual interruption of the machine's cycle.

Programming

Currently there are three methods of writing CNC programs: manually, conversationally, and using CAD/CAM software.

Manual programming requires the programmer to write out the program code for code and line by line. For simple parts, manual programming is efficient. However, as the complexity of the part increases, so does the time required to write the program. As part complexity increases, the amount of math required to calculate missing dimensions for coordinates increases. Even programmers who are proficient with geometry and trigonometry can spend a great deal of time on these calculations. When writing programs for complex parts, conversational programming and CAD/CAM (Computer Aided Drawing/Computer Aided Machining) programming have the advantage.

In order to write programs manually, the programmer must possess an intimate knowledge of the control, the machine, the tooling, the mathematics, and the actual codes used in the program. First, the programmer must decide on a process to machine the part. Next, the tools that will best suit the part and the process are selected. Then, the math needs to be done to calculate for missing dimensions on the part and to calculate the speed and feed values for the cutting conditions. The programmer can then write out the program line by line.

Most programmers who write manual programs use some form of editor on a computer to enter the actual program. If this is an editor designed for NC programming, there will usually be some built-in macros to speed up the data entry and to help with some of the math operations.

CAD/CAM programming minimizes the need for the programmer to perform complex math calculations. Due to the minimal amount of math required by the programmer, CAD/CAM programming tends to be more efficient than manual programming. Most product design is performed on computers using CAD software. By using the existing CAD drawing file, and importing that data into the CAM software, the programmer

only needs to decide on which process and tooling to employ. Tooling data and cutting conditions are also stored with the CAM system, so a programmer only needs to select them from a list.

Once the CAD drawing file is imported into the CAM software and the tooling is selected with the appropriate cutting conditions, the programmer selects which areas on the part are to be machined with each tool. The CAM software then calculates all of the moves necessary to machine the part. The CAM software will also check to be sure that the tool can physically cut the part with the specified tool geometry. Tool by tool, the programmer selects the area to machine and the tool to perform the operation. When complete, the data is output to an intermediate file that cannot be loaded into the machine tool. Another step is required to convert that intermediate data into machine specific code. That step is called postprocessing.

In an ideal world, CAD/CAM programming would be perfect. However, programmers must keep in mind that there is some potential for unintentionally entering incorrect data. This can occur when a product designer draws a part. Some features on the part may be exaggerated to make them more visually clear. An example of this would be a very small chamfer or a shallow taper. Tiny features such as these do not show up well on the printed drawing, and often the designer will exaggerate the features so that they appear more obvious. To compensate for this, the designer will make sure that the actual numeric dimensions are correct, even though the feature size is exaggerated on the drawing, so that the programmer's entries will be accurate if the printed numeric dimensions are used to produce the part. However, if this CAD drawing file is then loaded into the CAM software, the final part will be machined incorrectly, because the CAM software does not read the "printed" text describing the features. Instead, it uses the actual drawing geometry. When the design and programming departments are in close communication with each other this problem can be minimized.

Another problem with the CAD/CAM concept is that many times the original CAD drawing file is not available to the CNC programmer. A good example of this would be when a contract or job shop manufactures parts for another company. CAM software could still be used, but the drawing must be recreated using CAM software. For complex parts, this is still faster than manual programming. With simple parts, however, the extra time needed to recreate the drawing may take longer than writing the program manually.

Conversational programming refers to the special capability of some controls. Conversational programming is performed at the machine tool on its control, which will have a programming system built into it that is similar to CAM software. There is generally no provision to load a CAD file, which means the programmer must first recreate the drawing on the control. Once that step is completed, the programmer then selects a machining operation and a cutting tool. Using a built-in database of cutting conditions the control then creates the tool paths necessary to machine the part. Each operation is programmed in this fashion. Some of the conversational controls can then run the program, whereas other controls may require another step to postprocess the data for the machine.

Conversational controls are advantageous in certain applications, but, because the programming is performed on the shop floor, many distractions exist which can cause errors while programming. Another problem is that the machine may sit idle while the program is created. Although most conversational controls allow for programming while the machine is running, if the cycle time is very short there is little time left for writing programs.

Each of the three programming methods has advantages and disadvantages. Many shops today use all three methods.

As a precautionary step, many programmers run the CNC program on a CNC verification software package. CNC verification software reads and interprets each of the codes in the program. The resulting interpretation is then graphically animated on the computer screen, showing the tool motions and the material being removed from the part. By verifying the tool paths on a computer prior to running the program on an actual machine, time is saved during the setup. Safety is also increased as fewer errors in the program make it to the machine where mistakes can be very costly.

Nearly all CNC machines are programmed using a similar format. Although there is a published standard (EIA-274-D), many of the newer CNC controls follow the standard created by one of the largest producers of industrial controls. The programming codes and formats used with the popular CNC controlled equipment are very similar to the EIA-274-D standard.

Syntax. The NC programming language is a relatively simple language with strict rules. Each program must begin with a program label, which is the letter "O" followed by a four-digit number. Each code in the program begins with a letter followed by a number. In proper NC terminology, a letter is usually referred to as an address, but sometimes the whole code (word) is also referred to as an address. Programs must end with a program end code (M30), or stop code (M02).

On some controls, there can only be one M code per line. However, on some newer controls this rule is obsolete. Generally, when using the inch units, numbers can have a precision of four decimal places. With metric units there are only three decimal places of precision.

Program execution proceeds from top down. However, there are special codes that allow the programmer to cause program execution to jump to a certain line number.

Axes of Motion. Each machine tool has its own basic set of axes, such as X and Z for a turning center, and X, Y, and Z for a machining center. The diagrams below show these axes of motion for four different machine tools.

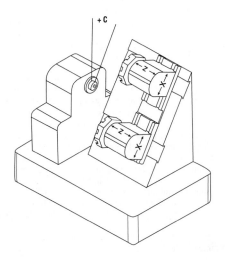

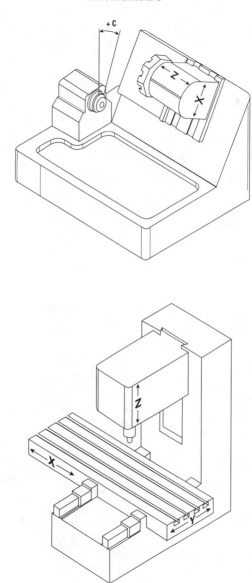

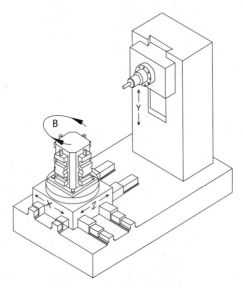

Virtually all machining centers and turning centers follow the basic coordinate system exactly as shown. Additional axes nomenclature tends to be more obscure.

When there is a second turret on a turning center that can only move along one axis, a different axes label will be used for this axis of motion. One machine tool builder may call this the B-axis while another will call it the Z-axis. Part of the reason for this has to do with how the machine tool is controlled. Some machines use two controls to handle the multiple turrets, whereas other machines only use one control to handle both turrets. On machines with two controls, the second turret's motion is referred to as the Z-axis, as is the case of a four-axis machine. Machine tools that only have one control will use the B address to represent the Z-axis for the second turret. As machine tools become more complex with many different axes of motion, there will probably be more deviation from the established standards.

Another consideration when dealing with axes of motion has to do with how an axis is programmed. For example, on a machining center the X-, Y-, and Z-axes are all programmed with linear coordinates. This means that if any axis moves, it will move in a straight line. Also, the specific distance of movement along the axis must be programmed. To move one inch requires a command to move one inch.

On a turning center, the Z-axis is considered linear. The X-axis, even though it moves in a straight line, is slightly different. The X-axis on a turning center moves the tool perpendicular to the centerline of the spindle. The tool is positioned relative to the diameter of the part. Although some machine tools use radius programming, most turning centers use diameter positioning for the X-axis. For a machine that programs using diameter programming, the command X1.5 will position the tool at a 1.5 diameter. Likewise, a machine that is programmed using radius coordinates uses the command X.75 to position the tool at a 1.5 diameter.

For rotary axes, like a C-axis on a turning center and the C- and B-axis on a machining center, the coordinate system is not a linear coordinate system but rather a rotary coordinate system using degrees. Although the angular measurement of degrees is fairly international, it still presents some problems. On blueprints and in engineering documents, angles are often expressed as Degrees - Minutes - Seconds. A minute is 1/60th of a degree, and a second is 1/60th of a minute. On most CNC machines, the programmer commands decimal parts of a degree. An angle of 47 degrees, 22 minutes would be expressed as 47.3666 (22 ÷ 60 = .3666).

Coordinate Systems. Nearly all CNC machines use one of two basic coordinate systems: incremental or absolute. Early in the history of NC and CNC, incremental coordinates were used exclusively. However, when absolute coordinates became available on CNC machines, and once they were understood, the majority of programmers switched over to the absolute coordinate system.

Explaining the incremental coordinate system is similar to giving someone driving directions. There is no singular point of reference and all positions are expressed as a distance and direction to move relative to the current position. The direction is expressed with the axis name along with a positive or negative value. The distance is specified as a numeric value using either inch or metric units. On a machining center, the following commands can be used to make a one unit square using the incremental coordinate system:

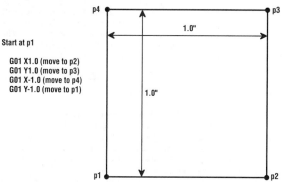

Each command is specifying a distance and a direction to move the tool. The first line commands the tool to move in the X-positive direction (right) a distance of one unit. The second line commands the tool to move in the Y-positive direction (up) a distance of one unit. The third line commands the tool to move in the X-negative direction (left) a distance of one unit. Finally, the fourth line commands the tool to move in the Y-negative direction (down) a distance of one unit.

With absolute coordinates, there is one reference point on the workpiece that is called program zero. This point on the workpiece is X zero, Y zero, and Z zero. All movements or positions are made relative to this origin or reference point. The diagram below shows how the absolute coordinate system works.

On the diagram, X and Y zero are in the lower left corner of the square. All positioning moves will be commanded as a position relative to that origin point. Another way to think about absolute coordinates is to think in terms of telling the tool what position to move to, NOT how far it should move.

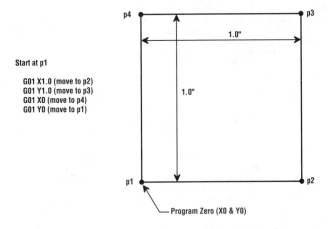

Start at p1

G01 X1.0 (move to p2)
G01 Y1.0 (move to p3)
G01 X0 (move to p4)
G01 Y0 (move to p1)

The first line commands the tool to move to the position of X1.0, which will place the tool one unit to the right of X zero. The second line commands the tool to move to the position of Y1.0, which will place the tool one unit above Y zero. The third line commands the tool to move back to the position of X0, which places the tool at X zero. The fourth line commands the tool to move back to the position of Y0, which places the tool at Y zero.

By comparing the two program examples, it can be seen that they both started with the same command. However, after the initial movement, the programs vary greatly. Many beginning programmers think that incremental coordinates are easier to use, so they try to write programs using incremental coordinates. However, as programs grow in size, it becomes more difficult to read and understand an incremental program. To read an incremental program, always start by reading from the very first line and follow each motion through the program. This is necessary because every command is relative to the tool's last position. With a program written in absolute coordinates, each command within the program shows exactly where the tool is positioned relative to program zero.

Although absolute coordinates dominate the programming style, incremental coordinates are still used in special cases, such as subprograms.

Types of Motion. CNC machines all have three basic types of motion: rapid motion, linear interpolation, and circular interpolation. To specify the type of motion, use the appropriate G code: G00 for rapid motion, G01 for linear interpolation, and G02 or G03 for circular interpolation.

Rapid motion (G00) is used to move the tool quickly into position for cutting. When rapid motion is commanded, the tool will move as fast as the machine will allow (there is no control, except with manual overrides). Because the speed at which the tool moves is not controlled, rapid motion cannot be used when cutting the workpiece.

When the tool needs to move along a workpiece to cut a straight line, linear interpolation (G01) is used. Linear interpolation causes the tool to move in a straight line motion along one or more axes at a controlled rate of feed. The rate of feed is infinitely variable up to the machine's maximum allowable feed rate. Because the tool can move in any direction and with any combination of axes, cutting any angle is also infinitely variable.

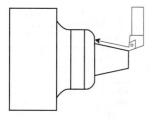
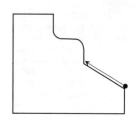

To cut a circle or radius on a part, use circular interpolation (G02 or G03). Circular interpolation will combine the movement of two axes to create a circular motion. Like linear interpolation, circular interpolation allows for infinite control of the rate of feed. Because there are two directions in which the tool can move around a radius, there are two G codes used for circular interpolation: G02 and G03. G02 moves the tool in a clockwise direction, whereas G03 moves the tool in a counterclockwise direction.

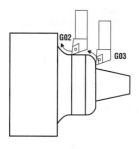
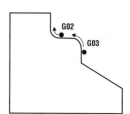

For all three types of motion, the programmer specifies the position to move to by inserting the appropriate axes address. For G00 and G01, only the destination point must be specified. However, with G02 and G03, additional information, which defines the radius of the arc or circle that is to be machined, must be specified. For many controls, this is simply done with the R address. For example, to mill a $3/4$" radius onto a part, the command G02 X3.0 Y.5 R.75 could be used. The X and Y coordinates specify the ending position of the tool, and the R code specifies the radius of the arc. The control creates a $3/4$" radius between the tool's starting and ending points.

Another type of motion, called helical interpolation, is used on machining centers. Helical interpolation is used primarily for milling threads onto a workpiece and is covered in depth in the Thread Milling section of this book. The motion itself is actually a combination of linear and circular interpolation. The X- and Y-axes move in a circular motion while the Z-axis moves in a linear motion. This combination of moves creates a helical motion that allows for threads to be milled onto a part. The milling cutters used for helical interpolation are called thread mills, which are similar in appearance to a tap, but they are ground very differently with regard to cutting edges and clearances. Prior to helical interpolation, it was extremely challenging to machine male threads or large diameter female threads with a machining center. With helical interpolation, it is possible to machine a thread on any diameter workpiece that will fit on the machine.

Rate of Feed. With linear and circular interpolation it is necessary to specify the velocity or speed at which the tool will move. For each type of tool and workpiece material, the programmer will also determine the correct rate of feed. This information can be found in the speeds and feeds section of this book, or in catalogs and flyers distributed by tooling vendors. In most cases a feed rate will be specified as a distance per revolution such as Inches Per Revolution (IPR or in./rev), Inches Per Tooth (IPT or in./t), Millimeters Per Revolution (MMPR or mm/rev), or Millimeters Per Tooth (MMPT or mm/t). Machining centers generally need to be programmed in Inches Per Minute (IPM or in./min) or Millimeters Per Minute (MMPM or mm/min) whereas turning centers use IPR or MMPR. In either case, the rate of feed is commanded with the F address. For example, F3.42 for inch units and F86.868 for metric units.

All tools that cut along the Z axis of motion will have their rate of feed recommendation specified as IPR or MMPR. Turning and boring feed rates are usually recommended in these units as well. This simple formula can be used to convert from IPR to IPM and MMPR to MMPM.

RPM × IPR = IPM	*or*	RPM × MMPR = MMPM
	600 RPM	
.0057 IPR	*or*	.145 MMPR
600 × .0057 = 3.42 IPM	*or*	600 × .14478 = 86.868 MMPM

Many tools used on machining centers cut sideways. That is to say, they cut with the side of the tool as is the case with milling. These types of tools will have their feed rate recommendations specified as IPT or MMPT. This formula can be used to convert from IPT or IPMM.

NT × IPT × RPM = IPM	*or*	NT × MMPT × RPM = MMPM
NT (number of teeth or flutes) 4 Flute end mill		
.0025 IPT	*or*	.0635 MMPT
	600 RPM	
4 × .0025 × 600 = 6 IPM	*or*	4 × .0635 × 600 = 152.4 MMPM

On most machine tools, the control feeds the tool at the specified feed rate, even if the tool cannot handle the chip load. In this situation, the tool will usually break. A special feature that is resident in some types of controls allows for the control to automatically override the programmed feed rate under certain conditions. This is called adaptive control. With an adaptive control, the programmer specifies the desired cutting parameters (speed, feed, power consumption, etc.). However, while the machine is executing the program, the control monitors the cutting load on the tool and automatically reduces the feed rate if the load becomes too high. Opinions vary on the use of adaptive controls when applied to metal cutting applications. There are cases where adaptive controls are advantageous. These include milling molds, wire EDM machining, and sinker type EDM machining. Many times these machines can perform their operations while unattended, and monitoring the feed rate with adaptive controls can be very beneficial.

Spindle Speed. The spindle speed on most CNC machine tools is infinitely variable within the limitations of the machine. Programming the spindle speed on a machining center only requires that it is programmed to run at a specific RPM. The programmer calculates the appropriate RPM for the cutting tool as applied to the specific machining operation. The S code is used to specify the spindle speed, and an M code is used to turn the spindle on. On most machine tools, M03 turns the spindle on in forward rotation, M04 turns the spindle on in reverse rotation, and M05 stops the spindle.

S750 M03 → Turns the spindle on in forward rotation at 750 RPM.

Turning centers can also be programmed in direct RPM mode. Because of the nature of turning, boring, and facing operations, the cut diameter is constantly changing. Because the cut diameter is constantly changing, turning centers have a special spindle speed mode that will keep the surface footage constant by varying the spindle speed, depending on the cutting diameter. In order to keep the surface footage constant while the cut diameter changes, the RPM must be changed dynamically during the cutting operation. The control handles all of the complexities of varying the RPM. The programmer need only specify the desired SFM or SMM.

Because a turning center's spindle can be programmed two different ways, a G code is used to set the machine tool into the desired mode. G97 is used to set the machine into RPM mode. G96 is used to set the machine into constant surface speed mode. The S code is used to specify the speed for both modes. Like a machining center, M03 turns the spindle on in forward rotation, M04 turns the spindle on in reverse rotation, and M05 stops the spindle.

G97 S600 M03 → Turns the spindle on in forward rotation at 600 RPM.

G96 S450 M03 → Turns the spindle on in forward rotation at 450 surface feet per minute (or 450 SMM if in metric mode).

Offsets and Tool Compensation. In a perfect world, the operator would load the program, part, and tools into the machine and run the cycle, which would result in a perfect part. However, in the real world tooling is not perfect. Tools are not all the same size, and they get dull. Offsets are the feature used to adjust for tool size and for wear.

A CNC machine offers several means of compensating for tool geometry and tool wear. A CNC machine also allows for compensation of the workpiece location, relative to the machine's home position. Even though offsets are similar for turning centers and machining centers, the explanations are best done separately.

Turning center offsets: Most modern CNC turning centers provide the operator with three types of offsets: wear, geometry, and work shift.

The wear and geometry offsets are activated when a tool is called up. Usually, this is done with a combination tool call up and offset call up code. For example, T1212 calls up tool twelve (the first two digits after T) with offset twelve (the last two digits). With this coding system, the programmer can use a different offset number for individual tools.

The offset number specified with the tool call up command does not actually specify an offset value. Instead, it specifies an offset register where the operator can place an offset value. Offset values are variable information, which means that they will change not only during a production run, but also from one setup to the next. Because offset values are temporary data, they do not belong inside the actual CNC program. The CNC control has an offset page or screen where the offset values are input. The diagram on the next page shows what an offset screen might look like.

When the program executes an offset command such as T0303, the offset becomes active. When the program commands the tool to move to any position, the final position will be the result of the programmed commands (with the X and or Z value), plus the offset value stored in offset register. The resulting position is offset from the programmed tool path by the amount specified in the offset command.

On most turning centers, there are actually two offsets that are activated with the tool call up: wear and geometry. The wear offset is used to compensate for the wear on the tool as it gets dull. It is also used to adjust the size of the part during setup and production.

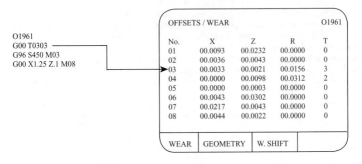

The geometry offset stores basic dimensional information about the tool such as how far the tool projects from the turret, the radius of the insert, and the orientation of the tool relative to the part. The diagram below shows one example of geometry offset measurements.

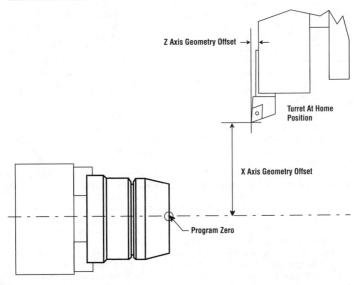

It should be noted that geometry offsets are used many different ways on turning centers. The explanation shown above is only one of many possible methods.

Another offset used on turning centers is the work shift offset. The work shift offset is used to shift the relative position of all the tools so that they line up with program zero on the workpiece.

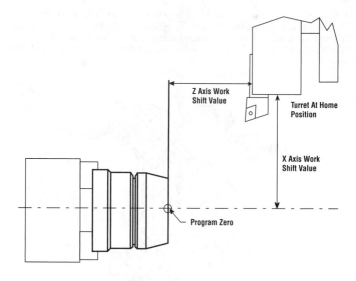

Another form of compensation used on turning centers is called tool nose radius compensation. All turning and boring tools have a small radius formed on them. When using standard carbide inserts, this radius is usually .0156", .0312", or .0469". When the tool is cutting parallel to an axis, there is no need to compensate for the tool's radius. However, when the tool is cutting an angle or a radius on a workpiece, compensation must be made for the radius of the tool. When a turning or boring tool is programmed on a turning center, the programmer treats the tool as if it has a sharp corner. All positioning moves are made relative to this sharp corner. In the diagram below, the theoretical sharp corner of the tool and the actual radius of the tool do not follow the same path.

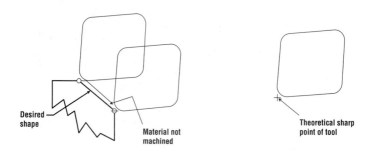

To compensate for this, the control needs to know the tool's radius and orientation. This information is entered into the geometry offset screen during the setup procedure. The actual radius value of the tool is entered as expected. The tool orientation is entered as a numeric code. Using the diagram below, the different tool orientations and their respective codes can be observed.

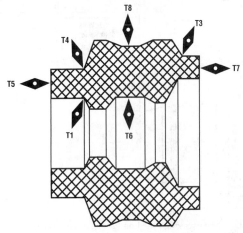

Within the program, a few additional codes need to be added to use tool nose radius compensation. G41 is used to turn tool nose radius compensation on when the tool is to the left of the programmed path. G42 is used to turn tool nose radius compensation on when the tool is to the right of the programmed path. G40 is used to turn the feature off.

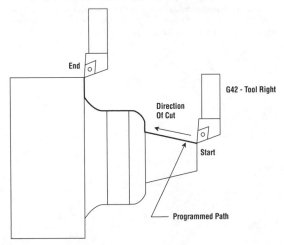

Machining center offsets: Most machining centers have two types of offsets that can be used to compensate for three distinct features: tool length, tool diameter, and fixture location.

It is nearly impossible to assemble all of the tools for a program to the exact same length. Furthermore, when rerunning a job, it is just as difficult to reassemble tools that are the same length as the tools used the last time it was run. Knowing this, CNC controls provide a feature, called tool length compensation, that compensates for the tool's length. During setup, the operator inputs the length of the tool into an offset register.

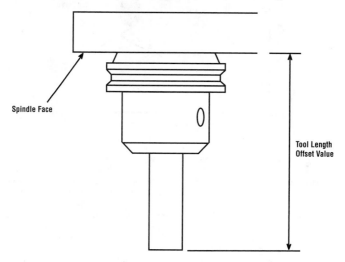

Spindle Face

Tool Length
Offset Value

The program accesses this tool length offset by activating the appropriate tool offset. The command G43 tells the machine to compensate for the tool's length in Z axis. An H code is used to specify which offset code to use. The whole command might look like this:

G00 G43 Z.1 H01.

This command would be used to bring the tool from the Z axis home position to a point .100" above the part while compensating for the tool length which is stored in offset register 1.

The second type of compensation and offset on a machining center deals with the tool's radius. When cutting with the side of an end mill while milling a contour, compensation must be allowed for the distance from the center of the tool to its cutting edge (the radius of the tool). This is important because when a contour is programmed, the actual dimensions of the part must be used for the coordinates. The programmer does not have to shift the coordinates to compensate for the tool's radius. Instead, the setup person inputs a value into the offset page that is equal to the tool's radius. The diagram below shows a simple contour example with the programmed coordinates and the tool's centerline path.

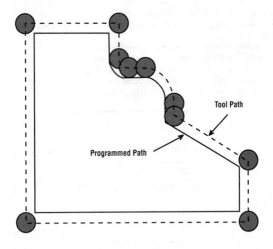

Tool Path

Programmed Path

The third and final type of compensation deals with the entire coordinate system. Most machine tools assume X, Y, and Z zero to be at the machine's home position. However, when a part is programmed, all coordinates used in the program are relative to one point on the part called "program zero." Obviously, program zero and the machine's home position are two entirely different positions. Work offsets are used to bring the machine's origin point and the program zero point on the workpiece into alignment. During setup, the setup person would use a probe or some form of indicator to measure the program zero location as it relates to machine home position. These measurements would then be input into the work offset page of the control.

Many controls today offer six work offsets as standard with an option for up to 48 additional offsets. This is because, on machining centers, it is very common to machine multiple parts per cycle. To call up the offset from within the program, the programmer commands the appropriate G code. The G codes: G54, G55, G56, G57, G58, and G59 call up and activate the basic six work offsets.

Canned Cycles. This section describes some important canned cycles found on many CNC machine tools. There are numerous others, but this listing will provide an overview of the advantages of using these programming time-savers.

Canned cycles for machining centers: Canned cycles, sometimes called fixed cycles, perform a specific machining operation with a minimum of input. In the case of machining centers, canned cycles are primarily used for hole machining such as drilling and tapping. With canned cycles on a machining center, the programmer initializes the canned cycle, then, on subsequent lines, lists the X and Y coordinates where the same operation is to be replicated. After the last repetitive operation is machined, the canned cycle can be cancelled with a G80 code.

Each canned cycle has a required number of variables that must be set. The following codes are universal to all canned cycles: X, Y, R, Z, and F. The X and Y codes specify the location where the machining operation will be performed. If the X and Y codes are not specified, the machine will perform the operation at its current location. The R code specifies the rapid plane, which is a position on Z-axis that the tool will rapid to. Generally the rapid plane is .050" to .100" above Z zero or the top surface of the hole. The Z code specifies the depth to which the machining operation shall be performed. The F code specifies the feed rate for the machining operation.

There are two cycles for spot/center drilling: G81 and G82. The G81 cycle feeds into the part then rapids out. The G82 cycle will dwell for a specified period of time at the Z depth. For the G81 cycle, the universal variables are used as described above.

G81 X_Y_R_Z_F_

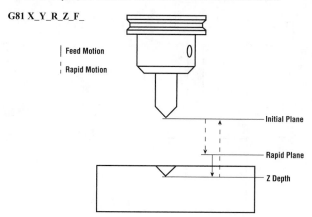

The G82 cycle, like the G81, needs minimal input. The only additional information needed is the P code, which specifies the duration of the dwell at the bottom of the hole.

G82 X_Y_R_Z_P_F_

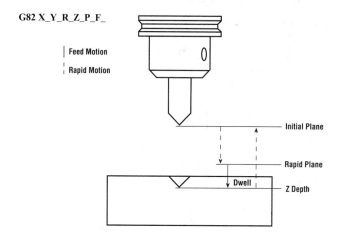

For drilling deep holes, there are two choices of cycles: G73 and G83. The G73 cycle drills to depth, and will also perform a chip breaking operation at specified intervals. The in-feed amount per chip-break is specified with the Q code. This cycle is also called "rapid peck drilling."

G73 X_Y_R_Z_Q_F_

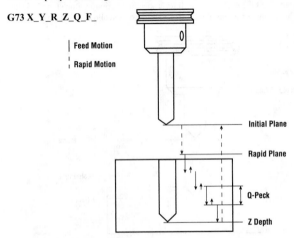

The G83 also performs a pecking cycle, but instead of backing up a small amount to break the chip, the tool retracts completely out of the hole. With this type of pecking, the chips are pulled out of the hole and fresh coolant can flood the tool for cooling. The Q code specifies the amount of in-feed per peck.

G83 X_Y_R_Z_Q_F_

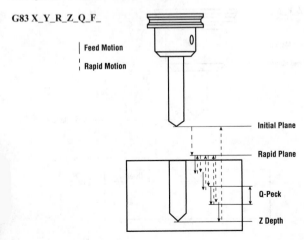

To perform tapping operations on a CNC machining center, use one of the tapping canned cycles: G74 or G84. Both tapping cycles operate in the same way, but G74 is used for left hand taps and G84 for right hand taps. The spindle should be running in the reverse direction prior to initializing the G74 tapping cycle.

G74 X_Y_R_Z_F_

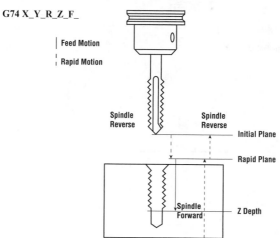

The G84 cycle is programmed identically to the G74 cycle with the exception of the G code. The spindle should be running in the forward direction prior to initializing the tapping cycle.

G84 X_Y_R_Z_F_

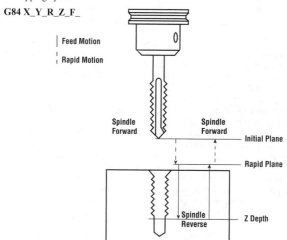

A special reaming cycle feeds the tool into and back out of the part. This cycle operates in a similar fashion to the tapping cycles, but the spindle does not change directions at the bottom of the hole.

G85 X_Y_R_Z_F_

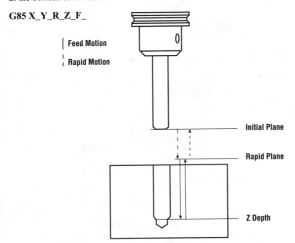

Canned cycles for turning centers: On a turning center, there are generally two types of canned cycles, simple and multiple repetitive. Simple canned cycles perform simple machining operations such as peck drilling or making a turning pass. The multiple repetitive cycles perform all of the roughing cuts necessary by initializing the roughing cycle and programming the finish shape of the part. Multiple repetitive cycles are also used to perform single point threading.

The diagrams on the next page illustrate several multiple repetitive cycles. What is not shown are the programmed coordinates that define the actual finished part. Common to the multiple repetitive cycles are the following codes: P, Q, U, W, D, and F. The P code specifies the first line label number that defines the finishing tool path. The Q code specifies the line label number of the last line of the finishing tool path. The U code specifies the amount of stock that will be left on the diameter of the part. The W code specifies the amount of stock to be left on the faces of the part for the finishing operation. The D code specifies the depth of cut per side. The F code specifies the roughing feed rate.

The G71 cycle is used to rough turn a part. The tool needs to be positioned at the stock diameter, which is how the control determines how much material to remove from the part. That position is called the convenient starting position. Based on the convenient starting position and the finish path definition, the control will then calculate each of the roughing passes needed to completely remove all of the waste material. When the cycle is completed, the part will be oversize in diameter by the value specified with the U code, and all of the faces will have material left on them as specified by the W code. Each rough pass depth of cut is controlled by the D code.

G71 P_Q_U_W_D_F_

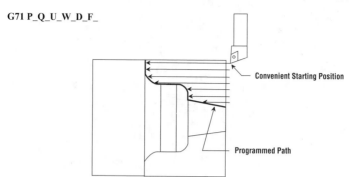

The G72 cycle works just like the G71 cycle, except that the cutting operation is that of facing as opposed to turning cuts.

G72 P_Q_U_W_D_F_

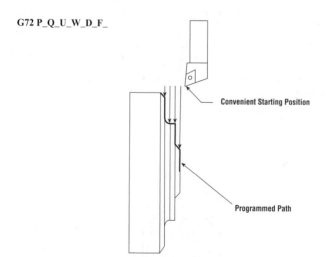

The G73 cycle is used to rough machine a casting or forging. Unlike the G71 and G72 cycles, the cutting motion is not parallel to an axis. Instead, the roughing passes will run parallel to the finish shape. The I and K codes specify the amount of material that is on the X- and Z-axis, respectively.

G73 P_Q_I_K_U_W_D_F_

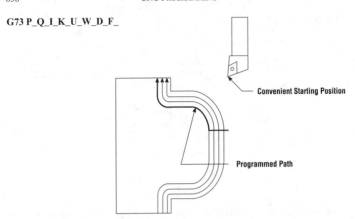

Convenient Starting Position

Programmed Path

Single point threading also requires a series of passes to properly and safely machine a thread onto a part. The X code specifies the minor diameter of the thread (major diameter in the case of a female thread). The Z code specifies the ending Z position of the thread. The K code specifies the first pass depth of cut. After the first pass, each successive depth of cut will be slightly smaller than the preceding one. As many passes as necessary are performed to machine the complete thread. The A code specifies the angle of the tool that is used to define the in-feed angle for each pass. The D code specifies the radial depth of the thread. The F code is the feed rate, which is also the lead of the thread.

G76 X_Z_K_A_D_F_

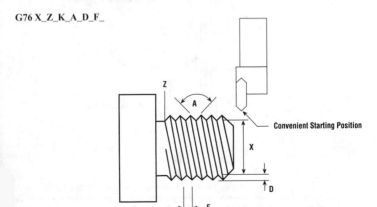

Convenient Starting Position

Subprograms. Subprograms can be powerful tools when writing programs. Some of the more typical uses for subprograms are to store the coordinates for hole locations or for a contour. Subprograms can save time when writing programs manually, as duplicate coordinates do not have to be written repeatedly. Another benefit of subprograms is that with proper use, program size is reduced. The CNC control has a limited amount of storage space. When that limitation is exceeded, subprograms are an efficient way to reduce the size of the program.

A sub program is an external program from the main program. It is called into execution one or more times by the main program with the following command: M98 P1962. The example command would call up program O1962 and execute it from the beginning. A subprogram ends with an M99 command instead of M30 as used in main programs. When the M99 command is executed, the main program is called up and program execution continues after the M98 P1962 command.

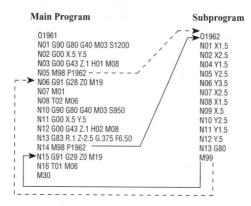

Main Program	Subprogram
O1961	O1962
N01 G90 G80 G40 M03 S1200	N01 X1.5
N02 G00 X.5 Y.5	N02 X2.5
N03 G00 Z1.1 H01 M08	N04 Y1.5
N05 M98 P1962	N05 Y2.5
N06 G91 G28 Z0 M19	N06 Y3.5
N07 M01	N07 X2.5
N08 T02 M06	N08 X1.5
N10 G90 G80 G40 M03 S950	N09 X.5
N11 G00 X.5 Y.5	N10 Y2.5
N12 G00 G43 Z1.1 H02 M08	N11 Y1.5
N13 G83 R1 Z-2.5 G.375 F6.50	N12 Y.5
N14 M98 P1962	N13 G80
N15 G91 G29 Z0 M19	M99
N16 T01 M06	
M30	

In the example above, the main program initializes a canned cycle for drilling holes and the first hole is drilled. Then the subprogram, which contains the remainder of the hole locations, is called up. When the subprogram ends at the M99 command, the main program is recalled and program execution continues on the line after the M98 command.

When used with incremental coordinates, subprograms can be very powerful. However, the most popular use for subprograms is to store simple absolute coordinates for contours and hole locations that are machined with multiple tools.

Popular Codes. Although there are published standards for CNC programming, each control manufacturer can, and usually does, create its own code set. One of the more popular controls manufacturers over the last 30 years has created their own "standard" of sorts. Today, many new controls now adopt that code set into their controls.

Addresses (codes)

Code	Description
O	Program label
N	Line label
G	Preparatory function
X, Y, Z, Q, R, I, J, K	Dimension word

F	Feed
S	Spindle speed
T	Tool function
M	Miscellaneous functions
X, P	Dwell
P	Program designation
H	Offset number
B	Second miscellaneous function

Preparatory Functions (G codes)

Code	Description	Application
G00	Rapid Motion	Lathe & Mill
G01	Linear Interpolation	Lathe & Mill
G02	Clockwise circular interpolation	Lathe & Mill
G03	Counterclockwise circular interpolation	Lathe & Mill
G04	Dwell	Lathe & Mill
G10	Data setting	Lathe & Mill
G11	Data setting mode cancel	Lathe & Mill
G15	Polar coordinates mode cancel	Lathe & Mill
G16	Polar coordinates mode on	Lathe & Mill
G17	X Y Plane selection	Lathe & Mill
G18	X Z Plane selection	Lathe & Mill
G19	Y Z Plane selection	Lathe & Mill
G20	Units - Inch	Lathe & Mill
G21	Units - Metric	Lathe & Mill
G27	Reference point return check	Lathe & Mill
G28	Return to reference point	Lathe & Mill
G30	Return to auxiliary reference point	Lathe & Mill
G31	Skip cutting	Lathe & Mill
G32	Thread cutting	Lathe
G34	Variable pitch thread cutting	Lathe
G40	Tool nose radius compensation cancel	Lathe
G40	Tool radius compensation cancel	Mill
G41	Tool nose radius compensation left	Lathe
G41	Tool radius compensation left	Mill
G42	Tool nose radius compensation right	Lathe
G42	Tool radius compensation right	Mill
G43	Tool length compensation + direction	Mill
G44	Tool length compensation - direction	Mill
G45	Tool offset increase	Mill
G46	Tool offset decrease	Mill
G47	Tool offset double increase	Mill
G48	Tool offset double decrease	Mill
G50	Coordinate system setting (Max RPM limit)	Turn
G50	Scaling on	Mill
G51	Scaling cancel	Mill
G52	Local coordinate system setting	Lathe & Mill
G54	Work offset #1	Lathe & Mill
G55	Work offset #2	Lathe & Mill
G56	Work offset #3	Lathe & Mill

G57	Work offset #4	Lathe & Mill
G58	Work offset #5	Lathe & Mill
G59	Work offset #6	Lathe & Mill
G60	Single direction positioning	Mill
G61	Exact stop mode	Mill
G62	Automatic corner override	Mill
G63	Tapping mode (special)	Mill
G64	Normal cutting mode	Mill
G65	Macro call	Lathe & Mill
G66	Modal Macro call	Lathe & Mill
G68	Coordinate rotation on	Mill
G69	Coordinate rotation cancel	Mill
G70	Finishing cycle	Lathe
G71	Rough turning cycle	Lathe
G72	Rough facing cycle	Lathe
G73	Pattern repeating cycle	Lathe
G73	Peck drilling cycle	Mill
G74	Peck drilling cycle	Lathe
G74	Left hand tapping	Mill
G75	Grooving cycle	Lathe
G76	Threading cycle	Lathe
G76	Fine boring cycle	Mill
G80	Canned cycle cancel	Mill
G81	Spot/center drilling cycle	Mill
G82	Spot/center drilling cycle W/dwell	Mill
G83	Deep peck drilling cycle	Mill
G84	Right hand tapping cycle	Mill
G85	Reaming cycle	Mill
G86	Boring cycle	Mill
G87	Back boring cycle	Mill
G88	Boring cycle	Mill
G89	Boring cycle	Mill
G90	Simple turn cycle	Lathe
G90	Absolute coordinates	Mill
G91	Incremental coordinates	Mill
G92	Simple thread cycle	Lathe
G92	Coordinate system setting	Mill
G94	Simple facing cycle	Lathe
G94	Feed per minute mode	Mill
G95	Feed per revolution mode	Mill
G96	Constant surface speed control	Lathe
G97	RPM spindle mode	Lathe
G98	Return to initial plane in canned cycle	Mill
G99	Return to rapid plane in canned cycle	Mill

It should be noted that as the capabilities of turning centers continue to increase with regard to milling, many of the mill-only codes are now also available for turning centers.

Miscellaneous Functions (M codes)

Code	Description
M00	Program stop
M01	Optional program stop
M02	Program end without rewind to start
M03	Spindle on - forward rotation
M04	Spindle on - reverse rotation
M05	Spindle stop
M06	Change tools - Mill
M07	Mist coolant on
M08	Flood coolant on
M09	Coolant off
M30	Program end with rewind to start
M98	Subprogram call
M99	Subprogram end

Machine tool builders generally follow these basic codes. However, the machine tool builder has complete control over the function of any and all M codes. Some machine tools will have only those codes listed, whereas other machine tools may have many more codes.

Minimizing Door-to-Door Time

Today, progressive machining operations are intensely focused on lead-time reduction. Driven by their customers, these shops realize that offering shorter lead-times is a critical competitive advantage, and in some cases the only way to retain business. Unfortunately, many shops are seeking to provide shorter delivery times in ways that add significant cost and complexity to the business, while only marginally reducing lead-time.

This section, authored by Miles Arnone, introduces the principles necessary to cost-effectively and continuously reduce the time required to deliver machined parts to customers. In so doing, the entire chain of events that takes place from the moment an order is received to when the finished product leaves the shipping dock ("in the door to out the door") must be considered. As will be discussed, it is counterproductive to speak of "minimizing machining times," "minimizing inspection times," etc., without first considering the entire system of which each of these operations is a part. It is also important to note that the principles discussed here do not apply exclusively to machining operations. But, to provide concrete examples and context, the following discussion examines the reduction of door-to-door time in the context of machining.

Throughput focus versus lead-time focus

Two terms need to be defined. *Lead-time* is the amount of elapsed time required to complete a specific set of tasks. In the context of this discussion, it is the total elapsed time to produce the part or product, measured from receipt of order to shipment. Minimizing door-to-door time is synonymous with reducing overall lead-time. The units of lead-time are minutes, hours, days, months, etc. *Throughput* is a measure of how many parts or products can be produced by a specific manufacturing operation, or the entire business, per unit time. It can be measured in terms of products produced per unit time or dollars (as measured by value to the customer, i.e., sale price) per unit time. Clearly, lead-time and throughput are related. But, it is important to remember that a reduction in lead-time does not inherently boost throughput. One can imagine an example where a part moves through a shop and every successive manufacturing operation is ordered to stop production of other parts immediately to accommodate the part in question—the classic expediting process. This would lead to a very short lead-time for the specific part, but lead-times for every other part in the shop would increase dramatically. The goal of a shop is to minimize lead-times in the context of increased throughput.

An extended example will be used as a means of exploring various methods (some valid and some misdirected) one might employ to reduce lead-time. Consider a machine shop that makes two families of products (A and B). *Figure 1* illustrates the process flow for each product. For each operation undertaken, the average cycle time required to complete that operation, including setup time, is indicated.[1] **Table 1** identifies the quantity of each resource (in hours) available per month.[2] It will be assumed that all of the shop's costs are fixed; a reasonable assumption in the short-term. The variable contribution for product A is $700—and for product B is $500.[3] Also assumed is that demand exceeds supply for both product A and product B.

The first thing one is likely to do, after having collected such information, is to calculate the theoretical lead-time for each product. According to the information available, which

---

[1] All cycle times are represented in terms of an 8-hour workday. Therefore, EDM, which requires 40 hours, takes place over 5 workdays, for example.

[2] Heat treat, which is outsourced, is assumed to have a high amount of excess capacity, and is therefore ignored for the purpose of this example.

[3] Variable contribution is defined, in this case, as the sale price less materials and consumables. All labor costs are considered fixed costs in the near-term.

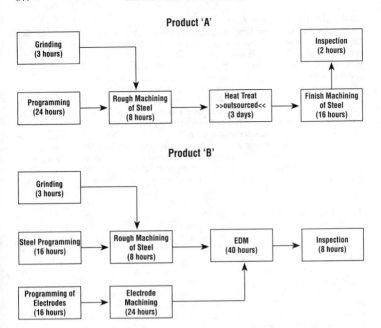

Figure 1. Process flow for each family of products.

Table 1. Resource Availability for Manufacturing System of Figure 1. Each resource unit is available eight hours per day, twenty days per month.

Resource Name	Units	Total Hours Available per Month
Programmers	3	480
Steel cutting machining centers	2	320
Graphite machining centers	1	160
Grinders	1	160
EDM	2	320
CMM (inspection)	1	160

was presumably summarized from time cards, one unit of product family 'A' can be made in an elapsed time of 9.25 business days, while one unit of product family 'B' can be produced in 11 business days. These times are calculated by simply adding up all of the operations that appear on the critical path for each product. Immediately, it can be seen that it is impossible to actually produce products in this time frame. Our model assumes that each resource required to make the product is available immediately as it is required. In other words, products 'A' and 'B' never wait in a queue between operations. Second, products 'A' and 'B' use some of the same resources. Therefore, the lead-time of either product cannot be predicted without knowing the exact product mix. Finally, unless the exact sequence and timing in which orders for product 'A' and product 'B' are received is known, it is not possible to determine the lead-time between receipt of order and shipment. If there are no products in the plant and an order is received for an 'A' it could, theoretically, be produced in 9.25 business days. But what if an order is received for an 'A' on Monday, two 'B's on Tuesday, and three 'A's and two more 'B's on Thursday? What if the sequence of orders was B, A, A, B, B, B, A, A instead? Very difficult indeed!

It can be seen, even in the context of this simple machining operation, that explicitly predicting the lead-time for a product is very difficult unless the product is being mass produced within a dedicated manufacturing system or cell. In fact, to make any headway on this problem whatsoever, one must focus on maximizing the shop's throughput rather than on reducing lead-time. Whereas lead-time focuses on the progression of a single manufactured good through the production process, throughput focuses on the performance of the manufacturing system as a whole. Only when this approach is taken can more products move through the factory in less time. In fact, by increasing throughput, lead-times are generally reduced. The only time this is not true is if the throughput is increased by adding resources in the same proportion as they currently exist, and at the same rate as the desired increase in output. For example, if the quantity of every resource is doubled, and the number of parts produced was also doubled, the throughput would double as well, but lead-time would remain the same.

The goal is to minimize the time it takes to get products to customers, while at the same time optimizing profitability. This condition must be added to the problem statement because, otherwise, the solution to minimize lead-time could very well bankrupt our fictional shop. Looking again at *Figure 1*, it can be seen that to optimize the throughput of our operation, a baseline throughput must first be established. To do this, the output of the shop in the context of a product mix must be evaluated. Having stated that demand for product families 'A' and 'B' exceeds supply, what product mix should be produced in one month to optimize the profits? A natural answer would be to produce as many of product 'A' as resources will allow, and then to produce as many of 'B' as possible with what is left over. Common sense suggests that this would maximize profits because product 'A' contributes $700 of profit per unit, while 'B' contributes only $500.

If this approach is taken, the production of our machine shop would be thirteen of 'A' and one of 'B' for a net profit of $9,600.[4] **Table 2** outlines the hours consumed for each resource to make the products, and how many hours are left over. One can see from **Table 2** that the first resource to run out of capacity will be the steel cutting machining centers. In this case, where thirteen 'A' and one 'B' have been made, the steel cutting machining centers are the *bottleneck*. In other words, our shop is limited from producing more, and from decreasing lead-time, by the amount of available steel cutting machining centers. To produce more parts in the same amount of elapsed time, the productive capacity of the bottleneck needs to increase.

Looking again at **Table 2**, it can be seen that there is significant excess capacity across many of the other resources in the shop, even as steel cutting capacity is exhausted. When one plays with the figures a bit, a more profitable production mix can be found. Surprisingly, if the shop produces eleven of 'A' and six of 'B' in a given month, a net profit of $10,700 is generated! **Table 3** details the hours consumed and the hours remaining per resource in this scenario. In this case, the capacity of steel cutting machining centers will run out if the shop tries to produce a fourteenth unit 'A.' But, if a seventh unit 'B,'

---

4   It is possible to build a simple spreadsheet with the information provided from the example to evaluate the throughput of the example machine shop, and to evaluate potential throughput increasing actions. A copy of the spreadsheet is available by request through the publisher.

Table 2. Resource Consumption to Produce (13A, 1B).

Resource Name	Total Hours Available per Month	Resources Consumed (hours, percentage)
Programmers	480	344, 72%
Steel cutting machining centers	320	320, 100%
Graphite machining centers	160	24, 15%
Grinders	160	42, 27%
EDM	320	40, 13%
CMM (inspection)	160	34, 21%

is produced, the shop will be short of both programming and graphite machining center resources. The bottleneck differs depending upon the route that is chosen, and it becomes clear that the nature of the bottleneck can shift depending upon the mixture of products moving through the shop. Further, as soon as a bottleneck is eliminated, there is always a new one. It is, in fact, highly undesirable from a profitability standpoint, and in terms of the need to control the manufacturing process, to operate without a bottleneck. If a shop has no bottleneck, it is probably suffering from overcapacity and should be taking in work at any price above its marginal costs or selling off capacity. What is most important about the example, as it has developed so far, is that in the case where the shop produces 11 of 'A' and 6 of 'B,' throughput is higher and lead-time is lower than when it produces 13 of 'A' and 1 of 'B,' because production is more ideally matched to fit the available resources.

Table 3. Resource Consumption to Produce (11A, 6B).

Resource Name	Total Hours Available per Month	Resources Consumed (hours, percentage)
Programmers	480	456, 95%
Steel cutting machining centers	320	312, 98%
Graphite machining centers	160	144, 90%
Grinders	160	51, 32%
EDM	320	240, 75%
CMM (inspection)	160	70, 44%

What if the problem of minimizing lead-times to customers is approached by focusing on the process flow for a product independent of the overall work plan for the shop? In other words, by focusing on lead-time rather than throughput. For example, if by examining the process flow for product 'A', it should be possible to reduce the overall lead-time by shortening the time required to undertake any step along the critical path. That is, if the shop can reduce the time required to complete any step, besides grinding, it should reduce overall lead-time. Eliminating three hours of programming time per product 'A', for example, would presumably reduce the lead-time of product 'A' to 8.875 days from 9.25 days. Actually, one finds that the shop can still produce no more than 11 of 'A' and 6 of 'B' because programming is not the bottleneck of the operation. In other words, saving time in programming contributes almost no improvement to shop performance. It only creates more idle time for the programmers and has a minimal effect on the shop's ability to ship more product in less time!

Figure 2. Five steps of production.

Table 4. Part Flow Through 5-step Production Cell. Beginning with unit 2, parts wait between steps B and C because step C requires 2 hours to complete. Steps D and E are often idle for the same reason.

Unit	STEP A		STEP B		STEP C		STEP D		STEP E		Lead Time
	In	Out	In	Out	In	Out	In	Out	In	Out	
1	8am	9	9	10	10	12pm	12	1	1	2pm	6 hrs.
2	9am	10	10	11	12pm	2	2	3	3	4pm	7
3	10am	11	11	12pm	2	4	4	5	5	6pm	8
4	11am	12pm	12	1	4	6	6	7	7	8pm	9
5	12pm	1	1	2	6	8	8	9	9	10pm	10

This point is made more clearly through another example. Assume that the production process consists of five steps (*Figure 2*). Steps A, B, D, and E take 1 hour each. Step C takes 2 hours. Thinking in terms of the flow of a single product through this process, it can be seen that the lead-time to produce one unit is 6 hours. But, as soon as one tries to produce more than one unit, the lead-time increases. This point is made clear in **Table 4.** Only the first unit can be produced in six hours. Every unit, thereafter, takes an additional hour to complete! This delay continues to build ad infinitum, because units "pile up" in front of operation C and must wait for C to complete work on the previous units. If the time to undertake operation A is reduced by 30 minutes, the resulting change in flow of product through the manufacturing process is shown in **Table 5.** Note that the total lead-time increases, and that, in fact, by reducing the time to undertake operation A, the lead-time quickly increases at a rate of 1.5 hours per unit. Although each unit is completed 30 minutes before they had been previously (i.e., unit three is completed at 5: 30 p.m., instead of 6 p.m.), each spends considerably more time in process. This increases the cost of carrying inventory, requires greater amounts of floor space, and increases the prospects for inadvertent mishandling of the product. If, rather than reducing the time for a nonbottleneck step, the bottleneck step C is reduced from 2 hours to 1.5 hours, significant benefits for lead-time and throughput accrue. **Table 6** outlines the resulting production process. Note that the overall lead-time for each unit is less than that of the original scenario (**Table 4**), and significantly better than when the cycle time of step A is reduced.

Table 5. Part Flow Through 5-step Production Cell Where the Time for Step A has been Reduced from 1 Hour to 0.5 Hours. Times for other steps are as before: {B,1}{C,2}{D,1} {E,1}. Delays (also WIP) are now present before both steps B and C.

Unit	STEP A		STEP B		STEP C		STEP D		STEP E		Lead Time
	In	Out	In	Out	In	Out	In	Out	In	Out	
1	8am	8:30	8:30	9:30	9:30	11:30	11:30	12:30	12:30	1:30	5.5 hrs.
2	8:30	9:00	9:30	10:30	11:30	1:30	1:30	2:30	2:30	3:30	7
3	9:00	9:30	10:30	11:30	1:30	3:30	3:30	4:30	4:30	5:30	8.5
4	9:30	10:00	11:30	12:30	3:30	5:30	5:30	6:30	6:30	7:30	10
5	10:00	10:30	12:30	1:30	5:30	7:30	7:30	8:30	8:30	9:30	11.5

Table 6. Part Flow Through 5-step Production Cell Where All Operations Take 1 Hour to Complete Except Step C, Which Takes 1.5 Hours.

Unit	STEP A		STEP B		STEP C		STEP D		STEP E		Lead Time
	In	Out	In	Out	In	Out	In	Out	In	Out	
1	8:00	9:00	9:00	10:00	10:00	11:30	11:30	12:30	12:30	1:30	5.5
2	9:00	10:00	10:00	11:00	11:30	1:00	1:00	2:00	2:00	3:00	6
3	10:00	11:00	11:00	12:00	1:00	2:30	2:30	3:30	3:30	4:30	6.5
4	11:00	12:00	12:00	1:00	2:30	4:00	4:00	5:00	5:00	6:00	7
5	12:00	1:00	1:00	2:00	4:00	5:30	5:30	6:30	6:30	7:30	7.5

The following summarizes our findings to this point.

- To minimize lead-time, one must evaluate the production system as a whole.
- Optimizing throughput is the key to achieving lower lead-times and higher profits.
- Focusing on lead-time reduction as an end in and of itself is not productive. Only by focusing on increasing throughput of the entire manufacturing system can meaningful lead-time reductions be achieved.
- To optimize throughput, all efforts should be placed on reducing the cycle time of the bottleneck step. Effort expended to reduce the cycle time of other steps is largely wasted.
- It is not practical to discuss lead-time without a solid understanding of the product mix, and the operational steps that each different product undergoes.

Common pitfalls to avoid

A number of common missteps that should be avoided by shops trying to minimize lead-times can now be identified.

Avoid local optima. Because equipment used in each operational step of a manufacturing process is usually supplied by a different vendor, it is easy to compartmentalize one's thinking and consider each step as a self-contained island. Vendors tend to promote the latest benefits of their technology independent of other steps in the manufacturing process. It is common, therefore, for shops to be drawn into seeking local optima for each operational step. In our machine shop example, if one considered the EDM process independently it may seem reasonable to consider retrofitting the machine with a better generator that reduced burn times. After all, it is the single longest process in the production of product 'B'. But, as has been seen, any investment in increasing the production rate of the EDM machine would be totally wasted! In fact, even if the cycle time for the EDM process increased from 40 hours to 50 hours, our total plant output would remain the same. Only if the system is considered as a whole, rather than as a series of independent elements, is this clear.

Utilization can be misleading. Many managers run their machining operations by monitoring utilization. If all the machines are running, then everything is perceived to be going well. If not, it is assumed that there is a problem. As our example makes evident, it is normal that most machines will not be running all of the time. In fact, the only process for which it is important to achieve a very high utilization is the bottleneck process. Examples of this problem can be found in many shops. The manager, walking around the machine shop, notices that the brand new graphite machining center he bought is not running. The manager is shocked. After all, he knows that this machine costs him $40 per hour, and he was advised to buy it because the machine it replaced couldn't keep up. So, the manager bursts into the CAM programmers' office and demands to know why the machine isn't running. The programmers scramble to get some programs out onto the machine, and it is soon making parts. But, what if it wasn't running because the programmers were optimizing the part programs to run 90 minutes per part instead of 150 minutes? Such a saving is quite possible if Z-level or flow-machining strategies are used instead of traditional "grid" tool paths. But, it might take an extra 30 minutes for the programmer to produce an optimized tool path for each part. In a case where the graphite machining center is the bottleneck, it is actually better to have the machine sitting idle for 30 minutes while the programmers optimized their work. The result would be a net saving of 30 minutes on the bottleneck process.

Technology traps. Many machine shops are fond of new technology. For example, high speed machining is one of the more popular technological developments being promoted by machine tool builders and professional societies today. But, too often, new technology is adopted because shops are enamored of it rather than because it addresses a fundamental need. New technology should only be adopted if it directly addresses a bottleneck operation. For example, assume the grinder in our example was 20 years old and took three times longer to produce parts as a new machine (i.e., 3 hours instead of 1). It would be a mistake to buy a new grinder believing that 66% of the cost of our grinding operation would evaporate, thereby increasing profits. Unless headcount can be reduced, or spending eliminated as a direct result of the new grinder, it would simply "shift" cost from one place to another. In fact, our costs would more likely increase due to the new machine's payments. On the other hand, it is advisable to embrace technology that increases the productivity of the bottleneck operation. So, if the steel cutting machining center is the bottleneck, one should evaluate the purchase of a more capable machine that produces parts faster with less scrap, automation to allow the machine to run 24×7, or a new CAM system that enables the programmers to create programs with shorter run times. It is not necessary that the technology in question be applied directly to the bottleneck operation, only that it directly affects the productivity of the bottleneck.

Practical steps to eliminate bottlenecks

As has been demonstrated, the process of identifying bottlenecks is not straightforward. Even in our highly simplified example, it was not readily apparent which operation was the bottleneck. And, in many operations, the bottleneck will regularly move from operation to operation as the product mix changes. To fully optimize throughput, a finite scheduling package is required. Such systems can be very expensive and complex, and are often best suited to operations with little or moderate change in product mix. Unfortunately, many machining operations are essentially job shops. In other words, they produce a wide range of different jobs, many of which are repeated on an infrequent basis, if at all. How, then, is the manager of a machining operation to proceed towards lower lead-times and higher throughput? The remainder of this discussion will outline specific actions and guidelines that can be undertaken to improve performance in these areas. The specific action items are as follows.

- Identify bottlenecks.
- Focus on increasing the throughput of the bottleneck.
  Eliminate nonvalue-added steps within the bottleneck process.
- Recognize and minimize the impact of process variability. Structure the operation to stabilize bottlenecks as best as possible—variability from other areas that would change the bottleneck must be eliminated and/or controlled.
- Position bottlenecks towards the front of the production process.
- Make sure that the bottleneck coincides with the firm's core competency.

Identify bottlenecks. This is the first step towards reducing lead-times. The process of identifying bottlenecks, on a rough-cut basis, requires that one lay out a process flow diagram for the manufacturing process. Each product type with significantly different processing requirements must be diagrammed independently. In our example, product types A and B were treated separately because they required different machinery and process steps. The process flows should identify the average cycle times for each processing step/resource. If the range of possible cycle times to process particular product type varies widely relative to the average, it is necessary to further refine the product type by splitting it into two or more categories.

After the process flow diagrams have been completed, the bottleneck is identified by loading the factory, as represented by the flow diagrams, with work until one of the process steps is fully consumed. This is best done iteratively, using a spreadsheet. For each process step, identify the number of hours available per unit time (for example, per month). Then, using the appropriate product mix, "load" the plant to see which resource is most fully consumed. If none of the process steps is fully consumed, continue to load the factory by adding additional production requirements until the first resource becomes overextended. Do not be misled into believing that one can easily intuit what the bottleneck is. Even if this is possible the first go-around, as one optimizes the bottleneck in later steps it will be increasingly important to formally revise the process flow diagrams to direct successive efforts. Also, it is useful to repeat this exercise several times with varying product mixes. If the bottleneck tends to shift from one work center to another based on the product mix, additional steps will need to be taken to stabilize the bottleneck.

Referring to our ongoing example, our factory was operating with a base product mix of 11 of 'A' and 6 of 'B.' In this case, it appears that there might be two bottlenecks, programming and steel cutting. Both are close to being fully utilized. But, assume that our business is expected to shift more towards product 'A,' at the expense of product 'B'. So, projecting forward to a mix of 12 of 'A' and 5 of 'B,' for example, it can be seen that steel cutting is the first resource to be overloaded. This holds true for the case of 12 'A' and 6 'B' as well. The resulting resource consumption is detailed in **Table 7** for these scenarios. Generally, it can be said that the limit to producing more units in less time is the steel cutting machining process. But, **Table 7** also makes clear that the process is fairly close to being constrained by the programmers generating tool path for the steel cutting machines. It is highly likely, because of process variability, that both areas will have to be addressed to boost throughput. Our example has focused exclusively on the manufacturing steps themselves. Support functions, such as order receipt and entry, purchasing of materials, and shipping of the finished product, have been excluded. A complete analysis would include these tasks, as they could contain a bottleneck.

Focus on increasing the throughput of the bottleneck. Once the bottleneck has been identified, steps must be taken to increase the rate at which it completes jobs. In our case, the focus is on increasing the throughput of the steel cutting work cell. The most

Table 7. Resource Consumption, in Hours, and as a Percentage of Available Resources, to Produce (12A, 5B) or (12A, 6B) Without any Changes to the Process of Figure 1.

Resource Name	(12A, 5B) Resource Consumption	(12A, 6B) Resource Consumption
Programmers	448, 93%	480, 100%
Steel cutting machining centers	328, 103%	336, 105%
Graphite machining centers	120, 75%	144, 90%
Grinders	51, 32%	54, 34%
EDM	200, 63%	240, 75%
CMM (inspection)	64, 40%	72, 45%

obvious way to go about this is to purchase another machine. This would increase available capacity by 50%, or 160 hours per month. Working through our process flow model, it can be determined that this would allow production to be boosted to 12 'A' and 6 'B,' or 13 'A' and 5 'B,' from 11 'A' and 6 'B' before the programming function became the bottleneck. This increases profitability by only $700 to $900 per month, before taking into consideration the additional machine payment and associated expenses.

Even if production could be boosted more dramatically, increasing throughput by purchasing additional capacity should be the method of last resort. It is expensive to bring new capacity on line, and, most important, it leaves unchanged the fundamental processes that cause bottlenecks in the first place.

The best method for increasing throughput of the bottleneck is to reorganize the work so as to gain more output with a minimum of additional expense. Begin by determining what percentage of the resource's time is used for processing parts versus setup. Eliminate as much nonvalue-added time as possible from the bottleneck resource. Redesign the work process at the resource to move inspections and setups to off-line stations so that they can be undertaken while the machine produces another set of parts. Determine how much time can be made available to the resource by improving its reliability, and therefore its up-time. For our steel cutting machining cell the following improvements might be considered.

1. Adopt quick-change fixturing to allow 'part zeros' to be established in an off-line jig. Time savings (avg.): 30 minutes per setup for roughing and finishing.
2. Purchase an off-line tool measurement station to measure all cutting tools before they enter the machine. Additional capacity created (avg.): 30 minutes per day per machine.
3. Schedule the machine so that a highly reliable job is always run at the end of the shift and into the off-shift (as opposed to by chance). Additional capacity created (avg.): 1 hour per day per machine.[5]
4. Retrofit the steel cutting machines with Ethernet to reduce down time spent waiting for programs to be transferred to machine using DNC. Time savings: 15 minutes per tool path (i.e., 30 minutes total across roughing and finishing operations for product A).
5. Rearrange the two machines and the above equipment into a cell so that a single machinist can monitor the machines. Purchase pneumatic tools, workbenches, power assisted lifting equipment, etc., to allow one machinist to reasonably support two machines without too much additional strain or stress. Rearrange the work schedules so that one machinist works from 7 a.m. to 4 p.m., the other from 9 a.m. to 6 p.m. Stagger their lunch hours and breaks accordingly. This adds approximately another 20 minutes per day per machine by eliminating lost time when the machines stop but the machinists are at lunch or on break.[6] It also adds another 4 hours of machining capacity per day, because the machines are both running from 7 a.m. to 6 p.m., rather than (for example) 8 a.m. to 5 p.m.

Our process model can now be revised to reflect the increase in machine capacity, and the reductions in cycle time. It is important to carefully consider whether an improvement should be classified as a reduction in cycle time for a specific job type, or as an increase in bottleneck capacity for all jobs. Improper classification of an improvement can cause

---

[5] We assume it is not always possible to schedule the machine in this way. Otherwise, additional capacity created would be much higher.

[6] On some days, significantly more than 20 minutes of additional capacity would be recovered, but there will be days when the machines would not stop for a work change-over or alarm condition during the lunch hour at all. Therefore, we estimate the average benefit as 20 minutes per day.

improper identification of the bottleneck. For example, above items 2, 3, and 5 will increase capacity of the steel cutting machines for any and all jobs. In the case of the tool setter, it can be assumed that previously each machine was loaded up with tools for the entire day in the morning and that the tool lengths were measured on the machine. By moving this off-line, the capacity to run any job is regained. Item 4, on the other hand, only provides a benefit for specific jobs. Assume, for example, that the roughing part programs for product B run on the steel cutting machine are relatively small. Thus, the installation of Ethernet would not positively affect the cycle time for product B as it related to the steel cutting machines. Item 4 is of benefit for steel machining of product A, because it requires longer programs that must be loaded into the CNC for each job due to differences in the final part geometry. Finally, item 1, adopted for all steel workpieces, would improve the cycle time for products A and B.

    Table 8 identifies the increases in capacity of the steel cutting cell, and the decrease in the steel cutting cycle time for product A and product B, as a result of the aforementioned changes. These changes, taken together, dramatically increase the available productive capacity of the steel cutting machining centers at a relatively modest cost. The steel cutting process is no longer our bottleneck. But, as would be the case if another machine was added, there is now significantly more capacity than can be used, because as soon as production is ramped up beyond (12A, 6B), a new bottleneck in the part programming process is created.

Table 8. Details of Capacity Increase for the Two Steel Cutting Machines, and Decrease in the Steel Cutting Cycle Time for Product A and Product B as a Result of Process Changes.

Original Steel Cutting Capacity = 320 hours	
Incremental Improvement	Additional capacity
Off-line tool setter	+ 20 hours
Improved job planning	+ 40 hours
Reorganize cell layout and working hours	+ 173.3 hours
New Steel Cutting Capacity = 553.3 hours	

Original Steel Cutting Cycle Time for Process 'A' = 24 hours	
Incremental Improvement	Reduction in cycle time
Quick change fixturing	- 1 hour (30 minutes per setup)
Ethernet	- 30 minutes (15 min. per operation)
New Cycle Time for Steel Cutting of 'A' = 22.5 hours	

Original Steel Cutting Cycle Time for Process 'B' = 8 hours	
Incremental Improvement	Reduction in cycle time
Quick change fixturing	- 30 minutes
New Cycle Time for Steel Cutting of 'B' = 7.5 hours	

    Alternately, if production of product 'B' ceases altogether, it would be possible to produce 20 'A' and 0 'B' for a profit of $14,000. This dramatic change, to where it no longer makes sense to manufacture unit 'B' at all, occurs because the bottleneck was shifted from steel cutting to the programming process. Now, for each unit of A, which creates a net profit contribution of $700, a full 24 hours of programming is needed. For each unit of B, which only contributes $500, a total of 32 hours of programming is required. Therefore, $29.17 of incremental profit per hour of the bottleneck is gained when producing unit A.

With product B, the earning is $15.63 per hour of the bottleneck. Previously, when steel machining was the bottleneck, a profit of $29 per hour of steel cutting was earned when making unit A, and $40 per hour of steel cutting with unit B. Profit of the enterprise is optimized when the profit generated per hour of bottleneck consumption is maximized. To continue forward, assume that due to market considerations (i.e., our customers demand both products or they won't buy either), in the near-term the shop produces 12 'A' and 6 'B' instead of 20 'A' and 0 'B'.[7] Also, despite the overabundance of steel machining resources these steps seem to create, we would still implement the aforementioned changes before shifting our focus to the programming process for reasons that will be discussed later.

After breaking the steel cutting bottleneck, the following steps could be undertaken to address the new bottleneck of the part programming process. Again, the goal is to improve throughput without hiring additional personnel. Note that it is true that as soon as one bottleneck is broken, a new one is created.

1. Provide programmers with computers and internet accounts they can use at home for any purpose at no charge if they put in an extra 8 hours of part programming per month from home.
2. Dedicate specific programmers to each product type to improve efficiency and process knowledge. Such a step might reduce the time required to create a tool path by 3%. It might also have the further benefit of leading to a cycle time reduction of approximately 5% for machining operations due to improved tool path quality.
3. Create a training/promotion program that allows machinists to take classes in CAM, and outfit them with a shop floor programming system that can be used to program the roughing tool paths for product 'A'. Increase pay to motivate them to take on this additional task, and create a promotion path into the CAM department. Assume that shop floor programming reduces the time required by the programming department by 3 hours per product 'A' produced, but adds back 4 hours of steel cutting time because the tool paths are less efficient.

Table 9 outlines the increases in part programming capacity, and **Table 10** details the resources now required to produce a unit of each product type. With these changes, and those previously undertaken in the steel cutting cell, the factory can now produce 14 'A' and 7 'B' for a monthly contribution of $13,300. This represents an almost 28% increase in product A from our baseline with relatively modest investment. Clearly, the lead-time that can be offered customers for product A will also drop significantly. Again, a more radical realignment to 21 'A' and 2 'B' (from 20 'A', 0 'B') can be undertaken for a variable contribution of $15,700. In either case, the lead-time and throughput for product B are not improved substantially because per unit of programming time, more profit is generated from a unit of A than a unit of B. For every 20.3 hours of programming time, there will be

Table 9. Details of Capacity Increase in Part Programming as a Result of Process Changes.

Original Part Programming Capacity = 480 hours	
Incremental Improvement	Additional Capacity
At home programming	+ 24 hours
Product focus efficiency	+ 14.4 hours
New Part Programming Capacity = 518.4 hours	

---

[7] It certainly would warrant discussion as to whether we could move our business away from production of product B altogether in the mid- to long-term given expected demand for product A versus product B.

Table 10. Changes to Resources Required to Produce One Product A or One Product B After Steel Cutting and Part Programming Process Changes.

Resource Name	Product A Hours Required	Notes
Programmers	20.3	3% efficiency improvement, shop floor programming
Steel cutting machining centers	25.3	Ethernet, improved fixturing, 5% improvement due to product focus by programmers. *Shop floor programming increases hours required, offsetting other benefits.*

Resource Name	Product B Hours Required	Notes
Programmers	31	3% efficiency improvement due to product focus
Steel cutting machining centers	7.1	Improved fixturing, 5% improvement due to product focus by programmers
Graphite machining centers	22.8	5% improvement due to product focus by programmers

an incremental profit of $700 with product A, whereas with product B an incremental $500 per 31 hours of programming time can be earned. So, it is logical to apply as much of our bottleneck resource to producing unit A rather than unit B.

Note that the third item undertaken to improve programming throughput represents a typical tradeoff that must often be made to boost throughput and profitability. In this case, three hours were cut out of the bottleneck, but four hours were added to a non bottleneck operation. Using conventional cost accounting, such a move would appear to be ill advised. The cost per hour for machining time is likely to be higher than for programming time, and so such a move would "increase" our costs. This distortion exists because of the way overhead is typically applied, based on labor hours. In reality, our total costs would remain stable because of an increase in consumption of a resource (steel machining) that was not fully utilized.[8] This distortion will exist only so far as the calculation of gross profit and gross margin are concerned, and is corrected as the complete profit and loss statement is compiled.

Recognize and minimize the impact of process variability. It is helpful to repeat this process of identifying and breaking the bottleneck again and again to boost system throughput and to reduce product lead-time. But, in addition to making deterministic changes such so these, the process variability must also be addressed if lead-time and throughput are to improve. To this point all of the analysis has worked exclusively with averages—average cycle time, average capacity, etc. This is acceptable so long as the variation surrounding these averages is relatively small. But, what if the variation around each of these processes is large? For example, what if the time to program product A is 21.3 hours on average, but varies between 4 hours and 30 hours? Clearly this would make it difficult to predict lead-time for any customer order of product A, and would make it very hard to predict the bottleneck. In reality, the cycle time for a process is not determinant. That is, it does not equal the average value every time. As a result, our analysis is forced to deal with process variability.

Process variability is a significant impediment to improving throughput and lead-times. It is much easier to optimize a production system with long cycle times that are determinant than one with short but highly variable cycle times. When faced with variability it is very difficult to predict the lead-time. As a result, one must quote customers longer lead-times

---

[8] Even if we pay the machinists a small differential for taking on some programming tasks, this is more than offset by the increase in throughput in the programming department.

than may be necessary to prevent late deliveries. Further, work-in-process inventories will be higher, and it will be much harder to isolate the bottleneck because it will shift from job to job due to variability within each process.

Every process has variability, and eliminating it is impossible. Even if the machining time, for example, is set definitively by the part program, occurrences such as broken tools, accidents, power failures, machine failures, and so on all contribute variability to the time required to complete a process. To optimize throughput and lead-time, variability must be minimized, particularly within the bottleneck step. Consider *Figure 3*, which illustrates the cumulative probability for cycle times of a machining process. In *Figure 3A*, the cycle time probability is evenly distributed between 1 hour and 4 hours, with an average of 2.5 hours. *Figure 3B* illustrates the cumulative probability for a normally distributed cycle time between 1 and 5 hours, with an average of 3 hours and standard deviation of 0.6 hours. Even though the average cycle time in *Figure 3B* is higher, this is a preferred distribution relative to that of *Figure 3A*. Because *Figure 3B* is normally distributed, the probability of achieving a cycle time near the average is significantly higher than that of *Figure 3A*. The process in *Figure 3B* is much better controlled than that of *Figure 3A*. For example, the probability of achieving a cycle time of [average ± 0.5 hour] is 33 % for the distribution of *Figure 3A*, and 60% for that of *Figure 3B*.

Cumulative Cycle Time Probability (linear distribution)

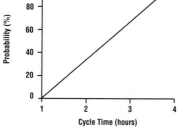

Cumulative Cycle Time Probability (normal distribution)

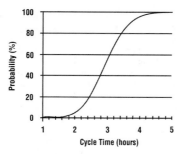

Figure 3A (top) and 3B (bottom). Cumulative probability for cycle times.

Once the concepts of process flow charting and bottleneck identification on a deterministic basis have been mastered, it is essential to identify and eliminate sources of variability. This is particularly true within the bottleneck operation, and those operations that take place before it in the manufacturing process. Variability downstream of the bottleneck (i.e., inspection) can be tolerated to a greater extent. Process variability affects overall system performance something akin to the domino effect—variation early in the process flow will negatively affect each subsequent operation. In our example, the excess capacity created in the steel machining operation, due to our bottleneck breaking activities, represents a safeguard against the negative effects of process variability within and before this operation. Because the machining step takes place early in the process flow, it is extremely important to 1) reduce variability as much as possible, and 2) safeguard against the variability that remains from negatively impacting downstream operations. Just as steel machining is insulated against variability within or prior to this process due to excess capacity, the programming process, which has a minimum of excess capacity, is currently susceptible to variation. Because the programming process is the bottleneck, and because it is the first process in the process flow, one must insure that variation is minimized, and that a buffer against that variation is created.

To discuss in detail all the countermeasures one can take against process variability is beyond the scope of this discussion. However, a general approach to addressing variability begins by charting the cycle time of each unit through the production step of interest to characterize the nature of the variability (i.e., its distribution and causes). Then, a root cause analysis of the variability sources can be undertaken. Begin by addressing the greatest cause of variation and proceed to those with a lesser impact. It may be that some causes of variation will be beyond the shop's control (i.e., customer changes to the product specification during manufacture). In such cases, it will be necessary to build enough capacity into the manufacturing system to minimize their negative impact.

Keep in mind, also, that "exceptions are the rule." In other words, variability is the norm not the exception. Variability in the flow of incoming orders, customer changes, machine failures, and so on, are a part of everyday life. They cannot be eliminated, but they can be planned for.

Position bottlenecks towards the front of the production process. In our example, the actions that were taken to increase throughput inadvertently moved the bottleneck to the front of the production process. This has a number of positive consequences. First, it eliminates the likelihood that variation in upstream processes could cause the bottleneck to be idle, waiting for work. This is critically important, because the bottleneck dictates overall throughput and lead-time. If the bottleneck process is waiting for work due to a breakdown in an upstream piece of equipment, for example, that productive time and output is lost forever—it cannot be made up! In our example, the bottleneck programming step is now exposed only to variation in the process of order entry and the receipt of the necessary information from the customer to undertake the job. When the bottleneck was steel cutting, problems or delays in the programming step, not just problems within the steel cutting process, could also contribute to lost productive capacity.

Second, a front-end bottleneck makes management of the manufacturing process much easier. By having the bottleneck at the "front of the line," management can more easily focus on ensuring that it is properly loaded, and that issues of capacity constraints or variability are readily addressed. Essentially, the operation at the front of the line has greater visibility because its operation is less clouded by issues associated with upstream operations. It is not always possible to position the bottleneck at the front of the line, nor is it always desirable, as will be discussed momentarily. But, whenever the bottleneck can be moved towards the front of the process, a more easily managed production system, with less susceptibility to variability, will be created. If one manages the bottleneck properly, downstream operations will, by definition, run more smoothly because they do not dictate the overall pace of production. If the bottleneck is at the beginning of the process, management's task is greatly simplified, and therefore the probability of error is also reduced.

Make sure that the bottleneck coincides with the firm's core competency. This last point is fairly short, but extremely important. A company that concentrates on reducing lead-times and increasing throughput (and profits) will spend a lot of time focused on improving the performance of its bottleneck. It therefore makes sense that the bottleneck should be structured, if at all possible, to coincide with the firm's core competency. In this way, improvements in quality and capability within the core competency will have a direct bearing upon the throughput of the firm by reducing process variability and cycle time.

A firm's core competency is that set of skills and capabilities around which the firm seeks to differentiate itself from the competition. In the production of molds and dies, for example, core competencies might be hard die machining techniques, or tool path generation to optimize accuracy of steel and graphite parts. The core competencies probably should not be order entry, grinding blocks to square, running a great web site, or producing simple base plates. The core competencies should be at most a handful of capabilities that provide competitive benefits relative to the competition as measured by the customer.

Note that "ultra-fast production of molds" has not been suggested as a possible core competency. If a firm adopts a strategy of achieving ultra-fast production as an end in and of itself, it is likely to end up with a relatively unfocused effort where it does many things fast but not particularly well. And, as has been seen, doing things other than the bottleneck quickly generally provides no benefit! High throughput, low lead-time production is best derived as a result of unifying a firm's bottlenecks and its core competencies and then relentlessly driving increases in the throughput of these operations. This necessarily reduces lead-times and increases product quality. Improved product quality is driven by a program to increase throughput, because without high quality, process variation will inflate lead-times and decrease throughput.

Referring to our ongoing example, if EDM or graphite machining were this company's core competencies, a reorganization of the business could be in order. As organized, the company's profitability is optimized when requirements for programming, followed by steel cutting, are highest. Currently, this company obtains its leverage through these processes. If the firm is better at EDM and graphite machining than part programming and steel cutting, its core competencies and bottlenecks would be at loggerheads. As a result, the firm is not as likely to successfully compete in the market for product B as it is in the market for product A. This is reflected in our analyses above, where the firm's profits were ultimately optimized at 21 'A' and 2 'B.' Clearly, this firm should get out of the product B market and focus on product family A, if possible.

Conclusion

Establishing shorter lead-times is best achieved by identifying and breaking process bottlenecks. By taking a systemwide approach to improving throughput, firms can continuously reduce lead-times and also increase product quality and profits. A throughput-centric approach can serve as a strong guiding principle for the management of a wide range of manufacturing operations, including machining.

Indexable Inserts

Insert classification by grade

It is important to be familiar with the performance characteristics of each cutting tool material. All inserts recommended in the speeds and feeds charts in the milling and turning sections of this book are identified by both insert material and ISO grades classification, factors that define the application range of the insert. Most insert manufacturers provide charts that position their insert grades on an ISO and/or ANSI chart for grade classification (an example of a manufacturer's chart is shown in *Figure 1*). By using the manufacturer's chart, the intended application of any insert grade can be determined. For instance, if the speeds and feeds tables in the milling or turning section recommend an "M40 PVD Insert," it can be seen in *Figure 1* that the manufacturer (in this instance, Kennametal) recommends its grade KC720 for the operation.

In addition to showing the manufacturer's suggested grade for specific machining operations, *Figure 1* also illustrates the ISO grade classification system from ISO Standard 513 and shows corresponding ANSI classifications for each grade. Low number ISO ranges (higher ANSI numbers) require cutting grades with high wear resistance, while toughness is a primary consideration for the higher ISO number (lower ANSI) ranges. The different ISO classifications, P, M, and K, represent the following materials and machining operations, and on most manufacturer's charts each class is represented by a designated color: P is blue, M is yellow, and K is red.

ISO P Class is designated by the color blue. This class is comprised of ferrous materials: most steels, cast steel, some stainless steel, and malleable cast iron. When machined, these materials are characterized by long chips. Group P01 designates machining processes on steel and steel castings such as finish turning and boring with no shocks, high cutting speed, and low feed and light depth of cut, producing high quality surface finishes. P10 includes finishing and semifinishing operations, such as threading, performed on steel and steel castings at high speeds and generating small to medium chip cross section. General machining operations such as turning and milling carried out on steel, steel castings, and malleable cast iron at low to medium speeds that create long medium to large chip cross sections are characteristic of class P20. Class P30 represents operations such as turning and milling performed under unfavorable machining conditions on steel, steel castings, and malleable cast iron. Chips are long with medium to large cross section. Class P40 incorporates operations performed under unfavorable cutting conditions, with low cutting speeds and heavy chiploads (such as roughing) on steel and difficult steel castings. An additional Class, P50, is sometimes included for very difficult operations at low speeds that create large chip sections, as performed, for example, on steel and steel castings with low tensile strength. There is no assigned ANSI equivalent of P50.

ISO M Class is represented by the color yellow. The M class includes difficult-to-machine materials that produce either long or short chips such as nickel based alloys, manganese steel, austenitic stainless steel, titanium, gray cast iron, and alloyed cast iron. As in the ISO P class, the lower numbers represent higher speed finishing operations. M10 grades are for turning at medium to high speeds with small to medium chip sections. M20 is for general machining operations at medium speeds and medium chip sections. M30 is for general machining operations on more difficult materials that produce medium to large chip sections. M40 operations include turning and parting-off operations on difficult materials. Although ANSI does not assign numbers to the "M" class, the equivalents shown in *Figure 1* can be used as guidelines for selecting cutting tool grades for this class.

ISO K Class is indicated by a red color. Materials in this class include aluminum, bronze, brass (and their alloys), chilled cast iron, and nonferrous materials such as abrasive

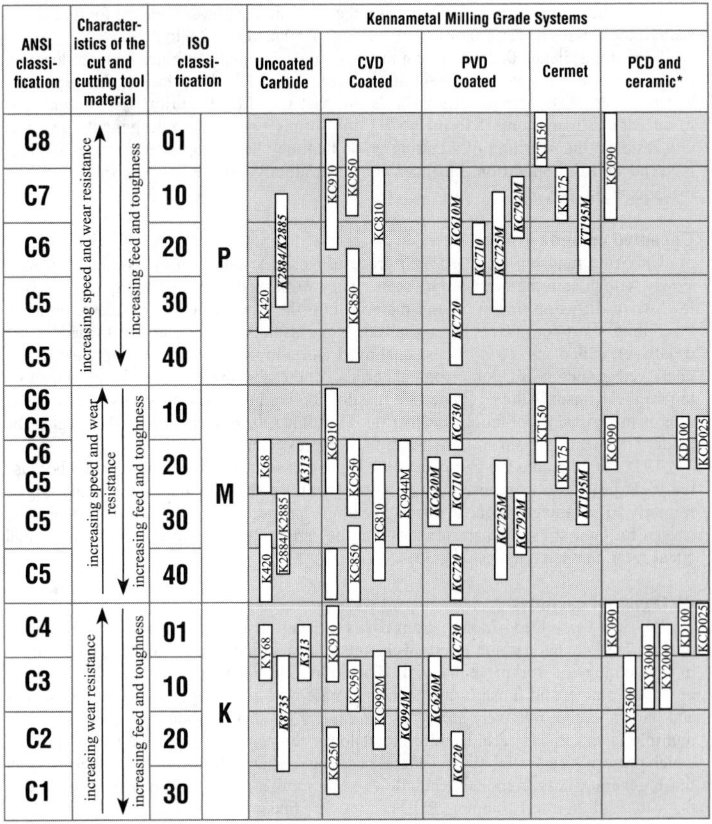

ANSI classi-fication	Characteristics of the cut and cutting tool material	ISO classi-fication		Kennametal Milling Grade Systems				
				Uncoated Carbide	CVD Coated	PVD Coated	Cermet	PCD and ceramic*
C8	↑ increasing speed and wear resistance ↓ increasing feed and toughness	01	P	K2884/K2885 (K420)	KC910 KC950 KC810 KC850	KC610M KC710 KC725M KC792M KC20	KT150 KT175 KT195M	KC090
C7		10						
C6		20						
C5		30						
C5		40						
C6 C5	↑ increasing speed and wear resistance ↓ increasing feed and toughness	10	M	K68 K313 K2884/K2885 K420	KC910 KC950 KC810 KC850 KC944M	KC730 KC620M KC710 KC725M KC792M KC20	KT150 KT175 KT195M	KC090 KD100 KCD025
C6 C5		20						
C5		30						
C5		40						
C4	↑ increasing wear resistance ↓ increasing feed and toughness	01	K	KY68 K313 K8735 K250	KC910 KC950 KC992M KC994M KC720	KC730 KC620M	KC090	KD100 KCD025 KY3000 KY2000 KY3500 KY1500
C3		10						
C2		20						
C1		30						

NOTE: Grades shown in **bold italic type** are recommended first choice grades. *Advanced materials recommended for high-speed milling applications.

Figure 1. Manufacturer's Grade Application Chart for milling operations. (Source, Kennametal Inc.)

plastics, ceramics, and hard and soft woods. Chips are short, ranging from near powder to small, self-breaking chips. K01 work materials include chilled and hard gray cast iron, high silicone aluminum alloys, ceramics, and abrasive plastics. Minimal removal operations such as finish turning, light milling, and scraping are representative. K10 indicates general machining on gray cast iron (over 220 BHN), hardened steel, aluminum and bronze alloys, and materials such as glass, porcelain, and stone. K20 covers general machining operations on copper, brass, aluminum, and gray cast iron. K30 indicates unfavorable machining conditions (large cutting angles, for example) on low tensile steel, cast iron, and plywood. Another class, K40, is sometimes included for unfavorable machining conditions on softer materials.

MACHINING

Using the softest K materials as a base, the comparative power required for machining each of the classes is: K= 1.0 to 1.50; P= 1.70 to 2.50; and M= 2.0 to 3.20.

Economics dictate that the cutting material grade selected combines high productivity (metal removal rate) with sufficient and consistent tool life. Part design, fixture design, and machine tool compatibility will also impact tool life. Rigidity is the foundation of successful machining. Sacrificing rigidity in any one of the components involved will result in the selection of a cutting grade that may be appropriate for the workpiece material and the operation, but may not be productive due to factors introduced by inadequate rigidity.

Uncoated carbide grades

Uncoated tungsten carbide (WC) grade usage, in standard insert configurations, is slowly fading from the metalcutting scene. They work well when surface speeds are very low, when diffusion of the cutting material into the workpiece is a concern, or when machining short run jobs. The cutting edges are usually sharp. Uncoated carbides are usually classified into two groups: unalloyed and alloyed. The primary application for unalloyed grades is for nonferrous materials, where abrasive wear is the primary tool failure mechanism. Alloyed grades are usually applied in ferrous materials, where crater wear is the primary tool failure mechanism. The higher the binder content, the tougher the grade. The finer the grain size, the better the wear resistance.

PVD coated grades are particularly well suited to replacing uncoated carbides because the PVD process allows inserts to be coated with up-sharp or unhoned cutting edges, formerly an advantage unique to uncoated carbide inserts. When used, uncoated grades are chosen for their strength characteristics, but they are more susceptible to flank, nose, and crater wear compared to coated grades.

PVD coated carbides

Physical Vapor Deposition coatings have technical advantages that make PVD titanium nitride (TiN) coated carbide inserts well suited for a variety of metalcutting applications in a wide range of workpiece materials. They offer outstanding performance in threading, grooving, cutoff, and finish turning. They operate well at low to medium cutting speeds and have a stable, relatively sharp cutting edge. PVD coatings are applied in single or multiple layers to the carbide substrate at lower temperatures (approximately 500° C), which preserves edge strength and enables coating of sharp edges. Their relatively smooth finish generates less frictional heat, allows lower cutting forces, and resists edge buildup that can result in insert chipping. PVD coatings are fine grained and very hard, so they are resistant to abrasive wear mechanisms. The cutting edges are sharp, or only slightly honed, resulting in less chatter, fewer burrs, finer surface finishes, and better dimensional control on the insert. Feed rates can be as low as 0.001 inch per revolution.

CVD coated carbides

Chemical Vapor Deposition coated carbide inserts are especially suited for most metalcutting applications in ferrous workpiece materials. These coatings are applied to the carbide substrate at high temperatures (approximately 1000° C), which results in interdiffusion of the coating with the substrate to assure a strong bond. CVD coated inserts generally have honed edges, and the process also permits deposition of multilayer coatings of titanium carbide (TiC), titanium nitride (TiN) and aluminum oxide (Al_2O_3) that combine different coating compounds to reduce both crater and flank wear. Cobalt enrichment is commonly used to increase strength edge and fatigue resistance while maintaining deformation resistance. CVD grades coated with aluminum oxide (Al_2O_3) permit higher cutting speeds and the coating also combats abrasive and crater wear. CVD coated carbide inserts require heavier hones on the cutting edge, and feed rates over 0.003 inch per revolution are recommended. Due to the high temperatures used in the CVD process, these inserts provide improved abrasion and crater wear resistance compared to PVD coated carbides, but residual tensile stress in the coatings causes toughness to be reduced from PVD levels.

Ceramic grades

Ceramic cutting grades can be divided into two basic families: alumina based (Al_2O_3), and silicon nitride based (Si_3N_4). The alumina based ceramics are produced in three forms: pure, mixed, and reinforced. Pure alumina based ceramics are white (if manufactured by cold pressing) or gray (if hot pressed), and offer excellent abrasion resistance for machining soft cast irons and steels to 330 BHN (35 R_C). Pure inserts have relatively low strength and toughness values and are subject to cutting edge fracture if improperly employed. Pure ceramics are commonly used for semi-roughing and roughing cast iron and steel. Mixed alumina based ceramics contain 20 to 40% titanium carbide (TiC) and titanium nitride (TiN) and are black in color. They have high toughness and thermal shock resistance for machining carbon steels, alloy steels, tool steels, and stainless steels to 653 BHN (60 R_C). Reinforced alumina based ceramics contain microscopic "whiskers" (approximately 30% by content) to reinforce the strength and thermal shock resistance, and are green in color. Reinforced ceramics are a relatively new development and are finding acceptance for interrupted cuts, hard turning, and for machining super alloys.

Compared to alumina based grades, silicon nitride based ceramics have improved toughness and thermal shock resistance. Silicon nitride based ceramics are especially suited for high productivity milling and turning of gray cast irons. Other uses include high-speed roughing of nickel-based alloys where scale or interruptions are present, and high-speed machining of high-temperature super alloys.

Guidelines for using ceramic inserts when machining high-temperature alloys. Notching (exaggerated abrasion wear concentrated at the depth of cut intersection of the insert) is the most common mode of ceramic tool failure. Excessive notching can lead to premature tool failure and insert breakage. The following steps can minimize notching.

- Use round inserts whenever possible.
- Use longest lead angle possible.

With round inserts, lead angle is optimized when depth of cut is 5% to 15% of the insert diameter. Maximum depth of cut is 25% of the insert diameter.

- Use flood coolant.
- Prechamfer the workpiece to eliminate stress points on the cutting edge of the insert. When using a separate prechamfering operation, feed the insert at a 90° angle to the chamfer to minimize insert notching during the prechamfer.

Guidelines for using ceramic inserts when machining carbon, alloy, tool, and stainless steels. The feed rate can be calculated from the width of the insert's T-land.

For turning or boring, the feed per revolution (fpr) is 1 to 1.5 % the width of the land. If width is 0.008" *or* 0.2032 mm:

fpr = 1 to 1.5 × 0.008"   *or*   0.2032 mm = 0.008–0.012" *or* 0.2032–0.2794 mm.

For milling, the feed per tooth (fpt) is 0.75 to 1.0 × the width of the land. If width is 0.008" *or* 0.2032 mm:

fpt = 0.75 to 1.0 × 0.008"   *or*   0.2032 mm = 0.006–0.008" *or* 0.1524–0.2032 mm.

• Coolant is normally not recommended for turning, boring, or milling of steels with ceramic cutting tools.

Guidelines for using ceramic inserts when machining cast irons. The feed rate can be calculated from the width of the insert's T-land.

For turning or boring, the feed per revolution (fpr) is 1 to 3 × the width of the land. If width is 0.008" *or* 0.2032 mm:

fpr = 1 to 3 × 0.008" *or* 0.2032 mm = 0.008–0.024" *or* 0.2032–0.0.6096 mm.

For milling, the feed per tooth (fpt) is 0.75 to 1.5 × the width of the land. If width is 0.008" *or* 0.2032 mm:

fpt = 0.75 to 1.5 × 0.008" *or* 0.2032 mm = 0.006–0.012" *or* 0.1524–0.2794 mm.

Cermet grades

Most cermet grades are composed of titanium carbonitride (TiCN) and titanium nitride (TiN) with a nickel or cobalt binder. Cermets (CERamics with METallic binders) are known for exceptional high wear resistance, which can be attributed to their chemical and thermal stability. Cermets hold a sharp cutting edge at high speeds and temperatures, allowing them to produce exceptionally good surface finishes under proper conditions. One of their primary uses is high-speed finish machining of most steels. They are less suited than carbides for heavy roughing or interrupted cutting conditions.

Carbon, alloy, stainless steels, and malleable (ductile) irons are the primary materials for TiC/TiN cermet applications. Free-machining aluminum and nonferrous alloys (copper, zinc, and brass) can also be machined successfully. Note that all of these materials produce a ductile chip. TiC/TiN cermets have metallic binders and do not exhibit the high hot hardness necessary for machining hardened workpiece materials with a hardness of 370 BHN (40 R_C) or better. For these materials, ceramic or CBN cutting tools are recommended.

As seen in **Table 1**, cermets offer the potential for increased productivity, primarily because they allow higher speeds.

Table 1. Cutting Speed of Cermets Compared to Uncoated Carbide.

Workpiece Material	Speed Potential (sfm *or* smm) vs. Uncoated Carbide
Carbon Steel, Stainless Steels, Ductile Irons	+ 30 to 50% (approx.)
Alloy Steels	+ 20% (approx.)

Compared to uncoated or coated carbides, cermets are more feed and fracture sensitive. The practical upper feed limit for a tough cermet is approximately 0.022–0.025 in./rev *or* 0.5588–0.6530 mm/rev when turning carbon steel. Most successful cermet applications (turning, boring, and milling) are at conventional carbide speeds and feeds. Success will come through extended tool life, superior surface finish, and economy of application. Since TiC/TiN cermets are more sensitive to thermal shock than carbide inserts, coolant is not recommended when rough turning, boring, or milling, but in very light finish turning and boring, coolant can be beneficial to workpiece surface finish.

Polycrystalline grades

Superhard polycrystalline cutting tool materials are divided into two basic groups: diamond (PCD) and cubic boron nitride (PCBN). Both are relatively expensive on a per unit basis, but can pay off by providing extraordinarily high productivity, resulting in very low per unit manufacturing costs. PCD tool life can actually be up to one hundred times higher than cemented carbide. In order to exploit these advantages, rigid machine tools and workholding fixtures are essential. Diamonds are the hardest cutting tool materials available, and are usually applied on nonferrous materials at very high speeds. They possess exceptional good abrasion resistance and strength and have high thermal conductivity to aid heat dissipation when used at very high speeds. PCD grades offer unsurpassed tool life when machining aluminum alloys and nonferrous and nonmetallic materials at high sfm/smm. PCD inserts are not suited for machining materials containing carbon (steel and other ferrous materials) because of a chemical reaction that causes rapid breakdown of the cutting material.

Cubic boron nitride groups can be divided into two groups: low PCBN content and high PCBN content. Low content grades have lower thermal conductivity and comparatively higher compressive strength. These features enable low content grades to promote self-induced hot cutting by combining high cutting speeds and negative rake tool geometries to soften the workpiece material, and effectively remove it from the part. These characteristics make low content grades ideal for finish machining hardened steels up to 480–740 BHN (50–65 R_C), with a maximum depth of cut of 0.60".

High PCBN content grades possess extremely high thermal conductivity and toughness, and can operate at high speeds and in severely interrupted cuts. High content grades are the cutting tool materials of choice for rough cutting hardened steel where severe mechanical edge loading is likely to occur. The very high hardness of these cutting tools results in excellent abrasion resistance. This is also a very good cutting tool material for fully pearlitic gray cast iron.

Edge preparation

Ceramics, cermets, and polycrystalline cutting tool materials have a relatively high material hardness. Because of this, the cutting edge is more brittle than conventional carbide cutting tools. To optimize the performance of these advanced materials, it is critical that the edge preparation be matched to the cutting tool material, the workpiece material, and the machining operation being performed. There are four major choices when considering edge preparation for advanced materials: T-land, hone, T-land plus hone, and up-sharp. The up-sharp alternative is not recommended for ceramic, cermet, or polycrystalline inserts as the resulting edge is highly susceptible to chipping.

T-lands (also called negative lands or K-lands) act to protect the cutting edge by eliminating a sharp cutting edge (thereby reducing edge chipping), and by redistributing the cutting forces back through the larger body cross section of the insert which makes the insert stronger. **Table 2** demonstrates the effect of T-lands and provides specifications for widths and angles.

Increasing either the width "T" or the angle "A" of the T-land strengthens the insert, but it also increases the cutting forces and is potentially detrimental to both the cutting tool and/or the workpiece material. The optimum situation is to have the minimum edge preparation (in terms of T-land size and angle) that will perform the desired operation and adequately protect the insert cutting edge, thereby achieving satisfactory tool life.

Hones are used to protect the cutting edge by eliminating the sharp edge and reducing edge chipping, but they do not allow the high level of protection afforded by a T-land. However, they are very satisfactory when using ceramics, cermets, or polycrystalline inserts in finishing operations such as when depth of cut and feed rates are kept at a minimum, and minimal cutting pressure is desired. Hones are also called "radius" or edge rounding.

In some situations, such as rough turning steel with ceramics, minute chipping can occur at the intersections of the T-land and the rake surface or flank surface of the insert.

This condition can be corrected by applying a radius hone to one or both intersections that effectively improves the performance of the edge preparation while keeping the T-land size and angle constant.

Table 2. Angle/Width Relationships of T-Lands. *(Source, Kennametal Inc.)*

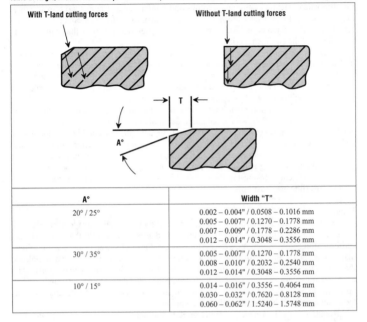

A°	Width "T"
20° / 25°	0.002 – 0.004" / 0.0508 – 0.1016 mm 0.005 – 0.007" / 0.1270 – 0.1778 mm 0.007 – 0.009" / 0.1778 – 0.2286 mm 0.012 – 0.014" / 0.3048 – 0.3556 mm
30° / 35°	0.005 – 0.007" / 0.1270 – 0.1778 mm 0.008 – 0.010" / 0.2032 – 0.2540 mm 0.012 – 0.014" / 0.3048 – 0.3556 mm
10° / 15°	0.014 – 0.016" / 0.3556 – 0.4064 mm 0.030 – 0.032" / 0.7620 – 0.8128 mm 0.060 – 0.062" / 1.5240 – 1.5748 mm

Figure 2. Radius hone (left). T-land plus hone (right).

Inserts and surface finish

In milling operations, surface finishes are generally a combination of roughness superimposed along a wave pattern on the workpiece surface. In turning, a single insert (or brazed single-point tool) produces the surface finish, while in milling, there may be many inserts that overlap as they feed across the workpiece to produce the final surface profile. Since the axial and radial location of all the inserts in an indexable milling cutter are never in perfect alignment, the machined surface finish has waves. The portion of the wave produced by the cutter can be exaggerated further if the machine spindle has runout. The wave produced by the machine spindle is easily observed in the part surface as a repeated mark or pattern dependent on the feed rate of the milling cutter. The type of insert selected can have a significant impact on part finish, as described below (also see the information on surface finish in the Turning, Grinding, and Metrology sections of this book).

Surface finish can be measured on a variety of scales. The most often used are the *Roughness Average* (R_a) and the *Roughness Depth* (R_{max}). The Roughness Average is the arithmetic average of the high points and the low points of the finish of a measured surface. This measurement is also called the "Arithmetic Average" (AA) or the "Centerline Average" (CLA). The Roughness Depth, on the other hand, represents the highest peak (the Maximum Roughness Height, H_{max}) and the lowest valley measured (the Maximum Roughness Depth D_{max}) on five samples of a measured surface. Another common, but officially obsolete, measurement is Root Square Mean (RMS). This is a measurement of a theoretical surface that would be created by removing the roughness peaks on a sample and then using these peaks to fill the remaining valleys. In general, an equivalent Roughness Average will be about 11% below that of a Root Mean Square value for the same surface: $1.11 \times R_a =$ RMS. Roughness is measured in micro-inches (1 μin = 1/1,000,000 inch, or 0.0254 μm) and micrometers (1 μm = 1/1,000,000 meter, or 39.37 μin). **Table 3** provides conversions between different measuring systems.

Table 3. Conversions (Approximate) Between Surface Measuring Systems.

Roughness Depth (R_{max}) to Roughness Average (R_a)				RMS μin to RMS μm	
R_{max}		R_a		RMS	
μm	μinch	μm	μinch	μm	μinch
0.025	1	0.006	0.20	0.007	0.20
0.033	1.3	0.008	0.30	0.009	0.30
0.046	1.8	0.010	0.40	0.011	0.40
0.057	2.3	0.012	0.50	0.013	0.50
0.076	3	0.015	0.60	0.018	0.70
0.101	4	0.020	0.80	0.022	0.90
0.127	5	0.025	1.0	0.027	1.1
0.152	6	0.03	1.2	0.033	1.3
0.203	8	0.04	1.6	0.044	1.8
0.254	10	0.05	2.0	0.055	2.2
0.305	12	0.06	2.4	0.066	2.6
0.406	16	0.08	3.2	0.088	3.5
0.508	20	0.10	4.0	0.11	4.4
0.584	23	0.12	4.8	0.13	5.2

(Continued)

Table 3. *(Continued)* **Conversions (Approximate) Between Surface Measuring Systems.**

Roughness Depth (R_{max}) to Roughness Average (R_a)				RMS min to RMS mm	
R_{max}		R_a		RMS	
μm	μinch	μm	μinch	μm	μinch
0.813	32	0.15	6.0	0.18	7.2
1.016	40	0.20	8.0	0.22	8.8
1.270	50	0.25	10	0.27	10.8
1.524	60	0.30	12.0	0.34	13.3
1.778	70	0.36	14.0	0.39	15.5
2.032	80	0.41	16.0	0.45	17.8
2.286	90	0.46	18.0	0.51	20.0
2.540	100	0.51	20.0	0.56	22.2
2.794	110	0.56	22.0	0.62	24.4
3.048	120	0.61	24.0	0.68	26.6
3.556	140	0.71	28.0	0.79	31.1
4.064	160	0.81	32.0	0.90	35.5
4.318	180	0.91	36.0	1.0	40.0
5.080	200	1.0	40.0	1.1	44.4
6.096	240	1.2	48.0	1.4	53.3
7.112	280	1.4	56.0	1.6	62.2
8.128	320	1.6	64.0	1.8	71.0
9.144	360	1.8	72.0	2.0	79.9
10.160	400	2.1	82.0	2.3	90.7
15.240	600	3.2	127.0	3.6	141.0
20.320	800	4.5	177.0	5.0	196.0
25.400	1000	5.8	230.0	6.5	255.0

Radiused inserts. Since the actual shape of the insert cutting edge will be machined into the part surface in both milling and turning operations, radiused inserts will leave grooves in the workpiece surface. The final surface is, therefore, dependent on the size of the radius and the feed per insert (assuming that all inserts are in the exact same plane both radially and axially). Since extremely accurate setups require lengthy setup time resulting in high cost, radiused inserts are typically restricted to roughing operations. The following formula can be used to calculate the groove depth on a workpiece finish that will be left by the path of a radiused insert in a milling operation. Since the formula assumes that the groove depth will be repeated consistently, the depth is theoretically equal to the average surface roughness (R_a) although, in reality, it is a measurement of maximum peak to valley roughness height, R_{max}. It must be remembered that the formula assumes perfect setup alignment and machine rigidity.

$$R_{max} = 125,000 \times ([fpt^2] \div r)$$

where R_{max} = Roughness (profile depth) in μinches
fpt = Feed per tooth in inch/tooth
r = Radius of the insert in decimal inches.

Example. Calculate the groove depth of a milled surface where the feed per tooth (or advance per tooth) was 0.008" and the radius of the insert is $^1/_{64}$" (0.0156"). Note that milling applications surface finish averages can range from a very rough 2000 μin. R_a (classified as "cleanup") to a "polished" finish of 8 μin. R_a. Average milling operations range from 250 μin. R_a ("medium") to 32 μin. R_a ("smooth").

$$R_{max} = 125,000 \times ([.008^2] , .0156) = 513 \text{ min.}$$

Chamfered inserts have small chamfers or wiping flats and are often used to achieve R_q ratings of 125 min. or less in milling. R_q is root mean square roughness, and establishes the mean (rather than the average) roughness depth of a finished surface. In turning, R_q (also referred to as RMS) measurements typically result in readings 15% to 25% higher than measurements achieved by the average roughness method. For milling, the range is higher, from 15% to 40%, because of the irregularities introduced by multiple inserts. The chamfer runs parallel to the workpiece surface, and the width of the chamfer must exceed the face mill's feed per revolution (or advance per revolution). When the feed per revolution exceeds the width of the chamfered flat, two or more inserts will produce the final surface finish profile. In this case, the final surface profile will be entirely dependent on the accuracy of the insert alignment in the axial plane. To assure surface finish numbers of 125 R_q, the advance per revolution should not exceed 80% of the flat width of the chamfer.

Wiper inserts are used in milling to obtain fine surface finishes when the advance per revolution greatly exceeds the width of the corner chamfer. In this situation, a periphery insert is installed in the cutter body to reduce waviness and minimize the peak to valley height in the surface profile. Wiper inserts have slight crowns so that differences in spindle camber or tilt from application to application will eliminate the typical saw-toothed profile associated with a lack of parallelism between the workpiece and the edge of the insert. The cutting depth of the wiper is normally ranges from 0.002" to a maximum of 0.004" below the other inserts in the cutter. The feed per revolution should be less than the width of the wiper, and in most instances should not exceed 50% of the wiper's length. Wipers in large diameter cutters should be aligned axially to insure proper overlap. Wipers are most effective on short chipping materials such as brass and cast iron. When used on steel, they can cause high axial forces and may introduce vibration due to the difficulty of forming a wide chip only 0.002" to 0.004" in height.

Wiper inserts can be used effectively in turning to improve surface finish. In milling wipers are used to reduce the depth of the grove created by other inserts in the cutter, but in turning their geometry can be used to advantage. When running at twice the feed rate as a normal insert, the R_{max} value will be equal to that of the normal insert. The radius of a wiper insert, when operating at the same feed as a normal insert, will produce a R_{max} value approximately one-half that of a normal insert.

To find the theoretical R_{max} value of a wiper insert used in a turning operation.

$$R_{max} = ([fpr^2 \times 10^6] \div [8 \times r]) \div 2$$

where R_{max} = Roughness (profile depth) in μinches
        fpr = Feed per revolution in inch/rev
        r = Radius of the insert in decimal inches.

Chip breakers

While most inserts have positive rake angles and, in relation to the toolholder, are negatively inclined to enhance cutting action and chip formation, the need for reliable chip control has led to the development of the chip breaker insert. In simplest terms, a chip breaker insert is manufactured with a groove that allows the insert to form chips in predetermined segments. The groove (or other irregularity–many inserts now contain intricate and almost artistic sinusoidal wave patterns for chipbreaking) encourages the chip to curl into short sections, eventually bending to the breaking point. Any artificial obstruction to normal chip flow on a cutting edge is, in essence, a chip breaker.

The key to chip breaking is to allow the chip to flow as freely as possible while breaking it as required for a specific machining operation. The force required to bend and break the chip consumes power, therefore the goal is to utilize the minimal power necessary to attain acceptable chip control, thereby breaking the chip to a manageable size without creating excess cutting force that could be detrimental to tool life. To produce chips of a desired size or configuration, feed and radial cutting force requirements can rise significantly. As forces rise, the ridge on the cutting edge of the insert will eventually erode or break away, possibly resulting in catastrophic tool failure. Because the shape of the insert's groove affects the radial and feed forces more than the tangential forces, using a chip breaker normally results in less tool force on a cutting operation and, therefore, less likelihood of vibration than would be experienced with a flat-top insert.

Chip control is also a function of the machinability of the workpiece material. Steels with carbon content of .20% and less are highly ductile in an annealed condition (less than 18 R_C) and, especially when outside diameter turning with shallow depths of cut on these steels, chip control problems are common. When these materials are machined in a heat treated condition (30-35 R_C), chip control is much more predictable. While increased hardness lowers the machinability rating of a material, the added hardness benefits chip handling and control up to the point where the additional hardness becomes obviously detrimental to tool life. When this point is reached, the chip will not increase in thickness as it passes over the top rake surface of the tool to the same degree as it would when machining the same material at a lower hardness level. Chips become thinner and harder to break for a given machine feed. The added hardness makes the chip stronger, with a higher resistance to bending, making it more difficult to break.

Unacceptable chip control can be the result of several factors. *Irregular*, or *ribbon shaped*, chips occur when the feed rate for an insert is too low for a specific depth of cut, or when the depth of cut is too shallow for a particular chip breaker design. Since the chip has relatively low strength, it will not curl and break predictably. *Flat spiral* chips are often caused by an extremely light depth of cut, generally very close to the nose radius of the insert. To correct this condition, choose a smaller nose radius, increase the depth of cut, increase the feed rate, or use a different chip breaker. *Helical* and *long spiral* chips are formed when using a chip groove that is too wide for a specific feed rate and depth of cut. To correct this condition, increase the feed rate or replace the insert for one with a narrower groove designed for lighter depths of cut. *Crumbling* chips are almost always the result of overfeeding the chip breaker.

Inserts are available with chip breakers on one or both sides of the insert. Most single-sided chip breaker geometries are recommended for roughing applications where they are subjected to high tangential cutting loads. Under these conditions, using a two-sided insert can result in the insert rocking under the cutting load, as fully 30 to 40% of the insert cutting edge is unsupported when compared to a single-sided design. The flat underside of a single-sided insert provides 100% support for the cutting edge. Under severe conditions, the use of two-sided inserts can lead to vibration or tool breakage.

Troubleshooting inserts

Troubleshooting should be performed in a systematic manner to identify and solve problems experienced when milling or turning with indexable inserts. The following guidelines are specifically directed to milling operations, but many of the causes and solutions also apply to turning and boring. These problems can be recognized as premature insert edge failure, unacceptable part appearance, machining noise, machine vibration, and cutter appearance. The five key areas of concern are:

1) cutting tool material   4) workpiece
2) cutter/adapter   5) setup/fixturing
3) machine

There are seven primary edge condition problems.

Chipping can appear to be normal flank wear to the untrained eye, but normal flank wear lands have a fine, smooth wear pattern while a land formed by chipping has a saw-toothed, uneven surface. If not detected early, it may be perceived as depth of cut notching. In most cases, the problem can be solved by changing to a stronger grade or different edge preparation such as a larger hone or T-land, or from a 0° cutter geometry to a lead angle cutter geometry. Another cause can be the recutting of chips.

Depth of cut notching appears when chipping or localized wear at the depth of cut line on the rake face and flank of the insert occurs. It is primarily the result of the condition of the workpiece material. The causes include abrasive skin or scale on workpiece, machining abrasive materials such as high temperature alloys, machining a work-hardened outer layer, or heat treated material above 560 BHN (55 R_C).

Thermal cracks run perpendicular to the insert's cutting edge and are caused by the extreme temperature variations involved in machining. These variations create heat stresses in the insert that can result in thermal cracks.

Built-up edge involves the adhesion of layers of workpiece material to the top surface of the insert. Hardened pieces of the adhered material periodically break free leaving an irregularly shaped depression along the cutting edge. Cutting forces will be increased by built-up edge.

Crater wear is seen as a relatively smooth, regular depression on the insert's rake face. It is caused either by workpiece material adhering to the insert's top edge being dislodged and carrying away minute fragments of the insert, or by frictional heat resulting from the flow of chips over the top surface of the insert. Eventually, the heat build-up softens the insert behind the cutting edge and carries away minute particles of the insert.

Flank wear, when uniform, is the preferred mode of insert failure because it can be predicted. Excessive flank wear increases cutting forces and results in a loss of surface finish. When flank wear occurs at an unacceptable rate or becomes unpredictable, the causes must be investigated. When roughing, inserts should be indexed when flank wear reaches 0.015–0.029" (0.3810–0.5080 mm). When finishing, inserts should be indexed when flank wear reaches 0.010–0.015" (0.2540–0.3810 mm) or sooner.

Multiple factors. When wear, chipping, thermal cracking, and breakage all occur at once, the machine operator must look beyond normal feed, speed, and depth of cut adjustments to find the problem. The system's rigidity should be closely inspected for loose or worn parts.

Insert identification

Indexable inserts are manufactured in a variety of shapes, sizes, and thicknesses. They are available with straight holes, countersunk holes, or no holes, with or without chip breakers, and in many other variations. The selection of the appropriate turning toolholder geometry accompanied by the correct insert shape and chip breaker geometry has a significant impact on the productivity and tool life of a specific operation. Proper insert selection is extremely important. Insert strength is one important factor in selecting the correct geometry for a workpiece material or hardness range. The general rule for

Table 4. Troubleshooting Insert Condition Problems. *(Source, Kennametal, Inc.)*

Cutting Tool Material	Problem	Solution	Comments
Carbide	chipping	• check system rigidity. • correct worn gibbs/bearings. • check for improper cutter mounting.	• these steps will reduce or cure chatter.
		• use largest hone or T-land available.	• strengthens edge.
		• use tougher grade.	• resists chipping.
		• increase speed.	• reduces or eliminates built-up edge.
		• reduce feed per tooth.	• allows chip evacuation.
		• choose cutter geometry with correct pitch. • use air blast or coolant.	• avoids recutting of chips.
	depth of cut notching	• change to 15°, 30°, or 45° lead cutters.	• improves cutter geometry.
		• change to more wear resistant grade.	• resists d.o.c. notching.
		• reduce feed per tooth.	• reduces cutting forces.
		• reduce speed.	• reduces cutting forces.
		• use hone or T-land insert.	• strengthens edge.
		• vary depth of cut on very abrasive materials.	• program alternating depths of cut to vary depth of cut line.
	thermal cracking	• reduce cutting speed and possibly feed per tooth.	• reduces cutting edge temperature.
		• assure coolant is properly applied or eliminate it altogether.	• temperature variations create stress.
		• use coated grade designed for wet milling.	• some grades resist thermal cracking.
	built-up edge	• increase cutting speed.	• higher speed reduces adhesion.
		• increase feed per tooth.	• larger chips less likely to adhere.
		• use mist or flood coolant when machining stainless and aluminum alloys.	• coolant reduces tendency of chips to adhere to the insert.
		• use diamond tipped or coated inserts on alloys.	• required at higher speeds.
		• use sharp edge, positive rake PVD or polished inserts.	• reduces tendency of material to adhere to inserts.
	crater wear	• use more wear-resistant grade.	• resists crater wear better.
		• reduce cutting speed.	• reduces frictional heat.
		• use smaller T-land or increase feed.	• feed should be correct for T-land.
	flank wear	• check that speed is correct.	• speed should be reduced without changing feed rate.
		• increase feed per tooth.	• feed should be high enough to avoid rubbing at shallow d.o.c.
		• use more wear resistant grade.	• if using uncoated grade, change to coated.
		• inspect insert/cutter combination.	• insert and cutter must be compatible.

(Continued)

Table 4. *(Continued)* **Troubleshooting Insert Condition Problems.** *(Source, Kennametal, Inc.)*

Cutting Tool Material	Problem	Solution	Comments
Carbide	multiple factors	• check for loose cutter mounting. • improve cutter and fixture rigidity. • check for worn hardware or improper insert installation.	• system must have sufficient rigidity.
		• reduce feed rate.	• reduces cutting forces.
		• if possible, use lead angle cutter.	• directs cutting forces away from insert nose.
		• use larger nose radius, T-land insert, or tougher grade.	• proper insert and grade will reduce wear.
Cermet	chipping	• reduce feed per insert.	• cermets have excellent resistance to built-up edge and are dry milling grades. Do not use coolant.
		• eliminate coolant.	
		• apply hone or T-land insert.	
	breakage (fracture)	• reduce depth of cut and chip load.	
		• increase speed.	
		• apply hone or T-land insert.	
Ceramic (alumina grade)	chipping & breakage	• check system rigidity.	• coolant is not recommended for milling. • use lead angle cutters where possible.
		• increase hone or T-land.	
		• switch to thicker insert.	
		• increase speed.	
		• reduce feed from heavy to moderate/light.	
	wear	• reduce speed.	
Ceramic (silicon nitride grade)	depth of cut notching	• reduce hone or size of T-land edge.	• grades intended for general purpose machining are more prone to d.o.c. notch than grades designed for roughing.
		• if possible, use lead angle cutter.	
		• prechamfer part to relieve stress points.	
		• vary depth of cut.	
	chipping	• change edge preparation.	• minor chipping is normal, especially on Inconel.
		• reduce chip load.	
	flank wear	• use 0.060" (1.5240 mm) as indexing criterion.	• for general purpose grade inserts only.
		• reduce speed, increase feed.	• silicon nitride grades can be used with or without coolant.
	fracture	• use thicker inserts.	
		• do not over-torque clamping device.	
Poly-crystalline Diamond	chipping & breaking	• check system rigidity.	• can be used with coolant.
		• reduce chip load, increase sfm/smm.	
CBN	chipping & breaking	• check system rigidity.	• use with coolant.
		• additional edge preparation may be required (hone or T-land).	

rating an insert's strength based on its size is "the larger the included (nose) angle on the insert corner, the greater the insert strength. The Turning section of this book contains additional information on inserts and insert holders, and the Single Point Threading section describes inserts used for threading on toolholders.

This section details the identification system for idexable inserts used in milling, turning, drilling, and boring operations. ISO and ANSI both provide standards for identifying inserts. The ISO standard may be used to quickly determine the dimensions of any standard insert, and all inserts are easily ordered by their ISO catalog number. In the ISO catalog number, the first letter, taken from the *Position 1* chart, identifies the insert's shape, the second letter identifies the clearance angle on the major cutting edge of the insert (see *Position 2*), the third letter denotes the tolerance class of the insert (see *Position 3*), and the fourth letter specifies the clamping system and the geometry of the insert as shown in *Position 4*. The initial four letters are the same for both the ANSI and ISO identification systems. Following the initial four letters in the ISO catalog number, are a series of four numbers that identify the physical characteristics of the insert. The first two letters provide the cutting edge length (see *Position 5*), the next two the thickness of the insert (see *Position 6*). The next two places will contain either two letters, designating the lead angle and corner relief angle, or two numbers indicating the corner radius (see *Position 7*). Two additional places are provided which can be used to specify the cutting edge form, and the cutting direction as shown in *Position 8* and *Position 9*.

Position 1: Insert shape. Enter the letter that correctly specifies the shape of the insert.

Letter Symbol	Shape	Nose Angle (Degrees)	Application
A	parallelogram	85	milling & turning
B	parallelogram	82	turning
C	rhomboid (diamond)	80	milling & turning
D	rhomboid (diamond)	55	turning
E	rhomboid (diamond)	75	milling & turning
F	rhomboid (diamond)	50	turning
H	hexagon	120	milling & turning
L	rectangle	90	milling & turning
M	rhomboid (diamond)	86	turning
N/K	parallelogram	55	turning
O	octagon	135	milling & turning
P	pentagon	108	turning
R	round	—	milling & turning
S	square	90	milling & turning
T	triangle	60	milling & turning
V	rhomboid (diamond)	35	turning
W	hexagon (trigon)	80	turning & drilling
X	manufacturer option		

Position 2: Clearance angle on the major cutting edge. **A** = 3°. **B** = 5°. **C** = 7°. **D** = 15°. **E** = 20°. **F** = 25°. **G** = 30°. **N** = 0°. **P** = 11°. Enter the letter that describes the clearance angle on the insert.

Position 3: Tolerance class. The top chart indicates tolerances by width ("A") of the insert. Enter the letter from the lower chart (Part II) that corresponds to the tolerance for the insert size.

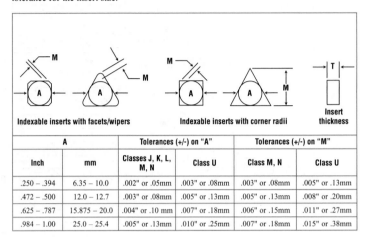

Indexable inserts with facets/wipers   **Indexable inserts with corner radii**   **Insert thickness**

A		Tolerances (+/-) on "A"		Tolerances (+/-) on "M"	
Inch	mm	Classes J, K, L, M, N	Class U	Class M, N	Class U
.250 – .394	6.35 – 10.0	.002" or .05mm	.003" or .08mm	.003" or .08mm	.005" or .13mm
.472 – .500	12.0 – 12.7	.003" or .08mm	.005" or .13mm	.005" or .13mm	.008" or .20mm
.625 – .787	15.875 – 20.0	.004" or .10 mm	.007" or .18mm	.006" or .15mm	.011" or .27mm
.984 – 1.00	25.0 – 25.4	.005" or .13mm	.010" or .25mm	.007" or .18mm	.015" or .38mm

Position 3: Part II.

Class		A	M	T	Class		A	M	T
A	inch	.001	.0002	.001	J	inch	.002 – .006*	.0002	.001
	mm	.025	.005	.025		mm	.05 – .15*	.005	.025
F	inch	.0005	.0002	.001	K	inch	.002 – .006*	.0005	.001
	mm	.013	.005	.025		mm	.05 – .15*	.013	.025
C	inch	.001	.0005	.001	L	inch	.002 – .006*	.001	.001
	mm	.025	.013	.025		mm	.05 – .15*	.025	.025
H	inch	.0005	.0005	.001	M	inch	.002 – .006*	.003 – .008*	.005
	mm	.013	.013	.025		mm	.05 – .15*	.08 – .20*	.130
E	inch	.001	.001	.001	N	inch	.002 – .006*	.003 – .008*	.001
	mm	.025	.025	.025		mm	.05 – .15*	.08 – .20*	.025
G	inch	.001	.001	.005	U	inch	.003 – .010*	.005 – .015*	.005
	mm	.025	.025	.130		mm	.08 – .25*	.13 – .38	.130

*See table above for tolerances according to insert size and class. All tolerances are (+/–).

Position 4: Geometry and clamping type. Letter symbol denotes the presence and shape of hole in the insert, if a chip breaker is used, and if the chip breaker appears on one or both sides of the insert.

Geometry and Clamping Type				
Symbol	Hole	Shape of Hole	Chip Breaker	Shape of Insert Section
N	none		without	
R			single sided	
F			double sided	
A	yes	cylindrical hole	without	
M			single sided	
G			double sided	
W		partly cylindrical hole 40–60° countersink	without	
T			single sided	
Q		partly cylindrical hole 40–60° double countersink	without	
U			double sided	
B		partly cylindrical hole 70–90° countersink	without	
H			single sided	
C		partly cylindrical hole 70–90° double countersink	without	
J			double sided	
X	special design			

Position 5: Cutting edge length. **Inch units:** Based on cutting edge length in increments of eights of an inch (3.18 mm). $^1/_4$" = 2, $^3/_8$" = 3, $^7/_{16}$" = 3.5, $^1/_2$" = 4, etc. **Metric Units:** Cutting edge length in millimeters.

Position 6: Thickness. Codes are specified for common thicknesses (see table).

Insert Thickness			
Measurement		**Symbol**	
inch	mm	inch	mm
1/32	.1	0.5 (1)*	–
.040	1.0	.6	T0
1/16	1.59	1 (2)*	01
5.64	1.98	1.2	T1
3/32	2.38	1.5 (3)*	02
1/8	3.18	2	03
5/32	3.97	2.5	T3
3/16	4.76	3	04
7/32	5.56	3.5	05
1/4	6.35	4	06
5/16	7.94	5	07
3/8	9.52	6	09
7/16	11.11	7	11
1/2	12.70	8	12

*Inch sizes in parenthesis are alternative symbols used for D and E inserts with under $1/4$" IC.

Position 7: Corner description. Two letters, the first specifies the lead angle as follows: **A** = 45°, **D** = 60°, **E** = 75°, **F** = 85°, **G** = 87°, **P** = 90°, **Z** = any other edge angle. The second letter identifies the facet clearance (corner relief angle): **A** = 3°, **B** = 5°, **C** = 7°, **D** = 15°, **E** = 20°, **F** = 25°, **G** = 30°, **N** = 0°, **P** = 11°, **Z** = any other angle. Alternatively, the numbers, from the chart below, may be used to identify the corner radius.

Corner Radius		Symbol	
inch	mm	ANSI	ISO
–	.051	–	00
.002	–	0	–
.004	.102	0.2	01
.008 (1/125)	.203	0.5	02
1/64	.397	1	04
1/32	.794	2	08
3/64	1.191	3	12
1/16	1.6	4	16
5/64	1.984	5	20
3/32	2.381	6	24
7/64	2.778	7	28
1/8	3.175	8	32
5/32	3.968	10	40
3/16	4.762	12	48
7/32	5.556	14	56
1/4	6.350	16	64
round insert (inch)		00	
round insert (metric)			M0

Position 8: Edge preparation. **A** = Honed edge 0.0005" to less than 0.003" (0.013 mm to less than 0.076 mm). **B** = Honed edge 0.003" to less than 0.005" (0.076 mm to less than 0.127 mm). **C** = Honed edge 0.005" to less than 0.007" (0.127 mm to less than 0.178 mm). **E** = Rounded edge. **F** = Sharp edge. **J** = Polished to 4-microinch AA, rake face only. **K** = Double chamfered cutting edge. **P** = Double chamfered and rounded cutting edge. **S** = Chamfered and rounded cutting edge (honed T-land). **T** = Chamfered cutting edge (T-land).

Position 9: Hand of insert. **R** = Right hand. **L** = Left hand. **N** = Either hand (neutral).

Position 10 is used for additional information. It is often used to designate facet size or special cutting edge conditions or control features.

Jig bushings

Jigs are special tools that not only secure the workpiece, but also contain guides (usually holes in the jig plate) that accurately position the cutting tools that pass through these guides and shape the workpiece. A fixture, on the other hand, simply holds the work during machining, and does not contain any special devices for positioning the tools that form the workpiece. Jigs are primarily used in drilling, tapping, or reaming hole-making operations. Jig bushings are used to protect expensive jig plates, and to assure repeatable accuracy. Rather than replacing a complete jig plate when guide holes become nonconcentric or oversize, bushings offer a fast and inexpensive repair that allows the original jig plate to be reused repeatedly.

Jig bushings can be obtained from several manufacturers in a vast number of sizes and configurations. Specialty companies are capable of producing one-off bushings for specialized operations, or they may be fabricated in the shop. The bushings listed in this section are commonly available and will meet most needs, but should by no means be considered a complete catalog of available sizes. Standardized bushings are identified with an ANSI identification system that allows for the dimensions of most bushings to be identified by their ANSI Number or Symbol. This system assigns a numerical prefix for each style of bushing, and the prefix is followed by two sets of numbers that provide the basic dimensions. The first numbers specify the outside diameter in $1/64$ths of an inch. For instance, a $3/4$" outside diameter would be expressed with the number 48, because $3/4$" equals $48/64$". The second number specifies the length of the bushing in $1/16$ths of an inch. This means that a 1" long bushing would be identified with the number 16. The ANSI "P" headless press fit bushing is the most widely used and inexpensive bushing, and it is commonly used for single stage drilling operations. A headless press fit bushing with an outside diameter of $3/4$", and a length of 1", would be identified with the ANSI Number P-48-16. It is very important to remember that, when specifying bushings, the height of the head of the bushing is not included in its length.

Bushings are also available in many different materials, including bronze, tool steels, tungsten carbide, and stainless steel, but most are supplied in high grade steel with a hardness ranging from 58 to 64 R_C. Selecting the hardness requirements and style of bushing for a specified operation may be influenced by the projected length of the production run. Other considerations are whether the bushing will be impacted from the top, or if heavy axial loads might eventually force a headless bushing out of the jig plate. In both instances, bushings with heads should be fitted. Both press fit (type H, **Table 1**, which also provides dimensions of type P press fit headless bushings) and renewable bushings (types S and SF, **Table 2** and **3**) are available with heads. Renewable bushings are used with a hardened liner press fit bushing (types HL and L, **Table 4**) that has been pressed into the jig plate. The liner bushing creates a precise location point for the renewable bushing, which is held in place by a lock screw (type S and type SF) or clamp (type SF only) fitted into a milled recess in the head of the bushing. This allows these bushings to be quickly changed, either to replace worn-out bushings, or to substitute another slip or slip fixed bushing with a different inside diameter in order to perform multiple tasks, such as drilling and reaming, with a single hole size. Tolerances for jig bushing dimensions are given in **Table 5**. Dimensions of lockscrews and clamps used to hold renewable bushings in place are given in **Tables 6** through **9**.

Another option when purchasing press fit bushings is to choose an "unfinished" bushing, usually designated with the letter U following the normal type designation. A type HU, for example, would be an unfinished head type press fit bushing. Generally, unfinished bushings with an outside diameter of up to $1/4$" are supplied with a maximum oversize of 0.010", above $1/4$" and up to $13/32$" the maximum oversize is 0.015", and above

$^1/_2$" the oversize is 0.020". Unfinished bushings are desirable for shops that maintain their own press fit limits.

Installation procedures

Press fit bushings

American Drill Bushing Co. recommends that press fit bushings (headless type P and type H with head) be installed according to the following guidelines. All proper installations begin by assuring that the mounting hole is round. This is best achieved by either jig boring or reaming the hole, since an ordinary twist drill seldom produces a perfectly round and accurately sized hole. Roundness is important because bushings will, sooner or later, assume the shape of their mounting hole. In any interference fit, material is displaced, so it is desirable to use the minimal interference necessary to retain the bushing. Otherwise, distortion of the jig plate, bushing distortion, or bushing seizure can result. Normally, diametral interference fits of 0.0005" to 0.0008" work well for headless press fit bushings or liners, and 0.0003" to 0.0005" is adequate for press fit bushings and liners with heads. The other variable is the depth of the mounting hole, and as plate thickness and bushing length increase, less interference is necessary.

Press fit bushings should be lubricated prior to installation. Bushings installed without lubricant are more likely to pick up metal and score the mounting hole during installation. Lubrication will also make removal easier. Whenever possible, a hand arbor should be used to press the bushing into the jig plate. A hammer should be used only as a last resort (and should never strike the bushing directly), as it may damage the bushing or compromise centerline perpendicularity.

High accuracy dictates that the bushing is placed in close proximity to the workpiece. The greater clearance between the bushing and workpiece, the higher the probability for error. However, there must be sufficient clearance to allow for the removal of chips, and the type of chip produced is an important consideration when setting the clearance. The recommended clearance when drilling materials such as cast irons, which produce small chips, is one-half of the drill diameter. For materials that produce long chips, such as cold-rolled steel, the clearance should be one to one and one-half times the diameter of the drill. Chip removal and accuracy can be maximized by substituting bushings of varying length based on the drill size and material being machined. For example, when drilling a $^3/_4$" hole in steel, select a bushing that will allow adequate clearance for chip removal. Then, after drilling, insert a bushing that allows very little clearance, but high accuracy, and ream the hole to the required tolerance.

Slip and fixed renewable bushings

Because these bushings are easily replaceable when properly installed in liner bushings, they rely on an outside locking mechanism—either a lockscrew or clamp— to retain them in place. Slip fit renewable bushings are held in place by a flange on the head of the bushing that allows for the bushing to be removed by simply twisting the head of the bushing and then lifting it from its liner, without removing the jig from the production line. Fixed renewable bushings require the lock screw or clamp to be removed before the bushing can be lifted from its liner. Both lock screws and lock clamps are designed to fit into the recess machined into the head of the bushing. The lock screws are tightened via a traditional slotted head, while the clamps are secured by socket head screws.

Other bushings

Special bushings are available for use with very thin materials. These bushings, commonly referred to as template bushings, are used when standard press fit bushings are

(Text continued on p. 890)

Table 1. Press Fit Jig Bushings, Type H (With Head) and Type P (Headless). *(Source, American Drill Bushing Co.)*

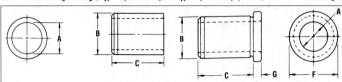

Hole Size A		Body Diameter B			Head Dimension—H Type		Radius R	Body Length L*	ANSI Number**
Drill	Decimal	Nom.	Max.	Min.	Dia. F	Height G			
#80 to 1/16	0.0135 through 0.0625	5/32"	0.1578	0.1575	0.2500	0.0938	0.016	0.250 0.312 0.375 0.500 0.750	P/H-10-4 P/H-10-5 P/H-10-6 P/H-10-8 PH-10-12
#55 to #39	0.0520 to 0.0995	3/16"	0.1890	0.1887	-	-	0.016	0.250 0.312 0.375 0.500 0.625 0.750	P-12-4 P-12-5 P-12-6 P-12-8 P-12-10 P-12-12
#55 to #39	0.0520 to 0.0995	13/64"	0.2046	0.2043	0.3125	0.0938	0.016	0.250 0.312 0.375 0.500 0.750 1.000 1.375	P/H-13-4 P/H-13-5 P/H-13-6 P/H-13-8 P/H-13-12 P/H-13-16 P-13-22
#40 to 9/64	0.0980 to 0.1406	1/4"	0.2516	0.2513	0.3750	0.0938	0.016	0.250 0.312 0.375 0.500 0.625 0.750 1.000 1.375 1.750	P/H-16-4 P/H-16-5 P/H-16-6 P/H-16-8 P/H-16-10 P/H-16-12 P/H-16-16 P/H-16-22 P-16-28
1/8 to #10	0.1406 to 0.1875	5/16"	0.1341	0.3138	0.4375	0.1250	0.031	0.250 0.312 0.375 0.500 0.750 1.000 1.375 1.750	P/H-20-4 P/H-20-5 P/H-20-6 P/H-20-8 P/H-20-12 P/H-20-16 P/H-20-22 P-20-28
#10 to #1	0.1935 to 0.2280	3/8"	0.3766	0.3763	0.5000	0.0938	0.031	0.250 0.312 0.375 0.500 0.625 0.750 1.000	P/H-24-4 P/H-24-5 P/H-24-6 P/H-24-8 P/H-24-10 P/H-24-12 P/H-24-16

(Continued)

Table 1. *(Continued)* **Press Fit Jig Bushings, Type H (With Head) and Type P (Headless).**
(Source, American Drill Bushing Co.)

Hole Size A		Body Diameter B			Head Dimension—H Type		Radius R	Body Length L*	ANSI Number**
Drill	Decimal	Nom.	Max.	Min.	Dia. F	Height G			
3/16 to F	0.1875 to 0.2570	13/32"	0.4078	0.0475	0.5312	0.1562	0.031	0.250 0.312 0.375 0.500 0.750 1.000 1.375 1.750	P/H-26-4 P/H-26-5 P/H-26-6 P/H-26-8 P/H-26-12 P/H-26-16 P/H-26-22 P/H-26-28
A To L	0.2340 to 0.2900	7/16"	0.4391	0.4388	0.5625	0.0938	0.031	0.250 0.312 0.375 0.500 0.625 0.750 1.000	P/H-28-4 P/H-28-5 P/H-28-6 P/H-28-8 P/H-28-10 P/H-28-12 P/H-28-16
3/16 to 5/16	0.1875 to 0.3125	1/2"	0.5017	0.5014	0.6250	0.2188	0.047	0.250 0.312 0.375 0.500 0.750 1.000 1.375 1.500 1.750 2.1250	P/H-32-4 P/H-32-5 P/H-32-6 P/H-32-8 P/H-32-12 P/H-32-16 P/H-32-22 P-32-24 P/H-32-28 P-32-34
M To 11/32	0.2950 to 0.3437	9/16"	0.5642	0.5639	0.6875	0.2188	0.047	0.250 0.312 0.375 0.500 0.625 0.750 1.000	P-36-4 P-36-5 P-36-6 P/H-36-8 P/H-36-10 P/H-36-12 P/H-36-16
5/16 to 7/16	0.3125 to 0.4375	5/8"	0.6267	0.6264	0.8125	0.2188	0.047	0.250 0.312 0.375 0.500 0.625 0.750 1.000 1.375 1.500 1.750 2.125	P/H-40-4 P/H-40-5 P/H-40-6 P/H-40-8 P/H-40-10 P/H-40-12 P/H-40-16 P/H-40-22 P/H-40-24 P/H-40-28 P/H-40-34
5/16 to 17/32	0.3125 to 0.5312	3/4"	0.7518	0.7515	0.9375	0.2188	0.062	0.250 0.312 0.375 0.500 0.625 0.750 1.000 1.250 1.375 1.500 1.750 2.125	P/H-48-4 P/H-48-5 P/H-48-6 P/H-48-8 P/H-48-10 P/H-48-12 P/H-48-16 P/H-48-20 P/H-48-22 P/H-48-24 P/H-48-28 P/H-48-34

(Continued)

Table 1. *(Continued)* **Press Fit Jig Bushings, Type H (With Head) and Type P (Headless).**
(Source, American Drill Bushing Co.)

Hole Size A		Body Diameter B			Head Dimension—H Type		Radius R	Body Length L*	ANSI Number**
Drill	Decimal	Nom.	Max.	Min.	Dia. F	Height G			
1/2 to 21/32	0.5156 to 0.6250	7/8"	0.8768	0.8765	1.1250	0.2500	0.062	0.312 0.375 0.500 0.625 0.750 1.000 1.250 1.375 1.500 1.750 2.125	P-56-5 P/H-56-6 P/H-56-8 P/H-56-10 P/H-56-12 P/H-56-16 P/H-56-20 P/H-56-22 P-56-24 P/H-56-28 P/H-56-34
1/2 to 49/64	0.5000 to 0.7656	1"	1.0018	1.0015	1.2500	0.3125	0.0938	0.500 0.625 0.750 1.000 1.250 1.375 1.500 1.750 2.125 2.500 3.000	P/H-64-8 P-64-10 P/H-64-12 P/H-64-16 P-64-20 P/H-64-22 P-64-24 P/H-64-28 P/H-64-34 P-64-40 P-64-48
5/8 to 1 1/32	0.7656 to 1.0312	1 3/8"	1.3772	1.3768	1.6250	0.3750	0.0938	0.500 0.625 0.750 1.000 1.250 1.375 1.500 1.750 2.000 2.125 2.500 3.000	P/H-88-8 P-88-10 P/H-88-12 P/H-88-16 P-88-20 P/H-88-22 P-88-24 P/H-88-28 P-88-32 P/H-88-34 P/H-88-40 P-88-48
1" to 1 25/64	1.0000 to 1.3906	1 3/4"	1.7523	1.7519	2.0000	0.3750	0.0938	0.750 1.000 1.375 1.500 1.750 2.125 2.500 3.000	P/H-112-12 P/H-112-16 P/H-112-22 P/H-112-24 P/H-112-28 P/H-112-34 P-112-40 P-112-48
1 3/8 to 1 49/64	1.3906 to 1.7500	2 1/4"	2.2525	2.2521	2.5000	0.3750	0.0938	1.000 1.375 1.500 1.750 2.125 2.500 3.000	P/H-144-16 P/H-144-22 P-144-24 P/H-144-28 P-144-34 P-144-40 P-144-48

All dimensions in inches. Diameter A to be concentric to body or head diameter to within 0.0005".

* Total height of Type H bushings is head height plus body length.

** Both styles are not available in all sizes. Check prefix for ANSI number for availability.

Table 2. Body Dimensions for Slip Type (S) and Fixed Type (SF) Renewable Wearing Bushings.
(For Head Dimensions, see Table 3.) *(Source, American Drill Bushing Co.)*

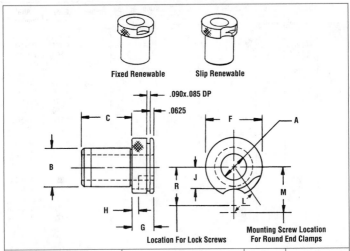

Fixed Renewable   Slip Renewable

.090x.085 DP

.0625

Location For Lock Screws

Mounting Screw Location
For Round End Clamps

Hole Size A			Body Diameter B			Body Length C	ANSI Number
Drill	Decimal	mm	Nom.	Max.	Min.		
#80 to #40	.0135 to .0980	.34 to 2.48	3/16"	0.1875	0.1873	0.250	S/SF-12-4
						0.312	S/SF-12-5
						0.375	S/SF-12-6
						0.500	S/SF-12-8
						0.625	S/SF-12-10
						0.750	S/SF-12-12
#39 to 9/64	.0995 to .1406	2.52 to 3.57	1/4"	0.2500	0.2498	0.250	S/SF-16-4
						0.312	SS/F-16-5
						0.375	S/SF-16-6
						0.500	S/SF-16-8
						0.625	S/SF-16-10
						0.750	S/SF-16-12
#55 to #10	0.0520 to 0.1935	1.32 to 4.91	5/16"	0.3125	0.3123	0.250	S/SF-20-4
						0.312	S/SF-20-5
						0.375	S/SF-20-6
						0.500	S/SF-20-8
						0.750	S/SF-20-12
						1.000	S/SF-20-16
						1.375	S/SF-20-22
#10 to #1	0.1935 to 0.2280	4.91 to 5.79	3/8"	0.3750	0.3748	0.500	S/SF-24-8
						0.625	S/SF-24-10
						0.750	S/SF-24-12
						1.000	S/SF-24-16

(Continued)

Table 2. *(Continued)* **Body Dimensions for Slip Type (S) and Fixed Type (SF) Renewable Wearing Bushings.**
(For Head Dimensions, see Table 3.) *(Source, American Drill Bushing Co.)*

Hole Size A			Body Diameter B			Body Length C	ANSI Number
Drill	Decimal	mm	Nom.	Max.	Min.		
A to L	0.2340 to 0.2900	5.94 to 7.36	7/16"	0.4375	0.4373	0.500 0.625 0.750 1.000	S/SF-28-8 S/SF-28-10 S/SF-28-12 S/SF-28-16
#27 to 11/32	0.1440 to 0.3437	3.65 to 8.72	1/2"	0.5000	0.4998	0.312 0.375 0.500 0.750 1.000 1.375 1.500 1.750 2.125	S/SF-32-5 S/SF-32-6 S/SF-32-8 S/SF-32-12 S/SF-32-16 S/SF-32-22 S/SF-32-24 S/SF-32-28 S/SF-32-34
S to 7/16	0.3480 to 0.4375	8.83 to 11.11	5/8"	0.6250	0.6248	0.500 0.625 0.750 1.000 1.250	S/SF-40-8 S/SF-40-10 S/SF-40-12 S/SF-40-16 S/SF-40-20
K to 9/16	0.2811 to 0.5625	7.13 to 14.28	3/4"	0.7500	0.7498	0.312 0.500 0.750 1.000 1.375 1.500 1.750 2.125 2.500 3.000	S/SF-48-5 S/SF-48-8 S/SF-48-12 S/SF-48-16 S/SF-48-22 S/SF-48-24 S/SF-48-28 S/SF-48-34 S/SF-48-40 S/SF-48-48
35/64 to 21/32	0.5469 to 0.6562	13.89 to 16.66	7/8"	0.8750	0.8748	0.500 0.625 0.750 1.000 1.250 1.500	S/SF-56-8 S/SF-56-10 S/SF-56-12 S/SF-56-16 S/SF-56-20 S/SF-56-24
15/32 to 25/32	0.4687 to 0.7812	11.90 to 19.84	1.00"	1.0000	0.9998	0.500 0.750 1.000 1.375 1.500 1.750 2.125 2.500 3.000	S/SF-64-8 S/SF-64-12 S/SF-64-16 S/SF-64-22 S/SF-64-24 S/SF-64-28 S/SF-64-34 S/SF-64-40 S/SF-64-48
23/32 to 1 1/16	0.7187 to 1.0625	18.25 to 26.98	1 3/8"	1.3750	1.3747	0.750 1.000 1.375 1.500 1.750 2.125 2.500 3.000	S/SF-88-12 S/SF-88-16 S/SF-88-22 S/SF-88-24 S/SF-88-28 S/SF-88-34 S/SF-88-40 S/SF-88-48

(Continued)

Table 2. *(Continued)* **Body Dimensions for Slip Type (S) and Fixed Type (SF) Renewable Wearing Bushings.** (For Head Dimensions, see Table 3.) *(Source, American Drill Bushing Co.)*

Hole Size A			Body Diameter B			Body Length C	ANSI Number
Drill	Decimal	mm	Nom.	Max.	Min.		
31/32 to 1 13/32	0.9687 to 1.4062	24.60 to 35.71	1 3/4"	1.7500	1.7497	0.750	S/SF-112-12
						1.000	S/SF-112-16
						1.375	S/SF-112-22
						1.500	S/SF-112-24
						1.750	S/SF-112-28
						2.125	S/SF-112-34
						2.500	S/SF-112-40
						3.000	S/SF-112-48
1 11/32 to 1 7/8	1.3437 to 1.8750	34.12 to 47.62	2 1/4"	2.2500	2.2496	0.750	S/SF-144-12
						1.000	S/SF-144-16
						1.375	S/SF-144-22
						1.500	S/SF-144-24
						1.750	S/SF-144-28
						2.125	S/SF-144-34
						2.500	S/SF-144-40
						3.000	S/SF-144-48

All dimensions in inches. Diameter A to be concentric to the head diameter within 0.0005". Total height of Type SF and S bushings is head height (G) from Table 3, plus body length (C).

Table 3. Head Dimensions for Type S and Type SF Renewable Bushings. (For explanation of symbols, see illustration at top of Table 2.) *(Source, American Drill Bushing Co.)*

B	F	G	H	J	L	M	R
3/16	5/16	3/16	3/32	3/32	55	-	17/64
1/4	7/16	1/4	1/8	9/64	30	-	23/64
5/16	35/64	1/4	1/8	11/64	65	5/8	1/2
3/8	5/8	1/4	1/8	15/64	30	-	29/64
7/16	5/8	1/4	1/8	15/64	30	-	29/64
1/2	51/64	1.4	1/8	19/64	65	3/4	5/8
5/8	7/8	1/4	1/8	23/64	30	-	37/64
3/4	1 3/64	1/4	1/8	27/64	50	7/8	3/4
7/8	1 1/4	3/8	3/16	31/64	30	-	53/64
1	1 27/64	3/8	3/16	19/32	35	1 7/64	59/64
1 3/8	1 51/64	3/8	3/16	25/32	30	1 19/64	1 7/64
1 3/4	2 19/64	3/8	3/16	1	30	1 41/64	1 25/64
2 1/4	2 51/64	3/8	3/16	1 1/4	25	1 57/64	1 41/64

Table 4. Liner Type Jig Bushings Type HL (with Head) and Type L (Headless). *(Source, American Drill Bushing Co.)*

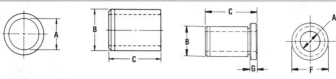

Inside Diameter A			Outside Diameter B			Head Dimension HL Only		Radius R	Body Length C*	ANSI Number**
Nom.	Max.	Min.	Nom.	Max.	Min.	Dia. F	Height G			
3/16"	0.1879	0.1876	5/16"	0.3141	0.3138	-	-	0.0312	0.250	L-20-4
									0.312	L-20-5
									0.375	L-20-6
									0.500	L-20-8
									0.625	L-20-10
									0.750	L-20-12
1/4"	0.2504	0.2501	7/16"	0.4392	0.4389	-	-	0.3750	0.250	L-28-4
									0.312	L-28-5
									0.375	L-28-6
									0.500	L-28-8
									0.625	L-28-10
									0.750	L-28-12
5/16"	0.3129	0.3126	1/2"	0.5017	0.5014	0.6250	0.0938	0.0469	0.250	L-32-4
									0.312	L/HL-32-5
									0.375	L/HL-32-6
									0.500	L/HL-32-8
									0.750	L/HL-32-12
									1.000	L/HL-32-16
									1.375	L/HL-32-22
5/16"	0.3129	0.3126	9/16"	0.5642	0.5639	-	-	0.0469	0.250	L-36-4
									0.312	L-36-5
									0.375	L-36-6
									0.375	L-36-8
									0.625	L-36-10
									0.500	L-36-12
									1.000	L-36-16
3/8"	0.3754	0.3751	5/8"	0.6267	0.6264	-	-	0.0469	0.500	L-40-8
									0.625	L-40-10
									0.750	L-40-12
									1.000	L-40-16
7/16"	0.4380	0.4377	5/8"	0.6267	0.6264	-	-	0.0625	0.500	L-40-8
									0.625	L-40-10
									0.750	L-40-12
									1.000	L-40-16
1/2"	0.5005	0.5002	3/4"	0.7518	0.7515	0.8750	0.0938	0.0625	0.312	L/HL-48-5
									0.375	L/HL-46-6
									0.500	L.HL-48-8
									0.750	L/HL-48-12
									1.000	L/HL-48-16
									1.375	L/HL-48-22
									1.500	L/LH-48-24
									1.750	L/HL-48-28
									2.125	L/LH-48-34

(Continued)

Table 4. *(Continued)* **Liner Type Jig Bushings Type HL (with Head) and Type L (Headless).**
(Source, American Drill Bushing Co.)

Inside Diameter A			Outside Diameter B			Head Dimension HL Only		Radius R	Body Length C*	ANSI Number**
Nom.	Max.	Min.	Nom.	Max.	Min.	Dia. F	Height G			
5/8"	0.6255	0.6252	7/8"	0.8768	0.8765	-	-	0.0625	0.500 0.625 0.750 1.000 1.250	L-56-8 L-56-10 L-56-12 L-56-16 L-56-20
3/4"	0.7506	0.7503	1"	1.0018	1.0015	1.1250	0.1250	0.0625	0.312 0.500 0.750 1.000 1.375 1.500 1.750 2.125 2.500 3.000	L-64-5 L/HL-64-8 L/HL-64-12 L/HL-64-16 L/HL-64-22 L/HL-64-24 L/HL-64-28 L/HL-64-34 L/HL-64-40 L/LH-64-48
7/8"	0.8757	0.8754	1 1/4"	1.2520	1.2517	-	-	0.0625	0.375 0.500 0.625 0.750 1.000 1.250 1.500	L-80-6 L-80-8 L-80-10 L-80-12 L-80-16 L-80-20 L-80-24
1"	1.0007	1.0004	1 3/8"	1.3772	1.3768	1.5000	0.1250	0.0938	0.500 0.750 1.000 1.375 1.5000 1.750 2.125 2.500 3.000	HL-88-8 L/HL-88-12 L.HL-88-16 L/HL-88-22 L/HL-88-24 L/HL-88-28 L/HL-88-34 L/HL-88-40 L/HL-88-48
1 1/4"	1.2600	1.2506	1 5/8"	1.6272	1.6268	-	-	0.0938	0.750 1.000 1.375 1.750 2.125 2.500 3.000	L/HL-104-12 L.HL-104-16 L/HL-104-22 L/HL-104-28 L/HL-104-34 L/HL-104-40 L/HL-104-48
1 3/8"	1.3760	1.3756	1 3/4"	1.7523	1.7519	1.8750	0.1875	0.0938	1.000 1.375 1.500 1.750 2.125 2.500 3.000	L/HL-112-12 L/HL-112-16 L/HL-112-22 L/HL-112-28 L/HL-112-34 L/HL-112-40 L/HL-112-48
1 1/2"	1.5010	1.5006	17/8"	1.8773	1.8769	-	-	0.0938	1.000 1.500 1.750 2.125 2.500 3.000	L/HL-120-12 L/HL-120-22 L/HL-120-28 L/HL-120-34 L/HL-120-40 L/HL-120-48

(Continued)

Table 4. *(Continued)* **Liner Type Jig Bushings Type HL (with Head) and Type L (Headless).**
(Source, American Drill Bushing Co.)

Inside Diameter A			Outside Diameter B			Head Dimension HL Only		Radius R	Body Length C*	ANSI Number**
Nom.	Max.	Min.	Nom.	Max.	Min.	Dia. F	Height G			
1 3/4"	1.7512	1.7508	2 1/4"	2.2525	2.2521	2.3750	0.1875	0.0938	1.000	L/HL-144-12
									1.375	L.HL-144-16
									1.500	L/HL-144-22
									1.750	L-144-28
									2.125	L-144-34
									2.500	L-144-40
									3.000	L-144-48
2 1/4"	2.2515	2.2510	2 3/4"	2.7526	2.7522	2.8750	0.1875	0.1250	1.000	L/HL-176-12
									1.375	L.HL-176-16
									1.500	L/HL-176-22
									1.750	L-176-28
									2.125	L-176-34
									2.500	L-176-40
									3.000	L-176-48

All dimensions in inches. Diameter A to be concentric to diameter F within 0.0005". Body diameters (B) shown are for finished bushings. Unfinished diameters are 0.015" to 0.020" larger and must be sized (ground) to fit jig plate holes.
* Total length of Type HL bushings is head height plus body length.
** Both styles are not available in all sizes. Check prefix for ANSI number for availability.

Table 5. Jig Bushing Tolerances.

Press Fit (H and P), Slip (S), and Fixed (SF) Jig Bushing		
Drill Sizes	**Inside Dia. (Bore)**	**Tolerance**
#80 to 1/4"	Nom.	+0.0001" to +0.0004"
Over 1/4" to 3/4"	Nom.	+0.0001" to +0.0005"
Over 3/4" to 1 1/2"	Nom.	+0.0002" to +0.0006"
Over 1 1/2"	Nom.	+0.0003" to +0.0007"
Press Fit Liner (HL and L) Jig Bushings		
Drill Sizes	**Inside Dia. (Bore)**	**Tolerance**
3/16" to 3/8"	Nom.	+0.0001" to +0.0004"
7/16" to 5/8"	Nom.	+0.0002" to +0.0005"
3/4"	Nom.	+0.0003" to +0.0006"
7/8" to 1"	Nom.	+0.0004" to +0.0007"
1 3/8"	Nom.	+0.0006" to +0.0010"
1 3/4"	Nom.	+0.0008" to +0.0012"
2 1/4"	Nom.	+0.0010" to +0.0015"
Reamer Bushing*		
Reamer Sizes	**Inside Dia.**	**Tolerance**
Up to 1/4"	Nom.	+0.0005" to +0.0008"
Over 1/4" to 1"	Nom.	+0.0006" to +0.0010"
Over 1"	Nom.	+0.0008" to +0.0012"

(Continued)

Table 5. *(Continued)* **Jig Bushing Tolerances.**

Concentricity	
Inside Dia.	Tolerance
1/8" to 1"	0.0003" T.I.R
Over 1"	0.0005" T.I.R.
Length	
Nom.	+/- 0.010"
Head Height	
Nom.	+/- 0.015"

*Reamer bushings are available as special sizes. Check supplier for availability. All dimensions in inches.

Table 6. Lock Screw Dimensions for Renewable Bushings.

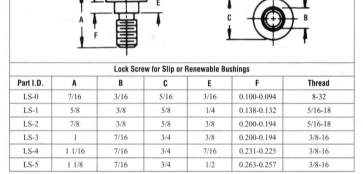

Lock Screw for Slip or Renewable Bushings						
Part I.D.	A	B	C	E	F	Thread
LS-0	7/16	3/16	5/16	3/16	0.100-0.094	8-32
LS-1	5/8	3/8	5/8	1/4	0.138-0.132	5/16-18
LS-2	7/8	3/8	5/8	3/8	0.200-0.194	5/16-18
LS-3	1	7/16	3/4	3/8	0.200-0.194	3/8-16
LS-4	1 1/16	7/16	3/4	7/16	0.231-0.225	3/8-16
LS-5	1 1/8	7/16	3/4	1/2	0.263-0.257	3/8-16

All dimensions in inches.

Table 7. Round Clamp Dimensions for Renewable Bushings.

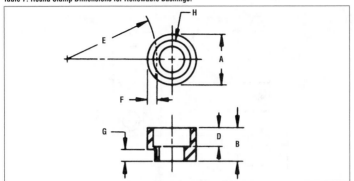

				Round Clamps for Renewable Bushings			
Part I.D.	A	B	D	E	F	G	H Socket Head Screw
RC-1	5/8	5/16	5/32	35/64	0.109	0.1250	5/16
RC-2	5/8	7/16	7/32	59/64	0.125	0.1875	5/16
RC-3	3/4	1/2	9/32	1 27/64	0.156	0.1875	3/8
RC-4	3/4	17/32	1/4	2 33/64	.156	0.2187	3/8
RE-5	3/4	9/16	1/4	2 33/64	.156	0.250	3/8

All dimensions in inches.

Table 8. Round End Clamp Dimensions for Renewable Bushings.

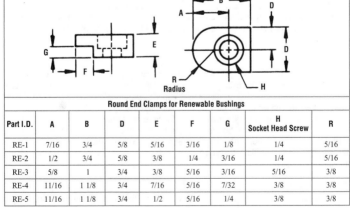

				Round End Clamps for Renewable Bushings				
Part I.D.	A	B	D	E	F	G	H Socket Head Screw	R
RE-1	7/16	3/4	5/8	5/16	3/16	1/8	1/4	5/16
RE-2	1/2	3/4	5/8	3/8	1/4	3/16	1/4	5/16
RE-3	5/8	1	3/4	3/8	5/16	3/16	5/16	3/8
RE-4	11/16	1 1/8	3/4	7/16	5/16	7/32	3/8	3/8
RE-5	11/16	1 1/8	3/4	1/2	5/16	1/4	3/8	3/8

All dimensions in inches.

Table 9. Flat Clamp Dimensions for Renewable Bushings.

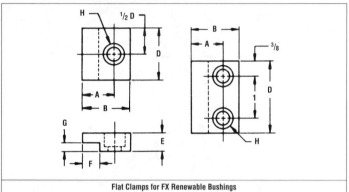

		Flat Clamps for FX Renewable Bushings						
Part I.D.	# Holes	A	B	D	E	F	G	H Socket Head Screw
FC-1	1	1/2	7/8	5/8	5/16	3/16	1/8	1/4
FC-2	1	5/8	1	1	3/8	1/4	3/16	5/16
FC-3	1	5/8	1 1/8	1	3/8	5/16	3/16	5/16
FC-4	1	51/64	1 11/64	1 1/4	7/16	27/64	7/32	3/8
FC-5	2	51/64	1 11/64	1 3/4	7/16	27/64	7/32	3/8
FC-6	1	51/64	1 11/64	1 1/4	1/2	27/64	1/4	3/8
FC-7	2	51/64	1 11/64	1 3/4	1/2	27/64	1/4	3/8

All dimensions in inches.

unsatisfactory, and they are widely available in thicknesses ranging from $1/16$ to $3/8$".

Bushings designed for installation in plastic laminate and other castable tooling materials, as well as wood, are available with knurled or serrated outer surfaces that may either be pressed into the jig plate or embedded and cast into the plate when it is manufactured.

Oil groove bushings

Bushings with oil grooves provide positive lubrication and longer life, and are available in all ANSI standard drill bushings. The grooves are normally $1/16$" wide and $1/32$" deep, with the oil holes being $1/4$" in diameter. "Wiper" versions are designed so that the oil groove does not break out at the wiper end of the bushing, which provides protection against dirt and chips. They are available in three configurations: headless, with the wiper end at the base of the bushing; head type, with the wiper end at the head of the bushing; and head type, with the wiper end at the base of the bushing. Sizes range from $3/8$" to 8" inside diameter, and $1/2$" to 8" long. Oil groove bushing configurations are shown in *Figure 1*.

Proper mounting

Sufficient clearance should be provided between the bushing and the workpiece to permit the removal of chips. The exception to this rule occurs in drilling operations

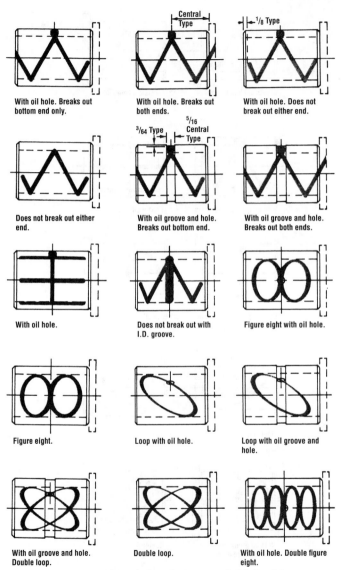

Figure 1. Oil groove bushings. (Source, American Drill Bushing Co.)

(Continued)

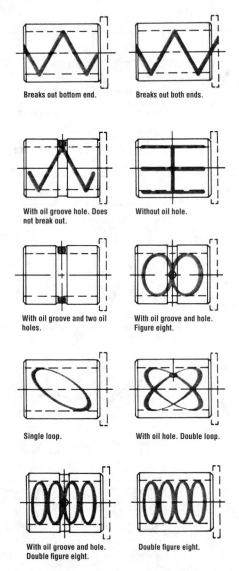

Figure 1. (Continued) Oil groove bushings. (Source, American Drill Bushing Co.)

requiring maximum precision where the bushing should be in direct contact with the workpiece. However, suitable chip clearance should be provided in most applications because the abrasive action of metal particles will accelerate bushing wear. Chip breaker bushings designed to break up long stringy chips are available.

The recommended chip clearance for metals that produce small chips (such as cast iron) is equal to one-half the bushing diameter. For other metals which produce long, stringy chips (such as cold-rolled steel), the chip clearance should be at least equal to, but not more than $1^{1}/_{2}$ times, the bushing inside diameter. Excessive chip clearance should be avoided. Because most cutting tools are slightly larger in diameter at the cutting end, excessive clearance reduces the guiding effect of the bushing and results in less accurate drilling.

When performing multiple operations, such as drilling and reaming, slip renewable bushings of different lengths can be used to obtain the combined advantages of adequate chip removal and precise accuracy. The slip renewable bushing should be short enough to provide proper chip clearance during the drilling operation, while the reamer bushing may be long enough to contact or closely approach the workpiece.

Burr clearance should be provided between the bushing and workpiece when drilling wiry metals such as copper. Metals of this type tend to produce secondary burrs around the top of the drilled holes that act to lift the jig from the workpiece, and cause difficulty in the removal of workpieces from side-loaded jigs. The recommended work clearance is one-half the inside diameter of the bushing.

For many applications requiring close center-to-center placement of bushings, extra thin-wall series bushings will prove helpful. However, for especially difficult close-hole patterns, it may be necessary to grind flats on the bushing outside diameter or head to achieve minimum spacing. When this technique is used, the bushing flats and mounting holes must be accurately machined.

When adapting bushings to suit applications involving irregular work surfaces, the ends of the bushings should be formed to the contour of the workpiece (see *Figure 2*). In many applications of this nature, the drill point does not enter perpendicular to the work surface, which induces a tendency to skid or wander. For this reason, the distance between the bushing and the workpiece must be held to a minimum so that the full guiding effect of the bushing can be obtained. The side load exerted by the drill in applications of this type is usually concentrated at a point near the drill exit end of the bushing, which causes accelerated bushing wear. Except in short production runs, the use of fixed renewable bushings should be considered to simplify the replacement of worn bushings and or facilitate proper orientation of the bushing with respect to the contour of the work surface.

Metric size jig bushings

Bushings to metric sizes are available in the following basic configurations.

Type PM. Plain (headless) press fit bushings. A headless bushing, identical in appearance to ANSI type P. Available in outside diameters of 4 mm to 125 mm, and in lengths 6 mm to 112 mm.

Type HM. Head type press fit bushing. Identical in appearance to ANSI type H. Available in outside diameters of 4 mm to 125 mm, and in lengths 6 mm to 112 mm.

Type SFM. Slip fixed renewable bushings. Available in outside diameters of 8 mm to 105 mm, and in lengths of 10 mm to 112 mm.

Type LM. Plain liners for slip fit renewable bushings. Available in outside diameters of 12 to 125 mm, lengths of 10 to 112 mm, and inside diameters from 8 mm to 105 mm.

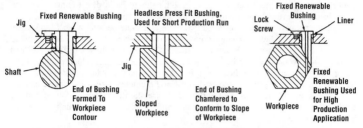

Figure 2. Adapting bushings to irregular work surfaces. (Source, American Drill Bushing Co.)

Type HLM. Head liners for slip fit renewable bushings. Available in the same dimensions as type LM.

Dimensions for lock screws and clamps used to position metric renewable bushings are given in **Tables 10** and **11**.

Inside diameter tolerances for metric drill bushings are to metric tolerance class G6. The tolerances, in millimeters, are given below.

Inside Diameter	Inside Diameter Tolerance
0.35 to 3.00	+0.002 to +0.008
3.01 to 6.00	+0.004 to +0.012
6.01 to 10.00	+0.005 to +0.014
10.01 to 18.00	+0.006 to +0.017
18.01 to 30.00	+0.007 to +0.020
30.01 to 50.00	+0.009 to +0.025
50.01 to 80.00	+0.010 to +0.029
80.01 to 105.00	+0.012 to +0.034

Outside diameter tolerances for press fit bushings are to metric tolerance class s6, and tolerance class h5 is applicable for metric slip fixed renewable bushings.

Table 10. Lock Screw Dimensions for Metric Renewable Bushings.

Part I.D.	A	F	B	C	Thread
LSM-1	15	3	7.5	13	M5 × 0.8
LSM-3	18	4	9.5	16	M6 × 1.0
LSM-5	22	5.5	12	20	M8 × 1.25
LSM-7	32	7	15	24	M10 × 1.5

All dimensions in millimeters.

Table 11. Round Clamp Dimensions for Metric Renewable Bushings.

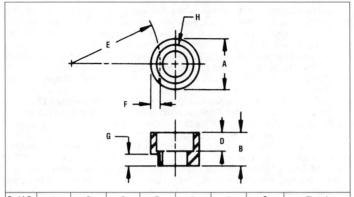

Part I.D.	A	B	C	D	E	F	G	Thread
RCM-1	13	8	10	4	9.5	3.7	3	M5 × 0.8
RCM-3	16	10	12	5	15	4.7	4	M6 × 1.0
RCM-5	20	12	15	5	30	6.2	5.5	M8 × 1.25
RCM-7	24	16	18	7	80	7.5	7	M10 × 1.5

All dimensions in millimeters.

Milling

Milling is a complex metal removal process that usually involves a clamped workpiece being fed in a linear direction into a multiple tooth rotating cutter. Due to the vast variety of machinery, controls, and tools that are available, milling is becoming the universal machining method of choice. It offers excellent machining efficiency, good surface finishes, unmatched flexibility, and a high degree of accuracy. While turning is frequently used to create round surfaces with a single cutting edge, milling normally uses several cutting edges in a single tool to create flat faces, shoulders, slots, and contoured surfaces. Coordinating the cutting edge cyclically entering and exiting the cut with a workpiece being fed into the cut requires the machine operator and tool engineer to possess knowledge of cutter geometry, cutting edge density, available power, and cutter and part stability and rigidity. *Unless otherwise credited, the technical material in this section is courtesy of Kennametal Inc., a leader in milling technology.*

There are many types of machine tools used to perform a milling operation, including manually controlled, numerically controlled, and special dedicated machines. The primary difference between machines dedicated to milling and multipurpose machines is the spindle orientation. Dedicated milling machines are designed to mill in one direction and have a spindle tilt to retard back striking. Dedicated machines include traditional ram-type machines, knee and column machines, and planer- or gantry-type milling machines.

Calculating machining parameters for milling

(See pages 969 – 983 for tables of suggested speeds and speeds for face milling, and pages 984 and 985 for suggested speeds and feeds for end milling). When selecting starting spindle speeds (rpm) and feeds (sfm–surface feet {or meters–smm} per minute, which is the rate at which the tool tip passes through the workpiece material due to the rotational speed of the cutter), the fundamental consideration when milling with tungsten carbide or other advanced cutting materials is controlling the amount of heat generated at the cutting edge. Considerations that affect the heat at the cutting edge include material composition, material hardness, and the cutter speed. The circumference of the cutter and the spindle rpm are used to calculate the cutting speed, but it is normal to refer to a standardized chart for a recommended cutting speed in sfm or smm to calculate the proper rpm for the cutter. Formulas for converting rpm to cutter speed (and cutter speed to rpm) are as follows.

to find	given	formula
sfm	cutter diameter in inches = D revolutions per minute = rpm	sfm = (D × rpm) ÷ 3.82*
rpm	surface feet per minute = sfm cutter diameter in inches = D	rpm = (sfm ÷ D) × 3.82*
smm	cutter diameter in mm = D revolutions per minute = rpm	smm = (D × rpm) ÷ 318.3**
rpm	surface meters per minute = smm cutter diameter in mm = D	rpm = (smm ÷ D) × 318.3**

* In inch units, 12 (one foot or twelve inches) divided by π (12 ÷ 3.1416 = 3.82).
** In metric units, 1000 (one meter or 1000 mm) divided by π (1000 ÷ 3.1416 = 318.3).

Example. Find the feed rate for a 6 in. (152.4 mm) cutter machining at 400 rpm.

  sfm = (6 × 400) ÷ 3.82 = 628   *note: 1 foot = 0.305 meters*
  smm = (152.4 × 400) ÷ 318.3 = 191.5   *1 meter = 3.281 feet*

Example. Find the rpm needed to maintain a cutting speed of 4000 sfm (1220 smm) using an 8" (203.2 mm) milling cutter.

rpm = (4000 ÷ 8) × 3.82 = 1910   *or*   rpm = (1220 ÷ 203.2) × 318.3 = 1911.

As cutting speed increases, so does the amount of heat produced at the cutting edge. Every cutting material (carbide, cermet, ceramic, etc.) has a specific heat range within which it will perform effectively. The goal is to mate all parameters including cutting material, workpiece material, cutting speed, and spindle speed so that the cutting edge does not exceed the effective heat range. Excessive temperatures shorten tool life and place limitations on cutter speed. When selecting a starting speed, it should be remembered that the harder and more difficult the material is to machine, the lower the starting speed should be. The example below demonstrates how spindle rpm should change due to increased hardness levels.

material	hardness	recommended sfm / smm	cutter diameter in. / mm	rpm
4340 steel (annealed)	248 BHN	500 / 150	8" / 203 mm	239
4340 steel (heat treated)	444 BHN	240 / 72	8" / 203 mm	115

Calculating chip load, advance per revolution, and advance per minute

Chip load is the feed load on each individual insert in a cutter. Even though the feed rate of a milling cutter is usually given in "inches per minute" or "millimeters per minute," the starting feed rate should be calculated as the chip load per effective tooth. Maintaining proper chip load (fpt–*feed per tooth*) is the principal method for dissipating heat buildup at the cutting edge. The temperature in the cutting zone is, to a large extent, the result of contact between the chip and the tool. Therefore, maintaining proper chip load will provide a chip thick enough to carry the majority of the heat away from the cutter and workpiece. As with speed, chip load has an effective range, and chips either too heavy or too light will result in temperature problems with the cutting edge. Chip load calculations should be adjusted according to the machine, fixture, and workpiece rigidity conditions. When the workpiece width plus the cutter diameter and pitch combination is such that two teeth are continuously engaged in the workpiece, the chip load (fpt) can be increased because of the stability of the cutting condition. When the radial width of cut allows only one tooth to be engaged in the cut at any time, the chip load should be decreased because the interrupted cut inherent to this condition will not allow for a continuous load on the cutter and spindle drive train. The minimum starting advance per tooth should rarely be below 0.003" per effective tooth per revolution (0.075 mm per effective tooth per revolution). The maximum advance per tooth will seldom exceed 0.012" per effective tooth per revolution (0.30 mm per effective tooth per revolution).

Once the starting chip load has been established, it is necessary to calculate the feed per revolution (fpr) in inches or millimeters. The feed (or advance) per revolution will increase by a factor of the number of effective teeth in the cutter. For example, a milling cutter with one insert, with a feed per tooth (fpt) and a feed per revolution (fpr) of 0.008" (0.20 mm), will advance 0.008" (0.20 mm) at the cutter centerline every time the cutter rotates one full revolution. The same style cutter with four inserts will advance 0.032" (0.80 mm) at the cutter centerline every time the cutter completes one revolution. As with the single insert cutter, the milling cutter with four inserts will advance only 0.008" (0.20 mm) per tooth, but with four teeth it will advance 0.032" (0.80 mm) per revolution. The following formulas are used to calculate the feed (advance) in inches and millimeters. The feed rate in inches per minute (in./min) or millimeters per minute (mm/min) can be calculated given sfm or smm and the feed per tooth or the feed per revolution.

to find	given	formula
fpt (feed per tooth or *chip load*) in inches *or* millimeters	number of effective teeth = nt in. *or* mm per minute = in./min *or* mm/min revolutions per minute = rpm	fpt = in./min *or* mm/min ÷ (nt × rpm)
fpt (feed per tooth or *chip load*) in inches *or* millimeters	number of effective teeth = nt feed per revolution (in. *or* mm) = fpr	fpt = fpr ÷ nt
dbt (distance between teeth) in inches *or* millimeters	π = 3.1416 cutter diameter (in. *or* mm) = D number of effective teeth = nt	dbt = (π × D) ÷ nt
fpr (feed per revolution) in inches *or* millimeters	in. *or* mm per minute = in./min *or* mm/min revolutions per minute = rpm	fpr = in./min *or* mm/min ÷ rpm
fpr (feed per revolution) in inches *or* millimeters	feed per tooth (in. *or* mm) = fpt number of effective teeth = nt	fpr = fpt × nt
rpm (revolutions per minute)	in. *or* mm per minute = in./min *or* mm/min feed per revolution (in. *or* mm) = fpr	rpm = in./min *or* mm/min ÷ fpr
rpm (revolutions per minute)	surface feet per minute = sfm cutter diameter in inches = D	rpm= (sfm ÷ D) × 3.82
rpm (revolutions per minute)	surface meters per minute = smm cutter diameter in millimeters = D	rpm = (smm ÷ D) × 318.3
rpm (revolutions per minute)	in. *or* mm per minute = in./min *or* mm/min number of effective teeth = nt feed per tooth (in. or mm) = fpt	rpm = in./min *or* mm/min ÷ (nt × fpt)
in./min or mm/min (feed)	feed per revolution (in. or mm) = fpr revolutions per minute = rpm	in./min *or* mm/min = fpr × rpm
in./min or mm/min (feed)	feed per tooth (in. or mm) = fpt revolutions per minute = rpm number of effective teeth = nt	in./min *or* mm/min = fpt × nt × rpm

Example. Find rpm, in./min, or mm/min, and fpr for a 6 in. (152.4 mm) diameter cutter with eight effective teeth, and feed rate of 628 sfm (191.5 smm) and fpt of 0.006" (.15 mm).

rpm = (628 sfm ÷ 6") × 3.82 = 400 rpm
  or rpm = (191.5 smm ÷ 152.4 mm) × 318.3 = 400 rpm

in./min = 0.006" fpt × 8 × 400 rpm = 19.2"/min
  or mm/min = .15 mm × 8 × 400 rpm = 480 mm/min

fpr = 19.2 in./min ÷ 400 rpm = 0.048"/revolution
  or fpr = 480 mm ÷ 400 rpm = 1.2 mm/revolution.

In calculations where the number of effective teeth are involved, it should be remembered that in some milling operations using staggered toothed slotted cutters, step mills, some inserted helical fluted end mills, and face mills loaded with serrated (patterned) roughing inserts, some teeth will not be effective. In such instances, feed per revolution should be calculated on the number of effective teeth in the cutter. Another consideration in all milling calculations involves the load conditions imposed by different machining operations. **Table 1** indicates how speed and feed tables can be interpreted for existing conditions. Adjustments should be made in ten percent increments while maintaining chip load until optimum settings are achieved.

Table 1. Machining Compensation for Various Operations.

condition	speed	feed
roughing	▼	▲
finishing	▲	▼
end milling	▲	▼
slotting	▲	▼
hard material	▼	●
soft material	▲	▲
scale	▼	▲
tool life	▼	▲
heavy d.o.c.	▼	▼

lower = ▼   same = ●
higher = ▲

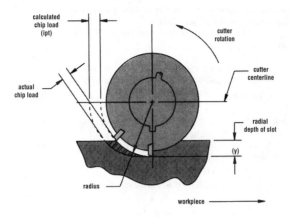

Figure 1. In slotting, chip thickness is calculated at the centerline of the cutter.

In slot milling (and in face milling and end milling when the centerline of the cutter is not located on the workpiece), a different formula which considers the chip load (fpt) to be at the centerline of the cutter can be used. (See *Figure 1.*) When the cutter is situated so that the centerline is not on the workpiece, in./min or mm/min can be increased to achieve the required chip load.

$$\text{fpt} = \frac{\left(\dfrac{\sqrt{(D - y) \times (y)}}{r}\right) \times \left(\dfrac{\text{in./min or mm/min}}{\text{rpm}}\right)}{\text{nt}}$$

$$\text{in./min } or \text{ mm/min} = \text{rpm} \times \text{nt} \times \left[\left(\frac{\text{fpt}}{\dfrac{\sqrt{(D - y) \times (y)}}{r}}\right)\right]$$

where: y = radial depth of slot
        D = diameter of cutter
        r = radius of cutter
        nt = number of effective teeth.

Calculating power requirements for milling

Milling cutters can consume significant amounts of power. Often, the lack of power is the limiting factor when deciding on a particular operation. When large diameter cutters or heavy stock removal is necessary, it is advantageous to first calculate the material removal rate (Q) in order to calculate necessary power requirements.

Use the following formula to find Q, the metal removal rate.

Q = d.o.c. × w.o.c. × feed rate (in./min) = in.³/min.

or Q = (d.o.c. [mm] × w.o.c. [mm] × feed rate [mm/min]) ÷ 1000 = cm³/min

or Q = (d.o.c. [cm] × w.o.c. [cm] × feed rate [cm/min]) = cm³/min.

Example. Where d.o.c. (depth of cut) = 0.200" or 5.08 mm (0.508 cm)
$\qquad$ w.o.c. (width of cut) = 1.64" or 42 mm (4.2 cm)
$\qquad$ feed = 19.2 in./min or 487.7 mm/min (48.77 cm/min)

$Q = 0.200 \times 1.64 \times 19.2 = 6.298 \ in.^3/min.$ $\qquad$ *note: 1 in.^{3} = 16.387 cm^3*
$\qquad\qquad\qquad\qquad\qquad\qquad\qquad\qquad\qquad\qquad\qquad$ *1 cm^3 = 0.061 in.^{3}*

$\quad$ *or* (5.08 mm × 42 mm × 487.7 mm/min) ÷ 1000 = 104 cm^3/min

$\quad$ *or* (0.508 cm × 4.2 cm × 48.77 cm/min) = 104 cm^3/min

Use the following formula to find HP$_c$, power at the cutter in English units (horsepower).

$\quad$ HP$_c$ = Q × K.

Example. To find HP$_c$, use the Q value from the above example and find the power constant, (K), from the "K" Factors for Face Milling chart. K represents the cubic inches of material per minute that can be removed by one horsepower. For this example, K = 0.641 (4140 steel, 220 BHN).

$\quad$ HP$_c$ = 6.298 × 0.641 = 4.04 HP at the cutter.

Use the following formula to find HP$_m$, power at the motor in English units.

$\quad$ HP$_m$ = HP$_c$ ÷ E.

$\quad$ To determine horsepower at the motor (HP$_m$), the spindle efficiency must be considered. Spindle efficiency "E" varies from 75% to 90% (E = 0.75 to 0.90). Using the above example,

$\quad$ HP$_m$ = 4.04 ÷ .8 = 5.0

Use the following formula to find kW$_c$, power at the cutter in metric units (kilowatts).

$\quad$ kW$_c$ = Q × K.

Example. To find kW$_c$, use the Q value from the above example and find the power constant, (K), from the "K" Factors for Face Milling chart. K represents the cubic centimeters of material per minute that can be removed by one kilowatt. For this example, K = 0.029 (4140 Steel, 220 BHN).

$\quad$ kW$_c$ = 104 × 0.029 = 3.0 $\qquad\qquad\qquad$ *note: 1 HP = 0.7457 kW*
$\qquad\qquad\qquad\qquad\qquad\qquad\qquad\qquad\qquad\qquad\qquad$ *1 kW = 1.341 HP*

Use the following formula to find kW$_m$, power at the motor in metric units.

$\quad$ kW$_m$ = kW$_c$ ÷ E.

Example. To determine power at the motor (kW$_m$), the spindle efficiency must be considered. Spindle efficiency "E" varies from 75% to 90% (E = 0.75 to 0.90).

$\quad$ kW$_m$ = 3.0 ÷ 0.8 = 3.75.

Tangential force, power, and torque calculations for positive/negative (shear) geometry cutters

$\quad$ When using high shear cutter geometry, tangential force (F$_t$, force at the spindle) accounts for the greatest portion of the machining power at the cutting tool. Using the tangential force formula is a quick way to determine the approximate forces that the fixtures, part wall sections, or spindle bearings will endure.

1. tangential force (F$_t$) in lb *or* kg
2. ultimate material strength (S) in lb/in.^{2} *or* kg/cm^2
3. cross sectional area of chip removed by the milling insert (A) in in.^{2} *or* cm^2
4. number of inserts in cut (Z$_c$)
5. machinability factor (C$_m$)
6. tool wear factor (C$_w$).

$F_t = S \times A \times Z_c \times C_m \times C_w$.

Find S. The approximate relationship between the ultimate material strength and hardness of the most commonly used work materials such as steels, irons, titanium alloys, can be expressed by the empirical formula:

$S = 500 \times BHN = lb/in.^2$ *note: 1 lb/in.² = 0.070307 kg/cm²*

or $S = 35.1535 \times BHN = kg/cm^2$ *1 kg/cm² = 14.22334 lb/in.²*

Note. BHN = Brinell hardness number. These numbers are obtained, primarily, at the 3000-kgf load. When testing soft metals such as aluminum alloys, the 500-kgf load is used. Hardness obtained at the 500-kgf load should be converted into the hardness equivalent of the 3000-kgf load by using the load factor of 1.15. For example, 130 BHN at the 500-kgf load is equivalent to 150 BHN at the 3000-kgf load ($130 \times 1.15 = 150$).

Example. For 4340 steel (heat treated), BHN = 444.

$S = 500 \times 444 = 222,000$ lb/in.²   *or* $S = 35.1535 \times 444 = 15,608$ kg/cm²

Find A, the cross sectional area of the chip.

$A = d.o.c. \times fpt = in.^2$ or cm^2

"K" Factors for Face Milling			
work material	**hardness BHN**	**"K" factor English units**	**"K" factor Metric units**
steels, free machining	110-200	.609	.0277
low-, medium-, and high carbon	201-253	.641	.0292
	254-286	.781	.0355
	287-330	.909	.0413
alloy steels, alloy and	135-200	.721	.0327
lower carbon tool steels	201-253	.794	.0361
	254-286	.856	.0389
	287-327	.940	.0427
	328-371	1.15	.0523
	372-481	1.45	.0659
	482-560	1.69	.0768
	561-700	1.86	.0845
stainless steels, wrought and cast	135-275	.65 - 1.31	.0296 - .0596
(ferritic, austenitic, martensitic)	286-421	1.35 - 2.0	.061 - .0909
	422-600	2.1 - 2.52	.0955 - .1146
stainless steels, precipitation hardened	150-375	.786 - 2.38	.0357 - .108
cast irons (gray, ductile, and malleable)	120-175	.439	.0199
	110-190	.5	.0227
	176-200	.529	.024
	201-250	.657	.0299
	251-300	.788	.0328
	301-320	.84	.0348
aluminum alloys, copper, brass, and zinc	50-150	.16 - .3	.007 - .0136
nickel alloys	140-300	1.10 - 1.89	.050 - .0859
	301-475	1.9 - 2.18	.0864 - .099
iron base, heat resistant alloys	135-320	1.10 - 1.89	.050 - .0859
high temperature alloys, cobalt base	200-360	1.21 - 2.08	.055 - .0946
titanium	250-375	.751 - 1.15	.034 - .0523
magnesium alloys	40-90	.10 - .15	.0045 - .0068
copper alloys	100-150	.3	.0136
	151-243	.5	.0227

Example. Where d.o.c. = 0.200" or 0.508 cm, and fpt = 0.006" or 0.01524 cm

A = 0.200 × 0.006 = 0.0012 in.^{2} *note: 1 in.^{2} =6.452 cm^2*

  or   A = 0.508 × 0.01524 = 0.00774 cm^2 *1 cm^2 = 0.155 in.^{2}*

Find Z_c. The number of inserts in the cut (simultaneously engaged with the work material) depends on the number of inserts in the cutter (Z) and the engagement angle (α). This relationship is:

  $Z_c = (Z \times \alpha°) \div 360°$.

The engagement angle (α) is relative to the width of the cut (W) and the cutter diameter (D). "Example One" below calculates α in cases where the w.o.c. exceeds the radius of the cutter (D/2). "Example Two" calculates α when the w.o.c. is less than the radius of the cutter.

Example One. Where W exceeds D/2. (See *Figure 2a*.)

$$Z_c = (Z \times \alpha) \div 360°$$
$$\alpha = 90° + \alpha_1$$
$$\sin_1 = \frac{AB}{OB} = \frac{(W - D/2)}{D/2} = \frac{(2[W - D/2])}{D} = \frac{(2W - D)}{D}$$
$$_1 = \text{arc sin} \frac{2W - D}{D}$$
$$Z_c = \frac{Z\left(90° + \text{arc sin} \frac{2W - D}{D}\right)}{360°}$$

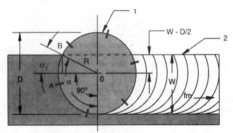

Figure 2a. Example One.

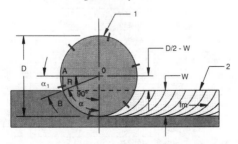

Figure 2b. Example Two.

Example Two. Where W is less than D/2. (See *Figure 2b*.)

$$Z_c = (Z \times \alpha) \div 360°$$

$$\alpha = 90° - \alpha_1$$

$$\sin \alpha_1 = \frac{AB}{OB} = \frac{(D/2 - W)}{D/2} = \frac{(2[D/2 - W])}{D} = \frac{(D - 2W)}{D}$$

$$\alpha_1 = \text{arc sin} \frac{D - 2W}{D}$$

$$Z_c = \frac{Z\left(90° - \text{arc sin} \frac{D - 2W}{D}\right)}{360°}$$

Where: D = cutter diameter
W = w.o.c.
α = engagement angle
α_1 = angle between cutter centerline and cutter radius to the peripheral point of exit or entry
Z = number of inserts on cutter
Z_c = number of inserts in cut.

Z_c values are based on the W/D ratio and Z. For example, if W equals D (W÷D = 1.0), α equals 180° and $Z_c = (Z \times 180°) \div 360° = 0.5Z$. If W is one-half D, α equals 90° and $Z_c = (Z \times 90°) \div 360° = 0.25Z$. The values of Z_c, depending on the given W/D ratios, are shown in **Table 2**.

Table 2. Z_c Values.

W/D	0.88	0.80	0.75	0.67	0.56	0.38	0.33	0.19	0.125
Z_c	0.38Z	0.35Z	0.33Z	0.30Z	0.27Z	0.21Z	0.20Z	0.14Z	0.12Z

C_m, the machinability factor, is used to indicate the degree of difficulty in machining various workpiece materials. The values are based on milling tests with a torque dynamometer. C_m factor values for some of the most common workpiece materials are provided in **Table 3**.

Table 3. C_m Values.

workpiece material	C_m
carbon and alloy steels	1.0 – 1.3
stainless steel	2.0 – 2.3
gray cast iron	1.0 – 1.3
titanium alloys	1.0 – 1.4
aluminum alloys	1.0 – 1.1

C_w, the tool wear factor, has a base of 1.0 which represents short time operation with sharp cutting tools. For longer operations (before the inserts are indexed) the following C_w values are representative:

light and heavy medium face milling $\quad C_w = 1.10 - 1.25$
heavy-duty face milling $\qquad\qquad\quad C_w = 1.30 - 1.60$.

With this information, the tangential force F_t can be calculated.

Example. Find F_t where:

S = 222,000 lb/in.² *or* 15,608 kg/cm²
A = 0.0012 in.² *or* 0.00774 cm²
Z_c = 2 (where Z = 8 and α = 90°)
C_m = 1.2 (from the table for carbon and alloy steel)
C_w = 1.25 (the wear factor for heavy medium faced milling)

$F_t = 222,000 \times 0.0012 \times 2 \times 1.2 \times 1.25 = 799.2$ lb $\qquad$ *note: 1 lb = 0.4536 kg*
or $F_t = 15,608 \times 0.00774 \times 2 \times 1.2 \times 1.25 = 362.42$ kg $\qquad$ *1 kg = 2.2046 lb*

Use the following formula to find T, torque at the cutter.

$$T = \frac{F_t \times D}{2} = \text{in.-lb}$$

note: 1 in.-lb = 0.11298 N-m
1 N-m = 8.8507 in.-lb
1 kg-m × 9.80665 = 1 N-m

$$or \quad T = \frac{F_t \times D}{2} \times 9.80665 = \text{N-m}$$

Example.

$$T = \frac{799.2 \times 6}{2} = 2397.6 \text{ in.-lb}$$

$$or \quad T = \frac{362.42 \times .1524}{2} \times 9.80665 = 270.82 \text{ N-m}.$$

Once the torque generated by tangential force is known, the machining power (HP or kW) at the cutter and at the motor can be calculated. These formulas assume sharp cutter edges and a spindle efficiency (E). All values are from previous examples in this section.

Use the follwing formula to find power at the cutter given tangential force and surface feed,

$$HP_c = \frac{F_t \times sfm}{33,000}$$

note: 33,000 is a constant for inch units
6120 is a constant for metric units

$$or \quad kW_c = \frac{F_t \times smm}{6120}$$

Example.

$$\frac{799 \times 628}{33,000} = 15.2 \text{ HP}_c$$

$$or \quad \frac{362.42 \times 191.5}{6120} = 11.3 \text{ kW}_c.$$

Use the following formula to find power at the cutter given torque and rpm,

$$HP_c = \frac{T \times rpm}{63,000}$$

note: 63,000 is a constant for inch units
9570 is a constant for metric units

$$or \quad kW_c = \frac{T \times rpm}{9570}$$

Example.

$$\frac{2397.6 \times 400}{63,000} = 15.2 \text{ HP}_c$$

$$or \quad \frac{270.82 \times 400}{9570} = 11.3 \text{ kW}_c.$$

Power at the motor uses the machine tool efficiency factor (E), which varies from 75% to 90% (E = 0.75 to 0.90).

$HP_m \ or \ kW_m = HP_c \ or \ kW_c \div E.$

Lead angles and chip thickness in face milling

Face mills are available with a variety of lead angles in order to meet part feature requirements and to provide geometry options for different machining situations and conditions. The lead angle is the angle formed by the leading edge of the insert and cutter body. Chip thickness, cutting forces, and tool life are all impacted by lead angle. Increasing lead angle reduces chip thickness for a given feed rate, and permits the cutting edge to gradually enter and exit the workpiece. The reduction in chip thickness results from the same amount of material being spread over a greater area of the insert cutting edge. Allowing the cutting area to gradually enter and exit the workpiece helps reduce radial pressure, protects the insert edge, and lessens the potential for breakout. When machining thin cross-section parts, it should be remembered that axial pressure increases caused by increased lead angles can cause deflection of the machined surface. Also, lead angle cutters cannot mill square shoulders. **Table 4** and examples below demonstrate how lead angle changes affect chip thickness even though the advance (feed) remains constant. (See *Figure 3*.)

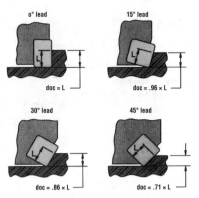

Figure 3. Lead angle effect on maximum depth of cut (doc).

Table 4. Lead Angle/Chip Thickness Relationship.

lead angle	feed per tooth (fpt)	actual chip thickness *
0°	A	A
10°	A	0.9848 × A
15°	A	0.9659 × A
20°	A	0.9397 × A
30°	A	0.8660 × A
45°	A	0.7071 × A
example		
0°	0.010"	0.010"
10°	0.010"	0.009848"
15°	0.010"	0.009659"
20°	0.010"	0.009397"
30°	0.010"	0.00866"
45°	0.010"	0.007071"

* Chip thickness (ct) is the cosine of the lead angle multiplied by the feed per tooth.
ct = cosine (of lead angle) × fpt.

Example. Using **Table 4**, find chip thickness when the lead angle is 20%, and fpt is 0.010" *or* 0.2540 mm.

0.9397 × 0.010" = 0.009397"   *or*   0.9397 × 0.2540 = 0.2387 mm.

Example. Calculate the difference in feed (in./min or mm/min) for a given chip thickness between a 0° lead angle cutter and a 45° lead angle face mill with eight effective teeth, 0.006" (0.15 mm) fpt, operating at 305 rpm. Note that the feed rate can be increased by 41% by changing only the lead angle.

For 0° lead angle: in./min = fpt × nt × rpm   *or*   mm/min = fpt × nt × rpm
$\qquad$ 0.006 × 8 × 305 = 14.64   *or*   0.15 × 8 × 305 = 366.

For 45° lead angle:

$$\text{in./min} = \frac{\text{fpt}}{0.707} \times 8 \times 305 \qquad or \qquad \text{mm/min} = \frac{\text{fpt}}{0.707} \times 8 \times 305.$$

$$\frac{0.006}{0.707} \times 8 \times 305 = 20.7 \qquad or \qquad \frac{0.15}{0.707} \times 8 \times 305 = 517.7.$$

Table 5 presents considerations for choosing a lead angle.

Table 5. Factors to Consider When Selecting a Lead Angle Cutter.

lead angle	advantages	disadvantages
0°	1. Use when 90° shoulder is required. 2. Good choice for thin-wall workpieces.	1. Highest radial cutting forces. 2. High entry shock load. 3. Increased potential for burr on insert exit side of part.
15° and 20°	1. For general milling applications and relatively rigid conditions. 2. Good relation of insert size and maximum depth of cut. 3. Reduced entry shock load.	1. Higher radial forces can cause problems in weak machine/workpiece/fixture conditions.
45°	1. Well balanced axial and radial cutting forces. 2. Less breakout on workpiece corner. 3. Entry shock minimized. 4. Less radial force directed into spindle bearings. 5. Higher feed rates possible.	1. Reduced maximum depth of cut due to lead angle. 2. Larger body diameter can cause fixture clearance problems.

Conventional milling versus climb milling

Conventional milling, or up-milling, occurs when the cutter rotation pushes against the direction of the workpiece feed. See *Figure 4*. The insert enters the workpiece at zero chip thickness and exits at maximum chip thickness. The insert, therefore, must be forced into the cut, causing high friction and excess heat. This phenomenon hardens the surface of the workpiece, and each successive insert has to enter the hardened layer which ultimately reduces tool life. Chip welding can also be a problem with conventional milling since any chips welded onto an

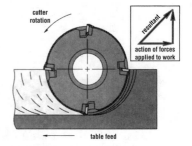

Figure 4. Conventional milling.

insert may be carried around to the entry point of the cut and become wedged between the workpiece and the insert, resulting in breakage. Conventional milling is recommended for materials with a hard scale or flame cut surface. Using this method, the inserts enter the workpiece below the part surface and avoid cutting through the hard scale which shortens tool life.

Climb milling, or down-milling, occurs when the cutter rotation (on either a face mill or an arbor mill) pulls in the same direction as the workpiece feed. See *Figure 5*. The insert enters the workpiece at near maximum chip thickness and exits the cut at zero chip thickness. Climb milling causes less heat to be generated and requires less feed drive power. Since there is no sliding action at the entry of the cut, tool life is better than with conventional milling. Climb milling is the preferred method for high-nickel alloys, titanium, and stainless steels in order to avoid the disastrous effects of workpiece hardening on tool life. It should be avoided only when the feed drive system has backlash, which will cause momentary fluctuations in the actual feed per insert. This condition will result in edge breakage due to increased chip thickness.

Angle of entry

The angle of entry of the insert to the workpiece is determined by the location of the cutter centerline in relation to the edge of the workpiece. A negative angle of entry assures insert contact with the workpiece at the strongest point on its cutting edge, reduces chipping of the insert, and permits heavier feeds on harder materials. If the center of the cutter is kept within the workpiece, a negative angle of entry is assured. With a positive entry angle, the insert makes contact with the workpiece at its weakest point, which can result in insert chipping. Although negative angle of entry is normally recommended, a positive entry angle can be used on some softer materials and cast irons. See *Figures 6a* and *6b*. Changing to more shock-resistant carbide grade, and using an insert with a negative land or hone, can help prevent cutting edge breakdown. When cutting a workpiece that contains cavities or holes, the angle of entry will change as the cutting edge exits and reenters the workpiece. In these conditions, the angle of entry cannot be controlled, and if problems are encountered, a tougher carbide insert or stronger insert geometry is recommended.

Face milling cutter selection and operation

When selecting a cutter, the machining characteristics of the workpiece material, the specifications of the machine tool, and the rigidity of the fixturing must all be considered. The cutter selection process should provide a cutter geometry that will provide maximum performance

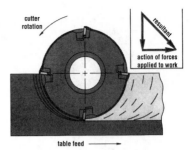

Figure 5. Climb milling.

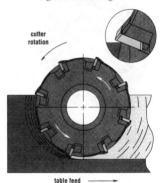

Figure 6a. Negative angle of entry.

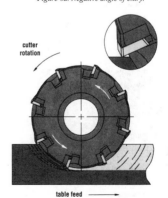

Figure 6b. Positive angle of entry.

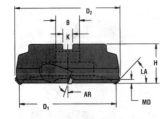

D₁ = effective cutting diameter
D₂ = maximum diameter
H = cutter body height
B = bore diameter
K = keyway width
LA = lead angle
AR = axial rake
RR = radial rake
BC = bolt circle
MD = maximum depth of cut
TRA = true rake angle

Figure 7. Face mill dimensions.

in relation to workpiece configuration, material composition, and machine tool capacity. *Figure 7* identifies the dimensional parameters that must be taken into account when choosing a face mill for a cutting operation. While it may not be possible to employ the optimum milling cutter for every application, sound judgement will improve productivity, reduce costs, and provide higher quality parts. There are four basic geometry types.

Shear angle geometry (also called positive/negative geometry) utilizes positive axial rake and negative radial rake inserts. This geometry is recommended for all steels and cast irons because of the free cutting action of the cutter. The positive axial rake shearing action directs chips up and away from the surface being machined, and the negative radial rake provides a strong cutting edge that allows large d.o.c. at high fpt rates. The large lead angle normally used with this geometry reduces cutting pressure and thins chips, and the geometry permits roughing and finishing in one pass while obtaining good surface finishes. This geometry does have limitations. It is not as effective as positive rake in cutting softer materials, and it can use only four cutting edges per insert. The shear angle requires more power than positive rake machines, but less than demanded by negative rake cutters. See *Figure 8.*

High shear angle offers the same

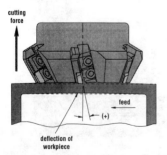

Figure 8a. Positive rake angle effect on cutting forces.

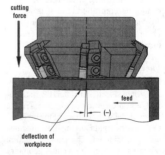

Figure 8b. Negative rake angle effect on cutting forces.

distinct advantages as shear angle cutters but with even more positive cutting action. The higher positive axial rake reduces cutting forces, thereby reducing power requirements and allowing metal removal rates to be increased.

Double positive geometry incorporates both positive axial rake and positive radial rake. This cutter rates highest in efficiency because it consumes less power, reduces entry shock, and is less likely to deflect thin workpieces. For each positive degree increase, power requirements are reduced by approximately 1.5%. The reduced cutting pressures provide lower cutting temperatures, and, in turn, reduced workhardening. They can be used effectively on aluminum, brass and ductile materials, free-machining steels, stainless steels, and cast iron, and the positive axial rake directs chips up and away from the machined surface. Positive cutters are recommended where there is considerable spindle extension, and they are more effective than negative rake on older machines with questionable spindle and feed mechanisms. Surface finishes are improved as a result of reduced cutting pressure, deflection, and vibration. The limitations of positive geometry are a weaker insert cutting edge, a tendency to lift the workpiece, and fewer cutting edges per insert.

Double negative geometry cutters are the logical choice when adequate power is available, when the point of contact at the workpiece requires a strong cutting edge, or when heavier feeds are desired. The workpiece should have sufficient cross section and strength to resist high cutting forces and entry shock. Steel and cast iron are recommended materials, and the high cutting temperatures may workharden the workpiece. Negative cutters offer excellent economy since the inserts have up to eight cutting edges. Negative geometry consumes substantial power, may cause workpiece deflection, and provides a compromised surface finish due to the deflection of chips onto the milled surface.

The cutter selection guide in **Table 6** for face milling identifies preferred cutters for machining different materials. When conditions permit, a lead angle cutter is preferable to a 0° cutter. The cutter geometry is identified by the rake angle of the insert. The rake angle is the angle between the face of the cutting tool and the workpiece. If the face of the tool is perpendicular to the work, it has a zero degree, or neutral, rake. If the angle of the insert face makes the cutting edge more acute, it is positive rake. When the angle is more blunt, it is negative rake.

Cutter pitch, or density, refers to the number of inserts in a cutter. Considerations in selecting the pitch include fpt, d.o.c., and chip clearance. There must be adequate space between inserts so that the chip can pass without restricting its formation. Consequently, cutters designed for heavy metal removal have fewer inserts to allow for maximum chip clearance. Ideally, a cutter should have at least two inserts, preferably three, in the cut at all times.

Course pitch is recommended for general purpose milling where adequate power is available and where maximum d.o.c. is required. Course pitch provides maximum chip clearance and is popular for roughing and finishing steel and also in machining conditions where vibration is not a threat to the success of the operation.

Medium, or close, pitch is recommended when moderate fpt is required, and where it is more advantageous to have more than one insert in the cut. Medium pitch reduces entry shock and cutting pressure while maintaining feed rates. Chip clearance is slightly smaller than with a course pitch cutter. Medium pitch is commonly used for cast iron milling.

Fine, or extra close, pitch is recommended for high production milling, to reduce breakout when machining thin wall sections, and where d.o.c. does not exceed 0.25" (6.35 mm). Chip clearance is considerably smaller than with a course pitch cutter. Fine pitch is useful when machining materials such as titanium where low cutting speeds are necessary.

Table 6. Cutter Selection Guide.

	Cutter Geometry											
	High shear		Shear		Positive			Double positive		High positive		Double negative
Lead Angle	0°	45°	0°	20°	0°	15°	45°	0°-1.5°	15°	0°	15°	0°/15°
Material												
free machining & carbon steels	○	●	●	●	○	○	○	○	○	—	—	○
alloy steels & 400, 500, PH stainless	●	●	●	●	○	○	○	○	○	—	—	○
gray cast iron	○	●	●	●	○	○	○	●	●	—	—	○
malleable & nodular iron	—	●	○	●	—	—	—	—	—	—	—	○
chilled cast iron	○	○	—	○	—	—	—	—	—	—	—	—
200, 300 stainless	—	●	●	●	—	—	—	○	○	○	○	—
ultra high strength steel	—	●	○	●	—	—	—	—	○	—	—	○
high-speed steel	○	●	○	●	—	—	—	—	—	○	○	○
hot work tool steel	○	●	○	●	—	—	—	—	—	—	—	○
cold work tool steel	○	●	○	●	—	—	—	—	—	—	—	—
high magnesium steels	○	●	●	●	—	—	—	—	○	—	—	—
magnesium alloys	—	—	—	—	—	—	●	○	○	—	●	—
aluminum alloys	●	○	—	—	●*	—	○	○	○	●	●	—
titanium, pure & alloyed	●	●	○	○	—	—	○	○	○	—	○	—
copper alloys	●	●	○	○	—	—	●	○	●	○	●	—
brass & bronze	●	●	○	○	—	—	●	○	●	○	○	—
zinc alloys	●	●	—	—	—	—	●	○	●	○	●	—
manganese	—	●	○	○	—	—	●	○	●	○	○	—
thermo-plastics	—	○	○	—	—	—	○	○	●	○	●	—
high-temperature alloys	—	●	●	—	—	—	●	○	●	—	—	—

Primary cutter recommendation = ●
(*Note:* If application permits, a lead angle cutter is always preferable to a 0° cutter.)
Alternative cutter recommendation = ○
Not recommended = —
* PCD cutter (polycrystalline diamond)

Differential pitch. A cutter with unequally spaced inserts is a differential pitch milling cutter. This feature is often standard on course pitch face mills where vibration can be a problem. This configuration breaks up harmonics that result from equally spaced inserts, greatly reducing the chance of vibration.

Cutter pitch affects fpr and the in./min or mm/min set or programmed at the machine. **Table 7** illustrates how feed is impacted by pitch on an 8" or 203 mm diameter cutter.

Table 7. Feed/Pitch Relationship in Face Milling.

8" (203 mm) D cutter @ 524 sfm (160 smm)	chip load (fpt) in./min or mm/min	cutter pitch	number of effective inserts	fpr in./rev or mm/rev	feed in./min or mm/min
250 rpm	.005"/.13 mm	coarse	10	.050"/1.27 mm	12.5"/317.5 mm
250 rpm	.005"/.13 mm	medium	14	.070"/1.78 mm	17.5"/444.5 mm
250 rpm	.005"/.13 mm	fine	22	.110"/2.79 mm	27.5"/698.5 mm

Determining what face mill *cutter diameter* to use is based on workpiece dimensions. Cutter ratio to part w.o.c. ratio should be approximately 3:2, or $1^1/_2$ times the part width. For instance, if the w.o.c. is *four inches*, choose a cutter diameter of six inches. If the workpiece is extremely wide, select a cutter diameter that matches the spindle capacity and multiple passes. For example, if the w.o.c. is twenty-four inches and the machine has a standard #50 taper spindle, use an eight inch diameter and take five passes at slightly less than five inches per pass, or four passes at six inches per pass. The wider cut will require extra power and rigidity. It is undesirable when the cutting diameter is approximately equal to the w.o.c. In this situation, the chip formed at the entrance and exit of the cut will be very thin. Thin chips do carry heat away as effectively as thick ones, and the result will be that excessive heat is transferred to the insert thereby causing premature edge failure and contributing to the possibility of workhardening. In operations where the ideal cutter diameter is not available, proper cutter positioning will provide positive results. The cutter should be positioned with approximately $^1/_4$ of the cutter body outside the workpiece and two passes should be made. The negative angle of entry produced is desirable, and increased tool life can be expected using this approach.

A step-by-step procedure should be followed when selecting a cutter. The steps below will aid in selecting and applying a cutter suited to a specific operation. The cutter selection guide above should be consulted for selecting a cutter style for the material being milled.

Step 1: Determine an objective and identify goals. Is production rate a factor? How important is finish? Are the number of passes required important? Can one pass machining be applied or will a rough and finish pass be necessary?

Step 2: Identify the characteristics of the machine to be used. Is it in good condition? Is it suited for the part to be milled? Does it have sufficient power for the planned production rate?

Step 3: Evaluate the workpiece. Is the part suited to be run across the selected machine? Can it be properly fixtured on the machine to provide adequate stability? What are the finish requirements?

Step 4: Select the proper cutter and carbide grade for the intended application. Tests should be conducted with coated and uncoated carbide grades. Before conducting tests, check the selected cutter for runout and be sure all inserts are securely locked in place and properly seated.

Step 5: Wipe the cutter mounting surface and bore to remove dirt and grease. Wipe the spindle face before mounting the cutter. Check for, and remove, any burrs on the cutter and spindle mounting surfaces.

Step 6: Select a starting speed and feed for the material being milled. Use the speed and feed charts on the following pages for proper starting speeds and feeds for normal cutting conditions.

Step 7: The final step is to take a trial cut. Observe chip flow and direction. Note the amount of heat being generated. Is the heat in the chip and not in the workpiece? Are chips being recut (caught between the workpiece and cutting tool after they are formed and then cut again)? Is the cutter running free and, above all, does it produce a smooth constant sound with no change in frequency? Additional cuts should be run at higher feed rates to determine the optimum rate. Closely observe all conditions encountered and make adjustments to maximize performance. The verification chart in **Table 8** will aid in solving and correcting any negative conditions encountered. It is important to make corrections for only one condition at a time so that results can be properly analyzed.

Table 8. Machining Correction Guidelines.

variables	excessive edge wear	built-up edge	edge chipping	cratering	blade breakage	chip welding
speed	too high	too low	too high on hard materials	too high		hard materials – too high soft materials – too low
feed	too low	too low	too high	too high	too high	too low
setup			not rigid		not rigid	not rigid
cutter			check insert seating		check insert seating and clamping	incorrect cutter for material
machine			not rigid		insufficient power	not rigid

Table 9 (Feed for 1 RPM) can be used to quickly identify the feed in in./min for multitooth cutters. For example, given a 24-tooth cutter operating at a feed rate of 0.002 per tooth, the cutter feed (from the table) for one rpm is 0.048 in./min. For 100 rpm, the value would be 100 × .048, or 4.8 in./min., and for 1000 rpm the value would be 48 in./min. To determine feed rates for cutters with more than 40 teeth, multiply the values on the chart to find the appropriate value. For example, for a 48-tooth cutter, simply double the values for the 24-tooth cutter used in the example above. The feed for 1 rpm would be 0.048 × 2, or 0.096 in./min. For 100 rpm and 1000 rpm, the feed rates would be 9.6 in./min and 96 in./min. To find the feed for 1 rpm when the feed per tooth in inches exceeds 0.020, combine the figures on the chart that combine to produce the required feed. For example, to find the feed per tooth on an 8-tooth cutter when the feed is 0.030, combine the value for 0.010 (0.080) and the value for 0.020 (0.160) to determine that the feed would be 0.240 in./min. Another means of arriving at this result would be to multiply the value for 0.150 by two: 2 × 0.120 = 0.240 in./min. These values can easily be converted to mm/min by multiplying the feed in in./min by 25.4.

Surface finish and workpiece troubleshooting – face milling

Indexable carbide cutters produce surface finishes within the range of 32 Ra to 150 Ra. The broad range is due to many variables such as work material, machine rigidity, spindle alignment, fixturing, insert nose geometry, insert wear, cutting feed and speed, heat generated chip wear, and chatter. Even in ideal conditions, surface finishes are generally a combination of roughness superimposed along a wave pattern on the workpiece surface. In milling, several inserts overlap as they feed across the workpiece to produce the final surface profile. Since the axial and radial location of all the inserts in an indexable milling cutter are never in perfect alignment, the final milled surface has waves. The portion of the wave produced by the cutter can be exaggerated even more if the machine spindle has

Table 9. Feed for 1 RPM

Feed Per Tooth

Inches	0.0002	0.0004	0.0005	0.0006	0.0008	0.001	0.002	0.004	0.005	0.006	0.008	0.010	0.012	0.015	0.020
No. of Teeth															
2	0.0004	0.0008	0.0010	0.0012	0.0016	0.002	0.004	0.008	0.010	0.012	0.016	0.020	0.024	0.030	0.040
3	0.0006	0.0012	0.0015	0.0018	0.0024	0.003	0.006	0.012	0.015	0.018	0.024	0.030	0.036	0.450	0.060
4	0.0008	0.0016	0.0020	0.0024	0.0032	0.004	0.008	0.016	0.020	0.024	0.032	0.040	0.048	0.600	0.080
6	0.0012	0.0024	0.0030	0.0036	0.0048	0.006	0.012	0.024	0.030	0.036	0.048	0.060	0.072	0.900	0.120
8	0.0016	0.0032	0.0040	0.0048	0.0064	0.008	0.016	0.032	0.040	0.048	0.064	0.080	0.096	0.120	0.160
10	0.0020	0.0040	0.0050	0.0060	0.0080	0.010	0.020	0.040	0.050	0.060	0.080	0.100	0.120	0.150	0.200
12	0.0024	0.0048	0.0060	0.0072	0.0096	0.012	0.024	0.048	0.060	0.072	0.096	0.120	0.144	0.180	0.240
14	0.0028	0.0056	0.0070	0.0084	0.0112	0.014	0.028	0.056	0.070	0.084	0.112	0.140	0.168	0.210	0.280
16	0.0032	0.0064	0.0080	0.0096	0.0128	0.016	0.032	0.064	0.080	0.096	0.128	0.160	0.192	0.240	0.320
18	0.0036	0.0072	0.0090	0.0108	0.0144	0.018	0.036	0.072	0.090	0.108	0.144	0.180	0.216	0.270	0.360
20	0.0040	0.0080	0.0100	0.0120	0.0160	0.020	0.040	0.080	0.100	0.120	0.160	0.200	0.240	0.300	0.400
22	0.0044	0.0088	0.0110	0.0132	0.0176	0.022	0.044	0.088	0.110	0.132	0.176	0.220	0.264	0.330	0.440
24	0.0048	0.0096	0.0120	0.0144	0.0192	0.024	0.048	0.096	0.120	0.144	0.192	0.240	0.288	0.360	0.480
26	0.0052	0.0104	0.0130	0.0156	0.0208	0.026	0.052	0.104	0.130	0.156	0.208	0.260	0.312	0.390	0.520
28	0.0056	0.0112	0.0140	0.0168	0.0224	0.028	0.056	0.112	0.140	0.168	0.224	0.280	0.336	0.420	0.560
30	0.0060	0.0120	0.0150	0.0180	0.0240	0.030	0.060	0.120	0.150	0.180	0.240	0.300	0.360	0.450	0.600
32	0.0064	0.0128	0.0160	0.0192	0.0256	0.032	0.064	0.128	0.160	0.192	0.256	0.320	0.384	0.480	0.640
36	0.0072	0.0144	0.0180	0.0216	0.0288	0.036	0.072	0.144	0.180	0.216	0.288	0.360	0.432	0.540	0.720
40	0.0080	0.0160	0.0200	0.0240	0.0320	0.040	0.080	0.160	0.200	0.240	0.320	0.400	0.480	0.600	0.800

runout. The wave produced by the machine spindle is easily identified on the part surface as a repeated mark or pattern dependent on the feed rate of the cutter.

The actual shape of an insert's cutting edge will be machined into the part surface. Since radiused inserts produce grooves in the workpiece, the final finish will be dependent on the size of the radius (r) and the feed per tooth (fpt) assuming that all inserts are in exactly the same plane both radially and axially. Because an exact setup requires significant time and expense, radiused inserts are used on roughing operations.

Inserts with small chamfers (or wiping flats) are often used when surface finish requirements are a magnitude of 125 micro-inches (3.150 micrometers or microns) or less. The chamfer runs parallel to the workpiece surface, and its width must exceed the feed per revolution (fpr) of the face mill. In instances where the fpr exceeds the width of the chamfered flat, two or more inserts will provide the final surface finish profile which will be totally dependent on their alignment in the axial plane.

Wiper inserts are used to obtain fine surface finishes when the fpr greatly exceeds the width of the corner chamfer. In this situation, a periphery wiping insert is installed in the cutter body to reduce waviness and minimize the peak to valley height of the surface profile. Wiper inserts have slight crowns, so differences in spindle camber or tilt from application to application will eliminate the saw-tooth profile associated with a lack of parallelism between the workpiece and the insert edge. The wiper cutting depth is normally 0.002" to 0.004" (0.0508 mm to 0.1016 mm) below the other inserts on the cutter. The fpr should be less than the length of the wiper, and in most cases should be 50% of the wiper length. Wipers in large diameter cutters should be aligned axially to ensure proper overlap. Cutters using wiper inserts should be used on short chipping materials such as brass and cast iron. Their use on steel often results in large axial forces and vibration due to the difficulty of forming a wide chip only 0.002" to 0.004" (0.0508 mm to 0.1016 mm) in height.

Increased surface quality normally means higher manufacturing costs, so the required design parameters should specify the necessary surface quality. The chart below suggests solutions for poor surface finish conditions.

cause	solution
cutter runout	Check for high insert, dirt in the pockets, dirty spindle, and cutter mounting face. Examine cutter for burrs and damage in cutter pocket.
worn or chipper insert	Index insert.
feed per revolution exceeds flat on wiper	Reduce feed rate or install wiper with greater effective insert facet width.
wiper insert set too high	Set wiper insert 0.0005" to 0.002" (0.01270 to 0.0508 mm) above highest insert.
chatter	Check machine and fixturing rigidity. Check arbor and spindle. Adjust feed rates, rpm, or reduce cut width. Consider cutter with fewer pockets.

Workpieces vary by size, hardness, condition, and composition. These variables can cause problems with surface finish and other machining parameters. Size and shape considerations include thin sections made of hard materials. Hardness can have a major impact on tool life and can cause tool deflection. Condition relates to the properties of the material being machined. For example, impurities can cause hard spots in castings that can break tools, and the surface scale on forgings can reduce tool life. The composition of the workpiece material affects tool life and machinability. The following charts provide methods for correcting specific workpiece problems.

Chatter is a vibration condition involving the milling machine and milling cutter. Once chatter arises, it tends to reside until the problem is corrected. It can be identified when lines or grooves, caused by teeth of the cutter vibrating in and out of the workpiece, appear at regular intervals on the workpiece. Spacing of the lines and grooves depends on the frequency of vibration.

problems	solutions
rigidity	Check system rigidity.
feed rate	Reduce feed rate. Check actual power consumption.
axial d.o.c.	Reduce axial d.o.c.
radial d.o.c.	Reduce radial d.o.c.
insert edge prep	Use sharp edge PVD coated inserts. Eliminate hone or T-land.

Burring is a condition whereby small slivers of alloy roll out over the shoulders and points of the workpiece as the tool exits the cut. Burring is most common when cutting soft, gummy materials.

problems	solutions
dull insert	Index inserts to sharp edge.
insert edge prep	Reduce T-land. Reduce or omit hone.
grade	Use sharp PVD coated inserts.
angle of entry	Change cutter angle of entry or exit.
lead angle	0° lead is the least desirable. Use 45°, 30°, or 15° lead angle.
chip load	Adjust chip load up or down. Prechamfer the workpiece.

Breakout describes the condition when an uneven break of material occurs as the cutter exits the part. It is most common when cutting cast iron and powdered-metal materials.

problems	solutions
geometry (lead angle)	Increase cutter lead angle.
insert corner geometry	Apply double facet inserts.
chip load per tooth	Lower the chip load.
angle of entry	Change cutter angle of entry or exit.

Back-striking marks are surface feed marks that are on a different height and run opposite in direction to normal feed marks. These marks are caused by a portion of the cutter's inserts that are not in the cut rubbing on the part during cutter rotation. Most machines designed specifically for milling have a spindle that is tilted slightly off parallel from the workpiece surface (approximately 0.0015" per foot or 0.125 mm per meter). Machining centers and other multi purpose machines have 0.00" spindle tilt, making back-striking unavoidable unless a ramp angle of 0.0015" per foot or 0.125 mm per meter is programmed. Programming this ramp angle will increase tool life as the inserts will no longer be rubbing. See *Figure 9*.

problems	solutions
dull inserts	Index inserts to a sharp edge. Use sharp PVD coated inserts.
machine spindle	Check spindle tilt on manual milling machines. Program slight ramp angle on CNC machines. Take advantage of any play in the slides by feeding in the opposite direction, thus raising the cutter heel.
cutter geometry	Use a more positive geometry cutter. Reduce either axial or radial d.o.c. cutter forces. Increase speed.
workpiece/fixturing	Improve workpiece support.

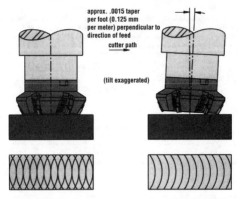

approx. .0015 taper
per foot (0.125 mm
per meter) perpendicular to
direction of feed

cutter path

(tilt exaggerated)

| Figure 9a. Back striking. | Figure 9b. No back striking. |

Surface finish in face milling is affected by the machine condition, cutter design and geometry, cutting data, workpiece, and stability of fixturing. When workpiece finish is not acceptable, the problem is usually identified as roughness (micro-finish), waviness, or the surface is not flat to specification.

problems	solutions
roughness	Check advance per revolution versus facet flat width. Apply wiper insert. Use PVD coated inserts. Check cutter for high tooth (improper preset). Check for built-up edge. Increase speed. Apply more lubricant to coolant.
waviness	Check system deflection of cutter, fixturing, and machine. Reduce chip load per insert.
flatness	Check system deflection of cutter, fixturing, and machine. Check for part deflection. Check for worn locators. Improve fixturing. Check spindle tilt. Apply a larger cutter. Apply a more positive rake cutter.

Surface finish in slotting or half-side mill operations present special problems because the finish produced by the outside diameter of the cutter will be different from the surface produced by the side of the cutter. The outside diameter will produce a surface similar to that produced by an end mill, while the finish on the side of the slot will be similar to that produced by a face mill. The closer to the centerline of the cutter, the higher the Ra, and the closer to the outside diameter of the cutter, the lower the Ra. This variation in finish produced by the chip load affects the taper of the slot or side condition in half-side milling.

problems	solutions
chip load	Reduce the advance per revolution.
insert	Apply special facet or wiper insert. Use sharp PVD coated cutters.
burring	Use a larger diameter cutter if rigidity permits.

Workpieces with thin-walled surfaces, composed of two or more pieces made from hard materials or with a box- or bowl-like shape, are particularly prone to chatter, size control, and burring problems.

problems	solutions
chatter	The workpiece can be dampened in several ways: wrap the workpiece with a rubber belt, a hose, or a long extension spring, or apply shot bags, sand, plaster of Paris, or plumber's lead.
size control	Fixturing must be rigid in order to absorb the thrust of the machining operation.
burring	Thin wall or soft materials may have a tendency to burr. See "burring" above.

End milling

End mills are rotary cutting tools with one or more cutting elements (teeth). They have straight or helical teeth on the periphery, which intermittently engage the workpiece and remove material by relative movement of the workpiece and cutter. *Figure 10* identifies the dimensions of conventional and indexable end mills. Precision Twist Drill Company provided the following descriptions and terminology.

2-flute end mills are general-purpose mills with two flutes for maximum chip volume. Used for plunge milling, roughing out slots, and peripheral milling, these multipurpose tools allow for high feed rates where part finish and dimensional accuracy are not critical. Best suited for use in low alloyed steels, cast iron, and nonferrous materials.

3-flute end mills are more rigid, with less cut interruption than the 2-flute. They have more chip volume area than the 4-flute, allowing higher feed rates. The 3-flute end mill has all the machining capabilities of a 2-flute end mill and is ideal for slotting applications. Improved part finish and dimensional accuracy can be achieved in a wider range of materials.

4-flute end mills are stronger than the 2- or 3-flute versions. The added rigidity allows higher feed rates with minimum deflection, allowing for improved workpiece finishes and dimensional accuracy. The inherent limited chip volume area restricts stock removal rates and deep plunge cutting capabilities. The 4-flute design is commonly used for finishing applications.

Ball nose end mills with full radius on end (ball nose) are held to within ± 0.002" of true radius blend. Material applications are the same as plain end mills. They are used for slotting and peripheral milling with a radius.

Indexable end mills are discussed later in this section.

End mill terminology

Axis: The line about which a cutter rotates.

Clearance: The additional space provided behind the relieved land of a cutter tooth to eliminate undesirable contact between the cutter and workpiece.

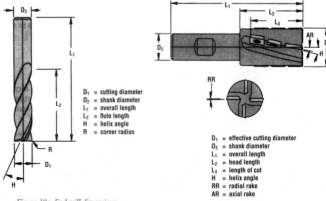

D_1 = cutting diameter
D_2 = shank diameter
L_1 = overall length
L_2 = flute length
H = helix angle
R = corner radius

Figure 10a. End mill dimensions.

D_1 = effective cutting diameter
D_2 = shank diameter
L_1 = overall length
L_2 = head length
L_3 = length of cut
H = helix angle
RR = radial rake
AR = axial rake

Figure 10b. Indexable end mill dimensions.

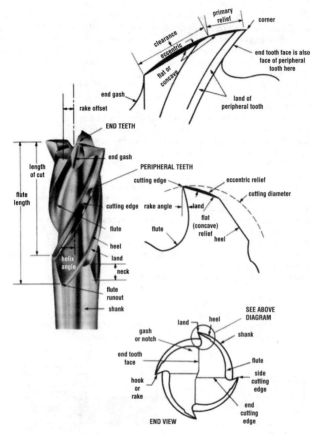

Figure 11. Physical characteristics of end mills. (Source, DoAll Company.)

Core Diameter: The diameter of a circle which is tangent to the bottom of the flutes.

Cutting Edge: The leading edge of the cutter tooth.

Cutting Edge Angle: The angle which a cutting edge makes with an axial plane at any given point. A constant lead will produce a constant cutting edge angle on a cylindrical cutter, and a varying cutting edge angle on a tapered cutter. A varying lead can be used to produce a constant cutting edge angle on a tapered cutter.

End Dish (concavity): Dish is defined as the angle at which the end teeth are ground from the end mill diameter to the center of the tool. It is measured from a plane perpendicular to the axis of the tool. The purpose of the dish on an end mill is to provide clearance for the end cut as the mill laterally traverses through the material.

Flute: The chip space between the back of one tooth and the face of the following tooth.

FPT: Feed per tooth in inches or millimeters.

Gash Angle: The angle of the gash that forms the end cutting teeth. It is measured from a plane perpendicular to the axis of the tool.

Hairline: A small retained portion of the circular land of the end mill after relieving.

Helical: A cutting edge or flute which progresses uniformly around a cylindrical surface in an axial direction.

Helix Angle: The cutting edge angle which a helical cutting edge makes with a plane containing the axis of a cylindrical cutter.

IPM: Feed rate in inches or millimeters (mm/min) per minute = FPT × NT × RPM.

Land: The narrow surface of a profile sharpened cutter tooth immediately behind the cutting edge.

Cylindrical: A narrow portion of the peripheral land, adjacent to the cutting edge, having no radial relief.

Relieved: The portion of the land adjacent to the cutting edge, which provides a relief.

Lead: The axial advance of a helical cutting edge in one turn around the axis.

Length of Cut: The effective axial length of a cutting edge; not including the cutter sweep.

NT: Number of teeth.

Rake: The angular relationship between the tooth face, or a tangent to the tooth face at a given point and a given reference plane or line.

Axial Rake: Applies to angular (not helical) flutes. The axial rake at a given point on the face of the flute is the angle between the tool axis and a tangent plane at the given point.

Hook: A concave condition of a tooth face. The rake of a hooked tooth face must be determined at a given point.

Negative Rake: Describes a tooth face in rotation whose cutting edge lags the surface of a tooth face.

Positive Rake: Describes a tooth face in rotation whose cutting edge leads the surface of the tooth face.

Radial Rake: The angle between the tooth face and a radial line passing through the cutting edge in a plane perpendicular to the cutter axis.

Relief: The relief of the removal of tool material behind or adjacent to the cutting edge to provide clearance and prevent rubbing.

Axial Relief: The relief measured in the axial direction between a plane perpendicular to the axis and the relieved surface. It can be measured by the amount of indicator drop at a given radius in a given amount of angular rotation.

Radial Relief: Relief in a radial direction measured in the plane of rotation. It can be measured by the amount of indicator drop at a given radius in a given amount of angular rotation.

Relief Angle: The angle formed between a relieved surface and a given plane tangent to a cutting edge or to a point on a cutting edge.

Shank: The projecting portion of a cutter which locates and drives the cutter from the machine spindle or adapter.

TD: Tool diameter in inches or millimeters.

Machining with end mills

See pages 984 and 985 for speed and feed information for end mills. When machining with end mills, the surface machined with the end of the cutter is perpendicular to the axis of rotation, and the surface machined by the tool periphery is parallel to the axis of rotation. The cutting edges are located on both the face and periphery and extend along its length parallel to the axis of rotation. It is normally used for intricate operations where limited clearance prohibits the use of a face mill or where slab milling is required. Solid carbide and high-speed steel end mills are suitable for a large variety of job shop and short run applications such as machining square shoulders, squaring ends, grooving operations, and in recess work for die making. Though economical, HSS end mills suffer from decreased hardness in the cutting edge as temperature increases when machining hard materials. Their most common application is small diameter end milling (less than 0.75" or 19 mm), and they have been produced in diameters of 0.15" or 3.81 mm, and even smaller. Solid carbide end mills are less prone to lose effectiveness at high temperature, and they are especially well suited for machining materials such as aluminum and zinc alloys which are low tensile strength nonferrous, nonmetallic, and highly abrasive materials. Their ability to withstand high cutting edge temperatures makes them appropriate for slotting, drilling, and side milling, and they are available in a wide variety of sizes from 0.375" to 2.00" (9.525 mm to 50.8 mm). Indexable end mills can be run at much higher speed and feed rates than HHS, and they are capable of machining a much wider range of materials. The axial depth of cut is restricted to the insert's edge length, and indexable end mills leave a slight scallop on the part wall at the point where inserts overlap.

Setup. Cutting efficiency and quality depend on a secure and rigid setup. Climb milling requires an extremely rigid setup.

Workpiece material. Use higher speeds on soft materials with low alloy content, and lower speeds on hard materials. Materials containing nickel, chromium, tungsten, molybdenum, or other alloys require lower cutting speeds even though they may be classified as soft.

End mill. Choose an end mill with the largest possible diameter, the shortest possible flute, and the shortest possible overall length. Large diameter end mills are stiffer and run cooler, allowing for the use of higher speeds and feeds. The work should be set up so that the spindle nose is positioned as close as possible to the workpiece. Use proper speeds and feeds for the application. Carbide end mills operate at much higher speeds and feeds than HSS. Increased cobalt content in HSS increases hot hardness, but decreases strength. Determine that the flute chip clearance is adequate for the workpiece material and take action to avoid recutting of chips which will damage or dull the cutting edge. Multiple flute end mills permit higher rates of production since feed per tooth is multiplied by the number of teeth, allowing for an increase in the feed rate. Condition of the end mill should be closely monitored until reasonable tool life data figures have been established. At the first sign of dulling, the end mill should be replaced or reground. Otherwise, the tool may fail beyond repair and damage the workpiece. Wear signs include: chip welding (increase the speed), changes in the sound of the cutting action, and changes in the shape or color of chips.

Tool life. High-speed steel end mills begin to destruct at 1050°F (565.5°C). Observe chip color, and reduce cutting speed if they become discolored by excessive heat.

Toolholder. Single angle collet chucks retain better concentricity between the tool and machine spindle than conventional end mill adapters. Cleanliness is essential with all types of end mill holders if runout is to be held to a minimum. A spindle wiper should be used periodically to remove chips, lint, oil, sludge, and coolants from the spindle taper to assure proper adjuster seating.

Machine tool. The machine tool must possess sufficient rigidity to minimize spindle deflection, and have sufficient power to perform at recommended speeds and feeds. Before increasing feed rate, the overall machine condition, workpiece rigidity, and cutting edge strength must be considered. Climb mill when possible for roughing cuts, particularly where the finishing allowance is small. Climb milling will permit higher cutting speeds and will also extend tool life.

Coolant is recommended when milling steel and high temperature alloys. Coolant or jets of air will direct chips away from the cutting tool and the workpiece and prevent recutting of chips. When machining titanium, coolant should be used in a heavy flow directed at the end mill cutting area to clear the end mill of chips and reduce the hazard of fire.

Circular and helical interpolation

Circular interpolation exists when a cutter rotating about its own axis travels in an orbiting motion about an inside diameter (ID) or outside diameter (OD) workpiece circumference without any vertical shift during the operation. This orbiting movement utilizes the X- and Y-axis.

Helical interpolation requires milling with three axes. The operation consists of a cutter rotating about its own axis traveling in an orbiting motion about an ID or OD workpiece in the X- and Y- planes, with a simultaneous linear movement in the Z-axis, creating a helical movement. For example, the path shown in *Figure 12* from point A to point B on the envelope of the cylinder combines a circular movement on the X- and Y-planes with a linear movement in the Z-direction. On most CNC systems, this function can be commanded with two different codes.

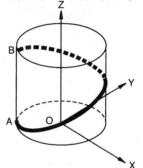

Figure 12. Helical interpolation.

- GO2: helical interpolation in a clockwise direction.
- GO3: helical interpolation in a counterclockwise direction.

On most CNC machines, the feed rate required for programming contour (circular or helical) milling is calculated based on the centerline of the tool. When programming linear tool movement, the feed rate at the cutting edge and centerline are identical, but with circular movement, this is not the case. The following formula is used to calculate the tool feed rate at the cutting edge.

to find	given	formula
F_1 (feed at cutting edge) in in./min *or* mm/min	feed per tooth (chip load) in inches *or* mm = fpt number of effective teeth = nt revolutions per minute = rpm	$F_1 = fpt \times nt \times rpm$

To calculate the feed rate at the tool centerline with circular tool movement about the ID of a workpiece (see *Figure 13a*), use the following formula.

to find	given	formula
F_2 (feed at tool centerline) in in./min *or* mm/min	feed at cutting edge in inches or mm = F_1 ID of workpiece in inches or mm = D cutter diameter in inches or mm = d_1 feed per tooth in inches or mm = fpt number of effective teeth = nt revolutions per minute = rpm	$F_2 = \dfrac{(F_1 \times [D - d_1])}{D}$

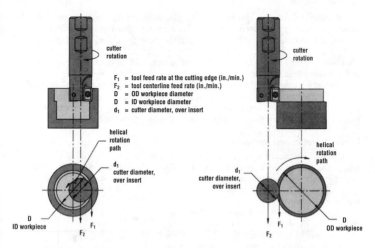

Figure 13a. Inside diameter helical interpolation.

Figure 13b. Outside diameter helical interpolation.

In ID contour applications, note that the tool centerline feed rate is always less than the cutting edge feed rate.

Example. Find the feed rate at the tool centerline when using circular or helical interpolation to contour mill the inside of a workpiece with a 4" (101.6 mm) inside diameter when using a 3" (76.2 mm) diameter cutter with 7 effective teeth operating at 637 rpm with feed per tooth of 0.008" (0.2032 mm).

First, calculate the feed rate at the cutting edge.

$F_1 = 0.008 \times 7 \times 637 = 35.672$ in./min   *or* $F_1 = 0.2032 \times 7 \times 637 = 906.07$ mm/min.

The feed rate at the tool centerline can now be calculated.

$$F_2 = \frac{(35.672 \times [4-3])}{4} = 8.92 \text{ in./min} \quad or \quad F_2 = \frac{(906.07 \times [101.6 - 76.2])}{101.6} = 226.52 \text{ mm/min.}$$

Therefore, to have the cutting edge feed rate (F_1) equal 35.672 in./min (906.07 mm), the machine tool must be programmed for the centerline cutting rate (F_2) of 8.92 in./min (226.52 mm/min). In this instance, the centerline cutting rate for ID circular and helical interpolation is approximately 75% less than the cutting edge feed rate.

To calculate the feed rate at the tool centerline with circular movement about the OD of a workpiece (see *Figure 13b*), use the following formula.

to find	given	formula
F_2 (feed at tool centerline) in in./min or mm/min	feed at the cutting edge in inches or mm = F_1 OD of the workpiece in inches or mm = D cutter diameter in inches or mm = d_1 feed per tooth in inches or mm = fpt number of effective teeth = nt revolutions per minute = rpm	$F_2 = \dfrac{(F_1 \times [d_1 + D])}{D}$

In OD contour applications, note that the tool centerline feed rate is always larger than the cutting edge feed rate.

Example. Find the feed rate at the tool centerline when using circular or helical interpolation to contour mill the outside of a workpiece with a 5" (127 mm) outside diameter when using a 2" (50.8 mm) diameter cutter with 5 effective teeth operating at 955 rpm with feed per tooth of 0.008" (0.2032 mm).

First, calculate the feed rate at the cutting edge.

$F_1 = 0.008 \times 5 \times 955 = 38.2$ in./min   *or* $F_1 = 0.2032 \times 5 \times 955 = 970.28$ mm/min.

The feed rate at the tool centerline can now be calculated.

$$F_2 = \frac{(38.2 \times [2 + 5])}{5} = 53.48 \text{ in./min} \qquad or \quad F_2 = \frac{(970.28 \times [50.8 + 127])}{127} = 1358.4 \text{ mm/min.}$$

Therefore, to have the cutting edge feed rate (F_1) equal 38.2 in./min (970.28 mm), the machine tool must be programmed for the centerline cutting rate (F_2) of 53.48 in./min (1358.4 mm/min). In this instance, the centerline cutting rate for OD circular and helical interpolation is approximately 40% more than the cutting edge feed rate.

Table 10. High-speed Steel End Mill Materials.

Material	Nominal Composition Percentage						Application
	C	W	Mo	Cr	V	Co	
tungsten high-speed steels							
T1	0.75	18.0		4.00	1.00	—	general purpose
T2	0.80	18.0		4.00	2.00	—	general purpose-higher strength
molybdenum high-speed steels							
M2	0.85	6.00	5.00	4.00	2.00	—	general purpose
M3, Class 1	1.05	6.00	5.00	4.00	2.40	—	fine edge tools
M7	1.00	1.75	8.75	4.00	2.00	—	fine edge tools-abrasion resistant
M10	0.85	—	8.00	4.00	2.00	—	general purpose-high strength
high-speed steels containing cobalt							
M6	0.80	4.00	5.00	4.00	1.50	12.0	heavy cuts-abrasion resistant
M30	0.80	2.00	8.00	4.00	1.25	5.00	heavy cuts-abrasion resistant
M42	1.10	1.50	9.50	3.75	1.15	8.00	heavy cuts-abrasion resistant
M43	1.20	2.75	8.00	3.75	1.60	8.25	heavy cuts-abrasion resistant
M47	1.10	1.50	9.50	3.75	1.25	5.00	heavy cuts-abrasion resistant
T4	0.75	18.0	—	4.00	1.00	5.00	heavy cuts
T5	0.80	18.0	—	4.00	2.00	8.00	heavy cuts-abrasion resistant
T6	0.80	20.0	—	4.50	1.50	12.0	heavy cuts-hard material
Century XL	1.30	6.25	10.5	3.75	2.00	—	general purpose-hard material
STC (Super Tuf-Cut)	1.50	10.0	5.25	3.75	3.00	9.00	extreme abrasion resistant
T15	1.50	12.0	—	4.00	5.00	5.00	extreme abrasion resistant

Table 11. End Mill Surface Treatments and Coatings.

Description	Characteristics	Application
Nitride Approx. Hardness 1200 HV Rc 72	Consists of a thin, hardened case 0.0005" to 0.002" deep on the surface of the tool to resist abrasion and reduce galling.	Can be used in most abrasive materials, both ferrous and nonferrous. Not recommended where chipping may be a problem.
Double Nitride Approx. Hardness 1400 HV Rc 74	Consists of a higher hardened case on the surface of the tool to resist abrasion and reduce galling. Prone to brittleness and chipping.	Can be used on nonmetallic, highly abrasive materials such as Bakelite, plastics, hard rubber and fibers.
Steam Oxide Approx. Hardness No change from Base Material	Consists of a layer of ferrous oxide on the surface of the tool which has good lubricant retaining properties. Improves toughness by relieving grinding stresses.	Can be used in low carbon, stainless, and free machining steels. Not recommended for use in soft, nonferrous materials where it may cause galling.
Nitride and Oxide Approx. Hardness 1200 HV Rc 72	A combination of two treatments which produces the favorable characteristics of both, resistance to abrasion, and galling.	Can be used in iron and cast iron, stainless and high tensile steels. Not recommended for use in nonferrous Materials where it may cause galling.
Chrome Plate Cr, Hard Chromium Approx. Hardness 1200 HV Rc 72	Consists of a very thin layer of hard chromium on the surface of the tool which reduces friction and prevents galling.	Can be used on most ferrous, nonferrous, and nonmetallic materials. While unlikely, it may cause galling in high chromium stainless steels.
Titanium Nitride TiN, PVD Process Approx. Hardness 2400 HV *Rc 86	Consists of a very hard coating on the surface of the tool which has outstanding wear resistance, reduces friction, and prevents galling.	Can be used on most ferrous, nonferrous, and nonmetallic materials. While unlikely, it may cause galling in titanium and titanium alloys.
Titanium Carbonitride TiCN, PVD Process Approx. Hardness 3000 HV *Rc 94	Consists of an extremely hard coating on the surface of the tool which has outstanding wear resistance, reduces friction, and prevents galling.	Can be used on most ferrous, nonferrous and abrasive materials. Very effective at higher speeds. While unlikely, it may cause galling in titanium and titanium alloys.
Chromium Carbide CrC, PVD Process Approx. Hardness 1850 HV Rc 80	Consists of a very hard coating on the surface of the tool which has excellent wear resistance, reduces friction, and prevents galling.	Can be used on titanium, titanium alloys, exotic materials, and die cast aluminum. Very effective at higher speeds and in many tapping applications. Under certain conditions it may cause galling in wrought aluminum.
Chromium Nitride CrN, PVD Process Approx. Hardness 1750 HV Rc 79	Consists of a very hard coating on the surface of the tool which has excellent wear resistance, reduces friction, and prevents galling.	Can be used on titanium, titanium alloys, nickel-base alloys, and copper alloys. Very effective at higher speeds and in many tapping applications. Under certain conditions, it may cause galling in wrought aluminum.
Titanium Aluminum Nitride TiAlN, PVD Process Approx. Hardness 2600 HV *Rc 89	Consists of an extremely hard coating on the surface of the tool which has outstanding wear resistance, reduces friction, and prevents galling. Forms an aluminum oxide layer at high speeds and elevated temperatures.	Can be used on titanium, titanium alloys, nickel-base alloys, stainless steel, and cast iron. Very effective at higher speeds and in some tapping applications. Not recommended for wrought aluminum, copper and brass.

*Theoretical values for approximate comparison to the Vickers Hardness values.

Note: While most surface treatments and coatings have antigalling properties, they may cause galling in materials composed of or containing identical base elements. Also, steam oxide and some coatings may cause galling in soft materials such as aluminum. *(Source; Fastcut Tool Corporation).*

Guidelines for resharpening general purpose end mills

The Weldon Tool Co. provided these guidelines to help answer the question "To what minimum diameter should an end mill be reground?" This is a difficult question because today's machinist has access to an almost endless combination of end mill and milling machine types, sizes, and styles. The issue is further complicated by the variety of workpiece shapes and materials, and machining conditions. There are additional factors that must be given consideration before the question can be answered. The first, and most obvious, is that there is a loss of end mill diameter as the peripheral edges are reground. This progressively reduces the cross-sectional area, permitting greater deflection under load (see *Figures 14a* and *14b*). With each regrind and reduction in diameter, there is a subsequent and significant reduction in radial rake, which is problematic as efficient metal removal is highly dependent on the end mill having sufficient rake angle for the material being machined. In conjunction with the reduction in diameter, the helix angle also slightly changes, which further affects the dynamic rake of the end mill as shown in *Figures 14a* and *14b*. As the diameter of the end mill is reduced, the end mill's tooth geometry causes a rapid increase of the width of the secondary land, thereby increasing regrinding time.

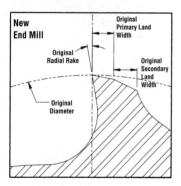

Figure 14a.

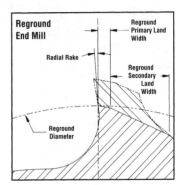

Figure 14b..

The flute depth is also reduced in proportion to the end mill diameter, and can reach a point where chip handling ability is impaired to such a degree that inefficient feed rates must be employed (see *Figure 15*). In fact, for every 1% reduction in end mill diameter (as a result of sharpening), there is a $1^1/_2$% to 2% decrease in cutting efficiency. The decrease is constant at this rate until the diameter reduction reaches 15%. At that point, each additional 1% reduction in diameter results in an additional efficiency loss of 3% to 4%. This pinpoints the fact that loss of cutting tooth rake, resulting in end mill diameter reduction through sharpening,

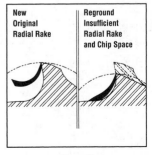

Figure 15

is the principal reason for cutting torque increase or loss of efficiency. For these reasons, it is advised that a 12% to 14% reduction of end mill diameter, based on standard tool diameter, is the economical limit beyond which an end mill cannot be ground and still perform efficiently.

Table 12. Permissible Reduction of Flute Diameter for "Regrinding" Purposes.

New Diameter	Amount of Reduction (Approximate Percent of the "New" Diameter)				
	2 Flutes	3 Flutes	4 Flutes	5 Flutes	6 Flutes
Under 3/8"	13%	13%	13%	—	—
3/8 – 5/8	14%	14%	13%	12%	—
11/16 – 1	14%	14%	14%	13%	—
Over 1	14%	14%	14%	13%	12%

Table 13. Resharpening Geometry for General Purpose End Mills. *(Source, Fastcut Tool Corp.)*

A	B	C	D	E	F	G
1/16	0.010	22°	32°	0.013	—	4°–5°
3/32	0.013	18°	30°	0.013	—	4°–5°
1/8	0.013	16°	28°	0.017	25°	3°
5/32	0.013	16°	26°	0.017	25°	3°
3/16	0.013	14°	26°	0.023	25°	3°
7/32	0.013	14°	24°	0.023	25°	3°
1/4	0.013	13°	22°	0.023	25°	2°
9/32	0.013	13°	22°	0.028	25°	2°
5/16	0.013	12°	22°	0.028	25°	2°
11/32	0.013	12°	20°	0.028	25°	2°
3/8	0.013	12°	20°	0.028	20°	2°
13/32	0.014	11°	20°	0.033	20°	2°
7/16	0.014	11°	20°	0.033	20°	2°
15/32	0.014	10°	20°	0.033	20°	2°
1/2	0.014	10°	18°	0.038	20°	2°
9/16	0.014	10°	18°	0.038	20°	1°
5/8	0.014	10°	18°	0.043	20°	1°

(Continued)

Table 13. *(Continued)* **Resharpening Geometry for General Purpose End Mills.**

A	B	C	D	E	F	G
11/16	0.014	9°	18°	0.062	20°	1°
3/4	0.014	9°	18°	0.062	20°	1°
13/16	0.023	9°	17°	0.062	20°	1°
7/8	0.023	9°	17°	0.062	20°	1°
15/16	0.023	8°	17°	0.062	20°	1°
1	0.023	8°	16°	0.075	20°	1°
1 1/8	0.028	8°	16°	0.094	20°	1°
1 1/4	0.028	7°	16°	0.094	20°	1°
1 1/2	0.028	7°	16°	0.094	18°	1°
2	0.033	6°	14°	0.100	18°	1/2°
3	0.043	5°	14°	0.125	18°	1/2°

NOTE: Increase relief angle approximately 20% when milling softer ranges of Aluminum, Copper, Magnesium, Nickel, Stainless Steel, Steel, Titanium, and Zinc. Decrease relief angle approximately 20% when milling harder materials such as Alloy Steels, Armor Plate, High Strength Steels, and Tool Steels.

Cutter clearance dimensions for conventional end mills

Table 14. **Conventional Inch End Mill Cutter Clearance Dimensions.** *(Source, Weldon Tool Co.)*

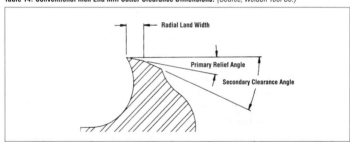

End Mill Diameter (Inch)	Primary Relief Angle	Secondary Clearance Angle	Primary Land Width (Inch)
1/16"	22°	32°	.005/.007
3/32"	18°	28°	.005/.007
1/8"	16°	28°	.005/.007
5/32"	15°	26°	.007/.009
3/16"	14°	26°	.007/.009
7/32"	13°	24°	.007/.009
1/4"	12°	22°	.007/.009
9/32"	12°	21°	.009/.012
5/16"	12°	21°	.009/.012
11/32"	11°	19°	.009/.012
3/8"	11°	19°	.009/.012
13/32"	11°	19°	.012/.016
7/16"	11°	19°	.012/.016

(Continued)

Table 14. *(Continued)* Conventional Inch End Mill Cutter Clearance Dimensions.

End Mill Diameter (Inch)	Primary Relief Angle	Secondary Clearance Angle	Primary Land Width (Inch)
15/32"	10°	19°	.012/.016
1/2"	10°	18°	.012/.016
9/16"	10°	18°	.016/.020
5/8"	10°	18°	.016/.020
11/16"	9°	18°	.016/.020
3/4"	9°	17°	.016/.020
13/16"	9°	17°	.020/.025
7/8"	9°	17°	.020/.025
15/16"	8°	16°	.020/.025
1"	8°	16°	.020/.025
1-1/8"	8°	14°	.025/.030
1-1/4"	7°	14°	.025/.030
1-3/8"	7°	13°	.025/.030
1-1/2"	7°	13°	.025/.030
1-5/8"	7°	13°	.030/.035
1-3/4"	6°	12°	.030/.035
1-7/8"	6°	12°	.030/.035
2"	6°	12°	.030/.035
2-1/4"	5°	11°	.035/.040
2-1/2"	5°	11°	.035/.040
2-3/4"	5°	11°	.035/.040
3"	5°	11°	.035/.040
3-1/2"	5°	11°	.035/.040
4"	4°	10°	.040/.045
5"	4°	10°	.040/.045
6"	4°	10°	.040/.045

Table 15. Metric End Mill Cutter Clearance Dimensions. *(Source, Weldon Tool Co.)*

End Mill Diameter (mm)	Primary Relief Angle	Secondary Clearance Angle	Primary Land Width (mm)
4	15°	26°	.18–.23
5	14°	26°	.18–.23
6–6.3	12°	22°	.18–.23
8	12°	21°	.23–.30
10	11°	19°	.30–.41
12	10°	19°	.30–.41
12.5	10°	18°	.30–.41
14	10°	18°	.41–.51
16	10°	18°	.41–.51
18	9°	18°	.41–.51
20	9°	17°	.41–.51
22	9°	17°	.51–.64
25	8°	16°	.51–.64
28–30	8°	14°	.64–.76

(Continued)

Table 15. *(Continued)* **Metric End Mill Cutter Clearance Dimensions.**

End Mill Diameter (mm)	Primary Relief Angle	Secondary Clearance Angle	Primary Land Width (mm)
32	7°	14°	.64–.76
36	7°	13°	.64–.76
38–40	7°	13°	.64–.76
45	6°	12°	.76–.89
50	6°	12°	.76–.89
56	5°	11°	.89–1.02
63	5°	11°	.89–1.02
70	5°	11°	.89–1.02
75	5°	11°	.89–1.02
80	5°	11°	.89–1.02
100	4°	10°	1.02–1.14
125	4°	10°	1.02–1.14
160	4°	10°	1.02–1.14

Table 16. Dimensions for End Mills Manufactured to NAS 986 Standards.

End Mill Diameter (Inch)	Primary Relief Angle	Secondary Clearance Angle	Primary Land Width (Inch)
1/2"	12°	20°	.013/.023
5/8"	12°	20°	.013/.023
3/4"	11°	20°	.015/.025
7/8"	11°	20°	.015/.025
1"	10°	20°	.020/.030
1-1/4"	10°	20°	.020/.030
1-1/2"	10°	20°	.020/.030
2"	9°	18°	.023/.033

Guidelines for resharpening roughing end mills

Roughing end mills are sharpened by grinding the radial face and fillet of the flute on a tool and grinding machine using a saucer-shaped grinding wheel. The wheel should be dressed to match the form of the flute face from the outside diameter to the fillet. The grinding head should be turned to an angle slightly greater than the helix angle of the end mill (1° to 3° extra). This allows the leading edge of the wheel to hollow grind the rake face (see *Figure 16*). Typically, the finished hook measures 8° to 12°. A tooth support finger is required to assure that the proper lead is maintained. The support finger rests on the back of the tooth being sharpened, located

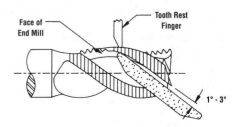

Figure 16. Grinding wheel location for end mill sharpening.
(Source, Weldon Tool Co.)

below the form (*Figure 17*). It is important to remove the entire wear land when sharpening tools. The tool can be spun-ground on the periphery to establish a guide for the amount of outside diameter metal to be removed. Regrinding can generate considerable heat, and care should be taken not to burn the cutting edge.

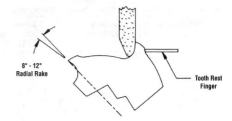

Figure 17. Location of support finger. (Source, Weldon Tool Co.)

Table 17. Light Duty Plain Milling Cutters. *(Source, Fastcut Tool Corp.)*

	All dimensions in inches (see tolerance information below)		
Cutter Dia. (D_1)	**Width of Face** (W)	**Hole Size** (D_3)	**# of Teeth***
2 1/2	3/16, 1/4, 5/16, 3/8, 1/2, 5/8	1	18
	3/4, 1, 1 1/2, 2, 3	1	14
	3/16, 1/4, 5/16, 3/8	1	18
3	1/2, 5/8	1 1/4	18
	3/4, 1, 1 1/4, 1 1/2, 2, 3	1 1/4	16
	1/4, 5/16, 3/8	1	24
4	3/8	1 1/4	20
	1/2, 5/8	1 1/4	24
	3/4, 1, 1 1/2, 2, 3, 4	1 1/4	20

* Number of teeth can vary by manufacturer.

Light duty plain milling cutters are designed for light cuts on plain surfaces and are used on horizontal milling machines.

Dimensions:
1. Cutters with less than $^3/_4$" width of face have straight teeth. Cutters $^3/_4$" and over have helix angle of 15° to 25°, dependent on manufacturer.
2. (ANSI/ASME B94.19-1985) specifies that the cutter diameter tolerance is ± 0.015". Tolerance on width of face up to 1 inch is ± 0.01", over 1" to 2" is +0.010", -0.000, over 2" is +0.020, -0.000.

Table 18. Heavy Duty Plain Milling Cutters. All dimensions in inches (see tolerance information below). *(Source, Fastcut Tool Corp.)*

Cutter Dia. (D_1)	**Width of Face** (W)	**Hole Size** (D_3)	**# of Teeth***
2 1/2	2, 4	1	8
3	2, 2 1/2, 3, 4, 6	1 1/4	8
4	2, 3, 4, 6	1 1/2	10

* Number of teeth can vary by manufacturer.

Heavy duty plain milling cutters provide good performance on heavy cuts due to the tooth rake and deep flutes.

Dimensions:
1. Heavy duty cutters have a helix angle of 25° to 45°, dependent on manufacturer.
2. (ANSI/ASME B94.19-1985) specifies that the cutter diameter tolerance is ± 0.015". Tolerance on width of face for cutters over 2" is +0.020, -0.000.

Table 19. High Helix Plain Milling Cutters. All dimensions in inches (see tolerance information below). *(Source, Fastcut Tool Corp.)*

Cutter Dia. (D₁)	Width of Face (W)	Hole Size (D₃)	# of Teeth*
3	4,6	1 1/4	8
4	8	1 1/2	6

* Number of teeth can vary by manufacturer.

High helix plain milling cutters give smooth workpiece finishes when run at high speeds and feeds. They cut with a shearing action because of tooth angle and high degree of helix.

Dimensions:

1. High helix cutters have a helix angle of 45° to 52°, dependent on manufacturer.
2. (ANSI/ASME B94.19-1985) specifies that the cutter diameter tolerance is ± 0.015". Tolerance on width of face for cutters over 2" is +0.020, -0.000.

Table 20. Single End, Short Length, Two- and Four-Flute Plain End Straight Shank End Mills. *(Source, Weldon Tool Co.)*

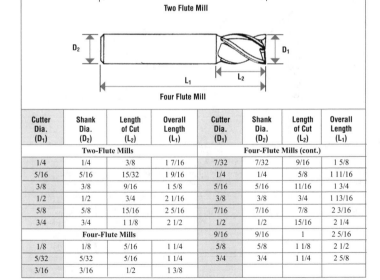

For general purpose milling, including slots, keyways, plunge cutting, and pockets.

Two Flute Mill

Four Flute Mill

Cutter Dia. (D₁)	Shank Dia. (D₂)	Length of Cut (L₂)	Overall Length (L₁)	Cutter Dia. (D₁)	Shank Dia. (D₂)	Length of Cut (L₂)	Overall Length (L₁)
Two-Flute Mills				**Four-Flute Mills (cont.)**			
1/4	1/4	3/8	1 7/16	7/32	7/32	9/16	1 5/8
5/16	5/16	15/32	1 9/16	1/4	1/4	5/8	1 11/16
3/8	3/8	9/16	1 5/8	5/16	5/16	11/16	1 3/4
1/2	1/2	3/4	2 1/16	3/8	3/8	3/4	1 13/16
5/8	5/8	15/16	2 5/16	7/16	7/16	7/8	2 3/16
3/4	3/4	1 1/8	2 1/2	1/2	1/2	15/16	2 1/4
Four-Flute Mills				9/16	9/16	1	2 5/16
1/8	1/8	5/16	1 1/4	5/8	5/8	1 1/8	2 1/2
5/32	5/32	5/16	1 1/4	3/4	3/4	1 1/4	2 5/8
3/16	3/16	1/2	1 3/8				

All dimensions in inches. Cutters right hand helix, right hand cut.

Table 21. Single Angle Milling Cutters with Weldon Shanks. *(Source, Weldon Tool Co.)*

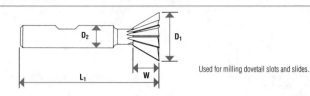

Used for milling dovetail slots and slides.

Single Angle	Cutter Dia. (D₁)	Width of Cutter (W)	Shank Dia. (D₂)	Overall Length (L₁)
45°	3/4	7/32	3/8	2 3/32
	1 3/8	3/8	5/8	2 11/16
	1 7/8	1/2	7/8	2 15/16
	2 1/4	11/16	1	3 9/16
60°	3/4	5/16	3/8	2 1/8
	1 3/8	9/16	5/8	2 7/8
	1 7/8	13/16	7/8	3 1/4
	2 1/4	1 1/16	1	3 3/4

All dimensions in inches. Straight flutes, right hand cut.

Table 22. Single and Double Angle, Arbor Type Milling Cutters. *(Source, Fastcut Tool Corp.)*

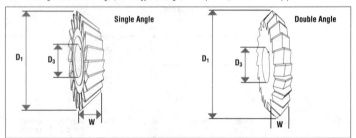

Single Angle*	Cutter Dia. (D₁)	Width of Cutter (W)	Hole Dia. (D₃)	Number of Teeth †
45°	2 3/4	1/2	1	20
	3	1/2	1 1/4	20
90°	2 3/4	1/2	1	20
	3	1/2	1 1/4	20
Double Angle	**Cutter Dia. (D)**	**Width of Cutter (W)**	**Hole Dia. (D₃)**	**Number of Teeth †**
45°	2 3/4	1/2	1	20
60°	2 3/4	1/2	1	20
90°	2 3/4	1/2	1	20

* Single angle cutters are straight flute, and are available with either right or left hand cut.

† Number ot teeth may vary by manufacturer. All dimensions in inches.

Table 23. Straight Tooth Side Milling Cutters. *(Source, Kennametal, Inc.)*

All dimensions in inches (see tolerance information below).

Cutter Dia. (D₁)	Width of Face (W)	Hole Dia. (D₃)	Number of Teeth*
2	3/16, 1/4, 3/8	1/2	14
	3/16, 1/4, 3/8	5/8	14
2 1/2	1/16, 3/32, 1/8	7/8	28
	1/4, 5/16, 3/8, 1/2	7/8	16
3	1/16, 3/32, 1/8, 5/32, 3/16, 7/32, 1/4	1	32
	3/16, 7/32, 1/4, 9/32, 5/16, 11/32, 3/8, 13/32, 7/16, 1/2, 9/16, 5/8, 11/16, 3/4, 13/16, 7/8,15/16, 1	1	20
4	1/16, 3/32, 1/8, 5/32, 3/16, 7/32, 1/4	1	36
	7/32, 1/4, 9/32, 5/16, 11/32, 3/8, 13/32, 7/16, 1/2, 9/16, 5/8, 11/16, 3/4, 13/16, 7/8, 15/16,1	1	24
	1/16, 3/32, 1/8, 5/32, 3/16, 7/32, 1/4	1 1/4	36
	7/32, 1/4, 9/32, 5/16, 11/32, 3/8, 13/32, 7/16, 1/2, 9/16, 5/8, 11/16, 3/4, 13/16, 7/8, 15/16, 1	1 1/4	24
5	1/16, 3/32, 1/8, 5/32, 3/16, 7/32, 1/4	1	40
	1/4, 9/32, 5/16, 11/32, 3/8, 13/32, 7/16, 1/2, 9/16, 5/8, 11/16, 3/4, 13/16, 7/8, 15/16, 1	1	26
	1/16, 3/32, 1/8, 5/32, 3/16, 7/32, 1/4	1 1/4	40
	1/4, 9/32, 5/16, 11/32, 3/8, 13/32, 7/16, 1/2, 9/16, 5/8, 11/16, 3/4, 13/16, 7/8, 15/16, 1	1 1/4	26
6	1/16, 3/32, 1/8, 5/32, 3/16, 7/32, 1/4, 3/8	1	42
	1/2, 3/4,	1	30
	1/16, 3/32, 1/8, 5/32, 3/16, 7/32, 1/4, 3/8	1 1/4	42
	1/4, 9/32, 5/16, 11/32, 3/8, 13/32, 7/16, 1/2, 9/16, 5/8, 11/16, 3/4, 13/16, 7/8, 15/16, 1	1 1/4	30
7	1/8, 3/16	1	48
	1/8, 3/16	1 1/4	48
	1/4, 5/16, 3/8, 7/16, 1/2, 9/16, 5/8, 11/16, 3/4, 13/16, 7/8, 15/16, 1	1 1/4	32
8	1/8	1	48
	1/8, 5/32, 3/16, 7/32, 1/4, 3/8	1 1/4	48
	1/4, 5/16, 3/8, 7/16, 1/2, 9/16, 5/8, 11/16, 3/4, 13/16, 7/8, 15/16, 1	1 1/4	34
	1/8, 5/32, 3/16, 7/32, 1/4, 3/8	1 1/2	48
9	1/4, 5/16, 3/8, 7/16, 1/2, 9/16, 5/8, 11/16, 3/4, 13/16, 7/8, 15/16, 1	1 1/4	36
10	3/16, 7/32, 1/4, 3/8	1 1/4	56
	3/16, 7/32, 1/4, 3/8	1 1/2	56
	1/4, 5/16, 3/8, 7/16, 1/2, 9/16, 5/8, 11/16, 3/4, 13/16, 7/8, 15/16, 1	1 1/2	38
12	1/4, 5/16, 3/8	1 1/2	64
	1/4, 5/16, 3/8, 7/16, 1/2, 9/16, 5/8, 11/16, 3/4, 13/16, 7/8, 15/16, 1	1 1/2	42

*Number of teeth may vary by manufacturer.

Side mills have straight flutes and side teeth on both sides. They are used for straddle milling, shallow slotting, and general tool rool work.

Dimensions:

1. (ANSI/ASME B94.19-1985) specifies that the cutter diameter tolerance is ± 0.015". Tolerance on width of face is +0.000 to -0.0005".

Table 24. Staggered Tooth Side Milling Cutters All dimensions in inches (see tolerance information below).
(Source, Kennametal, Inc.)

Cutter Dia. (D₁)	Width of Face (W)	Hole Dia. (D₃)	Number of Teeth*
2 1/8	3/16, 1/4, 5/16, 3/8	3/4	14
2 1/2	1/4, 5/16, 3/8, 1/2	7/8	16
2 3/4	1/4, 5/16, 3/8, 7/16, 1/2	1	16
3	3/32, 1/8, 5/32, 3/16, 7/32, 1/4	1	28
	3/16, 7/32, 1/4, 9/32, 5/16, 11/32, 3/8, 13/32, 7/16, 1/2, 9/16, 5/8, 11/16, 3/4, 13/16, 7/8, 15/16, 1, 1 1/4	1	16
	3/16, 7/32, 1/4, 9/32, 5/16, 11/32, 3/8, 13/32, 7/16, 1/2, 9/16, 5/8, 11/16, 3/4, 13/16, 7/8, 15/16, 1, 1 1/4	1 1/4	18
3 1/4	1/4, 3/8, 1/2, 3/4	1 1/4	18
3 1/2	7/32, 1/4, 9/32, 5/16, 11/32, 3/8, 13/32, 7/16, 1/2, 9/16, 5/8, 11/16, 3/4, 13/16, 7/8, 15/16, 1	1	18
	7/32, 1/4, 9/32, 5/16, 11/32, 3/8, 13/32, 7/16, 1/2, 9/16, 5/8, 11/16, 3/4, 13/16, 7/8, 15/16, 1	1 1/4	18
3 3/4	1/4, 3/8, 1/2, 3/4	1 1/4	18
4	3/32, 1/8, 5/32, 7/32, 1/4	1	32
	3/16, 7/32, 1/4, 9/32, 5/16, 11/32, 3/8, 13/32, 7/16, 1/2, 9/16, 5/8, 11/16, 3/4, 13/16, 7/8, 15/16, 1	1	18
	3/32, 1/8, 5/32, 3/16, 7/32, 1/4	1 1/4	32
	3/16, 7/32, 1/4, 9/32, 5/16, 11/32, 3/8, 13/32, 7/16, 1/2, 9/16, 5/8, 11/16, 3/4, 13/16, 7/8, 15/16, 1, 1 1/8, 1 1/4, 1 3/8, 1 1/2	1 1/4	18
4 1/4	1/4, 3/8, 1/2, 3/4	1 1/4	18
4 1/2	7/32, 1/4, 5/16, 3/8, 1/2, 5/8, 3/4, 1	1	18
	7/32, 1/4, 5/16, 3/8, 1/2, 5/8, 3/4, 1	1 1/4	18
5	3/32, 1/8, 5/32, 3/16, 7/32, 1/4	1	36
	3/16, 7/32, 1/4, 9/32, 5/16, 11/32, 3/8, 13/32, 7/16, 15/32, 1/2	1	24
	3/32, 1/8, 5/32, 3/16, 7/32, 1/4	1 1/4	36
	3/16, 7/32, 1/4, 9/32, 5/16, 11/32, 3/8, 13/32, 7/16, 15/32, 1/2, 17/32, 9/16, 19/32, 5/8, 11/16, 3/4, 13/16, 7/8, 15/16, 1	1 1/4	24
5 1/2	1/4, 3/8, 1/2, 3/4	1 1/4	24
6	1/8, 5/32, 3/16, 7/32, 1/4, 3/8	1	40
	3/16, 7/32, 1/4, 15/16, 11/32, 3/8, 13/32, 7/16, 15/32, 1/2	1	24
	1/8, 5/32, 3/16, 7/32, 1/4	1 1/4	40
	3/16, 7/32, 1/4, 9/32, 5/16, 11/32, 3/8, 13/32, 7/16, 15/32, 1/2, 17/32, 9/16, 5/8, 11/16, 3/4, 13/16, 7/8, 15/16, 1, 1 1/16, 1 1/8, 1 13/16, 1 1/4	1 1/4	24
	3/16, 7/32, 1/4, 9/32, 5/16, 11/32, 3/8, 13/32, 7/16, 15/32, 1/2, 9/16, 5/8, 11/16, 3/4, 13/16, 7/8, 15/16, 1	1 1/2	28
7	1/8, 3/16	1	44
	1/8, 3/16	1 1/4	44
	1/4, 5/16, 3/8, 7/16, 1/2, 9/16, 5/8, 11/16, 3/4, 13/16, 7/8, 15/16, 1	1 1/4	24
	1/4, 5/16, 3/8, 7/16, 1/2, 9/16, 5/8, 11/16, 3/4, 13/16, 7/8, 15/16, 1	1 1/2	28
8	1/8, 5/32, 3/16, 7/32, 1/4, 3/8	1 1/4	48
	3/16, 7/32, 1/4, 9/32, 5/16, 11/32, 3/8, 13/32, 7/16, 15/32, 1/2, 17/32, 9/16, 19/32, 5/8, 11/16, 3/4, 13/16, 7/8, 15/16, 1	1 1/4	28
	1/8, 5/32, 3/16, 7/32, 1/4, 3/8	1 1/2	48
	3/16, 7/32, 1/4, 9/32, 5/16, 11/32, 3/8, 13/32, 7/16, 15/32, 1/2, 17/32, 9/16, 19/32, 5/8, 11/16, 3/4, 13/16, 7/8, 15/16, 1	1 1/2	28
9	1/4, 5/16, 3/8, 7/16, 1/2, 9/16, 5/8, 11/16, 3/4, 13/16, 7/8, 15/16, 1	1 1/4	28
	1/4, 5/16, 3/8, 7/16, 1/2, 9/16, 5/8, 11/16, 3/4, 13/16, 7/8, 15/16, 1	1 1/2	28

(Continued)

Table 24. *(Continued)* **Staggered Tooth Side Milling Cutters**

Cutter Dia. (D₁)	Width of Face (W)	Hole Dia. (D₃)	Number of Teeth*
10	3/16, 7/32, 1/4, 3/8	1 1/4	56
	3/16, 7/32, 1/4, 3/8	1 1/2	56
	3/16, 7/32, 1/4, 9/32, 5/16, 11/32, 3/8, 13/32, 7/16, 15/32, 1/2, 17/32, 9/16, 5/8, 11/16, 3/4, 13/16, 7/8, 15/16, 1	1 1/2	32
12	1/4, 5/16, 3/8,	1 1/2	64
	1/4, 5/16, 11/32, 3/8, 13/32, 7/16, 15/32, 1/2, 17/32, 9/16, 5/8, 11/16, 3/4, 13/16, 7/8, 15/16, 1	1 1/2	36

*Number of teeth may vary by manufacturer.

Staggered tooth side cutters have alternate right hand and left hand helix and alternate side teeth.

Dimensions:

1. (ANSI/ASME B94.19-1985) specifies that the cutter diameter tolerance is ± 0.015". Tolerance on width of face up to $3/4$" is +0.000 to -0.0005". For face widths over $3/4$" to 1", tolerance is +0.000 to -0.0010".

Table 25. Half-Side Milling Cutters All dimensions in inches (see tolernce information below). *(Source, Fastcut Tool Corp.)*

Cutter Dia. (D₁)	Width of Face (W)	Hole Dia. (D₃)	Number of Teeth*
4	3/4	1 1/4	16
5	3/4	1 1/4	18
6	3/4	1 1/4	20

*Number of teeth may vary by manufacturer.

Half-side cutters have side teeth on one side only. They are available with either right hand cut or left hand cut. They are primarily used for straddle milling.

Dimensions:

1. (ANSI/ASME B94.19-1985) specifies that the cutter diameter tolerance is ± 0.015". Tolerance on width of face over $3/4$" to 1" is +0.000 to -0.0010".

Table 26. Corner Rounding End Mills with Weldon Shank. *(Source, Weldon Tool Co.)*

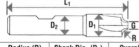

Used to form a radius between a horizontal and vertical surface. These are form relieving tools that can be sharpened without changing the form and size of the radius.

Radius (R)	Shank Dia. (D₂)	Overall Diameter (D₁)	Nose Width (G)	Overall Length (L₁)
1/16	3/8	7/16	1/4	2 1/2
3/32	3/8	1/2	1/4	2 1/2
1/8	1/2	5/8	1/4	3
5/32	1/2	3/4	5/16	3
3/16	1/2	7/8	5/16	2 15/16
1/4	1/2	1	3/8	3
5/16	1/2	1 1/8	3/8	3 1/4
3/8	1/2	1 1/4	3/8	3 1/2
3/16	3/4	7/8	5/16	3 1/8
1/4	3/4	1	3/8	3 1/4
5/16	7/8	1 1/8	3/8	3 1/2
3/8	7/8	1 1/4	3/8	3 3/4
7/16	1	1 3/8	3/8	4
1/2	1	1 1/2	3/8	4 1/8

All dimensions in inches. Straight flutes, right-hand cut.

Table 27. Heavy Duty, Medium Helix, Plain End, Single End, 2″ and 2 1/2″ Noncenter Cutting Combination Shank End Mills. *(Source, Fastcut Tool Corp.)*

Large diameter increases rigidity and reduces deflection. Combination shank fits either Weldon or Pin Drive Holders.

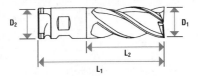

Heavy Duty, Four Flute

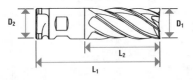

Heavy Duty, Multiflute

See Table 57 for Shank Dimensions.

Cutter Dia. (D_1)	Shank Dia. (D_2)	Length of Cut (L_2)	Overall Length (L_1)	No. of Flutes *
2	2	2	5 3/4	6
2	2	3	6 3/4	4, 6
2	2	4	7 3/4	4, 6
2	2	6	9 3/4	4, 6
2	2	8	11 3/4	4, 6
2	2	10	13 3/4	6
2	2	12	15 3/4	6
2 1/2	2	4	7 3/4	4, 6
2 1/2	2	6	9 3/4	4, 6
2 1/2	2	8	11 3/4	4, 6
2 1/2	2 1/2	4	8	6, 8
2 1/2	2 1/2	6	10	4, 6
2 1/2	2 1/2	8	12	4, 6
2 1/2	2 1/2	10	14	6
2 1/2	2 1/2	12	16	6
3	2 1/2	4	7 3/4	6, 8
3	2 1/2	6	9 3/4	6, 8
3	2 1/2	8	11 3/4	6, 8
3	2 1/2	12	15 3/4	6, 8

*Number of flutes may vary by manufacturer.

All dimensions in inches. End mills are noncenter cutting, right hand helix, right hand cut. Helix angle between 19° and 39°.

Table 28. Heavy Duty, Medium Helix, Plain and Ball End, Single End, 2" and 2 1/2" Center Cutting Combination Shank End Mills. *(Source, Fastcut Tool Corp.)*

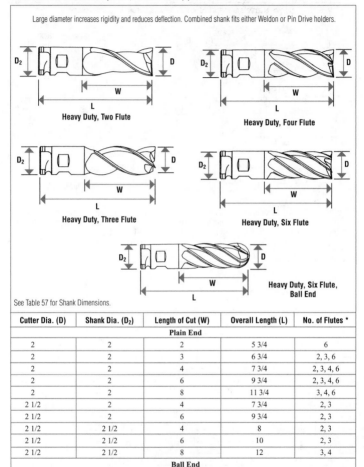

Large diameter increases rigidity and reduces deflection. Combined shank fits either Weldon or Pin Drive holders.

Heavy Duty, Two Flute

Heavy Duty, Four Flute

Heavy Duty, Three Flute

Heavy Duty, Six Flute

Heavy Duty, Six Flute, Ball End

See Table 57 for Shank Dimensions.

Cutter Dia. (D)	Shank Dia. (D₂)	Length of Cut (W)	Overall Length (L)	No. of Flutes *
		Plain End		
2	2	2	5 3/4	6
2	2	3	6 3/4	2, 3, 6
2	2	4	7 3/4	2, 3, 4, 6
2	2	6	9 3/4	2, 3, 4, 6
2	2	8	11 3/4	3, 4, 6
2 1/2	2	4	7 3/4	2, 3
2 1/2	2	6	9 3/4	2, 3
2 1/2	2 1/2	4	8	2, 3
2 1/2	2 1/2	6	10	2, 3
2 1/2	2 1/2	8	12	3, 4
		Ball End		
2	2	3	6 3/4	6
2	2	4	7 3/4	6
2	2	6	9 3/4	6
2	2	8	11 3/4	6

*Number of flutes may vary by manufacturer.

All dimensions in inches. End mills are center cutting, right hand helix, right hand cut. Helix angle between 19° and 39°.

Table 29. Single End Stub, Regular, Medium, and Long Length Plain End Two Flute End Mills with Weldon Shanks. *(Source, Fastcut Tool Corp.)*

For general purpose milling, including slots, keyways, plunge cutting, and pockets.

Cutter Dia. (D_1)	Shank Dia. (D_2)	Length of Cut (L_2)	Overall Length (L_1)	Cutter Dia. (D_1)	Shank Dia. (D_2)	Length of Cut (L_2)	Overall Length (L_1)	See Next Page
Stub Length				**Regular Length**				
1/8	3/8	3/16	2 1/8	7/8	7/8	1 1/2	3 3/4	
3/16	3/8	9/32	2 3/16	15/16	5/8	1 1/2	3 5/8	
1/4	3/8	3/8	2 1/4	15/16	3/4	1 1/2	3 3/4	
Regular Length				15/16	7/8	1 1/2	3 3/4	
1/8	3/8	3/8	2 5/16	1	5/8	1 1/2	3 5/8	
5/32	3/8	3/8	2 5/16	1	3/4	1 1/2	3 3/4	
3/16	3/8	7/16	2 5/16	1	7/8	1 1/2	3 3/4	
7/32	3/8	7/16	2 5/16	1	1	1 5/8	4 1/8	
1/4	3/8	1/2	2 5/16	1 1/8	7/8	1 5/8	3 7/8	
9/32	3/8	1/2	2 5/16	1 1/8	1	1 5/8	4 1/8	
5/16	3/8	9/16	2 5/16	1 1/4	7/8	1 5/8	3 7/8	
3/8	3/8	1/2	2 5/16	1 1/4	1	1 5/8	4 1/8	
7/16	3/8	13/16	2 1/2	1 1/4	1 1/4	1 5/8	4 1/8	
1/2	3/8	13/16	2 1/2	1 3/8	1	1 5/8	4 1/8	
1/2	1/2	1	3	1 1/2	1	1 5/8	4 1/8	
9/16	1/2	1 1/8	3 1/8	1 1/2	1 1/4	1 5/8	4 1/8	
5/8	1/2	1 1/8	3 1/8	1 5/8	1 1/4	1 5/8	4 1/8	
5/8	5/8	1 5/16	3 7/16	1 3/4	1 1/4	1 5/8	4 1/8	
11/16	1/2	1 5/16	3 5/16	1 7/8	1 1/4	1 5/8	4 1/8	
11/16	5/8	1 5/16	3 7/16	2	1 1/4	1 5/8	4 1/8	
3/4	1/2	1 5/16	3 5/16	**Medium Length**				
3/4	5/8	1 5/16	3 7/16	1/4	3/8	1	2 13/16	
3/4	3/4	1 5/16	3 9/16	5/16	3/8	1	2 3/4	
13/16	5/8	1 1/2	3 5/8	3/8	3/8	1	2 3/4	
13/16	3/4	1 1/2	3 5/8	1/2	1/2	1 1/2	3 1/2	
7/8	5/8	1 1/2	3 5/8	5/8	5/8	1 5/8	3 3/4	
7/8	3/4	1 1/2	3 3/4	3/4	3/4	1 3/4	4	

(Continued)

Table 29. *(Continued)* **Single End Stub, Regular, Medium, and Long Length Plain End Two Flute End Mills with Weldon Shanks.**

Cutter Dia. (D₁)	Shank Dia. (D₂)	Length of Cut (L₂)	Overall Length (L₁)	Cutter Dia. (D₁)	Shank Dia. (D₂)	Length of Cut (L₂)	Overall Length (L₁)	
Medium Length				**Long Length**				
7/8	7/8	2	4 1/4	1 1/4	1	3	5 1/2	
1	1	2 1/4	4 3/4	1 1/4	1 1/4	3	5 1/2	
1 1/8	1	2 1/4	4 3/4	1 3/8	1	3	5 1/2	
1 1/4	1	2 1/4	4 3/4	1 1/2	1 1/4	3	5 1/2	
1 1/4	1 1/4	2 1/4	4 3/4	1 5/8	1 1/4	3	5 1/2	
1 3/8	1	2 1/4	4 3/4	1 3/4	1 1/4	3	5 1/2	
1 1/2	1 1/4	2 1/4	4 3/4	1 7/8	1 1/4	3	5 1/2	
1 5/8	1 1/4	2 1/4	4 3/4	2	1 1/4	3	5 1/2	
1 3/4	1 1/4	2 1/4	4 3/4	**Extended Length**				**(B) ***
1 7/8	1 1/4	2 1/4	4 3/4	1/8	3/8	3/8	2 3/8	13/16
2	1 1/4	2 1/4	4 3/4	3/16	3/8	1/2	2 11/16	1 1/8
Long Length				1/4	3/8	5/8	3 1/16	1 1/2
1/4	3/8	1 1/2	3 5/16	5/16	3/8	3/4	3 5/16	1 3/4
5/16	3/8	1 1/2	3 1/4	3/8	3/8	3/4	3 5/16	1 3/4
3/8	3/8	1 1/2	3 1/4	1/2	1/2	1	4	2 7/32
1/2	1/2	2	4	5/8	5/8	1 3/8	4 5/8	2 23/32
5/8	5/8	2	4 1/8	3/4	3/4	1 5/8	5 3/8	3 11/32
3/4	¾	2 1/4	4 3/4	7/8	7/8	2	6 1/16	3 31/32
7/8	7/8	2 1/2	4 3/4	1	1	2 1/2	7 1/4	4 31/32
1	1	3	5 1/2	1 1/4	1 1/4	3	7 1/4	4 31/32
1 1/8	1	3	5 1/2					

*(B) is the length below the shank. The extended, unfluted shank section provides greater rigidity to reduce deflection. Extended length mills are intended for operations where a longer reach, but not a longer cutting length, is required.

All dimensions in inches.

Table 30. **Double End, Regular Length, Four Flute, Medium Helix, Center Cutting End Mills with Weldon Shanks.** *(Source, Fastcut Tool Corp.)*

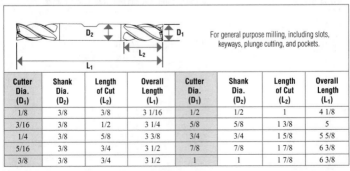

For general purpose milling, including slots, keyways, plunge cutting, and pockets.

Cutter Dia. (D₁)	Shank Dia. (D₂)	Length of Cut (L₂)	Overall Length (L₁)	Cutter Dia. (D₁)	Shank Dia. (D₂)	Length of Cut (L₂)	Overall Length (L₁)
1/8	3/8	3/8	3 1/16	1/2	1/2	1	4 1/8
3/16	3/8	1/2	3 1/4	5/8	5/8	1 3/8	5
1/4	3/8	5/8	3 3/8	3/4	3/4	1 5/8	5 5/8
5/16	3/8	3/4	3 1/2	7/8	7/8	1 7/8	6 3/8
3/8	3/8	3/4	3 1/2	1	1	1 7/8	6 3/8

All dimensions in inches. Cutters are center cutting, right hand helix, right hand cut. Helix angle between 19° and 39°.

Table 31. Double End Stub and Regular Length Plain End Two Flute End Mills with Weldon Shanks.
(Source, Fastcut Tool Corp.)

For general purpose milling, including slots, keyways, plunge cutting, and pockets.

Cutter Dia. (D_1)	Shank Dia. (D_2)	Length of Cut (L_2)	Overall Length (L_1)	Cutter Dia. (D_1)	Shank Dia. (D_2)	Length of Cut (L_2)	Overall Length (L_1)
Stub Length				Regular Length (cont.)			
1/8	3/8	3/16	2 3/4	21/64	3/8	9/16	3 1/8
5/32	3/8	1/4	2 3/4	11/32	3/8	9/16	3 1/8
3/16	3/8	9/32	2 3/4	23/64	3/8	9/16	3 1/8
7/32	3/8	11/32	2 7/8	3/8	3/8	9/16	3 1/8
1/4	3/8	3/8	2 7/8	25/64	1/2	13/16	3 3/4
Regular Length				13/32	1/2	13/16	3 3/4
7/64	3/8	3/8	3 1/16	27/64	1/2	13/16	3 3/4
1/8	3/8	3/8	3 1/16	7/16	1/2	13/16	3 3/4
9/64	3/8	7/16	3 1/8	29/64	1/2	13/16	3 3/4
5/32	3/8	7/16	3 1/8	15/32	1/2	13/16	3 3/4
11/64	3/8	7/16	3 1/8	31/64	1/2	13/16	3 3/4
3/16	3/8	7/16	3 1/8	1/2	1/2	13/16	3 3/4
13/64	3/8	1/2	3 1/8	9/16	5/8	1 1/8	4 1/2
7/32	3/8	1/2	3 1/8	5/8	5/8	1 1/8	4 1/2
15/64	3/8	1/2	3 1/8	11/16	3/4	1 5/16	5
1/4	3/8	1/2	3 1/8	3/4	3/4	1 5/16	5
17/64	3/8	9/16	3 1/8	13/16	7/8	1 9/16	5 1/2
9/32	3/8	9/16	3 1/8	7/8	7/8	1 9/16	5 1/2
19/64	3/8	9/16	3 1/8	15/16	1	1 5/8	5 1/2
5/16	3/8	9/16	3 1/8	1	1	1 5/8	5 1/2

All dimensions in inches. Cutters are center cutting, right hand helix, right hand cut. Helix angle between 19° and 39°.

Table 33. Double End Stub and Regular Length Ball End Two Flute End Mills with Weldon Shanks.
For general purpose milling, including slots, keyways, plunge cutting, and pockets. *(Source, Fastcut Tool Corp.)*

Cutter Dia. (D_1)	Shank Dia. (D_2)	Length of Cut (L_2)	Overall Length (L_1)	Cutter Dia. (D_1)	Shank Dia. (D_2)	Length of Cut (L_2)	Overall Length (L_1)
Stub Length				Regular Length (cont.)			
1/8	3/8	3/16	2 3/4	5/16	3/8	9/16	3 1/8
3/16	3/8	9/32	2 3/4	11/32	3/8	9/16	3 1/8
1/4	3/8	3/8	2 7/8	3/8	3/8	9/16	3 1/8
Regular Length				13/32	1/2	13/16	3 3/4
1/8	3/8	3/8	3 1/16	7/16	1/2	13/16	3 3/4
5/32	3/8	7/16	3 1/8	1/2	1/2	13/16	3 3/4
3/16	3/8	7/16	3 1/8	5/8	5/8	1 1/8	4 1/2
7/32	3/8	1/2	3 1/8	3/4	3/4	1 5/16	5
1/4	3/8	1/2	3 1/8	7/8	7/8	1 9/16	5 1/2
9/32	3/8	9/16	3 1/8	1	1	1 5/8	5 78

All dimensions in inches. Cutters are center cutting, right hand helix, right hand cut. Helix angle between 19° and 39°.

Table 34. Single and Double End Three Flute Medium Helix End Mills with Weldon Shanks.
(Source, Fastcut Tool Corp.)

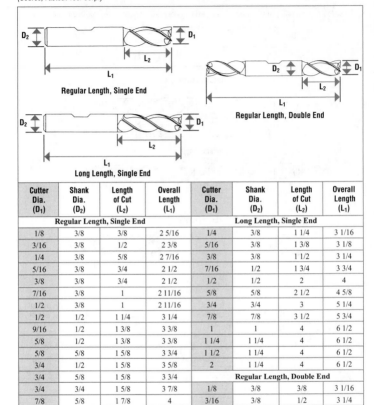

Cutter Dia. (D₁)	Shank Dia. (D₂)	Length of Cut (L₂)	Overall Length (L₁)	Cutter Dia. (D₁)	Shank Dia. (D₂)	Length of Cut (L₂)	Overall Length (L₁)
Regular Length, Single End				**Long Length, Single End**			
1/8	3/8	3/8	2 5/16	1/4	3/8	1 1/4	3 1/16
3/16	3/8	1/2	2 3/8	5/16	3/8	1 3/8	3 1/8
1/4	3/8	5/8	2 7/16	3/8	3/8	1 1/2	3 1/4
5/16	3/8	3/4	2 1/2	7/16	1/2	1 3/4	3 3/4
3/8	3/8	3/4	2 1/2	1/2	1/2	2	4
7/16	3/8	1	2 11/16	5/8	5/8	2 1/2	4 5/8
1/2	3/8	1	2 11/16	3/4	3/4	3	5 1/4
1/2	1/2	1 1/4	3 1/4	7/8	7/8	3 1/2	5 3/4
9/16	1/2	1 3/8	3 3/8	1	1	4	6 1/2
5/8	1/2	1 3/8	3 3/8	1 1/4	1 1/4	4	6 1/2
5/8	5/8	1 5/8	3 3/4	1 1/2	1 1/4	4	6 1/2
3/4	1/2	1 5/8	3 5/8	2	1 1/4	4	6 1/2
3/4	5/8	1 5/8	3 3/4	**Regular Length, Double End**			
3/4	3/4	1 5/8	3 7/8	1/8	3/8	3/8	3 1/16
7/8	5/8	1 7/8	4	3/16	3/8	1/2	3 1/4
7/8	7/8	1 7/8	4 1/8	1/4	3/8	5/8	3 3/8
7/8	3/4	1 7/8	4 1/8	5/16	3/8	3/4	3 1/2
1	5/8	1 7/8	4	3/8	3/8	3/4	3 1/2
1	7/8	1 7/8	4 1/8	7/16	1/2	1	4 1/8
1	3/4	1 7/8	4 1/8	1/2	1/2	1	4 1/8
1	1	2	4 1/2	9/16	5/8	1 3/8	5
1 1/8	1	2	4 1/2	5/8	5/8	1 3/8	5
1 1/4	1	2	4 1/2	3/4	3/4	1 5/8	5 5/8
1 1/4	1 1/4	2	4 1/2	7/8	7/8	1 7/8	6 1/8
1 1/2	1 1/4	2	4 1/2	1	1	1 7/8	6 3/8
1 3/4	1 1/4	2	4 1/2				
2	1 1/4	2	4 1/2				

All dimensions in inches. Cutters are center cutting, right hand helix, right hand cut. Helix angle between 19° and 39°.

Table 35. Single End, Regular, Medium, Long, and Extra Long Ball End Four Flute End Mills with Weldon Shanks. *(Source, Fastcut Tool Corp.)*

For contour or pocket milling where a ball end is required to produce a radius. Also can be used for general purpose milling.

Regular Length

Medium Length

Long Length

Extra Long Length

Cutter Dia. (D_1)	Shank Dia. (D_2)	Length of Cut (L_2)	Overall Length (L_1)	Cutter Dia. (D_1)	Shank Dia. (D_2)	Length of Cut (L_2)	Overall Length (L_1)
Regular Length				**Long Length**			
1/4	3/8	3/4	2 9/16	1/4	3/8	1 1/4	3 1/16
5/16	3/8	1	2 3/4	5/16	3/8	1 1/2	3 1/4
3/8	3/8	1	2 3/4	3/8	3/8	1 1/2	3 1/4
1/2	1/2	1	3	1/2	1/2	2	4
5/8	5/8	1 1/4	3 3/8	5/8	5/8	2 1/2	4 5/8
3/4	3/4	1 1/2	3 3/4	3/4	3/4	3	5 1/4
1	1 1/4	2	4 1/2	1	1	4	6 1/2
1 1/4	1 1/4	2	4 1/2	1 1/2	1 1/4	4	6 1/2
1 1/2	1 1/4	2	4 1/2	**Extra Long Length**			
Medium Length				3/8	3/8	2 1/2	4 5/16
1/4	3/8	1	2 13/16	1/2	1/2	3	5
1/2	1/2	1 1/2	3 1/2	5/8	5/8	3 3/4	5 7/8
5/8	5/8	1 3/4	3 7/8	3/4	3/4	4 1/2	6 3/4
3/4	3/4	2 1/4	4 1/2	1	1	6	8 1/2
1	1	3	5 1/2				
1 1/4	1 1/4	3	5 1/2				

All dimensions in inches. Cutters are center cutting, right hand helix, right hand cut. Helix angle between 19° and 39°.

Table 36. Miniature, Double End Two Flute Plain End and Ball End 3/16" Diameter Straight Shank End Mills.
(Source, Fastcut Tool Corp.)

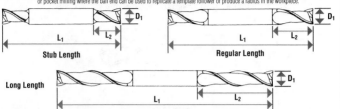

Plain ends for general purpose milling including profiling, slotting, keyways, pockets, and plunge cuts. Ball ends for general purpose milling and contour or pocket milling where the ball end can be used to replicate a template follower or produce a radius in the workpiece.

Cutter Dia. (D_1)	Stub Length		Regular Length		Long Length	
	Length of Cut (L_2)	Overall Length (L_1)	Length of Cut (L_2)	Overall Length (L_1)	Length of Cut (L_2)	Overall Length (L_1)
1/32	3/64	2	3/32	2 1/4	–	–
3/64	1/16	2	9/64	2 1/4	–	–
1/16	3/32	2	3/16	2 1/4	7/32	2 1/2
5/64	1/8	2	15/64	2 1/4	–	–
3/32	9/64	2	9/32	2 1/4	9/32	2 5/8
7/64	5/32	2	21/64	2 1/4	–	–
1/8	3/16	2	3/8	2 1/4	3/4	3 1/8
9/64	7/32	2	13/32	2 1/4	–	–
5/32	15/64	2	7/16	2 1/4	7/8	3 1/4
11/64	1/4	2	1/2	2 1/4	–	–
3/16	9/32	2	1/2	2 1/4	1	3 3/8

All dimensions are in inches. Cutters are center cutting, right hand helix, right hand cut. Helix angle between 19° and 39°.

Table 37. Miniature, Double End Four Flute Plain End and Ball End 3/16" Diameter Straight Shank End Mills.
(Source, Fastcut Tool Corp.)

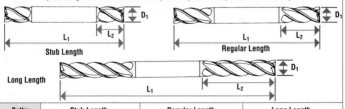

Plain ends for general purpose milling including profiling, slotting, keyways, pockets, and plunge cuts. Ball ends for general purpose milling and contour or pocket milling where the ball end can be used to replicate a template follower or produce a radius in the workpiece.

Cutter Dia. (D_1)	Stub Length		Regular Length		Long Length	
	Length of Cut (L_2)	Overall Length (L_1)	Length of Cut (L_2)	Overall Length (L_1)	Length of Cut (L_2)	Overall Length (L_1)
1/16	3/32	2	3/16	2 1/4	7/32	2 1/2
3/32	9/64	2	9/32	2 1/4	9/32	2 5/8
1/8	3/16	2	3/8	2 1/4	3/4	3 1/8
5/32	15/64	2	7/16	2 1/4	7/8	3 1/4
3/16	9/32	2	1/2	2 1/4	1	3 3/8

All dimensions are in inches. Cutters are center cutting, right hand helix, right hand cut. Helix angle between 19° and 39°.

Table 38. Single End Regular, and Extended Length Ball End Two Flute End Mills with Weldon Shanks.
(Source, Fastcut Tool Corp.)

For general purpose milling, including slots, keyways, plunge cutting and pockets.

Regular Length

Extended Length

Cutter Dia. (D_1)	Shank Dia. (D_2)	Length of Cut (L_2)	Overall Length (L_1)	Cutter Dia. (D_1)	Shank Dia. (D_2)	Length of Cut (L_2)	Overall Length (L_1)	
Regular Length				1 1/8	1	2 1/4	4 3/4	
1/8	3/8	3/8	2 5/16	1 1/4	1 1/4	2 1/2	5	
3/16	3/8	1/2	2 3/8	1 3/8	1 1/4	2 1/2	5	
1/4	3/8	5/8	2 7/16	1 1/2	1 1/4	2 1/2	5	
5/16	3/8	3/4	2 1/2	**Extended Length**				**(B)***
3/8	3/8	3/4	2 1/2					
7/16	1/2	1	3	1/8	3/8	3/8	2 3/8	13/16
1/2	1/2	1	3	3/16	3/8	1/2	2 11/16	1 1/8
9/16	1/2	1	3 1/16	1/4	3/8	5/8	3 1/16	1 1/2
5/8	1/2	1 3/8	3 3/8	5/16	3/8	3/4	3 5/16	1 3/4
5/8	5/8	1 3/8	3 1/2	3/8	3/8	3/4	3 5/16	1 3/4
3/4	1/2	1 5/8	3 5/8	7/16	1/2	1	3 11/16	1 7/8
3/4	3/4	1 5/8	3 7/8	1/2	1/2	1	4	2 7/32
7/8	7/8	2	4 1/4	5/8	5/8	1 3/8	4 5/8	2 23/32
7/8	3/4	2	4 1/4	3/4	3/4	1 5/8	5 3/8	3 11/32
13/16	3/4	2	4 1/4	7/8	7/8	2	6 1/16	3 31/32
1	1	2 1/4	4 3/4	1	1	2 1/2	7 1/4	4 31/32
1	3/4	2 1/4	4 1/2	1 1/4	1 1/4	3	7 1/4	4 31/32

*(B) is the length below the shank. The exended, unfluted shank provides greater rigidity to reduce deflection. Extended length mills are intended for operations where a longer reach, but not a longer cutting length, is required.

All dimensions in inches.

Table 39. Single End Stub, Regular, and Long Lengths, Multiple Flute, Medium Helix Non-Center Cutting End Mills with Weldon Shanks. *(Source, Fastcut Tool Corp.)*

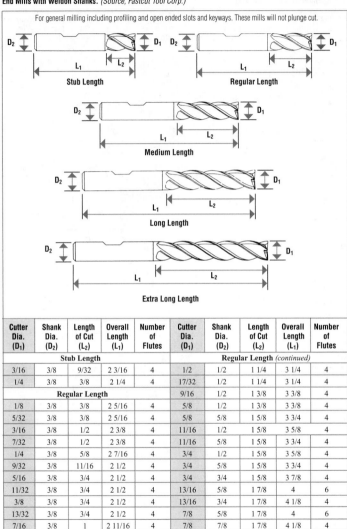

For general milling including profiling and open ended slots and keyways. These mills will not plunge cut.

Stub Length

Regular Length

Medium Length

Long Length

Extra Long Length

Cutter Dia. (D_1)	Shank Dia. (D_2)	Length of Cut (L_2)	Overall Length (L_1)	Number of Flutes	Cutter Dia. (D_1)	Shank Dia. (D_2)	Length of Cut (L_2)	Overall Length (L_1)	Number of Flutes
Stub Length					**Regular Length** *(continued)*				
3/16	3/8	9/32	2 3/16	4	1/2	1/2	1 1/4	3 1/4	4
1/4	3/8	3/8	2 1/4	4	17/32	1/2	1 1/4	3 1/4	4
Regular Length					9/16	1/2	1 3/8	3 3/8	4
1/8	3/8	3/8	2 5/16	4	5/8	1/2	1 3/8	3 3/8	4
5/32	3/8	3/8	2 5/16	4	5/8	5/8	1 5/8	3 3/4	4
3/16	3/8	1/2	2 3/8	4	11/16	1/2	1 5/8	3 5/8	4
7/32	3/8	1/2	2 3/8	4	11/16	5/8	1 5/8	3 3/4	4
1/4	3/8	5/8	2 7/16	4	3/4	1/2	1 5/8	3 5/8	4
9/32	3/8	11/16	2 1/2	4	3/4	5/8	1 5/8	3 3/4	4
5/16	3/8	3/4	2 1/2	4	3/4	3/4	1 5/8	3 7/8	4
11/32	3/8	3/4	2 1/2	4	13/16	5/8	1 7/8	4	6
3/8	3/8	3/4	2 1/2	4	13/16	3/4	1 7/8	4 1/8	4
13/32	3/8	3/4	2 1/2	4	7/8	5/8	1 7/8	4	6
7/16	3/8	1	2 11/16	4	7/8	7/8	1 7/8	4 1/8	4
15/32	3/8	1	2 11/16	4	7/8	3/4	1 7/8	4 1/8	4
1/2	3/8	1	2 11/16	4	15/16	5/8	1 7/8	4	6

(Continued)

Table 39. *(Continued)* Single End Stub, Regular, and Long Lengths, Multiple Flute, Medium Helix Noncenter Cutting End Mills with Weldon Shanks.

Cutter Dia. (D₁)	Shank Dia. (D₂)	Length of Cut (L₂)	Overall Length (L₁)	Number of Flutes	Cutter Dia. (D₁)	Shank Dia. (D₂)	Length of Cut (L₂)	Overall Length (L₁)	Number of Flutes
Regular Length *(continued)*					Long Length *(continued)*				
15/16	7/8	1 7/8	4 1/8	4	5/8	5/8	2 1/2	4 5/8	4
15/16	3/4	1 7/8	4 1/8	4	11/16	5/8	2 1/2	4 5/8	4
1	5/8	1 7/8	4	6	3/4	3/4	3	5 1/4	4
1	3/4	1 7/8	4 1/8	4	13/16	3/4	3	5 1/4	4
1	7/8	1 7/8	4 1/8	4	7/8	7/8	3 1/2	5 3/4	4
1	1	2	4 1/2	4	1	1	4	6 1/2	4
1 1/8	7/8	2	4 1/4	6	1 1/8	1	4	6 1/2	6
1 1/8	1	2	4 1/2	6	1 1/4	1	4	6 1/2	6
1 1/4	1	2	4 1/2	6	1 1/4	1 1/4	4	6 1/2	6
1 1/4	1 1/4	2	4 1/2	6	1 3/8	1	4	6 1/2	6
1 3/8	1	2	4 1/2	6	1 1/2	1	4	6 1/2	6
1 1/2	1	2	4 1/2	6	1 1/2	1 1/4	4	6 1/2	6
1 1/2	1 1/4	2	4 1/2	6	1 3/4	1 1/4	4	6 1/2	6
1 5/8	1 1/4	2	4 1/2	6	2	1 1/4	4	6 1/2	6
1 3/4	1 1/4	2	4 1/2	6	Extra Long Length				
1 7/8	1 1/4	2	4 1/2	6	1/4	3/8	1 3/4	3 9/16	4
2	1 1/4	2	4 1/2	6	5/16	3/8	2	3 3/4	4
Medium Length					3/8	3/8	2 1/2	4 1/4	4
1	1	3	5 1/2	4	1/2	1/2	3	5	4
1 1/4	1 1/4	3	5 1/2	4	9/16	1/2	3	5	4
1 1/2	1 1/4	3	5 1/2	4	5/8	5/8	4	6 1/8	4
1 3/4	1 1/4	3	5 1/2	6	11/16	5/8	4	6 1/8	4
2	1 1/4	3	5 1/2	6	3/4	3/4	4	6 1/4	4
Long Length					13/16	3/4	4	6 1/4	4
1/4	3/8	1 1/4	3 1/16	4	7/8	7/8	5	7 1/4	4
5/16	3/8	1 3/8	3 1/8	4	1	1	6	8 1/2	4
3/8	3/8	1 1/2	3 1/4	4	1 1/4	1 1/4	6	8 1/2	6
7/16	1/2	1 3/4	3 3/4	4	1 1/2	1 1/4	6	8 1/2	6
1/2	1/2	2	4	4	1 1/2	1 1/4	6	10 1/2	6
9/16	1/2	2	4	4					

All dimensions in inches. Cutters are noncenter cutting, right hand helix, right hand cut. Helix angle between 19° and 39°.

Table 40. Single End, Regular, Medium, Long, and Extra Long Length Multiple Flute Medium Helix Center Cutting End Mills with Weldon Shanks. *(Source, Fastcut Tool Corp.)*

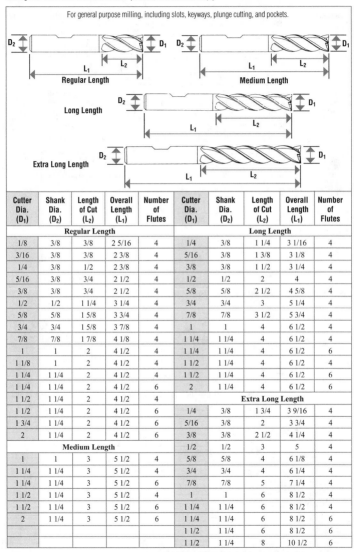

For general purpose milling, including slots, keyways, plunge cutting, and pockets.

Cutter Dia. (D₁)	Shank Dia. (D₂)	Length of Cut (L₂)	Overall Length (L₁)	Number of Flutes	Cutter Dia. (D₁)	Shank Dia. (D₂)	Length of Cut (L₂)	Overall Length (L₁)	Number of Flutes
Regular Length					**Long Length**				
1/8	3/8	3/8	2 5/16	4	1/4	3/8	1 1/4	3 1/16	4
3/16	3/8	3/8	2 3/8	4	5/16	3/8	1 3/8	3 1/8	4
1/4	3/8	1/2	2 3/8	4	3/8	3/8	1 1/2	3 1/4	4
5/16	3/8	3/4	2 1/2	4	1/2	1/2	2	4	4
3/8	3/8	3/4	2 1/2	4	5/8	5/8	2 1/2	4 5/8	4
1/2	1/2	1 1/4	3 1/4	4	3/4	3/4	3	5 1/4	4
5/8	5/8	1 5/8	3 3/4	4	7/8	7/8	3 1/2	5 3/4	4
3/4	3/4	1 5/8	3 7/8	4	1	1	4	6 1/2	4
7/8	7/8	1 7/8	4 1/8	4	1 1/4	1 1/4	4	6 1/2	4
1	1	2	4 1/2	4	1 1/4	1 1/4	4	6 1/2	6
1 1/8	1	2	4 1/2	4	1 1/2	1 1/4	4	6 1/2	4
1 1/4	1 1/4	2	4 1/2	4	1 1/2	1 1/4	4	6 1/2	6
1 1/4	1 1/4	2	4 1/2	6	2	1 1/4	4	6 1/2	6
1 1/2	1 1/4	2	4 1/2	4	**Extra Long Length**				
1 1/2	1 1/4	2	4 1/2	6	1/4	3/8	1 3/4	3 9/16	4
1 3/4	1 1/4	2	4 1/2	6	5/16	3/8	2	3 3/4	4
2	1 1/4	2	4 1/2	6	3/8	3/8	2 1/2	4 1/4	4
Medium Length					1/2	1/2	3	5	4
1	1	3	5 1/2	4	5/8	5/8	4	6 1/8	4
1 1/4	1 1/4	3	5 1/2	4	3/4	3/4	4	6 1/4	4
1 1/4	1 1/4	3	5 1/2	6	7/8	7/8	5	7 1/4	4
1 1/2	1 1/4	3	5 1/2	4	1	1	6	8 1/2	4
1 1/2	1 1/4	3	5 1/2	6	1 1/4	1 1/4	6	8 1/2	4
2	1 1/4	3	5 1/2	6	1 1/4	1 1/4	6	8 1/2	6
					1 1/2	1 1/4	6	8 1/2	6
					1 1/2	1 1/4	8	10 1/2	6

All dimensions in inches. Cutters are center cutting, right hand helix, right hand cut. Helix angle between 19° and 39°.

Table 41. Single End, Regular and Long Length Multiple Flute High Helix (Left-Hand) Noncenter Cutting End Mill. *(Source, Fastcut Tool Corp.)*

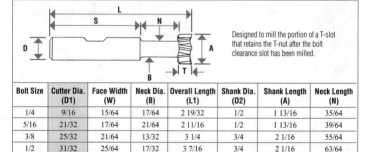

Designed to profile a wide range of materials. Left-hand helix produces downward force on the workpiece in the direction of the machining vibration usually associated with milling thin sections or stacks of thin parts.

Cutter Dia. (D₁)	Shank Dia. (D₂)	Length of Cut (L₂)	Overall Length (L₁)	Number of Flutes	Cutter Dia. (D₁)	Shank Dia. (D₂)	Length of Cut (L₂)	Overall Length (L₁)	Number of Flutes
Regular Length					Long Length				
1/4	3/8	5/8	2 7/16	4	1/4	3/8	1 1/4	3 1/16	4
5/16	3/8	3/4	2 1/2	4	5/16	3/8	1 3/8	3 1/8	4
3/8	3/8	3/4	2 1/2	4	3/8	3/8	1 1/2	3 1/4	4
7/16	3/8	1	2 11/16	4	7/16	1/2	1 3/4	3 3/4	4
1/2	1/2	1 1/4	3 1/4	4	1/2	1/2	2	4	4
5/8	5/8	1 5/8	3 3/4	4	5/8	5/8	2 1/2	4 5/8	4
3/4	3/4	1 3/4	4	4	3/4	3/4	3	5 1/4	4
1	1	2	4 1/2	4	1	1	4	6 1/2	4
1 1/4	1	2	4 1/2	6					
1 1/2	1	2	4 1/2	6					
2	1 1/4	2	4 1/2	8					

All dimensions in inches. Cutters are noncenter cutting, 45° left hand helix, right hand cut.

Table 42. T-Slot Cutters with Weldon Shanks. *(Source, Weldon Tool Co.)*

Designed to mill the portion of a T-slot that retains the T-nut after the bolt clearance slot has been milled.

Bolt Size	Cutter Dia. (D1)	Face Width (W)	Neck Dia. (B)	Overall Length (L1)	Shank Dia. (D2)	Shank Length (A)	Neck Length (N)
1/4	9/16	15/64	17/64	2 19/32	1/2	1 13/16	35/64
5/16	21/32	17/64	21/64	2 11/16	1/2	1 13/16	39/64
3/8	25/32	21/64	13/32	3 1/4	3/4	2 1/16	55/64
1/2	31/32	25/64	17/32	3 7/16	3/4	2 1/16	63/64
5/8	1 1/4	31/64	21/32	3 15/16	1	2 5/16	1 9/16
3/4	1 15/32	5/8	25/32	4 7/16	1	2 5/16	1 1/2
1	1 27/32	53/64	1 1/32	4 13/16	1 1/4	2 5/16	1 43/64
1 1/4	2 7/32	1 3/32	1 7/32	5 3/8	1 1/4	2 5/16	1 31/32
1 1/2	2 21/32	1 11/32	1 17/32	5 29/32	1 1/4	2 7/16	2 1/8

T-Slot cutters are also available with B.&S. Tapers in the following sizes and dimensions.
Bolt size: 1/2, Length 5, Taper No. 7. Bolt size: 5/8, Length 5 1/4, Taper No. 7.
Bolt size: 3/4, Length 6 7/8, Taper No. 9. Bolt size: 1, 7 1/4, Taper No. 9.
All dimensions in inches. Cutters are noncenter cutting, right hand cut.

Table 43. Multiflute, Medium Helix, Plain End, Noncenter Cutting, Weldon and Combination Shank Roughing End Mills. *(Source, Weldon Tool Co.)*

For general purpose rough milling including profiling and open ended slots.
A finishing pass with a conventional tool is normally required.

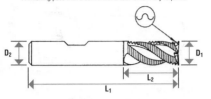

Combination shank not shown. See Table 57 for combination shank dimensions.

Cutter Dia. (D₁)	Shank Dia. (D₂)	Length of Cut (L₂)	Overall Length (L₁)	Number of Flutes*	Cutter Dia. (D₁)	Shank Dia. (D₂)	Length of Cut (L₂)	Overall Length (L₁)	Number of Flutes*
Weldon Shank					**Weldon Shank** *(continued)*				
1/4	3/8	5/8	2 7/16	3	1 1/2	3/4	4	6 1/4	6
5/16	3/8	3/4	2 1/2	3	1 1/2	1 1/4	2	4 1/2	6
3/8	3/8	3/4	2 1/2	4	1 1/2	1 1/4	3	5 1/2	6
1/2	1/2	1	3	4	1 1/2	1 1/4	4	6 1/2	6
1/2	1/2	1 1/4	3 1/4	4	1 1/2	1 1/4	6	8 1/2	6
1/2	1/2	2	4	4	1 3/4	1 1/4	2 1/2	5	6
5/8	5/8	1 5/8	3 3/4	4	1 3/4	1 1/4	4 1/2	7	6
5/8	5/8	2 1/2	4 5/8	4	2	1 1/4	2	4 1/2	8
3/4	3/4	1 5/8	3 7/8	4	2	1 1/4	4	6 1/2	8
3/4	3/4	3	5 1/4	4	**Combination Shank**				
7/8	3/4	1 7/8	4 1/8	5	2	2	3	6 3/4	8
7/8	3/4	3 1/2	5 3/4	5	2	2	4	7 3/4	8
1	3/4	3/4	3	5	2	2	5	8 3/4	8
1	3/4	1 7/8	4 1/8	5	2	2	6	9 3/4	8
1	3/4	4	6 1/4	5	2	2	7	10 3/4	8
1	1	2	4 1/2	5	2	2	8	11 3/4	8
1	1	3	5 1/2	5	2 1/2	2	4	7 3/4	8
1	1	4	6 1/2	5	2 1/2	2	6	9 3/4	8
1 1/8	1	2	4 1/2	6	2 1/2	2	8	11 3/4	8
1 1/4	3/4	2	4 1/2	6	2 1/2	2	10	13 3/4	8
1 1/4	3/4	4	6 1/4	6	2 1/2	2	12	15 3/4	8
1 1/4	1 1/4	2	4 1/2	6	3	2 1/2	4	7 3/4	10
1 1/4	1 1/4	3	5 1/2	6	3	2 1/2	6	9 3/4	10
1 1/4	1 1/4	4	6 1/2	6	3	2 1/2	8	11 3/4	10
1 1/2	3/4	3/4	3	6	3	2 1/2	10	13 3/4	10
1 1/2	3/4	2 1/4	4 1/2	6	3	2 1/2	12	15 3/4	10

* Number of flutes may vary by manufacturer.
All dimensions in inches. End mills are noncenter cutting, right-hand helix, right hand cut. Helix angle between 19° and 39°.

Table 44. Woodruff Keyseat Straight Shank Milling Cutters. *(Source, Fastcut Tool Corp.)*

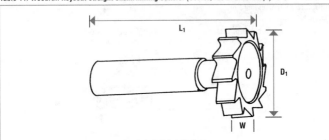

Shank size is ½ inch for all cutters.

Cutter No.*	Nominal Cutter Dia. (D₁)	Face Width (W)	Overall Length (L₁)	Number of Teeth†	Cutter No.*	Cutter Dia. (D₁)	Face Width (W)	Overall Length (L₁)	Number of Teeth†
202	1/4	1/16	2 1/16	8	708	1	7/32	2 7/32	12
202 1/2	5/16	1/16	2 1/16	8	1208	1	3/8	2 3/8	12
302 1/2	5/16	3/32	2 3/32	8	609	1 1/8	3/16	2 3/16	12
203	3/8	1/16	2 1/16	8	807	7/8	1/4	2 1/4	12
303	3/8	3/32	2 3/32	8	808	1	1/4	2 1/4	12
403	3/8	1/8	2 1/8	8	709	1 1/8	7/32	2 7/32	12
204	1/2	1/16	2 1/16	10	809	1 1/8	1/4	2 1/4	12
304	1/2	3/32	2 3/32	10	610	1 1/4	3/16	2 3/16	14
305	5/8	3/32	2 3/32	10	710	1 1/4	7/32	2 7/32	14
404	1/2	1/8	2 1/8	10	810	1 1/4	1/4	2 1/4	14
405	5/8	1/8	2 1/8	10	811	1 3/8	1/4	2 1/4	14
406	3/4	1/8	2 1/8	10	812	1 1/2	1/4	2 1/4	16
505	5/8	5/32	2 5/32	10	1008	1	5/16	2 5/16	12
605	5/8	3/16	2 3/16	10	1009	1 1/8	5/16	2 5/16	12
506	3/4	5/32	2 5/32	10	1010	1 1/4	5/16	2 5/16	14
806	3/4	1/4	2 1/4	10	1011	1 1/4	5/16	2 5/16	14
507	7/8	5/32	2 5/32	12	1012	1 1/2	5/16	2 5/16	16
606	3/4	3/16	2 3/16	10	1210	1 1/4	3/8	2 3/8	14
607	7/8	3/16	2 3/16	12	1211	1 3/8	3/8	2 3/8	14
707	7/8	7/32	2 7/32	12	1212	1 1/2	3/8	2 3/8	16
608	1	3/16	2 3/16	12					

* The cutter number indicates the nominal key dimensions or size cutter. The last two digits give the nominal diameter in 8ths of an inch, and the digits preceeding the last two give the nominal width in 32nds of an inch. For example, Cutter No. 204 indicates a size 2/32" × 4/8" (or 1/16" thick × 1/2" diameter). Cutter No. 1210 indicates a size 2/8" thick × 1 1/4" diameter). Keys and cutters correspond in size and number with the following exceptions:

For Key No. 121 - Use Cutter No. 807.
For Key No. 141 - Use Cutter No. 808.
For Key No. 131 - Use Cutter No. 1008.
For Key No. 161 - Use Cutter No. 1009.

† Number of teeth may vary by manufacturer.
All dimensions in inches. Cutters are straight flute, right hand cut.

Table 45. Woodruff Keyseat, Arbor Style Milling Cutters. *(Source, Fastcut Tool Corp.)*

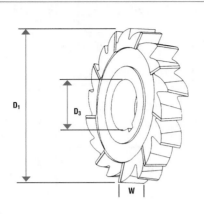

Cutter No.	For Use With Key No.	Nominal Cutter Dia. (D₁)	Face Width (W)	Hole Dia. (D3)	No. of Teeth *
617	126	2 1/8	3/16	3/4	16
817	127	2 1/8	1/4	3/4	16
1017	128	2 1/8	5/16	3/4	16
1217	129	2 1/8	3/8	3/4	16
822	RX	2 3/4	1/4	1	16
1022	SX	2 3/4	5/16	1	16
1222	TX	2 3/4	3/8	1	16
1422	UX	2 3/4	7/16	1	16
1622	VX	2 3/4	1/2	1	16
1228	30	3 1/2	3/8	1	20
1628	32	3 1/2	1/2	1	20
1828	33	3 1/2	9/16	1	18
2028	34	3 1/2	5/8	1	18
2428	36	3 1/2	3/4	1	18

* Number of teeth may vary by manufacturer.

All dimensions in inches. Arbor style Woodruff Keyseat cutters have staggered teeth and a slight relief on the side teeth. They are commonly made with diameters 1/32" larger than specified to allow for sharpening.

Table 46. Standard Keys and Keyways for Milling Cutters and Arbors. (*Source, ASA B5-3-1950.*)

Arbor and Keyseat

Cutter Bore and Keyway

Arbor and Key

Dia. of Arbor	Nom. Size Key	Arbor and Keyseat				Bore and Keyway					Arbor and Key			
		A Max.	A Min.	B Max.	B Min.	C Max.	C Min.	D* Min.	H Nom.	Cor. Rad.	E Max.	E Min.	F Max.	F Min.
1/2	3/32	.0947	.0937	.4531	.4481	.106	.099	.5578	3/64	.02	.0932	.0927	.5468	.5408
5/8	1/8	.126	.125	.5625	.5575	.137	.130	.6985	1/16	1/32	.1245	.1240	.6875	.6815
3/4	1/8	.126	.125	.6875	.6825	.137	.130	.8225	1/16	1/32	.1245	.1240	.8125	.8065
7/8	1/8	.126	.125	.8125	.8075	.137	.130	.9475	1/16	1/32	.1245	.1240	.9375	.9315
1	1/4	.251	.250	.8438	.8388	.262	.255	1.104	3/32	3/64	.2495	.2490	1.094	1.088
1 1/4	5/16	.3135	.3125	1.063	1.058	.325	.318	1.385	1/8	1/16	.3120	.3115	1.375	1.369
1 1/2	3/8	.376	.375	1.281	1.276	.410	.385	1.666	5/32	1/16	.3745	.3740	1.656	1.650
1 3/4	7/16	.4385	.4375	1.500	1.495	.473	.448	1.948	3/16	1/16	.4370	.4365	1.938	1.932
2	1/2	.501	.500	1.687	1.682	.535	.510	2.198	3/16	1/16	.4995	.4990	2.188	2.182
2 1/2	5/8	.626	.625	2.094	2.089	.660	.635	2.733	7/32	1/16	.6245	.6240	2.718	2.712
3	3/4	.751	.750	2.500	2.495	.785	.760	3.265	1/4	3/32	.7495	.7490	3.250	3.244
3 1/2	7/8	.876	.875	3.000	2.995	.910	.885	3.890	3/8	3/32	.8745	.8740	3.875	3.869
4	1	1.001	1.001	3.375	3.370	1.035	1.010	4.390	3/8	3/32	.9995	.9990	4.375	4.369
4 1/2	1 1/8	1.126	1.125	3.813	3.808	1.160	1.135	4.953	7/16	1/8	1.1245	1.1240	4.938	4.932
5	1 1/4	1.251	1.250	4.250	4.245	1.285	1.260	5.515	1/2	1/8	1.2495	1.2490	5.500	5.494

* D Max. = D Min. + 0.010″. All dimensions in inches.

Table 47. Single End, Regular Length, Two Flute Keyway End Mills with Weldon Shanks.
(Source, Fastcut Tool Corp.)

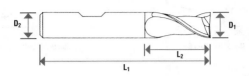

Used to mill keyways to close tolerances

Cutter Dia. (D_1)	Shank Dia. (D_2)	Length of Cut (L_2)	Overall Length (L_1)	Cutter Dia. (D_1)	Shank Dia. (D_2)	Length of Cut (L_2)	Overall Length (L_1)
.124	3/8	3/8	2 5/16	.624	5/8	1 5/16	3 7/16
.123	3/8	3/8	2 5/16	.623	5/8	1 5/16	3 7/16
.122	3/8	3/8	2 5/16	.622	5/8	1 5/16	3 7/16
.1865	3/8	7/16	2 5/16	.749	5/8	1 5/16	3 7/16
.1855	3/8	7/16	2 5/16	.748	5/8	1 5/16	3 7/16
.1845	3/8	7/16	2 5/16	.747	5/8	1 5/16	3 7/16
.249	3/8	1/2	2 5/16	.874	7/8	1 1/2	3 3/4
.248	3/8	1/2	2 5/16	.873	7/8	1 1/2	3 3/4
.247	3/8	1/2	2 5/16	.872	7/8	1 1/2	3 3/4
.3115	3/8	9/16	2 5/16	.999	1	1 5/8	4 1/8
.3105	3/8	9/16	2 5/16	.998	1	1 5/8	4 1/8
.3095	3/8	9/16	2 5/16	.997	1	1 5/8	4 1/8
.374	3/8	9/16	2 5/16	1.249	1 1/4	1 5/8	4 1/8
.373	3/8	9/16	2 5/16	1.248	1 1/4	1 5/8	4 1/8
.372	3/8	9/16	2 5/16	1.247	1 1/4	1 5/8	4 1/8
.499	1/2	1	3	1.499	1 1/4	1 5/8	4 1/8
.498	1/2	1	3	1.498	1 1/4	1 5/8	4 1/8
.497	1/2	1	3	1.497	1 1/4	1 5/8	4 1/8

All dimensions in inches.

Table 48. Metric Plain and Ball Nosed End Mills with Weldon Shanks. *(Source, Weldon Tool Co.)*

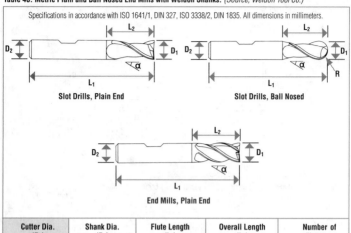

Specifications in accordance with ISO 1641/1, DIN 327, ISO 3338/2, DIN 1835. All dimensions in millimeters.

Slot Drills, Plain End

Slot Drills, Ball Nosed

End Mills, Plain End

Cutter Dia. (D_1)	Shank Dia. (D_2)	Flute Length (L_2)	Overall Length (L_1)	Number of Flutes
\multicolumn{5}{c}{Slot Drills, Plain End}
3	6	5	49	2 or 3
4	6	7	51	2 or 3
5	6	8	52	2 or 3
6	6	8	52	2 or 3
8	10	11	61	2 or 3
10	10	13	63	2 or 3
12	12	16	73	2 or 3
14	12	16	73	2 or 3
16	16	19	79	2 or 3
18	16	19	79	2 or 3
20	20	22	88	2 or 3
25	25	26	102	2 or 3
32	32	32	112	2
\multicolumn{5}{c}{Slot Drills, Ball Nosed}
3	6	5	49	2
4	6	7	51	2
5	6	8	52	2
6	6	8	52	2
8	10	11	61	2
10	10	13	63	2
12	12	16	73	2
14	12	16	73	2
16	16	19	79	2
18	16	19	79	2
20	20	22	88	2
25	25	26	102	2

(Continued)

Table 48. *(Continued)* **Metric Plain and Ball Nosed End Mills with Weldon Shanks.**

Cutter Dia. (D₁)	Shank Dia. (D₂)	Flute Length (L₂)	Overall Length (L₁)	Number of Flutes
End Mills, Plain End					
6	6	13	57	4
6	6	24	68	4
8	10	19	69	4
8	10	38	88	4
10	10	22	72	4
10	10	45	95	4
12	12	26	83	4
12	12	53	110	4
14	12	26	83	4
16	16	32	92	4
16	16	63	123	4
18	16	32	92	4
20	20	38	104	4
20	20	75	141	4
25	25	45	121	6
32	32	53	133	6
40	40	63	155	6

All dimensions in millimeters.

Table 49. Shell Type Mill. *(Source, Weldon Tool Co.)*

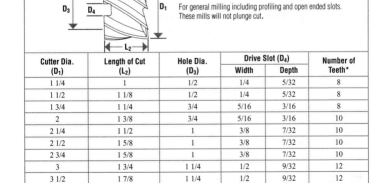

For general milling including profiling and open ended slots. These mills will not plunge cut.

Cutter Dia. (D₁)	Length of Cut (L₂)	Hole Dia. (D₃)	Drive Slot (D₄)		Number of Teeth*
			Width	Depth	
1 1/4	1	1/2	1/4	5/32	8
1 1/2	1 1/8	1/2	1/4	5/32	8
1 3/4	1 1/4	3/4	5/16	3/16	8
2	1 3/8	3/4	5/16	3/16	10
2 1/4	1 1/2	1	3/8	7/32	10
2 1/2	1 5/8	1	3/8	7/32	10
2 3/4	1 5/8	1	3/8	7/32	10
3	1 3/4	1 1/4	1/2	9/32	12
3 1/2	1 7/8	1 1/4	1/2	9/32	12
4	2 1/4	1 1/2	5/8	3/8	14
4 1/2	2 1/4	1 1/2	5/8	3/8	14
5	2 1/4	1 1/2	5/8	3/8	16
6	2 1/4	2	3/4	7/16	16

All dimensions in inches. Cutters have square corners, are noncenter cutting, 20° right hand helix, right-hand cut.

*Number of teeth may vary by manufacturer.

Table 50. Concave Radius, Convex Radius, and Corner Rounding Cutters. *(Source, Fastcut Tool Corp.)*

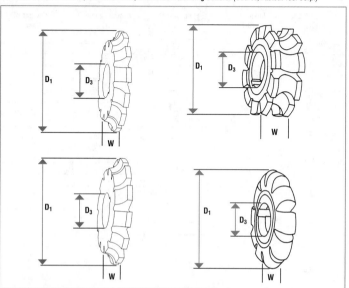

Concave and Convex Cutters					
Dia. (C) or Radius (R)	**Cutter Dia (D₁)**	**Cutter Width (W)**		**Hole Dia. (D₃)**	**Number of Teeth ***
		Concave	**Convex**		
1/8	2 1/4	1/4	1/8	1	16
3/16	2 1/4	3/8	3/16	1	16
1/4	2 1/2	7/16	1/4	1	14
5/16	2 3/4	9/16	5/16	1	14
3/8	2 3/4	5/8	3/8	1	12
7/16	3	3/4	7/16	1	12
1/2	3	13/16	1/2	1	12
5/8	3 1/2	1	5/8	1 1/4	12
3/4	3 3/4	1 13/16	3/4	1 1/4	12
7/8	4	1 3/8	7/8	1 1/4	12
1	4 1/4	1 9/16	1	1 1/4	10
Corner Rounding Cutters					
1/8	2 1/2	1/4		1	14
1/4	3	13/32		1	12
3/8	3 3/4	9/16		1 1/4	12
1/2	4 1/4	3/4		1 1/4	10
5/8	4 1/4	15/16		1 1/4	10

* Number of teeth may vary by manufacturer.

All dimensions in inches. Concave and Convex have straight flutes. Corner rounding cutters have straight flutes and are available in right hand and left hand cut.

Tapered end mills

Tapered end mills are extremely versatile cutting tools that are used extensively in die making and in the aircraft industry to produce "draft" or tapered walls on a workpiece. Weldon Tool Company provided the following guidelines for using tapered end mills which, despite their cone-shaped geometry, have much in common with straight end mills. One notable difference, however, is the "balanced compromise" which is a result of the number of flutes that can be designed for a particular tapered end mill. Two flutes might be proper for the small end, while four flutes would be more appropriate for the large end of the taper. A balanced compromise will settle on three flutes, and that is the way the tool is made. The same balanced compromise approach will influence how teeth are designed for a tapered end mill. For plunge cutting with the end teeth, two teeth would be ideal for chip flow. Four or more teeth, on the other hand, would give a much better feed-per-tooth condition, but can cause poor chip flow. The balanced compromise again settles on three teeth.

The primary relief on the small diameter of a tapered end mill is the same as on an equivalent-size straight diameter end mill. As the cutting diameter increased on a tapered tool, however, the relief angle also changes progressively to suit the change in diameter (see *Figure 18, a, b, and c*). The change in relief angle is necessary to obtain efficient cutting action all along the tool's cutting length. The change becomes even greater on "steeper tapered" tools—that is, tools that have a small point diameter plus a long flute length, where the cutting diameters are at a maximum variation.

Constant helix tapered end mills

Tapered end mills fall into two main design categories: the *constant helix with varying lead* type (*Figure 18b*), and the traditional *constant lead with varying helix* type (*Figure 18c*). The *constant helix* has enhanced cutting performance because tapered end mills made with a constant helix have more uniform radial rake or hook.

Constant helix tools, with their uniform hook angle, are also easier to sharpen because the relationship of the hooked tooth face and the support finger remains the same throughout the tool's cutting length. This constant relationship assures full control over the desired amount of primary relief angle. Constant hook and proper relief are essential if the tool is expected to provide the best cutting efficiency. Making tapered end mills with a constant helix, however, requires special equipment. This is an important difference to consider when selecting a tapered end mill.

As noted above, the primary relief

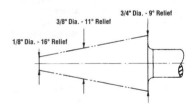

Figure 18a. Tapered end mill.

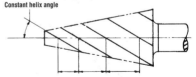

Figure 18b. Varying lead.

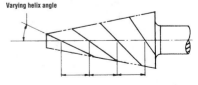

Figure 18c. Constant lead.

and secondary clearance angles are not consistent along the cutting length of tapered end mills. The primary relief angle, for example, should be greater at the small end of the tool for the best cutting performance, and the angle should decrease as the diameter increases. The amount of primary relief angle should be approximately equal, at any given point along the taper, to the relief angle of an equivalent diameter regular straight end mill.

Determining speed and feed for tapered end mills

When calculating the proper RPM speed for a tapered end mill, base the calculation on the largest diameter of the taper that will be contacting the workpiece. The table feed, on the other hand, is determined by the amount of feed-per-tooth that the smallest diameter of the taper can handle. If the tool's smallest diameter will not actually be engaged by the workpiece, use the smallest diameter of the tool that *will* be engaged. Since the point diameters of some tools are quite small, feed rates will be equally low—sometimes on the order of 0.0005" fpt or less—to prevent chip clogging and tool breakage. Extra care is required when feeding small, point-diameter tools that have been altered to a ball radius end because these tools are subject to higher cutting stresses and easier chip clogging.

Sharpening tapered end mills

Sharpening or resharpening a tapered end mill is rather complicated. To produce the correct taper on the outside of the mill, its centerline must first be swung away in a horizontal direction from the grinding wheel plane (*Figure 19,* angle "A"). At the same time, the centerline must also be tilted in a vertical plane to accomplish the required progressive change in the cutting edge angle (*Figure 19,* angle "B", and **Table 52**). While these manipulations are being accomplished, it may also be necessary to maintain the small end diameter to meet blueprint specification, or to accommodate a required ball radius. There are many other interrelated requirements in the sharpening setup that must be controlled as well.

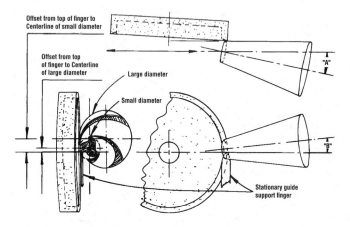

Figure 19. Correct angles for grinding tapered end mills.

One of the most critical requirements in the sharpening setup is the placement of the guide support finger. This finger is positioned and fixed securely adjacent to the corner of the grinding wheel. The vertical relationship of the finger point to the taper end mill centerline is called "offset," and establishes the amount of primary relief being ground on the cutting edge (refer again to *Figure 19*, and to **Tables 51, 52,** and **53**). The horizontal positioning of the finger adjacent to the grinding wheel corner permits placing a tooth face onto the finger before it contacts or after it leaves the grinding wheel, depending on whether the grinding is done from tip to shank or from shank to tip. The finger contact with a curved tooth face is made so that the cutting edge being ground is always slightly outside the finger in a radial direction (see *Figure 20*).

These conditions may require some compensation from the true theoretical "offset" position when the vertical relationship of the finger point to the centerline of the tool is being set (refer again to the tables). *Figure 20* shows the vertical, horizontal, and radial position of the support finger in the normal grinding setup.

Table 53 titled "Practical offset dimension…," is simply a modification of **Table 52**, labeled "Theoretical offset dimension…," to compensate for the effects of the support finger relationships, mentioned above, which cannot be accounted for in strictly mathematical terms. These modified values are for average constant helix conditions, but common variations can change even these values. Another critical condition is the need to know the exact size of the small end diameter that will properly accommodate a full ball radius. **Table 54** provides this information.

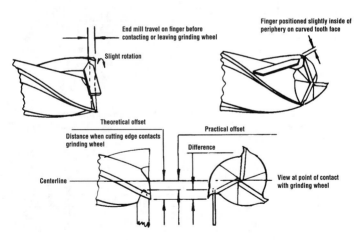

Figure 20. Position of support finger in normal grinding setup.

Table 51. Tilt Angle Settings for Sharpening Tapered End Mills.

Set Small Diameter Centerline Offset to Top of Finger—Tilt Up Rear at Angle Listed		Height of Block to Produce Angle of Tilt
Taper of end mill, one side	"B" Angle of tilt	5" Sine bar
1°	0° 10'	.015
2°	0° 20'	.029
3°	0° 30'	.044
4°	0° 37'	.054
5°	0° 45'	.065
6°	0° 52'	.076
7°	1° 00'	.087
8°	1° 10'	.102
9°	1° 20'	.116
10°	1° 30'	.131

All dimensions in inches.

Table 52. Theoretical Offset Dimension to Produce the Required Primary Angle on a Given Diameter.

Dia.	Prim.	Off.	Dia.	Prim.	Off.	Dia.	Prim.	Off.
1/16"	22°	.012	5/16"	12°	.033	5/8"	10°	.055
3/32	18°	.015	11/32	11°	.034	11/16	9°	.057
1/8	16°	.018	3/8	11°	.036	3/4	9°	.059
5/32	15°	.022	13/32	11°	.039	13/16	9°	.064
3/16	14°	.023	7/16	11°	.042	7/8	9°	.069
7/32	13°	.025	15/32	10°	.043	15/16	8°	.068
1/4	12°	.027	1/2	10°	.044	1	8°	.070
9/32	12°	.030	9/16	10°	.050	1-1/8	8°	.078

All dimensions in inches.

Table 53. Practical Offset Dimensions to Produce the Required Primary Angle on a Given Diameter.

Dia.	Prim.	Off.	Dia.	Prim.	Off.	Dia.	Prim.	Off.
1/16"	22°	.0025	5/16"	12°	.018	5/8"	10°	.042
3/32	18°	.0035	11/32	11°	.020	11/16	9°	.046
1/8	16°	.005	3/8	11°	.023	3/4	9°	.050
5/32	15°	.007	13/32	11°	.026	13/16	9°	.053
3/16	14°	.009	7/16	11°	.028	7/8	9°	.056
7/32	13°	.010	15/32	10°	.030	15/16	8°	.059
1/4	12°	.012	1/2	10°	.032	1	8°	.062
9/32	12°	.015	9/16	10°	.037	1-1/8	8°	.068

All dimensions in inches.

Table 54. Diameter Needed at Small End to Accommodate a Full Ball Radius.

Desired full ball radius	Degree of taper on one side						
	1/2°	1°	1-1/2°	2°	2-1/2°	3°	4°
1/32-.0312	.061	.061	.060	.060	.059	.059	.058
1/16-.0625	.123	.122	.121	.120	.119	.118	.116
1/8-.125	.247	.245	.243	.241	.239	.237	.233
3/16-.1875	.371	.368	.365	.362	.358	.355	.349
1/4-.250	.495	.491	.487	.482	.478	.474	.466
5/16-.312	.618	.613	.607	.602	.597	.592	.581
3/8-.375	.743	.736	.730	.724	.717	.711	.699
7/16-.437	.866	.855	.851	.843	.836	.829	.814
1/2-.500	.991	.982	.974	.965	.957	.948	.932
5/8-.625	1.0239	1.228	1.217	1.207	1.196	1.186	1.165
Constant	1.9825	1.9653	1.9482	1.9313	1.9144	1.8979	1.8649
Factor	57.295	28.645	19.094	14.318	11.452	9.540	7.150

Desired full ball radius	Degree of taper on one side						
	5°	6°	7°	8°	9°	10°	15°
1/32-.0312	.057	.056	.055	.054	.053	.052	.047
1/16-.0625	.114	.112	.110	.108	.106	.104	.095
1/8-.125	.229	.225	.221	.217	.213	.209	.191
3/16-.1875	.343	.337	.331	.325	.320	.314	.287
1/4-.250	.458	.450	.442	.434	.427	.419	.383
5/16-.312	.571	.561	.552	.542	.532	.523	.478
3/8-.375	.687	.675	.663	.651	.640	.629	.575
7/16-.437	.800	.786	.773	.759	.746	.733	.670
1/2-.500	.916	.900	.884	.869	.854	.839	.767
5/8-.625	1.145	1.125	1.105	1.086	1.067	1.048	.959
Constant	1.8326	1.8008	1.7694	1.7385	1.7082	1.6781	1.5347
Factor	5.715	4.757	4.072	3.557	3.156	2.835	1.866

To find small diameter for ball radius not listed, multiply full ball radius desired (decimal) by constant.

If small diameter is less than shown on chart, subtract it from chart diameter and multiply by factor to find length to cut off.

All dimensions in inches.

Table 55. Single End, Tapered, Center Cutting, Three Flute End Mills with Weldon Shanks.
(Source, Fastcut Tool Corp.)

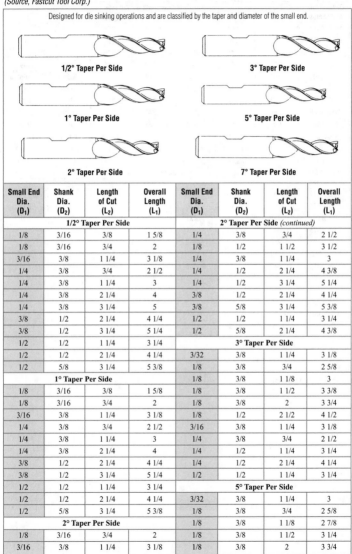

Designed for die sinking operations and are classified by the taper and diameter of the small end.

1/2° Taper Per Side

3° Taper Per Side

1° Taper Per Side

5° Taper Per Side

2° Taper Per Side

7° Taper Per Side

Small End Dia. (D_1)	Shank Dia. (D_2)	Length of Cut (L_2)	Overall Length (L_1)	Small End Dia. (D_1)	Shank Dia. (D_2)	Length of Cut (L_2)	Overall Length (L_1)
1/2° Taper Per Side				2° Taper Per Side *(continued)*			
1/8	3/16	3/8	1 5/8	1/4	3/8	3/4	2 1/2
1/8	3/16	3/4	2	1/8	1/2	1 1/2	3 1/2
3/16	3/8	1 1/4	3 1/8	1/4	3/8	1 1/4	3
1/4	3/8	3/4	2 1/2	1/4	1/2	2 1/4	4 3/8
1/4	3/8	1 1/4	3	1/4	1/2	3 1/4	5 1/4
1/4	3/8	2 1/4	4	3/8	1/2	2 1/4	4 1/4
1/4	3/8	3 1/4	5	3/8	5/8	3 1/4	5 3/8
3/8	1/2	2 1/4	4 1/4	1/2	1/2	1 1/4	3 1/4
3/8	1/2	3 1/4	5 1/4	1/2	5/8	2 1/4	4 3/8
1/2	1/2	1 1/4	3 1/4	3° Taper Per Side			
1/2	1/2	2 1/4	4 1/4	3/32	3/8	1 1/4	3 1/8
1/2	5/8	3 1/4	5 3/8	1/8	3/8	3/4	2 5/8
1° Taper Per Side				1/8	3/8	1 1/8	3
1/8	3/16	3/8	1 5/8	1/8	3/8	1 1/2	3 3/8
1/8	3/16	3/4	2	1/8	3/8	2	3 3/4
3/16	3/8	1 1/4	3 1/8	1/8	1/2	2 1/2	4 1/2
1/4	3/8	3/4	2 1/2	3/16	3/8	1 1/4	3 1/8
1/4	3/8	1 1/4	3	1/4	3/8	3/4	2 1/2
1/4	3/8	2 1/4	4	1/4	1/2	1 1/4	3 1/4
3/8	1/2	2 1/4	4 1/4	1/4	1/2	2 1/4	4 1/4
3/8	1/2	3 1/4	5 1/4	1/2	1/2	1 1/4	3 1/4
1/2	1/2	1 1/4	3 1/4	5° Taper Per Side			
1/2	1/2	2 1/4	4 1/4	3/32	3/8	1 1/4	3
1/2	5/8	3 1/4	5 3/8	1/8	3/8	3/4	2 5/8
2° Taper Per Side				1/8	3/8	1 1/8	2 7/8
1/8	3/16	3/4	2	1/8	3/8	1 1/2	3 1/4
3/16	3/8	1 1/4	3 1/8	1/8	3/8	2	3 3/4

(Continued)

Table 55. *(Continued)* **Single End, Tapered, Center Cutting, Three Flute End Mills with Weldon Shanks.**

Small End Dia. (D₁)	Shank Dia. (D₂)	Length of Cut (L₂)	Overall Length (L₁)	Small End Dia. (D₁)	Shank Dia. (D₂)	Length of Cut (L₂)	Overall Length (L₁)
5° Taper Per Side *(continued)*				7° Taper Per Side			
1/8	1/2	2 1/2	4 1/2	1/2	1/2	1 1/4	3 1/4
3/16	1/2	1 1/4	3 3/8	3/16	1/2	1 1/4	3 1/4
1/4	3/8	3/4	2 1/2	1/4	1/2	1 1/4	3 1/4
1/4	1/2	1 1/4	3 1/4	1/4	3/4	2 1/4	4 1/2
1/4	3/4	3 1/4	5 1/2	3/8	3/4	2 1/4	4 1/2
3/8	3/4	2 1/4	4 1/2	1/2	5/8	1 1/4	3 3/8
3/8	3/4	3 1/4	5 1/2				
1/2	1/2	1 1/4	3 1/4				

All dimensions in inches. End mills are center cutting, right hand spiral, right hand cut.

Table 56. Standard Weldon Shanks. *(Source, Weldon Tool Co.)*

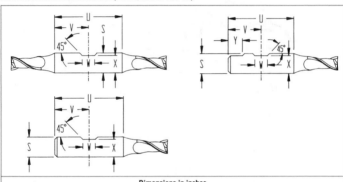

		Dimensions in inches					
Shank Dia. (S)	Length of Shank (U)	V	W		X	Y	
			Min.	Max.			
3/8	1 9/16	25/32	0.280	0.282	0.325	-	
1/2	1 25/32	57/64	0.330	0.332	0.440	-	
5/8	1 29/32	61/64	0.400	0.402	0.560	-	
3/4	2 1/32	1 1/64	0.455	0.457	0.675	-	
7/8	2 1/32	1 1/64	0.455	0.457	0.810	1/2	
1	2 9/32	1 9/64	0.515	0.517	0.925	1/2	
1 1/4	2 9/32	1 9/64	0.515	0.517	1.156	1/2	
1 1/2	2 11/16	1 3/16	0.515	0.517	1.406	9/16	
2	3 1/4	1 27/32	0.700	0.702	1.900	27/32	
2 1/2	3 1/2	1 15/16	0.700	0.702	2.400	27/32	

Tolerances in inches			
Element	Range	Direction	Tolerance
Dia. of Shank (S)	All Sizes	Minus	0.0001 to 0.0005
Length of Shank (U)	All Sizes	Plus or Minus	1/32
Dimension V	All Sizes	Plus or Minus	1/64
Dimension X	All Sizes	Minus	1/64
Dimension Y	7/8 to 2 1/2 incl.	Plus or Minus	1/32

Table 57. Combination Shanks for End Mills. *(Source, Weldon Tool Co.)*

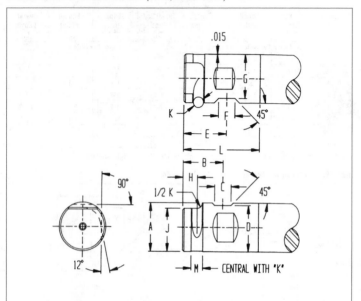

Dimensions in inches

Shank Dia. (D)	Length of Shank (L)	B	C	D	E	F
2	3 1/4	1 23/32	0.700	1.900	1 3/4	0.700
2 1/2	3 1/2	1 15/16	0.700	2.400	2	0.700
Shank Dia. (D)	**Length of Shank (L)**	**G**	**H**	**J**	**K**	**M**
2	3 1/4	1.809	5/8	1.772	0.440	1/2
2 1/2	3 1/2	2.312	3/4	2.245	0.503	9/16

Tolerances in inches

Element	Direction	Tolerance
Dia. of Shank (**S**)	Minus	0.0001 to 0.0005
Length of Shank (**L**)	Plus or Minus	1/32
Dimension **B**	Plus or Minus	1/64
Dimension **C**	Plus	0.002
Dimension **D**	Minus	1/64
Dimension **E**	Plus or Minus	1/64
Dimension **F**	Plus or Minus	0.005
Dimension **G**	Minus	1/64
Dimension **H**	Plus	1/64
Dimension **J**	Plus or Minus	0.002
Dimension **K**	Plus	0.003

Slotting mill selection

Slotting cutters are arbor mills that machine three surfaces simultaneously in a single machine pass. The slot produced by these tools has two sidewalls perpendicular to the axis of rotation and a bottom surface parallel to the axis of rotation. *Figure 21* shows the important dimensions of an indexable slotting cutter. A slotting mill has cutting edges that mill on two faces and its periphery. Arbor mills are generally more rigid and productive than end mills on slotting operations. These mills are selected for specific w.o.c. and d.o.c. requirements. Slotting cutters, half-side, and double-side cutters have a hole, or bore, through the center with one or more keyways for mounting on an arbor. See the section on

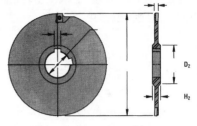

D_1 = effective cutting diameter
D_2 = hub diameter
H_1 = cutting width
H_2 = hub width
B = bore diameter
K = keyway width

Figure 21. Slotting cutters.

"Calculating chip load, advance per revolution, and advance per minute" on page 897 for calculations used in slot milling. The rigidity of the machine and the arbor supports have a direct bearing on cutter performance, surface finish, and tool life. The larger the arbor, the greater the rigidity as shown in the chart below which shows diameter-to-rigidity for four common arbor sizes.

1" ∅	1 1/4" ∅	1 1/2" ∅	2" ∅
basic rigidity	2.5 × basic rigidity	5 × basic rigidity	16 % basic rigidity

For best results, use the cutter with the largest bore that can mill to specifications. Place the cutter on the arbor as close to the spindle as the workpiece will permit, and position the overarm supports near the cutter. Proper arbor length is determined by the number of cutters mounted on it and the number of supports used. Arbor overhang beyond the outer support should be avoided as it can cause vibration (chatter) and poor performance, as well as present a safety hazard.

Metal slitting saws

Metal slitting saws are used for a variety of cutoff operations and for thin slotting applications. To assure maximum performance, sharp cutting edges must be maintained. Although available in many styles, each type of saw has been designed for specific operations.

Concave (plain) slitting saws have no side teeth and are generally used for shallow slotting or cutoff operations in all types of materials. Cutting edges are on the periphery, and the sides are ground with a slight dish (concavity) toward the arbor hole to avoid dragging in the cut.

Straight tooth saws with side teeth have both peripheral and side teeth. They are recommended for cutting thin-walled parts and for applications requiring a fine finish on the workpiece. The side tooth design improves chip flow and size control.

Staggered tooth saws with side teeth have peripheral teeth of alternate right- and left-hand helix and alternate side teeth. They are efficient for deep slotting applications.

Table 58. General Speed Range for Saws Under 0.100 inches (2.54 mm) in Width.

material being cut	sfm	smm
steel	50-130	15-40
cast iron	30-110	10-34
brass	200-800	61-244
copper	100-350	30-106
bronze	150-300	46-91
aluminum	500-700	152-213

Table 59. Concave (Plain) Metal Slitting Saws. *(Source, Kennametal, Inc.)*

Cutter Dia.	Face Width	Hole Dia.	# Teeth *
2 1/2	1/32, 3/64, 1/16, 3/32, 1/8	7/8	36
3	1/32, 3/64, 1/16, 3/32, 1/8, 5/32	1	36
4	1/32, 3/64, 1/16, 3/32, 1/8, 5/32, 3/16	1	40
4	1/32, 3/64, 1/16, 3/32, 1/8, 5/32, 3/16	1 1/4	40
5	1/32, 1/16, 3/32, 1/8, 5/32, 3/16	1	44
5	1/32, 1/16, 3/32, 1/8, 5/32, 3/16	1 1/4	44
6	1/32, 1/16, 3/32, 1/8, 5/32, 3/16	1	48
6	1/32, 1/16, 3/32, 1/8, 5/32, 3/16	1 1/4	48
8	1/8	1	58
8	1/8	1 1/4	58

* Number of teeth may vary by manufacturer.
All dimensions in inches.

Table 60. Straight Tooth Metal Slitting Saws with Side Teeth. *(Source, Kennametal, Inc.)*

Cutter Dia.	Face Width	Hole Dia.	# Teeth *
2 1/2	1/16, 3/32, 1/8	7/8	28
3	1/16, 5/64, 3/32, 7/64, 1/8, 9/64, 5/32, 11/64, 3/16, 13/64, 7/32, 15/64, 1/4	1	32
4	1/16, 5/64, 3/32, 7/64, 1/8, 9/64, 5/32, 11/64, 3/16, 13/64, 7/32, 15/64, 1/4	1	36
4	1/16, 5/64, 3/32, 7/64, 1/8, 9/64, 5/32, 11/64, 3/16, 13/64, 7/32, 15/64, 1/4	1 1/4	36
5	1/16, 3/32, 1/8, 5/32, 3/16, 7/32, 1/4	1	40
5	1/16, 3/32, 1/8, 5/32, 3/16, 7/32, 1/4	1 1/4	40
6	1/16, 5/64, 3/32, 7/64, 1/8, 9/64, 5/32, 11/64, 3/16, 13/64, 7/32, 15/64, 1/4, 17/64, 9/32, 19/64, 5/16, 21/64, 11/32, 23/64, 3/8	1	42
6	1/16, 5/64, 3/32, 7/64, 1/8, 9/64, 5/32, 11/64, 3/16, 13/64, 7/32, 15/64, 1/4, 17/64, 9/32, 19/64, 5/16, 21/64, 11/32, 23/64, 3/8	1 1/4	42
7	1/8, 3/16	1	48
7	1/8, 3/16	1 1/4	48
8	1/8	1	48
8	1/8, 5/32, 3/16, 7/32, 1/4, 3/8	1 1/4	48
8	1/8, 5/32, 3/16, 7/32, 1/4, 3/8	1 1/2	48
10	3/16, 7/32, 1/4, 3/8	1 1/4	56
10	3/16, 7/32, 1/4, 3/8	1 1/2	56
12	1/4, 5/16, 3/8	1 1/2	64

* Number of teeth may vary by manufacturer.
All dimensions in inches.

Table 61. Staggered Tooth Metal Slitting Saws with Side Teeth. *(Source, Kennametal, Inc.)*

Cutter Dia.	Face Width	Hole Dia.	# Teeth *
3	3/32, 1/8, 5/32, 3/16, 7/32, 1/4	1	28
4	3/32, 1/8, 5/32, 3/16, 7/32, 1/4	1	32
4	3/32, 1/8, 5/32, 3/16, 7/32, 1/4	1 1/4	32
5	3/32, 1/8, 5/32, 3/16, 7/32, 1/4	1	36
5	3/32, 1/8, 5/32, 3/16, 7/32, 1/4	1 1/4	36
6	1/8, 5/32, 3/16, 7/32, 1/4, 3/8	1	40
6	1/8, 5/32, 3/16, 7/32, 1/4, 3/8	1 1/4	40
7	1/8, 3/16	1	44
7	1/8, 3/16	1 1/4	44
8	1/8, 5/32, 3/16, 7/32, 1/4, 3/8	1 1/4	48
8	1/8, 5/32, 3/16, 7/32, 1/4, 3/8	1 1/2	48
10	3/16, 7/32, 1/4, 3/8	1 1/4	56
10	3/16, 7/32, 1/4, 3/8	1 1/2	56
12	1/4, 5/16, 3/8	1 1/2	64

* Number of teeth may vary by manufacturer.
All dimensions in inches.

Speeds and feeds for sawing are influenced by the same variables that affect milling cutters. Adjusting speeds and feeds for slitting saws must be done with great care due to the thin cross section design of the saws. Recommended chip loads range from 0.0002 to 0.002 in./tooth (0.005 to 0.05 mm/tooth). The speeds on **Table 58** are for high speed steel saws.

Screw slotting and jeweler's saws

Screw slotting saws are used for slotting screw heads and for making shallow slotting cuts. They have ground sides that slightly dish toward the arbor hole to avoid dragging in the cut. Cutoff operations on precious metals are performed with jeweler's saws. They are also ideal for very thin slotting cuts on small screw heads. When possible, the use of larger three-inch or four-inch saws with larger arbor holes is recommended to reduce deflection.

Table 62. Screw Slotting Saws. *(Source, Kennametal, Inc.)*

Cutter Dia.	Face Width	Hole Dia.	# Teeth *
1 3/4	.064, .057, .051, .045, .040, .036, .032, .028, 025, .023, .020, .018, .014, .013, .010	5/8	90
2 1/4	.072, .064, .057, .051, .045, .040, .036, .032, .028, 025, .023, .020, .018, .014, .013, .010	5/8	60
2 3/4	.144, .128, .144, .102, .091, .081, .072, .064, .057, 051, .045, 040, .036, .032, .028, .025, .023, .020, .018, .014, .013, .010	1	72

* Number of teeth may vary by manufacturer.
All dimensions in inches.

Table 63. Jeweler's Saws. *(Source, Kennametal, Inc.)*

Cutter Dia.	Face Width	Hole Dia.	# Teeth *
1	.032, .028, .025, .023, .020, .018, .016, .014	3/8	80
1	.012, .010	3/8	100
1 1/4	.032, .028, .025, .023, .020, .018, .016, .014, .012, .010	3/8	100
1 1/4	.012, .010	3/8	120
1 1/2	.028, .025, .023, .020, .018, .016, .014	1/2	110
1 1/2	.012, .010	1/2	140
1 3/4	.032, .028, .025, .023, .020, .018, .016, .014	1/2	130
1 3/4	.012, .010	1/2	160
2	.057, .051, .045, .040, .035, .032	1/2	110
2	.028, .025, .023, .020, .018, .016, .014, .012, .010	1/2	150
2 1/2	.057, .051, .045, .040, .035, .032	1/2	140
2 1/2	.028, .025, .023, .020, .018, .016, .014, .012	1/2	190
2 1/2	.010	1/2	240
3	.057, .051, .045, .040, .035, .032	1	170
3	.057, .051, .045, .040, .035, .032	1 1/2	170
3	.028, .025, .023, .020, .018, .016, .014	1	230
3	.028, .025, .023, .020, .018, .016, .014	1 1/2	230
3	.012, .010	1	280
3	.012, .010	1 1/2	280
4	.057, .051, .045, .040, .035, .032	1	220
4	.057, .051, .045, .040, .035, .032	1 1/2	220
4	.028, .025, .023, .020, .018, .016, .014	1	300
4	.028, .025, .023, .020, .018, .016, .014	1 1/2	300
4	.012, .010	1	380
4	.012, .010	1 1/2	380

Recommended Speeds and Feeds for Face Milling

The following charts provide recommended speeds and feeds for face milling operations. This data is the result of testing and experimentation carried out by Kennametal, Inc. Consult the Inserts section of this book (page 858) for information on how to interpret the recommended ISO Grade and Type cutting materials for proper insert selection.

FREE MACHINING AND LOW CARBON STEELS (Hardness of 110-225 BHN): AISI 1008, 1010, 1018, 1020, 1026, 1108, 1117, 1141, 1151, 12L13, 12L14.

Material characteristics	Possible result at the cutting edge	Steps to offset condition
Low carbon content, very soft material.	At low speeds, built-up edge forms on insert rake face making it difficult to obtain a good finish. Material burrs easily.	1. Increase speed and/or chip load. 2. Apply a lower force cutting geometry such as double positive or high shear. 3. Apply a low pressure mist coolant for lubricity.

ISO Grade / Type	Starting Speed		Starting Feed		Speed Range		Feed Range		Wet=W
	sfm	smm	in./t	mm/t	sfm	smm	in./t	mm/t	Dry=D
P15-P35 / UCC	430	131	.006	.15	250-600	76-183	.004-.011	.102-.28	D
P20-P35 / PVD	500	152	.006	.15	250-800	75-244	.003-.012	.075-.30	D
P30-P45 / PVD	300	91	.008	.20	200-500	61-152	.004-.013	.102-.33	D
P15-P35 / PVD	500	152	.007	.18	250-700	76-213	.003-.010	.075-.25	W/D [1]
P10-P20 / PVD	750	229	.006	.15	600-1150	76-351	.004-.009	.102-.23	D
P25-P45 / CVD	550	168	.008	.20	300-700	91-213	.005-.014	.13-.36	D
P10-P25 / CERMET	700	213	.004	.102	500-1200	152-366	.003-.011	.075-.28	D

[1] Dry is preferred to minimize thermal shock.

Notes (Free Machining and Low Carbon Steels):
1. Suggested starting conditions are based on a .100" axial depth of cut. Cutting speeds should be adjusted for greater or lesser depth.
2. Using a cutter with the recommended 3:2 cutter diameter to workpiece width ratio will minimize the burr produced on the exit side of the part (where the tooth leaves the cut).
3. The softer the material, the higher the cutting speed can be, and for more rigid parts, the chip load per tooth can go higher.
4. The more interrupted the surface being milled, the greater the requirement for a tool grade with good mechanical shock resistance.
5. The lower the carbon content, the higher the speed.

MEDIUM- AND HIGH-CARBON STEELS (Hardness of 180-330 BHN): AISI 1040, 1045, 1085, 1090, 1055, 1080, 1095, 1541, 1548, 1551.

Material characteristics	Possible result at the cutting edge	Steps to offset condition
Significant carbon content.	Increased levels of pearlite and cementite (abrasive) can cause rapid tool wear.	1. Reduce cutting speed and increase chip load to maintain productivity. 2. Apply most wear-resistant grade possible. 3. Annealing will improve tool life.

ISO Grade / Type	Starting Speed		Starting Feed		Speed Range		Feed Range		Wet=W
	sfm	smm	in./t	mm/t	sfm	smm	in./t	mm/t	Dry=D
P15-P35 / UCC	380	116	.006	.15	200-550	61-168	.003-.010	.075-.25	D
P20-P35 / PVD	430	131	.006	.15	220-700	67-123	.003-.011	.075-.28	D
P30-P45 / PVD	250	76	.008	.20	150-450	46-137	.004-.012	.102-.30	D
P15-P35 / PVD	390	119	.007	.17	200-600	61-183	.003-.009	.075-.23	W/D [1]
P10-P20 / PVD	700	213	.006	.15	550-1000	168-305	.004-.009	.102-.23	D
P25-P45 / CVD	400	122	.008	.20	250-550	76-168	.005-.013	.13-.33	D
P10-P25 / CERMET	600	183	.004	.102	350-1100	106-335	.003-.009	.075-.23	D

[1] Dry is preferred to minimize thermal shock.

Notes (Medium- and High- Carbon Steels):

1. Suggested starting conditions are based on a .100" axial depth of cut. Cutting speed should be adjusted for greater or lesser depth.
2. The higher the carbon content, the lower the speed.
3. For highly interrupted surfaces, or if two teeth cannot be maintained in the workpiece due to a small radial width of cut, a tougher grade, or an insert with a T-land should be used.
4. Climb milling is preferred to obtain maximum tool life if the machine tool condition permits.

ALLOY STEELS AND LOWER CARBON TOOL STEELS (Hardness of 135-700 BHN)

Material characteristics	Possible result at the cutting edge	Steps to offset condition
Alloyed (chromium, tungsten, molybdenum, vanadium, nickel).	Abrasive metal carbides formed. Can cause poor finish and low tool life.	1. Lower speed may be required to reduce abrasive wear. 2. Higher chip load may extend life due to less time in cut. 3. Use a more wear resistant grade at lighter chip loads.
Alloyed (chromium, nickel).	Workhardened part surfaces can cause chipping or notching of cutting edge.	1. Apply lead angle cutter. 2. Climb mill if machine tool condition permits. 3. Apply high shear (low force) cutter. 4. Consider single pass milling.
Very low alloy content.	Built-up edge on insert due to gumminess of steel.	1. Increase speed and/or chip load. 2. Apply high shear cutter for reduced forces. 3. Apply mist cooler for lubricity.

Alloy Steels and Lower Carbon Tool Steels (Hardness of 135-300 BHN): Alloy - AISI 1335, 4012, 4023, 4140, 4150, 4320, 4340, 4422, 5120, 52100, 8620, 8622, 8640. Lower Carbon Tool Steels - AISI P1, P6, P20, W1, W2, L6, H10, H11, H13, S5, S7.

ISO Grade / Type	Starting Speed		Starting Feed		Speed Range		Feed Range		Wet=W
	sfm	smm	in./t	mm/t	sfm	smm	in./t	mm/t	Dry=D
P15-P35 / UCC	380	116	.006	.15	200-500	61-152	.003-.010	.075-.25	D
P20-P35 / PVD	410	126	.005	.13	220-650	67-198	.003-.011	.075-.28	D
P15-P35 / PVD	380	116	.007	.18	200-580	61-177	.003-.009	.075-.23	W/D [1]
P10-P20 / PVD	680	207	.005	.13	500-950	152-289	.004-.009	.102-.23	D
P25-P45 / CVD	400	122	.008	.20	220-580	66-177	.005-.013	.13-.33	D
P10-P25 / CERMET	650	198	.004	.102	400-1000	122-304	.003-.009	.075-.23	D

[1] Dry is preferred to minimize thermal shock.

Alloy Steels and Lower Carbon Tool Steels (Hardness of 330-425 BHN): Alloy - AISI 1335, 4012, 4023, 4140, 4150, 4320, 4340, 4422, 5120, 52100, 8620, 8622, 8640. Lower Carbon Tool Steels - AISI H10, H11, H13.

ISO Grade / Type	Starting Speed		Starting Feed		Speed Range		Feed Range		Wet=W
	sfm	smm	in./t	mm/t	sfm	smm	in./t	mm/t	Dry=D
P30-P45 / PVD	250	76	.005	.13	150-400	46-122	.003-.008	.075-.20	D
P15-P35 / PVD	250	76	.005	.13	150-450	46-136	.003-.008	.075-.20	W/D [1]
M10-M20 / PVD [2]	350	106	.003	.075	280-500	85-152	.003-.006	.075-.15	D
P10-P20 / PVD	500	152	.004	.102	350-700	106-213	.003-.007	.075-.18	D
P01-P20 / CVD [2]	500	152	.004	.102	400-700	122-213	.003-.007	.075-.18	D
P10-P25 / CERMET	500	152	.003	.075	300-800	91-224	.003-.006	.075-.15	D

[1] Dry is preferred to minimize thermal shock.
[2] Applicable only in the higher hardness range at very light axial depths of cut.

Alloy Steels and Lower Carbon Tool Steels (Hardness of 425-700 BHN): Alloy - AISI 1335, 4140, 4150, 4340, 5046, 5140, 52100, 8625, 8640. Lower Carbon Tool Steels - AISI H10, H11, H12, H13, H14, H19.

ISO Grade / Type	Starting Speed		Starting Feed		Speed Range		Feed Range		Wet=W
	sfm	smm	in./t	mm/t	sfm	smm	in./t	mm/t	Dry=D
P01-P15 / CERAMIC (Al₂O₃)	550	168	.0025	.0635	400-1000	122-305	.002-.005	.0508-.13	D
P05-P20 / PCB	600	183	.004	.102	300-900	91-274	.003-.008	.075-.20	D

Notes (Alloy Steels and Lower Carbon Tool Steels):

1. A higher carbon content normally dictates a lower cutting speed or grade which is more resistant to abrasive wear.
2. The more rigid the part and fixture, the higher the chip load per tooth can go.
3. If two teeth cannot be maintained in the workpiece due to a small radial width of cut or the surface being milled has many interruptions such as drilled holes or cast cavities, reduce chip load or apply higher strength cutting grade.

Cutting speeds comparison for the tool steel material groups. Using the "W" series tool steel as the base alloy (100%), cutting speeds for the other tool steel groups, relative to "W," will be as follows: S=74%, O=59%, A=59%, D=40%, M2 and T1=48%, M3-1 and T4=43%, M15 and T15=35%, L=74%, F=65%, P1 and P6=92%, and P2 and P21=80%.

AUSTENITIC STAINLESS STEELS (200 AND 300 SERIES) INCLUDING DUPLEX (Hardness of 135-275 BHN): AISI 201, 209, 219, 302, 303, 304, 316, 321, 347, 329. ASTM XM-1, XM-5, XM-7, XM-21, CF-8M.

Material characteristics	Possible result at the cutting edge	Steps to offset condition
High workhardening rate due to high chromium content.	Depth of cut notch and chipping, rapid wear.	1. Maintain sufficient chip load (.003 minimum). 2. Climb mill only (machine condition permitting). 3. Apply cutter with greatest lead angle (45°). 4. Use high shear or double positive geometry cutter. 5. Consider one pass milling.
Gummy while being machined (nickel).	Built-up edge leading to poor part finish or edge chipping.	1. Increase cutting speed or increase chip load. 2. Apply low pressure mist coolant (lubricity). 3. Use cermet grade if conditions permit. 4. Apply flood coolant with proper grade.
Low thermal conductivity.	More heat goes into the insert causing rapid wear (low tool life) which in turn may generate a poor surface finish.	1. Maintain sufficient chip load. 2. Apply low pressure mist coolant. 3. Apply flood coolant with proper grade.

ISO Grade / Type	Starting Speed		Starting Feed		Speed Range		Feed Range		Wet=W
	sfm	smm	in./t	mm/t	sfm	smm	in./t	mm/t	Dry=D
M10-M20 / UCC	400	122	.005	.13	250-650	76-198	.003-.008	.075-.20	D
M25-M40 / PVD	450	137	.006	.15	150-600	46-183	.004-.012	.102-.30	W/D[1]
M25-M45 / PVD	500	152	.007	.18	250-800	77-244	.005-.011	.13-.28	W/D[1]
M05-M20 / PVD	600	183	.004	.102	200-800	61-244	.003-.007	.075-.18	D
M10-M25 / CERMET	700	213	.004	.102	300-1000	91-305	.003-.010	.075-.25	D

[1] Flood coolant heightens thermal shock. Proper chip load puts the heat in the chip. Low pressure mist coolant adds lubricity with less thermal shock.

Notes (Austenitic Stainless Steels (200 and 300 Series) Including Duplex):
1. Suggested starting conditions are based on a .100″ axial depth of cut. Cutting speeds should be adjusted for greater or lesser depth.
2. Due to the high tensile strength, high ductility, and toughness of this material, every effort must be made to achieve maximum rigidity in the fixture and machine tool.
3. A single pass at full axial depth of cut (face milling) may be more effective than multiple passes due to the workhardening nature of the material (rigidity and power permitting).
4. Duplex stainless materials are very tough and require a high strength cutting edge. Starting speeds should be reduced by approximately 30%.

FERRITIC-MARTENSITIC 400 & 500 SERIES STAINLESS STEELS (Hardness of 135-600 BHN)

Material characteristics	Possible result at the cutting edge	Steps to offset condition
Alloyed (chromium, molybdenum, vanadium, nickel).	Abrasive metal carbides formed. Can cause poor finish and low tool life.	1. Lower speed may be required to reduce wear. 2. Higher chip load may extend tool life due to less time in cut. 3. Apply more wear resistant grades at lighter chip loads.
Alloyed (chromium, nickel).	Workhardened part surfaces cause chipping or notching of the cutting edge.	1. Apply lead angle cutter. 2. Climb mill if machine tool condition permits. 3. Apply high shear (low force) cutter. 4. Verify sufficient chip load. 5. Consider single pass milling.

Ferritic-Martensitic 400 & 500 Series Stainless Steels (Hardness of 135-330 BHN): AISI 416, 416F, 416 Se, 420F, 430F, 439F Se, 502, 504.

ISO Grade / Type	Starting Speed		Starting Feed		Speed Range		Feed Range		Wet=W
	sfm	smm	in./t	mm/t	sfm	smm	in./t	mm/t	Dry=D
P15-P35 / UCC	380	116	.006	.15	200-500	61-152	.003-.010	.075-.25	D
P20-P35 / PVD	410	125	.005	.13	200-650	61-198	.003-.011	.075-.28	D
P15-P35 / PVD	380	116	.007	.18	200-580	61-177	.003-.009	.075-.23	W/D [2]
P10-P20 / PVD	680	207	.005	.13	500-950	152-289	.004-.009	.102-.23	D
M30 / CVD	400	122	.008	.20	220-580	67-177	.005-.013	.13-.33	D
P10-P25 / CERMET	650	198	.004	.102	400-1000	122-305	.003-.009	.075-.23	D

[1] See "Notes" below.
[2] Dry is preferred to minimize thermal shock.

Ferritic-Martensitic 400 & 500 Series Stainless Steels (Hardness of 330-450 BHN): AISI 416, 416F, 416 Se, 420F, 430F Se, 440C, 502, 504.

ISO Grade / Type	Starting Speed		Starting Feed		Speed Range		Feed Range		Wet=W
	sfm	smm	in./t	mm/t	sfm	smm	in./t	mm/t	Dry=D
P30-P45 / PVD	250	76	.005	.13	150-400	46-122	.003-.008	.075-.20	D
P15-P35 / PVD	250	76	.005	.13	150-450	46-137	.003-.008	.075-.20	D
M10-M20 / PVD[2]	350	106	.003	.075	280-500	82-152	.003-.006	.075-.15	D
P10-P20 / PVD	500	152	.004	.102	350-700	106-213	.003-.007	.075-.18	D
P10-P25 / CERMET	500	152	.003	.075	300-800	91-244	.003-.006	.075-.15	D

[1] See "Notes" below.
[2] For very light axial depth of cut and light chip loading applications only.

Martensitic 400 Series Stainless Steels (Hardness of 450-600 BHN)

ISO Grade / Type	Starting Speed		Starting Feed		Speed Range		Feed Range		Wet=W
	sfm	smm	in./t	mm/t	sfm	smm	in./t	mm/t	Dry=D
P01-P15 / CERAMIC (Al_2O_3)	550	167	.004	.102	400-1000	122-305	.003-.006	.075-.15	D
P05-P20 / PCB	600	183	.006	.15	300-900	91-274	.003-.010	.075-.25	D

Notes (Ferritic-Martensitic 400 & 500 Series Stainless Steels):

1. Suggested starting conditions for free-machining materials under 450 BHN are based on a .100" axial depth of cut. Cutting speed should be adjusted for greater or lesser depth and for nonfree machining materials.

2. Higher carbon content or hardness normally dictate the use of lower cutting speeds or a more wear resistant grade of insert.

3. AISI numbers for free-machining 400 series stainless steel are (ferritic) 430F, 430F Se, (martensitic) 416, 416 Se, 420F, 420F Se, 440F, and 440F Se.

4. The non-free-machining materials are (ferritic) 405, 409, 430, 434, 442, 446, (martensitic) 403, 410, 414, 420, 431, 440A, 440B, and 440C. Cutting speeds for nonfree-machining materials should be lower and tool life will be less.

5. The more rigid the part and fixture, the higher the chip load can be.

6. If two teeth cannot be maintained in the workpiece due to a small radial width of cut, reduce the chip load and use higher strength tool grades. An insert with a greater T-land may be preferable in this situation, and speeds can usually be increased due to the small percentage of time in cut per cutter revolution.

PRECIPITATION HARDENING STAINLESS STEELS (Hardness of 150-375 BHN): 15-5 PH, 17-4 PH, 17-7 PH; AM350, AM355; custom 450, custom 455; PH 13-8 Mo, PH 15-7 Mo.

Material characteristics	Possible result at the cutting edge	Steps to offset condition
High tensile strength, very tough, often having tough skin.	Insert chipping and short life.	1. Use inserts with T-land for maximum edge strength. 2. Apply lead angle cutter, 40 or 45°. 3. Use most rigid machine tool available.
High hardness can make material very abrasive.	Rapid flank wear results in short tool life.	1. Use greatest practical chip load per tooth to minimize time in cut per tooth. 2. Apply more wear resistant grades at lighter depths and chip loads.

ISO Grade / Type	Starting Speed		Starting Feed		Speed Range		Feed Range		Wet=W
	sfm	smm	in./t	mm/t	sfm	smm	in./t	mm/t	Dry=D
M20-M30 / CVD	350	106	.007	.18	200-550	61-167	.004-.012	.102-.30	W/D [2]
M25-M45 / PVD	300	91	.007	.18	250-500	76-152	.004-.010	.102-.25	W/D [2]
M10-M20 / PVD [1]	400	122	.004	.102	250-800	76-244	.003-.007	.075-.18	D
M20 / CERMET [1]	500	152	.004	.102	300-1000	91-305	.003-.008	.075-.20	D

[1] For finishing and semifinishing cuts only.
[2] Dry is preferred to minimize thermal shock.

Notes (Precipitation Hardening Stainless Steels):

1. Suggested starting conditions are based on .100″ axial depth of cut and milling of easier-to-machine material such as the 17-4 PH. Cutting speed should be adjusted for greater or lesser depth of cut and for more difficult materials.
2. A 45° lead angle cutter should be considered as a first choice when milling these materials.
3. Consider using a T-land for maximum edge strength.

GRAY CAST IRONS (Hardness of 120-320 BHN): ASTM A48: class 20, 25, 30, 35, 40, 45, 50, 55, 60. SAE J431: G1800, G3000, G3500, G4000.

Material characteristics	Possible result at the cutting edge	Steps to offset condition
Corners and thin sections can be chilled (very hard and brittle).	Edge breakout of the casting may occur whenever an insert exits the cut. A high hardness level can cause accelerated wear of the cutting edge.	1. Use a high lead angle (45°) cutter. 2. Prechamfer casting edge. 3. Reduce cutting speed if wear is a problem. 4. Apply a ground periphery sharp edged PVD coated or uncoated grade.
Unless cleaned prior to machining, sand can adhere to or be embedded in the cast surface.	Accelerated wear at the depth of cut line on the inserts can develop into depth of cut notch.	1. Shot blast to clean casting. 2. Take corrective action at the casting source. 3. Reduce cutting speed to minimize exposure to sand. 4. Set axial depth to ensure continuous engagement (get under the scale).
Contains abrasive elements.	Flank and nose wear on the insert.	1. Use aluminum oxide coated grades or silicon nitride based grades. 2. Increase chip load to lessen time in cut. 3. Reduce speed to minimize heat build-up.

ISO Grade / Type	Starting Speed		Starting Feed		Speed Range		Feed Range		Wet=W
	sfm	smm	in./t	mm/t	sfm	smm	in./t	mm/t	Dry=D
K20-K30 / UCC	400	122	.006	.15	300-500	91-152	.003-.012	.075-.30	W/D [1]
K10-K20 / PVD	450	137	.006	.15	300-600	91-183	.003-.008	.075-.20	D
K10-K25 / CVD	700	213	.009	.23	300-1000	91-305	.004-.014	.102-.36	W/D [1]
K05-K25 / CERMET	2500	762	.007	.17	1000-3800	305-1169	.004-.012	.102-.30	W/D [1]
K05-K20 / CERMET	2500	762	.006	.15	1200-3700	360-1139	.003-.010	.075-.25	D [1]

[1] Dry is preferred to minimize thermal shock. Coolant is acceptable and is used primarily to control cast iron dust.

Notes (Gray Cast Irons):
1. Starting speed and chip load suggestions are based on a .100" axial depth of cut and a hardness range of 190-220 BHN. Speeds may have to be reduced by as much as 50% on the higher hardness materials.
2. The elevated speeds at which the cermet grades operate will provide a very high metal removal rate. Make sure the machine tool selected has adequate horsepower and rigidity to allow effective use of these grades.
3. Negative rake cutters (higher cutting forces) are commonly applied on cast iron parts but shear angle geometry (lowest cutting forces) cutters can also be applied very effectively.
4. The greater the lead angle, the lower the possibility of edge breakout on the part.
5. If the cast iron part has numerous interruptions (cast holes, depressions, cavities), a fine pitch cutter should be used in order to maintain a constant load for stability (two teeth in the work at any one time).
6. Longer tool life can be obtained when milling fully annealed (ferritic) iron. Tool life decreases as the percentage of ferrite decreases and pearlite increases. The higher tensile strength materials require lower cutting speeds.

DUCTILE AND MALLEABLE CAST IRONS (Hardness of 120-320 BHN): Ductile Cast Iron: ASTM A536: 60-40-18, 65-45-12, 80-55-06, 100-70-03. SAE J434: D4018, D4512, D5506, D7003. **Malleable Cast Iron:** ASTM A47: grade 32510, 35018. ASTM A220: grade 45006, 50005, 60004. SAE J148: grade M3210, M4504, M5003.

Material characteristics	Possible result at the cutting edge	Steps to offset condition
Higher tensile strength than gray cast iron.	Possible edge chipping.	1. Negative rake inserts offer greater strength at the cutting edge. 2. Increase fixture rigidity. 3. Apply an insert with a T-land for greater edge strength.
Very abrasive.	Rapid flank and nose wear on the insert.	1. Use aluminum-oxide coated grades. 2. Increase chip load to minimize time in cut. 3. Reduce speed to reduce heat.

ISO Grade / Type	Starting Speed		Starting Feed		Speed Range		Feed Range		Wet=W
	sfm	smm	in./t	mm/t	sfm	smm	in./t	mm/t	Dry=D
P20-P30 / UCC	350	106	.006	.15	300-500	91-152	.003-.012	.075-.30	D
P15-P35 / PVD	400	122	.008	.20	250-800	76-244	.004-.012	.102-.30	W/D [1]
K10 / PVD	450	137	.005	.13	300-800	91-244	.003-.008	.075-.20	D
P10-P20 / PVD	500	152	.007	.18	300-800	91-244	.004-.011	.102-.28	D
K10-K25 / CVD	550	168	.008	.20	300-1000	91-305	.004-.012	.102-.30	W/D
P20-P35 / CVD	550	168	.009	.23	300-1200	91-365	.004-.014	.102-.36	W/D

[1] Dry is preferred to minimize thermal shock. Coolant is acceptable.

Notes (Ductile and Malleable Cast Irons):

1. Recommended starting conditions are based on a hardness range of 190-225 BHN and an axial depth of cut of .100". Adjust speed for greater or lesser depth of cut and higher hardness values.
2. The higher tensile strength of these materials dictates the need for maximum rigidity. The need for a high degree of strength at the cutting edge may mean that a T-land is desirable.
3. The greater the lead angle of the cutter, the lower the possibility of edge breakout on the part.
4. If the part has numerous interruptions (cast holes, depressions, cavities), a fine pitch cutter should be used in order to maintain a constant load on the cutter for stability (at least two teeth engaged in the work).

HARDENED AND CHILLED CAST IRONS (Hardness of 400-600 BHN): ASTM: A532, class I and class II (types A,B,C, & D). Class III, type E. Class 3, type A. UNS F45000 through F45009 (NiCr cast iron). Ni-HARD 2C, Meehanite.

Material characteristics	Possible result at the cutting edge	Steps to offset condition
Hard crust and scale.	Depth of cut notching in the insert cutting edge.	1. Use a lead angle cutter (45°) or round inserts that create a lead angle effect. 2. Apply an insert with a K-land or K-land plus a hone.
Brittle.	Material breaks out of the casting whenever an insert exits the cut.	1. Use a 45° lead angle cutter or round inserts. 2. Chamfer the edge of the casting prior to face milling. 3. Reduce chip load.
Highly abrasive.	Excessive flank wear on insert.	1. Increase depth of cut or chip load. 2. Increase cutting speed.

ISO Grade / Type	Starting Speed		Starting Feed		Speed Range		Feed Range		Wet=W
	sfm	smm	in./t	mm/t	sfm	smm	in./t	mm/t	Dry=D
KO2-KO9 / PCB	600	183	.008	.20	150-900	46-274	.005-.012	.13-.30	D
K01-K10 / CERAMIC (Al_2O_3)[1]	700	213	.004	.102	300-1100	91-335	.003-.008	.075-.20	D
K10-K20 / CERAMIC (Si_3N_4)	750	228	.006	.15	250-1100	76-335	.004-.010	.102-.25	D

[1] For very light axial depth of cut applications on white or chilled cast irons only.
[2] Dry is preferred to minimize thermal shock.

Notes (Hardened and Chilled Cast Irons):

1. Suggested starting speeds are based on a .040" axial depth of cut and a mid-range hardness of 500 BHN. Cutting speeds should be decreased for higher hardness and increased for lower hardness materials.
2. Starting speeds should be reduced by approximately 30% when machining Ni-HARD material and increased by 30% when machining Meehanite (.020" depth of cut).
3. As hardness increases, so does the need for rigidity. Any lack of rigidity is likely to result in chipping of the cutting edge.

ALUMINUM AND OTHER FREE-MACHINING, NONFERROUS METALS (Copper, Brass, Zinc, Magnesium).
Hardness of 50-150 BHN: ALUMINUM: AA; 2024-T4, 6061-T6, 7075-T6. BRASS: SAE J461, J463. ASTM B121, B453, B16, B124, B103, B30, B584, B271, B505.

Material characteristics	Possible result at the cutting edge	Steps to offset condition
Gummy material, low melting temperature.	Built-up edge on the rake face of the tool can cause a poor finish. Can also increase cutting forces generated.	1. Use high-positive, high-velocity cutter with inserts having a polished rake face (SEEN 12 03 08J, TEEN 16 03 08J, TECN 16 03 PERJW etc.). 2. Use PCD high-velocity face mill. 3. Use SEHT-12 04 AFN insert geometry with 45° high shear cutter. 4. Apply coolant.
Chips can workharden.	Workhardened chips deflect off of milled surface causing damage to the finish.	1. Use high positive axial rake cutter that directs chips away from the finished surface.

ISO Grade / Type	Starting Speed		Starting Feed		Speed Range		Feed Range		Wet=W
	sfm	smm	in./t	mm/t	sfm	smm	in./t	mm/t	Dry=D
K20-K30 / UCC	1800	550	.009	.23	1000-10000	300-3000	.004-.018	.102-.46	W/D
K10-K20 / PVD	2000	610	.009	.23	1200-10000	360-3000	.004-.012	.102-.30	W/D
K25-K35 / PVD	1500	457	.010	.25	1000-5000	300-1500	.005-.018	.13-.46	W/D
K02-K08 / PCD	3000	915	.006	.15	1000-8000	300-2400	.003-.009	.075-.23	W/D
K01-K10 / PCD	4000	1220	.008	.20	1000-10000	300-3000	.003-.015	.075-.38	W/D

[1] Coolant is acceptable and can help minimize built-up edge problems.

Notes (Aluminum and other Free-Machining, Nonferrous Metals):

1. For magnesium, the starting speed should be approximately 50% of the starting speed shown above. Use extreme caution—partially dry chips and sludge are extremely flammable and may ignite spontaneously. See National Fire Protection Association standard NFPA 480 for information.

2. Recommended starting conditions are based on .100" axial depth of cut. Speed may need to be adjusted for greater depth of cut.

3. High positive rake, high shear or positive rake geometry cutters clear chips away from the machined surface which is beneficial to maintaining a good surface finish.

4. Sharp-edged, ground periphery inserts will generally produce less of a burr than a molded, honed insert. Best possible finishes can be obtained with grade K01-K10 PCD.

HIGHER SILICON ALUMINUM/ALUMINUM BRONZE: AA number [1] 308.0 (5.5), A355.0 (5), 319.0 (6), A356.0 (7), 359.0 (9), A380.0 (8.5), 332.0 (9.5), A360.0 (9.5), 383.0 (10.5), A413.0 (12), 390.0 (17), 392.0 (19), 4004 (9.8), 4032 (12.2). **Aluminum Bronze** ASTM SB111, B359, B395, Ampco 21, Wearite 4-13.

[1] The number in parentheses after each material designation is the percentage of silicon in the aluminum.

Material characteristics	Possible result at the cutting edge	Steps to offset condition
The higher the silicon content, the more abrasive the material.	Rapid flank wear of the cutting edge causing burrs and therefore short tool life.	1. Use PCD grade K01-K10 or diamond film coated grade K02-K08. 2. Reduce cutting speed and/or increase chip load.
Chips can workharden.	Workhardened chips deflect off the milled surface causing damage to the finish.	1. Use high positive axial rake cutter that directs chips away from the finished surface.
(Bronze) resilient material.	Tool deflects away from work, especially in very light depths of cut.	1. Use K01-K10 (sharp edge). 2. Use ground periphery uncoated or PVD coated grade with up-sharp edge.

ISO Grade / Type	Starting Speed		Starting Feed		Speed Range		Feed Range		Wet=W
	sfm	smm	in./t	mm/t	sfm	smm	in./t	mm/t	Dry=D
K20-K30 / UCC	400	122	.012	.30	250-900	76-274	.004-.015	.102-.38	W/D
K10-K20 / PVD	500	152	.010	.25	300-1000	91-305	.003-.015	.075-.38	W/D
K02-K08 / Diamond film coated	1500	457	.008	.20	1000-5500	305-1667	.005-.012	.13-.30	W/D
K01-K10 / PCD	1500	457	.009	.23	1000-6000	305-1829	.004-.014	.102-.36	W/D

Mist coolant is often used to reduce the amount of fluid consumed.

Notes (Higher Silicon Aluminum/Aluminum Bronze):

1. Suggested starting speeds shown are for the materials containing about 17% to 19% silicon at a .150" axial depth of cut. Cutting speed can be increased as much as 400% for the lower silicon materials depending on hardness and silicon content.

2. For aluminum bronze materials, suggested starting speed should be about 50% of the value shown in the chart.

3. Grade K01-K10 PCD will provide the best possible finish.

NICKEL-BASE, HEAT-RESISTANT ALLOYS (Hardness of 140-475 BHN)

Material characteristics	Possible result at the cutting edge	Steps to offset condition
Workhardens.	A notch forms at the depth of cut line on the insert (depth of cut notch) causing high cutting forces and eventual failure of the insert. Also increases abrasive wear.	1. Use 45° lead angle cutter or round insert cutter that will simulate a lead angle cutter. 2. Increase hone level or K-land on insert to increase edge strength.
Workhardened burr forms at depth of cut line.	Same as above. Burr causes chipping of the insert edge above the depth of cut line.	1. Use cutter with greatest lead angle available (minimizes formation of burr at depth of cut line).
Material forms hardened burr whenever inserts exit the workpiece (over the edge of the part).	Causes notching and chipping of the cutting edge.	1. Establish cutter path that minimizes exits. 2. Use maximum lead angle cutter. 3. Use insert with a K-land for greater strength.

Nickel-Base, Heat-Resistant Alloys (Hardness of 140-300 BHN), typically annealed or solution treated: Hastelloy B, C-276, Inconel 601, Inconel 617, Nickel 200, Monel K500, Waspalloy, Inconel 625, 901, X-750, 718.

ISO Grade / Type	Starting Speed		Starting Feed		Speed Range		Feed Range		Wet=W
	sfm	smm	in./t	mm/t	sfm	smm	in./t	mm/t	Dry=D
M30 / PVD	100	31	.005	.13	60-350	18-106	.002-.008	.05-.20	D
M10 / PVD	120	37	.005	.13	80-400	24-122	.002-.008	.05-.20	D
M20-M30 / PVD	150	46	.005	.13	100-450	31-137	.002-.008	.05-.20	W/D [1]
M25-M35 / CVD	180	55	.005	.13	120-480	37-146	.002-.008	.05-.20	W/D [1]
K05-K20 / CERAMIC (Si_3N_4)	1000	305	.006	.15	600-1600	183-488	.004-.009	.102-.23	D

Nickel-Base, Heat-Resistant Alloys (Hardness of 300-475 BHN) typically annealed or solution treated: Inconel 718, X-750, Nimonic 80A, Waspaloy, Udimet 700, Rene 95, Inconel MA6000, Hastelloy C, ASTM grades HW and HX, IN100, MAR-M200.

ISO Grade / Type	Starting Speed		Starting Feed		Speed Range		Feed Range		Wet=W
	sfm	smm	in./t	mm/t	sfm	smm	in./t	mm/t	Dry=D
M30 / PVD	70	21	.004	.102	50-180	15-55	.002-.008	.05-.20	D
M10 / PVD	80	24	.004	.102	50-200	15-61	.002-.008	.05-.20	D
M20-M30 / PVD	90	27	.004	.102	50-220	15-67	.002-.008	.05-.20	W/D [1]
M25-M35 / CVD	120	37	.004	.102	80-260	24-79	.002-.008	.05-.20	W/D [1]
K05-K20 / CERAMIC (Si3N4)	800	244	.005	.13	500-1200	152-366	.004-.009	.102-.23	D

[1] Dry is preferred to minimize thermal shock. Low pressure mist coolant adds lubricity with less thermal shock.

Notes (Nickel-Base, Heat-Resistant Alloys):

1. Suggested starting conditions are based on a .100" axial depth of cut. Cutting speed should be adjusted for a greater or lesser depth of cut and condition of the workpiece surface (scale, pre-machined, etc.).

2. It is very important to maintain two teeth engaged in the workpiece to gain stability. Most insert chipping will occur at the beginning and end (entry and exit) of each milling pass.

3. As edge chipping is a common failure mode, an insert with a K-land will provide additional edge strength. Use the most rigid machine tool available.

4. A 45° lead angle or very large nose radius insert is the primary choice to minimize formation of a workhardened burr at the depth of cut line.

IRON BASE, HEAT-RESISTANT ALLOYS (Hardness of 135-320 BHN) Wrought: A-286, Discaloy, Incoloy 801, N-155, 16-25-6, 19-9DL. **Cast:** ASTM A297 grade HC. ASTM 351 grade HK-30, 40, and HT-30. ASTM A608 grade HD50, HF30, HN40. ASTM A297 grade HD, HE, HK, HP.

Material characteristics	Possible result at the cutting edge	Steps to offset condition
Workhardens.	A notch forms at the depth of cut line on the insert (depth of cut notch) causing high cutting forces and eventual failure of the insert. Also increases abrasive wear.	1. Use 45° lead angle cutter or round insert cutter that will simulate a lead angle cutter. 2. Increase hone level or K-land on insert to increase edge strength.
Workhardened burr forms at depth of cut line.	Same as above. Burr causes chipping of the insert edge above the depth of cut line.	1. Use cutter with greatest lead angle available (minimizes formation of burr at depth of cut line).
Material forms hardened burr whenever inserts exit the workpiece (over the edge of the part).	Causes notching and chipping of the cutting edge.	1. Establish cutter path that minimizes exits. 2. Use maximum lead angle cutter. 3. Use insert with a K-land for greater strength.

ISO Grade / Type	Starting Speed		Starting Feed		Speed Range		Feed Range		Wet=W
	sfm	smm	in./t	mm/t	sfm	smm	in./t	mm/t	Dry=D
M30 / PVD	200	61	.006	.15	60-350	18-106	.003-.010	.075-.25	D
M10 / PVD [1]	300	91	.004	.102	60-450	18-137	.002-.008	.05-.20	D
M20-M30 / PVD	280	85	.007	.18	60-400	18-122	.003-.011	.075-.28	W/D [2]
K05-K20 / CVD	330	99	.007	.18	80-500	24-152	.004-.012	.102-.30	D

Wrought materials (A286, Discaloy, etc.) are more difficult to machine than the cast materials. For these materials, reduce starting speed by 50% to 70%.

[1] For relatively light axial depth of cut and light chip load applications only.

[2] Dry is preferred to minimize thermal shock. Low pressure mist coolant adds lubricity with less thermal shock.

Notes (Iron Base, Heat Resistant Alloys):

1. Suggested starting conditions are based on a .100" axial depth of cut and hardness of approximately 185 BHN. Speed should be adjusted for greater or lesser depth and hardness.
2. Due to workhardening nature of this material, single-pass milling should be considered.
3. The need for a high strength cutting edge dictates that a negative rake cutter should be used. Workhardening characteristics of the material determine the need for a low force cutter. For optimum performance, consider a high shear, high lead angle cutter utilizing a K-land insert for maximum edge strength.
4. Machine tool and workpiece fixturing must be as rigid as possible.

COBALT BASE, HEAT-RESISTANT ALLOYS (Hardness of 150-425 BHN) Wrought: AiResist 213, Haynes 25 (L605), Haynes 188, J-1570, Stellite. **Cast:** AiResist 13, Haynes 21, MAR-M302, MAR-M509, NASA Co-W-Re, WI52.

Material characteristics	Possible result at the cutting edge	Steps to offset condition
Brittle material.	Part edge breaks out where inserts exit the workpiece. Can cause chipping on the insert.	1. Use 45° lead angle cutter or round insert cutter that will simulate a lead angle cutter. 2. Prechamfer exit edge of the workpiece.
Workhardens.	A notch forms at the depth of cut line on the insert (depth of cut notch) causing increased cutting forces and eventual insert failure. Also increases abrasive wear on the insert.	1. Use a ground periphery insert with an up-sharp edge. 2. Do not dwell in the workpiece (spindle running, no slide motion). 3. Maintain sufficient chip load. 4. Climb mill.

ISO Grade / Type	Starting Speed		Starting Feed		Speed Range		Feed Range		Wet=W
	sfm	smm	in./t	mm/t	sfm	smm	in./t	mm/t	Dry=D
M20 / UCC	60	18	.003	.075	20-150	6-46	.001-.008	.025-.20	D
M40 / PVD	50	15	.004	.102	10-120	3-37	.002-.009	.05-.23	D
M20 / PVD	80	24	.003	.075	30-200	9-61	.001-.008	.025-.20	D
M30 / PVD	80	24	.004	.102	20-200	6-61	.002-.010	.05-.25	W/D [1]

[1] Dry is preferred to minimize thermal shock.

Notes (Cobalt Base, Heat-Resistant Alloys):
1. Cast cobalt materials are more difficult to machine than the wrought materials. Suggested starting speeds should be reduced by 30% to 70% for the cast materials, depending on the hardness.
2. Climb milling is preferred. Machine tool and workpiece must be as rigid as possible. Sacrifices in rigidity will shorten tool life.
3. Maximum axial depth of cut is usually in the area of .060".

TITANIUM-ALLOYED: Hardness of 110-275 BHN: commercially pure Ti98.8, Ti99.9. **Hardness of 300-340 BHN:** alpha alloy, annealed Ti5Al2. **Hardness of 310-380 BHN:** alpha-beta alloy, annealed or solution treated and aged: Ti6Al-4V, Ti6Al-2Sn-4Zr-2Mo. **Hardness of 275-440 BHN:** beta alloy, annealed, solution treated, or solution treated and aged: Ti-3Al-8V-6Cr-4Mo-4Zr, Ti-10V-2Fe-3Al, Ti-13V-11Cr-3Al.

Material characteristics	Possible result at the cutting edge	Steps to offset condition
Chemically reactive.	Chip welding on the rake face of the insert is common. This leads to insert chipping and/or a galled finish on the part.	1. Apply a high shear geometry cutter and inserts with a 20° relief angle. 2. Use high positive axial rake.
Low heat conductivity.	More heat goes into the cutting edge, resulting in rapid wear.	1. Use lower cutting speeds and the highest practical chip load for setup conditions.
Workhardens rapidly.	Accelerates insert edge wear. Workhardened surfaces can cause tool deflection.	1. Do not dwell in the workpiece (spindle running, no slide motion). 2. Apply inserts with up-sharp cutting edges. 3. Maintain sufficient chip load. 4. Climb mill.
Low elastic modulus.	Material easily deflects away from dull cutting edge.	1. Material must be rigidly supported by fixture. 2. Apply ground tolerance sharp-edged inserts.

ISO Grade / Type	Starting Speed		Starting Feed		Speed Range		Feed Range		Wet=W
	sfm	smm	in./t	mm/t	sfm	smm	in./t	mm/t	Dry=D
M40 / PVD	130	40	.006	.15	50-500	15-152	.003-.010	.075-.25	W
M30 / PVD	150	46	.006	.15	50-550	15-167	.003-.010	.075-.25	W
M20 / PVD	130	40	.005	.13	50-600	15-184	.003-.008	.075-.20	W
M20 / UCC	120	37	.005	.13	50-500	15-152	.003-.008	.075-.20	W

Low pressure mist coolant adds lubricity with less thermal shock.

Notes (Titanium-Alloyed):

1. Starting speeds shown are for materials in the 300 to 350 BHN range at a .100" axial depth of cut. The starting speeds should be reduced by 40% to 60% for materials in the high hardness range (350-440 BHN).
2. Commercially pure titanium can be milled at speeds two to four times higher than those used to mill annealed alpha/beta material.
3. Climb milling is preferred because reduce it the possibility of built-up edge, extends tool life, and improves surface finish. Machine tools must not have any backlash in the slides.
4. Chip disposal is critical. Recutting chips will reduce tool life dramatically.
5. Program cutter path so that cutting forces are directed into fixed part locators or workstops and the more rigid workpiece cross-sectional areas.

Recommended Speeds and Feeds for End Milling

Recommended Starting Feeds and Speeds for High-Speed Steel End Mills for Peripheral Milling. (For slot milling, reduce speed value by 25%.)

workpiece material	speed		cutter diameter and feed per tooth (in./t or mm/t)								
			1/8" (3 mm)		5/32"-1/4" (3.5-6 mm)		5/16"-1/2" (7-13 mm)		9/16"-3/4" (14-20 mm)		13/16"-1" (21-25 mm)	
	sfm	smm	in./t	mm/t	in./t	mm/t	in./t	mm/t	in./t	mm/t	in./t	mm/t
tool steel annealed	85	26	.0006	.015	.0015	.038	.002	.051	.004	.102	.005	.127
P20 mold steel	85	26	.0005	.013	.0006	.015	.003	.076	.004	.102	.006	.152
die steel <390 BHN	35	11	.0005	.013	.0012	.03	.0018	.046	.003	.076	.005	.127
low carbon steel annealed	100	30	.0005	.013	.0012	.03	.0035	.089	.005	.127	.006	.152
medium carbon steel	80	24	.0005	.013	.001	.025	.003	.076	.004	.102	.0055	.140
hardened steel >450 BHN	30	9	.0002	.005	.0005	.013	.0015	.038	.0025	.063	.004	.102
stainless steel 400 series	140	42	.0006	.015	.001	.025	.0025	.063	.004	.102	.0055	.140
stainless steel 300 series	115	35	.0006	.015	.001	.025	.002	.051	.0035	.089	.005	.127
cast iron <221 BHN	90	27	.0008	.020	.001	.025	.003	.076	.004	.102	.0055	.140
cast iron >221 BHN	50	15	.0004	.010	.0008	.020	.002	.051	.003	.076	.004	.102
ductile iron	45	13	.0005	.013	.001	.025	.003	.076	.004	.102	.005	.127
malleable iron	85	26	.0005	.013	.001	.025	.003	.076	.0045	.115	.006	.152
aluminum [1]	375	112	.001	.025	.002	.051	.004	.102	.006	.152	.008	.203
high silicon aluminum [1]	220	66	.0005	.013	.001	.025	.003	.076	.005	.127	.006	.152
brass, bronze, copper alloy [1]	185	56	.0008	.020	.001	.025	.002	.051	.004	.102	.005	.127
magnesium [1]	350	105	.0008	.020	.001	.025	.0035	.089	.005	.127	.008	.203
plastics	375	112	.0015	.038	.0025	.063	.0055	.140	.010	.254	.015	.381
titanium alloy	33	10	.0003	.008	.0005	.013	.001	.025	.002	.051	.003	.076
nickel base high temp alloy	20	6	.0004	.010	.0008	.020	.001	.025	.0015	.038	.002	.051

[1] Recommended tool = Three flute, high helix end mill.

Notes (High-Speed Steel End Mill Feeds and Speeds):

1. Unless noted, figures given are for regular length four-flute, profiling end mill.
2. Axial depth of cut should not exceed 1.5 times the cutter diameter.

Recommended Starting Feeds and Speeds for Carbide End Mills for Peripheral Milling. (For slot milling, reduce speed value by 20%)

workpiece material	speed		cutter diameter and feed per tooth (in./t or mm/t)										
			1/16" (1.5 mm)		3/32"-1/8" (2-3 mm)		5/32"-1/4" (3.5-6mm)		5/16"-1/2" (7-13 mm)		9/16"-3/4" (14-20 mm)		13/16"-1" (25 mm)	
	sfm	smm	in./t	mm/t	in./t	mm/t	in./t	mm/t	in./t	mm/t	in./t	mm/t	in./t	mm/t
tool steel <480 BHN [1]	150-200	45-60	.0001	.002	.0002	.005	.0005	.013	.001	.025	.002	.051	.003	.076
tool steel annealed [1]	175-275	50-82	.0005	.013	.0006	.015	.0015	.038	.002	.051	.004	.102	.005	.127
P20 mold steel[1]	275-500	82-150	.0005	.013	.0006	.015	.001	.025	.003	.076	.005	.127	.008	.203
die steel <390 BHN [1]	125-200	38-60	.0002	.005	.0005	.013	.0015	.038	.002	.051	.004	.102	.006	.152
die steel >391 BHN [1]	50-90	15-28	.0002	.005	.0004	.010	.0005	.013	.001	.025	.002	.051	.003	.076
low carbon steel annealed [1]	250-400	75-120	.0004	.010	.0005	.013	.001	.025	.003	.076	.005	.127	.007	.178
medium carbon steel [1]	125-300	38-90	.0005	.013	.0006	.013	.0015	.038	.002	.051	.004	.102	.005	.127
hardened steel >450 BHN [1]	30-100	9-30	.0001	.002	.0002	.005	.0005	.013	.001	.025	.002	.051	.003	.076
stainless steel 400 series [2]	400-500	120-150	.0005	.013	.0008	.020	.001	.025	.002	.051	.004	.102	.006	.152
stainless steel 300 series [2]	300-400	90-120	.0005	.013	.0008	.020	.001	.025	.002	.051	.004	.102	.006	.152
cast iron <221 BHN [3]	250-600	75-180	.0005	.013	.001	.025	.002	.051	.003	.076	.006	.152	.008	.203
cast iron >221 BHN [3]	100-300	30-90	.0002	.005	.0004	.010	.0008	.020	.002	.051	.003	.076	.004	.102
ductile iron [3]	100-400	30-120	.0003	.008	.0005	.013	.001	.025	.002	.051	.004	.102	.006	.152
malleable iron [3]	250-500	75-180	.0003	.008	.0005	.013	.001	.025	.003	.076	.005	.127	.007	.178
aluminum [4]	1000+	300+	.0005	.013	.001	.025	.002	.051	.004	.102	.006	.152	.008	.203
high silicon aluminum [4]	500-900	150-270	.0005	.013	.001	.025	.002	.051	.004	.102	.006	.152	.008	.203
brass, bronze, copper alloy [4]	400-800	120-240	.0005	.013	.001	.025	.002	.051	.003	.076	.004	.102	.005	.127
magnesium [4]	900+	270+	.0005	.013	.001	.025	.002	.051	.004	.102	.006	.152	.010	.245
plastics [4]	1000+	300+	.001	.025	.0015	.038	.003	.076	.006	.152	.010	.254	.015	.381
titanium alloy [5]	100-150	30-45	.0002	.005	.0003	.008	.0005	.013	.001	.025	.002	.051	.004	.102
nickel base high temp alloy [4]	25-100	8-30	.0002	.005	.0004	.010	.0008	.020	.001	.025	.001	.025	.002	.051
Monel high nickel steel [2]	200-250	60-75	.0003	.008	.0005	.013	.001	.025	.002	.051	.003	.076	.004	.102

[1] Recommended tool = PVD coated 30° right-hand spiral end mill. [2] Recommended tool = PVD coated 60° high helix end mill. [3] Recommended tool = Uncoated carbide 30° right-hand spiral end mill. [4] Recommended tool = Uncoated carbide high shear end mill. [5] Recommended tool = Uncoated carbide 60° high helix end mill.

Notes (Carbide End Mill Feeds and Speeds):

A. For lighter radial depths of cut, use the higher range of speeds.

B. For greater radial depths of cut, use the lower range of speeds.

C. Axial depth of cut should not exceed 1.5 times the cutter diameter.

D. Reduce sfm/smm as material hardens. When milling steels up to 650 BHN (60 Rc), chip color indicates correct speed and feed. A tan color indicates proper operating parameters. If chips are blue or dark in color, reduce speed. If chips are white, increase speed.

External Turning

Turning involves the removal of material from the outer diameter of a rotating workpiece in order to create a circular shape through the use of a single point tool. In ideal conditions, the tool is used to shear metal from a workpiece in short, distinct chips that are easily reclaimed for recycling. Early turning tools were solid rectangular pieces of HSS with rake and clearance angles on one end. When the tool became dull, it was sharpened on a pedestal grinder and reused. HSS tools are still in use and are usually employed on older, less rigid machines running at lower cutting speeds. Carbide turning tools first appeared as small pieces or blanks that were brazed to a steel shank and ground with the same rake and clearance angles as those found on earlier HSS tools. The increased wear resistance and hardness of carbide provided improvements in productivity and tool life, but the expense and expertise required to resharpen these tools with diamond wheels made their widespread use impractical. Today's predominant turning tool features a toolholder with either an uncoated or coated carbide insert mechanically locked into place.

Early turning operations were performed with gear driven manual lathes. The modern lathe is CNC controlled and contains one or two machine turrets. Single turret machines have two axes (vertical and horizontal), while two turret machines have four axes (two on each turret). Unless otherwise credited, the technical material in this section is courtesy of Kennametal Inc., a leader in metalworking technology.

Calculating cutting speeds and cutting time in turning and boring operations

(See pages 1029-1048 for tables of suggested speeds and feeds for turning.) Since turning is a combination of linear (tool) and rotational (workpiece) machine movements, the cutting speed is defined as the rotational distance (sfm—surface feet {*or* meters—smm} per minute) traveled in one minute by a point on the part surface. The feed rate (in inches *or* millimeters per revolution) is the linear distance that the tool travels along or across the part surface. The following formulas are used to calculate the cutting speed.

to find	given	formula
sfm *or* smm	workpiece circumference = C revolutions per minute = rpm	sfm *or* smm = C × rpm
C (workpiece circumference in feet)	π = 3.1416 workpiece diameter in inches = D	C = (π ÷ 12) × D C = .262 × D
C (workpiece circumference in meters)	π = 3.1416 workpiece diameter in millimeters = D	C = (π ÷ 1000) × D C = 0.003146 × D
rpm	surface feet per minute = sfm workpiece diameter in inches = D	rpm = (sfm ÷ D) × (12 ÷ π) rpm = (sfm ÷ D) × 3.82
rpm	surface meters per minute = smm workpiece diameter in millimeters = D	rpm = (smm ÷ D) × (1000 ÷ π) rpm = (smm ÷ D) × 318.3

Example. Using the formulas above, find the cutting speed for finish turning a steel axle shaft to a 3.25 in. (82.55 mm) diameter at 1057 rpm.

sfm = (.262 × 3.25) × 1057 = 900   *note: 1 foot = 0.305 meters*
smm = (.003146 × 82.55) × 1057 = 274.5   *1 meter = 3.281 feet.*

Example. Find the rpm needed to maintain a cutting speed of 2000 sfm (609.57 smm) on a cast iron brake drum with a finished outer diameter of 7.2 in. (182.88 mm).

rpm = (2000 ÷ 7.2) × 3.82 = 1060   *or*   rpm = (609.57 ÷ 182.88) × 318.3 = 1060.

The prevalent use of CNC lathes with constant surface speed control has virtually eliminated the need to manually calculate surface speeds. The machine operator selects a surface speed, and the machine automatically determines and alters the rpm as the cutting tool transverses the different diameters along the outer profile of the workpiece. The ability to calculate cutting speeds, however, remains a requirement for determining cycle times for part processes. After the cutting speed has been selected for a particular workpiece material and condition, the appropriate feed rate must be determined. In roughing operations, the goal is to obtain the maximum metal removal rate possible with the available part rigidity and machine power. In finish turning operations, feed rates are established to produce the surface finish specified on the part blueprint. Feed per revolution (fpr) is measured in inches per revolution (in./rev) or millimeters per revolution (mm/rev), which represents the linear distance the tool travels for each revolution of the workpiece. Feed is also expressed as the linear distance the tool travels in a single minute (in./min or mm/min).

$$\text{in./min} = \text{in./rev} \times \text{rpm} \qquad or \qquad \text{mm/min} = \text{mm/rev} \times \text{rpm}.$$

Given the feed rate, the time required for cutting a given workpiece can be calculated.

where: T_m = time, in minutes
C = workpiece circumference in feet or meters
L = length of cut in inches or millimeters
fpr = feed rate per revolution in inches or millimeters
sfm or smm = cutting speed in feet or meters

$$T_m = (C \times L) \div (\text{fpr} \times \text{sfm}).$$

Example. Find the cutting time for one pass over a 24 inch (609.6 mm) long axle shaft that is 0.8515 feet (0.259 meters) in circumference when the cutting speed is 900 sfm (274.5 smm), the feed is 0.003 in./rev (0.0762 mm/rev), and machine speed is 1057 rpm.

$$T_m = (0.8518 \times 24) \div (0.003 \times 900) = 7.57 \text{ minutes (7 minutes, 34 seconds)}$$
$$or\ T_m = (0.259 \times 609.6) \div (0.0762 \times 274.5) = 7.56 \text{ minutes (7 minutes, 34 seconds)}$$

Cutting time can also be calculated using the machine rpm.

$$T_m = L \div (\text{fpr} \times \text{rpm})$$

Using the values from the above example, find the cutting time using rpm.

$$T_m = 24 \div (0.003 \times 1057) = 7.57 \text{ minutes}$$
$$or\ T_m = 609.6 \div (0.0762 \times 1057) = 7.57 \text{ minutes}.$$

Calculating power requirements for turning and boring

In turning and boring, the metal removal rate (Q) and necessary power requirements can be calculated with the following formulas.

Use the following formula to find Q, the metal removal rate.

$$Q = 12 \times \text{d.o.c. (inches)} \times \text{feed rate (in./rev)} \times \text{cutting speed (sfm)} = \text{in.}^3/\text{min}$$
$$or\ Q = \text{d.o.c. (mm)} \times \text{feed rate (mm/rev)} \times \text{cutting speed (smm)} = \text{cm}^3/\text{min}.$$

Example. Where   d.o.c. = 0.150" or 3.81 mm
feed rate = 0.008" or 0.2032 mm
cutting speed = 400 sfm or 122 smm

$$Q = 12 \times 0.15 \times 0.008 \times 400 = 5.76 \text{ in.}^3/\text{min} \qquad \textit{note:}\ 1 \text{ in.}^3 = 16.387 \text{cm}^3$$
$$or\ Q = (3.81 \times 0.2032 \times 122) = 94.45 \text{ cm}^3/\text{min} \qquad\qquad 1 \text{ cm}^3 = 0.061 \text{in.}^3$$

Using the metal removal rate and a power constant, the power required for a given operation can be calculated.

Use the following formula to find HP_s, the power at the spindle for English units (horsepower).

$$HP_s = Q \times K.$$

Example. To find HP_s use the Q value from the above example and find the power constant (K) from the "K" Factors for Turning and Boring chart. K represents the cubic inches of material per minute that can be removed by one horsepower. For this example, $K = 0.609$ (1108 steel, 160 BHN).

$HP_s = 5.76 \times 0.609 = 3.5$ HP at the spindle.

Use the following formula to find HP_m, the power at the motor in English units.

$HP_m = HP_s \div E$.

To determine horsepower at the motor (HP_m), the spindle efficiency must be considered. Spindle efficiency "E" varies from 75% to 90% (E= 0.75 to 0.90). Using the above example,

$HP_m = 3.5 \div 0.8 = 4.375$

Use the following formula to find kW_s, power at the spindle in metric units (kilowatts).

$kW_s = Q \times K$.

"K" Factors for Turning and Boring			
work material	hardness BHN	"K" factor English units	"K" factor Metric units
steels, free machining low-, medium-, and high carbon	110-200	.609	.0277
	201-253	.635	.0289
	254-286	.760	.0346
	287-330	.886	.0403
alloy steels, alloy and lower carbon tool steels	135-200	.721	.0327
	201-253	.785	.0357
	254-286	.847	.0385
	287-327	.940	.0427
	328-371	1.26	.0573
	372-481	1.61	.0732
	482-560	1.82	.0828
	561-700	2.0	.0909
stainless steels, wrought and cast (ferritic, austenitic, martensitic)	135-275	.602 - 1.20	.0274 - .0546
	286-421	1.21 - 1.86	.055 - .0846
	422-600	1.90 - 2.43	.0864 - .1105
stainless steels, precipitation hardened	150-375	.772 - 2.24	.0352 - .1019
cast irons (gray, ductile, and malleable)	120-175	.452	.0205
	110-190	.513	.0233
	176-200	.548	.025
	201-250	.720	.0327
	251-300	.819	.0372
	301-320	.872	.0396
aluminum alloys, copper, brass, and zinc	50-150	.121 - .286	.0055 - .013
nickel alloys	140-300	1.19 - 2.0	.0541 - .0909
	301-475	2.03 - 2.31	.0923 - .105
iron base, heat resistant alloys	135-320	1.10 - 1.89	.050 - .0859
high temperature alloys, cobalt base	200-360	1.20 - 2.11	.0546 - .0961
titanium	250-375	.722 - 1.0	.0328 - .0454
magnesium alloys	40-90	.10 - .152	.0045 - .0068
copper alloys	100-150	.315	.0143
	151-243	.520	.0236

Example. To find kW$_s$, use the Q value from the above example and find the power constant (K) from the "K" Factors for Turning and Boring chart. K represents the cubic centimeters of material per minute that can be removed by one kilowatt. For this example, $K = 0.027$ (1108 steel, 160 BHN).

kW$_s$ = 94.45 × 0.027 = 2.55   *note:* 1 HP = 0.7457 kW
                                        1kW = 1.341 HP.

Use the following formula to find kW$_m$, power at the motor in metric units.

kW$_m$ = kW$_s$ ÷ E.

To determine kW$_m$, the spindle efficiency must be considered. Spindle efficiency "E" varies from 75% to 90% (E= 0.75 to 0.90). Using the above example,

kW$_m$ = 2.55 ÷ 0 .8 = 3.19.

Tool geometry

The effectiveness of the cutting tool is affected by the angle of the tool in relation to the workpiece. The definitions in this section reference the insert to define cutting and relief angles, but are applicable to brazed single point tools as well.

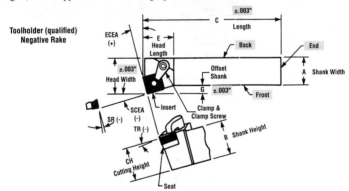

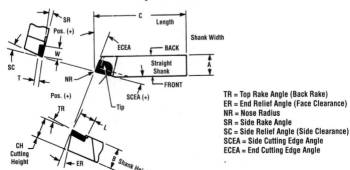

TR = Top Rake Angle (Back Rake)
ER = End Relief Angle (Face Clearance)
NR = Nose Radius
SR = Side Rake Angle
SC = Side Relief Angle (Side Clearance)
SCEA = Side Cutting Edge Angle
ECEA = End Cutting Edge Angle

Figure 1. Single Point Tool Terminology. (Source, Stellram USA.)

Top rake angle (also known as the back rake angle or angle of inclination) is the angle formed between the angle of inclination of the insert and the line perpendicular to the workpiece when viewing the tool from the side, front to back. The top rake is positive when it slopes downward from the cutting point and into the shank, and negative when it slopes upward from the cutting point and above the shank. It is neutral when the line formed from the top of the insert is parallel with the top of the shank (*Figure 2*). It should be noted that inserts are also categorized as being positive or negative. Positive inserts have angled sides and can be mounted in toolholders having positive top and side rake angles. Negative inserts have square sides (90° included angle) relative to the top face of the insert and are designed to be mounted with toolholders having negative top and side rake angles (*Figure 3*).

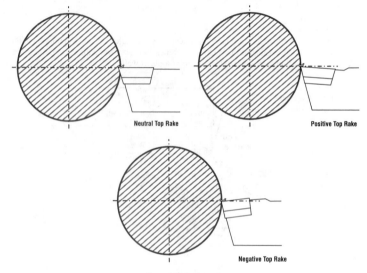

Figure 2. Top rake angles.

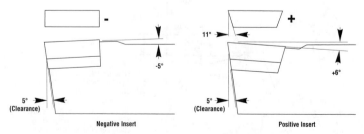

Figure 3. Negative and positive insert application.

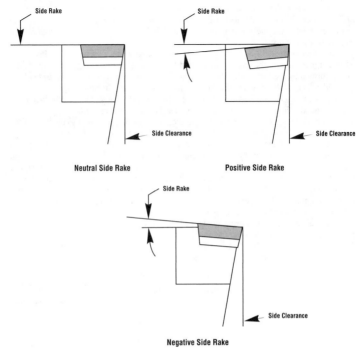

Figure 4. Side rake angles.

Side rake angle is formed between the face of the insert and the line perpendicular to the workpiece when viewing the end. It is positive if it slopes away from the cutting edge, and negative if it slopes upward. It is neutral when perpendicular to the cutting edge. The thickness of the tool behind the cutting edge is dependent on the side rake angle. A small angle allows for a thick tool for maximum strength, but will require higher cutting forces *(Figure 4)*.

Increased rake angles produce thinner chips, reducing cutting force requirements. However, once a maximum recommended angle is exceeded, the cutting edge is weakened and heat transfer is diminished. **Table 1** provides suggested top rake angles for selected workpiece materials.

Side cutting edge angle or *lead angle* is formed between the insert side cutting edge and the side of the tool shank. This angle leads the tool into the workpiece. Enlarging this angle produces wider chips (increasing inversely with the cosine of the angle).

End cutting edge angle is the angle formed between the insert cutting edge on the end of the tool and a line perpendicular to the back side of the tool shank. It determines the clearance between the cutting tool and the finished surface of the workpiece.

While the end and side cutting edge angles are absolute, the top rake angle is dependent on the geometry of the insert. For example, an insert with a positive ground or molded chipbreaker can change the effective top rake angle from negative to positive. Soft and ductile materials normally require larger clearance angles than hard and brittle materials. See the section on Inserts in this book for additional information on chipbreaker geometry.

End relief angle is formed below the end cutting edge by the end face of the insert and a line perpendicular to the base of the toolholder. The end clearance angle may be greater as it is formed by the end face of the toolholder and a line perpendicular to the base of the toolholder. Tip overhang will make the clearance angle exceed the relief angle.

Side relief angle is formed below the side cutting edge by the side face of the insert and a line perpendicular to the base of the toolholder. The side clearance angle may be greater as it is formed by the side face of the toolholder and a line perpendicular to the base of the toolholder. Tip overhang will make the clearance angle exceed the relief angle.

Relief and clearance angles on the end, as well as the side, of the tool are required to permit the tool to enter the cut. Without clearance, it would be impossible for chip formation to occur. Without adequate relief, the cutter will rub and produce heat. However, too large an angle results in a weak cutting edge. In general, the softer and more ductile the workpiece material, the greater the top rake of the turning tool. Ductile workpiece materials require high positive shearing angles while harder, tougher materials are best cut with more neutral or negative geometries. Single point turning tools also need end and side clearance angles in order to enter the cut. Tools without adequate clearance will push away from the workpiece and there will not be sufficient clearance to allow chip formation to take place. **Table 1** provides general recommendations for top rake angles and clearance angles for specific materials.

Table 1. Top Rake and Clearance Angles for Selected Workpiece Materials.

Material	Recommended Rake		Recommended clearance
	Carbide	HSS	
Aluminum	20° positive	25° positive	10°
Titanium, Inconel	10° positive	25° positive	6°
Low Carbon Steel, Ductile Brass	5° positive	25° positive	5°
Malleable Cast Iron <160 BHN	5° negative	18° positive	5°
High Carbon Steel, Cast Irons >160 BHN	0–5° negative	12° positive	4°
Steel Castings	10° negative	12° positive	6°

Tangential, radial, and axial cutting forces all act on the workpiece in a turning operation. The tangential force has the greatest effect on the power consumption of a turning operation, while the axial (feed) force exerts pressure through the part in a longitudinal direction. The radial (depth of cut) force tends to push the workpiece and toolbar apart. When these three components are added together, a resultant cutting force is established. The ratio of these forces is approximately 4:2:1 for a zero degree lead angle tool. This means that if the tangential force equals 1000 lbs (454 kgs), the axial (feed) force will be 500 lbs (227 kgs), and the radial (depth of cut) force 250 lbs (113.5 kgs). The magnitude of the radial and axial forces are altered as the toolholder lead angle is changed. In a turning operation, the greater the lead angle, the larger the radial cutting force and the smaller the axial force *(Figure 5)*.

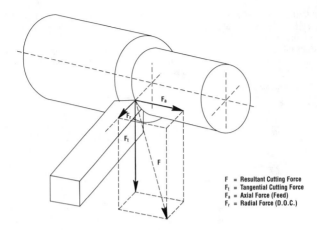

F = Resultant Cutting Force
F_t = Tangential Cutting Force
F_a = Axial Force (Feed)
F_r = Radial Force (D.O.C.)

Figure 5. Cutting forces in turning operations.

Side cutting edge angle (also known as lead angle or attack angle)

The primary consideration when selecting the side cutting edge angle (lead angle) of a turning tool is often the geometry of the workpiece or the material condition. See *Figure 6.* For example, when cutting through scale, interruptions, or a hardened surface, a lead angle tool will allow a reasonable rate of productivity without subjecting the cutting tool edge to severe shock. This benefit must be balanced against the possibility of increased part deflection or vibration due to the higher radial force created by the lead angle. The radial force increases as the lead angle increases. The majority of turning operations are most efficiently performed within a lead angle range of 10° to 30°. It should be noted that in countries using the metric system, lead angle is normally expressed differently.

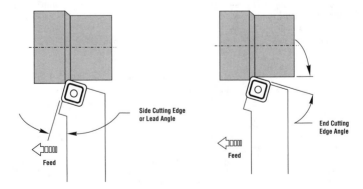

Side Cutting Edge
or Lead Angle

Feed

End Cutting
Edge Angle

Feed

Figure 6. Side and end cutting edge angles.

A 0° lead angle is referred to as a 90° lead angle, a 30° lead angle would be 60°, etc. The charts and formulas below are based on lead angles as they are commonly expressed in the United States.

Chip thickness is equal to feed when using zero degree lead angle cutting tools. However, when lead angle cutting tools are applied, chip thinning occurs. **Table 2** shows how lead angle changes affect chip thickness.

Table 2. Lead Angle/Chip Thickness Relationship.

Lead Angle	Feed Per Revolution fpr	Actual Chip Thickness *
0°	A	A
10°	A	0.9848 × A
15°	A	0.9659 × A
20°	A	0.9397 × A
30°	A	0.8660 × A
45°	A	0.7071 × A
Example		
0°	0.010"	0.010"
10°	0.010"	0.009848"
15°	0.010"	0.009659"
20°	0.010"	0.009397"
30°	0.010"	0.00866"
45°	0.010"	0.007071"

* Chip thickness (ct) is the cosine of the lead angle multiplied by the feed per revolution.

ct = cosine (of lead angle) × fpr.

Example. Where lead angle is 20°, and fpr is 0.010" *or* 0.2540 mm, find the chip thickness.

$$0.9397 \times 0.010 = 0.009397" \qquad or \qquad 0.9397 \times 0.2540 = 0.2387 \text{ mm}$$

Example. Calculate the difference in feed rate for a given chip thickness, between a 0° lead angle cutter and a 45° lead angle cutter. The feed rate is 0.003" (0.0762 mm), operating at 1057 rpm. (Note that the feed rate can be *increased* by 41% by changing only the lead angle.)

For 0° lead angle: in./min = fpr × rpm   *or*   mm/min = fpr × rpm
                 0.003 × 1057 = 3.171   *or*   0.076 × 1057 = 80.332

For 45° lead angle:
         in./min = (fpr ÷ 0.707) × 1057   *or*   mm/min = (fpr ÷ 0.707) × 1057
         (0.003 ÷ 0.707) × 1057 = 4.485   *or*   (0.0762 ÷ 0.707) × 1057 = 113.92

Chip thickness decreases with increases in the lead angle. However, for a given chip thickness, a 45° lead angle will increase both the chip width and feed rate by 41% over that achieved with a zero degree lead angle tool, all other parameters being equal.

Zero degree lead angle turning tools produce chips that are equal in width to the cutting depth of the turning operation. When a lead angle cutting tool is introduced, the effective cutting depth and corresponding chip width will exceed the actual cutting depth on the workpiece. **Table 3** calculates how different lead angles affect feed rate, chip thickness, and chip width in relation to increases in the lead angle.

Table 3. Selected Machining Parameters Influenced by Lead Angle.

Lead Angle	fpr	Actual Chip Thickness	Chip Width
0°	100%	100%	100%
10°	101.5%	98%	101.5%
15°	103.5%	97%	103.5%
20°	106.4%	94%	106.4%
30°	115.5%	87%	115.5%
45°	141.4%	71%	141.4%

Actual chip width for any depth of cut and lead angle can be calculated from the percentages in **Table 3**.

Example. Where lead angle is 30°, and d.o.c. is 0.20" (5.08 mm), find the chip width (c_w).

c_w = d.o.c. × c_w % expressed as a decimal

c_w = 0.20 × 1.155 = 0.231"   *or* c_w = 5.08 × 1.155 = 5.8674 mm.

Nose radius selection and surface finish

The surface finish produced on the part during a turning operation is dependent on the rigidity of the tool, the machine, and the workpiece. Once rigidity can be assured, the relationship between the machine feed (in./rev *or* mm/rev) and the nose contour of the insert or cutting tool can be used to determine the surface finish of the workpiece. The nose contour, expressed as a radius, has a direct effect on surface finish. Generally, the larger the radius, the better the surface finish. However, if too large, chatter may be induced. For a machining operation where a smaller than optimum radius is required, feed rate may have to be reduced to achieve a desired finish. *Figure 7* provides achievable surface finish results for prescribed feed rates and nose radii. For more information on surface finish and the use of wiper inserts in turning, see the Insert Selection section of this book.

Productivity

Provided that the appropriate level of power from the machine is available, productivity in a turning operation can be improved by increasing one or more of the following variables: depth of cut, feed rate, and speed.

Depth of cut. The easiest cutting parameter to alter is depth of cut. In turning, doubling the depth of cut will double productivity without increasing cutting temperature, cutting force per cubic inch or centimeter of material removed (specific cutting force), or the tensile strength of the chip. The power required from the machine will virtually double without any reduction in tool life, assuming the cutting edge can withstand the added tangential cutting force. However, increasing the depth of cut to increase productivity is an option only if there is sufficient material to be removed and there is sufficient power available from the machine.

Feed rate. Because feed rate is often easy to alter, it is commonly increased to gain added productivity. Doubling the feed rate results in a chip twice as thick, making it more difficult for the chip to curl and bend. If feed is doubled, the tangential cutting force, cutting temperature, and required power also increase, but they do not double. This occurs because the tool is cutting more efficiently and less power is wasted in heat generation. Tool life is reduced, but not halved. The additional force imposed on the cutting edge often

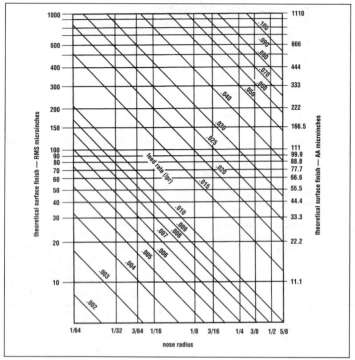

Nose radius and feed rate have the greatest impact on surface finish. To determine the nose radius required for a theoretical surface finish, use the following procedure and the chart above.

1. Locate the required surface finish (RMS or AA) on the vertical axis.
2. Follow the horizontal line corresponding to the desired theoretical finish to where it intersects the diagonal line corresponding to the intended feed rate.
3. Project a line downward to the nose radius scale and read the required nose radius.
4. If this line falls between two values, choose the larger value.
5. If no available nose radius will produce the required finish, feed rate must be reduced.
6. Reverse the procedure to obtain surface finish from a given nose radius.

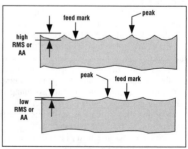

Note: Peaks produced with a small radii insert (top) compared to those produced with a large radius insert (bottom).

Figure 7. Nose radius selection and surface finish. (Source, Kennametal Inc.)

results in cratering of the top rake insert surface due to the increased temperature and friction generated during the cut. However, the cutting force, in relation to the amount of material removed (specific cutting force) actually decreases. The effect on the tool must be carefully monitored as feed is increased in order to avoid catastrophic failure of the insert when the chip becomes stronger than the cutting edge.

Speed. Increasing cutting speed imposes several detrimental effects on the cutting edge. In addition to a significant rise in cutting temperature, doubling the cutting speed requires more power and reduces tool life by more than half. The tangential and specific cutting forces actually diminish when speed is doubled as more power is consumed in heat generation per cubic inch *or* centimeter of workpiece material turned. The actual load on the top rake face of the insert is reduced, but cratering can still result due to higher cutting temperatures.

Troubleshooting

Insert wear is an accurate indicator of the success or failure of machining factors selected for any turning operation. **Table 4** identifies common insert failure modes and provides corrective measures for each.

Table 4. Insert Troubleshooting Chart for Turning. *(Source, Kennametal Inc.)*

Problem	Corrective Action	Problem	Corrective Action
Edge Wear*	Increase feed rate. Reduce speed (sfm or smm). Use more wear resistant grade. Apply coated grade.	Chipping	Utilize stronger grade. Consider edge preparation. Check rigidity of system. Increase lead angle.
Heat Deformation	Reduce feed rate. Reduce speed (sfm or smm). Reduce depth of cut. Use grade with higher hot hardness.	Depth of Cut Notching	Change lead angle. Consider edge preparation. Apply different grade. Adjust feed
Thermal Cracking	Properly apply coolant. Reduce speed (sfm or smm). Reduce feed rate. Apply coated grade.	Built-up Edge	Increase speed (sfm or smm). Increase feed rate. Apply PVD coated grades or cermets. Utilize coolant. Apply edge prep (smaller hone).
Cratering	Reduce feed rate. Reduce speed (sfm or smm). Apply coated grades or cermets. Utilize coolant.	Catastrophic Breakage	Utilize stronger grade/geometry. Reduce feed rate. Reduce depth of cut. Check rigidity of system. Examine edge prep/nose radius.

* Generally, inserts should be indexed when 0.030" (0.77 mm) flank wear is reached. For finishing operations, index at 0.015" (0.38 mm) or sooner.

Other areas of investigation are identified by unacceptable chips and workpiece and machine problems, which impact a wide variety of areas. **Table 5** suggests probable causes and solutions.

Table 5. Troubleshooting Turning Operations. *(Source, Kennametal Inc.)*

Possible Causes and Areas of Investigation	Unacceptable Chips		Workpiece Concerns		Machine Concerns	Insert Failure Modes							
	stringer/ribbons (light silver color)	corrugated/tight (dark blue or black color)	finish/rms tolerance	interrupted cuts	areas of investigation	edge wear	heat deformation (upset)	thermal cracking	crater	chipping	depth-of-cut notching	built-up edge	catastrophic breakage
speeds (sfm or smm)	P↑	•	P	P↑	•	P	P↓	•	P↓	•	•	P↑	•
feed	P↑	P↓	P	P↓	•	P	P↓	•	P↓	•	•	P↑	•
depth of cut	•	•	-	P↓	-	•	P↓	•	-	-	-	-	•
grade	-	-	•	•	-	P	•	P	•	P	•	P	•
coolant	•	•	•	-	•	•	•	P	•	•	-	•	•
rake angle	•	•	•	•	-	-	•	•	•	•	•	•	•
edge preparation	•	•	•	•	-	-	-	-	-	P	•	P	•
material (type/condition)	•	•	•	•	-	-	-	-	-	•	P	P	•
center height	-	-	•	-	•	•	-	-	-	•	-	•	-
geometry (insert)	P	P	-	•	-	-	-	-	•	•	-	•	•
insert finish	-	-	•	-	-	-	-	-	-	-	-	•	-
insert thickness	-	-	-	•	-	-	•	•	-	•	•	-	•
nose radius	•	•	P	•	-	-	•	-	-	•	-	-	•
lead angle	•	•	-	↑	-	-	-	-	-	•	P	-	•
holder (type/condition)	-	-	-	-	-	-	-	-	-	•	-	-	•
machine condition	-	-	•	•	•	-	-	-	-	•	-	-	•
chip flow direction	•	-	•	-	-	-	-	-	•	-	-	-	-
horsepower	-	-	-	•	•	-	-	-	-	•	-	-	-
excessive overhand	-	-	-	•	-	-	-	-	-	-	-	-	•
gibs (worn/out of adjustment)	-	-	-	•	•	-	-	-	-	-	-	-	-
spindle bearings	-	-	-	•	•	-	-	-	-	-	-	-	-
turret	-	-	-	-	•	-	-	-	-	-	-	-	-
head stock	-	-	-	-	•	-	-	-	-	-	-	-	-
machine level	-	-	-	-	•	-	-	-	-	-	-	-	-
machine anchored	-	-	-	-	•	-	-	-	-	-	-	-	-
workholding	-	-	-	•	•	-	-	-	-	•	-	-	•
leadscrew	-	-	-	-	•	-	-	-	-	-	-	-	-
rigidity	-	-	-	•	P	-	-	-	-	•	-	-	P
chatter	-	-	-	•	-	-	-	-	-	•	-	-	P

↑↓ Arrows indicate direction of adjustment.
"P" indicates areas of primary investigation.

Toolholder identification

The following paragraphs detail the identification system for mechanical clamping indexable insert holders as identified by ISO and ANSI standards. The nine sections of this system provide dimensional specifications, styles, and letter or numerical designations to identify the dimensions of the toolholder. The number or letter used in each section (or position) describes the characteristics of the holder for positive and consistent identification. It should be noted that manufacturers of toolholders might expand these classifications to include additional characteristics relevant to their products.

Section 1: Insert clamping method. Inserts are mechanically locked into place in square and rectangular shank tooling holders using a variety of methods. Cutting tool manufacturers have developed innovative variations of the four basic clamping systems for use with their shafts and inserts, but there remain four common holding methods, each identified by a letter (see *Figure 8*). **C**-clamping utilizes a top clamp and is employed on inserts that do not have a center hole. It relies entirely on friction to retain the insert, and it is normally used with positive rake inserts on medium to light turning and in boring operations. **M**-clamping is used on negative rake inserts with a center hole. The insert is forced against the pocket wall with a cam lockpin that secures the protective insert pocket shim. The top clamp holds the back of the insert and keeps it from lifting when a cutting load is applied to the nose of the insert. **M**-clamping is designed for medium to heavy turning operations. **S**-clamping uses a simple torx or allen head screw to secure the insert in the toolholder pocket. The inserts used with the S system must have either counterbores or countersinks. A disadvantage of this system is that the screws can sometimes seize when used in an operation that produces intense heat. Ideally, **S**-clamping is used for light to medium turning operations and in boring operations. **P**-clamping is the ISO standard for turning tools—the design principle is for negative rake inserts with holes. The insert is forced against the pocket walls using a pivoted lever that tilts as the adjustment screw is seated. This design is suited for medium to heavy turning operations but it does not prohibit the insert from lifting during the cut.

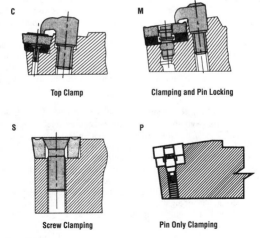

C

Top Clamp

M

Clamping and Pin Locking

S

Screw Clamping

P

Pin Only Clamping

Figure 8. Clamping methods.

Section 2: Insert shape. Each shape is identified by a letter *(see Figure 9)*. The section on Inserts contains additional detailed information on insert shapes and dimensions.

Section 3: Tool style or lead angle. **A** style = Straight shank, 0° side cutting edge angle. **B** style = Straight shank, 15° side cutting edge angle. **C** style = Straight shank, 0° end cutting edge angle. **D** style = Straight shank, 45° side cutting edge angle. **E** style = Straight shank, 30° side cutting edge angle. **F** style = Offset shank, 0° end cutting edge angle. **G** style = Offset shank, 0° side cutting edge angle. **H** style = Straight shank, 38° side cutting edge angle. **J** style = Offset shank, negative 3° side cutting edge angle. **K** style = Offset shank, 15° end cutting edge angle. **L** style = Offset shank, negative 5° side cutting angle. **M** style =

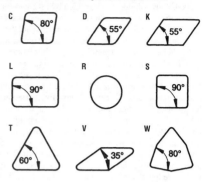

Figure 9. Insert shapes.

Straight shaft, 40° side cutting edge angle. **N** style = Straight shank, 27° side cutting edge angle. **P** style = Straight shank, 27°30' side cutting edge angle. **Q** style = Offset shank, 17°30' end cutting edge angle. **R** style = Offset shank, 15° side cutting edge angle. **S** style = Offset shank, 45° end cutting edge angle. **T** style = Offset shank, 30° side cutting edge angle. **U** style = Offset shank, negative 3° end cutting edge angle. **V** style = Straight shank, 17°30' side cutting edge angle. **W** style = Offset shank, 10° side cutting edge angle. **Y** style = Offset shank, 5° cutting edge angle (see *Figures 10 and 11*).

Section 4: Holder rake angle or insert clearance angle. For rake angle: **P** = Positive rake **N** = Negative rake **O** = Neutral rake. The rake angle is also called the angle of inclination or top rake angle. It is the angle formed between the angle of inclination of the insert and the line perpendicular to the workpiece when viewing the tool from the side, front to back. The rake is *positive* when the sum of the end clearance angle and wedge angle is less than 90°. It is *negative* when the sum of the end clearance angle and wedge angle exceeds 90°. It is *neutral* when the sum of the end clearance angle and the wedge angle is 90°. For insert clearance angle, the following letters have been assigned: **A** = 3°, **B** = 5°, **C** = 7°, **D** = 15°, **E** = 20°, **F** = 25°, **G** = 30°, **N** = 0°, **P** = 11°.

Section 5: Hand of tool. **R** = Right hand. **L** = Left hand. **N** = Either hand (neutral). A right hand tool cuts from right to left, a left hand tool cuts from left to right.

Sections 6 and 7: Holder shank dimensions. **Inch units**: (ANSI B94.45M-1979). The shank dimensions are identified by a significant two-digit number that indicates the holder cross section. For shanks under ⁵/₈" (15.88 mm) square, the number will represent the total number of sixteenths (1.59 mm) of width and height, and will be preceded by a zero. For shanks ⁵/₈" (15.88 mm) square and over, the numbers will represent the total number of sixteenths (1.59 mm) of width and height. For rectangular holders, the first digit represents the number of eighths (3.18 mm) of width and the second digit the number of quarters (6.35 mm) of height, except for a toolholder 1 ¹/₄ × 1 ¹/₂" (31.75 × 38.10 mm) which is given the number 91.

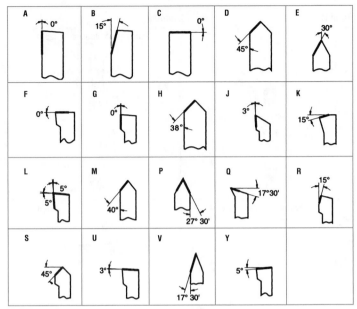

Figure 10. Tool style and lead angle designations.

$05 = \frac{5}{16} \times \frac{5}{16}$ $85 = 1\frac{1}{4} \times 1$

$06 = \frac{3}{8} \times \frac{3}{8}$ $20 = 1\frac{1}{4} \times 1\frac{1}{4}$

$08 = \frac{1}{2} \times \frac{1}{2}$ $24 = 1\frac{1}{2} \times 1\frac{1}{2}$

$10 = \frac{5}{8} \times \frac{5}{8}$ $86 = 1\frac{1}{2} \times 1$

$12 = \frac{3}{4} \times \frac{3}{4}$ $89 = 1 \times 1\frac{1}{4}$

$16 = 1 \times 1$ $32 = 2 \times 2$

Section 6: Tool height for rectangular shank cross sections as a two-digit number in millimeters, or as a one-digit number preceded by a zero. **Metric units** (ANSI B212.5-1986).

Section 7: Tool width for rectangular shank cross sections is recorded in the same manner as the height.

Section 8: Insert size based on inscribed circle. **Inch units** (ANSI B94.45M-1979). Based on inscribed circle in increments of eighths of an inch (3.18 mm). $\frac{1}{4}" = 2$, $\frac{3}{8}" = 3$, $\frac{1}{2}" = 4$, etc. **Metric units** (ANSI B212.5-1986). For rectangular and parallelogram shapes, the size shall be the length of the longest cutting edge in millimeters, disregarding decimals, and preceded by a zero if a one-digit number. For all other shapes, the side length (or the diameter for round inserts) in millimeters, disregarding decimals, and preceded by a zero if a one digit number. (**Note**: Insert size is *Section 9* of the metric standard.)

Section 9: Qualified surface and tool length. **Inch units** based on ANSI B94.45M-1979.

Qualified Back and End:
A = 4" (101.60 mm) long.
B = 4.5" (114.30 mm) long.

Qualified Front and End:
M = 4" (101.60 mm) long.
N = 4.5" (114.30 mm) long.

(Text continued on p. 1008)

Turning　　　　　　　　　　　　　　**Turning**

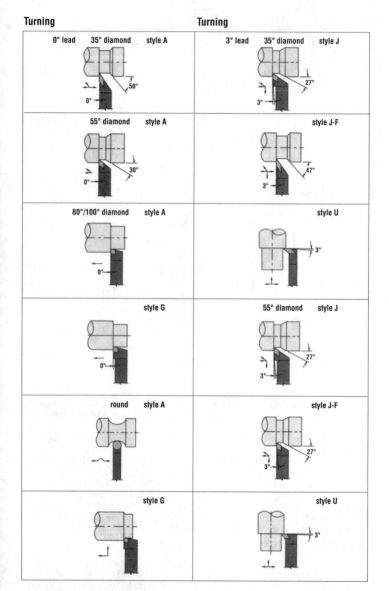

Figure 11. Toolholder/insert combinations for turning and profiling operations. (Source, Kennametal Inc.)

Turning   **Turning**

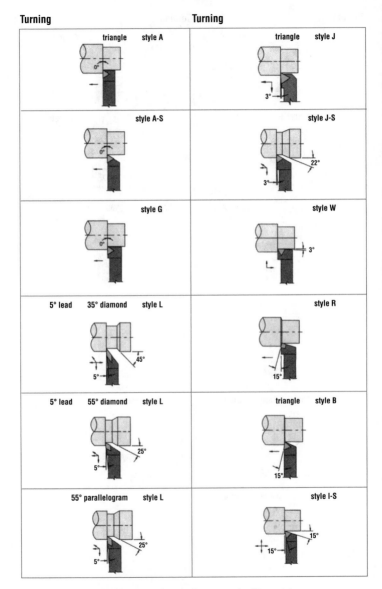

Figure 11. (Continued) Toolholder/insert combinations for turning and profiling operations. (Source, Kennametal Inc.)

Turning   Turning

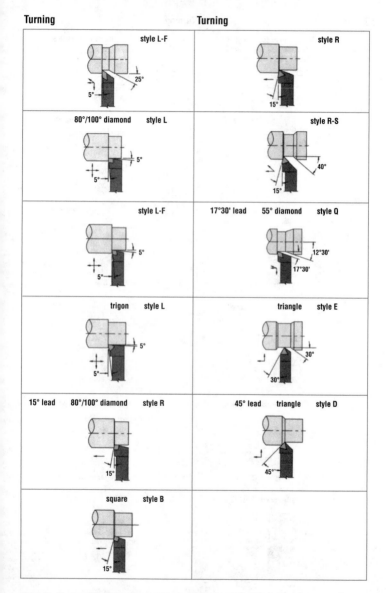

Figure 11. (Continued) Toolholder/insert combinations for turning and profiling operations.
(Source, Kennametal Inc.)

Facing

Profiling

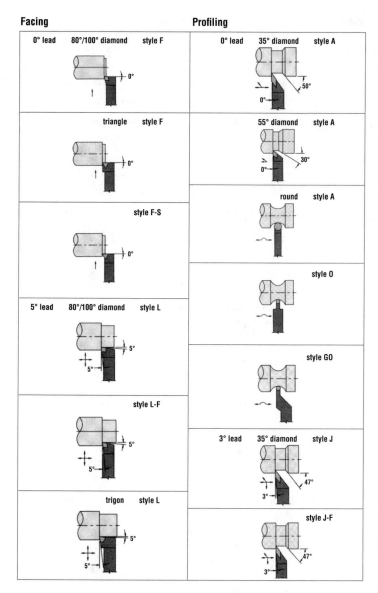

Figure 11. (Continued) Toolholder/insert combinations for turning and profiling operations. (Source, Kennametal Inc.)

Profiling   Profiling

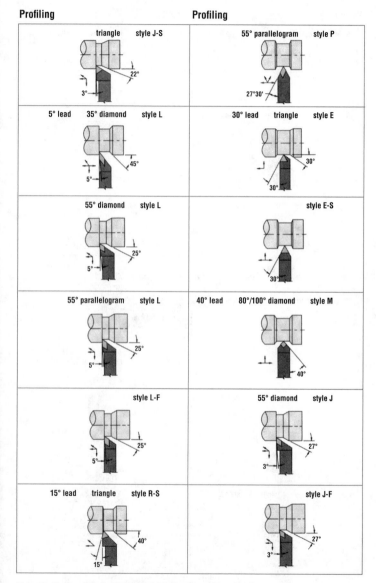

Figure 11. (Continued) Toolholder/insert combinations for turning and profiling operations. (Source, Kennametal Inc.)

Profiling

Profiling

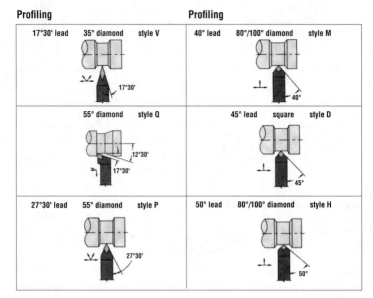

Chamfering

Plunging

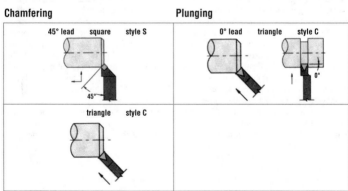

Figure 11. (Continued) Toolholder/insert combinations for turning and profiling operations. (Source, Kennametal Inc.)

C = 5" (127.00 mm) long.	P = 5" (127.00 mm) long.
D = 6" (152.40 mm) long.	R = 6" (152.40 mm) long.
E = 7" (177.80 mm) long.	S = 7" (177.80 mm) long.
F = 8" (203.20 mm) long.	T = 8" (203.20 mm) long.
G = 5.5" (139.70 mm) long.	U = 5.5" (139.70 mm) long.

Tool Length. **Metric units** based on ANSI B212.5-1986. (**Note**: Tool Length is *Section 8* of the metric standard.)

A = 32 mm	G = 90 mm	N = 160 mm	U = 350 mm
B = 40 mm	H = 100 mm	P = 170 mm	V = 400 mm
C = 50 mm	J = 110 mm	Q = 180 mm	W = 450 mm
D = 60 mm	K = 125 mm	R = 200 mm	X = to be specified
E = 70 mm	L = 140 mm	S = 250 mm	Y = 500 mm
F = 80 mm	M = 150 mm	T = 300 mm	

Section 10: Special Tolerances. The metric standard includes a provision for qualified holders that have tolerances of ±0.08 mm for back and end (**Q**), front and end (**F**), or back, front, and end (**B**) dimensions.

Maximum effective cutting edge length

The toolholder style, nose radius, and insert shape profoundly impact the effective cutting edge length in turning operations. **Table 6** provides the maximum effective cutting edge length for several toolholder/insert combinations. **Table 7** provides dimensional compensation required for various insert radius and toolholder combinations.

Brazed single-point tools

These tools are available in a variety of styles and sizes. Solid single-point tools may be made of high-speed steels, carbon steel, cobalt alloys, and carbides. Due to the expense of producing a complete tool from costly materials, solid tools have been largely replaced with brazed-tipped turning tools. Typically, brazed-tip tools are made with a body of relatively inexpensive material with a tip or blank of more costly cutting material brazed to the cutting point. Materials used for the tip include high-speed steel, carbides, and cubic boron nitride. They are available in standard styles designated by the letters A to G. Offset point styles A, B, E, F, and G are available as either right- or left-hand cutting tools. For square shanks, the number following the letter designation indicates the number of 1/16ths of an inch required to equal the height or width of the tool. For rectangular shanked tools, the first number is the total of 1/8ths of an inch in the shank width, and the second number is the total of 1/4ths of an inch in the shank height. The nose radius is related to shank size, and care should be taken to match the tool size to the finish requirements on each machining operation. **Tables 8** through **16** provide dimensions for single-point turning tools.

Specialized tools for boring

Boring is primarily used to finish large cored holes in castings or pierced holes in forgings. The operation is also referred to as internal turning, and the tools and processes used in boring are in many ways similar to turning. However, because boring is usually a finishing operation, and therefore chip flow is a major concern, the cutting angles employed are critical to success. Single point boring tools have been standardized (ANSI B212.1-1984) and have been assigned a simple identification system. The first letter of the identification denotes that the tool is either solid (S) or tipped (T). The second describes the shape, either square (S) or round (R). The third letter, A-H, refers to the side edge cutting angle as follows: A=0°, B=10°, C=30°, D=40°, E=45°, F=55°, G=90° with 0° rake,

(Text continued on p. 1020)

Table 6. Maximum Effective Cutting Edge Length for Selected Toolholder/Insert Combinations.
(Source, Stellram USA.)

Insert Shape				"T" 60° Triangular							
Tool Style				"A" "F" & "G"		"B" & "R"		"E"		"J"	
Inscribed Circle		Nose Radius		0°		15°		30°		3°	
inch	mm	inch	mm	inch	mm	inch	mm	inch	mm	inch	mm
1/4	6.35	1/125	.203	.413	10.49	.397	10.08	.355	9.02	.413	10.49
1/4	6.35	1/64	.397	.395	10.03	.378	9.60	.366	9.30	.395	10.03
1/4	6.35	1/32	.794	.356	9.04	.337	8.56	.297	7.54	.357	9.07
1/4	6.35	3/64	1.191	.317	8.05	.296	8.05	.258	6.55	.320	8.13
5/16	7.94	1/125	.203	.522	13.26	.502	12.75	.449	11.40	.521	13.23
5/16	7.94	1/64	.397	.503	12.78	.482	12.24	.430	10.92	.503	12.78
5/16	7.94	1/32	.794	.464	11.78	.441	11.20	.391	9.93	.465	11.81
5/16	7.94	3/64	1.191	.426	10.82	.401	10.18	.352	8.94	.428	10.87
3/8	9.53	1/125	.203	.630	16.00	.607	15.42	.543	13.79	.629	15.98
3/8	9.53	1/64	.397	.611	15.52	.587	14.91	.524	13.31	.611	15.52
3/8	9.53	1/32	.794	.572	14.53	.546	13.87	.484	12.29	.573	14.55
3/8	9.53	3/64	1.191	.534	13.56	.505	12.83	.445	11.30	.536	13.61
3/8	9.53	1/16	1.588	.496	12.60	.465	11.81	.406	10.31	.498	12.65
3/8	9.53	3/32	2.381	.418	10.62	.383	9.73	.328	8.33	.423	10.74
1/2	12.70	1/125	.203	.846	21.49	.816	20.73	.730	18.54	.846	21.49
1/2	12.70	1/64	.397	.828	21.03	.796	20.22	.711	18.06	.827	21.00
1/2	12.70	1/32	.794	.789	20.04	.755	19.18	.672	17.07	.790	20.07
1/2	12.70	3/64	1.191	.751	19.08	.714	18.14	.633	16.08	.752	19.10
1/2	12.70	1/16	1.588	.712	18.08	.674	17.12	.594	15.09	.714	18.14
1/2	12.70	3/32	2.381	.635	16.13	.592	15.04	.516	13.11	.639	16.23
1/2	12.70	1/8	3.175	.558	14.17	.511	12.98	.438	11.12	.564	14.33
5/8	15.88	1/64	.397	1.044	26.52	1.005	25.52	.899	22.83	1.044	26.52
5/8	15.88	1/32	.794	1.005	25.52	.964	24.48	.859	21.82	1.006	25.55
5/8	15.88	3/64	1.191	.967	24.56	.924	23.47	.820	20.83	.968	24.59
5/8	15.88	1/16	1.588	.929	23.60	.883	22.43	.781	19.84	.931	23.65
5/8	15.88	3/32	2.381	.851	21.61	.801	20.35	.703	17.86	.855	21.72
5/8	15.88	1/8	3.175	.775	19.68	.720	18.29	.625	16.56	.780	19.81
3/4	19.05	1/32	.794	1.222	31.04	1.173	29.79	1.047	26.59	1.222	31.04
3/4	19.05	3/64	1.191	1.184	30.07	1.133	28.78	1.008	25.60	1.184	30.07
3/4	19.05	1/16	1.588	1.145	29.08	1.092	27.74	.969	24.61	1.147	29.13
3/4	19.05	3/32	2.381	1.068	27.12	1.010	25.65	.891	22.63	1.072	27.22
3/4	19.05	1/8	3.175	.991	25.17	.929	23.60	.813	20.65	.996	25.30
1	25.40	3/32	2.381	1.501	38.13	1.429	36.30	1.266	32.16	1.504	44.20
1	25.40	1/8	3.175	1.424	36.17	1.347	34.21	1.188	30.18	1.429	36.30

(Continued)

Table 6. *(Continued)* **Maximum Effective Cutting Edge Length for Selected Toolholder/Insert Combinations.**
(Source, Stellram USA.)

Insert Shape				"S" Square				"V" 35° Diamond			
Tool Style				"D" & "S"		"R" & "K"		"J"		"P"	
Inscribed Circle		Nose Radius									
inch	mm	inch	mm	inch	mm	inch	mm	inch	mm	inch	mm
1/4	6.35	1/125	.203	.168	4.27	.232	5.89	.418	10.62	.376	9.55
1/4	6.35	1/64	.397	.159	4.04	.223	5.66	.402	10.21	.366	9.30
1/4	6.35	1/32	.794	.142	3.61	.204	5.18	.369	9.37	.344	8.74
1/4	6.35	3/64	1.191	.124	3.15	.186	4.72	.336	8.53	.323	8.20
5/16	7.94	1/125	.203	.212	5.38	.292	7.42	.527	13.38	.473	12.01
5/16	7.94	1/64	.397	.204	5.18	.283	7.19	.511	12.98	.462	11.74
5/16	7.94	1/32	.794	.186	4.72	.265	6.73	.478	12.14	.441	11.20
5/16	7.94	3/64	1.191	.168	4.27	.246	6.25	.445	11.30	.420	10.67
3/8	9.53	1/125	.203	.256	6.50	.353	8.97	.636	16.15	.569	14.45
3/8	9.53	1/64	.397	.248	6.30	.344	8.74	.620	15.75	.559	15.21
3/8	9.53	1/32	.794	.230	5.84	.325	8.25	.587	14.91	.538	13.67
3/8	9.53	3/64	1.191	.213	5.41	.306	7.77	.554	14.07	.517	13.13
3/8	9.53	1/16	1.588	.195	4.95	.288	7.32	.521	13.23	.496	12.60
3/8	9.53	3/32	2.381	.160	4.06	.251	6.38	.455	11.58	.453	11.51
1/2	12.70	1/125	.203	.345	8.76	.473	12.01	.854	21.69	.762	19.35
1/2	12.70	1/64	.397	.336	8.53	.464	11.78	.838	21.28	.752	19.10
1/2	12.70	1/32	.794	.319	8.10	.446	11.33	.804	20.42	.731	18.57
1/2	12.70	3/64	1.191	.301	7.64	.427	10.84	.771	19.58	.710	18.03
1/2	12.70	1/16	1.588	.284	7.21	.409	10.39	.738	18.74	.689	17.50
1/2	12.70	3/32	2.381	.248	6.30	.371	9.42	.672	17.07	.647	16.43
1/2	12.70	1/8	3.175	.213	5.41	.334	8.48	.606	15.39	.604	15.34
5/8	15.88	1/64	.397	.424	10.77	.585	14.60	1.055	26.80	.946	24.03
5/8	15.88	1/32	.794	.407	10.34	.566	14.38	1.022	25.96	.924	23.47
5/8	15.88	3/64	1.191	.389	9.88	.548	13.92	.989	25.12	.903	22.94
5/8	15.88	1/16	1.588	.372	9.45	.529	13.44	.956	24.28	.882	22.40
5/8	15.88	3/32	2.381	.337	8.56	.492	12.50	.890	22.61	.840	21.34
5/8	15.88	1/8	3.175	.302	7.67	.455	11.56	.824	20.93	.798	20.27
3/4	19.05	1/32	.794	.495	12.57	.687	17.45	1.240	31.50	1.118	28.39
3/4	19.05	3/64	1.191	.478	12.14	.669	16.99	1.207	30.66	1.097	27.86
3/4	19.05	1/16	1.588	.460	11.68	.650	16.51	1.174	29.82	1.075	27.30
3/4	19.05	3/32	2.381	.425	10.80	.613	15.57	1.107	28.12	1.033	26.24
3/4	19.05	1/8	3.175	.390	9.91	.576	14.63	1.041	26.44	.991	25.17
1	25.40	3/32	2.381	.602	15.29	.854	21.69	1.543	39.19	1.420	36.07
1	25.40	1/8	3.175	.567	14.40	.817	20.75	1.477	37.51	1.378	35.00

(Continued)

Table 6. *(Continued)* **Maximum Effective Cutting Edge Length for Selected Toolholder/Insert Combinations.** *(Source, Stellram USA.)*

Insert Shape				"D" 55° Diamond				"C" 80° Diamond		"W" 80° Trigon	
Tool Style				"J"		"P"		"F" & "G"		"P"	
Inscribed Circle		Nose Radius									
inch	mm	inch	mm	inch	mm	inch	mm	inch	mm	inch	mm
1/4	6.35	1/125	.203	.294	7.47	.258	6.55	.246	6.25	.148	3.76
1/4	6.35	1/64	.397	.283	7.19	.245	6.22	.238	6.04	.147	3.73
1/4	6.35	1/32	.794	.261	6.63	.220	5.59	.222	5.64	.145	3.68
1/4	6.35	3/64	1.191	.240	6.10	.194	4.93	.206	5.23	.144	3.66
5/16	7.94	1/125	.203	.370	9.40	.325	8.26	.309	7.85	.185	4.70
5/16	7.94	1/64	.397	.359	9.12	.313	7.95	.301	7.64	.184	4.67
5/16	7.94	1/32	.794	.338	8.58	.287	7.29	.285	7.24	.182	4.62
5/16	7.94	3/64	1.191	.316	8.03	.262	6.65	.269	6.83	.181	4.60
3/8	9.53	1/125	.203	.446	11.33	.393	9.98	.373	9.47	.222	5.64
3/8	9.53	1/64	.397	.436	11.07	.381	9.68	.365	9.27	.221	5.61
3/8	9.53	1/32	.794	.414	10.52	.355	9.02	.349	8.86	.220	5.59
3/8	9.53	3/64	1.191	.392	9.96	.330	8.38	.332	8.43	.218	5.54
3/8	9.53	1/16	1.588	.371	9.42	.304	7.72	.316	8.03	.216	5.49
3/8	9.53	3/32	2.381	.327	8.30	.253	6.43	.284	7.21	.213	5.41
1/2	12.70	1/125	.203	.599	15.21	.528	13.41	.500	12.70	.296	7.52
1/2	12.70	1/64	.397	.588	14.94	.516	13.11	.492	12.50	.295	7.49
1/2	12.70	1/32	.794	.566	14.38	.491	12.47	.475	12.06	.294	7.47
1/2	12.70	3/64	1.191	.545	13.84	.465	11.81	.459	11.66	.292	7.42
1/2	12.70	1/16	1.588	.523	13.28	.440	11.18	.443	11.25	.291	7.39
1/2	12.70	3/32	2.381	.480	12.19	.389	9.88	.411	10.44	.287	7.29
1/2	12.70	1/8	3.175	.436	11.07	.338	8.58	.379	9.63	.284	7.21
5/8	15.88	1/64	.397	.740	18.80	.651	16.54	.619	15.72	.369	9.37
5/8	15.88	1/32	.794	.719	18.26	.626	15.90	.602	15.29	.368	9.35
5/8	15.88	3/64	1.191	.697	17.70	.600	15.24	.586	14.88	.366	9.30
5/8	15.88	1/16	1.588	.675	17.14	.575	14.60	.570	14.48	.365	9.27
5/8	15.88	3/32	2.381	.632	16.05	.524	13.31	.538	13.66	.362	9.19
5/8	15.88	1/8	3.175	.589	14.96	.473	12.01	.506	12.85	.359	9.12
3/4	19.05	1/32	.794	.871	22.12	.761	19.33	.729	18.52	.442	11.23
3/4	19.05	3/64	1.191	.849	21.56	.736	18.70	.713	18.11	.441	11.20
3/4	19.05	1/16	1.588	.828	21.03	.710	18.03	.697	17.70	.439	11.15
3/4	19.05	3/32	2.381	.784	19.91	.660	16.76	.665	16.89	.436	11.07
3/4	19.05	1/8	3.175	.741	18.82	.609	15.47	.633	16.08	.433	11.00
1	25.40	3/32	2.381	1.089	27.66	.930	23.62	.919	23.34	.584	14.83
1	25.40	1/8	3.175	1.046	26.57	.879	22.33	.887	22.53	.581	14.76

Table 7. Insert Radius Compensation. *(Source, Stellram USA.)*

Triangle Profile				

A & G Style 0° SCEA

Rad.	L1	L2	D1	D2
1/64	.0114	.0271	.00	.0156
1/32	.0229	.0541	.00	.0312
3/64	.0343	.0812	.00	.0469
1/16	.0458	.1082	.00	.0625

B & R Style 15° SCEA

Rad.	L1	L2	D1	D2
1/64	.0146	.0302	.0081	.0039
1/32	.0291	.0604	.0162	.0078
3/64	.0437	.0906	.0243	.0117
1/16	.0582	.01207	.0324	.0156

C & F Style 0° ECEA

Rad.	L1	L2	D1	D2
1/64	.00	.0156	.0114	.0271
1/32	.00	.0312	.0229	.0541
3/64	.00	.0469	.0343	.0812
1/16	.00	.0625	.0458	.1082

E Style 30° SCEA

Rad.	L1	L2	D1	D2
1/64	.0156	.0312	.00	.0090
1/32	.0312	.0625	.00	.0180
3/64	.0469	.0938	.00	.0270
1/16	.0625	.1250	.00	.0361

J Style 3° SCEA

Rad.	L1	L2	D1	D2
1/64	.0106	.0262	.0014	.0170
1/32	.0212	.0524	.0028	.0340
3/64	.0318	.0786	.0042	.0511
1/16	.0423	.1048	.0056	.0681

Note: D1, L1 = Sharp Corner to Nose Radius. D2, L2 = Sharp Corner to Centerline Nose Radius.
D3, L3 = Sharp Corner to Preset Point. F = Head Width. C = Toolholder length.

(Continued)

Table 7. *(Continued)* **Insert Radius Compensation.** *(Source, Stellram USA.)*

Square Profile

B Style 15° SCEA

Rad.	L1	L2	D1	D2
1/64	.0035	.0191	.0110	.0009
1/32	.0070	.0383	.0221	.0019
3/64	.0105	.0574	.0331	.0028
1/16	.0140	.0765	.0552	.0038

D & S Style 45° ECEA & SCEA

Rad.	L1	L2	D1	D2
1/64	.0065	.0221	.00	.0065
1/32	.0129	.0442	.00	.0129
3/64	.0194	.0663	.00	.0194
1/16	.0259	.0884	.00	.0259

K Style 15° Reverse ECEA

Rad.	L1	L2	D1	D2
1/64	.0110	.0009	.0035	.0191
1/32	.0221	.0019	.0070	.0383
3/64	.0331	.0028	.0105	.0574
1/16	.0442	.0038	.0140	.0765

35° Diamond Profile

J Style 3° Reverse SCEA

Rad.	L1	L2	D1	D2
1/64	.0330	.0487	.0026	.0182
1/35	.0661	.0973	.0051	.0364
3/64	.0991	.1460	.0077	.0546
1/16	.1322	.1947	.0103	.0728

55° Diamond Profile

3° Reverse SCEA

Rad.	L1	L2	D1	D2
1/64	.0135	.0292	.0015	.0172
1/32	.0271	.0583	.0031	.0343
3/64	.0406	.0875	.0046	.0519
1/16	.0541	.1166	.0062	.0687

(Continued)

Table 7. *(Continued)* Insert Radius Compensation. *(Source, Stellram USA.)*

80° Diamond Profile

F Style 0° ECEA

Rad.	L1	L2	D1	D2
1/64	.00	.0156	.0030	.0186
1/32	.00	.0312	.0060	.0372
3/64	.00	.0469	.0090	.0559
1/16	.00	.0625	.0120	.0745

G Style 0° SCEA

Rad.	L1	L2	D1	D2
1/64	.0030	.0186	.00	.0156
1/32	.0069	.0312	.00	.0312
3/64	.0090	.0559	.00	.0469
1/16	.0120	.0745	.00	.0625

K Style 15° Reverse ECEA

Rad.	L1	L2	D1	D2
1/64	.0117	.0003	.0011	.0167
1/32	.0234	.0006	.0022	.0334
3/64	.0351	.0009	.0032	.0501
1/16	.0468	.0012	.0043	.0668

L Style 5° Reverse ECEA & SCEA

Rad.	L1	L2	D1	D2
1/64	.0016	.0172	.0016	.0172
1/32	.0031	.0344	.0031	.0344
3/64	.0047	.0516	.0047	.0516
1/16	.0062	.0688	.0062	.0688

100° Diamond Profile

B Style 15° SCEA

Rad.	L1	L2	D1	D2
1/64	.0011	.0167	.0117	.0003
1/32	.0022	.0384	.0234	.0006
3/64	.0032	.0501	.0351	.0009
1/16	.0043	.0668	.0468	.0012

Table 8. Style A (0° Side Cutting-edge Angle) Brazed Single-point Tools. For turning, facing, or boring to 90° square shoulder. *(Source, Kennametal Inc.)*

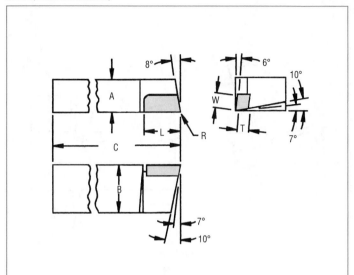

Style		Shank Dimensions			Tip Dimensions			(R) Nose Radius
Left	Right	Width (A)	Height (B)	Length (C)	Thickness (T)	Width (W)	Length (L)	
AL 4	AR 4	1/4	1/4	2	3/32	3/16	5/16	1/64
AL 5	AR 5	5/16	5/16	2 1/4	3/32	1/4	1/2	1/64
AL 6	AR 6	3/8	3/8	2 1/2	3/32	1/4	1/2	1/64
AL 7	AR 7	7/16	7/16	3	3/32	1/4	1/2	1/32
AL 8	AR 8	1/2	1/2	3 1/2	1/8	5/16	5/8	1/32
AL 10	AR 10	5/8	5/8	4	5/32	3/8	3/4	1/32
AL 12	AR 12	3/4	3/4	4 1/2	3/16	7/16	13/16	1/32
AL 16	AR 16	1	1	7*	1/4	9/16	1	1/32

*AL 16 and AR 16 also available in 6" length.

Table 9. Style B (15° Side Cutting-edge Angle) Brazed Single-point Tools. For turning, facing, or boring with lead angle in order to allow higher feeds with reduced tool pressure. *(Source, Kennametal Inc.)*

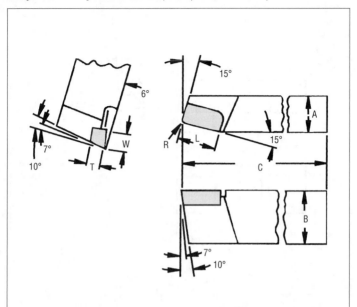

Style		Shank Dimensions			Tip Dimensions			(R) Nose Radius
Left	Right	Width (A)	Height (B)	Length (C)	Thickness (T)	Width (W)	Length (L)	
BL 4	BR 4	1/4	1/4	2	1/16	5/32	1/4	1/64
BL 5	BR 5	5/16	5/16	2 1/4	3/32	3/16	5/16	1/64
BL 6	BR 6	3/8	3/8	2 1/2	3/32	1/4	1/2	1/64
BL 7	BR 7	7/16	7/16	3	3/32	1/4	1/2	1/32
BL 8	BR 8	1/2	1/2	3 1/2	1/8	5/16	5/8	1/32
BL 10	BR 10	5/8	5/8	4	5/32	3/8	3/4	1/32
BL 12	BR 12	3/4	3/4	4 1/2	3/16	7/16	13/16	1/32
BL 16	BR 16	1	1	7*	1/4	9/16	1	1/32

*AL 16 and AR 16 also available in 6" length.

Table 10. Style C (0° End Cutting-edge Angle) Brazed Single-point Tools. For chamfering, facing, and plunge cutting. *(Source, Kennametal Inc.)*

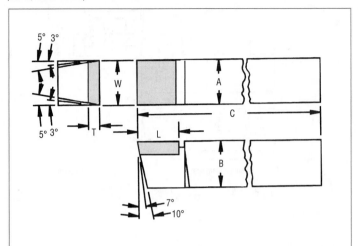

Style	Shank Dimensions			Tip Dimensions		
	Width (A)	Height (B)	Length (C)	Thickness (T)	Width (W)	Length (L)
C 4	1/4	1/4	2	1/16	1/4	5/16
C 5	5/16	5/16	2 1/4	3/32	5/16	3/8
C 6	3/8	3/8	2 1/2	3/32	3/8	3/8
C 7	7/16	7/16	3	3/32	7/16	1/2
C 8	1/2	1/2	3 1/2	1/8	1/2	1/2
C 10	5/8	5/8	4	5/32	5/8	5/8
C 12	3/4	3/4	4 1/2	3/16	3/4	3/4
C 16	1	1	7*	1/4	1	3/4
C 20	1 1/4	1 1/4	8**	5/16	1 1/4	3/4
C 44	1/2	1	7*	3/16	1/2	1/2
C 54	5/8	1	6	1/4	5/8	5/8
C 55	5/8	1 1/4	8**	1/4	5/8	5/8

* Also available in 6" length.
**Also available in 7" length.

Table 11. Style CT (Cut-off) Brazed Single-point Tools. For workpiece cut-off. *(Source, Kennametal Inc.)*

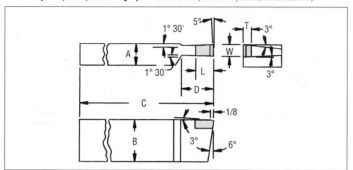

Style	Shank Dimensions				Tip Dimensions		
	Width (A)	Height (B)	O. A. Length (C)	Length (D)	Thickness (T)	Width (W)	Length (L)
CT 120	1/2	1	5	1	3/16	5/16	7/16
CT 121	1/2	1	5	1	1/8	1/4	3/4
CT 122	1/2	1	5	7/8	1/8	3/16	3/4
CT 130	5/8	1 1/4	5	1 1/4	3/16	3/8	1/2
CT 140	3/4	1 1/2	6	1 1/4	3/16	3/8	5/8

Table 12. Style D (80° Nose-angle) Brazed Single-point Tools.
For right-hand or left-hand turning, boring, or chamfering. *(Source, Kennametal Inc.)*

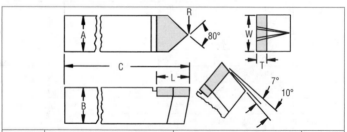

Style	Shank Dimensions			Tip Dimensions			(R) Nose Radius
	Width (A)	Height (B)	Length (C)	Thickness (T)	Width (W)	Length (L)	
D 4	1/4	1/4	2	1/16	1/4	5/16	1/64
D 5	5/16	5/16	2 1/4	3/32	5/16	3/8	1/64
D 6	3/8	3/8	2 1/2	3/32	3/8	1/2	1/64
D 7	7/16	7/16	3	3/32	7/16	1/2	1/32
D 8	1/2	1/2	3 1/2	1/8	1/2	1/2	1/32
D 10	5/8	5/8	4	5/32	5/8	5/8	1/32
D 12	3/4	3/4	4 1/2	3/16	3/4	3/4	1/32
D 16	1	1	7*	1/4	1	3/4	1/32

* D 16 also available in 6" length.

Table 13. Style E (60° Nose-angle) Brazed Single-point Tools. For 60° V-threading. *(Source, Kennametal Inc.)*

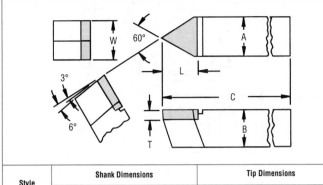

Style	Shank Dimensions			Tip Dimensions		
	Width (A)	Height (B)	Length (C)	Thickness (T)	Width (W)	Length (L)
E 5	5/16	5/16	2 1/4	3/32	5/16	3/8
E 6	3/8	3/8	2 1/2	3/32	3/8	1/2
E 8	1/2	1/2	3 1/2	1/8	1/2	1/2
E 10	5/8	5/8	4	5/32	5/8	5/8
E 12	3/4	3/4	4 1/2	3/16	3/4	3/4

Table 14. Style EL and ER (60° Nose-angle with Offset Point) Brazed Single-point Tools.
For 60° V-threading to shoulder. *(Source, Kennametal Inc.)*

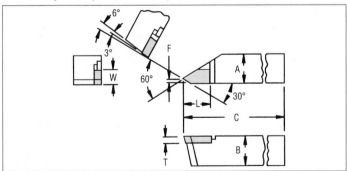

Style		Shank Dimensions				Tip Dimensions		
Left	Right	Width (A)	Height (B)	Length (C)	Offset (F)	Thickness (T)	Width (W)	Length (L)
EL 6	ER 6	3/8	3/8	2 1/2	1/16	3/32	1/4	3/8
EL 8	ER 8	1/2	1/2	3 1/2	3/32	1/8	5/16	5/8
EL 10	ER 10	5/8	5/8	4	3/32	5/32*	5/16	5/8
EL 12	ER 12	3/4	3/4	4 1/2	1/8	5/32	3/8	3/4

*Also available with 1/8" thick tip.

Table 15. Style F (0° End Cutting-edge Angle, Offset). Brazed Single-point Tools. For facing (tool shank 90° to the work axis) or turning and boring (tool shank parallel to work axis). *(Source, Kennametal Inc.)*

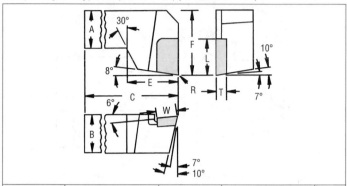

Style		Shank Dimensions			Head Dimensions		Tip Dimensions			(R) Nose Radius
Left	Right	Width (A)	Height (B)	Length (C)	Length (E)	Height (F)	Thickness (T)	Width (W)	Length (L)	
FL 8	FR 8	1/2	1/2	3 1/2	3/4	3/4	1/8	5/16	5/8	1/32
FL 10	FR 10	5/8	5/8	4	1	1	5/32	3/8	3/4	1/32
FL 12	FR 12	3/4	3/4	4 1/2	1 1/8	1 3/8	3/16	7/16	13/16	1/32
FL 16	FR 16	1	1	7*	1 3/8	1 3/4	1/4	9/16	1	1/32

*Also available in 6" length

and H=90° with 10° rake. The three letter code is followed by a hyphen and a number. On square tools SS and SR, the first number denotes the size in number of 1/32nds of an inch. The second number on SS and SR tools provides the length in number of 1/8ths of an inch. On round tools TS and TR, the first number denotes the size in numbers of 1/16ths of an inch. Therefore, a tool designated TRC-7 would be tipped, round, have a 30° side cutting edge angle, and be 7/16 inches in diameter. A tool designated SSE-610 would be solid, square, have a 45° side cutting edge angle, and be 3/16ths of an inch wide and 1 1/4 inch long. Complete dimensions for popular brazed carbide tipped boring tools are provided in **Tables 17** through **22**.

Boring bars for indexable inserts are available in eight common specifications, although custom boring bar systems exist in many different configurations. Standard bars are available in the following configurations.

Boring bar style or lead angle. **E** style = Straight shank, 30° side cutting edge angle. **G** style = Offset Shank, 0° side cutting edge angle. **F** style = Offset shank, 0° end cutting edge angle. **K** style = Offset shank, 15° end cutting edge angle. **L** style = Offset shank, negative 5° side cutting angle. **P** style = Straight shank, 27°30' side cutting edge angle. **Q** style = Offset shank, 17°30' end cutting edge angle. **S** style = Offset shank, 45° end cutting edge angle. **U** style = Offset shank, negative 3° end cutting edge angle. These bars.

These bars are normally fitted with one of the following insert shapes: **C** (80° diamond), **D** (55° diamond), **K** (55° parallelogram), **S**, (90° square), **T**, (60° triangle),

Table 16. Style G (0° Side Cutting-edge Angle, Offset) Brazed Single-point Tools. For turning a shoulder (tool shank 90° to the work axis). *(Source, Kennametal Inc.)*

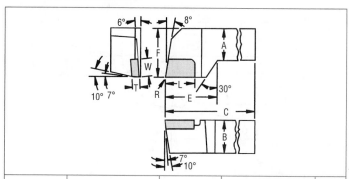

Style		Shank Dimensions			Head Dimensions		Tip Dimensions			(R) Nose Radius
Left	Right	Width (A)	Height (B)	Length (C)	Length (E)	Height (F)	Thickness (T)	Width (W)	Length (L)	
GL 8	GR 8	1/2	1/2	3 1/2	1 1/16	3/4	1/8	5/16	5/8	1/32
GL 10	GR 10	5/8	5/8	4	1 3/8	1	5/32	3/8	3/4	1/32
GL 12	GR 12	3/4	3/4	4 1/2	1 1/2	1 1/8	3/16	7/16	13/16	1/32
GL 16	GR 16	1	1	7*	1 11/16	1 1/2	1/4	9/16	1	1/32
GL 64	GR 64	3/4	1	6	1 7/16	1 1/4	1/4	1/2	3/4	1/32

*Also available in 6" length.

V, 35° diamond, and **W** (80° trigon). *Figure 12* illustrates the application of various bar/insert combinations.

Rigidity in boring

The key to productivity in boring operations is assuring that the boring tool is as rigid as possible. Since boring bars are often required to reach long distances into parts to remove stock, the rigidity of the operation can be compromised. The diameter of the hole and the need for added clearance for chip removal directly influence the maximum size of the boring bar that can be utilized. The practical overhang for steel boring bars is four times their shank diameter. When the tool overhang exceeds this limit, the metal removal rate of the boring bar is compromised significantly, due to loss of rigidity, and the resultant increased possibility of vibration. By envisioning the boring bar as a round beam, its deflection can be approximated with the following formula.

$$\delta = (64 \div 3\pi) \times (W \times L^3) \div (E \times D^4)$$

where: δ = Deflection of the beam
   D = The diameter of the beam in inches
   E = The modulus of elasticity of the material
   L = The length of the beam in inches
   W = The load on the beam in pounds

The beam deflection can be minimized for a given load by altering the beam's length,

(Text continued on p. 1028)

Table 17. Style TSA Brazed Single-point Boring Tools for 90° Boring Bar. *(Source, Kennametal Inc.)*

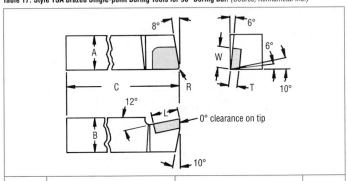

Style	Shank Dimensions			Tip Dimensions			(R) Nose Radius
	Width (A)	Height (B)	Length (C)	Thickness (T)	Width (W)	Length (L)	
TSA 5	5/16	5/16	1 1/2	3/32	3/16	5/16	1/64
TSA 6	3/8	3/8	1 3/4	3/32	3/16	5/16	1/64
TSA 8	1/2	1/2	2 1/4	1/8	5/16	7/16	1/32

Table 18. Style TSC Brazed Single-point Boring Tools for 60° Boring Bar. *(Source, Kennametal Inc.)*

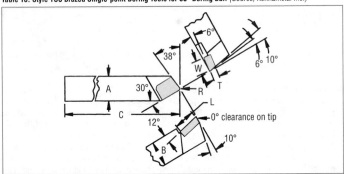

Style	Shank Dimensions			Tip Dimensions			(R) Nose Radius
	Width (A)	Height (B)	Length (C)	Thickness (T)	Width (W)	Length (L)	
TSC 5	5/16	5/16	1 1/2	3/32	3/16	5/16	1/64
TSC 6	3/8	3/8	1 3/4	3/32	3/16	5/16	1/64
TSC 8	1/2	1/2	2 1/4	1/8	5/16	7/16	1/32
TSC 10	5/8	5/8	3	5/32	3/8	9/16	1/32
TSC 12	3/4	3/4	3 1/2	3/16	7/16	5/8	1/32

Table 19. Style TSE Brazed Single-point Boring Tools for 45° Boring Bar. *(Source, Kennametal Inc.)*

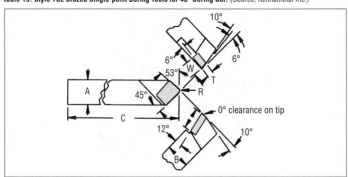

Style	Shank Dimensions			Tip Dimensions			(R) Nose Radius
	Width (A)	Height (B)	Length (C)	Thickness (T)	Width (W)	Length (L)	
TSE 5	5/16	5/16	1 1/2	3/32	3/16	5/16	1/64
TSE 6	3/8	3/8	1 3/4	3/32	3/16	5/16	1/64
TSE 8	1/2	1/2	2 1/2	1/8	5/16	7/16	1/32
TSE 10	5/8	5/8	3	5/32	3/8	9/16	1/32
TSE 12	3/4	3/4	3 1/2	3/16	7/16	5/8	1/32

Table 20. Style TRC Brazed Single-point Boring Tools for 60° Boring Bar. *(Source, Kennametal Inc.)*

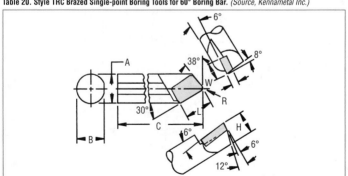

Style	Shank Dimensions				Tip Dimensions			(R) Nose Radius
	Dia. (A)	Width (B)	Height (H)	Length (C)	Thickness (T)	Width (W)	Length (L)	
TRC 5	5/16	19/64	7/32	1 1/2	1/16	3/16	1/4	1/64
TRC 6	3/8	11/32	9/32	1 3/4	3/32	3/16	5/16	1/64
TRC 8	1/2	15/32	3/2	2 1/2	3/32	1/4	3/8	1/32

Table 21. Style TRE Brazed Single-point Boring Tools for 45° Boring Bar. *(Source, Kennametal Inc.)*

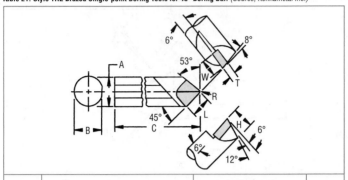

Style	Shank Dimensions				Tip Dimensions			(R) Nose Radius
	Dia. (A)	Width (B)	Height (H)	Length (C)	Thickness (T)	Width (W)	Length (L)	
TRE 5	5/16	19/64	7/32	1 1/2	1/16	3/16	1/4	1/64
TRE 6	3/8	11/32	9/32	1 3/4	1/16	3/16	1/4	1/64
TRE 8	1/2	15/32	3/2	2 1/2	3/32	5/16	3/8	1/32

Table 22. Style TRG Brazed Square End Boring Tools. *(Source, Kennametal Inc.)*

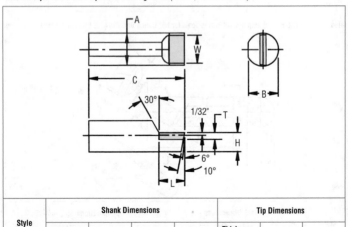

Style	Shank Dimensions				Tip Dimensions		
	Dia. (A)	Width (B)	Height (H)	Length (C)	Thickness (T)	Width (W)	Length (L)
TRG 5	5/16	19/64	3/16	1 1/2	1/16	1/4	1/4
TRG 6	3/8	11/32	7/32	1 3/4	1/16	5/16	1/4
TRG 8	1/2	15/32	9/32	2 1/2	3/32	3/8	3/8

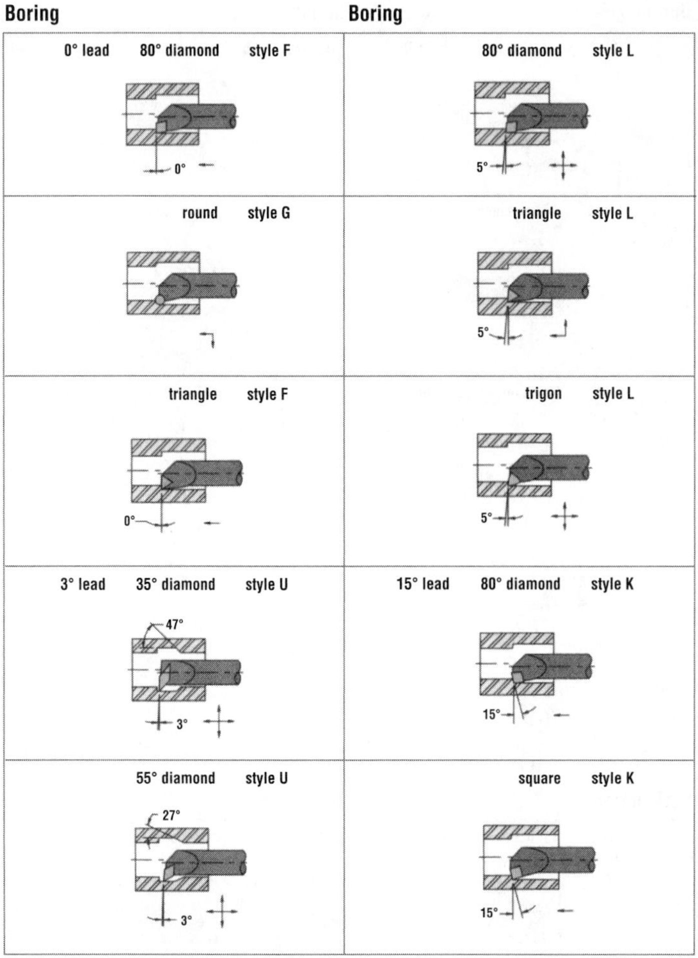

Boring

Boring

Figure 12. Toolholder/insert combinations for boring operations. (Source, Kennametal Inc.)

Boring   **Boring**

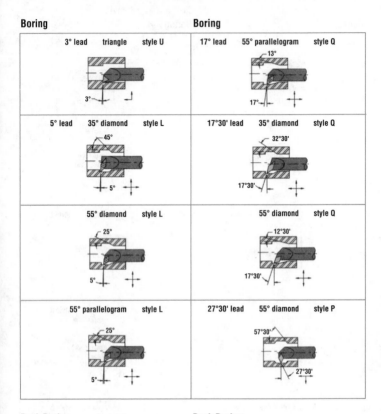

Back Boring   **Back Boring**

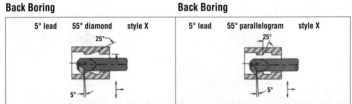

Figure 12. (Continued) Toolholder/insert combinations for boring operations. (Source, Kennametal Inc.)

Profiling   Profiling

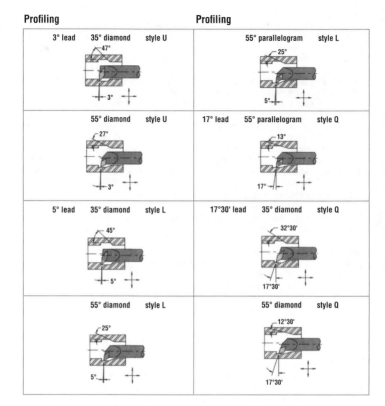

Figure 12. (Continued) Toolholder/insert combinations for boring operations. (Source, Kennametal Inc.)

its diameter, or its material. As can be seen by the formula, the beam diameter (to the fourth power) has the most significant impact on its stiffness or resistance to deflection. A 15% reduction in beam diameter will reduce rigidity by 48% and increase deflection by 92%. A 15% increase in length will reduce rigidity by only 34% and increase deflection by 52%. Therefore, it is obvious that, in terms of changes to a round beam (boring bar), a reduction in diameter will have a greater effect on rigidity than a similar increase in length. Increasing the diameter by 15% will result in reducing deflection by 43%, and shortening the bar by 15% will produce a 38% gain in deflection.

The boring bar's material also plays a significant role in the rate of deflection. This characteristic is known as the modulus of elasticity. The modulus of elasticity for some common boring bar materials are: Steel = 30,000,000 psi; Heavy Metal = 45,000,000 psi; Carbide = 90,000,000 psi. Therefore, a beam made of steel will deflect three times as much as one made of carbide under identical conditions, assuming that they have the same dimensions. Since the modulus of elasticity is a material constant, a characteristic of the material and independent of its hardness, heat treating a steel bar will have no effect on its resistance to deflection (stiffness). For most boring operations, steel shank boring bars offer satisfactory performance as long as the length/diameter ratio of the bar does not exceed 4:1. Tungsten carbide shank boring bars perform satisfactorily up to a length/diameter ratio of 6:1. When these ratios are exceeded, special materials should be employed to assure that rigidity requirements are met.

Cutting tool geometry in boring

As in turning, the radial and axial cutting forces in boring are greatly influenced by the lead angle. The increased axial force of a small lead angle can be useful in reducing vibration because the forces are directed along the boring bar. As the lead angle increases, radial forces increase, resulting in increased forces perpendicular to the direction of the cut. Radial forces contribute to vibration in boring operations. To control vibration, recommended lead angles are 0° to 15° (90° to 65° in countries using the metric system). At 15°, radial cutting forces are almost twice as high as with a 0° lead angle tool. In order to reduce tangential forces, positive rake angle cutting tools are preferred for most operations in order to reduce cutting forces. However, since a positive rake angle tool has smaller clearance angle than a negative angle tool, the possibility of contact between the tool and the workpiece must be taken into account. When boring smaller diameter holes in particular, caution must be taken to assure that an adequate clearance angle is maintained. Another consideration is the nose radius of the tool. Radial and tangential forces increase as the nose radius increases, but the relationship between the nose radius and radial force is affected by the lead angle. If the depth of cut is equal to or exceeds the nose radius, radial forces will be determined by the lead angle. If the nose radius exceeds the depth of cut, radial force will increase with cutting depth. For this reason, it is advantageous to select a nose radius that is smaller than the cutting depth.

Recommended Speeds and Feeds for Turning

The following charts provide recommended speeds and feeds for turning operations. These data are the result of testing and experimentation carried out by Kennametal Inc. Consult the Inserts section of this book (page 858) for information on how to interpret the recommended ISO Grade and Type cutting materials for proper insert selection.

FREE MACHINING AND LOW CARBON STEELS (Hardness 100-225 BHN): AISI 1008, 1010, 1020, 1026, 1108, 1117, 1141, 1151, 10L18, 10L45, 10L50, 11L44, 12L14.

Material characteristics	Resulting conditions	Steps to offset condition
Free Machining Steels are easy to machine but chip control can be difficult. High cutting speeds and high d.o.c. are achievable.	Torn finish (built-up edge).	1. Increase speed. 2. Use cermets. 3. Use positive rake inserts. 4. Increase coolant concentration.
	Burrs and sharp edges.	1. Change cutting path. 2. Use positive or high positive rake inserts.
Low Carbon Steels are soft and gummy, burrs and sharp edges and torn finish are common. High speeds are achievable.	Difficult chip control (low carbon steel).	1. Increase feed rate. 2. Change lead angle. 3. Reduce nose radius.
Interrupted cutting.		Increase speeds by 25 to 50%.

Machining conditions	Starting conditions		Roughing		Finishing	
d.o.c.-inches	050-.20		.10-.30		.010-.10	
d.o.c.-mm	1.3-5.0		2.5-7.6		.25-2.50	
feed rate-in./rev	.012		.010-.030		.006-.012	
feed rate-mm/rev	.30		.25-.75		.15-.30	
ISO Grade / Type	*surface speed*					
	sfm	smm	sfm	smm	sfm	smm
HSS	120	37	100-160	31-49	110-175	34-53
P01-P20 / CVD	950	290	-	-	800-1400	244-427
P15-P35 / CVD	750	229	650-1100	198-335	700-1100	213-335
P20-P40 / CVD	700	213	500-800	152-244	600-1100	183-335
P25-P45 / CVD	600	183	300-750	91-229	-	-
P30-P45 / CVD	600	183	300-750	91-229	-	-
P35-P50 / CVD	500	152	300-700	91-213	-	-

MEDIUM- AND HIGH-CARBON STEELS (Hardness of 180-330 BHN): AISI 1040, 1045, 1085, 1090, 1055, 1080, 1095, 1235, 1541, 1548, 1551, 1561, 1572.

Material characteristics	Resulting conditions	Steps to offset condition
Steel with *higher carbon* or *manganese* content is more abrasive, requires lower cutting speeds and feeds, and achieves surface finish easier.	Flank wear.	1. Use TiC/alumina ceramic grade. 2. Use aluminum oxide coated grade. 3. Increase feed or d.o.c. 4. Increase coolant concentration.
	Crater wear.	1. Use aluminum oxide coated grade. 2. Reduce speed or feed rate. 3. Use cermets. 4. Increase coolant pressure.
	Thermal deformation.	1. Use more wear resistant grade. 2. Reduce speed or feed rate. 3. Use generous amount of coolant.
Interrupted cutting.		Increase speeds by 25 to 50%.

Machining conditions	Starting conditions		Roughing		Finishing	
d.o.c.-inches	.050-.20		.10-.30		.010-.10	
d.o.c.-mm	1.3-5.0		2.5-7.6		.25-2.5	
feed rate-in./rev	.012		.010-.03		.006-.012	
feed rate-mm/rev	.30		.25-.75		.15-.30	
ISO Grade / Type	*surface speed*					
	sfm	smm	sfm	smm	sfm	smm
HSS	70	21	50-100	15-31	60-135	18-41
P01-P20 / CVD	900	274	650-1000	198-305	700-1200	213-366
P10-P20 / CVD	750	229	550-900	168-274	650-1000	198-305
P15-P35 / CVD	700	213	550-800	168-244	650-900	198-274
P20-P40 / CVD	650	198	500-750	152-229	600-800	183-244
P25-P45 / CVD	400	122	250-600	76-183	250-700	76-213
P30-P45 / CVD	500	152	300-600	91-183	-	-
P35-P50 / CVD	450	137	200-550	61-168	-	-

ALLOY AND TOOL STEELS (HARDNESS OF 200-325 BHN): Alloy - AISI 1300 series, 2000 series, 3000 series, 4012, 4023, 4140, 4150, 4320, 4422, 5120, E51100, E52100, 6118, 6150, 7000 series, 8620, 8622, 8640, 8822, E-9310, 94B15, and Rycut 40, Cr-Mo alloy and Super-Kut steels. Tool Steels - SAE M and T high speed, SAE A,D,H,O,and S hot and cold work, and wrought high carbon/low alloy W1, W2, L2, P1, P6, and P20.

Material characteristics	Resulting conditions	Steps to offset condition
Tool Steels are abrasive, workhardening, and produce tough-to-break chips.	Flank wear.	1. Use aluminum oxide coated grade. 2. Reduce speed. 3. Increase feed.
	Crater wear.	1. Use aluminum oxide coated grade. 2. Use cermets. 3. Reduce speed or feed rate.
	Dull surface finish.	1. Increase feed. 2. Use cermets. 3. Increase coolant concentration. 4. Use positive rake inserts.
Alloys: Higher carbon content results in increased abrasion but surface finish is easily achieved. Use lower speeds. *High carbon and alloy* content increases machining difficulty. *Higher chrome, nickel and moly* content increases tendency to workharden. *Higher nickel* content makes chip control more difficult.	Flank wear.	1. Use TiC/alumina ceramic grade. 2. Use aluminum oxide coated grade. 3. Increase feed. 4. Increase coolant concentration. 5. Increase depth of cut.
	Crater wear.	1. Use aluminum oxide coated grade. 2. Reduce speed or feed rate. 3. Use more free-cutting chip control. 4. Use cermets. 5. Increase coolant pressure.
	Thermal deformation.	1. Use more wear resistant grade. 2. Reduce speed or feed rate.
Interrupted cutting.		Increase speed 25 to 50%.

Machining conditions	Starting conditions		Roughing		Finishing	
d.o.c.-inches	.050-.20		.10-.30		.010-.10	
d.o.c.-mm	1.3-5.0		2.5-7.6		.25-2.5	
feed rate-in./rev	.012		.010-.030		.006-.012	
feed rate-mm/rev	.30		.25-.75		.15-.30	
ISO Grade / Type	*surface speed*					
	sfm	smm	sfm	smm	sfm	smm
HSS	70	21	40-90	12-27	50-120	15-37
P01-P10 / CVD	800	244	450-1000	137-305	550-1150	168-351
P10-P20 / CVD	700	213	300-900	91-274	500-950	152-290
P15-P35 / CVD	600	183	300-750	91-229	450-900	137-274
P20-P40 / CVD	550	168	300-750	91-229	400-850	122-259
P25-P45 / CVD	450	137	175-600	53-183	250-750	76-229
P30-P45 / CVD	500	152	250-650	76-198	-	-
P35-P50 / CVD	450	137	250-600	76-183	-	-

PREHEAT-TREATED ALLOY AND BEARING STEELS (Hardness 325-480 BHN): AISI 4330, 4140, 4150, 4340, 8640, and stress proof steel.

Material characteristics	Resulting conditions	Steps to offset condition
These steels are abrasive, workhardening, and produce tough, hard-to-break chips.	Flank wear.	1. Use aluminum oxide coated grade. 2. Increase feed. 3. Reduce speed.
	Crater wear.	1. Use aluminum oxide coated grade. 2. Reduce speed or feed rate.
	Dull surface finish.	1. Increase speed. 2. Use positive rake inserts. 3. Increase coolant concentration.
Interrupted cutting.		Increase speeds by 25 to 50%.

Machining conditions	Starting conditions		Roughing		Finishing	
d.o.c.-inches	.010-.08		.06-.25		.003-.060	
d.o.c.-mm	.25-2.0		1.5-6.3		.08-1.5	
feed rate-in./rev	.010		.006-.024		.003-.014	
feed rate-mm/rev	.25		.15-.60		.08-.35	
ISO Grade / Type	*surface speed*					
	sfm	smm	sfm	smm	sfm	smm
HSS	40	12	30-60	9-18	35-75	11-23
P10-P20 / CERAMIC (Al2O3)	850	259	-	-	600-1100	183-335
P05-20 / CVD	250	76	125-450	38-137	175-500	53-152
P20-P40 / CVD	200	61	125-400	38-122	175-400	53-122
P25-P45 / CVD	175	53	100-275	30-84	-	-
P30-P45 / CVD	250	76	150-350	46-107	-	-
P35-P50 / CVD	200	61	150-300	46-91	-	-

HARDENED IRONS, HEAT-TREATED ALLOY STEEL, AND TOOL STEEL (Hardness 480-750 BHN): AISI 4330, 300-M, ETD-180, 52100, 17-4 PH, and Hy Tuff steel. Die steels, hot and cold work tool steel.

Material characteristics	Resulting conditions	Steps to offset condition
Heat-treated alloy steels and *tool steels* are abrasive, workhardening, and produce tough, hard-to-break chips.	Flank wear.	1. Use aluminum oxide coated grade. 2. Reduce speed. 3. Increase feed.
	Crater wear.	1. Use aluminum oxide coated grade. 2. Use cermets. 3. Reduce speed or feed rate.
	Dull surface finish.	1. Increase speed. 2. Use cermets. 3. Increase coolant concentration. 4. Use positive rake (PVD) inserts.
Hardened irons are highly abrasive and require high cutting forces.	Excessive edge wear.	1. Apply alumina based ceramics or CBN grade. 2. Feed above .004 in./rev / .102 mm/rev. 3. Increase lead angle or use round inserts. 4. Check coolant application.
	Catastrophic breakage.	1. Increase lead angle or use round inserts. 2. Use large nose radii. 3. Repair toolholder. 4. Reduce feeds.
Interrupted cutting.		Increase speed 25 to 50%.

Machining conditions	Starting conditions	Roughing	Finishing
d.o.c.-inches	.015-.08	.060-.25	.003-.060
d.o.c.-mm	.40-2.0	1.5-6.3	.08-1.5
feed rate-in./rev	.015 (solid CBN) .009 (tipped CBN) .006 (ceramic)	.008-.024	.003-.014
feed rate-mm/rev	.38 (solid CBN) .23 (tipped CBN) .15 (ceramic)	.20-.60	.08-.35

ISO Grade / Type	surface speed					
	sfm	smm	sfm	smm	sfm	smm
K01-K03 / TIPPED CBN	350	107	-	-	220-650	67-198
K03-K05 / TIPPED CBN	300	91	-	-	100-550	30-168
K05-K15 / TIPPED CBN	300	91	-	-	100-450	30-137
K05-K15 / SOLID CBN	250	76	150-350	46-107	100-400	30-122
P01-P05 / CERAMIC (Al$_2$O$_3$)	300	91	-	-	100-450	30-137
P10-P20 / PVD	150	46	-	-	75-225	23-69

AUSTENITIC 200 AND 300 SERIES STAINLESS STEELS (Hardness 135-275): AISI 201, 302, 303, 316L, 321, 347, 384.

Material characteristics	Resulting conditions	Steps to offset condition
These materials are gummy when machined due to high nickel content. Small depths of cut are difficult, and chips are tough and stringy and buildup on the tool. Low thermal conductivity causes high heat at tool tip, and high chromium content encourages workhardening. Softer materials are more difficult to machine.	Depth of cut notch.	1. Increase lead angle. 2. Use tougher grade. 3. Set feed over .005 in. per rev / .13 mm per rev.
	Built-up edge.	1. Increase speed. 2. Use positive rake (PVD) inserts.
	Workpiece glazing.	1. Increase d.o.c. 2. Increase feed rate, decrease speed. 3. Reduce nose radius. 4. Use sharp PVD grade. 5. Maintain feed rate of at least .003 in. per rev / .08mm per rev.
	Dull surface finish.	1. Increase speed. 2. Reduce feed. 3. Increase positive rake.
Interrupted cutting.		Increase speeds by 25 to 50%.

Machining conditions	Starting conditions		Roughing		Finishing	
d.o.c.-inches	.060-.175		.10-.50		.02-.10	
d.o.c.-mm	1.5-4.4		2.5-12.7		.5-2.5	
feed rate-in./rev	.014		.006-.030		.003-.015	
feed rate-mm/rev	.35		.15-.75		.08-.40	
ISO Grade / Type	*surface speed*					
	sfm	smm	sfm	smm	sfm	smm
HSS	70	21	40-80	12-24	55-100	17-31
M15-M25 / CVD	550	168	300-650	91-198	400-800	122-244
M20-M35 / CVD	550	168	300-650	91-198	400-800	122-244
M30-M45 / CVD	500	152	250-500	76-152	375-750	114-229
M30-M40 / CVD	400	122	200-600	61-183	400-800	122-244
M35-M45 / CVD	400	122	200-600	61-183	400-800	122-244
M10-M25 / PVD	525	160	250-575	76-175	300-650	91-198
M30-M40 / PVD	275	84	175-325	53-99	200-400	61-122

FERRITIC-MARTENSITIC STAINLESS STEEL.

Material characteristics	Resulting conditions	Steps to offset condition
These materials require high cutting forces. Workhardening (especially PH stainless) and brittle, stringy chips are characteristic.	Depth of cut notch.	1. Increase lead angle. 2. Use tougher grade. 3. Set feed over .005 in./rev / .13 mm/rev.
	Built-up edge.	1. Increase speed. 2. Use positive rake inserts. 3. Increase coolant concentration.
	Workpiece glazing.	1. Increase d.o.c. 2. Reduce nose radius. 3. Increase coolant concentration.
	Torn or dull surface finish.	1. Increase speed. 2. Reduce feed. 3. Increase positive rake. 4. Increase coolant concentration.
Interrupted cutting.		Increase speeds by 25 to 50%.

Ferritic-Martensitic Stainless Steel (Hardness 135-300 BHN): AISI 410, 416, 416F, 420, 430, 440, 440C.

Machining conditions	Starting conditions		Roughing		Finishing	
d.o.c.-inches	.06-.175		.10-.50		.015-.10	
d.o.c.-mm	1.5-4.4		2.5-12.7		.38-2.5	
feed rate-in./rev	.015		.006-.017		.003-.015	
feed rate-mm/rev	.38		.15-.42		.08-.38	
ISO Grade / Type	surface speed					
	sfm	smm	sfm	smm	sfm	smm
HSS	75	23	65-85	20-26	70-100	21-31
P05-P20 / CVD	500	152	450-550	137-168	450-650	137-198
P20-P40 / CVD	450	137	400-550	122-168	400-600	122-183
P25-P45 / CVD	325	99	250-450	76-137	-	-
P30-P45 / CVD	300	91	250-400	76-122	-	-
P35-P50 / CVD	275	84	250-350	76-107	-	-

Ferritic-Martensitic Stainless Steel (Hardness 300-450 BHN): AISI 416, 416F, 416 Se, 440C, 502, 504.

Machining conditions	Starting conditions		Roughing		Finishing	
d.o.c.-inches	.06-.175		.10-.50		.015-.10	
d.o.c.-mm	1.5-4.4		2.5-12.7		.38-2.5	
feed rate-in./rev	.015		.006-.017		.003-.015	
feed rate-mm/rev	.38		.15-.42		.08-.38	
ISO Grade / Type	surface speed					
	sfm	smm	sfm	smm	sfm	smm
HSS	35	11	20-50	6-15	30-75	9-23
P05-P20 / CVD	450	137	400-500	122-152	400-550	122-168
P20-P40 / CVD	400	122	350-500	107-152	350-500	-
P25-P45 / CVD	280	85	180-350	55-107	-	-
P30-P45 / CVD	225	69	180-300	55-91	-	-
P35-P50 / CVD	225	69	180-300	55-91	-	-

GRAY CAST IRONS (Hardness 135-340 BHN): Class 20, 25, 30, 45, 55, 60 high tensile.

Material characteristics	Resulting conditions	Steps to offset condition
Abrasive, with a tendency to breakout during exit from cut. Potential for scale and inclusions and for chatter on thin wall sections. Corners and thin sections can be hard and brittle.	Excessive edge wear.	1. Use ceramics. 2. Increase feed to reduce in-cut time.
	Chipping.	1. Increase lead angle. 2. Use stronger grade. 3. Use prehoned inserts or inserts with edge preparation.
	Workpiece breakout.	1. Lower feed rate during exit. 2. Increase speed. 3. Increase lead angle. 4. Use positive geometry for finish cuts.
	Workpiece chatter.	1. Use positive rake inserts. 2. Increase lead angle. 3. Increase feed.
Interrupted cutting.		Increase speeds by 25 to 50%.

Machining conditions	Starting conditions		Roughing		Finishing	
d.o.c.-inches	.50-.25		.10-.50		.015-.10	
d.o.c.-mm	1.25-6.3		2.5-12.7		.38-2.5	
feed rate-in./rev	.015		.01-.03		.004-.015	
feed rate-mm/rev	.38		.25-.75		.10-.38	
ISO Grade / Type	*surface speed*					
	sfm	smm	sfm	smm	sfm	smm
HSS	70	21	55-90	17-27	65-120	20-37
K05-K15 / CERAMIC (Si$_3$N$_4$)	2400	732	700-3250	213-991	650-4000	198-1220
K10-K20 / CVD	750	229	400-1000	122-305	600-1400	183-427
K05-K20 / CVD	650	198	400-1000	122-305	400-1000	122-305
K15-K35 / CVD	600	183	300-750	91-229	400-900	122-274
K20-K30 / CVD	600	183	300-700	91-231	400-800	122-244
K30-K40 / CVD	500	152	250-600	76-183	-	-
K25-K35 / CVD	450	137	200-550	61-168	-	-

DUCTILE AND MALLEABLE CAST IRONS: Nodular/ductile, ferritic/pearlitic, and pearlitic/martensitic.

Material characteristics	Resulting conditions	Steps to offset condition
These materials are abrasive and difficult to machine. Hard spots may be caused by concentrations of carbide. Good surface finishes are difficult to achieve.	Edge wear.	1. Use harder grade insert. 2. Increase feed to reduce time in cut.
	Crater wear.	1. Reduce speed. 2. Increase coolant concentration.
	Torn or dull surface finish.	1. Increase lead angle. 2. Increase coolant concentration. 3. Use larger nose radius insert.
Interrupted cutting.		Increase speeds by 25 to 50%.

Ductile and Malleable Cast Irons (Hardness of 140-260 BHN).

Machining conditions	Starting conditions		Roughing		Finishing	
d.o.c.-inches	.05-.175		.10-.50		.015-.10	
d.o.c.-mm	1.25-4.40		2.5-12.7		.38-2.5	
feed rate-in./rev	.012		.01-.03		.003-.02	
feed rate-mm/rev	.30		.25-.75		.08-.50	
ISO Grade / Type	*surface speed*					
	sfm	smm	sfm	smm	sfm	smm
HSS	80	24	65-95	20-29	75-130	23-40
P01-P20 / CVD	750	229	500-900	152-274	500-950	152-290
P10-P20 / CVD	650	198	400-800	122-244	400-800	122-244
P15-P35 / CVD	550	168	300-700	91-213	300-750	91-229
P20-P40 / CVD	550	168	300-700	91-213	300-700	91-213
P30-P45 / CVD	500	152	300Ñ600	91-183	-	-
P35-P50 / CVD	450	137	300-550	91-168	-	-

Ductile and Malleable Cast Irons (Hardness of 270-400 BHN).

Machining conditions	Starting conditions		Roughing		Finishing	
d.o.c.-inches	.05-.175		.10-.50		.01-.10	
d.o.c.-mm	1.3-4.4		2.5-12.7		.25-2.5	
feed rate-in./rev	.01 (carbide) .006 (ceramic)		.01-.03		.003-.02	
feed rate-mm/rev	.25 (carbide) .15 (ceramic)		.25-.75		.08-.50	
ISO Grade / Type	*surface speed*					
	sfm	smm	sfm	smm	sfm	smm
HSS	30	9	15-45	6-14	25-65	8-20
P01-P20 / CVD	600	183	-	-	400-800	122-244
P10-P20 / CVD	500	152	350-600	107-183	300-600	91-183
P15-P35 / CVD	400	122	300-500	91-168	300-550	91-168
P20-P40 / CVD	400	122	250-500	76-152	300-550	91-168
P30-P45 / CVD	350	107	175-450	53-137	-	-
P35-P50 / CVD	300	91	150-400	46-122	-	-

FREE-MACHINING ALUMINUM ALLOYS AND HIGH-SILICON ALUMINUM (Hardness of 50-150 BHN):
Free-machining aluminum alloys: 1000, 1100, 1200, 1300 series, 2011 through 2024 series, 3000 series.
High-Silicon Aluminum: 4000, 5000, 6000, and 7000 series.

Material characteristics	Resulting conditions	Steps to offset condition
Free-Machining Aluminum is soft and gummy, has a low melting temperature, and tends to stick to the cutting tool. It possesses a low modulus of elasticity, high ductility, and tends to deform under pressure of the cutting tool. 2011-2024 series contains copper, manganese, and magnesium.	Built-up edge.	1. Increase speed. 2. Use positive rake PVD insert. 3. Use appropriate coolant.
	Chipping.	1. Avoid built-up edge. 2. Increase toolholder lead angle. 3. Ensure proper insert seating.
	Torn or dull surface finish.	1. Increase speed. 2. Reduce feed. 3. Use positive rake PVD insert. 4. Use appropriate coolant. 5. Use PCD grade.
High-Silicon Aluminum is abrasive and tough. High cutting forces are generated and chips are torn, rather than sheared, from the workpiece. Diamond is the most economical cutting tool. 5000 series contains chromium. 6000 series contains silicon, copper, magnesium, and chrome. 7000 series contains copper, magnesium, chrome, and zinc.	Edge wear.	1. Use coated grade. 2. Use PCD grade. 3. Increase coolant concentration.
	Built-up edge.	1. Increase speed. 2. Increase coolant concentration.
	Poor surface finish.	1. Use largest nose radius that is practical and does not cause chatter. 2. Check rigidity. 3. Minimize tool overhang.
	Chipping.	1. Avoid built-up edge. 2. Increase toolholder lead angle. 3. Decrease feed in interrupted cutting. 4. Ensure proper insert seating.

Machining conditions	Starting conditions	Roughing	Finishing
d.o.c.-inches	.06-.20	.15-.35	.01-.15
d.o.c.-mm	1.5-5.0	3.8-8.8	.25-3.8
feed rate-in./rev	.018	.008-.025	.003-.016
feed rate-mm/rev	.45	.20-.62	.08-.40

ISO Grade / Type	surface speed					
	sfm	smm	sfm	smm	sfm	smm
HSS	500	153	300-600	92-183	400-800	122-244
K03-K20 / PCD *	2500	762	-		1000-10000	305-3050
K05-K20 / PVD *	1800	549	700-2600	213-793	200-3000	61-915
M10 / CERMET *	1600	488	700-2300	213-701	900-2600	274-783
M10-M20 / UCC *	1600	488	650-2000	198-610	650-2200	198-671
K03-K20 / PCD **	1700	518	1200-2500	366-762	2000-3000	610-915

* Free machining aluminum alloys only.
** High-silicon aluminum only.

NICKEL BASED, HEAT-RESISTANT ALLOYS (Hardness of 125-450 BHN).

Material characteristics	Resulting conditions	Steps to offset condition
These materials are unusually abrasive (rather than hard), and workharden rapidly. Small d.o.c. is difficult, and tool life is relatively poor. High forces at the cutting edge result in high heat concentration at the cutting edge. High cutting speed may cause insert failure by plastic deformation.	Depth of cut notch.	1. Increase lead angle. 2. Use tougher grade insert. 3. Use .015" / .38 mm or greater d.o.c. 4. Feed over .005" / .13 mm per revolution. 5. Increase coolant concentration.
	Built-up edge.	1. Increase speed. 2. Use positive rake PVD insert. 3. Do not overrun cutting edge. 4. Increase coolant concentration.
	Workpiece glazing.	1. Increase d.o.c. or feed rate. 2. Reduce nose radius. 3. Use positive rake PVD insert.
	Torn or dull surface finish.	1. Increase speed. 2. Reduce feed. 3. Use positive rake PVD insert. 4. Increase coolant concentration.
Interrupted cutting.		Increase speeds by 25 to 50%.

Nickel Based, Heat-Resistant Alloys (Hardness of 125-250 BHN): Nickel 200, Monel, R405, Monel K500, Inconel 600, Inconel 625/901/X750/718, Waspalloy.

Machining conditions	Starting conditions		Roughing		Finishing	
d.o.c.-inches	.10		.06-.25		.01-.06	
d.o.c.-mm	2.5		1.5-6.3		.25-1.5	
feed rate-in./rev	.006		.004-.01		.004-.012	
feed rate-mm/rev	.15		.10-.25		.10-.30	
ISO Grade / Type	*surface speed*					
	sfm	smm	sfm	smm	sfm	smm
HSS	25	8	10-30	3-9	25-35	8-11
M10-M25 / CERAMIC (Si_3N_4)	700	213	400-750	122-229	-	-
M05-M20 / CERAMIC (Si_3N_4)	750	229	300-800	91-244	500-1200	152-366
M10-M30 / CERAMIC (Si_3N_4)	650	198	300-700	91-213	400-1000	122-305
M20-M35 / CVD	225	69	100-300	30-91	200-400	61-122
M30-M40 / CVD	225	69	100-300	30-91	200-400	61-122
M35-M45 / CVD	225	69	100-300	30-91	200-400	61-122
K01-K15 / CERAMIC (Al_2O_3)	850	259	-	-	600-1000	183-305
M10-M25 / PVD	175	53	-	-	90-300	27-91

Nickel Based, Heat-Resistant Alloys (Hardness of 200-450): Inconel 718, Hastelloy C, Rene 95, Waspalloy.

Machining conditions	Starting conditions		Roughing		Finishing	
d.o.c.-inches	.10		.06-.25		.01-.06	
d.o.c.-mm	2.5		1.5-6.3		.25-1.5	
feed rate-in./rev	.006		.004-.01		.004-.012	
feed rate-mm/rev	.15		.10-.25		.10-.30	
ISO Grade / Type	*surface speed*					
	sfm	smm	sfm	smm	sfm	smm
HSS	15	5	10-20	3-6	15-30	5-9
M10-M25 / CERAMIC (Si₃N₄)	700	213	300-700	91-213	-	-
M05-M20 / CERAMIC (Si₃N₄)	750	229	300-800	91-244	500-1200	152-366
M10-M30 / CERAMIC (Si₃N₄)	650	198	300-700	91-213	400-1000	122-305
M20-M35 / CVD	225	69	100-300	30-91	200-400	61-122
M30-M40 / CVD	225	69	100-300	30-91	200-400	61-122
M35-M45 / CVD	225	69	100-300	30-91	200-400	61-122
K01-K15 / CERAMIC (Al₂O₃)	850	259	-	-	600-1300	183-396
M10-M25 / PVD	175	53	-	-	90-275	27-84

IRON BASED, HEAT-RESISTANT ALLOYS (Hardness of 180-325 BHN): Waspalloy A286, 19-9 DL, Discaloy, 16-25-6, Incoloy 800.

Material characteristics	Resulting conditions	Steps to offset condition
These materials are unusually abrasive (rather than hard), and workharden rapidly. Small d.o.c. is difficult, and chips are tough and stringy. Tool life is relatively poor.	Depth of cut notch.	1. Increase lead angle. 2. Use tougher grade insert. 3. Use .015" / .38 mm or greater d.o.c. 4. Feed over .005" / .13 mm per revolution. 5. Increase coolant concentration.
	Built-up edge.	1. Increase speed. 2. Use positive rake PVD insert. 3. Do not overrun cutting edge. 4. Increase coolant concentration
	Workpiece glazing.	1. Increase d.o.c. or feed rate. 2. Reduce nose radius. 3. Use positive rake PVD insert.
	Torn or dull surface finish.	1. Increase speed. 2. Reduce feed. 3. Use positive rake PVD insert. 4. Increase coolant concentration.
Interrupted cutting.		Increase speeds by 25 to 50%.

Machining conditions	Starting conditions		Roughing		Finishing	
d.o.c.-inches	.060-.20		.10-.25		.01-.10	
d.o.c.-mm	1.5-5.0		2.5-6.3		.25-2.5	
feed rate-in./rev	.008		.006-.02		.003-.012	
feed rate-mm/rev	.20		.15-.50		.08-.30	
ISO Grade / Type	surface speed					
	sfm	smm	sfm	smm	sfm	smm
HSS	30	9	15-35	5-11	30-40	9-12
M10-M25 / CERAMIC (Si$_3$N$_4$)	500	152	225-600	69-183	-	-
M05-M20 / CERAMIC (Si$_3$N$_4$)	500	152	225-600	69-183	300-700	107-213
M10-M25 / PVD	150	46	100-250	30-76	-	-
M30-M40 / PVD	150	46	100-250	30-76	100-300	30-91
K01-K15 / CERAMIC (Al$_2$O$_3$)	600	183	-	-	325-750	99-229

COBALT BASED, HEAT-RESISTANT ALLOYS: Haynes 188, Stellite F, 5816, Haynes 25 (L605), J1650.

Material characteristics	Resulting conditions	Steps to offset condition
These materials are unusually abrasive (rather than hard), and workharden rapidly. Small d.o.c. is difficult, and tool life is relatively poor. High forces at the cutting edge result in high heat concentration at the cutting edge. High cutting speed may cause insert failure by plastic deformation.	Depth of cut notch.	1. Increase lead angle. 2. Use tougher grade insert. 3. Use .015" / .38 mm or greater d.o.c. 4. Feed over .005" / .13 mm per revolution. 5. Increase coolant concentration.
	Built-up edge.	1. Increase speed. 2. Use positive rake PVD insert. 3. Do not overrun cutting edge. 4. Increase coolant concentration.
	Workpiece glazing.	1. Increase d.o.c. or feed rate. 2. Reduce nose radius. 3. Use positive rake PVD insert.
	Torn or dull surface finish.	1. Increase speed. 2. Reduce feed. 3. Use positive rake PVD insert. 4. Increase coolant concentration.
Interrupted cutting.		Increase speeds by 25 to 50%.

Cobalt Based, Heat-Resistant Alloys (Hardness of 150-250 BHN).

Machining conditions	Starting conditions		Roughing		Finishing	
d.o.c.-inches	.10		.06-.25		.01-.06	
d.o.c.-mm	2.5		1.5-6.3		.25-1.5	
feed rate-in./rev	.005		.005-.01		.002-.008	
feed rate-mm/rev	.12		.12-.25		.05-.20	
ISO Grade / Type	surface speed					
	sfm	smm	sfm	smm	sfm	smm
M10-M25 / PVD	90	29	30-110	9-34	50-125	15-38
M30-M40 / PVD	60	18	25-90	8-27	-	-
M10-M20 / UCC	75	23	30-100	9-30	40-100	12-30

Cobalt Based, Heat-Resistant Alloys (Hardness of 250-450 BHN).

Machining conditions	Starting conditions		Roughing		Finishing	
d.o.c.-inches	.10		.06-.25		.01-.06	
d.o.c.-mm	2.5		1.5-6.3		.25-1.5	
feed rate-in./rev	.005		.005-.01		.002-.008	
feed rate-mm/rev	.12		.12-.25		.05-.20	
ISO Grade / Type	surface speed					
	sfm	smm	sfm	smm	sfm	smm
M10-M25 / PVD	80	24	25-100	8-30	40-120	12-36
M30-M40 / PVD	50	15	20-80	6-24	-	-
M10-M20 / UCC	65	20	25-80	8-24	35-100	11-30
K01-K15 / CERAMIC (Al_2O_3)	500	152	-	-	250-800	76-244
K05-K15 / PCBN (Solid)	500	152	-	-	275-725	84-221

Cobalt Based, Heat-Resistant Alloys (Hardness of 450-750 BHN).

Machining conditions	Starting conditions		Roughing		Finishing	
d.o.c.-inches	.10		.06-.25		.01-.06	
d.o.c.-mm	2.5		1.5-6.3		.25-1.5	
feed rate-in./rev	.005		.005-.01		.002-.008	
feed rate-mm/rev	.12		.12-.25		.05-.20	
ISO Grade / Type	surface speed					
	sfm	smm	sfm	smm	sfm	smm
M10-M25 / PVD	60	18	20-75	6-23	30-90	9-27
K01-K15 / CERAMIC (Al_2O_3)	300	91	190-550	60-168	200-650	61-198
K05-K15 / PCBN (Solid)	400	122	175-500	53-152	180-625	55-190

TITANIUM ALLOYS (Hardness of 250-450 BHN): Pure, alpha, alpha-beta, and beta titanium.

Material characteristics	Resulting conditions	Steps to offset condition
Titanium Alloys require lower cutting speeds than steels of equal hardness. Tool life is relatively poor, even at low cutting speeds, and high chemical reactivity causes chips to gall and weld to the cutting edge. These materials possess low thermal conductivity which increases cutting temperatures, and they are usually abrasive, tough, and stringy. Workhardening is common.	Depth of cut notch.	1. Increase lead angle. 2. Remove embrittled surface from mill stock. 3. Use .015" / .38 mm or greater d.o.c. 4. Feed over .005" / .13 mm per revolution. 5. Increase coolant concentration.
	Built-up edge.	1. Keep a sharp edge. 2. Use positive rake PVD insert. 3. Increase coolant concentration.
	Workpiece glazing.	1. Increase d.o.c. 2. Reduce nose radius. 3. Use positive rake PVD insert. 4. Index cutting edge.
	Torn or dull surface finish.	1. Increase speed. 2. Reduce feed. 3. Use positive rake PVD insert. 4. Index cutting edge. 5. Increase coolant concentration.
Interrupted cutting.		Increase speeds by 25 to 50%.

Titanium Alloys (Hardness of 250-350 BHN).

Machining conditions	Starting conditions		Roughing		Finishing	
d.o.c.-inches	.040-.15		.06-.25		.01-.06	
d.o.c.-mm	1.0-3.8		1.5-6.3		.25-1.5	
feed rate-in./rev	.006		.008-.015		.004-.01	
feed rate-mm/rev	.15		.20-.38		.10-.25	
ISO Grade / Type	surface speed					
	sfm	smm	sfm	smm	sfm	smm
HSS	40	12	30-50	9-15	35-60	11-18
M10-M25 / PVD	200	61	140-275	43-84	170-300	52-91
M30-M40 / PVD	150	46	90-200	27-61	120-325	36-99
M10-M20 / UCC	200	61	120-225	36-69	140-250	43-76

Titanium Alloys (Hardness of 350-450 BHN).

Machining conditions	Starting conditions		Roughing		Finishing	
d.o.c.-inches	.040-.15		.06-.25		.01-.06	
d.o.c.-mm	1.0-3.8		1.5-6.3		.25-1.5	
feed rate-in./rev	.006		.008-.015		.004-.01	
feed rate-mm/rev	.15		.20-.38		.10-.25	
ISO Grade / Type	surface speed					
	sfm	smm	sfm	smm	sfm	smm
M10-M25 / PVD	140	43	40-170	12-52	50-200	15-61
M30-M40 / PVD	100	30	30-150	9-46	40-175	12-53
M10-M20 / UCC	120	37	35-130	11-40	50-175	15-53

NONFERROUS MACHINING ALLOYS (Hardness of 50-150 BHN): Copper, brass, zinc.

Machining conditions	Starting conditions		Roughing		Finishing	
d.o.c.-inches	.06-.20		.10-.30		.015-.10	
d.o.c.-mm	1.5-5.0		2.5-7.5		.38-2.5	
feed rate-in./rev	.10		.008-.03		.006-.014	
feed rate-mm/rev	.25		.20-.75		.15-.35	
ISO Grade / Type	surface speed					
	sfm	smm	sfm	smm	sfm	smm
HSS	175	53	150-300	46-91	160-400	49-122
K10-K20 / CVD	1300	396	-	-	900-2500	274-762
K20-K30 / CVD	1100	335	500-1600	152-488	-	-
K05-K20 / PVD	900	274	375-1300	114-396	450-1500	137-457
K03-K15 / Diamond film coated	1700	518	900-2600	274-793	-	-
K03-K20 / PCD	1700	518	900-2600	274-793	1000-3000	305-915
K05-K20 / UCC	850	259	350-1450	107-442	375-1750	114-534

MAGNESIUM (Hardness of 50-90 BHN).

Machining conditions	Starting conditions		Roughing		Finishing	
d.o.c.-inches	.06-.20		.10-.30		.015-.10	
d.o.c.-mm	1.5-5.0		2.5-7.5		.38-2.5	
feed rate-in./rev	.10		.008-.03		.006-.014	
feed rate-mm/rev	.25		.20-.75		.15-.35	
ISO Grade / Type	surface speed					
	sfm	smm	sfm	smm	sfm	smm
HSS	500	152	-	-	250-1000	76-305
K03-K20 / PCD	2000	610	-	-	1000-3500	305-915
K05-K20 / PVD	1650	503	-	-	600-3000	183-915
K05-K20 UCC	1450	442	-	-	550-2600	168-793

ALUMINUM BASED METAL MATRIX COMPOSITES: SiC or Al_2O_3 particulates 20-40% of volume.

Machining conditions	Starting conditions		Roughing		Finishing	
d.o.c.-inches	.04		.04-.08		.01-.04	
d.o.c.-mm	1.0		2.5-7.5		.25-1.0	
feed rate-in./rev	.008		.008-.03		.003-.016	
feed rate-mm/rev	.20		.20-.75		.08-.40	
ISO Grade / Type	surface speed					
	sfm	smm	sfm	smm	sfm	smm
K03-K20 / PCD	1200	365	-	-	1000-2500	305-760
K03-K15 / Diamond film coated	1200	365	700-2000	210-610	1000-2500	305-760

TIN ALLOYS AS CAST: ASTM B23, alloys 1,2,3,11.

Machining conditions	Starting conditions		Roughing		Finishing	
d.o.c.-inches	.10		.05-.30		.01-.05	
d.o.c.-mm	2.5		1.2-7.6		.25-1.2	
feed rate-in./rev	.008		.005-.015		.006-.012	
feed rate-mm/rev	.20		.12-.38		.15-.30	
ISO Grade / Type	surface speed					
	sfm	smm	sfm	smm	sfm	smm
HSS	350	107	250-400	76-122	300-500	91-152
K05-K20 / PVD	700	213	625-775	190-236	750-900	229-274
K05-K20 / UCC	600	183	400-625	122-190	500-750	152-229

LEAD ALLOYS AS CAST: ASTM B23, alloys 7,8,13,15.

Machining conditions	Starting conditions		Roughing		Finishing	
d.o.c.-inches	.10		.10-.30		.025-.10	
d.o.c.-mm	2.5		2.5-7.6		.62-2.5	
feed rate-in./rev	.10		.10-.30		.025-.10	
feed rate-mm/rev	2.5		2.5-7.6		.62-2.5	
ISO Grade / Type	surface speed					
	sfm	smm	sfm	smm	sfm	smm
HSS	450	137	250-500	76-152	300-550	91-168
K05-K20 / PVD	850	259	750-900	229-274	850-1250	259-381
K05-K20 / UCC	800	244	700-900	213-274	750-1000	229-305

ZINC ALLOYS AS CAST (Hardness of 80-220 BHN).

Machining conditions	Starting conditions		Roughing		Finishing	
d.o.c.-inches	.008		.15-.20		.20-.15	
d.o.c.-mm	.20		3.8-.20		.50-3.8	
feed rate-in./rev	.008		.008-.018		.006-.012	
feed rate-mm/rev	.20		.20-.45		.15 / .30	
ISO Grade / Type	surface speed					
	sfm	smm	sfm	smm	sfm	smm
K05-K20 / PVD	1000	305	650-1050	198-320	750-1250	229-381
K05-K20 / UCC	950	290	625-1000	190-305	725-1200	221-366

TUNGSTEN ALLOYS.

Machining conditions	Starting conditions		Roughing		Finishing	
d.o.c.-inches	.10		.05-.20		.01-.05	
d.o.c.-mm	2.5		1.2-5.0		.25-1.2	
feed rate-in./rev	.008		.006-.018		.005-.01	
feed rate-mm/rev	.20		.15-.45		.12-.25	
ISO Grade / Type	surface speed					
	sfm	smm	sfm	smm	sfm	smm
K05-K20 / PVD	275	84	200-400	61-122	300-500	91-152
K05-K20 / UCC	250	76	175-375	53-114	225-400	69-122

COMPACTED GRAPHITE IRON (Hardness of 175-250 BHN).

Machining conditions	Starting conditions		Roughing		Finishing	
d.o.c.-inches	.06		.008-.024		.004-.014	
d.o.c.-mm	1.5		.20-.60		.10-.35	
feed rate-in./rev	.008		.008-.024		.004-.014	
feed rate-mm/rev	.20		.20-.60		.10-.35	
ISO Grade / Type	surface speed					
	sfm	smm	sfm	smm	sfm	smm
P01-P05 / CERAMIC (Al_2O_3)	1100	335	-	-	1000-5000	305-457
P01-P20 / CVD	500	152	350-675	107-206	400-750	122-229
P10-P20 / PVD	225	69	-	-	150-300	46-91
P20-P40 / CVD	450	137	260-590	79-180	300-650	91-198

MANGANESE (Hardness of 140-220 BHN): Wrought.

Machining conditions	Starting conditions		Roughing		Finishing	
d.o.c.-inches	.10		.05-.20		.01-.05	
d.o.c.-mm	2.5		1.2-5.0		.25-1.2	
feed rate-in./rev	.008		.006-.018		.005-.01	
feed rate-mm/rev	.20		.15-.45		.12-.25	
ISO Grade / Type	surface speed					
	sfm	smm	sfm	smm	sfm	smm
K05-K20 / PVD	200	61	125-300	38-91	150-325	49-99
K25-K35 / PVD	125	38	90-175	27-53	100-200	30-61
K05-K20 / UCC	110	34	90-200	27-61	125-250	38-76

DEPLETED URANIUM (Hardness of 150-180 BHN): Wrought.

Machining conditions	Starting conditions		Roughing		Finishing	
d.o.c.-inches	.08		.40-.20		.01-.04	
d.o.c.-mm	2.0		1.0-5.0		.25-1.0	
feed rate-in./rev	.006		.005-.015		.005-.013	
feed rate-mm/rev	.15		.12-.38		.12-.30	
ISO Grade / Type	surface speed					
	sfm	smm	sfm	smm	sfm	smm
K05-K20 / PVD	700	213	-	-	135-300	41-91
K05-K20 / UCC	110	34	75-150	23-46	125-225	38-69

ZIRCONIUM (HARDNESS OF 150-300 BHN).

Machining conditions	Starting conditions		Roughing		Finishing	
d.o.c.-inches	.05-.10		.05-.20		.01-.50	
d.o.c.-mm	1.2-2.5		1.2-5.0		.25-1.2	
feed rate-in./rev	.008		.01-.02		.005-.01	
feed rate-mm/rev	.20		.25-.50		.12-.25	
ISO Grade / Type	surface speed					
	sfm	smm	sfm	smm	sfm	smm
K05-K20 / PVD	225	69	160-300	49-91	180-320	55-98
K25-K35 / PVD	175	53	125-225	38-69	-	-

CARBON AND GRAPHITE COMPOSITES (Hardness of 270-400 BHN): Brush alloys, Kevlar, graphite.

Machining conditions	Starting conditions		Roughing		Finishing	
d.o.c.-inches	.075		.03-.25		.005-.03	
d.o.c.-mm	1.9		.75-6.3		.12-.75	
feed rate-in./rev	.008		.01-.06		.005-.014	
feed rate-mm/rev	.20		.25-1.5		.12-.35	
ISO Grade / Type	surface speed					
	sfm	smm	sfm	smm	sfm	smm
K03-K15 / Diamond film coated	2500	762	-	-	1800-4300	549-1311
K03-K20 / PCD	2500	762	-	-	1800-4300	549-1311
K05-K20 / PVD	650	198	350-850	107-259	425-950	130-290

GLASS AND CERAMICS (Hardness of 200-250 BHN).

Machining conditions	Starting conditions		Roughing		Finishing	
d.o.c.-inches	.05		-		.005-.05	
d.o.c.-mm	1.2		-		.12-1.2	
feed rate-in./rev	.008		-		.005-.01	
feed rate-mm/rev	.20		-		.12-.25	
ISO Grade / Type	surface speed					
	sfm	smm	sfm	smm	sfm	smm
K03-K15 / Diamond film coated	1650	503	-	-	750-3300	229-1006
K03-K20 / PCD	1650	503	-	-	750-3300	229-1006

NYLON, PLASTICS, RUBBERS, PHENOLICS, RESINS (Hardness of 0-200).

Machining conditions	Starting conditions		Roughing		Finishing	
d.o.c.-inches	.05		.02-.20		.005-.02	
d.o.c.-mm	1.2		.50-5.0		.12-.50	
feed rate-in./rev	.005		.006-.015		.003-.012	
feed rate-mm/rev	.12		.15-.38		.08-.30	
ISO Grade / Type	surface speed					
	sfm	smm	sfm	smm	sfm	smm
K03-K20 / PCD	1300	396	-	-	500-2400	152-732
K05-K20 / PVD	550	168	300-700	91-213	325-750	99-229

Twist Drills

Drilling is the process used to create a cylindrical hole in a solid section of workpiece material. When using mechanical methods and tools, drilling involves rotational and linear feed motions. Invented almost two centuries ago, today's twist drill is a general purpose roughing tool used to produce holes ranging in size from miniature (0.004" or 0.102 mm) to very large (4.0" or 101.6 mm). Normally twist drills can reliably produce hole depths equal to approximately five times their diameter. The point geometry of a conventional twist drill features a 118° included angle and a web thickness ranging from 18 to 30% of the drill's diameter. *Figure 1* identifies twist drill features and dimensions.

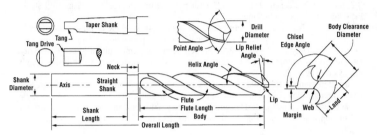

Figure 1. Drill terminology. (Source, Kennametal Inc.)

The vast majority of twist drills in use today are of the following types: screw machine length (a short length range), jobber length (medium length), or taper length (long length). For instance, general purpose $^3/_8$" drills of these types have the following dimensions: screw machine—1 $^7/_8$" flute, 3 $^1/_4$" overall; jobber length—3 $^5/_8$" flute, 5" overall; taper length—4 $^1/_4$" flute, 6 $^3/_4$" overall.

Screw machine length drills (also known as stub drills) were developed for use in screw machines where their short flutes and short overall length provide maximum rigidity. They are available in fractional sizes from $^3/_{64}$" to 1", number sizes from #60 through #1, every letter size, and most metric sizes from 1 mm to 25 mm, and in lengths from 1 $^3/_8$" (35 mm) to 6" (152 mm). These stout drills are commonly employed in portable drilling, often in sheet metal or other auto body work. They can be supplied in the following configurations.

1) General Purpose, with 118° point and either plain or tang shaft. Primarily for use in screw machines and designed to perform well in a wide range of low tensile strength materials.

2) General Purpose, with 118° point and left hand spiral. The left hand helix is for use in spindles that rotate counter-clockwise.

3) NAS 907 C, with 135° split point, a general- to medium-duty design that performs well in a broad range of materials in the iron and steel families.

4) Double Margin, 118° point. This design provides added support when used in drill bushings. For use in a broad range of materials in the iron and steel families.

5) Wide Land Parabolic Flute, 135° split point. This configuration is useful in deep hole drilling where the flute design promotes close tolerances due to improved chip and coolant flow and high torsional rigidity.

See **Tables 12** and **14** for straight shank screw machine length drill dimensions.

Jobber length drills are extremely versatile. They are available in every fractional size up to $^{11}/_{16}$", every number (wire gauge) size, every letter size, and metric sizes through 17.50 mm, and in lengths from $^3/_4$" (19 mm) to 7 $^5/_8$". They perform well in many high production operations. They have straight shanks and short flutes, and they are comparatively rigid due to their short overall length to diameter ratio. The name jobbers length, as implied, refers only to the drill length, and they are available with a wide variety of points and flute configurations including:

1) General Purpose, with 118° point and either plain or tang shaft. Commonly used for portable and machine drilling, and designed to perform well in a wide range of materials in the iron and steel families.

2) General Purpose, with 118° point and left hand spiral. The left hand helix is for use in spindles that rotate counter-clockwise.

3) NAS 907 A, with 118° split point. Conforming to National Aerospace Standard 907, these self centering drills are ideal for portable or machine drilling in mild steels and other light-duty applications.

4) NAS 907 B, with 135° split point. For portable or machine drilling of low tensile strength alloys, stainless steel, and other medium-duty applications.

5) Heavy Duty, 135° split point. This self centering point reduces thrust. Originally designed to drill cotter pin holes, this design resists torsional strain when drilling high tensile strength steels.

6) High Helix, 118° point. High helix and wide flute promote chip ejection in low tensile strength materials such as aluminum, zinc, and soft steels.

7) Low Helix, 118° point. Designed to reduce grabbing on entrance and breakthrough and to improve chip ejection. Well suited for drilling brass, plastics, and aluminum (shallow holes).

8) Half Round, 118° point. Designed to allow higher feed rates and improved hole finish in horizontal drilling of free machining brass, bronze, nylon, and similar materials. Using a centering drill to start the hole is recommended.

9) Wide Land Parabolic Flute, 135° split point. Very effective in deep hole drilling where the flute design promotes close tolerances due to improved chip and coolant flow and high torsional rigidity.

See **Table 12** for dimensions.

Taper length drills are popular for use in lathes and screw machines. They are available in the same sizes as screw machine length drills (see above), and in lengths from 2 $\frac{1}{4}$" (57 mm) to 11" (279 mm). These drills are commonly available with the following point and flute configurations.

1) General Purpose, with 118° point with plain shaft. Commonly used for drilling operations where a length longer than a Jobber Length is required, they perform well in a wide range of materials in the iron and steel families.

2) Automotive, taper shank, 118° point. Featuring a taper shank for ASA split sleeve drivers, these drills perform well in a broad range of materials in the iron and steel families.

3) Double Margin, 118° point. This design provides added support when used in drill bushings. Commonly used for drilling operations where a length longer than a jobber length is required.

4) High Helix, 118° point. Designed for improved chip ejection, this style is well suited for deep hole drilling in aluminum, magnesium, zinc, and low tensile strength steels.

5) Wide Land Parabolic Flute, 135° split point, with plain or taper shank. Effective in deep hole drilling where the flute design promotes close tolerances due to improved chip and coolant flow and high torsional rigidity.

6) Coolant Hole, 118° notched point, available with 23° or 34° helix. Ideal for high production, particularly when used horizontally while the workpiece rotates and the drill is stationary. Notch point aids penetration and coolant holes provide high pressure coolant flow to the point.

7) Core, 118° chamfer point, with three or four flutes. Used to size or enlarge cored,

punched, or other preformed holes. When using core drills, reduce regular speed by 25% and double the feed rate.

See **Tables 12** and **13** for straight shank taper length drill dimensions.

In addition to the three popular drill styles described above, there is a wide variety of specialized drilling tools available for specific operations. Specifications for many of these special varieties can be found in the tables in this section. In addition, drills are available with a variety of points for normal and specialized operations, as described below, courtesy of Kennametal, Inc.

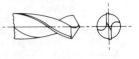

Conventional point drills with 118° included point angles are the most commonly used. They provide satisfactory results in a wide variety of materials. The straight chisel edge contributes to walking at the drill point, often making it necessary to spot the hole for improved accuracy.

Split point (or crankshaft point) drills were developed to produce deep holes in automotive crankshafts, or in any application where holes deeper than five to six times the drill diameter are specified. It is available in both 118° and 135° included angles. It also has two positive rake cutting edges extending to the center of the drill, which assist as

a chipbreaker to produce small, easily ejected, chips. Reduced thrust and the elimination of "walking" at the drill point are its main advantages, making it a good choice for portable drills or in applications where bushings cannot be used.

Notched points were developed for drilling tough alloys. As with the split point, the notched point can be incorporated on both 118° and 135° included angles, and it contains two additional positive rake cutting edges extending toward the center of the

drill. These secondary cutting lips, which extend no further than half the original cutting lip, can assist in chip control and reduce the torque required in drilling tough materials. They are suitable for drilling a wide variety of materials and they are commonly used on heavy web drills that allow the point to withstand the higher thrust loads required in drilling tough alloys.

Helical points feature an "S" contour with a radiused crown effect that has its highest point at the center of the drill axis. This crown contour creates a continuous cutting edge

from margin to margin across the web, resulting in a self-centering ability that allows the helical point drill to cut closer to actual drill diameter. Helical points are not available under $^1/_{16}$" diameter. This design is also known as a Winslow-Helical point.

Racon® points feature a continuously variable point angle which allows the lips and margins to blend together to form a smooth curve. The lips cut on a long, curved

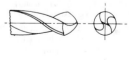

cutting edge which results in less load per unit area and less heat generated during the cut. Although this geometry is self-centering and must be used through a guide bushing, it can eliminate breakthrough burrs and it will provide increased tool life when drilling abrasive materials.

The *Bickford™ point* combines the self-centering feature of the Helical point with the long life, burr-free breakthrough, and higher feed capacity of the Racon®

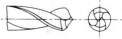

point. It is a good choice for producing accurate holes on NC machines without prior spot drilling. Bickford points should not be used for hand-held drilling operations because the point tends to grab the workpiece during breakthrough.

Double angle points were designed to reduce corner wear at the outer periphery of the cutting lip when drilling very abrasive materials such as medium and hard cast irons. The point is generated by first grinding a larger included angle and then a smaller included angle on the corner, providing,  in effect, a chamfer which reduces wear, improves hole size, acts as a chipbreaker, and reduces chipping corners of the lips when drilling hard materials. The length of the corner angle should be $1/3$ of the original cutting lip length.

Reduced rake points are produced by flattening or dubbing both cutting lips from the outer periphery to the chisel. This reduction in the effective rake angle produces a plowing (rather than shearing) action that prevents the drill point from grabbing in  low tensile strength materials such as brass, bronze, and plastics. A side benefit of the reduced rake is that it strengthens the cutting lip and assists in breaking chips.

Low angle points generally have an included angle of 60° or 90°, thereby reducing the effective rake at the outer periphery of the cutting lip. These points reduce cracking in plastics and grabbing on  breakthrough when drilling low tensile nonferrous materials.

Calculating speed and feed rates for drilling

The formulas for calculating speed and feed rates for drilling are the same as those used in milling. The speed of a drill is determined by the rate which the outer periphery of the tool rotates in relation to the material being cut. In general, the sfm *or* smm at which a drill will operate is based on the workpiece material, its condition and hardness, and the depth of the hole. **Table 11** provides cutting speeds for wire, fractional, and letter size drills. The formulas below can be used to calculate speeds for any diameter drill.

to find	given	formula
sfm	drill diameter in inches = D revolutions per minute = rpm	sfm = (D × rpm) ÷ 3.82*
rpm	surface feet per minute = sfm drill diameter in inches = D	rpm = (sfm ÷ D) × 3.82*
smm	drill diameter in mm = D revolutions per minute = rpm	smm = (D × rpm) ÷ 318.3**
rpm	surface meters per minute = smm drill diameter in mm = D	rpm = (smm ÷ D) × 318.3**

* In inch units, 12 (one foot or twelve inches) divided by π (12 ÷ 3.1416 = 3.82).
** In metric units, 1000 (one meter or 1000 mm) divided by π (1000 ÷ 3.1416 = 318.3).

Example. Find the feed rate for a #12 drill (0.1890" *or* 4.80 mm) rotating at 400 rpm.

sfm = (0.1890 × 400) ÷ 3.82 = 19.78   *note: 1 foot = 0.305 meters*
smm = (4.80 × 400) ÷ 318.3 = 6.03   *1 meter = 3.281 feet*

Example. Find the rpm needed to maintain a cutting speed of 40 sfm (12.2 smm) using a $1/2$" (12.70 mm) drill.

rpm = (40 ÷ 0.50) × 3.82 = 305.6   *or*   rpm = (12.2 ÷ 12.70) × 318.3 = 305.7

Once the cutting speed has been selected for a particular workpiece material and condition, the appropriate feed rate can be established. Drilling feed rates are selected to maximize productivity while maintaining chip control. The following formulas are used to calculate the feed (advance) in inches and millimeters. The feed rate in inches per minute (in./min) or millimeters per minute (mm/min) can be calculated given sfm or smm or the feed per revolution.

to find	given	formula
fpr (feed per revolution) in inches *or* millimeters	in. *or* mm per minute = in./min *or* mm/min revolutions per minute = rpm	fpr = in./min *or* mm/min ÷ rpm
rpm (revolutions per minute)	in. *or* mm per minute = in./min *or* mm/min feed per revolution (in. *or* mm) = fpr	rpm = in./min *or* mm/min ÷ fpr
rpm (revolutions per minute)	surface feet per minute = sfm cutter diameter in inches = D	rpm= (sfm ÷ D) × 3.82
rpm (revolutions per minute)	surface meters per minute = smm cutter diameter in millimeters = D	rpm = (smm ÷ D) × 318.3
in./min *or* mm/min (feed)	feed per revolution (in. *or* mm) = fpr revolutions per minute = rpm	in./min or mm/min = fpr × rpm

Example. Find rpm and in./min *or* mm/min for a $^1/_2$" (12.70 mm) drill with a feed rate of 30 sfm (9.14 smm) and a fpr of 0.009" *or* 0.2286 mm.

rpm = 30 sfm ÷ 0.5" × 3.82 = 229   *or*   rpm = 9.14 smm ÷ 12.70 mm × 318.3 = 229
in./min = 0.009" × 229 = 2.061   *or*   mm/min = 0.2286 mm × 229 = 52.35

Calculating time requirements in drilling

Drilling time for a specified hole depth can be calculated from rpm and fpr operating parameters.

cutting time (minutes) = (hole depth in inches *or* mm ÷ rpm) ÷ in./rev *or* mm/rev

Example. Calculate the time to required to drill a one inch (25.4 mm) deep hole at 400 rpm and a feed rate of 0.005 in./rev *or* 0.1270 mm/rev.

(1" ÷ 400) ÷ 0.005 = 0.50 minutes (30 seconds)
or (25.4mm ÷ 400) ÷ 0.1270 = 0.50 minutes (30 seconds)

Table 1 shows the time needed, in seconds, to drill a one inch hole at different speeds and feeds. The increased productivity that can be gained by using higher speed and/or feed rates is evident from this table.

Calculating power requirements in drilling

Drilling with large diameter drills can consume significant amounts of power. Often, the lack of power is the limiting factor when deciding on a particular operation. When large diameter drills or heavy stock removal is necessary, it is advantageous to first calculate the material removal rate (Q) in order to calculate necessary power requirements.

Find Q, the material removal rate.

$Q = (p × D^2 × fpr × rpm) ÷ 4 = in.^3/min$. Where feed rate is expressed in in./min.
or $Q = (p × D^2 × fpr × rpm) ÷ 4000 = cm^3/min$. Where feed rate is expressed in mm/min.

Example. Where   D (drill diameter) = 0.50" *or* 12.70 mm
fpr = 0.009" *or* 0.2286 mm
rpm = 229

Q = (3.1416 × 0.25 × 0.009 × 229) ÷ 4 = 0.4047 in³/min.   *note:* 1 in.³ = 16.387 cm³
or (3.1416 × 161.29 × 0.2286 × 229) ÷ 4000 = 6.6315 cm³/min   1 cm³ = 0.061 in.³

Find HP$_c$, power at the cutter in English units (horsepower).
HP$_c$ = $Q × K$.

Table 1. Drilling Time (in Seconds) Required to Penetrate 1" (25.4 mm) of Material.

						Feed per Revolution							
Inches:	0.0005	0.0010	0.002	0.003	0.004	0.005	0.007	0.01	0.015	0.02	0.03	0.05
Millimeters:	0.0127	0.0254	0.0508	0.0762	0.1016	0.127	0.1778	0.254	0.381	0.508	0.762	1.27
RPM												
10000	12.0	6.0	3.0	2.0	1.5	1.2						
9500	12.6	6.3	3.2	2.1	1.6	1.3						
9000	13.3	6.7	3.3	2.2	1.7	1.3	1.0					
8500	14.1	7.1	3.5	2.4	1.8	1.4	1.0					
8000	15.0	7.5	3.8	2.5	1.9	1.5	1.1					
7500	16.0	8.0	4.0	2.7	2.0	1.6	1.1					
7000	17.1	8.6	4.3	2.9	2.1	1.7	1.2					
6500	18.5	9.2	4.6	3.1	2.3	1.8	1.3					
6000	20.0	10.0	5.0	3.3	2.5	2.0	1.4	1.0				
5500	21.8	10.9	5.5	3.6	2.7	2.2	1.6	1.1				
5000	24.0	12.0	6.0	4.0	3.0	2.4	1.7	1.2				
4500	26.7	13.3	6.7	4.4	3.3	2.7	1.9	1.3				
4000	30.0	15.0	7.5	5.0	3.8	3.0	2.1	1.5	1.0			
3500	34.3	17.1	8.6	5.7	4.3	3.4	2.4	1.7	1.1			
3000	40.0	20.0	10.0	6.7	5.0	4.0	2.9	2.0	1.3	1.0		
2500	48.0	24.0	12.0	8.0	6.0	4.8	3.4	2.4	1.6	1.2		
2000	60.0	30.0	15.0	10.0	7.5	6.0	4.3	3.0	2.0	1.5	1.0	
1500	80.0	40.0	20.0	13.3	10.0	8.0	5.7	4.0	2.7	2.0	1.3	
1000	120.0	60.0	30.0	20.0	15.0	12.0	8.6	6.0	4.0	3.0	2.0	1.2
900	133.3	66.7	33.3	22.2	16.7	13.3	9.5	6.7	4.4	3.3	2.2	1.3
800	150.0	75.0	37.5	25.0	18.8	15.0	10.7	7.5	5.0	3.8	2.5	1.5

(Continued)

Table 1. *(Continued)* **Drilling Time (in Seconds) Required to Penetrate 1" (25.4 mm)**

	Feed per Revolution											
Inches:	0.0005	0.0010	0.002	0.003	0.004	0.005	0.007	0.01	0.015	0.02	0.03	0.05
Millimeters:	0.0127	0.0254	0.0508	0.0762	0.1016	0.127	0.1778	0.254	0.381	0.508	0.762	1.27
RPM												
700	171.4	85.7	42.9	28.6	21.4	17.1	12.2	8.6	5.7	4.3	2.9	1.7
600	200.0	100.0	50.0	33.3	25.0	20.0	14.3	10.0	6.7	5.0	3.3	2.0
500	240.0	120.0	60.0	40.0	30.0	24.0	17.1	12.0	8.0	6.0	4.0	2.4
400		150.0	75.0	50.0	37.5	30.0	21.4	15.0	10.0	7.5	5.0	3.0
300		200.0	100.0	66.7	50.0	40.0	28.6	20.0	13.3	10.0	6.7	4.0
200		300.0	150.0	100.0	75.0	60.0	42.9	30.0	20.0	15.0	10.0	6.0
150			200.0	133.3	100.0	80.0	57.1	40.0	26.7	20.0	13.3	8.0
125			240.0	160.0	120.0	96.0	68.6	48.0	32.0	24.0	16.0	9.6
100			300.0	200.0	150.0	120.0	85.7	60.0	40.0	30.0	20.0	12.0
75				266.7	200.0	160.0	114.3	80.0	53.3	40.0	26.7	16.0
50				400.0	300.0	240.0	171.4	120.0	80.0	60.0	40.0	24.0

Example. To find HP_c, use the Q value from the above example and find the power constant (K) from "K" Factors for Drilling chart. K represents the cubic inches of material per minute that can be removed by one horsepower. For this example, $K = 0.635$ (4140 steel, 220 BHN).

$HP_c = .4047 \times 0.635 = 0.257$ HP at the cutter.

Use the following formula to find HP_m, power at the motor in English units.

$HP_m = HP_c \div E$.

To determine horsepower at the motor (HP_m), the spindle efficiency must be considered. Spindle efficiency "E" varies from 75% to 90% (E = 0.75 to 0.90). Using the above example,

$HP_m = 0.257 \div .8 = 3.2$

Use the following formula to find kW_c, power at the cutter in metric units (kilowatts).

$kW_c = Q \times K$.

Example. To find kW_c, use the Q value from the above example and find the power constant (K) from the "K" Factors for Drilling chart. K represents the cubic centimeters of material per minute that can be removed by one kilowatt. For this example, $K = 0.0289$ (4140 Steel, 220 BHN).

$kW_c = 6.6315 \times 0.0289 = 0.19$ *note:* 1 HP = 0.7457 kW
                                              1 kW = 1.341 HP

Use the following formula to find kW_m, power at the motor in metric units.

$kW_m = kW_c \div E$.

Example. To determine power at the motor (kW_m), the spindle efficiency must be considered. Spindle efficiency "E" varies from 75% to 90% (E = 0.75 to 0.90).

$kW_m = .19 \div 0.8 = 0.238$

"K" Factors for Drilling			
work material	hardness BHN	"K" factor English units	"K" factor Metric units
steels, free machining low-, medium-, and high-carbon	110-200	.587	.0267
	201-253	.635	.0289
	254-286	.775	.0353
	287-330	.905	.0412
alloy steels, alloy, and lower carbon tool steels	135-200	.706	.0321
	201-253	.785	.0357
	254-286	.856	.0389
	287-327	.940	.0428
	328-371	1.15	.0523
	372-481	1.45	.066
	482-560	1.69	.0769
	561-700	1.86	.0846
stainless steels, wrought and cast (ferritic, austenitic, martensitic)	135-275	.593 – 1.08	.027 - .0491
	286-421	1.1– 1.72	.05 - .0782
	422-600	1.75 – 2.28	.0796 - .1037

(Continued)

"K" Factors for Drilling *(Continued)*			
work material	hardness BHN	"K" factor English units	"K" factor Metric units
stainless steels, precipitation hardened	150-375	.747 – 2.0	.034 - .091
cast irons (gray, ductile, and malleable)	120-175	.624	.0284
	110-190	.753	.0342
	176-200	.771	.035
	201-250	.867	.0394
	251-300	.938	.0426
	301-320	1.2	.0546
aluminum alloys, copper, brass, and zinc	50-150	.08 - .16	.0036 - .0073
nickel alloys	140-300	1.15 – 1.94	.0523 - .0882
	301-475	1.96 – 2.26	.089 - .1028
iron base, heat resistant alloys	135-320	.988 – 1.64	.045 - .0745
high temperature alloys, cobalt base	200-360	1.21 – 2.08	.055 - .0946
titanium	250-375	.751 – 1.15	.034 - .0523
magnesium alloys	40-90	.10 - .152	.0045 - .0068
copper alloys	100-150	.3	.0136

Surface treatment, tolerance specifications, and troubleshooting of twist drills

Most jobbers length, screw machine length, and taper length drills are available in cobalt steel, as carbide tipped, and in treated or untreated high-speed steel. Cobalt steel drills are especially heat resistant and especially well suited for drilling tough, high tensile strength materials such as PH stainless steel, titanium, and Inconel. Carbide tipped drills offer enhanced wear resistance and are most commonly used for drilling cast irons, nonferrous metals, and many composite materials. They are not the first choice for drilling steel. High-speed steel drills are often coated with a variety of surface treatments in order to increase their durability. The following treatments extend the versatility of HSS drills and are among the most widely available.

Oxide–Oxide is applied to finished tools and produces a thin black iron oxide surface coating. It provided additional tempering and stress relieving and reduces galling and chip welding, while increasing the ability of the tool to retain lubricants. Oxide coated drills are recommended for drilling iron and steel, but should not be used in nonferrous metals such as aluminum because the coating increases the loading tendencies of the tool.

Nitride–This treatment produces a hard case that is highly resistant to abrasion. It also retards the tendency of softer materials to cling, or load, on tools. Recommended uses are for drilling operations in ferrous, nonferrous, and nonmetallic materials which are abrasive and have loading characteristics.

Nitride and Oxide–The advantages of both coatings are combined when used together. Oxide improves lubricity, and nitriding enhances abrasion resistance. Well suited for abrasive ferrous applications but not recommended for soft materials such as aluminum, magnesium, or similar nonferrous materials.

Chrome Plating–An extremely thin layer of chrome is deposited on the tool's surface with this process. The layer of chrome reduces the coefficient of friction and resists chip weld and abrasion. Recommended for nonferrous and nonmetallic materials.

Titanium Nitride–This treatment improves tool life by acting as a wear resistant and thermal barrier. It also provides a low coefficient of friction and very high surface hardness. Other benefits are reduced friction and chip welding, and it acts as a thermal insulator between the chip and the tool. Tools treated with titanium nitride are very well suited for drilling in ferrous materials below R_c 40, and in nonferrous materials.

While drilling, unlike jig boring and reaming, is not expected to produce perfectly round holes, even though that is the objective, it is especially important that a drill is inspected before being used to assure that it is within prescribed dimensional tolerances for its size range. **Table 2** provides element and dimensional tolerances for twist drills, and **Table 3** gives straightness tolerances. Although not specifically listed in the table, metric sizes take the appropriate tolerance for the diameter range in which they fall. A guide for troubleshooting various problem that may be encountered while drilling with "in tolerance" tools is provided in **Table 4**.

Table 2. Element and Dimensional Tolerances for Two-, Three-, and Four-Flute Drills. Metric Sizes take the appropriate tolerance for the diameter range in which they fall. *(Source, Precision Twist Drill Co.)*

Specification	Drill Diameter Range				
	#97 to #81 .0059-.0130	Over #81 to 1/8" .0131 to .1250	1/16" to 1/8" .625 to .1250	1/16" to 1/2" .0625 to .5000	Over 1/8" to 1/4" .1251 to .2500
Dia. at Point	+.0002 to -.0002	+.0000 to -.0005			+.0000 to -.0007
Shank Dia.	+.0002 to -.0002	+.0000 to -.0025			-.0005 to -.0030
Back Taper (per inch)	None	.0000 to .0008			.0002 to .0008
Flute Length	+1/64" to -1/64"	+1/8" to -1/16"			+1/8" to -1/8"
Overall Length	+1/32" to -1/32"	+1/8" to -1/16"			
Included Point Angle				118° ± 5°	
Lip Height			TIV* = .0020		TIV* = .0030
Centrality of Web			TIV**= .0030		TIV**= .0040
Flute Spacing			TIV† = .0030‡		TIV† = .0060‡

Specification	Drill Diameter Range				
	Over 1/8" to 1/2" .1251 to .5000	Over 1/4" to 1/2" .2501 to .5000	Over 1/2" to 1" .5001 to 1.000	Over 1/2" to 1 1/2" .5001 to 1.500	Over 1/2" to 2" .5001 to 2.000
Dia. at Point		+.0000 to -.0010	+.0000 to -.0012		
Shank Dia.		-.0005 to -.0045			-.0005 to -.0030
Back Taper (per inch)		.0002 to .0009	.0003 to .0011		
Flute Length	+1/8" to -1/8"		+1/4" to -1/8"		
Overall Length	+1/8" to -1/8"		+1/4" to -1/8"		
Included Point Angle				118° ± 3°	
Lip Height		TIV* = .0040	TIV* = .0050		
Centrality of Web		TIV**= .0050	TIV**= .0070		
Flute Spacing		TIV† = .0100‡	TIV† = 0.140‡		

(Continued)

Table 2. *(Continued)* **Element and Dimensional Tolerances for Two-, Three-, and Four-Flute Drills.** Metric Sizes take the appropriate tolerance for the diameter range in which they fall. *(Source, Precision Twist Drill Co.)*

Specification	Drill Diameter Range			
	Over 1" to 2" 1.001 to 2.000	Over 1" to 3 1/2" 1.001 to 3.500	Over 1 1/2" to 3 1/2" 1.5001 to 3.500	Over 2" to 3 1/2" 2.001 to 3.500
Dia. at Point	+.0000 to -.0015			+.0000 to -.0020
Shank Dia.				
Back Taper (per inch)		.0004 to .0015		
Flute Length	+1/4" to -1/4"			+3/8" to -3/8"
Overall Length	+1/4" to -1/4"			+3/8" to -3/8"
Included Point Angle			118° ± 2°	
Lip Height		TIV* = .0060		
Centrality of Web		TIV**= .0100		
		TIV† = .0260‡		

* Total Indicator Variation. To measure: Rotate drill in vee block against a back end stop. Measure the cutting lip height variation on a comparator, or with an indicator set at a location approximately 75% of the distance from the center to the periphery of the drill.

** Total Indicator Variation. To measure: Rotate drill in a close fitting bushing. Record the difference in indicator readings of the web at the point as the drill is indexed 180°.

† Total Indicator Variation. To measure: Place drill in vee block against a back end stop, and rotate it against a radial finger stop. Take indicator reading at the leading edge of the margin on the opposite flute. Repeat for the other flute and note the difference between the two readings. The deviation in flute spacing is equal to one-half the difference between the two readings.

‡ The Actual Deviation is equal to one-half the TIV. See (†) above.

Table 3. Runout (Straightness) Tolerances for HSS Twist Drills. Metric Sizes take the appropriate tolerance for the diameter range in which they fall. *(Source, Precision Twist Drill Co.)*

Drill Diameter	Jobber Lengths[1]	Taper Lengths[2]	Screw Machine Lengths[3]	6" Aircraft Drills[4]	12" Aircraft Drills	Taper Shank Drills[5]
1/16"	.0080	.0200	.0070			
1/8"	.0070	.0180	.0060			
1/4"	.0060	.0120	.0050			
3/8"	.0050	.0090	.0050			
1/2"	.0050	.0090	.0050			
3/4"		.0080	.0050			
1"		.0070	.0040			
1 1/2"		.0060	.0040			
.0400-.0624				.0200	.0400	
.0625-.1249				.0150		
.0625-.1934					.0350	
.1250-.2499				.0120		
.1250-.3749				.0100		.0150
.1935-.5000					.0250	

(Continued)

Table 3. *(Continued)* **Runout (Straightness) Tolerances for HSS Twist Drills.** Metric Sizes take the appropriate tolerance for the diameter range in which they fall. *(Source, Precision Twist Drill Co.)*

Drill Diameter	Jobber Lengths[1]	Taper Lengths[2]	Screw Machine Lengths[3]	6" Aircraft Drills[4]	12" Aircraft Drills	Taper Shank Drills[5]
.2500-.3749						
.3750-.4999				.0070		
.3750-1.4999						.0100
.5000 and up				.0050		
1.500 and up						.0070

1. *Jobber Lengths.* For drills smaller than 0.03125" (up to $^1/_{32}$" or 0.80 mm), runout tolerances to be at manufacturer's discretion. For drills 0.03125" and larger, runout tolerances are to be calculated with the following formula:

   Maximum Runout (TIR) = (0.0001316 × [Overall Length ÷ Drill Diameter]) + 0.00368".

2. *Taper Lengths.* For drills smaller than 0.03125" (up to $^1/_{32}$" or 0.80 mm), runout tolerances to be at manufacturer's discretion. For drills 0.03125" and larger, runout tolerances are to be calculated with the following formula:

   Maximum Runout (TIR) = (0.000375 × [Overall Length ÷ Drill Diameter]) + 0.0027".

3. *Screw Machine Lengths.* For drills smaller than 0.0400" (up to #61 or 1.0 mm), runout tolerances to be at manufacturer's discretion. For drills 0.0400" and larger, runout tolerances are to be calculated with the following formula:

   Maximum Runout (TIR) = (0.0001316 × [Overall Length ÷ Drill Diameter]) + .00368".

4. *Taper Shank Tolerance Rate.* The rate of taper on taper shanks is held to 0.0020 inch per foot in the direction that increases the rate of taper.

Note: Suggested method of measuring twist drill straightness is as follows. *Straight shank drills* are to be checked with 1" of shank in a vee block with measurements taken on the margins at the point. If shank length is shorter than 1", use the full length of the shank. Taper shank drills are to be checked with the shank in a tapered vee block or a precision socket. Calculated values are to be rounded to the nearest thousandth of an inch.

Table 4. Troubleshooting Guide for Twist Drills. *(Source, Precision Twist Drill Co.)*

Problem	Probable Cause	Possible Solution
Outer corner breakdown	Improper speed and feed	Reduce speed, increase feed
	Insufficient coolant flow	Review and adjust
	Improper clearance	Resharpen or replace
	Chip congestion	Check geometry references
	Misalignment	Review and adjust
	Inconsistency in material	Review and adjust
Chipped cutting lips	Excessive clearance	Resharpen or replace
	Improper feed	Review and adjust
Chipping on margin	Misalignment	Review and adjust
	Oversize bushing	Replace
Broken drill	Chip congestion	Check geometry references
	Improper point geometry	Resharpen or replace
	Dull drill	Resharpen or replace
	Misalignment	Review and adjust
	Vibration and chatter	Review setup rigidity and adjust

(Continued)

Table 4. *(Continued)* **Troubleshooting Guide for Twist Drills.** *(Source, Precision Twist Drill Co.)*

Problem	Probable Cause	Possible Solution
Drill splits up center	Insufficient clearance	Resharpen or replace
	Web too thin	Resharpen or replace
	Improper feed	Reduce feed
Drill will not enter workpiece	Insufficient clearance	Resharpen or replace
	Web too thick	Resharpen or replace
	Dull drill	Resharpen or replace
	Chisel edge angle too high	Resharpen or replace
	Wrong rotation	Change rotation
Oversize hole	Improper point on drill	Resharpen or replace
	Chip congestion	Check geometry references
	Dull drill	Resharpen or replace
	Misalignment	Review and adjust
Rough hole	Dull drill	Resharpen or replace
	Improper feed	Reduce feed
	Improper point on drill	Resharpen or replace
	Insufficient coolant flow	Review and adjust
	Chip congestion	Check geometry references
Broken tang	Drill improperly seated in socket	Review and adjust

Drill reconditioning

(Precision Twist Drill Co. recommends the following procedures for reconditioning drills.) Since drills are "new" only once, the majority of a drill's life is spent in a reconditioned state. To realize the optimum holemaking benefits throughout a drill's life requires that it be reconditioned before excessive wear occurs, and to do so on a regular basis. The four most common drill points are the conventional, split point, notched, and helical. Each is reconditioned differently and a specific machine setup is required to assure successful reconditioning. Four separate steps are involved in drill reconditioning: removing worn drill portions, regrinding the point surface, web thinning, and inspecting the drill.

Removing worn portions. The outer corners are the first parts of a drill to wear (*Figure 2*). Wear begins as a slight rounding, then proceeds along the drill's margins. Wear causes a loss of size and a reverse taper of the drill body. Failure to remove all worn portions will cause the drill to cut improperly and eventually to break.

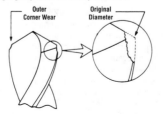

Figure 2. Left: Outer corner wear. Right: When resharpening, the worn area must be removed.

There are two procedures for removing worn portions. If wear is minimal, lightly regrinding the drill point surfaces will suffice. (Regularly recycled drills should not have more than $^1/_{16}$" inch [0.0625 mm] ground from their point surfaces.) For severe wear—when the drill body is worn back $^1/_4$" or more—cut away the worn areas with an abrasive cutoff wheel, then restore the original point edge.

Regrinding the point surface. In addition to removing the worn portion of the drill, the surfaces of the point must be reground. These two conical surfaces intersect with the faces of the flutes to form the cutting lips. They also intersect with each other to form the chisel edge (*Figure 3*).

As with any other cutting tool, the surface behind the cutting lips (heel) must not rub against the workpiece during the drilling operation. To prevent rubbing, the cutting lips must be relieved, which permits the chisel edge to penetrate. Without such relief, the drill will rub on the surface rather than penetrate material. Grinding clearance on the surface behind the cutting lips makes the heel lower than the cutting edge and permits the cutting lips to penetrate while drilling. This clearance (lip relief) is measured at the periphery of the drill, across the margin portion of the land as shown in *Figure 4*.

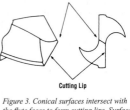

Figure 3. Conical surfaces intersect with the flute faces to form cutting lips. Surfaces meet at point to form the chisel edge.

The amount of lip relief varies, depending on the material drilled. Relief is lower for cutting difficult to machine materials because lower relief increases lip edge strength. The degree of lip relief increases toward the center of the drill. The relieved surfaces intersect at the drill's center and form the chisel edge angle. There is a relationship between the chisel edge angle and the lip relief angle. The higher the lip relief angle, the higher the chisel edge angle. An 8° to 10° lip angle = 125° chisel edge angle; 12° to 15° lip relief = 135° chisel edge angle; and when lip relief is greater than 15°, chisel edge angles may be as high as 145°.

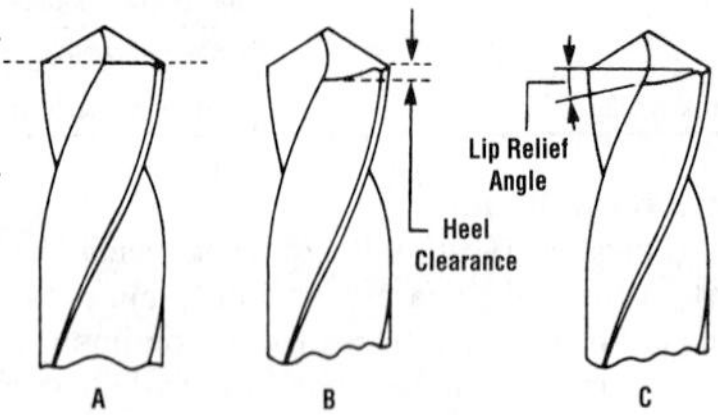

*Figure 4. A: Without lip relief, drill will not cut.
B: Grinding heel lower creates cutting tip at edges.
C: Lip relief angle is measured at periphery of drill.*

These relationships apply to conventional, split, and notched points. The helical point has chisel edge angles that are influenced not only by lip relief angles, but also by web thickness. The thicker the web, the lower the chisel angle (*Figure 5*).

Web thinning. For a drill to penetrate workpiece materials properly, the web at the point must be a certain thickness. Most drills are manufactured with the web increasing in thickness from the drill point to the back of the flute. Since removing worn portions of a drill results in a shorter flute area, the drill point will have a thicker web at the point, and a longer chisel edge, than it did prior to grinding. Although a drill's chisel edge does not actually cut, it does wedge material out of the way. Therefore, a longer chisel edge means that additional power will be

Figure 5. Helical point chisel angle.

required to drive the drill, resulting in extra heat, diminished tool life, and increased likelihood of tool breakage. Reconditioning a drill necessitates thinning the heavier web so that the chisel edge is restored to its correct length (*Figure 6*).

Drills for free-machining materials, such as aluminum and mild steel, have lighter webs than those designed for more difficult to machine materials like alloy steels. Unless unusual conditions prevail, repointed drills should be thinned to the specifications indicated in **Table 5**.

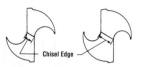

Figure 6. As a drill is reground, the chisel edge lengthens on tools with web increase.

Web thinning should be performed with a grinder designed specifically for the task. A web thinning grinder assures that equal stock is removed from both sides of the web, providing good web centrality, uniform web shape, and consistency from drill to drill. If a web thinning grinder is not available, the operation can be performed by an experienced operator. Whether done by machine or by hand, the grinding wheel used should be soft enough to remove stock without burning the cutting edges.

Table 5. Web-thinning Specifications for Repointed Drills.

Drill Diameter		Web Thickness at Point (% of Drill Diameter)
inches	mm	
.0135 to .0350	.35 to .85	30
.0350 to .0520	.90 to 1.30	25
.0520 to .1875	1.35 to 4.75	20
.1875 to .2500	4.80 to 6.30	17
.2500 to .6250	6.40 to 15.75	15
.6250 to 1.3750	16.0 to 34.50	13
1.3750 to 2.3750	35.0 and up	12
2.3750 and up		11

Web thinning should extend up the flute to one-third to one-half the drill's diameter. This assures that the thinning does not end abruptly and create a "chip pocket" that would prevent chips from flowing smoothly up the flutes, and could lead the drill to break.

The next step is to check and verify web centrality (*Figure 7*). A common method for web thinning conventional and notched points is to use a radiused grinding wheel. The direction of the grind should be 30° to the drill's center line, with a resulting rake angle equaling 8° to 12°. The secondary cutting edge that is created must be straight and at a 25° angle from the original cutting edge, as shown in *Figure 8*.

Figure 9 provides parameters for thinning the web on split point drills, which are most commonly available in 118° or 135° point angles. Since the latter is used most widely, this example will focus on 135° points. When grinding a 135° split point surface, the chisel edge angle should measure 105° to 120° (see **Table 6** for suggested lip relief angles). The two

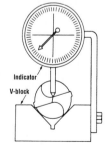

Figure 7. Measure the difference in readings from each side of the web to check centrality.

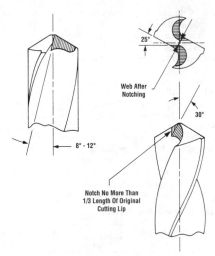

Figure 8. Notch thinning a web creates a secondary straight cutting edge at 25° from the original cutting edge.

Figure 9. Web thinning a split point drill provides positive rake. Lip relief angles are smaller for larger diameter drills.

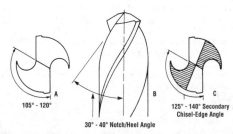

Figure 10. Notch/heel angle forms secondary chisel angle on split point.

notch/heel angles are ground to a 30° to 40° angle so that they form a secondary chisel angle of approximately 130° at the drill point center (*Figure 10*).

The two notching angles should not exceed 40°, as higher angles will create a chip pocket. The notch grinds should also produce a 3° to 8° positive rake angle in the face of the notches, thereby improving the drill's cutting action.

Table 6. Suggested Lip Relief Angles at Periphery.

Drill Diameters	General Purpose	Hard, Tough Materials	Soft, Free Machining Materials
#97 to #81	28°	22°	30°
#80 to #61	24°	20°	26°
#60 to #41	21°	18°	24°
#40 to #31	18°	16°	22°
#30 to 1/4"	16°	14°	20°
F to 11/32"	14°	12°	18°
S to 1/2"	12°	10°	16°
33/64" to 3/4"	10°	8°	14°
49/64" and over	8°	7°	12°

Helical drills are the exception to the web thinning rule. They do not require web thinning because their design incorporates a heavy web and a point with an S-shaped chisel angle that acts as a continuous cutting edge. This positive cutting action centers the drill, reduces thrust, and improves chip formation, allowing the drill to cut close to drill size.

Speeds and feeds for twist drilling

Tables 7–10 provide recommended speeds and feeds for twist drilling operations with high-speed steel tools. This data is the result of testing and experimentation carried out by Kennametal, Inc. See the sections on spade drilling and indexable insert drills for additional speeds and feeds data. **Table 11**, "Cutting Speeds: Wire, Fractional, and Letter Size Drills," can be used to convert surface feet per minute to revolutions per minute. For surface feet per minute requirements in excess of 100 feet per minute, combine the figures on rpm figures on the columns that add up to the required reading. For example, to find the rpm needed to operate a $1/32$" drill at 175 sfm, add the columns for 100 sfm (12,223), 60 sfm (7334), and 15 sfm (1833), which results in an indicated speed of 21,390 rpm.

Table 7. Recommended Surface Feet or Meters per Minute, and Coolant by Material Application.
(Source, Kennametal Inc.)

Ferrous Materials	BHN	SFM	SMM	Coolant
Low Carbon Steel	85 - 125	80 - 95	24 - 28	Soluble Oil
Medium Carbon Steel	125 - 175	70 - 85	21 - 26	Soluble Oil
High Carbon Steel	175 - 225	45 - 65	14 - 20	Soluble Oil
Steels - Alloyed	Under 200	60 - 90	18 - 27	Soluble Oil
	200 - 300	40 - 70	12 - 21	Soluble Oil
	Over 300	20 - 30	6 - 9	Soluble Oil
Steel Drop Forgings,	330 - 370	30 - 40	9 - 12	Cutting Oil
Heat Treated	370 - 420	20 - 30	6 - 9	Cutting Oil
	Over 420	10 - 20	3 - 6	Cutting Oil

(Continued)

Table 7. *(Continued)* **Recommended Surface Feet or Meters per Minute, and Coolant by Material Application.**
(Source, Kennametal Inc.)

Ferrous Materials	BHN	SFM	SMM	Coolant
Gray Cast Iron, Soft	125	75 - 150	22 - 45	Dry
Gray Cast Iron, Medium	120 - 200	50 - 100	15 - 30	Dry or Soluble Oil
Gray Cast Iron, Hard	Up to 350	25 - 40	7 - 12	Soluble Oil
Titanium Alloys Ti-75A	300 - 400	50 - 60	15 - 18	Cutting Oil
Ti-150A, RS-120	300 - 400	40 - 50	12 - 15	Cutting Oil
Ti-140A, RC-130B	300 - 400	30 - 40	9 - 12	Cutting Oil
Ti-6AL-4V	300 - 400	20 - 30	6 - 9	Cutting Oil
Stainless Steel, 300 Series	120 - 200	20 - 40	6 - 12	Cutting Oil
Stainless Steel, 400 Series	200 - 300	40 - 70	12 - 21	Cutting Oil
Martensitic 416, 420, F 416 Plus K, 400F, 4165SE, 440F	135 - 185	40 - 50	12 - 15	Cutting Oil
Precipitation Hardening	Up to 300	40 - 60	12 - 18	Cutting Oil
Precipitation Hardening	325 - 375	30	9	Cutting Oil
Stainless Steel, Cast	400 - 450	20	6	Cutting Oil
Heat Resisting Steels	175 - 225	10 - 25	3 - 7	Cutting Oil
Nimonic Alloys	100 - 300	10 - 20	3 - 6	Cutting Oil
Manganese, 12 - 14% min.	125 - 220	10 - 12	3 - 4	Cutting Oil
Spring Steels	402	15 - 30	5 - 9	Soluble Oil
Armor Plate	200 - 250	40	12	Soluble Oil
	250 - 300	35	10	Soluble Oil
	300 - 350	30	9	Cutting Oil
Nonferrous Materials				
Aluminum, Pure	140 - 350	130 - 200	41 - 60	Soluble Oil
Aluminum Alloys	140 - 330	150 - 300	45 - 90	Soluble Oil
Aluminum, Leaded	40 - 100	200 - 325	60 - 97	Soluble Oil
Aluminum, Silicon Alloy Die Cast	40 - 100	25 - 50	7 - 15	Soluble Oil
Brass	190 - 210	100 - 250	30 - 75	Cutting or Soluble Oil
Bronze	150 - 200	70 - 150	21 - 45	Soluble Oil
Copper, Nickel & Copper Tin Alloy	65 - 100	100 - 200	30 - 60	Cutting or Soluble Oil
Copper Aluminum Alloys	30 - 100	120 - 200	36 - 60	Cutting or Soluble Oil
Magnesium Alloys, Wrought	50 - 90	140 - 330	42 - 100	Cutting or Soluble Oil
Magnesium Alloys, Cast	50 - 90	140 - 365	42 - 100	Cutting or Soluble Oil
High Temperature Alloys, Cobalt, Nickel, or Iron Base	180-300	5 - 20	2 - 6	Cutting or Soluble Oil
Titanium Alloys	Over 250	20 - 45	6 - 13	Cutting Oil
Nickel Alloys, Wrought	80 - 170	70	21	Cutting or Soluble Oil
& Cast Monel	115 - 240	55	16	Cutting or Soluble Oil
Beryllium Nickel	200 - 250	12	4	Soluble Oil
Zinc Alloy	112 - 126	200 - 250	60 - 75	Soluble Oil
Plastic & Related Materials		100 - 200	30 - 60	Dry

Table 8. Recommended Drill Feeds per Revolution.

Diameter Range		FPR—Light Conditions		FPR—Medium Conditions		FPR—Heavy Conditions	
inch	mm	inch	mm	inch	mm	inch	mm
1/16-1/8	1.60-3.10	0.0005-0.001	0.013-0.025	0.001-0.002	0.025-0.05	0.002-0.004	0.05-0.10
1/8-1/4	3.20-6.30	0.001-0.003	0.025-0.075	0.003-0.005	0.075-0.13	0.004-0.006	0.10-0.15
1/4-3/8	6.40-9.50	0.003-0.005	0.075-0.13	0.005-0.007	0.13-0.18	0.006-0.010	0.15-0.25
3/8-1/2	9.60-12.70	0.004-0.006	0.10-0.15	0.005-0.008	0.13-0.20	0.008-0.012	0.20-0.30
1/2-3/4	12.70-19.0	0.005-0.007	0.13-0.18	0.007-0.010	0.18-0.25	0.009-0.014	0.23-0.36
3/4-1.0	19.50-25.0	0.007-0.010	0.18-0.25	0.009-0.014	0.23-0.36	0.014-0.020	0.36-0.50

Note: For inch units, RPM = (SFM ÷ Drill Diameter) × 3.82.
For metric units, RPM = (SMM ÷ Drill Diameter) × 318.3.

Speeds and feeds for deep-hole drilling

Holes that are three drill diameters or more deep are considered "deep holes," and speeds and feeds must be adjusted for conditions. The deeper the hole, the greater the tendency for chips to pack and clog the flutes of the drill. This leads to increased heat, and coolant flow is restricted. The buildup of heat at the point will eventually result in premature failure. Step drilling, or the process of drilling a short distance, then withdrawing the drill, will often reduce chip packing. The deeper the hole, the more frequent the drill must be retracted for this procedure to be effective. A reduction in speed and feed to reduce the amount of heat generated is generally required in most deep-hole applications where coolant cannot be effectively applied. **Table 9** provides correction factors for deep-hole drilling.

Table 9. Speeds and Feeds for Deep-hole Drilling—Correction Factors.

Hole Depth to Diameter Ratio (depth ÷ drill diameter)	Speed Reduction (in %)	Feed Reduction (in %)
3	10%	10%
4	20%	10%
5	30%	20%
6	35 - 40%	20%

Feed per revolution—parabolic drills

Parabolic drills enable improved chip flow, resulting in improved heat dissipation. It is important to maintain a constant, heavy feed rate regardless of depth of hole. **Table 10** provides feed rates and correction rates based on drill diameter and hole depth.

(Text continued on p. 1098)

Table 10. Speed and Feed Correction Factors for Parabolic Drills.

Drill Diameter Range		Feed Rate	
fraction	mm	inch	mm
1/16" to 1/8"	1.60 to 3.10	.0010 - .0040	.0254 - .01016
1/8" to 1/4"	3.20 to 6.30	.0030 - .0090	.0762 - .2286
1/4" to 3/8"	6.40 to 9.50	.0050 - .0100	.1270 - .2540
3/8" to 1/2"	9.60 to 12.70	.0070 - .0150	.1778 - .3810
1/2" to 3/4"	12.80 to 19.00	.0100 - .0160	.2540 - .4064
3/4" to 1"	19.50 to 25.0	.0150 - .0220	.3810 - .5588

Speed Correction Factor - Parabolic Drills	
Hole Depth-to-Diameter Ratio (depth ÷ drill diameter)	**Speed Reduction (in %)**
3	0
4	0
5	5%
6 to 8	10%
8 to 11	20%
11 to 14	30%
12 to 17	40%
17 to 20	50%

Example: For a $1/4$" drill drilling a $1\,1/2$" hole, the hole depth to diameter ratio would be 6 ($1\,1/2$" ÷ $1/4$" = 6).

Table 11. Cutting Speeds: Wire, Fractional, and Letter Size Drills. Wire Size #80 Through 3 1/2".

Drill Size	10'	12'	15'	20'	25'	30'	35'	40'	45'	50'	60'	70'	80'	90'	100'
							Surface Feet per Minute								
							Revolutions per Minute								
80	2829	3395	4244	5659	7074	8488	9903	11318	12732	14150	16980	19810	22640	25470	28300
79	2634	3161	3951	5269	6586	7903	9220	10537	11854	13170	15804	18438	21072	23706	26340
1/64	2245	2934	3664	4889	6112	7334	8556	9778	11001	12223	14668	17112	19557	22001	24446
78	2387	2865	3581	4775	5968	7162	8356	9549	10743	11935	14322	16709	19096	21483	23870
77	2122	2546	3183	4244	5305	6366	7427	8488	9549	10610	12732	14854	16976	19098	21220
76	1910	2292	2865	3820	4775	5730	6684	7639	8594	9550	11460	13370	15280	17190	19100
75	1819	2183	2728	3638	4547	5457	6366	7276	8185	9095	10914	12733	14552	16371	18190
74	1698	2037	2546	3395	4244	5093	5942	6791	7639	8510	10212	11914	13616	15318	17020
73	1592	1910	2387	3183	3979	4775	5570	6366	7162	7960	9552	11144	12736	14328	15920
72	1528	1833	2272	3056	3280	4584	5348	6112	6875	7640	9168	10696	12224	13752	15280
71	1469	1763	2204	2938	3673	4407	5142	5876	6611	7365	8838	10311	11784	13257	14730
70	1364	1637	2046	2728	3410	4093	4775	5457	6139	6820	8184	9548	10912	12276	13640
69	1308	1570	1962	2616	3270	3924	4578	5232	5887	6530	7836	9142	10488	11754	13060
68	1232	1479	1848	2464	3080	3696	4313	4929	5545	6160	7392	8624	9856	11088	12320
1/32	1222	1467	1833	2445	3056	3667	4278	4889	5500	6112	7334	8556	9778	11001	12223
67	1194	1432	1790	2387	2984	3581	4178	4775	5371	5970	7164	8358	9552	10746	11940
66	1157	1389	1736	2315	2894	3472	4051	4630	5207	5790	6948	8106	9264	10422	11580
65	1091	1310	1637	2183	2728	3274	3820	4365	4911	5455	6546	7637	8728	9819	10910
64	1061	1273	1592	2122	2653	3183	3714	4244	4775	5305	6366	7427	8488	9549	10610
63	1032	1239	1549	2065	2581	3097	3613	4129	4646	5160	6192	7224	8256	9288	10320
62	1005	1206	1508	2010	2513	3016	3518	4021	4523	5025	6030	7035	8040	9045	10050
61	979	1175	1469	1959	2449	2938	3428	3918	4407	4897	5876	6856	7835	8815	9794

(Continued)

Table 11. *(Continued)* Cutting Speeds: Wire, Fractional, and Letter Size Drills. Wire Size #80 Through 3 1/2".

Drill Size	Surface Feet per Minute														
	10'	12'	15'	20'	25'	30'	35'	40'	45'	50'	60'	70'	80'	90'	100'
	Revolutions per Minute														
60	955	1146	1432	1910	2387	2865	3342	3820	4297	4775	5729	6684	7639	8594	9549
59	932	1118	1397	1863	2329	2795	3261	3726	4192	4658	5590	6521	7453	8388	9316
58	909	1091	1364	1819	2274	2728	3183	3638	4093	4547	5456	6367	7275	8166	9095
57	888	1066	1332	1777	2221	2665	3109	3553	3997	4452	5342	6232	7122	8013	8903
56	821	986	1232	1643	2054	2464	2875	3286	3696	4108	4929	5751	6572	7394	8215
3/64	815	978	1222	1630	2037	2445	2852	3259	3667	4074	4889	5704	6519	7334	8149
55	735	881	1102	1469	1836	2204	2571	2938	3306	3673	4408	5142	5877	6611	7346
54	694	833	1042	1389	1736	2083	2431	2778	3125	3473	4167	4862	5556	6251	6945
53	642	770	963	1284	1605	1926	2247	2568	2889	3207	3848	4490	5131	5773	6414
1/16	611	733	917	1222	1528	1833	2139	2445	2750	3056	3667	4278	4889	5500	6112
52	602	722	902	1203	1504	1805	2105	2406	2707	3008	3609	4211	4812	5414	6015
51	570	684	855	1140	1425	1710	1995	2280	2565	2851	3421	3991	4561	5131	5701
50	546	655	819	1091	1364	1637	1910	2183	2456	2729	3274	3820	4533	4911	5457
49	523	628	785	1046	1308	1570	1831	2093	2355	2617	3140	3663	4186	4710	5233
48	503	603	754	1005	1256	1508	1759	2010	2262	2513	3016	3518	4021	4523	5026
5/64	489	587	733	978	1222	1467	1711	1956	2200	2445	2934	3422	3911	4400	4889
47	487	584	730	973	1216	1460	1703	1946	2190	2433	2920	3406	3893	4379	4866
46	472	566	707	943	1179	1415	1650	1886	2122	2358	2830	3301	3773	4244	4716
45	466	559	699	932	1165	1397	1630	1863	2096	2329	2795	3261	3726	4192	4658
44	444	533	666	888	1110	1332	1555	1777	1999	2221	2665	3109	3554	3999	4442
43	429	515	644	858	1073	1288	1502	1717	1931	2146	2575	3004	3434	3863	4292
42	409	490	613	817	1021	1226	1430	1634	1838	2043	2451	2860	3268	3677	4085

(Continued)

Table 11. *(Continued)* Cutting Speeds: Wire, Fractional, and Letter Size Drills. Wire Size #80 Through 3 1/2".

Drill Size	Surface Feet per Minute															
	10'	12'	15'	20'	25'	30'	35'	40'	45'	50'	60'	70'	80'	90'	100'
	Revolutions per Minute															
3/32	407	489	611	815	1019	1222	1426	1630	1833	2037	2445	2852	3259	3667	4074
41	398	477	597	796	995	1194	1393	1592	1790	1990	2387	2785	3183	3581	3979
40	390	468	585	780	974	1169	1364	1559	1754	1949	2339	2729	3118	3508	3898
39	384	461	576	768	960	1152	1344	1536	1728	1920	2303	2687	3071	3455	3839
38	376	452	564	753	941	1129	1317	1505	1693	1882	2258	2634	3010	3387	3763
37	367	441	551	735	918	1102	1285	1469	1653	1837	2204	2571	2938	3306	3673
36	359	430	538	717	897	1076	1255	1435	1514	1794	2152	2511	2870	3228	3587
7/64	349	419	524	698	873	1048	1222	1397	1572	1746	2095	2445	2794	3143	3492
35	347	417	521	694	868	1042	1215	1389	1563	1736	2083	2430	2778	3125	3472
34	344	413	516	688	860	1032	1204	1376	1549	1721	2065	2409	2753	3097	3442
33	338	406	507	676	845	1014	1183	1352	1521	1690	2028	2366	2704	3042	3380
32	329	395	494	659	823	988	1152	1317	1482	1647	1976	2305	2634	2964	3293
31	318	382	477	637	796	955	1114	1273	1432	1592	1910	2228	2546	2865	3183
1/8	306	367	458	611	764	917	1070	1222	1375	1528	1833	2139	2445	2750	3056
30	297	357	446	595	743	892	1040	1189	1338	1487	1784	2081	2378	2676	2973
29	281	337	421	562	702	843	983	1123	1264	1405	1685	1966	2247	2528	2809
28	272	326	408	544	680	816	952	1087	1223	1360	1631	1903	2175	2447	2719
9/64	272	326	407	543	679	815	951	1086	1222	1358	1630	1901	2173	2445	2716
27	265	318	398	531	663	796	928	1061	1194	1327	1592	1857	2122	2388	2653
26	260	312	390	520	650	780	909	1039	1169	1299	1559	1819	2078	2338	2598
25	255	307	383	511	639	766	894	1022	1150	1276	1533	1789	2044	2300	2555
24	251	302	377	503	628	754	880	1055	1131	1257	1508	1759	2010	2262	2513

(Continued)

Table 11. (Continued) Cutting Speeds: Wire, Fractional, and Letter Size Drills. Wire Size #80 Through 3 1/2".

Drill Size	Surface Feet per Minute														
	10'	12'	15'	20'	25'	30'	35'	40'	45'	50'	60'	70'	80'	90'	100'
	Revolutions per Minute														
23	248	298	372	496	620	744	868	992	1116	1240	1488	1736	1984	2232	2480
5/32	244	293	367	489	611	733	856	978	1100	1222	1467	1711	1956	2200	2445
22	243	292	365	487	608	730	852	973	1095	1217	1460	1703	1946	2190	2433
21	240	288	360	480	601	721	841	961	1081	1201	1441	1681	1922	2162	2402
20	237	285	356	474	593	712	830	949	1068	1186	1423	1660	1898	2135	2372
19	230	276	345	460	575	690	805	920	1035	1151	1381	1611	1841	2071	2301
18	225	270	338	451	563	676	789	901	1014	1130	1356	1582	1808	2034	2260
11/64	222	267	333	444	556	667	778	889	1000	1111	1333	1556	1778	2000	2222
17	221	265	331	442	552	662	773	883	994	1104	1325	1546	1766	1987	2208
16	216	259	324	432	540	647	755	863	971	1079	1295	1511	1726	1942	2158
15	212	255	318	424	531	637	743	849	955	1064	1276	1489	1702	1914	2127
14	210	252	315	420	525	630	735	839	944	1050	1259	1469	1679	1889	2099
13	206	248	310	413	516	619	723	826	929	1032	1239	1450	1652	1859	2065
3/16	204	244	306	407	509	611	713	815	917	1019	1222	1426	1630	1833	2037
12	202	243	303	404	505	606	707	808	909	1010	1213	1415	1617	1819	2021
11	200	240	300	400	500	600	700	800	900	1000	1200	1400	1600	1800	2000
10	197	237	296	395	494	592	691	790	888	987	1184	1382	1579	1777	1974
9	195	234	292	390	487	585	682	780	877	975	1169	1364	1559	1754	1949
8	192	230	288	384	480	576	672	768	864	960	1151	1343	1535	1727	1919
7	190	228	285	380	475	570	665	760	855	950	1140	1330	1520	1710	1900
13/64	188	226	282	376	470	564	658	752	846	940	1128	1316	1504	1692	1880
6	187	225	281	374	468	562	655	749	843	936	1123	1310	1498	1685	1872

(Continued)

Table 11. *(Continued)* Cutting Speeds: Wire, Fractional, and Letter Size Drills. Wire Size #80 Through 3 1/2".

Drill Size	Surface Feet per Minute																
	10'	12'	15'	20'	25'	30'	35'	40'	45'	50'	60'	70'	80'	90'	100'
	Revolutions per Minute														
5	186	223	279	372	465	558	651	743	836	930	1115	1310	1487	1673	1859
4	183	219	274	366	457	548	640	731	822	914	1097	1280	1462	1645	1828
3	179	215	269	359	448	538	628	717	807	897	1076	1255	1434	1614	1793
7/32	175	210	262	349	437	524	611	698	786	873	1048	1222	1397	1572	1746
2	173	207	259	346	432	519	605	691	778	864	1037	1210	1382	1555	1728
1	168	201	251	335	419	503	586	670	754	838	1005	1173	1340	1508	1675
A	163	196	245	326	408	490	571	653	735	818	982	1145	1309	1472	1636
15/64	163	196	244	326	407	489	570	652	733	815	978	1141	1304	1467	1630
B	160	193	241	321	401	481	562	642	722	803	963	1124	1284	1445	1605
C	158	189	237	316	395	473	552	631	710	789	947	1105	1262	1420	1578
D	155	186	233	311	388	466	543	621	699	778	934	1089	1245	1400	1556
1/4 E	153	183	229	306	382	458	535	611	687	764	917	1070	1222	1375	1528
F	149	178	223	297	372	446	520	595	669	743	892	1040	1189	1337	1486
G	146	176	220	293	366	439	512	585	659	732	878	1024	1170	1317	1463
H	144	172	215	287	359	431	503	574	646	718	862	1005	1149	1294	1436
I	140	169	211	281	351	421	492	562	632	702	842	983	1123	1264	1404
J	138	165	207	276	645	414	483	552	621	690	827	965	1103	1241	1379
K	136	163	204	272	340	408	476	544	612	680	815	951	1087	1223	1359
9/31	136	163	204	272	340	407	475	543	611	679	815	951	1086	1222	1358
L	132	158	198	263	329	395	461	527	593	659	790	922	1054	1185	1317
M	129	155	194	259	324	388	453	518	583	648	777	907	1036	1166	1295
N	126	152	190	253	316	379	442	506	569	633	759	886	1012	1139	1265

(Continued)

Table 11. (Continued) Cutting Speeds: Wire, Fractional, and Letter Size Drills. Wire Size #80 Through 3 1/2".

Drill Size	Surface Feet per Minute														
	10'	12'	15'	20'	25'	30'	35'	40'	45'	50'	60'	70'	80'	90'	100'
	Revolutions per Minute														
5/16	122	147	183	244	306	367	428	489	550	611	733	856	978	1000	1222
O	121	145	181	242	302	363	423	484	544	605	725	846	967	1088	1209
P	118	142	177	237	296	355	414	473	532	592	710	828	946	1065	1183
Q	115	138	173	230	288	345	403	460	518	575	690	805	920	1035	1150
R	113	135	169	225	282	338	394	451	507	564	676	789	902	1014	1127
11/32	111	133	167	222	278	333	389	444	500	556	667	778	889	1000	1111
S	110	132	165	220	274	329	384	439	494	549	659	769	878	988	1098
T	107	128	160	213	267	320	373	427	480	533	640	746	853	959	1066
U	104	125	156	208	259	311	363	415	467	519	623	727	830	934	1037
3/8	102	122	153	204	255	306	357	407	458	509	611	713	815	917	1019
V	101	122	152	203	253	304	355	405	456	507	608	709	810	912	1013
W	99	119	148	198	247	297	346	396	445	495	594	693	792	891	989
X	96	115	144	192	240	289	337	385	433	481	576	672	769	865	962
Y	95	113	142	189	236	284	331	378	425	473	567	662	756	851	945
13/32	94	113	141	188	235	282	329	376	423	470	564	658	752	846	940
Z	92	111	139	185	231	277	324	370	416	462	555	647	740	832	925
7/16	87	105	131	175	218	262	306	349	393	437	524	611	698	786	873
15/32	81	98	122	163	204	244	285	326	367	407	489	570	652	733	815
1/2	76	92	115	153	191	229	267	306	344	382	458	535	611	688	764
9/16	68	81	102	136	170	204	238	272	306	340	407	475	543	611	679
5/8	61	73	92	122	153	183	214	244	275	306	367	428	489	550	611
11/16	56	67	83	111	139	167	194	222	250	278	333	389	444	500	556

(Continued)

Table 11. (Continued) Cutting Speeds: Wire, Fractional, and Letter Size Drills. Wire Size #80 Through 3 1/2".

Drill Size	Surface Feet per Minute														
	10'	12'	15'	20'	25'	30'	35'	40'	45'	50'	60'	70'	80'	90'	100'
	Revolutions per Minute														
3/4	51	61	76	102	127	153	178	204	229	255	306	357	407	458	509
13/16	47	56	71	94	118	141	165	188	212	235	282	329	376	423	470
7/8	44	52	65	87	109	131	153	175	196	218	262	306	349	393	437
15/16	41	49	61	81	102	122	143	163	183	204	244	285	326	367	407
1.0	38	46	57	76	95	115	134	153	172	191	229	267	306	344	382
1 1/8	34	41	51	68	85	102	119	136	153	170	204	238	272	306	340
1 1/4	31	37	46	61	76	92	107	122	138	153	183	214	244	275	306
1 3/8	28	33	42	56	69	83	97	111	125	139	167	194	222	250	278
1 1/2	25	31	38	51	64	76	89	102	115	127	153	178	204	229	255
1 5/8	24	28	35	47	59	71	82	94	106	118	141	165	188	212	235
1 3/4	22	26	33	44	55	65	76	87	98	109	131	153	175	196	218
1 7/8	20	24	31	41	51	61	71	81	92	102	122	143	163	183	204
2.0	19	23	29	38	48	57	67	76	86	95	115	134	153	172	191
2 1/4	17	20	25	34	42	51	59	68	76	85	102	119	136	153	170
2 1/2	15	18	23	31	38	46	53	61	69	76	92	107	122	138	153
2 3/4	14	17	21	28	35	42	49	56	63	69	83	97	111	125	139
3.0	13	15	19	25	32	38	45	51	57	64	76	89	102	115	127
3 1/2	11	13	16	22	27	33	38	44	49	55	65	76	87	98	109

Table 12. Straight Shank Jobber Length, Taper Length, and Screw Machine Length Twist Drills. Jobber Length #97 Through 7.50 mm. Taper Length and Screw Machine Length 1.00 mm Through 1 Inch. *(Source, Kennametal Inc.)*

| Inch Sizes | Drill Diameter | | | Jobber Length | | | | Taper Length | | | | Screw Machine Length | | | |
| | mm Sizes | Equivalent | | Flute | | Overall | | Flute | | Overall | | Flute | | Overall | |
		inch	metric	inch	mm	inch	mm	inch	mm	inch	mm	inch	mm	inch	mm
97	.15	.0059	.15	1/16	1.5	3/4	19								
96	.16	.0063	.16	1/16	1.5	3/4	19								
95	.17	.0067	.17	1/16	1.5	3/4	19								
94	.18	.0071	.18	1/16	1.5	3/4	19								
93	.19	.0075	.19	1/16	1.5	3/4	19								
92	.20	.0079	.20	1/16	1.5	3/4	19								
91	.21	.0083	.21	5/64	2	3/4	19								
90	.22	.0087	.22	5/64	2	3/4	19								
89	.23	.0091	.23	5/64	2	3/4	19								
88	.24	.0095	.23	5/64	2	3/4	19								
	.25	.0098	.25	5/64	2	3/4	19								
87		.0100	.254	5/64	2	3/4	19								
	.26	.0102	.26	5/64	2	3/4	19								
86		.0105	.267	3/32	2.5	3/4	19								
	.27	.0106	.27	3/32	2.5	3/4	19								

Specifications begin below. First size is 1.0 mm.

(Continued)

Table 12. *(Continued)* **Straight Shank Jobber Length, Taper Length, and Screw Machine Length Twist Drills.** Jobber Length #97 Through 7.50 mm. Taper Length and Screw Machine Length 1.00 mm Through 1 Inch. *(Source, Kennametal Inc.)*

Inch Sizes	mm Sizes	Equivalent		Jobber Length				Taper Length				Screw Machine Length			
		inch	metric	Flute		Overall		Flute		Overall		Flute		Overall	
				inch	mm	inch	mm	inch	mm	inch	mm	inch	mm	inch	mm
85		.0110	.28	3/32	2.5	3/4	19								
	.28	.0114	.29	3/32	2.5	3/4	19								
84		.0115	.292	3/32	2.5	3/4	19								
	.29	.0118	.30	3/32	2.5	3/4	19								
83	.30	.0120	.305	3/32	2.5	3/4	19								
82		.0125	.318	3/32	2.5	3/4	19								
	.32	.0126	.32	3/32	2.5	3/4	19								
81		.0130	.33	3/32	2.5	3/4	19								
	.34	.0134	.34	1/8	3										
80		.0135	.343	1/8	3	3/4	19								
	.35	.0138	.35	1/8	3	3/4	19								
	.36	.0142	.36	1/8	3	3/4	19								
79		.0145	.368	1/8	3	3/4	19								
	.38	.0150	.38	3/16	5										
1/64		.0156	.396	3/16	5	3/4	19								
	.40	.0157	.40	3/16	5	3/4	19								
78		.0160	.406	3/16	5	3/4	19								
	.42	.0165	.42	3/16	5	3/4	19								
	.45	.0177	.45	3/16	5	3/4	19								
77		.0180	.457	3/16	5	3/4	19								

(Continued)

Table 12. *(Continued)* **Straight Shank Jobber Length, Taper Length, and Screw Machine Length Twist Drills.** Jobber Length #97 Through 7.50 mm. Taper Length and Screw Machine Length 1.00 mm Through 1 Inch. *(Source, Kennametal Inc.)*

Drill Diameter				Jobber Length					Taper Length				Screw Machine Length			
		Equivalent		Flute		Overall		Flute		Overall		Flute		Overall	
Inch Sizes	mm Sizes	inch	metric	inch	mm	inch	mm	inch	mm	inch	mm	inch	mm	inch	mm	inch	mm
	.46	.0181	.46	3/16	5	3/4	19								
	.48	.0189	.48	3/16	5	3/4	19								
	.50	.0197	.50	3/16	5	3/4	19								
76		.0200	.508	3/16	5	3/4	19								
75		.0210	.533	1/4	6	1	25								
	.55	.0217	.550	1/4	6	1	25								
74		.0225	.572	1/4	6	1	25								
	.60	.0236	.60	5/16	8	1 1/8	29								
73		.0240	.610	5/16	8	1 1/8	29								
72		.0250	.635	5/16	8	1 1/8	29								
	.65	.0256	.65	3/8	10	1 1/4	32								
71		.0260	.66	3/8	10	1 1/4	32								
	.70	.0276	.70	3/8	10	1 1/4	32								
70		.0280	.711	3/8	10	1 1/4	32								
69		.0292	.742	1/2	13	1 3/8	35								
	.75	.0295	.75	1/2	13	1 3/8	35								
68		.0310	.787	1/2	13	1 3/8	35								
1/32		.0312	.792	1/2	13	1 3/8	35								

(Continued)

Table 12. *(Continued)* **Straight Shank Jobber Length, Taper Length, and Screw Machine Length Twist Drills.** Jobber Length #97 Through 7.50 mm. Taper Length and Screw Machine Length 1.00 mm Through 1 Inch. *(Source, Kennametal Inc.)*

Drill Diameter				Jobber Length				Taper Length				Screw Machine Length			
Inch Sizes	mm Sizes	Equivalent inch	metric	Flute inch	mm	Overall inch	mm	Flute inch	mm	Overall inch	mm	Flute inch	mm	Overall inch	mm
	.80	.0315	.80	1/2	13	1 3/8	35								
67		.0320	.813	1/2	13	1 3/8	35								
66		.0330	.838	1/2	13	1 3/8	35								
	.85	.0335	.85	5/8	16	1 1/2	38								
65		.0350	.889	5/8	16	1 1/2	38								
	.90	.0354	.899	5/8	16	1 1/2	38								
64		.0360	.914	5/8	16	1 1/2	38								
63		.0370	.940	5/8	16	1 1/2	38								
	.95	.0374	.95	5/8	16	1 1/2	38								
62		.0380	.965	5/8	16	1 1/2	38								
61		.0390	.991	11/16	17	1 5/8	41								
	1.0	.0394	1	11/16	17	1 5/8	41	1 1/8	29	2 1/4	57	1/2	13	1 3/8	35
60		.0400	1.016	11/16	17	1 5/8	41	1 1/8	29	2 1/4	57	1/2	13	1 3/8	35
59		.0410	1.041	11/16	17	1 5/8	41	1 1/8	29	2 1/4	57	1/2	13	1 3/8	35
	1.05	.0413	1.05	11/16	17	1 5/8	41	1 1/8	29	2 1/4	57	1/2	13	1 3/8	35
58		.0420	1.067	11/16	17	1 5/8	41	1 1/8	29	2 1/4	57	1/2	13	1 3/8	35
57		.0430	1.092	3/4	19	1 3/4	44	1 1/8	29	2 1/4	57	1/2	13	1 3/8	35
	1.10	.0433	1.10	3/4	19	1 3/4	44	1 1/8	29	2 1/4	57	1/2	13	1 3/8	35
	1.15	.0453	1.15	3/4	19	1 3/4	44	1 1/8	29	2 1/4	57	1/2	13	1 3/8	35

(Continued)

Table 12. *(Continued)* **Straight Shank Jobber Length, Taper Length, and Screw Machine Length Twist Drills.** Jobber Length #97 Through 7.50 mm. Taper Length and Screw Machine Length 1.00 mm Through 1 Inch. *(Source, Kennametal Inc.)*

Drill Diameter				Jobber Length				Taper Length				Screw Machine Length			
Inch Sizes	mm Sizes	Equivalent inch	Equivalent metric	Flute inch	Flute mm	Overall inch	Overall mm	Flute inch	Flute mm	Overall inch	Overall mm	Flute inch	Flute mm	Overall inch	Overall mm
56		.0465	1.181	3/4	19	1 3/4	44	1 1/8	29	2 1/4	57	1/2	13	1 3/8	35
3/64		.0469	1.191	3/4	19	1 3/4	44	1 1/8	29	2 1/4	57	1/2	13	1 3/8	35
	1.20	.0472	1.20	7/8	22	1 7/8	48	1 3/4	44	3	76	5/8	16	1 5/8	41
	1.25	.0492	1.25	7/8	22	1 7/8	48	1 3/4	44	3	76	5/8	16	1 5/8	41
	1.30	.0512	1.30	7/8	22	1 7/8	48	1 3/4	44	3	76	5/8	16	1 5/8	41
55		.0520	1.321	7/8	22	1 7/8	48	1 3/4	44	3	76	5/8	16	1 5/8	41
	1.35	.0531	1.35	7/8	22	1 7/8	48	1 3/4	44	3	76	5/8	16	1 5/8	41
54		.0550	1.397	7/8	22	1 7/8	48	1 3/4	44	3	76	5/8	16	1 5/8	41
	1.40	.0551	1.40	7/8	22	1 7/8	48	1 3/4	44	3	76	5/8	16	1 5/8	41
	1.45	.0571	1.45	7/8	22	1 7/8	48	1 3/4	44	3	76	5/8	16	1 5/8	41
	1.50	.0591	1.50	7/8	22	1 7/8	48	1 3/4	44	3	76	5/8	16	1 5/8	41
53		.0595	1.511	7/8	22	1 7/8	48	1 3/4	44	3	76	5/8	16	1 5/8	41
	1.55	.0610	1.550	7/8	22	1 7/8	48	1 3/4	44	3	76	5/8	16	1 5/8	41
1/16		.0625	1.588	7/8	22	1 7/8	48	1 3/4	44	3	76	5/8	16	1 5/8	41
	1.60	.0630	1.60	7/8	22	1 7/8	48	1 3/4	44	3	76	5/8	16	1 5/8	41
52		.0635	1.613	7/8	22	1 7/8	48	2	51	3 3/4	95	11/16	17	1 11/16	43
	1.65	.0650	1.65	1	25	2	51	2	51	3 3/4	95	11/16	17	1 11/16	43
	1.70	.0669	1.70	1	25	2	51	2	51	3 3/4	95	11/16	17	1 11/16	43
51		.0670	1.702	1	25	2	51	2	51	3 3/4	95	11/16	17	1 11/16	43
	1.75	.0689	1.75	1	25	2	51	2	51	3 3/4	95	11/16	17	1 11/16	43

(Continued)

Table 12. (Continued) **Straight Shank Jobber Length, Taper Length, and Screw Machine Length Twist Drills.** Jobber Length #97 Through 7.50 mm. Taper Length and Screw Machine Length 1.00 mm Through 1 Inch. (Source, Kennametal Inc.)

Drill Diameter				Jobber Length						Taper Length						Screw Machine Length					
Inch Sizes	mm Sizes	Equivalent		Flute		Overall		Flute		Overall		Flute		Overall		Flute		Overall			
		inch	metric	inch	mm	inch	mm	inch	mm	inch	mm	inch	mm	inch	mm	inch	mm	inch	mm		
50		.70	1.778	1	25	2	51					2	51	3 3/4	95	11/16	17	1 11/16	43		
	1.80	.0709	1.80	1	25	2	51					2	51	3 3/4	95	11/16	17	1 11/16	43		
	1.85	.0728	1.85	1	25	2	51					2	51	3 3/4	95	11/16	17	1 11/16	43		
49		.0730	1.854	1	25	2	51					2	51	3 3/4	95	11/16	17	1 11/16	43		
	1.90	.0748	1.90	1	25	2	51					2	51	3 3/4	95	11/16	17	1 11/16	43		
48		.0760	1.930	1	25	2	51					2	51	3 3/4	95	11/16	17	1 11/16	43		
	1.95	.0768	1.95	1	25	2	51					2	51	3 3/4	95	11/16	17	1 11/16	43		
5/64		.0781	1.984	1	25	2	51					2	51	3 3/4	95	11/16	17	1 11/16	43		
47		.0785	1.994	1	25	2	51					2 1/4	57	4 1/4	108	11/16	17	1 11/16	43		
	2.0	.0787	2.0	1	25	2	51					2 1/4	57	4 1/4	108	11/16	17	1 11/16	43		
	2.05	.0807	2.050	1 1/8	29	2 1/8	54					2 1/4	57	4 1/4	108	3/4	19	1 3/4	44		
46		.0810	2.057	1 1/8	29	2 1/8	54					2 1/4	57	4 1/4	108	3/4	19	1 3/4	44		
45		.0820	2.083	1 1/8	29	2 1/8	54					2 1/4	57	4 1/4	108	3/4	19	1 3/4	44		
	2.10	.0827	2.10	1 1/8	29	2 1/8	54					2 1/4	57	4 1/4	108	3/4	19	1 3/4	44		
44	2.15	.0846	2.15	1 1/8	29	2 1/8	54					2 1/4	57	4 1/4	108	3/4	19	1 3/4	44		
		.0860	2.184	1 1/8	29	2 1/8	54					2 1/4	57	4 1/4	108	3/4	19	1 3/4	44		
	2.20	.0866	2.20	1 1/4	32	2 1/4	57					2 1/4	57	4 1/4	108	3/4	19	1 3/4	44		
	2.25	.0886	2.250	1 1/4	32	2 1/4	57					2 1/4	57	4 1/4	108	3/4	19	1 3/4	44		
43		.0890	2.261	1 1/4	32	2 1/4	57					2 1/4	57	4 1/4	108	3/4	19	1 3/4	44		
	2.30	.0906	2.30	1 1/4	32	2 1/4	57					2 1/4	57	4 1/4	108	3/4	19	1 3/4	44		

(Continued)

Table 12. *(Continued)* **Straight Shank Jobber Length, Taper Length, and Screw Machine Length Twist Drills.** Jobber Length #97 Through 7.50 mm. Taper Length and Screw Machine Length 1.00 mm Through 1 Inch. *(Source, Kennametal Inc.)*

Inch Sizes	mm Sizes	Equivalent		Jobber Length				Taper Length				Screw Machine Length			
		inch	metric	Flute inch	mm	Overall inch	mm	Flute inch	mm	Overall inch	mm	Flute inch	mm	Overall inch	mm
	2.35	.0925	2.35	1 1/4	32	2 1/4	57	2 1/4	57	4 1/4	108	3/4	19	1 3/4	44
42		.0935	2.375	1 1/4	32	2 1/4	57	2 1/4	57	4 1/4	108	3/4	19	1 3/4	44
3/32		.0938	2.383	1 1/4	32	2 1/4	57	2 1/4	57	4 1/4	108	3/4	19	1 3/4	44
	2.40	.0945	2.40	1 3/8	35	2 3/8	60	2 1/2	64	4 5/8	117	13/16	21	1 13/16	46
41		.0960	2.438	1 3/8	35	2 3/8	60	2 1/2	64	4 5/8	117	13/16	21	1 13/16	46
40	2.45	.0965	2.45	1 3/8	35	2 3/8	60	2 1/2	64	4 5/8	117	13/16	21	1 13/16	46
		.0980	2.489	1 3/8	35	2 3/8	60	2 1/2	64	4 5/8	117	13/16	21	1 13/16	46
39	2.50	.0984	2.50	1 3/8	35	2 3/8	60	2 1/2	64	4 5/8	117	13/16	21	1 13/16	46
		.0995	2.527	1 3/8	35	2 3/8	60	2 1/2	64	4 5/8	117	13/16	21	1 13/16	46
38		.1015	2.578	1 7/16	37	2 1/2	64	2 1/2	64	4 5/8	117	13/16	21	1 13/16	46
	2.60	.1024	2.60	1 7/16	37	2 1/2	64	2 1/2	64	4 5/8	117	13/16	21	1 13/16	46
37		.1040	2.642	1 7/16	37	2 1/2	64	2 1/2	64	4 5/8	117	13/16	21	1 13/16	46
	2.70	.1063	2.70	1 7/16	37	2 1/2	64	2 1/2	64	4 5/8	117	13/16	21	1 13/16	46
36		.1065	2.705	1 7/16	37	2 1/2	64	2 1/2	64	4 5/8	117	13/16	21	1 13/16	46
	2.75	.1083	2.75	1 7/16	37	2 1/2	64	-	-	-	-	-	-	-	-
7/64		.1094	2.779	1 1/2	38	2 5/8	67	2 1/2	64	4 5/8	117	13/16	21	1 13/16	46
35		.1100	2.794	1 1/2	38	2 5/8	67	2 3/4	70	5 1/8	130	7/8	22	1 7/8	48
	2.80	.1102	2.80	1 1/2	38	2 5/8	67	2 3/4	70	5 1/8	130	7/8	22	1 7/8	48
34		.1110	2.819	1 1/2	38	2 5/8	67	2 3/4	70	5 1/8	130	7/8	22	1 7/8	48

(Continued)

Table 12. (Continued) Straight Shank Jobber Length, Taper Length, and Screw Machine Length Twist Drills. Jobber Length #97 Through 7.50 mm. Taper Length and Screw Machine Length 1.00 mm Through 1 Inch. (Source, Kennametal Inc.)

Drill Diameter		Equivalent		Jobber Length				Taper Length				Screw Machine Length			
				Flute		Overall		Flute		Overall		Flute		Overall	
Inch Sizes	mm Sizes	inch	metric	inch	mm	inch	mm	inch	mm	inch	mm	inch	mm	inch	mm
33		.1130	2.870	1 1/2	38	2 5/8	67	2 3/4	70	5 1/8	130	7/8	22	1 7/8	48
	2.90	.1142	2.90	1 5/8	41	2 3/4	70	2 3/4	70	5 1/8	130	7/8	22	1 7/8	48
32		.1160	2.946	1 5/8	41	2 3/4	70	2 3/4	70	5 1/8	130	7/8	22	1 7/8	48
	3.0	.1181	3.0	1 5/8	41	2 3/4	70	2 3/4	70	5 1/8	130	7/8	22	1 7/8	48
31		.1200	3.048	1 5/8	41	2 3/4	70	2 3/4	70	5 1/8	130	7/8	22	1 7/8	48
	3.10	.1220	3.10	1 5/8	41	2 3/4	70	2 3/4	70	5 1/8	130	7/8	22	1 7/8	48
1/8		.1250	3.175	1 5/8	41	2 3/4	70	2 3/4	70	5 1/8	130	7/8	22	1 7/8	48
	3.20	.1260	3.20	1 5/8	41	2 3/4	70	3	76	5 3/8	137	15/16	24	1 15/16	49
	3.25	.1280	3.25	1 5/8	41	2 3/4	70	-	-	-	-	-	-	-	-
30		.1285	3.264	1 5/8	41	2 3/4	70	3	76	5 3/8	137	15/16	24	1 15/16	49
	3.30	.1299	3.30	1 3/4	44	2 7/8	73	3	76	5 3/8	137	15/16	24	1 15/16	49
	3.40	.1339	3.40	1 3/4	44	2 7/8	73	3	76	5 3/8	137	15/16	24	1 15/16	49
29		.1360	3.454	1 3/4	44	2 7/8	73	3	76	5 3/8	137	15/16	24	1 15/16	49
	3.50	.1378	3.50	1 3/4	44	2 7/8	73	3	76	5 3/8	137	15/16	24	1 15/16	49
28		.1405	3.569	1 3/4	44	2 7/8	73	3	76	5 3/8	137	15/16	24	1 15/16	49
9/64		.1406	3.571	1 3/4	44	2 7/8	73	3	76	5 3/8	137	15/16	24	1 15/16	49
	3.60	.1417	3.60	1 7/8	48	3	76	3	76	5 3/8	137	1	25	2 1/16	52
27		.1440	3.658	1 7/8	48	3	76	3	76	5 3/8	137	1	25	2 1/16	52
	3.70	.1457	3.70	1 7/8	48	3	76	3	76	5 3/8	137	1	25	2 1/16	52

(Continued)

Table 12. (Continued) Straight Shank Jobber Length, Taper Length, and Screw Machine Length Twist Drills. Jobber Length #97 Through 7.50 mm. Taper Length and Screw Machine Length 1.00 mm Through 1 Inch. (Source, Kennametal Inc.)

Drill Diameter				Jobber Length				Taper Length				Screw Machine Length			
Inch Sizes	mm Sizes	Equivalent		Flute		Overall		Flute		Overall		Flute		Overall	
		inch	metric	inch	mm	inch	mm	inch	mm	inch	mm	inch	mm	inch	mm
26		.1470	3.734	1 7/8	48	3	76	3	76	5 3/8	137	1	25	2 1/16	52
	3.75	.1476	3.75	1 7/8	48	3	76	-	-	-	-	-	-	-	-
25		.1495	3.797	1 7/8	48	3	76	3	76	5 3/8	137	1	25	2 1/16	52
	3.80	.1496	3.80	1 7/8	48	3	76	3	76	5 3/8	137	1	25	2 1/16	52
24		.1520	3.861	2.0	51	3 1/8	79	3.0	76	5 3/8	137	1	25	2 1/16	52
	3.90	.1535	3.90	2.0	51	3 1/8	79	3.0	76	5 3/8	137	1	25	2 1/16	52
23		.1540	3.912	2.0	51	3 1/8	79	3.0	76	5 3/8	137	1	25	2 1/16	52
5/32		.1562	3.967	2.0	51	3 1/8	79	3.0	76	5 3/8	137	1	25	2 1/16	52
22		.1570	3.988	2.0	51	3 1/8	79	3 3/8	86	5 3/4	146	1 1/16	27	2 1/8	54
	4.0	.1575	4.0	2 1/8	54	3 1/4	83	3 3/8	86	5 3/4	146	1 1/16	27	2 1/8	54
21		.1590	4.039	2 1/8	54	3 1/4	83	3 3/8	86	5 3/4	146	1 1/16	27	2 1/8	54
20		.1610	4.089	2 1/8	54	3 1/4	83	3 3/8	86	5 3/4	146	1 1/16	27	2 1/8	54
	4.10	.1614	4.10	2 1/8	54	3 1/4	83	3 3/8	86	5 3/4	146	1 1/16	27	2 1/8	54
	4.20	.1654	4.20	2 1/8	54	3 1/4	83	3 3/8	86	5 3/4	146	1 1/16	27	2 1/8	54
19		.1660	4.216	2 1/8	54	3 1/4	83	3 3/8	86	5 3/4	146	1 1/16	27	2 1/8	54
	4.25	.1673	4.25	2 1/8	54	3 1/4	83	-	-	-	-	-	-	-	-
	4.30	.1693	4.30	2 1/8	54	3 1/4	83	3 3/8	86	5 3/4	146	1 1/16	27	2 1/8	54
18		.1695	4.305	2 1/8	54	3 1/4	83	3 3/8	86	5 3/4	146	1 1/16	27	2 1/8	54
11/64		.1719	4.366	2 1/8	54	3 1/4	83	3 3/8	86	5 3/4	146	1 1/16	27	2 1/8	54

(Continued)

Table 12. *(Continued)* **Straight Shank Jobber Length, Taper Length, and Screw Machine Length Twist Drills.** Jobber Length #97 Through 7.50 mm. Taper Length and Screw Machine Length 1.00 mm Through 1 Inch. *(Source, Kennametal Inc.)*

Drill Diameter				Jobber Length						Taper Length						Screw Machine Length					
Inch Sizes	mm Sizes	Equivalent		Flute		Overall				Flute		Overall				Flute		Overall			
		inch	metric	inch	mm	inch	mm			inch	mm	inch	mm			inch	mm	inch	mm		
17		.1730	4.394	2 3/16	56	3 3/8	86			3 3/8	86	5 3/4	146			1 1/8	29	2 3/16	56		
	4.40	.1732	4.40	2 3/16	56	3 3/8	86			3 3/8	86	5 3/4	146			1 1/8	29	2 3/16	56		
16		.1770	4.496	2 3/16	56	3 3/8	86			3 3/8	86	5 3/4	146			1 1/8	29	2 3/16	56		
	4.50	.1772	4.50	2 3/16	56	3 3/8	86			3 3/8	86	5 3/4	146			1 1/8	29	2 3/16	56		
15		.1800	4.572	2 3/16	56	3 3/8	86			3 3/8	86	5 3/4	146			1 1/8	29	2 3/16	56		
	4.60	.1811	4.60	2 3/16	56	3 3/8	86			3 3/8	86	5 3/4	146			1 1/8	29	2 3/16	56		
14		.1820	4.623	2 3/16	56	3 3/8	86			3 3/8	86	5 3/4	146			1 1/8	29	2 3/16	56		
13	4.70	.1850	4.70	2 5/16	59	3 1/2	89			3 3/8	86	5 3/4	146			1 1/8	29	2 3/16	56		
	4.75	.1870	4.75	2 5/16	59	3 1/2	89			-	-	-	-			-	-	-	-		
3/16		.1875	4.762	2 5/16	59	3 1/2	89			3 3/8	86	5 3/4	146			1 1/8	29	2 3/16	56		
12	4.80	.1890	4.80	2 5/16	59	3 1/2	89			3 5/8	92	6	152			1 3/16	30	2 1/4	57		
11		.1910	4.851	2 5/16	59	3 1/2	89			3 5/8	92	6	152			1 3/16	30	2 1/4	57		
	4.90	.1929	4.90	2 7/16	62	3 5/8	92			3 5/8	92	6	152			1 3/16	30	2 1/4	57		
10		.1935	4.915	2 7/16	62	3 5/8	92			3 5/8	92	6	152			1 3/16	30	2 1/4	57		
9		.1960	4.978	2 7/16	62	3 5/8	92			3 5/8	92	6	152			1 3/16	30	2 1/4	57		
	5.0	.1969	5.0	2 7/16	62	3 5/8	92			3 5/8	92	6	152			1 3/16	30	2 1/4	57		
8		.1990	5.054	2 7/16	62	3 5/8	92			3 5/8	92	6	152			1 3/16	30	2 1/4	57		
	5.10	.2008	5.10	2 7/16	62	3 5/8	92			3 5/8	92	6	152			1 3/16	30	2 1/4	57		
7		.2010	5.105	2 7/16	62	3 5/8	92			3 5/8	92	6	152			1 3/16	30	2 1/4	57		
13/64		.2031	5.159	2 7/16	62	3 5/8	92			3 5/8	92	6	152			1 3/16	30	2 1/4	57		

(Continued)

Table 12. *(Continued)* **Straight Shank Jobber Length, Taper Length, and Screw Machine Length Twist Drills.** Jobber Length #97 Through 7.50 mm. Taper Length and Screw Machine Length 1.00 mm Through 1 Inch. *(Source, Kennametal Inc.)*

Drill Diameter				Jobber Length				Taper Length				Screw Machine Length			
Inch Sizes	mm Sizes	Equivalent		Flute		Overall		Flute		Overall		Flute		Overall	
		inch	metric	inch	mm	inch	mm	inch	mm	inch	mm	inch	mm	inch	mm
6		.2040	5.182	2 1/2	64	3 3/4	95	3 5/8	92	6	152	1 1/4	32	2 3/8	60
	5.20	.2047	5.20	2 1/2	64	3 3/4	95	3 5/8	92	6	152	1 1/4	32	2 3/8	60
5		.2055	5.220	2 1/2	64	3 3/4	95	3 5/8	92	6	152	1 1/4	32	2 3/8	60
	5.25	.2067	5.25	2 1/2	64	3 3/4	95	-	-	-	-	-	-	-	-
	5.30	.2087	5.30	2 1/2	64	3 3/4	95	3 5/8	92	6	152	1 1/4	32	2 3/8	60
4		.2090	5.309	2 1/2	64	3 3/4	95	3 5/8	92	6	152	1 1/4	32	2 3/8	60
	5.40	.2126	5.40	2 1/2	64	3 3/4	95	3 5/8	92	6	152	1 1/4	32	2 3/8	60
3		.2130	5.410	2 1/2	64	3 3/4	95	3 5/8	92	6	152	1 1/4	32	2 3/8	60
	5.50	.2165	5.50	2 1/2	64	3 3/4	95	3 5/8	92	6	152	1 1/4	32	2 3/8	60
7/32		.2188	5.558	2 1/2	64	3 3/4	95	3 5/8	92	6	152	1 1/4	32	2 3/8	60
	5.60	.2205	5.60	2 5/8	67	3 7/8	98	3 3/4	95	6 1/8	156	1 5/16	33	2 7/16	62
2		.2210	5.613	2 5/8	67	3 7/8	98	3 3/4	95	6 1/8	156	1 5/16	33	2 7/16	62
	5.70	.2244	5.70	2 5/8	67	3 7/8	98	3 3/4	95	6 1/8	156	1 5/16	33	2 7/16	62
	5.75	.2264	5.75	2 5/8	67	3 7/8	98	-	-	-	-	-	-	-	-
1		.2280	5.791	2 5/8	67	3 7/8	98	3 3/4	95	6 1/8	156	1 5/16	33	2 7/16	62
	5.80	.2283	5.80	2 5/8	67	3 7/8	98	3 3/4	95	6 1/8	156	1 5/16	33	2 7/16	62
	5.90	.2323	5.90	2 5/8	67	3 7/8	98	3 3/4	95	6 1/8	156	1 5/16	33	2 7/16	62
A		.2340	5.944	2 5/8	67	3 7/8	98	3 3/4	95	6 1/8	156	1 5/16	33	2 7/16	62
15/64		.2344	5.954	2 5/8	67	3 7/8	98	3 3/4	95	6 1/8	156	1 5/16	33	2 7/16	62
	6.0	.2362	6.0	2 3/4	70	4	102	3 3/4	95			1 3/8	35	2 1/2	64

(Continued)

Table 12. *(Continued)* **Straight Shank Jobber Length, Taper Length, and Screw Machine Length Twist Drills.** Jobber Length #97 Through 7.50 mm. Taper Length and Screw Machine Length 1.00 mm Through 1 Inch. *(Source, Kennametal Inc.)*

Drill Diameter				Jobber Length				Taper Length				Screw Machine Length			
Inch Sizes	mm Sizes	Equivalent		Flute		Overall		Flute		Overall		Flute		Overall	
		inch	metric	inch	mm	inch	mm	inch	mm	inch	mm	inch	mm	inch	mm
B		.2380	6.045	2 3/4	70	4	102	3 3/4	95	6 1/8	156	1 3/8	35	2 1/2	64
	6.10	.2402	6.10	2 3/4	70	4	102	3 3/4	95	6 1/8	156	1 3/8	35	2 1/2	64
C		.2420	6.147	2 3/4	70	4	102	3 3/4	95	6 1/8	156	1 3/8	35	2 1/2	64
	6.20	.2441	6.20	2 3/4	70	4	102	3 3/4	95	6 1/8	156	1 3/8	35	2 1/2	64
D		.2460	6.248	2 3/4	70	4	102	3 3/4	95	6 1/8	156	1 3/8	35	2 1/2	64
	6.25	.2461	6.25	2 3/4	70	4	102	-	-	-	-	-	-	-	-
	6.30	.2480	6.30	2 3/4	70	4	102	3 3/4	95	6 1/8	156	1 3/8	35	2 1/2	64
1/4 E		.2500	6.350	2 3/4	70	4	102	3 3/4	95	6 1/8	156	1 3/8	35	2 1/2	64
	6.40	.2520	6.40	2 7/8	73	4 1/8	105	3 7/8	98	6 1/4	159	1 7/16	37	2 5/8	67
	6.50	.2559	6.50	2 7/8	73	4 1/8	105	3 7/8	98	6 1/4	159	1 7/16	37	2 5/8	67
F		.2570	6.528	2 7/8	73	4 1/8	105	3 7/8	98	6 1/4	159	1 7/16	37	2 5/8	67
	6.60	.2598	6.60	2 7/8	73	4 1/8	105	-	-	-	-	1 7/16	37	2 5/8	67
G		.2610	6.629	2 7/8	73	4 1/8	105	3 7/8	98	6 1/4	159	1 7/16	37	2 5/8	67
	6.70	.2638	6.700	2 7/8	73	4 1/8	105	-	-	-	-	1 7/16	37	2 5/8	67
17/64		.2656	6.746	2 7/8	73	4 1/8	105	3 7/8	98	6 1/4	159	1 7/16	37	2 5/8	67
	6.75	.2657	6.75	2 7/8	73	4 1/8	105	-	-	-	-	-	-	-	-
H		.2660	6.756	2 7/8	73	4 1/8	105	3 7/8	98	6 1/4	159	1 1/2	38	2 11/16	68
	6.80	.2677	6.80	2 7/8	73	4 1/8	105	3 7/8	98	6 1/4	159	1 1/2	38	2 11/16	68
	6.90	.2717	6.90	2 7/8	73	4 1/8	105	-	-	-	-	1 1/2	38	2 11/16	68
I		.2720	6.909	2 7/8	73	4 1/8	105	3 7/8	98	6 1/4	159	1 1/2	38	2 11/16	68

(Continued)

Table 12. (Continued) Straight Shank Jobber Length, Taper Length, and Screw Machine Length Twist Drills. Jobber Length #97 Through 7.50 mm. Taper Length and Screw Machine Length 1.00 mm Through 1 Inch. (Source, Kennametal Inc.)

Inch Sizes	Drill Diameter			Jobber Length				Taper Length				Screw Machine Length			
	mm Sizes	Equivalent		Flute		Overall		Flute		Overall		Flute		Overall	
		inch	metric	inch	mm	inch	mm	inch	mm	inch	mm	inch	mm	inch	mm
J	7.0	.2756	7.0	2 7/8	73	4 1/8	105	3 7/8	98	6 1/4	159	1 1/2	38	2 11/16	68
		.2770	7.036	2 7/8	73	4 1/8	105	3 7/8	98	6 1/4	159	1 1/2	38	2 11/16	68
	7.10	.2795	7.10	2 15/16	75	4 1/4	108	-	-	-	-	1 1/2	38	2 11/16	68
K		.2810	7.137	2 15/16	75	4 1/4	108	3 7/8	98	6 1/4	159	1 1/2	38	2 11/16	68
9/32		.2812	7.142	2 15/16	75	4 1/4	108	3 7/8	98	6 1/4	159	1 1/2	38	2 11/16	68
	7.20	.2834	7.20	2 15/16	75	4 1/4	108	4	102	6 3/8	162	1 1/2	38	2 11/16	68
	7.25	.2854	7.25	2 15/16	75	4 1/4	108	-	-	-	-	-	-	-	-
	7.30	.2874	7.30	2 15/16	75	4 1/4	108	-	-	-	-	1 9/16	40	2 3/4	70
L		.2900	7.366	2 15/16	75	4 1/4	108	4	102	6 3/8	162	1 9/16	40	2 3/4	70
	7.40	.2913	7.40	3 1/16	78	4 3/8	111	-	-	-	-	1 9/16	40	2 3/4	70
M		.2950	7.493	3 1/16	78	4 3/8	111	4	102	6 3/8	162	1 9/16	40	2 3/4	70
	7.50	.2953	7.50	3 1/16	78	4 3/8	111	4	102	6 3/8	162	1 9/16	40	2 3/4	70
19/64		.2969	7.541	3 1/16	78	4 3/8	111	4	102	6 3/8	162	1 9/16	40	2 3/4	70
	7.60	.2992	7.60	3 1/16	78	4 3/8	111	-	-	-	-	1 5/8	41	2 13/16	71
N		.3020	7.671	3 1/16	78	4 3/8	111	4	102	6 3/8	162	1 5/8	41	2 13/16	71
	7.70	.0301	7.70	3 3/16	81	4 1/2	114	-	-	-	-	1 5/8	41	2 13/16	71
	7.75	.3051	7.75	3 3/16	81	4 1/2	114	-	-	-	-	-	-	-	-
	7.80	.3071	7.80	3 3/16	81	4 1/2	114	4	102	6 3/8	162	1 5/8	41	2 13/16	71
	7.90	.3110	7.90	3 3/16	81	4 1/2	114	-	-	-	-	1 5/8	41	2 13/16	71

(Continued)

Table 12. *(Continued)* Straight Shank Jobber Length, Taper Length, and Screw Machine Length Twist Drills. Jobber Length #97 Through 7.50 mm. Taper Length and Screw Machine Length 1.00 mm Through 1 Inch. *(Source, Kennametal Inc.)*

| Drill Diameter | | | | Jobber Length | | | | Taper Length | | | | Screw Machine Length | | | |
| Inch Sizes | mm Sizes | Equivalent | | Flute | | Overall | | Flute | | Overall | | Flute | | Overall | |
		inch	metric	inch	mm	inch	mm	inch	mm	inch	mm	inch	mm	inch	mm
5/16		.3125	7.938	3 3/16	81	4 1/2	114	4	102	6 3/8	162	1 5/8	41	2 13/16	71
	8.0	.3150	8.0	3 3/16	81	4 1/2	114	4 1/8	105	6 1/2	165	1 11/16	43	2 15/16	75
O		.3160	8.026	3 3/16	81	4 1/2	114	4 1/8	105	6 1/2	165	1 11/16	43	2 15/16	75
	8.10	.3189	8.10	3 5/16	84	4 5/8	117	-	-	-	-	1 11/16	43	2 15/16	75
	8.20	.3228	8.20	3 5/16	84	4 5/8	117	4 1/8	105	6 1/2	165	1 11/16	43	2 15/16	75
P		.3230	8.204	3 5/16	84	4 5/8	117	4 1/8	105	6 1/2	165	1 11/16	43	2 15/16	75
	8.25	.3248	8.25	3 5/16	84	4 5/8	117	-	-	-	-	-	-	-	-
	8.30	.3268	8.30	3 5/16	84	4 5/8	117	-	-	-	-	1 11/16	43	2 15/16	75
21/64		.3281	8.344	3 5/16	84	4 5/8	117	4 1/8	105	6 1/2	165	1 11/16	43	2 15/16	75
	8.40	.3307	8.40	3 7/16	87	4 3/4	121	-	-	-	-	1 11/16	43	3	76
Q		.3320	8.433	3 7/16	87	4 3/4	121	4 1/8	105	6 1/2	165	1 11/16	43	3	76
	8.50	.3346	8.50	3 7/16	87	4 3/4	121	4 1/8	105	6 1/2	165	1 11/16	43	3	76
	8.60	.3386	8.60	3 7/16	87	4 3/4	121	-	-	-	-	1 11/16	43	3	76
R		.3390	8.611	3 7/16	87	4 3/4	121	4 1/8	105	6 1/2	165	1 11/16	43	3	76
	8.70	.3425	8.70	3 7/16	87	4 3/4	121	-	-	-	-	1 11/16	43	3	76
11/32		.3438	8.733	3 7/16	87	4 3/4	121	4 1/8	105	6 1/2	165	1 11/16	43	3	76
	8.75	.3445	8.75					-	-	-	-	-	-	-	-
	8.80	.3465	8.80	3 1/2	89	4 7/8	124	4 1/4	108	6 3/4	171	1 3/4	44	3 1/16	78
S		.3480	8.839	3 1/2	89	4 7/8	124	4 1/4	108	6 3/4	171	1 3/4	44	3 1/16	78

(Continued)

Table 12. (Continued) **Straight Shank Jobber Length, Taper Length, and Screw Machine Length Twist Drills.** Jobber Length #97 Through 7.50 mm. Taper Length and Screw Machine Length 1.00 mm Through 1 Inch. (Source, Kennametal Inc.)

Inch Sizes	Drill Diameter — mm Sizes	Equivalent inch	Equivalent metric	Jobber Length — Flute inch	mm	Overall inch	mm	Taper Length — Flute inch	mm	Overall inch	mm	Screw Machine Length — Flute inch	mm	Overall inch	mm
	8.90	.3504	8.90	3 1/2	89	4 7/8	124	-	-	-	-	1 3/4	44	3 1/16	78
	9.0	.3543	9.0	3 1/2	89	4 7/8	124	4 1/4	108	6 3/4	171	1 3/4	44	3 1/16	78
T		.3580	9.093	3 1/2	89	4 7/8	124	4 1/4	108	6 3/4	171	1 3/4	44	3 1/16	78
	9.10	.3583	9.10	3 1/2	89	4 7/8	124	-	108	-	171	1 3/4	44	3 1/16	78
23/64		.3594	9.129	3 1/2	89	4 7/8	124	4 1/4	108	6 3/4	171	1 3/4	44	3 1/16	78
	9.20	.3622	9.20	3 5/8	92	5.0	127	-	108	-	171	1 13/16	46	3 1/8	79
	9.25	.3642	9.25	3 5/8	92	5.0	127	-	-	-	-	-	-	-	-
	9.30	.3661	9.30	3 5/8	92	5.0	127	-	-	-	-	1 13/16	46	3 1/8	79
U		.3680	9.347	3 5/8	92	5.0	127	4 1/4	108	6 3/4	171	1 13/16	46	3 1/8	79
	9.40	.3701	9.40	3 5/8	92	5.0	127	-	-	-	-	1 13/16	46	3 1/8	79
	9.50	.3740	9.50	3 5/8	92	5.0	127	4 1/4	108	6 3/4	171	1 13/16	46	3 1/8	79
3/8		.3750	9.525	3 5/8	92	5.0	127	4 1/4	108	6 3/4	171	1 13/16	46	3 1/8	79
V		.3770	9.576	3 5/8	92	5.0	127	4 3/8	111	7	178	1 7/8	48	3 1/4	83
	9.60	.3780	9.60	3 3/4	95	5 1/8	130	-	-	-	-				
	9.70	.3819	9.70	3 3/4	95	5 1/8	130	-	-	-	-	1 7/8	48	3 1/4	83
	9.75	.3839	9.75	3 3/4	95	5 1/8	130	-	-	-	-	1 7/8	48	3 1/4	83
	9.80	.3858	9.80	3 3/4	95	5 1/8	130	4 3/8	111	7	178	-	-	-	-
W		.3860	9.804	3 3/4	95	5 1/8	130	4 3/8	111	7	178	1 7/8	48	3 1/4	83
	9.90	.3898	9.90	3 3/4	95	5 1/8	130	-	-	-	-	1 7/8	48	3 1/4	83
25/64		.3906	9.921	3 3/4	95	5 1/8	130	4 3/8	111	7	178	1 7/8	48	3 1/4	83

(Continued)

Table 12. (Continued) Straight Shank Jobber Length, Taper Length, and Screw Machine Length Twist Drills. Jobber Length #97 Through 7.50 mm. Taper Length and Screw Machine Length 1.00 mm Through 1 Inch. (Source, Kennametal Inc.)

Drill Diameter				Jobber Length				Taper Length				Screw Machine Length			
Inch Sizes	mm Sizes	Equivalent		Flute		Overall		Flute		Overall		Flute		Overall	
		inch	metric	inch	mm	inch	mm	inch	mm	inch	mm	inch	mm	inch	mm
	10.0	.3937	10.0	3 3/4	95	5 1/8	130	4 3/8	111	7	178	1 15/16	49	3 5/16	84
X		.3970	10.084	3 3/4	95	5 1/8	130	4 3/8	111	7	178	1 15/16	49	3 5/16	84
	10.20	.4016	10.20	3 7/8	98	5 1/4	133	4 3/8	111	7	178	1 15/16	49	3 5/16	84
Y		.4040	10.262	3 7/8	98	5 1/4	133	4 3/8	111	7	178	1 15/16	49	3 5/16	84
	10.30	.4055	10.30	3 7/8	98	5 1/4	133	-	-	-	-	-	-	-	-
13/32		.4062	10.317	3 7/8	98	5 1/4	133	4 3/8	111	7	178	1 15/16	49	3 5/16	84
	10.40	.4094	10.40	3 7/8	98	5 1/4	133	-	-	-	-	-	-	-	-
Z		.4130	10.409	3 7/8	98	5 1/4	133	4 5/8	117	7 1/4	184	2	51	3 3/8	86
	10.50	.4134	10.50	3 7/8	98	5 1/4	133	4 5/8	117	7 1/4	184	2	51	3 3/8	86
	10.60	.4173	10.60	3 15/16	100	5 1/8	137	-	-	-	-	-	-	-	-
	10.70	.4213	10.70	3 15/16	100	5 1/8	137	-	-	-	-	-	-	-	-
27/64		.4219	10.716	3 15/16	100	5 1/8	137	4 5/8	117	7 1/4	184	2	51	3 3/8	86
	10.80	.4252	10.80	4 1/16	103	5 1/2	140	4 5/8	117	7 1/4	184	2 1/16	52	3 7/16	87
	10.90	.4291	10.90	4 1/16	103	5 1/2	140	-	-	-	-	-	-	-	-
	11.0	.4331	11.0	4 1/16	103	5 1/2	140	4 5/8	117	7 1/4	184	2 1/16	52	3 7/16	87
	11.10	.4370	11.10	4 1/16	103	5 1/2	140	-	-	-	-	-	-	-	-
7/16		.4375	11.112	4 1/16	103	5 1/2	140	4 5/8	117	7 1/4	184	2 1/16	52	3 7/16	87
	11.20	.4409	11.20	4 3/16	106	5 5/8	143	4 3/4	121	7 1/2	190	2 1/8	54	3 9/16	90

(Continued)

Table 12. *(Continued)* **Straight Shank Jobber Length, Taper Length, and Screw Machine Length Twist Drills.** Jobber Length #97 Through 7.50 mm. Taper Length and Screw Machine Length 1.00 mm Through 1 Inch. *(Source, Kennametal Inc.)*

Drill Diameter				Jobber Length				Taper Length				Screw Machine Length			
		Equivalent		Flute		Overall		Flute		Overall		Flute		Overall	
Inch Sizes	mm Sizes	inch	metric	inch	mm	inch	mm	inch	mm	inch	mm	inch	mm	inch	mm
	11.30	.4449	11.30	4 3/16	106	5 5/8	143	-	-	-	-	-	-	-	-
	11.40	.4488	11.40	4 3/16	106	5 5/8	143	-	-	-	-	-	-	-	-
	11.50	.4528	11.50	4 3/16	106	5 5/8	143	4 3/4	121	7 1/2	190	2 1/8	54	3 9/16	90
29/64		.4531	11.509	4 3/16	106	5 5/8	143	4 3/4	121	7 1/2	190	2 1/8	54	3 9/16	90
	11.60	.4567	11.60	4 5/16	110	5 3/4	146	-	-	-	-	-	-	-	-
	11.70	.4606	11.70	4 5/16	110	5 3/4	146	-	-	-	-	-	-	-	-
	11.80	.4646	11.80	4 5/16	110	5 3/4	146	4 3/4	121	7 1/2	190	2 1/8	54	3 5/8	92
	11.90	.4685	11.90	4 5/16	110	5 3/4	146	-	-	-	-	-	-	-	-
15/32		.4688	11.908	4 5/16	110	5 3/4	146	4 3/4	121	7 1/2	190	2 1/8	54	3 5/8	92
	12.0	.4724	12.0	4 3/8	111	5 7/8	149	4 3/4	121	7 3/4	197	2 3/16	56	3 11/16	94
	12.10	.4764	12.10	4 3/8	111	5 7/8	149	-	-	-	-	-	-	-	-
	12.20	.4803	12.20	4 3/8	111	5 7/8	149	4 3/4	121	7 3/4	197	2 3/16	56	3 11/16	94
	12.30	.4843	12.10	4 3/8	111	5 7/8	149	-	-	-	-	-	-	-	-
31/64		.4844	12.304	4 3/8	111	5 7/8	149	4 3/4	121	7 3/4	197	2 3/16	56	3 11/16	94
	12.40	.4882	12.40	4 3/8	111	5 7/8	149	-	-	-	-	-	-	-	-
	12.50	.4921	12.50	4 1/2	114	6	152	4 3/4	121	7 3/4	197	2 1/4	57	3 3/4	95
	12.60	.4961	12.60	4 1/2	114	6	152	-	-	-	-	-	-	-	-
1/2	12.70	.5000	12.70	4 1/2	114	6	152	4 3/4	121	7 3/4	197	2 1/4	57	3 3/4	95
	12.80	.5039	12.80	4 1/2	114	6	152	4 3/4	121	8	203	-	-	-	-

(Continued)

Table 12. (Continued) Straight Shank Jobber Length, Taper Length, and Screw Machine Length Twist Drills. Jobber Length #97 Through 7.50 mm. Taper Length and Screw Machine Length 1.00 mm Through 1 Inch. (Source, Kennametal Inc.)

| Drill Diameter | | Equivalent | | Jobber Length | | | | Taper Length | | | | Screw Machine Length | | | |
| | | | | Flute | | Overall | | Flute | | Overall | | Flute | | Overall | |
Inch Sizes	mm Sizes	inch	metric	inch	mm	inch	mm	inch	mm	inch	mm	inch	mm	inch	mm
	13.0	.5118	13.0	4 1/2	114	6	152	4 3/4	121	8	203	-	-	-	-
33/64		.5156	13.096	4 13/16	122	6 5/8	168	4 3/4	121	8	203	2 3/8	60	3 7/8	98
	13.20	.1597	13.20	-	-	-	-	4 3/4	121	8	203	-	-	-	-
17/32		.5312	13.492	4 13/16	122	6 5/8	168	4 3/4	121	8	203	2 3/8	60	3 7/8	98
	13.50	.5315	13.50	4 13/16	122	6 5/8	168	4 3/4	121	8	203	-	-	-	-
	13.80	.5433	13.80	-	-	-	-	4 7/8	124	8 1/4	210	-	-	-	-
35/64		.5469	13.891	4 13/16	122	6 5/8	168	4 7/8	124	8 1/4	210	2 1/2	64	4	102
	14.0	.5512	14.0	4 13/16	122	6 5/8	168	4 7/8	124	8 1/4	210	-	-	-	-
	14.25	.5610	14.25	-	-	-	-	4 7/8	124	8 1/4	210	-	-	-	-
9/16		.5625	14.288	4 13/16	122	6 5/8	168	4 7/8	124	8 1/4	210	2 1/2	64	4	102
	14.50	.5709	14.50	4 13/16	122	6 5/8	168	4 7/8	124	8 3/4	222	-	-	-	-
37/64		.5781	14.684	-	-	-	-	4 7/8	124	8 3/4	222	2 5/8	67	4 1/8	105
	14.75	.5807	14.75	-	-	-	-	4 7/8	124	8 3/4	222	-	-	-	-
	15.0	.5906	15.0	5 3/16	132	7 1/8	181	4 7/8	124	8 3/4	222	-	-	-	-
19/32		.5938	15.083	5 3/16	132	7 1/8	181	4 7/8	124	8 3/4	222	2 5/8	67	4 1/8	105
	15.25	.6004	15.25	-	-	-	-	4 7/8	124	8 3/4	222	-	-	-	-
39/64		.6094	15.479	5 3/16	132	7 1/8	181	4 7/8	124	8 3/4	222	2 3/4	70	4 1/4	108
	15.50	.6102	15.50	5 3/16	132	7 1/8	181	4 7/8	124	8 3/4	222	-	-	-	-
	15.75	.6201	15.75	-	-	-	-	4 7/8	124	8 3/4	222	-	-	-	-
5/8		.6250	15.875	5 3/16	132	7 1/8	181	4 7/8	124	8 3/4	222	2 3/4	70	4 1/4	108

(Continued)

Table 12. *(Continued)* **Straight Shank Jobber Length, Taper Length, and Screw Machine Length Twist Drills.** Jobber Length #97 Through 7.50 mm. Taper Length and Screw Machine Length 1.00 mm Through 1 Inch. *(Source, Kennametal Inc.)*

Drill Diameter				Jobber Length				Taper Length				Screw Machine Length			
Inch Sizes	mm Sizes	Equivalent inch	Equivalent metric	Flute inch	Flute mm	Overall inch	Overall mm	Flute inch	Flute mm	Overall inch	Overall mm	Flute inch	Flute mm	Overall inch	Overall mm
	16.0	.6299	16.0	5 3/16	132	7 1/8	181	5 1/8	130	9	228	-	-	-	-
	16.25	.6398	16.25	-	-	-	-	5 1/8	130	9	228	-	-	-	-
41/64		.6406	16.271	5 3/16	132	7 1/8	181	5 1/8	130	9	228	2 7/8	73	4 1/2	114
	16.50	.6496	16.50	5 3/16	132	7 1/8	181	5 1/8	130	9	228	-	-	-	-
21/32		.6562	16.669	5 3/16	132	7 1/8	181	5 1/8	130	9	228	2 7/8	73	4 1/2	114
	16.75	.6594	16.75	-	-	-	-	5 3/8	137	9 1/4	235	-	-	-	-
	17.0	.6693	17.0	5 5/8	143	7 5/8	194	5 3/8	137	9 1/4	235	2 7/8	73	4 1/2	114
43/64		.6719	17.006	5 5/8	143	7 5/8	194	5 3/8	137	9 1/4	235	2 7/8	73	4 1/2	114
	17.25	.6791	17.25	-	-	-	-	5 3/8	137	9 1/4	235	-	-	-	-
11/16		.6875	17.462	5 5/8	143	7 5/8	194	5 3/8	137	9 1/4	235	2 7/8	73	4 1/2	114
	17.50	.6890	17.50	5 5/8	143	7 5/8	194	5 3/8	137	9 1/4	235	-	-	-	-
45/64		.7031	17.859	-	-	-	-	5 5/8	143	9 1/2	241	3	76	4 3/4	121
	18.0	.7087	18.0	-	-	-	-	5 5/8	143	9 1/2	241	-	-	-	-
23/32		.7188	18.258	-	-	-	-	5 5/8	143	9 1/2	241	3	76	4 3/4	121
	18.50	.7283	18.50	-	-	-	-	5 7/8	149	9 3/4	247	-	-	-	-
47/64		.7344	18.654	-	-	-	-	5 7/8	149	9 3/4	247	3 1/8	79	5	127
	19.00	.7480	19.0	-	-	-	-	5 7/8	149	9 3/4	247	-	-	-	-
3/4		.750	19.050	-	-	-	-	6	152	9 7/8	251	3 1/8	79	5	127
49/64		.7656	19.446	-	-	-	-	6	152	9 7/8	251	3 1/4	83	5 1/8	130

(Continued)

Table 12. *(Continued)* **Straight Shank Jobber Length, Taper Length, and Screw Machine Length Twist Drills.** Jobber Length #97 Through 7.50 mm. Taper Length and Screw Machine Length 1.00 mm Through 1 Inch. *(Source, Kennametal Inc.)*

| Drill Diameter | | | | Jobber Length | | | | Taper Length | | | | Screw Machine Length | | | |
Inch Sizes	mm Sizes	Equivalent inch	metric	Flute inch	mm	Overall inch	mm	Flute inch	mm	Overall inch	mm	Flute inch	mm	Overall inch	mm
	19.50	.7677	19.50	-	-	-	-	6	152	9 7/8	251	-	-	-	-
25/32		.7812	19.845	-	-	-	-	6	152	9 7/8	251	3 1/4	83	5 1/8	130
	20.0	.7879	20.0	-	-	-	-	6 1/8	156	10	254	-	-	-	-
51/64		.7969	20.241	-	-	-	-	6 1/8	156	10	254	3 3/8	86	5 1/4	133
	20.50	.8071	20.50	-	-	-	-	6 1/8	156	10	254	-	-	-	-
13/16		.8125	20.638	-	-	-	-	6 1/8	156	10	254	3 3/8	86	5 1/4	133
	21.0	.8268	21.0	-	-	-	-	6 1/8	156	10	254	-	-	-	-
53/64		.8281	21.034	-	-	-	-	6 1/8	156	10	254	3 1/2	89	5 3/8	137
27/32		.8438	21.433	-	-	-	-	6 1/8	156	10	254	3 1/2	89	5 3/8	137
	21.50	.8465	21.50	-	-	-	-	6 1/8	156	10	254	-	-	-	-
55/64		.8594	21.829	-	-	-	-	6 1/8	156	10	254	3 1/2	89	5 3/8	137
	22.0	.8661	22.0	-	-	-	-	6 1/8	156	10	254	-	-	-	-
7/8		.8750	22.235	-	-	-	-	6 1/8	156	10	254	3 1/2	89	5 3/8	137
	22.50	.8858	22.50	-	-	-	-	6 1/8	156	10	254	-	-	-	-
57/64		.8906	22.621	-	-	-	-	6 1/8	156	10	254	3 5/8	92	5 5/8	143
	23.0	.9055	23.0	-	-	-	-	6 1/8	156	10	254	-	-	-	-
29/32		.9062	23.017	-	-	-	-	6 1/8	156	10	254	3 5/8	92	5 5/8	143
59/64		.9219	23.416	-	-	-	-	6 1/8	156	10 3/4	273	3 3/4	95	5 3/4	146
	23.50	.9252	23.50	-	-	-	-	6 1/8	156	10 3/4	273	-	-	-	-

(Continued)

Table 12. *(Continued)* **Straight Shank Jobber Length, Taper Length, and Screw Machine Length Twist Drills.** Jobber Length #97 Through 7.50 mm. Taper Length and Screw Machine Length 1.00 mm Through 1 Inch. *(Source, Kennametal Inc.)*

Drill Diameter		Equivalent		Jobber Length				Taper Length				Screw Machine Length			
Inch Sizes	mm Sizes			Flute		Overall		Flute		Overall		Flute		Overall	
		inch	metric	inch	mm	inch	mm	inch	mm	inch	mm	inch	mm	inch	mm
15/16		.9375	23.812	-	-	-	-	6 1/8	156	10 3/4	273	3 5/8	92	5 5/8	143
	24.0	.9449	24.0	-	-	-	-	6 3/8	162	11	279	-	-	-	-
61/64		.9531	24.209	-	-	-	-	6 3/8	162	11	279	3 7/8	98	5 7/8	149
	24.50	.9646	24.50	-	-	-	-	6 3/8	162	11	279	-	-	-	-
31/32		.9688	24.608	-	-	-	-	6 3/8	162	11	279	3 7/8	98	5 7/8	149
	25.0	.9843	25.0	-	-	-	-	6 3/8	162	11	279	-	-	-	-
63/64		.9844	25.004	-	-	-	-	6 3/8	162	11	279	4	102	6	152
1		1.0	25.40	-	-	-	-	6 3/8	162	11	279	4	102	6	152

Table 13. Dimensions for Taper Length Straight Shank Twist Drills. 1″ to 1 3/4″ and 25.50 mm to 31.50 mm in Diameter. *(Source, Precision Twist Drill Co.)*

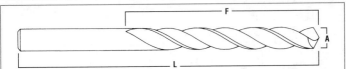

Drill Diameter				Lengths			
Fraction Sizes	mm Sizes	Equivalent		Flute (F)		Overall (L)	
		inch	metric	inch	mm	inch	mm
1		1.0	25.40	6 3/8	162	11	279
	25.50	1.0039	25.50	7 1/2	190	11 13/32	290
1 1/64		1.0156	25.796	6 1/2	165	11 1/8	282
	26.0	1.0236	26.0	7 1/2	190	11 13/32	290
1 1/32		1.0312	26.192	6 1/2	165	11 1/8	282
	26.50	1.0433	26.50	7 1/2	190	11 13/32	290
1 3/64		1.0469	26.591	6 5/8	168	11 1/4	286
1 1/16		1.0625	26.988	6 5/8	168	11 1/4	286
	27.0	1.0630	27.0	7 5/8	195	11 23/32	298
1 5/64		1.0781	27.384	6 7/8	175	11 1/2	292
	27.50	1.0827	27.50	7 5/8	195	11 23/32	298
1 3/32		1.0938	27.783	6 7/8	175	11 1/2	292
	28.0	1.1024	28.0	7 5/8	195	11 23/32	298
1 7/64		1.1094	28.179	7 1/8	181	11 3/4	298
	28.50	1.1220	28.50	7 29/32	201	12 1/8	307
1 1/8		1.1250	28.575	7 1/8	181	11 3/4	298
1 9/64		1.1406	28.971	7 1/4	184	11 7/8	301
	29.0	1.1417	29.0	7 29/32	201	12 1/8	307
1 5/32		1.1562	29.367	7 1/4	184	11 7/8	301
	29.50	1.1614	29.50	7 29/32	201	12 1/8	307
1 11/64		1.1719	29.766	7 3/8	187	12	305
	30.0	1.1811	30.0	7 29/32	201	12 1/8	307
1 3/16		1.1875	30.162	7 3/8	187	12	305
	30.50	1.2008	30.50	8 5/32	207	12 7/16	316
1 13/64		1.2031	30.559	7 1/2	190	12 1/8	308
1 7/32		1.2188	30.958	7 1/2	190	12 1/8	308
	31.0	1.2205	31.0	8 5/32	207	12 7/16	316

(Continued)

Table 13. *(Continued)* **Dimensions for Taper Length Straight Shank Twist Drills.** 1" to 1 3/4" and 25.50 mm to 31.50 mm in Diameter. *(Source, Precision Twist Drill Co.)*

Fraction Sizes	mm Sizes	Equivalent inch	metric	Flute (F) inch	mm	Overall (L) inch	mm
1 15/64		1.2344	31.354	7 7/8	200	12 1/2	317
	31.50	1.2402	31.50	8 5/32	207	12 7/16	316
1 1/4		1.250	31.750	7 7/8	200	12 1/2	317
1 9/32		1.2812	32.542	8 1/2	216	14 1/8	359
1 5/16		1.3125	33.338	8 5/8	219	14 1/4	362
1 11/32		1.3438	34.133	8 3/4	222	14 3/8	365
1 3/8		1.3750	34.925	8 7/8	225	14 1/2	368
1 13/32		1.4062	35.717	9.0	229	14 5/8	372
1 7/16		1.4375	36.512	9 1/8	232	14 3/4	375
1 15/32		1.4688	37.308	9 1/4	235	14 7/8	378
1 1/2		1.50	38.10	9 3/8	238	15	381
1 9/16		1.5625	39.688	9 5/8	244	15 1/4	387
1 5/8		1.6250	41.275	9 7/8	251	15 5/8	397
1 3/4		1.750	44.450	10 1/2	267	16 1/4	413

* *Note:* Taper length drills over 1" may be supplied with or without necks. Overall and flute lengths are the same for drills with or without necks. Neck lengths are 5/8" on drills up to 1 9/16" diameter, and 3/4" on larger sizes. Metric sizes are to DIN lengths.

Table 14. Dimensions for Reduced Shank Screw Machine Length Drills. 1" to 2" and 25.50 mm to 38.0 mm in Diameter.

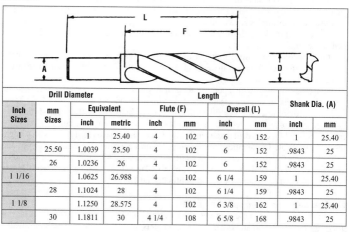

Inch Sizes	mm Sizes	Equivalent inch	metric	Flute (F) inch	mm	Overall (L) inch	mm	Shank Dia. (A) inch	mm
1		1	25.40	4	102	6	152	1	25.40
	25.50	1.0039	25.50	4	102	6	152	.9843	25
	26	1.0236	26	4	102	6	152	.9843	25
1 1/16		1.0625	26.988	4	102	6 1/4	159	1	25.40
	28	1.1024	28	4	102	6 1/4	159	.9843	25
1 1/8		1.1250	28.575	4	102	6 3/8	162	1	25.40
	30	1.1811	30	4 1/4	108	6 5/8	168	.9843	25

(Continued)

Table 14. *(Continued)* **Dimensions for Reduced Shank Screw Machine Length Drills.** 1″ to 2″ and 25.50 mm to 38.0 mm in Diameter.

Drill Diameter				Length				Shank Dia. (A)	
Inch Sizes	mm Sizes	Equivalent		Flute (F)		Overall (L)			
		inch	metric	inch	mm	inch	mm	inch	mm
1 3/16		1.1875	30.162	4 1/4	108	6 5/8	168	1	25.40
1 1/4		1.250	31.750	4 3/8	111	6 3/4	172	1	25.40
	32	1.2598	32	4 3/8	111	7	178	1.2402	31.50
1 5/16		1.3125	33.338	4 3/8	111	7	178	1.250	31.75
	34	1.3386	34	4 1/2	114	7 1/8	181	1.2402	31.50
1 3/8		1.3750	34.925	4 1/2	114	7 1/8	181	1.250	31.75
	36	1.4173	36	4 3/4	121	7 3/8	187	1.2402	31.50
1 7/16		1.4375	36.512	4 3/4	121	7 3/8	187	1.250	31.75
	38	1.4961	38	4 7/8	124	7 1/2	190	1.2402	31.50
1 1/2		1.50	38.10	4 7/8	124	7 1/2	190	1.250	31.75
1 9/16		1.5625	39.688	4 7/8	124	7 3/4	197	1.50	38.10
1 5/8		1.6250	41.275	4 7/8	124	7 3/4	197	1.50	38.10
1 11/16		1.6875	42.8620	5 1/8	130	8	203	1.50	38.10
1 3/4		1.750	44.450	5 1/8	130	8	203	1.50	38.10
1 13/16		1.8125	46.038	5 3/8	137	8 1/4	210	1.50	38.10
1 7/8		1.8750	47.625	5 3/8	137	8 1/4	210	1.50	38.10
1 15/16		1.9375	49.212	5 5/8	143	8 1/2	216	1.50	38.10
2		2	50.80	5 5/8	143	8 1/2	216	1.50	38.10

Twist drills with tangs

Many small to medium diameter straight shank drill sizes are available with tangs. The inclusion of a tang on a straight shank drill normally indicates that it is intended for heavy duty use. Some examples are as follows.

Jogger length drills–Split point (automotive or crankcase) drills for drilling in iron and steel.

Taper length drills–Split point drills (as above) and parabolic flute design drills which are effective for deep drilling of aluminum alloys and other low to medium tensile strength materials.

Straight shank drills with tapers are adapted to Morse Tapers by way of an intermediate drill driver. These drivers are essentially collets that fit inside an appropriate Morse Taper adapter to provide drive for the drill. Drill drivers come in five sizes (0–4), each fitting into the standard Morse Taper with the same numerical designation. A separate driver is, of course, required for each drill diameter. **Table 15** provides dimensions for tangs on straight shank drills. **Table 16** identifies the recommended drill diameters for each size driver, and **Table 17** gives full dimensions for standard Morse Tapers. Full dimensions for fractional Morse Taper shank drills are found in **Tables 18-20**, and dimensions for metric Morse Taper shanks and drills are provided in **Tables 21** and **22**.

Coolant hole, core, and specialty drills

Drills with coolant holes are designed for high production operations. They are of heavy duty construction and often have notch points to aid penetration. The holes, which

(Text continued on p. 1113)

Table 15. General Dimensions of Tangs for Straight Shank Drills. *(Source, Kennametal Inc.)*

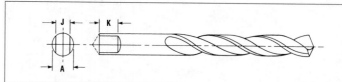

Nominal Diameter of Drill Shank (A)		Thickness of Tang (J)				Length of Tang (K)	
inch	mm	inch		mm		inch	mm
		max.	min.	max.	min.		
from 1/8 to 3/16	3.20 to 4.75	.0940	.0900	2.39	2.29	9/32	7.0
over 3/16 to 1/4	4.80 to 6.30	.1220	.1180	3.10	3.00	5/16	8.0
over 1/4 to 5/16	6.40 to 7.90	.1620	.1580	4.11	4.01	11/32	8.5
over 5/16 to 3/8	8.0 to 9.50	.2030	.1990	5.16	5.06	3/8	9.5
over 3/8 to 15/32	9.60 to 11.90	.2430	.2390	6.17	6.07	7/16	11.0
over 15/32 to 9/16	12.0 to 14.25	.3030	.2970	7.70	7.55	1/2	12.5
over 9/16 to 21/32	14.50 to 16.50	.3730	.3670	9.47	9.32	9/16	14.5
over 21/32 to 3/4	16.75 to 19.0	.4430	.4370	11.25	11.10	5/8	16.0
over 3/4 to 7/8	19.50 to 22.0	.5140	.5080	13.05	12.90	1 1/16	17.5
over 7/8 to 1	22.50 to 25.0	.6090	.6010	15.47	15.27	3/4	19.0
over 1 to 1 3/16	25.50 to 30.0	.7000	.6920	17.78	17.58	13/16	20.5
over 1 3/16 to 1 3/8	30.50 to 34.93	.8170	.8090	20.75	20.55	7/8	22.0

Concentricity of Tang			
Thickness of Tang (J)		Total Indicator Variation (TIV)	
inch	mm	inch	mm
from .0900 to .2430	from 2.29 to 6.17	.0060	0.16
over .2430 to .4430	over 6.17 to 11.25	.0070	0.18
over .4430 to .6090	over 11.25 to 15.47	.0080	0.21
over .6090 to .8170	over 15.47 to 20.75	.0090	0.23

Table 16. Recommended Drill Diameters for Split Sleeve, Collet Type, Drill Drivers.

Drill Size	First Choice Driver	Optional Driver	Second Optional
#61 to #53	Taper #1	Taper #0	-
1/16" to 3/16"	Taper #1	Taper #0	Taper #2
#12 to 1/4"	Taper #1	Taper #2	Taper #0
F to 3/8"	Taper #1	Taper #2	-
W to 9/16"	Taper #2	Taper #3	-
37/64" to 3/4"	Taper #3	Taper #4	-

Table 17. American National Standard Taper Shank Dimensions. *(Source, Precision Twist Drill Co.)*

	Drawing Symbol	Morse Taper Number							
Dimension		**0**	**1**	**2**	**3**	**4**	**4 1/2**	**5**	**6**	**7**
Plug small end dia.	D	.25200	.36900	.57200	.77800	1.02000	1.26600	1.47500	2.11600	2.75000
Dia. at end of socket	A	.35610	.47500	.70000	.93800	1.23100	1.50000	1.74800	2.49400	3.27000
Length	B	2 11/32	2 9/16	3 1/8	3 7/8	4 7/8	5 3/8	6 1/8	8 9/16	11 5/8
Depth	S	2 7/32	2 7/16	2 15/16	3 11/16	4 5/8	5 1/8	5 7/8	8 1/4	11 1/4
Depth of drilled hole	G	2 1/16	2 3/16	2 21/32	3 5/16	4 3/16	4 5/8	5 5/16	7 13/32	10 5/32
Depth of reamed hole	H	2 1/32	2 5/32	2 39/64	3 1/4	4 1/8	4 9/16	5 1/4	7 21/64	7 1/4
Standard plug depth	P	2	2 1/8	2 9/16	3 3/16	4 1/16	4 1/2	5 3/16	7 1/4	10

(Continued)

Table 17. *(Continued)* **American National Standard Taper Shank Dimensions.** *(Source, Precision Twist Drill)*

	Dimension	Drawing Symbol	Morse Taper Number								
			0	1	2	3	4	4 1/2	5	6	7
Tang	Thickness	t	5/32	13/64	1/4	5/16	15/32	9/16	5/8	3/4	1 1/8
	Length	T	1/4	3/8	7/16	9/16	5/8	11/16	3/4	1 1/8	1 3/8
	Radius	R	5/32	3/16	1/4	9/32	5/16	3/8	3/8	1/2	3/4
	Radius	a	3/64	3/64	1/16	5/64	3/32	1/8	1/8	5/32	3/16
Slot	Width	W	11/64	7/32	17/64	21/64	31/64	37/64	21/32	25/32	1 5/32
	Length	L	9/16	3/4	7/8	1 3/16	1 1/4	1 3/8	1 1/2	1 3/4	2 5/8
	End of socket to slot	K	1 15/16	2 1/16	2 1/2	3 1/16	3 7/8	4 5/16	4 15/16	7	9 1/2
	Taper per inch		.052050	.049882	.049951	.050196	.051938	.052000	.052626	.052138	.052000
	Taper per foot		.62460	.59858	.59941	.60235	.62326	.62400	.63151	.62565	.62400

Table 18. Morse Taper Shank Twist Drills - Fractional Sizes. *(Source, Kennametal, Inc.)*

Drill Dia.	Equivalent		Flute Length		Overall Length		Taper #
	Deci.	mm	inch	mm	inch	mm	
1/8	.1250	3.175	1 7/8	48	5 1/8	130	1
9/64	.1406	3.571	2 1/8	54	5 3/8	137	1
5/32	.1562	3.967	2 1/8	54	5 3/8	137	1
11/64	.1719	4.366	2 1/2	64	5 3/4	146	1
3/16	.1875	4.762	2 1/2	64	5 3/4	146	1
13/64	.2031	5.159	2 3/4	70	6	152	1
7/32	.2183	5.558	2 3/4	70	6	152	1
15/64	.2344	5.954	2 7/8	73	6 1/8	156	1
1/4	.250	6.350	2 7/8	73	6 1/8	156	1
17/64	.2656	6.746	3	76	6 1/4	159	1
9/32	.2812	7.142	3	76	6 1/4	159	1
19/64	.2969	7.541	3 1/8	79	6 3/8	162	1
5/16	.3125	7.938	3 1/8	79	6 3/8	162	1
21/64	.3281	8.334	3 1/4	83	6 1/2	165	1
11/32	.3438	8.733	3 1/4	83	6 1/2	165	1
23/64	.3594	9.129	3 1/2	89	6 3/4	171	1
3/8	.3750	9.525	3 1/2	89	6 3/4	171	1
25/64	.3906	9.921	3 5/8	92	7	178	1
13/32	.4062	1.320	3 5/8	92	7	178	1
27/64	.4219	1.716	3 7/8	98	7 1/4	184	1
7/16	.4375	11.112	3 7/8	98	7 1/4	184	1
29/64	.4531	11.509	4 1/8	105	7 1/2	190	1
15/32	.4688	11.906	4 1/8	105	7 1/2	190	1
31/64	.4844	12.304	4 3/8	111	8 1/4	210	2
1/2	.50	12.70	4 3/8	111	8 1/4	210	2
33/64	.5156	13.096	4 5/8	117	8 1/2	216	2
17/32	.5312	13.492	4 5/8	117	8 1/2	216	2
35/64	.5469	13.891	4 7/8	124	8 3/4	222	2
9/16	.5625	14.288	4 7/8	124	8 3/4	222	2
37/64	.5781	14.684	4 7/8	124	8 3/4	222	2
19/32	.5938	15.083	4 7/8	124	8 3/4	222	2
39/64	.6094	15.479	4 7/8	124	8 3/4	222	2

(Continued)

DRILLS

Table 18. (Continued) Morse Taper Shank Twist Drills - Fractional Sizes. (Source, Kennametal, Inc.)

Drill Dia.	Equivalent Deci.	Equivalent mm	Flute Length inch	Flute Length mm	Overall Length inch	Overall Length mm	Taper #
5/8	.6250	15.875	4 7/8	124	8 3/4	222	2
41/64	.6406	16.271	5 1/8	130	9	229	2
21/32	.6562	16.667	5 1/8	130	9	229	2
46/64	.6719	17.066	5 3/8	137	9 1/4	235	2
11/16	.6875	17.462	5 3/8	137	9 1/4	235	2
45/64	.7031	17.859	5 5/8	143	9 1/2	241	2
23/32	.7188	18.258	5 5/8	143	9 1/2	241	2
47/64	.7344	18.654	5 7/8	149	9 3/4	248	2
3/4	.750	19.050	5 7/8	149	9 3/4	248	2
49/64	.7656	19.446	6	152	9 7/8	251	2
25/32	.7812	19.843	6	152	9 7/8	251	2
51/64	.7969	2.241	6 1/8	156	10 3/4	273	3
13/16	.8125	2.638	6 1/8	156	10 3/4	273	3
53/64	.8281	21.034	6 1/8	156	10 3/4	273	3
27/32	.8438	21.433	6 1/8	156	10 3/4	273	3
55/64	.8594	21.829	6 1/8	156	10 3/4	273	3
7/8	.8750	22.225	6 1/8	156	10 3/4	273	3
57/64	.8906	22.621	6 1/8	156	10 3/4	273	3
29/32	.9062	23.017	6 1/8	156	10 3/4	273	3
59/64	.9219	23.416	6 1/8	156	10 3/4	273	3
15/16	.9375	23.813	6 1/8	156	10 3/4	273	3
61/64	.9531	24.209	6 3/8	162	11	279	3
31/32	.9688	24.608	6 3/8	162	11	279	3
63/64	.9844	25.004	6 3/8	162	11	279	3
1.0	1.0	25.40	6 3/8	162	11	279	3
1 1/64	1.0156	25.796	6 1/2	165	11 1/8	283	3
1 1/32	1.0312	26.192	6 1/2	165	11 1/8	283	3
1 3/64	1.0469	26.591	6 5/8	168	11 1/4	286	3
1 1/16	1.0625	26.988	6 5/8	168	11 1/4	286	3
1 5/64	1.0781	27.384	6 7/8	175	12 1/2	318	4
1 3/32	1.0938	27.783	6 7/8	175	12 1/2	318	4
1 7/64	1.1094	28.179	7 1/8	181	12 3/4	324	4
1 1/8	1.1250	28.575	7 1/8	181	12 3/4	324	4
1 9/64	1.1406	28.971	7 1/4	184	12 7/8	327	4
1 5/32	1.1562	29.367	7 1/4	184	12 7/8	327	4
1 11/64	1.1719	29.797	7 3/8	187	13	330	4
1 3/16	1.1875	3.162	7 3/8	187	13	330	4
1 13/64	1.2031	3.559	7 1/2	190	13 1/8	333	4
1 7/32	1.2188	3.958	7 1/2	190	13 1/8	333	4
1 15/64	1.2344	31.354	7 7/8	200	13 1/2	343	4
1 1/4	1.250	31.750	7 7/8	200	13 1/2	343	4
1 17/64	1.2656	32.146	8 1/2	216	14 1/8	359	4
1 9/32	1.2812	32.542	8 1/2	216	14 1/8	359	4
1 19/64	1.2969	32.941	8 5/8	219	14 1/4	362	4

(Continued)

Table 18. *(Continued)* Morse Taper Shank Twist Drills - Fractional Sizes. *(Source, Kennametal, Inc.)*

Drill Dia.	Equivalent Deci.	Equivalent mm	Flute Length inch	Flute Length mm	Overall Length inch	Overall Length mm	Taper #
1 5/16	1.3125	33.338	8 5/8	219	14 1/4	362	4
1 24/64	1.3281	33.734	8 3/4	222	14 3/8	365	4
1 11/32	1.3438	34.133	8 3/4	222	14 3/8	365	4
1 23/64	1.3594	34.529	8 7/8	225	14 1/2	368	4
1 3/8	1.3750	34.925	8 7/8	225	14 1/2	368	4
1 25/64	1.3906	35.321	9	229	14 5/8	371	4
1 13/32	1.4062	35.717	9	229	14 5/8	371	4
1 27/64	1.4219	36.116	9 1/8	232	14 3/4	375	4
1 7/16	1.4375	36.512	9 1/8	232	14 3/4	375	4
1 29/64	1.4531	36.909	9 1/4	235	14 7/8	378	4
1 15/32	1.4688	37.308	9 1/4	235	14 7/8	378	4
1 31/64	1.4844	37.704	9 3/8	238	15	381	4
1 1/2	1.50	38.10	9 3/8	238	15	381	4
1 17/32	1.5312	38.892	9 3/8	238	16 3/8	416	5
1 9/16	1.5625	39.688	9 5/8	244	16 5/8	422	5
1 19/32	1.5938	4.483	9 7/8	251	16 7/8	429	5
1 5/8	1.6250	41.275	10	254	17	432	5
1 21/32	1.6562	42.067	10 1/8	257	17 1/8	435	5
1 11/16	1.6875	42.862	10 1/8	257	17 1/8	435	5
1 23/32	1.1788	43.658	10 1/8	257	17 1/8	435	5
1 3/4	1.750	44.450	10 1/8	257	17 1/8	435	5
1 25/32	1.7812	45.242	10 1/8	257	17 1/8	435	5
1 13/16	1.8125	46.038	10 1/8	257	17 1/8	435	5
1 27/32	1.8434	46.833	10 1/8	257	17 1/8	435	5
1 7/8	1.8750	47.625	10 3/8	264	17 3/8	441	5
1 29/32	1.9062	48.417	10 3/8	264	17 3/8	441	5
1 15/16	1.9375	49.212	10 3/8	264	17 3/8	441	5
1 31/32	1.9688	5.008	10 3/8	264	17 3/8	441	5
2.0	2.0	5.80	10 3/8	264	17 3/8	441	5
2 1/32	2.0312	51.592	10 1/4	260	17 3/8	441	5
2 1/16	2.0625	52.388	10 1/4	260	17 3/8	441	5
2 3/32	2.0938	53.183	10 1/4	260	17 3/8	441	5
2 1/8	2.1250	53.975	10 1/4	260	17 3/8	441	5
2 5/32	2.1562	54.767	10 1/4	260	17 3/8	441	5
2 3/16	2.1875	55.563	10 1/4	260	17 3/8	441	5
2 7/32	2.2188	56.358	10 1/8	257	17 3/8	441	5
2 1/4	2.250	57.150	10 1/8	257	17 3/8	441	5
2 5/16	2.3125	58.738	10 1/8	257	17 3/8	441	5
2 3/8	2.3750	6.325	10 1/8	257	17 3/8	441	5
2 7/16	2.4375	61.912	11 1/4	286	18 3/4	476	5
2 1/2	2.50	63.50	11 1/4	286	18 3/4	476	5
2 9/16	2.5625	65.088	11 7/8	302	19 1/2	495	5
2 5/8	2.6250	66.675	11 7/8	302	19 1/2	495	5
2 11/16	2.6875	68.262	12 3/4	324	20 3/8	518	5

(Continued)

Table 18. *(Continued)* **Morse Taper Shank Twist Drills - Fractional Sizes.** *(Source, Kennametal, Inc.)*

Drill Dia.	Equivalent		Flute Length		Overall Length		Taper #	Drill Dia.	Equivalent		Flute Length		Overall Length		Taper #
	Deci.	mm	inch	mm	inch	mm			Deci.	mm	inch	mm	inch	mm	
2 3/4	2.750	68.850	12 3/4	324	20 3/8	518	5	3.0	3.0	76.20	14	356	21 3/4	552	5
2 316	2.8125	71.438	13 3/8	340	21 1/8	537	5	3 1/8	3.1250	79.375	14 5/8	371	24 1/2	622	6
2 7/8	2.8750	73.025	13 3/8	340	21 1/8	537	5	3 1/4	3.250	82.550	15 1/2	394	25 1/2	648	6
2 15/16	2.9375	74.612	14	356	21 3/4	552	5								

Table 19. Morse Taper Shank Twist Drills - Small Shank Fractional Series. *(Source, Kennametal Inc.)*

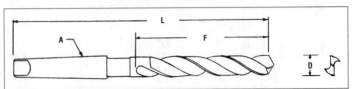

Drill Dia. A	Equivalent		Length				Taper #
			Flute (F)		Overall (L)		
	inch	mm	inch	mm	inch	mm	
31/64	0.4844	12.304	4 3/8	111	7 3/4	198	1
1/2	0.50	12.70	4 3/8	111	7 3/4	198	1
33/64	0.5156	13.096	4 5/8	117	8	203	1
17/32	0.5312	13.492	4 5/8	117	8	203	1
35/64	0.5469	13.891	4 7/8	124	8 1/4	210	1
13/16	0.8125	20.638	6 1/8	156	10	254	2
53/64	0.8281	21.034	6 1/8	156	10	254	2
27/32	0.8438	21.433	6 1/8	156	10	254	2
55/64	0.8594	21.829	6 1/8	156	10	254	2
7/8	0.8750	22.225	6 1/8	156	10	254	2
57/64	0.8906	22.621	6 1/8	156	10	254	2
29/32	0.9062	23.017	6 1/8	156	10	254	2
1 5/64	1.0781	27.384	6 7/8	175	11 1/2	292	3
1 3/32	1.0983	27.783	6 7/8	175	11 1/2	292	3
1 7/64	1.1094	28.179	7 1/8	181	11 3/4	298	3
1 1/8	1.1250	28.575	7 1/8	181	11 3/4	298	3
1 9/64	1.1406	28.971	7 1/4	184	11 7/8	302	3
1 5/32	1.1562	29.367	7 1/4	184	11 7/8	302	3
1 11/64	1.1719	29.797	7 3/8	187	12	305	3
1 3/16	1.1875	30.162	7 3/8	187	12	305	3
1 13/64	1.2031	30.559	7 1/2	190	12	1/8	3
1 7/32	1.2188	30.958	7 1/2	190	12	1/8	3
1 15/64	1.2344	31.354	7 7/8	200	12 1/2	318	3
1 1/4	1.250	31.750	7 7/8	200	12 1/2	318	3
1 33/64	1.5156	38.496	9 3/8	238	15	381	4
1 17/32	1.5312	38.892	9 3/8	238	15	381	4
1 35/64	1.5469	39.291	9 5/8	244	15 1/4	387	4
1 9/16	1.5625	39.688	9 5/8	244	15 1/4	387	4
1 37/64	1.5781	40.084	9 7/8	251	15 1/2	394	4
1 19/32	1.5938	40.483	9 7/8	251	15 1/2	394	4
1 39/64	1.6094	40.879	10	254	15 5/8	397	4

(Continued)

Table 19. *(Continued)* Morse Taper Shank Twist Drills - Small Shank Fractional Series. *(Source, Kennametal Inc.)*

Drill Dia. A	Equivalent		Length				Taper #
			Flute (F)		Overall (L)		
	inch	mm	inch	mm	inch	mm	
1 5/8	1.6250	41.275	10	254	15 5/8	397	4
1 41/64	1.6406	41.671	10 1/8	257	15 3/4	400	4
1 21/32	1.6562	42.067	10 1/8	257	15 3/4	400	4
1 43/64	1.6719	42.466	10 1/8	257	15 3/4	400	4
1 11/16	1.6875	42.862	10 1/8	257	15 3/4	400	4
1 45/64	1.7031	43.259	10 1/8	257	15 3/4	400	4
1 23/32	1.7188	43.658	10 1/8	257	15 3/4	400	4
1 47/64	1.7344	44.054	10 3/8	264	16 1/4	413	4
1 3/4	1.750	44.050	10 3/8	264	16 1/4	413	4
1 25/32	1.7812	45.242	10 3/8	264	16 1/4	413	4
1 13/16	1.8125	46.038	10 3/8	264	16 1/4	413	4
1 27/32	1.8438	46.833	10 3/8	264	16 1/4	413	4
1 7/8	1.8750	47.625	10 1/2	267	16 1/2	419	4
1 29/32	1.9062	48.417	10 1/2	267	16 1/2	419	4
1 15/16	1.9375	49.212	10 5/8	270	16 5/8	422	4
1 31/32	1.9688	50.008	10 5/8	270	16 5/8	422	4
2.0	2	50.80	10 5/8	270	16 5/8	422	4
3 1/8	3.1250	79.375	14 1/4	362	22	559	5
3 1/4	3.250	82.550	15 1/4	387	23	584	5
3 1/2	3.50	88.90	16 1/4	413	24	610	5

Table 20. Morse Taper Shank Twist Drills - Fractional Large Shank Series. *(Source, Kennametal Inc.)*

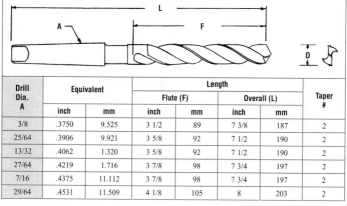

Drill Dia. A	Equivalent		Length				Taper #
			Flute (F)		Overall (L)		
	inch	mm	inch	mm	inch	mm	
3/8	.3750	9.525	3 1/2	89	7 3/8	187	2
25/64	.3906	9.921	3 5/8	92	7 1/2	190	2
13/32	.4062	1.320	3 5/8	92	7 1/2	190	2
27/64	.4219	1.716	3 7/8	98	7 3/4	197	2
7/16	.4375	11.112	3 7/8	98	7 3/4	197	2
29/64	.4531	11.509	4 1/8	105	8	203	2

(Continued)

Table 20. *(Continued)* **Morse Taper Shank Twist Drills - Fractional Large Shank Series.** *(Source, Kennametal Inc.)*

Drill Dia. A	Equivalent		Length				Taper #
			Flute (F)		Overall (L)		
	inch	mm	inch	mm	inch	mm	
15/32	.4688	11.906	4 1/8	105	8	203	2
41/64	.6406	16.271	5 1/8	130	9 3/4	248	3
21/32	.6562	16.667	5 1/8	130	9 3/4	248	3
43/64	.6719	17.066	5 3/8	137	10	254	3
11/16	.6875	17.462	5 3/8	137	10	254	3
45/64	.7031	17.859	5 5/8	143	10 1/4	260	3
23/32	.7188	18.258	5 5/8	143	10 1/4	260	3
47/64	.7344	18.654	5 7/8	149	10 1/2	267	3
3/4	.750	19.050	5 7/8	149	10 1/2	267	3
49/64	.7656	19.446	6	152	10 5/8	270	3
25/32	.7812	19.843	6	152	10 5/8	270	3
1.0	1.0	25.40	6 3/8	162	12	305	4
1 1/32	1.0312	26.192	6 1/2	165	12 1/8	308	4
1 1/16	1.0625	26.988	6 5/8	168	12 1/4	311	4

Table 21. Morse Taper Shank Dimensions (DIN 228, Form B and BK). *(Source ,Werkö GmbH.)*

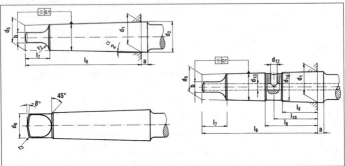

Symbol	Morse Taper Shaft, DIN 228						
	MK 0	MK 1	MK 2	MK 3	MK 4	MK 5	MK 6
a	3,0	3,5	5,0	5,0	6,5	6,5	8,0
b	3,9	5,2	6,3	7,9	11,9	15,9	19,0
d_1	9,045	12,065	17,780	23,825	31,267	44,399	63,348
d_2	9,2	12,2	18,0	24,1	31,6	44,7	63,8
d_5	6,1	9,0	14,0	19,1	25,2	36,5	52,4
d_6*	6,0	8,7	13,5	18,5	24,5	35,7	51,0
d_{12}			4,2	5,0	6,8	8,5	10,2

(Continued)

Table 21. *(Continued)* **Morse Taper Shank Dimensions (DIN 228, Form B and BK).** *(Source,Werkö GmbH.)*

Symbol	Morse Taper Shaft, DIN 228						
	MK 0	MK 1	MK 2	MK 3	MK 4	MK 5	MK 6
d_{13}			4,2	5,0	6,8	8,5	10,2
d_{14}**			15,0	21,0	28,0	40,0	56,0
L_6**	56,5	62,0	75,0	94,0	117,5	149,5	210,0
L_7*	10,5	13,5	16,0	20,0	24,0	29,0	40,0
L_8			20,0	29,0	39,0	51,0	81,0
L_9			34,0	43,0	55,0	69,0	99,0
L_{10}			27,0	36,0	47,0	60,0	90,0
r_2	4,0	5,0	6,0	7,0	8,0	10,0	13,0
r_3	1,0	1,2	1,6	2,0	2,5	3,0	4,0
$\alpha/2$	1°29'27"	1°25'43"	1°25'50"	1°26'16"	1°29'15"	1°30'26"	1°29'36"

All dimensions in millimeters.
*Maximum
** Tolerance +0/-0.1

DRILLS

Table 22. Morse Taper Shank Twist Drills - Metric Series. *(Source, Kennametal Inc.)*

Drill Dia. A mm	Deci.	Flute (F) inch	Flute (F) mm	Overall (L) inch	Overall (L) mm	Taper #
3.0	.1181	1 7/8	48	5 1/8	130	1
3.20	.1260	2 1/8	54	5 3/8	137	1
3.50	.1378	2 1/8	54	5 3/8	137	1
3.80	.1496	2 1/8	54	5 3/8	137	1
4.0	.1575	2 1/2	64	5 3/4	146	1
4.20	.1654	2 1/2	64	5 3/4	146	1
4.50	.1772	2 1/2	64	5 3/4	146	1
4.80	.1890	2 3/4	70	6	152	1
5.0	.1969	2 3/4	70	6	152	1
5.20	.2047	2 3/4	70	6	152	1
5.50	.2165	2 3/4	70	6	152	1
5.80	.2283	2 7/8	73	6 1/8	156	1
6.0	.2362	2 7/8	73	6 1/8	156	1
6.20	.2441	2 7/8	73	6 1/8	156	1
6.50	.2559	3	76	6 1/4	159	1
6.80	.2677	3	76	6 1/4	159	1
7.0	.2756	3	76	6 1/4	159	1
7.20	.2835	3 1/8	79	6 3/8	162	1
7.50	.2953	3 1/8	79	6 3/8	162	1
7.80	.3071	3 1/8	79	6 3/8	162	1
8.0	.3150	3 1/4	83	6 1/2	165	1
8.20	.3228	3 1/4	83	6 1/2	165	1
8.50	.3346	3 1/4	83	6 1/2	165	1
8.80	.3465	3 1/2	89	6 3/4	171	1
9.0	.3543	3 1/2	89	6 3/4	171	1
9.20	.3622	3 1/2	89	6 3/4	171	1
9.50	.3740	3 1/2	89	6 3/4	171	1
9.80	.3858	3 5/8	92	7	178	1
10.0	.3937	3 5/8	92	7	178	1
10.20	.4016	3 5/8	92	7	178	1
10.50	.4134	3 7/8	98	7 1/4	184	1
10.80	.4252	3 7/8	98	7 1/4	184	1

(Continued)

Table 22. *(Continued)* Morse Taper Shank Twist Drills - Metric Series. *(Source, Kennametal Inc.)*

Drill Dia. A		Length				Taper #
		Flute (F)		Overall (L)		
mm	Deci.	inch	mm	inch	mm	
11.0	.4331	3 7/8	98	7 1/4	184	1
11.20	.4409	4 1/8	105	7 1/2	190	1
11.50	.4528	4 1/8	105	7 1/2	190	1
11.80	.4646	4 1/8	105	7 1/2	190	1
12.0	.4724	4 3/8	111	8 1/4	210	2
12.20	.4803	4 3/8	111	8 1/4	210	2
12.50	.4921	4 3/8	111	8 1/4	210	2
12.80	.5039	4 5/8	117	8 1/2	216	2
13.0	.5118	4 5/8	117	8 1/2	216	2
13.20	.5197	4 5/8	117	8 1/2	216	2
13.50	.5315	4 5/8	117	8 1/2	216	2
13.80	.5433	4 7/8	124	8 3/4	222	2
14.0	.5512	4 7/8	124	8 3/4	222	2
14.25	.5610	4 7/8	124	8 3/4	222	2
14.50	.5709	4 7/8	124	8 3/4	222	2
14.75	.5709	4 7/8	124	8 3/4	222	2
15.0	.5906	4 7/8	124	8 3/4	222	2
15.25	.6004	4 7/8	124	8 3/4	222	2
15.50	.6102	4 7/8	124	8 3/4	222	2
15.75	.6201	4 7/8	124	8 3/4	222	2
16.0	.6299	5 1/8	130	9	229	2
16.25	.6398	5 1/8	130	9	229	2
16.50	.6496	5 1/8	130	9	229	2
16.75	.6594	5 3/8	137	9 1/4	235	2
17.0	.6693	5 3/8	137	9 1/4	235	2
17.25	.6791	5 3/8	137	9 1/4	235	2
17.50	.6890	5 5/8	143	9 1/2	241	2
18.00	.7087	5 5/8	143	9 1/2	241	2
18.50	.7283	5 7/8	149	9 3/4	248	2
19.0	.7480	5 7/8	149	9 3/4	248	2
19.50	.7677	6	152	9 7/8	251	2
20.0	.7874	6 1/8	156	10 3/4	273	3
20.50	.8071	6 1/8	156	10 3/4	273	3
21.0	.8268	6 1/8	156	10 3/4	273	3
21.50	.8465	6 1/8	156	10 3/4	273	3
22.0	.8661	6 1/8	156	10 3/4	273	3
22.50	.8858	6 1/8	156	10 3/4	273	3
23.0	.9055	6 1/8	156	10 3/4	273	3
23.50	.9252	6 1/8	156	10 3/4	273	3
24.0	.9449	6 3/8	162	11	279	3
24.50	.9646	6 3/8	162	11	279	3
25.0	.9843	6 3/8	162	11	279	3
25.50	1.0039	6 1/2	165	11 1/8	283	3
26.0	1.0236	6 1/2	165	11 1/8	283	3

(Continued)

Table 22. *(Continued)* **Morse Taper Shank Twist Drills - Metric Series.** *(Source, Kennametal Inc.)*

| Drill Dia. A | | Length | | | | Taper # |
mm	Deci.	Flute (F) inch	Flute (F) mm	Overall (L) inch	Overall (L) mm	
26.50	1.0433	6 5/8	168	11 1/4	286	3
27.0	1.0630	6 5/8	168	11 1/4	286	3
27.50	1.0827	6 7/8	175	12 1/2	318	4
28.0	1.1024	7 1/8	181	12 3/4	324	4
28.50	1.1220	7 1/8	181	12 3/4	324	4
29.0	1.1417	7 1/4	184	12 7/8	327	4
29.50	1.1614	7 3/8	187	13	330	4
30.0	1.1811	7 3/8	187	13	330	4
30.50	1.2008	7 1/2	190	13 1/8	333	4
31.0	1.2205	7 7/8	200	13 1/2	343	4
31.50	1.2402	7 7/8	200	13 1/2	343	4
32.0	1.2598	8 1/2	216	14 1/8	359	4
32.50	1.2795	8 1/2	216	14 1/8	359	4
33.0	1.2992	8 5/8	219	14 1/4	362	4
33.50	1.3189	8 3/4	222	14 3/8	365	4
34.0	1.3386	8 3/4	222	14 3/8	365	4
34.50	1.3583	8 7/8	225	14 1/2	368	4
35.0	1.3780	9	229	14 5/8	371	4
35.50	1.3976	9	229	14 5/8	371	4

| Drill Dia. A | | Length | | | | Taper # |
mm	Deci.	Flute (F) inch	Flute (F) mm	Overall (L) inch	Overall (L) mm	
36.0	1.4173	9 1/8	232	14 3/4	375	4
36.50	1.4370	9 1/8	232	14 3/4	375	4
37.0	1.4567	9 1/4	235	14 7/8	378	4
37.50	1.4764	9 3/8	238	15	381	4
38.0	1.4961	9 3/8	238	15	381	4
39.0	1.5354	9 5/8	244	16 5/8	422	5
40.0	1.5748	9 7/8	251	16 7/8	429	5
41.0	1.6142	10	254	17	432	5
42.0	1.6535	10 1/8	257	17 1/8	435	5
43.0	1.6929	10 1/8	257	17 1/8	435	5
44.0	1.7323	10 1/8	257	17 1/8	435	5
45.0	1.7717	10 1/8	257	17 1/8	435	5
46.0	1.8110	10 1/8	257	17 1/8	435	5
47.0	1.8504	10 3/8	264	17 3/8	441	5
48.0	1.8898	10 3/8	264	17 3/8	441	5
49.0	1.9291	10 3/8	264	17 3/8	441	5
50.0	1.9685	10 3/8	264	17 3/8	441	5
51.0	2.0079	10 3/8	264	17 3/8	441	5

pass through the interior of the drill and exit at the point, allow for high pressure coolant flow to the point for higher speeds, lower operating temperatures, and better chip ejection. These drills are often used horizontally while the drill is stationary and the workpiece rotates. Drills used in this manner normally have a smaller flute helix angle than rotating drills. Straight shank coolant holes over $1/2$" are normally tapped to securely attach coolant supply. Thread size and depths for these drills are shown in **Table 23**. Drills less that $1/2$" in diameter are normally cupped to aid coolant flow.

Table 23. Pipe Tap Thread for Straight Shank Coolant Hole Drills.

Drill Diameter	Thread	Depth
1/2" to 5/8"	1/8"– 27	3/4"
21/32" to 3/4"	1/8"– 27	1"
25/32" to 1 1/4"	1/4"– 18	1 1/4"
1 5/16" to 1 1/2"	3/8"– 18	1 1/2"

Morse Taper shank drills with coolant holes are fed coolant through a pipe tap or untapped hole in the recess between the drill's taper and its flutes. Straight shank drills are tapped directly into the rear of the drill, and the drill's shank is the same diameter as its body. **Tables 24** and **25** provide dimensions for popular sizes of coolant hole drills.

Core drills are used to enlarge cored, punched, or other preformed holes. They are not designed to drill from solid. Available in both three- and four-flute styles, core drills are able to operate at heavier feeds than two flute drills because chip loads are spread over a greater number of cutting edges. When using core drills, regular speed rates should be reduced by 25% and feed rates can be increased to up to double the regular rate. Dimensions for taper shank core drills are given in **Table 26**, and dimensions for straight shank core drills are found in **Table 27**.

Twist drills are available in almost unlimited variety, with the most uncommon usually designed for a specialized application. One of the most popular specialized drills is the *Silver and Deming reduced shank drill*. These relatively short (overall length of six inches) drills are widely used in machine shops and maintenance departments wherever the maximum chuck capacity is $1/2$". They are also in screw machines and machining centers when a short, rigid drill is required. Dimensions are given in **Table 28**.

(Text continued on p. 1124)

Table 24. Coolant Hole, 32° Helix, 118° Notched Point Drill-Taper Shank. Heavy duty construction with notched point for ease of penetration. Coolant holes allow for higher feeds and better chip ejection. *(Source, Precision Twist Drill Co.)*

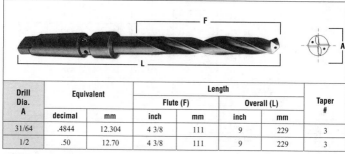

Drill Dia. A	Equivalent		Length				Taper #
			Flute (F)		Overall (L)		
	decimal	mm	inch	mm	inch	mm	
31/64	.4844	12.304	4 3/8	111	9	229	3
1/2	.50	12.70	4 3/8	111	9	229	3

(Continued)

Table 24. *(Continued)* **Coolant Hole, 32° Helix, 118° Notched Point Drill-Taper Shank.** Heavy duty construction with notched point for ease of penetration. Coolant holes allow for higher feeds and better chip ejection. *(Source, Precision Twist Drill Co.)*

Drill Dia. A	Equivalent		Length				Taper #
			Flute (F)		Overall (L)		
	decimal	mm	inch	mm	inch	mm	
17/32	.5312	13.492	4 5/8	117	9 1/4	235	3
9/16	.5625	14.288	4 7/8	124	9 1/2	241	3
19/32	.5938	15.083	4 7/8	124	9 1/2	241	3
5/8	.6250	15.875	4 7/8	124	9 1/2	241	3
21/32	.6562	16.667	5 1/8	130	9 3/4	248	3
11/16	.6875	17.462	5 3/8	137	10	254	3
23/32	.7188	18.258	5 5/8	143	10 1/4	260	3
3/4	.750	19.050	5 7/8	149	10 1/2	267	3
25/32	.7812	19.843	6	152	10 5/8	270	3
13/16	.8125	20.638	6 1/8	156	10 3/4	273	3
27/32	.8438	21.433	6 1/8	156	10 3/4	273	3
7/8	.8750	22.225	6 1/8	156	10 3/4	273	3
29/32	.9062	23.017	6 1/8	156	10 3/4	273	3
15/16	.9375	23.813	6 1/8	156	10 3/4	273	3
31/32	.9688	24.608	6 3/8	162	11	279	3
1"	1	25.40	6 3/8	162	11	279	3
1 1/32	1.0312	26.192	6 1/2	165	11 1/8	283	3
1 1/16	1.0625	26.988	6 5/8	168	11 1/4	286	3
1 3/32	1.0938	27.783	6 7/8	175	12 1/2	318	4
1 1/8	1.1250	28.575	7 1/8	181	12 3/4	324	4
1 5/32	1.1562	29.367	7 1/4	184	12 7/8	327	4
1 3/16	1.1875	30.162	7 3/8	187	13	330	4
1 7/32	1.2188	30.958	7 1/2	190	13 1/8	333	4
1 1/4	1.250	31.750	7 7/8	200	13 1/2	343	4
1 9/32	1.2812	32.542	8 1/2	216	14 1/8	359	4
1 5/16	1.3125	33.338	8 5/8	219	14 1/4	362	4
1 11/32	1.3438	34.133	8 3/4	222	14 3/8	365	4
1 3/8	1.3750	34.925	8 7/8	225	14 1/2	368	4
1 13/32	1.4062	35.717	9	229	14 5/8	371	4
1 7/16	1.4375	36.512	9 1/8	232	14 3/4	375	4
1 15/32	1.4688	37.308	9 1/4	235	14 7/8	378	4
1 1/2	1.50	38.10	9 3/8	238	15	381	4
1 9/16	1.5625	39.688	9 5/8	244	16 5/8	422	5
1 5/8	1.6250	41.275	10	254	17	432	5
1 11/16	1.6875	42.862	10 1/8	257	17 1/8	435	5
1 3/4	1.750	44.450	10 1/8	257	17 1/8	435	5

(Continued)

Table 24. *(Continued)* **Coolant Hole, 32° Helix, 118° Notched Point Drill-Taper Shank.** Heavy duty construction with notched point for ease of penetration. Coolant holes allow for higher feeds and better chip ejection. *(Source, Precision Twist Drill Co.)*

Drill Dia. A	Equivalent		Length				Taper #
			Flute (F)		Overall (L)		
	decimal	mm	inch	mm	inch	mm	
1 13/16	1.8125	46.038	10 1/8	257	17 1/8	435	5
1 7/8	1.8750	47.625	10 3/8	264	17 3/8	441	5
1 15/16	1.9375	49.212	10 3/8	264	17 3/8	441	5
2"	2	50.80	10 3/8	264	17 3/8	441	5

Table 25. Coolant Hole, 23° Helix, 118° Notched Point (Sizes $^3/_8$ - 1 $^1/_2$") and 34° Helix, 118° Notched Point (Sizes 3/8 - 1") Taper Length Drills. For high pressure coolant delivery. See thread dimensions below. *(Source, Precision Twist Drill Co.)*

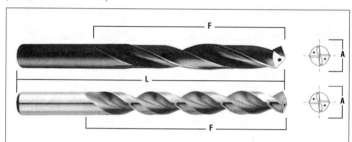

Drill Dia. A	Equivalent		Length			
			Flute (F)		Overall (L)	
	decimal	mm	inch	mm	inch	mm
3/8	.3750	9.525	4 1/4	108	6 3/4	171
13/32	.4062	10.317	4 3/8	111	7	178
7/16	.4375	11.112	4 5/8	117	7 1/4	184
15/32	.4688	11.908	4 7/8	124	7 1/2	190
1/2	.5000	12.70	5	127	7 3/4	197
17/32	.5312	13.492	5 1/4	133	8	203
9/16	.5625	14.288	5 3/8	137	8 1/4	210
19/32	.5938	15.083	5 5/8	143	8 1/2	216
5/8	.6250	15.75	5 3/4	146	8 3/4	222
21/32	.6562	16.669	5 7/8	149	9	228
11/16	.6875	17.462	6	152	9 1/4	235
23/32	.7188	18.258	6 3/16	157	9 1/2	241
3/4	.7500	19.050	6 3/8	162	9 3/4	248
25/32	.7812	19.845	6 1/2	165	9 7/8	251
13/16	.8125	20.638	6 5/8	168	10	254

(Continued)

Table 25. *(Continued)* **Coolant Hole, 23° Helix, 118° Notched Point (Sizes ³/₈ - 1 ¹/₂") and 34° Helix, 118° Notched Point (Sizes 3/8 - 1") Taper Length Drills.** For high pressure coolant delivery. See thread dimensions below. *(Source, Precision Twist Drill Co.)*

Drill Dia. A	Equivalent		Length			
			Flute (F)		Overall (L)	
	decimal	mm	inch	mm	inch	mm
27/32	.8438	21.433	6 3/4	171	10 1/4	260
7/8	.8750	22.235	7	178	10 1/2	267
29/32	.9062	23.017	7	178	10 5/8	270
15/16	.9375	23.812	7	178	10 3/4	273
31/32	.9688	24.608	7 1/8	181	10 7/8	276
1	1	25.4	7 3/16	183	11	279
1 1/32	1.0312	26.192	7 5/16	186	11 1/8	283
1 1/16	1.0625	26.988	7 3/8	187	11 1/4	286
1 3/32	1.0938	27.783	7 5/8	194	11 1/2	292
1 1/8	1.1250	28.575	7 7/8	200	11 3/4	298
1 5/32	1.1562	29.367	8	203	11 7/8	302
1 3/16	1.1875	30.162	8 1/8	206	12	305
1 7/32	1.2188	30.958	8 1/8	206	12 1/8	308
1 1/4	1.2500	31.750	8 1/2	216	12 1/2	318
1 5/16	1.3125	33.338	9 1/4	235	14 1/4	362
1 3/8	1.3750	34.925	9 1/2	241	14 1/2	368
1 7/16	1.4375	36.512	9 5/8	244	14 3/4	375
1 1/2	1.5000	38.10	9 7/8	250	15	381

Table 26. Taper Shank Core Drills, 118° Point, Fractional Sizes. Available in Three- and Four-Flute Styles. *(Source, Kennametal Inc.)*

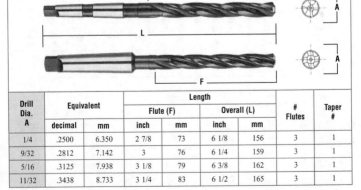

Drill Dia. A	Equivalent		Length				# Flutes	Taper #
			Flute (F)		Overall (L)			
	decimal	mm	inch	mm	inch	mm		
1/4	.2500	6.350	2 7/8	73	6 1/8	156	3	1
9/32	.2812	7.142	3	76	6 1/4	159	3	1
5/16	.3125	7.938	3 1/8	79	6 3/8	162	3	1
11/32	.3438	8.733	3 1/4	83	6 1/2	165	3	1

(Continued)

Table 26. *(Continued)* **Taper Shank Core Drills, 118° Point, Fractional Sizes.** Available in Three- and Four-Flute Styles. *(Source, Kennametal Inc.)*

Drill Dia. A	Equivalent		Length				# Flutes	Taper #
	decimal	mm	Flute (F)		Overall (L)			
			inch	mm	inch	mm		
3/8	.3750	9.525	3 1/2	89	6 3/4	171	3	1
13/32	.4062	10.319	3 5/8	92	7	178	3	1
7/16	.4375	11.112	3 7/8	98	7 1/4	184	3	1
15/32	.4688	11.908	4 1/8	105	7 1/2	190	3	1
1/2	.500	12.700	4 3/8	111	8 1/4	210	3 or 4	2
17/32	.5312	13.492	4 5/8	117	8 1/2	216	3 or 4	2
9/16	.5625	14.288	4 7/8	124	8 3/4	222	3 or 4	2
19/32	.5938	15.083	4 7/8	124	8 3/4	222	3 or 4	2
5/8	.6250	15.815	4 7/8	124	8 3/4	222	3 or 4	2
21/32	.6562	16.668	5 1/8	130	9	229	3 or 4	2
11/16	.6875	17.462	5 3/8	137	9 1/4	235	3 or 4	2
23/32	.7188	18.258	5 5/8	143	9 1/2	241	3 or 4	2
3/4	.7500	19.050	5 7/8	149	9 3/4	248	3 or 4	2
25/32	.7812	19.842	6	152	9 7/8	251	3 or 4	2
13/16	.8125	20.638	6 1/8	156	10 3/4	273	3 or 4	2
27/32	.8438	21.433	6 1/8	156	10 3/4	273	3 or 4	2
7/8	.8750	22.225	6 1/8	156	10 3/4	273	3 or 4	3
29/32	.9062	23.019	6 1/8	156	10 3/4	273	3 or 4	3
15/16	.9375	23.812	6 1/8	156	10 3/4	273	3 or 4	3
31/32	.9688	24.608	6 3/8	162	11	279	3 or 4	3
1	1.000	25.400	6 3/8	162	11	279	3 or 4	3
1 1/32	1.0312	26.192	6 1/2	165	11 1/8	283	3 or 4	3
1 1/16	1.0625	26.988	6 5/8	168	11 1/4	286	3 or 4	3
1 3/32	1.0938	27.783	6 7/8	175	12 1/2	318	3 or 4	4
1 1/8	1.1250	28.575	7 1/8	181	12 3/4	324	3 or 4	4
1 5/32	1.1562	29.367	7 1/4	184	12 7/8	327	3 or 4	4
1 13/16	1.1875	30.162	7 3/8	187	13	330	3 or 4	4
1 7/32	1.2188	30.958	7 1/2	190	13 1/8	333	3 or 4	4
1 1/4	1.2500	31.750	7 7/8	200	13 1/2	343	3 or 4	4
1 9/32	1.2812	32.542	8 1/2	216	14 1/8	359	4	4
1 5/16	1.3125	33.338	8 5/8	219	14 1/4	362	4	4
1 11/32	1.3438	34.133	8 3/4	222	14 3/8	365	4	4
1 3/8	1.3750	34.925	8 7/8	225	14 1/2	368	4	4
1 13/32	1.4062	35.717	9	229	14 5/8	371	4	4
1 7/16	1.4375	36.512	9 1/8	232	14 3/4	375	4	4
1 15/32	1.4688	37.3.6	9 1/4	235	14 7/8	378	4	4
1 1/2	1.500	38.100	9 3/8	238	15	381	4	4

(Continued)

Table 26. *(Continued)* **Taper Shank Core Drills, 118° Point, Fractional Sizes.** Available in Three- and Four-Flute Styles. *(Source, Kennametal Inc.)*

Drill Dia. A	Equivalent		Length				# Flutes	Taper #
	decimal	mm	Flute (F)		Overall (L)			
			inch	mm	inch	mm		
1 17/32	1.5312	38.892	9 3/8	238	16 3/8	416	4	5
1 9/16	1.5625	39.688	9 5/8	244	16 5/8	422	4	5
1 19/32	1.5938	40.483	9 7/8	251	16 7/8	429	4	5
1 5/8	1.6250	41.275	10	254	17	432	4	5
1 21/32	1.6562	42.067	10 1/8	257	17 1/8	435	4	5
1 11/16	1.6875	42.862	10 1/8	257	17 1/8	435	4	5
1 23/32	1.7188	43.658	10 1/8	257	17 1/8	435	4	5
1 3/4	1.7500	44.450	10 1/8	257	17 1/8	435	4	5
1 25/32	1.7812	45.244	10 1/8	257	17 1/8	435	4	5
1 13/16	1.8125	46.038	10 1/8	257	17 1/8	435	4	5
1 27/32	1.8438	46.833	10 1/8	257	17 1/8	435	4	5
1 7/8	1.8750	47.625	10 3/8	264	17 3/8	441	4	5
1 29/32	1.9062	48.417	10 3/8	264	17 3/8	441	4	5
1 15/16	1.9375	49.212	10 3/8	264	17 3/8	441	4	5
1 31/32	1.9688	50.008	10 3/8	264	17 3/8	441	4	5
2	2.000	50.800	10 3/8	264	17 3/8	441	4	5
2 1/8	2.1250	53.975	10 1/4	260	17 3/8	441	4	5
2 1/4	2.2500	57.150	10 1/8	257	17 3/8	441	4	5
2 3/8	2.3750	60.325	10 1/8	257	17 3/8	441	4	5
2 1/2	2.5000	63.500	11 1/4	286	18 3/4	476	4	5

Table 27. Straight Shank Core Drills, 118° Chamfer Point, Fractional Sizes. Available in Three- and Four-Flute Styles. *(Source, Precision Twist Drill Co.)*

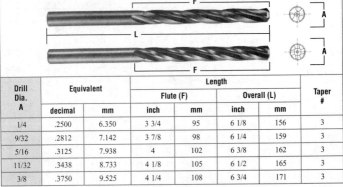

Drill Dia. A	Equivalent		Length				Taper #
	decimal	mm	Flute (F)		Overall (L)		
			inch	mm	inch	mm	
1/4	.2500	6.350	3 3/4	95	6 1/8	156	3
9/32	.2812	7.142	3 7/8	98	6 1/4	159	3
5/16	.3125	7.938	4	102	6 3/8	162	3
11/32	.3438	8.733	4 1/8	105	6 1/2	165	3
3/8	.3750	9.525	4 1/4	108	6 3/4	171	3

(Continued)

Table 27. *(Continued)* **Straight Shank Core Drills, 118° Chamfer Point, Fractional Sizes.** Available in Three- and Four-Flute Styles. *(Source, Precision Twist Drill Co.)*

Drill Dia. A	Equivalent		Length				Taper #
			Flute (F)		Overall (L)		
	decimal	mm	inch	mm	inch	mm	
13/32	.4062	10.319	4 3/8	111	7	178	3
7/16	.4375	11.112	4 5/8	117	7 1/4	184	3
15/32	.4688	11.908	4 3/4	121	7 1/2	190	3
1/2	.500	12.700	4 3/4	121	7 3/4	197	3 or 4
17/32	.5312	13.492	4 3/4	121	8	203	3 or 4
9/16	.5625	14.288	4 7/8	124	8 1/4	210	3 or 4
19/32	.5938	15.083	4 7/8	124	8 3/4	222	3 or 4
5/8	.6250	15.815	4 7/8	124	8 3/4	222	3 or 4
21/32	.6562	16.668	5 1/8	130	9	229	3 or 4
11/16	.6875	17.462	5 3/8	137	9 1/4	235	3 or 4
23/32	.7188	18.258	5 5/8	143	9 1/2	241	3 or 4
3/4	.7500	19.050	5 7/8	149	9 3/4	248	3 or 4
25/32	.7812	19.842	6	152	9 7/8	251	4
13/16	.8125	20.638	6 1/8	156	10	254	4
27/32	.8438	21.433	6 1/8	156	10	254	4
7/8	.8750	22.225	6 1/8	156	10	254	4
29/32	.9062	23.019	6 1/8	156	10	254	4
15/16	.9375	23.812	6 1/8	156	10 3/4	273	4
31/32	.9688	24.608	6 3/8	162	11	279	4
1	1	25.400	6 3/8	162	11	279	4

Table 28. Silver and Deming Reduced Shank Drills. Flute Length 3" (76.2 mm), Overall Length 6" (153 mm) for all diameters. *(Source, Precision Twist Drill Co.)*

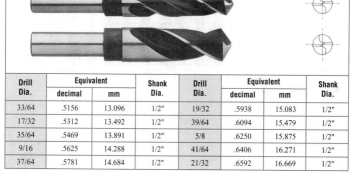

Drill Dia.	Equivalent		Shank Dia.	Drill Dia.	Equivalent		Shank Dia.
	decimal	mm			decimal	mm	
33/64	.5156	13.096	1/2"	19/32	.5938	15.083	1/2"
17/32	.5312	13.492	1/2"	39/64	.6094	15.479	1/2"
35/64	.5469	13.891	1/2"	5/8	.6250	15.875	1/2"
9/16	.5625	14.288	1/2"	41/64	.6406	16.271	1/2"
37/64	.5781	14.684	1/2"	21/32	.6592	16.669	1/2"

(Continued)

Table 28. *(Continued)* **Silver and Deming Reduced Shank Drills.** Flute Length 3" (76.2 mm), Overall Length 6" (153 mm) for all diameters. *(Source, Precision Twist Drill Co.)*

Drill Dia.	Equivalent decimal	Equivalent mm	Shank Dia.	Drill Dia.	Equivalent decimal	Equivalent mm	Shank Dia.
43/64	.6719	17.006	1/2"	1 19/64	1.2969	32.941	1/2"
11/16	.6875	17.462	1/2"	1 5/16	1.3125	33.338	1/2"
45/64	.7031	17.859	1/2"	1 21/64	1.3281	33.734	1/2"
23/32	.7188	18.258	1/2"	1 11/32	1.3438	34.133	1/2"
47/64	.7344	18.654	1/2"	1 23/64	1.3594	34.529	1/2"
3/4	.750	19.050	1/2"	1 3/8	1.3750	34.925	1/2"
49/64	.7656	19.446	1/2"	1 25/64	1.3906	35.321	1/2"
25/32	.7812	19.845	1/2"	1 13/32	1.4062	35.717	1/2"
51/64	.7969	20.241	1/2"	1 27/64	1.4219	36.116	1/2"
13/16	.8125	20.638	1/2"	1 7/16	1 4375	36.512	1/2"
53/64	.8281	21.034	1/2"	1 29/64	1.4531	36.909	1/2"
27/32	.8438	21.433	1/2"	1 15/32	1.4688	37.308	1/2"
55/64	.8594	21.829	1/2"	1 31/64	1.4844	37.704	1/2"
7/8	.8750	22.235	1/2"	1 1/2	1.50	38.10	1/2"
57/64	.8906	22.621	1/2"	1"	1.0	25.40	3/4"
29/32	.9062	23.017	1/2"	1 1/32	1.0312	26.192	3/4"
59/64	.9219	23.416	1/2"	1 1/16	1.0625	26.988	3/4"
15/16	.9375	23.812	1/2"	1 3/32	1.0938	27.783	3/4"
61/64	.9531	24.209	1/2"	1 1/8	1.1250	28.575	3/4"
31/32	.9688	24.608	1/2"	1 5/32	1.1562	29.367	3/4"
63/64	.9844	25.004	1/2"	1 3/16	1.1875	30.162	3/4"
1"	1.0	25.40	1/2"	1 7/32	1.2188	30.958	3/4"
1 1/32	1.0312	26.192	1/2"	1 1/4	1.250	31.750	3/4"
1 3/64	1.0469	26.591	1/2"	1 9/32	1.2812	32.542	3/4"
1 1/16	1.0625	26.988	1/2"	1 5/16	1.3125	33.338	3/4
1 5/64	1.0781	27.384	1/2"	1 11/32	1.3438	34.133	3/4"
1 3/32	1.0938	27.783	1/2"	1 3/8	1.3750	34.925	3/4"
1 7/64	1.1094	28.179	1/2"	1 13/32	1.4062	35.717	3/4"
1 1/8	1.1250	28.575	1/2"	1 7/16	1 4375	36.512	3/4"
1 9/64	1.1406	28.971	1/2"	1 15/32	1.4688	37.308	3/4"
1 5/32	1.1562	29.367	1/2"	1 1/2	1.50	38.10	3/4"
1 11/64	1.1719	29.797	1/2"	1 9/16	1.5625	39.688	3/4"
1 3/16	1.1875	30.162	1/2"	1 5/8	1.6250	41.275	3/4"
1 13/64	1.2031	30.559	1/2"	1 11/16	1.6875	42.862	3/4"
1 7/32	1.2188	30.958	1/2"	1 3/4	1.750	44.450	3/4"
1 15/64	1.2344	31.354	1/2"	1 13/16	1.8125	46.038	3/4"
1 1/4	1.250	31.750	1/2"	1 7/8	1.8750	47.625	3/4"
1 17/64	1.2656	32.146	1/2"	1 15/16	1.9375	49.212	3/4"
1 9/32	1.2812	32.542	1/2"	2"	2.0	50.8	3/4"

The Aerospace Industries Association has standardized a range of split pointed drills that are used in the aircraft and aerospace industries. The split point is self-centering and reduces thrust. They are known as *NAS 907 drills*, with a "type" letter following the standard number as shown in **Table 29**. Most NAS split point drills have a 135° point, but many varieties are available with a conventional 118° point. **Table 30** provides dimensions for NAS 907 extra length 135° split point drills.

Table 29. NAS 907 Drill Type Designations.

Type	Overall Length	Flute Length	Material	Application
NAS 907, Type A	Jobber	Jobber	HSS	Light Duty
NAS 907, Type B	Jobber	Jobber	HSS	Medium Duty
NAS 907, Type C	Screw Machine	Screw Machine	HSS	Medium Duty
NAS 907, Type D	Jobber	Short–Standard Helix	Cobalt Steel	Heavy Duty
NAS 907, Type E	Jobber	Short–Low Helix	Cobalt Steel	Heavy Duty
NAS 907, Type J	Jobber	Jobber	Cobalt Steel	Heavy Duty
NAS Extra Length	See **Table 30**	Jobber	HSS	Light–Medium

Spotting and centering drills are extremely rigid to reduce runout and eccentricity. They are used to locate a hole position for follow-up drilling by a larger drill, and they are ideally suited for shallow hole drilling, as they are designed to drill no further than the depth of their point, and for single operation centering and chamfering. Although shorter lengths offer superior rigidity. Spotting drills are available in short, regular, and long lengths. **Table 31** provides dimensions for 118° point spotting and centering drills, and **Table 32** shows sizes of spotting drills, available with either 90° or 120° included angle.

Table 30. NAS 907 Extra Length 135° Split Point Drills. *(Source, Precision Twist Drill Co.)*
All Sizes Available in 6" (152 mm) and 12" (304 mm) Overall Length.

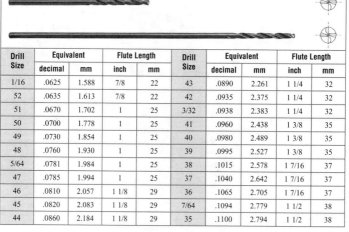

Drill Size	Equivalent		Flute Length		Drill Size	Equivalent		Flute Length	
	decimal	mm	inch	mm		decimal	mm	inch	mm
1/16	.0625	1.588	7/8	22	43	.0890	2.261	1 1/4	32
52	.0635	1.613	7/8	22	42	.0935	2.375	1 1/4	32
51	.0670	1.702	1	25	3/32	.0938	2.383	1 1/4	32
50	.0700	1.778	1	25	41	.0960	2.438	1 3/8	35
49	.0730	1.854	1	25	40	.0980	2.489	1 3/8	35
48	.0760	1.930	1	25	39	.0995	2.527	1 3/8	35
5/64	.0781	1.984	1	25	38	.1015	2.578	1 7/16	37
47	.0785	1.994	1	25	37	.1040	2.642	1 7/16	37
46	.0810	2.057	1 1/8	29	36	.1065	2.705	1 7/16	37
45	.0820	2.083	1 1/8	29	7/64	.1094	2.779	1 1/2	38
44	.0860	2.184	1 1/8	29	35	.1100	2.794	1 1/2	38

(Continued)

Table 30. *(Continued)* **NAS 907 Extra Length 135° Split Point Drills.** All Sizes Available in 6" (152 mm) and 12" (304 mm) Overall Length. *(Source, Precision Twist Drill Co.)*

Drill Size	Equivalent		Flute Length		Drill Size	Equivalent		Flute Length	
	decimal	mm	inch	mm		decimal	mm	inch	mm
34	.1110	2.819	1 1/2	38	7/32	.2188	5.558	2 1/2	64
33	.1130	2.870	1 1/2	38	2	.2210	5.613	2 5/8	67
32	.1160	2.946	1 5/8	41	1	.2280	5.791	2 5/8	67
31	.1200	3.048	1 5/8	41	A	.2340	5.944	2 5/8	67
1/8	.1250	3.175	1 5/8	41	15/64	.2344	5.954	2 5/8	67
30	.1285	3.264	1 5/8	41	B	.2380	6.045	2 3/4	70
29	.1360	3.454	1 3/4	44	C	.2420	6.147	2 3/4	70
28	.1405	3.569	1 3/4	44	D	.2460	6.248	2 3/4	70
9/64	.1406	3.571	1 3/4	44	1/4 E	.250	6.350	2 3/4	70
27	.1440	3.658	1 7/8	48	F	.2570	6.528	2 7/8	73
26	.1470	3.734	1 7/8	48	G	.2610	6.629	2 7/8	73
25	.1495	3.797	1 7/8	48	17/64	.2656	6.746	2 7/8	73
24	.1520	3.861	2	51	H	.2660	6.756	2 7/8	73
23	.1540	3.912	2	51	I	.2720	6.909	2 7/8	73
5/32	.1562	3.967	2	51	J	.2770	7.036	2 7/8	73
22	.1570	3.988	2	51	K	.2810	7.137	2 15/16	75
21	.1590	4.039	2 1/8	54	9/32	.2812	7.142	2 15/16	75
20	.1610	4.089	2 1/8	54	L	.290	7.366	2 15/16	75
19	.1660	4.216	2 1/8	54	M	.2950	7.493	3 1/16	78
18	.1695	4.305	2 1/8	54	19/64	.2969	7.541	3 1/16	78
11/64	.1719	4.366	2 1/8	54	N	.3020	7.671	3 1/16	78
17	.1730	4.394	2 3/16	56	5/16	.3125	7.938	3 1/16	78
16	.1770	4.496	2 3/16	56	O	.3160	8.026	3 1/16	78
15	.1800	4.572	2 3/16	56	P	.3230	8.204	3 5/16	84
14	.1820	4.623	2 3/16	56	21/64	.3281	8.344	3 5/16	84
13	.1850	4.70	2 5/16	59	Q	.3320	8.433	3 7/16	84
3/16	.1875	4.762	2 5/16	59	R	.3390	8.611	3 7/16	84
12	.1890	4.80	2 5/16	59	11/32	.3438	8.733	3 7/16	84
11	.1910	4.851	2 5/16	59	S	.3480	8.839	3 1/2	89
10	.1935	4.915	2 7/16	62	T	.3580	9.093	3 1/2	89
9	.1960	4.978	2 7/16	62	U	.3680	9.347	3 5/8	89
8	.1990	5.054	2 7/16	62	3/8	.3750	9.525	3 5/8	89
7	.2010	5.105	2 7/16	62	V	.3770	9.576	3 3/4	95
13/64	.2031	5.159	2 7/16	62	W	.3860	9.804	3 3/4	95
6	.2040	5.182	2 1/2	64	25/64	.3906	9.921	3 3/4	95
5	.2055	5.220	2 1/2	64	X	.3937	10.084	3 3/4	95
4	.2090	5.309	2 1/2	64	Y	.4040	10.262	3 7/8	98
3	.2130	5.410	2 1/2	64	13/32	.4062	10.317	3 7/8	98

(Continued)

Table 30. *(Continued)* **NAS 907 Extra Length 135° Split Point Drills.** All Sizes Available in 6" (152 mm) and 12" (304 mm) Overall Length. *(Source, Precision Twist Drill Co.)*

Drill Size	Equivalent		Flute Length		Drill Size	Equivalent		Flute Length	
	decimal	mm	inch	mm		decimal	mm	inch	mm
Z	.4130	10.409	3 7/8	98	15/32	.4688	11.908	4 5/16	110
27/64	.4219	10.716	3 5/16	100	31/64	.4844	12.304	4 3/8	111
7/16	.4375	11.112	4 1/6	103	1/2	.50	12.70	4 1/2	114
29/64	.4531	11.509	4 3/16	106					

Table 31. Spotting and Centering Drill, 118° point. *(Source, Kennametal Inc.)*

Drill Dia.	Equivalent		Flute Length		Overall Length	
	decimal	mm	inch	mm	inch	mm
1/16	.0625	1.588	3/4	19.050	1 1/4	31.750
3/32	.0938	2.383	3/4	19.050	1 1/4	31.750
1/8	.1250	3.175	3/4	19.050	1 1/4	31.750
5/32	.1562	3.967	1	25.40	1 1/2	38.10
3/16	.1875	4.762	1	25.40	1 1/2	38.10
7/32	.2188	5.558	1	25.40	1 1/2	38.10
1/4	.2500	6.350	1	25.40	1 1/2	38.10
5/16	.3125	7.938	1	25.40	1 1/2	38.10
3/8	.3750	9.525	1	25.40	2	50.80
1/2	.5000	12.70	1	25.40	2	50.80
5/8	.6250	15.875	1 1/8	28.575	2 1/4	57.150
3/4	.7500	19.050	1 1/8	28.575	2 1/4	57.150
1	1.0000	25.40	1 1/4	31.750	2 1/2	63.50

Table 32. Spotting Drill, 90° or 120° included angle. *(Source, Kennametal Inc.)*

Drill Dia.	Equivalent		Flute Length		Overall Length	
	decimal	mm	inch	mm	inch	mm
1/4	.2500	6.350	1	25.4	2 1/2	63.50
1/4	.2500	6.350	1	25.4	4	102
1/4	.2500	6.350	1	25.4	6	152
3/8	.3750	9.525	1 1/8	28.575	3 1/8	79
3/8	.3750	9.525	1 1/8	28.575	5	127
3/8	.3750	9.525	1 1/8	28.575	7	178
1/2	.5000	12.70	1 1/2	38.10	3 3/4	92
1/2	.5000	12.70	1 1/2	38.10	6	152
1/2	.5000	12.70	1 1/2	38.10	8	203
5/8	.6250	15.875	1 5/8	41.275	4 1/2	114
5/8	.6250	15.875	1 5/8	41.275	7 1/8	181
5/8	.6250	15.875	1 5/8	41.275	9	229
3/4	.7500	19.050	1 3/4	44.450	5	127

(Continued)

Table 32. *(Continued)* Spotting Drill, 90° or 120° included angle. *(Source, Kennametal Inc.)*

Drill Dia.	Equivalent		Flute Length		Overall Length	
	decimal	mm	inch	mm	inch	mm
3/4	.7500	19.050	1 3/4	44.450	8	203
3/4	.7500	19.050	1 3/4	44.450	10	254
1	1.000	25.40	1 3/4	44.450	6	152
1	1.000	25.40	1 3/4	44.450	8	203
1	1.000	25.40	1 3/4	44.450	10	254

Counterboring, countersinking, and spot-facing

Counterboring, countersinking, and spot-facing are drilling operations. Counterboring is the process of enlarging a hole for a portion of its length and leaving a square shoulder, at a right angle to the hole's axis, at the bottom of the enlarged portion. The most common use is to form a seat for the head of a sunken screw, but when used to a very shallow depth to provide a true surface for a bolt head, the operation is known as spot-facing. Counterbores are available with and without pilots, but pilotless tools must be guided by drill bushings. Speeds and feeds for counterboring are generally somewhat lower than for drilling. Dimensions for piloted counterbores are shown in **Tables 33** and **34**. The fasteners section of this handbook contains dimensions for 1960 series and 1936 series socket head set screws that are compatible with these tools.

Countersinking is used to drill a conical seat for screws with conical heads. The conical seat is concentric with the hole and is at an angle less than 90° to the hole. Multiple fluted countersinks should be operated at the same rpm as a reamer for the same size hole, and the maximum speed is normally between one-half and two-thirds the speed used for drilling the same material. The spacing of the flutes provides a centralizing action that works well in most materials. Fluted countersinks used for countersinking flat-head screws are available with included angles of 60° (for centering), 82° (for countersinking), 90° (for chamfering), and 100° (for deburring). Four-flute tools are also known as machine countersinks, and are designed for use with fixed machining centers. Three-flute tools are popular for use with portable equipment, and they provide superior tool life over single-flute tools. If multiflute countersinks are operated at higher speeds, chatter can occur. For this reason, when operating at higher speeds, a single-fluted tool should be applied. Single-fluted countersinks begin cutting in a smaller hole than multiflute tools because of the tool's high positive rake angle, and the single-flute cutting edge extends almost to the center of the tool. For this reason, they can be used to countersink much smaller holes than multiflute tools. The predrilled hole should be no smaller than 10% of the diameter of the countersink. See **Tables 35, 36,** and **37** for specifications of single and multifluted countersinks. **Table 38** shows the minimum and maximum diameter chamfers that can be produced with single and multifluted countersinks of different standard diameters.

Fluteless or chatterless countersinks are available in several configurations. **Table 39** provides specifications for these countersinking and deburring tools that have a single hole with a cutting edge that allows for chips to flow away from the workpiece. These tools are easily sharpened with a small mounted grinding wheel which is inserted into the hole. A *combined drill and countersink* is recommended to provide a smooth, accurate countersink when countersinks are required in abrasive, difficult to machine materials. They are available with 60° (for centering) and 120° (for chamfering) included angles (see **Table 40**).

Table 33. Straight Shank and Morse Taper Shank Piloted Counterbores for 1960 Series Socket Head Cap Screws. *(Source, Fastcut Tool Corp.)*

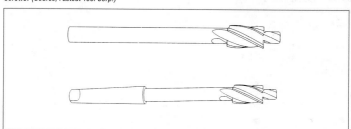

Pilot Dia.	Cutter Dia.	For U.S.S. Cap Screw	Shank Dia.	Morse Taper	Overall Length
5/16	15/32	5/16	7/16	1	6
21/64	31/64	5/16	7/16	1	6
11/32	1/2	5/16	7/16	1	6
7/16	21/32	7/16	1/2	2	7
29/64	43/64	7/16	1/2	-	7
15/32	11/16	7/16	1/2	2	7
5/8	15/16	5/8	3/4	3	8 1/4
21/32	31/32	5/8	3/4	3	8 1/4
11/16	1	5/8	3/4	3	8 1/4
3/4	1 1/8	3/4	1	-	8 13/16
25/32	1 5/32	3/4	1	3	8 13/16
13/16	1 3/16	3/4	1	3	8 13/16
29/32	1 11/32	7/8	-	3	8 13/16
1	1 1/2	1	1	-	8 13/16
1 1/32	1 17/32	1	-	3	8 13/16
1 1/16	1 9/16	1	1	3	8 13/16

All dimensions in inches. Cutters are 20°, right hand helix, right hand cut.

Table 34. Straight Shank and Morse Taper Shank Piloted Counterbores for 1936 Series Socket Head Cap Screws. *(Source, Fastcut Tool Corp.)*

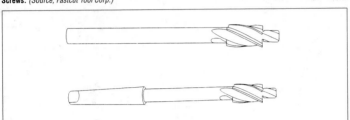

Pilot Dia.	Cutter Dia.	For U.S.S. Cap Screw	Shank Dia.	Morse Taper	Overall Length
.135	.230	6	7/32	-	3
.150	.242	6	7/32	-	3
.161	.274	8	1/4	-	3
.178	.286	8	1/4	-	3
.187	.316	10	5/16	-	3 1/2
.204	.328	10	5/16	-	3 1/2
.213	.348	12	11/32	-	3 1/2
.229	.360	12	11/32	-	3 1/2
1/4	3/8	1/4	3/8	1	5 3/4
17/64	25/64	1/4	3/8	1	5 3/4
9/32	13/32	1/4	3/8	1	5 3/4
5/16	7/16	5/16	7/16	1	6
21/64	29/64	5/16	7/16	1	6
11/32	15/32	5/16	7/16	1	6
3/8	9/16	3/8	1/2	1	6 1/2
25/64	37/64	3/8	1/2	1	6 1/2
13/32	19/32	3/8	1/2	1	6 1/2
7/16	5/8	7/16	1/2	2	7
29/64	41/64	7/16	1/2	-	7
15/32	21/32	7/16	1/2	2	7
1/2	3/4	1/2	1/2	2	7 1/4
33/64	49/64	1/2	1/2	2	7 1/4
17/32	25/32	1/2	1/2	2	7 1/4
9/16	13/16	9/16	3/4	2	7 1/2
19/32	27/32	9/16	3/4	-	7 1/2
5/8	7/8	5/8	3/4	-	8 1/4
21/32	29/32	5/8	3/4	3	8 1/4
11/16	15/16	5/8	3/4	-	8 1/4
3/4	1	3/4	1	3	8 13/16

(Continued)

Table 34. *(Continued)* **Straight Shank and Morse Taper Shank Piloted Counterbores for 1936 Series Socket Head Cap Screws.** *(Source, Fastcut Tool Corp.)*

Pilot Dia.	Cutter Dia.	For U.S.S. Cap Screw	Shank Dia.	Morse Taper	Overall Length
25/32	1 1/32	3/4	1	3	8 13/16
13/16	1 1/16	3/4	1	3	8 13/16
1	1 5/16	1	-	3	8 13/16

All dimensions in inches. Cutters are 20°, right-hand helix, right-hand cut.

Table 35. Single-Flute Countersink. Available in 60°, 82°, and 90° included angles. The predrilled hole should be no less than 10% of the body diameter of the countersink. *(Source, Kennametal Inc.)*

Drill Dia.	Equivalent		Shank Diameter		Overall Length	
	decimal	mm	inch	mm	inch	mm
1/4	.250	6.350	3/16	4.762	1 1/2	38
3/8	.375	9.525	1/4	6.350	1 3/4	44.5
1/2	.50	12.70	1/4	6.350	2	51
5/8	.625	15.875	3/8	9.525	2 1/4	57
3/4	.750	19.050	3/8	9.525	2 5/8	67
1.0	1	25.40	1/2	12.70	3 1/8	79.5

Table 36. Center Reamer Three-Flute Countersink. Available in 60°, 82°, 90°, and 100° included angles. Used for centering lathe work on predrilled holes or countersinking for screw heads. *(Source, Kennametal Inc.)*

Drill Dia.	Equivalent		Shank Diameter		Overall Length	
	decimal	mm	inch	mm	inch	mm
1/4	.250	6.350	3/16	4.762	1 1/2	38
3/8	.375	9.525	1/4	6.350	1 3/4	44.5
1/2	.50	12.70	3/8	9.525	2	51
5/8	.625	15.875	3/8	9.525	2 1/4	57
3/4	.750	19.050	1/2	12.70	2 5/8	67
1.0	1	25.40	1/2	12.70	3	76

Table 37. Four-Flute Countersink. Available in 60° and 80° included angles. For countersinking flat head screws. *(Source, Kennametal Inc.)*

Drill Dia.	Equivalent		Shank Diameter		Shank Length		Overall Length	
	decimal	mm	inch	mm	inch	mm	inch	mm
1/2	.50	12.70	1/2	12.70	2 1/4	57	3 23/32	94.458
5/8	.625	15.875	1/2	12.70	2 1/4	57	3 27/32	97.633
3/4	.750	19.050	1/2	12.70	2 1/4	57	3 29/32	99.217
7/8	.875	22.235	1/2	12.70	2 1/4	57	4	101.60
1.0	1	25.40	1/2	12.70	2 1/4	57	4 1/4	107.95

Table 38. Chamfer Dimensions for Countersinks. *(Source, Fastcut Tool Corp.)*

60°, 82°, 90° and Included Angles				
Body Dia.	Shank Dia.	No. Flutes	Min. Dia. Chamfer	Max Dia. Chamfer
1/4	1/4	1	.030	.250
1/4	1/4	3	.093	.250
3/8	1/4	1	.035	.375
3/8	1/4	3	.093	.375
1/2	3/8	1	.045	.500
1/2	3/8	3	.093	.500
5/8	3/8	1	.050	.625
5/8	3/8	3	.125	.625
3/4	1/2	1	.055	.750
3/4	1/2	3	.125	.750
1	1/2	1	.065	1.000
1	1/2	3	.125	1.000
1 1/2	3/4	1	.075	1.500
1 1/2	3/4	4	.156	1.500

All dimensions in inches.

Table 39. Fluteless Countersinks and Deburring Tools. *(Source, Weldon Tool Company.)*

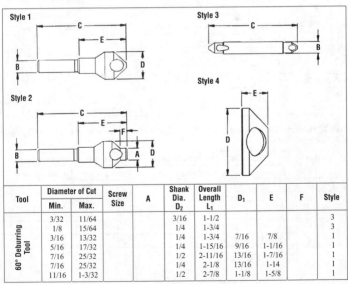

Tool	Diameter of Cut		Screw Size	A	Shank Dia. D_2	Overall Length L_1	D_1	E	F	Style
	Min.	Max.								
60° Deburring Tool	3/32	11/64			3/16	1-1/2				3
	1/8	15/64			1/4	1-3/4				3
	3/16	13/32			1/4	1-3/4	7/16	7/8		1
	5/16	17/32			1/4	1-15/16	9/16	1-1/16		1
	7/16	25/32			1/2	2-11/16	13/16	1-7/16		1
	7/16	25/32			1/4	2-1/8	13/16	1-14		1
	11/16	1-3/32			1/2	2-7/8	1-1/8	1-5/8		1

(Continued)

Table 39. *(Continued)* **Fluteless Countersinks and Deburring Tools.** *(Source, Weldon Tool Company.)*

Tool	Diameter of Cut Min.	Diameter of Cut Max.	Screw Size	A	Shank Dia. D_2	Overall Length L_1	D_1	E	F	Style
82° Countersink			1/4	1/4	3/8	2	33/64	1	5/32	2
			1/4	17/64	3/8	2	33/64	1	5/32	2
			5/16	5/16	3/8	2-3/16	41/64	1-3/16	5/32	2
			5/16	21/64	3/8	2-3/16	41/64	1-3/16	5/32	2
			3/8	3/8	1/2	2-5/8	49/64	1-3/8	5/32	2
			3/8	25/64	1/2	2-5/8	49/64	1-3/8	5/32	2
			7/16	7/16	1/2	2-3/4	53/64	1-1/2	7/32	2
			7/16	29/64	1/2	2-3/4	53/64	1-1/2	7/32	2
			1/2	1/2	1/2	2-13/16	15/16	1-9/16	1/4	2
			1/2	33/64	1/2	2-13/16	15/16	1-9/16	1/4	2
82° Pilotless Countersink	0.073	15/64			1/4	1-3/4				3
	3/16	17/32			1/4	1-7/8	9/16	1		1
	0.307	25/32			1/2	2-5/8	13/16	1-3/8		1
	0.307	25/32			1/4	2-1/16	13/16	1-1/4		1
	0.0431	7/8			1/2	2-13/16	1-1/8	1-9/16		1
	1/2	1-15/32			1/2	3-1/2	1-1/2	2-1/8		1
			Nos. 4,5		1/4	1-1/2	17/64	5/8		1
			No. 6		1/4	1-1/2	19/64	5/8		1
			No. 8		1/4	1-5/8	11/32	3/4		1
			No. 10		1/4	1-5/8	25/64	3/4		1
90° Deburring Tool	1/64	1/64			3/16	1-1/2				3
	1/16	11/64			3/16	1-1/2				3
	1/8	15/64			1/4	1-3/4				3
	5/32	13/32			1/4	1-11/16	7/16	13/16		1
	3/16	17/32			1/4	1-7/8	9/16	1		1
	5/16	25/32			1/2	2-19/32	13/16	1-3/8		1
	5/16	25/32			1/4	2-1/32	13/16	1-5/32		1
	9/16	1-3/32			1/2	2-13/16	1-1/8	1-9/16		1
	1/2	1-15/32			1/2	3-1/2	1-1/2	2-1/8		1
	1-1/16	2					2-1/8	7/8		4
	2	3					3-1/8	1		4
100° Countersink	3/32	1/4			1/4	1-1/2	17/64	5/8		1
	1/8	27/64			1/4	1-11/16	7/16	13/16		1
	3/16	35/64			1/4	1-7/8	9/16	1		1

All dimensions in inches.

Table 40. Combined Drills and Countersinks. Sizes 000 through 0 available in standard length with 60° countersink angle. Sizes 1 through 8 available in standard and extra lengths with a countersink angle of 60°. Sizes 11 through 20 available in standard length Bell type with a 60° countersink and a 120° chamfer. *(Source, Kennametal Inc.)*

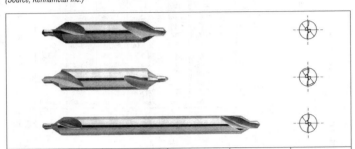

Size	Overall Diameter		Drill Diameter		Drill Length		Overall Length		Bell Diameter	
	inch	mm	inch	mm	inch	mm	inch	mm	inch	mm
000	1/8	3.175	.020	0.508	.030	.762	1 1/4	32	-	-
00	1/8	3.175	.025	0.640	.030	.762	1 1/8	29	-	-
0	1/8	3.175	1/32	0.794	.038	.965	1 1/8	29	-	-
1	1/8	3.175	3/64	1.191	3/64	1.191	1 1/4	32	-	-
1	1/8	3.175	3/64	1.191	3/64	1.191	3	76	-	-
1	1/8	3.175	3/64	1.191	3/64	1.191	4	102	-	-
2	3/16	4.762	5/64	1.984	5/64	1.984	1 7/8	48	-	-
2	3/16	4.762	5/64	1.984	5/64	1.984	4	102	-	-
2	3/16	4.762	5/64	1.984	5/64	1.984	5	127	-	-
3	1/4	6.350	7/64	2.778	7/64	2.778	2	51	-	-
3	1/4	6.350	7/64	2.778	7/64	2.778	4	102	-	-
3	1/4	6.350	7/64	2.778	7/64	2.778	5	127	-	-
4	5/16	7.938	1/8	3.175	1/8	3.175	2 1/8	54	-	-
4	5/16	7.938	1/8	3.175	1/8	3.175	4	102	-	-
4	5/16	7.938	1/8	3.175	1/8	3.175	5	127	-	-
4	5/16	7.938	1/8	3.175	1/8	3.175	6	152	-	-
5	7/16	11.112	3/16	4.762	3/16	4.762	2 3/4	70	-	-
5	7/16	11.112	3/16	4.762	3/16	4.762	5	127	-	-
5	7/16	11.112	3/16	4.762	3/16	4.762	6	152	-	-
6	1/2	12.70	7/32	5.556	7/32	5.556	3	76	-	-
6	1/2	12.70	7/32	5.556	7/32	5.556	5	127	-	-
6	1/2	12.70	7/32	5.556	7/32	5.556	6	152	-	-
7	5/8	15.875	1/4	6.350	1/4	6.350	3 1/4	83	-	-
7	5/8	15.875	1/4	6.350	1/4	6.350	6	152	-	-
8	3/4	19.050	5/16	7.938	5/16	7.938	3 1/2	89	-	-
8	3/4	19.050	5/16	7.938	5/16	7.938	6	152	-	-

(Continued)

Table 40. *(Continued)* **Combined Drills and Countersinks.** Sizes 000 through 0 available in standard length with 60° countersink angle. Sizes 1 through 8 available in standard and extra lengths with a countersink angle of 60°. Sizes 11 through 20 available in standard length Bell type with a 60° countersink and a 120° chamfer. *(Source, Kennametal Inc.)*

Size	Overall Diameter		Drill Diameter		Drill Length		Overall Length		Bell Diameter	
	inch	mm	inch	mm	inch	mm	inch	mm	inch	mm
11	1/8	3.175	3/64	1.191	3/64	1.191	1 1/4	32	0.10	2.540
12	3/16	4.762	1/16	1.588	1/16	1.588	1 7/8	48	0.150	3.810
13	1/4	6.350	3/32	2.383	3/32	2.383	2	51	0.20	5.080
14	5/16	7.938	7/64	2.778	7/64	2.778	2 1/8	54	0.250	6.350
15	7/16	11.112	5/32	3.969	5/32	3.969	2 3/4	70	0.35	8.890
16	1/2	12.70	3/16	4.762	3/16	4.762	3	76	0.40	10.16
17	5/8	15.875	7/32	5.556	7/32	5.556	3 1/4	83	0.50	12.70
18	3/4	19.050	1/4	6.350	1/4	6.350	3 1/2	89	0.60	15.24
19	7/8	22.225	5/16	7.938	5/16	7.938	3 5/8	92	0.70	17.78
20	1	25.40	3/8	9.525	3/8	9.525	3 3/4	95	0.80	20.32

Spade drilling

Spade drills are commonly used to produce deep holes, and larger holes ranging from 1" (25.4 mm) to 6" (152.4 mm) in diameter. Generally, they are not regarded as precision tools, and should not be used for finishing operations requiring tolerances below ± 0.010" (0.254 mm). Since they employ special holders (or shanks), blades may be quickly replaced on the machine, saving setup time on NC machines and allowing for many different diameter holes simply by interchanging blades in the holder. Advantages of spade drills over traditional twist drills include increased rigidity, lower initial cost, and their ability to drill deep holes. Chip removal can be problematic when vertical drilling, and it is import that adequate fluid is injected from the outside when using holders without coolant provision. Compared to twist drills, spade drills have a larger cutting angle (130° versus 118°) which shortens the cutting edge, and increases the depth of cut, chip thickness, and force required for cutting. These disadvantages are offset somewhat by the lower operating speeds required by spade drills. The cutting edge of the blade features chip breaker notches (also called chip "splitters") that divide what would naturally be wide, difficult to remove chips, into several smaller chips that can be disposed of more easily. The notches vary in size depending on blade diameter. Blades 1" to 2.31" in diameter have notches 0.050" in width. For blades 2.32" to 3.00", the notches are 0.070". For blades 3.01" to 4.00", the notches are 0.090", and for blades 4.00" and larger, they are 0.120" in width. For most applications, the clearance (or relief) face has a relief angle of 6 to 8°, but it can be increased to to 10 to 12° for drilling easily machined materials such as aluminum. Angles in excess of 12° are fragile and should only be used for the softest materials.

Feed selection is critical, as a feed rate that is too light can cause breakage at the drill point due to instability of the chisel edge. This instability causes the drill to "walk" or pull to one side of the hole. Also, starting holes can be difficult because the 130° point angle of the spade drill is incompatible with spotting done with a center drill. For large holes, a smaller spade drill can be used for centering by cutting a hole approximately 1/8" (3.175 mm) deep to guide the larger blade. Rigidity of the machine is critical to maximizing feed rates, and deeper holes will require reduced feeds. When the drill's overhang is greater than four times its diameter, feed should be reduced by 10%. If overhang is in excess of eight times diameter, reduce feed by 25%. High cutting feeds and speeds require sufficient coolant. Heat must be kept to a minimum and chips must be flushed from the hole. Except when flood cooling with short holders, coolant should be supplied through the holder, directly onto the spade blade cutting edge.

Speeds and feeds for spade drilling

Recommended speeds and feeds shown in **Table 41** are for machines of adequate rigidity. For heavy duty machines with adequate power, rates can be increased by 20%. For older machines, vertical machines such as radial arm presses, fragile setups, and light duty machines, feed rates should be reduced by as much as 20%. The specific energy factor (K) given on the table is used for estimating horsepower and thrust requirements later in this section.

Table 41. Recommended Speeds and Feeds for Spade Drilling with Coated Blades. *(Source, Kennametal Inc.)*

Material	Hardness (BHN)	Speed		Unit Power Factor (K)		Feed Rate in./rev *or* mm/rev			
		sfm	smm	inch	metric	1-2 1/2"	25-38 mm	2 1/2" +	39 mm +
Standard carbon steel and alloy steel	up to 225	110	33	0.80	0.036	.014	.356	.023	.559
	225–300	70	20	1.00	0.046	.012	.305	.019	.483
	300–350	60	17	1.20	0.055	.010	.254	.017	.432
Ferritic stainless steel	up to 150	100	31	0.90	0.041	.015	.381	.023	.584
	150–200	80	21	0.95	0.043	.012	.305	.020	.508
Austenitic stainless steel	up to 225	60	17	0.95	0.043	.015	.381	.024	.610
	225–300	50	15	1.00	0.046	.012	.305	.020	.508
Cast iron: nodular, ductile, or malleable	up to 150	150	45	0.50	0.022	.020	.508	.039	.991
	150–250	100	31	0.55	0.025	.016	.406	.028	.711
	250–350	80	21	1.10	0.050	.010	.254	.018	.457
Aluminum and magnesium	Up to 100	450	137	0.17	0.008	.020	.508	.038	.965
	100–150	390	119	0.21	0.010	.018	.457	.033	.838
High temperature alloys	up to 200	70	20	1.40	0.064	.012	.305	.018	.457
	200–350	50	15	1.45	0.066	.012	.305	.017	.432
Titanium alloys	up to 220	90	27	0.60	0.027	.007	.178	.013	.076
	220–400	40	12	0.75	0.034	.006	.152	.011	.279

Figures quoted are for titanium nitride coated spade blades. HSS blades are not recommended for many of the materials on this Table.

Power and thrust requirements for spade drilling

Spade drilling consumes more power than when using twist or indexable drills, and no operation should be performed without assuring that the machine and setup is sufficiently rigid, and that the machine can perform to the required limits for power, torque, and thrust. Power requirements at the spindle can be calculated using a percentage of tool sharpness, the unit power factor (K), the drill diameter, and the speed and feed factors.

Example. Find power requirements (in horsepower and kilowatts) at the spindle for a 2 1/4" (57.15 mm) spade drill cutting ferritic stainless steel (140 BHN) at 191 rpm (100 sfm *or* 31 smm) with a feed of 0.015 in./rev (0.381 mm/rev). From the above table, the K factor for this material is 0.9 in English units, 0.041 in metric units.

$$P_c = ([\pi \ D^2] \div 4) \times rpm \times feed \times K$$

$$or \quad kW_c = ([p \times D^2] \div 4000) \times rpm \times feed \times K$$

Where: D^2 = diameter of the drill in inches or millimeters, squared
feed = in./rev or mm/rev
K = unit power factor from Table 40

HP_c = ([3.1416 × 5.0625] ÷ 4) × 191 × 0.015 × 0.9 = 10.25 HP_c

or kW_c = ([3.1416 × 3266.12] ÷ 4000) × 191 × 0.381 × 0.041 = 6.65 kW_c.

Spade drilling is among the most power-consuming machining operations, and tool sharpness can have a dramatic effect on calculations. The above formulas are for sharp drills. For dull tools, the required power should be multiplied by 1.25. Power at the cutter can be converted to power at the motor by multiplying by the drive train factor of 1.1.

Approximate thrust requirements for drilling with a spade blade using the K factor and the following formula.

Thrust (lb) = 148,500 × D × feed × K   *note:* 1 lb = 0.4536 kg
or   thrust (kg) = 2292 × D × feed × K   1 kg = 2.2046 lb

Example. Using the operating parameters from the above example, find thrust in pounds and kilograms.

148,500 × 2.25 × 0.015 × 0.9 = 4511 lb   *note:* 148,500 is a constant for
                                                       inch units
or 2294.65 × 57.15 × 0.381 × 0.041 = 2048 kg   2294.65 is a constant for
                                                       metric units

For dull blades, thrust calculations should be multiplied by 1.25.

Troubleshooting spade blades

The high power and thrust requirements of spade drilling contribute to tool failure. Drill point wobble and drilling oversize holes can be caused by uneven webs, and care must be taken to remove equal amounts of stock from both webs when regrinding. Oversize holes can also be caused by an off-center drill that may have unequal cutting edge lengths, or a misaligned holder. Improper positioning of the blade in the grinding fixture when resharpening is often the cause. Insufficient chip removal or poor heat dissipation will result in galling and loading on the drill blade, caused by material annealing to the cutting edge and top rake angle. Changing to a better rake angle and improving coolant flow will solve the problem. Unequal blade angles will cause the drill to pull to one side, resulting in bell-mouth holes and accelerated dulling. Causes can be incorrect blade location in the holder, misalignment of the holder, or improper positioning of the blade in the grinding fixture when resharpening.

Drilling at excessive speed, or the continued use of a dull blade, will produce worn corners and discoloration. Excessive feed promotes abnormal end thrust that leads to a breakdown of the point and cutting edges and eventual tool failure. When experiencing premature dulling of the blade, the problem may be a too-large front clearance angle, or drilling at excessive speeds. A too-small a front clearance angle generates excessive heat and increases end thrust. In this condition, the blade will rub on the front clearance angle and produce work-hardened holes.

Indexable insert drills

Indexable insert drills use a complex design that is simple in concept. While most drills use two inserts, larger sizes have four. To perform the drilling operation, the inside insert cuts the inner portion of the hole while the outside insert, which is responsible for the final hole diameter, cuts the outer portion. The inside insert cuts a path that amounts to about

(Text continued on p. 1137)

Table 42. Long and Short Series Straight Shank Blade Holders With Coolant Provision.
(Source, Kennametal Inc.)

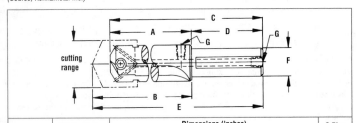

Series	Cutting Range (inches)	Dimensions (inches)						G Pipe Tap
		F	A	B	C	D	E	
AA	31/32 – 1 3/8	1	3	3.56	6.50	3.50	7.06	1/8
AA	31/32 – 1 3/8	1	3	3.56	6.50	3.50	7.06	1/8
AA	31/32 – 1 3/8	1	7.75	8.31	11.25	3.50	11.88	1/8
AA	31/32 – 1 3/8	1.50	7.75	8.31	11.25	3.50	11.88	1/8
A	31/32 – 1 3/8	1	3	3.56	6.50	3.50	7.06	1/8
A	31/32 – 1 3/8	1.50	3	3.56	6.50	3.50	7.06	1/8
A	31/32 – 1 3/8	1	7.75	8.31	11.25	3.50	11.88	1/8
A	31/32 – 1 3/8	1.50	7.75	8.31	11.25	3.50	11.88	1/8
B	1 1/4 – 1 3/4	1	3.50	4.09	7	3.50	7.59	1/4
B	1 1/4 – 1 3/4	1.50	3.50	4.09	7	3.50	7.59	1/4
B	1 1/4 – 1 3/4	1.250	8.12	8.71	11.62	3.50	12.21	1/4
B	1 1/4 – 1 3/4	1.50	8.12	8.71	11.62	3.50	12.21	1/4
C	1 1/2 – 2 3/8	1	4	5	8	4	9	1/4
C	1 1/2 – 2 3/8	1.25	4	5	8	4	9	1/4
C	1 1/2 – 2 3/8	1.50	4	5	8	4	9	1/4
C	1 1/2 – 2 3/8	1.50	8.50	9.50	12.50	4	13.50	1/4
D	2 – 2 7/8	1.50	4.50	5.50	8.50	4	9.50	1/4
D	2 – 2 7/8	1.750	4.50	5.50	8.50	4	9.50	1/4
D	2 – 2 7/8	1.750	9	10	13	4	14	1/4
E	2 1/2 – 3 3/8	1.750	5	6.06	9	4	10.06	1/2
E	2 1/2 – 3 3/8	2	5	6.06	9	4	10.06	1/2
E	2 1/2 – 3 3/8	2	10	11.06	14.0	4	15.06	1/2
F	3 – 3 7/8	2	5.50	6.75	9.50	4	10.75	1/2
F	3 – 3 7/8	2.50	5.50	6.75	9.50	4	10.75	1/2
F	3 – 3 7/8	2.50	11.50	12.75	15.50	4	16.75	1/2
G	3 1/2 – 4 1/2	3	6	7.44	11	5	12.44	1/2
G	3 1/2 – 4 1/2	3	13	14.43	18	5	19.43	1/2
H	4 – 6	3	7	8.56	13	6	14.56	1/2
H	4 – 6	3	15	16.56	21	6	22.56	1/2

All dimensions in inches.

Table 43. Short Series Morse Taper Shank Blade Holders Without Coolant Provision.
(Source, Kennametal Inc.)

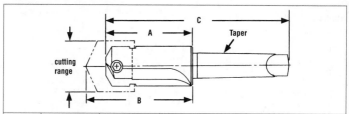

Series	Cutting Range (inches)	Morse Taper #	Dimensions (inches)		
			A	B	C
AA	31/32 – 1 3/8	3	3	3.75	6.87
A	31/32 – 1 3/8	3	3	3.75	6.87
B	1 1/4 – 1 3/4	4	3.50	4.34	8.37
C	1 1/2 – 2 3/8	4	4	5.25	8.87
D	2 – 2 7/8	4	4.50	5.75	9.37
E	2 1/2 – 3 3/8	5	5	6.31	11.12
F	3 – 3 7/8	5	5.50	7	11.62
G	3 1/2 – 4 1/2	5	6	7.68	12.12
H	4 – 6	6	7	8.87	15.56

All dimensions in inches.

Table 44. Extra Long, Long, and Short Series Morse Taper Shank Blade Holders With Coolant Provision.
(Source, Kennametal Inc.)

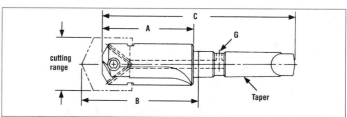

Series	Cutting Range (inches)	Morse Taper #	Dimensions (inches)			G Pipe Tap
			A	B	C	
AA	31/32 – 1 3/8	3	3	5.15	8.29	1/8
AA	31/32 – 1 3/8	4	7.75	10.25	14.31	1/4
AA	31/32 – 1 3/8	4	15	17.50	21.56	1/4
A	31/32 – 1 3/8	3	3	5.15	8.28	1/8
A	31/32 – 1 3/8	4	7.75	10.25	14.31	1/4
A	31/32 – 1 3/8	4	15.00	17.50	21.56	1/4
B	1 1/4 – 1 3/4	4	3.50	6.03	10.06	1/4
B	1 1/4 – 1 3/4	4	8.12	10.65	14.68	1/4
B	1 1/4 – 1 3/4	4	15	17.53	21.56	1/4

(Continued)

Table 44. *(Continued)* **Extra Long, Long, and Short Series Morse Taper Shank Blade Holders With Coolant Provision.** *(Source, Kennametal Inc.)*

Series	Cutting Range (inches)	Morse Taper #	Dimensions (inches)			G Pipe Tap
			A	B	C	
C	1 1/2 – 2 3/8	4	4	6.93	10.56	1/4
C	1 1/2 – 2 3/8	4	8.50	11.43	15.06	1/4
C	1 1/2 – 2 3/8	4	18.00	20.94	24.56	1/4
D	2 – 2 7/8	4	4.50	7.43	11.06	1/4
D	2 – 2 7/8	4	9	11.93	15.56	1/4
D	2 – 2 7/8	4	18	20.94	24.56	1/4
E	2 1/2 – 3 3/8	5	5	8.62	13.43	1/2
E	2 1/2 – 3 3/8	5	10	13.62	18.43	1/2
F	3 – 3 7/8	5	5.50	9.31	13.93	1/2
F	3 – 3 7/8	5	11.50	15.31	19.93	1/2
G	3 1/2 – 4 1/2	5	6	10	14.43	1/2
G	3 1/2 – 4 1/2	5	13	17	21.43	1/2
H	4 – 6	6	7	11.18	17.87	1/2
H	4 – 6	6	15.00	19.18	25.87	1/2

All dimensions in inches.

Table 45. Dimensions of Spade Drill Blades. Dimensions Apply to both Style "S" and Style "F" Blades. *(Source, Kennametal Inc.)*

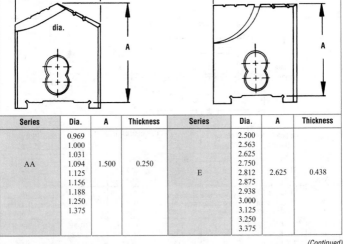

Series	Dia.	A	Thickness	Series	Dia.	A	Thickness
AA	0.969 1.000 1.031 1.094 1.125 1.156 1.188 1.250 1.375	1.500	0.250	E	2.500 2.563 2.625 2.750 2.812 2.875 2.938 3.000 3.125 3.250 3.375	2.625	0.438

(Continued)

Table 45. *(Continued)* Dimensions of Spade Drill Blades. Dimensions Apply to both Style "S" and Style "F" Blades. *(Source, Kennametal Inc.)*

Series	Dia.	A	Thickness	Series	Dia.	A	Thickness
A	0.969 1.000 1.031 1.063 1.125 1.188 1.219 1.250 1.313 1.375	2.000	0.188	F	3.000 3.125 3.250 3.375 3.500 3.625 3.750 3.875	3.312	0.500
B	1.250 1.281 1.344 1.375 1.469 1.500 1.563 1.625 1.688 1.750	1.500	0.281	G	3.500 3.625 3.750 3.875 4.000 4.125 4.250 4.375 4.000	3.625	0.625
C	1.500 1.563 1.594 1.625 1.719 1.750 1.813 1.875 1.969 2.000 2.125 2.375	2.000	0.312	H	4.000 4.125 4.250 4.375 4.500 4.625 4.750 4.875 5.000 5.250 5.375 5.500 5.625 5.750 5.875 6.000	3.750	0.688
D	2.000 2.063 2.125 2.188 2.250 2.313 2.375 2.438 2.500 2.625 2.750 2.875	2.250	0.375				

All dimensions in inches.

30% of the total area, the outside insert removes the remaining 70%. Insert placement varies with drill type, but two basic configurations are used. In *Figure 11*, Drill A uses square inserts, with the inside insert leading the cutting edge of the outside insert to cut a clearance path for the lead corner of the outside insert. In *Figure 12*, Drill B uses trigon inserts, with the outside insert leading the inside insert. In both cases, the overlapping asymmetrically placed inserts work together to provide high productivity when drilling larger diameter holes. Another feature of indexable insert drills is that different grades

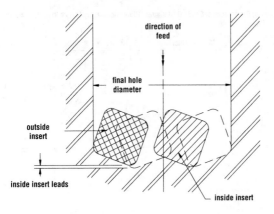

Figure 11. Drill A. Indexable insert drill with square inserts.

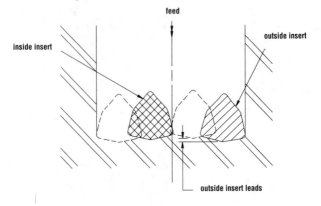

Figure 12. Drill B. Indexable insert drill with trigon inserts.

of inserts can be combined to optimize machining procedures. The illustrations in this section were provided by Kennametal Inc.

These drills reduce machine cycle time by producing higher penetration rates, lowering friction, and allowing for chip control. They were developed for use with NC machinery where these attributes allow the drill to make maximum use of the machine's available power. In addition to drilling applications, they can perform boring, facing, and turning operations (see *Figure 13*). As in all drilling operations, chip evacuation is enhanced with a steady supply of coolant. The coolant evacuates chips and dissipates heat, thus assuring adequate coolant pressure and volume will prolong tool life.

Indexable insert drills must be accurately aligned. To check alignment, insert a ground bar (without runout) into the toolholder. Mount a magnetic dial base indicator on the

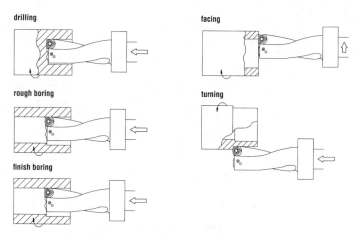

Figure 13. Operations that can be performed with drills with trigon inserts.

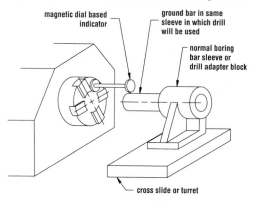

Figure 14. Indexable drill alignment.

spindle and take four readings to determine if the ground bar is above, below, left, or right of center. Also check angularity. On nonrotating applications, alignment must be within ± 0.005 and the angularity must be within ± 0.005 over a six-inch length. If the drill is rotating, place the dial indicator on the table and rotate the spindle. For rotating applications, runout should not exceed ± 0.005 tir. To check final alignment: 1) take four readings by rotating the chuck; 2) adjust cross slide or offsets (shimming the tool may be required); 3) take four more readings to check results and 4) replace ground bar with drill and make final check on the actual drill. See *Figure 14* for setup.

Drill misalignment (see *Figure 15a–c*) will result in cutting the wrong diameter hole. If the drill is to the left of center, both inserts will move to the left and the diameter of

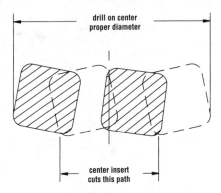

Figure 15a. Correctly centered drill.

If drill is to the left of center, both inserts move to the left.

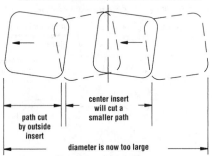

Figure 15b. Drill misaligned to left of center.

If drill is to the right of center, both inserts move to the right.

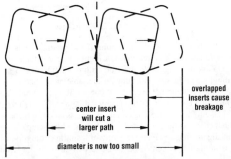

Figure 15c. Drill misaligned to right of center.

the hole will increase. The center insert will cut a smaller path and the gap between the outside and center insert can cause breakage. If the drill is to the right of center, both inserts will move to the right and the diameter of the hole will decrease. The center insert will cut a larger path, and the insert overlap could cause breakage. The smaller diameter could also cause rubbing on the outside diameter. **Table 46** provides troubleshooting guidelines for indexable drilling.

Table 46. Troubleshooting Guide for Indexable Drilling. *(Source, Kennametal Inc.)*

Problem	Source	Solution
Insert chipping (especially on inner insert) or breakage.	Off-center drill (misalignment).	Check alignment. Concentricity not to exceed ± 0.005 tir.
	Improper seating of tool in toolholder spindle or turret.	Check tool shank and socket for nicks and dirt. Check parting line between tool shank and socket with feeler gage. Make sure tool is locked tight.
	Deflection due to too much overhang and lack of rigidity.	Check with indicator to see if tool can be moved by hand. Check if tool can be held shorter.
	Improper seating of inserts in pocket.	Clean pockets whenever indexing or changing inserts. Check pockets for nicks and burrs. Make sure inserts rest completely on pocket bottoms.
	Damaged insert screws.	Check head and thread for nicks and burrs. Do not overtorque screws.
	Improper feeds and speeds.	Check recommended guidelines for given material.
	Insufficient coolant supply.	Check coolant flow.
	Improper carbide grade used on inboard station.	Use straight grade for multiple insert drills.
Grooving on back stroke. Drill body rubbing hole wall. Oversize or undersize holes.	Off-center drill.	Restore drill to proper alignment and concentricity. Check bottom of hole or disk for center stub.
	Deflection.	Check setup rigidity. Check speed/feed guidelines.
Rough cutting action (tool rumbles and deflects).	Too much thrust. Feed rate too high.	Lower feed and/or increase speed.
	Recutting chips.	Increase coolant flow. Add coolant grooves to tool.
Poor hole surface finish.	Vibration.	Check setup and tool rigidity. Check tool set in spindle or toolholder. Check speeds and feeds.
	Insufficient coolant pressure and volume.	Increase coolant pressure and flow. Check coolant flow for interruptions. Make sure coolant reaches inserts at all times.
	Recutting chips.	Increase coolant flow. Add coolant grooves to tool.
	Poor chip control. Chips trapped in hole.	Modify speed or feed.
	Chatter.	Modify feed rate.
Very short, thick, flat chips.	Feed rate too high in relation to cutting speed.	Lower feed rate and/or increase speed.
Long, stringy chips.	Feed rate too low in relation to cutting speed.	Increase feed rate and/or decrease speed, or use dimple inserts.
Can't loosen insert locking screws.	Seized threads due to insufficient coolant or high heat buildup.	Apply water- and heat-resistant lubricant to threads.

Speeds and feeds for indexable drilling

Table 47 provides recommended speeds and feeds for indexable drilling operations. This data is the result of testing and experimentation carried out by Kennametal, Inc.

Table 47. Recommended Speeds and Feeds for Indexable Drilling. *(Source, Kennametal Inc.)*

Material	Feed		Speed	
	in./rev	mm/rev	sfm	smm
Steel				
Armor plate	.005-.015	.1270-.381	125-175	38-53
Carbon steel resulfurized B-1112CD B-1113CD RSC-12613 (leaded) C-11L17 (leaded) C-1117CD	.0035-.007	.0889-.1778	700-1400	213-427
Low carbon C1137-C1145 cold drawn-hot rolled C1008-C1035 C10L45	.0035-.007	.0889-.1778	500-1200	152-366
Medium carbon C1045 1095 4140-4340 hot rolled 5145-5160 6150	.0035-.007	.0889-.1778	500-1000	152-305
High carbon 4340 cold drawn E-6180 nitralloy-135 52100 C1050 hot rolled	.007-.012	.1778-.3048	200-380	61-115
Cast steels	.005-.015	.1270-.381	240-500	73-152
High manganese steel	.015-.020	.381-.3048	50-125	15-38
High-speed steel	.005-.015	.1270-.381	200-350	61-107
Hot work diesteels	.005-.015	.1270-.381	250-350	76-107
Tool steel	.005-.015	.1270-.381	200-300	61-91
Stainless Steel				
Ferritic 405 430F 442 CC50	.004-.009 .003-.008 .003-.008	.1016-.2286 .0762-.2032 .0762-.2032	450-510 485-600 350-410	137-155 147-183 107-125
Martensitic 410 CA40 440	.003-.007 .004-.009 .003-.007	.0762-.1778 .1016-.2286 .0762-.1778	420-510 360-400 275-325	127-155 110-122 84-99
Precipitation hardening 350 17-7	.006-.009 .004-.009	.1524-.2286 .1016-.2286	125-175 140-200	38-53 42-61

(Continued)

Table 47. *(Continued)* **Recommended Speeds and Feeds for Indexable Drilling.** *(Source, Kennametal Inc.)*

Material	Feed		Speed	
	in./rev	mm/rev	sfm	smm
Aluminum				
Wrought				
2011-T3	.003-.008	.0762-.2032	2400-3000	731-915
3003-5052-5056	.003-.006	.0762-.1524	1800-2600	549-783
4032	.004-.008	.1016-.2032	800-1400	244-427
2014-2018-6151	.003-.008	.0762-.2032	1600-2200	488-671
6061-7075	.003-.008	.0762-.2032	1000-1600	305-488
Cast				
108-113-319	.003-.018	.0762-.4572	1600-2000	488-610
220-195-750	.003-.018	.0762-.4572	1800-2400	549-790
122-319-355-356	.004-.009	.1016-.2286	1200-1800	366-549
132	.004-.009	.1016-.2286	800-1400	244-427
Beryllium	.002-.005	.0508-.1270	200-250	61-76
Cast Iron				
Gray iron				
Class 30 ferritic	.006-.012	.1524-.3048	475-650	145-198
Class 40 pearlitic	.009-.012	.2286-.3048	290-380	88-116
Nodular				
60-45-10 ferritic	.006-.013	.1524-.3302	650-800	198-244
80-60-03 pearlitic	.006-.012	.1524-.3048	460-600	140-183
100-70-03 H.T.	.004-.010	.1016-.254	320-360	97-110
Meehanite 6A	.006-.013	.1524-.3302	650-850	143-259
Malleable				
53004 (85M)	.004-.011	.1016-.2794	400-520	122-157
35018 (86M)	.007-.013	.1778-.3302	600-750	183-229
Chilled				
60 shore	.007-.015	.1778-.3810	175-300	53-91
75 shore	.007-.016	.1778-.4064	125-225	38-69
Cobalt Base Alloys				
Stelite #6	.005-.008	.1270-.2032	30-40	9-12
Stelite #19	.003-.005	.0762-.1270	20-30	6-9
Stelite #21	.005-.008	.1270-.2032	50-65	15-19
Stelite #31	.005-.010	.1270-.254	50-80	15-24
Haynes alloy #25	.005-.010	.1270-.254	80-120	24-37
J-1570 (G.E.)	.003-.005	.0762-.1270	50-60	15-17
G-34 (Jessup)	.005-.008	.1270-.2032	30-50	9-15
Refractory 70	.005-.010	.1270-.254	90-130	27-40
Refractory 80	.003-.008	.0762-.2032	80-120	24-37
Rexalloy 33	.002-.003	.0508-.0762	25-40	7-12
S-816 (Allegheny)	.005-.010	.1270-.254	50-100	15-31
X-40	.005-.010	.1270-.254	50-100	15-31
Copper Alloys				
240 copper zinc	.004-.009	.1016-.2286	400-650	122-198
260	.006-.010	.1524-.254	500-880	152-268
268	.003-.009	.0762-.2286	600-900	183-274
280	.006-.010	.1524-.254	450-700	137-213
220	.003-.009	.0762-.2286	450-700	137-213
464	.003-.009	.0762-.2286	500-880	152-268
102 oxygen free	.0015-.006	.0381-.1524	1000-1600	305-488
172 beryllium copper	.0015-.006	.0381-.1524	600-900	183-274

(Continued)

Table 47. *(Continued)* **Recommended Speeds and Feeds for Indexable Drilling.** *(Source, Kennametal Inc.)*

Material	Feed		Speed	
	in./rev	mm/rev	sfm	smm
Leaded				
314 com. bronze	.003-.008	.0762-.2032	600-900	183-274
330 low brass	.003-.009	.0762-.2286	550-700	168-213
360 free cutting	.002-.006	.0508-.1524	1000-1600	305-488
353 high leaded	.002-.007	.0508-.1778	900-1200	274-366
366 Muntz metal	.003-.009	.0762-.2286	700-1000	213-305
High strength				
655 silicon bronze	.004-.009	.1016-.2286	450-700	137-137
675 Mn bronze	.002-.008	.0508-.2032	350-600	107-183
544 phos. bronze	.002-.008	.0508-.2032	550-800	168-268
614 Al bronze	.004-.009	.1016-.2286	300-550	91-168
620 Al bronze	.004-.009	.1016-.2286	285-470	86-143
Cast alloys				
red brass	.003-.009	.0762-.2286	800-1100	268-335
yellow brass	.003-.009	.0762-.2286	800-1100	268-335
bearing bronze	.002-.006	.0508-.1524	1100-1300	335-396
silicon bronze	.002-.006	.0508-.1524	1100-1300	335-396
Al bronze	.002-.006	.0508-.1524	1100-1300	335-396
Nickel Base Alloys				
Monel R	.004-.010	.1016-.254	250-375	76-119
Monel KR	.004-.010	.1016-.254	250-375	76-119
Monel K	.004-.010	.1016-.254	200-300	61-91
Monel S	.003-.008	.0762-.2032	150-250	45-76
Inconel	.004-.010	.1016-.254	160-200	47-51
Inconel 901	.004-.010	.1016-.254	125-160	38-48
Nimoric 75	.004-.010	.1016-.254	100-125	31-38
Inconel X	.004-.010	.1016-.254	60-90	17-27
Nimonic 80A	.004-.008	.1016-.2032	60-90	17-27
Hastelloy B	.005-.008	.1270-.2032	60-90	17-27
Nimonic 90	.005-.008	.1270-.2032	60-80	17-24
Hastelloy C	.005-.008	.1270-.2032	65-90	19-27
Udimet 500	.003-.008	.0762-.2032	75-120	22-37
Inconel 700	.003-.008	.0762-.2032	50-65	15-19
713	.003-.008	.0762-.2032	20-40	6-12
R-235	.004-.008	.1016-.2032	50-75	15-22
Rene 41	.004-.007	.1016-.1778	60-100	17-31
Astroloy	.004-.010	.1016-.254	60-75	17-22
Waspaloy	.003-.008	.0762-.2032	75-125	22-38
J-1500	.003-.008	.0762-.2032	60-100	17-31
Chromalloy	.004-.012	.1016-.3048	250-350	76-107
J-1300 (M308)	.005-.015	.1270-.381	100-140	31-43
Lapelloy	.004-.012	.1016-.3048	275-375	83-119
M-308	.004-.012	1016-.3048	90-150	27-45
Timken 16-25-6	.005-.015	.1270-.381	90-180	27-41
HS-21	.006-.009	.1524-.2286	75-120	22-36
HS-31	.006-.009	.1524-.2286	60-90	17-27
HS-30	.006-.009	.1524-.2286	70-100	20-31
Hastelloy X	.003-.015	.0762-.381	110-180	33-41
Hastelloy W	.003-.015	.0762-.381	90-150	27-45
Inconel 600	.003-.015	.0762-.381	160-200	47-61
K-42-B	.003-.015	.0762-.381	30-50	9-15
N-155	.006-.012	.1524-.3048	70-100	20-31
S-588	.006-.012	.1524-.3048	60-80	17-24

(Continued)

Table 47. (Continued) **Recommended Speeds and Feeds for Indexable Drilling.** (Source, Kennametal Inc.)

Material	Feed		Speed	
	in./rev	mm/rev	sfm	smm
19-9DL	.006-.012	.1524-.3048	170-220	51-67
AM 355	.004-.012	.1016-.3048	250-350	76-107
Discaloy	.004-.012	.1016-.3048	175-260	53-79
F-1570	.006-.009	.1524-.3048	55-100	16-31
J-1570	.005-.015	.1270-.381	60-100	17-31
J-1650	.006-.009	.1524-.2286	65-110	18-34
J-605	.006-.009	.1524-.2286	40-65	12-18
Tinadur	.004-.012	.1016-.3048	100-200	31-61
Refractory 26	.004-.012	.1016-.3048	80-120	21-36
A-286	.004-.012	.1016-.3048	120-180	36-41
S-816	.006-.009	.1524-.2286	70-120	20-36
M-252	.003-.010	.0762-.254	70-110	20-34
DCM	.003-.010	.0762-.254	60-75	16-22
Other Materials				
Magnesium				
AZ31B	.003-.010	.0762-.254	800-1400	244-427
AZ61A	.003-.010	.0762-.254	1000-1800	305-549
ZK60A	.003-.010	.0762-.254	1600-2400	488-790
AZ63A	.004-.012	.1016-.3048	750-1000	229-305
AZ91C	.003-.010	.0762-.254	800-1400	244-427
AZ93C	.004-.012	.1016-.3048	1200-2000	366-610
HK31A	.003-.010	.0762-.254	800-1200	244-366
Titanium	.005-.010	.1270-.254	90-150	27-45
Titanium alloys	.005-.010	.1270-.254	90-150	27-45
Tungsten	.002-.005	.0508-.1270	60-70	17-20
Tungsten rhenium alloy	.003-.005	.0762-.1270	80-100	21-31
Zirconium	.002-.003	.0508-.0762	225-250	68-76
Tantalum (pure)	.005-.007	.1270-.1778	400-800	122-268
Tantalum-tungsten	.010-.020	.254-.508	200-250	61-76
Thalium	.002-.005	.0508-.1270	150-200	45-61
Thorium	.002-.005	.0508-.1270	500-600	152-183
Tin-bismuth alloy	.002-.005	.0508-.1270	300-500	91-152
Manganese	.003-.005	.0762-.1270	50-65	15-18
Molybdenum	.003-.005	.0762-.1270	300-500	91-152
Magnetic alloys	.002-.008	.0508-.2032	75-120	22-37
Graphite	.005-.008	.1270-.2032	150-200	45-61
Lava (steatite)	.003-.005	.0762-.1270	500-750	152-229
Fiberglass	.005-.008	.1270-.2032	200-250	61-76
Plastic	.002-.010	.0508-.254	500-1500	152-457

Reaming and Broaching

Reaming is used to accurately and predictably enlarge and smooth the internal finish of a predrilled hole. To achieve the desired size and finish, the reamer removes a minimal amount of material (chips) from the wall of the hole while being rotated at speeds lower than those used for drilling. Although the reamer rotates in the same direction as a drill, its spiral flutes are the opposite hand to those of the drill used to make the hole, which allows the reamer to push away swarf from the cutting edge (unlike a drill, where swarf travels up the flutes). This ensures that the swarf does not interfere with the quality of the surface finish. Broaching, which is used both externally and internally to shape workpieces, is covered after reaming in this section.

Reamer nomenclature

The following descriptions and definitions were provided by Taylor and Jones Limited and Precision Twist Drill Co.

Axis: the longitudinal centerline of the reamer.

Back Taper: the reduction in diameter per inch of taper of reamer from the entering end toward the shank.

Bevel Lead: the angular cutting portion at the entering end that facilitates the reamer into the hole. It is not provided with a circular land. Also called the *Chamfer.*

Bevel Lead Angle: the angle formed by the cutting edges of the bevel lead and the reamer axis.

Bevel Lead Length: the length of the bevel lead, measured axially.

Body: the portion of the reamer extending from the entering end of the reamer to the commencement of the shank.

Circular Land: the cylindrically ground surface adjacent to the cutting edge, on the leading edge of the land. Also called the *Margin.*

Clearance: the space created by the relief behind the cutting edge or margin of a reamer.

Clearance Angles: the angles formed by the primary or secondary clearance and the tangent to the periphery of the reamer at the cutting edge (primary clearance angle) or behind the circular land (secondary clearance angle).

Core: the central portion of a reamer below the flutes which join the lands.

Core Diameter: the diameter at a given point along the axis of the largest circle that does not project into the flutes.

Cutting Edge: the edge formed by the intersection of the face and the circular land or the surface left by the provision of primary clearance.

Cutting Edge Length: the axial length of that portion of the fluted body provided with primary clearances or circular lands and including the taper and bevel lengths.

Diameter: the maximum cutting diameter of the reamer at the entering end (also see *Taper Reamers*).

Face: the portion of the flute edge adjacent to the cutting edge on which the chip impinges as it is cut from the workpiece.

Flutes: grooves in the body of the reamer that provide cutting edges, permit the removal of chips, and allow cutting fluid to reach the cutting edges.

Hand of Helix: (1) Right-hand helix: when viewed from either end, the flutes twist in a clockwise direction. (2) Left-hand helix: when viewed from either end, the flutes twist in a counterclockwise direction.

Heel: the edge formed by the intersection of the surface left by the provision of secondary clearance and the flute.

Helix Angle: the angle between the cutting edge and the reamer axis.

Land: the portion of the fluted body not cut away by the flutes; the surface or surfaces included between the cutting edge and the heel.

Lead of Helix: the distance, measured parallel to the reamer axis, between corresponding points on the leading edge of a land in one complete turn of a flute. Also called *Lead of Flute.*

Length of Square: the length of the squared portion at the extreme end of a parallel hand shank.

Overall Length: the length over the extreme ends of the reamer.

Primary Clearance: the portion of the land removed to provide clearance immediately behind the cutting edge.

Rake: the angular relationship between the cutting face, or a tangent to the cutting face at a given point, and a given reference plane or line.

> *Axial Rake:* angular (not helical or spiral) cutting faces. The angle between a plane containing the cutting face, or tangent to the cutting face at a given point, and the reamer axis.

> *Helical Rake:* helical and spiral cutting faces (not angular) only. The angle between a plane tangent to the cutting face at a given point on the cutting edge, and the reamer axis.

> *Hook:* a concave condition of a cutting face. The rake of a hooked cutting face must be determined at a given point.

> *Negative Rake:* a cutting face in rotation whose cutting edge lags the surface of the cutting face.

> *Positive Rake:* a cutting face in rotation whose cutting edge leads the surface of the cutting face.

> *Radial Rake Angle:* the angle in a transverse plane between a straight cutting face and a radial line passing through the cutting edge.

Recess: the portion of the body that is reduced in diameter below the cutting edges, pilot, or guide diameters.

Recess Length: the length of that portion of the body that is reduced in diameter below the cutting edges, pilot, or guide diameters.

Relief: the result of the removal of tool material behind or adjacent to the cutting edge to provide clearance and prevent rubbing.

> *Chamfer Relief:* the axial relief on the chamfer of the reamer.

> *Primary Relief:* The relief immediately behind the cutting edge or margin. Properly called "Relief."

> *Radial Relief:* relief in a radial direction measured in the plane of rotation. It can be measured by the amount of indicator drop at a given radius in a given amount of angular rotation.

Rotation of Cutting: (1) Right-hand cutting reamer: a reamer that rotates in a counterclockwise direction when viewed on the entering end of the reamer. (2) Left-hand cutting reamer: a reamer that rotates in a clockwise direction when viewed on the entering end of the reamer.

Secondary Clearance: the portion of the land removed to provide clearance behind the primary clearance or circular land. Also called *Secondary Relief.*

Shank: the portion of the reamer by which it is held and driven. (1) Straight or parallel shank for machine use is a cylindrical ground shank without square. (2) Parallel hand shank is a cylindrical ground shank with a square at its extreme for driving the reamer. (3) Taper shank is a shank of recognized standard taper for machine use.

Size of Square: the dimension across the flats of the squared portion at the extreme end of a parallel hand shank.

Taper Hole: defined for size by stating the large end diameter and the taper per foot on diameter, or the included angle of the taper, or as a taper ratio, e.g., 1 in 96 (included).

Taper Lead: the tapered cutting portion at the entering end that facilitates the entry of the reamer into the hole. It is not provided with a circular land.

Taper Lead Angle: the angle formed by the cutting edges of the taper lead and the reamer axis.

Taper Lead Length: the length of the taper head, measured axially.

Taper Reamers: (1) Large end diameter: the maximum diameter over the tapered cutting edges. (2) Small end diameter: the minimum diameter over the tapered cutting edges.

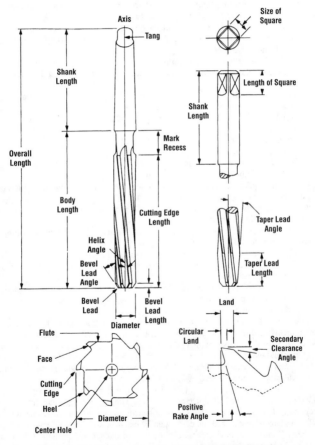

Figure 1. Defining dimensions for reamers. (Source, Taylor & Jones Limited.)

Hand reamers

Hand reamers always have a square shank for fitting a tap wrench. They also have an additional taper lead behind the 45° cutting bevel to help the reamer centralize in the hole if there is any misalignment. They may be either straight-cut or helical, and the helical is preferred in situations where there is an interruption to the cut, such as a keyway. Hand reamers are used in operations that require a small amount of material removal, such as moderately enlarging an existing hole or fitting bushings or bearings. See **Table 1** (fractional sizes) and **Table 3** (metric sizes) for dimensions of commonly available hand reamers.

Taper pin hand reamers have a taper angle of one in forty-eight (for Imperial- see **Table 4**) or one in fifty (for metric- see **Table 5**) and are intended to be used for reaming holes for the fitting of standard taper pins.

Table 1. Fractional Size Hand Reamers with Straight or Helical Flutes. See Table 2 for dimensions of squares. *(Source, Kennametal Inc.)*

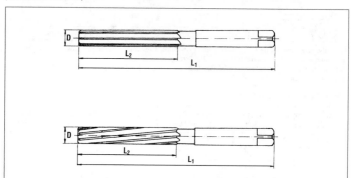

Diameter		Overall Length L_1	Flute Length L_2	Diameter		Overall Length L_1	Flute Length L_2
Fraction	Decimal			Fraction	Decimal		
1/8	.1250	3	1 1/2	11/32	.3438	4 3/4	2 3/8
9/64	.1406	3 1/4	1 5/8	23/64	.3594	5	2 1/2
5/32	.1562	3 1/4	1 5/8	3/8	.3750	5	2 1/2
11/64	.1719	3 1/2	1 3/4	25/64	.3906	5 1/4	2 5/8
3/16	.1875	3 1/2	1 3/4	13/32	.4062	5 1/4	2 5/8
13/64	.2031	3 3/4	1 7/8	27/64	.4219	5 1/2	2 3/4
7/32	.2188	3 3/4	1 7/8	7/16	.4375	5 1/2	2 3/4
15/64	.2344	4	2	29/64	.4531	5 3/4	2 7/8
1/4	.2500	4	2	15/32	.4688	5 3/4	2 7/8
17/64	.2656	4 1/4	2 1/8	31/64	.4844	6	3
9/32	.2812	4 1/4	2 1/8	1/2	.5000	6	3
19/64	.2969	4 1/2	2 1/4	17/32	.5312	6 1/4	3 1/8
5/16	.3125	4 1/2	2 1/4	9/16	.5625	6 1/2	3 1/4
21/64	.3281	4 3/4	2 3/8	19/32	.5938	6 3/4	3 3/8

(Continued)

Table 1. *(Continued)* **Fractional Size Hand Reamers with Straight or Helical Flutes.** See Table 2 for dimensions of squares. *(Source, Kennametal Inc.)*

Diameter		Overall Length L_1	Flute Length L_2	Diameter		Overall Length L_1	Flute Length L_2
Fraction	Decimal			Fraction	Decimal		
5/8	.6250	7	3 1/2	31/32	.9688	10 5/8	5 5/16
21/32	.6562	7 3/8	3 11/16	1	1.0000	10 7/8	5 7/16
11/16	.6875	7 ¾	3 7/8	1 1/16	1.0625	11 1/4	5 5/8
23/32	.7188	8 1/8	4 1/16	1 1/8	1.1250	11 5/8	5 13/16
3/4	.7500	8 3/8	4 3/16	1 3/16	1.1875	12	6
25/32	.7812	8 3/4	4 3/8	1 1/4	1.2500	12 1/4	6 1/8
13/16	.8125	9 1/8	4 9/16	1 5/16	1.3125	12 1/2	6 1/4
27/32	.8438	9 3/8	4 11/16	1 3/8	1.3750	12 5/8	6 5/16
7/8	.8750	9 3/4	4 7/8	1 7/16	1.4375	12 7/8	6 7/16
29/32	.9062	10	5	1 1/2	1.5000	13	6 1/2
15/16	.9375	10 1/4	5 1/8	All dimensions in inches.			

Typically, hand reamers up to $3/8$" have four to six flutes, sizes from $25/64$" to $3/4$" have six to eight flutes, sizes from $25/32$" to 1 $5/16$" have eight to twelve flutes, and sizes $1/38$" and above have ten to fourteen flutes. Check with individual manufacturers for specifications. Standard hand reamers are right-hand cut with either straight or left-hand helix flutes.

Table 2. Dimensions of Hand Reamer Squares.*

Reamer Dia.	Square Length	Square Size	Reamer Dia.	Square Length	Square Size	Reamer Dia.	Square Length	Square Size
1/8	5/32	.095	7/16	7/16	.330	3/4	3/4	.560
5/32	7/32	.115	15/32	7/16	.350	13/16	13/16	.610
3/16	7/32	.140	1/2	1/2	.375	7/8	7/8	.655
7/32	1/4	.165	17/32	1/2	.400	15/16	15/16	.705
1/4	1/4	.185	9/16	9/16	.420	1	1	.750
9/32	1/4	.210	19/32	9/16	.445	1 1/8	1	.845
5/16	5/16	.235	5/8	5/8	.470	1 1/4	1	.935
11/32	5/16	.255	21/32	5/8	.490	1 3/8	1	1.030
3/8	3/8	.280	11/16	11/16	.515	1 1/2	1 1/8	1.125
13/32	3/8	.305	23/32	11/16	.540	All dimensions in inches.		

*Square sizes are constant for fractional, adjustable, and expansion hand reamers.

Adjustable hand reamers are capable of reaming a series of odd-sized holes over a limited range, and they are handy for making a hole marginally larger. Typically, a modern small diameter $3/8$" adjustable hand reamer can expand about 0.31" (0.8 mm) over its minimum diameter, and a large 2 $3/4$" size can expand approximately 0.593" (15 mm). Adjustment is accomplished by raising or lowering adjusting nuts that raise or lower the cutting blades by moving the blades along tapered seatings. Replacement blades are available. See **Table 6** for dimensions.

Expansion hand reamers offer less range, but are a good choice for minimally increasing the size of a hole. An adjusting screw is used to minimally increase the cutter diameter. Typically, a small $1/4$" expansion reamer can expand about 0.006" (0.1524 mm), intermediate

Table 3. Metric Hand Reamers with Straight or Helical Flutes.

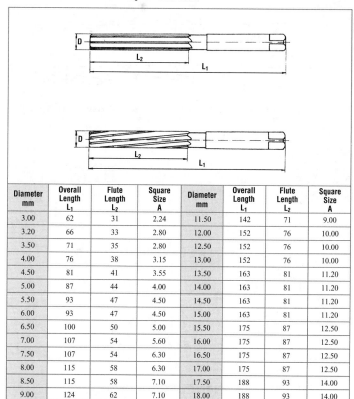

Diameter mm	Overall Length L_1	Flute Length L_2	Square Size A	Diameter mm	Overall Length L_1	Flute Length L_2	Square Size A
3.00	62	31	2.24	11.50	142	71	9.00
3.20	66	33	2.80	12.00	152	76	10.00
3.50	71	35	2.80	12.50	152	76	10.00
4.00	76	38	3.15	13.00	152	76	10.00
4.50	81	41	3.55	13.50	163	81	11.20
5.00	87	44	4.00	14.00	163	81	11.20
5.50	93	47	4.50	14.50	163	81	11.20
6.00	93	47	4.50	15.00	163	81	11.20
6.50	100	50	5.00	15.50	175	87	12.50
7.00	107	54	5.60	16.00	175	87	12.50
7.50	107	54	6.30	16.50	175	87	12.50
8.00	115	58	6.30	17.00	175	87	12.50
8.50	115	58	7.10	17.50	188	93	14.00
9.00	124	62	7.10	18.00	188	93	14.00
9.50	124	62	8.00	18.50	188	93	14.00
10.00	133	66	8.00	19.00	188	93	16.00
10.50	133	66	8.00	19.50	201	100	16.00
11.00	142	71	9.00	20.00	201	100	16.00

All dimensions in millimeters.

sizes ($^1/_2$" to $^{31}/_{32}$") expand up to 0.010", and large (1" and up) sizes can expand about 0.12" (3.048 mm). **Table 7** provides dimensions for expansion hand reamers.

Morse taper reamers and *Brown and Sharpe taper reamers* are available for producing (and maintaining) tapered sockets. As hand reamers, both reamers are used for reaming out existing sockets, but the Morse taper version is also available with a Morse taper shank for use on machines. Either style reamer should be ordered by specifying the desired Morse (No. 0 through No. 5) or Brown and Sharpe (No. 1 through No. 10) taper number.

Table 4. Fractional Size Taper Pin Hand Reamers with Straight or Helical Flutes.

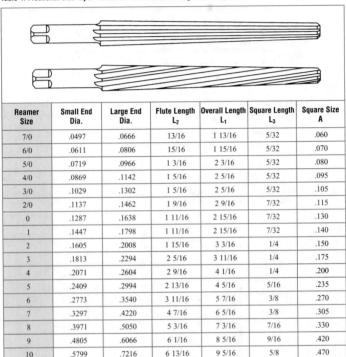

Reamer Size	Small End Dia.	Large End Dia.	Flute Length L_2	Overall Length L_1	Square Length L_3	Square Size A
7/0	.0497	.0666	13/16	1 13/16	5/32	.060
6/0	.0611	.0806	15/16	1 15/16	5/32	.070
5/0	.0719	.0966	1 3/16	2 3/16	5/32	.080
4/0	.0869	.1142	1 5/16	2 5/16	5/32	.095
3/0	.1029	.1302	1 5/16	2 5/16	5/32	.105
2/0	.1137	.1462	1 9/16	2 9/16	7/32	.115
0	.1287	.1638	1 11/16	2 15/16	7/32	.130
1	.1447	.1798	1 11/16	2 15/16	7/32	.140
2	.1605	.2008	1 15/16	3 3/16	1/4	.150
3	.1813	.2294	2 5/16	3 11/16	1/4	.175
4	.2071	.2604	2 9/16	4 1/16	1/4	.200
5	.2409	.2994	2 13/16	4 5/16	5/16	.235
6	.2773	.3540	3 11/16	5 7/16	3/8	.270
7	.3297	.4220	4 7/16	6 5/16	3/8	.305
8	.3971	.5050	5 3/16	7 3/16	7/16	.330
9	.4805	.6066	6 1/16	8 5/16	9/16	.420
10	.5799	.7216	6 13/16	9 5/16	5/8	.470

All dimensions in inches. Taper is $1/4$" per foot (one in forty-eight).

Table 5. Metric Size Taper Pin Hand Reamers with Straight or Helical Flutes.

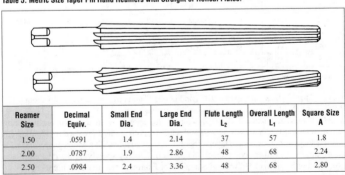

Reamer Size	Decimal Equiv.	Small End Dia.	Large End Dia.	Flute Length L_2	Overall Length L_1	Square Size A
1.50	.0591	1.4	2.14	37	57	1.8
2.00	.0787	1.9	2.86	48	68	2.24
2.50	.0984	2.4	3.36	48	68	2.80

(Continued)

Table 5. *(Continued)* **Metric Size Taper Pin Hand Reamers with Straight or Helical Flutes.**

Reamer Size	Decimal Equiv.	Small End Dia.	Large End Dia.	Flute Length L_2	Overall Length L_1	Square Size A
3.00	.1181	2.9	4.06	58	80	3.15
4.00	.1575	3.9	5.26	68	93	4.00
5.00	.1969	4.9	6.36	73	100	5.00
6.00	.2362	5.9	8.00	105	135	6.30
8.00	.3150	7.9	10.80	145	180	8
10.00	.3937	9.9	13.40	175	215	10
12.00	.4724	11.80	16.00	210	255	11.2
13.00	.5118	12.86	16.74	210	255	12.5
14.00	.5512	13.86	17.74	210	255	12.5
16.00	.6299	15.80	20.40	230	280	14
20.00	.7874	19.80	24.80	250	310	18

All dimensions in millimeters except "Decimal Equiv." column. Taper is one in fifty.

Table 6. Adjustable Hand Reamer Dimensional Sizes (Fractional and Metric).

Size	Size Range		Flute Length		Overall Length		Expansion Range
	inch	mm	inch	mm	inch	mm	inch
8/A	1/4-9/32	6.35-7.14	1 11/32	34.13	3 1/4	82.55	1/32
7/A	9/32-5/16	7.14-7.94	1 11/32	34.13	3 1/2	88.90	1/32
6/A	5/16-11/32	7.94-8.73	1 1/2	38.10	4 1/8	104.78	1/32
5/A	11/32-3/8	8.73-9.52	1 1/2	38.10	4 3/8	111.12	1/32
4/A	3/8-13/32	9.52-10.31	1 1/2	38.10	4 3/4	120.65	1/32
3/A	13/32-7/16	10.31-11.11	1 1/2	38.10	5	127.00	1/32
2/A	7/16-15/32	11.11-11.90	1 5/8	41.28	5 1/4	133.35	1/32
A	15/32-17/32	11.90-13.49	1 5/8	41.28	5 1/2	139.70	1/16
B	17/32-19/32	13.49-15.08	1 13/16	46.04	5 3/4	146.05	1/16
C	19/32-21/32	15.08-16.66	2 1/16	52.39	6 1/2	165.10	1/16
D	21/32-23/32	16.66-18.25	2 3/16	55.56	6 3/4	171.45	1/16
E	23/32-25/32	18.25-19.84	2 1/2	63.50	7	177.80	1/16
F	25/32-27/32	19.84-21.43	2 5/8	66.68	7 3/8	187.32	1/16
G	27/32-15/16	21.43-23.81	3	76.20	8	203.20	3/32
H	15/16-1 1/16	23.81-26.98	3 1/4	82.55	9	228.60	1/8
I	1 1/16-1 3/16	26.98-30.16	3 3/8	85.72	10	254.00	1/8

(Continued)

Table 6. *(Continued)* **Adjustable Hand Reamer Dimensional Sizes (Fractional and Metric).**

Size	Size Range		Flute Length		Overall Length		Expansion Range
	inch	mm	inch	mm	inch	mm	inch
J	1 3/16-1 11/32	30.16-34.13	3 7/8	98.42	11	279.40	5/32
K	1 11/32-1 1/2	34.13-38.10	4 1/4	107.95	12	304.80	5/32
L	1 1/2-1 13/16	38.10-46.03	4 7/16	112.71	14	355.60	5/16
M	1 13/16-2 7/32	46.03-56.35	4 7/8	123.83	16	406.40	13/32
N	2 7/32-2 3/4	56.35-69.85	5	127.00	18	457.20	17/32
O	2 3/4-3 11/32	69.85-84.93	5 1/4	133.35	20	508.00	19/32

Fractional Series. Metric measurements for reference only.

Table 7. Expansion Hand Reamer Dimensions.

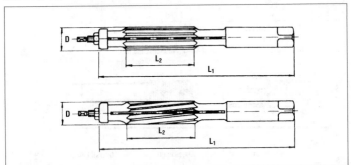

Nom. Dia.*	Flute Length	Overall Length	Nom. Dia.*	Flute Length	Overall Length
1/4	1 7/16	3 7/8	3/4	3 1/2	8
9/32	1 3/4	4	25/32	3 1/2	8
5/16	1 3/4	4	13/16	3 1/2	8
3/8	2	5	27/32	3 1/2	8
13/32	2	5	7/8	4	9
7/16	2	5	29/32	4	9
15/32	2	5	15/16	4	9
1/2	2	5 1/2	31/32	4	9
17/32	2	5 1/2	1	4 1/2	10
9/16	2 1/2	6	1 1/16	4 1/2	10
19/32	2 1/2	6	1 1/8	4 3/4	10 1/2
5/8	3	7	1 3/16	4 3/4	10 1/2
21/32	3	7	1 1/4	5	11
11/16	3	7	1 5/16	5	11
23/32	3	7	1 3/8	5 1/4	11 1/2

(Continued)

Table 7. *(Continued)* **Expansion Hand Reamer Dimensions.**

Nom. Dia.*	Flute Length	Overall Length	Nom. Dia.*	Flute Length	Overall Length
1 7/16	5 1/4	11 1/2	1 7/8	5 1/2	12 3/8
1 1/2	5 1/2	12	2	5 1/2	13 1/8
1 3/4	5 1/2	12 3/8	All dimensions in inches.		

*Recommended expansion limit is 0.01" times the nominal diameter. See text for maximum expansion limits for different diameters. Reamers are right-hand cut and are available with straight or spiral flutes.

Table 8. Taper Pipe Reamers. *(Source, Yankee Reamer.)*

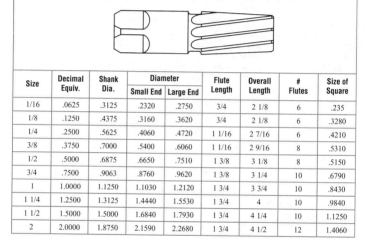

Size	Decimal Equiv.	Shank Dia.	Diameter		Flute Length	Overall Length	# Flutes	Size of Square
			Small End	Large End				
1/16	.0625	.3125	.2320	.2750	3/4	2 1/8	6	.235
1/8	.1250	.4375	.3160	.3620	3/4	2 1/8	6	.3280
1/4	.2500	.5625	.4060	.4720	1 1/16	2 7/16	6	.4210
3/8	.3750	.7000	.5400	.6060	1 1/16	2 9/16	8	.5310
1/2	.5000	.6875	.6650	.7510	1 3/8	3 1/8	8	.5150
3/4	.7500	.9063	.8760	.9620	1 3/8	3 1/4	10	.6790
1	1.0000	1.1250	1.1030	1.2120	1 3/4	3 3/4	10	.8430
1 1/4	1.2500	1.3125	1.4440	1.5530	1 3/4	4	10	.9840
1 1/2	1.5000	1.5000	1.6840	1.7930	1 3/4	4 1/4	10	1.1250
2	2.0000	1.8750	2.1590	2.2680	1 3/4	4 1/2	12	1.4060

All dimensions in inches. Taper is $3/4$" per foot and are used to ream holes to be tapped with American National Standard Taper Pipe Taps. Reamers have left hand flutes.

Taper pipe reamers are used to ream holes that are to be threaded with American Standard taper pipe threads. The taper is $3/4$" to the foot and they can be purchased in sizes ranging from large end diameters of $1/16$" to 2". Left hand spiral flutes are standard on high speed steel reamers. See **Table 8** for dimensions.

Machine reamers

Machine reamers have either straight or Morse taper shanks (or in some cases with DIN 1809 specification tangs), and they are available in high-speed steel, tungsten carbide tipped, and solid carbide in several configurations.

Chucking reamers have short flutes. The straight flute version has a 45° chamfer and is used on most materials. The helical flute version is available with either left or right helix and is primarily used for difficult to machine materials. Also, they are available in both fluted chucking and rose chucking configurations, with the rose chucking version having deeper flutes and no relief on the periphery. Chucking reamers are available in

Wire Gauge, Fractional, Letter, Decimal, and Metric sizes, so an ideal reamer can be obtained for almost any application. Expansion chucking reamers are available but they do not operate with the same flexibility enjoyed by the hand expansion reamers. Instead, they have the ability to maintain their basic diameter after repeated sharpening cycles. They are not designed to have their expansion screws loosened in order to reduce the diameter. **Table 9** gives dimensions of standard straight shank chucking reamers and standard jobber drill length chucking reamers. **Table 10** provides specifications for standard metric chucking reamers with Morse taper shanks, and **Table 11** gives dimensions for standard chucking reamers with Morse taper shanks. Expansion chucking reamers are especially useful for reaming abrasive materials because, as the diameter wears down, they can be expanded by tightening the adjusting screw and resharpening to the original diameter. They should not be considered adjustable reamers suitable for producing holes of different sizes. Dimensions for expansion chucking reamers are given in **Table 12**.

Jobber reamers are supplied with Morse tapers, have longer flutes than chucking reamers—approximately equivalent to those found on hand reamers-and have a sharpened 45° chamfer. They are commonly available in sizes ranging from $1/4$" to 1 $1/2$". Jobber reamers are detailed in **Table 13**.

Bridge reamers derive their name from their intended use, which is reaming pairs of holes that may be slightly out of alignment on structural iron and steel constructions such as bridges. They have a taper lead angle of one in ten on diameter on either straight or helical flutes, and can be obtained in diameters ranging from $3/8$" to 1 $9/16$" and in millimeter sizes of 8.40 mm to 40.00 mm. See **Table 14** for bridge reamer dimensions. A shorter version of the bridge reamer is the *car reamer* (see **Table 15**) that is intended for use in cramped situations where space is at a premium.

Taper pin reamers for use on machines, like the hand version, are used for reaming holes for standard tapers. For dimensions, see **Tables 4** and **5** (*machine taper pin* reamers have high spiral flutes, but they share dimensions with the hand taper pin reamer).

Shell reamers are mounted on either a taper or straight shank arbor that is tapered to fit the hole in the reamer, allowing several different sizes of shell reamers to be mounted on one size of arbor. They are essentially chucking reamers that are produced in diameters ranging from $3/4$" to 2 $1/4$" and ranging in overall length from 2 $1/4$" to 3 $3/4$". Due to their nature, these reamers are not as rigid as single piece reamers, but the ability to use as many as six different reamers on a single size arbor can offer economic advantages. **Table 16** gives dimensions for straight shank shell rose and rose chucking reamers, **Table 17** provides specifications for shell reamer arbors, and **Table 18** gives dimensions of driving slots and lugs for shell reamers. **Table 19** gives dimensions for DIN 219 metric shell reamers.

Stub screw-machine reamers, as their name implies, are used on automatic screw machines. Their standard configuration is heft-hand helix (right-hand cut), and they are produced in numbered sizes from #000 (0.030") to #31 (1.4421"). They are cross drilled in the shank to allow their use on pin drive floating holders, but they are more often fitted to standard holders. Their dimensions are given in **Table 20**.

Dowel pin reamers are similar in appearance to standard chucking reamers, but differ in taper and diameter tolerance. Dowel pin reamers are produced with increased back taper and a minus diameter tolerance (back taper = 0.0005" per 1 $1/2$" of length; tolerance = 0.000 - 0.002") whereas standard chucking reamers have very little back taper and a plus diameter tolerance. See **Table 21** for standard dimensions.

Yankee Reamer and Danly Machine Co. recommend the following guidelines for using dowel pin reamers for most applications.

1) Drill hole $^1/_{64}$" under dowel pin size in either cast iron or steel when hole is below $^1/_2$" in diameter. For holes over $^1/_2$", drill hole $^1/_{32}$" under dowel pin size.

2) Ream holes in cast iron 0.0005" undersize. For tool steel which is to be hardened, ream to size.

For high-grade tools or dies, they recommend the following.

1) Drill hole $^1/_{64}$" under dowel pin size when hole is below $^1/_2$" in diameter. For holes over $^1/_2$", drill hole $^1/_{32}$" under dowel pin size.

2) Ream hole 0.002" undersize.

3) Ream hole 0.0005" undersize.

4) Harden tool steel and lap holes to fit reamer.

5) Align and secure tool steel, use lapped hole as a jig and follow drill and reamer sequence in steps 1 and 2 above.

Combined drills and reamers are hybrid tools designed to combine the drilling and reaming operations. These tools are available in two configurations: step construction and subland construction. The step version has flutes that extend the entire length of the cutting edge, but the drill section is restricted to no more than the first quarter turn of the flute (1.5 to 2 times the diameter). After that point, the diameter of the tool increases slightly as the reamer section begins. The subland version has a constant diameter drill margin over the entire length of the cutting edge, and the reamer flutes are located behind the frill flutes. Due to the minimal length of the drilling section, only shallow holes can be drilled with combined drills and reamers, and blind holes cannot be cut and reamed to full depth. To operate properly, the drill section must pass entirely through the workpiece before the reamer section begins its pass.

It should be noted that several manufacturers produce reamers in nonstandardized configurations for specialized or even custom applications on modern machine tools. These reamers are commonly provided with tungsten carbide, titanium nitride, polycrystalline diamond, or other advanced cutting materials. Also, chucking reamers, in particular, are manufactured, or can be special ordered in virtually every conceivable decimal size diameter, and in metric sizes in increments of 0.01 mm. Consult manufacturers or suppliers for availability. *(Text continued on p. 1173)*

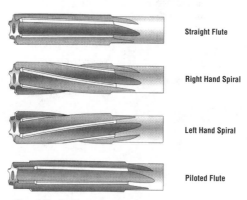

Straight Flute

Right Hand Spiral

Left Hand Spiral

Piloted Flute

Figure 2. Standard flute configurations for reamers. (Source, Alvord-Polk Tool Co.)

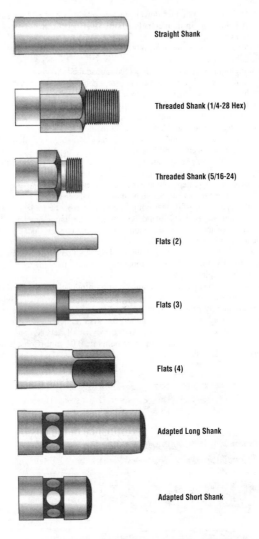

Straight Shank

Threaded Shank (1/4-28 Hex)

Threaded Shank (5/16-24)

Flats (2)

Flats (3)

Flats (4)

Adapted Long Shank

Adapted Short Shank

Figure 3. Standard shank configurations for reamers. (Source, Alvord-Polk Tool Co.)

Table 9. Straight Shank Chucking Reamer and Jobber Drill Length Chucking Reamer Dimensions. Fractional, Wire, Letter, and Decimal Sizes. *(Source, Kennametal Inc.)*

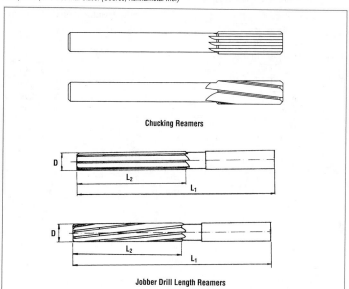

Chucking Reamers

Jobber Drill Length Reamers

Reamer Diameter			Shank Dia.		Chucking Reamer		Jobber Chucking Reamer	
Size	Equivalent		Max.	Min.	Overall Length	Flute Length	Overall Length	Flute Length
	Decimal	Metric						
1/32	.0312	.792	.0290	.0280	2 1/4	1/2	-	-
#67	.0320	.813	.0290	.0280	2 1/4	1/2	-	-
#66	.0330	.838	.0290	.0280	2 1/4	1/2	-	-
#65	.0350	.889	.0290	.0280	2 1/4	1/2	-	-
#64	.0360	.914	.0290	.0280	2 1/4	1/2	-	-
#63	.0370	.940	.0290	.0280	2 1/4	1/2	-	-
#62	.0380	.965	.0290	.0280	2 1/4	1/2	-	-
#61	.0390	.991	.0290	.0280	2 1/4	1/2	-	-
#60	.0400	1.016	.0390	.0380	2 1/2	1/2	1 5/8	1/2
#59	.0410	1.041	.0390	.0380	2 1/2	1/2	1 5/8	1/2
#58	.0420	1.067	.0390	.0380	2 1/2	1/2	1 5/8	1/2
#57	.0430	1.092	.0390	.0380	2 1/2	1/2	1 5/8	1/2
#56	.0465	1.181	.0455	.0445	2 1/2	1/2	1 3/4	1/2

Dimensions in inches except Metric Equivalent column. Reamers are available in straight flute, left-hand spiral, and right-hand spiral versions. Sizes to #31 have four flutes, sizes $1/8$" through $23/32$" have six flutes, sizes $3/4$" through $1 1/8$" have eight flutes, sizes $1 3/16$" through $1 5/16$" have ten flutes, and sizes $1 3/8$" and larger have twelve flutes. Number of flutes may vary with manufacturer.

(Continued)

Table 9. *(Continued)* **Straight Shank Chucking Reamer and Jobber Drill Length Chucking Reamer Dimensions.** Fractional, Wire, Letter, and Decimal Sizes. *(Source, Kennametal Inc.)*

Reamer Diameter			Shank Dia.		Chucking Reamer		Jobber Chucking Reamer	
Size	Equivalent		Max.	Min.	Overall Length	Flute Length	Overall Length	Flute Length
	Decimal	Metric						
3/64	.0469	1.191	.0455	.0445	2 1/2	1/2	1 3/4	1/2
#55	.0520	1.321	.0510	.0500	2 1/2	1/2	1 7/8	1/2
#54	.0550	1.397	.0510	.0500	2 1/2	1/2	1 7/8	1/2
#53	.0595	1.511	.0585	.0575	2 1/2	1/2	1 7/8	1/2
1/16	.0625	1.588	.0585	.0575	2 1/2	1/2	1 7/8	1/2
#52	.0635	1.613	.0585	.0575	2 1/2	1/2	1 7/8	1/2
#51	.0670	1.702	.0660	.0650	3	3/4	2	3/4
#50	.0700	1.778	.0660	.0650	3	3/4	2	3/4
#49	.0730	1.854	.0660	.0650	3	3/4	2	3/4
#48	.0760	1.930	.0720	.0710	3	3/4	2	3/4
5/64	.0781	1.984	.0720	.0710	3	3/4	2	3/4
#47	.0785	1.994	.0720	.0710	3	3/4	2	3/4
#46	.0810	2.057	.0771	.0761	3	3/4	2 1/8	3/4
#45	.0820	2.083	.0771	.0761	3	3/4	2 1/8	3/4
#44	.0860	2.184	.0810	.0800	3	3/4	2 1/8	3/4
#43	.0890	2.261	.0810	.0800	3	3/4	2 1/8	3/4
#42	.0935	2.375	.0880	.0870	3	3/4	2 1/4	3/4
3/32	.0938	2.383	.0880	.0870	3	3/4	2 1/4	3/4
#41	.0960	2.438	.0928	.0918	3 1/2	7/8	2 3/8	7/8
#40	.0980	2.489	.0928	.0918	3 1/2	7/8	2 3/8	7/8
#39	.0995	2.527	.0928	.0918	3 1/2	7/8	2 3/8	7/8
#38	.1015	2.578	.0950	.0940	3 1/2	7/8	2 1/2	7/8
#37	.1040	2.642	.0950	.0940	3 1/2	7/8	2 1/2	7/8
#36	.1065	2.705	.1030	.1020	3 1/2	7/8	2 1/2	7/8
7/64	.1094	2.779	.1030	.1020	3 1/2	7/8	2 5/8	7/8
#35	.1100	2.794	.1030	.1020	3 1/2	7/8	2 5/8	7/8
#34	.1110	2.819	.1055	.1045	3 1/2	7/8	2 5/8	7/8
#33	.1130	2.870	.1055	.1045	3 1/2	7/8	2 5/8	7/8
#32	.1160	2.946	.1120	.1110	3 1/2	7/8	2 3/4	7/8
#31	.1200	3.048	.1120	.1110	3 1/2	7/8	2 3/4	7/8
1/8	.1250	3.175	.1190	.1180	3 1/2	7/8	2 3/4	7/8
#30	.1285	3.264	.1190	.1180	3 1/2	7/8	2 ¾	7/8
#29	.1360	3.454	.1275	.1265	4	1	2 7/8	1
#28	.1405	3.569	.1350	.1340	4	1	2 7/8	1

Dimensions in inches except Metric Equivalent column. Reamers are available in straight flute, left-hand spiral, and right-hand spiral versions. Sizes to #31 have four flutes, sizes 1/8" through 23/32" have six flutes, sizes 3/4" through 1 1/8" have eight flutes, sizes 1 3/16" through 1 5/16" have ten flutes, and sizes 1 3/8" and larger have twelve flutes. Number of flutes may vary with manufacturer.

(Continued)

Table 9. *(Continued)* **Straight Shank Chucking Reamer and Jobber Drill Length Chucking Reamer Dimensions.** Fractional, Wire, Letter, and Decimal Sizes. *(Source, Kennametal Inc.)*

Reamer Diameter			Shank Dia.		Chucking Reamer		Jobber Chucking Reamer	
Size	Equivalent		Max.	Min.	Overall Length	Flute Length	Overall Length	Flute Length
	Decimal	Metric						
9/64	.1406	3.571	.1350	.1340	4	1	2 7/8	1
#27	.1440	3.658	.1350	.1340	4	1	3	1
#26	.1470	3.734	.1430	.1420	4	1	3	1
#25	.1495	3.797	.1430	.1420	4	1	3	1
#24	.1520	3.861	.1460	.1450	4	1	3 1/8	1
#23	.1540	3.912	.1460	.1450	4	1	3 1/8	1
5/32	.1562	3.967	.1510	.1500	4	1	3 1/8	1
#22	.1570	3.988	.1510	.1500	4	1	3 1/8	1
#21	.1590	4.039	.1530	.1520	4 1/2	1 1/8	3 1/4	1 1/8
#20	.1610	4.089	.1530	.1520	4 1/2	1 1/8	3 1/4	1 1/8
#19	.1660	4.216	.1595	.1585	4 1/2	1 1/8	3 1/4	1 1/8
#18	.1695	4.305	.1595	.1585	4 1/2	1 1/8	3 1/4	1 1/8
11/64	.1719	4.366	.1645	.1635	4 1/2	1 1/8	3 1/4	1 1/8
#17	.1730	4.394	.1645	.1635	4 1/2	1 1/8	3 3/8	1 1/8
#16	.1770	4.496	.1704	.1694	4 1/2	1 1/8	3 3/8	1 1/8
#15	.1800	4.572	.1755	.1745	4 1/2	1 1/8	3 3/8	1 1/8
#14	.1820	4.623	.1755	.1745	4 1/2	1 1/8	3 3/8	1 1/8
#13	.1850	4.70	.1805	.1795	4 1/2	1 1/8	3 1/2	1 1/8
3/16	.1875	4.762	.1805	.1795	4 1/2	1 1/8	3 1/2	1 1/8
#12	.1890	4.80	.1805	.1795	4 1/2	1 1/8	3 1/2	1 1/8
#11	.1910	4.851	.1860	.1850	5	1 1/4	3 1/2	1 1/4
#10	.1935	4.915	.1860	.1850	5	1 1/4	3 5/8	1 1/4
#9	.1960	4.978	.1895	.1885	5	1 1/4	3 5/8	1 1/4
#8	.1990	5.054	.1895	.1885	5	1 1/4	3 5/8	1 1/4
#7	.2010	5.105	.1945	.1935	5	1 1/4	3 5/8	1 1/4
13/64	.2031	5.159	.1945	.1935	5	1 1/4	3 5/8	1 1/4
#6	.2040	5.182	.1945	.1935	5	1 1/4	3 3/4	1 1/4
#5	.2055	5.220	.2016	.2006	5	1 1/4	3 3/4	1 1/4
#4	.2090	5.309	.2016	.2006	5	1 1/4	3 3/4	1 1/4
#3	.2130	5.5410	.2075	.2065	5	1 1/4	3 3/4	1 1/4
7/32	.2188	5.558	.2075	.2065	5	1 1/4	3 3/4	1 1/4
#2	.2210	5.613	.2173	.2163	6	1 1/2	3 7/8	1 1/2
#1	.2280	5.791	.2173	.2163	6	1 1/2	3 7/8	1 1/2
A	.2340	5.944	.2265	.2255	6	1 1/2	3 7/8	1 1/2

Dimensions in inches except Metric Equivalent column. Reamers are available in straight flute, left-hand spiral, and right-hand spiral versions. Sizes to #31 have four flutes, sizes $\frac{1}{8}$" through $\frac{23}{32}$" have six flutes, sizes $\frac{3}{4}$" through 1 $\frac{1}{8}$" have eight flutes, sizes 1 $\frac{3}{16}$" through 1 $\frac{5}{16}$" have ten flutes, and sizes 1 $\frac{3}{8}$" and larger have twelve flutes. Number of flutes may vary with manufacturer.

(Continued)

Table 9. *(Continued)* **Straight Shank Chucking Reamer and Jobber Drill Length Chucking Reamer Dimensions.** Fractional, Wire, Letter, and Decimal Sizes. *(Source, Kennametal Inc.)*

Size	Reamer Diameter — Equivalent Decimal	Reamer Diameter — Equivalent Metric	Shank Dia. Max.	Shank Dia. Min.	Chucking Reamer Overall Length	Chucking Reamer Flute Length	Jobber Chucking Reamer Overall Length	Jobber Chucking Reamer Flute Length
15/64	.2344	5.954	.2265	.2255	6	1 1/2	3 7/8	1 1/2
B	.2380	6.045	.2329	.2319	6	1 1/2	4	1 1/2
C	.2420	6.147	.2329	.2319	6	1 1/2	4	1 1/2
D	.2460	6.248	.2329	.2319	6	1 1/2	4	1 1/2
E 1/4	.2500	6.350	.2405	.2395	6	1 1/2	4	1 1/2
F	.2570	6.528	.2485	.2475	6	1 1/2	4 1/8	1 1/2
G	.2610	6.629	.2485	.2475	6	1 1/2	4 1/8	1 1/2
17/64	.2656	6.746	.2485	.2475	6	1 1/2	4 1/8	1 1/2
H	.2660	6.756	.2485	.2475	6	1 1/2	4 1/8	1 1/2
I	.2720	6.909	.2485	.2475	6	1 1/2	4 1/8	1 1/2
J	.2770	7.036	.2485	.2475	6	1 1/2	4 1/8	1 1/2
K	.2810	7.137	.2485	.2475	6	1 1/2	4 1/4	1 1/2
9/32	.2812	7.142	.2485	.2475	6	1 1/2	4 1/4	1 1/2
L	.2900	7.366	.2792	.2782	6	1 1/2	4 1/4	1 1/2
M	.2950	7.493	.2792	.2782	6	1 1/2	4 3/8	1 1/2
19/64	.2969	7.541	.2792	.2782	6	1 1/2	4 3/8	1 1/2
N	.3020	7.671	.2792	.2782	6	1 1/2	4 3/8	1 1/2
5/16	.3125	7.938	.2792	.2782	6	1 1/2	4 1/2	1 1/2
O	.3160	8.026	.2792	.2782	6	1 1/2	4 1/2	1 1/2
P	.3230	8.204	.2792	.2782	6	1 1/2	4 5/8	1 1/2
21/64	.3281	8.344	.2792	.2782	6	1 1/2	4 5/8	1 1/2
Q	.3320	8.433	.2792	.2782	6	1 1/2	4 3/4	1 1/2
R	.3390	8.611	.2792	.2782	6	1 1/2	4 3/4	1 1/2
11/32	.3438	8.733	.2792	.2782	6	1 1/2	4 3/4	1 1/2
S	.3480	8.839	.2792	.2782	7	1 3/4	4 7/8	1 3/4
T	.3580	9.093	.2792	.2782	7	1 3/4	4 7/8	1 3/4
23/64	.3594		.2792	.2782	7	1 3/4	4 7/8	1 3/4
U	.3680	9.347	.2792	.2782	7	1 3/4	5	1 3/4
3/8	.3750	9.525	.3105	.3095	7	1 3/4	5	1 3/4
V	.3770	9.576	.3105	.3095	7	1 3/4	5	1 3/4
W	.3860	9.804	.3105	.3095	7	1 3/4	5 1/8	1 3/4
25/64	.3906	9.921	.3105	.3095	7	1 3/4	5 1/8	1 3/4
X	.3970	10.084	.3105	.3095	7	1 3/4	5 1/8	1 3/4
Y	.4040	10.262	.3105	.3095	7	1 3/4	5 1/4	1 3/4

Dimensions in inches except Metric Equivalent column. Reamers are available in straight flute, left-hand spiral, and right-hand spiral versions. Sizes to #31 have four flutes, sizes 1/8" through 23/32" have six flutes, sizes 3/4" through 1 1/8" have eight flutes, sizes 1 3/16" through 1 5/16" have ten flutes, and sizes 1 3/8" and larger have twelve flutes. Number of flutes may vary with manufacturer.

(Continued)

Table 9. *(Continued)* Straight Shank Chucking Reamer and Jobber Drill Length Chucking Reamer Dimensions. Fractional, Wire, Letter, and Decimal Sizes. *(Source, Kennametal Inc.)*

Reamer Diameter			Shank Dia.		Chucking Reamer		Jobber Chucking Reamer	
Size	Equivalent		Max.	Min.	Overall Length	Flute Length	Overall Length	Flute Length
	Decimal	Metric						
13/32	.4062	10.317	.3105	.3095	7	1 3/4	5 1/4	1 3/4
Z	.4130	10.409	.3730	.3720	7	1 3/4	5 1/4	1 3/4
27/64	.4219	10.716	.3730	.3720	7	1 3/4	5 3/8	1 3/4
7/16	.4375	11.112	.3730	.3720	7	1 3/4	5 1/2	1 3/4
29/64	.4531	11.509	.3730	.3720	7	1 3/4	5 5/8	1 3/4
15/32	.4688	11.908	.3730	.3720	7	1 3/4	5 3/4	1 3/4
31/64	.4844	12.304	.4355	.4345	8	2	5 7/8	2
1/2	.5000	12.70	.4355	.4345	8	2	6	2
17/32	.5312	13.492	.4355	.4345	8	2	-	-
9/16	.5625	14.288	.4355	.4345	8	2	-	-
19/32	.5938	15.083	.4355	.4345	8	2	-	-
5/8	.6250	15.875	.5620	.5605	9	2 1/4	-	-
21/32	.6562	16.669	.5620	.5605	9	2 1/4	-	-
11/16	.6875	17.462	.5620	.5605	9	2 1/4	-	-
23/32	.7188	18.258	.5620	.5605	9	2 1/4	-	-
3/4	.7500	19.050	.6245	.6230	9 1/2	2 1/2	-	-
25/32	.7812	19.845	.6245	.6230	9 1/2	2 1/2	-	-
13/16	.8125	20.638	.6245	.6230	9 1/2	2 1/2	-	-
27/32	.8438	21.433	.6245	.6230	9 1/2	2 1/2	-	-
7/8	.8750	22.235	.7495	.7480	10	2 5/8	-	-
29/32	.9062	23.017	.7495	.7480	10	2 5/8	-	-
15/16	.9375	23.812	.7495	.7480	10	2 5/8	-	-
31/32	.9688	24.608	.7495	.7480	10	2 5/8	-	-
1	1.0000	25.40	.8745	.8730	10 1/2	2 3/4	-	-
1 1/16	1.0625	-	.8745	.8730	10 1/2	2 3/4	-	-
1 1/8	1.1250	-	.8745	.8730	11	2 7/8	-	-
1 3/16	1.1875	-	.9995	.9980	11	2 7/8	-	-
1 1/4	1.2500	-	.9995	.9980	11 1/2	3	-	-
1 5/16	1.3125	-	.9995	.9980	11 1/2	3	-	-
1 3/8	1.3750	-	.9995	.9980	12	3 1/4	-	-
1 7/16	1.4375	-	.9995	.9980	12	3 1/4	-	-
1 1/2	1.5000	-	1.2495	1.2480	12 1/2	3 1/2	-	-

Dimensions in inches except Metric Equivalent column. Reamers are available in straight flute, left-hand spiral, and right-hand spiral versions. Sizes to #31 have four flutes, sizes 1/8" through 23/32" have six flutes, sizes 3/4" through 1 1/8" have eight flutes, sizes 1 3/16" through 1 5/16" have ten flutes, and sizes 1 3/8" and larger have twelve flutes. Number of flutes may vary with manufacturer.

Table 10. Metric Chucking Reamers with Taper Shanks. *(Source, Werkö GmbH.)*

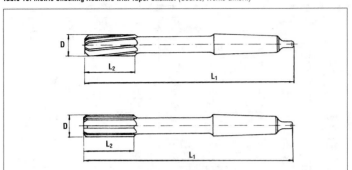

Size	Overall Length	Flute Length	Morse Taper	Size	Overall Length	Flute Length	Morse Taper
3.00	112	15	1	27.00	277	71	3
4.00	125	19	1	28.00	277	71	3
5.00	133	23	1	29.00	281	73	3
6.00	138	26	1	30.00	281	73	3
7.00	150	31	1	31.00	285	75	3
8.00	156	33	1	32.00	317	77	4
9.00	162	36	1	33.00	317	77	4
10.00	168	38	1	34.00	321	78	4
11.00	175	41	1	35.00	321	78	4
12.00	182	44	1	36.00	325	79	4
13.00	182	44	1	37.00	325	79	4
14.00	189	47	1	38.00	329	81	4
15.00	204	50	2	39.00	329	81	4
16.00	210	52	2	40.00	329	81	4
17.00	214	54	2	41.00	333	82	4
18.00	219	56	2	42.00	333	82	4
19.00	223	58	2	43.00	336	83	4
20.00	228	60	2	44.00	336	83	4
21.00	232	62	2	45.00	336	83	4
22.00	237	64	2	46.00	340	84	4
23.00	241	66	3	47.00	340	84	4
24.00	268	68	3	48.00	344	86	4
25.00	268	68	3	49.00	344	86	4
26.00	273	70	3	50.00	344	86	4

All dimensions in millimeters. Specifications in accordance with DIN 208.

Table 11. Chucking Reamers with Morse Taper Shanks. *(Source, Kennametal Inc.)*

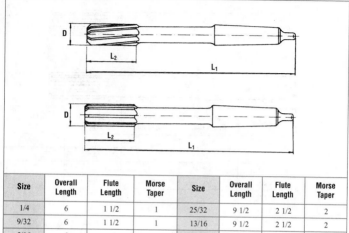

Size	Overall Length	Flute Length	Morse Taper	Size	Overall Length	Flute Length	Morse Taper
1/4	6	1 1/2	1	25/32	9 1/2	2 1/2	2
9/32	6	1 1/2	1	13/16	9 1/2	2 1/2	2
5/16	6	1 1/2	1	27/32	9 1/2	2 1/2	2
11/32	6	1 1/2	1	7/8	10	2 5/8	2
3/8	7	1 3/4	1	29/32	10	2 5/8	3
13/32	7	1 3/4	1	15/16	10	2 5/8	3
7/16	7	1 3/4	1	31/32	10	2 5/8	3
15/32	7	1 3/4	1	1	10 1/2	2 3/4	3
1/2	8	2	1	1 1/16	10 1/2	2 3/4	3
17/32	8	2	1	1 1/8	11	2 7/8	3
9/16	8	2	1	1 3/16	11	2 7/8	3
19/32	8	2	1	1 1/4	11 1/2	3	4
5/8	9	2 1/4	2	1 5/16	11 1/2	3	4
21/32	9	2 1/4	2	1 3/8	12	3 1/4	4
11/16	9	2 1/4	2	1 7/16	12	3 1/4	4
23/32	9	2 1/4	2	1 1/2	12 1/2	3 1/2	4
3/4	9 1/2	2 1/2	2	All dimensions in inches.			

Reamers are available with straight, left-hand spiral, and right-hand spiral flute, all right-hand cut. Sizes to $^{11}/_{32}$ have four to six flutes. Sizes from $^{7}/_{16}$ to $^{3}/_{4}$ have six to eight flutes. Sizes from $^{25}/_{32}$ to $^{31}/_{32}$ have eight to ten flutes, and sizes one inch and above have ten to twelve flutes.

Table 12. Expansion Chucking Reamers with Straight or Tapered Shanks. *(Source, Precision Twist Drill Co.)*

Size	Diameter	Straight Shank Dia.	Morse Taper No.	Flute Length	Overall Length
3/8	.3750	5/16	1	3/4	7
13/32	.4063	5/16	1	3/4	7
7/16	.4375	3/8	1	7/8	7
15/32	.4688	3/8	1	7/8	7
1/2	.5000	7/16	1	1	8
17/32	.5313	7/16	1	1	8
9/16	.5625	7/16	1	1 1/8	8
19/32	.5938	7/16	1	1 1/8	8
5/8	.6250	9/16	2	1 1/4	9
21/32	.6563	9/16	2	1 1/4	9
11/16	.6875	9/16	2	1 1/4	9
23/32	.7188	9/16	2	1 1/4	9
3/4	.7500	5/8	2	1 3/8	9 1/2
25/32	.7812	5/8	2	1 3/8	9 1/2
13/16	.8125	5/8	2	1 3/8	9 1/2
27/32	.8438	5/8	2	1 3/8	9 1/2
7/8	.8750	3/4	2	1 1/2	10
29/32	.9063	3/4	2	1 1/2	10
15/16	.9375	3/4	3	1 1/2	10
31/32	.9688	3/4	3	1 1/2	10
1	1.000	7/8	3	1 5/8	10 1/2
1 1/16	1.0625	7/8	3	1 5/8	10 1/2
1 1/8	1.1250	7/8	3	1 3/4	11
1 3/16	1.1875	1	3	1 3/4	11
1 1/4	1.2500	1	4	1 7/8	11 1/2
1 5/16	1.3125	1	4	1 7/8	11 1/2
1 3/8	1.3750	1	4	2	12
1 7/16	1.4375	1 1/4	4	2	12
1 1/2	1.5000	1 1/4	4	2 1/8	12 1/2

All dimensions in inches. Tolerances on diameters 0.3750" through 1.000" = +.0001/+.0005. For sizes above 1.000", tolerances are +.0002/+.0006. Sizes through $^{31}/_{32}$ have six flutes, sizes 1.0000 through 1 $^3/_8$ have eight flutes. Sizes 1 $^7/_{16}$ and above have ten flutes. Number of flutes may vary with manufacturer.

Table 13. Jobber Reamers with Straight Flutes and Morse Taper Shanks.

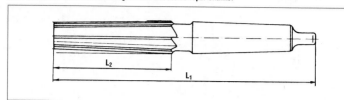

Size	Overall Length	Flute Length	Morse Taper	Size	Overall Length	Flute Length	Morse Taper
1/4	5 3/16	2	1	23/32	8	3 7/8	2
9/32	5 3/16	2	1	3/4	8 3/8	4 3/16	2
5/16	5 1/2	2 1/4	1	25/32	8 3/8	4 3/16	2
11/32	5 1/2	2 1/4	1	13/16	8 13/16	4 9/16	2
3/8	5 13/16	2 1/2	1	27/32	9 3/16	4 7/8	2
13/32	5 13/16	2 1/2	1	7/8	9 3/16	4 7/8	2
7/16	6 1/8	2 3/4	1	15/16	10	5 1/8	3
15/32	6 1/8	2 3/4	1	1	10 3/8	5 7/16	3
1/2	6 7/16	3	1	1 1/16	10 5/8	5 5/8	3
17/32	6 7/16	3	1	1 1/8	10 7/8	5 13/16	3
9/16	6 3/4	3 1/4	1	1 3/16	11 1/8	6	3
19/32	6 3/4	3 1/4	1	1 1/4	12 9/16	6 1/8	4
5/8	7 9/16	3 1/2	2	1 5/16	12 11/16	6 1/4	4
21/32	7 9/16	3 1/2	2	1 3/8	12 13/16	6 5/16	4
11/16	8	3 7/8	2	1 7/16	13	6 7/16	4
All dimensions in inches.				1 1/2	13 1/8	6 1/2	4

Sizes to $^{21}/_{32}$ may have six to eight flutes. Sizes from $^{11}/_{16}$ to $1\,^1/_8$ may have 8 to 10 flutes. Sizes $1\,^3/_{16}$ to $1\,^5/_{16}$ may have 8 to 12 flutes. Sizes $1\,^3/_8$ and above may have 10 to 12 flutes.

Table 14. Bridge Reamers with Taper Shanks and Straight or Helical Flutes.

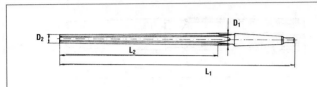

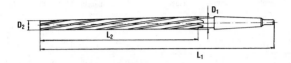

Size	Approx. Dia. at Point D_2		Flute Length	Overall Length	Morse
D_1	Fraction	Decimal	L_2	L_1	Taper
5/16	11/64	.1719	4 3/8	8 1/4	2
13/32	7/32	.2188	4 3/8	8 1/4	2
7/16	17/64	.2656	4 3/8	8 1/4	2
15/32	9/32	.2812	5 1/8	9	2
1/2	5/16	.3125	5 1/8	9	2
17/32	11/32	.3438	5 1/8	9	2
9/16	3/8	.3750	5 1/8	9	2
5/8	25/64	.3906	6 1/8	10	2
11/16	13/32	.4062	7 1/8	11 3/4	3
3/4	15/32	.4688	7 3/8	12	3
13/16	35/64	.5469	7 3/8	12	3
7/8	39/64	.6094	7 3/8	12	3
15/16	43/64	.6719	7 3/8	12	3
1	47/64	.7344	7 3/8	12	3
1 1/16	13/16	.8125	7 3/8	12	3
1 1/8	55/64	.8594	7 3/8	12	3
1 3/16	59/64	.9219	7 3/8	12	3
1 1/4	63/64	.9844	7 3/8	13	4
1 5/16	1 1/16	1.0625	7 3/8	13	4
1 3/8	1 1/8	1.1250	7 3/8	13	4
1 7/16	1 3/16	1.1875	7 3/8	13	4
1 1/2	1 1/4	1.2500	7 3/8	13	4
1 9/16	1 1/4	1.2500	7 3/8	13	4
1 5/8	1 3/8	1.3750	7 3/8	13	4
1 11/16	1 3/8	1.3750	7 3/8	13	4
1 3/4	1 7/16	1.4375	7 3/8	13	4

(Continued)

Table 14. *(Continued)* **Bridge Reamers with Taper Shanks and Straight or Helical Flutes.**

Size D_1	Approx. Dia. at Point D_2		Flute Length L_2	Overall Length L_1	Morse Taper
	Fraction	Decimal			
1 13/16	1 7/16	1.4375	7 3/8	13	4
1 7/8	1 1/2	1.5000	7 3/8	13	4
1 15/16	1 1/2	1.5000	7 3/8	13	4
2	1 3/4	1.7500	7 3/8	13	4

All dimensions in inches.

Table 15. Car Reamers with Morse Taper Shanks.

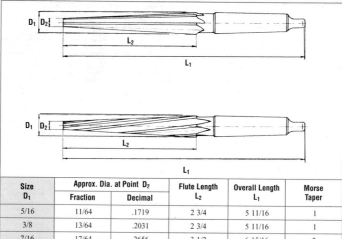

Size D_1	Approx. Dia. at Point D_2		Flute Length L_2	Overall Length L_1	Morse Taper
	Fraction	Decimal			
5/16	11/64	.1719	2 3/4	5 11/16	1
3/8	13/64	.2031	2 3/4	5 11/16	1
7/16	17/64	.2656	3 1/2	6 15/16	2
1/2	5/16	.3125	4	7 9/16	2
9/16	3/8	.3750	4	7 9/16	2
5/8	25/64	.3906	4 1/2	8 1/16	2
11/16	13/32	.4062	4 1/2	8 13/16	3
3/4	15/32	.4688	5	9 1/2	3
13/16	35/64	.5469	5	9 1/2	3
15/16	43/64	.6719	5	9 1/2	3
1	47/64	.7344	5	9 1/2	3
1 1/16	13/16	.8125	5	9 1/2	3

All dimensions in inches. Diameter at point may vary with manufacturer.

Table 16. Chucking and Rose Chucking Shell Reamers with Straight or Helical Flutes. See Table 17 for Arbor Dimensions.

Size D_1	Hole Dia. D_2	Overall Length L_2	Arbor #	Size D_1	Hole Dia. D_2	Overall Length L_2	Arbor #
3/4	3/8	2 1/4	4	2 7/16	1 1/4	3 3/4	9
13/16	1/2	2 1/2	5	2 1/2	1 1/4	3 3/4	9
7/8	1/2	2 1/2	5	2 9/16	1 1/2	4	10
15/16	1/2	2 1/2	5	2 5/8	1 1/2	4	10
1	1/2	2 1/2	5	2 11/16	1 1/2	4	10
1 1/16	5/8	2 3/4	6	2 3/4	1 1/2	4	10
1 1/8	5/8	2 3/4	6	2 13/16	1 1/2	4	10
1 3/16	5/8	2 3/4	6	2 7/8	1 1/2	4	10
1 1/4	5/8	2 3/4	6	2 15/16	1 1/2	4	10
1 5/16	3/4	3	7	3	1 1/2	4	10
1 3/8	3/4	3	7	3 1/16	1 3/4	4 1/2	11
1 7/16	3/4	3	7	3 1/8	1 3/4	4 1/2	11
1 1/2	3/4	3	7	3 3/16	1 3/4	4 1/2	11
1 9/16	3/4	3	7	3 1/4	1 3/4	4 1/2	11
1 5/8	3/4	3	7	3 5/16	1 3/4	4 1/2	11
1 11/16	1	3 1/2		3 3/8	1 3/4	4 1/2	11
1 3/4	1	3 1/2		3 7/16	1 3/4	4 1/2	11
1 13/16	1	3 1/2		3 1/2	1 3/4	4 1/2	11
1 7/8	1	3 1/2		3 9/16	2	5	12
1 15/16	1	3 1/2		3 5/8	2	5	12
2	1	3 1/2	8	3 11/16	2	5	12
2 1/16	1 1/4	3 3/4	9	3 3/4	2	5	12
2 1/8	1 1/4	3 3/4	9	3 13/16	2	5	12
2 3/16	1 1/4	3 3/4	9	3 7/8	2	5	12
2 1/4	1 1/4	3 3/4	9	3 15/16	2	5	12
2 5/16	1 1/4	3 3/4	9	4	2	5	12
2 3/8	1 1/4	3 3/4	9	All dimensions in inches.			

Notes: 1) Flute length (F_2) can be determined as follows. For sizes $3/4$" to $1\ 5/8$", $F_2 = F_1 - 3/4$". For sizes $1\ 9/16$" and larger, $F_2 = F_1 - 1.00$". 2) The number of flutes on the reamer will vary by manufacturer within the following ranges. Diameters up to 1" may have eight to ten flutes. Diameters from $1\ 1/16$" to $1\ 7/16$" may have eight to twelve flutes. Diameters from $1\ 1/2$" to $1\ 11/16$" may have ten to fourteen flutes. Diameters from $1\ 3/4$" to 2" may have twelve to fourteen flutes. Diameters from $2\ 1/16$" to $2\ 1/4$" may have twelve to sixteen flutes. Diameters from $2\ 5/16$" to $2\ 1/2$" may have fourteen to sixteen flutes. Diameters from $2\ 9/16$" to 3" may have sixteen to eighteen flutes. Diameters from $3\ 1/16$" to $3\ 1/2$" may have sixteen to twenty flutes. Diameters above $3\ 9/16$" may have eighteen to twenty flutes.

Table 17. Arbors for Shell Reamers - Straight and Morse Taper Shanks.

Arbor No.	Arbor Dia.	Taper Length	Fits Shell Reamer Dia.	Overall Length	Morse Taper Shank No.	Straight Shank Dia.
4	3/8	2 1/4	21/32 to 25/32	9	2	1/2
5	1/2	2 1/2	13/16 to 1 1/32	9 1/2	2	5/8
6	5/8	2 3/4	1 1/16 to 1 9/32	10	3	3/4
7	3/4	3	1 5/16 to 1 5/8	11	3	7/8
8	1	3 1/2	1 11/16 to 2	12	4	1 1/8
9	1 1/4	3 3/4	2 1/16 to 2 1/2	13	4	1 3/8
10	1 1/2	4	2 9/16 to 3	14	5	1 5/8
11	1 3/4	4 1/2	3 1/16 to 3 1/2	15	5	2
12	2	5	3 9/16 to 4	16	5	2 1/8

All dimensions in inches.

Table 18. Driving Slot and Lug Dimensions for Shell Reamers and Shell Reamer Arbors.

Arbor No.	Hole Dia. Large End	Fits Shell Reamer Dia.	Driving Slot		Lug on Arbor	
			Width	Depth	Width	Depth
4	3/8	21/32 to 25/32	5/32	3/16	9/64	5/32
5	1/2	13/16 to 1 1/32	3/16	1/4	11/64	7/32
6	5/8	1 1/16 to 1 9/35	3/16	1/4	11/64	7/32
7	3/4	1 5/16 to 1 5/8	1/4	5/16	15/64	9/32
8	1	1 11/16 to 2	1/4	5/16	15/64	9/32
9	1 1/4	2 1/16 to 2 1/2	5/16	3/8	19/64	11/32
10	1 1/2	2 9/16 to 3	5/16	3/8	19/64	11/32
11	1 3/4	3 1/16 to 3 1/2	3/8	7/16	23/64	13/32
12	2	3 9/16 to 4	3/8	7/16	23/64	13/32

All dimensions in inches.

Table 19. Shell reamers with Straight and Spiral Flutes, Metric Sizes.

Size D_1	Overall Length L_1	Flute Length L_2	Hole Size D_2	Size D_1	Overall Length L_1	Flute Length L_2	Hole Size D_2
20mm - 24mm	50	40	10	52mm - 60mm	71	50	27
25mm - 30mm	45	32	13	62mm - 70mm	80	56	32
31mm - 35mm	50	36	16	72mm - 85mm	90	63	40
36mm - 42mm	56	40	19	88mm - 100mm	100	71	50
43mm - 50mm	63	45	22	All dimensions in millimeters. Ref: DIN 219			

Table 20. Stub Screw Machine Reamers. *(Source, Yankee Reamer.)*

Size No.	Diameter Range	Flute Length L_2	Overall Length L_1	Shank Dia. D_2	# Flutes	Hole Dia. H
000	.0300 to .0599	1/2	2	1/8	4	1/46
00	.0600 to .0660	1/2	2	1/8	4	1/16
0	.0661 to .0740	1/2	2	1/8	4	1/16
1	.0741 to .0840	1/2	2	1/8	4	1/16
2	.0841 to .0960	1/2	2	1/8	4	1/16
3	.0961 to .1260	3/4	2	1/8	4	1/16
4	.1261 to .1580	1	2 1/4	1/4	4	3/32
5	.1581 to .1880	1	2 1/4	1/4	6	3/32
6	.1881 to .2190	1	2 1/4	1/4	6	3/32
7	.2191 to .2510	1	2 1/4	1/4	6	3/32
8	.2511 to .2820	1	2 1/4	3/8	6	1/8
9	.2821 to .3130	1	2 1/4	3/8	6	1/8
10	.3131 to .3440	1 1/4	2 1/2	3/8	6	1/8
11	.3441 to .3760	1 1/4	2 1/2	3/8	6	1/8
12	.3761 to .4070	1 1/4	2 1/2	1/2	6	3/16
13	.4071 to .4390	1 1/4	2 1/2	1/2	6	3/16
14	.4391 to .4700	1 1/4	2 1/2	1/2	6	3/16
15	.4701 to .5050	1 1/4	2 1/2	1/2	6	3/16
16	.5051 to .5670	1 1/2	3	5/8	8	1/4
17	.5671 to .6300	1 1/2	3	5/8	8	1/4
18	.6301 to .6920	1 1/2	3	5/8	8	1/4
19	.6921 to .7550	1 1/2	3	3/4	8	5/16
20	.7551 to .8170	1 1/2	3	3/4	8	5/16
21	.8171 to .8800	1 1/2	3	3/4	8	5/16
22	.8801 to .9420	1 1/2	3	3/4	8	5/16
23	.9421 to 1.010	1 1/2	3	3/4	8	5/16
24	1.0011 to 1.0700	1 1/2	3	3/4	8-10	5/16
25	1.0701 to 1.1320	1 1/2	3	3/4	8-10	5/16
26	1.1321 to 1.1940	1 1/2	3	3/4	8-10	5/16
27	1.1941 to 1.2560	1 1/2	3	3/4	8-10	5/16
28	1.2561 to 1.3180	1 1/2	3	3/4	8-10	5/16
29	1.3181 to 1.3800	1 1/2	3	3/4	8-10	5/16
30	1.3801 to 1.4420	1 1/2	3	3/4	8-10	5/16
31	1.4421 to 1.5312	1 1/2	3	3/4	8-10	5/16

All dimensions in inches.

Table 21. Dowel Pin Reamer Dimensions. *(Source, Yankee Reamer.)*

Reamer Dia.	For Dowel Pin	Tap Drill* Size	Shank Dia.	Flute Length	Overall Length	No. Flutes
.1230	1/8	7/64	.1190	7/8	3 1/2	4
.1247	1/8	7/64	.1190	7/8	3 1/2	4
.1855	3/16	11/64	.1805	1 1/8	4 1/2	6
.1870	3/16	11/64	.1805	1 1/8	4 1/2	6
.2480	1/4	15/64	.2405	1 1/2	6	6
.2495	1/4	15/64	.2405	1 1/2	6	6
.3105	5/16	19/64	.2792	1 1/2	6	6
.3120	5/16	19/64	.2792	1 1/2	6	6
.3730	3/8	23/64	.3105	1 3/4	7	6
.3745	3/8	23/64	.3105	1 3/4	7	6
.4355	7/16	27/64	.3730	1 3/4	7	6
.4370	7/16	27/64	.3730	1 3/4	7	6
.4980	1/2	15/32	.4355	2	8	6
.4995	1/2	15/32	.4355	2	8	6

* Recommended hole is $1/64$" under dowel pin diameter. See text for details. All dimensions in inches unless otherwise stated.

Tolerances for reamers

Reamers are precision tools that are manufactured to very tight tolerances. The United States Cutting Tool Institute specifies that dimensions must comply with the following tolerances.

Reamer Diameter: 1) Hand reamers with straight or helical flutes and squared shanks; 2) taper shank jobber reamers with straight flutes; 3) shell reamers with straight or helical flutes; 4) expansion chucking reamers, straight flute, straight or tapered shank; 5) chucking reamers with straight or helical flutes and tapered or straight shanks; and 6) stub screw machine reamers are toleranced by diameter ranges as follows.

No. 60 to $1/4$" inclusive (0.0400" to 0.2500")   +0.0001" to +0.0004"
Over $1/4$" to 1" inclusive (0.2501" to 1.0000")   +0.0001" to +0.0005"
Over 1" to 3" inclusive (1.0001" to 3.0000")   +0.0002" to +0.0006"

Flute Length: 1) Hand reamers with straight or helical flutes and squared shanks; 2) taper shank jobber reamers with straight flutes; 3) shell reamers with straight or helical flutes; 4) expansion chucking reamers, straight flute, straight or tapered shank; 5) chucking reamers with straight or helical flutes and tapered or straight shanks; 6) Stub screw machine reamers; and 7) expansion hand reamers with straight flute and squared shaft are toleranced by diameter ranges as follows.

No. 60 to 1" inclusive (0.0400" to 1.0000") $+1/16$" to $-1/16$"
Over 1" to 2" inclusive (1.0001" to 2.0000") $+3/32$" to $-3/32$"
Over 2" to 3" inclusive (2.0001" to 3.0000") $+1/8$" to $-1/8$"

Overall Length: 1) Hand reamers with straight or helical flutes and squared shanks; 2) taper shank jobber reamers with straight flutes; 3) shell reamers with straight or helical flutes; 4) expansion chucking reamers, straight flute, straight or tapered shank; 5) chucking reamers with straight or helical flutes, and tapered or straight shanks; 6) stub screw machine reamers; and 7) expansion hand reamers with straight flute and squared shaft are toleranced by diameter ranges as follows.

No. 60 to 1" inclusive (0.0400" to 1.0000") $+^1/_{16}$" to $-^1/_{16}$"
Over 1" to 2" inclusive (1.0001" to 2.0000") $+^3/_{32}$" to $-^3/_{32}$"
Over 2" to 3" inclusive (2.0001" to 3.0000") $+^1/_8$" to $-^1/_8$"

Shank Diameter: 1) Straight shank solid hand reamers with straight or helical flutes, and 2) straight shank expansion reamers with straight or helical flutes are toleranced by diameter ranges as follows.

From $^1/_8$" to 1" inclusive (0.1250" to 1.0000")   -0.0010" to -0.0050"
Over 1" to 1 $^1/_2$" inclusive (1.0001" to 1.5000")   -0.0015" to -0.0060"

- Straight shank solid chucking reamers with straight or helical flutes, and straight shank solid rose chucking reamers with straight or helical flutes are toleranced by diameter ranges as follows:
  From 0.0390" to 0.4355" inclusive   +0.0000" to -0.0010"
  Over 0.4355" to 1.2495" inclusive   +0.0000" to -0.0015"
- Taper pipe reamers with straight shank and straight flutes (carbon steel):
  From 0.4375" to 0.7000" inclusive   +0.0000" to -0.0070"
  Over 0.7000" to 1.1250" inclusive   +0.0000" to -0.0090"
- Taper pipe reamers with straight shank and spiral flutes (high speed steel):
  0.4375"   +0.0000" to -0.0070"
  Over 0.4375" to 1.1250" inclusive   +0.0000" to -0.0020"
  Over 1.1250" to 1.8750" inclusive   +0.0000" to -0.0030"
- Expansion chucking reamers:
  From 0.3125" to 1.7500" inclusive   -0.0005" to -0.0020"
- Stub screw machine reamers:
  From 0.1250" to 0.7500" inclusive   -0.0005" to -0.0020"
- Morse taper reamers:
  From 0.3125" to 1.5000" inclusive   -0.0050" to -0.0020"
- Brown and Sharpe taper reamers:
  From 0.2812" to 1.1250" inclusive   -0.0050" to -0.0020"
- Taper pin reamers with straight or spiral flutes:
  From 0.0781" to 0.6250" inclusive   -0.0010" to -0.0050"
- Taper pin reamer with high spiral flutes:
  From 0.0781" to 0.6250" inclusive   -0.0005" to -0.0020"

Determining hole size for reaming

Reamers require that a predetermined amount of stock be left in the workpiece in order to prolong tool life and produce a hole of the desired size and finish. If excessive material remains in a drilled hole to be removed by the reamer, problems with chip evacuation and potential damage to the reamer can occur. Too little stock will result in excessive wear. To determine the proper reamer and drill to be used for a given size hole, Yankee Reamer provides the following reamer and drill selection procedures.

Selecting the correct reamer and drill size for a given application requires three steps. The first is to determine the correct reamer diameter based on the total hole tolerance and the machinability of the workpiece material. The second is to determine the correct drilled hole diameter needed for proper reaming, and the third is to select the proper drill size to produce a hole of the required diameter.

Step 1) Calculate the best diameter reamer, based on the machinability of the material to be reamed by deducting the allowances in **Table 22** from the maximum finished hole size. For example, if the finished hole size is 0.500" with a -0.0005/+.0000" tolerance, and the material is 302 Stainless Steel with a machinability rating of 50, the Table shows that the range to deduct is 0.0004 to 0.0005". (See page 408 for chart of machinability ratings for a variety of materials.)

0.5000 - 0.0004 = 0.4996"

0.5000 - 0.0005 = 0.4995"

Therefore, the best obtainable reamer size is 0.4995".

Using a metric example, Step 1 is calculated as follows. If the finished hole size is 10 mm, with a -0.0762/+.0000 mm tolerance, and the material is 302 Stainless Steel with a machinability rating of 50, the Table shows that the range to deduct is 0.0229 to 0.0508 mm.

10 - 0.0229 = 9.9771 mm.

10 - 0.0508 = 9.9492 mm.

Therefore, the best obtainable reamer size is 9.50 mm.

Table 22. Total Hole Tolerances in Reaming. *(Source, Yankee Reamer.)*

Total Hole Tolerance		Machinability Range			
		Below 30 and Above 100		31-100	
inch	mm	inch	mm	inch	mm
0.0005	0.0127	0.0003-0.0004	0.0076-0.0102	0.0004-0.0005	0.0102-0.0127
0.0010	0.0254	0.0004-0.0006	0.0102-0.0152	0.0006-0.0008	0.0152-0.0203
0.0015	0.0381	0.0005-0.0008	0.0127-0.0203	0.0008-0.0010	0.0203-0.0254
0.0020	0.0508	0.0006-0.0010	0.0152-0.0254	0.0010-0.0014	0.0254-0.0356
0.0030	0.0762	0.0009-0.0020	0.0229-0.0508	0.0009-0.0020	0.0229-0.0508
0.0040	0.1016	0.0012-0.0026	0.0279-0.0660	0.0012-0.0026	0.0279-0.0660
0.0050	0.1270	0.0014-0.0032	0.0356-0.0813	0.0014-0.0032	0.0356-0.0813
0.0060	0.1524	0.0017-0.0038	0.0432-0.0965	0.0017-0.0038	0.0432-0.0965
0.0070	0.1778	0.0019-0.0044	0.0483-0.1118	0.0019-0.0044	0.0483-0.1118
0.0080	0.2032	0.0022-0.0050	0.0559-0.1270	0.0022-0.0050	0.0559-0.1270
0.0090	0.2286	0.0024-0.0056	0.0610-0.1422	0.0024-0.0056	0.0610-0.1422
0.0100	0.2540	0.0027-0.0062	0.0686-0.1575	0.0027-0.0062	0.0686-0.1575

Step 2) Using the diameter range for the reamer selected in Step 1, **Table 23** shows the amount of stock that should be removed by the reamer (leaving the recommended amount will ensure the best finish and tolerance). Using the decimal-inch example from Step 1, the table shows that the amount to deduct from the reamer size is 0.007 to 0.015".

0.4995 - 0.015 = 0.4845"

0.4995 - 0.007 = 0.4925".

Therefore, the best hole size for reaming is between 0.4845 and 0.4925".

Using the diameter range for the metric reamer selected in Step 1, **Table 23** shows that the amount to deduct from the reamer size is 0.1778 to 0.3810 mm.

9.50 - 0.3810 = 9.1190 mm

9.50 - 0.1778 = 9.3222 mm.

Therefore, the best hole size for reaming is between 9.1190 and 9.3222 mm.

Step 3) Using the hole diameters selected in Step 2, **Table 24** provides the oversize allowance that should be deducted from the hole size to determine the correct drill size. Using the decimal-inch example, the table shows that the amount to deduct from the hole diameters selected in Step 2 is 0.0048".

0.4845 - 0.0048 = 0.4797"

0.4925 - 0.0048 = 0.4877".

Table 23. Ideal Stock Removal Based on Reamer Diameter and Machinability. *(Source, Yankee Reamer.)*

Reamer Diameter Range		Machinability Range			
		30 or Less		31 and Up	
inch	mm	inch	mm	inch	mm
to 0.0625	to 1.5880	0.006-0.010	0.1524-0.2540	0.003-0.005	0.0762-0.1270
0.0626-0.1250	1.5900-3.1750	0.006-0.010	0.1524-0.2540	0.004-0.007	0.1016-0.1778
0.1251-0.2500	3.1775-6.3500	0.006-0.010	0.1524-0.2540	0.005-0.010	0.1270-0.2540
0.2501-0.5000	6.3525-12.700	0.012-0.020	0.3048-0.5080	0.007-0.015	0.1778-0.3810
0.5001-1.000	12.7025-25.40	0.012-0.020	0.3048-0.5080	0.010-0.020	0.2540-0.5080

Therefore, the best practical drill size is $^{31}/_{64}$ (0.4844) because it falls between 0.4797 and 0.4877".

Using the metric example, **Table 24** shows that the amount to deduct from the hole diameters selected in Step 2 is 0.1219 mm.

9.1190 - 0.1219 = 8.9971 mm

9.3222 - 0.1219 = 9.2003 mm.

Therefore, either a Letter "T" size (9.093 mm) or a 9.10 mm drill size can be used because both fall between 8.9971 and 9.2003 mm.

Table 24. Oversize Allowances for Drilled Holes. *(Source, Yankee Reamer.)*

Hole Diameter Range		Machinability Range			
		Below 30 and Above 100		31-100	
inch	mm	inch	mm	inch	mm
to 0.0625	to 1.5880	0.0005	0.0127	0.0015	0.0381
0.0626-0.1250	1.5900-3.1750	0.0006	0.0152	0.0028	0.0711
0.1251-0.1875	3.1775-4.7625	0.0008	0.0203	0.0036	0.0914
0.1876-0.2500	4.7650-6.3500	0.0008	0.0203	0.0042	0.1067
0.2501-0.5000	6.3525-12.700	0.0010	0.0254	0.0048	0.1219
0.5001-0.7500	12.7025-19.050	0.0012	0.0279	0.0052	0.1321
0.7501-1.0000	19.1525-25.400	0.0020	0.0508	0.0065	0.1651

Operating parameters for reamers

To achieve optimal performance, attention must be paid to the following.

1) The machine utilizing the tooling must have the necessary rigidity to minimize spindle deflection, and sufficient horsepower to perform at recommended speeds and feeds. As a general starting point, divide speeds used for drilling by one-half, and double the feeds used for drilling.

2) Holders and collets must provide good concentricity between tool and machine spindle.

3) The workpiece must be rigidly clamped and supported to minimize deflection.

4) Solid carbide reamers cannot tolerate chatter. Use a rigid setup, and make sure there is proper alignment between the reamer and hole.

5) Use as short a reamer as the application will permit—this will provide maximum tool rigidity.

6) Use the correct speeds and feeds to suit the application and material being machined.

7) Ensure that the proper amount of stock has been left in a hole before reaming. Insufficient stock will cause excessive reamer wear; excessive stock will cause problems with chips that may lead to the reamer freezing in the hole or possible breakage.

8) Coolant improves tool life. Direct the flow of coolant to the cutting edges. Insufficient or poorly directed coolant supply can cause hot spots, resulting in chipping or flaking of carbide. Reaming is not a heavy cutting operation, and a 40:1 dilution of soluble oil is normally satisfactory. For gray cast iron, air blasting is useful for cooling and chip removal.

9) If chatter is exhibited in spite of good setup conditions, reduce spindle RPM. If chatter persists, increase feed rate.

10) Use power feed when reaming materials up to Rc 50; hand feed on harder materials.

11) The reamer must be sharp. When indication of a dull reamer is present, resharpen or replace.

12) The use of guide bushings is recommended wherever possible. The bushing inside diameter should be approximately 0.0002" larger than the reamer's diameter.

13) The reamer should never be stopped while engaged in the hole.

14) If reamed holes are oversize, investigate the following potential causes: misalignment, vibration, or chatter; reamer is bottoming in blind hole; or dull reamer.

15) If reamed holes are undersize check the reamer diameter; excessive heat from improper speeds and feeds.

16) If reamed holes are bellmouthed, check for misalignment, vibration, or chatter; bent reamer shank; excessive tool overhang.

17) If the finish of a reamed hole is unsatisfactory, check for dull or damaged reamer; misalignment, vibration, or chatter; excessive stock removal; poor machinability of workpiece material; tool marks being left in hole.

18) Use the proper rake angle for the material being reamed. **Table 25** provides suggested rake angles for various materials.

Table 25. Suggested Rake Angles for Reaming. *(Source, Alvord-Polk Tool Co.)*

Material	Rake in Degrees
Aluminum, Magnesium, Stainless Steel	6 - 8
Brass, Malleable Iron, Monel, Steel, Titanium	3 - 5
Bronze	2 - 4
Cast Iron, Plastics	0 - 3

Speeds and feeds for reaming

Chatter is an indication of excessive speed. Feed rates need to be increased if the lower speeds do not eliminate the chatter. When reaming harder materials such as Rc 50 or Rc 60, speeds should be greatly reduced. When reaming nonferrous metals, the speeds may be increased. It is always preferable to hand feed when reaming materials that are harder than Rc 50. When a carbide reamer is engaged in a cut, the feed should never be stopped. The shock from a sudden stop can very easily chip the tool. **Table 26** gives recommended starting speeds and feeds in surface feet per minute (sfm) and inches per revolution (ipr) for the most commonly used materials. **Table 27** provides metric units in surface meters per minute (smm) and millimeters per revolution (mm/rev).

Table 26: Speeds and Feeds for Reaming in SFM and Inches. (Source, Precision Twist Drill Co.)

Material	Speed in SFM		Reamer Diameter in Inches					
	HSS	Carbide	To 1/16 IPR	.0626 to 1/8 IPR	.1250 to 1/4 IPR	.2501 to 1/2 IPR	.5001 to 1 IPR	Over 1 IPR
Aluminum	150-300	500-1000	.0005-.003	.002-.006	.004-.010	.006-.015	.010-.030	.020-.050
Brass/Bronze - Free Machining	125-200	250-400	.0005-.002	.002-.004	.004-.006	.006-.010	.010-.020	.020-.040
Brass/Bronze - Tough	75-125	150-250	.0005-.002	.002-.004	.004-.006	.006-.010	.010-.020	.020-.040
Cast Iron - Soft (Ferritic)	50-100	150-250	.001-.003	.003-.006	.006-.010	.010-.015	.015-.030	.030-.050
Cast Iron - Medium (Pearlitic)	25-50	75-150	.0002-.002	.001-.004	.002-.006	.004-.010	.006-.020	.010-.040
Cast Iron - Hard (Martensitic)	15-25	50-75	.0002-.001	.001-.002	.002-.004	.004-.006	.006-.010	.010-.020
Copper - Hard Bronze	50-75	100-150	.0002-.001	.001-.002	.002-.004	.004-.006	.006-.010	.010-.020
High Temp. Alloy - Nickel Base	10-20	40-70	.0002-.001	.001-.002	.002-.004	.004-.006	.006-.010	.010-.020
High Temp. Alloy - Cobalt Base	10-15	30-45	.0002-.001	.001-.002	.002-.004	.004-.006	.006-.010	.010-.020
Magnesium	200-400	500-1000	.0005-.003	.002-.006	.004-.010	.006-.015	.010-.030	.020-.050
Stainless Steel - Free Mach/400 Ann	40-60	150-250	.0005-.002	.002-.004	.004-.006	.006-.010	.010-.020	.020-.040
Stainless Steel - 300 Series	20-30	80-120	.0005-.002	.002-.004	.004-.006	.006-.010	.010-.020	.020-.040
Stainless Steel - PH & HT 400 Series	15-25	60-100	.0002-.002	.001-.004	.002-.006	.004-.010	.006-.010	.010-.040
Steel - Under 200 BHN	55-80	200-300	.0005-.003	.002-.006	.004-.010	.006-.015	.010-.030	.020-.050
Steel - 200 to 300 BHN	30-55	125-200	.0005-.002	.002-.004	.004-.006	.006-.010	.010-.020	.020-.040
Steel - 300 to 400 BHN	20-30	50-125	.0005-.002	.002-.004	.004-.006	.006-.010	.010-.020	.020-.040
Steel - 400 to 500 BHN	10-20	35-50	.0005-.002	.002-.004	.004-.006	.006-.010	.010-.020	.020-.040
Steel - Over 500 BHN	-	15-35	.0005-.002	.002-.004	.004-.006	.006-.010	.010-.020	.020-.040
Titanium - Pure	35-50	50-100	.0005-.002	.002-.004	.004-.006	.006-.010	.010-.020	.020-.040
Titanium - Alloys	10-20	35-50	.0002-.002	.001-.004	.002-.006	.004-.010	.006-.020	.010-.040

Note: RPM = (SFM ÷ Tool Diameter) × 3.82. The material operating parameters provided above are suggested starting points under ideal conditions. Adjustments to speeds and feeds may be required to meet individual job requirements. Preliminary cuts should be made at the lowest range and increased as conditions allow.

Table 27: Speeds and Feeds for Reaming in SMM and Millimeters. (Source, Precision Twist Drill Co.)

Material	Speed in SMM		Reamer Diameter in Millimeters					
	HSS	Carbide	To 1.6	1.61 to 3.2	3.21 to 6.4	6.41 to 12.7	12.71 to 25	Over 25
			mm/rev	mm/rev	mm/rev	mm/rev	mm/rev	mm/rev
Aluminum	46-91	152-305	.013-.075	.05-.15	.10-.25	.15-.38	.25-.76	.50-1.30
Brass/Bronze - Free Machining	38-61	76-122	.013-.05	.05-.10	.10-.15	.15-.25	.25-.50	.50-1.0
Brass/Bronze - Tough	22-38	46-76	.013-.05	.05-.10	.10-.15	.15-.25	.25-.50	.50-1.0
Cast Iron - Soft (Ferritic)	15-31	46-76	.025-.075	.075-.15	.15-.25	.25-.38	.38-.76	.76-1.3
Cast Iron - Medium (Pearlitic)	8-15	22-46	.005-.05	.005-.10	.05-.15	.10-.25	.15-.50	.25-1.0
Cast Iron - Hard (Martensitic)	4-8	15-22	.005-.025	.005-.05	.05-.15	.10-.15	.15-.25	.25-.50
Copper - Hard Bronze	15-22	31-46	.005-.025	.005-.05	.05-.10	.10-.15	.15-.25	.25-.50
High Temp. Alloy - Nickel Base	3-6	12-21	.005-.025	.005-.05	.05-.10	.10-.15	.15-.25	.25-.50
High Temp. Alloy - Cobalt Base	3-4	9-14	.005-.025	.005-.05	.05-.10	.10-.15	.15-.25	.25-.50
Magnesium	61-122	152-305	.013-.075	.05-.15	.10-.25	.15-.38	.25-.76	.50-1.30
Stainless Steel - Free Mach/400 Ann	12-17	46-76	.013-.05	.05-.10	.10-.15	.15-.25	.25-.50	.50-1.0
Stainless Steel - 300 Series	6-9	24-37	.013-.05	.05-.10	.10-.15	.15-.25	.25-.50	.50-1.0
Stainless Steel - PH & HT 400 Series	4-8	17-31	.005-.05	.005-.10	.05-.15	.10-.25	.15-.50	.25-1.0
Steel - Under 200 BHN	16-24	61-91	.013-.075	.05-.15	.10-.25	.15-.38	.25-.76	.50-1.30
Steel - 200 to 300 BHN	9-16	38-61	.013-.05	.05-.10	.10-.25	.15-.25	.25-.50	.50-1.0
Steel - 300 to 400 BHN	6-9	15-38	.013-.05	.05-.10	.10-.25	.15-.25	.25-.50	.50-1.0
Steel - 400 to 500 BHN	3-6	10-15	.013-.05	.05-.10	.10-.25	.15-.25	.25-.50	.50-1.0
Steel - Over 500 BHN	-	4-10	.013-.05	.05-.10	.10-.25	.15-.25	.25-.50	.50-1.0
Titanium - Pure	10-15	15-31	.013-.05	.05-.10	.10-.25	.15-.25	.25-.50	.50-1.0
Titanium - Alloys	3-6	10-15	.005-.05	.005-.10	.05-.15	.10-.25	.15-.50	.25-1.0

Note: RPM = (SMM ÷ Tool Diameter) × 318.3. The material operating parameters provided above are suggested starting points under ideal conditions. Adjustments to speeds and feeds may be required to meet individual job requirements. Preliminary cuts should be made at the lowest range and increased as conditions allow.

Broaching

A broach is a precision metal cutting tool that incorporates a series of rough, semi-finish, and finish teeth designed to remove successive portions of stock as it moves through or across a workpiece in a one-pass operation. Each tooth is calibrated to remove only a small amount of stock appropriate to the type of material broached, thereby permitting continuous clean chip removal. Broaching processes and tools are classified first as either surface or internal, then as standard or special. Each broaching tool is designed to fit the starting point, and the number and length of its teeth are set by the condition of the stock (machined, cast, or forged) and the amount of material to be removed. When properly applied, no other machining method can remove metal faster than broaching, while achieving the tolerances and accuracy that this technology provides. Much of the information in this section was provided courtesy of Hassay Savage Company, including *Figure 4* that graphically explains basic broach terminology terms.

Surface (or external) broaching is commonly used in place of milling or shaping operations on the surfaces of parts or components. Almost any shape can be broached—including flats, notches, slots, keyways, contoured surfaces, external gear teeth, and serrations—if the surface is in a straight line and unobstructed. Surface broaching is

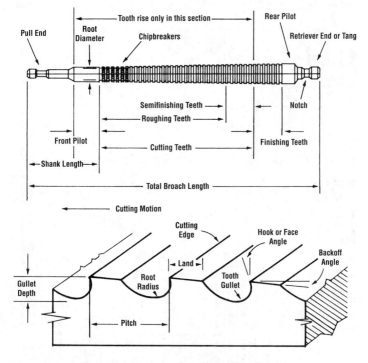

Figure 4. Broach terminology. (Source, Hassay Savage Company.)

especially effective when compared to other methods because it can produce highly finished surfaces of complex shapes with exacting dimensions at a rate much faster than can be achieved by milling. Internal broaching begins with a round-drilled hole and shapes it to practically any configuration from the simplest application (reaming) to the most intricate (rifling gun barrels). Almost all internal broaching is done by pull broaches because they remove significantly more stock than push broaches, even when the workpiece has a thin or irregular wall. Also, a pulled internal broach can handle longer lengths of cut. Internal broaching tools include round pull, round push, rotor cut style, straight spline and serration, and combination round and spline. Internal applications include tooling fixtures, gear and pulley keyways, keyways, rifling, and hole configurations in all configurations including square or hexagonal.

Three major factors enter into all broaching operations: the material to be broached, tool alignment, and lubrication. The hardness and composition of a material determines how effectively a part can be broached. Materials with a hardness of up to R_C 35 (HB 327) can be broached with excellent results. Harder materials are not recommended. Although some steels with hardnesses of up to R_C 42 can be broached, tool life suffers dramatically, surface finishes are often not as smooth as desired, and machining power requirements are very high. When broaching steels of 18 to 32 R_C, accuracies of ±0.0005" (±0.013 mm) with surface finishes of up to 32 to 100 μinch (0.8 to 2.5 μm) can be achieved. Workpiece material is also influential in determining the pitch of the broach. Pitch, expressed in number of teeth per inch, is the linear distance from the cutting edge of one tooth to the next (see *Figure 4*). For roughing, it will be coarser than for finishing, and all broaching operations require that at least two teeth are engaged at all times. A simple formula for determining the best pitch for short broaches is

$$P = 0.35 \sqrt{L} \quad \text{for inch units, and} \qquad P = 1.764 \sqrt{L} \quad \text{for metric units,}$$

where: P = pitch, in inches or millimeters

L = length of cut, in inches or millimeters.

Poor tool alignment will cause tool breakage and drifting that will produce a part outside of tolerance. To assure that the tool is aligned properly at all times, the following measures should be taken. 1) Assure that at least two (and preferably more) teeth are engaged at all times. 2) Use bushings below (or beyond) the workpiece to provide appropriate support for the broach. 3) In push broaching, center the tool below the ram so that the ram thrust forces drive the broach directly through the bushing to avoid drift. 4) Use a pilot hole that is appropriately dimensioned and accurately drilled to start the broach. Keep in mind that even when the above steps are followed, alignment will be adversely affected if the broach has become dulled through long use, or by incorrect use in materials that were initially too hard. Dull broaches produce cuts with incorrect tolerances, unacceptable finishes, and they tend to drift out of alignment. Broaches require resharpening when excessive pressure is required to run the broach, when tearing or poor surface finish is evident, when the broach drifts during machining, when chatter occurs, and when cutting sections show visible signs of wear.

Push broaching of keyways is done with a hydraulic press, and larger size keyways may require repeated passes using slightly larger shims for each successive pass. In order to achieve sufficient rigidity, push broaches must be of adequate cross section as to resist bending or deflection. Both American standard and metric keyway broaches and shims are available in a wide range of sizes (see **Tables 28** and **29**). Standard inch and metric square broaches and standard inch full square broach sizes are provided in **Table 30**, standard inch and metric hexagon broaches in **Table 31**, and standard round push broaches in **Table 32**. Pull broaches can be thin in cross section because pulling tends to straighten the hole or keyway. Inch pull-type keyway broach sizes are found in **Table 33**, and standard inch and metric pull-type keyseating broaches for Davis-type machines are shown in **Table 34**.

All cutting processes involve friction that causes heat, distortion, and vibration that can adversely affect the cut. Proper lubrication helps reduce friction and assists in chip removal. When broaching steel, use a good grade of cutting oil or a soluble coolant sprayed directly on the broach teeth. Hard steels, especially those containing nickel, may require a chlorine additive or a sulfur-based cutting oil. Brass may be broached dry or with a water soluble lubricant. Cast iron should be broached dry, but the bushings should be lubricated with oil. For copper and bronze, use oil or a water-soluble oil. Aluminum requires special lubricants, depending on the nature of the aluminum. Harder varieties may be treated as soft steel. In general, the alloy manufacturer's recommendation for low-speed machining operations should be followed. **Table 35** provides recommended machining parameters for broaching operations.

Troubleshooting broaching operations

Poor finishes may be caused by the workpiece material being too soft, by broach misalignment, by improperly applied or unsuitable coolant, or by a poorly sharpened broach. Chatter marks indicate that the broach may be too sharp, that the coolant is improperly applied or unsuitable, that speeds may be too high, that the tool is dull or poorly applied, or that the tooth spacing of the broach is not compatible with the job. A drifting broach can be caused by incorrect support or alignment, by the material being too soft, or by a dull or poorly sharpened broach. Excessive edge wear may be caused by abrasive workpiece materials, by an improperly applied or unsuitable lubricant, by incorrect speed rates, or by a broach made of unsuitable material. Chipping can be caused by too little or unsuitable lubricant, incorrect tooth spacing for the workpiece material, by the tool speed being too fast for effective chip removal, or by the chip load being too high.

Table 28. American Standard Keyway Broach and Shim Dimensions and Machining Information.
(Source, Hassay Savage Company.)

Broach Size	Width Tolerance	Dimensions	No. of Shims	Length of Cut Min.	Length of Cut Max.	Pressure for Max. l/c*
1/16	.0625 - .0635	1/8 × 5 1/4	0	13/64	1 1/8	270 lbs.
3/32	.0937 - .0947	1/8 × 5 1/4	0	13/64	1 1/8	810 lbs.
1/8	.1252 - .1262	1/8 × 5 1/4	1	13/64	1 1/8	720 lbs.
3/32	.0937 - .0947	3/16 × 7	0	19/64	1 11/16	990 lbs.
1/8	.1252 - .1262	3/16 × 7	1	19/64	1 11/16	780 lbs.
5/32	.1564 - .1574	3/16 × 7	1	19/64	1 11/16	1,560 lbs.
3/16	.1877 - .1887	3/16 × 7	1	19/64	1 11/16	2,100 lbs.
3/16	.1877 - .1887	3/8 × 11 7/8	1	13/32	2 1/2	1,440 lbs.
1/4	.2502 - .2512	3/8 × 11 7/8	1	13/32	2 1/2	2,580 lbs.
5/16	.3127 - .3137	3/8 × 11 7/8	1	13/32	2 1/2	4,200 lbs.
3/8	.3755 - .3765	3/8 × 11 7/8	2	13/32	2 1/2	3,600 lbs.
5/16	.3127 - .3137	9/16 × 13 7/8	1	3/4	6	9,250 lbs.
3/8	.3755 - .3765	9/16 × 13 7/8	2	3/4	6	8,000 lbs.
7/16	.4380 - .4390	9/16 × 13 7/8	2	3/4	6	11,750 lbs.
1/2	.5006 - .5016	9/16 × 13 7/8	3	3/4	6	11,000 lbs.
9/16	.5630 - .5640	9/16 × 13 7/8	3	3/4	6	11,100 lbs.
5/8	.6260 - .6270	3/4 × 15 1/4	4	3/4	6	11,500 lbs.
3/4	.7515 - .7525	3/4 × 15 1/4	5	3/4	6	13,500 lbs.
7/8	.8765 - .8775	1 × 19 1/4	6	3/4	6	11,000 lbs.
1	1.0015 - 1.0025	1 × 19 1/4	7	3/4	6	14,000 lbs.
1 1/8	1.1260 - 1.1270	1 1/8 × 23	8	3/4	6	15,750 lbs.
1 1/4	1.2510 - 1.2520	1 1/4 × 23	9	3/4	6	17,500 lbs.
1 3/8	1.3760 - 1.3770	1 3/8 × 23	10	3/4	6	19,250 lbs.
1 1/2	1.5020 - 1.5030	1 1/2 × 23	11	3/4	6	21,000 lbs.

Shims for Standard Keyway Broaches

Shim Size	Thickness	Length × Width	Shim Size	Thickness	Length × Width
1/8	.031	1 3/8 × 1/8	1/2	.062	6 1/4 × 9/16
1/8	.031	1 7/8 × 3/16	9/16	.062	6 1/4 × 9/16
5/32	.042	1 7/8 × 3/16	5/8	.062	6 3/4 × 3/4
3/16	.050	1 7/8 × 3/16	3/4	.062	6 3/4 × 3/4
3/16	.050	2 7/8 × 3/8	7/8	.062	6 3/4 × 1
1/4	.062	2 7/8 × 3/8	1	.062	6 3/4 × 1
5/16	.078	2 7/8 × 3/8	1 1/8	.062	7 × 1 1/8
3/8	.062	2 7/8 × 3/8	1 1/4	.062	7 × 1 1/4
7/16	.075	6 1/4 × 9/16			

* l/c = length of cut.

Broaches and bushings are sized for compatibility and several bushing sizes can be obtained for individual broach sizes up to 1". Required shims are normally supplied with broaches. Shim sizes correspond to broach size, not to bushing size. Consult supplier for details. All broach dimensions are in inches.

Table 29. ISO J-9 and ISO P9 Standard Tolerance Keyway Broaches. (Source, Hassay Savage Company.)

Broach Size mm	ISO J-9 Width Tolerance	ISO P9 Width Tolerance	Key Size (mm)	Dimensions (inches)	No. of Shims	Length of Cut (inches) Min.	Length of Cut (inches) Max.	Pressure for Max. l/c*
2	.0708 - .0792	.0778 - .0783	2 × 2	1/8 × 5 1/4	0	13/64	1 1/8	510 lbs.
3	.1181 - .1186	.1172 - .1177	3 × 3	1/8 × 5 1/4	1	13/64	1 1/8	720 lbs.
4	.1575 - .1580	.1563 - .1568	4 × 4	1/4 × 7	1	19/64	1 11/16	1,140 lbs.
5	.1968 - .1973	.1957 - .1962	5 × 5	1/4 × 7	1	19/64	1 11/16	2,040 lbs.
5	.1968 - .1973	.1957 - .1962	5 × 5	3/8 × 11 7/8	1	13/32	2 1/2	1,680 lbs.
6	.2362 - .2367	.2350 - .2355	6 × 6	3/8 × 11 7/8	1	13/32	2 1/2	1,890 lbs.
7	.2756 - .2761		7 × 7	3/8 × 11 7/8	1	13/32	2 1/2	2,985 lbs.
8	.3150 - .3155	.3137 - .3142	8 × 7	3/8 × 11 7/8	1	13/32	2 1/2	3,995 lbs.
10	.3937 - .3942	.3924 - .3929	10 × 8	9/16 × 13 7/8	2	3/4	6	8,100 lbs.
12	.4724 - .4730	.4710 - .4715	12 × 8	9/16 × 13 7/8	2	3/4	6	10,550 lbs.
14	.5512 - .5519	.5498 - .5503	14 × 9	9/16 × 13 7/8	2	3/4	6	11,090 lbs.
16	.6299 - .6306	.6280 - .6285	16 × 10	3/4 × 15 1/4	3	3/4	6	11,375 lbs.
18	.7087 - .7093	.7070 - .7075	18 × 11	3/4 × 15 1/4	3	3/4	6	12,000 lbs.
20	.7874 - .7883	.7855 - .7860	20 × 12	1 × 19 1/4	3	3/4	6	11,000 lbs.
22	.8661 - .8669	.8643 - .8648	22 × 14	1 × 19 1/4	4	3/4	6	11,200 lbs.
24	.9449 - .9458	.9430 - .9435	24 × 14	1 × 19 1/4	4	3/4	6	13,075 lbs.
25	.9842 - .9851	.9824 - .9829	25 × 14	1 × 19 1/4	4	3/4	6	13,275 lbs.
28	1.1030 - 1.1040	-	28 × 14	1 1/8 × 23	9	3/4	6	14,700 lbs.
30	1.1817 - 1.1827	-	30 × 14	1 1/4 × 23	9	3/4	6	15,700 lbs.
32	1.2605 - 1.2615	-	32 × 14	1 1/4 × 23	9	3/4	6	16,790 lbs.

(Continued)

Table 29. *(Continued)* **ISO J-9 and ISO P9 Standard Tolerance Keyway Broaches.** *(Source, Hassay Savage Company.)*

Shims for Standard Metric Keyway Broaches						
Shim Size	Thickness	Length × Width	Shim Size	Thickness	Length × Width
3mm	.031	1 3/8 × 1/8	16mm	.062	6 3/4 × 3/4
4mm	.038	1 7/8 × 1/4	18mm	.062	6 3/4 × 3/4
5mm	.050	1 7/8 × 1/4	20mm	.062	6 3/4 × 1
5mm	.047	2 7/8 × 3/8	22mm	.062	6 3/4 × 1
6mm	.062	2 7/8 × 3/8	24mm	.062	6 3/4 × 1
7mm	.062	2 7/8 × 3/8	25mm	.062	6 3/4 × 1
8mm	.078	2 7/8 × 3/8	28mm	.062	7 × 1
10mm	.056	6 1/4 × 9/16	30mm	.062	7 × 1
12mm	.056	6 1/4 × 9/16	32mm	.062	7 × 1 1/4
14mm	.062	6 1/4 × 9/16	36mm	.062	7 × 1 1/4

* l/c = length of cut.

Broaches and bushings are sized for compatibility, and several bushing sizes can be obtained for individual broach sizes up to 25 mm. Required shims are normally supplied with broaches. Shim sizes correspond to broach size, not to bushing size. Consult supplier for details. All broach and shim dimensions are in inches unless otherwise noted.

Table 30. Standard Inch Square and Metric Square Push-Type Broaches and Standard Inch Push-Type Full Square Broaches. *(Source, Hassay Savage Company.)*

Size Square	Tolerance	Dimensions Across Corners	Pilot Dia.	Drill Size	Broach Length	Length of Cut Min.	Length of Cut Max.	Pressure for Max. l/c*
1/8	.1250 - .1260	.1750 - .1770	.1280	#30	4 5/8	3/16	1/2	700 lbs.
5/32	.1565 - .1575	.2180 - .2200	.1585	#21	5 5/8	1/4	1/2	800 lbs.
4 mm	.1580 - .1585	.2080 - .2090	.1655	4.2 mm	5 9/16	5/16	5/8	-
3/16	.1880 - .1890	.2620 - .2640	.1930	#10	5 11/16	1/4	5/8	1,120 lbs.
3/16†	.1880 - .1890	.2628 - .2632	.1870	3/16	6 1/4	3/16	1/2	1,200 lbs.
5 mm	.1973 - .1983	.2645 - .2655	.2047	5.2 mm	6 7/8	3/8	3/4	-
7/32	.2190 - .2200	.3050 - .3070	.2275	#1	6 7/8	1/4	3/4	1,580 lbs.
6 mm	.2367 - .2377	.3295 - .3300	.250	6.35 mm	7	3/8	3/4	-

(Continued)

Table 30. (*Continued*) Standard Inch Square and Metric Square Push-Type Broaches and Standard Inch Push-Type Full Square Broaches. (*Source, Hassay Savage Company.*)

Size Square	Tolerance	Dimensions Across Corners	Pilot Dia.	Drill Size	Broach Length	Length of Cut		Pressure for Max. l/c*
						Min.	Max.	
1/4	.2505 - .2515	.3500 - .3520	.2651	17/64	7	1/4	3/4	1,740 lbs.
1/4†	.2505 - .2515	.3513 - .3517	.2495	1/4	8 5/8	1/4	5/8	1,600 lbs.
9/32	.2815 - .2825	.3930 - .3950	.2964	19/64	7 1/4	3/8	1	2,300 lbs.
5/16	.3130 - .3140	.4370 - .4390	.3276	21/64	8 1/8	3/8	1	2,630 lbs.
5/16†	.3120 - .3140	.4375 - .4380	.3120	5/16	10 1/4	1/4	3/4	2,300 lbs.
8 mm	.3155 - .3165	.4405 - .4410	.3267	8.3 mm	8 3/16	7/16	7/8	-
11/32	.3440 - .3450	.4810 - .4830	.3589	23/64	9 5/8	3/8	1 1/4	3,050 lbs.
3/8	.3755 - .3765	.5230 - .5250	.3901	25/64	9 7/8	3/8	1 1/4	3,625 lbs.
3/8†	.3755 - .3765	.5243 - .5247	.3745	3/8	12 1/4	3/8	1	3,750 lbs.
10 mm	.3942 - .3952	.5435 - .5445	.4057	10.3 mm	10 7/8	1/2	1	-
13/32	.4065 - .4075	.5680 - .5700	.4214	27/64	10 7/8	1/2	1 3/8	5,200 lbs.
7/16	.4385 - .4395	.6110 - .6130	.4526	29/64	11 3/8	1/2	1 3/8	5,700 lbs.
12 mm	.4729 - .4739	.6540 - .6550	.4921	12.5 mm	12 1/2	5/8	1 1/4	-
15/32	.4690 - .4700	.6560 - .6580	.4995	1/2	12 1/2	1/2	1 3/8	5,900 lbs.
1/2	.5005 - .5015	.6970 - .6990	.5307	17/32	12 5/8	1/2	1 3/8	6,000 lbs.
1/2†	.5005 - .5015	.6980 - .6990	.4995	1/2	14 7/16	3/8	1	5,000 lbs.
14 mm	.5517 - .5527	.7700 - .7710	.5905	15.0 mm	14 7/8	3/4	1 1/2	-
9/16	.5630 - .5640	.7860 - .7880	.5932	19/32	14 7/8	1/2	1 1/2	6,200 lbs.
5/8	.6260 - .6270	.8710 - .8730	.6557	21/32	16 13/16	5/8	1 5/8	7,100 lbs.
16 mm	.6310 - .6320	.8780 - .8790	.6693	17.0 mm	16 13/16	7/8	1 3/4	-
11/16	.6885 - .6895	.9610 - .9630	.7495	3/4	18 1/2	5/8	1 5/8	7,300 lbs.
18 mm	.7092 - .7102	.9880 - .9890	.7874	20.0 mm	18 1/2	7/8	1 3/4	-

(Continued)

Table 30. *(Continued)* Standard Inch Square and Metric Square Push-Type Broaches and Standard Inch Push-Type Full Square Broaches. *(Source, Hassay Savage Company.)*

Size Square	Tolerance	Dimensions Across Corners	Pilot Dia.	Drill Size	Broach Length	Length of Cut		Pressure for Max. l/c*
						Min.	Max.	
3/4	.7510 - .7520	1.0450 - 1.0470	.8120	13/16	18 7/8	5/8	1 5/8	8,600 lbs.
20 mm	.7879 - .7889	1.0990 - 1.1000	.8661	22.0 mm	18 7/8	7/8	1 3/4	-
22 mm	.8666 - .8676	1.2110 - 1.2120	.9448	24.0 mm	23 1/4	7/8	1 3/4	-
7/8	.8765 - .8775	1.2280 - 1.0300	.9370	15/16	22 3/4	1/2	2	12,000 lbs.
24 mm	.9454 - .9464	1.3170 - 1.3180	1.0236	26.0 mm	24 9/16	7/8	1 3/4	-
25 mm	.9848 - .9858	1.3730 - 1.3740	1.0630	27.0 mm	24 9/16	7/8	1 3/4	-
1"	1.0020 - 1.0030	1.4030 - 1.4050	1.0932	1 3/32	24 1/2	1/2	2	12,500 lbs.

* l/c = length of cut.

† Indicates Standard full square broach. All other inch dimension broaches are standard inch square.
All dimensions are in inches unless otherwise indicated.

Table 31. Standard Inch and Metric Push-Type Hexagon Broaches. *(Source, Hassay Savage Company.)*

Size Square	Tolerance	Dimensions Across Corners	Pilot Dia.	Drill Size	Broach Length	Length of Cut		Pressure for Max. l/c*
						Min.	Max.	
1/8	.1255 - .1260	.1450 - .1470	.1245	1/8	4 5/8	3/16	3/8	120 lbs.
5/32	.1565 - .1570	.1800 - .1820	.1557	5/32	5 1/2	1/4	1/2	340 lbs.
4 mm	.1580 - .1585	.1820 - .1830	.1575	4 mm	5 1/2	1/4	1/2	340 lbs.
3/16	.1880 - .1890	.2145 - .2155	.1870	3/16	5 9/16	1/4	5/8	520 lbs.
5 mm	.1973 - .1983	.2270 - .2280	.1968	5 mm	6	1/4	3/4	890 lbs.
7/32	.2190 - .2200	.2500 - .2520	.2182	7/32	6	1/4	3/4	890 lbs.
6 mm	.2367 - .2377	.2730 - .2740	.2362	6 mm	6 1/2	1/4	3/4	1,250 lbs.
1/4	.2505 - .2515	.2865 - .2875	.2495	1/4	6 1/2	1/4	3/4	1,250 lbs.
7 mm	.2760 - .2770	.3180 - .3190	.2756	7 mm	7 3/4	3/8	1	1,650 lbs.
9/32	.2815 - .2825	.3220 - .3240	.2807	9/32	7 3/4	3/8	1	1,650 lbs.
5/16	.3130 - .3140	.3580 - .3590	.3120	5/16	8 1/4	3/8	1	2,175 lbs.
8 mm	.3155 - .3165	.3634 - .3644	.3150	8 mm	8 1/4	3/8	1	2,175 lbs.
11/32	.3440 - .3450	.3950 - .3970	.3432	11/32	8 1/4	3/8	1 1/4	3,500 lbs.
3/8	.3755 - .3765	.4300 - .4310	.3745	3/8	9	3/8	1 1/4	3,700 lbs.

(Continued)

Table 31. (Continued) **Standard Inch and Metric Push-Type Hexagon Broaches.** (Source, Hassay Savage Company.)

Size Square	Tolerance	Dimensions Across Corners	Pilot Dia.	Drill Size	Broach Length	Length of Cut		Pressure for Max. l/c*
						Min.	Max.	
10 mm	.3942 - .3952	.4543 - .4553	.3937	10 mm	10	1/2	1 3/8	4,100 lbs.
13/32	.4065 - .4075	.4670 - .4690	.4057	13/32	10	1/2	1 3/8	4,100 lbs.
7/16	.4385 - .4395	.5020 - .5030	.4370	7/16	10 3/4	1/2	1 3/8	4,575 lbs.
15/32	.4690 - .4700	.5390 - .5410	.4682	15/32	12 3/8	1/2	1 3/8	4,700 lbs.
12 mm	.4729 - .4739	.5452 - .5462	.4724	12 mm	12 3/8	1/2	1 3/8	4,700 lbs.
1/2	.5005 - .5015	.5740 - .5750	.4995	1/2	12 1/2	1/2	1 3/8	5,300 lbs.
14 mm	.5517 - .5527	.6332 - .6342	.5512	14 mm	14 1/4	1/2	1 1/2	6,700 lbs.
9/16	.5630 - .5640	.6480 - .6500	.5620	9/16	14 1/4	1/2	1 1/2	6,700 lbs.
5/8	.6260 - .6270	.7170 - .7185	.6245	5/8	16 7/8	5/8	1 5/8	7,250 lbs.
16 mm	.6310 - .6320	.7248 - .7258	.6290	16 mm	16 7/8	5/8	1 5/8	7,250 lbs.
11/16	.6880 - .6890	.7930 - .7950	.6870	11/16	17	5/8	1 5/8	9,300 lbs.
18 mm	.7092 - .7102	.8150 - .8160	.7087	18 mm	17	5/8	1 5/8	9,400 lbs.
3/4	.7510 - .7520	.8610 - .8625	.7495	3/4	17 7/8	5/8	1 5/8	13,500 lbs.
20 mm	.7879 - .7889	.9049 - .9059	.7874	20 mm	17 7/8	5/8	1 5/8	14,150 lbs.
22 mm	.8666 - .8676	.9958 - .9968	.8661	22 mm	18 7/8	5/8	1 5/8	18,500 lbs.
7/8	.8755 - .8765	1.0060 - 1.0075	.8745	7/8	18 7/8	5/8	1 5/8	18,500 lbs.
24 mm	.9454 - .9464	1.0868 - 1.0878	.9449	24 mm	19 7/8	5/8	1 5/8	20,500 lbs.
25 mm	.9848 - .9858	1.1323 - 1.1333	.9842	25 mm	19 7/8	5/8	1 5/8	22,000 lbs.
1"	1.0020 - 1.0030	1.1520 - 1.1530	.9995	1	19 7/8	5/8	1 5/8	22,050 lbs.

* l/c = length of cut.
All dimensions are in inches unless otherwise indicated.

Table 32. Standard Round Push Broaches. *(Source, Hassay Savage Company.)*

Broach Dia.	Round Tolerance Size	Pilot Dia.	Drill Size	Overall Length	Length of Cut		Pressure for Max. l/c*
					Min.	**Max.**	
1/4	.2510 - .2505	.2344	15/64	6	5/16	3/4	840 lbs.
5/16	.3135 - .3130	.2969	19/64	6	5/16	3/4	1,050 lbs.
3/8	.3760 - .3755	.3594	23/64	7	3/8	1	1,440 lbs.
7/16	.4385 - .4380	.4219	27/64	7	3/8	1	1,620 lbs.
1/2	.5010 - .5005	.4844	31/64	9 3/16	1/2	1 1/4	2,240 lbs.
9/16	.5635 - .5630	.5469	35/64	9 3/16	1/2	1 1/4	2,490 lbs.
5/8	.6260 - .6255	.6094	39/64	9 3/16	1/2	1 1/4	2,500 lbs.
11/16	.6885 - .6880	.6719	43/64	9 3/16	1/2	1 1/4	2,730 lbs.
3/4	.7510 - .7505	.7344	47/64	10 1/4	5/8	1 1/2	3,720 lbs.
13/16	.8135 - .8130	.7969	51/64	10 1/4	5/8	1 1/2	4,025 lbs.
7/8	.8765 - .8760	.8594	55/64	10 1/4	5/8	1 1/2	4,340 lbs.
15/16	.9390 - .9380	.9219	59/64	10 3/4	5/8	1 3/4	4,620 lbs.
1	1.0015 - 1.0005	.9844	63/64	10 3/4	5/8	1 3/4	4,970 lbs.

* l/c = length of cut.

Round broaches are used for precision sizing, maintaining close tolerances, and producing reamer quality surface finishes. All dimensions are in inches.

Table 33. Inch Pull-Type Keyway Broaches with Screw Shanks and Notched Shanks (see Notes). *(Source, Hassay Savage Company.)*

| A | | Hole (min.) | Length of Cut | | B** | C | D*** | E | F | G | No. of Cuts | Thread |
Nom. Dia.	Decimal Dia.*		Min.	Max.								
1/16	.0635	3/8	3/8	1 1/4	.1552	20	.313	.271	7 13/16	.042	1	1/4-20
3/32	.0948	7/16	1/2	1 1/2	.1865	24	.367	.309	8 1/4	.058	1	5/16-18
3/32	.0948	5/8	5/8	2 1/2	.249	33	.491	.433	10	.058	1	3/8-16
1/8	.126	1/2	1/2	1 1/2	.249	30	.438	.364	9	.074	1	3/8-16
1/8	.126	7/8	5/8	2 1/2	.3115	36	.594	.520	10	.074	1	1/2-13
5/32	.1572	19/32	1/2	1 1/2	.249	30	.525	.436	9	.089	1	3/8-16
5/32	.1572	23/32	5/8	2 1/2	.3115	33	.625	.536	10	.089	1	1/2-13
3/16	.1885	11/16	5/8	2 1/2	.374	36	.581	.476	10	.105	1	1/2-13
3/16	.1885	15/16	11/16	3 1/2	.374	36	.796	.691	10 11/16	.105	1	1/2-13
7/32	.2198	11/16	5/8	2 1/2	.374	33	.557	.437	10	.120	1	1/2-13
11/32	.2198	15/16	11/16	3 1/2	.374	42	.813	.693	11 1/16	.120	1	1/2-13
1/4	.251	11/16	5/8	2 1/2	.374	36	.612	.476	10	.136	1	1/2-13

(Continued)

Table 33: (Continued) Inch Pull-Type Keyway Broaches with Screw Shanks and Notched Shanks (see Notes). (Source, Hassay Savage Company.)

| A | | Hole (min.) | Length of Cut | | B** | C | D*** | E | F | G | No. of Cuts | Thread |
Nom. Dia.	Decimal Dia.*		Min.	Max.								
1/4	.251	1	11/16	4	.499	45	.877	.741	11 13/16	.136	1	5/8-11
1/4	.251	1 7/16	7/8	6	.624	51	1.250	1.114	13 1/2	.136	1	3/4-10
9/32	.2828	7/8	11/16	4	.499	42	.716	.564	11 5/8	.152	1	5/8-11
9/32	.2828	1 1/4	7/8	6	.499	51	1.093	.941	13 1/2	.152	1	5/8-11
5/16	.314	1	11/16	4	.499	45	.908	.741	11 13/16	.167	1	5/8-11
5/16	.314	1 5/16	7/8	6	.499	51	1.158	.991	13 1/2	.167	1	5/8-11
3/8	.3765	1 1/16	11/16	4	.499	45	.938	.739	11 13/16	.199	1	5/8-11
3/8	.3765	1 5/16	7/8	6	.499	54	1.189	.990	13 1/2	.199	1	5/8-11
7/16	.439	1 9/16	11/16	4	.624	48	1.360	1.160	12	.230	1	3/4-10
7/16	.439	2	1	8	.624	48	1.611	1.496	15 5/8	.230	2	3/4-10
1/2	.5015	1 1/2	11/16	4	.624	48	1.312	1.051	12	.261	1	3/4-10
1/2	.5015	1 1/2	1	8	.624	48	1.377	1.246	16 1/2	.261	2	3/4-10
9/16	.5645	1 3/4	11/16	4	.6865	54	1.438	1.146	11 13/16	.292	1	1-8
9/16	.5645	1 5/8	1	8	.6865	51	1.391	1.245	16	.292	2	1-8
9/16	.5645	2 1/4	1 1/8	12	.874	60	1.641	1.495	20	.292	2	1-8
5/8	.627	1 7/8	11/16	4	.749	60	1.625	1.301	12 3/16	.324	2	1-8
5/8	.627	2 1/2	1	8	.874	54	1.657	1.495	16 3/8	.324	2	1-8
5/8	.627	2 1/4	1 1/8	12	.874	57	1.657	1.495	20	.324	2	1-8
3/4	.752	1 7/8	11/16	4	.874	60	1.625	1.239	12 3/16	.386	1	1-8
3/4	.752	2	1	8	.999	60	1.688	1.495	16 3/8	.386	2	1 1/4-7
3/4	.752	2 1/4	1 1/8	12	.999	57	1.688	1.560	20	.386	3	1 1/4-7
7/8	.877	2 1/4	11/16	4	1.124	63	1.875	1.426	12 3/8	.449	1	1 1/4-7

Table 33: (Continued) Inch Pull-Type Keyway Broaches with Screw Shanks and Notched Shanks (see Notes). (Source, Hassay Savage Company.)

A		Hole (min.)	Length of Cut		B**	C	D***	E	F	G	No. of Cuts	Thread
Nom. Dia.	Decimal Dia.*		Min.	Max.								
7/8	.877	2 1/4	1	8	1.124	63	1.719	1.494	15 3/4	.449	2	1 1/4-7
7/8	.877	2 1/4	1 1/8	12	1.124	63	1.719	1.569	20	.449	3	1 1/4-7
1	1.002	2 1/4	5/8	2 1/2	1.249	63	13750	1.239	10 1/2	.511	1	1 1/2-6
1	1.002	2 1/4	7/8	6	1.249	63	1.750	1.494	14 1/4	.511	2	1 1/2-6
1	1.002	2 1/4	1 1/8	12	1.249	60	1.750	1.580	20	.511	3	1 1/2-6

* Tolerance on Decimal Dimension "A" is ±0.0002" for all broaches with decimal diameters through 0.5015", and ±0.0003" for sizes 0.5645" through 1.002".
** Tolerance on Dimension "B" is +0.0000/-.0005".
*** Tolerance on Dimension "D" is 0.0005".
Note: Thread size is for threaded shanks only. Dimensions of notched shank are given in illustration above.
All dimensions in inches.

Table 34. Standard Inch and Metric Pull-Type Keyseating Broaches. *(Source, Hassay Savage Company.)*

Broach Size	Keyseat Width Tolerance	Body Dimensions Width × Height × Length	Length of Cut Max.	Length of Cut Min.
1/16	.0625 - .0635	3/16 × 3/8 × 16	5/8	1 7/8
2 mm	.0787 - .0792	3/16 × 3/8 × 16	5/8	1 7/8
3/32	.0937 - .0947	3/16 × 3/8 × 16	5/8	1 7/8
3 mm	.1181 - .1186	3/16 × 7/16 × 16	5/8	1 7/8
1/8	.1252 - .1262	3/16 × 7/16 × 16	5/8	1 7/8
5/32	.1564 - .1574	3/16 × 1/2 × 16	5/8	1 7/8
4 mm	.1575 - .1580	3/16 × 1/2 × 16	5/8	1 7/8
5 mm	.1968 - .1973	5 mm × 9/16 × 16	5/8	1 7/8
5 mm	.1968 - .1973	5 mm × 3/4 × 20	5/8	3 1/8
3/16	.1877 - .1887	3/16 × 9/16 × 16	5/8	1 7/8
3/16	.1877 - .1887	3/16 × 3/4 × 20	5/8	3 1/8
6 mm	.2362 - .2367	6 mm × 3/4 × 16	5/8	1 7/8
6 mm	.2362 - .2367	6 mm × 3/4 × 20	5/8	3 1/8
1/4	.2502 - .2512	1/4 × 3/4 × 16	5/8	1 5/8
1/4	.2502 - .2512	1/4 × 3/4 × 20	5/8	3 1/8
5/16	.3127 - .3137	5/16 × 7/8 × 16	1 1/16	3 3/16
5/16	.3127 - .3137	5/16 × 7/8 × 20	1 1/16	5 3/16
8 mm	.3150 - .3155	8 mm × 7/8 × 16	1 1/16	3 3/16
8 mm	.3150 - .3155	8 mm × 7/8 × 20	1 1/16	5 5/16
3/8	.3755 - .3765	3/8 × 7/8 × 16	1 1/16	3 3/16
3/8	.3755 - .3765	3/8 × 7/8 × 20	1 1/16	5 5/16
10 mm	.3937 - .3942	10 mm × 7/8 × 16	1 1/16	5 5/16
10 mm	.3937 - .3942	10 mm × 7/8 × 20	1 1/16	5 5/16
7/16	.4380 - .4390	7/16 × 1 × 16	1 1/16	3 3/16
7/16	.4380 - .4390	7/16 × 1 × 20	1 1/16	5 3/16
12 mm	.4724 - .4730	12 mm × 1 × 20	1 1/16	5 5/16
1/2	.5006 - .5016	1/2 × 1 × 20	1 1/16	5 3/16
14 mm	.5512 - .5519	14 mm × 1 × 20	1 1/16	5 5/16
9/16	.5630 - .5640	9/16 × 1 × 20	1 1/16	5 3/16
5/8	.6260 - .6270	5/8 × 1 × 20	1 1/16	5 3/16
16 mm	.6299 - .6306	16 mm × 1 × 20	1 1/16	5 5/16
18 mm	.7087 - .7093	18 mm × 1 × 20	1 1/16	5 5/16
3/4	.7515 - .7525	3/4 × 1 × 20	1 1/16	5 3/16
20 mm	.7874 - .7883	20 mm × 1 × 20	1 1/16	5 5/16
22 mm	.8661 - .8669	22 mm × 1 × 20	1 1/16	5 5/16
7/8	.8765 - .8775	7/8 × 1 × 20	1 1/16	5 3/16
24 mm	.9449 - .9458	24 mm × 1 × 20	1 1/16	5 5/16
25 mm	.9843 - .9853	25 mm × 1 × 20	1 1/16	5 5/16
1	1.0015 - 1.0025	1 × 1 × 20	1 1/16	5 3/16

* l/c = length of cut. Cutters of these dimensions are suitable for Davis AF Series cutters. All dimensions are in inches unless otherwise indicated.

Table 35. Machining Parameters for Broaching. *(Source, Hassay Savage Company.)*

Material	Broach Grade*	Rake Face Angle	Per Tooth Chip Load (max.)		Speed		Coolant
			inch	mm	fpm	mpm	
Monel Alloy	A	18°	.002	.05	12	3.7	Oil w/25% Sulfur Additive
Nickel Alloy	A	15°	.002	.05	10	3.0	Oil w/25% Sulfur Additive
Beta Alloy	A	18°	.002	.05	10	3.0	Oil w/25% Sulfur Additive
Titanium Alpha-Beta Alloy	A	14°-18°	.003	.08	8	2.4	Oil w/25% Sulfur Additive
Malleable Cast Iron	A	6°-8°	.002	.05	10	3.0	Oil, No Additive
Stainless Steel	A	18°-20°	.002	.05	8	2.4	Oil w/Sulfur Additive
Stainless Steel Cast	A	18°-20°	.002	.05	20	6.1	Oil w/Sulfur Additive
Stainless Steel Wrought	A	18°-20°	.003	.08	20	6.1	Oil w/Sulfur Additive
Armor & Aircraft Plate	A	12°-15°	.002	.05	15	4.5	Oil w/Sulfur Additive
Tool Steel	A	12°-15°	.002	.05	10	3.0	Oil, No Additive
High Strength Steel	B	15°-18°	.003	.08	20	6.1	Oil Base Water Soluble
Free Machining Alloy	B	12°-15°	.004	.10	30	9.1	Oil Base Water Soluble
Medium Carbon Steel	B	12°-15°	.003	.08	25	7.6	Oil Base Water Soluble
Low Carbon Steel	B	12°-15°	.004	.10	30	9.1	Oil Base Water Soluble
Aluminum Alloy T-5, T-6	B	(T-5) 6°-10° (T-6) 15°-18°	.006	.15	40	12.2	Oil Base Water Soluble
Magnesium Alloy	B	15°-18°	.006	.15	40	12.2	Oil Base Water Soluble
Copper Alloy	B	12°-15°	.005	.13	25	7.6	Oil Base Water Soluble
Brass	B	0°-3°	.006	.15	30	9.1	Oil Base Water Soluble
Bronze	B	0°-3°	.006	.15	30	9.1	None or water
Plastic	B	10°-12°	.010	.25	30	9.1	None or water
Wax	B	0°-3°	.008	.20	30	9.1	None or water

* Broach materials are identified by the following code letters: A = Tungsten Molybdenum (M4 or equivalent) or Tungsten Cobalt (T15 or equivalent) grade high speed steels. B = M2 high speed steel or equivalent.

Taps and Tapping

Taps are among the most important and widely used tools in manufacturing. The selection of the ideal tap for a given operation is essential for assuring a strong, reliable threaded hole. Tapping can be done either by hand or machine, using one of the three basic varieties of taps: taper, plug, and bottoming. All three are identical in thread form and size, but differ in the chamfered (tapered) section at the point of the tap as follows.

Tapered taps are tapered for approximately the first seven to ten threads.

Plug taps are tapered for approximately three to five threads.

Bottoming taps are only tapered for one to two threads. Some manufacturers also offer a "semi-bottoming" tap that has two to two and one-half threads on the chamfer. This configuration is not universal.

Taps are also identified by a style designation as Style 1, Style 2, or Style 3 taps. Style refers to the tap diameter and physical characteristics of the tap. This identification method, which appears to have originated with the United States Cutting Tool Institute (and is used by ASME/ANSI in their Standard ASME/ANSI B94.9), identifies Styles as follows.

Style 1 taps are Machine Screw sizes #0–12 and Metric sizes 1.6 mm to 5 mm. These taps have external centers on the thread and shank ends that can be removed on the thread end for bottoming taps.

Style 2 taps are in Fractional sizes $1/4" - 3/8"$ (0.250 to 0.3750") and Metric sizes 6 mm to 10 mm. They may have an external center on the thread end that can be removed for bottoming taps, and a partial cone on the shank end that is about one-quarter of the diameter of the shank.

Style 3 taps are all taps above the Fractional size $3/8"$ and the Metric size 10 mm. They have internal centers on both the thread and shank ends.

Dimensions for all Style 1, Style 2, and Style 3 Screw, Fractional, and Metric sizes, plus applicable tolerances, are given in **Table 1**.

Tap nomenclature *(Figure 1)*

Angle of Thread – The angle included between the flanks of the thread measured in an axial plane.

Axis – The longitudinal centerline of the tap.

Back Taper – A slight axial relief on the thread of the tap that makes the pitch diameter of the thread near the shank somewhat smaller than that of the chamfered end.

Basic – The theoretical or nominal standard size from which all variations are made.

Chamfer – The tapering of the threads at the front end of each land of a tap by cutting away and relieving the crest of the first few teeth to distribute the cutting action over several teeth. When the tapering amounts to 7 to 10 threads, the tap is called a "taper" tap; 3 to 5 threads, a "plug" tap; and 1 to 2 threads, a "bottoming" tap.

Chamfer Angle – The angle formed between the chamfer and the axis of the tap, measured in an axial plane at the cutting edge.

Chamfer Relief – The gradual decrease in land height from cutting edge to heel on the chamfered portion, to provide clearance for the cutting action as the tap advances.

Core – The diameter of the body of the tap between the flutes.

Crest – The top surface joining the two flanks of a thread. The crest of an external thread is at its major diameter, while the crest of an internal thread is at its minor diameter.

Cutting Face – The leading side of the land in the direction of rotation for cutting on which the chip impinges.

Dryseal – A pipe threaded fuel connection for both external and internal application designed for use where the assembled product must withstand high fluid or gas pressures without the use of a sealing compound, or where a sealer is functionally objectionable.

Flutes – The longitudinal channels formed in a tap to create cutting edges on the thread profile and to provide chip spaces and cutting fluid passages.

Heel – The back edge of the land.

Height of Thread – The distance between the crest and the base of a thread, measured normal to the axis.

Helical Flute – A flute with uniform axial lead and constant helix in a helical path around the axis of a cylindrical tap.

Helix – The angle at which the thread is ground to give the tap its advance. Also used to describe the angle of the flutes in Spiral Flute taps.

Hook Face – A concave cutting face, usually specified either as Chordal Hook or Tangential Hook.

Chordal Hook – The angle between the chord passing through the root and crest of a thread form at the cutting face, and a radial line through the crest at the cutting edge.

Tangential Hook Angle – The angle between a line tangent to a hook cutting face at the cutting edge and a radial line to the same point.

Interrupted Thread – A tap having an odd number of lands, with every other tooth along with the thread helix removed.

Land – The threaded section of the tap between the flutes.

Lead – The distance a screw thread advances axially in one complete turn. On a single lead screw or tap, the lead and pitch are identical. On a double lead screw or tap, the lead is twice the pitch, etc.

Pitch – The distance from any point on a screw or tap thread to a corresponding point on the next thread, measured parallel to the axis. The pitch equals one divided by the number of threads per inch.

Pitch Diameter – On a straight thread, the diameter of an imaginary co-axial cylinder, the surface of which would pass through the thread profile at such points as to make equal the width of the threads and the width of the spaces cut by the surface of the cylinder. On a taper thread, the diameter at a given distance from a reference plane perpendicular to the axis of an imaginary co-axial cone, the surface of which would pass through the thread profile at such points as to make equal the width of the threads and the width of the spaces cut by the surface of the cone.

Rake – Any deviation of a straight cutting face of the tooth from a radial line. Positive Rake means that the crest of the cutting face is angularly advanced ahead of the balance of the face of the tooth. Negative Rake means that the same point is angularly behind the balance of the cutting face of the tooth. Zero Rake means that the cutting face is directly on the centerline.

Relief – Back taper is a form of relief. This usually equals .0005" to .001" per inch of thread. See *Thread Relief.*

Root – The bottom surface joining the flanks of two adjacent threads. The root of an external thread is at its minor diameter, while the root of an internal thread is at its major diameter.

Spiral Point (Chip Driver) – A supplementary angular fluting cut in the cutting face of the land at the chamfer end. It is slightly longer than the chamfer on the tap and of the opposite hand to that of rotation.

Thread Relief – The clearance produced by removal of metal from behind the cutting edge. When the thread angle is relieved from the heel to cutting edge, the tap is said to have "eccentric" relief. If relieved from heel for only a portion of land width, the tap is said to have "con-eccentric" relief (a portion of the thread is concentric before the relief begins). This can be $1/3$ concentric and $2/3$ eccentric.

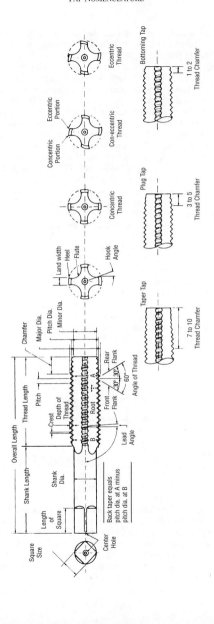

Figure 1. Tap terminology. (Source, Kennametal Inc.)

Types of taps

Standards for taps classify taps as having either "cut" or "ground" threads. Cut Thread Taps are created by turning and, for today's high accuracy and quality requirements, they cannot consistently produce threads to the same degree of excellence that can be achieved with Ground Thread Taps which are produced by grinding wheels to very exacting tolerances and have become the "standard" for taps. In fact, most manufacturers have abandoned Cut Thread Taps completely, and it is probable that it would be difficult to find one on the industrial market today. Even though Cut Thread Taps are uncommon, **Table 38** has been included for dimension limits for Unified threads produced with cut thread taps. Other than this table, all references and dimensions in this section refer to Ground Thread Taps. Thread limits and tolerances for Unified Thread and Metric Thread Ground Thread Taps are given in **Tables 35** through **37**. All Ground Thread Taps made to industry standards are marked with their nominal size, the tpi of the tap, and a symbol to identify the thread type (these symbols are identified in **Table 2**). In addition, left-hand threads are marked LH, and high-speed steel taps are marked HS.

Table 3 can be used for determining size and configuration availability of the first nine varieties of taps listed below. For example, #2 at the top of the Table refers to *Spiral Pointed Plug Taps*. From the Table, it can be seen that this tap is available in size 0-80 NF with an H2 Limit (see the section that immediately follows for a discussion of the H-Limit system). The "P" indicates that it is available only as a plug tap. The appearance of "TPB" in a box indicates that the tap size and type is available in all three configurations: taper, plug, and bottoming. The illustrations in this section were provided by Fastcut Tool Corporation.

1. Straight Fluted Taps. General purpose taps that can be used by hand or under power in both blind and through holes in most materials. The flutes retain chips during use, and for holes deeper than $1^1/2$ times the diameter three flute taps should be used. Special versions designed to produce small, powdery chips are available for tapping cast iron, cast brass, and other brasses. They are available in all Machine Screw Sizes, Fractional Sizes through $1^1/2$", and Metric sizes through M36, in taper, plug, and bottoming configurations. See Table 1 for dimensions, and Table 3 for size availability.

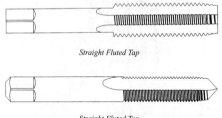

Straight Fluted Tap

Straight Fluted Tap

2. Spiral Pointed Plug Taps. Primarily intended for machine use, these taps can also be used by hand, although obtaining a true start in a hole can be difficult with any spiral pointed tap. Although designed for tapping through holes (and usually considered the first choice for this operation), they can be used in blind holes that are deep enough to allow for chip accumulation in the bottom of the hole. The spiral point forces chips ahead of the tap, which prevents clogging and recutting of the chips, and makes long holes, in excess of $1^1/2$ diameters, possible to cut. Suitable for most materials, but especially good performance can be obtained in those with high ductility. Available in Machine Screw sizes, Fractional sizes through $3/4$", and Metric sizes through M20. See Table 1 for dimensions, and Table 3 for size availability.

Spiral Pointed Plug Tap

3. Spiral Pointed Bottoming Taps. Short, blind holes that require full threads close to the hole's base can be cut with this tap. The notched front of the tap allows for additional area to accommodate cut chips; and the thicker chip, resulting from the short chamfer, tends to break up chips rather than leaving them long and stringy. These taps can be used by hand or under power in most materials, especially those with high ductility. Available in Machine Screw sizes and Fractional sizes through $5/16$". See Table 1 for dimensions, and Table 3 for size availability.

Spiral Pointed Bottoming Tap

4. Spiral Pointed Fluteless Plug Taps. Also called Spiral Point-only Taps, these taps are designed for tapping short through holes. The spiral point forces the chips ahead of the tap, which prevents clogging and recutting of the chips. They can be used by hand or under power in most materials. Available in Machine Screw sizes 4–12 and popular Fractional sizes to $1/2$". See Table 1 for dimensions, and Table 3 for size availability.

Spiral Pointed Fluteless Plug Tap

5. Spiral Fluted Taps. The helical flutes are the same hand of flute lead as the hand of cut. These taps can be used for most materials but are especially useful for tapping blind holes in mild steel and brass either by hand or under power. The spiral flutes draw the chips out of the hole, and also allow the tap to bridge a keyway or slot inside the hole without binding. The helix angle is 25° to 35°. Available in either plug or bottoming configurations in Machine Screw sizes 4–10 and most popular Fractional sizes through $1/2$". See Table 1 for dimensions, and Table 3 for size availability.

Spiral Fluted Tap

6. Fast Spiral Fluted Taps. Intended for tapping relatively deep blind holes, these taps have the same advantages as Spiral Fluted Taps but the helix angle is 45° to 60°. They perform best in mild steel, aluminum, magnesium, copper, and brass, and are available in both plug and bottoming configurations. Available in several Machine Screw sizes 3–12, popular Fractional sizes through $1/4$", and Metric sizes M3–M12. See Table 1 for dimensions, and Table 3 for size availability.

Fast Spiral Fluted Tap

7. Screw Thread Insert Taps. These taps produce threads with a pitch diameter that is slightly more than two thread heights larger than the basic pitch diameter, allowing for the installation of a screw thread insert. Inserts are primarily used in soft material prone to thread stripping, or to repair damaged threads. They can be used by hand or under power to produce either through or blind holes in most materials. They are available in plug and bottoming configurations in Machine Screw sizes 4–12 and Fractional sizes $1/4$"–$1/2$". See Table 1 for dimensions, Table 3 for size availability, and **Table 26** for recommended tap drill sizes for screw thread insert taps.

Screw Thread Insert Tap

8. Spiral Pointed Screw Thread Insert Plug Taps. The attributes of this tap are basically the same as the Screw Thread Insert Tap, but the spiral point, which forces chips ahead of the tap, allows for the tapping of long holes in excess of $1\frac{1}{2}$ times the diameter of the tap. Available in Machine Screw sizes 4–12 and Fractional sizes through $1/2$". See Table 1 for dimensions, Table 3 for size availability, and **Table 26** for recommended tap drill sizes for screw thread insert taps.

Spiral Pointed Screw Thread Insert Plug Tap

9. Thread Forming Taps. These taps do not cut threads, they cold form them by displacing material from the major diameter toward the minor diameter. They are extremely effective in ductile materials such as aluminum, copper, brass, leaded steels, low carbon steels, and stainless steels, but they will not function in materials that are brittle and have low ductility. Thread Forming Taps must be run under power and usually at speeds of 150% to 200% in excess of cutting taps. Because the tap displaces material above the mouth of the hole, countersinking before tapping is recommended. Also, since the hole diameter is actually reduced by the forming process, a larger tap drill is required for a forming tap than a cutting tap. See **Tables 18** through **21** for suggested drill sizes for forming taps. These taps are available in Machine Screw sizes and Fractional sizes through $^{3}/_{4}$" in both plug and bottoming configurations. See Table 1 for dimensions, and Table 3 for size availability.

Thread Forming Taps

10. Pulley Taps. These specialty taps are used to tap threads into pulley hubs. The various lengths allow the machine spindle to clear the pulley sheave. See **Table 27** for dimensions.

Pulley Tap

11. Nut Taps. Designed for tapping nuts, these taps have a long chamfer that produces low chip loads. After tapping a nut, the tap is not reversed. Instead, the nuts pass onto the long shank and accumulate until the shank becomes loaded and the tap is removed and the nuts are dumped. See **Table 28** for dimensions.

Nut Tap

12. Taper Pipe Taps. These taps are available for tapping NPT (Regular), NPTF (Dryseal), ANPT (Aeronautical), and PTF pipe threads. They are available in Continuous or Interrupted thread versions. Interrupted taps are designed to provide additional chip space, better coolant flow, and reduced drag. The thread size should be specified when ordering. See **Tables 29** through **31** and **Tables 34** and **35** for taper pipe tap specifications and information.

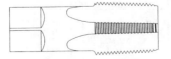

Taper Pipe Tap

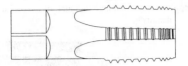

Interrupted Taper Pipe Tap

13. Straight Pipe Taps. These taps are available for tapping NPS (Regular) and NPSF (Dryseal) pipe threads. See **Tables 32**, **33**, and **34** for straight pipe tap specifications and information.

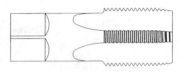

Straight Pipe Tap

14. Miniature Taps. Taps for cutting UNM threads are available in both thread forming and thread cutting varieties. Available tap sizes, recommended tap drill sizes, and information on their use can be found later in this "Taps and Tapping" section under the heading "Miniature taps." *(Text continued on p. 1215)*

Table 1. Dimensions and Tolerances for Standard Ground Thread Hand Taps. *(Source, New England Tap Corp.)*

Style 1 Sizes #0-12 Machine Screw Sizes 1.6-6.3 mm				**Style 2** Sizes #14 Machine Screw Sizes 7-10 mm				**Style 3** Sizes larger than 3/8" Sizes 12 mm or larger	

Nominal Dia. Range-In.		Machine Screw Size No.	Nominal Fractional Diameter (in.)	Nominal Metric Diameter (mm)	Style	Overall Length A	Thread Length B	Square Length C	Shank Diameter D	Size of Square E
Over	To (incl.)									
.052	.065	0	1/16	M1.6	1	1-5/8	5/16	3/16	.141	.110
.065	.078	1		M1.8	1	1-11/16	3/8	3/16	.141	.110
.078	.091	2		M2, M2.2	1	1-3/4	7/16	3/16	.141	.110
.091	.104	3	3/32	M2.5	1	1-13/16	1/2	3/16	.141	.110
.104	.117	4			1	1-7/8	9/16	3/16	.141	.110
.117	.130	5	1/8	M3, M3.5	1	1-15/16	5/8	3/16	.141	.110
.130	.145	6		M3.5	1	2	11/16	3/16	.141	.110
.145	.171	8	5/32	M4	1	2-1/8	3/4	1/4	.168	.131
.171	.197	10	3/16	M4.5, M5	1	2-3/8	7/8	1/4	.194	.152
.197	.223	12	7/32		1	2-3/8	15/16	9/32	.220	.165

(Continued)

Table 1. *(Continued)* **Dimensions and Tolerances for Standard Ground Thread Hand Taps.**

Nominal Dia. Range-In.		Machine Screw Size No.	Nominal Fractional Diameter (in.)	Nominal Metric Diameter (mm)	Style	Overall Length A	Thread Length B	Square Length C	Shank Diameter D	Size of Square E
Over	To (incl.)									
.223	.260	14	1/4	M6, M6.3	2	2-1/2	1	5/16	.255	.191
.260	.323		5/16	M7, M8	2	2-23/32	1-1/8	3/8	.318	.238
.323	.395		3/8	M10	2	2-15/16	1-1/4	7/16	.381	.286
.395	.448		7/16		3	3-5/32	1-7/16	13/32	.323	.242
.448	.510		1/2	M12, M12.5	3	3-3/8	1-21/32	7/16	.367	.275
.510	.573		9/16	M14	3	3-19/32	1-21/32	1/2	.429	.322
.573	.635		5/8	M16	3	3-13/16	1-13/16	9/16	.480	.360
.635	.709		11/16	M18	3	4-1/32	1-13/16	5/8	.542	.406
.709	.760		3/4		3	4-1/4	2	11/16	.590	.442
.760	.823		13/16	M20	3	4-15/32	2	11/16	.652	.489
.823	.885		7/8	M22	3	4-11/16	2-7/32	3/4	.697	.523
.885	.948		15/16	M24	3	4-29/32	2-7/32	3/4	.760	.570
.948	1.010		1	M25	3	5-1/8	2-1/2	13/16	.800	.600
1.010	1.073		1-1/16	M27	3	5-1/8	2-1/2	7/8	.896	.672
1.073	1.135		1-1/8		3	5-7/16	2-9/16	7/8	.896	.672
1.135	1.198		1 3/16	M30	3	5 7/16	2 9/16	1	1.021	766
1.198	1.260		1 1/4		3	5 3/4	2 9/16	1	1.021	66
1.260	1.323		1 5/16	M33	3	5 3/4	2 9/16	1 1/16	1.108	831
1.323	1.385		1 3/8		3	6 1/16	3	1 1/16	1.108	831
1.385	1.448		1 7/16	M36	3	6 1/16	3	1 1/8	1.2′	.925
1.448	1.510		1 1/2		3	6 3/8	3	1 1/8	1.′	.925

Tolerances

Element	Nominal Dia. Range Inches		Direction	Tolerance Inches Ground Thread
	Over	To (Incl.)		
Length Overall-A	.052	1.010	Plus or Minus	1/32
	1.010	1.510	Plus or Minus	1/16
Length of Thread-B	.052	.223	Plus or Minus	3/64
	.223	.510	Plus or Minus	1/16
	.510	1.510	Plus or Minus	3/32
Length of Square-C	.052	1.010	Plus or Minus	1/32
	1.010	1.510	Plus or Minus	1/16
Diameter of Shank-D	.052	.223	Minus	.0015
	.223	.635	Minus	.0015
	.635	1.010	Minus	.002
	1.010	1.510	Minus	.002
Size of Square-E	.052	.510	Minus	.004
	.510	1.010	Minus	.006
	1.010	1.510	Minus	.008

All dimensions in inches.

Table 2. Standard Tap Symbol Marking System. *(Source, Kennametal Inc.)*

Tap Marking	Thread Designations	Thread Series
ACME-C	ACME-C	Acme threads, centralizing
ACME-G	ACME-G	Acme threads, general purpose *(Also see STUB ACME)*
AMO	AMO	American Standard microscope objective threads
NPT	ANPT	Aeronautical National Form Taper pipe threads
BUTT	BUTT	Buttress Threads, pull type
PUSH-BUTT	PUSH-BUTT	Buttress Threads, push type
F-PTF	F-PTF	Dryseal fine taper pipe thread series
M	M	Metric Screw Threads—M Profile, with basic ISO 68 profile
MJ	MJ	Metric Screw Threads—MJ Profile, with rounded root of radius 0.1501 1P to 0.18042P
MJS	MJS	Metric Screw Threads—MJ Profile, special series
NC	NS5 IF	Class 5 interference fit Internal Threads entire ferrous material range
NC	NC5 INF	Class 5 interference fit Internal Threads entire nonferrous material range
NGO-RH or LH	NGO	National gas outlet threads
NGS	NGS	National gas straight threads
NGT	NGT	National gas taper threads *(Also see SGT)*
NH	NH	American Standard hose coupling threads of full form
NPS	NPSC	American Standard straight pipe threads in pipe couplings
NPSF	NPSF	Dryseal American Standard fuel internal straight pipe threads
NPSH	NPSH	American Standard straight hose coupling threads for joining to American Standard taper pipe threads
NPSI	NPSI	Dryseal American Standard intermediate internal straight pipe threads
NPSL	NPSL	American Standard straight pipe threads for loose fitting mechanical joints with locknuts
NPS	NPSM	American Standard straight pipe threads for free fitting mechanical joints for fixtures
NPT	NPT	American Standard taper pipe threads for general use
NPTF	NPTF	Dryseal American Standard taper pipe threads
NPTR	NPTR	American Standard taper pipe threads for railing joints
PTF SHORT	PTF—SAE SHORT	Dryseal SAE short taper pipe threads
PTF-SPL SHORT	PTF-SPL SHORT	Dryseal special short taper pipe threads
PTF-SPL EXTRA SHORT	PFT-SPL EXTRA SHORT	Dryseal special extra short taper pipe threads *(Also see SPL-PTF)*
M	S	ISO Miniature screw threads 0.25 to 1.4 mm inclusive
SGT	SGT	Special gas taper threads
SPL-PTF	SPL-PTF	Dryseal special taper pipe threads
STUB ACME	STUB ACME	Stub Acme threads
STUB ACME M1	STUB ACME M1	Stub Acme Modified Form 1
STUB ACME M2	STUB ACME M2	Stub Acme Modified Form 2

(Continued)

Table 2. *(Continued)* **Standard Tap Symbol Marking System.**

Tap Marking	Thread Designations	Thread Series
N	UN	Unified Inch Screw Thread, constant-pitch series
NC	UNC	Unified Inch Screw Thread, coarse pitch series
NF	UNF	Unified Inch Screw Thread, fine pitch series
NEF	UNEF	Unified Inch Screw Thread, extra-fine pitch series
N	UNJ	Unified Inch Screw Thread, constant-pitch series with rounded root of radius 0.15011P to 0.18042P, on external thread only
NC	UNJC	Unified Inch Screw Thread, coarse pitch series with rounded root of radius 0.15011P to 0.18042P. On external thread only Unified Inch Screw Thread, fine pitch series, with rounded root of radius 0.15011P to 0.18042P. On external thread only
NF	UNJF	Unified Inch Screw Thread, fine pitch series, with rounded root of radius 0.15011P. to 0.18042P. On external thread only
NEF	UNJEF	Unified Inch Screw Thread, extra-fine pitch series, with rounded root of radius 0.15011P to 0.18042P. On external thread only
N	UNR	Unified Inch Screw Thread, constant-pitch series, with rounded root of radius not less than 0.108P. (External thread only)
NC	UNRC	Unified Inch Screw Thread, course thread series, with rounded root of radius not less than 0.108P. (External thread only)
NF	UNRF	Unified Inch Screw Thread, fine pitch series, with rounded root of radius not less than 0.108P. (External thread only)
NEM	UNREF	Unified Inch Screw Thread, extra-fine pitch series, with rounded root of radius not less than 0.108P. (External thread only)
UNM	UNM	Unified miniature thread series
NS	UNS	Unified Inch Screw Thread, special diameter pitch, or length of engagement
Tap & Die Marking	**Thread Designations**	**British Screw Threads**
BA	BA	British Association
BSC	BSC	British Cycle
BSF	BSF	British Whitworth Fine
BSW	BSW	British Whitworth Course
BSPP (OLD)	BSPP	British Straight Pipe
BSPT (OLD)	BSPT	British Taper Pipe
WHIT	WHIT	British Whitworth Special
Tap Marking	**Thread Designations**	**British Pipe Threads**
G1/4 (BSPP)	G	British Internal Straight Pipe for Mechanical Joints
Rc1/4 (BSPT)	Rc	British Internal Taper Pipe for Pressure Tight Joints
Rp1/4 (BSPP)	Rp	British Internal Straight Pipe for Pressure Tight Joints
Die Marking	**Thread Designations**	**British Pipe Threads**
G1/4A (BSPP)	GxA	British External Straight Pipe for Mechanical Joints
G1/4B (BSPP)	GxB	British External Straight Pipe for Mechanical Joints
R1/4 (BSPT)	R	British External Taper Pipe for Pressure Tight Joints

Table 3. Commercially Available Ground Thread Tap Sizes for Taper, Plug, and Bottoming Taps. See Notes.

			Machine Screw Sizes								
Size	Thread	H Limit	1.	2.	3.	4.	5.	6.	7.	8.	9.
0-80	NF	H1	TPB	P	B	-	-	-	-	-	-
0-80	NF	H2	PB	P	B	-	-	-	-	-	PB
0-80	NF	H3	-	-	-	-	-	-	-	-	PB
1-64	NC	H1	TPB	P	-	-	-	-	-	-	-
1-64	NC	H2	P	P	B	-	-	-	-	-	PB
1-72	NF	H1	TPB	P	-	-	-	-	-	-	-
1-72	NF	H2	PB	P	B	-	-	-	-	-	B
2-56	NC	H1	TPB	P	B	-	-	-	-	-	-
2-56	NC	H2	TPB	P	B	-	-	-	-	-	PB
2-56	NC	H3	-	-	-	-	-	-	-	-	PB
2-64	NF	H1	-	P	-	-	-	-	-	-	-
2-64	NF	H2	TPB	P	-	-	-	-	-	-	B
2-64	NF	H3	-	-	-	-	-	-	-	-	B
3-48	NC	H1	P	P	-	-	-	-	-	-	-
3-48	NC	H2	TPB	P	B	-	-	PB	-	-	PB
3-48	NC	H3	-	-	-	-	-	-	-	-	B
3-56	NF	H1	-	P	-	-	-	-	-	-	-
3-56	NF	H2	TPB	P	B	-	-	-	-	-	PB
3-56	NF	H3	-	-	-	-	-	-	-	-	B
4-40	NC	H1	TPB	P	B	-	-	-	-	PB	P	-
4-40	NC	H2	TPB	P	B	P	PB	PB	-	-	-
4-40	NC	H3	-	-	-	-	-	-	-	-	PB
4-40	NC	H5	-	-	-	-	-	-	-	-	PB
4-48	NF	H1	P	P	-	-	-	-	-	-	-
4-48	NF	H2	TPB	P	B	-	-	-	-	-	-
4-48	NF	H3	-	-	-	-	-	-	-	-	PB
4-48	NF	H5	-	-	-	-	-	-	-	-	PB
4-36	NS	H2	TPB	P	-	-	-	-	-	-	-
5-40	NC	H1	PB	P	-	-	-	-	PB	P	-
5-40	NC	H2	TPB	P	B	P	PB	PB	-	-	-
5-40	NC	H3	-	-	-	-	-	-	-	-	PB
5-40	NC	H5	-	-	-	-	-	-	-	-	PB
5-44	NF	H2	TPB	P	B	-	-	-	-	-	-

Note: The three column heads on left indicate Tap size, Fine or Course Unified Thread, and H Limit of Tap. The numbered column heads identify a specific tap style as follows: 1 = Straight Fluted Taps. 2 = Spiral Pointed Plug Taps. 3 = Spiral Pointed Bottoming Taps. 4 = Spiral Pointed Fluteless Plug Taps. 5 = Spiral Fluted Taps. 6 = Fast Spiral Fluted Taps. 7 = Screw Thread Insert Taps. 8 = Spiral Pointed Screw Thread Insert Plug Taps. 9 = Thread Forming Taps. A "T," "P," or "B" in any box indicates that the indicated tap is available in Taper, Plug, or Bottoming configurations. All taps are ground thread and commercially available in high-speed steel with wear-resistant coatings (see text). Check with manufacturer for number of flutes. See Table 1 for tap dimensions.

(Continued)

Table 3. *(Continued)* **Commercially Available Ground Thread Tap Sizes for Taper, Plug, and Bottoming Taps.**
See Notes.

			Machine Screw Sizes								
Size	Thread	H Limit	1.	2.	3.	4.	5.	6.	7.	8.	9.
5-44	NF	H3	-	-	-	-	-	-	-	-	PB
5-44	NF	H5	-	-	-	-	-	-	-	-	PB
6-32	NC	H1	TPB	P	B	-	-	-	-	-	-
6-32	NC	H2	TPB	P	B	-	-	-	PB	P	-
6-32	NC	H3	TPB	P	B	P	PB	PB	PB	-	PB
6-32	NC	H5	-	-	-	-	-	-	-	-	PB
6-32	NC	H7	PB	P	B	-	-	-	-	-	PB
6-40	NF	H1	P	P	-	-	-	-	PB	P	-
6-40	NF	H2	TPB	P	B	-	-	-	-	-	-
6-40	NF	H3	-	-	-	-	-	-	-	-	PB
6-40	NF	H5	-	-	-	-	-	-	-	-	PB
8-32	NC	H1	TPB	P	B	-	-	-	-	-	-
8-32	NC	H2	TPB	P	B	-	-	-	PB	P	-
8-32	NC	H3	TPB	P	B	P	PB	PB	PB	-	PB
8-32	NC	H5	-	-	-	-	-	-	-	-	PB
8-32	NC	H7	PB	P	B	-	-	-	-	-	PB
8-36	NF	H1	P	P	-	-	-	-	-	-	-
8-36	NF	H2	TPB	P	B	-	-	-	-	-	-
8-36	NF	H3	-	-	-	-	-	-	-	-	PB
10-24	NC	H1	TPB	P	B	-	-	-	-	-	-
10-24	NC	H2	PB	P	B	-	-	-	PB	P	
10-24	NC	H2	TPB	-	-	-	-	-	-	-	-
10-24	NC	H3	TPB	P	B	P	PB	PB	PB	-	-
10-24	NC	H4	-	-	-	-	-	-	-	-	PB
10-24	NC	H6	-	-	-	-	-	-	-	-	PB
10-24	NC	H7	PB	P	-	-	-	-	-	-	-
10-32	NF	H1	PB	P	B	-	-	-	-	-	-
10-32	NF	H1	TPB	-	-	-	-	-	-	-	-
10-32	NF	H2	TPB	P	B	-	-	-	PB	P	-
10-32	NF	H3	TPB	P	B	P	PB	PB	PB	-	-
10-32	NC	H4	-	-	-	-	-	-	-	-	PB
10-32	NF	H6	-	-	-	-	-	-	-	-	PB
10-32	NF	H7	PB	P	-	-	-	-	-	-	-

Note: The three column heads on left indicate Tap size, Fine or Course Unified Thread, and H Limit of Tap. The numbered column heads identify a specific tap style as follows: 1 = Straight Fluted Taps. 2 = Spiral Pointed Plug Taps. 3 = Spiral Pointed Bottoming Taps. 4 = Spiral Pointed Fluteless Plug Taps. 5 = Spiral Fluted Taps. 6 = Fast Spiral Fluted Taps. 7 = Screw Thread Insert Taps. 8 = Spiral Pointed Screw Thread Insert Plug Taps. 9 = Thread Forming Taps. A "T," "P," or "B" in any box indicates that the indicated tap is available in Taper, Plug, or Bottoming configurations. All taps are ground thread and commercially available in high-speed steel with wear-resistant coatings (see text). Check with manufacturer for number of flutes. See Table 1 for tap dimensions.

(Continued)

Table 3. *(Continued)* **Commercially Available Ground Thread Tap Sizes for Taper, Plug, and Bottoming Taps.** See Notes.

Size	Thread	H Limit	1.	2.	3.	4.	5.	6.	7.	8.	9.
Machine Screw Sizes											
12-24	NC	H1	-	P	-	-	-	-	-	-	-
12-24	NC	H2	-	-	-	-	-	-	PB	P	-
12-24	NC	H3	TPB	P	B	P	-	PB	PB	-	-
12-24	NC	H4	-	-	-	-	-	-	-	-	PB
12-24	NC	H6	-	-	-	-	-	-	-	-	PB
12-28	NF	H1	P	-	-	-	-	-	-	-	-
12-28	NF	H3	TPB	P	-	-	-	-	-	-	-
12-28	NF	H4	-	-	-	-	-	-	-	-	PB
Fractional Sizes											
1/4-20	NC	H1	TPB	P	-	-	-	-	-	-	-
1/4-20	NC	H2	TPB	P	-	-	-	-	PB	P	-
1/4-20	NC	H3	TPB	P	B	P	PB	PB	PB	-	P
1/4-20	NC	H4	-	-	-	-	-	-	-	-	PB
1/4-20	NC	H5	PB	P	-	-	-	-	-	-	-
1/4-20	NC	H6	-	-	-	-	-	-	-	-	PB
1/4-28	NF	H1	PB	P	-	-	-	-	-	-	-
1/4-28	NF	H2	PB	P	-	-	-	-	PB	P	-
1/4-28	NF	H3	TPB	P	P	-	PB	PB	PB	-	-
1/4-28	NF	H4	PB	P	-	-	-	-	-	-	PB
1/4-28	NF	H6	-	-	-	-	-	-	-	-	PB
5/16-18	NC	H1	TPB	P	-	-	-	-	-	-	-
5/16-18	NC	H2	TPB	P	-	-	-	-	-	-	-
5/16-18	NC	H3	TPB	P	B	P	PB	PB	PB	P	-
5/16-18	NC	H5	PB	P	-	-	-	-	-	-	PB
5/16-18	NC	H7	-	-	-	-	-	-	-	-	PB
5/16-24	NF	H1	TPB	P	-	-	-	-	-	-	-
5/16-24	NF	H2	PB	P	-	-	-	-	PB	P	-
5/16-24	NF	H3	TPB	P	B	-	PB	PB	-	-	-
5/16-24	NF	H4	PB	P	-	-	-	-	-	-	-
5/16-24	NF	H5	-	-	-	-	-	-	-	-	PB
5/16-24	NF	H7	-	-	-	-	-	-	-	-	PB
3/8-16	NC	H1	PB	P	-	-	-	-	-	-	-
3/8-16	NC	H2	TPB	P	-	-	-	-	-	-	-

Note: The three column heads on left indicate Tap size, Fine or Course Unified Thread, and H Limit of Tap. The numbered column heads identify a specific tap style as follows: 1 = Straight Fluted Taps. 2 = Spiral Pointed Plug Taps. 3 = Spiral Pointed Bottoming Taps. 4 = Spiral Pointed Fluteless Plug Taps. 5 = Spiral Fluted Taps. 6 = Fast Spiral Fluted Taps. 7 = Screw Thread Insert Taps. 8 = Spiral Pointed Screw Thread Insert Plug Taps. 9 = Thread Forming Taps. A "T," "P," or "B" in any box indicates that the indicated tap is available in Taper, Plug, or Bottoming configurations. All taps are ground thread and commercially available in high-speed steel with wear-resistant coatings (see text). Check with manufacturer for number of flutes. See Table 1 for tap dimensions.

(Continued)

Table 3. *(Continued)* **Commercially Available Ground Thread Tap Sizes for Taper, Plug, and Bottoming Taps.** See Notes.

Fractional Sizes											
Size	Thread	H Limit	1.	2.	3.	4.	5.	6.	7.	8.	9.
3/8-16	NC	H5	PB	P	-	-	-	-	-	-	PB
3/8-16	NC	H7	-	-	-	-	-	-	-	-	PB
3/8-24	NF	H1	TPB	P	-	-	-	-	-	-	-
3/8-24	NF	H2	PB	P	-	-	-		PB	P	-
3/8-24	NF	H3	TPB	P	-	-	PB	PB	-	-	-
3/8-24	NF	H4	PB	P	-	-	-	-	-	-	-
3/8-24	NF	H5	-	-	-	-	-	-	-	-	PB
3/8-24	NF	H7	-	-	-	-	-	-	-	-	PB
7/16-14	NC	H1	PB	-	-	-	-	-	-	-	-
7/16-14	NC	H2	PB	P	-	-	-	-	-	-	-
7/16-14	NC	H3	TPB	P	-	-	PB	PB	PB	P	-
7/16-14	NC	H5	PB	P	-	-	-	-	-	-	PB
7/16-14	NC	H8	-	-	-	-	-	-	-	-	PB
7/16-20	NF	H1	PB	-	-	-	-	-	-	-	-
7/16-20	NF	H2	PB	P	-	-	-	-	-	-	-
7/16-20	NF	H3	TPB	P	-	-	PB	PB	PB	P	-
7/16-20	NF	H5	PB	P	-	-	-	-	-	-	PB
7/16-20	NF	H8	-	-	-	-	-	-	-	-	PB
1/2-13	NC	H1	PB	P	-	-	-	-	-	-	-
1/2-13	NC	H2	PB	P	-	-	-	-	-	-	-
1/2-13	NC	H3	TPB	P	-	P	PB	PB	PB	P	-
1/2-13	NC	H5	PB	P	-	-	-	-	-	-	PB
1/2-13	NC	H8	-	-	-	-	-	-	-	-	PB
1/2-20	NF	H1	TPB	P	-	-	-	-	-	-	-
1/2-20	NF	H2	PB	P	-	-	-	-	-	-	-
1/2-20	NF	H3	TPB	P	-	-	PB	PB	PB	P	-
1/2-20	NF	H5	PB	P	-	-	-	-	-	-	PB
1/2-20	NF	H8	-	-	-	-	-	-	-	-	PB
9/16-12	NC	H2	P	-	-	-	-	-	-	-	-
9/16-12	NC	H3	TPB	P	-	-	-	-	-	-	-
9/16-12	NC	H5	PB	-	-	-	-	-	-	-	-
9/16-12	NC	H7	-	-	-	-	-	-	-	-	PB
9/16-12	NC	H10	-	-	-	-	-	-	-	-	PB

Note: The three column heads on left indicate Tap size, Fine or Course Unified Thread, and H Limit of Tap. The numbered column heads identify a specific tap style as follows: 1 = Straight Fluted Taps. 2 = Spiral Pointed Plug Taps. 3 = Spiral Pointed Bottoming Taps. 4 = Spiral Pointed Fluteless Plug Taps. 5 = Spiral Fluted Taps. 6 = Fast Spiral Fluted Taps. 7 = Screw Thread Insert Taps. 8 = Spiral Pointed Screw Thread Insert Plug Taps. 9 = Thread Forming Taps. A "T," "P," or "B" in any box indicates that the indicated tap is available in Taper, Plug, or Bottoming configurations. All taps are ground thread and commercially available in high-speed steel with wear-resistant coatings (see text). Check with manufacturer for number of flutes. See Table 1 for tap dimensions.

(Continued)

Table 3. *(Continued)* **Commercially Available Ground Thread Tap Sizes for Taper, Plug, and Bottoming Taps.** See Notes.

Fractional Sizes											
Size	Thread	H Limit	1.	2.	3.	4.	5.	6.	7.	8.	9.
9/16-18	NF	H2	P	-	-	-	-	-	-	-	-	-
9/16-18	NF	H3	TPB	P	-	-	-	-	-	-	-	-
9/16-18	NF	H5	PB	-	-	-	-	-	-	-	-	-
9/16-18	NF	H7	-	-	-	-	-	-	-	-	PB
9/16-18	NF	H10	-	-	-	-	-	-	-	-	PB
5/8-11	NC	H1	P		-	-	-	-	-	-	-	-
5/8-11	NC	H2	P		-	-	-	-	-	-	-	-
5/8-11	NC	H3	TPB	P	-	-	-	-	-	-	-	-
5/8-11	NC	H5	PB	P	-	-	-	-	-	-	-	-
5/8-11	NC	H7	-	-	-	-	-	-	-	-	PB
5/8-11	NC	H10	-	-	-	-	-	-	-	-	PB
5/8-18	NF	H1	P	-	-	-	-	-	-	-	-
5/8-18	NF	H2	PB	-	-	-	-	-	-	-	-
5/8-18	NF	H3	TPB	P	-	-	-	-	-	-	-	-
5/8-18	NF	H5	PB	-	-	-	-	-	-	-	-	-
5/8-18	NF	H7	PB	-	-	-	-	-	-	-	PB
5/8-18	NF	H10	PB	-	-	-	-	-	-	-	PB
11/16-11	NS	H3	TPB	-	-	-	-	-	-	-	-
11/16-16	NS	H3	TPB	-	-	-	-	-	-	-	-
3/4-10	NC	H1	P	-	-	-	-	-	-	-	-
3/4-10	NC	H2	P	-	-	-	-	-	-	-	-
3/4-10	NC	H3	TPB	P	-	-	-	-	-	-	-	-
3/4-10	NC	H5	PB	P	-	-	-	-	-	-	-	-
3/4-10	NC	H7	-	-	-	-	-	-	-	-	PB
3/4-10	NC	H10	-	-	-	-	-	-	-	-	PB
3/4-16	NF	H1	P	-	-	-	-	-	-	-	-
3/4-16	NF	H2	P	-	-	-	-	-	-	-	-
3/4-16	NF	H3	TPB	P	-	-	-	-	-	-	-	-
3/4-16	NF	H5	PB	-	-	-	-	-	-	-	-	-
3/4-16	NF	H7	-	-	-	-	-	-	-	-	PB
3/4-16	NF	H10	-	-	-	-	-	-	-	-	PB
7/8-9	NC	H2	P	-	-	-	-	-	-	-	-
7/8-9	NC	H4	TPB	-	-	-	-	-	-	-	-

Note: The three column heads on left indicate Tap size, Fine or Course Unified Thread, and H Limit of Tap. The numbered column heads identify a specific tap style as follows: 1 = Straight Fluted Taps. 2 = Spiral Pointed Plug Taps. 3 = Spiral Pointed Bottoming Taps. 4 = Spiral Pointed Fluteless Plug Taps. 5 = Spiral Fluted Taps. 6 = Fast Spiral Fluted Taps. 7 = Screw Thread Insert Taps. 8 = Spiral Pointed Screw Thread Insert Plug Taps. 9 = Thread Forming Taps. A "T," "P," or "B" in any box indicates that the indicated tap is available in Taper, Plug, or Bottoming configurations. All taps are ground thread and commercially available in high-speed steel with wear-resistant coatings (see text). Check with manufacturer for number of flutes. See Table 1 for tap dimensions.

(Continued)

Table 3. *(Continued)* **Commercially Available Ground Thread Tap Sizes for Taper, Plug, and Bottoming Taps.** See Notes.

			Fractional Sizes									
Size	**Thread**	**H Limit**	**1.**	**2.**	**3.**	**4.**	**5.**	**6.**	**7.**	**8.**	**9.**
7/8-9	NC	H4	TPB	-	-	-	-	-	-	-	-
7/8-9	NC	H6	P	-	-	-	-	-	-	-	-
7/8-14	NF	H2	P	-	-	-	-	-	-	-	-
7/8-14	NF	H4	TPB	-	-	-	-	-	-	-	-
7/8-14	NF	H6	P	-	-	-	-	-	-	-	-
1-8	NC	H2	P	-	-	-	-	-	-	-	-
1-8	NC	H4	TPB	-	-	-	-	-	-	-	-
1-8	NC	H6	P	-	-	-	-	-	-	-	-
1-12	NF	H4	TPB	-	-	-	-	-	-	-	-
1-14	NS	H2	P	-	-	-	-	-	-	-	-
1-14	NS	H4	TPB	-	-	-	-	-	-	-	-
1-14	NS	H6	TPB	-	-	-	-	-	-	-	-
1-1/8-7	NC	H4	TPB	-	-	-	-	-	-	-	-
1-1/8-12	NF	H4	TPB	-	-	-	-	-	-	-	-
1-1/4-7	NC	H4	TPB	-	-	-	-	-	-	-	-
1-1/4-12	NF	H4	TPB	-	-	-	-	-	-	-	-
1-3/8-6	NC	H4	TPB	-	-	-	-	-	-	-	-
1-3/8-12	NF	H4	TPB	-	-	-	-	-	-	-	-
1-1/2-6	NC	H4	TPB	-	-	-	-	-	-	-	-
1-1/2-12	NF	H4	TPB	-	-	-	-	-	-	-	-

Note: The three column heads on left indicate Tap size, Fine or Course Unified Thread, and H Limit of Tap. The numbered column heads identify a specific tap style as follows: 1 = Straight Fluted Taps. 2 = Spiral Pointed Plug Taps. 3 = Spiral Pointed Bottoming Taps. 4 = Spiral Pointed Fluteless Plug Taps. 5 = Spiral Fluted Taps. 6 = Fast Spiral Fluted Taps. 7 = Screw Thread Insert Taps. 8 = Spiral Pointed Screw Thread Insert Plug Taps. 9 = Thread Forming Taps. A "T," "P," or "B" in any box indicates that the indicated tap is available in Taper, Plug, or Bottoming configurations. All taps are ground thread and commercially available in high-speed steel with wear-resistant coatings (see text). Check with manufacturer for number of flutes. See Table 1 for tap dimensions.

Table 3. *(Continued)* **Commercially Available Ground Thread Tap Metric Sizes for Taper, Plug, and Bottoming Taps.** See Notes.

				Metric Sizes					
Size	**Pitch**	**Class of Fit**	**Limit**	**1.**	**2.**	**5.**	**6.**	**9.**	**10.**
M1.6	0.35	6H	D3	TPB	P	-	-	-	-
M1.8	0.35	6H	D3	TPB	P	-	-	-	-
M2	0.40	6H	D3	TPB	P	-	-	-	-
M2.2	0.45	6H	D3	TPB	P	-	-	-	-
M2.5	0.45	6H	D3	TPB	P	-	-	-	-
M3	0.50	6H	D3	TPB	P	PB	PB	PB	-
M3.5	0.60	6H	D4	TPB	P	PB	PB	PB	-
M4	0.70	6H	D4	TPB	P	PB	PB	PB	-

(Continued)

Table 3. *(Continued)* **Commercially Available Ground Thread Tap Metric Sizes for Taper, Plug, and Bottoming Taps.** See Notes.

Metric Sizes									
Size	Pitch	Class of Fit	Limit	1.	2.	5.	6.	9.	10.
M4.5	0.75	6H	D4	TPB	P	PB	PB	PB	-
M5	0.80	6H	D4	TPB	P	PB	PB	PB	-
M6	1.0	6H	D5	TPB	P	PB	PB	PB	-
M7	1.0	6H	D5	TPB	P	PB	PB	PB	-
M8	1.25	6H	D5	TPB	P	PB	PB	PB	-
M8	1.0	6H	D5	TPB	P	PB	PB	PB	-
M10	1.50	6H	D6	TPB	P	PB	PB	PB	-
M10	1.25	6H	D5	TPB	P	PB	PB	PB	-
M10	1.0	-	H3	-	-	-	-	-	P
M12	1.75	6H	D6	TPB	P	PB	PB	PB	-
M12	1.25	6H	D5	TPB	P	PB	PB	PB	-
M14	2.0	6H	D7	TPB	P	-	-	PB	-
M14	1.50	6H	D6	TPB	P	-	-	PB	-
M14	1.25	-	H4	-	-	-	-	-	P
M16	2.0	6H	D7	TPB	P	-	-	PB	-
M16	1.50	6H	D6	TPB	P	-	-	-	-
M18	2.50	6H	D7	TPB	P	-	-	-	-
M18	1.50	6H	D6	TPB	-	-	-	PB	-
M18	1.50	-	H4	-	-	-	-	-	P
M20	2.50	6H	D7	TPB	P	-	-	-	-
M20	1.50	6H	D6	TPB	-	-	-	-	-
M22	2.50	6H	D7	TPB	-	-	-	-	-
M22	1.50	6H	D6	TPB	-	-	-	-	-
M24	3.0	6H	D8	TPB	-	-	-	-	-
M24	2.0	6H	D7	TPB	-	-	-	-	-
M27	3.0	6H	D8	TPB	-	-	-	-	-
M27	2.0	6H	D7	TPB	-	-	-	-	-
M30	3.50	6H	D9	TPB	-	-	-	-	-
M30	2.0	6H	D7	TPB	-	-	-	-	-
M33	2.0	6H	D7	TPB	-	-	-	-	-
M36	4.0	6H	D9	TPB	-	-	-	-	-
M36	3.0	6H	D8	TPB	-	-	-	-	-
M39	4.0	6H	D9	TPB	-	-	-	-	-
M39	3.0	6H	D8	TPB	-	-	-	-	-

Note: The four column heads on left indicate Tap size, Pitch of Thread, Class of Metric Thread, and Limit of Tap. The numbered column heads identify a specific tap style as follows: 1 = Straight Fluted Taps. 2 = Spiral Pointed Plug Taps. 5 = Spiral Fluted Taps. 6 = Fast Spiral Fluted Taps. 9 = Thread Forming Taps. 10 = Spark Plug Tap. A "T," "P," or "B" in any box indicates that the indicated tap is available in Taper, Plug, or Bottoming configurations. All taps are ground thread and commercially available in high-speed steel with wear-resistant coatings (see text). Check with manufacturer for number of flutes. See Table 1 for tap dimensions.

Pitch diameter limits identification system

Inch System ground thread screw machine and fractional size taps that are made to conform to standards for pitch diameter limits are marked with the letter "G" (for ground thread), plus the letter "H" or "L" and a pitch diameter limit number. "H" stands for high, or above basic size, and "L" indicates low, or under basic size. For taps under one inch in diameter, the pitch diameter limits are the amount the pitch diameter of the tap is over or under basic pitch diameter divided by .0005 as follows.

L 4 = basic minus .0015 to basic minus .002"   H 5 = basic plus .002 to basic plus .0025"
L 3 = basic minus .001 to basic minus .0015"   H 6 = basic plus .0025 to basic plus .003"
L 2 = basic minus .0005 to basic minus .001"   H 7 = plus basic.003 to basic plus .0035"
L 1 = basic to basic minus .0005"   H 8 = basic plus .0035 to basic plus .004"
H 1 = basic to basic plus .0005"   H 9 = basic plus .004 to basic plus .0045"
H 2 = basic plus .0005 to basic plus .001"   H 10 = basic plus .0045 to basic plus .005"
H 3 = basic plus .001 to basic plus .0015"   H 11 = basic plus .005 to basic plus .0055"
H 4 = basic plus .0015 to basic plus .002"   H 12 = basic plus .0055 to basic plus .006"

For fractional size taps over one inch in diameter to one and one-half inches in diameter, the following pitch diameter limits apply.

H 4 = basic plus .001 to basic plus .002"
H 6 = basic plus .002 to basic plus .003"
H 8 = basic plus .003 to basic plus .004"

Table 4 gives pitch diameters for H–1 through H–3 limit taps for Machine Screw size ground thread taps, and **Table 5** provides H–1 through H–6 limits for Fractional size ground thread taps. Recommended taps for Class 2, 2B, 3, and 3B Machine Screw and Fractional size threads for ground thread taps are given in **Table 6**. Forming tap recommendations for Class 2, 2B, and 3B Machine Screw and Fractional size threads are found in **Table 22**.

Metric System ground thread taps use the letter "D" for taps where the pitch diameter is over basic, and the letters "DU" for taps that have pitch diameters under basic. The D and DU pitch diameter limits are the amount the pitch diameter of the tap is over or under the basic pitch diameter divided by .013 mm. Since .013 mm equals .0005118", D limits are essentially equal to H limits for inch system taps, and DU limits are roughly equivalent to L limits. Pitch diameter limits (in inches) for recommended metric taps are as follows.

D 3 = basic plus .0009" to basic plus .0015"   D 7 = basic plus .0019" to basic plus .0035"
D 4 = basic plus .0012" to basic plus .002"   D 8 = basic plus .0024" to basic plus .004"
D 5 = basic plus .0015" to basic plus .0025"   D 9 = basic plus .0025" to basic plus .0045"
D 6 = basic plus .0018" to basic plus .003"

Table 7 gives pitch diameter limits for D–3 through D–9 limits for M Profile ground thread taps in both millimeters and inches, and **Table 8** shows recommended taps (with pitch diameter limits in both millimeters and inches) for Class 4H and 6H Metric Threads.

Calculating hole sizes for length of engagement and desired percentage of thread for ground thread cutting taps

Before drilling, reaming, and tapping a hole for a selected thread size, the percentage of thread engagement produced by the operation should be determined (see *Figure 2*). As the desired percentage of thread increases, the amount of strain on the tap's teeth increases, requiring more power and raising the risk of the tap breaking in the hole. Tap breakage is

(Text continued on p. 1221)

Table 4. Machine Screw Size Taps—Ground Thread—Unified Screw Threads. *(Source, Kennametal Inc.)*

Size	Threads Per Inch			Major Diameter			Basic Pitch Dia.	Pitch Diameter Limits					
	UNC	UNF	UNS	Basic	Min.	Max.		H1 Limit		H2 Limit		H3 Limit	
								Min.	Max.	Min.	Max.	Min.	Max.
0	-	80	-	.0600	.0605	.0616	.0519	.0519	.0524	.0524	.0529	-	-
1	64	-	-	.0730	.0736	.0750	.0629	.0629	.0634	.0634	.0639	-	-
1	-	72	-	.0730	.0736	.0748	.0640	.0640	.0645	.0645	.0650	-	-
2	56	-	-	.0860	.0867	.0883	.0744	.0794	.0749	.0749	.0754	-	-
2	-	64	-	.0860	.0866	.0880	.0759	-	-	.0764	.0769	-	-
3	48	-	-	.0990	.0999	.1017	.0855	.0855	.0860	.0860	.0865	-	-
3	-	56	-	.0990	.0997	.1013	.0874	.0874	.0879	.0879	.0884	-	-
4	-	-	36	.1120	.1135	.1156	.0940	-	-	.0945	.0950	-	-
4	40	-	-	.1120	.1133	.1152	.0958	.0958	.0963	.0963	.0968	-	-
4	-	48	-	.1120	.1129	.1147	.0985	.0985	.0990	.0990	.0995	-	-
5	40	-	-	.1250	.1263	.1282	.1088	.1088	.1093	.1093	.1098	-	-
5	-	44	-	.1250	.1263	.1280	.1102	-	-	.1107	.1112	-	-
6	32	-	-	.1380	.1401	.1421	.1177	.1177	.1182	.1182	.1187	.1187	.1192
6	-	40	-	.1380	.1393	.1412	.1218	.1218	.1223	.1223	.1228	-	-
8	32	-	-	.1640	.1661	.1681	.1437	.1437	.1442	.1442	.1447	.1447	.1452
8	-	36	-	.1640	.1655	.1676	.1460	.1460	.1465	.1465	.1470	-	-
10	24	-	-	.1900	.1927	.1954	.1629	.1629	.1634	.1634	.1639	.1639	.1644
10	-	32	-	.1900	.1921	.1941	.1697	.1697	.1702	.1702	.1707	.1707	.1712
12	24	-	-	.2160	.2187	.2214	.1889	.1889	.1894	-	-	.1899	.1904
12	-	28	-	.2160	.2183	.2206	.1928	.1928	.1933	-	-	.1938	.1942

All dimensions in inches.

Table 5. Fractional Size Taps—Ground Thread—Unified Screw Threads. *(Source, Fastcut Tool Corp.)*

Size	Threads Per Inch			Major Diameter			Basic Pitch Dia.	Pitch Diameter Limits			
	UNC	UNF	UNS	Basic	Min.	Max.		H1 Limit		H2 Limit	
								Min.	Max.	Min.	Max.
1/4	20	-	-	0.2500	0.2533	0.2565	0.2175	0.2175	0.2180	0.2180	0.2185
1/4	-	28	-	0.2500	0.2533	0.2546	0.2268	0.2268	0.2273	0.2273	0.2278
5/16	18	-	-	0.3125	0.3161	0.3197	0.2764	0.2764	0.2769	0.2769	0.2774
5/16	-	24	-	0.3125	0.3152	0.3179	0.2854	0.2854	0.2859	0.2859	0.2864
3/8	16	-	-	0.3750	0.3790	0.3831	0.3344	0.3344	0.3349	0.3349	0.3354
3/8	-	24	-	0.3750	0.3777	0.3804	0.3479	0.3479	0.3484	0.3484	0.3489
7/16	14	-	-	0.4375	0.4422	0.4468	0.3911	0.3911	0.3916	0.3916	0.3921
7/16	-	20	-	0.4375	0.4408	0.4440	0.4050	0.4050	0.4055	0.4055	0.4060
1/2	13	-	-	0.5000	0.5050	0.5100	0.4500	0.4500	0.4505	0.4505	0.4510
1/2	-	20	-	0.5000	0.5033	0.5065	0.4675	0.4675	0.4680	0.4680	0.4685
9/16	12	-	-	0.5625	0.5679	0.5733	0.5084	0.5084	0.4680	0.5089	0.5094
9/16	-	18	-	0.5625	0.5661	0.5697	0.5264	0.5264	0.5269	0.5269	0.5274
5/8	11	-	-	0.6250	0.6309	0.6368	0.5660	0.5660	0.5665	0.5665	0.5670
5/8	-	18	-	0.6250	0.6286	0.6322	0.5889	0.5889	0.5894	0.5894	0.5899
11/16	-	-	11	0.6875	0.6934	0.6993	0.6285	-	-	-	-
11/16	-	-	16	0.6875	0.6915	0.6956	0.6469	-	-	-	-
3/4	10	-	-	0.7500	0.7565	0.7630	0.6850	0.6850	0.6855	0.6855	0.6860
3/4	-	16	-	0.7500	0.7540	0.7581	0.7094	0.7094	0.7099	0.7099	0.7104
7/8	9	-	-	0.8750	0.8822	0.8894	0.8028	0.8028	0.8033	0.8033	0.8038
7/8	-	14	-	0.8750	0.8797	0.8843	0.8286	0.8286	0.8291	0.8291	0.8296

(Continued)

Table 5. (Continued) **Fractional Size Taps—Ground Thread—Unified Screw Threads.** (Source, Fastcut Tool Corp.)

Size	Threads Per Inch			Major Diameter			Basic Pitch Dia.	Pitch Diameter Limits			
	UNC	UNF	UNS	Basic	Min.	Max.		H1 Limit		H2 Limits	
								Min.	Max.	Min.	Max.
1	8	-	-	1.0000	1.0081	1.0162	0.9188	0.9188	0.9193	0.9193	0.9198
1	-	12	-	1.0000	1.0054	1.0108	0.9459	-	-	-	-
1	-	-	14	1.0000	1.0047	1.0093	0.9536	-	-	0.9541	0.9546
1 1/8	7	-	-	1.1250	1.1343	1.1436	1.0322	-	-	-	-
1 1/8	-	12	-	1.1250	1.1304	1.1358	1.0709	-	-	-	-
1 1/4	7	-	-	1.2500	1.2593	1.2686	1.1572	-	-	-	-
1 1/4	-	12	-	1.2500	1.2554	1.2608	1.1959	-	-	-	-
1 3/8	6	-	-	1.3750	1.3859	1.3967	1.2667	-	-	-	-
1 3/8	-	12	-	1.3750	1.3804	1.3858	1.3209	-	-	-	-
1 1/2	6	-	-	1.5000	1.5109	1.5217	1.3917	-	-	-	-
1 1/2	-	12	-	1.5000	1.5054	1.5108	1.4459	-	-	-	-

Size	Threads Per Inch			Pitch Diameter Limits							
	UNC	UNF	UNS	H3 Limit		H4 Limit		H5 Limit		H6 Limit	
				Min.	Max.	Min.	Max.	Min.	Max.	Min.	Max.
1/4	20	-	-	0.2185	0.2190	-	-	0.2195	0.2200	-	-
1/4	-	28	-	0.2278	0.2283	0.2283	0.2288	-	-	-	-
5/16	18	-	-	0.2774	0.2779	-	-	0.2784	0.2789	-	-
5/16	-	24	-	0.2864	0.2869	0.2869	0.2874	-	-	-	-
3/8	16	-	-	0.3354	0.3359	-	-	0.3364	0.3369	-	-
3/8	-	24	-	0.3489	0.3494	0.3494	0.3499	-	-	-	-
7/16	14	-	-	0.3921	0.3926	-	-	0.3931	0.3936	-	-
7/16	-	20	-	0.4060	0.4065	-	-	0.4070	0.4075	-	-
1/2	13	-	-	0.4510	0.4515	-	-	0.4520	0.4525	-	-
1/2	-	20	-	0.4685	0.4690	-	-	0.4695	0.4700	-	-
9/16	12	-	-	0.5094	0.5099	-	-	0.5104	0.5109	-	-
9/16	-	18	-	0.5274	0.5279	-	-	0.5284	0.5289	-	-
5/8	11	-	-	0.5670	0.5675	-	-	0.5680	0.5685	-	-
5/8	-	18	-	0.5899	0.5904	-	-	0.5909	0.5914	-	-
11/16	-	-	11	0.6295	0.6300	-	-	-	-	-	-
11/16	-	-	16	0.6479	0.6484	-	-	-	-	-	-
3/4	10	-	-	0.6860	0.6865	-	-	0.6870	0.6875	-	-
3/4	-	16	-	0.7104	0.7109	-	-	0.7114	0.7119	-	-
7/8	9	-	-	-	-	0.8043	0.8048	-	-	0.8053	0.8058
7/8	-	14	-	-	-	0.8301	0.8306	-	-	0.8311	0.8316
1	8	-	-	-	-	0.9203	0.9208	-	-	0.9213	0.9218
1	-	12	-	-	-	0.9474	0.9479	-	-	-	-
1	-	-	14	-	-	0.9551	0.9556	-	-	0.9561	0.9566
1 1/8	7	-	-	-	-	1.0332	1.0342	-	-	-	-
1 1/8	-	12	-	-	-	1.0719	1.0729	-	-	-	-
1 1/4	7	-	-	-	-	1.1582	1.1592	-	-	-	-
1 1/4	-	12	-	-	-	1.1969	1.1979	-	-	-	-
1 3/8	6	-	-	-	-	1.2677	1.2687	-	-	-	-
1 3/8	-	12	-	-	-	1.3219	1.3229	-	-	-	-
1 1/2	6	-	-	-	-	1.3927	1.3937	-	-	-	-
1 1/2	-	12	-	-	-	1.4469	1.4479	-	-	-	-

All dimensions in inches.

Table 6. Tap Recommendations for Classes 2, 3, 2B, and 3B Unified Machine Screw and Fractional Size Screw Threads. *(Source, New England Tap Corp.)*

Size	Threads Per Inch		Recommended Tap for Class of Thread				Pitch Diameter Limits				
	UNC	UNF	Class 2	Class 3	Class 2B	Class 3B	Min. All Classes (Basic)	Max. Class 2	Max. Class 3	Max. Class 2B	Max. Class 3B
0	-	80	G H1	G H1	G H2	G H1	.0519	.0536	.0532	.0542	.0536
1	64	-	G H1	G H1	G H2	G H1	.0629	.0648	.0643	.0655	.0648
1	-	72	G H1	G H1	G H2	G H1	.0640	.0658	.0653	.0665	.0659
2	56	-	G H1	G H1	G H2	G H1	.0744	.0764	.0759	.0772	.0765
2	-	64	G H1	G H1	G H2	G H1	.0759	.0778	.0773	.0786	.0779
3	48	-	G H1	G H1	G H2	G H1	.0855	.0877	.0871	.0885	.0877
3	-	56	G H1	G H1	G H2	G H1	.0874	.0894	.0889	.0902	.0895
4	40	-	G H2	G H1	G H2	G H2	.0958	.0982	.0975	.0991	.0982
4	-	48	G H1	G H1	G H2	G H1	.0985	.1007	.1001	.1016	.1008
5	40	-	G H2	G H1	G H2	G H2	.1088	.1112	.1105	.1121	.1113
5	-	44	G H1	G H1	G H2	G H1	.1102	.1125	.1118	.1134	.1126
6	32	-	G H2	G H1	G H3	G H2	.1177	.1204	.1196	.1214	.1204
6	-	40	G H2	G H1	G H2	G H2	.1218	.1242	.1235	.1252	.1243
8	32	-	G H2	G H1	G H3	G H2	.1437	.1464	.1456	.1475	.1465
8	-	36	G H2	G H1	G H2	G H2	.1460	.1485	.1478	.1496	.1487
10	24	-	G H3	G H1	G H3	G H3	.1629	.1662	.1653	.1672	.1661
10	-	32	G H2	G H1	G H3	G H2	.1697	.1724	.1716	.1736	.1726
12	24	-	G H3	G H1	G H3	G H3	.1889	.1922	.1913	.1933	.1922
12	-	28	G H3	G H1	G H3	G H3	.1928	.1959	.1950	.1970	.1959
1/4	20	-	G H3	G H2	G H5	G H3	.2175	.2211	.2201	.2223	.2211
1/4	-	28	G H3	G H1	G H4	G H3	.2268	.2299	.2290	.2311	.2300
5/16	18	-	G H3	G H2	G H5	G H3	.2764	.2805	.2794	.2817	.2803
5/16	-	24	G H3	G H1	G H4	G H3	.2854	.2887	.2878	.2902	.2890
3/8	16	-	G H3	G H2	G H5	G H3	.3344	.3389	.3376	.3401	.3387
3/8	-	24	G H3	G H1	G H4	G H3	.3479	.3512	.3503	.3528	.3516
7/16	14	-	G H5	G H3	G H5	G H3	.3911	.3960	.3947	.3972	.3957
7/16	-	20	G H3	G H1	G H5	G H3	.4050	.4086	.4076	.4104	.4091
1/2	13	-	G H5	G H3	G H5	G H3	.4500	.4552	.4537	.4565	.4548
1/2	-	20	G H3	G H1	G H5	G H3	.4675	.4711	.4701	.4731	.4717
9/16	12	-	G H5	G H3	G H5	G H3	.5084	.5140	.5124	.5152	.5135
9/16	-	18	G H3	G H2	G H5	G H3	.5264	.5305	.5294	.5323	.5308
5/8	11	-	G H5	G H3	G H5	G H3	.5660	.5719	.5702	.5732	.5714
5/8	-	18	G H3	G H2	G H5	G H3	.5889	.5930	.5919	.5949	.5934
3/4	10	-	G H5	G H3	G H5	G H5	.6850	.6914	.6895	.6927	.6907
3/4	-	16	G H3	G H2	G H5	G H3	.7094	.7139	.7126	.7159	.7143
7/8	9	-	G H6	G H4	G H6	G H4	.8028	.8098	.8077	.8110	.8089

(Continued)

Table 6. *(Continued)* Tap Recommendations for Classes 2, 3, 2B, and 3B Unified Machine Screw and Fractional Size Screw Threads. *(Source, New England Tap Corp.)*

Size	Threads Per Inch		Recommended Tap for Class of Thread				Pitch Diameter Limits				
	UNC	UNF	Class 2	Class 3	Class 2B	Class 3B	Min. All Classes (Basic)	Max. Class 2	Max. Class 3	Max. Class 2B	Max. Class 3B
7/8	-	14	G H4	G H2	G H6	G H4	.8286	.8335	.8322	.8356	.8339
1	8	-	G H6	G H4	G H6	G H4	.9188	.9264	.9242	.9276	.9254
1	-	12	G H4	G H2	G H6	G H4	.9459	.9515	.9499	.9535	.9516
1		14 NS	G H4	G H2	G H6	G H4	.9536	.9585	.9572	.9609	.9590
1 1/8	7	-	G H8	G H4	G H8	G H4	1.0322	1.0407	1.0381	1.0416	1.0393
1 1/8	-	12	G H4	G H4	G H6	G H4	1.0709	1.0765	1.0749	1.0787	1.0768
1 1/4	7	-	G H8	G H4	G H8	G H4	1.1572	1.1657	1.1631	1.1668	1.1644
1 1/4	-	12	G H4	G H4	G H6	G H4	1.1959	1.2015	1.1999	1.2039	1.2019
1 3/8	6	-	G H8	G H4	G H8	G H4	1.2667	1.2768	1.2738	1.2771	1.2745
1 3/8	-	12	G H4	G H4	G H6	G H4	1.3209	1.3265	1.3249	1.3291	1.3270
1 1/2	6	-	G H8	G H4	G H8	G H4	1.3917	1.4018	1.3988	1.4022	1.3996
1 1/2	-	12	G H4	G H4	G H6	G H4	1.4459	1.4515	1.4499	1.4542	1.4522

The above recommended taps normally produce the Class of Thread indicated in average materials when used with reasonable care. However, if the tap specified does not give a satisfactory gage fit in the work, a choice of some other limit tap will be necessary. All dimensions in inches.

Table 7. Metric Size Taps Dimensions and Pitch Diameter Limits M Profile Thread. *(Source, Kennametal Inc.)*

Dimensions										
Millimeters					Inches					
Nom. Dia.	Pitch	Basic Major Dia.	Min. Major Dia.	Max. Major Dia.	Basic Pitch Dia.	Pitch	Basic Major Dia.	Min. Major Dia.	Max. Major Dia.	Basic Pitch Dia.
1.6	0.35	1.600	1.628	1.653	1.373	.01378	.06299	.06409	.06508	.05406
2	0.4	2.000	2.032	2.057	1.740	.01575	.07874	.08000	.08098	.06850
2.5	0.45	2.500	2.536	2.561	2.208	.01772	.09843	.09984	.10083	.08693
3	0.5	3.000	3.040	3.065	2.675	.01969	.11811	.11969	.12067	.10531
3.5	0.6	3.500	3.548	3.573	3.110	.02362	.13780	.13969	.14067	.12244
4	0.7	4.000	4.056	4.097	3.545	.02756	.15748	.15969	.16130	.13957
4.5	0.75	4.500	4.560	4.601	4.013	.02953	.17717	.17953	.18114	.15799
5	0.8	5.000	5.064	5.105	4.480	.03150	.19685	.19937	.20098	.17638
6	1	6.000	6.080	6.121	5.350	.03937	.23622	.23937	.24098	.21063
7	1	7.000	7.080	7.121	6.350	.03937	.27559	.27874	.28035	.25000
8	1.25	8.000	8.100	8.164	7.188	.04921	.31496	.31890	.32142	.28299
10	1.5	10.000	12.120	10.184	9.026	.05906	.39370	.39843	.40094	.35535
12	1.75	12.000	12.140	12.204	10.863	.06890	.47244	.47795	.48047	.42768
14	2	14.000	14.160	14.224	12.701	.07874	.55118	.55748	.56000	.50004
16	2	16.000	16.160	16.224	14.701	.07874	.62992	.63622	.63874	.57878
20	2.5	20.000	20.200	20.264	18.376	.09843	.78740	.79528	.79780	.72346
24	3	24.000	24.240	24.340	22.051	.11811	.94488	.95433	.95827	.86815
30	3.5	30.000	30.280	30.380	27.727	.13780	1.18110	1.19213	1.19606	1.09161
36	4	36.000	36.360	36.420	33.402	.15748	1.41732	1.42992	1.43386	1.31504

(Continued)

Table 7. *(Continued)* **Metric Size Taps Dimensions and Pitch Diameter Limits M Profile Thread.**
(Source, Kennametal Inc.)

Pitch Diameter Limits											
Nom. Dia. mm	Pitch	D3 Limits mm		D3 Limits inch		Nom. Dia. mm	Pitch	D6 Limits mm		D6 Limits inch	
		Min.	Max.	Min.	Max.			Min.	Max.	Min.	Max.
1.6	0.35	1.397	1.412	.05500	.05559	10	1.5	9.073	9.104	.35720	.35843
2	0.4	1.764	1.779	.06945	.07004	12	1.75	10.901	10.941	.42953	.43075
2.5	0.45	2.232	2.247	.08787	.08846			**D7 Limits mm**		**D7 Limits inch**	
3	0.5	2.699	2.714	.10626	.10685			Min.	Max.	Min.	Max.
		D4 Limits mm		**D4 Limits inch**		14	2	12.751	12.792	.50201	.50362
		Min.	Max.	Min.	Max.	16	2	14.751	14.792	.58075	.58236
3.5	0.6	3.142	3.162	.12370	.12449	20	2.5	18.426	18.467	.72543	.72705
4	0.7	3.577	3.597	.14083	.14161			**D8 Limits mm**		**D8 Limits inch**	
4.5	0.75	4.045	4.065	.15925	.16004			Min.	Max.	Min.	Max.
5	0.8	4.512	4.532	.17764	.17843	24	3	22.114	22.155	.87063	.87224
		D5 Limits mm		**D5 Limits inch**				**D9 Limits mm**		**D9 Limits inch**	
		Min.	Max.	Min.	Max.			Min.	Max.	Min.	Max.
6	1	5.390	5.415	.21220	.21319	30	3.5	27.792	27.844	1.09417	1.09622
7	1	6.390	6.415	.25157	.25256	36	4	33.467	33.519	1.31760	1.31965
8	1.25	7.222	7.253	.28433	.28555						

Lead Tolerance
A maximum lead deviation of ±0.013 mm (± 0.0005") within any two threads not more than 25.4 mm (1") apart is permitted.

Angle Tolerance	
Pitch (mm)	**Deviation of Half Angle**
Over 0.25 to 2.5 Inclusive	±30'
Over 2.5 to 4 Inclusive	±25'
Over 4 to 6 Inclusive	±20'

Basic pitch diameter is the same as minimum pitch diameter of Internal Thread Class 6H—Table 21, ANSI B1.13M-1979.

Table 8. Tap Recommendations for Classes 4H and 6H Metric M Profile Screw Threads.
(Source, Fastcut Tool Corp.)

Threads per inch		Recommended Tap for Class of Thread		Pitch Diameter Limits for Class of Thread					
				Millimeters			Inches		
Nominal Diameter	Pitch	4H	6H	Min. All Classes (Basic)	Max. 4H	Max. 6H	Min. All Classes (Basic)	Max. 4H	Max. 6H
M1.6	0.35	D1	D3	1.373	1.426	1.458	.0541	.0561	.0574
M2	0.4	D1	D3	1.740	1.796	1.830	.0685	.0707	.0720
M2.5	0.45	D1	D3	2.208	2.268	2.303	.0869	.0893	.0907
M3	0.5	D1	D3	2.675	2.738	2.775	.1053	.1078	.1092
M3.5	0.6	D1	D4	3.110	3.181	3.222	.1224	.1252	.1268

(Continued)

Table 8. (Continued) **Tap Recommendations for Classes 4H and 6H Metric M Profile Screw Threads.**
(Source, Fastcut Tool Corp.)

Threads per inch		Recommended Tap for Class of Thread		Pitch Diameter Limits for Class of Thread					
				Millimeters			Inches		
Nominal Diameter	Pitch	4H	6H	Min. All Classes (Basic)	Max. 4H	Max. 6H	Min. All Classes (Basic)	Max. 4H	Max. 6H
M4	0.7	D2	D4	3.545	3.620	3.663	.1396	.1425	.1442
M4.5	0.75	D2	D4	4.013	4.088	4.131	.1580	.1609	.1626
M5	0.8	D2	D4	4.480	4.560	4.605	.1764	.1795	.1813
M6	1	D3	D5	5.350	5.445	5.500	.2106	.2144	.2165
M6	0.75	D2	D3	5.513	5.598	5.645	.2170	.2204	.2222
M7	1	D3	D5	6.350	6.445	6.500	.2500	.2537	.2559
M7	0.75	D2	D4	6.513	6.598	6.645	.2564	.2598	.2616
M8	1.25	D3	D5	7.188	7.288	7.348	.2830	.2869	.2893
M8	1	D3	D5	7.350	7.445	7.500	.2894	.2931	.2953
M10	1.5	D3	D6	9.026	9.138	9.206	.3554	.3598	.3624
M10	1.25	D3	D5	9.188	9.288	9.348	.3617	.3657	.3680
M12	1.75	D3	D6	10.863	10.988	11.063	.4277	.4326	.4356
M12	1.25	D3	D5	11.188	11.300	11.368	.4405	.4449	.4476
M14	2	D3	D7	12.701	12.833	12.913	.5000	.5052	.5084
M14	1.5	D3	D6	13.026	13.144	13.216	.5128	.5175	.5203
M16	2	D4	D7	14.701	14.833	14.913	.5788	.5840	.5871
M16	1.5	D3	D6	15.026	15.144	15.216	.5916	.5962	.5990
M18	2.5	D4	D7	16.376	16.516	16.600	.6447	.6502	.6535
M18	1.5	D3	D6	17.026	17.144	17.216	.6703	.6750	.6778
M20	2.5	D4	D7	18.376	18.516	18.600	.7235	.7290	.7323
M20	1.5	D3	D5	19.026	19.144	19.216	.7490	.7537	.7565
M24	3	D4	D8	22.051	22.221	22.316	.8681	.8748	.8786
M24	1.5	D3	D5	23.026	23.151	23.226	.9065	.9114	.9144
M27	3	D5	D8	25.051	25.221	25.316	.9863	.9930	.9967
M27	2	D5	D7	25.701	25.841	25.925	1.0118	1.0174	1.0207
M30	3.5	D5	D9	27.727	27.907	28.007	1.0916	1.0987	1.1026
M30	2	D5	D7	28.701	28.841	28.925	1.1300	1.1355	1.1388
M33	3.5	D5	D9	30.727	30.907	31.007	1.2097	1.2168	1.2207
M33	2	D5	D7	31.701	31.841	31.925	1.2481	1.2536	1.2569
M36	4	D5	D9	33.402	33.592	33.702	1.3150	1.3225	1.3268
M36	2	D5	D7	34.701	34.841	34.925	1.3662	1.3717	1.3750

The above recommended taps normally produce the Class of Thread indicated in average materials when used with reasonable care. However, if the tap specified does not give a satisfactory gage fit in the work, a choice of some other limit tap will be necessary. D1 Limit to have minus .0005 tolerance.

infrequent if the diameter of the tap is over $1/2$" or if the length of thread to be tapped is less than $1/2$". In tougher materials, tapping a successful hole becomes increasingly difficult, which further discourages attempts to achieve high percentages of thread of 70% or more, especially when using small diameter taps to produce holes for UNC threads. Normally, in the production of threads, it is considered impractical to tap a thread unless its nominal diameter is greater than six times the basic thread height. Therefore, when the ratio of diameter (D) is greater than six times the basic thread height (h), the use of a larger diameter, a finer pitch of thread, or both should be considered.

As a rule, when both sides of the assembly are to be concentric, the difference between the minimum major diameter of the external thread and the maximum minor diameter of the internal thread should never be less than twice the addendum of the external thread (0.75H). Another way to calculate is that the sum of the major diameter tolerance and allowance (if any) of the external thread plus the minor diameter tolerance of the internal thread should never be more than 4/3 of the addendum of the external thread (0.5H). This will provide for a minimum safe overlap of 50 percent thread engagement. When one side of the assembly is to be eccentric, the difference between the maximum pitch diameter of the internal thread and the minimum pitch diameter of the external thread should not exceed the basic thread height (0.625H). Otherwise stated, the sum of the pitch diameter tolerances of both threads and the allowance, if any, should not be greater than 0.625H. This will provide for an eccentric assembly condition equal to half the basic thread height (0.312H) and zero minimum overlap on one side. If the results from the limits of size selected violate the above rules, the tolerances should be reduced by using a closer class of tolerance, assuming tolerance consistent with manufacturing possibility, or using a coarser pitch to increase the amount of overlap. The major diameter tolerance of the external thread or minor diameter tolerance of the internal thread should not be less than the pitch diameter allowance of the respective thread to maintain thread form. Also, it should be noted that if the tolerance on the minor diameter of the internal thread must necessarily be large, the major diameter of the external thread must be held close to the maximum major diameter and vice versa.

An additional important consideration when selecting the desired percentage of thread is hole depth and length of engagement. For short lengths of engagement, the hole diameter required prior to threading should be held near the minimum limit to maximize thread height for maximum joint strength. As length of engagement increases, the hole diameter should be increased for more economical tapping with less risk of tap breakage. The following recommended hole sizes before threading (from FED-STD-H28/2B) are suggested to permit economical tapping. It will be noted that the difference between limits in each range is the same and equal to half of the minor diameter tolerance. However, the minimum differences for thread sizes below $^1/_4$" are equal to the minor diameter tolerances (tolerance is the difference between specified maximum and minimum limits) for lengths of engagement to and including 0.33 diameter. For sizes $^1/_4$" and larger, with lengths of engagement exceeding 0.33 diameter, the minimum values between maximum and minimum hole sizes should never be less than 0.0040".

Table 9 gives minor diameter limits and corresponding percentages of thread for Class 1B and 2B Unified thread sizes from $^1/_4$–20 UNC through 1 $^1/_8$–16 UN. **Table 10** provides the same information for Class 1B and 2B sizes 1 $^1/_8$–18 UNEF through 3 $^3/_4$–4 UNC.

Length of Engagement	Minimum Hole Size	Maximum Hole Size
Up to and including 0.33 diameter	Minimum Minor Diameter	Minimum Minor Diameter plus $^1/_2$ Minor Diameter Tolerance
Above 0.33 through 0.67 diameter	Minimum Minor Diameter plus $^1/_4$ Minor Diameter Tolerance	Minimum Minor Diameter plus $^3/_4$ Minor Diameter Tolerance
Above 0.67 through 1.50 diameter	Minimum Minor Diameter plus $^1/_2$ Minor Diameter Tolerance	Maximum Minor Diameter (Minimum Minor Diameter plus Minor Diameter Tolerance)
Above 1.5 through 3.0 diameter *	Minimum Minor Diameter plus $^3/_4$ Minor Diameter Tolerance	Maximum Minor Diameter plus $^1/_4$ Minor Diameter Tolerance *

* Recommended maximum hole size is outside standard minor diameter limits. Use of a minor diameter larger than standard will result in a reduction in shear area of the external threads of the mating part. If manufacturing process permits, maximum hole size before threading should be maintained at the high end of the standard minor diameter limits.

Table 11 gives this information for Class 3B Unified thread sizes 0–80 UNF through 1 $^{1}/_{8}$–16 UN. **Table 12** covers Class 3B sizes 1 $^{1}/_{8}$–18 UNEF through 3 $^{3}/_{4}$–4. These tables also list sizes of drills that may be expected to drill holes within or near the specified minor diameter limits. The resulting probable hole size and percentages of thread are tabulated. Since drills normally produce marginally oversize holes, the tabulated hole sizes have been derived from probable mean oversizes when drilling holes equal to 1.5 times drill diameter with measurement being taken at midpoint of the depth drilled. Due to the increase in amount oversize experienced with drills in excess of $^{3}/_{4}$", reaming of holes this size and larger is recommended. It should be noted that the Unified Thread profile allows for a maximum permissible percentage of thread of 83.3%, and that some of the indicated percentages on the Tables are in excess of this amount. This is either due to the allowances for drill sizes to cut oversize, or lack of availability of drills within specified minor diameter limits. A reading in excess of 83.3% therefore indicates that the drill size is smaller than the minimum minor diameter, and additional machining of the hole may be necessary to permit economical tapping.

    Table 13 gives recommended hole size limits for Unified Class 1B and 2B threads for three lengths of engagement (see chart above), and **Table 14** provides the same suggested hole sizes for Unified Class 3B threads. For both Tables, refer to Tables 11 and 12 for basic minimum and minor diameter and percent of thread specifications. Hole size limits for diameter-pitch combinations for sizes over 1.00 inch not included on these Tables can be derived provided that there is a diameter-pitch combination in the Tables that: 1) has the same pitch and 2) has a diameter that is less by an integral amount than the diameter of the diameter-pitch combination for which hole size values are desired.

Example. To obtain the values for a 4.00–8UN–3B thread, add 2.00 to the values for the 2.00–8UN–3B thread given in the table. These values would then become 3.8650, 3.8722, 3.8684, 3.8759, 3.8722, 3.8797. The percentages of thread will remain unchanged.

    Recommended tap drill sizes for Metric M Profile threads are given in **Table 15**, and **Table 16** provides pitch diameter limits for Class 4H and Class 6H threads. See **Table 17** for decimal inch equivalents of standard M Profile metric threads. *(Text continued p. 1255)*

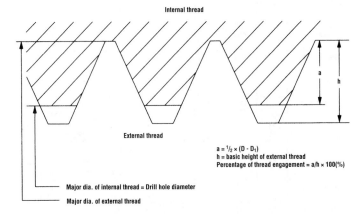

Internal thread

$a = \frac{1}{2} \times (D - D_1)$
h = basic height of external thread
Percentage of thread engagement = $a/h \times 100(\%)$

External thread

Major dia. of internal thread = Drill hole diameter

Major dia. of external thread

Figure 2. Percentage of thread engagement. External thread = basic profile. (Source, OSG Tap and Die, Inc.)

Table 9. Tap Drill Sizes and Percentage of Thread, Unified Threads Class 1B and 2B, Size $1/4$-20 UNC* Through 1 $1/8$-16 UN. *(Source, FED-STD-H28/2B.)*

Size, TPI, and Series*	Classes 1B and 2B Minor Dia.				Tap Drill and Percentage of Thread				
	D_1 (min.)	% of Thread*	D_1 (max.)	% of Thread*	Drill Size		% of Thread*	Probable Hole Size	% of Thread*
1/4-20 UNC	.196	83.1	.207	66.2	#9	.1960	83	.1998	77
					#8	.1990	79	.2028	73
					#7	.2010	75	.2048	70
					13/64	.2031	72	.2069	66
					#6	.2040	71	.2078	65
					#5	.2055	69	.2093	63
1/4-28 UNF	.211	84.1	.220	64.7	#3	.2130	80	.2168	72
					7/32	.2188	67	.2226	59
1/4-32 UNEF	.216	83.8	.224	64.0	7/32	.2188	77	.2226	67
					#2	.2210	71	.2248	62
1/4-36 UNS	.220	83.1	.226	66.5	#2	.2210	80	.2248	70
5/16-18 UNC	.252	83.8	.265	65.8	F	.2570	77	.2608	72
					G	.2610	71	.2651	66
5/16-20 UN	.258	83.9	.270	65.4	F	.2570	85	.2608	80
					G	.2610	79	.2651	73
					H	.2660	72	.2701	65
5/16-24 UNF	.267	84.1	.277	65.6	H	.2660	86	.2701	78
					I	.2720	75	.2761	67
					J	.2770	66	.2811	58
5/16-28 UN	.274	83.0	.282	65.7	J	.2770	77	.2811	68
					K	.2810	68	.2852	59
					9/32	.2812	67	.2854	58
5/16-32 UNEF	.279	82.5	.286	65.3	K	.2810	78	.2852	67
					9/32	.2812	77	.2854	67
5/16-36 UNS	.282	84.5	.289	65.1	7.25 mm	.2854	75	.2896	63
3/8-16 UNC	.307	83.8	.321	66.5	5/16	.3125	77	.3169	72
					O	.3160	73	.3204	67
3/8-20 UN	.321	83.1	.332	66.2	P	.3230	80	.3274	73
					Q	.3320	66	.3364	59
3/8-24 UNF	.330	83.1	.340	64.7	Q	.3320	79	.3364	71
					R	.3390	67	.3434	58
3/8-28 UN	.336	84.1	.345	64.7	R	.3390	78	.3434	68
					11/32	.3438	67	.3483	58
3/8-32 UNEF	.341	83.8	.349	64.0	11/32	.3438	77	.3483	66
					S	.3480	67	.3525	55
3/8-36 UNS	.345	83.1	.352	63.7	S	.3480	75	.3525	62
7/16-14 UNC	.360	83.5	.376	66.3	T	.3580	86	.3626	81
					23/64	.3594	84	.3640	79
7/16-16 UN	.370	83.1	.384	65.9	3/8	.3750	77	.3796	71
					V	.3770	75	.3816	69
7/16-20 UNF	.383	83.9	.395	65.4	W	.3860	79	.3906	72
					25/64	.3906	72	.3952	65
7/16-28 UNEF	.399	83.0	.407	65.7	Y	.4040	72	.4086	62

* For Machine Screw sizes, see **Table 11**: Tap Drill sizes and Percentage of Thread, Unified Threads Class 3B, Size 0-80 UNF Through 1 $1/8$-16 UN.

(Continued)

Table 9. *(Continued)* **Tap Drill Sizes and Percentage of Thread, Unified Threads Class 1B and 2B, Size $^1/_4$-20 UNC* Through 1 $^1/_8$-16 UN.** *(Source, FED-STD-H28/2B.)*

Size, TPI, and Series*	Classes 1B and 2B Minor Dia.				Tap Drill and Percentage of Thread				
	D_1 (min.)	% of Thread*	D_1 (max.)	% of Thread*	Drill Size		% of Thread*	Probable Hole Size	% of Thread*
7/16-32 UN	.404	82.5	.41	65.3	Y	.4040	83	.4086	71
					13/32	.4062	77	.4108	66
1/2-12 UNS	.410	83.1	.428	66.5	Z	.4130	80	.4177	76
					27/64	.4219	72	.4266	68
1/2-13 UNC	.417	83.1	.434	66.0	27/64	.4219	78	.4266	73
1/2-16 UN	.432	83.8	.446	66.5	7/16	.4375	77	.4422	71
1/2-20 UNF	.446	83.1	.457	66.2	29/64	.4531	72	.4578	65
1/2-28 UNEF	.461	84.1	.470	64.70	15/32	.4688	67	.4736	57
1/2-32 UN	.466	83.8	.474	64.0	15/32	.4688	77	.4736	65
9/16-12 UNC	.472	83.6	.490	67.0	15/32	.4688	87	.4736	82
					31/64	.4844	72	.4892	68
9/16-16 UN	.495	83.1	.509	65.9	1/2	.5000	77	.5048	71
					0.5062	.5062	69	.5110	63
9/16-18 UNF	.502	83.8	.515	65.8	1/2	.5000	87	.5048	80
					0.5062	.5062	78	.5110	71
9/16-20 UN	.508	83.9	.520	65.4	33/64	.5156	72	.5204	65
9/16-24 UNEF	.517	84.1	.527	65.6	33/64	.5156	87	.5204	78
					0.5203	.5203	78	.5251	69
9/16-28 UN	.524	83.0	.532	65.7	17/32	.5312	67	.5361	57
					0.5263	.5263	78	.5312	67
9/16-32 UN	.529	82.5	.536	65.3	17/32	.5312	77	.5361	65
5/8-11 UNC	.527	83.0	.546	66.9	17/32	.5312	79	.5361	75
5/8-12 UN	.535	83.1	.553	66.5	35/64	.5469	72	.5518	68
5/8-16 UN	.557	83.8	.571	66.5	9/16	.5625	77	.5674	71
					0.5687	.5687	69	.5736	63
5/8-18 UNF	.565	83.1	.578	65.1	9/16	.5625	87	.5674	80
					0.5687	.5687	78	.5736	71
5/8-20 UN	.571	83.1	.582	66.2	37/64	.5781	72	.5830	65
5/8-24 UNEF	.580	83.1	.590	64.7	37/64	.5781	87	.5830	78
					0.5828	.5828	78	.5877	69
5/8-28 UN	.586	84.1	.595	64.7	19/32	.5938	67	.5987	57
5/8-32 UN	.591	83.8	.599	64.0	19/32	.5938	77	.5987	65
11/16-12 UN	.597	83.6	.615	67.0	19/32	.5938	87	.5987	82
					39/64	.6094	72	.6143	68
11/16-16 UN	.620	83.1	.634	65.9	5/8	.6250	77	.6300	71
11/16-20 UN	.633	83.9	.645	65.4	41/64	.6406	72	.6456	65
11/16-24 UNEF	.642	84.1	.652	65.6	41/64	.6406	87	.6456	77
11/16-28 UN	.649	83.0	.657	65.7	21/32	.6562	67	.6612	57
11/16-32 UN	.654	82.5	.661	65.3	21/32	.6562	77	.6612	65

* For Machine Screw sizes, see **Table 11**: Tap Drill sizes and Percentage of Thread, Unified Threads Class 3B, Size 0-80 UNF Through 1 $^1/_8$-16 UN.

(Continued)

Table 9. *(Continued)* Tap Drill Sizes and Percentage of Thread, Unified Threads Class 1B and 2B, Size $^1/_4$-20 UNC* Through 1 $^1/_8$-16 UN. *(Source, FED-STD-H28/2B.)*

Size, TPI, and Series*	Classes 1B and 2B Minor Dia.				Tap Drill and Percentage of Thread				
	D_1 (min.)	% of Thread*	D_1 (max.)	% of Thread*	Drill Size		% of Thread*	Probable Hole Size	% of Thread*
3/4-10 UNC	.642	83.1	.663	67.0	41/64	.6406	84	.6456	80
					21/32	.6562	72	.6612	68
3/4-12 UN	.660	83.1	.678	66.5	21/32	.6562	87	.6612	82
					43/64	.6719	72	.6769	68
3/4-16 UNF	.682	83.8	.696	66.5	11/16	.6875	77	.6925	71
3/4-20 UNEF	.696	83.1	.707	66.2	45/64	.7031	72	.7082	64
3/4-28 UN	.711	84.1	.720	64.7	23/32	.7188	67	.7239	56
3/4-32 UN	.716	83.8	.724	64.0	23/32	.7188	77	.7239	64
13/16-12 UN	.722	83.6	.740	67.0	47/64	.7344	72	.7395	67
13/16-16 UN	.745	83.1	.759	65.9	3/4	.7500	77	.7552	71
13/16-20 UNEF	.758	83.9	.770	65.4	49/64	.7656	72	.7708	64
13/16-28 UN	.774	83.0	.782	65.7	25/32	.7812	67	.7864	56
13/16-32 UN	.779	82.5	.786	65.3	25/32	.7812	77	.7864	64
7/8-9 UNC	.755	83.1	.778	67.2	49/64	.7656	76	.7708	72
7/8-12 UN	.785	83.1	.803	66.5	25/32	.7812	87	.7864	82
					54/64	.7969	72	.8021	67
7/8-14 UNF	.798	83.0	.814	65.7	57/64	.7969	84	.8021	67
					0.8024	.8024	78	.8076	73
					13/16	.8125	67	.8177	62
7/8-16 UN	.807	83.8	.821	66.5	13/16	.8125	77	.8178	70
7/8-20 UNEF	.821	83.1	.832	66.2	53/64	.8281	72	.8335	64
7/8-28 UN	.836	84.1	.845	64.7	27/32	.8438	67	.8493	55
7/8-32 UN	.841	83.8	.849	64.0	27/32	.8438	77	.8493	63
15/16-12 UN	.847	83.6	.865	67.0	27/32	.8438	87	.8493	81
					55/64	.8594	72	.8650	67
15/16-16 UN	.870	83.1	.884	65.9	7/8	.8750	77	.8807	70
15/16-20 UNEF	.883	83.9	.895	65.4	57/64	.8906	72	.8965	63
15/16-28 UN	.899	83.0	.907	65.7	29/32	.9062	67	.9122	55
15/16-32 UN	.904	82.5	.911	65.3	29/32	.9062	77	.9122	62
1-8 UNC	.865	83.1	.890	67.7	55/64	.8594	87	.8653	83
					7/8	.8750	77	.8809	73
1-12 UNF	.910	83.1	.928	66.5	29/32	.9062	87	.9122	81
					59/64	.9219	72	.9279	67
1-14 UNS	.923	83.0	.938	66.8	59/64	.9219	84	.9279	78
					0.9274	.9274	78	.9335	72
1-16 UN	.923	83.8	.946	66.5	15/16	.9375	77	.9437	69
1-20 UNEF	.946	83.1	.957	66.2	61/64	.9531	72	.9594	63
1-28 UN	.961	84.1	.970	64.7	31/32	.9688	67	.9753	53

* For Machine Screw sizes, see **Table 11**: Tap Drill sizes and Percentage of Thread, Unified Threads Class 3B, Size 0-80 UNF Through 1 $^1/_8$-16 UN.

(Continued)

Table 9. *(Continued)* Tap Drill Sizes and Percentage of Thread, Unified Threads Class 1B and 2B, Size $^{1}/_{4}$-20 UNC* Through 1 $^{1}/_{8}$-16 UN. *(Source, FED-STD-H28/2B.)*

Size, TPI, and Series*	Classes 1B and 2B Minor Dia.				Tap Drill and Percentage of Thread				
	D_1 (min.)	% of Thread*	D_1 (max.)	% of Thread*	Drill Size		% of Thread*	Probable Hole Size	% of Thread*
1-32 UN	.966	83.8	.974	64.0	31/32	.9688	77	.9753	61
1 1/16-8 UN	.927	83.4	.952	68.0	59/64	.9219	87	.9279	83
					0.9274	.9274	83	.9335	79
					15/16	.9375	77	.9437	73
1 1/16-12 UN	.927	83.6	.990	67.0	31/32	.9688	87	.9753	81
					63/64	.9844	72	.9911	66
1 1/16-16 UN	.995	83.1	1.009	65.9	1	1.0000	77	1.0069	68
1 1/16-18 UNEF	1.002	83.8	1.015	65.8	1	1.0000	87	1.0069	77
1 1/16-20 UN	1.008	83.9	1.020	65.4	1 1/64	1.0156	72	1.0226	61
1 1/16-28 UN	1.024	83.0	1.032	65.7	1 1/32	1.0312	67	1.0383	52
1 1/8-7 UNC	.970	83.5	.998	68.4	31/32	.9688	84	.9750	81
					63/64	.9844	76	.9911	72
1 1/8-8 UN	.990	83.1	1.015	67.7	1	1.0000	77	1.0069	73
1 1/8-12 UNF	1.035	83.1	1.053	66.5	1 1/32	1.0312	87	1.0383	80
					1 3/64	1.0469	72	1.0541	65
1 1/8-16 UN	1.057	83.8	1.071	66.5	1 1/16	1.0625	77	1.0699	68

* For Machine Screw sizes, see **Table 11**: Tap Drill sizes and Percentage of Thread, Unified
Threads Class 3B, Size 0-80 UNF Through 1 $^{1}/_{8}$-16 UN.

Table 10. Tap Drill Sizes and Percentage of Thread, Unified Threads Class 1B and 2B, Size 1 $^{1}/_{8}$-18 UNEF Through 3 $^{3}/_{4}$-4 UNC. *(Source, FED-STD-H28/2B.)*

Size, TPI, and Series*	Classes 1B and 2B Minor Dia.				Tap Drill and Percentage of Thread		
	D_1 (min.)	% of Thread*	D_1 (max.)	% of Thread*	Drill Size		% of Thread*
1 1/8-18 UNEF	1.065	83.1	1.078	65.1	1 1/16	1.0625	87
					1 5/64	1.0781	65
1 1/8-20 UN	1.071	83.1	1.082	66.2	1 5/64	1.0781	72
1 1/8-28 UN	1.086	84.1	1.095	64.7	1 3/32	1.0938	67
1 3/16-8 UN	1.052	83.4	1.077	68.0	1 1/16	1.0625	77
1 3/16-12 UN	1.097	83.6	1.115	67.0	1 3/32	1.0938	87
1 3/16-16 UN	1.120	83.1	1.134	65.9	1 1/8	1.1250	77
1 3/16-18 UNEF	1.127	83.8	1.140	65.8	1 1/8	1.1250	87
					1 9/64	1.1406	65
1 3/16-20 UN	1.133	83.9	1.145	65.4	1 9/64	1.1406	72
1 3/16-28 UN	1.149	83.0	1.157	65.7	1 5/32	1.1562	67
1 1/8-7 UNC	1.095	83.5	1.123	68.4	1 3/32	1.0938	84
1 1/8-8 UN	1.115	83.1	1.140	67.7	1 1/8	1.1250	77
1 1/8-12 UNF	1.160	83.1	1.178	66.5	1 5/32	1.1562	87
					1 11/64	1.1719	72
1 1/8-16 UN	1.182	83.8	1.196	66.5	1 3/16	1.1875	77

(Continued)

Table 10. *(Continued)* Tap Drill Sizes and Percentage of Thread, Unified Threads Class 1B and 2B, Size 1 $^1/_8$-18 UNEF Through 3 $^3/_4$-4 UNC. *(Source, FED-STD-H28/2B.)*

Size, TPI, and Series*	Classes 1B and 2B Minor Dia.				Tap Drill and Percentage of Thread		
	D_1 (min.)	% of Thread*	D_1 (max.)	% of Thread*	Drill Size		% of Thread*
1 1/8-18 UNEF	1.190	83.1	1.203	65.1	1 3/16	1.1875	87
					1 13/64	1.2031	65
1 1/8-20 UN	1.196	83.1	1.207	66.2	1 13/64	1.2031	72
1 1/8-28 UN	1.211	84.1	1.220	64.7	1 7/32	1.2188	67
1 5/16-8 UN	1.177	83.4	1.202	68.0	1 11/64	1.1719	87
					1 3/16	1.1875	77
1 5/16-12 UN	1.222	83.6	1.240	67.0	1 7/32	1.2188	87
					1 15/64	1.2344	72
1 5/16-16 UN	1.245	83.1	1.259	65.9	1 1/4	1.2500	77
1 5/16-18 UNEF	1.252	83.8	1.265	65.8	1 1/4	1.2500	87
					1 17/64	1.2656	65
1 5/16-20 UN	1.258	83.9	1.270	65.4	1 17/64	1.2656	72
1 5/16-28 UN	1.274	83.0	1.282	65.7	1 9/32	1.2812	67
1 3/8-6 UNC	1.195	83.1	1.225	69.3	1 3/16	1.1875	87
					1 13/64	1.2031	79
					1 15/64	1.2188	72
1 3/8-8 UN	1.240	83.1	1.265	67.7	1 15/64	1.2344	87
					1 1/4	1.2500	77
1 3/8-12 UNF	1.285	83.1	1.303	66.5	1 9/32	1.2812	87
					1 19/64	1.2969	72
1 3/8-16 UN	1.307	83.8	1.321	66.5	1 5/16	1.3125	77
1 3/8-18 UNEF	1.315	83.1	1.328	65.1	1 5/16	1.3125	87
					1 21/64	1.3281	65
1 3/8-20 UN	1.321	83.1	1.332	66.2	1 21/64	1.3281	72
1 3/8-28 UN	1.336	84.1	1.345	64.7	1 11/32	1.3438	67
1 7/16-6 UN	1.257	83.4	1.288	69.1	1 17/64	1.2656	79
					1 9/32	1.2812	72
1 7/16-8 UN	1.302	83.4	1.327	68.0	1 19/64	1.2969	87
					1 5/16	1.3125	77
1 7/16-12 UN	1.347	83.6	1.365	67.0	1 11/32	1.3438	87
					1 23/64	1.3594	72
1 7/16-16 UN	1.370	83.1	1.384	65.9	1 3/8	1.3750	77
1 7/16-18 UNEF	1.377	83.8	1.390	65.8	1 3/8	1.3750	87
1 7/16-20 UN	1.383	83.9	1.395	65.4	1 25/64	1.3906	72
1 7/16-28 UN	1.399	83.0	1.407	65.7	1 13/32	1.4062	
1 1/2-6 UNC	1.320	83.1	1.350	69.3	1 5/16	1.3125	87
					1 21/64	1.3281	79
1 1/2-8 UN	1.365	83.1	1.390	67.7	1 23/64	1.3594	87
					1 3/8	1.3750	77
1 1/2-12 UNF	1.410	83.1	1.428	66.5	1 13/32	1.4062	87
					1 27/64	1.4219	72
1 1/2-16 UN	1.432	83.8	1.446	66.5	1 7/16	1.4375	77

(Continued)

Table 10. *(Continued)* **Tap Drill Sizes and Percentage of Thread, Unified Threads Class 1B and 2B, Size 1 1/8-18 UNEF Through 3 3/4-4 UNC.** *(Source, FED-STD-H28/2B.)*

Size, TPI, and Series*	Classes 1B and 2B Minor Dia.				Tap Drill and Percentage of Thread		
	D_1 (min.)	% of Thread*	D_1 (max.)	% of Thread*	Drill Size		% of Thread*
1 1/2-18 UNEF	1.440	83.1	1.452	66.5	1 7/16	1.4375	87
1 1/2-20 UN	1.446	83.1	1.457	66.2	1 29/64	1.4531	72
1 1/2-28 UN	1.461	84.1	1.470	64.7	1 15/32	1.4688	67
1 9/16-6 UN	1.382	83.4	1.413	69.1	1 25/64	1.3906	79
					1 13/32	1.4062	72
1 9/16-8 UN	1.427	83.4	1.452	68.0	1 27/64	1.4219	87
					1 7/16	1.4375	77
1 9/16-12 UN	1.472	83.6	1.490	67.0	1 15/32	1.4688	87
					1 31/64	1.4844	72
1 9/16-16 UN	1.495	83.1	1.509	65.9	1 1/2	1.5000	77
1 9/16-18 UNEF	1.502	83.8	1.515	65.8	1 1/2	1.5000	87
					1 33/64	1.5156	65
1 9/16-20 UN	1.508	83.9	1.520	65.4	1 33/64	1.5156	72
1 5/8-6 UN	1.445	83.1	1.475	69.3	1 29/64	1.4531	79
					1 15/32	1.4688	72
1 5/8-8 UN	1.490	83.1	1.515	67.7	1 31/64	1.4844	87
					1 1/2	1.5000	77
1 5/8-12 UN	1.535	83.1	1.553	66.5	1 17/32	1.5312	87
					1 35/64	1.5469	72
1 5/8-16 UN	1.557	83.8	1.571	66.5	1 9/16	1.5625	77
1 5/8-18 UNEF	1.565	83.1	1.578	65.1	1 9/16	1.5625	87
					1 37/64	1.5781	65
1 5/8-20 UN	1.571	83.1	1.582	66.2	1 37/64	1.5781	72
1 11/16-6 UN	1.507	83.4	1.538	69.1	1 1/2	1.5000	87
					1 33/64	1.5156	79
					1 17/32	1.5312	72
1 11/16-8 UN	1.552	83.4	1.577	68.0	1 9/16	1.5625	77
1 11/16-12 UN	1.597	83.6	1.615	67.0	1 19/32	1.5938	87
					1 39/64	1.6094	72
1 11/16-16 UN	1.620	83.1	1.634	65.9	1 5/8	1.6250	77
1 11/16-18 UNEF	1.627	83.8	1.640	65.8	1 5/8	1.6250	87
					1 41/64	1.6406	65
1 11/16-20 UN	1.633	83.9	1.645	65.4	1 41/64	1.6406	72
1 3/4-5 UNC	1.534	83.1	1.568	70.1	1 17/32	1.5312	84
					1 35/64	1.5469	78
1 3/4-6 UN	1.570	83.1	1.600	69.3	1 37/64	1.5781	79
					1 19/32	1.5938	72
1 3/4-8 UN	1.615	83.1	1.640	67.7	1 39/64	1.6094	87
					1 5/8	1.6250	77
					1 41/64	1.6406	67
1 3/4-12 UN	1.660	83.1	1.678	66.5	1 21/32	1.6562	87
					1 43/64	1.6719	72

(Continued)

Table 10. *(Continued)* **Tap Drill Sizes and Percentage of Thread, Unified Threads Class 1B and 2B, Size 1 $^1/_8$-18 UNEF Through 3 $^3/_4$-4 UNC.** *(Source, FED-STD-H28/2B.)*

Size, TPI, and Series*	Classes 1B and 2B Minor Dia.				Tap Drill and Percentage of Thread		
	D_1 (min.)	% of Thread*	D_1 (max.)	% of Thread*	Drill Size		% of Thread*
1 3/4-16 UN	1.682	83.8	1.696	66.5	1 11/16	1.6875	77
1 3/4-20 UN	1.696	83.1	1.707	66.2	1 45/64	1.7031	72
1 13/16-6 UN	1.632	83.4	1.663	69.1	1 5/8	1.6250	87
					1 41/64	1.6406	79
					1 21/32	1.6562	72
1 13/16-8 UN	1.677	83.4	1.702	68.0	1 43/64	1.6719	87
					1 11/16	1.6875	77
1 13/16-12 UN	1.722	83.6	1.740	67.0	1 23/32	1.7188	87
					1 47/64	1.7344	72
1 13/16-16 UN	1.745	83.1	1.759	65.9	1 3/4	1.7500	77
1 13/16-20 UN	1.758	83.9	1.770	65.4	1 49/64	1.7656	72
1 7/8-6 UN	1.695	83.1	1.725	69.3	1 45/64	1.7031	79
					1 23/32	1.7188	72
1 7/8-8 UN	1.740	83.1	1.765	67.7	1 3/4	1.7500	77
1 7/8-12 UN	1.785	83.1	1.803	66.5	1 25/32	1.7812	87
					1 51/64	1.7969	72
1 7/8-16 UN	1.807	83.8	1.821	66.5	1 13/16	1.8125	77
1 7/8-20 UN	1.821	83.1	1.832	66.2	1 53/64	1.8281	72
1 15/16-6 UN	1.757	83.4	1.788	69.1	1 49/64	1.7656	79
					1 25/32	1.7812	72
1 15/16-8 UN	1.802	83.4	1.827	68.0	1 51/64	1.7969	87
					1 13/16	1.8125	77
1 15/16-12 UN	1.847	83.6	1.865	67.0	1 27/32	1.8438	87
					1 55/64	1.8594	72
1 15/16-16 UN	1.870	83.1	1.884	65.9	1 7/8	1.8750	77
					1 25/32	1.7812	72
1 15/16-8 UN	1.802	83.4	1.827	68.0	1 51/64	1.7969	87
					1 13/16	1.8125	77
1 15/16-12 UN	1.847	83.6	1.865	67.0	1 27/32	1.8438	87
					1 55/64	1.8594	72
1 15/16-16 UN	1.870	83.1	1.884	65.9	1 7/8	1.8750	77
1 15/16-20 UN	1.883	83.9	1.895	65.4	1 57/64	1.8906	72
2-4.5 UNC	1.759	83.5	1.795	71.0	1 25/32	1.7812	76
2-6 UN	1.820	83.1	1.850	69.3	1 53/64	1.8281	79
					1 27/32	1.8438	72
2-8 UN	1.865	83.1	1.890	67.7	1 7/8	1.8750	77
2-12 UN	1.910	83.1	1.928	66.5	1 29/32	1.9062	87
					1 59/64	1.9219	72
2-16 UN	1.932	83.8	1.946	66.5	1 15/16	1.9375	77
2-20 UN	1.946	83.1	1.957	66.2	1 61/64	1.9531	72
2 1/16-16 UNS	1.995	83.1	2.009	65.9	2	2.0000	77

(Continued)

Table 10. *(Continued)* Tap Drill Sizes and Percentage of Thread, Unified Threads Class 1B and 2B, Size 1 $^1/_8$-18 UNEF Through 3 $^3/_4$-4 UNC. *(Source, FED-STD-H28/2B.)*

Size, TPI, and Series*	Classes 1B and 2B Minor Dia.				Tap Drill and Percentage of Thread		
	D_1 (min.)	% of Thread*	D_1 (max.)	% of Thread*	Drill Size		% of Thread*
2 1/8-6 UN	1.945	83.1	1.975	69.3	1 61/64	1.9531	79
					1 31/32	1.9688	72
2 1/8-8 UN	1.990	83.1	2.015	67.7	2	2.0000	77
2 1/8-12 UN	2.035	83.1	2.053	66.5	2 1/32	2.0312	87
2 1/8-16 UN	2.057	83.8	2.071	66.5	2 1/16	2.0625	77
2 1/8-20 UN	2.071	83.1	2.082	66.2	2 1/16	2.0625	96
2 3/16 UNS	2.120	83.1	2.134	65.9	2 1/8	2.1250	77
2 1/4-4.5 UNC	2.090	83.5	2.045	71.0	2	2.0000	87
					2 1/32	2.0312	76
2 1/4-6 UN	2.070	83.1	2.100	69.3	2 1/16	2.0625	87
2 1/4-8 UN	2.115	83.1	2.140	67.7	2 1/8	2.1250	77
2 1/4-12 UN	2.160	83.1	2.178	66.5	2 5/32	2.1562	87
2 1/4-16 UN	2.182	83.8	2.196	66.5	2 3/16	2.1875	77
2 1/4-20 UN	2.196	83.1	2.207	66.2	2 3/16	2.1875	96
2 5/16-16 UNS	2.245	83.1	2.259	65.9	2 1/4	2.2500	77
2 3/8-6 UN	2.195	83.1	2.226	68.8	2 3/16	2.1875	87
2 3/8-8 UN	2.240	83.1	2.265	67.7	2 1/4	2.2500	77
2 3/8-12 UN	2.285	83.1	2.303	66.5	58 mm	2.2835	85
2 3/8-16 UN	2.307	83.8	2.321	66.5	2 5/16	2.3125	77
2 3/8-20 UN	2.321	83.1	2.332	66.2	2 5/16	2.3125	96
2 7/16-16 UNS	2.370	83.1	2.384	65.9	2 3/8	2.3750	77
2 1/2-4 UNC	2.229	83.4	2.267	71.7	2 7/32	2.2188	87
					2 1/4	2.2500	77
2 1/2-6 UN	2.320	83.1	2.350	69.3	2 5/16	2.3125	87
2 1/2-8 UN	2.365	83.1	2.390	67.7	2 3/8	2.3750	77
2 1/2-12 UN	2.410	83.1	2.428	66.5	2 13/32	2.4062	87
2 1/2-16 UN	2.432	83.8	2.446	66.5	2 7/16	2.4375	77
2 1/2-20 UN	2.446	83.1	2.457	66.2	2 7/16	2.4375	96
2 5/8-4 UN	2.354	83.4	2.392	71.7	2 11/32	2.3438	87
					2 3/8	2.3750	77
2 5/8-6 UN	2.445	83.1	2.475	69.3	2 7/16	2.4375	87
2 5/8-8 UN	2.490	83.1	2.515	67.7	2 1/2	2.5000	77
2 5/8-12 UN	2.535	83.1	2.553	66.5	2 17/32	2.5312	87
2 5/8-16 UN	2.557	83.8	2.571	66.5	2 9/16	2.5625	77
2 5/8-20 UN	2.571	83.1	2.582	66.2	2 9/16	2.5625	96
2 3/4-4 UNC	2.479	83.4	2.517	71.7	2 1/2	2.5000	77
2 3/4-6 UN	2.570	83.1	2.600	69.3	2 9/16	2.5625	87
2 3/4-8 UN	2.615	83.1	2.640	67.7	2 5/8	2.6250	77

(Continued)

Table 10. *(Continued)* **Tap Drill Sizes and Percentage of Thread, Unified Threads Class 1B and 2B, Size 1 1/8-18 UNEF Through 3 3/4-4 UNC.** *(Source, FED-STD-H28/2B.)*

Size, TPI, and Series*	Classes 1B and 2B Minor Dia.				Tap Drill and Percentage of Thread		
	D_1 (min.)	% of Thread*	D_1 (max.)	% of Thread*	Drill Size		% of Thread*
2 3/4-12 UN	2.660	83.1	2.678	66.5	2 21/32	2.6562	87
2 3/4-16 UN	2.682	83.8	2.696	66.5	2 11/16	2.6875	77
2 3/4-20 UN	2.696	83.1	2.707	66.2	2 11/16	2.6875	96
2 7/8-4 UN	2.604	83.4	2.642	71.7	2 5/8	2.6250	77
2 7/8-6 UN	2.695	83.1	2.725	69.3	2 11/16	2.6875	87
2 7/8-8 UN	2.740	83.1	2.765	67.7	2 3/4	2.7500	77
2 7/8-12 UN	2.785	83.1	2.803	66.5	2 25/32	2.7812	87
2 7/8-16 UN	2.807	83.8	2.821	66.5	2 13/16	2.8125	77
2 7/8-20 UN	2.821	83.1	2.832	66.2	2 13/16	2.8125	96
3-4 UNC	2.729	83.4	2.767	71.7	2 3/4	2.7500	77
3-6 UN	2.820	83.1	2.850	69.3	2 13/16	2.8125	87
3-8 UN	2.865	83.1	2.890	67.7	2 7/8	2.8750	77
3-12 UN	2.910	83.1	2.928	66.5	74 mm	2.9134	80
3-16 UN	2.932	83.8	2.946	66.5	2 15/16	2.9375	77
3-20 UN	2.946	83.1	2.957	66.2	2 15/16	2.9375	96
3 1/4-4 UNC	2.979	83.4	3.017	71.7	3	3.0000	77
3 1/2-4 UNC	3.229	83.4	3.267	71.7	3 1/4	3.2500	77
3 3/4-4 UNC	3.479	83.4	3.517	71.7	3 1/2	3.5000	77

Table 11. Tap Drill Sizes and Percentage of Thread, Unified Threads Class 3B, Size 0-80 UNF Through 1 1/8-16 UN. *(Source, FED-STD-H28/2B.)*

Size, TPI, and Series*	Classes 3B Minor Dia.				Tap Drill and Percentage of Thread				
	D_1 (min.)	% of Thread*	D_1 (max.)	% of Thread*	Drill Size		% of Thread*	Probable Hole Size	% of Thread*
0-80 UNF	.0465	83.1	.0514	52.9	#56	.0465	83	.0480	74
					3/64	.0469	81	.0484	71
1-64 UNC	.0561	83.3	.0623	52.7	#54	.0550	89	.0565	81
					#53	.0595	67	.0610	59
1-72 UNF	.0580	83.1	.0635	52.7	#53	.0595	75	.0610	67
					1/16	.0625	58	.0640	50
2-56 UNC	.0667	83.2	.0737	53.0	#51	.0670	82	.0687	75
					#50	.0700	69	.0717	62
					#49	.0730	56	.0747	49
2-64 UNF	.0691	83.3	.0753	52.7	#50	.0700	79	.0717	70
					#49	.0730	64	.0747	56
3-48 UNC	.0764	83.5	.0845	53.6	#48	.0760	85	.0779	78
					5/64	.0781	77	.0800	70
					#47	.0785	76	.0804	69
					#46	.0810	67	.0829	60
					#45	.0820	63	.0839	56

(Continued)

Table 11. *(Continued)* Tap Drill Sizes and Percentage of Thread, Unified Threads Class 3B, Size 0-80 UNF Through 1 1/8-16 UN. *(Source, FED-STD-H28/2B.)*

Size, TPI, and Series*	Classes 3B Minor Dia.				Tap Drill and Percentage of Thread				
	D_1 (min.)	% of Thread*	D_1 (max.)	% of Thread*	Drill Size		% of Thread*	Probable Hole Size	% of Thread*
3-56 UNF	.0797	83.2	.0865	53.9	#46	.0810	78	.0829	69
					#45	.0820	73	.0839	65
					#44	.0860	56	.0879	48
4-40 UNC	.0849	83.4	.0939	55.7	#44	.0860	80	.0879	74
					#43	.0890	71	.0910	65
					#42	.0935	57	.0955	51
					3/32	.0938	56	.0958	50
4-48 UNF	.0894	83.5	.0968	56.2	#43	.0890	85	.0910	78
					#42	.0935	68	.0955	61
					3/32	.0938	67	.0958	60
					#41	.0960	59	.0980	52
5-40 UNC	.0979	83.4	.1062	57.9	#40	.0980	83	.1003	76
					#39	.0995	79	.1018	71
					#38	.1015	72	.1038	65
					#37	.1040	65	.1063	58
5-44 UNF	.1004	83.3	.1079	57.9	#38	.1015	80	.1038	72
					#37	.1040	71	.1063	63
					#36	.1065	63	.1088	55
6-32 UNC	.1040	83.8	.1140	59.1	#37	.1040	84	.1063	78
					#36	.1065	78	.1088	72
					7/64	.1094	70	.1120	64
					#35	.1100	69	.1126	63
					#34	.1110	67	.1136	60
					#33	.1130	62	.1156	55
6-40 UNF	.1110	83.1	.1186	59.7	#34	.1110	83	.1136	75
					#33	.1130	77	.1156	69
					#32	.1160	68	.1186	60
8-32 UNC	.1300	83.8	.1389	61.8	#29	.1360	69	.1389	62
8-36 UNF	.1340	83.1	.1416	62.1	#29	.1360	78	.1389	70
					#28	.1405	65	.1434	57
					9/64	.1406	65	.1435	57
10-24 UNC	.1450	83.1	.1555	63.7	#27	.1440	85	.1472	79
					#26	.1470	79	.1502	74
					#25	.1495	75	.1527	69
					#24	.1520	70	.1552	64
					#23	.1540	66	.1572	61
10-32 UNF	.1560	83.8	.1641	63.8	5/32	.1562	83	.1594	75
					#22	.1570	81	.1602	73
					#21	.1590	76	.1622	68
					#20	.1610	71	.1642	64
12-24 UNC	.1710	83.1	.1807	65.2	11/64	.1719	82	.1754	75
					#17	.1730	79	.1765	73
					#16	.1770	72	.1805	66
					#15	.1800	67	.1835	60
12-28 UNF	.1770	84.1	.1857	65.3	#16	.1770	84	.1805	77
					#15	.1800	78	.1835	70
					#14	.1820	73	.1855	66
					#13	.1850	67	.1885	59

(Continued)

Table 11. *(Continued)* **Tap Drill Sizes and Percentage of Thread, Unified Threads Class 3B, Size 0-80 UNF Through 1 1/8-16 UN.** *(Source, FED-STD-H28/2B.)*

Size, TPI, and Series*	Classes 3B Minor Dia.				Tap Drill and Percentage of Thread				
	D_1 (min.)	% of Thread*	D_1 (max.)	% of Thread*	Drill Size		% of Thread*	Probable Hole Size	% of Thread*
12-32 UNEF	.1820	83.8	.1895	65.3	#14	.1820	84	.1855	75
					#13	.1850	76	.1885	68
					3/16	.1875	70	.1910	62
					#12	.1890	67	.1925	58
1/4-20 UNC	.1960	83.1	.2067	66.7	#9	.1960	83	.1998	77
					#8	.1990	79	.2028	73
					#7	.2010	75	.2048	70
					13/64	.2031	72	.2069	66
					#6	.2040	71	.2078	65
					#5	.2055	69	.2093	63
1/4-28 UNF	.2110	84.1	.2190	66.8	#3	.2130	80	.2168	72
					7/32	.2188	67	.2226	59
1/4-32 UNEF	.2160	83.8	.2229	66.8	7/32	.2188	77	.2226	67
					#2	.2210	71	.2248	62
5/16-18 UNC	.2520	83.8	.2630	68.6	F	.2570	77	.2608	72
					G	.2610	71	.2651	66
5/16-20 UN	.2580	83.9	.2680	68.5	F	.2570	85	.2608	80
					G	.2610	79	.2651	73
					H	.2660	72	.2701	65
5/16-24 UNF	.2670	84.1	.2754	68.5	H	.2660	86	.2701	78
					I	.2720	75	.2761	67
5/16-28 UN	.2740	83.0	.2807	68.5	J	.2770	77	.2811	68
5/16-32 UNEF	.2790	82.5	.2847	68.5	K	.2810	78	.2852	67
					9/32	.2812	77	.2854	67
3/8-16 UNC	.3070	83.8	.3182	70.0	5/16	.3125	77	.3169	72
					O	.3160	73	.3204	67
3/8-20 UN	.3210	83.1	.3297	69.7	P	.3230	80	.3274	73
3/8-24 UNF	.3300	83.1	.3372	69.8	Q	.3320	79	.3364	71
3/8-28 UN	.3360	84.1	.3426	69.8	R	.3390	78	.3434	68
3/8-32 UNEF	.3410	83.8	.3469	69.2	11/32	.3438	77	.3483	68
7/16-14 UNC	.3600	83.5	.3717	70.9	T	.3580	86	.3626	81
					23/64	.3594	84	.3640	79
7/16-16 UN	.3700	83.1	.3800	70.8	3/8	.3750	77	.3796	71
					V	.3770	75	.3816	69
7/16-20 UNF	.3830	83.9	.3916	70.7	W	.3860	79	.3906	72
					25/64	.3906	72	.3952	65
7/16-28 UNEF	.3990	83.0	.4051	69.8	Y	.4040	72	.4086	62
7/16-32 UN	.4040	82.5	.4094	69.2	Y	.4040	83	.4086	71
					13/32	.4062	77	.4108	66
1/2-12 UNS	.4100	83.1	.4223	71.8	Z	.4130	80	.4177	76
					27/64	.4219	72	.4266	68
1/2-13 UNC	.4170	83.1	.4284	71.7	27/64	.4219	78	.4266	73
1/2-16 UN	.4320	83.8	.4419	71.6	7/16	.4375	77	.4422	71
1/2-20 UNF	.4460	83.1	.4537	71.3	29/64	.4531	72	.4578	65

(Continued)

Table 11. *(Continued)* Tap Drill Sizes and Percentage of Thread, Unified Threads Class 3B, Size 0-80 UNF Through 1 1/8-16 UN. *(Source, FED-STD-H28/2B.)*

Size, TPI, and Series*	Classes 3B Minor Dia.				Tap Drill and Percentage of Thread				
	D_1 (min.)	% of Thread*	D_1 (max.)	% of Thread*	Drill Size		% of Thread*	Probable Hole Size	% of Thread*
1/2-28 UNEF	.4610	84.1	.4676	69.8	11.8mm	.4646	76	.4693	66
1/2-32 UN	.4660	83.8	.4719	69.2	15/32	.4688	77	.4736	65
9/16-12 UNC	.4720	83.6	.4843	72.2	15/32	.4688	87	.4736	82
					31/64	.4844	72	.4892	68
9/16-16 UN	.4950	83.1	.5040	72.1	1/2	.5000	77	.5048	71
9/16-18 UNF	.5020	83.8	.5106	71.9	1/2	.5000	87	.5048	80
					0.5062	.5062	78	.5110	71
9/16-20 UN	.5080	83.9	.5162	71.3	33/64	.5156	72	.5204	65
9/16-24 UNEF	.5170	84.1	.5244	70.4	33/64	.5156	87	.5204	78
					0.5203	.5203	78	.5251	69
9/16-28 UN	.5240	83.0	.5301	69.8	0.5263	.5263	78	.5312	67
9/16-32 UN	.5290	82.5	.5344	69.2	17/32	.5312	77	.5361	65
5/8-11 UNC	.5270	83.0	.5391	72.7	17/32	.5312	79	.5361	75
5/8-12 UN	.5350	83.1	.5463	72.7	35/64	.5469	72	.5518	68
5/8-16 UN	.5570	83.8	.5662	72.4	9/16	.5625	77	.5674	71
5/8-18 UNF	.5650	83.1	.5730	72.1	9/16	.5625	87	.5674	80
					0.5687	.5687	78	.573	71
5/8-20 UN	.5710	83.1	.5787	71.3	37/64	.5781	72	.5830	65
5/8-24 UNEF	.5800	83.1	.5869	70.4	37/64	.5781	87	.5830	78
					0.5828	.5828	78	.5877	69
5/8-28 UN	.5860	84.1	.5926	69.8	0.5828	.5828	91	.5877	80
5/8-32 UN	.5910	83.8	.5969	69.2	19/32	.5938	77	.5987	65
11/16-12 UN	.5970	83.6	.6085	73.0	19/32	.5938	87	.5987	.82
11/16-16 UN	.6200	83.1	.6284	72.8	5/8	.6250	77	.6300	71
11/16-20 UN	.6330	83.9	.6412	71.3	41/64	.6406	72	.6456	65
11/16-24 UNEF	.6420	84.1	.6494	70.4	41/64	.6406	87	.6456	77
11/16-28 UN	.6490	83.0	.6551	69.8	16.5mm	.6496	82	.6546	71
11/16-32 UN	.6540	82.5	.6594	69.2	21/32	.6562	77	.6612	65
3/4-10 UNC	.6420	83.1	.6545	73.5	41/64	.6406	04	.6456	80
3/4-12 UN	.6600	83.1	.6707	73.3	21/32	.6562	87	.6612	82
3/4-16 UNF	.6820	83.8	.6908	72.9	11/16	.6875	77	.6925	71
3/4-20 UNEF	.6960	83.1	.7037	71.3	45/64	.7031	72	.7082	64
3/4-28 UN	.7110	84.1	.7176	69.8	18mm	.7087	89	.7138	78
3/4-32 UN	.7160	83.8	.7219	69.2	23/32	.7188	77	.7239	64
13/16-12 UN	.7220	83.6	.7329	73.5	18.5mm	.7283	78	.7334	73
13/16-16 UN	.7450	83.1	.7533	72.9	3/4	.7500	77	.7552	71
13/16-20 UNEF	.7580	83.9	.7662	71.3	49/64	.7656	72	.7708	64
13/16-28 UN	.7740	83.0	.7801	69.8	19.75mm	.7776	75	.7828	64
13/16-32 UN	.7790	82.5	.7844	69.2	25/32	.7812	77	.7864	64

(Continued)

Table 11. *(Continued)* Tap Drill Sizes and Percentage of Thread, Unified Threads Class 3B, Size 0-80 UNF Through 1 1/8-16 UN. *(Source, FED-STD-H28/2B.)*

Size, TPI, and Series*	Classes 3B Minor Dia.				Tap Drill and Percentage of Thread				
	D_1 (min.)	% of Thread*	D_1 (max.)	% of Thread*	Drill Size		% of Thread*	Probable Hole Size	% of Thread*
7/8-9 UNC	.7550	83.1	.7681	74.1	49/64	.7656	76	.7708	72
7/8-12 UN	.7850	83.1	.7952	73.7	25/32	.7812	87	.7864	82
7/8-14 UNF	.7980	83.0	.8068	73.5	51/64	.7969	84	.8021	79
					0.8024	.8024	78	.8076	73
7/8-16 UN	.8070	83.8	.8158	72.9	13/16	.8125	77	.8178	70
7/8-20 UNEF	.8210	83.1	.8287	71.3	53/64	.8281	72	.8335	64
7/8-28 UN	.8360	84.1	.8426	69.8	21.25mm	.8366	83	.8420	71
7/8-32 UN	.8410	83.8	.8469	69.2	27/32	.8438	77	.8493	63
15/16-12 UN	.8470	83.6	.8575	73.9	27/32	.8438	87	.8493	81
15/16-16 UN	.8700	83.1	.8783	72.9	7/8	.8750	77	.8807	70
15/16-20 UNEF	.8830	83.9	.8912	71.3	57/64	.8906	72	.8965	63
15/16-28 UN	.8990	83.0	.9051	69.8	22.75mm	.8957	90	.9017	77
15/16-32 UN	.9040	82.5	.9094	69.2	29/32	.9062	77	.9122	62
1.000-8 UNC	.8650	83.1	.8797	74.1	55/64	.8594	.87	.8653	83
					7/8	.8750	77	.8809	73
1.000-12 UNF	.9100	83.1	.9198	74.1	29/32	.9062	87	.9122	81
1.000-14 UNS	.9230	83.0	.9315	73.8	59/64	.9219	84	.9279	78
					0.9274	.9274	78	.9335	72
1.000-16 UN	.9320	83.8	.9408	72.9	15/16	.9375	77	.9437	69
1.000-20 UNEF	.9460	83.1	.9537	71.3	61/64	.9531	72	.9594	63
1.000-28 UN	.9610	84.1	.9676	69.8	24.5mm	.9645	77	.9709	63
1.000-32 UN	.9660	83.8	.9719	69.2	31/32	.9688	77	.9753	61
1 1/16-8 UN	.9270	83.4	.9422	74.1	59/64	.9219	87	.9279	83
					0.9274	.9274	83	.9335	79
					15/16	.9375	77	.9437	73
1 1/16-12 UN	.9720	83.6	.9823	74.1	31/32	.9688	87	.9753	81
1 1/16-16 UN	.9950	83.1	1.0033	72.9	1	1.0000	77	1.0069	68
1 1/16-18 UNEF	1.0020	83.8	1.0105	72.1	1	1.0000	87	1.0069	77
1 1/16-20 UN	1.0080	83.9	1.0162	71.3	1 1/64	1.0156	72	1.0226	61
1 1/16-28 UN	1.0240	83.0	1.0301	69.8	1 1/32	1.0312	67	1.0383	52
1 1/8-7 UNC	.9700	83.5	.9875	74.1	31/32	.9688	84	.9750	81
					63/64	.9844	76	.9911	72
1 1/8-8 UN	.9900	83.1	1.0047	74.1	1	1.0000	77	1.0069	73
1 1/8-12 UNF	1.0350	83.1	1.0448	74.1	1 1/32	1.0312	87	1.0383	80
1 1/8-16 UN	1.0570	83.8	1.0658	72.9	1 1/16	1.0625	77	1.0699	68

Table 12. Tap Drill Sizes and Percentage of Thread, Unified Threads Class 3B, Size 1 1/8-18 UNEF Through 3 3/4-4 UNC. (Source, FED-STD-H28/2B.)

Size, TPI, and Series*	Class 3B Minor Dia.				Tap Drill and Percentage of Thread		
	D_1 (min.)	% of Thread*	D_1 (max.)	% of Thread*	Drill Size		% of Thread*
1 1/8-18 UNEF	1.1650	83.1	1.0730	72.1	1 1/16	1.0625	87
1 1/8-20 UN	1.0710	83.1	1.0787	71.3	1 5/64	1.0781	72
1 1/8-28 UN	1.0860	84.1	1.0926	69.8	1 3/32	1.0938	67
1 3/16-8 UN	1.0520	83.4	1.0672	74.1	1 1/16	1.0625	77
1 3/16-12 UN	1.0970	83.6	1.1073	74.1	1 3/32	1.0938	87
1 3/16-16 UN	1.1200	83.1	1.1283	72.9	1 1/8	1.1250	77
1 3/16-18 UNEF	1.1270	83.8	1.1355	72.1	1 1/8	1.1250	87
1 3/16-20 UN	1.1330	83.9	1.1412	71.3	1 9/64	1.1406	72
1 3/16-28 UN	1.1490	83.0	1.1551	69.8	29.25mm	1.1516	77
1 1/4-7 UNC	1.0950	83.5	1.1125	74.1	1 3/32	1.0938	84
1 1/4-8 UN	1.1150	83.1	1.1297	74.1	1 1/8	1.1250	77
1 1/4-12 UNF	1.1600	83.1	1.1698	74.1	1 5/32	1.1562	87
1 1/4-16 UN	1.1820	83.8	1.1908	72.9	1 3/16	1.1875	77
1 1/4-18 UNEF	1.1900	83.1	1.1980	72.1	1 3/16	1.1875	87
1 1/4-20 UN	1.1960	83.1	1.2037	71.3	1 13/64	1.2031	72
1 1/4-28 UN	1.2110	84.1	1.2176	69.8	30.75mm	1.2106	85
1 5/16-8 UN	1.1770	83.4	1.1922	74.1	1 11/64 / 1 3/16	1.1719 / 1.1875	87 / 77
1 5/16-12 UN	1.2220	83.6	1.2323	74.1	1 7/32	1.2188	87
1 5/16-16 UN	1.2450	83.1	1.2533	72.9	1 1/4	1.2500	77
1 5/16-18 UNEF	1.2520	83.8	1.2605	72.1	1 1/4	1.2500	87
1 5/16-20 UN	1.2580	83.9	1.2662	71.3	1 17/64	1.2656	72
1 5/16-28 UN	1.2740	83.0	1.2801	69.8	32.5mm	1.2795	71
1 3/8-6 UNC	1.1950	83.1	1.2146	74.1	1 3/16 / 1 13/64	1.1875 / 1.2031	87 / 79
1 3/8-8 UN	1.2400	83.1	1.2547	74.1	1 15/64 / 1 1/4	1.2344 / 1.2500	87 / 77
1 3/8-12 UNF	1.2850	83.1	1.2948	74.1	1 9/32	1.2812	87
1 3/8-16 UN	1.3070	83.8	1.3158	72.9	1 5/16	1.3125	77
1 3/8-18 UNEF	1.3150	83.1	1.3230	72.1	1 5/16	1.3125	87
1 3/8-20 UN	1.3210	83.1	1.3287	71.3	1 21/64	1.3281	72
1 3/8-28 UN	1.3360	84.1	1.3426	69.8	34mm	1.3386	78
1 7/16-6 UN	1.2570	83.4	1.2771	74.1	1 17/64	1.2656	79
1 7/16-8 UN	1.3020	83.4	1.3172	74.1	1 19/64 / 1 5/16	1.2969 / 1.3125	87 / 77
1 7/16-12 UN	1.3470	83.6	1.3573	74.1	1 11/32	1.3438	87
1 7/16-16 UN	1.3700	83.1	1.3783	72.9	1 3/8	1.3750	77
1 7/16-18 UNEF	1.3770	83.8	1.3855	72.1	1 3/8	1.3750	87

(Continued)

Table 12. *(Continued)* **Tap Drill Sizes and Percentage of Thread, Unified Threads Class 3B, Size 1 1/8-18 UNEF Through 3 3/4-4 UNC.** *(Source, FED-STD-H28/2B.)*

Size, TPI, and Series*	Class 3B Minor Dia.				Tap Drill and Percentage of Thread		
	D_1 (min.)	% of Thread*	D_1 (max.)	% of Thread*	Drill Size		% of Thread*
1 7/16-20 UN	1.3830	83.9	1.3912	71.3	1 25/64	1.3906	72
1 7/16-28 UN	1.3990	83.0	1.4051	69.8	35.5mm	1.3976	86
1 1/2-6 UNC	1.3200	83.1	1.3396	74.1	1 5/16 1 21/64	1.3125 1.3281	87 79
1 1/2-8 UN	1.3650	83.1	1.3797	74.1	1 23/64 1 3/8	1.3594 1.3750	87 77
1 1/2-12 UNF	1.4100	83.1	1.4198	74.1	1 13/32	1.4062	87
1 1/2-16 UN	1.4320	83.8	1.4408	72.9	1 7/16	1.4375	77
1 1/2-18 UNEF	1.4400	83.1	1.4480	72.1	1 7/16	1.4375	87
1 1/2-20 UN	1.4460	83.1	1.4537	71.3	1 29/64	1.4531	72
1 1/2-28 UN	1.4610	84.1	1.4676	69.8	37mm	1.4567	93
1 9/16-6 UN	1.3820	83.4	1.4021	74.1	1 25/64	1.3906	79
1 9/16-8 UN	1.4270	83.4	1.4422	74.1	1 27/64 1 7/16	1.4219 1.4375	87 77
1 9/16-12 UN	1.4720	83.6	1.4823	74.1	1 15/32	1.4688	87
1 9/16-16 UN	1.4950	83.1	1.5033	72.9	1 1/2	1.5000	77
1 9/16-18 UNEF	1.5020	83.8	1.5105	72.1	1 1/2	1.5000	87
1 9/16-20 UN	1.5080	83.9	1.5162	71.3	1 33/64	1.5156	72
1 5/8-6 UN	1.4450	83.1	1.4646	74.1	1 29/64	1.4531	79
1 5/8-8 UN	1.4900	83.1	1.5047	74.1	1 31/64 1 1/2	1.4844 1.5000	87 77
1 5/8-12 UN	1.5350	83.1	1.5448	74.1	1 17/32	1.5312	87
1 5/8-16 UN	1.5570	83.8	1.5658	72.9	1 9/16	1.5625	77
1 5/8-18 UNEF	1.5650	83.1	1.5730	72.1	1 9/16	1.5625	87
1 5/8-20 UN	1.5710	83.1	1.5787	71.3	1 37/64	1.5781	72
1 11/16-6 UN	1.5070	83.4	1.5271	74.1	1 1/2 1 33/64	1.5000 1.5156	87 79
1 11/16-8 UN	1.5520	83.4	1.5672	74.1	1 9/16	1.5625	77
1 11/16-12 UN	1.5970	83.6	1.6073	74.1	1 19/32	1.5938	87
1 11/16-16 UN	1.6200	83.1	1.6283	72.9	1 5/8	1.6250	77
1 11/16-18 UNEF	1.6270	83.8	1.6355	72.1	1 5/8	1.6250	87
1 11/16-20 UN	1.6330	83.9	1.6412	71.3	1 41/64	1.6406	72
1 3/4-5 UNC	1.5340	83.1	1.5575	74.1	1 17/32 1 35/64	1.5312 1.5469	84 78
1 3/4-6 UN	1.5700	83.1	1.5896	74.1	1 9/16 1 37/64	1.5625 1.5781	87 79
1 3/4-8 UN	1.6150	83.1	1.6297	74.1	1 39/64 1 5/8	1.6094 1.6250	87 77
1 3/4-12 UN	1.6600	83.1	1.6698	74.1	1 21/32	1.6562	87

(Continued)

Table 12. *(Continued)* **Tap Drill Sizes and Percentage of Thread, Unified Threads Class 3B, Size 1 1/8-18 UNEF Through 3 3/4-4 UNC.** *(Source, FED-STD-H28/2B.)*

Size, TPI, and Series*	Class 3B Minor Dia.				Tap Drill and Percentage of Thread		
	D_1 (min.)	% of Thread*	D_1 (max.)	% of Thread*	Drill Size		% of Thread*
1 3/4-16 UN	1.6820	83.8	1.6908	72.9	1 11/16	1.6875	77
1 3/4-20 UN	1.6960	83.1	1.7037	71.3	1 45/64	1.7031	72
1 13/16-6 UN	1.6320	83.4	1.6521	74.1	1 5/8	1.6250	87
					1 41/64	1.6406	79
1 13/16-8 UN	1.6770	83.4	1.6922	74.1	1 43/64	1.6719	87
					1 11/16	1.6875	77
1 13/16-12 UN	1.7220	83.6	1.7323	74.1	1 23/32	1.7188	87
1 13/16-16 UN	1.7450	83.1	1.7533	72.9	1 3/4	1.7500	77
1 13/16-20 UN	1.7580	83.9	1.7662	71.3	1 49/64	1.7656	72
1 7/8-6 UN	1.6950	83.1	1.7146	74.1	1 45/64	1.7031	79
1 7/8-8 UN	1.7400	83.1	1.7547	74.1	1 3/4	1.7500	77
1 7/8-12 UN	1.7850	83.1	1.7948	74.1	1 25/32	1.7812	87
1 7/8-16 UN	1.8070	83.8	1.8158	72.9	1 13/16	1.8125	77
1 7/8-20 UN	1.8210	83.1	1.8287	71.3	1 53/64	1.8281	72
1 15/16-6 UN	1.7570	83.4	1.7771	74.1	1 49/64	1.7656	79
1 15/16-8 UN	1.8020	83.4	1.8172	74.1	1 51/64	1.7969	87
					1 13/16	1.8125	77
1 15/16-12 UN	1.8470	83.6	1.8573	74.1	1 27/32	1.8438	87
1 15/16-16 UN	1.8700	83.1	1.8783	72.9	1 7/8	1.8750	77
1 15/16-20 UN	1.8830	83.9	1.8912	71.3	1 57/64	1.8906	72
2-4.5 UNC	1.7590	83.5	1.7861	74.1	1 25/32	1.7812	76
2-6 UN	1.8200	83.1	1.8396	74.1	1 53/64	1.8281	79
2-8 UN	1.8650	83.1	1.8797	74.1	1 7/8	1.8750	77
2-12 UN	1.9100	83.1	1.9198	74.1	1 29/32	1.9062	87
2-16 UN	1.9320	83.8	1.9408	72.9	1 15/16	1.9375	77
2-20 UN	1.9460	83.1	1.9537	71.3	1 61/64	1.9531	72
2 1/16-16 UNS	1.9950	83.1	2.0033	72.9	2	2.0000	77
2 1/8-6 UN	1.9450	83.1	1.9646	74.1	1 61/64	1.9531	79
2 1/8-8 UN	1.9900	83.1	2.0047	74.1	2	2.0000	77
2 1/8-12 UN	2.0350	83.1	2.0448	74.1	2 1/32	2.0312	87
2 1/8-16 UN	2.0570	83.8	2.0658	72.9	2 1/16	2.0625	77
2 1/8-20 UN	2.0710	83.1	2.0787	71.3	2 1/16	2.0625	96
2 3/16-UNS	2.1200	83.1	2.1283	72.9	2 1/8	2.1250	77
2 1/4-4.5 UNC	2.0090	83.5	2.0361	74.1	2	2.0000	87
					2 1/32	2.0312	76
2 1/4-6 UN	2.0700	83.1	2.0896	74.1	2 1/16	2.0625	87
2 1/4-8 UN	2.1150	83.1	2.1297	74.1	2 1/8	2.1250	77
2 1/4-12 UN	2.1600	83.1	2.1698	74.1	2 5/32	2.1562	87

(Continued)

Table 12. *(Continued)* **Tap Drill Sizes and Percentage of Thread, Unified Threads Class 3B, Size 1 1/8-18 UNEF Through 3 3/4-4 UNC.** *(Source, FED-STD-H28/2B.)*

Size, TPI, and Series*	Class 3B Minor Dia.				Tap Drill and Percentage of Thread		
	D_1 (min.)	% of Thread*	D_1 (max.)	% of Thread*	Drill Size		% of Thread*
2 1/4-16 UN	2.1820	83.8	2.1908	72.9	2 3/16	2.1875	77
2 1/4-20 UN	2.1960	83.1	2.2037	71.3	2 3/16	2.1875	96
2 5/16-16 UNS	2.2450	83.1	2.2533	72.9	2 1/4	2.2500	77
2 3/8-6 UN	2.1950	83.1	2.2146	74.1	2 3/16	2.1875	87
2 3/8-8 UN	2.2400	83.1	2.2547	74.1	2 1/4	2.2500	77
2 3/8-12 UN	2.2850	83.1	2.2948	74.1	58mm	2.2835	85
2 3/8-16 UN	2.3070	83.8	2.3158	72.9	2 5/16	2.3125	77
2 3/8-20 UN	2.3210	83.1	2.3287	71.3	2 5/16	2.3125	96
2 7/16-16 UNS	2.3700	83.1	2.3783	72.9	2 3/8	2.3750	77
2 1/2-4 UNC	2.2290	83.4	2.2594	74.1	2 7/32 2 1/4	2.2188 2.2500	87 77
2 1/2-6 UN	2.3200	83.1	2.3396	74.1	2 5/16	2.3125	87
2 1/2-8 UN	2.3650	83.1	2.3797	74.1	2 3/8	2.3750	77
2 1/2-12 UN	2.4100	83.1	2.4198	74.1	2 13/32	2.4062	87
2 1/2-16 UN	2.4320	83.8	2.4408	72.9	2 7/16	2.4375	77
2 1/2-20 UN	2.4460	83.1	2.4537	71.3	2 7/16	2.4375	96
2 5/8-4 UN	2.3540	83.4	2.3844	74.1	2 11/32 2 3/8	2.3438 2.3750	87 77
2 5/8-6 UN	2.4450	83.1	2.4646	74.1	2 7/16	2.4375	87
2 5/8-8 UN	2.4900	83.1	2.5047	74.1	2 1/2	2.5000	77
2 5/8-12 UN	2.5350	83.1	2.5448	74.1	2 17/32	2.5312	87
2 5/8-16 UN	2.5570	83.8	2.5658	72.9	2 9/16	2.5625	77
2 5/8-20 UN	2.5710	83.1	2.5787	71.3	2 9/16	2.5625	96
2 3/4-4 UNC	2.4790	83.4	2.5094	74.1	2 1/2	2.5000	77
2 3/4-6 UN	2.5700	83.1	2.5896	74.1	2 9/16	2.5625	87
2 3/4-8 UN	2.6150	83.1	2.6297	74.1	2 5/8	2.6250	77
2 3/4-12 UN	2.6600	83.1	2.6698	74.1	2 21/32	2.6562	87
2 3/4-16 UN	2.6820	83.8	2.6908	72.9	2 11/16	2.6875	77
2 3/4-20 UN	2.6960	83.1	2.7037	71.3	2 11/16	2.6875	96
2 7/8-4 UN	2.6040	83.4	2.6344	74.1	2 5/8	2.6250	77
2 7/8-6 UN	2.6950	83.1	2.7146	74.1	2 11/16	2.6875	87
2 7/8-8 UN	2.7400	83.1	2.7547	74.1	2 3/4	2.7500	77
2 7/8-12 UN	2.7850	83.1	2.7948	74.1	2 25/32	2.7812	87
2 7/8-16 UN	2.8070	83.8	2.8158	72.9	2 13/16	2.8125	77

Table 13. Recommended Hole Size Limits Before Threading for Different Lengths of Engagement, Unified Class 1B and 2B Threads. *(Source, FED-STD-H28/2B.)*

Size, TPI, and Series	To and Including 0.33D		Above 0.33D to 0.67D		Above 0.67D through 1.5D	
	Min.	Max.	Min.	Max.	Min.	Max.
0-80 UNF	.0465	.0500	.0479	.0514	.0479	.0514
1-64 UNC	.0561	.0599	.0580	.0618	.0585	.0623
1-72 UNF	.0580	.0613	.0596	.0629	.0602	.0635
2-56 UNC	.0667	.0705	.0686	.0724	.0699	.0737
2-64 UNF	.0691	.0724	.0707	.0740	.0720	.0753
3-48 UNC	.0764	.0804	.0785	.0825	.0805	.0845
3-56 UNF	.0797	.0831	.0814	.0848	.0831	.0865
4-40 UNC	.0849	.0849	.0871	.0916	.0894	.0939
4-48 UNF	.0894	.0931	.0912	.0949	.0931	.0968
5-40 UNC	.0979	.1020	.1000	.1041	.1021	.1062
5-44 UNF	.1004	.1041	.1023	.1060	.1042	.1079
6-32 UNC	.104	.109	.106	.112	.109	.114
6-40 UNF	.111	.115	.113	.117	.115	.119
8-32 UNC	.130	.135	.132	.137	.134	.139
8-36 UNF	.134	.138	.136	.140	.138	.142
10-24 UNC	.145	.150	.147	.153	.150	.156
10-32 UNF	.156	.160	.158	.162	.160	.164
12-24 UNC	.171	.176	.173	.178	.176	.181
12-28 UNF	.177	.182	.179	.184	.181	.186
12-32 UNEF	.182	.186	.184	.188	.186	.190
1/4-20 UNC	.196	.202	.199	.204	.202	.207
1/4-28 UNF	.211	.216	.213	.218	.216	.220
1/4-32 UNEF	.216	.220	.218	.222	.220	.224
1/4-36 UNS	.220	.223	.221	.225	.222	.226
5/16-18 UNC	.252	.259	.256	.262	.259	.265
5/16-20 20UN	.258	.264	.261	.267	.264	.270
5/16-24 UNF	.267	.272	.270	.275	.272	.277
5/16-28 28UN	.274	.278	.276	.280	.278	.282
5/16-32 UNEF	.279	.282	.280	.284	.282	.286
5/16-36 UNS	.282	.286	.283	.287	.285	.289
3/8-16 UNC	.307	.314	.311	.318	.314	.321
3/8-20 20UN	.321	.327	.324	.330	.327	.332
3/8-24 UNF	.330	.335	.332	.337	.335	.340
3/8-28 28UN	.336	.340	.338	.343	.340	.345
3/8-32 UNEF	.341	.345	.343	.347	.345	.349
3/8-36 UNS	.345	.348	.346	.350	.348	.352
7/16-14 UNC	.360	.368	.364	.372	.368	.376
7/16-16 16UN	.370	.377	.373	.380	.377	.384

All dimensions in inches.   *(Continued)*

Table 13. *(Continued)* Recommended Hole Size Limits Before Threading for Different Lengths of Engagement, Unified Class 1B and 2B Threads. *(Source, FED-STD-H28/2B.)*

Size, TPI, and Series	To and Including 0.33D		Above 0.33D to 0.67D		Above 0.67D through 1.5D	
	Min.	Max.	Min.	Max.	Min.	Max.
7/16-20 UNF	.383	.389	.386	.392	.389	.395
7/16-28 UNEF	.399	.403	.401	.405	.403	.407
7/16-32 32UN	.404	.407	.405	.409	.407	.411
1/2-12 UNS	.410	.419	.414	.423	.419	.428
1/2-13 UNC	.417	.425	.421	.430	.425	.434
1/2-16 16UN	.432	.439	.436	.443	.439	.446
1/2-20 UNF	.446	.452	.449	.454	.452	.457
1/2-28 UNEF	.461	.466	.463	.468	.466	.470
1/2-32 32UN	.466	.470	.468	.472	.470	.474
9/16-12 UNC	.472	.481	.477	.486	.481	.490
9/16-16 16UN	.495	.502	.498	.505	.502	.509
9/16-18 UNF	.502	.509	.506	.512	.509	.515
9/16-20 20UN	.508	.514	.511	.517	.514	.520
9/16-24 UNEF	.517	.522	.520	.525	.522	.527
9/16-28 28UN	.524	.528	.526	.530	.528	.532
9/16-32 32UN	.529	.532	.530	.534	.532	.536
5/8-11 UNC	.527	.536	.532	.541	.536	.546
5/8-12 12UN	.535	.544	.539	.548	.544	.553
5/8-16 16UN	.557	.564	.561	.568	.564	.571
5/8-18UNF	.565	.571	.568	.574	.571	.578
5/8-20 20UN	.571	.577	.574	.580	.577	.582
5/8-24 UNEF	.580	.585	.582	.587	.585	.590
5/8-28 28UN	.586	.590	.588	.593	.590	.595
5/8-32 32UN	.591	.595	.593	.597	.595	.599
11/16-16-12 12UN	.597	.606	.602	.611	.606	.615
11/16-16-16UN	.620	.627	.623	.630	.627	.634
11/16-16-18UNS	.627	.634	.630	.637	.634	.640
11/16-20 20UN	.633	.639	.636	.642	.639	.645
11/16-24UNEF	.642	.647	.645	.650	.647	.652
11/16-28 28UN	.649	.653	.651	.655	.653	.657
11/16-32 32UN	.654	.657	.655	.659	.657	.661
3/4-10 UNC	.642	.652	.647	.658	.652	.663
3/4-12 12UN	.660	.669	.664	.673	.669	.678
3/4-16 UNF	.682	.689	.686	.693	.689	.696
3/4-18 UNS	.690	.696	.693	.699	.696	.703
3/4-20 UNEF	.696	.702	.699	.704	.702	.707
3/4-28 28UN	.711	.716	.713	.718	.716	.720
3/4-32 32UN	.716	.720	.718	.722	.720	.724

All dimensions in inches.

(Continued)

Table 13. *(Continued)* **Recommended Hole Size Limits Before Threading for Different Lengths of Engagement, Unified Class 1B and 2B Threads.** *(Source, FED-STD-H28/2B.)*

Size, TPI, and Series	To and Including 0.33D		Above 0.33D to 0.67D		Above 0.67D through 1.5D	
	Min.	Max.	Min.	Max.	Min.	Max.
13/16-12 12UN	.722	.731	.727	.736	.731	.740
13/16-16 16UN	.745	.752	.748	.755	.752	.795
13/16-18UNS	.752	.759	.756	.762	.759	.765
13/16-20UNEF	.758	.764	.761	.767	.764	.770
13/16-28 28UN	.774	.778	.776	.780	.778	.782
13/16-32 32UN	.779	.782	.780	.784	.782	.786
7/8-9UNC	.755	.766	.760	.772	.766	.778
7/8-12 12UN	.785	.794	.789	.798	.794	.803
7/8-14UNF	.798	.806	.802	.810	.806	.814
7/8-16 16UN	.807	.814	.811	.818	.814	.821
7/8-18UNS	.815	.821	.818	.824	.821	.828
7/8-20UNEF	.821	.827	.824	.830	.827	.832
7/8-28 28UN	.836	.840	.838	.843	.840	.845
7/8-32 32UN	.841	.845	.843	.847	.845	.849
15/16-12 12UN	.847	.856	.852	.861	.856	.865
15/16-16 16UN	.870	.877	.873	.880	.877	.884
15/16-20UNEF	.883	.889	.886	.892	.889	.895
15/16-28 28UN	.899	.903	.901	.905	.903	.907
15/16-32 32UN	.904	.907	.905	.909	.907	.911
1-8UNC	.865	.877	.871	.884	.877	.890
1-12UNF	.910	.919	.914	.923	.919	.928
1-14UNS	.923	.931	.927	.934	.931	.938
1-16 16UN	.932	.939	.936	.943	.939	.946
1-18UNS	.940	.946	.943	.949	.946	.953
1-20UNEF	.946	.952	.949	.954	.952	.957
1-28 28UN	.961	.966	.963	.968	.966	.970
1-32 32UN	.966	.970	.968	.972	.970	.974
1 1/16-8 8UN	.927	.940	.934	.946	.940	.952
1 1/16-12 12UN	.972	.981	.977	.986	.981	.990
1 1/16-14UNS	.985	.993	.989	.997	.993	1.001
1 1/16-16 16UN	.995	1.002	.998	1.005	1.002	1.009
1 1/16-18UNEF	1.002	1.009	1.006	1.012	1.009	1.015
1 1/16-20 20UN	1.008	1.014	1.011	1.017	1.014	1.020
1 1/16-28 28UN	1.024	1.028	1.026	1.030	1.028	1.032
1 1/8-7UNC	.970	.984	.977	.991	.984	.998
1 1/8 8UN	.990	1.002	.996	1.008	1.002	1.015
1 1/8-12UNF	1.035	1.044	1.039	1.048	1.044	1.053
1 1/8-16 16UN	1.057	1.064	1.061	1.068	1.064	1.071

All dimensions in inches.

(Continued)

Table 13. *(Continued)* **Recommended Hole Size Limits Before Threading for Different Lengths of Engagement, Unified Class 1B and 2B Threads.** *(Source, FED-STD-H28/2B.)*

Size, TPI, and Series	To and Including 0.33D		Above 0.33D to 0.67D		Above 0.67D through 1.5D	
	Min.	Max.	Min.	Max.	Min.	Max.
1 1/8-18UNEF	1.065	1.071	1.068	1.074	1.071	1.078
1 1/8-20 20UN	1.071	1.077	1.074	1.080	1.077	1.082
1 1/8-28 28UN	1.086	1.090	1.088	1.093	1.090	1.095
1 3/16-8 8UN	1.052	1.065	1.058	1.071	1.065	1.077
1 3/16-12 12UN	1.097	1.106	1.102	1.111	1.106	1.115
1 3/16-16 16UN	1.120	1.127	1.123	1.130	1.127	1.134
1 3/16-18UNEF	1.127	1.134	1.130	1.137	1.134	1.140
1 3/16-20 20UN	1.133	1.139	1.136	1.142	1.139	1.145
1 3/16-28 28UN	1.149	1.153	1.151	1.155	1.153	1.157
1 1/4-7UNC	1.095	1.109	1.102	1.116	1.109	1.123
1 1/4-8 8UN	1.115	1.127	1.121	1.134	1.127	1.140
1 1/4-12UNF	1.160	1.169	1.164	1.173	1.169	1.178
1 1/4-16 16UN	1.182	1.189	1.186	1.193	1.189	1.196
1 1/4-18UNEF	1.190	1.196	1.193	1.199	1.196	1.203
1 1/4-20 20UN	1.196	1.202	1.199	1.204	1.202	1.207
1 1/4-28 28UN	1.211	1.216	1.213	1.218	1.216	1.220
1 5/16-8 8UN	1.177	1.190	1.184	1.196	1.190	1.202
1 5/16-12 12UN	1.222	1.231	1.227	1.236	1.231	1.240
1 5/16-16 16UN	1.245	1.252	1.248	1.255	1.252	1.259
1 5/16-18 UNEF	1.252	1.259	1.256	1.262	1.259	1.265
1 5/16-20 20UN	1.258	1.264	1.261	1.267	1.264	1.270
1 5/16-28 28UN	1.274	1.278	1.276	1.280	1.278	1.282
1 3/8-6UNC	1.195	1.210	1.202	1.218	1.210	1.225
1 3/8-8 8UN	1.240	1.252	1.246	1.258	1.252	1.265
1 3/8-12UNF	1.285	1.294	1.289	1.298	1.294	1.303
1 3/8-16 16UN	1.307	1.314	1.311	1.318	1.314	1.321
1 3/8-18UNEF	1.315	1.321	1.318	1.324	1.321	1.328
1 3/8-20 20UN	1.321	1.327	1.324	1.330	1.327	1.332
1 3/8-28 28UN	1.336	1.340	1.338	1.343	1.340	1.345
1 7/16-6 6UN	1.257	1.272	1.265	1.280	1.272	1.288
1 7/16-8 8UN	1.302	1.315	1.308	1.321	1.315	1.327
1 7/16-12 12UN	1.347	1.356	1.352	1.361	1.356	1.365
1 7/16-16 16UN	1.370	1.377	1.373	1.380	1.377	1.384
1 7/16-18UNEF	1.377	1.384	1.380	1.387	1.384	1.390
1 7/16-20 20UN	1.383	1.389	1.386	1.392	1.389	1.395
1 7/16-28 28UN	1.399	1.403	1.401	1.405	1.403	1.407
1 1/2-6UNC	1.320	1.335	1.327	1.343	1.335	1.350
1 1/2-8 8UN	1.365	1.377	1.371	1.384	1.377	1.390

All dimensions in inches.   *(Continued)*

Table 13. *(Continued)* **Recommended Hole Size Limits Before Threading for Different Lengths of Engagement, Unified Class 1B and 2B Threads.** *(Source, FED-STD-H28/2B.)*

Size, TPI, and Series	To and Including 0.33D		Above 0.33D to 0.67D		Above 0.67D through 1.5D	
	Min.	Max.	Min.	Max.	Min.	Max.
1 1/2-12UNF	1.410	1.419	1.414	1.423	1.419	1.428
1 1/2-16 16UN	1.432	1.439	1.436	1.443	1.439	1.446
1 1/2-18UNEF	1.440	1.446	1.443	1.449	1.446	1.453
1 1/2-20 20UN	1.446	1.452	1.449	1.454	1.452	1.457
1 1/2-28 28UN	1.461	1.466	1.463	1.468	1.466	1.470
1 9/16-6 6UN	1.382	1.397	1.390	1.405	1.397	1.413
1 9/16-8 8UN	1.427	1.440	1.434	1.446	1.440	1.452
1 9/16-12 12UN	1.472	1.481	1.477	1.486	1.481	1.490
1 9/16-16 16UN	1.495	1.502	1.498	1.505	1.502	1.509
1 9/16-18UNEF	1.502	1.509	1.506	1.512	1.509	1.515
1 9/16-20 20UN	1.508	1.514	1.511	1.517	1.514	1.520
1 5/8-6 6UN	1.445	1.460	1.452	1.468	1.460	1.475
1 5/8-8 8UN	1.490	1.502	1.496	1.508	1.502	1.515
1 5/8-12 12UN	1.535	1.544	1.539	1.548	1.544	1.553
1 5/8-16 16UN	1.557	1.564	1.561	1.568	1.564	1.571
1 5/8-18UNEF	1.565	1.571	1.568	1.574	1.571	1.578
1 5/8-20 20UN	1.571	1.577	1.574	1.580	1.577	1.582
1 11/16-6 6UN	1.507	1.522	1.515	1.530	1.522	1.538
1 11/16-8 8UN	1.552	1.565	1.558	1.571	1.565	1.577
1 11/16-12 12UN	1.597	1.606	1.602	1.611	1.606	1.615
1 11/16-16 16UN	1.620	1.627	1.623	1.630	1.627	1.634
1 11/16-18UNEF	1.627	1.634	1.630	1.637	1.634	1.640
1 11/16-20 20UN	1.633	1.639	1.636	1.642	1.639	1.645
1 3/4-5UNC	1.534	1.550	1.542	1.559	1.550	1.568
1 3/4-6 6UN	1.570	1.585	1.577	1.593	1.585	1.600
1 3/4-8 8UN	1.615	1.627	1.6 21	1.634	1.627	1.640
1 3/4-12 12UN	1.660	1.669	1.664	1.673	1.669	1.678
1 3/4-16 16UN	1.682	1.689	1.686	1.693	1.689	1.696
1 3/4-20 20UN	1.696	1.702	1.699	1.704	1.702	1.707
1 13/16-6 6UN	1.632	1.647	1.640	1.655	1.647	1.663
1 13/16-8 8UN	1.677	1.690	1.684	1.696	1.690	1.702
1 13/16-12 12UN	1.722	1.731	1.727	1.736	1.731	1.740
1 13/16-16 16UN	1.745	1.752	1.748	1.755	1.752	1.759
1 13/16-20 20UN	1.7 58	1.764	1.761	1.767	1.764	1.770
1 7/8-6 6UN	1.695	1.710	1.702	1.718	1.710	1.725
1 7/8-8 8UN	1.740	1.752	1.746	1.758	1.752	1.765
1 7/8-12 12UN	1.785	1.794	1.789	1.798	1.794	1.803
1 7/8-16 16UN	1.807	1.814	1.811	1.818	1.814	1.821

All dimensions in inches. *(Continued)*

Table 13. *(Continued)* **Recommended Hole Size Limits Before Threading for Different Lengths of Engagement, Unified Class 1B and 2B Threads.** *(Source, FED-STD-H28/2B.)*

Size, TPI, and Series	To and Including 0.33D		Above 0.33D to 0.67D		Above 0.67D through 1.5D	
	Min.	Max.	Min.	Max.	Min.	Max.
1 7/8-20 20UN	1.821	1.827	1.824	1.830	1.827	1.832
1 15/16-6 6UN	1.757	1.772	1.765	1.780	1.772	1.788
1 15/16-8 8UN	1.802	1.815	1.808	1.821	1.815	1.827
1 15/16-12 12UN	1.847	1.856	1.852	1.861	1.856	1.865
1 15/16-16 16UN	1.870	1.877	1.873	1.880	1.877	1.884
1 15/16-20 20UN	1.883	1.889	1.886	1.892	1.889	1.895
2-4.5UNC	1.759	1.777	1.768	1.786	1.777	1.795
2-6 6UN	1.820	1.835	1.827	1.843	1.835	1.850
2-8 8UN	1.865	1.877	1.871	1.884	1.877	1.890
2-12 12UN	1.910	1.919	1.914	1.923	1.919	1.928
2-16 16UN	1.932	1.939	1.936	1.943	1.939	1.946
2-20 20UN	1.946	1.952	1.949	1.954	1.952	1.957
2 5/8-16UNS	1.995	2.002	1.998	2.005	2.002	2.009
2 1/8-6 6UN	1.945	1.960	1.952	1.968	1.960	1.975
2 1/8-8 8UN	1.990	2.002	1.996	2.008	2.002	2.015
2 1/8-12 12UN	2.035	2.044	2.039	2.048	2.044	2.053
2 1/8-16 16UN	2.057	2.064	2.061	2.068	2.064	2.071
2 1/8-20 20UN	2.071	2.077	2.074	2.080	2.077	2.082
2 3/16-16UNS	2.120	2.127	2.123	2.130	2.127	2.134
2 1/4-4.5UNC	2.009	2.027	2.018	2.036	2.027	2.045
2 1/4-6 6UN	2.070	2.085	2.077	2.093	2.085	2.100
2 1/2-4UNC	2.229	2.248	2.239	2.258	2.248	2.267
2 3/4-4UNC	2.479	2.498	2.489	2.508	2.498	2.517
3-4UNC	2.729	2.748	2.739	2.758	2.748	2.767
3 1/4-4UNC	2.979	2.998	2.989	3.008	2.998	3.017

All dimensions in inches.

Table 14. Recommended Hole Size Limits Before Threading for Different Lengths of Engagement, Unified Class 3B Threads. *(Source, FED-STD-H28/2B.)*

Size, TPI, and Series	To and Including 0.33D		Above 0.33D to 0.67D		Above 0.67D through 1.5D	
	Min.	Max.	Min.	Max.	Min.	Max.
0-80 UNF	.0465	.0500	.0479	.0514	.0479	.0514
1-64 UNC	.0561	.0599	.0580	.0618	.0585	.0623
1-72 UNF	.0580	.0613	.0596	.0629	.0602	.0635
2-56 UNC	.0667	.0705	.0686	.0724	.0699	.0737
2-64 UNF	.0691	.0724	.0707	.0740	.0720	.0753
3-48 UNC	.0764	.0804	.0785	.0825	.0805	.0845
3-56 UNF	.0797	.0831	.0814	.0848	.0831	.0865

All dimensions in inches.

(Continued)

Table 14. *(Continued)* **Recommended Hole Size Limits Before Threading for Different Lengths of Engagement, Unified Class 3B Threads.** *(Source, FED-STD-H28/2B.)*

Size, TPI, and Series	To and Including 0.33D		Above 0.33D to 0.67D		Above 0.67D through 1.5D	
	Min.	Max.	Min.	Max.	Min.	Max.
4-40 UNC	.0849	.0894	.0871	.0916	.0894	.0939
4-48 UNF	.0894	.0931	.0912	.0949	.0931	.0968
5-40 UNC	.0979	.1020	.1000	.1041	.1021	.1062
5-44 UNF	.1004	.1041	.1023	.1060	.1042	.1079
6-32 UNC	.1040	.1091	.1066	.1115	.1091	.1140
6-40 UNF	.1110	.1148	.1128	.1167	.1147	.1186
8-32 UNC	.1300	.1345	.1324	.1367	.1346	.1389
8-36 UNF	.1340	.1377	.1359	.1397	.1378	.1416
10-24 UNC	.1450	.1502	.1475	.1528	.1502	.1555
10-32 UNF	.1560	.1601	.1582	.1621	.1602	.1641
12-24 UNC	.1710	.1758	.1733	.1782	.1758	.1807
12-28 UNF	.1770	.1815	.1794	.1836	.1815	.1857
12-32 UNEF	.1820	.1858	.1841	.1877	.1859	.1895
1/4-20 UNC	.1960	.2013	.1986	.2040	.2013	.2067
1/4-28 UNF	.2110	.2152	.2131	.2171	.2150	.2190
1/4-32 UNEF	.2160	.2196	.2172	.2212	.2189	.2229
1/4-36 UNS	.2200	.2229	.2203	.2243	.2218	.2258
5/16-18 UNC	.2520	.2577	.2551	.2604	.2577	.2630
5/16-20 20UN	.2580	.2632	.2608	.2656	.2632	.2680
5/16-24 UNF	.2670	.2714	.2694	.2734	.2714	.2754
5/16-28 28UN	.2740	.2772	.2749	.2789	.2767	.2807
5/16-32 UNEF	.2790	.2817	.2792	.2832	.2807	.2847
5/16-36 UNS	.2820	.2850	.2823	.2863	.2837	.2877
3/8-16 UNC	.3070	.3127	.3101	.3155	.3128	.3182
3/8-20 20UN	.3210	.3253	.3231	.3275	.3253	.3297
3/8-24 UNF	.3300	.3336	.3314	.3354	.3332	.3372
3/8-28 28UN	.3360	.3395	.3370	.3410	.3386	.3426
3/8-32 UNEF	.3410	.3441	.3415	.3455	.3429	.3469
3/8-36 UNS	.3450	.3475	.3450	.3490	.3461	.3501
7/16-14 UNC	.3600	.3660	.3630	.3688	.3659	.3717
7/16-16 16UN	.3700	.3749	.3723	.3774	.3749	.3800
7/16-20 UNF	.3830	.3875	.3855	.3896	.3875	.3916
7/16-28 UNEF	.3990	.4020	.3995	.4035	.4011	.4051
7/16-32 32UN	.4040	.4066	.4040	.4080	.4054	.4094
1/2-12 UNS	.4100	.4161	.4129	.4192	.4160	.4223
1/2-13 UNC	.4170	.4225	.4196	.4254	.4226	.4284
1/2-16 16UN	.4320	.4371	.4347	.4395	.4371	.4419
1/2-20 UNF	.4460	.4498	.4477	.4517	.4497	.4537

All dimensions in inches.

(Continued)

Table 14. *(Continued)* **Recommended Hole Size Limits Before Threading for Different Lengths of Engagement, Unified Class 3B Threads.** *(Source, FED-STD-H28/2B.)*

Size, TPI, and Series	To and Including 0.33D		Above 0.33D to 0.67D		Above 0.67D through 1.5D	
	Min.	Max.	Min.	Max.	Min.	Max.
1/2-28 UNEF	.4610	.4645	.4620	.4660	.4636	.4676
1/2-32 32UN	.4660	.4691	.4665	.4705	.4679	.4719
9/16-12 UNC	.4720	.4783	.4753	.4813	.4783	.4843
9/16-16 16UN	.4950	.4994	.4971	.5017	.4994	.5040
9/16-18 UNF	.5020	.5065	.5045	.5086	.5065	.5106
9/16-20 20UN	.5080	.5123	.5102	.5142	.5122	.5162
9/16-24 UNEF	.5170	.5209	.5186	.5226	.5204	.5244
9/16-28 28UN	.5240	.5270	.5245	.5285	.5261	.5301
9/16-32 32UN	.5290	.5316	.5290	.5330	.5304	.5344
5/8-11 UNC	.5270	.5328	.5298	.5360	.5329	.5391
5/8-12 12UN	.5350	.5406	.5377	.5435	.5405	.5463
5/8-16 16UN	.5570	.5617	.5596	.5640	.5618	.5662
5/8-18 UNF	.5650	.5690	.5669	.5710	.5689	.5730
5/8-20 20UN	.5710	.5748	.5727	.5767	.5747	.5787
5/8-24 UNEF	.5800	.5834	.5811	.5851	.5829	.5869
5/8-28 28UN	.5860	.5895	.5870	.5910	.5886	.5926
5/8-32 32UN	.5910	.5941	.5915	.5955	.5929	.5969
11/16-12 12UN	.5970	.6029	.6001	.6057	.6029	.6085
11/16-16 16UN	.6200	.6241	.6219	.6262	.6241	.6284
11/16-18 UNS	.6270	.6315	.6294	.6335	.6314	.6355
11/16-20 20UN	.6330	.6373	.6352	.6392	.6372	.6412
11/16-24 UNEF	.6420	.6459	.6436	.6476	.6454	.6494
11/16-28 28UN	.6490	.6520	.6495	.6535	.6511	.6551
11/16-32 32UN	.6540	.6566	.6540	.6580	.6554	.6594
3/4-10 UNC	.6420	.6481	.6449	.6513	.6481	.6545
3/4-12 12UN	.6600	.6652	.6626	.6680	.6653	.6707
3/4-16 UNF	.6820	.6866	.6844	.6887	.6865	.6908
3/4-18 UNS	.6900	.6940	.6919	.6960	.6939	.6980
3/4-20 UNEF	.6960	.6998	.6977	.7017	.6997	.7037
3/4-28 28UN	.7110	.7145	.7120	.7160	.7136	.7176
3/4-32 32UN	.7160	.7191	.7165	.7205	.7179	.7219
13/16-12 12UN	.7220	.7276	.7250	.7303	.7276	.7329
13/16-16 16UN	.7450	.7491	.7469	.7512	.7490	.7533
13/16-18 UNS	.7520	.7565	.7544	.7585	.7564	.7605
13/16-20 UNEF	.7580	.7623	.7602	.7642	.7622	.7662
13/16-28 28UN	.7740	.7770	.7745	.7785	.7761	.7801
13/16-32 32UN	.7790	.7816	.7790	.7830	.7804	.7844
7/8-9 UNC	.7550	.7614	.7580	.7647	.7614	.7681

All dimensions in inches.   *(Continued)*

Table 14. *(Continued)* **Recommended Hole Size Limits Before Threading for Different Lengths of Engagement, Unified Class 3B Threads.** *(Source, FED-STD-H28/2B.)*

Size, TPI, and Series	To and Including 0.33D		Above 0.33D to 0.67D		Above 0.67D through 1.5D	
	Min.	Max.	Min.	Max.	Min.	Max.
7/8-12 12UN	.7850	.7900	.7874	.7926	.7900	.7952
7/8-14 UNF	.7980	.8022	.8000	.8045	.8023	.8068
7/8-16 16UN	.8070	.8116	.8094	.8137	.8115	.8158
7/8-18 UNS	.8150	.8190	.8169	.8210	.8189	.8230
7/8-20 UNEF	.8210	.8248	.8227	.8267	.8247	.8287
7/8-28 28UN	.8360	.8395	.8370	.8410	.8386	.8426
7/8-32 32UN	.8410	.8441	.8415	.8455	.8429	.8469
15/16-12 12UN	.8470	.8524	.8499	.8550	.8524	.8575
15/16-16 16UN	.8700	.8741	.8719	.8762	.8740	.8783
15/16-20 UNEF	.8830	.8873	.8852	.8892	.8872	.8912
15/16-28 28UN	.8990	.9020	.8995	.9035	.9011	.9051
15/16-32 32UN	.9040	.9066	.9040	.9080	.9054	.9094
1-8 UNC	.8650	.8722	.8684	.8759	.8722	.8797
1-12 UNF	.9100	.9148	.9123	.9173	.9148	.9198
1-14 UNS	.9230	.9271	.9249	.9293	.9271	.9315
1-16 16UN	.9320	.9366	.9344	.9387	.9365	.9408
1-18 UNS	.9400	.9440	.9419	.9460	.9439	.9480
1-20 UNEF	.9460	.9498	.9477	.9517	.9497	.9537
1-28 28UN	.9610	.9645	.9620	.9660	.9636	.9676
1-32 32UN	.9660	.9691	.9665	.9705	.9679	.9719
1 1/16-8 8UN	.9270	.9347	.9309	.9384	.9347	.9422
1 1/16-12 12UN	.9720	.9773	.9748	.9798	.9773	.9823
1 1/16-14 UNS	.9850	.9896	.9874	.9918	.9896	.9940
1 1/16-16 16UN	.9950	.9991	.9969	1.0012	.9990	1.0033
1 1/16-18 UNEF	1.0020	1.0065	1.0044	1.0085	1.0064	1.0105
1 1/16-20 20UN	1.0080	1.0123	1.0102	1.0142	1.0122	1.0162
1 1/16-28 28UN	1.0240	1.0270	1.0245	1.0285	1.0261	1.0301
1 1/8-7 UNC	.9700	.9790	.9747	.9833	.9789	.9875
1 1/8-8 8UN	.9900	.9972	.9934	1.0009	.9972	1.0047
1 1/8-12 UNF	1.0350	1.0398	1.0373	1.0423	1.0398	1.0448
1 1/8-16 16UN	1.0570	1.0616	1.0594	1.0637	1.0615	1.0658
1 1/8-18 UNEF	1.0650	1.0690	1.0669	1.0710	1.0689	1.0730
1 1/8-20 20UN	1.0710	1.0748	1.0727	1.0767	1.0747	1.0787
1 1/8-28 28UN	1.0860	1.0895	1.0870	1.0910	1.0886	1.0926
1 3/16-8 8UN	1.0520	1.0597	1.0559	1.0634	1.0597	1.0672
1 3/16-12 12UN	1.0970	1.1023	1.0998	1.1048	1.1023	1.1073
1 3/16-16 16UN	1.1200	1.1241	1.1219	1.1262	1.1240	1.1283
1 3/16-18 UNEF	1.1270	1.1315	1.1294	1.1335	1.1314	1.1355

All dimensions in inches. *(Continued)*

Table 14. *(Continued)* **Recommended Hole Size Limits Before Threading for Different Lengths of Engagement, Unified Class 3B Threads.** *(Source, FED-STD-H28/2B.)*

Size, TPI, and Series	To and Including 0.33*D*		Above 0.33*D* to 0.67*D*		Above 0.67*D* through 1.5*D*	
	Min.	Max.	Min.	Max.	Min.	Max.
1 3/16-20 20UN	1.1330	1.1373	1.1352	1.1392	1.1372	1.1412
1 3/16-28 28UN	1.1490	1.1520	1.1495	1.1535	1.1511	1.1551
1 1/4-7 UNC	1.0950	1.1040	1.0997	1.1083	1.1039	1.1125
1 1/4-8 8UN	1.1150	1.1222	1.1184	1.1259	1.1222	1.1297
1 1/4-12 UNF	1.1600	1.1648	1.1623	1.1673	1.1648	1.1698
1 1/4-16 16UN	1.1820	1.1866	1.1844	1.1887	1.1865	1.1908
1 1/4-18 UNEF	1.1900	1.1940	1.1919	1.1960	1.939	1.1980
1 1/4-20 20UN	1.1960	1.1998	1.1977	1.2017	1.1997	1.2037
1 1/4-28 28UN	1.2110	1.2145	1.2120	1.2160	1.2136	1.2176
1 5/16-8 8UN	1.1770	1.1847	1.1809	1.1884	1.1847	1.1922
1 5/16-12 12UN	1.2220	1.2273	1.2248	1.2298	1.2273	1.2323
1 5/16-16 16UN	1.2450	1.2491	1.2469	1.2512	1.2490	1.2533
1 5/16-18 UNEF	1.2520	1.2565	1.2544	1.2585	1.2564	1.2605
1 5/16-20 20UN	1.2580	1.2623	1.2602	1.2642	1.2622	1.2662
1 5/16-28 28UN	1.2740	1.2770	1.2745	1.2785	1.2761	1.2801
1 3/8-6 UNC	1.1950	1.2046	1.1996	1.2096	1.2046	1.2146
1 3/8-8 8UN	1.2400	1.2472	1.2434	1.2509	1.2472	1.2547
1 3/8-12 UNF	1.2850	1.2898	1.2873	1.2923	1.2898	1.2948
1 3/8-16 16UN	1.3070	1.3116	1.3094	1.3137	1.3115	1.3158
1 3/8-18 UNEF	1.3150	1.3190	1.3169	1.3210	1.3189	1.3230
1 3/8-20 20UN	1.3210	1.3248	1.3227	1.3267	1.3247	1.3287
1 3/8-28 28UN	1.3360	1.3395	1.3370	1.3410	1.3386	1.3426
1 7/16-6 6UN	1.2570	1.2671	1.2621	1.2721	1.2671	1.2771
1 7/16-8 8UN	1.3020	1.3097	1.3059	1.3134	1.3097	1.3172
1 7/16-12 12UN	1.3470	1.3523	1.3498	1.3548	1.3523	1.3573
1 7/16-16 16UN	1.3700	1.3741	1.3719	1.3762	1.3740	1.3783
1 7/16-18 UNEF	1.3770	1.3815	1.3794	1.3835	1.3814	1.3855
1 7/16-20 20UN	1.3830	1.3873	1.3852	1.3892	1.3872	1.3912
1 7/16-28 28UN	1.3990	1.4020	1.3995	1.4035	1.4011	1.4051
1 1/2-6 UNC	1.3200	1.3296	1.3246	1.3346	1.3296	1.3396
1 1/2-8 8UN	1.3650	1.3722	1.3684	1.3759	1.3722	1.3797
1 1/2-12 UNF	1.4100	1.4148	1.4123	1.4173	1.4148	1.4198
1 1/2-16 16UN	1.4320	1.4366	1.4344	1.4387	1.4365	1.4408
1 1/2-18 UNEF	1.4400	1.4440	1.4419	1.4460	1.4439	1.4480
1 1/2-20 20UN	1.4460	1.4498	1.4477	1.4517	1.4497	1.4537
1 1/2-28 28UN	1.4610	1.4645	1.4620	1.4660	1.4636	1.4676
1 9/16-6 6UN	1.3820	1.3921	1.3871	1.3971	1.3921	1.4021
1 9/16-8 8UN	1.4270	1.4347	1.4309	1.4384	1.4347	1.4422

All dimensions in inches.   *(Continued)*

Table 14. *(Continued)* **Recommended Hole Size Limits Before Threading for Different Lengths of Engagement, Unified Class 3B Threads.** *(Source, FED-STD-H28/2B.)*

Size, TPI, and Series	To and Including 0.33D		Above 0.33D to 0.67D		Above 0.67D through 1.5D	
	Min.	Max.	Min.	Max.	Min.	Max.
1 9/16-12 12UN	1.4720	1.4773	1.4748	1.4798	1.4773	1.4823
1 9/16-16 16UN	1.4950	1.4991	1.4969	1.5012	1.4990	1.5033
1 9/16-18 UNEF	1.5020	1.5065	1.5044	1.5085	1.5064	1.5105
1 9/16-20 20UN	1.5080	1.5123	1.5102	1.5142	1.5122	1.5162
1 5/8-6 6UN	1.4450	1.4546	1.4496	1.4596	1.4546	1.4646
1 5/8-8 8UN	1.4900	1.4972	1.4934	1.5009	1.4972	1.5047
1 5/8-12 12UN	1.5350	1.5398	1.5373	1.5423	1.5398	1.5448
1 5/8-16 16UN	1.5570	1.5616	1.5594	1.5637	1.5615	1.5658
1 5/8-18 UNEF	1.5650	1.5690	1.5669	1.5710	1.5689	1.5730
1 5/8-20 20UN	1.5710	1.5748	1.5727	1.5767	1.5747	1.5787
1 11/16-6 6UN	1.5070	1.5171	1.5121	1.5221	1.5171	1.5271
1 11/16-8 8UN	1.5520	1.5597	1.5559	1.5634	1.5597	1.5672
1 11/16-12 12UN	1.5970	1.6023	1.5998	1.6048	1.6023	1.6073
1 11/16-16 16UN	1.6200	1.6241	1.6219	1.6262	1.6240	1.6283
1 11/16-18 UNEF	1.6270	1.6315	1.6294	1.6335	1.6314	1.6355
1 11/16-20 20UN	1.6330	1.6373	1.6352	1.6392	1.6372	1.6412
1 3/4-5 UNC	1.5340	1.5455	1.5395	1.5515	1.5455	1.5575
1 3/4-6 6UN	1.5700	1.5796	1.5746	1.5846	1.5796	1.5896
1 3/4-8 8UN	1.6150	1.6222	1.6184	1.6259	1.6222	1.6297
1 3/4-12 12UN	1.6600	1.6648	1.6623	1.6673	1.6648	1.6698
1 3/4-16 16UN	1.6820	1.6866	1.6844	1.6887	1.6865	1.6908
1 3/4-20 20UN	1.6960	1.6998	1.6977	1.7017	1.6997	1.7037
1 13/16-6 6UN	1.6320	1.6421	1.6371	1.6471	1.6421	1.6521
1 13/16-8 8UN	1.6770	1.6847	1.6809	1.6884	1.6847	1.6922
1 13/16-12 12UN	1.7220	1.7273	1.7248	1.7298	1.7273	1.7323
1 13/16-16 16UN	1.7450	1.7491	1.7469	1.7512	1.7490	1.7533
1 13/16-20 20UN	1.7580	1.7623	1.7602	1.7642	1.7622	1.7662
1 7/8-6 6UN	1.6950	1.7046	1.6996	1.7096	1.7046	1.7146
1 7/8-8 8UN	1.7400	1.7472	1.7434	1.7509	1.7472	1.7547
1 7/8-12 12UN	1.7850	1.7898	1.7873	1.7923	1.7898	1.7948
1 7/8-16 16UN	1.8070	1.8116	1.8094	1.8137	1.8115	1.8158
1 7/8-20 20UN	1.8210	1.8248	1.8227	1.8267	1.8247	1.8287
1 15/16-6 6UN	1.7570	1.7671	1.7621	1.7721	1.7671	1.7771
1/15/16-8 8UN	1.8020	1.8097	1.8059	1.8134	1.8097	1.8172
1 15/16-12 12UN	1.8470	1.8523	1.8498	1.8548	1.8523	1.8573
1 15/16-16 16UN	1.8700	1.8741	1.8719	1.8762	1.8740	1.8783
1 15/16-20 20UN	1.8830	1.8873	1.8852	1.8892	1.8872	1.8912
2-4.5 UNC	1.7590	1.7727	1.7661	1.7794	1.7728	1.7861

All dimensions in inches.   *(Continued)*

Table 14. *(Continued)* **Recommended Hole Size Limits Before Threading for Different Lengths of Engagement, Unified Class 3B Threads.** *(Source, FED-STD-H28/2B.)*

Size, TPI, and Series	To and Including 0.33D		Above 0.33D to 0.67D		Above 0.67D through 1.5D	
	Min.	Max.	Min.	Max.	Min.	Max.
2-6 6UN	1.8200	1.8296	1.8246	1.8346	1.8296	1.8396
2-8 8UN	1.8650	1.8722	1.8684	1.8759	1.8722	1.8797
2-12 12UN	1.9100	1.9148	1.9123	1.9173	1.9148	1.9198
2-16 16UN	1.9320	1.9366	1.9344	1.9387	1.9365	1.9408
2-20 20UN	1.9460	1.9498	1.9477	1.9517	1.9497	1.9537
2 1/16-16 UNS	1.9950	1.9991	1.9969	2.0012	1.9990	2.0033
2 1/8-6 6UN	1.9450	1.9546	1.9496	1.9596	1.9546	1.9646
2 1/8-8 8UN	1.9900	1.9972	1.9934	2.0009	1.9972	2.0047
2 1/8-12 12UN	2.0350	2.0398	2.0373	2.0423	2.0398	2.0448
2 1/8-16 16UN	2.0570	2.0616	2.0594	2.0637	2.0615	2.0658
2 1/8-20 20UN	2.0710	2.0748	2.0727	2.0767	2.0747	2.0787
2 3/16-16 UNS	2.1200	2.1241	2.1219	2.1262	2.1240	2.1283
2 1/4-4.5 UNC	2.0090	2.0227	2.0161	2.0294	2.0228	2.0361
2 1/4-6 UN	2.0700	2.0796	2.0746	2.0846	2.0796	2.0896
2 1/2-4 UNC	2.2290	2.2444	2.2369	2.2519	2.2444	2.2594
2 3/4-4 UNC	2.4790	2.4944	2.4869	2.5019	2.4944	2.5094
3-4 UNC	2.7290	2.7444	2.7369	2.7519	2.7444	2.7594
3 1/4-4 UNC	2.9790	2.9944	2.9869	3.0019	2.9944	3.0094

All dimensions in inches.

Table 15. Drill Sizes for Metric Taps. *(Source, OSG Tap and Die, Inc.)*

Nom. size (mm)	Pitch (mm)	Basic Major Dia. (inches)	Tap Drill Size	Dec. Equiv. Tap Drill (inches)	Theo. % of Thread	Probable Oversize (inches)	Probable Hole Size (inches)	Prob. % of Thread
1.6	0.35	.0630	1.25mm	.0492	77	.0015	.0507	69
1.8	0.35	.0709	1.45mm	.0571	77	.0015	.0586	69
2	0.40	.0787	1/16	.0625	79	.0015	.0640	72
			1.60mm	.0630	77	.0017	.0647	69
			52	.0635	74	.0017	.0652	66
2.2	0.45	.0866	1.75mm	.0689	77	.0017	.0706	70
2.5	0.45	.0984	2.05mm	.0807	77	.0019	.0826	69
			46	.0810	76	.0019	.0829	67
			45	.0820	71	.0019	.0839	63
3	0.50	.1181	40	.0980	79	.0023	.1003	70
			2.5mm	.0984	77	.0023	.1007	68
			39	.0995	73	.0023	.1018	64

Note: Sizes with asterisk(*) are not standard drills.

(Continued)

Table 15. *(Continued)* **Drill Sizes for Metric Taps.** *(Source, OSG Tap and Die, Inc.)*

Nom. size (mm)	Pitch (mm)	Basic Major Dia. (inches)	Tap Drill Size	Dec. Equiv. Tap Drill (inches)	Theo. % of Thread	Probable Oversize (inches)	Probable Hole Size (inches)	Prob. % of Thread
3.5	0.60	.1378	33	.1130	81	.0026	.1156	72
			2.9mm	.1142	77	.0026	.1168	68
			32	.1160	71	.0026	.1186	63
4	0.70	.1574	3.25mm	.1280	82	.0029	.1309	74
			30	.1285	81	.0029	.1314	73
			3.3mm	.1299	77	.0029	.1328	69
4.5	0.75	.1772	3.7mm	.1457	82	.0032	.1489	74
			26	.1470	79	.0032	.1502	70
			3.75mm	.1476	77	.0032	.1508	69
			25	.1495	72	.0032	.1527	64
5	0.80	.1968	4.2mm	.1654	77	.0032	.1686	69
			19	.1660	75	.0032	.1692	68
6	1.00	.2362	10	.1935	84	.0038	.1973	76
			9	.1960	79	.0038	.1998	71
			5.0mm	.1968	77	.0038	.2006	70
			8	.1990	73	.0038	.2028	65
7	1.00	.2756	A	.2340	81	.0038	.2378	74
			15/64	.2344	81	.0038	.2382	73
			6.0mm	.2362	77	.0038	.2400	70
			B	.2380	74	.0038	.2418	66
8	1.25	.3150	6.7mm	.2638	80	.0041	.2679	74
			17/64	.2656	77	.0041	.2697	71
			6.75mm	.2657	77	.0041	.2698	71
			H	.2660	77	.0041	.2701	70
			6.8mm	.2677	74	.0041	.2718	68
8	1.00	.3150	7.0mm	.2756	77	.0041	.2797	69
			J	.2770	74	.0041	.2811	66
10	1.50	.3937	8.4mm	3307	82	.0044	.3351	76
			Q	.3320	80	.0044	.3364	75
			8.5mm	.3346	77	.0044	.3390	71
10	1.25	.3937	8.7mm	.3425	80	.0046	.3471	73
			11/32	.3438	78	.0046	.3483	71
			8.75mm	.3445	77	.0046	.3491	70
12	1.75	.4724	*10.25mm	.4035	77	.0047	.4082	72
			Y	.4040	76	.0047	.4087	71
			13/32	.4062	74	.0047	.4109	69
12	1.25	.4724	27/64*	.4219	79	.0047	.4266	72
			10.75mm	.4232	77	.0047	.4279	70
14	2	.5512	15/32	.4688	81	.0048	.4736	76
			12mm	.4724	77	.0048	.4772	72
14	1.5	.5512	12.5mm	.4921	77	.0048	.4969	71
16	2	.6299	35/64*	.5469	81	.0049	.5518	76
			14mm	.5512	77	.0049	.5561	72
16	1.5	.6299	*14.5mm	.5709	77	.0049	.5758	71
18	2.5	.7087	39/64*	.6094	78	.0050	.6144	74
			15.5mm	.6102	77	.0050	.6152	73
18	1.5	.7087	*16.5mm	.6496	77	.0050	.6546	70

Note: Sizes with asterisk(*) are not standard drills.

(Continued)

Table 15. *(Continued)* Drill Sizes for Metric Taps. *(Source, OSG Tap and Die, Inc.)*

Nom. size (mm)	Pitch (mm)	Basic Major Dia. (inches)	Tap Drill Size	Dec. Equiv. Tap Drill (inches)	Theo. % of Thread	Probable Oversize (inches)	Probable Hole Size (inches)	Prob. % of Thread
20	2.5	.7874	11/16*	.6875	78	.0050	.6925	74
			17.5mm	.6890	77	.0052	.6942	73
20	1.5	.7874	*18.5mm	.7283	77	.0052	.7335	70
22	2.5	.8661	49/64*	.7656	79	.0052	.7708	75
			19.5mm	.7677	77	.0052	.7729	73
22	1.5	.8661	*20.5mm	.8071	77	.0052	.8123	70
24	3	.9449	*21mm	.8268	77	.0059	.8327	73
			53/64	.8281	76	.0059	.8340	72
24	2	.9449	*22mm	.8661	77	.0059	.8720	71
27	3	1.0630	15/16	.9375	82	.0060	.9435	78
			*24mm	.9449	77	.0062	.9511	73
27	2	1.0630	*25mm	.9843	77	.0070	.9913	70
			63/64	.9844	77	.0070	.9914	70
30	3.5	1.1811	*26.5mm	1.0433	77			
			1-3/64	1.0469	75			
30	2	1.1811	*28mm	1.1024	77			
			1-7/64	1.1094	70			
33	3.5	1.2992	1-5/32	1.1562	80			
			*29.5mm	1.1614	77	Reaming		
			1-11/64	1.1719	71	Recommended		
33	2	1.2992	1-7/32	1.2188	79			
			*31mm	1.2205	77			
36	4	1.4173	1-1/4	1.2500	82			
			*32mm	1.2598	77			
36	3	1.4173	*33mm	1.2992	77			

Note: Sizes with asterisk(*) are not standard drills.

Table 16. Pitch Diameter Limits for Class 4H and 6H (Internal) M Profile Metric Threads.

Size (mm)	Pitch (mm)	Pitch Diameter Limits for Class of Thread		
		Min. (Basic)	Max. 4H	Max. 6H
M1.6	0.35	1.373	1.426	1.458
M1.8	0.35	1.573	1.626	1.658
M2	0.4	1.740	1.796	1.830
M2.2	0.45	1.908	1.968	2.003
M2.5	0.45	2.208	2.268	2.303
M3	0.5	2.675	2.738	2.775
M3.5	0.6	3.110	3.181	3.222
M4	0.7	3.545	3.620	3.663
M4.5	0.75	4.013	4.088	4.131
M5	0.8	4.480	4.560	4.605
M6	1	5.350	5.445	5.500
M7	1	6.350	6.445	6.500
M8	1.25	7.188	7.288	7.348
M8	1	7.350	7.445	7.500

All dimensions in millimeters.

(Continued)

Table 16. *(Continued)* Pitch Diameter Limits for Class 4H and 6H (Internal) M Profile Metric Threads.

Size (mm)	Pitch (mm)	Pitch Diameter Limits for Class of Thread		
		Min. (Basic)	Max. 4H	Max. 6H
M10	1.5	9.026	9.138	9.206
M10	1.25	9.188	9.288	9.348
M12	1.75	10.863	10.988	11.063
M12	1.25	11.188	11.300	11.368
M14	2	12.701	12.833	12.913
M14	1.5	13.026	13.144	13.216
M16	2	14.701	14.833	14.913
M16	1.5	15.026	15.144	15.216
M18	2.5	16.376	16.516	16.600
M18	1.5	17.026	17.144	17.216
M20	2.5	18.376	18.516	18.600
M20	1.5	19.026	19.144	19.216
M22	2.5	20.376	20.516	20.600
M22	1.5	21.026	21.144	21.216
M24	3	22.051	22.221	22.316
M24	2	22.701	22.841	22.925
M27	3	25.051	25.221	25.316
M27	2	25.701	25.841	25.925
M30	3.5	27.727	27.907	28.007
M30	2	28.701	28.841	28.925
M33	3.5	30.727	30.907	31.007
M33	2	31.701	31.841	31.925
M36	4	33.402	33.592	33.702
M36	3	34.051	34.221	34.316
M39	4	36.402	36.592	36.702
M39	3	37.051	37.221	37.316

All dimensions in millimeters.

Several different mathematical formulas have been advanced for predicting tap drill sizes for thread cutting taps that will result in a desired percentage of thread engagement. One thing that should always be kept in mind is that the drill will actually be forming the minor diameter of the thread. Therefore, under no circumstance should the drill diameter be in excess of the maximum minor diameter of the internal thread to be cut. Equation 1, which follows, is useful for finding the correct tap drill size for a desired percentage of thread, and Equation 2 can be used for determining the depth of thread that will be achieved with a given drill size. *One should always keep in mind* that the results predicted by these equations are based on a perfect hole, precisely the same size as the drill that produced it. Actual hole sizes will vary as a result of the condition of the drill, the machine doing the drilling, and the material being drilled. Actual percentage of thread engagement can only be determined by pin gaging the hole.

Equation 1) For ground thread taps only. Find the closest *tap drill size* for a desired percentage of thread for Unified and Metric M Profile threads. (**Table 12** in the Drilling section of this book provides a comprehensive listing of available drill sizes.)

Example for Unified Screw Threads:

Where: *S* is the size of the tap drill expressed in decimal inches

   D is the basic major diameter of the thread expressed in decimal inches

% is the desired percentage of thread engagement

tpi is the number of threads per inch.

$$S = D - ([.01299 \times \%] \div tpi)$$

Find the drill size needed for a $^1/_4$–20 tap to produce 75% of full thread.

.2500 – ([.01299 × 75] ÷ 20) = .20128" diameter hole.

Therefore, use a #7 (.2010") tap drill.

Example for Metric Screw Threads:

Where: *S* is the size of the tap drill expressed in millimeters

D is the basic major diameter of the thread expressed in millimeters

% is the desired percentage of thread engagement

P is the Pitch of the thread expressed in millimeters.

$$S = D - ([\% \times P] \div 76.98)$$

Find the drill size needed for an M5 × 0.80 tap to produce 75% of full thread.

5 – ([75 × .80] ÷ 76.98) = 4.22 mm

Therefore, use a #19 (4.216 mm / .1660") tap drill.

Equation 2) For ground thread taps only. Find the *percentage of thread engagement* that a specified tap drill will produce. (**Table 12** in the Drilling section of this book provides a comprehensive listing of available drill sizes.)

Example for Unified Screw Threads:

Where: % is the desired percentage of thread engagement

tpi is the number of threads per inch

D is the basic major diameter of the thread expressed in decimal inches

S is the size of the tap drill expressed in decimal inches.

$$\% = tpi \times ([D - S] \div .01299)$$

Find the percent of thread engagement that will result from using a #7 (.2010") drill to produce a hole for a $^1/_4$–20 tap.

20 × ([.2500 – .2010] ÷ .01299) = 75.44 %

Example for Metric Screw Threads:

Where: *S* is the size of the tap drill expressed in millimeters

D is the basic major diameter of the thread expressed in millimeters

% is the desired percentage of thread engagement

P is the pitch of the thread expressed in millimeters

$$\% = (76.980 \div P) \times (D - S)$$

Find the percent of thread engagement that will result from using a #19 (4.216 mm) drill to produce a hole for an M5 × 0.80 tap.

(76.980 ÷ .8) × (5 – 4.216) = 75.44 %

Calculating hole sizes for desired percentage of thread for forming taps

Also known as cold-forming taps, these taps do not remove metal by cutting. Instead, they displace material by plastic flow from the major diameter toward the minor diameter. Consequently, they do not produce chips, and they are extremely effective in ductile materials such as aluminum, copper, brass, leaded steels, low carbon steels, and stainless steels. However, they will not function in materials that are brittle and have low ductility. Threads created by these taps are stronger than those produced by cutting taps, and 65%

Table 17. Decimal Inch Equivalents of Standard M Profile Metric Thread Sizes.

Thread Designation	Dia. (Inches)	Pitch (Inches)	Thread Designation	Dia. (Inches)	Pitch (Inches)	Thread Designation	Dia. (Inches)	Pitch (Inches)
M1 × .25	.03937	.00984	M7 × 1.0	.27559	.03937	M18 × 2.5	.70866	.09483
M1.4 × .30	.05512	.01181	M8 × 1.25	.31496	.04921	M20 × 1.5	.78740	.05906
M1.6 × .35	.06299	.01378	M10 × 1.25	.39370	.04921	M20 × 2.5	.78740	.09483
M2 × .40	.07874	.01575	M10 × 1.5	.39370	.05906	M24 × 2	.94488	.07874
M2.5 × .45	.09843	.01772	M12 × 1.25	.47244	.04921	M24 × 3	.94488	.11811
M3 × .50	.11811	.01969	M12 × 1.75	.47244	.06890	M27 × 3	1.06299	.11811
M3.5 × .60	.13780	.02362	M14 × 1.5	.55118	.05906	M30 × 3.5	1.18110	.13780
M4 × .70	.15748	.02756	M14 × 2	.55118	.07874	M33 × 3.5	1.29921	.13780
M4.5 × .75	.17717	.02953	M16 × 1.5	.62992	.05906	M36 × 4	1.41732	.15748
M5 × .80	.19685	.03150	M16 × 2	.62992	.07874	mm × 0.03937 = inch equiv.		
M6 × 1.0	.23622	.03937	M18 × 1.5	.70866	.05906	inch × 25.4 = mm equiv.		

of thread is satisfactory for most applications. As can be seen in **Table 3**, thread forming taps for very coarse threads are not available. This is because of the difficulty in uniformly deforming the large volume of material required by coarse threads. Coarse threads down to 12 tpi can be reliably cut into most materials, but high ductility and low hardness are requirements for forming threads as coarse as 10 tpi.

Thread forming taps must be run under power and are usually effective when run at speeds of 150% to 200% in excess of thread cutting taps. Torque requirements are normally 125% more than with cutting taps, and power requirements can increase by 200% to 400%. Good lubrication and heat control can reduce these requirements. Because the tap displaces material above the mouth of the hole, countersinking before tapping is recommended. Also, since the hole diameter is actually reduced by the forming process, a larger tap drill is required for a forming tap than a cutting tap. Forming taps produce no chips, and tool life normally exceeds that of cutting taps. However, with forming taps, some imperfections can be expected at the minor diameter because the minor diameter, formed by displaced material that flows into the tap threads, is actually smaller than the diameter of the original drilled hole. With forming taps, the preferred tap drill size for a given thread designation will usually fall somewhere midway between the thread's specified major and minor diameters. These taps work best with fine pitches, and satisfactory threads coarser than 12 tpi are very difficult to achieve. **Tables 18** through **20** give recommended tap drill and forming tap sizes for Machine Screw and Fractional size threads, along with probable resulting percentage of thread. **Table 21** provides the same information for Metric M Profile threads. **Table 22** gives H limit forming tap recommendations for Machine Screw and Fractional size threads.

Since drilled holes for forming taps must be larger than holes drilled for cutting taps, the equation used for predicting percentage of full thread is unique to forming taps. Equation 3, which follows, is useful for finding the correct tap drill size for a desired percentage of thread, and Equation 4 can be used for determining the depth of thread that will be achieved with a given drill size. In all drilling operations, the true hole size will be affected by the condition of the drill and the material being machined. These equations can be used to predict percentage of thread for a given drill size, but the actual percentage of thread achieved can only be determined by pin gaging the drilled hole. The tables in this section are based on probable actual hole size and should be consulted for guidance in selecting a drill size for a particular thread.

Equation 3) For forming taps only. Find the closest *tap drill size* for a desired percentage of thread for Unified and Metric M Profile threads. (**Table 12** in the Drilling section of this book provides a comprehensive listing of available drill sizes.)

Example for Unified Screw Threads:

Where: S is the size of the tap drill expressed in decimal inches

D is the basic major diameter of the thread expressed in decimal inches

% is the desired percentage of thread engagement

tpi is the number of threads per inch.

$$S = D - ([.006799 \times \%] \div tpi)$$

Find the drill size needed for a $1/4$–20 tap to produce 70% of full thread.

.2500 – ([.006799 × .70] ÷ 20) = .2262" diameter hole.

Therefore, use a 5.75 mm (.2264") tap drill.

Example for Metric Screw Threads:

Where: S is the size of the tap drill expressed in millimeters

D is the basic major diameter of the thread expressed in millimeters

% is the desired percentage of thread engagement

P is the pitch of the thread expressed in millimeters.

$$S = D - ([\% \times P] \div 147.06)$$

Find the drill size needed for an M5 × 0.80 tap to produce 70% of full thread.

5 – ([70 × .8] ÷ 147.06) = 4.619 mm

Therefore, use a #14 (4.623 mm / .1820") tap drill.

Equation 4) For forming taps only. Find the *percentage of thread engagement* that a specified tap drill will produce. (**Table 12** in the Drilling section of this book provides a comprehensive listing of available drill sizes.)

Example for Unified Screw Threads:

Where: % is the desired percentage of thread engagement

tpi is the number of threads per inch

D is the basic major diameter of the thread expressed in decimal inches

S is the size of the tap drill expressed in decimal inches.

$$\% = tpi \times ([D - S] \div .006799)$$

Find the percent of thread engagement that will result from using a 5.75 mm drill to produce a hole for a $1/4$–20 tap.

5.75 mm = .2264"

20 × ([.2500 – .2264] ÷ .006799) = 69.42 %

Example for Metric Screw Threads:

Where: % is the desired percentage of thread engagement

D is the basic major diameter of the thread expressed in millimeters

S is the size of the tap drill expressed in millimeters

P is the pitch of the thread expressed in millimeters.

$$\% = ([D - S] \times 147.06) \div P$$

Find the percent of thread engagement that will result from using a #14 (4.623 mm) drill to produce a hole for an M5 × 0.80 tap.

([5 – 4.623] × 147.06) ÷ .8 = 69.30 %

(Text continued on p. 1265)

Table 18. Forming Tap Drill Sizes for Class 2B Fit Unified Threads. *(Source, New England Tap Corp.)*

Tap			75% Thread		65% Thread		55% Thread	
Size	UNC	UNF	Theor. Size	Near Drill	Theor. Size	Near Drill	Theor. Size	Near Drill
0		80	.0543	#54	.0549	#54	.0555	1.4mm
1	64		.0657	1.7mm	.0664	#51	.0672	#51
1		72	.0666	1.7mm	.0672	#51	.0679	#51
2	56		.0776	5/64	.0783	#47	.0792	#47
2		64	.0788	2mm	.0795	2.05mm	.0803	2.05mm
3	48		.0895	#43	.0904	2.3mm	.0914	2.3mm
3		56	.0905	2.3mm	.0913	2.3mm	.0922	2.35mm
4	40		.1000	#39	.1010	#38	.1023	2.6mm
4		48	.1021	2.6mm	.1030	2.6mm	.1040	#37
5	40		.1129	#33	.1140	2.9mm	.1152	2.9mm
5		44	.1141	2.9mm	.1149	2.9mm	.1161	#32
6	32		.1227	3.1mm	.1240	1/8	.1256	1/8
6		40	.1259	1/8	.1269	3.2mm	.1282	#30
8	32		.1486	#25	.1499	3.8mm	.1515	#24
8		36	.1504	#24	.1516	#24	.1530	#23
10	24		.1692	4.3mm	.1710	11/64	.1730	#17
10		32	.1746	4.4mm	.1760	#16	.1776	4.5mm
12	24		.1951	#9	.1969	5mm	.1990	#8
12		28	.1983	#8	.1998	5.1mm	.2016	#7
1/4	20		.2248	5.7mm	.2269	5.75mm	.2294	5.8mm
1/4		28	.2323	5.9mm	.2338	A	.2357	15/64
5/16	18		.2844	7.25mm	.2868	7.3mm	.2895	L
5/16		24	.2916	7.4mm	.2934	M	.2955	7.5mm
3/8	16		.3432	11/32	.3459	8.8mm	.3490	S
3/8		24	.3541	9mm	.3559	9mm	.3581	T
7/16	14		.4007	X	.4040	Y	.4076	13/32
7/16		20	.4122	Z	.4144	10.5mm	.4169	10.5mm
1/2	13		.4605	15/32	.4638	15/32	.4677	15/32
1/2		20	.4747	12mm	.4769	12mm	.4795	31/64
9/16	12		.5193	33/64	.5230	17/32	.5273	17/32
9/16		18	.5343	17/32	.5368	17/32	.5396	17/32
5/8	11		.5780	37/64	.5819	37/64	.5866	37/64
5/8		18	.5968	19/32	.5992	19/32	.6021	19/32
3/4	10		.6980	45/64	.7024	45/64	.7075	45/64
3/4		16	.7182	23/32	.7209	23/32	.7242	23/32
7/8	9		.818	13/16	.823	.823	.829	.823
7/8		14	.839	21.25mm	.843	27/32	.845	27/32

(Continued)

Table 18. *(Continued)* **Forming Tap Drill Sizes for Class 2B Fit Unified Threads.** *(Source, New England Tap Corp.)*

Tap			75% Thread		65% Thread		55% Thread	
Size	UNC	UNF	Theor. Size	Near Drill	Theor. Size	Near Drill	Theor. Size	Near Drill
1	8		.935	15/16	.942	15/16	.948	15/16
1		12	.959	.963	.963	.963	.967	.963

Note: Forming taps are very effective in ductile materials up to 250 BHN, such as aluminum, copper, low carbon steel, leaded steel, and some stainless steel. They must be power driven at speeds up to 150% to 200% of the speeds for cutting taps. Because forming taps reduce the drilled hole diameter by displacing material from the major diameter toward the minor diameter, a tap drill larger than that which is required for a cutting tap must be used. The tap drill sizes on this Table are for use with forming taps only, and should never be used for cutting taps.

Table 19. Forming Tap Drill Sizes for Class 3B Fit Unified Threads. *(Source, New England Tap Corp.)*

Tap			75% Thread		65% Thread		55% Thread	
Size	UNC	UNF	Theor. Size	Near Drill	Theor. Size	Near Drill	Theor. Size	Near Drill
0		80	.0540	1.35mm	.0546	#54	.0552	1.4mm
1	64		.0654	1.65mm	.0661	1.7mm	.0669	1.7mm
1		72	.0663	1.7mm	.0669	#51	.0676	#51
2	56		.0773	5/64	.0780	5/64	.0789	2mm
2		64	.0784	#47	.0791	2.05mm	.0799	2.05mm
3	48		.0886	2.25mm	.0895	#43	.0905	2.3mm
3		56	.0902	2.3mm	.0910	2.3mm	.0919	2.35mm
4	40		.0995	#39	.1005	#38	.1018	2.6mm
4		48	.1017	2.6mm	.1026	2.6mm	.1036	#37
5	40		.1125	#33	.1136	2.9mm	.1148	2.9mm
5		44	.1137	2.9mm	.1145	2.9mm	.1157	#32
6	32		.1222	3.1mm	.1235	3.1mm	.1251	1/8
6		40	.1255	1/8	.1265	3.2mm	.1278	3.25mm
8	32		.1481	3.8mm	.1494	#25	.1510	#24
8		36	.1500	#25	.1512	#24	.1526	#24
10	24		.1686	4.3mm	.1704	#18	.1724	#17
10		32	.1741	4.4mm	.1755	4.4mm	.1771	#16
12	24		.1941	5mm	.1959	#9	.1980	#8
12		28	.1978	#8	.1993	#8	.2011	#7
1/4	20		.2242	5.7mm	.2263	5.75mm	.2288	5.8mm
1/4		28	.2322	5.9mm	.2337	A	.2356	15/64
5/16	18		.2837	7.2mm	.2861	7.3mm	.2888	7.3mm
5/16		24	.2910	7.4mm	.2928	M	.2949	M
3/8	16		.3425	8.7mm	.3452	8.8mm	.3483	S
3/8		24	.3535	9mm	.3553	9mm	.3575	T
7/16	14		.3999	X	.4032	Y	.4068	13/32

(Continued)

Table 19. (Continued) Forming Tap Drill Sizes for Class 3B Fit Unified Threads. (Source, New England Tap Corp.)

Tap			75% Thread		65% Thread		55% Thread	
Size	UNC	UNF	Theor. Size	Near Drill	Theor. Size	Near Drill	Theor. Size	Near Drill
7/16		20	.4116	Z	.4138	10.5mm	.4163	10.5mm
1/2	13		.4596	15/32	.4629	15/32	.4668	15/32
1/2		20	.4740	12mm	.4762	12mm	.4788	31/64
9/16	12		.5185	33/64	.5222	17/32	.5265	17/32
9/16		18	.5335	17/32	.5360	17/32	.5388	17/32
5/8	11		.5771	37/64	.5810	37/64	.5857	37/64
5/8		18	.5961	19/32	.5985	19/32	.6014	19/32
3/4	10		.6970	45/64	.7014	45/64	.7065	45/64
3/4		16	.7174	23/32	.7201	23/32	.7234	23/32
7/8	9		.818	13/16	.823	.823	.829	.823
7/8		14	.839	21.25mm	.843	27/32	.845	27/32
1	8		.935	15/16	.942	15/16	.948	15/16
1		12	.959	.963	.963	.963	.967	.963

Note: Forming taps are very effective in ductile materials up to 250 BHN, such as aluminum, copper, low carbon steel, leaded steel, and some stainless steel. They must be power driven at speeds up to 150% to 200% of the speeds for cutting taps. Because forming taps reduce the drilled hole diameter by displacing material from the major diameter toward the minor diameter, a tap drill larger than that which is required for a cutting tap must be used. The tap drill sizes on this Table are for use with forming taps only, and should never be used for cutting taps.

Table 20. Forming Tap Drill Sizes for Class 2 Fit Unified Threads. (Source, New England Tap Corp.)

Tap			75% Thread		65% Thread		55% Thread	
Size	UNC	UNF	Theor. Size	Near Drill	Theor. Size	Near Drill	Theor. Size	Near Drill
0		80	.0540	1.35mm	.0546	#54	.0552	1.4mm
1	64		.0654	1.65mm	.0661	1.7mm	.0669	1.7mm
1		72	.0662	1.7mm	.0668	#51	.0675	#51
2	56		.0772	5/64	.0779	5/64	.0788	2mm
2		64	.0784	#47	.0791	2.05mm	.0799	2.05mm
3	48		.0886	2.25mm	.0895	#43	.0905	2.3mm
3		56	.0901	2.3mm	.0909	2.3mm	.0918	2.35mm
4	40		.0995	#39	.1005	#38	.1018	2.6mm
4		48	.1016	2.6mm	.1025	2.6mm	.1035	#37
5	40		.1124	#33	.1135	2.9mm	.1147	2.9mm
5		44	.1137	2.9mm	.1145	2.9mm	.1157	#32
6	32		.1222	3.1mm	.1235	3.1mm	.1251	1/8
6		40	.1254	1/8	.1264	3.2mm	.1277	3.25mm
8	32		.1481	3.8mm	.1494	#25	.1510	#24
8		36	.1499	#25	.1511	#24	.1525	#24
10	24		.1687	4.3mm	.1705	#18	.1725	#17
10		32	.1740	4.4mm	.1754	4.4mm	.1770	#16

(Continued)

Table 20. *(Continued)* **Forming Tap Drill Sizes for Class 2 Fit Unified Threads.** *(Source, New England Tap Corp.)*

Tap			75% Thread		65% Thread		55% Thread	
Size	UNC	UNF	Theor. Size	Near Drill	Theor. Size	Near Drill	Theor. Size	Near Drill
12	24		.1941	5mm	.1959	#9	.1980	#8
12		28	.1978	#8	.1993	#8	.2011	#7
1/4	20		.2242	5.7mm	.2263	5.75mm	.2288	5.8mm
1/4		28	.2317	5.9mm	.2332	A	.2351	15/64
5/16	18		.2838	7.2mm	.2862	7.3mm	.2889	7.3mm
5/16		24	.2909	7.4mm	.2927	M	.2948	M
3/8	16		.3426	8.7mm	.3453	8.8mm	.3484	S
3/8		24	.3533	9mm	.3551	9mm	.3573	T
7/16	14		.4001	X	.4034	Y	.4070	13/32
7/16		20	.4113	Z	.4135	10.5mm	.4160	10.5mm
1/2	13		.4598	15/32	.4631	15/32	.4670	15/32
1/2		20	.4737	12mm	.4759	12mm	.4785	.3464
9/16	12		.5187	33/64	.5224	17/32	.5267	17/32
9/16		18	.5334	17/32	.5359	17/32	.5387	17/32
5/8	11		.5774	37/64	.5813	37/64	.5860	37/64
5/8		18	.5959	19/32	.5983	19/32	.6012	19/32
3/4	10		.6973	45/64	.7017	45/64	.7068	45/64
3/4		16	.7172	23/32	.7199	23/32	.7232	23/32
7/8	9		.818	13/16	.823	.823	.829	.823
7/8		14	.839	21.25mm	.843	27/32	.845	27/32
1	8		.935	15/16	.942	15/16	.948	15/16
1		12	.959	.963	.963	.963	.967	24.5mm

Note: Forming taps are very effective in ductile materials up to 250 BHN, such as aluminum, copper, low carbon steel, leaded steel, and some stainless steel. They must be power driven at speeds up to 150% to 200% of the speeds for cutting taps. Because forming taps reduce the drilled hole diameter by displacing material from the major diameter toward the minor diameter, a tap drill larger than that which is required for a cutting tap must be used. The tap drill sizes on this Table are for use with forming taps only, and should never be used for cutting taps.

Table 21. Forming Tap Drill Sizes for Metric Screw Threads. *(Source, New England Tap Corp.)*

Tap Size	Tap Drill Size	Decimal Equiv. of Tap Drill (Inches)	Prob. % of Thread Engmt.
M2×.35	1.39mm	.0547	72
	1.41mm	.0555	64
	1.43mm	.0563	55
M2×.4	1.76mm	.0693	74
	50	.0700	67
	1.81mm	.0713	55
M3×.45	2.24mm	.0882	71
	43	.0890	65
	2.29mm	.0902	55
M3×.5	2.7mm	.1063	75
	2.75mm	.1083	61
M4×.6	3.15mm	.1240	75
	3.18mm	.1252	67
	3.22mm	.1268	57
M4×.7	3.6mm	.1417	74
	3.65mm	.1437	64
	3.68mm	.1449	57
M5×.75	4.06mm	.1598	77
	4.1mm	.1614	69
	4.15mm	.1634	59
M5×.8	4.55mm	.1791	73
	4.6mm	.1811	64
	4.65mm	.1831	55
M6×1	5.45mm	.2146	73
	5.5mm	.2165	66
	7/32	.2188	57

Tap Size	Tap Drill Size	Decimal Equiv. of Tap Drill (Inches)	Prob. % of Thread Engmt.
M7×1	6.45mm	.2539	72
	6.5mm	.2559	65
	6.55mm	.2579	58
M8×1.25	7.3mm	.2874	75
	L	.2900	67
	7.45mm	.2933	57
M10×1.25	9.3mm	.3661	74
	U	.3680	69
	9.45mm	.3720	56
M10×1.5	9.15mm	.3602	77
	9.25mm	.3642	67
	9.35mm	.3681	57
M12×1.25	11.3mm	.4449	75
	11.35mm	.4469	67
	11.4mm	.4488	57
M12×1.75	11.0mm	.4331	74
	7/16	.4375	64
	11.25mm	.4429	57
M14×1.5	13.2mm	.5197	77
	13.25mm	.5217	69
	13.3mm	.5236	59
M14×2	12.9mm	.5079	73
	13.0mm	.5118	64
	33/64	.5156	55

Tap Size	Tap Drill Size	Decimal Equiv. of Tap Drill (Inches)	Prob. % of Thread Engmt.
M16×1.5	15.1mm	.5945	79
	15.2mm	.5984	69
	15.3mm	.6024	60
M16×2	14.85mm	.5846	78
	15.0mm	.5906	67
	19/32	.5938	61
M18×1.5	17.2mm	.6772	69
	17.3mm	.6811	59
M18×2.5	21/32	.6563	73
	16.8mm	.6614	65
	16.9mm	.6654	59
M20×1.5	19.1mm	.7520	78
	19.2mm	.7559	68
	19.3mm	.7598	58
M20×2.5	18.6mm	.7323	76
	18.75mm	.7382	67
	18.9mm	.7441	58
M24×1.5	23.2mm	.9134	66
	23.25mm	.9154	61
M24×3	22.4mm	.8819	73
	22.5mm	.8858	68
	57/64	.8906	62

The percent of thread engagement in this table is based upon the probable hole size the drill will cut. The actual hole size may vary as a result of the condition of the drill, machine and material being drilled. The actual percent of thread engagement may be determined by pin gaging the hole.

Note: Forming taps are very effective in ductile materials up to 250 BHN, such as aluminum, copper, low carbon steel, leaded steel, and some stainless steel. They must be power driven at speeds up to 150% to 200% of the speeds for cutting taps. Because forming taps reduce the drilled hole diameter by displacing material from the minor diameter toward the major diameter, a tap drill larger than that which is required for a cutting tap must be used. The tap drill sizes on this Table are for use with forming taps only, and should never be used for cutting taps.

Table 22. Forming Tap Recommendations for Classes 2, 2B, and 3B Unified Machine Screw and Fractional Size Screw Threads. *(Source, Fastcut Tool Corp.)*

Size	Threads Per Inch		Recommended Forming Tap for Class of Thread			Pitch Diameter Limits			
	UNC	UNF	Class 2	Class 2B	Class 3B	Min. All Classes (Basic)	Max. Class 2	Max. Class 2B	Max. Class 3B
0		80	G H2	B H3	G H2	.0519	.0536	.0542	.0536
1	64		G H2	G H3	G H2	.0629	.0648	.0655	.0648
1		72	G H2	G H3	G H2	.0640	.0658	.0665	.0659
2	56		G H2	G H3	G H2	.0744	.0764	.0772	.0765
2		64	G H2	G H3	G H2	.0759	.0778	.0786	.0779
3	48		G H2	G H3	G H2	.0855	.0877	.0885	.0877
3		56	G H2	G H3	G H2	.0874	.0894	.0902	.0895
4	40		G H3	G H5	G H3	.0958	.0982	.0991	.0982
4		48	G H3	G H5	G H3	.0985	.1007	.1016	.1008
5	40		G H3	G H5	G H3	.1088	.1112	.1121	.1113
5		44	G H3	G H5	G H3	.1102	.1125	.1134	.1126
6	32		G H3	G H5	G H3	.1177	.1204	.1214	.1204
6		40	G H3	G H5	G H3	.1218	.1242	.1252	.1243
8	32		G H3	G H5	G H3	.1437	.1464	.1475	.1465
8		36	G H3	G H5	G H3	.1460	.1485	.1496	.1487
10	24		G H4	G H6	G H4	.1629	.1662	.1672	.1661
10		32	G H4	G H6	G H4	.1697	.1724	.1736	.1726
12	24		G H4	G H6	G H4	.1889	.1922	.1933	.1922
12		28	G H4	G H6	G H4	.1928	.1959	.1970	.1959
1/4	20		G H4	G H6	G H4	.2175	.2211	.2223	.2211
1/4		28	G H4	G H6	G H4	.2268	.2299	.2311	.2300
5/16	18		G H5	G H7	G H5	.2764	.2805	.2817	.2803
5/16		24	G H5	G H7	G H5	.2854	.2887	.2902	.2890
3/8	16		G H5	G H7	G H5	.3344	.3389	.3401	.3387
3/8		24	G H5	G H7	G H5	.3479	.3512	.3528	.3516
7/16	14		G H5	G H8	G H5	.3911	.3960	.3972	.3957
7/16		20	G H5	G H8	G H5	.4050	.4086	.4104	.4091
1/2	13		G H5	G H8	G H5	.4500	.4552	.4565	.4548
1/2		20	G H5	G H8	G H5	.4675	.4711	.4731	.4717
9/16	12		G H7	G H10	G H7	.5084	.5140	.5152	.5135
9/16		18	G H7	G H10	G H7	.5264	.5305	.5323	.5308
5/8	11		G H7	G H10	G H7	.5660	.5719	.5732	.5714
5/8		18	G H7	G H10	G H7	.5889	.5930	.5949	.5934
3/4	10		G H7	G H10	G H7	.6850	.6914	.6927	.6907
3/4		16	G H7	G H10	G H7	.7094	.7139	.7159	.7143
7/8	9		G H9	G H12	G H9	.8028	.8098	.8110	.8089
7/8		14	G H9	G H12	G H9	.8286	.8335	.8356	.8339

(Continued)

Table 22. *(Continued)* **Forming Tap Recommendations for Classes 2, 2B, and 3B Unified Machine Screw and Fractional Size Screw Threads.** *(Source, Fastcut Tool Corp.)*

Size	Threads Per Inch		Recommended Forming Tap For Class of Thread			Pitch Diameter Limits			
	UNC	UNF	Class 2	Class 2B	Class 3B	Min. All Classes (Basic)	Max. Class 2	Max. Class 2B	Max. Class 3B
1	8		G H9	G H12	G H9	.9188	.9264	.9276	.9254
1		12	G H9	G H12	G H9	.9459	.9515	.9535	.9516

The above recommended taps normally produce the Class of Thread indicated in average materials when used with reasonable care. However, if the tap specified does not give a satisfactory gage fit in the work, a choice of some other limit tap will be necessary. All of the H-Limits shown will produce a class 2B fit.

Percentage of thread in large holes

Even though most types of taps are readily available only up to 1 $\frac{1}{2}$" or 36 mm in diameter, larger sizes are available through tap manufacturers. Sizes up to 3–16 NS are readily obtainable in taper, plug, and bottoming configurations for use on large holes. Holes in excess of this diameter are usually threaded by chasing or by turning. Larger holes are usually bored, and **Table 23** provides recommended bore sizes and resulting percentage of thread engagement for Classes 1B, 2B, and 3B UNC and UNF Unified threads sized between 1 $\frac{1}{4}$–7 and 6–16.

Interference-fit threads

Recommended tap drill sizes for Class 5 interference-fit threads can be found in Table 10 on page 1461.

Miniature taps

Miniature taps require special attention during setup. Tap and target hole must be carefully aligned to prevent breakage. Once in the hole, the taps should be rotated forward only a fraction of a turn, then backed off to break and clear the chip. 75% of thread engagement should be considered the maximum allowable for these taps, and that should only be attempted when producing short threaded holes in soft materials. **Table 24** provides recommended tap drill sizes for three lengths of thread engagement for cut threads, and a suggested drill diameter when using forming taps. Dimensions of commonly available miniature plug and forming taps are given in **Table 25**.

Speeds and feeds for tapping

For additional information about speeds and feeds for forming taps, see the above section titled "Calculating hole sizes for desired percentage of thread for forming taps." As with all machining operations, the rigidity of the machine and workpiece is essential to assure repeatable performance in tapping. Since tapping, unlike drilling, is subject to error on two different planes, it is especially important to follow prescribed parameters. In tapping as in drilling, if the machine and workpiece are not absolutely rigid, hole size will be affected. However, in tapping, excessive force on the cutting tool can elongate and distort the pitch diameter of the thread. Hole depth also influences tapping speeds, and a recommended reduction of speed of 5% should be made for every 100% increase in hole depth. For example, if a $\frac{1}{4}$" deep hole being tapped at 40 sfm is increased in depth to 1", it is recommended that the speed should be decreased by 20% to 32 sfm. Achieving a desired percentage of thread engagement also affects speed. Obviously, with a given amount of power, tapping for 60% of thread can be done at higher speed than tapping for 80% of thread. These and other variables make tapping a challenging operation.

Table 39 provides tap speed ranges for various materials. The lower range should be adhered to when tapping deeper holes or when tapping for high percentage of thread engagement, and the higher ranges are to be used when tapping shallow holes with low percentage of thread requirements. Care should be exercised when exiting from holes. Often, the tap may be reversed at up to twice the cutting speed, but high speeds may result in the tap recutting as it is exited. This not only distorts the tapped thread, it also dramatically shortens the tap's life. The selection of the proper coolant is important for tool life and surface finish. RPM for achieving desired cutting speeds in sfm for most Unified and Metric taps can be found in **Tables 40** and **41**. Recommended coolants for a variety of workpiece materials are shown in **Table 42**, **Table 43** is a guide for selecting coatings that act as hardening treatments for high speed steel taps. The selection of the proper coating can greatly enhance tap performance in different workpiece materials. Different specialized coatings are offered by individual manufacturers, so it is advisable to consult with your supplier to determine which coatings are available. *(Text continued on p. 1290)*

Table 23. Recommended Bores for 1B, 2B, and 3B Unified Threads, 1 1/4" - 6".
(Source, Landis Threading Systems.)

Size and TPI	UNC and UNF Classes 1B, 2B				UNC and UNF Class 3B			
	Min. Hole Size	% of Thread[a]	Max. Hole Size[b]	% of Thread[a]	Min. Hole Size	% of Thread[a]	Max. Hole Size[b]	% of Thread[a]
1 1/4-7	1.095	83.5	1.123	68.4	1.0950	83.5	1.1125	74.1
1 1/4-8	1.115	83.1	1.140	67.7	1.1150	83.1	1.1297	74.1
1 1/4-12	1.160	83.1	1.178	66.5	1.1600	83.1	1.1698	74.1
1 1/4-16	1.182	83.8	1.196	66.5	1.1820	83.8	1.1908	72.9
1 5/16-12	1.222	83.6	1.240	67.0	1.2220	83.6	1.2323	74.1
1 5/16-16	1.245	83.1	1.259	65.9	1.2450	83.1	1.2533	72.9
1 5/16-18	1.252	83.8	1.265	65.8	1.2520	83.8	1.2605	72.1
1 3/8-6	1.195	83.1	1.225	69.3	1.1950	83.1	1.2146	74.1
1 3/8-8	1.240	83.1	1.265	67.7	1.2400	83.1	1.2547	74.1
1 3/8-12	1.285	83.1	1.303	66.5	1.2850	83.1	1.2948	74.1
1 3/8-16	1.307	83.8	1.321	66.5	1.3070	83.8	1.3158	72.9
1 7/16-12	1.347	83.6	1.365	67.0	1.3470	83.6	1.3573	74.1
1 7/16-16	1.370	83.1	1.384	65.9	1.3070	83.8	1.3158	72.9
1 1/2-6	1.320	83.1	1.350	69.3	1.3200	83.1	1.3396	74.1
1 1/2-8	1.365	83.1	1.390	67.7	1.3650	83.1	1.3797	74.1
1 1/2-12	1.410	83.1	1.428	66.5	1.4100	83.1	1.4198	74.1
1 1/2-16	1.432	83.8	1.446	66.5	1.4320	83.8	1.4408	72.9
1 9/16-16	1.495	83.1	1.509	65.9	1.4950	83.1	1.5033	72.9
1 5/8-8	1.490	83.1	1.515	67.7	1.4900	83.1	1.5047	74.1
1 5/8-12	1.535	83.1	1.553	66.5	1.5350	83.1	1.5448	74.1
1 5/8-16	1.557	83.8	1.571	66.5	1.5570	83.8	1.5658	72.9
1 11/16-16	1.620	83.1	1.634	65.9	1.6200	83.1	1.6283	72.9
1 3/4-5	1.534	83.1	1.568	70.1	1.5340	83.1	1.5575	74.1

[a]Based on values as rounded off in preceding column.
[b]Based on length of engagement equal to the nominal diameter.

(Continued)

Table 23. *(Continued)* **Recommended Bores for 1B, 2B, and 3B Unified Threads, 1 1/4" - 6".**
(Source, Landis Threading Systems.)

Size and TPI	UNC and UNF Classes 1B, 2B				UNC and UNF Class 3B			
	Min. Hole Size	% of Thread[a]	Max. Hole Size[b]	% of Thread[a]	Min. Hole Size	% of Thread[a]	Max. Hole Size[b]	% of Thread[a]
1 3/4-8	1.615	83.1	1.640	67.7	1.6150	83.1	1.6297	74.1
1 3/4-12	1.660	83.1	1.678	66.5	1.6600	83.1	1.6698	74.1
1 3/4-16	1.682	83.8	1.696	66.5	1.6820	83.8	1.6908	72.9
1 13/16-16	1.745	83.1	1.759	65.9	1.7450	83.1	1.7533	72.9
1 7/8-8	1.740	83.1	1.765	67.7	1.7400	83.1	1.7547	74.1
1 7/8-12	1.785	83.1	1.803	66.5	1.7850	83.1	1.7948	74.1
1 15/16-16	1.870	83.1	1.884	65.9	1.8700	83.1	1.8783	72.9
2-4 1/2	1.759	83.5	1.795	71.0	1.7590	83.5	1.7861	74.1
2-8	1.865	83.1	1.890	67.7	1.8650	83.1	1.8797	74.1
2-12	1.910	83.1	1.928	66.5	1.9100	83.1	1.9198	74.1
2-16	1.932	83.8	1.946	66.5	1.9320	83.8	1.9408	72.9
2 1/16-16	1.995	83.1	2.009	65.9	1.9950	83.1	2.0033	72.9
2 1/8-8	1.990	83.1	2.015	67.7	1.9900	83.1	2.0047	74.1
2 1/8-12	2.035	83.1	2.053	66.5	2.0350	83.1	2.0448	74.1
2 1/8-16	2.057	83.8	2.071	66.5	2.0570	83.8	2.0658	72.9
2 3/16-16	2.120	83.1	2.134	65.9	2.1200	83.1	2.1283	72.9
2 1/4-4 1/2	2.009	83.5	2.045	71.0	2.0090	83.5	2.0361	74.1
2 1/4-8	2.115	83.1	2.140	67.7	2.1150	83.1	2.1297	74.1
2 1/4-12	2.160	83.1	2.178	66.5	2.1600	83.1	2.1698	74.1
2 1/4-16	2.182	83.8	2.196	66.5	2.1820	83.8	2.1908	72.9
2 5/16-16	2.245	83.1	2.259	65.9	2.2450	83.1	2.2533	72.9
2 3/8-12	2.285	83.1	2.303	66.5	2.2850	83.1	2.2948	74.1
2 3/8-16	2.307	83.8	2.321	66.5	2.3070	83.8	2.3158	72.9
2 7/16-16	2.370	83.1	2.384	65.9	2.3700	83.1	2.3783	72.9
2 1/2-4	2.229	83.4	2.267	71.7	2.2290	83.4	2.2594	74.1
2 1/2-8	2.365	83.1	2.390	67.7	2.3650	83.1	2.3797	74.1
2 1/2-12	2.410	83.1	2.428	66.5	2.4100	83.1	2.4198	74.1
2 1/2-16	2.432	83.8	2.446	66.5	2.4320	83.8	2.4408	72.9
2 5/8-12	2.535	83.1	2.553	66.5	2.5350	83.1	2.5448	74.1
2 5/8-16	2.557	83.8	2.571	66.5	2.5570	83.8	2.5658	72.9
2 3/4-4	2.479	83.4	2.517	71.7	2.4790	83.4	2.5094	74.1
2 3/4-8	2.615	83.1	2.640	67.7	2.6150	83.1	2.6297	74.1
2 3/4-12	2.660	83.1	2.678	66.5	2.6600	83.1	2.6698	74.1
2 3/4-16	2.682	83.8	2.696	66.5	2.6820	83.8	2.6908	72.9
2 7/8-12	2.785	83.1	2.803	66.5	2.7850	83.1	2.7948	74.1

(Continued)

[a] Based on values as rounded off in preceding column.
[b] Based on length of engagement equal to the nominal diameter.

Table 23. *(Continued)* **Recommended Bores for 1B, 2B, and 3B Unified Threads, 1 1/4" - 6".**
(Source, Landis Threading Systems.)

Size and TPI	UNC and UNF Classes 1B, 2B				UNC and UNF Class 3B			
	Min. Hole Size	% of Thread[a]	Max. Hole Size[b]	% of Thread[a]	Min. Hole Size	% of Thread[a]	Max. Hole Size[b]	% of Thread[a]
2 7/8-16	2.807	83.8	2.821	66.5	2.8070	83.8	2.8158	72.9
3-8	2.865	83.1	2.890	67.7	2.8650	83.1	2.8797	74.1
3-12	2.910	83.1	2.928	66.5	2.9100	83.1	2.9198	74.1
3-16	2.932	83.8	2.946	66.5	2.9320	83.8	2.9408	72.9
3 1/8-12	3.035	83.1	3.053	66.5	3.0350	83.1	3.0448	74.1
3 1/8-16	3.057	83.8	3.071	66.5	3.0570	83.8	3.0658	72.9
3 1/4-4	2.979	83.4	3.017	71.7	2.9790	83.4	3.0094	74.1
3 1/4-8	3.115	83.1	3.140	67.7	3.1150	83.1	3.1297	74.1
3 1/4-12	3.160	83.1	3.178	66.5	3.1600	83.1	3.1698	74.1
3 1/4-16	3.182	83.8	3.196	66.5	3.1820	83.8	3.1908	72.9
3 3/8-12	3.285	83.1	3.303	66.5	3.285	83.1	3.303	66.5
3 3/8-16	3.307	83.8	3.321	66.5	3.307	83.8	3.321	66.5
3 1/2-4	3.229	83.4	3.267	71.7	3.229	83.4	3.267	71.7
3 1/2-8	3.365	83.1	3.390	67.7	3.365	83.1	3.390	67.7
3 1/2-12	3.410	83.1	3.428	66.5	3.410	83.1	3.428	66.5
3 1/2-16	3.432	83.8	3.446	66.5	3.432	83.8	3.446	66.5
3 5/8-12	3.535	83.1	3.553	66.5	3.535	83.1	3.553	66.5
3 5/8-16	3.557	83.8	3.571	66.5	3.557	83.8	3.571	66.5
3 3/4-4	3.479	83.4	3.517	71.7	3.479	83.4	3.517	71.7
3 3/4-8	3.615	83.1	3.640	67.7	3.615	83.1	3.640	67.7
3 3/4-12	3.660	83.1	3.678	66.5	3.660	83.1	3.678	66.5
3 3/4-16	3.682	83.8	3.696	66.5	3.682	83.8	3.696	66.5
3 7/8-12	3.785	83.1	3.803	66.5	3.785	83.1	3.803	66.5
3 7/8-16	3.807	83.8	3.821	66.5	3.807	83.8	3.821	66.5
4-4	3.729	83.4	3.767	71.7	3.729	83.4	3.767	71.7
4-8	3.865	83.1	3.890	67.7	3.865	83.1	3.890	67.7
4-12	3.910	83.1	3.928	66.5	3.910	83.1	3.928	66.5
4-16	3.932	83.8	3.946	66.5	3.932	83.8	3.946	66.5
4 1/4-8	4.115	83.1	4.140	67.7	4.115	83.1	4.140	67.7
4 1/4-12	4.160	83.1	4.178	66.5	4.160	83.1	4.178	66.5
4 1/4-16	4.182	83.8	4.196	66.5	4.182	83.8	4.196	66.5
4 1/2-8	4.365	83.1	4.390	67.7	4.365	83.1	4.390	67.7
4 1/2-12	4.410	83.1	4.428	66.5	4.410	83.1	4.428	66.5
4 1/2-16	4.432	83.8	4.446	66.5	4.432	83.8	4.446	66.5
4 3/4-8	4.615	83.1	4.640	67.7	4.615	83.1	4.640	67.7

[a]Based on values as rounded off in preceding column.
[b]Based on length of engagement equal to the nominal diameter.

(Continued)

Table 23. *(Continued)* **Recommended Bores for 1B, 2B, and 3B Unified Threads, 1 1/4" - 6".** *(Source, Landis Threading Systems.)*

Size and TPI	UNC and UNF Classes 1B, 2B				UNC and UNF Class 3B			
	Min. Hole Size	% of Thread[a]	Max. Hole Size[b]	% of Thread[a]	Min. Hole Size	% of Thread[a]	Max. Hole Size[b]	% of Thread[a]
4 3/4-12	4.660	83.1	4.678	66.5	4.660	83.1	4.678	66.5
4 3/4-16	4.682	83.8	4.696	66.5	4.682	83.8	4.696	66.5
5-8	4.865	83.1	4.890	67.7	4.865	83.1	4.890	67.7
5-12	4.910	83.1	4.928	66.5	4.910	83.1	4.928	66.5
5-16	4.932	83.8	4.946	66.5	4.932	83.8	4.946	66.5
5 1/4-8	5.115	83.1	5.140	67.7	5.115	83.1	5.140	67.7
5 1/4-12	5.160	83.1	5.178	66.5	5.160	83.1	5.178	66.5
5 1/4-16	5.182	83.8	5.196	66.5	5.182	83.8	5.196	66.5
5 1/2-8	5.365	83.1	5.390	67.7	5.365	83.1	5.390	67.7
5 1/2-12	5.410	83.1	5.428	66.5	5.410	83.1	5.428	66.5
5 1/2-16	5.432	83.8	5.446	66.5	5.432	83.8	5.446	66.5
5 3/4-8	5.615	83.1	5.640	67.7	5.615	83.1	5.640	67.7
5 3/4-12	5.660	83.1	5.678	66.5	5.660	83.1	5.678	66.5
5 3/4-16	5.682	83.8	5.696	66.5	5.682	83.8	5.696	66.5
6-8	5.865	83.1	5.890	67.7	5.865	83.1	5.890	67.7
6-12	5.901	83.1	5.928	66.5	5.901	83.1	5.928	66.5
6-16	5.932	83.8	5.946	66.5	5.932	83.8	5.946	66.5

[a]Based on values as rounded off in preceding column.
[b]Based on length of engagement equal to the nominal diameter.

Table 24. Tap Drill Sizes for UNM Miniature Screw Threads. *(Source, Miniature Thread Specialists [MTS].)*

Thread Size	For Thread Cutting Taps						For Forming Taps
	To 2/3D		2/3 to 1.5D		1.5 to 3XD		
	Min.	Max.	Min.	Max.	Min.	Max.	
1.40UNM	0.0439	0.0455	0.0450	0.0471	0.0460	0.0481	0.0498
1.20UNM	0.0379	0.0393	0.0388	0.0406	0.0397	0.0415	0.0438
1.10UNM	0.0340	0.0354	0.0349	0.0367	0.0358	0.0376	0.0397
1.00UNM	0.0300	0.0314	0.0309	0.0327	0.0319	0.0337	0.0358
0.90UNM	0.0270	0.0283	0.0279	0.0295	0.0287	0.0304	0.0318
0.80UNM	0.0241	0.0252	0.0248	0.0263	0.0256	0.0270	0.0283
0.70UNM	0.0211	0.0221	0.0217	0.0231	0.0224	0.0237	0.0248
0.60UNM	0.0181	0.0190	0.0187	0.0198	0.0193	0.0204	0.0212
0.55UNM	0.0170	0.0178	0.0176	0.0186	0.0181	0.0191	0.0195
0.50UNM	0.0150	0.0158	0.0156	0.0166	0.0161	0.0171	
0.45UNM	0.0141	0.0147	0.0145	0.0154	0.0149	0.0158	
0.40UNM	0.0121	0.0127	0.0125	0.0134	0.0130	0.0138	

Tap drill sizes in inches.   *(Continued)*

Table 24. *(Continued)* Tap Drill Sizes for UNM Miniature Screw Threads.
(Source, Miniature Thread Specialists [MTS].)

Thread Size	For Thread Cutting Taps						For Forming Taps
	To 2/3*D*		2/3 to 1.5*D*		1.5 to 3X*D*		
	Min.	Max.	Min.	Max.	Min.	Max.	
0.35UNM	0.0105	0.0111	0.0109	0.0117	0.0113	0.0121	
0.30UNM	0.0089	0.0095	0.0093	0.0100	0.0096	0.0104	
00-90	0.0358	0.0380	0.0374	0.0395	0.0385	0.0405	0.0426
00-96	0.0371	0.0385	0.0381	0.0400	0.0390	0.0409	0.0429
00-112	0.0386	0.0398	0.0394	0.0410	0.0402	0.0419	0.0435
000-120	0.0261	0.0273	0.0270	0.0280	0.0277	0.0293	0.0306
0000-160	0.0151	0.0160	0.0157	0.0170	0.0164	0.0176	0.0184

Tap drill sizes in inches.

Table 25. Dimensions of Miniature Taps. *(Source, Miniature Thread Specialists [MTS].)*

Size	Three Flute Plug Taps			Forming Taps		
	Overall Length	Shank Dia.	Thread Length	Overall Length	Shank Dia.	Thread Length
1.40 UNM	30.0	2.0	8.0	25.0	2.0	5.5
1.20 UNM	25.0	1.5	7.0	25.0	1.5	4.5
1.10 UNM	25.0	1.5	7.0	25.0	1.5	4.5
1.00 UNM	25.0	1.5	6.0	25.0	1.5	4.5
.90 UNM	25.0	1.5	5.0	25.0	1.5	3.0
.80 UNM	25.0	1.5	4.0	25.0	1.5	2.5
.70 UNM	25.0	1.5	4.0	25.0	1.5	2.5
.60 UNM	22.0	1.0	3.0	22.0	1.0	1.5
.50 UNM	22.0	1.0	3.0			
.45 UNM	22.0	1.0	2.0			
.40 UNM	22.0	1.0	2.0	Not available in Forming Taps		
.35 UNM	22.0	1.0	2.0			
.30 UNM	22.0	1.0	2.0			
00-90	25.0	1.5	6.0	25.0	1.5	4.5
00-96	25.0	1.5	6.0	25.0	1.5	4.5
00-112	25.0	1.5	6.0	25.0	1.5	4.5
00-120	25.0	1.5	5.0	25.0	1.5	3.0
000-160	25.0	1.5	4.0	22.0	1.0	1.5

Tap dimensions in millimeters.

Table 26. Tap Drill Sizes for Screw Thread Insert Taps. *(Source, Fastcut Tool Corp.)*

Tap Size	—Aluminum—				—Steel, Plastic, Magnesium—			
	Tap Drill Size	Decimal Equiv. of Tap Drill	Minor Dia. Limits (After Tapping)		Tap Drill Size	Decimal Equiv. of Tap Drill	Minor Dia. Limits (After Tapping)	
			Min.	Max.			Min.	Max.
4-40	#31	.1200	.116	.121	#31	.1200	.119	.124
5-40	#30	.1285	.128	.133	#29	.1360	.131	.136
6-32	#25	.1495	.144	.150	#25	.1495	.148	.154
6-40	#26	.1470	.144	.149	#25	.1495	.148	.153
8-32	#17	.1730	.170	.176	#16	.1770	.174	.180
10-24	13/64	.2031	.199	.205	#5	.2055	.203	.209
10-32	#7	.2010	.196	.202	13/64	.2031	.200	.206
12-24	#2	.2210	.221	.227	#1	.2280	.225	.231
1/4-20	17/64	.2656	.261	.267	17/64	.2656	.265	.271
1/4-28	G	.2610	.257	.264	17/64	.2656	.261	.268
5/16-18	Q	.3320	.328	.334	Q	.3320	.331	.337
5/16-24	21/64	.3281	.323	.330	Q	.3320	.327	.334
3/8-16	X	.3970	.390	.398	X	.3970	.396	.402
3/8-24	25/64	.3906	.385	.392	25/64	.3906	.389	.396
7/16-14	29/64	.4531	.453	.463	15/32	.4687	.461	.471
7/16-20	29/64	.4531	.450	.458	29/64	.4531	.453	.461
1/2-13	33/64	.5156	.515	.525	17/32	.5312	.523	.533

All dimensions in inches.

Table 27. Pulley Tap Dimensions. *(Source, Hanson-Whitney Co.)*

Thread Size	Thread Length	Ground Length	Overall Length	Shank Dia.	Neck Length	Square Length	Square Size
1/4-20 UNC	1	1 1/2	6, 8	.255	3/8	5/16	.191
5/16-18 UNC	1 1/8	1 9/16	6, 8	.318	3/8	3/8	.238
3/8-16 UNC	1 1/4	1 5/8	6, 8, 10	.381	3/8	7/16	.286
7/16-14 UNC	1 7/16	1 11/16	6, 8	.444	7/16	1/2	.333
1/2-13 UNC	1 21/32	1 11/16	6, 8, 10, 12	.507	1/2	9/16	.380
5/8-11 UNC	1 13/16	2	6, 8, 10, 12	.633	5/8	11/16	.475
3/4-10 UNC	2	2 1/4	10, 12	.759	3/4	3/4	.569

All taps to G H3 limit. Neck length may vary with manufacturer. All dimensions in inches.

Table 28. Nut Tap Dimensions. *(Source, Hanson-Whitney Co.)*

Thread Size	Thread Length	Overall Length	Shank Dia.	Square Length	Square Size
1/4-20 UNC	1 5/8	5	.185	5/16	.139
5/16-18 UNC	1 13/16	5 1/2	.240	5/8	.180
3/8-16 UNC	2	6	.294	11/16	.220
7/16-14 UNC	2 3/8	6 1/2	.235	3/4	.259
1/2-13 UNC	2 1/2	7	.400	7/8	.300

All taps to G H3 limits. All dimensions in inches.

Table 29. Limits and Tolerances for Taper Pipe Taps (NPT, ANPT, and NPTF Forms).
(Source, OSG Tap and Die, Inc.)

Thread Limits					
Nom. Size	TPI	Gage Measurement		Taper per Foot	
		Projection*	Tolerance ±	Min.	Max.
1/16	27	.312	1/16	23/32	25/32
1/8	27	.312	1/16	23/32	25/32
1/4	18	.459	1/16	23/32	25/32
3/8	18	.454	1/16	23/32	25/32
1/2	14	.579	1/16	23/32	25/32
3/4	14	.565	1/16	23/32	25/32
1	11 1/2	.678	3/32	23/32	25/32
1 1/4	11 1/2	.686	3/32	23/32	25/32
1 1/2	11 1/2	.699	3/32	23/32	25/32
2	11 1/2	.667	3/32	23/32	25/32
2 1/2	8	.925	3/32	47/64	25/32
3	8	.925	3/32	47/64	25/32
3 1/2	8	.938	1/8	47/64	25/32
4	8	.950	1/8	47/64	25/32

Lead Tolerance
Maximum allowable lead deviation within any two threads not more than one inch apart is ±.0005".

Angle Tolerance	
Threads per Inch	Deviation in Half Angle
8	±25°
11 1/2 to 27 Inclusive	±30°

Width of Flats at Tap Crest and Root					
TPI	Element	NPT		NPTF	
		Min.	Max.	Min.	Max.
27	Major Dia.	.0014	.0041	.0040	.0055
	Minor Dia.	-	.0041	-	.0040
18	Major Dia.	.0021	.0057	.0050	.0065
	Minor Dia.	-	.0057	-	.0050

* Projection is distance the small end of the tap protrudes through a taper thread ring gage. *(Continued)*
All dimensions in inches.

Table 29. *(Continued)* **Limits and Tolerances for Taper Pipe Taps (NPT, ANPT, and NPTF Forms).** *(Source, OSG Tap and Die, Inc.)*

		Width of Flats at Tap Crest and Root			
TPI	Element	NPT		NPTF	
		Min.	Max.	Min.	Max.
14	Major Dia.	.0027	.0064	.0050	.0065
	Minor Dia.	-	.0064	-	.0050
11 1/2	Major Dia.	.0033	.0073	.0060	.0083
	Minor Dia.	-	.0073	-	.0060
8	Major Dia.	.0048	.0090	.0080	.0103
	Minor Dia.	-	.0090	-	.0030

Formulas for NPT Pipe Form					
Min. Major Dia. = Measured Pitch Dia. + A below			Max. Major Dia. = Measured Pitch Dia. + B below.		
Min. Minor Dia. = Measured Pitch Dia. - B below.			Max. Minor Dia. = Measured Pitch Dia. - C below.		
TPI	A	B	C	D	E
27	.0267	.0296	.0257	.0234	.0251
18	.0408	.0444	.0401	.0377	.0395
14	.0535	.0571	.0525	.0515	.0533
11 1/2	.0658	.0696	.0647	.0614	.0649
8	.0966	.1000	.0946	-	-

Formulas for Dryseal NPTF Pipe Form					
Min. Major Dia. = Measured Pitch Dia. + D above.			Max. Major Dia. = Measured Pitch Dia. + E above.		
Min. Minor Dia. = Maximum or smaller.			Max. Minor Dia. = Measured Pitch Dia. - E above.		

Table 30. Reaming Data and Tap Drill Sizes for NPT, NPTF, and SAE Short Pipe Threads. *(Source, Fastcut Tool Corp.)*

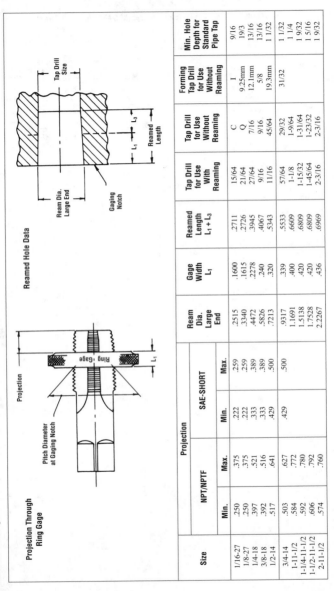

Size	Projection NPT/NPTF		Projection SAE-SHORT		Ream Dia. Large End	Gage Width L_1	Reamed Length $L_1 + L_3$	Tap Drill for Use With Reaming	Tap Drill for Use Without Reaming	Forming Tap Drill for Use Without Reaming	Min. Hole Depth for Standard Pipe Tap
	Min.	Max.	Min.	Max.							
1/16-27	.250	.375	.222	.259	.2515	.1600	.2711	15/64	C	I	9/16
1/8-27	.250	.375	.222	.259	.3340	.1615	.2726	21/64	Q	9.25mm	19/3
1/4-18	.397	.521	.333	.389	.4472	.2278	.3945	27/64	7/16	12.1mm	13/16
3/8-18	.392	.516	.333	.389	.5826	.240	.4067	9/16	9/16	5/8	13/16
1/2-14	.517	.641	.429	.500	.7213	.320	.5343	11/16	45/64	19.3mm	1 1/32
3/4-14	.503	.627	.429	.500	.9317	.339	.5533	57/64	29/32	31/32	1 1/32
1-11-1/2	.584	.772			1.1691	.400	.6609	1-1/8	1-9/64		1 1/4
1-1/4-11-1/2	.592	.780			1.5138	.420	.6809	1-15/32	1-31/64		1 9/32
1-1/2-11-1/2	.606	.792			1.7528	.420	.6809	1-45/64	1-23/32		1 5/16
2-11-1/2	.574	.760			2.2267	.436	.6969	2-3/16	2-3/16		1 9/32

All dimensions in inches.

Table 31. Recommended Bores for American National (NPT) and British Standard (BSTP) Taper Pipe Threads.*
(Source, Landis Threading Systems.)

NPT*			BSTP*		
Nominal Pipe Size	Thread Pitch	Bore Size	Nominal Pipe Size	Thread Pitch	Bore Size
1/8	27	11/32	1/8	28	5/16
1/4	18	7/16	1/4	19	27/64
3/8	18	19/32	3/8	19	9/16
1/2	14	23/32	1/2	14	11/16
3/4	14	15/16	3/4	14	29/32
1	11 1/2	1 5/32	1	11	1 1/8
1 1/4	11 1/2	1 1/2	1 1/4	11	1 15/32
1 1/2	11 1/2	1 23/32	1 1/2	11	1 25/32
2	11 1/2	2 3/16	2	11	2 5/32
2 1/2	8	2 5/8	2 1/2	11	2 25/32
3	8	3 1/4	3	11	3 9/32
3 1/2	8	3 3/4	3 1/2	11	3 3/4
4	8	4 1/4	4	11	4 1/4
5	8	5 5/16	5	11	5 1/4
6	8	6 3/8	6	11	6 1/4

*The above bore dimensions will result in a full thread. When using a collapsible tap, it is recommended that the hole be taper bored for best tool results. When boring, leave no more than .010" material per side below the minor diameter of the pipe thread to be tapped. All dimensions in inches.

Table 32. Limits and Tolerances for Straight Pipe Taps (NPS, NPSC, and NPSM Forms).
(Source, OSG Tap and Die, Inc.)

Thread Limits							
Nom. Size	TPI	Major Diameter			Pitch Diameter		
		Plug at Gaging Notch	Min. G	Max. H	Plug at Gaging Notch E	Min. K	Max. L
1/8	27	.3983	.4022	.4032	.3736	.3746	.3751
1/4	18	.5286	.5347	.5357	.4916	.4933	.4938
3/8	18	.6640	.8701	.6711	.6270	.6287	.6292
1/2	14	.8260	.8347	.8357	.7784	.7806	.7811
3/4	14	1.0364	1.0477	1.0457	.9889	.9906	.9916
1	11 1/2	1.2966	1.3062	1.3077	1.2386	1.2402	1.2412
1 1/4	11 1/2	1.6413	1.6507	1.6522	1.5384	1.5847	1.5862
1 1/2	11 1/2	1.8803	1.8897	1.8912	1.8223	1.8237	1.8252
2	11 1/2	2.3542	2.3639	2.3654	2.2963	2.2979	2.2294
2 1/2	8	2.8454	2.8604	2.8619	2.7622	2.7640	2.7660
3	8	3.4718	3.4868	3.4883	3.3885	3.3904	3.3924
3 1/2	8	3.9721	3.9872	3.9887	3.8888	3.8908	3.8928
4	8	4.4704	4.4855	4.4870	4.3871	4.3891	4.3911

(Continued)

Table 32. *(Continued)* **Limits and Tolerances for Straight Pipe Taps (NPS, NPSC, and NPSM Forms).**
(Source, OSG Tap and Die, Inc.)

Lead Tolerance	
Maximum allowable lead deviation within any two threads not more than one inch apart is ±.0005".	
Angle Tolerance	
Threads per Inch	**Deviation in Half Angle**
8	± 25'
11 1/2 to 27 Inclusive	± 30'
Minor Diameter	
Minor diameter is equal to actual measured pitch diameter minus the value below.	
TPI	**Value**
27	.0257
18	.0401
14	.0525
11 1/2	.0647
8	.0946

All dimensions in inches. Taps made to these specifications are marked NPS and used for NPS, NPSC, and NPSM straight pipe threads.

Table 33. Limits and Tolerances for Straight Dryseal Pipe Taps (NPSF Form). *(Source, OSG Tap and Die, Inc.)*

Thread Limits							
Nom. Size	**TPI**	**Major Diameter**			**Pitch Diameter**		**Minor Dia. Flat Max.**
		Plug at Gaging Notch E	**Min. G**	**Max. H**	**Min. K**	**Max. L**	
1/16	27	.2812	.3008	.3018	.2772	.2777	.004
1/8	27	.3736	.3932	.3942	.3696	.3701	.004
1/4	18	.4916	.5239	.5249	.4859	.4864	.005
3/8	18	.6270	.6593	.6603	.6213	.6218	.005
1/2	14	.7784	.8230	.8240	.7712	.7717	.005
3/4	14	.9889	1.0335	1.0345	.9817	.9822	.005
1	11 1/2	1.2386	1.2933	1.2943	1.2295	1.2305	.006
Lead Tolerance							
Maximum allowable lead deviation within any two threads not more than one inch apart is ±.0005".							
Angle Tolerance							
Threads per Inch				**Deviation in Half Angle**			
11 1/2 to 27 Inclusive				± 30'			
Minor Diameter							
Minor diameter is equal to actual measured pitch diameter minus the value below.							
TPI				**Value**			
27				.0251			
18				.0395			
14				.0533			
11 1/2				.0649			

All dimensions in inches. Taps made to these specifications are marked NPSF.

Table 34. Standard Pipe Tap Dimensions and Tolerances, Straight and Taper, Ground Thread.
(Source, Fastcut Tool Corp.)

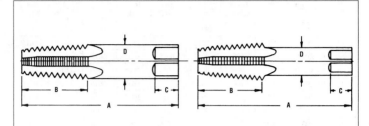

Dimensions					
Nominal Sizes Inches	Overall Length A	Length of Thread B	Length of Square C	Diameter of Shank D	Size of Square E
1/16	2-1/8	11/16	3/8	.3125	.234
1/8*	2-1/8	3/4	3/8	.3125	.234
1/8	2-1/8	3/4	3/8	.4375	.328
1/4	2-7/16	1-1/16	7/16	.5625	.421
3/8	2-9/16	1-1/16	1/2	.7000	.531
1/2	3-1/8	1-3/8	5/8	.6875	.515
3/4	3-1/4	1-3/8	11/16	.9063	.679
1	3-3/4	1-3/4	13/16	1.1250	.843
1-1/4	4	1-3/4	15/16	1.3125	.984
1-1/2	4-1/4	1-3/4	1	1.5000	1.125
2	4-1/2	1-3/4	1-1/8	1.8750	1.406
2-1/2	5-1/2	2-9/16	1-1/4	2.2500	1.687
3	6	2-5/8	1-3/8	2.6250	1.968
3-1/2	6-1/2	2-11/16	1-1/2	2.8125	2.108
4	6-3/4	2-3/4	1-5/8	3.0000	2.250

Tolerances			
Element	Range	Direction	Tolerance
Length Overall - A	1/16" to 3/4" incl.	Plus or Minus	1/32"
	1" to 4" incl.	Plus or Minus	1/16"
Length of Thread - B	1/16" to 3/4" incl.	Plus or Minus	1/16"
	1" to 1-1/4" incl.	Plus or Minus	3/32"
	1-1/2" to 4" incl.	Plus or Minus	1/8"
Length of Square - C	1/16" to 3/4" incl.	Plus or Minus	1/32"
	1" to 4" incl.	Plus or Minus	1/16"
Dia. Of Shank - D	1/16" to 1/8" incl.	Minus	.0015"
	1/4" to 1/2" incl.	Minus	.0020"
	3/4" to 1" incl.	Minus	.0020"
	1-1/4" to 4" incl.	Minus	.0030"
Size of Square - E	1/16" to 1/8" incl.	Minus	.0040"
	1/4" to 3/4" incl.	Minus	.0060"
	1" to 4" incl.	Minus	.0080"

*Small shank. All dimensions in inches.

Table 35. Tap Thread Limits and Tolerances, Unified Threads. See Notes Below. *(Source, OSG Tap and Die, Inc.)*

Threads Per Inch	A 0.130P	B	C			D			
			To 5/8" Inclusive	Over 5/8" to 2 1/2" Incl.	Over 2 1/2"	To 1" Incl.	Over 1" to 1 1/2" Incl.	Over 1 1/2" to 2 1/2" Incl.	Over 2 1/2"
80	.0016	.0011	.0005	.0010	.0015	.0005	.0010	.0010	.0015
72	.0018	.0012	.0005	.0010	.0015	.0005	.0010	.0010	.0015
64	.0020	.0014	.0005	.0010	.0015	.0005	.0010	.0010	.0015
56	.0023	.0016	.0005	.0010	.0015	.0005	.0010	.0010	.0015
48	.0027	.0018	.0005	.0010	.0015	.0005	.0010	.0010	.0015
44	.0030	.0017	.0005	.0010	.0015	.0005	.0010	.0010	.0015
40	.0032	.0019	.0005	.0010	.0015	.0005	.0010	.0010	.0015
36	.0036	.0021	.0005	.0010	.0015	.0005	.0010	.0010	.0015
32	.0041	.0020	.0010	.0010	.0015	.0005	.0010	.0010	.0015
28	.0046	.0023	.0010	.0010	.0015	.0005	.0010	.0010	.0015
24	.0054	.0027	.0010	.0010	.0015	.0005	.0010	.0015	.0015
20	.0065	.0032	.0010	.0010	.0015	.0005	.0010	.0015	.0015
18	.0072	.0036	.0010	.0010	.0015	.0005	.0010	.0015	.0015
16	.0081	.0041	.0010	.0010	.0015	.0005	.0010	.0015	.0020
14	.0093	.0046	.0010	.0015	.0015	.0005	.0010	.0015	.0020
13	.0100	.0050	.0010	.0015	.0015	.0005	.0010	.0015	.0020
12	.0108	.0054	.0010	.0015	.0015	.0005	.0010	.0015	.0020
11	.0118	.0059	.0010	.0015	.0020	.0005	.0010	.0015	.0020
10	.0130	.0065	-	.0015	.0020	.0005	.0010	.0015	.0020
9	.0144	.0072	-	.0015	.0020	.0005	.0010	.0015	.0020
8	.0162	.0081	-	.0015	.0020	.0005	.0010	.0015	.0020
7	.0186	.0093	-	.0015	.0020	.0010	.0010	.0020	.0025
6	.0217	.0108	-	.0015	.0020	.0010	.0010	.0020	.0025
5 1/2	.0236	.0118	-	.0015	.0020	.0010	.0015	.0020	.0025
5	.0260	.0130	-	.0015	.0020	.0010	.0015	.0020	.0025
4 1/2	.0289	.0144	-	.0015	.0020	.0010	.0015	.0020	.0025
4	.0325	.0162	-	.0015	.0020	.0010	.0015	.0020	.0025

Formulas for Unified Inch Screw Threads	
Max. Major Dia. = Basic + A	Min. Major Dia. = Max. Major Dia. - B
Max. Pitch Dia. = Min. Pitch Dia. + D	Min. Pitch Dia. = Basic Pitch Dia. + C

Lead Tolerance

Maximum allowable lead deviation within any two threads not more than one inch apart is ± .0005"

Angle Tolerance

Threads per Inch	Deviation in Half Angle
4 to 5 1/2 Inclusive	± 20'
6 to 9 Inclusive	± 25'
10 to 80 Inclusive	± 30'

A (Constant) = 0.130P for all Pitches. **B** = Major Diameter Tolerance as follows: 1) For 48 through 80 tpi, B = 0.087P. 2) For 36 through 47 tpi, B = 0.076P. 3) For 4 through 35 tpi, B = 0.065P. **C** = Amount over Basic for Minimum Pitch Diameter. **D** = Pitch Diameter Tolerance. *Notes:* 1) When the tap major diameter must be determined from a specified pitch diameter, the maximum major diameter equals the minimum specified pitch diameter minus **C**, plus 0.64952P, plus **A**. 2) Dimensions and constants on this table apply only to ground thread taps having a thread lead angle not in excess of 5°, unless otherwise specified.

Table 36. Tap Thread Limits and Tolerances, Metric Threads, in Inches. See Notes Below. (Source, OSG Tap and Die, Inc.)

| Pitch | | Basic Thread Height in Inches | W | X | Tap Limits for Metric Threads, W, X, Y, and Z. Values are in Inches | | | | | | | |
| | | | | | Y | | | | Z | | | |
mm	Inch Equiv.				M1.6 to M6.3 Incl.	Over M6.3 to M25 Incl.	Over M25 to M90 Incl.	Over M90	M1.6 to M6.3 Incl.	Over M6.3 to M25 Incl.	Over M25 to M90 Incl.	Over M90
0.3	.011811	.007671	.0009	.0010	.0015	.0015	.0020	.0020	.0006	.0006	.0008	.0008
0.35	.013779	.008950	.0011	.0010	.0015	.0015	.0020	.0020	.0006	.0006	.0008	.0008
0.4	.015748	.010229	.0013	.0010	.0015	.0020	.0020	.0020	.0006	.0006	.0008	.0010
0.45	.017716	.011507	.0014	.0010	.0015	.0020	.0020	.0020	.0006	.0008	.0008	.0010
0.5	.019685	.012786	.0016	.0010	.0015	.0020	.0020	.0025	.0006	.0008	.0010	.0010
0.6	.23622	.015343	.0019	.0010	.0020	.0020	.0025	.0025	.0008	.0008	.0010	.0010
0.7	.027559	.017900	.0022	.0016	.0020	.0020	.0025	.0025	.0008	.0008	.0010	.0010
0.75	.029527	.019178	.0024	.0016	.0020	.0025	.0025	.0025	.0008	.0010	.0010	.0012
0.8	.031496	.020457	.0025	.0016	.0020	.0025	.0025	.0030	.0008	.0010	.0010	.0012
0.9	.035433	.023014	.0028	.0016	.0020	.0025	.0025	.0030	.0008	.0010	.0010	.0012
1	.039370	.025572	.0032	.0016	.0025	.0025	.0025	.0030	.0010	.0010	.0012	.0012
1.25	.049212	.031964	.0039	.0025	.0025	.0025	.0030	.0035	.0010	.0012	.0012	.0016
1.5	.059055	.038357	.0047	.0025	.0025	.0030	.0030	.0035	.0010	.0012	.0012	.0016
1.75	.068897	.044750	.0055	.0025	-	.0030	.0035	.0040	-	.0012	.0016	.0016
2	.078740	.051143	.0063	.0025	-	.0035	.0035	.0040	-	.0016	.0016	.0016
2.5	.098425	.063929	.0079	.0025	-	.0035	.0040	.0045	-	.0016	.0016	.0020
3	.118110	.076715	.0095	.0039	-	.0040	.0040	.0050	-	.0016	.0020	.0020
3.5	.137795	.089501	.0110	.0039	-	.0040	.0045	.0050	-	.0016	.0020	.0020
4	.157480	.102286	.0126	.0039	-	.0040	.0045	.0055	-	.0020	.0020	.0025
4.5	.177165	.115072	.0142	.0039	-	-	.0050	.0055	-	.0020	.0020	.0025
5	.196850	.127858	.0158	.0039	-	-	.0050	.0060	-	-	.0025	.0025

(Continued)

Table 36. *(Continued)* Tap Thread Limits and Tolerances, Metric Threads, in Inches. See Notes Below. *(Source, OSG Tap and Die, Inc.)*

Pitch		Basic Thread Height in Inches	Tap Limits for Metric Threads, W, X, Y, and Z. Values are in Inches									
			W	X	Y				Z			
mm	Inch Equiv.				M1.6 to M6.3 Incl.	Over M6.3 to M25 Incl.	Over M25 to M90 Incl.	Over M90	M1.6 to M6.3 Incl.	Over M6.3 to M25 Incl.	Over M25 to M90 Incl.	Over M90
5.5	.216535	.140644	.0173	.0039	-	-	.0055	.0060	-	-	.0025	.0025
6	.236220	.153430	.0189	.0039	-	-	.0055	.0060	-	-	.0025	.0025

Formulas for Metric Screw Threads

Min. Major Dia. = Basic + W

Max. Major Dia. = Min. Major Dia. + X

Max. Pitch Dia. = Basic Pitch Dia. + Y

Min. Pitch Dia. = Max. Pitch Dia. - Z

Lead Tolerance

Maximum allowable lead deviation within any two threads not more than 25.4 mm apart is ± 0.013 mm

Angle Tolerance

Pitch in millimeters	Deviation in Half Angle
Over 0.25 to 2.5 Inclusive	± 30'
Over 2.5 to 4.0 Inclusive	± 25'
Over 4.0 to 6.0 Inclusive	± 20'

W = Constant to add to Basic Major Diameter (0.080*P*). **X** = Major Diameter Tolerance. **Y** = Amount over Basic for Maximum Pitch Diameter. **Z** = Pitch Diameter Tolerance.
Notes: 1) When the tap major diameter must be determined from a specified tap pitch diameter, the minimum major diameter equals the maximum specified tap pitch diameter minus **Y**, plus the basic single height of the thread, plus **W**. 2) Dimensions and constants on this table apply only to 60° metric threads with a P/8 flat at the major diameter of the basic thread form.

Table 37. Tap Thread Limits and Tolerances, Metric Threads, in Millimeters. See Notes Below. (Source, OSG Tap and Die, Inc.)

Pitch	Basic Thread Height	W	X	Y				Z			
				M1.6 to M6.3 Incl.	Over M6.3 to M25 Incl.	Over M25 to M90 Incl.	Over M90	M1.6 to M6.3 Incl.	Over M6.3 to M25 Incl.	Over M25 to M90 Incl.	Over M90
0.3	.194856	.024	.025	.039	.039	.052	.052	.015	.015	.020	.020
0.35	.227332	.028	.025	.039	.039	.052	.052	.015	.015	.020	.020
0.4	.259808	.032	.025	.039	.052	.052	.052	.015	.015	.020	.025
0.45	.292284	.036	.025	.039	.052	.052	.052	.015	.020	.020	.025
0.5	.324760	.040	.025	.039	.052	.052	.065	.015	.020	.025	.025
0.6	.389712	.048	.025	.052	.052	.065	.065	.020	.020	.025	.025
0.7	.454664	.056	.041	.052	.052	.065	.065	.020	.020	.025	.025
0.75	.487140	.060	.041	.052	.065	.065	.078	.020	.025	.025	.031
0.8	.519616	.064	.041	.052	.065	.065	.078	.025	.025	.025	.031
0.9	.584568	.072	.041	.065	.065	.078	.078	.025	.025	.025	.031
1	.649520	.080	.064	.065	.065	.078	.091	.025	.031	.031	.041
1.25	.811900	.100	.064	.065	.065	.078	.091	.025	.031	.031	.041
1.5	.974280	.120	.064	-	.078	.078	.091	.025	.031	.031	.041
1.75	1.13666	.140	.064	-	.078	.091	.104	-	.031	.031	.041
2	1.29904	.160	.064	-	.091	.091	.104	-	.041	.041	.041
2.5	1.62380	.200	.064	-	.091	.104	.117	-	.041	.041	.052
3	1.94856	.240	.100	-	.104	.104	.130	-	.041	.052	.052
3.5	2.27332	.280	.100	-	.104	.117	.130	-	.041	.052	.052
4	2.59808	.320	.100	-	.104	.117	.143	-	.052	.052	.064
4.5	2.92284	.360	.100	-	-	.130	.143	-	.052	.052	.064
5	3.24760	.400	.100	-	-	.130	.156	-	-	.064	.064

(Continued)

Table 37. *(Continued)* **Tap Thread Limits and Tolerances, Metric Threads, in Millimeters.** See Notes Below. *(Source, OSG Tap and Die, Inc.)*

Pitch	Basic Thread Height	W	X	Y				Z			
				M1.6 to M6.3 Incl.	Over M6.3 to M25 Incl.	Over M25 to M90 Incl.	Over M90	M1.6 to M6.3 Incl.	Over M6.3 to M25 Incl.	Over M25 to M90 Incl.	Over M90
5.5	3.57236	.440	.100	-	-	.143	.156	-	-	.064	.064
6	3.89712	.480	.100	-	-	.143	.156	-	-	.064	.064

Formulas for Metric Screw Threads

Min. Major Dia. = Basic + W
Max. Pitch Dia. = Basic Pitch Dia. + Y

Max. Major Dia. = Min. Major Dia. + X
Min. Pitch Dia. = Max. Pitch Dia. - Z

Lead Tolerance

Maximum allowable lead deviation within any two threads not more than 25.4 mm apart is ± 0.013 mm

Angle Tolerance

Pitch in millimeters	Deviation in Half Angle
Over 0.25 to 2.5 Inclusive	± 30'
Over 2.5 to 4.0 Inclusive	± 25'
Over 4.0 to 6.0 Inclusive	± 20'

W = Constant to add to Basic Major Diameter (0.080*P*). **X** = Major Diameter Tolerance. **Y** = Amount over Basic for Maximum Pitch Diameter. **Z** = Pitch Diameter Tolerance.
Notes: 1) When the tap major diameter must be determined from a specified tap pitch diameter, the minimum major diameter equals the maximum specified tap pitch diameter minus **Y**, plus the basic single height of the thread, plus **W**. 2) Dimensions and constants on this table apply only to 60° metric threads with a P/8 flat at the major diameter of the basic thread form.

Table 38. Cut Thread Tap Limits for Unified Threads. *(Source, Federal Standard GGG-T-60A.)*

Gage No. or Dia.	Threads per Inch			Major Diameter Limits			Pitch Diameter Limits		
	UNC	UNF	UNS	Basic	Min.	Max.	Basic	Min.	Max.
0	-	80	-	.0600	.0609	.0624	.0519	.0521	.0531
1	64	-	-	.0730	.0740	.0755	.0629	.0631	.0641
1	-	72	-	.0730	.0740	.0755	.0640	.0642	.0652
2	56	-	-	.0860	.0872	.0887	.0744	.0746	.0756
2	-	64	-	.0860	.0870	.0885	.0759	.0761	.0771
3	48	-	-	.0990	.1003	.1018	.0855	.0857	.0867
3	-	56	-	.0990	.1002	.1017	.0874	.0876	.0886
4	-	-	32	.1120	.1142	.1162	.0917	.0922	.0937
4	-	-	36	.1120	.1137	.1157	.0940	.0942	.0957
4	40	-	-	.1120	.1136	.1156	.0958	.0960	.0975
4	-	48	-	.1120	.1133	.1153	.0985	.0987	.1002
5	40	-	-	.1250	.1266	.1286	.1088	.1090	.1105
5	-	44	-	.1250	.1264	.1284	.1102	.1104	.1119
6	32	-	-	.1380	.1402	.1422	.1177	.1182	.1197
6	-	-	36	.1380	.1397	.1417	.1200	.1202	.1217
6	-	40	-	.1380	.1396	.1416	.1218	.1220	.1235
8	32	-	-	.1640	.1662	.1682	.1437	.1442	.1457
8	-	36	-	.1640	.1657	.1677	.1460	.1462	.1477
8	-	-	40	.1640	.1656	.1676	.1478	.1480	.1495
10	24	-	-	.1900	.1928	.1948	.1629	.1634	.1649
10	-	-	30	.1900	.1923	.1943	.1684	.1689	.1704
10	-	32	-	.1900	.1922	.1942	.1697	.1702	.1717
12	24	-	-	.2160	.2188	.2208	.1889	.1894	.1909
12	-	28	-	.2160	.2184	.2204	.1928	.1933	.1948
12	-	-	32	.2160	.2182	.2202	.1957	.1962	.1977
14	-	-	20	.2420	.2452	.2477	.2095	.2100	.2120
14	-	-	24	.2420	.2448	.2473	.2149	.2154	.2174
1/16	-	-	64	.0625	.0635	.0650	.0524	.0526	.0536
3/32	-	-	48	.0938	.0951	.0966	.0803	.0805	.0815
1/8	-	-	40	.1250	.1266	.1286	.1088	.1090	.1105
5/32	-	-	32	.1563	.1585	.1605	.1360	.1365	.1380
5/32	-	-	36	.1563	.1580	.1600	.1382	.1384	.1399
3/16	-	-	24	.1875	.1903	.1923	.1604	.1609	.1624
3/16	-	-	32	.1875	.1897	.1917	.1672	.1677	.1692
7/32	-	-	24	.2188	.2216	.2236	.1917	.1922	.1937
7/32	-	-	32	.2188	.2210	.2230	.1985	.1990	.2005
1/4	20	-	-	.2500	.2532	.2557	.2175	.2180	.2200
1/4	-	-	24	.2500	.2528	.2553	.2229	.2234	.2254

All dimensions in inches.   *(Continued)*

Table 38. *(Continued)* **Cut Thread Tap Limits for Unified Threads.** *(Source, Federal Standard GGG-T-60A.)*

Gage No. or Dia.	Threads per Inch			Major Diameter Limits			Pitch Diameter Limits		
	UNC	UNF	UNS	Basic	Min.	Max.	Basic	Min.	Max.
1/4	-	28	-	.2500	.2524	.2549	.2268	.2273	.2288
1/4	-	-	32	.2500	.2522	.2547	.2297	.2302	.2317
5/16	18	-	-	.3125	.3160	.3185	.2764	.2769	.2789
5/16	-	24	-	.3125	.3153	.3178	.2854	.2859	.2874
5/16	-	-	32	.3125	.3147	.3172	.2922	.2927	.2942
3/8	16	-	-	.3750	.3789	.3814	.3344	.3349	.3369
3/8	-	24	-	.3750	.3778	.3803	.3479	.3484	.3499
7/16	14	-	-	.4375	.4419	.4449	.3911	.3916	.3941
7/16	-	20	-	.4375	.4407	.4437	.4050	.4055	.4075
1/2	13	-	-	.5000	.5047	.5077	.4500	.4505	.4530
1/2	-	20	-	.5000	.5032	.5062	.4675	.4680	.4700
9/16	12	-	-	.5625	.5675	.5705	.5084	.5089	.5114
9/16	-	18	-	.5625	.5660	.5690	.5264	.5269	.5289
5/8	11	-	-	.6250	.6304	.6334	.5660	.5665	.5690
5/8	-	-	12	.6250	.6300	.6330	.5709	.5714	.5739
5/8	-	18	-	.6250	.6285	.6315	.5889	.5894	.5914
11/16	-	-	11	.6875	.6929	.6969	.6285	.6290	.6320
11/16	-	-	16	.6875	.6914	.6954	.6469	.6474	.6499
3/4	10	-	-	.7500	.7559	.7599	.6850	.6855	.6885
3/4	-	16	-	.7500	.7539	.7579	.7094	.7099	.7124
7/8	-	-	-	.8750	.8820	.8860	.8028	.8038	.8068
7/8	14	-	-	.8750	.8799	.8839	.8286	.8296	.8321
1	8	-	-	1.0000	1.0078	1.0118	.9188	.9198	.9228
1	-	12	-	1.0000	1.0055	1.0095	.9459	.9469	.9499
1	-	-	14	1.0000	1.0049	1.0089	.9536	.9546	.9571
1-1/8	7	-	-	1.1250	1.1337	1.1382	1.0322	1.0332	1.0367
1-1/8	-	12	-	1.1250	1.1305	1.1350	1.0709	1.0719	1.0749
1-1/4	7	-	-	1.2500	1.2587	1.2632	1.1572	1.1582	1.1617
1-1/4	-	12	-	1.2500	1.2555	1.2600	1.1959	1.1969	1.1999
1-3/8	6	-	-	1.3750	1.3850	1.3895	1.2667	1.2677	1.2712
1-3/8	-	12	-	1.3750	1.3805	1.3850	1.3209	1.3219	1.3249
1-1/2	6	-	-	1.5000	1.5100	1.5145	1.3917	1.3927	1.3962
1-1/2	-	12	-	1.5000	1.5055	1.5100	1.4459	1.4469	1.4499
1-3/4	5	-	-	1.7500	1.7602	1.7657	1.6201	1.6216	1.6256
2	4-1/2	-	-	2.0000	2.0111	2.0166	1.8557	1.8572	1.8612

All dimensions in inches.

Table 39. Recommended Speeds for Tapping Selected Workpiece Materials. *(Source, Weldon Co.)*

Material	Type	Hardness		Speed	
		Rc	BHN	SFM	SMM
Steel	Low Carbon Free Machining	<15	<180	20-65	6-19
	Medium to High Low Alloyed	<23	<240	20-50	6-15
	Castings and Forgings Heat Treatable Alloys	>24 <=38	>250 <=350	13-26	4-7
	Alloyed Tool Mold	>38 <=44	>350 <=420	6-13	2-4
Stainless Steel	Free Machining	<23	<240	5-15	1.5-4.5
	Heat and Corrosion Resistant	>24	<250	3-13	1-4
	Castings Precipitation Hardening	>38 <=44	>350 <=420	3-10	1-3
Cast Iron	Gray		<=220	20-65	6-19
	Nodular Chilled Meehanite Ductile		>=250	10-40	3-12
Aluminum	Pure Alloys			50-130	15-39
	Alloy Castings			30-65	9-19
Nickel Alloys	718 625 Hastelloy Monel Waspaloy Invar Incoloy	<=38	<=350	<13	<4
	718 Inconel A286	>38 <=44	>350 <=420	<13	<4
Titanium		<=38	<=350	<13	<4
Copper				25-50	7-15
Brass				30-65	9-19
Bronze		<44	<420	6-50	2-15
Zinc				25-50	7-15
Magnesium				30-65	9-19
Plastics	Thermoplastics			30-130	0-39
	Thermosetting Reinforced			15-40	4.5-12
Powder Metal		<44	<420	20-50	6-15
Graphite				15-35	4.5-10
Ceramics	Machinable Glasses			15-35	4.5-10
Special	Bronze Berrilium Copper			<13	<4
	Tungsten Ferro-Tic			5-15	1.5-4.5
	CPM10V			5-15	1.5-4.5

Table 40. Cutting Speeds: SFM Converted to RPM for Machine Screw, Fractional, and Pipe Tap Sizes. *(Source, Weldon Tool Co.)*

UNC/UNF Taps	NPT/NPTF Taps	Surface Feet per Minute																	
		5'	10'	15'	20'	25'	30'	40'	50'	60'	70'	80'	90'	100'	110'	120'	130'	140'	150'
		Revolutions per Minute																	
0		318	637	955	1273	1592	1910	2546	3183	3820	4456	5093	5729	6366	7003	7639	8276	8913	9549
1		273	546	819	1046	1308	1570	2093	2617	3140	3663	4186	4710	5233	5756	6279	6805	7326	7849
2		212	424	637	888	1110	1333	1777	2221	2665	3109	3554	3999	4442	4886	5330	5774	6218	6662
3		191	382	573	772	964	1157	1543	1929	2315	2701	3086	3472	3858	4244	4629	5015	5401	5787
4		174	347	521	682	853	1023	1364	1705	2046	2387	2728	3069	3411	3751	4092	4434	4775	5116
5		147	294	441	611	764	917	1222	1528	1833	2139	2445	2750	3056	3361	3667	3973	4278	4584
6		136	273	409	553	691	829	1106	1382	1659	1935	2212	2488	2766	3042	3318	3595	3871	4148
8		119	239	358	466	583	699	932	1165	1398	1631	1864	2097	2330	2563	2796	3029	3262	3495
10		101	201	302	402	502	603	804	1005	1205	1406	1607	1808	2009	2210	2411	2612	2813	3014
12		87	174	260	354	442	531	707	884	1061	1238	1415	1592	1769	1945	2122	2300	2476	2653
1/4		76	153	229	306	382	458	611	764	917	1070	1222	1375	1528	1681	1833	1986	2139	2292
5/16		62	123	185	245	306	367	489	611	733	856	978	1100	1222	1345	1467	1589	1711	1833
3/8		50	101	151	204	255	305	407	509	611	713	815	917	1019	1120	1222	1324	1426	1528
7/16	1/8	43	87	130	175	219	262	349	437	524	611	698	786	873	960	1048	1135	1222	1310
1/2	-	38	76	115	153	191	229	305	382	458	535	611	688	764	840	917	993	1070	1146
9/16	1/4	34	68	102	137	172	206	274	342	410	478	547	616	683	752	820	888	952	1020
5/8	-	32	64	96	122	153	183	244	306	367	428	489	550	611	672	733	794	856	917
11/16	3/8	28	55	83	111	138	167	222	278	333	389	444	500	556	611	667	722	778	833
3/4	-	25	51	76	102	128	153	203	255	305	357	407	458	509	560	611	662	713	764
7/8	1/2	22	43	65	87	109	131	175	218	262	306	350	392	437	480	524	568	611	655
1	-	19	38	57	76	96	115	153	191	230	268	305	344	382	420	458	497	535	573
1-1/8	3/4	17	34	51	68	84	102	136	170	204	238	272	306	340	373	407	441	475	509

(Continued)

Table 40. *(Continued)* Cutting Speeds: SFM Converted to RPM for Machine Screw, Fractional, and Pipe Tap Sizes. *(Source, Weldon Tool Co.)*

UNC/UNF Taps	NPT/NPTF Taps	Surface Feet per Minute																	
		5'	10'	15'	20'	25'	30'	40'	50'	60'	70'	80'	90'	100'	110'	120'	130'	140'	150'
		Revolutions per Minute																	
1-1/4	–	15	31	46	61	76	92	122	153	183	214	244	275	305	336	367	397	428	458
1-3/8	1	14	28	42	56	69	83	111	139	167	194	222	250	278	306	333	361	389	417
1-1/2	–	13	25	38	51	63	76	102	127	153	178	204	229	255	280	305	331	356	382
1-5/8		12	23	35	47	59	71	94	118	141	165	188	212	235	259	282	306	329	353
1-3/4		11	22	33	44	55	65	87	109	131	153	175	196	218	240	262	284	306	327
1-7/8		10	20	30	41	51	61	81	102	122	143	163	183	204	224	244	265	285	306
2		9	19	29	38	48	57	76	96	115	134	153	172	191	210	229	248	267	287

Table 41. Cutting Speeds: SFM Converted to RPM for Metric Tap Sizes. *(Source, Weldon Tool Co.)*

Tap Size	Surface Feet per Minute																	
	5'	10'	15'	20'	25'	30'	40'	50'	60'	70'	80'	90'	100'	110'	120'	130'	140'	150'
	Revolutions per Minute																	
M 1.0	490	979	1469	1959	2449	2938	3918	4897	5877	6856	7836	8815	9795	10774	11754	12733	13713	14692
M 2.0	242	484	725	967	1209	1451	1934	2418	2901	3385	3868	4352	4835	5319	5803	6286	6770	7253
M 3.0	162	324	486	647	809	971	1295	1619	1942	2266	2590	2914	3237	3561	3885	4208	4532	4856
M 3.5	138	277	415	554	692	830	1107	1384	1661	1938	2214	2491	2768	3045	3322	3599	3875	4152
M 4.0	122	243	365	487	608	730	973	1217	1460	1703	1946	2190	2433	2676	2920	3163	3406	3650
M 5.0	97	194	291	388	485	582	776	970	1163	1357	1551	1745	1939	2133	2327	2521	2715	2909
M 6.0	81	162	243	324	405	486	647	809	971	1133	1295	1457	1619	1781	1942	2104	2266	2428
M 7.0	69	138	208	277	346	415	554	692	830	969	1107	1246	1384	1522	1661	1799	1938	2076
M 8.0	61	121	182	243	303	364	485	606	728	849	970	1091	1213	1334	1455	1577	1698	1819
M 10.0	48	97	145	194	242	291	388	485	582	679	776	873	970	1067	1163	1260	1357	1454
M 12.0	40	81	121	162	202	243	324	405	486	567	647	728	809	890	971	1052	1133	1214
M 14.0	35	69	104	139	173	208	277	347	416	485	555	624	693	763	832	901	971	1040
M 16.0	30	61	91	121	152	182	243	303	364	424	485	546	606	667	728	788	849	910
M 18.0	27	54	81	108	135	162	216	269	323	377	431	485	539	593	647	700	754	808
M 20.0	24	49	73	97	121	146	194	243	291	340	388	437	485	534	582	631	680	728
M 22.0	22	44	66	88	110	132	176	221	265	309	353	397	441	485	529	573	618	662
M 24.0	20	40	61	81	101	121	162	202	243	283	323	364	404	445	485	526	566	606
M 27.0	18	36	54	72	90	108	144	180	216	252	287	323	359	395	431	467	503	539
M 30.0	16	32	49	65	81	97	129	162	194	226	259	291	323	356	388	420	453	485

Table 42. Material/Coolant Tapping Recommendations. *(Source, Landis Threading Systems.)*

Material	Recommended Coolant No. See Coolant Schedule
Aluminum	4
Bakelite	2
Brass-All	4
Plastic	2
Copper	4
Fiber	2
Iron	
Cast	3
Malleable	1-6
Wrought	1-6
Magnesium	4-6
Monel Metal	5-7
Nickel	5-7
Rubber	2
Steel-All	6

Coolant Schedule
1. Soluble Oil
2. Dry
3. Dry or Soluble Oil
4. Light Mineral Oil
5. Mineral Lard (20% lard)
6. Sulphur Base
7. Sulphurized Oil

Table 43. Surface Treatments and Coatings for High-Speed Steel Taps. *(Source, Fastcut Tool Corp.)*

Description	Characteristics	Application
Nitride Approx. Hardness 1200 HV Rc 72	Consists of a thin, hardened case 0.0005" to 0.002" deep on the surface of the tool to resist abrasion and reduce galling.	Can be used in most abrasive materials, both ferrous and nonferrous. Not recommended where chipping may be a problem.
Double Nitride Approx. Hardness 1400 HV Rc 74	Consists of a higher hardened case on the surface of the tool to resist abrasion and reduce galling. Prone to brittleness and chipping.	Can be used on nonmetallic, highly abrasive materials such as Bakelite, plastics, hard rubber, and fibers.
Steam Oxide Approx. Hardness No change from Base Material	Consists of a layer of ferrous oxide on the surface of the tool which has good lubricant retaining properties. Improves toughness by relieving grinding stresses.	Can be used in low carbon, stainless, and free machining steels. Not recommended for use in soft, nonferrous materials where it may cause galling.
Nitride and Oxide Approx. Hardness 1200 HV Rc 72	A combination of two treatments which produces the favorable characteristics of both resistance to abrasion and galling.	Can be used in iron and cast iron, stainless, and high tensile steels. Not recommended for use in nonferrous materials where it may cause galling.
Chrome Plate Cr, Hard Chromium Approx. Hardness 1200 HV Rc 72	Consists of a very thin layer of hard chromium on the surface of the tool which reduces friction and prevents galling.	Can be used on most ferrous, nonferrous and nonmetallic materials. While unlikely, it may cause galling in high chromium stainless steels.

(Continued)

Table 43. *(Continued)* Surface Treatments and Coatings for High Speed Steel Taps. *(Source, Fastcut Tool Corp.)*

Description	Characteristics	Application
Titanium Nitride TiN, PVD Process Approx. Hardness 2400 HV *Rc 86	Consists of a very hard coating on the surface of the tool which has outstanding wear resistance, reduces friction, and prevents galling.	Can be used on most ferrous, nonferrous and nonmetallic materials. While unlikely, it may cause galling in titanium and titanium alloys.
Titanium Carbonitride TiCN, PVD Process Approx. Hardness 3000 HV *Rc 94	Consists of an extremely hard coating on the surface of the tool which has outstanding wear resistance, reduces friction, and prevents galling.	Can be used on most ferrous, nonferrous and abrasive materials. Very effective at higher speeds. While unlikely, it may cause galling in titanium and titanium alloys.
Chromium Carbide CrC, PVD Process Approx. Hardness 1850 HV Rc 80	Consists of a very hard coating on the surface of the tool which has excellent wear resistance, reduces friction, and prevents galling.	Can be used on titanium, titanium alloys, exotic materials, and die cast aluminum. Very effective at higher speeds and in many tapping applications. Under certain conditions it may cause galling in wrought aluminum.
Chromium Nitride CrN, PVD Process Approx. Hardness 1750 HV Rc 79	Consists of a very hard coating on the surface of the tool which has excellent wear resistance, reduces friction, and prevents galling.	Can be used on titanium, titanium alloys, nickel-base alloys, and copper alloys. Very effective at higher speeds and in many tapping applications. Under certain conditions it may cause galling in wrought aluminum.
Titanium Aluminum Nitride TiAlN, PVD Process Approx. Hardness 2600 HV *Rc 89	Consists of an extremely hard coating on the surface of the tool which has outstanding wear resistance, reduces friction, and prevents galling. Forms an aluminum oxide layer at high speeds and elevated temperatures.	Can be used on titanium, titanium alloys, nickel-base alloys, stainless steel and cast iron. Very effective at higher speeds and in some tapping applications. Not recommended for wrought aluminum, copper, and brass.

*Theoretical values for approximate comparison to the Vickers Hardness values.

Note: While most surface treatments and coatings have antigalling properties, they may cause galling in materials composed of or containing identical base elements. Also, steam oxide and some coatings may cause galling in soft materials such as aluminum.

Solving tapped hole size problems

Tapped holes (internal threads) are inspected with *GO Plug* and *NOT GO Plug* Thread Gages (NOT GO gages are sometimes called NO GO gages). See *Figure 3*. To determine that a thread is acceptable for production, *GO* thread gages must be able to freely enter and pass through the length of the thread, thereby verifying all thread components except the minor diameter. *NOT GO* gages, on the other hand, must fail to pass through (or in some instances enter) the length of the thread. Therefore, a GO/NOT GO check is a pass/fail test only. The size of the part being inspected is never measured.

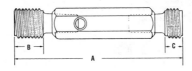

Figure 3. Typical thread plug gage. "A" is overall length, "B" indicates the GO member, and "C" shows the NOT GO member.

GO Plug gages check the extent of the tolerance, as applied to a specific screw thread, in the direction of the limit of maximum material and therefore represent the minimum limit of internal threads. The ideal thread will be a threaded counterpart of the GO Plug gage, made to precisely the same material limits. Thread Snap gages and Indicating gages have two significant advantages over GO Plug gages: 1) threads are inspected by applying the gage contacts within the internal thread at various discrete positions, and 2) threads can be inspected for roundness and taper my making additional checks along the thread length.

NOT GO Plug gages inspect the extent of tolerance in the direction of minimum material, which is the high limit of the internal thread. Separate gages are required to check pitch and major and minor diameter at minimum material. Thorough checking of the thread calls for a NOT GO Thread Snap gage. Two methods can be used for checking minimum material: functional diameter gaging or single element gaging.

Functional Diameter Gaging uses the NOT GO Plug gage when proper functioning of the thread assembly only requires control of the functional diameter of the thread at the minimum material limit. The gage should not engage for more than three threads. A NOT GO Thread Snap and Indicating gages are also applied at several points along the thread to assure compliance.

Single Element Gaging provides an economical control over the thread variables of lead, uniformity of helix, taper, roundness, and surface condition. Since Indicating gages give size value on production threads, their use alone provides reliable direction for adjusting any production irregularities.

Problems with thread and hole dimensions are indicated when GO gages fail to pass through the thread length, and when NOT GO gages do pass through, either partially or over the thread's full length. The following cases indicate trouble with tapped hole sizes. Figure 4 shows a typical thread design and the limits controlled by GO and NOT GO gages.

NOT GO gage enters full length. Obviously, this is an indication of an oversize hole, and the pitch diameter of the tap should be inspected to determine how much it is cutting oversize. If the amount is excessive, consider regrinding the tap to provide more hook or less width of land. This will require experimentation, and a more expeditious solution might to apply a different style of tap with a smaller pitch diameter to control the problem.

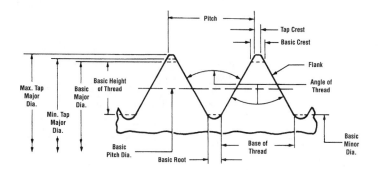

Figure 4. Thread dimensions controlled by thread plug gages. (Source, Kennametal Inc.)

Incorrect hook angle can also cause the tap to push, rather than cut, the thread, which will cause problems with the surface finish of the hole. If the hole is bell-mouthed, the likely cause is the alignment of the spindle and fixture. Spindle alignment can be especially troublesome when using a tapping head in a drill press. Another cause of tapping oversize holes is excessive feed pressure being applied to the tap. A lead screw can be employed or the pressure should be adjusted.

NOT GO gage partially enters the hole. In most cases, this indicates that the tap holder is not concentric with the spindle.

GO gage only partially enters the hole. This is an indication that the thread is tapered or the lead is out of specification. Taper normally indicates that the tap and the hole are not properly aligned. An error in the lead is caused by a defect in the tap and can only be corrected by replacement.

GO gage will not enter hole. Clearly, the first thing to check is the size of the tap and its pitch diameter. Check the tap's and the gage's markings to determine that the correct tap and gage is being used. The wall thickness of the workpiece may be causing the hole to distort. Try tapping the hole with roughing and finishing taps or a tap with more flutes. Another possibility is that the fixture is distorting the part. In this case, either reduce the clamping pressure or try using spiral point taps to remedy the problem.

Gages are available in several materials that can provide extra wear resistance. Those made of chrome plate and carbide are more resistant to wear than tool steel gages. GO gages, especially, are subject to wear through normal use and should be calibrated frequently. Calibration should be done at 68°F (20° C), and calibration schedules should be based on frequency of use, part abrasiveness, tolerance, and applicable quality procedures. With plug gages, GO is normally a plus tolerance and NOT GO is a minus tolerance. Gagemaker's tolerances for different classes of gages (XX, X, Y, Z, and ZZ are symbols for standard gage classes) are given in **Table 44**. See **Table 45** for GO and NOT GO Plug gage diameters for Class 2B and Class 3B Unified threads. **Table 46** gives GO and NOT GO Plug gage diameters for Class 6H Metric threads in both millimeters and inches.

Table 44. Gagemaker's Tolerance Chart for GO/NOT GO Plug Gages. *(Source, Vermont Gage Co.)*

Range	Class				
	XX	X	Y	Z	ZZ
Inch (Dimensions in Inches)					
.0009" to .8250"	.000020	.000040	.000070	.000100	.000200
.8251" to 1.5100"	.000030	.000060	.000090	.000120	.000240
1.5101" to 2.5100"	.000040	.000080	.000120	.000160	.000320
2.5101" to 4.5100"	.000050	.000100	.000150	.000200	.000400
4.5101" to 6.5100"	.000065	.000130	.000190	.000250	.000500
6.5101" to 9.0100"	.000080	.000160	.000240	.000320	.000640
9.0101" to 12.2600"	.000100	.000200	.000300	.000400	.000800
Metric (Dimensions in Millimeters)					
1.00mm to 21.00mm	.0005	.0010	.0018	.0030	.0050
21.01mm to 38.00mm	.0008	.0015	.0023	.0030	.0060
38.01mm to 64.00mm	.0010	.0020	.0030	.0040	.0080
64.01mm to 115.00mm	.0013	.0025	.0038	.0050	.0100
115.01mm to 165.00mm	.0017	.0033	.0048	.0060	.0130
165.01mm to 230.00mm	.0020	.0041	.0061	.0080	.0160
230.01mm to 300.00mm	.0025	.0051	.0076	.0100	.0200

Table 45. GO/NOT GO Thread Plug Gage Pitch Diameters for Class 2B and Class 3B Unified Threads.
(Source, Kennametal Inc.)

Nominal Size	GO Basic All Classes All Series	Unified Pitch Diameters				
		NOT GO		X Tol. GO = + NOT GO = -	X Lead Tol.	X ± Half Angle Tol.
		Class 2B	Class 3B			
#0-80 UNF	.0519	.0542	.0536	.0002	.0002	0°30'
#1-64 UNC	.0629	.0655	.0648	.0002	.0002	0°30'
#1-72 UNF	.0640	.0665	.0659	.0002	.0002	0°30'
#2-56 UNC	.0744	.0772	.0765	.0002	.0002	0°30'
#2-64 UNF	.0759	.0786	.0779	.0002	.0002	0°30'
#3-48 UNC	.0855	.0885	.0877	.0002	.0002	0°30'
#3-56 UNF	.0874	.0902	.0895	.0002	.0002	0°30'
#4-40 UNC	.0958	.0991	.0982	.0002	.0002	0°20'
#4-48 UNF	.0985	.1016	.1008	.0002	.0002	0°30'
#5-40 UNC	.1088	.1121	.1113	.0002	.0002	0°20'
#5-44 UNF	.1102	.1134	.1126	.0002	.0002	0°20'
#6-32 UNC	.1177	.1214	.1204	.0003	.0003	0°15'
#6-40 UNF	.1218	.1252	.1243	.0002	.0002	0°20'
#8-32 UNC	.1437	.1475	.1465	.0003	.0003	0°15'
#8-36 UNF	.1460	.1496	.1487	.0002	.0002	0°20'
#10-24 UNC	.1629	.1672	.1661	.0003	.0003	0°15'
#10-32 UNF	.1697	.1736	.1726	.0003	.0003	0°15'
#12-24 UNC	.1889	.1933	.1922	.0003	.0003	0°15'
#12-28 UNF	.1928	.1970	.1959	.0003	.0003	0°15'
#12-32 UNEF	.1957	.1998	.1988	.0003	.0003	0°15'
1/4-20 UNC	.2175	.2224	.2211	.0003	.0003	0°15'
1/4-28 UNF	.2268	.2311	.2300	.0003	.0003	0°15'
1/4-32 UNEF	.2297	.2339	.2328	.0003	.0003	0°15'
5/16-18 UNC	.2764	.2817	.2803	.0003	.0003	0°10'
5/16-24 UNF	.2854	.2902	.2890	.0003	.0003	0°15'
5/16-32 UNEF	.2922	.2964	.2953	.0003	.0003	0°15'
3/8-16 UNC	.3344	.3401	.3387	.0003	.0003	0°10'
3/8-24 UNF	.3479	.3528	.3516	.0003	.0003	0°15'
3/8-32 UNEF	.3547	.3591	.3580	.0003	.0003	0°15'
7/16-14 UNC	.3911	.3972	.3957	.0003	.0003	0°10'
7/16-20 UNF	.4050	.4104	.4091	.0003	.0003	0°15'
7/16-28 UNEF	.4143	.4189	.4178	.0003	.0003	0°15'
1/2-13 UNC	.4500	.4565	.4548	.0003	.0003	0°10'
1/2-20 UNF	.4675	.4731	.4717	.0003	.0003	0°15'
1/2-28 UNEF	.4768	.4816	.4804	.0003	.0003	0°15'
9/16 UNC	.5084	.5152	.5135	.0003	.0003	0°10'
9/16 UNF	.5264	.5323	.5308	.0003	.0003	0°10'
9/16-24 UNEF	.5354	.5405	.5392	.0003	.0003	0°15'
5/8-11 UNC	.5660	.5732	.5714	.0003	.0003	0°10'
5/8-18 UNF	.5889	.5949	.5934	.0003	.0003	0°10'
5/8-24 UNEF	.5979	.6031	.6018	.0003	.0003	0°15'
11/16-24 UNEF	.6604	.6656	.6643	.0003	.0003	0°15'
3/4-10 UNC	.6850	.6927	.6907	.0003	.0003	0°10'
3/4-16 UNF	.7094	.7159	.7143	.0003	.0003	0°10'
3/4-20 UNEF	.7175	.7232	.7218	.0003	.0003	0°15'
13/16-20 UNEF	.7800	.7857	.7843	.0003	.0003	0°15'

(Continued)

Table 45. *(Continued)* GO/NOT GO Thread Plug Gage Pitch Diameters for Class 2B and Class 3B Unified Threads. *(Source, Kennametal Inc.)*

Nominal Size	GO Basic All Classes All Series	Unified Pitch Diameters		X Tol. GO = + NOT GO = -	X Lead Tol.	X ± Half Angle Tol.
		NOT GO				
		Class 2B	Class 3B			
7/8-9 UNC	.8028	.8110	.8089	.0003	.0003	0°10'
7/8-14 UNF	.8286	.8356	.8339	.0003	.0003	0°10'
7/8-20 UNEF	.8425	.8482	.8468	.0003	.0003	0°15'
15/16-20 UNEF	.9050	.9109	.9094	.0003	.0003	0°15'
1"-8 UNC	.9188	.9276	.9254	.0004	.0004	0°05'
1"-12 UNF	.9459	.9535	.9516	.0003	.0003	0°10'
1"-14 UNS	.9536	.9609	.9590	.0003	.0003	0°10'
1"-20 UNEF	.9675	.9734	.9719	.0003	.0003	0°15'
1 1/16-18 UNEF	1.0315	1.0264	1.0310	.0003	.0003	0°10'
1 1/8-7 UNC	1.0322	1.0416	1.0393	.0004	.0004	0°05'
1 1/8-12 UNF	1.0709	1.0787	1.0768	.0003	.0003	0°10'
1 1/8-18 UNEF	1.0889	1.0951	1.0935	.0003	.0003	0°10'
1 3/16-18 UNEF	1.1514	1.1577	1.1561	.0003	.0003	0°10'
1 1/4-7 UNC	1.1572	1.1668	1.1644	.0004	.0004	0°05'
1 1/4-12 UNF	1.1959	1.2039	1.2019	.0003	.0003	0°10'
1 1/4-18 UNEF	1.2139	1.2202	1.2186	.0003	.0003	0°10'
1 5/16-18 UNEF	1.2764	1.2827	1.2811	.0003	.0003	0°10'
1 3/8-6 UNC	1.2667	1.2771	1.2745	.0004	.0004	0°05'
1 3/8-12 UNF	1.3209	1.3291	1.3270	.0003	.0003	0°10'
1 3/8-18 UNEF	1.3389	1.3452	1.3436	.0003	.0003	0°10'
1 7/16-18 UNEF	1.4014	1.4079	1.4062	.0003	.0003	0°10'
1 1/2-6 UNC	1.3917	1.4022	1.3996	.0004	.0004	0°05'
1 1/2-12 UNF	1.4459	1.4542	1.4522	.0003	.0003	0°10'
1 1/2-18 UNEF	1.4639	1.4704	1.4687	.0003	.0003	0°10'

Table 46. GO/NOT GO Thread Plug Gage Pitch Diameters for Class 6H Metric Threads.
(Source, Vermont Gage Co.)

Size	Pitch	Millimeters		Inch Equivalent	
		GO	NOT GO	GO	NOT GO
M1.6	.35	1.373	1.458	.05406	.05740
M1.8	.35	1.573	1.658	.06193	.06528
M2.0	.40	1.740	1.830	.06850	.07205
M2.2	.45	1.908	2.003	.07512	.07886
M2.5	.45	2.208	2.303	.08693	.09067
M3.0	.50	2.675	2.775	.10531	.10925
M3.5	.60	3.110	3.222	.12244	.12685
M4.0	.70	3.545	3.663	.13957	.14421
M4.5	.75	4.013	4.131	.15799	.16264
M5.0	.80	4.480	4.605	.17638	.18136
M5.5	.75	5.013	5.131	.19736	.20201
M6.0	1.00	5.350	5.500	.21063	.21654
M7.0	1.00	6.350	6.500	.25000	.25591

(Continued)

Table 46. *(Continued)* **GO/NOT GO Thread Plug Gage Pitch Diameters for Class 6H Metric Threads.** *(Source, Vermont Gage Co.)*

Size	Pitch	Millimeters		Inch Equivalent	
		GO	NOT GO	GO	NOT GO
M8.0	1.25	7.188	7.348	.28299	.28929
M9.0	1.25	8.188	8.348	.32236	.32866
M10.0	1.50	9.026	9.206	.35535	.36244
	1.25	9.188	9.348	.36173	.36803
	.75	9.513	9.645	.37453	.37972
M11.0	1.50	10.026	10.206	.39472	.40181
M12.0	1.75	10.863	11.063	.42768	.43555
	1.50	11.026	11.216	.43409	.44157
	1.25	11.188	11.368	.44047	.44756
M14.0	2.00	12.701	12.913	.50004	.50838
	1.50	13.026	13.216	.51283	.52031
M15.0	1.50	14.026	14.216	.55220	.55968
	1.00	14.350	14.510	.56496	.57126
M16.0	2.00	14.701	14.913	.57878	.58712
	1.50	15.026	15.216	.59157	.59905
M17.0	1.50	16.026	16.216	.63094	.63842
	1.00	16.350	16.510	.64370	.65000
M18.0	2.50	16.376	16.600	.64472	.65354
	1.50	17.026	17.216	.67031	.67779
M20.0	2.50	18.376	18.600	.72346	.73228
	1.50	19.026	19.216	.74905	.75653
	1.00	19.350	19.510	.76181	.76811
M22.0	2.50	20.376	20.600	.80220	.81102
	1.50	21.026	21.216	.82779	.83557
M24.0	3.00	22.051	22.316	.86815	.87858
	2.00	22.701	22.925	.89374	.90256
M25.0	2.00	23.701	23.925	.93311	.94193
	1.50	24.026	24.226	.94590	.95378
M26.0	1.50	25.026	25.226	.98527	.99315
	1.00	25.350	25.520	.99803	1.00472
M27.0	3.00	25.051	25.316	.98626	.99669
	2.00	25.701	25.925	1.01185	1.02067
M28.0	2.00	26.701	26.925	1.05122	1.06004
	1.50	27.026	27.226	1.06401	1.07189
	1.00	27.350	27.520	1.07677	1.08346
M30.0	3.50	27.727	28.007	1.09161	1.10264
	2.00	28.701	28.925	1.12996	1.13878
	1.50	29.026	29.226	1.14276	1.15063
M32.0	2.00	30.701	30.925	1.20870	1.21752
	1.50	31.026	31.226	1.22149	1.22937
M33.0	3.50	30.727	31.007	1.20972	1.22075
	2.00	31.701	31.925	1.24807	1.25689
M35.0	1.50	34.026	34.226	1.33960	1.34748
M36.0	4.00	33.402	33.702	1.31504	1.32685
	3.00	34.051	34.316	1.34059	1.35102
	2.00	34.701	34.925	1.36618	1.37500

(Continued)

Table 46. *(Continued)* GO/NOT GO Thread Plug Gage Pitch Diameters for Class 6H Metric Threads.
(Source, Vermont Gage Co.)

Size	Pitch	Millimeters		Inch Equivalent	
		GO	NOT GO	GO	NOT GO
M38.0	1.50	37.026	37.226	1.45771	1.46559
M39.0	4.00	36.402	36.702	1.43320	1.44500
	3.00	37.051	37.316	1.45870	1.46910
	2.00	37.701	37.925	1.48429	1.49311

Thread Rolling

Manufacturing male threads

Generally speaking, threads are manufactured either by rolling or cutting. In fact, over 90% of commercially available bolts up to 2 inches (51 millimeters) were manufactured by rolling. In addition, it is an economically efficient process. Since rolled threads are cold formed, no material is removed from the blank, resulting in significant material savings: for example, to cut a $^1/_2$ – 13 UNC bolt requires almost 20% more material than rolling the same bolt. The information in this handbook pertaining to thread rolling was provided by Landis Threading Systems, who have been producing threads by this method for almost 100 years.

Rolling versus cutting in thread production

Rolling is a chipless cold forming process capable of producing threads at high threading speeds with good tool life. Not all threads should be rolled. The following application considerations should be taken into account before beginning any rolling operation.

    1. In the majority of instances, materials that are well suited to rolling do not cut well, and those well suited to cutting do not roll well.

    2. In order to be suitable for rolling, the material should have an elongation factor of at least 12%. This is the characteristic that allows material to be plastically and permanently deformed. Rollability of materials will be covered in more detail later in this section.

    3. The design or end use of the workpiece may dictate the use of a certain material. Since cast iron, for example, does not flow, it is not a rollable material and its use would require cutting the threads.

    4. Since thread roll dies will accept only a specific volume of rolled material, using an oversize blank will result in excessive material flow that overfills the dies and can result in die breakage. Therefore, the blank diameter must be tightly controlled within specified limits. Recommended blank diameters for classes of threads are given later in this section.

Where finish is a prime consideration, rolling is superior to cutting. Some workpiece prints specify a finish on the thread, and rolling can be relied on to produce finishes 32 μin. (.813 μm) or less. Cutting, on the other hand, rarely produces threads better than 63 μin. (1.6 μm). Strength is another factor to consider. Because rolling does not cut across the grain structure, and also due to the high compressive forces created by the process, rolled threads are as much as 20% stronger than cut threads. Cutting shears the grain at every thread form, resulting in weakened threads (see *Figure 5a*).

When threading from the end of a workpiece, it is often necessary to roll close to a shoulder, and in such instances cutting is the preferred option because the cutting edge

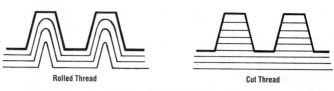

Rolled Thread **Cut Thread**

Figure 5a. Cross section of rolled thread and cut thread. (Source, Landis Threading Systems.)

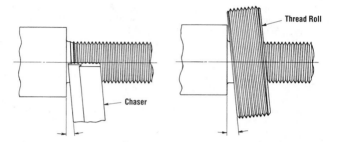

Thread Roll

Chaser

Figure 5b. Difference in proximity to shoulder in cutting (left) and rolling (right). (Source, Landis Threading Systems.)

lies at the helix angle on the center line of the work and tangentially stops at that point. The thread roll also lies at the helix angle, and contacts the workpiece on the center line; but instead of stopping at the point of contact, it must continue full circle. In doing so it is often unable to produce a thread close enough to the shoulder because the curve of the roll beyond the point of contact is beyond the center line and will strike the shoulder, as shown in *Figure 5b*.

On certain nonferrous metals (zinc and aluminum die castings, some aluminum and brass bars), thread cutting can be performed at relatively high speeds. However, on steels (especially the higher alloys), it is necessary to stay within the 5 to 60 sfm (1.5 to 19 smm) range. By comparison, a rolled thread can usually be initially cut at 100 sfm (30 smm) and in some instances taken as high as 400 sfm (120 smm). At higher speeds, caution should be exercised because the axial travel is very rapid, and short threads or threading going up to the shoulder can be difficult to control. Where the rolling head is being used on an automatic screw machine, and the threading operation does not govern the cycle time, using high speeds and having the head idle for some other operation is counterproductive. A speed of 100 sfm (30 smm) is usually sufficient for most operations.

When very deep, coarse, or multiple threads are required, cutting is preferred over rolling. It is possible when using cutting heads to make more than one pass without damage to the tool. This distributes the chip load over more cutting edges, thus improving both finish and tool life. This is not good practice in thread rolling because any tendency to workharden on the first pass will require the roll to work much harder on the second. Multiple threads introduce the problem of long throats and higher pressures than single starts. In cutting, the second pass can be adjusted so that it removes just enough metal to smooth up the thread if desired. On multiple start threads, especially Acme and trapezoidal, where the helix angle is high and the rear flank on the chaser has a negative rake, the throat teeth can be ground individually to provide positive rake for better cutting action. With a rolling head, the rolls cannot be turned to an angle great enough to accommodate many types of multiple threads.

Threading behind the shoulder can be done using a straddle type thread rolling attachment. This tool generally operates from the cross slide of an automatic screw machine simultaneously with other operations, and it is capable of rolling on the collet side of a shoulder even when the shoulder is larger than the thread. Attachments permit rolling very close to a shoulder or into relatively small relief. They can roll either straight or tapered threads, and have all the advantageous characteristics of other roll threading procedures.

Although thread cutting is acceptable for long runs, one of its advantages is that it requires less downtime than rolling in changeover for short runs. Also, there is a significant difference in the initial cost of a thread chaser compared to the greater cost of a roll. Many times, the workpiece for cut threads does not require the accuracy that a rolling blank does because the chaser can be made with a throat chamfer below the root of the new thread that acts as a hollow mill to remove excess metal.

Where it isn't necessary to have the full thread diameter on the workpiece back of the thread, such as with bent bolts, it is less expensive, from a blank material standpoint, to roll the threads. For instance, if rolling stock is purchased to blank diameter, there is approximately about a 20% saving in weight realized for $^1/_2$–13 UNC, and more than 15% for 1–8 UNC over the bar necessary for a cut thread.

Steel stock containing small amounts of lead is used, especially on automatic screw machines, for increasing machinability by as much as 25%. These steels do not work well for roll threading because the lead inclusions have a tendency to be squeezed out of the parent material as flakes, and will both foul the coolant and do serious damage to the surface finish. The same result occurs when rolling high sulfur steels.

Rollability of materials

A material's hardness and its percent of elongation factor indicate whether it can be plastically and permanently deformed. Generally, 32 R_c is the hardest material that is practical to roll. The elongation factor should be at least 12%. Other factors that influence a material's rollability are its yield point, microstructure, and the degree and speed at which it workhardens. Also, its modulus of elasticity, the nonmetallic content (in steel), and the workpiece diameter and pitch can be factors.

By its nature, thread rolling generates a great deal of pressure and heat, and coolant should be applied at the point of roll contact with the workpiece. Applying the coolant to the point of contact is especially important with revolving type heads which tend to sling the coolant away from the work. Cutting oil offers superior lubrication and works well for certain applications, but soluble oils are superior in carrying heat from the operation. See the "Lubricoolants" heading below for lubricant/coolant recommendations.

Basic open-hearth steels in the soft state are rollable providing carbon content does not exceed 1.5%. Sulfurized steels are rollable depending on the severity of cold working to be done and the percentage of sulfur content. While sulfur additives enhance the machinability of a cutting operation, they are not desirable for cold forming. Therefore, sulfur content should not exceed .13%, as higher rates cause extremely hard sulfur inclusions that require higher rolling pressures. In addition to shortening die life, inclusions resist cold forming, cause flaking, and can result in roll breakage.

Soft and malleable leaded steels are thought to have good rollability. However, while desirable for enhancing machinability, it is not conclusive that additives are compatible with rolling. Lead inclusions are softer than the base material, and lead inclusions result in intermittent and varying die loading that contributes to poor die life. Therefore, lead content should be no more than 0.1%.

Stainless steels are prone to workhardening. Most of the 400 Series steels are considered best for rolling, with 410 yielding the best results. Because of their extreme workhardening characteristics, the 300-Series should be avoided if possible even though their chemical analysis suggests that they would be appropriate.

To what degree workhardening will occur depends on the analysis of the material, its initial hardness, the severity of deformation, and the number of cold working cycles imposed on the workpiece. The greatest amount of deformation with the greatest hardness occurs in the root area. The least occurs on the flanks. Since the grain structure flows along the thread contour, the deeper and finer grain in the root reflects the high degree of hardness in that area. Workhardening resulting from rolling is superficial, and rolled materials retain core hardness 0.030" to 0.040" (0.76 mm to 1.02 mm) beneath the surface. Where a material has higher workhardening tendencies, hardening can extend as far as 0.150" (3.81 mm) below the surface. Die life will proportionally increase/decrease in direct ratio to any increase in workhardening. See **Tables 47** through **49**, which list the rollability factor and die life longevity for various materials.

Table 47. Rollability of Materials: Carbon and Alloy Steels. *(Source, Landis Threading Systems.)*

Material Designation	Thread Finish	Proportional Die Life				Remarks
		Soft	R_c 15-24	R_c 25-32	R_c 33+	
AISI 1008-1095	E	H	H-M	M	L	Excellent rollability
AISI 1108-1151 AISI 1211-1215 AISI B1111-B1113	G F F	H M H	H-M	M		These are free machining steels with high sulfur content. The highest sulfur materials (1110, 1144, and 1200) should be avoided when possible.
AISI 1330-1345 AISI 4012-4047	E E	H H-M	M M	M-L M-L	L L	These are medium alloy steels such as manganese, molybdenum, chrome, and nickel. Workhardening of material requires higher pressures, and some reduction in roll life over the 1000-1200 series will be experienced.
AISI 4118-4161	E	H-M	M	M-L	L	
AISI 4320-4340 AISI 4419 AISI 4615-4626 AISI 4718-4720 AISI 4815-4820	E	H-M	M	M-L	L	
AISI 5015-5060 AISI 5115-5160 AISI E51100-E52100	E	H-M	M	M-L	L	
AISI 6118-6150	E	H-M	M	M-L	L	
AISI 8615-8655 AISI 8720-8740 AISI 8822 AISI 9255-9260	E	H-M	M	M-L	L	
Stainless 302-304 309-317	E		M-L			These are nonhardenable austenitic steels containing higher quantities of nickel and chromium. High workhardening occurs with higher percent alloys. Also nonmagnetic. Recommend Carpenter #10 wherever possible. Material does not seam.
Stainless 305, 321, 347, 348 Carpenter & 12-20CB	E		M	L		
Stainless Carpenter #10	E			H-M	M-L	

(Continued)

Table 47. *(Continued)* **Rollability of Materials: Carbon and Alloy Steels.** *(Source, Landis Threading Systems.)*

Material Designation	Thread Finish	Proportional Die Life				Remarks
		Soft	R_c 15-24	R_c 25-32	R_c 33+	
Stainless 329-430F-446	E		M-L	L		Nonhardenable ferritic chromium stainless but magnetic. Lower work hardening but higher pressures required due to carbon.
Stainless 430-443	E		M	L		
Stainless 414-420F-440F	E		M-L	L		Hardenable martensitic chromium steels, magnetic. Most suited for rolling of the stainless grades, low workhardening.
Stainless 410-431-440C-501-502	E		H-M	L		
High Speed T1, M1, M2	E		M-L	L		Not generally recommended, however can be rolled under proper conditions.
Nitralloy 135-230	E		M	M-L	L	Not rollable after nitriding.

Letter designations for finish: E-excellent, G-good, F-fair, P-poor.
Letter designations for die life: H-high, M-medium, L-low.
Elongation factor: generally acceptable results can be achieved when percent elongation equals twelve (12) or more.

Table 48. Rollability of Materials: Wrought Copper and Copper Alloys. *(Source, Landis Threading Systems.)*

Material Designation		Alloy Name	Max. Hardness	Finish	Die Life	Remarks
SAE No.	ASTM No.					
CA102	B124 #12	Oxygen Free Copper	R_F 40	E	H	More than 90% copper.
CA110	B124 #12	Electrolytic Copper (ETP)	R_F 40	E	H	Excellent rollability.
CA122	B124 #12	Phosphorus Deoxidized (DHP)	R_F 45	E	H	
CA210	B36 #1	Gilding 95%	R_B 40	E	H	Copper-zinc alloys
CA220	B36 #2	Commercial Bronze 90%	R_B 42	E	H	basically good for
CA230	B36 #3	Red Brass	R_B 55	E	H	rolling except when zinc
CA240	B36 #4	Low Brass 80%	R_B 55	E	H	exceeds 30%. This tends
CA260	B36 #6	Cartridge Brass 70%	R_B 60	E	H	to produce poor finish
CA270	B36 #8	Yellow Brass	R_B 55	F	H	as indicated in
CA280	B135 #5	Muntz Metal	R_B 78	P	M	CA270-CA280.
CA314	B140-B	Leaded Comm. Bronze	R_B 65	P	M-H	Copper zinc alloys
CA335	B121 #2	Low Lead Brass	R_B 60	G	M-H	with lead added for
CA340	B121 #3	Medium Leaded Brass	R_B 60	G	M-H	improved machining
CA342	B121 #5	High Leaded Brass	R_B 55	P	M	characteristics. Poor to
CA345		Thread Rolling Brass	R_B 75	G	M-H	fair for rolling. Higher
CA353	B121 #5	High Leaded Brass 62%	R_B 55	P	M	lead produces poorer
CA356	B121 #6	Extra High Leaded Brass	R_B 55	P	L-M	thread finish and is not
CA360	B16	Free Cutting Brass	R_B 70	P	L	recommended for
CA365 to						rolling.
CA368	B171	Leaded Muntz Metal	R_B 70	P	M	
CA370	B135 #6	Free Cutting Muntz Metal	R_B 70	P	L	
CA377	B124 #2	Forging Brass	R_B 65	P	L	
CA385		Architectural Bronze	R_B 65	P	L	
CA443 to						Copper zinc alloy with 1.0% tin excellent
CA445	B171	Inhibited Admiralty	R_B 75	E	H	rollability.

(Continued)

Table 48. *(Continued)* **Rollability of Materials: Wrought Copper and Copper Alloys.**
(Source, Landis Threading Systems.)

Material Designation		Alloy Name	Max. Hardness	Finish	Die Life	Remarks
SAE No.	ASTM No.					
CA464 to CA467 CA485	B124 #3 B21-C	Naval Brass Leaded Naval Brass	R_B75 R_B80	P-F P	M L	Copper zinc alloy with lead and tin not conducive to good rolling characteristics. Alternate material should be used.
CA502 CA510 CA521 CA524 CA544	B105 B139-A B139-C B139-D B139-B2	Phosphor Bronze E Phosphor Bronze A Phosphor Bronze C Phosphor Bronze D Free Cutting Phosphor Bronze	R_B50 R_B65 R_B70 R_B70 R_B70	G G P-F P P	H H M-H M L	Copper tin alloy generally good for rolling but increasing tin content reduces rollability. CA544 contains some lead & zinc thereby reducing its rollability.
CA606 CA614 CA617 CA360 CA639	B150-3 B150-1 B150-2	Aluminum Bronze Aluminum Bronze D Aluminum Bronze Aluminum Bronze Aluminum Silicon Bronze	R_B70 R_B70 R_B70 R_B70 R_B75	G G P P G	M M L-M L L	Copper aluminum alloy fair to good rolling characteristics. Increased quantities of silicon nickel, introduce workhardening and reduce rollability.
CA651 CA655	B98-B B98-A	Low Silicon Bronze B High Silicon Bronze A	R_B70 R_B75	E FG	M M	Copper with silicon as basic alloy. Average rollability.
CA675	B138-A	Manganese Bronze A	R_B70	P	M	High zinc alloy, alternate material should be used.
CA706 CA715	B111 B111	Copper Nickel-10% Copper Nickel-30%	R_B70 R_P70	G G	M-H M	High nickel alloy reduces rollability proportionally.
CA745 CA752 CA754 CA757 CA770	B151-E B151-A B151-D B151-B	Nickel Silver 65-10 Nickel Silver 65-18 Nickel Silver 65-15 Nickel Silver 65-12 Nickel Silver 55-18	R_B70 R_B70 R_B70 R_P70 R_B70	E G-E E E G	H H H H M	Copper with zinc and nickel as alloy, rollability good to excellent. As alloy increases, rollability decreases.

Copper Casting Alloys, Annealed
Copper casting alloys, in the annealed condition are, for the most part, rated as poor rollability and poor die life. The copper alloys with basic quantities of tin, zinc, or silicon rate slightly better in die life with poor to fair finish. It is recommended that these materials be avoided where possible, and should only be considered for low production quantities.

Letter designations for finish: E-excellent, G-good, F-fair, P-poor.
Letter designations for die life: H-high, M-medium, L-low.
Elongation factor: generally acceptable results can be achieved when percent elongation equals twelve (12) or more.

Table 49. Rollability of Materials: Wrought Aluminum and Aluminum Alloys. *(Source, Landis Threading Systems.)*

Material Designation		Condition	Max. Hardness	% Elongation	Finish	Die Life	Remarks
SAE No.	ASTM No.						
1100-0	990A	Annealed	R_B23	45	E	H	99% aluminum
1100-H12	990A	1/4 Hard	R_B28	25	E	H	recommended for rolling.
110-H14	990A	1/2 Hard	R_B32	20	G-E	H	Workhardens very slowly,
1100-H16	990A	3/4 Hard	R_B38	17	G	H	cannot be heat treated.
1100-H18	990A	Full Hard	R_B44	15	F-G	M	Major alloy is silicon.
2011-T3	CB60A	Heat treated and cold worked	R_B95	15	FG	M-H	Lower quality finish is a result of lead and
2011-T6C	B60A	Heat treated and aged	R_B97	17	F	M-H	Bismuth alloys,
2011-T8	CB60A	Heat treated, cold worked & aged	R_B100	12	P-F	M-H	not generally recommended for rolling.
2014-0	CS41A	Annealed	R_B45	18	G	M-H	Copper, silicon,
2014-T4	CS41A	Heat treated and aged	R_B105	20	G-E	M-H	manganese major alloys,
2014-T6	CS41A	Heat treated and aged	R_B135	13	F	M-H	higher strength requires greater roll pressure.
2017-0	CM41A	Annealed	R_B45	22	E	H	Good rollability. Most
2017-T4	CM41A	Heat treated and aged	R_B105	22	E	H	commonly used for
2024-0	CG42A	Annealed	R_B47	22	E	H	rolling.
2024-T3	CG42A	Heat treated, cold worked	R_B120	18	E	H	
2024-T4	CG42A	Heat treated and aged	R_B120	18	E	H	
2117-T4	CG30A	Heat treated and aged	R_B70	27	E	H	
3003-0	MIA	Annealed	R_B28	40	E	H	99% aluminum,
3003-H12	MIA	1/4 Hard	R_B35	20	G-E	H	recommended for rolling.
3003-H14	MIA	1/2 Hard	R_B40	16	G	H	Work hardens very
3003-H16	MIA	3/4 Hard	R_B47	14	F	M	slowly, cannot be heat
3003-H18	MIA	Full Hard	R_B55	10	P-F	L-M	treated. Major alloy is manganese.
5052-0	CR20A	Annealed	R_B47	30	E	H	Fair to good rollability in
5052-H32	CR20A	1/4 Hard	R_B60	18	G	M	the lower hardness
5052-H34	CR20A	1/2 Hard	R_B68	14	F	M	condition, major alloy
5052-H36	CR20A	3/4 Hard	R_B73	10	P-F	L-M	manganese with
5053-H38	CR20A	Full Hard	R_B77	8	P	L	chromium.
5056-0	GM50A	Annealed	R_B65	35	E	H	Major alloy magnesium.
5056-H18	GM50A	Strain Hardened	R_B105	10	P	L-M	Recommend rolling in
5056-H38	GM50A	Strain Hardened and Stabilized	R_B100	15	P-F	L-M	annealed condition only.
6061-0	GS11A	Annealed	R_B30	30	E	H	Good to excellent
6061-T4	GS11A	Heat treated and aged	R_B25	65	G-E	H	rollability in conditions.
6061-T6	GS11A	Heat treated and aged	R_B17	95	G	H	
7075-0	ZG62A	Annealed	R_B60	16	F	H	Generally not
7075-T6	ZG62A	Heat treated and aged	R_B150	11	P	M	recommended for rolling.

Letter designations for finish: E-excellent, G-good, F-fair, P-poor.

Letter designations for die life: H-high, M-medium, L-low.

Elongation factor: generally acceptable results can be achieved when percent elongation equals twelve (12) or more.

Preparing and controlling blank diameters

The principles of rolling techniques will vary somewhat depending upon whether a rolling head, attachment, or machine is being used. However, the basic principles apply to all types of equipment. One requirement is the need to maintain the blank diameter within specific limits. Failure to do so can result in premature wear or early failure of the dies. Rolling dies can be compared to a vessel that holds liquid: they can only hold an amount of flowed metal equal to (but not more than) their displacement. While a vessel can overflow, rolling dies cannot. Since the roll life is much greater than that of a cutting tool, extra effort spent monitoring and controlling the blank to prevent breakage or premature wear is more than offset by the performance maintained. Unlike the cutting process which machines excess material away from the workpiece, rolling requires that the blank contain only enough material to fill the die cavity and form the thread. When too much material is present, the die cavity is overfilled. Conversely, too little material will result in an incorrectly formed thread.

An oversize blank exerts pressure against the dies, forcing them outward and damaging the die or producing an oversize thread. Oversize threads cannot be corrected by sizing the head smaller as that remedy would aggravate the condition and produce even greater pressure on the dies. The blank diameters given in **Tables 50** through **55** are recommended starting points. Depending on the material and its metallurgical makeup, the final blank diameter must finally be established by actual rolling. Although not always visible to the naked eye, rolled threads have a seam that is the result of material being flowed from the root area up to and folded in at the crest. Depending on the coarseness of the pitch and how tightly the thread is being rolled, the seam will be more visible on some threads than on others. Better die life will be realized by starting with the minimal acceptable blank diameter (which advantageously allows for growth) and accepting a larger seam. Seams have no bearing on thread strength, but they are often objected to on the basis of appearance. If appearance is a factor, threads can be rolled tighter (die life will be shortened) or the thread can be rolled with an oversize addendum which can be ground away to remove the seam.

With balanced threads such as the UN 60° included thread angle, the volume of the thread above the pitch diameter will approximate that below. The volume below is that which is flowed and displaced upward to form the addendum. Thus, it is apparent that, with a balanced form, the prepared blank's outside diameter will approximate the thread's pitch diameter. Starting at the minimum recommended blank diameter allows a certain amount of blank growth due to tool wear and allows the longest possible running time before the maximum allowable blank diameter is reached. Therefore, in practice, the blank diameter should be less than the maximum thread pitch diameter, and tolerances on the blank diameter should be as small as practical. When rolling shorter length threads, it may be necessary to increase the blank diameter slightly to compensate for endwise stretching.

Die thread form chipping can be minimized by beveling the end of the blank as shown in *Figure 6*. The small bevel diameter at the blank end should be equal to the thread minor diameter minus 0.10" (0.255 mm). Rolling will force the end threads outward and the finished bevel will be approximately 45°. Blanks should be as round and as straight as possible as rolling will not correct these inaccuracies. Avoid variations in diameter along the blank length. These cause uneven pressure distribution that can overload the rolls and result in premature failure. Maintain the smoothest finish possible on the blank, as it will influence the final thread smoothness. Finally, be aware that blank diameter may vary according to material hardness.

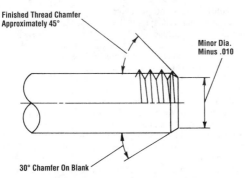

Figure 6: Proper beveling of the end of a blank. (Source, Landis Threading Systems.)

Lubricoolant recommendations for thread rolling

The pressure developed between the workpiece and the dies of a thread rolling machine can theoretically approach 500,000 psi (3,450 MPa) depending on the type of thread and workpiece hardness. During the rolling operation, the thread surfaces of the dies and workpiece are running together at a surface speed differential, because the largest surface of each die is in contact with and running on the smallest diameter of the thread. Accordingly, slippage under extreme pressure occurs between these two surfaces at diameters larger and smaller than the resolved, effective, rotating diameter which—for purposes of consideration—can be regarded as the pitch diameter. While this contributes to the heat generated by the rolling operation, the greatest amount of total heat is generated by the material being flowed or deformed to produce the thread.

Depending on the efficiency of the thread rolling machine, most of the input energy is converted to heat that must be dissipated by the dies and the lubricant, which acts as a coolant (hence, lubricoolant). The remaining heat in the workpiece is dissipated into the environment. It is desirable to remove as much heat as possible, as quickly as possible. Thus the lubricant with the best wetting and extreme pressure properties should be used.

Water, with a specific heat of 1.000, will absorb heat better than any other readily available substance. The specific heat of most oils ranges between .350 and .550, so water will clearly *absorb* twice as much heat as most oils. For this reason, it is desirable to use a water emulsion for thread rolling. Since the die spindle bearings in some thread rolling machines rely on the lubricant for cooling, the properties of the lubricoolant must have adequate bearing lubrication properties. However, cutting oils and emulsions should not contain active sulfur or chlorine, as both have been shown to have an erosive effect on thread rolling dies.

The following guidelines should be followed when selecting a lubricoolant.

1. The lubricant should be of the water emulsion or water miscible type for maximum heat dissipation.
2. Generally, a mix of 9 parts of water to 1 part concentrate works best. The mix may vary somewhat, but no combination should contain less than 4 parts of water to 1 part concentrate.

(Text continued on p. 1322)

Table 50. Recommended Steel Blank Diameters for Parallel Rolled UNF and UNC Class 2 Threads. *(Source, Landis Threading Systems.)*

Size		Steel											
		10-50 C Soft		30-50 C Soft		30-50 C or Alloy 15-25 R_c		30-50 C or Alloy 26-32 R_c		Stainless/Chrome Nickel 300 Series		Stainless/Chrome 400 Series*	
O.D.	Pitch	Min.	Max.	Min.	Max.	Min.	Max.	Min.	Max.	Min.	Max.	Min.	Max.
#0	80	0.0504	0.0498	0.0507	0.0501	0.0509	0.0503	0.0511	0.0505	0.0513	0.0507	0.0515	0.0509
#1	72	0.0623	0.0617	0.0627	0.0621	0.0629	0.0622	0.0631	0.0625	0.0633	0.0626	0.0635	0.0628
#1	64	0.0612	0.0605	0.0616	0.0609	0.0618	0.0611	0.0620	0.0613	0.0622	0.0615	0.0624	0.0617
#2	64	0.0742	0.0735	0.0746	0.0739	0.0748	0.0741	0.0750	0.0743	0.0752	0.0745	0.0754	0.0747
#2	56	0.0726	0.0719	0.0730	0.0723	0.0732	0.0725	0.0734	0.0728	0.0737	0.0730	0.0739	0.0732
#3	56	0.0855	0.0847	0.0859	0.0852	0.0861	0.0854	0.0863	0.0856	0.0866	0.0858	0.0868	0.0860
#3	48	0.0835	0.0827	0.0840	0.0832	0.0842	0.0834	0.0844	0.0837	0.0846	0.0839	0.0849	0.0841
#4	48	0.0964	0.0956	0.0969	0.0961	0.0971	0.0964	0.0974	0.0966	0.0976	0.0968	0.0979	0.0971
#4	40	0.0936	0.0928	0.0941	0.0933	0.0943	0.09335	0.0946	0.0938	0.0948	0.0940	0.0951	0.0943
#5	44	0.1081	0.1073	0.1086	0.1078	0.1088	0.1080	0.1091	0.1082	0.1093	0.1085	0.1096	0.1088
#5	40	0.1065	0.1057	0.1070	0.1062	0.1073	0.1064	0.1075	0.1067	0.1078	0.1070	0.1081	0.1072
#6	40	0.1195	0.1187	0.1200	0.1192	0.1203	0.1194	0.1205	0.1197	0.1208	0.1200	0.1211	0.1202
#6	32	0.1153	0.1144	0.1159	0.1149	0.1161	0.1152	0.1164	0.1155	0.1167	0.1158	0.1170	0.1161
#8	36	0.1436	0.1427	0.1442	0.1432	0.1444	0.1435	0.1447	0.1438	0.1450	0.1441	0.1453	0.1444
#8	32	0.1412	0.1402	0.1417	0.1408	0.1420	0.1411	0.1423	0.1414	0.1426	0.1416	0.1429	0.1419
#10	32	0.1671	0.1661	0.1677	0.1667	0.1680	0.1670	0.1683	0.1673	0.1686	0.1676	0.1689	0.1679
#10	24	0.1591	0.1589	0.1608	0.1596	0.1611	0.1599	0.1614	0.1602	0.1618	0.1606	0.1621	0.1609
#12	28	0.1900	0.1889	0.1907	0.1896	0.1910	0.1899	0.1913	0.1902	0.1916	0.1905	0.1920	0.1908
#12	24	0.1861	0.1848	0.1868	0.1855	0.1871	0.1859	0.1874	0.1862	0.1878	0.1865	0.1881	0.1869
1/4	20	0.2144	0.2131	0.2151	0.2138	0.2155	0.2142	0.2159	0.2146	0.2166	0.2153	0.2159	0.2146
1/4	28	0.2241	0.2228	0.2248	0.2235	0.2251	0.2238	0.2255	0.2242	0.2261	0.2248	0.2255	0.2242

(Continued)

Table 50. *(Continued)* Recommended Steel Blank Diameters for Parallel Rolled UNF and UNC Class 2 Threads. *(Source, Landis Threading Systems.)*

Size		Steel								Stainless/Chrome Nickel 300 Series		Stainless/Chrome 400 Series*	
		10-50 C Soft		30-50 C Soft		30-50 C or Alloy 15-25 R_c		30-50 C or Alloy 26-32R_c					
O.D.	Pitch	Min.	Max.	Min.	Max.	Min.	Max.	Min.	Max.	Min.	Max.	Min.	Max.
5/16	18	0.2729	0.2716	0.2737	0.2724	0.2741	0.2728	0.2745	0.2732	0.2753	0.2740	0.2745	0.2732
5/16	24	0.2823	0.2810	0.2830	0.2817	0.2834	0.2821	0.2837	0.2824	0.2845	0.2832	0.2837	0.2824
3/8	16	0.3306	0.3291	0.3315	0.3300	0.3320	0.3305	0.3324	0.3309	0.3333	0.3318	0.3324	0.3309
3/8	24	0.3448	0.3434	0.3455	0.3441	0.3459	0.3445	0.3463	0.3449	0.3471	0.3457	0.3463	0.3449
7/16	14	0.3871	0.3855	0.3880	0.3864	0.3885	0.3869	0.3890	0.3874	0.3899	0.3883	0.3890	0.3874
7/16	20	0.4012	0.3999	0.4021	0.4008	0.4025	0.4012	0.4029	0.4016	0.4037	0.4024	0.4029	0.4016
1/2	13	0.4458	0.4440	0.4468	0.4450	0.4473	0.4455	0.4478	0.4460	0.4488	0.4470	0.4478	0.4460
1/2	20	0.4637	0.4623	0.4646	0.4632	0.4650	0.4636	0.4655	0.4641	0.4663	0.4649	0.4655	0.4641
9/16	12	0.5039	0.5021	0.5050	0.5032	0.5055	0.5037	0.5060	0.5042	0.5070	0.5052	0.5060	0.5042
9/16	18	0.5225	0.5210	0.5234	0.5219	0.5238	0.5223	0.5243	0.5228	0.5252	0.5237	0.5243	0.5228
5/8	11	0.5614	0.5595	0.5625	0.5606	0.5630	0.5611	0.5636	0.5617	0.5647	0.5628	0.5636	0.5617
5/8	18	0.5850	0.5833	0.5859	0.5842	0.5864	0.5847	0.5869	0.5852	0.5878	0.5861	0.5869	0.5852
3/4	10	0.6799	0.6779	0.6811	0.6791	0.6817	0.6797	0.6823	0.6803	0.6834	0.6814	0.6823	0.6803
3/4	16	0.7052	0.7034	0.7062	0.7044	0.7067	0.7049	0.7072	0.7054	0.7082	0.7064	0.7072	0.7054
7/8	9	0.7972	0.7952	0.7985	0.7965	0.7991	0.7971	0.7998	0.7978	0.8010	0.7990	0.7998	0.7978
7/8	14	0.8240	0.8221	0.8251	0.8232	0.8257	0.8238	0.8262	0.8243	0.8273	0.8259	0.8262	0.8243
1	8	0.9131	0.9107	0.9144	0.9120	0.9151	0.9127	0.9158	0.9134	0.9172	0.9148	0.9158	0.9134
1	12	0.9408	0.9388	0.9420	0.9400	0.9426	0.9406	0.9432	0.9412	0.9443	0.9423	0.9432	0.9412
1-1/8	7	1.0262	1.0235	1.0276	1.0250	1.0283	1.0257	1.0290	1.0264	1.0304	1.0278	1.0290	1.0264
1-1/8	12	1.0657	1.0637	1.0669	1.0649	1.0675	1.0655	1.0681	1.0661	1.0693	1.0673	1.0681	1.0661
1-1/4	7	1.1510	1.1483	1.1525	1.1498	1.1533	1.1506	1.1540	1.1513	1.1555	1.1528	1.1540	1.1513

(Continued)

Table 50. *(Continued)* Recommended Steel Blank Diameters for Parallel Rolled UNF and UNC Class 2 Threads. *(Source, Landis Threading Systems.)*

Size		10-50 C Soft		30-50 C Soft		Steel							
						30-50 C or Alloy 15-25 R_c		30-50 C or Alloy 26-32R_c		Stainless/Chrome Nickel 300 Series		Stainless/Chrome 400 Series*	
O.D.	Pitch	Min.	Max.	Min.	Max.	Min.	Max.	Min.	Max.	Min.	Max.	Min.	Max.
1-1/4	12	1.1906	1.1885	1.1919	1.1898	1.1925	1.1904	1.1931	1.1910	1.1943	1.1922	1.1931	1.1910
1-3/8	6	1.2601	1.2571	1.2617	1.2587	1.2625	1.2595	1.2633	1.2603	1.2649	1.2619	1.2633	1.2603
1-3/8	12	1.3153	1.3133	1.3166	1.3146	1.3171	1.3152	1.3178	1.3158	1.3191	1.3171	1.13178	1.3158
1-1/2	6	1.3850	1.3820	1.3866	1.3836	1.3874	1.3844	1.3882	1.3852	1.3898	1.3868	1.3882	1.3851
1-1/2	12	1.4402	1.4382	1.4415	1.4395	1.4422	1.4402	1.4428	1.4408	1.4441	1.4421	1.4428	1.4408
1-3/4	5	1.6124	1.6094	1.6142	1.6112	1.6150	1.6120	1.6159	1.6129	1.6177	1.6147	1.6159	1.6129
2	4-1/2	1.8474	1.8442	1.8493	1.8461	1.8503	1.8471	1.8512	1.8480	1.8531	1.8499	1.8512	1.8480

Note: These dimensions are for set-up reference. Diameters must be finally established by actual rolling.

*Only certain grades rollable.

Table 51. Recommended Brass, Bronze, and Aluminum Blank Diameters for Parallel Rolled UNF and UNC Class 2 Threads. *(Source, Landis Threading Systems.)*

Size		Brass & Bronze		Aluminum Alloy			
				Soft		Hard	
O.D.	Pitch	Max.	Min.	Max.	Min.	Max.	Min.
#0	80	0.0507	0.0501	0.0509	0.0503	0.0507	0.0501
#1	72	0.0627	0.06321	0.0629	0.0622	0.0627	0.0621
#1	64	0.0616	0.0609	0.0618	0.0611	0.0616	0.0609
#2	64	0.0746	0.0739	0.0748	0.0741	0.0746	0.0739
#2	56	0.0730	0.0723	0.0732	0.0725	0.0730	0.0723
#3	56	0.0859	0.0852	0.0861	0.0854	0.0859	0.0852
#3	48	0.0840	0.0832	0.0842	0.0834	0.0840	0.0832
#4	48	0.0969	0.0961	0.0971	0.0964	0.0969	0.0961
#4	40	0.0941	0.0938	0.0943	0.0935	0.0941	0.0933
#5	44	0.1086	0.1078	0.1088	0.1080	0.1086	0.1078
#5	40	0.1070	0.1062	0.1073	0.1064	0.1070	0.1062
#6	40	0.1200	0.1192	0.1203	0.1194	0.1200	0.1192
#6	32	0.1159	0.1149	0.1161	0.1152	0.1159	0.1149
#8	36	0.1442	0.1432	0.1444	0.1435	0.1442	0.1432
#8	32	0.1417	0.1408	0.1420	0.1411	0.1417	0.1408
#10	32	0.1677	0.1667	0.1680	0.1670	0.1677	0.1667
#10	24	0.1608	0.1596	0.1611	0.1599	0.1608	0.1596
#12	28	0.1907	0.1896	0.1910	0.1899	0.1907	0.1896
#12	24	0.1868	0.1855	0.1871	0.1859	0.1868	0.1855
1/4	20	0.2151	0.2138	0.2155	0.2142	0.2151	0.2138
1/4	28	0.2248	0.2235	0.2251	0.2238	0.2248	0.2235
5/16	18	0.2737	0.2724	0.2741	0.2728	0.2737	0.2724
5/16	24	0.2830	0.2817	0.2834	0.2821	0.2830	0.2817
3/8	16	0.3315	0.3300	0.3320	0.3305	0.3315	0.3300
3/8	24	0.3455	0.3441	0.3459	0.3445	0.3455	0.3441
7/16	14	0.3880	0.3864	0.3885	0.3869	0.3880	0.3864
7/16	20	0.4021	0.4008	0.4025	0.4012	0.4021	0.4008
1/2	13	0.4468	0.4450	0.4473	0.4455	0.4468	0.4450
1/2	20	0.4646	0.4632	0.4650	0.4636	0.4646	0.4632
9/16	12	0.5050	0.5032	0.5055	0.5037	0.5050	0.5032
9/16	18	0.5234	0.5219	0.5238	0.5223	0.5234	0.5219
5/8	11	0.5625	0.5606	0.5630	0.5611	0.5625	0.5606
5/8	18	0.5859	0.5842	0.5864	0.5847	0.5859	0.5842
3/4	10	0.6811	0.6791	0.6817	0.6797	0.6811	0.6791
3/4	16	0.7062	0.7044	0.7067	0.7049	0.7062	0.7044
7/8	9	0.7985	0.7965	0.7991	0.7971	0.7985	0.7965
7/8	14	0.8251	0.8232	0.8257	0.8238	0.8251	0.8232

(Continued)

Table 51. *(Continued)* **Recommended Brass, Bronze, and Aluminum Blank Diameters for Parallel Rolled UNF and UNC Class 2 Threads.** *(Source, Landis Threading Systems.)*

Size		Brass & Bronze		Aluminum Alloy			
				Soft		Hard	
O.D.	Pitch	Max.	Min.	Max.	Min.	Max.	Min.
1	8	0.9144	0.9120	0.9151	0.9127	0.9144	0.9120
1	12	0.9420	0.9400	0.9426	0.9406	0.9420	0.9400
1-1/8	7	1.0276	1.0250	1.0283	1.0257	1.0276	1.0250
1-1/8	12	1.0669	1.0649	1.0675	1.0655	1.0669	1.0649
1-1/4	7	1.1525	1.1498	1.1533	1.1506	1.1525	1.1498
1-1/4	12	1.1919	1.1898	1.1925	1.1904	1.1919	1.1898
1-3/8	6	1.2617	1.2587	1.2625	1.2595	1.2617	1.2587
1-3/8	12	1.3166	1.3146	1.3172	1.3152	1.3166	1.3146
1-1/2	6	1.3866	1.3836	1.3874	1.3844	1.3866	1.3836
1-1/2	12	1.4415	1.4395	1.4422	1.4402	1.4415	1.4395
1-3/4	5	1.6142	1.6112	1.6150	1.6120	1.6142	1.6112
2	4-1/2	1.8442	1.8461	1.8503	1.8471	1.8442	1.8461

Note: These dimensions are for setup reference. Diameters must be finally established by actual rolling.

1310 THREAD ROLLING - BLANKS

Table 52. Recommended Steel Blank Diameters for Parallel Rolled UNF and UNC Class 3 Threads. *(Source, Landis Threading Systems.)*

Size		Steel											
O.D.	Pitch	10-50 C Soft		30-50 C Soft		30-50 C or Alloy 15-25 R_c		30-50 C or Alloy 26-32 R_c		Stainless/Chrome Nickel 300 Series		Stainless/Chrome 400 Series*	
		Min.	Max.	Min.	Max.	Min.	Max.	Min.	Max.	Min.	Max.	Min.	Max.
#0	80	0.0512	0.0507	0.0514	0.0510	0.0515	0.0511	0.0517	0.0513	0.0519	0.0515	0.0517	0.0513
#1	72	0.0632	0.0627	0.0635	0.0630	0.0637	0.0632	0.0638	0.0633	0.0641	0.0636	0.0638	0.0633
#1	64	0.0621	0.0616	0.0624	0.0619	0.0625	0.0620	0.0627	0.0622	0.0630	0.0625	0.0627	0.0622
#2	64	0.0751	0.0746	0.0754	0.0749	0.0755	0.0750	0.0757	0.0752	0.0760	0.0755	0.0757	0.0752
#2	56	0.0735	0.0730	0.0738	0.0733	0.0739	0.0734	0.0741	0.0736	0.0744	0.0739	0.0741	0.0736
#3	56	0.0865	0.0860	0.0868	0.0863	0.0869	0.0864	0.0871	0.0866	0.0874	0.0869	0.0871	0.0866
#3	48	0.0845	0.0840	0.0848	0.0843	0.0850	0.0845	0.0852	0.0847	0.0855	0.0850	0.0852	0.0847
#4	48	0.0975	0.0969	0.0978	0.0972	0.0980	0.0974	0.0982	0.0976	0.0986	0.0980	0.0982	0.0976
#4	40	0.0947	0.0941	0.0950	0.0945	0.0953	0.0947	0.0955	0.0949	0.0958	0.0952	0.0955	0.0949
#5	44	0.1091	0.1085	0.1095	0.1088	0.1097	0.1091	0.1099	0.1093	0.1102	0.1096	0.1099	0.1093
#5	40	0.1077	0.1071	0.1081	0.1075	0.1083	0.1077	0.1085	0.1079	0.1088	0.1082	0.1085	0.1079
#6	40	0.1207	0.1200	0.1211	0.1204	0.1213	0.1206	0.1215	0.1208	0.1219	0.1212	0.1215	0.1208
#6	32	0.1164	0.1157	0.1169	0.1162	0.1171	0.1164	0.1174	0.1167	0.1178	0.1171	0.1174	0.1167
#8	36	0.1448	0.1441	0.1452	0.1445	0.1454	0.1447	0.1457	0.1450	0.1461	0.1454	0.1457	0.1450
#8	32	0.1424	0.1417	0.1429	0.1422	0.1431	0.1424	0.1433	0.1426	0.1437	0.1430	0.1433	0.1426
#10	32	0.1683	0.1676	0.1688	0.1681	0.1690	0.1683	0.1693	0.1686	0.1697	0.1690	0.1693	0.1686
#10	24	0.1615	0.1607	0.1620	0.1612	0.1622	0.1614	0.1625	0.1617	0.1630	0.1622	0.1625	0.1617
#12	28	0.1913	0.1906	0.1918	0.1911	0.1921	0.1914	0.1923	0.1916	0.1928	0.1921	0.1923	0.1916
#12	24	0.1874	0.1866	0.1879	0.1871	0.1881	0.1873	0.1884	0.1876	0.1889	0.1881	0.1884	0.1876
1/4	20	0.2159	0.2150	0.2164	0.2155	0.2167	0.2158	0.2170	0.2161	0.2176	0.2167	0.2170	0.2161
1/4	28	0.2254	0.2246	0.2259	0.2251	0.2261	0.2253	0.2264	0.2256	0.2269	0.2261	0.2264	0.2256

(Continued)

Table 52. *(Continued)* Recommended Steel Blank Diameters for Parallel Rolled UNF and UNC Class 3 Threads. *(Source, Landis Threading Systems.)*

| Size | | Steel | | | | | | | | | | | |
| O.D. | Pitch | 10-50 C Soft | | 30-50 C Soft | | 30-50 C or Alloy 15-25 R_c | | 30-50 C or Alloy 26-32R_c | | Stainless/Chrome Nickel 300 Series | | Stainless/Chrome 400 Series* | |
		Min.	Max.	Min.	Max.	Min.	Max.	Min.	Max.	Min.	Max.	Min.	Max.
5/16	18	0.2724	0.2737	0.2753	0.2743	0.2756	0.2746	0.2759	0.2749	0.2765	0.2755	0.2759	0.2749
5/16	24	0.2839	0.2830	0.2845	0.2836	0.2847	0.2838	0.2850	0.2841	0.2855	0.2846	0.2850	0.2841
3/8	16	0.3326	0.3314	0.3333	0.3321	0.3336	0.3324	0.3340	0.3328	0.3346	0.3334	0.3340	.3328
3/8	24	0.3463	0.3453	0.3469	0.3459	0.3472	0.3462	0.3475	0.3465	0.3480	0.3470	0.3475	0.3465
7/16	14	0.3893	0.3880	0.3900	0.3887	0.3903	0.3890	0.3907	0.3894	0.3914	0.3901	0.3907	0.3894
7/16	20	0.4032	0.4022	0.4038	0.4028	0.4041	0.4031	0.4044	0.4034	0.4051	0.4041	0.4044	0.4034
1/2	13	0.4480	0.4467	0.4487	0.4474	0.4491	0.4478	0.4495	0.4482	0.4502	0.4489	0.4495	0.4482
1/2	20	0.4657	0.4646	0.4664	0.4653	0.4654	0.4643	0.4670	0.4659	0.4676	0.4665	0.4670	0.4659
9/16	12	0.5062	0.5049	0.5070	0.5057	0.5070	0.5057	0.5078	0.5065	0.5085	0.5072	0.5078	0.5065
9/16	18	0.5246	0.5233	0.5253	0.5240	0.5253	0.5240	0.5260	0.5247	0.5267	0.5254	0.5260	0.5247
5/8	11	0.5636	0.5623	0.5644	0.5631	0.5644	0.5631	0.5653	0.5640	0.5661	0.5648	0.5653	.5640
5/8	18	0.5871	0.5858	0.5878	0.5865	0.5878	0.5865	0.5885	0.5872	0.5892	0.5879	0.5885	0.5872
3/4	10	0.6825	0.6810	0.6834	0.6819	0.6834	0.6819	0.6843	0.6828	0.6852	0.6837	0.6843	0.6828
3/4	16	0.7073	0.7060	0.7080	0.7067	0.7080	0.7067	0.7088	0.7075	0.7096	0.7083	0.7088	0.7075
7/8	9	0.8002	0.7986	0.8010	0.7994	0.8011	0.7995	0.8021	0.8005	0.8030	0.8014	0.8021	0.8005
7/8	14	0.8262	0.8249	0.8270	0.8257	0.8270	0.8257	0.8279	0.8266	0.8287	0.8274	0.8279	0.8266
1	8	0.9160	0.9142	0.9170	0.9151	0.9170	0.9152	0.9180	0.9162	0.9191	0.9173	0.9180	0.9162
1	12	0.9434	0.9419	0.9443	0.9428	0.9428	0.9415	0.9452	0.9437	0.9461	0.9446	0.9452	0.9437
1-1/8	7	1.0292	1.0273	1.0303	1.0284	1.0303	1.0284	1.0314	1.0295	1.0325	1.0306	1.0314	1.0295
1-1/8	12	1.0684	1.0669	1.0693	1.0678	1.0693	1.0678	1.0693	1.0678	1.0711	1.0696	1.0693	1.0678
1-1/4	7	1.1542	1.1523	1.1553	1.1534	1.1553	1.1534	1.1553	1.1534	1.1575	1.1556	1.1553	1.1534

(Continued)

Table 52. (Continued) Recommended Steel Blank Diameters for Parallel Rolled UNF and UNC Class 3 Threads. (Source, Landis Threading Systems.)

Size		10-50 C Soft		30-50 C Soft		Steel							
						30-50 C or Alloy 15-25 R$_c$		30-50 C or Alloy 26-32R$_c$		Stainless/Chrome Nickel 300 Series		Stainless/Chrome 400 Series*	
O.D.	Pitch	Min.	Max.	Min.	Max.	Min.	Max.	Min.	Max.	Min.	Max.	Min.	Max.
1-1/4	12	1.1933	1.1918	1.1942	1.1927	1.1942	1.1927	1.1951	1.1936	1.1960	1.1945	1.1951	1.1936
1-3/8	6	1.2633	1.2613	1.2645	1.2625	1.2651	1.2631	1.2657	1.2637	1.2665	1.2649	1.2657	1.2637
1-3/8	12	1.3182	1.3167	1.3191	1.3176	1.3196	1.3181	1.3200	1.3185	1.3210	1.3195	1.3200	1.3185
1-1/2	6	1.3883	1.3862	1.3894	1.3874	1.3900	1.3880	1.3906	1.3886	1.3918	1.3898	1.3906	1.3886
1-1/2	12	1.4433	1.4416	1.4442	1.4425	1.4447	1.4430	1.4451	1.4434	1.4461	1.4444	1.4451	1.4434
1-3/4	5	1.6165	1.6141	1.6178	1.6154	1.6185	1.6161	1.6191	1.6167	1.6205	1.6181	1.6191	1.6167
2	4-1/2	1.8518	1.8493	1.8532	1.8507	1.8539	1.8514	1.8546	1.8521	1.8561	1.8536	1.8546	1.8521

Note: These dimensions are for setup reference. Diameters must be finally established by actual rolling.

Table 53. Recommended Brass, Bronze, and Aluminum Blank Diameters for Parallel Rolled UNF and UNC Class 3 Threads. *(Source, Landis Threading Systems.)*

| Size | | Brass & Bronze | | Aluminum Alloy | | | |
| O.D. | Pitch | | | Soft | | Hard | |
		Max.	Min.	Max.	Min.	Max.	Min.
#0	80	0.0514	0.0510	0.0515	0.0511	0.0514	0.0510
#1	72	0.0635	0.0630	0.0637	0.0632	0.0635	0.0630
#1	64	0.0624	0.0619	0.0625	0.0620	0.0624	0.0619
#2	64	0.0754	0.0749	0.0755	0.0750	0.0754	0.0749
#2	56	0.0738	0.0733	0.0739	0.0734	0.0738	0.0733
#3	56	0.0868	0.0863	0.0869	0.0864	0.0868	0.0863
#3	48	0.0848	0.0843	0.0850	0.0845	0.0848	0.0843
#4	48	0.0978	0.0972	0.0980	0.0974	0.0978	0.0972
#4	40	0.0950	0.0945	0.0953	0.0947	0.0950	0.0945
#5	44	0.1095	0.1088	0.1097	0.1091	0.1095	0.1088
#5	40	0.1081	0.1075	0.1083	0.1077	0.1081	0.1075
#6	40	0.1211	0.1204	0.1213	0.1206	0.1211	0.1204
#6	32	0.1169	0.1162	0.1171	0.1164	0.1169	0.1162
#8	36	0.1452	0.1445	0.1454	0.1447	0.1452	0.1445
#8	32	0.1429	0.1422	0.1431	0.1424	0.1429	0.1422
#10	32	0.1688	0.1681	0.1690	0.1683	0.1688	0.1681
#10	24	0.1620	0.1612	0.1622	0.1614	0.1620	0.1612
#12	28	0.1918	0.1911	0.1921	0.1914	0.1918	0.1911
#12	24	0.1879	0.1871	0.1881	0.1873	0.1879	0.1871
1/4	20	0.2164	0.2155	0.2167	0.2158	0.2164	0.2155
1/4	28	0.2259	0.2251	0.2261	0.2253	0.2259	0.2251
5/16	18	0.2753	0.2743	0.2756	0.2746	0.2753	0.2743
5/16	24	0.2845	0.2836	0.2847	0.2838	0.2845	0.2836
3/8	16	0.3346	0.3334	0.3340	0.3328	0.3333	0.3321
3/8	24	0.3469	0.3459	0.3472	0.3462	0.3469	0.3459
7/16	14	0.3900	0.3887	0.3903	0.3890	0.3900	0.3887
7/16	20	0.4038	0.4028	0.4041	0.4031	0.4038	0.4028
1/2	13	0.4487	0.4474	0.4491	0.4478	0.4487	0.4474
1/2	20	0.4664	0.4653	0.4654	0.4643	0.4664	0.4653
9/16	12	0.5070	0.5057	0.5070	0.5057	0.5070	0.5057
9/16	18	0.5253	0.5240	0.5253	0.5240	0.5253	0.5240
5/8	11	0.5644	0.5631	0.5644	0.5631	0.5644	0.5631
5/8	18	0.5878	0.5865	0.5878	0.5865	0.5878	0.5865
3/4	10	0.6834	0.6819	0.6834	0.6819	0.6834	0.6819
3/4	16	0.7080	0.7067	0.7080	0.7067	0.7087	0.7067
7/8	9	0.8010	0.7994	0.8011	0.7995	0.8010	0.7994
7/8	14	0.8270	0.8257	0.8270	0.8257	0.8270	0.8257

(Continued)

Table 53. *(Continued)* **Recommended Brass, Bronze, and Aluminum Blank Diameters for Parallel Rolled UNF and UNC Class 3 Threads.** *(Source, Landis Threading Systems.)*

Size		Brass & Bronze		Aluminum Alloy			
				Soft		Hard	
O.D.	Pitch	Max.	Min.	Max.	Min.	Max.	Min.
1	8	0.9170	0.9152	0.9170	0.9152	0.9170	0.9152
1	12	0.9443	0.9428	0.9428	0.9415	0.9443	0.9428
1-1/8	7	1.0303	1.0284	1.0303	1.0284	1.0303	1.0284
1-1/8	12	1.0693	1.0678	1.0693	1.0678	1.0693	1.0678
1-1/4	7	1.1553	1.1539	1.1553	1.1534	1.1553	1.1534
1-1/4	12	1.1942	1.1927	1.1942	1.1927	1.1942	1.1927
1-3/8	6	1.2645	1.2625	1.2651	1.2631	1.2645	1.2625
1-3/8	12	1.3191	1.3176	1.3196	1.3161	1.3191	1.3176
1-1/2	6	1.3894	1.3874	1.3900	1.3880	1.3894	1.3874
1-1/2	12	1.4442	1.4425	1.4447	1.4430	1.4442	1.4425
1-3/4	5	1.6178	1.6154	1.6185	1.6161	1.6178	1.6154
2	4-1/2	1.8532	1.8507	1.8539	1.8514	1.8532	1.8507

Note: These dimensions are for setup reference. Diameters must be finally established by actual rolling.

Table 54. Recommended Steel Blank Diameters for Straight, Rolled Metric Threads. *(Source, Landis Threading Systems.)*

Series Designation		Dia. in mm & inches	Steel												
			10-50 C Soft		30-50 C Soft		30-50 C or Alloy 15-25 R_c		30-50 C or Alloy 26-32 R_c		Stainless/Chrome Nickel 300 Series		Stainless/Chrome 400 Series*	
Size	Pitch		Min.	Max.	Min.	Max.	Min.	Max.	Min.	Max.	Min.	Max.	Min.	Max.
3	.35	mm	2.756	2.743	2.764	2.751	2.769	2.756	2.771	2.758	2.779	2.766	2.771	2.758
		inch	0.1085	0.1080	0.1088	0.1083	0.1090	0.1085	0.1091	0.1086	0.1094	0.1089	0.1091	0.1086
3	.5	mm	2.652	2.639	2.659	2.647	2.664	2.652	2.667	2.654	2.675	2.662	2.667	2.654
		inch	0.1044	0.1039	0.1047	0.1042	0.1049	0.1044	0.1050	0.1045	0.1053	0.1048	0.1050	0.1045
3.5	.35	mm	3.256	3.244	3.261	3.249	3.266	3.254	3.269	3.256	3.277	3.264	3.269	3.256
		inch	0.1282	0.1277	0.1284	0.1279	0.1286	0.1281	0.1287	0.1282	0.1290	0.1285	0.1287	0.1282
3.5	6	mm	3.084	3.071	3.094	3.081	3.096	3.084	3.101	3.089	3.109	3.096	3.101	3.089
		inch	0.1214	0.1209	0.1218	0.1213	0.1219	0.1214	0.1221	0.1216	0.1224	0.1219	0.1221	0.1216
4	.5	mm	3.653	3.640	3.660	3.647	3.663	3.650	3.668	3.655	3.675	3.663	3.668	3.655
		inch	0.1438	0.1433	0.1441	0.1436	0.1442	0.1437	0.1444	0.1439	0.1447	0.1442	0.1444	0.1439
4	.7	mm	3.520	3.505	3.528	3.513	3.533	3.518	3.538	3.523	3.548	3.533	3.538	3.523
		inch	0.1386	0.1380	0.1389	0.1383	0.1391	0.1385	0.1393	0.1387	0.1397	0.1391	0.1393	0.1387
4.5	.5	mm	4.153	4.140	4.161	4.148	4.163	4.150	4.168	4.155	4.176	4.163	4.168	4.155
		inch	0.1635	0.1630	0.1638	0.1633	0.1639	0.1634	0.1641	0.1636	0.1644	0.1639	0.1641	0.1636
5	.5	mm	4.651	4.638	4.661	4.648	4.663	4.651	4.666	4.653	4.676	4.663	4.666	4.653
		inch	0.1831	0.1826	0.1835	0.1830	0.1836	0.1831	0.1837	0.1832	0.1841	0.1836	0.1837	0.1832
5	.8	mm	4.455	4.437	4.463	4.448	4.468	4.453	4.473	4.458	4.483	4.465	4.473	4.463
		inch	0.1754	0.1747	0.1757	0.1751	0.1759	0.1753	0.1761	0.1755	0.1765	0.1758	0.1761	0.1755
6	.75	mm	5.484	5.469	5.494	5.479	5.499	5.484	5.504	5.489	5.512	5.496	5.504	5.489
		inch	0.2159	0.2153	0.2163	0.2157	0.2165	0.2159	0.2167	0.2161	0.2170	0.2164	0.2167	0.2161
6	1.0	mm	5.314	5.293	5.326	5.306	5.334	5.314	5.339	5.319	5.352	5.331	5.339	5.319
		inch	0.2092	0.2084	0.2097	0.2089	0.2100	0.2092	0.2102	0.2094	0.2107	0.2099	0.2102	0.2094
7	.75	mm	6.485	6.469	6.495	6.480	6.500	6.485	6.505	6.490	6.513	6.497	6.505	6.490
		inch	0.2553	0.2547	0.2557	0.2551	0.2559	0.2553	0.2561	0.2555	0.2564	0.2558	0.1561	0.2561

(Continued)

Table 54. (Continued) Recommended Steel Blank Diameters for Straight, Rolled Metric Threads. (Source, Landis Threading Systems.)

Series Designation		Dia. in mm & inches	Steel 10-50 C Soft		30-50 C Soft		30-50 C or Alloy 15-25 Rc		30-50 C or Alloy 26-32Rc		Stainless/Chrome Nickel 300 Series		Stainless/Chrome 400 Series*	
Size	Pitch		Min.	Max.	Min.	Max.	Min.	Max.	Min.	Max.	Min.	Max.	Min.	Max.
7	1.0	mm	6.314	6.294	6.327	6.307	6.335	6.314	6.340	6.320	6.353	6.332	6.340	6.320
		inch	0.2486	0.2478	0.2491	0.2483	0.2494	0.2486	0.2496	0.2488	0.2501	0.2493	0.2496	0.2488
8	1.0	mm	7.315	7.295	7.328	7.308	7.336	7.315	7.341	7.320	7.353	7.333	7.341	7.320
		inch	0.2880	0.2872	0.2885	0.2877	0.2888	0.2880	0.2890	0.2882	0.2895	0.2887	0.2890	0.2882
8	1.25	mm	7.150	7.130	7.163	7.142	7.168	7.148	7.176	7.155	7.188	7.168	7.176	7.155
		inch	0.2815	0.2807	0.2820	0.2812	0.2822	0.2814	0.2825	0.2817	0.2830	0.2822	0.2825	0.2817
9	1.0	mm	8.313	8.293	8.326	8.306	8.334	8.313	8.339	8.319	8.352	8.331	8.339	8.319
		inch	0.3273	0.3265	0.3278	0.3270	0.3281	0.3273	0.3283	0.3275	0.3288	0.3280	0.3283	0.3275
9	1.25	mm	8.151	8.131	8.164	8.143	8.169	8.148	8.176	8.156	8.189	8.169	8.176	8.156
		inch	0.3209	0.3201	0.3214	0.3206	0.3216	0.3208	0.3219	0.3211	0.3224	0.3216	0.3219	0.3211
10	1.0	mm	9.314	9.294	9.327	9.307	9.335	9.314	9.340	9.319	9.352	9.332	9.340	9.319
		inch	0.3667	0.3659	0.3672	0.3664	0.3675	0.3667	0.3677	0.3669	0.3682	0.3674	0.3677	0.3669
10	1.5	mm	8.984	8.956	8.999	8.971	9.007	8.979	9.014	8.989	9.030	9.004	9.014	8.989
		inch	0.3537	0.3526	0.3543	0.3532	0.3546	0.3535	0.3549	0.3539	0.3555	0.3545	0.3549	0.3539
11	1.5	mm	9.982	9.954	9.997	9.970	10.005	9.977	10.012	9.987	10.027	10.003	10.012	9.987
		inch	0.3930	0.3919	0.3936	0.3925	0.3939	0.3928	0.3942	0.3932	0.3948	0.3938	0.3942	0.3932
12	1.5	mm	10.983	10.955	10.998	10.970	11.006	10.978	11.013	10.988	11.029	11.003	11.013	10.988
		inch	0.4324	0.4313	0.4330	0.4319	0.4333	0.4322	0.4336	0.4326	0.4342	0.4332	0.4336	0.4326
12	1.75	mm	10.818	10.785	10.836	10.803	10.843	10.810	10.853	10.820	10.871	10.838	10.853	10.820
		inch	0.4259	0.4246	0.4266	0.4253	0.4269	0.4256	0.4273	0.4260	0.4280	0.4267	0.4273	0.4260
14	1.5	mm	12.982	12.954	12.997	12.969	13.005	12.977	13.012	12.987	13.028	13.002	13.012	12.987
		inch	0.5111	0.5100	0.5117	0.5106	0.5120	0.5109	0.5123	0.5113	0.5129	0.5119	0.5123	0.5113
14	2	mm	12.647	12.614	12.664	12.631	12.675	12.642	12.685	12.652	12.705	12.672	12.685	12.652
		inch	0.4979	0.4966	0.4986	0.4973	0.4990	0.4977	0.4994	0.4981	0.5002	0.4989	0.4994	0.4981

(Continued)

Table 54. *(Continued)* Recommended Steel Blank Diameters for Straight, Rolled Metric Threads. *(Source, Landis Threading Systems.).*

Series Designation		Dia. in mm & inches	Steel								Stainless/Chrome Nickel 300 Series		Stainless/Chrome 400 Series*	
			10-50 C Soft		30-50 C Soft		30-50 C or Alloy 15-25 Rc		30-50 C or Alloy 26-32 Rc					
Size	Pitch	mm / inch	Min.	Max.	Min.	Max.	Min.	Max.	Min.	Max.	Min.	Max.	Min.	Max.
16	1.5	mm	14.983	14.956	14.999	14.971	15.006	14.978	15.014	14.989	15.029	15.004	15.014	14.989
		inch	0.5899	0.5888	0.5905	0.5894	0.5908	0.5897	0.5911	0.5901	0.5917	0.5907	0.5911	0.5901
16	2	mm	14.648	14.615	14.666	14.633	14.676	14.643	14.686	14.653	14.707	14.674	14.686	14.653
		inch	0.5767	0.5754	0.5774	0.5761	0.5778	0.5765	0.5782	0.5769	0.5790	0.5777	0.5782	0.5769
18	1.5	mm	16.982	16.955	16.998	16.970	17.005	16.977	17.013	16.988	17.028	17.003	17.013	16.988
		inch	0.6686	0.6675	0.6692	0.6681	0.6695	0.6684	0.6698	0.6688	0.6704	0.6694	0.6698	0.6688
18	2.5	mm	16.312	16.274	16.335	16.297	16.345	16.307	16.358	16.320	16.380	16.342	16.358	16.320
		inch	0.6422	0.6407	0.6431	0.6416	0.6435	0.6420	0.6440	0.6425	0.6449	0.6434	0.6440	0.6425
20	1.5	mm	18.984	18.956	18.999	18.971	19.007	18.979	19.014	18.989	19.030	19.004	19.014	18.989
		inch	0.7474	0.7463	0.7480	0.7469	0.7483	0.7472	0.7486	0.7476	0.7492	0.7482	0.7486	0.7476
20	2.5	mm	18.313	18.275	18.336	18.298	18.346	18.308	18.359	18.321	18.382	18.344	18.359	18.321
		inch	0.7210	0.7195	0.7219	0.7204	0.7223	0.7208	0.7228	0.7213	0.7237	0.7222	0.7228	0.7213
22	1.5	mm	20.983	20.955	20.998	20.970	21.006	20.978	21.013	20.988	21.029	21.003	21.013	20.988
		inch	0.8261	0.8250	0.8267	0.8256	0.8270	0.8259	0.8273	0.8263	0.8279	0.8269	0.8273	0.8263
22	2.5	mm	20.312	20.274	20.335	20.297	20.345	20.307	20.358	20.320	20.381	20.343	20.358	20.320
		inch	0.7997	0.7982	0.8006	0.7991	0.8010	0.7995	0.8015	0.8000	0.8024	0.8009	0.8015	0.8000
24	2	mm	22.647	22.614	22.664	22.631	22.675	22.642	22.685	22.652	22.705	22.672	22.685	22.652
		inch	0.8916	0.8903	0.8923	0.8910	0.8927	0.8914	0.8931	0.8918	0.8939	0.8926	0.8931	0.8918
24	3	mm	21.979	21.930	22.004	21.958	22.017	21.971	22.029	21.984	22.055	22.007	22.029	21.984
		inch	0.8653	0.8634	0.8663	0.8645	0.8668	0.8650	0.8673	0.8655	0.8683	0.8664	0.8673	0.8655
27	2	mm	25.646	25.613	25.664	15.631	25.674	25.641	25.684	25.651	25.705	25.672	25.684	25.651
		inch	1.0097	1.0084	1.0104	1.0091	1.0108	1.0095	1.0112	1.0099	1.012	1.0107	1.0112	1.0099
27	3	mm	24.981	24.933	25.006	24.961	25.019	24.973	25.032	24.986	25.060	25.011	25.032	24.986
		inch	0.9835	0.9816	0.9845	0.9827	0.9850	0.9832	0.9855	0.9837	0.9866	0.9847	0.9855	0.9837

(Continued)

Table 54. (Continued) Recommended Steel Blank Diameters for Straight, Rolled Metric Threads. (Source, Landis Threading Systems.)

Size	Pitch	Dia. in mm & inches	Steel — 10-50 C Soft Min	Max	30-50 C Soft Min	Max	30-50 C or Alloy 15-25 Rc Min	Max	30-50 C or Alloy 26-32 Rc Min	Max	Stainless/Chrome Nickel 300 Series Min	Max	Stainless/Chrome 400 Series* Min	Max
30	2	mm	28.646	28.613	28.664	28.631	28.674	28.641	28.684	28.651	28.705	28.672	28.684	28.651
		inch	1.1278	1.1265	1.1285	1.1272	1.1289	1.1276	1.1293	1.1280	1.1301	1.1288	1.1293	1.1280
30	3.5	mm	27.643	27.595	27.673	27.623	27.689	27.638	27.704	27.653	27.732	27.683	27.704	27.653
		inch	1.0883	1.0864	1.0895	1.0875	1.0901	1.0881	1.0907	1.0887	1.0918	1.0899	1.0907	1.0887
33	2	mm	31.648	31.615	31.666	31.633	31.676	31.643	31.687	31.653	31.707	31.674	31.687	31.653
		inch	1.2460	1.2447	1.2467	1.2454	1.2471	1.2458	1.2475	1.2462	1.2483	1.2470	1.2475	1.2462
33	3.5	mm	30.643	30.594	30.673	30.622	30.688	30.637	30.704	30.653	30.731	30.683	30.704	30.653
		inch	1.2064	1.2045	1.2076	1.2056	1.2082	1.2062	1.2088	1.2068	1.2099	1.2080	1.2088	1.2068
36	3	mm	33.973	33.925	34.001	33.953	34.016	33.967	34.029	33.981	34.057	34.009	34.029	33.981
		inch	1.3375	1.3356	1.3386	1.3367	1.3392	1.3373	1.3397	1.3378	1.3408	1.3389	1.3397	1.3378
36	4	mm	33.316	33.265	33.346	33.295	33.361	33.311	33.377	33.326	33.407	33.356	33.377	33.326
		inch	1.3116	1.3096	1.3128	1.3108	1.3134	1.3114	1.3140	1.3120	1.3152	1.3132	1.3140	1.3120
39	3	mm	36.973	36.925	37.001	36.953	37.015	36.967	37.029	36.981	37.057	37.009	37.029	36.981
		inch	1.4556	1.4537	1.4567	1.4548	1.4573	1.4554	1.4578	1.4559	1.4589	1.4570	1.4578	1.4559
39	4	mm	36.316	36.265	36.346	36.295	36.361	36.311	36.377	36.326	36.407	36.356	36.377	36.326
		inch	1.4297	1.4277	1.4309	1.4289	1.4315	1.4295	1.4321	1.4301	1.4333	1.4313	1.4321	1.4301
42	3	mm	39.971	39.923	39.999	39.951	40.014	39.966	40.028	39.980	40.056	40.008	40.028	39.980
		inch	1.5737	1.5718	1.5748	1.5729	1.5754	1.5735	1.5759	1.5740	1.5770	1.5751	1.5759	1.5740
42	4.5	mm	38.987	38.928	39.020	38.961	39.036	38.978	39.053	38.994	39.086	39.027	39.053	38.994
		inch	1.5348	1.5325	1.5361	1.5338	1.5368	1.5345	1.5374	1.5351	1.5387	1.5364	1.5374	1.5351
45	3	mm	42.971	42.923	42.999	42.951	43.014	42.966	43.028	42.980	43.056	43.008	43.028	42.980
		inch	1.6918	1.6899	1.6929	1.6910	1.6935	1.6916	1.6940	1.6921	1.6951	1.6932	1.6940	1.6921
45	4.5	mm	41.987	41.928	42.020	41.961	42.036	41.978	42.053	41.994	42.086	42.027	42.053	41.994
		inch	1.6529	1.6506	1.6542	1.6519	1.6549	1.6526	1.6555	1.6532	1.6568	1.6545	1.6555	1.6532

(Continued)

Table 54. (Continued) Recommended Steel Blank Diameters for Straight, Rolled Metric Threads. (Source, Landis Threading Systems.)

Series Designation		Dia. in mm & inches	Steel											
			10-50 C Soft		30-50 C Soft		30-50 C or Alloy 15-25 R_c		30-50 C or Alloy 26-32R_c		Stainless/Chrome Nickel 300 Series		Stainless/Chrome 400 Series*	
Size	Pitch		Min.	Max.	Min.	Max.	Min.	Max.	Min.	Max.	Min.	Max.	Min.	Max.
48	3	mm	45.971	45.921	46.000	45.950	46.015	45.964	46.029	45.979	46.058	46.008	46.029	45.979
		inch	1.8099	1.8079	1.8111	1.8091	1.8117	1.8097	1.8123	1.8103	1.8135	1.8115	1.8123	1.8103
48	5	mm	44.655	44.592	44.691	44.628	44.709	44.645	44.727	44.663	44.762	44.699	44.727	44.663
		inch	1.7581	1.7556	1.7595	1.7570	1.7601	1.7577	1.7609	1.7584	1.7623	1.7598	1.7609	1.7584
52	3	mm	49.969	49.918	49.999	49.948	50.013	49.963	50.028	49.977	50.058	50.007	50.028	49.977
		inch	1.9673	1.9653	1.9685	1.9665	1.9691	1.9671	1.9697	1.9677	1.9709	1.9689	1.9697	1.9677
52	5	mm	48.643	48.592	48.678	48.628	48.696	48.645	48.714	48.663	48.749	48.699	48.714	48.663
		inch	1.9151	1.9131	1.9165	1.9145	1.9172	1.9152	1.9179	1.9159	1.9193	1.9173	1.9179	1.9159

Note: These dimensions are for setup reference. Diameters must be finally established by actual rolling.
*Only certain grades rollable.

1320 THREAD ROLLING - BLANKS

Table 55. Recommended Brass, Bronze, and Aluminum Blank Diameters for Straight, Rolled Metric Threads.
(Source, Landis Threading Systems.)

Series Designation		Dia. in mm & inches	Brass & Bronze		Aluminum Alloy			
					Soft		Hard	
Size	Pitch		Max.	Min.	Max.	Min.	Max.	Min.
3	.35	mm	2.764	2.751	2.769	2.756	2.764	2.751
		inch	0.1088	0.1083	0.1090	0.1085	0.1088	0.1083
3	.5	mm	2.659	2.647	2.664	2.652	2.659	2.647
		inch	0.1047	0.1042	0.1049	0.1044	0.1047	0.1042
3.5	.35	mm	3.261	3.249	3.266	3.254	3.261	3.249
		inch	0.1284	0.1279	0.1286	0.1281	0.1284	0.1279
3.5	6	mm	3.094	3.081	3.096	3.084	3.094	3.081
		inch	0.1218	0.1213	0.1219	0.1214	0.1218	0.1213
4	.5	mm	3.660	3.647	3.663	3.650	3.660	3.647
		inch	0.1441	0.1436	0.1442	0.1437	0.1441	0.1436
4	.7	mm	3.528	3.513	3.533	3.518	3.528	3.513
		inch	0.1389	0.1383	0.1391	0.1385	0.1389	0.1383
4.5	.5	mm	4.161	4.148	4.163	4.150	4.161	4.148
		inch	0.1638	0.1633	0.1639	0.1634	0.1638	0.1633
5	.5	mm	4.661	4.648	4.663	4.651	4.661	4.648
		inch	0.1835	0.1830	0.1836	0.1831	0.1835	0.1830
5	.8	mm	4.448	4.468	4.453	4.463	4.448	4.448
		inch	0.1757	0.1751	0.1759	0.1753	0.1757	0.1751
6	.75	mm	5.494	5.479	5.499	5.484	5.494	5.479
		inch	0.2163	0.2157	0.2165	0.2159	0.2163	0.2157
6	1.0	mm	5.326	5.306	5.334	5.314	5.326	5.306
		inch	0.2097	0.2089	0.2100	0.2092	0.2097	0.2089
7	.75	mm	6.495	6.480	6.500	6.485	6.495	6.480
		inch	0.2557	0.2551	0.2559	0.2553	0.2557	0.2551
7	1.0	mm	6.327	6.307	6.335	6.314	6.327	6.307
		inch	0.2491	0.2483	0.2494	0.2486	0.2491	0.2483
8	1.0	mm	7.328	7.308	7.336	7.315	7.328	7.308
		inch	0.2885	0.2877	0.2888	0.2880	0.2885	0.2877
8	1.25	mm	7.163	7.142	7.168	7.148	7.163	7.142
		inch	0.2820	0.2812	0.2822	0.2814	0.2820	0.2812
9	1.0	mm	8.326	8.306	8.334	8.313	8.326	8.306
		inch	0.3278	0.3270	0.3281	0.3273	0.3278	0.3270
9	1.25	mm	8.164	8.143	8.169	8.148	8.164	8.143
		inch	0.3214	0.3206	0.3216	0.3208	0.3214	0.3206
10	1.0	mm	9.327	9.307	9.335	9.314	9.327	9.307
		inch	0.3672	0.3664	0.3675	0.3667	0.3672	0.3664
10	1.5	mm	8.999	8.971	9.007	8.979	8.999	8.971
		inch	0.3543	0.3532	0.3546	0.3535	0.3543	0.3532
11	1.5	mm	9.997	9.970	10.005	9.977	9.997	9.970
		inch	0.3936	0.3925	0.3939	0.3928	0.3936	0.3925
12	1.5	mm	10.998	10.970	11.006	10.978	10.998	10.970
		inch	0.4330	0.4319	0.4333	0.4322	0.4330	0.4319
12	1.75	mm	10.836	10.803	10.843	10.810	10.836	10.803
		inch	0.4266	0.4253	0.4269	0.4256	0.4266	0.4253

(Continued)

Table 55. *(Continued)* **Recommended Brass, Bronze, and Aluminum Blank Diameters for Straight, Rolled Metric Threads.** *(Source, Landis Threading Systems.)*

Series Designation		Dia. in mm & inches	Brass & Bronze		Aluminum Alloy			
					Soft		Hard	
Size	Pitch		Max.	Min.	Max.	Min.	Max.	Min.
14	1.5	mm	12.997	12.969	13.005	12.977	12.997	12.969
		inch	0.5117	0.5106	0.5120	0.5109	0.5117	0.5106
14	2	mm	12.664	12.631	12.657	12.642	12.664	12.631
		inch	0.4986	0.4973	0.4990	0.4977	0.4986	0.4973
16	1.5	mm	14.999	14.971	15.006	14.978	14.999	14.971
		inch	0.5905	0.5894	0.5908	0.5897	0.5905	0.5894
16	2	mm	14.666	14.633	14.676	14.643	14.666	14.633
		inch	0.5774	0.5761	0.5778	0.5765	0.5774	0.5761
18	1.5	mm	16.998	16.970	17.005	16.977	16.998	16.970
		inch	0.6692	0.6681	0.6695	0.6684	0.6692	0.6681
18	2.5	mm	16.335	16.297	16.345	16.307	16.335	16.297
		inch	0.6431	0.6416	0.6435	0.6420	0.6431	0.6416
20	1.5	mm	18.999	18.971	19.007	18.979	18.999	18.971
		inch	0.7480	0.7469	0.7483	0.7472	0.7480	0.7469
20	2.5	mm	18.336	18.298	18.346	18.308	18.336	18.298
		inch	0.7219	0.7204	0.7223	0.7208	0.7219	0.7204
22	1.5	mm	20.998	20.970	21.006	20.978	20.998	20.970
		inch	0.8267	0.8256	0.8270	0.8259	0.8267	0.8256
22	2.5	mm	20.335	20.297	20.345	20.307	20.335	20.297
		inch	0.8006	0.7991	0.8010	0.7995	0.8006	0.7991
24	2	mm	22.664	22.631	22.675	22.642	22.664	22.631
		inch	0.8923	0.8910	0.8927	0.8914	0.8923	0.8910
24	3	mm	22.004	21.958	22.017	21.971	22.004	21.958
		inch	0.8663	0.8645	0.8668	0.8650	0.8663	0.8645
27	2	mm	25.664	25.631	25.674	25.641	25.664	25.631
		inch	1.0104	1.0091	1.0208	1.0095	1.0104	1.0091
27	3	mm	25.006	24.961	25.109	24.973	25.006	24.961
		inch	0.9845	0.9827	0.9850	0.9832	0.9845	0.9827
30	2	mm	28.664	28.631	28.674	28.641	28.664	28.631
		inch	1.1285	1.1272	1.1289	1.1276	1.1285	1.1272
30	3.5	mm	27.673	27.623	27.689	27.638	27.673	27.623
		inch	1.0895	1.0875	1.0901	1.0881	1.0895	1.0875
33	2	mm	31.666	31.633	31.676	31.643	31.666	31.633
		inch	1.2467	1.2454	1.2471	1.2458	1.2467	1.2454
33	3.5	mm	30.673	30.622	30.688	30.637	30.673	30.622
		inch	1.2076	1.2056	1.2082	1.2062	1.2076	1.2056
36	3	mm	34.001	33.953	34.015	33.967	34.001	33.953
		inch	1.3367	1.3392	1.3373	1.3386	1.3386	1.3367
36	4	mm	33.346	33.295	33.361	33.311	33.346	33.295
		inch	1.3128	1.3108	1.3134	1.3114	1.3128	1.3108
39	3	mm	37.001	36.953	37.015	36.967	37.001	36.953
		inch	1.4567	1.4548	1.4573	1.4554	1.4567	1.4548
39	4	mm	36.346	36.295	36.361	36.311	36.346	36.295
		inch	1.4309	1.4289	1.4315	1.4295	1.4309	1.4289

(Continued)

Table 55. *(Continued)* **Recommended Brass, Bronze, and Aluminum Blank Diameters for Straight, Rolled Metric Threads.** *(Source, Landis Threading Systems.)*

Series Designation		Dia. in mm & inches	Brass & Bronze		Aluminum Alloy			
					Soft		Hard	
Size	Pitch		Max.	Min.	Max.	Min.	Max.	Min.
42	3	mm	39.999	39.951	40.014	39.966	39.999	39.951
		inch	1.5860	1.5841	1.5865	1.5846	1.5860	1.5841
42	4.5	mm	39.020	38.961	39.036	38.978	39.020	38.961
		inch	1.5361	1.5338	1.5368	1.5345	1.5361	1.5338
45	3	mm	42.999	42.951	43.014	42.966	42.999	42.951
		inch	1.7041	1.7022	1.7046	1.7027	1.7041	1.7022
45	4.5	mm	42.020	41.961	42.036	41.978	42.020	41.961
		inch	1.6542	1.6519	1.6549	1.6526	1.6542	1.6519
48	3	mm	46.000	45.950	46.015	45.964	46.000	45.950
		inch	1.8111	1.8091	1.8117	1.8097	1.8111	1.8091
48	5	mm	44.691	44.628	44.709	44.645	44.691	44.628
		inch	1.7595	1.7570	1.7602	1.7577	1.7595	1.7570
52	3	mm	49.999	49.948	50.013	49.963	49.999	49.948
		inch	1.9685	1.9665	1.9691	1.9671	1.9685	1.9665
52	5	mm	48.678	48.628	48.696	48.645	48.678	48.628
		inch	1.9165	1.9145	1.9172	1.9152	1.9165	1.9145

Note: These dimensions are for setup reference. Diameters must be finally established by actual rolling.

3. The concentrate, when mixed with water, must form a thorough, stable mixture with no clotting or livering.
4. The mix should not foster the growth of fungi, mold, or bacteria, nor shall it cause dermatological effects.
5. The lubricoolant should not contain sulfur, chlorine, or other elements that may have a corrosive effect on iron, steel, copper, brass, aluminum, or other materials used for rolling.
6. The mix must remain stable in the presence of lime, aluminum-stearate, or other drawing compounds that may coat cold-drawn steel.
7. It should provide maximum rust preventative properties and have no tendency to remove paint.

Cutting speeds for thread rolling

Recommended rolling sfm recommendations range from 75 sfm (23 smm) to 250 sfm (75 smm), depending on the workpiece material, thread specification, and available machine power. With softer, more ductile material that is readily cold worked, rates of 250 sfm or even higher can be achieved. 100 sfm (30 smm) is considered a good reference point, and most threads are rolled in the 100 – 125 sfm (30 – 38 smm) range. **Table 56** gives recommended rolling head RPM based on a speed of 100 sfm (30 smm).

Maintenance and troubleshooting in thread rolling operations

Die Life Expectancy. As blank hardness increases, the difficulty in rolling and the degree of die life decreases at a proportionally faster rate due to the corresponding reduction in the material elongation (ductility) factor. **Table 57**, based on rolls for thread rolling machines, serves to illustrate the effect of hardness on die life. These

Table 56. RPM Chart for Thread Rolling Heads and Threading Machines. *(Source, Landis Threading Systems.)*

Thread Size	RPM*	Thread Size	RPM*	Thread Size	RPM*	Thread Size	RPM*
1/4–28	1680	1/2–20	820	7/8–14	460	1 3/8–12	289
5/16–18	1380	9/16–12	750	1–8	420	1 1/2–6	274
5/16–24	1340	9/16–18	730	1–12	400	1 1/2–12	264
3/8–16	1140	5/8–11	680	1 1/8–7	370	1 3/4–5	235
3/8–24	1100	5/8–18	650	1 1/8–12	360	1 3/4–12	225
7/16–14	975	3/4–10	560	1 1/4–7	330	2–4 1/2	206
7/16–20	940	3/4–16	540	1 1/4–12	320	2–12	196

* Based on speed of 100 surface feet per minute.

Table 57. Effect of Workpiece Hardness on Die Life. *(Source, Landis Threading Systems.)*

Workpiece Hardness	Yield (PSI)	Yield (MPa)	% Elongation	Pressure Factor	Average Die Life*
200 BHN (15 RC)	65,000	448	22.5	1.0	1,000,000
250 BHN (25 RC)	95,000	655	19.0	1.6	500,000
300 BHN (32 RC)	135,000	930	16.0	2.3	100,000
350 BHN (38 RC)	162,000	1117	14.0	2.8	10,000

* Average die life measured in number of pieces produced.

values are not absolute, but do serve as an indicator of the critical effect that only a few points increase of R_C has on die life.

Even though hardness and the elongation factor are regarded as indices of rollability, they are not absolute deciding factors. Certain high-temperature alloys may be comparatively soft, have high elongation and low yield considerations, yet develop a yield point that makes cold work impossible. Rolling affects the workhardening characteristics of certain materials to such a degree that the yield is increased to such a degree that permanent deformation cannot be accomplished. An example of a material with these characteristics is Haynes Alloy #25.

Head Maintenance. Thread rolling heads are generally equipped with multiple bushing type bearings. When extreme pressures are involved, such as when producing coarse pitches or threading heat treated or stainless steels, these bushings tend to mushroom or elongate endwise. Either of these conditions will sometimes cause the bearings to bind, causing the rolls to skip or stall. When this occurs, the bearings can be honed or filed to relieve the condition. An alternative is to substitute single bronze or carbide bushings for multiple bushing bearings. See **Table 58** for specific information to correct threading problems.

Prestart Checklist. The following checks should be made before beginning any rolling operation.

1. Is the blank diameter within specified tolerances? Start with the recommended minimum tolerance to assure maximum running time before tool wear results in the blank "growing" over the maximum allowable blank outside diameter.
2. Does the blank have the recommended chamfer angle?
3. Has the head been properly aligned with the work?
4. Are the rolls installed in the proper rotation? Install in 1,2,3 clockwise sequence

Table 58. Troubleshooting Chart for Thread Rolling. *(Source, Landis Threading Systems.)*

Problem	Cause	Correction
Excessive truncation on thread crest and oversize pitch diameter.	Material too hard or prone to workhardening, causing the head to deflect.	Use softer material or material less prone to workharden.
Taper at leave-off end of full threads.	Using improper helix angle bushings or rolls too low at rear end.	Check helix angle bushings. Confirm that helix angle is correct for threat being produced.
Highlights or marks on crests and flanks of threads where rolls leave work.	Overrolling. Thread is being piled and pushed ahead of the rolls.	If o.d. of blank is correct, size head to increase pitch diameter. If blank o.d. is oversize, reduce blank to recommended tolerances.
Enlarged and burnished radius in root, normally discernable with a comparator and thread chart.	Rolls not laying close enough to the helix angle of the thread.	Use correct helix angle bushings.
Variations in outside diameter of thread.	Variable blank diameter or variable hardness.	Confirm blank diameter is within recommended tolerances. Hold heat treatment to closer limits and assure uniformity.
Taper on diameter at beginning of the thread.	Axial flow of surface material toward end of work, leaving insufficient material to finish the thread.	Change to less ductile material or, if there is an unthreaded section in front of the thread, cut off the thread after rolling.
Taper on pitch diameter at beginning of thread. Usually on the first two or three threads.	Head deflecting due to material hardness.	Use softer material or use rolls with a longer throat angle.
Cupped end on work (first thread forced over the end of the work).	Material is flowing over an insufficient chamfer on the work.	Start with a chamfer angle of 30° from the centerline of the work. Some materials may require a chamfer angle of 15° to 12°.
Material flakes and adheres to thread flanks.	Material is either aluminum or steel that is leaded or high in sulfur content. Also caused by overrolling.	Change to a more suitable material. Decrease blank diameter.
Excessive truncation with correct pitch diameter.	More ductile materials may extrude excessively. Rough finish on the blank.	Increase blank diameter. Use blank with fine finish.
Rolls break and separate into washer-type segments several threads thick.	Overrolling or rolls laying at a higher-than-standard helix angle.	Reduce blank diameter. Use correct helix angle bushings for the application.
Premature thread crest breakdown and chipping.	Excessively hard material, incorrect helix angle, or incorrect starting lead.	Reduce workpiece hardness. Use suitable helix angle bushings. Use correct starting lead rate.
Head overheats.	Insufficient coolant supply or improper coolant application.	Increase coolant volume. Apply coolant to the workpiece at the point of roll contact, not to the rolls from the head periphery.
Thread eccentric to workpiece.	Improper alignment.	Check head and machine for misalignment.
Broken roll shafts.	Excessively hard material, striking the shoulder, or overrolling.	Use softer material. Reset head tripping point. Use recommended blank diameter.

for right-hand threading, the opposite for left-hand. For left-hand threading, install left-hand helix angle bushings.

5. Deliberately oversize the head for the first piece rolled. Then size down on subsequent pieces until the desired thread size is obtained. This will prevent overrolling and possible damage to the rolls.

Die Threading (Chasing)

Thread chasers are self-opening die heads and collapsing taps that are used to cut uniform internal or external helical threads. Die heads are capable of reproducing most thread forms including tapered pipe threads. A notable advantage of chasing threads is that, upon completion of cutting the thread form, the die head self-opens and retracts from the workpiece, thereby eliminating the need to be backed off the length of the thread. Die heads can be used on automatic screw machines, universal threading machines, and some varieties of lathes, and they may be either rotating or nonrotating.

Chaser types

External thread die chasers are fitted to either rotating or non-rotating heads, and are available in three styles: tangent, circular, or radial. *Tangential chasers* are flat and threaded on one side. They provide long life, free cutting action, and natural clearance. These chasers may either have straight "threads" ground across its length, or have the helix angle incorporated into the cutting edges. When setting straight cut chasers, refer to the helix angle Tables on pages 1360-1394. As shown in *Figure 7*, straight thread tangential chasers have two angles that are important to cutting operations. The *lead angle* is the angle created by the end of the chaser and the stamped edge. It varies with the helix angle at which the chaser is used, the type of die head, and whether or not a leadscrew feed is used. When cutting UNC (as well as metric and Whitworth coarse) standard diameter and

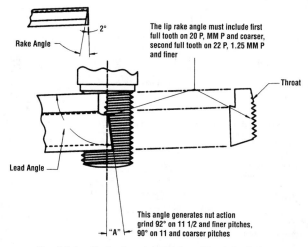

Figure 7. Rake and lead angles on tangential chasers. "Lip rake" angle shown.
(Source, Landis Threading Systems.)

Figure 8. Adjustable die head for tangential cutters. (Source, Greenfield Industries, Inc.)

pitch combinations, and UNF (as well as metric and BSF) standard threads of 11 pitch or coarser, the lead angle should be ground to 90°. For UNF and other fine threads 12 pitch and finer, the suggested lead angle is 92°. The *rake angle* establishes the cutting edge of the chaser. It varies with the machinability of the workpiece, and **Table 59** provides recommended rake angles for selected materials for tangential cutters. A third angle, the *throat angle* is set and does not require regrinding. It is predetermined to provide maximum chaser life and is important in determining thread finish quality. *Figure 8* shows a solid adjustable die head with four tangential cutters locked into place. This versatile head is available for tapping machines, Goss and De Leeuw machines, radial drills, drill presses, and boring mills. Chamfer angle is discussed below.

Although tangential cutters are generally considered to provide the highest accuracy of all thread chasers, *circular thread cutters* have several significant features. First, they can

Table 59. Rake Angles for Tangential Thread Chasing Cutters. *(Source, Landis Threading Systems.)*

Material	Rake Angle	Material	Rake Angle
Cast iron	15°	Aluminum alloy bars and shapes	25°
Wrought/gray cast iron Malleable/ductile iron	18°	Copper	28°
		Bronze or brass, cast	5° Neg.–0°
Low carbon steel free machining (B1112, C1117, etc.), and low carbon steel, non-free machining (C1010, C1018, etc.)	22°	Brass, forged or ruled, except free cutting	22°
		Brass, free cutting bars, forgings; bronze, rolled or forged	10°
Alloy steel (SAE 2000 to 6000 series, etc.), 160–200 BHN	25°	Manganese bronze	0°–10°
		Silicon bronze (Everdur)	22°
Alloy steel (SAE 2000 to 6000 series, etc.), 200–300 BHN	18°–22°	Aluminum bronze (Ampco)	18°–22°
		Naval bronze	0°–10°
Stainless steel	25°	Plastics and fibers	0°–35° Neg.
Aluminum bars and castings Aluminum alloy castings	10°	Bakelite	0°–10°
		Lucite	0°–15° Neg.

Note: All rake angles on this table are positive unless otherwise noted.

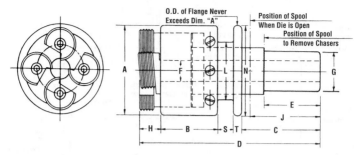

Figure 9. Revolving type, circular thread chaser head. (Source, Cleveland Twist Drill.)

be preset on chaser blocks with a micrometer, and then mounted directly on the machine tool, thereby eliminating setup time. Since they are circular, with a single flute to aid in chip removal, their cutting face can be ground through a full 270°, giving them long life and reducing replacement costs to a minimum. Their comparatively large bodies contribute to heat dissipation. Finally, since their thread grooves are annular (no lead is cut into the threading grooves), the helix angle is contained in the chaser holding block that mounts the cutter on the head. A front and profile view of a revolving style head with circular dies is shown in *Figure 9*. This head can cut straight, taper, and right- or left-hand threads. The pitch diameter is adjustable by turning opposing screws on the periphery, and when fitted with special dies, it can be used for end forming or hollow milling. It is available in a range of head diameters as small as 1.937 inches (49.20 mm) and as large as 18 inches (457.20 mm). Rake angles for circular cutters are, in most cases, the same as for tangential chasers. Chamfer angles for circular thread cutters are discussed below.

Radial thread chasers are available in several styles, with variations available from different manufacturers. In general, *milled radial chasers*, because their hardness—even though they are milled prior to heat treatment—is not equal to ground chasers, are used where misalignment and subsequent chipping due to worn machinery and/or inexperienced operators are expected. The helix angle is milled into the cutter, and the threads are resharpened on the chamfer. *Ground thread radial chasers* are held to tighter tolerances than miller chasers, and hold a much closer tolerance (within class 3 thread limits). Both of these chasers are made in sizes to fit most die heads and collapsing taps. *Insert chasers* allow interchangeability between die head and cutter. A single insert style cutter can be used in up to 100 different heads, and their small size permits economical storage—up to 3500 sets can be stored in one cubic foot. *Figure 10* provides nomenclature for insert chasers. Many of the terms and dimensions of insert chasers apply to all radial thread chasers. Rake angles for radial cutters are, in most cases, one-half that of tangential chasers. Chamfer angles for circular thread cutters are discussed below.

Chamfer angle

The chamfer angle (or throat length) on a chaser is specified to allow the chaser to thread close to the shoulder of a bolt head. Since chip size decreases, and chaser life and stress on the cutting edge increase with smaller chamfer angles, a minimal angle should be used unless threading as close to the shoulder as possible is required. A short chamfer (larger angle) is required to provide a thread close to the head, and good tool life and thread quality. **Tables 60** and **61** provide the amount of undercut for chamfer angles

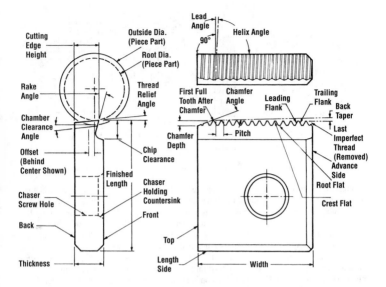

Figure 10. Insert chaser nomenclature. (Source, Cleveland Twist Drill.)

from 10° to 45°. However, it should be remembered that these distances do not allow for clearance for the die head's forward travel during opening at the completion of cutting. The following distances should be added to the undercuts given in the tables in order to assure that no damage is done from the chasers contacting the workpiece when the die head opens.

For 32–14 pitch chasers, add $1/32$" (.03125 inch/.7938 millimeter).

For 13–8 pitch chasers, add $3/64$" (.046875 inch/1.1906 millimeters).

For 7–4 pitch chasers, add $1/16$" (.0625 inch/1.5875 millimeters).

For .5–1.75 mm pitch chasers, add .8 millimeter.

For 2.0–3.0 mm pitch chasers, add 1.2 millimeters.

For 3.5–6.0 mm pitch chasers, add 1.5 millimeters.

Workpiece design should always allow for the widest relief possible, so that a chaser with the longest possible throat can be used. This will allow the chip to be divided over a longer cutting edge, which will enhance chaser life and produce a superior thread finish.

Cutting fluids for thread chasing

As in all machining operations, efficiency and quality can be maximized with the proper tools and fluids. The following cutting fluids are suggested by Cleveland Twist Drill for chasing operations.

Steel, Free Machining. B1112, B1113, C1117, C1118, and leaded high sulfur grades. Use noncorrosive sulfurized animal fatty oil; viscosity 125/175 SUS at 100° F; or water emulsifiable oil.

Carbon Steel. Carbon range of .10 to .40. Use chemically combined sulfurized

Table 60. Chamfer Length for Thread Chasers—Unified Threads. *(Source, Cleveland Twist Drill.)*

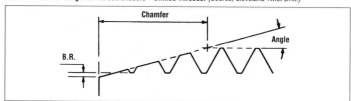

Threads Per Inch	Chamfer Angle							
	10°	15°	20°	25°	30°	35°	40°	45°
	.010" Below Root					.005" Below Root		
5	.797	.520	.383	.299	.241	.192	.160	.135
5-1/2	.727	.476	.350	.274	.221	.175	.146	.123
6	.674	.440	.324	.253	.204	.161	.134	.113
7	.585	.383	.281	.219	.177	.139	.116	.097
8	.519	.342	.252	.197	.159	.124	.103	.087
9	.467	.305	.224	.175	.141	.110	.091	.077
10	.426	.278	.205	.160	.129	.099	.083	.072
-	.008" Below Root					.003" Below Root		
11	.382	.252	.186	.145	.117	.089	.075	.062
11-1/2	.368	.240	.177	.138	.112	.085	.071	.059
12	.354	.233	.172	.134	.108	.082	.069	.058
13	.330	.216	.159	.124	.100	.076	.063	.053
14	.310	.203	.150	.117	.094	.071	.059	.049
16	.277	.180	.132	.103	.083	.062	.051	.043
-	.005" Below Root					.002" Below Root		
18	.234	.155	.114	.088	.072	.055	.046	.038
19	.222	.146	.107	.084	.068	.052	.043	.036
20	.213	.140	.103	.081	.065	.050	.041	.035
22	.196	.129	.095	.074	.060	.045	.038	.032
24	.183	.118	.087	.067	.055	.041	.034	.029
-	.002" Below Root					.001" Below Root		
26	.154	.101	.074	.058	.047	.037	.031	.026
27	.149	.097	.071	.056	.045	.036	.030	.025
28	.142	.094	.069	.054	.044	.035	.029	.024
30	.135	.088	.065	.050	.041	.032	.027	.023
32	.127	.085	.062	.048	.039	.031	.026	.021
36	.114	.075	.055	.043	.035	.027	.023	.019
40	.104	.067	.052	.039	.032	.025	.021	.017
44	.095	.063	.046	.036	.029	.023	.019	.016
48	.088	.058	.042	.033	.026	.021	.017	.014
56	.077	.051	.037	.029	.024	.018	.015	.013

All dimensions in inches.

Table 61. Chamfer Length for Thread Chasers—Metric Threads. *(Source, Cleveland Twist Drill.)*

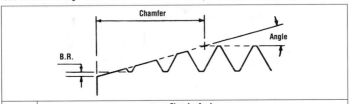

Metric Pitch	Chamfer Angle							
	10°	15°	20°	25°	30°	35°	40°	45°
	.25 Millimeter Below Root					.13 Millimeter Below Root		
6.00	23.06	15.19	11.18	8.71	7.04	5.64	4.70	3.94
5.50	21.26	14.00	10.31	8.05	6.50	5.18	4.32	3.63
5.00	19.46	12.80	9.42	7.37	5.94	4.72	3.94	3.30
4.50	17.65	11.63	8.56	6.68	5.38	4.27	3.56	3.00
4.00	15.85	10.44	7.70	5.99	4.85	3.81	3.18	2.67
3.50	14.05	9.25	6.81	5.31	4.29	3.35	2.79	2.36
3.00	12.25	8.08	5.94	4.62	3.73	2.90	2.41	2.03
2.50	10.46	6.88	5.05	3.96	3.20	2.44	2.06	1.73
-	.20 Millimeter Below Root					.08 Millimeter Below Root		
2.00	8.36	5.51	4.04	3.15	2.57	1.93	1.60	1.35
1.75	7.47	4.90	3.61	2.82	2.29	1.70	1.42	1.19
1.50	6.55	4.32	3.18	2.49	2.01	1.47	1.21	1.04
-	.13 Millimeter Below Root					.05 Millimeter Below Root		
1.25	5.23	3.43	2.54	1.98	1.60	1.22	1.02	.84
-	.05 Millimeter Below Root					.025 Millimeter Below Root		
1.00	3.89	2.57	1.88	1.47	1.19	.94	.79	.66
.80	3.18	2.08	1.55	1.19	.97	.76	.64	.53
.75	3.00	1.98	1.45	1.14	.91	.71	.61	.51
.70	2.82	1.85	1.37	1.07	.86	.66	.56	.48
.60	2.43	1.63	1.19	.91	.74	.58	.48	.41
.50	2.08	1.37	1.02	.79	.64	.48	.41	.36
.45	1.91	1.24	.91	.71	.58	.46	.38	.30
.40	1.73	1.14	.84	.66	.53	.41	.33	.28
.35	1.55	1.02	.76	.58	.48	.36	.30	.25
.30	1.37	.89	.66	.51	.41	.30	.25	.23

All dimensions in millimeters.

oil—active sulfur 1.5% min., plus animal fatty oil, 4% max.; viscosity 150 to 200 SUS at 100° F, or water emulsifiable oil.

Carbon Steel. Carbon range of .45 and higher. Use sulfur-chlorinated animal fatty oil—compounded in petroleum vehicle; viscosity 250 SUS at 100° F.

Alloy Steel. Carbon range of .10 to .35. Same as for carbon steel, .10 to .40 carbon.

Alloy Steel. Carbon range of .40 and higher. Use 10% to 30% sulfurized fatty oil, or sulfur-chlorinated fatty oil compounded in petroleum vehicle; viscosity 250 SUS at 100° F.

Stainless Steel, Type 300 series. Same as carbon steel, .10 to .40 carbon, or water emsulsifiable oil.

Stainless Steel, Type 400 series. Same as alloy steel, .40 carbon and higher, or water emsulsifiable oil.

Cast Iron, Malleable. Use water emulsifiable oil. For superior finish, use sulfurized animal fatty oil, active sulfur 0.5% to 1% in finish viscosity petroleum vehicle of 60 SUS at 100° F.

Cast Iron, Wrought. Same as carbon steel, .10 to .40 carbon.

Aluminum, Pure Alloy. Use petroleum oil containing 1%–2% animal fatty oil, viscosity of finished oil 40–60 SUS at 100° F, or use water emulsifiable oil.

Bronze, Regular Commercial. Use petroleum oil containing 15%–25% animal fatty oil in compound with viscosity 150–200 SUS at 100° F, or water emulsifiable oil.

Bronze, Silicon, Phosphor, or Aluminum. Use petroleum oil containing in compound sulfurized animal fatty oil (active sulfur 5% max.) or water emulsifiable oil.

Copper. Same as commercial bronze.

Brass, Free Cutting. Use water emulsifiable oil, or paraffin neutral oil, viscosity range 75–100 SUS at 100° F.

Brass, Forging. Same as silicon bronze.

Monel. Same as silicon bronze, but clean immediately with petroleum solvent.

Magnesium. Same petroleum oil as pure alloy aluminum. Do not use water. Tools should always be flooded.

Fiber. Dry.

Rubber. Use water emulsifiable oil, plus the addition of sodium carbonate in the amount of 25% total emulsion.

Power requirements for thread chasing

Calculating the power required for a thread chasing operation can be useful in evaluating machining methods, tools, tool life, and production rates. The method provided below is a simple way to estimate the horsepower needed to produce a given pitch (tpi) on a given workpiece material. Due to the variables involved, it is not 100% accurate, and the rating is for horsepower required at the tool. Due to machine inefficiency, the actual power required at the motor may be 10% to 20% higher than the values provided by the following formula. Therefore, an appropriate allowance, based on machine age and condition, should be made with regard to age and condition of the machine.

$$\text{P.P.V.} \times \text{M.P.F.} \times \text{sfm} \times \text{T.S.} = \text{HP}$$

where: P.P.V. = Pitch Power Value from **Table 62**
M.P.F. = Material Power Factor from **Table 63**
sfm = surface feet per minute
T.S. = Tool Sharpness factor.

Table 62 provides power requirements based on threading free machining steels B1111 and 1213. The P.P.V. value (in the second column) has already been factored into the values given in the Table, and is given for reference when machining other materials. While the values given in the Table are based on the Material Power Factor rating for free machining steel (1.0), they can be used for other materials by multiplying the chart value by the M.P.F. of the material being cut. For example, the M.P.F. factor for stainless steel (from **Table 63**) is 2.40. Thus the power required to cut 16 tpi in stainless steel at 30 sfm can be calculated as follows.

$$2.40 \times .405 = .972 \text{ HP at the tool,}$$

where 2.40 is the M.P.F. for stainless steel, and .405 is the HP needed to cut 16 tpi into free machining steel (from **Table 62**).

The Tool Sharpness (T.S.) is factored as 1.0 for a sharp tool; 1.5 for a medium sharp tool; and 2.0 for a dull tool.

Example. Find the power at the tool required to machine 14 tpi onto alloy steel (BHN 220) with a dull tool at 25 sfm.

Table 62. Power Requirements for Thread Chasing (See Notes). *(Source, Landis Threading Systems.)*

SFM		10	15	20	25	30	35	40	45	50
TPI	P.P.V.	Horsepower								
4	.120	1.800	2.700	3.600	4.500	5.400	6.300	7.200	8.100	9.000
4.5	.0951	1.427	2.1398	2.853	3.5663	4.2795	4.9928	5.706	6.4193	7.1325
5	.077	1.155	1.7325	2.310	2.8875	3.465	4.0425	4.620	5.1975	5.775
6	.0532	.7984	1.1976	1.5968	1.996	2.3952	2.7944	3.1936	3.5928	3.992
7	.0393	.589	.8835	1.178	1.4725	1.767	2.0615	2.356	2.6505	2.945
8	.030	.450	.675	.900	1.125	1.350	1.575	1.800	2.025	2.250
9	.0248	.3716	.5574	.7432	.929	1.1148	1.3006	1.4864	1.6722	1.858
10	.020	.300	.450	.600	.750	.900	1.050	1.200	1.350	1.500
11	.016	.240	.360	.480	.600	.720	.840	.960	1.080	1.200
11.5	.0155	.232	.349	.465	.581	.698	.814	.930	1.046	1.163
12	.015	.225	.3375	.450	.5625	.675	.7875	.900	1.0125	1.125
13	.012	.180	.270	.360	.450	.540	.630	.720	.810	.900
14	.0103	.1545	.2318	.309	.3863	.4635	.5408	.618	.6953	.7725
16	.009	.135	.2025	.270	.3375	.405	.4725	.540	.6075	.675
18	.007	.105	.1575	.210	.2625	.315	.3675	.420	.4725	.525
20	.006	.090	.135	.180	.225	.270	.315	.360	.405	.450
24	.004	.060	.090	.120	.150	.180	.210	.240	.270	.300
28	.003	.045	.0675	.090	.1125	.135	.1575	.180	.2025	.225
32	.002	.030	.045	.060	.075	.090	.105	.120	.135	.150
36	.0016	.0238	.0357	.0476	.0595	.0714	.0833	.0952	.1071	.119
40	.0013	.0192	.0288	.0384	.048	.0576	.0672	.0768	.0864	.0960
44	.0010	.0154	.0231	.0308	.0385	.0462	.0539	.0616	.0693	.077
48	.0009	.0131	.0197	.0262	.0328	.0393	.0459	.0524	.0590	.0655
56	.0007	.0100	.015	.020	.025	.030	.035	.040	.045	.050
64	.0005	.0077	.0116	.0154	.0193	.0231	.0270	.0308	.0347	.0385

Notes: For speeds in excess of 50 sfm, add the totals of existing columns. For example, to find the horsepower required to chase a 12 tpi thread at 75 sfm, add the totals from the columns for 50 sfm (1.125 HP) and 25 sfm (.5625 HP): the power required is 1.6875. To convert horsepower to kilowatts (kW), multiply HP by 0.7456999. To convert surface feet per minute (sfm) to surface meters per minute (smm), multiply sfm by 0.3048.

The values in this Table are for threading B1111 and 1213 free machining steel with a Material Power Factor (M.P.F.) of 1.0 with medium sharp tools having a Tool Sharpness (T.S.) rating of 1.5 (medium-sharp tool). P.P.V. stands for Pitch Power Factor. See text for full explanation of terms and derivations of factors used for calculating power requirements for thread chasing operations.

.0103 × 2.10 × 25 × 2.0 = 1.0815 HP

where: 0.0103 is the P.P.V. for 14 tpi (**Table 62**)
2.10 is the M.P.F. for alloy steel (BHN 220) from **Table 63**
25 is the speed in surface feet per minute
2.0 is the T.S. for a dull tool.

Cutting speeds for thread chasing

As in any cutting operation, tool rigidity, workpiece material, required tolerances, and desired finish are all instrumental in setting cutting speeds. In addition, tool balance can be critical at higher speeds when using thread chasing heads. Recommended surface speeds for high speed steel chasers cutting unified or metric threads are given in **Table 64**.

Collapsing taps

Collapsing taps are widely used for threading larger holes, especially for pipe threads. These taps quickly thread large diameter holes, and then collapse so that withdrawal does not require backing out by reverse threading. Since they are adjustable, a single tapping head can thread a large range of pipe sizes. They can be used for cutting standard 60° threads as well as taper pipe threads. These tools can be highly versatile, and *Figure 11* shows a collapsing tap fitted with rose reamer blades, facing blades, and threading chasers. This tool will ream a hole to tap drill size, face the workpiece, tap the hole, and collapse and rapidly extract from the workpiece. Collapsing taps are available in either stationary or revolving designs.

Table 63. Material Power Factors for Thread Chasing (See Notes). *(Source, Landis Threading Systems.)*

Material	Hardness BHN	Power Factor (M.P.F.)	Material	Hardness BHN	Power Factor (M.P.F.)
Aluminum	-	.28	Free Cutting Steel (111-1213)	140	1.00
Brass	-	.50			
Bronze	-	.50	Alloy Steel (1330-8642)	175	1.40
Cast Iron	-	.90		190	1.50
Copper	-	.62		200	1.60
Magnesium	-	.70		203	1.70
Malleable Iron	-	1.10		205	1.80
Stainless Steel	-	2.40		210	1.90
Titanium	-	2.15		215	2.00
Zinc	-	.50		220	2.10
	90	1.50		230	2.20
	110	1.60		240	2.30
	140	1.70		250	2.40
Carbon Steel (1008-1095)	170	1.80		330	3.20
	190	1.90		390	3.80
	200	2.00		470	4.50
	250	2.10	Structural Steel (A-36)	160	1.40

Notes: Hardness is expressed in Brinell hardness (BHN). M.P.F is Material Power Factor, with free cutting steel having a reference value M.P.F. of 1.0. See text for explanation of how to use M.P.F. in calculating power required for thread chasing operations.

Table 64. Recommended Surface Speeds for Thread Chasers (See Notes). *(Source, Cleveland Twist Drill.)*

Material	Threads per Inch								
	3-4 1/2	5-6	7-8	9-11 1/2	12-15	16-19	20-24	25-32	33-120
	Surface Feet per Minute								
Carbon Steels-Plain Carbon									
1052, 1064*, 1065*, 1066*, 1069*, 1070*, 1071*, 1074*, 1075*, 1080*, 1084*, 1085*, 1086*, 1090*, 1095*	5	6	7	9	11	13	14	15	16
1036, 1038, 1039, 1040, 1041, 1042, 1043, 1045, 1046, 1049, 1051, 1052, 1053, 1054*, 1055*, 1059*, 1060*, 1061*	8	11	14	17	20	22	24	25	27
1008, 1010, 1011, 1012, 1013, 1015, 1017, 1020, 1024, 1025, 1027, 1029, 1030, 1031, 1033, 1034, 1035, 1037, 1045*, 1050*	12	18	23	28	32	36	38	40	43
1016, 1018, 1019, 1021, 1022, 1023, 1026	17	26	35	43	49	54	58	61	66
Carbon Steels-Free Machining									
1137, 1140, 1141, 1145, 1146, 1151	12	18	23	28	32	36	38	40	43
1108, 1109, 1115, 1120, 1125, 1126, 1132, 1138, 1139, 1141*, 1144, 1145*, 1151*	17	26	35	43	49	54	58	61	66
1116, 1117, 1118, 1119, 1121	23	37	49	61	70	79	85	91	100
1112, 1113, 1119, 1212, 1213, 12L14	30	48	66	83	98	112	112	112	112
Magnesium Steels (1.60-1.90% Mn)									
1130*, 1335, 1340	8	11	14	17	20	22	24	25	27
Nickel Steel (3.50-5.00% Ni)									
2330									
Nickel Chromium Steels (1.25-3.50% Ni, 0.60-1.50% Cr)									
3140	8	11	14	17	20	22	24	25	27
3310*	5	6	7	9	11	13	14	15	16
3115, 3140*	12	18	23	28	32	36	38	40	43

(Continued)

Table 64. (Continued) Recommended Surface Speeds for Thread Chasers (See Notes). (Source, Cleveland Twist Drill.)

Material	Threads per Inch								
	3-4 1/2	5-6	7-8	9-11 1/2	12-15	16-19	20-24	25-32	33-120
	Surface Feet per Minute								
Molybdenum Steels									
4817*, 4820*	5	6	7	9	11	13	14	15	16
4063*, 4145*, 4147*, 4150*, 4320*, 4337*, 4340*, 4718, 4720, 4815*	8	11	14	17	20	22	24	25	27
4027, 4037*, 4042*, 4047*, 4130*, 4135*, 4137*, 4140*, 4142*, 4422, 4427, 4520, 4615, 4620, 4621	12	18	23	28	32	36	38	40	43
4023, 4024, 4118	17	26	35	43	49	54	58	61	66
Chromium Steels									
50100*, 51100*, 52100*	5	6	7	9	11	13	14	15	16
5150, 5150*, 5155*, 5160*, 51B60*	8	11	14	17	20	22	24	25	27
50B40*, 50B44*, 5046*, 50B46*, 50B50*, 5132*, 5135*, 5140*, 5145*, 5147*	12	18	23	28	32	36	38	40	43
5015, 5115, 5120	17	26	35	43	49	54	58	61	66
Chromium Vanadium Steels									
6120, 6150*	8	11	14	17	20	22	24	25	27
6118	12	18	23	28	32	36	38	40	43
Nickel Chromium Molybdenum Steels									
8625, 8627, 8645*, 86B45*, 8650*, 86B55*, 8660*, 8822, 9310*, 9840*	8	11	14	17	20	22	24	25	27
8115, 81B45*, 8615, 8617, 8620, 8622, 8630*, 8637*, 8640*, 8642*, 8720, 8735, 8740*, 8742*, 94B30	12	18	23	28	32	36	38	40	43
Silicon Manganese Steels									
9262*	5	6	7	9	11	13	14	15	16
9255*, 9260*, 9840*	8	11	14	17	20	22	24	25	27

(Continued)

Table 64. *(Continued)* Recommended Surface Speeds for Thread Chasers (See Notes). *(Source, Cleveland Twist Drill.)*

Material	Threads per Inch								
	3-4 1/2	5-6	7-8	9-11 1/2	12-15	16-19	20-24	25-32	33-120
	Surface Feet per Minute								
Nickel Chromium Molybdenum Manganese Steels									
94B15, 94B17, 94B40*	12	18	23	28	32	36	38	40	43
Stainless Steels									
200 Series, 302, 304, 321, 440A	5	6	7	9	11	13	14	15	16
303, 330, 410, 430	8	11	14	17	20	22	24	25	27
416, 430F	12	18	23	28	32	36	38	40	43
Aluminum Alloys									
1100	30	48	66	83	98	112	112	112	112
6061, Die Castings, Sand Castings	38	61	84	108	129	151	165	182	213
2011, 2017, 2024, 6262, 7075, 7178	57	96	136	186	237	300	332	370	418
Copper Alloys									
Red Brass, Manganese Bronze, Naval Brass, Electrolitic Tough Pitch Copper, Commercial Bronze, Phosphor Bronze, Al Bronze	17	26	35	43	49	54	58	61	66
Aluminum Silicon Bronze, Leaded Phosphor Bronze	23	37	49	61	70	79	85	91	100
Leaded Commercial Bronze, Forging Brass	30	48	66	83	98	112	112	112	112
Free Cutting Copper	38	61	84	108	129	151	165	182	213
Free Cutting Brass	47	77	106	140	170	208	232	259	302
Iron									
Wrought	8	11	14	17	20	22	24	25	27
Cast	17	26	35	43	49	54	58	61	66
Malleable	30	48	66	83	98	112	112	112	112
Magnesium	57	96	136	186	237	300	332	370	418

(Continued)

Table 64. *(Continued)* **Recommended Surface Speeds for Thread Chasers** (See Notes). *(Source, Cleveland Twist Drill.)*

Material	Threads per Inch									
	3-4 1/2	5-6	7-8	9-11 1/2	12-15	16-19	20-24	25-32	33-120
	Surface Feet per Minute								
Nickel Alloys									
Inconel	5	6	7	9	11	13	14	15	16
Monel and others	8	11	14	17	20	22	24	25	27
Plastics (Machinable)	47	77	106	140	170	208	232	259	302
Rubber (Machinable)	57	96	136	186	237	300	332	370	418
Zinc (Die Castings)	30	48	66	83	98	112	112	112	112

* Indicates steel has been annealed.

Notes: Reduce speeds by 25% when cutting NPT and NPTF threads. To convert surface feet per minute (sfm) values into surface meters per minute (smm), divide by 0.3048.

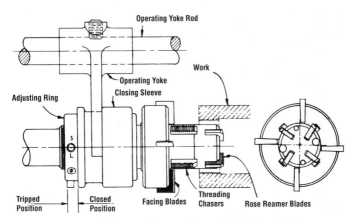

Figure 11. Multipurpose collapsing tap. (Source, Greenfield Industries Inc.)

Alignment is crucial and must be properly maintained between the tap and the hole being tapped. Misalignment will cause the tap to feed into the work at an angle that will spoil the threads and possibly damage the chasers. Rapid wearing or chipping on the tops of the chaser teeth is frequently caused by misalignment. It is advisable to use a lead screw or other positive feed to assure that the tap is fed into the work at the proper lead. If a positive feed is not used, the lead of the thread becomes largely dependent upon the operator.

The use of a proper lubricant assures the production of smooth and well formed threads. For tapping all grades of steel, including steel and wrought iron pipe, a good grade of sulfur base cutting oil is recommended. Cast and malleable iron can be tapped successfully with soluble oil as a lubricant. Cast brass, fiber, and hard rubber should be tapped dry, and kerosene works well for tapping aluminum.

Thread Grinding

While not widespread, thread grinding offers high precision and excellent surface finish. The most common use of thread grinding is for producing threads on taps and thread gages, and it is also often used to produce threads lead or adjusting screws used in measuring instruments. The process is also highly useful for threading materials that have a hardness in excess of R_C 36, and for materials with a hardness below R_C 17 that are difficult to machine cleanly with traditional threading processes. This makes threading of heat treated workpieces practical, thereby eliminating the possibility of distortion in a screw or bolt that is heat treated after threading. One of the primary benefits of grinding is that it is a gradual removal process, and stress cracks—that can cause failure under extreme pressure—are sometimes induced into sharp thread roots by traditional processes, but are eliminated by precision grinding. Ideally, ground threads should have a root radius of at least 0.003 in. (0.076 mm), and 0.010 in. (0.254 mm) is preferred. **Table 65** provides recommended grinding wheel specifications for internal and external threads. **Table 66** gives grain size requirements for grinding root widths from 0.002 to 0.010 in. (0.05 to 0.254 mm), and **Table 67** shows the coarsest grain sizes permissible for pitches ranging from 4 tpi to 80 tpi (0.3 to 6 mm pitch). The Norton Co. recommends the following cutting speeds for grinding: 9,000 to 10.500 sfm (2,750 to 3,200 smm) for resinoid wheels and 7,500 to 9,500 sfm (2,285 to 2,900 smm) for vitrified wheels.

Parts as small as 0.015 in. to as large as 16 in. (0.381 to 406 mm) can be ground with threads holding a 0.0002 in. (0.005 mm) pitch diameter tolerance limit, a lead tolerance of 0.0005 in./in. (0.0127 mm/mm), and a surface finish of 4 rms. Either center (cylindrical) or centerless grinding may be used, but centerless grinding provides increased efficiency because it allows the complete thread to be ground in one pass. Center grinding normally requires five to six passes when grinding from the solid. Centerless grinding uses

Table 65. Recommended Grinding Wheel Specifications for Thread Grinding. *(Source, Norton Co.)*

Application	Thread Type	TPI	Precision Quality	Commercial Quality
External Grinding				
Precision Screws	Fine Pitch	32-40 (from solid)	32A220-M9VG	-
		Finer than 40 (from solid)	A320-09B	-
Screws and Studs (Heat treated alloy steel)	American National Form Threads	8-12 (from solid)	-	23A100-R9BH
		14-20 (from solid)	-	23A120-R9BH
		24 and finer (from solid)	-	23A150-S9BH or
			-	23A150-T9BH or
			-	23A180-T9BH
Taps (High speed steel)	Acme Threads	8-18 (from solid)	-	23A100-R9BH
		20-26 (from solid)	-	23A150-S9BH
		27 and finer (from solid)	-	23A150-T9BH or
			-	23A180-T9BH
	American National Form Threads	4-12 (from solid)	32A100-K8VG	23A100-R9BH
		14-20 (from solid)	32A120-L9VG	23A120-R9BH
		24-36 (from solid)	32A180-N9VG	23A180-T9BH or
			-	A220-U9BH
	American Standard Pipe Threads	8-18 (from solid)	-	23A100-R9BH
		20-26 (from solid)	-	23A150-S9BH
		27 and finer (from solid)	-	23A150-T9BH or
			-	23A180-T9BH
Whitworth Form Threads	Whitworth Form Threads	8-16	32A120-K9VG	-
		18-24	32A150-L9VG	-
Worms and Lead Screws	Acme and Worm Threads	4-6 (precut)	32A80-J8VG	-
		8-12 (precut)	32A100-K8VG	-
		12-16 (from solid)	32A120-K9VG	23A100-R9BH
Internal Grinding				
Tool and Alloy Steel Hardened	Acme Threads	2-8 (precut)	38A80-J8VG	-
		10-20 (precut)	38A120-K9VG	-
	American National Form Threads	6-16 (precut)	38A120-L9VG	23A100-V9BH
		20-24 (precut)	38A220-M9VG	23A150-V9BH
		6-20 (from solid)	-	23A100-V9BH
		24 and finer (from solid)	-	23A150-V9BH

(Continued)

Table 65. *(Continued)* Recommended Grinding Wheel Specifications for Thread Grinding. *(Source, Norton Co.)*

Application	Thread Type	TPI	Precision Quality	Commercial Quality
		Multirib Grinding		
		Wheel Size ➤	Up to 14" Dia.	15" Dia. and Over
Hardened Steel	Centerless	5-10	-	32A120-Q9VG
	Cylindrical		38A120-N9VG	32A120-N9VG
	Surfacing		38A120-K9VG	32A120-K9VG
	Centerless	11-14	38A150-N9VG	38A150-Q10VG
	Cylindrical		38A150-N9VG	32A150-N9VG
	Surfacing		38A150-K9VG	32A150-K9VG
	Centerless	16-24	38A2203-N9VG	38A220-Q10VG
	Cylindrical		38A2203-N9VG	38A220-N9VG
	Surfacing		38A2203-K9VG	32A220-K9VG
	Centerless	24-32	38A320-N10VG	A320-Q11VG
	Cylindrical		38A320-N10VG	A320-N10VG
	Surfacing		38A320-K9VG	A320-K9VG

Table 66. Grain Size Requirements for Given Thread Roots. *(Source, Norton Co.)*

Root Width in./mm	Grain Size	
	Vitrified Wheel	**Resinoid Wheel**
0.002 / 0.05	–	180
0.003 / 0.076	220	180
0.004 / 0.10	180	150
0.005 / 0.127	150	120
0.006 / 0.15	120	120
0.007 / 0.178	120	100
0.008 / 0.20	100	100
0.009 / 0.229	100	90
0.010 / 0.254	90	90

Note: Even though resinoid wheels remove stock faster than vitrified wheels, they are considerably less rigid than vitrified wheels and care must be taken to assure that they do not deflect during cutting operations.

Table 67. Grain Size Requirements for Given Thread Pitches (Vitrified Wheels). (See Notes.)

Pitch tpi (mm)	Grain Size	Pitch tpi (mm)	Grain Size
4 (6)	70	16-20 (1.25-1.5)	120 / 220*
4 1/2–8 (3–5.5)	80	24-28 (.8-1)	150
5-10 (2.5-5)	120*	24-32 (.8-1)	320*
9-12 (2-2.5)	90	32-36 (.7-.75)	180
11-14 (1.75-2)	150*	40-64 (.4-.6)	220
13-14 (1.75-2)	100	72-80 (.3-.35)	240

Notes: *For multirib wheels. All other sizes are for single rib wheels. Metric pitch conversions are equivalent. All grain sizes are the coarsest allowable for pitch of thread.

multirib wheels and offers speed and economy. For example, production rates of 60 to 70 $1/4$–20 hardened socket set-screws can be ground from solid in less than a minute, with the grinding wheel having a useful life of up to 8 hours (up to 560 set-screws) before redressing.

Single-Rib Wheel. Single rib wheels have the thread profile on their cutting edge and must traverse, while inclined to the thread's helix angle, the length of the thread. Lead screw control and some other form of automatic infeed of the grinding machine is required, and multiple passes are normally required to finish a thread, with about two-thirds of the thread depth being removed on the first pass or passes, and the final finish cut removing the remaining one-third of thread. The exceptions are fine threads 12 tpi and finer that can often be cut in one pass. The maximum infeed for a single-rib wheel should be 0.040 in. (0.10 mm); 0.020 in. (0.05 mm) is commonly used for roughing, and 0.010 in. (0.025 mm) or less is used for achieving high accuracy. Grinding threads with a single-rib wheel is analogous to single-point threading on a lathe. It is the most common method of thread grinding and, with care, can be the most accurate.

Multirib Wheel. When production rate takes precedence over ultimate accuracy, multirib wheels are used. In instances where the thread is longer than the width of the wheel, traverse grinding is required, but it should be noted that the leading ribs on the wheel will experience accelerated wear and will eventually cut oversize threads while the trailing ribs have retained their accuracy. When using traverse grinding with multirib wheels, the thread pitch should not be in excess of one-eighth of the wheel's width, and the root width should be at least 0.007 in. (0.178 mm) wide due to the difficulty inherent in dressing the wheel for smaller root widths. In instances where the wheel width exceeds the thread length by 1.5 thread pitches, plunge grinding can be used to produce the complete thread in less than $1^{1}/_{2}$ revolutions of the workpiece, making this method the most productive type of thread grinding. In plunge grinding, the wheel is introduced to the work gradually over about one-half revolution and then cuts the complete thread and withdraws in just over a full revolution. In addition to conventional multirib wheels that contain several identical thread profiles, other varieties are the skip-rib (or alternate-rib) and three-rib

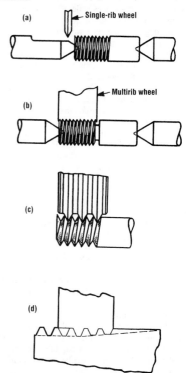

Figure 12. (a) Single-rib wheel. (b) Multirib wheel. (c) Skip-rib wheel. (d) Three-rib wheel with flattening rib.

wheels. The thread spacing on the skip-rib wheel is equal to twice the thread pitch, and approximately $2^1/_2$ workpiece revolutions are required to complete a thread. Due to the thread spacing, cutting fluids have better access to the cutting area, allowing for improved finish and making it easier to cut fine pitch threads. The three-rib design is named for its three individual purpose cutting ribs: a roughing rib that cuts to $^2/_3$ of the total thread depth, an intermediate rib that removes further material down to within approximately 0.005 in. (0.127 mm) of the total thread depth, and a finishing rib that completes the cut. An optional "flattening rib" can be used to finish grind the thread crests. Grinding wheel profiles are shown in *Figure 12*.

Thread Milling

Traditional thread milling is performed with either single- or multiform cutters. Single-form cutters approach the workpiece at an angle equal to the helix angle of the thread being cut, and advance and revolve around the workpiece at a preset rate to cut one thread at a time as the workpiece also rotates in the opposite direction of the cutter. While this method is capable of producing quality threads, it has grown out of favor due to the necessarily long setup and production time. It remains, however, a viable method for cutting large diameter workpieces or in instances where the thread length is too long for a multiform cutter. Using a multiform cutter reduces setup and production time. The complete thread form is contained on the tool which travels parallel to the workpiece which it circulates 1.1 revolutions before exiting. The most obvious limitation of the multiform cutter is that the length of the thread being cut cannot exceed the length of the cutter. Also, because the cutters are arbor or shank mounted, there are unavoidable size limitations when cutting internal threads. Most thread milling operations are now performed with full profile thread milling inserts that allow fast cutting and, as shown in **Table 68**, have been miniaturized to cut internal threads in bores as small as 0.374" (9.5 mm). **Table 69** provides troubleshooting solutions for CNC thread milling, and **Table 70** shows common tolerance classes for thread cutting inserts. The thread milling system that follows has been developed by Kennametal Inc. who produce a full range of milling holders and inserts.

When milling with inserts, the following points should be considered.

1) Inserts are designed to mill the full thread depth in one revolution.
2) When machining difficult materials, it may be desirable to make two passes. A 60% thread depth on the first pass, and a 40% thread depth on the second pass is recommended.
3) Thread relief grooves in blind holes are not necessary.
4) Thread milling large parts requires considerably less horsepower than other threading methods.
5) Thread milling produces short chips compared to stringy chips of other threading methods.
6) One holder is suitable for many different thread pitches.
7) PVD coated inserts provide maximum tool life for a wide variety of materials.

CNC thread milling

In order to perform the following thread milling operations with full profile inserts, a milling machine with three-axis control, capable of helical interpolation, is required. Helical interpolation is a CNC function producing tool movement along a helical path. Helical travel combines circular movement in one plane with a simultaneous linear motion in a plane perpendicular to the first. For example, the path from point A to point B [*Figure 13(a)*] on the envelope of the cylinder combines a circular movement in the X- and Y-plane with a linear movement in the Z-direction.

Table 68. Minimum Bore Diameters for Thread Milling with Thread Milling Inserts (UN-ISO-BSW). (Source, Kennametal Inc.)

tpi ▲	48	32	24	20	16	12	10	8
pitch (mm) ▲	0.5	0.75	1.0	1.25	1.5	2.0	2.5	3.0
Cutter Dia.	Minimum Bore Diameter inch/mm							
.35 / 8.9	.374 / 9.5	.394 / 10	.421 / 10.7	.449 / 11.4	-	-	-	-
.45 / 11.4	.472 / 12	.492 / 12.5	.520 / 13.2	.547 / 13.9	.571 / 14.5	-	-	-
.49 / 12.4	.512 / 13	.531 / 13.5	.559 / 14.2	.587 / 14.9	.610 / 15.5	-	-	-
.61 / 15.5	.630 / 16	.650 / 16.5	.667 / 16.9	.705 / 17.9	.728 / 18.5	.768 / 19.5	-	-
.67 / 17	.693 / 17.6	.717 / 18.2	.748 / 19	.772 / 19.6	.787 / 20	.827 / 21	-	-
.75 / 19	.776 / 19.7	.803 / 20.4	.827 / 21	.850 / 21.6	.866 / 22	.906 / 23	-	-
.79 / 19.3	.815 / 20.7	.843 / 21.4	.866 / 22	.890 / 22.6	.906 / 23	.945 / 24	-	-
.87 / 22.1	.893 / 22.7	.921 / 23.4	.945 / 24	.969 / 24.6	.984 / 25	1.024 / 26	-	-
1.02 / 25.9	1.051 / 26.7	1.079 / 27.4	1.102 / 28	1.130 / 28.7	1.154 / 29.4	1.193 / 30.3	1.441 / 36.6	1.535 / 39
1.18 / 30	1.209 / 29.7	1.236 / 31.4	1.260 / 32	1.291 / 32.8	1.319 / 33.5	1.362 / 34.6	1.732 / 44	1.830 / 47.2
1.46 / 37.1	1.496 / 38	1.520 / 38.6	1.555 / 39.5	1.591 / 40.4	1.614 / 41	1.654 / 42	1.929 / 49	2.047 / 52
1.65 / 42	1.701 / 43.2	1.724 / 43.8	1.772 / 45	1.811 / 46	1.831 / 46.5	1.866 / 47.4	-	-
tpi ▲	7	6		5		4.5		4
pitch (mm) ▲	3.5	4.0	4.5	5.0	5.5		6.0	
Cutter Dia.	Minimum Bore Diameter inch/mm							
1.18 / 30	1.654 / 42	1.772 / 45	1.890 / 48	-	-	-	-	-
1.46 / 37.1	1.929 / 49	2.047 / 52	2.185 / 55.5	-	-	-	-	-
1.65 / 42	2.146 / 54.5	2.268 / 57.6	2.401 / 56.9	-	-	-	-	-
1.38 / 35 (UN)	-	1.969 / 75.4	-	1.843 / 46.8	-	1.756 / 44.6	-	2.228 / 55.6
1.38 / 35 (ISO)	-	1.969 / 75.4	2.102 / 53.4	1.673 / 42.5	1.969 / 50	-	2.264 / 57.5	-
1.38 / 35 (BSW)	-	1.961 / 75.2	-	1.831 / 46.5	-	1.866 / 47.4	-	-

All diameters given as inch / millimeter. Millimeters are expressed to the nearest tenth.

Table 69. Troubleshooting Thread Milling Operations. *(Source, Kennametal Inc.)*

Problem	Possible Cause	Solution
Excessive insert flank wear	Cutting speed too high	Reduce cutting speed.
	Chip is too thin	Increase feed rate.
	Insufficient coolant	Increase coolant quality/pressure.
Chipping of cutting edge	Chip is too thick	Reduce feed rate. Use the tangential arc method of entering. Increase rpm.
	Vibration	Check rigidity.
Material buildup on cutting edge	Cutting speed too low	Increase cutting speed.
	Chip thickness too small	Increase feed rate.
Chatter / vibration	Feed rate too high	Reduce feed rate.
	Profile is too deep (coarse threads)	Execute two passes, each with increased cutting depth. Execute two passes, each cutting only half the thread length.
	Thread length is too long	Execute two passes, each cutting only half the thread length.
Insufficient thread accuracy	Tool deflection	Reduce feed rate. Execute a "zero" cut.

Table 70. Insert Tolerance Classes. *(Source, Kennametal Inc.)*

Thread Designations	Standard Designations	Tolerance Class
UN	ANSI B1.174	2A/2B
UNJ	MIL-S-8879A	3A/3B
ISO	R262 (DIN 13)	6g/6H
NPT	USAS B2.1:1968	Standard NPT
NPTF	ANSI B1.30.3-1976	Standard
BSW	B.S. 84:1956, DIN 259, ISO 228/1:1982	Medium Class A
BSPT	B.S. 21:1985	Standard BSPT
ACME	ANSI B1/5:1988	3G
PG	DIN 40430	Standard
TR	DIN 103	7e/7H

On most CNC systems, this function can be executed in two different ways:

GO2: helical interpolation in a *clockwise* direction, or

GO3: helical interpolation in a *counterclockwise* direction.

The thread milling operation shown in *Figure 13(b)* consists of a circular rotation of the tool about its own axis together with an orbiting motion along the bore or workpiece circumference. During one such orbit, the tool will move vertically one pitch length. These movements, combined with the insert geometry, create the required thread form.

Approaching the workpiece

There are three acceptable ways to approach the workpiece with the tool to initiate the

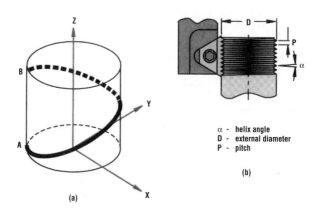

Figure 13. (a) Helical interpolation. (b) Thread milling. (Source, Kennametal Inc.)

thread: 1) along a tangential arc; 2) radially; and 3) along a tangential straight line. In all instances, climb milling is preferred.

Tangential Approach (Arc). With this method, shown in *Figure 14*, the tool enters and exits the workpiece smoothly. No marks are left on the workpiece, and there is no vibration, even with harder materials. Although the arc method requires slightly more complex programming than the radial approach discussed below, this method is recommended for machining the highest quality threads.

Radial Approach. Although this is the simplest method, there are two characteristics that should be noted about this approach. 1) A very small vertical mark that is of no significance to the thread itself may be left at the entry (and exit) point. 2) When using this method with very hard materials, there may be a tendency for the tool to vibrate as it approaches full cutting depth. When using this approach, radial feed during the entry to full profile depth should be only $1/3$ of the subsequent circular feed. *Figure 15* shows the radial approach.

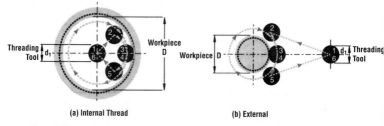

Figure 14. Tangential arc approach. (a) Internal thread. (b) External thread. (Source, Kennametal Inc.)
Points 1–2 show rapid approach path. 2–3 shows tool entry along tangential arc, with simultaneous feed along the Z-axis. 3–4 show the helical movement during one full orbit (360°). 4–5 show the tool exit path along the tangential arc, with continuing feed along the Z-axis. 5–6 show the rapid return path.

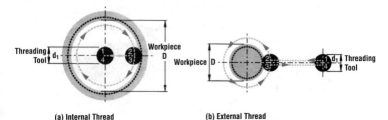

(a) Internal Thread   **(b) External Thread**

Figure 15. Radial approach. (a) Internal thread. (b) External thread. (Source, Kennametal Inc.)
Points 1–2 show the radial entry. 2–3 show the helical movement during one full orbit (360°).
3–4 show the return path.

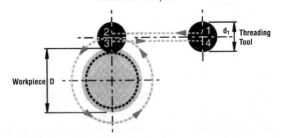

Figure 16. Tangential line approach. (Source, Kennametal Inc.)
Points 1–2 show the radial entry with simultaneous feed along the Z-axis. 2–3 show the helical movement during one full orbit (360°). 3–4 shows the radial exit path.

Tangential Approach (Line). This method, shown in *Figure 16*, is very simple, yet shares the advantages of the tangential arc method. However, it is applicable only to external threads.

Preparing a thread milling operation

The following description uses inch system units. However, the formulas for feed rate in metric units are given in the milling section of this book.

Calculating Feed Rate at the Cutting Edge. Before beginning the operation, the feed rates at both the cutting edge and the tool centerline must be calculated. The feed rate at the cutting edge can be determined with the following equation:

$$F_1 = \text{ipt} \times \text{nt} \times \text{rpm}$$

where F_1 = tool feed rate at the cutting edge
ipt = inch/tooth (feed rate)
nt = number of effective inserts in cutter
rpm = rotational speed (spindle rpm).

The rotational speed (rpm) is calculated with the following equation:

$$\text{rpm} = (12 \times \text{sfm}) \div (\pi \times d_1)$$

where sfm = cutting speed, surface feet per minute
d_1 = cutter diameter over insert.
π = 3.1416.

Calculating Feed Rate at the Tool Centerline. On most CNC machines, the feed rate required for programming is at the centerline of the tool. When dealing with linear tool movement, the feed rate at the cutting edge and centerline are identical, but, with circular tool movement, this is not the case. The following equations define the relationship between feed rates at the cutting edge and at the tool centerline.

For *internal threads.* (*Figure 17*)
$$F_2 = (F_1 \times [D - d_1]) \div D$$

and for *external threads.* (*Figure 18*)
$$F_2 = (F_1 \% [d_1 + D]) \div D$$

where F_2 = centerline feed rate (in./min)
   D = minor diameter, *internal thread*
   D = major diameter, *external thread*
   d_1 = cutter diameter over insert.

Calculating CNC program parameters

The following recommendations provided by Kennametal Inc. are a sure method for programming thread milling operations. For mass production, more complex methods exist. The equations in this section can be converted to metric units by substituting metric units (millimeters) for inch units.

Internal Threads (Climb Milling). The simplest way to perform an internal thread milling operation follows five programming steps. (See *Figure 19.*)

1) Rapidly move the tool to point "A." This point leaves a safe distance or clearance (C_L) between the cutter and the workpiece (usually about 0.02").

2) Move the cutter along the tool path (dashed line) from point "A" to point "B" so that within 90° (about the workpiece's centerline) it reaches maximum cutting depth, moving

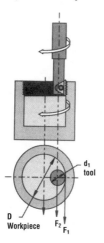

Figure 17. Calculating feed rate for internal threads. (Source, Kennametal Inc.)

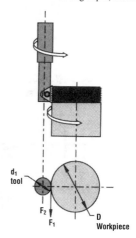

Figure 18. Calculating feed rate for external threads. (Source, Kennametal Inc.)

simultaneously $^1/_4$ pitch length in the Z-axis direction. This path along the entrance angle (β) is defined as radius R_e about a point "C" on the X-axis.

3) Machine the thread by completing one orbit around the circumference (starting and ending at point "B"), moving simultaneously one pitch length in the Z-axis direction.

4) Upon returning to point "B," exit the workpiece to point "D" along the tool path (dashed line), moving simultaneously $^1/_4$ pitch length in the Z-axis direction.

5) Rapidly return to point "O."

To program the path described above, six parameters much be calculated.

1) The entrance radius, R_e. The right triangle OAC enables R_e to be simply solved (see *Figure 20*).

$$OA = R_i - C_L$$
$$CA = R_e$$
$$OC = R_o - R_e$$
$$OA^2 + OC^2 = AC^2 \text{ (Pythagoras' Law)}$$

Replacing the actual values, $(R_i - C_L)^2 + (R_o - R_e)^2 = R_e^2$.

Therefore

$$R_e = ([R_i - C_L]^2 + R_o^2) \div 2 \times R_o$$

where $R_i = D \div 2$ (D = minor diameter, internal thread)
$R_o = D_o \div 2$ (D_o = nominal diameter, internal thread)
C_L = clearance between cutter and workpiece.

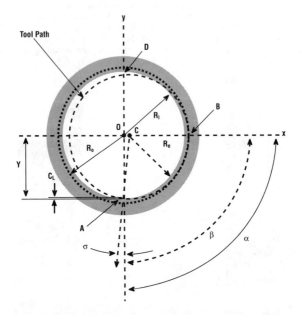

Figure 19. Machining internal threads with climb milling. (Source, Kennametal Inc.)

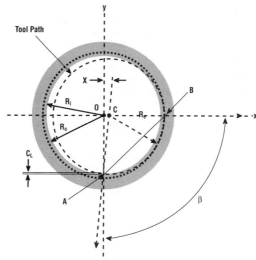

Figure 20. Solving for R_e. (Source, Kennametal Inc.)

2) The entrance angle about point "C," β (see *Figure 20*). β can be easily found using the same right triangle.

$$\beta = 90° + \sigma$$
$$\sin \sigma = OC \div CA = (R_o - R_e) \div R_e$$
$$\sigma = \text{arc sin } ([R_o - R_e] \div R_e).$$

Therefore,

$$\beta = 90° + \text{arc sin } ([R_o - R_e] \div R_e).$$

3) The vertical movement between points "A" and "B," Z_α.

$$Z_\alpha = P \times (\alpha \div 360°) = P \div 4 \text{ (since } \alpha = 90°)$$

where: P = pitch, in inches.

4) The values at the start of the entry approach, "X" and "Y" (see *Figure 19*).

$$X = 0$$
$$Y = -R_i + C_L$$

Note: in conventional milling, $Y = R_i - C_L$.

5) The Z-axis location at the start of the entry approach, Z.

$$Z = -(L + Z_\alpha)$$

where: L = thread length.

Note: in conventional milling, $Z = -L + P + Z_\alpha$.

6) The starting point, O.

$$X_a = 0$$
$$Y_a = 0.$$

External Threads (Climb Milling). External thread milling operations are similar to the internal application. It should be noted that, for climb milling, the orbit is clockwise, and the vertical displacement is in the positive direction. The simplest way to perform an external thread milling operation follows five programming steps. (See *Figure 21*).

1) Rapidly move the cutter to point "A." This point leaves a safe distance or clearance (C_L) between the cutter and the workpiece (usually about 0.02").

2) Move the cutter along the tool path (dashed line) to point "B" so that within 90° (about the workpiece's center) it reaches maximum cutting depth, moving simultaneously $1/4$ pitch length in the Z-axis direction. In this trajectory, the entrance angle (β) is defined by a radius R_e about a point "C" on the X-axis.

3) Machine the thread by completing one orbit around the circumference, moving simultaneously one pitch length in the Z-axis direction.

4) Upon returning to point "B," exit the workpiece along the tool path (dashed line) to point "D," moving simultaneously $1/4$ pitch length in the Z-axis direction.

5) Rapidly return to point "O."

To program the path described above, seven parameters much be calculated.

1) The entry radius, R_e. The right triangle OAC enables R_e to be simply solved (see *Figure 21*).

$$OA = R_o + C_L$$
$$CA = R_e$$
$$OC = R_e - R_i$$
$$OA^2 + OC^2 = AC^2 \text{ (Pythagoras' Law)}$$

Replacing the actual values, $(R_o + C_L)^2 + (R_e - R_i)^2 = R_e^2$.

Therefore,

$$R_e = ([R_o + C_L]^2 + R_i^2) \div 2 \times R_i$$

where: $R_i = D \div 2$ (D = minor diameter, external thread)
$R_o = D_o \div 2$ (D_o = nominal diameter, external thread)
C_L = clearance between cutter and workpiece.

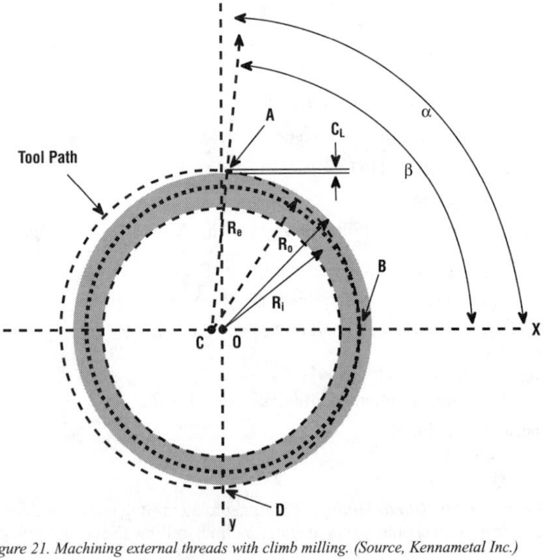

Figure 21. Machining external threads with climb milling. (Source, Kennametal Inc.)

2) The entrance angle about point "C," β (see *Figure 21*). β can be easily found using the same triangle.

$$\sin \beta = AO \div AC = (R_o + C_L) \div R_e$$
$$\beta = \text{arc sin } (R_o + C_L \div R_e).$$

3) The vertical movement between points "A" and "B," Z_α.

$$Z_\alpha = P \times (\alpha \div 360°) = P \div 4 \text{ (since } \alpha = 90°)$$

where P = pitch, in inches.

4) The values at the start of the entry approach, "X" and "Y" (see *Figure 21*).

$$X = 0$$
$$Y = -R_o + C_L.$$

Note: in conventional milling, $Y = -R_o - C_L$.

5) The Z-axis location at the start of the entry approach, Z.

$$Z = L - P - Z_\alpha$$

where L = thread length.

6) The tool's initiation point (prior to compensation).

$$X_a = 0$$
$$Y_a = (R_o + C_L) + R_t$$

where R_t = tool radius ($d_1 \div 2$).

Note: in conventional milling, $Y_a = -(R_o + C_L) - R_t$.

7) X_{CNC} and Y_{CNC} defined.

$$X_{CNC} = R_i$$
$$Y_{CNC} = R_o + C_L.$$

Step-by-step thread milling example

Machining parameters.

Thread: internal right hand $1^1/_4 \times 16$ UN-2B-RH(21)
Material: AISI 4140 (300 BHN)
Thread diameters: $D = 1.182"$ (minimum bore diameter)
                          $D_o = 1.25"$ (nominal diameter)
Thread length: 0.50 inch
Thread milling method: climb milling.

For the best thread quality, the cutter with the largest d_1 (cutter diameter) possible should be used. As can be seen in **Table 68**, any cutter with a diameter of 1.02" or less can be used for this operation. Even though a smaller cutter will perform the operation in less time, it will provide less tool rigidity and should be used with caution on tough materials. Given the material being used for this example, a cutter diameter of 0.79" will be selected. Next, an insert with the proper pitch and thread length must be mated to the cutter, and a feed rate must be specified based on the insert grade. For this example, a cutting speed of 500 sfm and a feed per tooth of 0.004" will be used.

1) Calculate the feed rate. Begin by finding the rpm.

$$\text{rpm} = (12 \times \text{sfm}) \div (\pi \times d_1) = (12 \times 500) \div (3.1416 \times .79) = 2418 \text{ rpm}.$$

Next, calculate F_1, the feed rate at the insert cutting edge.

$$F_1 = \text{ipt} \times \text{nt} \times \text{rpm} = 0.004 \times 1 \times 2418 = 9.67 \text{ in./min.}$$

Finally, calculate F_2 the feed rate at the cutter centerline.

$$F_2 = (F_1 \times [D - d_1]) \div D = (9.67 \times [1.182 - 0.79]) \div 1.182 = 3.207 \text{ in./min.}$$

2) Calculate the radius of the tangential arc, R_e.

$$R_e = ([R_i - C_L]^2 + R_o^2) \div 2 \times R_o = ([0.591 - 0.02]^2 + 0.625^2) \div 2 \times 0.625$$
$$= 0.573333 \text{ in.}$$

3) Calculate the angle β.

$\beta = 90° + $ arc sin $([R_o - R_e] \div R_e)$

$\beta = 90° + $ arc sin $([0.625 - 0.573333] \div 0.573333$

$\beta = 90° + 5.17° = 95.17° = 95° \ 10'$.

4) Calculate the movement, Z_α, along the Z-axis during the entry approach from point "A" to point "B."

$Z_\alpha = P \times (\alpha \div 360°) = P \div 4$ (since $\alpha = 90°) = 0.0625 \div 4 = 0.0156$ in.

5) Calculate the "X" and "Y" values at the start of the entry approach.

$X = 0$

$Y = - R_i + C_L = -0.591 + 0.02 = -0.571$ in.

6) Define the Z-axis location at the start of the entry approach.

$Z = -(L + Z_\alpha) = -(0.50 + 0.0156) = -0.5156$ in.

7) Define the starting point.

$X_a = 0$

$Y_a = 0$.

The CNC program (for Fanuc 11M) would be written as follows.

```
%
N10G90G00G57X0.000Y0.000
N20G43H10Z0.M3S2417
N30G91G00X0.Y0.Z-0.5156
N40G41D60X0.000Y-0.5710Z0.
N50G03X0.6250Y0.5710Z0.0156R0.5733F3.206
N60G03X0.Y0.Z0.0625I-0.625J0.
N70G03X-0.625Y0.5710Z0.0156R0.5733
N80G00G40X0.Y-0.5710Z0.
N90G49G57G00Z8.0M5
N100M30
%
```

Methods of thread milling

Climb milling results in lower cutting forces, better chip development, higher thread surface quality, and longer insert life. For these reasons, it is recommended whenever possible. However, in some cases of hardened materials or when milling certain difficult-to-machine exotic materials, conventional milling may be preferred. *Figure 22 a,b,c,d* illustrates four methods of external thread milling, both conventional and climb milling; and *Figure 23 a,b,c,d* shows how the same operations can be performed when milling internal threads.

Single Point Threading

In the earliest days of threadmaking, threads were often cut on engine, bench, and toolroom lathes, but today single point threading is usually restricted to CNC machines. While manual lathes are still used for very limited production threading jobs, the setup operation is very complicated and will not be detailed in this book. Suffice to say that the setup requires selecting a correct change gear for the required pitch and, of course, applying a brazed tool with the proper relief angle.

CNC, toolholder, and insert technology, when combined, have made single point threading much less troublesome than in the past. Careful selection of the toolholder, insert, and machining data is essential if the operation is to achieve optimum performance. Unlike conventional turning, only a limited number of machining parameters can be modified in threading. These include the thread form, which dictates the insert's physical size and shape. Surface speed, number of passes, depth of each pass, and infeed angle are all critical to successful single point threading.

Most standard 60° thread forms (UN and ISO) allow for root and crest truncation, so the selection of a full profile insert (as opposed to a full thread height insert) will increase

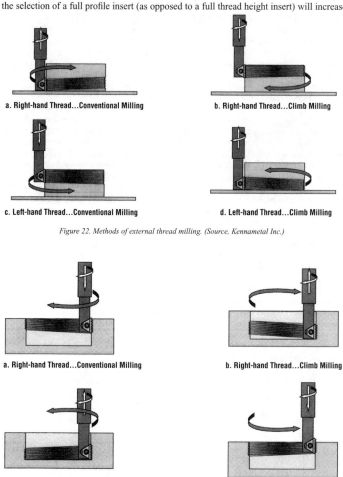

a. Right-hand Thread...Conventional Milling   b. Right-hand Thread...Climb Milling

c. Left-hand Thread...Conventional Milling   d. Left-hand Thread...Climb Milling

Figure 22. Methods of external thread milling. (Source, Kennametal Inc.)

a. Right-hand Thread...Conventional Milling   b. Right-hand Thread...Climb Milling

c. Left-hand Thread...Conventional Milling   d. Left-hand Thread...Climb Milling

Figure 23. Methods of internal thread milling. (Source, Kennametal Inc.)

edge strength and reduce the required number of passes. The main advantage of a full profile insert over a partial profile insert is that, in addition to producing an essentially burr free thread, less material is removed with a stronger tool and the resulting thread is stronger. *Figure 24* illustrates a partial and full 8UN thread. It can be seen that a partial profile insert would have to enter the workpiece a depth of 0.031" deeper than the full profile insert in order to achieve the same pitch diameter. There are various types of inserts for single point threading, as shown in *Figure 25*.

Infeed angle

In single point threading, the weakest part of the insert removes the greatest amount of material. There are four standard infeed angles, and each has specific advantages and disadvantages that should be considered before choosing the best method for a given operation. *Radial* angle infeed, as shown in *Figure 26*, appears to be a logical choice and is the common selection of inexperienced machinists. The obvious advantage is that it cuts on both sides of the thread form, placing all of the cutting edge in the cut—thereby minimizing the possibility of edge chipping. Unfortunately, this method concentrates frictional heat on the point of the insert, dramatically decreasing tool life. Additional disadvantages are that it forms a channel chip, which can be difficult to handle; burrs will be maximized; and there is an increased tendency to chatter because the entire cutting edge is engaged in the thread. *Flank* angle infeed feeds into the workpiece at the exact angle of the thread (see *Figure 27*), but for best results the angle should be slightly reduced. Since the leading edge of the tool is cutting, the chip flows out of the thread form area, reducing the possibility of burrs on the trailing edge of the tool. For best results when using this approach, the infeed angle should be 3° to 5° smaller than the angle of the thread (25° to 27° for UN and ISO threads) in order to avoid poor surface finish, chipping, or excessive flank wear from rubbing the insert's trailing edge. Problems to expect when using the flank angle are torn or poor surface finish, especially when cutting soft, gummy materials like low carbon steels, aluminum, and stainless steels. Also, the trailing edge of the insert may drag or rub, leading to a tendency to chip. *Modified flank* angle infeed (*Figure 28*) is similar to radial angle infeed—the tool cuts both sides of the thread form, offering some protection from chipping. And, even though a channel chip is produced, the uneven chip thickness assists in chip removal. *Alternating flank* angle infeed is difficult when using conventional machinery (*Figure 29*). Although some machine tools will require special programming techniques to use this method, it offers significantly increased tool life because both edges of the tool are used equally.

Laydown style inserts

Laydown triangular inserts (see *Figure 30*) are popular for single point threading, especially for use in small diameter bores. However, unlike other insert types, these inserts

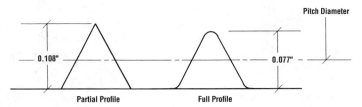

Figure 24. Partial- and full-profile for 8UN thread form. (Source, Duramet.)

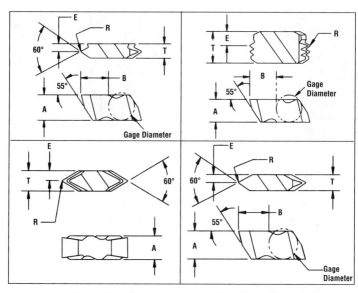

Figure 25. Single point thread cutting inserts. (Source, Kennametal Inc.)

require shims for thread helix angles. When setting up a thread cutting operation with laydown inserts, the first step is to determine the direction of the cut. If machining toward the chuck (when cutting a right hand thread with a right hand tool, or when cutting a left-hand thread with a left-hand tool), the standard helix method is used. If machining away from the chuck (when cutting a right-hand thread with a left-hand tool, or when cutting a left-hand thread with a right-hand tool), the reverse helix method is used. The part's shape and stability, and the flow of chips, are fundamental considerations when choosing to machine toward or away from the chuck.

Once the method of cutting has been established, the angle of the tool in the holder is adjusted, with shims or "anvils," to match the helix angle of the thread being cut. The correct helix angle is especially important when producing multistart and small-diameter threads in order to allow for adequate side clearance for the insert. Almost all toolholders, as shown in *Figures 31* and *32*, are supplied with a 1 $1/2$° lead angle, and shims are available in sizes to match helix angles from 4.5° to –1.5°. The helix angle of a UN or ISO 60° thread (and other threads) is a component of the pitch (or lead) and the effective pitch diameter of the thread (see *Figure 33*), and can be derived from the equation:

$$\tan \lambda = l \div (\pi \times d)$$

where: λ = helix angle
l = lead (1/tpi)
d = pitch diameter of thread.
Note: for multistart threads, l = 1/tpi × number of starts.

Helix angles for UN, Metric, and various Acme threads are provided in **Tables 75** through **80** at the end of this section.

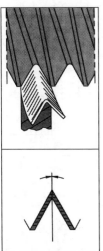

Figure 26.
Radial angle
infeed. (Source,
Kennametal Inc.)

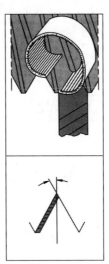

Figure 27.
Flank angle infeed.
(Source, Kennametal
Inc.)

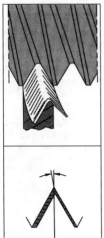

Figure 28.
Modified flank
angle infeed.
(Source,
Kennametal Inc.)

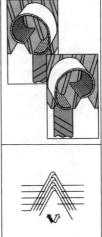

Figure 29.
Alternating flank
angle infeed.
(Source,
Kennametal Inc.)

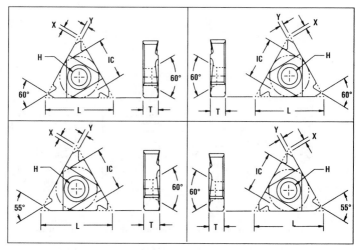

Figure 30. Laydown style inserts.

Infeed values

Single point threading requires making several passes, each with varying depths of cut, over the workpiece. Because a larger portion of the segment of the cutting edge of the insert is engaged in each successive pass, each pass should be made at a shallower depth than the preceding pass. Most CNC lathes have canned programs for threading operations, but **Table 71** and **Table 72** offer guidelines for externally and internally threading steel workpieces of <300 BHN. The depths of cut given are for full profile inserts. These depths are derived from a formula that assumes that the first pass will remove a recommended percentage of the final depth, and that each succeeding pass will remove that percentage of the total depth times the square root of the number of the pass, minus the initial depth of cut.

Example: Find the number of passes and recommended depths of cut for each pass for an 8 pitch UN external thread.

From the table, it can be seen that the total thread depth will be 0.0789" and the recommended depth of cut for the first pass is 0.0197", which is approximately 25% of the total thread depth. To determine the depth of cut for the second pass, multiply the depth of the first by the square root of 2 (for the second pass), and subtract the initial depth of cut from the total. For the third pass, multiply the depth of cut of the first pass by the square root of 3 (for the third pass), and subtract the initial depth of cut from the total. Continue until the accumulated cuts remove the total thread depth.

Second pass:	$0.0197 \times \sqrt{2} = 0.0278$
	$0.0278 - 0.0197 = 0.0082$ (d.o.c. for the second pass).
Third pass:	$0.0197 \times \sqrt{3} = 0.0341$
	$0.0341 - 0.0197 = 0.0063$ (d.o.c. for the third pass).
Fourth pass:	$0.0197 \times \sqrt{4} = 0.0394$
	$0.0394 - 0.0197 = 0.0053$ (d.o.c. for the fourth pass).

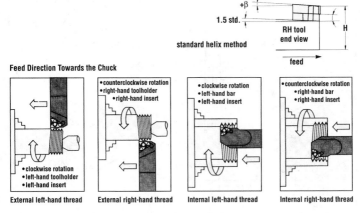

Figure 31. Standard helix toolholder and cutting diagrams for laydown inserts. (Source, Kennametal Inc.)

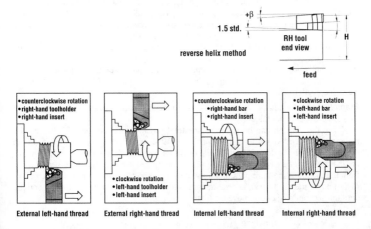

Figure 32. Reverse helix toolholder and cutting diagrams for laydown style inserts. (Source, Kennametal Inc.)

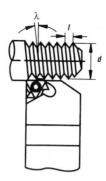

Figure 33. Values for determining helix angle of a thread. λ = helix angle; l = lead; and d = pitch diameter of thread. (Source, Kennametal Inc.)

If, when using this method, the final pass will remove excessive material, adjust the final pass so that the depth of cut exactly equals the amount needed to reach the total thread depth. In the case of the 8UN external thread on the table, the final depth of cut was reduced so that all sixteen passes resulted in the removal of precisely 0.0789" of material.

Surface speeds

The ideal surface speed will combine maximum tool life with required surface finish. In addition, in single point threading, side velocities must not exceed the capabilities of the machine, or thread quality will be adversely affected. Small diameter workpieces can also present problems as suggested surface feet per minute or surface meters per minute values may exceed practical tool parameters. To check whether maximum cutting speed recommendations exceed the maximum travel speed of the tool (as allowed by the machine in inches per minute or millimeters per minute), use the following equation.

Inch units: $\text{sfm (max)} = (D_3 \times \pi \times \text{tpi}) \times (\text{ipm (max)} \div 12)$
Metric units: $\text{smm (max)} = (D_3 \times \pi \times 1/\text{pitch}) \times (\text{mmpm (max)} \div 1000)$
where D_3 = part diameter in inches or millimeters.

The suggested speeds chart in **Table 73** provides speeds in sfm and smm for a variety of insert materials identified by their ISO grade classification. For an explanation of this system, see the Inserts section of this book.

Troubleshooting single point threading operations

Due to the large numbers of parameters that apply to single point threading, troubleshooting unsatisfactory results can be confusing. The solutions outlined in **Table 74** can help pinpoint solutions to a wide selection of problems.

Table 71. Infeed Values for Externally Threading Steel Workpieces (<300 BHN). (Source, Kennametal Inc.)

tpi (UN)	4	5	6	7	8	9	10	11	12	13
Thread Depth	.1578	.1262	.1052	.0902	.0789	.0701	.0631	.0574	.0526	.0485
1st pass	.0353	.0298	.0248	.0213	.0197	.0175	.0169	.0157	.0152	.0142
2nd pass	.0146	.0122	.0105	.0088	.0082	.0073	.0070	.0066	.0064	.0057
3rd pass	.0113	.0094	.0078	.0077	.0063	.0056	.0053	.0048	.0048	.0044
4th pass	.0095	.0079	.0067	.0059	.0053	.0047	.0045	.0041	.0042	.0037
5th pass	.0084	.0070	.0058	.0050	.0047	.0042	.0039	.0036	.0036	.0033
6th pass	.0076	.0063	.0052	.0045	.0043	.0037	.0036	.0031	.0032	.0030
7th pass	.0070	.0058	.0048	.0041	.0039	.0034	.0031	.0028	.0029	.0027
8th pass	.0065	.0054	.0045	.0038	.0036	.0032	.0030	.0026	.0027	.0025
9th pass	.0061	.0051	.0042	.0036	.0034	.0030	.0029	.0025	.0026	.0024
10th pass	.0057	.0048	.0040	.0034	.0032	.0028	.0028	.0024	.0025	.0023
11th pass	.0054	.0045	.0038	.0032	.0031	.0027	.0027	.0023	.0023	.0022
12th pass	.0052	.0043	.0036	.0031	.0029	.0026	.0026	.0022	.0022	.0021
13th pass	.0049	.0042	.0035	.0030	.0027	.0025	.0025	.0021	-	-
14th pass	.0048	.0041	.0034	.0029	.0026	.0024	.0024	.0020	-	-
15th pass	.0046	.0040	.0033	.0028	.0025	.0023	-	-	-	-
16th pass	.0044	.0039	.0032	.0027	.0025	.0022	-	-	-	-
17th pass	.0043	.0038	.0031	.0026	-	-	-	-	-	-
18th pass	.0042	.0037	.0030	.0025	-	-	-	-	-	-
19th pass	.0041	-	-	-	-	-	-	-	-	-
20th pass	.0039	-	-	-	-	-	-	-	-	-
tpi (UN)	**14**	**16**	**18**	**20**	**24**	**28**	**32**	**36**	**40**	**44**
Thread Depth	.0451	.0394	.0350	.0315	.0263	.0225	.0197	.0175	.0157	.0143
1st pass	.0136	.0125	.0124	.0119	.0118	.0112	.0098	.0087	.0078	.0073

(Continued)

Table 71. *(Continued)* **Infeed Values for Externally Threading Steel Workpieces (<300 BHN).** *(Source, Kennametal Inc.)*

tpi (UN)	14	16	18	20	24	28	32	36	40	44
Thread Depth	.0451	.0394	.0350	.0315	.0263	.0225	.0197	.0175	.0157	.0143
2nd pass	.0059	.0054	.0053	.0049	.0048	.0046	.0042	.0036	.0032	.0028
3rd pass	.0043	.0039	.0039	.0039	.0039	.0036	.0031	.0028	.0024	.0022
4th pass	.0036	.0034	.0033	.0032	.0031	.0031	.0026	.0024	.0020	.0020
5th pass	.0032	.0029	.0029	.0028	.0027	-	-	-	-	-
6th pass	.0029	.0026	.0026	.0025	-	-	-	-	-	-
7th pass	.0026	.0024	.0024	.0023	-	-	-	-	-	-
8th pass	.0024	.0022	.0022	-	-	-	-	-	-	-
9th pass	.0023	.0021	-	-	-	-	-	-	-	-
10th pass	.0022	.0020	-	-	-	-	-	-	-	-
11th pass	.0021	-	-	-	-	-	-	-	-	-

Note: These are nominal thread depths, in inches, for full profile inserts. When using partial profile inserts, reduce the initial depth of cut and increase the number of passes. When threading workhardening materials (e.g., stainless steel), the infeed should never be less than 0.003".

Table 72. Infeed Values for Internally Threading Steel Workpieces (<300 BHN). *(Source, Kennametal Inc.)*

tpi (UN)	4	5	6	7	8	9	10	11	12	13
Thread Depth	**.1353**	**.1082**	**.0902**	**.0773**	**.0676**	**.0601**	**.0541**	**.0492**	**.0451**	**.0416**
1st pass	.0303	.0255	.0213	.0183	.0169	.0150	.0145	.0132	.0131	.0120
2nd pass	.0125	.0105	.0090	.0076	.0073	.0062	.0064	.0055	.0054	.0050
3rd pass	.0096	.0083	.0069	.0058	.0053	.0047	.0046	.0044	.0041	.0038
4th pass	.0081	.0068	.0057	.0049	.0047	.0040	.0038	.0035	.0035	.0032
5th pass	.0071	.0060	.0050	.0043	.0041	.0035	.0034	.0031	.0031	.0028
6th pass	.0064	.0054	.0045	.0039	.0036	.0032	.0031	.0028	.0028	.0025
7th pass	.0059	.0050	.0041	.0036	.0033	.0029	.0028	.0026	.0026	.0023
8th pass	.0055	.0046	.0038	.0033	.0030	.0027	.0026	.0024	.0024	.0022
9th pass	.0052	.0043	.0036	.0031	.0028	.0025	.0024	.0022	.0022	.0021
10th pass	.0049	.0041	.0034	.0029	.0027	.0024	.0023	.0021	.0021	.0020
11th pass	.0046	.0039	.0032	.0028	.0026	.0023	.0022	.0020	.0020	.0019
12th pass	.0044	.0037	.0031	.0027	.0025	.0022	.0021	.0019	.0019	.0018
13th pass	.0042	.0036	.0030	.0026	.0024	.0021	.0020	.0018	-	-
14th pass	.0041	.0035	.0029	.0025	.0023	.0020	.0019	.0017	-	-
15th pass	.0040	.0034	.0028	.0024	.0022	.0019	-	-	-	-
16th pass	.0039	.0033	.0027	.0023	.0021	.0019	-	-	-	-
17th pass	.0038	.0032	.0026	.0022	-	-	-	-	-	-
18th pass	.0037	.0031	.0025	.0021	-	-	-	-	-	-
19th pass	.0036	-	-	-	-	-	-	-	-	-
20th pass	.0035	-	-	-	-	-	-	-	-	-
tpi (UN)	**14**	**16**	**18**	**20**	**24**	**28**	**32**	**36**	**40**	**44**
Thread Depth	**.0386**	**.0338**	**.0300**	**.0270**	**.0225**	**.0193**	**.0169**	**.0150**	**.0135**	**.0123**
1st pass	.0117	.0107	.0106	.0120	.0101	.0096	.0084	.0075	.0067	.0061

(Continued)

Table 72. *(Continued)* Infeed Values for Internally Threading Steel Workpieces (<300 BHN). *(Source, Kennametal Inc.)*

tpi (UN)	14	16	18	20	24	28	32	36	40	44
Thread Depth	.0386	.0338	.0300	.0270	.0225	.0193	.0169	.0150	.0135	.0123
2nd pass	.0048	.0043	.0044	.0042	.0042	.0039	.0035	.0031	.0029	.0025
3rd pass	.0037	.0034	.0033	.0032	.0032	.0033	.0027	.0023	.0021	.0019
4th pass	.0031	.0028	.0028	.0027	.0027	.0025	.0023	.0021	.0018	.0018
5th pass	.0027	.0025	.0025	.0024	.0023	-	-	-	-	-
6th pass	.0025	.0029	.0023	.0022	-	-	-	-	-	-
7th pass	.0023	.0021	.0021	.0021	-	-	-	-	-	-
8th pass	.0021	.0020	.0029	-	-	-	-	-	-	-
9th pass	.0020	.0019	-	-	-	-	-	-	-	-
10th pass	.0019	.0018	-	-	-	-	-	-	-	-
11th pass	.0018	-	-	-	-	-	-	-	-	-

Note: These are nominal thread depths, in inches, for full profile inserts. When using partial profile inserts, reduce the initial depth of cut and increase the number of passes. When threading workhardening materials (e.g., stainless steel), the infeed should never be less than 0.003".

Table 73. Recommended Grades and Speeds for Threading Various Workpiece Materials. (*Source, Kennametal Inc.*)

Workpiece Material		UCC* M20R†	PVD Coated P20R M15-25 K05-15	PVD Coated P 30-45 M40R K25-35	M 5-25 K 05-15	M 40R K 25-35	CVD Coated P10-20 M25-35	CVD Coated P25-45 M35>	CVD Coated P5-15 M10-25 K10	Cermet P10 M20	PCD M5-15 K01
Free-machining Carbon Steel	sfm	-	300-700	150-500	300-550	350-600	450-900	500-800	500-1000	500-1000	-
	smm	-	91-213	46-152	91-168	107-183	137-274	152-244	152-305	152-305	-
Plain Carbon Steel	sfm	-	250-600	150-400	250-500	300-500	300-750	400-700	400-800	450-900	-
	smm		76-183	46-122	76-152	91-152	91-229	122-213	122-244	137-274	
Alloy Steels 190-330 BHN	sfm	-	300-500	150-350	250-500	250-450	400-600	350-600	350-700	400-800	-
	smm	-	91-152	46-107	76-152	76-137	122-183	107-183	107-213	122-244	-
Alloy Steels 330-450 BHN	sfm	-	-	-	200-400	-	-	-	-	100-500	-
	smm	-	-	-	61-122	-	-	-	-	31-152	-
Gray Cast Iron 190-330 BHN	sfm	200-300	-	150-300	200-600	200-500	-	-	300-800	400-800	-
	smm	61-91	-	46-91	61-183	61-152	-	-	91-244	122-244	-
Gray Cast Iron 330-450 BHN	sfm	150-250	-	50-200	150-450	-	-	-	-	-	-
	smm	46-76	-	15-61	46-137	-	-	-	-	-	-
Alloy/Ductile Irons	sfm	150-250	300-500	100-400	150-500	250-500	-	300-600	300-750	400-800	-
	smm	46-76	91-152	31-122	46-152	76-152	-	91-183	91-229	122-244	-
Austenitic Stainless Steel*	sfm	200-350	-	150-350	200-500	300-500	-	400-650	-	500-850	-
	smm	61-107	-	46-107	61-152	91-152	-	122-198	-	152-259	-
Martensitic/ Ferritic Stainless	sfm	-	300-500	100-300	150-400	250-500	-	300-500	-	600-800	-
	smm	-	91-152	31-91	46-122	76-152	-	91-152	-	183-244	-
Hi-Temp Alloys** 200-260 BHN	sfm	80-120	-	50-150	80-300	-	-	-	-	600-800	-
	smm	24-35	-	15-46	-	-	-	-	-	183-344	-

(Continued)

Table 73. *(Continued)* **Recommended Grades and Speeds for Threading Various Workpiece Materials.** *(Source, Kennametal Inc.)*

Workpiece Material		UCC*	PVD Coated			CVD Coated				Cermet	PCD
		M20R†	P20R M15-25 K05-15	P 30-45 M40R K 25-35	M 5-25 K 05-15	M 40R K 25-35	P10-20 M25-35	P25-45 M35>	P5-15 M10-25 K10	P10 M20	M5-15 K01
Hi-Temp Alloys** 260-450 BHN	sfm	80-100	-	30-80	100-200	-	-	-	-	-	-
	smm	24-31	-	9-24	31-61	-	-	-	-	-	-
Titanium Alloys (Ti 6Al-4V)	sfm	110-180	-	-	110-250	-	-	-	-	-	-
	smm	34-55	-	-	34-76	-	-	-	-	-	-
Free-Machining Aluminum Alloys	sfm	400-800	-	-	400-1000	-	-	-	-	700-4000	Up to 4000
	smm	122-244	-	-	122-305	-	-	-	-	213-1220	Up to 1220
High-Silicon Aluminum Alloys	sfm	-	-	-	-	-	-	-	-	-	Up to 3000
	smm	-	-	-	-	-	-	-	-	-	Up to 915
Copper/Zinc/Brass	sfm	250-600	-	150-500	250-800	400-800	-	-	-	700-4000	Up to 3000
	smm	76-183	-	46-152	76-244	122-244	-	-	-	213-1220	Up to 915
Non-Metallics	sfm	400-1500	-	150-800	400-1500	-	-	-	-	700-4000	Up to 4500
	smm	122-457	-	46-244	122-457	-	-	-	-	213-1220	Up to 1220

*Uncoated carbide. **Example: Iconel 718. † ISO grades are expressed as the insert appears on the grade chart. "R" as in 20R indicates that the grade occupies the center of the 20 range (from 20 to 30, for example) on the chart. A rating of 25-35 indicates that the grade spans the center of the 20 range to the center of the 30 range on the chart. A rating of 30-40 indicates that the grade fills the entire 30 and 40 grids on the chart. The symbol ">" indicates that the grade rating begins at the position given and continues to the end of the chart section.

Table 74. Troubleshooting Single Point Threading Operations. *(Source, Kennametal Inc.)*

Quick Solution Guide

Problem ↓ / Solution →	Increase sfm/smm	Reduce sfm/smm	Increase chip load	Decrease chip load where failure occurs	Use tougher carbide grade	Use harder carbide grade	Apply coolant	Use coated carbide	Change infeed angle	Check for insert movement and reseat	Reduce tool overhang	Reseat shim	Reselect d.o.c.	Adjust center height	Begin cutting threads 0.5 inch before workpiece
Chatter	•			•						•	•			•	
Burr on crest	•														
Short tool life		•	•	•		•		•							
Chipped leading edge			•	•	•										
Chipped trailing edge			•		•				•						
Broken nose (first pass)	•														
Broken nose (after first pass)				•	•				•			•	•	•	
Build-up on cutting edge							•	•							•
Premature topping												•			
Splitting threads												•			

Detailed Solution Guide

Problem	Cause	Possible Solution
Thread with torn finish	• Burrs	• Use positive rake or full profile insert. • Increase coolant concentration.
	• Torn finish	• Alter infeed • Use PVD grade insert. • Use positive rake insert. • Increase speed.

(Continued)

Table 74. *(Continued)* **Troubleshooting Single Point Threading Operations.** *(Source, Kennametal Inc.)*

Problem	Cause	Possible Solution
		Detailed Solution Guide
Thread with torn finish	• Steps	• Check machine Z-axis travel. • Avoid cutting chips. • Check insert form. • Check for correct shim (laydown system).
Chatter	• Poor rigidity	• Minimize tool overhang • Check for workpiece deflection
	• Incorrect speed	• Adjust speed.
	• Wrong edge prep	• Adjust hone level.
	• Insert movement	• Check insert and clamp.
	• Improper infeed	• Use modified feed angle.
	• Off centerline	• Verify that tool cutting position is at workpiece centerline.
Built-up edge	• Speed too low	• Increase speed.
	• Insufficient coolant	• Increase coolant concentration and/or flow
	• Wrong edge prep	• Adjust hone size.
	• Chip load	• Adjust infeed angle. • Adjust depth of cut per pass.
Deformation of insert	• Speed too high	• Reduce speed.
	• Incorrect grade	• Use grade with a higher hot hardness.
	• Improper infeed angle	• Alter infeed angle.
Chipping of insert	• Light chip load	• Adjust chip load. • Increase or decrease number of passes.
	• Wrong grade	• Use tougher grade.
	• Improper infeed	• Alter infeed to modified flank

(Continued)

Table 74. (Continued) **Troubleshooting Single Point Threading Operations.** (Source, Kennametal Inc.)

Problem	Cause	Possible Solution
		Detailed Solution Guide
Chipping of insert	• Incorrect speed	• Increase speed if chipping on trailing edge. • Decrease speed if chipping on leading edge
	• Wrong edge prep	• Increase hone size.
	• Poor rigidity	• Minimize tool overhang. • Check for insert movement/check clamp. • Check for possible part deflection.
Broken nose on insert	• Heavy chip load	• Decrease chip load
	• Small nose radius	• Use larger nose radius if allowable.
	• Wrong grade	• Use tougher grade.
	• Wrong edge prep	• Increase hone size.
Flank wear on insert	• Wrong grade	• Use a more wear resistant grade.
	• Insufficient coolant	• Increase coolant flow.
	• Off centerline	• Check centerline height of the tool. (The smaller the diameter, the more critical the need for centerline accuracy.)

Table 75. Helix Angles for UN Threads #0 Through 1 5/16 Inch. (Source, Landis Threading Systems.)

Nom. Dia.	80	72	64	56	48	44	40	36	32	28	24	20	18	16
#0	4°23'	-	-	-	-	-	-	-	-	-	-	-	-	-
#1	3°31'	3°57'	4°31'	-	-	-	-	-	-	-	-	-	-	-
#2	2°55'	3°17'	3°45'	4°22'	-	-	-	-	-	-	-	-	-	-
#3	2°30'	2°49'	3°12'	3°43'	4°26'	-	-	-	-	-	-	-	-	-
#4	2°12'	2°28'	2°48'	3°14'	3°51'	4°15'	4°45'	-	-	-	-	-	-	-
#5	1°56'	2°11'	2°29'	2°52'	3°24'	3°45'	4°11'	4°43'	5°26'	-	-	-	-	-
#6	1°45'	1°58'	2°14'	2°34'	3°3'	3°22'	3°44'	4°13'	4°50'	5°39'	6°49'	-	-	-
#8	1°28'	1°38'	1°51'	2°8'	2°31'	2°46'	3°5'	3°28'	3°58'	4°37'	5°32'	-	-	-
#10	1°15'	1°24'	1°35'	1°49'	2°9'	2°22'	2°37'	2°57'	3°21'	3°54'	4°39'	-	-	-
#12	1°6'	1°13'	1°23'	1°35'	1°52'	2°4'	2°17'	2°33'	2°54'	3°22'	4°1'	-	-	-
3/16	-	-	1°36'	1°51'	2°11'	2°24'	2°40'	2°59'	3°24'	3°57'	4°44'	5°52'	6°40'	7°43'
1/4	-	-	1°11'	1°22'	1°36'	1°46'	1°57'	2°11'	2°29'	2°52'	3°24'	4°11'	4°44'	5°25'
5/16	-	-	0°57'	1°5'	1°16'	1°24'	1°32'	1°43'	1°57'	2°15'	2°40'	3°15'	3°40'	4°11'
3/8	-	-	-	0°54'	1°3'	1°9'	1°16'	1°25'	1°36'	1°51'	2°11'	2°40'	2°59'	3°24'
7/16	-	-	-	0°46'	0°54'	0°59'	1°5'	1°12'	1°22'	1°34'	1°51'	2°15'	2°31'	2°52'
1/2	-	-	-	0°40'	0°47'	0°51'	0°57'	1°3'	1°11'	1°22'	1°36'	1°57'	2°11'	2°29'
9/16	-	-	-	0°35'	0°42'	0°45'	0°50'	0°56'	1°3'	1°12'	1°25'	1°43'	1°55'	2°11'
5/8	-	-	-	0°32'	0°37'	0°41'	0°45'	0°50'	0°57'	1°5'	1°16'	1°32'	1°43'	1°57'
11/16	-	-	-	-	-	-	0°41'	0°45'	0°51'	0°59'	1°9'	1°24'	1°33'	1°46'
3/4	-	-	-	-	-	-	0°37'	0°42'	0°47'	0°54'	1°3'	1°16'	1°25'	1°36'
13/16	-	-	-	-	-	-	0°34'	0°38'	0°43'	0°50'	0°58'	1°10'	1°18'	1°29'
7/8	-	-	-	-	-	-	0°32'	0°35'	0°40'	0°46'	0°54'	1°5'	1°12'	1°22'
15/16	-	-	-	-	-	-	0°30'	0°33'	0°37'	0°43'	0°50'	1°0'	1°7'	1°16'

(Continued)

Table 75. *(Continued)* **Helix Angles for UN Threads #0 Through 1 5/16 Inch.** *(Source, Landis Threading Systems.)*

Nom. Dia.	Threads per Inch													
	16	18	20	24	28	32	36	40	44	48	56	64	72	80
1	1°11'	1°3'	0°57'	0°47'	0°40'	0°35'	0°31'	0°28'	-	-	-	-	-	-
1 1/16	1°7'	0°59'	0°53'	0°44'	0°38'	0°33'	0°29'	-	-	-	-	-	-	-
1 1/8	1°3'	0°56'	0°50'	0°42'	0°35'	0°31'	-	-	-	-	-	-	-	-
1 3/16	1°0'	0°53'	0°47'	0°39'	0°34'	0°29'	-	-	-	-	-	-	-	-
1 1/4	0°57'	0°50'	0°45'	0°37'	0°32'	0°28'	-	-	-	-	-	-	-	-
1 5/16	0°54'	0°48'	0°43'	0°35'	0°30'	0°26'	-	-	-	-	-	-	-	-

Nom. Dia.	Threads per Inch													
	3.5	4	4.5	5	5.5	6	7	8	9	10	11	12	13	14
5/16	-	-	-	-	-	-	-	-	-	-	-	5°52'	5°20'	4°53'
3/8	-	-	-	-	-	-	-	-	-	-	5°14'	4°44'	4°18'	3°57'
7/16	-	-	-	-	-	-	-	-	-	4°53'	4°22'	3°57'	3°37'	3°20'
1/2	-	-	-	-	-	-	-	-	4°43'	4°11'	3°45'	3°24'	3°7'	2°52'
9/16	-	-	-	-	-	-	-	5°26'	4°8'	3°40'	3°17'	2°59'	2°44'	2°31'
5/8	-	-	-	-	-	-	-	4°44'	3°40'	3°15'	2°56'	2°40'	2°26'	2°15'
11/16	-	-	-	-	-	-	4°53'	4°11'	3°17'	2°56'	2°38'	2°24'	2°12'	2°1'
3/4	-	-	-	-	-	5°14'	4°22'	3°45'	2°57'	2°40'	2°24'	2°11'	2°0'	1°51'
13/16	-	-	-	5°20'	4°46'	4°44'	3°57'	3°24'	2°44'	2°26'	2°12'	2°0'	1°50'	1°42'
7/8	-	-	-	4°53'	4°22'	4°16'	3°37'	3°7'	2°31'	2°15'	2°2'	1°51'	1°42'	1°34'
15/16	-	-	-	4°30'	4°2'	3°57'	3°20'	2°52'	2°20'	2°5'	1°53'	1°43'	1°35'	1°28'
1	-	5°26'	4°44'	4°11'	3°45'	3°40'	3°5'	2°40'	2°11'	1°57'	1°46'	1°36'	1°29'	1°22'
1 1/16	5°32'	5°3'	4°24'	3°54'	3°30'	3°24'	2°52'	2°29'	2°3'	1°50'	1°39'	1°30'	1°23'	1°17'
1 1/8	-	4°44'	4°8'	3°40'	3°17'	3°11'	2°41'	2°19'	1°55'	1°43'	1°33'	1°25'	1°18'	1°12'
1 3/16	5°11'	4°26'	3°53'	3°27'	3°6'	2°49'	2°23'	2°4'	1°49'	1°37'	1°28'	1°20'	1°14'	1°8'

(Continued)

Table 75. *(Continued)* Helix Angles for UN Threads #0 Through 1 $^5/_{16}$ Inch. *(Source, Landis Threading Systems.)*

Nom. Dia.	Threads per Inch													
	3.5	4	4.5	5	5.5	6	7	8	9	10	11	12	13	14
1 1/4	4°53'	4°11'	3°40'	3°15'	2°56'	2°40'	2°15'	1°57'	1°43'	1°32'	1°24'	1°16'	1°10'	1°5'
1 5/16	4°37'	3°58'	3°28'	3°5'	2°46'	2°31'	2°8'	1°51'	1°38'	1°28'	1°19'	1°12'	1°7'	1°2'

Nom. Dia.	Threads per Inch													
	7	8	9	10	11	12	13	14	16	18	20	24	28	32
1 3/8	2°2'	1°46'	1°33'	1°24'	1°16'	1°9'	1°4'	0°59'	0°51'	0°45'	0°41'	0°34'	0°29'	0°25'
1 7/16	1°56'	1°41'	1°29'	1°20'	1°12'	1°6'	1°1'	0°56'	0°49'	0°43'	0°39'	0°32'	0°28'	0°24'
1 1/2	1°51'	1°36'	1°25'	1°16'	1°9'	1°3'	0°58'	0°54'	0°47'	0°42'	0°37'	0°31'	0°26'	0°23'
1 9/16	1°46'	1°32'	1°22'	1°13'	1°6'	1°0'	0°56'	0°51'	0°45'	0°40'	0°36'	0°30'	0°25'	0°22'
1 5/8	1°42'	1°29'	1°18'	1°10'	1°4'	0°58'	0°53'	0°49'	0°43'	0°38'	0°34'	0°29'	0°24'	0°21'
1 11/16	1°38'	1°25'	1°15'	1°7'	1°1'	0°56'	0°51'	0°48'	0°42'	0°37'	0°33'	0°27'	0°23'	0°21'
1 3/4	1°34'	1°22'	1°12'	1°5'	0°59'	0°54'	0°49'	0°46'	0°40'	0°35'	0°32'	0°26'	0°23'	0°20'
1 13/16	1°31'	1°19'	1°10'	1°3'	0°57'	0°52'	0°48'	0°44'	0°39'	0°34'	0°31'	0°26'	0°22'	0°19'
1 7/8	1°28'	1°16'	1°7'	1°0'	0°55'	0°50'	0°46'	0°43'	0°37'	0°33'	0°30'	0°25'	0°21'	0°18'
1 15/16	1°25'	1°14'	1°5'	0°58'	0°53'	0°48'	0°45'	0°41'	0°36'	0°32'	0°29'	0°24'	0°20'	0°18'
2	1°22'	1°11'	1°3'	0°57'	0°51'	0°47'	0°43'	0°40'	0°35'	0°31'	0°28'	0°23'	0°20'	0°17'
2 1/8	1°17'	1°7'	0°59'	0°53'	0°48'	0°44'	0°41'	0°37'	0°33'	0°29'	0°26'	0°22'	0°19'	0°16'
2 1/4	1°12'	1°3'	0°56'	0°50'	0°45'	0°42'	0°38'	0°35'	0°31'	0°27'	0°25'	0°21'	0°17'	0°15'
2 3/8	1°8'	1°0'	0°53'	0°47'	0°43'	0°39'	0°36'	0°33'	0°29'	0°26'	0°23'	0°19'	0°17'	0°15'
2 1/2	1°5'	0°56'	0°50'	0°45'	0°41'	0°37'	0°34'	0°32'	0°28'	0°25'	0°22'	0°18'	0°16'	0°14'
2 5/8	1°2'	0°54'	0°48'	0°43'	0°39'	0°35'	0°33'	0°30'	0°26'	0°23'	0°21'	0°18'	0°15'	0°13'
2 3/4	0°59'	0°51'	0°45'	0°41'	0°37'	0°34'	0°31'	0°29'	0°25'	0°22'	0°20'	0°17'	0°14'	0°13'
2 7/8	0°56'	0°49'	0°43'	0°39'	0°35'	0°32'	0°30'	0°28'	0°24'	0°21'	0°19'	0°16'	0°14'	0°12'
3	0°54'	0°47'	0°42'	0°37'	0°34'	0°31'	0°29'	0°26'	0°23'	0°21'	0°18'	0°15'	0°13'	0°11'

(Continued)

Table 75. *(Continued)* Helix Angles for UN Threads #0 Through 1 5/16 Inch. *(Source, Landis Threading Systems.)*

Threads per Inch

Nom. Dia.	7	8	9	10	11	12	13	14	16	18	20	24	28	32
3 1/8	0°52'	0°45'	0°40'	0°36'	0°32'	0°30'	0°27'	0°25'	0°22'	0°20'	0°18'	0°15'	0°13'	0°11'
3 1/4	0°50'	0°43'	0°38'	0°34'	0°31'	0°29'	0°26'	0°24'	0°21'	0°19'	0°17'	0°14'	0°12'	0°11'
3 3/8	0°48'	0°42'	0°37'	0°33'	0°30'	0°27'	0°25'	0°24'	0°21'	0°18'	0°16'	0°14'	0°12'	0°10'
3 1/2	0°46'	0°40'	0°35'	0°32'	0°29'	0°26'	0°24'	0°23'	0°20'	0°18'	0°16'	0°13'	0°11'	0°10'
3 5/8	0°44'	0°39'	0°34'	0°31'	0°28'	0°26'	0°24'	0°22'	0°19'	0°17'	0°15'	0°13'	0°11'	0°9'
3 3/4	0°43'	0°37'	0°33'	0°30'	0°27'	0°25'	0°23'	0°21'	0°18'	0°16'	0°15'	0°12'	0°10'	0°9'
3 7/8	0°41'	0°36'	0°32'	0°29'	0°26'	0°24'	0°22'	0°20'	0°18'	0°16'	0°14'	0°12'	0°10'	0°9'
4	0°40'	0°35'	0°31'	0°28'	0°25'	0°22'	0°21'	0°20'	0°17'	0°15'	0°13'	0°11'	0°10'	0°9'
4 1/8	0°39'	0°34'	0°30'	0°27'	0°24'	0°22'	0°21'	0°19'	0°17'	0°15'	0°13'	0°11'	0°10'	0°8'
4 1/4	0°38'	0°33'	0°29'	0°26'	0°24'	0°22'	0°20'	0°18'	0°16'	0°14'	0°13'	0°11'	0°9'	0°8'
4 1/2	0°35'	0°31'	0°27'	0°25'	0°22'	0°21'	0°19'	0°18'	0°15'	0°14'	0°12'	0°10'	0°9'	0°8'

Threads per Inch

Nom. Dia.	3.5	4	4.5	5	5.5	6
1 3/8	4°22'	3°45'	3°17'	2°56'	2°38'	2°24'
1 7/16	4°9'	3°34'	3°8'	2°47'	2°31'	2°17'
1 1/2	3°57'	3°24'	2°59'	2°40'	2°24'	2°11'
1 9/16	3°47'	3°15'	2°51'	2°33'	2°18'	2°5'
1 5/8	3°37'	3°7'	2°44'	2°26'	2°12'	2°0'
1 11/16	3°28'	2°59'	2°37'	2°20'	2°7'	1°55'
1 3/4	3°20'	2°52'	2°31'	2°15'	2°2'	1°51'
1 13/16	3°12'	2°46'	2°26'	2°10'	1°57'	1°47'
1 7/8	3°5'	2°40'	2°20'	2°5'	1°53'	1°43'
1 15/16	2°58'	2°34'	2°15'	2°1'	1°49'	1°40'
2	2°52'	2°29'	2°11'	1°57'	1°46'	1°36'
2 1/8	2°41'	2°19'	2°3'	1°50'	1°39'	1°30'
2 1/4	2°31'	2°11'	1°55'	1°43'	1°33'	1°25'
2 3/8	2°23'	2°4'	1°49'	1°37'	1°28'	1°20'
2 1/2	2°15'	1°57'	1°43'	1°32'	1°24'	1°16'
2 5/8	2°8'	1°51'	1°38'	1°28'	1°19'	1°12'
2 3/4	2°2'	1°46'	1°33'	1°24'	1°15'	1°9'
2 7/8	1°56'	1°41'	1°29'	1°20'	1°12'	1°6'
3	1°51'	1°36'	1°25'	1°16'	1°9'	1°3'
3 1/8	1°46'	1°32'	1°22'	1°13'	1°6'	1°0'

(Continued)

Table 75. *(Continued)* Helix Angles for UN Threads #0 Through 1 5/16 Inch. *(Source, Landis Threading Systems.)*

Nom. Dia.	Threads per Inch					
	6	5.5	5	4.5	4	3.5
3 1/4	0°58'	1°4'	1°10'	1°18'	1°29'	1°42'
3 3/8	0°56'	1°1'	1°7'	1°15'	1°25'	1°38'
3 1/2	0°54'	0°59'	1°5'	1°12'	1°22'	1°34'
3 5/8	0°52'	0°57'	1°3'	1°10'	1°19'	1°31'
3 3/4	0°50'	0°55'	1°0'	1°7'	1°16'	1°28'

Nom. Dia.	Threads per Inch					
	6	5.5	5	4.5	4	3.5
3 7/8	0°48'	0°53'	0°58'	1°5'	1°14'	1°25'
4	0°47'	0°51'	0°57'	1°3'	1°11'	1°22'
4 1/8	0°45'	0°50'	0°55'	1°1'	1°9'	1°19'
4 1/4	0°44'	0°48'	0°53'	0°59'	1°7'	1°17'
4 1/2	0°41'	0°45'	0°50'	0°56'	1°3'	1°12'

Table 76. Helix Angles for ISO and Metric Threads 1 mm Through 24 mm. (Source, Landis Threading Systems.)

Nom. Dia.	Pitch in Millimeters													
	0.25	0.3	0.35	0.4	0.45	0.5	0.6	0.7	0.75	0.8	1	1.25	1.5	1.75
1	5°26'	6°46'	8°12'	-	-	-	-	-	-	-	-	-	-	-
1.1	4°51'	6°1'	7°17'	-	-	-	-	-	-	-	-	-	-	-
1.2	4°23'	5°26'	6°32'	-	-	-	-	-	-	-	-	-	-	-
1.4	3°41'	4°32'	5°26'	6°22'	7°22'	-	-	-	-	-	-	-	-	-
1.6	3°10'	3°53'	4°38'	5°26'	6°15'	-	-	-	-	-	-	-	-	-
1.7	2°58'	3°38'	4°20'	5°3'	5°49'	6°36'	-	-	-	-	-	-	-	-
1.8	2°47'	3°24'	4°3'	4°44'	5°26'	6°9'	-	-	-	-	-	-	-	-
2	2°29'	3°2'	3°36'	4°11'	4°48'	5°26'	6°46'	-	-	-	-	-	-	-
2.2	2°14'	2°44'	3°14'	3°45'	4°18'	4°51'	6°1'	-	-	-	-	-	-	-
2.3	2°8'	2°36'	3°5'	3°34'	4°5'	4°36'	5°43'	-	-	-	-	-	-	-
2.5	1°57'	2°22'	2°48'	3°15'	3°43'	4°11'	5°10'	6°13'	6°46'	-	-	-	-	-
2.6	1°52'	2°16'	2°41'	3°7'	3°33'	4°0'	4°56'	5°56'	6°27'	-	-	-	-	-
3	1°36'	1°57'	2°18'	2°40'	3°2'	3°24'	4°11'	5°0'	5°26'	5°52'	-	-	-	-
3.5	1°22'	1°39'	1°57'	2°15'	2°33'	2°52'	3°31'	4°11'	4°32'	4°53'	-	-	-	-
4	-	-	-	1°57'	2°13'	2°29'	3°2'	3°35'	3°53'	4°11'	5°26'	7°7'	8°58'	11°1'
4.5	-	-	-	1°43'	1°57'	2°11'	2°40'	3°9'	3°24'	3°40'	4°44'	6°9'	7°43'	9°24'
5	-	-	-	1°32'	1°45'	1°57'	2°22'	2°48'	3°2'	3°15'	4°11'	5°26'	6°46'	8°12'
5.5	-	-	-	1°24'	1°35'	1°46'	2°8'	2°32'	2°44'	2°56'	3°45'	4°51'	6°1'	7°17'
6	-	-	-	-	1°26'	1°36'	1°57'	2°18'	2°29'	2°40'	3°24'	4°23'	5°26'	6°32'
7	-	-	-	-	1°13'	1°22'	1°39'	1°57'	2°6'	2°15'	2°52'	3°41'	4°32'	5°26'
8	-	-	-	-	1°4'	1°11'	1°26'	1°41'	1°49'	1°57'	2°29'	3°10'	3°53'	4°38'
9	-	-	-	-	0°57'	1°3'	1°16'	1°30'	1°36'	1°43'	2°11'	2°47'	3°24'	4°3'
10	-	-	-	-	0°51'	0°57'	1°8'	1°20'	1°26'	1°32'	1°57'	2°29'	3°2'	3°36'

(Continued)

Table 76. *(Continued)* Helix Angles for ISO and Metric Threads 1 mm Through 24 mm. *(Source, Landis Threading Systems.)*

Nom. Dia.	Pitch in Millimeters													
	0.25	0.3	0.35	0.4	0.45	0.5	0.6	0.7	0.75	0.8	1	1.25	1.5	1.75
11	-	-	-	-	0°46'	0°51'	1°2'	1°13'	1°18'	1°24'	1°46'	2°14'	2°44'	3°14'
12	-	-	-	-	-	-	0°57'	1°6'	1°11'	1°16'	1°36'	2°2'	2°29'	2°56'
14	-	-	-	-	-	-	0°48'	0°57'	1°1'	1°5'	1°22'	1°44'	2°6'	2°29'
15	-	-	-	-	-	-	0°45'	0°53'	0°57'	1°0'	1°16'	1°36'	1°57'	2°18'
16	-	-	-	-	-	-	0°42'	0°49'	0°53'	0°57'	1°11'	1°30'	1°49'	2°9'
17	-	-	-	-	-	-	0°40'	0°46'	0°50'	0°53'	1°7'	1°24'	1°42'	2°1'
18	-	-	-	-	-	-	0°37'	0°44'	0°47'	0°50'	1°3'	1°20'	1°36'	1°54'
20	-	-	-	-	-	-	0°33'	0°39'	0°42'	0°45'	0°57'	1°11'	1°26'	1°42'
22	-	-	-	-	-	-	0°30'	0°36'	0°38'	0°41'	0°51'	1°5'	1°18'	1°32'
24	-	-	-	-	-	-	0°28'	0°33'	0°35'	0°37'	0°47'	0°59'	1°11'	1°24'

Nom. Dia.	Pitch in Millimeters										
	2	2.5	3	3.5	4	4.5	5	5.5	6	6.5	7
6	7°43'	-	-	-	-	-	-	-	-	-	-
7	6°22'	8°25'	-	-	-	-	-	-	-	-	-
8	5°26'	7°7'	-	-	-	-	-	-	-	-	-
9	4°44'	6°9'	-	-	-	-	-	-	-	-	-
10	4°11'	5°26'	6°46'	8°12'	-	-	-	-	-	-	-
11	3°45'	4°51'	6°1'	7°16'	8°37'	-	-	-	-	-	-
12	3°24'	4°23'	5°26'	6°32'	7°43'	8°58'	-	-	-	-	-
14	2°52'	3°41'	4°32'	5°26'	6°22'	7°22'	8°25'	-	-	-	-
15	2°40'	3°24'	4°11'	5°0'	5°52'	6°46'	7°43'	-	-	-	-
16	2°29'	3°10'	3°53'	4°38'	5°26'	6°15'	7°7'	8°1'	-	-	-
17	2°19'	2°58'	3°38'	4°20'	5°3'	5°49'	6°36'	7°26'	-	-	-

(Continued)

Table 76. *(Continued)* Helix Angles for ISO and Metric Threads 1 mm Through 24 mm. *(Source, Landis Threading Systems.)*

Nom. Dia. 18–24 — Pitch in Millimeters

Nom. Dia.	2	2.5	3	3.5	4	4.5	5	5.5	6	6.5	7
18	2°11'	2°47'	3°24'	4°3'	4°44'	5°26'	6°9'	6°55'	7°43'	-	-
20	1°57'	2°29'	3°2'	3°36'	4°11'	4°48'	5°26'	6°5'	6°46'	7°28'	-
22	1°46'	2°14'	2°44'	3°14'	3°45'	4°18'	4°51'	5°26'	6°1'	6°38'	-
24	1°36'	2°2'	2°29'	2°56'	3°24'	3°53'	4°23'	4°54'	5°26'	5°58'	6°32'

Nom. Dia. 25–56 — Pitch in Millimeters

Nom. Dia.	0.6	0.7	0.75	0.8	1	1.25	1.5	1.75	2	2.5	3	4	4.5	5
25	0°27'	0°31'	0°33'	0°36'	0°45'	0°57'	1°8'	1°20'	1°32'	1°57'	2°22'	3°15'	3°43'	4°11'
27	0°25'	0°29'	0°31'	0°33'	0°42'	0°52'	1°3'	1°14'	1°25'	1°48'	2°11'	2°59'	3°24'	3°50'
28	0°24'	0°28'	0°30'	0°32'	0°40'	0°50'	1°1'	1°11'	1°22'	1°44'	2°6'	2°52'	3°16'	3°41'
30	0°22'	0°26'	0°28'	0°30'	0°37'	0°47'	0°57'	1°6'	1°16'	1°36'	1°57'	2°40'	3°2'	3°24'
32	0°21'	0°24'	0°26'	0°28'	0°35'	0°44'	0°53'	1°2'	1°11'	1°30'	1°49'	2°29'	2°49'	3°10'
33	0°20'	0°24'	0°25'	0°27'	0°34'	0°42'	0°51'	1°0'	1°9'	1°27'	1°46'	2°24'	2°44'	3°4'
35	0°19'	0°22'	0°24'	0°25'	0°32'	0°40'	0°48'	0°57'	1°5'	1°22'	1°39'	2°15'	2°33'	2°52'
36	0°18'	0°22'	0°23'	0°25'	0°31'	0°39'	0°47'	0°55'	1°3'	1°20'	1°36'	2°11'	2°29'	2°47'
39	0°17'	0°20'	0°21'	0°23'	0°29'	0°36'	0°43'	0°51'	0°58'	1°13'	1°29'	2°0'	2°16'	2°33'
40	0°17'	0°19'	0°21'	0°22'	0°28'	0°35'	0°42'	0°49'	0°57'	1°11'	1°26'	1°57'	2°13'	2°29'
42	0°16'	0°18'	0°20'	0°20'	0°26'	0°33'	0°40'	0°47'	0°54'	1°8'	1°22'	1°57'	2°6'	2°21'
45	0°15'	0°17'	0°18'	0°20'	0°25'	0°31'	0°37'	0°44'	0°50'	1°3'	1°16'	1°43'	1°57'	2°11'
48	0°14'	0°16'	0°17'	0°18'	0°23'	0°29'	0°35'	0°41'	0°47'	0°59'	1°11'	1°36'	1°49'	2°2'
50	0°13'	0°15'	0°17'	0°18'	0°22'	0°28'	0°33'	0°39'	0°45'	0°57'	1°8'	1°32'	1°45'	1°57'
52	0°13'	0°15'	0°16'	0°17'	0°21'	0°27'	0°32'	0°38'	0°43'	0°54'	1°6'	1°29'	1°40'	1°52'
55	-	-	0°15'	0°16'	0°20'	0°25'	0°30'	0°36'	0°41'	0°51'	1°2'	1°24'	1°35'	1°46'
56	-	-	0°15'	0°16'	0°20'	0°25'	0°30'	0°35'	0°40'	0°50'	1°1'	1°22'	1°33'	1°44'

(Continued)

Table 76. *(Continued)* Helix Angles for ISO and Metric Threads 1 mm Through 24 mm. *(Source, Landis Threading Systems.)*

| | Pitch in Millimeters | | | | | | | | | | | | | |
Nom. Dia.	0.6	0.7	0.75	0.8	1	1.25	1.5	1.75	2	2.5	3	4	4.5	5
58	–	–	–	–	0°19'	0°24'	0°29'	0°34'	0°39'	0°49'	0°59'	1°19'	1°29'	1°40'
60	–	–	–	–	0°19'	0°23'	0°28'	0°33'	0°37'	0°47'	0°57'	1°16'	1°26'	1°36'
62	–	–	–	–	0°18'	0°22'	0°27'	0°31'	0°36'	0°45'	0°55'	1°14'	1°23'	1°33'
64	–	–	–	–	0°17'	0°22'	0°26'	0°30'	0°35'	0°44'	0°53'	1°11'	1°21'	1°30'
65	–	–	–	–	0°17'	0°21'	0°26'	0°30'	0°34'	0°43'	0°52'	1°10'	1°19'	1°29'
68	–	–	–	–	0°16'	0°20'	0°24'	0°29'	0°33'	0°41'	0°50'	1°7'	1°16'	1°24'
70	–	–	–	–	0°16'	0°20'	0°24'	0°28'	0°32'	0°40'	0°48'	1°5'	1°13'	1°22'
72	–	–	–	–	0°15'	0°19'	0°23'	0°27'	0°31'	0°39'	0°47'	1°3'	1°11'	1°20'
75	–	–	–	–	0°15'	0°18'	0°22'	0°26'	0°30'	0°37'	0°45'	1°0'	1°8'	1°16'
76	–	–	–	–	0°15'	0°18'	0°22'	0°26'	0°29'	0°37'	0°44'	1°0'	1°6'	1°15'
80	–	–	–	–	0°14'	0°17'	0°21'	0°24'	0°28'	0°35'	0°42'	0°57'	1°4'	1°11'
84	–	–	–	–	–	–	0°20'	0°23'	0°26'	0°33'	0°40'	0°54'	1°1'	1°8'
85	–	–	–	–	–	–	0°20'	0°23'	0°26'	0°33'	0°40'	0°53'	1°0'	1°7'
88	–	–	–	–	–	–	0°19'	0°22'	0°25'	0°32'	0°38'	0°51'	0°58'	1°5'
90	–	–	–	–	–	–	0°18'	0°22'	0°25'	0°31'	0°37'	0°50'	0°57'	1°3'

| Nom. Dia. | Pitch in Millimeters | | | | | |
---	5.5	6	6.5	7	7.5	8
25	4°40'	5°10'	5°41'	6°13'	–	–
27	4°16'	4°43'	5°11'	5°40'	6°9'	–
28	4°6'	4°32'	4°58'	5°26'	5°54'	–
30	3°47'	4°11'	4°35'	5°0'	5°26'	5°52'
32	3°31'	3°53'	4°16'	4°38'	5°2'	5°26'
33	3°24'	3°45'	4°7'	4°29'	4°51'	5°14'

| Nom. Dia. | Pitch in Millimeters | | | | | |
---	5.5	6	6.5	7	7.5	8
35	3°11'	3°31'	3°51'	4°11'	4°32'	4°53'
36	3°5'	3°24'	3°44'	4°3'	4°23'	4°43'
39	2°50'	3°7'	3°24'	3°42'	4°0'	4°19'
40	2°45'	3°2'	3°19'	3°36'	3°53'	4°11'
42	2°36'	2°52'	3°8'	3°24'	3°41'	3°58'
45	2°25'	2°40'	2°54'	3°9'	3°24'	3°40'

(Continued)

Table 76. *(Continued)* Helix Angles for ISO and Metric Threads 1 mm Through 24 mm. *(Source, Landis Threading Systems.)*

Nom. Dia.	Pitch in Millimeters					
	5.5	6	6.5	7	7.5	8
48	2°15'	2°29'	2°42'	2°56'	3°10'	3°24'
50	2°10'	2°22'	2°35'	2°48'	3°2'	3°15'
52	2°4'	2°17'	2°29'	2°41'	2°54'	3°7'
55	1°57'	2°8'	2°20'	2°32'	2°44'	2°56'
56	1°55'	2°6'	2°17'	2°29'	2°40'	2°52'
58	1°51'	2°1'	2°12'	2°23'	2°34'	2°46'
60	1°47'	1°57'	2°7'	2°18'	2°29'	2°40'
62	1°43'	1°53'	2°3'	2°13'	2°24'	2°34'
64	1°40'	1°49'	1°59'	2°9'	2°19'	2°29'
65	1°38'	1°47'	1°57'	2°7'	2°16'	2°26'

Nom. Dia.	Pitch in Millimeters					
	5.5	6	6.5	7	7.5	8
68	1°33'	1°42'	1°51'	2°1'	2°10'	2°19'
70	1°31'	1°39'	1°48'	1°57'	2°6'	2°15'
72	1°28'	1°36'	1°45'	1°54'	2°2'	2°11'
75	1°24'	1°32'	1°40'	1°49'	1°57'	2°5'
76	1°23'	1°31'	1°39'	1°47'	1°55'	2°4'
80	1°19'	1°26'	1°34'	1°41'	1°49'	1°57'
84	1°15'	1°22'	1°29'	1°36'	1°44'	1°51'
85	1°14'	1°21'	1°28'	1°35'	1°42'	1°50'
88	1°11'	1°18'	1°25'	1°32'	1°39'	1°46'
90	1°10'	1°16'	1°23'	1°30'	1°36'	1°43'

Table 77. Helix Angles for General Purpose Acme Threads 3/16 Through 1 1/2 Inch. *(Source, Landis Threading Systems.)*

Dia.	Threads per Inch											
	10	11	12	13	14	16	18	20	22	24	26	28
3/16	-	-	-	-	-	7°45'	6°44'	5°57'	5°22'	4°51'	4°25'	4°2'
1/4	9°31'	8°28'	7°26'	6°56'	6°22'	5°28'	4°47'	4°14'	3°49'	3°28'	3°11'	2°56'
5/16	7°11'	6°17'	5°42'	5°18'	4°53'	4°12'	3°41'	3°17'	2°58'	2°42'	2°29'	2°18'
3/8	5°43'	5°10'	4°41'	3°54'	3°57'	3°25'	2°59'	2°41'	2°25'	2°12'	2°2'	1°53'
7/16	4°49'	4°20'	3°56'	3°36'	3°19'	2°53'	2°31'	2°16'	2°3'	1°49'	1°43'	1°35'
1/2	4°9'	3°43'	3°23'	3°6'	2°52'	2°29'	2°12'	1°58'	1°46'	1°37'	1°30'	1°20'
9/16	3°39'	3°16'	2°59'	2°43'	2°31'	2°11'	1°56'	1°43'	1°34'	1°26'	1°18'	1°13'
5/8	3°12'	2°54'	2°39'	2°27'	2°15'	1°57'	1°43'	1°32'	1°24'	1°16'	1°11'	1°6'
11/16	2°54'	2°37'	2°23'	2°12'	2°3'	1°46'	1°34'	1°24'	1°16'	1°10'	1°4'	1°3'
3/4	2°39'	2°23'	2°11'	2°0'	1°51'	1°36'	1°25'	1°16'	1°4'	1°3'	59'	54'
13/16	2°25'	2°11'	2°0'	1°50'	1°43'	1°28'	1°18'	1°10'	59'	58'	54'	50'
7/8	2°14'	2°1'	1°51'	1°42'	1°34'	1°21'	1°12'	1°5'	55'	54'	50'	46'
15/16	2°5'	1°53'	1°43'	1°34'	1°28'	1°15'	1°7'	1°1'	51'	50'	46'	43'
1	1°56'	1°45'	1°36'	1°28'	1°22'	1°11'	1°3'	57'	51'	47'	43'	40'
1 1/16	1°49'	1°39'	1°30'	1°23'	1°18'	1°7'	1°0'	53'	48'	44'	40'	38'
1 1/8	1°45'	1°33'	1°25'	1°18'	1°13'	1°3'	56'	50'	45'	41'	38'	36'
1 3/16	1°38'	1°28'	1°19'	1°14'	1°9'	59'	53'	48'	43'	39'	36'	34'
1 1/4	1°32'	1°23'	1°16'	1°10'	1°5'	56'	50'	45'	41'	38'	34'	32'
1 5/16	1°27'	1°19'	1°12'	1°6'	1°2'	54'	48'	43'	39'	36'	32'	30'
1 3/8	1°23'	1°15'	1°9'	1°3'	59'	51'	46'	41'	37'	34'	31'	29'
1 7/16	1°19'	1°12'	1°6'	1°0'	57'	49'	44'	39'	36'	32'	30'	-
1 1/2	1°16'	1°9'	1°3'	58'	54'	47'	43'	37'	34'	31'	29'	-

(Continued)

Table 77. *(Continued)* Helix Angles for General Purpose Acme Threads $^3/_{16}$ Through 1 $^1/_2$ Inch. *(Source, Landis Threading Systems.)*

Dia.	Threads per Inch											
	9	8	7	6	5.5	5	4.5	4	3.5	3	2.5	2
1/4	10°52'	-	-	-	-	-	-	-	-	-	-	-
5/16	8°9'	9°25'	-	-	-	-	-	-	-	-	-	-
3/8	6°31'	7°30'	8°48'	10°40'	-	-	-	-	-	-	-	-
7/16	5°26'	6°14'	7°16'	8°45'	-	-	-	-	-	-	-	-
1/2	4°39'	5°19'	6°12'	7°25'	8°15'	-	-	-	-	-	-	-
9/16	4°5'	4°38'	5°24'	6°32'	7°9'	-	-	-	-	-	-	-
5/8	3°37'	4°7'	4°47'	5°42'	6°12'	7°2'	-	-	-	-	-	-
11/16	3°15'	3°41'	4°18'	5°6'	5°38'	6°17'	7°7'	-	-	-	-	-
3/4	2°57'	3°26'	3°54'	4°37'	5°5'	5°41'	6°27'	7°22'	8°40'	10°29'	-	-
13/16	2°43'	3°5'	3°33'	4°13'	4°40'	5°10'	5°51'	6°42'	7°50'	9°26'	-	-
7/8	2°29'	2°50'	3°17'	3°53'	4°17'	4°45'	5°23'	6°8'	7°13'	8°38'	10°49'	-
15/16	2°19'	2°37'	3°3'	3°36'	3°57'	4°24'	4°58'	5°40'	6°37'	7°56'	9°57'	-
1	2°10'	2°27'	2°53'	3°21'	3°41'	4°5'	4°37'	5°15'	6°8'	7°14'	9°10'	-
1 1/16	2°2'	2°18'	2°39'	3°8'	3°26'	3°49'	4°19'	4°54'	5°43'	6°49'	8°30'	-
1 1/8	1°55'	2°10'	2°29'	2°57'	3°13'	3°35'	4°2'	4°36'	5°21'	6°23'	7°56'	10°24'
1 3/16	1°48'	2°4'	2°22'	2°42'	3°3'	3°23'	3°49'	4°20'	5°2'	5°59'	7°25'	9°45'
1 1/4	1°42'	1°56'	2°14'	2°37'	2°53'	3°12'	3°36'	4°5'	4°45'	5°39'	6°59'	9°8'
1 5/16	1°37'	1°50'	2°7'	2°29'	2°44'	3°3'	3°24'	3°53'	4°29'	5°21'	6°34'	8°37'
1 3/8	1°33'	1°45'	2°1'	2°22'	2°36'	2°53'	3°14'	3°40'	4°15'	5°4'	6°13'	8°8'
1 7/16	1°29'	1°40'	1°56'	2°16'	2°29'	2°44'	3°5'	3°30'	4°3'	4°50'	5°55'	7°43'
1 1/2	1°25'	1°35'	1°50'	2°10'	2°22'	2°38'	2°57'	3°20'	3°51'	4°35'	5°38'	7°18'

(Continued)

Based on General Purpose Acme Basic Dimensions.

Table 77. *(Continued)* Helix Angles for General Purpose Acme Threads 1 9/16 Through 4 Inch. *(Source, Landis Threading Systems.)*

Dia.	Threads per Inch											
	10	11	12	13	14	16	18	20	22	24	26	28
1 9/16	1°13'	1°6'	1°0'	56'	52'	45'	40'	36'	32'	-	-	-
1 5/8	1°10'	1°3'	58'	54'	50'	43'	38'	35'	31'	-	-	-
1 3/4	1°5'	59'	54'	49'	46'	40'	35'	32'	-	-	-	-
1 7/8	1°0'	55'	50'	45'	43'	37'	33'	30'	-	-	-	-
2	56'	51'	47'	43'	40'	35'	31'	28'	-	-	-	-
2 1/8	53'	48'	45'	40'	-	-	-	-	-	-	-	-
2 1/4	50'	45'	42'	38'	-	-	-	-	-	-	-	-
2 3/8	47'	43'	40'	36'	-	-	-	-	-	-	-	-
2 1/2	45'	41'	37'	35'	-	-	-	-	-	-	-	-
2 5/8	43'	39'	35'	-	-	-	-	-	-	-	-	-
2 3/4	41'	37'	34'	-	-	-	-	-	-	-	-	-
2 7/8	39'	35'	32'	-	-	-	-	-	-	-	-	-
3	37'	34'	31'	-	-	-	-	-	-	-	-	-
3 1/8	36'	33'	-	-	-	-	-	-	-	-	-	-
3 1/4	34'	31'	-	-	-	-	-	-	-	-	-	-
3 3/8	33'	30'	-	-	-	-	-	-	-	-	-	-
3 1/2	32'	29'	-	-	-	-	-	-	-	-	-	-
3 5/8	31'	-	-	-	-	-	-	-	-	-	-	-
3 3/4	30'	-	-	-	-	-	-	-	-	-	-	-
3 7/8	29'	-	-	-	-	-	-	-	-	-	-	-
4	28'	-	-	-	-	-	-	-	-	-	-	-

(Continued)

Table 77. *(Continued)* Helix Angles for General Purpose Acme Threads 1 9/16 Through 4 Inch. *(Source, Landis Threading Systems.)*

Dia.	Threads per Inch											
	2	2.5	3	3.5	4	4.5	5	5.5	6	7	8	9
1 9/16	6°57'	5°24'	4°32'	3°42'	3°12'	2°49'	2°32'	2°16'	2°4'	1°46'	1°32'	1°21'
1 5/8	6°37'	5°9'	4°11'	3°32'	3°4'	2°42'	2°25'	2°10'	1°59'	1°41'	1°28'	1°18'
1 3/4	6°5'	4°43'	3°51'	3°15'	2°49'	2°30'	2°13'	2°0'	1°50'	1°33'	1°22'	1°12'
1 7/8	5°37'	4°21'	3°34'	3°1'	2°37'	2°19'	2°4'	1°52'	1°42'	1°27'	1°16'	1°8'
2	5°13'	4°4'	3°20'	2°48'	2°27'	2°10'	1°55'	1°44'	1°36'	1°21'	1°11'	1°3'
2 1/8	4°54'	3°48'	3°6'	2°38'	2°17'	2°2'	1°49'	1°38'	1°30'	1°17'	1°7'	59'
2 1/4	4°34'	3°33'	2°56'	2°29'	2°11'	1°55'	1°43'	1°32'	1°24'	1°13'	1°3'	56'
2 3/8	4°18'	3°21'	2°46'	2°21'	2°2'	1°48'	1°37'	1°28'	1°20'	1°9'	1°0'	53'
2 1/2	4°3'	3°10'	2°37'	2°13'	1°56'	1°43'	1°32'	1°23'	1°16'	1°5'	57'	50'
2 5/8	3°51'	3°1'	2°28'	2°7'	1°50'	1°37'	1°27'	1°19'	1°13'	1°2'	54'	47'
2 3/4	3°38'	2°52'	2°20'	2°1'	1°45'	1°33'	1°23'	1°15'	1°9'	59'	51'	45'
2 7/8	3°29'	2°44'	2°15'	1°55'	1°40'	1°29'	1°19'	1°12'	1°6'	56'	49'	43'
3	3°19'	2°37'	2°9'	1°50'	1°36'	1°25'	1°16'	1°9'	1°3'	53'	47'	41'
3 1/8	3°11'	2°30'	2°3'	1°46'	1°32'	1°21'	1°13'	1°6'	1°0'	52'	45'	40'
3 1/4	3°3'	2°24'	1°57'	1°41'	1°28'	1°18'	1°10'	1°3'	58'	50'	43'	38'
3 3/8	2°56'	2°19'	1°53'	1°37'	1°24'	1°15'	1°7'	1°1'	56'	48'	42'	37'
3 1/2	2°48'	2°13'	1°49'	1°33'	1°21'	1°12'	1°4'	59'	54'	46'	40'	36'
3 5/8	2°42'	2°8'	1°44'	1°30'	1°18'	1°9'	1°2'	56'	52'	45'	38'	35'
3 3/4	2°36'	2°3'	1°39'	1°27'	1°16'	1°7'	1°0'	54'	50'	43'	37'	33'
3 7/8	2°31'	1°59'	1°37'	1°24'	1°13'	1°5'	59'	53'	49'	42'	36'	32'
4	2°26'	1°55'	1°34'	1°21'	1°11'	1°3'	57'	51'	47'	41'	35'	30'

Based on General Purpose Acme Basic Dimensions.

Table 78. Helix Angles for General Purpose Double Start Acme Threads 3/16 Through 1 1/2 Inch. *(Source, Landis Threading Systems.)*

Dia.	Threads per Inch										
	28	20	18	16	14	12	11.5	11	10	9	8
3/16	7°51'	11°25'	12°52'	14°44'	17°51'						
1/4	5°43'	8°14'	9°14'	10°32'	12°15'	15°38'	15°22'	16°35'	18°32'	21°0'	
5/16	4°29'	6°26'	7°12'	8°12'	9°30'	11°33'	11°50'	12°25'	14°9'	15°59'	18°21'
3/8	3°41'	5°16'	5°54'	6°42'	7°45'	9°18'	9°37'	10°15'	11°20'	12°52'	14°45'
7/16	3°8'	4°28'	4°59'	5°40'	6°32'	7°50'	8°6'	8°37'	9°34'	10°46'	12°19'
1/2	2°44'	3°52'	4°20'	4°54'	5°39'	6°45'	6°59'	7°25'	8°16'	9°14'	10°33'
9/16	2°25'	3°25'	3°49'	4°19'	4°59'	5°57'	6°9'	6°31'	7°16'	8°6'	9°13'
5/8	2°10'	3°4'	3°25'	3°52'	4°27'	5°17'	5°29'	5°47'	6°23'	7°12'	8°11'
11/16	1°57'	2°46'	3°5'	3°30'	4°1'	4°46'	4°57'	5°13'	5°47'	6°29'	7°22'
3/4	1°47'	2°32'	2°49'	3°11'	3°40'	4°22'	4°31'	4°45'	5°17'	5°53'	6°42'
13/16	1°39'	2°20'	2°36'	2°56'	3°22'	4°0'	4°9'	4°22'	4°50'	5°25'	6°9'
7/8	1°32'	2°9'	2°24'	2°43'	3°7'	3°42'	3°50'	4°2'	4°28'	4°57'	5°39'
15/16	1°25'	2°0'	2°14'	2°32'	2°54'	3°26'	3°34'	3°46'	4°10'	4°37'	5°14'
1	1°20'	1°53'	2°6'	2°22'	2°43'	3°12'	3°20'	3°30'	3°52'	4°20'	4°53'
1 1/16	-	1°46'	1°58'	2°13'	2°33'	3°0'	3°7'	3°18'	3°38'	4°4'	4°36'
1 1/8	-	1°40'	1°51'	2°5'	2°24'	2°50'	2°57'	3°6'	3°30'	3°49'	4°20'
1 3/16	-	1°34'	1°45'	1°59'	2°16'	2°40'	2°47'	2°56'	3°16'	3°36'	4°8'
1 1/4	-	1°30'	1°40'	1°53'	2°9'	2°32'	2°38'	2°46'	3°4'	3°24'	3°52'
1 5/16	-	1°25'	1°35'	1°47'	2°3'	2°24'	2°30'	2°38'	2°54'	3°14'	3°40'
1 3/8	-	1°21'	1°30'	1°42'	1°57'	2°18'	2°23'	2°30'	2°46'	3°6'	3°30'
1 7/16	-	1°18'	1°26'	1°38'	1°52'	2°12'	2°17'	2°24'	2°38'	2°58'	3°20'
1 1/2	-	1°14'	1°23'	1°33'	1°47'	2°6'	2°11'	2°18'	2°32'	2°50'	3°10'

(Continued)

Based on General Purpose Acme Basic Dimensions.

Table 78. *(Continued)* Helix Angles for General Purpose Double Start Acme Threads 3/16 Through 1 1/2 Inch. *(Source, Landis Threading Systems.)*

Dia.	Threads per Inch									
	2	2.5	3	3.5	4	4.5	5	5.5	6	7
3/8	-	-	-	-	-	-	-	-	20°39'	17°12'
7/16	-	-	-	-	-	-	-	-	17°7'	14°18'
1/2	-	-	-	-	-	-	-	16°7'	14°36'	12°15'
9/16	-	-	-	-	-	-	-	14°5'	12°54'	10°43'
5/8	-	-	-	-	-	-	-	12°16'	11°17'	9°30'
11/16	-	-	-	-	-	-	-	11°10'	10°8'	8°33'
3/4	-	-	20°34'	16°57'	14°30'	12°41'	11°16'	10°5'	9°11'	7°46'
13/16	-	-	18°23'	15°23'	13°13'	11°34'	10°15'	9°16'	8°23'	7°4'
7/8	-	20°55'	17°2'	14°13'	12°8'	10°38'	9°34'	8°31'	7°43'	6°32'
15/16	-	19°20'	15°34'	13°4'	11°12'	9°50'	8°45'	7°52'	7°10'	6°5'
1	-	17°53'	14°16'	12°8'	10°25'	9°9'	8°9'	7°20'	6°41'	5°45'
1 1/16	-	16°38'	13°27'	11°19'	9°44'	8°33'	7°37'	6°50'	6°15'	5°17'
1 1/8	20°9'	15°35'	12°37'	10°37'	9°8'	8°1'	7°9'	6°25'	5°53'	4°57'
1 3/16	18°58'	14°36'	11°50'	9°59'	8°37'	7°33'	6°44'	6°5'	5°23'	4°43'
1 1/4	17°49'	13°46'	11°11'	9°26'	8°7'	7°9'	6°22'	5°45'	5°13'	4°28'
1 5/16	16°52'	12°58'	10°37'	8°55'	7°42'	6°46'	6°5'	5°27'	4°57'	4°14'
1 3/8	15°57'	12°17'	10°3'	8°27'	7°19'	6°26'	5°45'	5°11'	4°43'	4°2'
1 7/16	15°10'	11°43'	9°37'	8°4'	6°59'	6°8'	5°27'	4°57'	4°32'	3°52'
1 1/2	14°22'	11°9'	9°10'	7°40'	6°39'	5°51'	5°15'	4°43'	4°19'	3°40'

Based on General Purpose Acme Basic Dimensions.

(Continued)

Table 78. *(Continued)* Helix Angles for General Purpose Double Start Acme Threads 1 9/16 Through 4 Inch. *(Source, Landis Threading Systems.)*

Dia.	Threads per Inch									
	20	18	16	14	12	11.5	11	10	9	8
1 9/16	1°11'	1°19'	1°30'	1°43'	2°0'	2°6'	2°12'	2°26'	2°42'	3°3'
1 5/8	1°9'	1°16'	1°26'	1°39'	1°56'	2°1'	2°6'	2°20'	2°36'	2°56'
1 3/4	1°4'	1°11'	1°20'	1°31'	1°48'	1°52'	1°58'	2°10'	2°24'	2°44'
1 7/8	59'	1°6'	1°14'	1°25'	1°40'	1°44'	1°50'	2°0'	2°16'	2°32'
2	55'	1°2'	1°10'	1°20'	1°34'	1°37'	1°42'	1°52'	2°6'	2°22'
2 1/8	-	-	-	-	1°30'	1°32'	1°36'	1°46'	1°58'	2°14'
2 1/4	-	-	-	-	1°24'	1°26'	1°30'	1°40'	1°52'	2°6'
2 3/8	-	-	-	-	-	1°22'	1°26'	1°34'	1°46'	2°0'
2 1/2	-	-	-	-	-	1°18'	1°22'	1°30'	1°40'	1°54'
2 5/8	-	-	-	-	-	1°14'	1°18'	1°26'	1°34'	1°48'
2 3/4	-	-	-	-	-	1°10'	1°14'	1°22'	1°30'	1°42'
2 7/8	-	-	-	-	-	1°7'	1°10'	1°18'	1°26'	1°38'
3	-	-	-	-	-	1°4'	1°8'	1°14'	1°22'	1°34'
3 1/8	-	-	-	-	-	1°2'	1°6'	1°12'	1°20'	1°30'
3 1/4	-	-	-	-	-	59'	1°2'	1°8'	1°18'	1°26'
3 3/8	-	-	-	-	-	57'	1°0'	1°6'	1°14'	1°24'
3 1/2	-	-	-	-	-	55'	58'	1°4'	1°4'	1°20'
3 5/8	-	-	-	-	-	-	-	-	1°2'	1°16'
3 3/4	-	-	-	-	-	-	-	-	-	1°14'
3 7/8	-	-	-	-	-	-	-	-	-	1°12'
4	-	-	-	-	-	-	-	-	-	1°10'

Based on General Purpose Acme Basic Dimensions.

(Continued)

Table 78. (Continued) Helix Angles for General Purpose Double Start Acme Threads 1 9/16 Through 4 Inch. (Source, Landis Threading Systems.)

Dia.	Threads per Inch									
	2	2.5	3	3.5	4	4.5	5	5.5	6	7
1 9/16	13°42'	10°42'	8°45'	7°22'	6°23'	5°36'	5°3'	4°32'	4°8'	3°32'
1 5/8	13°4'	10°13'	8°19'	7°3'	6°6'	5°22'	4°50'	4°20'	3°58'	3°21'
1 3/4	12°2'	9°22'	7°40'	6°29'	5°37'	4°58'	4°25'	4°0'	3°40'	3°6'
1 7/8	11°8'	8°39'	7°6'	6°1'	5°14'	4°37'	4°8'	3°44'	3°24'	2°54'
2	10°21'	8°6'	6°39'	5°35'	4°48'	4°18'	3°50'	3°28'	3°12'	2°42'
2 1/8	9°44'	7°34'	6°11'	5°15'	4°33'	4°2'	3°34'	3°16'	3°0'	2°34'
2 1/4	9°5'	7°4'	5°51'	4°57'	4°22'	3°48'	3°26'	3°4'	2°48'	2°26'
2 3/8	8°33'	6°40'	5°31'	4°41'	4°4'	3°35'	3°14'	2°56'	2°40'	2°18'
2 1/2	8°4'	6°19'	5°12'	4°26'	3°52'	3°24'	3°4'	2°46'	2°32'	2°10'
2 5/8	7°40'	6°1'	4°56'	4°14'	3°40'	3°14'	2°55'	2°38'	2°26'	2°4'
2 3/4	7°14'	5°43'	4°40'	4°2'	3°30'	3°5'	2°46'	2°30'	2°18'	1°58'
2 7/8	6°57'	5°27'	4°30'	3°50'	3°20'	2°56'	2°39'	2°24'	2°12'	1°52'
3	6°37'	5°13'	4°18'	3°40'	3°12'	2°49'	2°32'	2°18'	2°6'	1°46'
3 1/8	6°21'	5°0'	4°6'	3°32'	3°4'	2°42'	2°26'	2°12'	2°0'	1°44'
3 1/4	6°5'	4°48'	3°54'	3°22'	2°56'	2°35'	2°20'	2°6'	1°56'	1°40'
3 3/8	5°51'	4°38'	3°46'	3°14'	2°48'	2°29'	2°14'	2°2'	1°52'	1°36'
3 1/2	5°35'	4°26'	3°38'	3°6'	2°42'	2°24'	2°8'	1°58'	1°48'	1°32'
3 5/8	5°23'	4°16'	3°29'	3°0'	2°36'	2°19'	2°4'	1°52'	1°44'	1°30'
3 3/4	5°11'	4°6'	3°18'	2°54'	2°32'	2°14'	2°0'	1°48'	1°40'	1°26'
3 7/8	5°2'	3°58'	3°14'	2°48'	2°26'	2°9'	1°57'	1°46'	1°38'	1°24'
4	4°51'	3°50'	3°8'	2°42'	2°22'	2°5'	1°54'	1°42'	1°34'	1°22'

Based on General Purpose Acme Basic Dimensions.

Table 79. Helix Angles for Standard Stub Acme Threads 3/16 Through 1 1/2 Inch. (Source, Landis Threading Systems.)

Table 79. Helix Angles for Standard Stub Acme Threads 3/16 Through 1 1/2 Inch. (Source, Landis Threading Systems.)

Dia.	Threads per Inch											
	28	26	24	22	20	18	16	14	12	11	10	9
3/16	3°47'	4°6'	4°28'	4°54'	5°26'	6°5'	6°56'	-	-	-	-	-
1/4	2°47'	3°0'	3°16'	3°35'	3°58'	4°26'	5°2'	5°48'	6°52'	7°45'	8°37'	9°43'
5/16	2°12'	2°22'	2°34'	2°49'	3°7'	3°29'	3°57'	4°33'	5°22'	6°0'	6°40'	7°29'
3/8	1°49'	1°57'	2°7'	2°19'	2°34'	2°52'	3°15'	3°44'	4°24'	4°54'	5°26'	6°5'
7/16	1°33'	1°40'	1°49'	1°59'	2°11'	2°26'	2°45'	3°10'	3°43'	4°8'	4°35'	5°8'
1/2	1°21'	1°27'	1°34'	1°43'	1°54'	2°7'	2°24'	2°45'	3°14'	3°35'	3°58'	4°26'
9/16	1°11'	1°17'	1°24'	1°31'	1°41'	1°52'	2°7'	2°26'	2°51'	3°9'	3°29'	3°54'
5/8	1°4'	1°9'	1°15'	1°22'	1°30'	1°41'	1°54'	2°11'	2°33'	2°49'	3°7'	3°29'
11/16	0°58'	1°3'	1°8'	1°14'	1°22'	1°31'	1°43'	1°58'	2°19'	2°33'	2°49'	3°9'
3/4	0°53'	0°57'	1°2'	1°8'	1°15'	1°23'	1°34'	1°48'	2°7'	2°20'	2°34'	2°52'
13/16	0°49'	0°53'	0°57'	1°3'	1°9'	1°17'	1°27'	1°39'	1°56'	2°8'	2°22'	2°38'
7/8	0°45'	0°49'	0°53'	0°58'	1°4'	1°11'	1°20'	1°32'	1°48'	1°59'	2°11'	2°26'
15/16	0°42'	0°46'	0°50'	0°54'	1°0'	1°6'	1°15'	1°26'	1°40'	1°50'	2°2'	2°16'
1	0°40'	0°43'	0°46'	0°51'	0°56'	1°2'	1°10'	1°20'	1°34'	1°43'	1°54'	2°7'
1 1/16	0°37'	0°40'	0°44'	0°48'	0°52'	0°58'	1°6'	1°15'	1°28'	1°37'	1°47'	1°59'
1 1/8	0°35'	0°38'	0°41'	0°45'	0°50'	0°55'	1°2'	1°11'	1°23'	1°31'	1°41'	1°52'
1 3/16	0°33'	0°36'	0°39'	0°43'	0°47'	0°52'	0°59'	1°7'	1°19'	1°26'	1°35'	1°46'
1 1/4	0°32'	0°34'	0°37'	0°40'	0°44'	0°49'	0°56'	1°4'	1°15'	1°22'	1°30'	1°41'
1 5/16	0°30'	0°32'	0°35'	0°38'	0°42'	0°47'	0°53'	1°1'	1°11'	1°18'	1°26'	1°36'
1 3/8	0°29'	0°31'	0°34'	0°37'	0°40'	0°45'	0°51'	0°58'	1°8'	1°14'	1°22'	1°31'
1 7/16	-	0°30'	0°32'	0°35'	0°39'	0°43'	0°48'	0°55'	1°5'	1°11'	1°18'	1°27'
1 1/2	-	0°28'	0°31'	0°34'	0°37'	0°41'	0°46'	0°53'	1°2'	1°8'	1°15'	1°23'

(Continued)

Based on General Purpose Acme Basic Dimensions.

Table 79. (Continued) Helix Angles for Standard Stub Acme Threads 3/16 Through 1 1/2 Inch. (Source, Landis Threading Systems.)

Dia.	Threads per Inch										
	2	2.5	3	3.5	4	4.5	5	5.5	6	7	8
3/8	-	-	-	-	-	-	-	-	9°34'	8°2'	6°56'
7/16	-	-	-	-	-	-	-	-	8°0'	6°45'	5°50'
1/2	-	-	-	-	-	-	-	7°34'	6°53'	5°48'	5°2'
9/16	-	-	-	-	-	-	-	6°38'	6°2'	5°6'	4°25'
5/8	-	-	-	-	-	-	6°33'	5°54'	5°22'	4°33'	3°56'
11/16	-	-	-	-	-	6°36'	5°53'	5°18'	4°50'	4°6'	3°33'
3/4	-	-	9°25'	7°55'	6°50'	6°0'	5°21'	4°50'	4°24'	3°44'	3°15'
13/16	-	-	8°35'	7°14'	6°15'	5°29'	4°54'	4°26'	4°2'	3°26'	2°59'
7/8	-	9°42'	7°54'	6°39'	5°45'	5°4'	4°31'	4°5'	3°43'	3°10'	2°45'
15/16	-	8°58'	7°18'	6°10'	5°20'	4°42'	4°12'	3°48'	3°28'	2°57'	2°34'
1	-	8°20'	6°48'	5°45'	4°58'	4°23'	3°55'	3°32'	3°14'	2°45'	2°24'
1 1/16	-	7°47'	6°21'	5°23'	4°39'	4°6'	3°40'	3°19'	3°2'	2°35'	2°15'
1 1/8	9°22'	7°18'	5°58'	5°3'	4°23'	3°52'	3°27'	3°7'	2°51'	2°26'	2°7'
1 3/16	8°48'	6°52'	5°37'	4°46'	4°8'	3°39'	3°16'	2°57'	2°42'	2°18'	2°0'
1 1/4	8°19'	6°29'	5°19'	4°30'	3°55'	3°27'	3°5'	2°48'	2°33'	2°11'	1°54'
1 5/16	7°52'	6°9'	5°3'	4°17'	3°43'	3°17'	2°56'	2°39'	2°26'	2°4'	1°48'
1 3/8	7°28'	5°50'	4°48'	4°4'	3°32'	3°7'	2°48'	2°32'	2°19'	1°58'	1°43'
1 7/16	7°6'	5°34'	4°34'	3°53'	3°22'	2°59'	2°40'	2°25'	2°12'	1°53'	1°38'
1 1/2	6°47'	5°19'	4°22'	3°42'	3°13'	2°51'	2°33'	2°19'	2°7'	1°48'	1°34'

(Continued)

Based on General Purpose Acme Basic Dimensions.

Table 79. (Continued) Helix Angles for Standard Stub Acme Threads 1 9/16 Through 4 Inch. (Source, Landis Threading Systems.)

Dia.	Threads per Inch										
	26	24	22	20	18	16	14	12	11	10	9
1 9/16	-	-	0°32'	0°35'	0°39'	0°44'	0°51'	0°59'	1°5'	1°12'	1°20'
1 5/8	-	-	0°31'	0°34'	0°38'	0°43'	0°49'	0°57'	1°3'	1°9'	1°17'
1 3/4	-	-	-	0°32'	0°35'	0°40'	0°45'	0°53'	0°58'	1°4'	1°11'
1 7/8	-	-	-	0°29'	0°33'	0°37'	0°42'	0°49'	0°54'	1°0'	1°6'
2	-	-	-	0°28'	0°31'	0°35'	0°40'	0°46'	0°51'	0°56'	1°2'
2 1/8	-	-	-	-	-	-	-	0°44'	0°48'	0°52'	0°58'
2 1/4	-	-	-	-	-	-	-	0°41'	0°45'	0°50'	0°55'
2 3/8	-	-	-	-	-	-	-	0°39'	0°43'	0°47'	0°52'
2 1/2	-	-	-	-	-	-	-	0°37'	0°40'	0°44'	0°49'
2 5/8	-	-	-	-	-	-	-	0°35'	0°38'	0°42'	0°47'
2 3/4	-	-	-	-	-	-	-	0°34'	0°37'	0°40'	0°45'
2 7/8	-	-	-	-	-	-	-	0°32'	0°35'	0°39'	0°43'
3	-	-	-	-	-	-	-	0°31'	0°34'	0°37'	0°41'
3 1/8	-	-	-	-	-	-	-	-	0°32'	0°35'	0°39'
3 1/4	-	-	-	-	-	-	-	-	0°31'	0°34'	0°38'
3 3/8	-	-	-	-	-	-	-	-	0°30'	0°33'	0°36'
3 1/2	-	-	-	-	-	-	-	-	0°29'	0°32'	0°35'
3 5/8	-	-	-	-	-	-	-	-	-	0°31'	0°34'
3 3/4	-	-	-	-	-	-	-	-	-	0°29'	0°33'
3 7/8	-	-	-	-	-	-	-	-	-	0°29'	0°32'
4	-	-	-	-	-	-	-	-	-	0°28'	0°31'

Based on General Purpose Acme Basic Dimensions.

(Continued)

Table 79. *(Continued)* Helix Angles for Standard Stub Acme Threads 1 9/16 Through 4 Inch. *(Source, Landis Threading Systems.)*

Dia.	Threads per Inch										
	2	2.5	3	3.5	4	4.5	5	5.5	6	7	8
1 9/16	6°29'	5°5'	4°11'	3°33'	3°5'	2°44'	2°27'	2°13'	2°1'	1°44'	1°30'
1 5/8	6°12'	4°52'	4°0'	3°24'	2°58'	2°37'	2°21'	2°8'	1°56'	1°39'	1°27'
1 3/4	5°43'	4°30'	3°42'	3°9'	2°44'	2°25'	2°10'	1°58'	1°48'	1°32'	1°20'
1 7/8	5°18'	4°10'	3°26'	2°56'	2°33'	2°15'	2°1'	1°50'	1°40'	1°26'	1°15'
2	4°57'	3°54'	3°13'	2°44'	2°23'	2°6'	1°53'	1°43'	1°34'	1°20'	1°10'
2 1/8	4°38'	3°39'	3°1'	2°34'	2°14'	1°59'	1°47'	1°37'	1°28'	1°15'	1°6'
2 1/4	4°21'	3°26'	2°50'	2°25'	2°6'	1°52'	1°40'	1°31'	1°23'	1°11'	1°2'
2 3/8	4°7'	3°15'	2°41'	2°17'	1°59'	1°46'	1°35'	1°26'	1°19'	1°7'	0°59'
2 1/2	3°54'	3°5'	2°33'	2°10'	1°53'	1°40'	1°30'	1°22'	1°15'	1°4'	0°56'
2 5/8	3°42'	2°55'	2°25'	2°4'	1°48'	1°35'	1°26'	1°18'	1°11'	1°1'	0°53'
2 3/4	3°31'	2°47'	2°18'	1°58'	1°43'	1°31'	1°22'	1°14'	1°8'	0°58'	0°51'
2 7/8	3°21'	2°39'	2°12'	1°53'	1°38'	1°27'	1°18'	1°11'	1°5'	0°55'	0°48'
3	3°13'	2°33'	2°6'	1°48'	1°34'	1°23'	1°15'	1°8'	1°2'	0°53'	0°46'
3 1/8	3°4'	2°26'	2°1'	1°43'	1°30'	1°20'	1°12'	1°5'	1°0'	0°51'	0°44'
3 1/4	2°57'	2°20'	1°56'	1°39'	1°27'	1°17'	1°9'	1°3'	0°57'	0°49'	0°43'
3 3/8	2°50'	2°15'	1°52'	1°35'	1°23'	1°14'	1°6'	1°1'	0°55'	0°47'	0°41'
3 1/2	2°44'	2°10'	1°48'	1°32'	1°20'	1°11'	1°4'	0°58'	0°53'	0°45'	0°40'
3 5/8	2°38'	2°5'	1°44'	1°29'	1°17'	1°9'	1°2'	0°56'	0°51'	0°44'	0°38'
3 3/4	2°32'	2°1'	1°40'	1°26'	1°15'	1°6'	1°0'	0°54'	0°49'	0°42'	0°37'
3 7/8	2°27'	1°57'	1°37'	1°23'	1°12'	1°4'	0°58'	0°52'	0°48'	0°41'	0°36'
4	2°22'	1°53'	1°34'	1°20'	1°10'	1°2'	0°56'	0°51'	0°46'	0°40'	0°35'

Based on General Purpose Acme Basic Dimensions.

Table 80. Helix Angles for Double Start Standard Stub Acme Threads ³/₁₆ Through 1 ½ Inch. *(Source, Landis Threading Systems.)*

Dia.	Threads per Inch										
	28	20	18	16	14	13	12	11.5	11	10	9
3/16	7°20'	10°28'	11°42'	13°16'	15°19'	-	-	-	-	-	-
1/4	5°26'	7°43'	8°38'	9°46'	11°15'	12°11'	13°16'	13°54'	14°34'	16°8'	18°5'
5/16	4°19'	6°7'	6°49'	7°43'	8°53'	9°37'	10°28'	10°57'	11°29'	12°42'	14°13'
3/8	3°35'	5°4'	5°39'	6°23'	7°20'	7°56'	8°37'	9°1'	9°27'	10°28'	11°42'
7/16	3°3'	4°19'	4°49'	5°26'	6°15'	6°45'	7°20'	7°40'	8°2'	8°53'	9°56'
1/2	2°40'	3°46'	4°11'	4°44'	5°26'	5°52'	6°23'	6°40'	6°59'	7°43'	8°37'
9/16	2°22'	3°20'	3°43'	4°11'	4°49'	5°12'	5°39'	5°54'	6°11'	6°49'	7°37'
5/8	2°8'	3°0'	3°20'	3°46'	4°19'	4°39'	5°4'	5°17'	5°32'	6°7'	6°49'
11/16	1°56'	2°43'	3°1'	3°25'	3°55'	4°13'	4°35'	4°47'	5°1'	5°32'	6°11'
3/4	1°46'	2°29'	2°46'	3°7'	3°35'	3°52'	4°12'	4°23'	4°35'	5°4'	5°39'
13/16	1°38'	2°17'	2°33'	2°53'	3°18'	3°33'	3°52'	4°2'	4°13'	4°39'	5°12'
7/8	1°31'	2°8'	2°22'	2°40'	3°3'	3°18'	3°35'	3°44'	3°55'	4°19'	4°49'
15/16	1°25'	1°59'	2°12'	2°29'	2°51'	3°4'	3°20'	3°29'	3°39'	4°1'	4°29'
1	1°19'	1°51'	2°4'	2°20'	2°40'	2°53'	3°7'	3°16'	3°25'	3°46'	4°11'
1 1/16	-	1°45'	1°57'	2°11'	2°30'	2°42'	2°56'	3°4'	3°12'	3°32'	3°56'
1 1/8	-	1°39'	1°50'	2°4'	2°22'	2°33'	2°46'	2°53'	3°1'	3°30'	3°43'
1 3/16	-	1°34'	1°44'	1°57'	2°14'	2°25'	2°37'	2°44'	2°52'	3°9'	3°31'
1 1/4	-	1°29'	1°39'	1°51'	2°8'	2°17'	2°29'	2°36'	2°43'	3°0'	3°20'
1 5/16	-	1°25'	1°34'	1°46'	2°1'	2°11'	2°22'	2°28'	2°35'	2°51'	3°10'
1 3/8	-	1°21'	1°30'	1°41'	1°56'	2°5'	2°15'	2°21'	2°28'	2°43'	3°2'
1 7/16	-	1°17'	1°26'	1°37'	1°51'	1°59'	2°9'	2°15'	2°21'	2°36'	2°53'
1 1/2	-	1°14'	1°22'	1°33'	1°46'	1°54'	2°4'	2°9'	2°15'	2°29'	2°46'

Based on General Purpose Acme Basic Dimensions.

(Continued)

Table 80. *(Continued)* Helix Angles for Double Start Standard Stub Acme Threads $3/16$ Through $1 1/2$ Inch. *(Source, Landis Threading Systems.)*

Dia.	Threads per Inch										
	8	7	6	5.5	5	4.5	4	3.5	3	2.5	2
5/16	16°9'	-	-	-	-	-	-	-	-	-	-
3/8	13°16'	15°19'	18°5'	-	-	-	-	-	-	-	-
7/16	11°15'	12°59'	15°19'	-	-	-	-	-	-	-	-
1/2	9°46'	11°15'	13°16'	14°34'	-	-	-	-	-	-	-
9/16	8°37'	9°56'	11°42'	12°51'	-	-	-	-	-	-	-
5/8	7°43'	8°53'	10°28'	11°29'	-	-	-	-	-	-	-
11/16	6°59'	8°2'	9°27'	10°22'	-	-	-	-	-	-	-
3/4	6°23'	7°20'	8°37'	9°27'	10°28'	11°42'	13°16'	15°19'	18°5'	-	-
13/16	5°52'	6°45'	7°56'	8°41'	9°37'	10°45'	12°11'	14°3'	16°36'	-	-
7/8	5°26'	6°14'	7°20'	8°2'	8°53'	9°56'	11°15'	12°59'	15°19'	18°39'	-
15/16	5°4'	5°49'	6°49'	7°28'	8°16'	9°14'	10°28'	12°4'	14°13'	17°18'	-
1	4°44'	5°26'	6°23'	6°59'	7°43'	8°37'	9°46'	11°15'	13°16'	16°9'	-
1 1/16	4°27'	5°6'	5°59'	6°33'	7°15'	8°5'	9°10'	10°33'	12°26'	15°7'	-
1 1/8	4°11'	4°49'	5°39'	6°11'	6°49'	7°37'	8°37'	9°56'	11°42'	14°13'	18°5'
1 3/16	3°58'	4°33'	5°20'	5°50'	6°27'	7°12'	8°9'	9°23'	11°3'	13°25'	17°4'
1 1/4	3°46'	4°19'	5°4'	5°32'	6°7'	6°49'	7°43'	8°53'	10°28'	12°42'	16°9'
1 5/16	3°35'	4°6'	4°49'	5°16'	5°49'	6°29'	7°20'	8°26'	9°56'	12°4'	15°19'
1 3/8	3°25'	3°55'	4°35'	5°1'	5°32'	6°11'	6°58'	8°2'	9°27'	11°29'	14°34'
1 7/16	3°16'	3°44'	4°23'	4°47'	5°17'	5°54'	6°40'	7°40'	9°1'	10°57'	13°54'
1 1/2	3°7'	3°35'	4°11'	4°35'	5°4'	5°39'	6°23'	7°20'	8°37'	10°28'	13°16'

Based on General Purpose Acme Basic Dimensions.

(Continued)

Table 80. *(Continued)* Helix Angles for Double Start Standard Stub Acme Threads 1 9/16 Through 4 Inch. *(Source, Landis Threading Systems.)*

Dia.	Threads per Inch										
	8	9	10	11	11.5	12	13	14	16	18	20
1 9/16	3°0'	2°39'	2°23'	2°10'	2°4'	1°59'	1°50'	1°42'	1°29'	1°19'	1°11'
1 5/8	2°53'	2°33'	2°17'	2°5'	1°59'	1°54'	1°45'	1°38'	1°25'	1°16'	1°8'
1 3/4	2°40'	2°22'	2°8'	1°56'	1°51'	1°46'	1°38'	1°31'	1°19'	1°10'	1°3'
1 7/8	2°29'	2°12'	1°59'	1°48'	1°43'	1°39'	1°31'	1°25'	1°14'	1°6'	0°59'
2	2°20'	2°4'	1°51'	1°41'	1°37'	1°33'	1°25'	1°19'	1°9'	1°2'	0°55'
2 1/8	2°11'	1°57'	1°45'	1°35'	1°31'	1°27'	1°20'	-	-	-	-
2 1/4	2°4'	1°50'	1°39'	1°30'	1°26'	1°22'	1°16'	-	-	-	-
2 3/8	1°57'	1°44'	1°34'	1°25'	1°21'	1°18'	1°12'	-	-	-	-
2 1/2	1°51'	1°39'	1°29'	1°21'	1°17'	1°14'	1°8'	-	-	-	-
2 5/8	1°46'	1°34'	1°25'	1°17'	1°14'	1°10'	-	-	-	-	-
2 3/4	1°41'	1°30'	1°21'	1°13'	1°10'	1°7'	-	-	-	-	-
2 7/8	1°37'	1°26'	1°17'	1°10'	1°7'	1°4'	-	-	-	-	-
3	1°33'	1°22'	1°14'	1°7'	1°4'	1°2'	-	-	-	-	-
3 1/8	1°29'	1°19'	1°11'	1°5'	1°2'	-	-	-	-	-	-
3 1/4	1°25'	1°16'	1°8'	1°2'	0°59'	-	-	-	-	-	-
3 3/8	1°22'	1°13'	1°6'	1°0'	0°57'	-	-	-	-	-	-
3 1/2	1°19'	1°10'	1°3'	0°58'	0°55'	-	-	-	-	-	-
3 5/8	1°17'	1°8'	-	-	-	-	-	-	-	-	-
3 3/4	1°14'	-	-	-	-	-	-	-	-	-	-
3 7/8	1°12'	-	-	-	-	-	-	-	-	-	-
4	1°9'	-	-	-	-	-	-	-	-	-	-

(Continued)

Based on General Purpose Acme Basic Dimensions.

Table 80. (*Continued*) Helix Angles for Double Start Standard Stub Acme Threads 1 9/16 Through 4 Inch. (*Source, Landis Threading Systems.*)

Dia.	Threads per Inch									
	2	2.5	3	3.5	4	4.5	5	5.5	6	7
1 9/16	12°42'	10°1'	8°16'	7°2'	6°7'	5°24'	4°51'	4°24'	4°3'	3°26'
1 5/8	12°11'	9°37'	7°56'	6°45'	5°52'	5°12'	4°39'	4°13'	3°52'	3°18'
1 3/4	11°15'	8°53'	7°20'	6°15'	5°26'	4°49'	4°19'	3°55'	3°35'	3°3'
1 7/8	10°28'	8°16'	6°49'	5°49'	5°4'	4°29'	4°1'	3°39'	3°20'	2°51'
2	9°46'	7°43'	6°23'	5°26'	4°44'	4°11'	3°46'	3°25'	3°7'	2°40'
2 1/8	9°10'	7°15'	5°59'	5°6'	4°27'	3°56'	3°32'	3°12'	2°56'	2°30'
2 1/4	8°37'	6°49'	5°39'	4°49'	4°11'	3°43'	3°20'	3°1'	2°46'	2°22'
2 3/8	8°9'	6°27'	5°20'	4°33'	3°58'	3°31'	3°9'	2°52'	2°37'	2°14'
2 1/2	7°43'	6°7'	5°4'	4°19'	3°46'	3°20'	3°0'	2°43'	2°29'	2°8'
2 5/8	7°20'	5°49'	4°49'	4°6'	3°35'	3°10'	2°51'	2°35'	2°22'	2°1'
2 3/4	6°59'	5°32'	4°35'	3°55'	3°25'	3°1'	2°43'	2°28'	2°15'	1°56'
2 7/8	6°40'	5°17'	4°23'	3°44'	3°16'	2°53'	2°36'	2°21'	2°9'	1°51'
3	6°23'	5°4'	4°11'	3°35'	3°7'	2°46'	2°29'	2°15'	2°4'	1°46'
3 1/8	6°7'	4°51'	4°1'	3°26'	3°0'	2°39'	2°23'	2°10'	1°59'	1°42'
3 1/4	5°52'	4°39'	3°52'	3°18'	2°53'	2°33'	2°17'	2°5'	1°54'	1°38'
3 3/8	5°39'	4°29'	3°43'	3°10'	2°46'	2°27'	2°12'	2°0'	1°50'	1°34'
3 1/2	5°26'	4°19'	3°35'	3°3'	2°40'	2°22'	2°8'	1°56'	1°46'	1°31'
3 5/8	5°14'	4°10'	3°27'	2°57'	2°35'	2°17'	2°3'	1°52'	1°42'	1°28'
3 3/4	5°4'	4°1'	3°20'	2°51'	2°29'	2°12'	1°59'	1°48'	1°39'	1°25'
3 7/8	4°53'	3°53'	3°13'	2°45'	2°24'	2°8'	1°55'	1°44'	1°36'	1°22'
4	4°44'	3°46'	3°7'	2°40'	2°20'	2°4'	1°51'	1°41'	1°33'	1°19'

Based on General Purpose Acme Basic Dimensions.

Screw Threads

Unified inch and ISO metric threads

The origin of the parallel helical thread has been lost to time, but what is known of its history suggests that early versions took on a wide variety of forms. The Englishman Sir Joseph Whitworth, in the early 1840s, is generally credited with being the first to standardize a thread: the 55° radiused root thread that bore his name became the British Standard thread and remained so for over a century. Other countries, of course, went their own way. Just three years after the introduction of the Whitworth thread, an American, William Sellers, contributed a new design with a 60° thread. Sellers found the 55° angle difficult to gauge, and his thread design was eventually accepted as the American Standard. To further confuse the issue, an international conference convened in Switzerland in 1898 and announced the S.I. (Systeme International) Metric thread. Various countries throughout Europe nationalized this thread, making it the standard on the Continent.

In 1948, in the interest of promoting standardized manufacturing—which had become a nagging problem for Allied efforts during the Second World War—the United States, Canada, and the United Kingdom agreed to adopt a new thread, the Unified screw thread. The Unified Standard th ad is the result, defined by American Standard B1.1-1949, which identifies a thread nfiguration that is essentially the same as the 60° American National thread that prece it. (There is still an "AN" series of fasteners able, but the abbreviation stands for the st letters of Air Force–Navy rather than can National, and AN fasteners use standa Unified threads.) Changes are most e' J the permitted allowances and tolerances fc ndividual sizes, and these modifi Jn ere introduced to provide greater fatigue strength, root clearance, and ease of as mb In addition, the new Unified Standard (as it came to be known) and the ISO metric thr ld also have the same basic configuration, but differ in diameters and pitch lengths for standard sizes. In instances where the tolerances for the two threads overlap, ISO metric and Unified inch threads are compatible, as shown in **Table 1a** which lists the six potentially compatible sizes. ISO metric and Unified size threads should never be combined, but this listing provides evidence of the basic compatibility of the two thread forms. Either series satisfies the need for a general purpose screw thread system for most applications. Allmetal Screw Products notes that there are 319 thread sizes in the Unified Thread Standard ranging in diameter from 0.06" (1.5 mm) to 6" (152 mm), and in pitches from 80 threads per inch (0.3175 mm pitch) to 4 tpi (6.35 mm pitch). Each is available in at least two classes of fit. There are even more metric sizes, 332 in all, ranging in diameter from 1 mm (0.03937") to 300 mm (11.81"), and in pitches from 0.2 mm (127 tpi) to 6 mm (4.23 tpi). Again, each size is available in more than a single tolerance grade.

Converting inch pitches to metric dimensions can be confusing, and **Table 1b** provides conversions for all popular metric pitches in millimeters to pitch and threads per inch values in inches.

Thread definitions

All threads share fundamental characteristics that, when defined, precisely identify all relevant dimensions of its shape. The following definitions were provided by Cleveland Twist Drill Co. *Figure 1* illustrates important screw thread nomenclature.

General terms relating to screw threads ·

Complete Thread. The complete or full form thread is the cross section of a threaded length having full form at both crest and root.

Incomplete Thread. An incomplete thread is a threaded profile having either crests

Table 1a. Potentially Compatible Unified and ISO Metric Threads. *(Source, Allmetal Screw Products.)*

Unified Thread Size	ISO Metric Thread Size
1–72 UNF Class 3A	1.8 × 0.35 (Coarse Pitch)
2–56 UNC Class 3A	2.2 × 0.45 (Coarse Pitch)
3–56 UNF Class 3A	2.5 × 0.45 (Coarse Pitch)
8–36 UNF Class 3A	4.0 × 0.70 (Coarse Pitch)
10–32 UNF Class 3A	5.0 × 0.80 (Coarse Pitch)
5/16–20 UN Class 3A	8.0 × 1.25 (Coarse Pitch)

Table 1b. Millimeters Pitch to Threads Per Inch Metric Conversion. *(Source, New England Tap Corp.)*

Pitch mm	Pitch inch	Threads per inch	Basic Height*	Pitch mm	Pitch inch	Threads per inch	Basic Height*
.25	0.00984	101.6000	0.00639	1.25	0.04921	20.3211	0.03196
.30	0.01181	84.6668	0.00767	1.50	0.05906	16.9316	0.03836
.35	0.01378	72.5689	0.00895	1.75	0.06890	14.5138	0.04475
.40	0.01575	63.4921	0.01023	2.00	0.07874	12.7000	0.05114
.45	0.01772	56.4334	0.01151	2.50	0.09843	10.1595	0.06393
.50	0.01969	50.8000	0.01279	3.00	0.11811	8.4667	0.07671
.60	0.02362	42.3370	0.01534	3.50	0.13780	7.2569	0.08950
.70	0.02756	36.2845	0.01790	4.00	0.15748	6.3500	0.10229
.75	0.02953	33.8639	0.01918	4.50	0.17717	5.6443	0.11508
.80	0.03150	31.7460	0.02046	5.00	0.19685	5.0800	0.12758
.90	0.03543	28.2247	0.02301	6.00	0.23622	4.2333	0.15344
1.00	0.03937	25.4000	0.02557	mm × 0.03937 = inch equiv. inch × 25.4 = mm equiv.			

* Calculated as twice the addendum, in inches.

or roots, or both crests and roots, not fully formed, resulting from their intersection with the cylindrical or end surface of the work or the vanish cone. It may occur at either end of the thread.

NOTE: Formerly in pipe thread terminology this was referred to as "the imperfect thread" but that is no longer considered desirable.

Lead Thread. The lead thread is the portion of the incomplete thread that is fully formed at root but not fully formed at crest which occurs at the entering end of either external or internal threads.

Vanish Thread (Partial Thread, Washout Thread or Thread Run-Out). A vanish thread is that part of the thread which is not fully formed at the root. It is produced by the chamfer at the starting end of the threading tool.

Effective Thread. The effective (or useful) thread includes the complete thread and those portions of the incomplete thread which are fully formed at the root but not at the crest (in taper pipe threads this includes the so-called black crest threads), thus excluding the vanish thread.

Total Thread. The total thread includes the complete and all of the incomplete thread, thus including the vanish thread.

Classes of Threads. Classes of threads are distinguished from each other by the amounts of tolerance or tolerance and allowance specified.

Thread Series. Thread series are groups of diameter-pitch combinations distinguished from each other by the number of threads per inch applied to specific diameters.

Single and Multiple Start Threads. A single thread is a thread made by forming one groove around a cylinder or inside a hole. Most threads are single threads. Double, triple, or other multiple threads have two, three, or more grooves formed around a cylinder or inside a hole.

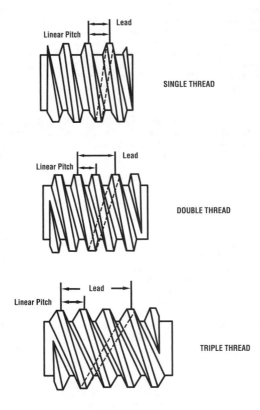

Screw Thread Forms

<u>Terms relating to size and fit</u>

Nominal Size. The nominal size is the designation that is used for the purpose of general information.

Basic Size. The basic size is the size from which the limits of size are derived by the application of allowances and tolerances.

Reference Size. A reference size is a size without tolerance used only for information purposes and does not govern manufacturing or inspection operations.

Design Size. The design size is the basic size with allowance applied, from which the limits of size are derived by the application of tolerances. If there is no allowance, the design size is the same as the basic size.

Actual Size. An actual size is a measured size.

Limits of Size. The limits of size are the applicable maximum and minimum sizes.

Maximum Material Limit. A maximum material limit is the limit of size that provides the maximum amount of material for the part. Normally it is the maximum limit of size of ʳal dimension or the minimum limit of size of an internal dimension.

Minᵢₘ al Material Limit. A minimum material limit is the limit of size that provides the minimum amount of material for the part. Normally it is the minimum limit of size of an external dimension or the maximum limit of size of an internal dimension.

NOTE: Examples of exceptions are: an exterior corner radius where the maximum radius is the minimum material limit and the minimum radius is the maximum material limit.

Allowance. An allowance is a prescribed difference between the maximum material limits of mating parts. It is the minimum clearance (positive allowance) or maximum interference (negative allowance) between such parts.

Tolerance. A tolerance is the total permissible variation of a size. The tolerance is the difference between the limits of size.

Unilateral Tolerance. A unilateral tolerance is a tolerance in which variation is permitted only in one direction from the design si

Bilateral Tolerance. A bilateral tolerance is a toleranᵤᵤ   h variation is permitted in both directions from the design size.

Terms relating to geometrical elements of screw threads

Axis of Thread. The axis of a thread is the axis of its pitch cylinder or cone.

Major Cylinder. The major cylinder is one that would bound the crests of an external straight thread or the roots of an internal straight thread.

Major Cone. The major cone is one that would bound the crests of an external taper thread or the roots of an internal taper thread.

Pitch Cylinder. The pitch cylinder is one of such diameter and location of its axis that its surface would pass through a straight thread in such a manner as to make the widths of the thread ridge and the thread groove equal and, therefore, is located equidistantly between the sharp major and minor cylinders of a given thread form. On a theoretically perfect thread these widths are equal to one-half of the basic pitch.

Pitch Cone. The pitch cone is one of such apex angle and location of its vertex and axis that its surface would pass through a taper thread in such a manner as to make the widths of the thread ridge and the thread groove equal and, therefore, is located equidistantly between the sharp major and minor cones of a given thread form. On a theoretically perfect taper thread these widths are equal to one-half of the basic pitch.

Minor Cylinder. The minor cylinder is one that would bound the roots of an external straight thread or the crests of an internal straight thread.

Minor Cone. The minor cone is one that would bound the roots of an external taper thread or the crests of an internal taper thread.

Pitch Line. The pitch line is a generator of the cylinder or cone specified in the definitions of pitch cylinder and pitch cone.

Form of Thread. The form of thread is its profile in an axial plane for a length of one pitch of the complete thread.

Basic Form of Thread. The basic form of a thread is the theoretical profile of the thread for a length of one pitch in an axial plane, from which the design forms of the threads for both the external and internal threads are developed.

Design Forms of Thread. The design forms for a thread are the maximum material forms permitted for the external and internal threads. In practice, however, the forms of roots are indeterminate roundings not encroaching on the maximum material form of the mating thread when assembled.

Terms relating to dimensions of screw threads

Pitch. The pitch of a thread having uniform spacing is the distance, measured parallel to its axis, between corresponding points on adjacent thread forms in the same side of the axis. The basic pitch is equal to the lead divided by the number of thread starts. See Table 4, No. 1.

Lead. When a threaded part is rotated about its axis with respect to a fixed mating, the lead is the axial distance moved by the part in relation to the amount of angular rotation. Lead is commonly specified as the distance to be moved in one complete rotation. It is necessary to distinguish measurement of lead from measurement of pitch, as uniformity of pitch measurements does not assure uniformity of lead. Variations in either lead or pitch cause the functional diameter of a thread to differ from the pitch diameter. See Table 4, No. 41.

Helix Variation. Helix variation of a thread is a wavy deviation from true helical advancement. The "helical path" includes the helix with its superimposed variation and is measured either as the maximum deviation from the true helix or as the "cumulative pitch." The cumulative pitch is the distance measured parallel to the axis of the thread between corresponding points on any two thread forms whether or not they are in the same axial plane.

Lead Angle. On a straight thread, the lead angle is the angle made by the helix of the thread at the pitch line with a plane perpendicular to the axis. On a taper thread, the lead angle at a given axial position is the angle made by the conical spiral of the thread, with the plane perpendicular to the axis, at the pitch line. See Table 4, No. 42.

Helix Angle. On a straight thread, the helix angle is the angle made by the helix of the thread at the pitch line with the axis. On a taper thread, the helix angle at a given axial position is the angle made by the conical spiral of the thread with the axis at the pitch line. The helix angle is the complement of the lead angle. See Table 4, No. 43.

NOTE: The helix angle was formerly defined in accordance with the present definition of lead angle.

Thickness of Thread Ridge. The thickness of thread ridge is the distance between the flanks of one thread ridge, normally measured parallel to the axis at the specified pitch radius. The thickness of thread ridge may be specified and measured parallel to the axis at any other specified radius.

NOTE: The pitch radius is equal to one-half of the pitch diameter.

Width of Thread Groove. The width of the thread groove is the distance between the flanks of adjacent thread ridges normally measured parallel to the axis at the specified pitch radius. The width of thread groove may be specified and measured parallel to the axis at any other specified radius.

Height of Thread. The height (or depth) of thread is the distance, measured radially between the major and minor cylinders or cones, respectively.

NOTE: In American practice the height of thread is often expressed as a percentage of three-fourths of the height of the fundamental triangle.

Addendum. The addendum of an external thread is the radial distance between the major and pitch cylinders or cones, respectively. The addendum of an internal thread is the radial distance between the minor and pitch cylinders or cones, respectively. See Table 4, No. 16 and No. 20.

Dedendum. The dedendum of an external thread is the radial distance between the pitch and minor cylinders or cones, respectively. The dedendum of an internal thread is the radial distance between the major and pitch cylinders or cones, respectively. See Table 4, No. 16–19.

Crest. The crest is the outermost surface of the thread form that joins the thread flanks. This is the farthest point on the thread from the cylinder or cone from which the thread projects. Hence, it is the point of the major diameter on external threads, and the minor diameter on internal threads. See Table 4, No. 34–36.

Crest Truncation. The crest truncation of a thread is the radial distance between the sharp crest (crest apex) and the cylinder or cone that would bound the crest. See Table 4, No. 20, 31, and 32.

Root. The root is the innermost surface of the thread form that joins the flanks, immediately adjacent to the cylinder or cone from which the thread projects. It is the point of the minor diameter on an external thread and the major diameter of an internal thread. See Table 4, No. 25–30.

Root Truncation. The root truncation of a thread is the radial distance between the sharp root (root apex) and the cylinder or cone that would bound the root. See Table 4, No. 20, 32, and 33.

Pitch Diameter. On a straight thread, the pitch diameter is the diameter of the pitch cylinder. On a taper thread, the pitch diameter at a given position on the thread axis is the diameter of the pitch cone at that position. On a single start thread of perfect form and lead, it is also the length between intercepts of a line that is perpendicular to the thread axis and intersects thread flanks on opposite sides of the thread axis. See Table 4, No. 12–15.

NOTE: When the crest of a thread is truncated beyond the pitch line, the pitch diameter, pitch cylinder, or pitch cone would be based on a theoretical extension of the thread flanks.

NOTE: Pitch diameter on the buttress casing thread is defined by the American Petroleum Institute, in API Standard 5B, as being midway between the major and minor diameters.

Major Diameter. On a straight thread, the major diameter is the diameter of the major cylinder. See Table 4, No. 3–5.

Minor Diameter. On a straight thread, the minor diameter is the diameter of the minor cylinder. See Table 4, No. 6–11.

Thread Groove Diameter (Simple Effective Diameter). On a straight thread, the thread groove diameter is the diameter of the coaxial cylinder, the surface of which would pass through the thread profiles at such points as to make the width of the thread groove equal to one-half of the basic pitch. It is the diameter yielded by measuring over or under cylinders (wires) or spheres (balls) inserted in the thread groove on opposite sides of the axis and computing the thread groove diameter as thus defined. On a taper thread, the thread groove diameter is the diameter at a given position on the thread axis of the coaxial cone, the surface of which would pass through the thread profiles at such points as to make the width of the thread groove (measured parallel to the axis) equal to one-half of the basic pitch. It is the diameter yielded by measuring over or under cylinders (wires) or spheres (balls) inserted in the thread groove on opposite sides of the axis and computing the thread groove diameter as thus defined.

Thread Ridge Diameter. On a straight thread, the thread ridge diameter is the diameter of the coaxial cylinder, the surface of which would pass through the thread profiles at such points as to make the thickness of the thread ridge equal to one-half of the basic pitch. On a taper thread, the thread ridge diameter is the diameter at a given position on the thread axis of the coaxial cone, the surface of which would pass through the thread profiles at such points as to make the thickness of the thread ridge (measured parallel to the axis) equal to one-half of the basic pitch.

Functional (Virtual) Diameter. The functional diameter of an external or internal thread is the pitch diameter of the enveloping thread of perfect pitch, lead, and flank angles, having full depth of engagement but clear at crests and roots, and of a specified length of engagement. It may be derived by adding to the pitch diameter in the case of an external thread, or subtracting from the pitch diameter in the case of an internal thread, the cumulative effects of deviations from specified profile, including variations in lead and flank angle over a specified length of engagement. The effects of taper, out-of-roundness, and surface defects may be positive or negative on either external or internal threads. (A perfect GO thread plug or ring gage, having a pitch diameter equal to that specified for the maximum material limit and having clearance at crest and root, is the enveloping thread corresponding to that limit.)

NOTE: Also called the Virtual Diameter, Effective Size, or Virtual Effective Diameter.

Length of Complete Thread. The length of complete thread is the axial length of a part where the thread section has full form at both crest and root; that is, the vanish threads are not included. However, on commercial fasteners where there are unfilled crests at the start of rolled threads or a chamfer at the start of a thread, not exceeding two pitches in length, this is traditionally included in the specified thread length.

NOTE: When designing threaded products, it is necessary to take cognizance of: (1) such permissible length of chamfer and (2) the first threads which by virtue of gaging practice may exceed the product limits and which may be included within the length of complete thread. However, when the application is such as to require a minimum or maximum number, or length, of complete threads, the specification shall so state. Similar specification is required for a definite length of engagement.

Length of Thread Engagement. The length of thread engagement of two mating threads is the axial distance over which two mating threads are designed to contact.

Depth of Thread Engagement. The depth (or height) of thread engagement between two coaxially assembled mating threads is the radial distance by which their thread forms overlap each other.

Unified threads by class

Unified threads have five fit classifications. The classes are intended to reflect the amount of tolerance specified. It should be remembered that these classes are entirely based on fit tolerances, and they should never be used as a determination of the grade or strength of an individual fastener. In all classes, the letter A applies to external threads, and the letter B applies to internal threads.

Class 1A and *1B* are very rarely used, although they are retained in the standards for Unified threads. Fasteners conforming to Class 1 are intended for conditions of low stress. Their loose fit is intended to allow for ease in assembly, even when the threads are slightly damaged or dirty. Only Class 1 external threads have an allowance.

Class 2A and *2B* are used for most applications. The specified allowance is intended to minimize galling and seizing in high cycle assembly, and to accommodate electroplated finishes. The 2A maximum diameters apply to an unplated part or to a part prior to plating, whereas the basic diameters (the 2A maximum diameter plus allowance) apply to a part

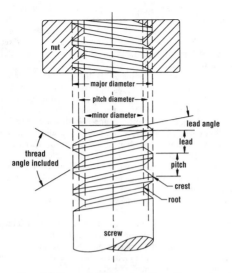

Figure 1. Basic screw thread dimensions. (Source, Kennametal, Inc.)

after plating. The minimum diameters of class 2B threads, whether or not plated or coated, are basic, affording no allowance or clearance in assembly at maximum material limits. This class is widely used for non-critical assembly with medium stress levels. Class 2A thread major diameter tolerances are equal to approximately twice their pitch diameter tolerances. The use of Class 2 threads provides for a small clearance between maximum material parts, except when the external thread is plated. Plated parts should be gaged with basic size GO gages.

Class 3A and *3B* are used where a close tolerance fit is important. Most aerospace fasteners are manufactured to Class 3 specifications. These threads require high quality assembly and inspection equipment. The maximum diameters of Class 3A threads and the minimum diameters of class 3B threads, whether or not plated or coated, are basic. This means that no allowance or clearance exists for assembly of maximum material components. However, since the tolerances on GO gages are within the limits of size of the product, the gages will assure a slight clearance when the fastener is made to maximum material limits. Note that thread formulas are generally based on specifications for Class 3 threads. For Class 1 and Class 2, the allowance stated in the tables must be subtracted.

Class 4 was designed for American National threads, and is no longer in use. Class 4 external threads have a negative allowance on the pitch diameter only.

Class 5 is an interference fit, designed for permanent installations. *Figure 2* illustrates the Class 5 thread at maximum and minimum material condition. Interference threads are intended to develop sufficient breakloose torques to prevent loosening of externally threaded fasteners. This is accomplished by precisely sizing the external thread larger than the internal thread so that, when torqued, either elastic compression or plastic movement of material causes both the internal and external components to become identical in

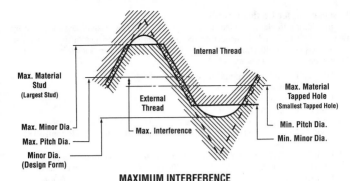

MAXIMUM INTERFERENCE

MINIMUM INTERFERENCE

Figure 2. Class 5 thread form. (Source, ANSI B1.12-1972 as published by the American Society of Mechanical Engineers.)

size. Class 5 fits are most often used for installing studs in hard materials or in similar situations where the mating of the components is intended to be permanent. Dimensions for Class 5 interference fit threads are given in **Table 10.** The subclass categories for Class 5 threads are as follows.

NC-5 HF   Ferrous material external threads for driving in hard ferrous material with hardness greater than 160 HB. The intended length of engagement should equal 1 $\frac{1}{4}$ times the diameter.

NC-5 CSF   Ferrous material external threads for driving in copper alloy and soft ferrous material with hardness of 160 HB or less. The intended length of engagement should equal 1 $\frac{1}{4}$ times the diameter.

NC-5 ONF   Ferrous material external threads for driving in non-ferrous materials other than copper alloys. The intended length of engagement should equal 2 $\frac{1}{2}$ times the diameter.

NC-5 IF   All ferrous material internal threads.

NC-5 INF   All non-ferrous material internal threads.

Designations for Unified threads

In addition to class, Unified threads are segmented by design, and identified by a two to four letter symbol intended to recognize the characteristics of the thread. Basic dimensions and specifications for UN and UNR thread forms are shown in *Figure 3*. The basic symbol designations are as follows.

UN	Unified constant-pitch thread.*
UNC	Unified coarse thread.*
UNF	Unified fine thread.*
UNEF	Unified extra-fine thread.*
UNS	Unified threads of special diameters, pitches, or lengths of engagement.*
UNR	Unified constant-pitch thread with a $0.108P$ to $0.144P$ controlled root radius.*
UNRC	Unified coarse thread with a $0.108P$ to $0.144P$ controlled root radius.*
UNRF	Unified fine thread with a $0.108P$ to $0.144P$ controlled root radius.*
UNREF	Unified extra-fine thread with a $0.108P$ to $0.144P$ controlled root radius.*
UNJ	Unified constant-pitch thread with a $0.15011P$ to a $0.18042P$ controlled root radius.**
UNJC	Unified coarse thread with a $0.15011P$ to a $0.18042P$ controlled root radius.**
UNJF	Unified fine thread with a $0.15011P$ to a $0.18042P$ controlled root radius.**
UNJEF	Unified extra-fine thread with a $0.15011P$ to a $0.18042P$ controlled root radius.**
UNM	Unified miniature thread.***

*For complete specifications, consult ANSI B1.1 Standard.

**For complete specifications, consult ANSI B1.15 Standard.

***For complete specifications, consult ANSI B1.10 Standard.

While UN and UNR threads are essentially identical, the root contour specified for external UN threads may be either flat (equal to $0.125P$) or rounded. For UN external threads, the nominal minor diameter is measured from the height of the flat root of the thread. The root contour (radius) for external UNR threads (both fine and course) is specified as continuous, and at a distance of not less than $0.625H$ below the basic major diameter. Radiused root contours (as opposed to flat) improve strength and fatigue life of the fastener. The majority of available socket headed cap screws are manufactured to UNR specifications. UNS and UNRS threads are intended for use only when standard series threads fail to conform to design standards.

The coarse thread forms, UNC and UNRC, are adaptable to rapid assembly and to conditions where corrosion or thread damage is likely to be encountered. Coarse threads are commonly specified in situations where the female thread is weaker than the male thread (for example, when threading steel bolts into aluminum castings). The coarse thread allows for the female thread to have maximum thread area and minimal minor diameter.

Fine thread forms, UNF and UNRF, provide more tensile strength than comparably sized coarse threads. Fine threads are specified where strength and fatigue resistance are important considerations, and in situations where the length of engagement is short. UNEF and UNREF forms enhance the positive characteristics of the basic fine thread, but require greater precision and increased time in assembly.

The limits of size for external and internal UN, UNF, UNC, UNS, and UNEF thread forms are provided in **Table 5** (external threads) and **Table 6** (internal threads). UNR form thread dimensions are identical to UN dimensions, with the exception of the minor

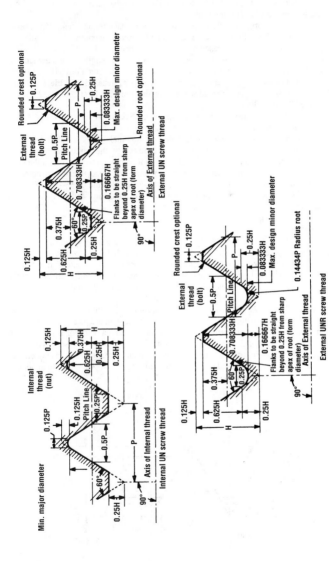

Figure 3. UN and UNR thread forms.

diameter of UNR external threads. To determine the minor diameter for an external UNR thread, use the constant for d_1 from **Table 2**.

UNJ threads were originally developed to meet aerospace requirements. The radius, larger than specified for UNR threads, was designed to provide maximum fatigue strength resulting primarily from their increased minor diameter. Bolts made to UNJ specifications normally have tensile strength of 160,000 psi or more. The length of engagement for UNJC, UNJF, and 8UNJ threads is equal to the basic major diameter. For UNJEF, 12UNJ, and 16UNJ threads, the length of engagement is equal to 9 pitches. See **Table 7** (external threads) and **Table 8** (internal threads) for specifications for UNJ threads. *Figure 4* illustrates the UNJ thread form.

UNM threads are discussed later in this section. See **Table 11** and **Table 12** for formulas and dimensions.

When designating a specific Unified fastener, the nominal size is stated first, usually as a fraction but in some cases as a decimal, and followed by the number of threads per inch, the series symbol, and the class symbol. If a number in parenthesis follows the class symbol, it is a reference to the gaging system number. A left-hand thread is denoted by "LH" at the end. An example of a Unified thread designation would be $^1/_2$–16 UN-2A. A designation of the same thread including a left-hand thread indicator and a gaging system number would be $^1/_2$–16 UN-2A-LH(21).

Gaging systems

UN, UNR, UNJ, M, and MJ are inspected by one of three gaging systems, depending on the engineering requirement of the threaded product. Selection of the appropriate inspection method is based on the features of most importance for the intended use of the thread. Consideration is given to such features as form, fit, function, and fabrication of the threaded product. Guidance for selection of an inspection method from the three standard gaging systems is given in FED-STD-H28/20B as follows.

System 21. System 21 provides for interchangeable assembly with functional size control at the maximum material limits within the length of standard gaging elements, and also control of characteristics identified as NOT GO functional diameters or as HI (internal) and LO (external) functional diameters. These functional gages provide some control at the minimum material limit when there is little variation in thread form characteristics such as lead, flank angle, taper, and roundness. System 21 is appropriate for use under one or more of the following conditions.

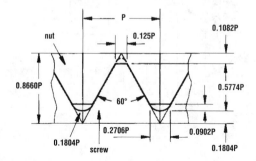

Figure 4. UNJ thread form. (Source, Kennametal, Inc.)

A) Where the threads of the product do not need specific mechanical strength properties, or where mechanical strength requirements are not specific for the product threads by either material strength and dimensional limits or by testing strength of the threads.

B) The threaded product has the mechanical properties specified, mechanical property testing of the threads is required, and the testing requires that the screw threads be subject to shear and beam loading by matching threads. If the threads have a locking element incorporated, locking torque values and tests must be specified and run on matching threads inspected by either System 22 or System 23.

C) For standard off-the-shelf general application fasteners when considered acceptable.

D) Internal thread is less than 0.190" (5 mm) nominal size.

System 22. System 22 provides for interchangeable assembly with functional size control at the maximum material limits within the length of standard gaging elements, and also control of the minimum material size limits over the length of the full thread. Other thread characteristics such as lead, flank angle, taper, and roundness variations are confined within these limits with no specific control of their magnitude. For UNJ and MJ external threads, control is also provided for the thread root radius and rounded root minor diameter. System 22 is recommended when none of the qualifying conditions for using either System 21 or System 23 is applicable.

System 23. System 23 provides for interchangeable assembly with functional size control at the maximum material limits within the length of standard gaging elements, and also control of the minimum material size limits over the length of the full thread. The magnitude of other thread characteristics such as lead, flank angle, taper, and roundness are further controlled within these limits. For UNJ and MJ external threads, control is also provided for the thread root radius and rounded root minor diameter. System 23 is recommended for use under one or more of the following conditions.

A) When thread element control is required to determine the extent of deviation in any of the elements of the thread, normally special applications.

B) For threaded product used in research investigations and testing to determine the effect that a specific thread element variation has on the attributes of the threaded product or the attributes of a threaded product's application.

C) The conductive of investigation and testing in analysis of thread failures.

Basic Unified thread dimensions

The basic pitch diameter (internal thread, specified as D_2 minimum) and the minor diameters of both external and internal Unified threads are all directly related to the basic major diameter of the internal thread (D). The basic major diameter is the exact equivalent of the decimal size of the fastener. For example, a $^3/_8$" nut has a basic major diameter of exactly 0.3750". These dimensions are constant for all Unified threads of the same pitch. For instance, from **Table 2** it can be seen that for all Unified threads, regardless of diameter, with 8 tpi, the basic (minimum) pitch diameter (D_2 min.) of the internal thread is equal to the basic major diameter (D) minus 0.0812". Also on the table, it can be seen that the minor diameter of an 8 tpi Class 3A external thread (d_1 basic) is equal to the basic major diameter minus 0.1488". Finally, from the same table, the minor diameter of an 8 tpi internal thread (D_1 min) is found to be equal to the basic major diameter minus 0.1353".

Theoretically, a 1 tpi internal Unified thread would have a minimum pitch diameter equal to the basic major diameter (D) minus 0.6496". The following formula, based on the constant 0.6496", can be used to find the minimum internal pitch diameter (D_2 min.) for any unified thread. Also, it should be noted that D_2 min. is *always equal* to the maximum pitch diameter for external (d_2 max.) Class 3A threads. For Class 1 and 2 threads, subtract the allowance for the specific thread as given in Table 5.

$$D_2 \text{ min.} = D - (0.6496 \div \text{tpi})$$

Similarly, the minor diameter (d_1) of a 1 tpi external Unified thread would be equal to the basic internal major diameter (D) minus 1.1908". Therefore, the following formula can be used to find d_1 for Class 3A threads. For other classes, subtract the allowance. It should be noted that this formula provides d_1 for UNR rounded radius threads (preferred).

$$d_1 = D - (1.1908 \div \text{tpi})$$

Finally, D_1, the internal thread minor diameter, for a 1 tpi nut would be equal to D minus 1.0824". Using this number, D_1 for any unified thread can be computed with the following formula.

$$D_1 = D - (1.0824 \div \text{tpi})$$

Table 2 can be used for quickly determining D_2 min., d_1, and D_1. To use the table, find the tpi in the left column, then subtract the number in any of the other columns from the basic internal major diameter. For example, to find D_1 for a 1 $^1/_{16}$–12 thread, subtract

Table 2. Constants for Determining Basic Dimensions for Unified Threads. (Subtract the constant from the major diameter of the internal thread.)

TPI	D_2 min.	d_1	D_1
4	0.1624	0.2977	0.2706
4.5	0.1444	0.2646	0.2405
5	0.1299	0.2382	0.2165
6	0.1083	0.1985	0.1804
7	0.0928	0.1701	0.1546
8	0.0812	0.1488	0.1353
9	0.0722	0.1323	0.1203
10	0.0650	0.1191	0.1082
11	0.0591	0.1082	0.0984
12	0.0541	0.0992	0.0902
13	0.0500	0.0916	0.0833
14	0.0464	0.0850	0.0773
16	0.0406	0.0744	0.0676
18	0.0361	0.0662	0.0601
20	0.0325	0.0595	0.0541
24	0.0271	0.0496	0.0451
27	0.0241	0.0441	0.0401
28	0.0232	0.0425	0.0386
32	0.0203	0.0372	0.0338
36	0.0180	0.0331	0.0301
40	0.0162	0.0298	0.0271
44	0.0148	0.0271	0.0246
48	0.0135	0.0248	0.0226

0.0902 from the major internal diameter (1.0625 − 0.0902 = 0.9723"). If the desired tpi is not on the table, use the universal formulas above.

Formulas for computing other dimensions are found in **Table 4**.

Unified threads by series

Unified threads are available in sizes that reflect the physical specifications of the individual thread forms. Larger diameters, naturally, tend to have fewer threads per inch, while smaller diameters have more. The "constant pitch" thread series, 4-UN/R, 6-UN/R, 8-UN/R, 8-UNJ, 12-UN/R, 12-UNJ, 16-UN/R, 16-UNJ, 20-UN/R, 28-UN/R, and 32-UN/R are commonly referred to as 8-thread series, 12-thread series, etc. The 8-, 12-, and 16-thread series are the preferred constant pitch sizes. The UN/UNR 8-series was designed for securing high-temperature, high-pressure joints, and is commonly used instead of UNC threads for applications where sizes in excess of one-inch are required. The UN/UNR 12-thread series was designed for use on boilers, but is commonly used in place of UNF threads for applications where diameters in excess of 1.5" are required. The UN/UNR 16-thread series is applied where fine-pitch fasteners 1 3/4" or larger are required. Available diameters, broken down by form and tpi, are shown in **Table 3**.

Table 3. Size/Pitch Combinations for Unified Threads. (See Note.)

Size*	Basic Dia.	4 UN	6 UN	8 UN/UNJ	12 UN/UNJ	16 UN/UNJ	20 UN	28 UN	32 UN	UNC/UNJC	UNF/ENJF	UNEF/UNJEF	UNS	NC-5**
0	0.0600										80			
1	0.0730									64	72			
2	0.0860									56	64			
3	0.0990									48	56			
4	0.1120									40	48			
5	0.1250									40	44			
6	0.1380								C	32	40			
8	0.1640								C	32	36			
10	0.1900								F	24	32		28, 36, 40, 48. 56	
12	0.1260							F	EF	24	28	32	36, 40,48, 56	
1/4	0.2500						C	F	EF	20	28	32	24, 27, 36, 40, 48, 56	20
5/16	0.3125						■	■	EF	18	24	32	27, 36, 40, 48	18
3/8	0.3750					C	■	■	EF	16	24	32	18, 27, 36, 40	16
7/16	0.4375					■	F	EF		14	20	28	18, 24, 27	14
1/2	0.5000					■	F	EF		13	20	28	12, 14, 18, 24, 27	13
9/16	0.5625				C	■	■	■		12	18	24	14, 27	12
5/8	0.6250				■	■	■	■		11	18	24	14, 27	11
11/16	0.6875				■	■	■	■				24		
3/4	0.7500				■	F	EF	■		10	16	20	14, 18, 24, 27	10
13/16	0.8125				■	■	EF	■				20		
7/8	0.8750				■	■	EF	■		9	14	20	10, 18, 24, 27	9
15/16	0.9375				■	■	EF	■				20		

(Continued)

Table 3. *(Continued)* **Size/Pitch Combinations for Unified Threads.** (See Note.)

Size*	Basic Dia.	4 UN	6 UN	8 UN/UNJ	12 UN/UNJ	16 UN/UNJ	20 UN	28 UN	32 UN	UNC/UNJC	UNF/ENJF	UNEF/UNJEF	UNS	NC-5**
1	1.0000			C	F	■	EF	■	■	8	12	20	10, 14, 18, 24, 27	8
1 1/16	1.0625			■	■	■	■					18		
1 1/8	1.1250			■	F	■	■			7	12	18	10, 14, 24	7
1 3/16	1.1875			■	■	■	■					18		
1 1/4	1.2500			■	F	■	■			7	12	18	10, 14, 24	7
1 5/16	1.3125			■	■	■	■					18		
1 3/8	1.3750		C	■	F	■	■			6	12	18	10, 14	6
1 7/16	1.4375		■	■	■	■	■					18		
1 1/2	1.5000		C	■	F	■	■			6	12	18	10, 14, 24	6
1 9/16	1.5625		■	■	■	■	■					18		
1 5/8	1.6250		■	■	■	■	■					18	10, 14, 24	
1 11/16	1.6875		■	■	■	■	■					18		
1 3/4	1.7500		■	■	■	■	■			5			10, 14, 18	
1 13/16	1.8125		■	■	■	■	■							
1 7/8	1.8750		■	■	■	■	■						10, 14, 18	
1 15/16	1.9375		■	■	■	■	■							
2	2.0000		■	■	■	■	■			4.5			10, 14, 18	
2 1/16	2.0625												16	
2 1/8	2.1250		■	■	■	■	■							
2 3/16	2.1875												16	
2 1/4	2.2500		■	■	■	■	■			4.5			14, 18	
2 5/16	2.3125												16	
2 3/8	2.3750		■	■	■	■	■							
2 7/16	2.4375												16	
2 1/2	2.5000	C	■	■	■	■				4			10, 14, 18	
2 5/8	2.6250	■	■	■	■	■								
2 3/4	2.7500	C	■	■	■	■				4			10, 14, 18	
2 7/8	2.8750	■	■	■	■	■								
3	3.0000	C	■	■	■	■				4			10, 14, 18	
3 1/8	3.1250	■	■	■	■	■								
3 1/4	3.2500	C	■	■	■	■				4			10, 14, 18	
3 3/8	3.3750	■	■	■	■	■								
3 1/2	3.5000	C	■	■	■	■				4			10, 14, 18	
3 5/8	3.6250	■	■	■	■	■								
3 3/4	3.7500	C	■	■	■	■				4			10, 14, 18	
3 7/8	3.8750	■	■	■	■	■								
4	4.0000	C	■	■	■	■				4			10, 14	

(Continued)

Table 3. *(Continued)* **Size/Pitch Combinations for Unified Threads.** *(See Note.)*

Size*	Basic Dia.	4 UN	6 UN	8 UN/UNJ	12 UN/UNJ	16 UN/UNJ	20 UN	28 UN	32 UN	UNC/UNJC	UNF/ENJF	UNEF/UNJEF	UNS	NC-5**
4 1/8	4.1250	■	■	■	■	■								
4 1/4	4.2500	■	■	■	■	■							10, 14	
4 3/8	4.3750	■	■	■	■	■								
4 1/2	4.5000	■	■	■	■	■							10, 14	
4 5/8	4.6250	■	■	■	■	■								
4 3/4	4.7500	■	■	■	■	■							10, 14	
4 7/8	4.8750	■	■	■	■	■								
5	5.0000	■	■	■	■	■							10, 14	
5 1/8	5.1250	■	■	■	■	■								
5 1/4	5.2500	■	■	■	■	■							10, 14	
5 3/8	5.3750	■	■	■	■	■								
5 1/2	5.5000	■	■	■	■	■							10, 14	
5 5/8	5.6250	■	■	■	■	■								
5 3/4	5.7500	■	■	■	■	■							10, 14	
5 7/8	5.8750	■	■	■	■	■								
6	6.0000	■	■	■	■	■							10, 14	

*Primary sizes are in normal type. Secondary sizes are in italics.
Note: A "C" in a box means that the constant pitch thread of that diameter is coarse. "F" designates fine, and "EF" extra fine. Dark boxes indicate availability. A blank box means that the thread combination is not available. A number in a box indicates the tpi of the thread of that diameter.

Formulas for 60° threads

Because of the similarities of the ISO Metric thread and the Unified thread, their dimensions can be determined with shared equations. Traditionally, metric thread equations have been stated as a fractional expression of the full height of the thread (the fundamental triangle) and Unified equations have been expressed as a decimal percentage of the pitch. Where applicable, both forms are given in **Table 4**.

Table 4. Basic Formulas for Calculating Unified and ISO 68 (1973) Thread Dimensions. *(See Notes Below.)*

	Symbol	Dimension	Equation	Remarks
1.	P	Pitch (number of threads per unit of length).	$P = 1 \div$ threads per inch, or $P = 1 \div$ threads per mm	To convert pitch in inches to metric pitch, multiply by 25.4. To convert metric pitch to pitch in inches, multiply by 0.03937.
2.	H	Height of fundamental triangle (sharp V-thread).	$H = 0.866025404P$ or $H = P \times \cos 30°$	The thread height if the crest and root were sharp, rather than having a slight flat.
3.	D	Major dia., int. thread.		Also called Basic Major Diameter. It is equal to the basic thread designation.

(Continued)

Table 4. *(Continued)* **Basic Formulas for Calculating Unified and ISO 68 (1973) Thread Dimensions.**
(See Notes Below.)

	Symbol	Dimension	Equation	Remarks
4.	d (max)	Major dia. (max.), ext. thread.	See Table 5	For Class 3A, equal to the basic thread designation. For Class 1A and 2A, equal to the basic thread dimension minus the allowance.
5.	d (min)	Major dia. (min.), ext. thread.	$d\,(\min) = d\,(\max) - 0.06 \times \sqrt[3]{P^2}$	This equation will not provide 100% accurate results for all thread classes. See Table 5 for exact dimensions.
6.	D_1 (min)	Minor dia. (min.), int. thread.	$D_1\,(\min) = D - 1.082532P$	The major diameter minus the depth of thread. Corresponds to recommended tap drill size. Also see Table 2.
7.	d_1 (max)	Minor dia. (max.), ext. thread.	$d_1\,(\max) = d - 1.226869P$	See Note [1] below.
8.	D_1 (max)	Minor dia. (max.), int. thread.	$D_1\,(\max) = D_1\,(\min) + [0.05 \times \sqrt[3]{P^2} + 0.03P \div D] - 0.002$	See Note [2] below.
9.	d_1 (min)	Minor dia. (min.), ext. thread.	$d_1\,(\min) = d_2\,(\min) - 0.56580P$	See Note [1] below.
10.	d_1 UNR	Minor dia., ext. UNR thread.	See Table 2	See Note [1] below.
11.	D_1 UNJ	Minor dia., int. UNJ and MJ thread.	$D_1 = D - 0.974279P$	Controlled root radius thread.
12.	D_2 (min)	Pitch dia. (min.), int. thread.	$D_2\,(\min) = D - 0.649519P$	Often estimated as $D - 2 \times h_{as}$. Also see Table 2.
13.	d_2 (max)	Pitch dia. (max.), ext. thread.	$d_2\,(\max) = d\,(\max) - 0.649519P$	Often estimated as $d - 2 \times h_{dn}$.
14.	D_2 (max)	Pitch dia. (max.), int. thread.	See Tables	
15.	d_2 (min)	Pitch dia. (min.), ext. thread.	See Tables	
16.	$h_{as}*$ $h_{dn}*$	Addendum, ext. thread. Dedendum, ext. and int. thread. UN, UNR (max.).	h_{as} or $h_{dn} = 3/8 \times H$, or h_{as} or $h_{dn} = 0.3247595P$	Twice $h_{as} = 3/4 \times H$, or $0.649519P$. One-half $h_{as} = 3H \div 16$, or $0.16238P$.
17.	h_{ds}	Dedendum, ext. thread. UNR (min.).	$h_{ds} = H/3$, or $h_{ds} = 0.288867P$	UNR only.
18.	h_{ds}	Dedendum, ext. thread. UNJ (max.).	$h_{ds} = 0.28290P$	UNJ only.
19.	h_{ds}	Dedendum, ext. thread. UNJ (min.).	$h_{ds} = 7/24 \times H$, or $h_{ds} = 0.252588P$	UNJ only.
20.	$h_{an}*$ $f_{cn}*$ $f_{rs}*$	Addendum, int. thread. Truncation, int. thread crest. Truncation, ext. thread root.	h_{an}, f_{cn}, or $f_{rs} = H/4$, or h_{an}, f_{cn}, or $f_{rs} = 0.216506351P$	f_{cn} is the radial distance from the apex of the fundamental triangle to the flat at the crest of the internal thread. Does not apply to UNJ and MJ threads. f_{rs} is the radial distance from the apex of the fundamental triangle to the flat at the root of the external thread.

(Continued)

Table 4. *(Continued)* **Basic Formulas for Calculating Unified and ISO 68 (1973) Thread Dimensions.**
(See Notes Below.)

	Symbol	Dimension	Equation	Remarks
21.	h_n* h_e* h_s* UN	Height of int. thread. Depth of thread engagement. Height of ext. UN and ISO metric thread.	h_n or $h_e = 5/8 \times H$, or h_n or $h_e = 0.541265877P$	Does not apply to UNJ and MJ threads. See separate listing. Twice $h_n = 5/4 \times H$, or $1.082532P$.
22.	h_s* UNR	Height of ext. UNR thread.	$h_s = 0.59539P$	UNR only.
23.	h_n* UNJ h_e* UNJ	Height of int. UNJ thread. Depth of UNJ thread engagement.	h_n or $h_e = 9H \div 16$, or h_n or $h_e = 0.48714P$	UNJ and MJ only.
24.	h_s* UNJ	Height of ext. UNJ thread.	$h_s = 2H \div 3$, or $h_s = 0.57735P$	UNJ and MJ only.
25.	r (max)	Root radius (max.), ext. thread.	$r = H \div 6$, or $r = 0.14434P$	Does not apply to UNJ and MJ threads.
26.	r (min)	Root radius/flat (min.), ext. thread.	$r = 0.125P$	Does not apply to UNR, UNJ and MJ threads.
27.	r UNR	Root radius (min.), ext. UNR thread.	$r = H \div 8$, or $r = 0.10825P$	UNR only.
28.	r UNR	Root radius (max.), ext. UNR thread.	$r = H \div 6$, or $r = 0.14434P$	UNR only.
29.	r UNJ	Root radius (min.), ext UNJ and MJ only.	$r = 0.15011P$	UNJ and MJ only.
30.	r UNJ	Root radius (max.), UNJ and MJ only.	$r = 5H \div 24$, or $r = 0.18042P$	Also height from sharp "V" to ext. thread root. UNJ and MJ only.
31.	f_{cn}* UNJ	Truncation, int. thread crest, UNJ thread.	$f_{cn} = 5H \div 16$, or $f_{cn} = 0.27063P$	UNJ and MJ only.
32.	f_{cs}* f_{rn}*	Truncation, ext. thread crest. Truncation, int. thread root.	f_{cs} or $f_{rn} = H \div 8$, or f_{cs} or $f_{rn} = 0.10825$	f_{cs} is the radial distance from the apex of the fundamental triangle to flat at the crest of the external thread. f_{rn} is the radial distance from the apex of the fundamental triangle to the flat at the root of the internal thread. Does not apply to UNR.
33.	f_{rn}* UNR	Truncation, int. thread root, UNR.	$f_{rn} = 0.16238P$	UNR only.
34.	F_{cs}* F_{rn}*	Flat, crest of ext. thread. Flat, root of int. thread.	F_{cs} or $F_{rn} = P \div 8$, or F_{cs}, or $F_{rn} = 0.125P$	
35.	F_{cn}*	Flat, int. thread crest.	$F_{cn} = 0.25P$	Does not apply to UNJ and MJ threads.
36.	F_{cn}* UNJ	Flat, int. thread crest.	$F_{cn} = 5P \div 16$, or $F_{cn} = 0.3125P$	UNJ and MJ only.
37.	NA	See remarks.	$11/12 \times H$, or $0.79386P$	Use to calculate the difference between D (max.) and D_2 (max.).

(Continued)

Table 4. *(Continued)* **Basic Formulas for Calculating Unified and ISO 68 (1973) Thread Dimensions.**
(See Notes Below.)

	Symbol	Dimension	Equation	Remarks
38.	NA	See remarks.	$0.711325H$, or $0.616025P$	Use to calculate the difference between the pitch dia. (min.) and minor dia. (min.) of external thread for $0.125P$ root radius.
39.	NA	See remarks. For UNJ and MJ only.	$0.6533H$, or $0.56580P$	Use to calculate the difference between the pitch dia. (min.) and minor dia. (min.) of external UNJ and MJ thread.
40.	NA	See remarks.	$H/2$, or $0.433013P$	Use to calculate the difference between the pitch dia. (max.) and minor dia. (max.) of external thread, and between the pitch dia. (min.) and minor dia. (min.) of internal thread.
41.	L *	Lead.	$1 \div 1/P$	$1/P$ is the number of turns per unit of length (threads per inch).
42.	λ *	Lead angle.	$\tan \lambda = L \div (\pi \times \text{pitch dia.})$	
43.	ψ *	Helix angle.	$\cot \psi = L \div (\pi \times \text{pitch dia.})$	Sometimes called the rake angle.
44.	NA	Area of minor dia., ext. thread.	$0.7854 \times d_1{}^2$	$\pi \div 4 = 0.7854$. Expressed in in.^{2} or mm^2.
45.	A_s	Tensile stress area.	$A_s = \pi \times ([d_2 \div 2] - [3H \div 16])^2$, or $A_s = 0.7854 \times (d - [0.9743 \div 1/P])^2$	$\pi \div 4 = 0.7854$. $1/P$ = number of threads per inch. Expressed in in.^{2} or mm^2.
46.	LE	Length of thread engagement.	$LE = 4\, A_s \div (\pi \times d_2\ {}_{BASIC})$	Based on combined shear failure of external and internal threads. See Note [3] below.
47.	AS_n	Shear area, internal threads.	$AS_n = \pi \times D_2\ {}_{BASIC} \times (3LE \div 4)$	For use only when d is greater than 0.250 inch. See Note [4] below.
48.	AS_s	Shear area, external threads.	$AS_s = \pi \times d_2\ {}_{BASIC} \times (5/8 \times LE)$	See Note [4] below.

Unless otherwise noted, formulas are for basic UN/ISO 68 thread form. Dimensions derived from these formulas for ISO 68 and M Profile threads do not compensate for tolerance grade variations.

* Symbol is traditional.

Note [1]: This specification has been stated in various ways in ANSI/ASME and Federal Standards. In FED-STD-H28, a single specification is provided for d_1 that is approximately the average of d_1 (max) and d_1 (min). These figures are given in the tables in this Handbook. In ANSI B1.1-1989, the specification given is stated as being for "UNR Minor Diameter (max)" which can be obtained by using the constant for d_1 that is specified in Table 2.

Note [2]: For 13 threads per inch or finer, this tolerance will not exceed $0.259809P$, nor shall it be less than $0.135315P$. For coarser threads, the tolerance will be equal to $0.120P$. This equation will not return values equal to the stated maximum and minimum figures provided in the tables, but they will be very close and will not exceed the values stated in the tables.

Note [3]: For a hollow part, subtract $0.7854\, d_h{}^2$ from A_s. d_h is the hole diameter on the externally threaded part.

Note [4]: Formulas given for AS_n and AS_s are simplified and are based on empirical data that vary from the minimum material shear areas.

Unified miniature threads

UNM threads are intended for use as general purpose fastening screws in watches, instruments, and miniature mechanisms. The diameter of UNM threads ranges from 0.30 mm to 1.40 mm (0.0118 to 0.0551 in.). Therefore, this thread form supplements the standard Unified series that begins at 0.060 in. The fourteen sizes are systematically distributed to provide a uniformly proportioned selection throughout the range. Primary sizes are recommended, but for restrictive conditions, secondary sizes may be specified. Since the standard is specified in metric units and then converted to inch units, the metric specifications take preference. Formulas for UNM threads are found in **Table 11**, and specifications are provided in **Table 12**. The thread form is illustrated in *Figure 5*. It should be noted that the maximum material limits given in Table 12 apply to both uncoated and coated threads. It is essential that measurable coatings remain with the stated dimensions since no allowance is provided between the limits of the internal and external threads. *(Text continued on p. 1464)*

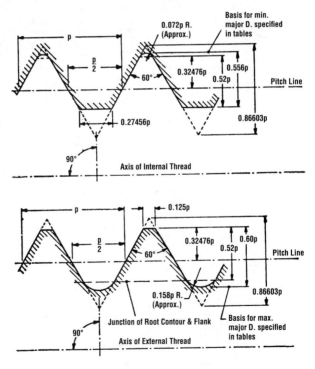

Figure 5. UN miniature thread form.

Table 5. Unified Screw Threads. External Dimensions. Classes 1A, 2A, and 3A.

Size, TPI, and Series	Class	Allowance	Major Diameter		Pitch Diameter		Minor Dia.
			max	min	max	min	
0-80 UNF	2A	.0005	.0595	.0563	.0514	.0496	.0442
	3A	-	.0600	.0568	.0519	.0506	.0447
1-64 UNC	2A	.0006	.0724	.0686	.0623	.0603	.0532
	3A	-	.0730	.0692	.0629	.0614	.0538
1-72 UNF	2A	.0006	.0724	.0689	.0634	.0615	.0554
	3A	-	.0730	.0695	.0640	.0626	.0560
2-56 UNC	2A	.0006	.0854	.0813	.0738	.0717	.0635
	3A	-	.0860	.0819	.0744	.0728	.0641
2-64 UNF	2A	.0006	.0854	.0816	.0753	.0733	.0662
	3A	-	.0860	.0822	.0759	.0744	.0668
3-48 UNC	2A	.0007	.0983	.0938	.0848	.0825	.0727
	3A	-	.0990	.0945	.0855	.0838	.0734
3-56 UNF	2A	.0007	.0983	.0942	.0867	.0845	.0764
	3A	-	.0990	.0949	.0874	.0858	.0771
4-40 UNC	2A	.0008	.1112	.1061	.0950	.0925	.0805
	3A	-	.1120	.1069	.0958	.0939	.0813
4-48 UNF	2A	.0007	.1113	.1068	.0978	.0954	.0857
	3A	-	.1120	.1075	.0985	.0967	.0864
5-40 UNC	2A	.0008	.1242	.1191	.1080	.1054	.0935
	3A	-	.1250	.1199	.1088	.1069	.0943
5-44 UNC	2A	.0007	.1243	.1195	.1095	.1070	.0964
	3A	-	.1250	.1202	.1102	.1083	.0971
6-32 UNC	2A	.0008	.1372	.1312	.1169	.1141	.0989
	3A	-	.1380	.1320	.1177	.1156	.0997
6-40 UNF	2A	.0008	.1372	.1321	.1210	.1184	.1065
	3A	-	.1380	.1329	.1218	.1198	.1073
8-32 UNC	2A	.0009	.1631	.1571	.1428	.1399	.1248
	3A	-	.1640	.1640	.1437	.1415	.1257
8-36 UNF	2A	.0008	.1632	.1577	.1452	.1424	.1291
	3A	-	.1640	.1585	.1460	.1439	.1299
10-24 UNC	2A	.0010	.1890	.1818	.1619	.1586	.1379
	3A	-	.1900	.1828	.1629	.1604	.1389
10-28 UNS	2A	.0010	.1890	.1825	.1658	.1625	.1452
10-32 UNF	2A	.0009	.1891	.1831	.1688	.1658	.1508
	3A	-	.1900	.1840	.1697	.1674	.1517
10-36 UNS	2A	.0009	.1891	.1836	.1711	.1681	.1550
10-40 UNS	2A	.0009	.1891	.1840	.1729	.1700	.1584
10-48 UNS	2A	.0008	.1892	.1847	.1757	.1731	.1636
10-56 UNS	2A	.0007	.1893	.1852	.1777	.1752	.1674
12-24 UNC	2A	.0010	.2150	.2078	.1879	.1845	.1639
	3A	-	.2160	.2088	.1889	.1863	.1849
12-28 UNF	2A	.0010	.2150	.2085	.1918	.1886	.1712
	3A	-	.2160	.2095	.1928	.1904	.1722
12-32 UNEF	2A	.0009	.2151	.2091	.1948	.1917	.1768
	3A	-	.2560	.2100	.1957	.1933	.1777

(Continued)

Table 6. Unified Screw Threads. Internal Dimensions. Classes 1B, 2B, and 3B.

Size, TPI, and Series	Class	Minor Diameter		Pitch Diameter		Major Dia.	Tap Drill*
		min	max	min	max		
0-80 UNF	2B	.0465	.0514	.0519	.0542	.0600	56
	3B	.0465	.0514	.0519	.0536	.0600	
1-64 UNC	2B	.0561	.0623	.0629	.0655	.0730	54
	3B	.0561	.0623	.0629	.0648	.0730	
1-72 UNF	2B	.0580	.0635	.0640	.0665	.0730	1.5mm
	3B	.0580	.0635	.0640	.0659	.0730	
2-56 UNC	2B	.0667	.0737	.0744	.0772	.0860	51
	3B	.0667	.0737	.0744	.0756	.0860	
2-64 UNF	2B	.0691	.0753	.0759	.0786	.0860	50
	3B	.0691	.0753	.0759	.0779	.0860	
3-48 UNC	2B	.0764	.0845	.0855	.0885	.0990	48
	3B	.0764	.0845	.0855	.0877	.0990	
3-56 UNF	2B	.0797	.0865	.0874	.0902	.0990	46
	3B	.0797	.0865	.0874	.0895	.0990	
4-40 UNC	2B	.0849	.0939	.0958	.0991	.1120	44
	3B	.0849	.0939	.0958	.0982	.1120	
4-48 UNF	2B	.0894	.0968	.0985	.1016	.1120	2.3mm
	3B	.0894	.0968	.0985	.1008	.1120	
5-40 UNC	2B	.0979	.1062	.1088	.1121	.1250	39
	3B	.0979	.1062	.1088	.1113	.1250	
5-44 UNC	2B	.1004	.1079	.1102	.1134	.1250	38
	3B	.1004	.1079	.1102	.1126	.1250	
6-32 UNC	2B	.104	.114	.1177	.1214	.1380	36
	3B	.1040	.1140	.1177	.1204	.1380	
6-40 UNF	2B	.111	.119	.1218	.1252	.1380	33
	3B	.1110	.1186	.1218	.1243	.1380	
8-32 UNC	2B	.130	.139	.1437	.1475	.1640	3.4mm
	3B	.1300	.1389	.1437	.1465	.1640	
8-36 UNF	2B	.134	.142	.1460	.1496	.1640	29
	3B	.1340	1416	.1460	.1487	.1640	
10-24 UNC	2B	.145	.156	.1629	.1672	.1900	3.7mm
	3B	.1450	.1555	.1629	.1661	.1900	
10-28 UNS	2B	.151	.160	.1668	.1711	.1900	
10-32 UNF	2B	.156	.164	.1697	.1736	.1900	5/32
	3B	.1560	.1641	.1697	.1726	.1900	
10-36 UNS	2B	.160	.166	.1720	.1759	.1900	
10-40 UNS	2B	.163	.169	.1738	.1775	.1900	
10-48 UNS	2B	.167	.172	.1765	.1799	.1900	
10-56 UNS	2B	.171	.175	.1784	.1816	.1900	
12-24 UNC	2B	.171	.181	.1889	.1933	.2160	17
	3B	.1710	.1807	.1889	.1922	.2160	
12-28 UNF	2B	.177	.186	.1928	.1970	.2160	15
	3B	.1770	.1857	.1928	.1959	.2160	
12-32 UNEF	2B	.182	.190	.1957	.1998	.2160	14
	3B	.1820	.1895	.1957	.1988	.2160	

(Continued)

Table 5. *(Continued)* Unified Screw Threads. External Dimensions. Classes 1A, 2A, and 3A.

Size, TPI, and Series	Class	Allowance	Major Diameter		Pitch Diameter		Minor Dia.
			max	min	max	min	
12-36 UNS	2A	.0009	.2151	.2096	.1971	.1941	.1810
12-40 UNS	2A	.0009	.2151	.2100	.1989	.1960	.1844
12-48 UNS	2A	.0008	.2152	.2107	.2017	.1991	.1896
12-56 UNS	2A	.0007	.2153	.2112	.2037	.2012	.1934
1/4-20 UNC	1A	.0011	.2489	.2367	.2164	.2108	.1876
	2A	.0011	.2489	.2408	.2164	.2127	.1876
	3A	-	.2500	.2419	.2175	.2147	.1887
1/4-24 UNS	2A	.0011	.2489	.2417	.2218	.2181	.1978
1/4-27 UNS	2A	.0010	.2490	.2423	.2249	.2214	.2036
1/4-28 UNF	1A	.0010	.2490	.2392	.2258	.2208	.2052
	2A	.0010	.2490	.2425	.2258	.2225	.2052
	3A	-	.2500	.2435	.2268	.2243	.2062
1/4-32 UNEF	2A	.0010	.2490	.2430	.2287	.2255	.2107
	3A	-	.2500	.2440	.2440	.2273	.2117
1/4-36 UNS	2A	.0009	.2491	.2436	.2311	.2280	.2150
1/4-40 UNS	2A	.0009	.2491	.2440	.2329	.2300	.2184
1/4-48 UNS	2A	.0008	.2492	.2447	.2357	.2330	.2236
1/4-56 UNS	2A	.0008	.2492	.2451	.2376	.2350	.2273
5/16-18 UNC	1A	.0012	.3113	.2982	.2752	.2691	.2431
	2A	.0012	.3113	.3026	.2752	.2712	.2431
	3A	-	.3125	.3038	.2764	.2734	.2443
5/16-20 UN	2A	.0012	.3113	.3032	.2788	.2748	.2500
	3A	-	.3125	.3044	.2800	.2770	.2512
5/16-24 UNF	1A	.0011	.3114	.3006	.2843	.2788	.2603
	2A	.0011	.3114	.3042	.2843	.2806	.2603
	3A	-	.3125	.3053	.2854	.2827	.2614
5/16-27 UNS	2A	.0010	.3115	.3048	.2874	.2839	.2661
5/16-28 UN	2A	.0010	.3115	.3050	.2883	.2849	.2677
	3A	-	.3125	.3060	.2893	.2867	.2687
5/16-32 UNEF	2A	.0010	.3115	.3055	.2912	.2880	.2732
	3A	-	.3125	.3065	.2922	.2898	.2742
5/16-36 UNS	2A	.0009	.3116	.3061	.2936	.2905	.2775
5/16-40 UNS	2A	.0009	.3116	.3065	.2954	.2925	.2809
5/16-48 UNS	2A	.0008	.3117	.3072	.2982	.2955	.2861
3/8-16 UNC	1A	.0013	.3737	.3595	.3331	.3266	.2970
	2A	.0013	.3737	.3643	.3331	.3287	.2970
	3A	-	.3750	.3656	.3344	.3311	.2983
3/8-18 UNS	2A	.0013	.3737	.3650	.3376	.3333	.3055
3/8-20 UN	2A	.0012	.3738	.3657	.3413	.3372	.3125
	3A	-	.3750	.3669	.3425	.3394	.3137
3/8-24 UNF	1A	.0011	.3739	.3631	.3468	.3411	.3228
	2A	.0011	.3739	.3667	.3468	.3430	.3228
	3A	-	.3750	.3678	.3479	.3450	.3239
3/8-27 UNS	2A	.0011	.3739	.3672	.3479	.3450	.3239

(Continued)

Table 6. *(Continued)* Unified Screw Threads. Internal Dimensions. Classes 1B, 2B, and 3B.

Size, TPI, and Series	Class	Minor Diameter		Pitch Diameter		Major Dia.	Tap Drill*
		min	max	min	max		
12-36 UNS	2B	.186	.192	.1980	.2019	.2160	
12-40 UNS	2B	.189	.195	.1998	.2035	.2160	
12-48 UNS	2B	.193	.198	.2025	.2059	.2160	
12-56 UNS	2B	.197	.201	.2044	.2076	.2160	
1/4-20 UNC	1B	.196	.207	.2175	.2248	.2500	9
	2B	.196	.207	.2175	.2223	.2500	
	3B	.1960	.2067	.2175	.2211	.2500	
1/4-24 UNS	2B	.205	.215	.2229	.2277	.2500	
1/4-27 UNS	2B	.210	.219	.2259	.2304	.2500	
1/4-28 UNF	1B	.211	.220	.2268	.2333	.2500	5.4mm
	2B	.211	.220	.2268	.2311	.2500	
	3B	.2110	.2190	.2268	.2300	.2500	
1/4-32 UNEF	2B	.216	.224	.2297	.2339	.2500	7/32
	3B	.2160	.2229	.2297	.2328	.2500	
1/4-36 UNS	2B	.220	.226	.2320	.2360	.2500	2
1/4-40 UNS	2B	.223	.229	.2320	.2360	.2500	
1/4-48 UNS	2B	.227	.232	.2365	.2401	.2500	
1/4-56 UNS	2B	.231	.235	.2384	.2417	.2500	
5/16-18 UNC	1B	.252	.265	.2764	.2843	.3125	F
	2B	.252	.265	.2764	.2817	.3125	
	3B	.2520	.2630	.2764	.2803	.3125	
5/16-20 UN	2B	.258	.270	.2800	.2852	.3125	G
	3B	.2580	.2662	.2800	.2839	.3125	
5/16-24 UNF	1B	.267	.277	.2854	.2925	.3125	H
	2B	.267	.277	.2854	.2902	.3125	
	3B	.2670	.2754	.2854	.2890	.3125	
5/16-27 UNS	2B	.272	.281	.2884	.2929	.3125	
5/16-28 UN	2B	.274	.282	.2893	.2937	.3125	J
	3B	.2740	.2801	.2893	.2926	.3125	
5/16-32 UNEF	2B	.279	.286	.2922	.2964	.3125	K
	3B	.2790	.2847	.2922	.2953	.3125	
5/16-36 UNS	2B	.282	.289	.2945	.2985	.3125	7.25mm
5/16-40 UNS	2B	.285	.291	.2963	.3001	.3125	
5/16-48 UNS	2B	.290	.295	.2990	.3026	.3125	
3/8-16 UNC	1B	.307	.321	.3344	.3429	.3750	5/16
	2B	.307	.321	.3344	.3401	.3750	
	3B	.3070	.3182	.3344	.3387	.3750	
3/8-18 UNS	2B	.315	.328	.3389	.3445	.3750	
3/8-20 UN	2B	.321	.332	.3425	.3479	.3750	P
	3B	.3210	.3287	.3425	.3465	.3750	
3/8-24 UNF	1B	.330	.340	.3479	.3553	.3750	Q
	2B	.330	.340	.3479	.3528	.3750	
	3B	.3300	.3372	.3479	.3516	.3750	
3/8-27 UNS	2B	.335	.344	.3509	.3556	.3750	

(Continued)

Table 5. *(Continued)* Unified Screw Threads. External Dimensions. Classes 1A, 2A, and 3A.

Size, TPI, and Series	Class	Allowance	Major Diameter		Pitch Diameter		Minor Dia.
			max	min	max	min	
3/8-28 UN	2A	.0011	.3739	.3674	.3507	.3471	.3301
	3A	-	.3750	.3685	.3518	.3491	.3312
3/8-32 UNEF	2A	.0010	.3740	.3680	.3537	.3503	.3357
	3A	-	.3750	.3690	.3547	.3522	.3367
3/8-36 UNS	2A	.0010	.3740	.3685	.3560	.3528	.3399
3/8-40 UNS	2A	.0009	.3741	.3690	.3579	.3548	.3434
7/16-14 UNC	1A	.0014	.4361	.4206	.3897	.3826	.3485
	2A	.0014	.4361	.4258	.3897	.3850	.3485
	3A	-	.4375	.4272	.3911	.3876	.3499
7/16-16 UN	2A	.0014	.4361	.4267	.3955	.3909	.3594
	3A	-	.4375	.4281	.3969	.3935	.3608
7/16-18 UNS	2A	.0013	.4362	.4275	.4001	.3958	.3680
7/16-20 UNF	1A	.0013	.4362	.4240	.4037	.3975	.3749
	2A	.0013	.4362	.4281	.4037	.3995	.3749
	3A	-	.4375	.4294	.4050	.4019	.3762
7/16-24 UNS	2A	.0011	.4364	.4292	.4093	.4055	.3853
7/16-27 UNS	2A	.0011	.4364	.4297	.4123	.4087	.3910
7/16-28 UNEF	2A	.0011	.4364	.4299	.4132	.4096	.3926
	3A	-	.4375	.4310	.4143	.4116	.3937
7/16-32 UN	2A	.0010	.4365	.4305	.4162	.4128	.3982
	3A	-	.4375	.4315	.4172	.4147	.3992
1/2-12 UNS	2A	.0016	.4984	.4870	.4443	.4389	.3962
	3A	-	.5000	.4886	.4459	.4419	.3978
1/2-13 UNC	1A	.0015	.4985	.4822	.4485	.4411	.4041
	2A	.0015	.4985	.4876	.4485	.4435	.4041
	3A	-	.5000	.4891	.4500	.4463	.4056
1/2-14 INS	2A	.0015	.4985	.4882	.4521	.4471	.4109
1/2-16 UN	2A	.0014	.4986	.4892	.4580	.4533	.4219
	3A	-	.5000	.4906	.4594	.4559	.4233
1/2-18 UNS	2A	.0013	.4987	.4900	.4626	.4582	.4305
1/2-20 UNF	1A	.0013	.4987	.4865	.4662	.4598	.4374
	2A	.0013	.4987	.4906	.4662	.4619	.4374
	3A	-	.5000	.4919	.4675	.4643	.4387
1/2-24 UNS	2A	.0012	.4988	.4916	.4717	.4678	.4477
1/2-27 UNS	2A	.0011	.4989	.4922	.4748	.4711	.4535
1/2-28 UNEF	2A	.0011	.4989	.4924	.4757	.4720	.4551
	3A	-	.5000	.4935	.4768	.4720	.4562
1/2-32 UN	2A	.0011	.4990	.4930	.4787	.4752	.4607
	3A	-	.5000	.4940	.4797	.4771	.4617
9/16-12 UNC	1A	.0016	.5609	.5437	.5068	.4990	.4587
	2A	.0016	.5609	.5495	.5068	.5016	.4587
	3A	-	.5625	.5511	.5084	.5045	.4603
9/16-14 UNS	2A	.0015	.5610	.5507	.5146	.5096	.4734
9/16-16 UN	2A	.0014	.5611	.5517	.5205	.5158	.4844
	3A	-	.5625	.5531	.5219	.5184	.4858

(Continued)

Table 6. *(Continued)* Unified Screw Threads. Internal Dimensions. Classes 1B, 2B, and 3B.

Size, TPI, and Series	Class	Minor Diameter		Pitch Diameter		Major Dia.	Tap Drill*
		min	max	min	max		
3/8-28 UN	2B	.336	.345	.3518	.3564	.3750	R
	3B	.3360	.3426	.3518	.3553	.3750	
3/8-32 UNEF	2B	.341	.349	.3547	.3591	.3750	11/32
	3B	.3410	.3469	.3547	.3580	.3750	
3/8-36 UNS	2B	.345	.352	.3570	.3612	.3750	S
3/8-40 UNS	2B	.348	.354	.3588	.3628	.3750	
7/16-14 UNC	1B	.360	.376	.3911	.4003	.4375	9.3mm
	2B	.360	.376	.3911	.3972	.4375	
	3B	.3600	.3717	.3911	.3957	.4375	
7/16-16 UN	2B	.370	.384	.3969	.4028	.4375	3/8
	3B	.3700	.3783	.3969	.4009	.4375	
7/16-18 UNS	2B	.377	.390	.4014	.4070	.4375	
7/16-20 UNF	1B	.383	.395	.4050	.4131	.4375	W
	2B	.383	.395	.4050	.4104	.4375	
	3B	.3830	.3916	.4050	.4091	.4375	
7/16-24 UNS	2B	.392	.402	.4104	.4153	.4375	
7/16-27 UNS	2B	.397	.406	.4134	.4181	.4375	
7/16-28 UNEF	2B	.399	.407	.4143	.4189	.4375	Y
	3B	.3990	.4051	.4143	.4178	.4375	
7/16-32 UN	2B	.404	.411	.4172	.4216	.4375	Y
	3B	.4040	.4094	.4172	.4205	.4375	
1/2-12 UNS	2B	.410	.428	.4459	.4529	.5000	Z
	3B	.4100	.4223	.4459	.4511	.5000	
1/2-13 UNC	1B	.417	.434	.4500	.4597	.5000	27/64
	2B	.417	.434	.4500	.4565	.5000	
	3B	.4170	.4284	.4500	.4548	.5000	
1/2-14 INS	2B	.423	.438	.4536	.4601	.5000	
1/2-16 UN	2B	.432	.446	.4594	.4655	.5000	7/16
	3B	.4320	.4408	.4594	.4640	.5000	
1/2-18 UNS	2B	.440	.453	.4639	.4697	.5000	
1/2-20 UNF	1B	.446	.457	.4675	.4759	.5000	11.4mm
	2B	.446	.457	.4675	.4731	.5000	
	3B	.4460	.4537	.4675	.4717	.5000	
1/2-24 UNS	2B	.455	.465	.4729	.4780	.5000	
1/2-27 UNS	2B	.460	.469	.4759	.4807	.5000	
1/2-28 UNEF	2B	.461	.470	.4768	.4816	.5000	15/32
	3B	.4610	.4676	.4768	.4804	.5000	
1/2-32 UN	2B	.466	.474	.4797	.4842	.5000	15/32
	3B	.4660	.4719	.4797	.4831	.5000	
9/16-12 UNC	1B	.472	.490	.5084	.5186	.5625	15/32
	2B	.472	.490	.5084	.5152	.5625	
	3B	.4720	.4843	.5084	.5135	.5625	
9/16-14 UNS	2B	.485	.501	.5161	.5226	.5625	
9/16-16 UN	2B	.495	.509	.5219	.5280	.5625	1/2
	3B	.4950	.5033	.5219	.5265	.5625	

(Continued)

Table 5. *(Continued)* **Unified Screw Threads. External Dimensions. Classes 1A, 2A, and 3A.**

Size, TPI, and Series	Class	Allowance	Major Diameter max	Major Diameter min	Pitch Diameter max	Pitch Diameter min	Minor Dia.
9/16-18 UNF	1A	.0014	.5611	.5480	.5250	.5182	.4929
	2A	.0014	.5611	.5524	.5250	.5205	.4929
	3A	-	.5625	.5538	.5264	.5230	.4943
9/16-20 UN	2A	.0013	.5612	.5531	.5287	.5245	.4999
	3A	-	.5625	.5544	.5300	.5268	.5012
9/16-24 UNEF	2A	.0012	.5613	.5541	.5342	.5303	.5102
	3A	-	.5625	.5553	.5354	.5325	.5114
9/16-27 UNS	2A	.0011	.5614	.5547	.5373	.5336	.5160
9/16-28 UN	2A	.0011	.5614	.5549	.5382	.5345	.5176
	3A	-	.5625	.5560	.5393	.5365	.5187
9/16-32	2A	.0010	.5615	.5555	.5412	.5377	.5232
	3A	-	.5625	.5565	.5422	.5396	.5242
5/8-11 UNC	1A	.0016	.6234	.6052	.5644	.5561	.5119
	2A	.0016	.6234	.6113	.5644	.5589	.5119
	3A	-	.6250	.6129	.5660	.5619	.5132
5/8-12 UN	2A	.0016	.6234	.6120	.5693	.5639	.5212
	3A	-	.6250	.6136	.5709	.5688	.5228
5/8-14 UNS	2A	.0015	.6235	.6132	.5771	.5720	.5359
5/8-16 UN	2A	.0014	.6236	.6142	.5830	.5782	.5469
	3A	-	.6250	.6156	.5844	.5808	.5483
5/8-18 UNF	1A	.0014	.6236	.6105	.5875	.5805	.5554
	2A	.0014	.6236	.6149	.5875	.5828	.5554
	3A	-	.6250	.6163	.5889	.5854	.5568
5/8-20 UN	2A	.0013	.6237	.6156	.5912	.5869	.5624
	3A	-	.6250	.6169	.5925	.5893	.5637
5/8-24 UNEF	2A	.0012	.6238	.6166	.5967	.5927	.5727
	3A	-	.6250	.6178	.5979	.5949	.5739
5/8-27 UNS	2A	.0011	.6239	.6172	.5998	.5960	.5785
5/8-28 UN	2A	.0011	.6239	.6174	.6007	.5969	.5801
	3A	-	.6250	.6185	.6018	.5990	.5812
5/8-32 UN	2A	.0011	.6239	.6179	.6036	.6000	.5856
	3A	-	.6250	.6190	.6047	.6020	.5867
11/16-12 UN	2A	.0016	.6859	.6745	.6318	.6264	.5837
	3A	-	.6875	.6761	.6334	.6293	.5853
11/16-16 UN	2A	.0014	.6861	.6767	.6455	.6407	.6094
	3A	-	.6875	.6781	.6469	.6433	.6108
11/16-20 UN	2A	.0013	.6862	.6871	.6537	.6494	.6249
	3A	-	.6875	.6794	.6550	.6518	.6262
11/16-24 UNEF	2A	.0012	.6863	.6791	.6592	.6552	.6352
	3A	-	.6875	.6803	.6604	.6574	.6364
11/16-28 UN	2A	.0011	.6864	.6799	.6632	.6594	.6426
	3A	-	.6875	.6810	.6643	.6615	.6437
11/16-32 UN	2A	.0011	.6864	.6804	.6661	.6625	.6481
	3A	-	.6875	.6815	.6672	.6645	.6492

(Continued)

Table 6. *(Continued)* **Unified Screw Threads. Internal Dimensions. Classes 1B, 2B, and 3B.**

Size, TPI, and Series	Class	Minor Diameter		Pitch Diameter		Major Dia.	Tap Drill*
		min	max	min	max		
9/16-18 UNF	1B	.502	.515	.5264	.5253	.5625	1/2
	2B	.502	.515	.5264	.5323	.5625	
	3B	.5020	.5106	.5264	.5308	.5625	
9/16-20 UN	2B	.508	.520	.5300	.5355	.5625	33/64
	3B	.5080	.5162	.5300	.5341	.5625	
9/16-24 UNEF	2B	.517	.527	.5354	.5405	.5625	33/64
	3B	.5170	.5244	.5354	.5392	.5625	
9/16-27 UNS	2B	.522	.531	.5384	.5432	.5625	
9/16-28 UN	2B	.524	.532	.5393	.5441	.5625	17/32
	3B	.5240	.5301	.5393	.5429	.5625	
9/16-32	2B	.529	.536	.5422	.5467	.5625	19/32
	3B	.5290	.5344	.5422	.5456	.5625	
5/8-11 UNC	1B	.527	.546	.5660	.5767	.6250	17/32
	2B	.527	.546	.5660	.5732	.6250	
	3B	.5270	.5391	.5660	.5714	.6250	
5/8-12 UN	2B	.535	.553	.5709	.5780	.6250	35/64
	3B	.5350	.5463	.5709	.5762	.6250	
5/8-14 UNS	2B	.548	.564	.5786	.5852	.6250	
5/8-16 UN	2B	.557	.571	.5844	.5906	.6250	9/16
	3B	.5570	.5658	.5844	.5890	.6250	
5/8-18 UNF	1B	.565	.578	.5889	.5980	.6250	9/16
	2B	.565	.578	.5889	.5949	.6250	
	3B	.5650	.5730	.5889	.5934	.6250	
5/8-20 UN	2B	.571	.582	.5925	.5981	.6250	37/64
	3B	.5710	.5787	.5925	.5967	.6250	
5/8-24 UNEF	2B	.580	.590	.5979	.6031	.6250	37/64
	3B	.5800	.5869	.5979	.6018	.6250	
5/8-27 UNS	2B	.585	.594	.6009	.6059	.6250	
5/8-28 UN	2B	.586	.595	.6018	.6067	.6250	19/32
	3B	.5860	.5936	.6018	.6055	.6250	
5/8-32 UN	2B	.591	.599	.6047	.6093	.6250	19/32
	3B	.5910	.5969	.6047	.6082	.6250	
11/16-12 UN	2B	.597	.615	.6334	.6405	.6875	19/32
	3B	.5970	.6085	.6334	.6387	.6875	
11/16-16 UN	2B	.620	.634	.6469	.6351	.6875	5/8
	3B	.6200	.6283	.6469	.6515	.6875	
11/16-20 UN	2B	.633	.645	.6550	.6606	.6875	41/64
	3B	.6330	.6412	.6550	.6592	.6875	
11/16-24 UNEF	2B	.642	.652	.6604	.6656	.6875	41/64
	3B	.6420	.6494	.6604	.6643	.6875	
11/16-28 UN	2B	.649	.657	.6643	.6692	.6875	21/32
	3B	.6490	.6551	.6643	.6680	.6875	
11/16-32 UN	2B	.654	.661	.6672	.6718	.6875	21/32
	3B	.6540	.6594	.6672	.6707	.6875	

(Continued)

Table 5. *(Continued)* **Unified Screw Threads. External Dimensions. Classes 1A, 2A, and 3A.**

Size, TPI, and Series	Class	Allowance	Major Diameter		Pitch Diameter		Minor Dia.
			max	min	max	min	
3/4-10 UNC	1A	.0018	.7482	.7288	.6832	.6744	.6255
	2A	.0018	.7482	.7353	.6832	.6773	.6773
	3A	-	.7500	.7371	.6850	.6806	.6273
3/4-12 UN	2A	.0017	.7483	.7369	.6942	.6887	.6461
	3A	-	.7500	.7386	.6959	.6918	.6478
3/4-14 UNS	2A	.0015	.7485	.7382	.7021	.6970	.6609
3/4-16 UNF	1A	.0015	.7485	.7343	.7079	.7004	.6718
	2A	.0015	.7485	.7391	.7079	.7029	.6718
	3A	-	.7500	.7406	.7094	.7056	.6733
3/4-18 UNS	2A	.0014	.7486	.7399	.7125	.7079	.6804
3/4-20 UNEF	2A	.0013	.7487	.7406	.7162	.7118	.6874
	3A	-	.7500	.7419	.7175	.7142	.6887
3/4-24 UNS	2A	.0012	.7488	.7416	.7217	.7176	.6977
3/4-27 UNS	2A	.0012	.7488	.7412	.7247	.7208	.7034
3/4-28 UN	2A	.0012	.7488	.7423	.7256	.7218	.7050
	3A	-	.7500	.7435	.7268	.7239	.7062
3/4-32 UN	2A	.0011	.7489	.7429	.7286	.7250	.7106
	3A	-	.7500	.7440	.7297	.7270	.7117
13/16-12 UN	2A	.0017	.8108	.7994	.7567	.7512	.7068
	3A	-	.8125	.8011	.7584	.7543	.7103
13/16-16 UN	2A	.0015	.8110	.8016	.7704	.7655	.7343
	3A	-	.8125	.8031	.7719	.7683	.7358
13/16-20 UNEF	2A	.0013	.8112	.8031	.7787	.7743	.7498
	3A	-	.8125	.8044	.7800	.7767	.7512
13/16-28 UN	2A	.0012	.8113	.8048	.7881	.7843	.7675
	3A	-	.8125	.8060	.7893	.7864	.7687
13/16-32 UN	2A	.0011	.8114	.8054	.7911	.7875	.7731
	3A	-	.8125	.8065	.7922	.7895	.7742
7/8-9 UNC	1A	.0019	.8731	.8523	.8009	.7914	7368
	2A	.0019	.8731	.8592	.8009	.7946	.7368
	3A	-	.8750	.8611	.8028	.7981	.7387
7/8-10 UNS	2A	.0018	.8732	.8603	.8082	.8022	.7505
7/8-12 UN	2A	.0017	.8733	.8619	.8192	.8137	.7711
	3A	-	.8750	.8636	.8209	.8168	.7728
7/8-14 UNF	1A	.0016	.8734	.8579	.8270	.8189	.7858
	2A	.0016	.8734	.8631	.8270	.8216	.7858
	3A	-	.8750	.8647	.8286	.8245	.7874
7/8-16 UN	2A	.0015	.8735	.8641	.8329	.8280	.7968
	3A	-	.8750	.8656	.8344	.8308	.7983
7/8-18 UNS	2A	.0014	.8736	.8649	.8375	.8329	.8054
7/8-20 UNEF	2A	.0013	.8737	.8656	.8412	.8368	.8124
	3A	-	.8750	.8669	.8425	.8392	.8137
7/8-24 UNS	2A	.0012	.8738	.8666	.8467	.8426	.8227
7/8-27 UNS	2A	.0012	.8738	.8671	.8497	.8458	.8284

(Continued)

Table 6. *(Continued)* Unified Screw Threads. Internal Dimensions. Classes 1B, 2B, and 3B.

Size, TPI, and Series	Class	Minor Diameter		Pitch Diameter		Major Dia.	Tap Drill*
		min	max	min	max		
3/4-10 UNC	1B	.642	.663	.6850	.6965	.7500	41/64
	2B	.642	.663	.6850	.6927	.7500	
	3B	.6420	.6545	.6850	.6907	.7500	
3/4-12 UN	2B	.660	.687	.6959	.7031	.7500	21/32
	3B	.6600	.6707	.6959	.7013	.7500	
3/4-14 UNS	2B	.673	.688	.7036	.7103	.7500	
3/4-16 UNF	1B	.682	.696	.7094	.7192	.7500	11/16
	2B	.682	.696	.7094	.7159	.7500	
	3B	.6820	.6908	.7094	.7143	.7500	
3/4-18 UNS	2B	.690	.703	.7139	.7199	.7500	
3/4-20 UNEF	2B	.696	.707	.7175	.7232	.7500	45/64
	3B	.6960	.7037	.7175	.7218	.7500	
3/4-24 UNS	2B	.705	.715	.7229	.7282	.7500	
3/4-27 UNS	2B	.710	.719	.7259	.7310	.7500	
3/4-28 UN	2B	.711	.720	.7268	.7318	.7500	23/32
	3B	.7110	.7176	.7268	.7305	.7500	
3/4-32 UN	2B	.716	.724	.7297	.7344	.7500	23/32
	3B	.7160	.7219	.7297	.7333	.7500	
13/16-12 UN	2B	.722	.740	.7584	.7656	.8125	47/64
	3B	.7220	.7329	.7584	.7638	.8125	
13/16-16 UN	2B	.745	.759	.7719	.7782	.8125	3/4
	3B	.7450	.7533	.7719	.7766	.8125	
13/16-20 UNEF	2B	.758	.770	.7800	.7857	.8125	49/64
	3B	.7580	.7662	.7800	.7843	.8125	
13/16-28 UN	2B	.774	.782	.7893	.7943	.8125	25/32
	3B	.7740	.7801	.7893	.7930	.8125	
13/16-32 UN	2B	.779	.786	.7922	.7969	.8125	25/32
	3B	.7790	.7844	.7922	.7958	.8125	
7/8-9 UNC	1B	.755	.778	.8028	.8151	.8750	49/64
	2B	.755	.778	.8028	.8110	.8750	
	3B	.7550	.7681	.8028	.8090	.8750	
7/8-10 UNS	2B	.767	.788	.8100	.8178	.8750	
7/8-12 UN	2B	.785	.803	.8209	.8281	.8750	25/32
	3B	.7850	.7948	.8209	.8263	.8750	
7/8-14 UNF	1B	.798	.814	.8286	.8392	.8750	51/64
	2B	.798	.814	.8286	.8356	.8750	
	3B	.7980	8068	.8286	.8339	.8750	
7/8-16 UN	2B	.807	.821	.8344	.8407	.8750	13/16
	3B	.8070	.8158	.8344	.8391	.8750	
7/8-18 UNS	2B	.815	.828	.8389	.8449	.8750	
7/8-20 UNEF	2B	.821	.832	.8425	.8482	.8750	53/64
	3B	.8210	.8287	.8425	.8468	.8750	
7/8-24 UNS	2B	.830	.840	.8479	.8532	.8750	
7/8-27 UNS	2B	.835	.844	.8509	.8560	.8750	

(Continued)

Table 5. *(Continued)* Unified Screw Threads. External Dimensions. Classes 1A, 2A, and 3A.

Size, TPI, and Series	Class	Allowance	Major Diameter		Pitch Diameter		Minor Dia.
			max	min	max	min	
7/8-28 UN	2A	.0012	.8738	.8673	.8506	.8468	.8300
	3A	-	.8685	.8685	.8518	.8489	.8312
7/8-32 UN	2A	.0011	.8739	.8679	.8536	.8500	.8356
	3A	-	.8750	.8690	.8547	.8520	.8367
15/16-12 UN	2A	.0017	.9358	.9244	.8817	.8760	.8336
	2B	-	.9375	.9261	.8834	.8793	.8353
15/16-16 UN	2A	.0015	.9360	.9266	.8954	.8904	.8593
	3A	-	.9375	.9281	.8969	.8932	.8608
15/16-20 UNEF	2A	.0014	.9361	.9280	.9036	.8991	.8748
	3A	-	.9375	.9294	.9050	.9016	.8762
15/16-28 UN	2A	.0012	.9363	.9298	.9131	.9091	.8925
	3A	-	.9375	.9310	.9143	.9113	.8937
15/16-32 UN	2A	.0011	.9364	.9304	.9161	.9123	.8981
	3A	-	.9375	.9315	.9172	.9144	.8992
1-8 UNC	1A	.0020	.9980	.9755	.9168	.9067	.8446
	2A	.0020	.9980	.9830	.9168	.9100	.8446
	3A	-	1.0000	.9850	.9188	.9137	.8466
1-10 UNS	2A	.0018	.9982	.9853	.9332	.9270	.8755
1-12 UNF	1A	.0018	.9982	.9810	.9441	.9353	.8960
	2A	.0018	.9982	.9868	.9441	.9382	.8960
	3A	-	1.0000	.9886	.9459	.9415	.8978
1-14 UNS	1A	.0017	.9983	.9828	.9519	.9435	.9107
	2A	.0017	.9983	.9880	.9519	.9463	.9107
	3A	-	1.0000	.9897	.9536	.9494	.9124
1-16 UN	2A	.0015	.9985	.9891	.9579	.9529	.9218
	3A	-	1.0000	.9906	.9594	.9557	.9233
1-18 UNS	2A	.0014	.9968	.9899	.9625	.9578	.9304
1-20 UNEF	2A	.0014	.9986	.9905	.9661	.9616	.9373
	3A	-	1.0000	.9919	.9675	.9641	.9387
1-24 UNS	2A	.0013	.9987	.9915	.9716	.9674	.9476
1-27 UNS	2A	.0012	.9988	.9921	.9747	.9707	.9534
1-28 UN	2A	.0012	.9988	.9923	.9756	.9716	.9550
	3A	-	1.0000	.9935	.9768	.9738	.9562
1-32 UN	2A	.0011	.9989	.9929	.9786	.9748	.9606
	3A	-	1.0000	.9940	.9797	.9769	.9617
1 1/16-8 UN	2A	.0020	1.0605	1.0455	.9793	.9725	.9071
	3A	-	1.0625	1.0475	.9812	.9762	.9091
1 1/16-12 UN	2A	.0017	1.0608	1.0494	1.0067	1.0010	.9586
	3A	-	1.0625	1.0511	1.0084	1.0042	.9603
1 1/16-16 UN	2A	.0015	1.0610	1.0516	1.0204	1.0154	.9843
	3A	-	1.0625	1.0531	1.0219	1.0182	.9858
1 1/16-18 UNEF	2A	.0014	1.0611	1.0524	1.0250	1.0203	.9929
	3A	-	1.0625	1.0538	1.0264	1.0228	.9943
1 1/16-20 UN	2A	.0014	1.0611	1.0530	1.0286	1.0241	.9998
	3A	-	1.0625	1.0544	1.0300	1.0266	1.0012

(Continued)

Table 6. *(Continued)* Unified Screw Threads. Internal Dimensions. Classes 1B, 2B, and 3B.

Size, TPI, and Series	Class	Minor Diameter		Pitch Diameter		Major Dia.	Tap Drill*
		min	max	min	max		
7/8-28 UN	2B	.836	.845	.8518	.8568	.8750	27/32
	3B	.8360	.8426	.8518	.8555	.8750	
7/8-32 UN	2B	.841	.849	.8547	.8594	.8750	27/32
	3B	.8410	.8469	.8547	.8583	.8750	
15/16-12 UN	2B	.847	.865	.8834	.8908	.9375	27/32
	3B	.8470	.8575	.8834	.8889	.9375	
15/16-16 UN	2B	.870	.884	.8969	.9034	.9375	7/8
	3B	.8700	.8783	.8969	.9018	.9375	
15/16-20 UNEF	2B	.883	.895	.9050	.9109	.9375	57/64
	3B	.8830	.8912	.9050	.9094	.9375	
15/16-28 UN	2B	.899	.907	.9143	.9195	.9375	29/32
	3B	.8990	.9061	.9143	.9182	.9375	
15/16-32 UN	2B	.904	.911	.9172	.9221	.9375	29/32
	3B	.9040	.9094	.9172	.9209	.9375	
1-8 UNC	1B	.865	.890	.9188	.9320	1.0000	55/64
	2B	.865	.890	.9188	.9276	1.0000	
	3B	.8650	.8797	.9188	.9254	1.0000	
1-10 UNS	2B	.892	.913	.9350	.9430	1.0000	
1-12 UNF	1B	.910	.928	.9459	.9573	1.0000	29/32
	2B	.910	.928	.9459	.9535	1.0000	
	3B	.9100	.9198	.9459	.9516	1.0000	
1-14 UNS	1B	.923	.938	.9536	.9645	1.0000	59/64
	2B	.923	.938	.9536	.9609	1.0000	
	3B	.9230	.9315	.9536	.9590	1.0000	
1-16 UN	2B	.932	.946	.9594	.9659	1.0000	15/16
	3B	.9320	.9408	.9594	.9643	1.0000	
1-18 UNS	2B	.940	.953	.9639	.9701	1.0000	
1-20 UNEF	2B	.946	.957	.9675	.9734	1.0000	61/64
	3B	.9460	.9537	.9675	.9719	1.0000	
1-24 UNS	2B	.955	.965	.9729	.9784	1.0000	
1-27 UNS	2B	.960	.969	.9759	.9811	1.0000	
1-28 UN	2B	.961	.970	.9768	.9820	1.0000	31/32
	3B	.9610	.9676	.9768	.9807	1.0000	
1-32 UN	2B	.966	.974	.9797	.9846	1.0000	31/32
	3B	.966	.9719	.9797	.9834	1.0000	
1 1/16-8 UN	2B	.927	.952	.9813	.9902	1.0625	59/64
	3B	.9270	.9422	.9813	.9880	1.0625	
1 1/16-12 UN	2B	.972	.990	1.0084	1.0158	1.0625	31/32
	3B	.9720	.9823	1.0084	1.0139	1.0625	
1 1/16-16 UN	2B	.995	1.009	1.0219	1.0284	1.0625	1
	3B	.9950	1.0033	1.0219	1.0268	1.0625	
1 1/16-18 UNEF	2B	1.002	1.015	1.0264	1.0326	1.0625	1
	3B	1.0020	1.0105	1.0264	1.0310	1.0625	
1 1/16-20 UN	2B	1.008	1.020	1.0300	1.0359	1.0625	1 1/64
	3B	1.0080	1.0162	1.0300	1.0344	1.0625	

(Continued)

Table 5. *(Continued)* Unified Screw Threads. External Dimensions. Classes 1A, 2A, and 3A.

Size, TPI, and Series	Class	Allowance	Major Diameter		Pitch Diameter		Minor Dia.
			max	min	max	min	
1 1/16-28 UN	2A	.0012	1.0613	1.0548	1.0381	1.0341	1.0175
	3A	-	1.0625	1.0560	1.0393	1.0363	1.0187
1 1/8-7 UNC	1A	.0022	1.1228	1.0982	1.0300	1.0191	.9475
	2A	.0022	1.1228	1.1064	1.0300	1.0228	.9475
	3A	-	1.1250	1.1086	1.0322	1.0268	.9497
1 1/8-8 UN	2A	.0021	1.1229	1.1079	1.0417	1.0348	.9695
	3A	-	1.1250	1.1100	1.0438	1.0386	.9716
1 1/8-10 UNS	2A	.0018	1.1232	1.1103	1.0582	1.0520	1.0005
1 1/8-12 UNF	1A	.0018	1.1232	1.1060	1.0691	1.0601	1.0210
	2A	.0018	1.1232	1.1118	1.0691	1.0631	1.0210
	3A	-	1.1250	1.1136	1.0709	1.0664	1.0228
1 1/8-14 UNS	2A	.0016	1.1234	1.1131	1.0770	1.0717	1.0358
1 1/8-16 UN	2A	.0015	1.1235	1.1141	1.0829	1.0779	1.0468
	3A	-	1.1250	1.1156	1.0844	1.0807	1.0483
1 1/8-18 UNEF	2A	.0014	1.1236	1.1149	1.0875	1.0828	1.0554
	3A	-	1.1250	1.1163	1.0889	1.0853	1.0568
1 1/8-20 UN	2A	.0014	1.1236	1.1155	1.0911	1.0866	1.0623
	3A	-	1.1250	1.1169	1.0925	1.0891	1.0637
1 1/8-24 UNS	2A	.0013	1.1237	1.1165	1.0966	1.0924	1.0726
1 1/8-28 UN	2A	.0012	1.1238	1.1173	1.1006	1.0966	1.0800
	3A	-	1.1250	1.1185	1.1018	1.0988	1.0812
1 3/16-8 UN	2A	.0021	1.1854	1.1704	1.1042	1.0972	1.0320
	3A	-	1.1875	1.1725	1.1063	1.1011	1.0341
1 3/16-12 UN	2A	.0017	1.1858	1.1744	1.1317	1.1259	1.0836
	3A	-	1.1875	1.1761	1.1334	1.1291	1.0853
1 3/16-16 UN	2A	.0015	1.1860	1.1766	1.1454	1.1403	1.1093
	3A	-	1.1875	1.1781	1.1469	1.1431	1.1108
1 3/16-18 UNEF	2A	.0015	1.1860	1.1773	1.1499	1.1450	1.1178
	3A	-	1.1875	1.1788	1.1514	1.1478	1.1193
1 3/16-20 UN	2A	.0014	1.1861	1.1780	1.1536	1.1489	1.1248
	3A	-	1.1875	1.1794	1.1550	1.1515	1.1262
1 3/16-28 UN	2A	.0012	1.1863	1.1798	1.1631	1.1590	1.1425
	3A	-	1.1875	1.1810	1.1643	1.1612	1.1437
1 1/4-7 UNC	1A	.0022	1.2478	1.2232	1.1550	1.1439	1.0725
	2A	.0022	1.2478	1.2314	1.1550	1.1476	1.0725
	3A	-	1.2500	1.2336	1.1572	1.1517	1.0747
1 1/4-8 UN	2A	.0021	1.2479	1.2329	1.6667	1.1579	1.0945
	3A	-	1.2500	1.2350	1.1688	1.1635	1.0966
1 1/4-10 UNS	2A	.0019	1.2481	1.2352	1.1831	1.1768	1.1254
1 1/4-12 UNF	1A	.0018	1.2481	1.2310	1.1941	1.1849	1.1460
	2A	.0018	1.2481	1.2368	1.1941	1.1879	1.1460
	3A	-	1.2500	1.2386	1.1959	1.1913	1.1478
1 1/4-14 UNS	2A	.0016	1.2484	1.2381	1.2020	1.1966	1.1608
1 1/4-16 UN	2A	.0015	1.2485	1.2391	1.2079	1.2028	1.1718
	3A	-	1.2500	1.2406	1.2094	1.2056	1.1733

(Continued)

Table 6. (Continued) Unified Screw Threads. Internal Dimensions. Classes 1B, 2B, and 3B.

Size, TPI, and Series	Class	Minor Diameter		Pitch Diameter		Major Dia.	Tap Drill*
		min	max	min	max		
1 1/16-28 UN	2B	1.024	1.032	1.0393	1.0445	1.0625	1 1/32
	3B	1.0240	1.0301	1.0393	1.0432	1.0625	
1 1/8-7 UNC	1B	.970	.998	1.0322	1.0463	1.1250	31/32
	2B	.970	.998	1.0322	1.0416	1.1250	
	3B	.9700	.9875	1.0322	1.0393	1.1250	
1 1/8-8 UN	2B	.990	1.015	1.0438	1.0528	1.1250	1
	3B	.990	1.0047	4.0438	1.0505	1.1250	
1 1/8-10 UNS	2B	1.017	1.0380	1.0600	1.0680	1.1250	
1 1/8-12 UNF	1B	1.035	1.053	1.0709	1.0826	1.1250	1 1/32
	2B	1.035	1.053	1.0709	1.0787	1.1250	
	3B	1.0350	1.0448	1.0709	1.0768	1.1250	
1 1/8-14 UNS	2B	1.048	1.064	1.0786	1.0855	1.1250	
1 1/8-16 UN	2B	1.057	1.071	1.0844	1.0909	1.1250	1 1/16
	3B	1.0570	1.0658	1.0844	1.0893	1.1250	
1 1/8-18 UNEF	2B	1.065	1.078	1.0889	1.0951	1.1250	1 1/16
	3B	1.0650	1.0730	1.0889	1.0935	1.1250	
1 1/8-20 UN	2B	1.071	1.082	1.0925	1.0984	1.1250	1 5/64
	3B	1.0710	1.0787	1.0925	1.0969	1.1250	
1 1/8-24 UNS	2B	1.080	1.090	1.0979	1.1034	1.1250	
1 1/8-28 UN	2B	1.086	1.095	1.1018	1.1070	1.1250	1 3/32
	3B	1.0860	1.0926	1.1018	1.1057	1.1250	
1 3/16-8 UN	2B	1.052	1.077	1.1063	1.1154	1.1875	1 1/16
	3B	1.0520	1.0672	1.1063	1.1131	1.1875	
1 3/16-12 UN	2B	1.097	1.115	1.1334	1.1409	1.1875	1 3/32
	3B	1.0970	1.1073	1.1334	1.1390	1.1875	
1 3/16-16 UN	2B	1.120	1.134	1.1469	1.1535	1.1875	1 1/8
	3B	1.1200	1.1283	1.1469	1.1519	1.1875	
1 3/16-18 UNEF	2B	1.127	1.140	1.1514	1.1577	1.1875	1 1/8
	3B	1.1270	1.1355	1.1514	1.1561	1.1875	
1 3/16-20 UN	2B	1.133	1.145	1.1550	1.1611	1.1875	1 9/64
	3B	1.1330	1.1412	1.1550	1.1595	1.1875	
1 3/16-28 UN	2B	1.149	1.157	1.1643	1.1696	1.1875	1 5/32
	3B	1.1490	1.1551	1.1643	1.1683	1.1875	
1 1/4-7 UNC	1B	1.095	1.123	1.1572	1.1716	1.2500	1 3/32
	2B	1.095	1.123	1.1572	1.1668	1.2500	
	3B	1.0950	1.1125	1.1572	1.1644	1.2500	
1 1/4-8 UN	2B	1.115	1.140	1.1688	1.1780	1.2500	1 1/8
	3B	1.1150	1.1297	1.1688	1.1757	1.2500	
1 1/4-10 UNS	2B	1.142	1.163	1.1850	1.1932	1.1250	
1 1/4-12 UNF	1B	1.160	1.178	1.1959	1.2079	1.2500	1 5/32
	2B	1.160	1.178	1.1959	1.2039	1.2500	
	3B	1.1600	1.1689	1.1959	1.2019	1.2500	
1 1/4-14 UNS	2B	1.173	1.188	1.2036	1.1206	1.2500	
1 1/4-16 UN	2B	1.182	1.196	1.2094	1.2160	1.2500	1 3/16
	3B	1.1820	1.1908	1.2094	1.2144	1.2500	

(Continued)

Table 5. *(Continued)* **Unified Screw Threads. External Dimensions. Classes 1A, 2A, and 3A.**

Size, TPI, and Series	Class	Allowance	Major Diameter		Pitch Diameter		Minor Dia.
			max	min	max	min	
1 1/4-18 UNEF	2A	.0015	1.2485	1.2398	1.2124	1.2075	1.1803
	3A	-	1.2500	1.2413	1.2139	1.2103	1.1818
1 1/4-20 UN	2A	.0014	1.2486	1.2405	1.2161	1.2114	1.1873
	3A	-	1.2500	1.2419	1.2175	1.2140	1.1887
1 1/4-24 UNS	2A	.0013	1.2487	1.2415	1.2216	1.2173	1.1976
1 1/4-28 UN	2A	.0012	1.2488	1.2423	1.2256	1.2215	1.2050
	3A	-	1.2500	1.2435	1.2268	1.2237	1.2062
1 5/16-8 UN	2A	.0021	1.3104	1.2954	1.2292	1.2221	1.1570
	3A	-	1.3125	1.2975	1.2313	1.2260	1.1591
1 5/16-12 UN	2A	.0017	1.3108	1.2994	1.2567	1.2509	1.2086
	3A	-	1.3125	1.3011	1.2584	1.2541	1.2103
1 5/16-16 UN	2A	.0015	1.3110	1.3016	1.2704	1.2653	1.2343
	3A	-	1.3125	1.3031	1.2719	1.2681	1.2358
1 5/16-18 UNEF	2A	.0015	1.3110	1.3023	1.2749	1.2700	1.2428
	3A	-	1.3125	1.3038	1.2764	1.2728	1.2443
1 5/16-20 UN	2A	.0014	1.3111	1.3030	1.2786	1.2739	1.2498
	3A	-	1.3125	1.3044	1.2800	1.2765	1.2512
1 5/16-28 UN	2A	.0012	1.3113	1.3048	1.2881	1.2840	1.2675
	3A	-	1.3125	1.3060	1.2893	1.2862	1.2687
1 3/8-6 UNC	1A	.0024	1.3726	1.3453	1.2643	1.2523	1.1681
	2A	.0024	1.3726	1.3544	1.2643	1.2563	1.1681
	3A	-	1.3750	1.3568	1.2667	1.2607	1.1705
1 3/8-8 UN	2A	.0022	1.3728	1.3578	1.2916	1.2844	1.2194
	3A	-	1.3750	1.3600	1.2938	1.2884	1.2216
1 3/8-10 UNS	2A	.0019	1.3731	1.3602	1.3081	1.3018	1.2504
1 3/8-12 UNF	1A	.0019	1.3731	1.3559	1.3190	1.3096	1.2709
	2A	.0019	1.3731	1.3617	1.3190	1.3127	1.2709
	3A	-	1.3750	1.3636	1.3209	1.3162	1.2728
3 1/8-14 UNS	2A	.0016	1.3734	1.3631	1.3270	1.3216	1.2858
1 3/8-16 UN	2A	.0015	1.3735	1.3641	1.3329	1.3278	1.2968
	3A	-	1.3750	1.3656	1.3344	1.3306	1.2983
1 3/8-18 UNEF	2A	.0015	1.3735	1.3648	1.3374	1.3325	1.3053
	3A	-	1.3750	1.3663	1.3389	1.3353	1.3068
1 3/8-20 UN	2A	.0014	1.3736	1.3655	1.3411	1.3364	1.3123
	3A	-	1.3750	1.3669	1.3425	1.3390	1.3137
1 3/8-28 UN	2A	.0012	1.3738	1.3673	1.3506	1.3465	1.3300
	3A	-	1.3750	1.3685	1.3518	1.3487	1.3312
1 7/16-6 UN	2A	.0024	1.4351	1.4169	1.3268	1.3188	1.2306
	3A	-	1.4375	1.4193	1.3292	1.3232	1.2330
1 7/16-8 UN	2A	.0022	1.4353	1.4203	1.3541	1.3469	1.2819
	3A	-	1.4375	1.4225	1.3563	1.3509	1.2841
1 7/16-12 UN	2A	.0018	1.4357	1.4243	1.3816	1.3757	1.3335
	3A	-	1.4375	1.4261	1.3834	1.3790	1.3353
1 7/16-16 UN	2A	.0016	1.4359	1.4265	1.3953	1.3901	1.3592
	3A	-	1.4375	1.4281	1.3969	1.3930	1.3608

(Continued)

Table 6. *(Continued)* **Unified Screw Threads. Internal Dimensions. Classes 1B, 2B, and 3B.**

Size, TPI, and Series	Class	Minor Diameter		Pitch Diameter		Major Dia.	Tap Drill*
		min	max	min	max		
1 1/4-18 UNEF	2B	1.190	1.203	1.2139	1.2202	1.2500	1 3/16
	3B	1.1900	1.1980	1.2139	1.2186	1.2500	
1 1/4-20 UN	2B	1.196	1.207	1.2175	1.2236	1.2500	1 13/64
	3B	1.1960	1.2037	1.2175	1.2220	1.2500	
1 1/4-24 UNS	2B	1.205	1.215	1.2229	1.2285	1.2500	
1 1/4-28 UN	2B	1.211	1.220	1.2268	1.2321	1.2500	1 7/32
	3B	1.2110	1.2176	1.2268	1.2308	1.2500	
1 5/16-8 UN	2B	1.177	1.202	1.2313	1.2405	1.3125	1 11/64
	3B	1.1770	1.1922	1.2313	1.2382	1.3125	
1 5/16-12 UN	2B	1.222	1.240	1.2584	1.2659	1.3125	1 7/32
	3B	1.2220	1.2323	1.2584	1.2640	1.3125	
1 5/16-16 UN	2B	1.245	1.259	1.2719	1.2785	1.3125	1 1/4
	3B	1.2450	1.2533	1.2719	1.2769	1.3125	
1 5/16-18 UNEF	2B	1.252	1.265	1.2764	1.2827	1.3125	1 1/4
	3B	1.2520	1.2605	1.2764	1.2811	1.3125	
1 5/16-20 UN	2B	1.258	1.270	1.2800	1.2861	1.3125	1 17/64
	3B	1.2580	1.2662	1.2800	1.2845	1.3125	
1 5/16-28 UN	2B	1.274	1.282	1.2893	1.2946	1.3125	1 9/32
	3B	1.2740	1.2801	1.2893	1.2933	1.3125	
1 3/8-6 UNC	1B	1.195	1.225	1.2667	1.2822	1.3750	1 13/16
	2B	1.195	1.225	1.2667	1.2771	1.3750	
	3B	1.1950	1.2146	1.2667	1.2745	1.3750	
1 3/8-8 UN	2B	1.240	1.265	1.2938	1.3031	1.3750	1 15/64
	3B	1.2400	1.2547	1.2938	1.3008	1.3750	
1 3/8-10 UNS	2B	1.267	1.288	1.3100	1.3182	1.3750	
1 3/8-12 UNF	1B	1.285	1.303	1.3209	1.3332	1.3750	1 9/32
	2B	1.285	1.303	1.3209	1.3291	1.3750	
	3B	1.2850	1.2948	1.3209	1.3270	1.3750	
1 3/8-14 UNS	2B	1.298	1.314	1.3286	1.3356	1.3750	
1 3/8-16 UN	2B	1.307	1.321	1.3344	1.3410	1.3750	1 5/16
	3B	1.3070	1.3158	1.3344	1.3394	1.3750	
1 3/8-18 UNEF	2B	1.315	1.328	1.3389	1.3452	1.3750	1 5/16
	3B	1.3150	1.3230	1.3389	1.3436	1.3750	
1 3/8-20 UN	2B	1.321	1.332	1.3425	1.3486	1.3750	1 21/64
	3B	1.3210	1.3287	1.3425	1.3470	1.3750	
1 3/8-28 UN	2B	1.336	1.345	1.3518	1.3571	1.3750	1 11/32
	3B	1.3360	1.3426	1.3518	1.3558	1.3750	
1 7/16-6 UN	2B	1.257	1.288	1.3292	1.3396	1.4375	1 17/64
	3B	1.2570	1.2771	1.3292	1.3370	1.4375	
1 7/16-8 UN	2B	1.302	1.327	1.3563	1.3657	1.4375	1 19/64
	3B	1.3020	1.3172	1.3563	1.3634	1.4375	
1 7/16-12 UN	2B	1.347	1.365	1.3834	1.3910	1.4375	1 11/32
	3B	1.3470	1.3573	1.3834	1.3891	1.4375	
1 7/16-16 UN	2B	1.370	1.384	1.3969	1.4037	1.4375	1 3/8
	3B	1.3700	1.3783	1.3969	1.4020	1.4375	

(Continued)

Table 5. *(Continued)* **Unified Screw Threads. External Dimensions. Classes 1A, 2A, and 3A.**

Size, TPI, and Series	Class	Allowance	Major Diameter		Pitch Diameter		Minor Dia.
			max	min	max	min	
1 7/16-18 UNEF	2A	.0015	1.4360	1.4273	1.3999	1.3949	1.3678
	3A	-	1.4375	1.4288	1.4014	1.3977	1.3693
1 7/16-20 UN	2A	.0014	1.4361	1.4280	1.4036	1.3988	1.3748
	3A	-	1.4375	1.4294	1.4050	1.4014	1.3762
1 7/16-28 UN	2A	.0013	1.4362	1.4297	1.4130	1.4088	1.3924
	3A	-	1.4375	1.4310	1.4143	1.4112	1.3937
1 1/2-6 UNC	1A	.0024	1.4976	1.4703	1.3893	1.3772	1.2931
	2A	.0024	1.4976	1.4794	1.3893	1.3812	1.2931
	3A	-	1.5000	1.4818	1.3917	1.3856	1.2955
1 1/2-8 UN	2A	.0022	1.4978	1.4828	1.4166	1.4093	1.3444
	3A	-	1.5000	1.4850	1.4188	1.4133	1.3466
1 1/2-10 UNS	2A	.0019	1.4981	1.4852	1.4331	1.4267	1.3754
1 1/2-12 UNF	1A	.0019	1.4981	1.4809	1.4440	1.4344	1.3959
	2A	.0019	1.4981	1.4867	1.4440	1.4376	1.3959
	3A	-	1.5000	1.4886	1.4459	1.4411	1.3978
1 1/2-14 UNS	2A	.0017	1.4983	1.4880	1.4519	1.4464	1.4107
1 1/2-16 UN	2A	.0016	1.4984	1.4890	1.4578	1.4526	1.4217
	3A	-	1.5000	1.4906	1.4594	1.4555	1.4233
1 1/2-18 UNEF	2A	.0015	1.4985	1.4898	1.4264	1.4574	1.4303
	3A	-	1.5000	1.4913	1.4639	1.4602	1.4318
1 1/2-20 UN	2A	.0014	1.4986	1.4905	1.4661	1.4613	1.4373
	3A	-	1.5000	1.4919	1.4675	1.4639	1.4387
1 1/2-24 UNS	2A	.0013	1.4987	1.4915	1.4716	1.4672	1.4476
1 1/2-28 UN	2A	.0013	1.4987	1.4922	1.4755	1.4713	1.4549
	3A	-	1.5000	1.4935	1.4768	1.4737	1.4562
1 9/16-6 UN	2A	.0024	1.5601	1.5419	1.4518	1.4436	1.3556
	3A	-	1.5625	1.5443	1.4542	1.4481	1.3580
1 9/16-8 UN	2A	.0022	1.5603	1.5453	1.4791	1.4717	1.4069
	3A	-	1.5625	1.5475	1.4813	1.4758	1.4091
1 9/16-12 UN	2A	.0018	1.5607	1.5493	1.5066	1.5007	1.4585
	3A	-	1.5625	1.5511	1.5084	1.5040	1.4603
1 9/16-16 UN	2A	.0016	1.5609	1.5515	1.5203	1.5151	1.4842
	3A	-	1.5625	1.5531	1.5219	1.5180	1.4858
1 9/16-18 UNEF	2A	.0015	1.5610	1.5523	1.5249	1.5199	1.4928
	3A	-	1.5625	1.5538	1.5264	1.5227	1.4943
1 9/16-20 UN	2A	.0014	1.5611	1.5530	1.5286	1.5238	1.4998
	3A	-	1.5625	1.5544	1.5300	1.5264	1.5014
1 5/8-6 UN	2A	.0025	1.6225	1.6043	1.5142	1.5060	1.4180
	3A	-	1.6250	1.6068	1.5167	1.5105	1.4205
1 5/8-8 UN	2A	.0022	1.6228	1.6078	1.5416	1.5342	1.4694
	3A	-	1.6250	1.6100	1.5438	1.5382	1.4716
1 5/8-10 UNS	2A	.0019	1.6231	1.6102	1.5581	1.5517	1.5004
1 5/8-12 UN	2A	.0018	1.6232	1.6118	1.5691	1.5632	1.5210
	3A	-	1.6250	1.6136	1.5709	1.5665	1.5228
1 5/8-14 UNS	2A	.0017	1.6233	1.6130	1.5769	1.5714	1.5357

(Continued)

Table 6. *(Continued)* Unified Screw Threads. Internal Dimensions. Classes 1B, 2B, and 3B.

Size, TPI, and Series	Class	Minor Diameter		Pitch Diameter		Major Dia.	Tap Drill*
		min	max	min	max		
1 7/16-18 UNEF	2B	1.377	1.390	1.4014	1.4079	1.4375	1 3/8
	3B	1.3770	1.3855	1.4014	1.4062	1.4375	
1 7/16-20 UN	2B	1.383	1.395	1.4050	1.4112	1.4375	1 25/64
	3B	1.3830	1.3912	1.4050	1.4096	1.4375	
1 7/16-28 UN	2B	1.399	1.407	1.4143	1.4198	1.4375	1 13/32
	3B	1.3990	1.4051	1.4143	1.4184	1.4375	
1 1/2-6 UNC	1B	1.320	1.350	1.3917	1.4075	1.5000	1 5/16
	2B	1.320	1.350	1.3917	1.4022	1.5000	
	3B	1.3200	1.3396	1.3917	1.3996	1.5000	
1 1/2-8 UN	2B	1.365	1.390	1.4188	1.4283	1.5000	1 23/64
	3B	1.3650	1.3797	1.4188	1.4259	1.5000	
1 1/2-10 UNS	2B	1.392	1.413	1.4350	1.4433	1.5000	
1 1/2-12 UNF	1B	1.410	1.428	1.4459	1.4584	1.5000	1 13/32
	2B	1.410	1.428	1.4459	1.4542	1.5000	
	3B	1.4100	1.4198	1.4459	1.4522	1.5000	
1 1/2-14 UNS	2B	1.423	1.438	1.4536	1.4608	1.5000	
1 1/2-16 UN	2B	1.432	1.446	1.4594	1.4662	1.5000	1 7/16
	3B	1.4320	1.4408	1.4594	1.4645	1.5000	
1 1/2-18 UNEF	2B	1.440	1.452	1.4639	1.4704	1.5000	1 7/16
	3B	1.4400	1.4480	1.4639	1.4687	1.5000	
1 1/2-20 UN	2B	1.446	1.457	1.4675	1.4737	1.5000	1 29/64
	3B	1.4460	1.4537	1.4675	1.4721	1.5000	
1 1/2-24 UNS	2B	1.455	1.465	1.4729	1.4787	1.5000	
1 1/2-28 UN	2B	1.461	1.470	1.4768	1.4823	1.5000	1 15/32
	3B	1.4610	1.4676	1.4768	1.4809	1.5000	
1 9/16-6 UN	2B	1.382	1.413	1.4542	1.4648	1.5625	1 25/64
	3B	1.3820	1.4021	1.4542	1.4622	1.5625	
1 9/16-8 UN	2B	1.427	1.452	1.4813	1.4909	1.5625	1 27/64
	3B	1.4270	1.4422	1.4813	1.4885	1.5625	
1 9/16-12 UN	2B	1.472	1.490	1.5084	1.5106	1.5625	1 15/32
	3B	1.4720	1.4823	1.5084	1.5141	1.5625	
1 9/16-16 UN	2B	1.495	1.509	1.5219	1.5287	1.5625	1 1/2
	3B	1.4950	1.5033	1.5219	1.5270	1.5625	
1 9/16-18 UNEF	2B	1.502	1.515	1.5264	1.5329	1.5625	1 1/2
	3B	1.5020	1.5105	1.5264	1.5312	1.5625	
1 9/16-20 UN	2B	1.508	1.520	1.5300	1.5362	1.5625	1 33/64
	3B	1.5080	1.5162	1.5300	1.5346	1.5625	
1 5/8-6 UN	2B	1.445	1.475	1.5167	1.5274	1.6250	1 29/64
	3B	1.4450	1.4646	1.5167	1.5247	1.6250	
1 5/8-8 UN	2B	1.490	1.515	1.5438	1.5535	1.6250	1 31/64
	3B	1.4900	1.5047	1.5438	1.5510	1.6250	
1 5/8-10 UNS	2B	1.517	1.538	1.5600	1.5683	1.6250	
1 5/8-12 UN	2B	1.535	1.553	1.5709	1.5785	1.6250	1 17/32
	3B	1.5350	1.5448	1.5709	1.5766	1.6250	
1 5/8-14 UNS	2B	1.548	1.564	1.5786	1.5858	1.6250	

(Continued)

Table 5. *(Continued)* **Unified Screw Threads. External Dimensions. Classes 1A, 2A, and 3A.**

Size, TPI, and Series	Class	Allowance	Major Diameter		Pitch Diameter		Minor Dia.
			max	min	max	min	
1 5/8-16 UN	2A	.0016	1.6234	1.6140	1.5828	1.5776	1.5467
	3A	-	1.6250	1.6156	1.5844	1.5805	1.5483
1 5/8-18 UNEF	2A	.0015	1.6235	1.6148	1.5874	1.5824	1.5553
	3A	-	1.6250	1.6163	1.5889	1.5852	1.5568
1 5/8-20 UN	2A	.0014	1.6236	1.6155	1.5911	1.5863	1.5623
	3A	-	1.6250	1.6169	1.5925	1.5889	1.5637
1 5/8-24 UNS	2A	.0013	1.6237	1.6165	1.5966	1.5922	1.5726
1 11/16-6 UN	2A	.0025	1.6850	1.6668	1.5767	1.5684	1.4805
	3A	-	1.6875	1.6693	1.5792	1.5730	1.4830
1 11/16-8 UN	2A	.0022	1.6853	1.6703	1.6041	1.5966	1.5319
	3A	-	1.6875	1.6725	1.6063	1.6007	1.5341
1 11/16-12 UN	2A	.0018	1.6857	1.6743	1.6316	1.6256	1.5835
	3A	-	1.6875	1.6761	1.6334	1.6289	1.5835
1 11/16-16 UN	2A	.0016	1.6859	1.6765	1.6453	1.6400	1.6092
	3A	-	1.6875	1.6781	1.6469	1.6429	1.6108
1 11/16-18 UNEF	2A	.0015	1.6860	1.6773	1.6499	1.6448	1.6178
	3A	-	1.6875	1.6788	1.6514	1.6476	1.6193
1 11/16-20 UN	2A	.0015	1.6860	1.6779	1.6535	1.6487	1.6247
	3A	-	1.6875	1.6794	1.6550	1.6514	1.6262
1 3/4-5 UNC	1A	.0027	1.7473	1.7165	1.6174	1.6040	1.5019
	2A	.0027	1.7473	1.7268	1.6174	1.6085	1.5019
	3A	-	1.7500	1.7295	1.6201	1.6134	1.5046
1 3/4-6 UN	2A	.0025	1.7475	1.7293	1.6392	1.6309	1.5430
	3A	-	1.7500	1.7318	1.6417	1.6354	1.5455
1 3/4-8 UN	2A	.0023	1.7477	1.7327	1.6665	1.6590	1.5943
	3A	-	1.7500	1.7350	1.6688	1.6632	1.5966
1 3/4-10 UNS	2A	.0019	1.7481	1.7352	1.6831	1.6766	1.6254
1 3/4-12 UN	2A	.0019	1.7482	1.7368	1.6941	1.6881	1.6460
	3A	-	1.7500	1.7386	1.6959	1.6914	1.6478
1 3/4-14 UNS	2A	.0017	1.7483	1.7380	1.7019	1.6963	1.6607
1 3/4-16 UN	2A	.0016	1.7484	1.7390	1.7078	1.7025	1.6717
	3A	-	1.7500	1.7406	1.7094	1.7054	1.6733
1 3/4-18 UNS	2A	.0015	1.7485	1.7398	1.7124	1.7073	1.6803
1 3/4-20 UN	2A	.0015	1.7485	1.7404	1.7160	1.7112	1.6872
	3A	-	1.7500	1.7419	7.7175	1.7139	1.6887
1 13/16-6 UN	2A	.0025	1.8100	1.7918	1.7017	1.6933	1.6055
	3A	-	1.8125	1.7943	1.7042	1.6979	1.6080
1 13/16-8 UN	2A	.0023	1.8102	1.7952	1.7290	1.7214	1.6568
	3A	-	1.8125	1.7975	1.7313	1.7256	1.6591
1 13/16-12 UN	2A	.0018	1.8107	1.7933	1.7566	1.7506	1.7085
	3A	-	1.8125	1.8011	1.7584	1.7539	1.7103
1 13/16-16 UN	2A	.0016	1.8109	1.8015	1.7703	1.7650	1.7342
	3A	-	1.8125	1.8031	1.7719	1.7679	1.7358
1 13/16-20 UN	2A	.0015	1.8110	1.8029	1.7785	1.7737	1.7497
	3A	-	1.8125	1.8044	1.7800	1.7764	1.7512

(Continued)

Table 6. *(Continued)* Unified Screw Threads. Internal Dimensions. Classes 1B, 2B, and 3B.

Size, TPI, and Series	Class	Minor Diameter		Pitch Diameter		Major Dia.	Tap Drill*
		min	max	min	max		
1 5/8-16 UN	2B	1.557	1.571	1.5844	1.5912	1.6250	1 9/16
	3B	1.5570	1.5658	1.5844	1.5895	1.6250	
1 5/8-18 UNEF	2B	1.565	1.578	1.5889	1.5954	1.6250	1 9/16
	3B	1.5650	1.5730	1.5889	1.5937	1.6250	
1 5/8-20 UN	2B	1.571	1.582	1.5925	1.5987	1.6250	1 37/64
	3B	1.5710	1.5787	1.5925	1.5971	1.6250	
1 5/8-24 UNS	2B	1.580	1.590	1.5979	1.6037	1.6250	
1 11/16-6 UN	2B	1.507	1.538	1.5792	1.5900	1.6875	1 1/2
	3B	1.5070	1.5271	1.5792	1.5873	1.6875	
1 11/16-8 UN	2B	1.552	1.577	1.6063	1.6160	1.6875	1 9/16
	3B	1.5520	1.5672	1.6063	1.6136	1.6875	
1 11/16-12 UN	2B	1.597	1.615	1.6334	1.6412	1.6875	1 19/32
	3B	1.5970	1.6070	1.6334	1.6392	1.6875	
1 11/16-16 UN	2B	1.620	1.634	1.6469	1.6538	1.6875	1 5/8
	3B	1.6200	1.6283	1.6469	1.6521	1.6875	
1 11/16-18 UNEF	2B	1.627	1.640	1.6514	1.6580	1.6875	1 5/8
	3B	1.6270	1.6355	1.6514	1.6563	1.6875	
1 11/16-20 UN	2B	1.633	1.645	1.6550	1.6613	1.6875	1 41/64
	3B	1.6330	1.6412	1.6550	1.6597	1.6875	
1 3/4-5 UNC	1B	1.534	1.568	1.6201	1.6375	1.7500	1 17/32
	2B	1.534	1.568	1.6201	1.6317	1.7500	
	3B	1.5340	1.5575	1.6201	1.6288	1.7500	
1 3/4-6 UN	2B	1.570	1.600	1.6417	1.6525	1.7500	1 9/16
	3B	1.5700	1.5896	1.6417	1.6498	1.7500	
1 3/4-8 UN	2B	1.615	1.640	1.6688	1.6786	1.7500	1 39/64
	3B	1.6150	1.6297	1.6688	1.6762	1.7500	
1 3/4-10 UNS	2B	1.642	1.663	1.6850	1.6934	1.7500	
1 3/4-12 UN	2B	1.660	1.678	1.6959	1.7037	1.7500	1 21/32
	3B	1.6600	1.6698	1.6959	1.7017	1.7500	
1 3/4-14 UNS	2B	1.673	1.688	1.7036	1.7109	1.7500	
1 3/4-16 UN	2B	1.682	1.696	1.7094	1.7163	1.7500	1 11/16
	3B	1.6820	1.6908	1.7094	1.7146	1.7500	
1 3/4-18 UNS	2B	1.690	1.703	1.7139	1.7205	1.7500	
1 3/4-20 UN	2B	1.696	1.707	1.7175	1.7328	1.7500	1 45/64
	3B	1.6960	1.7037	1.7175	1.7222	1.7500	
1 13/16-6 UN	2B	1.632	1.663	1.7042	1.7151	1.8125	1 45/64
	3B	1.6320	1.6521	1.7042	1.7124	1.8125	
1 13/16-8 UN	2B	1.677	1.702	1.7313	1.7412	1.8125	1 3/4
	3B	1.6770	1.6922	1.7313	1.7387	1.8125	
1 13/16-12 UN	2B	1.722	1.740	1.7584	1.7662	1.8125	1 23/32
	3B	1.7220	1.7320	1.7584	1.7642	1.8125	
1 13/16-16 UN	2B	1.745	1.759	1.7719	1.7788	1.8125	1 3/4
	3B	1.7450	1.7533	1.7719	1.7771	1.8125	
1 13/16-20 UN	2B	1.758	1.770	1.7800	1.7863	1.8125	1 49/64
	3B	1.7580	1.7662	1.7800	1.7847	1.8125	

(Continued)

Table 5. *(Continued)* Unified Screw Threads. External Dimensions. Classes 1A, 2A, and 3A.

Size, TPI, and Series	Class	Allowance	Major Diameter max	Major Diameter min	Pitch Diameter max	Pitch Diameter min	Minor Dia.
1 7/8-6 UN	2A	.0025	1.8725	1.8543	1.7642	1.7558	1.6680
	3A	-	1.8750	1.8568	1.7667	1.7604	1.6705
1 7/8-8 UN	2A	.0023	1.8727	1.8577	1.7915	1.7838	1.7193
	3A	-	1.8750	1.8600	1.7938	1.7881	1.7216
1 7/8-10 UNS	2A	.0019	1.8731	1.8602	1.8081	1.8016	1.7504
1 7/8-12 UN	2A	.0018	1.8732	1.8618	1.8191	1.8131	1.7710
	3A	-	1.8750	1.8636	1.8209	1.8164	1.7728
1 7/8-14 UNS	2A	.0017	1.8733	1.8630	1.8629	1.8213	1.7857
1 7/8-16 UN	2A	.0016	1.8734	1.8640	1.8328	1.8275	1.7967
	3A	-	1.8750	1.8656	1.8344	1.8304	1.7983
1 7/8-18 INS	2A	.0015	1.8735	1.8648	1.8374	1.8323	1.8053
1 7/8-20 UN	2A	.0015	1.8735	1.8654	1.8410	1.8362	1.8122
	3A	-	1.8750	1.8669	1.8425	1.8389	1.8137
1 15/16-6 UN	2A	.0026	1.9349	1.9167	1.8266	1.8181	1.7304
	3A	-	1.9375	1.9193	1.8292	1.8228	1.7330
1 15/16-8 UN	2A	.0023	1.9352	1.9202	1.8540	1.8463	1.7818
	3A	-	1.9375	1.9225	1.8563	1.8505	1.7841
1 15/16 -12 UN	2A	.0018	1.9357	1.9243	1.8816	1.8755	1.8335
	3A	-	1.9375	1.9261	1.8834	1.8789	1.8353
1 15/16-16 UN	2A	.0016	1.9359	1.9265	1.8953	1.8899	1.8592
	3A	-	1.9375	1.9281	1.8969	1.8929	1.8608
1 15/16-20 UN	2A	.0015	1.9360	1.9279	1.9035	1.8986	1.8747
	3A	-	1.9375	1.9294	1.9050	1.9013	1.8762
2-4 1/2 UNC	1A	.0029	1.9971	1.9641	1.8528	1.8385	1.7245
	2A	.0029	1.9971	1.9751	1.8528	1.8433	1.7245
	3A	-	2.0000	1.9780	1.8557	1.8486	1.7247
2-6 UN	2A	.0026	1.9974	1.9792	1.8891	1.8805	1.7929
	3A	-	2.0000	1.9818	1.8917	1.8853	1.7955
2-8 UN	2A	.0023	1.9977	1.9827	1.9165	1.9087	1.8443
	3A	-	2.0000	1.9850	1.9188	1.9130	1.8466
2-10 UNS	2A	.0020	1.9980	1.9851	1.9330	1.9265	1.8753
2-12 UN	2A	.0018	1.9982	1.9868	1.9441	1.9380	1.8960
	3A	-	1.0000	1.9886	1.9459	1.9414	1.8978
2-14 UNS	2A	.0017	1.9983	1.9880	1.9519	1.9462	1.9107
2-16 UN	2A	.0016	1.9984	1.9890	1.9578	1.9524	1.9217
	3A	-	2.0000	1.9906	1.9594	1.9554	1.9233
2-18 UNS	2A	.0015	1.9985	1.9898	1.9624	1.9573	1.9303
2-10 UN	2A	.0015	1.9985	1.9904	1.9660	1.9611	1.9372
	3A	-	2.0000	1.9919	1.9675	1.9638	1.9387
2 1/16-16 UNS	2A	.0016	2.0609	2.0515	2.0203	2.0149	1.9842
	3A	-	2.0625	2.0531	2.0219	2.0179	1.9858
2 1/8-6 UN	2A	.0026	2.1224	2.1042	2.0141	2.0054	1.9179
	3A	-	2.1250	2.1068	2.0167	2.0102	1.9205
2 1/8-8 UN	2A	.0024	2.1226	2.1076	2.0414	2.0335	1.9692
	3A	-	2.1250	2.1100	2.0438	2.0379	1.9716

(Continued)

Table 6. *(Continued)* Unified Screw Threads. Internal Dimensions. Classes 1B, 2B, and 3B.

Size, TPI, and Series	Class	Minor Diameter		Pitch Diameter		Major Dia.	Tap Drill*
		min	max	min	max		
1 7/8-6 UN	2B	1.695	1.725	1.7667	1.7777	1.8750	1 45/64
	3B	1.6950	1.7146	1.7667	1.7749	1.8750	
1 7/8-8 UN	2B	1.740	1.765	1.7938	1.8038	1.8750	1 3/4
	3B	1.7400	1.7547	1.7938	1.8013	1.8750	
1 7/8-10 UNS	2B	1.767	1.788	1.8100	1.8184	1.8750	
1 7/8-12 UN	2B	1.785	1.803	1.8209	1.8287	1.8750	1 25/32
	3B	1.7850	1.7948	1.8209	1.8287	1.8750	
1 7/8-14 UNS	2B	1.798	1.814	1.8286	1.8359	1.8750	
1 7/8-16 UN	2B	1.807	1.821	1.8344	1.8413	1.8750	1 13/16
	3B	1.8070	1.8158	1.8344	1.8396	1.8750	
1 7/8-18 INS	2B	1.815	1.828	1.8389	1.8455	1.8750	
1 7/8-20 UN	2B	1.821	1.832	1.8425	1.8488	1.8750	1 53/64
	3B	1.8210	1.8287	1.8425	1.8472	1.8750	
1 15/16-6 UN	2B	1.757	1.788	1.8292	1.8403	1.9375	1 49/64
	3B	1.7570	1.7771	1.8292	1.8375	1.9375	
1 15/16-8 UN	2B	1.802	1.827	1.8563	1.8663	1.9375	1 51/64
	3B	1.8020	1.8172	1.8563	1.8638	1.9375	
1 15/16-12 UN	2B	1.847	1.865	1.8834	1.8913	1.9375	1 27/32
	3B	1.8470	1.8570	1.8834	1.8893	1.9375	
1 15/16-16 UN	2B	1.870	1.884	1.8969	1.9039	1.9375	1 7/8
	3B	1.8700	1.8783	1.8969	1.9021	1.9375	
1 15/16-20 UN	2B	1.883	1.895	1.9050	1.9114	1.9375	1 57/64
	3B	1.8830	1.8912	1.9050	1.9098	1.9375	
2-4 1/2 UNC	1B	1.759	1.795	1.8557	1.8743	2.0000	1 25/32
	2B	1.759	1.795	1.8557	1.8681	2.0000	
	3B	1.7590	1.7861	1.8557	1.8650	2.0000	
2-6 UN	2B	1.820	1.850	1.8917	1.9028	2.0000	1 53/64
	3B	1.8200	1.8396	1.8917	1.9000	2.0000	
2-8 UN	2B	1.865	1.890	1.9188	1.9289	2.0000	1 7/8
	3B	1.8650	1.8797	1.9188	1.9264	2.0000	
2-10 UNS	2B	1.892	1.913	1.9350	1.9435	2.0000	
2-12 UN	2B	1.910	1.928	1.9459	1.9538	2.0000	1 29/32
	3B	1.9100	1.9198	1.9459	1.9518	2.0000	
2-14 UNS	2B	1.923	1.938	1.9536	1.9610	2.0000	
2-16 UN	2B	1.932	1.946	1.9594	1.9664	2.0000	1 15/16
	3B	1.9320	1.9408	1.9594	1.9646	2.0000	
2-18 UNS	2B	1.940	1.953	1.9639	1.9706	2.0000	
2-20 UN	2B	1.946	1.957	1.9675	1.9739	2.0000	1 61/64
	3B	1.9460	1.9537	1.9675	1.9723	2.0000	
2 1/16-16 UNS	2B	1.995	2.009	2.0219	2.0289	2.0625	2
	3B	1.9950	2.0033	2.0219	2.0271	2.0625	
2 1/8-6 UN	2B	1.945	1.975	2.0167	2.0280	2.1250	1 61/64
	3B	1.9450	1.9646	2.0167	2.0251	2.1250	
2 1/8-8 UN	2B	1.990	2.015	2.0438	2.0540	2.1250	2
	3B	1.9900	2.0047	2.0438	2.0515	2.1250	

(Continued)

Table 5. *(Continued)* Unified Screw Threads. External Dimensions. Classes 1A, 2A, and 3A.

Size, TPI, and Series	Class	Allowance	Major Diameter max	Major Diameter min	Pitch Diameter max	Pitch Diameter min	Minor Dia.
2 1/8-12 UN	2A	.0018	2.1232	2.1118	2.0691	2.0630	2.0210
	3A	-	2.1250	2.1136	2.0709	2.0664	2.0228
2 1/8-16 UN	2A	.0016	2.1234	2.1140	2.0828	2.0774	2.0467
	3A	-	2.1250	2.1156	2.0844	2.0803	2.0483
2 1/8-20 UN	3A	.0015	2.1235	2.1154	2.0910	2.0861	2.0622
	3A	-	2.1250	2.1169	2.0925	2.0888	2.0637
2 3/16-16 UNS	2A	.0016	2.1859	2.1765	2.1453	2.1399	2.1092
	3A	-	2.1875	2.1781	2.1469	2.1428	2.1108
2 1/4-4 1/2 UNC	1A	.0029	2.2471	2.2141	2.1028	2.0882	1.9745
	2A	.0029	2.2471	2.2251	2.1028	2.0931	1.9745
	3A	-	2.2500	2.2280	2.1057	2.0984	1.9774
2 1/4-6 UN	2A	.0026	2.2474	2.2292	2.1391	2.1303	2.0429
	3A	-	2.2500	2.2318	2.1417	2.1351	2.0455
2 1/4-8 UN	2A	.0024	2.2476	2.2326	2.1664	2.1584	2.0942
	3A	-	2.2500	2.2350	2.1688	2.1628	2.0966
2 1/4-10 UNS	2A	.0020	2.2480	2.2351	2.1830	2.1765	2.1253
2 1/4-12 UN	2A	.0018	2.2482	2.2368	2.1941	2.1880	2.1460
	3A	-	2.2500	2.2386	2.1959	2.1914	2.1478
2 1/4-14 UNS	2A	.0017	2.2483	2.2380	2.2019	2.1962	2.1607
2 1/4-16 UN	2A	.0016	2.2484	2.2390	2.2078	2.2024	2.1717
	3A	-	2.2500	2.2406	2.2094	2.2053	2.1733
2 1/4-18 UNS	2A	.0015	2.2485	2.2398	2.2124	2.2073	2.1803
2 1/4-20 UN	2A	.0015	2.2485	2.2404	2.2160	2.2111	2.1872
	3A	-	2.2500	2.2419	2.2175	2.2137	2.1887
2 5/16-16 UNS	2A	.0017	2.3108	2.3014	2.2702	2.2647	2.2341
	3A	-	2.3125	2.3031	2.2719	2.2678	2.2358
2 3/8-6 UN	2A	.0027	2.3723	2.3541	2.2640	2.2551	2.1678
	3A	-	2.3750	2.3568	2.2667	2.2601	2.1705
2 3/8-8 UN	2A	.0024	2.3726	2.3576	2.2914	2.2833	2.2192
	3A	-	2.3750	2.3600	2.2938	2.2878	2.2216
2 3/8-12 UN	2A	.0019	2.3731	2.3617	2.3190	2.3128	2.2709
	3A	-	2.3750	2.3636	2.3209	2.3163	2.2728
2 3/8-16 UN	2A	.0017	2.3733	2.3639	2.3327	2.3272	2.2966
	3A	-	2.3750	2.3656	2.3344	2.3303	2.2983
2 3/8-20 UN	2A	.0015	2.3735	2.3654	2.3410	2.3359	2.3122
	3A	-	2.3750	2.3669	2.3425	2.3387	2.3137
2 7/16-16 UNS	2A	.0017	2.4358	2.4264	2.3952	2.3897	2.3591
	3A	-	2.4375	2.4281	2.3969	2.3928	2.3608
2 1/2-4 UNC	1A	.0031	2.4969	2.4612	2.3345	2.3190	2.1902
	2A	.0031	2.4969	2.4731	2.3345	2.3241	2.1902
	3A	-	2.5000	2.4762	2.3376	2.3298	2.1933
2 1/2-6 UN	2A	.0027	2.4973	2.4791	2.3890	2.3800	2.2928
	3A	-	2.5000	2.4818	2.3917	2.3850	2.2955
2 1/2-8 UN	2A	.0024	2.4976	2.4826	2.4164	2.4082	2.3442
	3A	-	2.5000	2.4850	2.4188	2.4127	2.3466

(Continued)

Table 6. *(Continued)* **Unified Screw Threads. Internal Dimensions. Classes 1B, 2B, and 3B.**

Size, TPI, and Series	Class	Minor Diameter		Pitch Diameter		Major Dia.	Tap Drill*
		min	max	min	max		
2 1/8-12 UN	2B	2.035	2.053	2.0709	2.0788	2.1250	2 1/32
	3B	2.0350	2.0448	2.0709	2.0768	2.1250	
2 1/8-16 UN	2B	2.057	2.071	2.0844	2.0914	2.1250	2 1/16
	3B	2.0570	2.0658	2.0844	2.0896	2.1250	
2 1/8-20 UN	3B	2.071	2.082	2.0925	2.0989	2.1250	2 1/16
	3B	2.0710	2.0787	2.0925	2.0973	2.1250	
2 3/16-16 UNS	2B	2.120	2.134	2.1469	2.1539	2.1875	2 1/8
	3B	2.1200	2.1283	2.1469	2.1521	2.1875	
2 1/4-4 1/2 UNC	1B	2.009	2.045	2.1057	2.1247	2.2500	2
	2B	2.009	2.045	2.1057	2.1183	2.2500	
	3B	2.0090	2.0361	2.1057	2.1152	2.2500	
2 1/4-6 UN	2B	2.070	2.100	2.1417	2.1531	2.2500	2 1/16
	3B	2.0700	2.0896	2.1417	2.1502	2.2500	
2 1/4-8 UN	2B	2.115	2.140	2.1688	2.1792	2.2500	2 1/8
	3B	2.1150	2.1297	2.1688	2.1766	2.2500	
2 1/4-10 UNS	2B	2.142	2.163	2.1850	2.1935	2.2500	
2 1/4-12 UN	2B	2.160	2.178	2.1959	2.2038	2.2500	2 5/32
	3B	2.1600	2.1698	2.1959	2.2018	2.2500	
2 1/4-14 UNS	2B	2.173	2.188	2.2036	2.2110	2.2500	
2 1/4-16 UN	2B	2.182	2.196	2.0294	2.2164	2.2500	2 3/16
	3B	2.1820	2.1908	2.0294	2.2146	2.2500	
2 1/4-18 UNS	2B	2.190	2.203	2.2139	2.2206	2.2500	
2 1/4-20 UN	2B	2.196	2.207	2.2175	2.2239	2.2500	2 3/16
	3B	2.1960	2.2037	2.2175	2.2223	2.2500	
2 5/16-16 UNS	2B	2.245	2.259	2.2719	2.2791	2.3125	2 1/4
	3B	2.2450	2.2533	2.2719	2.2773	2.3125	
2 3/8-6 UN	2B	2.195	2.226	2.2667	2.2782	2.3750	2 3/16
	3B	2.1950	2.2146	2.2667	2.2753	2.3750	
2 3/8-8 UN	2B	2.240	2.265	2.2938	2.3043	2.3750	2 1/4
	3B	2.2400	2.2552	2.2938	2.3017	2.3750	
2 3/8-12 UN	2B	2.285	2.303	2.3209	2.3290	2.3750	58mm
	3B	2.2850	2.2948	2.3209	2.3269	2.3750	
2 3/8-16 UN	2B	2.307	2.321	2.3344	2.3416	2.3750	2 5/16
	3B	2.3070	2.3158	2.3344	2.3398	2.3750	
2 3/8-20 UN	2B	2.321	2.332	2.3425	2.3491	2.3750	2 5/16
	3B	2.3210	2.3287	2.3425	2.3475	2.3750	
2 7/16-16 UNS	2B	2.370	2.384	2.3969	2.4041	2.4375	2 3/8
	3B	2.3700	2.3783	2.3969	2.4023	2.4375	
2 1/2-4 UNC	1B	2.229	2.267	2.3376	2.3578	2.5000	2 7/32
	2B	2.229	2.267	2.3376	2.3511	2.5000	
	3B	2.2290	2.2594	2.3376	2.3477	2.5000	
2 1/2-6 UN	2B	2.320	2.350	2.3917	2.4033	2.5000	2 5/16
	3B	2.3200	2.3396	2.3917	2.4004	2.5000	
2 1/2-8 UN	2B	2.365	2.390	2.4188	2.4294	2.5000	2 3/8
	3B	2.3650	2.3797	2.4188	2.4268	2.5000	

(Continued)

Table 5. *(Continued)* Unified Screw Threads. External Dimensions. Classes 1A, 2A, and 3A.

Size, TPI, and Series	Class	Allowance	Major Diameter max	Major Diameter min	Pitch Diameter max	Pitch Diameter min	Minor Dia.
2 1/2-10 UNS	2A	.0020	2.4980	2.4851	2.4330	2.4263	2.3753
2 1/2-12 UN	2A	.0019	2.4981	2.4867	2.4440	2.4378	2.3959
	3A	-	2.5000	2.4886	2.4459	2.4413	2.3978
2 1/2-14 UNS	2A	.0017	2.4983	2.4880	2.4519	2.4461	2.4107
2 1/2-16 UN	2A	.0017	2.4983	2.4889	2.4577	2.4522	2.4216
	3A	-	2.5000	2.4906	2.4594	2.4553	2.4233
2 1/2-18 UNS	2A	.0016	2.4984	2.4897	2.4623	2.4570	2.4302
2 1/2-20 UN	2A	.0015	2.4985	2.4904	2.4660	2.4609	2.4372
	3A	-	2.5000	2.4919	2.4675	2.4637	2.4387
2 5/8-6 UN	2A	.0027	2.6223	2.6041	2.5140	2.5050	2.4187
	3A	-	2.6250	2.6068	2.5167	2.5099	2.4205
2 5/8-8 UN	2A	.0025	2.6225	2.6075	2.5413	2.5331	2.4691
	3A	-	2.6250	2.6100	2.5438	2.5376	2.4716
2 5/8-12 UN	2A	.0019	2.6231	2.6117	2.5690	2.5628	2.5209
	3A	-	2.6250	2.6136	2.5709	2.5663	2.5228
2 5/8-16 UN	2A	.0017	2.6233	2.6139	2.5827	2.5772	2.5644
	3A	-	2.6250	2.6156	2.5844	2.5803	2.5483
2 5/8-20 UN	2A	.0015	2.6235	2.6154	2.5910	2.5859	2.5622
	3A	-	2.6250	2.6169	2.5925	2.5887	2.5637
2 3/4-4 UNC	1A	.0032	2.7468	2.7111	2.5844	2.5686	2.4401
	2A	.0032	2.7468	2.7230	2.5844	2.5739	2.4401
	3A	-	2.7500	2.7262	2.5876	2.5797	2.4433
2 3/4-6 UN	2A	.0027	2.7473	2.7291	2.6390	2.6299	2.5428
	3A	-	2.7500	2.7318	2.6417	2.6349	2.5455
2 3/4-8 UN	2A	.0025	2.7475	2.7325	2.6663	2.6580	2.5941
	3A	-	2.7500	2.7350	2.6688	2.6625	2.5966
2 3/4-10 UNS	2A	.0020	2.7480	2.7351	2.6830	2.6763	2.6253
2 3/4-12 UN	2A	.0019	2.7481	2.7367	2.6940	2.6878	2.6459
	3A	-	2.7500	2.7386	2.6959	2.6913	2.6478
2 3/4-14 UNS	2A	.0017	2.7483	2.7380	2.7019	2.6961	2.6607
2 3/4-16 UN	2A	.0017	2.7483	2.7389	2.7077	2.7022	2.6716
	3A	-	2.7500	2.7406	2.7094	2.7053	2.6733
2 3/4-18 UNS	2A	.0016	2.7484	2.7397	2.7123	2.7070	2.6802
2 3/4-20 UN	2A	.0015	2.7485	2.7404	2.7160	2.7109	2.6872
	3A	-	2.7500	2.7419	2.7175	2.7137	2.6887
2 7/8-6 UN	2A	.0028	2.8722	2.8540	2.7639	2.7547	2.6677
	3A	-	2.8750	2.8568	2.7667	2.7598	2.6705
2 7/8-8 UN	2A	.0025	2.8725	2.8575	2.7913	2.7829	2.7191
	3A	-	2.8750	2.8600	2.7938	2.7875	2.7216
2 7/8-12 UN	2A	.0019	2.8731	2.8617	2.8190	2.8127	2.7709
	3A	-	2.8750	2.8636	2.8209	2.8162	2.7728
2 7/8-16 UN	2A	.0017	2.8733	2.8639	2.8327	2.8271	2.7966
	3A	-	2.8750	2.8656	2.8344	2.8302	2.7983
2 7/8-20 UN	2A	.0016	2.8734	2.8653	2.8409	2.8357	2.8121
	3A	-	2.8750	2.8669	2.8425	2.8386	2.8137

(Continued)

Table 6. *(Continued)* Unified Screw Threads. Internal Dimensions. Classes 1B, 2B, and 3B.

Size, TPI, and Series	Class	Minor Diameter		Pitch Diameter		Major Dia.	Tap Drill*
		min	max	min	max		
2 1/2-10 UNS	2B	2.392	2.413	2.4350	2.4437	2.5000	
2 1/2-12 UN	2B	2.41	2.428	2.4459	2.4540	2.5000	2 13/32
	3B	2.410	2.4198	2.4459	2.4519	2.5000	
2 1/2-14 UNS	2B	2.423	2.438	2.4536	2.4612	2.5000	
2 1/2-16 UN	2B	2.432	2.466	2.4594	2.4666	2.5000	2 7/16
	3B	2.4320	2.4408	2.4594	2.4648	2.5000	
2 1/2-18 UNS	2B	2.440	2.453	2.4639	2.4708	2.5000	
2 1/2-20 UN	2B	2.446	2.457	2.4675	2.4741	2.5000	2 1/16
	3B	2.4460	2.4537	2.4675	2.4725	2.5000	
2 5/8-6 UN	2B	2.445	2.475	2.5167	2.5285	2.6250	2 7/16
	3B	2.4450	2.4646	2.5167	2.5255	2.6250	
2 5/8-8 UN	2B	2.490	2.515	2.5438	2.5545	2.6250	2 1/2
	3B	2.4900	2.5052	2.5438	2.5518	2.6250	
2 5/8-12 UN	2B	2.535	2.553	2.5709	2.5790	2.6250	2 17/32
	3B	2.5350	2.5448	2.5709	2.5769	2.6250	
2 5/8-16 UN	2B	2.557	2.571	2.5844	2.5619	2.6250	2 9/16
	3B	2.5570	2.5658	2.5844	2.5898	2.6250	
2 5/8-20 UN	2B	2.571	2.582	2.5925	2.5991	2.6250	2 9/16
	3B	2.5710	2.5787	2.5925	2.5975	2.6250	
2 3/4-4 UNC	1B	2.479	2.517	2.5876	2.6082	2.7500	2 1/2
	2B	2.479	2.517	2.5876	2.6013	2.7500	
	3B	2.4790	2.5094	2.5876	2.5979	2.7500	
2 3/4-6 UN	2B	2.570	2.600	2.6417	2.6536	2.7500	2 9/16
	3B	2.5700	2.5869	2.6417	2.6506	2.7500	
2 3/4-8 UN	2B	2.615	2.640	2.6688	2.6796	2.7500	2 5/8
	3B	2.6150	2.6297	2.6688	2.6769	2.7500	
2 3/4-10 UNS	2B	2.642	2.663	2.6850	2.6937	2.7500	
2 3/4-12 UN	2B	2.660	2.678	2.6959	2.7040	2.7500	2 21/32
	3B	2.6600	2.6698	2.6959	2.7019	2.7500	
2 3/4-14 UNS	2B	2.673	2.688	2.7036	2.7112	2.7500	
2 3/4-16 UN	2B	2.682	2.696	2.7094	2.7166	2.7500	2 11/16
	3B	2.6820	2.6908	2.7094	2.7148	2.7500	
2 3/4-18 UNS	2B	2.690	2.703	2.7139	2.7208	2.7500	
2 3/4-20 UN	2B	2.696	2.707	2.7175	2.7241	2.7500	2 11/16
	3B	2.6960	2.7037	2.7175	2.7225	2.7500	
2 7/8-6 UN	2B	2.695	2.725	2.7667	2.7787	2.8750	2 11/16
	3B	2.6950	2.7146	2.7667	2.7757	2.8750	
2 7/8-8 UN	2B	2.740	2.765	2.7938	2.8048	2.8750	2 3/4
	3B	2.7400	2.7552	2.7938	2.8020	2.8750	
2 7/8-12 UN	2B	2.785	2.803	2.8209	2.8291	2.8750	2 25/32
	3B	2.7850	2.7948	2.8209	2.8271	2.8750	
2 7/8-16 UN	2B	2.807	2.821	2.8344	2.8417	2.8750	2 13/16
	3B	2.8070	2.8158	2.8344	2.8399	2.8750	
2 7/8-20 UN	2B	2.821	2.832	2.8425	2.8493	2.8750	2 13/16
	3B	2.8210	2.8287	2.8425	2.8476	2.8750	

(Continued)

Table 5. *(Continued)* **Unified Screw Threads. External Dimensions. Classes 1A, 2A, and 3A.**

Size, TPI, and Series	Class	Allowance	Major Diameter max	Major Diameter min	Pitch Diameter max	Pitch Diameter min	Minor Dia.
3-4 UNC	1A	.0032	2.9968	2.9611	2.8344	2.8183	2.6901
	2A	.0032	2.9968	2.9730	2.8344	2.8237	2.6901
	3A	-	-	2.9762	2.8376	2.8296	2.6933
3-6 UN	2A	.0028	2.9972	2.9790	2.8889	2.8796	2.7927
	3A	-	3.0000	2.9818	2.8917	2.8847	2.7955
3-8 UN	2A	.0026	2.9974	2.9824	2.9162	2.9077	2.8440
	3A	-	3.0000	2.9850	2.9188	2.9124	2.8466
3-10 UNS	2A	.0020	2.9980	2.9851	2.9330	2.9262	2.8753
3-12 UN	2A	.0019	2.9981	2.9867	2.9440	2.9377	2.8959
	3A	-	3.0000	2.9886	2.9459	2.9412	2.8978
3-14 UNS	2A	.0018	2.9982	2.9879	2.9518	2.9459	2.9106
3-16 UN	2A	.0017	2.9983	2.9889	2.9577	2.9521	2.9216
	3A	-	3.0000	2.9906	2.9594	2.9552	2.9233
3-18 UNS	2A	.0016	2.9984	2.9897	2.9623	2.9569	2.9302
3-20 UN	2A	.0016	2.9984	2.9903	2.9659	2.9607	2.9371
	3A	-	3.0000	2.9919	2.9675	2.9636	2.9387
3 1/8-6 UN	2A	.0028	3.1222	3.1040	3.0139	3.0045	2.9177
	3A	-	3.1250	3.1068	3.0167	3.0097	2.9205
3 1/8-8 UN	2A	.0026	3.1224	3.1074	3.0412	3.0326	2.9690
	3A	-	3.1250	3.1100	3.0438	3.0374	2.9716
3 1/8-12 UN	2A	.0019	3.1231	3.1117	3.0690	3.0627	3.0209
	3A	-	3.1250	3.1136	3.0709	3.0662	3.0228
3 1/8-16 UN	2A	.0017	3.1233	3.1139	3.0827	3.0771	3.0466
	3A	-	3.1250	3.1156	3.0844	3.0802	3.0483
3 1/4-4 UNC	1A	.0033	3.2467	3.2110	3.0843	3.0680	2.9400
	2A	.0033	3.2467	3.2229	3.0843	3.0734	2.9400
	3A	-	3.2500	3.2262	3.0876	3.0794	2.9433
3 1/4-6 UN	2A	.0028	3.2472	3.2290	3.1389	3.1294	3.0427
	3A	-	3.2500	3.2318	3.1417	3.1346	3.0455
3 1/4-8 UN	2A	.0026	3.2474	3.2324	3.1662	3.1575	3.0940
	3A	-	3.2500	3.2350	3.1688	3.1623	3.0966
3 1/4-10 UNS	2A	.0020	3.2480	3.2351	3.1830	3.1762	3.1253
3 1/4-12 UN	2A	.0019	3.2481	3.2367	3.1940	3.1877	3.1459
	3A	-	3.2500	3.2386	3.1959	3.1912	3.1478
3 1/4-14 UNS	2A	.0018	3.2482	3.2379	3.2018	3.1959	3.1606
3 1/4-16 UN	2A	.0017	3.2483	3.2389	3.2077	3.2021	3.1716
	3A	-	3.2500	3.2406	3.2094	3.2052	3.1733
3 1/4-18 UNS	2A	.0016	3.2484	3.2397	3.2123	3.2069	3.1802
3 3/8-6 UN	2A	.0029	3.3721	3.3539	3.2638	3.2543	3.1676
	3A	-	3.3750	3.3568	3.2667	3.2595	3.1705
3 3/8-8 UN	2A	.0016	3.3724	3.3574	3.2912	3.2824	3.2190
	3A	-	3.3750	3.3600	3.2938	3.2876	3.2216

(Continued)

Table 6. *(Continued)* Unified Screw Threads. Internal Dimensions. Classes 1B, 2B, and 3B.

Size, TPI, and Series	Class	Minor Diameter		Pitch Diameter		Major Dia.	Tap Drill*
		min	max	min	max		
3-4 UNC	1B	2.729	2.767	2.8376	2.8585	3.0000	2 3/4
	2B	2.729	2.767	2.8376	2.8515	3.0000	
	3B	2.7290	2.7954	2.8376	2.8480	3.0000	
3-6 UN	2B	2.820	2.850	2.8917	2.9038	3.0000	2 13/16
	3B	2.8200	2.8396	2.8917	2.9008	3.0000	
3-8 UN	2B	2.865	2.890	2.9188	2.9299	3.0000	2 7/8
	3B	2.8650	2.8797	2.9188	2.9271	3.0000	
3-10 UNS	2B	2.892	2.913	2.9350	2.9439	3.0000	
3-12 UN	2B	2.910	2.928	2.9459	2.9541	3.0000	74mm
	3B	2.9100	2.9198	2.9459	2.9521	3.0000	
3-14 UNS	2B	2.923	2.938	2.9536	2.9613	3.0000	
3-16 UN	2B	2.932	2.946	2.9594	2.9667	3.0000	2 15/16
	3B	2.9320	2.9408	2.9594	2.9649	3.0000	
3-18 UNS	2B	2.940	2.953	2.9639	2.9709	3.0000	
3-20 UN	2B	2.946	2.957	2.9675	2.9743	3.0000	2 15/16
	3B	2.9460	2.9537	2.9675	2.9726	3.0000	
3 1/8-6 UN	2B	2.945	2.975	3.0167	3.0289	3.0000	
	3B	2.9450	2.9646	3.0167	3.0259	3.0000	
3 1/8-8 UN	2B	2.990	3.015	3.0438	3.0550	3.0000	
	3B	2.9900	3.0052	3.0438	3.0522	3.0000	
3 1/8-12 UN	2B	3.035	3.053	3.0709	3.0791	3.0000	
	3B	3.0350	3.0448	3.0709	3.0771	3.0000	
3 1/8-16 UN	2B	3.057	3.071	3.0844	3.0917	3.0000	
	3B	3.0570	3.0658	3.0844	3.0899	3.0000	
3 1/4-4 UNC	1B	2.979	3.017	3.0876	3.1088	3.2500	
	2B	2.979	3.017	3.0876	3.1017	3.2500	
	3B	2.9790	3.0094	3.0876	3.0982	3.2500	
3 1/4-6 UN	2B	3.070	3.100	3.1417	3.1549	3.2500	
	3B	3.0700	3.0896	3.1417	3.1509	3.2500	
3 1/4-8 UN	2B	3.115	3.140	3.1688	3.1801	3.2500	
	3B	3.1150	3.1297	3.1688	3.1772	3.2500	
3 1/4-10 UNS	2B	3.412	3.163	3.1850	3.1939	3.2500	
3 1/4-12 UN	2B	3.160	3.178	3.1959	3.0241	3.2500	
	3B	3.1600	3.1698	3.1959	3.2021	3.2500	
3 1/4-14 UNS	2B	3.173	3.188	3.2036	3.2113	3.2500	
3 1/4-16 UN	2B	3.182	3.196	3.2094	3.2167	3.2500	
	3B	3.1820	3.1908	3.2094	3.2149	3.2500	
3 1/4-18 UNS	2B	3.190	3.203	3.2139	3.2209	3.2500	
3 3/8-6 UN	2B	3.195	3.225	3.2667	3.2791	3.3750	
	3B	3.1950	3.2146	3.2667	3.2760	3.3750	
3 3/8-8 UN	2B	3.240	3.265	3.2938	3.3052	3.3750	
	3B	3.2400	3.2552	3.2938	3.3023	3.3750	

Reaming Recommended

(Continued)

Table 5. *(Continued)* Unified Screw Threads. External Dimensions. Classes 1A, 2A, and 3A.

Size, TPI, and Series	Class	Allowance	Major Diameter max	Major Diameter min	Pitch Diameter max	Pitch Diameter min	Minor Dia.
3 3/8-12 UN	2A	.0019	3.3731	3.3617	3.3190	3.3126	3.2709
	3A	-	3.3750	3.3636	3.3209	3.3161	3.2728
3 3/8-16 UN	2A	.0017	3.3733	3.3639	3.3327	3.3269	3.2966
	3A	-	3.3750	3.3656	3.3344	3.3301	3.2983
3 1/2-4 UNC	1A	.0033	3.4967	3.4610	3.3343	3.3177	3.1900
	2A	.0033	3.4967	3.4729	3.3343	3.3233	3.1900
	3A	-	3.5000	3.4762	3.3376	3.3293	3.1933
3 1/2-6 UN	2A	.0029	3.4971	3.4789	3.3888	3.3792	3.2926
	3A	-	3.5000	3.4818	3.3917	3.3845	3.2955
3 1/2-8 UN	2A	.0026	3.4974	3.4824	3.4162	3.4074	3.4440
	3A	-	3.5000	3.4850	3.4188	3.3466	3.3466
3 1/2-10 UNS	2A	.0021	3.4979	3.4850	3.4329	3.4260	3.3752
3 1/2-12 UN	2A	.0019	3.4981	3.4867	3.4440	3.4376	3.3959
	3A	-	3.5000	3.4886	3.4459	3.4411	3.3978
3 1/2-14 UNS	2A	.0018	3.4982	3.4879	3.4518	3.4457	3.4106
3 1/2-15 UN	2A	.0017	3.4983	3.4889	3.4577	3.4519	3.4216
	3A	-	3.5000	3.4906	3.4594	3.4551	3.4233
3 1/2-18 UNS	2A	.0017	3.4983	3.4896	3.4622	3.4567	3.4301
3 5/8-6 UN	2A	.0029	3.6221	3.6039	3.5138	3.5041	3.4176
	3A	-	3.6250	3.6068	3.5167	3.5094	3.4205
3 5/8-8 UN	2A	.0027	3.6223	3.6073	3.5411	3.5322	3.4689
	3A	-	3.6250	3.6100	3.5438	3.5371	3.4716
3 5/8-12 UN	2A	.0019	3.6231	3.6117	3.5690	3.5626	3.5209
	3A	-	3.6250	3.6136	3.5709	3.5661	3.5228
3 5/8-16 UN	2A	.0017	3.6233	3.6139	3.5827	3.5769	3.5466
	3A	-	3.6250	3.6156	3.5844	3.5801	3.5483
3 3/4-4 UNC	1A	.0034	3.7466	3.7109	3.5842	3.5674	3.4399
	2A	.0034	3.7466	3.7228	3.5842	3.5730	3.4399
	3A	-	3.7500	3.7262	3.5876	3.5792	3.4433
3 3/4-6 UN	2A	.0029	3.7471	3.7289	3.6388	3.6290	3.5426
	3A	-	3.7500	3.7318	3.6417	3.6344	3.5455
3 3/4-8 UN	2A	.0027	3.7473	3.7323	3.6661	3.6571	3.5939
	3A	-	3.7500	3.7350	3.6688	3.6621	3.5966
3 3/4-10 UNS	2A	.0021	3.7479	3.7350	3.6829	3.6760	3.3652
3 3/4-12 UN	2A	.0019	3.7481	3.7367	3.6940	3.6876	3.6459
	3A	-	3.7500	3.7386	3.6959	3.6911	3.6478
3 3/4-14 UNS	2A	.0018	3.7482	3.7379	3.7018	3.6957	3.6606
3 3/4-16 UN	2A	.0017	3.7483	3.7389	3.7077	3.7019	3.6716
	3A	-	3.7500	3.7406	3.7094	3.7051	3.6733
3 3/4-18 UNS	2A	.0017	3.7483	3.7396	3.7122	3.7067	3.6801
3 7/8-6 UN	2A	.0030	3.8720	3.8538	3.7637	3.7538	3.6675
	3A	-	3.8750	3.8568	3.7667	3.7593	3.6705
3 7/8-8 UN	2A	.0027	3.8723	3.8573	3.7911	3.7820	3.7189
	3A	-	3.8750	3.8600	3.7938	3.7870	3.7216

(Continued)

Table 6. *(Continued)* **Unified Screw Threads. Internal Dimensions. Classes 1B, 2B, and 3B.**

Size, TPI, and Series	Class	Minor Diameter		Pitch Diameter		Major Dia.	Tap Drill*
		min	max	min	max		
3 3/8-12 UN	2B	3.285	3.303	3.3209	3.3293	3.3750	
	3B	3.2850	3.2948	3.3209	3.3272	3.3750	
3 3/8-16 UN	2B	3.307	3.321	3.3344	3.3419	3.3750	
	3B	3.3070	3.3158	3.3344	3.3400	3.3750	
3 1/2-4 UNC	1B	3.229	3.267	3.3376	3.3591	3.5000	
	2B	3.229	3.267	3.3376	3.3519	3.5000	
	3B	3.2290	3.2594	3.3376	3.3484	3.5000	
3 1/2-6 UN	2B	3.320	3.350	3.3917	3.4042	3.5000	
	3B	3.3200	3.3396	3.3917	3.4011	3.5000	
3 1/2-8 UN	2B	3.365	3.390	3.4188	3.4303	3.5000	
	3B	3.3650	3.3797	3.4188	3.4274	3.5000	
3 1/2-10 UNS	2B	3.392	3.413	3.4350	3.4440	3.5000	
3 1/2-12 UN	2B	3.410	3.428	3.4459	3.4543	3.5000	
	3B	3.4100	3.4198	3.4459	3.4522	3.5000	
3 1/2-14 UNS	2B	3.423	3.438	3.4536	3.4615	3.5000	
3 1/2-15 UN	2B	3.432	3.446	3.4594	3.4669	3.5000	
	3B	3.4320	3.4408	3.4594	3.4650	3.5000	
3 1/2-18 UNS	2B	3.440	3.453	3.4639	3.4711	3.5000	
3 5/8-6 UN	2B	3.445	3.475	3.5167	3.5293	3.6250	Reaming Recommended
	3B	3.4450	3.4646	3.5167	3.5262	3.6250	
3 5/8-8 UN	2B	3.490	3.515	3.5438	3.5554	3.6250	
	3B	3.4900	3.5052	3.5438	3.5525	3.6250	
3 5/8-12 UN	2B	3.535	3.553	3.5709	3.5793	3.6250	
	3B	3.5350	3.5448	3.5709	3.5772	3.6250	
3 5/8-16 UN	2B	3.557	3.571	3.5844	3.5919	3.6250	
	3B	3.5570	3.5658	3.5844	3.5900	3.6250	
3 3/4-4 UNC	1B	3.479	3.517	3.5876	3.6094	3.7500	
	2B	3.479	3.517	3.5876	3.6021	3.7500	
	3B	3.4790	3.5094	3.5876	3.5985	3.7500	
3 3/4-6 UN	2B	3.570	3.600	3.6417	3.6544	3.7500	
	3B	3.5700	3.5896	3.6417	3.6512	3.7500	
3 3/4-8 UN	2B	3.615	3.640	3.6688	3.6805	3.7500	
	3B	3.6150	3.6297	3.6688	3.6776	3.7500	
3 3/4-10 UNS	2B	3.642	3.663	3.6850	3.6940	3.7500	
3 3/4-12 UN	2B	3.660	3.678	3.6959	3.7043	3.7500	
	3B	3.6600	3.6698	3.6959	3.7022	3.7500	
3 3/4-14 UNS	2B	3.673	3.688	3.7036	3.7115	3.7500	
3 3/4-16 UN	2B	3.682	3.696	3.7094	3.7169	3.7500	
	3B	3.6820	3.6908	3.7094	3.7150	3.7500	
3 3/4-18 UNS	2B	3.690	3.703	3.7139	3.7211	3.7500	
3 7/8-6 UN	2B	3.695	3.725	3.7667	3.7795	3.8750	
	3B	3.6950	3.7146	3.7667	3.7763	3.8750	
3 7/8-8 UN	2B	3.740	3.765	3.7938	3.8056	3.8750	
	3B	3.7400	3.7552	3.7938	3.8026	3.8750	

(Continued)

Table 5. *(Continued)* **Unified Screw Threads. External Dimensions. Classes 1A, 2A, and 3A.**

Size, TPI, and Series	Class	Allowance	Major Diameter		Pitch Diameter		Minor Dia.
			max	min	max	min	
3 7/8-12 UN	2A	.0020	3.8730	2.8616	3.8189	3.8124	3.7708
	3A	–	3.8750	3.8636	3.8209	3.8160	3.7728
3 7/8-16 UN	2A	.0018	3.8732	3.8636	3.8326	3.8267	3.7965
	3A	–	3.8750	3.8656	3.8344	3.8300	3.7983
4-4 UNC	1A	.0034	3.9966	3.9609	3.8342	3.8172	3.6899
	2A	.0034	3.9966	3.9728	3.8342	3.8229	3.6899
	3A	.0034	4.0000	3.9762	3.8376	3.8291	3.6933
4-8 UN	2A	.0027	3.9973	3.9823	3.9161	3.9070	3.8439
	3A	.0000	1.0000	3.9850	3.9188	3.9120	3.8466
4-12 UN	2A	.0020	3.9980	3.9866	3.9439	3.9374	3.8958
	3A	.0000	4.0000	3.9906	3.9459	3.9410	3.8978
4-16 UN	2A	.0018	3.9982	3.9888	3.9576	3.9517	3.9215
	3A	.0000	4.0000	3.9906	3.9594	3.9550	3.9233
4 1/4-8 UN	2A	.0028	4.2472	4.2322	4.1660	4.1567	4.0938
	3A	.0000	4.2500	4.2350	4.1688	4.1618	4.0966
4 1/4-12 UN	2A	.0020	4.2480	4.2366	4.1939	4.1874	4.1458
	3A	.0000	4.2500	4.2386	4.1959	4.1910	4.1478
4 1/4-16 UN	2A	.0018	4.2482	4.2388	4.2076	4.2017	4.1715
	3A	.0000	4.2500	4.2406	4.2094	4.2050	4.1735
4 1/2-8 UN	2A	.0028	4.4972	4.4822	4.4160	4.4066	4.3438
	3A	.0000	4.5000	4.4850	4.4188	4.4117	4.3466
4 1/2-12 UN	2A	.0020	4.4980	4.4866	4.4439	4.4374	4.3958
	3A	.0000	4.5000	4.4886	4.4459	4.4410	4.3978
4 1/2-16 UN	2A	.0018	4.4982	4.4888	4.4576	4.4517	4.4215
	3A	.0000	4.5000	4.4906	4.4594	4.4550	4.4233
4 3/4-8 UN	2A	.0029	4.7471	4.7321	4.6659	4.6464	4.5937
	3A	.0000	4.7500	4.7350	4.6688	4.5616	4.5966
4 3/4-12 UN	2A	.0020	4.7480	4.7366	4.6939	4.6872	4.6458
	3A	.0000	4.7500	4.7386	4.6959	4.6909	4.6478
4 3/4-16 UN	2A	.0018	4.7482	4.7388	4.7076	4.7015	4.5715
	3A	.0000	4.7500	4.7406	4.7094	4.7049	4.6733
5-8 UN	2A	.0029	4.9971	4.9821	4.9159	4.9062	4.8137
	3A	.0000	5.0000	4.9850	4.9188	4.9116	4.8466
5-12 UN	2A	.0020	4.9980	4.9866	4.9439	4.9372	4.8958
	3A	.0000	5.0000	4.9886	4.9459	4.9409	1.8978
5-16 UN	2A	.0018	4.9982	4.9888	4.9576	4.9515	4.9215
	3A	.0000	5.0000	4.9906	4.9594	4.9549	4.9233
5 1/4-8 UN	2A	.0029	5.2471	5.2321	5.1659	5.1561	5.0937
	3A	.0000	5.2500	5.2350	5.1688	5.1615	5.0966
5 1/4-12 UN	2A	.0020	5.2480	5.2366	5.1939	5.1872	5.1458
	3A	.0000	5.2500	5.2386	5.1959	5.1909	5.1478
5 1/4-16 UN	2A	.0018	5.2482	5.2388	5.2076	5.2015	5.1715
	3A	.0000	5.2500	5.2406	5.2094	5.2049	5.1733
5 1/2-8 UN	2A	.0030	5.4970	5.4820	5.4158	5.4059	5.3436
	3A	.0000	5.5000	5.4850	5.4188	5.4114	5.3466

(Continued)

Table 6. *(Continued)* **Unified Screw Threads. Internal Dimensions. Classes 1B, 2B, and 3B.**

Size, TPI, and Series	Class	Minor Diameter		Pitch Diameter		Major Dia.	Tap Drill*
		min	max	min	max		
3 7/8-12 UN	2B	3.785	3.803	3.8209	3.8294	3.8750	
	3B	3.7850	3.7948	3.8209	3.8273	3.8750	
3 7/8-16 UN	2B	3.807	3.821	3.8344	3.8420	3.8750	
	3B	3.8070	3.8158	3.8344	3.8401	3.8750	
4-4 UNC	1B	3.729	3.767	3.8376	3.8597	4.000	
	2B	3.729	3.767	3.8376	3.8523	4.000	
	3B	3.7290	3.7594	3.8376	3.8487	4.000	
4-8 UN	2B	3.865	3.890	3.9188	3.9307	4.000	
	3B	3.8650	3.8797	3.9188	3.9277	4.000	
4-12 UN	2B	3.910	3.928	3.9459	3.9544	4.000	
	3B	3.9100	3.9198	3.9459	3.9523	4.000	
4-16 UN	2B	3.932	3.946	3.9594	3.9670	4.000	
	3B	3.9320	3.9408	3.9594	3.9651	4.000	
4 1/4-8 UN	2A	4.115	4.140	4.1688	4.1809	4.2500	
	3A	4.1150	4.1297	4.1688	4.1778	4.2500	
4 1/4-12 UN	2A	4.160	4.178	4.1959	4.2044	4.2500	
	3A	4.1600	4.1689	4.1959	4.2023	4.2500	
4 1/4-16 UN	2A	4.182	4.196	4.2094	4.2170	4.2500	
	3A	4.1820	4.1908	4.2094	4.2151	4.2500	
4 1/2-8 UN	2A	4.365	4.390	4.4188	4.4310	4.5000	Reaming Recommended
	3A	4.3650	4.3797	4.4188	4.4280	4.5000	
4 1/2-12 UN	2A	4.410	4.420	4.4459	4.4544	4.5000	
	3A	4.4100	4.4198	4.4459	4.4523	4.5000	
4 1/2-16 UN	2A	4.432	4.446	4.4594	4.4670	4.5000	
	3A	4.4320	4.4408	4.4594	4.4651	4.5000	
4 3/4-8 UN	2A	4.615	4.640	4.6688	4.6812	4.7500	
	3A	4.6150	4.6297	4.6688	4.6781	4.7500	
4 3/4-12 UN	2A	4.660	4.678	4.6959	4.7046	4.7500	
	3A	4.6600	4.6698	4.6959	4.7025	4.7500	
4 3/4-16 UN	2A	4.682	4.696	4.7094	4.7133	4.7500	
	3A	4.6820	4.6908	4.7094	4.7153	4.7500	
5-8 UN	2A	4.865	4.890	4.9188	4.9314	5.0000	
	3A	4.8650	4.8797	4.9188	4.9282	5.0000	
5-12 UN	2A	4.910	4.928	4.9459	4.9546	5.0000	
	3A	4.9100	4.9198	4.9459	4.9525	5.0000	
5-16 UN	2A	4.932	4.946	4.9594	4.9673	5.0000	
	3A	4.9320	4.9408	4.9594	4.9653	5.0000	
5 1/4-8 UN	2A	5.115	5.140	5.1688	5.1815	5.2500	
	3A	5.1150	5.1297	5.1688	5.1783	5.2500	
5 1/4-12 UN	2A	5.160	5.178	5.1959	5.2046	5.2500	
	3A	5.1600	5.1698	5.1959	5.2025	5.2500	
5 1/4-16 UN	2A	5.182	5.196	5.2094	5.2173	5.2500	
	3A	5.1820	5.1908	5.2094	5.2153	5.2500	
5 1/2-8 UN	2A	5.365	5.390	5.4188	5.4317	5.5000	
	3A	5.3650	5.3797	5.4188	5.4285	5.5000	

(Continued)

Table 5. *(Continued)* **Unified Screw Threads. External Dimensions. Classes 1A, 2A, and 3A.**

Size, TPI, and Series	Class	Allowance	Major Diameter		Pitch Diameter		Minor Dia.
			max	min	max	min	
5 1/2-12 UN	2A	.0020	5.4980	5.4866	5.4439	5.4372	5.3958
	3A	.0000	5.5000	5.4886	5.4459	5.4409	5.3978
5 1/2-16 UN	2A	.0018	5.4982	5.4888	5.4576	5.4515	5.4215
	3A	.0000	5.5000	5.4906	5.4594	5.4549	5.4233
5 3/4-8 UN	2A	.0030	5.7470	5.7320	5.6658	5.6558	5.5936
	3A	.0000	5.7500	5.7350	5.6688	5.6613	5.5966
5 3/4-12 UN	2A	.0021	5.7479	5.7365	5.6938	5.5869	5.6457
	3A	.0000	5.7500	5.7386	5.6959	5.6907	5.6478
5 3/4-16 UN	2A	.0019	5.7481	5.7386	5.7075	5.7013	5.6714
	3A	.0000	5.7500	5.7406	5.7094	5.7047	5.6733
6-8 UN	2A	.0030	5.9970	5.9820	5.9158	5.9056	5.8436
	3A	.0000	6.0000	5.9850	5.9188	5.9112	5.8466
6-12 UN	2A	.0021	5.9979	5.9865	5.9438	5.9369	5.8957
	3A	.0000	6.0000	5.9886	5.9459	5.9407	5.8978
6-16 IN	2A	.0019	5.9981	5.9887	5.9575	5.9513	5.9214
	3A	.0000	6.0000	5.9906	5.9594	5.9547	5.9233

Notes: Source, National Bureau of Standards H28 and TC 9-524. Dimensions are generally in accordance with ANSI B1.1-1974 (See Note 2, Table 4). UNS series threads are to be used only if standard series threads do not meet requirements. Class 2A threads with an additive finish increase by the allowance to Class 3A dimensions.

Table 6. *(Conclusion)* **Unified Screw Threads. Internal Dimensions. Classes 1B, 2B, and 3B.**

Size, TPI, and Series	Class	Minor Diameter		Pitch Diameter		Major Dia.	Tap Drill*
		min	max	min	max		
5 1/2-12 UN	2A	5.410	5.428	5.4459	5.4546	5.5000	
	3A	5.4100	5.4198	5.4459	5.4525	5.5000	
5 1/2-16 UN	2A	5.432	5.446	5.4594	5.4673	5.5000	
	3A	5.4320	5.4408	5.4594	5.4653	5.5000	
5 3/4-8 UN	2A	5.615	5.640	5.6688	5.6818	5.7500	
	3A	5.6150	5.6297	5.6688	5.6786	5.7500	
5 3/4-12 UN	2A	5.660	5.678	5.6959	5.7049	5.7500	Reaming Recommended
	3A	5.6600	5.6698	5.6959	5.7026	5.7500	
5 3/4-16 UN	2A	5.682	5.696	5.7094	5.7175	5.7500	
	3A	5.6820	5.6908	5.7094	5.7155	5.7500	
6-8 UN	2A	5.865	5.890	5.9188	5.9320	6.0000	
	3A	5.8650	5.8797	5.9188	5.9287	6.0000	
6-12 UN	2A	5.910	5.928	5.9459	5.9549	6.0000	
	3A	5.9100	5.9198	5.9459	5.9526	6.0000	
6-16 IN	2A	5.932	5.946	5.9594	5.9675	6.0000	
	3A	5.9320	5.9408	5.9594	5.9655	6.0000	

* Tap drill recommendations are for maximum percent of thread engagement (Source, FED-STD-H28/2B). See Tapping Section for alternative sizes. Reaming is suggested for larger sizes.

Notes: Source, National Bureau of Standards H28 and TC 9-524. Dimensions are generally in accordance with ANSI B1.1-1974 (See Note 2, Table 4). UNS series threads are to be used only if standard series threads do not meet requirements.

Table 7. UNJ Series External Threads, Class 3A. Secondary Sizes in italics. *(Source, MIL-S-8879C.)*

Size, TPI, and Series	Major Diameter		Pitch Diameter		Minor Diameter		Root Radius	
	max	min	max	min	max	min	max	min
0-80 UNJF	0.0600	0.0568	0.0519	0.0506	0.0456	0.0435	0.0023	0.0019
1-64 UNJC	0.0730	0.0692	0.0629	0.0614	0.0550	0.0526	0.0028	0.0023
1-72 UNJF	0.0730	0.0695	0.0640	0.0626	0.0570	0.0547	0.0025	0.0021
2-56 UNJC	0.0860	0.0819	0.0744	0.0728	0.0654	0.0627	0.0032	0.0027
2-64 UNJF	0.0860	0.0822	0.0759	0.0744	0.0680	0.0656	0.0028	0.0023
3-48 UNJC	0.0990	0.0945	0.0855	0.0838	0.0750	0.0720	0.0038	0.0031
3-56 UNJF	0.0990	0.0949	0.0874	0.0858	0.0784	0.0757	0.0032	0.0027
4-40 UNJC	0.1120	0.1069	0.0958	0.0939	0.0832	0.0798	0.0045	0.0038
4-48 UNJF	0.1120	0.1075	0.0985	0.0967	0.0880	0.0849	0.0038	0.0031
5-40 UNJC	0.1250	0.1199	0.1088	0.1069	0.0962	0.0928	0.0045	0.0038
5-44 UNJF	0.1250	0.1202	0.1102	0.1083	0.0987	0.0954	0.0041	0.0034
6-32 UNJC	0.1380	0.1320	0.1177	0.1156	0.1019	0.0979	0.0056	0.0047
6-40 UNJF	0.1380	0.1329	0.1218	0.1198	0.1092	0.1057	0.0045	0.0038
8-32 UNJC	0.1640	0.1580	0.1437	0.1415	0.1279	0.1238	0.0056	0.0047
8-36 UNJF	0.1640	0.1585	0.1460	0.1439	0.1320	0.1282	0.0050	0.0042
10-24 UNJC	0.1900	0.1828	0.1629	0.1604	0.1418	0.3368	0.0075	0.0063
10-32 UNJF	0.1900	0.1840	0.1679	0.1674	0.1539	0.1497	0.0056	0.0047
12-24 UNJC	0.2160	0.2088	0.1889	0.1863	0.1678	0.1527	0.0075	0.0063
12-28 UNJF	0.2160	0.2095	0.1928	0.1904	0.1748	0.1702	0.0064	0.0054
12-32 UNJEF	0.2160	0.2100	0.1957	0.1933	0.1799	0.1756	0.0056	0.0047
1/4-20 UNJC	0.2500	0.2419	0.2175	0.2147	0.1922	0.1864	0.0090	0.0075
1/4-28 UNJF	0.2500	0.2435	0.2268	0.2243	0.2088	0.2041	0.0064	0.0054
1/4-32 UNJEF	0.2500	0.2440	0.2297	0.2273	0.2139	0.2096	0.0056	0.0047
5/16-18 UNJC	0.3125	0.3038	0.2764	0.2734	0.2483	0.2420	0.0100	0.0083
5/16-24 UNJF	0.3125	0.3053	0.2854	0.2827	0.2644	0.2591	0.0075	0.0063
5/16-32 UNJEF	0.3125	0.3065	0.2922	0.2898	0.2764	0.2721	0.0056	0.0047
3/8-16 UNJC	0.3750	0.3656	0.3344	0.3311	0.3028	0.2957	0.0113	0.0094
3/8-24 UNJF	0.3750	0.3678	0.3479	0.3450	0.3268	0.3214	0.0075	0.0063
3/8-32 UNJEF	0.3750	0.3690	0.3547	0.3522	0.3389	0.3345	0.0056	0.0047
7/16-14 UNJC	0.4375	0.4272	0.3911	0.3876	0.3550	0.3472	0.0129	0.0107
7/16-16 UNJ	0.4375	0.4281	0.3969	0.3935	0.3653	0.3581	0.0113	0.0094
7/16-20 UNJF	0.4375	0.4294	0.4050	0.4019	0.3797	0.3736	0.0090	0.0075
7/16-28 UNJEF	0.4375	0.4310	0.4143	0.4116	0.3963	0.3914	0.0064	0.0054
1/2-13 UNJC	0.5000	0.4891	0.4500	0.4463	0.4111	0.4028	0.0139	0.0115
1/2-16 UNJ	0.5000	0.4906	0.4594	0.4559	0.4278	0.4205	0.0113	0.0094
1/2-20 UNJF	0.5000	0.4919	0.4675	0.4643	0.4422	0.4360	0.0090	0.0075
1/2-28 UNJEF	0.5000	0.4935	0.4768	0.4740	0.4588	0.4538	0.0064	0.0054
9/16-12 UNJC	0.5625	0.5511	0.5084	0.5045	0.4663	0.4574	0.0150	0.0125
9/16-16 UNJ	0.5625	0.5531	0.5219	0.5184	0.4903	0.4830	0.0113	0.0094
9/16-18 UNJF	0.5625	0.5538	0.5264	0.5230	0.4983	0.4916	0.0100	0.0083
9/16-24 UNJEF	0.5625	0.5553	0.5354	0.5325	0.5144	0.5089	0.0075	0.0063
5/8-11 UNJC	0.6250	0.6129	0.5660	0.5619	0.5201	0.5105	0.0164	0.0136
5/8-12 UNJ	0.6250	0.6136	0.5709	0.5668	0.5288	0.5196	0.0150	0.0125
5/8-16UNJ	0.6250	0.6156	0.5844	0.5808	0.5528	0.5454	0.0113	0.0094
5/8-18 UNJF	0.6250	0.6163	0.5889	0.5854	0.5608	0.5540	0.0100	0.0083
5/8-24 UNJEF	0.6250	0.6178	0.5979	0.5949	0.5768	0.5713	0.0075	0.0063
11/16-12 UNJ	0.6875	0.6761	0.6334	0.6293	0.5913	0.5822	0.0150	0.0125
11/16-16 UNJ	0.6875	0.6781	0.6469	0.6433	0.6153	0.6079	0.0113	0.0094
11/16-24 UNJEF	0.6875	0.6803	0.6604	0.6574	0.6394	0.6338	0.0075	0.0063

(Continued)

Table 8. UNJ Series Internal Threads, Class 3B. Secondary Sizes in italics. *(Source, MIL-S-8879C.)*

Size, TPI, and Series	Minor Diameter		Pitch Diameter		Major Diameter
	min	max	min	max	min
0-80 UNJF	0.0479	0.0511	0.0519	0.0536	0.0600
1-64 UNJC	0.0578	0.0619	0.0629	0.0648	0.0730
1-72 UNJF	0.0595	0.0631	0.0640	0.0659	0.0730
2-56 UNJC	0.0686	0.0732	0.0744	0.0765	0.0860
2-64 UNJF	0.0708	0.0749	0.0759	0.0779	0.0860
3-48 UNJC	0.0787	0.0862	0.0855	0.0877	0.0990
3-56 UNJF	0.0816	0.0862	0.0874	0.0895	0.0990
4-40 UNJC	0.0877	0.0942	0.0958	0.0982	0.1120
4-48 UNJF	0.0917	0.0971	0.0985	0.1008	0.1120
5-40 UNJC	0.1007	0.1072	0.1088	0.1113	0.1250
5-44 UNJF	0.1029	0.1088	0.1102	0.1126	0.1250
6-32 UNJC	0.1076	0.1157	0.1177	0.1204	0.1380
6-40 UNJF	0.1137	0.1202	0.1218	0.1243	0.1380
8-32 UNJC	0.1336	0.1417	0.1437	0.1465	0.1640
8-36 UNJF	0.1370	0.1442	0.1460	0.1487	0.1640
10-24 UNJC	0.1494	0.1600	0.1629	0.1166	0.1900
10-32 UNJF	0.1596	0.1675	0.1697	0.1726	0.1900
12-24 UNJC	0.1754	0.1852	0.1889	0.1922	0.2160
12-28 UNJF	0.1812	0.1896	0.1928	0.1959	0.2160
12-32 UNJEF	0.1856	0.1929	0.1957	0.1988	0.2160
1/4-20 UNJC	0.2013	0.2121	0.2175	0.2211	0.2500
1/4-28 UNJF	0.2152	0.2229	0.2268	0.2300	0.2500
1/4-32 UNJEF	0.2196	0.2263	0.2297	0.2328	0.2500
5/16-18 UNJC	0.2584	0.2690	0.2764	0.2803	0.3125
5/16-24 UNJF	0.2719	0.2799	0.2854	0.2890	0.3725
5/16-32 UNJEF	0.2820	0.2880	0.2922	0.2953	0.3125
3/8-16 UNJC	0.3142	0.3251	0.3344	0.3387	0.3750
3/8-24 UNJF	0.3344	0.3418	0.3479	0.3516	0.3750
3/8-32 UNJEF	0.3446	0.3501	0.3547	0.3580	0.3750
7/16-14 UNJC	0.3680	0.3795	0.3911	0.3957	0.4375
7/16-16 UNJ	0.3767	0.3869	0.3969	0.4014	0.4375
7/16-20 UNJF	0.3888	0.3970	0.4050	0.4091	0.4375
7/16-28 UNJEF	0.4027	0.4086	0.4143	0.4178	0.4375
1/2-13 UNJC	0.4251	0.4368	0.4500	0.4548	0.5000
1/2-16 UNJ	0.4392	0.4488	0.4594	0.4640	0.5000
1/2-20 UNJF	0.4513	0.4591	0.4675	0.4717	0.5000
1/2-28 UNJEF	0.4652	0.4708	0.4768	0.4804	0.5000
9/16-12 UNJC	0.4814	0.4914	0.5084	0.5135	0.5625
9/16-16 UNJ	0.5017	0.5109	0.5219	0.5265	0.5625
9/16-18 UNJF	0.5084	0.5166	0.5264	0.5308	0.5625
9/16-24 UNJEF	0.5219	0.5281	0.5354	0.5392	0.5625
5/8-11 UNJC	0.5365	0.5474	0.5660	0.5714	0.6250
5/8-12 UNJ	0.5439	0.5539	0.5709	0.5762	0.6250
5/8-16UNJ	0.5642	0.5731	0.5844	0.5890	0.6250
5/8-18 UNJF	0.5708	0.5788	0.5889	0.5934	0.6250
5/8-24 UNJEF	0.5844	0.5904	0.5929	0.6013	0.6250
11/16-12 UNJ	0.6064	0.6164	0.6334	0.6387	0.6875
11/16-16 UNJ	0.6267	0.6353	0.6469	0.6515	0.6875
11/16-24 UNJEF	0.6469	0.6547	0.6604	0.6643	0.6875

(Continued)

Table 7. *(Continued)* **UNJ Series External Threads, Class 3A.** Secondary Sizes in italics. *(Source, MIL-S-8879C.)*

Size, TPI, and Series	Major Diameter		Pitch Diameter		Minor Diameter		Root Radius	
	max	min	max	min	max	min	max	min
3/4-10 UNJC	0.7500	0.7371	0.6850	0.6806	0.6345	0.6240	0.0180	0.0150
3/4-12 UNJ	0.7500	0.7386	0.6959	0.6918	0.6538	0.6446	0.0150	0.0125
3/4-16 UNJF	0.7500	0.7406	0.7094	0.7056	0.6778	0.6702	0.0113	0.0094
3/4-20 UNJEF	0.7500	0.7419	0.7175	0.7142	0.6922	0.6859	0.0090	0.0075
13/16-12 UNJ	0.8125	0.8011	0.7584	0.7543	0.7163	0.7072	0.0150	0.0125
13/16-16 UNJ	0.8125	0.8031	0.7719	0.7683	0.7403	0.7329	0.0113	0.0094
13/16-20 UNJEF	0.8125	0.8044	0.7800	0.7767	0.7547	0.7484	0.0090	0.0075
7/8-9 UNJC	0.8750	0.8611	0.8028	0.7981	0.7467	0.7352	0.0200	0.0167
7/8-12 UNJ	0.8750	0.8636	0.8209	0.8168	0.7788	0.7696	0.0150	0.0125
7/8-14 UNJF	0.8750	0.8647	0.8286	0.8245	0.7925	0.7841	0.0129	0.0107
7/8-16 UNJ	0.8750	0.8656	0.8344	0.8308	0.8028	0.7954	0.0113	0.0094
7/8-20 UNJEF	0.8750	0.8669	0.8425	0.8392	0.8172	0.8109	0.0090	0.0075
15/16-12 UNJ	0.9375	0.9261	0.8834	0.8793	0.8413	0.8320	0.0150	0.0125
15/16-16 UNJ	0.9375	0.9281	0.8969	0.8932	0.8653	0.8578	0.0113	0.0094
15/16-20 UNJEF	0.9375	0.9294	0.9050	0.9016	0.8797	0.8733	0.0090	0.0075
1-8 UNJC	1.0000	0.9850	0.9188	0.9137	0.8556	0.8430	0.0226	0.0188
1-12 UNJF	1.0000	0.9886	0.9459	0.9415	0.9038	0.8944	0.0150	0.0125
1-16 UNJ	1.0000	0.9906	0.9594	0.9557	0.9278	0.9203	0.0113	0.0094
1-20 UNJEF	1.0000	0.9919	0.9675	0.9641	0.9422	0.9358	0.0090	0.0075
1 1/16-8 UNJ	1.0625	1.0475	09813	0.9762	0.9182	0.9055	0.0226	0.0188
1 1/16-12 UNJ	1.0625	1.0511	1.0084	1.0042	0.9663	0.9570	0.0150	0.0125
1 1/16-16 UNJ	1.0625	1.0531	1.0219	1.0182	0.9903	0.9828	0.0113	0.0094
1 1/16-18 UNJEF	1.0625	1.0538	1.0264	1.0228	0.9983	0.9914	0.0100	0.0083
1 1/8-7 UNJC	1.1250	1.1086	1.0322	1.0268	0.9600	0.9460	0.0258	0.0214
1 1/8-7 UNJC	1.2500	1.2336	1.1572	1.1517	1.0850	1.0709	0.0218	0.0214
1 1/8-8 UNJ	1.1250	1.1100	1.0438	1.0386	0.9806	0.9661	0.0226	0.0188
1 1/8-12 UNJF	1.1250	1.1136	1.0709	1.0664	1.0288	1.0192	0.0150	0.0125
1 1/8-16 UNJ	1.1250	1.1156	1.0844	1.0807	1.0528	1.0453	0.0113	0.0094
1 1/8-18 UNJEF	1.1250	1.1163	1.0889	1.0853	1.0608	1.0539	0.0100	0.0083
1 3/16-8 UNJ	1.1875	1.1725	1.1063	1.1011	1.0432	1.0304	0.0226	0.0188
1 3/16-12 UNJ	1.1875	1.1761	1.1334	1.1291	1.0913	1.0820	0.0150	0.0125
1 3/16-16 UNJ	1.1875	1.1781	1.1469	1.1431	1.1153	1.1077	0.0113	0.0094
1 3/16-18 UNJEF	1.1875	1.1788	1.1514	1.1478	1.1233	1.1164	0.0100	0.0083
1 1/4-8 UNJ	1.2500	1.2350	1.1688	1.1635	1.1056	1.0928	0.0226	0.0188
1 1/4-12 UNJF	1.2500	1.2386	1.1959	1.1913	1.1538	1.1442	0.0150	0.0125
1 1/4-16 UNJ	1.2500	1.2406	1.2094	1.2056	1.1778	1.1702	0.0113	0.0094
1 1/4-18 UNJEF	1.2500	1.2413	1.2139	1.2103	1.1858	1.1789	0.0100	0.0083
1 5/16-8 UNJ	1.3125	1.2975	1.2313	1.2260	1.1682	1.1553	0.0226	0.0188
1 5/16-12 UNJ	1.3125	1.3011	1.2584	1.2541	1.2163	1.2070	0.0150	0.0125
1 5/16-16 UNJ	1.3125	1.3031	1.2719	1.2681	1.2403	1.2327	0.0113	0.0094
1 5/16-18 UNJEF	1.3125	1.3038	1.2764	1.2728	1.2483	1.2414	0.0100	0.0083
1 3/8-6 UNJC	1.3750	1.3568	1.2667	1.2607	1.3825	1.1664	0.0301	0.0250
1 3/8-8 UNJ	1.3750	1.3600	1.2938	1.2884	1.2306	1.2177	0.0226	0.0188
1 3/8-12 UNJF	1.3750	1.3636	1.3209	1.3162	1.2788	1.2690	0.0150	0.0125
1 3/8-16 UNJ	1.3750	1.3656	1.3344	1.3306	1.3028	1.2952	0.0113	0.0094
1 3/8-18 UNJEF	1.3750	1.3663	1.3389	1.3353	1.3108	1.3039	0.0100	0.0083
1 7/16-8 UNJ	1.4375	1.4225	1.3563	1.3509	1.2932	1.2802	0.0226	0.0188
1 7/16-12 UNJ	1.4375	1.4261	1.3834	1.3790	1.3413	1.3318	0.0150	0.0125
1 7/16-16 UNJ	1.4375	1.4281	1.3969	1.3930	1.3653	1.3576	0.0113	0.0094
1 7/16-18 UNJEF	1.4375	1.4288	1.4014	1.3977	1.3733	1.3663	0.0100	0.0083

(Continued)

Table 8. *(Continued)* **UNJ Series Internal Threads, Class 3B.** Secondary Sizes in italics. *(Source, MIL-S-8879C.)*

Size, TPI, and Series	Minor Diameter		Pitch Diameter		Major Diameter
	min	max	min	max	min
3/4-10 UNJC	0.6526	0.6646	0.6850	0.6907	0.7500
3/4-12 UNJ	0.6689	0.6789	0.6959	0.7013	0.7500
3/4-16 UNJF	0.6892	0.6977	0.7094	0.7143	0.7500
3/4-20 UNJEF	0.7013	0.7081	0.7175	0.7218	0.7500
13/16-12 UNJ	0.7314	0.7414	0.7584	0.7638	0.8125
13/16-16 UNJ	0.7517	0.7602	0.7719	0.7766	0.8125
13/16-20 UNJEF	0.7638	0.7706	0.7800	0.7843	0.8125
7/8-9 UNJC	0.7668	0.7801	0.8028	0.8089	0.8750
7/8-12 UNJ	0.7939	0.8039	0.8209	0.8263	0.8750
7/8-14 UNJF	0.8055	0.8152	0.8286	0.8339	0.8750
7/8-16 UNJ	0.8142	0.8227	0.8344	0.8391	0.8750
7/8-20 UNJEF	0.8263	0.8331	0.8425	0.8468	0.8750
15/16-12 UNJ	0.8564	0.8664	0.8834	0.8889	0.9375
15/16-16 UNJ	0.8767	0.8852	0.8969	0.9018	0.9375
15/16-20 UNJEF	0.8888	0.8956	0.9050	0.9094	0.9375
1-8 UNJC	0.8783	0.8933	0.9188	0.9254	1.0000
1-12 UNJF	0.9189	0.9298	0.9459	0.9516	1.0000
1-16 UNJ	0.9392	0.9477	0.9594	0.9643	1.0000
1-20 UNJEF	0.9513	0.9581	0.9675	0.9719	1.0000
1 1/16-8 UNJ	0.9408	0.9558	0.9813	0.9880	1.0625
1 1/16-12 UNJ	0.9814	0.9914	1.0084	1.0139	1.0625
1 1/16-16 UNJ	1.0017	1.0102	1.0219	1.0268	1.0625
1 1/16-18 UNJEF	1.0084	1.0159	1.0264	1.0310	1.0625
1 1/8-7 UNJC	0.9859	1.0030	1.0322	0.0393	1.1250
1 1/8-8 UNJ	1.0033	1.0183	1.0438	1.0305	1.1250
1 1/8-12 UNJF	1.0439	1.0539	1.0709	1.0768	1.1250
1 1/8-16 UNJ	1.0642	1.0727	1.0844	1.0893	1.1250
1 1/8-18 UNJEF	1.0709	1.0784	1.0889	1.0935	1.1250
1 3/16-8 UNJ	1.0658	1.0808	1.1063	1.1131	1.1875
1 3/16-12 UNJ	1.1064	1.1164	1.1334	1.1390	1.1875
1 3/16-16 UNJ	1.1267	1.1352	1.1469	1.1519	1.1875
1 3/16-18 UNJEF	1.1334	1.1409	1.1514	1.1561	1.1875
1 1/4-8 UNJ	1.1283	1.1433	1.1688	1.1757	1.2500
1 1/4-12 UNJF	1.1689	1.1789	1.1959	1.2019	1.2500
1 1/4-16 UNJ	1.1892	1.1977	1.2094	1.2144	1.2500
1 1/4-18 UNJEF	1.1959	1.2034	1.2139	1.2186	1.2500
1 5/16-8 UNJ	1.1908	1.2058	1.2313	1.2382	1.3125
1 5/16-12 UNJ	1.2314	1.2414	1.2584	1.2640	1.3125
1 5/16-16 UNJ	1.2517	1.2602	1.2719	1.2769	1.3125
1 5/16-18 UNJEF	1.2584	1.2659	1.2754	1.2811	1.3125
1 3/8-6 UNJC	1.2127	1.2327	1.2667	1.2745	1.3750
1 3/8-8 UNJ	1.2533	1.2683	1.2938	1.3008	1.3750
1 3/8-12 UNJF	1.2939	1.3039	1.3209	1.3270	1.3750
1 3/8-16 UNJ	1.3142	1.3227	1.3344	1.3394	1.3750
1 3/8-18 UNJEF	1.3209	1.3284	1.3389	1.3436	1.3750
1 7/16-8 UNJ	1.3158	1.3308	1.3563	1.3634	1.4375
1 7/16-12 UNJ	1.3564	1.3664	1.3834	1.3891	1.4375
1 7/16-16 UNJ	1.3767	1.3852	1.3969	1.4020	1.4375
1 7/16-18 UNJEF	1.3834	1.3909	1.4014	1.4062	1.4375

(Continued)

Table 7. *(Continued)* **UNJ Series External Threads, Class 3A.** Secondary Sizes in italics. *(Source, MIL-S-8879C.)*

Size, TPI, and Series	Major Diameter		Pitch Diameter		Minor Diameter		Root Radius	
	max	min	max	min	max	min	max	min
1 1/2-6 UNJC	1.5000	1.4818	1.3917	1.3856	1.3075	1.2913	0.0301	0.0250
1 1/2-8 UNJ	1.5000	1.4850	1.4188	1.4133	1.3556	1.3426	0.0226	0.0188
1 1/2-12 UNJF	1.5000	1.4886	1.4459	1.4411	1.4038	1.3940	0.0150	0.0125
1 1/2-16 UNJ	1.5000	1.4906	1.4594	1.4555	1.4278	1.4201	0.0113	0.0094
1 1/2-18 UNJEF	1.5000	1.4913	1.4639	1.4602	1.4358	1.4288	0.0100	0.0083
1 9/16-8 UNJ	1.5625	1.5475	1.4813	1.4758	1.4182	1.4051	0.0226	0.0188
1 9/16-12 UNJ	1.5625	1.5511	1.5084	1.5040	1.4663	1.4568	0.0150	0.0125
1 9/16-16 UNJ	1.5625	1.5531	1.5219	1.5180	1.4903	1.4826	0.0113	0.0094
1 9/16-18 UNJEF	1.5625	1.5538	1.5264	1.5227	1.4983	1.4913	0.0100	0.0083
1 5/8-8 UNJ	1.6250	1.6100	1.5438	1.5382	1.4806	1.4675	0.0226	0.0188
1 5/8-12 UNJ	1.6250	1.6136	1.5709	1.5665	1.5288	1.5194	0.0150	0.0125
1 5/8-16 UNJ	1.6250	1.6156	1.5844	1.5805	1.5528	1.5451	0.0113	0.0094
1 5/8-18 UNJEF	1.6250	1.6163	1.5889	1.5852	1.5608	1.5538	0.0100	0.0083
1 11/16-8 UNJ	1.6875	1.6725	1.6063	1.5007	1.5432	1.5300	0.0226	0.0188
1 11/16-12 UNJ	1.6875	1.6761	1.6334	1.6289	1.5913	1.5818	0.0150	0.0125
1 11/16-16 UNJ	1.6875	1.6781	1.6469	1.6429	1.6153	1.6075	0.0113	0.0094
1 11/16-18 UNJEF	1.6875	1.6788	1.6514	1.6476	1.6233	1.6162	0.0100	0.0083
1 3/4-5 UNJC	1.7500	1.7295	1.6201	1.6134	1.5191	1.5002	0.0361	0.0300
1 3/4-8 UNJ	1.7500	1.7350	1.6688	1.5632	1.6056	1.5924	0.0226	0.0188
1 3/4-12 UNJ	1.7500	1.7386	1.6959	1.6914	1.6538	1.6442	0.0150	0.0125
1 3/4-16 UNJ	1.7500	1.7406	1.7094	1.7054	1.6778	1.6700	0.0113	0.0094
1 13/16-8 UNJ	1.8125	1.7975	1.7313	1.7256	1.6682	1.6549	0.0226	0.0188
1 13/16-12 UNJ	1.8125	1.8011	1.7584	1.7539	1.7163	1.7068	0.0150	0.0125
1 13/16-16 UNJ	1.8125	1.8031	1.7719	1.7679	1.7403	1.7325	0.0113	0.0094
1 7/8-8 UNJ	1.8750	1.8600	1.7938	1.7881	1.7306	1.7174	0.0226	0.0188
1 7/8-12 UNJ	1.8750	1.8636	1.8209	1.8164	1.7788	1.7692	0.0150	0.0125
1 7/8-16 UNJ	1.8750	1.8656	1.8344	1.8304	1.8028	1.7950	0.0113	0.0094
1 15/16-8 UNJ	1.9375	1.9225	1.8563	1.8505	1.7932	1.7798	0.0226	0.0188
1 15/16-12 UNJ	1.9375	1.9261	1.8834	1.8789	1.8413	1.8318	0.0150	0.0125
1 15/16-16 UNJ	1.9375	1.9281	1.8969	1.8929	1.8653	1.8575	0.0113	0.0094
2-4 1/2 UNJC	2.0000	1.9780	1.8557	1.8486	1.7434	1.7229	0.0401	0.0334
2-8 UNJ	2.0000	1.9850	1.9188	1.9130	1.8556	1.8423	0.0226	0.0188
2-12 UNJ	2.0000	1.9886	1.9459	1.9414	1.9038	1.8942	0.0150	0.0125
2-16 UNJ	2.0000	1.9906	1.9594	1.9554	1.9278	1.9200	0.0113	0.0094
2 1/8-8 UNJ	2.1250	2.1100	2.0438	2.0379	1.9806	1.9672	0.0226	0.0188
2 1/8-12 UNJ	2.1250	2.1136	2.0709	2.0664	2.0288	2.0192	0.0150	0.0125
2 1/8-16 UNJ	2.1250	2.1156	2.0844	2.0803	2.0528	2.0450	0.0113	0.0094
2 1/4-4 1/2 UNJC	2.2500	2.2280	2.1057	2.0984	1.9934	1.9727	0.0401	0.0334
2 1/4-8 UNJ	2.2500	2.2350	2.1688	2.1628	2.1056	2.0921	0.0226	0.0188
2 1/4-12 UNJ	2.2500	2.2386	2.1959	2.1914	2.1538	2.1442	0.0150	0.0125
2 1/4-16 UNJ	2.2500	2.2406	2.2094	2.2053	2.1778	2.1700	0.0113	0.0094
2 3/8-8 UNJ	2.3750	2.3600	2.2938	2.2878	2.2306	2.2971	0.0226	0.0188
2 3/8-12 UNJ	2.3750	2.3636	2.3209	2.3163	2.2788	2.2692	0.0150	0.0125
2 3/8-16 UNJ	2.3750	2.3656	2.3344	2.3303	2.3028	2.2949	0.0113	0.0094
2 1/2-4 UNJC	2.5000	2.4762	2.3376	2.3298	2.2113	2.1884	0.0401	0.0375
2 1/2-8 UNJ	2.5000	2.4850	2.4188	2.4127	2.3556	2.3420	0.0226	0.0188
2 1/2-12 UNJ	2.5000	2.4886	2.4459	2.4413	2.4038	2.3942	0.0150	0.0125
2 1/2-16 UNJ	2.5000	2.4906	2.4594	2.4553	2.4278	2.4199	0.0113	0.0094

(Continued)

Table 8. *(Continued)* **UNJ Series Internal Threads, Class 3B.** Secondary Sizes in italics. *(Source, MIL-S-8879C.)*

Size, TPI, and Series	Minor Diameter		Pitch Diameter		Major Diameter
	min	max	min	max	min
1 1/2-6 UNJC	1.2127	1.2327	1.2667	1.2745	1.3750
1 1/2-8 UNJ	1.3783	1.3933	1.4188	1.4259	1.5000
1 1/2-12 UNJF	1.1489	1.4289	1.4459	1.4522	1.5000
1 1/2-16 UNJ	1.4392	1.4477	1.4594	1.4645	1.5000
1 1/2-18 UNJEF	1.4459	1.4534	1.4639	1.4687	1.5000
1 9/16-8 UNJ	1.4408	1.4558	1.4813	1.4885	1.5625
1 9/16-12 UNJ	1.4814	1.4914	1.5084	1.5141	1.5625
1 9/16-16 UNJ	1.5017	1.5102	1.5219	1.5270	1.5625
1 9/16-18 UNJEF	1.5084	1.5159	1.5264	1.5312	1.5625
1 5/8-8 UNJ	1.5033	1.5183	1.5438	1.5510	1.6250
1 5/8-12 UNJ	1.5439	1.5539	1.5709	1.5766	1.6250
1 5/8-16 UNJ	1.5642	1.5727	1.5844	1.5895	1.6250
1 5/8-18 UNJEF	1.5709	1.5784	1.5889	1.5937	1.6250
1 11/16-8 UNJ	1.5658	1.5803	1.6063	1.6136	1.6875
1 11/16-12 UNJ	1.6064	1.6164	1.6334	1.6392	1.6875
1 11/16-16 UNJ	1.6267	1.6352	1.6469	1.6521	1.6875
1 11/16-18 UNJEF	1.6334	1.6409	1.6514	1.6563	1.6875
1 3/4-5 UNJ	1.5552	1.5792	1.6201	1.6288	1.7500
1 3/4-8 UNJ	1.6283	1.6433	1.6688	1.6762	1.7500
1 3/4-12 UNJ	1.6689	1.6789	1.6959	1.7017	1.7500
1 3/4-16 UNJ	1.6892	1.6977	1.7094	1.7146	1.7500
1 13/16-8 UNJ	1.6908	1.7058	1.7313	1.7387	1.8125
1 13/16-12 UNJ	1.7314	1.7414	1.7584	1.7642	1.8125
1 13/16-16 UNJ	1.7517	1.7602	1.7719	1.7771	1.8125
1 7/8-8 UNJ	1.7533	1.7683	1.7938	1.8013	1.8750
1 7/8-12 UNJ	1.7939	1.8039	1.8209	1.8267	1.8750
1 7/8-16 UNJ	1.8142	1.8227	1.8344	1.8396	1.8750
1 15/16-8 UNJ	1.8158	1.8308	1.8563	1.8638	1.9375
1 15/16-12 UNJ	1.8564	1.8664	1.8834	1.8893	1.9375
1 15/16-16 UNJ	1.8767	1.8852	1.8969	1.9021	1.9375
2-4 1/2 UNJC	1.7835	1.8102	1.8557	1.8650	2.0000
2-8 UNJ	1.8783	1.8933	1.9188	1.9264	2.0000
2-12 UNJ	1.9189	1.9289	1.9459	1.9518	2.0000
2-16 UNJ	1.9392	1.9477	1.9594	1.9646	2.0000
2 1/8-8 UNJ	2.0033	2.0183	2.0438	2.0515	2.1250
2 1/8-12 UNJ	2.0439	2.0539	2.0709	2.0768	2.1250
2 1/8-16 UNJ	2.0642	2.0727	2.0844	2.0896	2.1250
2 1/4-4 1/2 UNJC	2.0335	2.0602	2.1057	2.1152	2.2500
2 1/4-8 UNJ	2.1283	2.1433	2.1688	2.1766	2.2500
2 1/4-12 UNJ	2.1689	2.1789	2.1959	2.2018	2.2500
2 1/4-16 UNJ	2.1892	2.1977	2.2094	2.2146	2.2500
2 3/8-8 UNJ	2.2533	2.2683	2.2938	2.3017	2.3750
2 3/8-12 UNJ	2.2939	2.3039	2.3209	2.3269	2.3750
2 3/8-16 UNJ	2.3142	2.3227	2.3344	2.3398	2.3750
2 1/2-4 UNJC	2.2565	2.2865	2.3376	2.3477	2.5000
2 1/2-8 UNJ	2.3783	2.3933	2.4188	2.4268	2.5000
2 1/2-12 UNJ	2.4189	2.4289	2.4459	2.4519	2.5000
2 1/2-16 UNJ	2.4392	2.4477	2.4594	2.4648	2.5000

(Continued)

Table 7. *(Continued)* **UNJ Series External Threads, Class 3A.** Secondary Sizes in italics. *(Source, MIL-S-8879C.)*

Size, TPI, and Series	Major Diameter		Pitch Diameter		Minor Diameter		Root Radius	
	max	min	max	min	max	min	max	min
2 5/8-8 UNJ	2.6250	2.6100	2.5438	2.5376	2.4669	2.5438	0.0226	0.0188
2 5/8-12 UNJ	2.6250	2.6136	2.5709	2.5663	2.5288	2.5192	0.0150	0.0125
2 5/8-16 UNJ	2.6250	2.6156	2.5844	2.5803	2.5528	2.5449	0.0113	0.0094
2 3/4-4 UNJC	2.7500	2.7262	2.5876	2.5797	2.4613	2.4382	0.0401	0.0375
2 3/4-8 UNJ	2.7500	2.7350	2.6688	2.6625	2.6056	2.5918	0.0226	0.0188
2 3/4-12 UNJ	2.7500	2.7386	2.6959	2.6913	2.6538	2.6442	0.0150	0.0125
2 3/4-16 UNJ	2.7500	2.7406	2.7094	2.7053	2.6778	2.6699	0.0113	0.0094
2 7/8-8 UNJ	2.8750	2.8600	2.7938	2.7875	2.7306	2.7168	0.0226	0.0188
2 7/8-12 UNJ	2.8750	2.8636	2.8209	2.8162	2.7788	2.7690	0.0150	0.0125
2 7/8-16 UNJ	2.8750	2.8656	2.8344	2.8302	2.8028	2.7948	0.0113	0.0094
3-4 UNJC	3.0000	2.9762	2.8376	2.8296	2.7113	2.6882	0.0451	0.0375
3-8 UNJ	3.0000	2.9850	2.9188	2.9124	2.8556	2.8417	0.0226	0.0188
3-12 UNJ	3.0000	2.9886	2.9459	2.9412	2.9038	2.8940	0.0150	0.0125
3-16 UNJ	3.0000	2.9906	2.9594	2.9552	2.9278	2.9198	0.0113	0.0094
3 1/8-8 UNJ	3.1250	3.1100	3.0438	3.0374	2.9806	2.9667	0.0226	0.0188
3 1/8-12 UNJ	3.1250	3.1136	3.0709	3.0662	3.0288	3.0190	0.0150	0.0125
3 1/8-16 UNJ	3.1250	3.1156	3.0844	3.0802	3.0528	3.0448	0.0113	0.0094
3 1/4-4 UNJC	3.2500	3.2262	3.0876	3.0794	2.9613	2.9380	0.0451	0.0375
3 1/4-8 UNJ	3.2500	3.2350	3.1688	3.1623	3.1056	3.0916	0.0226	0.0188
3 1/4-12 UNJ	3.2500	3.2386	3.1959	3.1912	3.1538	3.1440	0.0150	0.0125
3 1/4-16 UNJ	3.2500	3.2406	3.2094	3.2052	3.1778	3.1698	0.0113	0.0094
3 3/8-8 UNJ	3.3750	3.3600	3.2938	3.2872	3.2306	3.1765	0.0226	0.0188
3 3/8-12 UNJ	3.3750	3.3636	3.3209	3.3161	3.2788	3.2690	0.0150	0.0125
3 3/8-16 UNJ	3.3750	3.3656	3.3344	3.3301	3.3028	3.2947	0.0113	0.0094
3 1/2-8 UNJ	3.5000	3.4850	3.4188	3.4122	3.3556	3.3415	0.0226	0.0188
3 1/2-12 UNJ	3.5000	3.4886	3.4459	3.4411	3.4038	3.3940	0.0150	0.0125
3 1/2-16 UNJ	3.5000	3.4909	3.4594	3.4551	3.4278	3.4197	0.0113	0.0094
3 5/8-8 UNJ	3.6250	3.6100	3.5438	3.5371	3.4806	3.4664	0.0226	0.0188
3 5/8-12 UNJ	3.6250	3.6136	3.5709	3.5661	3.5288	3.5190	0.0150	0.0125
3 5/8-16 UNJ	3.6250	3.6156	3.5844	3.5801	3.5528	3.5447	0.0113	0.0094
3 3/4-8 UNJ	3.7500	3.7350	3.6688	3.6621	3.6056	3.5914	0.0226	0.0188
3 3/4-12 UNJ	3.7500	3.7386	3.6959	3.6911	3.6538	3.6440	0.0150	0.0125
3 3/4-16 UNJ	3.7500	3.7406	3.7094	3.7051	3.6778	3.6697	0.0113	0.0094
3 7/8-8 UNJ	3.8750	3.8600	3.7938	3.7870	3.7306	3.7163	0.0226	0.0188
3 7/8-12 UNJ	3.8750	3.8636	3.8209	3.8160	3.7788	3.7688	0.0150	0.0125
3 7/8-16 UNJ	3.8750	3.8656	3.8344	3.8300	3.8028	3.7946	0.0113	0.0094
4-8 UNJ	4.000	3.9850	3.9188	3.9120	3.8556	3.8413	0.0226	0.0188
4-12 UNJ	4.000	3.9886	3.9459	3.9410	3.7038	3.8938	0.0150	0.0125
4-16 UNJ	4.000	3.9906	3.9594	3.9550	3.9278	3.9196	0.0113	0.0094
4 1/8-12 UNJ	4.1250	4.1136	4.0709	4.0660	4.0288	4.0188	0.0150	0.0125
4 1/8-16 UNJ	4.1250	4.1156	4.0844	4.0800	4.0528	4.0446	0.0113	0.0094
4 1/4-12 UNJ	4.2500	4.2386	4.1959	4.1910	4.1538	4.1438	0.0150	0.0125
4 1/4-16 UNJ	4.2500	4.2406	4.2094	4.2050	4.1778	4.1696	0.0113	0.0094
4 3/8-12 UNJ	4.3750	4.3636	4.3209	4.3160	4.2788	4.2688	0.0150	0.0125
4 3/8-16 UNJ	4.3750	4.3656	4.3344	4.3300	4.3028	4.2946	0.0113	0.0094
4 1/2-12 UNJ	4.5000	4.4886	4.4459	4.4410	4.4038	4.3938	0.0150	0.0125
4 1/2-16 UNJ	4.5000	4.4906	4.4594	4.4550	4.4278	4.4196	0.0113	0.0094

(Continued)

Table 8. *(Continued)* **UNJ Series Internal Threads, Class 3B.** Secondary Sizes in italics. *(Source, MIL-S-8879C.)*

Size, TPI, and Series	Minor Diameter min	Minor Diameter max	Pitch Diameter min	Pitch Diameter max	Major Diameter min
2 5/8-8 UNJ	2.5033	2.5183	2.5438	2.5518	2.6250
2 5/8-12 UNJ	2.5439	2.5539	2.5709	2.5769	2.6250
2 5/8-16 UNJ	2.5642	2.5727	2.5844	2.5898	2.6250
2 3/4-4 UNJC	2.5065	2.5365	2.5876	2.5979	2.7500
2 3/4-8 UNJ	2.6283	2.6433	2.6688	2.6769	2.7500
2 3/4-12 UNJ	2.6689	2.6789	2.6959	2.7019	2.7500
2 3/4-16 UNJ	2.6892	2.6977	2.7094	2.7148	2.7500
2 7/8-8 UNJ	2.7533	2.7683	2.7938	2.8020	2.8750
2 7/8-12 UNJ	2.7939	2.8039	2.8209	2.8271	2.8750
2 7/8-16 UNJ	2.8142	2.8227	2.8344	2.8399	2.8750
3-4 UNJC	2.7565	2.7865	2.8376	2.8480	3.0000
3-8 UNJ	2.8783	2.8933	2.9188	2.9271	3.0000
3-12 UNJ	2.9189	2.9289	2.9459	2.9521	3.0000
3-16 UNJ	2.9392	2.9477	2.9594	2.9649	3.0000
3 1/8-8 UNJ	3.0033	3.0183	3.0438	3.0522	3.1250
3 1/8-12 UNJ	3.0439	3.0539	3.0709	3.0771	3.1250
3 1/8-16 UNJ	3.0642	3.0727	3.0844	3.0899	3.1250
3 1/4-4 UNJ	3.0065	3.0365	3.0876	3.0982	3.2500
3 1/4-8 UNJ	3.1283	3.1433	3.1688	3.1773	3.2500
3 1/4-12 UNJ	3.1689	3.1789	3.1959	3.2021	3.2500
3 1/4-16 UNJ	3.1892	3.1977	3.2094	3.2149	3.2500
3 3/8-8 UNJ	3.2533	3.2683	3.2938	3.3023	3.3750
3 3/8-12 UNJ	3.2939	3.3039	3.3209	3.3272	3.3750
3 3/8-16 UNJ	3.3142	3.3227	3.3344	3.3400	3.3750
3 1/2-8 UNJ	3.3783	3.3933	3.4188	3.4274	3.5000
3 1/2-12 UNJ	3.4189	3.4289	3.4459	3.4522	3.5000
3 1/2-16 UNJ	3.4392	3.4477	3.4594	3.4650	3.5000
3 5/8-8 UNJ	3.5033	3.5183	3.5438	3.5525	3.6250
3 5/8-12 UNJ	3.5439	3.5539	3.5709	3.5772	3.6250
3 5/8-16 UNJ	3.5642	3.5727	3.5844	3.5900	3.6250
3 3/4-8 UNJ	3.6283	3.6433	3.6688	3.6776	3.7500
3 3/4-12 UNJ	3.6689	3.6789	3.6959	3.7022	3.7500
3 3/4-16 UNJ	3.6892	3.6977	3.7094	3.7150	3.7500
3 7/8-8 UNJ	3.7533	3.7683	3.7938	3.8026	3.8750
3 7/8-12 UNJ	3.7939	3.8039	3.8209	3.8273	3.8750
3 7/8-16 UNJ	3.8142	3.8227	3.8344	3.8401	3.8750
4-4 UNJC	3.7565	3.7865	3.8376	3.8487	4.0000
4-8 UNJ	3.8783	3.8933	3.9188	3.9277	4.0000
4-12 UNJ	3.9189	3.9289	3.9459	3.9523	4.0000
4-16 UNJ	3.9392	3.9477	3.9594	3.9651	4.0000
4 1/8-12 UNJ	4.0439	4.0539	4.0709	4.0773	4.1250
4 1/8-16 UNJ	4.0642	4.0727	4.0844	4.0901	4.1250
4 1/4-12 UNJ	4.1689	4.1789	4.1959	4.2023	4.2500
4 1/4-16 UNJ	4.1892	4.1977	4.2094	4.2151	4.2500
4 3/8-12 UNJ	4.2939	4.3039	4.3209	4.3273	4.3750
4 3/8-16 UNJ	4.3142	4.3227	4.3344	4.3401	4.3750
4 1/2-12 UNJ	4.4189	4.4289	4.4459	4.4523	4.5000
4 1/2-16 UNJ	4.4392	4.4477	4.4594	4.4651	4.5000

(Continued)

Table 7. *(Continued)* **UNJ Series External Threads, Class 3A.** Secondary Sizes in italics. *(Source, MIL-S-8879C.)*

Size, TPI, and Series	Major Diameter		Pitch Diameter		Minor Diameter		Root Radius	
	max	min	max	min	max	min	max	min
4 5/8-12 UNJ	4.6250	4.6136	4.5709	4.5659	4.5288	4.5188	0.0150	0.0125
4 5/8-16 UNJ	4.6250	4.6156	4.5844	4.5799	4.5528	4.5445	0.0113	0.0094
4 3/4-12 UNJ	4.7500	4.7386	4.6959	4.6909	4.6538	4.6438	0.0150	0.0125
4 3/4-16 UNJ	4.7500	4.7406	4.7094	4.7049	4.6778	4.6695	0.0113	0.0094
4 7/8-12 UNJ	4.8750	4.8636	4.8209	4.8159	4.7788	4.7688	0.0150	0.0125
4 7/8-16 UNJ	4.8750	4.8656	4.8344	4.8299	4.8028	4.7945	0.0113	0.0094
5-12 UNJ	5.0000	4.9886	4.9459	4.9409	4.9038	4.8938	0.0150	0.0125
5-16 UNJ	5.0000	4.9906	4.9594	4.9549	4.9278	4.9195	0.0113	0.0094
5 1/8-12 UNJ	5.1250	5.1136	5.0709	5.0659	5.0288	5.0188	0.0150	0.0125
5 1/8-16 UNJ	5.1250	5.1156	5.0844	5.0799	5.0528	5.0445	0.0113	0.0094
5 1/4-12 UNJ	5.2500	5.2386	5.1959	5.1909	5.1538	5.1438	0.0150	0.0125
5 1/4-16 UNJ	5.2500	5.2406	5.2094	5.2049	5.1778	5.1695	0.0113	0.0094
5 3/8-12 UNJ	5.3750	5.3636	5.3209	5.3159	5.2788	5.2688	0.0150	0.0125
5 3/8-16 UNJ	5.3750	5.3656	5.3344	5.3299	5.3028	5.2945	0.0113	0.0094
5 1/2-12 UNJ	5.5000	5.4886	5.4459	5.4409	5.4038	5.3938	0.0150	0.0125
5 1/2-16 UNJ	5.5000	5.4906	5.4594	5.4549	5.4278	5.4195	0.0113	0.0094
5 5/8-12 UNJ	5.6250	5.6136	5.5709	5.5657	5.5288	5.5186	0.0150	0.0125
5 5/8-16 UNJ	5.6250	5.6156	5.5844	5.5797	5.5528	5.5443	0.0113	0.0094
5 3/4-12 UNJ	5.7500	5.7386	5.6959	5.6907	5.6538	5.6436	0.0150	0.0125
5 3/4-16 UNJ	5.7500	5.7406	5.7094	5.7047	5.6778	5.6693	0.0113	0.0094
5 7/8-12 UNJ	5.8750	5.836	5.8209	5.8157	5.7788	5.7686	0.0150	0.0125
5 7/8-16 UNJ	5.8750	5.8656	5.8344	5.8297	5.8028	5.7943	0.0113	0.0094
6-12 UNJ	6.0000	5.9886	5.9459	5.9407	5.9038	5.8936	0.0150	0.0125
6-16 UNJ	6.0000	5.9906	5.9594	5.9547	5.9278	5.9193	0.0113	0.0094

All dimensions in inches. Source, MIL-S-8879C.

Table 8. *(Continued)* **UNJ Series Internal Threads, Class 3B.** Secondary Sizes in italics. *(Source, MIL-S-8879C.)*

Size, TPI, and Series	Minor Diameter		Pitch Diameter		Major Diameter
	min	max	min	max	min
4 5/8-12 UNJ	4.5439	4.5539	4.5709	4.5775	4.6250
4 5/8-16 UNJ	4.5642	4.5727	4.5844	4.5903	4.6250
4 3/4-12 UNJ	4.6689	4.6789	4.6959	4.7025	4.7500
4 3/4-16 UNJ	4.6892	4.6977	4.7094	4.7153	4.7500
4 7/8-12 UNJ	4.7939	4.8039	4.8209	4.8275	4.8750
4 7/8-16 UNJ	4.8142	4.8227	4.8344	4.8403	4.8750
5-12 UNJ	4.9189	4.9289	4.9459	4.9525	5.0000
5-16 UNJ	4.9392	4.9477	4.9594	4.9653	5.0000
5 1/8-12 UNJ	5.0439	5.0539	5.0709	5.0775	5.1250
5 1/8-16 UNJ	5.0642	5.0727	5.0844	5.0903	5.1250
5 1/4-12 UNJ	5.1689	5.1789	5.1959	5.2025	5.2500
5 1/4-16 UNJ	5.1892	5.1977	5.2094	5.2153	5.2500
5 3/8-12 UNJ	5.2939	5.3039	5.3209	5.3275	5.3750
5 3/8-16 UNJ	5.3142	5.3227	5.3344	5.3403	5.3750
5 1/2-12 UNJ	5.4189	5.4289	5.4459	5.4525	5.5000
5 1/2-16 UNJ	5.4392	5.4477	5.4594	5.4653	5.5000
5 5/8-12 UNJ	5.5439	5.5539	5.5709	5.5776	5.6250
5 5/8-16 UNJ	5.5642	5.5727	5.5844	5.5905	5.6250
5 3/4-12 UNJ	5.6689	5.6789	5.6959	5.7026	5.7500
5 3/4-16 UNJ	5.6892	5.6977	5.7094	5.7155	5.7500
5 7/8-12 UNJ	5.7939	5.8039	5.8209	5.8276	5.8750
5 7/8-16 UNJ	5.8142	5.8227	5.8344	5.8405	5.8750
6-12 UNJ	5.9189	5.9289	5.9459	5.9526	6.0000
6-16 UNJ	5.9392	5.9477	5.9594	5.9655	6.0000

All dimensions in inches. Source, MIL-S-8879C.

Table 9. Thread Form for UN, UNR, and UNJ Threads. *(Source, Cleveland Twist Drill .)*

TPI	Pitch P	Height of External Thread		Addendum	Crest Flat	Root Flat (Basic Profile)	Root Radii			
		UN, UNR	UNJ	UN, UNR, UNJ	UN, UNR, UNJ	UN	UN		UNJ	
							Max.	Min.	Max.	Min.
		0.70833H	0.66667H	0.375H	0.125P	0.250P	0.14434P	0.10825P	0.18042P	0.15011P
80	0.125	.0077	.0072	.0041	.0016	.0031	.0018	.0014	.0023	.0019
72	.0139	.0085	.0080	.0045	.0017	.0035	.0020	.0015	.0025	.0021
64	.0156	.0096	.0090	.0051	.0020	.0039	.0023	.0017	.0028	.0023
56	.0179	.0110	.0103	.0058	.0022	.0045	.0026	.0019	.0032	.0027
48	.0208	.0128	.0120	.0068	.0026	.0052	.0030	.0023	.0038	.0031
44	.0227	.0139	.0131	.0074	.0028	.0057	.0033	.0025	.0041	.0034
40	.0250	.0153	.0144	.0081	.0031	.0062	.0036	.0027	.0045	.0038
36	.0278	.0170	.0160	.0090	.0035	.0069	.0040	.0030	.0050	.0042
32	.0312	.0192	.0180	.0101	.0039	.0078	.0045	.0034	.0056	.0047
28	.0357	.0219	.0206	.0116	.0045	.0089	.0052	.0039	.0064	.0054
27	.0370	.0227	.0214	.0120	.0046	.0093	.0053	.0040	.0067	.0056
24	.0417	.0256	.0241	.0135	.0052	.0104	.0060	.0045	.0075	.0063
20	.0500	.0307	.0289	.0162	.0062	.0125	.0072	.0054	.0090	.0075
18	.0556	.0341	.0321	.0180	.0069	.0139	.0080	.0060	.0100	.0088
16	.0625	.0383	.0361	.0203	.0078	.0156	.0090	.0068	.0113	.0094
14	.0714	.0438	.0412	.0232	.0089	.0179	.0103	.0077	.0129	.0107
13	.0770	.0472	.0444	.0250	.0096	.0193	.0111	.0083	.0139	.0115
12	.0833	.0511	.0481	.0271	.0104	.0208	.0120	.0090	.0150	.0125
11-1/2	.08696	.0533	.0502	.0282	.0109	.0217	.0126	.0094	.0157	.0130
11	.0909	.0558	.0525	.0295	.0114	.0227	.0131	.0098	.0164	.0136
10	.100	.0613	.0577	.0325	.0125	.0250	.0144	.0108	.0180	.0150
9	.1111	.0682	.0642	.0361	.0139	.0278	.0160	.0120	.0201	.0167
8	.125	.0767	.0722	.0406	.0156	.0312	.0180	.0135	.0226	.0188
7	.1429	.0876	.0825	.0464	.0179	.0357	.0206	.0155	.0258	.0214
6	.1667	.1022	.0962	.0541	.0208	.0417	.0241	.0180	.0301	.0250
5	.200	.1227	.1155	.0650	.0250	.0500	.0239	.0216	.0361	.0300
4-1/2	.225	.1363	.1283	.0722	.0278	.0556	.0321	.0241	.0401	.0334
4	.250	.1534	.1443	.0812	.0312	.0625	.0361	.0271	.0451	.0375

All dimensions in inches.

Table 10. Unified Class 5 Interference-Fit Thread Dimensions.

	External Threads						
Class	NC-5 HF		NC-5 CSF NC-5 ONF		All Classes-Diameters		
Size	Major Dia.		Major Dia.		Pitch		Minor
	Max.	Min.	Max.	Min.	Max.	Min.	Max.
1/4-20	.2470	.2408	.2470	.2408	.2230	.2204	.1932
5/16-18	.3080	.3020	.3090	.3030	.2829	.2799	.2508
3/8-16	.3690	.3626	.3710	.3646	.3414	.3382	.3053
7/16-14	.4305	.4233	.4330	.4258	.3991	.3955	.3579
1/2-13	.4920	.4846	.4950	.4876	.4584	.4547	.4140
9/16-12	.5540	.5460	.5580	.5495	.5176	.5136	.4695
5/8-11	.6140	.6056	.6195	.6111	.5758	.5716	.5233
3/4-10	.7360	.7270	.7440	.7350	.6955	.6910	.6378
7/8-9	.8600	.8502	.8685	.8587	.8144	.8095	.7503
1-8	.9835	.9727	.9935	.9827	.9316	.9262	.8594
1 1/8-7	1.1070	1.0952	1.1180	1.1062	1.0465	1.0406	.9640
1 1/4-7	1.2320	1.2200	1.2430	1.2312	1.1715	1.1656	1.0890
1 3/8-6	1.3560	1.3410	1.3680	1.3538	1.2839	1.2768	1.1877
1 1/2-6	1.4810	1.4670	1.4930	1.4788	1.4089	1.4018	1.3127

	Internal Threads						
Class	NC-5 IF			NC-5 INF		Both Classes	
Size	Minor Dia.		Tap Drill	Minor Dia.		Tap Drill	Pitch Dia.		Major
	Min.	Max.		Min.	Max.		Min.	Max.	Min.
1/4-20	.1960	.2060	.2031	.1960	.2060	.2031	.2175	.2201	.2532
5/16-18	.2520	.2630	.2610	.2520	.2630	.2610	.2764	.2794	.3161
3/8-16	.3070	.3180	.3160	.3070	.3180	.3160	.3344	.3376	.3790
7/16-14	.3740	.3810	.3750	.3600	.3720	.3680	.3911	.3947	.4421
1/2-13	.4310	.4400	.4331	.4170	.4290	.4219	.4500	.4537	.5050
9/16-12	.4880	.4970	.4921	.4720	.4850	.4844	.5084	.5124	.5679
5/8-11	.5440	.5540	.5469	.5270	.5400	.5313	.5660	.5702	.6309
3/4-10	.6670	.6780	.6719	.6420	.6550	.6496	.6850	.6895	.7565
7/8-9	.7770	.7890	.7812	.7550	.7690	.7656	.8028	.8077	.8822
1-8	.8900	.9040	.8906	.8650	.8800	.8750	.9188	.9242	1.0081
1 1/8-7	1.0000	1.0150	1.0000	.9700	.9860	.9844	1.0322	1.0381	1.1343
1 1/4-7	1.1250	1.1400	1.1250	1.0950	1.1110	1.1094	1.1572	1.1631	1.2593
1 3/8-6	1.2290	1.2470	1.2344	1.1950	1.2130	1.2031	1.2667	1.2738	1.3858
1 1/2-6	1.3540	1.3720	1.3594	1.3200	1.3380	1.3281	1.3917	1.3988	1.5108

All dimensions in inches. For descriptions of each class of interference-fit thread, refer to "Unified threads by class" earlier in this section. Source, ASA B1.12-1972 as published by the American Society of Mechanical Engineers.

Table 11. Basic Formulas for Calculating UNM Thread Dimensions and Tolerances.

Symbol	Dimension	Equation
P	Pitch (number of threads per unit of length).	$P = 1 \div$ threads per inch, or $P = 1 \div$ threads per mm
H	Height of fundamental triangle.	$H = 0.866025404P$
D	Major dia. (basic).	
D	Major dia., int. thread.	$D + 0.072P$
d	Minor dia., ext. thread.	$D - 1.20P$
D_1	Minor dia., int. thread.	$D - 1.04P$
E	Pitch dia. (basic).	$D - 0.64952P$
h	Thread height (basic).	$0.556P$
h_c	Depth of engagement.	$0.52P$
r_n	Int. root radius.	$0.072P$
r_s	Ext. root radius.	$0.158P$

Other dimensions in accordance with formulas for UN threads in Table 4.

Tolerances on Design Dimensions*		
External Thread (−)	Dimension	Internal Thread (+)
$0.12P + 0.006$	Major Diameter	$0.168P + 0.008$
$0.08P + 0.008$	Pitch Diameter	$0.08P + 0.008$
$0.16P + 0.008$	Minor Diameter	$0.32P + 0.012$

*Tolerances are applied to the dimensions specified in Table 12.

Table 12. Unified Miniature Thread Series. *(Source, FED-STD-H28/5.)*

UNM - External Threads							
Size (External)	Pitch mm	Major Dia. mm		Pitch Dia. mm		Minor Dia. mm	
		Max.	Min.	Max.	Min.	Max.	Min.
0.30 UNM	0.080	0.300	0.284	0.248	0.234	0.204	0.183
0.35 UNM	0.090	0.350	0.333	0.292	0.277	0.242	0.220
0.40 UNM	0.100	0.400	0.382	0.335	0.319	0.280	0.256
0.45 UNM	0.100	0.450	0.432	0.385	0.369	0.330	0.306
0.50 UNM	0.125	0.500	0.479	0.419	0.401	0.350	0.322
0.55 UNM	0.125	0.550	0.529	0.469	0.451	0.400	0.372
0.60 UNM	0.150	0.600	0.576	0.503	0.483	0.420	0.388
0.70 UNM	0.175	0.700	0.673	0.586	0.564	0.490	0.454
0.80 UNM	0.200	0.800	0.770	0.670	0.646	0.560	0.520
0.90 UNM	0.225	0.900	0.867	0.754	0.728	0.630	0.586
1.00 UNM	0.250	1.000	0.964	0.838	0.810	0.700	0.652
1.10 UNM	0.250	1.100	1.064	0.938	0.910	0.800	0.752
1.20 UNM	0.250	1.200	1.164	1.038	1.010	0.900	0.852
1.40 UNM	0.300	1.400	1.358	1.205	1.173	1.040	0.984
Size (External)	Pitch tpi	Major Dia. inch		Pitch Dia. inch		Minor Dia. inch	
		Max.	Min.	Max.	Min.	Max.	Min.
0.30 UNM	318	0.0118	0.0112	0.0098	0.0092	0.0080	0.0072
0.35 UNM	282	0.0138	0.0131	0.0115	0.0109	0.0095	0.0086

(Continued)

Table 12. *(Continued)* **Unified Miniature Thread Series.** *(Source, FED-STD-H28/5.)*

UNM - External Threads							
Size (External)	**Pitch** tpi	**Major Dia. inch**		**Pitch Dia. inch**		**Minor Dia. inch**	
		Max.	**Min.**	**Max.**	**Min.**	**Max.**	**Min.**
0.40 UNM	254	0.0157	0.0150	0.0132	0.0126	0.0110	0.0101
0.45 UNM	254	0.0177	0.0170	0.0152	0.0145	0.0130	0.0120
0.50 UNM	203	0.0197	0.0189	0.0165	0.0158	0.0138	0.0127
0.55 UNM	203	0.0217	0.0208	0.0185	0.0177	0.0157	0.0146
0.60 UNM	169	0.0236	0.0227	0.0198	0.0190	0.0165	0.0153
0.70 UNM	145	0.0276	0.0265	0.0231	0.0222	0.0193	0.0179
0.80 UNM	127	0.0315	0.0303	0.0264	0.0254	0.0220	0.0205
0.90 UNM	113	0.0354	0.0341	0.0297	0.0287	0.0248	0.0231
1.00 UNM	102	0.0394	0.0380	0.0330	0.0319	0.0276	0.0257
1.10 UNM	102	0.0433	0.0419	0.0369	0.0358	0.0315	0.0296
1.20 UNM	102	0.0472	0.00458	0.0409	0.0397	0.0354	0.0335
1.40 UNM	85	0.0551	0.0535	0.0474	0.0462	0.0409	0.0387
UNM - Internal Threads							
Size (Internal)	**Pitch** mm	**Minor Dia. mm**		**Pitch Dia. mm**		**Major Dia. mm**	
		Min.	**Max.**	**Min.**	**Max.**	**Min.**	**Max.**
0.30 UNM	0.080	0.217	0.254	0.248	0.262	0.306	0.327
0.35 UNM	0.090	0.256	0.297	0.292	0.307	0.356	0.380
0.40 UNM	0.100	0.296	0.340	0.335	0.351	0.407	0.432
0.45 UNM	0.100	0.346	0.390	0.385	0.401	0.457	0.482
0.50 UNM	0.125	0.370	0.422	0.419	0.437	0.509	0.538
0.55 UNM	0.125	0.420	0.472	0.469	0.487	0.559	0.588
0.60 UNM	0.150	0.444	0.504	0.503	0.523	0.611	0.644
0.70 UNM	0.175	0.518	0.586	0.586	0.608	0.713	0.750
0.80 UNM	0.200	0.592	0.668	0.670	0.694	0.814	0.856
0.90 UNM	0.225	0.666	0.750	0.754	0.780	0.916	0.962
1.00 UNM	0.250	0.740	0.832	0.838	0.866	1.018	1.068
1.10 UNM	0.250	0.840	0.932	0.938	0.966	1.118	1.168
1.20 UNM	0.250	0.940	1.032	1.038	1.066	1.218	1.268
1.40 UNM	0.300	1.088	1.196	1.205	1.237	1.422	1.480
Size (Internal)	**Pitch** tpi	**Minor Dia. inch**		**Pitch Dia. inch**		**Major Dia. inch**	
		Min.	**Max.**	**Min.**	**Max.**	**Min.**	**Max.**
0.30 UNM	318	0.0085	0.0100	0.0098	0.0104	0.0120	0.0129
0.35 UNM	282	0.0101	0.0117	0.0115	0.0121	0.0140	0.0149
0.40 UNM	254	0.0117	0.0134	0.0132	0.0138	0.0160	0.0170
0.45 UNM	254	0.0136	0.0154	0.0152	0.0158	0.0180	0.0190
0.50 UNM	203	0.0146	0.0166	0.0165	0.0172	0.0200	0.0212
0.55 UNM	203	0.0165	0.0186	0.0185	0.0192	0.0220	0.0231
0.60 UNM	169	0.0175	0.0198	0.0198	0.0206	0.0240	0.0254

(Continued)

Table 12. *(Continued)* **Unified Miniature Thread Series.** *(Source, FED-STD-H28/5.)*

Size (Internal)	Pitch tpi	Minor Dia. inch		Pitch Dia. inch		Major Dia. inch	
		Min.	Max.	Min.	Max.	Min.	Max.
0.70 UNM	145	0.0204	0.0231	0.0231	0.0240	0.0281	0.0295
0.80 UNM	127	0.0233	0.0263	0.0264	0.0273	0.0321	0.0337
0.90 UNM	113	0.0262	0.0295	0.0297	0.0307	0.0361	0.0379
1.00 UNM	102	0.0291	0.0327	0.0330	0.0341	0.0401	0.0420
1.10 UNM	102	0.0331	0.0367	0.0369	0.0380	0.0440	0.0460
1.20 UNM	102	0.0370	0.0406	0.0409	0.0420	0.0480	0.0499
1.40 UNM	85	0.0428	0.0471	0.0474	0.0487	0.0560	0.0583

All dimensions are provided in both millimeters and inches, as stated in FED-STD-H28/5.
Preferred sizes are in normal type. Secondary sizes are in italics.

60° Stub thread

The stub version of the 60° thread was actually designed for use as an Acme style thread for producing transverse movement. Shear strength is higher than with the Acme form, but load transmission efficiency is lower. The angle between the flanks of the thread is 60°, and the threads are truncated top and bottom. The width of flat at the screw crest is equal to $0.250P$ and the flat at the root equals $0.227P$. Threads have a basic height of $0.433P$, a basic thickness at pitch line of $0.50P$, and are symmetrical about a line perpendicular to the axis of the screw. A typical thread designation is 1 1/2–9 SPL 60° FORM, followed by limits of size. *Figure 6* illustrates the thread form. FED-STD-H28/19 contains information on specifications, allowances, and tolerances.

ISO Metric threads

As described earlier, the ISO Metric and ANSI/ASME M Profile thread shares the basic profile of the Unified inch screw thread. The tolerances in this system have a heritage that

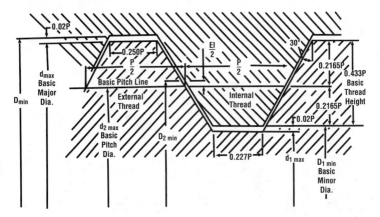

Figure 6. 60° Stub thread form.

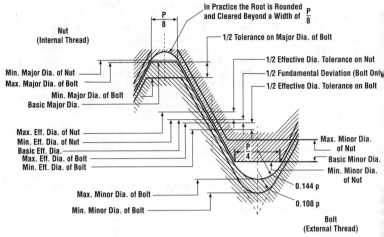

In Practice the Root is Rounded and Cleared Beyond a Width of $\frac{P}{8}$

Figure 7. ISO Metric/M Profile thread form.

can be traced to the ISO system of limits and fits, and thread designations are intended to provide full tolerance information about a thread, much as the Class 1–5 system identifies the basic fit of Unified threads. The metric identification system first identifies the diameter in millimeters, then places a multi sign before the pitch. A dash then separates the thread dimensions from the tolerance information. Tolerance grades are numbers, and the number is followed by a tolerance position letter. A typical metric thread designation would be M10 × 1.5-6g. To indicate a gaging system number and a left hand thread, the designation would be M10 × 1.5-6g – LH (21). *Figure 7* illustrates the ISO Metric thread form. See the heading "Gaging systems" earlier in this section for details about System 21, System 22, and System 23.

The ISO Metric screw thread tolerance system provides for tolerance grades and tolerance positions (allowances) for the pitch diameter and crest diameter. For internal threads, the tolerance position letter is capitalized, and for external threads it is lower case. Internal threads have five grades of tolerance, 4,5,6,7,8. Grade 6 is the most commonly used grade, and lower numbers have tighter tolerances. There are two tolerance position letters for internal threads, H and G. "H" indicates that the minimum internal thread is at the basic pitch diameter size, and it is preferred for general use. "G" indicates that the minimum internal thread is above the pitch diameter, thereby creating an allowance. External threads have seven tolerance grades, 3,4,5,6,7,8,9, with the tolerance increasing with the numbers. "6" is the most commonly used grade. There are three tolerance position letters for external threads, h, g, and e, with the first two being the most widely used. "h" indicates no allowance, "g" provides a small allowance, and "e" has a larger allowance that does not find many applications. Metric threads can have two tolerance references. For instance, a thread designated 4g6g, has two tolerance grade and two tolerance position references. This thread will have its pitch diameter toleranced to 4g and its major diameter will have a 6g tolerance. The first reference is always to pitch diameter, and the second refers to the major diameter of the external thread of the minor diameter of the internal thread. Two factors determine the selection of a suitable tolerance class: the length of thread engagement, and the quality requirement. **Table 13** provides preferred ISO metric tolerance classes.

Since tolerance Grade 6 is by far the most common thread, the other tolerance grades are commonly measured in relation to Grade 6. The approximate ratios of other grades to Grade 6 are as follows: 3 = 0.5; 4 = 0.63; 5 = 0.8; 7 = 1.25; 8 = 1.6; and 9 = 2.0. The tolerances for Grade 6 threads are given in **Table 14** and **Table 15**. To find the approximate tolerance for other grades, multiply the appropriate Grade 6 tolerance by the ratio for the desired tolerance. For example, to find the external major diameter (d) tolerance for a Grade 4 thread with a pitch of 1.25, consult Table 13. From the table, it will be seen that the Grade 6 tolerance is 0.212 mm. To find the Grade 4 tolerance, multiple by the ratio given above (0.63): $0.212 \times 0.63 = 0.133$.

Metric grade tolerances can be derived from the tables of dimensions for 6H and 6g threads (**Table 16** for coarse external threads, **Table 17** for coarse internal threads, **Table 18** provides calculated data for coarse 6g threads, **Table 19** for fine external threads, and **Table 20** for fine internal threads). In consulting the tables it will be seen that the difference between the maximum and minimum specifications for all of the four dimensions specified in Table 14 and Table 15 (internal pitch diameter, external pitch diameter, internal minor diameter, and external major diameter) are equal to the tolerances specified in these tables. It should be noted that although fine threads enhance static strength and in some instances can be used without locking mechanisms because they are less prone to vibration, coarse threads are recommended for most applications. Therefore, it is suggested that

Table 13. Preferred Tolerance Classes for Metric Screw Threads. *(Source, OSG Tap and Die, Inc.)*

Quality Requirement	Tolerance Position "e" (Large Allowance)			External Thread					
				Tolerance Position "g" (Small Allowance)			Tolerance Position "h" (No Allowance)		
	Length of Thread Engagement			Length of Thread Engagement			Length of Thread Engagement		
Fine Close Fit							3h-4h	4h	5h-4h
Medium General Purpose		6e	7e-6e	5g-6g	6g	7g-6g	5h-6h	6h	7h-6h
Coarse Difficult Manufacturing Applications					8g	9g-8g			

Quality Requirement	Tolerance Position "G" (Small Allowance)			Internal Thread		
				Tolerance Position "H" (No Allowance)		
	Length of Thread Engagement			Length of Thread Engagement		
Fine Close Fit				4H	5H	6H
Medium General Purpose	5G	6G	7G	5H	6H	7H
Coarse Difficult Manufacturing Applications		7G	8G		7H	8H

Note: Tolerance Position "e" is not to be applied to pitches finer than 0.5 mm. Tolerance Classes 6g and 6H are for commercial screw, bolt, and nut threads.

coarse thread fasteners be employed unless a particular application requires the functional properties of fine threads. The Metric position allowance (also called the fundamental deviation) can also be derived from the tables. As noted above there is no allowance for positions H and h. The allowance for 6g is given in Table 16, and explained in the notes in the bottom of the table. G allowances are mirror images of g allowances. As can be seen in Table 16, the allowance for any diameter "g" thread with a 1.25 pitch will be 0.028. That number is derived from the three basic formulas at the bottom of the table. For any "G" thread, the allowance will be the exact same, but a positive (in other words, the allowance for the 'g' thread will be expressed as –0.028, and for the 'G' thread it will be +0.028).

Generally, M profile 6g/6H metric designation can be equated to a Class 2A/2B Unified designation, and 4h/5H will provide a close fit similar to Class 3A/3B. British metric threads are defined in British Standard 3643, which is basically in accordance with the ISO and ANSI/ASME standards. For dimensions, consult Tables 16–20.

The **MJ profile** is used where 3A/3B or UNJ threads are required. MJ threads should be used where efficient use of material (maximum loads with minimum material), fatigue life, and stress levels take priority over cost considerations. FED-STD-H28/21B states that "The MJ profile provides a mandatory-controlled rounded root, for external thread, with an increased minor diameter. The rounded roots greatly reduce the concentration of stress, hence increase the fatigue life of the part. External threads produced by rolling after heat treatment and with rounded roots more than double part fatigue life. Its large core diameter improves the static tensile strength." As can be seen in Table 4, many specifications for MJ threads can be calculated with the same formulas as used for determining UNJ dimensions. For full specifications of MJ threads, consult ANSI B1.21M. See **Table 21** for additional information on the M and MJ thread forms.

Table 14. Internal (D_2) and External (d_2) Pitch Diameter Tolerances for Grade 6 Metric Thread.

Major Diameter		Pitch	D_2 Tolerance	d_2 Tolerance	Major Diameter		Pitch	D_2 Tolerance	d_2 Tolerance
Over	To				Over	To			
1.5	2.8	0.2	–	0.050	22.4	45	1	0.170	0.125
		0.25	–	0.056			1.5	0.200	0.150
		0.35	0.085	0.063			2	0.224	0.170
		0.4	0.090	0.067			3	0.265	0.200
		0.45	0.095	0.071			3.5	0.280	0.212
							4	0.300	0.224
							4.5	0.315	0.236
2.8	5.6	0.35	0.090	0.067	45	90	1.5	0.212	0.160
		0.5	0.100	0.075			2	0.236	0.180
		0.6	0.112	0.085			3	0.280	0.212
		0.7	0.118	0.090			4	0.315	0.236
		0.75	0.118	0.090			5	0.335	0.250
		0.8	0.125	0.095			5.5	0.355	0.265
							6	0.375	0.280
5.6	11.2	0.75	0.132	0.100	90	180	2	0.250	0.190
		1	0.150	0.112			3	0.300	0.224
		1.25	0.160	0.118			4	0.335	0.250
		1.5	0.180	0.132			6	0.400	0.300
11.2	22.4	1	0.160	0.118	180	355	3	0.335	0.250
		1.25	0.180	0.132			4	0.375	0.280
		1.5	0.190	0.140			6	0.425	0.315
		1.75	0.200	0.150					
		2	0.212	0.160					
		2.5	0.224	0.170					

All dimensions in millimeters.

Table 15. Internal Minor Diameter (D_1) and External Major Diameter (d) Tolerances for Grade 6 Metric Thread.

Pitch	D_1 Tolerance	d Tolerance	Pitch	D_1 Tolerance	d Tolerance
0.2	–	0.056	1.25	0.265	0.212
0.25	–	0.067	1.5	0.300	0.236
0.3	0.085	0.075	1.75	0.335	0.265
0.35	0.100	0.085	2	0.375	0.280
0.4	0.112	0.095	2.5	0.450	0.335
0.45	0.125	0.100	3	0.500	0.375
0.5	0.140	0.106	3.5	0.560	0.425
0.6	0.160	0.125	4	0.600	0.475
0.7	0.180	0.140	4.5	0.670	0.500
0.75	0.190	0.140	5	0.710	0.530
0.8	0.200	0.150	5.5	0.750	0.560
1	0.236	0.180	6	0.800	0.600

All dimensions in millimeters.

Table 16. External Metric Thread - Coarse Pitch - M Profile - Tolerance Class 6g. All dimensions in millimeters.

Thread Designation	Major Diameter (d)		Pitch Diameter (d_2)		Minor Diameter (d_1)		d_2 (max.) minus d_1 (max.)	d_2 (min.) minus d_1 (min.)	Allowance*
	Max.	Min.	Max.	Min.	Max.	Min.			
1 × 0.25	1.000	0.933	0.838	0.785	0.730	0.631	0.108	0.154	-
1.1 × 0.25	1.100	1.033	0.938	0.885	0.830	0.731	0.108	0.154	-
1.2 × 0.25	1.200	1.133	1.038	0.985	0.930	0.831	0.108	0.154	-
1.4 × 0.30	1.400	1.325	1.205	1.149	1.075	0.964	0.130	0.185	-
1.6 × 0.35	1.581	1.496	1.354	1.291	1.202	1.075	0.152	0.216	0.019
1.8 × 0.35	1.781	1.696	1.554	1.491	1.402	1.275	0.152	0.216	0.019
2 × 0.40	1.981	1.886	1.721	1.654	1.548	1.408	0.173	0.246	0.019
2.2 × 0.45	2.180	2.080	1.888	1.817	1.693	1.540	0.195	0.277	0.020
2.5 × 0.45	2.480	2.380	2.188	2.117	1.993	1.840	0.195	0.277	0.020
3 × 0.50	2.980	2.874	2.655	2.580	2.438	2.272	0.217	0.308	0.020
3.5 × 0.60	3.479	3.354	3.089	3.004	2.829	2.634	0.260	0.370	0.021
4 × 0.70	3.978	3.838	3.523	3.433	3.220	3.002	0.303	0.431	0.022
4.5 × 0.75	4.478	4.338	3.991	3.901	3.666	3.439	0.325	0.462	0.022
5 × 0.80	4.976	4.826	4.456	4.361	4.110	3.868	0.346	0.493	0.024
6 × 1.00	5.974	5.794	5.325	5.212	4.892	4.596	0.433	0.616	0.026
7 × 1.00	6.974	6.794	6.324	6.212	5.891	5.596	0.433	0.616	0.026
8 × 1.25	7.972	7.760	7.160	7.042	6.619	6.272	0.541	0.770	0.028
9 × 1.25	8.972	8.760	8.160	8.042	7.619	7.272	0.541	0.770	0.028
10 × 1.50	9.968	9.732	8.994	8.862	8.344	7.938	0.650	0.924	0.032
11 × 1.50	10.968	10.732	9.994	9.862	7.344	8.938	0.650	0.924	0.032

(Continued)

Table 16. *(Continued)* **External Metric Thread - Coarse Pitch - M Profile - Tolerance Class 6g.**
All dimensions in millimeters.

Thread Designation	Major Diameter (d)		Pitch Diameter (d₂)		Minor Diameter (d₁)		d_2 (max.) minus d_1 (max.)	d_2 (min.) minus d_1 (min.)	Allowance*
	Max.	Min.	Max.	Min.	Max.	Min.			
12 × 1.75	11.966	11.701	10.829	10.679	10.071	9.601	0.758	1.078	0.034
14 × 2.00	13.962	13.682	12.663	12.503	11.797	11.271	0.866	1.232	0.038
16 × 2.00	15.962	15.682	14.663	14.503	13.797	13.271	0.866	1.232	0.038
18 × 2.50	17.958	17.623	16.334	16.164	15.251	14.624	1.083	1.540	0.042
20 × 2.50	19.958	19.623	18.334	18.164	17.251	16.624	1.083	1.540	0.042
22 × 2.50	21.958	21.623	20.334	20.164	19.251	18.624	1.083	1.540	0.042
24 × 3.00	23.952	23.577	22.003	21.803	20.704	19.955	1.299	1.848	0.048
27 × 3.00	26.952	26.577	25.003	24.803	23.704	22.955	1.299	1.848	0.048
30 × 3.50	29.947	29.522	27.674	27.462	26.158	25.306	1.516	2.156	0.053
33 × 3.50	32.947	32.522	30.674	30.462	29.158	28.306	1.516	2.156	0.053
36 × 4.00	35.940	35.465	33.342	33.118	31.610	30.654	1.732	2.464	0.060
39 × 4.00	38.940	38.465	36.342	36.118	34.610	33.654	1.732	2.464	0.060
42 × 4.50	41.937	41.437	39.014	38.778	37.065	36.006	1.949	2.772	0.063
45 × 4.50	44.937	44.437	42.014	41.778	40.065	39.006	1.949	2.772	0.063
48 × 5.00	47.929	47.399	44.681	44.431	42.516	41.351	2.165	3.080	0.071
52 × 5.00	51.929	51.399	48.681	48.431	46.516	45.351	2.165	3.080	0.071
56 × 5.50	55.925	55.365	52.353	52.088	49.971	48.700	2.382	3.388	0.075
60 × 5.50	59.925	59.365	56.353	56.088	53.971	52.700	2.382	3.388	0.075
64 × 6.00	63.920	63.320	60.023	59.743	57.425	56.047	2.598	3.696	0.080
68 × 6.00	67.920	67.320	64.023	63.743	61.425	60.047	2.598	3.696	0.080
72 × 6.00	71.920	71.320	68.023	67.743	65.425	64.047	2.598	3.696	0.080
80 × 6.00	79.920	79.320	76.023	75.743	76.065	74.787	2.598	3.696	0.080
90 × 6.00	89.920	89.320	86.023	85.743	83.425	82.047	2.598	3.696	0.080
100 × 6.00	99.920	99.320	96.023	95.723	93.425	92.027	2.598	3.696	0.080

* The external thread allowance, or upper deviation, is derived from the following formulas: **Dmin - dmax; D₂min - d₂max;** and **D₁min - d₁max.** The thread specifications for metric M series coarse threads in this table are in accordance with ANSI/ASME B1.13 M and ISO 68, Tolerance Class 6g. Formulas used to calculate minor diameter dimensions in this table are: d_1 (max.) = d (max.) - 1.082532P and d_1 (min.) = d_2 (min.) - 0.616025P. *(Sources, Maryland Metrics, Metric & Multistandard Components Corp., and Landis Threading Systems.)*

Table 17. Internal Metric Thread - Coarse Pitch - M Profile - Tolerance Class 6H. All dimensions in millimeters.

Nominal Size ISO M	Pitch (P)	Minor Diameter (D₁)		Pitch Diameter (D₂)		Major Diameter (D)		D_2 (min.) minus D_1 (min.)	D (max.) minus D_2 (max.)	Tap Drill Size*
		Max.	Min.	Max.	Min.	Max.	Min.			
1	.25	0.785	0.729	0.894	0.838	1.092	1.000	0.109	0.198	.75mm
1.1	.25	0.885	0.829	0.994	0.938	1.192	1.100	0.109	0.198	.85mm
1.2	.25	0.985	0.929	1.094	1.038	1.292	1.200	0.109	0.198	.95mm
1.4	.30	1.142	1.075	1.265	1.205	1.503	1.400	0.130	0.238	1.1mm

(Continued)

Table 17. *(Continued)* **Internal Metric Thread - Coarse Pitch - M Profile - Tolerance Class 6H.**
All dimensions in millimeters.

Nominal Size ISO M	Pitch (P)	Minor Diameter (D₁)		Pitch Diameter (D₂)		Major Diameter (D)		D₂ (min.) minus D₁ (min.)	D (max.) minus D₂ (max.)	Tap Drill Size*
		Max.	Min.	Max.	Min.	Max.	Min.			
1.6	.35	1.321	1.221	1.458	1.373	1.736	1.600	0.152	0.278	1.25mm
1.8	.35	1.521	1.421	1.658	1.573	1.936	1.800	0.152	0.278	1.45mm
2	.40	1.679	1.567	1.830	1.740	2.148	2.000	0.173	0.318	1.6mm
2.2	.45	1.838	1.713	2.003	1.908	2.360	2.200	0.195	0.357	1.75mm
2.5	.45	2.138	2.013	2.303	2.208	2.660	2.500	0.195	0.357	2.05mm
3.0	.50	2.599	2.459	2.775	2.675	3.172	3.000	0.216	0.397	2.5mm
3.5	.60	3.010	2.850	3.222	3.110	3.698	3.500	0.260	0.476	2.9mm
4	.70	3.422	3.242	3.663	3.545	4.219	4.000	0.303	0.556	3.3mm
4.5	.75	3.878	3.688	4.131	4.013	4.726	4.500	0.325	0.595	26
5	.80	4.334	4.134	4.605	4.480	5.240	5.000	0.346	0.635	19
6	1.00	5.153	4.917	5.550	5.350	6.294	6.000	0.433	0.794	5mm
7	1.00	6.153	5.917	6.500	6.350	7.294	7.000	0.433	0.794	6.0mm
8	1.25	6.912	6.647	7.348	7.188	8.340	8.000	0.541	0.992	17/64
9	1.25	7.912	7.647	8.348	8.188	9.340	9.000	0.541	0.992	7.75mm
10	1.50	8.676	8.376	9.206	9.026	10.396	10.000	0.650	1.190	8.5mm
11	1.50	9.676	9.376	10.206	10.026	11.396	11.000	0.650	1.190	9.5mm
12	1.75	10.441	10.106	11.063	10.863	12.452	12.000	0.757	1.389	Y
14	2.00	12.210	11.835	12.913	12.701	15.501	14.000	0.866	1.588	12.0mm
16	2.00	14.210	13.835	14.913	14.701	16.501	16.000	0.866	1.588	14.0mm
18	2.50	15.744	15.294	16.660	16.376	18.645	18.000	1.082	1.985	39/64
20	2.50	17.744	17.294	18.600	18.376	20.585	20.000	1.082	1.985	17.5mm
22	2.50	19.744	19.294	20.600	20.376	22.585	22.000	1.082	1.985	49/64
24	3.00	21.252	20.752	22.316	22.051	24.698	24.000	1.299	2.382	21.0mm
27	3.00	24.252	23.752	25.316	25.051	27.698	27.000	1.299	2.382	24.0mm
30	3.50	26.771	26.211	28.007	27.727	30.785	30.000	1.516	2.778	26.5mm
33	3.50	29.771	29.211	31.007	30.727	33.785	33.000	1.516	2.778	29.5mm
36	4.00	32.270	31.670	33.702	33.402	36.877	36.000	1.732	3.175	32mm
39	4.00	35.270	34.670	36.702	36.402	39.877	39.000	1.732	3.175	35mm
42	4.50	37.799	37.129	39.392	39.077	42.964	42.000	1.948	3.572	1 15/32
45	4.50	40.799	40.129	42.392	42.077	45.964	45.000	1.948	3.572	40.5mm
48	5.00	43.297	42.587	45.087	44.752	49.056	48.000	2.165	3.969	1 11/16
52	5.00	47.297	46.587	49.087	48.752	53.056	52000	2.165	3.969	47mm
56	5.50	50.796	50.046	52.783	52.428	57.149	56.000	2.382	4.366	50.5mm
60	5.50	54.796	54.046	56.783	56.428	61.149	60.000	2.382	4.366	54.5mm
64	6.00	58.305	57.505	60.478	60.103	65.241	64.000	2.598	4.763	57.5mm
68	6.00	62.305	61.505	64.478	64.103	69.241	68.000	2.598	4.763	62mm
72	6.00	66.305	65.505	68.478	68.103	73.241	72.000	2.598	4.763	70mm

(Continued)

Table 17. *(Continued)* **Internal Metric Thread - Coarse Pitch - M Profile - Tolerance Class 6H.**
All dimensions in millimeters.

Nominal Size ISO M	Pitch (P)	Minor Diameter (D_1)		Pitch Diameter (D_2)		Major Diameter (D)		D_2 (min.) minus D_1 (min.)	D (max.) minus D_2 (max.)	Tap Drill Size*
		Max.	Min.	Max.	Min.	Max.	Min.			
80	6.00	74.305	73.505	76.478	76.103	81.241	80.000	2.598	4.763	74mm
90	6.00	84.305	83.505	86.478	86.103	91.241	90.000	2.598	4.763	84mm
100	6.00	94.305	93.505	96.503	96.103	101.266	100.00	2.598	4.763	94mm

* See Tapping section for alternative sizes. The pitch diameter tolerance, not given in the table, is equal to the difference between the maximum and minimum pitch diameters: **D_2 max - D_2 min.** The thread specifications for metric M series coarse threads in this table are in accordance with ANSI/ASME B1.13 M and ISO 68, Tolerance Class 6H. *(Sources, Maryland Metrics, Metric & Multistandard Components Corp., Landis Threading Systems, and Kennametal Inc.)*

Table 18. Calculated Data for M Profile - Coarse Pitch - Class 6g Threads.

Nominal Dia. ISO M	Pitch (P)	Root Radius (.125 P)	Lead Angle	Minor Dia. Section mm²*	Stress Area mm²**
1	0.25	0.03125	5°25'	0.4185	0.4498
1.1	0.25	0.03125	4°51'	0.5411	0.5766
1.2	0.25	0.03125	4°23'	0.6793	0.7190
1.4	0.30	0.04350	4°31'	0.9076	0.9635
1.6	0.35	0.04375	4°38'	1.1347	1.2082
1.8	0.35	0.04375	4°30'	1.5438	1.6294
2	0.40	0.05000	4°11'	1.8820	1.9883
2.2	0.45	0.05625	4°18'	2.2512	2.3830
2.5	0.45	0.05625	3°43'	3.1196	3.2745
3	0.50	0.06250	3°24'	4.6683	4.8798
3.5	0.60	0.07500	3°31'	6.2857	6.5786
4	0.70	0.08750	3°25'	8.1433	8.5306
4.5	0.75	0.09375	3°24'	10.5554	11.0296
5	0.80	0.10000	3°15'	13.2671	13.8293
6	1.00	0.12500	3°24'	18.7959	19.6369
7	1.00	0.12500	2°52'	27.2564	28.2672
8	1.25	0.15625	3°10'	34.4093	35.8277
9	1.25	0.15625	2°47'	45.5918	47.2224
10	1.50	0.18750	3°20'	54.6814	56.8368
11	1.50	0.18750	2°44'	68.5735	70.9848
12	1.75	0.21875	2°56'	78.6592	82.6880
14	2.00	0.25000	2°52'	109.3035	113.3519
16	2.00	0.25000	2°29'	149.5066	154.2350
18	2.50	0.31250	2°47'	182.6785	189.2308
20	2.50	0.31250	2°29'	233.7327	241.1367
22	2.50	0.31250	2°14'	291.0700	299.3257

(Continued)

Table 18. *(Continued)* **Calculated Data for M Profile - Coarse Pitch - Class 6g Threads.**

Nominal Dia. ISO M	Pitch (P)	Root Radius (.125 P)	Lead Angle	Minor Dia. Section mm² *	Stress Area mm² **
24	3.00	0.37500	2°29'	336.6661	347.3095
27	3.00	0.37500	2°11'	441.3003	453.4738
30	3.50	0.43750	2°18'	537.4029	553.1026
33	3.50	0.43750	2°50'	667.7384	685.2258
36	4.00	0.50000	2°11'	784.7655	806.4106
39	4.00	0.50000	2°00'	940.7930	955.5608
42	4.50	0.56250	2°60'	1078.994	1107.568
45	4.50	0.56250	1°57'	1260.727	1291.600
48	5.00	0.62500	2°20'	1419.697	1456.071
52	5.00	0.62500	1°52'	1699.400	1739.174
56	5.50	0.68750	1°55'	1961.223	2008.271
60	5.50	0.68750	1°47'	2287.767	2338.558
64	6.00	0.75000	1°49'	2589.959	2648.872
68	6.00	075000	1°42'	2963.338	3026.332
72	6.00	0.75000	1°36'	3361.850	3428.924
80	6.00	0.75000	1°26'	4544.233	4309.508
90	6.00	0.75000	1°16'	5466.172	5551.609
100	6.00	0.75000	1°80'	6855.152	6950.791

All dimensions in millimeters. Note: 645.16 mm² = 1 in.².
* At maximum material condition. Area = .7854 × d_1 (max.)².
** A_s = ($[d_2 ÷ 2]$ - $[3H ÷ 16])^2$ or A_s = .7854 (d - $[.9743 ÷ n]$).

Table 19. External Metric Thread - Fine Pitch - M Profile - Tolerance Class 6g. All dimensions in millimeters.

Nominal Dia. ISO MF	Pitch (P)	Major Diameter (d)		Pitch Diameter (d2)		Root Radius (.125 P)	Minor Diameter (d1)		d2 (max.) minus d1 (max.)	d2 (min.) minus d1 (min.)	Allowance*
		Max.	Min.	Max.	Min.		Max.	Min.			
8	1.00	7.974	7.794	7.324	7.212	0.12500	6.891	6.596	0.433	0.616	0.026
10	0.75	9.978	9.838	9.491	9.391	0.09375	9.166	8.929	0.325	0.462	0.022
10	1.25	9.972	9.760	9.160	9.042	0.15625	8.619	8.272	0.541	0.770	0.028
12	1.00	11.974	11.794	11.324	11.206	0.12500	10.891	10.590	0.433	0.616	0.026
12	1.25	11.972	11.760	11.160	11.028	0.15625	10.619	10.258	0.541	0.777	0.028
12	1.50	11.968	11.732	10.994	10.854	0.18750	10.344	9.930	0.650	0.924	0.032
14	1.50	13.968	13.732	12.994	12.854	0.18750	12.344	11.930	0.650	0.924	0.032
15	1.00	14.974	14.794	14.324	14.206	0.12500	13.891	13.590	0.433	0.616	0.026
16	1.50	15.968	15.732	14.994	14.854	0.18750	14.344	13.930	0.650	0.924	0.032
17	1.00	16.974	16.794	16.324	16.206	0.12500	15.891	15.590	0.433	0.616	0.026
18	1.50	17.968	17.732	16.994	16.854	0.18750	16.344	15.930	0.650	0.924	0.032
20	1.00	19.974	19.794	19.324	19.206	0.12500	18.891	18.590	0.433	0.616	0.026
20	1.50	19.968	19.732	18.994	18.854	0.18750	18.344	17.930	0.650	0.924	0.032
22	1.50	21.968	21.732	20.994	20.854	0.18750	20.344	19.930	0.650	0.924	0.032
24	2.00	23.962	23.682	22.663	22.493	0.25000	21.797	21.261	0.866	1.232	0.038
25	1.50	24.968	24.732	23.994	23.844	0.18750	23.344	22.920	0.650	0.924	0.032
27	2.00	26.962	26.682	25.663	25.493	0.25000	24.797	24.261	0.866	1.232	0.038
30	1.50	29.968	29.732	28.994	28.844	0.18750	28.344	27.920	0.650	0.924	0.032
30	2.00	29.962	29.682	28.663	28.493	0.25000	27.797	27.261	0.866	1.232	0.038
33	2.00	32.962	32.682	31.663	31.493	0.25000	30.797	30.261	0.866	1.232	0.038
35	1.50	34.968	34.732	33.994	33.844	0.18750	33.344	32.920	0.650	0.924	0.032
36	2.00	35.952	35.577	34.003	33.803	0.25000	33.137	32.571	0.866	1.232	0.038

(Continued)

Table 19. *(Continued)* **External Metric Thread - Fine Pitch - M Profile - Tolerance Class 6g.** All dimensions in millimeters.

Nominal Dia. ISO MF	Pitch (P)	Major Diameter (d)		Pitch Diameter (d2)		Root Radius (.125 P)	Minor Diameter (d1)		d2 (max.) minus d1 (max.)	d2 (min.) minus d1 (min.)	Allowance*
		Max.	Min.	Max.	Min.		Max.	Min.			
39	2.00	38.952	38.577	37.003	36.803	0.25000	36.137	35.571	0.866	1.232	0.038
40	1.50	35.698	39.732	38.994	38.844	0.18750	38.344	37.920	0.650	0.924	0.032
42	2.00	41.965	41.682	40.663	40.493	0.25000	39.797	39.261	0.866	1.232	0.038
45	1.50	44.968	44.732	43.994	43.844	0.18750	43.344	42.920	0.650	0.924	0.032
48	2.00	47.692	47.682	46.663	46.483	0.25000	45.797	45.251	0.866	1.232	0.038
50	1.50	49.968	49.732	48.994	48.834	0.18750	48344	47.910	0.650	0.924	0.032
55	1.50	54.968	54.732	53.994	53.834	0.18750	53.344	52.910	0.650	0.924	0.032
56	2.00	55.962	55.682	54.663	54.483	0.25000	53.797	53.251	0.866	1.234	0.038
60	1.50	55.968	59.732	58.994	58.834	0.18750	58.344	57.910	0.650	0.924	0.032
65	1.50	64.968	64.732	63.944	63.834	0.18750	63.344	62.910	0.650	0.924	0.032
70	1.50	69.968	69.732	68.994	68.834	0.18750	68.344	67.910	0.650	0.924	0.032
75	1.50	74.968	74.732	73.994	73.834	0.18750	73.344	72.910	0.650	0.924	0.032
80	1.50	79.968	79.732	78.994	78.834	0.18750	78.344	77.910	0.650	0.924	0.032

* The external thread allowance, or upper deviation, is derived from the following formulas: **Dmin − dmax**; **D2min − d2max**; and **D1min − d1max**. The thread specifications for metric M series coarse threads in this table are in accordance with ANSI/ASME B1.13 M and ISO 68, Tolerance Class 6g. *(Sources, Maryland Metrics, Metric & Multistandard Components Corp., and Landis Threading Systems.)*

Table 20. Internal Metric Thread - Fine Pitch - M Profile - Tolerance Class 6H. All dimensions in millimeters.

Nominal Size ISO MF	Pitch (P)	Minor Diameter (D₁)		Pitch Diameter (D₂)		Major Diameter (D)		D₂ (min.) minus D₁ (min.)	D (max.) minus D₂ (max.)	Tap Drill Size*
		Max.	Min.	Max.	Min.	Max.	Min.			
8	1.00	7.153	6.917	7.500	7.350	8.294	8.00	0.433	0.794	7.0mm
10	0.75	9.378	9.188	9.645	9.513	10.240	10.00	0.325	0.595	9.2mm
10	1.25	8.912	8.647	9.348	9.188	10.340	10.00	0.541	0.992	11/32
12	1.00	11.153	10.917	11.510	11.350	12.304	12.00	0.433	0.794	11.0mm
12	1.25	10.912	10.647	11.368	11.188	12.360	12.00	0.541	0.992	27/64
12	1.50	10.676	10.376	11.216	11.026	12.406	12.00	0.650	1.190	10.4mm
14	1.50	12.676	12.376	13.216	13.026	14.406	14.00	0.650	1.190	12.5mm
15	1.00	14.153	13.917	14.510	14.350	15.304	15.00	0.433	0.794	14.0mm
16	1.50	14.676	14.376	15.216	15.026	16.406	16.00	0.650	1.190	14.5mm
17	1.00	16.153	15.917	16.510	16.350	17.304	17.00	0.433	0.794	16.0mm
18	1.50	16.676	16.376	17.216	17.026	18.406	18.00	0.650	1.190	16.5mm
20	1.00	19.153	18.917	19.510	19.350	20.304	20.00	0.433	0.794	19.0mm
20	1.50	18.676	18.376	19.216	19.026	20.406	20.00	0.650	1.190	18.5mm
22	1.50	20.676	20.376	21.216	21.026	22.406	22.00	0.650	1.190	20.5mm
24	2.00	22.210	21.835	22.925	22.701	24.513	24.00	0.866	1.588	22mm
25	1.50	23.676	23.376	24.226	24.026	25.416	25.00	0.650	1.190	23.5mm
27	2.00	25.210	24.835	25.925	25.701	27.513	27.00	0.866	1.588	25.0mm
30	1.50	28.676	28.376	29.226	29.026	30.416	30.00	0.650	1.190	28.5mm
30	2.00	28.210	27.835	28.925	28.701	30.513	30.00	0.866	1.588	28.0mm
33	2.00	31.210	30.835	31.925	31.701	33.513	33.00	0.866	1.588	1 7/32
35	1.50	33.676	33.376	34.226	34.026	35.416	35.00	0.650	1.190	33.5mm
36	2.00	34.210	33.835	34.925	34.701	36.513	36.00	0.866	1.588	34mm
39	2.00	37.210	36.835	37.925	37.701	39.513	39.00	0.866	1.588	1 29/64
40	1.50	38.676	38.376	39.226	39.026	40.416	40.00	0.650	1.190	38.5mm
42	2.00	40.210	39.835	40.925	40.701	42.513	42.00	0.866	1.588	40mm
45	1.50	43.676	43.376	44.026	43.676	45.416	45.00	0.650	1.190	43.5mm
48	2.00	46.210	45.835	46.937	46.701	48.525	48.00	0.866	1.588	46mm
50	1.50	48.676	48.376	49.238	49.026	50.428	50.00	0.650	1.190	1 29/32
55	1.50	53.676	53.376	54.238	54.026	55.428	55.00	0.650	1.190	53.5mm
56	2.00	54.210	53.835	54.937	54.701	56.525	56.00	0.866	1.588	2 1/8
60	1.50	58.676	58.376	59.238	59.026	60.428	60.00	0.650	1.190	58.5mm
65	1.50	63.676	63.376	64.238	64.026	65.428	65.00	0.650	1.190	63.5mm
70	1.50	68.676	68.376	69.238	69.026	70.428	70.00	0.650	1.190	68.5mm
75	1.50	73.676	73.376	74.238	74.026	75.428	75.00	0.650	1.190	73.5mm
80	1.50	78.676	78.376	79.238	79.026	80.428	80.00	0.650	1.190	78.5mm

* See Tapping section for alternative sizes. The pitch diameter tolerance, not specified in the table, is equal to the difference between the maximum and minimum pitch diameters: **D₂ max - D₂ min**. The thread specifications for metric M series fine threads in this table are in accordance with ANSI/ASME B1.13 M and ISO 68, Tolerance Class 6H. *(Sources, Maryland Metrics, Metric & Multistandard Components Corp., Landis Threading Systems,* and *Kennametal Inc.)*

Table 21. Thread Form for M and MJ Threads. *(Source, Cleveland Twist Drill.)*

Pitch P	Height of External Thread		Addendum	Root Flat (Basic Profile)	Root Radii		
	M	MJ	M, MJ	M	M (Min.)	MJ	
						Min.	Max.
	0.69717H	0.66667H	0.375H	0.250P	0.125P	0.18042P	0.15011P
0.2	.1208	.1155	.0650	.0500	.0250	.0361	.0300
0.25	.1509	.1443	.0812	.0625	.0313	.0451	.0375
0.3	.1811	.1732	.0974	.0750	.0375	.0541	.0450
0.35	.2113	.2021	.1137	.0788	.0438	.0631	.0525
0.4	.2415	.2309	.1299	.1000	.0500	.0722	.0600
0.45	.2717	.2598	.1461	.1125	.0563	.0812	.0675
0.5	.3019	.2887	.1624	.1250	.0625	.0902	.0751
0.6	.3623	.3464	.1949	.1500	.0750	.1083	.0901
0.7	.4226	.4041	.2273	.1750	.0875	.1263	.1051
0.75	.4528	.4330	.2436	.1875	.0938	.1353	.1126
0.8	.4830	.4619	.2598	.2000	.1000	.1443	.1201
1	.6038	.5774	.3248	.2500	.1250	.1804	.1501
1.25	.7547	.7217	.4059	.3125	.1563	.2255	.1876
1.5	.9057	.8660	.4871	.3750	.1875	.2706	.2252
1.75	1.0566	1.0104	.5683	.4375	.2188	.3157	.2627
2	1.2075	1.1547	.6495	.5000	.2500	.3608	.3002
2.5	1.5094	1.4434	.8119	.6250	.3125	.4511	.3753
3	1.8113	1.7321	.9743	.7500	.3750	.5413	.4503
3.5	2.1132	2.0207	1.1367	.8750	.4375	.6315	.5254
4	2.4151	2.3094	1.2990	1.0000	.5000	.7217	.6004
4.5	2.7170	2.5981	1.4614	1.1250	.5625	.8119	.6755
5	3.0188	2.8868	1.6238	1.2500	.6250	.9021	.7506
5.5	3.3207	3.1754	1.7862	1.3750	.6875	.9923	.8256
6	3.6226	3.4641	1.9486	1.5000	.7500	1.0825	.9007
8	4.8301	4.6188	2.5981	2.0000	1.0000	1.4434	1.2009

All dimensions in millimeters.

B.S.F., B.S.W. Whitworth, and other British thread forms

Although the Whitworth thread form had been superceded by the Unified thread and the British Standard metric thread in Great Britain, its widespread usage over time assures a continued demand for replacing worn or damaged fasteners. There are two designations, B.S.W. (British Standard Whitworth) which indicates the coarse series, and B.S.F., the fine series. All Whitworth threads have a 55° included angle, a radius equal to $0.137329P$, and a thread height equal to $0.640327P$. The height of the sharp V thread is equal to $0.96049P$. The Whitworth thread form is shown in *Figure 8*, and basic dimensions are given in **Table 22** (coarse threads) and **Table 23** (fine threads). It should be noted that, by reducing the maximum bolt limits below basic size for medium and free fit dimensions, an allowance is created for plating.

The British Standard Taper Pipe (BSTP) thread for pressure tight joints and the British

(Text continued on p. 1483)

Table 22. British Standard Whitworth (B.S.W. - Coarse) Thread Dimensions. *(Source, Landis Threading Systems.)*

Size and TPI	Fit	Major Diameter Max.	Min.	Plated	Effective Diameter Max.	Min.	Plated	Minor Diameter Max.	Min.	Plated	Drill Size
1/8-40	Close	0.1250	0.1215	-	0.1090	0.1071	-	0.0930	0.0890	-	2.50mm
	Med.	0.1238	0.1193	0.1250	0.1078	0.1049	0.1090	0.0918	0.0857	0.0930	
	Free	0.1238	0.1179	0.1250	0.1078	0.1035	0.1090	0.0918	0.0843	0.0930	
3/16-24	Close	0.1875	0.1832	-	0.1608	0.1585	-	0.1341	0.1291	-	3.70mm
	Med.	0.1863	0.1808	0.1875	0.1596	0.1561	0.1608	0.1329	0.1253	0.1341	
	Free	0.1863	0.1791	0.1875	0.1596	0.1544	0.1608	0.1329	0.1236	0.1341	
1/4-20	Close	0.2500	0.2452	-	0.2180	0.2154	-	0.1860	0.1805	-	5.10mm
	Med.	0.2488	0.2427	0.2500	0.2168	0.2129	0.2180	0.1848	0.1764	0.1860	
	Free	0.2488	0.2408	0.2500	0.2168	0.2110	0.2180	0.1848	0.1764	0.1860	
5/16-18	Close	0.3125	0.3073	-	0.2769	0.2741	-	0.2413	0.2354	-	6.50mm
	Med.	0.3112	0.3046	0.3125	0.2756	0.2714	0.2769	0.2400	0.2311	0.2413	
	Free	0.3112	0.3025	0.3125	0.2756	0.2693	0.2769	0.2400	0.2290	0.2413	
3/8-16	Close	0.3750	0.3695	-	0.3350	0.3320	-	0.2950	0.2888	-	7.90mm
	Med.	0.3736	0.3666	0.3750	0.3336	0.3291	0.3350	0.2936	0.2841	0.2950	
	Free	0.3736	0.3646	0.3750	0.3336	0.3268	0.3350	0.2936	0.2818	0.2950	
7/16-14	Close	0.4375	0.4316	-	0.3918	0.3886	-	0.3461	0.3394	-	9.20mm
	Med.	0.4360	0.4285	0.4375	0.3903	0.3855	0.3918	0.3446	0.3345	0.3461	
	Free	0.4360	0.4260	0.4375	0.3903	0.3830	0.3918	0.3446	0.3320	0.3461	
1/2-12	Close	0.5000	0.4937	-	0.4466	0.4432	-	0.3932	0.3860	-	10.40mm
	Med.	0.4985	0.4904	0.5000	0.4451	0.4399	0.4466	0.3917	0.3807	0.3932	
	Free	0.4985	0.4879	0.5000	0.4451	0.4374	0.4466	0.3917	0.3782	0.3932	
5/8-11	Close	0.6250	0.6183	-	0.5668	0.5631	-	0.5086	0.5010	-	13.40mm
	Med.	0.6233	0.6147	0.6250	0.5651	0.5595	0.5668	0.5069	0.4953	0.5086	
	Free	0.3233	0.6119	0.6250	0.5651	0.5567	0.5668	0.5069	0.4925	0.5086	
3/4-10	Close	0.7500	0.7428	-	0.6860	0.6820	-	0.6220	0.6139	-	16.25mm
	Med.	0.7482	0.7390	0.7500	0.6842	0.6782	0.6860	0.6202	0.6079	0.6220	
	Free	0.7482	0.7360	0.7500	0.6842	0.6752	0.6860	0.6202	0.6049	0.6220	

An allowance is provided on bolts in the "Medium" and "Free" classes by reducing the maximum bolt limits below basic size. This reduction allows for plating, the plating falling within the allowance. It also facilitates easy assembly of maximum metal bolts and nuts.

(Continued)

Table 22. *(Continued)* British Standard Whitworth (B.S.W. - Coarse) Thread Dimensions. *(Source, Landis Threading Systems.)*

Size and TPI	Fit	Major Diameter			Effective Diameter			Minor Diameter			Drill Size
		Max	Min	Plated	Max	Min	Plated	Max	Min	Plated	
7/8-9	Close	0.8750	0.8674	-	0.8039	0.7996	-	0.7328	0.7242	-	19.25mm
	Med.	0.8750	0.8653	-	0.8039	0.7975	-	0.7328	0.7197	-	
	Free	0.8750	0.8621	-	0.8039	0.7943	-	0.7328	0.7165	-	
1-8	Close	1.0000	0.9920	-	0.9200	0.9155	-	0.8400	0.8309	-	22.00mm
	Med.	1.0000	0.9897	-	0.9200	0.9132	-	0.8400	0.8261	-	
	Free	1.0000	0.9863	-	0.9200	0.9098	-	0.8400	0.8227	-	
1 1/8-7	Close	1.1250	1.1164	-	1.0335	1.0287	-	0.9420	0.9323	-	24.50mm
	Med.	1.1250	1.1140	-	1.0335	1.0263	-	0.9420	0.9272	-	
	Free	1.1250	1.1105	-	1.0335	1.0228	-	0.9420	0.9237	-	
1 1/4-7	Close	1.2500	1.2413	-	1.1585	1.1536	-	1.0670	1.0572	-	27.25mm
	Med.	1.2500	1.2388	-	1.1585	1.1511	-	1.0670	1.0520	-	
	Free	1.2500	1.2351	-	1.1585	1.1474	-	1.0670	1.0483	-	
1 1/2-6	Close	1.5000	1.4906	-	1.3933	1.3880	-	1.2866	1.2760	-	33.50mm
	Med.	1.5000	1.4879	-	1.3933	1.3853	-	1.2866	1.2704	-	
	Free	1.5000	1.4839	-	1.3933	1.3813	-	1.2866	1.2664	-	
1 3/4-5	Close	1.7500	1.7398	-	1.6219	1.6162	-	1.4938	1.4823	-	38.50mm
	Med.	1.7500	1.7369	-	1.6219	1.6133	-	1.4938	1.4763	-	
	Free	1.7500	1.7326	-	1.6219	1.6090	-	1.4938	1.4720	-	
2-4.5	Close	2.0000	1.9892	-	1.8577	1.8516	-	1.7154	1.7032	-	44.50
	Med.	2.0000	1.9862	-	1.8577	1.8486	-	1.7154	1.6969	-	
	Free	2.0000	1.9816	-	1.8577	1.8440	-	1.7154	1.6923	-	
2 1/4-4	Close	2.2500	2.2368	-	2.0899	2.0838	-	1.9298	1.9169	-	50.00mm
	Med.	2.2500	2.2354	-	2.0899	2.0803	-	1.9298	1.9102	-	
	Free	2.2500	2.2306	-	2.0899	2.0755	-	1.9298	1.9054	-	
2 1/2-4	Close	2.5000	2.4884	-	2.3399	2.3333	-	2.1798	2.1667	-	56.00mm
	Med.	2.5000	2.4850	-	2.3399	2.3299	-	2.1798	2.1598	-	
	Free	2.5000	2.4801	-	2.3399	2.3250	-	2.1798	2.1549	-	

(Continued)

An allowance is provided on bolts in the "Medium" and "Free" classes by reducing the maximum bolt limits below basic size. This reduction allows for plating, the plating falling within the allowance. It also facilitates easy assembly of maximum metal bolts and nuts.

Table 22. (Continued) British Standard Whitworth (B.S.W. - Coarse) Thread Dimensions. (Source, Landis Threading Systems.)

Size and TPI	Fit	Major Diameter			Effective Diameter			Minor Diameter			Drill Size
		Max.	Min.	Plated	Max.	Min.	Plated	Max.	Min.	Plated	
2 3/4-3.5	Close	2.7500	2.7377	-	2.5670	2.5600	-	2.3840	2.3701	-	61.50mm
	Med.	2.7500	2.7343	-	2.5670	2.5566	-	2.3840	2.3629	-	
	Free	2.7500	2.7290	-	2.5670	2.5513	-	2.3840	2.3576	-	
3-3.5	Close	3.0000	2.9875	-	2.8170	2.8098	-	2.6340	2.6199	-	68.00mm
	Med.	3.0000	2.9839	-	2.8170	2.8062	-	2.6340	2.6125	-	
	Free	3.0000	2.9786	-	2.8170	2.8009	-	2.6340	2.6072	-	
3 1/2-3.25	Close	3.5000	3.4868	-	3.3030	3.2954	-	3.1060	3.0912	-	73.75mm
	Med.	3.5000	3.4830	-	3.3030	3.2916	-	3.1060	3.0835	-	
	Free	3.5000	3.4773	-	3.3030	3.2859	-	3.1060	3.0778	-	
3 3/4-3	Close	3.7500	3.7364	-	3.5366	3.5287	-	3.3232	3.3078	-	85.50mm
	Med.	3.7500	3.7324	-	3.5366	3.5248	-	3.3232	3.2998	-	
	Free	3.7500	3.7265	-	3.5366	3.5189	-	3.3232	3.2939	-	
4-3	Close	4.0000	3.9862	-	3.7866	3.7786	-	3.5732	3.5577	-	92.00mm
	Med.	4.0000	3.9822	-	3.7866	3.7745	-	3.5732	3.5496	-	
	Free	4.0000	3.9761	-	3.7866	3.7685	-	3.5732	3.5436	-	
4 1/2-2.875	Close	4.5000	4.4857	-	4.2773	4.2689	-	4.0546	4.0385	-	104.20mm
	Med.	4.5000	4.4815	-	4.2773	4.2647	-	4.0546	4.0302	-	
	Free	4.5000	4.4752	-	4.2773	4.2584	-	4.0546	4.0239	-	
5-2.75	Close	5.0000	4.9852	-	4.7672	4.7584	-	4.5344	4.5178	-	116.50mm
	Med.	5.0000	4.9808	-	4.7672	4.7541	-	4.5344	4.5092	-	
	Free	5.0000	4.9743	-	4.7672	4.7475	-	4.5344	4.5026	-	
5 1/2-2.875	Close	5.5000	5.4847	-	5.2561	5.2470	-	5.0122	4.9951	-	128.50mm
	Med.	5.5000	5.4802	-	5.2561	5.2424	-	5.0122	4.9862	-	
	Free	5.5000	5.4733	-	5.2561	5.2356	-	5.0122	4.9794	-	
6-2.5	Close	6.0000	5.9842	-	5.7439	5.7345	-	5.4878	5.4701	-	141.00mm
	Med.	6.0000	5.9795	-	5.7439	5.7298	-	5.4878	5.4601	-	
	Free	6.0000	5.9725	-	5.7439	5.7227	-	5.4878	5.4539	-	

An allowance is provided on bolts in the "Medium" and "Free" classes by reducing the maximum bolt limits below basic size. This reduction allows for plating, the plating falling within the allowance. It also facilitates easy assembly of maximum metal bolts and nuts.

Table 23. British Standard Fine (B.S.F.) Thread Dimensions. (Source, Landis Threading Systems.)

Size and TPI	Fit	Major Diameter			Effective Diameter			Minor Diameter			Drill Size
		Max.	Min.	Plated	Max.	Min.	Plated	Max.	Min.	Plated	
3/16-32	Close	0.1875	0.1835	-	0.1675	0.1653	-	0.1475	0.1430	-	4mm
	Med.	0.1864	0.1813	0.1875	0.1664	0.1631	0.1675	0.1464	0.1396	0.1475	
	Free	0.1864	0.1796	0.1875	0.1664	0.1614	0.1675	0.1464	0.1379	0.1475	
7/32-28	Close	0.2188	0.2145	-	0.1959	0.1935	-	0.1730	0.1681	-	4.60mm
	Med.	0.2177	0.2122	0.2188	0.1948	0.1912	0.1959	0.1719	0.1645	0.1730	
	Free	0.2177	0.2105	0.2188	0.1948	0.1895	0.1959	0.1719	0.1628	0.1730	
1/4-26	Close	0.2500	0.2455	-	0.2254	0.2229	-	0.2008	0.1958	-	5.30mm
	Med.	0.2489	0.2432	0.2500	0.2243	0.2206	0.2254	0.1997	0.1921	0.2008	
	Free	0.2489	0.2413	0.2500	0.2243	0.2187	0.2254	0.1997	0.1902	0.2008	
5/16-22	Close	0.3125	0.3077	-	0.2834	0.2807	-	0.2543	0.2488	-	6.80mm
	Med.	0.3113	0.3051	0.3125	0.2822	0.2781	0.2834	0.2531	0.2447	0.2543	
	Free	0.3113	0.3030	0.3125	0.2822	0.2760	0.2834	0.2531	0.2426	0.2543	
3/8-20	Close	0.3750	0.3699	-	0.3430	0.3401	-	0.3110	0.3052	-	8.30mm
	Med.	0.3737	0.3671	0.3750	0.3417	0.3373	0.3430	0.3097	0.3008	0.3110	
	Free	0.3737	0.3649	0.3750	0.3417	0.3351	0.3430	0.3097	0.2986	0.3110	
7/16-18	Close	0.4375	0.4320	-	0.4019	0.3988	-	0.3663	0.3601	-	9.70mm
	Med.	0.4361	0.4290	0.4375	0.4005	0.3958	0.4019	0.3649	0.3555	0.3663	
	Free	0.4361	0.4267	0.4375	0.4005	0.3935	0.4019	0.3649	0.3532	0.3663	
1/2-16	Close	0.5000	0.4942	-	0.4600	0.4567	-	0.4200	0.4135	-	11.10mm
	Med.	0.4985	0.4910	0.5000	0.4585	0.4535	0.4600	0.4158	0.4085	0.4200	
	Free	0.4985	0.4886	0.5000	0.4585	0.4511	0.4600	0.4185	0.4061	0.4200	
9/16-16	Close	0.5625	0.5566	-	0.5225	0.5191	-	0.4825	0.4759	-	12.70mm
	Med.	0.5610	0.5533	0.5625	0.5210	0.5158	0.5225	0.4810	0.4708	0.825	
	Free	0.5610	0.5508	0.5625	0.5210	0.5133	0.5225	0.4810	0.4683	0.4825	
5/8-14	Close	0.6250	0.6187	-	0.5793	0.5757	-	0.5336	0.5265	-	14.00mm
	Med.	0.6234	0.6153	0.6250	0.5777	0.5723	0.5793	0.5320	0.5213	0.5336	
	Free	0.6234	0.6126	0.6250	0.5777	0.5696	0.5793	0.5320	0.5186	0.5336	

An allowance is provided on bolts in the "Medium" and "Free" classes by reducing the maximum bolt limits below basic size. This reduction allows for plating, the plating falling within the allowance. It also facilitates easy assembly of maximum metal bolts and nuts.

(Continued)

Table 23. *(Continued)* British Standard Fine (B.S.F.) Thread Dimensions. *(Source, Landis Threading Systems.)*

Size and TPI	Fit	Major Diameter			Effective Diameter			Minor Diameter			Drill Size
		Max.	Min.	Plated	Max.	Min.	Plated	Max.	Min.	Plated	
3/4-12	Close	0.7500	0.7432	-	0.6966	0.6927	-	0.6432	0.6355	-	16.75mm
	Med.	0.7482	0.7394	0.7500	0.6948	0.6889	0.6966	0.6414	0.6297	0.6432	
	Free	0.7482	0.7365	0.7500	0.6948	0.6860	0.6966	0.6414	0.6268	0.6432	
7/8-11	Close	0.8750	0.8678	-	0.8168	0.8126	-	0.7586	0.7505	-	19.75mm
	Med.	0.8750	0.8658	-	0.8168	0.8106	-	0.7586	0.7464	-	
	Free	0.8750	0.8627	-	0.8168	0.8075	-	0.7586	0.7433	-	
1-10	Close	1.0000	0.9924	-	0.9360	0.9316	-	0.8720	0.8635	-	22.75mm
	Med.	1.0000	0.9902	-	0.9360	0.9294	-	0.8720	0.8591	-	
	Free	1.0000	0.9869	-	0.9360	0.9261	-	0.8720	0.8558	-	
1 1/8-9	Close	1.1250	1.1171	-	1.0539	1.0493	-	0.9828	0.9739	-	25.50mm
	Med.	1.1250	1.1148	-	1.0539	1.0470	-	0.9828	0.9692	-	
	Free	1.1250	1.1113	-	1.0539	1.0435	-	0.9828	0.9657	-	
1 1/4-9	Close	1.2500	1.2419	-	1.1789	1.1741	-	1.1078	1.0987	-	28.50mm
	Med.	1.2500	1.2395	-	1.1789	1.1717	-	1.1078	1.0939	-	
	Free	1.2500	1.2359	-	1.1789	1.1681	-	1.1078	1.0903	-	
1 1/2-8	Close	1.5000	1.4913	-	1.4200	1.4148	-	1.3400	1.3302	-	34.50mm
	Med.	1.5000	1.4888	-	1.4200	1.4123	-	1.3400	1.3252	-	
	Free	1.5000	1.4849	-	1.4200	1.4084	-	1.3400	1.3213	-	
1 3/4-7	Close	1.7500	1.7407	-	1.6585	1.6530	-	1.5670	1.5566	-	40.50mm
	Med.	1.7500	1.7380	-	1.6585	1.6502	-	1.5670	1.5512	-	
	Free	1.7500	1.7338	-	1.6585	1.6461	-	1.5670	1.5470	-	
2-7	Close	2.0000	1.9905	-	1.9085	1.9027	-	1.8170	1.8063	-	47.00mm
	Med.	2.0000	1.9876	-	1.9085	1.8998	-	1.8170	1.8008	-	
	Free	2.0000	1.9832	-	1.9085	1.8955	-	1.8170	1.7965	-	
2 1/4-6	Close	2.2500	2.2398	-	2.1433	2.1372	-	2.0366	2.0252	-	53.00mm
	Med.	2.2500	2.2368	-	2.1433	2.1341	-	2.0366	2.0193	-	
	Free	2.2500	2.2322	-	2.1433	2.1296	-	2.0366	2.0147	-	

An allowance is provided on bolts in the "Medium" and "Free" classes by reducing the maximum bolt limits below basic size. This reduction allows for plating, the plating falling within the allowance. It also facilitates easy assembly of maximum metal bolts and nuts.

(Continued)

Table 23. *(Continued)* **British Standard Fine (B.S.F.) Thread Dimensions.** *(Source, Landis Threading Systems.)*

Size and TPI	Fit	Major Diameter			Effective Diameter			Minor Diameter			Drill Size
		Max.	Min.	Plated	Max.	Min.	Plated	Max.	Min.	Plated	
2 1/2-6	Close	2.2500	2.4896	-	2.3933	2.3870	-	2.2866	2.2750	-	59.00
	Med.	2.2500	2.4864	-	2.3933	2.3838	-	2.2866	2.2689	-	
	Free	2.2500	2.4817	-	2.3933	2.3791	-	2.2866	2.2642	-	
2 3/4-6	Close	2.7500	2.7394	-	2.6433	2.6368	-	2.5366	2.5247	-	-
	Med.	2.7500	2.7361	-	2.6433	2.6355	-	2.5366	2.5186	-	
	Free	2.7500	2.7312	-	2.6433	2.6286	-	2.5366	2.5137	-	
3-5	Close	3.0000	2.9887	-	2.8719	2.8650	-	2.7438	2.7311	-	-
	Med.	3.0000	2.9852	-	2.8719	2.8616	-	2.7438	2.7245	-	
	Free	3.0000	2.9801	-	2.8719	2.8564	-	2.7438	2.7194	-	

An allowance is provided on bolts in the "Medium" and "Free" classes by reducing the maximum bolt limits below basic size. This reduction allows for plating, the plating falling within the allowance. It also facilitates easy assembly of maximum metal bolts and nuts.

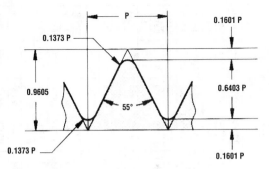

Figure 8. Whitworth (B.S.W.) thread form.

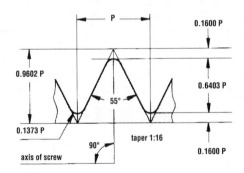

Figure 9. British Standard Taper Pipe form.

Standard Parallel Pipe (BSPP) for non-pressure type joints are also based on the Whitworth form (see *Figure 9*). Both BSTP and BSPP threads share major, pitch (effective), and minor diameters. Full specifications for BSTP threads are given in **Table 24**. (The Japanese parallel pipe thread–JIS PF–shares dimensions with the BSPP thread.)

The British Association (B.A.) Miniature Thread's origin dates back to the mid-1880s, and it was widely used for screws in instruments and clocks. It has an included angle of 47.5°, a radius equal to $0.2617P$, and a thread height equal to $0.6000P$. The height of the sharp V thread is equal to $1.1363365P$. Tolerances are based on two classes of fit (close and normal), and provide an allowance for plating sizes 0 through 10, and are given in **Table 25**. Dimensions for the B.A. thread are given in **Table 26**.

Table 24. Specifications for British Standard Taper Pipe (BSTP) Thread for Pressure Fit. (Source, Landis Threading Systems.)

| | Nominal Size | Sizes 1/8 - 3" | | | | | | | | | | |
		1/8	1/4	3/8	1/2	3/4	1	1 1/4	1 1/2	2	2 1/2	3
Pipe O.D.		13/32	17/32	11/16	27/32	1 1/16	1 11/32	1 11/16	1 29/32	2 3/8	3	3 1/2
TPI		28	19	19	14	14	11	11	11	11	11	11
Pitch		0.03571	0.05263	0.05263	0.07143	0.07143	0.09091	0.09091	0.09091	0.09091	0.09091	0.09091
Thread Height		0.0229	0.0337	0.0337	0.0457	0.0457	0.0582	0.0582	0.0582	0.0582	0.0582	0.0582
Basic Dia. at Gage Plane	Major Gage Dia.	0.3830	0.5180	0.6560	0.8250	1.0410	1.3090	1.6500	1.8820	2.3470	2.9600	3.4600
	Pitch Dia.	0.3601	0.4843	0.6223	0.7793	0.9953	1.2508	1.5918	1.8238	2.2888	2.9018	3.4018
	Minor Dia.	0.3372	0.4506	0.5886	0.7336	0.9496	1.1926	1.5336	1.7656	2.2306	2.8436	3.3436
Gage Length	Basic - Turns	4 3/8	4 1/2	4 3/4	4 1/2	5 1/4	4 1/2	5 1/2	5 1/2	6 7/8	7 9/16	8 15/16
	Basic - Inches	0.1563	0.2367	0.2500	0.3214	0.3750	0.4091	0.5000	0.5000	0.6250	0.6875	0.8125
Length	Tol. (+ and -) Turns	1	1	1	1	1	1	1	1	1	1	1 1/2
	Tol. (+ and -) Inches	0.0357	0.0526	0.0526	0.0714	0.0714	0.0909	0.0909	0.0909	0.0909	0.1364	0.1364
	Max. - Turns	5 3/8	5 1/2	5 3/4	5 1/2	6 1/4	5 1/2	6 1/2	6 1/2	7 7/8	9 1/16	10 7/16
	Max. - Inches	0.1920	0.2893	0.3026	0.3928	0.4464	0.5000	0.5909	0.5909	0.7159	0.8239	0.9489
	Min. - Turns	3 3/8	3 1/2	3 3/4	3 1/2	4 1/2	3 1/2	4 1/2	4 1/2	5 7/8	6 1/16	7 7/16
	Min. - Inches	0.1206	0.1841	0.1974	0.2500	0.3036	0.3182	0.4091	0.4091	0.5341	0.5511	0.6761
Length of Useful Thread at Pipe End (Minimum)	For Basic Gage - Turns	7 1/8	7 1/4	7 1/2	7 1/4	8	7 1/4	8 1/4	8 1/4	10 1/8	11 9/16	12 15/16
	For Basic Gage - Inches	0.2545	0.3814	0.3947	0.5178	0.5714	0.6591	0.7500	0.7500	0.9204	1.0511	1.1761
	For Max Gage - Turns	8 1/8	8 1/4	8 1/2	8 1/4	9	8 1/4	9 1/4	9 1/4	11 1/8	13 1/16	14 7/16
	For Max Gage - Inches	0.2902	0.4340	0.4473	0.5892	0.6428	0.7500	0.8409	0.8409	1.0113	1.1875	1.3125
	For Min Gage - Turns	6 1/8	6 1/4	6 1/2	6 1/4	7	6 1/4	7 1/4	7 1/4	9 1/8	10 1/16	11 7/16
	For Min Gage - Inches	0.2188	0.3288	0.3421	0.4464	0.5000	0.5682	0.6591	0.6591	0.8295	0.9147	0.0397

(Continued)

Table 24. (Continued) Specifications for British Standard Taper Pipe (BSTP) Thread for Pressure Fit. (Source, Landis Threading Systems.)

Nominal Size		3 1/2	4	5	6	7	8	9	10	11	12
					Sizes 3 1/2 - 12"						
Pipe O.D.		4	4 1/2	5 1/2	6 1/2	7 1/2	8 1/2	9 1/2	10 5/8	11 5/8	12 5/8
TPI		11	11	11	11	10	10	10	10	8	8
Pitch		0.09091	0.09091	0.09091	0.09091	0.1000	0.1000	0.1000	0.1000	0.1250	0.1250
Thread Height		0.0582	0.0582	0.0582	0.0582	0.0640	0.0640	0.0640	0.0640	0.8000	0.8000
Basic Dia. at Gage Plane	Major Gage Dia.	3.9500	4.4500	5.4500	6.4500	7.4500	8.4500	9.4500	10.4500	11.4500	12.4500
	Pitch Dia.	3.8918	4.3918	5.3918	6.3918	7.3860	8.3860	9.3860	10.3860	11.3700	12.3700
	Minor Dia.	3.8336	4.3336	5.3336	6.3336	7.3220	8.3220	9.3220	10.3220	11.2900	12.2900
Gage Length	Basic - Turns	9 5/8	11	12 3/8	12 3/8	13 3/4	15	15	16 1/4	13	13
	Basic - Inches	0.8750	1.000	1.1250	1.1250	1.3750	1.5000	1.5000	1.6250	1.6250	1.6250
	Tol. (+ and -) Turns	1 1/2	1 1/2	1 1/2	1 1/2	2	2	2	2	2	2
	Tol. (+ and -) Inches	0.1364	0.1364	0.1364	0.1364	0.2000	0.2000	0.2000	0.2000	0.2000	0.2000
	Max. - Turns	11 1/8	12 1/2	13 7/8	13 7/8	15 3/4	17	17	18 1/4	15	15
	Max - Inches	0.0114	1.1364	1.2614	1.2614	1.5750	1.7000	1.7000	1.8250	1.8750	1.8750
	Min. - Turns	8 1/8	9 1/2	10 7/8	10 7/8	11 3/4	13	13	14 1/4	11	11
	Min. - Inches	0.7386	0.8636	0.9886	0.9886	1.1750	1.3000	1.3000	1.4250	1.3750	1.3750
Length of Useful Thread at Pipe End (Minimum)	For Basic Gage -Turns	13 5/8	15 1/2	17 3/8	17 3/8	19 1/4	20 1/2	20 1/2	21 3/4	18 1/2	18 1/2
	For Basic Gage - Inches	1.2386	1.4091	1.5795	1.5795	1.9250	2.0500	2.0500	2.1750	2.3125	2.3125
	For Max Gage - Turns	15 1/8	17	18 7/8	18 7/8	21 1/4	22 1/2	22 ½	23 3/4	20 1/2	20 1/2
	For Max Gage - Inches	1.3750	1.5455	1.7519	1.7519	2.1250	2.2500	2.2500	2.3750	2.5625	2.5625
	For Min Gage - Turns	12 1/8	14	15 7/8	15 7/8	17 1/4	18 1/2	18 ½	19 3/4	16 1/2	16 1/2
	For Min Gage - Inches	1.1022	1.2727	1.4431	1.4431	1.7250	1.8500	1.8500	1.9750	2.0625	2.0625

Table 25. Tolerances for B.A. Screw Threads.

Thread and Fit		Pitch Dia.	Minor Dia.	Major Dia.
Ext. Thread (minus)	Close Fit	$0.08P + 0.02$mm	$0.16P + 0.04$mm	$0.15P$
	Normal Fit	$0.10P + 0.025$mm	$0.020P + 0.05$mm	Size 0–10 = $0.20P$
				Size 11–16 = $0.025P$
Int. Thread (+)	All	$0.12P + 0.03$mm	$0.375P$	–

Table 26. B.A. Standard (Miniature) Thread Dimensions.

Size #	Major Dia. mm	Pitch mm	TPI	Pitch Dia. mm	Minor Dia. mm	Thread Height mm	Tap Drill mm
0	6.000	1.0000	25.40	5.400	4.800	0.600	5.00
1	5.300	0.9000	28.22	4.760	4.220	0.540	4.40
2	4.700	0.8100	31.35	4.215	3.730	0.485	3.90
3	4.100	0.7300	34.79	3.660	3.220	0.440	3.40
4	3.600	0.6600	38.48	3.205	2.810	0.395	2.95
5	3.200	0.5900	43.05	2.845	2.490	0.355	2.60
6	2.800	0.5300	47.92	2.480	2.160	0.320	2.30
7	2.500	0.4800	52.92	2.210	1.920	0.290	2.00
8	2.200	0.4300	59.07	1.940	1.680	0.260	1.80
9	1.900	0.3900	65.12	1.665	1.430	0.235	1.50
10	1.700	0.3500	72.57	1.490	1.280	0.210	1.35
11	1.500	0.3100	38.93	1.315	1.130	0.185	1.20
12	1.300	0.2800	90.71	1.130	0.960	0.170	1.00
13	1.200	0.2500	101.60	1.050	0.900	0.150	0.95
14	1.000	0.2300	110.40	0.860	0.720	0.140	0.75
15	0.900	0.2100	121.00	0.775	0.650	0.125	–
16	0.790	0.1900	134.00	0.675	0.560	0.115	–

Acme screw threads

The form for Acme threads dates back to the last decade of the 19th century, when the form was designed to replace square and other configured thread forms used for the purpose of producing transverse motion on mechanical mechanisms. Today, their use has expanded to a wide variety of applications wherever heavy loads must be transmitted. There are three standardized forms of Acme threads: General Purpose, Centralizing, and Stub, and each has an included angle of 29°. The maximum major diameter of the external thread is basic and equal to the nominal major diameter. The minimum pitch diameter of the internal thread is basic and equal to the basic major diameter minus the basic height of the thread ($0.5P$). The basic minor diameter is equal to the basic major diameter minus twice the basic thread height. The minimum minor diameter of the internal General Purpose thread is basic. The minimum minor diameter of the internal Centralizing thread is $0.1P$ above basic. Tolerances specified are applicable to lengths of engagement not exceeding twice the nominal diameter. The descriptions in this section are from FED-STD-H28/12, FED-STD-H28/13. ANSI/ASME B1.5 and ANSI/ASME B1.8 also provide dimensional data for Acme threads. The original Acme thread standards were issued in 1952 as ASA B1.5-1952 and ASA B1.8-1952. The form remains basically unaltered from the initial standard.

There are three classes of *General Purpose* threads (2G, 3G, 4G, with the G standing for General Purpose), each with clearances on all diameters for free movement so that they may be used in assemblies with the internal thread rigidly fixed and the external thread moving perpendicular to its axis limited by its bearing or bearings. Class 2G is preferred, but either of the other classes should be selected if excessive backlash or endplay is desired. The three classes of *Centralizing* threads (2C, 3C, and 4C, with the C standing for Centralizing) have a limited clearance on the major diameters of the internal and external threads, so that a bearing at the major diameter maintains approximate alignment of the thread axis and prevents wedging on the flanks of the thread. Previously, classes 5C and 6C were recognized, but they are no longer recommended. Class 2C is preferred; and while selecting one of the other classes can reduce excessive backlash or endplay, the result may be that the tolerances will be difficult to maintain. For both General Purpose and Centralizing threads, it is recommended that external and internal threads of the same class be used together. Centralizing threads are intended to prevent wedging on the flanks of the thread. The *Stub* Acme thread originated in the early 1900s, and its use is generally confined to those unusual applications where a coarse pitch thread of shallow depth is required due to mechanical or metallurgical considerations. Class 2G, general-purpose tolerances and allowances, are recommended for Stub Acme threads, with 3G and 4G being alternative selections where excessive backlash and endplay are encountered (see the following section).

Acme threads are designated by nominal diameter expressed as a fraction or a decimal, tpi, and class. A typical designation would be 7/16–12 ACME-2G. The same thread, with a left-hand thread, would be designated 7/16–12 Acme-2G-LH. Stub Acme threads do not utilize a class and the designation would read 7/16–12 Stub Acme, or 7/16–12 Stub Acme-LH.

All classes of General Purpose external and internal threads may be used interchangeably. The requirement for a Centralizing fit is that the sum of the major diameter tolerance plus the major diameter allowance on the internal thread, and the major diameter tolerance on the external thread shall equal or be less than the pitch diameter allowance on the external thread. A class 2C external thread that has a larger pitch diameter allowance than either

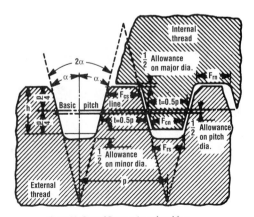

Figure 10. General Purpose Acme thread form.

a class 3C or 4C external thread can be used interchangeably with classes 2C, 3C, or 4C internal threads and fulfill this requirement. Similarly, a class 3C external thread can be used interchangeably with class 3C and 4C internal threads, but only a class 4C internal thread can be used with a class 4C external thread. *Figure 10* shows the General Purpose form, *Figure 11* the Centralizing form, and the Stub form is illustrated in *Figure 12*.

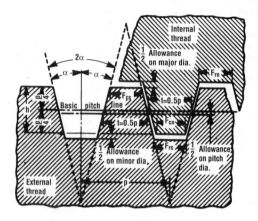

Figure 11. Centralizing Acme thread form.

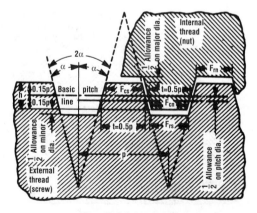

Figure 12. Stub Acme thread form.

Table 27. Basic Formulas for Calculating Acme Thread Dimensions.

Dimension	General Purpose Acme Thread	Centralizing Acme Thread	Stub Acme Threads *	Remarks
Basic Height of Thread	$H = 0.5P$	$H = 0.5P$	$H = 0.3P$	See Note [1]
Basic Thickness of Thread**	$t = 0.5P$	$t = 0.5P$	$t = 0.5P$	
Basic Width of Flat of Crest of Internal Thread	$F_{cn} = 0.3707P$	$F_{cn} = 0.3707P + 0.259 \times$ minor dia. allowance, int. thread	$F_{cn} = 0.4224P + 0.259 \times$ minor dia. allowance, int. thread	See Note [2]
Basic Width of Flat of Crest of External Thread	$F_{cs} = 0.3707P - 0.259 \times$ pitch dia. allowance, ext. thread	$F_{cs} = 0.3707P - 0.259 \times$ pitch dia. allowance, ext. thread	$F_{cn} = 0.4224P + 0.259 \times$ pitch dia. allowance, ext. thread	See Table 28 for pitch diameter allowances.
Basic Flat at Root of Internal Thread	$F_{rn} = 0.0307P - 0.259 \times$ major dia. allowance, int. thread	$F_{rn} = 0.0307P - 0.259 \times$ major dia. allowance, int. thread	$F_{rm} = 0.4224P - 0.259 \times$ major dia. allowance, int. thread	
Basic Flat at Root of External Thread	$F_{rs} = 0.0307P - 0.259 \times$ (minor dia. allowance, ext. thread − pitch dia. allowance, ext. thread)	$F_{rs} = 0.0307P - 0.259 \times$ (minor dia. allowance, ext. thread − pitch dia. allowance, ext. thread)	$F_{rs} = 0.4224P - 0.259 \times$ (minor dia. allowance, ext. thread − pitch dia. allowance, ext. thread)	See Table 28 for pitch diameter allowances.

* Basic Stub Acme form.

** The basic thickness of the thread measured at a diameter smaller by one-half the pitch than the basic major diameter (i.e., at the pitch diameter line).

Note [1]: For Modified Form 1 Stub Acme, $H = 0.375P$. For Modified Form 2 Stub Acme, $H = 0.025P$.

Note [2]: For Modified Form 1 Stub Acme, $F_{cs} = 0.4030P$. For Modified Form 2 Stub Acme, $F_{cs} = 0.4353P$.

Allowances, tolerances, and dimensions for Acme threads

For General Purpose Acme threads and Stub Acme threads, the allowance (minimum clearance) at major and minor diameters is established as follows: a minimal diametrical clearance is provided at the minor diameter of all external threads by establishing the maximum minor diameter of 0.020 in. below the basic minor diameter for 10 tpi and coarser, and 0.010 in. below the basic minor diameter for finer pitches. A minimal diametrical clearance at the major diameter is obtained by establishing the minimum major diameter of the internal thread 0.020 in. above the basic minor diameter for 10 tpi and coarser, and 0.010 in. above the basic minor diameter for finer pitches. General Purpose Acme external threads may have the crest corners chamfered at an angle of 45° with the axis, to a maximum depth of $0.667P$. This corresponds to a maximum width of chamfer flat of $0.945P$.

For Centralizing Acme threads, a minimal diametrical clearance is provided at the minor diameter of all external threads by establishing the maximum minor diameter 0.020 in. below the basic minor for 10 tpi and coarser, and 0.010 in. for finer pitches. A minimum diametrical clearance for the fillet is provided at the minor diameter by establishing the minimum minor diameter of the internal thread $0.1P$ greater than the basic minor diameter. A minimum diametrical clearance at the major diameter is obtained by establishing the minimum major diameter of the internal thread at $0.001\sqrt{D}$ above the basic major diameter. Centralizing Acme external threads have the crest corners chamfered to an angle of 45° with the axis, to a minimum depth of $0.05P$ and a maximum depth of $0.667P$, corresponding to a minimum width of chamfer flat of $0.0707P$ and a maximum width of $0.0945P$. External threads may also have a fillet at the minor diameter not greater than $0.1P$. Internal threads may have a fillet at the major diameter not greater than $0.06P$.

Pitch diameter allowances for General Purpose and Centralizing Acme Threads are given in **Table 28**. Dimensions and tolerances for General Purpose Acme Threads are given in **Tables 29–31**, for Centralizing Acme Threads see **Tables 30–32**, and for Stub Acme Threads see **Table 33**.

Table 28. Pitch Diameter Allowances for Acme Threads. *(Source, FED-STD H28/12.)*

Nominal Size Range		Pitch Diameter Allowances on External Threads, General Purpose, Centralizing, and Stub Acme*		
Above	To and Including	Class 2G and 2C and Stub Acme $0.008\sqrt{D}$	Class 3G and 3C $0.006\sqrt{D}$	Class 4G and 4C $0.004\sqrt{D}$
0	3/16	0.0024	0.0018	0.0012
3/16	5/16	0.0040	0.0030	0.0020
5/16	7/16	0.0049	0.0037	0.0024
7/16	9/16	0.0057	0.0042	0.0028
9/16	11/16	0.0063	0.0047	0.0032
11/16	13/16	0.0069	0.0052	0.0035
13/16	15/16	0.0075	0.0056	0.0037
15/16	1 1/16	0.0080	0.0060	0.0040
1 1/16	1 3/16	0.0085	0.0064	0.0042
1 3/16	1 5/16	0.0089	0.0067	0.0045
1 5/16	1 7/16	0.0094	0.0070	0.0047

(Continued)

Table 28. *(Continued)* Pitch Diameter Allowances for Acme Threads. *(Source, FED-STD H28/12.)*

Nominal Size Range		Pitch Diameter Allowances on External Threads, General Purpose, Centralizing, and Stub Acme*		
Above	To and Including	Class 2G and 2C and Stub Acme $0.008\sqrt{D}$	Class 3G and 3C $0.006\sqrt{D}$	Class 4G and 4C $0.004\sqrt{D}$
1 7/16	1 9/16	0.0098	0.0073	0.0049
1 9/16	1 7/8	0.0105	0.0079	0.0052
1 7/8	2 1/8	0.0113	0.0085	0.0057
2 1/8	2 3/8	0.0120	0.0090	0.0060
2 3/8	2 5/8	0.0126	0.0095	0.0063
2 5/8	2 7/8	0.0133	0.0099	0.0066
2 7/8	3 1/4	0.0140	0.0105	0.0070
3 1/4	3 3/4	0.0150	0.0112	0.0075
3 3/4	4 1/4	0.0160	0.0120	0.0080
4 1/4	4 3/4	0.0170	0.0127	0.0085
4 3/4	5 1/2	0.0181	0.0136	0.0091

All dimensions in inches.
*An increase in 10% of the allowance is recommended for each inch, or fraction thereof, that the length of engagement exceeds two diameters.

Modified square thread

Another example of an Acme type thread is the modified square thread that was designed for translation applications and for resisting axial loads where only small radial loads are permitted. The angle between the flanks of the thread is 10°, and the threads are truncated top and bottom. The thread had a basic height of 0.5P, a basic thickness at the pitch line of 0.5P, and is symmetrical about a line perpendicular to the axis of the external thread. The 10° angle results in a thread that is practically the equivalent of a square thread, and it is therefore capable of being produced economically. A typical Modified Square thread designation would be 1 3/4–6 SPL 10° FORM. *Figure 13* shows the thread form. FED-STD-H28/19A provides additional information on dimensions, allowances, and tolerances.

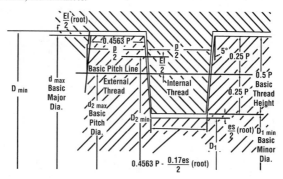

Figure 13. Modified Square thread form.

Table 29. Dimensions and Tolerances for Class 2G General Purpose Acme Screw Threads.
(Source, Landis Threading Systems.)

Nominal Size and TPI	Major Diameter			Minor Diameter		Pitch Diameter		
	Max.	Min.	Tol.	Max.	Min.	Max.	Min.	Tol.
External Threads - Class 2G								
1/4-16	0.2500	0.2450	0.0050	0.1775	0.1618	0.2148	0.2043	0.0105
5/16-14	0.3125	0.3075	0.0050	0.2311	0.2140	0.2728	0.2614	0.0114
3/8-12	0.3750	0.3700	0.0050	0.2817	0.2632	0.3284	0.3161	0.0123
7/16-12	0.4375	0.4325	0.0050	0.3442	0.3253	0.3909	0.3783	0.0126
1/2-10	0.5000	0.4950	0.0050	0.3800	0.3594	0.4443	0.4306	0.0137
5/8-8	0.6250	0.6188	0.0062	0.4800	0.4570	0.5562	0.5408	0.0154
3/4-6	0.7500	0.7417	0.0083	0.5633	0.5371	0.6598	0.6424	0.0174
7/8-6	0.8750	0.8667	0.0083	0.6883	0.6615	0.7842	0.7663	0.0179
1-5	1.0000	0.9900	0.0100	0.7800	0.7509	0.8920	0.8726	0.0194
1 1/8-5	1.1250	1.1150	0.0100	0.9050	0.8753	1.0165	0.9967	0.0198
1 1/4-5	1.2500	1.2400	0.0100	1.0300	0.9998	1.1411	1.1210	0.0201
1 3/8-4	1.3750	1.3625	0.0125	1.1050	1.0719	1.2406	1.2185	0.0220
1 1/2-4	1.5000	1.4875	0.0125	1.2300	1.1965	1.3652	1.3429	0.0223
1 3/4-4	1.7500	1.7375	0.0125	1.4800	1.4456	1.6145	1.5916	0.0229
2-4	2.0000	1.9875	0.0125	1.7300	1.6948	1.8637	1.8402	0.0235
2 1/4-3	2.2500	2.2333	0.0167	1.8967	1.8572	2.0713	2.0450	0.0263
2 1/2-3	2.5000	2.4833	0.0167	2.1467	2.1065	2.3207	2.2939	0.0268
2 3/4-3	2.7500	2.7333	0.0167	2.3967	2.3558	2.5700	2.5427	0.0273
3-2	3.0000	2.9750	0.2500	2.4800	2.4326	2.7360	2.7044	0.0316
3 1/2-2	3.5000	3.4750	0.2500	2.9800	2.9314	3.2350	3.2026	0.0324
4-2	4.0000	3.9750	0.2500	3.4800	3.4302	3.7340	3.7008	0.0332
4 1/2-2	4.5000	4.4750	0.2500	3.9800	3.9201	4.2330	4.1991	0.0339
5-2	5.0000	4.9750	0.2500	4.4800	4.4281	4.7319	4.6973	0.0346
Internal Threads - Class 2G								
1/4-16	0.2600	0.2700	0.0100	0.1875	0.1925	0.2188	0.2293	0.0105
5/16-14	0.3225	0.3325	0.0100	0.2411	0.2461	0.2768	0.2882	0.0114
3/8-12	0.3850	0.3950	0.0100	0.2917	0.2967	0.3333	0.3456	0.0123
7/16-12	0.4475	0.4575	0.0100	0.3542	0.3592	0.3958	0.4084	0.0126
1/2-10	0.5200	0.5400	0.0200	0.4000	0.4050	0.4500	0.4637	0.0137
5/8-8	0.6450	0.6650	0.0200	0.5000	0.5062	0.5625	0.5779	0.0154
3/4-6	0.7700	0.7900	0.0200	0.5833	0.5916	0.6667	0.6841	0.0174
7/8-6	0.8950	0.9150	0.0200	0.7083	0.7166	0.7917	0.8096	0.0179
1-5	1.0200	1.0400	0.0200	0.8000	0.8100	0.9000	0.9194	0.0194
1 1/8-5	1.1450	1.1650	0.0200	0.9250	0.9350	1.0250	1.0448	0.0198
1 1/4-5	1.2700	1.2900	0.0200	1.0500	1.0600	1.1500	1.1701	0.0201

(Continued)

Table 29. *(Continued)* **Dimensions and Tolerances for Class 2G General Purpose Acme Screw Threads.**
(Source, Landis Threading Systems.)

Nominal Size and TPI	Major Diameter			Minor Diameter			Pitch Diameter		
	Max.	Min.	Tol.	Max.	Min.	Tol.	Max.	Min.	Tol.
				Internal Threads - Class 2G					
1 3/8-4	1.3950	1.4150	0.0200	1.1250	1.1375	0.0125	1.2500	1.2720	0.0220
1 1/2-4	1.5200	1.5400	0.0200	1.2500	1.2625	0.0125	1.3750	1.3973	0.0223
1 3/4-4	1.7700	1.7900	0.0200	1.5000	1.5125	0.0125	1.6250	1.6479	0.0229
2-4	2.0200	2.0400	0.0200	1.7500	1.7625	0.0125	1.8750	1.8985	0.0235
2 1/4-3	2.2700	2.2900	0.0200	1.9167	1.9334	0.0167	2.0833	2.1096	0.0263
2 1/2-3	2.5200	2.5400	0.0200	2.1667	2.1834	0.0167	2.3333	2.3601	0.0268
2 3/4-3	2.7700	2.7900	0.0200	2.4167	2.4334	0.0167	2.5833	2.6106	0.0273
3-2	3.0200	3.0400	0.0200	2.5000	2.5250	0.0250	2.7500	2.7816	0.0316
3 1/2-2	3.5200	3.5400	0.0200	3.0000	3.0250	0.0250	3.2500	3.2824	0.0324
4-2	4.0200	4.0400	0.0200	3.5000	3.5250	0.0250	3.7500	3.7832	0.0332
4 1/2-2	4.5200	4.5400	0.0200	4.0000	4.0250	0.0250	4.2500	4.2839	0.0339
5-2	5.0200	5.0400	0.0200	4.5000	4.5250	0.0250	4.7500	4.7846	0.0346

All dimensions in inches.

Table 30. Dimensions and Tolerances for Class 3G General Purpose Acme Screw Threads.
(Source, Landis Threading Systems.)

Nominal Size and TPI	Major Diameter			Minor Diameter			Pitch Diameter		
	Max.	Min.	Tol.	Max.	Min.		Max.	Min.	Tol.
				External Threads - Class 3G					
1/4-16	0.2500	0.2450	0.0050	0.1775	0.1702		0.2158	0.2109	0.0049
5/16-14	0.3125	0.3075	0.0050	0.2311	0.2231		0.2738	0.2685	0.0053
3/8-12	0.3750	0.3700	0.0050	0.2817	0.2730		0.3296	0.3238	0.0058
7/16-12	0.4375	0.4325	0.0050	0.3442	0.3354		0.3921	0.3862	0.0059
1/2-10	0.5000	0.4950	0.0050	0.3800	0.3704		0.4458	0.4394	0.0064
5/8-8	0.6250	0.6188	0.0062	0.4800	0.4693		0.5578	0.5506	0.0072
3/4-6	0.7500	0.7417	0.0083	0.5633	0.5511		0.6615	0.6534	0.0081
7/8-6	0.8750	0.8667	0.0083	0.6883	0.6758		0.7861	0.7778	0.0083
1-5	1.0000	0.9900	0.0100	0.7800	0.7664		0.8940	0.8849	0.0091
1 1/8-5	1.1250	1.1150	0.0100	0.9050	0.8912		1.0186	1.0094	0.0092
1 1/4-5	1.2500	1.2400	0.0100	1.0300	1.0159		1.1433	1.1339	0.0094
1 3/8-4	1.3750	1.3625	0.0125	1.1050	1.0896		1.2430	1.2327	0.0103
1 1/2-4	1.5000	1.4875	0.0125	1.2300	1.2144		1.3677	1.3573	0.0104
1 3/4-4	1.7500	1.7375	0.0125	1.4800	1.4640		1.6171	1.6064	0.0107
2-4	2.0000	1.9875	0.0125	1.7300	1.7136		1.8665	1.8555	0.0110
2 1/4-3	2.2500	2.2333	0.0167	1.8967	1.8783		2.0743	2.0620	0.0123
2 1/2-3	2.5000	2.4833	0.0167	2.1467	2.1279		2.3238	2.3113	0.0125
2 3/4-3	2.7500	2.7333	0.0167	2.3967	2.3776		2.5734	2.5607	0.0127

(Continued)

Table 30. *(Continued)* **Dimensions and Tolerances for Class 3G General Purpose Acme Screw Threads.** *(Source, Landis Threading Systems.)*

External Threads - Class 3G								
Nominal Size and TPI	Major Diameter			Minor Diameter		Pitch Diameter		
	Max.	Min.	Tol.	Max.	Min.	Max.	Min.	Tol.
3-2	3.0000	2.9750	0.2500	2.4800	2.4579	2.7395	2.7248	0.0147
3 1/2-2	3.5000	3.4750	0.2500	2.9800	2.9574	3.2388	3.2237	0.0151
4-2	4.0000	3.9750	0.2500	3.4800	3.4568	3.7380	3.7225	0.0155
4 1/2-2	4.5000	4.4750	0.2500	3.9800	3.9563	4.2373	4.2215	0.0158
5-2	5.0000	4.9750	0.2500	4.4800	4.4558	4.7364	4.7202	0.0162

Internal Threads - Class 3G								
Nominal Size and TPI	Major Diameter			Minor Diameter			Pitch Diameter		
	Max.	Min.	Tol.	Max.	Min.	Tol.	Max.	Min.	Tol.
1/4-16	0.2600	0.2700	0.0100	0.1875	0.1925	0.0050	0.2188	0.2237	0.0049
5/16-14	0.3225	0.3325	0.0100	0.2411	0.2461	0.0050	0.2768	0.2821	0.0053
3/8-12	0.3850	0.3950	0.0100	0.2917	0.2967	0.0050	0.3333	0.3391	0.0058
7/16-12	0.4475	0.4575	0.0100	0.3542	0.3592	0.0050	0.3958	0.4017	0.0059
1/2-10	0.5200	0.5400	0.0200	0.4000	0.4050	0.0050	0.4500	0.4564	0.0064
5/8-8	0.6450	0.6650	0.0200	0.5000	0.5062	0.0062	0.5625	0.5697	0.0072
3/4-6	0.7700	0.7900	0.0200	0.5833	0.5916	0.0083	0.6667	0.6748	0.0081
7/8-6	0.8950	0.9150	0.0200	0.7083	0.7166	0.0083	0.7917	0.8000	0.0083
1-5	1.0200	1.0400	0.0200	0.8000	0.8100	0.0100	0.9000	0.9091	0.0091
1 1/8-5	1.1450	1.1650	0.0200	0.9250	0.9350	0.0100	1.0250	1.0342	0.0092
1 1/4-5	1.2700	1.2900	0.0200	1.0500	1.0600	0.0100	1.1500	1.1594	0.0094
1 3/8-4	1.3950	1.4150	0.0200	1.1250	1.1375	0.0125	1.2500	1.2603	0.0103
1 1/2-4	1.5200	1.5400	0.0200	1.2500	1.2625	0.0125	1.3750	1.3854	0.0104
1 3/4-4	1.7700	1.7900	0.0200	1.5000	1.5125	0.0125	1.6250	1.6357	0.0107
2-4	2.0200	2.0400	0.0200	1.7500	1.7625	0.0125	1.8750	1.8860	0.0110
2 1/4-3	2.2700	2.2900	0.0200	1.9167	1.9334	0.0167	2.0833	2.0956	0.0123
2 1/2-3	2.5200	2.5400	0.0200	2.1667	2.1834	0.0167	2.3333	2.3458	0.0125
2 3/4-3	2.7700	2.7900	0.0200	2.4167	2.4334	0.0167	2.5833	2.5960	0.1270
3-2	3.0200	3.0400	0.0200	2.5000	2.5250	0.0250	2.7500	2.7647	0.1470
3 1/2-2	3.5200	3.5400	0.0200	3.0000	3.0250	0.0250	3.2500	3.2651	0.1510
4-2	4.0200	4.0400	0.0200	3.5000	3.5250	0.0250	3.7500	3.7655	0.1550
4 1/2-2	4.5200	4.5400	0.0200	4.0000	4.0250	0.0250	4.2500	4.2658	0.1580
5-2	5.0200	5.0400	0.0200	4.5000	4.5250	0.0250	4.7500	4.7662	0.1620

All dimensions in inches.

Table 31. Dimensions and Tolerances for Class 4G General Purpose Acme Screw Threads.
(Source, Landis Threading Systems.)

Nominal Size and TPI	External Threads - Class 4G								
	Major Diameter			Minor Diameter			Pitch Diameter		
	Max.	Min.	Tol.	Max.	Min.	Max.	Min.	Tol.	
1/4-16	0.2500	0.2450	0.0050	0.1775	0.1722	0.2168	0.2133	0.0035	
5/16-14	0.3125	0.3075	0.0050	0.2311	0.2254	0.2748	0.2710	0.0038	
3/8-12	0.3750	0.3700	0.0050	0.2817	0.2755	0.3309	0.3268	0.0041	
7/16-12	0.4375	0.4325	0.0050	0.3442	0.3379	0.3934	0.3892	0.0042	
1/2-10	0.5000	0.4950	0.0050	0.3800	0.3731	0.4472	0.4426	0.0046	
5/8-8	0.6250	0.6188	0.0062	0.4800	0.4723	0.5593	0.5542	0.0051	
3/4-6	0.7500	0.7417	0.0083	0.5633	0.5546	0.6632	0.6574	0.0058	
7/8-6	0.8750	0.8667	0.0083	0.6883	0.6794	0.7880	0.7820	0.0060	
1-5	1.0000	0.9900	0.0100	0.7800	0.7703	0.8960	0.8895	0.0065	
1 1/8-5	1.1250	1.1150	0.0100	0.9050	0.8951	1.0208	1.0142	0.0066	
1 1/4-5	1.2500	1.2400	0.0100	1.0300	1.0199	1.1455	1.1388	0.0067	
1 3/8-4	1.3750	1.3625	0.0125	1.1050	1.0940	1.2453	1.2380	0.0073	
1 1/2-4	1.5000	1.4875	0.0125	1.2300	1.2188	1.3701	1.3627	0.0074	
1 3/4-4	1.7500	1.7375	0.0125	1.4800	1.4685	1.6198	1.6122	0.0076	
2-4	2.0000	1.9875	0.0125	1.7300	1.7183	1.8693	1.8615	0.0078	
2 1/4-3	2.2500	2.2333	0.0167	1.8967	1.8835	2.0773	2.0685	0.0088	
2 1/2-3	2.5000	2.4833	0.0167	2.1467	2.1333	2.3270	2.3181	0.0089	
2 3/4-3	2.7500	2.7333	0.0167	2.3967	2.3831	2.5767	2.5676	0.0091	
3-2	3.0000	2.9750	0.2500	2.4800	2.4642	2.7430	2.7325	0.0105	
3 1/2-2	3.5000	3.4750	0.2500	2.9800	2.9638	3.2425	3.2317	0.0108	
4-2	4.0000	3.9750	0.2500	3.4800	3.4634	3.7420	3.7309	0.0111	
4 1/2-2	4.5000	4.4750	0.2500	3.9800	3.9631	4.2415	4.2302	0.0113	
5-2	5.0000	4.9750	0.2500	4.4800	4.4627	4.7409	4.7294	0.0115	
Internal Threads - Class 4G									
Nominal Size and TPI	Major Diameter			Minor Diameter			Pitch Diameter		
	Max.	Min.	Tol.	Max.	Min.	Tol.	Max.	Min.	Tol.
1/4-16	0.2600	0.2700	0.0100	0.1875	0.1925	0.0050	0.2188	0.2223	0.0035
5/16-14	0.3225	0.3325	0.0100	0.2411	0.2461	0.0050	0.2768	0.2806	0.0038
3/8-12	0.3850	0.3950	0.0100	0.2917	0.2967	0.0050	0.3333	0.3374	0.0041
7/16-12	0.4475	0.4575	0.0100	0.3542	0.3592	0.0050	0.3958	0.4000	0.0042
1/2-10	0.5200	0.5400	0.0200	0.4000	0.4050	0.0050	0.4500	0.4546	0.0046
5/8-8	0.6450	0.6650	0.0200	0.5000	0.5062	0.0062	0.5625	0.5676	0.0051
3/4-6	0.7700	0.7900	0.0200	0.5833	0.5916	0.0083	0.6667	0.6725	0.0058
7/8-6	0.8950	0.9150	0.0200	0.7083	0.7166	0.0083	0.7917	0.7977	0.0060
1-5	1.0200	1.0400	0.0200	0.8000	0.8100	0.0100	0.9000	0.9065	0.0065
1 1/8-5	1.1450	1.1650	0.0200	0.9250	0.9350	0.0100	1.0250	1.0316	0.0066
1 1/4-5	1.2700	1.2900	0.0200	1.0500	1.0600	0.0100	1.1500	1.1567	0.0067

(Continued)

Table 31. *(Continued)* **Dimensions and Tolerances for Class 4G General Purpose Acme Screw Threads.**
(Source, Landis Threading Systems.)

Nominal Size and TPI	Major Diameter			Minor Diameter			Pitch Diameter		
	Max.	Min.	Tol.	Max.	Min.	Tol.	Max.	Min.	Tol.
1 3/8-4	1.3950	1.4150	0.0200	1.1250	1.1375	0.0125	1.2500	1.2573	0.0073
1 1/2-4	1.5200	1.5400	0.0200	1.2500	1.2625	0.0125	1.3750	1.3824	0.0074
1 3/4-4	1.7700	1.7900	0.0200	1.5000	1.5125	0.0125	1.6250	1.6326	0.0076
2-4	2.0200	2.0400	0.0200	1.7500	1.7625	0.0125	1.8750	1.8828	0.0078
2 1/4-3	2.2700	2.2900	0.0200	1.9167	1.9334	0.0167	2.0833	2.0921	0.0088
2 1/2-3	2.5200	2.5400	0.0200	2.1667	2.1834	0.0167	2.3333	2.3422	0.0089
2 3/4-3	2.7700	2.7900	0.0200	2.4167	2.4334	0.0167	2.5833	2.5924	0.0091
3-2	3.0200	3.0400	0.0200	2.5000	2.5250	0.0250	2.7500	2.7605	0.0105
3 1/2-2	3.5200	3.5400	0.0200	3.0000	3.0250	0.0250	3.2500	3.2608	0.0108
4-2	4.0200	4.0400	0.0200	3.5000	3.5250	0.0250	3.7500	3.7611	0.0111
4 1/2-2	4.5200	4.5400	0.0200	4.0000	4.0250	0.0250	4.2500	4.2613	0.0113
5-2	5.0200	5.0400	0.0200	4.5000	4.5250	0.0250	4.7500	4.7615	0.0115

All dimensions in inches.

Table 32. Dimensions and Tolerances for Class 2C Centralizing Acme Screw Threads.
(Source, Landis Threading Systems.)

Nominal Size and TPI	Major Diameter			Minor Diameter			Pitch Diameter		
	Max.	Min.	Tol.	Max.	Min.	Max.	Min.	Tol.
1/2-10	0.500	0.4975	0.0025	0.3800	0.3594	0.4443	0.4306	0.0137
5/8-8	0.6250	0.6222	0.0028	0.4800	0.4570	0.5562	0.5408	0.0154
3/4-6	0.7500	0.7470	0.0030	0.5633	0.5371	0.6598	0.6424	0.0174
7/8-6	0.8750	0.8717	0.0033	0.6883	0.6615	0.7842	0.7663	0.0179
1-5	1.0000	0.9965	0.0035	0.7800	0.7509	0.8920	0.8726	0.0194
1 1/8-5	1.1250	1.1213	0.0037	0.9050	0.8753	1.0165	0.9967	0.0198
1 1/4-5	1.1250	1.2461	0.0039	1.0300	0.9998	1.1411	1.1210	0.0201
1 3/8-4	1.3750	1.3709	0.0041	1.1050	1.0719	1.2406	1.2185	0.0220
1 1/2-4	1.5000	1.4957	0.0043	1.2300	1.1965	1.3652	1.3429	0.0223
1 3/4-4	1.7500	1.7454	0.0046	1.4800	1.4456	1.6145	1.5916	0.0229
2-4	2.0000	1.9951	0.0049	1.7300	1.6948	1.8637	1.8402	0.0235
2 1/4-3	2.2500	2.2448	0.0052	1.8967	1.8572	2.0713	2.0450	0.0263
2 1/2-3	2.5000	2.4945	0.0055	2.1467	2.1065	2.3207	2.2939	0.0268
2 3/4-3	2.7500	2.7442	0.0058	2.3967	2.3558	2.5700	2.5427	0.0273
3-2	3.0000	2.9939	0.0061	2.4800	2.4326	2.7360	2.7044	0.0316
3 1/2-2	3.5000	3.4935	0.0065	2.9800	2.9314	3.2350	3.2026	0.0324
4-2	4.0000	3.9930	0.0070	3.4800	3.4302	3.7340	3.7008	0.0332
4 1/2-2	4.5000	4.4926	0.0074	3.9800	3.9201	4.2330	4.1991	0.0339
5-2	5.0000	4.9922	0.0078	4.4800	4.4281	4.7319	4.6973	0.0346

(Continued)

Table 32. *(Continued)* **Dimensions and Tolerances for Class 2C Centralizing Acme Screw Threads.**
(Source, Landis Threading Systems.)

Nominal Size and TPI	Major Diameter			Minor Diameter			Pitch Diameter		
	Max.	Min.	Tol.	Max.	Min.	Tol.	Max.	Min.	Tol.
						Internal Threads - Class 2C			
1/2-10	0.50007	0.5032	0.0025	0.4100	0.4150	0.0050	0.4500	0.4637	0.0137
5/8-8	0.6258	0.6286	0.0026	0.5125	0.5187	0.0062	0.5625	0.5779	0.0154
3/4-6	0.7509	0.7539	0.0030	0.6000	0.6083	0.0083	0.6667	0.6841	0.0174
7/8-6	0.8759	0.8792	0.0033	0.7250	0.7333	0.0083	0.7917	0.8096	0.0179
1-5	1.0010	1.0045	0.0035	0.8200	0.8300	0.0100	0.9000	0.9194	0.0194
1 1/8-5	1.1261	1.1298	0.0037	0.9450	0.9550	0.0100	1.0250	1.0448	0.0198
1 1/4-5	1.2511	1.2550	0.0039	1.0700	1.0800	0.0100	1.1500	1.1701	0.0201
1 3/8-4	1.3762	1.3803	0.0041	1.1500	1.1625	0.0125	1.2500	1.2720	0.0220
1 1/2-4	1.5012	1.5055	0.0043	1.2750	1.2875	0.0125	1.3750	1.3973	0.0223
1 3/4-4	1.7513	1.7559	0.0046	1.5250	1.5375	0.0125	1.6250	1.6479	0.0229
2-4	2.0014	2.0063	0.0049	1.7750	1.7875	0.0125	1.8750	1.8985	0.0235
2 1/4-3	2.2515	2.2567	0.0052	1.9500	1.9667	0.0167	2.0833	2.1096	0.0263
2 1/2-3	2.5016	2.5071	0.0055	2.2000	2.2167	0.0167	2.3333	2.3601	0.0268
2 3/4-3	2.7517	2.7575	0.0058	2.4500	2.4667	0.0167	2.5833	2.6106	0.0273
3-2	3.0017	3.0078	0.0061	2.5500	2.5750	0.0250	2.7500	2.7816	0.0316
3 1/2-2	3.5019	3.5084	0.0065	3.0500	3.0750	0.0250	3.2500	3.2824	0.0324
4-2	4.0020	4.0090	0.0070	3.5500	3.5750	0.0250	3.7500	3.7832	0.0332
4 1/2-2	4.5021	4.5095	0.0074	4.0500	4.0750	0.0250	4.2500	4.2839	0.0339
5-2	5.0022	5.0100	0.0078	4.5500	4.5750	0.0250	4.7500	4.7846	0.0346

All dimensions in inches.

Table 33. Dimensions and Tolerances for Class 3C Centralizing Acme Screw Threads.
(Source, Landis Threading Systems.)

Nominal Size and TPI	Major Diameter			Minor Diameter		Pitch Diameter		
	Max.	Min.	Tol.	Max.	Min.	Max.	Min.	Tol.
				External Threads - Class 3C				
1/2-10	0.500	0.4989	0.0011	0.3800	0.3594	0.4443	0.4306	0.0137
5/8-8	0.6250	0.6238	0.0012	0.4800	0.4570	0.5562	0.5408	0.0154
3/4-6	0.7500	0.7487	0.0013	0.5633	0.5371	0.6598	0.6424	0.0174
7/8-6	0.8750	0.8736	0.0014	0.6883	0.6615	0.7842	0.7663	0.0179
1-5	1.0000	0.9985	0.0015	0.7800	0.7509	0.8920	0.8726	0.0194
1 1/8-5	1.1250	1.1234	0.0016	0.9050	0.8753	1.0165	0.9967	0.0198
1 1/4-5	1.1250	1.2483	0.0017	1.0300	0.9998	1.1411	1.1210	0.0201
1 3/8-4	1.3750	1.3732	0.0018	1.1050	1.0719	1.2406	1.2185	0.0220
1 1/2-4	1.5000	1.4982	0.0019	1.2300	1.1965	1.3652	1.3429	0.0223
1 3/4-4	1.7500	1.7480	0.0020	1.4800	1.4456	1.6145	1.5916	0.0229
2-4	2.0000	1.9979	0.0021	1.7300	1.6948	1.8637	1.8402	0.0235
2 1/4-3	2.2500	2.2478	0.0022	1.8967	1.8572	2.0713	2.0450	0.0263

(Continued)

Table 33. *(Continued)* **Dimensions and Tolerances for Class 3C Centralizing Acme Screw Threads.**
(Source, Landis Threading Systems.)

	External Threads - Class 3C								
Nominal Size and TPI	**Major Diameter**			**Minor Diameter**			**Pitch Diameter**		
	Max.	Min.	Tol.	Max.	Min.	Max.	Min.	Tol.	
2 1/2-3	2.5000	2.4976	0.0024	2.1467	2.1065	2.3207	2.2939	0.0268
2 3/4-3	2.7500	2.7475	0.0025	2.3967	2.3558	2.5700	2.5427	0.0273
3-2	3.0000	2.9974	0.0026	2.4800	2.4326	2.7360	2.7044	0.0316
3 1/2-2	3.5000	3.4972	0.0028	2.9800	2.9314	3.2350	3.2026	0.0324
4-2	4.0000	3.9970	0.0030	3.4800	3.4302	3.7340	3.7008	0.0332
4 1/2-2	4.5000	4.4968	0.0032	3.9800	3.9201	4.2330	4.1991	0.0339
5-2	5.0000	4.9966	0.0034	4.4800	4.4281	4.7319	4.6973	0.0346

	Internal Threads - Class 3C								
Nominal Size and TPI	**Major Diameter**			**Minor Diameter**			**Pitch Diameter**		
	Max.	Min.	Tol.	Max.	Min.	Tol.	Max.	Min.	Tol.
1/2-10	0.50007	0.5032	0.0025	0.4100	0.4150	0.0050	0.4500	0.4637	0.0137
5/8-8	0.6258	0.6286	0.0028	0.5125	0.5187	0.0062	0.5625	0.5779	0.0154
3/4-6	0.7509	0.7539	0.0030	0.6000	0.6083	0.0083	0.6667	0.6841	0.0174
7/8-6	0.8759	0.8792	0.0033	0.7250	0.7333	0.0083	0.7917	0.8096	0.0179
1-5	1.0010	1.0045	0.0035	0.8200	0.8300	0.0100	0.9000	0.9194	0.0194
1 1/8-5	1.1261	1.1298	0.0037	0.9450	0.9550	0.0100	1.0250	1.0448	0.0198
1 1/4-5	1.2511	1.2550	0.0039	1.0700	1.0800	0.0100	1.1500	1.1701	0.0201
1 3/8-4	1.3762	1.3803	0.0041	1.1500	1.1625	0.0125	1.2500	1.2720	0.0220
1 1/2-4	1.5012	1.5055	0.0043	1.2750	1.2875	0.0125	1.3750	1.3973	0.0223
1 3/4-4	1.7513	1.7559	0.0046	1.5250	1.5375	0.0125	1.6250	1.6479	0.0229
2-4	2.0014	2.0063	0.0049	1.7750	1.7875	0.0125	1.8750	1.8985	0.0235
2 1/4-3	2.2515	2.2567	0.0052	1.9500	1.9667	0.0167	2.0833	2.1096	0.0263
2 1/2-3	2.5016	2.5071	0.0055	2.2000	2.2167	0.0167	2.3333	2.3601	0.0268
2 3/4-3	2.7517	2.7575	0.0058	2.4500	2.4667	0.0167	2.5833	2.6106	0.0273
3-2	3.0017	3.0078	0.0061	2.5500	2.5750	0.0250	2.7500	2.7816	0.0316
3 1/2-2	3.5019	3.5084	0.0065	3.0500	3.0750	0.0250	3.2500	3.2824	0.0324
4-2	4.0020	4.0090	0.0070	3.5500	3.5750	0.0250	3.7500	3.7832	0.0332
4 1/2-2	4.5021	4.5095	0.0074	4.0500	4.0750	0.0250	4.2500	4.2839	0.0339
5-2	5.0022	5.0100	0.0078	4.5500	4.5750	0.0250	4.7500	4.7846	0.0346

All dimensions in inches.

Table 34. Dimensions and Tolerances for Class 4C Centralizing Acme Screw Threads.
(Source, Landis Threading Systems.)

	External Threads - Class 4C							
Nominal Size and TPI	**Major Diameter**			**Minor Diameter**			**Pitch Diameter**	
	Max.	Min.	Tol.	Max.	Min.	Max.	Min.	Tol.
1/2-10	0.5000	0.4993	0.0007	0.3800	0.3594	0.4443	0.4306	0.0137
5/8-8	0.6250	0.6242	0.0008	0.4800	0.4570	0.5562	0.5408	0.0154
3/4-6	0.7500	0.7491	0.0009	0.5633	0.5371	0.6598	0.6424	0.0174
7/8-6	0.8750	0.8741	0.0009	0.6883	0.6615	0.7842	0.7663	0.0179
1-5	1.0000	0.9990	0.0010	0.7800	0.7509	0.8920	0.8726	0.0194
1 1/8-5	1.1250	1.1239	0.0011	0.9050	0.8753	1.0165	0.9967	0.0198
1 1/4-5	1.1250	1.2489	0.0011	1.0300	0.9998	1.1411	1.1210	0.0201
1 3/8-4	1.3750	1.3738	0.0012	1.1050	1.0719	1.2406	1.2185	0.0220
1 1/2-4	1.5000	1.4988	0.0012	1.2300	1.1965	1.3652	1.3429	0.0223
1 3/4-4	1.7500	1.7487	0.0013	1.4800	1.4456	1.6145	1.5916	0.0229
2-4	2.0000	1.9986	0.0014	1.7300	1.6948	1.8637	1.8402	0.0235
2 1/4-3	2.2500	2.2485	0.0015	1.8967	1.8572	2.0713	2.0450	0.0263
2 1/2-3	2.5000	2.4984	0.0016	2.1467	2.1065	2.3207	2.2939	0.0268
2 3/4-3	2.7500	2.7483	0.0017	2.3967	2.3558	2.5700	2.5427	0.0273
3-2	3.0000	2.9983	0.0017	2.4800	2.4326	2.7360	2.7044	0.0316
3 1/2-2	3.5000	3.4981	0.0019	2.9800	2.9314	3.2350	3.2026	0.0324
4-2	4.0000	3.9980	0.0020	3.4800	3.4302	3.7340	3.7008	0.0332
4 1/2-2	4.5000	4.4979	0.0021	3.9800	3.9201	4.2330	4.1991	0.0339
5-2	5.0000	4.9978	0.0022	4.4800	4.4281	4.7319	4.6973	0.0346
	Internal Threads - Class 4C							
Nominal Size and TPI	**Major Diameter**			**Minor Diameter**			**Pitch Diameter**	
	Max.	Min.	Tol.	Max.	Min.	Tol.	Max.	Min.	Tol.
1/2-10	0.50007	0.5021	0.0014	0.4100	0.4150	0.0050	0.4500	0.4637	0.0137
5/8-8	0.6258	0.6274	0.0016	0.5125	0.5187	0.0062	0.5625	0.5779	0.0154
3/4-6	0.7509	0.7526	0.0017	0.6000	0.6083	0.0083	0.6667	0.6841	0.0174
7/8-6	0.8759	0.8778	0.0019	0.7250	0.7333	0.0083	0.7917	0.8096	0.0179
1-5	1.0010	1.0030	0.0020	0.8200	0.8300	0.0100	0.9000	0.9194	0.0194
1 1/8-5	1.1261	1.1282	0.0021	0.9450	0.9550	0.0100	1.0250	1.0448	0.0198
1 1/4-5	1.2511	1.2533	0.0022	1.0700	1.0800	0.0100	1.1500	1.1701	0.0201
1 3/8-4	1.3762	1.3785	0.0023	1.1500	1.1625	0.0125	1.2500	1.2720	0.0220
1 1/2-4	1.5012	1.5036	0.0024	1.2750	1.2875	0.0125	1.3750	1.3973	0.0223
1 3/4-4	1.7513	1.7539	0.0026	1.5250	1.5375	0.0125	1.6250	1.6479	0.0229
2-4	2.0014	2.0042	0.0028	1.7750	1.7875	0.0125	1.8750	1.8985	0.0235
2 1/4-3	2.2515	2.2545	0.0030	1.9500	1.9667	0.0167	2.0833	2.1096	0.0263
2 1/2-3	2.5016	2.5048	0.0032	2.2000	2.2167	0.0167	2.3333	2.3601	0.0268
2 3/4-3	2.7517	2.7550	0.0033	2.4500	2.4667	0.0167	2.5833	2.6106	0.0273
3-2	3.0017	3.0052	0.0035	2.5500	2.5750	0.0250	2.7500	2.7816	0.0316

(Continued)

Table 34. *(Continued)* **Dimensions and Tolerances for Class 4C Centralizing Acme Screw Threads.**
(Source, Landis Threading Systems.)

Internal Threads - Class 4C								
Nominal Size and TPI	Major Diameter			Minor Diameter			Pitch Diameter		
	Max.	Min.	Tol.	Max.	Min.	Tol.	Max.	Min.	Tol.
3 1/2-2	3.5019	3.5056	0.0037	3.0500	3.0750	0.0250	3.2500	3.2824	0.0324
4-2	4.0020	4.0060	0.0040	3.5500	3.5750	0.0250	3.7500	3.7832	0.0332
4 1/2-2	4.5021	4.5063	0.0042	4.0500	4.0750	0.0250	4.2500	4.2839	0.0339
5-2	5.0022	5.0067	0.0045	4.5500	4.5750	0.0250	4.7500	4.7846	0.0346

All dimensions in inches.

Table 35. Dimensions and Tolerances for Stub Acme Screw Threads. *(Source, FED-STD-H28/13.)*

External Threads - Stub Acme								
Nominal Size and TPI	Major Diameter			Pitch Diameter			Minor Diameter	
	Max.	Min.	Tol.	Max.	Min.	Tol.	Max.	Min.
1/4-16	0.2500	0.2469	0.0050	0.2272	0.2167	0.0105	0.2024	0.1919
5/16-14	0.3125	0.3089	0.0050	0.2871	0.2757	0.0114	0.2597	0.2483
3/8-12	0.3750	0.3708	0.0050	0.3451	0.3328	0.0123	0.3150	0.3027
7/16-12	0.4375	0.4333	0.0050	0.4076	0.3950	0.0126	0.3775	0.3649
1/2-10	0.5000	0.4950	0.0050	0.4643	0.4506	0.0137	0.4200	0.4063
5/8-8	0.6250	0.6188	0.0062	0.5812	0.5658	0.0154	0.5300	0.5146
3/4-6	0.7500	0.7417	0.0083	0.6931	0.6757	0.0174	0.6300	0.6126
7/8-6	0.8750	0.8667	0.0083	0.8175	0.7996	0.0179	0.7550	0.7371
1-5	1.0000	0.9900	0.0100	0.9320	0.9126	0.0194	0.8600	0.8406
1 1/8-5	1.1250	1.1150	0.0100	1.0565	1.0367	0.0198	0.9850	0.9652
1 1/4-5	1.2500	1.2400	0.0100	1.1811	1.1610	0.0201	1.1100	1.0899
1 3/8-4	1.3750	1.3625	0.0125	1.2906	1.2686	0.0220	1.2050	1.1830
1 1/2-4	1.500	1.4875	0.0125	1.4152	1.3929	0.0223	1.3300	1.3077
1 3/4-4	1.7500	1.7375	0.0125	1.6645	1.6416	0.0229	1.5800	1.5571
2-4	2.0000	1.9875	0.0125	1.9137	1.8902	0.0235	1.8300	1.8065
2 1/4-3	2.2500	2.2333	0.0167	2.1380	2.1117	0.0263	2.0300	2.0037
2 1/2-3	2.500	2.4833	0.0167	2.3874	2.3606	0.0268	2.2800	2.2532
2 3/4-3	2.7500	2.7333	0.0167	2.6367	2.6094	0.0273	2.5300	2.5027
3-2	3.0000	2.9750	0.2500	2.8360	2.8044	0.0316	2.6800	2.6484
3 1/2-2	3.5000	3.4750	0.2500	3.3350	3.3026	0.0324	3.1800	3.1476
4-2	4.0000	3.9750	0.2500	3.8340	3.8008	0.0332	3.6800	3.6468
4 1/2-2	4.5000	4.4750	0.2500	4.3330	4.2991	0.0339	4.1800	4.1461
5-2	5.0000	4.9750	0.2500	4.8319	4.7973	0.0346	4.6800	4.6454
Internal Threads - Stub Acme								
Nominal Size and TPI	Major Diameter			Pitch Diameter			Minor Diameter		
	Min.	Max.	Tol.	Min.	Max.	Tol.	Min.	Max.	Tol.
1/4-16	0.2600	0.2750	0.0100	0.2312	0.2417	0.0105	0.2125	0.2156	0.0050

(Continued)

Table 35. *(Continued)* Dimensions and Tolerances for Stub Acme Screw Threads. *(Source, FED-STD-H28/13.)*

Nominal Size and TPI	Major Diameter			Pitch Diameter			Minor Diameter		
	Min.	Max.	Tol.	Min.	Max.	Tol.	Min.	Max.	Tol.
5/16-14	0.3225	0.3339	0.0100	0.2911	0.3025	0.0114	0.2696	0.2732	0.0050
3/8-12	0.3850	0.3973	0.0100	0.3500	0.3623	0.0123	0.3250	0.3292	0.0050
7/16-12	0.4475	0.4601	0.0100	0.4125	0.4251	0.0126	0.3875	0.3917	0.0050
1/2-10	0.5200	0.5337	0.0200	0.4700	0.4837	0.0137	0.4400	0.4450	0.0050
5/8-8	0.6450	0.6604	0.0200	0.5875	0.6029	0.0154	0.5500	0.5562	0.0062
3/4-6	0.7700	0.7874	0.0200	0.7000	0.7174	0.0174	0.6500	0.6583	0.0083
7/8-6	0.8950	0.9129	0.0200	0.8250	0.8429	0.0179	0.7750	0.7833	0.0083
1-5	1.0200	1.0394	0.0200	0.9400	0.9594	0.0194	0.8800	0.8900	0.0100
1 1/8-5	1.1450	1.1648	0.0200	1.0650	1.0848	0.0198	1.0050	1.0150	0.0100
1 1/4-5	1.2700	1.2901	0.0200	1.1900	1.2101	0.0201	1.1300	1.1400	0.0100
1 3/8-4	1.3950	1.4170	0.0200	1.3000	1.3220	0.0220	1.2250	1.2375	0.0125
1 1/2-4	1.5200	1.5423	0.0200	1.4250	1.4473	0.0223	1.3500	1.3625	0.0125
1 3/4-4	1.7700	1.7929	0.0200	1.6750	1.6979	0.0229	1.6000	1.6125	0.0125
2-4	2.0200	2.0435	0.0200	1.9250	1.9485	0.0235	1.8500	1.8625	0.0125
2 1/4-3	2.2700	2.2963	0.0200	2.1500	2.1763	0.0263	2.0500	2.0667	0.0167
2 1/2-3	2.5200	2.5468	0.0200	2.4000	2.4268	0.0268	2.3000	2.3167	0.0167
2 3/4-3	2.7700	2.7973	0.0200	2.6500	2.6773	0.0273	2.5500	2.5667	0.0167
3-2	3.0200	3.0516	0.0200	2.8500	2.8816	0.0316	2.7000	2.7250	0.0250
3 1/2-2	3.5200	3.5524	0.0200	3.3500	3.3824	0.0324	3.2000	3.2250	0.0250
4-2	4.0200	4.0532	0.0200	3.8500	3.8832	0.0332	3.7000	3.7250	0.0250
4 1/2-2	4.5200	4.5539	0.0200	4.3500	4.3839	0.0339	4.2000	4.2250	0.0250
5-2	5.0200	5.0546	0.0200	4.8500	4.8846	0.0346	4.7000	4.7250	0.0250

Internal Threads - Stub Acme

Buttress screw thread

The American Buttress thread form is capable of withstanding exceptionally high unidirectional stress parallel to the axis of the thread. This strength is derived from the basic thread form, which provides a pressure flank at a nearly right angle to the force, fortified by a "buttress" leaning into the flank at an angle of 45°, commonly called a 7°/45° thread form. Buttress threads have a variety of applications where tubular members are screwed together, ranging from the common automobile hose clamp to the breech assemblies of heavy artillery. The initial standardization of the thread came with the issue of ASA B1.9 in 1953, which defined the thread profile and the dimensional formulas. The present standard is ANSI B1.9, which was issued in 1973.

The standard Buttress thread as shown in *Figure 14* has a load flank angle of 7°, measured in the axial plane. The height of the sharp V thread is $0.89064P$, the thread height is $0.66271P$, and the basic height of thread engagement is $0.6P$. At maximum material condition, the root radius equals $0.07141P$, and the root truncation is $0.08261P$. At minimum material condition, root radius is equal to $0.0357P$, and the root truncation is $0.0413P$. Crest width equals $0.16316P$, and crest truncation is $0.14532P$. **Table 36** provides basic dimensions for preferred sizes of American Buttress screw threads in maximum material condition. The dimensions in the table include the recommended allowance

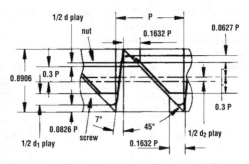

Figure 14. American Buttress 7°/45° thread form. (Source, Kennametal Inc.)

(clearance) specified for each diameter and pitch. Allowances are provided for external threads only. Internal threads are basic.

Pull-type Buttress threads are identified by their nominal diameter, stated as a decimal rather than a fraction, followed by the tpi and the letters BUTT. For push-type threads, the letters PUSH BUTT follow the diameter and tpi. The designation for a one and one-half inch, 10 tpi pull-type thread would be 1.5–10 BUTT. Class 2 Buttress threads are considered "Standard," and Class 3 are "Precision Grade." Both classes share identical allowances on their external threads, but Class 3 threads have slightly smaller tolerances (0.0019" on the smallest thread, and 0.0075" on the largest) on the major diameter of the external thread, the pitch diameter of both internal and external threads, and the minor diameter of the internal thread.

The British Standard 7°/45° Buttress thread shares its basic angular dimensions with the American form, but is not interchangeable. It has a basic thread height of $0.50586P$ and a root radius of $0.12055P$. A second Standard British Buttress thread has a pressure flank angle of 0° and a flank angle of 52°. Known as a 0°/52°, it has a basic thread height of $0.49298P$ and a root radius of $0.09298P$. Because of the 0° angle, this form offers economy of production. British Buttress threads are specified by British Standard B.S. 1657:1950.

Table 36. American Buttress Inch Screw Threads Basic Dimensions at Maximum Material Condition.

Nominal Size	TPI	External Thread			Internal Thread			Crest Width	Height of Thread	Height of Engagement
		Major Dia.	Pitch Dia.	Minor Dia.	Minor Dia.	Pitch Dia.	Major Dia.			
0.50	20	0.4963	0.4663	0.4300	0.4400	0.4700	0.5063	0.0082	0.0331	0.0300
0.50	16	0.4960	0.4585	0.4132	0.4250	0.4625	0.5078	0.0102	0.0414	0.0375
0.50	12	0.4956	0.4456	0.3852	0.4000	0.4500	0.5104	0.0136	0.0552	0.5000
0.65	20	0.6463	0.6163	0.5800	0.5900	0.6200	0.6563	0.0082	0.0331	0.0300
0.65	16	0.6460	0.6085	0.5632	0.5750	0.6125	0.6578	0.0102	0.0414	0.0375
0.65	12	0.6456	0.5956	0.5352	0.5500	0.6000	0.6604	0.0136	0.0552	0.0500
0.75	16	0.7458	0.7083	0.6630	0.6750	0.7125	0.7578	0.0102	0.0414	0.0375
0.75	12	0.7454	0.6954	0.6350	0.6500	0.7000	0.7604	0.0136	0.0552	0.0500
0.75	10	0.7451	0.6851	0.6126	0.6300	0.6900	0.7625	0.0163	0.0663	0.0600

(Continued)

Table 36. *(Continued)* American Buttress Inch Screw Threads Basic Dimensions at Maximum Material Condition.

Nominal Size	TPI	External Thread			Internal Thread			Crest Width	Height of Thread	Height of Engagement
		Major Dia.	Pitch Dia.	Minor Dia.	Minor Dia.	Pitch Dia.	Major Dia.			
0.875	16	0.8708	0.8333	0.7880	0.8000	0.8375	0.8828	0.0102	0.0414	0.0375
0.875	12	0.8704	0.8204	0.7600	0.7750	0.8250	0.8854	0.0136	0.0552	0.0500
0.875	10	0.8701	0.8101	0.7376	0.7550	0.8150	0.8875	0.0163	0.0663	0.0600
1.0	16	0.9958	0.9583	0.9130	0.9250	0.9625	1.0078	0.0102	0.0414	0.0375
1.0	12	0.9954	0.9454	0.8850	0.9000	0.9500	1.0104	0.0136	0.0552	0.0500
1.0	10	0.9951	0.9351	0.8626	0.8800	0.9400	1.0125	0.0163	0.0663	0.0600
1.25	12	1.2452	1.1952	1.1348	1.1500	1.2000	1.2604	0.0136	0.0552	0.0500
1.25	10	1.2449	1.1849	1.1124	1.1300	1.1900	1.2625	0.0163	0.0663	0.0600
1.25	8	1.2445	1.1695	1.0788	1.1000	1.1750	1.2657	0.0204	0.0828	0.0750
1.375	12	1.3702	1.3202	1.2598	1.2750	1.3250	1.3854	0.0136	0.0552	0.0500
1.375	10	1.3699	1.3099	1.2374	1.2550	1.3150	1.3875	0.0163	0.0663	0.0600
1.375	8	1.3695	1.2945	1.2038	1.2250	1.3000	1.3907	0.0204	0.0828	0.0750
1.50	12	1.4952	1.4452	1.3848	1.4000	1.4500	1.5104	0.0136	0.0552	0.0500
1.50	10	1.4949	1.4349	1.3624	1.3800	1.4400	1.5125	0.0163	0.0663	0.0600
1.50	8	1.4939	1.4189	1.3282	1.3500	1.4250	1.5157	0.0204	0.0828	0.0750
1.75	10	1.7447	1.6847	1.6122	1.6300	1.6900	1.7625	0.0163	0.0663	0.0600
1.75	8	1.7442	1.6692	1.5785	1.6000	1.6750	1.7657	0.0204	0.0828	0.0750
1.75	6	1.7436	1.6436	1.5227	1.5500	1.6500	1.7709	0.0272	0.1105	0.1000
2.0	10	1.9947	1.9347	1.8622	1.8800	1.9400	2.0125	0.0163	0.0663	0.0600
2.0	8	1.9942	1.9192	1.8285	1.8500	1.9250	2.0157	0.0204	0.0828	0.0750
2.0	6	1.9936	1.8936	1.7727	1.8000	1.9000	2.0209	0.0272	0.1105	0.1000
2.25	10	2.2447	2.1847	2.1122	2.1300	2.1900	2.2625	0.0163	0.0663	0.0600
2.25	8	2.2442	2.1692	2.0785	2.1000	2.1750	2.2657	0.0204	0.0828	0.0750
2.25	6	2.2436	2.1436	2.0227	2.0500	2.1500	2.2709	0.0272	0.1105	0.1000
2.50	10	2.4947	2.4347	2.3622	2.3800	2.4400	2.5125	0.0163	0.0663	0.0600
2.50	8	2.4942	2.4192	2.3285	2.3500	2.4250	2.5157	0.0204	0.0828	0.0750
2.50	6	2.4936	2.3936	2.2727	2.3000	2.4000	2.5209	0.0272	0.1105	0.1000
2.75	8	2.7439	2.6689	2.5782	2.6000	2.6750	2.7657	0.0204	0.0828	0.0750
2.75	6	2.7433	2.6433	2.5224	2.5500	2.6500	2.7709	0.0272	0.1105	0.1000
2.75	5	2.7429	2.6229	2.4778	2.5100	2.6300	2.7751	0.0326	0.1325	0.1200
3.0	8	2.9939	2.9189	2.8282	2.8500	2.9250	3.0157	0.0204	0.0828	0.0750
3.0	6	2.9933	2.8933	2.7724	2.8000	2.9000	3.0209	0.0272	0.1105	0.1000
3.0	5	2.9929	2.8729	2.7278	2.7600	2.8800	3.0251	0.0326	0.1325	0.1200
3.50	8	3.4939	3.4189	3.3282	3.3500	3.4250	3.5157	0.0204	0.0828	0.0750
3.50	6	3.4933	3.3933	3.2724	3.3000	3.4000	3.5209	0.0272	0.1105	0.1000
3.50	5	3.4929	3.3729	3.2278	3.2600	3.3800	3.5251	0.0326	0.1325	0.1200
4.0	8	3.9939	3.9189	3.8282	3.8500	3.9250	4.0157	0.0204	0.0828	0.0750
4.0	6	3.9933	3.8933	3.7724	3.8000	3.9000	4.0209	0.0272	0.1105	0.1000

(Continued)

Table 36. *(Continued)* **American Buttress Inch Screw Threads Basic Dimensions at Maximum Material Condition.**

Nominal Size	TPI	External Thread			Internal Thread			Crest Width	Height of Thread	Height of Engagement
		Major Dia.	Pitch Dia.	Minor Dia.	Minor Dia.	Pitch Dia.	Major Dia.			
4.0	5	3.9929	3.8729	3.7278	3.7600	3.8800	4.0251	0.0326	0.1325	0.1200
4.5	6	4.4930	4.3930	4.2721	4.3000	4.4000	4.5209	0.0272	0.1105	0.1000
4.5	5	4.4926	4.3726	4.2275	4.2600	4.3800	4.5251	0.0326	0.1325	0.1200
4.5	4	4.4920	4.3420	4.1607	4.2000	4.3500	4.5314	0.0408	0.1657	0.1500
5.0	6	4.9930	4.8930	4.7721	4.8000	4.9000	5.0209	0.0272	0.1105	0.1000
5.0	5	4.9926	4.8726	4.7275	4.7600	4.8800	5.0251	0.0326	0.1325	0.1200
5.0	4	4.9920	4.8420	4.6607	4.7000	4.8500	5.0314	0.0408	0.1657	0.1500
5.50	6	5.4930	5.3930	5.2721	5.3000	5.4000	5.5209	0.0272	0.1105	0.1000
5.50	5	5.4926	5.3726	5.2275	5.2600	5.3800	5.5251	0.0326	0.1325	0.1200
5.50	4	5.4920	5.3420	5.1607	5.2000	5.3500	5.5314	0.0408	0.1657	0.1500
6.0	6	5.9930	5.8930	5.7721	5.8000	5.9000	6.0209	0.0272	0.1105	0.1000
6.0	5	5.9926	5.8726	5.7275	5.7600	5.8800	6.0251	0.0326	0.1325	0.1200
6.0	4	5.9920	5.8420	5.6607	5.7000	5.8500	6.0314	0.0408	0.1657	0.1500
7.0	5	6.9922	6.8722	6.7271	6.7600	6.8800	7.0251	0.0326	0.1325	0.1200
7.0	4	6.9916	6.8416	6.6603	6.7000	6.8500	7.0314	0.0408	0.1657	0.1500
7.0	3	6.9907	6.7907	6.5489	6.6000	6.8000	7.0418	0.0544	0.2209	0.2000
8.0	5	7.9922	7.8722	7.7271	7.7600	7.8800	8.0251	0.0326	0.1325	0.1200
8.0	4	7.9916	7.8416	7.6603	7.7000	7.8500	8.0314	0.0408	0.1657	0.1500
8.0	3	7.9907	7.7907	7.5489	7.6000	7.8000	8.0418	0.0544	0.2209	0.2000
9.0	5	8.9922	8.8722	8.7271	8.7600	8.8800	9.0251	0.0326	0.1325	0.1200
9.0	4	8.9916	8.8416	8.6603	8.7000	8.8500	9.0314	0.0408	0.1657	0.1500
9.0	3	8.9907	8.7907	8.5489	8.6000	8.8000	9.0418	0.0544	0.2209	0.2000
10.0	5	9.9922	9.8722	9.7271	9.7600	9.8800	10.0251	0.0326	0.1325	0.1200
10.0	4	9.9916	9.8416	9.6603	9.7000	9.8500	10.0314	0.0408	0.1657	0.1500
10.0	3	9.9907	9.7907	9.5489	9.6000	9.8000	10.0418	0.0544	0.2209	0.2000
12.0	4	11.9911	11.8411	11.6598	11.7000	11.8500	12.0314	0.0408	0.1657	0.1500
12.0	3	11.9902	11.7902	11.5484	11.6000	11.8000	12.0418	0.0544	0.2209	0.2000
12.0	2.5	11.9896	11.7496	11.4594	11.5200	11.7600	12.0502	0.0653	0.2651	0.2400
14.0	4	13.9911	13.8411	13.6598	13.7000	13.8500	14.0314	0.0408	0.1657	0.1500
14.0	3	13.9902	13.7902	13.5484	13.6000	13.8000	14.0418	0.0544	0.2209	0.2000
14.0	2.5	13.9896	13.7496	13.4594	13.5200	13.7600	14.0502	0.0653	0.2651	0.2400
16.0	4	15.9911	15.8411	15.6598	15.7000	15.8500	16.0314	0.0408	0.1657	0.1500
16.0	3	15.9902	15.7902	15.5484	15.6000	15.8000	16.0418	0.0544	0.2209	0.2000
16.0	2.5	15.9896	15.7496	15.4594	15.5200	15.7600	16.0502	0.0653	0.2651	0.2400
18.0	3	17.9897	17.7897	17.5479	17.6000	17.8000	18.0418	0.0544	0.2209	0.2000
18.0	2.5	17.9891	17.7491	17.4589	17.5200	17.7600	18.0502	0.0653	0.2651	0.2400
18.0	2	17.9882	17.6882	17.3255	17.4000	17.7000	18.0627	0.0816	0.3314	0.3000
20.0	3	19.9897	19.7897	19.5479	19.6000	19.8000	20.0418	0.0544	0.2209	0.2000

(Continued)

Table 36. *(Continued)* American Buttress Inch Screw Threads Basic Dimensions at Maximum Material Condition.

Nominal Size	TPI	External Thread			Internal Thread			Crest Width	Height of Thread	Height of Engagement
		Major Dia.	Pitch Dia.	Minor Dia.	Minor Dia.	Pitch Dia.	Major Dia.			
20.0	2.5	19.9891	19.7491	19.4589	19.5200	19.7600	20.0502	0.0653	0.2651	0.2400
20.0	2	19.9882	19.6882	19.3255	19.4000	19.7000	20.0627	0.0816	0.3314	0.3000
22.0	3	21.9897	21.7897	21.5479	21.6000	21.8000	22.0418	0.0544	0.2209	0.2000
22.0	2.5	21.9891	21.7491	21.4589	21.5200	21.7600	22.0502	0.0653	0.2651	0.2400
22.0	2	21.9882	21.6882	21.3255	21.4000	21.7000	22.0627	0.0816	0.3314	0.3000
24.0	3	23.9897	23.7897	23.5479	23.6000	23.8000	24.0418	0.0544	0.2209	0.2000
24.0	2.5	23.9891	23.7491	23.4589	23.5200	23.7600	24.0502	0.0653	0.2651	0.2400
24.0	2	23.9882	23.6882	23.3255	23.4000	23.7000	24.0627	0.0816	0.3314	0.3000

All dimensions in inches.

Pipe threads

The modern American pipe form has been through several evolutions since it originated as the Briggs Standard in the last century. In 1913, the "Committee on the Standardization of Pipe Threads," jointly sponsored by the American Gas Association and the American Society of Mechanical Engineers, began six years of work that resulted in a standard that was published by the National Screw Thread Commission (NSTC). The NSTC continued work on the project, as did the AGA and the ASME. Since then, the standards have been reviewed and fine-tuned regularly, but the basic thread dimensions can be traced back to their earliest origins. There are two general categories of pipe threads: those that provide a pressure-tight fit, and those that produce a mechanical joint. Pressure fit threads may be designed to perform when assembled dry (NPTF for tapered fittings, NPSF for straight pipe fuel fittings, and NPSI for standard straight pipe fittings), or with a sealer (NPT series and NPSC for pipe couplings). Pipe threads that are used for joining, but are not pressure tight, are designed as parallel rather than taper threads. These include NPSM for free fitting mechanical joints, and NPSL for loose fitting mechanical joints. See Table 23 for specifications for the British Standard Taper Pipe Thread (BSTP).

The NPT (American National Standard Pipe Thread) form is the most widely used pipe thread. A taper is always present on the external thread, and in high pressure applications, the internal thread is also tapered. The basic form is shown in *Figures 15* and *16*. It utilizes a 60° included angle thread, and the height of the sharp V thread equals $0.866025P$. The basic thread height is $0.80000P$, and both the crest and root are truncated a minimum of $0.033P$. The maximum truncation is graduated by pitch as follows: 27 tpi = $0.0972P$; 18 tpi = $0.0882P$; 14 tpi = $0.0784P$; 11.5 tpi = $0.0725P$; and 8 tpi = $0.0624P$. The minimum width of the flat at the truncation $0.038P$. The maximum width of the flat at the truncation is graduated by pitch as follows: 27 tpi = $0.1107P$; 18 tpi = $0.1026P$; 14 tpi = $0.0896P$, 11.5 tpi = $0.0839P$, and 8 tpi = $0.0720P$. The taper on the NPT form is 0.75" per foot, producing a taper angle of 1°47'. The tolerance for the NPT thread is "plus or minus one turn (length of one pitch)." NPT thread specifications and formulas are contained in MIL-P-7105B (Aeronautical National Taper Pipe Thread Form, Symbol ANPT), ANSI B2.1-1968, and ANSI/ASME B1.20.1-1983. **Table 37** provides dimensions.

The NPSC thread is used on pipe couplings, and it exists only in an internal thread configuration. When assembled wrench-tight with an external NPT thread, a sealer must be used and the coupling should only be used for low-pressure applications. Dimensions for NPSC threads, which are the same form as NPT threads, are given in **Table 38**. ANSI

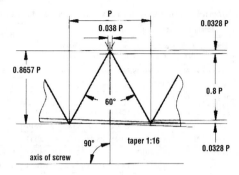

Figure 15. Basic NPT thread form.

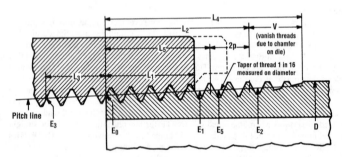

Figure 16. NPT assembly. Refer to Table 37 for dimensions.

B2.1-1968 and ANSI/ASME B1.20.1-1983 contain full information on the form.

NPSM pipe threads are used on free-fitting mechanical joints used for mechanical assembly in situations where there is no internal pressure. The form uses a 60° included angle thread with thread height (sharp V thread) equal to $0.866025P$. The internal thread height is truncated and is equal to $0.54126P$. The internal thread crest equals $0.2500P$. Tolerances are the same as Unified 2A/2B, and the minimum pitch diameter of the internal thread is equal to E_1 of the NPT thread form. **Table 39** provides dimensions for the NPSM thread form. NSPL pipe threads are intended for use in loose fitting mechanical joints with locknuts. Both NPSM and NPSL forms are covered by standards ANSI B2.2-1968 and ANSI/ASME B1.20.1-1983.

Dryseal pipe threads work through deformation when under load. Wrench-tight threads are "crushed," bringing the flanks into contact and thereby erecting a barrier to escaping fluids or gases. The dimensions of NPTF dryseal taper pipe threads are similar to those of NPT threads, except for the amount of truncation in the crest and root. **Table 40** provides formulas for determining both the truncation and the width of the flat at the crest and root. Although Class 1 and 2 NPTF threads are theoretically identical, Class 2 threads have been inspected to assure accuracy of the root and crest truncation. Therefore, when a pressure tight seal is critical, only Class 2 threads should be used. See **Table 41** for dimensions of

Table 37. American Standard Taper Pipe Threads NPT Basic Dimensions. *(Source, Landis Threading Systems.)*

Nominal Pipe Size		Sizes 1/16 - 3"											
		1/16	1/8	1/4	3/8	1/2	3/4	1	1 1/4	1 1/2	2	2 1/2	3
Pipe O.D.		0.3125	0.4050	0.5400	0.6750	0.8400	1.0500	1.3150	1.6600	1.9000	2.3750	2.8750	3.5000
TPI		27	27	18	18	14	14	11 1/2	11 1/2	11 1/2	11 1/2	8	8
Pitch P		0.03704	0.03704	0.05556	0.05556	0.07143	0.07143	0.08696	0.08696	0.08696	0.08696	0.12500	0.12500
E_0 *		0.27118	0.36351	0.47739	0.61201	0.75843	0.96768	1.21363	1.55713	1.79609	2.26902	2.71953	3.34062
Hand Tight Engagement	Inch L_1 *	0.1600	0.1615	0.2278	0.2400	0.3200	0.3390	0.4000	0.4200	0.4200	0.4360	0.682	0.766
	Threads	4.32	4.36	4.10	4.32	4.48	4.75	4.60	4.83	4.83	5.01	5.46	6.13
	E_1 *	0.28118	0.37360	0.49163	0.62701	0.77843	0.98887	1.23863	1.58338	1.82234	2.29627	2.76216	3.38850
Length of Full Thread	Inch L_2 *	0.2611	0.2639	0.4018	0.4078	0.5337	0.5457	0.6828	0.7068	0.7235	0.7565	1.1375	1.2000
	Threads	7.05	7.12	7.23	7.34	7.47	7.64	7.85	8.13	8.32	8.70	9.10	9.60
	E_2 *	0.28750	0.3800	0.50250	0.63750	0.79179	1.00179	1.25630	1.60130	1.84130	2.31630	2.79062	3.41562
Wrench Makeup Length, Internal Thread	Inch L_3 *	0.1111	0.1111	0.1667	0.1667	0.2143	0.2143	0.2609	0.2609	0.2609	0.2609	0.2500	0.2500
	Threads	3	3	3	3	3	3	3	3	3	3	2	2
	E_3 *	0.26424	0.35656	0.46697	0.60160	0.74504	0.95429	1.19733	1.54083	1.77978	2.25272	2.70391	3.32500
Vanish Threads	Inch	0.1285	0.1285	0.1928	0.1928	0.2478	0.2478	0.3017	0.3017	0.3017	0.3017	0.4337	0.4337
	Threads	3.47	3.47	3.47	3.47	3.47	3.47	3.47	3.47	3.47	3.47	3.47	3.47
Overall Length Ext. Threads	L_4 *	0.3896	0.3924	0.5946	0.6006	0.7815	0.7935	0.9845	1.0085	1.0252	1.0582	1.5712	1.6337
Nominal Perfect Ext Thread	Length L_5 *	0.1870	0.1898	0.2907	0.2967	0.3909	0.4029	0.5089	0.5329	0.5496	0.5826	0.8875	0.9500
	Dia.	0.28287	0.37537	0.49556	0.63056	0.78286	0.99286	1.24543	1.59043	1.83043	2.30543	2.77500	3.40000
Height of Thread		0.02963	0.2963	0.04444	0.04444	0.05714	0.05714	0.06957	0.06957	0.06957	0.06957	0.1000	0.1000
Increase in Dia. per Thread		0.00231	0.00231	0.00347	0.00347	0.00446	0.00446	0.00543	0.00543	0.00543	0.00543	0.00781	0.00781
Basic Minor Dia. at Small End		0.2416	0.3339	0.4329	0.5676	0.7013	0.9105	1.1441	1.4876	1.7265	2.1995	2.6195	3.2406

(Continued)

Table 37. (Continued) **American Standard Taper Pipe Threads NPT Basic Dimensions.** (Source, Landis Threading Systems.)

Nominal Pipe Size		Sizes 3 1/2 - 24 OD"											
		3 1/2	4	5	6	8	10	12	14 OD	16 OD	18 OD	20 OD	24 OD
Pipe O.D.		0.4000	4.5000	5.5630	6.6250	8.6250	10.7500	12.7500	14.0000	16.0000	18.0000	20.0000	24.0000
TPI		8	8	8	8	8	8	8	8	8	8	8	8
Pitch P		0.12500	0.12500	0.12500	0.12500	0.12500	0.12500	0.12500	0.12500	0.12500	0.12500	0.12500	0.12500
E_0 *		3.83750	4.33438	5.39073	6.44609	8.43359	10.54531	12.53281	13.77500	15.76250	17.7500	19.73750	23.71250
Hand Tight Engagement	Inch L_1 *	0.8210	0.8440	0.9370	0.9580	1.0630	1.2100	1.3600	1.5620	1.8120	2.0000	2.125	2.375
	Threads	6.57	6.75	7.50	7.66	8.50	9.68	10.88	12.50	14.50	16.00	17.00	19.00
	E_1 *	3.88881	4.38712	5.44929	6.50597	8.50003	10.62094	12.61781	13.87262	15.87575	17.8750	19.87031	23.86094
Length of Full Thread	Inch L_2 *	1.2500	1.3000	1.4063	1.5125	1.7125	1.9250	2.1250	2.2500	2.4500	2.6500	2.8500	3.2500
	Threads	10.00	10.40	11.25	12.10	13.70	15.40	17.00	18.90	19.60	21.20	22.80	26.00
	E_2 *	3.91562	4.41562	5.47862	6.54062	8.54062	10.66562	12.66562	13.91562	15.91562	17.91562	19.91562	23.91562
Wrench Makeup Length, Internal Thread	Inch L_3 *	0.2500	0.2500	0.2500	0.2500	0.2500	0.2500	0.2500	0.2500	0.2500	0.2500	0.2500	0.2500
	Threads	2	2	2	2	2	2	2	2	2	2	2	2
	E_3 *	3.82188	4.31875	5.37511	6.43047	8.41797	10.52969	12.51719	13.75938	15.74688	17.73438	19.72188	23.69688
Vanish Threads	Inch	0.4337	0.4337	0.4337	0.4337	0.4337	0.4337	0.4337	0.4337	0.4337	0.4337	0.4337	0.4337
	Threads	3.47	3.47	3.47	3.47	3.47	3.47	3.47	3.47	3.47	3.47	3.47	3.47
Overall Length Ext. Threads	L_4 *	1.6837	1.7337	1.8400	1.9462	2.1462	2.3587	2.5587	2.6837	2.8837	3.0837	3.2837	3.6837
Nominal Perfect Ext Thread	Length L_5 *	1.000	1.0500	1.1563	1.2625	1.4625	1.6750	1.8750	2.0000	2.2000	2.4000	2.6000	3.0000
	Dia.	3.90000	4.40000	5.46300	6.52500	8.52500	10.65000	12.65000	13.90000	15.90000	17.90000	19.90000	23.90000
Height of Thread		0.1000	0.1000	0.1000	0.1000	0.1000	0.1000	0.1000	0.1000	0.1000	0.1000	0.1000	0.1000
Increase in Dia. per Thread		0.00781	0.00781	0.00781	0.00781	0.00781	0.00781	0.00781	0.00781	0.00781	0.00781	0.00781	0.00781
Basic Minor Dia. at Small End		3.7375	4.2344	5.2907	6.3461	8.3336	10.4453	12.4328	13.6750	15.6625	17.6500	19.6375	23.6125

Notes: E_0 = The Pitch Diameter at the small end of the external thread = $D - (0.05D + 1.1)P$. E_1 = The Pitch Diameter at the large end of the Internal thread = $E_0 + 0.0625 L_1$. E_2 = Pitch Diameter at large end of the external tread = $E_1 + 0.0625 L_2$. E_3 = Pitch Diameter at small end of internal thread = $E_0 - 0.1875P$. L_1 = Normal engagement by hand between external and internal threads. L_2 = Effective length of external thread = $(0.80D + 6.8)P$. L_3 = Overall length of the internal thread. Note that MIL-P-7105B differs from this chart by specifying the wrench makeup to be 3 threads for all sizes up to 3'. L_4 = Overall length of the external thread. $L_1 + L_3$ = Overall length of complete external thread = $L_2 - 2P$.

Table 38. NPSC Internal Pipe Coupling Thread Dimensions.

Nominal Pipe Size	TPI	Pipe O.D.	Minor Dia. (Min.)	Pitch Diameter Min.	Pitch Diameter Max.
1/8	27	0.405	0.340	0.3701	0.3771
1/4	18	0.540	0.442	0.4864	0.4968
3/8	18	0.675	0.577	0.6218	0.6322
1/2	14	0.840	0.715	0.7717	0.7851
3/4	14	1.050	0.925	0.9822	0.9956
1	11 1/2	1.315	1.161	1.2305	1.2468
1 1/4	11 1/2	1.660	1.506	1.5752	1.5915
1 1/2	11 1/2	1.900	1.745	1.8142	1.8305
2	11 1/2	2.375	2.219	2.2881	2.3044
2 1/2	8	2.875	2.650	2.7504	2.7739
3	8	3.500	3.277	3.3768	3.4002
3 1/2	8	4.000	3.777	3.8771	3.9005
4	8	4.500	4.275	4.3754	4.3988

All dimensions in inches.

Table 39. Dimensions for NPSM Threads for Free-Fitting Mechanical Joints.

Nominal Size	TPI	Pipe O.D.	Allowance	Class 2A External Thread Major Dia. Max.	Min.	Pitch Dia. Max.	Min.	Class 2B Internal Thread Minor Dia. Min.	Max.	Pitch Dia. Min.	Max.
1/8	27	0.405	0.0011	0.397	0.039	0.3725	0.3689	0.358	0.364	0.3736	0.3783
1/4	18	0.540	0.0013	0.526	0.517	0.4903	0.4859	0.468	0.481	0.4916	0.4974
3/8	18	0.675	0.0014	0.662	0.653	0.6256	0.6211	0.603	0.612	0.6270	0.6329
1/2	14	0.840	0.0015	0.823	0.813	0.7769	0.7718	0.747	0.759	0.7784	0.7851
3/4	14	1.050	0.0016	1.034	1.024	0.9873	0.9820	0.958	0.970	0.9889	0.9958
1	11.5	1.315	0.0017	1.293	1.281	1.2369	1.2311	1.201	1.211	1.2386	1.2462
1 1/4	11.5	1.660	0.0018	1.638	1.626	1.5816	1.5756	1.546	1.555	1.5834	1.5912
1 1/2	11.5	1.900	0.0018	1.877	1.865	1.8205	1.8144	1.785	1.794	1.8223	1.8302
2	11.5	2.375	0.0019	2.351	2.339	2.9444	2.2882	2.259	2.268	2.2963	2.3044
2 1/2	8	2.875	0.0022	2.841	2.826	2.7600	2.7526	2.708	2.727	2.7622	2.7720
3	8	3.500	0.0023	3.467	3.452	3.3862	3.3786	3.334	3.353	3.3885	3.3984
3 1/2	8	4.000	0.0023	3.968	3.953	3.8865	3.8788	3.835	3.848	3.8888	3.8988
4	8	4.500	0.0023	4.466	4.451	4.3848	4.3711	4.333	4.346	4.3871	4.3971
5	8	5.563	0.0024	5.528	5.513	5.4469	5.4390	5.395	5.408	5.4493	5.4598
6	8	6.625	0.0024	6.585	6.570	6.5036	6.4955	6.452	6.464	6.5060	6.5165

All dimensions in inches.

Table 40. Formulas for Determining Root and Crest Truncation on NPTF Threads.

TPI	Pitch P	Truncation				Equivalent Width of Flat			
		Crest		Root		Crest		Root	
		Max.	Min.	Max.	Min.	Max.	Min.	Max.	Min.
27	0.03704	0.094 P	0.047 P	0.140 P	0.094 P	0.108 P	0.054 P	0.162 P	0.108 P
18	0.05556	0.078 P	0.047 P	0.109 P	0.078 P	0.090 P	0.054 P	0.126 P	0.090 P
14	0.07143	0.060 P	0.036 P	0.085 P	0.060 P	0.070 P	0.042 P	0.098 P	0.070 P
11.5	0.08696	0.060 P	0.040 P	0.090 P	0.060 P	0.069 P	0.046 P	0.103 P	0.069 P
8	0.12500	0.055 P	0.042 P	0.076 P	0.055 P	0.064 P	0.048 P	0.088 P	0.064 P

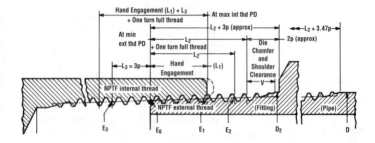

Figure 17. NPTF assembly. Refer to Table 38 for dimensions.

NPTF threads. The NPSF, or Standard Fuel Internal Straight Pipe Thread, is internal only. It is designed to be used with external NPTF threads and its major advantage is economy of manufacture because it is a straight rather than tapered thread. It is not as efficient as the internal NPTF thread, but the form provides an effective seal by allowing for root and crest interference when assembled. *Figure 17* shows the NPTF form.

Although not technically a pipe thread, the NPSH and NH (National Hose coupling Threads and National Fire Hose Coupling Threads) form is used on threaded sections of hose couplings, valves, nozzles, and all other fittings used in direct connection with hose intended for fire protection or for domestic, industrial, and general service. NH threads are standard hose threads ranging in size from 0.75" to 6" created by cutting or rolling. NPSH threads are straight hose couplings that can be mated to NPT threads using a gasket for sealing. They are available in sizes from 0.50" to 4". Another form of the thread, the NHR, is for garden hose connections and is used on thin-walled materials that are formed to profile. The dimensions for these threads are given in FED-STD-H28/10. This is a 60° included angle thread with a flat at the crest and root equal to 0.125P. Thread height is equal to 0.649519P. Tolerances are as follows. The pitch diameter tolerance provided for a mating nipple (external thread) and coupling (internal thread) are the same. Pitch diameter tolerances include lead and angle variations. The tolerance on the major diameter is twice that of the pitch diameter. The tolerance on the minor diameter of the external thread is equal to the tolerance on the pitch diameter plus two-ninths of the basic thread height. The minimum minor diameter of the external thread is such as to result in a flat equal to one-third of the basic flat at the root when the pitch diameter of the external thread

Table 41. NPTF Dryseal Pipe Threads Basic Dimensions. *(Source, Landis Threading Systems.)*

Nominal Pipe Size	1/16	1/8	1/4	3/8	1/2	3/4	1	1 1/4	1 1/2	2	2 1/2	3
Pipe O.D.												
TPI	27	27	18	18	14	14	11 1/2	11 1/2	11 1/2	11 1/2	8	8
Pitch P	0.03704	0.3704	0.0556	0.0556	0.07143	0.07143	0.08696	0.08696	0.08696	0.08696	0.12500	0.12500
E_0 *	0.27118	0.36351	0.47739	0.61201	0.75843	0.96768	1.21363	1.55713	1.79606	2.26902	2.71953	3.34062
E_1 *	0.28118	0.37360	0.49160	0.62701	0.77843	0.98887	1.23863	1.58338	1.82234	2.29267	2.76216	3.38850
E_2 *	0.28750	0.38000	0.50250	0.63750	0.79179	1.00179	1.25630	1.60130	1.84130	2.31630	2.79062	3.41562
E_3 *	0.26424	0.35656	0.46697	0.60160	0.74504	0.95429	1.19733	1.54083	1.77978	2.25272	2.70391	3.32500
Hand Tight Engagement — Inch L_1	0.1600	0.1615	0.2278	0.2400	0.3200	0.3390	0.4000	0.4200	0.4200	0.4360	0.6820	0.7660
Hand Tight Engagement — Threads	4.32	4.36	4.10	4.32	4.48	4.75	4.60	4.83	4.83	5.01	5.46	6.13
Length of Full Thread — Inch L_2	0.2611	0.2639	0.4018	0.4078	0.5337	0.5457	0.6828	0.7068	0.7235	0.7565	1.1375	1.2000
Length of Full Thread — Threads	7.05	7.12	7.23	7.34	7.47	7.64	7.85	8.13	8.32	8.70	9.10	9.60
Vanish Threads — Inch	0.1139	0.1112	0.1607	0.1547	0.2163	0.2043	0.2547	0.2620	0.2765	0.2747	0.3781	0.3781
Vanish Threads — Threads	3.075	3.072	2.892	2.791	3.028	2.860	2.929	3.013	3.180	3.159	3.025	3.025
Shoulder Length	0.3750	0.3750	0.5625	0.5625	0.7500	0.7500	0.9375	0.9688	1.0000	1.0312	1.5156	1.5781
External Thread for Draw — Inch	0.1011	0.1024	0.1740	0.1678	0.2137	0.2067	0.2828	0.2868	0.3035	0.3205	0.4555	0.4340
External Thread for Draw — Threads	2.73	2.76	3.13	3.02	2.99	2.89	3.25	3.30	3.49	3.69	3.64	3.47
Basic Internal Full Thread Length — Inch	0.2711	0.2726	0.3945	0.4067	0.5343	0.5533	0.6609	0.6809	0.6809	0.6969	1.0570	1.1410
Basic Internal Full Thread Length — Threads	7.32	7.36	7.10	7.32	7.48	7.75	7.60	7.83	7.83	8.01	8.46	9.13
Outside Dia. of Fitting	0.315	0.407	0.546	0.681	0.850	1.060	1.327	1.672	1.912	2.387	2.893	3.518
Outside Dia. of Pipe	0.3125	0.405	0.540	0.675	0.840	1.050	1.315	1.660	1.900	2.375	2.875	3.500

*Notes: E_0 = The Pitch Diameter at the small end of the external thread = $D - (0.05D) = 1.037P$. E_1 = The Pitch Diameter at the large end of the Internal thread = $E_0 + 0.0625\,L_1$. E_2 = Pitch Diameter at large end of the external tread. E_3 = Pitch Diameter at small end of internal thread. $L_2 = (0.8\,D + 5.8)P$.

is at its minimum value. The maximum minor diameter is basic. The tolerance on the major diameter of the internal thread is equal to the tolerance of the pitch diameter plus two-ninths of the basic thread height. The minimum major diameter of the internal thread is such as to result in a basic flat of $P/8$ ($0.125P$) when the pitch diameter of the internal thread is at its minimum value. The maximum major diameter of the internal thread is that corresponding to a flat equal to one-third of the basic flat. The tolerance on the minor diameter of the internal thread is twice the tolerance on the pitch diameter of the internal thread. The minimum minor diameter of the internal thread is such as to result in a basic flat of $P/8$ at the crest when the pitch diameter of the internal thread is at its minimum value. See **Table 42** basic dimensions and *Figure 18* for the thread form.

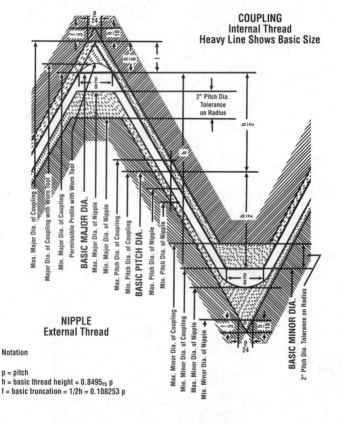

Figure 18. NPSH and NH thread form.

Table 42. Basic Dimensions of American National Hose Nipple and Coupling Threads, NPSH and NH.

Nominal Size*	TPI	Symbol	Service	Pitch	Allowance	External Thread Max.			Internal Thread Max.		
						Major Dia.	Pitch Dia.	Minor Dia.	Minor Dia.	Pitch Dia.	Major Dia.
1/2, 5/8, 3/4	11.5	.75-11.5 NHR	Garden Hose	0.08696	0.0100	1.0625	1.0060	0.9495	0.9595	1.0160	1.0725
3/4	8	.75-8 NH	Chemical	0.1250	0.0120	1.3750	1.2938	1.2126	1.2246	1.3058	1.3870
1 1/2	9	1.5-9 NH	Fire Hose	0.1111	0.0120	1.9900	1.9178	1.8457	1.8577	1.9298	2.0020
1/2	14	.5-14 NPSH	Steam, air, water and all other connections with standard pipe threads	0.07143	0.0075	0.8248	0.7784	0.7320	0.7395	0.7859	0.8323
3/4	14	.75-14 NSPH		0.07143	0.0075	1.0353	0.9889	0.9425	0.9500	0.9964	1.0428
1	11.5	1-11.5 NSPH		0.08696	0.0100	1.2951	1.2386	1.1821	1.1921	1.2486	1.3051
1 1/4	11.5	1.25-11.5 NSPH		0.08696	0.0100	1.6399	1.5834	1.5629	1.5369	1.5934	1.6499
1 1/2	11.5	1.5-11.5 NSPH		0.08696	0.0100	1.8788	1.8223	1.7658	1.7758	1.8323	1.8888
2	11.5	2-11 NSPH		0.08696	0.0100	2.3528	2.2963	2.2398	2.2498	2.3063	2.3628
2 1/2	7.5	2.5-7.5 NH		0.13333	0.0150	3.0686	2.9820	2.8954	2.9104	2.9970	3.0836
3	6	3-6 NH		0.16667	0.0150	3.6239	3.5156	3.4073	3.4223	3.5306	3.6389
3 1/2	6	3.5-6 NH		0.16667	0.0200	4.2439	4.1356	4.0273	4.0473	4.1556	4.2639
4	6	4-6 NH (SPL) *	Fire Hose	0.16667	0.0250	4.9082	4.7999	4.6916	4.7117	4.8200	4.9283
4	4	4-4 NH		0.2500	0.0250	5.0109	4.8485	4.6861	4.7111	4.8735	5.0359
4 1/2	4	4.5-4 NH		0.2500	0.0250	5.7609	5.5985	5.4361	5.4611	5.6235	5.7859
5	4	5-4 NH		0.2500	0.0250	6.2600	6.0976	5.9352	5.9602	6.1226	6.2850
6	4	6-4 NH		0.2500	0.0250	7.0250	6.8626	6.7002	6.7252	6.8876	7.0500

* 4-6 NH (SPL) thread is used extensively aboard ship by the Navy Dept. All dimensions in inches.

Tightening and Tensioning Threaded Fasteners

Bolted Joints

Axial Loaded Joints

(Federal Standard MIL-HDBK-60 contains information on calculating tension and torque in threaded fasteners. Much of the material in this section was derived from this standard and from FED-STD-H28/2B.) Applied preloads (or tension stress) for bolts, screws, and studs in *axial loaded* joints should be sufficient to assure that joint members remain in contact and in compression. Loss of joint compression can result in 1) leakage of pressurized fluids past compression gaskets, 2) loosening of the fastener, and 3) reduced fastener fatigue life. *Figure 1* shows that for a joint with no preload, the axial load on the joint members is transmitted directly to the fastener and the bolt load equals the joint load. In multiple bolted joints, each fastener will share a proportion of the joint load (shown as line OAB in Figure 1). However, when a preload (P_{B1} in the figure) is applied, the axial joint load is partially absorbed by reduction of the compression of the joint members. Therefore, bolt load increase is lower than in the nonpreloaded joint (shown as line $P_{B1}A$, with point A representing the separation of joint members with resulting loss of compression so that further loading beyond point A is direct along line AB). When an axial joint load in the range between lines P_a and P_b is applied to a nonpreloaded joint, the resulting bolt load will vary between P_{Ba} and P_{Bb}. When preload P_{B1} is applied, the resultant bolt load varies between $P_{Ba'}$ and $P_{Bb'}$. This condition considerably reduces cyclic load and results in enhanced fatigue life for the fastener. Preloading can increase fastener life from only a few hundred cycles to an essentially unlimited fatigue life. One laboratory study tested a bolt tightened to 1,420 pounds of tension (29,000 PSI of residual stress) by subjecting it to an alternating (tension and compression) load of 9,000 pounds. It experienced failure after 6,000 cycles. An identical bolt, tightened to 8,420 pounds of tension (80,000 PSI of residual stress), was cycled at the same 9,000 pound load for 4,650,000 cycles before failure.

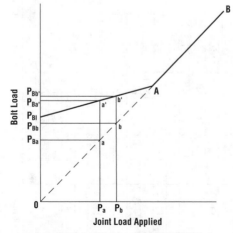

Figure 1. Bolt load in joint with applied axial load.

Shear Loaded Joints

Fasteners in *shear loaded* joints can be separated into two categories. First, in the case where the joint members are designed to slide, joint members transmit shear loads to the fasteners in the joint. The preload must be sufficient to keep the joint members in contact. Secondly, in the case where the joint members do not slide and there is no relative motion between the joint members, shear loads are transmitted within the joint by frictional force. Therefore, the preload must be sufficient to provide a frictional force greater than the applied shear forces. In both instances, shear stresses induced by preloading must be considered in the bolt design. The U.S. Navy Bureau of Aeronautics dictates the following fit tolerances for bolts loaded in shear.

Fit	Tolerance	Application
Class 1	± 0.001"	For use when one or two bolts are subjected to a reversal of loads in a critical joint assembly. Holes are to be reamed.
Class 2	± 0.002"	For use when there are more than two through primary structures, where vibration and reversal of loads are expected, or where joint rigidity is required. Holes are to be reamed.
Class 3	± 0.005 to ± 0.010"	For use where a large number of bolts not subjected to load reversals are used.

Combined Shear and Axial Loaded Joints

These joints are subjected to the sum of forces of axial loaded joints and shear loaded joints, and must be analyzed carefully to assure reliability.

Determining Preloads

Fasteners are normally designed for maximum utilization of material, and a maximum safe preload can usually be determined. However, if for standardization, logistical, or other reasons a lower strength fastener is replaced by one of higher strength, the preload values should not be increased. Preload should be based on calculated joint requirements. For joints subjected to cyclic loading and for joints using high strength fasteners where yield strain is near the point of fracture, preloads should be maintained below the yield point. Maximum preloads specified are generally within the following ranges.

50 – 80% of the maximum tensile ultimate strength,
75 – 90% of the minimum tensile yield strength or proof load, or
100% of the observed proportional limit or onset of yield.

For joints with primarily static loading using fasteners of ductile material, where yield strain is relatively far from the strain at fracture, preloads above yield are often used. Preload should not exceed the strength of the components (fastener head, juncture of head and shank, etc.), and there must be sufficient thread engagement to prevent stripping.

Torsion Load

When preloads are applied by turning nuts, bolts, or screws, a torsion load component is added to the desired axial bolt load. In fact, the highest load imposed on a fastener will be experienced during tightening. The combined loading increases the tensile stress on the fastener, a consideration often ignored by designers on the assumption that the majority of the torsional force rapidly dissipates after the driving force is removed. This may be true in joints tensioned near or beyond fastener yield, but for critical joints where fastener tension must be maintained below yield it is important to adjust the axial tension load requirements to include the effects of preload tension. For this adjustment, the combined tensile stress, f_{tc} (also known as "von Mises stress"), may be calculated from the following equation. (See **Table 1** for explanation of symbols for all equations found in this section.)

Table 1. Symbols Used in Equations in this Section.

A_B = Effective fastener* cross sectional area in in.^{2} or mm^2.

A_G = Gasket area in in.^{2} or mm^2.

A_J = Effective compressed area in the joint in in.^{2} or mm^2.

A_s = Tensile stress area in in.^{2} or mm^2.

d = External thread nominal diameter.

d_2 = External thread pitch diameter in inches.

E_B = Fastener* modulus of elasticity in PSI or MPa.

E_G = Gasket modulus of elasticity in PSI or MPa.

E_J = Joint material modulus of elasticity in PSI or MPa.

e = Coefficient of linear thermal expansion (in./in. % °F).

F = Maximum load stress in pounds or Newtons.

f_s = Shear stress caused by torsional load application in PSI or MPa.

f_t = Axial tensile stress applied in PSI or MPa.

L = Lead of thread helix in inches or millimeters.

L_B = Effective fastener* length in inches or millimeters.

n = Number of threads per inch.

P = Axial load applied to the joint in pounds or Newtons.

P (italicized) = Thread pitch in inch dimensions.

P_B = Axial fastener* load in pounds or Newtons.

P_G = Gasket compression load in pounds or Newtons.

P_J = Joint compression load in pounds or Newtons.

t = temperature in °F.

t_G = Gasket thickness in inches or millimeters.

t_0 = Operating temperature in °F.

t_J = Effective joint thickness in inches or millimeters.

Y = Proof load (stress) in PSI or MPa.

δ_B = Fastener* elongation in inches or millimeters.

δ_G = Gasket compression in inches or millimeters.

δ_J = Joint material compression in inches or millimeters.

μ = Coefficient of friction in threads.

Δ = Change caused by application of load "P."

θ = Amount of turn of a nut in degrees.

T (subscript) = Indicates applicability to total joint including gasket.

1 (subscript) = Indicates condition with applied joint fastener* preload only.

Relevant Conversions
1 PSI = 0.006894757 N/mm^2
1 ksi = 6.894757 N/mm^2
1 N/mm^2 = 145.0377 PSI
1 N/mm^2 = 0.1450377 ksi
1 MPa = 145.0377 PSI
1 N/mm^2 = 1 MPa
1 pound (force) = 4.448222 Newton
1 Newton = 0.2248089 pound (force)
1 in.^{2} = 645.16 mm^2
1 mm^2 = 0.001550003 in.^{2}

*The term "Fastener" as used in this section is interchangeable with threaded bolts, screws, and studs.

$$f_{tc} = \sqrt{f_t^2 + 3f_s^2}$$

A simple calculation for the amount of tensile stress on a fastener takes into consideration the force acting upon the fastener and its diameter.

$$f_t = F \div A_s$$

Maximum loading stresses for threaded fasteners are determined by their known proof yield capability and tensile stress area. Maximum load in pounds can be determined by the equation

$$F = Y \times A_s.$$

For single start Unified Threads and M Profile Metric threads, the following formula can be used to determine combined tensile stress. Tensile stress increase due to preload tension increases in significance as friction coefficients increase. (See **Table 2** for coefficient of friction values to be used in the following equations.)

$$f_{tc} = f_t \times \sqrt{1 + (3 \times [\{1.96 + 2.31\} \div \{1 - 0.325P \div d_2\} - 1.96]^2)}$$

For single start UNJ with a thread stress diameter equal to the pitch diameter, the following equation applies.

$$f_{tc} = f_t \times \sqrt{1 + (3 \times [\{0.637P \div d_2\} + 2.31\ ]^2)}$$

Table 2. Coefficient of Friction (μ)Values for Threaded Fasteners.

Bolt/Nut Material	Lubricant	$\mu \pm 20\%$
Steel, Carbon and Low Alloy	Graphite in Petrolatum or Oil	0.07
Steel, Carbon and Low Alloy	Molybdenum Disulfide Grease	0.11
Steel, Carbon and Low Alloy	Machine Oil	0.15
Steel, Cadmium Plated	None	0.12
Steel, Zinc Plated	None	0.17
Steel/Bronze	None	0.15
Steel, Corrosion Resistant Steel or Nickel Base Alloys. Silver Plated Materials	None	0.14
Titanium/Steel	Graphite in Petrolatum	0.08
Titanium	Molybdenum Disulfide Grease	0.10

As a further example of the relationship of torque to lubrication, *Figure 2* is a chart based on the results obtained with a 9/16–18 steel bolt screwed into an aluminum casting. Without lubrication, a torque of 115 to 125 lb./in. was required to develop tension of 800 to 1,400 lb. When lubricated, torque values were just 75 to 85 lb./in. to develop tension in the range of 1,000 to 1,250 pounds. Torque values are affected in various ways by different types of lubricants.

Calculating Tensile Stress Area

Tests referenced in FED-STD-H28/2B have shown that externally threaded parts fail in tension at loads corresponding to those of unthreaded parts with diameters midway between the pitch diameter and minor diameter. The following formulas for calculation of tensile stress area provide stress area based upon a diameter approximately midway between the minimum pitch diameter and the minimum minor diameter. These formulas have been successfully applied to steel and other metals with ultimate strengths up to 180,000 PSI. In these formulas, A_s is tensile stress area, d_2 is the pitch diameter,

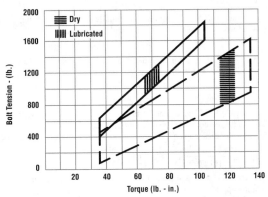

Figure 2. The effect of lubrication on the torque-tension relationship.(Source, Skidmore-Wilhiem Mfg. Co.)

d is major diameter, H is the height of the fundamental triangle, and n is the number of threads per inch ($1/P$).

$$A_s = \pi ([d_2 \div 2] - [3H \div 16])^2$$
$$\text{or } A_s = 0.7854 \times (d - [0.9743 \div n])^2$$

The grade designation of a bolt or nut can be used as a guide in the selection. Federal Standard FF-B-584F gives mechanical and chemical properties of carbon and alloy steel bolts, Grades 1 – 8, as provided in **Table 3**. Ultimate tensile strengths for standard Unified threads are given in **Table 4**, and **Table 5** gives ultimate tensile strengths for UNJ 3A threads. Ultimate shear strengths for standard Unified threads are shown in **Table 6**. **Table 7** gives mechanical properties for bolts, studs, and hex cap screws for fasteners made of corrosion resistant steels, aluminum, and copper and nickel alloys. In *Figure 3*, tensile strength/tensile load relationships for various size fasteners are shown. **Table 8** gives mechanical properties of various nut grades. **Table 9**, provided by American Fastener Technologies Corp., contains ASTM, SAE, and ISO markings and mechanical properties for steel fasteners. Tensile stress areas of Unified Thread fasteners are given in **Table 10**. The three most commonly used fastener grades are Grade 2, Grade 5, and Grade 8. Grade 2 fasteners are made of mild low or medium carbon steel, Grade 5 fasteners are made of medium hard steel, and Grade 8 fasteners are made from medium carbon alloy steel, quenched and tempered. ASTM standards ASTM 193 and ASTM 194 should be consulted for full material and chemical properties of hex head cap screws, studs, bolts, and nuts made of steel and stainless steel. For further information on grades, see SAE standard SAE J429.

Tensile stress areas for coarse M Profile Metric threads can be found in **Table 11**, and **Table 12** gives mechanical and chemical properties for standard classes of M Profile Metric bolts, screws, and studs. Metric Class (or "Property Class") ratings for externally threaded fasteners contain two numbers separated by a period. The first number, which may be one or two digits, represents 1/100 of the nominal tensile strength of the fastener, expressed in Newtons/mm^2 (one N/mm^2 = 145.0377 PSI, and one PSI = 0.006894757 N/mm^2). The number following the period is equal to ten times the ratio between the fastener's nominal yield stress and its nominal tensile strength. Multiplying the two numbers provides a value roughly equivalent to 1/10 of the nominal yield strength in

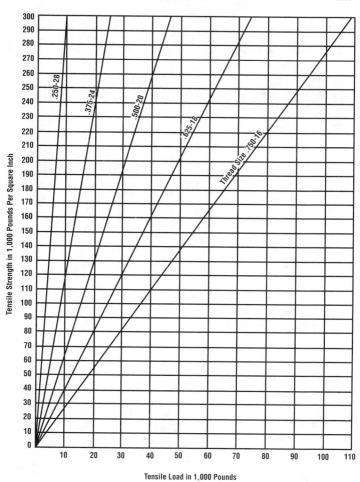

Figure 3. Tensile strength versus tensile load for various threaded fastener sizes. (Source, MIL-STD-1251A.)

N/mm². Additional information on external Metric fastener classes and properties can be found in ISO standard ISO 898. The four most commonly used classes are as follows. Class 4.8 indicates a low or medium carbon steel fastener commonly used for machine screws, machine bolts, and similar applications. Class 8.8 fasteners are usually hex cap screws and hex socket screws made of medium hard carbon steel, quenched and tempered. Class 10.9 is usually a quenched and tempered alloy steel used for hex cap screws. Finally,

(Text continued on p. 1541)

Table 3. Properties of Carbon and Alloy Steel Bolts, Screws, and Studs. (Source, FF-B-548F.)

Mechanical Properties

Grade	Nominal Dia. (inches)	Full Size Fastener		Machine Test Specimen				Core Hardness Rockwell	
		Proof Load ksi	Tensile Strength Min. ksi	Yield Strength Min. ksi	Tensile Strength Min. ksi	Elongation %	Reduction in Area %	Min.	Max.
1	1/4" - 1.5"	33	60	36	60	18	35	B70	B100
2	1/4" - 3/4"	55	74	57	74	18	35	B80	B100
	Over 3/4" to 1.5"	33	60	36	60	18	35	B70	B100
4	1/4" - 1.5"	-	115	100	115	10	35	C22	C32
5	1/4" - 1"	85	120	92	120	14	35	C25	C34
	Over 1" to 1.5"	74	105	81	105	14	35	C19	C30
5.1	No. 6" - 3/8"	85	120	-	-	-	-	C25	C40
5.2	1/4" - 1"	85	120	92	120	14	35	C26	C36
7	1/4" - 1.5"	105	133	115	133	12	35	C28	C34
8	1/4" - 1.5"	120	150	130	150	12	35	C33	C39
8.1	1/4" - 1.5"	120	150	130	150	10	35	C32	C38
8.2	1/4" - 1"	120	150	130	150	10	35	C35	C42

Chemical Properties

Grade	Material & Treatment	Element %					
		Carbon Min.	Carbon Max.	Manganese Min.	Phosphorus Max.	Sulpher Max.	Boron Min.
1	Low or Medium Carbon Steel	-	0.55	-	0.048	0.058	-
2	Low or Medium Carbon Steel	-	0.28	-	0.048	0.058	-
4	Medium Carbon Cold Drawn Steel	-	0.55	-	0.048	0.058	-
5	Medium Carbon Steel, Quenched and Tempered	0.28	0.55	-	0.048	0.058	-

(Continued)

Table 3. *(Continued)* **Properties of Carbon and Alloy Steel Bolts, Screws, and Studs.** *(Source, FF-B-548F.)*

Grade	Material & Treatment	Chemical Properties Element %								
		Carbon		Manganese	Phosphorus	Sulpher	Boron			
		Min.	Max.	Min.	Max.	Max.	Min.			
5.1	Low or Medium Carbon Steel, Quenched and Tempered	0.15	0.30	-	0.048	0.058	-			
5.2	Low Carbon Martensite Steel, Fully Killed, Fine Grain, Quenched and Tempered	0.15	0.25	0.74	0.048	0.058	0.0005			
7	Medium Carbon Alloy Steel, Quenched and Tempered	0.28	0.55	-	0.040	0.045	-			
8	Medium Carbon Alloy Steel, Quenched and Tempered	0.28	0.55	-	0.040	0.045	-			
8.1	Elevated Temperature Drawn Steel-Medium Carbon Alloy or SAE 1541 (formerly SAE 1041)	0.28	0.55	-	0.048	0.058	-			
8.2	Low Carbon Martensite Steel, Fully Killed, Fine Grain, Quenched and Tempered	0.15	0.25	0.74	0.048	0.058	0.0005			

Table 4. Ultimate Tensile Strengths of Class 2A and Class 3A Unified Threaded Fasteners. *(Source, MIL-HDBK-5H.)*

Fastener Size		Nominal Minor Area	Tensile Stress of Fastener (ksi)						
Dia. (inch)	See Note 3		Ultimate Tensile Strength in Pounds (See Notes 1, 2)						
			55	62	62.5	125	140	160	180
0.112	4-40	0.0050896	280	316	318	636	713	814	916
0.138	6-32	0.0076821	423	476	480	960	1075	1225	1380
0.164	8-32	0.012233	673	758	765	1525	1710	1955	2200
0.190	10-32	0.018074	994	1120	1130	2255	2530	2890	3250
0.250	1/4-28	0.033394	1835	2070	2085	4170	4680	5340	6010
0.312	5/16-24	0.053666	2950	3325	3350	6710	7510	8590	9660
0.375	3/8-24	0.082397	4530	5110	5150	10300	11500	13150	14800
0.438	7/16-20	0.11115	6110	6890	6950	13850	15550	17750	20000
0.500	1/2-20	0.15116	8310	9370	9450	18900	21150	24150	27200
0.562	9/16-18	0.19190	10550	11900	11950	23950	26850	30700	34500
0.625	5/8-18	0.24349	13350	15100	15200	30400	34050	38950	43800
0.750	3/4-16	0.35605	19550	22050	22250	44500	49800	57000	64100
0.875	7/8-14	0.48695	26750	30150	30400	60900	68200	77900	87700
1.000	1-12	0.63307	34800	39250	39550	79100	88600	101000	114000
1.125	1-1/8-12	0.82162	45200	50900	51400	102500	115000	131500	147500
1.250	1-1/4-12	1.0347	56900	64200	64700	129000	144500	165500	186000
1.375	1-3/8-12	1.2724	70000	78900	79500	159000	178000	203500	229000
1.500	1-1/2-12	1.5345	84400	95100	95900	191500	214500	245500	276000

1) Values shown for 0.112, 0.138, and 0.164 diameters are for 2A threads. All other values are for 3A threads.
2) Nuts and fastener heads designed to develop the ultimate tensile strength of the fastener are required to develop the tabulated tension loads.
3) Fractional equivalent or number of threads per inch.
Nominal area dimensions are in inches.

Table 5. Ultimate Tensile Strength of UNJ Unified Threaded Fasteners. *(Source, MIL-HDBK-5H.)*

Tensile Stress of Fastener (ksi)			160	180	220	260
Fastener Size		Nominal Minor Area	Ultimate Tensile Strength in Pounds (See Notes 1, 2)			
Dia. (inch)	See Note 3					
0.112	4-40	0.0054367	869	979	1195	1410
0.138	6-32	0.0081553	1305	1465	1790	2120
0.164	8-32	0.012848	2055	2310	2825	3340
0.190	10-32	0.018602	2975	3345	4090	4840
0.250	1/4-28	0.034241	5480	6160	7530	8900
0.312	5/16-24	0.054905	8780	9880	12050	14250
0.375	3/8-24	0.083879	13400	15100	18450	21800
0.438	7/16-20	0.11323	18100	20350	24900	29400
0.500	1/2-20	0.15358	24550	27600	33750	39900
0.562	9/16-18	0.19502	31200	35100	42900	50700
0.625	5/8-18	0.24700	39500	44500	54300	64200
0.750	3/4-16	0.36082	57700	64900	79400	93800
0.875	7/8-14	0.49327	78900	88800	108500	128000
1.000	1-12	0.64156	102500	115500	141000	166500
1.125	1-1/8-12	0.83129	133000	149500	182500	216000
1.250	1-1/4-12	1.0456	167000	188000	230000	271500
1.375	1-3/8-12	1.2844	205500	231000	282500	333500
1.500	1-1/2-12	1.5477	247500	278500	340500	402000

1) Values are for 3A threads.
2) Nuts and fastener heads designed to develop the ultimate tensile strength of the fastener are required to develop the tabulated tension loads.
3) Fractional equivalent or number and threads per inch.
4) The tension fastener allowables above are based on the maximum minor diameter thread area for UNJ threads as defined in MIL-S-8879. Maximum minor area is given in inches.

Table 6. Ultimate Shear Strength of Unified Threaded Fasteners. (Source, MIL-HDBK-5H.)

Fastener Dia. (inch)	Basic Shank Area	35	38	75	90	108	125	145	156
Shear Stress of Fastener (ksi)				Ultimate Single Shear Strength, in Pounds					
0.112	0.0098520	345	374	739	887	1060	1230	1425	1535
0.125	0.012272	430	466	920	1105	1325	1530	1775	1910
0.138	0.014957	523	568	1120	1345	1615	1870	2165	2330
0.156	0.019175	671	729	1435	1725	2070	2395	2780	2990
0.164	0.021124	739	803	1580	1900	2280	2640	3060	3295
0.188	0.027612	966	1045	2070	2485	2980	3450	4005	4310
0.190	0.028353	992	1075	2125	2550	3060	3540	4110	4420
0.216	0.036644	1280	1390	2745	3295	3955	4580	5315	5720
0.219	0.037582	1315	1425	2815	3380	4060	4700	5445	5860
0.250	0.049087	1715	1865	3680	4420	5300	6140	7115	7660
0.312	0.076699	2680	2915	5750	6900	8280	9590	11100	11950
0.375	0.11045	3865	4200	8280	9935	11900	13800	16000	17200
0.438	0.15033	5260	5710	11250	13500	16200	18750	21750	23450
0.500	0.19635	6870	7460	14700	17650	21200	24500	28450	30600
0.562	0.24850	8700	9440	18600	22350	26800	31050	36000	38750
0.625	0.30680	10700	11650	23000	27600	33100	38350	44500	47900
0.750	0.44179	15450	16750	33100	39750	47700	55200	64000	68900
0.875	0.60132	21050	22850	45100	54100	64900	75200	87200	93800
1.000	0.78540	27450	29850	58900	70700	84800	98200	113500	122500
1.125	0.99402	34750	37750	74600	89500	107000	124000	144000	155000
1.250	1.2272	43000	46600	92000	110000	132500	153000	177500	191000
1.375	1.4849	52000	56400	111000	133500	160000	185500	215000	231500
1.500	1.7671	61800	67100	132500	159000	190500	220500	256000	275500

Basic shank area dimensions are in inches.

Table 7. Mechanical Properties for Stainless Steel and Nonferrous Fasteners.

| Grade | Description & Treatment (See Note 1) | Bolts, Screws & Studs | | | | | | Nuts | |
| | | Full Size Fastener | | Machine Test Specimen | | | | | |
		Yield Strength Min. ksi	Tensile Strength Min. ksi	Yield Strength Min. ksi	Tensile Strength Min. ksi	Elongation Min. %	Hardness Rockwell	Proof Load Stress	Hardness Rockwell
303A	Austenitic Stainless Steel, SA	30	75	30	75	20	B75	75	B75
304-A	Austenitic Stainless Steel, SA	30	75	30	75	20	B75	75	B75
304	Austenitic Stainless Steel, CW	50	90	45	85	20	B85	90	B85
304-SH	Austenitic Stainless Steel, SH	See Note 2	See Note 2	See Note 2	See Note 2	15	C25	See Note 2	C20
305-A	Austenitic Stainless Steel, SA	30	75	30	75	20	B70	75	B70
305	Austenitic Stainless Steel, CW	50	90	45	85	20	B85	90	B85
305-SH	Austenitic Stainless Steel, SH	See Note 2	See Note 2	See Note 2	See Note 2	15	C25	See Note 2	C20
316-A	Austenitic Stainless Steel, SA	30	75	30	75	20	B70	75	B70
316	Austenitic Stainless Steel, CW	50	90	45	85	20	B85	90	B85
316-SH	Austenitic Stainless Steel, SH	See Note 2	See Note 2	See Note 2	See Note 2	15	C25	See Note 2	C20
XM7-A	Austenitic Stainless Steel, SA	30	75	30	75	20	B70	75	B70
XM7	Austenitic Stainless Steel, CW	50	90	45	85	20	B85	90	B85

(Continued)

Table 7. *(Continued)* Mechanical Properties for Stainless Steel and Nonferrous Fasteners.

Grade	Description & Treatment (See Note 1)	Bolts, Screws & Studs								Nuts		
		Full Size Fastener		Machine Test Specimen								
		Yield Strength Min. ksi	Tensile Strength Min. ksi	Yield Strength Min. ksi	Tensile Strength Min. ksi	Elongation Min. %	Hardness Rockwell			Proof Load Stress	Hardness Rockwell	
384-A	Austenitic Stainless Steel, SA	30	75	30	75	20	B70			75	B70	
384	Austenitic Stainless Steel, CW	50	90	45	85	20	B85			90	B85	
410-H	Martensitic Stainless Steel, HT	95	125	95	125	20	C22			125	C22	
410-HT	Martensitic Stainless Steel, HT	135	180	135	180	12	C36			180	C36	
416-H	Martensitic Stainless Steel, HT	95	125	95	125	20	C22			125	C22	
416-HT	Martensitic Stainless Steel, HT	135	180	135	180	12	C36			180	C36	
430	Ferritic Stainless Steel	40	70	40	70	20	B75			70	B75	
464-HF	Naval Brass	15	52	14	50	25	B56			52	B56	
464	Naval Brass	27	60	25	57	25	B65			60	B65	
462	Naval Brass	27	52	24	50	20	B65			52	B65	
642	Aluminum Bronze	35	72	35	72	15	B75			72	B75	
630	Aluminum Bronze	50	105	50	105	10	B90			105	B90	
614	Aluminum Bronze	40	75	40	75	30	B70			75	B70	
510	Phosphor Bronze	35	60	35	60	15	B60			60	B60	
675	Manganese Bronze	22	55	22	55	20	B60			55	B60	

(Continued)

Table 7. *(Continued)* Mechanical Properties for Stainless Steel and Nonferrous Fasteners.

Grade	Description & Treatment (See Note 1)	Bolts, Screws & Studs								Nuts		
		Full Size Fastener		Machine Test Specimen								
		Yield Strength Min. ksi	Tensile Strength Min. ksi	Yield Strength Min. ksi	Tensile Strength Min. ksi	Elongation Min. %	Hardness Rockwell			Proof Load Stress	Hardness Rockwell
655-HF	Silicon Bronze	20	52	18.5	50	20	B60			52	B60
655	Silicon Bronze	38	70	36	68	15	B75			70	B75
651	Silicon Bronze	45	75	42.5	72	8	B75			75	B75
661	Silicon Bronze	38	70	38	70	15	B75			70	B75
NICU-A-HF	Nickel-Copper Alloy A	25	70	25	70	20	B70			70	B70
NICU-A	Nickel-Copper Alloy A	40	80	40	80	20	B80			80	B80
NICU-B	Nickel-Copper Alloy B	40	80	40	80	20	B80			80	B80
NICU-K (7)	Nickel-Copper Aluminum Alloy	90	130	90	130	20	C24			130	C24
2024-T4	Aluminum Alloy	40	55	40	55	14	B70			55	B70
6061-T6	Aluminum Alloy	35	42	35	42	12	B50			42	B50

1) Abbreviations used for material treatment are: S.A. = Solution Annealed. CW = Cold Worked. SH = Strain Hardened. HT = Hardened and Tempered.
2) Austenitic stainless steel, strain hardened bolts, screws, studs, and nuts shall have the following strength per properties.

(Continued)

Table 7. *(Continued)* **Mechanical Properties for Stainless Steel and Nonferrous Fasteners.**

Product Size	Bolts, Screws & Studs				Nuts
	Tested Full Size		Machine Test Specimen		
	Yield Strength Min. ksi	Tensile Strength Min. ksi	Yield Strength Min. ksi	Tensile Strength Min. ksi	Proof Load Stress ksi
To 5/8 inch	100	125	90	115	125
Over 5/8 to 1 inch	70	105	65	100	105
Over 1 to 1.5 inch	50	90	45	85	90

Source, American Fastener Technologies Corp., and MIL–S–1222H.

Table 8. Mechanical Properties for Nuts. (MIL-S-1222H.)

Material	Grade	Heat Treatment or Conditions	Nominal Size (inch)	Proof Stress for Hex Nut* (psi)	Rockwell Hardness
Carbon and Alloy Steels	2	-	1/4 and over	90,000	B80 min
	2H	Hardened and Tempered	1/4 and over	150,000	C24-38
	4, 7	Hardened and Tempered	1/4 and over	150,000	C24-38
	5	Hardened and Tempered	1/4 thru 1	120,000	C23-32
			Over 1	105,000	C19-32
	8	Hardened and Tempered	1/4 and over	150,000	C24-38
Corrosion Resistant Steels	303, 303 Se, 304, 305, 316, 321, 347, 384	Annealed	1/4 and over	75,000	B65-95
		Cold Worked	1/4 and over	90,000	B95 min
	410, 416, 416, Se, 431	Hardened	1/4 and over	125,000	C25-34
		Hardened and Tempered	1/4 and over	180,000	C38-47
	630	Annealed and Age Hardened	1/4 and over	135,000	C28-38
Aluminum Alloys	2024	T4	1/4 and over	55,000	B70 min
	6061	T6	1/4 and over	40,000	B40 min
	7075	T73	1/4 and over	52,000	B60 min
Copper Alloys	462, 464, 482	-	1/4 and over	55,000	-
	510, 544	-	1/4 and over	60,000	-
	632	-	1/4 and over	90,000	-
	655, 661	-	1/4 and over	65,000	-
	670, 675	-	1/4 and over	55,000	-
Nickel Alloys	400, 405	-	1/4 and over	80,000	B80 min
	500	Annealed and Age Hardened	1/4 and over	130,000	C24 min
Titanium Alloys	T7	Annealed	1/4 and over	120,000	C26 max
		Solution Treated and Aged	1/4 and over	140,000	C26 min

Table 9. ASTM, SAE, and ISO Grade Markings and Mechanical Properties for Steel Fasteners.
(Source, American Fasteners Technologies Corp.)

Identification Grade Mark	Specification	Fastener Description	Material	Nominal Size Range	Mechanical Properties (ksi)		
					Proof Load	Yield Strength (min)	Tensile Strength (min)
No Grade Mark	SAE J429 Grade 1	Bolts, Screws, Studs	Low or Medium Carbon Steel	1/4 thru 1 1/2	33	36	60
	ASTM A307 Grades A&B		Low Carbon Steel	1/4 thru 4	-	-	60
	SAE J429 Grade 2		Low or Medium Carbon Steel	1/4 thru 3/4	55	57	74
				3/4 thru 1 1/2	33	36	60

(Continued)

Table 9. *(Continued)* **ASTM, SAE, and ISO Grade Markings and Mechanical Properties for Steel Fasteners.**
(Source, American Fasteners Technologies Corp.)

Identification Grade Mark	Specification	Fastener Description	Material	Nominal Size Range	Mechanical Properties (ksi)		
					Proof Load	Yield Strength (min)	Tensile Strength (min)
No Grade Mark	SAE J429 Grade 4	Studs	Medium Carbon Cold Drawn Steel	1/4 thru 1 1/2	-	100	115
B5	ASTM A193 Grade B5		AISI 501	1/4 thru 4	-	80	100
B6	ASTM A193 Grade B6		AISI 410			85	110
B7	ASTM A193 Grade B7		AISI 4140, 4142, or 4105	1/4 thru 2 1/2 / Over 2 1/2 thru 4 / Over 4 thru 7	- / - / -	105 / 95 / 75	125 / 115 / 100
B16	ASTM A193 Grade B16	Bolts, Screws, Studs for High Temp. Service	CrMoVa Alloy Steel			105 / 95 / 85	125 / 115 / 100
B8	ASTM A193 Grade B8		AISI 304				
B8C	ASTM A193 Grade B8C		AISI 347	1/4 and larger	-	30	75
B8M	ASTM A193 Grade B8M		AISI 316				
B8T	ASTM A193 Grade B8T		AISI 321	1/4 and larger	-	30	75

(Continued)

Table 9. *(Continued)* **ASTM, SAE, and ISO Grade Markings and Mechanical Properties for Steel Fasteners.**
(Source, American Fasteners Technologies Corp.)

Identification Grade Mark	Specification	Fastener Description	Material	Nominal Size Range	Mechanical Properties (ksi)		
					Proof Load	Yield Strength (min)	Tensile Strength (min)
B8	ASTM A193 Grade B8	Bolts, Screws, Studs for High Temp. Service	AISI 304 Strain Hardened	1/4 thru 3/4 Over 3/4 thru 1 Over 1 thru 1 1/4 Over 1 1/4 thru 1 1/2		100 80 65 50	125 115 105 100
B8C	ASTM A193 Grade B8C		AISI 347 Strain Hardened		- - - -		
B8M	ASTM A193 Grade B8M		AISI 316 Strain Hardened			95 80 65 50	110 100 95 90
B8T	ASTM A193 Grade B8T		AISI 321 Strain Hardened			100 80 65 50	125 115 105 100
L7	ASTM A320 Grade L7	Bolts, Screws, Studs for Low Temp. Service	AISI 4140, 4142 or 4145	1/4 thru 2 1/2	-	105	125
L7A	ASTM A320 Grade L7A		AISI 4037				
L7B	ASTM A320 Grade L7B		AISI 4137				
L7C	ASTM A320 Grade L7C		AISI 8740				
L43	ASTM A320 Grade L43		AISI 4340	1/4 thru 4	-	105	125

(Continued)

Table 9. *(Continued)* **ASTM, SAE, and ISO Grade Markings and Mechanical Properties for Steel Fasteners.**
(Source, American Fasteners Technologies Corp.)

Identification Grade Mark	Specification	Fastener Description	Material	Nominal Size Range	Mechanical Properties (ksi)		
					Proof Load	Yield Strength (min)	Tensile Strength (min)
B8	ASTM A320 Grade B8		AISI 304				
B8C	ASTM A320 Grade B8C		AISI 347				
B8T	ASTM A320 Grade B8T		AISI 321	1/4 and larger	-	30	75
B8F	ASTM A320 Grade B8F		AISI 303 or 303Se				
B8M	ASTM A320 Grade B8M	Bolts, Screws, Studs for Low Temp. Service	AISI 316				
B8	ASTM A320 Grade B8		AISI 304				
B8C	ASTM A320 Grade B8C		AISI 347	1/4 thru 3/4	-	100	100
B8F	ASTM A320 Grade B8F		AISI 303 or 303Se	Over 3/4 thru 1	-	80	80
				Over 1 thru 1 1/4	-	65	65
B8M	ASTM A320 Grade B8M		AISI 316	Over 1 1/4 thru 1 1/2		50	50
B8T	ASTM A320 Grade B8T		AISI 321				

(Continued)

Table 9. *(Continued)* **ASTM, SAE, and ISO Grade Markings and Mechanical Properties for Steel Fasteners.**
(Source, American Fasteners Technologies Corp.)

Identification Grade Mark	Specification	Fastener Description	Material	Nominal Size Range	Mechanical Properties (ksi)		
					Proof Load	Yield Strength (min)	Tensile Strength (min)
	SAE J429 Grade 5	Bolts, Screws, Studs	Medium Carbon Steel, Quenched and Tempered	1/4 thru 1 Over 1 to 1 1/2	85 74	92 81	120 105
	ASTM A449			1/4 thru 1 Over 1 to 1 1/2 Over 1 1/2 thru 3	85 74 55	92 81 58	120 105 90
	SAE J429 Grade 5.1	Sems	Low or Medium Carbon Steel, Quenched and Tempered	No. 6 thru 3/8	85	-	120
	SAE J429 Grade 5.2	Bolts, Screws, Studs	Low Carbon Martensitic Steel, Quenched and Tempered	1/4 thru 1	85	92	120
A325	ASTM A325 Type 1	High Strength Structural Bolts	Medium Carbon Steel, Quenched and Tempered	1/2 thru 1 1 1/8 thru 1 1/2	85 74	92 81	120 105
A325	ASTM A325 Type 2		Low Carbon Martensitic Steel, Quenched and Tempered	1/2 thru 1	85	92	120
A325	ASTM A325 Type 3		Atmospheric Corrosion Resisting Steel, Quenched and Tempered	1/2 thru 1 1 1/8 thru 1 1/2	85 74	92 81	120 105
BB	ASTM A354 Grade BB	Bolts, Studs	Alloy Steel, Quenched and Tempered	1/4 thru 2 1/2 2 3/4 thru 4	80 75	83 78	105 100
BC	ASTM A354 Grade BC				105 95	109 99	125 115
	SAE J429 Grade 7	Bolts, Screws	Medium Carbon Alloy Steel, Quenched and Tempered [4]	1/4 thru 1 1/2	105	115	133

(Continued)

Table 9. *(Continued)* **ASTM, SAE, and ISO Grade Markings and Mechanical Properties for Steel Fasteners.** *(Source, American Fasteners Technologies Corp.)*

Identification Grade Mark	Specification	Fastener Description	Material	Nominal Size Range	Mechanical Properties (ksi)		
					Proof Load	Yield Strength (min)	Tensile Strength (min)
⬡ (No Grade Mark shown)	SAE J429 Grade 8	Bolts, Screws, Studs	Medium Carbon Alloy Steel, Quenched and Tempered	1/4 thru 1 1/2	120	130	150
	ASTM A354 Grade BD		Alloy Steel, Quenched and Tempered [4]				
⬡ No Grade Mark	SAE J429 Grade 8.1	Studs	Medium Carbon Alloy or SAE 1041 Modified Elevated Temperature Drawn Steel	1/4 thru 1 1/2	120	130	150
⬡ A490	ASTM A490	High Strength Structural Bolts	Alloy Steel, Quenched and Tempered	1/2 thru 1 1/2	120	130	150 min. 170 max.
⬡ No Grade Mark	ISO R898 Class 4.6		Medium Carbon Steel, Quenched and Tempered		33	36	60
⬡ No Grade Mark	ISO R898 Class 5.8				55	57	74
8.8 ⬡ or ⬡ 88	ISO R898 Class 8.8	Bolts, Screws, Studs	Alloy Steel, Quenched and Tempered	All sizes thru 1 1/2	85	92	120
10.9 ⬡ or ⬡ 109	ISO R898 Class 10.9				120	130	150

(Continued)

Table 9. *(Continued)* **ASTM, SAE, and ISO Grade Markings and Mechanical Properties for Steel Fasteners.**
(Source, American Fasteners Technologies Corp.)

Identification Grade Mark	Specification	Fastener Description	Material	Nominal Size Range	Mechanical Properties (ksi)		
					Proof Load	Yield Strength (min)	Tensile Strength (min)
12.9 or 129	ISO 898 Class 12.9	Bolts, Screws, Studs	Alloy Steel, Quenched and Tempered	All sizes thru 1 1/2	140	175	200

Identification Grade Mark	Specification	Material	Nominal Size Range	Proof Load Stress (ksi)	Rockwell Hardness		See Note
					Min.	Max.	
No mark	ASTM A563-Grade 0	Carbon Steel	1/4 thru 1 1/2	69	B55	C32	3, 4
	ASTM A563-Grade A	Carbon Steel	1/4 thru 1 1/2	90	B68	C32	3, 4
	ASTM A563-Grade B	Carbon Steel	1/4 thru 1	120	B69	C32	3, 4
			Over 1 thru 1 1/2	105			
	ASTM A563-Grade C	Carbon Steel may be Quenched and Tempered	1/4 thru 4	144	B78	C38	5
	ASTM A563-Grade C3	Atmospheric Corrosion Resistant Steel may be Quenched and Tempered	1/4 thru 4	144	B78	C38	5, 9
	ASTM A563-Grade D	Carbon Steel may be Quenched and Tempered	1/4 thru 4	150	B84	C38	6
	ASTM A563-Grade DH	Carbon Steel, Quenched and Tempered	1/4 thru 4	175	C24	C38	6
	ASTM A563-Grade DH3	Atmospheric Corrosion Resistant Steel, Quenched and Tempered	1/4 thru 4	175	C24	C38	5, 9

(Continued)

Table 9. *(Continued)* **ASTM, SAE, and ISO Grade Markings and Mechanical Properties for Steel Fasteners.**
(Source, American Fasteners Technologies Corp.)

Identification Grade Mark	Specification	Material	Nominal Size Range	Proof Load Stress (ksi)	Rockwell Hardness Min.	Rockwell Hardness Max.	See Note
	ASTM A194-Grade 1	Carbon Steel	1/4 thru 4	130	B70	-	7
	ASTM A194-Grade 2	Medium Carbon Steel	1/4 thru 4	150	159	352	7, 8
	ASTM A194-Grade 2H	Medium Carbon Steel, Quenched and Tempered	1/4 thru 4	175	C24	C38	7
	ASTM A194-Grade 2HM	Medium Carbon Steel, Quenched and Tempered	1/4 thru 4	150	159	237	7, 8
	ASTM A194-Grade 4	Medium Carbon Alloy Steel, Quenched and Tempered	1/4 thru 4	175	C24	C38	7
	ASTM A194-Grade 7	Medium Carbon Alloy Steel, Quenched and Tempered	1/4 thru 4	175	C24	C38	7
	ASTM A194-Grade 7M	Medium Carbon Alloy Steel, Quenched and Tempered	1/4 thru 4	150	159	237	7
See Note 1, 2	See 10						

Notes:

1) In addition to the indicated grade marking, all grades, except A563 grades O, A and B, must be marked for manufacturer identification.
2) The markings shown for all grades of A194 nuts are for cold formed and hot forged nuts. When nuts are machined from bar stock the nut must be additionally marked with the letter 'B'.
3) Nuts are not required to be marked unless specified by the purchaser. When marked, the identification marking shall be the grade letter O, A, or B.
4) Properties shown are those of nonplated or noncoated coarse thread hex nuts.
5) Properties shown are those of coarse thread heavy hex bolts.
6) Properties shown are those of coarse thread heavy hex nuts.
7) Properties shown are those of coarse 8-pitch thread heavy hex nuts.
8) Hardnesses are Brinell Hardness Numbers.
9) The nut manufacturer, as an option, may add other markings to indicate the use of atmospheric corrosion resistant steel.
10) Specifications-ASTM A563-Carbon and Alloy Steel Nuts. ASTM A194/A194M-Carbon and Alloy Steel Nuts for bolts for high pressure and high temperature service.

Table 10. Tensile Stress Areas for Unified Thread Fasteners. *(Source, MIL-S-122H.)*

Nominal Dia. (inches)	Coarse Threads (UNC)		Fine Threads (UNF)		8-Thread Series (8UN)	
	Threads per inch	Stress Area (in.^{2})	Threads per inch	Stress Area (in.^{2})	Threads per inch	Stress Area (in.^{2})
1/4	20	0.0318	28	0.0364	-	-
5/16	18	0.0524	24	0.0580	-	-
3/8	16	0.0775	24	0.0878	-	-
7/16	14	0.1063	20	0.1187	-	-
1/2	13	0.1419	20	0.1599	-	-
9/16	12	0.182	18	0.203	-	-
5/8	11	0.226	18	0.256	-	-
3/4	10	0.334	16	0.373	-	-
7/8	9	0.462	14	0.509	-	-
1	8	0.606	12	0.663	8	0.606
1-1/8	7	0.763	12	0.856	8	0.790
1-1/4	7	0.969	12	1.073	8	1.000
1-3/8	6	1.155	12	1.315	8	1.233
1-1/2	6	1.405	12	1.581	8	1.492
1-5/8	-	-	-	-	8	1.78
1-3/4	5	1.90	-	-	8	2.08
1-7/8	-	-	-	-	8	2.41
2	4-1/2	2.50	-	-	8	2.77
2-1/4	4-1/2	3.25	-	-	8	3.56
2-1/2	4	4.00	-	-	8	4.44
2-3/4	4	4.93	-	-	8	5.43
3	4	5.97	-	-	8	6.51
3-1/4	4	7.10	-	-	8	7.69
3-1/2	4	8.33	-	-	8	8.96
3-3/4	4	9.66	-	-	8	10.34
4	4	11.08	-	-	8	11.81

Table 11. Tensile Stress Areas for M Profile Metric Fasteners - Tolerance Class 6g.

Basic Dia. mm	Coarse Series		Fine Series	
	Pitch	Stress Area mm^2	Pitch	Stress Area mm^2
1	.25	.4498	-	-
1.1	.25	.5766	-	-
1.2	.25	.7190	-	-
1.4	.30	.9635	-	-
1.6	.35	1.208	-	-
1.8	.35	1.629	-	-
2	.40	1.988	-	-

(Continued)

Table 11. *(Continued)* Tensile Stress Areas for M Profile Metric Fasteners - Tolerance Class 6g.

Basic Dia. mm	Coarse Series		Fine Series	
	Pitch	Stress Area mm²	Pitch	Stress Area mm²
2.2	.45	2.383	-	-
2.5	.45	3.274	-	-
3	.50	4.878	-	-
3.5	.60	6.579	-	-
4	.70	8.531	-	-
4.5	.75	11.030	-	-
5	.80	13.829	-	-
6	1.00	19.637	0.75	22.03
7	1.00	28.267	-	-
8	1.25	35.828	1.00	39.167
9	1.25	47.222	-	-
10	1.50	56.837	1.25	61.199
11	1.50	70.985	-	-
12	1.75	82.688	1.25	92.075
14	2.00	113.35	1.50	124.54
16	2.00	154.24	1.50	167.25
18	2.50	189.23	1.50	216.23
20	2.50	241.14	1.50	271.50
22	2.50	299.32	1.50	333.13
24	3.00	347.31	2.00	384.42
27	3.00	453.47	2.00	495.74
30	3.50	553.10	2.00	621.20
33	3.50	685.23	2.00	760.80
36	4.00	806.41	2.00	914.54
39	4.00	955.56	2.00	1082.41
42	4.50	1107.57	2.00	1264.42
45	4.50	1291.60	1.50	1492.51
48	5.00	1456.07	2.00	1670.85
52	5.00	1739.17	-	-
56	5.50	2008.27	2.00	2300.72
60	5.50	2338.56	1.50	2696.36
64	6.00	2648.87	2.00	3031.13
68	6.00	3026.33	-	-
72	6.00	3428.92	2.00	3862.06
80	6.00	4309.51	2.00	4793.53
90	6.00	5551.61	2.00	6099.23
100	6.00	6950.79	2.00	7562.02

Table 12. Mechanical and Chemical Properties of M Profile Carbon and Alloy Steel Bolts, Screws, and Studs.

Mechanical Properties

Class		4.6	4.8	5.6	5.8	6.6	6.8	6.9	8.8	10.9	12.9	14.9
Designation	Formerly	4D	4S	5D	5S	6D	6S	6G	8G	10K	12K	-
	U.S. Grade	1	2	2	2	3	3	3	5	8	-	-
Brinell	Min.	110		140		170			225	280	330	390
	Max.	170		215		245			300	365	425	-
Rockwell B	Min.	63		78		88						
	Max.	88		97		102						
Rockwell C	Min.								18	27	34	40
	Max.								31	38	44	49
Yield Point	ksi Min.		45		56			76	91	128	153	180
Tensile Strength	ksi Min.		56		70		85	85.4	113.8	142.2	170	200
	ksi Max.		78		100		113	99.6	128	170.7	200	230

Chemical Properties

Class	Material & Treatment	Element %			
		Carbon		**Phosphorus**	**Sulphur**
		Min.	Max.	Max.	Max.
4.6	Carbon Steel	-	0.55	0.05	0.06
4.8	Carbon Steel	-	0.55	0.05	0.06
5.6	Carbon Steel	0.15	0.55	0.05	0.06
5.8	Carbon Steel	-	0.55	0.05	0.06
6.8	Carbon Steel	-	0.55	0.05	0.06
8.8	Carbon Steel with Additives (B, Mn, or Cr), Quenched and Tempered	0.15	0.40	0.035	0.035
	Carbon Steel, Quenched and Tempered	0.25	0.55	0.035	0.035

(Continued)

Table 12. *(Continued)* Mechanical and Chemical Properties of M Profile Carbon and Alloy Steel Bolts, Screws, and Studs.

Class	Material & Treatment	Chemical Properties			
		Element %			
		Carbon		Phosphorus	Sulphur
		Min.	Max.	Max.	Max.
10.9	Carbon Steel, Quenched and Tempered	0.25	0.55	0.035	0.035
	Carbon Steel with Additives (B, Mn, or Cr), Quenched and Tempered	0.20	0.55	0.035	0.035
	Alloy Steel, Quenched and Tempered	0.20	0.55	0.035	0.035
<u>10.9</u>*	Carbon Steel with Additives (B, Mn, or Cr), Quenched and Tempered	0.15	0.35	0.035	0.035
12.9	Alloy Steel, Quenched and Tempered	0.20	0.50	0.035	0.035

* Products to this specification have the Property Class underlined.

Table 13. Mechanical Properties of Steel Metric Nuts.

Nominal Size mm		Class 5		Class 8		Class 10		Class 12	
Over	Up To	Proof load N/mm²	Hardness Rockwell C	Proof load N/mm²	Hardness Rockwell C	Proof load N/mm²	Hardness Rockwell C	Proof load N/mm²	Hardness Rockwell C
-	4	520	30	800	30	1040	28-38	1150	31-38
4	7	580	30	810	30	1040	28-38	1150	31-38
7	10	590	30	830	30	1040	28-38	1160	31-38
10	16	610	30	840	30	1050	28-38	1190	31-38
16	39	630	30	920	38	1060	28-38	1200	31-38

Note: Hexagon nuts size M 5 and larger, and all nuts Class 8 and higher are marked either on the bearing surface or on the side on one flat. Left-hand thread nuts size M 6 and larger are marked either on the bearing surface with a left-hand arrow, or by a groove on the corners halfway up the nut height.

Class 12.9 fasteners are usually hex socket cap screws and set screws made of quenched and tempered alloy steel. Mechanical properties of internally threaded Metric nuts are presented in **Table 13**. Property Classes of nuts differ from bolts as they have only one designation number. ISO 898 Part 2 covers the properties of metric, and the values given in **Table 13** are in accord with that standard.

Preload Relaxation

When preloads are initially applied to a fastener in a joint, local yielding takes place due to excess bearing stress under nut and bolt heads, local high spots, rough surface finish, and lack of perfect squareness of the bearing surfaces. Also, loads are not distributed evenly on each thread in a joint, which may lead to thread deformation. These conditions can result in loss of the intended preload over time. As a general guideline, an allowance for 10% loss of preload may be anticipated when designing a joint. Increasing the resilience of a joint may increase its resistance to local yielding. If practical, a ratio of joint length to bolt diameter of four or higher is recommended. Use of through bolts, farside tapped holes, spacers, and washers are design options that can be used to increase the ratio. Over time, the prescribed preload may be reduced or completely lost due to vibration, temperature cycling, creep joint load, etc. Use of a thread-locking solution that prevents relative motion within a joint may solve these problems. **Table 14** provides operating parameters for four types of thread-locking liquids produced by the Loctite Corporation.

Table 14. Common Thread-locking Liquid Solutions. *(Source, Loctite Corp.)*

Product	Fastener Size	Break/Prevail Torque	Temperature Range	Recommended Uses	Key Specifications
Low Strength Thread-locker (Purple)	#2 to 1/4" 2.2 to 6 mm	62/27	−65° to 300°F −54° to 149°C	Set screws, adjustment screws, calibration screws meters and gages	Conforms to MIL-S-46163A
Removable Thread-locker (Blue)	1/4" to 3/4" 6 to 20 mm	115/53	−65° to 300°F −54° to 149°C	Machine tools and presses, pumps and compressors, mounting bolts, gear boxes	Conforms to MIL-S-46163A. NSF/ANSI 61 Approved
High Strength Thread-locker (Red)	3/8" to 1" 9.5 to 25 mm	85/250	−65° to 300°F −54° to 149°C	Heavy equipment, suspension bolts, motor and pump mounts, bearing cap bolts and studs	Conforms to MIL-S-46163A
Penetrating Thread-locker (Green)	#2 to 1/2" 2.2 to 12 mm	85/250	−65° to 300°F −54° to 149°C	Preassembled fasteners, instrumentation screws, electrical connectors	Conforms to MIL-S-46163A. NSF/ANSI 61 Approved

Bolted joint behavior under applied load

The following derivation of a general formula describing behavior of a bolted joint subjected to an applied axial load covers only the condition where the axial load (P in pounds or Newtons) applied to the joint is applied over the same effective joint thickness (t_J in inches or millimeters) that is subjected to the joint compression load (P_J in pounds or Newtons). In many instances, the actual loading planes reside within the joint, leading to a complex loading situation. The condition covered in this section predicts somewhat higher bolt tensions prior to joint separation than exists when loading planes are within the joint material and joint separation occurring at a higher applied joint load. Eccentric loading on a joint causes joint separation and at a lower load than axial loading, and is not covered by the following examples.

Preload Application

Tightening a fastener in a joint compresses the joint materials and puts the fastener in tension. The preload on the fastener (P_B in pounds or Newtons) is equal to the compressive load on the joint material (P_J) – see *Figure 4*. The Modulus of Elasticity (E_B), or Young's Modulus, is a ratio between the stress applied to a material and the elastic strain produced by the stress. For carbon and alloy steel, E_B = 28,500,000 PSI; for stainless steel, E_B = 27,600,000 PSI; and for aluminum alloys, E_B has a range of 9,900,000 to 10,300,000 PSI.

Stretch of a fastener under preload in this configuration is equal to
$$\delta_{B1} = (P_{B1} \times L_B) \div (A_B \times E_B).$$

Compression of the joint material under preload is equal to
$$\delta_{J1} = (P_{J1} \times t_J) \div (A_J \times E_J).$$

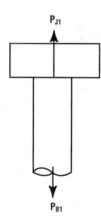

Figure 4. Fastener forces under preload.

Axial Load Applied and Joint Materials in Compression

When an axial load (P) is applied to a joint, fastener stretch is increased by $\Delta\delta_B$ and fastener load increases by ΔP_B. The increased (stretched) length of the fastener reduces the compression of the joint materials by $\Delta\delta_J$ and the compressive loading of the joint materials by ΔP_J. See *Figure 5*.

The increased amount of *stretch* of a fastener under preload in this configuration ($\Delta\delta_B$) and the amount of fastener load increase (ΔP_B) are equal to

$$\Delta\delta_B = (\Delta P_B \times L_B) \div (A_B \times E_B), \text{ and}$$

$$\Delta P_B = (\Delta\delta_B \times A_B \times E_B) \div L_B.$$

The reduction in the compression of joint materials ($\Delta\delta_J$) and the compressive loading (ΔP_J) are equal to

$$\Delta\delta_J = (\Delta P_J \times t_J) \div (A_J \times E_J), \text{ and}$$

$$\Delta P_J = (\Delta\delta_J \times A_J \times E_J) \div t_J.$$

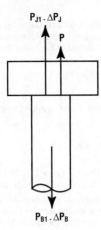

Figure 5. Forces inherent in a loaded joint fastener.

Therefore (see *Figure 5*), $P + P_{J1} - \Delta P_J = P_{B1} + \Delta P_B$. From *Figure 4* it can be seen that $P_{J1} = P_{B1}$. The amount of additional fastener stretch equals the reduction in joint material compression, i.e., $\Delta \delta_B = \Delta \delta_J = \Delta \delta$. Using these relationships and substituting the values for ΔP_B and ΔP_J, the following formulas can be derived.

$$\Delta P_B = (A_B \times E_B \div L_B) \div ([A_B \times E_B \div L_B] + [A_J \times E_J \div t_J]), \text{ and}$$
$$P_B = P_{B1} + \Delta P_B = P_{B1} + P \div (1 + [L_B \times A_J \times E_J \div t_J \times A_B \times E_B]).$$

Axial Load Applied, but Joint Separated

Whenever the applied axial load is sufficiently large to open the joint, there is no longer any load applied to the fastener by the joint materials. As shown in *Figure 6*, the load (P) becomes equal to the fastener load (P_B).

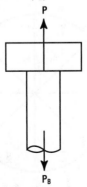

Figure 6. Forces applied to an open joint fastener.

Joint Separation

Referring to Figure 1, line $P_{B1}A$ is the condition discussed above in *Axial Load Supplied and Joint Materials in Compression*. Line AB in the same figure is the condition described immediately above in *Axial Load Applied but Joint Separated*. The point of joint separation, A, occurs when the applied load, P, is described by both conditions.

$$P = P_{B1} \times (1 + [t_J \times A_B \times E_B \div L_B \times A_J \times E_J]).$$

Gasketed Joints

When preload is applied to a joint, gasket compression load is equal to the load on the rest of the joint.

Gasket compression is
$$\delta_{G1} = P_{G1} \times t_G \div A_G \times E_G.$$

Total joint compression is
$$\delta_{JT1} = \delta_{J1} + \delta_{G1} = (P_{J1} \times t_J \div A_J \times E_J) + (P_{G1} \times t_G \div A_G \times E_G)$$
$$= P_{B1} \times ([t_J \div A_J \times E_J] + [t_G \div A_G \times E_G]) \text{ since } P_{J1} = P_{G1} = P_{B1}.$$

With the axial load, P, applied to the joint as in *Axial Load Applied and Joint Materials in Compression* (above) fastener loading remains unchanged, but compression in the total joint is reduced by $\Delta\delta_{JT}$ and compression load is reduced by ΔP_{JT}.

$$\Delta\delta_{JT} = \Delta P_{JT} ([t_J \div A_J \times E_J] + [t_G \div A_G \times E_G]), \text{ and}$$
$$\Delta P_{JT} = \Delta\delta_{JT} \div ([t_J \div A_J \times E_J] + [t_G \div A_G \times E_G]).$$

Therefore, $\Delta P_B = P \div (1 + K)$ where
$$K = ([L_B \div \{A_B \times E_B\}] \div [t_J \div \{A_J \times E_J\}] + [t_G \div A_G \times E_G]),$$
and $P_B = P_{B1} + \Delta P_B = P_{B1} + (P \div [1 + K]).$

It is important to apply approximately equal tension to each of the fasteners in a flange joint, especially if the joint contains a gasket. Unequal tension will result in leakage past the gasket. The recommended procedure is to tighten the fasteners in stages, following a pattern of tightening a pair of opposite fasteners, then a pair of opposite fasteners 90° away, followed by pairs of opposite fasteners in between as shown in *Figure 7*. This should be done in two or three steps in order to allow the shear stress to relax between steps. After tightening, it is advisable to retighten after a few hours to help assure that preload relaxation will not cause the joint to leak.

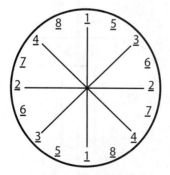

Figure 7. Correct tightening order for pairs of flange bolts.

Controlling bolt tension

There are numerous methods for controlling fastener tension, and **Table 15** provides the expected accuracy and relative costs for widely used preloading techniques as provided by SPS Technologies. Under ideal conditions, bolt elongation control using an ultrasonic detector can produce an accuracy as good as the strain gage method and at comparable cost. Tightening methods using power drive are similar in accuracy to manual methods, but costs are higher.

Table 15. Estimated Accuracy and Relative Costs for Manual Preloading Methods. *(Source, SPS Technologies.)*

Preload Method	Load Accuracy %	Relative Cost
Feel (Sensitivity to touch)	± 35	1
Torque Wrench	± 25	1.5
Turn of Nut	± 15	3
Preload Indicating Washer	± 10	7
Sensor Wrench (Computer controlled) —Below yield using turn of nut	± 15	8
—Yield point sensing	± 8	8
Bolt Elongation	± 3 – 5	15
Strain Gage	± 1	20

Elongation Measurement

Bolt elongation increases in the same proportion as stress increases. If both ends of a bolt are accessible, as in *Figure 8,* it is quite simple to measure the bolt with a micrometer before and after tension is applied to assure proper axial stress. The following formula applies:

$$\delta_B = (f_t \times L_B) \div E_B.$$

For more complex bolt geometry, elongation is equal to the sum of the elongations for each section, also taking into consideration traditional stresses in bolt head height and nut engagement length. Measurement by micrometer is most accurate when the bolt is threaded its entire length, or it has few threads in the bolt grip area so that elongation will be practically uniform throughout its length.

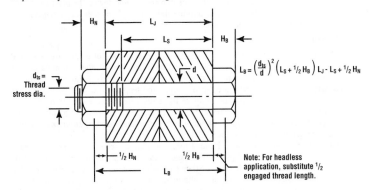

$$L_B = \left(\frac{d_{ts}}{d}\right)^2 \left(L_S + \frac{1}{2}H_B\right) L_J - L_S + \frac{1}{2}H_N$$

Note: For headless application, substitute $\frac{1}{2}$ engaged thread length.

Figure 8. Effective length applicable in elongation formula.

In situations where both ends of a fastener are not accessible for measurement, it is sometimes possible to drill the fastener axially and take measurements with a micrometer depth gage as the fastener is tightened. *Figure 9* shows a cross section of a drilled fastener. A drilled fastener can also be used in another way to measure elongation. A pin placed through the length of the hole, and secured at the bottom, will move in relation to its reference surface as the fastener is tightened. If, for example, 2 mm of the pin protruded from the top of the fastener prior to tightening, the fastener will have elongated 1 mm over its length when the pin's protrusion is reduced to 1 mm during tightening.

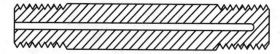

Figure 9. Axial hole in fastener for elongation measurement when one end is not accessible.

The *"turn of the nut"* method for measuring elongation uses the lead of the thread helix to determine the amount of elongation. Accuracy is affected by elastic deformation of the threads, by roughness of the bearing surfaces, and by the difficulty in establishing the starting point for measuring the angle. Generally, the nut is tightened to firmly seat the contacting surfaces and then it is loosened just enough to release bolt tension and twisting. This is the starting point for measuring the angle. Although nut turn angle will differ with bolt size, material, length, and thread lead, the following guidelines are useful for estimating the number of turns past snug for most bolt length/diameter relationships.

Bolt Length (diameters)	Amount of Rotation Past Snug
Length ≤ 4	One-third turn
4 < Length ≤ 8	One-half turn
8 < Length ≤ 12	Two-thirds turn

It should be kept in mind that unless the deformation of the nut and joint materials under load is negligible compared to that of the bolt, a significant portion of the nut rotation will be absorbed through means other than elongation of the bolt. In such instances, the following equation will not provide reliable results and the nut turn angle should be determined empirically with a tension measuring device. In most cases, the following equations can be used to determine the turn angle of the nut in degrees that will result in a given elongation.

$$\delta_B = (\theta \times L) \div 360 \quad \text{or} \quad \theta = 360 \times ([f_t \times L_B] \div [E \times L]).$$

The *ultrasonic method* of measuring elongation uses sound pulses that are generated from one end of a bolt and travel at the speed of sound through the bolt material and bounce off the other end and return to the point of origin. The time required to travel the bolt's length and return is measured to determine the length of the bolt. Using various inputs, the ultrasonic system can compute the stress, load, and elongation of a bolt at any time by comparing pulse travel time in loaded and unstressed conditions. The equipment requires a high degree of sophistication as the speed of sound varies with material, temperature, and stress level.

Strain Gages use a thin wire bonded to a bolt that stretches as the bolt elongates under load. The stretch causes a change in electrical resistance of the wire that can be measured and correlated with bolt load. Strain gage wires or groups of wires can be bonded either to the outside of a fastener or to the inside surface of a small axial hole drilled through the fastener. An additional mounting method uses the recess on the outside of a plastic tube that is bonded to the inside of a small axial hole in the bolt.

Temperature can also be used to induce tension and elongation. When a hot bolt and nut in a joint are air cooled, the bolt will shrink as it cools and tension will be developed. The following formula can be used to calculate the temperature necessary to induce the required axial tensile stress in a bolt when the stress is below the elastic limit.

$$t = (f_t \div E_B \times e) + t_o.$$

To provide bolt tension, the bolt and nut should be heated to a temperature slightly higher than calculated temperature to allow for cooling prior to nut tightening. The nut should be tightened snugly and the assembly should then be allowed to cool. Tension is induced by cooling. This method can only be applied where the bolt is accessible for heating and where the heating will not degrade joint materials. Accuracy is affected by the difficulty of controlling bolt temperature and by any significant deformity of the joint materials.

Temperature can also be used to control the expansion of a bolt. Using a thickness gage inserted under the nut, the elongation of the bolt can be measured as temperature is applied. When the desired length is achieved, the nut is tightened snugly. After cooling, the axial expansion in the bolt should be approximately equal to the desired stress level.

Torque control

Unified Threads

Suggested preloads and tightening torques for any size fastener can be calculated from the tables found in this section. The suggested preload, in most cases is 90% of the maximum clamping load for permanent connections, and 75% maximum clamping load for reusable connections. As shown earlier, the following equation provides maximum clamping load:

$$F = Y \times A_s.$$

Values for Y have been provided in Tables 3 and 9, and tensile stress areas for standard fasteners have been given in Tables 10 and 11. For materials without a given proof load (Y), the value can be estimated as 85% of the minimum yield strength. Once the value of 90% F or 75% F is known for a specific grade and size of fastener, the tightening torque (T) can be determined by multiplying F times the coefficient of friction (from Table 2) and the nominal fastener's nominal diameter.

$$T = \%F \times \mu \times d.$$

The following example shows how to estimate clamping force and desired tightening torque for a Grade 5, $5/8 - 11$ UNC bolt to be used for a permanent connection. From Table 3, it can be seen that the proof load for this fastener is 85,000 PSI, and from Table 10 it can be seen that the tensile stress area is 0.226 in.[2]. Therefore, F = 85000 × 0.226 = 19,210 pounds. Since the connection is designed to be permanent, the desired clamping force will be 90% of the maximum clamping load, or 17,289 pounds. The desired torque can now be found by multiplying the clamping load times the coefficient of friction from Table 2 (15%), and the bolt diameter (0.6250). 17289 × .15 × .625 = 1621 lb in. or 135 lb ft.

Metric Threads

The equations above can be used, with metric measurements, to solve the following example for a Class 8.8, 16 × 2.00 fastener used in a reusable connection. From Table 9 it can be seen that the proof load is 85,000 PSI; 85000 × .006894757 = 586 MPa. The tensile stress area, from Table 11, for this fastener is 154.235 mm[2]. To find F multiply,

586 × 154.235 = 90,382 Newtons. Next, multiply the clamping force by 75% to determine the amount of force desirable for a reusable connection. The desired clamping force is therefore 67,786 Newtons. From Table 2, obtain the coefficient of friction. The bolt diameter is 16 mm. Tightening torque can now be found with the following formula: 67786 × .15 × 16 = 162,686 N/mm or 162.69 Nm.

Table 16 provides clamping load and tightening torque recommendations for bolts, screws, and studs with Unified threads, and **Table 17** provides this information for M Profile metric threads. Seating torques for Unified thread socket screws are given in **Table 18** and **Table 19**, and **Table 20** provides these values for metric thread socket screws. The higher values in these tables represent 100% of the allowable load and torque and the lower values are 75% of the maximum. The values were obtained from the formulas discussed immediately above and assume a 0.15% coefficient of friction. Proof load values to determine the clamping load and tightening torque were obtained from Table 3 and Table 10. It should be noted that, depending on the source of the data, proof loads, tensile strength, and yield strength values will vary by as much as 10%. This can be attributed to independent testing on different samples that provide different results, to averaging several samples, and to discrepancies in the calibration of test equipment. The varying values for grades and classes is sufficient reason not to tension fasteners beyond 90% of their proof load rating.

UNJ Threads. As stated earlier, calculation of tensile stress is based on the diameter approximately midway between the minimum pitch diameter and the minimum minor diameter. However, for UNJ threads in accordance with MIL-S-8879, tensile stress area is considered to be at the basic pitch diameter. Therefore, tightening torque requirements for UNJ threaded fasteners is higher than for an equally stressed Unified threaded fastener in an equivalent joint. To convert the Unified torque to the appropriate UNJ torque requirement, use the following formula.

$$\text{UNJ}_{\text{Torque}} = \text{Unified}_{\text{Torque}} \times ([d_{\text{Basic}} \times n] \div [d_{\text{Basic}} \times n])^2.$$

Table 16. Clamp Load and Tightening Torque Values for Unified Thread Bolts, Screws, and Studs.
(See Notes for Dimensional Information.)

Nominal Size and tpi	Grade 2		Grade 5		Grade 8	
	Clamp Load Range [1]	Torque Range [2]	Clamp Load Range [1]	Torque Range [2]	Clamp Load Range [1]	Torque Range [2]
4-40	249 - 332	4 - 6	385 - 513	6 - 9	544 - 725	9 - 12
4-48	273 - 364	5 - 6	421 - 562	7 - 9	595 - 793	10 - 13
5-40	375 - 500	7 - 9	579 - 773	11 - 14	818 - 1091	15 - 20
5-44	342 - 457	6 - 9	529 - 706	10 - 13	747 - 996	14 - 19
6-32	371 - 495	8 - 10	574 - 765	12 - 16	810 - 1080	17 - 22
6-40	419 - 558	9 - 12	647 - 863	13 - 18	914 - 1218	19 - 25
8-32	578 - 770	14 - 19	893 - 1190	22 - 29	1260 - 1680	31 - 41
8-36	608 - 811	15 - 20	940 - 1253	23 - 31	1327 - 1769	33 - 44
10-24	722 - 963	21 - 27	1116 - 1488	32 - 42	1575 - 2100	45 - 60
10-32	825 - 1100	24 - 31	1275 - 1700	36 - 48	1800 - 2400	51 - 68
12-24	998 - 1331	32 - 43	1543 - 2057	50 - 67	2178 - 2904	71 - 94
12-28	1064 - 1419	34 - 46	1645 - 2193	53 - 71	2322 - 3096	75 - 100
1/4-20	1312 - 1749	4 - 5	2027 - 2703	6 - 8	2862 - 3816	9 - 12

(Continued)

Table 16. *(Continued)* Clamp Load and Tightening Torque Values for Unified Thread Bolts, Screws, and Studs. *(See Notes for Dimensional Information.)*

Nominal Size and tpi	Grade 2		Grade 5		Grade 8	
	Clamp Load Range [1]	Torque Range [2]	Clamp Load Range [1]	Torque Range [2]	Clamp Load Range [1]	Torque Range [2]
1/4-28	1502 - 2002	5 - 6	2321 - 3094	7 - 10	3276 - 4368	10 - 14
5/16-18	2162 - 2882	8 - 11	3341 - 4454	13 - 17	4716 - 6288	18 - 25
5/16-24	2393 - 3190	9 - 12	3698 - 4930	14 - 19	5220 - 6960	20 - 27
3/8-16	3197 - 4263	15 - 20	4941 - 6588	23 - 31	6975 - 9300	33 - 44
3/8-24	3622 - 4829	17 - 23	5597 - 7463	26 - 35	7902 - 10536	37 - 49
7/16-14	4385 - 5847	24 - 32	6777 - 9036	37 - 49	9567 - 12756	52 - 70
7/16-20	4896 - 6529	27 - 36	7567 - 10090	41 - 55	10683 - 14244	58 - 78
1/2-13	5853 - 7805	37 - 49	9046 - 12062	57 - 75	12771 - 17028	80 - 106
1/2-20	6596 - 8795	41 - 55	10194 - 13592	64 - 85	14391 - 19188	90 - 120
9/16-12	7508 - 10010	53 - 70	11603 - 15470	82 - 109	16380 - 21840	115 - 153
9/16-18	8374 - 11165	59 - 78	12941 - 17255	91 - 121	18270 - 24360	128 - 171
5/8-11	9323 - 12430	73 - 97	14408 - 19210	113 - 150	20340 - 27120	159 - 212
5/8-18	10560 - 14080	83 - 110	16320 - 21760	128 - 170	23040 - 30720	180 - 240
3/4-10	13778 - 18370	129 - 172	21293 - 28390	200 - 266	30060 - 40080	282 - 376
3/4-16	15386 - 20515	144 - 192	23779 - 31705	223 - 297	33570 - 44760	315 - 420
7/8-9	11435 - 15246	125 - 167	29453 - 39270	322 - 430	41580 - 55440	455 - 606
7/8-14	12598 - 16797	138 - 184	32449 - 43265	355 - 473	45810 - 61080	501 - 668
1-8	14999 - 19998	187 - 250	33633 - 44844	420 - 561	54540 - 72720	682 - 909
1-12	16409 - 21879	205 - 273	36797 - 49062	460 - 613	59670 -79560	746 - 995
1-1/8-7	18884 - 25179	266 - 354	42347 - 56462	595 - 794	68670 - 91560	966 - 1288
1-1/8-8	19553 - 26070	275 - 367	43845 - 58460	617 - 822	71100 - 94800	1000 - 1333
1-1/8-12	21186 -28248	298 - 397	47508 - 63344	668 - 891	77040 - 102720	1083 - 1445
1-1/4-7	23983 - 31977	375 - 500	53780 - 71706	840 - 1120	87210 - 116280	1363 - 1817
1-1/4-8	24750 - 33000	387 - 516	55500 - 74000	867 - 1156	90000 - 120000	1406 - 1875
1-1/4-12	26557 - 35409	415 - 553	59552 - 79402	930 - 1241	96570 - 128760	1509 - 2012
1-3/8-6	28586 - 38115	491 - 655	64103 - 85470	1102 - 1469	103950 - 138600	1787 - 2382
1-3/8-8	30517 - 40689	525 - 699	68432 - 91242	1176 - 1568	110970 - 147960	1907 - 2543
1-3/8-12	32546 - 43395	559 - 746	72983 - 97310	1254 - 1673	118350 - 157800	2034 - 2712
1-1/2-6	34774 - 46365	652 - 869	77978 - 103970	1462 - 1949	126450 - 168600	2371 - 3161
1-1/2-8	36927 - 49236	692 - 923	82806 - 110408	1553 - 2070	134280 - 179040	2518 - 3357
1-1/2-12	39130 - 52173	734 - 978	87746 - 116994	1645 - 2194	142290 - 189720	2668 - 3557
1-5/8-8	44055 - 58740	895 - 1193	98790 - 131720	2007 - 2676	160200 - 213600	3254 - 4339
1-3/4-5	47025 - 62700	1029 - 1372	105450 - 140600	2307 - 3076	171000 - 228000	3741 - 4988
1-3/4-8	51480 - 68640	1126 - 1502	115440 - 153920	2525 - 3367	187200 - 249600	4095 - 5460
1-7/8-8	59648 - 79530	1398 - 1864	133755 - 178340	3135 - 4180	216900 - 289200	5084 - 6778
2-4.5	61875 - 82500	1547 - 2063	138750 - 185000	3469 - 4625	225000 - 300000	5625 - 7500

Notes: 1) Lower value equals 75% of maximum clamping load in pounds. Higher value equals maximum (100%) clamping load in pounds.

2) For sizes through 12-28 (shaded on the Table), tightening torque is given in lb in. For sizes $^1/_4$-20 and larger, tightening torques are given in lb ft. Lower value equals 75% of maximum tightening torque. Higher value equals maximum (100%) tightening torque.

Table 17. Clamp Load and Tightening Torque Values for M Profile Metric Bolts, Screws, and Studs.
(See Notes for Dimensional Information.)

Nominal Size and Pitch mm	Grade 4.8			Grade 8.8		
	Clamp Load Range [1]	Torque Range [2] Nm	Torque Range [3] lb ft	Clamp Load Range [1]	Torque Range [2] Nm	Torque Range [3] lb ft
1 × 0.25	105 - 139	0.02 - 0.02	0.14 - 0.19	202 - 270	0.03 - 0.04	0.27 - 0.36
1.1 × 0.25	134 - 179	0.02 - 0.03	0.20 - 0.26	259 - 346	0.04 - 0.06	0.38 - 0.51
1.2 × 0.25	167 - 223	0.03 - 0.04	0.27 - 0.36	324 - 431	0.06 - 0.08	0.52 - 0.69
1.4 × 0.3	224 - 299	0.05 - 0.06	0.42 - 0.56	434 - 578	0.09 - 0.12	0.81 - 1.1
1.6 × 0.35	281 - 374	0.07 - 0.09	0.60 - 0.80	544 - 725	0.13 - 0.17	1.2 - 1.5
1.8 × 0.35	379 - 505	0.10 - 0.14	0.9 - 1.2	733 - 977	0.20 - 0.26	1.8 - 2.3
2.0 × 0.4	462 - 616	0.14 - 0.18	1.2 - 1.6	895 - 1193	0.27 - 0.36	2.4 - 3.2
2.2 × 0.45	554 - 739	0.18 - 0.24	1.6 - 2.2	1072 - 1430	0.35 - 0.47	3.1 - 4.2
2.5 × 0.45	761 - 1015	0.29 - 0.38	2.5 - 3.4	1473 - 1964	0.55 - 0.74	4.9 - 7
3 × 0.5	1134 - 1512	0.51 - 0.68	4.5 - 6	2195 - 2927	0.99 - 1.3	9 - 12
3.5 × 0.6	1530 - 2039	0.80 - 1.1	7 - 9	2961 - 3947	1.6 - 2.1	14 - 18
4 × 0.7	1983 - 2645	1.2 - 1.6	11 - 14	3839 - 5119	2.3 - 3.1	20 - 27
4.5 × 0.75	2564 - 3419	1.7 - 2.3	15 - 20	4964 - 6618	3.4 - 4.5	30 - 40
5 × 0.8	3215 - 4287	2.4 - 3.2	1.8 - 2.4	6223 - 8297	4.7 - 6	3.4 - 4.6
6 × 1	4566 - 6087	4.1 - 5.5	3.0 - 4.0	8837 - 11782	8 - 11	6 - 8
6 × 0.75	5122 - 6829	4.6 - 6	3.4 - 4.5	9914 - 13218	9 - 12	7 - 9
7 × 1	6572 - 8763	7 - 9	5.1 - 7	12720 - 16960	13 - 18	10 - 13
8 × 1.25	8330 - 11107	10 - 13	7 - 10	16123 - 21497	19 - 26	14 - 19
8 × 1	9106 - 12142	11 - 15	8 - 11	17625 - 23500	21 - 28	16 - 21
9 × 1.25	10979 - 14639	15 - 20	11 - 15	21250 - 28333	29 - 38	21 - 28
10 × 1.5	13215 - 17619	20 - 26	15 - 19	25577 - 34102	38 - 51	28 - 38
10 × 1.25	14229 - 18972	21 - 28	16 - 21	27540 - 36719	41 - 55	30 - 41
11 × 1.5	16504 - 22005	27 - 36	20 - 27	31943 - 42591	53 - 70	39 - 52
12 × 1.75	19225 - 25633	35 - 46	26 - 34	37210 - 49613	67 - 89	49 - 66
12 × 1.25	21407 - 28543	39 - 51	28 - 38	41434 - 55245	75 - 99	55 - 73
14 × 2	26354 - 35139	55 - 74	41 - 54	51008 - 68010	107 - 143	79 - 105
14 × 1.5	28956 - 38607	61 - 81	45 - 60	56043 - 74724	118 - 157	87 - 116
16 × 2	35861 - 47814	86 - 115	63 - 85	69408 - 92544	167 - 222	123 - 164
16 × 1.5	38886 - 51848	93 - 124	69 - 92	75263 - 100350	181 - 241	133 - 178
18 × 2.5	43996 - 58661	119 - 158	88 - 117	85154 - 113538	230 - 307	170 - 226
18 × 1.5	50273 - 67031	136 - 181	100 - 133	97304 - 129738	263 - 350	194 - 258
20 × 2.5	56065 - 74753	168 - 224	124 - 165	108513 - 144684	326 - 434	240 - 320
20 × 1.5	63124 - 84165	189 - 252	140 - 186	122175 - 162900	367 - 489	270 - 360
22 × 1.5	69592 - 92789	230 - 306	169 - 226	134694 - 179592	444 - 593	328 - 437
22 × 1.5	77453 - 103270	256 - 341	189 - 251	149909 - 199878	495 - 660	365 - 486
24 × 3	80764 - 107685	291 - 388	214 - 286	156317 - 208422	563 - 750	415 - 553
24 × 2	89378 - 119170	322 - 429	237 - 316	172989 - 230652	623 - 830	459 - 612

(Continued)

Table 17. *(Continued)* Clamp Load and Tightening Torque Values for M Profile Metric Bolts, Screws, and Studs. *(See Notes for Dimensional Information.)*

Nominal Size and Pitch mm	Grade 4.8			Grade 8.8		
	Clamp Load Range [1]	Torque Range [2] Nm	Torque Range [3] lb ft	Clamp Load Range [1]	Torque Range [2] Nm	Torque Range [3] lb ft
27 × 3	105432 - 140576	427 - 569	315 - 420	204062 - 272082	826 - 1102	610 - 813
27 × 2	115260 - 153679	467 - 622	344 - 459	223083 - 297444	903 - 1205	666 - 889
30 ×3.5	128596 - 171461	579 - 772	427 - 569	248895 - 331860	1120 - 1493	826 - 1101
30 × 2	144406 - 192541	650 - 866	479 - 639	279495 - 372660	1258 - 1677	928 - 1237
33 × 3.5	159316 - 212421	789 - 1051	582 - 776	308354 - 411138	1526 - 2035	1126 - 1501
33 × 2	176886 - 235848	876 - 1167	646 - 861	342360 - 456480	1695 - 2260	1250 - 1667
36 × 4	187490 - 249987	1012 - 1350	747 - 996	362885 - 483846	1960 - 2613	1445 - 1927
36 × 2	212631 - 283507	1148 - 1531	847 - 1129	411543 - 548724	2222 - 2963	1639 - 2185
39 × 4	222168 - 296224	1300 - 1733	959 - 1278	430002 - 573336	2516 - 3354	1855 - 2474
39 × 2	251660 - 335547	1472 - 1963	1086 - 1448	487085 - 649446	2849 - 3799	2102 - 2802

Nominal Size and Pitch mm	Grade 10.9			Grade 12.9		
	Clamp Load Range [1]	Torque Range [2] Nm	Torque Range [3] lb ft	Clamp Load Range [1]	Torque Range [2] Nm	Torque Range [3] lb ft
1 × 0.25	280 - 373	0.04 - 0.06	0.37 - 0.50	327 - 436	0.05 - 0.07	0.43 - 0.58
1.1 × 0.25	359 - 479	0.06 - 0.08	0.52 - 0.70	419 - 559	0.07 - 0.09	0.61 - 0.82
1.2 × 0.25	448 - 597	0.08 - 0.11	0.71 - 0.95	523 - 697	0.09 - 0.13	0.83 - 1.11
1.4 × 0.3	600 - 800	0.13 - 0.17	1.1 - 1.5	701 - 935	0.15 - 0.20	1.30 - 1.74
1.6 × 0.35	752 - 1003	0.18 - 0.24	1.6 - 2.1	879 - 1172	0.21 - 0.28	1.87 - 2.49
1.8 × 0.35	1014 - 1352	0.27 - 0.37	2.4 - 3.2	1185 - 1580	0.32 - 0.43	2.83 - 3.78
2.0 × 0.4	1238 - 1650	0.37 - 0.50	3.3 - 4.4	1446 - 1928	0.43 - 0.58	3.84 - 5.12
2.2 × 0.45	1483 - 1978	0.49 - 0.65	4.3 - 5.8	1734 - 2312	0.57 - 0.76	5.06 - 6.75
2.5 × 0.45	2038 - 2717	0.76 - 1.0	6.8 - 9.0	2382 - 3176	0.89 - 1.19	7.91 - 10.54
3 × 0.5	3037 - 4049	1.4 - 1.8	12 - 16	3549 - 4732	1.60 - 2.13	14.13 - 18.85
3.5 × 0.6	4095 - 5461	2.2 - 2.9	19 - 25	4786 - 6382	2.51 - 3.35	22.24 - 29.65
4 × 0.7	5311 - 7081	3.2 - 4.2	28 - 38	6206 - 8275	3.72 - 4.97	32.96 - 43.94
4.5 × 0.75	6866 - 9155	4.6 - 6	41 - 55	8024 - 10699	5.42 - 7.22	47.94 - 63.92
5 × 0.8	8609 - 11478	6 - 9	5 - 6	10061 - 13414	7.55 - 10.06	5.57 - 7.42
6 × 1	12224 - 16299	11 - 15	8 - 11	14286 - 19048	12.86 - 17.14	9.48 - 12.64
6 × 0.75	13714 - 18285	12 - 16	9 - 12	16027 - 21369	14.42 - 19.23	10.64 - 14.18
7 × 1	17596 - 23462	18 - 25	14 - 18	20564 - 27419	21.59 - 28.79	15.93 - 21.23
8 × 1.25	22303 - 29737	27 - 36	20 - 26	26065 - 34753	31.28 - 41.70	23.07 - 30.76
8 × 1	24381 - 32509	29 - 39	22 - 29	28494 - 37992	34.19 - 45.59	25.22 - 33.63
9 × 1.25	29396 - 39194	40 - 53	29 - 39	34354 - 45805	46.38 - 61.84	34.21 - 45.61
10 × 1.5	35381 - 47175	53 - 71	39 - 52	41349 - 55132	62.02 - 82.70	45.75 - 60.99
10 × 1.25	38096 - 50795	57 - 76	42 - 56	44522 - 59363	66.78 - 89.04	49.26 - 65.68
11 × 1.5	44188 - 58918	73 - 97	54 - 72	51642 - 68855	85.21 - 113.61	62.85 - 83.80
12 × 1.75	51473 - 68631	93 - 124	68 - 91	60156 - 80207	108.28 - 144.37	79.86 - 106.48
12 × 1.25	57317 - 76422	103 - 138	76 - 101	66985 - 89313	120.57 - 160.76	88.93 - 118.57

(Continued)

Table 17. *(Continued)* Clamp Load and Tightening Torque Values for M Profile Metric Bolts, Screws, and Studs. *(See Notes for Dimensional Information.)*

Nominal Size and Pitch mm	Grade 10.9			Grade 12.9		
	Clamp Load Range [1]	Torque Range [2] Nm	Torque Range [3] lb ft	Clamp Load Range [1]	Torque Range [2] Nm	Torque Range [3] lb ft
14 × 2	70560 - 94081	148 - 198	109 - 146	82462 - 109950	173.17 - 230.89	127.72 - 170.30
14 × 1.5	77526 - 103368	163 - 217	120 - 160	90603 - 120804	190.27 - 253.69	140.33 - 187.11
16 × 2	96014 - 128019	230 - 307	170 - 227	112210 - 149613	269.30 - 359.07	198.63 - 264.84
16 × 1.5	104113 - 138818	250 - 333	184 - 246	121674 - 162233	292.02 - 389.36	215.38 - 287.18
18 × 2.5	117796 - 157061	318 - 424	235 - 313	137665 - 183553	371.70 - 495.59	274.15 - 365.53
18 × 1.5	134603 - 179471	363 - 485	268 - 357	157307 - 209743	424.73 - 566.31	313.26 - 417.69
20 × 2.5	150110 - 200146	450 - 600	332 - 443	175429 - 233906	526.29 - 701.72	388.17 - 517.56
20 × 1.5	169009 - 225345	507 - 676	374 - 499	197516 - 263355	592.55 - 790.07	437.04 - 582.72
22 × 1.5	186327 - 248436	615 - 820	454 - 605	217755 - 290340	718.59 - 958.12	530.01 - 706.68
22 × 1.5	207373 - 276498	684 - 912	505 - 673	242352 - 323136	799.76 - 1066.35	589.87 - 786.50
24 × 3	216238 - 288317	778 - 1038	574 - 766	252712 - 336949	909.76 - 1213.02	671.01 - 894.67
24 × 2	239301 - 319069	861 - 1149	635 - 847	279666 - 372887	1006.80 - 1342.39	742.57 - 990.10
27 × 3	282285 - 376380	1143 - 1524	843 - 1124	329899 - 439866	1336.09 - 1781.46	985.45 - 1313.94
27 × 2	308598 - 411464	1250 - 1666	922 - 1229	360651 - 480868	1460.64 - 1947.51	1077.31 - 1436.41
30 ×3.5	344305 - 459073	1549 - 2066	1143 - 1524	402380 - 536507	1810.71 - 2414.28	1335.51 - 1780.68
30 × 2	386635 - 515513	1740 - 2320	1283 - 1711	451850 - 602467	2033.33 - 2711.10	1499.70 - 1999.61
33 × 3.5	426556 - 568741	2111 - 2815	1557 - 2076	498505 - 664673	2467.60 - 3290.13	1820.01 - 2426.68
33 × 2	473598 - 631464	2344 - 3126	1729 - 2305	553482 - 737976	2739.74 - 3652.98	2020.73 - 2694.30
36 × 4	501990 - 669320	2711 - 3614	1999 - 2666	586663 - 782218	3167.98 - 4223.98	2336.58 - 3115.44
36 × 2	569301 - 759068	3074 - 4099	2267 - 3023	665328 - 887104	3592.77 - 4790.36	2649.89 - 3533.19
39 × 4	594836 - 793115	3480 - 4640	2567 - 3422	695170 - 926893	4066.74 - 5422.33	2999.48 - 3999.30
39 × 2	673800 - 898400	3942 - 5256	2907 - 3876	787453 - 1049938	4606.60 - 6142.14	3397.66 - 4530.21

Notes: 1) Lower value equals 75% of maximum clamp load in Newtons. Higher value equals maximum (100%) clamp load in Newtons.

2) Lower value equals 75% of maximum tightening torque in Nm. Higher value equals maximum (100%) tightening torque.

3) For sizes through 4.5 × 0.75 (shaded in the Table), tightening torque is given in lb in. For sizes 5 × 0.8 and larger, tightening torques are given in lb ft. Lower value equals 75% of maximum tightening torque. Higher value equals maximum (100%) tightening torque.

Table 18. Recommended Seating Torques for Socket Head Cap and Flat Head Screws. *(See Notes.)*

Thread Size	Alloy Steel Socket Head Cap Screws[1]		Stainless Steel Socket Head Cap Screws[2]		Alloy Steel Flat Head Socket Screws[3]		Stainless Steel Flat Head Socket Screws[4]	
Nom	UNRC	UNRF	UNRC	UNRF	UNRC	UNRF	UNRC	UNRF
#0	-	3	-	1.3	-	1.5	-	1
#1	5	5	2	2.3	205	2.5	1.7	1.8
#2	7	8	3.8	4	4.5	4.5	2.8	3
#3	12	13	5.7	6	7	7	4.3	4.6
#4	18	19	8	9	8	8	6.0	6.6
#5	24	25	12	14	12	13	8.9	9.3

(Continued)

Table 18. *(Continued)* **Recommended Seating Torques for Socket Head Cap and Flat Head Screws.** *(See Notes.)*

Thread Size	Alloy Steel Socket Head Cap Screws[1]		Stainless Steel Socket Head Cap Screws[2]		Alloy Steel Flat Head Socket Screws[3]		Stainless Steel Flat Head Socket Screws[4]	
Nom	**UNRC**	**UNRF**	**UNRC**	**UNRF**	**UNRC**	**UNRF**	**UNRC**	**UNRF**
#6	34	36	15	17	15	17	11	12
#8	59	60	28	29	30	31	20	21
#10	77	91	40	45	40	45	30	34
1/4	200	240	95	110	100	110	71	81
5/16	425	475	170	190	200	220	123	136
3/8	750	850	300	345	350	400	218	247
7/16	1,200	1,350	485	545	560	625	349	388
1/2	1,850	2,150	750	850	850	1,000	532	600
9/16	2,500	2,700	920	1,050	1,200	1,360	767	856
5/8	3,400	3,820	1,270	4,450	1,700	1,900	1,060	1,200
3/4	6,000	6,800	2,260	2,520	3,000	3,200	1,880	2,100
7/8	8,400	9,120	3,790	4,180	5,000	5,400	3,030	3,340
1	12,500	13,200	5,690	6,230	7,200	7,600	4,550	5,000
1 1/8	14,900	16,600						
1 1/4	25,000	27,000						
1 3/8	33,000	35,000						
1 1/2	43,500	47,000						
1 3/4	71,500	82,500						
2	108,000	125,000						
2 1/4	155,000	186,000						
2 1/2	215,000	248,000						
2 3/4	290,000	330,000						
3	375,000	430,000						

Notes: Caution: Even when using the recommended seating torques in this Table, the induced loads obtained may vary by as much as ± 25%, depending on uncontrolled variables such as mating material, lubrication, surface finish, hardness, etc. Torque values are in inch pounds. To convert inch pounds (in.-lbs) to Newton meters (N-m), divide the values on this Table by 8.88. Seating torques given are for screws made to the specifications referenced, as follows.

1) Alloy Steel Socket Head Cap Screws: ASTM A574. Hardness: RC 38-43. Tensile Strength: 190,000 PSI thru 1/2" size, 180,000 PSI over 1/2" size. Yield Strength: 170,000 PSI thru 1/2" size, 155,000 PSI over 1/2" size.

2) Stainless Steel Socket Head Cap Screws: ASTM F837. Hardness: RC 33. Tensile Strength: 95,000 PSI. Yield Strength 30,000 PSI.

3) Alloy Steel Flat Head Socket Screws: ASTM F835. Hardness: RC 38-43. Tensile Strength: 160,000 PSI min.

4) Stainless Steel Flat Head Socket Screws: ASTM F879. Hardness: RC 33. Tensile Strength 90,000 PSI min.

Source, Unbrako/SPS Technologies, Inc.

Table 19. Recommended Seating Torques for Socket Button Head, Low Head Cap, Shoulder, and Set Screws. (See Notes.)

Thread Size Nom	Alloy Steel Button Head Cap Screws[1]		Stainless Steel Button Head Cap Screws[2]		Low Head Cap Screws[3]	Socket Head Shoulder Screws[4]	Alloy Steel Socket Set Screws[5]	Stainless Steel Socket Set Screws[6]
	UNRC	UNRF	UNRC	UNRF				
#0	-	1.5	-	1			1	0.4
#1	2.5	2.5	1.7	1.8			1.8	1.2
#2	4.5	4.5	2.8	3			1.8	1.2
#3	7	7	4.3	4.6			5	4
#4	8	8	6.0	6.6			5	4
#5	12	13	8.9	9.3			10	7
#6	15	17	11	12			10	7
#8	30	31	20	21	25		20	16
#10	40	45	30	34	35		36	26
1/4	100	110	71	81	80	45	87	70
5/16	200	220	123	136	157	112	165	130
3/8	350	400	218	247	278	230	290	230
7/16	560	625					430	340
1/2	850	1,000	532	600	226	388	620	500
9/16	1,200	1,360					620	500
5/8	1,700	1,900	1,060	1,200		990	1,325	980
3/4						1,975	2,400	1,700
7/8							3,600	3,000
1						3,490	5,000	4,000
1 1/8							7,200	5,600
1 1/4						5,610	9,600	7,700
1 3/8							9,600	7,700
1 1/2						12,000	11,300	9,100
1 3/4						16,000		
2						30,000		

Notes: Caution: Even when using the recommended seating torques in this Table, the induced loads obtained may vary by as much as ± 25%, depending on uncontrolled variables such as mating material, lubrication, surface finish, hardness, etc. Torque values are in inch pounds. To convert inch pounds (in.-lbs) to Newton meters (N-m), divide the values on this Table by 8.88. Seating torques given are for screws made to the specifications referenced, as follows.

1) Alloy Steel Button Head Socket Screws: ASTM F835. Hardness: RC 38-43. Tensile Strength: 160,000 PSI min.

2) Stainless Steel Button Head Socket Screws: ASTM F879. Hardness: RC 33. Tensile Strength: 90,000 PSI min.

3) Low Head Cap Screws: ASTM A574. Hardness: RC 38-43. Tensile Strength: 170,000 PSI min. Yield Strength: 150,000 PSI min.

4) For Socket Head Shoulder Screws, the nominal dimension given is for the shoulder diameter of the screw. ASTM A574. Hardness: RC 36-43. Tensile Strength: 160,000 PSI.

5) Alloy Steel Socket Set Screws: ASTM F912. Hardness: RC 45-53.

6) Stainless Steel Socket Set Screws: ASTM F880. Hardness: RC 33.

Source, Unbrako/SPS Technologies, Inc.

Table 20. Recommended Seating Torques for Metric Socket Head Screws. *(See Notes.)*

Thread Size[1]	Socket Head Cap Screws[2]		Flat Head Cap Screws[3]		Button Head Cap Screws[4]		Socket Head Shoulder Screws[5]		Low Head Cap Screws[6]		Socket Set Screws[7]	
Nom	N-m	in.-lbs	N-m	in.-lbs	N-m	in.-lbs	N-m	in.-lbs	N-m	in.-lbs	N-m	in.-lbs
M1.6	0.29	2.6									0.09	0.8
M2	0.60	5.3									0.21	1.8
M2.5	1.21	11									0.57	5
M3	2.1	19	1.2	11	1.2	11					0.92	8
M4	4.6	41	2.8	25	2.8	25			4.5	40	2.2	19
M5	9.5	85	5.5	50	5.5	50			8.5	75	4.0	35
M6	16	140	9.5	85	9.5	85	7	60	14.5	130	7.2	64
M8	39	350	24	210	24	210	12	105	35	310	17	150
M10	77	680	47	415	47	415	29	255	70	620	33	290
M12	135	1,200	82	725	82	725	57	500	120	1,060	54	480
M14	215	1,900					100	885				
M16	330	2,900	205	1,800	205	1,800	240	2,125	300	2,650	134	1,190
M20	650	5,700	400	3,550			470	4,160	575	5,100	237	2,100
M24	1,100	9,700	640	5,650							440	3,860
M30	2,250	19,900										
M36	3,850	34,100										
M42	6,270	19,900										
M48	8,560	75,800										

Notes: Caution: Even when using the recommended seating torques in this Table, the induced loads obtained may vary by as much as ± 25%, depending on uncontrolled variables such as mating material, lubrication, surface finish, hardness, etc. 1) All threads are Class 4g 6g. Seating torques given are for screws made to the specifications referenced, as follows. 2) Socket Head Cap Screws: ASTM A574M, DIN912 alloy steel. Hardness: RC 38-43. Tensile Stress: 1300 MPa thru M16 size, 1250 MPa over M16 size. Yield Stress: 1170 MPa thru M16 size, 1125 over M16 size. 3) Flat Head Cap Screws: ASTM F835M. Hardness: RC 38-43. Tensile Stress: 1040 MPa. Yield Stress: 945 MPa. 4) Button Head Cap Screws: ASTM F835M. Hardness: RC 38-43. Tensile Stress: 1040 MPa. Yield Stress: 945 MPa. 5) Socket Head Shoulder Screws: ASTM A574. Hardness: RC 36-43. Tensile Stress: 1100 MPa based on minimum thread neck area. Shear Stress: 660 MPa. 6) Low Head Cap Screws: ASTM A574M. Hardness: RC 33-39. Tensile Stress: 1040 MPa. Yield Stress: 940 MPa. 7) Socket Set Screws: ASTM F912M. Hardness: Hardness: RC 45-53. *Source, Unbrako/SPS Technologies, Inc.*

Dimensions for UNJ threads are included in the Screw Threads section of this book.

Bolt torque and axial tension relationship

Axial Load Torque (Figure 10)

In order to achieve a desired axial load, in a fastener, torque is applied. The torque must be sufficient to produce the axial load and also overcome friction in the threads and friction under the nut or bolt head (bearing surface). Axial load (P_B) is a component of the normal force developed between threads. The normal force component that is perpendicular to the thread's helix is the thread helix force ($P_N\lambda$), and the final component of this force is the torque load ($P_B \tan \lambda$). Assuming that the force is applied at the pitch diameter of the thread, the torque (T_1) required to develop the axial load is

$$T_1 = P_B \times \tan \lambda \times (d_2 \div 2).$$

As can be seen in *Figure 9*, $\tan \lambda = L \div (\pi \times d_2)$; therefore,

$$T_1 = (P_B \times L) \div 2\pi.$$

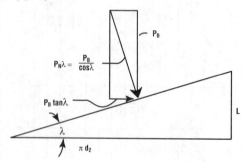

Figure 10. Thread helix forces.

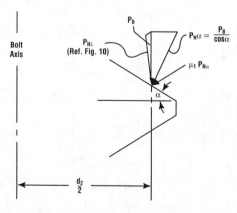

Figure 11. Thread friction force.

Thread Friction Torque (Figure 11)

As shown in the figure, the normal force component perpendicular to the thread flanks is $P_N\alpha$. With a coefficient of friction (μ_1) between the threads, the total friction load is equal to $\mu_1 \times P_N\alpha$, or $\mu_1 \times P_N\alpha \div \cos \alpha$. Assuming that the force is applied at the pitch diameter of the thread, the amount of torque required to overcome thread friction is

$$T_2 = (d_2 \times \mu_1 \times P_B) \times 2 \cos \alpha.$$

Nut or Bolt Underhead Friction Torque (Figure 12)

Given a coefficient of friction of μ_2 between the bolt head washer face and the joint, the friction load equals $\mu_2 \times P_B$. Assuming that the force is applied midway between the nominal diameter (d) and the washer face diameter (b), the torque (T_3) necessary to overcome the nut or bolt underhead friction can be calculated as

$$T_3 = ([d + b] \div 4) \times (\mu_2 \times P_B).$$

Torque-Tension Relation

Total torque (T) required to develop axial bolt load (P_B) is equal to the sum of T_1, T_2, and T_3 above.

$$T = P_B ([L \div 2\pi] + [d_2 \times \mu_1 \div 2 \cos \alpha] + [\{d + b \times \mu_2\} \div 4]).$$

For fasteners with 60° threads, $\mu = 30°$, and d_2 is approximately $0.92d$. If no loose washer is used under the rotated nut or bolt head, b = approximately $1.5d$.

$$T = P_B ([0.159 \times L] + [d \times \{\mu_1 + 0.625 \mu_2\}]).$$

If, in addition to these conditions, μ_1 is approximately equal to μ_2, then $\mu_1 = \mu$, then $T = P_B \times ([0.159 \times L] + [0.625 \times \mu d]).$

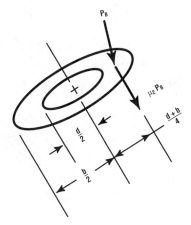

Figure 12. Nut or bolt head friction force.

Shear area

Minimum Material Condition

The geometric shear area of an internal thread at minimal material condition is equal to the area of that thread that is intersected by a cylinder with a diameter equal to the minimum major diameter of the mating external thread over the length of engagement (see *Figure 13*). Similarly, the geometric shear area of an external thread at minimum material condition is equal to the area of that thread that is intersected by a cylinder with a diameter equal to the maximum minor diameter of the mating internal thread (also shown in *Figure 13*).

Basic Size

The geometric shear area of an external thread at basic size is equal to the area of a basic size thread that is intersected by a cylinder with a diameter equal to the basic minor diameter. The geometric shear area of an internal thread at basic size is not normally used for calculation.

Formulas for Determining Shear Area

In order to determine stress areas with the formulas given below, it is first necessary to choose a length of engagement. Commonly, length of engagement, based on requirements, varies from a low of one-third of the major diameter of the internal thread to 1.5 times the major diameter of the internal thread. In these formulas, n = number of threads per inch; d is the minor diameter of the external thread, d_2 is the pitch diameter of the external thread; D_1 is the minor diameter of the internal thread; D_2 is the pitch diameter of the internal

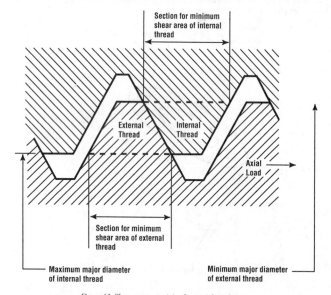

Figure 13. Shear areas at minimal material condition.

thread; Td, Td_2, TD_1, and TD_2 are the tolerances on d, d_2, D_1, and D_2, respectively; es is external thread allowance; AS_n is the shear area of the internal thread; AS_s is the shear area of the external thread; and LE is the length of engagement.

Shear area of internal threads when both external and internal threads are at minimal material condition:

$$AS_n \text{ (min)} = \pi \div n \times LE \times d_{min} \times ([1 \div 2n] + 0.57735 \times [d_{min} - D_{2\,max}]), \text{ or}$$
$$AS_n \text{ (min)} = \pi \times d_{min} \times (0.875 - 0.57735 \times n \times [Td + TD_2 + es] \times LE.$$

Shear area of internal thread (simplified), when d is 0.25" or greater:

$$AS_n = \pi \times D_{2\,basic} \times (3 \times LE \div 4).$$

Shear area of external threads when both external and internal threads are at minimal material condition:

$$AS_s \text{ (min)} = \pi \times n \times LE \times D_{1\,max} \times ([1 \div 2n + 0.57735] \times [d_{2\,min} - D_{1\,max}]), \text{ or}$$
$$AS_s \text{ (min)} = \pi \times D_{1\,max} ([0.75 - 0.57735 \times n] \times (TD_1 + Td_2 + es] \times LE.$$

Shear area of external thread (simplified):

$$AS_s = \pi \times d_{2\,basic} \times (^5/_8 \times LE).$$

Shear area of external threads when both external and internal threads are at basic size:

$$AS_s \text{ (max)} = \pi \times D_{1\,basic} \times (^3/_4 \times LE).$$

Shear area of combined failure:

$$AS = \pi \times D_{2\,basic} \times (LE \div 2).$$

The shear stress area ratio (R_1) can be found by $AS_{s\,max}$ to $AS_{n\,min}$.

Length of engagement

The length of engagement for UNC, UNF, and 8UN series threads, upon which their specified tolerances are based, is equal to basic major diameter. These tolerances are applicable for lengths of engagement for these threads of 1.0 to 1.5 times the basic major diameter. The length of engagement for UNEF, 12UN, and 16UN series threads, upon which their specified tolerances are based, is equal to 9 pitches. These tolerances are applicable for lengths of engagement for these threads of 5 to 15 pitches. For miniature thread series, all tolerances governing limits of size are based on pitch only and apply to lengths of engagement from 0.67 to 1.5 times the nominal diameter.

When mating internal and external threads are on parts manufactured from materials of approximately equal strength, failure will usually take place in both threads simultaneously. However, it is generally more economical if the externally threaded part should break, rather than either the internal or external part strip. In other words, the length of thread engagement (LE) should be sufficient to develop the full strength of the screw. Thus, the length of the internal thread and its dimensions (particularly its minor diameter) should be such that, taking into account a possible difference in strength of material of the internal and external threads, the externally threaded part will break before either the external or internal threads strip. Due to this situation, formulas for length of engagement are derived from shear formulas with tensile stress area (A_s) replaced by $2A_s$ because the required area in shear is twice the tensile stress area in order to fully develop the full strength of the externally threaded part (and $4A_s$ when calculating the combined full strength of both the external and internal threads). This relationship is based on experiments made by the National Bureau of Standards, in which it was found that for hot-rolled and cold-rolled steel, and brass screws and nuts, this factor varied from 1.7 to 2.0. The effect of combined stress is not often taken into consideration in calculating length of engagement because the added shear load affects both the tensile and shear stresses in approximately the same proportion.

To find length of engagement, based on shear of the external thread, requires tensile stress area and shear stress area values derived from formulas introduced above.

$$LE = 2 A_s \div AS_{s\ min}.$$

Length of engagement based on developing full tensile stress of the external thread with threads at basic size:

$$LE_{max} = 2A_s \div AS_{s\ max}.$$

Length of engagement based on combined shear failure of external and internal threads (assumes combination stress failure of both threads):

$$LE = 4A_s\ \pi \times d_{2\ basic}.$$

For a hollow part, subtract $0.7854d_h^2$ from A_s in the above formulas. d_h is the hole diameter on the externally threaded part.

Length of engagement based on shear of the internal thread is equal to the shear area stress ratio (R_1 – see above) and the ultimate material strength ratio (R_2), which is found by dividing the ultimate tensile strength of the internally threaded part by the ultimate tensile strength of the externally threaded part.

$$LE_{max} = R_1 \div R_2.$$

The above formulas for length of engagement yield approximate values because they are based in part on shear stress areas that are inexact due to nut dilation that varies with geometry, friction forces, and material properties.

Stripping strength of tapped holes

The Unbrako division of SPS Technologies has done research into obtaining minimum thread engagement based on applied load, material, type of thread, and bolt diameter. The results are reported in this section and shown graphically in *Figures 14 through 19*.

Knowledge of the thread stripping strength of tapped holes is necessary to develop full tensile strength of a bolt or, for that matter, the minimum engagement needed for any lesser load. Conversely, if only limited length of engagement is available, the data can help determine the maximum load that can be safely applied without stripping the threads of the tapped hole.

Attempts to compute lengths of engagement and related factors by formula have not been entirely satisfactory—mainly because of subtle differences between various materials. Therefore, strength data was empirically developed from a series of tensile tests of tapped specimens for seven commonly used metals including steel, aluminum, brass, and cast iron.

This design data is summarized in the charts shown in Figures 14 through 19, and covers a range of screw thread sizes from #0 to one-inch in diameter for both coarse and fine threads. Though developed from tests of Unbrako socket head cap screws having minimum tensile strength (depending on their diameter) from 190,000 to 180,000 PSI, these stripping strength values are valid for all other screws or bolts of equal or lower strength having a standard thread form. Data are based on static loading only.

In the test program, bolts threaded into tapped specimens of the metal under study were stressed in tension until the threads stripped. Load at which stripping occurred and the length of engagement of the specimen were noted. Conditions of the tests, all of which are met in a majority of industrial bolt applications, were as follows.

Tapped holes had a basic thread depth within the range of 65 to 80 percent. Threads of tapped holes were Class 2B or better.

Minimum amount of metal surrounding the tapped hole was 2.5 times the major diameter.

Test loads were applied slowly in tension to screws having standard Class 3A threads. (Data, though, will be equally applicable to Class 2A external threads as well.

Study of the test results revealed certain factors that greatly simplified the compilation of thread stripping strength data.

Stripping strengths are almost identical for loads applied either by pure tension or by screw torsion. These data are, therefore, equally valid for either condition of application.

Stripping strength values vary with diameter of screw. For a given load and material, larger diameter bolts require greater engagement.

Minimum length of engagement (as a percent of screw diameter) is a straight line function of load. This permits easy interpolation of test data for any intermediate load condition.

When engagement is plotted as a percentage of bolt diameter, it is apparent that stripping strengths for a wide range of screw sizes are close enough to be grouped in a single curve. Thus, in the accompanying charts, data for size #0 through #12 have been represented by a single set of curves.

With this data, it becomes a simple matter to determine stripping strengths and lengths of engagement for any condition of application, as can be seen from the four examples that follow.

Example 1. Calculate length of thread engagement to develop the minimum ultimate tensile strength (190,000 PSI) of a $^1/_2$-13 (National Coarse) Unbrako cap screw in cast iron having an ultimate shear strength of 30,000 PSI. *Figure 14* is for screw sizes from #0 through #10. *Figures 15 and 16* are for sizes from $^1/_4$ inch through $^5/_8$ inch. *Figures 17 and 18* are for sizes $^3/_4$ inch through 1 inch, and *Figure 19* is for screws over one inch in diameter. For this example, we can see from Figure 15 that a value of 1.4D is reached at the spot where the Cast Iron line intersects the ultimate tensile strength value of the screw. Multiplying the nominal bolt diameter (0.500 inch) by 1.4 gives a minimum length of engagement of 0.725 inch.

Example 2. Calculate the length of engagement for the above conditions if only 140,000 PSI is to be applied (which is the equivalent of using a bolt with a maximum tensile strength of 140,000 PSI). From Figure 15 obtain the value, which is 1.06D. Minimum length of engagement is therefore 1.06 × 0.5 = 0.530 inch.

Example 3. Suppose in Example 1 that minimum length of engagement to develop full tensile strength was not available because the thickness of the metal allowed a tapped hole only 0.600 inch deep. Hole depth in terms of bolt diameter equals 0.600 ÷ 0.500 = 1.20D. By working backwards in Figure 15, maximum hole load that can be carried is approximately 159,000 PSI.

Example 4. Suppose that the hole in Example 1 is in steel having an ultimate tensile shear strength of 65,000 PSI. Although there is no curve for this steel in Figure 15, a design value can be obtained by taking a point midway between the curves for the 80,000 PSI and 50,000 PSI steels that are listed. Under the conditions of the example, a length of engagement of either 0.825D or 0.413 inch will be obtained.

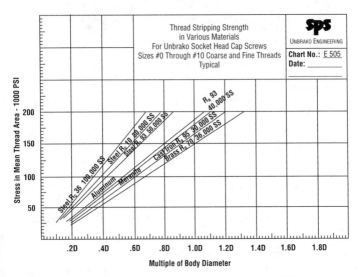

Figure 14. Thread stripping strength for nominal sizes #0 through #10.

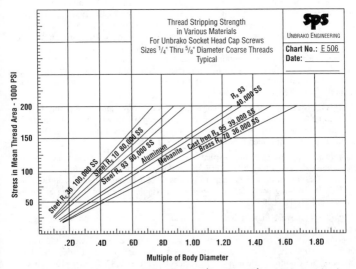

Figure 15. Thread stripping strength for nominal sizes $1/4$ inch through $5/8$ inch with coarse threads.

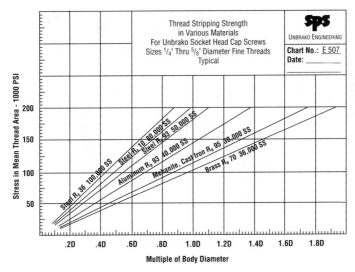

Figure 16. Thread stripping strength for nominal sizes $^1/_4$ inch through $^5/_8$ inch with fine threads.

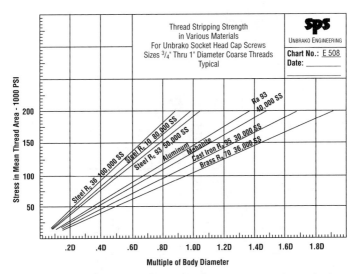

Figure 17. Thread stripping strength for nominal sizes $^3/_4$ inch through 1 inch with coarse threads.

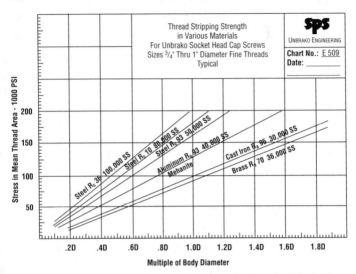

Figure 18. Thread stripping strength for nominal sizes ³/₄ inch through 1 inch with fine threads.

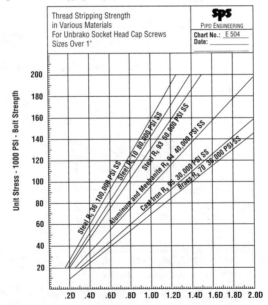

Figure 19. Thread stripping strength for nominal sizes over 1 inch.

Plating and coating threaded parts

Because they can change the diameters of a fastener, platings or coatings must be applied under strict quality regulations. The thickness of a coating can severely impact the torque and preload requirements, so special care should be taken to assure that coated fasteners remain with diameter limits. Plating and coating thickness can vary from as little as 0.00003" for gold flash to as much as 0.0005" for solder plate over nickel plate. However, most finishes range in thickness from 0.0002" to 0.0003". MIL-S-7742D provides the following guidelines for dimensioning threads for coating/plating.

Material Limits for Coated/Plated External Threads

When externally threaded parts are to be coated/plated, the threads should not be undercut more than 0.001" on all threads for which the specified pitch diameter tolerance does not exceed 0.0035". For threaded parts with a pitch diameter tolerance in excess of 0.0035", the specified minimum pitch diameter may be reduced by an amount equal to 0.3 times the pitch diameter tolerance, up to a maximum of 0.0015". To determine before coating/plating gaging limits for a uniformly coated/plated external thread, make the following reductions. Reduce the tabulated maximum pitch diameter by an amount equal to four times the maximum plating/coating thickness. Reduce the tabulated minimum pitch diameter, minimum major diameter, and minimum minor diameter by an amount equal to two times the minimum coating/plating thickness. Reduce the tabulated maximum major diameter and maximum minor diameter by an amount equal to two times the maximum coating/plating thickness. All thread elements must be in tolerance, as modified above, before coating/plating. Unless otherwise specified, coated Class 2A threads should meet the dimensional limits for Class 3A threads.

The following formulas, where d_2 is pitch diameter and t_c is coating thickness, can be used as a quick guide for predicting alteration in pitch diameter caused by coating/plating.

$$d_2 \text{ max before plating} = d_2 \text{ max after plating} - (t_c \text{ max} \times 4)$$
$$d_2 \text{ min before plating} = d_2 \text{ min after plating} - (t_c \text{ min} \times 4).$$

Material Limits for Coated/Plated Internal Threads

Internal threads may be overcut to allow for coating/plating thickness. Unless otherwise specified, before plating/coating gaging limits for a uniformly coated/plated internal thread should be increased as follows. Increase the tabulated minimum pitch diameter by an amount equal to four times the maximum coating/plating thickness. Increase the tabulated minimum major diameter and the minimum minor diameter by an amount equal to two times the maximum coating/plating thickness. Increase the tabulated maximum pitch diameter, maximum major diameter, and maximum minor diameter by an amount two times the minimum coating/plating thickness. All thread elements must be in tolerance, as modified above, before coating/plating.

Military specifications and requirements for fasteners

Mil-B-6812E *Military Specification, Bolts, Aircraft* provides stringent tensile strength requirements for bolts. As defined by the standard, the test specifies minimum ultimate tensile strength and minimum double shear strength for noncorrosion resistant steel bolts, corrosion resistant steel bolts, and aluminum alloy bolts. These strength requirements are shown in **Table 21**.

An important consideration for many fastener applications, especially the aerospace industry, is the strength to weight ratio of the fastener material. Strength to weight ratio is defined as the ratio of tensile strength to density. In applications where weight is an important

Table 21. Strength Requirements for Aircraft Bolts. *(Source, MIL-B-6812E.)*

Size	Ultimate Tensile Strength[1] (min. in pounds)				Double Shear Strength[2] (min. in pounds)	
	Steel			Al Alloy	Steel	Al Alloy
	Eyebolt	Hexagon Head Bolts			All Types of Bolts	
		Fine Thread[3]	Coarse Thread[3]	Fine Thread		
No. 6	-	-	-	-	2,120	1,080
No. 8	-	-	-	-	3,000	1,570
No. 10	1,150	2,210	1,800	-	4,250	2,092
1/4	2,450	4,080	3,360	1,100	7,360	3,650
5/16	3,910 (AN44)	6,500	5,660	2,030	11,500	5,700
5/16[4]	5,290 (AN45)	6,500	5,660	3,220	11,500	5,700
3/8	7,015	10,100	8,470	3,220	16,560	8,250
7/16	9,200	13,600	11,680	5,020	22,500	11,200
1/2	14,375	18,500	15,730	6,750	29,400	14,600
9/16	20,125	23,600	20,300	9,180	37,400	18,500
5/8	-	30,100	25,100	11,700	46,000	22,800
3/4	-	44,000	37,800	14,900	66,300	33,000
7/8	-	60,000	-	21,800	90,100	45,000
1	-	80,700	-	29,800	117,800	58,500
1-1/8	-	101,800	-	40,000	147,500	74,000
1-1/4	-	130,200	-	50,500	182,100	91,000

1) The values shown for the ultimate tensile strength are for minimum values and are based on:
    a. 125,000 PSI for noncorrosion-resistant and corrosion-resistant steel
    b. 62,000 PSI for aluminum alloy.
    The strength values shown for the eyebolts are based on the strength of the eye. The root area of the thread is the basis of calculation for the tensile strength of hexagon head bolts.
    Clevis bolts shall have tensile strengths equal to one-half of the requirements for hexagon-head bolts when used with AN320 or MS21083 nuts. Clevis bolts are intended primarily for use in shear applications.
2) Ultimate shear strengths are computed on the basis of 60 percent of the ultimate tensile strengths.
3) Class of thread is as specified on the applicable standard drawing.
4) Different from size 5/16 above in the design of the eye section.

consideration, the strength to weight ratio becomes a factor (see **Table 22**).

All standard studs, bolts, hex cap screws, socket head screws, and nuts in diameters of 1/4 inch and larger, "for critical applications where a high degree of reliability is required," that are procured by any branch of the military are required to meet the material specifications laid out in MIL-S-1222H *Military Specification, Studs, Bolts, Hex Cap Screws, Socket Head Cap Screws, and Nuts.* The grade, chemical, and identification requirements from that standard are shown in **Table 23**.

Washers

(The published informational catalogs of West Coast Lockwasher (WCL) Company were the source of much of the material in this section.)

Table 22. Strength to Weight Ratios for Various Fastener Materials. *(Source, ITT Harper.)*

Material	Tensile Strength ksi	Density lb./in.^{3}	Strength to Weight Ratio
Martensitic Stainless Steel (410, 416)	180	0.280	6.4
Aluminum (2002-T4)	60	0.098	6.1
Austenitic Stainless Steel (18-8) Strain Hardened	125	0.290	4.3
Titanium (Commercially Pure)	50	0.163	3.1
Nylon	12	0.041	2.9
Austenitic Stainless Steel (18-8) Annealed	80	0.290	2.8
Monel 400	80	0.319	2.5
Silicon Bronze	75	0.308	2.4
Brass	60	0.308	2.0
Mild Steel	50	0.282	1.8

Table 23. Material Grades, Chemical, and Identification Requirements for Fasteners. *(Source, MIL-S-1222H.)*

Material Type		Material Grade	Fastener Type	Chemical Requirement	I.D. Marking
Carbon and Alloy Steel		2	Stud, bolt, hex cap screw	SAE J 429	See Note 1
		5	Stud, bolt, hex cap screw	SAE J 429	See Note 2
		8	Stud, bolt, hex cap screw	SAE J 429	See Note 2
		2	Nut	SAE J 995	See Note 1
		5	Nut	SAE J 995	See Note 2
		8	Nut	SAE J 995	See Note 2
Alloy Steels		B7 See Note 5	Stud, bolt, hex cap screw	ASTM A 193	B7
		B16	Stud, bolt, hex cap screw	ASTM A 193	B16
		A574	Socket head cap screw	ASTM A 574	See Note 1
		4340	Socket head cap screw	ASTM A 574	4340
		L7	Stud, bolt, hex cap screw	ASTM A 320	L7
		L43	Stud, bolt, hex cap screw	ASTM A 320	L43
		2H	Nut	ASTM A 194	2H
		4	Nut	ASTM A 194	4
		7 See Note 9	Nut	ASTM A 194	7
Corrosion Resistant Steels	Austenitic	303	Stud, bolt, hex cap screw, socket head cap screw	ASTM A 194	303 See Notes 3 & 4
		303 Se See Note 6	Stud, bolt, hex cap screw, socket head cap screw	ASTM F 593	303 Se See Notes 3 & 4
		304	Stud, bolt, hex cap screw, socket head cap screw	ASTM F 593	304 See Notes 3 & 4
		305	Stud, bolt, hex cap screw, socket head cap screw	ASTM F 593	305 See Notes 3 & 4
		316	Stud, bolt, hex cap screw, socket head cap screw	ASTM F 593	316 See Notes 3 & 4

(Continued)

Table 23. *(Continued)* Material Grades, Chemical, and Identification Requirements for Fasteners.
(Source, MIL-S-1222H.)

Material Type		Material Grade	Fastener Type	Chemical Requirement	I.D. Marking
Corrosion Resistant Steels	Austenitic	321	Stud, bolt, hex cap screw, socket head cap screw	ASTM F 593	321 See Notes 3 & 4
		347	Stud, bolt, hex cap screw, socket head cap screw	ASTM F 593	347 See Notes 3 & 4
		384	Stud, bolt, hex cap screw, socket head cap screw	ASTM F 593	384 See Notes 3 & 4
	Martensitic	410 See Note 6	Stud, bolt, hex cap screw, socket head cap screw	ASTM F 593	410H, 410HT See Note 3
		416, 416 Se	Stud, bolt, hex cap screw, socket head cap screw	ASTM F 593	416H, 416HT, 416seH, 416SeHT See Note 3
		431	Stud, bolt, hex cap screw, socket head cap screw	ASTM F 593	431H, 431HT See Note 3
	Age Hardened	630	Stud, bolt, hex cap screw, socket head cap screw	ASTM F 593	630 See Note 3
	Austenitic	303	Nut	ASTM F 594	303 See Notes 3 & 4
		303 Se See Note 6	Nut	ASTM F 594	303Se See Notes 3 & 4
		304	Nut	ASTM F 594	304 See Notes 3 & 4
		305	Nut	ASTM F 594	305, See Notes 3 & 4
		384	Nut	ASTM F 594	384 See Notes 3 & 4
		316	Nut	ASTM F 594	316 See Notes 3 & 4
		321	Nut	ASTM F 594	321 See Notes 3 & 4
		347	Nut	ASTM F 594	347 See Notes 3 & 4
	Martensitic	410 See Note 6	Nut	ASTM F 594	410H, 410HT See Note 3
		416	Nut	ASTM F 594	416H, 416HT See Note 3
		416 Se	Nut	ASTM F 594	416seH, 416seHT See Note 3
		431	Nut	ASTM F 594	431H, 431HT See Note 3
	Age Hardened	630	Nut	ASTM F 594	630, See Note 3
Nickel Alloys	Ni-Cu	400	Stud, bolt, hex cap screw, socket head cap screw	QQ-N-281, class A	NC
	Ni-Cu-Al	500	Stud, bolt, hex cap screw, socket head cap screw	QQ-N-286, class A	K, See Note 8

(Continued)

Table 23. *(Continued)* **Material Grades, Chemical, and Identification Requirements for Fasteners.**
(Source, MIL-S-1222H.)

Material Type		Material Grade	Fastener Type	Chemical Requirement	I.D. Marking
Nickel Alloys	Ni-Cu	400 See Note 7	Nut	QQ-N-281, class A	NC
		405	Nut	QQ-N-281, class A	NC-R
	Ni-Cu-Al	500	Nut	QQ-N-286, class A	K
Copper Alloys	N	462 See Note 6	Stud, bolt, hex cap screw	ASTM F 468	462, See Note 3
		464	Stud, bolt, hex cap screw	ASTM F 468	464, See Note 3
		482	Stud, bolt, hex cap screw	QQ-B-637	482, See Note 3
	P	510 See Note 6	Stud, bolt, hex cap screw	ASTM F 468	510, See Note 3
		544	Stud, bolt, hex cap screw	ASTM B 139	544, See Note 3
	Ni-Al Bronze	632	Stud, bolt, hex cap screw	QQ-C-465	632, See Note 3
	Silicon Bronze	655 See Note 6	Stud, bolt, hex cap screw	ASTM F 468	655, See Note 3
		661	Stud, bolt, hex cap screw	ASTM F 468	661, See Note 3
	M	670 See Note 6	Stud, bolt, hex cap screw	ASTM B 138	670, See Note 3
		675	Stud, bolt, hex cap screw	ASTM F 468	675, See Note 3
	N	462	Nut	ASTM F 467	462, See Note 3
		464	Nut	ASTM F 467	464, See Note 3
	P	510	Nut	ASTM F 467	510, See Note 3
	S	655	Nut	ASTM F 467	655, See Note 3
		661	Nut	ASTM F 467	661, See Note 3
	M	675	Nut	ASTM F 467	675, See Note 3
Titanium Alloy		T-7	Stud, bolt, hex cap screw	MIL-T-9047	T7
Aluminum Alloys		2024	Stud, bolt, hex cap screw	ASTM F 468	2024, See Note 3
		6061	Stud, bolt, hex cap screw	ASTM F 468	6061, See Note 3
		7075	Stud, bolt, hex cap screw	ASTM F 468	7075, See Note 3
		2024	Nut	ASTM F 467	2024, See Note 3
		6061	Nut	ASTM F 467	6061, See Note 3
		7075	Nut	ASTM F 467	7075, See Note 3

Notes:

1) Individual fastener marking to identify the material grade is not required.
2) Grade 5 shall be marked with three radial lines equally spaced for externally threaded fasteners and three circumferential dashes equally spaced for nuts. Grade 8 shall be marked with six radial lines equally spaced for externally threaded fasteners and six circumferential dashes equally spaced for nuts.
3) When specified each fastener shall be marked with the material grade.
4) In addition to the material grade, the marking shall include the heat treatment symbol "An" when the fastener is machined from annealed stock or when the fastener is annealed after forming, or the identification symbol "SH" when the fastener is cold headed and rolled from strain hardened stock thus acquiring a degree of cold work.
5) Suitable nuts for B7 externally threaded fasteners are grades 2H, 4, and 7, and suitable nuts for B16 externally threaded fasteners are grades 4 and 7.

(Continued)

Table 23. *(Continued)* **Material Grades, Chemical, and Identification Requirements for Fasteners.**
(Source, MIL-S-1222H.)

6) The following groups of material are considered equivalent and interchangeable. When approved by the contracting activity, the contractor may select any alloy within each group:
   a. 303,303 Se, 304, 305, 384
   b. 32L, 347
   c. 410, 416, 416 Se, 431
   d. 462, 464, 482
   e. 510, 544
   f. 655, 661
   g. 670, 675
7) The following groups of material are considered equivalent and interchangeable. When approved by the contracting activity, the contractor may select any alloy within each group:
   a. 400, 405
8) Heat or lot number and the manufacturer's symbol are required for fasteners $^1/_2$ inch in diameter and larger. Only the material symbol shall be required for fasteners less than $^1/_2$ inch in diameter. Numbers used for

Plain Flat Washers

Flat washers are intended to act as bearing surfaces that prevent bolt heads and nuts from damaging work surfaces. In addition, since their surface is known to be smooth and flat, they allow more accuracy in achieving installation torque or strain measurements. Several standards issuing organizations, in countries throughout the world, have specified preferred and alternative sizes for internal and external dimensions of flat washers. One manufacturer claims to keep over 7,000 different flat washers in inventory. In addition, it should be remembered that manufacturers of washers offer custom size washers in almost any conceivable dimension and material. For instance, one leading company offers sizes ranging from the smallest at 0.022" internal diameter and 0.040" outside diameter to the largest with dimensions of 2.515" internal diameter to 3.250" outside diameter. Each of the hundreds of available sizes is available in seventeen different thicknesses, eighteen different materials, and dozens of finishes. Typically these custom washers are available with the following tolerances: external diameter = +0.020" to –0.005"; internal diameter = ±0.010"; and thickness = ±0.003" for washers up to 0.032" thick, ±0.006" for washers from 0.033" to 0.062" thick, ±0.009" for washers from 0.063" to 0.093" thick; and ±0.010" for washers 0.094" thick and over. More exact tolerances are available on special order. Therefore, should a washer size be required that is not included in a printed standard of suggested or recommended sizes, that washer will almost certainly be available from one of several manufacturers or suppliers.

The inside diameter of conventional flat washers have three distinct profiles that are a result of the punch process that is used to produce the part. The punch, pushing through the washer, creates a rounded corner on the entry side. The center section has approximately parallel sides as the punch passes through, and the exit side is tapered resulting from punch breakout. The internal dimension of the washer is measured across the approximately parallel flats. The diameter of the breakout taper may exceed the specified internal diameter on the washer by an amount equal to 25% of the washer thickness (see *Figure 20*).

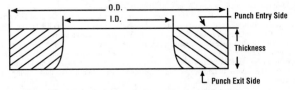

Figure 20. Profile view of a flat washer. (Source, West Coast Lockwasher Company.)

Helical Spring Lockwashers

Helical spring lockwashers, which are among the most widely used antiloosening assembly components available, have been specified in countless assembly applications because of their reliable performance. They enhance the security of general industrial assemblies by 1) providing greater bolt tension per unit of applied torque, 2) providing hardened bearing surfaces that create more uniform torque control, 3) allowing more uniform load distribution through controlled radii (section) cutoff, 4) providing protection against looseness resulting from vibration and corrosion, and 5) providing optimum locking performance in applications with hardened faying or bearing surfaces.

Helical spring lockwashers are slightly thicker on their inside diameter than on their outside diameter (see *Figure 21*). Therefore, the design thickness is, in theory, situated midway across the surface of the washer and may be determined with the formula

$$T = (t_o + t_i) \div 2$$

where T = specified thickness
t_o = outside thickness
t_i = inside thickness.

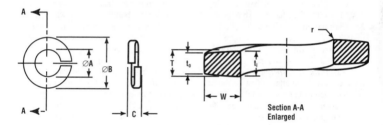

Figure 21. Sectional view of a helical spring lockwasher.

The "split" in the helical spring lockwasher absorbs initial driving torque and visually closes under nominal bolt loading. When tension in the assembly is reduced and loosening occurs, it provides resistance to the backoff rotation of the screw or nut. Helical washers are trapezoidal in section. After the single-coil spring closes to the flat condition, further loading results in additional deformation of the washer caused by a complex twisting of the trapezoidal section and a slight increase in the diameter of the washer under load (see *Figure 22*). The spring rate developed by the final deformation is very high and provides a reactive load that is equivalent to a significant increase in effective bolt length. Bolts stretch under load. The longer the effective length of a bolt, the more it can stretch. A

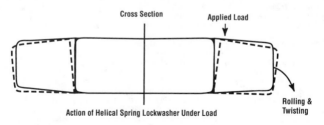

Action of Helical Spring Lockwasher Under Load

Figure 22. Action of a helical spring lockwasher under load. (Source, West Coast Lockwasher Company.)

hardened steel bolt, stressed at 60,000 PSI, will elongate approximately 0.002" per inch of effective length. Therefore, a long bolt can become a very effective spring that, like a spring when stretched, will attempt to return to its original length. This provides a clamping or tightening force to the assembly. Two terms are used to define and measure these forces: spring rate and effective bolt length.

Spring rate is the ratio of load to deflection in the loaded part. It is deflection related to load and is expressed in terms of the amount of load required to achieve specific levels of deflection. Optimum joint performance is achieved when the spring elements of the fastening system have a spring rate that is low enough to assure that any yielding of the joint members in compression will not significantly reduce the designed tensile stress of the fastener. To obtain this optimal condition, the conventional solution is to utilize the spring characteristics of long bolts, but an auxiliary spring element, such as a helical spring lockwasher, is an effective alternative.

Tests run at Lawrence Technological University have shown that the typical helical spring lockwasher exhibits a spring rate after flattening that is approximately 70% more effective than a flat washer of the same thickness. This means that the effective bolt length in the joint is increased by the thickness of the flattened helical spring lockwasher, plus the equivalent length provided by the spring rate derived from the visually flattened washer. Tested at 75% of the hardened bolt proof load, the equivalent bolt length is shown in **Table 24**. The total contribution of a helical spring lockwasher to the integrity of an assembly, in addition to the commonly recognized frictional resistance to backoff rotation, includes the reactive length added to the bolt by the washer thickness and by the tension of spring rate generated by its compression.

Testing helical spring lockwashers. In preparing a helical spring lockwasher for hardness testing, it should first be twisted to remove the helix so that it forms a near flat surface. Due to the trapezoidal section of the washer, both sides must be filed or ground flat to assure accurate readings (see *Figure 23*). Care must be taken to assure that surface temperature does not exceed 250° F (121° C) during this operation. Essential requirements

Table 24. Equivalent Bolt Length. *(Source, West Coast Lockwasher Company.)*

Washer Size (inches)	Additional Spring Rate Contributed by Compressed Washer	
	Regular Lockwasher	Heavy Lockwasher
3/8	65.6%	74.1%
7/16	68.6%	72.2%
1/2	73%	80.3%

Comparison of Washerless Joint to Joint with Lockwasher			
Joint without Washer		Joint with 3/8 inch Helical Spring Lockwasher	
Bolt head height	0.24"	Bolt head height	0.24"
Two assembled 1/8" plates	0.25"	Two assembled 1/8" plates	0.25"
Nut thickness	0.32"	Helical washer thickness	0.10"
		Nut thickness	0.32"
Total thickness	0.81"	**Total thickness**	0.91"
Half of head height (deflected)	0.12"	Half of head height (deflected)	0.12"
Total thickness of assembled plates	0.025"	Total thickness of assembled plates	0.025"
Half of nut thickness (deflected)	0.16"	Total thickness of washer	0.10"
		Additional spring rate contributed by compressed washer	0.656"
		Half of nut thickness (deflected)	0.16"
Equivalent bolt length	0.53"	**Equivalent bolt length**	1.286"

HARDNESS

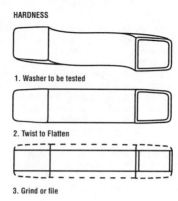

1. Washer to be tested

2. Twist to Flatten

3. Grind or file

Figure 23. Steps required for preparing a helical spring lockwasher for testing.
(Source, West Coast Lockwasher Company.)

of the Rockwell test are that the penetrator be perpendicular to the surface of the test piece, and that the test piece must not move, even in the slightest degree, when the test load is applied. Since one point of hardness represents a depth of only 0.00008", a movement of only 0.001" could cause an error of over 10 Rockwell numbers.

Penetration should be made in the center of the washer's flattened surface, and readings should not be taken too close together. If the indentation is made too close to the edge of the test washer, the material will yield, giving incorrect low readings. Also, since the area surrounding an indent will be cold worked, subsequent indentations made in these areas will give incorrect readings—usually higher than virgin material. The penetration test should be performed on only one side of the washer.

Applicable hardness requirements are as follows.

Material	HR_C	HV
Carbon Steel	38-46	372-458
Austenitic Stainless Steel	35-43	345-423
Monel K500	33-40	327-392

	HR_B	HV
Aluminum Alloy	75-97	137-222
Phosphor Bronze	90 min.	185 min.
Silicon Bronze	90 min.	185 min.

In addition, carbon steel spring washers, tested in accordance with ANSI B18.21.1, should meet the following limits for decarburization.

Dia. of Section of Washer	Max. Depth of Free Ferrite	Max. Total Affected Depth
Up to 0.140" incl.	0.002"	0.006"
0.140" to 0.250" incl.	0.003"	0.008"
0.250" to 0.375" incl.	0.004"	0.010"
0.375" to 0.500" incl.	0.006"	0.015"

Spring Washers

Spring washers are specifically designed to provide a compensating spring force and sustain a load or absorb a shock. Many design variations have evolved to best serve one or the other of these two basic functions or to optimize both functions in a single part within specific I.D./O.D. limits. Two principal factors continually increase the requirements for spring washers: 1) the continuing effort to downsize many end products, relative to both weight and cost, creates a need for small, multifunctional assembly components such as washers that support a load, span a hole, or both, while providing a compensating spring force; and 2) automated assembly requires some "play" or tolerance in the "fit" of components. Spring tension is needed to compensate for these tolerances.

Recognizing these two broad areas of influence, it can be stated that the more common applications for spring washers are 1) to take up "play" in assemblies due to cumulative tolerances, 2) to compensate for small dimensional changes in assembled components, 3) to eliminate end-play or rattles, 4) to maintain fastener tension, 5) to compensate for expansion and contraction or cold flow of material, and 6) to absorb intermittent shock loads and function as working springs capable of providing controlled reaction under dynamic loads.

Design considerations. Load and deflection are the key characteristics of a spring washer as it must be determined how much the washer will deflect under a given load and at what point it will flatten. These values are commonly expressed in Load/Deflection (L/D) curves with load (applied force) measured on one axis and washer deflection on the other. A typical L/D curve for a simple coil spring is shown in *Figure 24*. Point "A" on the curve represents a spring upon which no force has been applied and no deflection has taken place. The vertical axis (Load) is measured in increments of ten pounds of pressure, and the horizontal axis (Deflection) represents deflection in one-inch increments. It can be seen that with a load of 20 pounds, the spring deflects two inches (point "C"). At point "E", the spring is fully compressed. The graph represents a linear curve, or constant spring rate, where the sample spring deflects one inch for every increase of ten pounds applied to the spring. Most spring washers do not perform this consistently.

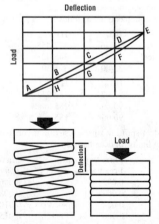

Figure 24. Load and deflection curve for a spring washer. (Source, West Coast Lockwasher Company.)

As the force is gradually released, it does not return to the full extension it had before deflection. This can be seen on the return at points "F," "G," and "H." At point "F," for example, at a deflection of three inches, the spring is supplying just under thirty pounds of reactive force. This results from the spring having used some of its stored energy to overcome friction caused by the bearing surface at both ends. Some stored energy is also lost through the increased temperature resulting from intermolecular friction caused by the initial deflection.

When considering washer design, the obvious relationship between washer thickness and load bearing characteristics should be kept in mind. Likewise, there is an inverse relationship between washer thickness and spring compensation or deflection. By varying either of these characteristics, the performance capabilities of a washer can be predicted by the designer.

Load/deflection characteristics. When specifying a spring washer, it is important to analyze both functional and physical requirements. In considering load requirements, two basic types (static and dynamic) must be recognized in addition to the amount of load. In a static load environment, the basic function of the spring washer is to retain load. In such an environment, the elastic load of the material may be exceeded. In a dynamic environment, the washer functions as a regularly flexing spring, and the elastic limits of the material must not be exceeded. Loading a washer beyond its yield strength will result in permanent distortion of its crown height. The type and magnitude of the load to which a spring washer will be subjected, and the reactive force it will be required to exert, are the primary factors that determine the type of spring washer best suited to a specific application. This range of deflection, or "spring travel," is an important element in spring washer design. Values for maximum load and maximum deflection for both single and multiple wave washers can be determined mathematically. The following equations, while useful as guidelines, can produce values that can vary up to 35% from actual test results. If load and deflection values are exceeded, a spring washer cannot perform within its elastic limits. Therefore, to avoid overstressing the washer, a general design rule when using these formulas is to select a washer that has twice the required deflection.

Estimating load (P) in pound-feet and deflection (f) in inches for single wave washers:

$$P = (S \times [D - d] \times t^2) \div 6 \times D$$
$$f = (S \times D^2) \div (6 \times [E \times t]).$$

Estimating load (P) in pound-feet and deflection (f) in inches for multiple wave washers:

$$P = (S \times N^2 \times t^2 \times [D - d]) \div (.75 \times [D + d])$$
$$f = (S \times 2D^2) \div (12\, E \times t \times N^2)$$

where: N = number of waves
P = load in pound-feet
E = modulus of elasticity in PSI (30,000,000 PSI for steel)
t = material thickness in inches
f = deflection in inches
d = inside diameter in inches
D = outside diameter in inches
S = max. allowable stress in PSI (200,000 PSI for steel).

Space envelope. Wave washers and conical washers increase in diameter as they are compressed. Therefore, allowable inside and outside diameter limits are an important consideration when specifying spring washers. The overall space occupied by a spring washer can be described as a hollow cylinder as shown in *Figure 25*. This cylinder of space

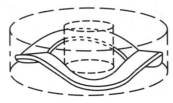

Figure 25. Space cylinder (envelope) for a spring washer. (Source, West Coast Lockwasher Company.)

must be recognized in assembly design considerations which may restrict the acceptable dimensions of the washer. Generally, the larger the washer outside diameter, the greater the load that can be supported and distributed.

Operational environment. The environment in which a spring washer operates can affect the anticipated performance of the washer in terms of load bearing and reactive characteristics. Temperature and exposure to corrosive agents are important environmental considerations, and washer materials specifications must respect these factors. The limitations imposed by even relatively low ambient temperatures on various spring washer materials are shown below. The indicated temperature limits can, of course, be exceeded, but this will result in increased relaxation of the washer under load.

Material	Recommended Operating Limit (°F)
SAE 1050 Steel	250°
SAE 1065 Steel	250°
425 Phosphor Bronze	225°
Beryllium Copper	225°
Spring Brass	150°

Belleville Conical Disc Spring Washers

Disc springs are among the most widely used tension generating washers. Common uses include spanning alignment holes, distributing bearing loads, and generating and sustaining the tension needed to hold assemblies together. Belleville disc spring washers (see *Figure 26*) combine high spring force for short movement with high-energy storage capacity. In true Belleville washers, the ratio of material thickness to rim width is held to about one in five, and crown height should not exceed 40% of the material thickness to assure that, when loaded to flat, yield strength is not exceeded and the washer will return to full crown height when compression is removed. However, commercial springs are not held to specific outside diameter/crown height/thickness ratio, and the crown height to thickness ratio is traditionally much greater for commercial disc springs. Therefore, when loaded to flat, their yield point may be exceeded. These washers are often used in applications where they function entirely within their elastic range. When loaded beyond their yield point, they will act in a consistent manner over a reduced crown height although they have, in effect, become deformed.

By carrying the thickness/crown height relationship, it is possible to meet a wide range of load/deflection requirements with disc spring washers. A crown height to thickness ratio ranging from 0.4 to 0.8, for example, produces a fairly constant spring rate (*Figure 27a*), while increasing the thickness ratio up to 1.4 will provide a positive rate of increase in the load up to 100% deflection (*Figure 27b*). Once a ratio of 1.4 is reached, the disc spring will show a constant load over a fairly large deflection, making it useful in applications where

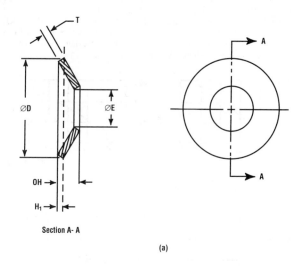

(a)

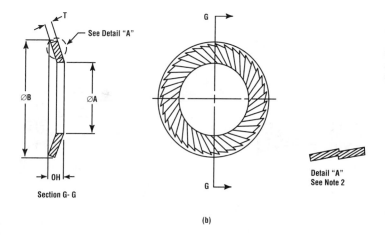

(b)

Figure 26. (a) Belleville spring washer in cross section. (b) Serrated Belleville spring washer. Serrated washers have sawtooth surfaces top and bottom with a rake angle of 30° +5°.

extreme wear conditions must be absorbed (*Figure 27c*). Finally, when the crown height to thickness ratio exceeds 1.4, yielding will occur (*Figure 27d*).

Disc springs can be stacked to enhance performance characteristics. When stacked in parallel (*Figure 28a*), load bearing characteristics are enhanced. When stacked in series (*Figure 28b*), greater travel can be achieved. Combination stacking, in both parallel and series, can be used to increase both load bearing and deflection.

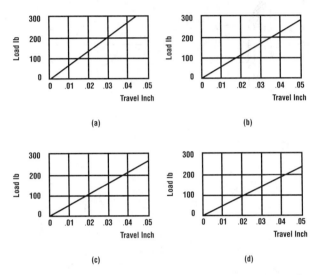

Figure 27. Effect of thickness/crown height ratio on disc springs (see text).
(Source, West Coast Lockwasher Company.)

Figure 28. (a) Belleville washers stacked in parallel. (b) Belleville washers stacked in series.

Abrasive Machining

The grinding process

Grinding is a process that removes material by the application of abrasives that are bonded to form a rotating tool, usually in the form of a wheel. As the rotating wheel contacts a workpiece, the abrasive particles act as tiny cutting tools—each removes a small chip as it passes over the workpiece. The Grinding Wheel Institute describes the grinding process as follows. Abrasive grains are miniature cutting tools with millions of teeth biting into the work and removing metal. During the grinding process, dull grains are torn from the bond and new, sharp grains take their places. During grinding, the abrasive is subjected to two types of wear that take place simultaneously: mechanical and chemical. The mechanical process is the mechanism by which the wheel becomes self-sharpening because the grains have been torn away and replaced by fresh grains. The other is chemical. At the high speed of grinding, the temperature at the point of contact between the abrasive grain and the workpiece is extremely high and chemical reaction between the abrasive and workpiece, or the surrounding atmosphere, can occur rapidly. The heat generated by dull abrasives weakens the bond, and the abrasives are torn loose. The dulling of the grain necessitates wheel dressing; also, the abrasive cutting edges dull like any cutting tool.

IMPORTANT NOTE: All grinding operations should follow the guidelines provided by ANSI B7.1 "Safety Requirements for the Use, Care and Protection of Abrasive Wheels." All grinding wheels and other abrasive machining tools should be operated within their intended limits, as stated by the manufacturer.

Principles of grinding

All grinding processes use many cutting edges (randomly located) to remove material. The chips formed in grinding are very small relative to turning or milling operations. They are almost microscopic. Taking just small amounts of material from the workpiece is how grinding can maintain generally superior surface finishes and closer tolerances than other chip making processes. Single point cutters are less effective in creating very small chips. This section, the result of an interview with Dr. K. Subramanian, Director of the World Grinding Center at Norton Company, is based on an article that appeared in *Modern Machine Shop*. It investigates the grinding cutting zone to identify what's happening when an abrasive cutter contacts a workpiece. Understanding this microscopic interaction is key to understanding the grinding process.

In the zone

A grinding cutting tool is a combination of an abrasive grain and the bond that holds it. It is analogous to a single point cutter and toolholder. There are three fundamental interactions that occur between an abrasive tool and the workpiece. These are cutting, plowing, and sliding. Because there are so many cutting tools at work in a grinding operation, all of these interactions occur to some degree at all times during the process. Grinding process control is an effort to balance these interactions.

Cutting a clean chip is interaction number one. Ideally, the abrasive grain is sufficiently exposed to penetrate the workpiece material and curl a good chip. In this condition, there is sufficient clearance between the grain, bond, and workpiece for the chip to escape by being flushed out with coolant or thrown away from the cutting zone by wheel action. Process parameters that contribute to this condition include a proper choice of abrasive and work combination reflecting the right size and shape of abrasive grains coupled with correct settings of feed, speed, and depth to enable the cut chip to evacuate the cutting zone.

Due to the dynamic nature of the grinding process, other interactions also occur in the cut zone. Plowing of the workpiece material can occur in the cut zone. In this cutting interaction, the abrasive grain is unable to get sufficient penetration into the material to lift a chip. Instead, it pushes the material ahead of the abrasive edge, in effect, plowing it.

The third cutting zone interaction is sliding. This condition can occur for several reasons. A lack of cut depth can cause the abrasive grain to slide across the workpiece surface. Insufficient clearance between the abrasive grit and the workpiece can trap the chip, causing it to slide on either the grinding wheel or the workpiece. In cases where a grit stays bonded to the wheel too long, and it doesn't break off soon enough, the binder can come into contact with the workpiece and create slide marks on the surface.

Every grinding operation is an effort to balance these cutting zone interactions. They occur in every grinding process including centerless, internal, cylindrical, and even snag grinding.

Understanding grinding

Understanding the basic cutting interactions gives shops a vocabulary to discuss different grinding processes. The three cut-zone interactions are involved, to differing degrees, in all grinding processes. The degree to which each of the three interactions occurs determines the kind of grinding being done. Commercial grinding processes break down into three general categories. These are rough grinding, precision grinding, and high or ultra-precision grinding.

The differentiating factor for each of these categories is the amount of metal removed. Metal removal is balanced against the desired tolerance or finish. In grinding, like turning and milling, high metal removal rates are generally in inverse proportion to close tolerances. Which is why shops use roughing and finishing passes.

In rough grinding, the desired workpiece/wheel interaction is focused on cutting. In these applications, maximum metal removal is the goal. Cutting off billets, snagging gates and risers from castings, or grinding weld beads smooth are all processes where the maximum amount of metal removal is the goal. Precise control of the size and surface finish is a secondary consideration. To create size and surface finish control for high metal removal—the precision grinding application—roughing passes are generally followed by finish passes. During rough grinding operations, stress is often loaded into the machine tool structure.

To alleviate this load, a spark-out pass is often programmed. This final pass is designed to take out the built-up stress in the machine tool structure and impart a good surface finish and size tolerance to the workpiece. A spark-out pass often results in some plowing of the workpiece material, because the depth of cut is shallow. An ideal spark-out pass occurs when the abrasive grain completes the chip making on the workpiece coincidental with the machine tool's return to a relaxed state from being stress loaded.

Precision grinding applications combine high metal removal with good part size control. Creep feed grinding, high efficiency deep grinding, and slot grinding are processes developed to increase the chip making interaction during the grind, while delivering very good tolerance and part geometry.

In ultra-precision grinding operations, little or no actual cutting is done. Instead, the workpiece surface is in effect rubbed clean primarily by sliding action from very fine abrasive grains. Ultra-precision grinding is the surface finishing of a very precisely sized workpiece. Most surface finishing processes generally fall into this category, including lapping and polishing.

Other variables

Interaction between the workpiece and the abrasive cutter defines the grinding process. Moving away from the cutting zone brings other influences into the process. How successfully a shop can control what happens in the grind is dependent on hundreds of other variables that can be categorized into just four process inputs: the machine tool, workpiece material, wheel selection, and operational factors.

Machine tool factors. A machine's design deals specifically with its rigidity, precision, and dynamic stability. A machine that cannot mechanically hold the close relationship between a grinding wheel and the workpiece is less likely to deliver consistent, predictable results. Other machine tool factors that influence the grind are machine controls, power/speed capability, slide movements, truing, and dressing mechanisms. These control how closely the grinding wheel and workpiece can be positioned on a repeatable basis. Leaving the workzone to true the grinding wheel and then returning precisely is an important machine tool factor.

The capability of the machine tool's coolant delivery system is also a factor in the successful grind by maintaining interface temperature, providing lubricity, and flushing of chips. Pump capacity, coolant type, pressure ranges, the filtration system, and coolant flow all play a large part in the cutting zone.

Work material factors. Defining those aspects of the workpiece that influence the grind is especially critical to successful cutting. Workpiece characteristics include the material's mechanical properties, its machinability, thermal stability, abrasion resistance, microstructure, and chemical resistance. For example, steel can't be cut with diamond when the abrasive/workpiece contact pressure is high because of the chemical reaction between the carbon in both the abrasive and the workpiece. However, for lower pressure grinding operations such as honing, steel and diamond can be used together.

Geometry, or determining how well the grinding wheel and workpiece conform to each other, is another work material factor for grinding. Workpiece shape can restrict coolant flow to the workzone, and sharp edges and very tight radii are special considerations that should be included in the analysis of the workpiece.

Part quality requirements are critical factors to the successful grinding operation. These include feature tolerances and surface finish requirements. The consistency of these tolerances and finishes gage the success of the application process.

Wheel selection factors. Obviously, selecting the abrasive is an important factor in grinding. Grain types, properties, size, distribution, and concentration should be identified.

The "toolholder" for these grains (bond) is an equally important factor. Bond characteristics can be classified by their type, hardness, stiffness, porosity, and thermal conductivity. How the grinding wheel is constructed influences the cut. Shape and size of the wheel should be matched to the workpiece geometry. Core material of the wheel and its form or profile need to coexist with the other selection factors.

Operational factors. These factors include fixturing, wheel balancing, frequency of truing and dressing, application of coolant, and whether the part is gaged in-process or off-line.

To achieve maximum influence on the grinding process, the grinding process should be considered as a whole. It's more effective to look at all of the factors (machine tool, work material, wheel selection, and operation) in terms of their contribution to the grinding system. Understanding how each of these factors influences the cutting zone helps predict behavior within the grind. This approach permits a shop to focus on a few critical factors in each of the four categories instead of trying to deal with a large number of variables on a hit-or-miss basis.

Evaluating the process

Because the cutting zone interactions between grinding grains and workpieces occur on a microscopic level, monitoring what's happening in the cut must be done indirectly. Each of the three cutter/workpiece interactions—cutting, plowing, and sliding—generates a signature that can be measured and monitored. Based on the desired workpiece finish requirements, the optimum combination of cutting, plowing, and sliding that produces those results becomes a baseline measurement. Deviation from that base can be measured and monitored using force, power consumption on the machine tool, and, on occasion, temperature measurements. Wheel wear, patterns, workpiece geometry, finish, and surface quality also serve as evidence about the interactions taking place in the cutting zone.

In application, using the four categories of machine, material, wheel, and operational factors, a shop can set up an initial part run that fits within the known parameters of all four categories. Changing one or more categories will produce results either positive to the process or negative. Making workpieces that meet or exceed their technical and economic standards is the ultimate result of the grinding process.

Grinding wheel fundamentals

This section on Grinding Wheel Fundamentals, authored by Joe Sullivan of Norton Company, originally appeared in *Modern Machine Shop* magazine.

Abrasives–grits and grains

Grinding wheels and other bonded abrasives have two major components: the abrasive *grains* that do the actual cutting, and the *bond* that holds the grains together and supports them while they cut. The percentages of grain and bond, and their spacing in the wheel, determine the wheel's structure.

The particular abrasive used in a wheel is chosen based on the way it will interact with the work material. The ideal abrasive has the ability to stay sharp with minimal point dulling. When dulling begins, the abrasive fractures, creating new cutting points. Each abrasive type is unique with distinct properties for hardness, strength, fracture toughness, and resistance to impact.

Grinding wheels are generally labeled with a maximum safe operating speed. Don't exceed this speed limit. The safest course is not even to mount a given wheel on any grinder fast enough to exceed this limit.

Aluminum oxide is the most common abrasive used in grinding wheels. It is usually the abrasive chosen for grinding carbon steel, alloy steel, high speed steel, annealed malleable iron, wrought iron, and bronzes and similar metals. There are many different types of aluminum oxide abrasives, each specially made and blended for particular types of grinding jobs. Each abrasive type carries its own designation—usually a combination of a letter and a number. These designations vary by manufacturer.

Zirconia alumina is another family of abrasives, each one made from a different percentage of aluminum oxide and zirconium oxide. The combination results in a tough, durable abrasive that works well in rough grinding applications, such as cut-off operations, on a broad range of steels and steel alloys. As with aluminum oxide, there are several different types of zirconia alumina from which to choose.

Silicon carbide is an abrasive used for grinding gray iron, chilled iron, brass, soft bronze, and aluminum, as well as stone, rubber, and other nonferrous materials.

Ceramic aluminum oxide is the newest major development in abrasives. This is a high-purity grain manufactured in a gel sintering process. The result is an abrasive with the ability to fracture at a controlled rate at the submicron level, constantly creating thousands of new cutting points. This abrasive is exceptionally hard and strong. It is primarily used for precision grinding in demanding applications on steels and alloys that are the most difficult to grind. The abrasive is normally blended in various percentages with other abrasives to optimize its performance for different applications and materials.

Superabrasives make up a special category of bonded abrasives designed for grinding the hardest, most challenging work materials. Because carbides, high speed steels, PCD, PCBN, ceramics, and some other materials used to make cutting tools can be nearly as hard as conventional abrasives, the job of sharpening them falls to a special class of abrasives—diamond and CBN, the superabrasives.

These materials offer extreme hardness, but they are more expensive than conventional abrasives (silicon carbide and aluminum oxide). Therefore, superabrasive grinding wheels have a different construction than conventional abrasive wheels. Where a conventional abrasive product is made up of abrasive all the way through, superabrasive wheels have abrasive on the cutting edge of the wheel that is bonded to a core material, which forms the shape of the wheel and contributes to the grinding action.

Superabrasive wheels are supplied in the same standard grit range as conventional wheels (typically 46 through 2,000 grit). Like other types of wheels, they can be made in a range of grades and concentrations (the amount of diamond in the bond) to fit the operation.

Once the grain is known, the next question relates to *grit size*. Every grinding wheel has a number designating this characteristic. Grit size is the size of individual abrasive grains in the wheel. It corresponds to the number of openings per linear inch in the final screen size used to size the grain. In other words, higher numbers translate to smaller openings in the screen the grains pass through. Lower numbers (such as 10, 16, or 24) denote a wheel with coarse grain. The coarser the grain, the larger the size of the material removed. Coarse grains are used for rapid stock removal where finish is not important. Higher numbers (such as 70, 100, and 180) are fine grit wheels. They are suitable for imparting fine finishes, for small areas of contact, and for use with hard, brittle materials. See the subsection below on *grit sizes* in the Grinding Wheel Specifications section for specific information on dimensions and nomenclature.

Bond materials

To allow the abrasive in the wheel to cut efficiently, the wheel must contain the proper bond. The bond is the material that holds the abrasive grains together so they can cut effectively. It must also wear away as the abrasive grains wear and are expelled so new sharp grains are exposed. There are three principal types of bonds used in conventional grinding wheels. Each is capable of giving distinct characteristics to the grinding action of the wheel. The type of bond selected depends on such factors as the wheel operating speed, the type of grinding operation, the precision required, and the material to be ground.

Vitrified bonds. Most grinding wheels are made with vitrified bonds, which consist of a mixture of carefully selected clays. At the high temperatures produced in the kilns where grinding wheels are made, the clays and the abrasive grain fuse into a molten glass condition. During cooling, the glass forms a span that attaches each grain to its neighbor and supports the grains while they grind.

Grinding wheels made with vitrified bonds are very rigid, strong, and porous. They remove stock material at high rates and grind to precise requirements. They are not affected by water, acid, oil, or variations in temperature. Vitrified bonds are very hard, but at the same time they are brittle like glass. They are broken down by the pressure of grinding.

Organic bonds. Some bonds are made of *organic* substances. These bonds soften under the heat of grinding. The most common organic bond type is the *resinoid* bond, which is made from synthetic resin. Wheels with resinoid bonds are good choices for applications that require rapid stock removal, as well as those where better finishes are needed. They are designed to operate at higher speeds, and they are often used for wheels in fabrication shops, foundries, billet shops, and for saw sharpening and gumming.

Another type of organic bond is *rubber*. Wheels made with rubber bonds offer a smooth grinding action. Rubber bonds are often found in wheels used where a high quality of finish is required, such as ball bearing and roller bearing races. They are also frequently used for cut-off wheels where burr and burn must be held to a minimum.

The strength of a bond is designated in the *grade* of the grinding wheel. The bond is said to have a hard grade if the spans between each abrasive grain are very strong and retain the grains well against the grinding forces tending to pry them loose. A wheel is said to have a soft grade if only a small force is needed to release the grains. It is the relative amount of bond in the wheel that determines its grade or hardness.

Hard grade wheels are used for longer wheel life, for jobs on high-horsepower machines, and for jobs with small or narrow areas of contact. Soft grade wheels are used for rapid stock removal, for jobs with large areas of contact, and for hard materials such as tool steels and carbides.

There are four different bonds used in superabrasive wheels. Resinoid bond wheels are exceptionally fast and cool cutting. They are well suited to sharpening multitooth cutters and reamers, and for all precision grinding operations. Resin is the "workhorse" bond, most commonly used and most forgiving. Vitrified bond wheels combine fast cutting with a resistance to wear. They are often used in high-volume production operations. Metal bond wheels are used for grinding and cutting nonmetallic materials such as stone, reinforced plastics, and semiconductor materials that cannot be machined by other cutting tools. Single-layer plated wheels are used when the operation requires both fast stock removal and the generation of a complex form.

Wheel shapes

The wheel itself is available in a variety of shapes. The product typically envisioned when one thinks of a grinding wheel is the *straight wheel*. The grinding *face*—the part of the wheel that addresses the work—is on the periphery of a straight wheel. A common variation of the straight wheel design is the *recessed wheel*, so called because the center of the wheel is recessed to allow it to fit on a machine spindle flange assembly.

On some wheels, the cutting face is on the *side* of the wheel. These wheels are usually named for their distinctive shapes, as in *cylinder wheels*, *cup wheels*, and *dish wheels*. Sometimes bonded abrasive sections of various shapes are assembled to form a continuous or intermittent side grinding wheel. These products are called *segments*. Wheels with cutting faces on their sides are often used to grind the teeth of cutting tools and other hard-to-reach surfaces.

Mounted wheels are small grinding wheels with special shapes, such as cones or plugs, which are permanently mounted on a steel mandrel. They are used for a variety of off-hand and precision internal grinding operations.

Further considerations

A number of factors must be considered in order to select the best grinding wheel for the job at hand. The first consideration is the material to be ground. This determines the kind of abrasive needed in the wheel. For example, aluminum oxide or zirconia alumina should be used for grinding steels and steel alloys. For grinding cast iron, nonferrous metals, and nonmetallic materials, select a silicon carbide abrasive.

Hard, brittle materials generally require a wheel with a fine grit size and a softer grade. Hard materials resist the penetration of abrasive grains and cause them to dull quickly. Therefore, the combination of finer grit and softer grade lets abrasive grains break away as they become dull, exposing fresh, sharp cutting points. On the other hand, wheels with the coarse grit and hard grade should be chosen for materials that are soft, ductile, and easily penetrated.

The amount of stock to be removed is also a consideration. Coarser grits provide rapid stock removal since they are capable of greater penetration and heavier cuts. However, if the work material is hard to penetrate, a slightly finer grit wheel will cut faster because there are more cutting points to do the work.

Wheels with vitrified bonds provide fast cutting. Resin, rubber, or shellac bonds should be chosen if a smaller amount of stock is to be removed, or if the finish requirements are higher.

Another factor that affects the choice of wheel bond is the wheel speed in operation. Usually vitrified wheels are used at speeds less than 6,500 surface feet per minute. At higher speeds, the vitrified bond may break. Organic bond wheels are generally the choice between 6,500 and 9,500 surface feet per minute. Working at higher speeds usually requires specially designed wheels for high speed grinding.

In any case, do not exceed the safe operating speed shown on the wheel or its blotter. This might be specified in either rpm or sfm.

The next factor to consider is the area of grinding contact between the wheel and the workpiece. For a broad area of contact, use a wheel with coarser grit and softer grade. This ensures a free, cool cutting action under the heavier load imposed by the size of the surface to be ground. Smaller areas of grinding contact require wheels with finer grits and harder grades to withstand the greater unit pressure.

Next, consider the severity of the grinding action. This is defined as the pressure under which the grinding wheel and the workpiece are brought and held together. Some abrasives have been designed to withstand severe grinding conditions when grinding steel and steel alloys.

Grinding machine horsepower must also be considered. In general, harder grade wheels should be used on machines with higher horsepower. If horsepower is less than wheel diameter, a softer grade wheel should be used. If horsepower is greater than wheel diameter, choose a harder grade wheel.

Grinding wheel care

Grinding wheels must be handled, mounted, and used with the right amount of precaution and protection. They should always be stored so they are protected from banging and gouging. The storage room should not be subjected to extreme variations in temperature and humidity because these can damage the bonds in some wheels.

Immediately after unpacking, all new wheels should be closely inspected to be sure they have not been damaged in transit. All used wheels returned to the storage room should also be inspected.

Wheels should be handled carefully to avoid dropping and bumping, since this may lead to damage or cracks. Wheels should be carried to the job, not rolled. If the wheel is too heavy to be carried safely by hand, use a hand truck, wagon, or forklift truck with cushioning provided to avoid damage.

Before mounting a vitrified wheel, ring test it as explained in the American National Standards Institute's B7.1 Safety Code for the Use, Care, and Protection of Grinding Wheels. The ring test is designed to detect any cracks in a wheel. Never use a cracked wheel.

A wise precaution is to be sure the spindle rpm of the machine you're using does not exceed the maximum safe speed of the grinding wheel.

Always use a wheel with a center hole size that fits snugly yet freely on the spindle without forcing it. Never attempt to alter the center hole. Use a matched pair of clean, recessed flanges at least one-third the diameter of the wheel. Flange bearing surfaces must be flat and free of any burrs or dirt buildup.

Tighten the spindle nut only enough to hold the wheel firmly without overtightening. If mounting a directional wheel, look for the arrow marked on the wheel itself and be sure it points in the direction of spindle rotation.

Always make sure that all wheel and machine guards are in place, and that all covers are tightly closed before operating the machine. After the wheel is securely mounted and the guards are in place, turn on the machine, step back out of the way, and let it run for at least one minute at operating speed before starting to grind.

Grind only on the *face* of a straight wheel. Grind only on the *side* of a cylinder, cup, or segment wheel. Make grinding contact gently, without bumping or gouging. Never force grinding so that the motor slows noticeably or the work gets hot. The machine ampmeter can be a good indicator of correct performance.

If a wheel breaks during use, make a careful inspection of the machine to be sure that protective hoods and guards have not been damaged. Also, check the flanges, spindle, and mounting nuts to be sure they are not bent, sprung, or otherwise damaged.

Grinding wheel specifications

As explained above, grinding wheels are identified by their characteristics. While certain characteristics differ by manufacturer, others are standardized by ANSI, Coated Abrasives Manufacturer's Institute (CAMI), or Federation of European Producers of Abrasives (FEPA).

Grinding wheel marking systems

Grinding wheel marking systems for most wheels (excluding superabrasives) are set forth in ANSI B74.13-1990, but even with these standardized recommendations, the marking system can still vary by manufacturer. This is most noticeable in the "abrasive type" category, which often reflects a particular manufacturer's unique abrasive. However, with certain possible exceptions, a typical marking might look like the one below.

–	1	2	3	4	5	6
Prefix	Abrasive Type	Grain Size	Grade	Structure	Bond Type	Manufacturer's Record
51	A	36	L	5	V	23

In this example, each station of the marking is read as follows.

Prefix. Unnumbered and optional. This space is for the manufacturer's inclusion of information indicating the exact type of abrasive. In most instances, this space is not used.

Abrasive Type. Station 1. The standard provides three specific letters for this station. The letter "A" indicates aluminum oxide; "C" indicates silicon carbide; and "Z" indicates aluminum zirconium. In reality, the entry in this station is more likely to have an entry such as "5SG," which is Norton Company's nomenclature for their ceramic aluminum oxide abrasive with the highest ceramic content. "53A" would indicate a general-purpose blend of light and dark aluminum oxide abrasive. In order to understand many of the possible entries in Station 1, a manufacturer's catalog is required.

Grain Size. Station 2. On the example above, the grain size (36) indicates a coarse medium grain. There are three grain sizes, and each is identified by a series of numbers.

In each category, lower numbers are coarser than the higher numbers. Coarse grain: 8, 10, 12, 14, 16, 20, 24. Medium Grain: 30, 36, 46, 54, 60. Fine: 70, 80, 90, 100, 120, 150, 180. Very Fine: 220, 240, 280, 320, 400, 500, 600. Grain (or grit) size is discussed in more detail below.

Grade. Station 3. Any of the twenty-six letters of the alphabet may be used in this Station. "A" is the softest, and "Z" is the hardest. Usually, these are unofficially subclassified as follows. Soft: A–G. Medium: I–P. Hard: Q–Z. See further explanation below.

Structure. Station 4. Optional, and often not used. Numbers 1–16 are intended to indicate the grain spacing on the wheel. A "1" wheel has very closely grouped grains. A "16" has a more open structure with wider spaced grains. Closed coats (low number rating) permit the greatest degree of stock removal and provide the longest product life. Open coats (high number rating) have only 50 to 75% of their surface coated by abrasive grain. Evenly spaced voids between the particles of grain reduce the effect of loading caused by metal particles.

Bond Type. Station 5. Up to nine different letters or letter combinations are assigned to identify the bond used on the wheel. B = Resinoid (synthetic resins). BF = Resinoid Reinforced. E = Shellac. O = Oxychloride. P = Plastic. R = Rubber. RF = Rubber Reinforced. S = Silicate. V = Vitrified.

Manufacturer's Record. Station 6. This station is reserved for a manufacturer's private marking to indicate pertinent information, such as a variation to the standard bond.

Explanation of grain (grit) size

Grain sizes are standardized by both CAMI and FEPA, and the numbers for each are to a large degree simultaneous up to #240. In other words, a CAMI rating of 24 and a FEPA rating of P24 are approximately the same; and the grain size, in microns (μm), will be between 715 and 740 μm. Similarly, a CAMI rating of 240 and an FEPA rating of P240 have similar grain sizes (53.5 versus 58.5 μm). However, as the grain size becomes finer, the two ranking systems diverge significantly and should not be interpreted as equal. A CAMI rating of 600, for example, has a grain size of 16 μm, while an FEPA rating of P600 has a size of 25.75 μm. The scales also cover different ranges: CAMI's scale is from 12 to 1,200, while FEPA ranges from P12 to P2500. **Table 1** provides a comparison chart for CAMI and FEPA ratings, and includes Japanese JIS grades for comparison.

Importance of wheel grade rating

The wheel grade rating given in Station 3 of the standardized marking system is not precisely defined by testing criteria. In general, wheels with high bond content are "hard" wheels, and those with a small amount of bond will be "soft." Ideally, the wheel chosen for a given operation should be hard enough to prevent it from breaking down too fast, but soft enough to allow worn grain to break away to prevent glazing.

Hard wheels are capable of removing large amounts of material at rapid rates, and are useful when precise shapes must be machined. Care must be taken when using a "hard" wheel to avoid excessive heat buildup. Burning can take place if the wheel glazes and contacts the workpiece, which can raise the temperature sufficiently to vaporize the workpiece surface. Glazed wheels have a characteristic shiny appearance caused by the grain wearing flat, rather than being discarded by grinding forces. Glazing also leads to chatter. Heat checking, which causes fine cracks in the workpiece, is normally caused by inadequate or inappropriate grinding fluid, or by improper abrasive selection for the material being ground.

Soft wheels break down more quickly than hard wheels, but glazing is less likely because grain separation from the wheel is accelerated. This rapid grain evacuation can result in

Table 1. Grit Comparison Chart. *(Source, Norton Company.)*

Average Particle Size (Microns)	Grading Systems			Emery Products	Average Particle Size (Microns)	Grading Systems			Emery Products
	CAMI	FEPA	JIS			CAMI	FEPA	JIS	
5.0	-	-	-	-	64.0	-	-	-	2
6.0	-	-	3000	-	65.0	-	P220	220	-
6.5	1200	-	2500	4/0	66.0	220	-	-	-
8.4	-	P2500	-	-	78.0	180	P180	180	-
8.5	-	-	2000	-	79.0	-	-	-	3
9.2	1000	-	-	3/0	93.0	150	-	150	-
10.3	-	P2000	-	-	95.0	-	-	-	FINE
10.5	-	-	1500	-	97.0	-	P150	-	-
12.2	800	-	1200	-	116.0	120	-	120	-
12.6	-	P1500	-	-	127.0	-	P120	100	-
15.0	-	-	-	-	136.0	-	-	-	MEDIUM
15.3	-	P1200	1000	-	141.0	100	-	-	-
16.0	600	-	-	2/0	156.0	-	P100	-	-
18.3	-	P1000	800	-	189.0	-	-	-	COURSE
19.7	500	-	-	0	192.0	80	-	80	-
20.0	-	-	-	-	197.0	-	P80	-	-
21.8	-	P800	-	-	260.0	-	P60	-	-
23.6	400	-	600	-	268.0	60	-	60	-
25.0	-	-	-	-	326.0	-	P50	50	-
25.75	-	P600	-	-	341.0	-	-	-	EX. CRS.
28.8	360	-	500	-	351.0	50	-	-	-
30.0	-	P500	-	-	412.0	-	P40	40	-
35.0	-	P400	400	-	428.0	40	-	-	-
36.0	320	-	-	-	524.0	-	P36	-	-
40.0	-	-	360	-	535.0	36	-	36	-
40.5	-	P360	-	-	622.0	-	P30	-	-
44.0	280	-	-	1	638.0	30	-	30	-
45.0	-	-	-	-	715.0	24	-	-	-
46.2	-	P320	320	-	740.0	-	P24	24	-
50.0	-	-	-	-	905.0	20	-	-	-
52.5	-	P280	280	-	984.0	-	P20	20	-
53.5	240	-	-	-	1320.0	16	-	16	-
55.0	-	-	-	-	1324.0	-	P16	-	-
58.5	-	P240	-	-	1764.0	-	P12	-	-
60.0	-	-	240	-	1842.0	12	-	-	-

loose grain becoming caught in the cut and causing scratches on the workpiece.

To make a wheel act softer:	To make a wheel act harder:
Increase work speed	Decrease work speed
Increase traverse speed	Decrease traverse speed
Increase infeed	Decrease infeed
Decrease wheel speed	Increase wheel speed
Increase dresser traverse rate	Dress at slower traverse rate
Increase depth of dress per traverse	Decrease depth of dress per traverse.
Dress more often.	

Grinding wheel shapes

Grinding wheels can be broken down into two basic types: *peripheral wheels,* in which the outer rim contacts the workpiece; and *side,* or "face," wheels. In addition, a small number of available cone and plug shapes have also been standardized. These wheel, cone, and plug profiles are standardized by ANSI B74.2-1982, but wheel manufacturers often develop their own profiles or variations of the standardized forms. The standard forms have, in most instances, remained unchanged for decades, and what evolution, or revolution, that has occurred in this field can be attributed to developments in grinding with CBN and diamond "superabrasive" wheels and disks. The basic profiles of traditional abrasive wheels are shown in *Figures 1* and *2*. Profiles for superabrasive wheels will be covered later in this section. It should be remembered that the forms shown in the figure are meant to be representative. The Type 01 straight wheel, for example, is available from Norton Company with a vitrified bond in diameters as large as fourteen inches and as small as three inches, and in thicknesses ranging from one-quarter inch to two inches.

The edge or face of the wheel that contacts the workpiece can have many different contours depending on the operation. When ordering wheels, a 90° straight face will be delivered unless a specific contour is specified. A straight face can also be dressed to assume a required profile. See *Figure 3.*

Grinding wheel speeds – definitions

Although speeds for grinding wheels are measured in sfm or smm, the wheels themselves are commonly rated in rpm. Wheels should never be operated in excess of their stated speed, and, for safety, it is always good policy to use a wheel rated at a higher rpm than the machine that it is mounted on is capable of reaching. **Table 2** provides recommended wheel speeds for vitrified and resinoid bonded wheels of several shapes. **Table 3** converts sfm into actual rpm for grinding wheels from 1 inch to 36 inches in diameter.

As grinding wheel speeds increase, each grain on the wheel actually cuts less, which results in less wear on the wheel because as speeds increase the physical characteristics of the cut make the wheel act like it has a harder grade rating. Conversely, as speeds decrease, each grain remains in the cut longer and abrasive grain wear increases (as with a softer wheel).

Work speed refers to the speed at which a grinding wheel passes over a workpiece or rotates about a center. As in other forms of metal cutting, a high work speed can be advantageous because the faster a cutter moves across the workpiece, less heat is retained and the potential of thermal damage is reduced. Higher work speeds, or reducing the diameter of the wheel, results in increased grain depth of cut, and enables the wheel to perform in the manner of a softer grade wheel. If the work speed decreases, or wheel diameter increases, grain depth of cut decreases, and the wheel tends to act like a harder grade wheel.

(Text continued on p. 1594)

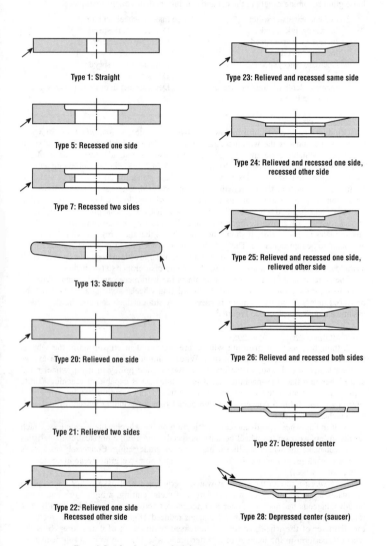

Figure 1. Peripheral grinding wheel shapes. Arrows point to grinding face(s).

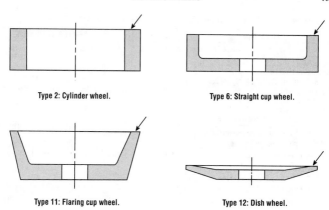

Type 2: Cylinder wheel.

Type 6: Straight cup wheel.

Type 11: Flaring cup wheel.

Type 12: Dish wheel.

Figure 2. Side, or face, grinding wheel shapes. Arrows point to grinding face.

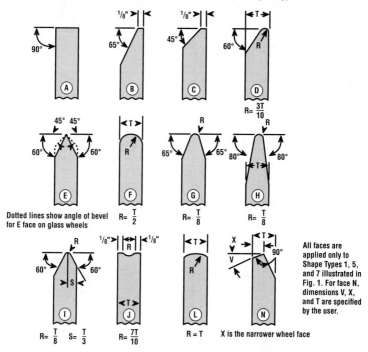

Figure 3. Standard grinding wheel face profiles. R = radius, S = length of cylindrical section, and T = overall thickness.

Table 2. Maximum Wheel Speeds for Grinding Wheels. *(Source, Pacific Grinding Wheel Co.)*

Wheel Type	SFM - Vitrified			SFM or RPM - Resinoid		
Wheel Hardness	A - D	E - G	H+	A - G	H - K	L+
Type 1 straight wheels Type 4 tapered wheels Types 5, 7, 20, 21, 22, 23, 24, 25, 26, dove tailed and relieved Type 12 dish wheels Type 13 saucer wheels Type 17, 17R, 18, & 19 plug and cone wheels	5,500	6,000	6,500	6,500	8,000	9,500
Type 2 cylinder wheels, plate mounted, nut inserted and projecting stud	5,000	5,500	6,000	5,000	6,000	7,000
Types 6 & 11 cup-shaped tool room wheels, fixed base machines	4,500	5,000	6,000	6,000	7,550	8,500
NOTE: Type 11 7" & 8", 3, 600		6,000	7,000	6,000	7,550	8,500
Types 6 & 11 portable snagging wheels.	4,500	5,500	6,500	6,000	8,000	9,500
Type 1 nut inserted & plate mounted abrasive discs	5,500	6,000	6,500	5,500	7,000	8,500
Type 1 FC (reinforced) snagging wheels: · Maximum diameter 4", maximum thickness 1/4" · All other diameters thicker than 1/4"				9,500 9,500	12,500 12,500	14,200 12,500
Type 1 high-speed wheels (BH4)				4" dia. 16,000 rpm 5" dia. 12,987 rpm 6" dia. 10,824 rpm 8" dia. 8,000 rpm		
Type 11 high-speed, all fine thread				4" dia. 12,000 rpm 5" dia. 9,000 rpm 6" dia. 8,000 rpm		
Type 1 cut-off (including reinforced) · 16" diameter or less · Over 16"				9,500 9,500	12,000 12,000	16,000 14,200
Thread and flute grinding	8,000	10,000	12,000	8,000	10,000	12,000
Crankshaft and camshaft grinding	5,500	8,000	8,500	6,500	8,000	9,500
Epoxy bonded wheels				6,500	7,000	8,000
Type 27 and 28, 3/8" thick and thinner, BFC and Honeycomb Flex				3" dia. 20,000 rpm 4" dia., 3/8" arbor 15,000 rpm 4" dia., 5/8" arbor 15,000 rpm 4 1/2" dia. 13,300 rpm 5" dia. 13,300 rpm 6" dia. 9,000 rpm 7" dia. 8,500 rpm 8" dia. 6,500 rpm 9" dia. 6,000 rpm		
Type 27 high speed (FC 1), 1/4" thick and thin				6" dia., 5/8"-18 thd. 12,000 rpm* 7" dia., 3/4"-16 thd. 9,000 rpm 9" dia., 7/8"-14 thd. 8,000 rpm*		
Type 27 and 28, 1/2" thick and thicker				6" dia. 7,000 rpm 7" dia. 7,000 rpm* 9" dia. 4,670 rpm		
Type 27, 3/8" thick and thinner				7" dia. 8,000 rpm 9" dia 6,000 rpm		

* Denotes special speed.

Table 3. Grinding Wheel Speeds (sfm) for Indicated Revolutions Per Minute. *(Source, Norton Company.)*

Wheel Diameter (inches)	Surface Speed in Feet Per Minute								
	4,000	4,500	5,000	5,500	6,000	6,500	7,000	7,500	8,000
	Revolutions Per Minute								
1	15,279	17,189	19,098	21,008	22,918	24,828	26,737	28,647	30,558
2	7,639	8,594	9,549	10,504	11,459	12,414	13,368	14,328	15,278
3	5,093	5,729	6,366	7,003	7,639	8,276	8,913	9,549	10,186
4	3,820	4,297	4,775	5,252	5,729	6,207	6,685	7,162	7,640
5	3,056	3,438	3,820	4,202	4,584	4,966	5,348	5,730	6,112
6	2,546	2,865	3,183	3,501	3,820	4,138	4,456	4,775	5,092
7	2,183	2,455	2,728	3,001	3,274	3,547	3,820	4,092	4,366
8	1,910	2,148	2,387	2,626	2,865	3,103	3,342	3,580	3,820
9	1,698	1,910	2,122	2,334	2,546	2,758	2,970	3,182	3,396
10	1,528	1,719	1,910	2,101	2,292	2,483	2,674	2,865	3,056
12	1,273	1,432	1,591	1,751	1,910	2,069	2,228	2,386	2,546
14	1,091	1,228	1,364	1,500	1,637	1,773	1,910	2,046	2,182
16	955	1,074	1,194	1,313	1,432	1,552	1,672	1,791	1,910
18	849	955	1,061	1,167	1,273	1,379	1,485	1,591	1,698
20	764	859	955	1,050	1,146	1,241	1,337	1,432	1,528
22	694	781	868	955	1,042	1,128	1,215	1,302	1,388
24	637	716	796	875	955	1,034	1,115	1,194	1,274
26	588	661	734	808	881	955	1,028	1,101	1,176
28	546	614	682	750	818	887	955	1,023	1,092
30	509	573	637	700	764	828	891	955	1,018
36	424	477	530	583	637	690	742	795	848
Wheel Diameter (inches)	Surface Speed in Feet Per Minute								
	8,500	9,000	9,500	10,000	12,000	12,500	14,200	16,000	16,500
	Revolutions Per Minute								
1	32,467	34,377	36,287	38,196	45,836	47,745	54,240	61,116	63,025
2	16,238	17,188	18,143	19,098	22,918	23,875	27,120	30,558	31,510
3	10,822	11,459	12,096	12,732	15,278	15,915	18,080	20,372	21,010
4	8,116	8,595	9,072	9,549	11,459	11,940	13,560	15,278	15,755
5	6,494	6,876	7,258	7,640	9,168	9,550	10,850	12,224	12,605
6	5,411	5,729	6,048	6,366	7,639	7,960	9,040	10,186	10,505
7	4,638	4,911	5,183	5,456	6,548	6,820	7,750	8,732	9,005
8	4,058	4,297	4,535	4,775	5,729	5,970	6,780	7,640	7,880
9	3,606	3,820	4,032	4,244	5,092	5,305	6,030	6,792	7,000
10	3,247	3,438	3,629	3,820	4,584	4,775	5,425	6,112	6,300
12	2,705	2,864	3,023	3,183	3,820	3,980	4,520	5,092	5,250
14	2,319	2,455	2,592	2,728	3,274	3,410	3,875	4,366	4,500
16	2,029	2,149	2,268	2,387	2,865	2,985	3,390	3,820	3,940
18	1,803	1,910	2,016	2,122	2,546	2,655	3,015	3,396	3,500

(Continued)

Table 3. *(Continued)* **Grinding Wheel Speeds (sfm) for Indicated Revolutions Per Minute.**
(Source, Norton Company.)

Wheel Diameter (inches)	Surface Speed in Feet Per Minute								
	8,500	9,000	9,500	10,000	12,000	12,500	14,200	16,000	16,500
	Revolutions Per Minute								
20	1,623	1,719	1,814	1,910	2,292	2,390	2,715	3,056	3,150
22	1,476	1,562	1,649	1,736	2,084	2,170	2,465	2,776	2,865
24	1,353	1,433	1,512	1,591	1,910	1,990	2,260	2,546	2,625
26	1,248	1,322	1,395	1,468	1,762	1,840	2,090	2,352	2,425
28	1,159	1,228	1,296	1,364	1,637	1,705	1,940	2,182	2,250
30	1,082	1,146	1,210	1,274	1,528	1,595	1,810	2,056	2,100
36	902	954	1,007	1,061	1,273	1,330	1,510	1,698	1,750

Notes: SFM = 0.262 × wheel diameter (inches). To convert SFM to meters per second, divide SFM by 196.85.

Traverse distance or speed (also called crossfeed) refers to the distance the workpiece is moved across the face of the wheel, and it affects surface finish and productivity. Good surface finishes are more easily achieved when the traverse distance is no more than one-quarter of the wheel width. As the traverse distance increases to one-half, or more, of the wheel width, surface finish steadily diminishes. On the other hand, increasing the ratio between traverse distance and wheel width also increases productivity; the relationship between surface finish and productivity rate must be compromised based on the requirements of the operation.

Calculations used in grinding

For evaluation purposes, the work material removal per unit of width (Z_W') can be calculated. The following formulas calculate Z_W' for internal grinding, surface grinding, external plunge grinding, and external traverse grinding. Z_W' is measured in U.S. customary units in inches cubed per minute inch [in.^{3}/(min•in.)], and in metric units as millimeters cubed per millimeter second [mm^3/(mm•s)].

Internal grinding (ID plunge grinding)

U.S. customary units $\qquad Z_W' = \pi \times D_W \times V_W \times 60$

Metric units $\qquad Z_W' = \pi \times D_W \times V_W$

where $\quad \pi = 3.14159$
$D_W =$ the initial inner diameter of the work, in inches or millimeters
$V_W =$ the radial feed of the wheel into the workpiece, in in./s. or mm/s.

Surface grinding

$Z_W' = a \times V_W$

where $\quad a =$ depth of cut in inches or millimeters
$V_W =$ the workpiece speed in in./min or mm/s.

External grinding (OD plunge grinding)

U.S. customary units $\qquad Z_W' = \pi \times D_W \times V_W \times 60$

Metric units $Z_W{}' = \pi \times D_W \times V_W$

where $\pi = 3.14159$

$D_W =$ the initial outer diameter of the work, in inches or millimeters

$V_W =$ the radial feed of the wheel into the workpiece, in in./s. or mm/s.

External grinding (OD traverse)

For proper evaluation in this mode, both $Z_W{}'$ and Z_W (total work material removal) are calculated to illustrate the effect of work speed and traverse speed, respectively. Z_W is measured in units of in.³/s and mm³/s, and in this mode $Z_W{}'$ has units of in.³/(min•[in./rev]) or mm³/(s•[mm/rev]).

U.S. customary or metric units $Z_W = \pi \times D_W \times V_W \times a$

U.S. customary units $Z_W{}' = \pi \times D_W \times N_W \times a$

Metric units $Z_W{}' = \pi \times D_W \times N_W \times a \times 0.01667$

where $\pi = 3.14159$

$D_W =$ the initial outer diameter of the work, in inches or millimeters

$V_W =$ the axial traverse feed rate of the wheel along the workpiece, in in./s or mm/s.

$N_W =$ the workpiece speed in revolutions per minute

$a =$ the radial depth of cut in inches or millimeters.

Grinding ratio, or g-ratio, is the ratio of the volume of work material removed to the volume of wheel consumed. Economically, it is better to remove a lot of material and consume very little wheel. This means that a higher g-ratio is better. It is expressed as

G-ratio = Volume of work removed ÷ Volume of wheel worn.

Cylindrical grinding

In cylindrical grinding operations, both the wheel and workpiece rotate. The workpiece is either fixed and driven between centers, or driven by a revolving chuck or collet while supported in a center. There are two basic methods of cylindrical grinding: traverse and plunge. In traverse grinding, the wheel traverses axially across the part. In plunge grinding, the wheel is thrust, or "plunged," into the part. Type 01 straight wheels are most commonly used in cylindrical grinding.

There are several kinds of cylindrical grinding machines including plain cylindrical (or roll) grinders, inside- outside-diameter grinders, and centerless grinders (covered in more detail below).

The following guidelines, as published by Norton Company, can be used to determine recommended machining parameters in traverse grinding. They are intentionally conservative but provide excellent guidelines for initiating an operation.

Wheel speed. Typical speeds are 6,500, 8,500 or 12,000 sfm.

Speed is not to exceed manufacturer's recommendation.

Suggested infeed rate

rough grinding = 0.002"

finish grinding = 0.0005".

Width of cut (wheel width per work revolution)

for materials 50 R_C or softer.

rough grinding: width of cut = $^1/_2$ of wheel thickness.

finish grinding: width of cut = $^1/_6$ of wheel thickness.

for materials harder than 50 R_C.

rough grinding: width of cut = $1/4$ of wheel thickness.

finish grinding: width of cut = $1/8$ of wheel thickness.

Work speed in sfm

for materials 50 R_C or softer.

work speed (sfm) = wheel speed (sfm) ÷ 60.

for materials harder than 50 R_C.

work speed (sfm) = wheel speed (sfm) ÷ 100.

Work speed in rpm. (work speed {sfm} = [rpm × 0.2618] × workpiece diameter)

rpm = (work speed [sfm] ÷ 0.2618) ÷ workpiece diameter.

Traverse rate (inches per minute)

wheel thickness × width of cut × work speed in rpm.

The above formulas for traverse grinding apply to operations where the wheel passes over the workpiece. In plunge grinding, where the wheel is fed directly into the workpiece, the following parameters and formulas apply. As with the above formulas, they are generally conservative.

Wheel speed. Typical speeds are 6,500, 8,500 or 12,000 sfm.

Speed is not to exceed manufacturer's recommendation.

Infeed per revolution of the workpiece

steel (soft)

rough grinding = 0.0005"

finish grinding = 0.0002".

plain carbon steel (hardened)

rough grinding = 0.0002"

finish grinding = 0.00005".

alloy and tool steel (hardened)

rough grinding = 0.0001"

finish grinding = 0.000025".

Work speed in sfm

for materials 50 R_C or softer.

work speed (sfm) = wheel speed (sfm) ÷ 60.

for materials harder than 50 R_C.

work speed (sfm) = wheel speed (sfm) ÷ 100.

Internal cylindrical grinding

Internal diameter grinding of bores and holes is performed with cylindrical grinding. The process is capable of generating size and concentricity within millionths of an inch. Due to the necessity of having to use a small wheel that will fit within a bore, wheels typically range in diameter from one-half inch to three inches, and Type 01 Straight wheels and Type 05 Recessed One Side wheels are most often used. The necessarily small size of the wheel leads to rapid wear, which has led to the popularity of CBN and diamond wheels in crush dressable and vitrified form for internal grinding. The diameter of the wheel selected should be 66 to 75% of the diameter of the hole being ground, and the wheel width should not exceed the diameter of the hole.

The following guidelines can be used to determine parameters for internal grinding.

Work speed = 150 to 200 sfm for most applications.

Wheel speed ([rpm × 0.2618] × workpiece diameter)
    general applications = 5,000 to 6,500 sfm
    production applications = 6,500 to 12,000 sfm.

Wheel size = 66% to 75% of hole diameter to be ground.

Dressing recommendations
    choose a lead between 0.006" and 0.012" per wheel revolution
    rotate tool frequently to maintain a sharp point
    traverse rate per minute = rpm × choice of lead.

Troubleshooting cylindrical grinding operations. *(Source, Norton Company.)*

Problem	Possible Cause	Suggested Correction
Burn or cracks on workpiece	Wheel too hard	Try softer wheel or make wheel act softer.
	Dress is too fine	Dress faster.
	Nozzle not properly directing coolant	Redirect flow.
	Work speed too slow, or wheel speed too fast	Adjust speeds.
	Wheel feed too slow, causing excess contact time	Increase feed rate.
Chatter marks on workpiece	Loose spindle, end play, or bearing failure	Check wheel spindle on assembly.
	Work drive faults, out of true plate	Check gear action, truth, drive on pin pressure
	Wheel too hard, poor truing, poor mounting	Use softer grade, dress wheel correctly, secure mounting.
Poor quality finish	Poor quality dressing	Reduce final depth of dress on wheel.
	Diamond insecurely held	Check and replace diamond, and secure properly.
	Misalignment of head and tail stocks	Check for properly aligned workpiece.
	Dirty coolant	Clean coolant frequently, use efficient filter.
	Scored workpiece centers	Check and lap center bolts as required

Surface grinding

The term surface grinding is usually used to refer to the process of grinding a plane surface by feeding the workpiece beneath a rotating grinding wheel. The workpiece may travel traversely under the wheel and reciprocate (move back and forth) beneath a grinding wheel mounted on a horizontal spindle, or move in circles on a rotary table beneath a vertical spindle cutting on the face of the grinding wheel or grinding "segment." The surface may be ground perfectly flat, or grooves may be introduced by grinding straight channels into the workpiece. The result is, of course, essentially the same as would be achieved by milling the surface, but in certain instances grinding can offer significant advantages over milling. The most obvious example of a requirement lending itself to grinding is surface finish—grinding can produce the finest finishes with the highest dimensional accuracy. Other advantages are economic: tooling can be less expensive, and required contours can be dressed into the profile of the wheel. Fixturing is also less expensive as it is normally less complex than the fixturing required for milling. Very hard or abrasive surfaces are often more easily and inexpensively machined with grinding than milling. As in all machining operations, accuracy and surface finish depend on machine rigidity, proper setup, and sharp and true cutting tools.

Grinding "segments" used on Blanchard style vertical spindle surface grinding machines provide high material removal rates on flat surfaces. In appearance, segments look like the tip of a recessed grinding wheel. When mounted on a vertical spindle above a rotary table, they are capable of high infeed rates capable of removing large amounts of stock in a very short time. When vertical grinding with segments, production rates can be maximized by choosing the hardest grade that will not overload the machine, along with a grit size as coarse as possible in order to remove stock quickly in large chips. For soft and mild steels, 24 and 30 grit are recommended. For hard steels, select 36 or 46 grit. If glazing is encountered, it is likely that the table speed is too slow or the grade is too hard.

The following guidelines can be used to determine recommended machining parameters in horizontal spindle surface grinding. They are intentionally conservative but provide excellent guidelines for initiating an operation.

Wheel speed. Recommended wheel speeds are 5,500 to 6,500 sfm.
Speed is not to exceed manufacturer's recommendation.

Table feed. Recommended table speed is 50 to 100 fpm.

Down feed per pass
plain carbon steel
rough grinding = 0.003"
finish grinding = 0.0005"

alloy steel, 50 R_C or softer
rough grinding = 0.003"
finish grinding = 0.001"

alloy steel, harder than 50 R_C
rough grinding = 0.003"
finish grinding = 0.0005"

tool steel
rough grinding = 0.002"
finish grinding = 0.0005"

cast steel, gray iron, or ductile iron 50 R_C or softer
rough grinding = 0.003"
finish grinding = 0.001"

cast steel, harder than 50 R_C
rough grinding = 0.003"
finish grinding = 0.0005"

martensitic stainless steel 250 HRB or softer
rough grinding = 0.002"
finish grinding = 0.0005"

martensitic stainless steel, harder than 250 HRB
rough grinding = 0.001"
finish grinding = 0.0005"

aluminum alloys
rough grinding = 0.003"
finish grinding = 0.001".

Crossfeed per pass should be 20 to 30% of wheel width for materials 50 R_C or softer, and 10 to 15% for harder materials.

Troubleshooting horizontal surface grinding operations. *(Source, Norton Company.)*

Problem	Possible Cause	Suggested Correction
Workpiece not flat or not parallel	Too heavy a cut	Take lighter cut, especially on finishing pass.
	Burning on the workpiece	Choose a softer grade wheel.
	Dirty magnetic chuck	Clean magnetic chuck before final grind.
Scratching of workpiece	Dirty coolant	Clean coolant tank or replace grinding fluid.
	Wheel too soft	Choose harder grade and/or finer grit wheel.
	Wheel too coarse	Dress wheel "closed" before final pass.
Burning or checking of workpiece	Wheel too hard	Choose softer grade and/or coarser grit wheel.
		Choose more open structure wheel.
	Table speed too slow	Increase table speed.
	Wheel dressed too fine	Dress wheel coarser, more open.
	Depth of cut too deep	Take lighter cuts and dress more frequently.
Chatter marks on workpiece	Worn spindle bearings	Check and correct for truth and end play.
	Wheel out of true	Redress wheel and check mounting.
	Poorly clamped wheel	Check tightness of mounting nuts.
	Glazed wheel face	Redress wheel with sharp dressing tool.
	Wheel too hard	Use softer wheel, perhaps with open structure.
Wheel loading or glazing	Wheel too fine	Use coarser grit or softer grade wheel.
	Wheel too finely dressed	Dress wheel very open.
	Depth of cut too small	Increase feed and traverse speed.
	Poor coolant quality	Change coolant or use high-detergent type.
	Wheel structure too closed	Choose a more open structure wheel.
	Diamond dressing tool worn	Replace diamond, and dress coarse (open).

Centerless grinding

Centerless grinding is used to produce cylindrical forms to extremely close tolerances. The process eliminates the need for center holding the workpiece (as in cylindrical grinding) by supporting the work at three separate points: the grinding wheel, a work support blade, and a feed wheel. (See *Figure 4.*) The wheels maintain a constant "grip"

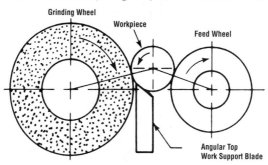

Figure 4. Proper setup for centerless grinding with angular topped blade. (Source, Norton Company.)

on the workpiece. However, because nothing actually clamps the workpiece in place, each piece can float freely for continuous production (also known as "throughfeed centerless grinding").

In centerless grinding, the grinding wheel normally rotates at 6,500 sfm, but sometimes speeds up to 12,000 sfm are used with qualified wheels. The feed wheel turns in the same direction as the grinding wheel, but only at 36 to 900 sfm. The workpiece rides between the two wheels and rotates in the opposite direction. Although cutting pressure forces the workpiece downward, it is held in place against the feed wheel by the work support blade. Feed wheels are usually rubber bond abrasive wheels, and rely on good frictional characteristics to revolve the workpiece at the same peripheral speed as the feed wheel.

It can be seen that the position of the work blade governs the position of the workpiece. If the support is flat-topped and positioned so that the centerline of the workpiece is in the same plane as the centers of the grinding and feed wheels, a constant diameter cross section will be ground but it will not necessarily be cylindrical because any high spot on the surface of the workpiece, as it comes in contact with the feed wheel, will cause the workpiece to be pushed toward the grinding wheel, thus grinding a low spot on the opposite side (see *Figure 5*). By having a common centerline on work and both wheels, the error is accentuated with every revolution.

By raising the blade to project the centerline of the workpiece above the common centerline of the wheels, the problem is partially solved. Now, when a high spot on the workpiece contacts the feed wheel, the corresponding low spot ground will not be directly opposite. Instead of making the error greater with each revolution, it will be progressively diminished. The final step to assuring consistent roundness is to make the top of the work blade angular rather than flat. By raising the work blade, and making the top of the blade angular, the shape of the space occupied by the workpiece is changed and effects of high-spot errors are effectively dampened out. Generally speaking, the higher the work is raised above the common wheel centerline, the faster a cylindrical shape will be obtained.

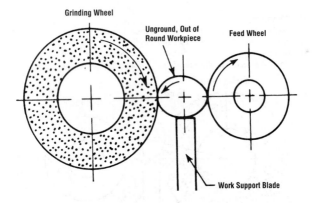

Figure 5. Problems caused by centerless grinding with a flat-top blade and having the wheels and workpiece on common center. (Source, Norton Company.)

Throughfeed centerless grinding. This process is used on straight cylindrical workpieces without interfering shoulders or projections. The workpiece is fed by the offset axis feed wheel, past the grinding wheel that performs the shaping, and through to a discharge position. The diameter of the workpiece is limited only by machine capacity, and its length can be any practical dimension. Workpieces can be fed continuously through the machine. To provide continuous throughfeed, the axis of the feed wheel is tilted slightly (3° to 8°) relative to the axis of the grinding wheel. **Table 4** shows the approximate theoretical feed rate, in inches per minute, that a workpiece will feed past the grinding wheel given the diameter, speed of rotation of the feed wheel, and the angle of inclination of the feed wheel. For example, if the feed wheel diameter is 10.5 inches, the feed wheel speed is 40 rpm, and the angle of inclination of the feed wheel is 4°, it can be seen from the Table that the traverse rate of the workpiece will be 92 inches per minute.

The following guidelines can be used for starting specifications for throughfeed centerless grinding.

Vitrified wheel

Grit size	Surface finish (RMS μ in.)	Part outside dia.	Wheel grade
46	60–70	3" >	K, L
60	32	2–2.50"	L
80	15–20	.50–.75"	L, M

Organic wheel

Grit size	Surface finish (RMS μ in.)	Part outside dia.	Wheel grade
54	32–50	> 1"	Q
54	32–50	1" & <	S
90	10–15	All	Q

Infeed (plunge) centerless grinding. When a workpiece has projections, irregular shapes, varying diameters, or shoulders, infeed grinding is commonly used. To accommodate the irregularities, the grinding wheel (or wheels) is trued to form the various part diameters and lengths required. In infeed grinding, the workpiece can be fed to the wheels from above, with no lateral movement of the piece while it is being ground. This makes the process well suited for profiles and multi-diameter workpieces. *Figure 6* illustrates profiled workpiece being ground with infeed grinding.

(Text continued on p. 1607)

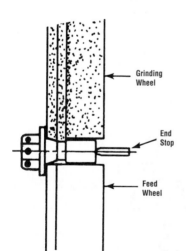

Grinding Wheel

End Stop

Feed Wheel

Figure 6. Position of feed wheel and workpiece in infeed grinding. (Source, Norton Company.)

Table 4. Throughfeed Work Traverse Rates. See Notes. (Source, Norton Company.)

Feed Wheel Angle	Feed Wheel Dia. (inches)	Feed Wheel Speeds										
		10 RPM	15 RPM	20 RPM	25 RPM	30 RPM	40 RPM	50 RPM	70 RPM	90 RPM	125 RPM	160 RPM
		Work Feed Rates (Inches Per Minute)										
1°	8.0	4.4	6.6	8.8	11.0	13.2	17.5	21.9	30.7	39.5	54.8	70.2
	8.5	4.7	7.0	9.3	11.6	14.0	18.6	23.3	32.6	41.9	58.2	74.6
	9.0	4.9	7.4	9.9	12.3	14.8	19.7	24.7	34.5	44.4	61.7	78.9
	9.5	5.2	7.8	10.4	13.0	15.6	20.8	26.0	36.5	46.9	65.1	83.3
	10.0	5.5	8.2	11.0	13.7	16.4	21.9	27.4	38.4	49.3	68.5	87.7
	10.5	5.8	8.6	11.5	14.4	17.3	23.0	28.8	40.3	51.8	72.0	92.1
	11.0	6.0	9.0	12.1	15.1	18.1	24.1	30.2	42.2	54.3	75.4	96.5
	11.5	6.3	9.5	12.6	15.8	18.9	25.2	31.5	44.1	56.7	78.8	100.9
	12.0	6.6	9.9	13.2	16.4	19.7	26.3	32.9	46.0	59.2	82.2	105.3
	12.5	6.9	10.3	13.7	17.1	20.5	27.4	34.2	47.9	61.6	85.6	109.7
	13.0	7.1	10.7	14.3	17.8	21.3	28.5	35.6	49.8	64.1	89.0	114.1
	13.5	7.4	11.1	14.8	18.5	22.1	29.6	36.9	51.7	66.5	92.4	118.5
	14.0	7.6	11.5	15.4	19.2	22.9	30.7	38.3	53.6	69.0	95.8	122.9
2°	8.0	8.8	13.2	17.5	21.9	26.3	35.1	43.9	61.4	78.9	109.6	140.3
	8.5	9.3	14.0	18.6	23.3	28.0	37.3	46.6	65.2	83.9	116.5	149.1
	9.0	9.9	14.8	19.7	24.7	29.6	39.5	49.3	69.1	88.8	123.3	157.9
	9.5	10.4	15.6	20.8	26.0	31.2	41.7	52.1	72.9	93.7	130.2	166.7
	10.0	11.0	16.4	21.9	27.4	32.9	43.9	54.8	76.7	98.7	137.1	175.4
	10.5	11.5	17.3	23.0	28.8	34.5	46.0	57.6	80.6	103.6	143.9	184.2
	11.0	12.1	18.1	24.1	30.2	36.2	48.2	60.3	84.4	108.5	150.8	193.0
	11.5	12.6	18.9	25.2	31.5	37.8	50.4	63.0	88.3	113.5	157.6	201.7

(Continued)

Table 4. *(Continued)* **Throughfeed Work Traverse Rates.** See Notes. *(Source, Norton Company.)*

Feed Wheel Angle	Feed Wheel Dia. (inches)	Feed Wheel Speeds										
		10 RPM	15 RPM	20 RPM	25 RPM	30 RPM	40 RPM	50 RPM	70 RPM	90 RPM	125 RPM	160 RPM
		Work Feed Rates (Inches Per Minute)										
2°	12.0	13.2	19.7	26.3	32.9	39.5	52.6	65.8	92.1	118.4	164.5	210.5
	12.5	13.7	20.5	27.4	33.2	41.1	54.8	68.5	96.0	123.4	171.3	219.2
	13.0	14.3	21.3	28.5	33.6	42.8	57.0	71.3	99.8	128.3	174.2	228.5
	13.5	14.8	22.1	29.6	33.9	44.4	59.2	74.0	103.7	133.3	177.0	236.7
	14.0	15.4	22.9	30.7	34.3	46.1	61.4	76.8	107.5	138.2	183.9	245.5
3°	8.0	13.2	19.7	26.3	32.9	39.5	52.6	65.8	92.1	118.4	164.4	210.5
	8.5	14.0	21.0	28.0	34.9	41.9	55.9	69.9	97.8	125.8	174.7	223.6
	9.0	14.8	22.2	29.6	37.0	44.4	59.2	74.0	103.6	133.2	185.0	236.8
	9.5	15.6	23.4	31.2	39.1	46.9	62.5	78.1	109.3	140.6	195.3	249.9
	10.0	16.4	24.7	32.9	41.1	49.3	65.8	82.2	115.1	148.0	205.5	263.1
	10.5	17.3	25.9	34.5	43.2	51.8	69.1	86.3	120.9	155.4	215.8	276.2
	11.0	18.1	27.1	36.2	45.2	54.3	72.3	90.4	126.6	162.8	226.1	289.4
	11.5	18.9	28.4	37.8	47.3	56.7	75.6	94.5	132.4	170.2	236.4	302.6
	12.0	19.7	29.6	39.5	49.3	59.2	78.9	98.7	138.1	177.6	246.6	315.7
	12.5	20.5	30.9	41.1	51.4	61.6	82.2	102.8	143.9	185.0	256.9	328.8
	13.0	21.3	32.1	42.8	53.4	64.1	85.5	106.9	149.6	192.4	267.2	341.9
	13.5	22.1	33.4	44.4	55.5	66.5	88.8	111.0	155.4	199.8	277.5	355.0
	14.0	22.9	34.6	46.1	57.5	69.0	92.1	115.1	161.1	207.2	287.8	368.1
4°	8.0	17.5	26.3	35.1	43.8	52.6	70.1	87.7	122.7	157.8	219.2	280.5
	8.5	18.6	27.9	37.3	46.6	55.9	74.5	93.1	130.4	167.7	232.9	298.1
	9.0	19.7	29.6	39.4	49.3	59.2	78.9	98.6	138.1	177.5	246.6	315.6

(Continued)

Table 4. (Continued) Throughfeed Work Traverse Rates. See Notes. (Source, Norton Company.)

Feed Wheel Angle	Feed Wheel Dia. (inches)	Feed Wheel Speeds										
		10 RPM	15 RPM	20 RPM	25 RPM	30 RPM	40 RPM	50 RPM	70 RPM	90 RPM	125 RPM	160 RPM
		Work Feed Rates (Inches Per Minute)										
4°	9.5	20.8	31.2	41.6	52.1	62.5	83.3	104.1	145.7	187.4	260.3	333.1
	10.0	21.9	32.9	43.8	54.8	65.7	87.7	109.6	153.4	197.2	273.9	350.7
	10.5	23.0	34.5	46.0	57.5	69.0	92.0	115.1	161.1	207.1	287.6	368.2
	11.0	24.1	36.2	48.2	60.3	72.3	96.4	120.5	168.8	217.0	301.3	385.7
	11.5	25.2	37.8	50.4	63.0	75.6	100.8	126.0	176.4	226.8	315.0	403.3
	12.0	26.3	39.4	52.6	65.7	78.9	105.2	131.5	184.1	236.7	328.7	420.8
	12.5	27.4	41.0	54.8	68.5	82.2	109.6	136.9	191.7	246.5	342.4	438.4
	13.0	28.5	42.6	57.0	71.3	85.5	114.0	142.4	199.4	256.4	356.1	456.0
	13.5	29.6	44.2	59.2	74.1	88.8	118.4	147.8	207.0	266.2	369.8	473.6
	14.0	30.7	45.8	61.4	76.9	92.1	122.8	153.3	214.7	276.1	383.5	491.2
5°	8.0	21.9	32.9	43.8	54.8	65.7	87.6	109.5	153.3	197.2	273.8	350.5
	8.5	23.3	34.9	46.5	58.2	69.8	93.1	116.4	162.9	209.5	290.9	372.4
	9.0	24.6	37.0	49.3	61.6	73.9	98.6	123.2	172.5	221.8	308.0	394.3
	9.5	26.0	39.0	52.0	65.0	78.0	104.1	130.1	182.1	234.0	325.2	416.2
	10.0	27.4	41.1	54.8	68.5	82.1	109.5	136.9	191.7	246.4	342.3	438.1
	10.5	28.8	43.1	57.5	71.9	86.3	115.0	143.8	201.3	258.8	359.4	460.0
	11.0	30.1	45.2	60.2	75.3	90.4	120.5	150.6	210.8	271.1	376.5	481.9
	11.5	31.5	47.2	63.0	78.7	94.5	126.0	157.4	220.4	283.4	393.6	503.8
	12.0	32.9	49.3	65.7	82.1	98.6	131.4	164.3	230.0	295.7	410.7	525.7
	12.5	34.3	51.3	68.5	85.5	102.7	136.9	171.2	239.6	308.0	427.8	547.6
	13.0	35.7	53.4	71.2	88.9	106.8	142.4	178.1	249.2	320.3	444.9	569.5

(Continued)

Table 4. *(Continued)* **Throughfeed Work Traverse Rates.** See Notes. *(Source, Norton Company.)*

Feed Wheel Angle	Feed Wheel Dia. (inches)	Feed Wheel Speeds										
		10 RPM	15 RPM	20 RPM	25 RPM	30 RPM	40 RPM	50 RPM	70 RPM	90 RPM	125 RPM	160 RPM
		Work Feed Rates (Inches Per Minute)										
5°	13.5	37.1	55.4	74.0	92.3	110.9	147.9	185.0	258.8	332.6	462.0	591.4
	14.0	38.5	57.5	76.7	95.7	115.0	153.4	191.9	268.4	344.9	479.1	613.3
6°	8.0	26.3	39.4	52.5	65.7	78.8	105.1	131.4	183.9	236.4	328.4	420.3
	8.5	27.9	41.9	55.8	69.8	83.7	111.7	139.6	195.4	251.2	348.9	446.6
	9.0	29.6	44.3	59.1	73.9	88.7	118.2	147.8	206.9	266.0	369.4	472.9
	9.5	31.2	46.8	62.4	78.0	93.6	124.8	156.0	218.4	280.8	390.0	499.2
	10.0	32.8	49.3	65.7	82.1	98.5	131.4	164.2	229.9	295.6	410.5	525.4
	10.5	34.5	51.7	69.0	86.2	103.4	137.9	172.4	241.4	310.3	431.0	551.7
	11.0	36.1	54.2	72.2	90.3	108.4	144.5	180.6	252.9	325.1	451.5	578.0
	11.5	37.8	56.6	75.5	94.4	113.3	151.1	188.8	264.4	339.9	472.1	604.2
	12.0	39.4	59.1	78.8	98.5	118.2	157.6	197.0	275.8	354.7	492.6	630.5
	12.5	41.0	61.6	82.1	102.6	123.1	164.3	205.2	287.4	369.5	513.1	656.8
	13.0	42.6	64.1	85.4	106.7	128.0	170.9	213.4	214.4	384.3	533.6	683.1
	13.5	44.2	66.6	88.7	110.8	132.9	177.5	221.6	227.8	399.1	554.1	709.4
	14.0	45.8	69.1	92.0	114.9	137.8	184.1	229.8	241.2	413.9	574.6	735.7
7°	8.0	30.6	45.9	61.3	76.6	91.9	122.5	153.1	214.4	275.7	382.9	490.1
	8.5	32.5	48.8	65.1	81.4	97.6	130.2	162.7	227.8	292.9	406.8	520.7
	9.0	34.5	51.7	68.9	86.1	103.4	137.8	172.3	241.2	310.1	430.7	551.3
	9.5	36.4	54.6	72.7	90.9	109.1	145.5	181.9	254.6	327.4	454.7	582.0
	10.0	38.3	57.4	76.6	95.7	114.9	153.1	191.4	268.0	344.6	478.6	612.6
	10.5	40.2	60.3	80.4	100.5	120.6	160.8	201.0	281.4	361.8	502.5	643.2

(Continued)

Table 4. (Continued) Throughfeed Work Traverse Rates. See Notes. (Source, Norton Company.)

Feed Wheel Angle	Feed Wheel Dia. (inches)	Feed Wheel Speeds										
		10 RPM	15 RPM	20 RPM	25 RPM	30 RPM	40 RPM	50 RPM	70 RPM	90 RPM	125 RPM	160 RPM
		Work Feed Rates (Inches Per Minute)										
7°	11.0	42.1	63.2	84.2	105.3	126.3	168.5	210.6	294.8	379.0	526.4	673.8
	11.5	44.0	66.0	88.1	110.1	132.1	176.1	220.1	308.2	396.3	550.4	704.5
	12.0	45.9	68.9	91.9	114.9	137.8	183.8	229.7	321.6	413.5	574.3	735.1
	12.5	47.8	71.8	95.7	119.7	143.5	191.5	239.3	335.0	430.7	598.2	765.7
	13.0	49.7	74.7	99.5	124.5	149.2	199.2	248.9	348.4	447.9	622.1	796.3
	13.5	51.6	77.6	103.3	129.3	154.9	206.9	258.5	361.8	465.1	646.0	826.9
	14.0	53.5	80.5	107.1	134.1	160.6	214.6	268.1	375.2	482.3	669.9	857.5
8°	8.0	35.0	52.5	70.0	87.4	104.9	139.9	174.9	244.8	314.8	437.2	559.6
	8.5	37.2	55.7	74.3	92.9	111.5	148.7	185.8	260.1	334.5	464.5	594.6
	9.0	39.3	59.0	78.7	98.4	118.0	157.4	196.7	275.4	354.1	491.9	629.6
	9.5	41.5	62.3	83.1	103.8	124.6	166.1	207.7	290.7	373.5	519.2	664.6
	10.0	43.7	65.6	87.4	109.3	131.2	174.9	218.6	306.1	393.5	546.5	699.5
	10.5	45.9	68.9	91.8	114.8	137.7	183.6	229.5	321.4	413.2	573.8	734.5
	11.0	48.1	72.1	96.2	120.2	144.3	192.4	240.5	336.7	432.8	601.2	769.5
	11.5	50.3	75.4	100.6	125.7	150.8	201.1	251.4	352.0	452.5	628.5	804.5
	12.0	52.5	78.7	104.9	131.2	157.4	209.9	262.3	367.3	472.2	655.8	839.5
	12.5	54.7	82.0	109.3	136.7	164.0	218.6	273.2	382.6	491.9	683.1	874.5
	13.0	56.9	85.3	113.6	142.2	170.6	227.4	284.1	397.9	511.6	710.4	909.5
	13.5	59.1	88.6	118.0	147.7	177.2	236.1	295.0	413.2	531.3	737.7	944.5
	14.0	61.3	91.9	122.3	153.2	183.8	244.9	305.9	428.5	551.0	765.0	979.5

Notes: In centerless grinding, the speed that the work feeds past the grinding wheel on throughfeed jobs depends on the diameter, speed of rotation, and angle of inclination of the feed wheel.

The following guidelines can be used for starting specifications for infeed centerless grinding.

Wheel speed. Typical wheel speeds are 6,500, 8,500, 12,000 sfm.
Speed is not to exceed manufacturer's recommendation.

Wheel speed to work ratio
for materials 50 R_C and softer
work speed (sfm) = wheel speed (sfm) ÷ 60.

for materials harder than 50 R_C
work speed (sfm) = wheel speed (sfm) ÷ 100.

Infeed per revolution of the workpiece
steel (soft)
rough grinding = 0.0005"
finish grinding = 0.0002"

plain carbon steel (hardened)
rough grinding = 0.0002"
finish grinding = 0.00005"

alloy and tool steel (hardened)
rough grinding = 0.0001"
finish grinding = 0.000025".

Endfeed centerless grinding. This operation is used to grind conically tapered cylindrical sections, such as shanks on "A" and "B" taper drill bits. The feed wheel, grinding wheel, and work blade are set up in a fixed relationship to each other, and then the two wheels are dressed to a shape to match the desired taper of the workpiece, and the work support blade is ground to the same taper. The workpiece is fed from the front of the machine and is ground until it reaches an end stop.

Work support blade

The work support blade performs three functions. 1) It positions the workpiece correctly in the space between the grinding and feed wheels to obtain the desired part diameter and roundness. 2) It supports and guides the workpiece during the grinding operation. 3) It keeps the workpiece in continuous contact with the feed wheel.

The length of the support blade is determined by the face width of the grinding wheel. It should be at least as long as the wheel is wide. In throughfeed operations, it should be long enough to provide support during loading and unloading as well. Its thickness depends on the diameter of the workpiece. The following chart provides recommended thicknesses.

Workpiece Diameter mm/inch	Recommended Blade Thickness mm/inch
6 / 0.250	3 / 0.125
6–7 / 0.3125	4 / 0.1875
7–9 / 0.375	5 / 0.250
9–11 / 0.4375	7 / 0.3125
11–13 / 0.500	9 / 0.375
13–16 / 0.625	11 / 0.4375
16 and over / 0.625 and over	14 / 0.5625

In most instances, a 30° top angle on the blade will be satisfactory. However, when broad faced grinding wheels call for a correspondingly long work blade, the angle may be reduced to add strength to the support. In such cases, an angle of 20°, or in some cases as shallow as 20°, may work satisfactorily.

Blade material is important because the workpiece rotates on the blade, so the potential for scratching or otherwise damaging the workpiece exists. Sintered carbide blades have extremely good wear characteristics, making them suitable for long production runs, and especially desirable when workpiece materials are oil- or case-hardened steel, cast iron, or stainless steel. High speed steel blades are commonly used with nonferrous metals, or small diameter parts. Dense cast iron, Meehanite, and hard bronze are sometimes used when scoring of the workpiece is a potential problem.

Feed wheels

Feed wheels must have three important characteristics. 1) A high coefficient of friction in order to provide the best possible control of work travel rate. 2) The ability to absorb vibrations that can generate geometric defects such as chatter marks. 3) Minimal wear and form alteration under operation. **Table 5** provides common shapes and dimensions of feed wheels.

Table 5. Shapes and Dimensions of Standard Feed Wheels. *(Source, Norton Company.)*

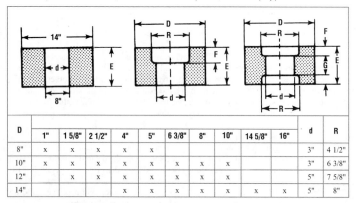

D	1"	1 5/8"	2 1/2"	4"	5"	6 3/8"	8"	10"	14 5/8"	16"	d	R
8"	x	x	x	x	x						3"	4 1/2"
10"	x	x	x	x	x	x	x	x			3"	6 3/8"
12"		x	x	x	x	x	x	x			5"	7 5/8"
14"				x	x	x	x	x	x	x	5"	8"

Troubleshooting centerless grinding operations. *(Source, Norton Company.)*

Problem	Possible Cause	Suggested Correction
Chatter	Wheel out of balance	Heavy dress or replace wheel.
	Wheel not tight on spindle	Tighten flanges.
	Blade angle too steep	Change work blade.
	Work too high above center	Readjust work blade.
	Work blade too thin or warped	Replace blade.
	Wheel too hard or too fine	Replace wheel or make wheel act softer.
	Glazed or loaded wheel	Redress.
	Blade not properly clamped	Readjust blade.
	Too heavy a cut	Decrease stock removal rate.
	Vibration transferred to machine	Check for vibration.

(Continued)

Troubleshooting centerless grinding operations. *(Continued) (Source, Norton Company.)*

Problem	Possible Cause	Suggested Correction
Feed lines	Grinding wheel not relieved on exit side	Dress wheel so that it stops grinding 1/2" from exit edge.
	Work guides improperly set	Readjust guides.
	Wheel not uniform across face	Redress.
Scored work	Pick-up on blade (dirt, chips, abrasive)	Use coolant with greater lubricity or change work blade to softer material.
Burned work	Improper coolant flow	Readjust nozzle.
	Dull diamond glazing the wheel	Replace diamond or rotate.
	Wheel too hard	Replace wheel or make wheel act softer.
	Dress rate too slow	Increase diamond traverse rate.
Fish tails	Dirty coolant	Replace coolant or filters.
	Loose grit in wheel	Redress wheel.
Poor finish	Improper wheel grade	Proper adjustments are self-evident.
	Dirty coolant	
	Dress rate too fast	
Poor finish	Jumpy, erratic dressing traverse rate	Proper adjustments are self-evident.
	Dull diamond	
	Diamond loose	
	Blade not tight	
	Blade angle too steep	
	Pick-up on blade	
	Oil or loading on wheel face	
	Feed wheel loose	
	Feed wheel speed too high	
	Work too high above center	
Out-of-round work	Work not high enough above center	Readjust work blade.
	Workpiece bowed	Prestraighten workpiece before grinding.
	Feed wheel not round	Properly redress feed wheel.
	Blade angle not steep enough	Change work blade.
	Feed wheel too slow	Increase feed wheel speed.
Barrel-shaped work (low ends)	Work guides on entrance and exit sides deflected toward feed wheel	Readjust entrance and exit guides.
	Face of feed wheel not straight at contact point with workpiece	Properly redress feed wheel. Be sure truing fixture is properly set.
Concave work (low center)	Work guides on entrance side deflected toward the grinding wheel	Readjust the entrance and exit guides.
Taper on front end of work	Work guides on entrance side deflected toward the feed wheel	Readjust entrance guides.

(Continued)

Troubleshooting centerless grinding operations. *(Continued) (Source, Norton Company.)*

Problem	Possible Cause	Suggested Correction
Taper on rear end of work	Work guides on exit side deflected toward the feed wheel	Readjust exit guides.
Erratic work sizes	Gibs not adjusted properly on upper and lower slides	Readjust gibs.
	Wear in infeed screw and nut	Repair infeed mechanism.
	Infeed lever not held tight	Tighten nut.
	Insufficient lubrication on slides	Provide proper lubrication.
	Work blade loose	Tighten blade.
	Parts too hot	Change wheel, adjust coolant flow, or make wheel cut more freely.
	Feed wheel camming	Tighten collet screws, replace feed wheel bearing.
Work not straight	Work not straightening sufficiently before grinding	Prestraighten workpiece before grinding.
	Too much stock removal on a previous pass	Adjust stock removal rates on prior pass.
	Parts too hot	Change wheel, adjust control flow, or make wheel cut more freely.
Straight taper on work	Work guides improperly set	Readjust entrance and exit guides.
		Properly redress feed wheel.
	Blade not level	Readjust blade.
	Loose gibs	Adjust and tighten gibs.
	Parts not end-to-end in through feed	Adjust feed rate.
	Parts too hot	Change wheel, adjust coolant flow, or make wheel cut more freely.

Creep feed grinding

This form of grinding gained initial popularity in Europe and Japan and has now become an established technique in North America. It offers high productivity, and its cost effectiveness has encouraged wide acceptance. Because creep feed is a "one-pass" operation that makes a deep cut of up to one inch in steel materials, conventional grinding machines are not suitable for the process. Table drives on creep feed machines are mechanical rather than hydraulic, and horsepower requirements are high. The grinding wheel is formed, and the form is transmitted onto the workpiece. The machine must be absolutely rigid.

Creep feed grinding is a plunge operation, and the wheel makes only one or two passes at very low table speeds (0.5 to 1 inch per minute). A heavy flow of cutting fluid is applied close to the nip in order to remove chips and cool the work, and refrigeration is sometimes used to lower operating temperatures. A second pass, when required, is typically no more than 0.002" deep and is used after a dressing cycle to "clean up" the workpiece. Vitrified aluminum oxide wheels, usually in grades D through G, are popular because their softness and open structure provides good chip clearance and allows for the application of cutting fluids. Diamond roll dressing is preferred.

Continuous dressing of the wheel with a diamond roll during grinding, in amounts on the order of 20 to 60 millionths of an inch per revolution, can assure that the wheel remains sharp throughout the operation, and dramatically reduce cutting times. Modern creep feed grinding machines automatically compensate for this removal of material from the wheel, and this has led to the process becoming popular for fully automated machining.

Creep feed wheels are normally purchased in an unfinished size and then altered by dressing for a specific application. The following chart provides alteration parameters for standard size wheels.

Alteration parameters for creep feed grinding wheels—all dimensions in inches			
Unfinished size* D × T × H	Finished size		
	Outside diameter (min.)	Width (min.)	Bore (max.)
16 × 1 × 5	14	1/2	8
16 × 2 1/2 × 5	14	1 1/4	8
16 × 4 × 5	14	2 3/4	8
20 × 1 × 8	17	1/2	12
20 × 2 1/2 × 8	17	2 1/4	12
20 × 4 × 8	17	2 3/4	12
24 × 2 × 8	20	3/4	12
24 × 4 × 8	20	2 3/4	12

Snagging

Snagging is a rough grinding application used to remove unwanted metal. Since surface finish is not critical, durable wheels are used. Typical applications include removing unwanted metal on castings; removing flaws and cracks; removing gates, risers, and parting lines; rough beveling; grinding down heavy welds; and preparing surfaces for cleaning or painting. Straight (Type 1) and straight cup (Type 06) wheels are commonly used in either horizontal or straight shaft grinding machines, or flaring cup (Type 11) wheels are used in right angle grinders. Various round and square tipped cones and plugs are also used.

Troubleshooting snagging operations. *(Source, Norton Company.)*

Problem	Possible Cause	Suggested Correction
Poor stock removal	Insufficient pressure applied	Increase pressure to use all available power.
	Wheel too coarse or too hard	Use finer grit and/or softer grinding wheel.
Grinding cost too high	Wheel acting too soft	Use harder and/or coarser wheel.
Wheel loading or glazing	Grade too hard	Try softer grade.
	Grit too coarse	Try finer grit.
Wheel "dusty"	Wheel too soft	Try harder grade.
	Bond too soft	Try harder bond.
Wheel doesn't hold corner, or is spauling	Wheel too coarse	Use finer grit, or 2 grit combination.
	Wheel too soft	Use harder grade.

(Continued)

Troubleshooting snagging operations. *(Continued) (Source, Norton Company.)*

Problem	Possible Cause	Suggested Correction
Burning the workpiece	Wheel too hard	Use softer grade.
	Wheel glazed or loaded	Dress the wheel face.
	Grinding pressure too low	Increase grinding force.
	Grinding pressure too high	Reduce grinding force.
Surface finish too rough	Wheel too coarse	Try finer grit.
	Wheel speed too slow	Try higher speed machine.
	Worn machine bearings	Rebuild machine.
Vibration	Wheel worn out of round	Try dressing the wheel.
	Faulty flanges	Check flanges for flatness and burrs (see ANSI B7.1).
	Bent machine spindle	Check spindle run-out.
	Worn machine bearings	Rebuild machine.
Wheel stalls or slows	Grinding force too high	Reduce pressure or contact area.
	Wheel too hard	Use softer grade wheel.
	Air supply inadequate	Check air pressure, CFPM air flow, and air line size.
	Machine power too low	Replace machine with higher power grinder.

Cutting-off

Cutting-off operations can be done by several methods, including the use of cut-off abrasive wheels. Other methods include the laser, the abrasive water jet, the metal saw, the friction saw, and oxyacetylene and plasma arc torches. All of these methods can be used on ferrous materials, but studies by the Norton Company suggest that the abrasive wheel is capable of outperforming all other methods. For cutting off nonferrous materials such as aluminum, brass, copper, and bronze, the selection is usually between the abrasive wheel and metal saw.

While the metal saw continues to be the acceptable method of cutting off large cross sections (such as steel beams and plate), the abrasive wheel can often perform the same operation more rapidly and at less expense. A careful study of costs made by one manufacturer showed the following.

The ratios of cutting time of the saw compared with the abrasive wheel ranged from 12:1 to 20:1. This decrease in cutting time with the abrasive cut-off wheel allowed the operator to be more productive.

The operating cost, including the cost of cutting tools and power, plus the labor cost to produce the same amount of work, was, in the majority of cases, less with the abrasive wheel than with the saw.

The reason cited for the superior performance of the abrasive wheel is that the wheel provides a vastly greater number of cutting points compared to the saw. The typical cut-off wheel presents a multitude of sharp abrasive points of teeth on its cutting face, whereas the saw has only a limited few. And abrasive cutting teeth, when rotated at a speed of two to three miles a minute, actually cut their way through the work as truly as a saw.

Cut-off wheels should be run at the highest possible speed marked on the wheel and cuts should be made as quickly as possible. Further, it is recommended that a cut-off machine should have one horsepower for every inch of wheel diameter. If less power is available, a softer specification wheel is recommended.

Types of cut-off wheels

Nonreinforced wheels. These wheels are designed for use on "fixed base" types of cut-off machines where the work is securely clamped, the wheel is adequately guarded, and the machine is designed so the wheel operates in a controlled cutting plane. Although some of these wheels have a reinforced zone around the hole, they are not considered to be truly reinforced. These wheels have resin, rubber, or shellac bonds. Most production jobs use resin on dry applications, but rubber wheels are also used dry, and either rubber or resinoid wheels can be used for wet applications. Shellac wheels are generally supplied in smaller diameters for either wet or dry applications, and they are predominately used in applications where extreme versatility and quality of cut are major considerations.

The hardest applicable grade nonreinforced wheel should be used for maximum durability, but softer grades tend to resharpen more readily and should be used on more difficult to cut materials for better cutting quality, achieved at the expense of higher wheel wear. Wheels with higher structure numbers have greater durability but are slower cutting. The effect of going higher in structure number is somewhat similar to using harder letter grades.

Reinforced wheels. These wheels contain strengthening fabric, or filaments, that make them resistant to breakage caused by cross bending. Portable cut-off, swingframe, locked head push-through, and foundry chop stroke operations require reinforced cut-off wheels. Wheels are Type 01 Straight configuration and are commonly available in diameters ranging from $^3/_4$" to 72", although the most popular sizes range from 3" to 30" in diameter. Wheel thickness is determined by grit size, construction, and wheel diameter, but maximum thickness is under $^1/_4$". Bonds are either resinoid for dry cutting, or rubber for wet cutting.

The hardest applicable grade reinforced wheel should be used for maximum durability, but softer grades tend to resharpen more readily and should be used on more difficult to cut materials for better cutting quality, achieved at the expense of higher wheel wear. Wheels with higher structure numbers have greater durability but are slower cutting. The effect of going higher in structure number is somewhat similar to using harder letter grades. It

Diameter	Spindle Speed (RPM)	SFM
6	10,185	16,000
7	8,730	16,000
8	7,640	16,000
10	6,115	16,000
12	5,095	16,000
14	4,365	16,000
16	3,820	16,000
18	3,015	14,200
20	2,715	14,200
24	2,260	14,200
26	2,090	14,200
30	1,810	14,200
34	1,595	14,200

should be noted that diamond blades are also available for cutting glass, quartz, germanium, silicon, ceramics, refractories, and cemented carbides.

Reinforced wheels up to 16" in diameter are approved for use as high as 16,000 sfm. For wheels over 16" in diameter, the maximum rating is 14,200 sfm. The following chart provides the maximum spindle speed (rpm) allowed for popular diameters of reinforced cut-off wheels.

Troubleshooting cut-off operations (Source, Norton Company)

When a cut-off wheel is of proper specification for an operation and is being used properly, the face of the wheel will develop one of three contours: 1) convex, slightly rounded, correct for dry applications; 2) square, correct for wet applications; 3) concave, indented with a constant radius, correct when mild specifications are used wet with light pressure as when cutting thin-wall tubing, or wet cutting titanium.

Improperly specified or used cut-off wheels will develop one of the following contours on the face of the wheel: 1) pointed, indicating that the wheel is too hard acting or that feed rates are too slow; 2) chisel, resulting from unequal coolant distribution, possibly from a plugged or misaligned nozzle.

Problem	Possible Cause	Selected Correction
Nonsquare cuts	Work not properly clamped	Check clamp and clean to remove swarf.
	Misaligned spindle bearings	Check for bearing truth and alignment.
	Poor coolant distribution	Ensure equal volume of coolant to each side of the wheel.
	Wheel is too hard	Wheel is too hard.
Workpiece burn	Insufficient feed rate	Work machine to maximum power available.
	Poor coolant flow	Increase volume and direct at cutting point.
	Wheel is too coarse	Use finer wheel or more machine power.
	Wheel is too hard	Use softer grade wheel.
	Wheel is running out	Check spindle and flanges.
	Wheel speed is too slow	Ensure tight flanges and maximum speed is being used.
Poor finish	Too much burr	Use finer grit or softer wheel.
	Wheel too coarse	Use finer grit.

Grinding tool steels

Improper grinding can affect the performance of tool steels not only by the formation of grinding cracks that can lead to eventual breakage, but by the development of a softened surface zone. As explained in an article prepared by Timken Latrobe Steel ("The Effect of Improper Grinding Technique on the Surface Structure of High Speed Steels"), the visual appearance of the softened zone is deceptive because the amount of visible burning often appears very slight, but the microscopic effect can be shown to progress a considerable distance below the surface with consequent negative results on tool performance. This fact is of particular importance when subsequent manufacturing operations to remove the visible burn color are either a chemical treatment or a very shallow grinding pass.

In order to illustrate that the effects of abusive grinding can penetrate well below the ground surface, Timken Latrobe's engineers performed the following experiment on a $7/16$" sample of round, hardened, and tempered high speed steel with a hardness rating tested at R_C 64.5/65. The sample was clamped on an adjustable vise and ground on one side at

30° to the longitudinal axis with a "soft" 32C46F8VP wheel. The sample was then rotated 180° so the wheel would produce a similar flat having a 60° included angle with the first. For the cut on this side, a badly glazed "hard" wheel (220 grit Alundum, vitrified bond) was used. The side ground with the soft wheel appeared to be visibly smoother in 125× magnification, and no appearance of discoloration was evident. On the side ground with the hard wheel, however, a rehardened zone is evident at the same magnification. At much higher magnification (1500%), distortion in the grains can be seen, indicating that the steel was plastic in the surface zone during grinding with the hard wheel.

Generally, wheel burning is believed to cause a high "temper" in the surface zones. However, it is also true that excessive wheel burn can generate sufficient heat to cause a rehardening of the surface. Therefore, it is possible to have either a dark etching or light etching microstructure, depending on the degree of wheel burn. In this experiment, the burning caused partial rehardening. The resulting structure changed to a light etching appearance, but the surface hardness that developed was not that of a fully hardened material. Underlying the rehardened zone, the normal "overtempered" zone existed, but it was of lower hardness than the unaffected metal. The microhardness results were:

Depth below surface (inches)	Hardness R_C
0.0006	61.1
0.0025	62.2
0.0045	63.8
0.0067	63.5
0.0087	64.5
interior	64.5

The depth and hardness of the burned zone varies considerably within the affected area owing to mechanical conditions of grinding such as coolant flow, rate of wheel breakdown, chatter, etc.

Selecting grinding wheels for tool steels

Tool steels have been broadly broken down into four classes to indicate their grindability, and wheel recommendations are commonly made by class of material. In **Table 6**, the four classes are explained and the tool steels assigned to each class, by AISI/SAE designation, are provided. In **Table 7**, a grinding wheel selection guide for both surface and cylindrical grinding of tool steels gives specific operational guidelines. **Table 8** provides first and second choices for wheels for particular operations, and wheel configurations by class of steel being ground.

Superabrasive Machining

CBN wheels

Grinding wheels or mounted points using either CBN or diamond as an abrasive are commonly referred to as superabrasives. CBN stands for cubic boron nitride, which was initially conceived through the efforts of General Electric researchers who were attempting to develop artificial diamond materials. Today, CBN is the workhorse material for grinding very hard steels to precise profiles and finishes. The one obvious drawback of CBN materials is that they do appear to react to water at high temperatures, which results in the dissolution of the protective boron oxide layer. Also, at normal speeds, it is not the abrasive of choice for grinding soft, gummy material, as the wheels will become clogged with the material. At higher speeds, however, this tendency disappears and soft materials can be ground successfully. Generally speaking, CBN is used to grind ferrous materials such as high speed tool steels, die steel, hardened carbon steel, alloy steels, aerospace alloys, and abrasion resistant ferrous materials.

Table 6. Tool Steel Classes Rated by Grindability.

Class Details	AISI/SAE Designation
Class 1 Tool Steels These steels are considered easy to grind. On a scale of 1 to 100, steels in this class have a grindability rating between 12 and 100.	A6, A10; H10, H11, H12, H13, H14, H19, H21; L2, L3, L6; All "O" series tool steels; All "S" series tool steels; All "W" series tool steels.
Class 2 Tool Steels These steels are considered average to grind. On a scale of 1 to 100, steels in this class have a grindability rating between 3 and 12.	D5; H22, H23, H24, H26; M1, M2, M8, M42; P6, P20, P21; T1, T2.
Class 3 Tool Steels These steels are considered difficult to grind. On a scale of 1 to 100, steels in this class have a grindability rating between 1 and 3.	D2, D3; M2 (high carbon), M7, M10; T3, T6, T5.
Class 4 Tool Steels These steels are extremely difficult to grind. On a scale of 1 to 100, steels in this class have a grindability rating of less than 1.0.	D7; M3; T15.

Table 7. Grinding Wheel Selection Guide for Tool Steels. *(Source, Bill Bryson.)*

		Class 1 Tool Steels	Class 2 Tool Steels	Class 3 Tool Steels	Class 4 Tool Steels
Surface Grinding					
Abrasive		Semi-friable	Semi-friable	Friable	Friable
Grit Size		36-60	36-60	46-60	80-100
Hardness		H - J	G - H	F - G	F
Bond		Vitrified	Vitrified	Vitrified	Vitrified
Wheel Speed - rpm		5,500 - 6,500	5,500 - 6,500	5,500 - 6,500	3,000
Table Speed	ft/m	50 - 75	50 - 75	Maximum	Maximum
	m/m	15.2 - 22.9	15.2 - 22.9		
Cross Feed	inch	1/32 - 1/16	1/32 - 1/16	1/64 - 1/32	Wheel Width
	mm	0.794 - 1.588	0.794 - 1.588	0.397 - 0.794	
Infeed	inch	0.001 - 0.005	0.001 - 0.003	0.001 - 0.005	0.0005 - 0.001
	mm	0.0254 - 0.1270	0.0254 - 0.0762	0.0254 - 0.1270	0.0127 - 0.0254
Cylindrical Grinding					
Abrasive		Semi-friable	Semi-friable	Friable	Friable
Grit Size		60-80	60-80	60-80	80-100
Hardness		K - L	J - K	H - J	F - H
Bond		Vitrified	Vitrified	Vitrified	Vitrified
Wheel Speed - rpm		5,500 - 6,500	5,500 - 6,500	5,500 - 6,500	3,000
Table Speed	ft/m	90 - 150	90 - 150	100 - 150	100 - 150
	m/m	27.4 - 45.7	27.4 - 45.7	30.45 - 45.7	30.45 - 45.7
Cross Feed	inch	1/32 - 1/16	1/32 - 1/16	1/32 - 1/16	1/16 - 1/8
	mm	0.794 - 1.588	0.794 - 1.588	0.794 - 1.588	1.599 - 3.175
Infeed	inch	0.001 - 0.002	0.001 - 0.002	0.001 - 0.002	0.0005 - 0.001
	mm	0.0254 - 0.0508	0.0254 - 0.0508	0.0254 - 0.0508	0.0127 - 0.0254

Note: Table reproduced from *Heat Treatment, Selection, and Application of Tool Steels* by Bill Bryson (Hanser Gardner Publications).

Table 8. Recommended Wheels for Grinding Hardened Tool Steels. See Notes. *(Source, Norton Company.)*

Operation		Wheel Dia.	Wet or Dry	First Choice	Second Choice
Class 1 Tool Steels					
Production Grinding	Centerless	-	Wet	5SG60-MVS	53A60-M8VCN
	Cylindrical	14" and Smaller	Wet	5SG60-LVS	32A60-L5VBE
		16" and Larger	Wet	5SG60-KVS	53A60-L8VCN
	Internal	Under 1/2"	Wet	5SG80-NVS	32A80-N8VBE
		1/2" to 1"	Wet	5SG60-MVS	32A60-N6VBE
		1" to 3"	Wet	5SG60-LVS	5SG60-J12VSP
		Over 3"	Wet	5SG45-LVS	5SG46-K12VSP
Production and Toolroom Grinding	Surface Grinding-Straight Wheel	14" and Smaller	Wet	5SG46-G12VSP	5SG46-IVS
			Dry	5SG46-IVS	5SG46-H12VSP
		Over 14"	Wet	5SG46-IVS	5SG46-H12VSP
	Segments	1 1/2" Rim	Wet	32A36-E12VBEP	5SG36-E12VSP
	Cylinders	1 1/2" Rim	Wet	5SG30-F12VSP	32A36-F12VBEP
	Cups	3/4" Rim	Wet	5SG46-JVS	32A46-J5VBE
Toolroom Grinding	Internal	Under 1/2"	Dry	5SG80-MVS	32A80-M8VBE
		1/2" to 1"	Dry	5SG60-LVS	32A60-M6VBE
		1" to 3"	Dry	5SG60-KVS	5SG60-I12VSP
		Over 3"	Dry	5SG46-KVS	5SG46-J12VSP
	Cutter Sharpening-Straight Wheel	-	Wet	5SG46-KVS	32A46-K8VBE
		-	Dry	5SG46-JVS	32A46-K5VBE
	Dish Shape	-	Dry	5SG60-JVS	32A60-L5VBE
	Cup Shape	-	Dry	5SG60-KVS	32A60-L5VBE
		-	Wet	5SG60-LVS	32A60-L5VBE
	Form Tool	8" and Smaller	Wet	5SG60-LVS	32A60-L5VBE
			Dry	5SG60-KVS	32A60-L5VBE
		10" and Larger	Wet	5SG60-LVS	32A60-L8VBE
Class 2 Tool Steels					
Production Grinding	Centerless	-	Wet	5SG60-MVS	53A60-L8VCN
	Cylindrical	14" and Smaller	Wet	5SG60-LVS	32A60-L5VBE
		16" and Larger	Wet	5SG60-KVS	53A60-K8VCN
	Internal	Under 1/2"	Wet	5SG80-LVS	32A80-L6VBE
		1/2" to 1"	Wet	5SG70-KVS	32A60-L6VBE
		1" to 3"	Wet	5SG60-JVS	5SG60-I12VSP
		Over 3"	Wet	5SG60-JVS	5SG46-I12VSP
Production and Toolroom Grinding	Surface Grinding-Straight Wheel	14" and Smaller	Wet	5SG46-G12VSP	5SG46-IVS
			Dry	5SG46-IVS	5SG46-G12VSP
		Over 14"	Wet	5SG46-IVS	5SG46-H12VSP
	Segments	1 1/2" Rim	Wet	32A36-E12VBEP	5SG36-E12VSP
	Cylinders	1 1/2" Rim	Wet	5SG30-E12VSP	32A36-F12VBEP
	Cups	3/4" Rim	Wet	5SG46-JVS	32A46-J5VBE

(Continued)

Table 8. *(Continued)* **Recommended Wheels for Grinding Hardened Tool Steels.** See Notes.
(Source, Norton Company.)

Operation		Wheel Dia.	Wet or Dry	First Choice	Second Choice
Class 2 Tool Steels					
Toolroom Grinding	Internal	Under 1/2"	Dry	5SG80-KVS	32A80-K6VBE
		1/2" to 1"	Dry	5SG70-JVS	32A60-K6VBE
		1" to 3"	Dry	5SG60-IVS	5SG60-H12VSP
		Over 3"	Dry	5SG60-IVS	5SG46-H12VSP
	Cutter Sharpening-Straight Wheel	-	Wet	5SG46-LVS	32A46-J8VBE
		-	Dry	5SG46-JVS	32A46-J5VBE
	Dish Shape	-	Dry	5SG60-JVS	32A60-K5VBE
	Cup Shape	-	Dry	5SG60-KVS	32A60-K5VBE
		-	Wet	5SG60-LVS	32A60-K5VBE
	Form Tool	8" and Smaller	Wet	5SG60-KVS	32A60-K5VBE
			Dry	5SG80-KVS	32A60-K5VBE
		10" and Larger	Wet	5SG60-LVS	32A60-K8VBE
Class 3 Tool Steels					
Production Grinding	Centerless	-	Wet	5SG60-LVS	53A60-K8VCN
	Cylindrical	14" and Smaller	Wet	5SG60-LVS	32A60-K5VBE
		16" and Larger	Wet	5SG60-KVS	53A60-K8VCN
	Internal	Under 1/2"	Wet	5SG90-LVS	32A80-L6VBE
		1/2" to 1"	Wet	5SG80-LVS	32A60-L6VBE
		1" to 3"	Wet	5SG70-KVS	53G60-I12VSP
		Over 3"	Wet	5SG60-JVS	5SG60-I12VSP
Production and Toolroom Grinding	Surface Grinding-Straight Wheel	14" and Smaller	Wet	5SG46-G12VSP	5SG46-IVS
			Dry	5SG60-IVS	5SG60-H12VSP
		Over 14"	Wet	5SG46-IVS	5SG46-H12VSP
	Segments	1 1/2" Rim	Wet	32A36-E12VBEP	5SG46-E12VSP
	Cylinders	1 1/2" Rim	Wet	5SG46-E12VSP	32A46-F12VBEP
	Cups	3/4" Rim	Wet	5SG46-IVS	32A46-I8VBE
Toolroom Grinding	Internal	Under 1/2"	Dry	5SG90-KVS	32A80-K6VBE
		1/2" to 1"	Dry	5SG80-KVS	32A60-K6VBE
		1" to 3"	Dry	5SG70-JVS	5SG60-H12VSP
		Over 3"	Dry	5SG60-IVS	5SG60-H12VSP
	Cutter Sharpening-Straight Wheel	-	Wet	5SG46-JVS	32A46-I8VBE
		-	Dry	5SG46-IVS	32A46-I8VBE
	Dish Shape	-	Dry	5SG60-IVS	32A60-J5VBE
	Cup Shape	-	Dry	5SG60-IVS	32A60-J5VBE
		-	Wet	5SG60-JVS	32A60-J5VBE
	Form Tool	8" and Smaller	Wet	5SG80-KVS	32A80-K8VBE
			Dry	5SG80-JVS	32A80-K8VBE
		10" and Larger	Wet	5SG80-JVS	32A80-K8VBE

(Continued)

Table 8. *(Continued)* **Recommended Wheels for Grinding Hardened Tool Steels.** See Notes. *(Source, Norton Company.)*

Operation		Wheel Dia.	Wet or Dry	First Choice	Second Choice
Class 4 Tool Steels					
Production Grinding	Centerless	-	Wet	5SG60-JVS	32A60-J8V127
	Cylindrical	14" and Smaller	Wet	5SG80-JVS	5SG60-KVS
		16" and Larger	Wet	5SG80-IVS	53A80-J8V127
	Internal	Under 1/2"	Wet	5SG100-LVS	32A100-L6VBE
		1/2" to 1"	Wet	5SG90-KVS	32A80-L6VBE
		1" to 3"	Wet	5SG80-JVS	5SG80-I12VSP
		Over 3"	Wet	5SG80-IVS	5SG80-I12VSP
Production and Toolroom Grinding	Surface Grinding-Straight Wheel	14" and Smaller	Wet	5SG46-G12VSP	5SG60-IVS
			Dry	5SG60-IVS	5SG60-G12VSP
		Over 14"	Wet	5SG46-IVS	5SG46-G12VSP
	Segments	1 1/2" Rim	Wet	32A36-E12VBEP	5SG46-E12VSP
	Cylinders	1 1/2" Rim	Wet	5SG60-E12VSP	5SG46-E12VSP
	Cups	3/4" Rim	Wet	5SG60-IVS	5SG60-G12VSP
Toolroom Grinding	Internal	Under 1/2"	Dry	5SG100-KVS	32A100-K6VBE
		1/2" to 1"	Dry	5SG90-JVS	32A80-K6VBE
		1" to 3"	Dry	5SG80-IVS	5SG80-H12VSP
		Over 3"	Dry	5SG80-HVS	5SG80-H12VSP
	Cutter Sharpening-Straight Wheel	-	Wet	5SG60-IVS	5SG60-G12VSP
		-	Dry	5SG60-IVS	32A46-H8VBE
	Dish Shape	-	Dry	5SG80-IVS	5SG60-IVS
	Cup Shape	-	Dry	5SG80-IVS	5SG60-IVS
		-	Wet	5SG60-IVS	5SG80-JVS
	Form Tool	8" and Smaller	Wet	5SG80-JVS	32A80-J8VBE
			Dry	5SG80-JVS	32A80-J8VBE
		10" and Larger	Wet	5SG80-JVS	32A80-J8VBE

Notes: See Table 6 for an explanation of Tool Steel Classes as used in this Table. Grinding Wheels identification numbers are for wheels produced by the Norton Company. See the text for details of Norton's designations, and for information on the ANSI Standard grinding wheel marking system.

CBN wears at a significantly slower rate than either aluminum oxide or silicon carbide, and provides a much longer life between dressing than either of these materials. This long life, together with the ability to remove large amounts of stock at high spindle speeds, somewhat offsets the higher cost of CBN abrasives.

The following identification is used by Norton Company on their CBN grinding and cut-off wheels.

Abrasive Type	Grit Size	Grade	Bond Type	Abrasive Depth
CB	120	T	B99	$1/8$

In this example, the codes indicate the following.

Abrasive Type. CB indicates coated cubic boron nitride crystal.

Grit Size. There are seven possible entries for this field. 100 indicates roughing. 120 is for roughing or cut-off. 150 can be used for roughing or finishing. 180 is used to improve finish. 220, 320, and 400 are all for finish grinding. **Table 9** provides an approximate comparison of various grit size grading systems for CBN and diamond and CBN, and **Table 10** gives expected surface finish by grit size.

Grade. Three grades are available. Q is used for a broad area of contact. T is the first choice, and used for dry grinding and reshaping. W is the most durable grade, used with high volume coolant.

Bond Type. B99 is a resin bond that can be used wet or dry. Other bonds are Aztec III, recommended for dry tool resharpening, and Aztec .007 for increased feed rates and heavier cuts.

Abrasive Depth. The depth of the usable abrasive. Possible depths are $1/16$", $1/8$", $7/32$", $1/4$", and solid.

Diamond wheels

Synthetic diamond abrasives are most commonly used to grind nonferrous materials. This is because diamond is a carbon based material that reacts with iron. This results in dissolution of the diamond into the carbon of the steel, which accelerates wear. On the other hand, diamond abrasives are the first choice for grinding tungsten carbide, cermets, glass, ceramics, fiberglass, natural stones, and plastics.

Diamond wheels should be used with a full flow coolant properly directed toward the grinding zone—water with a rust inhibitor is recommended. Vitrified wheels, in particular, should not be used without a coolant.

The following identification is used by Norton Company on their diamond grinding and cut-off wheels.

Abrasive Type	Grit Size	Grade	Concentration	Bond Type	Abrasive Depth
ASD	120	R	75	B99	$1/8$

In this example, the codes indicate the following.

Abrasive Type. ASD can be used wet or dry, D is micro-sized diamond for finishing and polishing, and SD is free cutting and can be used wet or dry. ASD, D, and SD are all resin bond wheels. M4D is a metal bond wheel, used on glass, ceramics, refractories, and other nonmetallics. Vitrified bond wheels are RMD and SD.

Grit Size. There are seven possible entries for this field. 100 indicates roughing. 120 is for roughing or cut-off. 150 can be used for roughing or finishing. 180 is used to improve finish. 220, 320, and 400 are all for finish grinding. **Table 9** provides an approximate comparison of various grit size grading systems for diamond and CBN abrasives, and **Table 10** gives expected surface finish by grit size.

Grade. R and N are resin bond wheels, N and L are metal bond wheels, and P and R are vitrified bond wheels.

Concentration. 50 is for large area of contact grinding. 75 is freer cutting than 100. When durability is required, 100 is recommended under flood coolant conditions. 125 is selected when precise form-holding grinding is required.

Bond Type. Resin bonds are B99, which can be used wet or dry, and B105 for dry toolroom reconditioning operations. Metal bonds are M99, for cut-off and grinding glass and ceramics, and MSL for dry reconditioning of carbide tools. V99 is a vitrified bond, used for finishing, or plunge grinding of carbide tools.

Superabrasive wheel identification system

CBN and diamond grinding and cut-off wheels are available in a variety of standardized

Table 9. Comparison of Grit and Micron Sizes of Standardized Grading Systems. *(Source, Norton Company.)*

Norton Company Grit Size		U.S. Standard	European (FEPA) Size	Japanese (JIS) Size	Average Particle Size (Inches)	Particles Per Caret
Diamond	CBN	Mesh				
16	-	16/20	1182	-	0.046	33
24	24	20/30	892	20	0.033	-
36	36	30/40	602	30	0.023	282
-	-	35/40	501	35	-	-
-	46	40/50	427	40	0.017	770
50	-	50/60	427	40	0.017	770
-	60	50/60	301	50	0.012	2M
60	-	60/70	301	50	0.012	2M
-	80C	50/80	-	-	-	-
-	80	60/80	252	60	0.0098	3M
80	-	70/80	-	-	-	-
100	-	80/100	252	60	0.0098	3M
105	100	80/100	181	80	0.007	10M
100S	120C	80/120	-	-	0.0064	14M
110	120	100/120	151	100	0.006	17M
120	150	120/140	126	120	0.005	21M
150	180	140/170	107	140	0.004	49M
180	220	170/200	91	170	0.0035	83M
220	230	200/230	76	200	0.003	140M
240	240	230/270	64	230	0.0025	252M
320	320	270/325	54	270	0.0018	284M
400	400	325/400	46	325	0.0015	660M
Micron Size Comparison						
Nominal Micron Size	Norton Grit Size*	Approximate European (FEPA) Size	Japanese (JIS) Size	Average Particle Size (Inches)	Particles Per Caret	
40/60	-	M63	-	0.0014	-	
30/40	500	M40	500	0.0012	-	
20/30	-	M25	700	0.00087	-	
15/25	-	-	800	0.0006	-	
10/20	600	M16	1000	0.00049	17MM	
8/16	-	-	1500	0.00037	-	
6/12	800	M10	2000	0.0003	79MM	
4/8	-	M6.3	2500	0.0002	262MM	
3/6	-	-	4000	0.00018	-	
2/4	-	M4.0	5000	0.00011	205MM	
0/2	-	M1.0	15000	0.000039	62MMM	

* Used for metal coated diamond and CBN only. (Not all micron sizes are available with metal coatings.

Table 10. Expected Surface Finish by Grit Size. See Note. *(Source, Norton Company.)*

Diamond		
Grit Size	**Expected Finish (Micro inch AA)**	**Maximum Depth of Cut Per Pass**
100	24 to 32	0.001" to 0.002"
120	16 to 18	0.001" to 0.002"
150	14 to 16	0.001" to 0.002"
180	12 to 14	0.001" to 0.002"
220	10 to 12	0.001" to 0.002"
320	8	0.0004" to 0.0006"
400	7 to 8	0.0003" to 0.0005"
Cubic Boron Nitride (CBN)		
Grit Size	**RMS Finish with Oscillation**	**Expected Finish Plunge**
100	35 to 40	40 to 45
120	30 to 35	35 to 40
150	25 to 30	30 to 35
180	20 to 25	25 to 30
220	15 to 20	20 to 25
320	10 to 15	15 to 20
400	4 to 8	5 to 10

Note: This Table is a guide to expected surface finishes. Surface finish is affected by a number of variables including machine type and condition, type of material ground, coolant, wheel speed, bond system, etc.

shapes, and can be custom ordered for unique applications. Common shapes, and the application for each, are shown in *Figure 7.* Note that superabrasive and standard abrasive grinding wheels share many of the same basic core shape profiles and number designations as shown in *Figures 1* and *2*, but that the abrasives on superabrasive wheels are applied to discrete areas as shown by dark patches on the wheels in Figure 7. The first identifying number of a superabrasive wheel refers to the core profile of the wheel.

The second code is a letter, or pair of letters, that identifies the shape of the cross section of the diamond or CBN cutting section. There are 29 different codes, and the shapes and assigned code letters are shown in *Figure 8.*

The third code used is a number, 1 through 10, to show the location of the CBN or diamond section on a grinding wheel. Nine of the ten positions are shown in *Figure 9.* Position 7, not shown and not commonly used, positions the abrasive on the side of the wheel, not extending to the periphery.

A series of letter codes are added to the three basic numerical and letter designations to provide additional information about modifications to the wheel. The assigned modification designations for superabrasive wheels are as follows. *B*, Drill and Counterbore: holes drilled and counterbored in the wheel's core. *C*, Drill and Countersink: holes drilled and countersunk in the wheel's core. *H*, Plain Hole: straight hole drilled in the core of the wheel. *M*, Holes Plain or Thread: core of wheel contains mixed holes, some plain and some threaded. *P*, Relieved One Side: the core has been relieved on one side—the thickness of the core is less than the wheel thickness. *Q*, Diamond or CBN Insert: three surfaces of the abrasive are partially or entirely enclosed by the core. *R*, Relieved Two Sides: core of

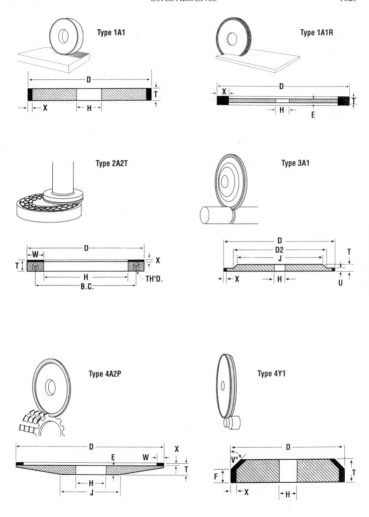

Figure 7. CBN and diamond grinding wheel shapes. Dimensional letters are BC = bolt diameter, D = diameter, D2 = large hub diameter, d = diameter of mounting holes, E = back thickness, H = hole diameter, J = hub diameter, K = inside diameter, S = face angle, T = thickness, THD = mounting thread size, U = abrasive thickness, V = insert length, W = rim width, X = abrasive depth (Source, Norton Company.)

(Continued)

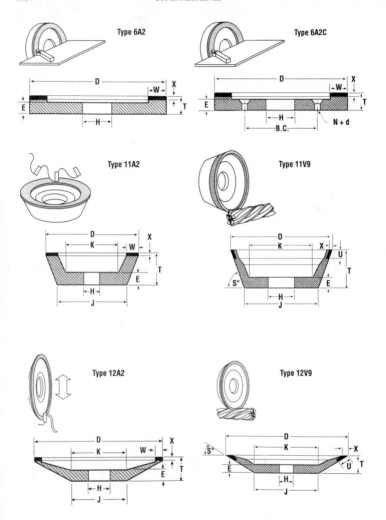

Figure 7. (Continued) CBN and diamond grinding wheel shapes. Dimensional letters are BC = bolt diameter, D = diameter, D2 = large hub diameter, d = diameter of mounting holes, E = back thickness, H = hole diameter, J = hub diameter, K = inside diameter, S = face angle, T = thickness, THD = mounting thread size, U = abrasive thickness, V = insert length, W = rim width, X = abrasive depth (Source, Norton Company.)

(Continued)

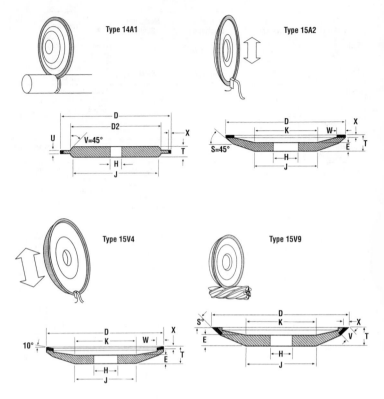

Figure 7. (Continued) CBN and diamond grinding wheel shapes. Dimensional letters are BC = bolt diameter, D = diameter, D2 = large hub diameter, d = diameter of mounting holes, E = back thickness, H = hole diameter, J = hub diameter, K = inside diameter, S = face angle, T = thickness, THD = mounting thread size, U = abrasive thickness, V = insert length, W = rim width, X = abrasive depth (Source, Norton Company.)

wheel relieved on both sides—the thickness of the core is less than the wheel thickness. S, Segmented Diamond or CBN Section: abrasive is mounted in segmented sections on the wheel's core. SS, Segmented and Slotted: wheel has separated (by slots) segments, and abrasive is mounted on segments. V, Diamond or CBN Inverted: abrasive surface of wheel is concave in shape. These modifications are shown in *Figure 10.*

Using the above designations and letters, superabrasive wheels can be fully identified. A wheel with the designation 6A2C, for example, is a straight cup (6) core, rectangular diamond section (A) mounted on its side (2), modified with a drilled and countersunk hole (C) as shown in *Figure 11.*

The core material of the wheel has a direct influence on its performance characteristics, and should be selected to meet operational requirements. The following chart provides the characteristics of core materials used for superabrasive wheels.

Core Material	Vibration Dampening	Heat Dissipation	Mechanical Strength	Remarks
Steel	Very Poor	Excellent	Excellent	Used predominately on cut-off wheels.
Aluminum Solid	Poor	Excellent	Excellent	May not be suitable for all dry grinding operations.
Resaloy	Good	Good	Good	Made of aluminum powder and resin. May be used wet or dry.
Resaloy Self-dressing	Good	Good	Poor	–
Copper	Good	Excellent	Good	Made of copper powder and resin.
Bakelite	Good	Poor	Good	Made of phenolic resin with a glass filler.
Composition	Good	Poor	Fair	Made of abrasive and resin.

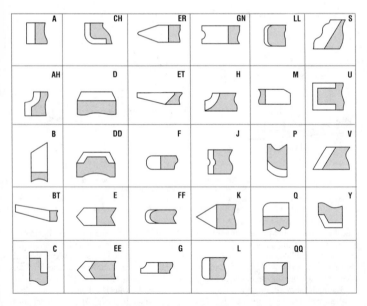

Figure 8. Designation letters for the shape of the diamond or CBN section. Shaded area is core of wheel.

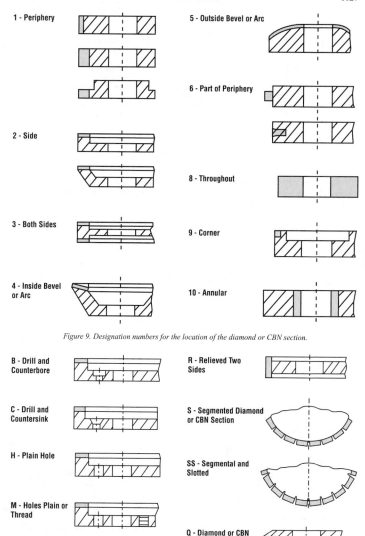

1 - Periphery

5 - Outside Bevel or Arc

6 - Part of Periphery

2 - Side

8 - Throughout

3 - Both Sides

9 - Corner

4 - Inside Bevel or Arc

10 - Annular

Figure 9. Designation numbers for the location of the diamond or CBN section.

B - Drill and Counterbore

R - Relieved Two Sides

C - Drill and Countersink

S - Segmented Diamond or CBN Section

H - Plain Hole

SS - Segmental and Slotted

M - Holes Plain or Thread

Q - Diamond or CBN Insert

P - Relieved One Side

V - Diamond or CBN Inverted

Figure 10. Designation letters for modifications made to the core of the wheel.

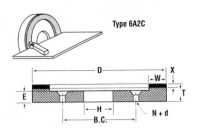

Type 6A2C

D - Diameter
E - Back Thickness
H - Hole
BC - Bolt Circle Diameter
T - Thickness
N - Number or Mounting Holes
W - Rim Width
X - Abrasive Depth
d - Diameter of Mounting Holes

Figure 11. A 6A2C grinding wheel.

Table 11 provides recommended wheels and materials for superabrasive grinding operations. To interpret the wheel specifications, refer to the superabrasive wheel identification system for CBN and diamond wheels provided above.

(Text continued on p. 1635)

Table 11. Recommended Wheels and Materials for Superabrasive Grinding Operations.
(Source, Norton Company.)

Application or Common Machine Type	Common Wheel Size, Type, and Blueprint	Variables	Better	Best
Tool Room Grinding				
Chip Breaker/Clearance	Various sizes Type 1A1	Dry Wet	CB180-TBB CB180-WB99E	CB80-TBD Refer to P.E.
Cutter Sharpening (Milling Cutters, Broaches, Reamers, etc.)	3 3/4" × 1 1/2 × 1 1/4" Type 11V9	Dry Wet	Aztec 111 120T-1/8 CB120-TB99-1/8	Aztec .007-100-1/8 CB120-WB99-1/8
	6 × 1 × 1 1/4" Type 12A2	Wet or Dry	CB120-TB99-1/16	CB120-WB99-1/16
	6 × 3/4 × 1 1/4" Type 12V9	Wet or Dry	CB120-TB99-1/8	CB150-WB99-1/8
	6 × 3/4 × 1 1/4" Type 15V9 or 1FF1	Wet or Dry	CB100-WB99-1/16	CB120-WBB-1/8
Cylindrical Grinding	12 × 1/2 × 3" Type 1A1		CB150-TB99E-1/8	CB150-TBA-1/8
Internal Grinding Tools	1/2" or thicker Type 1A1	Wet Dry	CB120-TB99E	CB120-WBA CB120-WBB
	Thinner than 1/2" Type DW	Wet or Dry	CB100-WB99	CB120-WBB
Saw Sharpening	4" to 8" diameter Type 4A2P, 12A2 or 15A2	Wet Dry	CB180-TBB-1/8 Aztec 111 150T-1/8	CB150-TBA-1/8 Aztec .007-150-1/8
Single-Point	6 × 3/4 × 1 1/4" Type 6A2	Wet or Dry	CB150-QB99-1/16	CB150-TBB-1/16
Surface Grinding	10 × 1/2 × 3" Type 1A1		CB100-TB99-1/16 CB120-TBA	CB120-TBB

(Continued)

Table 11. *(Continued)* **Recommended Wheels and Materials for Superabrasive Grinding Operations.** *(Source, Norton Company.)*

Application or Common Machine Type	Common Wheel Size, Type, and Blueprint	Variables	Better	Best
Tool Steels (R$_c$ 50+)				
Centerless Grinding	Various sizes Type 1A1	Water base or oil coolant	CB150-TBA	
Cylindrical Grinding	10 × 1/2 × 3" Type 1A1	Water base coolant, under 16" diameter	CB100-TB99-1/16	CV5SG100-1/8
		Water base coolant, under 16" diameter and over	CB120-T99-1/8	CB120-TBA-1/8
		Oil coolant	CB120-TB99E-1/8	B120-H100VX222C-1/8
		No coolant	CB100-B99-1/8	CB120-TBD-1/8
Internal Grinding	Various sizes Type DW	Wet	B120-EP	CB120-TBB
		Dry	CB120-TBB	B180-J125VX223C
	Type 1A8 Under 1/2"	Wet	CB120-TBA	B180-H150VX322C
	1/2" and over	Wet	CB120-TBA	B180-H150VX322C
	Type 1A1 trued with a Shank Tool 1/2" to 3/4"	Wet	CB150-TBA-1/8	B180-H150VX322C
	3/4" to 1"	Wet	CB120-TBA-1/8	B180-H200VX322C
	Type 1A1 trued with Rotary Diamond 1/2" to 3/4"	Wet	B180-H150VX322C	B240-I200VX322C
	3/4" to 1"	Wet	B180-H200VX322C	B240-I200VX322C
Slotting	7 × 1/4 × 1 1/4" Type 1A1R or 1A1RN	Water base coolant	CB120-WBB-1 1/16	C3B120-WMN-1/8
		Oil coolant	C3B120-WMN	B120-H125VX224C-1/8
Surface Grinding	6 × 1/4 × 1 1/4"	Water base coolant	CB100-TB99-1/16	CVSG-100-1/8
	6 × 3/8 × 1 1/4"	Oil coolant	CB120-TB99-1/8	B120-J100VX222C-1/8
	6 × 1/2 × 1 1/4" Type 1A1	No coolant	Aztec 111-100W-1/8	Aztec .007-100-1/16

Application or Common Machine Type	Common Wheel Size, Type, and Blueprint	Variables	Starting Specifications
Aerospace Materials (Inconel, Rene, etc.)			
Creep feed Grinding	Various sizes Type 1A1 or special shapes	Water base coolant Better Best	B150-H100VXL224C 3B150-H125VXL224C
		Oil coolant Best	B150-H125VX222C-1/8

(Continued)

Table 11. *(Continued)* **Recommended Wheels and Materials for Superabrasive Grinding Operations.**
(Source, Norton Company.)

Application or Common Machine Type		Common Wheel Size, Type, and Blueprint	Variables	Starting Specifications
Aerospace Materials (Inconel, Rene, etc.)				
Cylindrical Grinding		Various sizes Type 1A1	Water base or oil coolant: Better Best	B120-H125VX222C 5B120-J125VX222C
Rotary Tool Manufacturing				
Broaches-Straight (Huffman or Walter Machines)		Various sizes Various types	Wet or Dry	CB150-WBB
Circular Broaches		Various sizes Type 1FF1	Wet or Dry	CB120-WBB
Drill Clearing		Various sizes Type 1A1	Wet - straight oil	CB150-WBB CB180-TBB CB180-WBB
Drill Fluting from Grid (Hertlein or Normac Machines)		Various sizes Various types	High oil pressure	C5B120-WBE
Drill Pointing		Various sizes Various types	High oil pressure	CB150-WBB
End Mill	Ball Nose Gashing Healing	Various sizes Types 1V+B1P	Wet - straight oil	CB100-WBE CB100-TBE
	Gashing (Huffman Machine)	12 × .775 × 5" Type 4V9P	Wet - straight oil Production grinding	CB120-WBB
		12 × 3/4 × 5" Type 4V9P	Wet - straight oil Large end mills	C5B120-WB98
	End Feature (Huffman Machine)	11 × 1 1/2 × 5" Type 3A1P	Wet - straight oil	CB120-TBB
	Fluting from Solid (Huffman, Normac or ITM)	Various sizes, Type 1A1 or 14V1RN Steel Core	High oil pressure Low oil pressure	CB120-WBE CB120-WB98 or CB120-WBA
	Fluting Polishing (Huffman Machine)	12 × 3/4 × 5" Type 4V9P	Wet - straight oil Production Resharpening	CB150-WBB CB150-WBB
	Outside Diameter Primary and Secondary (Huffman Machine)	12 × 1/4 × 3/8 × 5" Type 3A1	Wet - straight oil	CB180-WBB
Reamers		Various sizes Various types	Wet or dry	CB120-WBB
Tap Fluting (Hertlein or Normac Machines)		Various sizes. Type 1FF1RN Steel Core recommended	High oil pressure	CB150-WBE
Tap Camfering		Various sizes Type 1A1	Wet - straight oil	CB150-WBB

(Continued)

Table 11. *(Continued)* **Recommended Wheels and Materials for Superabrasive Grinding Operations.**
(Source, Norton Company.)

Application or Common Machine Type		Common Wheel Size, Type, and Blueprint	Variables	Starting Specifications
Glass Grinding				
Glass Flat	Cutting off	Various sizes Type 1A1R		M4D100-N50M
	Pencil Edging-Automotive	Various sizes Type 1F6Y double spindle	Single spindle Roughing Finishing	M4DR180-35MX872D M4DR150-40MX872D M4DR220-30MX872D
	Pencil Edging-Mirror/ Furniture	Various sizes Type 1F6Y intermediate	Roughing Finishing	M4DR120-40M173 M4DR150-40M173 M4D220-40M173
	Square Edging	Various sizes Type 1A1 or 6A2		M4D180-50M146
	Glassline	7 × 5/8 × 2 1/2" Type 1FF6YH Type 1FF16YH Type 1FF6YH	Station #1 Station #2 Station #3	M4D120-30M175-3/16 M4D150-30M175-3/16 M4D240-30M175-3/16
	Hermitec	7 × 5/8 × 22 mm Type 1FF6YH		M4D180-35M173-1/8
	Somaca	10 × 5/8 × 8" Type 1F6YB+T	Station #1 Durable Free Cut Station #2 Durable Free Cut	M4D120-25M173-1/8 M4D120-75M57-1/8 M4D220-25M173-1/8 M4D220-75M57-1/8
Sun Tool	8" model	8 × 5/8 × 5.500" Type 1F6YT	Single station	M4DR180-30MX872D
	9 3/4 and 10" models	9 3/4 × 5/8 × 7.531" 10 × 5/8 × 7.531" Type 1F6YT	Single station	M4DR180-35MX872D
Glass Lenses (generating with coolant)		Various sizes Type 2FF2	Roughing Finishing	M4D105-30MX219C M4D30/40MIC-25M173
		Type 1A1, 1EE1V, 12Y1V	Edging finishing	M4D240-50M146
Glass Prisms (surfacing)		Various sizes Type 1A1	Roughing Finishing	M4D120-N50M M4D40/60MIC-N50M
Glass Tubing (cutting)		Various sizes Type 1A1R	Durable Free cut	M4D100-N25M SD150-R75B69
			Large tubing Over 3" dia.	ASD180-R50B56
			Thin wall Durable Free cut	M4D180-L25M SD220-R75BX107
			Very thin wall Free cut	SD220-R75B69

(Continued)

Table 11. *(Continued)* **Recommended Wheels and Materials for Superabrasive Grinding Operations.**
(Source, Norton Company.)

	Application or Common Machine Type	Common Wheel Size, Type, and Blueprint	Variables	Starting Specifications
Ceramics Grinding and Miscellaneous Applications				
Bonded Abrasives	Cutting Off	Various sizes Type 1A1R	Coarse (soft) Fine (hard)	M4D80-N75M99 M4D100-N75M99
	Surface Grinding Disc and Blanchard	Various sizes and designs Grinding segments and buttons	Free cutting, maximum material removal	M4D24-50M24
Coated Abrasives	Skiving	Various sizes Type 1A1 or 6A2	(Dry) Coarse grit Fine grit	M4D46-N100M M4D80-N100M
Agate -Gems	Cutting Off	Various sizes Type 1A1R		M4D60-L25M
	Edging	Various sizes Type 1A1 or 6A2	Roughing (Wet) Finishing	M4D100-N25M D6/12MIC-N25M
Aluminum Oxide		See Advanced Ceramics		
Boron Carbide	Cutting Off	Various sizes Type 1A1R	(Wet) Durable Free cut	M4D100-N50M ASD100-R100B56 ASD100-R100B99
	Internal Grinding	Various sizes Type 1A1 or 1A8	(Wet) Roughing	ASD100-R100B56 ASD100-R100B99
			Finishing	ASD320-R100B56 ASD320-R100B99
	Outside Diameter Grinding	Various sizes Type 1A1	(Wet) Roughing	ASD100-R100B54
			Semi-finishing	ASD220-R100B54
			Finishing	SD320-R100B69
	Surface Grinding	Various sizes Type 1A1, 2A2 or 6A2	(Wet) Roughing Durable	M4D100-25MX156
			Free cut	ASD100-R100B56
Molded Brake Lining	Cutting Off	Various sizes Type 1FF1RSS	(Dry) 10,000 SFPM-12,000 SFPM	D30/40-H-MSL
	Surface Grinding	Various sizes Type 1A1 or 6A2	8,500 SFPM-10,000 SFPM	D20/30-H-MSL
Carbon	Soft Cutting Off	Various sizes Type 1A1R	(Wet or dry) Durable Free cut	M4D46-N50M D80/100-H-MSL
	Surface Grinding	Various sizes Type 1A1	Durable Free cut	M4D60-N100M D80/100-H-MSL
	Hard Cutting Off	Various sizes Type 1A1R	(Wet) Durable Free cut	M4D60-L50M D80/100-H-MSL

(Continued)

Table 11. *(Continued)* **Recommended Wheels and Materials for Superabrasive Grinding Operations.** *(Source, Norton Company.)*

Application or Common Machine Type		Common Wheel Size, Type, and Blueprint	Variables	Starting Specifications
Ceramics Grinding and Miscellaneous Applications				
Advanced Ceramics[1]	Cutting Off	Various sizes Type 1A1R	(Wet) Durable Free cut	M4D220-N25M ASD220-R75BX615C
	Internal Grinding	Various sizes Type 1A1 or 1A8	(Wet) Roughing	SD320-R100B80
			Finishing	D15/25MIC-R100B80 D10/25MIC-L6V10
			Ultra-finishing	D4/8MIC-R100BX619C D2/4MIC-R100BX619C
	Outside Diameter Grinding	Various sizes Type 1A1 or 3A1	(Wet) Roughing	SD320-R100B80 SD320-N6V10
			Finishing	D15/25MIC-R100B80 D15/25MIC-R100BX619C D10/20MIC-L6VX37C
			Ultra-finishing	D4/8MIC-R100BX619C D2/4MIC-R100BX619C
	Surface Grinding	Various sizes Type 1A1, 2A2 or 6A2	(Wet) Roughing	SD220-R75BX620C SD220-L4VX223C SD320-R100B80
			Finishing	D15/25MIC-R100BX619C B10-20MIC-L6VX37C
			Ultra-finishing	D4/8MIC-R100BX619C D2/4MIC-R100BX619C
Technical Ceramics[2]	Cutting Off Outside Diameter Grinding	Various sizes Type 1A1 or 3A1	(Wet) Roughing	M4DC120-N75M ASD120-R75BX615C
			Finishing	SD320-R100B80 SD320-R100B69
	Surface Grinding	Various sizes Type 1A1, 2A2 or 6A2	(Wet) Roughing	M4DC120-N50M SD120-L5VX223C
			Finishing	SD320-R100B69 SD320-R100B80
Chrome Honing		Various sizes Type HHA2 or SR		M4D100-75M175
Easy to Grind [3]		Various sizes and types Type 1A1R discs, wheels, Centerless wheels	Maximum Material Removal Polished Surface	ASD4/6MIC-N50M M4D100-L50M

(Continued)

Table 11. *(Continued)* **Recommended Wheels and Materials for Superabrasive Grinding Operations.**
(Source, Norton Company.)

Application or Common Machine Type		Common Wheel Size, Type, and Blueprint	Variables	Starting Specifications
Ceramics Grinding and Miscellaneous Applications				
Ferrite (Magnets)	Arc Magnets	Various sizes and designs Type 9FF1		PD60-EP
	Cutting Off	Various sizes Type 1A1R		M4D150-N25M
	Disc Grinding	Various sizes and designs Type 2Y2		M4D100-N50M9
	Outside Diameter Grinding	Various sizes Type 1A1	Durable Free Cut	M4D120-N75M9 PD80-EP
	Surface and Internal Grinding (Blanchard)	Various sizes and designs Type 1A1 Type 2A2		M4D150-N75M9 M4D150-N50M9
Fiberglass Cutting Off		Various sizes Type 1FF1RSS	(Dry) 12,000 SFPM (Wet) 6,000 SFPM	PD46-EP PD46-EP
Minerals (Lapidary) Cutting Off		Various sizes Type 1A1R	Moh Hardness 7 or less Moh Hardness over 7 Durable Free Cut	M4D100-N25M M4D100-L25M PD240-EP
Norbide		See Ceramic-Boron Carbide		
PCB (Polycrystalline Borazon)		Various sizes Type 6A2 or 11A2	(Wet) Roughing	AD30/40MIC-UMXL7519
PCD (Polycrystalline Diamond)		Various sizes	(Wet) Roughing	AD30/40MIC-UMXL7519
Ewag Type Machines		Type 6A2 or 11A2		
Plastics Cutting off and surface grinding		Various sizes Type 1A1, 1FF1R or 1FF1RSS	(Dry) 10,000 SFPM - 12,000 SFPM (Wet) 6,000 SFPM	
Porcelain	Cutting Off	Various sizes Type 1A1R		M4D100-N25M
	Outside Diameter Grinding	Various sizes Type 1A1		M4D100-N75M
	Surface Grinding	Various sizes Type 2A2SS		
Refractories	Cutting Off	Various sizes Type 1A1 or 1A1R	Roughing	M4D36-N75M M4D60-N100M
			Finishing	M4D120-N100M ASD150-R75BX615C SD120-R100B69
	Surface Grinding	Type 2A2 or 6A2	Roughing	M4D36-25M77
			Finishing	M4D120-N100M ASD150-R75BX615C

(Continued)

Table 11. *(Continued)* **Recommended Wheels and Materials for Superabrasive Grinding Operations.**
(Source, Norton Company.)

Application or Common Machine Type		Common Wheel Size, Type, and Blueprint	Variables	Starting Specifications
Ceramics Grinding and Miscellaneous Applications				
Rokide (Plasma Spray)		Various sizes Type 1A1	(Wet) Roughing	SD320-R50B69
			Finishing	D4/8MIC-R50BX619C
Steatite	Cutting off	Various sizes Type 1A1R		M4D120-N25M
	Surface Grinding	Various sizes Type 1A1		ASD100-R75B56 ASD100-R75B99
Tile	Cutting Off	Various sizes Type 1A1R Type 1FF1R	(Dry) Offhand	M4D36-N25M PD36-EP

Notes: [1] Advanced Ceramics include Alumina-Oxide (high density), Alumina Nitride, Alumina Titanate, Boron Carbide, Boron Nitride, Sialon, Silicon Carbide, Silicon Nitride, Titanium-Diboride, Zirconium-Oxide.
[2] Technical Ceramics include Aluminum-Oxide (porous 90), Aluminum-Silicate, Beryllia, Magnesia, Mulite.
[3] Easy to Grind materials include Corderites, Ferrites/Magnets (Aluminum-Oxide), Porous Alumina (90×), Silicate (1200 Knoop hardness) Thermal Sprays (Rokide).

Superabrasive grinding parameters

The following guidelines, as published by Norton Company, can be used to determine recommended machining parameters for CBN and diamond grinding wheels. Note that the wheel speeds given are not intended to be maximum operating speeds. ANSI B7.1 should be consulted to confirm maximum speeds. Speeds on the following chart are given in surface feet per minute (sfm) and meters per second (m/s).

$$sfm = \text{wheel diameter} \times rpm \times 0.262$$

$$m/s = sfm \div 196.85 \quad sfm = m/s \times 196.85.$$

Type of Bond	Wet Grinding		Dry Grinding	
	Cup Wheels	Peripheral Wheels	Cup Wheels	Peripheral Wheels
CBN Grinding Wheels				
Resin Bond Wheels	5,906 to 9,843 sfm 30 to 50 m/s	5,906 to 9,843 sfm 30 to 50 m/s	2,953 to 5,906 sfm 15 to 30 m/s	2,953 to 5,906 sfm 15 to 30 m/s
Diamond Grinding Wheels				
Resin Bond Wheels	4,921 to 7,874 sfm 25 to 40 m/s	4,921 to 7,874 sfm 25 to 40 m/s	2,756 to 3,543 sfm 14 to a8 m/s	2,756 to 3,543 sfm 14 to a8 m/s
Metal Bond Wheels	–	3,937 to 5,906 sfm 20 to 30 m/s	–	–
Vitrified Bond Wheels	2,953 to 5,906 sfm 15 to 30 m/s	2,953 to 5,906 sfm 15 to 30 m/s	–	–

Troubleshooting superabrasive grinding operations. *(Source, Norton Company.)*

Problem	Possible Cause	Suggested Correction
Dry Grinding		
Burning (excessive heat)	Wheel loaded or glazed	Dress wheel with a dressing stick.
	Excessive feed rate	Reduce infeed of wheel or workpiece.
	Wheel too durable	Use freer cutting specification or slow down wheel speed
Poor finish	Grit size too coarse	Select a finer grit size.
	Excessive feed rate	Reduce infeed of wheel or workpiece.
Chatter	Wheel out of true	True wheel and ensure that it is not slipping on mount.
Wet Grinding		
Burning (excessive heat)	Wheel loaded or glazed	Redress wheel.
	Poor coolant placement	Apply coolant directly to wheel/workpiece interface.
	Excessive material removal rate	Reduce downfeed and/or crossfeed.
Poor finish	Excessive dressing	Use lighter dressing pressure. Stop dressing as soon as wheel begins to consume stick rapidly.
	Grit size too coarse	Select a finer grit size.
	Poor coolant flow or location	Apply heavy flood so it reaches wheel/workpiece interface.
Chatter	Wheel out of true	True wheel and ensure that it is not slipping on mount.
Wheel will not cut	Glazed by truing	Dress lightly until wheel opens up.
	Wheel loaded	Dress lightly until wheel opens up. Increase coolant flow to keep wheel surface clean. Never run wheel with coolant turned off.
Slow cutting	Low speeds and feeds	Increase feed rate and/or increase wheel speed.
Short wheel life	Incorrect coolant flow	Apply coolant to flood wheel/work surface.
	Low wheel speed	Increase wheel speed.
	Excessive dressing	Use lighter dressing pressure.
	Wheel too soft or too hard	Change grit or grade; use higher concentration.

Dressing grinding wheels

Conventional abrasive wheels

Grinding wheels should be trued and dressed before being put into service. Truing is the process of making the periphery of the wheel run concentric to the axis of rotation. Truing can also include altering the wheel by putting a special form or contour into its face. Dressing is the process of altering the cutting action of the wheel. The hard diamond point of a dressing tool breaks the bond posts of a vitrified wheel and fractures the grains, removing dull abrasive grains entirely and producing a new surface with sharp, abrasive particles. Additionally, dressing removes tiny pieces of material from the pores of the wheel face. These particles are picked up during grinding, leading to "loading" of the wheel which slows the grinding process and can result in burn or chatter marks on the workpiece. With conventional abrasive wheels, truing and dressing occur simultaneously, but with superabrasive wheel applications, truing is done with a tool or roll, and dressing occurs with a vitrified dressing stick.

It is important not to take too deep a cut when dressing and truing a wheel. Too heavy an infeed can damage the diamond dressing tool, waste abrasive from the wheel surface, dull the abrasive particles, and even damage the bond post in the wheel. Suggested infeed rates per pass are

Single point and cone point tools	0.001" maximum
Chisel, HPB blade tools	0.002" maximum
Multipoint (Nib) and cluster tools	0.005" maximum.

Single point tools should approach the grinding wheel at a 10° to 15° "drag" angle to create a sharpening effect for the tool. Multipoint (impregnated) tools do not require sharpening, and should approach the wheel with full-face contact.

It is not uncommon in external grinding to find the wheel face worn 0.015" to 0.020" on the edges, with a high spot in the center. In these cases, it is best to contact the highest point before traversing in order to prevent excessive penetration and possible damage to the dressing tool. For high finish grinding, however, contact should not be made without traverse, as the diamond may leave a mark on the wheel that will require excessive dressing to remove. In such cases, it is best to bring the diamond almost into contact with the high spot. Start the coolant, then the traverse, and feed in 0.001" to 0.002" at the end of each pass until contact is made. Then proceed in the usual way.

The traverse rate is critical to obtaining the desired part finish and metal removal rate. A slow traverse rate tends to "close up" a wheel, allowing for better part finishes but lower metal removal rates. Too slow a traverse rate can close a wheel too much, causing parts to be burned. A faster traverse rate creates a more "open" wheel face, allowing for better metal removal rates and less chance for burn. Fast traverse rates will generally contribute to rougher part finishes. Traverse rate is calculated as follows.

wheel speed × traverse per wheel revolution = traverse rate per minute.

Recommended traverse rates for selected tool types are

Tool Type	Traverse Rate
Single point, cone point, chisel, and multicut (blade type)	0.002" (fine finish) to 0.010" (coarse finish) per revolution
Multipoint (impregnated) and cluster type	0.006" (fine finish) to 0.030" (coarse finish) per revolution.

Wheels may be dressed either wet or dry, but the dressing operation should always be carried out under the same conditions as when grinding. When *grinding wet, dress wet* as follows. The coolant nozzle should be arranged to flood the entire wheel face or should follow the diamond across. It is best to filter the coolant, as dirt or chips carried to the work will not only affect the quality of finish, but also tend to load the wheel and necessitate frequent redressing. When *grinding dry, dress dry* but allow frequent intervals for the diamond to cool. Otherwise, burning or fracture will occur.

Guidelines for using diamond dressing tools. The diamond dresser is a cutting tool, which means that its usefulness is in direct proportion to the definition of its point. Dull diamonds force worn abrasive particles into the wheel pores, and glaze the wheel face to produce a dull wheel. To maintain a well-defined cutting point at all times, longer wheel life, and greater wheel efficiency, the single point diamond should be turned one-quarter turn at regular intervals. Frequency of turning depends on use, but they should be rotated a minimum of once a day. Care should be taken not to bump the tool against the wheel as such shocks can result in flattening or fracturing the diamond point.

Guidelines for using multipoint dressing tools. To assure optimum performance, it is essential that the tool and wheel are in full face contact. To achieve this, four or five passes at 0.005" infeed is recommended on initial installation. The high infeed will expose the diamond particles in the tool's matrix. Coolant should be applied generously at the wheel/tool point of contact at all times during the dressing cycle. To obtain normal rate of wear of both the wheel and the tool, no more than 0.002" infeed for rough grinding should be taken. For finish grinding, 0.0005" per pass with no infeed on the last pass for extremely fine finishes. Also for fine finish requirements, dress at a traverse rate of 0.006" to 0.015" per wheel revolution. For an open free-cutting wheelface, use a traverse rate of 0.016" to 0.030" per wheel revolution. When dressing a badly worn or new wheel face, locate the high point by manual infeed, then use mechanical infeed.

Superabrasive wheels

Truing. Prior to truing the wheel, run a wax crayon over the wheel face (use of the crayon is safe, but liquid base inks should never be applied to wheel faces). Mount the wheel on a brake controlled truing device (always use the brake controlled truing device dry). Bring the diamond/CBN wheel and the truing wheel together until they almost touch, then run the diamond/CBN wheel to normal speed. Start the truing wheel in the same direction and bring the two spinning wheels together until they touch. Traverse the wheel back and forth at 30 to 60 inches per minute, and downfeed 0.0005" to 0.001" at the end of each traverse. At the end of truing, the diamond/CBN wheel should be smooth and true. Any crayon remaining on the wheel face after truing will reveal untrued areas.

Dressing. Even though superabrasive wheels seldom require dressing, the procedure is accomplished by applying an aluminum oxide or silicon carbide dressing stick directly to a turning wheel. Aluminum oxide sticks are white, and even though they are designed for dressing CBN wheels, they also work well with diamond wheels. Silicon carbide sticks are black and, in finer grades, are used for dressing diamond wheels. High performance norbide and boron carbide dressing sticks are also available–they are the hardest sticks available. For resinoid and vitrified bond wheels, select a dressing stick that is one or two grit sizes finer than the abrasive in the diamond/CBN wheel. For metal bond wheels, choose a stick with the same grit or one grit coarser than the wheel abrasive. *Figure 12* shows the features of a properly dressed diamond/CBN wheel face.

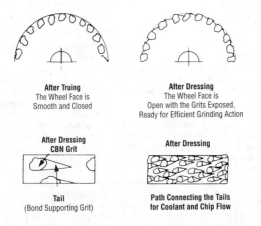

After Truing
The Wheel Face is
Smooth and Closed

After Dressing
The Wheel Face is
Open with the Grits Exposed,
Ready for Efficient Grinding Action

After Dressing
CBN Grit

Tail
(Bond Supporting Grit)

After Dressing

Path Connecting the Tails
for Coolant and Chip Flow

Figure 12. Properly dressed diamond/CBN wheel face. (Source, Norton Company.)

Superfinishing

Grinding is a high surface temperature operation that produces a series of peaks and valleys in the workpiece—even using the finest grit will not eliminate them entirely. A ground surface is also minutely wavey and may contain chatter marks. Even though these marks may not be easily visible without magnification, they can become destructive during high speed rolling or sliding. To remove surface finish peaks and to provide a good load-bearing surface of undisturbed base metal, hardness, and structure, superfinishing is required. Typical examples of parts requiring superfinishing are roller bearings, shock absorber rods, sliding vane pumps, piston pins, crankshaft journals, and cam lobes.

Superfinishing is a high-precision operation for removing minute amounts of surface material on dimensionally finished parts. The process uses fine grit adhesive particles bonded in stick form for cylindrical work, and in cup or cylindrical form for spherical work. Because superfinishing is a low-temperature abrading process that exposes the base material structure, it produces a smooth, long-wearing bearing surface on rotating, sliding, and sealing surfaces.

Superfinishing is known around the world by other terms such as microfinishing, superhoning, short stroke honing, or simply as polishing. It should not, however, be confused with honing, which is primarily a metal removal process that produces a crosshatch pattern on the part. Superfinishing usually takes place after grinding. It can remove as little as 0.5 micron (0.00002"), or as much as 20 microns (0.008") or more of material to further refine the surface quality and finish.

In superfinishing, a stick oscillates rapidly with a very short stroke while the workpiece rotates. The stick may transverse a long piece of material, or the workpiece may throughfeed under the stone. As the abrasive oscillates, the parts rotate or oscillate under the cup or cylinder, producing micro-fine chips. For stick applications on spherical shapes, motions include workpiece rotation, infeed pressure, and stick oscillation. The stick may also be applied in a plunge mode where narrow areas are involved. Typical process parameters are

Contact pressure: 1 to 20 kg/cm^2 (14 to 300 PSI) on stick area

Part rotation: 12 to 400 m/min (36 to 1,200 sfm)

Reciprocation amplitude: 1.0 to 5.0 mm (0.040" to 0.200")

Reciprocation frequency: 200 to 3,000 cycles per minute

Traverse/throughfeed: up to 5 m/min (200 in./min).

The process is also done with rotating cups and cylinders for spherical and flat surfaces. Typical operating parameters for cup or cylinder superfinishing include

Contact pressure: 0.7 to 4.2 kg/cm^2 (10 to 60 PSI) on wheel face area

Part rotation: 15 to 60 m/min (50 to 200 sfm)

Abrasive rotation: 150 to 600 m/min (500 to 2,000 sfm), but on multistation machines the roughing station may rotate at up to 2,000 m/min (6,000 sfm) and remove as much as 0.2 mm (0.008") stock.

Superfinishing specification marking systems

Superfinishing materials are identified by a coding system that specifically denotes the characteristics of the material. There are six sections in the identification system, as follows.

Abrasive Type. Available abrasives include

SG: seeded gel aluminum oxide blend available in grits ranging from 240 to 1,200, and in grades from D through O. Used on bearing steels, chrome plated parts, and hardened and tool steels.

NLA: calcified alumina with an average particle size of 3 to 8 microns, available in grades C through M. Typically used for inner and outer bearing races.

NSA: high purity white aluminum oxide available in grits ranging from 240 to 1,200, and in grades B through P. Used at first-stations, and is usually first choice for single-station machines finishing steel materials.

SGG: a high performance seeded gel blend available in grits ranging from 240 to 1,200, and in grades from E through O. Used on cast and nodular iron engine components, crankshaft journals, cam lobes, and other applications where silicon carbide is the abrasive choice.

NVCA: green silicon carbide used for finishing and polishing with negligible stock removal. Used in last-station finishing. Available in 1,200 grit and grades F through K.

NMVC: green silicon carbide developed to retard loading. Available in grits ranging from 240 to 1,500, and in grades from C through Q. Used on nonferrous and cast iron materials.

Grit Size. Available in FEPA grit sizes 240 through 1,500. Grit 320 is considered "fine," and 1,200 and 1,500 are "very fine."

Grade. Indicates bond strength. Designations are B through Q, with B being the softest and Q the hardest. Use a softer grade for heavy stock removal, for large contact areas, when using light pressure, with finer grits, for harder materials, and in roughing stations. Use a harder grade for rough surfaces, when extra fine finish is required, for small contact areas, at single-station machines, and in finishing stations.

Manufacturer Number. Assigned at manufacturer's option to denote a special feature or characteristic.

Bond Type. Either vitrified or resin. Almost all superfinishing products utilize a vitrified bond. Wear occurs from pressure and dressing action.

Treatment. Most aluminum oxide superfinishing products, and most single-station and intermediate multistation silicon carbide products are sulfur treated. The treatment is designated by the letter "S." Sulfur fills the pores of vitrified bonded products and reinforces the abrasive particles. Benefits are longer life, better finish, and it retards loading. The treatment also makes the product act harder and slows the cut rate. The other common treatment is wax, designated by the letter "W." Wax performs similarly to sulfur in soft and medium grades where pressure is moderate. It is not as rigid as sulfur and loses its benefits at higher pressures.

Table 12 provides guidelines for single-station superfinishing, using the specification marking system described above, and **Table 13** gives similar information for multistation operations.

Product availability

Superfinishing materials are available in stick form, as segments and blanks, and as cups and cylinders. The following tolerances and dimensions are for superfinishing materials as manufactured by Norton Company.

Sticks. Superfinishing sticks are used to produce the final finish and surface characteristics of a wide variety of metal parts. They require very close tolerances on both thickness and width in order to conform to the part and to fit the stick securely in its holder. When ordering superfinishing sticks, sizes are specified by thickness, width, and length, and tolerances are as follows:

	Metric Units	U.S. Customary units
Thickness	+ 0.00 mm, – 0.10 mm	+ 0.000", – 0.004"
Width	+ 0.00 mm, – 0.10 mm	+ 0.000", – 0.004"
Length	+ 0.00 mm, – 2.00 mm	+ 0.000", – 0.080"

Segments. Segments are used in sets and chucked in a ring-type holder to form a cylinder that is normally mounted on a vertical spindle or oscillator in a manner similar to using segments on Blanchard style surface grinding systems. Superfinishing segments are typically used on large flat or spherical surfaces. Since segments are often beveled on their sides and one end, it is important to include a blueprint or sketch

Table 12. Specification Guidelines for Single-station Superfinishing. *(Source, Norton Company.)*

Material Type	Abrasive Type	Grit Size	Grade	Treatment*
Aluminum Light Metal Brass Bronze	NMVC	400 - 600	D - H	Wax
Ball Bearing Tracks 100 Cr6 (52100)	SG	600 - 1200	G - L	Sulfur
	NLA		D - I	Sulfur
	NMVC	600 - 1000	E - K	Sulfur or Wax
	NSA	600 - 1200	D - I	Sulfur or Wax
Camshaft Lobes	NMVC	600 - 800	L - O	Sulfur
	SGG	600 - 800	L - O	Sulfur
Cast Iron Gray	NMVC	600 - 1000	J - M	Sulfur
	SGG	600 - 1000	J - M	Sulfur
Chrome Plate Hard	SG	400 - 800	G - I	Sulfur
	NSA	800 - 1000	D - F	Sulfur
Crankshafts	NMVC	600 - 800	L - O	Sulfur
	SGG	600 - 800	L - O	Sulfur
Crankshaft Radii	NMVC	600 - 800	L - O	Sulfur
	SGG	600 - 800	L - O	Sulfur
Roller Bearing Tracks	SG	400 - 800	G - J	Sulfur
	NSA	500 - 800	D - H	Sulfur or Wax
Stainless Steel 300 Series	NMVC	400 - 1000	F - J	Sulfur or Wax
	NSA	500 - 800	D - F	Sulfur or Wax
Stainless Steel 400 Series	NSA	400 - 800	F - H	Sulfur or Wax
Steel-Hardened HRc 58-62	SG	400 - 1000	G - L	Sulfur
	NSA	600 - 1000	D - J	Sulfur
	NMVC	600 - 1000	F - K	Sulfur
Steel-Hardened HRc 62-64	SG	400 - 1000	F - K	Sulfur
	NSA	600 - 1000	D - I	Sulfur
	NMVC	600 - 1000	E - J	Sulfur
Steel-Nitrided	SG	600 - 1000	E - J	Sulfur
	NSA	800 - 1000	D - F	Sulfur
	NMVC	800 - 1000	E - G	Sulfur
Steel-Unhardened	NSA	400 - 800	F - J	Sulfur
	NMVC	400 - 800	G - O	Sulfur

*** *Note:* Where sulfur treatment is not acceptable, Wax Treatment or No Treatment may be appropriate. Consult with manufacturer for recommendations.

Table 13. Specification Guidelines for Multistation Superfinishing. (Source, Norton Company.)

Material Type	Up to 4 Superfinishing Stations				Up to 8 Superfinishing Stations			
	Abrasive Type	Grit Size	Grade	Treatment*	Abrasive Type	Grit Size	Grade	Treatment*
Ball Bearing Races	SG	400 - 800	F - I	Sulfur or Wax	Norton SG	320 - 600	E - G	Sulfur or Wax
	NSA	600 - 800	D - F	Sulfur or Wax	NSA	400 - 600	D - F	Sulfur or Wax
	NMVC	800 - 1200	F - H	Sulfur or Wax	NSA	800 - 1000	D - F	Sulfur or Wax
	NVCA	1200	F - H		NMVC	1000/1200	G - I/H - J	Sulfur or Wax
					NVCA	1200	F - H	
Chrome-Hard	SG	400 - 800	F - I		Norton SG	400 - 600	E - G	
	NSA	400 - 800	D - F	Sulfur	NSA	400 - 800	D - F	Sulfur
	NMVC	600 - 1000	E - K	Sulfur	NMVC	800 - 1000	E - G	Sulfur
	NSA	600 - 1200	D - I	Sulfur	NVCA	1200	G - I	Sulfur
						1200	F - H	
Gudgeon/Piston Pins-Hardened	SG	400 - 800	F - I		Norton SG	400 - 800	E - H	
	NMVC	800 - 1000	F - I	Sulfur	NSA	600 - 800	D - F	Sulfur
	NVCA	1200	F - H	Sulfur	NMVC	800 - 1200	F - H	Sulfur
					NVCA	1200	F - H	Sulfur
Rollers	SG	500 - 800	E - H	Sulfur or Wax	Norton SG	400 - 800	E - H	Sulfur or Wax
	NMVC	600 - 800	E - H	Sulfur or Wax	NSA	500 - 800	D - F	Sulfur or Wax
	NMVC	1000 - 1200	G - I	Sulfur or Wax	NMVC	600 - 1000	G - I	Sulfur or Wax
					NMVC	1000 - 1200	H - J	Sulfur or Wax
					NVCA	1200	F - H	Sulfur or Wax
Needle Rollers	SG	500 - 800	G - J	Sulfur or Wax	Norton SG	400 - 600	F - H	Sulfur or Wax
	NMVC	600 - 800	F - I	Sulfur or Wax	NSA	400 - 600	F - I	Sulfur or Wax
	NMVC	800 - 1200	H - J	Sulfur or Wax	NMVC	600 - 800	E - G	Sulfur or Wax
	NVCA	1200	F - H	Sulfur or Wax	NMVC	1000	G - I	Sulfur or Wax
					NVCA	1200	F - H	Sulfur or Wax

(Continued)

Table 13. *(Continued)* **Specification Guidelines for Multistation Superfinishing.** *(Source, Norton Company.)*

Material Type	Up to 4 Superfinishing Stations				Up to 8 Superfinishing Stations			
	Abrasive Type	Grit Size	Grade	Treatment*	Abrasive Type	Grit Size	Grade	Treatment*
Shock/Strut Rods before chrome					NMVC NMVC NMVC NVCA	320 - 500 600 - 800 600 - 800 1000	I - K G - I G - I F - H	Sulfur or Wax Sulfur or Wax Sulfur or Wax
after chrome	The application involves 5 or more stations.				Norton SG NSA NSA NVCA	400 - 600 400 - 600 600 - 800 1200	E - H D - F D - F F - K	Sulfur or Wax Sulfur or Wax Sulfur or Wax
Steel-Hardened	SG NSA NMVC NVCA	400 - 600 600 - 800 800 - 1200 1200	F - I E - G E - H F - H	Sulfur Sulfur Sulfur	Norton SG NSA NMVC NVCA	400 - 600 400 - 600 600 - 800 1200	F - I F - I F - H F - K	Sulfur Sulfur Sulfur
Steel-Unhardened	NMVC NMVC NVCA	600 - 800 1000 - 1200 1200	F - I E - H F - H	Sulfur Sulfur	NMVC NMVC NMVC NVCA	400 - 600 400 - 600 800 - 1000 1200	E - G E - G E - G F - K	Sulfur Sulfur Sulfur

* *Note:* Where sulfur treatment is not acceptable, Wax Treatment or No Treatment may be appropriate. Consult with manufacturer for recommendations.

when ordering, and the number of segments in a set should be specified. Tolerances for segments are as follows:

	Metric Units	U.S. Customary Units
Thickness	+ 0.00 mm, − 0.4 mm	+ 0.000", − 0.016"
Width	+ 0.00 mm, − 0.4 mm	+ 0.000", − 0.016"
Length	± 0.8 mm	± 0.032"

Blanks: Blanks are normally supplied "as molded," but finished blanks are available molded and ground to very precise thicknesses. Standard thicknesses for blanks are 26 mm for grits 500 and coarser, and 22 mm for grits 600 and finer. Standard width is 105 mm, and standard length is 155 mm. Tolerances for finished and unfinished blanks are as follows:

	Metric Units	U.S. Customary Units
Thickness, finished	+ 0.00 mm, − 0.40 mm	+ 0.000", − 0.016"
Thickness, unfinished	+ 1.5 mm, − 0.8 mm	+ $^1/_{16}$", − $^1/_{32}$"
Width, both	+ 6.0 mm, − 1.5 mm	+ $^1/_4$", − $^1/_{16}$"
Length, both	+ 06.0 mm, − 1.6 mm	+ $^1/_4$", − $^1/_{16}$"

Cups and cylinders. Superfinishing cups and cylinders are mechanically mounted on a spindle and are used on spherical, concave, convex, and flat surfaces. Cups should be ordered by the dimensions shown in *Figure 13*, and may be finished as one piece or with attached metal, plastic, or abrasive back plates. Cylinders should be ordered by specifying the dimensions shown in *Figure 14*. Tolerances for cups and cylinders are as follows:

	Metric Units	U.S. Customary Units
Diameter	+ 0.00 mm, − 0.25 mm	+ 0.000", − 0.010"
Thickness	± 0.50 mm	± 0.020"
Rim thickness	+ 0.00 mm, − 0.25 mm	+ 0.000", − 0.010"

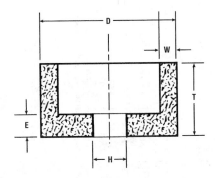

Figure 13. Superfinishing cup dimensions. D = diameter, T = thickness, H = hole diameter, W = rim thickness, and E = back thickness. (Source, Norton Company.)

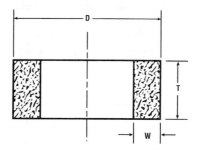

Figure 14. Superfinishing cylinder dimensions. D = diameter, T = thickness, and W = rim thickness. (Source, Norton Company.)

Surface finish in superfinishing

Superfinishing operations remove minute amounts on surface material on a dimensionally finished part. The primary purpose for superfinishing a ground part is to improve surface finish. Because the roughness heights of the part vary, as shown in *Figure 15*, it is necessary to average these heights to have a meaningful comparison. This measurement of the average value of the amplitude of peaks and valleys that form the roughness texture of a surface finish is known as "Ra." (**Table 14** provides terms and definitions used in identifying surface finish parameters, and should be referred to for additional information and explanations, and **Table 15** compares surface roughness values.)

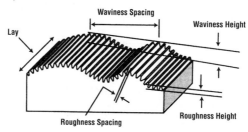

Figure 15. Roughness and waviness factors in surface finishing. (Source, Norton Company.)

Two techniques are used to trace surfaces for measurement and evaluation: skid and skidless stylus types (both are shown in *Figure 16*). A surface texture traced with a skid type stylus does not produce a graph that is representative of the actual surface. It may amplify or nullify the waviness of the surface depending on its geometry as related to the skid–stylus spacing. However, if a roughness assessment that eliminates the factors of waviness and error of form is desired, a skid stylus should be used. When using a skidless stylus, a more accurate profile of the surface can be obtained, including a representation of waviness and form.

Table 14. Surface Finish Parameters: Terms and Definitions. *(Source, Norton Company.)*

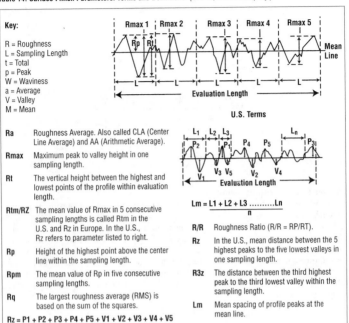

Key:

R = Roughness
L = Sampling Length
t = Total
p = Peak
W = Waviness
a = Average
V = Valley
M = Mean

U.S. Terms

Ra	Roughness Average. Also called CLA (Center Line Average) and AA (Arithmetic Average).
Rmax	Maximum peak to valley height in one sampling length.
Rt	The vertical height between the highest and lowest points of the profile within evaluation length.
Rtm/RZ	The mean value of Rmax in 5 consecutive sampling lengths is called Rtm in the U.S. and Rz in Europe. In the U.S., Rz refers to parameter listed to right.
Rp	Height of the highest point above the center line within the sampling length.
Rpm	The mean value of Rp in five consecutive sampling lengths.
Rq	The largest roughness average (RMS) is based on the sum of the squares.

$$Rz = \frac{P1 + P2 + P3 + P4 + P5 + V1 + V2 + V3 + V4 + V5}{10}$$

$$Lm = \frac{L1 + L2 + L3 \ldots\ldots\ldots Ln}{n}$$

R/R	Roughness Ratio (R/R = RP/RT).
Rz	In the U.S., mean distance between the 5 highest peaks to the five lowest valleys in one sampling length.
R3z	The distance between the third highest peak to the third lowest valley within the sampling length.
Lm	Mean spacing of profile peaks at the mean line.

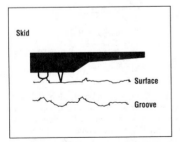

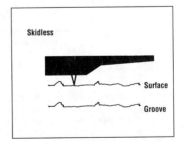

Figure 16. Skid (left) and skidless (right) tracing styluses. (Source, Norton Company.)

Table 15. Comparison Chart of Surface Roughness Values. *(Source, Norton Company.)*

Rt	Rtm	Ra/CLA/AA		Rq/RMS/Rs	
μm	μm	μm	μin.	μm	μin.
0.06	0.03	0.006	0.2	0.007	0.2
0.08	0.04	0.008	0.3	0.009	0.3
0.1	0.05	0.01	0.4	0.011	0.4
0.12	0.06	0.012	0.5	0.013	0.5
0.15	0.08	0.015	0.6	0.018	0.7
0.2	0.1	0.002	0.8	0.022	0.9
0.25	0.12	0.025	1.0	0.027	1.1
0.3	0.15	0.03	1.2	0.033	1.3
0.4	0.2	0.04	1.6	0.044	1.8
0.5	0.25	0.05	2.0	0.055	2.2
0.6	0.3	0.06	2.4	0.066	2.6
0.8	0.4	0.08	3.2	0.088	3.5
0.9	0.5	0.10	4.0	0.11	4.4
1.0	0.6	0.12	4.8	0.13	5.2
1.2	0.8	0.15	6.0	0.18	7.2
1.6	1.0	0.20	8.0	0.22	8.8
2.0	1.2	0.25	10	0.27	10.8
2.5	1.6	0.3	12	0.33	13.2
3.0	2.0	0.4	16	0.44	17.6
4.0	2.5	0.5	20	0.55	22
5.0	3.0	0.6	24	0.66	26
10	6.0	1.2	48	1.3	52
15	10	2.5	100	2.7	108
30	20	5.0	200	5.5	220
50	40	10	400	11	440
100	80	20	800	22	880
200	160	40	1600	44	1760
400	320	80	3200	88	3500

Note: Due to the nature of measuring surface finishes, comparison values may vary up to 25%.
1 μm = 0.001 mm = 0.000040" = 1 micron.   1 μin. = 0.000001" = 0.254 μm.

Figure 17. Two surface finishes with equal Ra values.

Using the roughness average, Ra, can sometimes be misleading. The two surfaces shown in *Figure 17*, for example, have dramatically different profiles, but both have the same Ra. A less common means of measuring roughness is the roughness root mean square, Rq or RMS, which provides more weight to the highest peaks. Rq is determined in a fashion similar to Ra, but the numbers are squared when averaging, as shown in *Figure 18*.

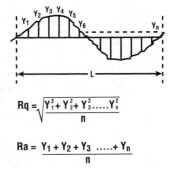

$$Rq = \sqrt{\frac{Y_1^2 + Y_2^2 + Y_3^2 \ldots\ldots Y_n^2}{n}}$$

$$Ra = \frac{Y_1 + Y_2 + Y_3 \ldots\ldots + Y_n}{n}$$

Figure 18. Method for measuring Rq (RMS), the roughness root mean square. (Source, Norton Company.)

Sawing

Sawing operations—usually involving the cutting off or slotting of sheet, tube, or bars—are normally carried out with three different machines: hacksaws, bandsaws, or circular saws.

Hacksawing

Although hacksawing by hand, especially on thin-walled materials, is still common, hacksaw machines are relatively low in cost and require little maintenance, and they are able to perform most cutting operations on a wide range of materials with fair to good accuracy. They cut by drawing the blade toward the motor end of the machine. At the completion of this movement—called the draw stroke—the hacksaw blade lifts slightly in order to clear the workpiece, and then moves an equal distance in the opposite direction (the return stroke) before lowering again to begin another stroke. The base of a hacksaw machine usually contains a coolant reservoir and a pump for conveying coolant to the work. The reservoir contains baffles that cause the chips to settle to the bottom of the tank. Horizontal machines are the most common, and are so named because of the orientation of the blade, which is fed downward during cutting. On vertical machines, the blade is oriented vertically and is fed axially into the workpiece to perform the cut. Generally, most materials, with the exception of cast iron and copper, will require a lubricant during cutting.

The cutting speed of hacksawing machines is measured in strokes per minute. Therefore, the length of the stroke is an important consideration. Since a longer stroke at a given speed will cut faster than a shorter stroke at the same speed, the length of the stroke must be specified when designating speeds. The length of the stroke on most machines is between four and ten inches (102 to 254 millimeters) depending on the size of the machine. On machines with an adjustable stroke, the wider the stock being cut, the shorter the stroke in order to prevent the blade holders from striking the workpiece. On most power hacksaws, the stroke length is adjustable within two or three inches, and on advanced machines, more than one speed can be selected. On single speed machines, the speed must be regulated by changing the stroke. Doubling the stroke will result in the machine cutting twice as fast, and reducing it by one-half will result in the machine cutting half as fast. This proportion can be applied to any fraction to increase or decrease the speed of the machine. Speeds given in **Table 1** are presented for guidance only. The correct speeds for cutting various materials will depend on the machine and blade being used. If a recommended speed cannot be approximated either by changing the stroke or the speed, the feed can be decreased to prevent undue wear on the hacksaw blade. Power hacksaw machines having a mechanical feed can usually be regulated to feed the saw blade downward from 0.001" (0.0254 mm) to 0.025" (0.635 mm) per stroke. On these machines, a device to stop the speed when hard spots are encountered is usually incorporated into the design. The feed of machines using gravity feed is regulated by the weight of the saw's frame and any additional weights of springs that may be connected to the frame to increase or decrease the downward force of the blade. Minimum and maximum blade pressures are determined by the machine's manufacturer and should be adhered to. The following rules apply for selecting proper feeds for hacksawing machines.

1) The feed should be very light when starting a cut and may be increased after the cut is started.
2) Hard materials require a lighter feed than soft materials. Reduce the feed when welds or hard spots are encountered.
3) Wide material requires a heavier feed than narrow material because the pressure is distributed over a larger surface area.

4) Sharp blades will cut well with lighter feeds. A dull blade will require heavier feeds.

Table 1. Power Hacksawing Machine Speeds and Feeds.

Material	4" to 6" Stroke			8" to 10" Stroke		
	Speed (Strokes/Minute)	Feed		Speed (Strokes/Minute)	Feed	
		inch	mm		inch	mm
Aluminum	135	0.005	0.127	65	0.016	0.406
Brass	135	0.005	0.127	65	0.016	0.406
Bronze	90	0.004	0.102	45	0.014	0.356
Cast Iron	90	0.004	0.102	45	0.014	0.356
Copper	135	0.005	0.127	65	0.016	0.406
Steel, Alloy	90	0.004	0.102	45	0.014	0.356
Steel, High Speed	60	0.003	0.076	30	0.013	0.330
Steel, Machine	135	0.005	0.127	65	0.016	0.406
Steel, Stainless	60	0.004	0.102	30	0.013	0.330
Steel, Tool (Annealed)	90	0.004	0.102	45	0.014	0.356
Steel, Tool (Unannealed)	60	0.003	0.076	30	0.013	0.330

Hacksaw Blades

Blades for hacksawing are classified as Grade A or Grade B. Grade A designates that the material used in manufacture is high-speed steel that (in the hardened and tempered condition) can be reheated to 1100° F (593° C) for five hours, followed by cooling to room temperature, and retain a hardness of Rockwell C of 60 or greater. Grade A, Class 1 (all hard) blades should be marked "high speed" or "HSS," and Grade A, Class 2 (flexible back) blades should be marked "high speed edge" or "HSS edge." Blades marked Grade A are used for machine hacksawing. Grade B blades are for hand use and are manufactured from heat treated high carbon or low tungsten steel. **Table 2** provides dimensions for both hand and machine hacksaw blades.

Straightness tolerance requires that blades must be free from crimps, buckles, or warpage. When a straightedge is placed against the side of a blade at the toothed edge and at the back edge, any concave or convex deviation in the contacting edge must not exceed the maximum deviations shown below.

Blade Length (inches)	Maximum Deviation (inches)
10	3/64 (0.046875)
12 through 18	1/16 (0.0625)
21 through 24	5/64 (0.078125)
30	3/32 (0.09375)

The blade ends should be heat-treated in such a manner as to preclude the pinholes pulling out or the ends breaking when a blade is tensioned for standardized testing.

On all saw blades, pitch is the distance from the tip of one tooth to the tip of the next tooth on a blade. It is more common to identify blades by the number of teeth per inch, which can vary from 2.5 to 18 teeth per inch for power saws, and to refer to that number as the blade's pitch. In order to keep at least three consecutive teeth within the cut at all times to reduce the possibility of snagging a tooth on the workpiece, blades with six teeth

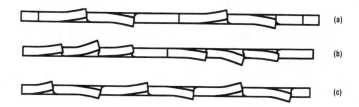

Figure 1. Tooth set on hacksaw blades: (a) raker, (b) wavy, (c) alternative.

per inch are usually used for average cutting conditions on materials from 1 to 5 inches in diameter. For cutting parts from 5 to 8 inches, four teeth per inch are recommended, and larger parts can be cut with blades having three teeth per inch. When cutting soft materials, a coarse blade is recommended to provide adequate space between the teeth for removal of chips. Hard materials require a finer blade in order to evenly distribute the cutting pressure to as many teeth as possible in order to reduce wear on the blade. The "tooth set" on blades identifies the configuration of the teeth when the blade is viewed from below where the angles of the individual teeth can best be observed. It is necessary for the tips of the teeth to bend left or right, in a pattern, in order to create clearance for the blade as it moves through a cut. Popular tooth sets are shown in *Figure 1*. The *raker set* is a three-tooth sequence—left, right, straight—that is repeated throughout the blade length. The raker set is preferred for heavy cutting. The *wavy set*, which is especially useful for cutting thin-walled materials with hand blades, has groups of teeth set to each side in a wave pattern. Wavy set blades are usually fine pitch and may have as many as 32 teeth per inch. The *alternative set* alternates the direction of every tooth—left, right, left, right, etc. Finally, tooth form is the appearance of the blade when viewed in profile. For further information on tooth form, see the section on Bandsaws.

Bandsawing

Bandsaw machines use a continuous blade (band) that runs over a driver wheel and one or more idler wheels to perform straight, radii, or contour cuts with repeatable accuracy of ±0.002" (0.051mm) per inch of cut. Vertical Bandsaw Machines, also called Contouring Machines, are versatile cutting tools with tilt tables that allow angle cuts and special contour cuts. Feed may be manual, gravity, or hydraulic, and the tilt frame can allow miter cutting (45°) on some machines. When combined with CNC, modern bandsaws are capable of producing three-dimensional shapes not normally associated with sawing. Horizontal Bandsaw Machines are used primarily for cutoff operations. In essence, it performs the same tasks as a power hacksaw, but it carries it out more efficiently.

Bandsaw Blades

For production work, *carbon steel blades* are available for low initial cost, and even though they will wear out faster than bi-metal blades they are commonly used to cut mild steels and other softer materials such as aluminum, brass, and copper. They are available with standard tooth forms (see below) and with embedded tungsten or diamond (commonly called "grit" blades) that offer smooth, fast cutting on some hard, abrasive materials such as fiberglass, composite materials, carbon and graphites, aramid fiber, plastics, white or

Table 2. Hacksaw Blade Dimensions. *(Source, Fed. Std. GGG-B-451d.)*

			Hand Blades				
Nominal Length	**Teeth per inch**	**Width**	**Thickness**		**Overall Length**	**Center-to-Center for Pinholes**	**Pinhole Dia.**
			HSS	**Standard Steel**			
10	18, 24, 32	1/2	0.025	0.025	10 3/8	9 7/8	0.158
12	14, 18, 24, 32	1/2	0.025	0.025	12 3/8	11 7/8	0.158
12	18/24*	1/2	–	0.025	12 3/8	11 7/8	0.158

		Power Blades					
Nominal Length	**Teeth per inch**	**Width**		**Thickness ±0.003**	**Overall Length ± 1/16**	**Center-to-Center for Pinholes ± 1/16**	**Pinhole Dia. ±0.010**
		Class 1 ± 1/64	**Class 2**				
12	14, 18	5/8	5/8 (+1/16, –1/64)	0.032	12 1/2	11 7/8	3/16
12	10, 14	1	1 (+3/32, –1/64)	.050	12 3/4	11 7/8	9/32
14	10, 14	1	1 (+3/32, –1/64)	.050	14 3/8	13 1/2	9/32
14	4, 6, 10	1 1/4	1 1/4 (+1/8, –1/64)	.062	14 1/2	13 1/2	9/32
14	4, 6	1 1/2	1 1/2 (+1/8, –1/64)	.075	14 1/2	13 1/2	9/32
17	10, 14	1	1 (+3/32, –1/64)	.050	17 3/8	16 1/2	9/32
17	4, 6, 10	1 1/4	1 1/4 (+1/8, –1/64)	.062	17 1/2	16 1/2	9/32
18	6, 10	1 1/4	1 1/4 (+1/8, –1/64)	.062	18 1/2	17 1/2	9/32
18	4, 6	1 1/2	1 1/2 (+1/8, –1/64)	.075	18 1/2	17 1/2	9/32
18	4, 6	1 3/4	1 3/4 (+1/8, –1/64)	.088	18 3/4	17 1/2	9/32
21	4, 6	1 3/4	1 3/4 (+1/8, –1/64)	.088	22 1/4	21	9/32
24	4, 6	1 3/4	1 3/4 (+1/8, –1/64)	.088	25 1/4	24	9/32
24	4	2	2 (+1/8, –1/64)	.100	25 1/4	24	25/64
30	4	2 1/2	2 1/2 (+1/8, –1/64)	.100	32	30	25/64
36	2 1/2	4 1/2	4 1/2 (+1/8, –1/64)	.125	38	36	1/2

All dimensions in inches.
Notes: Class 1 power blades are one piece, all hard. Class 2 power blades are welded with high-speed edge. Tolerances on hand blades are: Width ±1/64; Thickness (HSS) ±0.003; Thickness (Standard Steel) +0.001, –0.003; Overall Length ±1/16; Pinhole center-to-center ±1/16; Pinhole Dia. ±0.010.

chilled cast iron, satellites, high hardness tool steels, and super alloys. Carbon steel blades are also available in "hard back" and "flexible back" versions—hard back blades offer more support which results in straighter, faster cuts in similar materials. Recommended tension for flexible back blades is 15,000 – 20,000 PSI, and tension for hard back is 20,000 – 25,000 psi. Flexible back blades are ideal for cutting materials like aluminum that can be cut at high speed because the flexible back means that the blade experiences less fatigue while making rapid turns around the drive and idler wheels at higher speeds. Avoiding the extra costs of hardening allows flexible back blades to be the preferred economic choice for operations such as this. When using carbon steel blades, it is desirable to have 6 – 18 teeth in the cut at all times. *Tungsten carbide tipped blades* have tungsten carbide teeth welded to an alloy steel blade back. These blades are especially effective for cutting modern space-age heat-resistant alloys, nickel based alloys, cobalt alloys, titanium alloys, and cast tensile steels with tensile strength over 142,000 psi, and for nonferrous materials such as aluminum castings, fiberglass, graphite, and nonmetallic materials. These blades have the highest heat and wear resistance available and offer long life and fast and quiet cutting. *Bi-metal blades* have high-speed steel (normally M2 or M42) or matrix cobalt steel teeth electron beam welded to a spring steel blade back. These blades are most often used to cut low carbon steel, mild steel, some alloy steels and tool steels, stainless steels, and aluminum, monel, and inconel. Bi-metal blades cut truer than carbon steel blades, because their strength allows them to be tensioned higher, and they provide long life which results in lower costs per cut on most materials than can be achieved with carbon steel blades. These blades are the choice for automatic machines or where high performance takes precedence over expense. For tool and die contouring, narrow width bi-metal blades outperform carbon blades by ten to one. When using these blades, at least three teeth should be kept in the cut, with 6 to 12 being optimum. Recommended tension for bi-metal blades is 30,000 – 35,000 PSI.

Tooth forms for bandsaws vary with manufacturer, but standard, variable, skip, and hook forms are widely available (see *Figure 2*). *Standard form* blades have evenly spaced teeth with deep gullets, and a 0° rake angle. They offer good chip carrying ability and are favored for cutting a wide range of materials. Because they have no rake angle, they are especially suited for cutting bundles or stacks of thin or small diameter material. *Variable tooth* blades are similar in appearance to standard form blades and have 0° rake angle, but they have variable spacing between the teeth to reduce vibrations—especially when cutting irregularly shaped materials—and enhance the finish of the cut material. The *variable pitch, positive rake form* has a positive rake angle that encourages chip formation, provides improved tooth penetration, and increases blade life. The *skip form* also has 0° rake angle, wide gullets, and evenly spaced teeth, and it provides for coarse pitch on a narrow band. It is popular for cutting softer metals and for nonmetallic cutting including wood, plastics, and composition materials. It is also available in a *skip positive* form that has a positive rake angle to improve tooth penetration and chip formation. *Hook form* blades feature wide gullets, evenly spaced teeth, and positive rake angle. This form permits coarse teeth on a narrow band and provides good chip carrying capacity in nonmetallic applications. It is favored for cutting metals with discontinuous chips such as cast iron, and for wood, plastic, cork, and composition material.

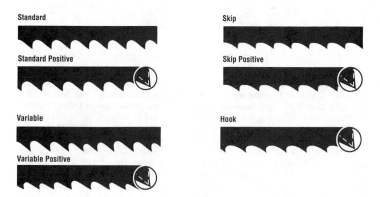

Figure 2. Bandsaw tooth forms. (Source, American Saw and Mfg. Company.)

Figure 3. Variable pitch raker set.

Tooth set was covered earlier in the discussion on hacksaw blades (see *Figure 1*), but more varieties are available for bandsaw blades. In bandsawing, the *raker set* offers fast cutting and a good finish. The *wavy set* is most often used for cutting tubing, and the *alternative set* is favored for fast cutting where surface finish is less important than speed. In addition to raker, wavy, and alternative tooth sets, the *variable pitch raker set* is a multi-tooth sequence that varies with pitch. It provides an improved surface finish over the standard raker set.

Blades for contour sawing may be carbon steel, bi-metal, or tungsten carbide, and a primary selection criterion is the radius that is to be cut. The widest blade possible should be used, so it is very important to review the planned cut carefully in order to determine the smallest radius to be cut. **Table 3** provides minimum radii for various widths of contour bandsaw blades.

Table 3. Radius Chart for Bandsaw Width Selection.

Blade Width		Minimum Radius	
inch	mm	inch	mm
1/16	1.6	Can Cut Square Radius	
3/32	2.4	1/16	1.6
1/8	3.2	1/8	3.2
3/16	4.8	5/16	8
1/4	6.4	5/8	16
3/8	9.5	1 7/16	37
1/2	12.7	2 1/2	64
5/8	16	3 3/4	95
3/4	19	5 7/16	138
1	25.4	7 1/4	184

Blade selection

Choosing a blade for a particular cutting operation requires consideration of many variables in addition to determining the minimum allowable radius. These include pitch (teeth per inch), tooth form, tooth set, and material. The American Saw and Manufacturing Company has devised a selection chart that is printed in the catalogs of several saw manufacturers that helps in the determination of how many teeth per inch should be used on a workpiece of a given shape and diameter. The selection chart is intended for use with variable tooth blades. As reproduced in *Figure 4*, the guide has three columns, each with a part shape on the top, and with numbers that indicate the number of teeth recommended. The column on the left is for square parts, the center for round parts, and the one on the right is for hollow tubing and pipe, and other structural shapes. The numbers on each side of the chart refer to length of the cut to be made—with inches on the left and millimeters on the right. To determine the number of blade teeth per inch for a cut, first select the column that represents the shape of the part being cut, then read down either side to the

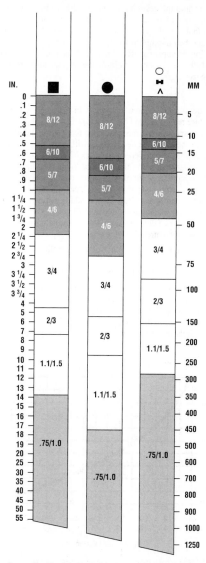

Figure 4. Tooth selection chart for variable tooth blades. (Source, Amada Cutting Technologies, Inc.)

desired length of cut to find the desired number of teeth for the saw. For instance, when cutting a 7" (178 mm) round bar, select the center column and follow it down to the 7" level which indicates that a blade with $^2/_3$ teeth per inch should be used for the cut. The $^2/_3$ variable tooth blade has maximum tooth spacing of approximately $^1/_2$ inch, and a minimum tooth spacing of approximately $^1/_3$ inch. When using the right-hand column for piping, tubes, and structural shapes, multiply the wall thickness by three to determine the length of cut to be used.

Break-in procedures

New bandsaw blades must be broken in properly in order to assure the longest possible blade life. The procedure recommended by American Saw and Manufacturing Co. begins with setting the bandsaw speed to the recommended feet per minute for the material being cut. The material should be the same as the band will be cutting during its product life. Next, select the starting feed rate based on band speed shown in *Figure 5* and begin the cut. After cutting to a distance equal to the width of the blade, slightly increase the feed. As the blade reaches the halfway point of the cut, increase the feed (again) lightly. Finish the cut at this speed. Start a fresh cut at the same speed and increase the feed rate again when approaching halfway in the cut. Repeat with additional cuts until the blade has cut the required number of square inches as indicated in *Figure 6*. The following formulas can be used to predict the number of cuts required on a given diameter workpiece, the area of a single cut round, and the recommended time of cut.

Total Area Required (*Figure 6*) ÷ Area of Workpiece = No. of Cuts Required

Diameter^2 × .7854 = Area of a Round

Area of Workpiece ÷ Cutting Rate (in.^{2}/min *or* cm^2/min – *Figure 6*) = Time of Cut

When using a new blade to cut steel, inspect the condition of the chip to determine that the blade is operating efficiently. Correct band speed, feed pressure, and tooth selection should produce chips that look like silvery spirals. Too many teeth and/or insufficient feed pressure will cut thin and powdery filings rather than spiral chips. To correct, try increasing blade pressure. If that fails, try a coarser blade. Large, heavy chips that appear burned indicate that feed pressure is too high, blade speed too low, or both.

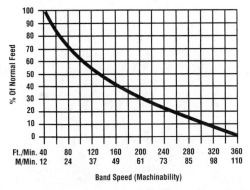

Figure 5. Starting feed rate for break-in. (Source, American Saw and Mfg. Co.)

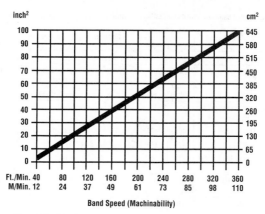

Figure 6. Total break-in area required. (Source, American Saw and Mfg. Co.)

Power Requirements

An interesting feature of bandsawing is that machines with minimal horsepower are capable of performing the same operations as much more powerful machines. This is because even though higher feed forces are needed for harder materials, horsepower requirements are lower because they are relative to chip removal rate, which is small. Therefore, bandsaws with up to 2 horsepower (1.5 kW) are able to cut hard materials such as stainless and alloy steels. To accomplish this, band speed must be low and the process time consuming. Paradoxically, low power machines are not as well suited for cutting softer materials like aluminum because the high chip removal rate at higher band speeds requires higher power.

Bandsawing Speeds for Cutting

Table 4 provides recommended band speeds for cutting a 4" (100 mm) workpiece of various materials with a $^{3}/_{4}$" variable pitch tooth bi-metal saw, using a cutting fluid. The *notes* on the table provide conversion factors for cutting materials of other diameters without fluids and with carbon steel blades.

Troubleshooting Bandsaw Cutting Operations. **Table 5** identifies several common problems that are experienced during bandsaw operations. Probable causes and solutions are identified. The manufacturer should be contacted for problems not covered by this chart.

Friction sawing

Many bandsaw machines can be used for friction sawing, a process that uses heat generated by friction to soften the workpiece material so that it can be removed quickly and economically. The bandsaw is ideally suited for this process because its long band length in relation to work thickness allows the cutting edge to dissipate the heat before reentering the cut. When friction sawing, the band must travel at high velocity (blades are commonly rated for 20,000 fpm although the process is usually performed at speeds between 6,000 and 15,000 fpm) while exerting light pressure on the work. This pressure generates tremendous heat immediately in front of the cutting edge, raising the temperature

Table 4. Recommended Bandsaw Speeds for Cutting (See Notes). *(Source, American Saw and Mfg. Co.)*

Material	Designation	Band Speed fpm	Band Speed mpm	Material	Designation	Band Speed fpm	Band Speed mpm
Carbon Steels				**Alloy Steels**			
Resulfurized Carbon Steel	1117	340	104	Ni-Cr-Mo Alloy Steel	4340, 8630, E4340	220	67
	1137	290	88		8615, 8620	240	73
	1141, 1144, 1144 HiStress	280	85		8640	200	67
	1213, 1215	400	122		E9310	175	53
	12L14	425	130	Cr-Mo Alloy Steel	4130	270	82
Carbon Steel	1008, 1015, 1020, 1025	320	98		4140, 4142	250	76
	1018, 1021, 1022, 1513	300	91		4145	210	64
	1030	330	101		4150	200	61
	1035	310	95		41L40	275	84
	1040, A36 (shapes)	270	82	Cr Alloy Steel	5160	220	67
	1042, 1541	250	76		6150	210	64
	1044, 1045	220	67		E52100	160	49
	1060	200	61	Mn Alloy Steel	1330	220	67
	1095	185	56		1345	210	64
Tool Steels					300M, HY-100, HY-80	160	49
Shock Resisting	S-1	200	61		A242 (1) COR-TEN A	300	91
	S-5	140	43		A242 (2) COT-TEN B	270	82
	S-7	125	38		HP 9-4-20, HP 9-4-25	105	32
Cold Working	A-10	160	49	**Mold Steel**	P-20	230	70
	A-2	180	55	**Stainless Steels**			
	A-6	200	61	Cr Stainless	410, 502	140	43
	D-2	90	27		414	115	35
Hot Working	H-11, H-12, H-13	190	58		420	190	58
Oil Hardening	O-1	200	61		430	150	46
	O-6	190	58		431	95	29
High-Speed	M-1	110	34		440 C	80	24
	M-2, M-42, T-1	100	30	Austenitic Stainless Steels	18-18-2, 309	90	27
	T-15	70	21		23-13-5, Nitronic 50, Nitronic 60	60	18
Low Alloy	L-6	190	58		304	120	37
Water Hardening	W-1	225	69		304L	115	35

(Continued)

Table 4. *(Continued)* **Recommended Bandsaw Speeds for Cutting** (See Notes).
(Source, American Saw and Mfg. Co.)

Material	Designation	Band Speed		Material	Designation	Band Speed	
		fpm	mpm			fpm	mpm
Stainless Steels *(Continued)*				**Cast Irons** *(Continued)*			
Austenitic Stainless Steels (Continued)	310	80	24	Ductile Cast Iron	A536 (100-70-03)	185	56
	316, 316L, 324	100	30		A536 (120-90-02)	120	37
	347	110	34		A536 (60-40-18)	360	110
Martensitic	Greek Ascoloy	95	29		A536 (80-55-06)	240	73
Ferritic-Austenitic Stainless Steel	2205	80	24	Malleable	A47 (50 & 53 ksi)	300	91
	Ferralium 255	90	27	Ductile, Austenitic	A439 (D-2)	80	24
Free Machining Stainless Steel	203 EZ, 430 F	150	46		A439 (D-2B)	60	18
	303, 303 PB, 303 SE	140	43	**Alloys**			
	416	190	58	Aluminum Alloys	1100, 2011, 2017, 2024, 3003, 5052, 5086, 6061, 6063, 6101, 6262, 7075	500	152
	440 F	160	49	Copper Alloys	101, 102, 110, 122, 172, 1751, 182, 220, 510, 625, 706, 715	235	72
Precipitation Hardening Stainless Steel	15-5 PH, 17-4 PH, 17-7 PH, AM 350, AM 355, PH 13-8 MO, PH 14-8 MO	80	24		1452, 187,	375	114
	PH 15-7 MO	75	23		623, 624	265	81
Custom Stainless Steel	Custom 450, Custom 455	80	24		173, 932, 934	315	96
Cast Irons					630	230	70
Pearlitic Malleable	A220 (50005)	240	73		811	215	66
	A220 (60004)	170	52		230, 260, 272, 280, 464, 632, 655	245	75
	A220 (70003)	145	44		330, 365	285	87
	A220 (80002)	125	38		353	400	122
	A220 (90001)	100	30		356, 360	450	137
Gray Iron, Austenitic	A436 (1B)	150	46		380, 544	350	107
	A436 (2)	140	43	Titanium Alloys	Ti-2Al-11Sn-5Zr-1Mo, Ti-5Al-2-5Sn, Ti-6Al-2Sn-4Zr-2Mo	75	23
	A436 (2B)	125	38		Ti-3Al-8V-6Cr-4Mo-4Zr, Ti-8Mo-8V-2Fe-3Al	80	24
Gray Iron	A48 (20 ksi)	230	70		Ti-6Al-4V	70	21
	A48 (40 ksi)	180	55		Ti-7Al-4Mo, Ti-8Mo-8V-2Fe-3Al	65	20
	A48 (60 ksi)	100	30	Magnesium	AZ318	900	274

(Continued)

Table 4. *(Continued)* **Recommended Bandsaw Speeds for Cutting** (See Notes).
(Source, American Saw and Mfg. Co.)

Material	Designation	Band Speed		Material	Designation	Band Speed	
		fpm	mpm			fpm	mpm
Alloys *(Continued)*				**Alloys** *(Continued)*			
Nickel Alloy	Nickel 200, Nickel 201, Nickel 205	85	26	Nickel Base Alloys	Inconol 625	100	30
Nickel Base Alloys	20CB3	80	24		Monel 400, Monel 401	75	23
	Hastelloy B, Hastelloy B-2, Hastelloy C, Hastelloy C-22, Hastelloy C-276, Hastelloy C-4, Hastelloy F, Hastelloy G, Hastelloy G-2, Hastelloy G-3, Hastelloy N, Hastelloy G-30, Hastelloy S, Hastelloy W, Hastelloy X, Incoloy 825, Incoloy 925, Incoloy 751, Incoloy X750, Waspaloy	70	21		Monel K500	65	20
					Monel R405	85	26
					Mimonic 75	50	15
				Iron Based Super Alloys	A286, Incoloy 800, Incoloy 801	90	27
	Incoloy 802, Incoloy 804,	90	27		PYROMET X-15	120	37
	Incoloy 901, Incoloy 903, Incoloy 600, Incoloy 718, NI-SPAN-C 902, Nimonic 263, RENE 41, UDIMET 500	60	18	Cobalt Based Alloys	ASTROLOY M	60	18
					WF-11	65	20

Notes: The material names Inconel, Monel, and Nimonic are registered.

The recommended speeds in this table are based on cutting a 4 " (100 mm) workpiece with a $^3/_4$ bi-metal variable tooth saw, using cutting fluid.

When cutting dry, reduce speeds 30% to 50%.

When cutting with carbon steel blades, reduce speeds by 50%.

For other workpiece thicknesses:

    Increase band speed:

        15% when cutting $^1/_4$" (6 mm) material with a 10/14 variable tooth bandsaw.

        12% when cutting a $^3/_4$" (19 mm) material with a 6/10 variable tooth bandsaw.

        10% when cutting a 1 $^1/_4$" (32 mm) material with a 5/8 variable tooth bandsaw.

        5% when cutting 2 $^1/_2$" (64 mm) material with a 4/6 variable tooth saw.

    Decrease Band Speed:

        12% when cutting 8" (200 mm) material with a 2/3 variable tooth saw.

Table 5. Troubleshooting Bandsaw Cutting Operations. *(Source, M. K. Morse Company.)*

Problem	Probable Cause	Likely Solution
Premature Blade Breakage (Straight break indicates fatigue)	1) Incorrect blade (teeth too coarse) 2) Blade tension too tight 3) Side guides too tight 4) Damaged or misadjusted blade guides 5) Excessive feed 6) Incorrect cutting fluid 7) Wheel diameter too small for blade 8) Blade rubbing on wheel flanges 9) Teeth in contact with work before starting saw 10) Incorrect blade velocity	1) Use finer tooth pitch. 2) Reduce blade tension (see machine operator's manual). 3) Check side guide clearance (see machine manual). 4) Check all guides for alignment and damage. 5) Reduce feed pressure. 6) Check coolant. 7) Use thinner blade. 8) Adjust wheel alignment. 9) Allow 1/2" clearance before starting cut. 10) Increase or decrease blade speed.
Premature Dulling of Teeth	1) **Teeth pointing in wrong direction (blade mounted backwards)** 2) **Improper or no blade break-in** 3) **Hard spots in material** 4) **Material work hardened** 5) **Improper coolant** 6) **Improper coolant concentration** 7) **Speed too high** 8) **Feed too light**	1) **Install blade correctly. If teeth are facing in the wrong direction, flip blade inside out.** 2) **Break in blade properly (see recommended procedures).** 3) **Check material for hardness or hard spots like scale or flame cut areas.** 4) **Increase feed pressure.** 5) **Check coolant type.** 6) **Check coolant mixture.** 7) **Check recommended blade speed.** 8) **Increase feed pressure.**
Inaccurate Cut	1) Tooth set damage 2) Excessive feed pressure 3) Improper tooth size 4) Cutting fluid not applied evenly 5) Guides worn or loose 6) Insufficient blade tension	1) Check for worn set on one side of blade. 2) Reduce feed pressure. 3) Check cutting chart. 4) Check coolant nozzles. 5) Tighten or replace guides, check for proper alignment. 6) Adjust to recommend tension.
Band Leading In Cut	1) Overfeed 2) Insufficient blade tension 3) Tooth set damage 4) Guide arms loose or set too far apart 5) Chips not being cleaned from gullets	1) Reduce feed force. 2) Adjust recommended tension. 3) Check material for hard inclusions. 4) Position arms as close to work as possible. Tighten arms. 5) Check chip brush.
Chip Welding	1) Insufficient coolant flow 2) Wrong coolant concentration 3) Excessive speed and/or pressure 4) Tooth size too small 5) Chip brush not working	1) Check coolant level and flow. 2) Check coolant ratio. 3) Reduce speed and/or pressure. 4) Use coarser tooth pitch. 5) Repair or replace chip brush.
Teeth Fracture Back of Tooth (Indicates work spinning in clamps)	1) Incorrect speed and/or feed 2) Incorrect blade pitch 3) Saw guides not adjusted properly 4) Chip brush not working	1) Check cutting chart. 2) Check cutting chart. 3) Adjust or replace saw guides. 4) Repair or replace chip brush.
Irregular Break (Indicates material movement)	1) Indexing out of sequence 2) Material loose in vise	1) Check proper machine movement. 2) Check vise or clamp.

(Continued)

Table 5. *(Continued)* **Troubleshooting Bandsaw Cutting Operations.** *(Source, M. K. Morse Company.)*

Problem	Probable Cause	Likely Solution
Teeth Stripping	1) Feed pressure too high 2) Tooth stuck in cut 3) Improper or insufficient coolant 4) Tooth pitch too large 5) Hard spots in material 6) Work spinning in vise (loose nest or bundle) 7) Blade speed too slow 8) Blade teeth running backwards 9) Chip brush not working	1) Reduce feed pressure. 2) Do not enter old cut with a new blade. 3) Check coolant flow and concentration. 4) Use finer tooth pitch. 5) Check material for hard inclusions. 6) Check clamping pressure (be sure work is held firmly). 7) Increase blade speed (see cutting chart). 8) Reverse blade (turn inside out). 9) Repair or replace chip brush.
Wear on Back of Blades	1) Excessive feed pressure 2) Insufficient blade tension 3) Back-up guide roll frozen, damaged, or worn 4) Blade rubbing on wheel flange	1) Decrease feed pressure. 2) Increase blade tension and readjust guides. 3) Repair or replace back-up roll or guide. 4) Realign wheel.
Rough Cut (Washboard surface, vibration, and/or chatter)	1) Dull or damaged blade 2) Incorrect speed or feed 3) Insufficient blade support 4) Incorrect tooth pitch 5) Insufficient coolant	1) Replace with new blade. 2) Increase speed or decrease feed. 3) Move guide arms as close as possible to the work. 4) Use finer pitch blade. 5) Check coolant flow.
Wear Lines, Loss of Set	1) Saw guide inserts or pulley are riding on teeth 2) Insufficient blade tension 3) Hard spots in material 4) Back-up guide worn	1) Check machine manual for correct blade width. 2) Tension blade properly. 3) Check material for inclusions. 4) Replace.
Twisted Blade (Profile sawing)	1) Blade binding in cut 2) Side guides too tight 3) Radius too small for blade width 4) Work not firmly held 5) Erratic coolant flow 6) Excessive blade tension	1) Decrease feed pressure. 2) Adjust side guide gap. 3) Use narrower blade. 4) Check clamping pressure. 5) Check coolant nozzles. 6) Decrease blade tension.
Blade Wear (Teeth blued)	1) Incorrect blade 2) Incorrect feed or speed 3) Improper or insufficient coolant	1) Use coarser tooth pitch. 2) Increase feed or decrease speed. 3) Check coolant flow.

to the point where it is softened but not melted and allowing the blade to easily and cleanly remove material. Friction sawing is normally used on ferrous materials hardened to R_C 38 or higher. While traditional sawing could be used to shape these materials, friction sawing can perform the operation faster and without the distortion (making the process popular for cutting many types of tubing) or rough edges that would be caused by torch cutting. Secondary machining operations to clean edges or resize a workpiece are not normally required. Blades used in friction sawing are normally one piece and have the standard tooth form.

Since this process relies on softening the work material before it is actually cut, it is ideal for most ferrous alloys up to $3/4''$ (19 mm) thickness. Other materials that can be cut are cast steel and gray and malleable cast iron. However, most cast irons and some steels containing tungsten cannot be friction sawed because the grains break off before softening. Aluminum and brass alloys are also inappropriate because of their high degree of thermal conductivity that does not allow heat to concentrate immediately in front of the cut. Stacking sheets before sawing is not recommended as the heat will fuse the layers together. See **Table 6** for recommended cutting speeds for friction sawing.

Table 6. Recommended Speeds for Friction Sawing. (Source, DoAll Company.)

Material	Thickness	Velocity		Linear Cutting Rate	
		fpm	mpm	in./min.	mm/min.
Carbon Steel (SAE) 1010 – 1095	To 1/4" (6.35 mm)	6,000	1830	60	1525
	1/2" (12.7 mm)	9,000	2745	30	760
	3/4" (19 mm)	12,500	3810	8	205
Manganese Steel T1330 – 1350	To 1/4" (6.35 mm)	6,000	1830	55	1400
	1/2" (12.7 mm)	9,000	2745	25	635
	3/4" (19 mm)	12,500	3810	7	180
Free Machining Steel X1112 – X1340	To 1/4" (6.35 mm)	6,000	1830	60	1525
	1/2" (12.7 mm)	9,000	2745	25	635
	3/4" (19 mm)	12,500	3810	7	180
Nickel Steel 2015 – 2515	To 1/4" (6.35 mm)	8,000	2440	45	1145
	1/2" (12.7 mm)	12,000	3660	20	510
	3/4" (19 mm)	14,000	4265	5	130
Nickel Chromium Steel 3115 – 3415	To 1/4" (6.35 mm)	8,000	2440	45	1145
	1/2" (12.7 mm)	12,000	3660	18	460
	3/4" (19 mm)	14,000	4265	5	130
Molybdenum Steel 4023 – 4820	To 1/4" (6.35 mm)	8,000	2440	50	1270
	1/2" (12.7 mm)	12,000	3660	20	510
	3/4" (19 mm)	14,000	4265	6	150
Chromium Steel 5120 – 5150	To 1/4" (6.35 mm)	8,000	2440	45	1145
	1/2" (12.7 mm)	12,000	3660	18	460
	3/4" (19 mm)	14,000	4265	5	130
Chromium Steel 51210 – 52100	To 1/4" (6.35 mm)	8,000	2440	40	1015
	1/2" (12.7 mm)	12,000	3660	18	460
	3/4" (19 mm)	14,000	4265	5	130
Chromium Vanadium 6115 – 6195	To 1/4" (6.35 mm)	8,000	2440	40	1015
	1/2" (12.7 mm)	12,000	3660	15	380
	3/4" (19 mm)	14,000	4265	5	130
Tungsten Steel 7260 – 71360	To 1/4" (6.35 mm)	8,000	2440	38	965
	1/2" (12.7 mm)	12,000	3660	12	305
	3/4" (19 mm)	14,000	4265	4	100
Silicon Steel 9255 – 9260	To 1/4" (6.35 mm)	8,000	2440	45	1145
	1/2" (12.7 mm)	12,000	3660	16	405
	3/4" (19 mm)	14,000	4265	4	100
Armor Plate	To 1/4" (6.35 mm)	7,000	2135	60	1525
	1/2" (12.7 mm)	10,000	3050	30	760
	3/4" (19 mm)	13,500	4115	6	150
Stainless Steel	To 1/4" (6.35 mm)	7,000	2135	60	1525
	1/2" (12.7 mm)	10,000	3050	30	760
	3/4" (19 mm)	13,500	4115	8	205

(Continued)

Table 6. *(Continued)* **Recommended Speeds for Friction Sawing.** *(Source, DoAll Company.)*

Material	Thickness	Velocity		Linear Cutting Rate	
		fpm	mpm	in./min.	mm/min.
Cast Steel	To 1/4" (6.35 mm)	7,000	2135	50	1270
	1/2" (12.7 mm)	10,000	3050	16	405
	3/4" (19 mm)	13,500	4115	7	180
Gray Cast Iron	To 1/4" (6.35 mm)	7,000	2135	55	1400
	1/2" (12.7 mm)	10,000	3050	20	510
	3/4" (19 mm)	13,500	4115	7	180
Malleable Cast Iron	To 1/4" (6.35 mm)	7,000	2135	55	1400
	1/2" (12.7 mm)	10,000	3050	20	510
	3/4" (19 mm)	13,500	4115	7	180
Meehanite Castings	To 1/4" (6.35 mm)	7,000	2135	55	1400
	1/2" (12.7 mm)	10,000	3050	20	510
	3/4" (19 mm)	13,500	4115	7	180

Circular sawing

Circular sawing is closely related to milling, with the saw's teeth being the equivalent of flutes or inserts in a milling cutter. Each saw tooth engages the workpiece and removes and carries away a curled chip. Metal slitting saws, used on milling machine arbors, differ from circular saws in that they are only made in diameters of up to 8" (203 mm) while one piece circular saw blades are usually made from high speed steel and are designed for cutting larger workpieces (usually up to 4" [100 mm]) and are available in diameters up to 16" (406 mm). Circular sawing is also known as cold sawing to differentiate the process from sawing operations in mills and forge shops. For larger cuts, multi-piece saws with segmented blades comprised of short sections—usually about four teeth—that are slotted and fit snugly over a central tongue, are riveted in place to produce saws that range in size from 12" (305 mm) to 120" (3050 mm). Because segmented blades offer greater shock resistance, and because segments rather than entire blades can be replaced if teeth are broken or otherwise damaged, they are widely used for reasons of both economy and efficiency. Their main drawback is that they are thicker and cannot cut to the same accuracy as solid blades. Carbide tipped segmented blades have found acceptance for sawing large forged steel billets, where they can be reground as many as nine times before retipping is required.

Due to the solid circular saw's high rigidity, high levels of accuracy and surface finish can be achieved. For large diameter blades, length tolerances of ±0.004" (0.10 mm) per inch, and squareness tolerances of ±0.0015" (0.04 mm) per diameter inch, can be achieved with surface finishes ranging (in steel) from 60 μinch to 125 μinch (1.5 μm to 3.2 μm). Surface finish numbers in aluminum and other soft materials can be expected to be approximately one-sixth those achieved in steel and other hard metals. A further advantage of circular saw machining is that most cuts can be expected to be virtually free of burrs, thereby making further finishing operations unnecessary in most operations.

When cutting most steels, cutting speeds for high speed steel circular saws range from 30 to 100 fpm (9 to 31 mpm) with feed rates of 3 to 8 ipm (76 to 200 mm/min). For stainless steel, recommended speeds are 15 to 40 fpm (4.5 to 12 mpm) together with feed rates of 1 to 4 ipm (25 to 102 mm/min). Cast irons can be cut at roughly the same speeds as steel, but with marginally higher feed rates. Much higher speeds, 300 to 15,000 fpm (90 to 4500 m/m) with feed rates of 8 to 80 ipm (200 to 2000 mm/min) can be used for cutting aluminum, copper, bronze, and their alloys. With the exception of cutting brass and cast iron, cutting fluid should be used in all circular saw operations.

Friction Sawing. This process, as performed with the circular saw, requires very high speeds of 18,000 to 25,000 fpm (5490 to 7600 mpm) and heavy feed pressure. This process is much the same as described above for friction sawing with a bandsaw, but the size of the circular saw limits the workpiece diameter to around 4" (100 mm). Additional factors are the high feed pressure which requires increased power, cutting tool heat which cannot be dissipated as easily as with a band saw, the creation of burrs that may make a finishing operation necessary, and the fact that the process works satisfactorily only with ferrous materials as nonferrous metals tend to adhere to the saw blade.

Welding, Brazing, and Soldering

Much of the material in this section was supplied by Ray Miller, an instructor of welding, and a sculptor who uses welding in his art.

The joining of production components falls into one of three main categories: 1) welding, brazing, and soldering; 2) adhesives; and 3) mechanical fastening. In discussing welding, brazing, and soldering, it should be understood that industry has only recently begun to standardize the terminology and symbols of welding and welding-related processes. The American Welding Society is the recognized technical body responsible for governing the bulk of the technical issues of welding-related processes. Unless otherwise noted, when referring to the general terminology of the welding, brazing, and soldering processes, the term "welding" is used in this section to encompass all three processes.

Given the great diversity of welding processes, and their applications, this section will focus on the most commonly applied operations. **Table 1** provides a comprehensive listing of welding processes classified by type.

Energy sources: electricity and chemical energy

The most basic element of any welding process is the source of the concentrated energy used to change the physical state of the materials involved. Welding processes are either fusion or nonfusion procedures. This definition refers to the parent or base materials involved. In fusion processes, the base materials are melted and fused together into a common substance. In nonfusion processes, a filler material is typically melted and used to complete the bond between the parent materials.

Energy sources used to accomplish this melting are either electrical or chemical. Within these two categories of energy source there are a wide variety of application technologies, ranging from the most basic (arc welding) to many advanced applications such as high frequency, laser, electron beam, inertia, and friction welding.

Within chemical energy based processes there is a wide range of dynamics to be considered. The most common chemical energy source is the oxygen fuel torch that is used in a variety of applications, including the most ancient and widely used welding/brazing procedures. Modern technology has allowed for increases in chemically generated energy, resulting in more dramatic processes like explosion and cadmium welding.

Fluxes and weld shielding

For the most part, welding and joining processes involve a change in the physical state of the materials being joined. In a vast number of these applications, it involves the melting of metals, and in this molten state oxidation can have a severe negative impact on the joint created. To isolate the joint during its "curing" or solidification, many processes incorporate a shielding medium, typically in the form of an inert gas. The inert gases used for this are argon, carbon dioxide, helium, and composite mixtures determined to have beneficial effects on the joint.

Fluxes can also contribute to the shielding function, in addition to providing material preparation characteristics. Solid coatings on welding rods are a widely used means of integrating flux and shielding into a procedure. As the welding electrode is melted during the arc welding process, the flux coating melts. This molten material can perform functions such as forming a liquid shield over the molten puddle, coalescing contaminants in the molten puddle, and imparting beneficial components to the molten material. The fluxing properties needed in some processes include precleaning and etching the surface to improve bonding through increased surface area. Solid flux coatings are used on coated electrodes and filler rods, but fluxes can also be in liquid, powder, and paste forms.

Prior to initiating welding or joining operations, there are many design characteristics that must be defined. These include the physical weld joint, joint preparation, material selection, loads or stresses involved, manufacturing or assembly requirements, quality requirements, operator or equipment limits, procedures, and, perhaps the most important governing characteristic, economics. Many of these decisions will depend on existing codes and standards that govern welding operations.

Codes and standards

ISO, QFD, AWS, ASME, DIN, and ANSI are acronyms that represent the organizations that issue the technical standards and quality codes applied to industrial processes to assure repeatability, quality, and safety in products. There are a vast number of governing codes, standards, organizations, industry associations, and governmental agencies involved in developing, implementing, and policing compliance the applicable codes and standards, including the following.

American Welding Society (AWS). Governs structural, sheet metal, welder certification, inspector certification.

American Society of Mechanical Engineers (ASME). Governs power piping, process piping, etc.

Nation Fire Protection Association (NFPA). Sprinkler piping, explosive gases.

American Petroleum Institute (API). Pipelines, tanks storage facilities.

American National Standards Institute (ANSI). National standards development for a variety of applications.

State basic building codes. Each state defines its own building code requirements which typically includes references to code welding applications, such as structural steel and piping systems.

In addition, many industries have developed their own governing body, and, in conjunction with that body, there is a document that defines standards of practice, and procedures for applying welding and joining technology. Consequently, the first consideration that needs to be investigated before undertaking any welding operation is whether the work being planned is covered by a code or standard. Application specific welding and joining codes fall into several categories.

Structural steel—governed by AWS D1.1
Power piping systems—governed by ASME B31.1
Process Piping Systems—governed by ASME B31.3
Petroleum Pipelines—API 1104.

Fabrications, machines, transportation equipment, airplanes, and forklift trucks are also examples of products governed by codes and standards. These are generated by government agencies, such as the FAA, and by industry associations.

Other code issues also involve welder certification, equipment calibration and maintenance, inspection requirements and documentation, and record keeping. For many companies these requirements are new, and highlight the need for the development of company specific weld procedures, welder certification programs, inspection and documentation guidelines, and a welding/quality control program.

Weld process specification

In the definition of welding procedures, the fundamental document required is a weld process specification (WPS). A WPS is a detailed process description that includes information associated with essential and nonessential variables associated with the

Table 1. Master Chart of Welding and Allied Processes. *(Source, American Welding Society.)*

gas metal arc welding	GMAW
—pulsed arc	GMAW-P
—short circuiting arc	GMAW-S
gas tungsten arc welding	GTAW
—pulsed arc	GTAW-P
plasma arc welding	PAW
shielded metal arc welding	SMAW
stud arc welding	SW
submerged arc welding	SAW
—series	SAW-S

block brazing	BB
diffusion brazing	DB
dip brazing	DB
exothermic brazing	EXB
flow brazing	FLOW
furnace brazing	FB
induction brazing	IB
infrared brazing	IRB
resistance brazing	RB
torch brazing	TB
twin carbon brazing	TCAB

electron beam welding	EBW
—high vacuum	EBW-HV
—medium vacuum	EBW-MV
—nonvacuum	EBW-NV
electroslag welding	ESW
flow welding	FLOW
induction welding	IW
laser beam welding	LBW
percussion welding	PEW
thermit welding	TW

air acetylene welding	AAW
oxyacetylene welding	OAW
oxyhydrogen welding	OHW
pressure gas welding	PGW

(Continued)

atomic hydrogen welding	AHW
bare metal arc welding	BMAW
carbon arc welding	CAW
—gas	CAW-G
—shielded	CAW-S
—twin	CAW-T
electrogas welding	EGW
flux cored arc welding	FCAW

coextrusion welding	CEW
cold welding	CW
diffusion welding	DFW
explosion welding	EXW
forge welding	FOW
friction welding	FRW
hot pressure welding	HPW
roll welding	ROW
ultrasonic welding	USW

dip soldering	DS
furnace soldering	FS
induction soldering	IS
infrared soldering	IRS
iron soldering	INS
resistance soldering	RS
torch soldering	TS
wave soldering	WS

flash welding	FW
projection welding	PW
resistance seam welding	RSEW
—high frequency	RSEW-HF
—induction	RSEW-I
resistance spot welding	RSW
upset welding	UW
—high frequency	UW-HF
—induction	UW-I

ARC WELDING (AW)

BRAZING (B)

OTHER WELDING

OXYFUEL GAS WELDING (OFW)

WELDING PROCESSES

ALLIED PROCESSES

SOLID-STATE WELDING (SSW)

SOLDERING (S)

RESISTANCE WELDING (RW)

continued next page

Table 1. *(Continued)* **Master Chart of Welding and Allied Processes.** *(Source, American Welding Society.)*

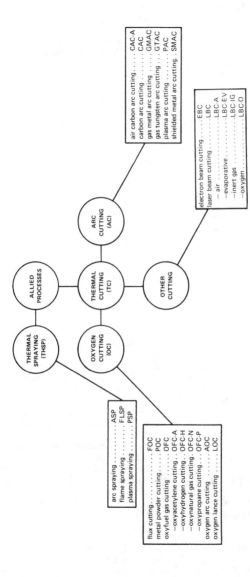

air carbon arc cutting . . . . CAC-A
carbon arc cutting . . . . CAC
gas metal arc cutting . . . . . . GMAC
gas tungsten arc cutting . . . . GTAC
plasma arc cutting . . . . PAC
shielded metal arc cutting. . SMAC

electron beam cutting . . . . . EBC
laser beam cutting . . . . LBC
 — air . . . . LBC-A
 — evaporative . . . . LBC-EV
 — inert gas . . . . LBC-IG
 — oxygen . . . . LBC-O

ARC CUTTING (AC)

ALLIED PROCESSES

THERMAL CUTTING (TC)

OTHER CUTTING

THERMAL SPRAYING (THSP)

OXYGEN CUTTING (OC)

arc spraying . . . . ASP
flame spraying . . . . FLSP
plasma spraying . . . . . . PSP

flux cutting . . . . FOC
metal powder cutting . . . . . POC
oxyfuel gas cutting . . . . . . OFC
 — oxyacetylene cutting . . OFC-A
 — oxyhydrogen cutting . . OFC-H
 — oxynatural gas cutting . OFC-N
 — oxypropane cutting . . . OFC-P
oxygen arc cutting . . . . . . AOC
oxygen lance cutting . . . . . LOC

process. Essential variables are aspects of the process which, if changed, will affect the resultant weld. Examples include material, position, and the process chosen for the operation.

The use of a WPS helps provide process control. It provides the operator with the information needed to produce a conforming weld. Typical WPS information includes the following information, but may be expanded to incorporate additional variables as required by the specific operation.

Description of the process—SMAW, GMAW, GTAW, SAW, etc.
Allowable materials classifications—P1, P8, etc. (P1, carbon steels; P8, stainless steels)
Current and voltage settings
Electrodes allowed, size, type
Joint preparation
Preheat requirements
Postheat treatments
Inspection requirements
Weld positions permitted
Shielding gases
Prepurge/postpurge requirements.

The following sections will cover the fundamentals of a WPS, defining weld procedure, applications, joint preparation, weld joints, inspection requirements, welder certifications, shielding gases, fluxes, etc. *Figure 1* illustrates a wide variety of joint preparations, weld joint configurations, plus provides an overview of standardized welding-related drawing symbols.

Welding positions

In producing weldments, the operator is likely to encounter a variety of welding positions. Each position presents unique requirements.

The most welder-friendly position is flat welding. The flat position consists of the weld joint being presented to the welder in a true horizontal configuration, meaning that the components which comprise the weldment are laid flat.

The next position is horizontal. The main difference between the flat and horizontal positions is that gravity can pull the weld puddle down, out of the joint, as the weldment components are arranged at right (or near 90°) angles to the operator with the weld joint in either a vertical or horizontal orientation.

The most challenging position is welding overhead. In this configuration, the weld joint is directly over the welder's head. Weld direction travel is not critical. In this position, gravity tends to pull the weld puddle out of the joint.

Following overhead in degree of difficulty are the vertical positions: vertical up and vertical down. In these positions the weld joint is aligned vertically and the direction of travel is either up or down. With vertical up, the weld puddle is "pulled" up hill and the operator must take action to prevent the puddle from falling out of the joint. With vertical down, the weld point must stay ahead of the puddle and the operator must assure that the puddle does not fall out of the joint, and that there is complete fusion to the parent metal.

Weld procedures

It is always recommended to use a WPS for each welding process. When performing code welds, a WPS is required. When performing noncode welds, a WPS can provide a process quality control function, although it is not required.

The most common welding processes are electric arc techniques. With these processes, current is applied to an electrode holder and workpiece through cables from a current/voltage supply. A transformer is used when welding with alternating current. In welding with direct current, the operation requires rectifiers, generators, or frequency converters.

Procedure qualification

Once a WPS is either developed or identified for a specific application, it needs to be qualified. This can be through destructive testing means, where a welder performs the WPS on a sample coupon and the coupon is then destructively tested by means outlined by the appropriate code or standard. If the WPS passes the testing requirements, it is then a qualified WPS and the welder who performed the qualification weld is automatically certified. The documentation of the qualification test for a WPS is a Procedure Qualification record, or PQR. Another source of WPSs are the prequalified WPSs incorporated into certain codes and standards. For example, in AWS D1.1 there is a prequalified WPS that can be used for structural steel welding applications. A company can then test their welders to that WPS. This leads to the next topic of discussion, welder certification.

Welder certification

As a matter of compliance with most state building codes, a company performing code welding needs certified welders. To perform code welding, the welders need to complete a test coupon using the appropriate qualified WPS. These coupons are then destructively tested. Successful completion of this test will certify the welder in this specific WPS. For many industries, welders are required to be certified in several WPSs. In designing a company's weld quality program, the testing and certification of welders is an important issue. This requires documented coupon materials, metered welding equipment, and destructive testing, usually by an independent lab.

A basic aspect of welder certification is vision testing for welders prior to certification. Certifications also expire, depending on codes and standards. AWS D1.1 structural steel WPS certifies a welder for life. However, state building codes require periodic use of the WPS to maintain certification, and periodic retesting to verify that a welder maintains the proper skill level.

Welding Processes

Shielded metal arc welding (SMAW)

Shielded metal arc welding (SMAW) is the most widely used weld process. It is also known as MMA (manual metal arc), electric arc welding, and stick-electrode welding. The process is primarily used for the welding of ferrous alloys (although electrodes suitable for nonferrous alloys such as aluminum and certain copper and nickel-based alloys also exist). Many large and small manufacturing industries use SMAW, and it is widely employed in shipbuilding, pipeline construction, fabricating pressure vessels, and structural steelwork. The capital costs are inexpensive when compared to other methods, although its deposition rate is relatively low, which makes SMAW a slow process when compared to other welding operations (deposition rate is the rate of material deposited per unit time). SMAW is used when more mechanized methods cannot be applied due to cost or process restrictions. SMAW is often used for short stitch welds or welding heavy gage materials where mobility is important. It is typically a fully manual process.

The arc is started by striking the electrode directly against the workpiece. The electrode is held at an angle to the joint so the welder may see the weld puddle forming below the arc along the joint line. Using the arc, the welder must ensure that the molten metal flows ahead of the electrode at all times, and that both sides of the joint are uniformly melted.

Basic Welding Symbols and Their Location Significance

Location Significance	Fillet	Plug or Slot	Spot or Projection	Stud	Seam	Back or Backing	Surfacing	Flange Corner	Flange Edge
Arrow Side						Groove weld symbol			
Other Side				Not used		Groove weld symbol	Not used	Not used	Not used
Both Sides	Not used	Not used	Not used	Not used	Not used	Not used	Not used	Not used	Not used
No Arrow Side or Other Side Significance		Not used		Not used		Not used	Not used	Not used	Not used

Location Significance	Groove								
	Square	V	Bevel	U	J	Flare-V	Flare-Bevel		Scarf for Brazed Joint
Arrow Side									
Other Side									
Both Sides			Not used	Not used	Not used	Not used	Not used		
No Arrow Side or Other Side Significance		Not used	Not used	Not used	Not used	Not used	Not used		Not used

(Continued)

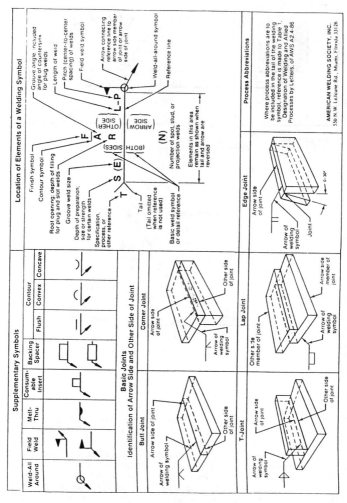

Figure 1. Welding symbols and basic joint configurations. (Source, American Welding Society.)

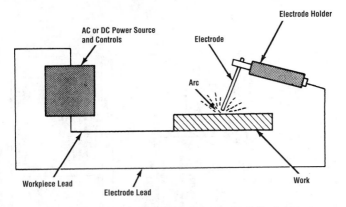

Figure 2a. Typical circuit for SMAW welding. (Source, American Welding Society.)

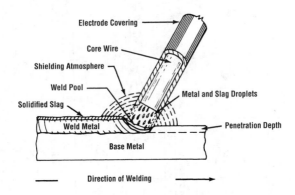

Figure 2b. The SMAW process. (Source, American Welding Society.)

Selection of the proper electrode is essential. Typical (these examples are for E60xx electrodes) electrode diameters are between $3/32$"–$1/4$" (2.5–6.4 mm), and typical length is approximately 18" (460 mm). They are normally replaced when the electrode shortens to approximately 2" (50 mm). Welding currents range between 50–350 amps for arc voltages of magnitude 30–35 volts. The current rate varies by electrode diameter, with $3/32$" (2.5 mm) electrodes having a range of 50–90 amps; $1/8$" (3.2 mm) electrodes having a range of 65–130 amps; $5/32$" (4.0 mm) electrodes having a range of 110–180 amps; $3/16$" (5.0 mm) electrodes having a range of 150–250 amps; and $1/4$" (6.4 mm) electrodes having a range of 200–350 amps. The best deposition rate will result from using as large an electrode as possible while still maintaining a conforming weld. Small diameter electrodes are used for thinner materials and for the root pass in multipass welds in order to reach the bottom of the joint. Many classes of electrodes are available. The variables in an SMAW electrode include material, flux coating, weld position, current flow settings, and diameter.

Electrodes come printed with a 5 digit AWS/ANSI alphanumeric code. They typically begin with an "E" (which stands for low-carbon steel, metal arc welding electrode) then 4 digits. These codes are as follows.

> Exx10   DC+ (DC reverse or DCEP) electrode positive
> Exx11   AC or DC- (DC straight or DCEN) electrode negative
> Exx12   AC or DC-
> Exx13   AC or DC+ or DC-
> Exx14   AC or DC+ or DC-
> Exx15   DC+
> Exx16   AC or DC+
> Exx18   AC or DC+ or DC-
> Exx20   AC or DC+ or DC-
> Exx24   AC or DC+ or DC-
> Exx27   AC or DC+ or DC-

where the xx represents the tensile strength of the filler material:
> E60xx have a tensile strength of 60,000 psi
> E70xx would be 70,000 psi, etc.

Weld position is defined by the third digit:
> Exx1x is for use in all positions
> Exx2x is for use in flat or horizontal positions
> Exx3x is for flat welding only.

The type of flux coating is represented by the combination of third and fourth digits. Due to the variety of codes, it is recommended that the electrode supplier's reference material be consulted for specific information before selecting an electrode for an operation.

It should be noted that alternate codes are used for electrode identification. The ISO 2560 standard provides an eleven digit code which is read as follows. First digit = same as AWS/ANSI. Digits two through five = tensile strength, yield stress, ductility, and Charpy value. Digit six = type of covering: A = acid, C = cellulose, R = rutile, and B = basic. Digits seven through nine indicate electrode efficiency. Digits ten and eleven provide operating characteristics. In the United Kingdom, the corresponding standard is BS 639. Canadian (CSA), German (DIN), and Japanese (JIS) standards issuing agencies also provide their own electrode classifications.

Gas metal arc welding (GMAW) and flux cored arc welding (FCAW)

Gas Metal Arc Welding (GMAW) utilizes an inert shielding gas, carbon dioxide, helium, argon, or a combination and a semiautomatic wire fed filler wire/electrode. This is also known as metal inert-gas (MIG) welding, metal active-gas (MAG) welding, metal arc gas-shielded (MAGS) welding, and CO_2 welding. Some of the shielding gases are not inert gases and have some effect on the weld metal. Common gases used in GMAW are argon, argon mixed with 5% oxygen, argon mixed with 10 to 25% carbon dioxide, and pure carbon dioxide. Amperage requirements are 60 to 500 amps, with voltage ranges between 14 and 22 volts, but operations are typically performed within a voltage range of 16 to 18 volts. The process is very similar to Flux Cored Arc Welding, except that no residual slag is left from the welding operation.

In GMAW, the continuous solid metal or metal cored electrode/filler wire is fed by drive system, through a cable, to the weld control gun. An arc burns between the workpiece

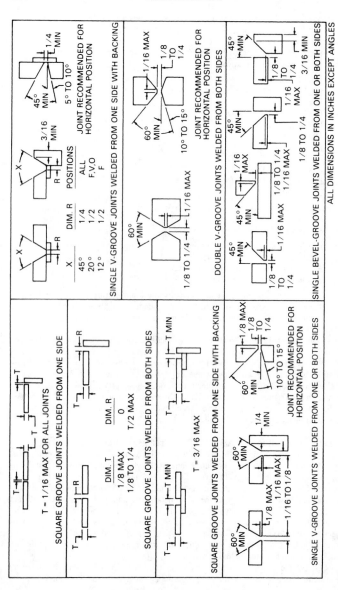

Figure 3a. Joint geometries for SMAW welding of steel. (Source, American Welding Society.)

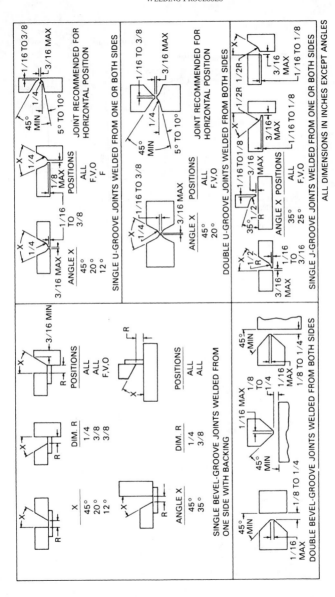

Figure 3b. Joint geometries for SMAW welding of steel. (Source, American Welding Society.)

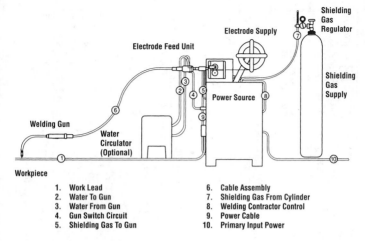

1. Work Lead
2. Water To Gun
3. Water From Gun
4. Gun Switch Circuit
5. Shielding Gas To Gun
6. Cable Assembly
7. Shielding Gas From Cylinder
8. Welding Contractor Control
9. Power Cable
10. Primary Input Power

Figure 4a. GMAW welding equipment. (Source, American Welding Society.)

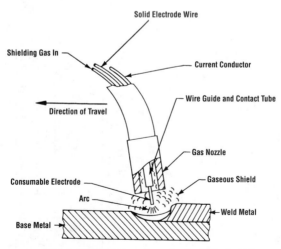

Figure 4b. The GMAW process. (Source, American Welding Society.)

and the electrode, which is melted as it is fed. The most popular electrode diameters are 0.035" (1.0 mm), commonly used for welding material thicknesses ranging from 20 gauge (0.036" or 1 mm) to $^1/_4$" (6.4 mm), and 0.045" (1.2 mm) which is used for materials ranging from $^3/_{16}$" (5 mm) to $^7/_{16}$" (11 mm). Material from the molten electrode end is transferred in fine droplets to the workpiece where it builds filler in the joint. Shielding gas flows out of the nozzle, around the wire, so that the welding area and the heated metal are protected from oxidation. The arc length is self-regulating since the current source has constant voltage control. The current source usually consists of a transformer with a rectifier since the welding commonly occurs with direct current and a positive electrode.

Gas metal arc welding is becoming more widely used as equipment becomes more affordable. Examples include most major industrial products and an expanded role in consumer products, such as automobiles, with the advent of CNC robot welders. It may be used as a semiautomatic method but may also be fully automated, allowing for many uses generating a lower welding cost compared to using coated electrodes. The method is very important in sheet metal and robot welding. GMAW is a semiautomatic process. The operator sets the feed rate and current and initiates the arc just by pressing a trigger. The power source controls the voltage.

When the welding gun is triggered (in semiautomatic welding), the shielding gas begins to flow out of the nozzle and the electrode is fed out. The arc is lit when the electrode strikes the workpiece. The welding gun is then guided along the joint assuring even melting of both sides of the joint and that an adequate amount of filler is applied.

Short arc welding is used in welding of thin sheet metals. The arc length is decreased by reducing the voltage on the current source, so that the metal droplets on the electrode end short-circuit immediately with the workpiece. When the droplet is incorporated into the puddle, the arc is lit again and the sequence is repeated at high frequency. Larger arc voltages are used for thicker metal, and one commonly refers to spray arc welding with finely dispersed, nonshort-circuiting droplets. In spray arc welding with high currents, one may obtain very deep penetration into the base material.

A popular variation of GMAW is flux cored arc welding (FCAW). Even though FCAW is identified as a unique weld process, there are two versions that have many similarities to GMAW and SMAW. Both FCAW and GMAW use virtually identical welding equipment, with two significant changes. First, the welding wire for FCAW is a hollow tube filled with welding flux—similar to those used in SMAW—as opposed to the solid wire used in GMAW. A second major difference is that the polarity for FCAW is reversed, due to the nature of the flux cored wire. The resultant welds produced by FCAW also have a slag covering that must be removed, while GMAW welds are slag free. Many references consider FCAW a very distinct process. There are two distinct versions of FCAW: gas shielded and self-shielded. The gas shielded method uses carbon dioxide or a carbon dioxide and argon mixture to protect the molten metal from oxygen and nitrogen in the air, and is normally used in the same applications as GMAW. The self-shielded method uses vaporized flux ingredients to protect the molten weld pool and is used in the same applications as SMAW.

Gas tungsten arc welding (GTAW)

Gas Tungsten Arc Welding (GTAW) or TIG welding (an abbreviation for Tungsten Inert Gas) is a method that uses the inert gases helium and argon as shielding gases. It is also known as tungsten arc gas-shielded (TAGS) welding and heliarc welding. This process also uses a solid tungsten nonconsumable electrode and manually fed filler rods when required. (Some new products on the market include "cold

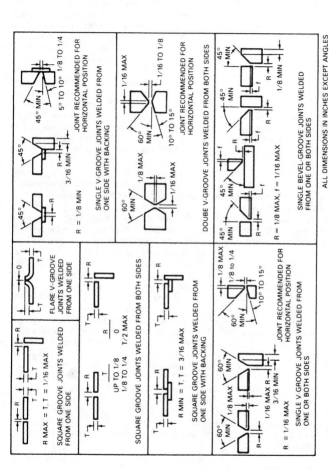

Figure 5a. Joint geometries for GMAW. (Source, American Welding Society.)

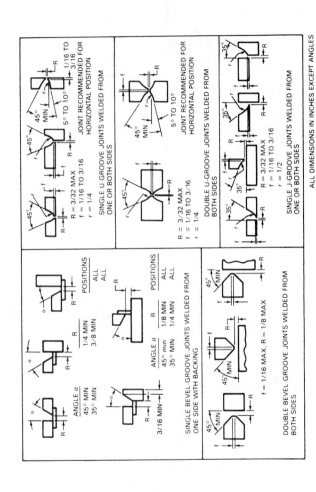

ALL DIMENSIONS IN INCHES EXCEPT ANGLES

SINGLE U-GROOVE JOINTS WELDED FROM
ONE OR BOTH SIDES

R = 3/32 MAX
f = 1/16 TO 3/16
r = 1/4

JOINT RECOMMENDED FOR
HORIZONTAL POSITION

5° TO 10°

45°
MIN

1/16 TO
3/16

45°

DOUBLE U-GROOVE JOINTS WELDED FROM
BOTH SIDES

R = 3/32 MAX
f = 1/16 TO 3/16
r = 1/4

JOINT RECOMMENDED FOR
HORIZONTAL POSITION

5° TO 10°

45°
MIN

45°

SINGLE J-GROOVE JOINTS WELDED FROM
ONE OR BOTH SIDES

R = 3/32 MAX
f = 1/16 TO 3/16
r = 1/2

35°

35°

35°

SINGLE BEVEL-GROOVE JOINTS WELDED FROM
ONE SIDE WITH BACKING

ANGLE α	R	POSITIONS
45° MIN	1/4 MIN	ALL
35° MIN	3/8 MIN	ALL

ANGLE α	R	POSITIONS
45° min	1/8 MIN	ALL
35° MIN	1/4 MIN	ALL

3/16 MIN

DOUBLE BEVEL GROOVE JOINTS WELDED FROM
BOTH SIDES

f = 1/16 MAX, R = 1/8 MAX

45°
MIN

45°
MIN

45°
MIN

Figure 5b. Joint geometries for GMAW. (Source, American Welding Society.)

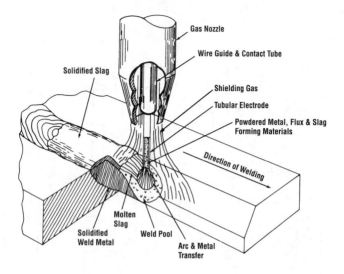

Figure 6a. Gas shielded flux cored arc welding. (Source, American Welding Society.)

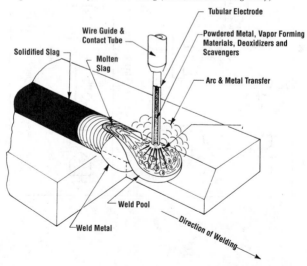

Figure 6b. Self-shielded flux cored arc welding. (Source, American Welding Society.)

wire feeders" which would allow the GTAW/TIG process to function almost identically to GMAW.) Current requirements are in the range of 10 to 300 amps.

The shielding gas (argon or argon plus helium is most commonly used) flows from a nozzle around the electrode and protects the weld puddle, the heated metal, and the electrode from oxidation. In order to obtain the best arc stability, the tungsten electrode has traditionally been alloyed with 1-2% thorium. (Today, environmental, health, and safety concerns have raised questions about the dangers of cumulative exposure to radioactive thorium dust from grinding points into the tungsten electrodes.) Safer alternatives are becoming available from suppliers. The current source usually consists of a transformer with rectifier allowing both direct current (negative electrode) and alternating current

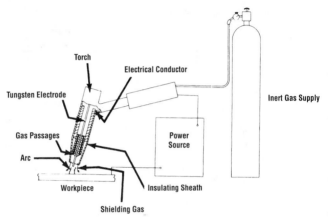

Figure 7a. GTAW welding equipment. (Source, American Welding Society.)

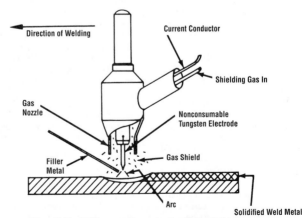

Figure 7b. The GTAW process. (Source, American Welding Society.)

welding. Electrodes that include thorium can be readily identified by their "Th" suffix, followed by a one-digit number indicating the percentage of the substance in the electrode. EWTh-2, for instance, is an electrode (E) consisting of 2% thorium and 98% tungsten (W, for wolfram, is the chemical symbol for tungsten). Zirconium (Zr) is the other element found in tungsten electrode composition. The equipment usually includes a high frequency source for arc starting and time relays which control shielding gas flow for a few seconds after switching off the arc in order to protect the electrode and weld puddle from oxidation. Welding torches for currents of over 100 amps are typically water-cooled, lower current rated torches are air-cooled.

Typical maximum currents for tungsten electrodes are shown in **Table 2**.

Table 2. Recommended Maximum Currents for Tungsten Electrodes.

Electrode Diameter		Maximum Amperage	
inch	mm	Thorium (DC)	Zirconium (AC)
3/64	1.2	75	45
1/16	1.6	150	55
3/32	2.4	245	90
1/8	3.2	390	160
5/32	4.0	450	225

GTAW is well suited to joining thin materials, exotic materials, or small parts where excessive heat buildup in the workpiece cannot be allowed. The process produces very high quality welds for a number of different steels and metal types, and can be automated. It is commonly used for precision welding for the aircraft, nuclear, and instrumentation industries. It is a very important method for the butt welding of stainless steel and nonferrous metals, even on large bore piping systems where root pass cleanliness and full fusion is critical. GTAW is typically a fully manual process.

The arc is started via the welder's remote foot pedal (manual welding), and uses a high frequency arc to bridge the gap between the electrode and the workpiece. In welding with DC, the high frequency is switched off automatically as soon as the arc is stable. However, with AC, the high frequency remains on in order to maintain the arc. Starting the arc by striking the electrode against the workpiece may lead to electrode damage and tungsten inclusions in the weld. The welding torch is held still a few seconds after starting the arc until a weld puddle is formed. The actual welding occurs by controlling the weld puddle along the joint using filler material if required. GTAW is usually performed using DC straight polarity (electrode negative). Aluminum welding must be carried out with AC welding, as it tends to build a heavy oxide coating. Other nonferrous materials such as copper and brass also require the use of AC current.

When using filler material, it is important to use compatible alloys to the base metal and to make sure that the filler will produce a weld which meets the strength, corrosion, and quality requirements of the base material and WPS. The WPS will define the filler requirements. Some processes can be completed without the use of a filler wire, while others require a filler for a sound weld.

Laser welding (LSW)

Laser welding (LSW) equipment uses a focused beam of light (actually a beam of coherent electro-magnetic radiation) on a small point as the source of the concentrated energy needed to melt the base material. The energy melts, and sometimes even vaporizes, the metal. The molten weld hardens very quickly when it is no longer subjected to the beam,

Table 3. Recommended Current, Tungsten Electrode, and Shielding Gases for GTAW of Selected Metals. (*Source, American Welding Society.*)

Type of Metal	Thickness	Type of Current	Electrode*	Shielding Gas
Aluminum	All	Alternating	Pure or Zirconium	Argon or Argon-helium
	Over 1/8"	DCEN	Thoriated	Argon-helium or Argon
	Under 1/8"	DCEP	Thoriated or Zirconium	Argon
Copper, Copper Alloys	All	DCEN	Thoriated	Helium
	Under 1/8"	Alternating	Pure or Zirconium	Argon
Magnesium Alloys	All	Alternating	Pure or Zirconium	Argon
	Under 1/8"	DCEP	Zirconium or Thoriated	Argon
Nickel Alloys	All	DCEN	Thoriated	Argon
Plain Carbon, Low Alloy Steels	All	DCEN	Thoriated	Argon or Argon-helium
	Under 1/8"	Alternating	Pure or Zirconium	Argon
Stainless Steel	All	DCEN	Thoriated	Argon or Argon-helium
	Under 1/8"	Alternating	Pure or Zirconium	Argon
Titanium	All	DCEN	Thoriated	Argon

* Where thoriated electrodes are recommended, ceriated or lanthanated electrodes may also be used.

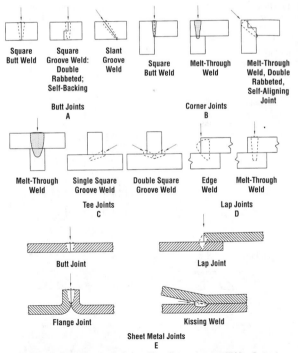

Square Butt Weld

Square Groove Weld: Double Rabbeted; Self-Backing

Slant Groove Weld

Square Butt Weld

Melt-Through Weld

Melt-Through Weld, Double Rabbeted, Self-Aligning Joint

Butt Joints
A

Corner Joints
B

Melt-Through Weld

Single Square Groove Weld

Double Square Groove Weld

Edge Weld

Melt-Through Weld

Tee Joints
C

Lap Joints
D

Butt Joint

Lap Joint

Flange Joint

Kissing Weld

Sheet Metal Joints
E

Figure 8. Joint geometries for laser welding. (Source, American Welding Society.)

since the area of the heated surface is not widespread. The light source is permanently rigged, and the light beam is controlled by lenses and mirrors to the different welding points. The welding is commonly carried out in overlapping joints without filler. There is, however, a large range of fillers that may be obtained, primarily for prewelding in the preparation of surfaces that are to achieve special properties. The high-energy concentration enables the heat-affected zone to be small and the deformation minimal. With laser welding, the investment cost of the system can be very expensive, and production requirements need to be high in order justify the investment. Since lasers can produce deep, narrow welds over contoured surfaces, they offer an excellent alternative for repetitive, high production manufacturing. LSW is an automatic process.

Plasma arc welding (PAW)

The plasma arc welding (PAW) method is the next generation of development of the GTAW process. In plasma welding, an arc of light usually burns between the workpiece and a nonmelting tungsten electrode. A plasma gas (commonly argon) flows down along the electrode and is thereby forced through a small hole, precisely in front of the electrode tip, via a nozzle, together with the electric arc. The plasma arc obtained in this manner has a very high and concentrated energy content. A shielding gas, usually being the same as the plasma gas, flows out of an outer nozzle on the welding torch in order to protect the molten weld and warm metal from oxidation. When the arc burns directly between the workpiece and electrode, the term "transmitted electrode" is used, which is the most common method of operation. The alternative is to utilize a "nontransmitted electrode," in which the arc burns between the electrode and the plasma nozzle. In the latter case, a very hot gas flows out of the nozzle's outlet at high speed. The hot gas may be used to weld, for example, nonelectrically conductive materials. Welding commonly takes place with direct current and negative electrode, and the current source may consist of a transformer with rectifier or converter.

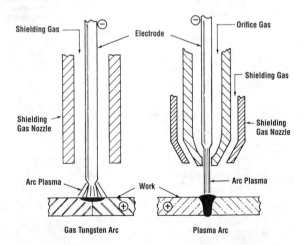

Figure 9. Comparison of GTAW and PAW processes. (Source, American Welding Society.)

A high frequency unit is also part of the equipment and is used to initiate the arc. A time relay enables the plasma and shielding gases to continue to flow a few seconds after having switched the arc off in order to protect the warm electrode and weld from oxidation. Should filler be required, it is usually added separately from the side.

Plasma welding is often used for surface welding of stainless steel onto ordinary steel. This application allows for a close control of the penetration into the base material. PAW can be a manual, semiautomatic, or fully automatic process.

Submerged arc welding (SAW)

Submerged arc welding (SAW) is very similar to GMAW. In SAW the heat is also generated by an arc that burns between the workpiece and a continuously fed wire electrode. The differences are that the arc burns with complete protection under a weld flux powder, which forms a slag. This slag protects the molten material and weld joint from oxidation. As the arc melts the joint and electrode wire tip, the molten electrode material will be transferred in molten droplets into the joint forming the bead. Simultaneously, the closest welding flux also melts forming a slag that protects the molten material and weld joint from oxidation. SAW is also a fully automatic process. An operator sets the electrode feed rate, current, flux deposition rate, and workpiece indexing speed. SAW is limited to processes where the workpiece can be moved either on a rotating turntable or a linearly translating platen.

The flux, which does not melt, is collected and reused. The length of the arc is maintained with the aid of a control unit that increases or decreases the electrode's feeding speed based on current. This method is almost exclusively used in automated welding applications. The welding torch, wire feeder unit, wire holder, flux container, and electrical control units are all integrated into a complete package. The equipment can also be

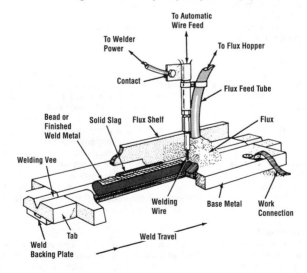

Figure 10. The SAW process. (Source, American Welding Society.)

permanently mounted onto a welding gantry over a moving workpiece. SAW may be carried out with either DC (positive or negative electrode) or AC. Therefore, transformers, rectifiers, or converters can be current sources.

SAW is used primarily for the welding of large fabrications and castings as well as common structural steel. Stainless steel and certain metals such as copper, aluminum, and titanium alloys may also be welded with SAW. Its main use is in industries that require long, continuous welds in thick sheets or plates, such as shipbuilding, beam and pressurized vessel industry, as well as in welding overlays. SAW is suited to flat position welds only due to the limitations of the sand flux. The most extensive application is for welding stainless steel overlays onto carbon steel, for example, nuclear reactor vessels.

The arc is started by feeding the electrode towards the workpiece and having it short-circuit. To establish the proper weld settings, the process needs to be prequalified and a WPS generated prior to use for production welding. The deepest penetration is obtained when using DCEP. DCEN produces minor penetration but a greater electrode deposition rate.

The electrode diameter is set according to current, deposition rate, penetration, and prequalification results. Due to the automated equipment used for SAW, large diameter electrodes can be used, allowing for high current settings. These high current settings, along with the high deposition rates, make SAW very attractive for the fabrication of large industrial weldments and castings. Also, since the flux actually covers the arc, SAW is very unobtrusive. Flash and sparks are not visible, and smoke and fumes are reduced to a minimum.

A large array of fluxes and filler materials also gives SAW the ability to be used in many diverse industries, including the weld repair of large steam turbine components. In welding of stainless steels, the wire should exhibit the same properties as the base material. Filler in the form of bands is available for the surface welding of stainless steel onto common steel. Powders of molten, agglomerated, and sintered types are available in different chemical compositions and grain sizes. Many powders (granulated fluxes) are sensitive to humidity.

Gas welding

Gas welding is also known as fusion welding or oxyacetylene welding. In gas welding the joint sides are melted by a flame generated by the combustion of a fuel gas, acetylene, propane, or MAPP and aided by pure oxygen. If filler is required, it is applied from the side. Fuel gas burned with the aid of oxygen produces a high enough flame temperature (approximately 5600°F [3100°C]) to achieve satisfactory weld results. Gas welding is characterized by being very flexible and adaptable to many application scenarios. The method may be used to weld most materials. Gas welding has decreased in importance through the years due to the development of new and more effective methods, although the method is still widely used in the assembly welding of pipes with thicknesses of up to 0.236" (5 mm) such as in the plumbing sector. Gas welding may, in such cases, be carried out in I-seams with very reliable welding results. Flux must be used with materials that easily oxidize. This is particularly important in the welding of stainless steel and noniron metals. The flux is applied onto the seam surfaces before welding.

Resistance (spot) welding

Resistance, or spot, welding is applied only to lap joints. Two copper electrodes, sometimes water-cooled, press the metal sheets together, one from each side. Once the electrode force has reached a given value, an electric current is applied between the electrodes. Heat is generated at the meeting point between the plates by the highly

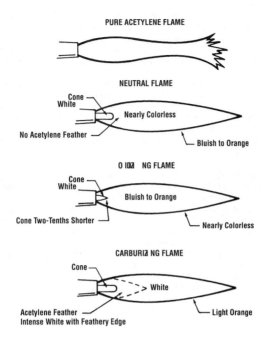

PURE ACETYLENE FLAME

NEUTRAL FLAME

Cone
White

Nearly Colorless

No Acetylene Feather

Bluish to Orange

O IDZ NG FLAME

Cone
White

Bluish to Orange

Cone Two-Tenths Shorter

Nearly Colorless

CARBURIZ NG FLAME

Cone

> White

Acetylene Feather
Intense White with Feathery Edge

Light Orange

Figure 11. Oxyacetylene flames. (Source, American Welding Society.)

concentrated current flow at that point. The heat quickly reaches the melting point of the base metal. The current is then switched off, and the electrodes continue to press with full force for a short period of time in order to cool down the puddle and allow the weld to form. Spot welding machines consist of a transformer that lowers the value of the main voltage to 5-10 V, creating the high current values. The on- and off-switching of the currents normally occurs with the aid of a controller. This controller sets current, duration of current flow, and dwell time when the joint is still clamped with no current flow. One of the electrodes is commonly fixed while the other one may move with the aid of pneumatic cylinders.

Spot welding is used for relatively thin sheet/plate thicknesses of $3/64$" to $5/64$" (1-2 mm). However, there are commercially available units that, in extreme cases, can weld up to 1" (25.4 mm). Very high welding speeds may be achieved in high production rate environments. It is, therefore, commonly used with industrial robots, a very common method in all sheet metal production, such as the automobile industry and the manufacture of household appliances. Carbon steel with low carbon levels, as well as stainless steel, are compatible with spot welding, but other metals may also be spot welded with good results.

Seam welding (SW)

Seam welding is essentially the same process as spot welding. The major difference being that the electrodes on a spot welding machine have been replaced by electrode rollers.

One of the rollers is motor driven which entails that the overlapping seam will be continuously fed forward between the rollers while the welding current is, normally, switched on and off in pulses. A welding point is created for each pulse. The distance between the points decreases as the pulsing frequency increases. In extreme cases, the points may be so closely placed that a continuous weld seam is obtained.

Seam welding is, as with spot welding, very common within the sheet metal industry, where the opportunity to obtain continuous weld seams can be exploited by SW. Sheet thicknesses of up to $^1/_8$" (3 mm) are normally seam welded. Examples of products which are seam welded include: tin cans, radiators, stainless steel wash basins, and plumbing/pipe fittings.

Electroslag welding (ESW)

Electroslag welding is used for welding thick plates in a vertical position. The welding is performed in thicknesses from around 1" up to 40" (25.4–1000 mm). The most common case being butt welding in I-seams. The sheets, or plates, are placed in a vertical position with a gap width of around $^3/_4$"–1-$^1/_4$" (20–30 mm). The welding is started at the bottom and carried out upwards. The weld puddle is held in place, in the joint, by water-cooled copper chucks. The slag shields the weld by floating on top of the molten metal. Filler is applied from above, through the gap, in the form of a continuous filler wire. No arc burns in the process, and the wire melts when it comes into contact with the floating slag due to the resistance heat generated when the welding current from the electrode passes through the slag. The electrode is sometimes driven into the melt by a guide tube, fixed in the joint. As the weld proceeds upwards, the wire melts and builds filler. ESW is used for manufacturing thick-walled pressure vessels and machine frames.

Electrogas welding (EGW)

Electrogas welding is carried out much in the same manner as electroslag welding. The main difference being that in EGW, the heat is generated by an arc that burns between the filler wire and the weld melt. Carbon dioxide, argon, or a mixture of these is used as shielding gas. This process is seldom used for thicknesses of over 4" (100 mm).

Stud welding

Stud welding is a method for welding bolts, screws, pins, and similar items onto a metal surface. The welding is carried out with the aid of a special gun in which the bolt or other item to be affixed to the surface is placed. An electric arc is formed between the bolt end and the workpiece. The bolt is automatically inserted into the puddle formed from an arc between the stud and the workpiece. Large steel bolts or studs for foundation work may be welded. The method is very rapid and simple to use and is applied, for example, for adding foundation studs in structural applications, for insulating studs in power generation equipment, boiler casings, and turbine shells.

Electron beam welding (EBW)

In electron beam welding (EBW), the energy source is a high-velocity electron beam. This beam hits the metal surface causing localized melting. An electron beam welding machine consists of an electron beam source, a vacuum chamber, and control equipment. The workpiece must be placed under vacuum during welding. The electron beam's large energy concentration allows for very high productivity rates. The process is well suited for certain metal combinations, such as unalloyed steel with stainless steel, low alloy steel with high-speed steel, and unalloyed steel with copper. It provides fast, deep, narrow welds with minimum distortion. Electron beam welding equipment has a high cost and therefore has been limited to more exotic production settings such as jet engine components, where finite control and low heat input is critical.

Friction (FRW) and inertia (IW) welding

In friction welding the heat is generated when one of the workpieces, with circular cross-section, rotates under pressure and at high speed against the other, usually static, workpiece. The rotation stops when the temperature has reached near melting point, and then the workpieces are butted together. Many materials and material combinations, which are usually difficult to weld with conventional welding methods, may be easily welded with this method. Inertia welding is very similar, in that it translates rotational energy into concentrated heat to cause welding at the joint. In IW, like FRW, one workpiece is rotated while the other remains stationary. When the programmed inertial energy is "stored" in the rotating element, it is forced into the stationary part and kept under pressure. The inertial energy is transferred into melting at the joint.

Ultrasonic welding

Ultrasonic welding is often used in the spot welding of thin metals, such as foils, and thin plastics in overlapping seams. The welding is carried out with the aid of an electrode that is pressed against the upper part of the workpiece and simultaneously transmits ultrasonic waves. The ultrasonic waves create a localized breakdown of the oxide film that produces a rise in temperature. This results in a material connection, or weld, with minimal deformation of the base material. Aluminum foil is easily welded with this method, but other metals and alloys are possible. This process is typically used to weld thin foil into a thicker material.

Explosive welding

In explosive welding, the workpieces are welded by the rapid application of pressure caused by the use of explosives. An example of the application of explosive welding is the welding together of stainless steel sheets onto ordinary steel in the production of compound steel sheets.

Brazing

Brazing is distinguished from gas in welding in two main areas. First of all, it is performed at a lower temperature (above 800° F [425° C]) than gas welding. The low temperature is an important consideration where distortion is a factor, or when maintaining the lowest possible temperature is essential. A nonferrous filler is used that is distributed between the closely fitted surfaces of the joint by capillary attraction. In brazing, there is no melting of the metals being joined, and the bond between the brazing alloy and base metal is obtained by slight diffusion of the brazing alloy into the heated base metal, by surface alloying of the base metal with the brazing alloy, or by a combination of both.

There are three common forms of brazing: 1) torch brazing, which uses air-acetylene, oxyacetylene, or other oxy-fuel gases; 2) furnace brazing which is popular for brazing preassembled or prejigged materials; and 3) induction brazing which lends itself to large production runs of single items.

Soldering

Soldering is a joining process wherein coalescence between metal parts is produced by heating to suitable temperatures (generally below 800° F [425° C]) and by using nonferrous filler metals (solder) having melting temperatures below those of the base metals. The solder is usually distributed between the properly fitted surfaces of the joints by capillary attraction.

The most common solders are composed of tin and lead in varying amounts. Solders with special properties may also include antimony, silver, bismuth, or indium. Most common solders are included in ASTM B-32 and ASTM B284. The main function of a flux is to promote good wetting action between the solder and the base metal. Fluxes may be highly corrosive, depending on the ingredients, and they are available in liquid and paste forms.

Welding, Brazing, and Soldering Low Carbon and Low Alloy Steels

Gas welding (oxyacetylene) welding can be used for joining low carbon steels, low alloy steels, and wrought iron. For welding, carbon steels containing more than 0.35% carbon are generally considered high carbon and require special care to preserve their mechanical properties. Low alloy steels with air hardening tendencies will be discussed below. Thin sheet (up to $^3/_{16}$ inch thick) can be completely melted by the flame without edge preparation. Material from $^3/_{16}$ to $^1/_4$ inch thick requires a slight root opening for complete penetration, and filler must be added to compensate for the opening. The joint edges of material $^1/_4$ inch or thicker should be beveled. The angle of the bevel for oxyacetylene welding of wrought iron and steel varies from 35 to 45°, which equates to an included angle of 70 to 90°. A root face of $^1/_{16}$ inch is recommended. Material $^3/_4$ inch thick and greater should be double beveled, if possible, and the joint should be welded from both sides. The root face can vary from 0 to $^1/_8$ inch. *Figure 12* shows suggested bevels and root faces.

A bad characteristic of oxyacetylene welding is the steep temperature gradient that is produced across the weld joint and in the surrounding weld joint. This often results in distortion unless precautions are taken. By first tack welding the assembly, and then

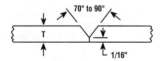

T = Thickness from 1/4" up to, but not including, 3/4" thickness

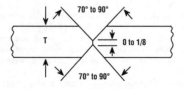

T = 3/4" thickness and over

Figure 12. Edge configurations for oxyacetylene welding.

beginning the weld at the point on the joint least subject to distortion, the problem can be minimized. If a steel contains sufficient carbon, and if it is allowed to cool too fast after welding, the weld metal and the metal in the heat affected zone will harden. This should be avoided because the affected material will lack ductility and be susceptible to cracking. In most hardenable steels, this can be avoided by playing the torch over the weld and the adjacent area for a short period after the weld is completed, or by preheating the material before welding. Air hardening steels may require a postheat treatment, such as a stress relief or annealing treatment, to eliminate the hardened zone.

Generally, the composition of the weld deposit should approach that of the base metal, especially if the weldment will be heat treated after welding. Steel rods and wire for oxyacetylene welding are designed to accommodate the base metal—rods meeting the ASM-ATSM Specifications for Iron and Steel Gas Welding Rods are suitable.

Shielded metal arc welding (SMAW) of wrought iron is best performed when the welding speed and current are slightly below those used for mild steel of the same thickness. The slower speed allows entrapped slag and gases to float out of the weld puddle, and the lower current minimizes the possibility of burning through thinner sections. Electrodes used for wrought iron should meet AWS-ASTM Specifications for Mild Steel Arc Welding Electrodes.

When the SMAW process is used on carbon steels containing up to 0.30% carbon, and less than one-inch in thickness, pre- or postheating is not necessary, and special electrodes are not required. However, when section thickness exceeds one-inch, preheating to 200° F, maintaining a minimum interpass temperature of 200° F, and postweld stress relief are advisable. When carbon content of the steel exceeds 0.30%, special precautions are required to prevent cracking in the weld deposit or in the adjacent heat affected zone. Preheating to, and maintaining a minimum interpass temperature of, 300 to 400° F is advisable for all thicknesses of material, and a stress relief treatment should follow welding. Carbon steels having a carbon content of over 0.50% are seldom welded, but, if special precautions are taken, satisfactory welds can be achieved.

In SMAW operations, the selection of the electrode type and the size of electrode to use in welding carbon steels depends on the weld strength requirements, the composition of the steel being welded, the thickness of the steel, the joint configuration, and the welding position. Electrodes of the E60XX series of AWS-ASTM Specifications for Mild Steel Arc Welding are suitable for low carbon steels, and the AWS-ASTM E70XX series are recommended for medium and higher carbon steels.

Low alloy steels should be preheated—400 to 750° F, depending on hardenability—prior to welding to avoid cracking. The following guidelines should be observed.

Carbon Content %	Hardenability Equivalency	Preheat and Interpass Temp.
0.16 through 0.34	AISI 4130 and steels of equivalent hardenability	400° F
0.35 through 0.43	Steels with hardenability higher than above	500 to 750° F

It is generally advantageous to stress relieve the low alloy steels after welding. In addition to relieving residual stresses in the welded structure, the temperatures involved in stress relieving are such that any hardened material adjacent to the weld will be tempered, with a corresponding improvement in ductility.

For low alloy steels, electrodes meeting the requirements of AWS-ASTM E70XX, E80XX, E90XX, and E100XX classifications are widely used. The electrodes are classified as low-hydrogen types, which generally have less than 0.2% moisture in the coating to avoid introducing hydrogen to the weld zone. Usually, they are packaged in hermetically sealed containers that should not be opened until time of use. After opening the electrodes, storing them in a holding oven at a temperature recommended by the manufacturer is recommended.

Gas metal arc welding (GMAW) and gas tungsten arc welding (GTAW) can be used to weld carbon steels. GTAW is usually limited to welding thinner sections with or without the addition of filler material. GMAW, because of its high deposition rate, is generally used to weld heavier section thicknesses. Quite often, the root pass of a multipass weld is made with the GTAW process, while the remaining passes are made with some other welding process. This procedure provides a clean, sound root bead.

In general, preheating practices should apply to the recommendations described in the SMAW section (above). A number of welding wires are available which are specifically designed for GMAW welding of low and medium carbon steels. These wires usually contain deoxidizers or other scavenging agents to prevent porosity or other damage to the weld metal from the oxygen, nitrogen, or hydrogen that may be in the shielding gas or which may reach the weld from the surrounding atmosphere. The deoxidizing agents commonly used are manganese, silicon, and aluminum.

Both GMAW and GTAW are used for welding low alloy steels, but GTAW is generally selected when optimal mechanical properties are required for highly stressed parts. Joint design and preparation, and preheating practices are similar to that used for SMAW (above). Several compositions of filler wire have been developed for using the GMAW process on low alloy steels. Selection is usually based on composition of the steel and the weld strength requirements. Generally, the composition of the wire is similar to the base metal except that, in some cases, the carbon content is lowered in order to obtain improved ductility in the weld.

Submerged arc welding (SAW) can be used for wrought iron with the same fluxes and electrodes used for welding low carbon steels. Slower welding speeds are recommended to allow the greater volume of gases generated during welding to escape before the molten metal solidifies.

SAW is one of the leading processes for joining low and medium carbon steels. Moderately thick sections with a carbon content ranging up to 0.35% can be welded without precautionary measures such as pre- or postheating. When the carbon content exceeds 0.35%, preheating is generally necessary.

The low alloy steels are easily welded with SAW, but precautions are necessary to avoid cracking in the weld and heat affected zone. The heat treatable low alloy steels may be welded provided that the weld area is preheated and the cooling rate is slow. An important factor in welding these steels is the proper selection of filler metals and fluxes that will deposit weld metal capable of response to heat treatment after welding. It is advisable to follow the manufacturer's recommendations in the choice of fluxes and wires.

In welding certain quenched and tempered structural steels, the heat input must be closely controlled to assure the retention of strength and notch toughness in the heat affected zone. Any practice that will retard cooling, such as preheating or high welding heat inputs, should be avoided whenever possible.

Electroslag welding (ESW) is primarily used on low tensile strength (45,000 to 72,000 PSI) structural steels, the medium tensile strength structural steels (65,000 to 85,000 PSI), and the high tensile strength structural steels (75,000 to 100,000 PSI). These include both high and low carbon steels, and several high strength heat treatable low alloy steels such as AISI 41XX and 43XX.

Stud welding can be used to weld carbon steels with up to 0.30% carbon without preheating. For higher carbon content steel, particularly in heavy sections, preheating is advisable in order to prevent cracking in the heat affected zone. Low alloy steels can be satisfactorily stud welded without preheating provided the carbon content is held to

0.12% maximum. Higher carbon content low alloy steels should be preheated. The heat treatable high strength low alloy structural steels require more attention because they are sufficiently hardenable to form martensite in the heat affected zone. Preheating to about 700° F is recommended.

Resistance (spot) welding of carbon steels is usually limited to low carbon steels with a maximum of 0.15% carbon content. Higher carbon contents require the use of special techniques involving a machine preheating cycle to avoid undesirable weld structures, cracking, and extremely low ductility in the weld region. Spot welding low alloy and high strength steels requires special techniques to reduce the hardness and avoid excessive brittleness in the weld area. Some of the low alloy high strength structural steels can be welded without preheating. In such cases, welding is followed by tempering in a furnace at about 1100° F to eliminate any undesirable weld zone microstructures and to improve weld ductility.

Brazing carbon and low alloy steels with copper or copper zinc alloys causes temperature induced grain growth in the base metal. When necessary, the grain structure and mechanical properties of the material can be improved by reheating the brazed assembly above the critical range of the steel, providing the brazing alloy used has a melting temperature above the austenitizing temperature of the steel. See **Table 4** for recommended brazing alloys for carbon and low alloy steels.

Table 4. Brazing Alloys for Carbon and Low Alloy Steels.

Brazing Alloy	Brazing Temperature Range °F/°C	Remarks
BCu (Copper)	200 to 2100 / 93 to 1150	Lap and butt joints commonly used.
RBCu-Zn-A (Copper Zinc)	1670 to 1750 / 910 to 955	Overheating should be avoided, as voids may form in joints as a result of entrapped zinc vapors.
BAg-1 (Silver base)	1145 to 1440 / 620 to 782	Flows freely into capillary joints.
BAg-1A (Silver base)	1175 to 1400 / 635 to 760	Low temperature applications. Free flowing.
BAg-2 (Silver base)	1295 to 1550 / 700 to 845	Good "bridging" characteristics. Forms fillets readily.
BAg-5 (Silver base)	1370 to 1550 / 745 to 845	Cadmium free. Applications involving "step" brazing.
BAg-18 (Silver base)	1325 to 1550 / 720 to 845	Good "wetting" in controlled atmosphere or vacuum brazing without flux.

Residual stresses in welded steels

When two pieces of steel are joined by welding, the weld metal and adjacent heat affected base metal undergo considerable expansion on heating, and contraction upon cooling to room temperature. If the two pieces are left free to move, they will be drawn closer together during contraction of the weld metal. If conditions are such that the pieces cannot move freely toward each other during contraction, they are said to be "under restraint." The forces set up by the restraining action create residual or "locked in" stresses in the weld region. Generally, these stresses are the highest in the metal near the center of the weld, which is the last to cool.

Although not always the case, residual stresses in structural steel are considered to be detrimental. There are two practical methods of reducing or eliminating residual stresses in welded structures: thermally, through stress relief heat treatment, or by peening (hammer blows on the weld deposit). Peening is frequently used to relieve residual stresses but its effectiveness is questionable because the method is difficult to control. Stress relief by thermal means, on the other hand, is closely controlled by heating the entire welded structure in a furnace or by locally heating the weld with a gas torch. In general, the stress relief of low carbon and low alloy steel welded structures is accomplished at temperatures ranging from 1100 to 1200° F. Usual soaking time at temperature is one hour for each inch of thickness, with a one hour minimum. Cooling should be as gradual as possible.

Much can be done to control residual stresses during welding, including the following.

1) The assembly and welding sequences should permit movement of the detail parts during welding. Joints of maximum fixity should be welded first.
2) An intermittent sequence of welding, such as "skip" or "backstep" welding, can be advantageous.
3) Introducing bending stresses during welding should be avoided.
4) Avoid excessively large welds ("overwelding").
5) Peening of all weld beads, except root and last bead, in multiple bead weld joints may be helpful.

Cracking of welded joints is caused by localized stress that, at some point, exceeds the ultimate strength of the material. When cracks occur during welding, usually little or no deformation is present. Perhaps the most troublesome form of cracking is termed "hot cracking." Factors leading to the formation of hot cracks in the weld deposit are high joint rigidity, the contour of the weld bead, a high carbon content, and a high sulfur or silicon content. This type of cracking usually appears in the last metal to freeze, and is intergranular in nature. A convex bead is less susceptible to cracking than a concave bead. A stress concentration is developed at the thinnest and hottest part of a concave bead during cooling from the welding temperature. The thinner center section of the convex bead makes it less likely to crack.

Small star shaped cracks are a type of hot crack associated with weld craters. Unless care is taken, there is usually a tendency to form a weld crater when the welding operation is interrupted. Crater cracks may be starting cracks for longitudinal weld cracks, especially when they form in a crater at the end of a weld bead. To avoid crater cracks, it is necessary to fill the crater with additional metal before breaking off the welding operation. Preheating usually minimizes any tendency toward hot cracking by reducing stress in the area.

A basic problem encountered in welding many of the low alloy steels has been termed "underbead cracking." Underbead cracking is invariably located under the fusion zone of the heat affected zone of alloy steels. It is attributable to the affects of dissolved hydrogen released from the austenite as it transforms. It can be virtually eliminated by preheating and the use of a low hydrogen welding process. Low carbon steels show no tendency to underbead cracking.

Cracks may also occur within the heat affected zone of the base metal. They are usually longitudinal in nature and are almost always associated with the hardened zone developed in hardenable steels (such as medium carbon and low alloy steels). Hardness and brittleness in the heat affected zone of the base metal are metallurgical effects produced by the heat of welding and are the chief factors that tend to cause cracking during welding. When base metal cracking is encountered with hardenable steels, improvements can be obtained by using a suitable preheat, increasing heat input that will tend to retard the cooling rate, and selecting the most suitable electrode for the steel being welded.

Welding, Brazing, and Soldering Aluminum and Aluminum Alloys

The welding of aluminum is common practice in industry because it is fast, easy, and relatively inexpensive. It is especially useful in making leakproof joints in thick or thin metal, and it can be employed with either wrought or cast aluminum, or a combination of each. Welding of aluminum is done by fusing the molten parent material together, with or without the use of filler material, or by using pressure (with or without heat generated by electrical resistance) to fuse the materials. **Tables 5** and **6** provide information on the suitability of several aluminum alloys for welding. See page 528 for a full description of the temper designations used on these tables.

Before welding, the corrosive resistant oxide film that protects aluminum must be removed and prevented from reforming by mechanical, chemical, or electric methods. If the film is allowed to remain in place, the wetting process that is required for coalescence of the metals during welding cannot take place. It can be removed by sanding, or with stainless steel (mechanical), with fluxes that dissolve and float the oxides away (chemical), or the application of reverse polarity current (electrical).

The good thermal conductivity of aluminum allows the heat of welding to spread rapidly from the weld zone, which can result in a loss in strength in work hardened or heat treated alloys through annealing. It can also cause buckling or total collapse of the parent material if the metal is not supported properly during welding. The good electrical conductivity of aluminum necessitates the use of higher currents in resistance welding. The low melting point of aluminum, in the range of 900 to 1216° F (482 to 658° C), increases the need for care. Since aluminum gives no visual indication of having obtained

Table 5. Weldability Ratings for Wrought Aluminum Alloys.

Alloy		Relative Suitability for Welding, Brazing, and Soldering						
		Gas Weld	Arc Weld (Flux)	Arc Weld (Inert Gas)	Resistance Weld	Pressure Weld	Soldering	Brazing
Wrought Aluminum Alloy	1100	A	A	A	B	A	A	A
	2011	D	D	D	D	D	D	D
	2014	D	B	B	B	D	D	D
	2014-Clad	D	B	B	A	C	D	D
	2017	D	B	B	B	D	D	D
	2018	D	B	B	B	D	D	D
	2024	D	B	B	B	C	D	D
	2024-Clad	D	B	B	A	C	D	D
	3003	A	A	A	A-B	A	A	A
	4032	D	B	B	C	C	D	D
	5052	A	A	A	A-B	A-B	C	C
	6061	A	A	A	A	C	B	A
	6151	A	A	A	A	C	B	B
	7075	D	D	D	B	D	D	D

When using filler wire, the wire should contain less than 0.0008% beryllium to avoid toxic fumes.

A = Generally weldable by all commercial procedures and methods.

B = Weldable with special techniques or for special applications that justify preliminary trials or testing to develop welding procedure and weld performance.

C = Limited weldability because of crack sensitivity or loss in resistance to corrosion and mechanical properties.

D = No commonly used welding methods have been developed.

Table 6. Weldability Ratings of Cast Aluminum Alloys. *(Source, MIL-HDBK-5H.)*

Alloy	Weldability[1]	
	Inert Gas Metal or Tungsten Arc	Resistant Spot
A201.0	C	C
354.0	B	B
355.0	B	B
C355.0	B	B
356.0	A	A
A356.0	A	A
A357.0	A	B
D357.0	A	A
359.0	A	B

[1] Weldability related to joining a casting to another part of the same composition. The weldability ratings are not applicable to minor weld repairs. When using filler wire, the wire should contain less than 0.0008% beryllium to avoid toxic fumes.

A = Generally weldable by all commercial procedures and methods.

B = Weldable with special techniques or for special applications that justify preliminary trials or testing to develop welding procedure and weld performance.

C = Limited weldability because of crack sensitivity or loss in resistance to corrosion and mechanical properties.

D = No commonly used welding methods have been developed.

welding temperature, the temperature must be measured by the physical condition of the material rather than by its appearance.

Where a considerable amount of general heating can be tolerated, and where an easily finished bead is desired, gas welding is preferred. *Oxyacetylene gas welding* works well on butt, lap, and fillet welds on material from 0.40 inch to one-inch in thickness. However, where minimum general heating, absence of flux, and very good properties are requirements, a variety of inert gas shielded arc welding should be used. See **Table 7** for tensile strengths of typical gas welded joints.

Table 7. Typical Tensile Strength of Gas Welded Joints on Aluminum Alloys. *(Source, MIL-HDBK-694A[MR].)*

Alloy			Thickness (inches)	Tensile Strength PSI
Type	Designation	Temper		
Sand Cast	B443.0	– F	0.50	12,000
	514.0	– F	0.50	12,000
Wrought	1100	– H14	0.249	11,000
	3003	– H14	0.249	14,000
	5052	– H34	0.249	27,000

See Materials section for explanation of tempers.

Shielded metal arc welding (SMAW) is especially suitable for heavy material. Welds in plate 2.5 inches thick can be performed satisfactorily by this method. Unsound joints are likely to appear in metal arc welded material that is less than 5/64 inch thick, and weld soundness and surface smoothness are not as good as with other arc welding methods. These factors, plus the necessity of a flux, have been responsible for the decrease in popularity of this process.

Gas tungsten arc welding (GTAW) has two distinct advantages over other forms of fusion welding: no flux is needed, and welds can be made with almost equal facility in the flat, vertical, or overhead positions. It can be used for either manual or automatic welding on aluminum materials 0.05 inch thick or thicker.

Resistance (spot) welding is especially useful for joining high strength aluminum alloy sheet with practically no loss of strength. Using resistance welding, these alloys can be spot welded, seam (line) welded, or butt (flash) welded. Spot welding is widely used to replace riveting, seam welding is essentially spot welding with the spots spaced so closely together that they overlap to produce a gas tight joint, and flash welding heats the aluminum by establishing an arc between the two pieces to be joined. **Table 8** provides typical shear strengths for spot welds, and **Table 9** gives tensile strengths of butt welded joints. See the Materials section for descriptions of the temper designations for aluminum and aluminum alloys.

Brazing will provide a joint equivalent to that of the metal in the annealed condition. However, in some instances where an age hardening alloy is used, the mechanical properties of the metal can be enhanced by treatment. For example, alloy 6061, when quenched from the brazing operation and when artificially aged, will exhibit a tensile strength of approximately 45,000 PSI, a yield strength of 40,000 PSI, and an elongation of 9% in two

Table 8. Typical Shear Strengths of Spot Welds on Aluminum Alloys. *(Source, MIL-HDBK-694A[MR].)*

Alloy	Temper	Thinnest Sheet in Joint (inches)									
		0.016	0.020	0.025	0.032	0.040	0.051	0.064	0.081	0.102	0.125
		Shear Strength (minimum), Pounds per Spot									
1100	– H14 to – H18	40	55	70	110	150	205	280	420	520	590
3003 5052	– H12 or – H18 to – O	70	100	145	210	300	410	565	775	950	1000
5052 6061	– H32 Or – H38 to – T4 or – T6	98	132	175	235	310	442	625	865	1200	1625
2024 (clad) 2024 7075 (clad) 7075	– T3 – T3 or – T6 – T6	108	140	185	260	345	480	690	1050	1535	2120

See Materials section for explanation of tempers.

Table 9. Typical Tensile Strengths of Butt Welded Joints on Aluminum Alloys. *(Source, MIL-HDBK-694A[MR].)*

Designation			Tensile Strength Across Weld (PSI)	
Base Metal		Filler Metal Alloy	As Welded	After Heat Treat And Aging
Alloy	Temper			
1100		1100	13.5	–
3003		1100	16	–
5052		5052	28	–
2014	– T6	4043	34	51,000
6061	– T6	4043	27	43,000
6063	– T5, T6	4043	20	–

See Materials section for explanation of tempers.

inches. Brazeable alloys are available in plate, sheet, tube, rod, bar, wire, and shapes. The most popular brazeable alloys are 1100, 3003, and 6061.

Soldering can be used to join aluminum to aluminum and other solderable metals. Using a soldering iron or torch is suitable for such applications as indoor electrical joints, but it is not recommended for structural members. Soldering aluminum is somewhat more difficult than soldering other materials because of the high thermal conductivity of aluminum and the presence of a tough oxide film. Aluminum solder melts at 550 to 700° F (288 to 371° C), as compared with 375 to 400° F (190 to 204° C) for other solders. This higher melting rate is required because of the difficulty of maintaining sufficient heat in the working area to melt the solder. Thus, only small parts that can be preheated are suitable for soldering with an iron. Larger parts require the use of a torch to concentrate sufficient heat.

Welding, Brazing, and Soldering Copper and Copper Alloys

Copper's high thermal conductivity must be taken into account during welding and brazing. Preheating and use of high heat inputs may be required to ensure successful welds. The relatively high thermal expansion coefficient of copper alloy must also be taken into account when fixturing parts for weld preparation. Oxygen containing coppers can become embrittled if the weld is contaminated with hydrogen. Often called "hydrogen sickness," this condition can be avoided by utilizing deoxidizing or oxygen-free coppers such as C12200. Copper alloys are sufficiently deoxidized and are not subject to this phenomenon.

Aluminum bronzes, silicon bronzes, and copper-nickels are also commonly welded, but brasses, manganese bronzes, and nickel silvers are more difficult to weld and generally must be brazed instead. Leaded copper alloys cannot tolerate the high heat of welding without suffering mechanical damage, and they should be brazed or soldered. All copper alloys can be soldered, but some of the alloys have a protective oxide film that can be troublesome and may require an aggressive flux.

For optimum results, brazing or soldering must be performed on properly prepared surfaces. Mechanical cleaning, as with abrasives, can be beneficial, although fluxes are used in most situations. Commercial fluxes are available in a range of chemical activities, or degrees of aggressiveness. Good practice calls for selecting the least aggressive flux commensurate with the type of work to be done since flux residues, if not completely removed, can lead to corrosion problems later on. Most copper alloys respond satisfactorily to mildly aggressive fluxes, with the notable exceptions of the aluminum bronzes and beryllium coppers which tend to resist fluxing on account of the refractory alumina and beryllia contents, respectively, of their oxide layers. Sometimes more aggressive fluxes may be required in these cases. After joining, the beryllium copper should be cleaned with a 15 to 20% aqueous solution of nitric acid, followed by a water rinse. **Table 10** provides joining characteristics of selected wrought copper alloys, and **Table 11** gives this information for casting alloys.

Weld Quality

In previous pages we have discussed the requirements of codes and standards and their effects on defining weld quality guidelines. To facilitate compliance with these codes and standards, we need to understand some simple aspects of weld quality.

Once a company has identified the applicable codes and standards associated with its industry and products, it needs to develop a weld quality control process. This process is a documented description of the requirements, how they are to be met internally, quality controls, identification of responsible parties, WPS's, procedure qualification

(Text continued on p. 1706)

Table 10. Joining Characteristics of Selected Wrought Copper Alloys. See Notes.
(Source, Copper Development Association.)

Copper Alloy No.	SMAW	GTAW & GMAW	Oxyacetylene Welding	Resistance Seam Welding	Resistance Spot Welding	Butt Welding	Electron Beam	Brazing	Soldering
C10100	D	B	C	D	D	B	A	A	A
C10200	D	B	C	D	D	B	A	A	A
C10300	D	A	C	D	D	B	A	A	A
C10400	D	B	C	D	D	B	A	A	A
C10500	D	B	C	D	D	B	A	A	A
C10700	D	B	C	D	D	B	A	A	A
C10800	D	A	B	D	D	B	A	A	A
C11000	D	C	D	D	D	B	A	B	A
C11300	D	C	D	D	D	B	A	B	A
C11400	D	C	D	D	D	B	A	B	A
C11500	D	C	D	D	D	B	A	B	A
C11600	D	C	D	D	D	B	A	B	A
C12000	D	A	C	D	D	B	A	A	A
C12100	D	A	C	D	D	B	A	A	A
C12200	D	A	B	D	D	B	A	A	A
C12900	D	C	D	D	D	B	-	B	A
C14500	D	C	C	D	D	C	-	B	A
C14520	D	C	C	D	D	C	-	B	A
C14700	D	D	D	D	D	B	-	A	A
C15000	D	D	D	D	D	B	-	B	A
C15715	D	D	D	C-D	C-D	C	-	B	A
C15725	D	D	D	C-D	C-D	C	-	B	A
C15760	D	D	D	C-D	C-D	C	-	B	A
C16200	D	B	B	D	D	B	-	A	A
C16500	D	C	D	D	D	A	-	A	A
C17000	D	B	C	B	B	B	A	B	A
C17200	D	B	C	B	B	B	A	B	A
C17300	D	C	C	C	C	C	C	D	A
C17410	D	B	C	B	B	C	B	B	A
C17500	D	B	C	B	B	C	B	B	A
C17510	D	B	C	B	B	C	B	B	A
C18135	D	B	D	D	D	B	-	A	A
C18200	D	B	D	D	D	C	-	B	A
C18400	D	B	D	D	D	C	-	B	A
C18700	D	D	D	D	D	C	-	B	A

(Continued)

Table 10. *(Continued)* **Joining Characteristics of Selected Wrought Copper Alloys.** See Notes.
(Source, Copper Development Association.)

Copper Alloy No.	SMAW	GTAW & GMAW	Oxyacetylene Welding	Resistance Seam Welding	Resistance Spot Welding	Butt Welding	Electron Beam	Brazing	Soldering
C19100	D	C	C	C-D	C-D	C-D	-	B	A
C22000	D	B	B	D	D	B	-	A	A
C22600	D	B	B	D	C	B	-	A	A
C26130	D	C	B	D	B	B	-	A	A
C26200	D	C	B	D	B	B	-	A	A
C26800	D	C	B	D	B	B	-	A	A
C27000	D	C	B	D	B	B	-	A	A
C28000	D	C	B	D	B	B	-	A	A
C31400	D	D	D	D	D	B	-	B	A
C31600	D	D	D	D	D	C	-	B	A
C32000	D	D	D	D	D	C	-	B	A
C33500	D	C	C	D	C	C	-	B	A
C34000	D	D	D	D	D	C	-	B	A
C34200	D	D	D	D	D	C	-	B	A
C35000	D	D	D	D	D	C	-	B	A
C35300	D	D	D	D	D	C	-	B	A
C35600	D	D	D	D	D	C	-	B	A
C36000	D	D	D	D	D	C	-	B	A
C37700	D	D	D	D	D	C	-	B	A
C38500	D	D	D	D	D	C	-	B	A
C46400	D	C	B	C	B	B	-	A	A
C48200	D	D	D	D	D	C	-	B	A
C48500	D	D	D	D	D	C	-	B	A
C51000	C	B	C	C	B	A	-	A	A
C52100	C	B	C	C	B	A	-	A	A
C54400	D	D	D	D	D	C	-	B	A
C61000	B	B	D	C	B	B	-	D	C
C61300	B	A	D	B	B	B	-	C	D
C61400	B	B	D	B	B	B	-	C	D
C61800	B	B	D	B	B	B	-	B	C
C62300	B	B	D	B	B	B	-	C	D
C62400	B	B	D	B	B	B	-	C	D
C62500	B	B	D	C	C	C	-	C	D
C63000	B	B	D	B	B	B	-	C	D
C63200	B	A	D	C	B	B	-	C	C

(Continued)

Table 10. *(Continued)* **Joining Characteristics of Selected Wrought Copper Alloys.** See Notes. *(Source, Copper Development Association.)*

Copper Alloy No.	SMAW	GTAW & GMAW	Oxyacetylene Welding	Resistance Seam Welding	Resistance Spot Welding	Butt Welding	Electron Beam	Brazing	Soldering
C63600	C	C	D	C	C	C	-	D	D
C64200	C	C	D	C	C	C	-	C	D
C65100	C	A	B	B	A	A	-	A	A
C65500	C	A	B	A	A	A	-	A	B
C67400	D	C	D	B	B	B	-	B	C
C67500	D	C	B	C	B	B	-	A	A
C69400	-	-	B	B	B	B	-	A	A
C70600	B	A	C	B	B	A	-	A	A
C71500	A	A	B	A	A	A	-	A	A
C74500	D	C	B	C	B	B	-	A	A
C75200	D	C	B	C	B	B	-	A	A
C75400	D	C	B	C	B	B	-	A	A
C75700	D	C	B	C	B	B	-	A	A

Notes:
A = Excellent   B = Good   C = Fair   D= Not Recommended

Table 11. Joining Characteristics of Selected Copper Casting Alloys. *(Source, Copper Development Association.)*

Copper Alloy No.	SMAW	GTAW & GMAW	Oxyacetylene Welding	Carbon Arc Welding	Brazing	Soldering
C80100	C	D	D	C	A	A
C81100	C	D	D	C	A	A
C81300	C	C	D	C	B	A
C81400	C	C	D	C	B	A
C81500	C	D	D	C	B	B
C82000	C	C	D	C	B	B
C82200	C	C	D	C	B	B
C82400	C	C	D	C	C	C
C82500	C	C	D	C	C	C
C82600	C	C	D	C	C	C
C82700	C	C	D	C	C	C
C82800	C	C	D	C	C	C
C83300	C	D	D	D	B	A

(Continued)

Table 11. *(Continued)* **Joining Characteristics of Selected Copper Casting Alloys.**
(Source, Copper Development Association.)

Copper Alloy No.	SMAW	GTAW & GMAW	Oxyacetylene Welding	Carbon Arc Welding	Brazing	Soldering
C83400	C	D	C	D	A	A
C83600	D	C	D	D	B	A
C83800	D	C	D	D	B	A
C84200	D	C	D	D	B	A
C84400	D	C	D	D	B	A
C84500	D	C	D	D	B	A
C84800	D	C	D	D	B	A
C85200	D	D	C	D	C	A
C85400	D	D	C	D	A	A
C85500	D	D	D	D	C	B
C85700	D	D	D	D	C	B
C85800	D	D	D	D	B	B
C86100	C	B	B	D	D	D
C86200	C	B	B	D	D	D
C86300	D	B	D	D	D	D
C86400	D	D	D	D	C	C
C86500	D	D	D	D	C	C
C86700	D	D	D	D	C	C
C86800	D	D	D	D	C	C
C87400	C	D	C	D	C	D
C87500	C	D	C	D	C	D
C87600	C	C	B	D	C	D
C87800	D	D	D	D	C	D
C90200	C	C	C	C	B	A
C90300	C	C	C	C	B	A
C90500	C	C	C	C	B	A
C90700	C	C	C	C	B	A
C90900	C	C	C	C	B	A
C91000	C	C	C	C	B	A
C91100	C	C	C	C	B	A
C91300	C	C	C	C	B	A
C91600	C	C	C	C	B	A
C91700	C	C	C	C	B	A
C92200	D	D	D	D	A	A
C92300	D	D	D	D	B	A

(Continued)

Table 11. *(Continued)* **Joining Characteristics of Selected Copper Casting Alloys.**
(Source, Copper Development Association.)

Copper Alloy No.	SMAW	GTAW & GMAW	Oxyacetylene Welding	Carbon Arc Welding	Brazing	Soldering
C92500	D	D	D	D	B	A
C92600	D	D	D	D	B	A
C92700	D	D	D	D	B	A
C92800	D	D	D	D	B	A
C92900	D	D	D	D	B	A
C93200	D	D	D	D	B	A
C93400	D	D	D	D	B	B
C93500	D	D	D	D	B	B
C93700	D	D	D	D	B	B
C93800	D	D	D	D	D	B
C93900	D	D	D	D	D	B
C94300	D	D	D	D	D	B
C94400	D	D	D	D	B	B
C94500	D	D	D	D	D	B
C94700	B	B	C	D	A	A
C94800	D	D	D	D	B	A
C95200	A	B	D	C	B	B
C95300	A	B	D	C	B	B
C95400	A	B	D	C	B	B
C95500	B	B	D	D	C	B
C95600	B	C	D	C	B	B
C95700	A	B	D	B	B	B
C95800	B	B	D	D	C	B
C96200	D	C	D	D	A	A
C96300	C	C	D	D	A	A
C96400	B	B	D	D	A	A
C96600	C	C	B	C	A	A
C97300	D	D	D	D	A	A
C97400	D	D	D	D	A	A
C97600	D	D	D	D	A	A
C97800	D	D	D	D	A	A
C97300	B	B	D	D	B	D
C99700	-	-	B	-	B	B
C99750	C	B	D	D	B	B

Notes: A = Excellent   B = Good   C = Fair   D = Not Recommended

records, welder certification, inspection practices, and documentation and record keeping requirements.

We have already had some discussion of codes and standards and how they apply. Given the expertise required to develop and manage a weld quality assurance/control program, an organization needs to assess its ability to perform this function well and weigh the value of utilizing outside resources to meet their needs.

Many professional and industry societies and associations have local chapters that may provide resources to help companies understand and manage their welding programs. In addition to these groups there are also professional consultants, the American Welding Society itself, and the Hobart Institute of Welding Technology in Troy, Ohio. The Hobart Institute is a not-for-profit school created by Hobart Brothers welding company for the advancement of welding technology and welding education.

Weld defects and inspection

As part of every code and standard associated with welding, a variety of quality requirements are defined. These definitions include prior topics such as WPS qualification, welder certification, and documentation. They also define specific weld quality characteristics and allowable limits. Some fundamental terms associated with weld defects are discussed below. It is acknowledged that every weld made has defects of some kind. Just as with any other item, there are allowable limits of defects. The presence of defects does not make a finished product defective, or scrap. Each industry has developed, through testing, experience, and research, allowable levels of defects in their respective products. These levels distinguish pieces with defects from defective pieces. Some of the defects associated with welding are as follows.

Inclusions—Foreign matter that becomes incorporated into a finished weld.

Undercut—Gaps or valleys along the weld joint in the parent metal. This is a result of excessive heat melting more metal from the parent material, and that metal becoming incorporated into the weld puddle.

Pinholes—Voids in a finished weld. These voids can be the result of gaseous products of combustion from contaminants in either the weld material itself, or from surface contamination. Pinholes or bubbles can also result from oxidation caused by either insufficient shielding gas or flux, or excessive heat.

Cold joints—Result from incomplete fusion of the filler and parent materials. Outward appearance can be very deceiving in these defects.

Toe cracks—Cracking created along the toe (line between the filler material and the parent material) from weld shrinkage.

Crater cracks—Cracks that can result from abrupt interruption of the weld arc at the end of a joint.

Excess reinforcement—When the welder builds up the filler material above the thickness of the base workpiece. Typically there should be no more than $1/8$" of reinforcement.

Incomplete fusion—In the root pass of a weld, the filler material is not completely fused into the base material.

As mentioned, various codes and standards define allowable limits of these and other defects.

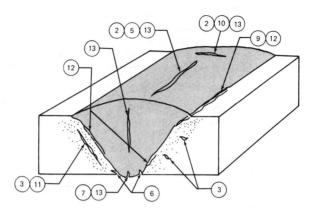

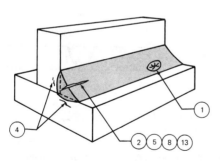

LEGEND
1. Crater Crack
2. Face Crack
3. Heat-Affected Zone Crack
4. Lamellar Tear
5. Longitudinal Crack
6. Root Crack
7. Root Surface Crack
8. Throat Crack
9. Toe Crack
10. Transverse Crack
11. Underbead Crack
12. Weld Interface Crack
13. Weld Metal Crack

Figure 13. Crack types.

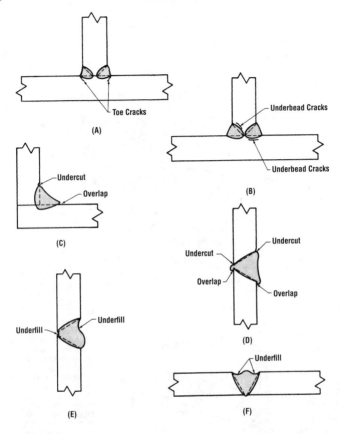

Figure 14. Weld flaws.

AWS Certification

One of the functions of the AWS has been to develop standardization in the welding industry. As a result, the AWS has developed a body of codes and standards that cover many topics associated with welding. These topics include structural steel codes, inspection and quality standards, and certification procedures to establish an individual's skill level as a welder, a weld inspector, or welding educator.

One of the fundamental certification skills is welder certification. The intention of the certification is to demonstrate and document a welder's ability to successfully complete a fairly generic WPS. A welder who has completed the AWS certification can be assumed to be a qualified welder. A qualified welder is someone who can reliably perform a variety of weld procedures. A certified welder is someone who has successfully completed a certification WPS. As mentioned, the AWS has a certification WPS. Although a solid measure of one's ability, performing a certification WPS does not certify an individual to perform any code welding. To perform code welding, the welder must be certified to a company-specific WPS. These certification tests involve destructive testing of the coupons completed by the welder. Once a welder's coupon is passed, his certification documentation is usually required to be filed with the state, just like a doctor, lawyer, or professional engineer.

Certified weld inspector/senior CWI

Inspection is a part of any welding program related to code welding. To standardize on the qualifications required to perform inspection, the AWS has developed the Certified Weld Inspector program. This program requires an individual to have specific education and experience, as well as pass a full day practical exam. This certification is valid for three years and can be renewed twice without reexamination. As this program has expanded in recent years, the designation of Senior CWI has been added to recognize those professionals with distinguished expertise, experience, and knowledge.

Certified weld educator

Education is the key to providing a complete approach to quality welding and welding technology. To communicate the skill level and experience of those professionals involved in welding education, the AWS has also developed the Certified Weld Educator qualification. This qualification can build on the knowledge of a CWI. A requirement has been included which requires that the CWE be individually certified in the processes being taught.

Company-Specific WPS's

As mentioned above, company-specific WPSs are required by most state building codes. These state codes typically refer to industrial codes and standards for technical requirements. We have also discussed prequalified WPSs. These are the only exceptions to a company-specific WPS, and are only related to the AWS codes such as D1.1 for structural steel. For the most part, all of the other codes and standards require companies to develop their own WPSs, perform PQRs (see below), and certify their welders to these company-specific WPSs. Once a welder has completed these certification tests, the welder is allowed to perform code welding for that company.

If a welder changes employers, this certification is not transportable. The fact that the welder was certified should, however, give his new employer an indication of the welder's ability to achieve recertification.

PQR

As discussed earlier in this section, the PQR, or Procedure Qualification Record, in the documentation is evidence that a company-specific WPS has passed the required testing, and the WPS is sound for use on code applications.

WQR

This is the documentation of the welder's certification (Welder Qualification Record). A welder may hold several WQRs, one for each WPS.

Inspection

To verify the quality of a completed weld joint, inspection is the final control. Weld quality begins with the use of a qualified WPS, a certified welder, and the use of the proper materials. Once the weld has been completed it is then necessary, as part of any weld quality control program, to incorporate inspection. For production welding and code welding, there are a wide variety of nondestructive examinations (NDE or sometimes referred to as NDT [nondestructive testing]) means. For WPS qualification, or for periodic process verification, destructive testing is utilized.

NDE methods range from the most commonly used and also the most cost effective method, visual inspection, to elaborate technical testing apparatus such as eddy current, radiography, and magnetic resonance imaging.

Nondestructive examination methods

RT. Radiographic testing, commonly referred to as X-raying welds, is a popular and cost effective method of verifying internal weld quality. Defects such as incomplete root fusion, inclusions, porosity, and cracking can be detected using this method. RT can only be performed by a certified NDE technician, as radioactive materials are used. Many codes require RT films to be on file to document weld quality.

VT. Visual inspection is the most commonly used and most cost effective means of verification of weld quality. Codes and standards identify visually measurable characteristics of welds, and the allowable limits for defects in these characteristics.

MP. Magnetic particle inspection takes advantage of lines of a magnetic field in a ferrous workpiece. The operator or technician will magnetize the workpiece and then dust the surface with iron powder, or spray on a developer that consists of iron powder in a suspension. MP is of limited use for cracking along the lines of magnetic field created. Technicians will vary the field to look for parallel indications. Perpendicular indications become obvious.

FP. Fluorescent penetrant inspection is a variation of MP. In FP, an ultraviolet sensitive media is used in addition to the magnetic indicator. Then, using a "black light," the technician can see finer detailed defects.

EC. Eddy current testing is another variation of MP. EC is used on nonmagnetic materials. Instead of the lines of the magnetic field used as the indicator for defects, a standing electric field is developed in the workpiece and the lines of the electric field become the indicator. This, of course, is limited to electrically conductive materials.

UT. Ultrasonic testing uses high frequency sound waves sent through the material to locate internal defects. UT cannot be applied to detected surface flaws.

LP. Liquid penetrant testing is the most widely used method to complement visual inspection. FP is used to detect surface defects by taking advantage of the capillary action of liquids being drawn into cracks, pinholes, etc. The part is cleaned and degreased, then the liquid dye is applied. The part is then cleaned of excess dye and a white powder developer is sprayed on the surface. Any cracks will show up as lines of dye in the developer and pinholes will show up as dots.

MR. Magnetic resonance imaging, as used for the health care industry, is also being applied to part quality inspection. A strong magnetic field is applied to excite the molecules of the part, and computer imaging can distinguish variations in material density.

Destructive testing methods

Tensile testing. A fundamental destructive test, the tensile test verifies that the sample being tested maintains the desired tensile strength. A standardized test coupon is cut from a part and then put into a tensile testing machine. This device puts the sample under tensile load and will chart the yield point, failure load, and percent elongation. This data will give the inspector access to the brittleness of the part, and the tensile strength.

Bend test. This is used to give an indication of the integrity of a welder's test weld joint. When a welder or WPS is being qualified, one of the methods used is the bend test. A sample strap is cut from the weld joint, across the joint. The strap is then put into a defined bending jig and bent at the joint. Some joint separation will occur. The applicable codes and standards will define allowable limits.

Charpy impact test. This unique test is used to measure the brittleness of the sample part. The lower the value, the more brittle or less impact resistant. Higher values indicate a more malleable material.

Weld Design and Economics

So far this section has discussed performing welding on an already defined joint, part, and process. Prior to performing the work, the joint needs to be defined based on the specific application. Issues such as dynamic load, cyclic fatigue, fabrication costs, and material properties all factor into definition of the joint design.

A fundamental distinction that determines a material's ability to withstand the loads applied is a review of the stresses present, and the material's tensile strength. In the discussion of Shielded Metal Arc Welding, the meaning of the characters in the electrode code were explained. For example, an E6010 had a filler material with a tensile strength of 60,000 psi. This indicates the filler material can withstand a load of 60,000 pounds force per square inch of cross sectional area. This is symbolized by the formula $P = F/A$, where P is the pressure applied, F is the force in pounds, and A is the cross sectional or surface area. When analyzing the loads applied, they can be considered as compressive loads, tensile loads, shear forces, and hoop stresses.

Production factors

In determining which welding processes are most applicable to a specific operation, it is important to understand many production related factors. Production rates for a process are related to the following variables.

Duty cycle, or the length of time, within a ten-minute period, that the equipment can operate at its rated current.

Deposition rate, which is the amount of filler material a process can typically deposit in the weld joint per unit of time. Usually expressed in pounds per hour.

Deposition efficiency, the percentage of filler material deposited which will be permanently incorporated into the weld joint. Efficiency is affected by weld spatter and grinding between beads.

Design requirements

Application requirements affect process selection. These include the following.

Cyclic fatigue. In the designed application, what types of loads will the workpiece experience? Some applications expose the workpiece to impact loads, static loads, and dynamic loads. How frequently these loads are applied determines the cycles, and therefore the lifespan or mean time to failure of the part. Many industries find it necessary run cyclic fatigue tests on components to establish warranty periods.

Tensile strength. The loads applied must be within the strength limits of the materials used and the design characteristics of the joint.

Yield strength. The point at which the material starts to permanently deform, and therefore begins to fail, identifies its yield strength.

Design for assembly. DFA, as it is referred to in industry, keeps the assembly requirements in the forefront of part design. Material costs should not be sacrificed for assembly costs.

Cost estimating

Accurate product pricing is based on effective cost estimating, which requires an understanding of several fundamental concepts.

Nesting patterns. The ability to layout part profiles on the raw material in such a way as to minimize scrap can cut costs even before the operation is performed.

Material estimating. Not only the raw materials for the parts, but also the number of pounds of weld material, shielding gases, electrodes, grinding discs, etc., all contribute to the cost of the weld.

Production rate estimating. Incorporating the production factors discussed above, as well as productivity rates, overhead costs, and material handling capabilities. These are variable costs that will be different for individual welders and companies.

Jig and fixture costs. When designing components for manufacture, it is necessary to design part holding fixtures, assembly jigs, match lines, and reference dimensioning hardware. The production of these items can cost more than the welding operation, but if properly designed and implemented, they can be essential for producing quality and repeatable welds.

Environmental Health and Safety Issues

As more information is assembled about the health risks associated with the materials we come in contact with on a daily basis, it becomes an even higher priority to provide a safe and healthy workplace. Several federal, state, and local agencies provide information to keep workers and employers informed of the risks associated with certain working conditions, materials, and processes. The results of these efforts include Right to Know legislation, an important means of giving workers access to detailed information about the health and safety risks of the materials they use in their jobs. The MSDS, or Material Safety Data Sheet, is a universal means to communicate the health and safety characteristics of materials. The MSDS contains recommendations for personal protective equipment such as safety glasses, special gloves, hearing protection, etc. There are no MSDS sheets for processes. When incorporating new processes, it is essential to review manufacturer recommendations, industry standards, and federal, state, and local safety guidelines.

Given the inherent characteristics of welding, appropriate eye protection to allow safe exposure to intense UV and IR radiation must be used. In addition, high noise levels are typical in most welding processes, indicating the value of hearing protection. The metal fumes and gases associated with welding processes also highlight the need for proper ventilation and respiratory protection.

Joining with Adhesives

Adhesive Bonding

Adhesive bonding is the process of uniting materials with the aid of an adhesive. An adhesive is any substance capable of holding such materials together by surface attachment. There are two principal types of adhesive bonding: structural and nonstructural. Structural adhesive bonding occurs when elements of a structure are fastened together with an adhesive, and structural adhesive bonds must be capable of transmitting structural stress without loss of structural integrity within design limits. Resistance to the service environment is a corollary property. Properties of structural adhesives used for bonding metals are shown in **Table 1**.

Modern adhesives have several ingredients, based on intended use. These components include the following.

Base or *binder* is the primary component of an adhesive as it holds the substrates together. The binder is usually the component from which the name of the adhesive is derived.

Hardeners are added to adhesives to promote the curing reaction by taking part in it. Two-part adhesive systems generally have one part that is the base, and a second part that is the hardener. Upon mixing the base with the hardener in metered amounts, a chemical reaction takes place that causes the adhesive to solidify.

Catalysts are sometimes incorporated into an adhesive formulation to speed up the reaction between the base and the hardener. Sometimes a catalyst takes the form of an *activator*. An activator is not mixed into the adhesive, but is applied in a thin layer to one substrate, while a thicker layer of the base system is applied to the other substrate, and then the two are joined.

Solvents are sometimes needed to thin the adhesive to a spreadable consistency. Solvents used with synthetic resins and elastomers are generally organic, and often a mixture of solvents is required to achieve the desired properties.

Diluents are ingredients added to an adhesive to reduce the concentration of a binder. They are primarily used to reduce the viscosity and modify the processing conditions of some adhesives. Reactive diluents react with the binder during cure, becoming part of the product, and do not evaporate, as does a solvent.

Fillers are relatively nonadhesive substances added to the adhesive to improve its working properties, strength, permanence, conductivity, and other qualities. They are also added to reduce material cost.

Carriers or *reinforcements* are usually a thin fabric or paper used to support the semicured adhesive composition.

The primary function of adhesives is to fasten parts together by transmitting stresses from one member to another in a manner that distributes the stresses much more uniformly than can be achieved with conventional mechanical fasteners. Consequently, adhesives often permit the fabrication of structures that are lower in cost and weight, but are mechanically equivalent to, or stronger than, conventional assemblies. Dissimilar materials are easily joined by adhesives, provided that proper surface treatments are used. Adhesives can be conductors or nonconductors of electricity, flexible to allow for thermal expansion changes between dissimilar materials, sealants for attached components, and mechanical dampers to improve fatigue resistance of structures. **Table 2** shows important advantages and disadvantages of adhesive bonds.

Adhesive characteristics by category

Synthetic adhesives are classified into four major categories: thermoplastic adhesives, thermosetting adhesives, elastomeric adhesives, and adhesive alloys. The characteristics of each are shown in **Table 3**.

Table 1. Properties of Structural Adhesives Used to Bond Metals. (Source, MIL-HDBK-691B.)

Adhesive	Service Temp.		Shear Strength		Peel Strength	Impact Strength	Creep Resistance	Solvent Resistance	Moisture Resistance	Type of Bond
	Max.	Min.	PSI	MPa						
Epoxy-amino	150° F 66° C	-50° F -46° C	3,000-5,000	20.7-34.5	Poor	Poor	Good	Good	Good	Rigid
Epoxy-polyamide	150° F 66° C	-60° F -51° C	2,000-4,000	13.8-27.6	Medium	Good	Good	Good	Medium	Tough and moderately flexible
Epoxy-anhydride	300° F 149° C	-60° F -15.6° C	3,000-5,000	20.7-34.9	Poor	Medium	Good	Good	Good	Rigid
Epoxy-phenolic	350° F 177° C	-423° F -253° C	3,200	22.1	Poor	Poor	Good	Good	Good	Rigid
Epoxy-nylon	180° F 82° C	-423° F -253° C	6,500	44.8	Very Good	Good	Good	Good	Poor	Tough
Epoxy-polysulfide	150° F 66° C	-100° F -73.3° C	3,000	20.7	Good	Medium	Medium	Good	Good	Flexible
Nitrile-phenolic	300° F 149° C	-100° F -73.3° C	3,000	20.7	Good	Good	Good	Good	Good	Tough and moderately flexible
Vinyl-phenolic	225° F 107° C	-60° F -51° C	2,000-5,000	13.8-34.5	Very Good	Good	Medium	Medium	Good	Tough and moderately flexible
Neoprane-phenolic	200° F 95° C	-70° F -56.7° C	3,000	20.7	Good	Good	Good	Good	Good	Tough and moderately flexible
Polyimide	600° F 316° C	-423° F -253° C	3,000	20.7	Poor	Poor	Good	Good	Medium	Rigid
Polybenzimidazole	500° F 260° C	-423° F -253° C	2,000-5,000	13.8-20.7	Poor	Poor	Good	Good	Good	Rigid
Polyurethane	150° F 66° C	-423° F -253° C	5,000	34.5	Good	Good	Good	Medium	Poor	Flexible
Acrylate acid diester	200° F 95° C	-60° F -51° C	2,000-4,000	13.8-27.6	Poor	Medium	Good	Poor	Poor	Rigid

(Continued)

Table 1. *(Continued)* Properties of Structural Adhesives Used to Bond Metals. *(Source, MIL-HDBK-691B.)*

Adhesive	Service Temp.		Shear Strength		Peel Strength	Impact Strength	Creep Resistance	Solvent Resistance	Moisture Resistance	Type of Bond
	Max.	Min.	PSI	MPa						
Cyanoacrylate	150° F 66° C	-60° F -51° C	2,000	13.8	Poor	Poor	Good	Poor	Poor	Rigid
Phenoxy	180° F 82° C	-70° F -56.7° C	2,500	17.2	Medium	Good	Good	Poor	Good	Tough and moderately flexible
Thermosetting acrylic	250° F 121° C	-60° F -51° C	3,000-4,000	20.7-27.6	Poor	Poor	Good	Good	Good	Rigid

Table 2. Advantages and Disadvantages of Adhesive Bonds.

Advantages	Disadvantages
Provide uniform distribution of stress, and increase stress bearing area.	Visual examination of the bond area is usually impossible.
Join thin or thick materials of any shape.	Require careful surface preparation, often with corrosive chemicals.
Join similar or dissimilar materials.	
Minimize or prevent electrochemical (galvanic) corrosion between dissimilar materials.	Long cure times may be required, particularly when high cure temperatures are not used.
Resist fatigue and cyclic loads.	Holding fixtures, presses, ovens, or autoclaves may be necessary.
Provide joints with smooth contours.	
Seal joints against a variety of environments.	The upper service temperature is approximately 350 °F (177° C), except for special adhesives
Insulate against heat and electrical conductance.	allowing limited use to 700° F (371° C).
Heat required is minimal, and in most instances too low to reduce the strength of metal parts.	Rigid process control, including emphasis on cleanliness, is required for most adhesives.
Dampen vibration and absorb shock.	The useful life of a joint depends on the environment to which it is exposed.
Provide good strength to weight ratio.	
Often faster or cheaper than mechanical fastening.	Natural or vegetable origin materials are subject to attack by bacteria, mold, rodents, or vermin.

Table 3. Characteristics of Adhesives by Major Type. *(Source, MIL-HDBK-691B.)*

Classification	Thermoplastic	Thermosetting	Elastomeric	Alloys
Types within group	Cellulose acetate, cellulose acetate butyrate, cellulose nitrate, polyvinyl acetate, vinyl vinylidene, polyvinyl acetals, polyvinyl alcohol, polyamide, acrylic, phenoxy	Cyanoacrylate, polyester urea formaldehyde, melamine formaldehyde, resorcinol and phenol-resorcinol formaldehyde, epoxy, polyimide, polybenzimidazole acrylic, acrylate acid diester	Natural rubber reclaimed rubber, butyl, polyisobutylene, nitrile, styrene-butadiene, polyurethane, polysulfide, silicone, neoprene	Epoxy-phenolic, epoxy-polysulfide, epoxy-nylon nitrile-phenolic, neoprene-phenolic, vinyl-phenolic
Most used form	Liquid, some dry film	Liquid, but all forms common	Liquid, some film	Liquid, paste, film
Common further classification	By vehicle (most are solvent dispersion or water emulsions)	By cure requirements (heat and/or pressure most common but some are catalyst types)	By cure requirements (all are common); also by vehicle (most are solvent dispersions or water emulsions)	By cure requirements (usually heat and pressure except some epoxy types); by vehicle (most are solvent dispersions or 100% solids); and by type of adherends or end-service conditions
Bond characteristics	Good to 150-200°F; (66-93°C) poor creep strength; fair peel strength	Good to 200-500°F; (93-260°C) good creep strength; fair peel strength	Good to 150-400°F; (66-204°C) never melt completely; low strength; high flexibility	Balanced combination of properties of other chemical groups depending on formulation; generally higher strength over wide temperature range

(Continued)

Table 3. *(Continued)* **Characteristics of Adhesives by Major Type.** *(Source, MIL-HDBK-691B.)*

Classification	Thermoplastic	Thermosetting	Elastomeric	Alloys
Major type of use	Unstressed joints; designs with caps, overlaps, stiffeners	Stresses joints at slightly elevated temperature	Unstressed joints on lightweight materials; joints in flexure	Where highest and strictest end-service conditions must be met; sometimes regardless of cost, as military uses
Materials most commonly bonded	Formulation range covers all materials, but emphasis on nonmetallics-especially wood, leather, cork, paper, etc.	For structural uses of most materials	Few used "straight" for rubber, fabric, foil, paper, leather, plastics, films; also as tapes. Most modified with synthetic resins	Metals, ceramics, glass thermosetting plastics; nature of adherends often not as vital as design or end-service conditions (i.e., high strength, temperature)

Thermoplastic Adhesives

Thermoplastics do not cross-link during cure, so they may be resoftened after heating. They are single-component systems that harden upon cooling from a melt, or by evaporating from a solvent or water vehicle. Hot-melt adhesives commonly used in packaging applications are examples of a solid thermoplastic that is applied in a molten state, and develops adhesive properties as it solidifies on cooling. Some wood glues, such as polyvinyl acetate emulsions, are thermoplastic emulsions that are common household items that harden by evaporation of water from an emulsion, but are capable of resoftening at high temperatures. Thermoplastic adhesives have a more limited temperature range than thermosetting types, and they should not be used at temperatures above 150° F (65° C). Their physical properties vary over a wide range because many polymers are used in a single adhesive formulation.

Thermosetting Adhesives

Thermosetting adhesives cannot be heated and softened repeatedly after the initial cure. These adhesives are cured by chemical reaction at room or elevated temperatures. Substantial pressure may be required with some thermosetting adhesives, and others are capable of providing strong bonds with only contact pressure. These adhesives are sometimes provided in a solvent medium to facilitate application, but they are also available as solventless liquids, pastes, and solids. They may be supplied as multiple- or single-part systems. Generally, the single-part adhesives require curing at elevated temperatures and have a limited shelf life. Multiple-part adhesives have longer shelf lives, and some can be cured at room temperature (or more rapidly at elevated temperature). However, they must be metered and mixed before application, and once the components are mixed the working life is limited. Because molecules of thermosetting resins are heavily cross linked, they have good resistance to heat and solvents, and show little elastic deformation under load at elevated temperatures.

Elastomeric Adhesives

Elastomerics are based on synthetic or naturally occurring polymers with outstanding toughness and elongation. They may be supplied as solvent solutions, latex cements, dispersions, pressure-sensitive types, and single- and multiple-part solventless liquids or pastes. The curing requirements vary with the type and form of elasto-meric adhesive. These adhesives can be formulated for a wide variety of applications, but they are generally known for their high degree of flexibility and good peel strength.

Adhesive Alloys

Adhesive alloys are made by combining thermosetting, thermoplastic, and elastomeric adhesives, and using the most useful properties of each constituent type. This means, in turn, that the adhesive alloy is usually as weak as its weakest constituent. Thermosetting resin additions provide a very strong cross-linked structure. Impact, bending, and peel strength are provided by the addition of thermoplastic or elastomeric materials. Adhesive alloys are commonly available as solvent solutions or as supported or unsupported film.

Adhesive families and selection guidelines

There are several popular methods of classifying adhesives into families or groups, and each is confused by the continual combination of two or more types of adhesives (alloying) to satisfy application requirements. The following discussion will classify specific adhesives into six groups so the designer can, to some degree, narrow the choices based on characteristics of each group.

Chemically Reactive Adhesives

Catalytic plural components–chemical cure. Adhesives of this type are normally provided in twin tubes or cans and are mixed immediately before use. In large operations, the components may be mixed in bulk to supply a work shift, or may be mixed directly in the application nozzle. Chemically reactive adhesives will harden at room temperature, but application of heat gives a better quality bond and faster cure. Reactive adhesives, including some varieties of epoxies, are available premixed, degassed, and quick frozen in small squeeze tubes and syringes. They come ready to use, packaged in dry ice, and must be retained in storage at $-40°$ F ($-40°$ C) or lower. This type of adhesive offers advantages in improved quality control, and eliminates messy two-part mixing.

Epoxy adhesives are relatively simple to use and they produce reliable bonds. Industrial epoxy adhesives may be similar to their household equivalents and come in twin tubes or cans, or they may be quite different. Epoxies are versatile, and they are particularly useful for joining dissimilar materials. They are usually chosen for specialized service where high strength is required, and a moderately high cost is acceptable. Peel strength and flexibility are low for unmodified epoxies, but they have excellent tensile-shear strength and resistance to moisture and solvents. In order to form a permanent adhesive bond, epoxy resins must be mixed with a hardener (catalyst) that chemically converts the liquid epoxy resin to a solid resin. The most important advantages of the epoxy class are that they solidify without evolution of volatile materials, and with minimal shrinkage. A wide range of properties can be obtained by use of different epoxy-hardener adhesive systems. In general, epoxy adhesives are the most versatile adhesives available, and they have excellent tensile-shear strength, and resistance to moisture and solvents. Their peel strength is, however, poor. Epoxy adhesives are used on metals, plastics, glass, wood, rubber, and ceramics.

Phenolic adhesives. Phenol-formaldehyde resins were the earliest synthetic fibers to attain commercial prominence. Although they are primarily used to mold plastic products, they are also widely used as a wood adhesive, mainly in the production of plywood. Since phenolics have good high temperature stability, they are used in blends with other resins to create alloys for bonding to metals. These bonds are exceptionally stable to their environment. Limited shelf stability before usage and the release of volatiles during cure are potential disadvantages. Phenolics are available either as a one-component heat curable liquid solution or powder, or as a liquid solution to which a catalyst must be added.

Polyurethanes are frequently used to bond "difficult" plastics, usually to dissimilar materials such as metal. Urethanes, as adhesives, resemble epoxies in cost, method of handling, and ability to bond to most surfaces. Cured polyurethane resembles a thermoplastic, as it is flexible, and rubbery, in contrast to rigid epoxies. This flexibility, combined with good adhesion, assures good bonding to flexible plastics where peel strength is important. The outstanding characteristic of urethanes is strength at severely cold (cryogenic) temperatures, which is better than for any other adhesive. The lap-shear strength, for instance, for a typical urethane adhesive with aluminum is only 1,900 PSI (12.4 MPa) at room temperature, but increases to 5,000 PSI (34.4 MPa) at $-100°$ F ($-73°$ C) and below.

"Second generation" acrylics. Older (first generation) acrylics used solutions of polymers in methacrylate monomers that were polymerized in the presence of reinforcing resin. In second generation acrylics, the adhesive is based on a combination of modifying polymers and a surface activator. The modifying polymer reinforces and toughens the bond and provides a reactive chemical site that acts as a catalyst in the presence of special activators. The system consists of two components, each being 100% solids, in fluid form, that react to form the adhesive film. Functionally, these adhesives resemble the epoxies and polyurethanes, but they cure by free-radical reaction rather than ionic polymerization. The properties of these adhesives, contrasted with other rapid thermosetting adhesives, are shown in **Table 4**. These adhesives also feature low shrinkage during curing and good environmental resistance. They are not resistant to concentrated acids, alkalies, or acetone.

Silicones have good heat stability, chemical inertness, and surface-active properties. These adhesives fall into four groups: primers, or coupling agents, used to solve difficult industrial bonding problems; adhesives and sealants; pressure sensitive adhesives; and heat cured adhesives. The use of silicones as adhesives is limited mainly by their high cost, but they are normally employed by necessity when organic materials (carbon based) are unable to withstand harsh environmental conditions, where superior reliability is required, or when the durability of silicones provides an economic advantage. The retention of flexibility and some degree of strength at temperatures ranging from cryogenic to over $500°$ F ($260°$ C) is very unusual, but the room temperature strength properties of silicones are quite low in comparison to the typical polymer. Silicone peel strength, on the other hand, is quite good.

Catalytic plural components–moisture cure. Adhesives of this type are supplied as one-component formulations, usually in collapsible tubes, to assure good protection from atmospheric moisture before use. The second requisite component, moisture, is always available—either from the surrounding air or as absorbed moisture from the surface of the substrates. The most troublesome situation where insufficient moisture may be present is in the bonding of nonporous substrates such as metals.

Silicones. RTV (room temperature vulcanizing) silicones work by reacting with atmospheric moisture and releasing acetic acid vapor which crosslinks the viscous paste into an elastomer (such as in silicone rubber adhesives).

Cyanoacrylate adhesives cure very rapidly through surface moisture, and may be used as quick-set adhesives to bond metal to metal. Lap shear strengths of 2,000 PSI (13.7 MPa) have been reported, but the resistance to moisture of these adhesives is only fair. Simple alkyl cyanoacrylates are widely distributed as household adhesives. Modern versions of these adhesives that are based on higher cyanoacrylic acid homologues

Table 4. Comparison of Properties of Acrylic and Other Adhesive Types (Room Temperature Curing). (Source, MIL-HDBK-691B.)

Property	First Generation Acrylic	Second Generation Acrylic	Epoxy	Urethane	Cyanoacrylate	Anaerobic
Lap Shear Strength[1]						
Oily, nonblasted steel	2,300-2,800 psi 15.8-19.3 MPa	2,400-2,600 psi 16.5-17.9 MPa	750-1,000 psi 5.17-6.89 MPa	0	NA	600-800 psi 4.13-5.52 MPa
Oily, nonblasted steel 350°F (177°C) 30 min. heat cycle	650-900 psi 4.48-6.20 MPa	2,800-3,000 psi 19.3-20.6 MPa	1,000-1,200 psi 6.89-8.27 MPa	700-1,000 psi 4.82-6.89 MPa	NA	NA
Solvent-cleaned Nonblasted steel	1,750-2,200 psi 12.1-15.2 MPa	2,400-3,000 psi 16.5-20.6 MPa	1,800-2,100 psi 12.4-14.5 MPa	550-650 psi 3.79-4.48 MPa	200 psi 1.38 MPa	NA
Solvent-cleaned Nonblasted steel	3,000-3,700 psi 20.6-25.5 MPa	3,300-3,500 psi 22.7-24.1 MPa	1,800-2,100 psi 12.4-14.5 MPa	650-1,050 psi 4.48-7.24 MPa	4000 psi 27.6 MPa	2,150-2,550 psi 14.8-17.6 MPa
Etched aluminum	3,700-4,000 psi 25.5-27.5 MPa	3,400-3,600 psi 23.4-24.8 MPa	2,000-2,400 psi 13.8-16.5 MPa	1,750-2,050 psi 12.1-14.1 MPa	3000 psi 20.0 MPa	NA
Solvent-cleaned aluminum	900-1,800 psi 6.21-12.4 MPa	2,000-2,200 psi 13.8-15.2 MPa	1,600-1,800 psi 11.0-12.4 MPa	375-400 psi 2.58-2.76 MPa	1000 psi 6.89 MPa	950-1,500 psi 6.55-10.7 MPa
Impact Strength[2]						
Room temperature	10-20 ft.-lb./in.^{2} 21,010-42,040 J/m^2	14-16 ft.-lb./in.^{2} 29,430-33,630 J/m^2	10-12 ft.-lb./in.^{2} 21,010-25,220 J/m^2	5-11 ft.-lb./in.^{2} 10,510-23,120 J/m^2	3-5 ft.-lb./in.^{2} 6,306-11,000 J/m^2	3-5 ft.-lb./in.^{2} 6,306-11,000 J/m^2
-20°F (-29°C)	3.5-4.5 ft.-lb./in.^{2} 7,357-9,459 J/m^2	10-14 ft.-lb./in.^{2} 21,020-29,430 J/m^2	3-5 ft.-lb./in.^{2} 6,306-10,510 J/m^2	2.5-5.3 ft.-lb./in.^{2} 5,255-11,140 J/m^2	NA	NA
0°F (-18°C)	5-8 ft.-lb./in.^{2} 10,510-16,016 J/m^2	11-14 ft.-lb./in.^{2} 23,000-29,428 J/m^2	5-6 ft.-lb./in.^{2} 10,510-12,612 J/m^2	2.5-3.5 ft.-lb./in.^{2} 5,255-7,357 J/m^2	NA	NA
180°F (82°C)	10-20 ft.-lb./in.^{2} 21,000-42,000 J/m^2	11-13 ft.-lb./in.^{2} 23,000-27,000 J/m^2	5-6 ft.-lb./in.^{2} 11,000-13,000 J/m^2	4-5.5 ft.-lb./in.^{2} 8,400-12,000 J/m^2	NA	NA

(Continued)

Table 4. (*Continued*) Comparison of Properties of Acrylic and Other Adhesive Types (Room Temperature Curing). (*Source, MIL-HDBK-691B.*)

Property	First Generation Acrylic	Second Generation Acrylic	Epoxy	Urethane	Cyanoacrylate	Anaerobic
			T-Peel Strength[3]			
Solvent-cleaned steel	24-26 lb/in 4,200-4,550 N/m	25-30 lb/in 4,375-5,250 N/m	5-10 lb/in 875-1,750 N/m	29-30 lb/in 5,075-5,250 N/m	NA	1.5 lb/in 262 N/m
Etched aluminum	8-6 lb/in 1,400-2,800 N/m	25-30 lb/in 4,375-5,250 N/m	15-20 lb/in 2,620-3,510 N/m	36-38 lb/in 6,300-6,650 N/m	NA	1-2 lb/in 175-350 N/m
			Set Time			
Room temperature	120 minutes	2-4 minutes	120-180 minutes	15 minutes	0.5 minutes	1 minute

[1] Lap shear test as defined in ASTM D1002. [2] Impact test as defined in ASTM D950. [3] T-peel test as defined in ASTM D1876.

such as ethyl, propyl, and butyl ethers offer very fast bond times (30 seconds or less) on a variety of materials including metal, high bond strength, no mixing, bonding at room temperature with contact pressure, and very little shrinkage. Although they are relatively high in price, only a few milligrams are needed for each bond. The key step in the successful application of a cyanoacrylate adhesive is the formation of a thin adhesive film between two well-mated surfaces—the thinner the film, the faster the rate of bond formation and the higher the bond strength. These adhesives have good resistance to various solvents and good bond aging properties, but when compared with other structural adhesives they have relatively poor impact resistance due to the fact that the bond is very thin and inflexible. Shear strengths of selected substrates bonded with methyl 2-cyanoacrylate are shown in **Table 5**.

Heat activated (one-part) system adhesives eliminate the need to mix metered amounts of the basic components, and may be supplied as solid films for uniform bonding thickness. These systems do require elevated temperature cures and may have short shelf lives.

Table 5. Shear Strengths of Selected Substrates Bonded with Methyl 2-Cyanoacrylate Adhesive.
(Source, MIL-HDBK-691.)

Substrate	Shear Strength	
	PSI[1]	MPa[1]
Aluminum-aluminum	2700	18.6
Steel-steel (cold rolled)	3000 [2]	20.7 [2]
Aluminum-steel	1750 [3]	12.2 [3]
Acrylic-acrylic	590 [4]	4.04 [4]
Rigid vinyl-rigid vinyl	900 [4]	6.21 [4]
Polyester glass-polyester glass	1100	7.59
Phenolic-phenolic	770 [4]	5.31 [4]
Nylon 6.6-nylon 6.6	650 [4]	4.52 [4]
Nylon 6.6-aluminum	500	3.44
Phenolic-aluminum	650	4.46
Butyl rubber-butyl rubber	70 [4]	0.48 [4]
SBR rubber-SBR rubber	90 [4]	0.62 [4]
Neoprene-neoprene	50 [4]	0.34 [4]
Natural rubber-natural rubber	50 [4]	0.34 [4]
Neoprene-phenolic	160 [4]	1.10 [4]
Butyl rubber-acrylic	170 [4]	1.10 [4]
Butyl rubber-polyester glass	100 [4]	0.69 [4]
SBR-polyester	110 [4]	0.759 [4]
Butyl rubber-steel	50 [4]	0.34 [4]
Butyl rubber-aluminum	70 [4]	0.486 [4]

Notes:
[1] Average lap-shear strength of 5 specimens aged for 24 hours at 73° F (23° C) following bond formation.
[2] With surface activator, ten minute cure.
[3] Without surface activator, ten minute cure.
[4] Substrate failure.

Polybenzimidazoles (PBI) were developed to fill the need for high temperature organic polymers, and *polymides (PI)* are analogous polymers competing for the same applications. PBIs are thermoplastics, even though their thermoplastic nature is not evident below 700° F (371° C), and they are relatively stable to air to 500° F (288° C) for short term applications. They can be successfully employed on stainless steels, beryllium, and titanium, including the assembly of honeycomb structures for space vehicles. In addition to high cost, they also have some moisture sensitivity at room temperature, and lap-shear strength drops gradually on heating to 600° F (316° C), and more rapidly above this temperature. PIs have superior long term strength retention qualities, and excellent properties for shear strength joints exposed to elevated temperatures.

Epoxies. If the hardener component of an epoxy adhesive is a meltable solid rather than a liquid, it may, in some cases, be included in a formulation supplied as a latent one-component adhesive. Dicyandiamide (DICY) is the standard hardener supplied in this manner, and it has a cure schedule of one hour at 250° F (121° C). A second hardener of this type is boron trifluoride-amine, which is not as latent or as effective in curing as DICY. Typically, one-component DICY epoxy polymers can achieve tensile strength of 11,000 PSI (76 MPa), shear strength of 7,000 PSI (8.3 MPa), and flexure strength of 13,000 PSI (90 MPa).

Nylons form the high molecular weight members of the family of polyamide resins. Although these polymers have excellent properties, with tensile strengths approaching 12,000 PSI (83 MPa), and elongation of 300% maximum, their high softening points and high melt viscosities limit their use. However, some specialty nylon resins having lower melting temperatures are used quite successfully for application by extrusion for high strength metal-to-metal bonding. Because epoxies and nylon are not truly compatible at room temperature and must be "forced" together in formulating the adhesive, the products are normally supplied as flexible films.

Polyvinyl acetal and *polyvinyl formal*, when blended with phenolics, form structural adhesives that are useful in flexibilizing thermoset resins to obtain structural adhesives for metals.

Evaporation or Diffusion Adhesives

Solvent based systems. Solvent based adhesives are supplied by coating the resin solution on a substrate, followed by some combination of solvent evaporation and contact with the second adherend, which may or may not be coated. With porous materials, such as wood or paper, the contact usually precedes the final drying step, and the last traces of solvent are absorbed by the substrate. Metals and nonporous plastics require essentially complete removal of solvent, so that good contact will be maintained during the bonding process, during which heat and pressure are used to activate the adhesive coating. The advantages of solvent systems are good water resistance, high heat strength, and ease of wetting some substrates. Disadvantages are toxicity, and they present a fire hazard.

Natural rubber and reclaimed rubber are suitable for pressure sensitive applications and for bonding nonmetallics such as leather, textile, and paper. Pressure sensitive applications include bonding nonmetallic trim to metal automotive bodies. These adhesives are susceptible to attack by solvents, oils, and chemicals.

Synthetic rubbers

Nitrile rubber is a copolymer of butadiene and acrylonitrile that offers good resistance to oil and grease. It is used to bond vinyl, other elastomers, and fabrics where good wear and oil and water resistance are important. Nitrile rubber is alloyed with phenolics and used to bond aluminum and stainless steel laminates. In bonding metals, tensile strengths up to 4,000 PSI (28 MPa) at 70° F (21° C), and 1,000 PSI (7 MPa) at 230° F (160° C) are achieved with nitrile-phenolic adhesives.

Neoprene (polychloroprene) is used extensively to bond aluminum. Although neoprene is similar in properties to natural rubber, it generally provides higher strength, better aging, and higher temperature resistance. For structural applications, neoprene is usually combined with a phenolic resin.

Phenolics. The combination of phenolics with synthetic rubber is discussed immediately above. For wood bonding (plywood), dilution of the phenol with an organic solvent is advantageous for handling, mixing, and coating. Other uses of solvent solutions of phenolic resins are in the preparation of wood particle boards, the bonding of glass fiber, and in aircraft honeycomb construction.

Urethanes may be diluted with dry solvents such as methylene for easy application. The adhesive can be coated on the substrate and the solvent allowed to evaporate, and this process works well for bonding rubber to fibers, rubber to metals, and for wood assemblies. Thermoplastic urethanes are also available for use in solvent based adhesives.

Vinyl resins. Several vinyl monomers (monomers are the individual units of molecules that, when linked together, form a polymer chain) are used to prepare thermoplastics for adhesive applications. Thermoplastic resins do not cure when heated. Instead, they develop joint strength by solvent evaporation, diffusion, or fusion. The most important vinyl resins for adhesives are polyvinyl acetate, the polyvinal acetals (butyral and formal), and polyvinyl alkyl ethers. Polyvinyl acetate resins can be dissolved in alcohols, ketones, and esters for coating on substrates. Polyvinyl acetal resins, in the form of polyvinyl butyral, are used for laminating safety glass. Acetal-phenol mixtures (commonly called vinyl phenolics) are used as structural adhesives for metal honeycomb, and for metal-to-wood laminates.

Acrylic resins are used for bonding cloth, plastics, leather, and metal foils. The advantages of acrylics over vinyls for adhesive use are that they are stable in sunlight, some remain flexible at low temperatures, some are oil resistant, and they are readily modified with crosslinkable monomers so that heat cured adhesives with improved solvent resistance can be obtained.

Water based systems. Water solutions, dispersions, or emulsions are widely used for application and bonding of various substrates. Water systems are most frequently used when at least one surface to be bonded is porous, allowing the water to escape by permeation. When compared to organic solvents, water's major advantage is cost. Other positive characteristics are nonflammability, low toxicity, and the wide range of viscosity and solids contents that may be obtained. Disadvantages include poor moisture resistance, slow drying, a tendency to freeze, and shrinkage of fabric substrates.

Natural rubber is isolated as a water dispersion known as latex, which may be used directly as an adhesive. Usually, latex is blended with other adhesive raw materials such as phenolics, casein, or natural tackifiers such as Kayara gum to improve bonding.

Synthetic rubbers, discussed above under "solvent based systems," are also available as emulsions in water. These adhesives are primarily used to bond paper, textile, and leather.

Vinyl resin. Polyvinyl acetate, the most widely used vinyl resin in water dispersion, is the basis for common household "white" glue. It can be used on paper, wood, leather, and cloth.

Acrylics are prepared commercially by emulsion polymerization to produce acrylic latex products.

Hot Melt Types

Hot melt adhesives, when contrasted to older techniques, offer increased production rates and simplicity in handling the adhesive. A hot melt may be defined as a 100% solids thermoplastic adhesive polymer. Hot melt materials are (ideally) solid up to 175° F (70° C) or higher, then melt sharply to give a low viscosity fluid that is readily applied and capable of wetting the surface to be bonded, followed by rapid setting upon cooling. Application temperatures for hot melts range from about 300 to 500° F (149 to 280° C). Despite the fact that all hot melts are based on only a few polymers, available commercial products number in the hundreds, making comparisons between competing brands difficult. From a designer's standpoint, one of the most important characteristics of hot melt adhesives is service temperature. Because they have relatively low melting temperatures, service temperatures are also low. Another limitation, also derived from their thermoplastic nature, is that they creep under load and with time.

Ethylene-vinyl acetate (EVA) and *polyolefin resins*. Hot melts based on these resins are the lowest cost materials. These are general purpose, low molecular weight adhesives for bonding substrates of paper, cardboard, wood, fabric, and similar materials. Their strength is adequate at temperatures ranging from –30 to +120° F (–34 to +49° C), and compounded versions are available that are suitable for nonloadbearing applications to about 160° F (71° C).

Polyamide (nylon) and *polyester resins* represent the next step up in strength and general service. Compounds of these materials, which are generally referred to as "high performance hot melts," are used to assemble products made from plastics, glass, hardboard, wood, fabric, foam materials, leather, hard rubber, and some metals. Various compounds made from these resins maintain adequate bonding characteristics in the temperature range of –40° F (–40° C) to above 180° F (62° C), and some formulations can be used over 200° F (93° C) and in no-load applications up to 300° F (149° C). Polyester based hot melts are generally stronger and more rigid than the nylon compounds. This is because the polyesters have high crystallinity, which gives them sharp melting points, often coupled with high tensile strength and good elongation. Both nylon and polyester hot melts are sensitive to moisture during application, but each in a different manner. Nylon compounds have the advantage of good strength and flexibility, but if not stored in a dry area, they may pick up moisture. Then, during application, the absorbed moisture causes foaming which, in turn, produces voids in the applied layer. Moisture affects polyester melts by entering their molecular structure. Then the longer the adhesive is heated, the more the molecular weight is reduced and the lower the viscosity becomes. The consequences make polyesters unsuitable for dispensing by means of a reservoir style system.

Thermoelastic elastomer hot melts are made up of polyurethanes and styrene-butadiene-styrene and styrene-isoprene-styrene polymers. They are widely used in pressure sensitive applications but, as a class, they are not as strong as the polyesters, but they are stronger than conventional thermosetting rubbers. They offer good flexibility and toughness for applications requiring dynamic endurance, vibration endurance, and vibration resistance, and they have good wetting characteristics. As a class they are very viscous, with high molecular weights, which often makes them difficult to apply unless formulated with other ingredients.

Delayed Tack Adhesives

A delayed tack adhesive differs from both a hot melt adhesive and a hot seal adhesive. A hot melt will be nontacky when cooled, while a hot seal adhesive will seal under heat and pressure, but does not remain tacky for any period of time after cooling. A delayed tack system, after it is heat activated and cooled, will remain tacky for anywhere from several minutes to several days over a wide temperature range. To obtain the delayed tack characteristics, a solid plasticizer, which also contributes to nontackiness before heat activation, must be used. The most common such plasticizers are dicyclohexyl phthalate, diphenyl phthalate, N-cyclohexyl-p-toluene-sulfonamide, and o/p-toluenesul-fonamide. Adhesives with different heat activation temperatures are possible because of the range of melting points, and the most commonly used resins are styrene-butadiene copolymers, polyvinyl acetate, polystyrene, and polyamides.

Film and Tape Adhesives

Bonds produced by a given adhesive will usually have similar properties, whether the adhesive is applied as a one-compound paste or as a solid film. The advantages of using the adhesive in film form are that it assures a uniform, controlled glue line thickness; it is fast and easy to apply (a clean, solvent free operation is permitted); and two-sided films can be prepared (these are used in lightweight sandwich construction—the honeycomb side of the adhesive provides good filleting, the skin side will provide high peel strength; if one side of the film is tacky, it will be easier to align the assembly to be bonded). Sometimes the term film adhesive is applied to the high molecular weight polymers, normally elastomers, blended with curing agents, fillers, and other ingredients, and then extruded, calendered, or cast into 5 to 15 mil unsupported thick films. When the mixture is cast or calendered onto a mesh support, such as a woven or nonwoven mesh of glass or other fibers, the result is a tape adhesive. Films and tapes may be either soft and tacky, or stiff and dry. They may be room temperature storable or require refrigeration. Film and tape adhesives differ from paste and liquid adhesives in that the former contain a high proportion of a high molecular weight polymer. The 100% solids paste and liquid adhesives contain only low molecular weight resins because that is the only way they can remain fluid and useable. The best of the tape and film types have higher peak values and broader surface temperature ranges than the best of the 100% solids. See *Figure 1* for typical tensile shear strength at temperature ratings for 100% solids adhesives, and *Figure 2* to compare those values with similar data for tape and film adhesives. Tape and film adhesives dominate the airframe market for structural adhesives, where it is sometimes combined with mechanical fastening. The fact that these adhesives seal as well as fasten makes possible air- and liquid-tight pressurized cabins, fuel, and hydraulic systems. **Table 6** provides chemical and strength properties for popular film and tape adhesives.

Vinyl-phenolics. The "vinyl" in the name is somewhat misleading, as it refers to either polyvinyl formal (PVF) or polyvinyl butyral (PVB), one of which forms the backbone polymer. These adhesives have generally excellent durability, both in water and in other adverse environments. Curing takes place at 350° F (177° C). Newer adhesives have lower cure temperatures, which has led to a trend away from vinyl-phenolics. These adhesives are also available in solvent solutions and emulsions, liquid, and coreacting powder form.

Epoxy-phenolics account for a small fraction of the present usage of structural adhesives, but are important for service applications including operational temperatures of 300 and 500° F (149 and 260° C). They are still considered to be among the best adhesives for long term use at these temperatures, but other adhesives are superior to 350° F (177° C). These adhesives are also available as a two-part paste.

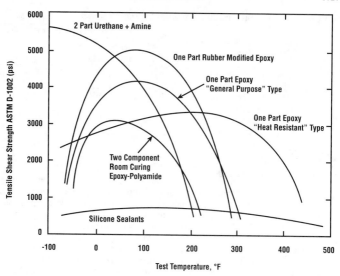

Figure 1. Typical tensile shear strength at temperature data for paste and liquid (100% solids) adhesives.

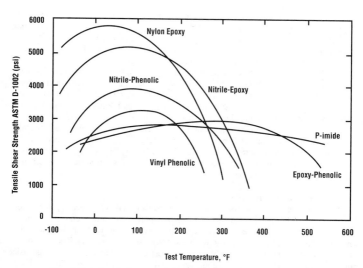

Figure 2. Typical tensile shear strength data for film, tape, and solvent based adhesives.

Table 6. Composition and Physical Properties of Selected Film and Tape Adhesives. *(Source, MIL-HDBK-691.)*

Type	Backbone Polymer	Crosslinking Resin	Catalyst	High Pressure Cure?	Tensile Shear Strength at Room Temperature		T-Peel Strength at Room Temperature	
					KSI	MPa	lb/in.	N/m
Nylon-epoxy	Soluble nylon	Liquid epoxy	DICY type	No	5.5–7.2	39–49	80–130	14,000–22,750
Elastomer-epoxy	Nitrile rubber	Liquid epoxy	DICY type	No	3.7–6.0	26–41	22–90	3,850–15,750
Nitrile-phenolic	Nitrile rubber	Phenolic novolac	Hexa-sulfur	Yes	3.0–4.5	21–31	15–60	2,625–10,500
Vinyl-phenolic	PVB or PVF	Resol phenolic	Acid	Yes	3.0–4.5	21–31	15–35	2,625–6,065
Epoxy-phenolic	Solid epoxy	Resol phenolic	Acid	Yes	2.0–3.2	14–22	6–12	1,050–2,100

Nitril-phenolics are made by blending a nitrile rubber with a phenolic novalac resin, plus other compounding ingredients. Their major uses are in the automotive industry in bonding brake shoes and clutch disks, but they are also used in aircraft assembly and other applications primarily due to their low cost and high bond strengths at temperatures up to 250° F (121° C). They have exceptional bond durability on steel and aluminum, and superior durability after extended salt spray, water immersion, and in other corrosive environments. Their major disadvantages are the need for high pressure cure (up to 200 PSA [1.38 MPa]), when the trend is toward lower pressures, and the need for high temperature (300° F [149° C]) long term cure, when the trend is toward lower temperature curing. These adhesives are also available in solvent solutions.

Nylon-epoxies first became prominent in the early 1960s because of the excellent properties they exhibited for that time. However, they have several drawbacks that limit their use, including loss in peel strength at low temperatures; poor creep resistance; and extreme sensitivity to moisture. Considerable efforts, often involving primers, have been made to solve the moisture problem, but without great success.

Elastomer-epoxies. These tape or film adhesives are toughened with an elastomer (usually a nitrile rubber). This feature as well as short and low temperature curing properties have made these adhesives popular in the aircraft industry. The bond durability of these high-peel strength elastomer epoxies is satisfactory, but does not match the durability of the older vinyl-phenolic or nitrile-phenolic types.

Neoprene-phenolics have high resistance to creep and to most service environments, plus excellent fatigue and impact strengths. Curing is under heat and pressure, and normal service temperatures are −70 to 200° F (−57 to 93° C). Shear strength, however, is lower than that of the other modified phenolic adhesives. These adhesives are also available in solvent solutions.

Pressure Sensitive Adhesives (PSAs)

A pressure sensitive adhesive is a substance capable of holding objects together when their surfaces are brought into contact under briefly applied pressure at room temperature. Generally, they are composed of a rubbery type polymer that has been modified with several additives such as plasticizers, tactifiers, fillers, solvents, and stabilizers. In order of decreasing volume and, generally, increasing price, the following are used as basic polymeric raw materials: natural rubber, styrene-butadiene rubber, reclaimed rubber, butyl rubber and butadiene-acrylonitrile rubber, polyacrylates, and polyvinyl alkyl ethers. A disadvantage of using the rubbers in PSAs is poor aging, which results in a loss of elasticity and tack, and a yellowing of the adhesive layer. When rigid, high performance standards are required for a PSA, polyvinyl alkyl ethers or polyacrylates are the polymers of choice.

Care and application of adhesives

Storage. Manufacturer's recommendations must be followed carefully when storing adhesives. Many adhesives must be stored in the dark or in opaque containers, while others must be maintained at low temperature. For example, one epoxy-phenolic has the following storage schedule for newly prepared adhesive film. At $-10°$ F ($-23°$ C), useful life is 180 days. At $30°$ F ($-1°$ C), useful life is 75 days. At $85-$ F ($29°$ C), useful life is 3 days.

Basic resins and curing agents for thermosetting adhesives should be kept apart so that accidental container breakage will not lead to contamination problems. Containers for solvent based adhesives should be sealed as soon as possible after use to prevent solvent evaporation or the escape of toxic or flammable vapors.

Preparation. Adhesive preparation requires careful attention. After cold storage, the adhesive must be warmed to the correct temperature for application, which is normally room temperature but, in some cases—as with hot melt adhesives—may be considerably higher. Where component mixing is required, it is important to measure the proportions correctly in order to obtain an optimum bond. This is especially true with catalytic reactions, such as amine curing agents for epoxy resins, where insufficient catalyst will result in incomplete polymerization of the binder resin, and too much catalyst can lead to brittleness in the cured material. Weighed components of multipart adhesives must be mixed thoroughly—mixing should be continued until no color streaks or density stratifications are noticeable. Overagitation should be avoided as it may introduce air into the mixture and lead to foaming during the heat cure. For large scale bonding operations, hand mixing is costly, and mixing repeatability is difficult to assure.

Application. The selection of the application method depends on the form of the adhesive, the size and shapes of the parts to be joined, the areas where the adhesive is to be applied, and production volume and rate.

Liquid adhesives offer the largest variety of application methods, but brushing, flowing, spraying, and roll coating are the methods most often employed. Brushing is appropriate for adherends with complex shapes. It requires a minimum of equipment and little waste, but it is not well suited to rapid assembly work, and the thickness of the adhesive is difficult to control. Flowing is especially useful when joining irregularly shaped flat shapes. The adhesive is fed under pressure through a simple nozzle (flow gun) or brush (flow brush) and is applied in a simple smooth sweep. It provides better thickness control and higher production rates than brushing. Spraying is the method of choice for applying adhesive to large areas with nonuniform contours. It allows for high production rates, and uniform thickness, but the solids content of the adhesive must be closely monitored. Applicators are similar to paint spray guns, so potentially toxic solvent mists are possible. Adequate ventilation must be provided. Roll coating is analogous to painting with a roller, and it works best on flat surfaces.

Paste adhesives are the simplest and most reproducible adhesives to apply. Little operator skill is required, application is usually clean, and little waste is experienced. Pastes may be applied either with a device such as a caulking gun, or with spatulas and trowels. Spatulas work well on hard-to-spread adhesives, and trowels are useful for controlling the thickness of the applied paste.

Powder adhesives are generally one-part epoxy based systems that require heat and pressure to cure. They may be applied by sifting onto a preheated substrate; by dipping the preheated substrate into the adhesive; or by conventional means if the powder is first melted into a liquid or paste.

Film adhesives are restricted to use on flat surfaces or simple curves, and application requires a degree of care to resist wrinkling, and to remove the separation sheets. They are supplied in heat activated, pressure sensitive, or solvent activated forms, and application will vary with each. Solvent activated types are made tacky and pressure sensitive by wiping them with solvent, and are best suited for contoured or irregularly shaped parts. Film adhesives provide many advantages, including ease of handling, minimal waste, high repeatability, and excellent physical properties.

Hot melts are normally applied through two basic delivery systems: melt reservoirs and progressive feed systems. Melt reservoir systems consist of a melting pot, a pump, feed hose, and an extrusion gun or applicator wheel. One of the main advantages of reservoir systems is that they use adhesives in their cheapest form. Granular or pellet adhesives in bulk quantities normally cost no more than one-third the price of the same adhesives in cartridges. Drawbacks of the reservoir system are that it may be difficult to maintain a uniform temperature due to the addition of fresh solid adhesive at periodic intervals. Progressive feed systems heat only a small amount of adhesive at a time. These systems may use a small hopper that transmits melted adhesive to a heated gun, or a self-contained gun that melts the adhesives within the gun.

Adhesive thickness. It is desirable to have a uniform bond line with a thickness ranging from 0.002" to 0.010" (0.05 mm to 0.25 mm). Starved adhesive joints, where some areas have no adhesive, will result in poor bonds, so steps must be taken to control the thickness. Mechanical shims, which may be removed after curing, will assure thickness and allow the flow of uncured adhesives. In some instances, it may be possible to design stops into the joint, which will serve the same function as shims. Employing a film adhesive that becomes highly viscous during the cure cycle will prevent excessive adhesive flow out of the joint. With supported films, the adhesive carrier itself can act as a shim. With film adhesives, the bond-line thickness can normally be determined by the original thickness of the adhesive film. With most adhesives, trial and error can be used to establish the proper pressure-adhesive-viscosity factors for producing the desired bond thickness.

Casting Processes

Selecting the casting process is an important element in the design cycle even though, in some cases, it is a decision that is left to the foundry. More often than not, the process to be used emerges logically from the product's size, shape, and technical requirements. This section, most of which was prepared by the Copper Development Association, provides guidelines for selecting a casting process to meet part requirements.

Among the more important factors that influence the choice of casting method are:

- the number of castings to be made
- the size and/or weight of the casting
- the shape and intricacy of the product
- the amount and quality of finish machining needed
- the required surface finish
- the prescribed level of internal soundness (pressure tightness) and/or the type and level of inspection to be performed
- the permissible variation of dimensional accuracy for a single part, and part-to-part consistency through the production run, and
- the casting characteristics of the copper alloy specified.

Other considerations, such as code requirements, can also play a role in selecting the casting process, but it is primarily the number and size of castings required, along with the alloy chosen, that determine how a casting will be made.

That is not to say that the designer has little choice; in fact, quite the opposite can be true. For example, small parts made in moderate to large quantities frequently lend themselves to several processes, in which case factors such as surface finish, soundness, or mechanical properties will bear strongly on the choice of method used. These parameters are set by the designer.

It is convenient to classify the casting processes as being applicable either to general shapes of more or less any form or to specific and usually rather simple shapes. In addition, several new special processes have been commercialized in recent years, one of which is described below.

Processes for general shapes

Sand casting

Sand casting is relatively inexpensive, acceptably precise, and, above all, highly versatile. It can be utilized for castings ranging in size from a few ounces to many tons. Further, it can be applied to simple shapes as well as castings of considerable complexity, and it can be used with all of the copper casting alloys.

Sand casting imposes few restrictions on product shape. The only significant exceptions are the draft angles that are always needed on flat surfaces oriented perpendicular to the parting line.

Dimensional control and consistency in sand castings ranges from about ± 0.030 to ± 0.125 in (± 0.8 to 3.2 mm). Within this range, the more generous tolerances apply across the parting line. Surface finish ranges between approximately 300 to 500 μin (7.7 – 12.9 μm) rms. With proper choice of molding sands and careful foundry practice, surprisingly intricate details can be reproduced. There are a number of variations on the sand casting process.

In *green sand* casting—still the most widely used process—molds are formed in unbaked (green) sand, which is most often silica, SiO_2, bonded with water and a small amount of a clay to develop the required strength. The clay minerals (montmorillonite, kaolinite) absorb water and form a natural bonding system that holds the sand particles together. Various sands and clays may be blended to suit particular casting situations.

The mold is made by ramming prepared sand around a *pattern*, held in a *flask*. The patterns are withdrawn, leaving the *mold cavity* into which metal will be poured. Molds are made in two halves, an upper portion, the *cope*, and a lower portion, the *drag*. The boundary between cope and drag is known as the *parting line*.

Cores, made from sand bonded with resins and baked to give sufficient strength, may be supported within the mold cavity to form the internal structure of hollow castings. *Chills* of various designs may be embedded in the mold cavity wall to control the solidification process.

Risers are reservoirs of molten metal used to ensure that all regions of the casting are adequately fed until solidification is complete. Risers also act as heat sources and thereby help promote directional solidification. Molten metal is introduced into the mold cavity through a sprue and distributed through a system of *gates* and *runners*. *Figure 1* shows the sequence of steps used to make a typical sand casting. Note how the gates, runners, and, risers are situated to ensure complete and even filling of the mold.

Bench molding operations are performed by hand. Quality and part-to-part consistency depend largely on the skill of the operator. The labor-intensive nature of bench molding usually restricts it to prototypes or short production runs. Patterns are another significant cost factor, especially if their cost cannot be amortized over a large number of castings. Still, bench molding remains the most economical method when only a few castings must be produced.

The *machine molding* method is automated and therefore faster than bench molding, but the casting process is essentially similar. Molding machines sling, ram, jolt, or squeeze sand onto patterns, which in this case may consist of several parts arranged on a mold board. Dimensional control, surface finish, and product consistency are better than those achievable with bench molding. Favorable costs can be realized from as few as several dozen castings. Machine-molded sand casting is therefore the most versatile process in terms of production volume.

Waterless molding aims to eliminate the sometimes detrimental effects of moisture in the molding sand. Clays are treated to react with oils rather than water to make them bond to the sand particles. The hot strength of the waterless-bonded sand is somewhat lower than that of conventional green sands. This reduces the force needed to displace the sand as the casting shrinks during solidification, which in turn reduces the potential for hot tearing. On the other hand, sands with low hot strength have a greater tendency to be damaged by hot metal flowing into the mold.

For large castings, molds may be baked or partially dried to increase their strength. The surfaces of skin-dried molds are treated with organic binders, then dried by means of torches or heaters. To make dry sand molds, simple organic bonding agents such as molasses are dissolved in the bonding water when making up the green sand mixture. The entire mold is then baked to develop the desired hot strength. Besides hardening the mold, removing water also reduces the chance for blowholes and other moisture-related casting defects. Baking and skin drying are expensive operations and the dry sand methods are rapidly being replaced by a variety of *no-bake processes*, described below.

There are three general types of low-temperature-curing, chemical binders. *Cement* has traditionally been used as a bonding agent in the extremely large molds used to cast marine propellers and similar products. Cement-bonded molds are extremely strong and durable, but they must be designed carefully since their inability to yield under solidification shrinkage stresses may cause hot tearing in the casting.

Organic binders utilize resins that cure by reaction with acidic catalysts. Furan-, phenolic-, and urethane-base systems are the most popular of the large variety of currently

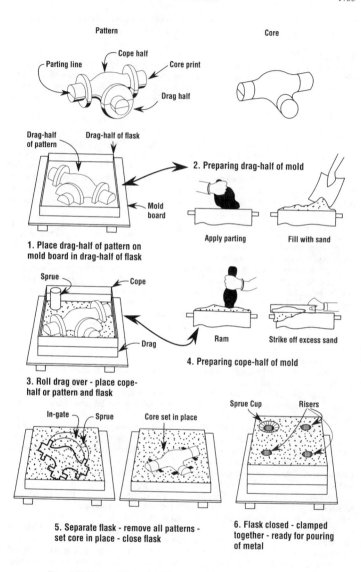

Figure 1. Making a mold for sand casting. (Source, Copper Development Association.)

available bonding agents. Of the *inorganic binders*, the well-known liquid sodium silicate-CO_2 process is most widely used for copper alloy castings.

No bake (air set)

In this process, silica sand is mixed with a resin that hardens when exposed to the atmosphere. The process requires no water. It can be used for molds as well as cores. It is applicable to products as small as 20 lb (9 kg), although it is mainly used for large castings weighing up to 20,000 lb (9,100 kg). The no-bake process has become very popular in the past 10 years.

Shell molding

Resin-bonded sand systems are also used in the shell molding process, in which prepared sand is contacted with a heated metal pattern to form a thin, rigid shell as shown in *Figure 2*. As in sand casting, two mating halves of the mold are made to form the mold cavity. Common shell-molding binders include phenolformaldehyde resins, furan or phenolic resins and baking oils similar to those used in cores. Nonbaking resins (furans, phenolics, urethanes) are also available; these can claim lower energy costs because they do not require heated patterns.

The shell molding process is capable of producing quite precise castings and nearly rivals metal-mold and investment casting in its ability to reproduce fine details and maintain dimensional consistency. Surface finish, about 125 μin (3.2 μm) rms, is considerably better than that from green sand casting.

Shell molding is best suited to small-to-intermediate size castings. Relatively high pattern costs (pattern halves must be made from metal) favor long production runs. On the other hand, the fine surface finishes and good dimensional reproducibility can, in many instances, reduce the need for costly machining. While still practiced extensively, shell molding has declined somewhat in popularity, mostly because of its high energy costs compared with no-bake sand methods; however, shell-molded cores are still very widely used.

Plaster molding

Plaster molds can be used to produce precision products of near-net shape. Plaster-molded castings are characterized by surface finishes as smooth as 32 μin rms and dimensional tolerances as close as ± 0.005 in. (± 0.13 mm), and typically require only minimal finish machining. In some cases, rubber patterns can be used. These have the advantage of permitting reentrant angles and zero-draft faces in the casting's design.

Gypsum plaster ($CaSO_4$) is normally mixed with refractory or fibrous compounds for strength and specific mechanical properties. The plaster must be made slightly porous to allow the escape of gases as the castings solidify. This can be achieved by *autoclaving* the plaster molds in steam, a technique known as the Antioch process. This produces very fine cast surfaces suitable for such precision products as tire molds, pump impellers, plaques and artwork. It is relatively costly.

Foaming agents produce similar effects at somewhat lower costs. Labor cost remains relatively high, however. Foamed plaster molds produce very fine surface finishes with good dimensional accuracy, but they are better suited to simple shapes.

Most plaster mold castings are now made using the Copaco process, which utilizes conventional wood or metal patterns and gypsum-fibrous mineral molding compounds. The process is readily adapted to automation; with low unit costs, it is the preferred plaster-mold method for long production runs. On the whole, however, plaster molding accounts for a very small fraction of the castings market.

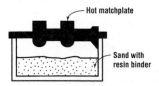

1. A heated metal matchplate is placed over a box containing sand mixed with thermoplastic resin.

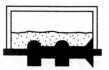

2. Box and matchplate are inverted for short time. Heat melts resin next to matchplate.

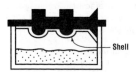

3. When box and matchplate are righted, a thin shell of resin bonded sand is retained on the matchplate.

4. Shell is removed from the matchplate.

5. Steps 1 through 4 are repeated using the other side of matchplate.

6. The shells are placed in oven and "heat treated" to thoroughly set resin bond.

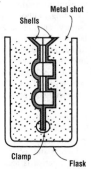

7. Shells are clamped together and placed in a flask. Metal shot or coarse sand is packed around the shell, and mold is ready to receive molten metal.

Figure 2. The shell molding process. (Source, Copper Development Association.)

Refractory molds

Of the several refractory-mold-based methods, the Shaw process is probably the best known. Here, the wood or metal pattern halves are dipped into an aggregate slurry containing a methyl silicate binder, forming a shell. After stripping the pattern, the shell is fired at a high temperature to produce a strong refractory mold. Metal is introduced into the mold while it is still hot. This aids feeding but it also produces the relatively show cooling rates and course-grained structures that are typical of the process.

Dimensional accuracy as good as ± 0.003 in. (± 0.08 mm) is attainable in castings smaller than about one inch (25 mm), while tolerances as close as ± 0.045 in. (± 1.1 mm) are claimed in castngs larger than 15 in (630 mm) in cross section. Additional allowances of about 0.010-0.020 in (0.25-0.5 mm) must be included across the parting line. Surface finishes are typically better than 80 μin (2 μm) rms in nonferrous castings.

Very fine surface finishes and excellent reproduction of detail are characteristic of the investment casting, or lost wax, process. The process was practiced by several ancient cultures and has survived virtually without modification for the production of artwork, statuary, and fine jewelry. Today, the process's most important commercial application is in the casting of complex, net shape precision industrial products such as impellers and gas turbine blades.

The process first requires the manufacture of an intricate metal die with a cavity in the shape of the finished product (or parts of it, if the product is to be assembled from several castings). It also requires special wax. Plastic or a low-melting alloy is cast into the die, then removed and carefully finished using heated tools. Clusters of wax patterns are dipped into a refractory/plaster slurry, which is allowed to harden as a shell or as a monolithic mold.

The mold is first heated to melt the wax (or volatilize the plastic), then fired at a high temperature to vitrify the refractory. Metal is introduced into the mold cavity and allowed to cool at a controlled rate. The sequence of steps involved in the investment method is illustrated in *Figure 3*.

Investment casting is capable of maintaining very high dimensional accuracy in small castings, although tolerances increase somewhat with casting size. Dimensional consistency ranks about average among the casting methods; however, surface finishes can be as fine as 60 μin (1.5 μm) rms, and the process is unsurpassed in its ability to reproduce intricate detail.

Investment casting is better suited to castings under 100 lbs (45 kg) in weight. Because of its relatively high tooling costs and higher-than-average total costs, the process is normally reserved for relatively large production runs of precision products, and is not often applied to copper alloys.

Metal-mold processes

Reusable or metal-mold processes are used more extensively for copper alloys in Europe and England than in North America; however, they are gaining recognition here as equipment and technology become increasingly available. Permanent mold casting in North America is identified as gravity die casting or simply die casting in Europe and the U.K. The process called die casting in North America is known as pressure die casting abroad.

Permanent mold casting utilizes a metallic mold. The mold is constructed such that it can be opened along a conveniently located parting line. Hot metal is poured through a sprue to a system of gates arranged so as to provide even, low-turbulence flow to all parts of the cavity. Baked sand cores can be provided just as they would be with conventional sand castings. Chills are unnecessary since the metal mold provides very good heat

1. Wax or plaster is injected into die to make a pattern.

2. Patterns are gated to a central sprue.

INVESTMENT FLASK CASTING

INVESTMENT SHELL CASTING

3. A metal flask is placed around the pattern cluster.

4. Flask is filled with investment mold slurry.

3. Pattern clusters are dipped in ceramic slurry.

4. Refractory grain is sifted onto coated patterns, steps 3 and 4 are repeated several times to obtain desired shell thickness.

5. After mold material has set and dried, patterns are melted out of mold.

6. Hot molds are filled with metal by gravity, pressure vacuum or centrifugal force.

5. After mold material has set and dried, patterns are melted out of mold.

6. Hot molds are filled with metal by gravity, pressure vacuum or centrifugal force.

7. Mold material is broken away from castings.

8. Castings are removed from sprue, and gate stubs ground off.

7. Mold material is broken away from castings.

Figure 3. Investment casting processes. (Source, Copper Development Association.)

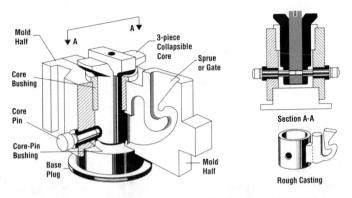

Figure 4. The permanent mold casting process. (Source, Copper Development Association.)

transfer. The nature of the process necessitates adequate draft angles along planar surfaces oriented perpendicular to the parting line. Traces of the parting line may be visible in the finished casting and there may be some adherent flashing, but both are easily removed during finishing.

Permanent mold castings are characterized by good part-to-part dimensional consistency and very good surface finishes (about 70 μin., 1.8 μm). Any traces of metal flow lines on the casting surface are cosmetic rather than functional defects. Permanent mold castings exhibit good soundness. There may be some microshrinkage, but mechanical properties are favorably influenced by the castings' characteristically fine grain size. The ability to reproduce intricate detail is only moderate, however, and for products in which very high dimensional accuracy is required, plaster mold or investment processes should be considered instead. *Figure 4* illustrates the process.

Permanent mold casting is more suitable for simple shapes in mid-size castings than it is for very small or very large products. Die costs are relatively high, but the absence of molding costs makes the overall cost of the process quite favorable for medium to large production volumes.

Die casting

Die casting involves the injection of liquid metal into a multipart die under high pressure. Pneumatically actuated dies make the process almost completely automated. Die casting is best known for its ability to produce high quality products at very low unit costs. Very high production rates offset the cost of the complex heat-resisting tooling required; and with low labor costs, overall casting costs are quite attractive.

Highly intricate products can be made by die casting (although investment casting is even better in this regard). Dimensional accuracy and part-to-part consistency are unsurpassed in both small (<1 in., 25 mm) and large castings. The attainable surface finish, often as good as 30 μin. (0.76 μm) rms, is better than with any other casting process. Extremely rapid cooling rates (dies are normally water cooled) results in very fine grain sizes and good mechanical properties. Die casting is ideally suited to the mass production of small parts. The process is illustrated in *Figure 5*.

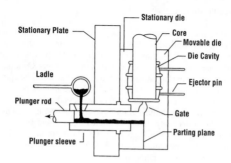

1. Metal poured into sleeve

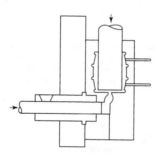

2. Plunger forces metal into die

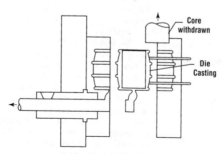

3. Die separated and casting ejected

Figure 5. The die casting process. (Source, Copper Development Association.)

Processes for specific shapes

Continuous casting

Continuous casting uses a mold cavity whose graphite or water-cooled metal side walls are fixed, while the bottom wall—also cooled—is free to move in the axial direction as molten metal is poured in from the top, as shown in *Figure 6*. The process is used to produce bearing blanks and other long castings with uniform cross sections. Continuous casting is the principal method used for the large-tonnage production of semifinished products such as cast rods, tube rounds, gear and bearing blanks, slabs, and custom shapes.

The extremely high cooling and solidification rates attending continuous casting can, depending on the alloy, produce columnar grains. The continuous supply of molten metal at the solidification interface effectively eliminates microshrinkage and produces high quality, sound products with very good mechanical properties. With its simple die construction, relatively low equipment cost, high production rate, and low labor requirements, continuous casting is a very economical production method.

Centrifugal casting

This casting process has been known for several hundred years, but its evolution into a sophisticated production method for other than simple shapes has taken place only in this century. Today, very high quality castings of considerable complexity are produced using this technique.

To make a centrifugal casting, molten metal is poured into a spinning mold. The mold may be oriented horizontally or vertically, depending on the casting's aspect ratio. Short, squat products are cast vertically while long tubular shapes are cast horizontally. In either case, centrifugal force holds the molten metal against the mold wall until it solidifies. Carefully weighted charges insure that just enough metal freezes in the mold to yield the desired wall thickness. The process is shown in *Figure 7*. In some cases, dissimilar alloys can be cast sequentially to produce a composite structure.

Molds for copper alloy castings and similar materials are usually made from carbon steel coated with a suitable refractory mold wash. Molds can be costly if ordered to custom dimensions, but the larger centrifugal foundries maintain sizeable stocks of molds in diameters ranging from a few inches to several feet.

The inherent quality of centrifugal castings is based on the fact that most nonmetallic impurities in castings are less dense than the metal itself. Centrifugal force causes impurities (dross, oxides) to concentrate at the casting's inner surface. This is usually machined away, leaving only clean metal in the finished product. Because freezing is rapid and completely directional, centrifugal castings are inherently sound and pressure tight. Mechanical properties can be somewhat higher than those of statically cast products.

Centrifugal castings are made in sizes ranging from approximately 2 in. to 12 ft. (50 mm to 3.7 m)

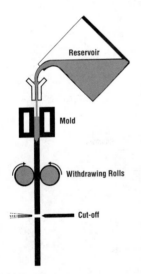

Figure 6. The continuous casting process. (Source, Copper Development Association.)

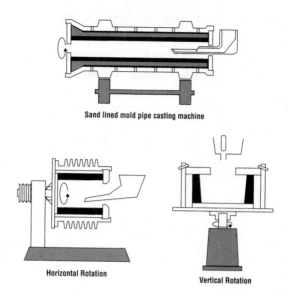

Sand lined mold pipe casting machine

Horizontal Rotation

Vertical Rotation

Figure 7. The centrifugal casting process. (Source, Copper Development Association.)

in diameter and from a few inches to many yards in length. Size limitations, if any, are likely not based on the foundry's melt shop capacity. Simple-shaped centrifugal castings are used for items such as pipe flanges and valve components, while complex shapes can be cast by using cores and shaped molds. Pressure-retaining centrifugal castings have been found to be mechanically equivalent to more costly forgings and extrusions.

In a related process called centrifuging, numerous small molds are arranged radially on a casting machine with their feed sprues oriented toward the machine's axis. Molten metal is fed to the spinning mold, filling the individual cavities. The process is used for small castings such as jewelry and dental bridgework, and is economically viable for both small and large production quantities. Several molding methods can be adapted to the process, and the unit costs of centrifuged castings will depend largely on the type of mold used.

Special casting processes

Recent years have seen the introduction of a number of new casting processes, often aimed at specific applications. While these techniques are still to some extent under development, and while they are certainly not available at all job shop foundries, their inherent advantages make them valuable additions to the designer's list of options.

Squeeze casting

This interesting process aims to improve product quality by solidifying the casting under a metallostatic pressure head sufficient to 1) prevent the formation of shrinkage defects, and 2) retain dissolved gases in solution until freezing is complete. This method was originally developed in Russia and has undergone considerable improvement in the U.S. It is carried out in metal molds resembling the punch and die sets used in sheet metal forming.

After introducing a carefully metered charge of molten metal, the upper die assembly is lowered into place, forming a tight seal. The "punch" portion of the upper die is then forced into the cavity, displacing the molten metal under pressure until it fills the annular space between the die halves.

Proponents of squeeze casting claim that it produces very low gas entrapment and that castings exhibit shrinkage volumes approximately one-half those seen in sand castings. Very high production rates, comparable to die casting but with considerably lower die costs, are also claimed.

The process produces the high quality surfaces typical of metal mold casting, with good reproduction of detail. Rapid solidification results in a fine grain size, which in turn improves mechanical properties. It is claimed that squeeze casting can be applied to many of the copper alloys, although die and permanent mold casting alloys should be favored.

Selecting a casting process

A product's shape, size, and physical characteristics often limit the choice of casting method to a single casting process, in which case the task simply becomes one of selecting a reliable foundry offering a fair price. If there is a choice of casting methods, it may be worthwhile to consult a trusted foundry, since the foundryman's experience can be a source of cost-saving ideas. In any event, it is advantageous to limit selection of the casting method to a few choices early in the design process so that the design and the casting method meet each other's requirements.

Making the selection is not inherently difficult, although it should be emphasized that the help of a skilled foundryman can be invaluable at this point. The factors listed at the beginning of this chapter determine the most suitable and economical process.

Casting design principles

The many factors involved in proper casting design are discussed in a number of excellent texts, including those published by the Non-Ferrous Founders' Society and the American Foundrymen's Association. It must be emphasized that successful casting design is a cooperative process involving all the parties involved. The advice of a skilled foundryman and patternmaker is invaluable, and the earlier in the design process such consultation is sought, the better.

The most important point to bear in mind when designing a casting is that the design doesn't simply set the shape of the product, it also determines the way that the casting will solidify—to the extent this is independent of foundry practice. It may be helpful to review the material presented earlier on the freezing behavior of the various copper alloys.

Short-freezing alloys, such as pure copper, high copper alloys and yellow brasses, aluminum bronzes, and copper-nickels solidify from the mold walls inward and tend to form shrinkage cavities in regions where the last remaining liquid metal solidifies.

Long-freezing alloys, such as tin bronzes, leaded alloys, and the red and semi-red brasses solidify by going through a mushy stage more or less uniformly throughout the casting's volume. They tend to form internal porosity that cannot always be avoided, but can often be tolerated.

There is a spectrum of freezing behaviors between the short- and long-freezing alloys. Exactly how a casting solidifies depends on alloy composition, casting shape, pouring temperature, and the rate of heat extraction. For both short- and long-freezing alloys, however, it is important to ensure that the metal freezes in a directional manner such that the last metal to solidify within the mold cavity (not including the metal left in risers) is adequately fed by liquid metal until solidification is complete. No partially liquid region of the casting should be shut off from a supply of molten metal. Hot spots should be avoided since these tend to remain liquid longest. Approximate pouring temperatures for several metal alloys are given in **Table 1**.

The simplest way to ensure proper solidification is by the placement of risers. These reservoirs of liquid metal are placed either where they can feed relatively thin sections that might otherwise freeze off and isolate adjacent regions of the casting, or where they can, with their high sensible heat, help bring about directional solidification.

Risers must be large enough to remain liquid well after the casting has solidified. Risers are more important in the casting of short-freezing alloys, where feeding takes place over considerable distances. In long-freezing alloys, risers are less helpful in promoting directional solidification and are used instead to ensure uniform solidification rates.

The design and placement of risers is beyond the scope of this alloy selection guide, but the designer should recognize their importance. Since the need for risers may affect the shape or layout of a casting, it is best to consult with the foundryman about riser placement before committing to a final configuration.

It is also important to take into consideration the shrinkage stresses a casting may be subjected to as it solidifies. The ability of a casting to resist such stresses without cracking depends on the alloy's structure, solidification behavior, and elevated-temperature properties. The presence of a second phase, particularly beta, tends to improve strength and ductility at high temperatures, and this reduces the tendency for restrained sections to tear as the metal solidifies and shrinks. The type of molding material is also important. Properly made sand molds can accommodate shrinkage, while permanent or plaster molds cannot. **Table 2** provides solidification shrinkage information of several common foundry alloys, and **Table 3** gives shrinkage allowances in inches and millimeters per foot.

Table 1. Approximate Pouring Temperatures for Metal Alloys. *(Source, Cost-Effective Design by Michael A. Gwyn.)*

Alloy	°F	°C
Solder	~450	~230
Tin	~600	~315
Lead	~650	~345
Zinc Alloys	650 - 850	345 - 455
Aluminum Alloys	1150 - 1350	620 - 735
Magnesium Alloys	1150 - 1350	620 - 735
Copper-base Alloys	1650 - 2150	900 - 1180
Cast Iron: Gray, Ductile, Malleable	2450 - 2700	1340 - 1480
High Manganese Steel	2550 - 2650	1400 - 1455
Monel (70 Ni, 30 Cu)	2500 - 2800	1370 - 1540
Nickel-base Superalloys	2600 - 2800	1430 - 1540
High Alloy Steels	2700 - 2900	1480 - 1600
High Alloy Irons	2800 - 3000	1540 - 1650
Carbon and Low Alloy Steels	2850 - 3100	1565 - 1700
Titanium Alloys	3100 - 3300	1700 - 1820
Zirconium Alloys	3350 - 3450	1845 - 1900

Table 2. Characteristics of Common Foundry Alloys. *(Source, Cost-Effective Design by Michael A. Gwyn.)*

Alloy Group	Type*	Fluid Life	Slag or Dross
Ferrous			
Gray Iron	Eutectic	Excellent	Little
Ductile Iron	Eutectic	Good	Little-Moderate
Carbon and Low Alloy Steel	Directional	Poor	Moderate
High Alloy Steel	Directional	Fair	Moderate
Nonferrous			
Zinc Alloys	Eutectic	Excellent	Little
	Equiaxed	Good	Little
Aluminum Alloys	Eutectic	Excellent	Moderate
	Directional		
	Equiaxed	Fair	Large
Magnesium Alloys	Eutectic	Excellent	Little
	Directional		Moderate
Red and Yellow Brasses, Bronzes	Eutectic	Very Good	Little
	Directional		
	Equiaxed	Fair	Moderate
Aluminum Bronzes	Equiaxed	Good Fair	Moderate Large

* Type is based on how the alloy solidifies metallurgically. The definitions below are important to the selection of casting geometry. Proper solidification is a combination of metallurgy and heat transfer patterns that come from geometry.

Eutectic. Being, in fact, eutectic alloys or behaving like them, these alloys remain liquid in the mold for a brief period, cooling; then they solidify very quickly all over. This phenomenon minimizes internal shrinkage and the need for risers.

Directional. These alloys begin solidifying quickly, perpendicular to the mold walls. Solidification "direction" and pathways are perpendicular from casting geometry and thermal patterns in the mold walls. Without proper pathway geometry, isolated thermal shrinkage can result.

Equiaxed. These alloys not only begin solidifying perpendicular to mold walls but also begin to solidify in the midst of liquid, forming *equiaxed islands of solid*. Solidified pathways are interrupted by the equiaxed islands, making these alloys difficult to bend.

As an example of the interplay between metal and molding material in the choice of a casting process, consider the alpha-beta structure of yellow brasses. The alloys' good high temperature ductility, along with their relatively short freezing range, suit them to the permanent mold and die casting processes.

Design Fundamentals

Observing a few simple rules will go a long way toward avoiding the most prevalent design-based casting defects. It should become apparent that these rules are based on the solidification behaviors described above.

- Avoid abrupt changes in section thickness. Taper the larger section such that it blends into the thinner section, *Figure 8*.
- Always avoid sharp internal corners. Use generous fillets and rounded corners whenever possible to avoid the formation of hot spots.

Table 3. Foundry Alloy Shrinkage Allowances. *(Source, Cost-Effective Design by Michael A. Gwyn.)*

Alloy	Inches per Foot	Millimeters per Meter
Aluminum	0.156	13.025
Aluminum Bronze	0.313	26.050
Brass and Bronze	0.188 - 0.156	15.682 - 13.025
Iron, Gray *	0.318	26.443
Iron, Ductile **	0 - 0.063	0 - 5.249
Iron, Malleable	0.318	26.443
Lead	0.313	26.050
Monel	0.250	20.833
Nickel	0.250	20.833
Ni-Resist 1 and 2	0.188	15.682
Steel, Carbon	0.250	20.833
Stainless Alloys CA-15, CA-40 CF-8, CF-8M CN-7M (Alloy 20)	0.250 0.281 0.281	20.833 23.425 23.425

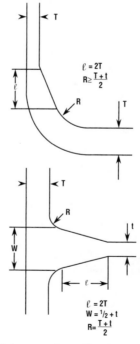

Figure 8. Wall thickness design recommendations. (Source, Copper Development Association.)

- Minimize the use of L intersections, avoid X intersections, and take care in designing T intersections. Use rounded corners insofar as possible, and substitute two Ts for each X intersection whenever possible.
- Visualize how the metal will solidify, and design the casting to take this into account. Consider the type of freezing the candidate alloy will undergo and use this understanding to avoid shrinkage cavities or porosity.
- Identify the constraints the solidifying and cooling casting will undergo, and formulate the design accordingly to avoid hot tearing.
- Do not hesitate to add metal (padding) to facilitate the feeding of thin sections and remove metal where it creates an abrupt change in section size. Removing metal may, in fact, strengthen the casting.
- If there is a possibility that shrinkage stresses will demand some degree of flexibility during solidification, use curved members in place of straight sections whenever possible.

Following these guidelines will help ensure that the casting design process will begin correctly, and that the need for changes later on–when they may be more expensive–will be minimized.

Fabricating with sheet metal

This section provides basic information on metal bending and shearing. Sheet metal thicknesses are based on the Manufacturer's Standard Gage for Sheet Steel, which is based on a weight of 41.82 pounds per square foot per inch of thickness. Thickness in both inches and millimeters, plus weights in pounds and kilograms, for standard Manufacturer's Gage sizes are given in **Table 1**. Gages and weights for galvanized sheet steel are based on a different system, using a standard weight of 501.9 lb/ft^3 for galvanized steel. These gage sizes and weights by gage number are provided in **Table 2**. In the United Kingdom, sheet metal sizes are normally based on the Imperial Wire Gage, also known as the British Standard Wire Gage. Dimensions for this standard are given later in this section in **Table 9** titled "Standards for Wire Gages."

Thickness specifications for most materials other than iron and steel are commonly designated in fractions of an inch, rather than by gage size. Weights of aluminum and copper for fractional sizes in $1/16$ths of an inch are given in **Table 3**.

Bending metal

Fabrication often requires the bending of metal by mechanical means. To be predictable, the bending process must be uniform along a straight axis. After bending, the inside surface of the bend is in compression and the outer surface in tension, and the strains and stresses that take place in the bending plane are influenced by the angle of the bend and the radius of the bend. While most annealed metals can be bent to any desired angle, the practical radius of the bend is influenced and restricted by the thickness of the metal. In general, most metals can be bent to a radius equal to their thickness, but there are exceptions, of course, including malleable aluminums, brasses, and other nonferrous materials. Suggested curvature factor constants that can be used to calculate minimum bending radii for several metals are given in **Table 4.**

When laying out work for bending, or deep drawing, allowances must be made for dimensional changes that occur in the bend or in stretching the material. For bending, this change in length is known as the *bend allowance* and it can be calculated (for most bends) with the aid of a constant that locates the neutral axis (where no stretching or compression takes place) of the bend. Normally, the value used for the constant (k) is 0.5, which locates the neutral axis in the center of the metal, but in fact it varies with thickness and material from 0.5 to as low as 0.33 for mild tempered sheet metals. The working formula for estimating the bend allowance (BA) is

$$BA = (\alpha_A \div 360) \times 2\,\pi\,(r + k\,t)$$

where α_A = bend angle, in degrees
r = radius of the bend
k = constant
t = thickness of the material.

To determine the amount of bend allowance introduced per degree of angle of bend, a somewhat simpler formula can be used.

$$BA \text{ for } 1° = (\pi \div 180) \times (r + k\,t).$$

Another variable that must often be calculated before bending is the amount of *setback* that will be generated for a sheet of given thickness when the radius is known. As can be seen in *Figure 1*, setback is a measurement of the distance from a line projected beyond the mold line on one leg of a bent sheet or plate to the point that it intersects with an identical line projected from the other leg. It can accurately be calculated through approximately 90° of bend (45° to 135°). When the bend angle is less than 43°, setback cannot be calculated accurately with this formula.

$$SB = 2(r + t) - ([\alpha_B \div 360] \times [2\pi\{r + kt\}])$$

where α_B = setback angle, in degrees (see *Figure 1*).

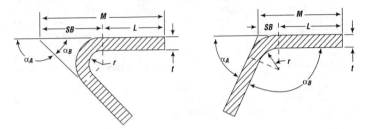

Figure 1. Bending terminology. M = mold line dimension. SB = setback. L = leg of part. t = thickness of part. r = radius of bend. α_A = bend angle. α_B = setback angle.

Table 5 provides bend allowance and setback for bends of 90°, plus bend allowance per degree, for several common radii and sheet thicknesses. In the calculation of this Table, k was assumed to be 0.5.

Springback is caused by residual stress in a material after bending that encourages it to return to its original shape. In fact, a purely elastic material will return to its original shape after bending, and the partial-plastic nature of most metals makes it difficult to calculate the exact shape a part will assume after being removed from the press. Springback is calculated by determining a factor (the springback ratio–K) which, unfortunately, can be assigned only through trial and error (see *Figure 2*).

$$K = \alpha_1 \div \alpha_2 = (r_1 + 0.5t) \div (r_2 + 0.5t)$$

where α_1 = angle of part, after bending, in degrees
        α_2 = angle of the die, in degrees
        r_1 = inside radius of die
        r_2 = inside radius of part, after bending
        t = thickness of part.

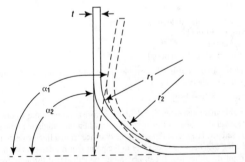

Figure 2. Springback. α_1 = angle in die. α_2 = angle on part after removal. r_1 = inside radius on die. r_2 = inside radius on part.

(Text continued on p. 1758)

Table 1. Sheet Steel Weight by Manufacturer's Standard Gage Number.

Gage Number	Thickness		Weight			
	inch	millimeter	Pounds/ft.^{2}	Pounds/in.^{2}	kg/cm^2	kg/m^2
3	0.2391	6.0731	9.999162	0.069439	0.004882	48.820200
4	0.2242	5.6947	9.376044	0.065111	0.004578	45.777870
5	0.2092	5.3137	8.748744	0.060755	0.004272	42.715120
6	0.1943	4.9352	8.125626	0.056428	0.003967	39.672790
7	0.1793	4.5542	7.498326	0.052072	0.003661	36.610040
8	0.1644	4.1758	6.875208	0.047745	0.003357	33.567710
9	0.1495	3.7973	6.252090	0.043417	0.003053	30.525390
10	0.1345	3.4163	5.624790	0.039061	0.002746	27.462640
11	0.1196	3.0378	5.001672	0.034734	0.002442	24.420310
12	0.1046	2.6568	4.374372	0.030378	0.002136	21.357560
13	0.0897	2.2784	3.751254	0.026050	0.001832	18.315230
14	0.0747	1.8974	3.123954	0.021694	0.001525	15.252480
15	0.0673	1.7094	2.814486	0.019545	0.001374	13.741530
16	0.0598	1.5189	2.500836	0.017367	0.001221	12.210150
17	0.0538	1.3665	2.249916	0.015624	0.001099	10.985060
18	0.0478	1.2141	1.998996	0.013882	0.000976	9.759956
19	0.0418	1.0617	1.748076	0.012139	0.000853	8.534857
20	0.0359	0.9119	1.501338	0.010426	0.000733	7.330176
21	0.0329	0.8357	1.375878	0.009555	0.000672	6.717627
22	0.0299	0.7595	1.250418	0.008683	0.000611	6.105077
23	0.0269	0.6833	1.124958	0.007812	0.000549	5.492528
24	0.0239	0.6071	0.999498	0.006941	0.000488	4.879978
25	0.0209	0.5309	0.874038	0.006070	0.000427	4.267428
26	0.0179	0.4547	0.748578	0.005198	0.000365	3.654879
27	0.0164	0.4166	0.685848	0.004763	0.000335	3.348604
28	0.0149	0.3785	0.623118	0.004327	0.000304	3.042329
29	0.0135	0.3429	0.564570	0.003921	0.000276	2.756473
30	0.0120	0.3048	0.501840	0.003485	0.000245	2.450198
31	0.0105	0.2667	0.439110	0.003049	0.000214	2.143923
32	0.0097	0.2464	1.405654	0.002817	0.000198	1.980577
33	0.0090	0.2286	0.376380	0.002614	0.000184	1.837649
34	0.0082	0.2083	0.342924	0.002381	0.000167	1.674302
35	0.0075	0.1905	0.313650	0.002178	0.000153	1.531374
36	0.0067	0.1702	0.280194	0.001946	0.000137	1.368027
37	0.0064	0.1626	0.267648	0.001859	0.000131	1.306772
38	0.0060	0.1524	0.250920	0.001743	0.000123	1.225099

Based on a weight of 41.82 pounds per square foot per inch of thickness, the Manufacturer's Standard Gage for Sheet Steel.

Table 2. Galvanized Sheet Steel Weight by U.S. Standard Gage Number.

Gage Number	Thickness		Weight		
	inch	millimeter	Pounds/ft.^{2}	Pounds/in.^{2}	kg/m^2
8	0.1681	4.270	0.0488	7.031	34.33
9	0.1532	3.891	0.0445	6.406	31.28
10	0.1382	3.510	0.0401	5.781	28.23
11	0.1233	3.132	0.0358	5.156	25.18
12	0.1084	2.753	0.0315	4.531	22.12
13	0.0934	2.372	0.0271	3.906	19.07
14	0.0785	1.994	0.0228	3.281	16.02
15	0.0710	1.803	0.0206	2.969	14.49
16	0.0635	1.613	0.0184	2.656	12.97
17	0.0575	1.461	0.0167	2.406	11.75
18	0.0516	1.311	0.0150	2.156	10.53
19	0.0456	1.158	0.0132	1.906	9.307
20	0.0396	1.006	0.0115	1.656	8.087
21	0.0366	0.930	0.0106	1.531	7.476
22	0.0336	0.853	0.0098	1.406	6.866
23	0.0306	0.077	0.0089	1.281	6.256
24	0.0276	0.701	0.0080	1.156	5.645
25	0.0247	0.627	0.0072	1.031	5.035
26	0.0217	0.551	0.0063	0.906	4.425
27	0.0202	0.513	0.0059	0.844	4.120
28	0.0187	0.475	0.0054	0.781	3.814
29	0.0172	0.437	0.0050	0.719	3.509
30	0.0157	0.399	0.0046	0.656	3.204
31	0.0142	0.361	0.0041	0.594	2.899
32	0.0134	0.340	0.0039	0.563	2.746

Note: Weights based on standard of 501.9 lb/ft^3.

Table 3. Weight of Aluminum and Copper Sheet and Plate.

Thickness			Aluminum				Copper			
Fraction	Decimal	mm	lb./ft.²	lb./in.²	kg/cm²	kg/m²	lb./in.²	lb./ft.²	kg/cm²	kg/m²
1/64	.015625	0.396875	0.214844	.001492	.000105	1.048959	.005027	0.723958	.000353	3.249301
1/32	.031250	0.793750	0.429688	.002984	.000210	2.097919	.010055	1.447917	.000707	6.498602
3/64	.046875	1.190625	0.644531	.004476	.000315	3.146878	.015082	2.171875	.001060	9.747903
1/16	.062500	1.587500	0.859375	.005968	.000420	4.195837	.020110	2.895833	.001414	12.99720
5/64	.078125	1.984375	1.074219	.007460	.000524	5.244797	.025137	3.619792	.001767	16.24650
3/32	.093750	2.381250	1.289063	.008952	.000629	6.293756	.030165	4.343750	.002121	19.49581
7/64	.109375	2.778125	1.503906	.010444	.000734	7.342715	.035192	5.067708	.002474	22.74511
1/8	.125000	3.175000	1.718750	.011936	.000839	8.391675	.04022	5.791667	.002828	25.99441
9/64	.140625	3.571875	1.933594	.013428	.000944	9.440634	.045247	6.515625	.003181	29.24371
5/32	.156250	3.968750	2.148438	.014920	.001049	10.48959	.050275	7.239583	.003535	32.49301
11/64	.171875	4.365625	2.363281	.016412	.001154	11.53855	.055302	7.963542	.003888	35.74231
3/16	.187500	4.762500	2.578125	.017904	.001259	12.58751	.060330	8.687500	.004242	38.99161
13/64	.203125	5.159375	2.792969	.019396	.001364	13.63647	.065357	9.411458	.004595	42.24091
7/32	.218750	5.556250	3.007813	.020888	.001469	14.68543	.070385	10.13542	.004949	45.49021
15/64	.234375	5.953125	3.222656	.022380	.001573	15.73439	.075412	10.85938	.005302	48.73951
1/4	.250000	6.350000	3.437500	.023872	.001678	16.78335	.080440	11.58333	.005655	51.98881
17/64	.265625	6.746875	3.652344	.025363	.001783	17.83231	.085467	12.30729	.006009	55.23811
9/32	.281250	7.143750	3.867188	.026855	.001888	18.88127	.090495	13.03125	.006362	58.48742
19/64	.296875	7.540625	4.082031	.028347	.001993	19.93023	.95522	13.75521	.006716	61.73672
5/16	.312500	7.937500	4.296875	.029839	.002098	20.97919	.100550	14.47917	.007069	64.98602
21/64	.328125	8.334375	4.511719	.031331	.002203	22.02815	.105577	15.20313	.007423	68.23532
11/32	.343750	8.731250	4.726563	.032823	.002308	23.07711	.110605	15.92708	.007776	71.48462
23/64	.359375	9.128125	4.941406	.034315	.002413	24.12607	.115632	16.65104	.008130	74.73392
3/8	.375000	9.525000	5.156250	.035807	.002518	25.17502	.120660	17.37500	.008483	77.98322

(Continued)

Table 3. *(Continued)* **Weight of Aluminum and Copper Sheet and Plate.**

Thickness			Aluminum				Copper			
Fraction	Decimal	mm	lb./ft.²	lb./in.²	kg/cm²	kg/m²	lb./ft.²	lb./in.²	kg/cm²	kg/m²
25/64	.390625	9.921875	5.371094	.037299	.002622	26.22398	18.09896	.125687	.008837	81.23252
13/32	.406250	10.31875	5.585938	.038791	.002727	27.27294	18.82292	.130715	.009190	84.48182
27/64	.421875	10.71563	5.800781	.040283	.002832	28.32190	19.54688	.135742	.009544	87.73112
7/16	.437500	11.11250	6.015625	.041775	.002937	29.37086	20.27083	.140770	.009897	90.98042
29/64	.453125	11.50938	6.230469	.043267	.003042	30.41982	20.99479	.145797	.010251	94.22972
15/32	.468750	11.90625	6.445313	.044759	.003147	31.46878	21.71875	.150825	.010604	97.47903
31/64	.484375	12.30313	6.660156	.046251	.003252	32.51774	22.44271	.155852	.010957	100.7283
1/2	.500000	12.70000	6.875000	.047743	.003357	33.56670	23.16667	.160880	.011311	103.9776
33/64	.515625	13.09688	7.089844	.049235	.003462	34.61566	23.89063	.165907	.011664	107.2269
17/32	.531250	13.49375	7.304688	.050727	.003566	35.66462	24.61458	.170935	.012018	110.4762
35/64	.546875	13.89063	7.519531	.052219	.003671	36.71358	25.33854	.175962	.012371	113.7255
9/16	.562500	14.28750	7.734375	.053711	.003776	37.76254	26.06250	.180990	.012725	116.9748
37/64	.578125	14.68438	7.949219	.055203	.003881	38.81150	26.78646	.186017	.013078	120.2241
19/32	.593750	15.08125	8.164063	.056695	.003986	39.86046	27.51042	.191045	.013432	123.4734
39/64	.609375	15.47813	8.378906	.058187	.004091	40.90941	28.23438	.196072	.013785	126.7227
5/8	.625000	15.87500	8.593750	.059679	.004196	41.95837	28.95833	.201100	.014139	129.972
41/64	.640625	16.27188	8.808594	.061171	.004301	43.00733	29.68229	.206127	.014492	133.2213
21/32	.656250	16.66875	9.023438	.062662	.004406	44.05629	30.40625	.211154	.014845	136.4706
43/64	.671875	17.06563	9.238281	.064155	.004511	45.10525	31.13021	.216182	.015199	139.7199
11/16	.687500	17.46250	9.453125	.065647	.004615	46.15421	31.85417	.221209	.015553	142.9692
45/64	.703125	17.85938	9.667969	.067139	.004720	47.20317	32.57813	.226237	.015906	146.2185
23/32	.718750	18.25625	9.882813	.068631	.004825	48.25213	33.30208	.231264	.016260	149.4678
47/64	.734375	18.65313	10.097660	.070123	.004930	49.30109	34.02604	.236292	.016613	152.7171
3/4	.750000	19.05000	10.312500	.071615	.005035	50.35005	34.75000	.241319	.016966	155.9664

Notes: Weights are based on: Aluminum 165 lb/ft³; Copper 556 lb/ft³.

Table 4. Minimum Bending Radius.

The curvature factors given below can be used to estimate minimum bending radius by applying the formula

$$r_{min} = t \, c$$

where t is the sheet thickness of the workpiece, and c is the curvature factor from the Table, which applies only to bending operations traverse to the direction of rolling. When bending lines are parallel to the direction of rolling, the above formula should be modified as follows.

$$r_{min} = (0.5 \, t) \times c$$

Material	c factor	Material	c factor
Sheet Steel	0.6	AlMg 9-soft	2.2
Deep Drawing Sheet Steel	0.5	AlMg 9-half-hard	5.0
Stainless Steel		AlMgSi-soft	1.2
Martensitic/ferritic	0.8	AlMgSi-recrystallization hardened	2.5
Austenitic	0.5	AlSi-soft	0.8
Copper	0.25	AlSi-hard	6.0
Tin Bronze	0.6	AlMn-soft	1.0
Aluminum Bronze	0.5	AlMn-strain hardened	1.2
CuZn 28	0.3	AlMn-hard	1.2
CuZn 40	0.35	AlCu-soft	1.0
Zinc	0.4	AlCu- recrystallization hardened	3.0
Aluminum-soft	0.6	AlCuMg-soft	1.2
Aluminum-half-hard	0.9	AlCuMg-strain hardened	1.5
Aluminum-hard	2.0	AlCuMg- recrystallization hardened	3.0
AlMg 3-soft	1.0	AlCuNi-annealed	1.4
AlMg 3-half-hard	1.3	AlCuNi-unannealed	3.5
AlMg 7-soft	2.0	MgMn	5.0
AlMg 7-half-hard	3.0	MgAl 6	3.0

Source, *Handbook of Technical Formulas,* Karl-Friedrich Fischer, ed. Hanser Gardner Publications.

Table 5. Bend Allowances and Setback Lengths for Sheet Metal. BA = Bend Allowance. SB = Setback. (See Notes.)

Radius	Condition	0.0097" 32 Gage	0.0149" 28 Gage	0.0179" 26 Gage	0.0209" 25 Gage	0.0239" 24 Gage	0.0299" 22 Gage	0.0359" 20 Gage	0.0478" 18 Gage
1/32 0.03125"	BA for 1°	0.00063	0.00068	0.00070	0.00073	0.00075	0.00081	0.00086	0.00096
	BA for 90°	0.0567	0.0608	0.0631	0.0655	0.0679	0.0726	0.0773	0.0866
	SB for 90°	0.0252	0.0315	0.0352	0.0388	0.0424	0.0497	0.0570	0.0715
1/16 0.0625"	BA for 1°	0.00118	0.00122	0.00125	0.00127	0.00130	0.00135	0.00140	0.00151
	BA for 90°	0.1058	0.1099	0.1122	0.1146	0.1169	0.1217	0.1264	0.1357
	SB for 90°	0.0386	0.0449	0.0486	0.0522	0.0559	0.0631	0.0704	0.0849
3/32 0.09375"	BA for 1°	0.00172	0.00177	0.00179	0.00182	0.00184	0.00190	0.00195	0.00205
	BA for 90°	0.1549	0.1590	0.1613	0.1637	0.1660	0.1707	0.1755	0.1848
	SB for 90°	0.0520	0.0583	0.0620	0.0656	0.0693	0.0766	0.0838	0.0983
1/8 0.125"	BA for 1°	0.00227	0.00231	0.00234	0.00236	0.00239	0.00244	0.00249	0.00260
	BA for 90°	0.2040	0.2081	0.2104	0.2128	0.2151	0.2198	0.2245	0.2339
	SB for 90°	0.0654	0.0717	0.0754	0.0790	0.0827	0.0900	0.0973	0.1117
5/32 0.15625"	BA for 1°	0.00281	0.00286	0.00288	0.00291	0.00294	0.00299	0.00304	0.00314
	BA for 90°	0.2531	0.2571	0.2595	0.2619	0.2642	0.2689	0.2736	0.2830
	SB for 90°	0.0788	0.0852	0.0888	0.0924	0.0961	0.1034	0.1107	0.1251
3/16 0.1875"	BA for 1°	0.00336	0.00340	0.00343	0.00345	0.00348	0.00353	0.00359	0.00369
	BA for 90°	0.3021	0.3062	0.3086	0.3109	0.3133	0.3180	0.3227	0.3321
	SB for 90°	0.0923	0.0986	0.1022	0.1059	0.1095	0.1168	0.1241	0.1385
8/32 0.21875"	BA for 1°	0.00390	0.00395	0.00397	0.00400	0.00403	0.00408	0.00413	0.00424
	BA for 90°	0.3512	0.3553	0.3577	0.3600	0.3624	0.3671	0.3718	0.3812
	SB for 90°	0.1057	0.1120	0.1156	0.1193	0.1229	0.1302	0.1375	0.1519

Column headers span: **Sheet Metal Thickness**

Notes: Gage thicknesses are manufacturer's gage. For thicknesses of U.S. Standard gage, see Table "Sheet Steel Weight by U.S. Standard Gage Number." All dimensions in inches.

(Continued)

Table 5. *(Continued)* Bend Allowances and Setback Lengths for Sheet Metal. BA = Bend Allowance. SB = Setback. (See Notes.)

Radius	Condition	Sheet Metal Thickness							
		0.0097" 32 Gage	0.0149" 28 Gage	0.0179" 26 Gage	0.0209" 25 Gage	0.0239" 24 Gage	0.0299" 22 Gage	0.0359" 20 Gage	0.0478" 18 Gage
1/4 0.25"	BA for 1°	0.00445	0.00449	0.00452	0.00455	0.00457	0.00462	0.00468	0.00478
	BA for 90°	0.4003	0.4044	0.4068	0.4091	0.4115	0.4162	0.4209	0.4302
	SB for 90°	0.1191	0.1254	0.1290	0.1327	0.1363	0.1436	0.1509	0.1654
9/32 0.28125"	BA for 1°	0.00499	0.00504	0.00506	0.00509	0.00512	0.00517	0.00522	0.00533
	BA for 90°	0.4494	0.4535	0.4558	0.4582	0.4606	0.4653	0.4700	0.4793
	SB for 90°	0.1325	0.1388	0.1425	0.1461	0.1497	0.1570	0.1643	0.1788
5/16 0.3125"	BA for 1°	0.00554	0.00558	0.00561	0.00564	0.00566	0.00572	0.00577	0.00587
	BA for 90°	0.4985	0.5026	0.5049	0.5073	0.5096	0.5144	0.5191	0.5284
	SB for 90°	0.1459	0.1522	0.1559	0.1595	0.1632	0.1704	0.1777	0.1922
11/32 0.34375"	BA for 1°	0.00608	0.00613	0.00616	0.00618	0.00621	0.00626	0.00631	0.00642
	BA for 90°	0.5476	0.5517	0.5540	0.5564	0.5587	0.5634	0.5682	0.5775
	SB for 90°	0.1593	0.1656	0.1693	0.1729	0.1766	0.1839	0.1911	0.2056
3/8 0.375"	BA for 1°	0.00663	0.00668	0.00670	0.00673	0.00675	0.00681	0.00686	0.00696
	BA for 90°	0.5967	0.6008	0.6031	0.6055	0.6078	0.6125	0.6172	0.6266
	SB for 90°	0.1727	0.1790	0.1827	0.1863	0.1900	0.1973	0.2046	0.2190
7/16 0.4375"	BA for 1°	0.00772	0.00777	0.00779	0.00782	0.00784	0.00790	0.00795	0.00805
	BA for 90°	0.6948	0.6989	0.7013	0.7036	0.7060	0.7107	0.7154	0.7248
	SB for 90°	0.1996	0.2059	0.2095	0.2132	0.2168	0.2241	0.2314	0.2458
1/2 0.5"	BA for 1°	0.00881	0.00886	0.00888	0.00891	0.00894	0.00899	0.00904	0.00914
	BA for 90°	0.7930	0.7971	0.7995	0.8018	0.8042	0.8089	0.8136	0.8229
	SB for 90°	0.2264	0.2327	0.2363	0.2400	0.2436	0.2509	0.2582	0.2727

Notes: Gage thicknesses are manufacturer's gage. For thicknesses of U.S. Standard gage, see Table "Sheet Steel Weight by U.S. Standard Gage Number." All dimensions in inches.

(Continued)

1756 FABRICATION PROCESSES

Table 5. (*Continued*) Bend Allowances and Setback Lengths for Sheet Metal. BA = Bend Allowance. SB = Setback. (See Notes.)

Radius	Condition	Sheet Metal Thickness							
		0.0598" 16 Gage	0.0747" 14 Gage	0.0897" 13 Gage	0.1196" 11 Gage	0.1345" 10 Gage	0.1495" 9 Gage	0.1793" 7 Gage	0.2391" 3 Gage
1/32 0.03125"	BA for 1°	0.00107	0.00120	0.00133	0.00159	0.00172	0.00185	0.00211	0.00263
	BA for 90°	0.0961	0.1078	0.1195	0.1430	0.1547	0.1665	0.1899	0.2369
	SB for 90°	0.0860	0.1041	0.1224	0.1587	0.1768	0.1950	0.2312	0.3038
1/16 0.0625"	BA for 1°	0.00161	0.00174	0.00187	0.00213	0.00226	0.00240	0.00266	0.00318
	BA for 90°	0.1451	0.1568	0.1686	0.1921	0.2038	0.2156	0.2390	0.2860
	SB for 90°	0.0995	0.1176	0.1358	0.1721	0.1902	0.2084	0.2446	0.3172
3/32 0.09375"	BA for 1°	0.00216	0.00229	0.00242	0.00268	0.00281	0.00294	0.00320	0.00372
	BA for 90°	0.1942	0.2059	0.2177	0.2412	0.2529	0.2647	0.2881	0.3351
	SB for 90°	0.1129	0.1310	0.1492	0.1855	0.2036	0.2218	0.2580	0.3306
1/8 0.125"	BA for 1°	0.00270	0.00283	0.00296	0.00323	0.00336	0.00349	0.00375	0.00427
	BA for 90°	0.2433	0.2550	0.2668	0.2903	0.3020	0.3138	0.3372	0.3841
	SB for 90°	0.1263	0.1444	0.1626	0.1989	0.2170	0.2352	0.2714	0.3441
5/32 0.15625"	BA for 1°	0.00325	0.00338	0.00351	0.00377	0.00390	0.00403	0.00429	0.00481
	BA for 90°	0.2924	0.3041	0.3159	0.3394	0.3511	0.3629	0.3863	0.4332
	SB for 90°	0.1397	0.1578	0.1760	0.2123	0.2304	0.2486	0.2848	0.3575
3/16 0.1875"	BA for 1°	0.00379	0.00392	0.00406	0.00432	0.00445	0.00458	0.00484	0.00536
	BA for 90°	0.3415	0.3532	0.3650	0.3885	0.4002	0.4119	0.4353	0.4823
	SB for 90°	0.1531	0.1712	0.1894	0.2257	0.2438	0.2621	0.2983	0.3709
8/32 0.21875"	BA for 1°	0.00434	0.00447	0.00460	0.00486	0.00499	0.00512	0.00538	0.00590
	BA for 90°	0.3906	0.4023	0.4141	0.4375	0.4492	0.4610	0.4844	0.5314
	SB for 90°	0.1665	0.1846	0.2028	0.2392	0.2573	0.2755	0.3117	0.3843

Notes: Gage thicknesses are manufacturer's gage. For thicknesses of U.S. Standard gage, see Table "Sheet Steel Weight by U.S. Standard Gage Number." All dimensions in inches.

(*Continued*)

Table 5. *(Continued)* Bend Allowances and Setback Lengths for Sheet Metal. BA = Bend Allowance. SB = Setback. (See Notes.)

Radius	Condition	Sheet Metal Thickness							
		0.0598" 16 Gage	0.0747" 14 Gage	0.0897" 13 Gage	0.1196" 11 Gage	0.1345" 10 Gage	0.1495" 9 Gage	0.1793" 7 Gage	0.2391" 3 Gage
1/4 0.25"	BA for 1°	0.00489	0.00502	0.00515	0.00541	0.00554	0.00567	0.00593	0.00645
	BA for 90°	0.4397	0.4514	0.4632	0.4866	0.4983	0.5101	0.5335	0.5805
	SB for 90°	0.1799	0.1980	0.2162	0.2526	0.2707	0.2889	0.3251	0.3977
9/32 0.28125"	BA for 1°	0.00543	0.00556	0.00569	0.00595	0.00608	0.00621	0.00647	0.00700
	BA for 90°	0.4888	0.5005	0.5122	0.5357	0.5474	0.5592	0.5826	0.6296
	SB for 90°	0.1933	0.2114	0.2297	0.2660	0.2841	0.3023	0.3385	0.4111
5/16 0.3125"	BA for 1°	0.00598	0.00611	0.00624	0.00650	0.00663	0.00676	0.00702	0.00754
	BA for 90°	0.5378	0.5495	0.5613	0.5848	0.5965	0.6083	0.6317	0.6787
	SB for 90°	0.2068	0.2249	0.2431	0.2794	0.2975	0.3157	0.3519	0.4245
11/32 0.34375"	BA for 1°	0.00652	0.00665	0.00678	0.00704	0.00717	0.00730	0.00756	0.00809
	BA for 90°	0.5869	0.5986	0.6104	0.6339	0.6456	0.6574	0.6808	0.7278
	SB for 90°	0.2202	0.2383	0.2565	0.2928	0.3109	0.3291	0.3653	0.4379
3/8 0.375"	BA for 1°	0.00707	0.00720	0.00733	0.00759	0.00772	0.00785	0.00811	0.00863
	BA for 90°	0.6360	0.6477	0.6595	0.6830	0.6947	0.7065	0.7299	0.7768
	SB for 90°	0.2336	0.2517	0.2699	0.3062	0.3243	0.3425	0.3787	0.4514
7/16 0.4375"	BA for 1°	0.00816	0.00829	0.00842	0.00868	0.00881	0.00894	0.00920	0.00972
	BA for 90°	0.7342	0.7459	0.7577	0.7812	0.7929	0.8046	0.8280	0.8750
	SB for 90°	0.2604	0.2785	0.2967	0.3330	0.3511	0.3694	0.4056	0.4782
1/2 0.5"	BA for 1°	0.00925	0.00938	0.00951	0.00977	0.00990	0.01003	0.01029	0.01081
	BA for 90°	0.8324	0.8441	0.8559	0.8793	0.8910	0.9028	0.9262	0.9732
	SB for 90°	0.2872	0.3053	0.3235	0.3599	0.3780	0.3962	0.4324	0.5050

Notes: Gage thicknesses are manufacturer's gage. For thicknesses of U.S. Standard gage, see Table "Sheet Steel Weight by U.S. Standard Gage Number." All dimensions in inches.

Bending pressures

Calculations for estimated pressures for bending specific materials are based on the ultimate tensile strength of the material, the ratio of the die opening to the material thickness, the length of the bent part, and the width of the channel in the die. Ultimate tensile strengths for most materials can be found in the appropriate Materials section of this book. The formula for calculating required bending force (F_B), in tons, is as follows.

$$F_B = (T\,k\,l\,t^2) \div W_V$$

where T = ultimate tensile strength, in tons per in.^{2}
   k = die opening factor (see below)
   l = length of bent part, in inches
   t = thickness of part, in inches
   W_V = width of channel in lower die.

The k factor, or die opening-to-material thickness ratio, is rated from 16-times material thickness to 8-times material thickness as follows.

Die opening-to-material thickness	k factor
16 times	1.20
15 times	1.21625
14 times	1.2325
13 times	1.24875
12 times	1.265
11 times	1.28125
10 times	1.2975
9 times	1.31375
8 times	1.33

Example. Using the above formula, calculate the pressure necessary to bend a 0.250" thick piece of AISI 1020 annealed steel. The width of the channel in the lower die is 3.5", the length of the bend is 5", and the die opening is 2.5" (or 10 times material thickness). From Table 7 of the Ferrous Materials section of this book (page 430), it can be seen that the tensile strength for AISI 1020 (annealed) is 57,250 PSI, which is equivalent to 28.6 tons/in.^{2}. From the list of k factors above, it can be seen that the correct factor for 10-times thickness is 1.2975.

$$F_B = (28.6 \times 1.2975 \times 5 \times 0.25^2) \div 3.5 = 3.31 \text{ tons.}$$

Drawing

Drawing is a complex bending process that transforms a flat metal blank into a hollow vessel. A ram is used to flow the metal into a preformed die cavity. When properly done, the finished workpiece will retain adequate thickness throughout, and will not exhibit visible stress signs such as wrinkling. Produced shapes may be round, conical, box-shape, oval, etc. When the depth of the die cavity equals or exceeds the minimum part width, the process is known as *deep drawing*. The basic dimensions and forces of a drawn part are shown in *Figure 3*.

As stated in the caption in the figure, the maximum drawing ratio (β) of the original blank diameter to the diameter of the ram is 2.0 for most materials, but, depending on acceptable quality requirements, ratios of up to 2.2 can be achieved on some steel and brass sheets. The failure mode is premature cracking in the bottom radius of the drawn part. Some ferritic stainless steels have a drawing ratio of no more than 1.55, and should be drawn at elevated temperatures. For most carbon steels, a ratio of 1.8 to 2.0 is recommended.

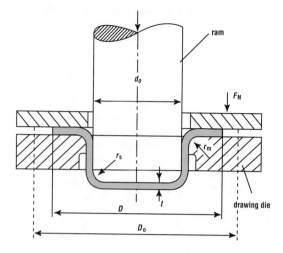

Figure 3. Drawing. D_O is the original diameter of the blank. D is the instantaneous size after drawing. d_o is the diameter of the ram. r_s is the radius of the ram. r_m is the radius of the drawing die. F_N is the hold-down force. The ram diameter is usually less than half of the original blank diameter; the radius of the die is approximately $1.5 \times d_o$, and the radius of the drawing die should be approximately eight times the thickness of the base of the drawn part. In order to avoid wrinkling, the hold-down force should range from 145 to 1,450 PSI (1 to 10 MPa).

In some operations, redrawing may be required to satisfy part requirements. This requires a second, or even a third, drawing operation to deepen the cavity formed by the first draw. Although this can often be accomplished without annealing, the drawing ratio should be dramatically reduced, usually to about 66% of β_{MAX}. After annealing, redrawing of most metals can be successfully done at up to 75% of β_{MAX}.

Determining blank size. For most drawing operations, the size of the blank required can be determined with a simple formula that can be used on cylindrical parts without a rounded bottom.

$$D_O = \sqrt{d^2 + 4dh}$$

where D_O = blank diameter
d = shell diameter
h = height.

This formula works well when the shell diameter-to-corner radius ratio (d/r) is 20 or more, but for lower ratios, the formula should be modified as follows. For d/r of 15 to 20, subtract $0.5\,r$ from the answer. For d/r of 10 to 15, subtract r from the answer. **Table 6** gives blank diameter formulas for other rotationally symmetrical drawn parts.

Shearing operations
Blanking and punching

Blanking is the process of using a press to cut a part out of metal sheet or plate (or other material). It is, in effect, the opposite of punching, where the sheet or plate is

Table 6. Blank Diameter Formulas for Rotationally Symmetrical Drawn Parts.

	$D_0 =$
	$\sqrt{d^2 + 4dh}$
	$\sqrt{d_2^2 + 4d_1 h}$
	$\sqrt{d_2^2 + 4(d_1 h_1 + d_2 h_2)}$
	$\sqrt{d_3^2 + 4(d_1 h_1 + d_2 h_2)}$
	$\sqrt{d_1^2 + 4d_1 h + 2f(d_1 + d_2)}$
	$\sqrt{d_2^2 + 4(d_1 h_1 + d_2 h_2) + 2f(d_2 + d_3)}$
	$\sqrt{2d^2} = 1,414d$
	$\sqrt{d_1^2 + d_2^2}$
	$1,414\sqrt{d_1^2 + f(d_1 + d_2)}$
	$1,414\sqrt{d^2 + 2dh}$
	$\sqrt{d_1^2 + d_2^2 + 4d_1 h}$
	$1,414\sqrt{d_1^2 + 2d_1 h + f(d_1 + d_2)}$
	$\sqrt{d_2^2 + 4h^2}$

(Continued)

Table 6. *(Continued)* **Blank Diameter Formulas for Rotationally Symmetrical Drawn Parts.**

	$D_0 =$
	$\sqrt{d_2^2 + 4h^2}$
	$\sqrt{d_2^2 + 4(h_1^2 + d_1 h_2)}$
	$\sqrt{d^2 + 4(h_1^2 + d h_2)}$
	$\sqrt{d_1^2 + 4h^2 + 2f(d_1 + d_2)}$
	$\sqrt{d_1^2 + 4\left[h_1^2 + d_1 h_2 + \dfrac{f}{2}(d_1 + d_2)\right]}$
	$\sqrt{d_1^2 + 2s(d_1 + d_2)}$
	$\sqrt{d_1^2 + 2s(d_1 + d_2) + d_3^2 - d_2^2}$
	$\sqrt{d_1^2 + 2\left[s(d_1 + d_2) + 2d_2 h\right]}$
	$\sqrt{d_1^2 + 6,28 r d_1 + 8r^2}$ or $\sqrt{d_2^2 + 2,28 r d_2 - 0,56 r^2}$
	$\sqrt{d_1^2 + 6,28 r d_1 + 8r^2 + d_3^2 - d_2^2}$ or $\sqrt{d_3^2 + 2,28 r d_2 - 0,56 r^2}$
	$\sqrt{d_1^2 + 6,28 r d_1 + 8r^2 + 4d_2 h + d_3^2 - d_2^2}$ or $\sqrt{d_3^2 + 4d_2(0,57r + h) - 0,56 r^2}$

(Continued)

Table 6. *(Continued)* **Blank Diameter Formulas for Rotationally Symmetrical Drawn Parts.**

	$D_0 =$
	$\sqrt{d_1^2 + 6,28rd_1 + 8r^2 + 2f(d_2 + d_3)}$ or
	$\sqrt{d_2^2 + 2,28rd_2 + 2f(d_2 + d_3) - 0,56r^2}$
	$\sqrt{d_1^2 + 6,28rd_1 + 8r^2 + 4d_2h + 2f(d_2 + d_3)}$ or
	$\sqrt{d_2^2 + 4d_2\left(0,57r + h + \dfrac{f}{2}\right) + 2d_3f - 0,56r^2}$
	$\sqrt{d_1^2 + 4(1,57rd_1 + 2r^2 + hd_2)}$ or
	$\sqrt{d_2^2 + 4d_2(h + 0,57r) - 0,56r^2}$

Source, *Handbook of Technical Formulas,* Karl-Friedrich Fischer, ed. Hanser Gardner Publications.

the part, and the material that is removed by the punch becomes scrap. In blanking, the outer contour of the part is produced in a single press stroke. Fine-blanking is a more refined evolution of the process that uses a triple action press (closing force, blanking force, and counterforce to eject the part) to produce highly accurate parts (to 0.001") to net or near-net shapes.

Due to the nature of blanking, layout considerations are important in order to reduce scrap. Blanking layout can be compared in some ways to maximizing the number of cookies that can be cut from a sheet of dough. The goal is to produce as many parts as possible from the available sheet. For most metals, blanks can be safely pierced from sheets as near as $3/4$" from the edge of the sheet, and up to $1/2$" from other blanks.

Pressures required for blanking and punching can be calculated with the following formula.

$$F = S_H \times l \times t$$

where F = force (or pressure) in lb

        S_H = shear strength of material in PSI

        l = length of the cutting line

        t = thickness of the material (inches).

This formula is simplified somewhat for cutting round holes, which is a common application for punch work.

$$F = S_H \times \pi D \times t$$

where D = the diameter of the hole, in inches.

The shear strength of a material is a measure of its ability to resist puncture, and absolute values can be difficult to calculate for a specific process. Factors that influence the shearing resistance include the curvature of the cutter, the clearance between the ram and the die, and sharpness of the tool–a blunt cutting edge can increase pressure requirements by as much as 1.6 times normal. Generally, shear strength is related to tensile strength, and for most operations a reasonable starting estimate for most materials is that the shear strength of a material will be approximately 75% of its tensile strength.

Table 7 provides pressure requirements for blanking (or punching) several materials of fractional thicknesses. The values in the Table show requirements per lineal inch of cut, and can be converted for larger or smaller cuts by multiplying the value by the number of lineal inches actually in the cut. **Table 8** provides pressure, in tons, required to punch holes of various diameters in standard gage material. Again, pressure requirements for holes of larger diameters can be estimated by multiplying values on the Table (see Note at end of the Table).

Wire gages

A variety of wire gage sizing standards have been developed both by the U.S.A. and abroad, and the most widely encountered are shown in **Table 9.** Due to the proliferation of gage titles, it is recommended that wire be ordered by decimal size rather than gage number, and both inch and millimeter decimal equivalents are provided in the Table. In the United States, the American Steel Wire (or Washburn and Moen) is generally considered the "standard" gage to specify, and in the United Kingdom, the British Imperial (or Standard Wire Gage [S.W.G.]) is the first choice. Weights of 1000 feet of copper, steel, and aluminum wire, by gage number, are given in **Table 10.**

Table 7. Pressure Required (in Tons) Per Lineal Inch for Blanking Various Materials. (See Notes.)

Thickness (inch)	Steel 0.25% Carbon 55,000	Steel 0.75% Carbon 80,000	Steel 1% Carbon 85,000	Iron Wrought 40,000	Aluminum 32 BHN 12,000	Aluminum 45 BHN 18,000	Aluminum 63 BHN 23,000	Copper Cast 25,000	Copper Rolled 28,000	Bronze Cast 35,000	Fiber Hard 20,000
1/64	0.4297	0.6250	0.6641	0.3125	0.0938	0.1406	0.1797	0.1953	0.2188	0.2734	0.1563
1/32	0.8594	1.2500	1.3281	0.6250	0.1875	0.2813	0.3594	0.3906	0.4375	0.5469	0.3125
3/64	1.2891	1.8750	1.9922	0.9375	0.2813	0.4219	0.5391	0.5859	0.6563	0.8203	0.4688
1/16	1.7188	2.5000	2.6563	1.2500	0.3750	0.5625	0.7188	0.7813	0.8750	1.0938	0.6250
5/64	2.1484	3.1250	3.3203	1.5625	0.4688	0.7031	0.8984	0.9766	1.0938	1.3672	0.7813
3/32	2.5781	3.7500	3.9844	1.8750	0.5625	0.8438	1.0781	1.1719	1.3125	1.6406	0.9375
7/64	3.0078	4.3750	4.6484	2.1875	0.6563	0.9844	1.2578	1.3672	1.5313	1.9141	1.0938
1/8	3.4375	5.0000	5.3125	2.5000	0.7500	1.1250	1.4375	1.5625	1.7500	2.1875	1.2500
9/64	3.8672	5.6250	5.9766	2.8125	0.8438	1.2656	1.6172	1.7578	1.9688	2.4609	1.4063
5/32	4.2969	6.2500	6.6406	3.1250	0.9375	1.4063	1.7969	1.9531	2.1875	2.7344	1.5625
11/64	4.7266	6.8750	7.3047	3.4375	1.0313	1.5469	1.9766	2.1484	2.4063	3.0078	1.7188
3/16	5.1563	7.5000	7.9688	3.7500	1.1250	1.6875	2.1563	2.3438	2.6250	3.2813	1.8750
13/64	5.5859	8.1250	8.6328	4.0625	1.2188	1.8281	2.3359	2.5391	2.8438	3.5547	2.0313
7/32	6.0156	8.7500	9.2969	4.3750	1.3125	1.9688	2.5156	2.7344	3.0625	3.8281	2.1875
15/64	6.4453	9.3750	9.9609	4.6875	1.4063	2.1094	2.6953	2.9297	3.2813	4.1016	2.3438
1/4	6.8750	10.0000	10.6250	5.0000	1.5000	2.2500	2.8750	3.1250	3.5000	4.3750	2.5000
17/64	7.3047	10.6250	11.2891	5.3125	1.5938	2.3906	3.0547	3.3203	3.7188	4.6484	2.6563
9/32	7.7344	11.2500	11.9531	5.6250	1.6875	2.5313	3.2344	3.5156	3.9375	4.9219	2.8125
19/64	8.1641	11.8750	12.6172	5.9375	1.7813	2.6719	3.4141	3.7109	4.1563	5.1953	2.9688
5/16	8.5938	12.5000	13.2813	6.2500	1.8750	2.8125	3.5938	3.9063	4.3750	5.4688	3.1250
21/64	9.0234	13.1250	13.9453	6.5625	1.9688	2.9531	3.7734	4.1016	4.5938	5.7422	3.2813

Notes: Values given are for one lineal inch, but required pressures for any hole can be determined by multiplying values on the Table by the number of lineal inches in the cut. 1 ton = 2,000 lb = 907.2 kg. 1,000 PSI = 6.895 MPa.

(Continued)

Table 7. *(Continued)* Pressure Required (in Tons) Per Lineal Inch for Blanking Various Materials. (See Notes.)

Thickness (inch)	Steel 0.25% Carbon 55,000	Steel 0.75% Carbon 80,000	Steel 1% Carbon 85,000	Iron Wrought 40,000	Aluminum 32 BHN 12,000	Aluminum 45 BHN 18,000	Aluminum 63 BHN 23,000	Copper Cast 25,000	Copper Rolled 28,000	Bronze Cast 35,000	Fiber Hard 20,000
11/32	9.4531	13.7500	14.6094	6.8750	2.0625	3.0938	3.9531	4.2969	4.8125	6.0156	3.4375
23/64	9.8828	14.3750	15.2734	7.1875	2.1563	3.2344	4.1328	4.4922	5.0313	6.2891	3.5938
3/8	10.3125	15.0000	15.9375	7.5000	2.2500	3.3750	4.3128	4.6875	5.2500	6.5625	3.7500
25/64	10.7422	15.6250	16.6016	7.8125	2.3438	3.5156	4.4922	4.8828	5.4688	6.8359	3.9063
13/32	11.1719	16.2500	17.2656	8.1250	2.4375	3.6563	4.6719	5.0781	5.6875	7.1094	4.0625
27/64	11.6016	16.8750	17.9297	8.4375	2.5313	3.7969	4.8516	5.2734	5.9063	7.3828	4.2188
7/16	12.0313	17.5000	18.5938	8.7500	2.6250	3.9375	5.0313	5.4688	6.1250	7.6563	4.3750
29/64	12.4609	18.1250	19.2578	9.0625	2.7188	4.0781	5.2109	5.6641	6.3438	7.9297	4.5313
15/32	12.8906	18.7500	19.9219	9.3750	2.8125	4.2188	5.3906	5.8594	6.5625	8.2031	4.6875
31/64	13.3203	19.3750	20.5859	9.6875	2.9063	4.3594	5.5703	6.0547	6.7813	8.4766	4.8438
1/2	13.7500	20.0000	21.2500	10.0000	3.0000	4.5000	5.7500	6.2500	7.0000	8.7500	5.0000
33/64	14.1797	20.6250	21.9141	10.3125	3.0938	4.6406	5.9297	6.4453	7.2188	9.0234	5.1563
17/32	14.6094	21.2500	22.5781	10.6250	3.1875	4.7813	6.1094	6.6406	7.4375	9.2969	5.3125
35/64	15.0391	21.8750	23.2422	10.9375	3.2813	4.9219	6.2891	6.8359	7.6563	9.5703	5.4688
9/16	15.4688	22.5000	23.9063	11.2500	3.3750	5.0625	6.4688	7.0313	7.8750	9.8438	5.6250
37/64	15.8984	23.1250	24.5703	11.5625	3.4688	5.2031	6.6484	7.2266	8.0938	10.1172	5.7813
19/32	16.3281	23.7500	25.2344	11.8750	3.5625	5.3438	6.8281	7.4219	8.3125	10.3906	5.9375
39/64	16.7578	24.3750	25.8984	12.1875	3.6563	5.4844	7.0078	7.6172	8.5313	10.6641	6.0938
5/8	17.1875	25.0000	26.5625	12.5000	3.7500	5.6250	7.1875	7.8125	8.7500	10.9375	6.2500
41/64	17.6172	25.6250	27.2266	12.8125	3.8438	5.7656	7.3672	8.0078	8.9688	11.2109	6.4063
21/32	18.0469	26.2500	27.8906	13.1250	3.9375	5.9063	7.5469	8.2031	9.1875	11.4844	6.5625

Notes: Values given are for one lineal inch, but required pressures for any hole can be determined by multiplying values on the Table by the number of lineal inches in the cut. 1 ton = 2,000 lb = 907.2 kg. 1,000 PSI = 6.895 MPa.

(Continued)

Table 7. *(Continued)* Pressure Required (in Tons) Per Lineal Inch for Blanking Various Materials. (See Notes.)

Thickness (inch)	Steel			Iron	Aluminum			Copper		Bronze	Fiber
	0.25% Carbon 55,000	0.75% Carbon 80,000	1% Carbon 85,000	Wrought 40,000	32 BHN 12,000	45 BHN 18,000	63 BHN 23,000	Cast 25,000	Rolled 28,000	Cast 35,000	Hard 20,000
43/64	18.4766	26.8750	28.5547	13.4375	4.0313	6.0469	7.7266	8.3984	9.4063	11.7578	6.7188
11/16	18.9063	27.5000	29.2188	13.7500	4.1250	6.1875	7.9063	8.5938	9.6250	12.0313	6.8750
45/64	19.3359	28.1250	29.8828	14.0625	4.2188	6.3281	8.0859	8.7891	9.8438	12.3047	7.0313
23/32	19.7656	28.7500	30.5469	14.3750	4.3125	6.4688	8.2656	8.9844	10.0625	12.5781	7.1875
47/64	20.1953	29.3750	31.2109	14.6875	4.4063	6.6094	8.4453	9.1797	10.2813	12.8516	7.3438
3/4	20.6250	30.0000	31.8750	15.0000	4.5000	6.7500	8.6250	9.3750	10.5000	13.1250	7.5000

Notes: Values given are for one lineal inch, but required pressures for any hole can be determined by multiplying values on the Table by the number of lineal inches in the cut. 1 ton = 2,000 lb = 907.2 kg. 1,000 PSI = 6.895 MPa.

Table 8. Pressure Required (in Tons) to Punch Round Holes in Sheet and Plate Steel (50,000 PSI Shear Strength). (See Notes.)

Mfg. Gage	Thickness (inches)	Hole Diameter (inches)							
		1/16	1/8	3/16	1/4	5/16	3/4	1/2	5/8
32	.0097	0.0476	0.0952	0.1428	0.1905	0.2381	0.2857	0.3809	0.4761
31	.0105	0.0515	0.1031	0.1546	0.2062	0.2577	0.3093	0.4123	0.5154
30	.0120	0.0589	0.1178	0.1767	0.2356	0.2945	0.3534	0.4712	0.5891
29	.0135	0.0663	0.1325	0.1988	0.2651	0.3313	0.3976	0.5301	0.6627
28	.0149	0.0731	0.1463	0.2194	0.2926	0.3657	0.4388	0.5851	0.7314
27	.0164	0.0805	0.1610	0.2415	0.3220	0.4025	0.4830	0.6440	0.8050
26	.0179	0.0879	0.1757	0.2636	0.3515	0.4393	0.5272	0.7029	0.8787
25	.0209	0.1026	0.2052	0.3078	0.4104	0.5130	0.6156	0.8207	1.0259
24	.0239	0.1173	0.2346	0.3520	0.4693	0.5866	0.7039	0.9386	1.1732
23	.0269	0.1320	0.2641	0.3961	0.5282	0.6602	0.7923	1.0564	1.3205
22	.0299	0.1468	0.2935	0.4403	0.5871	0.7339	0.8806	1.1742	1.4677
21	.0329	0.1615	0.3230	0.4845	0.6460	0.8075	0.9690	1.2920	1.6150
20	.0359	0.1762	0.3524	0.5287	0.7049	0.8811	1.0573	1.4098	1.7622
19	.0418	0.2052	0.4104	0.6156	0.8207	1.0259	1.2311	1.6415	2.0519
18	.0478	0.2346	0.4693	0.7039	0.9386	1.1732	1.4078	1.8771	2.3464
17	.0538	0.2641	0.5282	0.7923	1.0564	1.3205	1.5845	2.1127	2.6409
16	.0598	0.2935	0.5871	0.8806	1.1742	1.4677	1.7613	2.3483	2.9354
15	.0673	0.3304	0.6607	0.9911	1.3214	1.6518	1.9822	2.6429	3.3036
14	.0747	0.3667	0.7334	1.1001	1.4667	1.8334	2.2001	2.9335	3.6668
13	.0897	0.4403	0.8806	1.3209	1.7613	2.2016	2.6419	3.5225	4.4031
12	.1046	0.5135	1.0269	1.5404	2.0538	2.5673	3.0807	4.1076	5.1346
11	.1196	0.5871	1.1742	1.7613	2.3483	2.9354	3.5225	4.6967	5.8709
10	.1345	0.6602	1.3205	1.9807	2.6409	3.3011	3.9614	5.2818	6.6023

(Continued)

Table 8. *(Continued)* Pressure Required (in Tons) to Punch Round Holes in Sheet and Plate Steel (50,000 PSI Shear Strength). (See Notes.)

Mfg. Gage	Thickness (inches)	Hole Diameter (inches)							
		1/16	1/8	3/16	1/4	5/16	3/4	1/2	5/8
9	.1495	0.7339	1.4677	2.2016	2.9354	3.6693	4.4031	5.8709	7.3386
8	.1644	0.8070	1.6140	2.4210	3.2280	4.0350	4.8420	6.4560	8.0700
7	.1793	0.8801	1.7603	2.6404	3.5206	4.4007	5.2808	7.0411	8.8014
6	.1943	0.9538	1.9075	2.8613	3.8151	4.7689	5.7226	7.6302	9.5377
5	.2092	1.0269	2.0538	3.0807	4.1076	5.1346	6.1615	8.2153	10.2691
4	.2242	1.1005	2.2011	3.3016	4.4022	5.5027	6.6033	8.8043	11.0054
3	.2391	1.1737	2.3474	3.5210	4.6947	5.8684	7.0421	9.3895	11.7368

Mfg. Gage	Thickness (inches)	Hole Diameter (inches)							
		3/4	7/8	1	1 1/4	1 1/2	1 3/4	2	2 1/2
32	.0097	0.5714	0.6666	0.7618	0.9523	1.1428	1.3332	1.5237	1.9046
31	.0105	0.6185	0.7216	0.8247	1.0308	1.2370	1.4432	1.6493	2.0617
30	.0120	0.7069	0.8247	0.9425	1.1781	1.4137	1.6493	1.8850	2.3562
29	.0135	0.7952	0.9278	1.0603	1.3254	1.5904	1.8555	2.1206	2.6507
28	.0149	0.8777	1.0240	1.1702	1.4628	1.7554	2.0479	2.3405	2.9256
27	.0164	0.9660	1.1270	1.2881	1.6101	1.9321	2.2541	2.5761	3.2201
26	.0179	1.0544	1.2301	1.4059	1.7573	2.1088	2.4603	2.8117	3.5147
25	.0209	1.2311	1.4363	1.6415	2.0519	2.4622	2.8726	3.2830	4.1037
24	.0239	1.4078	1.6425	1.8771	2.3464	2.8157	3.2849	3.7542	4.6928
23	.0269	1.5845	1.8486	2.1127	2.6409	3.1691	3.6973	4.2255	5.2818
22	.0299	1.7613	2.0548	2.3483	2.9354	3.5225	4.1096	4.6967	5.8709
21	.0329	1.9380	2.2610	2.5840	3.2300	3.8759	4.5219	5.1679	6.4599
20	.0359	2.1147	2.4671	2.8196	3.5245	4.2294	4.9343	5.6392	7.0490
19	.0418	2.4622	2.8726	3.2830	4.1037	4.9245	5.7452	6.5659	8.2074

(Continued)

Table 8. *(Continued)* Pressure Required (in Tons) to Punch Round Holes in Sheet and Plate Steel (50,000 PSI Shear Strength). (See Notes.)

Mfg. Gage	Thickness (inches)	Hole Diameter (inches)							
		3/4	7/8	1	1 1/4	1 1/2	1 3/4	2	2 1/2
18	.0478	2.8157	3.2849	3.7542	4.6928	5.6313	6.5699	7.5084	9.3855
17	.0538	3.1691	3.6973	4.2255	5.2818	6.3382	7.3945	8.4509	10.5636
16	.0598	3.5225	4.1096	4.6967	5.8709	7.0450	8.2192	9.3934	11.7417
15	.0673	3.9643	4.6250	5.2857	6.6072	7.9286	9.2500	10.5715	13.2144
14	.0747	4.4002	5.1336	5.8669	7.3337	8.8004	10.2671	11.7339	14.6673
13	.0897	5.2838	6.1644	7.0450	8.8063	10.5676	12.3288	14.0901	17.6126
12	.1046	6.1615	7.1884	8.2153	10.2691	12.3229	14.3767	16.4306	20.5382
11	.1196	7.0450	8.2192	9.3934	11.7417	14.0901	16.4384	18.7868	23.4835
10	.1345	7.9227	9.2432	10.5636	13.2045	15.8454	18.4864	21.1273	26.4091
9	.1495	8.8063	10.2740	11.7417	14.6772	17.6126	20.5480	23.4835	29.3543
8	.1644	9.6840	11.2980	12.9120	16.1400	19.3680	22.5960	25.8240	32.2799
7	.1793	10.5617	12.3219	14.0822	17.6028	21.1233	24.6439	28.1644	35.2056
6	.1943	11.4452	13.3528	15.2603	19.0754	22.8905	26.7056	30.5206	38.1508
5	.2092	12.3229	14.3767	16.4306	20.5382	24.6459	28.7535	32.8611	41.0764
4	.2242	13.2065	15.4076	17.6087	22.0108	26.4130	30.8152	35.2173	44.0217
3	.2391	14.0842	16.4315	18.7789	23.4736	28.1684	32.8631	37.5578	46.9473

Notes: Larger hole sizes, or values for sizes not given on the table, can be determined by multiplying or adding values on the Table. For instance, it can be seen that the pressure needed to punch a two-inch hole is equal to twice the amount needed to punch a one-inch hole. The amount needed to punch a three-inch hole would be three times the amount needed for a one-inch hole. Likewise, it can be seen that the amount of pressure needed to punch a 1 ¹/₂" hole is equal to the sum of the amount needed to punch a one-inch hole plus a one-half inch hole. Likewise, the Table can be used to find pressures required for materials with other than 50,000 PSI (345 MPa) shear strength. For instance, for materials with 100,000 PSI shear strength, double the values on the Table; for materials with 25,000 PSI shear strength, halve the values on the Table. 1 ton = 2,000 lb = 907.2 kg. 1,000 PSI = 6.895 MPa.

Table 9. Standards for Wire Gages.

Wire Gage Number	U.S. Steel or Washburn & Moen		British Imperial Standard Wire Gage		American or Browne & Sharpe		Birmingham or Stubbs Iron Wire		Stubbs Steel Wire		Music Wire (MWG)	
	inch	mm	inch	mm	inch	mm	inch	mm	inch	mm	inch	mm
00000000 (8/0)	-	-	-	-	.731429	18.5783	-	-	-	-	-	-
0000000 (7/0)	.49000	12.4460	.50000	12.7000	.651356	16.5444	-	-	-	-	-	-
000000 (6/0)	.46150	11.7221	.46400	11.7856	.580049	14.7333	-	-	-	-	.004	.102
00000 (5/0)	.43050	10.9347	.43200	10.9728	.516549	13.1203	-	-	-	-	.005	.127
0000 (4/0)	.39380	10.0025	.40000	10.1600	.460000	11.6840	.4540	11.532	-	-	.006	.152
000 (3/0)	.36250	9.2075	.37200	9.4488	.409642	10.4049	.4250	10.795	-	-	.007	.178
00 (2/0)	.33100	8.4074	.34800	8.8392	.364797	9.2658	.3800	9.652	-	-	.008	.203
0	.30650	7.7851	.32400	8.2296	.324861	8.2515	.3400	8.636	-	-	.009	.229
1	.28300	7.1882	.30000	7.6200	.289279	7.3481	.3000	7.620	.227	5.7658	.010	.254
2	.26250	6.6675	.27600	7.0104	.257626	6.5437	.2840	7.214	.219	5.5626	.011	.279
3	.24370	6.1900	.25200	6.4008	.229423	5.8273	.2590	6.579	.212	5.3848	.012	.305
4	.22530	5.7226	.23200	5.8928	.204307	5.1894	.2380	6.045	.207	5.2578	.013	.330
5	.20700	5.2578	.21200	5.3848	.181941	4.6213	.2200	5.588	.204	5.1816	.014	.356
6	.19200	4.8768	.19200	4.8768	.162023	4.1154	.2030	5.156	.201	5.1054	.016	.406
7	.17700	4.4958	.17600	4.4704	.144285	3.6649	.1800	4.572	.199	5.0546	.018	.457
8	.16200	4.1148	.16000	4.0640	.128490	3.2636	.1650	4.191	.197	5.0038	.020	.508
9	.14830	3.7668	.14400	3.6576	.114424	2.9064	.1480	3.759	.194	4.9276	.022	.559
10	.13500	3.4290	.12800	3.2512	.101897	2.5882	.1340	3.404	.191	4.8514	.024	.610
11	.12050	3.0607	.11600	2.9464	.090742	2.3048	.1200	3.048	.188	4.7752	.026	.660
12	.10550	2.6797	.10400	2.6416	.080808	2.0525	.1090	2.769	.185	4.6990	.029	.737
13	.09150	2.3241	.09200	2.3368	.071962	1.8278	.0950	2.413	.182	4.6228	.031	.787

(Continued)

Table 9. *(Continued)* **Standards for Wire Gages.**

Wire Gage Number	U.S. Steel or Washburn & Moen		British Imperial Standard Wire Gage		American or Browne & Sharpe		Birmingham or Stubbs Iron Wire		Stubbs Steel Wire		Music Wire (MWG)	
	inch	mm	inch	mm	inch	mm	inch	mm	inch	mm	inch	mm
14	.08000	2.0320	.08000	2.0320	.064084	1.6277	.0830	2.108	.180	4.5720	.033	.838
15	.07200	1.8288	.07200	1.8288	.057068	1.4495	.0720	1.829	.178	4.5212	.035	.889
16	.06250	1.5875	.06400	1.6256	.050821	1.2908	.0650	1.651	.175	4.4450	.037	.940
17	.05400	1.3716	.05600	1.4224	.045257	1.1495	.0580	1.473	.172	4.3688	.039	.991
18	.04750	1.2065	.04800	1.2192	.040303	1.0237	.0490	1.245	.168	4.2672	.041	1.041
19	.04100	1.0414	.04000	1.0160	.035891	0.9116	.0420	1.067	.164	4.1656	.043	1.092
20	.03480	0.8839	.03600	0.9144	.031961	0.8118	.0350	0.889	.161	4.0894	.045	1.143
21	.03175	0.8065	.03200	0.8128	.028462	0.7229	.0320	0.813	.157	3.9878	.047	1.194
22	.02860	0.7264	.02800	0.7112	.025347	0.6438	.0280	0.711	.155	3.9370	.049	1.245
23	.02580	0.6553	.02400	0.6096	.022572	0.5733	.0250	0.635	.153	3.8862	.051	1.295
24	.02300	0.5842	.02200	0.5588	.020101	0.5106	.0220	0.559	.151	3.8354	.055	1.397
25	.02040	0.5182	.02000	0.5080	.017900	0.4547	.0200	0.508	.148	3.7592	.059	1.499
26	.01810	0.4597	.01800	0.4572	.015941	0.4049	.0180	0.457	.146	3.7084	.063	1.600
27	.01730	0.4394	.01640	0.4166	.014196	0.3606	.0160	0.406	.143	3.6322	.067	1.702
28	.01620	0.4115	.01480	0.3759	.012641	0.3211	.0140	0.356	.139	3.5306	.071	1.803
29	.01500	0.3810	.01360	0.3454	.011258	0.2859	.0130	0.330	.134	3.4036	.075	1.905
30	.01400	0.3556	.01240	0.3150	.010025	0.2546	.0120	0.305	.127	3.2258	.080	2.032
31	.01320	0.3353	.01160	0.2946	.008928	0.2268	.0100	0.254	.120	3.0480	.085	2.159
32	.01280	0.3251	.01080	0.2743	.007950	0.2019	.0090	0.229	.115	2.9210	.090	2.286
33	.01180	0.2997	.01000	0.2540	.007080	0.1798	.0080	0.203	.112	2.8448	.095	2.413
34	.01040	0.2642	.00920	0.2337	.006305	0.1601	.0070	0.178	.110	2.7940	.100	2.5400

(Continued)

Table 9. (Continued) Standards for Wire Gages.

Wire Gage Number	U.S. Steel or Washburn & Moen		British Imperial Standard Wire Gage		American or Browne & Sharpe		Birmingham or Stubbs Iron Wire		Stubbs Steel Wire		Music Wire (MWG)	
	inch	mm	inch	mm	inch	mm	inch	mm	inch	mm	inch	mm
35	.00950	0.2413	.00840	0.2134	.005615	0.1426	.0050	0.127	.108	2.7432	.106	2.6924
36	.00900	0.2286	.00760	0.1930	.005000	0.1270	.0040	0.102	.106	2.6924	.112	2.8448
37	.00850	0.2159	.00680	0.1727	.004453	0.1131	-	-	.103	2.6162	.118	2.9972
38	.00800	0.2032	.00600	0.1524	.003965	0.1007	-	-	.101	2.5654	.124	3.1496
39	.00750	0.1905	.00520	0.1321	.003531	0.0897	-	-	.099	2.5146	.130	3.3020
40	.00700	0.1778	.00480	0.1219	.003145	0.0799	-	-	.097	2.4638	.138	3.5052
41	.00660	0.1676	.00440	0.1118	.002800	0.0711	-	-	.095	2.4130	.146	3.7048
42	.00620	0.1575	.00400	0.1016	.002490	0.0633	-	-	.092	2.3368	.154	3.9116
43	.00600	0.1524	.00360	0.0914	.002220	0.0564	-	-	.088	2.2352	.162	4.1148
44	.00580	0.1473	.00320	0.0813	.001980	0.0502	-	-	.085	2.1590	.170	4.3180
45	.00550	0.1397	.00280	0.0711	.001760	0.0447	-	-	.081	2.0574	.180	4.5720
46	.00520	0.1321	.00240	0.0610	.001570	0.0398	-	-	.079	2.0066	-	-
47	.00500	0.1270	.00200	0.0508	.001400	0.0355	-	-	.077	1.9558	-	-
48	.00480	0.1219	.00160	0.0406	.001240	0.0316	-	-	.075	1.9050	-	-
49	.00460	0.1168	.00120	0.0305	.001110	0.0281	-	-	.072	1.8288	-	-
50	.00440	0.1118	.00100	0.0254	.000990	0.0251	-	-	.069	1.7526	-	-
51	-	-	-	-	.000880	0.0223	-	-	.066	1.6764	-	-
52	-	-	-	-	.000780	0.0199	-	-	.063	1.6002	-	-
53	-	-	-	-	.000700	0.0177	-	-	.058	1.4732	-	-
54	-	-	-	-	.000620	0.0158	-	-	.055	1.3970	-	-
55	-	-	-	-	.000550	0.0140	-	-	.050	1.2700	-	-
56	-	-	-	-	.000490	0.0125	-	-	.045	1.1430	-	-

Table 10. Weight per 1000 Linear Feet (304.8 Linear Meters) of Copper, Steel, and Aluminum Wire.

American Steel Wire Gage			Copper		Steel		Aluminum	
Number	inch	mm	lb	kg	lb	kg	lb	kg
0000000 (7/0)	.49000	12.4460	728.1073	330.2694	657.3918	298.1929	220.0036	99.79365
000000 (6/0)	.46150	11.7221	645.8722	292.9676	583.1436	264.5139	195.1556	88.5226
00000 (5/0)	.43050	10.9347	562.0171	254.9309	507.4327	230.1715	169.8181	77.02949
0000 (4/0)	.39380	10.0025	470.278	213.3181	424.6035	192.6001	142.0984	64.45582
000 (3/0)	.36250	9.2075	398.4916	180.7558	359.7892	163.2004	120.4075	54.61686
00 (2/0)	.33100	8.4074	332.2456	150.7066	299.9771	136.0696	100.3907	45.53724
0	.30650	7.7851	284.8814	129.2222	257.2131	116.6719	86.07928	39.04556
1	.28300	7.1882	242.8712	110.1664	219.283	99.46678	73.38555	33.28769
2	.26250	6.6675	208.9594	94.78396	188.6647	85.57833	63.1388	28.63976
3	.24370	6.1900	180.1002	81.69346	162.6085	73.7592	54.41877	24.68436
4	.22530	5.7226	153.9308	69.82302	138.9807	63.04165	46.51147	21.0976
5	.20700	5.2578	129.9403	58.94092	117.3202	53.2i644	39.26254	17.80949
6	.19200	4.8768	111.7907	50.70826	100.9333	45.78336	33.77848	15.32192
7	.17700	4.4958	95.00571	43.09459	85.77854	38.90915	28.70676	13.02139
8	.16200	4.1148	79.58537	36.09992	71.85585	32.59382	24.04738	10.90789
9	.14830	3.7668	66.69381	30.25231	60.21635	27.31414	20.15209	9.140986
10	.13500	3.4290	55.26762	25.06939	49.8999	22.63459	16.69957	7.574924
11	.12050	3.0607	44.0329	19.97332	39.75632	18.03347	13.30491	6.035105
12	.10550	2.6797	33.75267	15.31021	30.47453	13.82325	10.19865	4.626107
13	.09150	2.3241	25.38899	11.51644	22.92315	10.39794	7.671493	3.479789
14	.08000	2.0320	19.40811	8.803517	17.52315	7.948499	5.86432	2.660056
15	.07200	1.8288	15.72057	7.130849	14.19375	6.438284	4.750099	2.154645
16	.06250	1.5875	11.84577	5.37324	10.69528	4.851379	3.579297	1.623569
17	.05400	1.3716	8.842819	4.011103	7.983984	3.621535	2.671931	1.211988
18	.04750	1.2065	6.842116	3.103584	6.177594	2.802157	2.067402	0.937773
19	.04100	1.0414	5.097661	2.312299	4.602564	2.087723	1.5403	0.69868
20	.03480	0.8839	3.672499	1.665846	3.315817	1.504055	1.109676	0.503349
21	.03175	0.8065	3.056966	1.38664	2.760067	1.251966	0.923688	0.418985
22	.02860	0.7264	2.480477	1.125145	2.239568	1.015868	0.749497	0.339972
23	.02580	0.6553	2.018564	0.915621	1.822517	0.826694	0.609926	0.276662
24	.02300	0.5842	1.604201	0.727666	1.448398	0.656993	0.484723	0.21987
25	.02040	0.5182	1.262012	0.572449	1.139443	0.516851	0.381327	0.17297
26	.01810	0.4597	0.993483	0.450644	0.896993	0.406876	0.300189	0.136166
27	.01730	0.4394	0.907602	0.411688	0.819454	0.371704	0.274239	0.124395
28	.01620	0.4115	0.795854	0.360999	0.718559	0.325938	0.240474	0.109079
29	.01500	0.3810	0.682316	0.309499	0.616048	0.279439	0.206168	0.093518
30	.01400	0.3556	0.594373	0.269608	0.536646	0.243423	0.179595	0.081464
31	.01320	0.3353	0.528386	0.239676	0.477068	0.216398	0.159656	0.07242

(Continued)

Table 10. *(Continued)* **Weight per 1000 Linear Feet (304.8 Linear Meters) of Copper, Steel, and Aluminum Wire.**

American Steel Wire Gage			Copper		Steel		Aluminum	
Number	inch	mm	lb	kg	lb	kg	lb	kg
32	.01280	0.3251	0.496848	0.22537	0.448593	0.203482	0.150127	0.068097
33	.01180	0.2997	0.422248	0.191532	0.381238	0.17293	0.127586	0.057873
34	.01040	0.2642	0.327997	0.148779	0.296141	0.13433	0.099107	0.44955
35	.00950	0.2413	0.273685	0.124143	0.247104	0.112086	0.082696	0.037511
36	.00900	0.2286	0.245634	0.11142	0.221777	0.100598	0.07422	0.033666
37	.00850	0.2159	0.219099	0.099383	0.19782	0.089731	0.066203	0.03003
38	.00800	0.2032	0.194081	0.088035	0.175231	0.79485	0.058643	0.026601
39	.00750	0.1905	0.170579	0.077375	0.154012	0.06986	0.051542	0.023379
40	.00700	0.1778	0.148593	0.067402	0.134162	0.060856	0.044899	0.020366
41	.00660	0.1676	0.132096	0.059919	0.119267	0.054099	0.039914	0.018105
42	.00620	0.1575	0.11657	0.052876	0.105248	0.047741	0.035223	0.015977
43	.00600	0.1524	0.109171	0.04952	0.098568	0.04471	0.032987	0.014963

Note: Weights based on: Copper 556 lb/ft^3; Steel 502 lb/ft^3; Aluminum 168 lb/ft^3.

Precision Machining with Lasers [1]

Introduction

Lasers are one of the most important tools of the microtechnology age. Their use has become so widespread that they are now found in everything from small consumer products such as compact disc players to large, heavy engineering plants for cutting and welding. Furthermore, the range of parts produced using lasers is even larger and growing ever more rapidly.

Since the first demonstration of laser action in 1960 by Ted Maiman at Hughes Research Laboratories in the USA, there has been an incredibly rapid development in laser technology and its applications. Lasers now exist in all kinds of different media, covering wavelengths from X-rays to the far infrared, ranging in pulse duration from a few femtoseconds to continuous wave and possessing powers which cover a span of some 30 orders of magnitude. As the functionality of lasers has increased, so has the range of disciplines in which lasers have had a major impact, and many new products or devices would not be possible without the use of lasers.

This section provides a brief introduction to lasers and their applications, with an emphasis on developments in laser micromachining. Since this is a vast field, only an overall guide to the relevant topics can be provided, but the aim is to present the reader with an introductory overview. The demands of such a short treatment necessitate the omission of many aspects; but the fundamentals of modern laser micromachining, particularly for industrial applications, are introduced. In addition, the respective properties and uses of the three most commonly used and important lasers, in this area—solid-state, carbon dioxide, and excimer lasers—are examined.

Laser basics

Lasers are systems for the generation and amplification of light, where "light" can refer to any electromagnetic wavelength, e.g., X-rays, ultraviolet, visible, infrared. The term "laser" is an acronym for *Light Amplification by Stimulated Emission of Radiation*. The geometry, functions, materials, output characteristics, and many other details can vary greatly in lasers, but the main properties of most lasers are, generally speaking, very similar.

In its simplest form, a laser consists of a gain medium, a method of pumping the gain medium, and an optical cavity (also known as the resonator). The gain medium lends its name to the type of laser—for example, ruby, carbon dioxide, helium-neon, rhodamine dye—and the properties of the laser medium largely determine the output wavelength characteristics of the system. The output wavelength (or wavelengths) of the laser arise from electronic transitions in the gain medium, and many thousands of transitions have been demonstrated so far. The optical cavity of the laser determines many of the other properties of the laser including how the laser beam propagates and what coherence characteristics the laser possesses.

Spontaneous and stimulated emission

A simple atomic system of two energy levels E_1 and E_2 is shown in *Figure 1*. An electron can change from the lower level to a higher one by absorbing an incoming photon where the energy of the photon is $h\nu = E_2 - E_1$. When the electron is in the higher energy state E_2, there is a finite probability per unit time that it will undergo a transition to the lower energy state E_1 and, in the process, emit a photon of energy $h\nu = E_2 - E_1$.

---

[1] This section, Precision Machining with Lasers, was contributed by Nadeem Rizvi of Exitech Limited.

This process, which takes place without the influence of an external field, is called *spontaneous emission*. The amount of time, on average, that the electron spends in the upper level once it has been excited there from the lower level depends on the atomic system under consideration, i.e., it depends on the laser gain medium. This time is also known as the upper state lifetime.

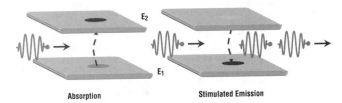

Absorption   **Stimulated Emission**

Figure 1. Photon absorption and emission.

If an electron is occupying the upper level E_2 and is then subjected to a photon with a frequency $v = (E_2-E_1)/h$, it can make a transition to the lower level E_1. This change of level by the electron is accompanied by the emission of another photon with an energy $hv = E_2-E_1$ and this process is termed *stimulated emission*. The photon, which is emitted by stimulated emission, is identical to the inducing photon in its frequency, polarization, phase, and direction. Stimulated emission is the fundamental process by which laser light is generated. Another important characteristic of stimulated emission is that the induced rate of emission is proportional to the intensity of the external field.

Population inversion

One necessary condition for laser action to occur is that the upper level from which the transitions are to take place should contain more electrons than the lower level—this is termed *population inversion*. This arises from the fact that the probability of stimulated emission between two levels is exactly the same as the probability of stimulated absorption between those two levels and hence the dominant process—emission or absorption—is governed by whichever level has the highest population number. Neither emission nor absorption can be preferentially chosen over the other process. Thus, if there is more population in the lower level, the light will, on average, be absorbed and the incoming signal reduced; if the upper level has the highest population, then the light will, on average, be amplified. The stimulated emission resulting from a population inversion thus provides the central optical amplification present in lasers.

Most practical laser systems use a four-level scheme in terms of their electronic transitions. *Figure 2* represents a simplified diagram of a four-level laser, and the important general factors to note are as follows.

- The pumping of the laser gain medium excites some of the population from the ground-state E_0 into an upper level E_3. In practice, E_3 may be a series of closely spaced levels instead of a single level, which helps in maximizing the absorption.
- The excited atoms cascade down to the upper laser level E_2. This is ideally a nonradiative process so that no photon emissions occur during the filling of the upper laser level. The atoms then drop down to the lower laser level accompanied by photon emission as already described.
- The atoms in the lower laser level then relax down to the ground state once more, ready for subsequent pumping.

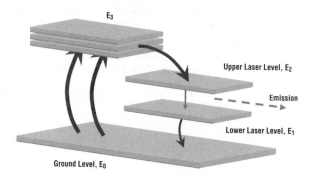

Figure 2. Representation of energy levels of a four-level laser.

One of the reasons why four-level lasers are generally more efficient than three-level systems is that in a four-level system, once some of the excited species are in level E_2, a population inversion automatically exists since level E_1 is not usually populated.

The necessary condition for obtaining population inversion is that the rate of relaxation from the lower laser level to the ground state should be faster than the rate of relaxation from the upper laser level into the lower laser level. One of the methods for ensuring this is that the relaxation time of the upper laser level—the upper state lifetime—should be long. In many practical lasers, the upper laser levels are metastable, which means that they can have lifetimes of many hundreds of microseconds. If such long upper state lifetimes are combined with short lifetimes of the lower laser level, then population inversion can be ensured if there is efficient pumping into the upper laser level.

Different laser systems have widely varying details of relaxation rates, lifetimes, efficiencies, etc., so the exact nature of specific lasers depends on their particular characteristics.

Resonators

A working practical laser is obtained by adding some feedback to the stimulated emission taking place from the pumped gain medium. The simplest laser resonator consists of the gain medium and two mirrors.

One mirror is usually highly reflecting to the laser wavelength and so reflects all of the light back into the resonator, or laser cavity. The other mirror is normally partially transmissive to allow some output from the laser. Hence, the feedback provided by the resonator allows a rapid buildup in the intracavity photon number whereas the partially reflecting output coupler enables the laser to have a usable output.

There are many different possible configurations of laser cavities, and lasers can be constructed in many forms; these may include, for example:

- a large number of mirrors,
- the layout of the cavity may follow a complicated path,
- there may be many foci inside the resonator,
- the cavity may include elements other than the laser medium (e.g., nonlinear crystals, prisms, semiconductor devices), and
- the cavity mirrors may be formed on the ends of the laser medium itself.

The exact design of the laser cavity is dependent, among other criteria, on the laser medium, the pumping geometry, how the laser is to be used, and the wavelength coverage required. The resonator geometry and design largely determine the output beam characteristics of the laser, and the main properties that are affected are beam divergence, spatial coherence, temporal coherence, focusability, and polarization. The contents of the resonator also affect the optical characteristics of the laser such as wavelength and linewidth (sometimes also called bandwidth).

One way of treating the propagation of light in laser resonators is to use matrix methods. In this method, each cavity element (e.g., mirror, lens, prism, nonlinear crystal) is represented by a matrix and the effects of successive cavity elements can be calculated to obtain the modes of light that propagate inside the resonator. This method provides a way of calculating the effects of particular cavity geometries on the optical modes inside the laser and is useful in evaluating how cavities behave and what the beam characteristics are.

Stability

A resonator may be said to be unstable if the beams do not reproduce themselves after a large number of roundtrips in the cavity. Mode analysis, for example, using matrix methods, is required to determine whether the optical field inside a particular resonator is stable or not.

The resonator geometry affects the propagation and coherence aspects of the output laser beam and so is sometimes an important condition to consider. Of the lasers detailed in this review, only excimer lasers are commonly used with unstable resonators, and even then their use is for special applications such as lithography or holography.

Resonator modes

There are two types of modes in a laser cavity: *transverse* modes and *longitudinal* modes. Transverse modes govern the spatial characteristics of the beam, and the longitudinal modes determine the spectral content of the laser emission.

The transverse mode structure is determined by the cavity geometry and by other optional apertures or cavity constraints that may be included. Many transverse modes can oscillate at one time although, in practice, the laser resonator is usually constrained to operate with just one single transverse mode. This is usually the lowest order mode, which is termed TEM_{00}. The terminology for TEM (transverse electromagnetic) modes derives from the behavior of the electric and magnetic fields at the boundaries of the laser resonator, and the two subscripts designate the number of nulls in the fields, e.g., TEM_{21} has two nulls in one axis and one null in the orthogonal axis.

The TEM_{00} mode is often called the fundamental mode of the laser. It is the smallest possible mode for a particular laser and hence can easily be selected by the placement of a suitable aperture in the cavity. Since the other transverse modes are larger in size, the aperture imparts a higher loss to those modes and thus only the fundamental mode propagates with sufficient gain to maintain oscillation. One drawback for the TEM_{00} is that, due to its small mode volume, it extracts less laser power than other transverse modes and hence high-order TEM modes often produce higher powers than the fundamental mode. Despite the lower power available, however, most solid-state and carbon dioxide lasers operate with the TEM_{00} mode due to its superior propagation and focusing characteristics.

In the case of excimer lasers, the nature of the resonator design and gas discharge mean that the output beam is highly multimode in nature, containing many thousands of spatial modes. This makes the excimer laser output very spatially incoherent, but, as will be seen later, this can be useful in beam homogenization and mask projection systems.

As already explained, a laser resonator consists of (at least) two mirrors separated by a distance L. The propagation of two travelling waves in opposite directions inside the laser produces a standing wave, and each mode of this standing wave is a longitudinal mode of the cavity. The longitudinal modes define the spectral content of the laser.

The criterion for the longitudinal modes is that an integral number of half-wavelengths fit exactly into the cavity. The frequency spacing of the modes is calculated as follows. The condition for an integral number of half-wavelengths in the resonator gives:

$$L = \frac{n\lambda}{2}$$

where n = an integer
L = the separation of the mirrors
λ = the wavelength of light.

In terms of frequency of the nth mode, f_n:

where c = the speed of light.

Since the frequency of the f_{n+1} mode is $(n+1)(c/2L)$, it can be seen that the frequency separation Δf is given by:

$$\Delta f = \frac{c}{2L}$$

It can be seen that the frequency spacing is independent of the laser wavelength but only depends on the spacing between the resonator mirrors. Hence, for a cavity that is 1 m in length, the frequency separation of longitudinal modes is 150 MHz—thus many thousands of spectral (or longitudinal) modes can be contained under the laser gain profile.

The number of longitudinal modes which fall under the laser gain profile and which are above the losses of the resonator are the only ones which are allowed to oscillate, as shown in *Figure 3*. The combination of the gain and the modes determines the spectral output of the laser as shown.

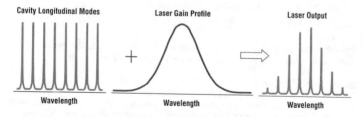

Figure 3. Longitudinal modes of a laser.

Coherence

Some applications of lasers require the laser source to have a particular set of optical properties, and one of the most important properties is that of coherence. There are two types of coherence—spatial coherence and temporal coherence.

Spatial coherence. The laser beam has coherence transverse to its direction of propagation. This is termed spatial coherence (or sometimes lateral or transverse coherence). Spatial coherence is the property of two distant points *across* a laser beam front to have a fixed phase relationship. This length, in effect, gives a distance over which a part of a beam front can interfere with another part of the same beam front. The spatial coherence length L_{SP}, is given by:

$$L_{SP} \approx \frac{\lambda}{\Theta_D} \approx \pi \omega_0$$

where Θ_D = the full-angle divergence and ω_0 is the fundamental beam waist diameter of the laser beam.

Spatial coherence is an important property in applications that require the interference of beams, for example, in holographic setups.

Temporal coherence. Temporal coherence results from the spectral output of the laser source and is inversely proportional to the laser linewidth (or bandwidth): the smaller the laser linewidth, the larger the temporal coherence. The temporal coherence time is defined as:

$$\tau_c \approx \frac{1}{\Delta v_L}$$

where Δv_L = the laser linewidth.

The term temporal coherence length, L_C, is defined as the maximum separation between two points *along* the beam propagation direction for a fixed phase relationship to be maintained between those points. The relationship between the coherence time and temporal coherence length is given by:

$$L_C = c\tau_c$$

One way of looking at temporal coherence is that if one is attempting to interfere two beams from the same laser source (e.g., a beam having been split by a beam-splitter and then recombined in, for example, an interferometric setup), then the maximum path length difference that there can be between the two beams for effective interference is given by the temporal coherence length.

Since $\tau_c \approx 1/\Delta v_L$, the temporal coherence length, is given by:

$$L_c \approx \frac{c}{\Delta v_L}$$

and it is noted that this also provides a measure of the extent of monochromaticity of the laser, i.e., very narrow linewidth lasers have a long temporal coherence length.

Applications of Coherence

There are uses of lasers where the spectral or spatial content of the laser beam can have significant effects on the desired results and so coherence effects have to be addressed carefully. A laser application that can require spatially or temporally coherent laser beams is in telecommunications—the production of gratings in optical fibers, or fiber Bragg gratings (FBGs) as they are more commonly known.

There are two main methods for producing FBGs. One is termed interferometric and the other is the proximity phase mask method. In the interferometric method, the laser beam is split into two beams and the two beams are then recombined at the fiber, producing an interference pattern resulting in the formation of the grating. The angle of the two beams determines the period of the grating. In this method, the temporal coherence length is important in that the optical path length difference between the two beams needs to be less than the temporal coherence length of the laser, otherwise efficient interference does not take place.

In the proximity phase mask method, the laser beam is diffracted by a phase mask, and the resulting diffracted beams interfere in the fiber to form the grating. The spatial coherence is more important in this method because the spatial coherence length, i.e., the separation of parts of the beam which can take part in the interference, effectively defines how far the fiber can be from the phase mask in order for effective interference to occur. Therefore, spatially coherent lasers are used in this type of FBG production and these lasers include unstable resonator excimer lasers and fundamental mode argon-ion lasers.

Line-narrowing

The two most commonly used excimer laser transitions for micromachining applications—KrF and ArF—have relatively broadband emissions with a gain bandwidth of the order of 300-400pm. In certain applications, the multiwavelength nature of these lasers can be an issue and techniques are sometimes employed to reduce the range of spectral outputs. This is termed line-narrowing.

Line-narrowing alters the gain profile of the laser by only allowing a very small number of wavelengths to lase, thereby reducing the spectral content of the output. This reduction in the laser linewidth is accompanied by an increase in the temporal coherence length of the laser, as already described. There are different methods for line-narrowing of various lasers but all of them require the insertion of some elements inside the laser cavity which can discriminate against certain wavelengths of the laser. In the case of excimer lasers, the two most commonly used methods for line-narrowing involve etalons or gratings. Prisms can also be used as intracavity devices for spectral selection but suffer from problems similar to that of etalons in that they are made from transmissive optical materials and so their use with UV light induces defects or color-centers over time. Although gratings do not, in general, provide as much dispersion as prisms, their reflective nature means they are often preferred for intracavity use.

An etalon usually consists of two plane-parallel plates separated by a distance d (see *Figure 4*). The plates normally have optical coatings on them that give each plate a particular reflectivity R_1 and R_2. Normally, it is the case that $R_1 = R_2 = R$. The operation

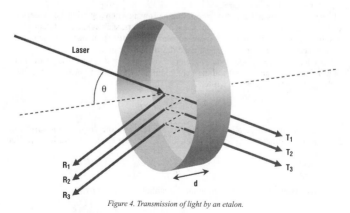

Figure 4. Transmission of light by an etalon.

of the etalon relies on constructive and destructive interference between the multiple reflections off the two plates as light passes through the etalon. Only light of certain wavelengths has the correct phase after the multiple reflections off the plates to interfere constructively and be passed by the etalon. Other wavelengths do not interfere constructively and hence are not transmitted.

The output spectrum of an etalon consists of a series of peaks that are the result of multiple reflections from the two etalon surfaces and their subsequent interference with each other. The separation of the peaks is called the *free spectral range* (*FSR*) and is purely a function of the distance d between the two plates, the refractive index of the medium between the plates n, and the angle of incidence θ, and is given by:

$$FSR = \frac{1}{2nd \cos \theta}$$

If d is given in centimeters, then the FSR is in terms of wavenumbers (cm^{-1}). The *FSR* can also be given in terms of frequency and wavelength as follows:

$$FSR = \frac{c}{2nd \cos \theta} \quad (Hz) \quad \text{or} \quad FSR = \frac{\lambda^2}{2nd \cos \theta} \quad (m)$$

The finesse F of the etalon is the ratio of the separation of the peaks to their width (*FWHM*) and is given by:

$$F = \frac{FSR}{FWHM}$$

The finesse is only a function of the reflectivity of the plates and is given by:

$$F = \frac{\pi \sqrt{R}}{1 - R}$$

By using etalons of different free spectral range and finesse inside a laser cavity, very narrow spectral output can be ensured. The *FSR* needs to be large enough such that only one transmission peak of the etalon falls within the gain profile of the laser (otherwise more than one line will lase) and the finesse needs to be high enough that a sufficiently narrow linewidth emission occurs.

In general, etalons give a high degree of line-narrowing without altering the shape of the laser beam. Their main disadvantage for use with UV lasers, such as excimer lasers, is that they suffer from the build-up of UV-induced color-centers after prolonged exposure to the laser light and their performance can degrade significantly with use.

If a reflection grating (see *Figure 5*), which has a spacing d between its grooves, has light of λ incident at an angle θ and the light is diffracted into the nth order at an angle ϕ, then the grating equation can be written as:

$$n\lambda = d(\sin\theta + \sin \phi)$$

In the Littrow configuration, the grating is rotated so the angle of the diffracted order is made identical to the angle of incidence, i.e., $\theta = \phi$, and the grating equation simplifies to:

$$n\lambda = 2d \sin\theta$$

The dispersion D_L for a Littrow-mounted grating is then given by:

$$D_L = \frac{2 \tan \theta}{\lambda}$$

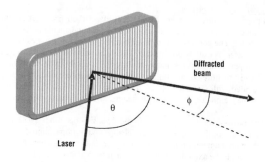

Figure 5. Reflection of light by a grating.

The dispersion of the grating limits the minimum linewidth that can be achieved for light incident in one pass on the grating:

$$\delta\lambda_D = \frac{\lambda\delta\theta}{2\tan\theta} = \frac{\lambda w}{2L\tan\theta}$$

where w = the width of the beam (limited by an intracavity aperture or slit)
L = the distance of the grating from the aperture.

The *resolution* for the Littrow configuration is given by:

$$R_L = wD_L$$

The resolution of the grating also limits the minimum linewidth that can be achieved for light incident in one pass on the grating:

$$\delta\lambda_R = \frac{\lambda^2}{2w\tan\theta}$$

The actual linewidth achieved in a single pass on a grating in the Littrow configuration is a combination of the limits set by the grating dispersion and resolution and the amount of line-narrowing can be obtained from :

$$\delta\lambda = (\delta\lambda_D{}^2 + \delta\lambda_R{}^2)^{1/2} \approx \frac{\lambda}{2\tan\theta}\left[\left(\frac{w}{L}\right)^2 + \left(\frac{\lambda}{w}\right)^2\right]^{1/2}$$

Because of the multiple bounces for light incident on a grating when used as an end reflector in a laser cavity, the actual linewidth produced by the laser is usually less than that calculated. It can be seen that the incident angle of illumination of a grating primarily determines the minimum linewidth achievable in the diffracted order.

Due to the $(1/\tan\theta)$ dependence of the line narrowing, θ should be made as large as possible to achieve minimum linewidth provided that the grating still diffracts efficiently. In general, the more grazing the angle of incidence, the higher the resolution produced by a grating but at the sacrifice of diffraction efficiency. Using a grating that has grooves with a saw-toothed profile—a *blazed* grating—is a convenient method for achieving high diffraction efficiencies at relatively high angles of incidence.

Propagation of Gaussian beams

The important features of laser beams that affect their propagation and focusing properties—modes and divergence, for example—arise fundamentally from the design of the laser resonator. The confinement of laser beams inside laser cavities for solid-state or carbon dioxide lasers is usually governed by Gaussian beam optics and this treatment then allows other properties such as resonator stability and focusability to be obtained. Excimer lasers, as already mentioned, are highly multimode in nature and their propagation is not usually treated with Gaussian optics. The main issue to keep in mind with excimer lasers is that they emit highly divergent beams, which have asymmetric divergences in each axis.

The beam cross-section at any point in the Gaussian beam has a Gaussian intensity profile but the width of the intensity profile changes with propagation distance along the axis. The minimum diameter of the beam $2\omega_0$ occurs at the point where the phase front of the beam is planar. If the axial distance z is taken as zero at this waist, then the variation in the size of the beam with distance z can be expressed as:

$$\omega_z = \omega_0 \left[1 + \left(\frac{\lambda z}{\pi \omega_0^2} \right)^2 \right]^{1/2}$$

where λ = the wavelength of light and z is the axial distance.

Hence, knowing the size and position of the Gaussian beam waist allows the beam size to be determined at any propagation distance.

The important features of this form of propagation are shown in the *Figure 6*. The full-angle divergence of the beam as shown in the figure is given by:

$$\Theta_D = \frac{2\lambda}{\pi \omega_0}$$

and this is the smallest spread that a beam with a waist size of $2\omega_0$ can have.

A useful parameter is the "M^2" value of the beam, which gives a measure of the deviation of the beam from the ideal, diffraction-limited propagation. The M^2 factor is given by:

$$M^2 = \frac{\pi \omega_0 \Theta_D}{2\lambda}$$

If M^2 is equal to unity, then the beam is said to "diffraction limited." Other values of M^2, which are always greater than unity, mean that the beam is M^2 times diffraction limited, and so relate to how the beam will propagate and can be focused.

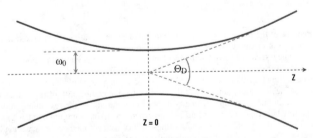

Figure 6. Propagation of a Gaussian beam.

Focusing of Gaussian beams

If a thin lens of focal length f is used to focus a Gaussian beam (which has a divergence Θ_D as shown above), then the focal spot diameter d is given by $d = f\Theta_D$:

$$d = \frac{4f\lambda}{\pi D}$$

where $D = 2\omega_0$, the beam diameter of the beam at the lens.

The *Rayleigh Range* for a Gaussian beam is defined as the distance over which the beam size increases by a factor of $\sqrt{2}$. The Rayleigh Range can be used to give a measure of the depth of focus of a focused beam and it is given by:

$$Z_R = \pm \frac{\pi\omega_0^2}{\lambda}$$

Terminology

Some common terms used in laser technology are defined in **Table 1**. Assume that a laser emits pulses at a repetition rate R (Hz) and at a wavelength λ (m), with single-pulse energy E (J), and pulse duration $\Delta\tau$ (s), and is then imaged or focused to a spot size with area A (cm^2).

Table 1. Common Laser Terminology and Symbols.

Property	Definition	Units
Average Power, P_{AV}	$E \cdot R$	W
Peak Power, P_P	$E/\Delta\tau$	W
Fluence, F	E/A	J/cm^2
Peak Intensity, I	$E/(\Delta\tau \cdot A)$	W/cm^2

Types of lasers

Many different types of lasers are in use today. These range from low-power devices like diode lasers for pattern writing or printing applications, to multi-kilowatt systems for welding and cutting of thick metals. The three most commonly used lasers for micromachining applications are carbon dioxide lasers, solid-state lasers, and excimer lasers. Details of their use and operation follow.

Carbon dioxide (CO$_2$) lasers

The carbon dioxide laser is a gas laser that emits light in the infrared region at different lines around a wavelength of 10 microns. Carbon dioxide lasers are widely used in many industrial laser applications, particularly where high continuous powers are required. These applications are most commonly ones that involve cutting or drilling of thick metals, but carbon dioxide lasers have also recently begun to be used in other research areas and for medical applications.

The carbon dioxide laser relies on the transitions of the CO$_2$ molecule. The transitions occur as a result of the vibrational and rotational modes of the molecule. Since many different vibrational and rotational modes can exist, CO$_2$ lasers can emit on many lines but the two main wavelengths are around 10.6 mm and 9.6 mm. Apart from carbon dioxide, the CO_2 laser also typically contains other gases such as helium or nitrogen. The nitrogen helps in the efficient transfer of the energy from the electrical discharge (which pumps the laser) into the CO_2 molecule, and the helium helps in the depopulation of the lower laser level.

Pumping of the CO_2 laser can be achieved by an electric discharge or with RF excitation, and optimized systems can have efficiencies of up to 20%. Industrial lasers exist that can emit continuous output powers of tens of kilowatts, so CO_2 lasers are among the highest power systems and are already highly developed, rugged industrial tools.

There are different types of CO_2 lasers that exhibit different output properties. The main types that are used in microengineering applications are as follows.

Sealed tube. A simple design where a tube filled with the laser and buffer gases is excited by electrical discharge. The cavity is formed by two mirrors on the ends of the tube and the output scales with the length of the tube at around 50 W/m. One problem with sealed tube lasers is that the heat generated by the discharge needs to be removed from the gases for efficient operation, and this heat extraction sets the practical limit on the power available.

Flowing gas. By making the gas flow, or circulate, the heat extraction can be increased dramatically, since it takes place by convection rather than conduction. Designs where a low pressure gas is flowed slowly can provide power extraction of around 100 W/m, but fast-axial flow lasers, where the gases are flowed out of the tube, cooled, and then reinjected, can produce output powers of many kilowatts.

Transverse flow. These systems make the gases flow transversely to the laser axis and the discharge is also orthogonal to the laser beam. Both these points make the heat extraction very efficient, and about 10 kW/m can be obtained for the output power, although at the expense of inferior beam quality.

TEA. The types of lasers listed above operate in continuous mode and the discharge properties limit the internal gas pressures to a few hundred millibars. In pulsed operation mode, this restriction is not important and much higher pressures can be used, thereby increasing greatly the gas density and hence the available energy. These lasers are called transversely excited atmospheric, or TEA, lasers and operate at internal pressures of 1 atmosphere or above. At elevated pressures (about 10 atmospheres or so), the broadening of the emission lines allows continuous tuning of the output wavelength.

Slab discharge. This is a recent development where the discharge volume is made rectangular in cross-section. The electrodes are quite close together (~2 mm) but are 50-100 mm wide and about 1 m in length. The output scales with the electrode area and is of the order of 1 W/cm². Output powers in the kilowatt range have been achieved. Another advantage of the slab-type lasers is that the output beam is of good quality through novel resonator designs, even though the powers are high.

The high output powers offered by CO_2 lasers are commonly used in many areas of industry, especially for cutting and drilling applications. Although many metals reflect strongly at 10 mm, certain metals such as titanium do have sufficient absorption to be cut well at these infrared wavelengths. Since titanium is widely used in manufacturing and is also very difficult to machine with conventional techniques, the CO_2 laser cutting of titanium is now a standard practice in industry. Steels are also routinely cut at rates of a few millimeters per minute with high power lasers. Other materials such as ceramics, textiles, rubbers, wood, and plastics are also very well cut using CO_2 lasers.

Other areas where CO_2 lasers are used include the following.

Marking. Moderate powers of a few tens of watts are sufficient for a wide variety of product marking applications. The main advantages of laser marking over printing or ink-based systems are that the marks are more permanent, the systems require less maintenance, and they are faster.

Heat Treatment. Many materials can be thermally cycled with IR lasers such as high power carbon dioxide lasers. The thermal effects can also be used in stripping paint off surfaces, where the removal can take place quickly, efficiently, and in parts difficult to access conventionally.

Remote Sensing. IR absorption of the laser light by contaminants, pollutants or other species of interest can be used for spectroscopic analysis of the atmosphere or other media, and low to high power CO_2 lasers are commonly used in many sensing roles.

Medical Applications. The continuous CO_2 laser beam is useful in surgery for cutting since the IR light cauterizes as it cuts. Localized heat treatments can also be applied.

Decontamination. This means the removal of contamination and coatings from industrial machinery and parts, including molds.

Carbon dioxide lasers are extremely widespread in industrial applications but they are usually not used for work where very high feature resolution is required, due to their relatively long operating wavelengths.

Solid-State Lasers

Numerous solid-state lasers have been developed, but the most commonly used one in machining applications is the neodymium yttrium aluminium garnet, or Nd:YAG, laser. This laser is currently used very widely in areas such as drilling, dicing, welding, soldering, trimming, cutting, and scribing.

The Nd:YAG is an infrared laser whose fundamental laser emissions are around wavelengths of 1 micron with the strongest output being at a wavelength of 1064 nm. One of the reasons why Nd:YAG lasers are so widely used is because of their versatility—they can operate continuously, Q-switched or mode-locked, giving pulse durations from a few picoseconds to milliseconds. Output pulse energies can vary from millijoules to many hundreds of joules, and average powers of thousands of watts have been demonstrated. Other features of Nd:YAG lasers include the ability to be diode-pumped (thus providing a very efficient and compact laser source) and to be harmonically converted to shorter wavelengths to give outputs in the visible and the UV.

The Nd:YAG laser relies on the transitions of Nd^{3+} ions which are housed in the YAG ($Y_3Al_5O_{12}$) host. The Nd^{3+} ions replace some of the ytterbium ions in the lattice and typical levels of the neodymium concentration are about 1% by weight or about 10^{20} atoms/cm^3. The Nd:YAG laser is a four-level system.

Efficient pumping of the neodymium ions is usually accomplished in pump bands around 730 nm and 800 nm. This is achieved commonly by flashlamps but laser diodes are also developing rapidly for pumping applications. After the Nd^{3+} ions have been excited into the higher energy levels from the ground state, they decay nonradiatively into the upper laser level from which the stimulated transition of the laser takes place. The upper laser level has a relatively long lifetime of about 270 msec. The lower laser level decays nonradiatively quickly back down to the ground state. The main transition between the upper and lower laser levels occurs, as has been mentioned, at 1064 nm but these levels can split into different energy bands, thus allowing different laser lines to lase.

The Nd^{3+} ion has been used in a variety of hosts to produce lasing transitions in the near-infrared spectral region, both in crystals and glasses. These hosts include phosphate glass, gadolinium scandium gallium garnet (GSGG), yttrium lithium fluoride (YLF), and vanadate (VYO_4). The optical properties of the laser in terms of the laser wavelength, the emission cross-section, the upper state lifetime, and lasing linewidth are determined by a combination of the properties of the neodymium ion and host, whereas the thermal and mechanical properties of the laser are largely dependent on the host medium. Hence, for example, Nd:YAG is inferior to Nd:YLF in its thermal lensing properties, so YLF is preferred over YAG when considering a laser with lower thermal problems.

In general, the parameters of the laser system such as laser material, doping, host material, etc., can be varied to suit the application; for example, low doping concentrations of around 0.5% are typically used in Nd:YAG lasers for continuous-wave applications while high power systems (which need to be Q-switched) require a larger doping concentration of around 1-2%.

The Nd:YAG laser operates at room temperature at its strongest transition of 1064 nm. Other wavelength outputs can be obtained by inserting a spectrally selective element inside the resonator and by controlling the temperature. The spectral selection can use a dispersive element like a prism inside the cavity or a resonant reflector (which is a device which operates using interference) as an output coupler, for example.

There are various types of Nd lasers available, and here, the discussion shall not be restricted to just Nd:YAG lasers but will cover most solid-state lasers that use the neodymium transition. These lasers provide:

- output powers from milliwatts to multiple kilowatts (depending on laser medium, pumping method, and geometry);
- pulse durations from picoseconds to microseconds (depending on the laser medium, pumping scheme, and intracavity energy modulation method);
- different laser linewidths ranging from 0.5 nm to tens of nanometers (depending primarily on the host medium);
- different wavelengths from the infrared to the UV, using tunability of the fundamental laser and harmonic conversion with nonlinear crystals.

The peak powers available from neodymium lasers can be very high (since the pulse durations are small); and this, combined with the wavelength, allows these lasers to be particularly suited to the microdrilling of many metals. The Nd-based systems are in fact preferred to the longer wavelength carbon dioxide laser since many metals reflect the far infrared radiation of the carbon dioxide laser. Nd-based lasers are also widely used in soldering, resistor trimming, and mask repair, for example, in the microelectronics industries.

Neodymium-doped lasers can be pumped either longitudinally or transversely. In longitudinal pumping, the pump source—usually another laser—is focused into the laser rod from one of its polished ends. In transverse pumping, the pump light enters the laser medium from the sides. Various pump sources can be used such as: high intensity broadband sources, continuous emission tungsten arc lamps, gas-discharge pulsed flashlamps, and diode lasers.

Transverse pumping is most commonly used with flashlamps while either transverse or longitudinal pumping can be used with diode lasers, depending on the power requirements of the laser. There are different designs for the pumping configurations using flashlamps. A lamp can be placed close to the rod for transverse pumping or a more-efficient pumping cavity can be designed for housing the lamp and rod. An example of this is an elliptical cavity where the rod is at one focus and the lamp is at the other so that the light from the lamp is focused into the rod. Other designs also exist which try and maximize the amount of light that the rod can absorb.

Diode laser pumping of neodymium lasers is inherently more efficient than lamp pumping since the pumping lines of the diodes can be made to overlap the absorption bands of the laser medium, thus optimizing the absorption. This has two major impacts.

1. The amount of heat generated in the gain medium is much lower, so less cooling is required.
2. The efficiency of conversion of electrical energy (needed for the pump source) into laser light can be as high as 10-15% with diode laser pumping, as opposed to about 1%, at best, for lamp pumping.

Neodymium lasers are most commonly encountered as systems where the laser medium is in the shape of a rod or cylinder. These are typically about a few millimeters in diameter and many tens of milimeters in length. Slabs or disks of material are also sometimes used as the gain medium.

The two main design criteria that have to be addressed in a solid-state laser system are the extraction of maximum power and the maintenance of good beam quality. There are, of course, other considerations but if these two criteria are met, then the laser is efficient, has high output power, and good focusability.

There is always a tradeoff between high extraction of power and a good optical beam. A single transverse mode of a laser does not extract energy from all of a rod, yet a multimode beam is not as focussable as a fundamental mode beam, so the user requirements of the particular system mostly determine the route that is taken.

Recent developments in the manufacture of diode lasers have also helped to make available very efficient pump sources. Diode lasers are now available that can provide many watts of average power with extremely high conversion efficiencies. Diode lasers can be used for pumping in both longitudinal and transverse geometries, and this also means that very compact, high power lasers can be constructed.

The main developments that have been under way for the past couple of years are centered around the miniaturization and ruggedization of solid-state lasers. Using diode-pumping, ever more efficient and compact systems are being demonstrated and improvements in the manufacture and design of these lasers has yielded high power, compact, efficient lasers at many wavelengths. Intracavity frequency conversion and a wider availability of pulse durations have also increased the choice of lasers for the user. Along with these developments, the cost of ownership and operation of many solid-state lasers has also dropped dramatically.

Excimer lasers

Excimer lasers are now routinely used in many varied applications of lasers such as the drilling of ink-jet printer nozzles, corrective eye surgery, annealing of flat-panel displays, production of solar panels, and production of biomedical devices. This expansion in the usage of excimer lasers has largely been governed by the very high performance offered by the excimer laser processing techniques—high resolution, high speed, reproducibility, flexibility, simplicity, etc.—but has also been accelerated in recent years by advances in the technology of the laser systems themselves. Commercial excimer lasers are available that operate at repetition rates of many kilohertz and with average powers of hundreds of watts.

Excimer lasers are molecular gas lasers that operate in the ultraviolet (UV) region of the spectrum. The term "excimer" originates from "excited dimer" because the laser transitions are related to the electronic states of excited diatomic molecules. The word "dimer" refers to a diatomic molecule of the same element, e.g., N_2 or O_2. The first excimer laser was demonstrated in 1971 when stimulated emission from Xe_2 around

170 nm was shown, but today excimers are generally used to describe species such as krypton fluoride (KrF), argon fluoride (ArF), xenon fluoride (XeF), or xenon chloride (XeCl) even though, strictly speaking, these should be called "exciplex" lasers—from excited complex molecules. Nonetheless, the term excimer is now universally accepted to cover the rare gas halides such as KrF, ArF, and XeCl. The first demonstration of one of today's common excimer lasers was in 1975 when lasing in XeBr was produced. The same year saw the first laser emission from KrF. Since then, many other transitions have also been demonstrated.

The so-called rare, or "inert," gases—helium, neon, argon, krypton, and xenon—normally have all their permissible electron orbits filled and therefore do not tend to react with other atoms. If a rare gas atom is electronically excited, however, either by altering the electronic orbit or removing an electron altogether, then the atom can become highly reactive. Hence, positively charged rare gas atoms (e.g., Kr^+, Ar^+, Xe^+) can bind very strongly with a negatively charged halogen atoms (e.g., F^-, Cl^-).

The excited level of the molecule, which is normally achieved by intense electrical pumping, is very short lived with a typical lifetime of $\sim 10 \times 10^{-9}$ s (10 nanoseconds, or 10 ns). After these few nanoseconds, an ultraviolet photon is emitted which takes the molecule back to its ground state. As the unexcited atoms are inert to each other (and may even repel), they disperse in fractions of picoseconds. Hence the ground state contains a negligible population and so makes for an ideal lower laser level, i.e., the population inversion—the difference between the upper and lower laser levels—is large. All practical excimer lasers are pulsed devices, with the laser output pulses lasting typically for 10-30 ns. The electrical pumping of the gas to produce the excited discharge is also achieved in short (50-100 ns) high voltage (35-50 kV) bursts, allowing peak current densities of ~ 1 kA/cm^2 to be sustained.

The four most common excimer lasers are listed in **Table 2** with their output spectral characteristics. It will be noticed that the krypton fluoride and argon fluoride lasers have a relatively broad emission over a range of ~ 300-400 pm, whereas the xenon fluoride and xenon chloride systems display discrete output emission lines. This difference often needs to be borne in mind when considering the use of excimer lasers, particularly if the spectral properties are of major importance.

Table 2. Spectral Properties of Common Excimer Lasers.

Laser Type	Main Output Wavelength (nm)	Emission Linewidth (pm)
ArF	193.3	370
KrF	248.5	310
XeF	348.75 (~15%) 353.05 (~27%) 351.05, 351.15, 351.3 (~58%)	~50 (for individual transitions)
XeCl	307.96 (~60%) 308.16, 308.21 (~40%)	~30 (for individual transitions)

Even though other lasers such as Nd:YAG and CO_2 can produce average powers of tens of kilowatts, the short pulse nature of excimer lasers together with their reasonably high average powers can produce unique interactions with materials, allowing for very efficient micromachining.

The majority of excimer laser cavities are formed by a plane parallel resonator, where two flat optics (usually a high reflector and an output coupler) provide feedback to the gas discharge. The material for the resonator optics is typically fused silica, calcium fluoride, or magnesium fluoride, all of which are highly transmissive at ultraviolet wavelengths. The reflector mirror is a substrate with a high-reflectivity coating on its rear surface that reflects all the light back into the cavity, and the output coupler is usually an uncoated window that permits some fraction of the light to leave the cavity. Both these optics are normally directly fixed to the ends of the gas discharge section and, for this reason, the coating of the rear reflector is on the surface not exposed to the excimer gas environment since this would be too corrosive for the coating. This high reflecting coating is either aluminium or made up of a multilayer dielectric stack.

Any transmissive optic used in excimer laser systems can suffer from increased absorption after extended exposure to UV light, especially if the energy densities of the laser pulses are significant. This degradation in performance usually manifests itself as the formation of color-centers in the bulk of the material, which then absorb light in the UV. This problem is most severe at shorter wavelengths, so that ArF lasers are more prone to this effect than XeCl, for example.

Resonator designs and construction. Most excimer lasers have cavities consisting of plane parallel optics that form a stable resonator. The discharge aperture for most excimer lasers can be as large as 3 cm^2; and with cavity lengths of 1000 mm-1500 mm, the high spectral gain bandwidth of excimer lasers permits many hundreds of thousands of axial and transverse modes to be supported. This results in excimer lasers having poor spatial and temporal coherence properties—in this respect, excimer lasers behave more like broadband light sources such as lamps than conventional lasers.

Stable resonators of excimer lasers typically produce output beams that have a rectangular cross-section (due to the discharge region), usually 5-10 mm in width by 20-30 mm in length. Due to the nature of the stable cavity from which it is produced, however, the output beam has a high beam divergence of 2-10 mrad. This relatively high divergence means that it is not readily focusable.

Excimer lasers are now routinely used in many areas that require UV laser light sources. One alternative, however, which is now increasingly becoming more attractive, is the use of UV solid-state lasers. These systems offer many features such as high "wall-plug" efficiency, compact design, low operating costs, and better beam quality. The main differences in the use of UV excimer lasers and UV solid-state lasers are in the output beam parameters—excimers output rectangular, divergent, multimode beams (ideal for mask projection systems), whereas solid-state lasers produce round, low-divergence TEM$_{00}$ beams ideal for focusing to small spots.

Laser micromachining

The applications considered here concern laser ablation, where the laser energy penetrates the material and leads to some transformation. This penetration can range from a few nanometers in depth to many hundreds of microns, or even millimeters.

Laser ablation

Laser ablation is the direct removal of material resulting from the interaction of the laser light with the sample. Most commonly, ablation relies on the laser light being absorbed in the material, leading to material interaction. As such, the matching of the laser wavelength with the absorption properties of the sample is an important parameter for efficient ablation.

Due to their short wavelengths in the UV, excimer lasers have relatively high photon energies (around 3.5-7.9 eV). This means that the UV photons have sufficient energy to directly break some common molecular bonds, particularly in materials such as polymers.

Infrared lasers such as Nd:YAG or carbon dioxide lasers, however, have much lower photon energies so they cannot directly break such molecular bonds. These IR lasers have to rely on the absorption of multiple photons, which then heat up the material, leading to a molten phase and subsequent evaporation of the material.

In the case of excimer laser ablation, where single photons directly break molecular bonds, the process can be viewed as follows. The excimer laser illuminates an area of the sample that absorbs the UV photons in a shallow depth of a few microns or less. As the UV light breaks the molecular bonds, the particle number density increases rapidly in the exposed volume and this in turn leads to a rapid rise in the pressure in this region. This increase in the internal pressure causes the dissociated material to be ejected at very high speeds from the exposure site, leaving the machined site clean.

Excimer laser ablation is generally a process where the influence of thermal effects is not significant. This is partly due to the short-pulse nature of the excimer lasers but also because the high absorption of the UV photons restricts the interaction area to that of the "footprint" of the exposed area. Hence, excimer laser machining produces a negligible heat-affected zone (HAZ) and can be used for the high quality processing of many materials such as plastics, glasses, ceramics, composite materials, and biological samples.

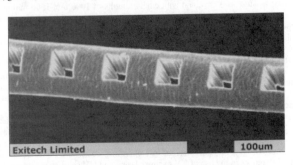

Figure 7. Human hair with holes drilled using an excimer laser.

Figure 7 shows a very good example of the lack of thermal damage for an excimer laser machined sample. The picture shows a scanning electron micrograph of a human hair of 100 mm diameter. This has been machined with an excimer laser to produce 40 mm square holes. As can be clearly seen, the holes have been precisely machined with no HAZ in the surrounding areas.

With infrared wavelength lasers, and with those lasers having longer pulse durations, the absorption in the material may not be as efficient, and the production of a molten phase in the interaction region can lead to a significant HAZ. The quality of machining and extent of the HAZ are also affected by how the machining is conducted (in terms of the precise details of the processing conditions) but high quality results are nonetheless possible.

One exception to the above model of ablation is that of the recently developed femtosecond laser micromachining. In this case, the linear absorption of the laser in the material is not necessary for efficient ablation to occur, and so even materials that are transparent to the laser radiation (dielectrics, for example) can be micromachined. In general, femtosecond systems used for micromachining applications operate in the near infrared region of the spectrum (750-1000 nm) and are based on titanium sapphire lasers. The short, femtosecond pulses from these lasers provide excellent quality of machining in virtually all materials, unlike their longer-pulse counterparts. This is due to a number of reasons: the deposition of energy into the sample is extremely rapid, leading to the speedy formation of vapor and plasma phases; interaction with the material is on timescales much shorter than the thermal diffusion times, which results in the absence of heat-affected zones; and there is no molten phase which leads to a reduction in redeposited material around the machined site.

In materials that are transparent to the laser light, the absorption occurs through nonlinear processes via laser-induced optical breakdown caused by multiphoton ionization. Dielectric materials do not absorb the laser light at low intensities because the bound valence electrons have bandgaps that are larger than the photon energies. Free conduction electrons, which are always present, can, however, gain sufficient energy through collisions with the lattice and bound electrons to cause the creation of further free electrons. The number density of free electrons rises and the avalanche process continues until sufficient bound electrons have been ionized to form a critical density plasma, which then breaks down the material and initiates absorption. At high electric field strengths (such as those generated by intense femtosecond pulses), however, a bound electron can absorb multiple photons and can ionize directly, creating large numbers of free electrons. Therefore, multiphoton ionization can provide the seed electrons which start the breakdown process, but because the process is strong due to the high field strengths, the breakdown is much more predictable and its threshold very precise for femtosecond laser pulses.

Femtosecond lasers have been shown to micromachine metals, glasses, ceramics, biological samples, and semiconductors, among other materials, with excellent quality and little or no thermal damage. They are likely to become increasingly important in applications where other lasers cannot be used, particularly in areas such as medicine.

In most laser micromachining applications, the material removal rate is usually relatively small and can be of the order of only tenths of a micron per pulse. This small etch rate may, at first sight, seem to be a disadvantage in that large amounts of materials cannot be ablated but, in practice, the small etch rate is used as the main tool with which to control the depth of the laser microstructure. The depth control which can be achieved for any laser machining which is undertaken is simply the etch rate value at the laser fluence used—for example, at 248 nm and ~1 J/cm^2, the depth can be controlled in polyimide to ~0.25 μm since that is the amount of material which a single pulse would remove. Even though the single pulse removal rate may be relatively small, the use of lasers with multikilohertz repetition rates can be used to produce rapid material machining.

Laser ablation techniques can be used to control the geometry of the microstructure being produced in two or three dimensions: the lateral (XY) dimensions are mainly a function of the beam size (primarily governed by the focussing or imaging optics) and the depth (Z) is largely dependent on the processing conditions used.

Processing techniques

The use of lasers is now extremely widespread, both on the commercial market (e.g., CD players, printers, light shows, pointers, range-finders, scanners, etc.) and in high-tech R&D and industrial applications (e.g., production of memory chips, flat-panel displays, solar panels, via holes in printed circuit boards, MEMS devices, etc.).

Lasers can offer a multitude of advantages including high precision and accuracy, ultrahigh resolution, noncontact processing, ability to process a wide range of materials, high speed processing, direct machining, avoiding photolithography steps, flexibility, and rapid prototyping capabilities.

Many criteria need to be considered regarding the use of lasers for a particular application. No single sequence can be defined to determine all the important factors that need to be addressed, but some of the main features in the use of lasers include:

- choice of material, which determines which lasers can be used;
- the resolution of the features required, which influences the optical system and processing method;
- the dimensions of the features to be machined, which affects the optics and the workpiece handling; and
- any issues of processing speed that may impact on the laser repetition rate.

Once a laser has been chosen for a particular application, then important parameters to optimize are the laser fluence at the sample, the number of shots used, the nature of processing (e.g., scanning or static machining), the repetition rate of the laser (to reduce thermal effects that may be present), and the use of any assist gases. The appropriate combination of such parameters is vital in obtaining optimum results.

Direct writing

One of the simplest ways to micromachine with a laser is to focus the beam with a lens and then to move the sample around under the focused spot to produce a pattern (as shown in *Figure 8a*). This is termed direct writing or serial scribing. Serial scribing is most commonly used with Nd:YAG and CO_2 lasers but it is not, however, well suited for use with excimer lasers since the optical properties of the excimer beam do not lend themselves to well-controlled direct focusing.

A type of serial scribing involves the use of a system where galvanometer-controlled mirrors deflect the laser beam in the X-Y plane (see *Figure 8b*). This is called scanning, and the motion required for the pattern to be produced is obtained from the beam deflection from the scanner. A lens can either be placed before or after the galvanometer scanner to focus the beam and this lens is usually a "flat-field" lens, where the focal plane of the lens is at a fixed distance from the lens, irrespective of the angle of the beam through the lens. The extent over which the beam can be deflected by the scanner/lens is called the field of the scanner and it is typically about 100 mm × 100 mm or so. To machine sample areas larger than the field size requires the sample to be stepped sideways and the scanning repeated, but this is only possible for regular patterns where adjacent fields can be "stitched" together. Where the pattern to be machined is irregular (i.e., nonrepeating) and larger than the field size, the sample can be moved with XY motion while the scanner is deflecting the beam, as long as the beam position from the scanner can be synchronized with the sample motion and the control for the scanner deflection updated continuously.

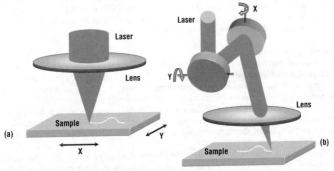

Figure 8. (a) Direct writing and (b) beam scanning techniques.

Mask projection

The vast majority of excimer laser applications use the technique of mask projection. This is partly due to the large sizes of most excimer beams (~2 cm² in area) and because of the mode characteristics. This method is very flexible and can be tailored to fit many requirements, and it has been developed extensively in recent years. The principle of mask projection is depicted in *Figure 9*. In general, the excimer laser beam is shaped and homogenized to improve the energy distribution of the beam. This is done since the unaltered output distribution of an excimer laser is not uniform and machining with a nonuniform leads to variations in the ablation depth.

A mask is placed at the plane of best uniformity of the homogenized beam. This plane—the mask plane—thus contains a pattern (the mask), which is illuminated with the uniform laser beam. A high-resolution projection lens images the mask plane onto the sample or workpiece to produce a replica of the mask pattern there. This is usually accomplished with a demagnification as well so that the image is smaller than the mask pattern. The exact optical configuration is not usually as in the figure, which is only meant to display the overall concept of the projection technique.

The mask projection technique has several attractions, such as the following.

- The mask is distant from the sample and so damage from debris can be avoided.
- Demagnification allows the mask manufacture constraints to be less severe.
- The system has flexibility since only the mask has to be changed to produce another pattern.

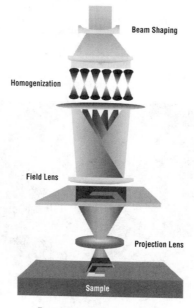

Beam Shaping

Homogenization

Field Lens

Projection Lens

Sample

Figure 9. Principle of mask projection.

- The laser fluence is higher at the sample than at the mask due to demagnification, and hence the mask can be used at well below its laser damage threshold. This means that lifetime issues of the mask can, in general, be ignored.
- Having independent control of the mask and the workpiece allows complicated processing techniques to be used.
- Nonplanar substrates can still be processed even with a planar mask.

Machining with Static Mask and Workpiece. The basic system for excimer laser mask projection, where both the mask and workpiece are static, usually includes some beam shaping and beam homogenization to improve the illumination uniformity of the mask pattern. The mask is then imaged onto the workpiece with a high-resolution projection lens.

The lens and the optical system define the maximum resolution achievable at the workpiece and the field size of the lens restricts the maximum size of the image. Since neither the mask nor the workpiece move, the size of the image is thus fundamentally limited to the lens field size. In practice, the mask pattern is much smaller than the lens field size (and is, obviously, also smaller than the beam size at the mask since otherwise the beam size and not the mask would define the size of the pattern imaged).

The practical limitation of this technique with a static mask and workpiece is that only patterns smaller than the beam size can be used, and hence its applicability is restricted to only a few areas. One of the most common uses of this technique is in the drilling of holes where a single hole or a small array of holes is required on a component.

Figure 10 shows a scanning electron micrograph of a hole drilled with a static mask projection system where in a hole has been drilled in a PVC tube which forms part of a medical bilumen catheter. This hole was drilled with a KrF laser and the hole diameter is 500 mm. This single hole can be drilled repetitively on successive devices with a static projection system.

In general, polymers machine very well with excimer lasers due primarily to the high absorption of most polymers in the UV. This leads to machined samples having the following attributes: excellent edge quality, lack of debris, absence of thermal damage, good depth control, and vertical walls.

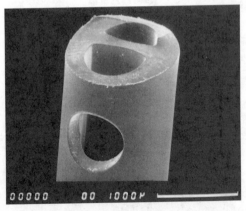

Figure 10. Hole in medical catheter drilled with excimer laser.

Refinements of the mask projection technique exist that offer the ability to machine large and complicated features into samples. These techniques rely on moving the mask and/or the sample in synchronism with the firing of the laser and allowing the benefits of mask projection to be extended over large (100's cm^2) areas.

Many applications that have evolved over the past few years use mask projection. These include the production of nozzles (printers), microchannels (biomedical or microfluidics), large area patterning (printing or displays), and manufacture of microoptics (photonics).

Laser beam shaping

It is often desirable to alter the shape, size, or energy distribution of the laser beam. The method by which this is usually done depends on the type of laser and the required properties for the beam. Traditionally, beam sizes have been altered by using combinations of positive and negative lenses in a variety of beam telescope configurations. The redistribution of the laser beam energy—beam homogenization—is sometimes necessary because high uniformity of ablation may be required for particular applications.

The simplest method for improving the uniformity of a beam is to use an aperture, which selects the most uniform portion of the beam. Although this is not strictly beam homogenization (since the energy of the beam is not being redistributed), it can provide a convenient way to eliminate those parts of the beam that are not desirable. Other common methods of beam homogenization require optics such as lenses or prisms to act on the beam.

Biprisms. Biprisms can be used to homogenize beams by spatially separating and then recombining two halves of a beam. The two halves of the beam travel at different angles but overlap at a certain plane, at which point the outer and inner parts of one half combine with the inner and outer parts of the other half, respectively, giving an increased beam uniformity.

The use of biprisms is again a simple technique and does not require complex optics, but the level of uniformity that can be obtained is limited.

Lenses. Combinations of lenses can be used to redistribute the energy in laser beams. These are normally used with lasers with Gaussian-like beam profiles that need to be converted into a more "top-hat" profiles. Common methods include using a pair of spherical lenses to provide differing amounts of spherical aberration or the use of aspheric lenses to alter the distribution of the propagating light.

Homogenizer Arrays. Arrays of small lenses are most commonly used with excimer lasers to provide an extremely high level of beam homogenization and give a very controllable method for beam smoothing. These homogenisers contain a few tens of small lenses ("lenslets") that are arranged in a close-packed geometry. These lenslets break the beam into small beamlets and the homogenization occurs by the superposition of these beamlets at one plane. The level of homogeneity increases with increasing numbers of lenslets since more beams take part in the beam uniformity averaging.

Homogenizer arrays can either contain cylindrical lenses (for homogenization of one axis) or spherical lenses (for homogenization of both axes), but the principle of operation is the same. *Figure 11* shows the setup for a double homogenizer system. When used with excimer lasers and mask projection systems, the plane of optimum uniformity is made coincident with the mask plane, which ensures that the mask is uniformly illuminated. The use of homogenizers with excimer lasers is effective since the output beams from standard excimer lasers contain many thousands of spatial modes and so interference (speckle) between the overlapping beamlets is not present.

If the size of each lenslet in *Figure 11* is h, and the focal lengths of the arrays and lens

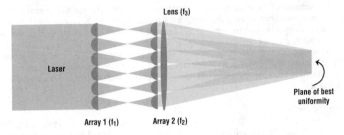

Figure 11. Principle of beam homogenization using a pair of homogenizer arrays.

are as shown, then the size of the beam at the plane of best uniformity d is given by:

$$d = \frac{hf_3}{f_2}$$

Diffractive Optical Elements (DOEs). An alternative method that has received much interest in the past couple of years is the use of diffractive optical elements (DOEs), even though DOEs have been around for many decades (in the form of Fresnel zone plates, for example). These are transmissive devices that alter the laser beam distribution by diffraction and are usually made of glasses, silica, or polymers (depending on the laser with which they are to be used). The main attraction of using DOEs in micromachining applications is that the light can be distributed precisely into predefined patterns (e.g., an array of holes or an annulus), which allows for a more efficient use of the light without aperturing or masking. The performance of DOEs is particularly dependent on the precision of manufacturing, since small deviations in the depths of the DOE surface lead to reduced efficiency of diffraction, with the main drawback being the presence of significant light in the zero order (i.e., light which passes straight through the DOE without being diffracted).

Industrial machining systems

Various laser applications have matured in recent years and transferred successfully into manufacturing environments. An example of this is the drilling of high-density micro-via holes for printed circuit boards, where alternative mechanical drilling or etching techniques cannot match the size, quality, and speed of laser drilling. As industries grow and demand higher specification products, the demands on the laser techniques also increase but laser applications show no sign of not being able to meet these new requirements. Even when other viable techniques exist, laser-based options usually allow a greater flexibility and scope for development than nonlaser processes, and this is also an important factor in the uptake of laser technology.

Manufacturing applications

The following is a brief introduction to certain industrial areas that have benefited and are continuing to develop rapidly as a result of laser processing techniques.

Printer Nozzles. The growth in the use of personal computers has also driven the technology of printers into new areas: smaller physical size, higher resolution, full color capability, and higher printing speeds. Typical desktop printers contain print heads that include many hundreds of holes through which the ink is discharged onto the paper. The demand for higher resolution (dots per inch or dpi), in particular, has moved the manufacturing of printer heads from EDM or etching into laser drilling. The sizes of typical printer holes are of the order of a few tens of microns but the requirements for accuracy of positioning, roundness, centricity, and reproducibility mean that laser drilling (usually using excimer lasers) is now the most common method for the manufacture of printer nozzles.

The production of ink-jet printer heads uses excimer laser mask projection where a mask pattern containing an entire nozzle head is projected onto a polymer tape. The laser production system drills a complete nozzle head containing many dozens of holes in a single pass, with each nozzle head taking 1-2 seconds to drill.

Vias and Interconnects. The markets for printed circuit boards have increased as the range of personal, domestic, and commercial electronic products has multiplied. The drive in these consumer products such as mobile telephones, digital cameras, personal computers, CD players, etc., is to decrease in size and increase in speed and functionality. This has meant that the sizes of the electronics inside the products have to keep shrinking, which in turn leads to a higher density of packing of microelectronic components. One result of this is that the interconnections between components on and between PCBs reduces and requires smaller interconnects and via holes. The use of mechanical drilling limits the smallest size of via hole to around 100 mm or so, and although other methods such as plasma etching can obtain smaller hole sizes, the simplicity and ultimate capabilities of lasers have proved to be irresistible. Lasers can drill holes at extremely high speeds of up to many hundreds or thousands of holes per second (depending on the material bring drilled) and have emerged as the preferred way of producing the micro-via holes for many consumer devices. The most commonly used lasers for PCB drilling are third-harmonic solid-state lasers (such as Nd:YAG or Nd:vanadate) operating at 355 nm, and carbon dioxide lasers operating around 9-10 mm. More than 90% of micro-via holes drilled currently are produced using laser drilling.

Most production machines for the drilling of micro-via holes use beam-scanning techniques (using galvanometer scanners) in conjunction with sample motion. This is done to enable ultrahigh speed drilling to be performed, which is not possible by simply using mechanical movement of the sample.

Semiconductors and Lithography. As the speed and memory demands of computers grow, the dimensions of the critical features inside the memory and processing chips diminish. The manufacturing of the silicon-based chips is reliant on the process of lithography, with the feature geometries being defined by optical exposure followed by various etching steps. Lasers have replaced the use of mercury lamps (as the exposure element) in all high specification lithography systems, and increasingly shorter laser wavelengths are being deployed to reduce the feature dimensions further. Semiconductor production facilities are currently using 248 nm excimer lasers and the planned reduction in features will use 193 nm excimer lasers, 157 nm fluorine lasers and, ultimately, X-ray wavelengths. All excimer laser-based lithography production tools use the technique of mask projection with extremely high levels of beam homogenization, since the uniformity of exposure is a critical issue for semiconductor chip manufacturing. The level of homogeneity required is usually achieved by using light pipes/guides, which provide a very high level of beam averaging, rather than just homogenizer arrays. It should be kept in mind that in contrast to many other laser machining applications, optical lithography relies on the *exposure* of photoresist as opposed to *ablation*, and so some of the constraints of machining are absent, only to be replaced by other demanding resolution and processing control issues.

Telecommunications. The use of lasers in applications related to telecommunications and photonics is also now very widespread. In particular, the requirements of high speed, high capacity digital communications systems have been driven by the availability of laser-based methods that have enabled ultrahigh specification devices to be fabricated. One example of such a development is the production of gratings in optical fibers whose uses include optical filters, stabilizers for diode lasers, and dispersion compensators. All these fiber-based gratings are produced using ultraviolet lasers (either frequency-doubled argon-ion lasers or 248 nm excimer lasers), which are used to induce changes of refractive index in the fibers, thereby leading to a controllable change in their spectral properties. Other uses of lasers in this area include laser tuning of waveguides (for example to control the properties of optical receiver or transmitter devices), laser trimming of the resonance properties of optical modulators by the controlled ablation of electrical conductors, and the laser cutting or dicing of optical materials.

MEMS/MOEMS. Micro-electro-mechanical structures (MEMS) and micro-opto-electro-mechanical structures (MOEMS) are multifunctional units which have been developed in the past few years for a variety of applications. MEMS and MOEMS systems are typically used as miniature devices to provide capabilities for micropositioning, microactuation, micromanipulation, sensing, light guiding, or other precise tasks, usually as part of a larger system or product. Commercial examples of MEMS/MOEMS devices include automotive air-bag sensors, pressure sensors, and optical switches.

The majority of MEMS and MOEMS have been produced using conventional etching techniques, but the increasing need for these devices to be made from a diverse selection of materials (for example, in biomedical applications) has opened up research into laser-produced MEMS devices. The benefits of laser micromachining already discussed can be applied to MEMS structures, and it is likely that more laser-produced MEMS devices will emerge, particularly where the geometries involved cannot be easily met with etching or conventional processing.

Electrical Discharge Machining (EDM)

Since its introduction in the mid-1950s, EDM has grown from a "nontraditional" machining method to the fourth most popular machining process, with EDM machine sales outpaced only by milling, turning, and grinding machines. In fact, the roots of EDM can be traced back to 1770, when the erosive effects of electrical discharges were first noted and recorded by Joseph Priestly, an English scientist. The modern "discovery" of EDM is attributed to B. Lazarenko and N. Lazarenko who harnessed this erosive process for machining purposes in the early 1940s.

Developments in EDM have accelerated rapidly in the past decade, and today the EDM process is often the method of choice for a growing variety of operations. In the early 1980s, CNC units first appeared, adding great versatility, including reliable untended operation, to the machines. Further advances in EDM technology have enhanced both surface finishes and machining times. The EDM process utilizes thousands of electrical discharges per second between a wire or electrode and an electrically conductive workpiece that is placed in a circulating, nonconductive dielectric fluid. The fluid acts as an insulator until a specific gap and voltage is achieved. At that point, the dielectric fluid ionizes and becomes an electrical conductor that permits the current (spark, or discharge) to flow through the fluid to the workpiece. The spark, with a temperature ranging from 14,430 to 21,630° F (8,000 to 12,000 °C), causes the workpiece to erode as it melts or vaporizes. In order to generate the spark, and allow for the molten material, or "swarf," to evacuate the workpiece, the current must be switched on and off at very high speeds. The relationship between on-time and off-time is known as the duty cycle, and while the goal is to minimize the off-time (during which no melting can take place) and maximize the on-time (to maximize cutting speed), there must be a reasonable balance that allows enough off-time for the dielectric fluid to flush the swarf from the workpiece.

One of the most appealing aspects of EDM is that the process does not recognize the hardness of the workpiece material. Since the material is "vaporized" rather than cut, and the tool never contacts the work, workpiece hardness is not a factor. For materials above R_C 38, EDM can be the machining method of choice.

Recent developments allow EDM machines to monitor each spark, and "adapt" the duty-cycle to optimum efficiency. This ability, plus the growing demand for precision part manufacturing in the aerospace, electronic, and medical industries has widened EDM's appeal. There are two different methods of EDM machining—ram and wire. Both use the same basic principle and both are capable of producing intricate components to tolerances within millionths of an inch, but there are notable differences in the two processes.

For a complete description of the technology and techniques of EDM machining, "The EDM Handbook" by E. Bud Guitrau (Hanser Gardner Publications) should be consulted.

Ram EDM

Ram electrode materials

Also known as vertical or "die sinker" EDMs (because early machines were widely used for removing broken taps and drills from workpieces), ram machines use an electrode, produced from an electrically conductive material, as the cutting tool in material removal operations. Depending on the surface finish required, metal removal rates for ram EDM can be as low as 0.001 in.^{3} /hr (16 mm^3 /hr) to as much as 25 in.^{3} /hr (408 cm^3 /hr) on either annealed or hardened materials. The shape of the cutting area of the electrode is a mirror image of the finished form or shape (cavity) that it produces in the workpiece. Much consideration must be given to electrode design because the size of the cavity cut by the electrode will actually be marginally larger than the electrode's surface area. This is called the *overcut*, a condition influenced by many factors including the amount of power applied and the purity of the dielectric fluid, and it normally amounts to 0.002 to 0.020" *per side* (0.005 to 0.51 mm/side). Although brass was favored for electrodes in the early days of EDM, copper and graphite are now the electrode materials of choice although other exotic materials capable of electrical conductivity (such as tantalum, tungsten, nickel, and molybdenum) are also used. Copper, graphite, and infiltrated graphite are estimated to make up 95% of all electrodes used in the USA, so we will focus on these materials.

Copper electrodes. Copper possesses fine granular structure that can be polished to an extremely smooth finish, which is why copper electrodes are selected for producing "glass-like" part surface finishes as smooth as 2 μinch (0.05 μm) Ra in small cavities. In spite of its ability to produce smooth workpiece surfaces, copper has several negative characteristics that restrict its application. While well suited for small parts, its sensitive thermal expansion characteristics make it less desirable for large part production. Even though copper electrodes can be machined quickly, especially with wire EDM, in most cases they must be deburred before use. Two additional disadvantages of copper are that it leads to oxidation by contaminating the dielectric oil, and it is toxic to living tissue. Since copper has a relatively moderate melting point of 1980° F (1083° C), a vaporization temperature of 4675° F (2580° C), and low resistance to wear, it is often combined with tungsten to expand its usefulness as an electrode material. Copper electrodes are normally favored for machining low melting temperature alloys such as aluminum, brass, and copper as well as for carbon and stainless steels. Copper tungsten (which is a more expensive material) electrodes contain up to 70% tungsten and are strong and resistant to damage, making them desirable for high precision work involving intricate details. The material is much more difficult to machine to shape than copper, and often requires carbide tooling. In addition to being used for carbon and stainless steels, copper tungsten is usually selected for machining tungsten carbide workpieces.

Graphite electrodes. Graphite is by far the most widely used electrode material in the United States, and is gaining popularity in other countries. Graphite is a form of carbon, and it features high heat resistance (its vaporization temperature is in the region of 6700° F / 3980° C), excellent electrical conductivity, ideal thermal expansion characteristics, and very good machinability. It can be machined to shape much faster than copper (though care must be taken to avoid chipping), and the resulting electrode will be burr free. However, care must be taken during machining, especially grinding, as the dust created is both abrasive and conductive, making it a potential nuisance to both machine tool parts and electrical components. Because graphite is produced from furnace sintered carbon powder and coal pitch, it is less dense than copper and its grain structure is too large for it to economically produce surface finishes lower than 20 μinch (0.51 μm) Ra, even when using high-grade graphite with fine (0.0005" / 0.13 mm) particle structure. Electrodes made from large (0.008" / 0.20 mm) particle graphite are popular for high volume production when neither a smooth finish nor fine details are required—surface roughness in these circumstances falling between 250 and 500 μinch (6.3 to 12.5 μm) Ra. Copper can be combined with graphite to produce infiltrated graphite electrode material that combines the positive characteristics of both materials. Graphite is used for machining high temperature alloys and steel, while fine particle infiltrated materials are often selected for roughing or finishing tungsten carbide.

Variables in ram EDM machining

All electrodes experience thermal damage during operation. Factors impacting wear include frequency, polarity, the duty cycle, and the properties of the workpiece and electrode. *Spark frequency* is the number of times in a given period that the current is switched on and off. Leaving the current on for longer periods—low frequency—for roughing operations increases the amount of workpiece material removed but sacrifices surface finish to increase higher material removal rates. Graphite electrodes have better wear resistance characteristics than copper electrodes at low frequencies. High frequency minimizes the duration of the on-time and is used for finishing operations. It generates more individual material removing cycles per second, but surface finish is improved because less material is removed with each cycle. As the frequency increases, graphite, in particular, exhibits increasing wear rates.

Polarity can be alternated in EDM operations to meet machining requirements. In most operations, the electrode represents the positive side of the spark gap and is sometimes referred to as the anode. Being positive helps protect the electrode from wear and in so doing assures that it retains dimensional accuracy for as long as possible. However, when polarity is reversed, and the electrode becomes the negative side of the spark gap (or cathode), metal removal rates using graphite electrodes can increase dramatically—as much as 50%—unfortunately to the detriment of the electrode which experiences wear rates up to 40% higher than with positive polarity. Negative polarity is commonly used with graphite electrodes when machining steel, high temperature alloys, carbide, titanium, and copper. Speed increases of this magnitude are not possible with copper electrodes, but negative polarity is preferred when machining titanium, carbide, and copper with copper electrodes.

The *duty cycle*, as described earlier, is the relationship between on-time to off-time. Since all cutting is done during the on-time, long on-times and short off-times are desirable for maximizing the material removal rate. Determining the amount of power desirable during the on-time depends on two variables: electrode size and the operation to be performed. The surface area of the electrode that contacts the workpiece is the first guideline for the amount of *amperage* required for the operation, with recommended amperage being from 50 to a maximum of 65 amperes per square inch. For example, an electrode with a surface area of 0.5 in.^{2} (3.23 cm^2) can utilize 25 to 32.5 amperes, while one with a 2 in.^{2} (12.9 cm^2) surface requires 100 to 130 amperes. Normally, the maximum overall rating for copper electrodes of any size is 100 amperes, while graphite can accept up to 600 amperes. Higher amperage is required for roughing operations and electrodes with large surface areas, while lower amperages are necessary for small electrodes and finishing operations with fine surface details.

Under very select conditions, modern controlled pulse power machines can be set to run in what is referred to as "no-wear" mode that arrests electrode wear to less than 1% of workpiece wear. While not possible with all materials, no-wear machining can be carried out with either copper or graphite electrodes when machining steel or aluminum. Operating in no-wear mode involves running with positive electrode polarity and controlling the arc duration at the point where the maximum amount of workpiece material is removed while sacrificing the minimal amount of electrode. The first step is to calculate the cutting area of the electrode to determine the maximum amperage that can be generated at the beginning of the cut. The second calculation is to determine the amount of amperage desirable for the final cut in order to meet surface finish requirements. The cut begins at maximum amperage, which is progressively reduced to coincide the final cut with the desired minimal amperage. Running in no-wear mode requires trial and error machining on manual machines, but modern adaptive controls can automatically select the ideal parameters.

Flushing ram EDM operations

Dielectric fluid must satisfactorily perform three primary duties for the EDM operation to be successful. 1) It must act as an insulator and conductor with equal success. 2) It must cool the workpiece and electrode. 3) It must "flush" machined particles away from the spark gap, which is normally between 0.0005 and 0.005 inches (0.013 to 0.13 mm) in width. The dielectric fluid used with ram EDM machines is usually a paraffin or naptha-based mineral oil that possesses a high flash point—at least 180° F / 82° C—and very low viscosity that allows it to flow freely through gaps as small as 0.0005" (0.013 mm). The fluid must be in constant circulation through the spark gap, and it must be constantly filtered and recirculated. With regular filter changes, it should last for 18 to 36 months of regular use. To confirm the condition of the fluid, it should be tested for acidity, hydrocarbons, oxidation, and other contaminants by an independent laboratory.

Without adequate flushing, the EDM process cannot be successful. There are four distinct methods of flushing on ram machines that will allow the dielectric fluid to adequately flow through the spark gap. *Normal flow* flushing is performed by directing pressurized fluid, introduced through passages in the electrode or workpiece, into the spark gap. All machines are equipped with outlets for pressurized flushing, making this option the most widely adopted method of flushing. Care must be taken before designing the electrode or workpiece to assure that a constant flow of fluid can be maintained through the spark gap during machining. Ideally, holes should be placed where the cut will be the deepest, and they should be as uniformly spaced as possible. An often undesirable side effect of pressurized flushing is that it unavoidably introduces a taper, or at least a rough finish, to the sidewall of the cut. This is caused by errant sparks, which are intended to travel from the end of the electrode to the workpiece, sometimes traveling off the side of the electrode as the tool advances into the workpiece. With normal flow flushing this is unavoidable due to the fact that the spark will always choose the shortest path, which, because warmed dielectric heats the sidewall and causes it to thermally expand, is sometimes off the side, rather than the end, of the tool.

Vacuum or suction flushing, where dielectric fluid enters the electrode at the cutting end and is sucked out the spindle end (or through a hole in the workpiece), can be used to eliminate sidewall tapering and pitting. Since the fluid is cool as it passes over the finished sidewall on its path to the spark area, it does not thermally expand the workpiece sidewalls, thereby assuring that the shortest gap for the spark to travel is always at the cutting end of the tool, not on its sides. The growing popularity of EDM is reflected in the recent availability of specialized drilling machines and drills being marketed for only one purpose: precision drilling EDM electrodes for normal flow of vacuum flushing. Other flushing methods do not require electrode or workpiece holes, and are normally chosen as a compromise. Flushing from one side, known as *jet or side* flushing, with an aimed spray or jet of fluid is not the most desirable method of flushing, but it is sometimes necessary because the size and shape of the electrode or workpiece will not permit either the normal or vacuum method. *Immersion* flushing can provide sufficient cooling and particle removal when the cut is to be very shallow or through a very thin section. The process is improved when either the electrode or the workpiece can be axially pulsed into and out of the cut, forcing fresh fluid to be repeatedly pumped into the cutting area.

Mold machining with ram EDM

One of the primary applications of ram EDM is the machining of molds. While the process is superior to milling molds in many ways, it has one disadvantage—"recast" and heat affected layers caused by the high heat of EDM machining—that will be discussed as necessary here and in detail later in this section. Most larger molds are produced with graphite electrodes, which cannot match the superior finishes of smaller copper electrodes but can, under proper circumstances, produce very satisfactory finishes down to 10 μinch

(0.25 μm) Ra without polishing. To reach this degree of smoothness, it is often necessary to rely on adding special chrome or silicon powders to the dielectric fluid. One such additive, introduced by Makino, uses a patented conductive powder that claims to break the spark into multiple sparks to produce superior finishes and significantly reduce the depth of the heat affected layer. When using additives, special filtration and other enhancements may be required, and the dielectric system must be structured to assure that the additive remains suspended and evenly distributed in the spark gap.

Another factor favoring EDM machining for mold making is that the time required to mill an electrode is less than the time needed to mill the form in metal, and the difference grows as the milling operation increases in sophistication. This ease in machining a difficult form onto an electrode has been one of the primary reasons for the continuing expansion of EDM machining. As the complexity of the geometry increases, the benefits of using EDM rather than milling become readily apparent. Sharp detail work also favors EDM, especially sharp inside corners that are impossible to cut perfectly with a milling machine. Another situation best handled by EDM is deep cutting, where the tool length-to-diameter ratio is particularly high. Tolerances down to 0.0001" can be realized with slow metal removal rates with EDM, but when even tighter tolerances are specified, milling is the preferred machining method. A final reason for selecting EDM for mold machining is untended cutting. EDM is an easier process to automate because it is more predictable than milling. It can use robots to load electrodes and workpieces, making it possible to run an EDM machine almost continuously. While a manual ram EDM may run six hours during a workday, a realistic target for a CNC machine is at least sixteen hours per day during one tended and two untended shifts.

Tapping with ram EDM

Since EDM operates with equal speed and efficiency on both annealed and heat treated materials, it is able to produce threaded holes in hardened materials and in tough aerospace alloys. *Modern Machine Shop* reported on one shop that drills and taps holes (in one operation) as large as $5/8$" and as small as $6/32$" in diameter, often in workpieces with thin walls or ribs as thin as 0.02" (0.51 mm) that would be prone to cracking if conventional drilling or tapping were attempted. In operation, the tapping head works in this manner. The electrode feeds through a threaded bushing or lead nut that keeps it straight and centered, as well as sets the pitch of the thread. Since pitch is variable, a split nut design that can be used to accommodate threads of different diameters is used. With the electrode positioned, a collapsible collar tightens around the lower end of the nut until the electrode cannot move. This automatically sets the orientation of the lead nut's internal threads to match those of the electrode. The nut is then slightly loosened and the electrode threads through as it feeds. It is important to keep the energy concentrated on the tip of the electrode and away from the sidewalls. The leading threads round off, but the following threads retain their shape and clean out the shallow threads that result from tip wear. The tip can then be filed or cut off to extend electrode life. Solid hole tapping electrodes are available in sizes as small as 0–80 UNF, and comparable metric sizes.

A more conventional method of EDM tapping can be performed on a CNC EDM with circular interpolation, which is the ability to program a circular motion that coincides with motion in the Z-axis. The resulting helical path permits the tapping electrode to create a preprogrammed thread form on the sidewall of a preexisting hole. In this case, the electrode is not a full size tap, but a narrower tapping electrode that orbits within the hole to cut the thread form. This process is generally slower than tapping from solid as described above because the electrical energy is dispersed across the entire length of the electrode engaged in the workpiece, and because the orbiting electrode is smaller than a corresponding thread size electrode used for solid hole tapping. There is, in effect, more work to do and less electrode area to perform the work. However, it does not require the expense of a tapping head, and the use of a predrilled hole can, in some instances, offset the time penalty. Orbiting style tapping electrodes are available in sizes as small as 2–56 UNC, and comparable metric sizes.

EDM control systems

Adapting linear motors on machine tools has created a lot of interest in recent years, but they are not a new invention. Rather, they are an innovative adaptation of an existing technology. It is, in fact, the very same technology that propels roller coasters and powers bullet trains to record speeds in Japan. What makes the current linear motor units so intriguing is their entry into mainstream machining operations. Many industry observers believe that linear motors will have a dramatic impact on the design of machine tool axis motion, outmoding the present technology much like CNC did to control systems a few years ago. The following article by Todd Laffin of Sodick, Inc., appeared in *Modern Machine Shop* and fully explains the impact of linear motors on EDM.

Linear motors

Linear motors are rather simple. On a CNC ram EDM, two series of magnetic plates are mounted on the Z-axis quill, along with fixed magnetic coils on each side of the axis. A linear scale with very fine resolution is mounted to the Z-axis quill in order to detect axis movement location. As the control signals the Z-axis destination, electrical current is introduced into the copper coils, which are adjacent to the magnet plates. The resulting difference in polarity between the plates propels them in opposite directions. Because the one set of plates is fixed, the other set is driven rapidly along its path. The more current that is introduced, the faster the moving axis will travel. The Z-axis can travel more than 1,400 ipm, or nearly 22 times faster than a traditional ballscrew equipped EDM.

No backlash

With conventional motors, before any movement is realized, the electrical motion (rotation) must be converted into mechanical (linear) motion through the use of belts, gearboxes or ballscrews. All of these conversions introduce issues of mass, inertia, backlash, lag-time, overshoot, friction, and heat. Even in the case of direct-drive systems (where the motor-shaft is mounted directly to the ballscrew), it must first overcome the mass, inertia, and friction of the ballscrew mechanism before it encounters the mass, inertia, and friction of the table and the workpiece weight. Then, of course, to stop this motion, the same amount of time and energy is required. With linear motors, most of these issues are reduced or eliminated because no conversion from rotational to linear motion takes place.

Although a linear drive produces less torque at low speed than conventional drive systems, EDM machines don't have high-torque, high-load requirements as chip cutting machines do. Therefore, this characteristic is not an issue for EDM. Because EDM is a noncontact machining process, it is a perfect fit for linear motor technology. Unlike machining centers that take advantage of the table speeds of linear motors, EDM uses linear technology's speed for the Z-axis (ram stroke) in order to create its own flushing capability. EDM also uses the speed of linear motors to react to changes in the spark gap. Linear motor EDMs will excel in difficult-to-flush applications.

Figure 1 compares a typical rotary ballscrew system to a typical linear motor system. In the ballscrew system, position and velocity are controlled by encoder signals. The ballscrew transfers the rotary movement into linear movement. During machining, the control needs to maintain the desired spark gap between the electrode and the workpiece, which is a distance of 3 to 50 microns. This oscillation is repeated up to 500 times per second. The inevitable mechanical twist and "backlash" resulting from the ballscrew's movement will ultimately reduce machining accuracy.

Backlash is a phenomenon that occurs in machine tools when the rotary motion of the ballscrew is converted into linear motion for the bed or table. The conversion is generally carried out via a geared feed mechanism, which requires a certain amount of clearance between the driving and driven gears. Although the clearance does not cause problems when gears are driven at a constant speed or direction, when the drive direction is reversed, there is a short delay before the driving gear teeth mesh with the driven teeth. This lag time, known as backlash, directly affects machining accuracy. As the driving mechanism wears over time, the gear tolerances degrade, which also contributes to a reduction in machining quality.

On the linear motor system, the motor is in effect the only moving part, so the travel distance measured by the linear scale is sent directly to the motor. This allows for a simpler control mechanism without the effects of backlash. Because the electrode is directly connected to the motor assembly, the movement of the electrode and the motor are in unison. This direct connection lets the system operate at very high speeds without mechanical vibration.

Therefore, voltage feedback of the spark gap is precisely followed up with an extremely fast response rate, resulting in faster machining speeds and improved machining accuracy.

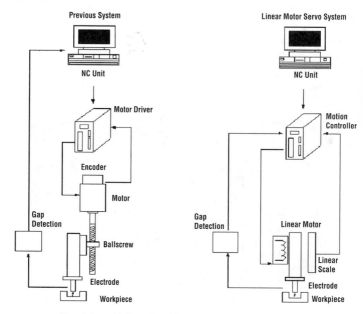

Figure 1. A typical ballscrew drive (left) compared to a linear motor axis drive.

Applying linear motors

Certain considerations have to be addressed by machine designers when creating a linear motor ram EDM. These include designing and building an extremely rigid yet lightweight head, balancing magnetic forces to eliminate distortion, and counterbalancing the head for accurate high speed machining.

Linear motors do generate a great deal of heat, which, if not dissipated, can reduce the life of the linear motor's magnets. To overcome this problem, some sort of system must be added throughout the magnets to circulate chilled coolant in order to maintain optimum operating temperatures.

Because the Z-axis column must travel at high speeds, a conventional head counterbalancing weight system is inadequate. The counterweight on Sodick's design, for example, was eliminated and replaced by an air cylinder system that could offset the forces generated by the high speeds and high acceleration rates at which the Z axis travels.

Similarly, the "ramp speed," or acceleration, of a linear motor system from command to actual position is more than 30 percent faster than a rotary system given the same command. A linear motor control must be able to process information and change the spark gap faster than a traditional ballscrew-driven unit, which has to send data through the NC unit to a motor driver, an encoder, and down to a ballscrew. The linear control sends gap detection data directly to the linear motor, which responds at a much faster rate to the command.

When set in the automatic jump cycle mode, a linear motor EDM can accelerate up to 1.22 G and reach a Z-axis speed of 36 m/min. (1,440 ipm). The maximum servo speed of ballscrew models is about 1.5 m/min. (60 ipm) with 0.05 G acceleration. With linear motors, the combination of speed and acceleration results in machining accuracy.

Natural flushing

This combination of speed and accuracy results in the ability to machine shapes up to (but not limited to) 4 inches deep without auxiliary flushing. Because the electrode is moving so fast, it creates its own natural flushing currents within the spark gap. Yet at this speed, motion is smooth enough to allow a nickel to be balanced upright on the chuck while

Figure 2. Ballscrew ram EDMs may require auxiliary flushing, as shown on the left. With the Z-axis traveling more than 1,400 ipm, a linear motor ram EDM creates its own natural flushing and evenly disperses the eroded particles from the spark gap (right).

moving at 1,400 ipm. The high-speed jump of the Z-axis creates a uniform and minimum spark gap while reducing electrode wear and improving cutting accuracy, as shown in *Figure 2*. The need for flushing ports and forced flushing operations is eliminated.

By creating its own natural flushing for large electrodes, a linear motor EDM produces an even, uniform surface before the finishing process takes place. Because there is less material to remove, time needed for the finishing pass is reduced. Likewise, polishing time is reduced.

Linear motor ram EDMs are proving to be easier to operate than ballscrew driven EDMs because dealing with auxiliary flushing is eliminated. Adding just the right amount of flushing is something many EDM operators learn only after years of experience. Too much or too little flushing can create uneven surfaces. With linear motor EDMs, it is unnecessary to determine where to drill flush holes through the electrode, where to set up auxiliary flush ports, or how to set the right amount of flushing pressure.

Eliminating the need for flushing is ideal for untended operations that use automatic electrode changers and robots. With only one main moving part, a linear motor EDM requires less maintenance and less downtime.

Linear motors for wire EDMs

While CNC ram EDMs have benefited the most from linear motor technology, their wire EDM counterparts have also posted impressive gains in surface finish, machining time, and accuracy from the application of this technology. The improved servo response and sensitivity that linear motors provide result in quicker and more accurate wire alignments and touch-off routines.

The superior response and virtually vibration-free movement of linear motors improves the corner accuracy, positioning accuracy, and roundness capabilities versus traditional ballscrew machines. The faster servo response and increased sensitivity of linear motors result in fewer wire breaks and lost time during rethreading operations. Because linear technology is combined with a higher level of control technology, more power can be put into the wire without breaking.

Cutting speeds on a linear motor driven system increase by 20 percent when compared to a ballscrew model. The combination of high speed roughing and improved discharge frequencies has been found to consistently deliver a surface roughness of 18 rms after only one skim pass. Eliminating multiple skim passes can reduce machining time by as much as 60 percent on certain wire EDM jobs.

Wire EDMs lend themselves to linear motors for another reason. Because the X- and Y-axes on a wire machine do not move with the same high speed as the Z-axis on a CNC ram machine, less heat is generated. Cooling lines are therefore not required on wire EDMs with linear motors.

Beyond EDM

In the Japanese machine tool market, linear motor ram EDMs are reportedly taking back work that had been moved to machining centers. This trend is attributed to the increased machining speeds seen on linear motor EDMs, the elimination of flushing, and the fact that no more operator intervention is required after the process begins. With linear motor technology, ram EDMs have fewer variables and are easier to operate, while the linear motor wire EDMs attain higher accuracy and better finishes with fewer skim passes.

Wire EDM

In concept, a wire EDM machine is simple to explain. A strand of wire feeds from a supply spool to the workpiece, via a wire drive system. The wire, energized by electrical contacts, passes through the workpiece at a determined velocity as required for a specific operation. A stream or bath of deionized water surrounds the wire, and the electrical current in the wire is pulsed on and off, creating sparks that cross the gap between wire and workpiece. Each spark erodes a minute particle of the workpiece as sparks are continuously emitted along the length of the wire. By moving the wire in relation to the workpiece, the channel eroded by sparking action follows a path that can be highly intricate. Spent wire exiting the workpiece from below is gathered onto a take-up reel or chopped up for disposal.

In reality, the wire EDM machine is much more complex. The movement of the wire from supply spool to disposal is manipulated by a series of rollers that pinch the wire and draw it downward while other rollers apply a precise amount of tension to keep it straight and taut. A set of upper and lower wire guides, one above and one below the workpiece, position the wire on its path through the workpiece.

Wire EDM can trace its roots to the final years of the 1960s, but it did not become a factor until the following decade. Since then, it has progressed rapidly in several important areas. In the earliest days, maximum cutting speed was only 1 in.² /hour (645 mm² / hr), while today it is approaching 30 in.² /hour (194 cm² /hr), and average cutting speeds have also increased almost as much due to improvement in machine techniques such as higher flushing pressure and advances in machine technology such as solid-state circuitry, and wire materials. Wire EDM's range of applications have also increased. Early machines were primarily used to create stamping dies and punches, and workpiece size rarely exceeded 4 inches. As dimensions in all axes increased, so did workpiece size. *Figure 3* illustrates how increases in axis travel and speed have increased during recent years, while machine price, adjusted for inflation, declined almost 75%.

Another area where wire EDM machines have shown great improvement is in the degree of taper they can achieve. While early machines were limited to a 1° die relief to 4 inches (100 mm), the maximum taper on contemporary wire EDMs is 30° to 16 inches (406 mm). Accuracy has similarly improved, from 0.001" (0.025 mm) on early machines to 0.00005" (0.0013 mm) today. Untended operation has been influential in the growth of

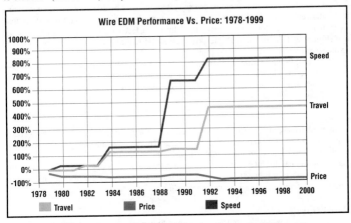

Figure 3. Increases in axis travel and speed have contributed to wire EDM's success.
(Source, Charmilles Technology Corp.)

EDM. For wire machines, this was boosted by reliable automatic wire threading systems that permit different diameter wires to be loaded without outside assistance. Other features now found on machines include automatic fluid pressure adjustment, automatic wire tensioning, and automatic slug removal from workpieces.

EDM wire selection

When selecting a basic wire, tensile strength, hardness, and thickness should be considered and an appropriate wire must be selected based on the requirements of the operation. If a wire with insufficient tensile strength is selected, breakage can be expected. If a wire is too hard—a measurement of its ability to elongate without breaking—it may be brittle and break under sudden shock. Due to their ability to bend and follow contours, softer wires are desirable for high-taper cutting operations. Choosing the proper thickness will depend on the intricacies and dimensions of the cut. The most commonly used range of wire diameters for normal EDM operations is 0.010" to 0.012" (0.254 to 0.30 mm). Fine wires can be obtained in diameters of 0.004" (0.10 mm) down to 0.001" (0.0254 mm). When used on appropriate machines, 0.001" diameter wire can hold tolerances to ±0.00012" (3 μm) within a work envelope of 13.8 × 9.8 × 10 inches (350 × 249 × 254 mm) in the X-, Y-, and Z-axis. Special wire guides are required for the finest wires. Much of the following information on EDM wires was provided by GISCO Equipment Inc.

Copper was initially the material of choice for EDM wire, primarily because of its excellent electrical conductivity (100% IACS [International Annealed Copper Standard] rating) and thermal conductivity. However, plain copper wire also possesses negative physical characteristics including low tensile strength (about 35,000 PSI *or* 241 MPa) and a high percentage of elongation (>20%) that commonly lead to breakage of the wire. Perhaps even more undesirable is the fact that pure copper wire does not flush as well as copper alloy wires.

Brass wire is today's most widely used EDM wire, and the base wires normally contain a Cu/Zn ratio of either 65/35 or 63/37. The superiority of brass over copper for a wire material is due to the lower vaporization temperature of brass, allowing it to flush much better than copper. Tensile strength for premium "63/37" brass wire can reach a maximum of 130,000 PSI (896 MPa), and its percentage of elongation is in the range of >15%<2%. IACS rating for thermal conductivity is 20%. The addition of 2% aluminum to solution harden the brass has a major effect on brass wire. Tensile strength increases to a maximum of 173,000 PSA (1193 MPa), and elongation increases to >20%<2%. The GISCO designation for this wire is CuZn33Al2, meaning that it has 33% zinc, 2% aluminum, and the remaining material is copper. Copper grades are available in even size diameters ranging from 0.002" (0.10 mm) to 0.014" (0.356 mm). CuZn wire is suitable for automatic wire threading machines.

Molybdenum wire is useful when cutting intricate workpieces with small internal radii, but its expense, very low vapor pressure, and relatively poor flushability (even when graphite coated) makes it impractical for all EDM operations. At 280,000 PSI (1930 MPa), its tensile strength is unmatched, but molybdenum is phased out by newer grades of wire.

Steel clad wire is a composite wire with a steel core to provide good tensile strength (135,000 PSI [930 MPa]), a coating of copper, and a final coating of brass. For certain extreme cutting operations, such as machining tall workpieces or parts with extreme flushing limitations, this wire can be a good choice. The steel core provides strength, the layer of copper improves conductivity, allowing for higher power settings and faster cutting, and the zinc enriched brass coating improves flushability because it vaporizes rapidly. A major advantage of composite steel core wire is fracture toughness, which is the wire's ability to withstand the formation of craters or pits on its surface—defects that are the leading cause of wire-breaking fractures.

In 1979, an alternative to wires that are composed of a mix of Cu/Zn was introduced. *Zinc coated copper core wires* offered additional advantages and the ability to cut a wide range of materials. Zinc has a relatively low vaporization temperature, allowing it to boil off rapidly during the cutting process. The low vaporization point of zinc enhances flushing by creating smaller debris particles when it is quenched by the dielectric. Smaller particles are removed more efficiently by the flushing system, which means that higher currents can be used and cutting speeds can be improved with less threat of wire breakage. These wires have a maximum tensile strength of 65,000 PSI (448 MPa) and elongation in the range of <3%.

The next generation zinc coated copper wire featured a high conductivity copper core, with 4% magnesium added to improve performance. A second improvement was the use of a zinc oxide coating that improved flushing performance. The tensile strength of these wires is 65,000 PSI (448 MPa), and their elongation is >3%. The benefit to the user can be seen in their ability to cut high workpieces, and especially in cutting graphite and aluminum alloys. It also features a high resistance to breakage, but it is not suitable for automatic wire threading or skim cutting.

Zinc coated brass wire has a heat treated core with a tensile strength of 70,000 PSI (483 MPa), providing greater accuracy in the cut than copper wire, and the coating has an oxide layer that permits the wire to slide through the guides without flaking the zinc coating. An additional quality of the oxide layer is that it is semiconductive. Therefore, when the wire is subjected to small vibrations against the workpiece that are induced by the mechanical transport of the wire through the guides, the flushing forces, and the discharge of electrical energy, resulting in numerous minute electrical "shorts" that can slow the machining process, the oxide actually masks the sensitivity of the machine servo system which allows more "go forward" signals and fewer "go backward, there is a short circuit." Zinc coated brass wires are suitable for high taper cuts, thanks to its high elongation (>20%), and for cutting exotic composite and man-made materials. It has limited suitability for automatic threaders.

The development of automatic wire threaders in the 1980s led to the development of an evolutionary zinc coated brass wire with a much higher tensile strength (130,000 PSI [896 MPa]) and very high precision. These high tensile strength wires feature excellent mechanical straightness, and they are popular where greater speed and excellent finish are required. Varieties of this wire are available with a zinc oxide coating for Japanese and some European machines using round diamond guides to prevent flaking, and without zinc oxide for machines using an open "vee" guide. These wires are very suitable for cuts involving 0.003", 0.004", and 0.006" (0.0762 mm, 0.1016 mm, and 0.1524 mm) diameter wire, and for taper cuts up to 7° on machines with conventional fixed nontilting guides. The elongation is <2%.

Silver coated brass wire was recently developed to improve the transfer of electrical energy from the carbide power contacts to the wire itself. Because silver has a 105% IACS conductivity rating versus 20% for plain brass wire, only an extremely thin coating of silver is required. The tensile strength is 130,000 PSI (896 MPa), and the elongation is <2%. This type of wire offers excellent price/performance value.

Diffusion annealed wires were developed to enhance wire characteristics in order to match advances in machine technology. Coated copper and brass wires were unable to keep up with cutting speeds of 16 in.²/hour, because the zinc coating had vaporized off after one inch of cutting height. Fortunately, it was discovered that a surface coating of 50% Cu/50% Zn was a workable solution. Available with either a 100% Cu or a CuZn20 base material, these composite wires have a thick coating (0.0014" [0.03556 mm]) of 50% copper and 50% zinc. Next, the coating undergoes a heat treatment process and becomes diffused. Finally, the coating undergoes a drawing operation that compresses (through cold working) the diffused layer into the core material. The diffused wire is porous at the surface, which aids in flushing.

The copper core diffusion annealed wire is suitable for precision and high speed cutting, and for tapers up to 7°. It has limited suitability for automatic threading. This wire has a tensile strength of 72,000 PSI (531 MPa), and elongation of <3%. The brass wire is suitable for automatic wire threaders (its tensile strength is 115,000 PSI [792 MPa]), and is capable of delivering cutting speeds up to 30 in.^{2}/hr.

EDM wires are available in uniform spool sizes, as shown in **Table 1**.

Table 1. Standard EDM Wire Spool Sizes.

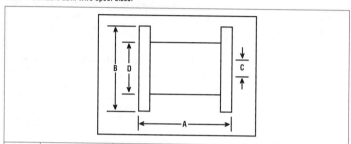

Spool Type	A		B		C		D	
	inch	mm	inch	mm	inch	mm	inch	mm
P-1	3.5	90	4.0	100	0.80	20	2.0	50
P-3	4.4	110	5.2	130	0.80	20	3.2	82
P-5	4.5	114	6.4	160	0.80	20	3.5	90
P-10	5.3	137	7.9	200	0.95	25	3.5	90

Flushing wire EDM operations

While flushing is as important to wire EDM as it is to ram EDM, it is concentrated along the section of the wire that is in contact with the workpiece. Consequently, the primary concern is to see that a continuous flow of deionized water (the dielectric for wire EDM) over this area is maintained. While this sometimes requires pressurized jets, it is more often accomplished by "pressure flush cup" method which positions a flush cup containing dielectric fluid above the cut. The wire and dielectric exit the cup through a nozzle at the bottom of the cup, introducing the dielectric directly to the cutting area. For best results, the nozzle should be placed as close as permissible to the cut, with the distance not exceeding 0.005" (0.13 mm). Contemporary wire EDMs offer flushing pressures up to 300 PSI (21 bar) that, when combined with the proper nozzle size (opening), will provide adequate flushing for most cutting operations. For difficult cutting, such as round surfaces that can deflect the dielectric, the workpiece can be submerged. While this eliminates the possibility of pressure flushing, it does assure that the entire cut area will be bathed in dielectric.

Adequate filtration of the dielectric is important to successful flushing, but a concern that cannot be controlled by simply removing particles from the dielectric is electrolysis.

Deionized water should be chemically pure, but normal EDM operations introduce chemical changes that cannot be filtered out and therefore remain and accumulate in the solution. When the current interacts with this contamination, usually through stray sparks attracted by the contaminants, surface degradation (noticeably rust on ferrous materials) begins, and the depth of the heat affected layer is increased. Other signs are bluing of titanium workpieces, cobalt binder depletion in carbides, anodic oxidation of aluminum, and microcracking of all materials. To overcome these problems, antielectrolysis power supplies have been developed that use voltage modulation in combination with transistorized pulse circuitry and sensors to significantly reduce—or even eliminate—the stray sparks that cause electrolysis.

Special drills are available for creating flushing holes. They feature short flute lengths, and a fast spiral designed to more efficiently draw chips from the cut. In addition, they have 135° points (rather than the conventional 118° point) that spread the chip load across the entire cutting edge. Flushing drills are available in diameters ranging from $1/4$" to $1/2$", and in lengths of 2, 4, and 6 inches.

EDM surface finish

While both ram and wire EDM, under properly controlled conditions, are capable of providing extremely fine surface finishes, all EDMed surfaces are, to some degree, affected by a recast (or "white") layer and a heat affected layer or zone (sometimes called the HAZ) below the recast layer. The recast is made up of material that was melted by the EDM spark but failed to leave the parent material due to poor flushing, gravitational attraction, or other factors. This once molten material cools on the surface of the workpiece to form a hardened layer with a depth of between 0.00015" (0.0038 mm) and 0.005" (0.127 mm). Shallower recast layers are produced during low metal removal rate operations using high frequency and weak sparks. Thicker layers result from low frequency/hot spark machining such as is employed in roughing operations. The HAZ can continue beneath the recast layer to a depth of 0.015" (0.38 mm), and is caused by surface hardening during material removal.

Timken Latrobe Steel performed an interesting study of recast and HAZ over a period of sixteen months that revealed that the condition is "readily evident on heat treated material subsequently cut by EDM techniques, but tools cut in the annealed state and then heat treated did not contain the white layer." In examining a sample of Badger material with an R_C of 57, they found that the EDM spark had increased the hardness to R_C 64, indicating that surface temperatures during machining had reached close to $1450 - 1500°$ F ($788 - 816°$ C) and that the dielectric fluid had liquid quenched the surface layer. It was decided that the recast is a highly stressed layer that could detract from the performance capabilities of the part or tool, particularly under extreme loading. Their conclusion was that the recast layer is inevitable on heat treatable steels, and the following "considerations" should be reviewed when using EDM on heat treated tools. "1) Use one or more light EDM finishing passes (high frequency spark) to remove the effect of heavier cuts and to minimize the depth of the rehardened layer. 2) Use some other finishing technique such as lapping or grinding to totally remove the rehardened zone. 3) Follow steps 1 & 2 by retempering the tool, at the highest temperature permitted by the hardness requirement, to minimize the stress level." Recast can be removed by various mechanical and chemical methods including electrochemical machining which is covered below.

Electrochemical Machining (ECM)

ECM is a process whereby metal is removed by electrochemical dissolution. The basic principles of the process date back to Faraday's work around 1824, although its first application in industry arose from the needs of the aerospace industry in the late 1950s to accommodate super alloys then being developed for more efficient turbine engines. ECM allows the manufacture of parts with stress- and burr-free surfaces that cannot be practically produced with traditional machining methods. Hard and soft materials can be machined with equal speed and precision, and with virtually no tool wear. For example, ECM can produce airfoils with thickness tolerances of ±0.002" (0.05 mm), and complex shapes and contours can be machined in one process rather than in several operations. Typical applications for ECM are turbine blades and vanes, artillery projectiles, bearing cages, molds and dies, surgical implants, and prostheses. It can be used to perform a wide variety of operations including shaping, broaching, sawing, trepanning, surfacing, drilling, and deburring. This section on ECM was contributed by John Adam Fussner.

The electrolytic cell is the heart of ECM. It consists of three parts: the anode, the cathode, and the electrolyte. The rest of the ECM machine is dedicated to maintaining control of the electrolyte and the electric flow to the electrolytic cell. The *anode*, which is positive in polarity, is the workpiece and can be any conductive material. In the aerospace industry, it is typically stainless steel, titanium, or nickel alloy. For blades and vanes, the dimensions may be as small as $0.5 \times 0.4"$ (12.7×10.2 mm), and the maximum size that can be machined is only limited by the amount of electricity that can be utilized economically, and by the design of the product. The *cathode*, or tool, is negative in polarity and, as in EDM, is the inverse shape of the part being produced. Popular materials for the cathode are copper, brass, bronze, stainless steel, and titanium because they have minimal electrical resistance but sufficient strength to withstand electrolyte pressures without deformation, and they are resistant to chemical attack by the electrode. Since there is no tool-to-workpiece contact, tool wear in ECM is not a significant factor—the primary causes are sparks and arcs resulting from short circuits, which have been essentially eliminated by modern solid state electronic controls. The *electrolyte* is the conductive fluid that completes the circuit between the anode and cathode. It is usually a sodium chloride or sodium nitrate solution that flows past the cathode and anode at high pressure (100 to 200 ft/sec *or* 30 to 60 m/sec). For stainless steel turbine blades, an aqueous solution of sodium chloride with a concentration of one pound per gallon, or a solution of sodium nitrate with a concentration of 1.5 to 2.6 pounds per gallon, is used.

In electrochemical machining, the anode, cathode, and electrolyte interact through electrolysis, which can be thought of as an accelerated corrosion process. Electrolysis chemically decomposes the anode into the inverse shape of the cathode. This takes place in two steps. 1) Transfer of electrons from the cathode to the electrode. 2) Transfer of electrons from the electrolyte to the anode (see *Figure 1*). In step one, a chemical reaction occurs near the surface of the cathode when current is applied to the electrolyte and the water within the electrolyte breaks down into hydrogen gas (H_2) and two hydrogen ions ($2OH^-$). The negatively charged ions are attracted to the positively charged anode, which sets up stage two of the process. In an electrolyte solution, ions tend to try to neutralize themselves by attaching to another ion with the opposite charge. One way these ions are formed is if the hydroxide ion finds an atom with a higher valance (valance is the measure of how easily an electron gives up its electrons). Metals are composed of high valance atoms, so they enthusiastically give up electrons and form a positive ion. As this occurs, the metal is dissolved at the surface of the anode, and a metal hydroxide is formed. ECM is sometimes described as a "reverse plating" operation because of the nature of the dissolution of the workpiece.

The combination of these two chemical reactions results in the dissolution of the atoms at the surface of the anode. As the metal atoms are pulled off the surface by the hydroxide ions in order to neutralize themselves, the atoms give off electrons that travel into the

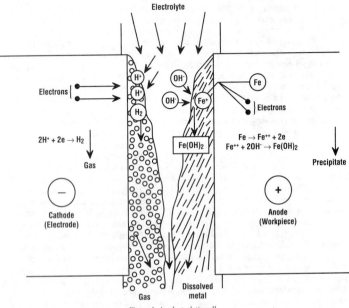

Figure 1. An electrolytic cell.

metal. The hydroxyl ions and metal ions form a metal hydroxide *precipitate* (also called ECM sludge) that is swiftly removed by the electrolyte.

The ECM machine

In order for the electrolytic cell (also called the *flow box*) to properly machine parts, it must maintain the electrochemical process. This requires that the anode is rigidly held in place, and that the cathode is advanced on a CNC controlled ram at the proper rate (normally between 0.01 and 0.75 in./min (0.25 to 19 mm/min). Voltage is supplied by a DC power supply that regulates the prescribed voltage to within ±0.5%. Should the voltage vary beyond ±0.5%, part tolerances will become unacceptable. During the machining process, the sludge and other impurities in the electrolyte must be constantly removed through an extensive cleaning system consisting of a dirty electrolyte tank, a clean electrolyte tank, a clarification system, and the pump and its associated plumbing. In the dirty electrolyte tank, circulating electrolyte is collected before being routed to the clarifier for the removal of foreign particles. The cleansed electrolyte is then sent to the clean tank where it is again filtered and cooled or heated as necessary. It is then returned to the machine for reuse.

Process parameters

The rate of material removal in ECM is a function of current density. The primary variables that affect the current density and removal rate are voltage, feed rate, electrolyte conductivity, electrolyte composition, electrolyte flow, and workpiece material. In most operations, the voltage is controlled at a constant value, usually between 5 and 30 volts. Current requirements range from 50 to 40,000 amperes, and current density ranges between 50 to 3,000 A/in.^{2} (8 to 465 A/cm^2). Feed rate must be closely monitored, as the cutting gap is inversely proportional to the feed rate. The frontal gap varies with feed rate because feeding the cathode at a higher rate decreases the gap and reduces the resistance. Lowering the resistance increases the current, resulting in an increase in the material removal rate. At slower speeds, the opposite is true. As the gap increases, resistance rises and the current drops, resulting in a decrease of the material removal rate. Feed rate is also directly proportional to the current; if the feed rate is doubled, the current will double. The feed rate can be calculated by dividing the metal removal rate by the projected area.

$$V_f = (MRR \times h) \div A_W$$

where V_f = feed rate
         MMR = material removal rate
         h = efficiency of the electrochemical reaction
         A_W = estimated frontal area.

The value of h is usually between 80 and 90%, and MMR can be calculated with the following formula.

$$\text{specific removal rate} = \text{atomic weight} \div (\text{valence} \times \text{density} \times 96{,}494)$$

In this equation, 96,494 represents a constant equal to the amount of electricity required to remove one gram of material. See *Figure 2* for an illustration of the process parameters. The feed rate is also related to the type of operation being performed and the workpiece material. **Table 1** provides feed rates for various ECM machining operations on Inconel 718.

Table 1. Feed Rates for ECM.

Material - Inconel 718	Feed Rate	
Operation	in./min.	mm/min.
Round holes	0.085	2.2
Simple cavities	0.085	2.2
Broaching (angled ECM)	0.3	7.6

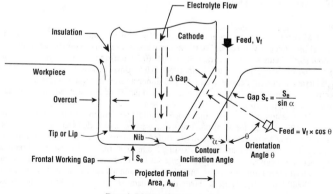

Figure 2. ECM machining parameters.

Electrical conductivity affects the resistance across the cutting gap. Factors affecting conductivity include concentration and temperature—if either concentrations or temperature increase in the electrolyte, conductivity will rise. To obtain the best electrolyte composition for a specific operation, pretesting is recommended. Temperature is directly impacted by electrolyte flow, which also affects the surface finish.

Cathode design

As in EDM, the tool (cathode) shape is usually a mirror image of the shape to be produced. Conventional milling techniques are normally employed to create the cathode shape. The design of the cathode must allow for a smooth flow of electrolyte within a small gap. Positive pressure is necessary to eliminate striation and sparking, and the design must allow for ECM overcut, which can range from 0.001 to 0.05" (0.025 to 1.3 mm). Flow considerations in cathode design are so important that an understanding of hydrodynamics will be helpful to the designer. Flow must be constant and uniform, and quantities of 0.25 gal/min (0.95 L/min) over the machining gap should be maintained at all times. Changes in the direction of flow may cause stagnation in the electrolyte, creating ridges on the workpiece surface.

Two primary methods of cathode design are used: trial and error, and equilibrium gap. As the name suggests, trial and error requires fabricating a cathode shape according to the part to be produced, machining the part, measuring the part, plotting the variation between the part and the part drawing dimensions, and then modifying the cathode according to the variation. This process must be repeated until the variations are acceptable. The equilibrium method takes into consideration the frontal equilibrium gap and the angle of the workpiece contour with respect to the feed direction, based on the following equation (see *Figure 2* for graphic representations of the unknowns).

$$S_C = S_e \div \sin \alpha$$

where S_C = the normal gap
S_e = the frontal equilibrium gap
α = the angle between the feed direction and the tangent of the contour surface.

If α is larger than 40°, the above equation will not yield reliable results. Most cathode designs are variations of the equilibrium gap method and the trial and error method. More complex designs can be made with CAD to factor in such variations as velocity of flow and the electrical field.

Tolerances and surface finishes with ECM

ECM offers precision accuracy, fine surface finishes, and superior dimensional repeatability. Accuracy of ±0.001" (0.03 mm) can be achieved on one dimensional cuts, and contour surface cuts with an accuracy of ±0.004" (0.10 mm) and better are typical. Surface finishes normally run from 16 to 32 µin. (0.41 to 0.81 µm) Ra. The finish is dependent on such factors as current density, feed rate, gap dimensions, workpiece microstructure, and electrolyte composition, viscosity, temperature, and flow. Under ideal conditions, a surface finish of 5 to 15 µin. (0.13 to 0.38 µm) Ra can be achieved.

In 1988, General Electric undertook a study to determine the most effective manufacturing method for axial compressor one-piece bladed disks known as "blisks." It is interesting to review the results in order to compare ECM to more traditional machining methods. Compared to the existing 5-axis DNC milling process, ECM reduced manufacturing labor by 28 hours per engine set (over 50% savings). The ECM process also allowed other indirect savings, including cutter and abrasive costs that amounted to over $200,000 annually. **Table 2** shows the comparative process capabilities for 5-axis milling and ECM for manufacturing blisks.

Another study of blisk manufacturing carried out at Sermatech-Lehr showed that switching from CNC milling to ECM reduced dimensional deviations from 14% to 6%. Also, it was found that blades could be spaced 10% to 15% closer using ECM.

Table 2. ECM versus 5-axis Milling for Blisk Production.

Dimension	Milling		ECM	
	inch	metric	inch	metric
Airfoil thickness	±0.008	±0.20 mm	±0.003	±0.076 mm
Chord	+0.009/−0.011	+0.23/−0.28 mm	±0.007	±0.18 mm
Contour	±0.002	±0.05 mm	±0.002	±0.05 mm
Airfoil finish Ra	32 µin. axial 75 µin. radial	1.04 µm axial 1.88 µm radial	24 µin. before peening 32 µin. after peening	0.61 µm before peening 0.81 µm after peening
Hub finish Ra	40 µin. bench prior to abrasive flow	1.02 µm bench prior to abrasive flow	32 µin. after peening	0.81 µm after peening
Radial position	0.005±	±0.13 mm	±0.005	±0.013 mm

ECM machining considerations

ECM sparks occur when a short circuit exists between the cathode and anode. Should a spark jump (or arc) between the cathode and anode, it can result in damage to the workpiece and require repair or replacement of the cathode. There are two primary reasons for undesired sparks: electrolyte contamination and improper electrolyte flow. Electrolyte contamination consists of either conductive or nonconductive particles, and either can result in spark damage. Should a conductive metal sliver from the ECM system become lodged in the machining gap, it will create an arc between the anode and cathode that produces an intense electrical discharge over a small area. Nonconductive particles may also become lodged between the cathode and anode, which will slow the electrolytic current and result in the slowing of workpiece dissolution. Under these conditions, when the cathode continues to advance at a predetermined rate but workpiece dissolution has slowed, the cathode will eventually contact the anode, resulting in a spark that will damage both the cathode and anode.

The intrinsic structure of some materials may lead to the stray pitting, or to intergranular attack (when etching occurs at the boundary layers within the microstrucure). These conditions are caused by passive films that exist on different phases of a material's microstructure. ECM rates vary due to the properties of the individual films, resulting in nonuniform dissolution of the anode that will cause protrusions to develop on its surface. These protrusions represent a reduction of the spark gap that leads to subsequent etching at the boundary layers.

One interesting aspect of ECM is that, unlike mechanical processes such as milling or turning, it does not induce compressive stresses onto the workpiece that can increase tensile strength and hardness. Consequently, the fatigue strength of an ECM'ed part, though identical in appearance to a milled part, may be as much as 10 to 25% less than the mechanically produced part. The ECM process produces parts with stress free surfaces, a definite advantage, that can be strengthened with peening or vapor honing if strength is critical.

Waterjet Cutting Technology

Waterjets are used in high production applications across the globe. They complement other technologies such as milling, laser, EDM, plasma, and routers. No noxious gases or liquids are used in waterjet cutting, and waterjets do not create hazardous materials or vapors. No heat affected zones or mechanical stresses are left on a waterjet cut surface. The waterjet has shown that it can do things that other technologies simply cannot. From cutting whisper thin details in stone, glass, and metals; to rapid hole drilling of titanium; to cutting of food, to the killing of pathogens in beverages and dips, the waterjet has proven itself unique. This section on waterjet cutting technology was provided by Chip Burnham of Flow International Corporation.

Dr. Norman Franz is regarded as the father of the waterjet. He was the first person who studied the use of ultrahigh-pressure (UHP) water as a cutting tool. The term UHP is defined as more than 30,000 pounds per square inch (PSI). Dr. Franz, a forestry engineer, wanted to find new ways to slice thick trees into lumber. In the 1950s, Franz first dropped heavy weights onto columns of water, forcing that water through a tiny orifice. He obtained short bursts of very high pressures (often many times higher than are currently in use), and was able to cut wood and other materials.

Mechanics of the waterjet machine

All waterjet machines have a pump that pressurizes the water and delivers it continuously, so that a cutting head can convert the pressurized water into a supersonic waterjet stream. Two types of pump can be used for waterjet applications—a direct drive based pump, or an intensifier based pump. The direct drive pump operates in the same manner as a low-pressure "pressure washer." It is a triplex pump that actuates three plungers directly from the electric motor. These pumps have been available since the late 1980s, and are gaining acceptance in the waterjet industry due to their simplicity. In their present state of development, direct drive pumps can deliver a maximum continuous operating pressure 10 to 25% lower than intensifier pump units (20,000-PSI to 50,000-PSI for direct drive, 40,000-PSI to 60,000-PSI for intensifiers). In metric units, PSI is measured as bar, and one bar equals 14.50377 PSI.

Though direct drive pumps are used in some industrial applications, the vast majority of all ultrahigh pressure pumps in the waterjet world today are intensifier based. The typical intensifier pump contains two fluid circuits, the water circuit and the hydraulic circuit. The water circuit consists of the inlet water filters, booster pump, intensifier, and shock attenuator. Ordinary tap water is filtered by the inlet water filtration system, which is usually comprised of a 1- and a 0.45-micron cartridge filter. The filtered water then travels to the booster pump, where the inlet water pressure is maintained at approximately 90-PSI to ensure that the intensifier is never "starved for water." The filtered water is then sent to the intensifier pump and pressurized to up to 60,000-PSI. Before the water leaves the pump unit to travel through the plumbing to the cutting head, it first passes through the shock attenuator, which dampens the pressure fluctuations to ensure that the water exiting the cutting head is steady and consistent. Without the attenuator, the water stream would visibly and audibly pulse, leaving marks on the workpiece surface.

The hydraulic circuit consists of an electric motor (25 to 200 HP), a hydraulic pump, an oil reservoir, a manifold, and a piston biscuit/plunger. The electric motor powers the hydraulic pump. The hydraulic pump pulls oil from the reservoir and pressurizes it to 3,000-PSI. This pressurized oil is sent to the manifold where the manifold's valves create the stroking action of the intensifier by alternating the injection of hydraulic oil to one side of the biscuit/plunger assembly, then the other. The 3,000-PSI hydraulic oil pushes against a piston biscuit. A plunger with a face area of 20 times less than the biscuit pushes against the water. Therefore, the 3,000-PSI oil pressure is "intensified" twenty times, yielding 60,000-PSI water pressure. The intensifier is a reciprocating pump, in that the biscuit/plunger assembly reciprocates back and forth, delivering high-pressure water out one side of the intensifier while low-pressure water fills the other side. The hydraulic oil is then cooled during recirculation to the reservoir.

The "intensification principle" varies the area component of the pressure equation to intensify, or increase, the pressure.

$$\text{Pressure} = \text{Force} \div \text{Area}$$

If Force = 20, Area = 20, then Pressure = 1. If the Force is held constant and the Area is greatly reduced, the Pressure will increase. For example, if the Area is reduced from 20 down to 1, the Pressure will increase from 1 to 20. In *Figure 1*, the small arrows denote the 3,000-PSI of oil pressure pushing against a biscuit face that has 20 times more area than the face of the plunger. The intensification ratio, therefore, is 20:1.

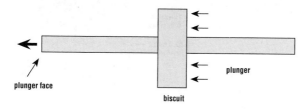

Figure 1. The plunger face/biscuit area ratio determines pressure.

Sophisticated check valves ensure the low-pressure and high-pressure water is only allowed to travel in one direction. The high-pressure cylinders and end caps that encase the plunger and biscuit assembly are specially designed to withstand the enormous force and the constant pressure cycles. The entire unit is designed for long life, and also designed to fail in a safe, gradual manner, rather than instantaneously. The seals and connections begin to leak slowly through specially designed weep holes. The operator or maintenance person can see a drip escaping from a weep hole. The location of the drip and the amount of water indicate when maintenance should be performed. Usually, the maintenance person can schedule the periodic maintenance of seals and check valves out 1 to 2 weeks into the future by simply monitoring the gradual weeping. Warning and shutdown sensors also cover the pumping unit to further safeguard against pump damage. See *Figure 2*.

Once the high-pressure pump has created the water pressure, water is delivered to the cutting head via high-pressure plumbing. In addition to transporting the high-pressure water, the plumbing also provides freedom of movement to the cutting head. The most common type of high-pressure plumbing is special stainless steel tubing that is available in different sizes for different purposes.

1/4 inch steel tubing: Because it is flexible, this tubing is typically used to plumb the motion equipment. It is not used to bring high-pressure water over long distances (for example, from pump to base of motion equipment). Long lengths of 10 to 20 feet are used to provide X, Y, and Z movement (called a high-pressure whip). It is easily bent, even into a coil (coils provide greater flexibility over short distances).

3/8 inch steel tubing: Typically, this tubing is used to deliver water from the pump to the base of the motion equipment. It can be bent, but it is not normally used to plumb the motion equipment.

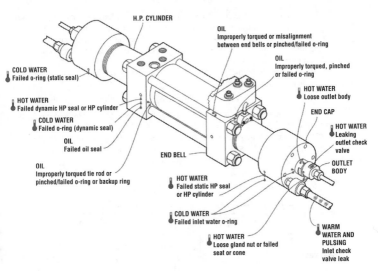

Figure 2. Troubleshooting an intensifier based pump: hot water dripping from a weep hole indicates a high-pressure leak, cold water indicates low-pressure. (Source, Flow International Corp.)

$^9/_{16}$ *inch steel tubing:* This tubing is commonly used to transport high-pressure water over long distances. The large internal diameter reduces pressure loss. When very large pumps are present, this tubing is especially beneficial (the larger the volume of high-pressure water needed to be transported, the larger the potential pressure loss). It cannot be bent, so fittings are required to create corners.

Pure waterjet machines

Pure waterjet is the original water cutting method. The first commercial applications were in the early to mid-1970s, and involved the cutting of corrugated cardboard. The largest commercial uses for pure waterjet cutting are disposable diapers, tissue paper, and automotive interiors. When cutting tissue paper and disposable diapers, the waterjet process creates less moisture on the material than touching or breathing on it. Pure waterjet attributes:

Very thin stream (0.004" to 0.010" in diameter is the most common range)
Extremely detailed geometry
Very little material loss due to cutting
Nonheat cutting
Cut very thick
Cut very thin
Usually cuts very quickly
Able to cut soft, light materials (e.g., fiberglass insulation up to 24" thick)
Extremely low cutting forces
Simple fixturing
24 hour per day operation

Pure waterjet cutting heads

In waterjet cutting, the material removal operation can be described as a supersonic erosion process. It is not pressure but stream velocity that tears away microscopic pieces or grains of material. Pressure and velocity are two distinct forms of energy. To achieve high stream velocity, the pressurized water passes through a tiny hole in a jewel (sapphire, ruby, or diamond) that is affixed to the end of the plumbing tubing. Pure waterjet orifice diameters range from 0.004" to 0.010" for typical cutting. At approximately 40,000-PSI, the resulting stream that passes out of the orifice is traveling at Mach 2. At 60,000-PSI the speed is in excess of Mach 3. For special applications, such as waterblasting concrete with a nozzle traversing back and forth on a tractor, a single large orifice of up to 0.10" is often used.

Each of the three common types of orifice materials (sapphire, ruby, diamond) has its own unique attributes. Sapphire is the most common orifice material in use. It is a man-made, single crystal jewel that produces a fairly good quality stream. Sapphire orifices can be expected to last, when used with good quality water, approximately 50 to 100 cutting hours. In abrasive waterjet applications the sapphire's life is one-half that of pure waterjet applications. Sapphire orifices typically cost between $15 and $30 each.

Ruby is often used in abrasive waterjet applications. The stream characteristics are well suited for abrasive jets, but are not well suited for pure waterjet cutting. The cost is approximately the same as the sapphire.

Diamond has considerably longer run life (800 to 2,000 hours) but is 10 to 20 times more costly. Diamond is especially useful where 24 hour per day operation is required, such as tissue paper, auto interior, and disposable diaper cutting. Diamonds, unlike the other orifice types, can sometimes be ultrasonically cleaned and reused. **Table 1** summarizes the characteristics of common orifice materials.

Table 1. Characteristics of Waterjet Orifice Materials. *(Source, Flow International Corp.)*

Material	Life	Use	Comments
Sapphire	50 to 100 hours	Pure waterjet	General purpose, though life reduces by one-half for abrasive waterjet applications
Ruby	50 to 100 hours	Abrasive waterjet	Stream not suitable for pure waterjet applications
Diamond	800 to 2,000 hrs	Primarily waterjet	10 to 20 times more expensive than ruby or sapphire

Abrasive waterjet machines

The abrasive waterjet differs from the pure waterjet in a few significant ways. In pure waterjet, the supersonic stream erodes the workpiece material. In the abrasive waterjet, the waterjet stream accelerates abrasive particles, and it is these particles, rather than the water, that erode the workpiece material. The abrasive waterjet is hundreds, if not thousands, of times more powerful than a pure waterjet. Where the pure waterjet cuts soft materials, the abrasive waterjet cuts hard materials such as metals, stone, composites, and ceramics. Abrasive waterjets using standard parameters can cut materials with hardness up to and slightly beyond that of aluminum oxide ceramic (often called alumina, AD 99.9). Abrasive waterjet attributes include:

Extremely versatile process
No heat affected zones
No mechanical stresses
Easy to program
Thin stream (0.020" to 0.050" in diameter)
Extremely detailed geometry

Thin material cutting
10 inch thick cutting
Stack cutting
Little material loss due to cutting
Simple to fixture
Low cutting forces (under 1 lb. while cutting)
One jet setup for nearly all abrasive jet jobs
Easily switched from single to multi-head use
Quickly switched from pure waterjet to abrasive waterjet
Reduced secondary operations
Little or no burr

Abrasive waterjet cutting heads

Within every abrasive waterjet is a pure waterjet. Abrasive is added after the pure waterjet stream is created. Then the abrasive particles are accelerated down the mixing tube. The abrasive used in abrasive waterjet cutting is hard sand that is specially screened and sized. The most common abrasive is garnet. Garnet is hard, tough, and inexpensive. Different mesh sizes are used for different jobs.

120 Mesh - produces smooth surface
80 Mesh - most common, general purpose
50 Mesh - cuts a little faster than 80, with slightly rougher surface.

The mixing tube acts like a rifle barrel to accelerate the abrasive particles. They, like the orifice, come in many different sizes and various replacement lives. Mixing tubes are approximately 3 inches long, 0.025" in diameter, and have internal diameters ranging from 0.020" to 0.060", with the most common being 0.040".

Although the abrasive waterjet machine is typically considered simple to operate and "bullet proof," the mixing tube does require operator attention. A major technological advancement in the waterjet was the invention of truly long-life mixing tubes. Unfortunately, the longer life tubes are far more brittle than their predecessors, which were tungsten carbide tubes. If the cutting head comes in contact with clamps, weights, or the target material, the tube may be broken. Broken tubes cannot be repaired and must be replaced. Today's most advanced systems incorporate collision detection to avoid crashes with the mixing tube. See **Table 2** for mixing tube material characteristics.

The standoff distance between the mixing tube and the target material is typically 0.010" to 0.200". Larger standoff (greater than 0.080") can cause a frosting to appear atop the cut edge of the part. Many abrasive waterjet systems reduce or eliminate frosting by cutting under water or using other techniques. The consumable items in an abrasive

Table 2. Characteristics of Abrasive Waterjet Mixing Tube Materials. *(Source, Flow International Corp.)*

Material	Life	Comments
Standard Tungsten Carbide	4 to 6 hours	These were the original mixing tubes. They are no longer used, due to their poor performance and cost per hour ratio. They tend to wear out of round, require very frequent replacement.
Low cost composite carbide	35 to 60 hours	Good for rough cutting or when training a new operator.
Mid-life composite carbide	80 to 90 hours	A good all-around tube.
Premium composite carbide	100 to 150 hours	The top-of-the-line. This popular tube exhibits the most concentric and predictable wear. Used for both precision work and everyday work.

waterjet are the water, abrasive, orifice (usually ruby), and mixing tube. The abrasive and mixing tube are exclusive to the abrasive waterjet. The other consumables are also found in the pure waterjet.

Motion equipment

There are many different styles and configurations of waterjet motion equipment (machine tools). In addition to providing motion, the machine tool must also include some means of holding the workpiece material catching the jet and collecting the water and debris created by the machining operation.

Stationary and 1-dimension machines: The simplest of machines is the stationary waterjet. Looking much like a bandsaw, it is usually used in the aerospace industry to trim composites. The operator feeds the material through the stream much like a bandsaw. After the material has been cut, the catcher collects the stream and debris. Usually outfitted with a pure waterjet, some stationary waterjet machines are equipped with abrasive waterjets. Another version of a stationary machine is a slitter. Here a product such as paper is fed through the machine, and the waterjet slits the product into specific widths. A cross cutter is another example of a machine that moves in one axis. Although it is not truly stationary, this simple machine often works in conjunction with a slitter. Where the slitter cuts product to specific widths, the cross cutter cuts across a product that is fed beneath it. Often the slitter and cross cutter work together to create a grid pattern in materials such as vending machine brownie cakes.

It is generally not recommended to use an abrasive waterjet manually (either moving the material by hand or the cutting head by hand). It is very difficult to manually move the object being cut at a specific and consistent speed. Most manufacturers will not recommend or quote a manually operated abrasive waterjet. Only in special cases where operator safety is not in question should an abrasive waterjet be used in a manual mode.

XY tables for 2-dimension cutting: XY tables, sometimes called "flatstock" machines, are the most common forms of waterjet motion equipment. These machines are used with pure waterjets to cut gaskets and plastics, rubber, and foam. Abrasive waterjets utilize these tables to cut metals, composites, glass stone, and ceramics. Abrasive waterjet and pure waterjet tables may be as small as 2 × 4 ft or as large as 30 × 100 ft. The basic components of an XY are:

  Control by either CNC or PC
  Servo motors, usually with closed-loop feedback, to ensure position and velocity
    integrity
  Base unit with linear ways, bearing blocks, and ball screw drive
  Bridge unit also with ways, blocks, and ball screw
  Catcher tank with material support

Many different machine styles are available, but two distinct styles dominate the industry: the mid-rail gantry and the cantilever. The mid-rail gantry machines have two base rails and a bridge. The cantilever system has one base and a rigid bridge. All machine types will have some form of adjustability for the head height (the head height is controlled by the Z-axis). The Z-axis adjustability can be in the form of a manual crank, motorized screw, or a fully programmable servo screw.

The catcher tanks on flatstock machines are usually water filled tanks that incorporate grating or slats to support the workpiece. These supports are slowly consumed during the cutting process. Catcher tanks can either be self-cleaning, where the waste is deposited into a container, or manual, where the tank is periodically shoveled empty.

Five + axis machines for 3-dimensional cutting: Many man-made items, like airplanes,

Table 3. Waterjet Machine Tool Motion Specifications. *(Source, Flow International Corp.)*

Envelope Size	The length of travel found in each axis of movement. The most common sizes for flatstock cutting on a waterjet or abrasive waterjet machine are 2 × 3 m × 0.3 m, or approximately 6 × 10 × 1 ft. The catcher tank is usually at least 6 inches larger than the travel length and width, aiding in heavy plate loading, allowing for clamping, and allowing for variability in raw sheet size.
Linear Positional Accuracy	Measures how accurately the machine can move to a point. One axis is measured at a time from one point to another. Speed is not a consideration here, as the measurement is taken once the machine has reached a full stop.
Machine Repeatability	Ability of the machine to return to a point.
Rapid Traverse Speed	Rapid traverse is the top speed a machine can move without cutting. The control system simply sends a signal to the drive motors saying, "go as fast as you can in that direction." The accuracy of machine motion is usually compromised during rapid traverse. Rapid traverse is used to move from one cut path (e.g., a hole cutout), to another cut path (e.g., the part perimeter).
Contour Speed	The top speed that the machine can move while maintaining all the accuracy specifications (i.e., accuracy, repeatability, velocity). This is a critical specification as it relates to part production cycle times and part accuracy.

have few flat surfaces on them. Also, advances in complex 3D composites and metal forming technologies suggest that there will be fewer flat parts in the future. Thus, the need for 3-dimension cutting increases each year. Waterjets are easily adapted to 3D cutting. The lightweight heads and low kickback forces during cutting give design engineers freedom not found when designing for the high loads found in milling and routing. And the thin, high-pressure plumbing, featuring excellent flexibility achieved through the use of swivels, provides freedom of movement.

The simplest of the 3-dimensional cutting systems is the Universal WaterRouter. This device is moved by hand, and is only intended for pure waterjet cutting of thin materials (e.g., aircraft interiors and thin composites). The handheld gun is counterweighted, and provides all degrees of freedom. This device was popular in the 1980s as a superior method of trimming composites. As a replacement for the router, the operator would press a special nozzle against the template, turn on the jet with dual thumb triggers, and trace the template with the nozzle while walking around the part. The jet would shoot into a point catcher after cutting the material. Many of these safe and effective tools are used today in thin aerospace composite cutting and other applications.

The need for enhanced production and the desire to eliminate the expensive templates required for routers and Universal WaterRouters lead to the use of fully programmable 5-axis machines. With these machines, a programmer creates the tool path in an office and downloads the program to the operator's machine control system that then cuts the material.

Even with the improvement of advanced 3-D offline programming software, 3-D cutting (contouring a shape in space) proves more complex than 2-D (flat plate cutting), regardless of whether the cutting process is waterjet, router, or another process. An extreme example of the complexity of cutting in 3-D is the composite tail components of the Boeing 777 aircraft. Many steps are taken to trim the part. First, the programs for the cut path and the flexible "pogostick" tooling are downloaded. Then the material is flown in via overhead crane while the pogosticks unscrew to their preprogrammed height. After the part has been roughly located and the pogostick vacuum tops have secured the part, a special Z-axis (not used for cutting) brings a touch probe into location to precisely locate the part in space. The touch probe samples a number of points to ensure the part elevation and orientation is known. Then the program part transformation takes place. Here the program is reoriented to match the actual location of the part. Finally the touch probe Z is retracted and the cutting head Z swings into action. Now the part is ready to cut. The 3-D program is accessed and run, controlling all 5 axes to hold the cutting head normal to the surface being cut, while traveling at the precise speed required.

The cutting of relatively thick composites (>0.05") or any metal requires the use of abrasives, which means that a method has to be found for stopping the 50 horsepower jet from cutting up the pogosticks and tooling bed after the cut has been made through the material. The only effective way known to date is to catch the jet in a very special point catcher. A steel ball point catcher can stop the full 50 HP in under 6 inches, then the slurry is vacuumed away to a waste handling tank. A "C" shaped frame connects the catcher to the Z-axis wrist. This C-frame can rotate to allow the head to trim the entire circumference of the wing part.

The point catcher consumes steel balls at a rate of approximately one-half to 1 pound per hour. The jet is actually stopped by the dispersion of kinetic energy. As the jet enters the small container of steel balls, the balls spin. The spinning balls then rub their neighbors and they spin. The spinning balls in the point catcher will consume the energy of the jet, leaving the cutting slurry to harmlessly leak out the screened catcher bottom. These point catchers have proven so effective that they can run horizontally, and even sometimes completely upside-down.

The complexities of locating the part in space, adjusting the part program, and accurately cutting a part are magnified as the size of the part increases. Many shops effectively use 3-D machines for simple 2-D cutting and complex 3-D cutting every day. Although software continues to get easier and machines continue to get more advanced, parts continue to get more complex.

Characteristics of part accuracy

A distinct difference exists between part accuracy and the accuracy that a machine can achieve in its movements. Even though a machine may have an accuracy of 0.00000000000001", with perfect dynamic motion, perfect velocity control, and dead-on repeatability, it does not mean it will cut perfect parts. Finished part accuracy is a combination of process error (the waterjet) + machine error (the XY performance) + workpiece stability (fixturing, flatness, stable with temperature).

Table 4 describes part errors that would occur even if the waterjet machine motion is perfect. The waterjet beam has characteristics that greatly affect part accuracy. Controlling these characteristics has been the focus of waterjet suppliers for many years. Simply put, a highly accurate and repeatable machine may eliminate machine motion from the part accuracy equation, but it does not eliminate other part errors (such as fixturing errors and inherent waterjet stream errors).

When cutting materials under 1 inch thick, a waterjet machine typically cuts parts with an accuracy from ±0.003" to 0.015" (.07 to .4 mm). For materials over 1 inch thick, the machines will produce parts from ±0.005" to 0.100" (.12 to 2.5 mm). A high performance XY table is designed to have an accuracy of about 0.005" linear positional accuracy or better.

The effects of power

A common misconception in waterjet and especially abrasive waterjet cutting is that it is best to use as little power, as little pressure, as little abrasive as possible to get the job done. Nothing could be further from the truth. The key is to cut as fast as possible. For most applications, the operating cost increase when running a system "flat out" is far outweighed by the money saved by producing more parts in a given time period. The

Table 4. Waterjet Cutting Part Errors. *(Source, Flow International Corp.)*

Part Errors	Description
Beam Deflection or "Stream Lag"	When the waterjet (or other beam type cutters like laser or plasma) are cutting through the material, the stream will deflect backwards (opposite direction of travel) when cutting power begins to drop. This problem causes: increased taper, inside corner problems, and sweeping out of arcs. Reduce lag error by increasing cutting power or slowing down the cut speed.
Increased Taper	A "V" shaped taper is created when cutting at high speeds. Taper can be minimized or eliminated by slowing down the cut path or increasing cutting power. Image is exaggerated for descriptive purposes.
Inside Corner Problems	When cutting an inside corner at high speed, the stream digs into the part as it comes out of the corner. This image is of the hole that is left when cutting a square cutout, viewed from the exit (or bottom) side. The image is exaggerated for descriptive purposes. No such stream lag error will be seen on the part top (the entrance side).
Sweeping Out of Arcs	When cutting at high speed around an arc or circle the stream lag sweeps out a cone. Image is exaggerated to illustrate the error. As with the inside corner problem, the sweeping to arcs error will not be seen on the top side of the part.
Fixturing	Even though the waterjet delivers under one-half pound of vertical force when cutting a high quality part, and under 5 pounds when rough cutting, proper fixturing is required to produce accurate parts. The part must not move during cutting or piercing, and it must not vibrate. To minimize these errors try to butt the workpiece up against the edge of the catcher or a solid bar stop secured to the table slats. Look for material vibration or movement during cutting.
Material Instability	Some materials, like plastics, can be very sensitive to temperature changes. Called thermal expansion, these materials may expand when heated or shrink when cooled. During waterjet cutting, the material does not get hot, but it can get warm. The temperature of the a typical abrasive waterjet is approximately 180°F (82.2°C), and the water filled catcher tank may warm enough to distort very heat sensitive workpiece material. Be especially careful of air gaps in cast material, as the stream tends to open up in air gaps. The abrasive water jet will not induce warpage in sheet material. It will, however, relieve stresses. When working with a sheared material <0.125" thick and starting the cut path off the part, enter into the part, and then cut the part, you may see the material twist and warp. Avoid this warpage whenever possible by beginning cut paths from within the material (pierce a hole and begin cutting) as opposed to beginning from outside the material.

(Continued)

Table 4. *(Continued)* **Waterjet Cutting Part Errors.** *(Source, Flow International Corp.)*

Part Errors	Description
Pump Issues	Beyond the obvious pump issues such as ensuring that the pump is delivering water at the set pressure, other issues can also impact part accuracy. If the pump has two or more intensifiers, do the intensifiers always stroke at the same time? If so, look on the part for vertical marks on the cut edge that match in frequency with the stroking. Check valves should be in good working order to assure that no loss of pressurized water is experienced.
Water Pressure at the Nozzle	Cut speed can be lost if excessive pressure drops (greater than 2,500-PSI) exist in the high-pressure plumbing run from pump to head. Ensure the in-line filter, usually located near the cutting head, is free of excessive buildup. If any changes have been made to the plumbing run (change of route, replacement of a large line with a smaller replacement line, etc.) then ensure that larger pressure drops have not been created. Any loss of pressure between the pump is to be minimized.
Cutter Compensation Error	Cutter compensation is the value entered into the control system that takes into account the width of cut of the jet; in effect, it sets the amount by which the cut path is enlarged so that the final part comes out to proper size. Before performing any high precision work where finished part tolerances are better than ±0.005", cut a test coupon and ensure that the cutter compensation is properly set. Also, make sure the operator knows at what frequency the cutter compensation must be adjusted to in order to compensate for mixing tube wear. Low quality mixing tubes may exhibit excessive wear and require compensation adjustments every few hours. Many good drawings have been cut wrong because the operator did not take the time to establish the best cutter compensation value.
Programming Error	Often the most difficult of all part accuracy errors to find is a programming error where the dimension of the part program does not perfectly match the dimension of the original CAD or hand-drawn drawing. Part programs that appear graphically on the screen of an XY control typically do not display dimensions. Therefore, this error can go undetected. When all else fails, double check that the dimensions on the part program match exactly those of the original drawing.
Abrasive Mesh Size	Typical abrasive mesh sizes are 120, 80, and 50 (similar to sandpaper abrasive used for woodworking). The different mesh sizes have only a minor impact on part accuracy. They have a greater impact on surface finish and overall cut speed. Finer abrasives (larger mesh number) produce slower cuts and smoother surfaces.
Machine Motion	**The positional accuracy and dynamic motion characteristics of a machine have an impact on part accuracy. Two of the many aspects to machine motion performance are:** backlash in the mechanical unit (changes in direction, is there slop in the gears or screw when the motor changes from clockwise to counterclockwise?); and repeatability, will the machine come back close to the same point over and over? Servo tuning is important. Improper tuning will cause backlash, squareness, repeatability errors, and can cause the machine to chatter (wiggle at high frequency) when moving. Position accuracy is important, as well as straightness, flatness, and parallelism of the linear rails. Small parts, under 12" in length and width, do not demand as much from the XY table as larger parts. A large part measuring, for example, 4 × 4 ft, will be greatly impacted by machine performance. A small part will not see position accuracy, or rail straightness as a major impact on finished part tolerance simply because the small part masks machine errors. Large parts expose such errors more evidently.

(Continued)

Table 4. *(Continued)* **Waterjet Cutting Part Errors.** *(Source, Flow International Corp.)*

Part Errors	Description
Machine Motion (Continued)	Remember that a machine motion characteristic does not directly correspond to finished part tolerance. An expensive super-precision machine (linear position accuracy of, for example, ±0.001" over full travel), will not automatically generate a finished part of ±0.001" — other part accuracy factors are still there (see above).

generic curves shown in *Figure 3* have been generated by countless Universities, waterjet manufacturers, waterjet users, and research companies. The curves always show the same tendency: as abrasive flow rate is increased from zero, cut speed goes up and cost per inch goes down until a peak point is reached, a point where the cut speed and cost per inch are both at their optimum. Of course, with every "rule of thumb" there are exceptions. However, for virtually all cutting in the world today, fastest cut speed = lowest cost per inch.

To cut as fast as possible, the system should be operated using the maximum horsepower available. If the machine has a 60,000-PSI pump with 50 HP, then whenever possible all 50 HP should be used. On a 100 HP system that can only effectively run a cutting head that consumes 50 HP, consider running two heads. Suggested cutting speeds and flow rates for efficient cutting are shown in **Table 5**, and water flow rate as a function of pressure and orifice size, both in gallons and liters, is provided in **Table 6** and **Table 7**.

Abrasive constitutes two-thirds of the machine operating cost of the equipment. Machine operating cost does not include labor, lease, depreciation, facilities, or other overhead costs. It does include power, water, air, seals, check valves, orifice, mixing tube, abrasive, inlet water filters, and long term spares (hydraulic pump, high-pressure cylinders, etc.). Basically, these are all the items that need replacement regularly or over the life of the waterjet system.

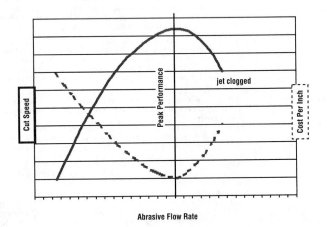

Abrasive Flow Rate

Figure 3. For any given set of parameters (pressure, orifice size, etc.), the cut speed goes up and costs go down as the abrasive flow rate is increased. Eventually, peak performance is reached.

Table 5. Waterjet Cutting Speeds, Horsepower Requirements, and Flow Rates for Various Materials. (Source, Flow International Corp.)

	Cutting Speed									
	55,000-PSI/3.793-BAR						40,000-PSI/2,759-BAR			
	Imperial	Metric	Imperial	Metric	Imperial	Metric	Imperial	Metric	Imperial	Metric
Orifice inch/mm	.010/.030	.254/.762	.014/.040	.356/1.016	.016/.040	.406/1.016	.010/.030	.254/.762	.013/.040	.330/1.016
Water gpm/liter/m	.5	2.27	1.0	4.55	1.24	5.64	.42	1.91	.71	3.23
Abrasive lb/m/kg/m	.8	.363	1.3	.590	1.5	.680	.6	.272	1.0	.454
Power HP/kW	25	18.6	50	37.3	60	44.7	11	8.2	25	18.6
Material	in./m*	mm/m*	in./m*	mm/m*	in./m*	mm/m*	in./m*	mm/m*	in./m*	mm/m*
Aluminum 1/4"	46.0	1168	67.1	1704	74.7	1897	27.1	688	39.6	1006
1/2"	21.8	554	31.8	808	35.4	899	12.8	325	18.8	478
1"	9.7	246	14.1	358	15.7	399	5.7	145	8.3	211
Granite 1/4"	81.2	2062	118.4	3007	131.7	3345	47.7	1212	69.9	1775
1/2"	38.5	978	56.1	1425	62.4	1585	22.6	574	33.1	841
1"	17.1	434	24.9	632	27.7	704	10.0	254	14.7	373
Graphite/Epoxy 1/4"	124.9	3172	182.2	4628	202.8	5151	73.4	1864	107.6	2733
1/2"	59.2	1504	86.3	2192	96.1	2441	34.8	884	60.0	1524
1"	26.3	668	38.4	975	42.7	1085	15.5	394	22.6	574
Inconel 1/4"	15.5	394	22.7	577	25.2	640	9.1	231	13.4	340
1/2"	7.4	188	10.7	272	11.9	302	4.3	109	6.3	160
1"	3.3	84	4.8	122	5.3	135	1.9	48	2.8	71
Marble 1/4"	95.7	2431	139.5	3543	155.2	3942	56.2	1427	82.3	2090
1/2"	45.3	1151	66.1	1679	73.5	1867	26.6	676	39.0	991
1"	20.1	511	29.4	747	32.7	831	11.8	300	17.3	439
Glass 1/4"	88.3	2243	128.8	3272	143.3	3640	51.9	1318	76.0	1930
1/2"	41.8	1062	61.0	1549	67.9	1725	24.6	625	36.0	914
1"	18.6	472	27.1	688	30.2	767	10.9	277	16.0	406
Steel (mild) 1/4"	18.3	465	26.6	676	29.6	752	10.7	272	15.7	399
1/2"	8.7	221	12.2	310	14.0	356	5.1	130	7.4	188
1"	3.8	97	5.6	142	6.2	157	2.3	58	3.3	84

(Continued)

Table 5. *(Continued)* Waterjet Cutting Speeds, Horsepower Requirements, and Flow Rates for Various Materials. *(Source, Flow International Corp.)*

	Cutting Speed									
	55,000-PSI/3,793-BAR						40,000-PSI/2,759-BAR			
	Imperial	Metric	Imperial	Metric	Imperial	Metric	Imperial	Metric	Imperial	Metric
Orifice inch/mm	.010/.030	.254/.762	.014/.040	.356/1.016	.016/.040	.406/1.016	.010/.030	.254/.762	.013/.040	.330/1.016
Water gpm/liter/m	.5	2.27	1.0	4.55	1.24	5.64	.42	1.91	.71	3.23
Abrasive lb/m kg/m	.8	.363	1.3	.590	1.5	.680	.6	.272	1.0	.454
Power HP/kW	25	18.6	50	37.3	60	44.7	11	8.2	25	18.6
Material	in./m*	mm/m*	in./m*	mm/m*	in./m*	mm/m*	in./m*	mm/m*	in./m*	mm/m*
Steel (stnls) 1/4"	17.0	432	24.8	630	27.6	701	10.0	254	14.7	373
1/2"	8.1	206	11.8	300	13.1	333	4.7	119	6.9	175
1"	3.6	91	5.2	132	5.8	147	2.1	53	3.1	79
Titanium 1/4"	22.2	564	32.3	820	36.0	914	13.0	330	19.1	485
1/2"	10.5	267	15.3	389	17.0	432	6.2	157	9.0	229
1"	4.7	119	6.8	173	7.6	193	2.7	69	4.0	102

* Speeds are for a separation (rough) cut measured in linear inches per minute (in./min) or linear millimeters per minute (mm/min). For a quality cut, multiply the speed by 0.4 (40% of the maximum speed).

Table 6. Water Flow Rate in Gallons Per Minute as a Function of Pressure and Orifice Size.
(Source, Flow International Corp.)

Orifice (Inch)	Pressure (PSI)							
	20,000	25,000	30,000	35,000	40,000	45,000	50,000	55,000
	Gallons per Minute							
.003	.026	.029	.032	.035	.037	.039	.042	.044
.004	.047	.053	.058	.062	.067	.071	.075	.078
.005	.074	.082	.090	.098	.104	.111	.117	.122
.006	.106	.119	.130	.140	.150	.159	.168	.176
.007	.145	.162	.177	.191	.204	.217	.229	.240
.008	.189	.211	.231	.250	.267	.283	.299	.313
.009	.239	.267	.293	.316	.338	.358	.378	.396
.010	.295	.330	.361	.390	.417	.443	.466	.489
.011	.357	.399	.437	.472	.505	.535	.564	.592
.012	.425	.475	.520	.562	.601	.637	.672	.704
.013	.499	.557	.611	.660	.705	.748	.788	.827
.014	.578	.646	.708	.765	.818	.867	.914	.959
.015	.664	.742	.813	.878	.939	.996	1.049	1.101
.016	.755	.844	.925	.999	1.068	1.133	1.194	1.252
.017	.853	.953	1.044	1.128	1.206	1.279	1.348	1.414
.018	.956	1.069	1.171	1.264	1.352	1.434	1.511	1.585
.019	1.065	1.191	1.304	1.409	1.506	1.597	1.684	1.766
.020	1.180	1.319	1.445	1.561	1.669	1.770	1.866	1.957
.021	1.301	1.455	1.573	1.721	1.840	1.951	2.057	2.157
.022	1.428	1.596	1.749	1.889	2.019	2.142	2.258	2.368
.023	1.561	1.745	1.911	2.064	2.207	2.341	2.467	2.588

Table 7. Water Flow Rate in Liters Per Minute as a Function of Pressure and Orifice Size.
(Source, Flow International Corp.)

Orifice (mm)	Pressure (BAR)							
	1378	1724	2069	2414	2759	3103	3448	3793
	Liters per Minute							
.076 mm	.098	.110	.121	.132	.140	.148	.159	.167
.100 mm	.18	.20	.22	.23	.25	.27	.28	.295
.127 mm	.28	.31	.34	.37	.39	.42	.44	.46
.152 mm	.40	.45	.49	.59	.57	.60	.63	.67
.178 mm	.55	.61	.67	.72	.77	.82	.87	.91
.203 mm	.71	.80	.87	.95	1.01	1.07	1.13	1.18
.229 mm	.90	1.01	1.11	1.20	1.28	1.35	1.43	1.49
.254 mm	1.12	1.25	1.36	1.48	1.58	1.68	1.76	1.85
.279 mm	1.351	1.51	1.65	1.79	1.91	2.02	2.13	2.24
.305 mm	1.608	1.80	1.97	2.13	2.27	2.41	2.54	2.66

(Continued)

Table 7. *(Continued)* **Water Flow Rate in Liters Per Minute as a Function of Pressure and Orifice Size.**
(Source, Flow International Corp.)

Orifice (mm)	Pressure (BAR)							
	1378	1724	2069	2414	2759	3103	3448	3793
	Liters per Minute							
.330 mm	1.888	2.11	2.31	2.50	2.67	2.83	2.98	3.13
.356 mm	2.187	2.44	2.68	2.90	3.10	3.28	3.46	3.63
.381 mm	2.51	2.81	3.08	3.32	3.55	3.77	3.97	4.17
.406 mm	2.86	3.19	3.50	3.78	4.04	4.29	4.52	4.74
.432 mm	3.23	3.61	3.95	4.27	4.56	4.84	5.10	5.35
.457 mm	3.62	4.05	4.43	4.78	5.12	5.43	5.72	5.99
.483 mm	4.03	4.51	4.94	5.33	5.70	6.04	6.37	6.68
.508 mm	4.47	4.99	5.47	5.91	6.32	6.70	7.06	7.41
.533 mm	4.92	5.51	5.95	6.51	6.96	7.38	7.79	8.16
.559 mm	5.40	6.04	6.62	7.15	7.64	8.11	8.55	8.96
.584 mm	5.91	6.60	7.23	7.81	8.35	8.86	9.34	9.80

In abrasive waterjet cutting it is often thought that to reduce the abrasive flow rate saves money. On the contrary, it wastes money. There is a peak performance point, shown in *Figure 3*, for abrasive waterjet operation. When all overhead is included, the cheapest cutting is always found at the fastest possible speed. This fact is independent of the material you are cutting, or the power of the system.

A variety of pump sizes are available. Some manufacturers offer just a few sizes, while others make the full range available. The full range is shown in **Table 8**.

In **Table 8**, the most common pump size currently in use is the 50 HP pump powering one head. From there, the order of popularity follows 100 HP, 25 HP, and 150 HP. Over 60% of all pumps produced today are 50 or 100 HP.

Since 1999, there has been a steady increase in the number of multi-head systems using 100 HP and even higher. The upgrade to 100 HP costs a little more ($30,000 to $50,000 at current prices), but these systems are more productive than 50 HP single head systems.

Table 8. Available Pump and Multiple-head Options for Waterjet Systems. *(Source, Flow International Corp.)*

Pump Power	Output (gallons per minute)	Maximum single orifice (diameter) it can power at full pressure	Multi-head options
25 HP	0.5 gpm	0.010" diameter	Multi-head with this pump is highly unusual as heads become small.
50 HP	1.0 gpm	0.014"	2 each 0.010"
75 HP	1.5 gpm	0.017"	3 each 0.010"
100 HP	2.0 gpm	0.21" (seldom use full 2.0 gpm because largest abrasive jet heads are 0.016" or possibly 0.018")	4 each 0.010" 2 each 0.014"
150 HP	3.0 gpm	0.28" (seldom use all 2.0 gpm because largest abrasive jet heads are 0.016" or possibly 0.018")	2 each 0.017" 3 each 0.014" 6 each 0.010"

Note: Simplified pump capacity table. Although other pump sizes exist from some manufacturers (40 HP, 60 HP, 200 HP, etc.), those in the above table are most common.

Ten important things to remember about abrasive waterjet cutting

1. When cutting material under 0.100" thick, little is gained by using medium (50 HP) or large (60 to 80 HP) cutting heads. Consider using a smaller sized head (25 HP) for these thin material cutting operations.

2. Avoid cutting through air gaps greater than 0.020". The jet tends to open up in the gap and cut the lower layer roughly. When stack cutting, keep the sheets together.

3. Smaller abrasive grains (120 mesh or smaller) will cut at a slightly slower rate but will produce a slightly smoother surface (as compared to 80 or 50 mesh).

4. Productivity is cost per inch, not cost per hour. It matters very little how much it costs per hour to run an abrasive waterjet. What matters is how many parts can be produced in a given length of time. Some users make the mistake of trying to reduce operating cost by minimizing the abrasive flow rate. Even though abrasive is two-thirds of the abrasive waterjet operating cost, parts must be produced quickly in order to consume overhead (labor, facilities, lease payment) expenses. Cut as fast as possible, using all available horsepower and the peak abrasive flow rate.

5. For waterjets used to pierce composites, glass, and stone on a regular basis, ensure the system has the ability to have the water pressure lowered and raised by the controller. Also, investigate vacuum assist or other techniques to improve probability of successfully piercing these brittle or laminated materials without damage to the target material.

6. Most machines do not employ material handling automation, such as shuttles. Only when material handling constitutes a significant portion of part production cost should automation be considered. 90% of all abrasive waterjet machines are loaded and unloaded either by hand or with the aid of simple overhead cranes, jib cranes, or fork lifts. Approximately 50% of all waterjet machines utilize material handling automation. Waterjet typically cuts thin, light material at very high speed. The time it takes to cut an entire sheet is fairly low, and the handling component of part production cost is high enough to justify the added capital expense.

7. Ordinary tap water is used to feed the waterjet systems. 90% of all waterjet and abrasive waterjet users require only water softening prior to sending that water through the pump's inlet water filters and then to the intensifier. Reverse Osmosis (RO) and De-Ionizers tend to make the water so pure that it becomes "ion starved." This aggressive water seeks to satisfy its ion starvation by taking ions from surrounding materials, such as the metals in the pump and high-pressure plumbing lines. RO and De-I can greatly extend orifice life, while simultaneously performing very expensive damage to the intensifier and plumbing. Orifices are rather inexpensive. High-pressure cylinders, check valves, and end caps damage will far outweigh orifice life improvements.

8. Cutting underwater reduces surface frosting or "hazing" found on the top edge of an abrasive waterjet cut. Cutting underwater also greatly reduces jet noise and workplace mess. The only negative is that operators cannot see the jet clearly during cutting. If the operator objects to underwater cutting, consider electronic performance monitoring. These monitors will detect deviation from peak cutting performance and stop the system prior to part damage.

9. For systems that often use different abrasive mesh sizes for different jobs, consider adding additional abrasive storage and metering for the mesh sizes commonly run. A small (100 lb) or large (500 to 2,000 lb) bulk transfer and associated metering valve will allow for quick changeover from one mesh size to the next, thereby reducing downtime and nuisance while increasing production.

10. Break-out tabs can prove effective for cutting of materials under 0.3" thick. Although break-out tabs generally guarantee that a secondary operation of grinding off the tabs will be required, their usage allows material handling to be performed faster. Simply unload a cut sheet with the cut parts still intact. The harder the material, the smaller the break-out tab should be. Consult the machine manufacturer for detailed suggestions, or experiment with different tab sizes and styles.

Threaded Fasteners

This section primarily contains dimensions and specifications for threaded fasteners. Explanations of standard fastener grades, plus the ratings and head and nut markings for each grade, were provided in the "Bolted Joints" section. Also, because washers are used in conjunction with threaded fasteners, they are included in this section so that they are easily accessible for review when considering bolt/screw/washer combinations. As anyone who has specified a fastener for a specific application is aware, there are vast numbers of new, innovative styles of threaded fasteners for almost any need. It is impossible to chronicle each of these ingenious devices in this book, as they have neither been standardized nor used in sufficient numbers to make them universally available. With this in mind, creative designers should, by consulting manufacturer's catalogs, be able to find a nearly ideal threaded fastener for almost any requirement.

After this introduction containing brief descriptions of the general types of threaded fasteners, Tables follow with dimensions for the most widely used English/U.S. Customary unit fasteners: Hexagon head and other bolt forms; hex nuts; socket head screws; machine, tapping, and other screws. Tables providing Metric measurement threaded fasteners are then grouped similarly. It is hoped that this arrangement will minimize confusion and make locating specific fasteners more efficient and less time consuming. For full information on specific fasteners, the following standards should be consulted.

English/U.S. Customary Unit Threaded Fasteners and Washers

ANSI/ASME B18.2.1. American National Standard Square Bolts, Hex Bolts, and Hex Screws.

ANSI/ASME B18.2.2. American National Standard Hex Nuts.

ANSI/ASME B18.3. American National Standard Socket Head Screws.

ANSI/ASME B18.5. American National Standard Round Head, Countersunk Head, and Special Head Bolts.

ANSI/ASME B18.6.2. American National Standard Slotted Head Cap Screws and Slotted and Square Head Set Screws.

ANSI/ASME B18.6.3. American National Standard Machine Screws and Machine Screw Nuts.

ANSI/ASME B18.6.4. American National Standard Self-Tapping and Metallic Drive Screws.

ANSI B18.17. American National Standard Wing Nuts and Screws, and Thumb Screws.

ANSI/ASME B18.21.1. American National Standard Lock Washers.

ANSI/ASME B18.22.1. American National Standard Plain Washers.

Metric Unit Fasteners and Washers*

ANSI B18.2.3.1M. Metric Hexagon Cap Screws.

ISO/DIN 4017 and DIN 933. Hexagon Head Screw, product grades A and B.

ISO/DIN 4018 and DIN 558. Hexagon Head Screw, product grade C.

ISO 8676 and DIN 961. Hexagon Head Screw with Fine Thread.

ANSI B18.2.3.4M. Metric Hexagon Flange Screws.

ISO 8100/04 and DIN 6921. Hexagon Flange Bolt.

ANSI B18.2.3.5M. Metric Hexagon Bolts.

ISO/DIN 4014 and DIN 931. Hexagon Head Bolt, product grades A and B.

ISO/DIN 4016 and DIN 601. Hexagon Head Bolt, product grade C.

ISO 8765 and DIN 960. Hexagon Head Bolt with Fine Thread.

ANSI B18.2.3.7M. Metric Heavy Hexagon Structural Bolts.

FASTENERS

DIN 9614. Hexagon Bolt for High Tensile Structural Bolting.

ANSI B18.2.3-8M. Metric Hex Lag Screws.

ANSI B18.3.1M. Metric Socket Head Cap Screws.

ISO 4762 and DIN 912. Socket Head Cap Screw.

ANSI B18.2.4.1M and B18.2.4.2M. Metric Hexagon Nuts.

ISO/DIN 4032 and DIN 934. Hexagon Nut, style 1, product grades A and B.

ISO/DIN 4034 and DIN 555. Hexagon Nut, style 1, product grade C.

ANSI B18.2.4.3M. Metric Slotted Hexagon Nuts.

DIN 935. Hexagon Castle Nut.

DIN 979. Hexagon Thin Castle Nut.

ANSI B18.2.4.4M. Metric Hexagon Flange Nuts.

ISO 4161 and DIN 6923. Hexagon Flange Nut.

ANSI B19.2.4.5M and 2.4.6M. Metric Hex Jam Nuts and Heavy Hexagon Nuts.

ANSI B18.3.3M. Metric Socket Head Shoulder Screws.

ISO 7379 and DIN 9841. Socket Head Shoulder Screws.

ANSI B18.3.4M. Metric Socket Button Head Cap Screws.

ANSI B18.3.5M. Metric Socket Flat Head Cap Screws.

DIN 7991. Hexagon Socket Countersunk Head Screw.

ANSI B18.3.6M. Metric Socket Set Screws.

ISO 4026, 4027, 4028, 4029, and DIN 913, 914, 915, 915. Socket Set Screws.

ANSI B18.5.2.2M. Metric Round Head Square Neck Bolts.

ANSI B18.18.3M-1998. Metric Prevailing Torque Hex Flange Nuts.

ISO 7043 and DIN 6926. Prevailing Torque Hex Nut with Nylon Insert.

ISO 7044 and DIN 6927. Prevailing Torque Hex Nut–All Metal.

ANSI B18.22M. Metric Plain Washers.

ISO 7089/90 and DIN-125-T1. Hex Head Bolt and Nut Washers up to Hardness 250 HV.

ISO 7089/90 and DIN-125-T2. Hex Head Bolt and Nut Washers from Hardness 300 HV.

ISO 7092 and DIN 433-T-1. Cheese Head Bolt and Nut Washers up to Hardness 250 HV.

ISO 7092 and DIN 433-T-2. Cheese Head Bolt and Nut Washers from Hardness 300 HV.

ISO 1207 and DIN 84. Slotted Cheese Head Screws.

ISO/DIN 1479. Hexagon Head Tapping Screws.

ISO/DIN 1481. Slotted Pan Head Tapping Screws.

ISO/DIN 1482. Slotted Countersunk Flat Head Tapping Screws.

ISO/DIN 1483. Slotted Raised Countersunk Oval Head Tapping Screws.

ISO/DIN 1580 and DIN 85. Slotted Pan Head Screws.

ISO/DIN 7045. Cross Recessed Pan Head Screws.

ISO/DIN 7046. Cross Recessed Countersunk Flat Head Screws.

ISO/DIN 7047. Cross Recessed Raised Countersunk Head Screws.

ISO/DIN 7049 and DIN 7981. Cross Recessed Pan Head Tapping Screws.

ISO/DIN 7050 and DIN 7982. Cross Recessed Countersunk Flat Head Tapping Screws.

ISO/DIN 7051 and DIN 7983. Cross Recessed Raised Countersunk Oval Head Tapping Screws.

* Grade A and B Metric fasteners are minimum property class 8.8 for bolts, and minimum property class 8 for nuts. Bolts adhering to these Grades, in diameters of 5 mm or higher, have their property class stamped on the head of the bolt. Nuts adhering to

these grades in nominal sizes of 5 mm or higher have their property class on one side flat (preferred), or on the clock face. Grade C (commercial) Metric bolts conform to property class 4.6 or better, and nuts of this Grade conform to property class 4 or better.

The Metric threaded fastener Tables reference appropriate ISO/DIN and DIN standard numbers. They should be consulted for additional standard reference numbers. The values in these tables are generally nominal values based on the relevant standards, but they, as with all the Tables in this section, should not be taken as absolute measurements. Appropriate standards should be consulted when dimensional references are critical.

ISO tolerances for Metric threaded fasteners are provided in **Table 75**, and ISO tolerances for socket screws are given in **Table 124**.

Bolts and screws

Even though individual bolts and screws may be interchangeable, a bolt is intended to be used with a nut. A screw, on the other hand, mates with preexisting threads, or, in some configurations such as self-tapping screws, a screw can make its own threads. Screws and bolts are both externally threaded fasteners that come in a variety of head/shoulder/shank/thread combinations, and are used to join separate elements, or to fasten something into place. A stud is used for similar applications, but does not have a head that can be used to torque it into place.

As seen in the following material provided by Unbrako/SPS Technologies, Inc. fabrication methods can impact the quality of a fastener. Producing threads on bolts and screws is covered in detail elsewhere in this book, but bolt and screw heads are produced by entirely different operations. There are two general methods of making bolt heads—forging and machining. The economy and grain flow resulting from forging make it the preferred method.

The temperature of forging can vary from room temperature to 2000° F (1094° C). By far, the greatest number of parts are cold upset on forging machines known as headers, or boltmakers. For materials lacking sufficient formability for cold forging, hot forging must be used. Hot forging is also used for bolts too large for cold upsetting due to machine capacity. In fact, the largest cold forming machines can make bolts up to 1 $1/2$ inches in diameter. For large quantities of bolt heads, hot forging is, unfortunately, more expensive than cold forging.

Some materials, such as stainless steel, are warm forged at temperatures up to 1,000° F (538° C). This heating provides two benefits. First, lower forging pressures are necessary due to the reduced yield strength of the material, and, second, work hardening rates are reduced.

Machining is used for very large diameters or small production runs. The disadvantage of machining is that the process cuts the metal grain flow, thus creating planes of weakness at the critical head-to-shank fillet area. This can result in reduced tension performance resulting from fracture planes.

The head-to-shank transition (fillet) represents a sizable change in cross section at a critical area of bolt performance. It is important that this notch effect be minimized. A generous radius in the fillet reduces the notch effect, but a compromise is necessary because too much radius will reduce the load bearing area under the head. Composite fillets, such as elliptical fillets, maximize curvature on the shank side of the fillet, and minimize it on the head side—thereby reducing loss of bearing area on the load bearing surface. The head-shank fillet must not be restricted or bound by the bolt hole. A sufficient chamfer or radius on the edge of the hole will prevent interference that could seriously reduce fatigue life. Also, if the bolt should seat on an unchamfered edge, there might be serious loss of preload if the edge breaks under load.

Bolt forms

Hexagon head bolts are available in two basic configurations: standard hex bolts, and heavy hex bolts/heavy hex structural bolts. The threads may be Class 2A Unified Coarse, Fine, or 8 thread series. Heavy hex bolts have head dimensions that are approximately $1/8$ inch wider than standard hex bolts throughout the range, thereby usefully increasing the bearing surface area. The material specification for standard and heavy Grade A bolts is ASTM A307. Heavy hex structural bolts have generally the same dimensions as heavy hex bolts, but the thread length is shorter (0.25 inch shorter on $1/2$ inch nominal size bolts, and 1 inch shorter on $1\,1/2$ inch nominal size bolts). Structural bolts are made to ASTM A325 and ASTM A490 specifications, and the head should be stamped with the appropriate specification. Since hex bolts and hex head cap screws essentially share dimensions other than head height (the head on screws is not as high as on bolts), many suppliers will substitute one for the other in smaller diameters up to $3/4$ inch. Therefore, if head height is critical, orders should specify bolt or screw. Also, different material standards (SAE J429, ASTM A449, or ASTM A354) are specified for screws.

Many round head and flat head bolts are in wide use, and the most common varieties are shown in the Tables.

Metric standard hex bolts and standard hex cap screws are almost dimensionally identical, but the screw is normally specified with a washer face while the bearing surface is often flat on the bolt.

Dimensions of English/U.S. Customary unit bolts are given in **Tables 1** through **15**, and dimensions for Metric bolts are given in **Tables 76** through **85**.

Nuts

The most common shapes for nuts are square and hex. Square nuts are not normally made to exacting tolerances and their use is generally restricted to lighter-duty applications. Hex nuts come in many configurations, in standard and heavy versions, and in thinner sizes known as jam nuts. Jam nuts are used as locking nuts. Normally, a jam nut is tensioned to seating torque on a screw or stud, and then held stationary while a full nut is torqued to full preload on top of the jam nut. This results in the two nuts bearing in opposite directions and "jamming" the threads to serve as a locking nut. Standard hex and jam nuts have slightly smaller dimensions than heavy hex and heavy jam nuts: up to $9/16$, the standard nut is $1/16$ inch narrower across flats; above $9/16$, standard nuts are $1/8$ inch narrower across flats. In addition, jam nuts are always thinner than hex nuts of the same nominal size. For standard nuts and jam nuts, the difference in thickness ranges from $1/16$ inch to $7/16$ inch, and for heavy nuts and heavy jam nuts the difference in thickness ranges from $2/16$ inch to $1\,3/4$ inch. Hex nuts are available in many other varieties including locking nuts with nylon inserts, flanged nuts, and castle (or slotted) nuts. Hex nuts for high temperature, high pressure service should be specified as follows. For moderate temperatures and pressures, specify ASTM A194 Grade 2 nuts. For quenched nuts suitable for high temperature and high pressure conditions, specify ASTM A194 Grade 2H nuts. For severe conditions, specify ASTM A194 Grades 4 or 7 for heat treated carbon molybdenum and chromium molybdenum nuts.

Hex nuts in English/U.S. Customary units are available in a wider variety of sizes than metric hex nuts, but metric versions of most hex shaped nuts are available including jam nuts, nylon insert locking nuts, flanged nuts, and castle nuts.

Dimensions of English/U.S. Customary unit nuts are given in **Tables 16** through **31**, and dimensions for Metric nuts are given in **Tables 86** through **95**.

Washers

Due to the fact that washers are manufactured in almost every conceivable width, internal diameter, and thickness, only standardized washers are included in this book. Companies such as WCL (West Coast Lockwasher) and ASM (Accurate Screw Machine) publish extensive lists of available sizes and materials and should be contacted for dimensions of flat, curved, Belleville, and wave form washers. See the section in this book on "Bolted Joints" for more information about these washer forms.

Dimensions of selected English/U.S. Customary unit washers are given in **Tables 32** (which provides comparative dimensions of both inch unit and Metric unit washers), **33, 34,** and **35,** and dimensions for Metric washers are given in **Tables 96** through **107.**

Common screws, and machine screw head and thread configurations

Hexagon head cap screws are the most common screws in use. They are briefly compared to hexagon head bolts in the section on bolts (above). There are many basic and standardized head configurations for machine screws. In addition, there are probably twice as many different "drive" systems (Phillips head, slotted head, Torx, etc.), and each of the more popular head styles can most likely be obtained with any of the drive systems. With the exception of slotted, cross recessed or Phillips, and socket head screws, most drive systems have not been standardized and their intended use can be very specialized. Therefore, before specifying a screw or drive system, it is often a good idea to shop around for the one best suited to a particular application. With the exception of the hex head screw, all machine screws have round heads when viewed from above.

Binding Head. This design is only available with a flat bearing surface. In appearance, it resembles the fillister head, but is wider and lower. Due to their low head height, binding head screws are available only with a slotted drive.

Cheese Head. This design is more popular with Metric screws in Europe than it is in the U.S. It resembles a tall fillister head, but has a flat or recessed slotted top surface.

Fillister Head. The fillister head is easily identified by its relatively high sides and its rounded top surface.

Flat Countersunk Head. There are two established designs for this head: one with a head angle of approximately 82° (usually preferred), and a less common head with an angle of approximately 100° (the 100° thread is, however, used exclusively in the miniature series). When installed correctly, this head design sits flush with the surrounding surface. A "flat countersunk trim head" provides either a head one size larger (large trim head) or two sizes smaller (small trim head) than a conventionally sized flat countersunk head. Trim heads are available only with cross-recessed drives.

Hex Head. Identified by six flat sides, a flat bearing surface, and a flat or slightly indented top surface, this design offers the choice of fastening or releasing either by wrenching or by a slotted driver.

Hex Washer Head. This is a refined version of the hex head that incorporates a round, flat bearing surface that extends beyond the raised head of the screw. It may be either wrenched or tightened/removed with a slotted driver.

Oval Countersunk Head. This head surface is rounded and the head angle is approximately 82°. An "oval countersunk trim head" provides either a head one size larger (large trim head) or two sizes smaller (small trim head) than a conventionally sized oval countersunk head. Trim heads are available only with cross-recessed drives.

Pan Head. The top surface of a pan head screw is flat if the screw is slotted, but rounded if it is cross-recessed. The pan head is generally preferred over the round head screw.

Round Head. This is the only semi-elliptical machine screw head. It has been to some degree replaced by the pan head, which provides a slot or recess of almost uniform depth across the head, and protrudes less from the bearing surface.

Truss Head. The truss head is the second widest head available (after the binding head). It is low and rounded on top, allowing for very little depth between the bearing surface and slot or recess bottom. Therefore, it is weak by design and should only be used with low driving torques.

Except for the smallest sizes, machine screws are available with UNC and UNF Class 2A threads, or UNRC and UNRF series threads. Thread lengths are:

Size	Thread Length
No. 5 and smaller, nominal length three diameters or less	Threads extend to at least one pitch of bearing surface
Nominal lengths greater than three diameters, to 1 $^1/_8$ inches	Threads extend to at least two pitches of bearing surface
No. 6 and larger, nominal length three diameters or less	Threads extend to at least one pitch of bearing surface
Nominal lengths greater than three diameters, to 2 inches	Threads extend to at least two pitches of bearing surface
Over 2 inches	1 $^1/_2$ inches, minimum.

Ironically, there are machine screw nuts as well. Although the use of a nut transforms the screw into a "bolt," hex nuts are available in standard plus small pattern sizes, and square nuts are available as well.

The Metric dimensions for machine screw lengths are based on a minimal screw length, which is equal to three diameters, and governs the unthreaded portion of the screw beneath the head. For the minimal lengths, this unthreaded portion is equal to approximately 20% of the diameter. For mid-range sizes (up to 30 mm for M × 2 and M × 2.5, and up to 50 mm for all other sizes), the length of the unthreaded area is approximately one-third of the diameter. For longer screws, the full form thread length is a minimum of 25 mm on M × 2 and M × 2.5 diameters, and 30 mm on larger diameters.

Hex head cap screws and machine screws are covered in **Tables 36** through **47**, and instrument (miniature) screws are covered in **Tables 48** through **50** (English/U.S. Customary units). Metric hex head and machine screw dimensions are given in **Tables 108** through **116**.

Self-tapping and thread cutting screws

For many years, the Type A point was the most widely specified thread forming tapping screw style. Its thread was very coarse (25% fewer threads per inch in the $^3/_8$ or #24 nominal diameter than the "AB" thread), and generally it fell out of favor and was relegated to the "not recommended" category. The AB thread has generally replaced the A, and it shares general dimensions with the DIN 7940 and ISO 1478 Type ST tapping screw thread. The Type AB thread combines the 45° gimlet point of the Type A, and the pitches of the Type B. (The ISO designation for this point is also "AB," but the DIN nomenclature is Type B.) The Type AB can be used for most applications, and recommended core hole sizes are provided in the Tables for tapping screws.

The Type B thread forming tapping screw is used for sheet metal, nonferrous castings, plywood, and plastics. It is essentially a flat end Type AB screw, and the Metric version is covered by DIN 7940 and ISO 1478. (The ISO standard refers to the flat end as a Type B point, while the DIN standard calls it a Type BZ.) The Type BT, also known as the Type 25, shares the threads and point of the Type B, but has a cutting groove on the point. It is commonly used in plastics.

Thread Type D (also known as Type 1), Type F, and Type T (also known as Type 23) are thread cutting screws, but they have finer threads approximately equivalent to those found on machine screws. They also feature blunt points and tapered entering threads with chip cavities. These screws are commonly used in sheet metal, aluminum, zinc, and lead die castings, cast iron, and plastics.

The Type U has a smooth, round, unslotted head. It is inserted with pressure and is not intended to be removed once in position.

Tapping and thread forming screws use many of the same head styles covered above in the discussion of machine screws, and most commonly with slotted or Phillips recessed drives.

Recommended hole sizes for English/U.S. Customary unit tapping and thread forming screws are given in **Tables 59** through **62**.

Tapping and thread cutting screws are covered in **Tables 55** through **58** (English/U.S. Customary units), and **Tables 117** through **120** (Metric screws).

Sems

Sems are screws with captive washers. While most commonly available as slotted or Phillips pan head screws available in size from No. 2 to $1/2''$ nominal diameter, they may also have round, flat, oval, fillister, truss, binding, socket, hex, or hex washer heads. Thread styles can be specified to meet specific needs, as sems can be obtained off the shelf with machine screw threads, or the following tapping threads: Type AB, Type B, Type D/Type1, Type F, and Type T/Type 23. Lock washer styles include internal tooth, external tooth, square cone, helical, and conical, and plain flat washer versions are also available. New versions of sems-type captive washer screws are constantly being introduced, so several suppliers should be contacted if an unusual thread/washer combination is desirable for a specific application.

Typically, sems are available in thread diameters of 0.0470'' (size 00) to $1/2''$, and Metric diameters of M1 through M12. Lengths range from $1/8$ to 3 inches, and 3 to 75 millimeters.

Socket screws

Socket screws were developed for applications with limited space, and their cylindrical head and internal wrenching features allow their use in locations where externally wrenched fasteners would be impractical. As originally designed, consistent relationships were not maintained among the nominal shank diameter, head diameter, and socket size throughout the range of sizes. However, in 1960, industry standards were issued that provided consistent relationships throughout the size range. Screws made to this standardized design are commonly referred to as "1960 Series" socket head cap screws. The older design, termed the "1936 Series" after the year of their design, are no longer in production. The designations "1936" and "1960" have caused some confusion because the numbers are sometimes thought to refer to the materials used in the screw's construction. The terms refer only to the dates of acceptance for dimensional relationships, and do not in any way define the mechanical properties of a screw.

The International Organization for Standardization standard ISO 898 contains a system of property classes that perform a function similar to the Society of Automobile Engineers SAE J429 grade designations. These property classes differ in material, and have up to 100% difference in strength. The two most pertinent classes, for socket head screws, are 10.9 and 12.9. The numerals used in property class designations refer to the nominal ultimate tensile strength and nominal percent yield strength. For example, a Property Class 10.9 fastener has 1,000 MPa nominal ultimate strength, and a yield of 90 percent of ultimate. The strength of Property Class 10.9 allows the use of plain carbon steels, while Grade 8 calls for only alloy steels.

The U.S. industry material standard for Metric, alloy steel socket head cap screws, ASTM A574M, specifies one strength level, which is equivalent to ISO Property Class 12.9 and is similar to the strength level for inch unit socket heads. As noted, the U.S. standard specifies only one strength level for all socket head cap screws, metric socket screws are manufactured around the world to various standards, many of which are not equal to ASTM A547M. Therefore, metric socket screws may look similar but may be Property Class 8.8, Property Class 10.9, or Property Class 12.9. U.S. standards call for all socket head cap screws to meet ASTM A574M, which is equal to Property Class 12.9 (1,220 MPa [177,000 PSI]) ultimate tensile strength.

Socket head set screws have less bearing area than hex head screws of the same nominal diameter, and a hardened and ground washer should be used with socket head set screws when the bolt is stressed in tension. Otherwise, there is the risk that the head will sink into the washer and relieve the residual stress within the screw.

Low head socket hex screws. These screws are normally used in parts too thin for standard height socket cap screws, and in applications with limited clearance.

Socket head shoulder screws. Shoulder screws have an undercut portion between the thread and shoulder, allowing for a close fit. They are used in many punch and die operations, such as the location and retention of stripper plates, and act as a guide in blanking and forming presses. Other applications include bearing pins for swing arms, links, and levers; shafts for cam rolls and other rotating parts, pivots, and stud bolts. They are sometimes called "stripper bolts," resulting from their use with stripper plates and springs.

Flat head and button head socket screws. These screws were designed to be used for moderate fastening applications such as machining guards, hinges, covers, etc. They are not recommended for use in critical high strength applications where socket head cap screws should be used.

Set screw points. These points are common to both socket and square head set screws.

Knurled cup. Designed for quick and permanent location of gears, collars, pulleys, knobs, or shafts. This style resists loosening even during severe vibration.

Plain cup. Use against hardened shafts, in zinc, die castings, and other soft materials where high tightening torques are impractical.

Flat. Designed for applications where parts must be frequently reset. Flat socket screws cause minimal damage to the surface they bear against. They can be used with hardened shafts, and as adjusting screws. Preferred for thin walled thicknesses and on soft plugs.

Oval. Like the flat point, oval points are popular for applications where frequent adjustment without deforming the part is required. Also for seating against an angular surface.

Cone. For permanent location of parts. Deep penetration gives highest axial and holding power. In material over R_C 15, point should be spotted to half its length to develop shear strength across point. Used for pivots and fine adjustment.

Dog/half dog. Used for permanent location of one part to another. Point is spotted in hole drilled in shaft or against flat (milled). Often used to replace dowel pins. Works well against hardened members or hollow tubing.

Torsional and axial holding power of English/U.S. Customary unit cup point socket set screws can be found in **Table 71**.

Dimensions for English/U.S. Customary socket screws are given in **Tables 63** through

71. Metric socket screws are covered in **Tables 121** through **130**. Grip and body lengths for English/U.S. Customary unit socket head cap screws can be found in **Table 64**, and hole and counterbore sizes for these screws are given in **Table 65**. The same information for Metric socket head cap screws is provided in **Tables 122** and **123**.

Retained nuts and speed nuts

These are essentially "clip on" internally threaded fasteners that offer several benefits for assembly operations. Since they are, essentially, floating nuts, they do not require drilling in the fixture being attached, and they do not require drilling or tapping. This system of floating alignment does not require special tools, skills, or other equipment. Retained nuts actually use a threaded square nut mounted on a retaining clip, while speed nuts provide a hole that accommodates a tapping nut or sem. Dimensions for Type-J and Type-U clip on fasteners in both English/U.S. Customary units and Metric units are given in **Tables 25–28**.

Studs

There are three basic configurations of studs in wide use. The most common is the continuous thread stud, which is available in two types. Type 1 continuous thread studs are measured from end to end, and threads are UNRC-2A. Type 2 continuous thread studs are measured from first thread to first thread, and the points are flat and chamfered. Type 2 studs are made to standards specified in ANSI B16.5, and threads are UNRC-2A for sizes one inch and under, and 8UNR-2A for longer sizes (for high temperature/high pressure service).

Tap-end studs have different types of thread on each end, and an unthreaded section (the "body") in the center. The end with the shorter threaded section is the tap-end, with Class NC5 interference fit threads or Class UNRC-3A threads. The tap-end has a chamfered point and is designed to be threaded into a tapped hole. The opposite end (with the longer thread) has UNRC-2A threads and either a chamfered or round point, and receives the nut. Length is measured overall. Type 1 tap-end studs are unthreaded blanks. Type 2 tap-end studs may have either an undersize body (rolled threads) or a full size body (cut threads). The body is finished to either a minimum Class 2A pitch diameter, or the maximum basic diameter of the nut-end thread. Type 3 tap-end studs have body tolerances equal to the major diameter of Class 2A threads, and Type 4 tap-end studs have a body milled to tolerances specified by the user.

Double-end studs are designed to receive nuts on both ends. The body separates Class-2A threaded sections of equal length, and the points are chamfered or rounded. Length is measured overall.

Table 1. Dimensions of Standard Hexagon Head Bolts.

Bolt Size		E Body Dia. (Max.)	F Width Across Flats			G Width Across Corners		H Head Height			R Fillet Radius		L_T Thread Length	
Nom.	Basic		Basic	Max.	Min.	Max.	Min.	Basic	Max.	Min.	Max.	Min.	<6 in. (Basic)	>6 in. (Basic)
1/4	0.2500	0.260	7/16	0.438	0.425	0.505	0.498	11/64	0.188	0.150	0.03	0.01	0.750	1.000
5/16	0.3125	0.324	1/2	0.500	0.484	0.577	0.552	7/32	0.235	0.195	0.03	0.01	0.875	1.125
3/8	0.3750	0.388	9/16	0.562	0.544	0.650	0.620	1/4	0.268	0.226	0.03	0.01	1.000	1.250
7/16	0.4375	0.452	5/8	0.625	0.603	0.722	0.687	19/64	0.316	0.272	0.03	0.01	1.125	1.375
1/2	0.5000	0.515	3/4	0.750	0.725	0.866	0.826	11/32	0.364	0.302	0.03	0.01	1.250	1.500
5/8	0.6250	0.642	15/16	0.938	0.906	1.083	1.033	27/64	0.444	0.378	0.06	0.02	1.500	1.750
3/4	0.7500	0.768	1 1/8	1.125	1.088	1.299	1.240	1/2	0.524	0.455	0.06	0.02	1.750	2.000
7/8	0.8750	0.895	1 5/16	1.312	1.269	1.516	1.447	37/64	0.604	0.531	0.06	0.02	2.000	2.250
1	1.0000	1.022	1 1/2	1.500	1.450	1.732	1.653	43/64	0.700	0.591	0.09	0.03	2.250	2.500
1 1/8	1.1250	1.149	1 11/16	1.688	1.631	1.949	1.859	3/4	0.780	0.658	0.09	0.03	2.500	2.750
1 1/4	1.2500	1.277	1 7/8	1.875	1.812	2.165	2.066	27/32	0.876	0.749	0.09	0.03	2.750	3.000
1 3/8	1.3750	1.404	2 1/16	2.062	1.994	2.382	2.273	29/32	0.940	0.810	0.09	0.03	3.000	3.250

(Continued)

All dimensions in inches. Source: Extracted from ANSI B18.2.1-1972, published by the American Society of Mechanical Engineers.

Table 1. *(Continued)* Dimensions of Standard Hexagon Head Bolts.

Bolt Size		E Body Dia. (Max.)	F Width Across Flats				G Width Across Corners			H Head Height			R Fillet Radius		L_T Thread Length	
Nom.	Basic		Basic	Max.	Min.		Max.	Min.	Basic	Max.	Min.		Max.	Min.	<6 in. (Basic)	>6 in. (Basic)
1 1/2	1.5000	1.531	2 1/4	2.250	2.175		2.598	2.480	1	1.036	0.902		0.09	0.03	3.250	3.500
1 3/4	1.7500	1.785	2 5/8	2.625	2.538		3.031	2.893	1 5/32	1.196	1.054		0.12	0.04	3.750	4.000
2	2.0000	2.039	3	3.000	2.900		3.464	3.306	1 11/32	1.388	1.175		0.12	0.04	4.250	4.500
2 1/4	2.2500	2.305	3 3/8	3.375	3.262		3.897	3.719	1 1/2	1.548	1.327		0.19	0.06	4.750	5.000
2 1/2	2.5000	2.559	3 3/4	3.750	3.625		4.330	4.133	1 21/32	1.708	1.479		0.19	0.06	5.250	5.500
2 3/4	2.7500	2.827	4 1/8	4.125	3.988		4.763	4.546	1 13/16	1.869	1.632		0.19	0.06	5.750	6.000
3	3.0000	3.081	4 1/2	4.500	4.350		5.196	4.959	2	2.060	1.815		0.19	0.06	6.250	6.500
3 1/4	3.2500	3.335	4 7/8	4.875	4.712		5.629	5.372	2 3/16	2.251	1.936		0.19	0.06	6.750	7.000
3 1/2	3.5000	3.589	5 1/4	5.250	5.075		6.062	5.786	2 5/16	2.380	2.057		0.19	0.06	7.250	7.500
3 3/4	3.7500	3.858	5 5/8	5.625	5.437		6.495	6.198	2 1/2	2.572	2.241		0.19	0.06	7.750	8.000
4	4.0000	4.111	6	6.000	5.800		6.928	6.612	2 11/16	2.764	2.424		0.19	0.06	8.250	8.500

All dimensions in inches. Source: Extracted from ANSI B18.2.1-1972, published by the American Society of Mechanical Engineers.

Table 2. Dimensions of Heavy Hexagon Head Structural Bolts.

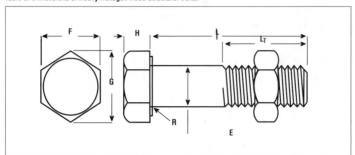

Bolt Size		E Body Dia.		F Width Across Flats			G Width Across Corners	
Nom.	Basic	Max.	Min.	Basic	Max.	Min.	Max.	Min.
1/2	0.5000	0.515	0.482	7/8	0.875	0.850	1.010	0.969
5/8	0.6250	0.642	0.605	1 1/16	1.062	1.031	1.227	1.175
3/4	0.7500	0.768	0.729	1 1/4	1.250	1.212	1.443	1.383
7/8	0.8750	0.895	0.852	1 7/16	1.438	1.394	1.660	1.589
1	1.0000	1.022	0.976	1 5/8	1.625	1.575	1.876	1.796
1 1/8	1.1250	1.149	1.098	1 13/16	1.812	1.756	2.093	2.002
1 1/4	1.2500	1.277	1.223	2	2.000	1.938	2.309	2.209
1 3/8	1.3750	1.404	1.345	2 3/16	2.188	2.119	2.526	2.416
1 1/2	1.5000	1.531	1.470	2 3/8	2.375	2.300	2.742	2.622

Bolt Size		H Height			R Fillet Radius		L_T Thread Length	V Transition Thread Length	Runout of Bearing Surface, FIR
Nom.	Basic	Basic	Max.	Min.	Max.	Min.	Basic	Max.	Max.
1/2	0.5000	5/16	0.323	0.302	0.031	0.009	1.00	0.19	0.016
5/8	0.6250	25/64	0.403	0.378	0.062	0.021	1.25	0.22	0.019
3/4	0.7500	15/32	0.483	0.455	0.062	0.021	1.38	0.25	0.022
7/8	0.8750	35/64	0.563	0.531	0.062	0.031	1.50	0.28	0.025
1	1.0000	39/64	0.627	0.591	0.093	0.062	1.75	0.31	0.028
1 1/8	1.1250	11/16	0.718	0.658	0.093	0.062	2.00	0.34	0.032
1 1/4	1.2500	25/32	0.813	0.749	0.093	0.062	2.00	0.38	0.035
1 3/8	1.3750	27/32	0.878	0.810	0.093	0.062	2.25	0.44	0.038
1 1/2	1.5000	15/16	0.974	0.902	0.093	0.062	2.25	0.44	0.041

All dimensions in inches. Source: Extracted from ANSI B18.2.1-1972, published by the American Society of Mechanical Engineers.

Table 3. Dimensions of Square Head Bolts.

Bolt Size		E Body Dia. (Max.)	F Width Across Flats			G Width Across Corners		H Head Height			R Filet Radius		L_T Thread Length	
Nom.	Basic		Basic	Max.	Min.	Max.	Min.	Basic	Max.	Min.	Max.	Min.	<6 in. (Basic)	>6 in. (Basic)
1/4	0.2500	0.260	3/8	0.375	0.362	0.530	0.498	11/64	0.188	0.156	0.03	0.01	0.750	1.000
5/16	0.3125	0.324	1/2	0.500	0.484	0.707	0.665	13/64	0.220	0.186	0.03	0.01	0.875	1.125
3/8	0.3750	0.388	9/16	0.562	0.544	0.795	0.747	1/4	0.268	0.232	0.03	0.01	1.000	1.250
7/16	0.4375	0.452	5/8	0.625	0.603	0.884	0.828	19/64	0.316	0.278	0.03	0.01	1.125	1.375
1/2	0.5000	0.515	3/4	0.750	0.725	1.061	0.995	21/64	0.348	0.308	0.03	0.01	1.250	1.500
5/8	0.6250	0.642	15/16	0.938	0.906	1.326	1.244	27/64	0.444	0.400	0.06	0.02	1.500	1.750
3/4	0.7500	0.768	1 1/8	1.125	1.088	1.591	1.494	1/2	0.524	0.476	0.06	0.02	1.750	2.000
7/8	0.8750	0.895	1 5/16	1.312	1.269	1.856	1.742	19/32	0.620	0.568	0.06	0.02	2.000	2.250
1	1.0000	1.022	1 1/2	1.500	1.450	2.121	1.991	21/32	0.684	0.628	0.09	0.03	2.250	2.500
1 1/8	1.1250	1.149	1 11/16	1.688	1.631	2.386	2.239	3/4	0.780	0.720	0.09	0.03	2.500	2.750
1 1/4	1.2500	1.277	1 7/8	1.875	1.812	2.652	2.489	27/32	0.876	0.812	0.09	0.03	2.750	3.000
1 3/8	1.3750	1.404	2 1/16	2.062	1.994	2.917	2.738	29/32	0.940	0.872	0.09	0.03	3.000	3.250
1 1/2	1.5000	1.531	2 1/4	2.250	2.175	3.182	2.986	1	1.036	0.964	0.09	0.03	3.250	3.500

All dimensions in inches. Source: Extracted from ANSI B18.2.1-1972, published by the American Society of Mechanical Engineers.

Table 4. Dimensions of 12-Point Flange Screws. (See Notes.)

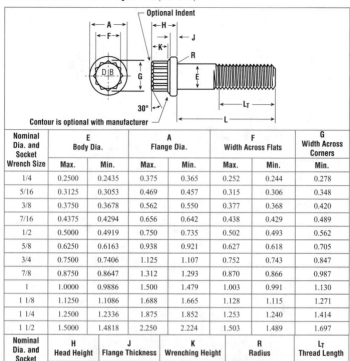

Nominal Dia. and Socket Wrench Size	E Body Dia.		A Flange Dia.		F Width Across Flats		G Width Across Corners
	Max.	Min.	Max.	Min.	Max.	Min.	Min.
1/4	0.2500	0.2435	0.375	0.365	0.252	0.244	0.278
5/16	0.3125	0.3053	0.469	0.457	0.315	0.306	0.348
3/8	0.3750	0.3678	0.562	0.550	0.377	0.368	0.420
7/16	0.4375	0.4294	0.656	0.642	0.438	0.429	0.489
1/2	0.5000	0.4919	0.750	0.735	0.502	0.493	0.562
5/8	0.6250	0.6163	0.938	0.921	0.627	0.618	0.705
3/4	0.7500	0.7406	1.125	1.107	0.752	0.743	0.847
7/8	0.8750	0.8647	1.312	1.293	0.870	0.866	0.987
1	1.0000	0.9886	1.500	1.479	1.003	0.991	1.130
1 1/8	1.1250	1.1086	1.688	1.665	1.128	1.115	1.271
1 1/4	1.2500	1.2336	1.875	1.852	1.253	1.240	1.414
1 1/2	1.5000	1.4818	2.250	2.224	1.503	1.489	1.697

Nominal Dia. and Socket Wrench Size	H Head Height	J Flange Thickness	K Wrenching Height	R Radius		L_T Thread Length
	Max.	Min.	Min.	Max.	Min.	Min.
1/4	0.250	0.058	0.15	0.014	0.009	1.000
5/16	0.312	0.074	0.18	0.017	0.012	1.125
3/8	0.375	0.095	0.21	0.020	0.015	1.250
7/16	0.438	0.109	0.26	0.023	0.018	1.375
1/2	0.500	0.129	0.29	0.026	0.020	1.500
5/8	0.625	0.166	0.36	0.032	0.024	1.750
3/4	0.750	0.200	0.44	0.039	0.030	2.000
7/8	0.875	0.234	0.51	0.044	0.034	2.250
1	1.000	0.268	0.60	0.050	0.040	2.500
1 1/8	1.125	0.310	0.66	0.055	0.045	2.750
1 1/4	1.250	0.350	0.73	0.060	0.050	3.000
1 1/2	1.500	0.433	0.87	0.070	0.060	3.500

Notes: 12-point flange screws and bolts are made to numerous dimensions as specified by Military (MS) and Aerospace (NAS) Standards. Material specifications will vary. The above screw is typical and produced from alloy steel with a hardness of RC 37-43, tensile strength of 170,000 PSI (min), yield strength of 153,000 PSI (min), and proof load of 140,000 PSI (min). Threads are UNRC/UNRF Class 3A for sizes through one inch, and UNRC/UNRF Class 2A for sizes 1 1/8 to 1 1/2 inch. Source: Darling Bolt Company.

Table 5. Dimensions of Round-Head Bolts.

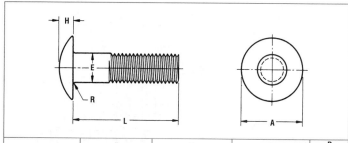

Bolt Size		E Body Dia.		A Head Dia.		H Head Height		R Fillet Radius
Nom.	Basic	Max.	Min.	Max.	Min.	Max.	Min.	Max.
No. 10	0.1900	0.199	0.182	0.469	0.438	0.114	0.094	0.031
1/4	0.2500	0.260	0.237	0.594	0.563	0.145	0.125	0.031
5/16	0.3125	0.324	0.298	0.719	0.688	0.176	0.156	0.031
3/8	0.3750	0.388	0.360	0.844	0.782	0.208	0.188	0.031
7/16	0.4375	0.452	0.421	0.969	0.907	0.239	0.219	0.031
1/2	0.5000	0.515	0.483	1.094	1.032	0.270	0.250	0.031
5/8	0.6250	0.642	0.605	0.344	1.219	0.344	0.313	0.062
3/4	0.7500	0.768	0.729	1.594	1.469	0.406	0.375	0.062

All dimensions in inches. Source: Extracted from ANSI B18.5-1971, published by the American Society of Mechanical Engineers.

Table 6. Dimensions of Round-Head Square-Neck Bolts.

Bolt Size		E Body Dia.		A Head Dia.		H Head Height		O Square Width		P Square Depth		Q Corner Radius	R Fillet Radius
Nom.	Basic	Max.	Min.	Max.	Min.	Max.	Min.	Max.	Min.	Max.	Min.	Max.	Max.
No. 10	0.1900	0.199	0.182	0.469	0.438	0.114	0.094	0.199	0.185	0.125	0.094	0.031	0.031
1/4	0.2500	0.260	0.237	0.594	0.563	0.145	0.125	0.260	0.245	0.156	0.125	0.031	0.031
5/16	0.3125	0.324	0.298	0.719	0.688	0.176	0.156	0.324	0.307	0.187	0.156	0.031	0.031
3/8	0.3750	0.388	0.360	0.844	0.782	0.208	0.188	0.388	0.368	0.219	0.188	0.047	0.031
7/16	0.4375	0.452	0.421	0.969	0.907	0.239	0.219	0.452	0.431	0.250	0.219	0.047	0.031
1/2	0.5000	0.515	0.483	1.094	1.032	0.270	0.250	0.515	0.492	0.281	0.250	0.047	0.031
5/8	0.6250	0.642	0.605	1.344	1.219	0.344	0.313	0.642	0.616	0.344	0.313	0.078	0.062
3/4	0.7500	0.768	0.729	1.594	1.469	0.406	0.375	0.768	0.741	0.406	0.375	0.078	0.062
7/8	0.8750	0.895	0.852	1.844	1.719	0.469	0.438	0.895	0.865	0.469	0.438	0.094	0.062
1	1.0000	1.022	0.976	2.094	1.969	0.531	0.500	1.022	0.990	0.531	0.500	0.094	0.062

All dimensions in inches. Source: Extracted from ANSI B18.5-1971, published by the American Society of Mechanical Engineers.

Table 7. Dimensions of Round-Head Fin-Neck Bolts.

| Bolt Size | | E Body Dia. | | A Head Dia. | | H Head Height | | M Fin Thickness | | N Distance Across Fins | | U Fin Depth | | R Fillet Radius |
Nom.	Basic	Max.	Min.	Max.	Min.	Max.	Min.	Max.	Min.	Max.	Min.	Max.	Min.	Max.
No. 10	0.1900	0.199	0.182	0.469	0.438	0.114	0.094	0.098	0.078	0.395	0.375	0.088	0.078	0.031
1/4	0.2500	0.260	0.237	0.594	0.563	0.145	0.125	0.114	0.094	0.458	0.438	0.104	0.094	0.031
5/16	0.3125	0.324	0.298	0.719	0.688	0.176	0.156	0.145	0.125	0.551	0.531	0.135	0.125	0.031
3/8	0.3750	0.388	0.360	0.844	0.782	0.208	0.188	0.161	0.141	0.645	0.625	0.151	0.141	0.031
7/16	0.4375	0.452	0.421	0.969	0.907	0.239	0.219	0.192	0.172	0.739	0.719	0.182	0.172	0.031
1/2	0.5000	0.515	0.483	1.094	1.032	0.270	0.250	0.208	0.188	0.833	0.813	0.198	0.188	0.031

All dimensions in inches. Source: Extracted from ANSI B18.5-1971, published by the American Society of Mechanical Engineers.

Table 8. Dimensions of Round-Head Ribbed-Neck Bolts.

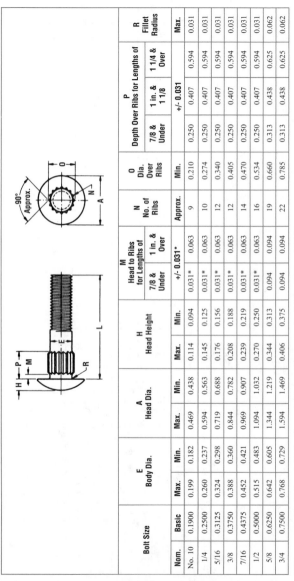

Bolt Size		E Body Dia.		A Head Dia.		H Head Height		M Head to Ribs for Lengths of		N No. of Ribs	O Dia. Over Ribs	P Depth Over Ribs for Lengths of			R Fillet Radius
								7/8 & Under	1 in. & Over			7/8 & Under	1 in. & 1 1/8	1 1/4 & Over	
								+/- 0.031*				+/- 0.031			
Nom.	Basic	Max.	Min.	Max.	Min.	Max.	Min.			Approx.	Min.				Max.
No. 10	0.1900	0.199	0.182	0.469	0.438	0.114	0.094	0.031*	0.063	9	0.210	0.250	0.407	0.594	0.031
1/4	0.2500	0.260	0.237	0.594	0.563	0.145	0.125	0.031*	0.063	10	0.274	0.250	0.407	0.594	0.031
5/16	0.3125	0.324	0.298	0.719	0.688	0.176	0.156	0.031*	0.063	12	0.340	0.250	0.407	0.594	0.031
3/8	0.3750	0.388	0.360	0.844	0.782	0.208	0.188	0.031*	0.063	12	0.405	0.250	0.407	0.594	0.031
7/16	0.4375	0.452	0.421	0.969	0.907	0.239	0.219	0.031*	0.063	14	0.470	0.250	0.407	0.594	0.031
1/2	0.5000	0.515	0.483	1.094	1.032	0.270	0.250	0.031*	0.063	16	0.534	0.250	0.407	0.594	0.031
5/8	0.6250	0.642	0.605	1.344	1.219	0.344	0.313	0.094	0.094	19	0.660	0.313	0.438	0.625	0.062
3/4	0.7500	0.768	0.729	1.594	1.469	0.406	0.375	0.094	0.094	22	0.785	0.313	0.438	0.625	0.062

* Tolerance on the No. 10 through 1/2 sizes for nominal lengths 7/8 in. and under shall be +0.031 and -0.000.
All dimensions in inches. Source: Extracted from ANSI B18.5-1971, published by the American Society of Mechanical Engineers.

Table 9. Dimensions of Step Bolts.

Bolt Size		E Body Dia.		A Head Dia.		H Head Height		O Square Width		P Square Depth		Q Corner Radius	R Fillet Radius
Nom.	Basic	Max.	Min.	Max.	Min.	Max.	Min.	Max.	Min.	Max.	Min.	Max.	Max.
No. 10	0.1900	0.199	0.182	0.656	0.625	0.114	0.094	0.199	0.185	0.125	0.094	0.031	0.031
1/4	0.2500	0.260	0.237	0.844	0.813	0.145	0.125	0.260	0.245	0.156	0.125	0.031	0.031
5/16	0.3125	0.324	0.298	1.031	1.000	0.176	0.156	0.324	0.307	0.187	0.156	0.031	0.031
3/8	0.3750	0.388	0.360	1.219	1.188	0.208	0.188	0.388	0.268	0.219	0.188	0.047	0.031
7/16	0.4375	0.452	0.421	1.406	1.375	0.239	0.219	0.452	0.431	0.250	0.219	0.047	0.031
1/2	0.5000	0.515	0.483	1.594	1.563	0.270	0.250	0.515	0.492	0.281	0.250	0.047	0.031

All dimensions in inches. Source: Fastbolt Corp.

Table 10. Dimensions of Countersunk Head Square-Neck Plow Bolts.

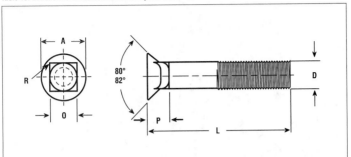

Bolt Size		A Head Dia.			K Feed Thickness	P Depth of Square and Head		O Width of Square		R Radius on Corners
		Max.	Min. Sharp	Abs. Min. With Flat	Max.	Max.	Min.	Max.	Min.	Max.
Nom.	Basic									
5/16	0.3125	0.605	0.578	0.563	0.025	0.269	0.243	0.325	0.313	1/32
3/8	0.3750	0.708	0.671	0.656	0.031	0.312	0.281	0.387	0.375	3/64
7/16	0.4375	0.826	0.781	0.766	0.036	0.364	0.328	0.450	0.438	3/64
1/2	0.2500	0.945	0.890	0.875	0.042	0.417	0.375	0.515	0.500	3/64
*9/16	0.5625	1.045	1.000	0.969	0.045	0.461	0.416	0.578	0.563	5/64
5/8	0.6250	1.147	1.094	1.063	0.050	0.506	0.456	0.640	0.625	5/64
3/4	0.7500	1.303	1.250	1.219	0.050	0.541	0.491	0.765	0.750	5/64
7/8	0.8750	1.512	1.469	1.406	0.063	0.626	0.563	0.906	0.875	3/32
1	1.0000	1.700	1.656	1.594	0.063	0.690	0.627	1.031	1.000	3/32
Repair †										
5/16	0.3125	0.556	0.531	0.516	0.020	0.232	0.212	0.325	0.313	1/32
3/8	0.3750	0.659	0.624	0.609	0.025	0.272	0.247	0.387	0.375	3/64
7/16	0.4375	0.779	0.734	0.719	0.030	0.324	0.294	0.450	0.438	3/64
1/2	0.2500	0.898	0.843	0.828	0.035	0.375	0.340	0.515	0.500	3/64
*9/16	0.5625	1.002	0.953	0.922	0.040	0.423	0.383	0.578	0.563	5/64
5/8	0.6250	1.096	1.047	1.016	0.040	0.458	0.418	0.640	0.625	5/64
3/4	0.7500	1.252	1.203	1.172	0.040	0.493	0.453	0.765	0.750	5/64
7/8	0.8750	1.465	1.422	1.359	0.050	0.573	0.523	0.906	0.875	3/32
1	1.0000	1.653	1.609	1.547	0.050	0.637	0.587	1.031	1.000	3/32

† The letter "R" shall be shown on top of the repair head, to distinguish it from the regular head bolt.
* This size is not recommended.
If the method of manufacture permits, it is recommended that the same radius be maintained on each of all four corners of the square.
All dimensions in inches. Source: Extracted from ANSI B18.9-1958, published by the American Society of Mechanical Engineers.

Table 11. Dimensions of Countersunk Bolts and Slotted Countersunk Bolts.

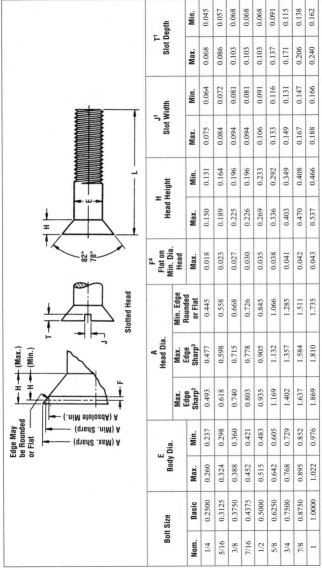

| Bolt Size | | E Body Dia. | | A Head Dia. | | | F⁴ Flat on Min. Dia. Head | H Head Height | | J¹ Slot Width | | T¹ Slot Depth | |
Nom.	Basic	Max.	Min.	Max. Edge Sharp²	Max. Edge Sharp³	Min. Edge Rounded or Flat	Max.	Max.	Min.	Max.	Min.	Max.	Min.
1/4	0.2500	0.260	0.237	0.493	0.477	0.445	0.018	0.150	0.131	0.075	0.064	0.068	0.045
5/16	0.3125	0.324	0.298	0.618	0.598	0.558	0.023	0.189	0.164	0.084	0.072	0.086	0.057
3/8	0.3750	0.388	0.360	0.740	0.715	0.668	0.027	0.225	0.196	0.094	0.081	0.103	0.068
7/16	0.4375	0.452	0.421	0.803	0.778	0.726	0.030	0.226	0.196	0.094	0.081	0.103	0.068
1/2	0.5000	0.515	0.483	0.935	0.905	0.845	0.035	0.269	0.233	0.106	0.091	0.103	0.068
5/8	0.6250	0.642	0.605	1.169	1.132	1.066	0.038	0.336	0.292	0.133	0.116	0.137	0.091
3/4	0.7500	0.768	0.729	1.402	1.357	1.285	0.041	0.403	0.349	0.149	0.131	0.171	0.115
7/8	0.8750	0.895	0.852	1.637	1.584	1.511	0.042	0.470	0.408	0.167	0.147	0.206	0.138
1	1.0000	1.022	0.976	1.869	1.810	1.735	0.043	0.537	0.466	0.188	0.166	0.240	0.162

(Continued)

Table 11. *(Continued)* Dimensions of Countersunk Bolts and Slotted Countersunk Bolts.

Bolt Size		E Body Dia.		A Head Dia.			F⁴ Flat on Min. Dia. Head	H Head Height		J¹ Slot Width		T¹ Slot Depth	
Nom.	Basic	Max.	Min.	Max. Edge Sharp²	Max. Edge Sharp³	Min. Edge Rounded or Flat	Max.	Max.	Min.	Max.	Min.	Max.	Min.
1 1/8	1.1250	1.149	1.098	2.104	2.037	1.962	0.043	0.604	0.525	0.196	0.178	0.257	0.173
1 1/4	1.2500	1.277	1.223	2.337	2.262	2.187	0.043	0.671	0.582	0.211	0.193	0.291	0.197
1 3/8	1.3750	1.404	1.345	2.571	2.489	2.414	0.043	0.738	0.641	0.226	0.208	0.326	0.220
1 1/2	1.5000	1.531	1.470	2.804	2.715	2.640	0.043	0.805	0.698	0.258	0.240	0.360	0.244

[1] Head shall be unslotted, unless otherwise specified. Slot dimensions are same as Slotted Flat Countersunk Head Cap Screws in American National Standard, ANSI B18.6.2.

[2] Maximum head height calculated on maximum sharp head diameter, basic bolt diameter and 78° head angle.

[3] Minimum head height calculated on minimum sharp head diameter, basic bolt diameter and 80° head angle.

[4] Flat on minimum diameter head calculated on minimum sharp and absolute minimum head diameters and 82° head angle.

All dimensions in inches. Source: Extracted from ANSI B18.5-1971, published by the American Society of Mechanical Engineers.

Table 12. Dimensions of 114-Degree Countersunk Neck Bolts.

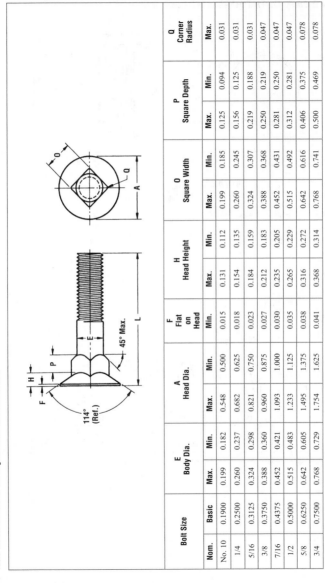

Bolt Size		E Body Dia.		A Head Dia.		F Flat on Head	H Head Height		O Square Width		P Square Depth		Q Corner Radius
Nom.	Basic	Max.	Min.	Max.	Min.	Min.	Max.	Min.	Max.	Min.	Max.	Min.	Max.
No. 10	0.1900	0.199	0.182	0.548	0.500	0.015	0.131	0.112	0.199	0.185	0.125	0.094	0.031
1/4	0.2500	0.260	0.237	0.682	0.625	0.018	0.154	0.135	0.260	0.245	0.156	0.125	0.031
5/16	0.3125	0.324	0.298	0.821	0.750	0.023	0.184	0.159	0.324	0.307	0.219	0.188	0.031
3/8	0.3750	0.388	0.360	0.960	0.875	0.027	0.212	0.183	0.388	0.368	0.250	0.219	0.047
7/16	0.4375	0.452	0.421	1.093	1.000	0.030	0.235	0.205	0.452	0.431	0.281	0.250	0.047
1/2	0.5000	0.515	0.483	1.233	1.125	0.035	0.265	0.229	0.515	0.492	0.312	0.281	0.047
5/8	0.6250	0.642	0.605	1.495	1.375	0.038	0.316	0.272	0.642	0.616	0.406	0.375	0.078
3/4	0.7500	0.768	0.729	1.754	1.625	0.041	0.368	0.314	0.768	0.741	0.500	0.469	0.078

All dimensions in inches. Source: Extracted from ANSI B18.5-1971, published by the American Society of Mechanical Engineers.

Table 13. Dimensions of Flat Countersunk Head Elevator Bolts.

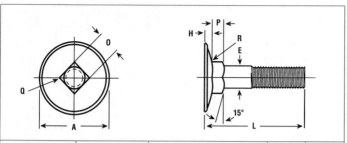

Bolt Size		E Body Dia.		A Head Dia.			C Head Angle, Degrees (Ref.)	F Flat on Min. Dia. Head
				Edge Sharp	Edge Sharp	Edge Flat		
Nom.	Basic	Max.	Min.	Max.	Min.	Min.		
No. 10	0.1900	0.199	0.182	0.790	0.750	0.740	9	0.025
1/4	0.2500	0.260	0.237	1.008	0.969	0.938	9	0.035
5/16	0.3125	0.324	0.298	1.227	1.188	1.157	9	0.035
3/8	0.3750	0.388	0.360	1.352	1.312	1.272	11	0.040
7/16	0.4375	0.452	0.421	1.477	1.438	1.397	13	0.040
1/2	0.5000	0.515	0.483	1.602	1.562	1.522	12	0.040

Bolt Size		H Head Height		O Square Width		P Square Depth		Q Corner Radius	R Fillet Radius
Nom.	Basic	Max.	Min.	Max.	Min.	Max.	Min.	Max.	Max.
No. 10	0.1900	0.082	0.062	0.210	0.185	0.125	0.094	0.031	0.031
1/4	0.2500	0.098	0.078	0.280	0.245	0.219	0.188	0.031	0.031
5/16	0.3125	0.114	0.094	0.342	0.307	0.250	0.219	0.031	0.031
3/8	0.3750	0.145	0.125	0.405	0.368	0.250	0.219	0.047	0.031
7/16	0.4375	0.176	0.156	0.468	0.431	0.281	0.250	0.047	0.031
1/2	0.5000	0.176	0.156	0.530	0.492	0.281	0.250	0.047	0.031

All dimensions in inches. Source: Fastbolt Corp.

Table 14. Type 1 Regular-Pattern Eyebolt.

Nominal Size		A Shank Dia.	B Shank Length	C Eye ID	D Nominal Eye OD	Eye Sect. Dia.	Overall Length	Min. Length Full Thread	Thread Size UNC - 2A	Safe Working Load, lb. at 0° [1]	Safe Working Load, lb. at 45° [1]	Safe Working Load, lb. at 90° [1]
1/4*	0.25	0.25 0.28	1.00 1.06	0.69 0.81	1.19	0.19 0.25	2.06 2.38	0.75	1/4 - 20 or .250 - 20	400	80	60
5/16*	0.31	0.31 0.34	1.12 1.19	0.81 0.94	1.44	0.25 0.31	2.44 2.75	0.81	5/16 - 18 or .3125 - 18	800	160	120
3/8*	0.38	0.38 0.41	1.25 1.38	0.94 1.06	1.69	0.31 0.38	2.81 3.19	0.88	3/8 - 16 or .375 - 16	1400	280	210
7/16	0.44	0.44 0.47	1.38 1.50	1.00 1.12	1.81	0.34 0.41	3.06 3.44	1.00	7/16 - 14 or .4375 - 14	2000	400	300
1/2*	0.50	0.50 0.53	1.50 1.62	1.12 1.25	2.12	0.44 0.50	3.50 3.88	1.12	1/2 - 13 or .500 - 13	2600	520	390

(Continued)

Table 14. *(Continued)* Type 1 Regular-Pattern Eyebolt.

Nominal Size		A Shank Dia.	B Shank Length	C Eye ID	D Nominal Eye OD	Eye Sect. Dia.	Overall Length	Min. Length Full Thread	Thread Size UNC - 2A	Safe Working Load, lb. at 0° [1]	Safe Working Load, lb. at 45° [1]	Safe Working Load, lb. at 90° [1]
9/16	0.56	0.56 / 0.59	1.62 / 1.75	1.19 / 1.31	2.25	0.47 / 0.53	3.75 / 4.12	1.25	9/16 - 12 or .5625 - 12	3000	600	450
5/8*	0.62	0.62 / 0.66	1.75 / 1.88	1.31 / 1.44	2.56	0.56 / 0.62	4.19 / 4.56	1.38	5/8 - 11 or .625 - 11	4000	800	600
3/4*	0.75	0.75 / 0.78	2.00 / 2.12	1.44 / 1.56	2.81	0.62 / 0.69	4.69 / 5.06	1.62	3/4 - 10 or .750 - 10	6000	1200	900
7/8*	0.88	0.88 / 0.91	2.25 / 2.38	1.56 / 1.69	3.19	0.75 / 0.81	5.31 / 5.69	1.81	7/8 - 9 or .875 - 9	6600	1300	1000
1*	1.00	1.00 / 1.06	2.50 / 2.62	1.69 / 1.81	3.56	0.88 / 0.94	5.94 / 6.31	2.06	1 - 8 or 1.000-8	8000	1600	1200
1 1/8	1.12	1.12 / 1.19	2.75 / 2.88	1.94 / 2.06	4.06	1.00 / 1.06	6.69 / 7.06	2.31	1 1/8 - 7 or 1.125 - 7	10000	2000	1500
1 1/4*	1.25	1.25 / 1.34	3.00 / 3.12	2.12 / 2.25	4.44	1.09 / 1.16	7.31 / 7.69	2.50	1 1/4 - 7 or 1.250 - 7	15000	3000	2250
1 1/2*	1.50	1.50 / 1.59	3.50 / 3.62	2.44 / 2.56	5.19	1.31 / 1.38	8.56 / 8.94	3.00	1 1/2 - 6 or 1.500 - 6	18000	3600	2700
1 3/4	1.75	1.75 / 1.84	3.75 / 3.88	2.75 / 3.00	6.00	1.50 / 1.62	9.50 / 10.12	3.19	1 3/4 - 5 or 1.750-5	22000	4400	3300
2*	2.00	2.00 / 2.09	4.00 / 4.25	3.06 / 3.44	6.88	1.75 / 1.88	10.56 / 11.44	3.38	2 - 4 1/2 or 2.000 - 4.50	26000	5200	3900
2 1/2*	2.50	2.50 / 2.62	5.00 / 5.25	3.81 / 4.19	8.50	2.19 / 2.31	13.19 / 14.06	4.25	2 1/2 - 4 or 2.500 - 4	32000	6400	4800

* Preferred. [1] These safe working loads are based on the following percentages of minimum proof loads shown in ASTM A 489; 0°—66.6 percent; 45°—13.3 percent, 90°—10.0 percent. All dimensions in inches. Source: Extracted from ANSI B18.15-1969, published by the American Society of Mechanical Engineers.

Table 15. Type 2 Shoulder-Pattern Eyebolt.

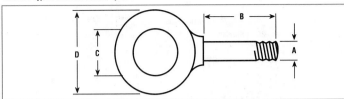

Nominal Size		A Shank Dia.	B Shank Length	C Eye ID	D Nominal Eye OD	Eye Sect. Dia.	Overall Length	Min. Length Full Thread
1/4*	0.25	0.25-0.28	1.00-1.06	0.69-0.81	1.19	0.19-0.25	2.22-2.53	0.75
5/16*	0.31	0.31-0.34	1.12-1.19	0.81-0.94	1.44	0.25-0.31	2.66-2.97	0.81
3/8*	0.38	0.38-0.41	1.25-1.38	0.94-1.06	1.69	0.31-0.38	3.09-3.47	0.88
7/16	0.44	0.44-0.47	1.38-1.50	1.00-1.12	1.81	0.34-0.41	3.41-3.78	1.00
1/2*	0.50	0.50-0.53	1.50-1.62	1.12-1.25	2.12	0.44-0.50	3.81-4.19	1.12
9/16	0.56	0.56-0.59	1.62-1.75	1.19-1.31	2.25	0.47-0.53	4.19-4.56	1.25
5/8*	0.62	0.62-0.66	1.75-1.88	1.31-1.44	2.56	0.56-0.62	4.56-4.94	1.38
3/4*	0.75	0.75-0.78	2.00-2.12	1.44-1.56	2.81	0.62-0.69	5.06-5.50	1.62
7/8*	0.88	0.88-0.91	2.25-2.38	1.56-1.69	3.19	0.75-0.81	5.75-6.19	1.81
1*	1.00	1.00-1.06	2.50-2.62	1.69-1.81	3.56	0.88-0.94	6.44-6.88	2.06
1 1/8	1.12	1.12-1.19	2.75-2.88	1.94-2.06	4.06	1.00-1.06	7.31-7.75	2.31
1 1/4*	1.25	1.25-1.34	3.00-3.12	2.12-2.25	4.44	1.09-1.16	8.00-8.44	2.50
1 1/2*	1.50	1.50-1.59	3.50-3.62	2.44-2.56	5.19	1.31-1.38	9.22-9.72	3.00
1 3/4	1.75	1.75-1.84	3.75-3.88	2.75-3.00	6.00	1.50-1.62	10.50-11.12	3.19
2*	2.00	2.00-2.09	4.00-4.25	3.06-3.44	6.88	1.75-1.88	11.53-12.22	3.38

Nominal Size		Thread Size UNC-2A	C_L Eye to Shoulder	Shoulder Dia.	Shoulder Height	Radius Under Shoulder	Safe Working Load, lb. at [1]		
							0°	45°	90°
1/4*	0.25	1/4 - 20 or .250 - 20	0.69 0.81	0.50 0.56	0.12 0.19	0.015 0.025	400	100	80
5/16*	0.31	5/16 - 18 or .3125 - 18	0.88 1.00	0.56 0.32	0.12 0.19	0.015 0.025	800	200	160
3/8*	0.38	3/8 - 16 or .375 - 16	1.06 1.19	0.62 0.69	0.12 0.19	0.015 0.025	1400	350	280
7/16	0.44	7/16 - 14 or .4375 - 14	1.19 1.31	0.75 0.81	0.19 0.25	0.015 0.025	2000	500	400
1/2*	0.50	1/2 - 13 or .500 - 13	1.31 1.44	0.88 0.94	0.19 0.25	0.015 0.025	2600	650	520
9/16	0.56	9/16 - 12 or .5625 - 12	1.50 1.62	0.94 1.00	0.22 0.28	0.020 0.045	3000	750	600
5/8*	0.62	5/8 - 11 or .625 - 11	1.59 1.72	1.00 1.06	0.25 0.31	0.020 0.045	4000	1000	800

(Continued)

Table 15. *(Continued)* **Type 2 Shoulder-Pattern Eyebolt.**

Nominal Size		Thread Size UNC-2A	C_L Eye to Shoulder	Shoulder Dia.	Shoulder Height	Radius Under Shoulder	Safe Working Load, lb. at [1]		
							0°	45°	90°
3/4*	0.75	3/4 - 10 or .750 - 10	1.72 1.91	1.12 1.25	0.25 0.31	0.020 0.045	6000	1500	1200
7/8*	0.88	7/8 - 9 or .875 - 9	2.03 2.22	1.31 1.44	0.31 0.38	0.040 0.065	6600	1670	1330
1*	1.00	1 - 8 or 1.000 - 8	2.22 2.41	1.50 1.62	0.38 0.44	0.060 0.095	8000	2000	1600
1 1/8	1.12	1 1/8 - 7 or 1.125 - 7	2.59 2.78	1.69 1.81	0.44 0.50	0.060 0.095	10000	2500	2000
1 1/4*	1.25	1 1/4 - 7 or 1.250 - 7	2.84 3.03	1.88 2.00	0.50 0.56	0.060 0.095	15000	3750	3000
1 1/2*	1.50	1 1/2 - 6 or 1.500 - 6	3.19 3.44	2.12 2.25	0.50 0.62	0.060 0.095	18000	4500	3600
1 3/4	1.75	1 3/4 - 5 or 1.750 - 5	3.88 4.12	2.50 2.62	0.50 0.62	0.060 0.095	22000	5500	4400
2*	2.00	2 - 4 1/2 or 2.000 - 4.50	4.25 4.50	2.88 3.00	0.62 0.75	0.060 0.095	26000	6500	5200

* Preferred. [1] Applies if shoulder is properly seated. Otherwise safe working load of Type 1 applies. These safe working loads are based on the following percentages of minimum proof loads shown in ASTM A 489; 0°—66.6 percent, 45°—16.7 percent, 90°—13.3 percent. All dimensions in inches. Source: Extracted from ANSI B18.15-1969, published by the American Society of Mechanical Engineers.

Table 16. Dimensions of Hex Nuts and Hex Jam Nuts.

Nut Size		F Width Across Flats			G Width Across Corners		H Thickness Hex Nuts			H₁ Thickness Hex Jam Nuts		
Nom.	Basic	Basic	Max.	Min.	Max.	Min.	Basic	Max.	Min.	Basic	Max.	Min.
1/4	0.2500	7/16	0.438	0.428	0.505	0.488	7/32	0.226	0.212	5/32	0.163	0.150
5/16	0.3125	1/2	0.500	0.489	0.577	0.557	17/64	0.273	0.258	3/16	0.195	0.180
3/8	0.3750	9/16	0.562	0.551	0.650	0.628	21/64	0.337	0.320	7/32	0.227	0.210
7/16	0.4375	11/16	0.688	0.675	0.794	0.768	3/8	0.385	0.365	1/4	0.260	0.240
1/2	0.5000	3/4	0.750	0.736	0.866	0.840	7/16	0.448	0.427	5/16	0.323	0.302
9/16	0.5625	7/8	0.875	0.861	1.010	0.982	31/64	0.496	0.473	5/16	0.324	0.301
5/8	0.6250	15/16	0.938	0.922	1.083	1.051	35/64	0.559	0.535	3/8	0.387	0.363
3/4	0.7500	1 1/8	1.125	1.088	1.299	1.240	41/64	0.665	0.617	27/64	0.446	0.398

(Continued)

Table 16. *(Continued)* Dimensions of Hex Nuts and Hex Jam Nuts.

Nut Size		F Width Across Flats			G Width Across Corners		H Thickness Hex Nuts			H₁ Thickness Hex Jam Nuts		
Nom.	Basic	Basic	Max.	Min.	Max.	Min.	Basic	Max.	Min.	Basic	Max.	Min.
7/8	0.8750	1 5/16	1.312	1.269	1.516	1.447	3/4	0.776	0.724	31/64	0.510	0.458
1	1.0000	1 1/2	1.500	1.450	1.732	1.653	55/64	0.887	0.831	35/64	0.575	0.519
1 1/8	1.1250	1 11/16	1.688	1.631	1.949	1.859	31/32	0.999	0.939	39/64	0.639	0.579
1 1/4	1.2500	1 7/8	1.875	1.812	2.165	2.066	1 1/16	1.094	1.030	23/32	0.751	0.687
1 3/8	1.3750	2 1/16	2.062	1.994	2.382	2.273	1 11/64	0.206	1.138	25/32	0.815	0.747
1 1/2	1.5000	2 1/4	2.250	2.175	2.598	2.480	1 9/32	1.317	1.245	27/32	0.880	0.808

All dimensions in inches. Source: Fastbolt Corp.

Table 17. Dimensions of Hex Nuts and Hex Jam Nuts.

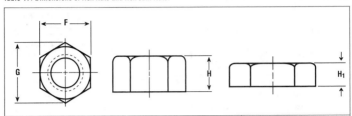

Nut Size		F Width Across Flats			G Width Across Corners		H Thickness Hex Nuts		
Nom.	Basic	Basic	Max.	Min.	Max.	Min.	Basic	Max.	Min.
1/4	0.2500	7/16	0.438	0.428	0.505	0.488	7/32	0.226	0.212
5/16	0.3125	1/2	0.500	0.489	0.577	0.557	17/64	0.273	0.258
3/8	0.3750	9/16	0.562	0.551	0.650	0.628	21/64	0.337	0.320
7/16	0.4375	11/16	0.688	0.675	0.794	0.768	3/8	0.385	0.365
1/2	0.5000	3/4	0.750	0.736	0.866	0.840	7/16	0.448	0.427
9/16	0.5625	7/8	0.875	0.861	1.010	0.982	31/64	0.496	0.473
5/8	0.6250	15/16	0.938	0.922	1.083	1.051	35/64	0.559	0.535
3/4	0.7500	1 1/8	1.125	1.088	1.299	1.240	41/64	0.665	0.617
7/8	0.8750	1 5/16	1.312	1.269	1.516	1.447	3/4	0.776	0.724
1	1.0000	1 1/2	1.500	1.450	1.732	1.653	55/64	0.887	0.831
1 1/8	1.1250	1 11/16	1.688	1.631	1.949	1.859	31/32	0.999	0.939
1 1/4	1.2500	1 7/8	1.875	1.812	2.165	2.066	1 1/16	1.094	1.030
1 3/8	1.3750	2 1/16	2.062	1.994	2.382	2.273	1 11/64	0.206	1.138
1 1/2	1.5000	2 1/4	2.250	2.175	2.598	2.480	1 9/32	1.317	1.245

Nut Size		H₁ Thickness Hex Jam Nuts			Hex Nuts, Specified Proof Load		Runout of Bearing Face, FIR Max
					Up to 150,000 psi	150,000 psi & Greater	Jam Nuts, All Strength Levels
Nom.	Basic	Basic	Max.	Min.			
1/4	0.2500	5/32	0.163	0.150	0.015	0.010	0.015
5/16	0.3125	3/16	0.195	0.180	0.016	0.011	0.016
3/8	0.3750	7/32	0.227	0.210	0.017	0.012	0.017
7/16	0.4375	1/4	0.260	0.240	0.018	0.013	0.018
1/2	0.5000	5/16	0.323	0.302	0.019	0.014	0.019
9/16	0.5625	5/16	0.324	0.301	0.020	0.015	0.020
5/8	0.6250	3/8	0.387	0.363	0.021	0.016	0.021
3/4	0.7500	27/64	0.446	0.398	0.023	0.018	0.023
7/8	0.8750	31/64	0.510	0.458	0.025	0.020	0.025

(Continued)

Table 17. *(Continued)* Dimensions of Hex Nuts and Hex Jam Nuts.

Nut Size		H_1 Thickness Hex Jam Nuts			Hex Nuts, Specified Proof Load		Runout of Bearing Face, *FIR* Max
Nom.	Basic	Basic	Max.	Min.	Up to 150,000 psi	150,000 psi & Greater	Jam Nuts, All Strength Levels
1	1.0000	35/64	0.575	0.519	0.027	0.022	0.027
1 1/8	1.1250	39/64	0.639	0.579	0.030	0.025	0.030
1 1/4	1.2500	23/32	0.751	0.687	0.033	0.028	0.033
1 3/8	1.3750	25/32	0.815	0.747	0.036	0.031	0.036
1 1/2	1.5000	27/32	0.880	0.808	0.039	0.034	0.039

All dimensions in inches. Source: ANSI B18.2.2-1972, as published by the American Society of Mechanical Engineers.

Table 18. Dimensions of Thick Pattern and Thin Pattern Nylon Insert Lock Nuts.

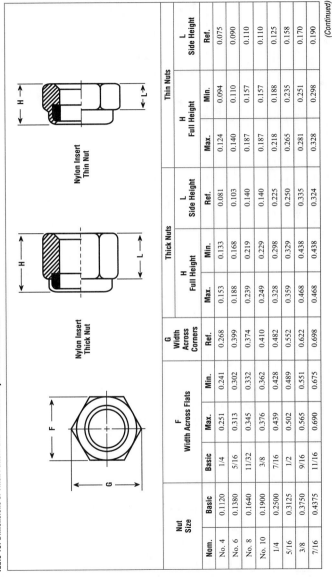

Nylon Insert Thick Nut

Nylon Insert Thin Nut

Nut Size		F Width Across Flats			G Width Across Corners	Thick Nuts				Thin Nuts			
						H Full Height		L Side Height		H Full Height		L Side Height	
Nom.	Basic	Basic	Max.	Min.	Ref.	Max.	Min.	Ref.		Max.	Min.	Ref.	
No. 4	0.1120	1/4	0.251	0.241	0.268	0.153	0.133	0.081		0.124	0.094	0.075	
No. 6	0.1380	5/16	0.313	0.302	0.399	0.188	0.168	0.103		0.140	0.110	0.090	
No. 8	0.1640	11/32	0.345	0.332	0.374	0.239	0.219	0.140		0.187	0.157	0.110	
No. 10	0.1900	3/8	0.376	0.362	0.410	0.249	0.229	0.140		0.187	0.157	0.110	
1/4	0.2500	7/16	0.439	0.428	0.482	0.328	0.298	0.225		0.218	0.188	0.125	
5/16	0.3125	1/2	0.502	0.489	0.552	0.359	0.329	0.250		0.265	0.235	0.158	
3/8	0.3750	9/16	0.565	0.551	0.622	0.468	0.438	0.335		0.281	0.251	0.170	
7/16	0.4375	11/16	0.690	0.675	0.698	0.468	0.438	0.324		0.328	0.298	0.190	

(Continued)

Table 18. *(Continued)* Dimensions of Thick Pattern and Thin Pattern Nylon Insert Lock Nuts.

Nut Size		F Width Across Flats			G Width Across Corners	Thick Nuts			Thin Nuts		
						H Full Height		L Side Height	H Full Height		L Side Height
Nom.	Basic	Basic	Max.	Min.	Ref.	Max.	Min.	Ref.	Max.	Min.	Ref.
1/2	0.5000	3/4	0.752	0.736	0.837	0.609	0.579	0.464	0.328	0.298	0.225
9/16	0.5625	7/8	0.877	0.861	0.978	0.656	0.626	0.469	-	-	-
5/8	0.6250	15/16	0.940	0.922	1.051	0.765	0.735	0.593	0.406	0.376	0.264
3/4	0.7500	1 1/8	1.127	1.088	1.191	0.890	0.860	0.742	0.421	0.391	0.288
7/8	0.8750	1 5/16	1.315	1.269	1.430	0.999	0.969	0.790	-	-	-
1	1.0000	1 1/2	1.502	1.450	1.615	1.078	1.061	0.825	-	-	-
1 1/8	1.1250	1 11/16	1.690	1.631	1.826	1.203	1.141	0.930	-	-	-
1 1/4	1.2500	1 7/8	1.877	1.812	2.038	1.422	1.360	1.125	-	-	-
1 1/2	1.5000	2 1/4	2.252	2.175	2.444	1.640	1.578	1.313	-	-	-

All dimensions in inches.

Table 19. Dimensions of Hex Slotted Nuts.

.016 Approx.

Nut Size		F Width Across Flats			G Width Across Corners		H Thickness			T Unslotted Thickness		S Width of Slot		Runout of Bearing Surface, *F/R*
Nom.	Basic	Basic	Max.	Min.	Max.	Min.	Basic	Max.	Min.	Max.	Min.	Max.	Min.	Max.
1/4	0.2500	7/16	0.438	0.428	0.505	0.488	7/32	0.226	0.212	0.14	0.12	0.10	0.07	0.015
5/16	0.3125	1/2	0.500	0.489	0.577	0.557	17/64	0.273	0.258	0.18	0.16	0.12	0.09	0.016
3/8	0.3750	9/16	0.562	0.551	0.650	0.628	21/64	0.337	0.320	0.21	0.19	0.15	0.12	0.017
7/16	0.4375	11/16	0.688	0.675	0.794	0.768	3/8	0.385	0.365	0.23	0.21	0.15	0.12	0.018
1/2	0.5000	3/4	0.750	0.736	0.866	0.840	7/16	0.448	0.427	0.29	0.27	0.18	0.15	0.019
9/16	0.5625	7/8	0.875	0.861	1.010	0.982	31/64	0.496	0.473	0.31	0.29	0.18	0.15	0.020
5/8	0.6250	15/16	0.938	0.922	1.083	1.051	35/64	0.559	0.535	0.34	0.32	0.24	0.18	0.021
3/4	0.7500	1 1/8	1.125	1.088	1.299	1.240	41/64	0.665	0.617	0.40	0.38	0.24	0.18	0.023
7/8	0.8750	1 5/16	1.312	1.269	1.516	1.447	3/4	0.776	0.724	0.52	0.49	0.24	0.18	0.025
1	1.0000	1 1/2	1.500	1.450	1.732	1.653	55/64	0.887	0.831	0.59	0.56	0.30	0.24	0.027
1 1/8	1.1250	1 11/16	1.688	1.631	1.949	1.859	31/32	0.999	0.939	0.64	0.61	0.33	0.24	0.030
1 1/4	1.2500	1 7/8	1.875	1.812	2.165	2.066	1 1/16	1.094	1.030	0.70	0.67	0.40	0.31	0.033
1 3/8	1.3750	2 1/16	2.062	1.994	2.382	2.273	1 11/64	1.206	1.138	0.82	0.78	0.40	0.31	0.036
1 1/2	1.5000	2 1/4	2.250	2.175	2.598	2.480	1 9/32	1.317	1.245	0.86	0.82	0.46	0.37	0.039

All dimensions in inches. Source: Extracted from ANSI B18.2.2-1972, published by the American Society of Mechanical Engineers.

Table 20. Dimensions of Hex Thick Slotted Nuts.

Nut Size		F Width Across Flats			G Width Across Corners		H Thickness			T Unslotted Thickness		S Width of Slot		Runout of Bearing Surface, FIR
Nom.	Basic	Basic	Max.	Min.	Max.	Min.	Basic	Max.	Min.	Max.	Min.	Max.	Min.	Max.
1/4	0.2500	7/16	0.438	0.428	0.505	0.488	9/32	0.288	0.274	0.20	0.18	0.10	0.07	0.015
5/16	0.3125	1/2	0.500	0.489	0.577	0.557	21/64	0.336	0.320	0.24	0.22	0.12	0.09	0.016
3/8	0.3750	9/16	0.562	0.551	0.650	0.628	13/32	0.415	0.398	0.29	0.27	0.15	0.12	0.017
7/16	0.4375	11/16	0.688	0.675	0.794	0.768	29/64	0.463	0.444	0.31	0.29	0.15	0.12	0.018
1/2	0.5000	3/4	0.750	0.736	0.866	0.840	9/16	0.573	0.552	0.42	0.40	0.18	0.15	0.019
9/16	0.5625	7/8	0.875	0.861	1.010	0.982	39/64	0.621	0.598	0.43	0.41	0.18	0.15	0.020
5/8	0.6250	15/16	0.938	0.922	1.083	1.051	23/32	0.731	0.706	0.51	0.49	0.24	0.18	0.021
3/4	0.7500	1 1/8	1.125	1.088	1.299	1.240	13/16	0.827	0.798	0.57	0.55	0.24	0.18	0.023
7/8	0.8750	1 5/16	1.312	1.269	1.516	1.447	29/32	0.922	0.890	0.67	0.64	0.24	0.18	0.025
1	1.0000	1 1/2	1.500	1.450	1.732	1.653	1	1.018	0.982	0.73	0.70	0.30	0.24	0.027
1 1/8	1.1250	1 11/16	1.688	1.631	1.949	1.859	1 5/32	1.176	1.136	0.83	0.80	0.33	0.24	0.030
1 1/4	1.2500	1 7/8	1.875	1.812	2.165	2.066	1 1/4	1.272	1.228	0.89	0.86	0.40	0.31	0.033
1 3/8	1.3750	2 1/16	2.062	1.994	2.382	2.273	1 3/8	1.399	1.351	1.02	0.98	0.40	0.31	0.036
1 1/2	1.5000	2 1/4	2.250	2.175	2.598	2.480	1 1/2	1.526	1.474	1.08	1.04	0.46	0.37	0.039

All dimensions in inches. Source: Extracted from ANSI B18.2.2-1972, published by the American Society of Mechanical Engineers.

Table 21. Dimensions of Hex Castle Nuts.

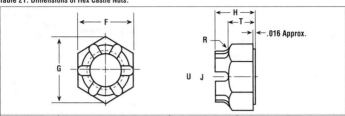

Nut Size		F Width Across Flats			G Width Across Corners		H Thickness		
Nom.	Basic	Basic	Max.	Min.	Max.	Min.	Basic	Max.	Min.
1/4	0.2500	7/16	0.438	0.428	0.505	0.488	9/32	0.228	0.274
5/16	0.3125	1/2	0.500	0.489	0.577	0.557	21/64	0.336	0.320
3/8	0.3750	9/16	0.562	0.551	0.650	0.628	13/32	0.415	0.398
7/16	0.4375	11/16	0.688	0.675	0.794	0.768	29/64	0.463	0.444
1/2	0.5000	3/4	0.750	0.736	0.866	0.840	9/16	0.573	0.552
9/16	0.5625	7/8	0.875	0.861	1.010	0.982	39/64	0.621	0.598
5/8	0.6250	15/16	0.938	0.922	1.083	1.051	23/32	0.731	0.706
3/4	0.7500	1 1/8	1.125	1.088	1.299	1.240	13/16	0.827	0.798
7/8	0.8750	1 5/16	1.312	1.269	1.516	1.447	29/32	0.922	0.890
1	1.0000	1 1/2	1.500	1.450	1.732	1.653	1	1.018	0.982
1 1/8	1.1250	1 11/16	1.688	1.631	1.949	1.859	1 5/32	1.176	1.136
1 1/4	1.2500	1 7/8	1.875	1.812	2.165	2.066	1 1/4	1.272	1.228
1 3/8	1.3750	2 1/16	2.062	1.994	2.382	2.273	1 3/8	1.399	1.351
1 1/2	1.5000	2 1/4	2.250	2.175	2.598	2.480	1 1/2	1.526	1.474

Nut Size		T Unslotted Thickness & Height of Flats			J Width of Slot		R Radius of Fillet	U Dia. of Cylindrical Part	Runout of Bearing Surface, FIR
Nom.	Basic	Nom.	Max.	Min.	Max.	Min.	+/- 0.010	Min.	Max.
1/4	0.2500	3/16	0.20	0.18	0.10	0.07	0.094	0.371	0.015
5/16	0.3125	15/64	0.24	0.22	0.12	0.09	0.094	0.425	0.016
3/8	0.3750	9/32	0.29	0.27	0.15	0.12	0.094	0.478	0.017
7/16	0.4375	19/64	0.31	0.29	0.15	0.12	0.094	0.582	0.018
1/2	0.5000	13/32	0.42	0.40	0.18	0.15	0.125	0.637	0.019
9/16	0.5625	27/64	0.43	0.41	0.18	0.15	0.156	0.744	0.020
5/8	0.6250	1/2	0.51	0.49	0.24	0.18	0.156	0.797	0.021
3/4	0.7500	9/16	0.57	0.55	0.24	0.18	0.188	0.941	0.023
7/8	0.8750	21/32	0.67	0.64	0.24	0.18	0.188	1.097	0.025
1	1.0000	23/32	0.73	0.70	0.30	0.24	0.188	0.254	0.027

(Continued)

Table 21. *(Continued)* **Dimensions of Hex Castle Nuts.**

Nut Size		T Unslotted Thickness & Height of Flats			J Width of Slot		R Radius of Fillet	U Dia. of Cylindrical Part	Runout of Bearing Surface, *FIR*
Nom.	Basic	Nom.	Max.	Min.	Max.	Min.	+/- 0.010	Min.	Max.
1 1/8	1.1250	13/16	0.83	0.80	0.33	0.24	0.250	1.411	0.030
1 1/4	1.2500	7/8	0.89	0.86	0.40	0.31	0.250	1.570	0.033
1 3/8	1.3750	1	1.02	0.98	0.40	0.31	0.250	1.726	0.036
1 1/2	1.5000	1 1/16	1.08	1.04	0.46	0.37	0.250	1.881	0.039

All dimensions in inches. Source: Extracted from ANSI B18.2.2-1972, published by the American Society of Mechanical Engineers.

Table 22. Dimensions of Serrated Flange Lock Nuts.

Nut Size		Width Across Flats F		Width Across Corners G		Flange Dia. A		Height H		Wrenching Length K	Flange Thickness S
Nom.	Basic	Max.	Min.	Max.	Min.	Max.	Min.	Max.	Min.	Min.	Max.
No. 6	0.1380	0.312	0.302	0.361	0.342	0.422	0.406	0.171	0.156	0.10	0.02
No. 8	0.1640	0.344	0.334	0.397	0.381	0.469	0.452	0.203	0.187	0.13	0.02
No. 10	0.1900	0.375	0.365	0.433	0.416	0.500	0.480	0.219	0.203	0.13	0.03
No. 12	0.216	0.438	0.428	0.505	0.488	0.594	0.574	0.236	0.222	0.14	0.04
1/4	0.2500	0.438	0.428	0.505	0.488	0.594	0.574	0.236	0.222	0.14	0.04
5/16	0.3125	0.500	0.489	0.577	0.557	0.680	0.660	0.283	0.268	0.17	0.04
3/8	0.3750	0.562	0.551	0.650	0.628	0.750	0.728	0.347	0.330	0.23	0.04
7/16	0.4375	0.688	0.675	0.794	0.768	0.937	0.910	0.395	0.375	0.26	0.04
1/2	0.5000	0.750	0.736	0.866	0.840	1.031	1.000	0.458	0.437	0.31	0.05
5/8	0.6250	0.938	0.922	1.083	1.051	1.281	1.248	0.569	0.545	0.40	0.05
3/4	0.7500	1.125	1.088	1.299	1.240	1.500	1.460	0.675	0.627	0.46	0.06

All dimensions in inches.

Table 23. Dimensions of Square Nuts.

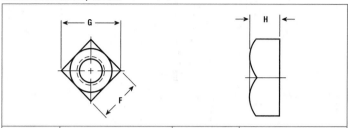

Nut Size		F Width Across Flats			G Width Across Corners		H Thickness		
Nom.	Basic	Basic	Max.	Min.	Max.	Min.	Basic	Max.	Min.
1/4	0.2500	7/16	0.438	0.425	0.619	0.584	7/32	0.235	0.203
5/16	0.3125	9/16	0.562	0.547	0.795	0.751	17/64	0.283	0.249
3/8	0.3750	5/8	0.625	0.606	0.884	0.832	21/64	0.346	0.310
7/16	0.4375	3/4	0.750	0.728	1.061	1.000	3/8	0.394	0.356
1/2	0.5000	13/16	0.812	0.788	1.149	1.082	7/16	0.458	0.418
5/8	0.6250	1	1.000	0.969	1.414	1.330	35/64	0.569	0.525
3/4	0.7500	1 1/8	1.125	1.088	1.591	1.494	21/32	0.680	0.632
7/8	0.8750	1 5/16	1.312	1.269	1.856	1.742	49/64	0.792	0.740
1	1.0000	1 1/2	1.500	1.450	2.121	1.991	7/8	0.903	0.847
1 1/8	1.1250	1 11/16	1.688	1.631	2.386	2.239	1	1.030	0.970
1 1/4	1.2500	1 7/8	1.875	1.812	2.652	2.489	1 3/32	1.126	1.062
1 3/8	1.3750	2 1/16	2.062	1.994	2.917	2.738	1 13/64	1.237	1.169
1 1/2	1.5000	2 1/4	2.250	2.175	3.182	2.986	1 5/16	1.348	1.276

All dimensions in inches. Source: Extracted from ANSI B18.2.2-1972, published by the American Society of Mechanical Engineers.

Table 24. Dimensions of Square and Hex Machine Screw Nuts. (See Notes.)

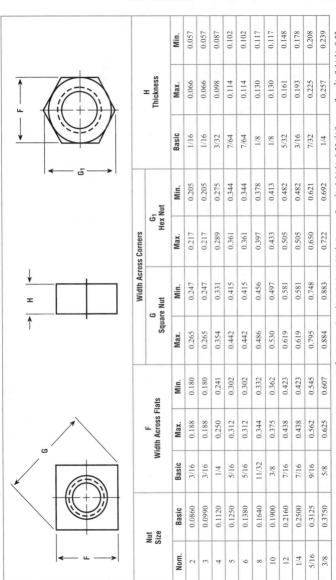

Nut Size		F Width Across Flats			Width Across Corners				H Thickness		
					G Square Nut		G₁ Hex Nut				
Nom.	Basic	Basic	Max.	Min.	Max.	Min.	Max.	Min.	Basic	Max.	Min.
2	0.0860	3/16	0.188	0.180	0.265	0.247	0.217	0.205	1/16	0.066	0.057
3	0.0990	3/16	0.188	0.180	0.265	0.247	0.217	0.205	1/16	0.066	0.057
4	0.1120	1/4	0.250	0.241	0.354	0.331	0.289	0.275	3/32	0.098	0.087
5	0.1250	5/16	0.312	0.302	0.442	0.415	0.361	0.344	7/64	0.114	0.102
6	0.1380	5/16	0.312	0.302	0.442	0.415	0.361	0.344	7/64	0.114	0.102
8	0.1640	11/32	0.344	0.332	0.486	0.456	0.397	0.378	1/8	0.130	0.117
10	0.1900	3/8	0.375	0.362	0.530	0.497	0.433	0.413	1/8	0.130	0.117
12	0.2160	7/16	0.438	0.423	0.619	0.581	0.505	0.482	5/32	0.161	0.148
1/4	0.2500	7/16	0.438	0.423	0.619	0.581	0.505	0.482	3/16	0.193	0.178
5/16	0.3125	9/16	0.562	0.545	0.795	0.748	0.650	0.621	7/32	0.225	0.208
3/8	0.3750	5/8	0.625	0.607	0.884	0.883	0.722	0.692	1/4	0.257	0.239

Notes: Small pattern nuts are available in the following sizes: 4–40 (³⁄₁₆″ across flats, ¹⁄₁₆″ thick); 6–32 (¹⁄₄″ across flats, ¹⁄₁₆″ thick); 8–32 (³⁄₁₆″ across flats, ³⁄₃₂″ thick); 8–32 (⁵⁄₁₆″ across flats, ⁷⁄₆₄″ thick); 10–32 (⁵⁄₁₆″ across flats, ⁷⁄₆₄″ thick). All dimensions in inches.

Table 25. Dimensions of J-Type Nut Retainers. (See Notes.)

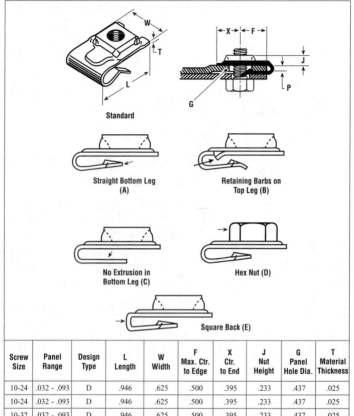

Standard

Straight Bottom Leg (A)

Retaining Barbs on Top Leg (B)

No Extrusion in Bottom Leg (C)

Hex Nut (D)

Square Back (E)

Screw Size	Panel Range	Design Type	L Length	W Width	F Max. Ctr. to Edge	X Ctr. to End	J Nut Height	G Panel Hole Dia.	T Material Thickness
10-24	.032 - .093	D	.946	.625	.500	.395	.233	.437	.025
10-24	.032 - .093	D	.946	.625	.500	.395	.233	.437	.025
10-32	.032 - .093	D	.946	.625	.500	.395	.233	.437	.025
M6×1	*2.29-3.56*	*D*	*26.29*	*19.05*	*12.70*	*11.69*	*8.74*	*13.47*	*.710*
1/4-20	.032 - .093	D	.946	.625	.500	.395	.233	.437	.025
1/4-20	.090 - .140	D	1.035	.750	.500	.460	.344	.531	.028
1/4-28	.032 - .093	D	.946	.625	.500	.395	.233	.437	.025
5/16-18	.050 - .090	D	1.05	.750	.546	.460	.344	.531	.028
5/16-18	.130-.140	ACDE	.820	.687	.400	.385	.330	.343	.031
3/8-16	.090-.140	D	1.035	.750	.500	.460	.344	.531	.028
3/8-16	.093-.203	Std.	1.48	.750	.750	.440	.230	.437	.028

Notes: Dimensions given are for Tinnerman brand fasteners. Dimensions for inch screw sizes are in inches, and dimensions for metric sizes (shown in italics) are in millimeters. Source: Tinnerman Fasteners.

Table 26. Dimensions of U-Type Nut Retainers. (See Notes.)

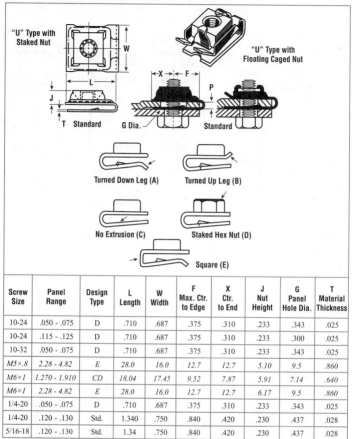

Screw Size	Panel Range	Design Type	L Length	W Width	F Max. Ctr. to Edge	X Ctr. to End	J Nut Height	G Panel Hole Dia.	T Material Thickness
10-24	.050 - .075	D	.710	.687	.375	.310	.233	.343	.025
10-24	.115 - .125	D	.710	.687	.375	.310	.233	.300	.025
10-32	.050 - .075	D	.710	.687	.375	.310	.233	.343	.025
M5×.8	*2.28 - 4.82*	*E*	*28.0*	*16.0*	*12.7*	*12.7*	*5.10*	*9.5*	*.860*
M6×1	*1.270 - 1.910*	*CD*	*18.04*	*17.45*	*9.52*	*7.87*	*5.91*	*7.14*	*.640*
M6×1	*2.28 - 4.82*	*E*	*28.0*	*16.0*	*12.7*	*12.7*	*6.17*	*9.5*	*.860*
1/4-20	.050 - .075	D	.710	.687	.375	.310	.233	.343	.025
1/4-20	.120 - .130	Std.	1.340	.750	.840	.420	.230	.437	.028
5/16-18	.120 - .130	Std.	1.34	.750	.840	.420	.230	.437	.028

Notes: Dimensions given are for Tinnerman brand fasteners. Dimensions for inch screw sizes are in inches, and dimensions for metric sizes (shown in italics) are in millimeters. Source: Tinnerman Fasteners.

Table 27. Dimensions of J-Type Speed Nuts. (See Notes.)

Straight Upper Leg (A)

Corners Turned Up (B)

No Extrusion on Lower Leg (C)

Relief Notch (D)

Corners Cut Off (E)

Impressions Turned 90° (F)

Sides Turned Up (G)

Corner Turned Down Lower Leg (J)

Barbs on Top Leg (K)

Standard

Corners Turned Down Top Leg (M)

Keyhole Impression (N)

Barbs on Lower Leg (P)

Screw Size	Panel Range	Design Type	L Length	W Width	F Max. Ctr. to Edge	X Ctr. to End	G Panel Hole Dia.	T Material Thickness
6-32	.042 - .062	ADF	.470	.500	.218	.210	.187	.017
M4.7×1.59	*2.50 - 2.50*	*N*	*14.10*	*12.70*	*7.13*	*6.40*	*8.53*	*.640*
8-32	.045 - .062	ADFN	.520	.500	.234	.250	.218	.017
8A	.051 - .064	ADF	.500	.250	.210	.220	.218	.022
8A or 8B	.080 - .105	ACF	.790	.630	.500	.250	.218	.028
10-24	.045 - .062	AEN	.970	.380	.562	.350	.250	.022
10A	.025 - .040	ADFN	.580	.630	.250	.270	.250	.031
10A or 10B	.168 - .198	DF	.570	.630	.281	.270	.250	.031
1/4-20	.040 - .051	ADK	.910	.562	.375	.430	.312	.025
14A or 14B	.100 - .140	ACE	1.09	.630	.580	.450	.250	.037
14B	.028 - .056	AE	.900	.560	.421	.400	.312	.037
5/16-10	.032 - .064	AE	.900	.630	.500	.430	.437	.044

Notes: Dimensions given are for Tinnerman brand fasteners. Dimensions for inch screw sizes are in inches, and dimensions for metric sizes (shown in italics) are in millimeters. Source: Tinnerman Fasteners.

Table 28. Dimensions of U-Type Speed Nuts. (See Notes.)

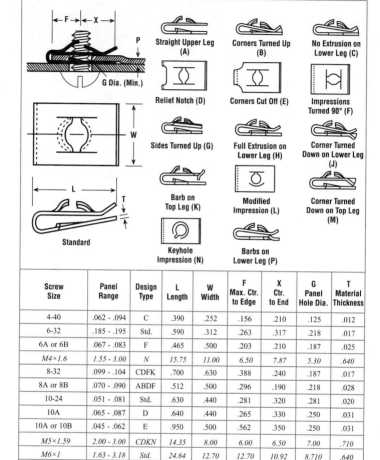

Screw Size	Panel Range	Design Type	L Length	W Width	F Max. Ctr. to Edge	X Ctr. to End	G Panel Hole Dia.	T Material Thickness
4-40	.062 - .094	C	.390	.252	.156	.210	.125	.012
6-32	.185 - .195	Std.	.590	.312	.263	.317	.218	.017
6A or 6B	.067 - .083	F	.465	.500	.203	.210	.187	.025
M4×1.6	*1.55 - 3.00*	*N*	*15.75*	*11.00*	*6.50*	*7.87*	*5.30*	*.640*
8-32	.099 - .104	CDFK	.700	.630	.388	.240	.187	.017
8A or 8B	.070 - .090	ABDF	.512	.500	.296	.190	.218	.028
10-24	.051 - .081	Std.	.630	.440	.281	.320	.281	.020
10A	.065 - .087	D	.640	.440	.265	.330	.250	.031
10A or 10B	.045 - .062	E	.950	.500	.562	.350	.250	.031
M5×1.59	*2.00 - 3.00*	*CDKN*	*14.35*	*8.00*	*6.00*	*6.50*	*7.00*	*.710*
M6×1	*1.63 - 3.18*	*Std.*	*24.64*	*12.70*	*12.70*	*10.92*	*8.710*	*.640*
1/4-20	.051 - .109	H	.790	.500	.312	.390	.343	.025
14A	.028 - .056	E	.930	.560	.421	.430	.312	.037
14B	.115 - .125	C	.970	.500	.500	.420	.281	.037
5/16-18	.112 - .140	Std.	1.035	.630	.562	.395	.375	.028

Notes: Dimensions given are for Tinnerman brand fasteners. Dimensions for inch screw sizes are in inches, and dimensions for metric sizes (shown in italics) are in millimeters. Source: Tinnerman Fasteners.

Table 29. Dimensions of Type A Wing Nuts.

Nut Size		Threads per inch	Series	Nut Blank Size (Ref.)	A Wing Spread		B Wing Height		C Wing Thickness		D Between Wings		E Boss Diameter		G Boss Height	
Nom.	Basic				Max.	Min.	Max.	Min.	Max.	Min.	Max.	Min.	Max.	Min.	Max.	Min.
No. 3	0.0990	48 & 56	Heavy	AA	0.72	0.59	0.41	0.28	0.11	0.07	0.21	0.17	0.33	0.29	0.14	0.10
No. 4	0.1120	40 & 38	Heavy	AA	0.72	0.59	0.41	0.28	0.11	0.07	0.21	0.17	0.33	0.29	0.14	0.10
No. 5	0.1250	40 & 44	**Light**	**AA**	**.072**	**0.59**	**0.41**	**0.28**	**0.11**	**0.07**	**0.21**	**0.17**	**0.33**	**0.29**	**0.14**	**0.10**
			Heavy	A	0.91	0.78	0.47	0.34	0.14	0.10	0.27	0.22	0.43	0.39	0.18	0.14
No. 6	0.1380	32 & 40	Light	AA	0.72	0.59	0.41	0.28	0.11	0.07	0.21	0.17	0.33	0.29	0.14	0.10
			Heavy	**A**	**0.91**	**0.78**	**0.47**	**0.34**	**0.14**	**0.10**	**0.27**	**0.22**	**0.43**	**0.39**	**0.18**	**0.14**
No. 8	0.1640	32 & 36	**Light**	**A**	**0.91**	**0.78**	**0.47**	**0.34**	**0.14**	**0.10**	**0.27**	**0.22**	**0.43**	**0.39**	**0.18**	**0.14**
			Heavy	B	1.10	0.97	0.57	0.43	0.18	0.14	0.33	0.26	0.50	0.45	0.22	0.17
No.10	0.1900	24 & 32	**Light**	**A**	**0.91**	**0.78**	**0.47**	**0.43**	**0.14**	**0.10**	**0.27**	**0.22**	**0.43**	**0.39**	**0.18**	**0.14**
			Heavy	B	1.10	0.97	0.57	0.43	0.18	0.14	0.33	0.26	0.50	0.45	0.22	0.17
No.12	0.2160	24 & 28	**Light**	**B**	**1.10**	**0.97**	**0.57**	**0.43**	**0.18**	**0.14**	**0.33**	**0.26**	**0.50**	**0.45**	**0.22**	**0.17**
			Heavy	C	1.25	1.12	0.66	0.53	0.21	0.17	0.39	0.32	0.58	0.51	0.25	0.20

(Continued)

Table 29. *(Continued)* **Dimensions of Type A Wing Nuts.**

Nut Size Nom.	Nut Size Basic	Threads per inch	Series	Nut Blank Size (Ref.)	A Wing Spread Max	A Wing Spread Min	B Wing Height Max	B Wing Height Min	C Wing Thickness Max	C Wing Thickness Min	D Between Wings Max	D Between Wings Min	E Boss Diameter Max	E Boss Diameter Min	G Boss Height Max	G Boss Height Min
1/4	0.2500	20 & 28	Light	B	1.10	0.97	0.57	0.43	0.18	0.14	0.39	0.26	0.50	0.45	0.22	0.17
			Regular	C	1.25	1.12	0.66	0.53	0.21	0.17	0.39	0.32	0.58	0.51	0.25	0.20
			Heavy	D	1.44	1.31	0.79	0.65	0.24	0.20	0.48	0.42	0.70	0.64	0.30	0.26
5/16	0.3125	18 & 24	Light	C	1.25	1.12	0.66	0.53	0.21	0.17	0.39	0.32	0.58	0.51	0.25	0.20
			Regular	D	1.44	1.31	0.79	0.65	0.24	0.20	0.48	0.42	0.70	0.64	0.30	0.26
			Heavy	E	1.94	1.81	1.00	0.87	0.33	0.26	0.65	0.54	0.93	0.86	0.39	0.35
3/8	0.3750	16 & 24	Light	D	1.44	1.31	0.79	0.65	0.24	0.20	0.48	0.42	0.70	0.64	0.30	0.26
			Regular	E	1.94	1.81	1.00	0.87	0.33	0.26	0.65	0.54	0.93	0.86	0.39	0.35
7/16	0.4375	14 & 20	Light	E	1.94	1.81	1.00	0.87	0.33	0.26	0.65	0.54	0.93	0.86	0.39	0.35
			Heavy	F	2.76	2.62	1.44	1.31	0.40	0.34	0.90	0.80	1.19	1.13	0.55	0.51
1/2	0.5000	13 & 20	Light	E	1.94	1.81	1.00	0.87	0.33	0.26	0.65	0.54	0.93	0.86	0.39	0.35
			Heavy	F	2.76	2.62	1.44	1.31	0.40	0.34	0.90	0.80	1.19	1.13	0.55	0.51
9/16	0.5825	12 & 18	Heavy	F	2.76	2.62	1.44	1.31	0.40	0.34	0.90	0.80	1.19	1.13	0.55	0.51
5/8	0.6250	11 & 18	Heavy	F	2.76	2.62	1.44	1.31	0.40	0.34	0.90	0.80	1.19	1.13	0.55	0.51
3/4	0.7500	10 & 16	Heavy	F	2.76	2.62	1.44	1.31	0.40	0.34	0.90	0.80	1.19	1.13	0.55	0.51

All dimensions in inches. Source: Extracted from ANSI B18.17-1968, published by the American Society of Mechanical Engineers.

Table 30. Dimensions of Type D, Style 2 Wing Nuts.

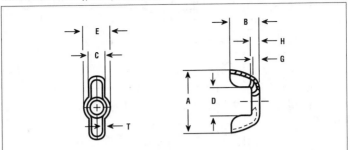

Nut Size		Threads per Inch	Series	Nut Blank Size (Ref.)	A Wing Spread		B Wing Height		C Wing Thickness	
Nom.	Basic				Max.	Min.	Max.	Min.	Max.	Min.
No. 5	0.1250	40	Regular	A	1.03	0.97	0.25	0.19	0.19	0.13
No. 6	0.1380	32	Regular	A	1.03	0.97	0.25	0.19	0.19	0.13
No. 8	0.1640	32	Regular	A	1.03	0.97	0.25	0.19	0.19	0.13
No. 10	0.1900	24	Regular	B	1.40	1.34	0.34	0.28	0.25	0.18
		32	Heavy	C	1.21	1.16	0.28	0.26	0.31	0.25
No. 12	0.2160	24	Regular	C	1.21	1.16	0.28	0.26	0.31	0.25
1/4	0.2500	20	Regular	C	1.21	1.16	0.28	0.26	0.31	0.25

Nut Size		Threads per Inch	Series	D Between Wings	E Boss Diameter		G Boss Height	H Wall Height	T Stock Thickness	
Nom.	Basic			Min.	Max.	Min.	Min.	Min.	Max.	Min.
No. 5	0.1250	40	Regular	0.30	0.40	0.34	0.07	0.09	0.04	0.03
No. 6	0.1380	32	Regular	0.30	0.40	0.34	0.08	0.09	0.04	0.03
No. 8	0.1640	32	Regular	0.30	0.40	0.34	0.08	0.09	0.04	0.03
No. 10	0.1900	24	Regular	0.32	0.53	0.47	0.09	0.16	0.05	0.04
		32	Heavy	0.60	0.61	0.55	0.09	0.13	0.05	0.04
No. 12	0.2160	24	Regular	0.60	0.61	0.55	0.11	0.13	0.05	0.04
1/4	0.2500	20	Regular	0.60	0.61	0.55	0.11	0.13	0.05	0.04

All dimensions in inches. Source: Extracted from ANSI B18.17-1968, published by the American Society of Mechanical Engineers.

Table 31. Dimensions of Type D, Style 3 Wing Nuts.

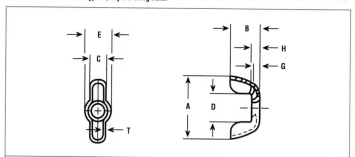

Nominal Size or Basic Major Dia. of Thread	Threads per Inch	Series	Nut Blank Size (Ref.)	A Wing Spread		B Wing Height		C Wing Thickness	
				Max.	Min.	Max.	Min.	Max.	Min.
10 - 0.1900	24 32	Light Regular	A C	1.31 1.40	1.25 1.34	0.48 0.53	0.42 0.47	0.29 0.25	0.23 0.19
12 - 0.2160	24	Regular	B	1.28	1.22	0.40	0.34	0.23	0.17
1/4 - 0.2500	20	Light Regular Heavy	B E D	1.28 1.78 1.47	1.22 1.72 1.40	0.40 0.66 0.50	0.34 0.60 0.44	0.23 0.31 0.37	0.17 0.25 0.31
5/16 - 0.3125	18	Regular Heavy	E D	1.78 1.47	1.72 1.40	0.66 0.50	0.60 0.44	0.31 0.37	0.25 0.31

Nominal Size or Basic Major Dia. of Thread	Threads per Inch	Series	Nut Blank Size (Ref.)	D Between Wings	E Boss Diameter		G Boss Height	H Wall Height	T Stock Thickness	
				Min.	Max.	Min.	Min.	Min.	Max.	Min.
10 - 0.1900	24 32	Light Regular	A C	0.47 0.50	0.65 0.75	0.59 0.69	0.08 0.08	0.12 0.14	0.04 0.04	0.03 0.03
12 - 0.2160	24	Regular	B	0.59	0.73	0.67	0.11	0.12	0.04	0.03
1/4 - 0.2500	20	Light Regular Heavy	B E D	0.59 0.70 0.66	0.73 1.03 1.03	0.67 0.97 0.97	0.11 0.14 0.14	0.12 0.17 0.14	0.04 0.06 0.08	0.03 0.04 0.06
5/16 - 0.3125	18	Regular Heavy	E D	0.70 0.66	1.03 1.03	0.97 0.97	0.14 0.14	0.17 0.14	0.06 0.08	0.04 0.06

All dimensions in inches. Source: Extracted from ANSI B18.17-1968, published by the American Society of Mechanical Engineers.

Table 32. Dimensions of Selected Metallic Flat Washers. (See Notes.)

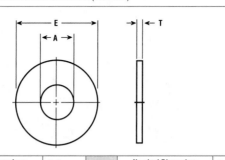

Size inches	Nominal Dimensions			Reference (see notes)	Size inches	Nominal Dimensions			Reference (see notes)
	A	E	Thickness			A	E	Thickness	
000	.040	.090	.010	-	M2.5	.106	.197	.020	DIN 433
M1	.043	.099	.012	DIN 433	M2.5	.106	.256	.020	DIN 125 A
M1	.043	.126	.012	DIN 125A	M2.5	.106	.315	.032	DIN 9021B
M1.2	.051	.118	.012	DIN 433	3	.109	.219	.025	ANS B-N
M1.2	.051	.149	.012	DIN 125A	3	.109	.312	.032	ANS B-R
00	.055	.156	.010	-	3	.109	.406	.040	ANS B-W
0	.065	.099	.016	NAS 620	M3	.119	.236	.032	DIN 1440
0	.065	.125	.025	MS1496	M2.5	.121	.231	.030	ANSI-N
M1.6	.067	.140	.012	DIN 433	M2.5	.121	.308	.030	ANSI-R
M1.6	.067	.158	.012	DIN 125A	M2.5	.121	.387	.030	ANSI-W
0	.068	.125	.025	ANS B-N	4	.125	.250	.022	MS15795
0	.068	.188	.025	ANS B-R	4	.125	.250	.032	ANS B-N
0	.068	.250	.025	ANS B-W	4	.125	.312	.032	ANS A-P
M1.8	.080	.152	.024	ANSI-N	4	.125	.375	.040	ANS B-R
M1.8	.080	.191	.024	ANSI-R	4	.125	.438	.040	ANS B-W
M1.8	.080	.230	.030	ANSI-W	M3	.126	.236	.020	DIN 433
1	.084	.156	.025	ANS B-N	M3	.126	.276	.020	DIN 125A
1	.084	.219	.025	ANS B-R	M3	.126	.355	.030	DIN 9021B
1	.084	.281	.032	ANS B-W	M3	.126	.355	.040	DIN 7349
M2	.086	.177	.012	DIN 433	M3	.126	.269	.030	ANSI-N
M2	.086	.197	.020	DIN 125A	M3	.141	.387	.040	ANSI-R
2	.094	.188	.025	ANS B-N	M3	.141	.464	.047	ANSI-W
2	.094	.250	.020	ANS A-P	5	.141	.281	.032	ANS B-N
2	.094	.250	.032	ANS B-R	5	.141	.406	.040	ANS B-R
2	.094	.344	.035	ANS B-W	5	.141	.500	.040	ANS B-W
M2	.101	.191	.030	ANS B-N	6	.156	.312	.032	ANS B-N
M2	.101	.230	.030	ANS B-R	6	.156	.375	.049	ANS A-P
M2	.101	.308	.030	ANS B-R	6	.156	.438	.040	ANS B-R

(Continued)

Table 32. *(Continued)* **Dimensions of Selected Metallic Flat Washers.** (See Notes.)

Size inches	Nominal Dimensions			Reference (see notes)	Size inches	Nominal Dimensions			Reference (see notes)
	A	E	Thickness			A	E	Thickness	
6	.156	.562	.040	ANS B-W	3/8	.406	.812	.065	ANS A-P
M4	.158	.315	.032	DIN 1440	3/8	.406	1.250	.100	ANS B-W
M3.5	.161	.348	.040	ANSI-N	M10	.414	.709	.063	DIN 433
M3.5	.161	.583	.058	ANSI-W	M10	.414	.827	.079	DIN 125A
M4	.169	.315	.020	DIN 433	M10	.433	1.093	.095	ANSI-N
M4	.169	.355	.032	DIN 125A	M10	.433	1.524	.118	ANSI-W
M4	.169	.473	.040	DIN 9021B	3/8	.438	1.000	.083	ANS A-P
8	.188	.375	.040	ANS B-N	7/16	.469	.922	.062	ANS A-P
8	.188	.375	.049	ANS A-P	7/16	.469	1.125	.063	ANS B-R
8	.188	.625	.063	ANS B-W	7/16	.469	1.469	.100	ANS B-W
10	.203	.406	.040	ANS B-N	7/16	.500	1.625	.083	ANS A
10	.203	.469	.049	ANS A	M12	.512	.788	.079	DIN 433
10	.203	.734	.063	ANS B-W	M12	.512	.946	.099	DIN 125A
M5	.209	.374	.040	DIN 433	M12	.529	.994	.095	ANSI-N
M5	.209	.394	.040	DIN 125A	M12	.529	1.721	.118	ANSI-W
M5	.209	.591	.063	DIN 9021B	1/2	.531	1.062	.095	ANS A-P
10	.219	.500	.049	ANS A-P	1/2	.531	1.250	.100	ANS B-R
M5	.224	.425	.047	ANSI-N	1/2	.531	1.750	.100	ANS B-W
M5	.224	.778	.077	ANSI-W	1/2	.562	1.250	.109	ANS A-P
12	.234	.438	.040	ANS B-N	9/16	.594	1.125	.063	ANS B-N
12	.234	.875	.063	ANS B-W	9/16	.594	1.156	.095	ANS A-P
12	.250	.562	.049	ANS A-P	9/16	.594	2.000	.100	ANS B-W
12	.250	.562	.065	ANS A-P	9/16	.625	1.469	.109	ANS A-P
M6	.253	.493	.062	DIN 125A	M16	.630	1.103	.118	DIN 1440
M6	.253	.670	.118	DIN 7349	5/8	.656	1.312	.095	ANS A-P
M6	.260	.493	.062	DIN 126	5/8	.656	1.750	.100	ANS B-R
1/4	.266	.625	.049	ANS A	5/8	.656	2.250	.160	ANS B-W
1/4	.281	.500	.063	ANS B-N	M16	.670	1.064	.079	DIN 433
1/4	.281	.625	.062	ANS A-P	M16	.670	1.182	.118	DIN 125A
1/4	.281	1.000	.063	ANS B-W	5/8	.688	1.750	.134	ANS A-P
1/4	.312	.734	.062	ANS A-P	M16	.709	1.182	.118	DIN 126
1/4	.312	.875	.062	ANS A	3/4	.812	1.469	.134	ANS A-P
M8	.331	.591	.062	DIN 433	3/4	.812	2.000	.100	ANS B-R
M8	.331	.670	.062	DIN 125A	3/4	.812	2.000	.148	ANS A-P
5/16	.344	.625	.063	ANS B-N	3/4	.812	2.500	.160	ANS B-W
5/16	.344	1.125	.063	ANS B-W	M20	.827	1.300	.099	DIN 433
5/16	.375	.734	.065	ANS A	M20	.867	1.458	.118	DIN 125
3/8	.390	1.125	.062	ANS A	M20	.865	1.154	.138	ANSI
3/8	.344	.625	.063	ANS B-N	M22	.906	1.537	.118	DIN 125A

(Continued)

Table 32. *(Continued)* **Dimensions of Selected Metallic Flat Washers.** (See Notes.)

Size inches	Nominal Dimensions			Reference (see notes)	Size inches	Nominal Dimensions			Reference (see notes)
	A	E	Thickness			A	E	Thickness	
7/8	.938	1.750	.134	ANS A-P	1 1/2	1.625	3.500	.181	ANS A-P
7/8	.938	2.000	.165	ANS A	1 5/8	1.750	3.000	.160	ANS B-N
7/8	.938	2.250	.165	ANS A-P	1 5/8	1.750	3.750	.180	ANS A-P
7/8	.938	2.750	.160	ANS B-W	1 5/8	1.750	4.250	.266	ANS B-W
1"	1.062	1.750	.100	ANS B-N	M42	1.773	3.073	.276	DIN 126
1"	1.062	2.000	.134	ANS A-P	1 3/4	1.875	3.250	.160	ANS B-N
1"	1.062	2.250	.165	ANS A	1 3/4	1.875	4.000	.180	ANS A-P
1"	1.062	2.500	.165	ANS A-P	1 3/4	1.875	4.500	.250	ANS B-W
1 1/16	1.078	1.812	.032	AN 960	1 3/4	1.890	2.625	.090	AN 960
1 1/8	1.188	2.000	.100	ANS B-N	1 7/8	2.000	3.500	.250	ANS B-N
1 1/8	1.188	2.750	.160	ANS B-R	1 7/8	2.000	4.250	.180	ANS A-P
1 1/8	1.188	3.325	.160	ANS B-W	1 7/8	2.000	4.750	.250	ANS B-W
M30	1.221	2.206	.158	DIN 125A	2"	2.125	3.750	.250	ANS B-N
1 1/8	1.250	2.840	.165	ANS A	2"	2.125	4.500	.180	ANS A-P
M30	1.300	2.206	.158	DIN 126	2"	2.125	5.000	.250	ANS B-W
1 1/4	1.312	2.250	.160	ANS B-N	2 1/4	2.265	3.000	.032	AN 960
1 1/4	1.312	3.500	.250	ANS B-W	2 1/4	2.375	4.750	.220	ANS A-P
1 5/16	1.375	2.500	.165	ANS A-P	2 1/2	2.515	3.250	.090	AN 960
1 5/16	1.375	3.000	.165	ANS A-P	2 1/2	2.625	5.000	.238	ANS A-P
1 3/8	1.500	3.250	.181	ANS A-P	2 3/4	2.875	5.250	.259	ANS A-P
1 1/2	1.515	1.250	.090	NAS 1149	2 7/8	2.968	3.385	.010	SHIM
M36	1.537	2.600	.058	DIN 126	3"	3.125	5.500	.284	ANS A-P
1 1/2	1.562	2.750	.160	ANS B-N	3"	3.155	3.500	.025	-
1 1/2	1.562	4.000	.250	ANS B-W					

All dimensions in inches. See text for information about available sizes, materials, and tolerances for flat washers. This table was compiled from information published by West Coast Lockwasher (WCL) Company and referenced standards.

Notes: Washers to DIN standards conform to German national standards. Full specifications for metric washers to ANSI specifications can be found in ANSI B18.22M. American National Standard washers are designated above as ANS, with suffixes A and B identifying the two standard series; and N designates narrow, R is regular, and W is wide. P indicates preferred series. For complete specifications for inch unit national standard washers, consult ANSI B 18.22.1. Air Force-Navy (AN) 960 washers are steel washers made to military specifications. This standard is scheduled to be replaced by NAS 1149. Multipurpose military washers conform to Mil-Std 15795. Mil-Std 15496 washers are made from stainless steel to reduced diameters.

Table 33. Dimensions of Regular (Medium) and High Collar Helical Spring Lock Washers.

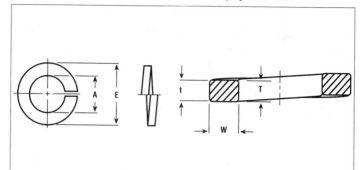

Nom. Washer Size	A Inside Dia.		Regular Lock Washer			High Collar Lock Washer		
			E Outside Dia.	T Mean Thickness[1]	W Section Width	E Outside Dia.	T Mean Thickness[1]	W Section Width
	Max.	Min.	Max.	Min.	Min.	Max.	Min.	Min.
#2	0.094	0.088	0.172	0.020	0.035	-	-	-
#3	0.107	0.101	0.195	0.025	0.040	-	-	-
#4	0.120	0.114	0.209	0.025	0.040	0.173	0.022	0.022
#5	0.133	0.127	0.236	0.031	0.047	0.202	0.030	0.030
#6	0.148	0.141	0.250	0.031	0.047	0.216	0.030	0.030
#8	0.174	0.167	0.293	0.040	0.055	0.267	0.047	0.042
#10	0.200	0.193	0.334	0.047	0.062	0.294	0.047	0.042
#12	0.227	0.220	0.377	0.056	0.070	-	-	-
1/4	0.262	0.254	0.489	0.062	0.109	0.365	0.078	0.047
5/16	0.326	0.317	0.586	0.078	0.125	0.460	0.093	0.062
3/8	0.390	0.380	0.683	0.094	0.141	0.553	0.125	0.076
7/16	0.455	0.443	0.779	0.109	0.156	0.647	0.140	0.090
1/2	0.518	0.506	0.873	0.125	0.171	0.737	0.172	0.103
9/16	0.582	0.570	0.971	0.141	0.188	-	-	-
5/8	0.650	0.635	1.079	0.156	0.203	0.923	0.203	0.125
11/16	0.713	0.698	1.176	0.172	0.219	-	-	-
3/4	0.775	0.760	1.271	0.188	0.234	1.111	0.218	0.154
13/16	0.843	0.824	1.367	0.203	0.250	-	-	-
7/8	0.905	0.887	1.464	0.219	0.266	1.296	0.234	0.182
15/16	0.970	0.950	1.560	0.234	0.281	-	-	-

(Continued)

Table 33. *(Continued)* Dimensions of Regular (Medium) and High Collar Helical Spring Lock Washers.

Nom. Washer Size	A Inside Dia.		Regular Lock Washer			High Collar Lock Washer		
			E Outside Dia.	T Mean Thickness[1]	W Section Width	E Outside Dia.	T Mean Thickness[1]	W Section Width
	Max.	Min.	Max.	Min.	Min.	Max.	Min.	Min.
1	1.042	1.017	1.661	0.250	0.297	1.483	0.250	0.208
1 1/16	1.107	1.080	1.756	0.266	0.312	-	-	-
1 1/8	1.172	1.144	1.853	0.281	0.328	1.669	0.313	0.236
1 3/16	1.237	1.208	1.950	0.297	0.344	-	-	-
1 1/4	1.302	1.271	2.045	0.312	0.359	1.799	0.313	0.236
1 5/16	1.366	1.334	2.141	0.328	0.375	-	-	-
1 3/8	1.432	1.398	2.239	0.344	0.391	2.041	0.375	0.292
1 7/16	1.497	1.462	2.334	0.359	0.406	-	-	-
1 1/2	1.561	1.525	2.430	0.375	0.422	2.170	0.375	0.292
1 3/4	1.811	1.775	-	-	-	2.602	0.469	0.383
2	2.061	2.025	-	-	-	2.852	0.469	0.383
2 1/4	2.311	2.275	-	-	-	3.352	0.508	0.508
2 1/2	2.561	2.525	-	-	-	3.602	0.508	0.508
2 3/4	2.811	2.775	-	-	-	4.102	0.633	0.633
3	3.061	3.025	-	-	-	4.352	0.633	0.633

Notes: [1] Mean thickness is inside diameter thickness (T) plus outside diameter thickness (t), divided by two. All dimensions in inches.

Table 34. Dimensions of External Tooth Lock Washers.

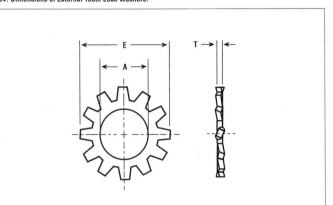

Nom. Washer Size	A Inside Dia.		E Outside Dia.		T Thickness
	Max.	Min.	Max.	Min.	
#4	0.123	0.116	0.255	0.245	0.015
#6	0.150	0.142	0.317	0.306	0.020
#8	0.176	0.168	0.381	0.370	0.020
#10	0.204	0.195	0.406	0.395	0.022
1/4	0.267	0.256	0.506	0.494	0.025
5/16	0.326	0.320	0.601	0.588	0.029
3/8	0.398	0.384	0.695	0.680	0.035
7/16	0.463	0.448	0.760	0.740	0.035
1/2	0.529	0.512	0.898	0.882	0.039
9/16	0.594	0.576	0.985	0.965	0.039
5/8	0.659	0.640	1.070	1.045	0.045
3/4	0.790	0.768	1.260	1.240	0.050
7/8	0.927	0.897	1.140	1.380	0.055
1	1.060	1.025	1.620	1.590	0.062
1 1/4	1.142	1.128	1.827	1.797	0.062

Source: MS35335.

Table 35. Dimensions of Internal Tooth Lock Washers.

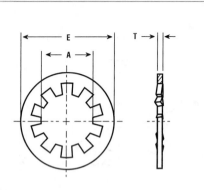

Nom. Washer Size	A Inside Dia.		E Outside Dia.		T Thickness
	Max.	Min.	Max.	Min.	
#2	0.095	0.089	0.185	0.175	0.013
#4	0.123	0.116	0.265	0.255	0.015
#6	0.150	0.142	0.288	0.278	0.017
#8	0.176	0.168	0.336	0.325	0.020
#10	0.204	0.195	0.381	0.370	0.022
1/4	0.267	0.256	0.478	0.466	0.025
5/16	0.332	0.320	0.607	0.594	0.029
3/8	0.398	0.384	0.692	0.678	0.035
7/16	0.463	0.448	0.789	0.774	0.035
7/16	0.480	0.472	0.607	0.593	0.017
1/2	0.529	0.512	0.883	0.867	0.039
9/16	0.572	0.565	0.692	0.678	0.017
5/8	0.659	0.640	1.071	1.053	0.045
3/4	0.795	0.769	1.245	1.220	0.050
3/4	0.785	0.775	1.077	1.047	0.022
7/8	0.918	0.894	1.386	1.364	0.055
1	1.043	1.019	1.637	1.613	0.062
1 1/4	1.280	1.260	1.950	1.921	0.062

Source: MS35333.

Table 36. Dimensions of Hexagon Head Cap Screws (Finished Hex Bolts).

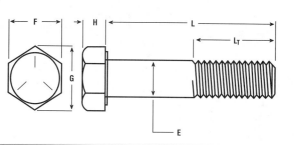

Screw/Bolt Size		E Body Dia.		F Width Across Flats			G Width Across Corners		Wrenching Height
Nom.	Basic	Max.	Min.	Basic	Max.	Min.	Max.	Min.	Min.
1/4	0.2500	0.2500	0.2450	7/16	0.438	0.428	0.505	0.488	0.106
5/16	0.3125	0.3125	0.3065	1/2	0.500	0.489	0.577	0.557	0.140
3/8	0.3750	0.3750	0.3690	9/16	0.562	0.551	0.650	0.628	0.160
7/16	0.4375	0.4375	0.4305	5/8	0.625	0.612	0.722	0.698	0.195
1/2	0.5000	0.5000	0.4930	3/4	0.750	0.736	0.866	0.840	0.215
9/16	0.5625	0.5625	0.5545	13/16	0.812	0.798	0.938	0.910	0.250
5/8	0.6250	0.6250	0.6170	15/16	0.938	0.922	1.083	1.051	0.269
3/4	0.7500	0.7500	0.7410	1-1/8	1.125	1.100	1.299	1.254	0.324
7/8	0.8750	0.8750	0.8660	1-5/16	1.312	1.285	1.516	1.465	0.378
1	1.0000	1.0000	0.9900	1-1/2	1.500	1.469	1.732	1.675	0.416
1-1/8	1.1250	1.1250	1.1140	1-11/16	1.688	1.631	1.949	1.859	0.461
1-1/4	1.2500	1.2500	1.2390	1-7/8	1.875	1.812	2.165	2.066	0.530
1-3/8	1.3750	1.3750	1.3630	2-1/16	2.062	1.994	2.382	2.273	0.569
1-1/2	1.5000	1.5000	1.4880	2-1/4	2.250	2.175	2.598	2.480	0.640
1-3/4	1.7500	1.7500	1.7380	2-5/8	2.625	2.538	3.031	2.893	0.748
2	2.0000	2.0000	1.9880	3	3.000	2.900	3.464	3.306	0.825
2-1/4	2.2500	2.2500	2.2380	3-3/8	3.375	3.262	3.897	3.719	0.933
2-1/2	2.5000	2.5000	2.4880	3-3/4	3.750	3.625	4.330	4.133	1.042
2-3/4	2.7500	2.7500	2.7380	4-1/8	4.125	3.988	4.763	4.546	1.151
3	3.0000	3.0000	2.9880	4-1/2	4.500	4.350	5.196	4.959	1.290

(Continued)

Table 36. *(Continued)* **Dimensions of Hexagon Head Cap Screws (Finished Hex Bolts).**

Screw/Bolt Size		H Height			L_T Thread Length for Screw Lengths		Transition Thread Lengths for Screw Lengths		Runout of Bearing Surface FIR
					6 in. and shorter	Over 6 in.	6 in. and shorter	Over 6 in.	
Nom.	Basic	Basic	Max.	Min.	Basic	Basic	Max.	Max.	Max.
1/4	0.2500	5/32	0.163	0.150	0.750	1.000	0.400	0.650	0.010
5/16	0.3125	13/64	0.211	0.195	0.875	1.125	0.417	0.667	0.011
3/8	0.3750	15/64	0.243	0.226	1.000	1.250	0.438	0.688	0.012
7/16	0.4375	9/32	0.291	0.272	1.125	1.375	0.464	0.714	0.013
1/2	0.5000	5/16	0.323	0.302	1.250	1.500	0.481	0.731	0.014
9/16	0.5625	23/64	0.371	0.348	1.375	1.625	0.750	0.750	0.015
5/8	0.6250	25/64	0.403	0.378	1.500	1.750	0.773	0.773	0.017
3/4	0.7500	15/32	0.483	0.455	1.750	2.000	0.800	0.800	0.020
7/8	0.8750	35/64	0.563	0.531	2.000	2.250	0.833	0.833	0.023
1	1.0000	39/64	0.627	0.591	2.250	2.500	0.875	0.875	0.026
1-1/8	1.1250	11/16	0.718	0.658	2.500	2.750	0.929	0.929	0.029
1-1/4	1.2500	25/32	0.813	0.749	2.750	3.000	0.929	0.929	0.033
1-3/8	1.3750	27/32	0.878	0.810	3.000	3.250	1.000	1.000	0.036
1-1/2	1.5000	15/16	0.974	0.902	3.250	3.500	1.000	1.000	0.039
1-3/4	1.7500	1-3/32	1.134	1.054	3.750	4.000	1.100	1.100	0.046
2	2.0000	1-7/32	1.263	1.175	4.250	4.500	1.167	1.167	0.052
2-1/4	2.2500	1-3/8	1.423	1.327	4.750	5.000	1.167	1.167	0.059
2-1/2	2.5000	1-17/32	1.583	1.479	5.250	5.500	1.250	1.250	0.065
2-3/4	2.7500	1-11/16	1.744	1.632	5.750	6.000	1.250	1.250	0.072
3	3.0000	1-7/8	1.935	1.815	6.250	6.500	1.250	1.250	0.079

All dimensions in inches. Source: Extracted from ANSI B18.2.1-1972, published by the American Society of Mechanical Engineers.

Table 37. Dimensions of Slotted Fillister Head Cap Screws.

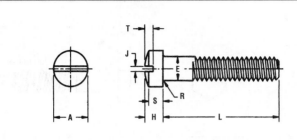

Screw/Bolt Size		E Body Dia.		A Head Dia.		S Head Side Height		H Total Head Height	
Nom.	Basic	Max.	Min.	Max.	Min.	Max.	Min.	Max.	Min.
1/4	0.2500	0.2500	0.2450	0.375	0.363	0.172	0.157	0.216	0.194
5/16	0.3125	0.3125	0.3070	0.437	0.424	0.203	0.186	0.253	0.230
3/8	0.3750	0.3750	0.3690	0.562	0.547	0.250	0.229	0.314	0.284
7/16	0.4375	0.4375	0.4310	0.625	0.608	0.297	0.274	0.368	0.336
1/2	0.5000	0.5000	0.4930	0.750	0.731	0.328	0.301	0.413	0.376
9/16	0.5625	0.5625	0.5550	0.812	0.792	0.375	0.346	0.467	0.427
5/8	0.6250	0.6250	0.6170	0.875	0.853	0.422	0.391	0.521	0.478
3/4	0.7500	0.7500	0.7420	1.000	0.976	0.500	0.466	0.612	0.566
7/8	0.8750	0.8750	0.8660	1.125	1.098	0.594	0.556	0.720	0.668
1	1.0000	1.0000	0.9900	1.312	1.282	0.656	0.612	0.803	0.743

Screw/Bolt Size		J Slot Width		T Slot Depth		R Fillet Radius	
Nom.	Basic	Max.	Min.	Max.	Min.	Max.	Min.
1/4	0.2500	0.075	0.064	0.097	0.077	0.031	0.016
5/16	0.3125	0.084	0.072	0.115	0.090	0.031	0.016
3/8	0.3750	0.094	0.081	0.142	0.112	0.031	0.016
7/16	0.4375	0.094	0.081	0.168	0.133	0.047	0.016
1/2	0.5000	0.106	0.091	0.193	0.153	0.047	0.016
9/16	0.5625	0.118	0.102	0.213	0.168	0.047	0.016
5/8	0.6250	0.133	0.116	0.239	0.189	0.062	0.031
3/4	0.7500	0.149	0.131	0.283	0.223	0.062	0.031
7/8	0.8750	0.167	0.147	0.334	0.264	0.062	0.031
1	1.0000	0.188	0.166	0.371	0.291	0.062	0.031

Note: A slight rounding on the edges at the periphery of the head is permissible if the diameter of the bearing circle is equal to no less than 90% of the specified minimum head diameter. All dimensions in inches. Source: Extracted from ANSI B16.2-1972, published by the American Society of Mechanical Engineers.

Table 38. Dimensions of Square Head and Hexagon Head Lag Screws.

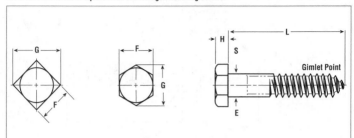

Screw Size		E Body or Shoulder Dia.		F Width Across Flats			S Shoulder Length	Radius of Fillet	
Nom.	Inch	Max.	Min.	Basic	Max.	Min.	Min.	Max.	Min.
No. 10	0.1900	0.199	0.178	9/32	0.281	0.271	0.094	0.003	0.001
1/4*	0.2500	0.260	0.237	3/8	0.375	0.362	0.094	0.003	0.001
1/4 †	0.2500	0.260	0.237	7/16	0.438	0.425	0.094	0.003	0.001
5/16	0.3125	0.324	0.298	1/2	0.500	0.484	0.125	0.003	0.001
3/8	0.3750	0.388	0.360	9/16	0.562	0.544	0.125	0.003	0.001
7/16	0.4375	0.452	0.421	5/8	0.625	0.603	0.156	0.003	0.001
1/2	0.5000	0.515	0.482	3/4	0.750	0.725	0.156	0.003	0.001
5/8	0.6250	0.642	0.605	15/16	0.938	0.906	0.312	0.006	0.002
3/4	0.7500	0.768	0.729	1 1/8	1.125	1.088	0.375	0.006	0.002
7/8	0.8750	0.895	0.852	1 5/16	1.312	1.269	0.375	0.006	0.002
1	1.0000	1.022	0.976	1 1/2	1.500	1.450	0.625	0.009	0.003
1 1/8	1.1250	1.149	1.098	1 11/16	1.688	1.631	0.625	0.009	0.003
1 1/4	1.2500	1.277	1.223	1 7/8	1.875	1.812	0.625	0.009	0.003

Screw Size		H Height of Square Head			G Width Across Square Corners		H Height of Hex Head			G Width Across Hex Corners	
Nom.	Inch	Basic	Max.	Min.	Max.	Min.	Basic	Max.	Min.	Max.	Min.
No. 10	0.1900	1/8	0.140	0.110	0.398	0.372	1/8	0.140	0.110	0.323	0.309
1/4	0.2500	11/64	0.188	0.156	0.530	0.498	11/64	0.188	0.150	0.505	0.484
5/16	0.3125	13/64	0.220	0.186	0.707	0.665	7/32	0.235	0.195	0.577	0.552
3/8	0.3750	1/4	0.268	0.232	0.795	0.747	1/4	0.268	0.226	0.650	0.620
7/16	0.4375	19/64	0.316	0.278	0.884	0.828	19/64	0.316	0.272	0.722	0.687
1/2	0.5000	11/32	0.348	0.308	1.061	0.995	11/32	0.364	0.302	0.866	0.826
5/8	0.6250	27/64	0.444	0.400	0.326	1.244	27/64	0.444	0.378	1.083	1.033
3/4	0.7500	1/2	0.524	0.476	1.591	1.494	1/2	0.524	0.455	1.299	1.240
7/8	0.8750	37/64	0.620	0.568	1.856	1.742	37/64	0.604	0.531	1.516	1.447
1	1.0000	43/64	0.684	0.628	2.121	1.991	43/64	0.700	0.591	1.732	1.653
1 1/8	1.1250	3/4	0.780	0.720	2.386	2.239	3/4	0.780	0.658	1.949	1.859
1 1/4	1.2500	27/32	0.876	0.812	2.652	2.489	27/32	0.876	0.749	2.165	2.066

* Square head lag screw. † Hex head lag screw. All dimensions in inches.
Source: Extracted from ANSI B18.2-1-1972, published by the American Society of Mechanical Engineers.

Table 39. Dimensions of Slotted Binding Head Machine Screws.

Screw Size	A Head Dia.		H Head Height		F Oval Height		J Slot Width		T Slot Width		U[1] Undercut Dia.		X[1] Undercut Depth	
	Max.	Min.	Max.	Min.	Max.	Min.	Max.	Min.	Max.	Min.	Max.	Min.	Max.	Min.
0000	0.046	0.040	0.014	0.009	0.006	0.003	0.008	0.004	0.009	0.005	-	-	-	-
000	0.073	0.067	0.021	0.015	0.008	0.005	0.012	0.006	0.013	0.009	-	-	-	-
00	0.098	0.090	0.028	0.023	0.011	0.007	0.017	0.010	0.018	0.012	-	-	-	-
0	0.126	0.119	0.032	0.026	0.012	0.008	0.023	0.016	0.018	0.009	0.098	0.086	0.007	0.002
1	0.153	0.145	0.041	0.035	0.015	0.011	0.026	0.019	0.024	0.014	0.120	0.105	0.008	0.003
2	0.181	0.171	0.050	0.043	0.018	0.013	0.031	0.023	0.030	0.020	0.141	0.124	0.010	0.005
3	0.208	0.197	0.059	0.052	0.022	0.016	0.035	0.027	0.036	0.025	0.162	0.143	0.011	0.006
4	0.235	0.223	0.068	0.061	0.025	0.018	0.039	0.031	0.042	0.030	0.184	0.161	0.012	0.007
5	0.263	0.249	0.078	0.069	0.029	0.021	0.043	0.035	0.048	0.035	0.205	0.180	0.014	0.009
6	0.290	0.275	0.087	0.078	0.032	0.024	0.048	0.039	0.053	0.040	0.226	0.199	0.015	0.010
8	0.344	0.326	0.105	0.095	0.039	0.029	0.054	0.045	0.065	0.050	0.269	0.236	0.017	0.012

(Continued)

Table 39. *(Continued)* Dimensions of Slotted Binding Head Machine Screws.

Screw Size	A Head Dia.		H Head Height		F Oval Height		J Slot Width		T Slot Width		U[1] Undercut Dia.		X[1] Undercut Depth	
	Max.	Min.	Max.	Min.	Max.	Min.	Max.	Min.	Max.	Min.	Max.	Min.	Max.	Min.
10	0.399	0.378	0.123	0.112	0.045	0.034	0.060	0.050	0.077	0.060	0.312	0.274	0.020	0.015
12	0.454	0.430	0.141	0.130	0.052	0.039	0.067	0.056	0.089	0.070	0.354	0.311	0.023	0.018
1/4	0.525	0.498	0.165	0.152	0.061	0.046	0.075	0.064	0.105	0.084	0.410	0.360	0.026	0.021
5/16	0.656	0.622	0.209	0.194	0.077	0.059	0.084	0.072	0.134	0.108	0.513	0.450	0.032	0.027
3/8	0.788	0.746	0.253	0.235	0.094	0.071	0.094	0.081	0.163	0.132	0.615	0.540	0.039	0.034

Note: [1] Unless otherwise specified, screws will not be undercut. All dimensions in inches. Source: Fastbolt Corp.

Table 40. Dimensions of Regular Hex Head and Hex Washer Head Machine Screws, Tapping Screws, and Self-Drilling Screws.

Screw Size	F Width Across Flats		G Width Across Corners	H Head Height		D Washer Dia.		U Washer Thickness		J Slot Width		T Slot Depth	
	Max.	Min.	Min.	Max.	Min.	Max.	Min.	Max.	Min.	Max.	Min.	Max.	Min.
4	0.187	0.181	0.202	0.060	0.049	0.243	0.225	0.019	0.011	0.039	0.031	0.042	0.025
6	0.250	0.244	0.272	0.093	0.080	0.328	0.302	0.025	0.015	0.048	0.039	0.053	0.033
8	0.250	0.244	0.272	0.110	0.096	0.348	0.322	0.031	0.019	0.054	0.045	0.074	0.052
10	0.312	0.305	0.340	0.120	0.105	0.414	0.384	0.031	0.019	0.060	0.050	0.080	0.057
12	0.312	0.305	0.340	0.155	0.139	0.432	0.398	0.039	0.022	0.067	0.056	0.103	0.077
14	0.375	0.367	0.409	0.190	0.172	0.520	0.480	0.050	0.030	0.075	0.064	0.111	0.083
1/4	0.375	0.367	0.409	0.190	0.172	0.520	0.480	0.050	0.030	0.075	0.064	0.111	0.083
5/16	0.500	0.489	0.545	0.230	0.208	0.676	0.624	0.055	0.035	0.084	0.072	0.134	0.100

All dimensions in inches. Source: Fastbolt Corp.

Table 41. Dimensions of Slotted and Phillips Fillister Head Machine Screws.

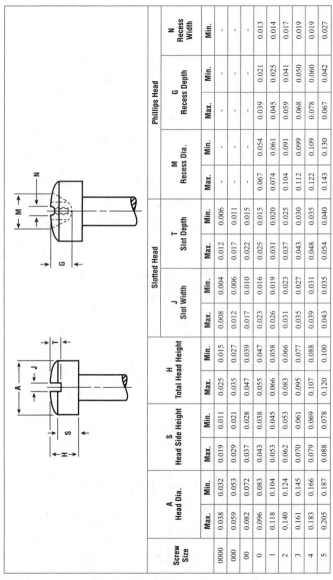

Screw Size	A Head Dia.		S Head Side Height		H Total Head Height		Slotted Head J Slot Width		T Slot Depth		Phillips Head M Recess Dia.		G Recess Depth		N Recess Width
	Max.	Min.	Max.	Min.	Max.	Min.	Max.	Min.	Max.	Min.	Max.	Min.	Max.	Min.	Min.
0000	0.038	0.032	0.019	0.011	0.025	0.015	0.008	0.004	0.012	0.006	-	-	-	-	-
000	0.059	0.053	0.029	0.021	0.035	0.027	0.012	0.006	0.017	0.011	-	-	-	-	-
00	0.082	0.072	0.037	0.028	0.047	0.039	0.017	0.010	0.022	0.015	-	-	-	-	-
0	0.096	0.083	0.043	0.038	0.055	0.047	0.023	0.016	0.025	0.015	0.067	0.054	0.039	0.021	0.013
1	0.118	0.104	0.053	0.045	0.066	0.058	0.026	0.019	0.031	0.020	0.074	0.061	0.045	0.025	0.014
2	0.140	0.124	0.062	0.053	0.083	0.066	0.031	0.023	0.037	0.025	0.104	0.091	0.059	0.041	0.017
3	0.161	0.145	0.070	0.061	0.095	0.077	0.035	0.027	0.043	0.030	0.112	0.099	0.068	0.050	0.019
4	0.183	0.166	0.079	0.069	0.107	0.088	0.039	0.031	0.048	0.035	0.122	0.109	0.078	0.060	0.019
5	0.205	0.187	0.088	0.078	0.120	0.100	0.043	0.035	0.054	0.040	0.143	0.130	0.067	0.042	0.027

(Continued)

Table 41. *(Continued)* Dimensions of Slotted and Phillips Fillister Head Machine Screws.

Screw Size	A Head Dia.		S Head Side Height		H Total Head Height		J Slot Width		T Slot Depth		M Recess Dia.		G Recess Depth		N Recess Width
	Max.	Min.	Max.	Min.	Max.	Min.	Max.	Min.	Max.	Min.	Max.	Min.	Max.	Min.	Min.
6	0.226	0.208	0.096	0.086	0.132	0.111	0.048	0.039	0.060	0.045	0.166	0.153	0.091	0.066	0.028
8	0.270	0.250	0.113	0.102	0.156	0.133	0.054	0.045	0.071	0.054	0.182	0.169	0.108	0.082	0.030
10	0.313	0.292	0.130	0.118	0.180	0.156	0.060	0.050	0.083	0.064	0.199	0.186	0.124	0.100	0.031
12	0.357	0.334	0.148	0.134	0.205	0.178	0.067	0.056	0.094	0.074	0.259	0.246	0.141	0.115	0.034
1/4	0.414	0.389	0.170	0.155	0.237	0.207	0.075	0.064	0.109	0.087	0.281	0.268	0.161	0.135	0.036
5/16	0.518	0.490	0.211	0.194	0.295	0.262	0.084	0.072	0.137	0.110	0.322	0.309	0.203	0.177	0.042
3/8	0.622	0.590	0.253	0.233	0.355	0.315	0.094	0.081	0.164	0.133	0.389	0.376	0.233	0.210	0.065
7/16	0.625	0.589	0.265	0.242	0.368	0.321	0.094	0.081	0.170	0.135	0.413	0.400	0.259	0.234	0.068
1/2	0.750	0.710	0.297	0.273	0.412	0.362	0.106	0.091	0.190	0.151	0.435	0.422	0.280	0.255	0.071
9/16	0.812	0.768	0.336	0.308	0.466	0.410	0.118	0.102	0.214	0.172	0.470	0.442	0.312	0.288	0.076
5/8	0.875	0.827	0.375	0.345	0.521	0.461	0.133	0.116	0.240	0.193	0.587	0.564	0.343	0.314	0.081
3/4	1.000	0.945	0.441	0.406	0.612	0.542	0.149	0.131	0.281	0.226	0.633	0.610	0.382	0.355	0.086

All dimensions in inches. Source: Fastbolt Corp.

Table 42. Dimensions of Slotted and Phillips Flat Countersunk Head Machine Screws, Tapping Screws, and Wood Screws.

| Screw Size | A Head Dia. | | H Head Height | Slotted Head | | | | Phillips Head | | | | N Recess Width |
| | Max.¹ | Min.² | Ref. | J Slot Width | | T Slot Depth | | M Recess Dia. | | Recess Depth | | |
				Max.	Min.	Max.	Min.	Max.	Min.	Max.	Min.	Min.
0000	0.043	0.037	0.011	0.008	0.004	0.007	0.003	-	-	-	-	-
000	0.064	0.058	0.016	0.011	0.007	0.009	0.005	-	-	-	-	-
00	0.092	0.076	0.028	0.017	0.010	0.014	0.009	-	-	-	-	-
0	0.119	0.099	0.035	0.023	0.016	0.015	0.010	0.069	0.056	0.043	0.027	0.014
1	0.146	0.123	0.043	0.026	0.019	0.019	0.012	0.077	0.064	0.051	0.035	0.015
2	0.172	0.147	0.051	0.031	0.023	0.023	0.015	0.102	0.089	0.063	0.047	0.017
3	0.199	0.171	0.059	0.035	0.027	0.027	0.017	0.107	0.094	0.068	0.052	0.018
4	0.225	0.195	0.067	0.039	0.031	0.030	0.020	0.128	0.115	0.089	0.073	0.018
5	0.252	0.220	0.075	0.043	0.035	0.034	0.022	0.154	0.141	0.086	0.063	0.027
6	0.279	0.244	0.083	0.048	0.039	0.038	0.024	0.174	0.161	0.106	0.083	0.029

(Continued)

Table 42. (Continued) Dimensions of Slotted and Phillips Flat Countersunk Head Machine Screws, Tapping Screws, and Wood Screws.

Screw Size	A Head Dia.		H Head Height	Slotted Head				Phillips Head				
				J Slot Width		T Slot Depth		M Recess Dia.		Recess Depth		N Recess Width
	Max.[1]	Min.[2]	Ref.	Max.	Min.	Max.	Min.	Max.	Min.	Max.	Min.	Min.
8	0.332	0.292	0.100	0.054	0.045	0.045	0.029	0.189	0.176	0.121	0.098	0.030
10	0.385	0.340	0.116	0.060	0.050	0.053	0.034	0.204	0.191	0.136	0.113	0.032
12	0.438	0.389	0.132	0.067	0.056	0.060	0.039	0.268	0.255	0.156	0.133	0.035
1/4	0.507	0.452	0.153	0.075	0.064	0.070	0.046	0.283	0.270	0.171	0.148	0.036
5/16	0.635	0.568	0.191	0.084	0.072	0.088	0.058	0.365	0.352	0.216	0.194	0.061
3/8	0.762	0.685	0.230	0.094	0.081	0.106	0.070	0.393	0.380	0.245	0.223	0.065
7/16	0.812	0.723	0.223	0.094	0.081	0.103	0.066	0.409	0.396	0.261	0.239	0.068
1/2	0.875	0.775	0.223	0.106	0.091	0.103	0.065	0.424	0.411	0.276	0.254	0.069
9/16	1.000	0.889	0.260	0.118	0.102	0.120	0.077	0.454	0.431	0.300	0.278	0.073
5/8	1.125	1.002	0.298	0.133	0.116	0.137	0.088	0.576	0.553	0.342	0.316	0.079
3/4	1.375	1.230	0.372	0.149	0.131	0.171	0.111	0.640	0.617	0.406	0.380	0.087

Notes: [1] Maximum head diameter is to top of sharp edge. [2] Minimum head diameter is to rounded edge or flat. All dimensions in inches. Source: Fastbolt Corp.

Table 43. Dimensions of Slotted and Phillips Undercut Flat Countersunk Head Machine Screws.

Screw Size	These Lengths or Shorter are Undercut	A Head Dia. Max.[1]	A Head Dia. Min.[2]	H Head Height Max.	H Head Height Min.	J Slot Width Max.	J Slot Width Min.	T Slot Depth Max.	T Slot Depth Min.	M Recess Dia. Max.	M Recess Dia. Min.	G Recess Depth Max.	G Recess Depth Min.	N Recess Width Min.
						Slotted Head				Phillips Head				
0	1/8	0.119	0.099	0.025	0.018	0.023	0.016	0.011	0.007	0.069	0.056	0.043	0.027	0.014
1	1/8	0.146	0.123	0.031	0.023	0.026	0.019	0.014	0.009	0.077	0.064	0.051	0.035	0.015
2	1/8	0.172	0.147	0.036	0.028	0.031	0.023	0.016	0.011	0.095	0.082	0.056	0.040	0.017
3	1/8	0.199	0.171	0.042	0.033	0.035	0.027	0.019	0.012	0.102	0.089	0.063	0.047	0.018
4	3/16	0.225	0.195	0.047	0.038	0.039	0.031	0.022	0.014	0.117	0.104	0.078	0.062	0.018
5	3/16	0.252	0.220	0.053	0.043	0.043	0.035	0.024	0.016	0.128	0.115	0.089	0.073	0.018
6	3/16	0.279	0.244	0.059	0.048	0.048	0.039	0.027	0.017	0.146	0.133	0.078	0.055	0.025
8	1/4	0.332	0.292	0.070	0.058	0.054	0.045	0.032	0.021	0.174	0.161	0.106	0.083	0.029
10	5/16	0.385	0.340	0.081	0.068	0.060	0.050	0.037	0.024	0.189	0.176	0.121	0.098	0.030

(Continued)

Table 43. *(Continued)* Dimensions of Slotted and Phillips Undercut Flat Countersunk Head Machine Screws.

Screw Size	These Lengths or Shorter are Undercut	A Head Dia.		Slotted Head						Phillips Head				
				H Head Height		J Slot Width		T Slot Depth		M Recess Dia.		G Recess Depth		N Recess Width
		Max.¹	Min.²	Max.	Min.	Max.	Min.	Max.	Min.	Max.	Min.	Max.	Min.	Min.
12	3/8	0.438	0.389	0.092	0.078	0.067	0.056	0.043	0.028	0.233	0.220	0.121	0.098	0.030
1/4	7/16	0.507	0.452	0.107	0.092	0.075	0.064	0.050	0.032	0.250	0.237	0.136	0.113	0.032
5/16	1/2	0.635	0.568	0.134	0.116	0.084	0.072	0.062	0.041	0.317	0.304	0.168	0.146	0.053
3/8	9/16	0.762	0.685	0.161	0.140	0.094	0.081	0.075	0.049	0.365	0.352	0.216	0.194	0.061
7/16	5/8	0.812	0.723	0.156	0.133	0.094	0.081	0.072	0.045	0.393	0.380	0.245	0.223	0.065
1/2	3/4	0.875	0.775	0.156	0.130	0.106	0.091	0.072	0.046	0.409	0.396	0.261	0.242	0.068

Notes: ¹ Maximum head diameter is to top of sharp edge. ² Minimum head diameter is to rounded edge or flat. All dimensions in inches. Source: Fastbolt Corp.

Table 44. Dimensions of Slotted and Phillips Undercut Oval Head Countersunk Machine Screws, Tapping Screws, and Wood Screws.

Screw Size	A Head Dia.		S Head Side Height	H Total Head Height		Slotted Head				Phillips Head				
						J Slot Width		T Slot Depth		M Recess Dia.		G Recess Depth		N Recess Width
	Max.[1]	Min.[2]	Ref.	Max.	Min.	Max.	Min.	Max.	Min.	Max.	Min.	Max.	Min.	Min.
00	0.093	0.083	0.028	0.042	0.034	0.017	0.010	0.023	0.016	-	-	-	-	-
0	0.119	0.099	0.035	0.056	0.041	0.023	0.016	0.030	0.025	0.074	0.061	0.045	0.027	0.014
1	0.146	0.123	0.043	0.068	0.052	0.026	0.019	0.038	0.031	0.077	0.064	0.048	0.030	0.015
2	0.172	0.147	0.051	0.080	0.063	0.031	0.023	0.045	0.037	0.112	0.099	0.069	0.052	0.018
3	0.199	0.171	0.059	0.092	0.073	0.035	0.027	0.052	0.043	0.124	0.111	0.081	0.064	0.019
4	0.225	0.195	0.067	0.104	0.084	0.039	0.031	0.059	0.049	0.136	0.123	0.094	0.077	0.019
5	0.252	0.220	0.075	0.116	0.095	0.043	0.035	0.067	0.055	0.158	0.145	0.085	0.061	0.028
6	0.279	0.244	0.083	0.128	0.105	0.048	0.039	0.074	0.060	0.178	0.165	0.105	0.080	0.030
8	0.332	0.292	0.100	0.152	0.126	0.054	0.045	0.088	0.072	0.192	0.179	0.119	0.095	0.031

(Continued)

Table 44. *(Continued)* Dimensions of Slotted and Phillips Undercut Oval Head Countersunk Machine Screws, Tapping Screws, and Wood Screws.

Screw Size	A Head Dia.		S Head Side Height	H Total Head Height		Slotted Head				Phillips Head				
						J Slot Width		T Slot Depth		M Recess Dia.		G Recess Depth		N Recess Width
	Max.[1]	Min.[2]	Ref.	Max.	Min.	Max.	Min.	Max.	Min.	Max.	Min.	Max.	Min.	Min.
10	0.385	0.340	0.116	0.176	0.148	0.060	0.050	0.103	0.084	0.209	0.196	0.137	0.113	0.033
12	0.438	0.389	0.132	0.200	0.169	0.067	0.056	0.117	0.096	0.270	0.257	0.152	0.128	0.038
1/4	0.507	0.452	0.153	0.232	0.197	0.075	0.064	0.136	0.112	0.290	0.277	0.173	0.148	0.040
5/16	0.635	0.568	0.191	0.290	0.249	0.084	0.072	0.171	0.141	0.381	0.368	0.226	0.202	0.064
3/8	0.762	0.685	0.230	0.347	0.300	0.094	0.081	0.206	0.170	0.400	0.387	0.245	0.221	0.066
7/16	0.812	0.723	0.223	0.345	0.295	0.094	0.081	0.210	0.174	0.410	0.397	0.257	0.233	0.068
1/2	0.875	0.775	0.223	0.354	0.299	0.106	0.091	0.216	0.176	0.422	0.409	0.269	0.245	0.070
9/16	1.000	0.889	0.260	0.410	0.350	0.118	0.102	0.250	0.207	-	-	-	-	-
5/8	1.125	1.002	0.298	0.467	0.399	0.133	0.116	0.285	0.235	-	-	-	-	-
3/4	1.375	1.230	0.372	0.578	0.497	0.149	0.131	0.353	0.293	-	-	-	-	-

Notes: [1] Maximum head diameter is to top of sharp edge. [2] Minimum head diameter is to rounded edge or flat. All dimensions in inches. Source: Fastbolt Corp.

Table 45. Dimensions of Slotted and Phillips Pan Head Machine Screws, Tapping Screws, and Self-drilling Screws.

| Screw Size | A Head Dia. | | Slotted Head | | | | | | Phillips Head | | | | | | N Recess Width |
| | Max. | Min. | H Head Height | | J Slot Width | | T Slot Depth | | H₁ Head Height | | M Recess Dia. | | G Recess Depth | | Min. |
			Max.	Min.	Max.	Min.	Max.	Min.	Max.	Min.	Max.	Min.	Max.	Min.	
0000	0.042	0.036	0.016	0.010	0.008	0.004	0.008	0.004	-	-	-	-	-	-	-
000	0.066	0.060	0.023	0.017	0.012	0.008	0.012	0.008	-	-	-	-	-	-	-
00	0.090	0.082	0.025	0.032	0.017	0.010	0.016	0.010	-	-	-	-	-	-	-
0	0.116	0.104	0.039	0.031	0.023	0.016	0.022	0.014	0.044	0.036	0.067	0.054	0.039	0.021	0.013
1	0.142	0.130	0.046	0.038	0.026	0.019	0.027	0.018	0.053	0.044	0.074	0.061	0.045	0.025	0.014
2	0.167	0.155	0.053	0.045	0.031	0.023	0.031	0.022	0.062	0.053	0.104	0.091	0.059	0.041	0.017
3	0.193	0.180	0.060	0.051	0.035	0.027	0.036	0.026	0.071	0.062	0.112	0.099	0.068	0.050	0.019
4	0.219	0.205	0.068	0.058	0.039	0.031	0.040	0.030	0.080	0.070	0.122	0.109	0.078	0.060	0.019
5	0.245	0.231	0.075	0.065	0.043	0.035	0.045	0.034	0.089	0.079	0.158	0.145	0.083	0.057	0.028

(Continued)

Table 45. *(Continued)* Dimensions of Slotted and Phillips Pan Head Machine Screws, Tapping Screws, and Self-drilling Screws.

Screw Size	A Head Dia.		Slotted Head						Phillips Head							
			H Head Height		J Slot Width		T Slot Depth		H₁ Head Height		M Recess Dia.		G Recess Depth		N Recess Width
	Max.	Min.	Max.	Min.	Max.	Min.	Max.	Min.	Max.	Min.	Max.	Min.	Max.	Min.	Min.
6	0.270	0.256	0.082	0.072	0.048	0.039	0.050	0.037	0.097	0.087	0.166	0.153	0.091	0.066	0.028
8	0.322	0.306	0.096	0.085	0.054	0.045	0.058	0.045	0.115	0.105	0.182	0.169	0.108	0.082	0.030
10	0.373	0.357	0.110	0.099	0.060	0.050	0.068	0.053	0.133	0.122	0.199	0.186	0.124	0.100	0.031
12	0.425	0.407	0.125	0.112	0.067	0.056	0.077	0.061	0.151	0.139	0.259	0.246	0.141	0.115	0.034
1/4	0.492	0.473	0.144	0.130	0.075	0.064	0.087	0.070	0.175	0.162	0.281	0.268	0.161	0.135	0.036
5/16	0.615	0.594	0.178	0.162	0.084	0.072	0.106	0.085	0.218	0.203	0.350	0.337	0.193	0.169	0.059
3/8	0.740	0.716	0.212	0.195	0.094	0.081	0.124	0.100	0.261	0.244	0.389	0.376	0.233	0.210	0.065
7/16	0.863	0.837	0.247	0.228	0.094	0.081	0.142	0.116	0.305	0.284	0.413	0.400	0.259	0.234	0.068
1/2	0.987	0.958	0.281	0.260	0.106	0.091	0.161	0.131	0.348	0.325	0.435	0.422	0.280	0.255	0.071
9/16	1.041	1.000	0.315	0.293	0.118	0.102	0.179	0.146	0.391	0.366	0.470	0.447	0.312	0.288	0.076
5/8	1.172	1.125	0.350	0.325	0.133	0.116	0.197	0.162	0.434	0.406	0.587	0.564	0.343	0.314	0.081
3/4	1.435	1.375	0.419	0.390	0.149	0.131	0.234	0.192	0.521	0.488	0.633	0.610	0.382	0.355	0.086

All dimensions in inches. Source: Fastbolt Corp.

Table 46. Dimensions of Slotted and Phillips Round Head Machine Screws and Wood Screws.

Screw Size	A Head Diameter		H Head Height		Slotted Head						Phillips Head					
	Max.	Min.	Max.	Min.	J Slot Width		T Slot Depth		M Recess Diameter		G Recess Depth		N Recess Width
					Max.	Min.	Max.	Min.	Max.	Min.	Max.	Min.	Min.
0000	0.041	0.035	0.022	0.016	0.008	0.004	0.017	0.013	-	-	-	-	-
000	0.062	0.056	0.031	0.025	0.012	0.008	0.018	0.012	-	-	-	-	-
00	0.089	0.080	0.045	0.036	0.017	0.010	0.026	0.018	-	-	-	-	-
0	0.113	0.099	0.053	0.043	0.023	0.016	0.039	0.029	0.073	0.060	0.042	0.022	0.014
1	0.138	0.122	0.061	0.051	0.026	0.019	0.044	0.033	0.082	0.069	0.052	0.033	0.015
2	0.162	0.146	0.069	0.059	0.031	0.023	0.048	0.037	0.100	0.087	0.053	0.034	0.017
3	0.187	0.169	0.078	0.067	0.035	0.027	0.053	0.040	0.109	0.096	0.062	0.042	0.018
4	0.211	0.193	0.086	0.075	0.039	0.031	0.058	0.044	0.118	0.105	0.072	0.053	0.019
5	0.236	0.217	0.095	0.083	0.043	0.035	0.063	0.047	0.154	0.141	0.074	0.046	0.027

(Continued)

Table 46. (Continued) Dimensions of Slotted and Phillips Round Head Machine Screws and Wood Screws.

Screw Size	A Head Diameter		H Head Height		Slotted Head				Phillips Head				
					J Slot Width		T Slot Depth		M Recess Diameter		G Recess Depth		N Recess Width
	Max.	Min.	Max.	Min.	Max.	Min.	Max.	Min.	Max.	Min.	Max.	Min.	Min.
6	0.260	0.240	0.103	0.091	0.048	0.039	0.068	0.051	0.162	0.149	0.084	0.156	0.027
8	0.309	0.287	0.120	0.107	0.054	0.045	0.077	0.058	0.178	0.165	0.101	0.075	0.030
10	0.359	0.334	0.137	0.123	0.060	0.050	0.087	0.065	0.195	0.182	0.119	0.093	0.031
12	0.408	0.382	0.153	0.139	0.067	0.056	0.096	0.073	0.249	0.236	0.125	0.099	0.032
1/4	0.472	0.443	0.175	0.160	0.075	0.064	0.109	0.082	0.268	0.255	0.147	0.121	0.034
5/16	0.590	0.557	0.216	0.198	0.084	0.072	0.132	0.099	0.308	0.295	0.187	0.161	0.040
3/8	0.708	0.670	0.256	0.237	0.094	0.081	0.155	0.117	0.387	0.374	0.228	0.202	0.064
7/16	0.750	0.707	0.328	0.307	0.094	0.081	0.196	0.148	0.402	0.389	0.241	0.216	0.066
1/2	0.813	0.766	0.355	0.332	0.106	0.091	0.211	0.159	0.416	0.403	0.256	0.231	0.068
9/16	0.938	0.887	0.410	0.385	0.118	0.102	0.242	0.183	0.459	0.436	0.292	0.265	0.075
5/8	1.000	0.944	0.438	0.411	0.133	0.116	0.258	0.195	0.554	0.531	0.318	0.277	0.077
3/4	1.250	1.185	0.547	0.516	0.149	0.131	0.320	0.242	0.654	0.631	0.418	0.379	0.088

All dimensions in inches. Source: Fastbolt Corp.

Table 47. Dimensions of Slotted and Phillips Truss Head Machine Screws and Tapping Screws.

Screw Size	A Head Diameter		H Head Height		Slotted Head J Slot Width		T Slot Depth		Phillips Head M Recess Diameter		G Recess Depth		N Recess Width
	Max.	Min.	Max.	Min.	Max.	Min.	Max.	Min.	Max.	Min.	Max.	Min.	Min.
0000	0.049	0.043	0.014	0.010	0.009	0.005	0.009	0.005	-	-	-	-	-
000	0.077	0.071	0.022	0.018	0.013	0.009	0.013	0.009	-	-	-	-	-
00	0.106	0.098	0.030	0.024	0.017	0.010	0.018	0.012	-	-	-	-	-
0	0.131	0.119	0.037	0.029	0.023	0.016	0.022	0.014	0.063	0.050	0.037	0.019	0.013
1	0.164	0.149	0.045	0.037	0.026	0.019	0.027	0.018	0.071	0.058	0.045	0.027	0.014
2	0.194	0.180	0.053	0.044	0.031	0.023	0.031	0.022	0.104	0.091	0.059	0.041	0.018
3	0.226	0.211	0.061	0.051	0.035	0.027	0.036	0.026	0.110	0.097	0.066	0.049	0.018
4	0.257	0.241	0.069	0.059	0.039	0.031	0.040	0.030	0.112	0.099	0.069	0.051	0.018
5	0.289	0.272	0.078	0.066	0.043	0.035	0.045	0.034	0.128	0.115	0.085	0.067	0.019

(Continued)

Table 47. *(Continued)* Dimensions of Slotted and Phillips Truss Head Machine Screws and Tapping Screws.

Screw Size	A Head Diameter		H Head Height		Slotted Head J Slot Width		T Slot Depth		Phillips Head M Recess Diameter		G Recess Depth		N Recess Width
	Max.	Min.	Max.	Min.	Max.	Min.	Max.	Min.	Max.	Min.	Max.	Min.	Min.
6	0.321	0.303	0.086	0.074	0.048	0.039	0.050	0.037	0.158	0.145	0.084	0.059	0.027
8	0.384	0.364	0.102	0.088	0.054	0.045	0.058	0.045	0.173	0.160	0.099	0.074	0.029
10	0.448	0.425	0.118	0.103	0.060	0.050	0.068	0.053	0.188	0.175	0.115	0.090	0.030
12	0.511	0.487	0.134	0.118	0.067	0.056	0.077	0.061	0.248	0.235	0.128	0.103	0.032
1/4	0.573	0.546	0.150	0.133	0.075	0.064	0.087	0.070	0.263	0.250	0.143	0.118	0.033
5/16	0.698	0.666	0.183	0.162	0.084	0.072	0.106	0.085	0.352	0.339	0.193	0.168	0.059
3/8	0.823	0.787	0.215	0.191	0.094	0.081	0.124	0.100	0.383	0.370	0.226	0.202	0.063
7/16	0.948	0.907	0.248	0.221	0.094	0.081	0.142	0.116	0.414	0.401	0.257	0.232	0.068
1/2	1.073	1.028	0.280	0.250	0.106	0.091	0.161	0.131	0.444	0.431	0.288	0.263	0.072
9/16	1.198	1.149	0.312	0.279	0.118	0.102	0.179	0.146	0.451	0.428	0.302	0.278	0.074
5/8	1.323	1.269	0.345	0.309	0.133	0.116	0.196	0.162	0.559	0.536	0.322	0.289	0.077
3/4	1.573	1.511	0.410	0.368	0.149	0.131	0.234	0.182	0.620	0.597	0.384	0.352	0.085

All dimensions in inches. Source: Fastbolt Corp.

Table 48. Dimensions of Binding Head Instrument (Miniature) Screws. [1]

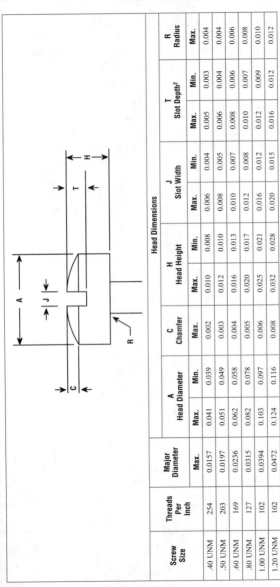

Screw Size	Threads Per Inch	Major Diameter Max.	Head Diameter Max.	Head Diameter Min.	C Chamfer Max.	Head Height Max.	Head Height Min.	Slot Width Max.	Slot Width Min.	Slot Depth[2] Max.	Slot Depth[2] Min.	R Radius Max.
			A	A		H	H	J	J	T	T	
.40 UNM	254	0.0157	0.041	0.039	0.002	0.010	0.008	0.006	0.004	0.005	0.003	0.004
.50 UNM	203	0.0197	0.051	0.049	0.003	0.012	0.010	0.008	0.005	0.006	0.004	0.004
.60 UNM	169	0.0236	0.062	0.058	0.004	0.016	0.013	0.010	0.007	0.008	0.006	0.006
.80 UNM	127	0.0315	0.082	0.078	0.005	0.020	0.017	0.012	0.008	0.010	0.007	0.008
1.00 UNM	102	0.0394	0.103	0.097	0.006	0.025	0.021	0.016	0.012	0.012	0.009	0.010
1.20 UNM	102	0.0472	0.124	0.116	0.008	0.032	0.028	0.020	0.015	0.016	0.012	0.012

Notes: [1] For screw lengths four times the maximum major diameter or less, thread length (Lt) shall extend to within two threads of the head bearing surface. Screws of greater length shall have complete threads for a minimum of four maximum major diameters. [2] Slot depth measured from bearing surface. All dimensions except screw size are in inches. Source: MS21271.

Table 49. Dimensions of Fillister Head Instrument (Miniature) Screws. [1]

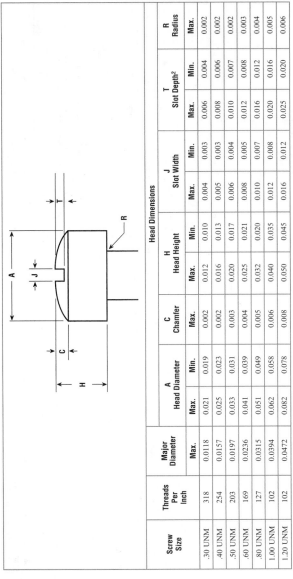

Screw Size	Threads Per Inch	Major Diameter	Head Dimensions													
		Max.	A Head Diameter		C Chamfer	H Head Height		J Slot Width		T Slot Depth[2]		R Radius
			Max.	Min.	Max.	Max.	Min.	Max.	Min.	Max.	Min.	Max.
.30 UNM	318	0.0118	0.021	0.019	0.002	0.012	0.010	0.004	0.003	0.006	0.004	0.002
.40 UNM	254	0.0157	0.025	0.023	0.002	0.016	0.013	0.005	0.003	0.008	0.006	0.002
.50 UNM	203	0.0197	0.033	0.031	0.003	0.020	0.017	0.006	0.004	0.010	0.007	0.002
.60 UNM	169	0.0236	0.041	0.039	0.004	0.025	0.021	0.008	0.005	0.012	0.008	0.003
.80 UNM	127	0.0315	0.051	0.049	0.005	0.032	0.020	0.010	0.007	0.016	0.012	0.004
1.00 UNM	102	0.0394	0.062	0.058	0.006	0.040	0.035	0.012	0.008	0.020	0.016	0.005
1.20 UNM	102	0.0472	0.082	0.078	0.008	0.050	0.045	0.016	0.012	0.025	0.020	0.006

Notes: [1] For screw lengths four times the maximum major diameter or less, thread length (L1) shall extend to within two threads of the head bearing surface. Screws of greater length shall have complete threads for a minimum of four maximum major diameters. [2] Slot depth measured from bearing surface. All dimensions except screw size are in inches. Source: MS21268.

Table 50. Dimensions of 100° Flat Head Instrument (Miniature) Screws. [1]

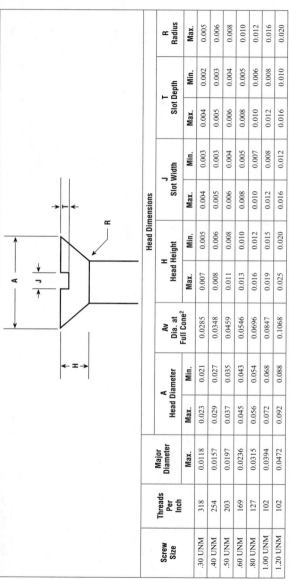

Screw Size	Threads Per Inch	Major Diameter Max.	A Head Diameter Max.	A Head Diameter Min.	Av Dia. at Full Cone[2]	Head Dimensions H Head Height Max.	H Head Height Min.	J Slot Width Max.	J Slot Width Min.	T Slot Depth Max.	T Slot Depth Min.	R Radius Max.
.30 UNM	318	0.0118	0.023	0.021	0.0285	0.007	0.005	0.004	0.003	0.004	0.002	0.005
.40 UNM	254	0.0157	0.029	0.027	0.0348	0.008	0.006	0.005	0.003	0.005	0.003	0.006
.50 UNM	203	0.0197	0.037	0.035	0.0459	0.011	0.008	0.006	0.004	0.006	0.004	0.008
.60 UNM	169	0.0236	0.045	0.043	0.0546	0.013	0.010	0.008	0.005	0.008	0.005	0.010
.80 UNM	127	0.0315	0.056	0.054	0.0696	0.016	0.012	0.010	0.007	0.010	0.006	0.012
1.00 UNM	102	0.0394	0.072	0.068	0.0847	0.019	0.015	0.012	0.008	0.012	0.008	0.016
1.20 UNM	102	0.0472	0.092	0.088	0.1068	0.025	0.020	0.016	0.012	0.016	0.010	0.020

Notes: [1] For screw lengths four times the maximum major diameter or less, thread length (Lt) shall extend to within two threads of the head bearing surface. Screws of greater length shall have complete threads for a minimum of four maximum major diameters. [2] "Av" derived from D (Max), H (Max), and mean angle. All dimensions except screw size are in inches. Source: MS21270.

Table 51. Dimensions of Square Head Set Screws.

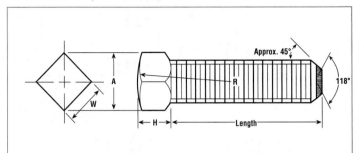

Screw Size		TPI Class 2A	W Width Across Flats		A Width Across Corners		H Height of Head		R Head Radius
Nom.	Basic	2A	Max.	Min.	Max.	Min.	Max.	Min.	Nom.
No. 10	0.1900	24	0.188	0.180	0.265	0.247	0.148	0.134	31/64
1/4	0.2500	20	0.250	0.241	0.354	0.331	0.196	0.178	5/8
5/16	0.3120	18	0.312	0.302	0.442	0.415	0.245	0.224	25/32
3/8	0.3750	16	0.375	0.362	0.530	0.497	0.293	0.270	15/16
7/16	0.4370	14	0.437	0.423	0.619	0.581	0.341	0.315	1-3/32
1/2	0.5000	13	0.500	0.484	0.707	0.665	0.389	0.361	1-1/4
9/16	0.5620	12	0.562	0.545	0.795	0.748	0.437	0.407	1-13/32
5/8	0.6520	11	0.625	0.606	0.884	0.833	0.485	0.452	1-9/16
3/4	0.7500	10	0.750	0.729	1.060	1.001	0.582	0.544	1-7/8
7/8	0.8750	9	0.875	0.852	1.237	1.170	0.678	0.635	2-3/16
1	1.000	8	1.000	0.974	1.414	1.337	0.774	0.726	2-1/2
1-1/8	1.125	7	1.125	1.096	1.591	1.505	0.870	0.817	2-13/16
1-1/4	1.250	7	1.250	1.219	1.768	1.674	0.966	0.908	3-1/8
1-3/8	1.375	6	1.375	1.342	1.945	1.843	1.063	1.000	3-7/16
1-1/2	1.500	6	1.500	1.464	2.121	2.010	1.159	1.091	3-3/4

Length Tolerances for Square Head Set Screws			
Size	Up to 1" incl.	Over 1" to 2" incl.	2" and over
No. 10 thru 5/8"	- 0.03	- 0.06	- 0.09
3/4" and over	- 0.06	- 0.12	- 0.18

All dimensions in inches. Source: Unbrako/SPS Technologies, Inc.

Table 52. Point Dimensions for Square Head Set Screws.

Plain Cup   Flat   Dog and Half Dog   Cone   Oval

Screw Size		C Point Dia. Cup and Flat		P Point Dia. Dog and Half Dog		Q Point Length Dog		Q Point Length Half Dog		R Radius Oval Point[1]	Y Cone Point Angle[2]
Nom.	Basic	Max.	Min.	Max.	Min.	Max.	Min.	Max.	Min.		
No. 10	0.1900	0.102	0.088	0.127	0.120	0.095	0.085	0.050	0.040	0.142	1/4
1/4	0.2500	0.132	0.118	0.156	0.149	0.130	0.120	0.068	0.058	0.188	5/16
5/16	0.3125	0.172	0.156	0.203	0.195	0.161	0.151	0.083	0.073	0.234	3/8
3/8	0.3750	0.212	0.194	0.250	0.241	0.193	0.183	0.099	0.089	0.281	7/16
7/16	0.4375	0.252	0.232	0.297	0.287	0.224	0.214	0.114	0.104	0.328	1/2
1/2	0.5000	0.291	0.270	0.344	0.334	0.255	0.245	0.130	0.120	0.375	9/16
9/16	0.5625	0.332	0.309	0.391	0.379	0.287	0.275	0.146	0.134	0.422	5/8

(Continued)

Table 52. *(Continued)* **Point Dimensions for Square Head Set Screws.**

Screw Size		C Point Dia. Cup and Flat		P Point Dia. Dog and Half Dog		Q Point Length Dog		Q Point Length Half Dog		R Radius Oval Point[1]	Y Cone Point Angle[2]
Nom.	Basic	Max.	Min.	Max.	Min.	Max.	Min.	Max.	Min.		
5/8	0.6250	0.371	0.347	0.469	0.456	0.321	0.305	0.164	0.148	0.469	3/4
3/4	0.7500	0.450	0.425	0.562	0.549	0.383	0.367	0.196	0.180	0.562	7/8
7/8	0.8750	0.530	0.502	0.656	0.642	0.446	0.430	0.227	0.211	0.656	1
1	1.0000	0.609	0.579	0.750	0.734	0.510	0.490	0.260	0.240	0.750	1 1/8
1 1/8	1.1250	0.689	0.655	0.844	0.826	0.572	0.552	0.291	0.271	0.844	1 1/4
1 1/4	1.2500	0.767	0.733	0.938	0.920	0.635	0.615	0.323	0.303	0.938	1 1/2
1 3/8	1.3750	0.848	0.808	1.031	1.011	0.698	0.678	0.354	0.334	1.031	1 5/8
1 1/2	1.5000	0.926	0.886	1.125	1.105	0.760	0.740	0.385	0.365	1.125	1 3/4

Notes: Point angle "X" for flat and cup points is 45° +5°/-0° for screws of nominal lengths equal to or longer than those listed in column Y, and 30° minimum for screws of shorter nominal lengths. The extent of rounding or flat at the apex of the cone point is not to exceed an amount equivalent to 10% of the basic screw diameter. [1] Radius of oval point value in column R is +0.031 / -0.000. [2] Cone point angle is 90° ± 2° for the nominal lengths shown in column Y, and for longer lengths. Cone point angle is 118° ± 2° for lengths shorter than those shown in column Y. All dimensions in inches. Source: Extracted from ANSI B18.6.2-1972, published by the American Society of Mechanical Engineers.

Table 53. Dimensions of Type A, Regular Thumb Screws.

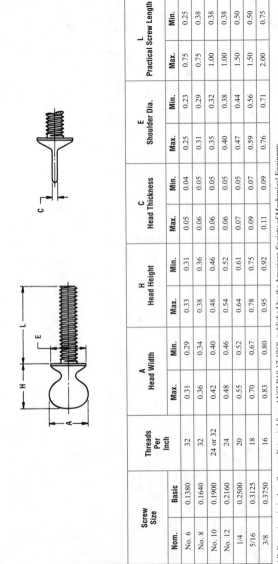

Screw Size		Threads Per Inch	A Head Width		H Head Height		C Head Thickness		E Shoulder Dia.		L Practical Screw Length	
Nom.	Basic		Max.	Min.	Max.	Min.	Max.	Min.	Max.	Min.	Max.	Min.
No. 6	0.1380	32	0.31	0.29	0.33	0.31	0.05	0.04	0.25	0.23	0.75	0.25
No. 8	0.1640	32	0.36	0.34	0.38	0.36	0.06	0.05	0.31	0.29	0.75	0.38
No. 10	0.1900	24 or 32	0.42	0.40	0.48	0.46	0.06	0.05	0.35	0.32	1.00	0.38
No. 12	0.2160	24	0.48	0.46	0.54	0.52	0.06	0.05	0.40	0.38	1.00	0.38
1/4	0.2500	20	0.55	0.52	0.64	0.61	0.07	0.05	0.47	0.44	1.50	0.50
5/16	0.3125	18	0.70	0.67	0.78	0.75	0.09	0.07	0.59	0.56	1.50	0.50
3/8	0.3750	16	0.83	0.80	0.95	0.92	0.11	0.09	0.76	0.71	2.00	0.75

All dimensions in inches. Source: Extracted from ANSI B18.17-1968, published by the American Society of Mechanical Engineers.

Table 54. Dimensions of Type A, Heavy Thumb Screws.

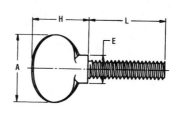

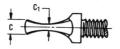

Screw Size		Threads per Inch	A Head Width		H Head Height		C Head Thickness	
Nom.	Basic		Max.	Min.	Max.	Min.	Max.	Min.
No. 10	0.1900	24	0.89	0.83	0.84	0.72	0.18	0.16
1/4	0.2500	20	1.05	0.99	0.94	0.81	0.24	0.22
5/16	0.3125	18	1.21	1.15	1.00	0.88	0.27	0.25
3/8	0.3750	16	1.41	1.34	1.16	1.03	0.30	0.28
7/16	0.4375	14	1.59	1.53	1.22	1.09	0.36	0.34
1/2	0.5000	13	1.81	1.72	1.28	1.16	0.40	0.38

Screw Size		Threads per Inch	C_1 Head Thickness		E Shoulder Dia.		L Practical Screw Length	
Nom.	Basic		Max.	Min.	Max.	Min.	Max.	Min.
No. 10	0.1900	24	0.10	0.08	0.33	0.31	2.00	0.50
1/4	0.2500	20	0.10	0.08	0.40	0.38	3.00	0.50
5/16	0.3125	18	0.11	0.09	0.46	0.44	4.00	0.50
3/8	0.3750	16	0.11	0.09	0.55	0.53	4.00	0.50
7/16	0.4375	14	0.13	0.11	0.71	0.69	2.50	1.00
1/2	0.5000	13	0.14	0.12	0.83	0.81	3.00	1.00

All dimensions in inches. Source: Extracted from ANSI B18.17-1968, published by the American Society of Mechanical Engineers.

Table 55. Dimensions of Type A and Type AB Self-Tapping Screw Threads. (See Notes.)

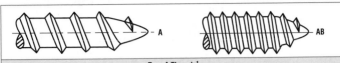

Type A Threads[1]							
Screw Size		Threads per Inch	Major Dia.		Minor Dia.		Length[2]	
Nom.	Basic		Max.	Min.	Max.	Min.	90° Head	Csk Head[3]
#0	0.0600	40	0.060	0.057	0.042	0.039	1/8	3/16
#1	0.0730	32	0.075	0.072	0.051	0.048	1/8	3/16
#2	0.0860	32	0.088	0.084	0.061	0.056	5/32	3/16
#3	0.0990	28	0.101	0.097	0.076	0.071	3/16	7/32
#4	0.1120	24	0.114	0.110	0.083	0.078	3/16	1/4
#5	1.1250	20	0.130	0.126	0.095	0.090	3/16	1/4
#6	0.1380	18	0.141	0.136	0.102	0.096	1/4	5/16
#7	0.1510	16	0.158	0.152	0.114	0.108	5/16	3/8
#8	0.1640	15	0.168	0.162	0.123	0.116	3/8	7/16
#10	0.1900	12	0.194	0.188	0.133	0.126	3/8	1/2
#12	0.2160	11	0.221	0.215	0.162	0.155	7/16	9/16
#14	0.2420	10	0.254	0.248	0.185	0.178	1/2	5/8
#16	0.2680	10	0.280	0.274	0.197	0.189	9/16	3/4
#18	0.2940	9	0.306	0.300	0.217	0.209	5/8	13/16
#20	0.3200	9	0.333	0.327	0.234	0.226	11/16	13/16
#24	0.3720	9	0.390	0.383	0.291	0.282	3/4	1
Type AB Threads							
Screw Size		Threads per Inch	Major Dia.		Minor Dia.		Minimum Length	
Nom.	Basic		Max.	Min.	Max.	Min.	90° Head	Csk Head[3]
#0	0.0600	48	0.060	0.057	0.036	0.033	3/32	7/64
#1	0.0730	42	0.075	0.072	0.049	0.046	1/8	9/64
#2	0.0860	32	0.088	0.084	0.064	0.060	9/64	11/64
#3	0.0990	28	0.101	0.097	0.075	0.071	11/64	3/16
#4	0.1120	24	0.114	0.110	0.086	0.082	3/16	7/32
#5	0.1250	20	0.130	0.126	0.094	0.090	3/16	1/4
#6	0.1380	20	0.139	0.135	0.104	0.099	7/32	17/64
#7	0.1510	19	0.154	0.149	0.115	0.109	17/64	5/16
#8	0.1640	18	0.166	0.161	0.122	0.116	9/32	21/64
#10	0.1900	16	0.189	0.183	0.141	0.135	21/64	3/8
#12	0.2160	14	0.215	0.209	0.164	0.157	3/8	13/32
1/4	0.2500	14	0.246	0.240	0.192	0.185	13/32	15/32
	0.3125	14	0.315	0.308	0.244	0.236	17/32	19/32

Notes: [1] Although still widely available, the A thread is generally considered obsolete and is no longer recommended. [2] Type A threads in these lengths or shorter should not be used. Substitute AB screws. [3] Csk = Countersunk Head. All dimensions in inches.

Table 56. Dimensions of Type B, Type BF, Type BP, and Type BT/Type 25 Self-Tapping Screw Threads.

Type B

Type 25

Screw Size		Threads per Inch	D Major Dia.		d Minor Dia.		P Point Dia.	
Nom.	Basic		Max.	Min.	Max.	Min.	Max.	Min.
#0	0.0600	48	0.060	0.057	0.036	0.033	0.031	0.027
#1	0.0730	42	0.075	0.072	0.049	0.046	0.044	0.040
#2	0.0860	32	0.088	0.084	0.064	0.060	0.058	0.054
#3	0.0990	28	0.101	0.097	0.075	0.071	0.068	0.063
#4	0.1120	24	0.114	0.110	0.086	0.082	0.079	0.074
#5	0.1250	20	0.130	0.126	0.094	0.090	0.087	0.082
#6	0.1380	20	0.139	0.135	0.104	0.099	0.095	0.089
#7	0.1510	19	0.154	0.149	0.115	0.109	0.105	0.099
#8	0.1640	18	0.166	0.161	0.122	0.116	0.112	0.106
#10	0.1900	16	0.189	0.183	0.141	0.135	0.130	0.123
#12	0.2160	14	0.215	0.209	0.164	0.157	0.152	0.145
1/4	0.2500	14	0.246	0.240	0.192	0.185	0.179	0.171
5/16	0.3125	12	0.315	0.308	0.244	0.236	0.230	0.222

(Continued)

Table 56. (Continued) Dimensions of Type B, Type BF, Type BP, and Type BT/Type 25 Self-Tapping Screw Threads.

Screw Size Nom.	Basic	Threads per Inch	D Major Dia. Max.	D Major Dia. Min.	d Minor Dia. Max.	d Minor Dia. Min.	P Point Dia. Max.	P Point Dia. Min.
3/8	0.3750	12	0.380	0.371	0.309	0.299	0.293	0.285
7/16	0.4375	10	0.440	0.431	0.359	0.349	0.343	0.335
1/2	0.5000	10	0.504	0.495	0.423	0.413	0.407	0.399

Nom. Size	S Point Taper Length				L Type B, Type BF, Type BT				L Type BP	
	Short Screws		Long Screws		Determine Length for Point Taper[1]		Minimum Practical Screw Length		Determine Min. Practical Length[2]	
	Max.	Min.	Max.	Min.	90° Head	Csk Head[3]	90° Head	Csk Head[3]	90° Head	Csk Head[3]
#0	0.042	0.031	0.052	0.042	5/64	1/8	5/64	3/32	5/32	13/64
#1	0.048	0.036	0.060	0.048	5/64	5/32	5/64	1/8	11/64	1/4
#2	0.062	0.047	0.078	0.062	7/64	3/16	7/64	5/32	13/64	9/32
#3	0.071	0.054	0.089	0.071	9/64	7/32	9/64	3/16	1/4	21/64
#4	0.083	0.063	0.104	0.083	3/16	1/4	9/64	3/16	5/16	3/8
#5	0.100	0.075	0.125	0.100	3/16	1/4	5/32	3/16	21/64	25/64
#6	0.100	0.075	0.125	0.100	1/4	5/16	11/64	1/4	25/64	15/32
#7	0.105	0.079	0.132	0.105	5/16	3/8	3/16	1/4	15/32	17/32
#8	0.111	0.083	0.139	0.111	5/16	7/16	3/16	1/4	31/64	39/64
#10	0.125	0.094	0.156	0.125	3/8	1/2	15/64	5/16	9/16	11/16
#12	0.143	0.107	0.179	0.143	7/16	9/16	9/32	3/8	21/32	25/32
1/4	0.143	0.107	0.179	0.143	1/2	5/8	9/32	3/8	3/4	7/8
5/16	0.167	0.125	0.208	0.167	1/2	5/8	5/16	7/16	53/64	61/64
3/8	0.167	0.125	0.208	0.137	1/2	5/8	5/16	7/16	29/32	1 1/32

(Continued)

Table 56. *(Continued)* Dimensions of Type B, Type BF, Type BP, and Type BT/Type 25 Self-Tapping Screw Threads.

| Nom. Size | S Point Taper Length | | | | L Type B, Type BF, Type BT | | | | | | L Type BP | | | |
| | Short Screws | | Long Screws | | Determine Length for Point Taper[1] | | Minimum Practical Screw Length | | | | Determine Min. Practical Length[2] | | | |
	Max.	Min.	Max.	Min.	90° Head	Csk Head[3]	90° Head	Csk Head[3]			90° Head	Csk Head[3]		
7/16	0.200	0.150	0.250	0.200	5/8	3/4	15/32	5/8			1 7/64	1 15/64		
1/2	0.200	0.150	0.250	0.200	5/8	3/4	15/32	5/8			1 3/16	1 5/16		

Notes: [1] Screws of the stated nominal length or shorter have point taper length specified for short screws. Longer lengths use the point taper length for long screws.
[2] Type BP screws of the stated nominal length or shorter have point taper length specified for short screws. Longer lengths use the point taper length for long screws.
[3] Csk = Countersunk Head. All dimensions in inches.

Table 57. Dimensions of Type D/Type 1, Type F, and Type T/Type 23 Thread Cutting Tapping Screws.

Type 1   Type F   Type 23

Screw Size		TPI[1]	D Major Dia.		P Point Dia.		S Point Taper Length		Determine Length for Point Taper[2]		Min. Practical Screw Length	
Nom.	Basic		Max.	Min.	Max.	Min.	Short Screws	Long Screws	90° Head	Csk Head[3]	90° Head[3]	Csk Head[3]
#2	0.0860	56	0.0860	0.0820	0.067	0.061	0.062	0.080	9/64	3/16	1/8	5/32
#2	0.0860	64	0.0860	0.0822	0.070	0.064	0.055	0.070	1/8	11/64	1/8	5/32
#3	0.0990	48	0.0990	0.0946	0.077	0.070	0.052	0.094	11/64	7/32	1/8	3/16
#3	0.0990	56	0.0990	0.0950	0.080	0.074	0.045	0.080	9/64	3/16	1/8	3/16
#4	0.1120	40	0.1120	0.1072	0.086	0.077	0.062	0.112	13/64	1/4	1/8	3/16
#4	0.1120	48	0.1120	0.1076	0.090	0.083	0.052	0.094	11/64	7/32	1/8	3/16
#5	0.1250	40	0.1250	0.1202	0.099	0.090	0.062	0.112	13/64	9/32	1/8	3/16
#5	0.1250	44	0.1250	0.1204	0.101	0.093	0.057	0.102	3/16	1/4	1/8	3/16
#6	0.1380	32	0.1380	0.1326	0.106	0.095	0.078	0.141	1/4	5/16	3/16	1/4
#6	0.1380	40	0.1380	0.1332	0.112	0.103	0.062	0.112	13/64	17/64	3/16	1/4

(Continued)

Table 57. *(Continued)* Dimensions of Type D/Type 1, Type F, and Type T/Type 23 Thread Cutting Tapping Screws.

Screw Size		TPI¹	D Major Dia.		P Point Dia.		S Point Taper Length				L			
							Short Screws		Long Screws		Determine Length for Point Taper²		Min. Practical Screw Length	
Nom.	Basic		Max.	Min.	Max.	Min.					90° Head	Csk Head³	90° Head	Csk Head³
#8	0.1640	32	0.1640	0.1586	0.132	0.121	0.109	0.078	0.109	0.141	1/4	21/64	3/16	1/4
#8	0.1640	36	0.1640	0.1590	0.135	0.125	0.097	0.069	0.097	0.125	15/64	19/64	3/16	1/4
#10	0.1900	24	0.1900	0.1834	0.147	0.133	0.146	0.104	0.146	0.188	11/32	27/64	15/64	5/16
#10	0.1900	32	0.1900	0.1846	0.158	0.147	0.109	0.178	0.109	0.141	1/4	11/32	15/64	5/16
#12	0.2160	24	0.2160	0.2094	0.173	0.159	0.146	0.104	0.146	0.188	11/32	7/16	17/64	3/8
#12	0.2160	28	0.2160	0.2098	0.179	0.167	0.125	0.089	0.125	0.161	19/64	25/64	17/64	3/8
1/4	0.2500	20	0.2500	0.2428	0.198	0.181	0.175	0.125	0.175	0.225	13/32	33/64	17/64	3/8
1/4	0.2500	28	0.2500	0.2438	0.213	0.201	0.125	0.089	0.125	0.161	19/64	13/32	17/64	3/8
5/16	0.3125	18	0.3125	0.3043	0.255	0.236	0.194	0.139	0.194	0.250	29/64	19/32	9/32	7/16
5/16	0.3125	24	0.3125	0.3059	0.269	0.255	0.146	0.104	0.146	0.188	11/32	15/32	9/32	7/16
3/8	0.3750	16	0.3750	0.3660	0.310	0.289	0.219	0.156	0.219	0.281	1/2	47/64	9/32	7/16
3/8	0.3750	24	0.3750	0.3684	0.332	0.318	0.146	0.104	0.146	0.188	11/32	1/2	9/32	7/16

Notes: ¹ TPI = threads per inch. ² Screws of the stated nominal length or shorter have point taper length specified for short screws. Longer lengths use the point taper length for long screws. ³ Csk = Countersunk Head. All dimensions in inches.

Table 58. Dimensions of Type U Round Head Drive Screws.

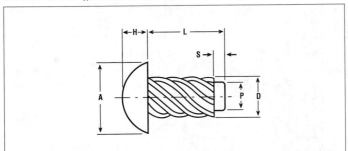

Nom. Screw Size	No of Thread Starts	D Body Dia.		A Head Dia.		H Head Height		P Pilot Dia.		Hole Size	
		Max.	Min.	Max.	Min.	Max.	Min.	Max.	Min.	Drill Size	Hole Dia.
00	6	0.060	0.057	0.099	0.090	0.034	0.026	0.049	0.046	#55	0.052
0	6	0.075	0.072	0.127	0.118	0.049	0.041	0.063	0.060	#51	0.067
2	8	0.100	0.097	0.162	0.146	0.069	0.059	0.083	0.080	#41	0.086
4	7	0.116	0.112	0.211	0.193	0.086	0.075	0.096	0.092	#37	0.104
6	7	0.140	0.136	0.260	0.240	0.103	0.091	0.116	0.112	#31	0.120
7	8	0.154	0.150	0.285	0.264	0.111	0.099	0.126	0.122	#29	0.136
8	8	0.167	0.162	0.309	0.287	0.120	0.107	0.136	0.132	#27	0.144
10	8	0.182	0.177	0.359	0.334	0.137	0.123	0.150	0.146	#20	0.161
12	8	0.212	0.206	0.408	0.382	0.153	0.139	0.177	0.173	#11	0.191
14	9	0.236	0.242	0.457	0.429	0.170	0.155	0.202	0.198	#2	0.221

Nom. Screw Size (L)	1/8	3/16	1/4	5/16	3/8	1/2	5/8	3/4	≥ 1
Pilot Length (S)	0.047	0.047	0.047	0.047	0.062	0.062	0.078	0.078	0.125

All dimensions in millimeters. Source: Fastbolt Corp.

Table 59. Hole Sizes for Types AB, B, and BP Tapping and Thread Forming Screws.

Screw Size	Sheet Metal Thickness	Steel, Stainless Steel, Monel, Brass Sheet Metal			Aluminum Alloy Steel Metal		
		Pierced or Extruded Hole	Drilled or Clean Punched Hole		Pierced or Extruded Hole	Drilled or Clean Punched Hole	
		Hole Dia.	Hole Dia.	Drill Size	Hole Dia.	Hole Dia.	Drill Size
2	.015		.064	52			
	.018		.064	52			
	.024		.067	51		.064	52
	.030		.070	50		.064	52
	.036		.073	49		.064	52
	.048		.073	49		.067	51
	.060		.076	48		.070	50
4	.015	.086	.086	44			
	.018	.086	.086	44			
	.024	.098	.089	43	.086		
	.030	.098	.094	42	.086	.086	44
	.036	.098	.094	42	.086	.086	44
	.048		.096	41	.086	.086	44
	.060		.100	39		.089	43
	.075		.102	38		.089	43
	.105					.094	42
6	.015	.111	.104	37			
	.018	.111	.104	37			
	.024	.111	.106	36	.111		
	.030	.111	.106	36	.111	.104	37
	.036	.111	.110	35	.111	.104	37
	.048	.111	.111	34	.111	.104	37
	.060		.116	32		.106	36
	.075		.120	31		.110	35
	.105		.128	30		.111	34
	.128 to .250					.120	31
8	.018	.136					
	.024	.136					
	.030	.136	.125	1/8	.136	.116	32
	.036	.136	.125	1/8	.136	.120	31
	.048	.136	.125	1/8	.136	.128	30
	.060		.128	30	.136	.136	29
	.075		.136	29		.140	28
	.105		.140	28		.147	26
	.125		.150	25		.147	26
	.135		.150	25		.149	25
	.162 to .375		.152	24		.152	24
10	.018	.157					
	.024	.157					
	.030	.157	.144	27	.157		
	.036	.157	.144	27	.157	.144	27
	.048	.157	.147	26	.157	.144	27
	.060		.152	24	.157	.144	27
	.075		.152	24		.147	26
	.105		.157	22		.147	26
	.125		.161	20		.154	23
	.135		.170	18		.154	23
	.164		.170	18		.159	21
	.200 to .375		.173	17		.166	19

(Continued)

Table 59. *(Continued)* Hole Sizes for Types AB, B, and BP Tapping and Thread Forming Screws.

Screw Size	Sheet Metal Thickness	Steel, Stainless Steel, Monel, Brass Sheet Metal			Aluminum Alloy Steel Metal		
		Pierced or Extruded Hole	Drilled or Clean Punched Hole		Pierced or Extruded Hole	Drilled or Clean Punched Hole	
		Hole Dia.	Hole Dia.	Drill Size	Hole Dia.	Hole Dia.	Drill Size
	.030	.209	.194	10			
	.036	.209	.194	10			
	.048	.209	.194	10			
	.060		.199	8		.199	8
	.075		.204	6		.201	7
1/4	.105		.209	4		.204	6
	.125		.228	1		.209	4
	.135		.228	1		.209	4
	.164		.234	15/64		.213	3
	.187		.234	15/64		.213	3
	.194		.234	15/64		.221	2
	.200 to .375					.228	1

Notes: Hole sizes for metal thicknesses 0.105 inch and above apply to Types B and BP only. All dimensions in inches. Source: MS24632.

Table 60. Approximate Hole Sizes for Types AB, B, and BP Tapping and Thread Forming Screws.

Screw Size	Cast Metals - Aluminum, Magnesium, Zinc, Brass, Bronze[1]			Resin Impregnated Plywood				
	Hole Dia.	Drill Size	Min. Penetration in Blind Hole	Hole Dia.	Drill Size	Min. Material Thickness	Penetration in Blind Hole	
							Min.	Max.
#2	0.078	#47	0.125	0.073	#49	0.125	0.188	0.500
#4	0.104	#37	0.188	0.100	#39	0.188	0.250	0.625
#6	0.128	#30	0.250	0.125	1/8	0.188	0.250	0.625
#8	0.152	#24	0.250	0.144	#27	0.188	0.250	0.750
#10	0.177	#16	0.250	0.173	#17	0.250	0.312	1.000
1/4	0.234	15/64	0.312	0.228	#1	0.312	0.375	1.000

Plastics[1]						
Screw Size	Phenol Formaldehyde			Cellulose Acetate and Nitrate, Acrylic and Styrene Resin		
	Hole Dia.	Drill Size	Min. Penetration in Blind Hole	Hole Dia.	Drill Size	Min. Penetration in Blind Hole
#2	0.078	#47	0.188	0.078	#47	0.188
#4	0.100	#39	0.250	0.094	#42	0.250
#6	0.128	#30	0.250	0.120	#31	0.250
#8	0.150	#25	0.312	0.144	#27	0.312
#10	0.177	#16	0.312	0.170	#18	0.312
1/4	0.214	15/64	0.375	0.221	#2	0.375

Notes: [1] These Tables refer to Type B and Type BP only. All dimensions in inches. Source: MS24632.

Table 61. Approximate Hole Sizes for Type BF and Type BT/Type 25 Self-Tapping Screws.

Screw Size	Die Cast Zinc and Aluminum			Screw Size	Die Cast Zinc and Aluminum		
	Material Thickness	Hole Dia.	Drill Size		Material Thickness	Hole Dia.	Drill Size
#2	0.060	0.073	#49	#8	0.125	0.149	#25
	0.083	0.073	#49		0.140	0.149	#25
	0.109	0.076	#48		0.188	0.149	#25
	0.125	0.076	#48		0.250	0.152	#24
	0.140	0.076	#48		0.312	0.152	#24
#4	0.109	0.098	#40	#10	0.125	0.166	#19
	0.125	0.100	#39		0.140	0.166	#19
	0.140	0.100	#39		0.188	0.166	#19
	0.188	0.100	#39		0.250	0.170	#18
	0.250	0.102	#38		0.312	0.172	11/64
					0.375	0.172	11/64
#6	0.125	0.120	#31	1/4	0.125	0.221	#2
	0.140	0.120	#31		0.140	0.221	#2
	0.188	0.120	#31		0.188	0.221	#2
	0.250	0.125	1/8		0.250	0.228	#1
	0.312	0.125	1/8		0.312	0.228	#1
					0.375	0.228	#1

Plastics[1]							
Screw Size	Phenol Formaldehyde				Cellulose Acetate and Nitrate, Acrylic and Styrene Resin		
	Hole Dia.	Drill Size	Penetration		Hole Dia.	Drill Size	Penetration
			Min.	Max.			Min.	Max.
#2	0.078	5/64	0.094	0.250	0.076	48	0.094	0.250
#4	0.104	#37	0.125	0.312	0.100	#39	0.125	0.312
#6	0.125	1/8	0.188	0.375	0.120	#31	0.188	0.375
#8	0.147	#26	0.250	0.500	0.144	#27	0.250	0.500
#10	0.170	#18	0.312	0.625	0.166	#19	0.312	0.625
1/4	0.228	#1	0.375	0.750	0.221	#2	0.375	0.750

All dimensions in inches. Source: MS24633.

Table 62. Approximate Hole Sizes for Type D/Type 1, Type F, and Type T/Type 23 "Machine Screw" Thread Cutting Tapping Screws.

Hardest Material Tapped	Screw Size No.	Thickness of Hardest Material Tapped									
		0.015 to 0.018 incl.	0.018 to 0.040 incl.	0.041 to 0.056 incl.	0.057 to 0.071 incl.	0.072 to 0.089 incl.	0.090 to 0.110 incl.	0.103 to 0.115 incl.	0.116 to 0.131 incl.	0.132 to 0.148 incl.	Over 0.148
Steel — Carbon Mild Quarter Hard 55,000 PSI Min. Tensile Strength	4-40	0.086	0.086	0.086	0.093	0.093	-	0.096	0.096	0.096	-
	6-32	0.104	0.104	0.104	0.113	0.113	-	0.113	0.120	0.120	-
	8-32	0.128	0.128	0.128	0.140	0.140	-	0.140	0.144	0.144	-
	10-24	0.149	0.149	0.149	0.154	0.154	-	0.154	0.161	0.161	-
	1/4-20	0.199	0.199	0.199	0.209	0.209	-	0.209	0.213	0.213	-
Chrome-Molybdenum Normalized [1] 90,000 PSI Min. Tensile Strength	4-40	0.189	0.189	0.189	0.089	0.096	0.096	-	0.098	-	-
	6-32	0.110	0.110	0.110	0.110	0.116	0.120	-	0.120	-	-
	8-32	0.136	0.136	0.136	0.136	0.140	0.144	-	0.144	-	-
	10-24	0.157	0.157	0.157	0.157	0.159	0.161	-	0.161	-	-
	1/4-20	0.209	0.209	0.209	0.209	0.213	0.218	-	0.218	0.218	-
Corrosion Resistant Half Hard [1] 150,000 PSI Min. Tensile Strength	4-40	0.093	0.093	0.096	-	-	-	-	-	-	-
	6-32	0.110	0.110	0.110	0.116	-	-	-	-	-	-
	8-32	0.136	0.136	0.140	0.140	0.144	-	-	-	-	-
	10-24	-	-	0.159	0.161	0.161	0.166	-	-	-	-
	1/4-20	-	-	0.205	0.205	0.209	0.213	0.218	0.221	-	-

Notes: [1] Lubricate with heavy cutting oil. All dimensions in inches. Source: AND10326.

(Continued)

Table 62. *(Continued)* Approximate Hole Sizes for Type D/Type 1, Type F, and Type T/Type 23 "Machine Screw" Thread Cutting Tapping Screws.

Hardest Material Tapped	Screw Size No.	Thickness of Hardest Material Tapped									
		0.015 to 0.018 incl.	0.018 to 0.040 incl.	0.041 to 0.056 incl.	0.057 to 0.071 incl.	0.072 to 0.089 incl.	0.090 to 0.110 incl.	0.103 to 0.115 incl.	0.116 to 0.131 incl.	0.132 to 0.148 incl.	Over 0.148
Al-Mg-Cr Half Hard 34,000 PSI Min. Tensile Strength	4-40	0.086	0.086	0.086	0.086	0.089	-	-	0.093	-	-
	6-32	0.104	0.104	0.104	0.104	0.106	-	-	0.106	-	-
	8-32	0.128	0.128	0.128	0.128	0.136	-	-	0.140	-	-
	10-24	0.149	0.149	0.149	0.149	0.152	-	-	0.154	-	-
	1/4-20	0.193	0.193	0.199	0.199	0.204	-	-	0.204	-	-
Al-Cu-Mg-Mn Heat Treated 52,000 PSI Min. Tensile Strength	4-40	0.086	0.086	0.086	0.086	0.089	0.089	-	0.093	-	-
	6-32	0.104	0.104	0.104	0.104	0.106	0.106	-	0.106	-	-
	8-32	0.128	0.128	0.128	0.128	0.136	0.140	-	0.140	-	-
	10-24	0.149	0.149	0.149	0.149	0.152	0.152	-	0.154	-	-
	1/4-20	0.199	0.199	0.199	0.199	0.204	0.204	-	0.204	-	-
Al-Cu-Mg (1.5% Mn) Heat Treated 62,000 PSI Min. Tensile Strength	4-40	0.086	0.086	0.086	0.089	0.089	-	-	0.	-	-
	6-32	0.104	0.104	0.104	0.109	0.109	-	-	0.	-	-
	8-32	0.129	0.129	0.0129	0.136	0.136	-	-	0.	-	-
	10-24	0.149	0.149	0.149	0.152	0.152	-	-	0.	-	-
	1/4-20	0.199	0.199	0.199	0.203	0.203	-	-	0.	-	-
Castings Aluminum Alloy or Aluminum	4-40	-	-	-	0.096	0.096	0.096	0.096	0.098	0.098	0.098
	6-32	-	-	-	0.116	0.116	0.116	0.116	0.120	0.120	0.120
	8-32	-	-	-	0.144	0.144	0.144	0.144	0.147	0.147	0.147
	10-24	-	-	-	0.161	0.161	0.161	0.161	0.166	0.166	0.166
	1/4-20	-	-	-	0.221	0.221	0.221	0.221	0.228	0.228	0.228

Aluminum Alloy

Notes: [1] Lubricate with heavy cutting oil. All dimensions in inches. Source: AND10326.

Table 63. Dimensions of Socket Head Cap Screws, Hexagon Recess, 1960 Series.

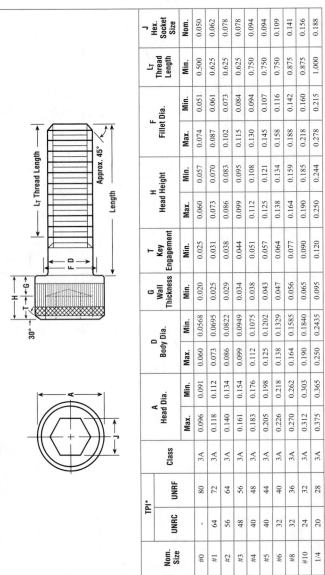

Nom. Size	TPI* UNRC	TPI* UNRF	Class	A Head Dia. Max	A Head Dia. Min	D Body Dia. Max	D Body Dia. Min	G Wall Thickness Min	T Key Engagement Min	H Head Height Max	H Head Height Min	F Fillet Dia. Max	F Fillet Dia. Min	L_T Thread Length Min	J Hex. Socket Size Nom.
#0	-	80	3A	0.096	0.091	0.060	0.0568	0.020	0.025	0.060	0.057	0.074	0.051	0.500	0.050
#1	64	72	3A	0.118	0.112	0.073	0.0695	0.025	0.031	0.073	0.070	0.087	0.061	0.625	0.062
#2	56	64	3A	0.140	0.134	0.086	0.0822	0.029	0.038	0.086	0.083	0.102	0.073	0.625	0.078
#3	48	56	3A	0.161	0.154	0.099	0.0949	0.034	0.044	0.099	0.095	0.115	0.084	0.625	0.078
#4	40	48	3A	0.183	0.176	0.112	0.1075	0.038	0.051	0.112	0.108	0.130	0.094	0.750	0.094
#5	40	44	3A	0.205	0.198	0.125	0.1202	0.043	0.057	0.125	0.121	0.145	0.107	0.750	0.094
#6	32	40	3A	0.226	0.218	0.138	0.1329	0.047	0.064	0.138	0.134	0.158	0.116	0.750	0.109
#8	32	36	3A	0.270	0.262	0.164	0.1585	0.056	0.077	0.164	0.159	0.188	0.142	0.875	0.141
#10	24	32	3A	0.312	0.303	0.190	0.1840	0.065	0.090	0.190	0.185	0.218	0.160	0.875	0.156
1/4	20	28	3A	0.375	0.365	0.250	0.2435	0.095	0.120	0.250	0.244	0.278	0.215	1.000	(0.188)

(Continued)

Table 63. *(Continued)* Dimensions of Socket Head Cap Screws, Hexagon Recess, 1960 Series.

| Nom. Size | TPI* | | Class | A Head Dia. | | D Body Dia. | | G Wall Thickness | T Key Engagement | H Head Height | | F Fillet Dia. | | Lt Thread Length | J Hex. Socket Size |
	UNRC	UNRF		Max.	Min.	Max.	Min.	Min.	Min.	Max.	Min.	Max.	Min.	Min.	Nom.
5/16	18	24	3A	0.469	0.457	0.3125	0.3053	0.119	0.151	0.312	0.306	0.347	0.273	1.125	0.250
3/8	16	24	3A	0.562	0.550	0.375	0.3678	0.143	0.182	0.375	0.368	0.415	0.331	1.250	0.312
7/16	14	20	3A	0.656	0.642	0.4375	0.4294	0.166	0.213	0.437	0.430	0.484	0.388	1.375	0.375
1/2	13	20	3A	0.750	0.735	0.500	0.4919	0.190	0.245	0.500	0.492	0.552	0.446	1.500	0.375
9/16	12	18	3A	0.843	0.827	0.5625	0.5538	0.214	0.265	0.562	0.554	0.6185	0.525	1.625	0.438
5/8	11	18	3A	0.938	0.921	0.625	0.6163	0.238	0.307	0.625	0.616	0.689	0.562	1.750	0.500
3/4	10	16	3A	1.125	1.107	0.750	0.7406	0.285	0.370	0.750	0.740	0.828	0.681	2.000	0.625
7/8	9	14	3A	1.312	1.293	0.875	0.8647	0.333	0.432	0.875	0.864	0.963	0.798	2.250	0.750
1	8	12	3A	1.500	1.479	1.000	0.9886	0.380	0.495	1.000	0.988	1.100	0.914	2.500	0.750
1 **	-	14 *	3A	1.500	1.479	1.000	0.9886	0.380	0.495	1.000	0.988	1.100	0.914	2.500	0.750
1 1/8	7	12	2A	1.688	1.665	1.125	1.1086	0.428	0.557	1.125	1.111	1.235	1.023	2.812	0.875
1 1/4	7	12	2A	1.875	1.852	1.250	1.2336	0.475	0.620	1.250	1.236	1.370	1.148	3.125	0.875
1 3/8	6	12	2A	2.062	2.038	1.375	1.3568	0.523	0.682	1.375	1.360	1.505	1.256	3.437	1.000
1 1/2	6	12	2A	2.250	2.224	1.500	1.4818	0.570	0.745	1.500	1.485	1.640	1.381	3.750	1.000
1 3/4	5	12	2A	2.625	2.597	1.750	1.7295	0.665	0.870	1.750	1.734	1.910	1.609	4.375	1.250
2	4 1/2	12	2A	3.000	2.970	2.000	1.9780	0.760	0.995	2.000	1.983	2.180	1.843	5.000	1.500
2 1/4	4 1/2	12	2A	3.375	3.344	2.250	2.2280	0.855	1.120	2.250	2.232	2.450	2.093	5.625	1.750
2 1/2	4	12	2A	3.750	3.717	2.500	2.4762	0.950	1.245	2.500	2.481	2.720	2.324	6.250	1.750
2 3/4	4	12	2A	4.125	4.090	2.750	2.7262	1.045	1.370	2.750	2.730	2.990	2.574	6.875	2.000
3	4	12	2A	4.500	4.464	3.000	2.9762	1.140	1.495	3.000	2.979	3.260	2.824	7.500	2.250

Notes: * Threads per inch. ** 1 inch size also available in 1-14 UNRS (special) thread form.

(Continued)

Table 63. (Continued) Dimensions of Socket Head Cap Screws, Hexagon Recess, 1960 Series.

Size	Length tolerances for head cap screws, hexagon recess, 1960 series				
	Up to 1" incl.	Over 1" to 2 1/2" incl.	Over 2 1/2" to 6" incl.	Over 6"	
#0 thru 3/8"	- 0.03	- 0.04	- 0.06	- 0.12	
7/16" thru 3/4"	- 0.03	- 0.06	- 0.08	- 0.12	
7/8" thru 1 1/2"	- 0.05	- 0.10	- 0.14	- 0.20	
Over 1 1/2"	-	- 0.18	- 0.20	- 0.24	

All dimensions in inches. Source: Unbrako/SPS Technologies, Inc.

Table 64. Grip (L_G) and Body (L_B) Lengths for 1960 Series Socket Head Cap Screws. (See Notes.)

Length	#1		#2		#3		#4		#5		#6	
	L_G	L_B	L_G	L_B	L_G	L_B	L_G	L_B	L_G	L_B	L_G	L_B
7/8	.250	.172	.250	.161	.250	.146						
1	.250	.172	.250	.161	.250	.146						
1 1/4	.625	.547	.625	.536	.625	.521	.250	.125	.250	.125		
1 1/2	.875	.797	.875	.786	.875	.771	.250	.125	.250	.125	.500	.344
1 3/4			1.125	1.036	1.125	1.021	.750	.625	.750	.625	.500	.344
2					1.125	1.021	.750	.625	.750	.625	1.000	.844
2 1/4						1.271	1.250	1.125	1.250	1.125	1.000	.844
2 1/2									1.250	1.125	1.500	1.344
2 3/4									1.750	1.625	1.500	1.344
											2.000	1.844

(Continued)

Table 64. *(Continued)* Grip (L_G) and Body (L_B) Lengths for 1960 Series Socket Head Cap Screws. (See Notes.)

Length	#8 L_G	#8 L_B	#10 L_G	#10 L_B	1/4 L_G	1/4 L_B	5/16 L_G	5/16 L_B	3/8 L_G	3/8 L_B	7/16 L_G	7/16 L_B
1 1/4	.375	.219	.375	.167	.500	.250						
1 1/2	.375	.219	.375	.167	.500	.250						
1 3/4	.875	.719	.875	.667	1.000	.750	.625	.347	.500	.187	.625	.268
2	.875	.719	.875	.667	1.000	.750	.625	.347	.500	.187	.625	.268
2 1/4	1.375	1.219	1.375	1.167	1.500	1.250	1.125	.847	1.000	.687	1.125	.768
2 1/2	1.375	1.219	1.375	1.167	1.500	1.250	1.125	.847	1.000	.687	1.125	.768
2 3/4	1.875	1.719	1.875	1.667	2.000	1.750	1.625	1.347	1.500	1.187	1.625	1.268
3	1.875	1.719	1.875	1.667	2.000	1.750	1.625	1.347	1.500	1.187	1.625	1.268
3 1/4	2.375	2.219	2.375	2.167	2.500	2.250	2.125	1.847	2.000	1.687	2.125	1.768
3 1/2			2.375	2.167	2.500	2.250	2.125	1.847	2.000	1.687	2.125	1.768
3 3/4			2.875	2.667	3.000	2.750	2.625	2.347	2.500	2.187	2.625	2.268
4					3.000	2.750	2.625	2.347	2.500	2.187	2.625	2.268
4 1/4					3.500	3.250	3.125	2.847	3.000	2.687	3.125	2.768
4 1/2					3.500	3.250	3.125	2.847	3.000	2.687	3.125	2.768
4 3/4					4.000	3.750	3.625	3.347	3.500	3.187	3.625	3.268
5					4.000	3.750	3.625	3.347	3.500	3.187	3.625	3.268
5 1/4							4.125	3.847	4.000	3.687	4.125	3.768
5 1/2							4.125	3.847	4.000	3.687	4.125	3.768
5 3/4							4.625	4.347	4.500	4.187	4.625	4.268
6							4.625	4.347	4.500	4.187	4.625	4.268
6 1/4							5.125	4.847	5.000	4.687	5.125	4.768
6 1/2									5.000	4.687	5.125	4.768
6 3/4									5.500	5.187	5.625	5.268

(Continued)

Table 64. *(Continued)* Grip (L_G) and Body (L_B) Lengths for 1960 Series Socket Head Cap Screws. (See Notes.)

Length	#8 L_G	#8 L_B	#10 L_G	#10 L_B	1/4 L_G	1/4 L_B	5/16 L_G	5/16 L_B	3/8 L_G	3/8 L_B	7/16 L_G	7/16 L_B
7									5.500	5.187	5.625	5.268
7 1/4									6.000	5.687	5.625	5.268
7 1/2											6.125	5.768
7 3/4											6.125	5.768
8											6.625	6.268
8 1/2											7.125	6.768
9											7.625	7.268

Length	1/2 L_G	1/2 L_B	9/16 L_G	9/16 L_B	5/8 L_G	5/8 L_B	3/4 L_G	3/4 L_B	7/8 L_G	7/8 L_B	1 L_G	1 L_B
2 1/4	.750	.365										
2 1/2	.750	.365	.875	.458	.750	.295						
2 3/4	.750	.365	.875	.458	.750	.295						
3	1.500	1.115	.875	.458	.750	.295	1.000	.500				
3 1/4	1.500	1.115	1.625	1.208	1.500	1.045	1.000	.500	1.000	.444		
3 1/2	1.500	1.115	1.625	1.208	1.500	1.045	1.000	.500	1.000	.444	1.000	.375
3 3/4	2.250	1.865	1.625	1.208	1.500	1.045	1.000	.500	1.000	.444	1.000	.375
4	2.250	1.865	2.375	1.958	2.250	1.795	2.000	1.500	1.000	.444	1.000	.375
4 1/4	2.250	1.865	2.375	1.958	2.250	1.795	2.000	1.500	2.000	1.444	1.000	.375
4 1/2	3.000	2.615	2.375	1.958	2.250	1.795	2.000	1.500	2.000	1.444	2.000	1.375
4 3/4	3.000	2.615	3.125	2.708	3.000	2.545	2.000	1.500	2.000	1.444	2.000	1.375
5	3.000	2.615	3.125	2.708	3.000	2.545	3.000	2.500	2.000	1.444	2.000	1.375
5 1/4	3.750	3.365	3.125	2.708	3.000	2.545	3.000	2.500	3.000	2.444	2.000	1.375
5 1/2	3.750	3.365	3.875	3.458	3.750	3.295	3.000	2.500	3.000	2.444	3.000	2.375

(Continued)

Table 64. *(Continued)* Grip (L_G) and Body (L_B) Lengths for 1960 Series Socket Head Cap Screws. (See Notes.)

Length	1/2		9/16		5/8		3/4		7/8		1	
	L_G	L_B	L_G	L_B	L_G	L_B	L_G	L_B	L_G	L_B	L_G	L_B
5 3/4	3.750	3.365	3.875	3.458	3.750	3.295	3.000	2.500	3.000	2.444	3.000	2.375
6	4.500	4.115	3.875	3.458	3.750	3.295	4.000	3.500	3.000	2.444	3.000	2.375
6 1/4	4.500	4.115	4.625	4.208	4.500	4.045	4.000	3.500	4.000	3.444	3.000	2.375
6 1/2	4.500	4.115	4.625	4.208	4.500	4.045	4.000	3.500	4.000	3.444	4.000	3.375
6 3/4	5.250	4.865	4.625	4.208	4.500	4.045	4.000	3.500	4.000	3.444	4.000	3.375
7	5.250	4.865	5.375	4.958	5.250	4.795	5.000	4.500	4.000	3.444	4.000	3.375
7 1/4	5.250	4.865	5.375	4.958	5.250	4.795	5.000	4.500	5.000	4.444	5.000	4.375
7 1/2	6.000	5.615	5.375	4.958	5.250	4.795	5.000	4.500	5.000	4.444	5.000	4.375
7 3/4	6.000	5.615	6.125	5.708	6.000	5.545	5.000	4.500	5.000	4.444	5.000	4.375
8	6.000	5.615	6.125	5.708	6.000	5.545	6.000	5.500	5.000	4.444	5.000	4.375
8 1/2	7.000	6.615	6.875	6.458	6.750	6.295	6.000	5.500	6.000	5.444	6.000	5.375
9	7.000	6.615	6.875	6.458	6.750	6.295	7.000	6.500	6.000	5.444	6.000	5.375
9 1/2	8.000	7.615	7.625	7.208	7.750	7.295	7.000	6.500	7.000	6.444	7.000	6.375
10	8.000	7.615	7.625	7.208	7.750	7.295	8.000	7.500	7.000	6.444	7.000	6.375
11			9.125	8.708	9.250	8.795	9.000	8.500	8.000	7.444	8.000	7.375
12			10.125	9.708	10.250	9.795	10.000	9.000	9.000	8.444	9.000	8.375

Notes: L_G is the maximum grip length and is the distance from the bearing surface to the first complete thread. L_B is the minimum body length and is the length of the unthreaded cylindrical portion of the shank. Note that 3/4", 7/8", and 1 inch diameters are available in lengths up to 20 inches. For sizes larger than 1 inch, the minimum complete thread length shall be equal to the basic thread length, and the total thread length including imperfect threads shall be basic thread length plus five pitches. Lengths too short to apply this formula shall be threaded to the head. All dimensions in inches. Source: Unbrako/SPS Technologies, Inc.

Table 65. Drill and Counterbore Sizes for 1960 Series Socket Head Cap Screws.

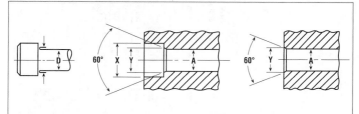

Nom. Size	Basic Screw Dia.	A				X Counter-bore Dia.	Y Counter-sink Dia.[3]	Hole Dimensions	
		Drill Size for Hole A						Tap Drill Dia.	
		Close Fit[1]		Normal Fit[2]				UNRC	UNRF
		Nom.	Decimal	Nom.	Decimal				
0	0.0600	#51	0.0670	#49	0.0730	1/8	0.074	-	3/64
1	0.0730	#46	0.0810	#43	0.0890	5/32	0.087	1.5mm	#53
2	0.0860	3/32	0.0937	#36	0.1065	3/16	0.102	#50	#50
3	0.0990	#36	0.1065	#31	0.1200	7/32	0.115	#47	#45
4	0.1120	1/8	0.1250	#29	0.1360	7/32	0.130	#43	#42
5	0.1250	9/64	0.1406	#23	0.1540	1/4	0.145	#38	#38
6	0.1380	#23	0.1540	#18	0.1695	9/32	0.158	#36	#33
8	0.1640	#15	0.1800	#10	0.1935	5/16	0.188	#29	#29
10	0.1900	#5	0.2055	#2	0.2210	3/8	0.218	#25	#21
1/4	0.2500	17/64	0.2656	9/23	0.2812	7/16	0.278	#7	#3
5/16	0.3125	21/64	0.3281	11/32	0.3437	17/32	0.346	F	I
3/8	0.0375	25/64	0.3906	13/32	0.4062	5/8	0.415	5/16	Q
7/16	0.4375	29/64	0.4531	15/32	0.4687	23/32	0.183	U	25/64
1/2	0.5000	33/64	0.5156	17/32	0.5312	13/16	0.552	27/64	29/64
5/8	0.6250	41/64	0.6406	21/32	0.6562	1	0.689	35/64	14.5mm
3/4	0.7500	49/64	0.7656	25/32	0.7812	1 3/16	0.828	21/32	11/16
7/8	0.8750	57/64	0.8906	29/32	0.9062	1 3/8	0.963	49/64	20.5mm
1	1.0000	1 1/64	1.0156	1 1/32	1.0312	1 5/8	1.100	7/8	59/64
1 1/4	1.2500	1 9/32	1.2812	1 5/32	1.3125	2	1.370	1 7/64	1 11/64
1 1/2	1.5000	1 17/32	1.5312	1 9/16	1.5625	2 3/8	1.640	34mm	36mm

Notes: [1] Close fit is limited to holes for those lengths of screws threaded to the head in assemblies in which: A) only one screw is used; or B) two or more screws are used and the mating holes are produced at assembly or by matched and coordinated tooling. [2] Normal fit is intended for: A) screws of relatively long length; or B) assemblies that involve two or more screws and where the mating holes are produced by conventional tolerancing methods. This fit provides the maximum allowable eccentricity of the longest standard screws and for certain deviations in the parts being fastened. [3] It is considered good practice to chamfer or break the edges of holes that are smaller than the screw diameter plus twice the fillet height (D Max. + 2F Max.) in parts in which the hardness approaches, equals, or exceeds the screw hardness. Dimension "C" on the Table provides this value. All dimensions are in inches. Source: Unbrako/SPS Technologies. Inc.

Table 66. Dimensions of Low Head Socket Head Cap Screws, Hexagon Recess.

Nom. Size	TPI* UNRC	TPI* UNRF	Class	A Head Dia. Max	A Head Dia. Min	D Body Dia. Max	D Body Dia. Min	T Key Engagement Min	H Head Height Max	H Head Height Min	Fillet Extension Max	Fillet Extension Min	J Hex. Socket Size Nom.
#4	40	48	3A	0.183	0.178	0.1120	0.1075	0.038	0.059	0.053	0.009	0.005	0.0500
#5	40	44	3A	0.205	0.200	0.1250	0.1202	0.044	0.065	0.059	0.010	0.006	0.0625
#6	32	40	3A	0.226	0.221	0.1380	0.1329	0.050	0.072	0.066	0.010	0.006	0.0625
#8	32	36	3A	0.270	0.265	0.1640	0.1585	0.060	0.085	0.079	0.012	0.007	0.0781
#10	24	32	3A	0.312	0.307	0.1900	0.1840	0.072	0.098	0.092	0.014	0.009	0.0938
1/4	20	28	3A	0.375	0.369	0.2500	0.2435	0.094	0.127	0.121	0.014	0.009	0.1250
5/16	18	24	3A	0.437	0.431	0.3125	0.3053	0.110	0.158	0.152	0.017	0.012	0.1562
3/8	16	24	3A	0.562	0.556	0.3750	0.3678	0.115	0.192	0.182	0.020	0.015	0.1875
1/2	13	20	3A	0.750	0.743	0.5000	0.4919	0.151	0.254	0.244	0.026	0.020	0.2500
5/8	11	18	3A	0.875	0.867	0.6250	0.6163	0.225	0.316	0.306	0.032	0.024	0.3125
3/4	10	16	3A	1.000	0.987	0.7500	0.7406	0.247	0.376	0.366	0.039	0.030	0.3750
7/8	9	14	3A	1.125	1.111	0.8750	0.8647	0.304	0.438	0.428	0.044	0.034	0.5000
1	8	12	3A	1.312	1.297	1.0000	0.9886	0.359	0.500	0.489	0.050	0.040	0.5620

* TPI = Threads per inch. All dimensions in inches. Source: SPS Technologies, Inc.

Table 67. Dimensions of Button Head, Hexagon Recess Screws.

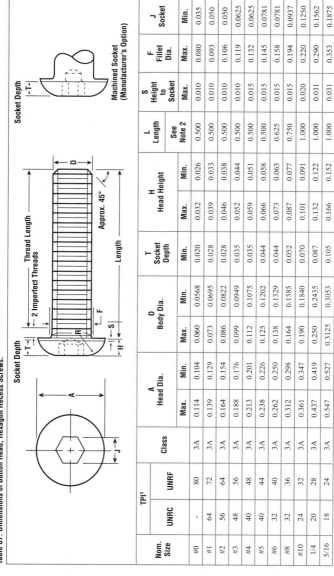

Nom. Size	TPI[1] UNRC	TPI[1] UNRF	Class	A Head Dia. Max.	A Head Dia. Min.	D Body Dia. Max.	D Body Dia. Min.	T Socket Depth Min.	H Head Height Max.	H Head Height Min.	L Length See Note 2	S Height to Socket Max.	F Fillet Dia. Max.	J Socket Min.
#0	–	80	3A	0.114	0.104	0.060	0.0568	0.020	0.032	0.026	0.500	0.010	0.080	0.035
#1	64	72	3A	0.139	0.129	0.073	0.0695	0.028	0.039	0.033	0.500	0.010	0.093	0.050
#2	56	64	3A	0.164	0.154	0.086	0.0822	0.028	0.046	0.038	0.500	0.010	0.106	0.050
#3	48	56	3A	0.188	0.176	0.099	0.0949	0.035	0.052	0.044	0.500	0.010	0.119	0.0625
#4	40	48	3A	0.213	0.201	0.112	0.1075	0.035	0.059	0.051	0.500	0.015	0.132	0.0625
#5	40	44	3A	0.238	0.226	0.125	0.1202	0.044	0.066	0.058	0.500	0.015	0.145	0.0781
#6	32	40	3A	0.262	0.250	0.138	0.1329	0.044	0.073	0.063	0.625	0.015	0.158	0.0781
#8	32	36	3A	0.312	0.298	0.164	0.1585	0.052	0.087	0.077	0.750	0.015	0.194	0.0937
#10	24	32	3A	0.361	0.347	0.190	0.1840	0.070	0.101	0.091	1.000	0.020	0.220	0.1250
1/4	20	28	3A	0.437	0.419	0.250	0.2435	0.087	0.132	0.122	1.000	0.031	0.290	0.1562
5/16	18	24	3A	0.547	0.527	0.3125	0.3053	0.105	0.166	0.152	1.000	0.031	0.353	0.1875

(Continued)

Table 67. *(Continued)* Dimensions of Button Head, Hexagon Recess Screws.

Nom. Size	TPI[1]		Class	A Head Dia.		D Body Dia.		T Socket Depth	H Head Height			L Length See Note 2	S Height to Socket	F Fillet Dia.	J Socket
	UNRC	UNRF		Max.	Min.	Max.	Min.	Min.	Max.	Min.		Max.	Max.	Min.
3/8	16	24	3A	0.656	0.636	0.375	0.3678	0.122	0.199	0.185	1.250	0.031	0.415	0.2187
1/2	13	20	3A	0.875	0.851	0.500	0.4919	0.175	0.265	0.245	2.000	0.046	0.560	0.3125
5/8	11	18	3A	1.000	0.970	0.625	0.6163	0.210	0.331	0.311	2.000	0.062	0.685	0.3750

Notes: [1] TPI = Threads per inch. [2] Screw lengths equal to or shorter than listed in column L will be threaded to head. For screws longer than listed, the minimum useable thread length shall be equal to twice the diameter + 0.500 inch. [3] Imperfect threads not to enter into fillet area.

Length tolerances for button head, hexagon recess screws		
Up to 1" incl.	Over 1" to 2" incl.	Over 2"
- 0.03	- 0.04	- 0.06

All dimensions in inches. Source: Unbrako/SPS Technologies, Inc.

Table 68. Dimensions of 82° Flat Countersunk Head, Hexagon Recess Screws.

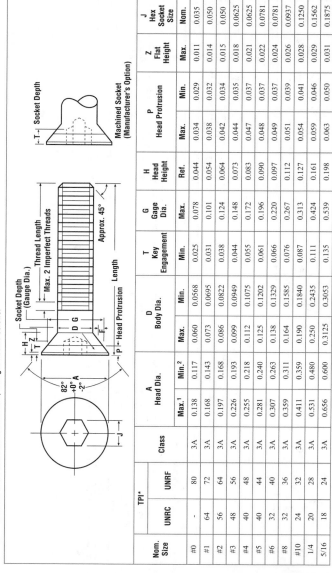

Nom. Size	TPI* UNRC	TPI* UNRF	Class	A Head Dia. Max.[1]	A Head Dia. Min.[2]	D Body Dia. Max.	D Body Dia. Min.	T Key Engagement Min.	G Gage Dia. Max.	H Head Height Ref.	P Head Protrusion Max.	P Head Protrusion Min.	Z Flat Height Max.	J Hex Socket Size Nom.
#0	–	80	3A	0.138	0.117	0.060	0.0568	0.025	0.078	0.044	0.034	0.029	0.011	0.035
#1	64	72	3A	0.168	0.143	0.073	0.0695	0.031	0.101	0.054	0.038	0.032	0.014	0.050
#2	56	64	3A	0.197	0.168	0.086	0.0822	0.038	0.124	0.064	0.042	0.034	0.015	0.050
#3	48	56	3A	0.226	0.193	0.099	0.0949	0.044	0.148	0.073	0.044	0.035	0.018	0.0625
#4	40	48	3A	0.255	0.218	0.112	0.1075	0.055	0.172	0.083	0.047	0.037	0.021	0.0625
#5	40	44	3A	0.281	0.240	0.125	0.1202	0.061	0.196	0.090	0.048	0.037	0.022	0.0781
#6	32	40	3A	0.307	0.263	0.138	0.1329	0.066	0.220	0.097	0.049	0.037	0.024	0.0781
#8	32	36	3A	0.359	0.311	0.164	0.1585	0.076	0.267	0.112	0.051	0.039	0.026	0.0937
#10	24	32	3A	0.411	0.359	0.190	0.1840	0.087	0.313	0.127	0.054	0.041	0.028	0.1250
1/4	20	28	3A	0.531	0.480	0.250	0.2435	0.111	0.424	0.161	0.059	0.046	0.029	0.1562
5/16	18	24	3A	0.656	0.600	0.3125	0.3053	0.135	0.539	0.198	0.063	0.050	0.031	0.1875

(Continued)

Table 68. (Continued) Dimensions of 82° Flat Countersunk Head, Hexagon Recess Screws.

Nom. Size	TPI* UNRC	TPI* UNRF	Class	A Head Dia. Max.¹	A Head Dia. Min.²	D Body Dia. Max.	D Body Dia. Min.	T Key Engagement Min.	G Gage Dia. Max.	H Head Height Ref.	P Head Protrusion Max.	P Head Protrusion Min.	Z Flat Height Max.	J Hex Socket Size Nom.
3/8	16	24	3A	0.781	0.720	0.375	0.3678	0.159	0.653	0.234	0.069	0.056	0.033	0.2187
7/16	14	20	3A	0.844	0.781	0.4375	0.4294	0.159	0.690	0.234	0.084	0.071	0.034	0.2500
1/2	13	20	3A	0.937	0.872	0.500	0.4919	0.172	0.739	0.251	0.110	0.096	0.034	0.3125
5/8	11	18	3A	1.188	1.112	0.625	0.6163	0.220	0.962	0.324	0.123	0.108	0.040	0.3750
3/4	10	16	3A	1.438	1.355	0.750	0.7406	0.220	1.186	0.396	0.136	0.121	0.040	0.5000
7/8	9	14	3A	1.688	1.605	0.875	0.8647	0.248	1.411	0.468	0.149	0.134	0.040	0.5625
1	8	12	3A	1.938	1.855	1.000	0.9886	0.297	1.635	0.540	0.162	0.146	0.040	0.6250

Notes: ¹ Maximum to theoretical sharp corners. ² Minimum–Absolute with a flat.

Length tolerances for 82° countersunk head, hexagon recess screws

Size	Up to 1" incl.	Over 1" to 2 1/2" incl.	Over 2 1/2" to 6" incl.	Over 6"
#0 thru 3/8"	- 0.03	- 0.04	- 0.06	- 0.12
7/16" thru 3/4"	- 0.03	- 0.06	- 0.08	- 0.12
7/8" and up	- 0.05	- 0.10	- 0.14	- 0.20

All dimensions in inches. Source: Unbrako/SPS Technologies, Inc.

Table 69. Dimensions for Hexagon Recess Socket Head Shoulder Bolts.

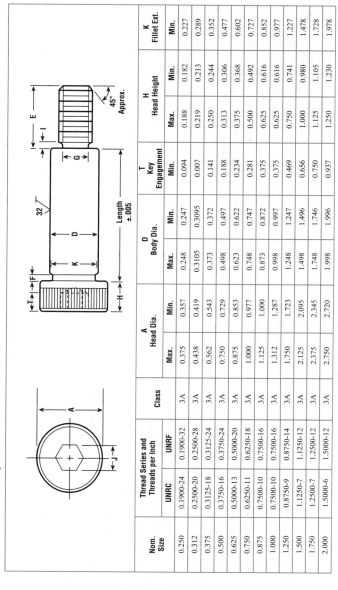

Nom. Size	Thread Series and Threads per Inch		Class	A Head Dia.		T Key Engagement	D Body Dia.		H Head Height		K Fillet Ext.
	UNRC	UNRF		Max.	Min.	Min.	Max.	Min.	Max.	Min.	Min.
0.250	0.1900-24	0.1900-32	3A	0.375	0.357	0.094	0.248	0.247	0.188	0.182	0.227
0.312	0.2500-20	0.2500-28	3A	0.438	0.419	0.007	0.3105	0.3095	0.219	0.213	0.289
0.375	0.3125-18	0.3125-24	3A	0.562	0.543	0.141	0.373	0.372	0.250	0.244	0.352
0.500	0.3750-16	0.3750-24	3A	0.750	0.729	0.188	0.498	0.497	0.313	0.306	0.477
0.625	0.5000-13	0.5000-20	3A	0.875	0.853	0.234	0.623	0.622	0.375	0.368	0.602
0.750	0.6250-11	0.6250-18	3A	1.000	0.977	0.281	0.748	0.747	0.500	0.492	0.727
0.875	0.7500-10	0.7500-16	3A	1.125	1.000	0.375	0.873	0.872	0.625	0.616	0.852
1.000	0.7500-10	0.7500-16	3A	1.312	1.287	0.375	0.998	0.997	0.625	0.616	0.977
1.250	0.8750-9	0.8750-14	3A	1.750	1.723	0.469	1.248	1.247	0.750	0.741	1.227
1.500	1.1250-7	1.1250-12	3A	2.125	2.095	0.656	1.498	1.496	1.000	0.980	1.478
1.750	1.2500-7	1.2500-12	3A	2.375	2.345	0.750	1.748	1.746	1.125	1.105	1.728
2.000	1.5000-6	1.5000-12	3A	2.750	2.720	0.937	1.998	1.996	1.250	1.230	1.978

(Continued)

Table 69. *(Continued)* Dimensions for Hexagon Recess Socket Head Shoulder Bolts.

Nom. Size	Thread Series and Threads per Inch		Class	G Thread Neck Dia.				F Shoulder Neck Width	I Thread Neck Width	E Thread & Neck Length	J Hex Socket Size
	UNRC	UNRF		UNRC		UNRF					
				Max.	Min.	Max.	Min.	Max.	Max.	See Note 1	Nom.
0.250	0.1900-24	0.1900-32	3A	0.142	0.133	0.156	0.147	0.093	0.083	0.375	0.125
0.312	0.2500-20	0.2500-28	3A	0.193	0.182	0.210	0.201	0.093	0.100	0.437	0.156
0.375	0.3125-18	0.3125-24	3A	0.249	0.237	0.265	0.256	0.093	0.111	0.500	0.187
0.500	0.3750-16	0.3750-24	3A	0.304	0.291	0.327	0.318	0.093	0.125	0.625	0.250
0.625	0.5000-13	0.5000-20	3A	0.414	0.397	0.443	0.432	0.093	0.154	0.750	0.312
0.750	0.6250-11	0.6250-18	3A	0.521	0.502	0.561	0.549	0.093	0.182	0.875	0.375
0.875	0.7500-10	0.7500-16	3A	0.638	0.616	0.678	0.665	0.093	0.200	1.000	0.500
1.000	0.7500-10	0.7500-16	3A	0.638	0.616	0.678	0.665	0.093	0.200	1.000	0.500
1.250	0.8750-9	0.8750-14	3A	0.750	0.726	0.796	0.778	0.125	0.222	1.125	0.625
1.500	1.1250-7	1.1250-12	3A	0.964	0.934	1.022	1.014	0.125	0.286	1.500	0.875
1.750	1.2500-7	1.2500-12	3A	1.089	1.059	1.147	1.139	0.125	0.286	1.750	1.000
2.000	1.5000-6	1.5000-12	3A	1.307	1.277	1.397	1.389	0.125	0.333	2.000	1.250

Notes: [1] Dimension is +0.000 / −0.020.

* TPI = Threads per inch.

All dimensions in inches. Source: SPS Technologies, Inc.

Table 70. Dimensions for Socket Hexagon Recess Set Screws. (See Notes.)

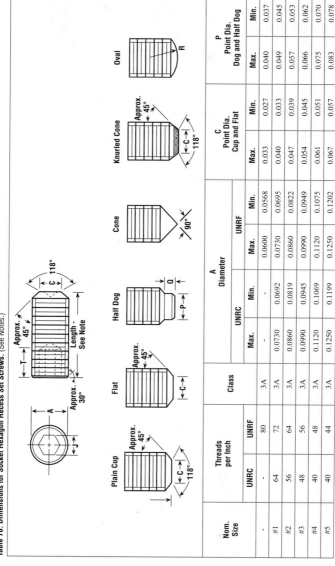

Nom. Size	Threads per Inch		Class	A Diameter						C Point Dia. Cup and Flat		P Point Dia. Dog and Half Dog	
	UNRC	UNRF		UNRC		UNRF							
				Max.	Min.	Max.	Min.			Max.	Min.	Max.	Min.
–	–	80	3A	–	–	0.0600	0.0568			0.033	0.027	0.040	0.037
#1	64	72	3A	0.0730	0.0692	0.0730	0.0695			0.040	0.033	0.049	0.045
#2	56	64	3A	0.0860	0.0819	0.0860	0.0822			0.047	0.039	0.057	0.053
#3	48	56	3A	0.0990	0.0945	0.0990	0.0949			0.054	0.045	0.066	0.062
#4	40	48	3A	0.1120	0.1069	0.1120	0.1075			0.061	0.051	0.075	0.070
#5	40	44	3A	0.1250	0.1199	0.1250	0.1202			0.067	0.057	0.083	0.078

(Continued)

Table 70. (Continued) Dimensions for Socket Hexagon Recess Set Screws. (See Notes.)

Nom. Size	Threads per Inch		Class	A Diameter				C Point Dia. Cup and Flat		P Point Dia. Dog and Half Dog	
	UNRC	UNRF		UNRC		UNRF					
				Max.	Min.	Max.	Min.	Max.	Min.	Max.	Min.
#6	32	40	3A	0.1380	0.1320	0.1380	0.1329	0.074	0.064	0.092	0.087
#8	32	36	3A	0.1640	0.1580	0.1640	0.1585	0.087	0.076	0.109	0.103
#10	24	32	3A	0.1900	0.1828	0.1900	0.1840	0.102	0.088	0.127	0.120
1/4	20	28	3A	0.2500	0.2419	0.2500	0.2435	0.132	0.118	0.156	0.149
5/16	18	24	3A	0.3125	0.3038	0.3125	0.3053	0.172	0.156	0.203	0.195
3/8	16	24	3A	0.3750	0.3656	0.3750	0.3678	0.212	0.194	0.250	0.241
7/16	14	20	3A	0.4375	0.4272	0.4375	0.4294	0.252	0.232	0.296	0.287
1/2	13	20	3A	0.5000	0.4891	0.5000	0.4919	0.291	0.270	0.343	0.334
9/16	12	18	3A	0.5625	0.5511	0.5625	0.5538	0.332	0.309	0.390	0.379
5/8	11	18	3A	0.6250	0.6129	0.6250	0.6163	0.371	0.347	0.468	0.456
3/4	10	16	3A	0.7500	0.7371	0.7500	0.7406	0.450	0.425	0.562	0.549
7/8	9	14	3A	0.8750	0.8611	0.8750	0.8647	0.530	0.502	0.656	0.642
1	8	12	3A	1.0000	0.9850	1.0000	0.9897	0.609	0.579	0.750	0.734
1 1/8	7	12	3A	1.1250	1.1086	1.1250	1.1136	0.689	0.655	0.843	0.826
1 1/4	7	12	3A	1.2500	1.2336	1.2500	1.2386	0.767	0.733	0.937	0.920
1 3/8	6	12	3A	1.3750	1.3568	1.3750	1.3636	0.848	0.808	1.031	1.011
1 1/2	6	12	3A	1.5000	1.4818	1.5000	1.4886	0.926	0.886	1.125	1.105

(Continued)

Table 70. (Continued) Dimensions for Socket Hexagon Recess Set Screws. (See Notes.)

Nom. Size	Threads per Inch		Class	Q Point Length Half Dog		Point Length Dog (Not Shown)		R Radius Oval Point	T¹ Socket Depth	J Hex Socket Size
	UNRC	UNRF		Max.	Min.	Max.	Min.	Basic	Min.	Nom.
–	–	80	3A	0.017	0.013	0.033	0.027	0.045	0.035	0.028
#1	64	72	3A	0.021	0.017	0.040	0.034	0.055	0.035	0.035
#2	56	64	3A	0.024	0.020	0.047	0.039	0.064	0.035	0.035
#3	48	56	3A	0.027	0.023	0.054	0.046	0.074	0.060	0.050
#4	40	48	3A	0.030	0.026	0.060	0.052	0.084	0.075	0.050
#5	40	44	3A	0.033	0.027	0.064	0.056	0.094	0.075	0.0625
#6	32	40	3A	0.038	0.032	0.074	0.066	0.104	0.075	0.0625
#8	32	36	3A	0.043	0.037	0.084	0.076	0.123	0.075	0.0781
#10	24	32	3A	0.049	0.041	0.095	0.085	0.142	0.105	0.0937
1/4	20	28	3A	0.0665	0.0585	0.130	0.120	0.188	0.105	0.125
5/16	18	24	3A	0.082	0.074	0.164	0.148	0.234	0.140	0.1562
3/8	16	24	3A	0.0987	0.0887	0.1955	0.1795	0.281	0.140	0.1875
7/16	14	20	3A	0.114	0.104	0.2267	0.2107	0.328	0.190	0.2187
1/2	13	20	3A	0.130	0.120	0.260	0.240	0.375	0.210	0.250
9/16	12	18	3A	0.1456	0.1356	0.291	0.271	0.422	0.265	0.250
5/8	11	18	3A	0.164	0.148	0.3225	0.3025	0.469	0.265	0.3125
3/4	10	16	3A	0.1955	0.1795	0.385	0.365	0.562	0.330	0.375
7/8	9	14	3A	0.2267	0.2107	0.4475	0.4275	0.656	0.450	0.500
1	8	12	3A	0.260	0.240	0.510	0.490	0.750	0.550	0.5625
1 1/8	7	12	3A	0.291	0.271	0.5775	0.5475	0.844	0.650	0.5625
1 1/4	7	12	3A	0.3225	0.3025	0.640	0.610	0.938	0.700	0.625

(Continued)

Table 70. (Continued) Dimensions for Socket Hexagon Recess Set Screws. (See Notes.)

Nom. Size	Threads per Inch		Class	Q Point Length Half Dog		Point Length Dog (Not Shown)		R Radius Oval Point	T¹ Socket Depth	J Hex Socket Size
	UNRC	UNRF		Max.	Min.	Max.	Min.	Basic	Min.	Nom.
1 3/8	6	12	3A	0.3537	0.3337	0.7025	0.6725	1.031	0.700	0.625
1 1/2	6	12	3A	0.385	0.365	0.765	0.735	1.125	0.750	0.750

Notes: When cone cup set screw length equals nominal diameter or less, included angle is 118°; for other lengths the included angle is 90°. When plain cup set screw length equals nominal diameter or less, included angle is 130°; for other lengths the included angle is 118°.

¹ Values shown in column T are for minimum stock length cup point screws. Screws shorter than nominal minimum length shown do not have sockets deep enough to utilize full key capacity which can result in failure of socket, key, or mating threads.

Length tolerance table for socket hexagon recess set screws			
Up to 0.630"	Over 0.630" to 2.0"	Over 2.0" to 6.0"	Over 6.0"
± 0.01	± 0.02	± 0.03	± 0.06

All dimensions in inches. Source: Unbrako/SPS Technologies, Inc.

Table 71. Torsional and Axial Holding Power of Cup Point Socket Set Screws. (See Notes.)

Nom. Size	Seating Torque (in.-lb)	Axial Holding Power (Pounds)	Shaft Diameter (shaft hardness Rc 15 to Rc 35) Torsional Holding Power (in.-lb)							
			1/16	3/32	1/8	5/32	3/16	7/32	1/4	5/16
#0	1.0	50	**1.5**	2.3	3.1	3.9	4.7	5.4	6.2	
#1	1.8	65	2.0	**3.0**	4.0	5.0	6.1	7.1	8.1	10.0
#2	1.8	85	2.6	4.0	**5.3**	6.6	8.0	9.3	10.6	13.2
#3	5	120	3.2	5.6	7.5	**9.3**	11.3	13.0	15.0	18.7
#4	5	160		7.5	10.0	12.5	**15.0**	17.5	20.0	25.0
#5	10	200			12.5	15.6	18.7	**21.8**	**25.0**	31.2
#6	10	250				19	23	27	31	**39**
#8	20	385				30	36	42	48	60
#10	36	540					51	59	68	84
1/4	87	1,000							125	156
5/16	165	1,500								234

Nom. Size	Seating Torque (in.-lb)	Axial Holding Power (Pounds)	Shaft Diameter (shaft hardness Rc 15 to Rc 35) Torsional Holding Power (in.-lb)							
			3/8	7/16	1/2	9/16	5/8	3/4	7/8	1
#2	1.8	85	16.0							
#3	5	120	22.5	26.3						
#4	5	160	30.0	35.0	40.0					
#5	10	200	37.5	43.7	50.0	56.2	62			
#6	10	250	47	55	62	70	78	94	109	
#8	20	385	**72**	84	96	108	120	144	168	192
#10	36	540	101	**118**	135	152	169	202	236	270
1/4	87	1,000	187	218	**250**	281	312	357	437	500

(Continued)

Table 71. *(Continued)* Torsional and Axial Holding Power of Cup Point Socket Set Screws. (See Notes.)

Nom. Size	Seating Torque (in.-lb)	Axial Holding Power (Pounds)	Shaft Diameter (shaft hardness R_C 15 to R_C 35)							
			3/8	7/16	1/2	9/16	5/8	3/4	7/8	1
			Torsional Holding Power (in.-lb)							
5/16	165	1,500	280	327	375	421	**468**	**562**	656	750
3/8	290	2,000	375	437	500	562	625	750	**875**	1000
7/16	430	2,500		545	625	702	780	937	1095	1250
1/2	620	3,000			750	843	937	1125	1310	1500
9/16	620	3,500				985	1090	1310	1530	1750
5/8	1,325	4,000					1250	1500	1750	2000
3/4	2,400	5,000						1875	2190	2500
7/8	3,600	5,600							2620	3000
1	5,000	6,500								3500

Nom. Size	Seating Torque (in.-lb)	Axial Holding Power (Pounds)	Shaft Diameter (shaft hardness R_C 15 to R_C 35)							
			1 1/4	1 1/2	1 3/4	2	2 1/2	3	3 1/2	4
			Torsional Holding Power (in.-lb)							
#10	36	540	338							
1/4	87	1,000	625	750						
5/16	165	1,500	937	1125	1310	1500				
3/8	290	2,000	1250	1500	1750	2000				
7/16	430	2,500	1560	**1875**	2210	2500	3125			
1/2	620	3,000	1875	**2250**	2620	3000	3750	4500		
9/16	620	3,500	2190	2620	**3030**	3500	4370	5250	6120	
5/8	1,325	4,000	2500	3000	3750	**4000**	5000	6000	7000	8000
3/4	2,400	5,000	3125	3750	4500	5000	**6250**	**7500**	8750	10000

(Continued)

Table 71. *(Continued)* **Torsional and Axial Holding Power of Cup Point Socket Set Screws.** (See Notes.)

Nom. Size	Seating Torque (in.-lb)	Axial Holding Power (Pounds)	Shaft Diameter (shaft hardness R_C 15 to R_C 35)							
			1 1/4	1 1/2	1 3/4	2	2 1/2	3	3 1/2	4
			Torsional Holding Power (in.-lb)							
7/8	3,600	5,600	3750	4500	5250	6000	7500	9000	10500	12000
1	5,000	6,500	4375	5250	6120	7000	8750	10500	12250	14000

Notes: Values are for cup point socket screws seated at recommended installation torques, using Class 3A threads in Class 2A tapped holes. Holding power was defined as the minimum load to produce 0.010 inch relative movement of shaft and collar. Tabulated axial and torsional holding powers are typical strengths and should be used accordingly, with specific safety factors appropriate to the given application and load conditions. Values in bold type indicate recommended set screw sizes on the basis that screw diameter should be roughly one-half shaft diameter. Source: Unbrako/SPS Technologies, Inc.

Table 72. Dimensions of Hexagon Keys.

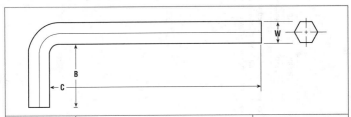

W Key Size		C Length of Long Arm				B Length of Short Arm	
		Short Series		Long Series			
Max.	Min.	Max.	Min.	Max.	Min.	Max.	Min.
0.0280	0.0275	1.312	1.125	2.688	2.500	0.312	0.125
0.0350	0.0345	1.312	1.125	2.766	2.578	0.438	0.250
0.0500	0.0490	1.750	1.562	2.938	2.750	0.625	0.438
1/16	0.0615	1.844	1.656	3.094	2.906	0.656	0.469
5/64	0.0771	1.969	1.781	3.281	3.094	0.703	0.516
3/32	0.0927	2.094	1.906	3.469	3.281	0.750	0.562
7/64	0.1079	2.219	2.031	3.656	3.469	0.797	0.609
1/8	0.1235	2.344	2.156	3.844	3.656	0.844	0.656
9/64	0.1391	2.469	2.281	4.031	3.844	0.891	0.703
5/32	0.1547	2.594	2.406	4.219	4.031	0.938	0.750
3/16	0.1860	2.844	2.656	4.594	4.406	1.031	0.844
7/32	0.2172	3.094	2.906	4.969	4.781	1.125	0.938
1/4	0.2485	3.344	3.156	5.344	5.156	1.219	1.031
5/16	0.3110	3.844	3.656	6.094	5.906	1.344	1.156
3/8	0.3735	4.344	4.156	6.844	6.656	1.469	1.281
7/16	0.4355	4.844	4.656	7.594	7.406	1.594	1.406
1/2	0.4975	5.344	5.156	8.344	8.156	1.719	1.531
9/16	0.5600	5.844	5.656	9.094	8.906	1.844	1.656
5/8	0.6225	6.344	6.156	9.844	9.656	1.969	1.781
3/4	0.7470	7.344	7.156	11.344	11.156	2.219	2.031
7/8	0.8720	8.344	8.156	12.844	12.656	2.469	2.281
1	0.9970	9.344	9.156	14.344	14.156	2.719	2.531
1-1/4	1.2430	11.500	11.000	-	-	3.250	2.750
1-1/2	1.4930	13.500	13.000	-	-	3.750	3.250
1-3/4	1.7430	15.500	15.000	-	-	4.250	3.750
2	1.9930	17.500	17.000	-	-	4.750	4.250

All dimensions in inches. Source: Unbrako/SPS Technologies, Inc.

Table 73. Dimensions of Spline Sockets.

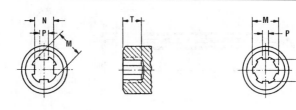

Nom. Socket and Key Size	No. of Teeth	M Socket Major Dia.		N Socket Minor Dia.		P Width of Tooth	
		Max.	Min.	Max.	Min.	Max.	Min.
0.033	4	0.0350	0.0340	0.0260	0.0255	0.0120	0.0115
0.048	6	0.050	0.049	0.041	0.040	0.011	0.010
0.060	6	0.062	0.061	0.051	0.050	0.014	0.013
0.072	6	0.074	0.073	0.064	0.063	0.016	0.015
0.096	6	0.098	0.097	0.082	0.080	0.022	0.021
0.111	6	0.115	0.113	0.098	0.096	0.025	0.023
0.133	6	0.137	0.135	0.118	0.116	0.030	0.028
0.145	6	0.149	0.147	0.128	0.126	0.032	0.030
0.168	6	0.173	0.171	0.150	0.147	0.036	0.033
0.183	6	0.188	0.186	0.163	0.161	0.039	0.037
0.216	6	0.221	0.219	0.190	0.188	0.050	0.048
0.251	6	0.256	0.254	0.221	0.219	0.060	0.058
0.291	6	0.298	0.296	0.254	0.252	0.068	0.066
0.372	6	0.380	0.377	0.319	0.316	0.092	0.089
0.454	6	0.463	0.460	0.386	0.383	0.112	0.109
0.595	6	0.604	0.601	0.509	0.506	0.138	0.134
0.620	6	0.631	0.627	0.535	0.531	0.149	0.145
0.698	6	0.709	0.705	0.604	0.600	0.168	0.164
0.790	6	0.801	0.797	0.685	0.681	0.189	0.185

All dimensions in inches. Source: Extracted from ANSI B18.3-1969, published by the American Society of Mechanical Engineers.

Table 74. Dimensions of Spline Keys.

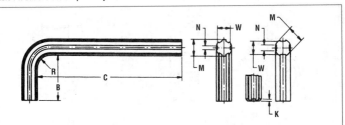

Nom. Socket and Key Size	M Major Dia.		W Minor Dia.		N Width of Indent		B Length of Short Arm	
	Max.	Min.	Max.	Min.	Max.	Min.	Max.	Min.
0.033	0.0330	0.0320	0.0250	0.0240	0.0140	0.0130	0.312	0.125
0.048	0.0480	0.0470	0.0390	0.0380	0.0130	0.0120	0.438	0.250
0.060	0.0600	0.0590	0.0490	0.0480	0.0160	0.0150	0.625	0.438
0.072	0.0720	0.0710	0.0620	0.0610	0.0190	0.0180	0.656	0.469
0.096	0.0960	0.0950	0.0790	0.0775	0.0240	0.0230	0.703	0.516
0.111	0.1110	0.1100	0.0940	0.0925	0.0280	0.0270	0.750	0.562
0.133	0.0330	0.1310	0.1140	0.1120	0.0340	0.0320	0.797	0.609
0.145	0.0450	0.1435	0.1240	0.1225	0.0355	0.0340	0.844	0.656
0.168	0.1680	0.1660	0.1440	0.1420	0.0410	0.0390	0.891	0.703
0.183	0.1830	0.1815	0.1580	0.1565	0.0440	0.0425	0.938	0.750
0.216	0.2160	0.2145	0.1840	0.1825	0.0550	0.0535	1.031	0.844
0.251	0.2510	0.2495	0.2140	0.2125	0.0655	0.0640	1.125	0.938
0.281	0.2910	0.2895	0.2460	0.2445	0.0775	0.0760	1.219	1.031
0.372	0.3720	0.3705	0.3100	0.3085	0.0975	0.0960	1.344	1.156
0.454	0.4540	0.4525	0.3770	0.3755	0.1185	0.1170	1.469	1.281
0.595	0.5950	0.5935	0.5000	0.4975	0.1460	0.1445	1.719	1.531
0.620	0.6200	0.6175	0.5240	0.5215	0.1615	0.1590	1.844	1.656
0.698	0.6980	0.6955	0.5930	0.5905	0.1805	0.1780	1.844	1.656
0.790	0.7900	0.7875	0.6740	0.6715	0.1975	0.1950	1.969	1.781

Nom. Key and Socket Size	C Length of Long Arm				Number of Splines	R Radius of Bend	K Chamfer
	Short Series		Long Series				
	Max.	Min.	Max.	Min.		Min.	Max.
0.033	1.312	1.125	-	-	4	0.062	0.003
0.048	1.312	1.125	-	-	6	0.062	0.004
0.060	1.750	1.562	-	-	6	0.062	0.006
0.072	1.844	1.656	-	-	6	0.062	0.008
0.096	1.969	1.781	-	-	6	0.078	0.008

(Continued)

Table 74. *(Continued)* Dimensions of Spline Keys.

Nom. Key and Socket Size	C Length of Long Arm				Number of Splines	R Radius of Bend	K Chamfer
	Short Series		Long Series				
	Max.	Min.	Max.	Min.		Min.	Max.
0.111	2.094	1.906	-	-	6	0.094	0.009
0.133	2.219	2.031	3.656	3.469	6	0.125	0.014
0.145	2.344	2.156	3.844	3.656	6	0.125	0.015
0.168	2.469	2.281	4.031	3.844	6	0.156	0.016
0.183	2.594	2.406	4.219	4.031	6	0.156	0.016
0.216	2.844	2.656	4.594	4.406	6	0.188	0.022
0.251	3.094	2.906	4.969	4.781	6	0.219	0.024
0.281	3.344	3.156	5.344	5.156	6	0.250	0.030
0.372	3.844	3.656	6.094	5.906	6	0.312	0.032
0.454	4.344	4.156	6.844	6.656	6	0.375	0.044
0.595	5.344	5.156	8.344	8.156	6	0.500	0.050
0.620	5.844	5.656	9.094	8.906	6	0.500	0.053
0.698	5.844	5.656	-	-	6	0.562	0.055
0.790	6.344	6.156	-	-	6	0.625	0.070

All dimensions in inches. Source: Extracted from ANSI B18.3-1960, published by the American Society of Mechanical Engineers.

Table 75. ISO Tolerances for Metric Fasteners.

Nom. Dia.		Tolerance Zone (External Threads)								
Over	To	h6	h8	h10	h11	h13	h14	h15	h16	js14
0	1	0 / -0.006	0 / -0.014	0 / -0.040	0 / -0.060	0 / -0.14	-	-	-	-
1	3	0 / -0.006	0 / -0.014	0 / -0.040	0 / -0.060	0 / -0.14	0 / -0.25	0 / -0.40	0 / -0.60	±0.125
3	6	0 / -0.008	0 / -0.018	0 / -0.048	0 / -0.075	0 / -0.18	0 / -0.30	0 / -0.48	0 / -0.75	±0.15
6	10	0 / -0.009	0 / -0.022	0 / -0.058	0 / -0.090	0 / -0.22	0 / -0.36	0 / -0.58	0 / -0.90	±0.18
10	18	0 / -0.011	0 / -0.027	0 / -0.070	0 / -0.110	0 / -0.27	0 / -0.43	0 / -0.70	0 / -1.10	±0.215
18	30	0 / -0.030	0 / -0.033	0 / -0.084	0 / -0.130	0 / -0.33	0 / -0.52	0 / -0.84	0 / -1.30	±0.26
30	50	-	-	-	-	0 / -0.39	0 / -0.62	0 / -1.00	0 / -1.60	±0.31
50	80	-	-	-	-	0 / -0.46	0 / -0.74	0 / -1.20	0 / -1.90	±0.37
80	120	-	-	-	-	0 / -0.54	0 / -0.87	0 / -1.40	0 / -2.20	±0.435

(Continued)

Table 75. *(Continued)* ISO Tolerances for Metric Fasteners.

Nom. Dia.		Tolerance Zone (External Threads)				Tolerance Zone (Internal Threads)					
Over	To	js15	js16	js17	m6	H7	H8	H9	H11	H13	H14
0	1	-	-	-	+0.002 / +0.008	+0.010 / 0	+0.014 / 0	+0.025 / 0	+0.060 / 0	+0.14 / 0	-
1	3	±0.20	±0.30	±0.50	+0.002 / +0.008	+0.010 / 0	+0.014 / 0	+0.025 / 0	+0.060 / 0	+0.14 / 0	+0.25 / 0
3	6	±0.24	±0.375	±0.60	+0.004 / +0.012	+0.012 / 0	+0.018 / 0	+0.030 / 0	+0.075 / 0	+0.18 / 0	+0.30 / 0
6	10	±0.29	±0.45	±0.075	+0.006 / +0.015	+0.015 / 0	+0.022 / 0	+0.036 / 0	+0.090 / 0	+0.22 / 0	+0.36 / 0
10	18	±0.35	±0.55	±0.90	+0.007 / +0.018	+0.018 / 0	+0.027 / 0	+0.043 / 0	+0.110 / 0	+0.27 / 0	+0.43 / 0
18	30	±0.42	±0.65	±1.05	+0.008 / +0.021	+0.021 / 0	+0.033 / 0	+0.052 / 0	+0.130 / 0	+0.33 / 0	+0.52 / 0
30	50	±0.50	±0.80	±1.25	-	-	-	-	-	+0.39 / 0	+0.62 / 0
50	80	±0.60	±0.95	±1.50	-	-	-	-	-	+0.46 / 0	+0.74 / 0
80	120	±0.70	±1.10	±1.75	-	-	-	-	-	+0.54 / 0	+0.87 / 0

All dimensions in millimeters. References: ISO R286, ISO 4759/I, ISO 4759/II, ISO 4759/III. Source: Unbrako/SPS Technologies, Inc.

Table 76. Dimensions of Metric Hexagon Cap Screw (ISO/DIN 4014, DIN 931).

Bolt Size	E Body Dia.		F Width Across Flats		G Width Across Corners	H Head Height		L_T Thread Length		
	Max.	Min.	Max.	Min.	Min.	Max.	Min.	L_T to 125	L_T to 200	L_T over 200
M4	4.0	3.82	7.00	6.78	7.66	2.50	2.40	14.00	–	–
M5	5.0	4.82	8.00	7.78	8.79	3.65	3.35	16.00	–	–
M6	6.0	5.82	10.00	9.78	11.05	4.15	3.85	18.00	24.00	–
M8	8.0	7.78	13.00	12.73	14.38	5.45	5.15	22.00	28.00	–
M10	10.0	9.78	16.00	15.73	18.90	6.58	6.22	26.00	32.00	45.00
M12	12.0	11.73	18.00	17.73	21.10	7.68	7.32	30.00	36.00	49.00
M14	14.0	13.73	21.00	20.67	24.49	8.98	8.62	34.00	40.00	53.00
M16	16.0	15.73	24.00	23.67	26.75	10.18	9.82	38.00	44.00	57.00
M18	18.0	17.73	27.00	26.67	30.14	11.72	11.28	42.00	48.00	61.00

(Continued)

Table 76. *(Continued)* Dimensions of Metric Hexagon Cap Screw (ISO/DIN 4014, DIN 931).

Bolt Size	E Body Dia.		F Width Across Flats		G Width Across Corners	H Head Height		L_T Thread Length			
	Max.	Min.	Max.	Min.	Min.	Max.	Min.	L_T to 125	L_T to 200	L_T over 200
M20	20.0	19.67	30.00	29.67	33.53	12.72	12.28	46.00	52.00	65.00
M22	22.0	21.67	34.00	33.38	35.72	14.22	13.78	50.00	56.00	69.00
M24	24.0	23.67	36.00	35.38	39.98	15.22	14.78	54.00	60.00	73.00
M27	27.0	26.67	41.00	40.38	45.20	17.05	16.35	60.00	66.00	79.00
M30	30.0	29.67	46.00	45.00	50.85	19.12	18.28	66.00	72.00	85.00
M33	33.0	32.61	50.00	49.00	55.37	20.92	20.08	72.00	78.00	91.00
M36	36.0	35.61	55.00	53.80	60.79	22.92	22.08	78.00	84.00	97.00

All dimensions in millimeters.

Table 77. Nominal Dimensions of Metric Fine Pitch Hexagon Head Bolt (DIN 960).

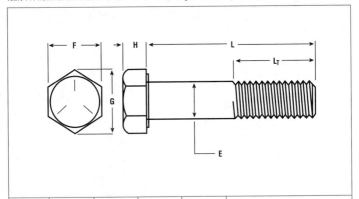

Bolt Size	E Body Dia.	F Width Across Flats	G Width Across Corners	H Head Height	L_T Thread Length		
					L_T to 125	L_T to 200	L_T over 200
M8×1	8.00	13.00	14.38	5.30	22.00	28.00	-
M10×1	10.00	17.00	18.90	6.40	26.00	32.00	-
M10×1.25	10.00	17.00	18.90	6.40	26.00	32.00	-
M12×1.25	12.00	19.00	21.10	7.50	30.00	36.00	-
M12×1.50	12.00	19.00	21.10	7.50	30.00	36.00	-
M14×1.50	14.00	22.00	24.49	8.80	34.00	40.00	-
M16×1.50	16.00	24.00	26.75	10.00	38.00	44.00	-
M18×1.50	18.00	27.00	30.14	11.50	42.00	48.00	-
M20×1.50	20.00	30.00	33.53	12.50	46.00	52.00	-
M20×2.00	20.00	30.00	33.53	12.50	46.00	52.00	-
M24×2.00	24.00	36.00	39.98	15.00	54.00	60.00	-
M27×2.00	27.00	41.00	45.20	17.00	60.00	66.00	79.00
M30×2.00	30.00	46.00	50.85	18.70	66.00	72.00	58.00
M33×2.00	33.00	50.00	55.37	21.00	72.00	78.00	91.00
M36×3.00	36.00	55.00	60.79	22.50	78.00	84.00	97.00

All dimensions in millimeters.

Table 78. Nominal Dimensions of Metric Fine Pitch Hexagon Head Bolt (DIN 961).

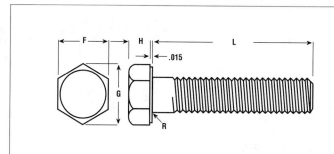

Bolt Size	Body Dia.	F Width Across Flats	G Width Across Corners	H Head Height
M8	8.00	13.00	14.38	5.30
M10	10.00	17.00	18.90	6.40
M12	12.00	19.00	21.10	7.50
M14	14.00	22.00	24.49	8.80
M16	16.00	24.00	26.75	10.00
M18	18.00	27.00	30.14	11.50
M20	20.00	30.00	33.53	12.50
M22	22.00	32.00	35.72	14.00
M24	24.00	36.00	39.98	15.00
M27	27.00	41.00	45.63	17.00
M30	30.00	46.00	51.28	19.00
M33	33.00	50.00	55.80	21.00
M36	36.00	55.00	61.31	23.00
M39	39.00	60.00	66.96	25.00
M42	42.00	65.00	72.61	26.00
M45	45.00	70.00	78.29	28.00
M48	48.00	75.00	83.91	30.00
M52	52.00	80.00	89.56	33.00

All dimensions in inches.

Table 79. Dimensions of Metric Heavy (Structural) Hexagon Head Bolt (DIN 6914). (See Notes.)

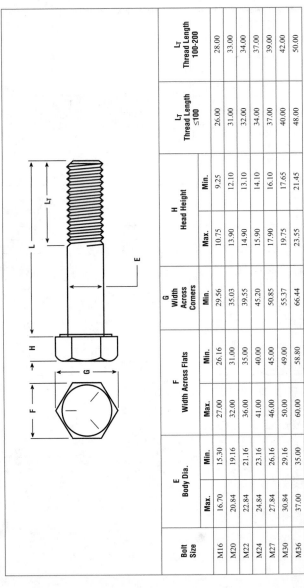

Bolt Size	E Body Dia.		F Width Across Flats		G Width Across Corners	H Head Height		L_T Thread Length ≤100	L_T Thread Length 100-200
	Max.	Min.	Max.	Min.	Min.	Max.	Min.		
M16	16.70	15.30	27.00	26.16	29.56	10.75	9.25	26.00	28.00
M20	20.84	19.16	32.00	31.00	35.03	13.90	12.10	31.00	33.00
M22	22.84	21.16	36.00	35.00	39.55	14.90	13.10	32.00	34.00
M24	24.84	23.16	41.00	40.00	45.20	15.90	14.10	34.00	37.00
M27	27.84	26.16	46.00	45.00	50.85	17.90	16.10	37.00	39.00
M30	30.84	29.16	50.00	49.00	55.37	19.75	17.65	40.00	42.00
M36	37.00	35.00	60.00	58.80	66.44	23.55	21.45	48.00	50.00

Notes: These hexagon bolts have increased width across flats for high tensile structural bolting. Use with hex nut specified in DIN 6915 and washer specified in DIN 6916. All dimensions in millimeters.

Table 80. Nominal Dimensions of Metric Hexagon Flange Bolts (DIN 6921).

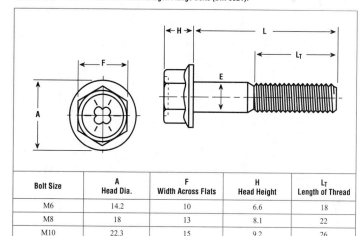

Bolt Size	A Head Dia.	F Width Across Flats	H Head Height	L$_T$ Length of Thread
M6	14.2	10	6.6	18
M8	18	13	8.1	22
M10	22.3	15	9.2	26
M12	26.6	16	11.5	30

All dimensions in millimeters.

Table 81. Maximum Grip Gaging Lengths and Minimum Body Lengths for Metric Hexagon Head Bolts.

Nom. Bolt Length	Nominal Bolt Diameter and Pitch									
	M6 × 1		M8 × 1.25		M10 × 1.5		M12 × 1.75		M14 × 2	
	Max. Grip Length	Min. Body Length	Max. Grip Length	Min. Body Length	Max. Grip Length	Min. Body Length	Max. Grip Length	Min. Body Length	Max. Grip Length	Min. Body Length
30	12.0	7.0	13.0	6.8						
35	17.0	12.0	13.0	6.08						
40	17.0	12.0	23.0	16.8	14.0	6.5				
45	27.0	22.0	23.0	16.8	19.0	11.5	15.0	6.2		
50	27.0	22.0	33.0	26.8	19.0	11.5	15.0	6.2	16.0	6.0
55	37.0	32.0	33.0	26.8	29.0	21.5	25.0	16.2	21.0	11.0
60	37.0	32.0	43.0	36.8	29.0	21.5	25.0	16.2	21.0	11.0
65			43.0	36.8	39.0	31.5	35.0	26.2	31.0	21.0
70			53.0	46.8	39.0	31.5	35.0	26.2	31.0	21.0
75			53.0	46.8	49.0	41.5	45.0	36.2	41.0	31.0
80					49.0	41.5	45.0	36.2	41.0	31.0
85					59.0	51.5	55.0	46.2	51.0	41.0
90					59.0	51.5	55.0	46.2	51.0	41.0
100					74.0	66.5	70.0	61.2	66.0	56.0
110							80.0	71.2	76.0	66.0
120							90.0	81.2	86.0	76.0
130									90.0	80.0
140									100.0	90.0

(Continued)

Table 81. (Continued) Maximum Grip Gaging Lengths and Minimum Body Lengths for Metric Hexagon Head Bolts.

Nom. Bolt Length	M16 × 2		M20 × 2.5		M24 × 3		M30 × 3.5	
	Max. Grip Length	Min. Body Length	Max. Grip Length	Min. Body Length	Max. Grip Length	Min. Body Length	Max. Grip Length	Min. Body Length
55	17.0	7.0						
60	17.0	7.0						
65	27.0	17.0	19.0	6.5				
70	27.0	17.0	19.0	6.5				
75	37.0	27.0	29.0	16.5				
80	37.0	27.0	29.0	16.5	26.0	11.0		
85	47.0	37.0	39.0	26.5	31.0	16.0		
90	47.0	37.0	39.0	26.5	36.0	21.0		
100	62.0	52.0	54.0	41.5	46.0	31.0	34.0	16.5
110	72.0	62.0	64.0	51.5	56.0	41.0	44.0	26.5
120	82.0	72.0	74.0	61.5	66.0	51.0	54.0	36.5
130	86.0	76.0	78.0	65.5	70.0	55.0	58.0	40.5
140	96.0	86.0	88.0	75.5	80.0	65.0	68.0	50.5
150	106.0	96.0	98.0	85.5	90.0	75.0	78.0	60.5
160	116.0	106.0	108.0	95.5	100.0	85.0	88.0	70.5
170			118.0	105.5	110.0	95.0	98.0	80.5
180			128.0	115.5	120.0	105.0	108.0	90.5
190			138.0	125.5	130.0	115.0	118.0	100.5
200			148.0	135.5	140.0	125.0	128.0	110.5
220					147.0	132.0	135.0	117.5
240					167.0	152.0	155.0	137.5

(Continued)

Table 81. *(Continued)* **Maximum Grip Gaging Lengths and Minimum Body Lengths for Metric Hexagon Head Bolts.**

Nom. Bolt Length	Nominal Bolt Diameter and Pitch							
	M16 × 2		M20 × 2.5		M24 × 3		M30 × 3.5	
	Max. Grip Length	Min. Body Length	Max. Grip Length	Min. Body Length	Max. Grip Length	Min. Body Length	Max. Grip Length	Min. Body Length
260							175.0	157.5
280							195.0	177.5

Notes: Bolts shorter than the sizes shown are threaded full length. All dimensions in millimeters. Source: DOD-B-70331.

Table 82. Dimensions of Metric Round (or Cup) Head Square Neck "Carriage" Bolts (DIN 603).

Bolt Size	E Body Dia.	A Head Dia.	H Head Height	P Square Depth	O Square Width	L_T Thread Length			
		Max.	Max.	Max.	Max.	L_T to 125	L_T to 200	L_T over 200	
M5	5.00	13.55	3.30	4.10	5.48	16.00	-	-	
M6	6.00	16.55	3.88	4.60	6.48	18.00	24.00	-	
M8	8.00	20.65	4.88	5.60	8.58	22.00	28.00	-	
M10	10.00	24.65	5.38	6.60	10.58	26.00	32.00	45.00	
M12	12.00	30.65	6.95	8.75	12.70	30.00	36.00	49.00	
M16	16.00	38.8	8.95	12.90	16.70	38.00	44.00	57.00	
M20	20.00	46.8	11.05	15.90	-	-	-	46.00	

All dimensions in millimeters.

Table 83. Nominal Dimensions of Metric Flat Countersunk Nib Bolts (DIN 604).

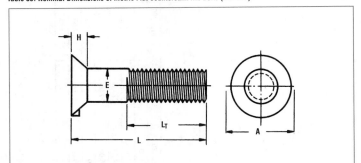

E Bolt Size	A Head Dia.	H Head Height	α Head Angle	L_T Thread Length
M8	16.00	5.00	90°	22.00
M10	19.00	5.50	90°	26.00
M12	24.00	7.00	90°	30.00
M16	32.00	9.00	90°	38.00
M20	32.00	11.50	60°	46.00
M24	38.80	13.00	60°	54.00

All dimensions in millimeters.

Table 84. Dimensions of Metric Flat Countersunk Square Neck Bolts - Long Square (DIN 605).

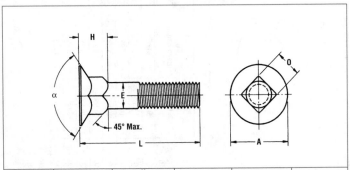

E Bolt Size	A Head Dia. Max.	H Head/Square Height Max.	α Head Angle	O Square Width	L_T Thread Length
M6	16.55	7.45	120°	6.48	18.00
M8	20.65	9.45	120°	8.58	22.00
M10	24.65	11.55	120°	10.58	26.00

All dimensions in millimeters.

Table 85. Nominal Dimensions of Metric Cup Head Nib Bolts (DIN 607).

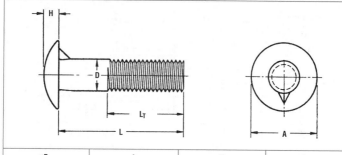

D Bolt Size	A Head Dia.	H Head Height	L_T Thread Length
M8	16.55	6.38	22.00
M10	19.65	7.45	26.00
M12	24.65	9.65	30.00
M16	30.65	11.75	38.00

All dimensions in millimeters.

Table 86. Nominal Dimensions of Metric Hexagon Nuts (ISO/DIN 4032, DIN 934).

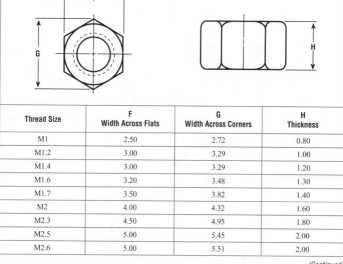

Thread Size	F Width Across Flats	G Width Across Corners	H Thickness
M1	2.50	2.72	0.80
M1.2	3.00	3.29	1.00
M1.4	3.00	3.29	1.20
M1.6	3.20	3.48	1.30
M1.7	3.50	3.82	1.40
M2	4.00	4.32	1.60
M2.3	4.50	4.95	1.80
M2.5	5.00	5.45	2.00
M2.6	5.00	5.51	2.00

(Continued)

Table 86. *(Continued)* Nominal Dimensions of Metric Hexagon Nuts (ISO/DIN 4032, DIN 934).

Thread Size	F Width Across Flats	G Width Across Corners	H Thickness
M3	5.50	6.01	2.40
M3.5	6.00	6.01	2.80
M4	7.00	7.66	3.20
M5	8.00	8.76	4.00
M6	10.00	11.05	5.00
M7	11.00	12.12	5.50
M8	13.00	14.38	6.50
M10	17.00	18.90	8.00
M12	19.00	21.10	10.00
M14	22.00	24.49	11.00
M16	24.00	26.75	13.00
M18	27.00	29.56	15.00
M20	30.00	32.95	16.00
M22	32.00	35.03	18.00
M24	36.00	39.55	19.00
M27	41.00	45.20	22.00
M30	46.00	50.85	24.00
M33	50.00	55.37	26.00
M36	55.00	60.79	29.00
M39	60.00	66.44	31.00
M42	65.00	71.30	34.00
M45	70.00	76.95	36.00
M48	75.00	82.60	38.00
M52	80.00	88.25	42.00
M56	85.00	95.07	45.00
M60	90.00	100.72	48.00
M64	95.00	106.37	51.00
M68	100.00	112.02	54.00
M72	105.00	117.67	58.00
M76	110.00	123.32	61.00
M80	115.00	128.97	64.00
M85	120.00	134.62	68.00
M90	130.00	145.77	72.00
M95	135.00	151.42	76.00
M100	145.00	162.72	80.00
M105	150.00	168.37	84.00
M110	155.00	174.02	88.00
M115	165.00	185.32	92.00
M120	170.00	190.97	96.00

(Continued)

Table 86. *(Continued)* **Nominal Dimensions of Metric Hexagon Nuts (ISO/DIN 4032, DIN 934).**

Thread Size	F Width Across Flats	G Width Across Corners	H Thickness
M125	180.00	202.27	100.00
M130	185.00	207.75	104.00
M135	190.00	213.40	108.00
M140	200.00	224.70	112.00
M145	210.00	236.00	116.00
M150	210.00	236.00	120.00

All dimensions in millimeters.

Table 87. Dimensions of Metric Hexagon Thin Nuts (DIN 439B).

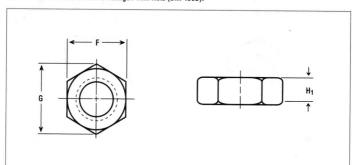

Thread Size	F Width Across Flats		G Width Across Corners	H Thickness	
	Max.	Min.	Min.	Max.	Min.
M2	4.00	-	4.32	1.20	-
M3	5.50	-	6.01	1.80	-
M4	7.00	-	7.66	2.20	-
M5	8.00	7.78	8.79	2.70	2.45
M6	10.00	9.78	11.05	3.20	2.90
M8	13.00	12.73	14.38	4.00	3.70
M10	17.00	15.73	18.90	5.00	4.70
M12	19.00	17.73	21.10	6.00	5.70
M14	22.00	-	24.49	7.00	-
M16	24.00	23.67	26.75	8.00	7.42
M18	27.00	-	29.56	9.00	-
M20	30.00	29.16	32.95	10.00	9.10
M22	32.00	-	35.03	11.00	-
M24	36.00	35.00	39.55	12.00	10.90

(Continued)

Table 87. *(Continued)* **Dimensions of Metric Hexagon Thin Nuts (DIN 439B).**

Thread Size	F Width Across Flats		G Width Across Corners	H Thickness	
	Max.	Min.	Min.	Max.	Min.
M27	41.00	40.00	45.20	13.50	12.40
M30	46.00	45.00	50.85	15.00	13.90
M33	50.00	49.00	55.37	16.50	15.40
M36	55.00	53.80	60.79	18.00	16.90

All dimensions in millimeters.

Table 88. Nominal Dimensions of Metric Structural Hexagon Nuts (DIN 6915). (See Notes.)

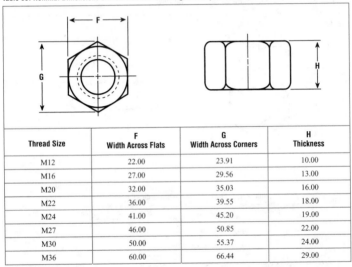

Thread Size	F Width Across Flats	G Width Across Corners	H Thickness
M12	22.00	23.91	10.00
M16	27.00	29.56	13.00
M20	32.00	35.03	16.00
M22	36.00	39.55	18.00
M24	41.00	45.20	19.00
M27	46.00	50.85	22.00
M30	50.00	55.37	24.00
M36	60.00	66.44	29.00

Notes: These hexagon nuts have increased width across flats for high tensile structural bolting. Designed to be used with bolt specified in DIN 6914 and washer specified in DIN 6916. All dimensions in millimeters.

Table 89. Nominal Dimensions of Metric Hexagon Flange Nuts (DIN 6923).

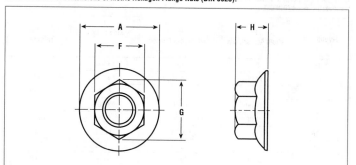

Thread Size	F Width Across Flats	G Width Across Corners	H Thickness	A Width of Flange
M5	8.00	8.79	5.00	11.80
M6	10.00	11.05	6.00	14.20
M8	13.00	14.38	8.00	17.90
M10	15.00	16.64	10.00	21.80
M12	18.00	20.03	12.00	26.00
M14	21.00	23.36	14.00	29.90
M16	24.00	26.75	16.00	34.50
M20	30.00	32.95	20.00	42.80

All dimensions in millimeters.

Table 90. Nominal Dimensions of Metric High-Style Prevailing Torque Hexagon Nut with Nonmetallic Insert (DIN 982).

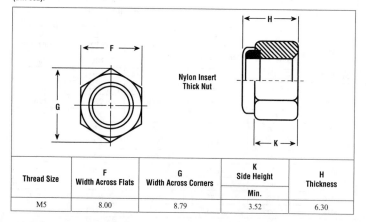

Thread Size	F Width Across Flats	G Width Across Corners	K Side Height Min.	H Thickness
M5	8.00	8.79	3.52	6.30

Table 90. *(Continued)* **Nominal Dimensions of Metric High-Style Prevailing Torque Hexagon Nut with Nonmetallic Insert (DIN 982).**

Thread Size	F Width Across Flats	G Width Across Corners	K Side Height Min.	H Thickness
M6	10.00	11.05	3.92	8.00
M8	13.00	14.38	5.15	9.50
M10	17.00	18.90	6.43	11.50
M12	19.00	21.10	8.30	14.00
M14	22.00	24.49	9.68	16.00
M16	24.00	26.75	11.28	18.00
M20	30.00	32.95	13.52	22.00
M24	36.00	39.55	16.16	28.00

All dimensions in millimeters.

Table 91. Nominal Dimensions of Metric Low Style Prevailing Torque Hexagon Nuts with Nonmetallic Insert. (DIN 985).

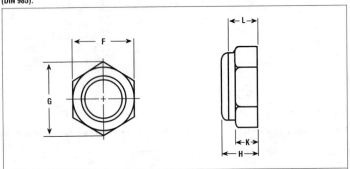

Thread Size	F Width Across Flats	G Width Across Corners	H Thickness	L Side Height Min.	K Thread Length Min.
M2.5/M2.6	5.00	5.51	3.80	2.00	-
M3	5.50	6.01	4.00	2.40	1.65
M4	7.00	7.66	5.00	2.90	2.20
M5	8.00	8.79	5.00	3.20	2.75
M6	10.00	11.05	6.00	4.00	3.30
M8	13.00	14.38	8.00	5.50	4.40
M10	17.00	18.90	10.00	6.50	5.50
M12	19.00	21.10	12.00	8.00	6.60
M14	22.00	24.49	14.00	9.50	7.70
M16	24.00	26.75	16.00	10.50	8.80
M18	27.00	29.56	18.50	13.00	9.90

(Continued)

Table 91. *(Continued)* **Nominal Dimensions of Metric Low Style Prevailing Torque Hexagon Nuts with Nonmetallic Insert. (DIN 985).**

Thread Size	F Width Across Flats	G Width Across Corners	H Thickness	L Side Height Min.	K Thread Length Min.
M20	30.00	32.95	20.00	14.00	11.00
M22	32.00	35.03	22.00	15.00	12.20
M24	36.00	39.55	24.00	15.00	13.20
M27	41.00	45.02	27.00	17.00	14.80
M30	46.00	50.85	30.00	19.00	16.50
M33	50.00	55.37	33.00	22.00	18.20
M36	55.00	60.79	36.00	25.00	19.80
M39	60.00	66.44	39.00	27.00	21.50
M42	65.00	72.09	42.00	29.00	23.10
M45	70.00	76.95	45.00	32.00	24.80
M48	75.00	82.60	48.00	36.00	26.50

All dimensions in millimeters.

Table 92. Nominal Dimensions of Metric Prevailing Torque Hexagon Nuts (DIN 980 Form V).

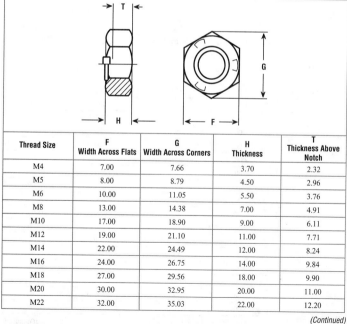

Thread Size	F Width Across Flats	G Width Across Corners	H Thickness	T Thickness Above Notch
M4	7.00	7.66	3.70	2.32
M5	8.00	8.79	4.50	2.96
M6	10.00	11.05	5.50	3.76
M8	13.00	14.38	7.00	4.91
M10	17.00	18.90	9.00	6.11
M12	19.00	21.10	11.00	7.71
M14	22.00	24.49	12.00	8.24
M16	24.00	26.75	14.00	9.84
M18	27.00	29.56	18.00	9.90
M20	30.00	32.95	20.00	11.00
M22	32.00	35.03	22.00	12.20

(Continued)

Table 92. *(Continued)* Nominal Dimensions of Metric Prevailing Torque Hexagon Nuts (DIN 980 Form V).

Thread Size	F Width Across Flats	G Width Across Corners	H Thickness	T Thickness Above Notch
M24	36.00	39.55	24.00	13.20
M27	41.00	45.02	27.00	14.80
M30	46.00	50.85	30.00	16.50
M33	50.00	55.37	33.00	18.20
M36	55.00	60.79	36.00	19.80

All dimensions in millimeters.

Table 93. Nominal Dimensions of Metric Hexagon Castle Nuts (DIN 935).

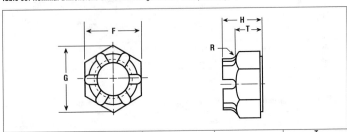

Thread Size	F Width Across Flats	G Width Across Corners	H Thickness	T Unslotted Thickness Min.
M4	7.00	7.74	5.00	3.20
M5	8.00	8.87	6.00	4.00
M6	10.00	11.05	7.50	5.00
M7	11.00	12.12	8.00	5.50
M8	13.00	14.38	9.50	6.50
M10	17.00	18.90	12.00	8.00
M12	19.00	21.10	15.00	10.00
M14	22.00	24.49	16.00	11.00
M16	24.00	26.75	19.00	13.00
M18	27.00	29.56	21.00	15.00
M20	30.00	32.95	22.00	16.00
M22	32.00	35.03	26.00	18.00
M24	36.00	39.55	27.00	19.00
M27	41.00	45.20	30.00	22.00
M30	46.00	50.85	33.00	24.00
M33	50.00	55.37	35.00	26.00
M36	55.00	60.79	38.00	29.00
M39	60.00	66.44	40.00	31.00
M42	65.00	71.30	46.00	34.00

(Continued)

Table 93. *(Continued)* Nominal Dimensions of Metric Hexagon Castle Nuts (DIN 935).

Thread Size	F Width Across Flats	G Width Across Corners	H Thickness	T Unslotted Thickness Min.
M45	70.00	78.26	48.00	36.00
M48	75.00	82.60	50.00	38.00
M52	80.00	88.25	54.00	42.00
M56	85.00	95.07	57.00	45.00
M58	90.00	100.72	63.00	48.00
M60	90.00	100.72	63.00	48.00
M64	95.00	106.37	66.00	51.00
M68	100.00	112.02	69.00	54.00
M72	105.00	117.67	73.00	58.00
M76	110.00	123.32	76.00	61.00
M80	115.00	128.97	79.00	64.00
M85	120.00	134.62	88.00	68.00
M90	130.00	145.77	92.00	72.00
M95	135.00	151.42	96.00	76.00
M100	145.00	162.72	100.00	80.00

Table 94. Nominal Dimensions of Metric Domed Cap Hexagon Nuts (DIN 1587).

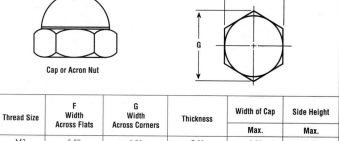

Cap or Acron Nut

Thread Size	F Width Across Flats	G Width Across Corners	Thickness	Width of Cap Max.	Side Height Max.
M3	5.50	6.01	7.00	5.50	2.40
M4	7.00	7.66	8.00	6.50	3.20
M5	8.00	8.79	10.00	7.50	4.00
M6	10.00	11.05	12.00	9.50	5.00
M8	13.00	14.38	15.00	12.50	6.50
M10	17.00	18.90	18.00	16.00	8.00
M12	19.00	21.10	22.00	18.00	10.00
M14	22.00	24.49	25.00	21.00	11.00

(Continued)

Table 94. *(Continued)* **Nominal Dimensions of Metric Domed Cap Hexagon Nuts (DIN 1587).**

Thread Size	F Width Across Flats	G Width Across Corners	Thickness	Width of Cap Max.	Side Height Max.
M16	24.00	26.75	28.00	23.00	13.00
M20	30.00	33.53	34.00	28.00	16.00
M24	36.00	39.98	42.00	34.00	19.00

All dimensions in millimeters.

Table 95. Nominal Dimensions of Metric Wing Nuts (DIN 315).

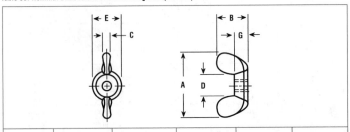

D Thread Size	A Wing Spread	B Wing Height	C Wing Thickness	E Boss Dia.	G Wall Height
M4	18.00	8.50	1.70	6.00	3.20
M5	24.00	11.00	2.30	8.00	4.00
M6	30.00	15.00	2.70	10.00	5.00
M8	36.00	18.00	3.60	13.00	6.50
M10	48.00	23.00	4.60	17.00	8.00
M12	62.00	31.00	5.60	20.00	10.00
M16	70.00	35.00	6.50	26.00	13.00
M20	86.00	44.00	7.00	32.00	16.00
M24	106.00	53.50	8.60	41.00	20.00

All dimensions in millimeters.

Table 96. Nominal Dimensions of Medium (DIN 125) and Grade C (DIN 126) Metric Washers for Hexagon Bolts.

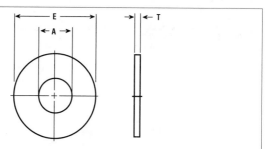

A Inside Dia. DIN 125	A Inside Dia. DIN 126	For Screw Size	E External Dia.	T Thickness
1.7	-	M1.6	4.00	0.30
2.2	-	M2	5.00	0.30
2.7	-	M2.5	6.00	0.50
3.2	-	M3	7.00	0.50
3.7	-	M3.5	8.00	0.50
4.3	-	M4	9.00	0.80
5.3	5.5	M5	10.00	1.00
6.4	6.6	M6	12.00	1.60
7.4	7.6	M7	14.00	1.60
8.4	9.0	M8	16.00	1.60
10.5	11.0	M10	20.00	2.00
13.0	13.5	M12	24.00	2.50
15.0	15.5	M14	28.00	2.50
17.0	17.5	M16	30.00	3.00
19.0	-	M18	34.00	3.00
21.0	22.0	M20	37.00	3.00
23.0	-	M22	39.00	3.00
25.0	26.0	M24	44.00	4.00
28.0	30.0	M27	50.00	4.00
31.0	33.0	M30	56.00	4.00
34.0	36.0	M33	60.00	5.00
37.0	39.0	M36	66.00	5.00
40.0	42.0	M39	72.00	6.00
43.0	45.0	M42	78.00	7.00
-	48.0	M45	85.00	7.00
50.0	52.0	M48	92.00	8.00

(Continued)

Table 96. *(Continued)* **Nominal Dimensions of Medium (DIN 125) and Grade C (DIN 126) Metric Washers for Hexagon Bolts.**

A Inside Dia. DIN 125	A Inside Dia. DIN 126	For Screw Size	E External Dia.	T Thickness
54.0	56.0	M52	98.00	8.00
-	62	M56	105.00	9.00
-	66	M60	110.00	9.00
-	70	M64	115.00	9.00

All dimensions in millimeters.

Table 97. Nominal Dimensions of Metric Split Lock Washers (DIN 127, Form A and B). (See Notes.)

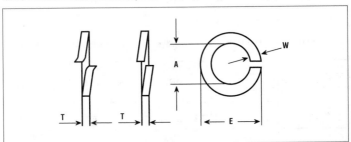

A Inside Dia.	E External Dia.	For Screw Size	W Width	T Thickness
2.1	M2	4.40	0.90	0.50
2.6	M2.5	5.10	1.00	0.60
3.1	M3	6.20	1.30	0.80
3.6	M3.5	6.70	1.30	0.80
4.1	M4	7.60	1.50	0.09
5.1	M5	9.20	1.80	1.20
6.1	M6	11.80	2.50	1.60
8.1	M8	14.80	3.00	2.00
10.2	M10	18.10	3.50	2.20
12.2	M12	21.10	4.00	2.50
14.2	M14	24.10	4.50	3.00
16.2	M16	27.40	5.00	3.50
18.2	M18	29.40	5.00	3.50
20.2	M20	33.60	6.00	4.00
22.5	M22	35.90	6.00	4.00
24.5	M24	40.00	7.00	5.00
27.5	M27	43.00	7.00	5.00
30.5	M30	48.20	8.00	6.00

(Continued)

Table 97. *(Continued)* **Nominal Dimensions of Metric Split Lock Washers (DIN 127, Form A and B).** (See Notes.)

A Inside Dia.	E External Dia.	For Screw Size	W Width	T Thickness
33.5	M33	55.20	10.00	6.00
36.5	M36	58.20	10.00	6.00
39.0	M38	61.20	10.00	6.00
42.5	M42	68.20	12.00	7.00
45.5	M45	71.20	12.00	7.00
49.0	M48	75.00	12.00	7.00
53.0	M52	86.00	14.00	8.00

Notes: Form A washers have tangs. Form B washers have square ends. All dimensions in millimeters.

Table 98. Nominal Dimensions of External Toothed (Form A) and Internal Toothed (Form J) Metric Lock Washers (DIN 6797).

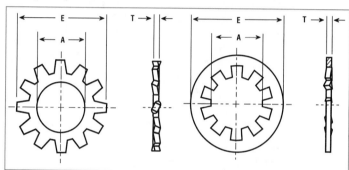

A Inside Dia.	For Screw Size	E External Dia.	T Thickness
2.20	M2	4.50	0.30
2.50	M2.3	5.00	0.40
2.80	M2.5	5.50	0.40
3.20	M3	6.00	0.40
3.70	M3.5	7.00	0.50
4.30	M4	8.00	0.50
5.10	M5	9.00	0.50
5.30	M5	10.00	0.60
6.40	M6	11.00	0.70
8.20	M8	14.00	0.80
8.40	M8	15.00	0.80
10.50	M10	18.00	0.90
13.00	M12	20.50	1.00

(Continued)

Table 98. *(Continued)* **Nominal Dimensions of External Toothed (Form A) and Internal Toothed (Form J) Metric Lock Washers (DIN 6797).**

A Inside Dia.	For Screw Size	E External Dia.	T Thickness
15.00	M14	24.00	1.00
17.00	M16	26.00	1.20
21.00	M20	33.00	1.40
25.00	M24	38.00	1.50
31.00	M30	48.00	1.60

All dimensions in millimeters.

Table 99. Nominal Dimensions of External (Form A) and Internal (Form J) Serrated Metric Lock Washers (DIN 6798).

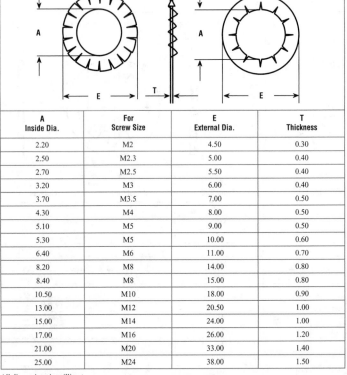

A Inside Dia.	For Screw Size	E External Dia.	T Thickness
2.20	M2	4.50	0.30
2.50	M2.3	5.00	0.40
2.70	M2.5	5.50	0.40
3.20	M3	6.00	0.40
3.70	M3.5	7.00	0.50
4.30	M4	8.00	0.50
5.10	M5	9.00	0.50
5.30	M5	10.00	0.60
6.40	M6	11.00	0.70
8.20	M8	14.00	0.80
8.40	M8	15.00	0.80
10.50	M10	18.00	0.90
13.00	M12	20.50	1.00
15.00	M14	24.00	1.00
17.00	M16	26.00	1.20
21.00	M20	33.00	1.40
25.00	M24	38.00	1.50

All dimensions in millimeters.

Table 100. Nominal Dimensions of Curved (Form A) Spring Metric Lock Washers (DIN 128).

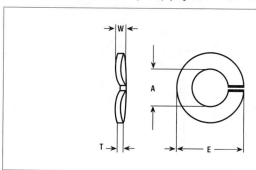

A Inside Dia.	For Screw Size	E External Dia.	T Thickness	W Uncompressed Thickness	
				Max.	Min.
2.1	M2	4.40	0.50	0.90	0.70
2.6	M2.5	5.10	0.60	1.10	0.90
3.1	M3	6.20	0.70	1.30	1.10
3.6	M3.5	6.70	0.70	1.30	1.10
4.1	M4	7.60	0.80	1.40	1.20
5.1	M5	9.20	1.00	1.70	1.50
6.1	M6	11.80	1.30	2.20	2.00
8.1	M8	14.80	1.60	2.75	2.45
10.2	M10	18.10	1.80	3.15	2.85
12.2	M12	21.10	2.10	3.65	3.35
14.2	M14	24.10	2.40	4.30	3.90
16.2	M16	27.40	2.80	5.10	4.50
18.2	M17	29.40	2.80	5.10	4.50
20.2	M20	33.60	3.20	5.90	5.10
22.2	M21	35.90	3.20	5.90	5.10
24.5	M24	40.00	4.00	7.50	6.50
30.5	M30	48.20	6.00	10.50	9.50

All dimensions in millimeters.

Table 101. Nominal Dimensions of Metric Curved (Form A) Spring Washers (DIN 137).

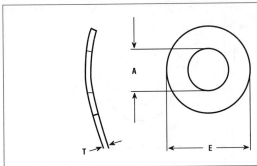

A Inside Dia.	For Screw Size	E External Dia.	T Thickness
1.10	M1	2.50	0.20
1.30	M1.2	3.00	0.20
1.50	M1.4	3.00	0.25
1.80	M1.6/M1.7	4.00	0.25
2.20	M2	4.50	0.30
2.80	M2.5	5.50	0.30
3.20	M3	6.00	0.40
3.70	M3.5	7.00	0.40
4.30	M4	8.00	0.50
5.30	M5	10.00	0.50
6.40	M6	11.00	0.50
7.40	M7	12.00	0.50
8.40	M8	15.00	0.50
10.50	M10	18.00	0.80

All dimensions in millimeters.

Table 102. Nominal Dimensions of Metric Waved (Form B) Spring Washers (DIN 137).

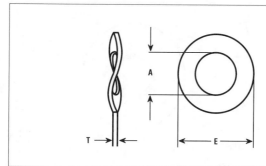

A Inside Dia.	For Screw Size	E External Dia.	T Thickness
3.20	M3	8.00	0.50
3.70	M3.5	8.00	0.50
4.30	M4	9.00	0.50
5.30	M5	11.00	0.50
6.40	M6	12.00	0.50
7.40	M7	14.00	0.80
8.40	M8	15.00	0.80
10.50	M10	21.00	1.00
13.00	M12	24.00	1.20
15.00	M14	28.00	1.60
17.00	M16	30.00	1.60
19.00	M18	34.00	1.60
21.00	M20	36.00	1.60
23.00	M22	40.00	1.80
25.00	M24	44.00	1.80
28.00	M27	50.00	2.00
31.00	M30	56.00	2.20
34.00	M33	60.00	2.20
37.00	M36	68.00	2.50

All dimensions in millimeters.

Table 103. Nominal Dimensions of Metric Disc Spring Washers (DIN 2093).

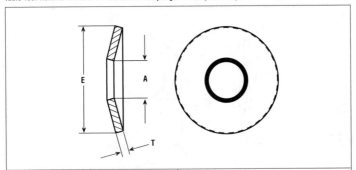

A Inside Dia.	E External Dia.	T Thickness
3.20	8.00	0.20, 0.30, 0.40
4.20	8.00	0.20, 0.30, 0.40
3.20	10.00	0.30, 0.40, 0.50
4.20	10.00	0.40, 0.50
5.20	10.00	0.25, 0.40, 0.50
4.20	12.00	0.40, 0.50, 0.60
5.20	12.00	0.50, 0.60
6.20	12.00	0.50, 0.60
5.20	12.50	0.50
6.20	12.50	0.35, 0.50, 0.70
7.20	14.00	0.35, 0.50, 0.80
5.20	15.00	0.40, 0.50, 0.60, 0.70
6.20	15.00	0.50, 0.60, 0.70
8.20	15.00	0.70, 0.80
8.20	16.00	0.40, 0.60, 0.70, 0.80, 0.90
6.20	18.00	0.40, 0.50, 0.60, 0.70, 0.80
8.20	18.00	0.50, 0.70, 0.80, 1.00
9.20	18.00	0.45, 0.70, 1.00
8.20	20.00	0.60, 0.70, 0.80, 0.90, 1.00
10.20	20.00	0.50, 0.80, 0.90, 1.00, 1.10, 1.25, 1.50
11.20	22.50	0.60, 0.80, 1.25
8.20	23.00	0.70, 0.80, 0.90, 1.00
10.20	23.00	0.90, 1.00, 1.25
12.20	23.00	1.00, 1.25, 1.50
10.20	25.00	1.00
12.20	25.00	0.70, 0.90, 1.00, 1.25, 1.50

(Continued)

Table 103. *(Continued)* Nominal Dimensions of Metric Disc Spring Washers (DIN 2093).

A Inside Dia.	E External Dia.	T Thickness
10.20	28.00	0.80, 1.00, 1.25, 1.50
12.20	28.00	1.00, 1.25, 1.50
14.20	28.00	0.80, 1.00, 1.25, 1.50
12.20	31.50	1.00, 1.25, 1.50
16.30	31.50	0.80, 1.25, 1.50, 1.75, 2.00
12.30	34.00	1.00, 1.25, 1.50
14.30	34.00	1.25, 1.50
16.30	34.00	1.50, 2.00
18.30	35.50	0.90, 1.25, 2.00
14.30	40.00	1.25, 1.50, 2.00
16.30	40.00	1.50, 2.00
18.30	40.00	2.00
20.40	40.00	1.00, 1.50, 2.00, 2.25, 2.50
22.40	45.00	1.25, 1.75, 2.50
18.40	50.00	1.25, 1.50, 2.00, 2.50, 3.00
20.40	50.00	2.00, 2.50
22.40	50.00	2.00, 2.50
25.40	50.00	1.25, 1.50, 2.00, 2.50, 3.00
28.50	56.00	1.50, 2.00, 3.00
20.50	60.00	2.00, 2.50, 3.00
25.50	60.00	2.50, 3.00
30.50	60.00	2.50, 3.00, 3.50
31.00	63.00	1.80, 2.50, 3.00, 3.50
25.50	70.00	2.00
30.50	70.00	2.50, 3.00
35.50	70.00	3.00, 4.00
40.50	70.00	4.00, 5.00
36.00	71.00	2.00, 2.50, 4.00
31.00	80.00	2.50, 3.00, 4.00
36.00	80.00	3.00, 4.00
41.00	80.00	2.25, 3.00, 4.00, 5.00
46.00	90.00	2.50, 3.50, 5.00
41.00	100.00	4.00, 5.00
51.00	100.00	2.70, 3.50, 4.00, 5.00, 6.00
57.00	112.00	3.00, 4.00, 6.00
41.00	125.00	4.00
51.00	125.00	4.00, 5.00, 6.00
61.00	125.00	5.00, 6.00, 8.00

(Continued)

Table 103. *(Continued)* **Nominal Dimensions of Metric Disc Spring Washers (DIN 2093).**

A Inside Dia.	E External Dia.	T Thickness
64.00	125.00	3.50, 5.00, 8.00
71.00	125.00	6.00, 8.00, 10.00
72.00	140.00	3.80, 5.00, 8.00
61.00	150.00	5.00, 6.00
71.00	150.00	6.00, 8.00
81.00	150.00	8.00, 10.00

All dimensions in millimeters.

Table 104. Nominal Dimensions of Metric Heavy Structural Steel Bolting Washer (DIN 6916). (See Notes.)

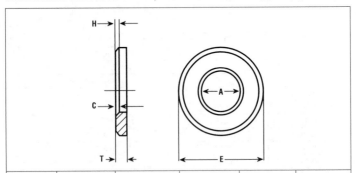

A Inside Dia.	For Thread Size	E External Dia.	T Thickness	H Chamfer Height	C Chamfer Width
13.00	M12	24.00	3.00	1.60	0.50
17.00	M16	30.00	4.00	1.60	1.00
21.00	M20	37.00	4.00	2.00	1.00
23.00	M22	39.00	4.00	2.00	1.00
25.00	M24	44.00	4.00	2.00	1.00
28.00	M27	50.00	5.00	2.50	1.00
31.00	M30	56.00	5.00	2.50	1.00
37.00	M36	66.00	6.00	3.00	1.50

Notes: These washers are intended for high tensile structural bolting. Use with hexagon bolt specified in DIN 9614 and hex nut specified in DIN 6915. All dimensions in millimeters.

Table 105. Nominal Dimensions of Large Size Metric Fender Washers (DIN 9021).

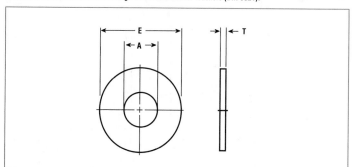

A Inside Dia.	For Screw Size	E External Dia. Max.	T Thickness
2.50	M2.3	7.00	0.80
2.70	M2.5	8.00	0.80
3.20	M3	9.00	0.80
4.30	M4	12.00	1.00
5.30	M5	15.00	1.20
6.40	M6	18.00	1.60
8.40	M8	24.00	2.00
10.50	M10	30.00	2.50
13.00	M12	37.00	3.00
15.00	M14	44.00	3.00
17.00	M16	50.00	3.00
20.00	M18	56.00	4.00
22.00	M20	60.00	4.00

All dimensions in millimeters.

Table 106. Nominal Dimensions of Metric Washers for Cheese Head Screws (DIN 433).

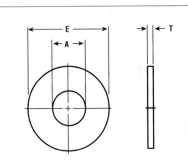

A Inside Dia.	For Screw Size	E External Dia.	T Thickness
1.10	M1	2.50	0.30
1.30	M1.3	3.00	0.30
1.50	M1.5	3.00	0.30
1.90	M1.8	4.00	0.30
2.20	M2	4.50	0.30
2.70	M2.5	5.00	0.50
3.20	M3	6.00	0.50
3.70	M3.5	7.00	0.50
4.30	M4	8.00	0.50
5.30	M5	9.00	1.00
6.40	M6	11.00	1.60
8.40	M8	15.00	1.60
10.50	M10	18.00	1.60
13.00	M12	20.00	2.00
15.00	M14	24.00	2.50
17.00	M16	28.00	2.50
19.00	M18	30.00	2.50
21.00	M20	34.00	3.00

All dimensions in millimeters.

Table 107. Nominal Dimensions of Medium Metric Washers for Clevis Pins (DIN 1440).

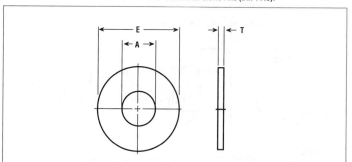

For Pin Size	E External Dia.	T Thickness
M3	6.00	0.80
M4	8.00	0.80
M5	10.00	0.80
M6	12.00	1.60
M7	14.00	1.60
M8	16.00	2.00
M10	20.00	2.50
M12	25.00	3.00
M13	25.00	3.00
M14	28.00	3.00
M16	28.00	3.00
M18	30.00	4.00
M20	32.00	4.00
M22	34.00	4.00
M23	36.00	4.00
M25	40.00	4.00
M26	40.00	5.00
M28	42.00	5.00
M30	45.00	5.00
M32	50.00	5.00
M33	50.00	5.00
M35	52.00	6.00
M36	52.00	6.00
M40	58.00	6.00
M45	62.00	7.00
M50	68.00	8.00

All dimensions in millimeters.

Table 108. Nominal Dimensions of Metric Hexagon Cap Screw (DIN 933).

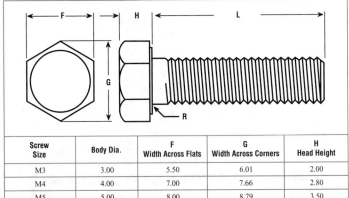

Screw Size	Body Dia.	F Width Across Flats	G Width Across Corners	H Head Height
M3	3.00	5.50	6.01	2.00
M4	4.00	7.00	7.66	2.80
M5	5.00	8.00	8.79	3.50
M6	6.00	10.00	11.05	4.00
M8	8.00	13.00	14.38	5.30
M10	10.00	17.00	18.90	6.40
M12	12.00	19.00	21.10	7.50
M14	14.00	22.00	24.49	8.80
M16	16.00	24.00	26.75	10.00
M18	18.00	27.00	30.14	11.50
M20	20.00	30.00	33.53	12.50
M22	22.00	32.00	35.72	14.00
M24	24.00	36.00	39.98	15.00
M27	27.00	41.00	45.20	17.00
M30	30.00	46.00	50.85	18.70
M33	33.00	50.00	55.37	21.00
M36	36.00	55.00	60.79	22.50
M42	42.00	65.00	71.30	26.00
M48	48.00	75.00	82.60	30.00

All dimensions in millimeters.

Table 109. Dimensions of Metric Lag (or Coach) Screws (DIN 571). (See Notes.)

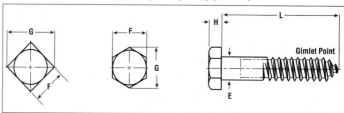

Screw Size	E Body Dia.		F Width Across Flats		G Width Across Corners	H Head Height	
	Max.	Min.	Max.	Min.	Min.	Max.	Min.
M6	6.48	5.52	10.0	9.64	10.89	4.38	3.62
M8	8.58	7.42	13.0	12.57	14.20	5.68	4.92
M10	10.58	9.42	16.0	15.57	18.72	6.85	5.95
M12	12.70	11.30	18.0	17.57	20.88	7.95	7.05
M16	16.70	15.30	24.0	23.16	26.17	10.75	9.25

Notes: L_T (length of thread) = 0.61L mm (minimum). All dimensions in millimeters.

Table 110. Nominal Dimensions of Metric Cross Recessed Raised Cheese Head Screws (DIN 7985).

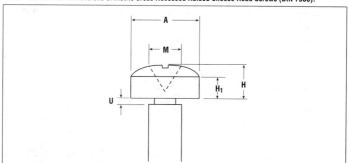

Screw Size	A Head Dia.	H Head Height	H_1 Flat Height	U Neck Relief Width	M Recess Width
	Max.		Approx.		Approx.
M3	6.00	2.40	1.60	1.00	3.10
M4	8.00	3.10	2.00	1.40	4.60
M5	10.00	3.80	2.50	1.60	5.30
M6	12.00	4.60	3.00	2.00	6.80
M8	16.00	6.00	3.70	2.50	9.00

All dimensions in millimeters.

Table 111. Nominal Dimensions of Metric Slotted Cheese Head Screws (DIN 84).

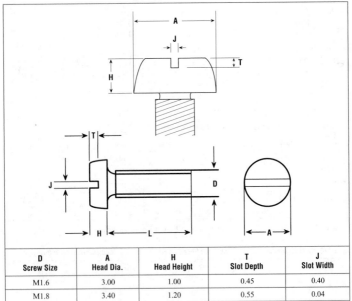

D Screw Size	A Head Dia.	H Head Height	T Slot Depth	J Slot Width
M1.6	3.00	1.00	0.45	0.40
M1.8	3.40	1.20	0.55	0.04
M2	3.80	1.30	0.60	0.50
M2.5	4.50	1.60	0.70	0.60
M3	5.50	2.00	0.85	0.80
M3.5	6.00	2.40	1.00	1.00
M4	7.00	2.60	1.10	1.20
M5	8.50	3.30	1.30	1.20
M6	10.00	3.90	1.60	1.60
M8	13.00	5.00	2.00	2.00
M10	16.00	6.00	2.40	2.50
M12	18.00	7.00	2.40	2.50

All dimensions in millimeters.

Table 112. Nominal Dimensions of Metric Slotted Pan Head Screws (DIN 85).

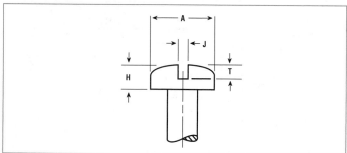

Screw Size	A Head Dia.	H Head Height	T Slot Depth	J Slot Width
M2.5/M2.6	5.00	1.50	0.60	0.60
M3	6.00	1.80	0.70	0.80
M4	8.00	2.40	1.00	1.20
M5	10.00	3.00	1.20	1.20
M6	12.00	3.60	1.40	1.60
M8	16.00	4.80	1.90	2.00
M10	20.00	6.00	2.40	2.50

All dimensions in millimeters.

Table 113. Nominal Dimensions of Metric Slotted Flat Head Countersunk Head Screws (DIN 963).

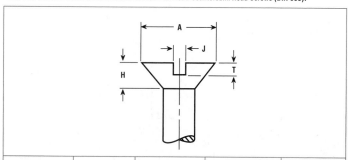

Screw Size	A Head Dia.	H Head Height	T Slot Depth Min.	J Slot Width
M1	1.90	0.60	0.20	0.25
M1.2	2.30	0.72	0.25	0.30
M1.4	2.60	0.84	0.28	0.30
M1.6	3.00	0.96	0.32	0.40

(Continued)

Table 113. *(Continued)* Nominal Dimensions of Metric Slotted Flat Head Countersunk Head Screws (DIN 963).

Screw Size	A Head Dia.	H Head Height	T Slot Depth Min.	J Slot Width
M1.8	3.40	1.08	0.35	0.40
M2	3.80	1.20	0.40	0.50
M2.5	4.70	1.50	0.50	0.60
M3	5.60	1.65	0.60	0.80
M3.5	6.50	1.93	0.70	0.80
M4	7.50	2.20	0.80	1.00
M5	9.20	2.50	1.00	1.20
M6	11.00	3.00	1.20	1.60
M8	14.50	4.00	1.60	2.00
M10	18.00	5.00	2.00	2.50
M12	22.00	6.00	2.40	3.00
M14	25.00	7.00	2.80	3.00
M16	29.00	8.00	3.20	4.00
M18	33.00	9.00	3.60	4.00
M20	36.00	10.00	4.00	5.00

All dimensions in millimeters.

Table 114. Nominal Dimensions of Metric Slotted Oval Countersunk Head Screws (DIN 964).

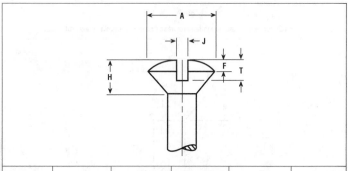

Screw Size	A Head Dia.	H Head Height	F Raised Height	T Slot Depth Min.	J Slot Width
M1	1.90	0.60	0.25	0.40	0.25
M1.2	2.30	0.72	0.30	0.50	0.30
M1.4	2.60	0.84	0.35	0.52	0.30
M1.6	3.00	0.96	0.40	0.65	0.40
M2	3.80	1.20	0.50	0.80	0.50

(Continued)

Table 114. *(Continued)* Nominal Dimensions of Metric Slotted Oval Countersunk Head Screws (DIN 964).

Screw Size	A Head Dia.	H Head Height	F Raised Height	T Slot Depth Min.	J Slot Width
M2.5	4.70	1.50	0.60	1.00	0.60
M3	5.60	1.65	0.75	1.20	0.80
M3.5	6.50	1.93	0.90	1.40	0.80
M4	7.50	2.20	1.00	1.60	1.00
M5	9.20	2.50	1.25	2.00	1.20
M6	11.00	3.00	1.50	2.40	1.60
M8	14.50	4.00	2.00	3.20	2.00
M10	18.00	5.00	2.50	4.00	2.50

All dimensions in millimeters.

Table 115. Nominal Dimensions of Metric Cross Recessed Countersunk Flat Head Screws (DIN 965).

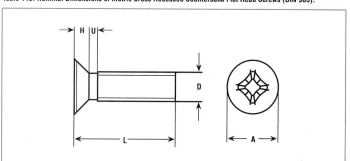

D Screw Size	A Head Dia.	H Head Height	U Neck Relief Width
M1.6	3.00	0.96	0.70
M1.8	3.40	1.08	0.70
M2	3.80	1.20	0.80
M2.5	4.70	1.50	0.90
M3	5.60	1.65	1.00
M3.5	6.50	1.93	1.20
M4	7.50	2.20	1.40
M5	9.20	2.50	1.60
M6	11.00	3.00	2.00
M8	14.50	4.00	2.50
M10	18.00	5.00	3.00

All dimensions in millimeters.

Table 116. Nominal Dimensions of Metric Cross Recessed Raised Countersunk Flat Head Screws (DIN 966).

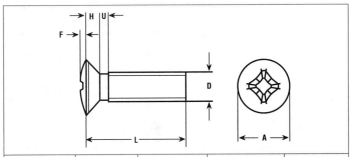

D Screw Size	A Head Dia.	H Head Height	F Raised Height	U Neck Relief Width
M1.6	3.00	0.96	0.40	0.70
M1.8	3.40	1.08	0.45	0.70
M2	3.80	1.20	0.50	0.80
M2.5	4.70	1.50	0.60	0.90
M3	5.60	1.65	0.75	1.00
M3.5	6.50	1.93	0.90	1.20
M4	7.50	2.20	1.00	1.40
M5	9.20	2.50	1.25	1.50
M6	11.00	3.00	1.50	2.00
M8	14.50	4.00	2.00	2.50
M10	18.00	5.00	2.50	3.00

All dimensions in millimeters.

Table 117. Nominal Dimensions of Metric Recessed Pan Head Sheet Metal Tapping Screws (ISO/DIN 7049, DIN 7981).

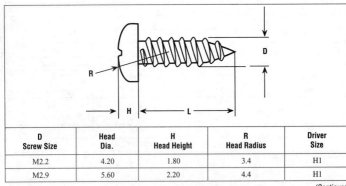

D Screw Size	Head Dia.	H Head Height	R Head Radius	Driver Size
M2.2	4.20	1.80	3.4	H1
M2.9	5.60	2.20	4.4	H1

(Continued)

Table 117. *(Continued)* **Nominal Dimensions of Metric Recessed Pan Head Sheet Metal Tapping Screws (ISO/DIN 7049, DIN 7981).**

D Screw Size	A Head Dia.	H Head Height	R Head Radius	Driver Size
M3.5	6.90	2.60	5.4	H2
M3.9	7.50	2.80	5.8	H2
M4.2	8.20	3.05	6.2	H2
M4.8	9.50	3.55	7.2	H2
M5.5	10.80	3.95	8.2	H3
M6.3	12.50	4.55	9.5	H3
M8	15.80	5.25	-	-

All dimensions in millimeters.

Table 118. Dimensions of Metric Recessed Flat Head Countersunk Sheet Metal Tapping Screws (ISO/DIN 7050, DIN 7982).

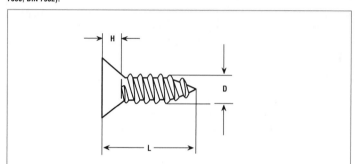

D Screw Size	Head Dia.	H Head Height	Neck Relief Width
M2.2	4.30	1.30	H1
M2.9	5.50	1.70	H1
M3.5	6.80	2.10	H2
M3.9	7.50	2.30	H2
M4.2	8.10	2.50	H2
M4.8	9.50	3.00	H2
M5.5	10.80	3.40	H3
M6.3	12.40	3.80	H3
M8	15.80	4.65	-

All dimensions in millimeters.

Table 119. Nominal Dimensions of Metric Recessed Oval Head Countersunk Sheet Metal Tapping Screws (ISO/DIN 7051, DIN 7983).

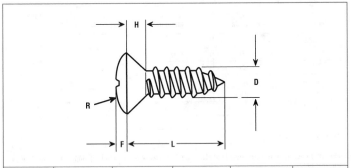

D Screw Size	Head Dia.	F Oval Height	H Countersink Depth	R Head Radius	Driver Size
M2.2	4.30	0.70	1.30	3.8	H1
M2.9	5.50	0.90	1.70	4.6	H1
M3.5	6.80	1.20	2.10	5.4	H2
M3.9	7.50	1.30	2.30	6	H2
M4.2	8.10	1.40	2.50	6.5	H2
M4.8	9.50	1.50	3.00	8.2	H2
M5.5	10.80	1.70	3.40	9.4	H3
M6.3	12.40	2.00	3.80	11.1	H3
M8	15.80	-	4.65	-	-

All dimensions in millimeters.

Table 120. Nominal Dimensions of Metric Square Head Set Screw with Short Dog Point (DIN 479).

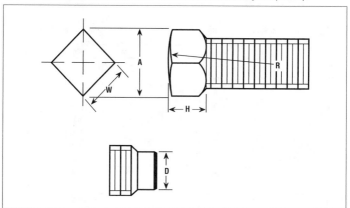

D Screw Size	W Width Across Flats	A Width Across Corners	D Point Dia.	H Head Height
M6	6.00	8.00	4.00	6.00
M8	8.00	10.00	5.50	8.00
M10	10.00	13.00	7.00	10.00
M12	13.00	17.00	8.50	12.00
M16	16.00	21.00	12.00	16.00

All dimensions in millimeters.

Table 121. Dimensions of Metric Socket Head Cap Screws.

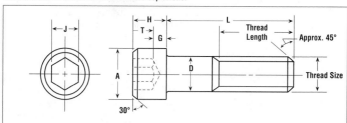

Nom. Thread Size	Pitch	A Head Dia.	D Body Dia.	H Head Height	J Socket Size	G Wall Thickness	T Socket Depth
		Max.	Max.	Max.	Nom.	Min.	Min.
M1.6	0.35	3.00	1.60	1.60	1.50	0.54	0.80
M2	0.40	3.80	2.00	2.00	1.50	0.68	1.00
M2.5	0.45	4.50	2.50	2.500	2.00	0.85	1.25

(Continued)

Table 121. *(Continued)* **Dimensions of Metric Socket Head Cap Screws.**

Nom. Thread Size	Pitch	A Head Dia.	D Body Dia.	H Head Height	J Socket Size	G Wall Thickness	T Socket Depth
		Max.	Max.	Max.	Nom.	Min.	Min.
M3	0.50	5.50	3.00	3.00	2.50	1.02	1.50
M4	0.70	7.00	4.00	4.00	3.00	1.52	2.00
M5	0.80	8.50	5.00	5.00	4.00	1.90	2.50
M6	1.00	10.00	6.00	6.00	5.00	2.28	3.00
M8	1.25	13.00	8.00	8.00	6.00	3.20	4.00
M10	1.50	16.00	10.00	10.00	8.00	4.00	5.00
M12	1.75	18.00	12.00	12.00	10.00	4.80	6.00
M14	2.00	21.00	14.00	14.00	12.00	5.60	7.00
M16	2.00	24.00	16.00	16.00	14.00	6.40	8.00
M20	2.50	30.00	20.00	20.00	17.00	8.00	10.00
M24	3.00	36.00	24.00	24.00	19.00	9.60	12.00
M30	3.50	45.00	30.00	30.00	22.00	12.00	15.00
M36	4.00	54.00	36.00	36.00	27.00	14.40	18.00
M42	4.50	63.00	42.00	42.00	32.00	16.80	21.00
M48	5.00	72.00	48.00	48.00	36.00	19.20	24.00

Length tolerances for socket head cap screws				
Size	Up to 16 mm incl.	Over 16 to 50 mm incl.	Over 50 to 120 mm incl.	Over 120 to 200 mm incl.	Over 200 mm
M1.6 thru M10	± 0.3	± 0.4	± 0.7	± 1.0	± 2.0
M12 thru M20	± 0.3	± 0.4	± 1.0	± 1.5	± 2.5
Over M20	-	± 0.7	± 1.5	± 2.0	± 3.0

All dimensions in millimeters. Source: Unbrako/SPS Technologies, Inc.

Table 122. Grip (L_G) and Body (L_B) Lengths for Metric Socket Head Cap Screws. (See Notes.)

Length	M1.6		M2		M2.5		M3		M4		M5		M6	
	L_G	L_B	L_G	L_B	L_G	L_B	L_G	L_B	L_G	L_B	L_G	L_B	L_G	L_B
20	4.8	3.0	4.0	2.0										
25	9.8	8.0	9.0	7.0	8.0	5.7	7.0	4.5						
30	14.8	13.0	14.0	12.0	13.0	10.7	12.0	9.5	10.0	6.5				
35			19.0	17.0	18.0	15.7	17.0	14.5	15.0	11.5	13.0	9.0	11.0	6.0
40			24.0	22.0	23.0	20.7	22.0	19.5	20.0	16.5	18.0	14.0	16.0	11.0
45					28.0	25.7	27.0	24.5	25.0	21.5	23.0	19.0	21.0	16.0
50					33.0	30.7	32.0	29.5	30.0	26.5	28.0	24.0	26.0	21.0
55							37.0	34.5	35.0	31.5	33.0	29.0	31.0	26.0
60							42.0	39.5	40.0	36.5	38.0	34.0	36.0	31.0
65							47.0	44.5	45.0	41.5	43.0	39.0	41.0	36.0
70									50.0	46.5	48.0	44.0	46.0	41.0
80									60.0	56.5	58.0	54.0	56.0	51.0
90											68.0	64.0	66.0	61.0

(Continued)

Table 122. (Continued) Grip (L_G) and Body (L_B) Lengths for Metric Socket Head Cap Screws. (See Notes.)

Length	M1.6 L_G	M1.6 L_B	M2 L_G	M2 L_B	M2.5 L_G	M2.5 L_B	M3 L_G	M3 L_B	M4 L_G	M4 L_B	M5 L_G	M5 L_B	M6 L_G	M6 L_B
100											78.0	74.0	76.0	71.0
110													86.0	81.0
120													96.0	91.0

Length	M8 L_G	M8 L_B	M10 L_G	M10 L_B	M12 L_G	M12 L_B	M14 L_G	M14 L_B	M16 L_G	M16 L_B	M20 L_G	M20 L_B	M24 L_G	M24 L_B
45	17.0	10.7		10.5										
50	22.0	15.7	18.0	15.5										
55	27.0	20.7	23.0	20.5										
60	32.0	25.7	28.0	25.5	24.0	15.2								
65	37.0	30.7	33.0	30.5	29.0	20.2	25.0	15.0						
70	42.0	35.7	38.0	35.5	34.0	25.2	30.0	20.0	26.0	16.0				
80	52.0	45.7	48.0	40.5	44.0	35.2	40.0	30.0	36.0	26.0				
90	62.0	55.7	58.0	50.5	54.0	45.2	50.0	40.0	46.0	36.0	38.0	25.5		
100	72.0	65.7	68.0	60.5	64.0	55.2	60.0	50.0	56.0	46.0	48.0	35.5	40.0	25.0
110	82.0	75.7	78.0	70.5	74.0	65.2	70.0	60.0	66.0	56.0	58.0	45.5	50.0	35.0
120	92.0	85.7	88.0	80.5	84.0	75.2	80.0	70.0	76.0	66.0	68.0	55.5	60.0	45.0
130	102.0	95.7	98.0	90.5	94.0	85.2	90.0	80.0	86.0	76.0	78.0	65.5	70.0	55.0
140	112.0	105.7	108.0	100.5	104.0	95.2	100.0	90.0	96.0	86.0	88.0	75.5	80.0	65.0
150	122.0	115.7	118.0	110.5	114.0	105.2	110.0	100.0	106.0	96.0	98.0	85.5	90.0	75.0
160	132.0	125.7	128.0	120.5	124.0	115.2	120.0	110.0	116.0	106.0	108.0	95.5	100.0	85.0
180			148.0	140.5	144.0	135.2	140.0	130.0	136.0	126.0	128.0	115.5	120.0	105.0
200			168.0	160.5	164.0	155.2	160.0	150.0	156.0	146.0	148.0	135.5	140.0	125.0
220					184.0	175.2	180.0	170.0	176.0	166.0	168.0	155.5	160.0	145.0

(Continued)

Table 122. *(Continued)* Grip (L_G) and Body (L_B) Lengths for Metric Socket Head Cap Screws. (See Notes.)

Length	M8		M10		M12		M14		M16		M20		M24	
	L_G	L_B	L_G	L_B	L_G	L_B	L_G	L_B	L_G	L_B	L_G	L_B	L_G	L_B
240					204.0	195.2	200.0	190.0	196.0	186.0	188.0	175.5	180.0	165.0
260							220.0	210.0	216.0	206.0	208.0	195.5	200.0	185.0
300									256.0	246.0	248.0	235.5	240.0	225.0

Notes: L_G is the maximum grip length and is the distance from the bearing surface to the first complete thread. L_B is the minimum body length and is the length of the unthreaded cylindrical portion of the shank. All dimensions in millimeters. Source: Unbrako/SPS Technologies, Inc.

Table 123. Drill and Counterbore Sizes for Metric Socket Head Cap Screws.

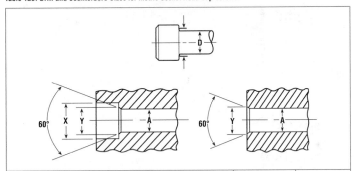

Nom. Size	D Basic Screw Dia.	A Drill Size for Hole A		X Counter-bore Dia.	Y Counter-sink Dia.[3]
		Close Fit[1]	Normal Fit[2]		
M1.6	1.60	1.80	1.95	3.50	2.0
M2	2.00	2.20	2.40	4.40	2.6
M2.5	2.50	2.70	3.00	5.40	3.1
M3	3.00	3.40	3.70	6.50	3.6
M4	4.00	4.40	4.80	8.25	4.7
M5	5.00	5.40	5.80	9.75	5.7
M6	6.00	6.40	6.80	11.25	6.8
M8	8.00	8.40	8.80	14.25	9.2
M10	10.00	10.50	10.80	17.25	11.2
M12	12.00	12.50	12.80	19.25	14.2
M14	14.00	14.50	14.75	22.25	16.2
M16	16.00	16.50	16.75	25.50	18.2
M20	20.00	20.50	20.75	31.50	22.4
M24	24.00	24.50	24.75	37.50	26.4
M30	30.00	30.75	31.75	47.50	33.4
M36	36.00	37.00	37.50	56.50	39.4
M42	42.00	43.00	44.00	66.00	45.6
M48	48.00	49.00	50.00	75.00	52.6

Notes: [1] Close fit is limited to holes for those lengths of screws threaded to the head in assemblies in which: A) only one screw is used; or B) two or more screws are used and the mating holes are produced at assembly or by matched and coordinated tooling. [2] Normal fit is intended for: A) screws of relatively long length; or B) assemblies that involve two or more screws and where the mating holes are produced by conventional tolerancing methods. This fit provides the maximum allowable eccentricity of the longest standard screws and for certain deviations in the parts being fastened. [3] It is considered good practice to chamfer or break the edges of holes that are smaller than the screw diameter plus twice the fillet height (D Max. + 2F Max.) in parts in which the hardness approaches, equals, or exceeds the screw hardness. Dimension "Y" on the Table provides this value. All dimensions in millimeters. Source: Unbrako/SPS Technologies, Inc.

Table 124. ISO Tolerances for Socket Screws.

Nom. Dia.		Tolerance Zone										
over	to	C13	C14	D9	D10	D11	D12	EF8	E11	E12	Js9	K9
0	3	+0.20 +0.06	+0.31 +0.06	+0.045 +0.020	+0.060 +0.020	+0.080 +0.020	+0.12 +0.02	+0.024 +0.010	+0.074 +0.014	+0.100 +0.014	± 0.0125	0 -0.025
3	6	+0.24 +0.06	+0.37 +0.07	+0.060 +0.030	+0.078 +0.030	+0.115 +0.030	+0.15 +0.03	+0.028 +0.014	+0.095 +0.020	+0.140 +0.020	± 0.015	0 -0.030
6	10	-	-	-	-	+0.130 +0.040	+0.19 +0.40	+0.040 +0.018	+0.115 +0.025	+0.115 +0.025	± 0.018	0 -0.036
10	18	-	-	-	-	-	+0.2 +0.05	-	+0.142 +0.032	+0.212 +0.032	-	-
18	30	-	-	-	-	-	+0.275 +0.065	-	-	-	-	-
30	50	-	-	-	-	-	+0.33 +0.08	-	-	-	-	-
50	80	-	-	-	-	-	+0.40 +0.10	-	-	-	-	-

All dimensions in millimeters. References: ISO R286, ISO 4759/I, ISO 4759/II, ISO 4759/III. Source: Unbrako/SPS Technologies, Inc.

Table 125. Dimensions of Metric Socket Flat Head Cap Screws.

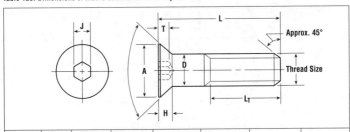

Nom. Thread Size	Pitch	A Head Dia.	D Body Dia.	H Head Height	J Socket Size	L_T Thread Length	T Socket Depth
		Max.	Max.	Ref.	Nom.	Min.	Min.
M3	0.5	6.72	3.00	1.70	2.00	18.00	1.10
M4	0.7	8.96	4.00	2.30	2.50	20.00	1.55
M5	0.8	11.20	5.00	2.80	3.00	22.00	2.05
M6	1.0	13.44	6.00	3.30	4.00	24.00	2.25
M8	1.25	17.92	8.00	4.40	5.00	28.00	3.20
M10	1.50	22.40	10.00	5.50	6.00	32.00	3.80
M12	1.75	26.88	12.00	6.50	8.00	36.00	4.35
M16	2.00	33.60	16.00	7.50	10.00	44.00	4.89
M20	2.50	40.32	20.00	8.50	12.00	52.00	5.45
M24	3.00	40.42	24.00	14.0	14.00	60.00	10.15

Length tolerances for socket flat head cap screws			
Size	Up to 16 mm incl.	Over 16 to 60 mm incl.	Over 60 mm
M3 thru M24	± 0.3	± 0.5	± 0.8

All dimensions in millimeters. Source: Unbrako/SPS Technologies, Inc.

Table 126. Dimensions of Metric Socket Button Head Cap Screws.

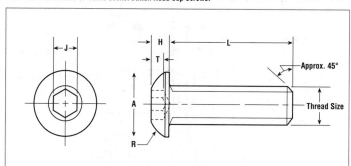

Screw Size	A Head Dia.	Oval Height	H Head Height	T Socket Depth	J Socket Size
M3	0.5	5.70	1.65	1.05	2.0
M4	0.7	7.60	2.20	1.35	2.5
M5	0.8	9.50	2.75	1.92	3.0
M6	1.0	10.50	3.30	2.08	4.0
M8	1.285	14.00	4.40	2.75	5.0
M10	1.50	18.00	5.50	3.35	6.0
M12	1.75	21.00	6.60	4.16	8.0
M16	2.0	28.00	8.60	5.20	10.0

Length tolerances for socket flat head cap screws			
Size	Up to 16 mm incl.	Over 16 to 60 mm incl.	Over 60 mm
M3 thru M24	± 0.3	± 0.5	± 0.8

All dimensions in millimeters. Source: Unbrako/SPS Technologies, Inc.

Table 127. Dimensions of Metric Socket Head Shoulder Screws.

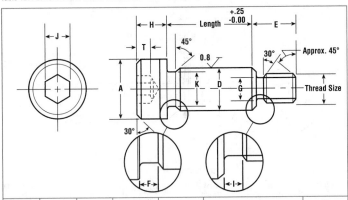

Nom. Size	Thread Size	Pitch	A Head Dia. Max.	T Socket Depth Min.	D Body Dia. Max.	D Body Dia. Min.	K Neck Dia. Min.
6	M5	0.80	10.00	2.40	6.00	5.982	5.42
8	M6	1.00	13.00	3.30	8.00	7.978	7.42
10	M8	1.25	16.00	4.20	10.00	9.978	9.42
12	M10	1.50	18.00	4.90	12.00	11.973	11.42
16	M12	1.75	24.00	6.60	16.00	15.973	15.42
20	M16	2.00	30.00	8.80	20.00	19.967	19.42
24	M20	2.50	36.00	10.00	24.00	23.967	23.42

Nom. Size	Thread Size	H Head Height Max.	G Thread Neck Dia. Min.	F Neck Width Max.	I Thread Neck Width Max.	E Thread/Neck Length Max.	J Socket Size Nom.
6	M5	4.50	3.68	2.50	2.40	9.75	3.00
8	M6	5.50	4.40	2.50	2.60	11.25	4.00
10	M8	7.00	6.03	2.50	2.80	13.25	5.00
12	M10	8.00	7.69	2.50	3.00	16.40	6.00
16	M12	10.00	9.35	2.50	4.00	18.40	8.00
20	M16	14.00	12.96	2.50	4.80	22.40	10.00
24	M20	16.00	16.30	3.00	5.60	27.40	12.00

All dimensions in millimeters. Source: Unbrako/SPS Technologies, Inc.

Table 128. Dimensions of Metric Low Head Cap Screws.

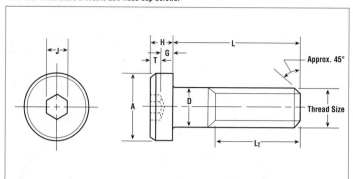

Nom. Thread Size	Pitch	A Head Dia.	D Body Dia.	G Wall Thickness	T Socket Depth	H Head Height	L_T Thread Length	J Socket Size
		Max.	Max.	Min.	Min.	Max.	Min.	Nom.
M4	0.7	7.00	4.00	1.06	1.48	2.8	20.00	3.0
M5	0.8	8.50	5.00	1.39	1.85	3.5	22.00	4.0
M6	1.0	10.00	6.00	1.65	2.09	4.0	24.00	5.0
M8	1.25	13.00	8.00	2.24	2.48	5.0	28.00	6.0
M10	1.5	16.00	10.00	2.86	3.36	6.5	32.00	8.0
M12	1.75	18.00	12.00	3.46	4.26	8.0	36.00	10.0
M16	2.0	24.00	16.00	4.91	4.76	10.0	44.00	12.0
M20	2.5	30.00	20.00	6.10	6.07	12.5	52.00	14.0

All dimensions in millimeters. Source: Unbrako/SPS Technologies, Inc.

Table 129. Dimensions of Metric Socket Set Screws - Knurled Cup Point and Plain Cup Point.

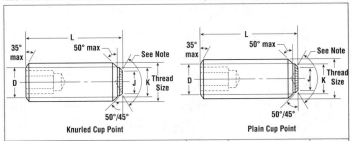

Knurled Cup Point   Plain Cup Point

Nom. Thread Size	Pitch	D Body Dia. Max.	J Point End Dia. Max		K Point Base Dia. Max.	L Length		Socket Size Nom.
			Plain Cup	Knurl Cup		Plain Cup	Knurl Cup	
M1.6	0.35	1.00	0.80	-	-	2.0	-	0.7
M2	0.40	1.32	1.00	-	-	2.5	-	0.9
M2.5	0.45	1.75	1.25	-	-	3.0	-	1.3
M3	0.50	2.10	1.50	1.40	2.06	3.0	3.0	1.5
M4	0.70	2.75	2.00	2.10	2.74	3.0	3.0	2.0
M5	0.80	3.70	2.50	2.50	3.48	4.0	4.0	2.5
M6	1.00	4.35	3.00	3.30	4.14	4.0	5.0	3.0
M8	1.25	6.00	5.00	5.00	5.62	5.0	6.0	4.0
M10	1.50	7.40	6.00	6.00	7.12	6.0	8.0	5.0
M12	1.75	8.60	8.00	8.00	8.58	8.0	10.0	6.0
M16	2.00	12.35	10.00	10.00	11.86	12.0	14.0	8.0
M20	2.50	16.00	14.00	14.00	14.83	16.0	18.0	10.0
M24	3.00	18.95	16.00	16.00	17.80	20.0	20.0	12.0

Note: The cap angle is 135° max. for screw lengths equal to or smaller than screw diameter. For longer lengths, the cup angle will be 124° max.

Length tolerances for Metric socket set screws with plain cup and knurled cup point			
Size	Up to 12 mm incl.	Over 12 to 50 mm incl.	Over 50 mm
M1.6 thru M24	± 0.3	± 0.5	± 0.8

All dimensions in millimeters. Source: Unbrako/SPS Technologies, Inc.

Table 130. Dimensions of Metric Socket Set Screws - Flat Point, Cone Point, and Dog Point. (See Notes.)

Nom. Thread Size	Pitch	D Body Dia.	W Socket Size Nom.	Flat Point J Point Dia. Max.	Flat Point L Length[1] Min.	Cone Point J Point Dia. Max.	Cone Point L Length[1] Min.	Dog Point H Point Length Half Dog	Dog Point H Point Length Full Dog	Dog Point L Length[1] Min.	V Point Dia. Max.
M3	0.5	2.10	1.5	2.0	3.0	0.3	4.0	0.75	1.5	5.0	2.00
M4	0.7	2.75	2.0	2.5	3.0	0.4	4.0	1.00	2.0	5.0	2.50
M5	0.8	3.70	2.5	3.5	4.0	0.5	5.0	1.25	2.5	6.0	3.50

(Continued)

Table 130. *(Continued)* Dimensions of Metric Socket Set Screws - Flat Point, Cone Point, and Dog Point. (See Notes.)

Nom. Thread Size	Pitch	D Body Dia.	W Socket Size Nom.	Flat Point J Point Dia. Max.	Flat Point L Length¹ Min.	Cone Point J Point Dia. Max.	Cone Point L Length¹ Min.	Dog Point H Point Length Half Dog	Dog Point H Point Length Full Dog	Dog Point L Length¹ Min.	Dog Point V Point Dia. Max.
M6	1.00	4.25	3.0	4.0	4.0	1.5	6.0	1.50	3.0	6.0	4.00
M8	1.25	6.00	4.0	5.5	5.0	2.0	6.0	2.00	4.0	8.0	5.50
M10	1.50	7.40	5.0	7.0	6.0	2.5	8.0	2.50	5.0	8.0	7.00
M12	1.75	8.60	6.0	8.5	8.0	3.0	10.0	3.00	6.0	12.0	8.50
M16	2.00	12.35	8.0	12.0	12.0	4.0	14.0	4.00	8.0	16.0	12.00
M20	2.50	16.00	10.0	15.0	14.0	6.0	18.0	5.00	10.0	20.0	15.00
M24	3.00	18.95	12.0	18.0	20.0	8.0	20.0	6.00	12.0	22.0	18.00

Notes: The cap angle is 135° max. for screw lengths equal to or smaller than screw diameter. For longer lengths, the cup angle will be 124° max. ¹ Minimum length is preferred.

Length tolerances for Metric socket set screws with flat point, cone point and dog point

Size	Up to 12 mm, incl.	Over 12 to 50 mm, incl.	Over 50 mm
M1.6 thru M24	± 0.3	± 0.5	± 0.8

All dimensions in millimeters. Source: Unbrako/SPS Technologies, Inc.

Table 131. Metric Hexagon Key Application Chart.

W Size	Socket Head Cap Screws	Flat Head Cap Screws	Button Head Cap Screws	Socket Head Shoulder Screws	Socket Set Screws	Low Head Cap Screws
0.7	-	-	-	-	M1.6	-
0.9	-	-	-	-	M2	-
1.3	-	-	-	-	M2.5	-
1.5	M1.6, M2	-	-	-	M3	-
2.0	M2.5	M3	M3	-	M4	-
2.5	M3	M4	M4		M5	
3.0	M4	M5	M5	M6	M6	M4
4.0	M5	M6	M6	M8	M8	M5
5.0	M6	M8	M8	M10	M10	M6
6.0	M8	M10	M10	M12	M12	M8
8.0	M10	M12	M12	M16	M16	M10
10.0	M12	M16	M16	M20	M20	M12
12.0	M14	M20	-	M24	M24	M16
14.0	M16	M24	-	-	-	M20
17.0	M20	-	-	-	-	-
19.0	M24	-	-	-	-	-
22.0	M30	-	-	-	-	-
27.0	M36	-	-	-	-	-
32.0	M42	-	-	-	-	-
36.0	M48	-	-	-	-	-

All dimensions in millimeters. Source: Unbrako/SPS Technologies, Inc.

Table 132. Dimensions and Mechanical Properties of Metric Hexagon Keys.

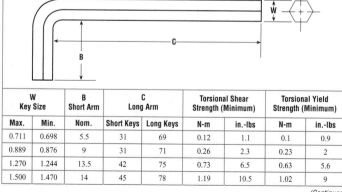

W Key Size		B Short Arm	C Long Arm		Torsional Shear Strength (Minimum)		Torsional Yield Strength (Minimum)	
Max.	Min.	Nom.	Short Keys	Long Keys	N-m	in.-lbs	N-m	in.-lbs
0.711	0.698	5.5	31	69	0.12	1.1	0.1	0.9
0.889	0.876	9	31	71	0.26	2.3	0.23	2
1.270	1.244	13.5	42	75	0.73	6.5	0.63	5.6
1.500	1.470	14	45	78	1.19	10.5	1.02	9

(Continued)

Table 132. *(Continued)* Dimensions and Mechanical Properties of Metric Hexagon Keys.

W Key Size		B Short Arm	C Long Arm		Torsional Shear Strength (Minimum)		Torsional Yield Strength (Minimum)	
Max.	Min.	Nom.	Short Keys	Long Keys	N-m	in.-lbs	N-m	in.-lbs
2.000	1.970	16	50	83	2.9	26	2.4	21
2.500	2.470	18	56	90	5.4	48	4.4	39
3.000	2.960	20	63	100	9.3	82	8	71
4.000	3.960	25	70	106	22.2	196	18.8	166
5.000	4.960	28	80	118	42.7	378	36.8	326
6.000	5.960	32	90	140	74	655	64	566
8.000	7.950	36	100	160	183	1,620	158	1,400
10.000	9.950	40	112	170	345	3,050	296	2,620
12.000	11.950	45	125	212	634	5,610	546	4,830
14.000	13.930	55	140	236	945	8,360	813	7,200
17.000	16.930	60	160	250	1,690	15,000	1,450	12,800
19.000	18.930	70	180	280	2,360	20,900	2,030	18,000
22.000	21.930	80	200	335	3,670	32,500	3,160	28,000
24.000	23.930	90	224	375	4,140	36,600	3,560	31,500
27.000	26.820	100	250	500	5,870	51,900	5,050	44,700
32.000	31.820	125	315	630	8,320	73,600	7,150	63,300
36.000	35.820	140	355	710	11,800	104,000	10,200	90,300

All dimensions in millimeters. Source: Unbrako/SPS Technologies, Inc.

Gears and Gearing

Gear definitions and nomenclature

Addendum. The point or portion of the tooth of a gear wheel that lies outside the pitch circle. In bevel gears, it is the portion of a tooth that projects above the pitch cone.

Arc of Action. The arc of the pitch circle through which a gear tooth travels during the period in which it is in contact with a mating tooth.

Axial Pitch. The distance parallel to the axis between corresponding sides of adjacent teeth. Sometimes referred to as linear pitch.

Axial Plane. In a single gear, any plane containing the axis and any given point. In a pair of gears, the plane that contains the axis of both gears.

Backlash. The amount by which the width of a tooth space exceeds the thickness of the engaging tooth at the pitch circle. Measured by holding one gear fixed and rotating the mating gear, and reading the total movement on a dial indicator.

Base Circle. The circle from which the involute tooth profiles are generated.

Bevel Gear. A conical shaped gear with the teeth cut on an angle in order to transmit power in any desired direction. Teeth on a bevel gear may be straight, or curved as found on spiral bevel, Zerol bevel, and hypoid gears.

Center Distance. The perpendicular distance between the axis of two mating gears.

Chordal Pitch. The distance from one point on a gear tooth to the corresponding point on the next tooth, measured as a chord of the pitch circle.

Chordal Thickness. The chordal distance between the sides of a gear tooth, at the pitch circle, as measured by a gear tooth caliper.

Circular Pitch. The distance along the circumference at the pitch circle between corresponding points on the profiles of adjacent teeth.

Circular Thickness. The length of arc between the two sides of a gear tooth, measured at the pitch circle.

Clearance. The distance between the outside circle of one gear, and the root circle of its mate, or the amount by which the dedendum of one gear exceeds the addendum of its mate.

Crowned Teeth. Teeth that have been shortened in height to prevent contact with a mating gear.

Dedendum. The portion of a gear tooth between the root circle and the pitch circle. It is equal to the addendum plus the clearance.

Diametral Pitch. The ratio between the number of teeth on a gear and its pitch diameter. If, for example, a gear has 30 teeth and the pitch diameter is 10, the diametral pitch is 3. Hence, the diametral pitch represents the number of teeth per inch of pitch diameter.

Fillet Radius or *Curve*. The concave portion of a tooth's profile that occurs where it joins the bottom of the tooth space.

Gear Ratio. The relation between the number of teeth on the drive gear and the driven gear.

Gear Train. Two or more gears connecting the driving and driven parts of a machine.

Helical Gear. A cylindrical gear with teeth cut at some angle other than a right angle to the gear face. Helical gears have two teeth engaged at all times and usually operate more quietly and are smoother than spur gears.

Helix Angle. The angle a helical gear tooth makes at the pitch circle in a plane perpendicular to the gear axis.

Herringbone Gear. A gear on which the teeth slope both ways from the center line on the gear face (having, in one gear, the characteristics of both a left- and a right-hand spiral gear). Often used on heavy machinery. Sometimes called a split or double helical gear.

Hypoid Bevel Gear. A cross between a spiral bevel gear and a worm gear in which the pinion is located above or below the center of the mating gears. This layout allows for higher ratios than are practical with other bevel gears, and allows for the pinion shaft to extend beyond the gear for additional support.

Intermediate Gear. A gear between the driving and the driven gears.

Intermittent Gear. A gear that has teeth on part, but not all, of the circumference. When the teeth are in mesh, the driven gear turns. When the plain surfaces between the teeth are in contact, the driven gear is stationary.

Internal Gear. A gear that engages with teeth set in the internal circumference of a ring. Almost always a spur gear.

Involute. A curve traced by a point on a line as it unrolls from a fixed curve.

Involute Gear. A gear whose tooth design is developed after the involute system, as distinguished from the cycloidal type. (The shape of a cycloidal curve is generated by a point on the circumference of a circle as it rolls along a straight line.) The involute form is popular for gear systems because it is easily produced by several machining methods, and it allows meshed gears to transmit smooth, uniform angular velocity.

Lead. The axial advance of a helix in one complete turn of a gear wheel.

Line of Action. A straight line generated by the path of contact between meshing gears that is tangent to both base circles and passes through the pitch point.

Miter Gear. Mating bevel gears having an equal number of teeth and a shaft angle of 90°.

Module. In the inch system, the ratio of the pitch diameter in inches to the number of teeth. In the metric system, the ratio of the pitch diameter in millimeters to the number of teeth (pitch diameter in millimeters divided by the number of teeth). Tooth proportions for module gears are the same as for those made for the inch (diametral pitch) system.

Outside Diameter. On a cylindrical or worm gear, the diameter of the circle that contains the tops of the teeth. On a bevel gear, the diameter of the crown circle.

Pinion. In mated gear sets, the gear with the smaller number of teeth.

Pitch. The distance from a point on one gear tooth to the corresponding point on the next tooth measured along a given curve.

Pitch Circle. A circle through the pitch points of each tooth on a gear wheel, having its center at the axis of the gear.

Pitch Cone. In gears with intersecting axes, a cone with its apex at the intersection of the axis and its base forming the pitch circle.

Pitch Diameter. The diameter of the pitch circle of a gear wheel.

Pitch Point. The point of tangency of two pitch circles. The pitch point of a gear tooth occurs at the point that it intersects the pitch circle.

Plane of Rotation. Any plane perpendicular to a gear axis.

Pressure Angle. The angle between the tooth profile and the plane that is tangent to the pitch surface, usually measured at the pitch point.

Rack. A gear with teeth spaced along a straight line.

Runout. The total difference of the off-center relation of the axis of the tooth profiles with respect to the rotational axis.

Spiral. A curve formed by a fixed point moving about a center and constantly increasing the distance from it.

Spiral Bevel Gear. A bevel gear with spiral teeth that are curved at an angle to the pitch cone element. The form allows for gradual engagement of the mating teeth along their entire length.

Spiral Gear. A gear of which each tooth constitutes a part of a helix (also called a screw gear).

Spur Gear. The most common form of gear wheel. Its teeth are cut parallel to the axes, and it transmits drive between parallel shafts.

Straight Bevel Gear. Bevel gears with straight teeth that, if extended, would intersect at the axis.

Stub-gear Tooth. A type of involute tooth with a pressure angle of 20°. The addendum and dedendum are both shorter than in the standard tooth. Most commonly used in automobile drives.

Stud Gear. Another name for an intermediate gear.

Transverse Plane. A plane perpendicular to both the axial plane and to the pitch plane. The transverse plane and the plane of rotation coincide in gears with parallel axes.

Undercut. When any part of the fillet radius lies inside of a line drawn tangent to the working profile at the point of juncture with the fillet, the tooth is said to be undercut. Often done intentionally to allow for finishing operations.

Whole Depth. The total depth of a tooth space, equal to the addendum plus the dedendum.

Working Depth. The depth of engagement of two standard gears on standard center distances. Equal to the sum of the addendums of the mating gears.

Worm. A gear with one or more teeth in the form of screw threads. The worm mates with a worm wheel and is usually used to transmit power between shafts situated at right angles.

Zerol Bevel Gears. A curved tooth bevel gear whose centerline is tangent to the pitch cone element at mid-face. (Zerol is a registered trademark of Gleason Gear Works.)

Gear manufacturing

Helical, spur, and worm gears

Helical and spur gears are among the most basic gear forms and can be produced with a variety of machining and forming methods. Actual production methods are usually selected based on economy of run, as all the methods briefly described below can produce acceptable accuracy and finish.

Broaching. Producing gear teeth by broaching is among the fastest methods available, but the speed advantage is offset by high costs. In broaching, a single pass of a full profile full-form broach both roughs and finishes the tooth. Among the most common applications is cutting internal helical gears, due to the difficulty of producing this form with other methods.

Grinding. Grinding is also used in gear production, with the dressed grinding wheel taking on the role of the hob, milling cutter, or shaper. The process is expensive, and is usually reserved for applications where the workpiece is too hard for other methods (48 R_C or higher), or when absolute accuracy is required.

Hobbing. A hob is a tool that is usually cylindrical in shape and greater in length than diameter, with teeth arranged along a helical thread. In hobbing, the hob and gear blank are rotated in timed relationship to each other, and teeth are generated on the blank as the hob is successively fed axially or tangentially across, or radially into, the workpiece. As stated, the hob "generates" teeth on the workpiece, meaning that the tooth profile of the hob is conjugate to, but different than, the profile of the cut tooth. Hobs usually have either an arbor or shank, and are designed to cut in either a left- or right-hand direction, depending on the direction of the hob's helix. Hobbing can be used to generate most tooth forms that are equally spaced about a cylindrical axis. It is an accurate, fast, and low cost method of cutting many gear forms. It should be noted that multiple thread milling cutters differ from hobs in that their teeth are not arranged along a helical thread.

Milling. Gear milling cutters "form" the teeth on a gear blank. By forming, it is meant that the resultant shape on the workpiece will, as exactly as possible, copy the shape of the cutter. Milling cutters traverse axially across a stationary workpiece, cutting one tooth at a time. CNC controls or, on less sophisticated milling machines, an indexing head advances the cutter until the complete gear wheel has been machined. Bevel, helical, and spur gears can be formed with the milling process. Form milling can be an attractive method of cutting a small run of gears, but the process is time consuming as only one tooth is cut at a time.

Roll Forming. Rolling is a cold forming process used to produce spur and helical gears in the same way in which splines and screw threads are rolled. The process is fast and provides excellent surface finishes. There are some limitations on rolling gears: spur gears with fewer than eighteen teeth cannot be rolled satisfactorily, and rolling gears with pressure angles less than 20° produces insufficient root and crest flats. Rolling is also used to form worms to engage with worm wheels.

Shaping. Shaping differs from hobbing and milling in that the shaping tool reciprocates, rather than revolves, across the workpiece. As the shaper and blank rotate in "mesh" the shaper also reciprocates, and this action cuts the gear form into the workpiece. The process can be used for both internal and external spur and helical gears. Since the cutter shape is usually slightly modified from the final gear form in order to enhance cutting, shaping is a "generating-forming" process.

Shaving. Shaving is used to enhance gear tooth accuracy by removing small amounts from hobbed or shaped teeth. Internal and external spur and helical teeth can be shaved. There are two basic methods of shaving: with a serrated tooth rotary tool in the form of a helical gear, or with a serrated tooth tool in the form of a rack. In rotary shaving, the workpiece and cutter are rotated in tight mesh and the shaver finishes the gear to final size. The rack method allows for the rotating gear to pass over a stationary rack.

Bevel gears

While cutting straight bevel gears can sometimes be done with traditional tools, those with curved teeth often require specialized processes. In straight bevel gears, an imaginary line drawn along the sides of the teeth will converge at the axis of the gear, forming the pitch cone apex. Teeth that converge in this manner are known by their registered name, Coniflex (a Gleason trade name), and almost all such gears must be generated with twin cutters that can be brought closer together as they travel toward the axis of the gear. This can be accomplished by having the two tools travel on reciprocating slides so that they straddle the tooth and cut both sides of a tooth simultaneously, or by using interlocking cutters, with each cutting one side of opposing teeth along the root plane. Tooth thickness is not determined by the cutting tool, as it is with spur gears, but by the adjusting the machine to allow the gear teeth to be cut teeth to a predetermined, and optimally desirable, thickness.

Hypoid, spiral, and Zerol gears are either generated using the process described for straight gears, or they are produced by either the Formate or Helixform process. These "nongenerating" processes can be used whenever the gear-to-pinion ratio is 2.1:1 or greater.

The Formate process uses a face milling style cutter, and two variations of the process are used. The first is a two-step roughing-finishing process, performed on two different machines—the first machine makes a roughing pass and the second finishes the workpiece using a cutter in which each blade is slightly longer and wider than the preceding blade, allowing it to act in the fashion of a circular broach as the gear is fed into the cutter or vice versa. The second variation is a one-step "completing" operation that doubles cutting speed and reduces feed to produce an adequate surface finish for most applications in a single pass.

In the Helixform method, the cutter both rotates and advances axially, allowing it to cut a true helical form. Otherwise, cutting is done in much the same manner and with the same tools as used in the Formate process.

Face mill cutters used in machining bevel gears are classified as inserted, integral, or segmental, and all three forms can be used for both roughing and finishing. Inserted blade cutters have blades that are either bolted or clamped to a slotted head along their periphery; integral (or solid) are one piece and are usually used for fine pitch gears that are smaller in diameter (less than 5"); and segmented cutters have blades bolted to the periphery of the individual sections that form the cutter.

AGMA gear quality class numbers

The American Gear Manufacturer's Association has assigned quality numbers ranging from 3 to 15 to identify accuracy levels required for gear teeth based on their application. These ratings include a vast array of tolerances based on several classes within each quality number. Companies producing power transmission gears should acquire the appropriate AGMA/ANSI standards relating to their work.

Spur gears

Gear systems are often classified by the relationship on the driving axis to the driven axis. Gears mounted on parallel axes include internal and external spur gears, helical gears, and herringbone gears. Spur gears are the most widely applied gears in the world, and seven basic rack forms have been standardized. (A basic rack form portrays the gear on a plane pitch surface.) Basic gear geometry is illustrated in *Figure 1*. The formulas for finding tooth and gear dimensions are found in **Tables 1–3**.

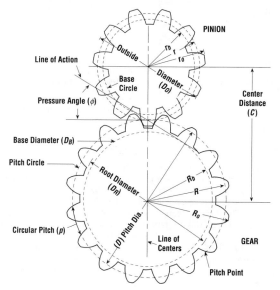

Figure 1. Basic gear geometry and terms.

14 $1/2°$ composite tooth form. This form is among the original configurations approved by the American Standards Association in 1927, and was standardized in ASA B6.1-1932. It was designed to be cut with formed milling cutters. Mating gears have equal addendums, operate on standardized centers, and are conjugate (meaning they can transmit uniform rotary motion) to a basic rack. Officially, this form is not recommended for new designs as it is obsolete.

Addendum = $1/P$ or $0.3183\,p$ Minimum Dedendum = $1.157/P$ or $0.3683\,p$
Working Depth = $2/P$ or $0.6366\,p$ Minimum Whole Depth = $2.157/P$ or $0.6866\,p$
Tooth Thickness = $1.5708/P$ or $p/2$ Minimum Clearance = $0.157/P$ or $0.05\,p$

14 $1/2°$ involute tooth form. Often called a 14 $1/2°$ full depth tooth, this form can be produced by gear hobs and pinion style shaping cutters. Also standardized in ASA B6.1-1932, and is now also considered obsolete. Tooth proportions are the same as for the composite tooth form.

Table 1. General Formulas for Finding Dimensions of Spur Gears.

To Get	Formula	To Get	Formula
C	$(N_P \times [m_G + 1]) \div 2P$	h_t	$a + b$
	$(D_P + D_G) \div 2$	m_G	$N_G \div N_P$
	$(N_P + N_G) \div 2P$	π	$P \times p$
	$([N_P + N_G] \times p) \div 2p$	N	$P \times D$
D	$N \div P$		$(\pi \times D) \div p$
	$NP \div \pi$		$\pi \div p$
D_B	$D \times \cos ø$	P	$N \div D$
D_O	$D + 2a$		$(N_P [m_G + 1]) \div 2C$
	$(N + 2) \div P$	p	$(\pi \times D) \div N$
	(full depth tooth)		$\pi \div P$
h_k	$a_G + a_P$	p_b	$p \times \cos ø$

ø = Pressure Angle	D_O = Outside Diameter
π = Pi (3.1416)	h_k = Working Depth of Tooth
a = Addendum	h_t = Whole Depth of Tooth
a_G = Addendum (Gear)	m_G = Gear Ratio
a_P = Addendum (Pinion)	N = Number of Teeth
b = Dedendum	N_G = Number of Teeth (Gear)
C = Center Distance	N_P = Number of Teeth on (Pinion)
D = Pitch Diameter	P = Diametral Pitch
D_B = Base Circle Diameter	p = Circular Pitch
D_G = Pitch Diameter (Gear)	p_b = Base Pitch

20° and 25° full-depth involute tooth form. The 20° pressure angle is specified if there are 18 or more teeth in the pinion, or 36 or more in the pair. This is currently the most widely adopted spur gear form in use. The 25° form can be used when there are 12 or more teeth in the pinion, or 24 or more in the pair, and is often specified when additional strength is required. This form is commonly called the coarse pitch spur gear, and was originally standardized in ASA B6.1-1932. It is currently fully described in ANSI B6.1-1968, R1974. Preferred diametral pitches for coarse pitch gears are 2, 2 $1/4$, 2 $1/2$, 3, 4, 6, 8, 10, 12, and 16. Tooth dimensions for this form are given in **Table 4**, and the basic rack form is shown in *Figure 2*.

(Text continued on p. 2031)

Table 2. Formulas for Finding Tooth Dimensions of 20° and 25° (Coarse Pitch) Full Depth Spur Gears.

To Get	Given P	Given p	To Get	Given P	Given p
a	$1 \div P$	$0.3183 \times p$	D_R	$(N - 2.5) \div P$ (standard)	$0.3183 \times (N - 2.5) p$ (standard)
b	$1.250 \div P$ (standard)	$0.3979 \times p$ (standard)		$(N - 2.7) \div P$ (shaved or ground teeth)	$0.3183 \times (N - 2.7) p$ (shaved or ground teeth)
	$1.350 \div P$ (shaved or ground teeth)	$0.4297 \times p$ (shaved or ground teeth)	h_k	$2 \div P$	$0.6366 \times p$
c	$0.250 \div P$ (standard)	$0.0796 \times p$ (standard)	h_t	$2.250 \div P$ (standard)	$0.7162 \times p$ (standard)
	$0.350 \div P$ (shaved or ground teeth)	$0.1114 \times p$ (shaved or ground teeth)		$2.350 \div P$ (shaved or ground tooth)	$0.7480 \times p$ (shaved or ground tooth)
D	$N \div P$	$0.3183 \times N_P$	r_f	$0.300 \div P$	$0.0955 \times p$
D_O	$(N + 2) \div P$	$0.3183 \times (N + 2) p$	t	$1.5708 \div P$	$p \div 2$

a = Addendum	h_t = Whole Depth of Tooth
b = Dedendum	N = Number of Teeth
c = Clearance	N_P = Number of Teeth on (Pinion)
D = Pitch Diameter	P = Diametral Pitch
D_O = Outside Diameter	p = Circular Pitch
D_R = Root Diameter	r_f = Fillet Radius
h_k = Working Depth of Tooth	t = Circular Thickness

Table 3. Formulas for Finding Tooth Dimensions of 14 1/2° , 20° , and 25° (Fine Pitch) Spur Gears.

To Get	Given P	Given p	To Get	Given P	Given p
a	$1 \div P$	$0.3183 \times p$	D_R	$([N - 2.2] \div P) - 0.004$ (standard)	$0.3183 \times (N - 2.2) p$ (standard)
b	$(1.200 \div P) + 0.002$ (standard)	$(0.3820 \times p) + 0.002$ (standard)		$([N - 2.35] \div P) - 0.004$ (shaved or ground teeth)	$0.3183 \times (N - 2.35) p - 0.004$ (shaved or ground teeth)
	$(1.350 \div P) + 0.002$ (shaved or ground teeth)	$(0.4297 \times p) + 0.002$ (shaved or ground teeth)	h_k	$2 \div P$	$0.6366 \times p$
c	$(0.200 \div P) + 0.002$ (standard)	$(0.0637 \times p) + 0.002$ (standard)	h_t	$(2.200 \div P) + 0.002$ (standard)	$(0.7003 \times p) + 0.002$ (standard)
	$(0.350 \div P) + 0.002$ (shaved or ground teeth)	$(0.1114 \times p) + 0.002$ (shaved or ground teeth)		$(2.350 \div P) + 0.002$ (shaved or ground tooth)	$(0.7480 \times p) + 0.002$ (shaved or ground tooth)
D	$N \div P$	$0.3183 \times N_P$	t	$1.5708 \div P$	$p \div 2$
D_O	$(N + 2) \div P$	$0.3183 \times (N + 2) p$			

a = Addendum	h_t = Whole Depth of Tooth
b = Dedendum	N = Number of Teeth
c = Clearance	N_P = Number of Teeth on (Pinion)
D = Pitch Diameter	P = Diametral Pitch
D_O = Outside Diameter	p = Circular Pitch
D_R = Root Diameter	t = Circular Thickness
h_k = Working Depth of Tooth	

Table 4. Dimensions of Coarse Pitch Full Depth Involute Spur Gear Teeth—20° and 25° Pressure Angles.

Diametral Pitch	Circular Pitch	Circular Thickness	Addendum (Std.)	Dedendum (Std.)	Dedendum (Special)	Working Depth	Whole Depth (Std.)	Whole Depth (Special)	Clearance (Std.)	Clearance (Special)	Fillet Radius
P	p	t	a	b	b	h_k	h_t	h_t	c	c	r_f
0.31416	10	5	3.18310	3.97888	4.29719	6.36620	7.16198	7.48029	0.79578	1.11409	0.95493
0.33069	9.5	4.75	3.02395	3.77993	4.08233	6.04789	6.80388	7.10627	0.75599	1.05838	0.90718
0.34907	9	4.5	2.86479	3.58099	3.86747	5.72958	6.44578	6.73226	0.71620	1.00268	0.85944
0.36960	8.5	4.25	2.70564	3.38205	3.65261	5.41127	6.08768	6.35825	0.67641	0.94697	0.81169
0.39270	8	4	2.54648	3.18310	3.43775	5.09296	5.72958	5.98423	0.63662	0.89127	0.76394
0.41888	7.5	3.75	2.38733	2.98416	3.22289	4.77465	5.37148	5.61022	0.59683	0.83556	0.71620
0.44880	7	3.5	2.22817	2.78521	3.00803	4.45634	5.01338	5.23620	0.55704	0.77986	0.66845
0.48332	6.5	3.25	2.06902	2.58627	2.79317	4.13803	4.65529	4.86219	0.51725	0.72416	0.62070
0.52360	6	3	1.90986	2.38733	2.57831	3.81972	4.29719	4.48817	0.47747	0.66845	0.57296
0.57120	5.5	2.75	1.75071	2.18838	2.36345	3.50141	3.93909	4.11416	0.43768	0.61275	0.52521
0.62832	5	2.5	1.59155	1.98944	2.14859	3.18310	3.58099	3.74014	0.39789	0.55704	0.47747
0.69813	4.5	2.25	1.43240	1.79049	1.93373	2.86479	3.22289	3.36613	0.35810	0.50134	0.42972
0.78540	4	2	1.27324	1.59155	1.71887	2.54648	2.86479	2.99212	0.31831	0.44563	0.38197
0.89760	3.5	1.75	1.11409	1.39261	1.50402	2.22817	2.50669	2.61810	0.27852	0.38993	0.33423
1	3.14159	1.57080	1.00000	1.25000	1.35000	2.00000	2.25000	2.35000	0.25000	0.35000	0.30000
1.25	2.51327	1.25664	0.80000	1.00000	1.08000	1.60000	1.80000	1.88000	0.20000	0.28000	0.24000
1.5	2.09439	1.04720	0.66667	0.83333	0.90000	1.33333	1.50000	1.56667	0.16667	0.23333	0.20000
1.75	1.79519	0.89760	0.57143	0.71429	0.77143	1.14286	1.28571	1.34286	0.14286	0.20000	0.17143
2	1.57080	0.78540	0.50000	0.62500	0.67500	1.00000	1.12500	1.17500	0.12500	0.17500	0.15000
2.25	1.39626	0.69813	0.44444	0.55556	0.60000	0.88889	1.00000	1.04444	0.11111	0.15556	0.13333
2.5	1.25664	0.62832	0.40000	0.50000	0.54000	0.80000	0.90000	0.94000	0.10000	0.14000	0.12000

(Continued)

Table 4. *(Continued)* Dimensions of Coarse Pitch Full Depth Involute Spur Gear Teeth—20° and 25° Pressure Angles.

Diametral Pitch	Circular Pitch	Circular Thickness	Addendum (Std.)	Dedendum (Std.)	Dedendum (Special)	Working Depth	Whole Depth (Std.)	Whole Depth (Special)	Clearance (Std.)	Clearance (Special)	Fillet Radius
P	p	t	a	b	b	h_k	h_t	h_t	c	c	r_f
2.75	1.14240	0.57120	0.36364	0.45455	0.49091	0.72727	0.81818	0.85455	0.09091	0.12727	0.10909
3	1.04720	0.52360	0.33333	0.41667	0.45000	0.66667	0.75000	0.78333	0.08333	0.11667	0.10000
3.25	0.96664	0.48332	0.30769	0.38462	0.41538	0.61538	0.69231	0.72308	0.07692	0.10769	0.09231
3.5	0.89760	0.44880	0.28571	0.35714	0.38571	0.57143	0.64286	0.67143	0.07143	0.10000	0.08571
3.75	0.83776	0.41888	0.26667	0.33333	0.36000	0.53333	0.60000	0.62667	0.06667	0.09333	0.08000
4	0.78540	0.39270	0.25000	0.31250	0.33750	0.50000	0.56250	0.58750	0.06250	0.08750	0.07500
4.5	0.69813	0.34907	0.22222	0.27778	0.30000	0.44444	0.50000	0.52222	0.05556	0.07778	0.06667
5	0.62832	0.31416	0.20000	0.25000	0.27000	0.40000	0.45000	0.47000	0.05000	0.07000	0.06000
5.5	0.57120	0.28560	0.18182	0.22727	0.24545	0.36364	0.40909	0.42727	0.04545	0.06364	0.05455
6	0.52360	0.26180	0.16667	0.20833	0.22500	0.33333	0.37500	0.39167	0.04167	0.05833	0.05000
6.5	0.48332	0.24166	0.15385	0.19231	0.20769	0.30769	0.34615	0.36154	0.03846	0.05385	0.04615
7	0.44880	0.22440	0.14286	0.17857	0.19286	0.28571	0.32143	0.33571	0.03571	0.05000	0.04286
7.5	0.41888	0.20944	0.13333	0.16667	0.18000	0.26667	0.30000	0.31333	0.03333	0.04667	0.04000
8	0.39270	0.19635	0.12500	0.15625	0.16875	0.25000	0.28125	0.29375	0.03125	0.04375	0.03750
8.5	0.36960	0.18480	0.11765	0.14706	0.15882	0.23529	0.26471	0.27647	0.02941	0.04118	0.03529
9	0.34907	0.17453	0.11111	0.13889	0.15000	0.22222	0.25000	0.26111	0.02778	0.03889	0.03333
9.5	0.33069	0.16535	0.10526	0.13158	0.14211	0.21053	0.23684	0.24737	0.02632	0.03684	0.03158
10	0.31416	0.15708	0.10000	0.12500	0.13500	0.20000	0.22500	0.23500	0.02500	0.03500	0.03000
11	0.28560	0.14280	0.09091	0.11364	0.12273	0.18182	0.20455	0.21364	0.02273	0.03182	0.02727
12	0.26180	0.13090	0.08333	0.10417	0.11250	0.16667	0.18750	0.19583	0.02083	0.02917	0.02500
13	0.24166	0.12083	0.07692	0.09615	0.10385	0.15385	0.17308	0.18077	0.01923	0.02692	0.02308

(Continued)

Table 4. *(Continued)* Dimensions of Coarse Pitch Full Depth Involute Spur Gear Teeth—20° and 25° Pressure Angles.

Diametral Pitch	Circular Pitch	Circular Thickness	Addendum (Std.)	Dedendum (Std.)	Dedendum (Special)	Working Depth	Whole Depth (Std.)	Whole Depth (Special)	Clearance (Std.)	Clearance (Special)	Fillet Radius
P	p	t	a	b	b	h_k	h_t	h_t	c	c	r_f
14	0.22440	0.11220	0.07143	0.08929	0.09643	0.14286	0.16071	0.16786	0.01786	0.02500	0.02143
15	0.20944	0.10472	0.06667	0.08333	0.09000	0.13333	0.15000	0.15667	0.01667	0.02333	0.02000
16	0.19635	0.09817	0.06250	0.07813	0.08438	0.12500	0.14063	0.14688	0.01563	0.02188	0.01875
17	0.18480	0.09240	0.05882	0.07353	0.07941	0.11765	0.13235	0.13824	0.01471	0.02059	0.01765
18	0.17453	0.08727	0.05556	0.06944	0.07500	0.11111	0.12500	0.13056	0.01389	0.01944	0.01667
19	0.16535	0.08267	0.05263	0.06579	0.07105	0.10526	0.11842	0.12368	0.01316	0.01842	0.01579
20	0.15708	0.07854	0.05000	0.06250	0.06750	0.10000	0.11250	0.11750	0.01250	0.01750	0.01500

Note: Preferred diametral pitches in bold type.

14 $\frac{1}{2}°$, 20°, and 25° fine pitch tooth form. This form was introduced by ANSI B6.7-1967, and emphasizes a 20° pressure angle with diametral pitches ranging from 20 to 120 (with preferred sizes being 20, 24, 32, 48, 64, 72, 80, 96, and 120), although 150 and 200 are also in general use. The standard provides specifications for the tooth form in both spur and helical gears. In addition to the 20° form, additional specifications included in the standard allow for 14 $\frac{1}{2}°$ and 25° pressure angle forms having the same tooth proportions as the 20° form. The 14 $\frac{1}{2}°$ pressure angle form is preferred for applications requiring unusually close control of backlash. The 25° pressure angle form is intended to provide additional strength, and to ease the production of precision cast or sintered gears. Tooth dimensions for this form are given in **Table 5**.

20° stub tooth involute form. As with other forms described above, this form was standardized in ASA B6.1-1932, and is now considered obsolete but is applied in certain situations, including plastic gearing.. It was designed to provide high strength for heavily loaded applications, especially at low speeds. The basic rack form of this style is shown in *Figure 2*. Basic dimensions, differing from those shown in **Table 1**, are as follows.

Addendum = 0.8/P or 0.2546 p Minimum Dedendum = 1.00/P or 0.3183 p
Working Depth = 1.6/P or 0.5092 p Minimum Whole Depth = 1.8/P or 0.5092 p
Outside Diameter = (N + 1.6)/P Root Diameter = (N – 2.0)/P
Minimum Clearance = 0.2/P or 0.0637 p Center Distance = (N_P + N_G)/2P
Tooth Thickness = 1.5708/P or p/2

Fellows 20° stub tooth form. A form introduced by Fellows Gear Shaper Company that is not in common use. It uses two diametrical pitches—one to determine the circular pitch, the number of teeth for a given pitch diameter, tooth thickness, and the center distance, and a second to determine the addendum, dedendum, and clearance.

Fellows 20° full depth involute form. With one exception, this form uses the same dimensional formulas as the standard 20° full depth involute system. The difference is that, for pitches 20P and finer, the clearance is 0.2/P + 0.002".

Table 6 provides pitch diameters for preferred diametral pitches for coarse gears, and **Table 7** provides pitch diameters for preferred diametral pitches for fine gears.

(Text continued on p. 2048)

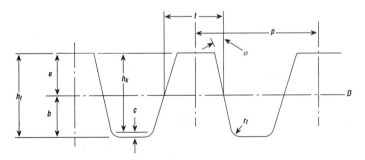

Figure 2. Basic rack form for 14 $\frac{1}{2}°$, 20°, and 25° full depth involute systems. Symbols: a = addendum, b = dedendum, c = clearance, D = pitch diameter, h_k = working depth, h_t = whole depth, p = circular pitch, t = circular thickness, r_f = fillet radius, ø = pressure angle.

Table 5. Dimensions of Fine Pitch Gear Teeth—14 1/2°, 20° and 25° Pressure Angles.

Diametral Pitch	Circular Pitch	Circular Thickness	Addendum (Std.)	Dedendum (Std.)	Dedendum (Special)	Working Depth	Whole Depth (Std.)	Whole Depth (Special)	Clearance (Std.)	Clearance (Special)
P	p	t	a	b	b	h_k	h_t	h_t	c	c
20	.15708	.07854	.05000	.06200	.06950	.10000	.11200	.11950	.01200	.01950
22	.14280	.07140	.04545	.05655	.06336	.09091	.10200	.10882	.01109	.01791
24	.13090	.06545	.04167	.05200	.05825	.08333	.09367	.09992	.01033	.01658
26	.12083	.06042	.03846	.04815	.05392	.07692	.08662	.09238	.00969	.01546
28	.11220	.05610	.03571	.04486	.05021	.07143	.08057	.08593	.00914	.01450
30	.10472	.05236	.03333	.04200	.04700	.06667	.07533	.08033	.00867	.01367
32	.09817	.04909	.03125	.03950	.04419	.06250	.07075	.07544	.00825	.01294
34	.09240	.04620	.02941	.03729	.04171	.05882	.06671	.07112	.00788	.01229
36	.08727	.04363	.02778	.03533	.03950	.05556	.06311	.06728	.00756	.01172
38	.08267	.4134	.02632	.03358	.03753	.05263	.05989	.06384	.00726	.01121
40	.07854	.03927	.02500	.03200	.03575	.05000	.05700	.06075	.00700	.01075
44	.07140	.03570	.02273	.02927	.03268	.04545	.05200	.05541	.00655	.00995
48	.06545	.03273	.02083	.02700	.03013	.04167	.04783	.05096	.00617	.00929
52	.06042	.03021	.01923	.02508	.02796	.03846	.04431	.04719	.00585	.00873
56	.05610	.02805	.01786	.02343	.02611	.03571	.04129	.04396	.00557	.00825
60	.05236	.02618	.01667	.02200	.02450	.03333	.03867	.04117	.00533	.00783
64	.04909	.02454	.01563	.02075	.02309	.03125	.03638	.03872	.00513	.00747
68	.04620	.02310	.01471	.01965	.02185	.02941	.03435	.03656	.00494	.00715
72	.04363	.02182	.01389	.01867	.02075	.02778	.03256	.03464	.00478	.00686
76	.04134	.02067	.01316	.01779	.01976	.02632	.03095	.03292	.00463	.00661
80	.03927	.01964	.01250	.01700	.01888	.02500	.02950	.03138	.00450	.00638

(Continued)

Table 5. *(Continued)* Dimensions of Fine Pitch Gear Teeth—14 1/2°, 20° and 25° Pressure Angles.

Diametral Pitch	Circular Pitch	Circular Thickness	Addendum (Std.)	Dedendum (Std.)	Dedendum (Special)	Working Depth	Whole Depth (Std.)	Whole Depth (Special)	Clearance (Std.)	Clearance (Special)
P	p	t	a	b	b	h_k	h_t	h_t	c	c
88	.03570	.01785	.01136	.01564	.01734	.02273	.02700	.02870	.00427	.00598
96	.03272	.01636	.01042	.01450	.01606	.02083	.02492	.02648	.00408	.00565
104	.03021	.01510	.00962	.01354	.01498	.01923	.02315	.02460	.00392	.00537
112	.02805	.01403	.00893	.01271	.01405	.01786	.02164	.02298	.00379	.00513
120	.02618	.01309	.00833	.01200	.01325	.01667	.02033	.02158	.00367	.00492

Note: Preferred diametral pitches in bold type.

Table 6. Pitch Diameters for Coarse Tooth Preferred Diametral Gear Sizes. See Notes.

Number of Teeth	Diametral Pitch										
	2	2.25	2.50	3	4	6	8	10	12	16
	Pitch Diameter										
10	5.0000	4.4444	4.0000	3.3333	2.5000	1.6667	1.2500	1.0000	0.8333	0.6250
11	5.5000	4.8889	4.4000	3.6667	2.7500	1.8333	1.3750	1.1000	0.9167	0.6875
12	6.0000	5.3333	4.8000	4.0000	3.0000	2.0000	1.5000	1.2000	1.0000	0.7500
13	6.5000	5.7778	5.2000	4.3333	3.2500	2.1667	1.6250	1.3000	1.0833	0.8125
14	7.0000	6.2222	5.6000	4.6667	3.5000	2.3333	1.7500	1.4000	1.1667	0.8750
15	7.5000	6.6667	6.0000	5.0000	3.7500	2.5000	1.8750	1.5000	1.2500	0.9375
16	8.0000	7.1111	6.4000	5.3333	4.0000	2.6667	2.0000	1.6000	1.3333	1.0000
17	8.5000	7.5556	6.8000	5.6667	4.2500	2.8333	2.1250	1.7000	1.4167	1.0625
18	9.0000	8.0000	7.2000	6.0000	4.5000	3.0000	2.2500	1.8000	1.5000	1.1250
19	9.5000	8.4444	7.6000	6.3333	4.7500	3.1667	2.3750	1.9000	1.5833	1.1875
20	10.0000	8.8889	8.0000	6.6667	5.0000	3.3333	2.5000	2.0000	1.6667	1.2500
21	10.5000	9.3333	8.4000	7.0000	5.2500	3.5000	2.6250	2.1000	1.7500	1.3125
22	11.0000	9.7778	8.8000	7.3333	5.5000	3.6667	2.7500	2.2000	1.8333	1.3750
23	11.5000	10.2222	9.2000	7.6667	5.7500	3.8333	2.8750	2.3000	1.9167	1.4375
24	12.0000	10.6667	9.6000	8.0000	6.0000	4.0000	3.0000	2.4000	2.0000	1.5000
25	12.5000	11.1111	10.0000	8.3333	6.2500	4.1667	3.1250	2.5000	2.0833	1.5625
26	13.0000	11.5556	10.4000	8.6667	6.5000	4.3333	3.2500	2.6000	2.1667	1.6250
27	13.5000	12.0000	10.8000	9.0000	6.7500	4.5000	3.3750	2.7000	2.2500	1.6875
28	14.0000	12.4444	11.2000	9.3333	7.0000	4.6667	3.5000	2.8000	2.3333	1.7500
29	14.5000	12.8889	11.6000	9.6667	7.2500	4.8333	3.6250	2.9000	2.4167	1.8125
30	15.0000	13.3333	12.0000	10.0000	7.5000	5.0000	3.7500	3.0000	2.5000	1.8750
31	15.5000	13.7778	12.4000	10.3333	7.7500	5.1667	3.8750	3.1000	2.5833	1.9375

(Continued)

Table 6. *(Continued)* **Pitch Diameters for Coarse Tooth Preferred Diametral Gear Sizes.** See Notes.

Number of Teeth	Diametral Pitch									
	2	2.25	2.50	3	4	6	8	10	12	16
	Pitch Diameter									
32	16.0000	14.2222	12.8000	10.6667	8.0000	5.3333	4.0000	3.2000	2.6667	2.0000
33	16.5000	14.6667	13.2000	11.0000	8.2500	5.5000	4.1250	3.3000	2.7500	2.0625
34	17.0000	15.1111	13.6000	11.3333	8.5000	5.6667	4.2500	3.4000	2.8333	2.1250
35	17.5000	15.5556	14.0000	11.6667	8.7500	5.8333	4.3750	3.5000	2.9167	2.1875
36	18.0000	16.0000	14.4000	12.0000	9.0000	6.0000	4.5000	3.6000	3.0000	2.2500
37	18.5000	16.4444	14.8000	12.3333	9.2500	6.1667	4.6250	3.7000	3.0833	2.3125
38	19.0000	16.8889	15.2000	12.6667	9.5000	6.3333	4.7500	3.8000	3.1667	2.3750
39	19.5000	17.3333	15.6000	13.0000	9.7500	6.5000	4.8750	3.9000	3.2500	2.4375
40	20.0000	17.7778	16.0000	13.3333	10.0000	6.6667	5.0000	4.0000	3.3333	2.5000
41	20.5000	18.2222	16.4000	13.6667	10.2500	6.8333	5.1250	4.1000	3.4167	2.5625
42	21.0000	18.6667	16.8000	14.0000	10.5000	7.0000	5.2500	4.2000	3.5000	2.6250
43	21.5000	19.1111	17.2000	14.3333	10.7500	7.1667	5.3750	4.3000	3.5833	2.6875
44	22.0000	19.5556	17.6000	14.6667	11.0000	7.3333	5.5000	4.4000	3.6667	2.7500
45	22.5000	20.0000	18.0000	15.0000	11.2500	7.5000	5.6250	4.5000	3.7500	2.8125
46	23.0000	20.4444	18.4000	15.3333	11.5000	7.6667	5.7500	4.6000	3.8333	2.8750
47	23.5000	20.8889	18.8000	15.6667	11.7500	7.8333	5.8750	4.7000	3.9167	2.9375
48	24.0000	21.3333	19.2000	16.0000	12.0000	8.0000	6.0000	4.8000	4.0000	3.0000
49	24.5000	21.7778	19.6000	16.3333	12.2500	8.1667	6.1250	4.9000	4.0833	3.0625
50	25.0000	22.2222	20.0000	16.6667	12.5000	8.3333	6.2500	5.0000	4.1667	3.1250
51	25.5000	22.6667	20.4000	17.0000	12.7500	8.5000	6.3750	5.1000	4.2500	3.1875
52	26.0000	23.1111	20.8000	17.3333	13.0000	8.6667	6.5000	5.2000	4.3333	3.2500
53	26.5000	23.5556	21.2000	17.6667	13.2500	8.8333	6.6250	5.3000	4.4167	3.3125

(Continued)

Table 6. (*Continued*) Pitch Diameters for Coarse Tooth Preferred Diametral Gear Sizes. See Notes.

Number of Teeth	Diametral Pitch										
	2	2.25	2.50	3	4	6	8	10	12	16
	Pitch Diameter										
54	27.0000	24.0000	21.6000	18.0000	13.5000	9.0000	6.7500	5.4000	4.5000	3.3750
55	27.5000	24.4444	22.0000	18.3333	13.7500	9.1667	6.8750	5.5000	4.5833	3.4375
56	28.0000	24.8889	22.4000	18.6667	14.0000	9.3333	7.0000	5.6000	4.6667	3.5000
57	28.5000	25.3333	22.8000	19.0000	14.2500	9.5000	7.1250	5.7000	4.7500	3.5625
58	29.0000	25.7778	23.2000	19.3333	14.5000	9.6667	7.2500	5.8000	4.8333	3.6250
59	29.5000	26.2222	23.6000	19.6667	14.7500	9.8333	7.3750	5.9000	4.9167	3.6875
60	30.0000	26.6667	24.0000	20.0000	15.0000	10.0000	7.5000	6.0000	5.0000	3.7500
61	30.5000	27.1111	24.4000	20.3333	15.2500	10.1667	7.6250	6.1000	5.0833	3.8125
62	31.0000	27.5556	24.8000	20.6667	15.5000	10.3333	7.7500	6.2000	5.1667	3.8750
63	31.5000	28.0000	25.2000	21.0000	15.7500	10.5000	7.8750	6.3000	5.2500	3.9375
64	32.0000	28.4444	25.6000	21.3333	16.0000	10.6667	8.0000	6.4000	5.3333	4.0000
65	32.5000	28.8889	26.0000	21.6667	16.2500	10.8333	8.1250	6.5000	5.4167	4.0625
66	33.0000	29.3333	26.4000	22.0000	16.5000	11.0000	8.2500	6.6000	5.5000	4.1250
67	33.5000	29.7778	26.8000	22.3333	16.7500	11.1667	8.3750	6.7000	5.5833	4.1875
68	34.0000	30.2222	27.2000	22.6667	17.0000	11.3333	8.5000	6.8000	5.6667	4.2500
69	34.5000	30.6667	27.6000	23.0000	17.2500	11.5000	8.6250	6.9000	5.7500	4.3125
70	35.0000	31.1111	28.0000	23.3333	17.5000	11.6667	8.7500	7.0000	5.8333	4.3750
71	35.5000	31.5556	28.4000	23.6667	17.7500	11.8333	8.8750	7.1000	5.9167	4.4375
72	36.0000	32.0000	28.8000	24.0000	18.0000	12.0000	9.0000	7.2000	6.0000	4.5000
73	36.5000	32.4444	29.2000	24.3333	18.2500	12.1667	9.1250	7.3000	6.0833	4.5625
74	37.0000	32.8889	29.6000	24.6667	18.5000	12.3333	9.2500	7.4000	6.1667	4.6250
75	37.5000	33.3333	30.0000	25.0000	18.7500	12.5000	9.3750	7.5000	6.2500	4.6875

(*Continued*)

Table 6. *(Continued)* Pitch Diameters for Coarse Tooth Preferred Diametral Gear Sizes. See Notes.

Number of Teeth	Diametral Pitch									
	2	2.25	2.50	3	4	6	8	10	12	16
	Pitch Diameter									
76	38.0000	33.7778	30.4000	25.3333	19.0000	12.6667	9.5000	7.6000	6.3333	4.7500
77	38.5000	34.2222	30.8000	25.6667	19.2500	12.8333	9.6250	7.7000	6.4167	4.8125
78	39.0000	34.6667	31.2000	26.0000	19.5000	13.0000	9.7500	7.8000	6.5000	4.8750
79	39.5000	35.1111	31.6000	26.3333	19.7500	13.1667	9.8750	7.9000	6.5833	4.9375
80	40.0000	35.5556	32.0000	26.6667	20.0000	13.3333	10.0000	8.0000	6.6667	5.0000
81	40.5000	36.0000	32.4000	27.0000	20.2500	13.5000	10.1250	8.1000	6.7500	5.0625
82	41.0000	36.4444	32.8000	27.3333	20.5000	13.6667	10.2500	8.2000	6.8333	5.1250
83	41.5000	36.8889	33.2000	27.6667	20.7500	13.8333	10.3750	8.3000	6.9167	5.1875
84	42.0000	37.3333	33.6000	28.0000	21.0000	14.0000	10.5000	8.4000	7.0000	5.2500
85	42.5000	37.7778	34.0000	28.3333	21.2500	14.1667	10.6250	8.5000	7.0833	5.3125
86	43.0000	38.2222	34.4000	28.6667	21.5000	14.3333	10.7500	8.6000	7.1667	5.3750
87	43.5000	38.6667	34.8000	29.0000	21.7500	14.5000	10.8750	8.7000	7.2500	5.4375
88	44.0000	39.1111	35.2000	29.3333	22.0000	14.6667	11.0000	8.8000	7.3333	5.5000
89	44.5000	39.5556	35.6000	29.6667	22.2500	14.8333	11.1250	8.9000	7.4167	5.5625
90	45.0000	40.0000	36.0000	30.0000	22.5000	15.0000	11.2500	9.0000	7.5000	5.6250
91	45.5000	40.4444	36.4000	30.3333	22.7500	15.1667	11.3750	9.1000	7.5833	5.6875
92	46.0000	40.8889	36.8000	30.6667	23.0000	15.3333	11.5000	9.2000	7.6667	5.7500
93	46.5000	41.3333	37.2000	31.0000	23.2500	15.5000	11.6250	9.3000	7.7500	5.8125
94	47.0000	41.7778	37.6000	31.3333	23.5000	15.6667	11.7500	9.4000	7.8333	5.8750
95	47.5000	42.2222	38.0000	31.6667	23.7500	15.8333	11.8750	9.5000	7.9167	5.9375
96	48.0000	42.6667	38.4000	32.0000	24.0000	16.0000	12.0000	9.6000	8.0000	6.0000
97	48.5000	43.1111	38.8000	32.3333	24.5000	16.1667	12.1250	9.7000	8.0833	6.0625

(Continued)

Table 6. *(Continued)* Pitch Diameters for Coarse Tooth Preferred Diametral Gear Sizes. See Notes.

Number of Teeth	Diametral Pitch									
	2	2.25	2.50	3	4	6	8	10	12	16
	Pitch Diameter									
98	49.0000	43.5556	39.2000	32.6667	24.5000	16.3333	12.2500	9.8000	8.1667	6.1250
99	49.5000	44.0000	39.6000	33.0000	24.7500	16.5000	12.3750	9.9000	8.2500	6.1875
100	50.0000	44.4444	40.0000	33.3333	25.0000	16.6667	12.5000	10.0000	8.3333	6.2500
101	50.5000	44.8889	40.4000	33.6667	25.2500	16.8333	12.6250	10.1000	8.4167	6.3125
102	51.0000	45.3333	40.8000	34.0000	25.5000	17.0000	12.7500	10.2000	8.5000	6.3750
103	51.5000	45.7778	41.2000	34.3333	25.7500	17.1667	12.8750	10.3000	8.5833	6.4375
104	52.0000	46.2222	41.6000	34.6667	26.0000	17.3333	13.0000	10.4000	8.6667	6.5000
105	52.5000	46.6667	42.0000	35.0000	26.2500	17.5000	13.1250	10.5000	8.7500	6.5625
106	53.0000	47.1111	42.4000	35.3333	26.5000	17.6667	13.2500	10.6000	8.8333	6.6250
107	53.5000	47.5556	42.8000	35.6667	26.7500	17.8333	13.3750	10.7000	8.9167	6.6875
108	54.0000	48.0000	43.2000	36.0000	27.0000	18.0000	13.5000	10.8000	9.0000	6.7500
109	54.5000	48.4444	43.6000	36.3333	27.2500	18.1667	13.6250	10.9000	9.0833	6.8125
110	55.0000	48.8889	44.0000	36.6667	27.5000	18.3333	13.7500	11.0000	9.1667	6.8750
111	55.5000	49.3333	44.4000	37.0000	27.7500	18.5000	13.8750	11.1000	9.2500	6.9375
112	56.0000	49.7778	44.8000	37.3333	28.0000	18.6667	14.0000	11.2000	9.3333	7.0000
113	56.5000	50.2222	45.2000	37.6667	28.2500	18.8333	14.1250	11.3000	9.4167	7.0625
114	57.0000	50.6667	45.6000	38.0000	28.5000	19.0000	14.2500	11.4000	9.5000	7.1250
115	57.5000	51.1111	46.0000	38.3333	28.7500	19.1667	14.3750	11.5000	9.5833	7.1875
116	58.0000	51.5556	46.4000	38.6667	29.0000	19.3333	14.5000	11.6000	9.6667	7.2500
117	58.5000	52.0000	46.8000	39.0000	29.2500	19.5000	14.6250	11.7000	9.7500	7.3125
118	59.0000	52.4444	47.2000	39.3333	29.5000	19.6667	14.7500	11.8000	9.8333	7.3750
119	59.5000	52.8889	47.6000	39.6667	29.7500	19.8333	14.8750	11.9000	9.9167	7.4375

(Continued)

Table 6. (Continued) Pitch Diameters for Coarse Tooth Preferred Diametral Gear Sizes. See Notes.

Number of Teeth	Diametral Pitch									
	2	2.25	2.50	3	4	6	8	10	12	16
	Pitch Diameter									
120	60.0000	53.3333	48.0000	40.0000	30.0000	20.0000	15.0000	12.0000	10.0000	7.5000
121	60.5000	53.7778	48.4000	40.3333	30.2500	20.1667	15.1250	12.1000	10.0833	7.5625
122	61.0000	54.2222	48.8000	40.6667	30.5000	20.3333	15.2500	12.2000	10.1667	7.6250
123	61.5000	54.6667	49.2000	41.0000	30.7500	20.5000	15.3750	12.3000	10.2500	7.6875
124	62.0000	55.1111	49.6000	41.3333	31.0000	20.6667	15.5000	12.4000	10.3333	7.7500
125	62.5000	55.5556	50.0000	41.6667	31.2500	20.8333	15.6250	12.5000	10.4167	7.8125
126	63.0000	56.0000	50.4000	42.0000	31.5000	21.0000	15.7500	12.6000	10.5000	7.8750
127	63.5000	56.4444	50.8000	42.3333	31.7500	21.1667	15.8750	12.7000	10.5833	7.9375
128	64.0000	56.8889	51.2000	42.6667	32.0000	21.3333	16.0000	12.8000	10.6667	8.0000
129	64.5000	57.3333	51.6000	43.0000	32.2500	21.5000	16.1250	12.9000	10.7500	8.0625
130	65.0000	57.7778	52.0000	43.3333	32.5000	21.6667	16.2500	13.0000	10.8333	8.1250
131	65.5000	58.2222	52.4000	43.6667	32.7500	21.8333	16.3750	13.1000	10.9167	8.1875
132	66.0000	58.6667	52.8000	44.0000	33.0000	22.0000	16.5000	13.2000	11.0000	8.2500
133	66.5000	59.1111	53.2000	44.3333	33.2500	22.1667	16.6250	13.3000	11.0833	8.3125
134	67.0000	59.5556	53.6000	44.6667	33.5000	22.3333	16.7500	13.4000	11.1667	8.3750
135	67.5000	60.0000	54.0000	45.0000	33.7500	22.5000	16.8750	13.5000	11.2500	8.4375
136	68.0000	60.4444	54.4000	45.3333	34.0000	22.6667	17.0000	13.6000	11.3333	8.5000
137	68.5000	60.8889	54.8000	45.6667	34.2500	22.8333	17.1250	13.7000	11.4167	8.5625
138	69.0000	61.3333	55.2000	46.0000	34.5000	23.0000	17.2500	13.8000	11.5000	8.6250
139	69.5000	61.7778	55.6000	46.3333	34.7500	23.1667	17.3750	13.9000	11.5833	8.6875
140	70.0000	62.2222	56.0000	46.6667	35.0000	23.3333	17.5000	14.0000	11.6667	8.7500
141	70.5000	62.6667	56.4000	47.0000	35.2500	23.5000	17.6250	14.1000	11.7500	8.8125

(Continued)

Table 6. *(Continued)* **Pitch Diameters for Coarse Tooth Preferred Diametral Gear Sizes.** See Notes.

Number of Teeth	Diametral Pitch									
	2	2.25	2.50	3	4	6	8	10	12	16
	Pitch Diameter									
142	71.0000	63.1111	56.8000	47.3333	35.5000	23.6667	17.7500	14.2000	11.8333	8.8750
143	71.5000	63.5556	57.2000	47.6667	35.7500	23.8333	17.8750	14.3000	11.9167	8.9375
144	72.0000	64.0000	57.6000	48.0000	36.0000	24.0000	18.0000	14.4000	12.0000	9.0000
145	72.5000	64.4444	58.0000	48.3333	36.2500	24.1667	18.1250	14.5000	12.0833	9.0625
146	73.0000	64.8889	58.4000	48.6667	36.5000	24.3333	18.2500	14.6000	12.1667	9.1250
147	73.5000	65.3333	58.8000	49.0000	36.7500	24.5000	18.3750	14.7000	12.2500	9.1875
148	74.0000	65.7778	59.2000	49.3333	37.0000	24.6667	18.5000	14.8000	12.3333	9.2500
149	74.5000	66.2222	59.6000	49.6667	37.2500	24.8333	18.6250	14.9000	12.4167	9.3125
150	75.0000	66.6667	60.0000	50.0000	37.5000	25.0000	18.7500	15.0000	12.5000	9.3750
151	75.5000	67.1111	60.4000	50.3333	37.7500	25.1667	18.8750	15.1000	12.5833	9.4375
152	76.0000	67.5556	60.8000	50.6667	38.0000	25.3333	19.0000	15.2000	12.6667	9.5000
153	76.5000	68.0000	61.2000	51.0000	38.2500	25.5000	19.1250	15.3000	12.7500	9.5625
154	77.0000	68.4444	61.6000	51.3333	38.5000	25.6667	19.2500	15.4000	12.8333	9.6250
155	77.5000	68.8889	62.0000	51.6667	38.7500	25.8333	19.3750	15.5000	12.9167	9.6875
156	78.0000	69.3333	62.4000	52.0000	39.0000	26.0000	19.5000	15.6000	13.0000	9.7500
157	78.5000	69.7778	62.8000	52.3333	39.2500	26.1667	19.6250	15.7000	13.0833	9.8125
158	79.0000	70.2222	63.2000	52.6667	39.5000	26.3333	19.7500	15.8000	13.1667	9.8750
159	79.5000	70.6667	63.6000	53.0000	39.7500	26.5000	19.8750	15.9000	13.2500	9.9375
160	80.0000	71.1111	64.0000	53.3333	40.0000	26.6667	20.0000	16.0000	13.3333	10.0000

Notes: All pitch diameters are theoretical. The outside diameter of a standard full depth gear is equal to the pitch diameter of a gear of the same diametral pitch with two additional teeth.

Table 7. Pitch Diameters for Fine Tooth Preferred Diametral Gear Sizes. See Notes.

Number of Teeth	Diametral Pitch									
	20	24	32	48	64	72	80	96	120	150
	Pitch Diameter									
10	0.5000	0.4167	0.3125	0.2083	0.1563	0.1389	0.1250	0.1042	0.0833	0.0667
11	0.5500	0.4583	0.3438	0.2292	0.1719	0.1528	0.1375	0.1146	0.0917	0.0733
12	0.6000	0.5000	0.3750	0.2500	0.1875	0.1667	0.1500	0.1250	0.1000	0.0800
13	0.6500	0.5417	0.4063	0.2708	0.2031	0.1806	0.1625	0.1354	0.1083	0.0867
14	0.7000	0.5833	0.4375	0.2917	0.2188	0.1944	0.1750	0.1458	0.1167	0.0933
15	0.7500	0.6250	0.4688	0.3125	0.2344	0.2083	0.1875	0.1563	0.1250	0.1000
16	0.8000	0.6667	0.5000	0.3333	0.2500	0.2222	0.2000	0.1667	0.1333	0.1067
17	0.8500	0.7283	0.5313	0.3542	0.2656	0.2361	0.2125	0.1771	0.1417	0.1133
18	0.9000	0.7500	0.5625	0.3750	0.2813	0.2500	0.2250	0.1875	0.1500	0.1200
19	0.9500	0.7917	0.5938	0.3958	0.2969	0.2639	0.2375	0.1979	0.1583	0.1267
20	1.0000	0.8333	0.6250	0.4167	0.3125	0.2778	0.2500	0.2083	0.1667	0.1333
21	1.0500	0.8750	0.6563	0.4375	0.3281	0.2917	0.2625	0.2188	0.1750	0.1400
22	1.1000	0.9167	0.6875	0.4583	0.3438	0.3056	0.2750	0.2292	0.1833	0.1467
23	1.1500	0.9583	0.7188	0.4792	0.3594	0.3194	0.2875	0.2396	0.1917	0.1533
24	1.2000	1.0000	0.7500	0.5000	0.3750	0.3333	0.3000	0.2500	0.2000	0.1600
25	1.2500	1.0417	0.7813	0.5208	0.3906	0.3472	0.3125	0.2604	0.2083	0.1667
26	1.3000	1.0833	0.8125	0.5417	0.4063	0.3611	0.3250	0.2708	0.2167	0.1733
27	1.3500	1.1250	0.8438	0.5625	0.4219	0.3750	0.3375	0.2813	0.2250	0.1800
28	1.4000	1.1667	0.8750	0.5833	0.4375	0.3889	0.3500	0.2917	0.2333	0.1867
29	1.4500	1.2083	0.9063	0.6042	0.4531	0.4028	0.3625	0.3021	0.2417	0.1933
30	1.5000	1.2500	0.9375	0.6250	0.4688	0.4167	0.3750	0.3125	0.2500	0.2000
31	1.5500	1.2917	0.9688	0.6458	0.4844	0.4306	0.3875	0.3229	0.2583	0.2067

(Continued)

Table 7. *(Continued)* Pitch Diameters for Fine Tooth Preferred Diametral Gear Sizes. See Notes.

Number of Teeth	Diametral Pitch									
	20	24	32	48	64	72	80	96	120	150
	Pitch Diameter									
32	1.6000	1.3333	1.0000	0.6667	0.5000	0.4444	0.4000	0.3333	0.2667	0.2133
33	1.6500	1.3750	1.0313	0.6875	0.5156	0.4583	0.4125	0.3438	0.2750	0.2200
34	1.7000	1.4167	1.0625	0.7083	0.5313	0.4722	0.4250	0.3542	0.2833	0.2267
35	1.7500	1.4583	1.0938	0.7292	0.5469	0.4861	0.4375	0.3646	0.2917	0.2333
36	1.8000	1.5000	1.1250	0.7500	0.5625	0.5000	0.4500	0.3750	0.3000	0.2400
37	1.8500	1.5417	1.1563	0.7708	0.5781	0.5139	0.4625	0.3854	0.3083	0.2467
38	1.9000	1.5833	1.1875	0.7917	0.5938	0.5278	0.4750	0.3958	0.3167	0.2533
39	1.9500	1.6250	1.2188	0.8125	0.6094	0.5417	0.4875	0.4063	0.3250	0.2600
40	2.0000	1.6667	1.2500	0.8333	0.6250	0.5556	0.5000	0.4167	0.3333	0.2667
41	2.0500	1.7083	1.2813	0.8542	0.6406	0.5694	0.5125	0.4271	0.3417	0.2733
42	2.1000	1.7500	1.3125	0.8750	0.6563	0.5833	0.5250	0.4375	0.3500	0.2800
43	2.1500	1.7917	1.3438	0.8958	0.6719	0.5972	0.5375	0.4479	0.3583	0.2867
44	2.2000	1.8333	1.3750	0.9167	0.6875	0.6111	0.5500	0.4583	0.3667	0.2933
45	2.2500	1.8750	1.4063	0.9375	0.7031	0.6250	0.5625	0.4688	0.3750	0.3000
46	2.3000	1.9167	1.4375	0.9583	0.7188	0.6389	0.5750	0.4792	0.3833	0.3067
47	2.3500	1.9583	1.4688	0.9792	0.7344	0.6528	0.5875	0.4896	0.3917	0.3133
48	2.4000	2.0000	1.5000	1.0000	0.7500	0.6667	0.6000	0.5000	0.4000	0.3200
49	2.4500	2.0417	1.5313	1.0208	0.7656	0.6806	0.6125	0.5104	0.4083	0.3267
50	2.5000	2.0833	1.5625	1.0417	0.7813	0.6944	0.6250	0.5208	0.4167	0.3333
51	2.5500	2.1250	1.5938	1.0625	0.7969	0.7083	0.6375	0.5313	0.4250	0.3400
52	2.6000	2.1667	1.6250	1.0833	0.8125	0.7222	0.6500	0.5417	0.4333	0.3467
53	2.6500	2.2083	1.6563	1.1042	0.8281	0.7361	0.6625	0.5521	0.4417	0.3533

(Continued)

Table 7. (Continued) Pitch Diameters for Fine Tooth Preferred Diametral Gear Sizes. See Notes.

Number of Teeth	Diametral Pitch										
	20	24	32	48	64	72	80	96	120	150
	Pitch Diameter										
54	2.7000	2.2500	1.6875	1.1250	0.8438	0.7500	0.6750	0.5625	0.4500	0.3600
55	2.7500	2.2917	1.7188	1.1458	0.8594	0.7639	0.6875	0.5729	0.4583	0.3667
56	2.8000	2.3333	1.7500	1.1667	0.8750	0.7778	0.7000	0.5833	0.4667	0.3733
57	2.8500	2.3750	1.7813	1.1875	0.8906	0.7917	0.7125	0.5938	0.4750	0.3800
58	2.9000	2.4167	1.8125	1.2083	0.9063	0.8056	0.7250	0.6042	0.4833	0.3867
59	2.9500	2.4583	1.8438	1.2292	0.9219	0.8194	0.7375	0.6146	0.4917	0.3933
60	3.0000	2.5000	1.8750	1.2500	0.9375	0.8333	0.7500	0.6250	0.5000	0.4000
61	3.0500	2.5417	1.9063	1.2708	0.9531	0.8472	0.7625	0.6354	0.5083	0.4067
62	3.1000	2.5833	1.9375	1.2917	0.9688	0.8611	0.7750	0.6458	0.5167	0.4133
63	3.1500	2.6250	1.9688	1.3125	0.9844	0.8750	0.7875	0.6563	0.5250	0.4200
64	3.2000	2.6667	2.0000	1.3333	1.0000	0.8889	0.8000	0.6667	0.5333	0.4267
65	3.2500	2.7083	2.0313	1.3542	1.0156	0.9028	0.8125	0.6771	0.5417	0.4333
66	3.3000	2.7500	2.0625	1.3750	1.0313	0.9167	0.8250	0.6875	0.5500	0.4400
67	3.3500	2.7917	2.0938	1.3958	1.0469	0.9306	0.8375	0.6979	0.5583	0.4467
68	3.4000	2.8333	2.1250	1.4167	1.0625	0.9444	0.8500	0.7083	0.5667	0.4533
69	3.4500	2.8750	2.1563	1.4375	1.0781	0.9583	0.8625	0.7188	0.5750	0.4600
70	3.5000	2.9167	2.1875	1.4583	1.0938	0.9722	0.8750	0.7292	0.5833	0.4667
71	3.5500	2.9583	2.2188	1.4792	1.1094	0.9861	0.8875	0.7396	0.5917	0.4733
72	3.6000	3.0000	2.2500	1.5000	1.1250	1.0000	0.9000	0.7500	0.6000	0.4800
73	3.6500	3.0417	2.2813	1.5208	1.1406	1.0139	0.9125	0.7604	0.6083	0.4867
74	3.7000	3.0833	2.3125	1.5417	1.1563	1.0278	0.9250	0.7708	0.6167	0.4933
75	3.7500	3.1250	2.3438	1.5625	1.1719	1.0417	0.9375	0.7813	0.6250	0.5000

(Continued)

Table 7. *(Continued)* Pitch Diameters for Fine Tooth Preferred Diametral Gear Sizes. See Notes.

Number of Teeth	Diametral Pitch									
	20	24	32	48	64	72	80	96	120	150
	Pitch Diameter									
76	3.8000	3.1667	2.3750	1.5833	1.1875	1.0556	0.9500	0.7917	0.6333	0.5067
77	3.8500	3.2083	2.4063	1.6042	1.2031	1.0694	0.9625	0.8021	0.6417	0.5133
78	3.9000	3.2500	2.4375	1.6250	1.2188	1.0833	0.9750	0.8125	0.6500	0.5200
79	3.9500	3.2917	2.4688	1.6458	1.2344	1.0972	0.9875	0.8229	0.6583	0.5267
80	4.0000	3.3333	2.5000	1.6667	1.2500	1.1111	1.0000	0.8333	0.6667	0.5333
81	4.0500	3.3750	2.5313	1.6875	1.2656	1.1250	1.0125	0.8438	0.6750	0.5400
82	4.1000	3.4167	2.5625	1.7083	1.2813	1.1389	1.0250	0.8542	0.6833	0.5467
83	4.1500	3.4583	2.5938	1.7292	1.2969	1.1528	1.0375	0.8646	0.6917	0.5533
84	4.2000	3.5000	2.6250	1.7500	1.3125	1.1667	1.0500	0.8750	0.7000	0.5600
85	4.2500	3.5417	2.6563	1.7708	1.3281	1.1806	1.0625	0.8854	0.7083	0.5667
86	4.3000	3.5833	2.6875	1.7917	1.3438	1.1944	1.0750	0.8958	0.7167	0.5733
87	4.3500	3.6250	2.7188	1.8125	1.3594	1.2083	1.0875	0.9063	0.7250	0.5800
88	4.4000	3.6667	2.7500	1.8333	1.3750	1.2222	1.1000	0.9167	0.7333	0.5867
89	4.4500	3.7083	2.7813	1.8542	1.3906	1.2361	1.1125	0.9271	0.7417	0.5933
90	4.5000	3.7500	2.8125	1.8750	1.4063	1.2500	1.1250	0.9375	0.7500	0.6000
91	4.5500	3.7917	2.8438	1.8958	1.4219	1.2639	1.1375	0.9479	0.7583	0.6067
92	4.6000	3.8333	2.8750	1.9167	1.4375	1.2778	1.1500	0.9583	0.7667	0.6133
93	4.6500	3.8750	2.9063	1.9375	1.4531	1.2917	1.1625	0.9688	0.7750	0.6200
94	4.7000	3.9167	2.9375	1.9583	1.4688	1.3056	1.1750	0.9792	0.7833	0.6267
95	4.7500	3.9583	2.9688	1.9792	1.4844	1.3194	1.1875	0.9896	0.7917	0.6333
96	4.8000	4.0000	3.0000	2.0000	1.5000	1.3333	1.2000	1.0000	0.8000	0.6400
97	4.8500	4.0417	3.0313	2.0208	1.5156	1.3472	1.2125	1.0104	0.8083	0.6467

(Continued)

Table 7. *(Continued)* Pitch Diameters for Fine Tooth Preferred Diametral Gear Sizes. See Notes.

Number of Teeth	Diametral Pitch										
	20	24	32	48	64	72	80	96	120	150
	Pitch Diameter										
98	4.9000	4.0833	3.0625	2.0417	1.5313	1.3611	1.2250	1.0208	0.8167	0.6533
99	4.9500	4.1250	3.0938	2.0625	1.5469	1.3750	1.2375	1.0313	0.8250	0.6600
100	5.0000	4.1667	3.1250	2.0833	1.5625	1.3889	1.2500	1.0417	0.8333	0.6667
101	5.0500	4.2083	3.1563	2.1042	1.5781	1.4028	1.2625	1.0521	0.8417	0.6733
102	5.1000	4.2500	3.1875	2.1250	1.5938	1.4167	1.2750	1.0625	0.8500	0.6800
103	5.1500	4.2917	3.2188	2.1458	1.6094	1.4306	1.2875	1.0729	0.8583	0.6867
104	5.2000	4.3333	3.2500	2.1667	1.6250	1.4444	1.3000	1.0833	0.8667	0.6933
105	5.2500	4.3750	3.2813	2.1875	1.6406	1.4583	1.3125	1.0938	0.8750	0.7000
106	5.3000	4.4167	3.3125	2.2083	1.6563	1.4722	1.3250	1.1042	0.8833	0.7067
107	5.3500	4.4583	3.3438	2.2292	1.6719	1.4861	1.3375	1.1146	0.8917	0.7133
108	5.4000	4.5000	3.3750	2.2500	1.6875	1.5000	1.3500	1.1250	0.9000	0.7200
109	5.4500	4.5417	3.4063	2.2708	1.7031	1.5139	1.3625	1.1354	0.9083	0.7267
110	5.5000	4.5833	3.4375	2.2917	1.7188	1.5278	1.3750	1.1458	0.9167	0.7333
111	5.5500	4.6250	3.4688	2.3125	1.7344	1.5417	1.3875	1.1563	0.9250	0.7400
112	5.6000	4.6667	3.5000	2.3333	1.7500	1.5556	1.4000	1.1667	0.9333	0.7467
113	5.6500	4.7083	3.5313	2.3542	1.7656	1.5694	1.4125	1.1771	0.9417	0.7533
114	5.7000	4.7500	3.5625	2.3750	1.7813	1.5833	1.4250	1.1875	0.9500	0.7600
115	5.7500	4.7917	3.5938	2.3958	1.7969	1.5972	1.4375	1.1979	0.9583	0.7667
116	5.8000	4.8333	3.6250	2.4167	1.8125	1.6111	1.4500	1.2083	0.9667	0.7733
117	5.8500	4.8750	3.6563	2.4375	1.8281	1.6250	1.4625	1.2188	0.9750	0.7800
118	5.9000	4.9167	3.6875	2.4583	1.8438	1.6389	1.4750	1.2292	0.9833	0.7867
119	5.9500	4.9583	3.7188	2.4792	1.8594	1.6528	1.4875	1.2396	0.9917	0.7933

(Continued)

Table 7. (Continued) Pitch Diameters for Fine Tooth Preferred Diametral Gear Sizes. See Notes.

Number of Teeth	Diametral Pitch										
	20	24	32	48	64	72	80	96	120	150
	Pitch Diameter										
120	6.0000	5.0000	3.7500	2.5000	1.8750	1.6667	1.5000	1.2500	1.0000	0.8000
121	6.0500	5.0417	3.7813	2.5208	1.8906	1.6806	1.5125	1.2604	1.0083	0.8067
122	6.1000	5.0833	3.8125	2.5417	1.9063	1.6944	1.5250	1.2708	1.0167	0.8133
123	6.1500	5.1250	3.8438	2.5625	1.9219	1.7083	1.5375	1.2813	1.0250	0.8200
124	6.2000	5.1667	3.8750	2.5833	1.9375	1.7222	1.5500	1.2917	1.0333	0.8267
125	6.2500	5.2083	3.9063	2.6042	1.9531	1.7361	1.5625	1.3021	1.0417	0.8333
126	6.3000	5.2500	3.9375	2.6250	1.9688	1.7500	1.5750	1.3125	1.0500	0.8400
127	6.3500	5.2917	3.9688	2.6458	1.9844	1.7639	1.5875	1.3229	1.0583	0.8467
128	6.4000	5.3333	4.0000	2.6667	2.0000	1.7778	1.6000	1.3333	1.0667	0.8533
129	6.4500	5.3750	4.0313	2.6875	2.0156	1.7917	1.6125	1.3438	1.0750	0.8600
130	6.5000	5.4167	4.0625	2.7083	2.0313	1.8056	1.6250	1.3542	1.0833	0.8667
131	6.5500	5.4583	4.0938	2.7292	2.0469	1.8194	1.6375	1.3646	1.0917	0.8733
132	6.6000	5.5000	4.1250	2.7500	2.0625	1.8333	1.6500	1.3750	1.1000	0.8800
133	6.6500	5.5417	4.1563	2.7708	2.0781	1.8472	1.6625	1.3854	1.1083	0.8867
134	6.7000	5.5833	4.1875	2.7917	2.0938	1.8611	1.6750	1.3958	1.1167	0.8933
135	6.7500	5.6250	4.2188	2.8125	2.1094	1.8750	1.6875	1.4063	1.1250	0.9000
136	6.8000	5.6667	4.2500	2.8333	2.1250	1.8889	1.7000	1.4167	1.1333	0.9067
137	6.8500	5.7083	4.2813	2.8542	2.1406	1.9028	1.7125	1.4271	1.1417	0.9133
138	6.9000	5.7500	4.3125	2.8750	2.1563	1.9167	1.7250	1.4375	1.1500	0.9200
139	6.9500	5.7917	4.3438	2.8958	2.1719	1.9306	1.7375	1.4479	1.1583	0.9267
140	7.0000	5.8333	4.3750	2.9167	2.1875	1.9444	1.7500	1.4583	1.1667	0.9333
141	7.0500	5.8750	4.4063	2.9375	2.2031	1.9583	1.7625	1.4688	1.1750	0.9400

(Continued)

Table 7. *(Continued)* **Pitch Diameters for Fine Tooth Preferred Diametral Gear Sizes. See Notes.**

Number of Teeth	Diametral Pitch										
	20	24	32	48	64	72	80	96	120	150
	Pitch Diameter										
142	7.1000	5.9167	4.4375	2.9583	2.2188	1.9722	1.7750	1.4792	1.1833	0.9467
143	7.1500	5.9583	4.4688	2.9792	2.2344	1.9861	1.7875	1.4896	1.1917	0.9533
144	7.2000	6.0000	4.5000	3.0000	2.2500	2.0000	1.8000	1.5000	1.2000	0.9600
145	7.2500	6.0417	4.5313	3.0208	2.2656	2.0139	1.8125	1.5104	1.2083	0.9667
146	7.3000	6.0833	4.5625	3.0417	2.2813	2.0278	1.8250	1.5208	1.2167	0.9733
147	7.3500	6.1250	4.5938	3.0625	2.2969	2.0417	1.8375	1.5313	1.2250	0.9800
148	7.4000	6.1667	4.6250	3.0833	2.3125	2.0556	1.8500	1.5417	1.2333	0.9867
149	7.4500	6.2083	4.6563	3.1042	2.3281	2.0694	1.8625	1.5521	1.2417	0.9933
150	7.5000	6.2500	4.6875	3.1250	2.3438	2.0833	1.8750	1.5625	1.2500	1.0000
151	7.5500	6.2917	4.7188	3.1458	2.3594	2.0972	1.8875	1.5729	1.2583	1.0067
152	7.6000	6.3333	4.7500	3.1667	2.3750	2.1111	1.9000	1.5833	1.2667	1.0133
153	7.6500	6.3750	4.7813	3.1875	2.3906	2.1250	1.9125	1.5938	1.2750	1.0200
154	7.7000	6.4167	4.8125	3.2083	2.4063	2.1389	1.9250	1.6042	1.2833	1.0267
155	7.7500	6.4583	4.8438	3.2292	2.4219	2.1528	1.9375	1.6146	1.2917	1.0333
156	7.8000	6.5000	4.8750	3.2500	2.4375	2.1667	1.9500	1.6250	1.3000	1.0400
157	7.8500	6.5417	4.9063	3.2708	2.4531	2.1806	1.9625	1.6354	1.3083	1.0467
158	7.9000	6.5833	4.9375	3.2917	2.4688	2.1944	1.9750	1.6458	1.3167	1.0533
159	7.9500	6.6250	4.9688	3.3125	2.4844	2.2083	1.9875	1.6563	1.3250	1.0600
160	8.0000	6.6667	5.0000	3.3333	2.5000	2.2222	2.0000	1.6667	1.3333	1.0667

Notes: All pitch diameters are theoretical. The outside diameter of a standard full depth gear is equal to the pitch diameter of a gear of the same diametral pitch with two additional teeth.

Metric module

Table 8 provides equivalent diametral pitch and circular pitch for preferred metric modules. It should be noted that no preferred modules are equivalent to any preferred diametral pitch. See the definitions section for an explanation of the metric module system. Dimensions for the ISO basic rack form for the module system gear form are given in **Table 9**.

Formulas for spur gear layout

Standard center distance. The theoretical center distance between two mating gears is equal to one-half of the sum of the pitch diameter of the two mating gears.

$$C = (D_G + D_P) \div 2$$

The above formula, as well as the others given for center distance in **Table 1**, provides a theoretically ideal center distance, but backlash can be introduced, or removed, by making minute changes in center distance. When calculating center distance for $14\,^1/_2°$ pressure angle gears, for example, the resulting backlash introduced, or reduced, per 0.0001" change in center distance is approximately 0.00005". For 20° pressure angle gears, the resulting backlash introduced, or reduced, per 0.0001" change in center distance is approximately 0.00007". For 25° pressure angle gears, the resulting backlash introduced, or reduced, per 0.0001" change in center distance is approximately 0.00009". Caution must be used when reducing backlash in order to avoid the introduction of interference between the mating gears. The amount of backlash that is introduced by a given change in center distance can be calculated with the following equation.

$$\Delta B = 2 \tan ø \times \Delta C$$

In some instances, the center distance is rigidly set and the gears must be selected to exactly fit the specified distance. In these situations, sometimes it is impossible to find a set of gears with the required ratio and a standard diametral. In such cases, it is possible to select gears based on their circular pitch using the following formula, in which N is the combined number of teeth on both gears.

$$p = (C \times 2\pi) \div N$$

Gear tooth strength

The maximum load that a gear tooth can safely transmit can be estimated if the tooth's dimensions and material, plus the diametral pitch of the gear, are known. The following equation is used for finding the maximum strength of gears working in clean, well-lubricated environments.

$$W_t = (S \times F \times Y) \div P$$

where W_t = Maximum transmittable load, in pounds or Newtons

S = Maximum tooth strength, equal to one-third the tensile strength of the gear material, in PSI or MPa

F = Face width of the gear, in inches or millimeters

Y = Lewis form factor, which is dimensionless (see below)

P = Diametral pitch.

For calculations using the metric module, the Lewis factor is converted to a value based on circular pitch.

$$Y_p = Y \div \pi$$

where Y_p = Lewis factor based on circular pitch, which is dimensionless.

Table 8. Preferred Metric Modules and Their American Equivalents.

Module	Diametral Pitch	Circular Pitch (inch)	Circular Pitch (mm)
50	0.5080	6.1842	157.080
42	0.6048	5.1948	131.947
36	0.7056	4.4527	113.097
30	0.8467	3.7105	92.248
24	1.0583	2.9685	75.398
20	1.2700	2.4737	62.832
18	1.4111	2.2263	56.548
16	1.5875	1.9790	50.267
14	1.8143	1.7316	43.983
12	2.1167	1.4842	37.699
10	2.5400	1.2368	31.415
8	3.1750	0.9895	25.133
7	3.6286	0.8658	21.991
6.5	3.9078	0.8039	20.420
6	4.2333	0.7421	18.850
5.5	4.6182	0.6803	17.279
5	5.0802	0.6184	15.707
4.75	5.3474	0.5875	14.923
4.5	5.6444	0.5566	14.138
4	6.3500	0.4947	12.565
3.75	6.7733	0.4638	11.781
3.5	7.2571	0.4329	10.996
3.25	7.8154	0.4020	10.211
3	8.4667	0.3711	9.426
2.75	9.2364	0.3401	8.639
2.5	10.1600	0.3092	7.854
2.25	11.2889	0.2783	7.069
2	12.7000	0.2474	6.284
1.75	14.5143	0.2164	5.497
1.5	16.9333	0.1855	4.712
1.25	20.3200	0.1546	3.927
1	25.4000	0.1237	3.142
0.9	28.2222	0.1113	2.827
0.8	31.7500	0.0989	2.513
0.75	33.8667	0.0928	2.357
0.7	36.2857	0.0865	2.199
0.6	42.3333	0.0742	1.885
0.5	50.8000	0.0618	1.571
0.4	63.5000	0.0495	1.256
0.3	84.6667	0.0371	0.943
0.2	127.0000	0.0247	0.628

Table 9. Formulas for Module Spur Gears (m = Module).

To Get	Formula	To Get	Formula
a	a is equal to m	D_O	$D + 2\,m$
b	$1.25\,m$		$m\,(N + 2)$
B	$2 \times \Delta C \times \tan \phi$	D_R	$D - 2.5\,m$
	B is equal to Δt	m	$25.4 \div P$
B_{LA}	$B \cos \phi$	N	$D \div m$
C	$(m \times [N_G + N_P]) \div 2$	p	$m \times \pi$
D	$m \times N$		$(D \div N) \times p$
D_B	$D \times \cos \phi$		$\pi \div P$

ϕ = Pressure Angle	D_O = Outside Diameter
π = Pi (3.1416)	D_R = Root Distance
a = Addendum	m = Module
b = Dedendum	N = Number of Teeth
B = Backlash	N_G = Number of Teeth (Gear)
B_{LA} = Backlash Along Line of Action	N_P = Number of Teeth on (Pinion)
C = Center Distance	P = Diametral Pitch
D = Pitch Diameter	p = Circular Pitch
D_B = Base Circle Diameter	

It is now possible to solve for W_t using the metric module.

$$W_t = S \times F \times Y_p \times m \times \pi$$

where m = Metric module, in millimeters.

The Lewis factor is named for Wilfred Lewis, who (in the early 1890s) introduced constants for estimating tooth beam strength. The Lewis factor values that apply to $14\,^1/_2°$ and $20°$ full depth involute gears are shown in **Table 10**. The equation used to calculate the factor values is

$$Y = (t^2 \times P) \div 6l$$

where t^2 = Tooth thickness (at root curvature) squared, in inches or millimeters
$\quad\quad\; l$ = Gear (beam) length, from thickness measurement point to top of tooth, in inches or millimeters.

Contact ratio. The ratio between the arc of action and the circular pitch is known as the contact ratio. It is often thought of as the number of pairs of teeth in contact at any time, and in spur gears this ratio should always be at least 1:4 to assure smooth and reliable operation. In the diametral pitch system, it can be found with the formula

$$m_f = (\sqrt{D_O{}^2 - D_B{}^2} + \sqrt{d_O{}^2 - d_B{}^2} - C \sin \phi) \div p \cos \phi$$

and in the module system the formula is

$$m_f = (\sqrt{D_O{}^2 - D_B{}^2} + \sqrt{d_O{}^2 - d_B{}^2} - C \sin \phi) \div m\,\pi \cos \phi$$

where m_f = Contact ratio
$\quad\quad D_O$ = Outside diameter of gear
$\quad\quad D_B$ = Base diameter of gear
$\quad\quad d_O$ = Outside diameter of pinion
$\quad\quad d_B$ = Base diameter of pinion
$\quad\quad C$ = Center diatance
$\quad\quad p$ = Circular pitch
$\quad\quad \phi$ = Pressure angle
$\quad\quad m$ = Module.

Table 10. Lewis Factor Values for 14 $\frac{1}{2}$° and 20° Full Depth Involute Gear Teeth.

14 $\frac{1}{2}$° Teeth			
Number of Teeth	Lewis Factor	Number of Teeth	Lewis Factor
10	0.176	34	0.325
11	0.192	36	0.329
12	0.210	38	0.332
13	0.223	40	0.336
14	0.236	45	0.340
15	0.245	50	0.346
16	0.255	55	0.352
17	0.264	60	0.355
18	0.270	65	0.358
19	0.277	70	0.360
20	0.283	75	0.361
22	0.292	80	0.363
24	0.302	90	0.366
26	0.308	100	0.368
28	0.314	150	0.375
30	0.318	200	0.378
32	0.322	300	0.382
20° Teeth			
Number of Teeth	Lewis Factor	Number of Teeth	Lewis Factor
10	0.201	34	0.370
11	0.226	36	0.377
12	0.245	38	0.383
13	0.264	40	0.389
14	0.276	45	0.399
15	0.289	50	0.408
16	0.295	55	0.415
17	0.302	60	0.421
18	0.308	65	0.425
19	0.314	70	0.429
20	0.320	75	0.433
22	0.330	80	0.436
24	0.337	90	0.442
26	0.344	100	0.446
28	0.352	150	0.458
30	0.358	200	0.463
32	0.364	300	0.471

Helical gears

The teeth on helical gears are cut along a helix on the pitch cylinder, rather than parallel to the axis as on a spur gear. On mating parallel gears, one helical gear will have a right helix, the other will have a left helix.

On gears at right angles to each other, the hands of the helix on both gears will be the same, and the sum of the helix angles will be 90°. When the shafts are at 45° to each other, and each helix is less than 45°, the helix will be of the same hand on each gear and the sum of the helix angles will be 45°. However, if either angle is greater than 45°, the gears will be of opposite hand, and the difference between the helix angles will be 45°.

If the shafts of the two gears are at any angle other than 45° or 90°, and each helix is less than the shaft angle, the gears will be of the same hand and the sum of the helix angles will be equal to the angles of the shafts. However, if either helix angle is greater than the shaft angle, the gears will be of opposite hand and the difference between the shaft angles will be equal to the shaft angle.

Helical gears use the same basic rack form (14 $\frac{1}{2}$° full depth involute) as used for spur gears, so many of the formulas for dimensions are interchangeable. However, when compared to spur gears of the same size and number of teeth, helical gears will run more smoothly and quietly, and will have more load carrying capacity. Most of these benefits are derived from the fact that there is increased axial tooth overlap. Unlike spur gears, helical gear teeth generate end thrust that can increase bearing loads and reduce the life of the gear.

In addition to having diametral and circular pitch, helical gears also have two related pitches: normal circular pitch (p_n), and normal diametral pitch (P_n). The normal pitches, as well as the axial pitch, are measured along a line perpendicular to the face of the teeth known as the plane of action (see *Figure 3*). Traverse pitches, which are equivalent to the standard pitches on a spur gear, are measured along a plane parallel to the axle. The difference between the traverse planes and the normal planes is directly related to the helix angle (ψ).

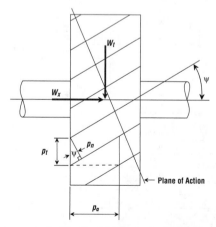

Figure 3. Helical gear geometry. Symbols: p_a = axial pitch, p_n = normal diametral pitch, p_t = transverse circular pitch, W_t = transverse load, Wx = axial load, ψ = helix angle.

The normal circular pitch is the distance between corresponding points on adjacent teeth, measured on the pitch circles at right angles to the face of the tooth. It is determined by the helix angle. The relationship between the normal circular pitch (p_n) and transverse circular pitch (p_t) can be seen in the following formulas.

$$p_t = \pi \div P_t \qquad p_n = p_t \times \cos \psi$$

The relationship between the normal diametral pitch (P_n), which is also influenced by the helix angle and the transverse diametral pitch (P_t), is expressed in the following formulas.

$$P_t = N \div D \qquad P_n = P_t \div \cos \psi$$

The normal tooth thickness on helical gears can be derived from the transverse tooth thickness measured on the plane of rotation using the following formula.

$$t_n = t \cos \psi$$

The axial pitch (p_a) is also measured across the helix, rather than along the axial plane, and is expressed with the following formula.

$$p_a = p_t \times \cot \psi$$

Note: The cosine of 45° is 1.4142, so gears with this helix angle have normal to transverse ratios of 1.4142:1.

The relation of transverse to normal pressure angles can also be calculated. The normal pressure angle will always be smaller than the transverse pressure angle. For example, a helical gear with a 45° helix angle and a 14 $^1/_2$° transverse pressure angle will have a normal pressure angle of approximately 10° 22 minutes.

$$\tan \phi_n = \tan \phi \times \cos \psi$$

Dimensional formulas for helical gears appear in **Table 11**.

Other helical gear formulas

Checking formula. The following checking formula can be used for confirming calculations of helical gear pitch diameters and center distances. The equations are balanced, and the sum on one side of the equation equals the sum of the other side. For shafts at 90° angles to each other, the formula is

$$N_G + (N_P + \tan \psi_P) = 2\,C \times P_{nP} \times \sin \psi_P ,$$

and for angles between shafts other than 90°, the following formula is used.

$$N_G + ([N_P \times \cos \psi_G] \div [\cos \psi_P]) = 2\,C \times P_{nP} \times \cos \psi_G$$

where the subscript p refers to the driving gear, and G refers to the following gear.

Gear strength. The Lewis equation formulas for strength based on tooth form (given earlier for spur gears) also apply to helical gears, but the formula is changed somewhat. Instead of using transverse diametral pitch, normal diametral pitch is substituted. In addition, the Y values are different. These values, for helical gears with 14 $^1/_2$° pressure angle and 45° helix angle, are given in **Table 12**. The contact ratio formula for helical gears on parallel shafts is also similar to the one used for spur gears, but the face width (F) is added. For identification of the other notations given in the formula, refer to the formula for spur gears, which was given earlier.

$$m_f = ([\sqrt{D_O{}^2 - D_B{}^2} + \sqrt{d_O{}^2 - d_B{}^2} - C \sin \phi] \div p \cos \phi) + (F \sin \psi \div p_n)$$

Angle formula for skew axis. When skew axis helical gears operate at a center distance other than standard, the active profile of the gears becomes altered, and determining the altered profile becomes necessary (see *Figure 4*). The solution is based on a premise

Table 11. Formulas for Dimensions of Helical Gear Teeth and Gears.

To Get	Formula	To Get	Formula
a	$1 \div P_n$	D_O	$D + 2_a$
b	$1.250 \div P_n$	D_P	$N_P \div \cos \psi$
$\cos \psi$	$p_n \div p_t$	D_R	$D - 2_b$
C	$(D_P + D_G) \div 2$	h_t	$2.157 \div P_n$
	$(N_P + N_G) \div (2P_n \times \cos \psi)$	L	$\pi D \times \cot \psi$
C *	$(P_n) \times ([N_G \div \cos \psi_G] + [N_G \div \cos \psi_P])$	t_n	$1.5708 \div P_n$
C **	$(D_G - D_P) \div 2$	$\tan \o_n$	$\tan \o \cos \psi$
D	$N \div (P_n \times \cos \psi)$	W_r	$W_N \times \sin \o_n$
D_B	$D \times \cos \o_t$	W_t	$W_N \times \cos \o_n \times \cos \psi$
D_G	$N_G \div \cos \psi$	W_x	$W_N \times \cos \o_n \times \sin \psi$

$\o$ = Pressure Angle	D_P = Pitch Diameter (Pinion)
$\o_G$ = Pressure Angle (Gear)	D_R = Root Diameter
$\o_P$ = Pressure Angle (Pinion)	h_t = Whole Depth of Tooth
$\o_n$ = Normal Pressure Angle	L = Lead
ψ = Helix Angle	N = Number of Teeth
ψ_G = Helix Angle (Gear)	N_G = Number of Teeth (Gear)
ψ_P = Helix Angle (Pinion)	N_P = Number of Teeth on (Pinion)
π = Pi (3.1416)	P_n = Normal Diametral Pitch
a = Addendum	p_n = Normal Circular Pitch
b = Dedendum	p_t = Transverse Circular Pitch
C = Center Distance	t_n = Normal Circular Thickness
D = Pitch Diameter	W_N = Load, Normal to Surface
D_B = Base Circle Diameter	W_r = Radial Load
D_G = Pitch Diameter (Gear)	W_t = Transverse Load
D_O = Outside Diameter	W_x = Axial Load

Notes: * Formulas identified with an asterisk are for nonparallel shafts only. ** Internal gears only. See text for other formulas for finding dimensions of helical gears.

Table 12. Lewis Factor Values for 14 $1/2°$ Pressure Angle, 45° Helix Angle Helical Gear Teeth.

Number of Teeth	Lewis Factor	Number of Teeth	Lewis Factor
10	0.314	30	0.364
12	0.327	32	0.365
15	0.339	36	0.367
16	0.342	40	0.370
18	0.345	48	0.372
20	0.352	50	0.373
24	0.358	60	0.374
25	0.361	72	0.377

similar to that of parallel axis gearing: the path of contact is a common internal tangent to the base diameters of the gear and its mate, regardless of the center distance. An added consideration is that the path of contact does not lie in a transverse plane of either gear, but rather in a plane containing the pitch point, and perpendicular to the helices described on the pitch cylinders of both gears. The pitch cylinders and the corresponding helix angles vary directly with the center distance. Consequently, the shaft angle varies, as the sum of the helix angles of the pitch cylinders are equal to the shaft angle when the direction of the helices is common.

$$\tan \psi_{r1} = (Y + \sqrt{Y^2 + 4XZ}) \div 2X$$

where $X = (D_{b1} \times \sin \phi_n \times \sin \Sigma) \div \sin \phi_1$

$Y = C_r \times \sin \psi_1 \times \cos \phi_n \times \sin \Sigma - ([D_{b1} \times \sin \phi_n \times \cos \Sigma] \div \sin \phi_1) - ([D_{b2} \times \sin \phi_n] \div \sin \phi_2)$

$Z = C_r \times \sin \psi_1 \times \cos \phi_n \times \sin \Sigma$

$\psi_{r2} = \Sigma - \psi_{r1}$

$\phi_{r1} = \arccos ([\tan \psi_1 \times \cos \phi_1] \div \tan \psi_{r1})$

$\phi_{r2} = \arccos ([\tan \psi_2 \times \cos \phi_2] \div \tan \psi_{r2})$

where ψ_r = Operating helix angle

ϕ_r = Operating pressure angle

D_b = Pitch circle at base

Σ = Shaft angle.

Formulas for module system helical gear dimensions are given in **Table 13**.

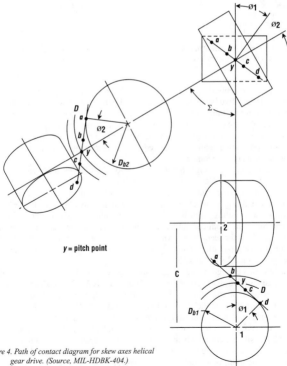

y = pitch point

Figure 4. Path of contact diagram for skew axes helical gear drive. (Source, MIL-HDBK-404.)

Table 13. Formulas for Module Helical Gears (*m* = Normal Module, *M* = Real Module).

To Get	Formula	To Get	Formula
a	a is equal to m		$D \div m$
b	$1.166\,m$	N	$(D \times \pi) \div p$
D	$(m \times N) \div \cos\psi$		$(D \times \cos\psi) \div m$
D_O	$D + 2m$		$M \times \pi$
	$m\,([N \div \cos\psi] + 2)$	p	$p_n \div \cos\psi$
h_t	$2.166\,m$		$m\,\pi \div \cos\psi$
L	$P \times \pi \times \cot\psi$		$m \times \pi$
	$N \times M \times \pi \times \cot\psi$	p_n	$p \times \cos\psi$
m	$M \times \cos\psi$		$(P \times \pi \times \cos\psi) \div N$
	$(p\cos\psi) \div \pi$	$\cos\psi$	$m \div M$
M	$m \div \cos\psi$		$(N \times m) \div P$
	$p \div \pi$		

ψ = Helix Angle	L = Lead
π = Pi (3.1416)	m = Module
a = Addendum	M = Real Module
b = Dedendum	N = Number of Teeth
D = Pitch Diameter	P = Diametral Pitch
D_O = Outside Diameter	p = Circular Pitch
h_t = Whole Depth of Tooth	p_n = Normal Circular Pitch

Notes: For helical gears, *m*, the normal module is measured at right angles to the tooth face, and equates with "normal" as used in the diametral pitch system. The real module, *M*, is measured along the pitch circle.

Bevel gears

A bevel gear is essentially a truncated cone with a number of cogs or teeth generated on the cone's surface. Two bevel gears, when rotated in respective nonparallel intersecting axes, should transmit uniform rotary motion. While in mesh, the gears roll on cones called pitch cones, and the apex of this cone for each of the two mating gears is at the point of intersection of the axes. Bevel gears are most commonly mounted on shafts that intersect at a 90° angle, but may actually be mounted at any intersecting angle. *Figure 5* shows the bevel gear form and its geometry.

The contemporary 20° *straight bevel gears* for 90° shaft angles are based on a system developed by the Gleason Gear Works, and use their registered trademark, Coniflex, to identify the form. (Those requiring detailed information about bevel gear design should acquire "Gleason Straight Bevel Gear Design" which is available from the Gleason Corporation.) The Coniflex tooth form is produced with a circular cutter and provides excellent control of the tooth contact length. High production rates can be achieved with the system, and since the teeth are crowned in a lengthwise direction, minor adjustments can be made to the gears after assembly. Other pressure angles sometimes applied to straight bevel gears are 14 $^1/_2$° and 16°.

Dimensions of bevel gears are taken at the large end of the gear, and formulas for finding dimensions of straight bevel gears are given in **Table 14**. **Table 15** provides similar information for bevel pinions. Straight bevel gears are the most widely used type of bevel gear, partly because they are the simplest to cut. However, at higher speeds (above 1,000 sfm) they produce more noise and vibration than spiral or hypoid bevel gears.

Spiral bevel gears are capable of operation at much higher speeds than straight gears. Speeds of 12,000 sfm are commonly achieved, and with special care in manufacturing this velocity can be at least doubled. Spiral gears are much stronger, largely due to the fact that more than one tooth is in contact at all times, and their normal spiral angle is 35°.

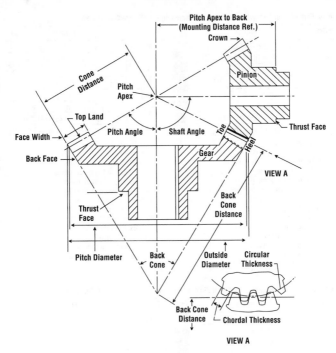

Figure 5. Bevel gear geometry.

Formulas for establishing the dimensions of spiral bevel gears are given in **Table 16**, and **Table 17** provides formulas for pinions.

Hypoid bevel gears are primarily used for automotive applications. The offset pinion used on hypoids found early favor with auto makers because it allowed the driveshaft to run lower, and protrude less into the passenger area than would have been possible with an intersecting shaft gear. Hypoid gears also provide smooth operation and produce relatively little noise. Pressure angles for hypoid gears normally range from 18° to 22.5°.

Assembling bevel gears

To assure proper assembly of bevel gears, the backlash should be measured and a bearing check should be administered. When checking backlash, measurements should be made with a dial gage at no fewer than four combinations of mating teeth to assure that the backlash is consistent throughout the revolution of the wheel and pinion. From the dial gage reading, which is the transverse backlash, the measurement can be converted to normal backlash with the following equation.

Backlash = Indicator reading × cos pressure angle × cos spiral angle, or

B = indicator reading × cos ø × cos ψ

Table 14. Formulas for Straight Bevel Gears. Pressure Angle = 20°, Shaft Angle = 90°.

To Get	Formula	To Get	Formula
a_{CG}	$a_G + ([t_G{}^2 \times \cos \Gamma_G] \div 4D_G)$	h_k	$2 \div P$
a_G	$(0.54 \div P) + (0.46 \div [P \times m_G{}^2])$	h_t	$(2.188 \div P) + 0.002$
A_O	$D_G \div (2 \sin \Gamma_G)$	p	$\pi \div P$
b_G	$(2.188 \div P) - a_G + 0.002$	Γ_G	$90° - \Gamma_P$
δ_G	$\tan^{-1}(b_G \div A_O)$	Γ_{OG}	$\Gamma_G + \delta_P$
c	$h_t - h_k$	Γ_{RG}	$\Gamma_G - \delta_G$
D	$N_G \div P$	t_{CG}	$t_G - (t_G{}^2 \div 6\,D_G) - 0.5\,B$
D_{OG}	$D_G + (2\,a_G \times \cos \Gamma_G)$	t_G	$0.5\,p - ([a_P - a_G] \times [\tan ø - \{K \div P\}])$

ø = Pressure Angle
δ_G = Dedendum Angle (Gear)
δ_P = Dedendum Angle of (Pinion)
Γ_G = Pitch Angle (Gear)
Γ_P = Pitch Angle (Pinion)
Γ_{OG} = Face Angle of Gear Blank (Gear)
Γ_{RG} = Root Angle (Gear)
π = Pi (3.1416)
a_{CG} = Chordal Addendum (Gear)
a_G = Addendum (Gear)
a_P = Addendum (Pinion)
A_O = Outer Cone Distance
b_G = Dedendum
B = Backlash

c = Clearance
D = Pitch Diameter
D_G = Pitch Diameter (Gear)
D_{OG} = Outside Diameter (Gear)
h_k = Working Depth of Tooth
h_t = Whole Depth of Tooth
K = Load Intensity Factor
m_G = Gear Ratio
N_G = Number of Teeth (Gear)
P = Diametral Pitch
p = Circular Pitch
t_{CG} = Chordal Thickness (Gear)
t_G = Circular Thickness (Gear)

Table 15. Formulas for Straight Bevel Gear Pinions. Pressure Angle = 20°, Shaft Angle = 90°.

To Get	Formula	To Get	Formula
a_{CP}	$a_P + ([t_P{}^2 \times \cos \Gamma_P] \div 4D)$	h_k	$2 \div P$
a_P	$h_k - a_G$	h_t	$(2.188 \div P) + 0.002$
A_O	$D_G \div (2 \sin \Gamma_G)$	p	$\pi \div P$
b_P	$(2.188 \div P) - a_P$	Γ_P	$\tan^{-1} \times (N_P \div N_G)$
δ_P	$\tan^{-1} \times (b_P \div A_O)$	Γ_{OP}	$\Gamma_P + \delta_G$
c	$h_t - h_k$	Γ_{RP}	$\Gamma_P - \delta_P$
D	$N_P \div P$	t_{CP}	$t_P - (t_P{}^2 \div 6\,D_P{}^2) - 0.5\,B$
D_{OP}	$D_P + (2\,a_P \times \cos \Gamma_P)$	T_P	$P - t_G$

δ_G = Dedendum Angle (Gear)
δ_P = Dedendum Angle of (Pinion)
Γ_G = Pitch Angle (Gear)
Γ_P = Pitch Angle (Pinion)
Γ_{OP} = Face Angle of Gear Blank (Pinion)
Γ_{RP} = Root Angle (Pinion)
π = Pi (3.1416)
a_{CP} = Chordal Addendum (Pinion)
a_G = Addendum (Gear)
a_P = Addendum (Pinion)
A_O = Outer Cone Distance
b_P = Dedendum (Pinion)
B = Backlash
c = Clearance

D = Pitch Diameter
D_B = Base Circle Diameter
D_G = Pitch Diameter (Gear)
D_{OP} = Outside Diameter (Pinion)
D_P = Pitch Diameter (Pinion)
h_k = Working Depth of Tooth
h_t = Whole Depth of Tooth
N_G = Number of Teeth (Gear)
N_P = Number of Teeth on (Pinion)
P = Diametral Pitch
p = Circular Pitch
t_{CP} = Chordal Thickness (Pinion)
t_G = Circular Thickness (Gear)
t_P = Circular Thickness of (Pinion)

Table 16. Formulas for Spiral Bevel Gears. Pressure Angle = 20°, Shaft Angle = 90°.

To Get	Formula	To Get	Formula
a_G	$(0.460 \div P) + (0.390 \div [P \times m_G^2])$	h_k	$1.700 \div P$
A_O	$D_G \div (2 \sin \Gamma_G)$	h_t	$1.888 \div P$
b_G	$h_t - a_G$	p	$\pi \div P$
δ_G	$\tan^{-1} \times (b_G \div A_O)$	Γ_G	$90° - \Gamma P$
c	$h_t - h_k$	Γ_{OG}	$\Gamma G + \delta_P$
D	$N_G \div P$	Γ_{RG}	$\Gamma G - \delta_G$
D_{OG}	$D_G + (2\, a_G \times \cos \Gamma_G)$	t_G	$0.5\, p \times ([a_P - a_G] \times [\tan ø \div \cos ø] - [K \div P])$

ø = Pressure Angle	D = Pitch Diameter
δ_G = Dedendum Angle (Gear)	D_G = Pitch Diameter (Gear)
δ_P = Dedendum Angle of (Pinion)	D_{OG} = Outside Diameter (Gear)
Γ_G = Pitch Angle (Gear)	h_k = Working Depth of Tooth
Γ_P = Pitch Angle (Pinion)	h_t = Whole Depth of Tooth
Γ_{OG} = Face Angle of Gear Blank (Gear)	K = Load Intensity Factor
Γ_{RG} = Root Angle (Gear)	m_G = Gear Ratio
π = Pi (3.1416)	N_G = Number of Teeth (Gear)
a_G = Addendum (Gear)	P = Diametral Pitch
a_P = Addendum (Pinion)	p = Circular Pitch
b_G = Dedendum	t_G = Circular Thickness (Gear)
c = Clearance	

Table 17. Formulas for Spiral Bevel Gear Pinions. Pressure Angle = 20°, Shaft Angle = 90°.

To Get	Formula	To Get	Formula
a_P	$h_k - a_G$	h_k	$1.700 \div P$
A_O	$D_G \div (2 \sin \Gamma_G)$	h_t	$1.888 \div P$
b_P	$h_t - a_P$	p	$\pi \div P$
δ_P	$\tan^{-1} \times (b_P \div A_O)$	Γ_P	$\tan^{-1} \times (N_P \div N_G)$
c	$h_t - h_k$	Γ_{OP}	$\Gamma P + \delta_G$
D	$N_P \div P$	Γ_{RP}	$\Gamma P - \delta_G$
D_{OP}	$D_P + (2\, a_P \times \cos \Gamma_P)$	t_P	$p - t_G$

δ_G = Dedendum Angle (Gear)	D_G = Pitch Diameter (Gear)
δ_P = Dedendum Angle of (Pinion)	D_{OP} = Outside Diameter (Pinion)
Γ_G = Pitch Angle (Gear)	D_P = Pitch Diameter (Pinion)
Γ_P = Pitch Angle (Pinion)	h_k = Working Depth of Tooth
Γ_{OP} = Face Angle of Gear Blank (Pinion)	h_t = Whole Depth of Tooth
Γ_{RP} = Root Angle (Pinion)	N_G = Number of Teeth (Gear)
π = Pi (3.1416)	N_P = Number of Teeth on (Pinion)
a_P = Addendum (Pinion)	P = Diametral Pitch
A_O = Outer Cone Distance	p = Circular Pitch
b_P = Dedendum (Pinion)	t_G = Circular Thickness (Gear)
c = Clearance	t_P = Circular Thickness of (Pinion)
D = Pitch Diameter	

The tooth bearing check is done by first painting the pinion or gear with a nonabrasive marking compound, and then rotating the gears in mesh using a minimal turning load. The teeth of the uncoated member are then observed to evaluate the bearing setting. *Figures 6* through *8* provide guidelines for interpreting the marks on the gear and pinion teeth that may indicate that bearing adjustments are required, or that an error has been made in mounting the gear. *Figure 6* shows the possible tooth bearing marks on a spiral, Zerol, or Coniflex pinion, and *Figure 7* shows marks for the gear member. *Figure 8* shows marks as they appear on straight bevel (non-Coniflex) gear teeth.

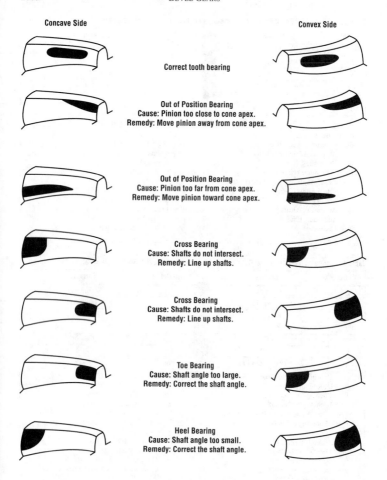

Figure 6. Tooth bearing marks on pinion member of Zerol, Spiral, and Coniflex bevel gears.

Bearing adjustments. If the bearing is too high on the pinion and too low on the gear, the pinion will be too near the axis of the gear member. The problem can be corrected by increasing the pinion mounting distance. If the bearing is too low on the pinion and too high on the gear, the pinion will be too far from the axis of the gear member. The problem can be corrected by decreasing the pinion mounting distance.

Mounting errors. When contact is too close on one end of a tooth on one side of a gear, and too close on the opposite end of a corresponding tooth on the other side, a cross-bearing mounting error is indicated. When this occurs, the shafts are noncoplanar (not mounted in the same plane) and must be realigned.

When contact is too far forward on corresponding teeth on both sides of a gear, the shaft angle is inaccurate. If the bearing area is too close to the top (small end), the shaft angle is too large. If the bearing area is too close to the heel (large end), the shaft angle is too small (see *Figures 6* and *7*).

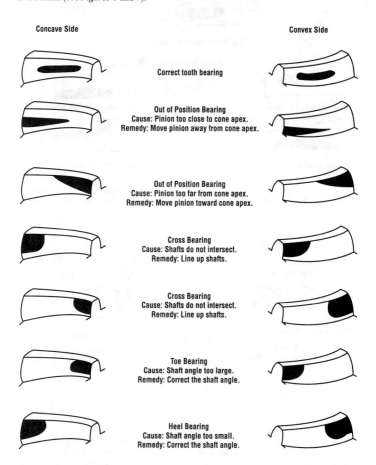

Concave Side   **Convex Side**

Correct tooth bearing

Out of Position Bearing
Cause: Pinion too close to cone apex.
Remedy: Move pinion away from cone apex.

Out of Position Bearing
Cause: Pinion too far from cone apex.
Remedy: Move pinion toward cone apex.

Cross Bearing
Cause: Shafts do not intersect.
Remedy: Line up shafts.

Cross Bearing
Cause: Shafts do not intersect.
Remedy: Line up shafts.

Toe Bearing
Cause: Shaft angle too large.
Remedy: Correct the shaft angle.

Heel Bearing
Cause: Shaft angle too small.
Remedy: Correct the shaft angle.

Figure 7. Tooth bearing marks on gear member of Zerol, Spiral, and Coniflex bevel gears.

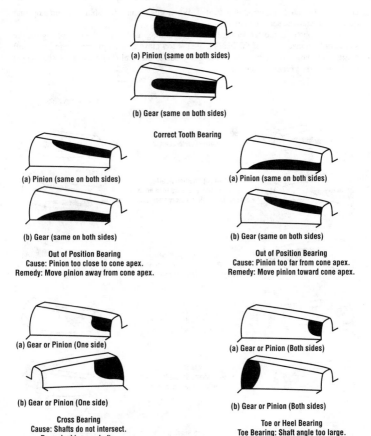

(a) Pinion (same on both sides)

(b) Gear (same on both sides)

Correct Tooth Bearing

(a) Pinion (same on both sides)

(b) Gear (same on both sides)

Out of Position Bearing
Cause: Pinion too close to cone apex.
Remedy: Move pinion away from cone apex.

(a) Pinion (same on both sides)

(b) Gear (same on both sides)

Out of Position Bearing
Cause: Pinion too far from cone apex.
Remedy: Move pinion toward cone apex.

(a) Gear or Pinion (One side)

(b) Gear or Pinion (One side)

Cross Bearing
Cause: Shafts do not intersect.
Remedy: Line up shafts.

(a) Gear or Pinion (Both sides)

(b) Gear or Pinion (Both sides)

Toe or Heel Bearing
Toe Bearing: Shaft angle too large.
Heel Bearing: Shaft angle too small.
Remedy: Correct the shaft angle.

Figure 8. Tooth bearing marks on straight bevel gears and pinions.

Worm gears

Worm gears are commonly used transmit motion, rather than power, through shafts at right angles to each other. In a worm gear set, the worm resembles a large screw (usually but not always having a single thread), and the worm wheel is essentially a helical gear. When worm and wheel shafts intersect at 90°, the lead angle of the worm is equal to the helix angle of the wheel. When used for speed reduction, the worm is the driving element. *Figure 9* shows basic worm and worm wheel geometry.

Because the worm is a thread around a cylinder, the wheel is usually concave at the point of contact to maximize the contact area, and is known as a single-enveloping worm set. Contact area can be increased by relieving the worm concavely along its length, so that both the worm and the wheel envelop each other. This is known as a double-enveloping worm set, and this form allows for higher transfer loads. However, this form is not in wide use, mainly due to difficulty in mounting.

The velocity ratio is the ratio of the number of teeth on the wheel to the number of thread starts on the worm. For example, a 20-tooth wheel meshing with a single thread worm will have a ratio of 1:20, meaning that one turn of the wheel will result in 20 turns of the worm, or that five turns of the worm occur when the wheel makes one-quarter revolution. Thus, it can be seen that large speed reductions result when the worm is the driven element, but that speed can be increased by the velocity ratio when the wheel drives the worm.

Efficiency is closely related to the helix angle of the worm, and a worm helix angle of 30° normally provides good results. The following formula can be used to calculate efficiency.

$$\tan \psi_W = L_W \div \pi \, D_W$$

Formulas for finding dimensions of worms are provided in **Table 18**, and **Table 19** gives dimensions for worm wheels.

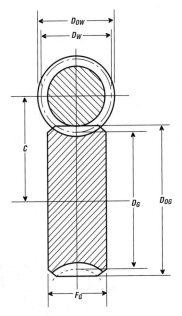

Figure 9. Worm/worm wheel geometry for single-enveloping worm set. Symbols: C = center distance, D_G = pitch diameter of wheel, D_{OG} = outside diameter of wheel, D_{OW} = outside diameter of worm, D_W = pitch diameter of worm, F_G = face width of wheel.

Table 18. Formulas for Finding Dimensions of Worms.

To Get	Formula	To Get	Formula
a_W *	$0.3183 \times p_n$	h_t *	$0.6866 \times p_n$
a_W **	$0.2865 \times p_n$	h_t **	$0.623 \times p_n$
c	$h_t - h_k$	L_W	$N_W \times p_x$
C	$(D_W + D_D) \div 2$	m_G	$N_G \div N_W$
	$(p_n \div 2\,\pi) \times ([N_G \div \cos \lambda] + [N_W \div \sin \lambda])$	p_n	$p_x \times \cos \lambda$
	$2C - D_G$		p
D_W	$D_{OW} - 2\,a_W$	p_x	$L_W \div N_W$
	$(N_W \times p_n) \div p \sin \lambda$		$p_n \div \cos \lambda$
D_{OW}	$D_W + 2\,a_W$	t	$p_n \div 2$
h_k	$0.6366\, p_n$	$\tan \lambda$	$L_W \div (\pi\, D_W)$

λ = Lead Angle	h_t = Whole Depth of Tooth
π = Pi (3.1416)	L_W = Lead (Worm)
a_W = Addendum	m_G = Ratio
c = Clearance	N_G = Number of Teeth (Wheel)
C = Center Distance	N_W = Number of Thread Starts (Worm)
D_G = Pitch Diameter (Wheel)	p = Circular Pitch (Wheel)
D_W = Pitch Diameter (Worm)	p_n = Circular Pitch of Worm (Normal)
D_{OW} = Outside Diameter (Worm)	p_x = Axial Circular Pitch (Worm)
h_k = Working Depth of Tooth	t = Tooth Thickness

* For single and double thread worms. ** For triple and quadruple thread worms.

Table 19. Formulas for Finding Dimensions of Worm Gear Wheels.

To Get	Formula	To Get	Formula
a_G *	$0.3183 \times p_n$	h_k	$0.6366\, p_n$
a_G **	$0.2865 \times p_n$	h_t *	$0.6866 \times p_n$
c	$h_t - h_k$	h_t **	$0.623 \times p_n$
C	$(D_W + D_D) \div 2$	m_G	$N_G \div N_W$
	$(p_n \div 2\,\pi) \times ([N_G \div \cos \lambda] + [N_W \div \sin \lambda])$		$p \times \cos \psi$
D_G	$(N_G \times p_x) \div \pi$	p_n	$p_x \times \cos \lambda$
	$(N_G \times p_n) \div p \cos \lambda$		$([2C - D_W] \times \pi) \div N_G$
D_{OG}	$2C - D_W + 2\,a_G$	p_x	$L \div N_W$
F_G	$\sqrt{(D_W + h_k)^2 - D_W{}^2}$	t	$p_n \div 2$

ψ = Helix Angle	h_k = Working Depth of Tooth
λ = Lead Angle	h_t = Whole Depth of Tooth
π = Pi (3.1416)	L_G = Lead (Wheel)
a_G = Addendum	m_G = Ratio
c = Clearance	N_G = Number of Teeth (Wheel)
C = Center Distance	N_W = Number of Thread Starts (Worm)
D_G = Pitch Diameter (Wheel)	p = Circular Pitch (Wheel)
D_W = Pitch Diameter (Worm)	p_n = Circular Pitch of Worm Gear (Normal)
D_{OG} = Outside Diameter (Wheel)	p_x = Axial Circular Pitch (Worm)
F_G = Face Width (Wheel)	t = Tooth Thickness

* For single and double thread worms. ** For triple and quadruple thread worms.

Designing plastic gears

Specialized plastics have been used for spur, helical, bevel, and worm gears in certain applications for the past two decades. With new developments in materials, their use has become widespread to the point that in many situations they can be considered as replacements for metal gears—especially those made from nonferrous metals, cast iron, and unhardened steel. Although nylon has significantly lower strength than steel, reduced friction and inertia, along with the resilience (bending) of thermoplastic gear teeth, make direct substitution possible in many instances. Quadrant Engineering Plastics Products offer the following guidelines for replacing existing metal gears, and for estimating requirements for new applications.

Spur gears

The methods described below were developed from Quadrant Engineering's test data and from maximum allowable bending stress analysis of plastic gear teeth (see *Figure 10*). In addition, four correctional factors were allowed for: 1) material strength and the presence and absence of lubrication; 2) pitch line velocity; 3) required service life; and 4) ambient temperature under service conditions.

Replacing an existing metal gear. The correction factors values for replacing an existing metal gear are given in the following figures and Tables.

C_S = Service/Lifetime correction factor. See *Figure 11*.
C_V = Velocity (at pitch line) correction factor. See *Figure 12*.
C_M = Material Strength correction factor. See **Table 20**.
C_T = Temperature correction factor:
   For ambient temperature $\leq 100°$ F: $C_T = 1$
   For $> 100°$ F but $< 200°$ F: $C_T =$
   $1 \div (1 + \alpha [\text{Temp} - 100° \text{F}])$

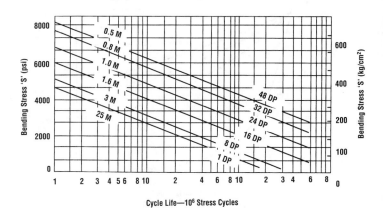

Figure 10. Maximum tooth bending stresses vs. cycle life for nylon gears.
(Source, Quadrant Engineering Plastic Parts.)

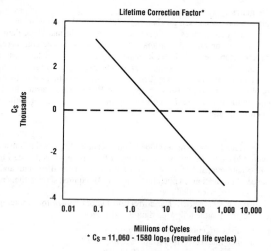

Figure 11. Service/Lifetime correction factor chart for plastic gears
(Source, Quadrant Engineering Plastic Parts.)

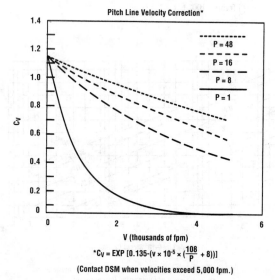

Figure 12. Pitch line velocity correction chart for plastic gears. (Source, Quadrant Engineering Plastic Parts.)

Table 20. Material Strength Correction Factor. *(Source, Quadrant Engineering Plastic Parts.)*

Material	Operating Conditions		
	No Lubrication	Periodic Lubrication	Continuous Lubrication
Nylon 6 and Nylon 66 materials	0.49 - 1.00	0.94 - 1.00	1.20 - 1.26
Acetel (POM)	-	-	1.04
Phenolic	-	0.96	1.13
UHMW Polyethylene (PE)	-	-	0.75

where $\alpha = 0.022$ for Nylon Type 6 materials such as Nylatron® GSM and NSM, and MC® 901/907 that have a continuous service temperature of 200° F and a compressive strength of 14,000 PSI

$\alpha = 0.004$ for Nylon Type 66 materials such as Nylatron® GS and NS, and Polypenco® 101 that have a continuous service temperature of 210 to 220° F, and a compressive strength ranging from 12,000 to 16,000 PSI

$\alpha = 0.010$ for Acetal materials such as Acetron® GP, which has a continuous service temperature of 180° F, and a compressive strength of 15,000 PSI.

For ambient temperatures 200° or higher, consult with manufacturers for specific applications.

Using the available correction factors, it is possible to calculate the maximum torque (T_{MAX}) and maximum horsepower (H_{MAX}) the plastic replacement gear can safely transmit by using the following Lewis equations.

$$T_{MAX} = (D\,F\,Y\,[1985 + 64.1P + C_S]\,C_V\,C_T\,C_M) \div 2P$$

$$H_{MAX} = (D\,F\,Y\,R\,[1985 + 64.1P + C_S]\,C_V\,C_T\,C_M) \div 126{,}000\,P$$

where D = Pitch diameter (inch)
F = Face width (inch)
Y = Lewis Factor for tooth form (see **Table 21**)
R = Gear RPM
P = Diametral pitch

Finally, compare the maximum torque (T_{MAX}) and horsepower (H_{MAX}) values for the theoretical plastic gear with the known torque (T_1) and horsepower (H_1) values transmitted through the existing metal gear. For the plastic gear to be an acceptable substitute, T_1 must be less than or equal to T_{MAX}, and H_1 must be less than or equal to H_{MAX}.

Designing a new plastic gear. When designing a new gear, first select the pitch diameter. The next step is to estimate the required diametral pitch. In the following equations, it is accepted that C_S, the service/lifetime correction factor, is equal to 10,000,000 cycles, that C_T is equal to 1 (ambient temperature $\leq 100°$ F), and C_V is equal to 2,000 sfm.

If torque is known or estimated

$$P = 1985 \div ([4\,T \div D\,F] - 64.1).$$

If horsepower is known or estimated

$$P = 1985\,([252{,}000\,H \div D\,F\,R] - 64.1).$$

The value of P derived from the above formulas should then be rounded to the nearest whole number. It is now possible to determine the number of teeth on the gear.

$$N = D \times P$$

Table 21. Tooth Form Factor for Plastic Gearing. *(Source, Quadrant Engineering Plastic Parts.)*

Number of Teeth	Pressure Angle/Form		
	14 1/2°	20° Full Depth	20° Stub
14	-	-	0.540
15	-	-	0.566
16	-	-	0.578
17	-	0.512	0.587
18	-	0.521	0.603
19	-	0.534	0.616
20	-	0.544	0.628
22	-	0.559	0.648
24	0.509	0.572	0.664
26	0.522	0.588	0.678
28	0.535	0.597	0.688
30	0.540	0.606	0.698
34	0.553	0.628	0.714
38	0.566	0.651	0.729
43	0.575	0.672	0.739
50	0.588	0.694	0.758
60	0.604	0.713	0.774
75	0.613	0.735	0.792
100	0.622	0.757	0.808
150	0.635	0.779	0.830
300	0.650	0.801	0.855
Rack	0.660	0.823	0.881

Now that the number of teeth on the gear is known, select the tooth form factor (Y) from **Table 21** based on the pressure angle specified for the gear. Next select a plastic material and identify the appropriate material correction factor (C_M) from **Table 20**, the service/lifetime correction factor (C_S) from *Figure 11*, the temperature correction factor (C_T), and the velocity correction factor (C_V) from *Figure 12*.

Using the information obtained so far, use the formulas for T_{MAX} and H_{MAX} given above to calculate the maximum torque and horsepower. If the gear is inadequate to meet the actual requirements, recalculate using a different material, an optional pitch diameter, or a different face width.

Helical gears

To determine the stress on plastic helical gears, spur gear design equations for T_{MAX} and H_{MAX} must be modified to compensate for the differing tooth contact angles and forces. The tooth form factor (Y) is calculated from the formative number (N_r) of teeth, rather than the actual number (N).

$$N_r = N \div \cos \psi$$

where ψ = Helix angle of the gear.

In addition, the normal diametral pitch (P_n) is used rather than the transverse diametral pitch (P), and is calculated with the formula

$$P_n = P \cos \psi.$$

It can be seen from these equations that the effect is more pronounced as the helix increases. Metal helical gears are often specified (rather than spur gears) in order to reduce noise and vibration, and it should be noted that equivalent plastic spur gears will often achieve these reductions more efficiently.

Using the formative number of teeth to select the Y value, and substituting the normal diametral pitch for the transverse diametral pitch, the T_{MAX} and H_{MAX} formulas introduced above for spur gears can be modified for helical gears.

Bevel gears

When calculating T_{MAX} and H_{MAX} for bevel gears, modifications similar to those made for helical gears are required. First, find the formative number (N_r) of teeth in order to select a value for Y.

$$N_r = N \div \cos \Gamma$$

where Γ = Pitch angle.

Next, the sum of the T_{MAX} and H_{MAX} equations are multiplied by a factor to allow for the geometry of bevel gears.

$$T_{MAX} = ([D \, F \, Y \, \{1985 + 64.1P + C_S\} \, C_V \, C_T \, C_M] \div 2P) \times ([A_O - F] \div A_O)$$

$$H_{MAX} = ([D \, F \, Y \, R \, \{1985 + 64.1P + C_S\} \, C_V \, C_T \, C_M] \div 126{,}000 \, P) \times ([A_O - F] \div A_O)$$

where A_O = Outer cone distance.

Assembly of plastic gears

Gears are commonly fastened to shafts using a variety of techniques, including:

Press fit over splined or knurled shafts (for gears transmitting low torques)
Set screws for an economical low torque gear
Bolting a metal hub through the gear width, which is suitable for drive gears produced in intermediate quantities
Machined keyways for gears carrying higher torques.

The use of machined keyways is preferred over square cornered keyways in order to reduce the stress concentration at the corners. The minimum corners keyway area can be determined with the following formula.

$$A = 63{,}000 \, H \div (R \, r \, S_K)$$

where H = Horsepower
R = Gear RPM
r = Mean radius of keyway
S_K = Material maximum permissible keyway stress from **Table 22**.

Table 22. Maximum Permissible Keyway Stress for Continuously Running Gears.
(Source, Quadrant Engineering Plastic Parts.)

Material	Keyway Stress (Sk) PSI
Nylon 6	2,000
Nylon 66	1,500
Acetal	2,000
UHMW Polyethylene	300

If the keyway size predicted by the above formula is impractical, and multiple keyways cannot be used, then a keyed flanged hub and check plate bolted through the gear should be considered. The required number of bolts and their diameters at a particular pitch circle radius can be calculated from a modified form of the formula.

Minimum number of bolts required = $63{,}000\ H \div (R\ r_1\ A_1\ S_K)$

where r_1 = Pitch circle radius of bolts

A_1 = Projected area of bolts (bolt diameter × gear width in contact with bolts)

Fractional answers should be raised to the next higher number of bolts. The bolts should not be tightened excessively during assembly in order to avoid the risk of gear distortion or bolt shearing due to material expansion during normal running. Consequently, the use of cup washers or similar types are recommended where practical, and nylon washers provide a satisfactory alternative.

Design considerations with plastic gears

Heat dissipation (and therefore performance) is optimized by running plastic gears against metal gears. When running an all-plastic gear system, dissimilar materials are suggested.

Sufficient backlash is required for plastic gears to accommodate the greater thermal expansion of plastic versus metal due to frictional heating and changes in ambient conditions. The suggested backlash can be calculated using the following formula.

$$B = 0.10 \div D$$

Compared to a 14 $^1/_2°$ pressure angle, a 20° pressure angle with a full root radius maximizes the bending strength of gear teeth. With the 20° angle, the load capacity is increased by 15%, and service life is increased 3.5 times at the same load.

When design permits, select the smallest tooth that will carry the load required. This will minimize heat buildup from higher teeth sliding velocities.

For higher torque capability, consider gear blanks directly cast over machine steel inserts.

The wear on a plastic gear is largely determined by the counterface or opposing gear. In general, it is best to avoid making both driven and driving gears from similar plastics, and it should be remembered that most plastic gears run well against metal gears. However, a surface finish of 12 to 16 μin. is recommended on metal gears running against plastic gears.

Involute Splines – Fit and Function

Splines used in industry take many forms. Involute splines, which are splines with gradually sloping rather than straight sides, trace their beginnings to the aircraft industry which employed early versions of 20° pressure angle involute splines. The 20° spline is not commonly used today, having been replaced by 30°, 37.5°, and 45° pressure angles primarily as a result of these angles becoming standardized by SAE and ASA standards. Existing standards for involute splines can trace their existence back to AS 84, which is the basis for the more recently released version of ANSI B92.1-1970, R 1993. Much of the material referenced in this section is derived from material in these standards that has remained timely in present editions. Metric sizes are covered in ANSI B92.2M-1980, Rev. 1989.

The standards consider both side fit and major diameter fit for 30° pressure angle splines, but only side fit for the 37.5° and 45° versions. Clearance dimensions are given for major and minor diameters. Mating splines contact at the major diameter (i.e., the diameter of the major circle, or the circle formed by the outermost surface of the spline—for the external spline this is the tooth-tip circle, and for the internal spline it is the root circle) to centralize the shaft, and at the sides of the teeth to drive the shaft. Clearance dimensions are given for the minor diameter (i.e., the diameter of the minor circle, or the circle formed by the innermost surface of the spline—for the external spline this is the root circle, and for the internal spline it is the tooth-tip circle). There are four classes of tolerances on space width and tooth thickness. Originally, only one tolerance class, Class 5, was specified, but a later version of the standard added three additional classes (all based on Class 5) as follows:

Class 4 Tolerance, which is equal to Class 5 × 0.71
Class 6 Tolerance, which is equal to Class 5 × 1.40
Class 7 Tolerance, which is equal to Class 5 × 2.00.

In involute splines, pitch is determined by a direct relationship between the diametral pitch (the number of splines per inch of pitch diameter) and the stub pitch (the radial distance from the pitch circle to the major circle of the external spline, or from the minor circle of the internal spline). Stub pitch is always equal to diametral pitch × 2. Thus, pitch is always expressed as two numbers such as 8/16, 32/64, or 48/96. Circular pitch, on the other hand, is measured along the pitch circle, and is the distance between corresponding points of adjacent spline teeth. *Figure 1* illustrates the terminology applied to an internal

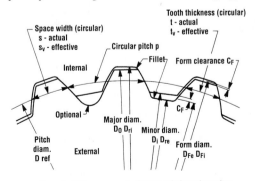

Figure 1. A 30° pressure angle, flat root, side fit involute spline.

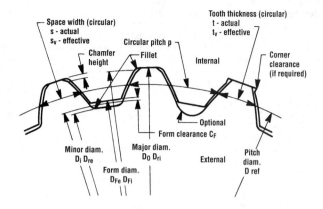

Figure 2. A 30° pressure angle, flat root, major-diameter fit involute spline.

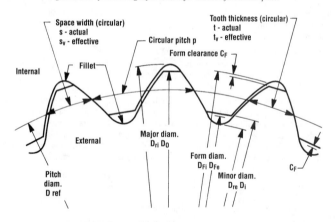

Figure 3. A 37.5° pressure angle, fillet root, side fit involute spline.

and external 30° pressure angle involute spline with flat root side fit, and *Figure 2* shows the same details for a 30° pressure angle slide with flat root major diameter fit. It can be seen that the flat root side fit allows for full contact with one side of each tooth, but the "form clearance" denies contact between the internal and external major diameters of the spline. In the flat root major diameter fit, the major diameters are in contact. These forms of fit are commonly used where space dimensions restrict the use of wall diameters that would allow the use of fillet roots (see below). When using major-diameter fits, corner clearance (as shown in the illustration) should always be machined into the spline.

Fillet root splines have an arch shaped root that joins the sides of adjacent teeth. They are preferred for heavy loads because the fillets reduce stress concentrations. *Figure 3* shows a 37.5° pressure angle fillet root, side fit spline. An internal spline with a fillet root can only be used for side fit applications.

The basic dimensions of involute splines are given in **Table 1**. The minimum effective space width (see *Figures*) dimension on the Table is basic, and refers to the dimension of the internal spline. It is equal to the circular tooth thickness on the pitch circle of an imaginary perfect external spline that would fit an internal spline without looseness of interference considering engagement of the entire axial length of the spline. Fit variations are obtained by varying the tooth thickness of the *external* spline. **Table 2** provides Class 5 tolerances for space width and tooth thickness. These can be converted to other Class tolerances by multiplying by the conversion factors provided earlier in this section. Both machining tolerances and variation allowances are given in this Table.

Variations in involute spline dimensions take many forms, with some being common enough to be specifically identified.

Index variations are variations in tooth dimensions that result in clearances varying from one set of mating teeth to another. This, of course, results in looseness or overtightness in the teeth with variations.

Profile variations occur at the point that is used to determine actual space width or tooth thickness—either the pitch point or contact point. Positive variations in the direction of space either reduce the effective clearance or increase the interference. Negative variations are in the direction of the tooth and reduce contact area but do not affect the fit.

Parallelism variations are measured along the reference axis of the spine and occur when a spline tooth is not parallel with other teeth on the spline.

Table 1. Basic Dimensions of Involute Splines.

Pitch	Circular Pitch	Minimum Effective Space Width (Basic)		
		30° Pressure Angle	37.5° Pressure Angle	45° Pressure Angle
2.5 / 5	1.2566	0.6283	0.6683	-
3 / 6	1.0472	0.5236	0.5569	-
4 / 8	0.7854	0.3927	0.4177	-
5 / 10	0.6283	0.3142	0.3342	-
6 / 12	0.5236	0.2618	0.2785	-
8 / 16	0.3927	0.1963	0.2088	-
10 / 20	0.3142	0.1571	0.1671	0.1771
12 / 24	0.2618	0.1309	0.1392	0.1476
16 / 32	0.1963	0.0982	0.1044	0.1107
20 / 40	0.1571	0.0785	0.0835	0.0885
24 / 48	0.1309	0.0654	0.0696	0.0738
32 / 64	0.0982	0.0491	0.0522	0.0553
40 / 80	0.0785	0.0393	0.0418	0.0443
48 / 96	0.0654	0.0327	0.0348	0.0369
64 / 128	0.0491	-	-	0.0277
80 / 160	0.0393	-	-	0.0221
128 / 256	0.0246	-	-	0.0138

All dimensions in inches. See text for explanation of terms.

Table 2. Maximum Tolerances for Space Width and Tooth Thickness for Class 5 Involute Splines.
(Dimensions given in ten thousandths of an inch: 15.8 = 0.00158.)

Pitch ⯈	2.2/5 and 3/6	4/8 and 5/10	6/12 and 8/16	10/20 and 12/24	16/32 and 20/40	24/48 and 48/96	64/128 and 80/160	128/256
No. of Teeth	Machining Tolerance							
10	15.8	14.5	12.5	12.0	11.7	11.7	9.6	9.5
20	17.6	16.0	14.0	13.0	12.4	12.4	10.2	10.0
30	18.4	17.5	15.5	14.0	13.1	13.1	10.8	10.5
40	21.8	19.0	17.0	15.0	13.8	13.8	11.4	-
50	23.0	20.5	18.5	16.0	14.5	14.5	-	-
60	24.8	22.0	20.0	17.0	15.2	15.2	-	-
70	-	-	-	18.0	15.9	15.9	-	-
80	-	-	-	19.0	16.6	16.6	-	-
90	-	-	-	20.0	17.3	17.3	-	-
100	-	-	-	21.0	18.0	18.0	-	-
	Variation Allowance							
10	23.5	20.3	17.0	15.7	14.2	12.2	11.0	9.8
20	27.0	22.6	19.0	17.4	15.4	13.4	12.0	10.6
30	30.5	24.9	21.0	19.1	16.6	14.6	13.0	11.4
40	34.0	27.2	23.0	21.6	17.8	15.8	14.0	-
50	37.5	29.5	25.0	22.5	19.0	17.0	-	-
60	41.0	31.8	27.0	24.2	20.2	18.2	-	-
70	-	-	-	25.9	21.4	19.4	-	-
80	-	-	-	27.6	22.6	20.6	-	-
90	-	-	-	29.3	23.8	21.8	-	-
100	-	-	-	31.0	25.0	23.0	-	-
	Profile Variation							
All	+7	+6	+5	+4	+3	+2	+2	+20
	-10	-8	-7	-6	-5	-4	-4	-4
	Lead Variation							
L_g * in Variation	0.3	0.5	1	2	3	4	5	6
	2	3	4	5	6	7	8	9

* L_g is length of engagement.

Alignment variation is when the effective spline axis is not parallel to the reference axis. In other words, the teeth are parallel to each other, but not to the reference axis of the spline.

Lead variation includes both parallelism and alignment variations, and is the variation of a spline's intended direction from the reference axis of the spline.

Most basic dimensions for involute splines can be determined with published formulas. In **Table 3**, the formulas for the most commonly required dimensions are

Table 3. Formulas for Determining Basic Dimensions of Involute Splines. (See explanation of symbols at end of Table.)

Symbol and Term	30° Pressure Angle			37.5° Pressure Angle	45° Pressure Angle
	Flat Root Side Fit — 2.5/5 to 32/64 Pitch	Flat Root Major Dia. Fit — 3/6 to 16/32 Pitch	Fillet Root Side Fit — 2.5/5 to 48/96 Pitch	Fillet Root Side Fit — 2.5/5 to 48/96 Pitch	Fillet Root Side Fit — 10/20 to 128/256 Pitch
c_F Form Clearance	0.001 D, with a maximum of 0.010 and a minimum of 0.002				
D Pitch Diameter	$N \div P$	$N \div P$	$N \div P$	$N \div P$	$N \div P$
D_b Base Diameter	$D \cos 30°$	$D \cos 30°$	$D \cos 30°$	$D \cos 37.5°$	$D \cos 45°$
D_{Fe} Form Dia. External Spline	$([N-1] \div P) + 2c_F$	$([N-1] \div P) + 2c_F$	$([N-1] \div P) + 2c_F$	$([N-0.8] \div P) + 2c_F$	$([N-0.6] \div P) + 2c_F$
D_{Fi} Form Dia. Internal Spline	$([N+1] \div P) + 2c_F$	$([N+0.8] \div P - 0.004) + 2c_F$	$([N+1] \div P) + 2c_F$	$([N+1] \div P) + 2c_F$	$([N+1] \div P) + 2c_F$
D_i Minor Dia. Internal Spline	$N - 1 \div P$	$N - 1 \div P$	$N - 1 \div P$	$N - 0.8 \div P$	$N - 0.6 \div P$
D_o Major Dia. External Spline	$N + 1 \div P$	$N + 1 \div P$	$N + 1 \div P$	$N + 1 \div P$	$N + 1 \div P$
D_{re} Minor Dia. Ext. Spl. (Root)	$N - 1.35 \div P$			$N - 1.8 \div P$	—
				$N - 2 \div P$	$N - 1$
D_{ri} Major Dia. Int. Spl. (Root)	$N + 1.35 \div P$	$N + 1 \div P$	$N + 1.8 \div P$	$N + 1.6 \div P$	$N + 1.4 \div P$
p Circular Pitch	$\pi \div P$	$\pi \div P$	$\pi \div P$	$\pi \div P$	$\pi \div P$
P_s Stub Pitch	$2P$	$2P$	$2P$	$2P$	$2P$
s Actual Space Width	$s_v + (m + \lambda)$	$s_v + (m + \lambda)$	$s_v + (m + \lambda)$	$s_v + (m + \lambda)$	$s_v + (m + \lambda)$

(Continued)

Table 3. *(Continued)* **Formulas for Determining Basic Dimensions of Involute Splines.** (See explanation of symbols at end of Table.)

Symbol and Term	30° Pressure Angle			37.5° Pressure Angle	45° Pressure Angle
	Flat Root Side Fit	Flat Root Major Dia. Fit	Fillet Root Side Fit	Fillet Root Side Fit	Fillet Root Side Fit
	2.5/5 to 32/64 Pitch	3/6 to 16/32 Pitch	2.5/5 to 48/96 Pitch	2.5/5 to 48/96 Pitch	10/20 to 128/256 Pitch
s_V Effective Space Width (Min.)	$\pi \div 2P$	$\pi \div 2P$	$\pi \div 2P$	$(0.5\,\pi + 0.1) \div 2P$	$(0.5\,\pi + 0.2) \div 2P$
t Actual Tooth Thickness (min.)	t_V max.$- (m + \lambda)$	t_V max.$- (m + \lambda)$	t_V max.$- (m + \lambda)$	t_v max.$- (m + \lambda)$	t_V max.$- (m + \lambda)$
t_V Effective Tooth Thickness (max.)	$\pi \div 2P$	$(\pi \div 2P)\ c_V$ min.	$\pi \div 2P$	$(0.5\,\pi + 0.10) \div P$	$(0.5\,\pi + 0.20) \div P$

Symbols used in formulas but not identified in the left column are as follows.

N = Number of Teeth
P = Diametral Pitch
m = Machining Tolerance
λ = Variation Tolerance
c_V = Effective Clearance, λ

Although not used on the Table, the symbol Ø is commonly used to indicate pressure angle.

provided. Tolerances for certain dimensions, such as actual space width, circular (s max), and actual tooth thickness, circular (t min) can be found by multiplying sum of the variations allowance and machining tolerance by the constants for each Tolerance Class (provided above).

Straight Splines

Due to the fact that splines were first widely applied to industrial use in the automobile industry, early standards were issued by the SAE. The dimensions are intended to apply to soft broached holes. Radii at the corners of the spline should not be in excess of 0.015 inch. Formulas for finding dimensions of 4, 6, 10, and 16 tooth spline fittings are found in **Table 4**, and **Tables 5** through **8** provide values for important dimensions of splined fittings. The root width of splineways on splined shafts, not provided in the Tables, can be found with the following formula (see *Figure 4* for references).

$$\sin ([360° \div N] - 2\alpha) \div 2 \times d = W$$

where N = Number of splines
            d = Diameter of shaft at its root
            W = Root width of splineway.

The angle, α, can be found with the following formula.

$$\sin \alpha = {}^1\!/_2\, T \div {}^1\!/_2\, d, \qquad \text{or} \qquad \sin \alpha = T \div d$$

where T = Width of the spline.

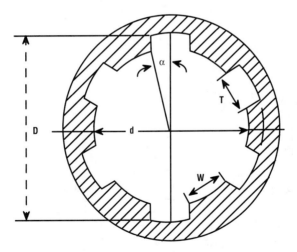

Figure 4. Six spline fitting.

Table 4. Constants for Calculating Spline Dimensions.

No. of Splines	W For All Fits	A (Permanent Fit)		B (Slide—Not Under Load)		C (Slide Under Load)	
		h	d	h	d	h	d
4	0.241D*	0.075D	0.850D	0.125D	0.750D	-	-
6	0.250D	0.050D	0.900D	0.075D	0.850D	0.100D	0.800D
10	0.156D	0.045D	0.910D	0.070D	0.860D	0.095D	0.810D
16	0.198D	0.045D	0.910D	0.070D	0.850D	0.095D	0.810D

Radii on corners of splines not to exceed 0.015 inch.

Splines shall not be more than 0.006 inch per foot out of parallel with respect to the axis of the shaft.

No allowance is made for radii on corners or for clearance. Dimensions are intended to apply to only the soft-broached hole. Allowance must be made for machining.

* Four splines, for fits A and B only.

Table 5. Four Spline Fittings.

| Nom. Dia. | For All Fits | | | | 4A Permanent Fit | | | | | 4B To Slide—Not Under Load | | | | |
| | D | | W | | d | | h | | T* | d | | h | | T* |
	Min.	Max.	Min.	Max.	Min.	Max.	Min.	Max.		Min.	Max.	Min.	Max.	
3/4	0.749	1.750	0.179	0.181	0.636	0.637	0.055	0.056	78	0.561	0.562	0.093	0.094	123
7/8	0.874	0.875	0.209	0.211	0.743	0.744	0.065	0.066	107	0.655	0.656	0.108	0.109	167
1	0.999	1.000	0.239	0.241	0.849	0.850	0.074	0.075	139	0.749	0.750	0.124	0.125	219
1 1/8	1.124	1.125	0.269	0.271	0.955	0.956	0.083	0.084	175	0.843	0.844	0.140	0.141	277
1 1/4	1.249	1.250	0.299	0.301	1.061	1.062	0.093	0.094	217	.0936	0.937	0.155	0.156	341
1 3/8	1.374	1.375	0.329	0.331	1.168	1.169	0.102	0.103	262	1.030	1.031	0.171	0.172	414
1 1/2	1.499	1.500	0.359	0.361	1.274	1.275	0.111	0.112	311	1.124	1.125	0.186	0.187	491
1 5/8	1.624	1.625	0.389	0.391	1.380	1.381	0.121	0.122	367	1.218	1.219	0.202	0.203	577
1 3/4	1.749	1.750	0.420	0.422	1.486	1.487	0.130	0.131	424	1.311	1.312	0.218	0.219	670
2	1.998	2.000	0.479	0.482	1.698	1.700	0.148	0.150	555	1.498	1.500	0.248	0.250	875
2 1/4	2.248	2.250	0.539	0.542	1.910	1.912	0.167	0.169	703	1.685	1.687	0.279	0.281	1,106
2 1/2	2.498	2.500	0.599	0.602	2.123	2.125	0.185	0.187	865	1.873	1.875	0.310	0.312	1,365
3	2.998	3.000	0.720	0.723	2.548	2.550	0.223	0.225	1,249	2.248	2.250	0.373	0.375	1,969

* The formula for torque provides a value in inch-pounds per inch of bearing length and at 1,000 pounds pressure per square inch.
Torque formula: $T = 1,000 \times 4 \times$ mean radius $\times h \times L$.

Table 6. Six Spline Fittings.

Nom. Dia. Min.	For All Fits				6A Permanent Fit			6B Slide—Not Under Load			6C Slide Under Load		
	D		W		d		T*	d		T*	d		T*
	Min.	Max.	Min.	Max.	Min.	Max.		Min.	Max.		Min.	Max.	
3/4	0.749	0.750	0.186	0.188	0.674	0.675	80	0.637	0.638	117	0.599	0.600	152
7/8	0.874	0.875	0.217	0.219	0.787	0.788	109	0.743	0.744	159	0.699	0.700	207
1	0.999	1.000	0.248	0.250	0.899	0.900	143	0.849	0.850	208	0.799	0.800	270
1 1/8	1.124	1.125	0.279	0.281	1.012	1.013	180	0.955	0.956	263	0.899	0.900	342
1 1/4	1.249	1.250	0.311	0.313	1.124	1.125	223	1.062	1.063	325	0.999	1.000	421
1 3/8	1.374	1.375	0.342	0.344	1.237	1.238	269	1.068	1.169	393	1.099	1.100	510
1 1/2	1.499	1.500	0.373	0.375	1.349	1.350	321	1.274	1.275	468	1.199	1.200	608
1 5/8	1.624	1.625	0.404	0.406	1.462	1.463	376	1.380	1.381	550	1.299	1.300	713
1 3/4	1.749	1.750	0.436	0.438	1.574	1.575	436	1.487	1.488	637	1.399	1.400	827
2	1.998	2.000	0.497	0.500	1.798	1.800	570	1.698	1.700	833	1.598	1.600	1,080
2 1/4	2.248	2.250	0.560	0.563	2.023	2.025	721	1.911	1.913	1,052	1.798	1.800	1,367
2 1/2	2.498	2.500	0.622	0.625	2.248	2.250	891	2.123	2.125	1,300	1.998	2.000	1,688
3	2.998	3.000	0.747	0.750	2.698	2.700	1,283	2.548	2.550	1,873	2.398	2.400	2,430

* The formula for torque provides a value in inch-pounds per inch of bearing length and at 1,000 pounds pressure per square inch. Torque formula: $T = 1,000 \times 6 \times$ mean radius $\times h \times L$.

Table 7. Ten Spline Fittings.

Nom. Dia.	For All Fits				10A Permanent Fit			10B Slide—Not Under Load			10C Slide Under Load		
	D		W		d			d			d		
	Min.	Max.	Min.	Max.	Min.	Max.	T*	Min.	Max.	T*	Min.	Max.	T*
3/4	0.749	0.750	0.115	0.117	0.682	0.683	120	0.644	0.645	183	0.607	0.608	241
7/8	0.874	0.875	0.135	0.137	0.795	0.796	165	0.752	0.753	248	0.708	0.709	329
1	0.999	1.000	0.154	0.156	0.909	0.910	215	0.859	0.860	326	0.809	0.810	430
1 1/8	1.124	1.125	0.174	0.176	1.023	1.024	271	0.967	0.968	412	0.910	0.911	545
1 1/4	1.249	1.250	0.193	0.195	1.137	1.138	336	1.074	1.075	508	1.012	1.013	672
1 3/8	1.374	1.375	0.213	0.215	1.250	1.251	406	1.182	1.183	614	1.113	1.114	813
1 1/2	1.499	1.500	0.232	0.234	1.364	1.365	483	1.289	1.290	732	1.214	1.215	967
1 5/8	1.624	1.625	0.252	0.254	0.478	1.479	566	1.397	1.398	860	1.315	1.316	1,135
1 3/4	1.749	1.750	0.271	0.273	1.592	1.593	658	1.504	1.505	997	1.417	1.418	1,316
2	1.998	2.000	0.309	0.312	1.818	1.820	860	1.718	1.720	1,302	1.618	1.620	1,720
2 1/4	2.248	2.250	0.348	0.351	2.046	2.048	1,088	1.933	1.935	1,647	1.821	1.823	2,176
2 1/2	2.498	2.500	0.387	0.390	2.273	2.275	1,343	2.148	2.150	2,034	2.023	2.025	2,688
3	2.998	3.000	0.465	0.468	2.728	2.730	1,934	2.578	2.580	2,929	2.428	2.430	3,869
3 1/2	3.497	3.500	0.543	0.546	3.182	3.185	2,632	3.007	3.010	3,987	2.832	2.835	5,266
4	3.997	4.000	0.621	0.624	3.637	3.640	3,438	3.437	3.440	5,208	3.237	3.240	6,878
4 1/2	4.497	4.500	0.699	0.702	4.092	4.095	4,351	3.867	3.870	6,591	3.642	3.645	8,705
5	4.997	5.000	0.777	0.780	4.547	4.550	5,371	4.297	4.300	8,137	4.047	4.050	10,746
5 1/2	5.497	5.500	0.855	0.858	5.002	5.005	6,500	4.727	4.730	9,846	4.452	4.455	13,003
6	5.997	6.000	0.933	0.936	5.457	5.460	7,735	5.157	5.160	11,718	4.857	4.860	15,475

* The formula for torque provides a value in inch-pounds per inch of bearing length and at 1,000 pounds pressure per square inch.
Torque formula: $T = 1,000 \times 10 \times$ mean radius $\times h \times L$.

Table 8. Sixteen Spline Fittings.

Nom. Dia.	For All Fits				16A Permanent Fit			16B Slide—Not Under Load			16C Slide Under Load		
	D		W		d		T	d		T	d		T
	Min.	Max.	Min.	Max.	Min.	Max.		Min.	Max.		Min.	Max.	
2	1.997	2.000	0.193	0.196	1.817	1.820	1,375	1.717	1.720	2,083	1.617	1.620	2,751
2 1/2	2.497	2.500	0.242	0.245	2.273	2.275	2,149	2.147	2.150	3,255	2.022	2.025	4,299
3	2.997	3.000	0.291	0.294	2.727	2.730	3,094	2.577	2.580	4,687	2.427	2.430	6,190
3 1/2	3.497	3.500	0.340	0.343	3.182	3.185	4,212	3.007	3.010	6,378	2.832	2.835	8,426
4	3.997	4.000	0.389	0.392	3.637	3.640	5,501	3.437	3.440	8,333	3.237	3.240	11,005
4 1/2	4.497	4.500	0.438	0.441	4.092	4.095	6,962	3.867	3.870	10,546	3.642	3.645	13,928
5	4.997	5.000	0.487	0.490	4.547	4.550	8,595	4.297	4.300	13,020	4.047	4.050	17,195
5 1/2	5.497	5.500	0.536	0.539	5.002	5.005	10,395	4.727	4.730	15,754	4.452	4.455	20,806
6	5.997	6.000	0.585	0.588	5.547	5.460	12,377	5.157	5.150	18,749	4.857	4.860	24,760

* The formula for torque provides a value in inch-pounds per inch of bearing length and at 1,000 pounds pressure per square inch.
Torque formula: $T = 1{,}000 \times 16 \times$ mean radius $\times h \times L$.

Square Shaft Fits and Taper Fittings

Square shafts are sometimes called on to do the duty of more expensive splined shafts in noncritical fittings. There are two basic classes of fittings—permanent and sliding and dimensions for each are given in **Table 9**. These suggested dimensions have remained unchanged for decades and have not been standardized, probably because square shafts are rarely considered for high precision applications. Tapered shafts with fitted nuts, on the other hand, remain in wide use. They were standardized many years ago by the SAE, originally with both plain and slotted nuts. More recently, the use of plain nuts has been and discouraged and slotted nuts with cotter-pins are specified. The centerline of the cotter-pin hole is positioned 90° from the position of the keyway. Dimensions of taper shaft ends and their slotted nuts are given in **Table 10**.

Table 9. Dimensions and Fits of Square Shafts.

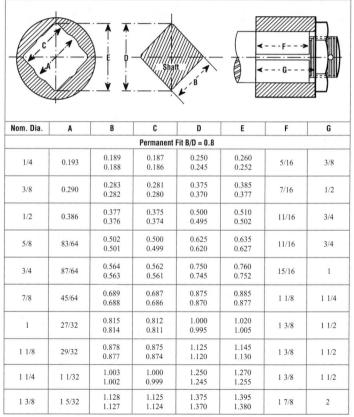

Nom. Dia.	A	B	C	D	E	F	G
			Permanent Fit B/D = 0.8				
1/4	0.193	0.189 0.188	0.187 0.186	0.250 0.245	0.260 0.252	5/16	3/8
3/8	0.290	0.283 0.282	0.281 0.280	0.375 0.370	0.385 0.377	7/16	1/2
1/2	0.386	0.377 0.376	0.375 0.374	0.500 0.495	0.510 0.502	11/16	3/4
5/8	83/64	0.502 0.501	0.500 0.499	0.625 0.620	0.635 0.627	11/16	3/4
3/4	87/64	0.564 0.563	0.562 0.561	0.750 0.745	0.760 0.752	15/16	1
7/8	45/64	0.689 0.688	0.687 0.686	0.875 0.870	0.885 0.877	1 1/8	1 1/4
1	27/32	0.815 0.814	0.812 0.811	1.000 0.995	1.020 1.005	1 3/8	1 1/2
1 1/8	29/32	0.878 0.877	0.875 0.874	1.125 1.120	1.145 1.130	1 3/8	1 1/2
1 1/4	1 1/32	1.003 1.002	1.000 0.999	1.250 1.245	1.270 1.255	1 3/8	1 1/2
1 3/8	1 5/32	1.128 1.127	1.125 1.124	1.375 1.370	1.395 1.380	1 7/8	2

(Continued)

Table 9. *(Continued)* **Dimensions and Fits of Square Shafts.**

Nom. Dia.	A	B	C	D	E	F	G
Permanent Fit B/D = 0.8							
1 1/2	1 5/32	1.253 1.252	1.250 1.249	1.500 1.495	1.520 1.505	1 7/8	2
1 3/4	1 27/64	1.378 1.377	1.375 1.374	1.750 1.745	1.770 1.755	2 1/8	2 1/4
2	1 35/64	1.504 1.503	1.500 1.498	2.000 1.995	2.020 2.005	2 7/8	3
2 1/4	1 13/16	1.754 1.753	1.750 1.748	2.250 2.245	2.270 2.255	2 7/8	3
2 1/2	2 1/16	2.004 2.003	2.000 1.998	2.500 2.495	2.520 2.505	3 3/8	3 1/2
2 3/4	2 5/16	2.254 2.253	2.250 2.248	2.750 2.745	2.770 2.755	3 3/8	3 1/2
3	2 37/64	2.504 2.503	2.500 2.498	3.000 2.995	3.020 3.005	3 7/8	4
3 1/2	2 55/64	2.754 2.753	2.750 2.748	3.500 3.495	3.520 3.505	4 3/8	4 1/2
4	3 23/64	3.254 3.253	3.250 3.248	4.000 3.995	4.020 4.005	5 3/8	5 1/2

Nom. Dia.	A		B	C		D	E
Sliding Fits B/D = 0.73							
1/4	0.257		0.248 0.247	0.250 0.249		0.344 0.339	0.354 0.346
3/8	0.386		0.373 0.372	0.375 0.374		0.516 0.511	0.526 0.518
1/2	33/64		0.498 0.497	0.500 0.499		0.687 0.682	0.697 0.689
5/8	41/64		0.623 0.622	0.625 0.624		0.844 0.839	0.854 0.846
3/4	49/64		0.748 0.747	0.750 0.749		1.031 1.026	1.051 1.036
7/8	29/32		0.873 0.872	0.875 0.874		1.187 1.182	1.207 1.192
1	1 1/32		0.998 0.997	1.000 0.999		1.375 1.370	1.395 1.380
1 1/8	1 5/32		1.123 1.122	1.125 1.124		1.562 1.557	1.582 1.567
1 1/4	1 9/32		1.248 1.247	1.250 1.249		1.687 1.682	1.707 1.692

(Continued)

Table 9. *(Continued)* **Dimensions and Fits of Square Shafts.**

Nom. Dia.	A	B	C	D	E
Sliding Fits B/D = 0.73					
1 3/8	1 27/64	1.373 1.372	1.375 1.374	1.875 1.870	1.895 1.880
1 1/2	1 35/64	1.498 1.497	1.500 1.499	2.062 2.057	2.082 2.067
1 3/4	1 13/16	1.748 1.747	1.750 1.749	2.375 2.370	2.395 2.380
2	2 1/16	1.997 1.996	2.000 1.998	2.750 2.745	2.770 2.755
2 1/4	2 5/16	2.247 2.246	2.250 2.248	3.062 3.057	3.082 3.067
2 1/2	2 37/64	2.497 2.496	2.500 2.498	3.437 3.432	3.457 3.442
2 3/4	2 55/64	2.747 2.746	2.750 2.748	3.750 3.745	3.770 3.755
3	3 3/32	2.997 2.996	3.000 2.998	4.125 4.120	4.145 4.130
3 1/2	3 39/64	3.497 3.496	3.500 3.498	4.750 4.745	4.770 4.755
4	4 1/8	3.997 3.996	4.000 3.998	5.500 5..495	5.520 5.505

Table 10. Taper Shaft and Nut Dimensions. (Taper per foot is 1.500 ±0.002 inch.)

Nom. Dia.	Shaft Dia. D_s		Hole Dia. D_h		L_c	L_a	L_h	L_t	T_s	T_p
	Max.	Min.	Max.	Min.						
1/4	0.250	0.249	0.248	0.247	9/16	5/16	3/8	5/16	7/32	9/64
3/8	0.375	0.374	0.373	0.372	47/64	7/16	1/2	23/64	17/64	3/16
1/2	0.500	0.499	0.498	0.497	63/64	11/16	3/4	23/64	17/64	3/16
5/8	0.625	0.624	0.623	0.622	1 3/32	11/16	3/4	17/32	7/16	1/4
3/4	0.750	0.749	0.748	0.747	1 11/32	15/16	1	17/32	7/16	1/4
7/8	0.875	0.874	0.873	0.872	1 11/16	1 1/8	1 1/4	11/16	1/2	5/16
1	1.001	0.999	0.997	0.995	1 15/16	1 3/8	1 1/2	11/16	1/2	5/16
1 1/8	1.126	1.124	1.122	1.120	1 15/16	1 3/8	1 1/2	11/16	1/2	5/16
1 1/4	1.251	1.249	1.247	1.245	1 15/16	1 3/8	1 1/2	11/16	1/2	5/16
1 3/8	1.376	1.374	1.372	1.370	2 7/16	1 7/8	2	11/16	1/2	5/16
1 1/2	1.501	1.499	1.497	1.495	2 7/16	1 7/8	2	11/16	1/2	5/16

(Continued)

Table 10. *(Continued)* **Taper Shaft and Nut Dimensions.** (Taper per foot is 1.500 ±0.002 inch.)

Nom. Dia.	Shaft Dia. D_s		Hole Dia. D_h		L_c	L_a	L_h	L_t	T_s	T_p
	Max.	Min.	Max.	Min.		Min.				
1 5/8	1.626	1.624	1.622	1.620	2 13/16	2 1/8	2 1/4	13/16	5/8	7/16
1 3/4	1.751	1.749	1.747	1.745	2 13/16	2 1/8	2 1/4	13/16	5/8	7/16
1 7/8	1.876	1.874	1.872	1.870	3 1/16	2 3/8	2 1/2	13/16	5/8	7/16
2	2.001	1.999	1.997	1.995	3 9/16	2 7/8	3	13/16	5/8	7/16
2 1/4	2.252	2.248	2.245	2.242	3 9/16	2 7/8	3	13/16	5/8	7/16
2 1/2	2.502	2.498	2.495	2.492	4 9/32	3 3/8	3 1/2	1 1/4	1	5/8
2 3/4	2.752	2.748	2.745	2.742	4 9/32	3 3/8	3 1/2	1 1/4	1	5/8
3	3.002	2.998	2.995	2.992	4 25/32	3 7/8	4	1 1/4	1	5/8
3 1/4	3.252	3.248	3.245	3.242	5 1/32	4 1/8	4 1/4	1 1/4	1	5/8
3 1/2	3.502	3.498	3.495	3.492	5 7/16	4 3/8	4 1/2	1 3/8	1 1/8	3/4
4	4.002	3.998	3.995	3.992	6 7/16	5 3/8	5 1/2	1 3/8	1 1/8	3/4

Nom. Dia.	Nut Flat Width	Threads per in.	Keyway W		Keyway H		Square Key		D_t	C
			Max.	Min.	Max.	Min.	Max.	Min.		
1/4	5/16	40	0.0625	0.0615	0.037	0.033	0.0635	0.0625	0.1935	5/64
3/8	1/2	32	0.0937	0.0927	0.053	0.049	0.0947	0.0937	0.3125	5/64
1/2	1/2	32	0.1250	0.1240	0.069	0.065	0.1260	0.1250	0.3125	5/64
5/8	3/4	28	0.1562	0.1552	0.084	0.080	0.1572	0.1562	0.5000	1/8
3/4	3/4	28	0.1875	0.1865	0.100	0.096	0.1885	0.1875	0.5000	1/8
7/8	15/16	24	0.2500	0.2490	0.131	0.127	0.2510	0.2490	0.6250	5/32
1	1 1/16	20	0.2500	0.2490	0.131	0.127	0.2510	0.2490	0.7500	5/32
1 1/8	1 1/4	20	0.3125	0.3115	0.162	0.158	0.3135	0.3125	0.8750	5/32
1 1/4	1 7/16	20	0.3125	0.3115	0.162	0.158	0.3135	0.3125	1.0000	5/32
1 3/8	1 7/16	20	0.3750	0.3740	0.194	0.190	0.3760	0.3750	1.0000	5/32

(Continued)

Table 10. *(Continued)* Taper Shaft and Nut Dimensions. (Taper per foot is 1.500 ±0.002.)

Nom. Dia.	Nut Flat Width	Threads per in.	Keyway W		Keyway H		Square Key		D_t	c
			Max.	Min.	Max.	Min.	Max.	Min.		
1 1/2	1 7/16	20	0.3750	0.3740	0.194	0.190	0.3760	0.3750	1.0000	5/32
1 5/8	2 3/16	18	0.4375	0.4365	0.225	0.221	0.4385	0.4375	1.2500	5/32
1 3/4	2 3/16	18	0.4375	0.4365	0.225	0.221	0.4385	0.4375	1.2500	5/32
1 7/8	2 3/16	18	0.4375	0.4365	0.225	0.221	0.4385	0.4375	1.2500	5/32
2	2 3/16	18	0.5000	0.4990	0.256	0.252	0.5010	0.5000	1.2500	5/32
2 1/4	2 3/8	18	0.5625	0.5610	0.287	0.283	0.5640	0.5625	1.5000	5/32
2 1/2	3 1/8	16	0.6250	.6235	0.319	0.315	0.6265	0.6250	2.000	7/32
2 3/4	3 1/8	16	0.6875	0.6860	0.350	0.346	0.6890	0.6875	2.000	7/32
3	3 1/8	16	0.7500	0.7485	0.381	0.377	0.7515	0.7500	2.000	7/32
3 1/4	3 1/8	16	0.7500	0.7485	0.381	0.377	0.7515	0.7500	2.000	7/32
3 1/2	3 7/8	16	0.8750	0.8735	0.444	0.440	0.8765	0.8750	2.5000	9/32
4	3 7/8	16	1.0000	0.9985	0.506	0.502	1.0015	1.0000	2.5000	9/32

All dimensions in inches.

Drive Belts

Belts of several types are used to transmit power in mechanical systems. The basic design of flat, v-style, and cogged belts is included in this section.

To find: Diameter of driving pulley, when speed and diameter of driven pulley, and the speed of the driving pulley, are known.

$$D = d \times \text{rpm}_d \div \text{rpm}_D$$

where D = diameter of driving pulley
 d = diameter of driven pulley
 rpm_D = speed of driving pulley
 rpm_d = speed of driven pulley.

To find: Speed of driving pulley, when the diameter and speed of the driven pulley, and the diameter of the driving pulley, are known.

$$\text{rpm}_D = d \times \text{rpm}_d \div D$$

To find: Diameter of the driven pulley, when the diameter and speed of the driving pulley, and the speed of the driven pulley, are known.

$$d = D \times \text{rpm}_D \div \text{rpm}_d$$

To find: Speed of the driven pulley, when the diameter and speed of the driving pulley, and the diameter of the driven pulley, are known.

$$\text{rpm}_d = D \times \text{rpm}_D \div d$$

To find: Effective belt length (L) of flat or v-belt for two sheave drives.

$$L = 2C + ([\pi \div 2] \% [D + d]) + ([D - d]^2 \div 4C)$$

where C = distance between centerpoint of sheave axles
 D = diameter of larger sheave
 d = diameter of smaller sheave.

To find: Distance (L_2) between two sheave axles, given length of belt and pulley diameters.

$$L_2 = p + \sqrt{p^2 - q}$$

where $p = (L \div 4) - ([\pi \times \{D + d\}] \div 8)$
 $q = (D - d)^2 \div 8$

To find: Arc of contact (β) on small sheave.

$$\cos \beta \div 2 = (D - d) \div 2C$$

To find: Torque acting on a driving belt (in in.-lb), when the amount of force transmitted (F – in pounds) and the pulley diameter (in inches) is known.

$$T = F (d \div 2)$$

To find: Horsepower acting on a driving belt, when torque and belt speed are known.

$$HP = T \times \text{rpm} \div 396{,}000$$

V-belt drives

Most v-belts are covered in standards issued by ANSI and the Rubber Manufacturers' Association (RMA). Only the most common of the wide variety of available belts are covered in this section. ANSI/RMA heavy-duty belt forms that are included are classical (sizes A, B, C, D, and E), and narrow (sizes 3 VX, 5VX, 5V, and 8V) belts. Belts exclusively intended for automobile use, which are covered by SAE standards, are also discussed.

Effective length of v-belts. Effective belt length measurements are made on a tensioned belt mounted on two equal diameter sheaves, and it is a measurement of the pitch length (not inside length) of the belt. The distance between the centerpoint of the two sheaves is then measured. Effective belt length is then calculated as twice center distance, plus the outside circumference of one of the sheaves. The formula for determining effective belt length (L) is given above.

Classical heavy-duty belts

Dimensions for classical belts are as follows. Size A and AX (belts designated with a single letter have slightly rounded tops, and belts designated "X" have square tops) = $1/2$" wide across top, $11/32$" in height. Size B and BX = $21/32$" wide across top, $7/16$" in height. Size C and CX = $7/8$" wide across top, $17/32$" in height. Size D and DX = $1 1/4$" wide across top, $3/4$" in height. Size E = $1 1/2$" wide across top, 1" in height.

Belt length for classical belts is often stated as a "datum" length, which is synonymous with "pitch" length. The substitution of datum for pitch was initiated by the RMA in order to bring domestic measurements in accordance with those of ISO. Thus, the center length of a belt may be referred to as interchangeably as its datum length, its pitch length, or its effective length. For sheaves, the terms outside diameter and pitch diameter are also interchangeable for standard sheaves, but not for deep groove or combination sheaves. The datum diameter of a sheave refers to the point on the sheave where a measuring ball or rod of predetermined size will contact the sides of a sheave.

A series belts are generally available in effective lengths of 27.3 to 129.3 inches; B series belts are generally available in lengths of 36.8 to 300.3 inches; C series belts are generally available in lengths of 53.9 to 420.9 inches; D series belts are generally available in lengths of 123.3 to 660.8 inches; and E series belts are generally available in lengths of 184.5 to 661 inches. Classic series belts retain an obsolete standard length designation that is the length of the inside diameter of the belt, and therefore does not exactly specify the length of the belt at pitch diameter. In fact, pitch length may be as much as 4.5 inches longer than the inside length, which is used for the designation. An example of the designation system is C75, which is a C series belt that actually has an effective length of 77.9.

Sheave angles and depths for classical belts are given in **Table 1**. The recommended maximum rim speed for classic v-belts is 6,000 ft/min.

Narrow heavy-duty belts

Dimensions of narrow heavy-duty v-belts are as follows: 3VX = $3/8$" across top, $5/16$" in height. Size 5VX and 5V (belts designated 'VX' have sharp corners at the sides of the top and bottom surfaces and have notched cross sections, and belts designated "V" have rounded shoulders) = $5/8$" wide across top, $17/32$" in height. Size 8V = 1" wide across top, $7/8$" in height.

The datum system for narrow belts, briefly described above, has been used since inception on narrow belts. Therefore, listings for belt datum length and effective length should coincide. 3V belts are generally available in effective lengths of 25 to 140 inches; 5V belts are generally available in lengths of 50 to 355 inches; and 8V belts are generally available in lengths of 100 to 425 inches. When ordering belts, the belt size should be specified first, followed by the standard length designation, which is the belt length without a final decimal point. For example, when ordering a 3V belt of 31.5 inches in length, the belt designation would be 3V315. When ordering an 8V belt of 315 inches in length, the belt designation would be 8V3150.

Table 1. Groove Dimensions for Standard Classical V-belt Sheaves.

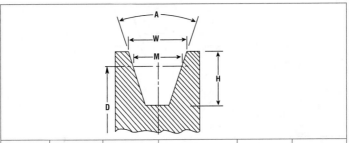

Belt Size	Width at Point of Measurement M	Depth of Groove (Min.) H	Diameter at Point of Measurement D	Angle of Groove[1] A	Top Width of Groove W
A, AX	0.418	0.460	Thru 5.4 incl.	34	0.494 ±0.005
			Over 5.4	38	0.504 ±0.005
B, BX	0.530	0.550	Thru 7 incl.	34	0.637 ±0.006
			Over 7	38	0.650 ±0.006
C, CX	0.757	0.750	Thru 7.99 incl.	34	0.879 ±0.007
			Over 7.99 to 12 incl.	36	0.887 ±0.007
			Over 12	38	0.895 ±0.007
D	1.076	1.020	Thru 12.99 incl.	34	12.59 ±0.008
			Over 12.99 to 17 incl.	36	1.271 ±0.008
			Over 17	38	1.283 ±0.008

[1] Groove angle tolerance is ±0.33°. All dimensions in inches.

Sheave angles and depths for narrow belts are given in **Table 2**. The recommended maximum rim speed for narrow belts it is 6,500 ft/min.

Automotive belts

V-belts for automotive use are available in nine sizes, each of which is the measurement of belt width across its top. Belts are available in one-half inch increments up to 80 inches in length, and in 1 inch increments in longer lengths. **Table 3** provides belt sizes and basic sheave angles and depths for automotive v-belts.

Synchronous belt drives

Often referred to as timing or positive drive belts, synchronous belts have teeth that mesh with the pulleys for predictable power transmission. These belts are typically reinforced with carbon steel, Kevlar, rayon, fiberglass, or stainless steel to assure that their length remains constant under power. Due to the range of strengthening components used in these belts, it is not advisable to base horsepower ratings of a belt on its width, even though formulas for calculating recommended maximum horsepower ratings are provided below. Supplier catalogs should be consulted to determine the actual performance parameters of individual belts.

Standard construction synchronous belts can normally be used with confidence

Table 2. Groove Dimensions for Standard Narrow V-belt Sheaves.

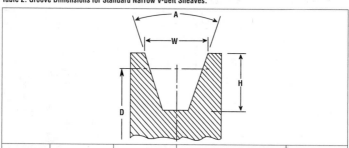

Belt Size	Top Width of Groove W	Depth of Groove (Min.) H	Diameter D	Angle of Groove[1] A
3V	0.350 ±0.005	0.340	Thru 3.49 incl.	36
			Over 3.49 thru 6 incl.	38
			Over 6 thru 12 incl.	40
			Over 12	42
5V	0.600 ±0.005	0.590	Thru 9.99 incl.	38
			Over 9.99 thru 16 incl.	40
			Over 16	42
8V	1.000 ±0.005	0.990	Thru 15.99 incl	38
			Over 15.99 thru 22.4 incl	40
			Over 22.4	42

[1] Groove angle tolerance is ±0.25°. All dimensions in inches.

Table 3. Groove Dimensions for Automotive V-belt Sheaves.

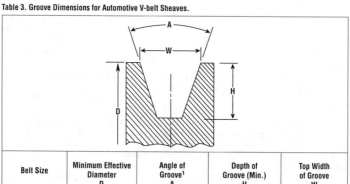

Belt Size	Minimum Effective Diameter D	Angle of Groove[1] A	Depth of Groove (Min.) H	Top Width of Groove W
0.250	2.25	36	0.276	0.248
0.318	2.25	36	0.354	0.315

(Continued)

Table 3. *(Continued)* Groove Dimensions for Automotive V-belt Sheaves.

Belt Size	Minimum Effective Diameter D	Angle of Groove[1] A	Depth of Groove (Min.) H	Top Width of Groove W
0.380	2.40	36	0.433	0.380
0.440	2.75	36	0.512	0.441
0.500	3.00	36	0.551	0.500
11/16	3.00	34	0.551	0.597
	Over 4.00	36		
	Over 6.00	38		
3/4	3.00	34	0.630	0.660
	Over 4.00	36		
	Over 6.00	38		
7/8	3.50	34	0.709	0.785
	Over 4.50	36		
	Over 6.00	38		
1	4.00	34	0.827	0.910
	Over 6.00	36		
	Over 8.00	38		

[1] Tolerance on groove angle is ±0.5°. All dimensions in inches.

within a temperature range of –30 to 185° F (–34 to 85° C), but belts are available that extend the operational range to –65 to 230° F (–54 to 110° C). Again, it is important to confirm that a belt selected for extreme temperatures will adequately perform in its operational environment.

Positive drive belts are also susceptible to chemical breakdown, as the base material is usually neoprene. Belts made from urethane provide good resistance to several chemicals, as well as better abrasion resistance than neoprene belts. Ideally, belts should have a clean environment, and they should never be lubricated or encased in oil.

Pulley alignment is critical for a synchronous drive system, and misalignment in excess of $^1/_4$° per foot of center distance should be corrected. At least one pulley should be adjustable to allow for installation and tensioning of the belt. If an idler is used to tension the belt, it must be placed on the slack span of the belt. Once installed and properly tensioned, positive drive belts are less critical to tension maintenance than v-belts. Pulleys (or sprockets) are available from belt manufacturers in a wide range of diameters for all belt styles and widths.

As with gear or chain drives, pitch must be considered in all positive drive belt applications. On the belt, pitch is the distance between the same location (usually measured from the centerpoint of the teeth as some belts have rounded, or curvilinear, "teeth") on adjacent teeth, and it is measured on the pitch line of the belt (see *Figure 1*). On the pulley, pitch is the distance between groove centers, measured on the pulley's pitch circle. The pitch circle on a positive drive belt pulley (from which the pitch diameter is measured) coincides with the pitch line of the belt mating with it, as seen in *Figure 2*. Positive drive belts must be run with pulleys of the same pitch. The following formula can be used to find the pitch diameter of positive drive pulleys.

$$PD = (N \times p) \div \pi$$

where N = number of teeth on pulley
p = pitch (spacing of teeth).

The pitch diameter times the number of teeth in the belt equals the belt length. Effective belt length may also be calculated with the formula for L that was provided in the section on v-belts. An additional formula, that includes the pitch of the belt, is as follows.

$$L_T = 2C + ([p \div 2] \times [D_T + d_T]) + ([p \div \pi] \times [D_T - d_T]^2 \div 4C)$$

where L_T = length of toothed belt
C = distance between centerpoint of pulley axles
p = belt pitch
D_T = pitch diameter of larger pulley
d_T = pitch diameter of smaller pulley.

The pitch of the belt can also be used to calculate the center distance between pulley axles.

$$C = ([b + \sqrt{b^2 - 32 \times \{ D_T - d_T \}^2}] \div 16)$$

where $b = 4 p - 6.28318 (D_T + d_T)$.

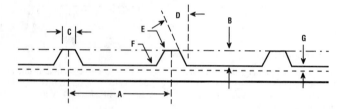

Figure 1. Trapezoidal tooth positive drive belt. A = pitch of teeth on belt. B = tooth depth.
C = tooth width at bottom of tooth. D = pressure angle. E = radius at bottom of tooth.
F = radius at top of tooth. G = pitch line differential.

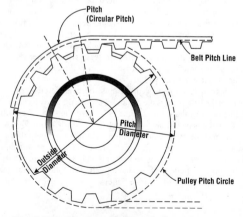

Figure 2. Nomenclature for positive drive pulley dimensions.

There are six conventional pitches for trapezoidal tooth belts, and five popular pitches for curvilinear "HTD" belts. Several belt manufacturers have developed proprietary "modified curvilinear" tooth profiles that are claimed to outperform the HTD style belts. Manufacturers should be consulted for details. Pitch for conventional and HTD type belts is given below.

| Trapezoidal Tooth Belts | | Curvilinear HTD Tooth Belts |
Designation	Pitch	Designation / Pitch
MXL – Mini Extra Light	0.080"	3 mm
XL – Extra Light	0.020"	5 mm
L – Light	0.375"	8 mm
H – Heavy	0.500"	14 mm
XH – Extra Heavy	0.875"	20 mm
XXH – Double Extra Heavy	1.25"	

The pitch diameter of the pulley minus the pitch line differential is equal to the outside diameter of the pulley. Pitch line differentials for common belt pitches are provided in the far right hand column (Flat Belt Dia. a × 2) of **Table 4**. Additional information about the dimensions of trapezoidal belt teeth can also be found in the Table.

Belt frequency refers to the number of belt teeth that engage and exit the pulley grooves in a given unit of time. It can be calculated with the following formula.

$$F = (N_G \times \text{rpm}) \div 60$$

where F = meshing frequency (teeth/sec or cycles/sec)
  N_G = number of grooves in pulley.

Horsepower ratings

Formulas for approximating maximum horsepower ratings for trapezoidal tooth belts with a minimum of six teeth in mesh are provided below. The belt widths given are representative of the full range of widths normally available for each belt style. Ratings for MXL belts are not given below because of their low capacities. In the formulas, d_T is the pitch diameter of the smaller pulley, and s is the rpm of the faster pulley divided by 1000.

Belt type	Width	Formula
XL	0.25	$(d_T \times s) \times (0.0916 - 7.07 \times 10^{-5} [d_T \times s]^2) \times 0.62$
XL	0.38	$(d_T \times s) \times (0.0916 - 7.07 \times 10^{-5} [d_T \times s]^2)$
L	0.50	$(d_T \times s) \times (0.436 - 3.01 \times 10^{-4} [d_T \times s]^2) \times 0.45$
L	0.75	$(d_T \times s) \times (0.436 - 3.01 \times 10^{-4} [d_T \times s]^2) \times 0.72$
L	1.00	$(d_T \times s) \times (0.436 - 3.01 \times 10^{-4} [d_T \times s]^2)$
H	0.75	$(d_T \times s) \times (3.73 - 1.41 \times 10^{-3} [d_T \times s]^2) \times 0.21$
H	1.00	$(d_T \times s) \times (3.73 - 1.41 \times 10^{-3} [d_T \times s]^2) \times 0.29$
H	1.50	$(d_T \times s) \times (3.73 - 1.41 \times 10^{-3} [d_T \times s]^2) \times 0.45$
H	2.00	$(d_T \times s) \times (3.73 - 1.41 \times 10^{-3} [d_T \times s]^2) \times 0.63$
H	3.00	$(d_T \times s) \times (3.73 - 1.41 \times 10^{-3} [d_T \times s]^2)$
XH	2.00	$(d_T \times s) \times (7.21 - 4.68 \times 10^{-3} [d_T \times s]^2) \times 0.45$
XH	3.00	$(d_T \times s) \times (7.21 - 4.68 \times 10^{-3} [d_T \times s]^2) \times 0.72$
XH	4.00	$(d_T \times s) \times (7.21 - 4.68 \times 10^{-3} [d_T \times s]^2)$
XXH	2.00	$(d_T \times s) \times (11.4 - 4.81 \times 10^{-3} [d_T \times s]^2) \times 0.35$
XXH	3.00	$(d_T \times s) \times (11.4 - 4.81 \times 10^{-3} [d_T \times s]^2) \times 0.56$
XXH	4.00	$(d_T \times s) \times (11.4 - 4.81 \times 10^{-3} [d_T \times s]^2) \times 0.78$
XXH	5.00	$(d_T \times s) \times (11.4 - 4.81 \times 10^{-3} [d_T \times s]^2)$

Table 4. Dimensions of Trapezoidal Tooth Positive Drive Belts.

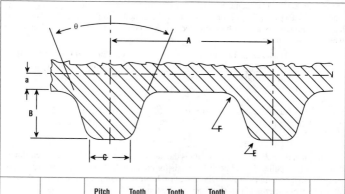

Belt Designation	Number of Grooves	Pitch Length A	Tooth Angle θ	Tooth Depth B	Tooth Width C	Radius E	Radius F	Flat Belt Dia. a × 2
		±.0001	±.25	+.002/-.000	+.002/-.000	±.001	±.001	
MXL	10 thru 23	.0800	56	.025	.024	.012	.009	.020
	24 and over		40		.0265			
XL	10 and over	.2000	50	.055	.050	.024	.024	.020
L	10 and over	.3750	40	.084	.122	.034	.021	.030
H	14 thru 19	.5000	40	.102	.167	.058	.041	.054
	Over 19						.056	
XH	18 and over	.8750	40	.271	.299	.079	.076	.110
XXH	18 and over	1.2500	40	.405	.457	.106	.111	.120

All dimensions in inches.

Roller Chain Drives

Roller chains are assembled from link plates, pins, bushings, and rollers. Links are classified as either roller links, which are the narrow links on the chain, or connecting (sometimes called pin links) links, which make up the wide, or outer, links on the chain. Each link of a roller chain contains up to ten individual pieces, making a roller chain a relatively complex medium for transmitting power. Modern chains transmit power with acceptably low noise and high efficiency. Links of the chain are connected using connecting links secured in place by either rivets, cotter pins, or, on smaller chains, spring clips. Chains secured by pins or clips are considered detachable. Roller chains built to U.S. standards are identified with numbers assigned many years ago by the American Standards Association. A two or three digit identification number indicates a single strand chain. Multistrand chains are identified with a hyphen, plus a number indicating the number of strands on the chain. For example, a 140-2 chain has two strands, and a 160-3 chain has three strands.

Dimensions and load capacities for single and double strand chain can be found in **Table 1**. Sprockets have likewise been standardized. Tooth dimensions are provided in **Table 2**, and outside sprocket dimensions for any pitch chain can be derived from the

Table 1. Dimensions of Single and Double Strand Standard Roller Chain.

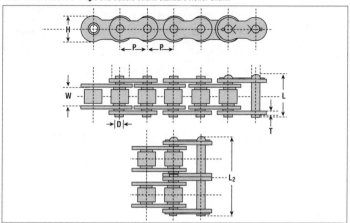

Chain Number		Pitch	Feet per Link	Links per Foot	Roller Width	Roller Dia.	Pin Dia.	Plate Height	Plate Thickness
Single Chain	Double Chain	P			W		D	H	T
25	-	0.250	0.0208	48.0000	0.125	0.130	0.090	0.230	0.030
35	35-2	0.375	0.0313	32.0000	0.188	0.200	0.141	0.354	0.050
41	-	0.500	0.0417	24.0000	0.250	0.306	0.141	0.385	0.050
40	40-2	0.500	0.0521	19.2000	0.312	0.312	0.156	0.472	0.060
50	50-2	0.625	0.0625	16.0000	0.375	0.400	0.200	0.590	0.080
60	60-2	0.750	0.0729	13.7143	0.500	0.469	0.234	0.712	0.094
80	80-2	1.000	0.0833	12.0000	0.625	0.625	0.312	0.948	0.125
100	100-2	1.250	0.1042	9.6000	0.750	0.750	0.375	1.185	0.156
120	120-2	1.500	0.1146	8.7273	1.000	0.875	0.437	1.425	0.187
140	140-2	1.750	0.1250	8.0000	1.000	1.000	0.500	1.660	0.219
160	160-2	2.000	0.1354	7.3846	1.250	1.125	0.562	1.898	0.250
180	180-2	2.250	0.1458	6.8571	1.406	1.406	0.687	2.134	0.281
200	200-2	2.500	0.1563	6.4000	1.500	1.562	0.781	2.373	0.312
240	240-2	3.000	0.1667	6.0000	1.875	1.875	0.937	2.850	0.375

Chain Number		Single Strand Chain				Double Strand Chain			
		Pin Length L		Max. Allowed Load	Minimum Tensile Strength	Pin Length L_2		Max. Allowed Load	Minimum Tensile Strength
Single Chain	Double Chain	Riveted	Cottered			Riveted	Cottered		
		LR	LC	lb	lb	LR	LC	lb	lb
25	-	0.300	0.339	130	780	-	-	-	-
35	35-2	0.460	0.500	495	1,760	0.858	0.896	920	3,520
41	-	0.520	0.579	550	1,500	-	-	-	-

(Continued)

Table 1. *(Continued)* **Dimensions of Single and Double Strand Standard Roller Chain.**

Chain Number		Single Strand Chain				Double Strand Chain			
		Pin Length L		Max. Allowed Load	Minimum Tensile Strength	Pin Length L_2		Max. Allowed Load	Minimum Tensile Strength
Single Chain	Double Chain	Riveted	Cottered			Riveted	Cottered		
		LR	LC	lb	lb	LR	LC	lb	lb
40	40-2	0.650	0.716	820	3,125	1.216	1.280	1,460	6,250
50	50-2	0.812	0.878	1,440	4,880	1.524	1.595	2,440	9,760
60	60-2	1.012	1.087	2,020	7,030	1.910	1.985	3,540	14,060
80	80-2	1.280	1.396	3,300	12,500	2.434	2.550	5,660	25,000
100	100-2	1.556	1.678	5,020	19,530	2.968	3.090	8,640	39,060
120	120-2	1.960	2.116	6,910	28,125	3.750	3.905	11,680	56,250
140	140-2	2.118	2.308	9,160	38,280	4.044	4.230	15,500	76,560
160	160-2	2.508	2.704	12,100	50,000	4.814	5.010	20,500	100,000
180	180-2	2.808	3.074	13,800	63,280	5.414	5.674	23,500	126,560
200	200-2	3.070	3.299	16,300	78,125	5.894	6.120	27,400	156,250
240	240-2	3.772	4.070	22,300	112,500	7.236	7.530	38,000	225,000

For transverse pitch of multiple strand chains, see **Table 3**. Dimensions, except where indicated, are in inches.

Table 2. Basic Roller Chain Sprocket Tooth Dimensions.

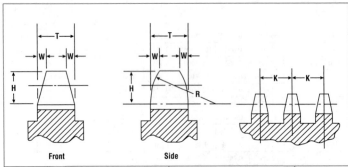

Chain Number	Tooth Section Dimensions			Sprocket Flange Thickness (Max.) T			Transverse Pitch for Multistrand Chains K
	Depth of Chamfer H	Width of Chamfer W	Maximum Radius R	Single Strand Chain	Two and Three Strands	Four or More Strands	
25	1/8	1/32	0.265	0.110	0.106	0.096	0.252
35	3/16	3/64	0.398	0.169	0.163	0.150	0.399
41	1/4	1/16	0.531	0.226	-	-	-
40	1/4	1/16	0.531	0.284	0.275	0.256	0.566
50	5/16	5/64	0.664	0.343	0.332	0.310	0.713
60	3/8	3/32	0.796	0.459	0.444	0.418	0.897

(Continued)

Table 2. *(Continued)* **Basic Roller Chain Sprocket Tooth Dimensions.**

Chain Number	Tooth Section Dimensions			Sprocket Flange Thickness (Max.) T			Transverse Pitch for Multistrand Chains K
	Depth of Chamfer H	Width of Chamfer W	Maximum Radius R	Single Strand Chain	Two and Three Strands	Four or More Strands	
80	1/2	1/8	1.062	0.575	0.556	0.526	1.153
100	5/8	5/32	1.327	0.692	0.669	0.633	1.408
120	3/4	3/16	1.593	0.924	0.894	0.848	1.789
140	7/8	7/32	1.858	0.924	0.894	0.848	1.924
160	1	1/4	2.124	1.156	1.119	1.063	2.305
180	1-1/8	9/32	2.392	1.302	1.259	1.198	2.592
200	1-1/4	5/16	2.654	1.389	1.344	1.278	2.817
240	1-1/2	3/8	3.187	1.738	1.682	1.602	3.458

All dimensions in inches.

values provided in **Table 3**—see the notes at the bottom of the Table for instructions. Pitch diameter values in this Table were obtained with the following formula, with p being the chain pitch, and N the number of teeth on the sprocket.

$$PD = p \div \sin (180° \div N).$$

The formula for finding the outside diameter, when the pitch (p) and the number of teeth on the sprocket are known, is:

$$OD = p \times (0.6 + \cot [180° \div N]).$$

The maximum hub diameter of the sprocket can be found with the following formula.

$$HD_{max} = p \, (\cot [180° \div N] - 1) - 0.030.$$

Formulas for obtaining the bottom diameters of a sprocket with an even number of teeth, and the caliper diameter of a sprocket with an odd number of teeth, are provided as a note at the end of the Table.

Chain lengths can be determined by multiplying the number of links in a chain by its pitch. Length in feet is equal to number of links times pitch divided by 12. **Table 1** includes linear measurement in feet per length, and links per foot, for each standard chain size.

The length of chain required, measured in pitches, for the center distance, also in pitches, between two sprockets can be obtained with the following formula. It should be noted that the tolerance on the value for L_P provided by the formula is ± one-half pitch. In addition, an even number of pitches is recommended, as an odd number of pitches will require an offset link in the chain, which can reduce its strength.

$$L_P = 2C_P + ([N + n] \div 2) + ([\{N - n\} \div 2\pi]^2 \div C_P)$$

where L_P = length of chain, in number of pitches (any fraction of L_P is considered one pitch)

C_P = center distance between two sprockets, in number of pitches (preferred C_P is 30 to 50 pitches, and the minimum is one-half the sum of the outside diameters of both sprockets)

N = number of teeth on large sprocket

n = number of teeth on small sprocket

Table 3. Roller Chain Sprocket Diameters for 1-inch Pitch Roller Chain. (See Notes.)

No. Teeth	Pitch Diameter	Outside Diameter	No. Teeth	Pitch Diameter	Outside Diameter	No. Teeth	Pitch Diameter	Outside Diameter
6	2.0000	2.332	46	14.6536	15.219	86	27.3807	27.962
7	2.3048	2.677	47	14.9717	15.538	87	27.6989	28.281
8	2.6131	3.014	48	15.2898	15.857	88	28.0171	28.599
9	2.9238	3.347	49	15.6079	16.176	89	28.3354	29.918
10	3.2361	3.678	50	15.9260	16.495	90	28.6536	29.236
11	3.5495	4.006	51	16.2441	16.813	91	28.9718	29.554
12	3.8637	4.332	52	16.5619	17.132	92	29.2900	29.873
13	4.1785	4.657	53	16.8803	17.451	93	29.6082	30.191
14	4.4940	4.982	54	17.1984	17.769	94	29.9264	30.510
15	4.8097	5.305	55	17.5166	18.088	95	30.2446	30.828
16	5.1259	5.627	56	17.8347	18.406	96	30.5628	31.146
17	5.4423	5.950	57	18.1529	18.725	97	30.8811	31.465
18	5.7588	6.271	58	18.4710	19.044	98	31.1994	31.783
19	6.0756	6.593	59	18.7892	19.363	99	31.5177	32.102
20	6.3925	6.914	60	19.1073	19.681	100	31.8360	32.420
21	6.7095	7.235	61	19.4255	20.000	101	32.1543	32.739
22	7.0266	7.555	62	19.7437	20.318	102	32.4726	33.057
23	7.3439	7.875	63	20.0618	20.637	103	32.7909	33.376
24	7.6613	8.196	64	20.3800	20.955	104	33.1091	33.694
25	7.9787	8.516	65	20.6982	21.274	105	33.4274	34.012
26	8.2962	8.836	66	21.0164	21.593	106	33.7457	34.331
27	8.6138	9.156	67	21.3346	21.911	107	34.0640	34.649
28	8.9315	9.475	68	21.6528	22.230	108	34.3823	34.968
29	9.2491	9.795	69	21.9710	22.548	109	34.7006	35.286
30	9.5668	10.114	70	22.2892	22.867	110	35.0189	35.605
31	9.8845	10.434	71	22.6074	23.185	111	35.3371	35.923
32	10.2023	10.753	72	22.9256	23.504	112	35.655	36.241
33	10.5201	11.072	73	23.2438	23.822	113	35.974	36.560
34	10.8380	11.392	74	23.5620	24.141	114	36.292	36.878
35	11.1558	11.711	75	23.8802	24.459	115	36.610	37.197
36	11.4737	12.030	76	24.1984	24.778	116	36.929	37.515
37	11.7917	12.349	77	24.5166	25.096	117	37.247	37.833
38	12.1096	12.668	78	24.8349	25.415	118	37.565	38.152
39	12.4275	12.987	79	25.1531	25.733	119	37.883	38.470
40	12.7455	13.306	80	25.4713	26.052	120	38.201	38.788
41	13.0635	13.625	81	25.7895	26.370	121	38.519	39.106
42	13.3815	13.944	82	26.1078	26.689	122	38.837	39.425
43	13.6995	14.263	83	26.4260	27.007	123	39.156	39.743
44	14.0175	14.582	84	26.7443	27.326	124	39.475	40.062
45	14.3356	14.901	85	27.0625	27.644	125	39.794	40.381

(Continued)

Table 3. *(Continued)* **Roller Chain Sprocket Diameters for 1-inch Pitch Roller Chain.** (See Notes.)

Notes. For pitches other than one inch, multiply the values on the Table by the required pitch. To obtain bottom diameters for sprockets with even number of teeth, subtract the diameter of the chain roller from the pitch diameter. To obtain bottom caliper diameters for sprockets with an odd number of teeth, use the following formula: Pitch diameter × cos (90° ÷ Number of teeth on sprocket) – roller diameter.

To establish an approximate center distance in pitches given a chain length in pitches and the number of teeth in both sprockets, use the following formula. When using this formula, it is recommended that an even number of chain pitches be used in order to avoid the use of an offset link.

$$C_P = 1p \div 8 \, (2L_P - N - n + \sqrt{[2L_P - N - n]^2 - 0.811 \, [N - n]^2}).$$

The previous two formulas refer to the necessity of adding an offset, or connecting link, if an odd number of pitches are contained in a chain. Although the terms offset and connecting link are sometimes used interchangeably, a connecting link is actually a detachable pin link that may be either a press or slip fit. Connecting links are secured by spring or cotter pins, and may be found on roller chains as a device allowing the length of the chain to be modified. An offset link, on the other hand, can be used on a chain with an odd number of pitches to give it an even number. A two pitch offset link is a combination roller link and offset link connected with a riveted pin that allows a connecting link to be attached to either end. A one pitch offset link can only be slip fitted, and therefore may well become the weakest link in the chain. One pitch links should be confined to applications where lighter loads, lower speeds, and minimal impact loads will be encountered.

Horsepower ratings for roller chains take several factors into consideration, including link plate fatigue, roller impact fatigue, bushing impact fatigue, and galling. The standard reference for horsepower ratings is ANSI B29.1, which grew out of data provided by the American Chain Association in 1966. A selection of these ratings for ANSI pitch chains is provided in **Table 4**. The values in the Table are for single strand chains in a smooth running application. To convert to more severe applications, and multiple strand chains, the following formula is used, which is followed by the necessary drive service factors (F) and multiple strand factors (F_S).

Required HP rating = (hp × F) ÷ F_S

where   hp = horsepower to be transmitted (from **Table 4**)

Drive service factors (F):

Type of Driven Load	Electric Motor or Turbine	Internal Combustion Engine, with Hydraulic Drive	Internal Combustion Engine, without Hydraulic Drive
Smooth	1.0	1.0	1.2
Some Impact	1.3	1.2	1.4
Large Impact	1.5	1.4	1.7

Multiple strand factors (F_S), with the number of roller chain strands followed by the factor in italics: 2 strands = *1.7*; 3 strands = *2.5*; 4 strands = *3.3*; 5 strands = *3.9*; and 6 strands = *4.6*.

Example: Find the number of desired teeth on the small sprocket for a double strand No. 50 (5/8 inch pitch) roller chain transmitting 28 horsepower at 1200 rpm. The chain operates on a hydraulic IC engine that experiences moderate impact. From the drive service factor table, it can be seen that F is 1.2, and the multiple strand factor F_S is 1.7. Therefore, the formula is:

Required HP rating = (28 × 1.2) ÷ 1.7 = 19.76.

From **Table 4**, it can be seen that a small sprocket with 26 teeth, operating at 1200 rpm, has a rating of 19.9 and is therefore the best candidate for this application.

The Table also provides the recommended method of lubrication for given horsepower and rpm. The shaded areas on the Table indicate that bath or slinger disk lubrication is recommended. Values to the left of the shaded area, which are printed on a white background, denote that the lubrication method should be manual or drip. Values to the right of the shaded area, again printed on white, denote that oil stream lubrication is recommended.

After determining the number of teeth required on the small sprocket for a given horsepower and chain pitch from **Table 4**, it is advisable to consult **Table 3** to confirm that the diameter of the necessary sprocket will meet the requirements of the application. If satisfactory, the number of teeth needed on the large sprocket can be found by a desired speed ratio, which should be a maximum of 7:1 under normal operating conditions. (The speed ratio is determined by one of three methods: dividing the pitch diameter of the large sprocket by the pitch diameter of the small sprocket; dividing the number of teeth on the large sprocket by the number of teeth on the small sprocket; or by dividing the speed in rpm of the small socket by the speed of the large socket.) Chain speed in ft/min can be found with the following formula.

$$S = (n_{rpm} \, \pi \, d) \div 12$$

where n_{rpm} = small sprocket rpm
d = small sprocket diameter (inches).

Given the speed of the chain and the horsepower to be transmitted, the suggested chain tension in lbs can be calculated.

$$T = (hp \times 33000) \div S$$

Generally, roller chains are installed with some slack. While too much slack can induce vibration that can be damaging, slack amounting to 4% of the center span between the sprockets is considered adequate. However, slack should be limited to 2% in the following circumstances: when the center distance between sprockets exceeds $3 \, ^1/_3$ feet; when severe loads and repeated starting are encountered; when the transmission of power is vertical rather than horizontal; and in installations where sudden reversal of direction is encountered. Since chains elongate up to 0.1% of their length after being in operation for some time, it is desirable to check it periodically, or rely on a tensioner to eliminate vibration.

Dimensions of BS/ISO roller chain are provided in **Table 5**.

Silent chain

Sometimes called an "inverted tooth" chain, silent chain is used for power transmission and conveying applications. Sprocket dimensions, pitches, and other design specifications are to ANSI standard B29.2M, which should be consulted for further information. Widths vary dramatically from $^1/_2$ inch to 30 inches. Some of these chains are capable of running at speeds up to 7,000 ft/min, and the more robust versions have a horsepower capacity of up to 2,000 HP.

Contrasted to roller chains, proponents of silent chains claim that in addition to quieter operation, they provide higher operating speeds, smoother power transmission, higher efficiency, and longer sprocket life.

Table 4. Horsepower Ratings of Standard Pitch Single Strand Roller Chains.

No. of Teeth in Small Sprocket	Small Sprocket Revolutions per Minute							
	500	1200	3000	4000	5000	6000	7000	8000
	1/4 inch Pitch No. 25 Roller Chain							
11	0.23	0.50	1.15	1.38	0.99	0.75	0.60	0.49
12	0.25	0.55	1.26	1.57	1.12	0.86	0.68	0.56
13	0.27	0.60	1.38	1.77	1.27	0.96	0.77	0.63
14	0.30	0.65	1.49	1.93	1.42	1.08	0.86	0.70
15	0.32	0.70	1.61	2.08	1.57	1.20	0.95	0.78
16	0.34	0.76	1.72	2.23	1.73	1.32	1.05	0.86
17	0.37	0.81	1.84	2.38	1.90	1.44	1.14	0.94
18	0.39	0.86	1.96	2.63	2.07	1.57	1.25	1.02
19	0.41	0.91	2.07	2.69	2.24	1.70	1.35	1.11
20	0.44	0.96	2.19	2.84	2.42	1.84	1.46	1.20
21	0.46	1.01	2.31	2.99	2.60	1.98	1.57	1.29
22	0.48	1.07	2.43	3.15	2.79	2.12	1.69	1.38
23	0.51	1.12	2.55	3.30	2.98	2.27	1.80	1.47
24	0.53	1.17	2.67	3.46	3.18	2.42	1.92	1.57
25	0.56	1.22	2.79	3.61	3.38	2.57	2.04	1.67
26	0.58	1.28	2.91	3.77	3.59	2.73	2.17	1.77
28	0.63	1.38	3.15	4.09	4.01	3.05	2.42	1.98
30	0.68	1.49	3.40	4.40	4.45	3.38	2.68	2.20
32	0.73	1.60	3.64	4.72	4.90	3.73	2.96	2.42
35	0.80	1.76	4.01	5.20	5.60	4.26	3.38	2.77
40	0.92	2.03	4.64	6.00	6.85	5.21	4.13	3.38
45	1.05	2.31	5.26	6.82	8.17	6.21	4.93	4.04
	3/8 inch Pitch No. 35 Roller Chain							
11	0.77	1.70	2.94	1.91	1.37	1.04	0.82	0.67
12	0.85	1.87	3.35	2.17	1.56	1.18	0.94	0.77
13	0.93	2.04	3.77	2.45	1.75	1.33	1.06	0.87
14	1.01	2.21	4.22	2.74	1.96	1.49	1.18	0.97
15	1.08	2.38	4.68	3.04	2.17	1.65	1.31	1.07
16	1.16	2.55	5.16	3.35	2.40	1.82	1.45	1.18
17	1.24	2.73	5.64	3.67	2.62	2.00	1.58	1.30
18	1.32	2.90	6.16	3.99	2.86	2.17	1.73	1.41
19	1.40	3.07	6.67	4.33	3.10	2.36	1.87	1.53
20	1.48	3.25	7.20	4.68	3.35	2.55	2.02	1.65
21	1.56	3.42	7.75	5.03	3.60	2.74	2.17	1.78
22	1.64	3.60	8.21	5.40	3.86	2.94	2.33	1.91

Notes: The shaded area on the Table indicates that oil bath or slinger disk lubrication is recommended. Values to the left of the shaded area denote that these applications are less severe, and lubrication can be done either manually or by oil drip. Values to the right of the shaded area are for more severe applications, and oil stream lubrication is recommended.

(Continued)

Table 4. *(Continued)* Horsepower Ratings of Standard Pitch Single Strand Roller Chains.

No. of Teeth in Small Sprocket	Small Sprocket Revolutions per Minute							
	500	1200	3000	4000	5000	6000	7000	8000
	3/8 inch Pitch No. 35 Roller Chain							
23	1.72	3.78	8.62	5.77	4.13	3.14	2.49	2.04
24	1.80	3.96	9.02	6.15	4.40	3.35	2.66	2.17
25	1.88	4.13	9.43	6.54	4.68	3.56	2.82	2.31
26	1.96	4.31	9.84	6.93	4.96	3.77	3.00	2.45
28	2.12	4.67	10.7	7.75	5.55	4.22	3.35	2.74
30	2.29	5.03	11.6	8.59	6.15	4.68	3.71	3.04
32	2.45	5.40	12.3	9.47	6.77	5.15	4.09	3.35
35	2.70	5.95	13.6	10.8	7.75	5.90	4.68	3.83
40	3.12	6.87	15.7	13.2	9.47	7.20	5.72	4.68
45	3.55	7.80	17.8	15.8	11.3	8.59	6.82	-

No. of Teeth in Small Sprocket	Small Sprocket Revolutions per Minute							
	100	200	500	1200	3000	4000	5000	6000
	1/2 inch Pitch No. 41 Lightweight Roller Chain							
11	0.24	0.44	1.01	1.71	0.43	0.28	0.20	0.15
12	0.26	1.49	1.11	1.95	0.49	0.32	0.23	0.17
13	0.28	0.53	1.21	2.20	0.56	0.36	0.26	0.20
14	0.31	0.57	1.31	2.46	0.62	0.40	0.29	0.22
15	0.33	0.62	1.41	2.73	0.69	0.45	0.32	0.24
16	0.36	0.66	1.51	3.01	0.76	0.49	0.35	0.27
17	0.38	0.71	1.61	3.29	0.83	0.54	0.39	0.29
18	0.40	0.75	1.72	3.59	0.91	0.59	0.42	0.32
19	0.43	0.80	1.82	3.89	0.98	0.64	0.46	0.35
20	0.45	0.84	1.92	4.20	1.06	0.69	0.49	0.38
21	0.48	0.89	2.03	4.46	1.14	0.74	0.53	0.40
22	0.50	0.93	2.13	4.69	1.23	0.80	0.57	0.43
23	0.53	0.98	2.24	4.92	1.31	0.85	0.61	0.46
24	0.55	1.03	2.34	5.15	1.40	0.91	0.65	0.49
25	0.57	1.07	2.45	5.38	1.49	0.96	0.69	0.53
26	0.60	1.12	2.55	5.61	1.58	1.02	0.73	0.56
28	0.65	1.21	2.77	6.08	1.76	1.14	0.82	0.62
30	0.70	1.31	2.98	6.55	1.95	1.27	0.91	0.69
32	0.75	1.40	3.20	7.03	2.15	1.40	1.00	-
35	0.83	1.54	3.52	7.74	2.46	1.60	1.14	-
40	0.96	1.78	4.07	8.94	3.01	1.95	1.40	-
45	1.08	2.02	4.62	10.20	3.59	2.33	-	-

Notes: The shaded area on the Table indicates that oil bath or slinger disk lubrication is recommended. Values to the left of the shaded area denote that these applications are less severe, and lubrication can be done either manually or by oil drip. Values to the right of the shaded area are for more severe applications, and oil stream lubrication is recommended.

(Continued)

Table 4. *(Continued)* Horsepower Ratings of Standard Pitch Single Strand Roller Chains.

No. of Teeth in Small Sprocket	Small Sprocket Revolutions per Minute							
	100	200	500	1200	3000	4000	5000	6000
	1/2 inch Pitch No. 40 Roller Chain							
11	0.43	0.80	1.83	4.63	2.17	1.41	1.01	0.77
12	0.47	0.88	2.01	5.09	2.47	1.60	1.15	0.87
13	0.52	0.96	2.20	5.55	2.79	1.81	1.29	0.98
14	0.56	1.04	2.38	6.01	3.11	2.02	1.45	1.10
15	0.60	1.12	2.66	6.47	3.46	2.24	1.60	1.22
16	0.65	1.20	2.75	6.94	3.80	2.47	1.77	1.34
17	0.69	1.29	2.93	7.41	4.17	2.71	1.94	1.47
18	0.73	1.37	3.12	7.88	4.54	2.95	2.11	1.60
19	0.78	1.45	3.31	8.36	4.92	3.20	2.29	1.74
20	0.82	1.53	3.50	8.83	5.31	3.45	2.47	1.88
21	0.87	1.62	3.69	9.31	5.72	3.71	2.66	2.02
22	0.91	1.70	3.88	9.79	6.13	3.98	2.85	2.17
23	0.96	1.78	4.07	10.30	6.55	4.26	3.05	2.32
24	1.00	1.87	4.26	10.80	6.99	4.54	3.25	2.47
25	1.05	1.95	4.46	11.20	7.43	4.82	3.45	2.63
26	1.09	2.04	4.64	11.70	7.88	5.12	3.66	2.79
28	1.18	2.20	5.03	12.70	8.80	5.72	4.09	3.11
30	1.27	2.38	5.42	13.70	9.76	6.34	4.54	3.45
32	1.36	2.55	5.81	14.70	10.80	6.99	5.00	-
35	1.50	2.81	6.40	16.20	12.30	7.99	5.76	-
40	1.74	3.24	7.39	18.70	15.00	9.76	6.99	-
45	1.97	3.68	8.40	21.20	17.90	11.70	-	-

No. of Teeth in Small Sprocket	Small Sprocket Revolutions per Minute							
	100	500	900	1200	1800	3000	3500	4000
	5/8 inch Pitch No. 50 Roller Chain							
11	0.84	3.57	6.06	7.85	5.58	2.59	2.06	1.68
12	0.92	3.92	6.65	8.62	6.35	2.95	2.34	1.92
13	1.00	4.27	7.25	9.40	7.16	3.33	2.64	3.16
14	1.09	4.63	7.86	10.20	8.01	3.72	2.95	2.42
15	1.17	4.99	8.47	11.00	8.88	4.13	3.27	2.68
16	1.26	5.35	9.08	11.80	9.78	4.56	3.61	2.95
17	1.34	5.71	9.69	12.60	10.70	4.98	3.95	3.23
18	1.43	6.07	10.30	13.40	11.70	5.42	4.30	3.52
19	1.51	6.44	10.90	14.20	12.70	5.88	4.67	3.82
20	1.60	6.80	11.50	15.00	13.70	6.35	5.04	4.13

Notes: The shaded area on the Table indicates that oil bath or slinger disk lubrication is recommended. Values to the left of the shaded area denote that these applications are less severe, and lubrication can be done either manually or by oil drip. Values to the right of the shaded area are for more severe applications, and oil stream lubrication is recommended.

(Continued)

Table 4. *(Continued)* **Horsepower Ratings of Standard Pitch Single Strand Roller Chains.**

No. of Teeth in Small Sprocket	Small Sprocket Revolutions per Minute							
	100	500	900	1200	1800	3000	3500	4000
	5/8 inch Pitch No. 50 Roller Chain							
21	1.69	7.17	12.20	15.80	14.70	6.84	5.42	4.44
22	1.77	7.54	12.80	16.60	15.80	7.39	5.82	4.76
23	1.86	7.91	13.40	17.40	16.90	7.84	6.22	5.09
24	1.95	8.29	14.10	18.20	18.00	8.35	6.33	5.42
25	2.03	8.66	14.70	19.00	19.10	8.88	7.05	5.77
26	2.12	9.03	15.30	19.90	20.30	9.42	7.47	6.12
28	2.30	9.79	16.60	21.50	22.60	10.50	8.35	6.84
30	2.49	10.50	17.90	23.20	25.10	11.70	9.26	7.58
32	2.66	11.30	19.20	24.90	27.70	12.90	10.20	8.35
35	2.93	12.50	21.10	27.40	31.60	14.70	11.70	9.55
40	3.38	14.40	24.40	31.60	38.70	18.00	14.30	-
45	3.84	16.30	27.70	35.90	46.10	21.40	-	-
	3/4 inch Pitch No. 60 Roller Chain							
11	1.44	6.13	10.40	11.90	6.45	3.00	2.38	1.95
12	1.58	6.74	11.40	13.50	7.35	3.42	2.71	2.22
13	1.73	7.34	12.50	15.20	8.29	3.85	3.06	2.50
14	1.87	7.96	13.50	17.00	9.26	4.31	3.42	2.80
15	2.01	8.57	14.50	18.80	10.30	4.77	3.79	3.10
16	2.16	9.19	15.60	20.20	11.30	5.26	4.17	3.42
17	2.31	9.81	16.70	21.60	12.40	5.76	4.57	3.74
18	2.45	10.40	17.70	22.90	13.50	6.28	4.98	4.08
19	2.60	11.10	18.80	24.30	14.60	6.81	5.40	4.42
20	2.75	11.70	19.80	25.70	15.80	7.35	5.83	-
21	2.90	12.30	20.90	27.10	17.00	7.91	6.28	-
22	3.05	13.00	22.00	28.50	18.20	8.48	6.73	-
23	3.19	13.60	23.10	29.90	19.50	9.07	7.19	-
24	3.35	14.20	24.10	31.30	20.80	9.66	7.67	-
25	3.50	14.90	25.30	32.70	22.10	10.30	8.15	-
26	3.65	15.50	26.40	34.10	23.40	10.90	8.65	-
28	3.95	16.80	28.50	37.00	26.20	12.20	-	-
30	4.26	18.10	30.80	39.80	29.10	13.50	-	-
32	4.56	19.40	33.00	42.70	32.00	14.90	-	-
35	5.03	21.40	36.30	47.10	36.60	17.0	-	-
40	5.81	23.70	42.00	54.40	44.70	-	-	-
45	6.60	28.10	47.70	61.70	53.40	-	-	-

Notes: The shaded area on the Table indicates that oil bath or slinger disk lubrication is recommended. Values to the left of the shaded area denote that these applications are less severe, and lubrication can be done either manually or by oil drip. Values to the right of the shaded area are for more severe applications, and oil stream lubrication is recommended.

(Continued)

Table 4. *(Continued)* **Horsepower Ratings of Standard Pitch Single Strand Roller Chains.**

No. of Teeth in Small Sprocket	Small Sprocket Revolutions per Minute							
	50	100	300	500	1000	1200	1400	2000
	1 inch Pitch No. 80 Roller Chain							
11	1.80	3.36	9.04	14.30	19.60	14.90	11.80	6.93
12	1.98	3.69	9.93	15.70	22.30	17.00	13.50	7.90
13	2.16	4.03	10.80	17.10	25.20	19.20	15.20	8.91
14	2.34	4.36	11.70	18.60	28.20	21.40	17.00	9.96
15	2.52	4.70	12.60	20.00	31.20	23.80	18.90	11.00
16	2.70	5.04	13.60	21.50	34.40	26.20	20.80	12.20
17	2.88	5.38	14.60	22.90	37.70	28.70	22.70	13.30
18	3.07	5.72	15.40	24.40	41.10	31.20	24.80	14.50
19	3.25	6.07	16.30	25.80	44.50	33.90	26.90	15.70
20	3.44	6.41	17.20	27.30	48.10	36.60	29.00	17.00
21	3.62	6.76	18.20	28.80	51.70	39.40	31.20	18.30
22	3.81	7.11	19.10	30.30	55.50	42.20	33.50	19.60
23	4.00	7.46	20.10	31.80	59.30	45.10	35.80	21.00
24	4.19	7.81	21.00	33.20	62.00	48.10	38.20	22.30
25	4.37	8.16	21.90	34.70	64.80	51.10	40.60	23.80
26	4.58	8.52	22.90	36.20	67.60	54.20	43.00	25.20
28	4.94	9.23	24.80	39.30	73.30	60.60	48.10	28.20
30	5.33	9.94	26.70	42.30	78.90	67.20	53.30	31.20
32	5.71	10.70	28.60	45.40	84.60	74.00	58.70	34.40
35	6.29	11.70	31.60	50.00	93.30	84.70	67.20	39.40
40	7.27	13.60	36.40	57.70	108	103	82.10	48.10
45	8.25	15.40	41.40	65.60	122	123	98.00	54.10
	1 1/4 inch Pitch No. 100 Roller Chain							
11	3.45	6.44	17.30	27.40	23.40	17.80	14.20	8.29
12	3.79	7.08	19.00	30.10	26.70	20.30	16.10	9.45
13	4.13	7.72	20.70	32.80	30.10	22.90	18.20	10.60
14	4.48	8.38	22.50	35.80	33.70	26.60	20.30	11.90
15	4.83	9.01	24.20	38.30	37.30	28.40	22.50	13.20
16	5.17	9.66	26.00	41.10	41.10	31.30	24.80	14.60
17	5.52	10.30	27.70	43.90	45.00	34.30	27.20	15.90
18	5.88	11.00	29.50	46.70	49.10	37.30	29.60	17.40
19	6.23	11.60	31.20	49.50	53.20	40.50	32.10	18.80
20	6.58	12.30	33.00	52.30	57.50	43.70	34.70	20.30
21	6.94	13.00	34.80	55.10	61.80	47.00	37.30	21.90
22	7.30	13.60	36.60	58.00	66.30	50.40	40.00	23.40
23	7.66	14.30	38.40	60.80	70.90	53.90	42.80	25.10

Notes: The shaded area on the Table indicates that oil bath or slinger disk lubrication is recommended. Values to the left of the shaded area denote that these applications are less severe, and lubrication can be done either manually or by oil drip. Values to the right of the shaded area are for more severe applications, and oil stream lubrication is recommended.

(Continued)

Table 4. *(Continued)* **Horsepower Ratings of Standard Pitch Single Strand Roller Chains.**

No. of Teeth in Small Sprocket	Small Sprocket Revolutions per Minute							
	50	100	300	500	1000	1200	1400	2000
	1 1/4 inch Pitch No. 100 Roller Chain							
24	8.02	15.00	40.20	63.70	75.60	57.50	45.60	26.70
25	8.38	15.80	42.00	66.60	80.30	61.10	48.50	28.40
26	8.74	16.30	43.80	69.40	85.20	64.80	51.40	30.10
28	9.47	17.70	47.50	75.20	95.20	72.40	57.50	33.70
30	10.20	19.00	51.20	81.00	106	80.30	63.70	10.00
32	10.90	20.40	54.90	86.90	116	88.50	70.20	-
35	12.00	22.50	60.40	95.70	133	101	80.30	-
40	13.90	26.00	69.80	111	163	124	98.10	-
45	15.80	29.50	79.30	126	194	148	117	-

Notes: The shaded area on the Table indicates that oil bath or slinger disk lubrication is recommended. Values to the left of the shaded area denote that these applications are less severe, and lubrication can be done either manually or by oil drip. Values to the right of the shaded area are for more severe applications, and oil stream lubrication is recommended.

Table 5. Dimensions of Single and Double Strand European Standard (BS 228, ISO 606, DIN 8187) Roller Chain.

Chain Number		Pitch	Inside Width	Roller Dia.	Pin Dia.	Plate Height	Plate Thickness	
Single Chain	Double Chain						Outside Pl.	Inside Pl.
		P	W		D	H	T	S
06B	06B-2	0.375	0.225	0.250	0.129	0.325	0.050	0.040
08B	08B-2	0.500	0.305	0.335	0.175	0.465	0.060	0.060
10B	10B-2	0.625	0.380	0.400	0.200	0.580	0063	0.063
12B	12B-2	0.750	0.460	0.475	0.225	0.635	0.071	0.071
16B	16B-2	1.000	0.670	0.625	0.326	0.827	0.157	0.122
20B	20B-2	1.250	0.770	0.750	0.401	1.032	0.177	0.138
24B	24B-2	1.500	1.000	1.000	0.576	1.307	0.217	0.197
28B	28B-2	1.750	1.220	1.100	0.626	1.456	0.276	0.236
32B	32B-2	2.000	1.220	1.150	0.701	1.614	0.276	0.236
40B	40B-2	2.500	1.500	1.550	0.901	2.083	0.346	0.312
48B	48B-2	3.000	1.800	1.900	1.151	2.432	0.488	0.456

Chain Number		Single Strand Chain			Double Strand Chain		
		Pin Length L		Min. Breaking Load	Pin Length L_2		Min. Breaking Load
Single Chain	Double Chain	Riveted	Cottered		Riveted	Cottered	
		LR	LC	lb	LR	LC	lb
06B	06B-2	0.531	0.661	2,050	0.937	1.067	3,600
08B	08B-2	0.670	0.823	4,100	1.220	1.374	7,200
10B	10B-2	0.772	0.933	5,110	1.425	1.587	9,000

(Continued)

Table 5. *(Continued)* **Dimensions of Single and Double Strand European Standard (BS 228, ISO 606, DIN 8187) Roller Chain.**

Chain Number		Single Strand Chain			Double Strand Chain		
		Pin Length L		Min. Breaking Load	Pin Length L_2		Min. Breaking Load
Single Chain	Double Chain	Riveted	Cottered		Riveted	Cottered	
		LR	LC	lb	LR	LC	lb
12B	12B-2	0.894	1.075	6,640	1.661	1.842	12,000
16B	16B-2	1.421	1.634	13,050	2.677	2.890	28,850
20B	20B-2	1.701	1.941	21,400	3.110	3.378	38,250
24B	24B-2	2.102	2.290	38,200	3.976	4.267	63,000
28B	28B-2	2.563	2.854	45,000	4.882	5.201	81,000
32B	32B-2	2.654	2.965	58,500	4.961	5.272	101,250
40B	40B-2	3.090	3.319	79,800	6.063	6.363	141,750
48B	48B-2	3.772	4.071	126,000	7.874	8.214	225,000

For identification of dimensional symbols, see **Table 1**. Dimensions, except where indicated, are in inches.

Bearings

Bearing terminology

Aligning ball or roller thrust bearing. A bearing that, by virtue of the shape of the seat washer, is capable of sustaining considerable misalignment.

Annular bearing. An antifriction bearing primarily designed to support load perpendicular to the shaft axis.

Ball bearings. Antifriction bearings that use balls as a rolling element. Examples of deep groove, angular contact, and thrust ball bearings are shown in *Figure 1*.

Ball retainer. Also called the "ball cage," this is the element of a ball bearing assembly that separates the balls and maintains their symmetrical radial spacing. They are made from a variety of materials including steel, phenolic, Teflon, and bronze.

Cam follower ball bearings. Special service ball bearings with an extra heavy outer ring.

Cylindrical roller bearings. A bearing in which the roller surface is perpendicular to the bearing axis.

DN value. Bearings are rated for many parameters. The DN value is a practical guide to judging if a bearing is capable of the speed needed for a particular application. DN = bore (mm) × speed (rpm). Typically, single-row ball bearings have a maximum DN value of 200,000 with grease lubrication and 300,000 with oil. Double-row ball bearings have a maximum value of 160,000 on grease, 220,000 on oil. Cylindrical roller bearings have a maximum value of 150,000 on grease, 200,000 on oil. Spherical roller bearings have a maximum value of 120,000 on grease, 170,000 on oil.

Double row ball or roller bearing. A bearing with two rows of rolling elements. This configuration allows for higher radial and thrust loads while increasing the width of the bearing from 60 – 80%.

Duplex ball bearings. Two single row angular contact bearings selected dimensionally to be a matched pair or set.

Flanged bearing. Generally descriptive of antifriction bearings with outer ring or cup flanged on outside diameter.

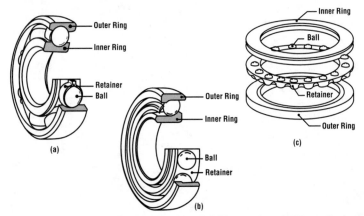

Figure 1. Deep groove (a), angular contact (b), and thrust (c) ball bearings. (Source, NTN Bearing Corp.)

Floating bearing. A bearing designed or mounted to permit axial displacement between shaft and housing.

Journal bearings. These bearings are of simple construction with a flanged or unflanged sleeve or bushing acting as the outer member, and rotation occurring between the sleeve and a structure connecting shaft or pin. They can be used without lubrication in static joints and provide an economical, easily replaced design element.

Maximum roller bearing. A cageless roller bearing with maximum number of rollers.

Miniature ball bearing. A ball bearing with outside diameter dimensions that are below and not including $3/8$" or 9 mm.

Multirow ball or roller bearing. A bearing with more than two rows of rolling elements.

Needle roller bearings. A radial needle roller bearing is a radial cylindrical roller bearing having a large ratio of pitch diameter to roller diameter and a large ratio of roller length to roller diameter. A typical needle bearing is shown in *Figure 2.*

Pillow block bearings. These bearings are sometimes used for mounting equipment and mechanisms. They consist of a bolt-down housing with a variety of types of cartridge bearings.

Plain bearings. Also called "sliding surface bearings," these include all types of bearings in which surfaces slide relative to one another, whether separated by a lubricant or in contact. Included are cylindrical sleeves, or bushings, used in static joint applications. Plain bearings are generally divided into four classes, based on design geometry and application: spherical, cylindrical, rod end, and special bearings.

Preload. Preload commonly refers to an internal loading characteristic in a bearing which is dependent of any external radial or axial load carried by the bearing.

Pure radial load. The load resulting from a single force acting through the center of the bearing at right angles to the bearing axis. Also see *radial load.*

Pure thrust load. The load resulting from a single force applied to the direction coaxial with the bearing axis.

Race. The inner or outer ring of a cylindrical or needle roller bearing.

Raceway depth. In a ball bearing, the vertical dimension from the bottom of the raceway to inner ring outside diameter, or outer ring inside diameter.

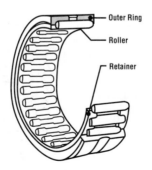

Figure 2. A typical needle roller bearing. (Source, NTN Bearing Corp.)

Radial bearings. Rolling element bearings that are primarily designed to support loads perpendicular to the shaft axis. Generally, radial bearings have a contact angle of less than 45°.

Radial internal clearance. The radial internal clearance of a single row radial contact ball bearing is the average outer ring raceway diameter, minus the average inner ring raceway diameter, minus twice the ball diameter. The raceway diameters are taken at the bottom of the raceway. The clearance is also called "radial play."

Radial load. The load that may result from a single force, or the "resultant" of several forces acting in a direction at right angles to the bearing axis.

Rod end bearings. These bearings are intended for use on push-pull rods as end fittings to permit functional attachments to mechanisms and structures. They consist of a self-aligning bearing in housing with an integral shank for attachment to the rod.

Roller bearings. Bearings that use rollers (not balls) as rolling elements. They are classified according to the shape of their rollers as either cylindrical, needle, taper, or spherical. Typical roller bearings are shown in *Figure 3*.

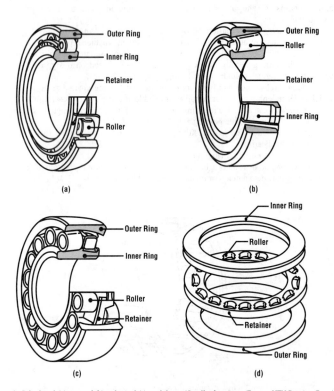

Figure 3. Cylindrical (a), tapered (b), spherical (c), and thrust (d) roller bearings. (Source, NTN Bearing Corp.)

Self-aligning ball or roller radial bearing. A bearing that, by virtue of the raceway or outer ring construction, is capable of considerable misalignment. Examples are spherical radial roller bearings and spherical roller thrust bearings.

Spherical bearings. A bearing composed of a spherical shaped piece with a hole for a shaft and employed as an inner race and spherically concave single piece outer member conforming to the shape of the inner race. The outer member may be permanently swaged around the outer portion of the inner member, or slotted to accept the inner member. *Metal-to metal* spherical bearings require lubrication, while *lined type* spherical bearings use sacrificial liners. *Figure 3* includes an example of a spherical roller bearing.

Straight roller bearing. A roller bearing of the radial type having cylindrical rollers.

Thrust bearings. Thrust bearings are rolling element bearings that have a contact area greater than 45° and have higher axial load capacity than radial bearings.

Bearing selection–ball, roller, and needle bearings

Bearing selection is based on many parameters, as shown in **Table 1**. Given the application of the bearing, the characteristics of each bearing type can be examined as follows in order to make a selection.

Friction and torque

As can be seen in **Table 1**, the running coefficient of friction for antifriction bearings is considered to be as follows. Ball bearings: 0.0015. Needle bearings: 0.0025. Roller bearings: 0.002. Sliding element bearings: 0.03 – 0.20. The coefficient of friction of a bearing is an indirect measure of the bearing's resistance to rotation. As such, the designer needs certain information on bearing internal geometry to make necessary calculations. This can be done most directly by calculating torsional resistance, or torque.

At normal temperatures bearings introduce relatively little frictional torque. The running torque of a grease lubricated antifriction bearing under load at moderate temperature can be estimated with the following formula:

$$T = \mu_f \times r \times F_r$$

where T = torque (inch pounds)
   μ_f = running coefficient of friction (see above)
   r = radius of bearing bore (inches)
   F_r = radial load (pounds).

Ball bearings

Because the point of contact where the ball meets the race is smaller in ball than in tapered roller bearings, it is normally necessary to use a larger ball bearing than tapered roller bearing for a given load capacity. The small contact area offers lower frictional resistance, which often makes them a first choice for high-speed applications with low torque and vibration requirements. They are most often used where size, load capacity, and radial capacity are not as important as low cost and ease of assembly.

"Deep groove" ball bearings (also known as "Conrad" type) have permanently fixed races and are provided fully assembled. Consequently, the radial clearance is set and cannot be adjusted. "Angular contact" ball bearings will accept low amounts of side thrust, and, compared to similar size deep groove bearings, load ratings are higher. "Thrust" ball bearings carry axial loads exclusively.

Static load capacity. Static load capacity is concerned with the strength of a bearing and its ability to resist significant deformation and fracture. The radial static load rating for a ball bearing is based on the following formula:

$$L_s = k \times n \times D^2$$

Table 1. Bearing Selection Parameters. *(Source, MIL-HDBK-1599A.)*

Operating Parameter	Rolling Element Bearings			Sliding Element Bearings	
	Ball Type	Needle Type	Roller Type	TFE Lined	Metal-to-Metal
Load Capacity					
(For a given envelope size)	Low	Medium	Medium	Medium	High
Radial (Static)	Low	Medium	Medium	Medium	High
Axial (Static)	Low	Medium	Low	Medium	High
Shock	Low	Medium	Medium (Preloading Helps)	Medium	High
Vibration	Rapid Fretting	Moderate Fretting	Rapid Fretting (Preloading Helps)	Medium	High
Type of Operation					
Small Oscillation	Possible Fretting	Best of Rolling Element	Possible	Handle Well	Reversed Loading Required
Oscillation	Good	Good	Good	Good	Reversed Loading Required
Slow Rotation	Good	Good	Good	Limited by Surface Speed	Excellent
Speed					
High Speed Oscillation (Continuous)	Excellent with Cages	Excellent with Cages	Excellent with Cages	Excellent with Small Angles	Poor
High Speed Oscillation (Intermittent)	Excellent	Excellent	Excellent	Excellent with Small Angles	Requires Average Loading
High Speed Rotation (Continuous)	Excellent with Cages	Not Intended	Excellent with Cages	Do Not Use	Not Recommended
High Speed Rotation (Intermittent)	Excellent with Cages	Not Intended	Excellent with Cages	Not Recommended	Not Recommended
High Acceleration	Requires Preload	Not Intended	Requires Preload	Good	Good
Duty Cycle	No Effect	No Effect	No Effect	May Cause Heat	May Cause Heat
Temperature					
Continuous	Depends on Seals and Material. Range -65° to 350° F (-54° to 177° C)	Depends on Seals and Material. Range -65° to 350° F (-54° to177° C)	Depends on Seals and Material. Range -65° to 350° F (-54° to 177° C)	-65° to 350° F	Depends on Material and Lube
Maximum				375° F (190° C)	Depends on Material
Minimum				Increase in Friction	Good
Misalignment	Decrease in Capacity	Cannot Tolerate	Good	Good	Good
Coefficient of Friction	0.0015	0.0025	0.002	0.03 to 0.10	0.2

where L_s = radial static load (pounds)
k = design factor (see below)
n = number of balls
D = ball diameter (inch).

Applicable k factors for calculating radial static load ratings for ball bearings are as follows. Deep groove bearing: 10,000. Single row self-aligning bearing: 4,800. Double row self-aligning bearing: 3,800. Rod end bearings: 3,200. The static load limit can be applied to the bearing for a short period of time without affecting the smooth operation or endurance under normal loads. The minimum static fracture load (where an actual breakage of the bearing occurs) is not less than 1.5 times the static load. Axial static capacity varies from approximately 50 – 60% of radial capacity for nonself-aligning ball bearings, and 13 – 20% for self-aligning types. A bearing should not be expected to take full radial and axial limit loads simultaneously.

Dynamic load capacity. The basic dynamic load rating is an expression of the load capacity that a bearing can sustain through one million revolutions. It is an essential calculation in estimating bearing life.

The basic dynamic capacity of a ball bearing is the constant radial load at which 90% of a group of identical bearings running at the same speed, temperature, and angle of oscillation will meet or exceed a fatigue life of 2,000 revolutions. An oscillatory cycle, through an angle of at least the ball spacing, is considered equivalent to a revolution. If a bearing life of more than 2,000 cycles or revolutions is required, the applied constant radial load must be reduced to a value below the basic dynamic capacity of the bearing. The dynamic capacity of the bearing is related to the cycle life by the following formula:

$$C = C_o \div L$$

where C = dynamic load capacity required
C_o = basic dynamic capacity at L_{10} = 2,000 revs. (average life of 10,000 cycles)
L = life factor, based on the formula

$$L = (n \div 2000)^{1/3.6}$$

where n = desired bearing life (number of cycles). See *Figure 4.*

The basic dynamic capacity is based on the inner race moving and the outer race stationary. If the reverse is true, the basic dynamic capacity should be reduced by dividing by a factor of 1.20.

In instances where the magnitude of the load applied to the bearing varies greatly at different speeds or cycles, a constant mean load can be computed to permit proper sizing of the bearing. The following rules apply.

If load magnitude varies less than 30%, use the higher load value.

If greater variation in load magnitude occurs, use an estimated load somewhat higher than the arithmetic average, taking into account the relative percentage of total life at each significant load level.

To obtain a more exact figure, the variable dynamic load magnitudes can be stepped off to a number of incremental load values associated with a finite proportion of total bearing life and used in the following equation:

$$P_e = ([N_1(P_1)^{3.6} + N_2(P_2)^{3.6} + \ldots N_n(P_n)^{3.6}] \div N)^{1/3.6}$$

where P_e = equivalent dynamic load to result in the same life as the variable loads
$N = N_1 + N_2 + \ldots N_n$ = total cycles or revolutions

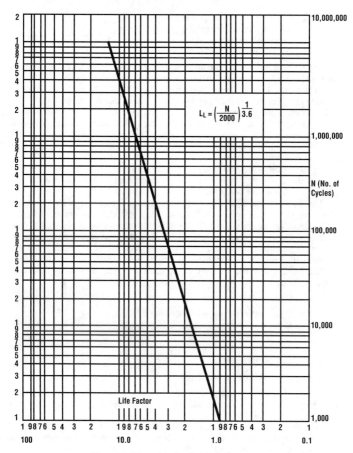

$$L_L = \left(\frac{N}{2000}\right)^{\frac{1}{3.6}}$$

Figure 4. Life factor (L) values for ball bearings.

P_1 = load acting during N_1 revolution
P_2 = load acting during N_2 revolution
P_n = load acting during N_n revolution.

Another instance where obtaining an equivalent load value becomes necessary is when a bearing is subjected to both axial and radial loads. It is then necessary to convert these loads into an equivalent load. An approximation for combining both radial and thrust loads for either static or dynamic conditions can be obtained with the following formula:

$$P_e = F_r + Y F_t$$

where P_e = equivalent load
F_r = applied radial load
F_t = applied thrust (axial) load
Y = thrust factor = radial load rating ÷ thrust load rating.

If the applied thrust load does not exceed 10% of the applied radial load, the equivalent load should be equal to the applied radial load. The calculated equivalent load must never be higher than the bearing rating in either the static or dynamic condition.

Roller bearings

Roller bearings have a larger load carrying capacity than ball bearings of similar size, and are often used in applications where long life and endurance, even when encountering shock loads, are essential. Roller bearings may be tapered, cylindrical, spherical, or needle in shape (needle bearings are covered in more detail below), and may contain single or multiple rows.

While roller bearings are not capable of the ultimate speeds of ball bearings, standard tapered roller bearings can operate at speeds up to about 6,000 ft/min (30 m/s), and, in some configurations, can achieve speeds of over 40,000 ft/min (200 m/s). Tapered roller bearings are a popular choice for many applications as their design allows them to operate with both radial and axial loads, and also provide minimal bearing size for a given load capacity.

Cylindrical roller bearings are commonly specified when it is necessary for the shaft to "float" axially in relation to the housing. The length-to-diameter ratio for the rollers ranges from 1:1 to 1:3.

Spherical roller bearings are self-aligning and have good radial load capacity, but under very heavy loads they do not have the inherent ability to distribute stress evenly, which limits their speed capability.

Static load capacity. The basic static load capacity of a roller bearing is termed the limit load rating. This is the maximum load that the bearing can carry without significantly affecting performance to the rated life at the dynamic load rating. The static load rating of both single- and double-row concave and convex bearings is given by the following formula:

$$L_s = n \times D \times l \times \cos \alpha$$

where L_s = static load rating (pounds)
n = number of rollers
D = mean roller diameter (inch)
l = roller contact length (inch)
α = roller inclination angle to the bearing bore axis.

The axial load capacity of roller bearings ranges from 10% of the radial capacity for single row bearings, to 72% for some wide series double-row bearings. The ultimate load rating is defined as the maximum load that can be applied and held for three minutes without structural failure of the bearing. In actual practice, brinelling will occur on the race surface if subjected to a load equal to the ultimate load rating.

Dynamic load capacity. The dynamic load rating of a roller bearing is defined on the basis of a unidirectional load that will result in an average bearing life (L_{10}) of 2,000 cycles at 90° oscillation before evidence of contact fatigue (spalling) occurs. Average bearing life (L_{10}) signifies that in a given group of bearings under test, 90% are expected to survive their hourly rating. For the test, the angle of oscillation is defined at 180° of angular travel within an included angle of 90°. In most cases, the applied (dynamic) operating load is much less than the rated load for the bearing. The load/life equation below was developed by ISO to calculate bearing life at any load. It should be noted that most bearing manufacturers use a modified, and more rigorous, version of this formula for rating their bearings.

$$L_{10} = (C \div P)^{10/3} \times (10^6 \div 60)$$

where L_{10} = rated life, in hours
C = radial rating of bearing, in pound-force or Newtons
P = applied operating load, or equivalent radial load in pound-force or Newtons.

Since bearing applications rarely involve a constant, unidirectional load, either radial or axial, the equivalent mean load should be calculated to aid in bearing selection. The following procedure can be used to determine equivalent loads when applied loads are of varying magnitude and direction.

First, the equivalent radial dynamic load can be calculated with the formula:

$$P_e = F_r + 2.5 \, F_a$$

where P_e = equivalent radial dynamic load
F_r = applied maximum radial dynamic load
F_a = applied maximum axial dynamic load.

This formula combines the thrust and radial loads. Single-row bearings can carry momentary or intermittent thrust loads, but are not designed for continuous thrust. Therefore, for single-row bearings, $P_e = F_r$. Multiple operating bearings should be considered separately.

Next, the equivalent radial dynamic load at 90° oscillations can be determined.

$$P = K_o \, P_o$$

where P = equivalent dynamic radial load at 90° oscillation
K_o = oscillation factor (from *Figure 5*)
P_o = equivalent radial dynamic load representing the various loads in the spectrum (using the following formula)
$$P_o = (f_1[P_1]^{3.67} + f_2[P_2]^{3.67} + \ldots f_n[P_n]^{3.67})^{.273}$$

where $f_1, f_2, \ldots f_n$ = number of cycles at one condition divided by the total spectrum cycles

$P_1, P_2, \ldots P_n$ = individual equivalent radial dynamic load representing one condition in the spectrum.

This equation is used only when various load levels exist. If only one dynamic load level occurs, then P becomes P_e.

The required radial dynamic load rating, assuming an average life of L_{10}, is then calculated:

$$C \geq (P_e \times [C_y \div 2{,}000])^{.273}$$

where C = radial dynamic load rating required
C_y = total cycles required.

A bearing can now be selected, based on the dynamic load rating required. The load limit on the bearing selected must be equal to or larger than the maximum applied static loads. Both radial and axial limit load ratings must be considered.

Needle bearings

While actually a form of roller bearing, needle bearings combine a very low profile cross section with a high radial load capacity. Generally speaking, to be classified as a

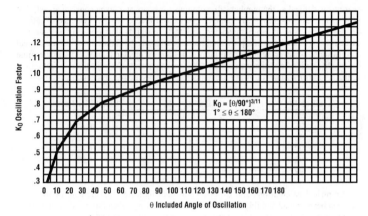

Figure 5. Oscillation angle factors for roller bearings.

needle bearing, the roller length must be at least four times its diameter. Their ability to sustain axial loads is low, however.

Static load capacity. A load that damages the bearing is considered the ultimate load, and it can be calculated with the following formula. It should be noted that this formula was developed for aerospace applications where maximum reliability is a minimum requirement. Hence, the *ASC* in the formula stands for "aircraft static capacity."

$$ASC = 1200 \times i \times (z - 3) \times l \times D$$

where ASC = aircraft static capacity (pound)
        i = number of rows of rollers
        z = number of rollers per row
        l = minimum roller contact (inch)
        D = minimum roller diameter (inch).

The above formula does not apply to "drawn cup" style extra-light duty needle bearings as defined by MS24462. For those bearings, the formula is as follows.

$$ASC = 15800 \times dm \times l$$

where dm = pitch diameter of the roller complement (inch).

Dynamic load capacity. With adequate lubrication, the mode of failure for needle bearings is considered to be surface fatigue. The basic dynamic rating, C_r, for a housed needle bearing is the calculated, constant radial load that a group of apparently identical bearings with a stationary outer ring can theoretically endure for a rating life of one million revolutions of the inner ring. The basic dynamic load rating is a reference value only, the base value of one million revolutions rating life having been chosen for ease of calculation. Since applied loading as great as the basic dynamic load rating tends to cause plastic deformation of the rolling surfaces, it is not anticipated that such heavy loading would normally be applied. The life of a housed needle bearing is expressed as the oscillations or cycles that a bearing will complete before failure. Life varies, of course, to some degree from one bearing to another, but stabilizes into a predictable pattern when considering a large group of the same size and type of bearings. The L_{10} rated life of a group of such bearings is defined as the number of revolutions (or hours at a given constant speed) that 90% of the tested bearings will complete or exceed before the first evidence of fatigue develops. Therefore, it can be predicted with 90% reliability that a bearing will meet or exceed its calculated L_{10} life, providing normal fatigue is the failure mode. Some critical applications, however, require definition of reliability greater than 90%. To determine bearing life with reliabilities greater than 90%, the L_{10} life must be adjusted by a factor (a_1), so that $L_1 = a_1 L_{10}$. The following life adjustment factors are recommended.

Reliability %	L_n Rating Life	a_1 Reliability Factor
90	L_{10}	1
95	L_5	0.62
96	L_4	0.53
97	L_3	0.44
98	L_2	0.33
99	L_1	0.21

Empirical calculations and experimental data indicate a predictable relationship between bearing load and life. The life of a housed bearing, when failure is due to surface fatigue, can be determined from the basic dynamic load rating and the equivalent load. This relationship is expressed in the following formula, in which the bearing life is found to vary inversely as the applied load to an exponential power.

$$L_L = C_r \div P_e$$

where C_r = basic dynamic load rating
P_e = equivalent load
L_L = life factor (see *Figure 6*).

The above relationship holds for bearings with a race hardness of at least R_C 58. If the race hardness is less than R_C 58, the formula is:

$$L_L = C_r \div (P_e \times H_F)$$

where H_F = the hardness factor, as follows.

Raceway Hardness R_C	Hardness Factor	Raceway Hardness R_C	Hardness Factor
57	1.02	48	2.00
56	1.04	47	2.24
55	1.07	46	2.50
54	1.12	45	2.76
53	1.19	44	3.06
52	1.28	43	3.39
51	1.41	42	3.77
50	1.59	41	4.16
49	1.78	40	4.55

Equivalent load (P_e) calculations for needle bearings are necessary when the bearing is subject to various load levels during its duty cycle. It can be determined with the following formula:

$$P_e = (\Sigma f [F]^{10/3})^{3/10}$$

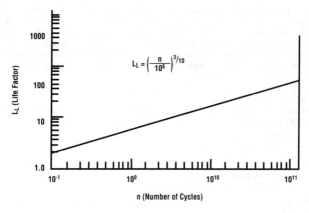

Figure 6. Life factors for housed needle bearings.

where f = proportion of total life at which particular load levels are applied
F = radial load level at F_1, F_2, F_n.

The equivalent load should never exceed the indicated load limit for the bearing.

Bearing selection–plain bearings

Friction and torque

The formula for calculating the torque of plain bearings is the same as for ball and roller bearings ($T = \mu_f \times r \times F_r$), which was given above. However, for spherical bearings turning on ball, r is equal to one-half ball diameter—for plain bearings and spherical bearings turning on bore, r equals one-half bore diameter. The friction coefficient value for metal-to-metal plain bearings, both spherical and journal, varies between 0.10 and 0.30, with the lower figure relating to reverse loading and considerable motion, and the higher value occurring with heavy, unidirectional loading and relatively slow motion. PTFE fabric lined bearings may be considered to have a coefficient of 0.03 to 0.10, with the coefficient decreasing with increases in temperature and load. If a design requires lower friction than is indicated above, it may be necessary to use special bearings with special variations in material, lubricants, and geometry.

PTFE lined plain spherical bearings

PTFE (or TFE, for polytetrafluoroethylene) lined plain spherical bearings have radial and axial static load limit definitions as follows.

The radial static limit load is the load that, when applied for two minutes to the bearing, will not cause a permanent set of more than 0.003" to the bearing.

The axial static load limit is the load that, when applied for two minutes to the bearing, will not cause a permanent set of more than 0.005" to the bearing.

The ultimate load is 1.5 times the radial or axial static load limit. When this load is applied, ball or race fracture, or ball push-out, will not occur.

Dynamic radial and axial loads require more stringent testing. Variables in these tests include type of loading (unidirectional or reversing), angle of oscillation, oscillation rate, temperature, and contamination. Bearing manufacturers have produced considerable test data resulting in modifying factors or curves that may be used to aid in bearing selection. However, since the effects of the variables on PTFE lined bearings are relatively intangible, both previous experience and performance play an important part in the selection process. An example of bearing selection follows.

Procedure. To select a bearing for L_N cycles of oscillatory life at a given oscillation angle (θ) and load (P).

To determine the equivalent life (L_E) of the bearing, the following formula (which is intended to show the relationship between the various factors that contribute to failure) can be used:

$$L_E = (L_N \times F_A \times F_C) \div K_\theta$$

where F_A = application factor (if more than one condition applies, see below)
F_C = contamination factor (see below)
K_θ = oscillation angle factor (from *Figure 7*).

Condition	F_A	Degree of Contamination	F_C
Steady load	1	None	1
Reversing load	2	Occasional	2
Vibration	3	Continuous	6
Impact load	4		
−65° F to 0°F, and			
175° F to 350° F	2		
Continuous misalignment	2		

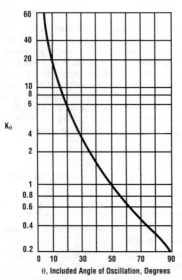

Figure 7. Oscillation angle factor (K_θ) for PTFE lined plain spherical bearings.

Similarly, the required rated dynamic capacity (P_R) of the bearing is determined from the following relationship.

$$P_R = P \div F_L$$

where F_L = load factor (see *Figure 8*).

When the design load varies during the bearing duty cycle, the equivalent load (P_e) can be calculated based on the following formula.

$$P_e = (\Sigma f[F]^{3.6})^{1/3.6}$$

where f = proportion of total life at which particular load levels are applied
        F = radial load level at F_1, F_2, F_n.

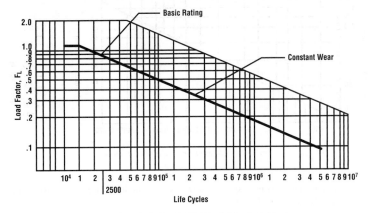

Figure 8. Load factor (F_L) values for PTFE lined plain spherical bearings.

Dry-film or grease lubricated plain spherical bearings

The static load capacity for these bearings is a load at which a specific amount of permanent set is not exceeded. The values selected are such that subsequent operation of the bearing is not adversely affected. Ultimate load capacity is 1.5 times the load limit capacity. At ultimate capacity, no fracture of the ball or race, or ball push-out, will occur.

Dry-film lubricated bearings have a high dynamic capacity of up to 50,000 PSI (345 MPa) unit loading, but life is difficult to predict due to wear through the thin film coating. Therefore, these bearings are best used in joints that are relatively static under load with only incidental rotation or misalignment motion. Bronze race bearings have been found to give essentially unlimited service life when properly lubricated and operating at 10 CPM maximum at loads up to 13,000 PSI (90 MPa). For shorter, finite life at higher loads, stress values given in *Figure 9* can be used with the bearing race projected area.

Bearings using beryllium copper balls are rated at 30,000 to 40,000 PSI (207 to 276 MPa), depending on whether rotation is at the spherical surface or in the bore, and whether the ball is clamped or free to rotate on the pin.

PTFE lined journal bearings

The *static limit* load capacity for these bearings is considered to be as follows.

$$T = T_{40} \times F_T$$

If T_{40} is not known

where
- T = Track capacity for a specific bearing, lb.
- T_{40} = Capacity of R_C = 40 track (standard MS sheets), lb.
- F_T = Track Strength factor from graph

where
- T = Track capacity for a specific bearing, lb.
- D = OD of track roller, In.
- L = Constant length of track roller, In.
- K = Constant, psi, from graph

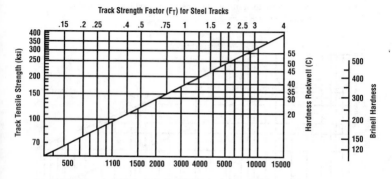

Figure 9. Strength factors for tracks of different hardness.

For aluminum bearings, static load (L_s) is equal to 50,000 times bore (B), times bearing length (L) minus 0.1, or

$$L_s = 50,000 \, B \, (L - 0.1).$$

For corrosion resistant steel bearings, static load (L_s) is equal to 78,500 times bore (B), times bearing length (L) minus 0.1, or

$$L_s = 78,500 \, B \, (L - 0.1).$$

Due to pin bending and edge loading, static capacity is not significantly improved by increasing bearing length beyond a dimension equal to bore diameter.

The *dynamic limit* load capacity is defined as follows.

Dynamic load (C_r) is equal to 37,500 times bore (B), times bearing length (L) minus 0.1, or

$$C_r = 37,500 \, B \, (L - 0.1).$$

Figure 10 is a nomogram relating bearing life, applied load, angle of oscillation, and the projected area of the bushing. It is an aid in sizing a bearing for a particular application. The performance shown in the figure will be obtained only with properly designed pin, shaft, or bolt as a mating member, and minimum hardness of R_C 40.

Metallic journal bearings

Metal journals are used primarily to handle static loads and can be loaded to values that are approximately one-half of the compressive yield strength of the material. Using

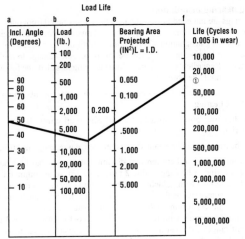

① **Example: A plain bearing with 0.3 IN² projected area, operating at ±25° oscillation and 5,000 lbs load, will give 26,000 cycles of life before 0.005 IN wear occurs**

Figure 10. Load-life characteristics for PTFE lined journal bearings.

the projected area of the bearing (length × diameter), and the following compressive yield strength values, the loaded capacity can be determined.

Material	Yield Strength		Temperature (Max.)	
	KSI	MPa	°F	°C
4130 Steel (180 ksi U.T.S.)	115	793	350	177
17-4 Steel (AMS 5643)	90	620	500	260
Beryllium Copper (QQ-C-530)	90	620	350	177
Al-Ni-Bronze (AMS 4640 & 488)	60	414	350	177
Al-Bronze (QQ-C-465)	40	275	350	177

Bearing cages

The rolling element retainer in a ball or roller bearing separates the rollers and enforces uniform spacing of the balls or rollers. These retainers are commonly referred to as cages, and their construction and material are influential to bearing performance. Cages are manufactured from a variety of materials, including bronze, steel, aluminum, and several synthetic materials.

Pressed steel cages are most commonly used for deep groove ball bearings, and they are light and compact. They are not, however, particularly suited for high speeds, especially when compared to molded phenolic laminate and reinforced nylon retainers that are often used in precision angular contact ball bearings. Cages made of these synthetic materials are capable of operating at temperatures up to 275 – 300° F (135 – 150° C), and they are quiet running. High performance synthetic cages made from materials such as porous polyimide and reinforced polyphenelyene sulfide are rated to temperatures of 450° F (230° C), and sometimes higher. Bronze cages are often preferred for very heavy-duty applications, and can be found in ball, cylindrical, spherical, and tapered roller bearings.

Rolling bearing installation

Bearings are intended to permit a load to be transmitted from one structural member to another, while at the same time permitting relative movement between the two members. Incorrect bearing installation and retention can result in a reduction of useful bearing life, an increase in bearing operating torque, or irreparable damage to the housing or bearing components. Regardless of the type of bearing being installed, the following preinstallation precautions should be observed.

The installation area should be clean, so that contaminants do not come into contact with the bearing.

The housing bore should be free of all metal chips, filings, and other foreign material.

The bearing should be retained in its protective wrapping until time of installation.

Galvanic corrosion between two dissimilar materials can be a source of damage to either the bearing or housing. **Table 2** provides a listing of common bearing and housing and shaft materials, with guidelines for preventing galvanic corrosion.

Another consideration that must be examined when selecting bearing and housing/shaft materials is the expansion coefficients of the materials involved. Care must be taken to avoid mating materials with radically different coefficients. **Table 3** provides expansion coefficient ranges for most materials likely to be encountered. In general, the materials grouped into the same class in the Table can be used together without expansion mismatch problems if the bearing diameters are small, if operating temperatures are not extreme, or if radial play requirements are not critical.

Table 2. Recommended Treatments to Prevent Galvanic Corrosion. *(Source, MIL-HDBK-1599A.)*

Bearing Material (Bore and OD Surface)	Housing or Shaft Material					
	Aluminum Alloy	Low Alloy Steel	Titanium	Corrosion Resistant Steel	Super Alloys	Refractory Metals
Aluminum Alloys	A	A, C	A	A	I	I
Bronze and Brass	I	I	G	S	S	I
Bronze and Brass Cadmium Plated	A	C	I	I	I	I
52100 and Low Alloy Steels	A, C	C	N	I	I	I
440C Stainless Steel	A	I	G	S	S	I
440C with Wet Primer	A	C	G	S	I	I
Corrosion Resistant Steels	A	I	G	S	S	I
300 Series Steels (17-4, etc.)	A	C	G	S	I	I
Superalloys-Rene (41, etc.)	I	I	G	S	S	O
Cermets (LT-2, TiC, etc.)	I	I	G	S	S	O
Ceramics Al_2O_3, ZrO_2	I	I	G	S	S	O

Legend for letter symbols.
I = Incompatible
A = Anodize aluminum per MIL-A-8625, type 1 or type 2
C = Cadmium plate low alloy steels per Fed-Spec QQ-P-416, type II, class 2
G = Use dry film on bearing OD and in titanium bore
N = Nickel plate per MIL-C-26074, grade C, class I or equivalent
O = Oxidation-resistant coating normally used will provide galvanic protection
S = Satisfactory for use with no surface treatment
Note: If plating or anodizing is required, it will be necessary to reestablish the housing tolerance.

Table 3. Expansion Classes of Bearing/Shaft/Housing Materials. *(Source, MIL-HDBK-1599A.)*

Bearing Materials	Shaft Materials	Housing Materials
Low Expansion Class (4.5 × 10⁻⁶ in./in./°F)		
Stainless Steels Tool Steels 400 Series Stainless Steel Low Alloy Steels Titanium Alloys	Stainless Steels Tool Steels 400 Series Stainless Steel Titanium Alloys	Stainless Steels 400 Series Stainless Steel Titanium Alloys
High Expansion Class (6.2 × 10⁻⁶ to 15 × 10⁻⁶ in./in./°F)		
A-286 Rene 41 Stellites, 3, 19, and 25 M-252 Waspaloy Inconels Other Superalloys Aluminum Magnesium	A-286 Rene 41 Other Superalloys Inconels	A-286 Rene 41 Inconels Aluminum Magnesium
Copper Based Alloys - Bearing Material Only		
Beryllium Copper 9.5 × 10⁻⁶ in./in./°F Aluminum Bronze 9.5 × 10⁻⁶ in./in./°F Aluminum Nickel Bronze 9.0 × 10⁻⁶ in./in./°F Aluminum Silicon Bronze 10.0 × 10⁻⁶ in./in./°F		

Care must also be taken in selecting rolling element materials that vary in expansion coefficient from the race materials, as a change in bearing radial play will become evident as the temperature is raised or lowered. The usefulness of the interference fit method for mounting bearings to be used in extreme temperatures depends on the difference between the expansion coefficients of the bearing and the shaft or housing. The practical maximum that can be tolerated is a difference in expansion coefficient of approximately 1.0×10^{-6} in./in./°F. This expansion difference could introduce a relief of 0.0005 in./in. of diameter in a press fit at an increase in operating temperature of 600° F, and 0.0018 in./in. of diameter at a change in operating temperature of 1,900° F.

Figure 11 shows the change in mounting tolerances introduced by different changes in operating temperature. The figure shows, for example, that if the difference between the housing/shaft and bearing coefficient is equal to 6.0×10^{-6} in./in./°F, and the increase or decrease in operating temperature is 1,000° F, then the gap or interference between the housing/shaft and the bearing will increase or decrease by 0.006 in./in. of diameter. At elevated or cryogenic temperatures, the expansion coefficients are instrumental in bearing selection.

Proper fit of a rolling element bearing is critical to the life of the bearing. Most bearings have sufficient internal clearance to permit an interference on either shaft or housing. However, an interference fit in both shaft and housing may induce too much preload into the bearing. A slight amount of preload can improve bearing performance, as it tends to improve the load distribution within the bearing. Too much preload, on the other hand, can reduce bearing life. In most applications, bearings are customarily mounted with a clearance fit between the shaft or pin (which remain stationary) and the bore of the inner ring. The outer ring must fit closely with the housing (which revolves), and the bearing is retained in the axial direction by mechanical means. In most instances, the fit with the housing is a hand push, or light tapping fit.

The quality and predictability of bearing fit into a housing or onto a shaft is

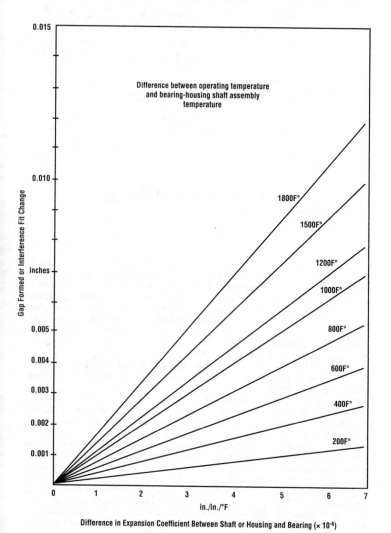

Figure 11. Effects of expansion coefficient on the fit between housing/shaft and bearing.

directly influenced by the tolerances of the bearing. The AFBMA (Antifriction Bearing Manufacturer's Association) has standardized tolerances for five classes of ball bearings (ABEC-1, ABEC-3, ABEC-5, ABEC-7, and ABEC-9), and three classes of cylindrical roller bearings (RBEC-1, RBEC-3, and RBEC-5) in ANSI/AFBMA Std 20. The lower number classes have the largest tolerance ranges, which progressively grow tighter in the higher numbered classes. In addition to ball bearing tolerances for bore, outside diameter, radial runout, and ring width found in all classes, ABEC-5, -7, and -9 also include tolerances for side runout and parallelism of the bearing sides. While ANSI/AFBMA Std. 20 tolerances are widely referenced in catalogs published by bearing manufacturers, there are several other international standards that are widely adhered to. JIS B 1514 and DIN 620 cover all types of bearings, ISO 492 for radial bearings, and ISO 199 for thrust ball bearings all have the same specification level, but are close to, but not in conformance with, ANSI/AFBMA Std. 20. In addition, these international standards assign their own class numbers, which are compared—to the extent possible—in **Table 4**.

The base classes ABEC-1 and RBEC-1 should be used in situations where high precision resulting from extremely close tolerances is not a requirement. ABEC-3 and RBEC-3 offer slightly closer tolerances, but, in general, the two base classes are best reserved for applications with light to normal load requirements. For precision fit and higher load requirements, ABEC-7 tolerances are normally recommended. Many manufacturers label their bearings that conform to ABEC-7 as "superprecision" bearings. When installing these bearings, the shaft size and tolerance can be considered to be the same as the bearing bore. For example, if the bore size is 2.1654 "max" and 2.1652 "min," then the shaft

Table 4. Standards for Bearing Tolerances, with Classes Equated to ANSI/AFBMA Std. 20. (See Notes.)

ANSI/AFBMA Std. 20 Tolerance Class ➤	ABEC-1 RBEC-1	ABEC-3 RBEC-3	ABEC-5 RBEC-5	ABEC-7	ABEC-9
▼ **International Standard** ▼	**Nearest Equivalent Class of International Standards**				
JIS B 1514 (All Types of Bearings)	Class 0 Class 6X	Class 6	Class 5	Class 4	Class 2
DIN 620 (All Types of Bearings)	P0	P6	P5	P4	P2
ISO 199 (Thrust Ball Bearings)	"Normal"	Class 6	Class 5	Class 4	Class 2
ISO 492 (Radial Bearings)	"Normal" Class 6X	Class 6	Class 5	Class 4	Class 2
ISO 578 (Inch Tapered Roller Bearings)	Class 4	-	Class 3	Class 0	Class 00
ISO 1224 (Instrument Bearings)	-	-	Class 5A	Class 4A	-
▼ **Other ANSI/AFBMA Standards** ▼	**Nearest Equivalent Class**				
ANSI/AFBMA Std. 21.1 (Metric Ball Instrument Bearings) ANSI/AFBMA Std. 12.2 (Inch Ball Instrument Bearings)	-	Class 3P	Class 5P Class 5T	Class 7P Class 7T	Class 9P
ANSI B 3.19 and AFBMA Std. 19 (Inch Tapered Roller Bearings)	Class 4	Class 2	Class 3	Class 0	Class 00
ANSI/AFBMA Std. 19.1 (Metric Tapered Roller Bearings)	Class K	Class N	Class C	Class B	Class A

Notes: ANSI/AFBMA Std. 20 provides tolerances for all radial bearings except tapered roller bearings and precision instrument bearings. This Table is intended to show *nearest equivalent classes* or several standards—it should not be assumed that nearest equivalent classes in different standards will be dimensionally identical. The actual Standard or the manufacturer's dimensional data should be consulted. This Table was compiled from information published by NTN Bearing Corporation.

diameter size should have the same maximum and minimum dimensions. The resulting bore to shaft mounting fit will be either 0.0002" tight or loose, or "line-to-line" if both shaft and bore are mutually ad maximum or minimum dimension. The minimum housing bore dimensions for ABEC-7 bearings should be the same as the maximum outside diameter of the bearing. If, for example, the maximum outside diameter tolerance is 0.0003", then the maximum bore diameter should be 0.003" larger than the minimum bore diameter. Again using 0.003" as an example, this will result in the resulting outside diameter to housing fit being loose at 0.0006", or tight at 0.0000", and the "average" fit will be 0.003" loose. As would be expected, increased interference is recommended for heavy bearing loads, or in conditions where shocks or vibration are to be expected.

Other bearing standards. ANSI/AFBMA Std. 18.1 provides tolerance limits for metric series drawn cup (without inner ring) needle roller bearings, Types NB, NBM, NY, NYM, NH, and NHM. ANSI/AFBMA Std. 18.2 provides tolerance limits for inch series drawn cup (without inner ring) needle roller bearings, Types NIB, NIBM, NIY, NIYM, NIH, and NIHM. The same standard covers cage, machined ring type NIA, and inner-ring type NIR. ANSI/AFBMA Std. 24.1 provides tolerance limits for metric series single direction thrust ball (Type TA) and Roller (Type TS) bearings. ANSI/AFBMA Std. 24.2 provides tolerance limits for cylindrical roller thrust bearings, and categorizes them into four classes: Type TP, in three series: extra light, light, and heavy, and Type TPC.

Lubrication of rolling bearings

Lubricants prevent metal-to-metal contact in bearings by providing a thin film of oil or grease between the contact surfaces. In addition to reducing friction and wear, a properly selected lubricant will also accelerate heat dissipation and prevent rust and corrosion.

Greases

Grease is by far the most commonly used lubricant for rolling bearings, primarily because application is simple and normally lasts for extended periods, and because only rudimentary sealing devices are required to retain the lubricant in the bearing. Greases contain a thickener, typically a lithium, sodium, or calcium based soap, or a nonsoap based thickener such as bentone, silica, polyurea, or carbon black, plus a base oil. The oil may be either a natural base mineral oil, or a synthetic lubricant such as Diester or silicone oil. The base oil is the most important ingredient, because it is largely responsible for the grease's lubricating characteristics. Low viscosity base oils will produce a lubricant that provides optimal performance in low temperatures, and also allows higher bearing speeds. High viscosity bases have advantages in high load applications.

Although a mineral and synthetic grease may share similar lubricating qualities, greases with different bases should not be mixed. If a synthetic lubricant is replacing a natural lubricant in a bearing, the bearing should be thoroughly cleaned before the substitution is made. Similarly, it is advisable not to mix greases from different manufacturers, even if the base oil is identical in both, because the different additives they contain may not be compatible. Mixing greases can cause the grease to become softer, and the temperature range of the resultant mixed grease may be significantly altered.

When lubricating a bearing, it is generally advisable to fill the housing to 30 – 60% of capacity, favoring the lesser amount in applications where high speeds are required and increases in temperature need to be kept to a minimum. Overfilling the housing can cause temperature rise that may result in softening of the grease and, ultimately, leakage of the lubricant.

Oil

For high speed and high temperature applications, oil lubrication is generally superior to grease but, unfortunately, oil can only be used in very controlled environments because it cannot be sealed inside the bearing but must be introduced through some mechanical process. Common varieties of oil lubrication include oil bath, splash, drip, mist, jet, and circulating.

Oil baths work well when speeds are low to moderate. When at rest, the oil level for bearings on horizontal shafts should be maintained at the center of the lowest bearing element. For vertical shaft applications, the elements can be submerged to 50 – 80% of their width. Multirow bearings are, therefore, unsuitable for vertical applications.

Oil splash requires a shaft mounted impeller or similar appendage to gather oil from a reservoir and distribute it onto the surface of the bearing. This method is effective for high speed applications but requires considerable engineering expertise to design a pick-up system that will distribute a manageable splash or spray throughout the speed range.

Drip lubrication offers a comparatively simple solution for continuous lubrication of bearings operating at relatively high speeds under moderate load. An oiling mechanism must be placed above the bearing and regulated to emit an intermittent drip of oil. When the dripping oil strikes the rotating bearing, it turns to mist.

Oil mist lubrication is similar to oil drip, but the mist is generated by pressurizing and atomizing the oil before it reaches the bearing. It requires a small amount of oil and the mist can be directed to several bearings at once. This is a very effective medium for high speed applications, but since it requires a pressurizing device for implementation, the system is bulky and can be too expensive except for very rigorous applications.

Oil jet sprays a continuous stream of oil directly onto the bearing. For severe operating conditions, it offers the most reliable and predictable means of lubrication. Since it requires pressure, it is only used where necessary, such as in lubricating main bearings in jet engines and gas turbines.

Circulating lubrication allows for oil to circulate through several bearings and return to a central location for cleaning and cooling. Since it is both automatic and continuous, it provides reliable lubrication for most conditions.

Viscosity

Oils are commonly referred to by their viscosity rating. While there are a great many other factors in choosing a lubricating oil, including the additive content, viscosity rating is an indication of a lubricant's ability to resist a shearing force, which is an indication of its ability to resist a certain level of force before experiencing breakdown. In fact, viscosity is a measurement of a fluid's resistance to flow due to molecular cohesion.

Dynamic viscosity (also called absolute viscosity) is usually expressed in $lb\text{-}s/ft^2$ in Customary U.S. units, and gram/cm-s, which is known as a "poise," in Metric units. One poise is equal to 0.1 kg/m-s. A centipoise is equal to one-hundredth of a poise. Dynamic viscosity (μ) is the ratio of the viscous stress to viscous strain rate.

$$\mu = \tau \div (d_v \div d_y)$$

where τ = viscous resistance per unit area
d_v = velocity change in given distance
d_y = distance, perpendicular to direction of flow.

Kinematic viscosity, on the other hand, is not a measurement of a property, but a ratio between the dynamic viscosity of a fluid and its specific gravity. The term used to express kinematic viscosity is the "stoke," which is equal to a cm^2/s, and is usually rated in centistokes, which is one-hundredth of a stoke. One stoke is equal to 10^{-4} m^2/s. Kinematic viscosity (v) is determined by the following formula:

$$v = \mu \div p$$

where p = specific gravity of the fluid.

Viscometers determine the viscosity of a fluid by measuring the time needed for a specific volume to flow through a capillary tube at a specific temperature. In the U.S., the Saybolt Universal Viscometer is most commonly used for measuring low to medium viscosities, which are expressed as SSU (Seconds Saybolt Universal) units. High viscosities are normally measured by the Saybolt Furol Viscometer, which measures in SSF (Seconds Saybolt Furol) units. In the U.K., the Redwood viscometer is frequently used, and readings are given in Redwood No. 1 Seconds for low and moderate viscosities, and Redwood No. 2 Seconds for more viscous fluids. **Table 5** is a conversion chart that can be used to equate values expressed in Centistokes, SSU, SSF, and Redwood No. 1 Seconds. **Table 6** provides viscosities for several lubricants, in SSU and Centistokes.

Table 5. Viscosity Conversion Table.

Seconds Saybolt Universal SSU	Seconds Saybolt Furol SSF	Kinematic Viscosity Centistokes cSt	Seconds Redwood 1	Seconds Saybolt Universal SSU	Seconds Saybolt Furol SSF	Kinematic Viscosity Centistokes cSt	Seconds Redwood 1
31	-	1.00	29	400	41.90	87.6	353
35	-	2.56	32.1	450	46.8	97.4	397
40	-	4.30	36.2	500	51.60	109	441
45	-	5.90	40.6	550	56.50	119	485
50	-	7.40	44.7	600	61.40	131	529
55	-	8.90	49.1	650	66.20	141	573
60	-	10.3	53.9	700	71.10	153	617
65	-	11.8	57.9	750	76.0	163	661
70	12.95	13.1	62.3	800	81.0	174	705
75	13.20	14.5	67.6	850	86.0	184	749
80	13.70	15.7	71.0	900	91.0	196	793
85	14.10	17.0	75.1	950	95.8	206	837
90	14.44	18.2	79.6	1000	100.70	218	882
100	15.24	20.6	88.4	1200	120	260	1058
120	16.59	25.0	105.9	1400	140	302	1234
140	17.95	29.8	123.6	1600	160	347	1411
150	19.30	32.1	132.4	1800	180	390	1537
160	20.70	34.3	141.1	2000	200	436	1763
180	22.10	38.8	158.8	2500	250	545	2204
200	23.50	43.2	176.4	3000	300	655	2646
220	25.0	47.5	194.0	4000	400	872	3526
240	26.50	51.9	212	5000	500	1090	4408
250	28.0	54.0	220	6000	600	1310	5290
260	29.50	56.5	229	7000	700	1525	6171
280	31.0	60.5	247	8000	800	1750	7053
300	32.50	64.9	265	9000	900	1970	7934
350	37.20	75.8	309	10000	1000	2200	8816

Table 6. Viscosity Ratings of Selected Fluids.

Liquid	Temperature		Viscosity	
	°F	°C	SSU	Centisokes
Automotive Lubrication Oil, Paraffin Base				
SAE 10W	0	-17.8	5000 to 10000	1100 to 2200
SAE 20W	0	-17.8	10000 to 40000	2200 to 8800
SAE 10	100	37.8	165 to 240	35.4 to 51.9
	130	54.4	90 to 120	18.2 to 25.3
SAE 20	100	37.8	240 to 400	51.9 to 86.6
	130	54.4	120 to 185	25.3 to 39.9
SAE 30	100	37.8	400 to 580	86.8 to 125.5
	130	54.4	185 to 255	39.9 to 55.1
SAE 40	100	37.8	580 to 950	125 to 206.6
	130	54.4	255 to 800	55.1 to 173.2
SAE 50	100	37.8	950 to 1600	205.6 to 352
	210	98.9	80 to 105	15.6 to 21.6
SAE 60	100	37.8	1600 to 2300	352 to 507
	210	98.9	105 to 125	21.6 to 26.2
Automotive Transmission Oil				
SAE 80	0	-17.8	100000 max.	22000 max.
SAE 90	100	37.8	800 to 1500	173.2 to 324.7
	130	54.4	300 to 500	64.5 to 108.2
SAE 140	130	54.4	950 to 2300	205.6 to 507
	210	98.9	120 to 200	25.1 to 42.9
Machine Lubricating Oil, Paraffin Base				
Fed. Spec. No. 8	100	37.8	112 to 160	23.4 to 34.3
	130	54.4	70 to 90	13.1 to 18.4
Fed. Spec. No. 10	100	37.8	160 to 235	34.3 to 50.8
	130	54.4	90 to 120	18.2 to 25.3
Fed. Spec. No. 20	100	37.8	235 to 385	50.8 to 83.4
	130	54.4	120 to 185	25.3 to 39.9
Fed. Spec. No. 30	100	37.8	385 to 550	83.4 to 119
	130	54.4	185 to 255	39.9 to 55.1

Based on figures published by the Hydraulic Institute in the Pipe Friction Manual, 1961.

Retaining Rings

Spiral retaining rings

Spiral retaining rings are used to provide continuous uniform retaining shoulders for positioning and retaining machine components on shafts. They are available in many configurations, but the most widely used were established and specified in MIL-R-27426, which provides material requirements and specifications for both internal and external medium and heavy duty rings. Spiral rings are available in various steel grades and several super alloys. **Table 1** provides engineering information on many of the more common materials.

Load capacity calculations

Calculating allowable loads requires an examination of both ring shear and groove deformation (both in pounds), and the design limitation will necessarily be the lesser of the two. Thrust load capacity of the ring can be calculated with the following formula, given the following conditions: 1) square corners at retained part and groove; 2) minimum clearance between retained part and shaft/housing; and 3) static loading conditions.

$$P_S = (D_O \times t \times S_S \times \pi) \div k$$

where P_S = Allowable thrust load capacity (lbs), based on shear strength of material
D_O = Shaft or housing diameter, in inches
t = Ring thickness, in inches
S_S = Shear strength of ring material, in PSI
k = Factor of safety.

The factor of safety (k) can be either 2 or 3. In order to provide a conservative estimate, most ring manufacturers assign a factor of 3 when providing ratings for their rings. Shear strength (S_S) for most materials can be taken from **Table 1**. It should be noted that ring shear is most often a potential problem when the groove material is hardened steel. Ring shear capacity (based on ring material shear strength of 120,000 PSI and a safety factor of 3) is provided in **Tables 3** through **6**.

The capacity of the groove wall can also be calculated. Permanent groove deformation can take many configurations, but axial deflection is the most frequent problem, and is caused by a ring twisting—causing it to grow in diameter and become dished (see *Figure 1*), and ultimately roll (or extrude) out of the groove. Allowable capacity of the groove wall can be calculated with the following formula.

$$P_T = (D_O \times D_I \times S_Y \times \pi) \div k$$

where P_T = Allowable load capacity (lbs), based on groove deformation
D_I = Groove depth, in inches
S_Y = Yield strength of the groove material, in PSI.

Again, the factor of safety (k) can be either 2 or 3, but most ring manufacturers calculate allowable thrust load using a safety factor of 2. The yield strength varies, but typical groove material yield strengths are as follows.

Hardened Steel (8620)	110,000 PSI
Cold Drawn Steel (1018)	70,000 PSI
Hot Rolled Steel (1018)	45,000 PSI
Aluminum (2017)	40,000 PSI
Cast Iron	10–40,000 PSI

Groove yield capacity (based on groove material yield strength of 45,000 PSI and a safety factor of 2) is provided in **Tables 3** through **6**.

Table 1. Properties of Materials Commonly Used for Spiral Retaining Rings, U.S. Customary Units.
(Source, Smalley Steel Ring Co.)

Material	Thickness	Min. Tensile Strength	Shear Strength	Temperature Limit [3]	Modulus of Elasticity
	inch	ksi	ksi	°F	ksi
Carbon Spring Temper Steel SAE 1070 - 1090	0.006-0.014	269	153	250	30,000
	0.0141-0.021	255	145		
	0.0211-0.043	221	126		
	0.0431 & larger	211	120		
AISI 302 AMS-5866	0.008-0.022	210	119	400	28,000
	0.0221-0.047	200	114		
	0.0471-0.062	185	105		
	0.0621-0.074	175	100		
	0.0741-0.089	165	94		
	0.0891-0.095	155	88		
AISI 316 ASTM A313 [1]	0.008-0.023	195	111	400	28,000
	0.0231-0.048	190	108		
	0.0481-0.061	175	99		
	0.0611 & larger	170	97		
17-7 PH Condition C/CH-900 AMS-5529	-	240 [2]	137 [2]	650	29,500
Beryllium Copper Alloy No. 25 ASTM B197 [1]	-	185 [2]	128 [2]	400	18,500
Phosphor Bronze Grade A QQ-B-750 [1]	0.010-0.025	106	61	200	16,000
	0.0251-0.036	95	54		
	0.0361 & larger	90	51		
Inconel X-750 * AMS-5699 [4,6] AMS-5699 [1] AMS-5699 [5] AMS-5698	-	220 RC 50 max. [2]	125 [2]	700	31,000
	-	136 RC 35 max	77	700	
	-	150 [2]	85 [2]	1300	
	-	155 [7]	88 [2]	1000	
A286 AMS-5810	-	180 [2]	105 [2]	1000	31,000
Elgiloy * AMS-5876 [1,4,5]	0.0043-0.025	280 [2]	160 [2]	800	30,000
	0.0251-0.047	255 [2]	145 [2]		
	0.0471-0.075	205 [2]	117 [2]		
	0.0751-0.100	160 [2]	91 [2]		
Monel K-500 QQ-N-286 1	-	RC 25 min	68	550	26,000
	-	170 [2]	97 [2]		

Notes: [1] Referenced for chemical composition only. [2] Values obtained after precipitation hardening.
[3] Exceeding these temperatures will cause increased relaxation. [4] Use in H_2S service. Conforms to NACE Standard MR-01-75. [5] High temperature service. [6] Spring temper. [7] #1 temper.
*Elgiloy is a registered trademark of the Elgiloy Company. Inconel X-750 is a registered trademark of Inco Alloys Int'l.

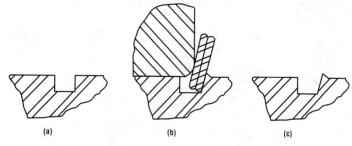

Figure 1. Axial deflection: (a) normal; (b) ring shown in deflection; (c) resulting deformation of groove.

The allowable deflection of the ring can also be calculated. For internal rings, the formula is as follows.

$$\text{Ø}_N = 114.6 \times R_G \left([\sigma + C_I \times M \times E] \div [\{1 - C_I\} \times D_N] \div [L \times M \times T_I] \right)$$

where Ø_N = Allowable angle of internal ring deflection

R_G = Mean groove radius, in inches

σ = Allowable working stress, in PSI

C_I = Percent change in ring diameter from free state to installed state, in inches

M = Ring material modulus of elasticity, in PSI

W_T = Ring radius wall, in inches

D_N = Neutral diameter of ring, in inches $(D_O - E)$

L = Number of turns on ring

T_I = Mean ring material thickness $(t \div 2)$.

In the above equation, the calculated value for Ø_N should be used if below 18°. However, if Ø_N is equal to 18° or better, use 18°. Material thickness refers to the thickness of a single ply of a ring. In the above formula, $T_I = (t \div 2)$, indicating that the ring has two turns.

The formula for allowable ring deflection for external rings is:

$$\text{Ø}_X = 114.6 \times R_G \left([\sigma - C_I \times M \times E] \div [\{1 + C_I\} \times D_N] \div [M \times T_I] \right)$$

As with internal rings, the calculated value for Ø_X should be used if below 18°. However, if Ø_X is equal to 18° or better, use 18°.

The allowable thrust load can be calculated using the allowable angle of deflection. This formula assumes that the load is applied through a retained part that applies the load very close to the shaft or bore diameter and where the load is of a static nature.

$$P_T = (A \times D_I \times S_Y \times \tan \text{Ø}) \div 0.073\, k$$

Centrifugal capacity

Proper functioning of an external retaining ring requires that it remain seated on the groove bottom. Centrifugal loading can overcome this state. The formula below is used by Smalley Steel Ring Company to predict the speed at which the force holding the ring tight on the groove (cling) becomes zero.

$$\text{rpm}_{max} = \left([3600 \times V \times M \times I \times g] \div [4\pi \times Y \times \gamma \times A \times R_M^5] \right)^{1/2}$$

where $V =$ Cling/2 (groove diameter minus free inside diameter, divided by 2), in inches

$M =$ Modulus of elasticity, in PSI

$I =$ moment of inertia $(t \times W_T) \div 12$, in units of in.^{4}

$g =$ gravitational acceleration (386.4 in./sec^2)

$Y =$ Multiple turn factor ($Y = 1.909$ for one turn; 3.407 for two turns; 4.958 for three turns; and 6.520 for four turns)

$\gamma =$ Material density, in lb/in.^{3}

$A =$ Cross sectional area $(t \times W_T) - .12 \times t^2$

$R_M =$ Mean free radius (free inside diameter plus W_T, divided by 2), in inches.

As will be noted, the calculation required to determine the maximum allowable rpm is relatively complex. Therefore, recommended maximum rpm for medium and heavy duty external rings is provided in **Table 2**.

Groove geometry

The groove, as well as the retained components, must have square corners.

MIL-R-27426 states that shafts or housings for medium and heavy duty external retaining rings up to 0.946" and less free diameter, the maximum radius on the groove bottom is 0.005". Those larger in diameter should have a maximum radius on the groove of 0.010". For plated rings, 0.002" upper thickness tolerance is allowed.

For internal rings, the standard states that medium and heavy duty internal retaining rings up to 1.063" free diameter, the maximum bottom radius is 0.005". Those larger in diameter should have a maximum radius on the groove of 0.010". For plated rings, 0.002" upper thickness tolerance is allowed.

The minimum distance between the outer groove wall and end of the shaft should be no less than three times groove depth.

Spiral retaining ring nomenclature

Crimp refers to a bend in the ring material at the gap, allowing for flat and parallel surfaces. It is also referred to as *offset*.

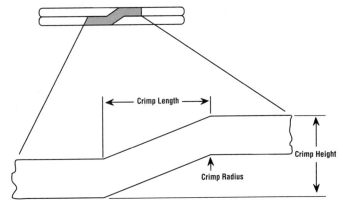

Crimp (section enlarged).

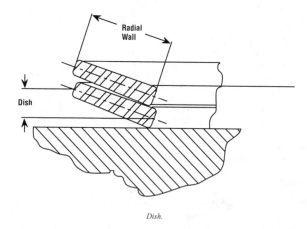

Dish.

Dish is the height difference in the ring cross section's axis of symmetry between the inside diameter and outside diameter. Maximum allowable dish ranges from 0.001" for a ring radial wall of 0.001–0.039", to 0.007" for a ring radial wall of 0.195–0.224", to 0.014" for a radial wall of 0.650–1.000".

Left hand wound refers to a counterclockwise winding direction. Normal wind is clockwise, or right hand.

Misalignment is a radial mismatch between the adjacent leaves of a retaining spring in free state.

Nominal Housing/Shaft Diameter	Maximum Misalignment
0.50 – 1.00"	0.015"
1.01 – 2.00"	0.025"
2.01 – 6.00"	0.045"
6.01 – 10.00"	0.070"
10.01 – 15.00"	0.100"

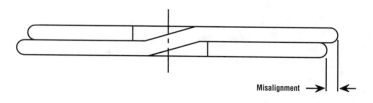

Misalignment.

Table 2. Maximum Allowable RPM for Medium and Heavy Duty Spiral Retaining Rings.
(Source, Smalley Steel Ring Co.)

Shaft Diameter S		Medium Duty Ring	Heavy Duty Ring	Shaft Diameter S		Medium Duty Ring	Heavy Duty Ring
inch	mm			inch	mm		
0.469	11.9	-	21,450	1.500	38.1	4,940	6,540
0.500	12.7	24,650	20,600	1.562	39.7	5,343	6,110
0.531	13.5	21,280	-	1.575	40.0	5,240	-
0.551	14.0	19,440	18,260	1.625	41.3	4,880	5,750
0.562	14.3	18.520	17,400	1.687	42.8	4,930	5,260
0.594	15.1	17,290	15,390	1.750	44.5	4,510	4,970
0.625	15.9	19,500	14,730	1.771	45.0	4,410	4,990
0.656	16.7	16,270	-	1.812	46.1	4,290	4,720
0.669	17.0	16,510	13,860	1.875	47.6	4,240	4,540
0.687	17.4	15,470	13,510	1.938	49.2	4,020	-
0.718	18.2	13,050	-	1.969	50.0	3,860	4,730
0.750	19.1	14,290	12,190	2.000	5038	3,740	4,560
0.781	19.8	12,960	11,110	2.062	52.4	3,550	3,810
0.812	20.6	12,470	10,150	2.125	54.0	3,400	3,560
0.843	21.4	10,770	-	2.156	54.8	3,490	3,450
0.875	22.2	10,570	10,340	2.165	55.0	3,370	-
0.906	23.0	9,180	-	2.188	55.6	3,290	-
0.937	23.8	9,400	8,760	2.250	57.2	3,220	3,240
0.968	24.6	8,920	-	2.312	58.7	3,020	3,040
0.984	25.0	9,530	8,640	2.362	60.0	2,870	-
1.000	25.4	9,160	8,940	2.375	60.3	2,890	3,380
1.023	26.0	9,070	8,500	2.437	61.9	2,920	3,180
1.031	26.2	8,080	-	2.500	63.5	2,750	3,090
1.062	27.0	8,610	11,260	2.559	65.0	2,600	2,920
1.093	27.8	7,350	-	2.562	65.1	2,600	-
1.125	28.6	7,470	9,820	2.625	66.7	2,500	2,750
1.156	29.4	6,700	-	2.688	68.3	2,470	2,680
1.188	30.2	7,350	9,040	2.750	69.9	2,340	2,790
1.218	30.9	6,340	-	2.813	71.5	2,380	-
1.250	31.8	6,750	8,042	2.875	73.0	2,260	2,590
1.281	32.5	5,860	-	2.937	74.6	2,140	2,460
1.312	33.3	6,310	8,280	2.952	75.0	2,160	-
1.343	34.1	5,960	-	3.000	76.2	2,080	2,410
1.375	34.9	6,110	7,430	3.062	77.8	2,020	2,290
1.406	35.7	5,580	-	3.125	79.4	1,980	2,240
1.437	36.5	5,490	6,700	3.149	80.0	1,980	-
1.468	37.3	5,020	-	3.156	80.2	-	2,190

(Continued)

Table 2. *(Continued)* **Maximum Allowable RPM for Medium and Heavy Duty Spiral Retaining Rings.**
(Source, Smalley Steel Ring Co.)

Shaft Diameter S		Medium Duty Ring	Heavy Duty Ring	Shaft Diameter S		Medium Duty Ring	Heavy Duty Ring
inch	mm			inch	mm		
3.187	80.9	1,930	-	5.125	130.2	1,020	-
3.250	82.6	1,870	2,100	5.250	133.4	970	1,210
3.312	84.1	1,840	-	5.375	136.5	900	-
3.343	84.9	1,840	1,960	5.500	139.7	890	1,120
3.375	85.7	1,790	-	5.511	140.0	870	-
3.437	87.3	1,750	1,880	5.625	142.9	840	-
3.500	88.9	1,700	2,090	5.750	146.1	820	1,030
3.543	90.0	1,730	2,080	5.875	149.2	780	-
3.562	90.5	1,680	1,970	5.905	150.0	760	-
3.625	92.1	1,660	1,890	6.000	152.4	750	950
3.687	93.6	1,600	1,890	6.125	155.6	880	-
3.740	95.0	1,520	-	6.250	158.8	860	1,090
3.750	95.3	1,530	1,860	6.299	160.0	830	-
3.182	96.8	1,470	-	6.375	161.9	810	-
3.875	98.4	1,500	1,750	6.500	165.1	790	960
3.938	100.0	1,510	1,690	6.625	168.3	650	-
4.000	101.6	1,470	1,660	6.750	171.5	732	980
4.063	103.2	1,400	-	6.875	174.6	700	-
4.125	104.8	1,350	-	7.000	177.8	680	920
4.134	105.0	1,380	-	7.125	181.0	650	-
4.188	106.4	1,360	-	7.250	184.2	640	800
4.250	108.0	1,360	1,440	7.375	187.3	610	-
4.312	109.5	1,300	-	7.500	190.5	580	760
4.331	110.0	1,300	-	7.625	193.7	570	-
4.375	111.1	1,290	1,360	7.750	196.9	560	720
4.437	112.7	1,230	-	7.875	200.0	540	-
4.500	114.3	1,270	1,300	8.000	203.2	520	680
4.562	115.9	1,280	-	8.250	209.6	490	640
4.625	117.5	1,240	-	8.500	215.9	550	610
4.687	119.0	1,220	-	8.750	222.3	520	570
4.724	120.0	1,180	-	9.000	228.6	490	610
4.750	120.7	1,180	1,180	9.250	235.0	460	580
4.812	122.2	1,140	-	9.500	241.3	440	560
4.875	123.8	1,120	-	9.750	247.7	420	520
4.937	125.4	1,090	-	10.000	254.0	400	500
5.000	127.0	1,050	1,080	10.250	260.4	380	480
5.118	130.0	1,000	-	10.500	266.7	360	460

(Continued)

Table 2. *(Continued)* **Maximum Allowable RPM for Medium and Heavy Duty Spiral Retaining Rings.** *(Source, Smalley Steel Ring Co.)*

Shaft Diameter S		Medium Duty Ring	Heavy Duty Ring	Shaft Diameter S		Medium Duty Ring	Heavy Duty Ring
inch	mm			inch	mm		
10.750	273.1	340	440	13.000	330.2	-	390
11.000	279.4	330	420	13.250	336.6	-	380
11.250	285.8	-	420	13.500	342.9	-	360
11.500	292.1	-	430	13.750	349.3	-	350
11.750	298.5	-	410	14.000	355.6	-	340
12.000	304.8	-	390	14.250	362.0	-	330
12.250	311.2	-	380	14.500	368.3	-	350
12.500	317.5	-	370	14.750	374.7	-	340
12.750	323.9	-	350	15.000	381.0	-	330

Tolerances for spiral retaining rings

External Medium Duty Spiral Retaining Rings. Specifications for medium duty external spiral rings are provided in **Table 3.** Although the Table provides dimensions of rings only up to shaft sizes of six inches, most manufacturers produce rings for shafts up to 11 inches in diameter. Tolerances that may be applied to the Table are as follows (all dimensions in inches).

Free Inside Diameter (G). For shaft sizes 0.500 to 1.031 incl.: +0.000, –0.013. For sizes 1.0620 to 1.500 incl.: +0.000, –0.015. For sizes 1.562 to 2.125 incl.: +0.000, –0.020. For sizes 2.156 to 2.688 incl.: +0.000, –0.025. For sizes 2.750 to 2.437 incl.: +0.000, –0.030. For sizes 3.500 to 5.125 incl.: +0.000, –0.040. For sizes 5.250 to 6.125 incl.: +0.000, –0.050. For sizes 6.250 to 7.375 incl.: +0.000, –0.060. For sizes 7.500 to 11.000 incl.: +0.000, –0.070.

Thickness (F). For shaft sizes 0.500 to 1.500 incl.: ±0.002. For sizes 1.562 to 4.500 incl.: ±0.003. For sizes 4.562 to 11.000 incl.: ±0.004.

Radial Wall (E). For shaft sizes 0.500 to 2.750 incl.: ±0.004. For sizes 2.183 to 6.000 incl.: ±0.005. For sizes 6.125 to 11.000 incl.: ±0.006.

Groove Diameter (C). For shaft sizes 0.500 to 0.562 incl.: ±0.002. For sizes 0.594 to 1.031 incl.: ±0.003. For sizes 1.062 to 1.500 incl.: ±0.004. For sizes 1.562 to 2.000 incl.: ±0.005. For sizes 2.062 to 5.125 incl.: ±0.006. For sizes 5.250 to 6.000 incl.: ±0.007. For sizes 6.125 to 11.000 incl.: ±0.008.

Groove Width (D). For shaft sizes 0.500 to 1.156 incl.: +0.003, –0.000. For sizes 1.188 to 2.952 incl.: +0.004, –0.000. For sizes 3.000 to 6.000 incl.: +0.005, –0.000. For sizes 6.125 to 11.000 incl.: +0.006, –0.000.

Metric external retaining rings are commonly available for shaft diameters ranging from 12 mm to 280 mm, and DIN series rings are available for shaft diameters ranging from 13 mm to 400 mm.

Table 3. External Medium Duty (Type A, Class 1) Spiral Retaining Ring Dimensions. See Text for Tolerances. *(Source, MIL27426 and Smalley Steel Ring Co.)*

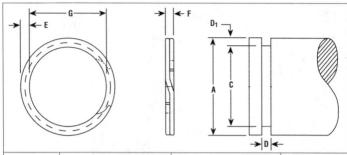

Shaft Diameter A		Ring Dimensions			Groove Dimensions			Thrust Capacity	
		Diameter G	Thickness F	Wall E	Diameter C	Width D	Depth D_1	Ring Shear[1]	Groove Yield[2]
inch	mm	Nom.	Nom.	Nom.	Nom.	Nom.	Nom.	lbs.	lbs.
0.500	12.7	0.467	0.025	0.045	0.474	0.030	0.013	2,000	460
0.531	13.5	0.498	0.025	0.045	0.505	0.030	0.013	2,130	490
0.551	14.0	0.518	0.025	0.045	0.525	0.030	0.013	2,210	510
0.562	14.3	0.529	0.025	0.045	0.536	0.030	0.013	2,250	520
0.594	15.1	0.561	0.025	0.045	0.569	0.030	0.013	2,380	550
0.625	15.9	0.585	0.025	0.055	0.594	0.030	0.016	2,500	710
0.656	16.7	0.617	0.025	0.055	0.625	0.030	0.016	2,630	740
0.669	17.0	0.629	0.025	0.055	0.638	0.030	0.016	2,680	760
0.687	17.5	0.647	0.025	0.055	0.656	0.030	0.016	2,750	780
0.718	18.3	0.679	0.025	0.055	0.687	0.030	0.016	2,880	810
0.750	19.0	0.710	0.031	0.065	0.719	0.036	0.016	3,360	850
0.781	19.8	0.741	0.031	0.065	0.750	0.036	0.016	3,500	880
0.812	20.6	0.771	0.031	0.065	0.781	0.036	0.016	3,640	920
0.843	21.4	0.803	0.031	0.065	0.812	0.036	0.016	3,780	950
0.875	22.2	0.828	0.031	0.065	0.838	0.036	0.019	3,920	1,180
0.906	23.0	0.860	0.031	0.065	0.869	0.036	0.019	4,060	1,220
0.937	23.8	0.889	0.031	0.065	0.900	0.036	0.019	4,200	1,260
0.968	24.4	0.916	0.037	0.075	0.925	0.042	0.021	5,180	1,440
0.984	25.0	0.930	0.037	0.075	0.941	0.042	0.021	5,260	1,460
1.000	25.4	0.946	0.037	0.075	0.957	0.042	0.021	5,350	1,480
1.023	26.0	0.968	0.037	0.075	0.980	0.042	0.021	5,470	1,520
1.031	26.3	0.978	0.037	0.075	0.988	0.042	0.021	5,510	1,530
1.062	27.0	1.007	0.037	0.075	1.020	0.042	0.021	5,680	1,580
1.093	27.8	1.040	0.037	0.075	1.051	0.042	0.021	5,840	1,620
1.125	29.0	1.070	0.037	0.075	1.083	0.042	0.021	6,020	1,670

(Continued)

Table 3. *(Continued)* **External Medium Duty (Type A, Class 1) Spiral Retaining Ring Dimensions.**
See Text for Tolerances. *(Source, MIL27426 and Smalley Steel Ring Co.)*

Shaft Diameter A		Ring Dimensions			Groove Dimensions			Thrust Capacity	
		Diameter G	Thickness F	Wall E	Diameter C	Width D	Depth D_1	Ring Shear[1]	Groove Yield[2]
inch	mm	Nom.	Nom.	Nom.	Nom.	Nom.	Nom.	lbs.	lbs.
1.156	39.3	1.102	0.037	0.075	1.114	0.042	0.021	6,180	1,720
1.188	30.2	1.127	0.043	0.085	1.140	0.048	0.024	7,380	2,020
1.218	31.0	1.159	0.043	0.085	1.170	0.048	0.024	7,570	2,070
1.250	31.7	1.188	0.043	0.085	1.202	0.048	0.024	7,770	2,120
1.281	32.6	1.221	0.043	0.085	1.233	0.048	0.024	7,960	2,170
1.312	33.3	1.251	0.043	0.095	1.264	0.048	0.024	8,150	2,230
1.343	34.1	1.282	0.043	0.095	1.295	0.048	0.024	8,350	2,280
1.375	34.9	1.308	0.043	0.095	1.323	0.048	0.026	8,540	2,530
1.406	35.8	1.340	0.043	0.095	1.354	0.048	0.026	8,740	2,580
1.437	36.5	1.370	0.043	0.095	1.385	0.048	0.026	8,930	2,640
1.468	37.3	1.402	0.043	0.095	1.416	0.048	0.026	9,120	2,700
1.500	38.1	1.433	0.043	0.095	1.448	0.048	0.026	9,320	2,760
1.562	39.6	1.490	0.049	0.108	1.507	0.056	0.028	10,100	3,090
1.575	40.0	1.503	0.049	0.108	1.520	0.056	0.028	10,190	3,120
1.625	41.7	1.549	0.049	0.108	1.566	0.056	0.030	10,510	3,450
1.687	42.8	1.610	0.049	0.118	1.628	0.056	0.030	10,910	3,580
1.750	44.4	1.673	0.049	0.118	1.691	0.056	0.030	11,310	3,710
1.771	44.9	1.690	0.049	0.118	1.708	0.056	0.032	11,450	4,010
1.813	46.0	1.730	0.049	0.118	1.749	0.056	0.032	11,720	4,100
1.875	47.6	1.789	0.049	0.128	1.808	0.056	0.034	12,120	4,510
1.938	49.2	1.844	0.049	0.128	1.861	0.056	0.034	12,530	4,660
1.969	50.0	1.882	0.049	0.128	1.902	0.056	0.034	12,730	4,730
2.000	50.8	1.909	0.049	0.128	1.929	0.056	0.035	12,930	4,950
2.062	52.3	1.971	0.049	0.128	1.992	0.056	0.035	13,330	5,100
2.125	53.9	2.029	0.049	0.128	2.051	0.056	0.037	13,740	5,560
2.156	54.7	2.060	0.049	0.138	2.082	0.056	0.037	13,940	5,640
2.165	55.0	2.070	0.049	0.138	2.091	0.056	0.037	14,000	5,660
2.188	55.5	2.092	0.049	0.138	2.113	0.056	0.037	14,150	5,720
2.250	57.1	2.153	0.049	0.138	2.176	0.056	0.037	14,550	5,890
2.312	58.7	2.211	0.049	0.138	2.234	0.056	0.039	14,950	6,370
2.362	59.9	2.261	0.049	0.138	2.284	0.056	0.039	15,270	6,510
2.375	60.3	2.273	0.049	0.138	2.297	0.056	0.039	15,360	6,550
2.437	61.9	2.331	0.049	0.148	2.355	0.056	0.041	15,760	7,060
2.500	63.5	2.394	0.049	0.148	2.418	0.056	0.041	16,160	7,250
2.559	64.9	2.449	0.049	0.148	2.473	0.056	0.043	16,550	7,780
2.562	65.1	2.452	0.049	0.148	2.476	0.056	0.043	16,560	7,790

(Continued)

Table 3. *(Continued)* **External Medium Duty (Type A, Class 1) Spiral Retaining Ring Dimensions.**
See Text for Tolerances. *(Source, MIL27426 and Smalley Steel Ring Co.)*

Shaft Diameter A		Ring Dimensions			Groove Dimensions			Thrust Capacity	
		Diameter G	Thickness F	Wall E	Diameter C	Width D	Depth D_1	Ring Shear[1]	Groove Yield[2]
inch	mm	Nom.	Nom.	Nom.	Nom.	Nom.	Nom.	lbs.	lbs.
2.625	66.6	2.514	0.049	0.148	2.539	0.056	0.043	16,970	7,980
2.688	68.2	2.572	0.049	0.158	2.597	0.056	0.045	17,380	8,550
2.750	69.8	2.635	0.049	0.158	2.660	0.056	0.045	17,780	8,750
2.813	71.4	2.696	0.049	0.168	2.722	0.056	0.045	18,190	8,950
2.875	73.0	2.755	0.049	0.168	2.781	0.056	0.047	18,590	9,550
2.937	74.5	2.817	0.049	0.168	2.843	0.056	0.047	18,990	9,760
2.952	74.9	2.831	0.049	0.168	2.858	0.056	0.047	19,090	9,810
3.000	76.1	2.877	0.061	0.168	2.904	0.068	0.048	24,150	10,180
3.062	77.7	2.938	0.061	0.168	2.966	0.068	0.048	24,650	10,390
3.125	79.3	3.000	0.061	0.178	3.027	0.068	0.049	25,150	10,820
3.149	79.9	3.023	0.061	0.178	3.051	0.068	0.049	25,350	10,910
3.187	81.0	3.061	0.061	0.178	3.089	0.068	0.049	25,650	11,040
3.250	82.5	3.121	0.061	0.178	3.150	0.068	0.050	26,160	11,490
3.312	84.4	3.180	0.061	0.188	3.208	0.068	0.052	26,660	12,170
3.343	84.9	3.210	0.061	0.188	3.239	0.068	0.052	26,910	12,290
3.375	85.8	3.242	0.061	0.188	3.271	0.068	0.052	27,170	12,410
3.437	87.2	3.301	0.061	0.188	3.331	0.068	0.053	27,660	12,880
3.500	88.8	3.363	0.061	0.188	3.394	0.068	0.053	28,170	13,110
3.543	89.9	3.402	0.061	0.198	3.433	0.068	0.055	28,520	13,770
3.562	90.5	3.422	0.061	0.198	3.452	0.068	0.055	28,670	13,850
3.625	92.0	3.483	0.061	0.198	3.515	0.068	0.055	29,180	14,090
3.687	93.6	3.543	0.061	0.198	3.575	0.068	0.056	29,680	14,600
3.740	95.0	3.597	0.061	0.198	3.628	0.068	0.056	30,100	14,800
3.750	95.2	3.606	0.061	0.198	3.638	0.068	0.056	30,180	14,840
3.812	97.1	3.668	0.061	0.198	3.700	0.068	0.056	30,680	15,090
3.875	98.3	3.724	0.061	0.208	3.757	0.068	0.059	31,190	16,160
3.938	99.9	3.784	0.061	0.208	3.820	0.068	0.059	31,700	16,420
4.000	101.5	3.842	0.061	0.218	3.876	0.068	0.062	32,200	17,530
4.063	103.2	3.906	0.061	0.218	3.939	0.068	0.062	32,700	17,810
4.125	104.8	3.967	0.061	0.218	4.000	0.068	0.062	33,200	18,080
4.134	105.0	3.975	0.061	0.218	4.010	0.068	0.062	33,270	18,120
4.188	106.5	4.030	0.061	0.218	4.058	0.068	0.062	33,710	19,240
4.250	107.9	4.084	0.061	0.228	4.120	0.068	0.065	34,210	19,530
4.312	109.5	4.147	0.061	0.228	4.182	0.068	0.065	34,710	19,810
4.331	110.0	4.164	0.061	0.228	4.200	0.068	0.065	34,860	19,900
4.375	111.0	4.208	0.061	0.228	4.245	0.068	0.065	35,210	20,100

(Continued)

Table 3. *(Continued)* **External Medium Duty (Type A, Class 1) Spiral Retaining Ring Dimensions.**
See Text for Tolerances. *(Source, MIL27426 and Smalley Steel Ring Co.)*

Shaft Diameter A		Ring Dimensions			Groove Dimensions			Thrust Capacity	
		Diameter G	Thickness F	Wall E	Diameter C	Width D	Depth D_1	Ring Shear[1]	Groove Yield[2]
inch	mm	Nom.	Nom.	Nom.	Nom.	Nom.	Nom.	lbs.	lbs.
4.437	112.7	4.271	0.061	0.228	4.307	0.068	0.065	35,710	20,390
4.500	114.2	4.326	0.061	0.238	4.364	0.068	0.068	36,220	21,630
4.562	116.0	4.384	0.072	0.250	4.422	0.079	0.070	43,340	22,570
4.625	117.4	4.447	0.072	0.250	4.485	0.079	0.070	43,940	22,890
4.687	119.0	4.808	0.072	0.250	4.547	0.079	0.070	44,530	23,190
4.724	120.0	4.546	0.072	0.250	4.584	0.079	0.070	44,880	23,370
4.750	120.6	4.571	0.072	0.250	4.610	0.079	0.070	45,130	23,500
4.812	122.2	4.633	0.072	0.250	4.672	0.079	0.070	45,720	23,810
4.875	123.8	4.695	0.072	0.250	4.735	0.079	0.070	46,310	24,120
4.937	125.4	4.757	0.072	0.250	4.797	0.079	0.070	46,900	24,430
5.000	126.9	4.820	0.072	0.250	4.856	0.079	0.072	47,500	25,450
5.118	130.0	4.934	0.072	0.250	4.974	0.079	0.072	48,620	26,050
5.125	130.1	4.939	0.072	0.250	4.981	0.079	0.072	48,690	26,080
5.250	133.2	5.064	0.072	0.250	5.107	0.079	0.072	49,880	26,720
5.375	136.5	5.187	0.072	0.250	5.228	0.079	0.074	51,060	28,120
5.500	139.9	5.308	0.072	0.250	5.353	0.079	0.074	52,250	28,770
5.511	140.0	5.320	0.072	0.250	5.364	0.079	0.074	52,360	28,830
5.625	142.8	5.433	0.072	0.250	5.478	0.079	0.074	53,440	29,420
5.750	145.9	5.550	0.072	0.250	5.597	0.079	0.077	54,630	31,300
5.875	149.2	5.674	0.072	0.250	5.722	0.079	0.077	55,810	31,980
5.905	150.0	5.705	0.072	0.250	5.752	0.079	0.077	56,100	32,140
6.000	152.3	5.798	0.072	0.250	5.847	0.079	0.077	57,000	32,660

[1] Based on a ring material shear strength of 120,000 PSI and a safety factor of 3.
[2] Based on a groove material yield of 45,000 PSI and a safety factor of 2.

External Heavy Duty Spiral Retaining Rings. Specifications for heavy duty external spiral rings are provided in **Table 4**. Although the Table provides dimensions of rings only up to shaft sizes of six inches, most manufacturers produce rings for shafts up to 15 inches in diameter. Tolerances that may be applied to the Table are as follows (all dimensions in inches).

Free Inside Diameter (G). For shaft sizes 0.469 to 1.500 incl.: +0.000, –0.013. For sizes 1.562 to 2.000 incl.: +0.000, –0.020. For sizes 2.062 to 2.687 incl.: +0.000, –0.025. For sizes 2.750 to 3.437 incl.: +0.000, –0.030. For sizes 3.500 to 5.000 incl.: +0.000, –0.035. For sizes 5.250 to 6.000 incl.: +0.000, –0.050. For sizes 6.250 to 7.000 incl.: +0.000, –0.060. For sizes 7.250 to 10.000 incl.: +0.000, –0.070. For sizes 10.250 to 12.500 incl.: +0.000, –0.090. For sizes 12.750 to 15.000 incl.: +0.000, –0.110.

Thickness (F). For shaft sizes 0.469 to 1.500 incl.: ±0.002. For sizes 1.562 to 5.000 incl.: ±0.003. For sizes 5.250 to 6.000 incl.: ±0.004. For sizes 6.250 to 15.000 incl.: ±0.005.

Radial Wall (E). For shaft sizes 0.469 to 2.000 incl.: ±0.004. For sizes 2.062 to 5.000 incl.: ±0.005. For sizes 5.250 to 6.000 incl.: ±0.006. For sizes 6.250 to 11.250 incl.: ±0.007. For sizes 11.500 to 15.000 incl.: +0.015, −0.010.

Groove Diameter (C). For shaft sizes 0.469 to 0.562 incl.: ±0.002. For sizes 0.594 to 1.023 incl.: ±0.003. For sizes 1.062 to 1.500 incl.: ±0.004. For sizes 1.562 to 2.000 incl.: ±0.005. For sizes 2.062 to 5.000 incl.: ±0.006. For sizes 5.250 to 6.000 incl.: ±0.007. For sizes 6.250 to 10.000 incl.: ±0.008. For sizes 10.250 to 12.500 incl.: ±0.010. For sizes 12.750 to 15.000 incl.: ±0.012.

Groove Width (D). For shaft sizes 0.469 to 1.023 incl.: +0.003, −0.000. For sizes 1.062 to 2.000 incl.: +0.004, −0.000. For sizes 2.062 to 5.000 incl.: +0.005, −0.000. For sizes 5.250 to 6.000 incl.: +0.006, −0.000. For sizes 6.025 to 15.000 incl.: +0.008, −0.000.

Table 4. External Heavy Duty (Type A, Class 2) Spiral Retaining Ring Dimensions. See Text for Tolerances. *(Source, MIL27426 and Smalley Steel Ring Co.)*

Shaft Diameter A		Ring Dimensions			Groove Dimensions			Thrust Capacity	
		Diameter G	Thickness F	Wall E	Diameter C	Width D	Depth D_1	Ring Shear[1]	Groove Yield[2]
inch	mm	Nom.	Nom.	Nom.	Nom.	Nom.	Nom.	lbs.	lbs.
0.469	11.9	0.439	0.025	0.045	0.443	0.029	0.013	1,880	430
0.500	12.7	0.464	0.035	0.050	0.468	0.039	0.016	2,530	570
0.551	14.0	0.514	0.035	0.050	0.519	0.039	0.016	2,790	620
0.562	14.3	0.525	0.035	0.050	0.530	0.039	0.016	2,840	640
0.594	15.1	0.554	0.035	0.050	0.559	0.039	0.018	3,000	760
0.625	15.9	0.583	0.035	0.055	0.588	0.039	0.019	3,160	840
0.669	17.0	0.623	0.035	0.055	0.629	0.039	0.020	3,380	950
0.688	17.5	0.641	0.042	0.065	0.646	0.046	0.021	4,180	1,020
0.750	19.0	0.698	0.042	0.065	0.704	0.046	0.023	4,550	1,220
0.781	19.8	0.727	0.042	0.065	0.733	0.046	0.024	4,740	1,330
0.812	20.6	0.756	0.042	0.065	0.762	0.046	0.025	4,930	1,440
0.875	22.2	0.814	0.042	0.075	0.821	0.046	0.027	5,310	1,670
0.938	23.8	0.875	0.042	0.075	0.882	0.046	0.028	5,690	1,860
0.984	25.0	0.919	0.042	0.085	0.926	0.046	0.029	5,970	2,020
1.000	25.4	0.932	0.042	0.085	0.940	0.046	0.030	6,070	2,120
1.023	26.0	0.953	0.042	0.085	0.961	0.046	0.031	6,210	2,240
1.062	27.0	0.986	0.050	0.103	0.998	0.056	0.032	7,010	2,400
1.125	28.6	1.047	0.050	0.103	1.059	0.056	0.033	7,420	2,620
1.188	30.2	1.105	0.050	0.103	1.118	0.056	0.035	7,840	2,940
1.250	31.7	1.163	0.050	0.103	1.176	0.056	0.037	8,250	3,270
1.312	33.3	1.218	0.050	0.118	1.232	0.056	0.040	8,660	3,710
1.375	34.9	1.277	0.050	0.118	1.291	0.056	0.042	9,070	4,080
1.438	36.5	1.336	0.050	0.118	1.350	0.056	0.044	9,490	4,470
1.500	38.1	1.385	0.050	0.118	1.406	0.056	0.047	9,900	4,980
1.562	39.6	1.453	0.062	0.128	1.468	0.068	0.047	12,780	5,190

(Continued)

Table 4. *(Continued)* External Heavy Duty (Type A, Class 2) Spiral Retaining Ring Dimensions.
See Text for Tolerances. *(Source, MIL27426 and Smalley Steel Ring Co.)*

Shaft Diameter A		Ring Dimensions			Groove Dimensions			Thrust Capacity	
		Diameter G	Thickness F	Wall E	Diameter C	Width D	Depth D_1	Ring Shear[1]	Groove Yield[2]
inch	mm	Nom.	Nom.	Nom.	Nom.	Nom.	Nom.	lbs.	lbs.
1.625	41.7	1.513	0.062	0.128	1.529	0.068	0.048	13,290	5,510
1.687	42.8	1.573	0.062	0.128	1.589	0.068	0.049	13,800	5,840
1.750	44.4	1.633	0.062	0.128	1.650	0.068	0.050	14,320	6,190
1.771	44.9	1.651	0.062	0.128	1.669	0.068	0.051	14,490	6,380
1.812	46.0	1.690	0.062	0.128	1.708	0.068	0.052	14,820	6,660
1.875	47.6	1.751	0.062	0.158	1.769	0.068	0.053	15,340	7,020
1.969	50.0	1.838	0.062	0.158	1.857	0.068	0.056	16,110	7,790
2.000	50.8	1.867	0.062	0.158	1.886	0.068	0.057	16,360	8,060
2.062	52.3	1.932	0.078	0.168	1.946	0.086	0.058	21,220	8,450
2.125	53.9	1.989	0.078	0.168	2.003	0.086	0.061	21,870	9,160
2.156	54.7	2.018	0.078	0.168	2.032	0.086	0.062	22,190	9,450
2.250	57.1	2.105	0.078	0.168	2.120	0.086	0.065	23,160	10,340
2.312	58.7	2.163	0.078	0.168	2.178	0.086	0.067	23,800	10,950
2.375	60.3	2.223	0.078	0.200	2.239	0.086	0.068	24,440	11,420
2.437	61.9	2.283	0.078	0.200	2.299	0.086	0.069	25,080	11,890
2.500	63.5	2.343	0.078	0.200	2.360	0.086	0.070	25,730	12,370
2.559	64.9	2.402	0.078	0.200	2.419	0.086	0.070	26,340	12,660
2.625	66.6	2.464	0.078	0.200	2.481	0.086	0.072	27,020	13,360
2.687	68.2	2.523	0.078	0.200	2.541	0.086	0.073	27,660	13,870
2.750	69.8	2.584	0.093	0.225	2.602	0.103	0.074	32,140	14,390
2.875	73.0	2.702	0.093	0.225	2.721	0.103	0.077	33,600	15,650
2.937	74.5	2.750	0.093	0.225	2.779	0.103	0.079	34,320	16,400
3.000	76.1	2.818	0.093	0.225	2.838	0.103	0.081	35,060	17,180
3.062	77.7	2.878	0.093	0.225	2.898	0.103	0.082	35,790	17,750
3.125	79.3	2.936	0.093	0.225	2.957	0.103	0.084	36,520	18,560
3.156	80.1	2.965	0.093	0.225	2.986	0.103	0.085	36,880	18,960
3.250	82.5	3.054	0.093	0.225	3.076	0.103	0.087	37,980	19,990
3.344	84.9	3.144	0.093	0.225	3.166	0.103	0.089	39,080	21,040
3.437	87.2	3.234	0.093	0.225	3.257	0.103	0.090	40,170	21,870
3.500	88.8	3.293	0.111	0.270	3.316	0.120	0.092	48,820	22,760
3.543	89.9	3.333	0.111	0.270	3.357	0.120	0.093	49,420	23,290
3.625	92.0	3.411	0.111	0.270	3.435	0.120	0.099	50,560	24,340
3.687	93.6	3.469	0.111	0.270	3.493	0.120	0.097	51,430	25,280
3.750	95.2	3.527	0.111	0.270	3.552	0.120	0.095	52,310	26,240
3.875	98.3	3.647	0.111	0.270	3.673	0.120	0.101	54,050	27,670
3.938	99.9	3.708	0.111	0.270	3.734	0.120	0.102	54,930	28,390

(Continued)

Table 4. *(Continued)* **External Heavy Duty (Type A, Class 2) Spiral Retaining Ring Dimensions.**
See Text for Tolerances. *(Source, MIL27426 and Smalley Steel Ring Co.)*

Shaft Diameter A		Ring Dimensions			Groove Dimensions			Thrust Capacity	
		Diameter G	Thickness F	Wall E	Diameter C	Width D	Depth D_1	Ring Shear[1]	Groove Yield[2]
inch	mm	Nom.	Nom.	Nom.	Nom.	Nom.	Nom.	lbs.	lbs.
4.000	101.5	3.765	0.111	0.270	3.792	0.120	0.104	55,800	29,410
4.250	107.9	4.037	0.111	0.270	4.065	0.120	0.093	59,280	27,940
4.375	110.0	4.161	0.111	0.270	4.190	0.120	0.093	61,030	28,760
4.500	114.2	4.280	0.111	0.270	4.310	0.120	0.095	62,770	30,220
4.750	120.6	4.518	0.111	0.270	4.550	0.120	0.100	66,260	36,930
5.000	126.9	4.756	0.111	0.270	4.790	0.120	0.105	69,740	37,110
5.250	133.2	4.995	0.127	0.350	5.030	0.139	0.110	83,790	40,820
5.500	140.0	5.228	0.127	0.350	5.265	0.139	0.118	87,780	45,880
5.750	145.9	5.466	0.127	0.350	5.505	0.139	0.123	91,770	49,990
6.000	152.3	5.705	0.127	0.350	5.745	0.139	0.128	95,760	54,290

[1] Based on a ring material shear strength of 120,000 PSI and a safety factor of 3.
[2] Based on a groove material yield of 45,000 PSI and a safety factor of 2.

Internal Medium Duty Spiral Retaining Rings. Specifications for medium duty internal spiral rings are provided in **Table 5.** Although the Table provides dimensions of rings only up to housing sizes of six inches, most manufacturers produce rings for housings up to 11 inches in diameter. Tolerances that may be applied to the Table are as follows (all dimensions in inches).

Free Outside Diameter (G). For housings sizes 0.500 to 1.031 incl.: +0.013, −0.000. For sizes 1.0620 to 1.500 incl.: +0.015, −0.000. For sizes 1.562 to 2.047 incl.: +0.020, −0.000. For sizes 2.062 to 3.000 incl.: +0.025, −0.000. For sizes 3.062 to 4.063 incl.: +0.030, −0.000. For sizes 4.125 to 5.125 incl.: +0.035, −0.000. For sizes 5.250 to 6.125 incl.: +0.045, −0.000. For sizes 6.250 to 7.125 incl.: +0.055, −0.000. For sizes 7.250 to 11.000 incl.: +0.0650, −0.000.

Thickness (F). For housing sizes 0.500 to 1.500 incl.: ±0.002. For sizes 1.562 to 4.562 incl.: ±0.003. For sizes 4.625 to 11.000 incl.: ±0.004.

Radial Wall (E). For housing sizes 0.500 to 2.813 incl.: ±0.004. For sizes 2.834 to 6.000 incl.: ±0.005. For sizes 6.125 to 11.000 incl.: ±0.006.

Groove Diameter (C). For housing sizes 0.500 to 0.750 incl.: ±0.002. For sizes 0.777 to 1.031 incl.: ±0.003. For sizes 1.062 to 1.500 incl.: ±0.004. For sizes 1.562 to 2.0047 incl.: ±0.005. For sizes 2.062 to 5.125 incl.: ±0.006. For sizes 5.250 to 6.000 incl.: ±0.007. For sizes 6.125 to 11.000 incl.: ±0.008.

Groove Width (D). For housing sizes 0.500 to 1.156 incl.: +0.003, −0.000. For sizes 1.188 to 2.952 incl.: +0.004, −0.000. For sizes 3.000 to 6.000 incl.: +0.005, −0.000. For sizes 6.125 to 11.000 incl.: +0.006, −0.000.

Metric internal retaining rings are commonly available for housing diameters ranging from 12 mm to 280 mm, and DIN series rings are available for housing diameters ranging from 13 mm to 400 mm.

Table 5. Internal Medium Duty (Type B, Class 1) Spiral Retaining Ring Dimensions. See Text for Tolerances. *(Source, MIL27426 and Smalley Steel Ring Co.)*

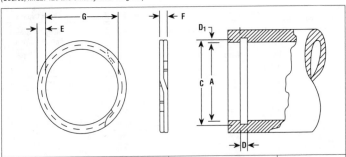

Housing Diameter A		Ring Dimensions			Groove Dimensions			Thrust Capacity	
		Diameter G	Thickness F	Wall E	Diameter C	Width D	Depth D_1	Ring Shear[1]	Groove Yield[2]
inch	mm	Nom.	Nom.	Nom.	Nom.	Nom.	Nom.	lbs.	lbs.
0.500	12.7	0.467	0.025	0.045	0.474	0.030	0.013	2,000	460
0.531	13.5	0.498	0.025	0.045	0.505	0.030	0.013	2,130	490
0.551	14.0	0.518	0.025	0.045	0.525	0.030	0.013	2,210	510
0.562	14.3	0.529	0.025	0.045	0.536	0.030	0.013	2,250	520
0.594	15.1	0.561	0.025	0.045	0.569	0.030	0.013	2,380	550
0.625	15.9	0.585	0.025	0.055	0.594	0.030	0.016	2,500	710
0.656	16.7	0.617	0.025	0.055	0.625	0.030	0.016	2,630	740
0.669	17.0	0.629	0.025	0.055	0.638	0.030	0.016	2,680	760
0.687	17.5	0.647	0.025	0.055	0.656	0.030	0.016	2,750	780
0.718	18.3	0.679	0.025	0.055	0.687	0.030	0.016	2,880	810
0.750	19.0	0.710	0.031	0.065	0.719	0.036	0.016	3,360	850
0.781	19.8	0.741	0.031	0.065	0.750	0.036	0.016	3,500	880
0.812	20.6	0.771	0.031	0.065	0.781	0.036	0.016	3,640	920
0.843	21.4	0.803	0.031	0.065	0.812	0.036	0.016	3,780	950
0.875	22.2	0.828	0.031	0.065	0.838	0.036	0.019	3,920	1,180
0.906	23.0	0.860	0.031	0.065	0.869	0.036	0.019	4,060	1,220
0.937	23.8	0.889	0.031	0.065	0.900	0.036	0.019	4,200	1,260
0.968	24.4	0.916	0.037	0.075	0.925	0.042	0.021	5,180	1,440
0.984	25.0	0.930	0.037	0.075	0.941	0.042	0.021	5,260	1,460
1.000	25.4	0.946	0.037	0.075	0.957	0.042	0.021	5,350	1,480
1.023	26.0	0.968	0.037	0.075	0.980	0.042	0.021	5,470	1,520
1.031	26.3	0.978	0.037	0.075	0.988	0.042	0.021	5,510	1,530
1.062	27.0	1.007	0.037	0.075	1.020	0.042	0.021	5,680	1,580
1.093	27.8	1.040	0.037	0.075	1.051	0.042	0.021	5,840	1,620
1.125	29.0	1.070	0.037	0.075	1.083	0.042	0.021	6,020	1,670

(Continued)

Table 5. *(Continued)* **Internal Medium Duty (Type B, Class 1) Spiral Retaining Ring Dimensions.**
See Text for Tolerances. *(Source, MIL27426 and Smalley Steel Ring Co.)*

Housing Diameter A		Ring Dimensions			Groove Dimensions			Thrust Capacity	
		Diameter G	Thickness F	Wall E	Diameter C	Width D	Depth D_1	Ring Shear[1]	Groove Yield[2]
inch	mm	Nom.	Nom.	Nom.	Nom.	Nom.	Nom.	lbs.	lbs.
1.156	29.3	1.210	0.037	0.075	1.198	0.042	0.021	6,180	1,720
1.188	30.2	1.249	0.043	0.085	1.236	0.048	0.024	7,380	2,020
1.218	31.0	1.278	0.043	0.085	1.266	0.048	0.024	7,570	2,070
1.250	31.7	1.312	0.043	0.085	1.298	0.048	0.024	7,770	2,120
1.281	32.6	1.342	0.043	0.085	1.329	0.048	0.024	7,960	2,170
1.312	33.3	1.374	0.043	0.085	1.360	0.048	0.024	8,150	2,230
1.343	34.1	1.408	0.043	0.085	1.395	0.048	0.026	8,350	2,470
1.375	34.9	1.442	0.043	0.095	1.427	0.048	0.026	8,540	2,530
1.406	35.8	1.472	0.043	0.095	1.458	0.048	0.026	8,740	2,580
1.437	36.5	1.504	0.043	0.095	1.489	0.048	0.026	8,930	2,640
1.456	37.0	1.523	0.043	0.095	1.508	0.048	0.026	9,050	2,680
1.468	37.3	1.535	0.043	0.095	1.520	0.048	0.026	9,120	2,700
1.500	38.1	1.567	0.043	0.095	1.552	0.048	0.026	9,320	2,760
1.562	39.6	1.634	0.049	0.108	1.617	0.056	0.028	10,100	3,090
1.574	40.0	1.649	0.049	0.108	1.633	0.056	0.030	10,180	3,340
1.625	41.7	1.701	0.049	0.108	1.684	0.056	0.030	10,510	3,350
1.653	42.0	1.730	0.049	0.108	1.712	0.056	0.030	10,690	3,510
1.687	42.8	1.768	0.049	0.118	1.750	0.056	0.031	10,910	3,700
1.750	44.4	1.834	0.049	0.118	1.813	0.056	0.031	11,310	3,840
1.813	46.0	1.894	0.049	0.118	1.875	0.056	0.031	11,720	3,970
1.850	47.0	1.937	0.049	0.118	1.917	0.056	0.034	11,960	4,450
1.875	47.6	1.960	0.049	0.118	1.942	0.056	0.034	12,120	4,510
1.938	49.2	2.025	0.049	0.118	2.005	0.056	0.034	12,530	4,660
2.000	50.8	2.091	0.049	0.128	2.071	0.056	0.035	12,930	4,950
2.047	52.0	2.138	0.049	0.128	2.118	0.056	0.035	13,240	5,060
2.062	52.3	2.154	0.049	0.128	2.132	0.056	0.035	13,330	5,100
2.125	53.9	2.217	0.049	0.128	2.195	0.056	0.035	13,740	5,260
2.165	55.0	2.260	0.049	0.138	2.239	0.056	0.037	14,000	5,660
2.188	55.5	2.284	0.049	0.138	2.262	0.056	0.037	14,150	5,720
2.250	57.1	2.347	0.049	0.138	2.324	0.056	0.037	14,550	5,890
2.312	58.7	2.413	0.049	0.138	2.390	0.056	0.039	14,950	6,370
2.375	60.3	2.476	0.049	0.138	2.453	0.056	0.039	15,360	6,550
2.437	61.9	2.543	0.049	0.148	2.519	0.056	0.041	15,760	7,060
2.440	62.0	2.546	0.049	0.148	2.522	0.056	0.041	15,780	7,070
2.500	63.5	2.606	0.049	0.148	2.582	0.056	0.041	16,160	7,250
2.531	64.2	2.641	0.049	0.148	2.617	0.056	0.043	16,360	7,690

(Continued)

Table 5. (Continued) **Internal Medium Duty (Type B, Class 1) Spiral Retaining Ring Dimensions.**
See Text for Tolerances. (Source, MIL27426 and Smalley Steel Ring Co.)

Housing Diameter A		Ring Dimensions			Groove Dimensions			Thrust Capacity	
		Diameter G	Thickness F	Wall E	Diameter C	Width D	Depth D₁	Ring Shear¹	Groove Yield²
inch	mm	Nom.	Nom.	Nom.	Nom.	Nom.	Nom.	lbs.	lbs.
2.562	65.1	2.673	0.049	0.148	2.648	0.056	0.043	16,560	7,790
2.625	66.6	2.736	0.049	0.148	2.711	0.056	0.043	16,970	7,980
2.677	68.0	2.789	0.049	0.158	2.767	0.056	0.045	17,310	8,520
2.688	68.2	2.803	0.049	0.158	2.778	0.056	0.045	17,380	8,550
2.750	69.8	2.865	0.049	0.158	2.841	0.056	0.045	17,780	8,750
2.813	71.4	2.929	0.049	0.158	2.903	0.056	0.045	18,190	8,950
2.834	71.9	2.954	0.049	0.168	2.928	0.056	0.047	18,320	9,520
2.875	73.0	2.995	0.049	0.168	2.969	0.056	0.047	18,590	9,550
2.937	74.5	3.058	0.049	0.168	3.031	0.056	0.047	18,990	9,760
2.952	75.0	3.073	0.049	0.168	3.046	0.056	0.047	19,090	9,810
3.000	76.1	3.122	0.061	0.168	3.096	0.068	0.048	24,150	10,180
3.062	77.7	3.186	0.061	0.168	3.158	0.068	0.048	24,650	10,390
3.125	79.3	3.251	0.061	0.178	3.223	0.068	0.048	25,150	10,600
3.149	80.0	3.276	0.061	0.178	3.247	0.068	0.048	25,350	10,680
3.187	81.0	3.311	0.061	0.178	3.283	0.068	0.048	25,650	10,810
3.250	82.5	3.379	0.061	0.178	3.350	0.068	0.050	26,160	11,490
3.312	84.4	3.446	0.061	0.188	3.416	0.068	0.052	26,660	12,170
3.346	84.9	3.479	0.061	0.188	3.450	0.068	0.052	26,930	12,300
3.375	85.7	3.509	0.061	0.188	3.479	0.068	0.052	27,170	12,410
3.437	87.2	3.574	0.061	0.188	3.543	0.068	0.053	27,660	12,880
3.500	88.8	3.636	0.061	0.188	3.606	0.068	0.053	28,170	13,110
3.543	89.9	3.684	0.061	0.198	3.653	0.068	0.055	28,520	13,770
3.562	90.5	3.703	0.061	0.198	3.672	0.068	0.055	28,670	13,850
3.625	92.0	3.769	0.061	0.198	3.737	0.068	0.056	29,180	14,350
3.687	93.6	3.832	0.061	0.198	3.799	0.068	0.056	29,680	14,600
3.740	95.0	3.885	0.061	0.198	3.852	0.068	0.056	30,100	14,800
3.750	95.2	3.894	0.061	0.198	3.862	0.068	0.056	30,180	14,840
3.812	96.8	3.963	0.061	0.208	3.930	0.068	0.059	30,680	15,900
3.875	98.3	4.025	0.061	0.208	3.993	0.068	0.059	31,190	16,160
3.938	99.9	4.089	0.061	0.208	4.056	0.068	0.059	31,700	16,420
4.000	101.5	4.157	0.061	0.218	4.124	0.068	0.062	32,200	17,530
4.063	103.2	4.222	0.061	0.218	4.187	0.068	0.062	32,700	17,810
4.125	104.8	4.284	0.061	0.218	4.249	0.068	0.062	33,200	18,080
4.188	106.5	4.347	0.061	0.218	4.311	0.068	0.062	33,710	18,350
4.250	107.9	4.416	0.061	0.228	4.380	0.068	0.065	34,210	19,530
4.312	109.7	4.479	0.061	0.228	4.442	0.068	0.065	34,710	19,810

(Continued)

Table 5. *(Continued)* **Internal Medium Duty (Type B, Class 1) Spiral Retaining Ring Dimensions.**
See Text for Tolerances. *(Source, MIL27426 and Smalley Steel Ring Co.)*

Housing Diameter A		Ring Dimensions			Groove Dimensions			Thrust Capacity	
		Diameter G	Thickness F	Wall E	Diameter C	Width D	Depth D_1	Ring Shear[1]	Groove Yield[2]
inch	mm	Nom.	Nom.	Nom.	Nom.	Nom.	Nom.	lbs.	lbs.
4.330	109.9	4.497	0.061	0.228	4.460	0.068	0.065	34,850	19,900
4.375	111.0	4.543	0.061	0.228	4.505	0.068	0.065	35,210	20,100
4.437	112.7	4.611	0.061	0.238	4.573	0.068	0.069	35,710	21,330
4.500	114.2	4.674	0.061	0.238	4.636	0.068	0.069	36,220	21,630
4.527	115.0	4.701	0.061	0.238	4.663	0.068	0.069	36,440	21,760
4.562	116.0	4.737	0.061	0.238	4.698	0.068	0.069	36,720	21,930
4.625	117.4	4.803	0.072	0.250	4.765	0.079	0.070	43,940	22,890
4.687	119.0	4.867	0.072	0.250	4.827	0.079	0.070	44,530	23,190
4.724	120.0	4.903	0.072	0.250	4.864	0.079	0.070	44,880	23,370
4.750	120.6	4.930	0.072	0.250	4.890	0.079	0.070	45,130	23,500
4.812	122.2	4.993	0.072	0.250	4.952	0.079	0.070	45,720	23,810
4.875	123.8	5.055	0.072	0.250	5.015	0.079	0.070	46,310	24,120
4.921	125.0	5.102	0.072	0.250	5.061	0.079	0.070	46,750	24,350
4.937	125.4	5.122	0.072	0.250	5.081	0.079	0.072	46,900	25,130
5.000	126.9	5.185	0.072	0.250	5.144	0.079	0.072	47,500	25,450
5.118	130.0	5.304	0.072	0.250	5.262	0.079	0.072	48,620	26,050
5.125	130.1	5.311	0.072	0.250	5.269	0.079	0.072	48,690	26,100
5.250	133.2	5.436	0.072	0.250	5.393	0.079	0.072	49,880	26,720
5.375	136.5	5.566	0.072	0.250	5.522	0.079	0.074	51,060	28,120
5.500	139.7	5.693	0.072	0.250	5.647	0.079	0.074	52,250	28,770
5.511	140.0	5.703	0.072	0.250	5.658	0.079	0.074	52,360	28,830
5.625	142.8	5.818	0.072	0.250	5.772	0.079	0.074	53,440	29,400
5.708	145.0	5.909	0.072	0.250	5.861	0.079	0.077	54,230	31,070
5.750	145.9	5.950	0.072	0.250	5.903	0.079	0.077	54,630	31,300
5.875	149.2	6.077	0.072	0.250	6.028	0.079	0.077	55,810	31,980
5.905	150.0	6.106	0.072	0.250	6.058	0.079	0.077	56,100	32,140
6.000	152.3	6.202	0.072	0.250	6.153	0.079	0.077	57,000	32,660

[1] Based on ring material shear strength of 120,000 PSI and a safety factor of 3.
[2] Based on a groove material yield of 45,000 PSI and a safety factor of 2.

Internal Heavy Duty Spiral Retaining Rings. Specifications for heavy duty internal spiral rings are provided in **Table 6.** Although the Table provides dimensions of rings only up to housing sizes of six inches, most manufacturers produce rings for housings up to 15 inches in diameter. Tolerances that may be applied to the Table are as follows (all dimensions in inches).

Free Outside Diameter (G). For housing sizes 0.500 to 1.500 incl.: +0.013, −0.000. For sizes 1.562 to 2.000 incl.: +0.020, −0.000. For sizes 2.062 to 2.531 incl.: +0.025, −0.000. For sizes 2.562 to 3.000 incl.: +0.030, −0.000. For sizes 3.062 to 5.000

incl.: +0.035, –0.000. For sizes 5.250 to 6.000 incl.: +0.050, –0.000. For sizes 6.250 to 7.000 incl.: +0.055, –0.000. For sizes 7.250 to 10.500 incl.: +0.070, –0.000. For sizes 10.750 to 12.750 incl.: +0.012, –0.000. For sizes 13.000 to 15.000 incl.: +0.140, –0.000.

Thickness (F). For housing sizes 0.500 to 1.500 incl.: ±0.002. For sizes 1.562 to 5.000 incl.: ±0.003. For sizes 5.250 to 6.000 incl.: ±0.004. For sizes 6.250 to 15.000 incl.: ±0.005.

Radial Wall (E). For housing sizes 0.500 to 2.000 incl.: ±0.004. For sizes 2.062 to 3.625 incl.: ±0.005. For sizes 3.750 to 7.000 incl.: ±0.006. For sizes 7.250 to 11.250 incl.: ±0.007. For sizes 11.500 to 15.000 incl.: +0.015, –0.010.

Groove Diameter (C). For housing sizes 0.500 to 0.750 incl.: ±0.002. For sizes 0.777 to 1.023 incl.: ±0.003. For sizes 1.062 to 1.500 incl.: ±0.004. For sizes 1.562 to 2.000 incl.: ±0.005. For sizes 2.062 to 5.000 incl.: ±0.006. For sizes 5.250 to 6.000 incl.: ±0.007. For sizes 6.250 to 10.500 incl.: ±0.008. For sizes 10.750 to 12.500 incl.: ±0.010. For sizes 12.750 to 15.000 incl.: ±0.012.

Groove Width (D). For housing sizes 0.500 to 1.023 incl.: +0.003, –0.000. For sizes 1.062 to 2.000 incl.: +0.004, –0.000. For sizes 2.062 to 5.000 incl.: +0.005, –0.000. For sizes 5.250 to 6.000 incl.: +0.006, –0.000. For sizes 6.025 to 15.000 incl.: +0.008, –0.000.

Table 6. Internal Heavy Duty (Type B, Class 2) Spiral Retaining Ring Dimensions. See Text for Tolerances. *(Source, MIL27426 and Smalley Steel Ring Co.)*

Housing Diameter A		Ring Dimensions			Groove Dimensions			Thrust Capacity	
		Diameter G	Thickness F	Wall E	Diameter C	Width D	Depth D_1	Ring Shear[1]	Groove Yield[2]
inch	mm	Nom.	Nom.	Nom.	Nom.	Nom.	Nom.	lbs.	lbs.
0.500	12.7	0.538	0.035	0.045	0.530	0.039	0.015	2,530	530
0.512	13.0	0.550	0.035	0.045	0.542	0.039	0.015	2,590	540
0.562	14.3	0.605	0.035	0.055	0.596	0.039	0.017	2,840	680
0.625	15.9	0.675	0.035	0.055	0.665	0.039	0.020	3,160	880
0.688	17.5	0.743	0.035	0.065	0.732	0.039	0.022	3,480	1,070
0.750	19.0	0.807	0.035	0.065	0.796	0.039	0.023	3,790	1,220
0.777	19.7	0.836	0.042	0.075	0.825	0.046	0.024	4,720	1,320
0.812	20.6	0.873	0.042	0.075	0.862	0.046	0.025	4,930	1,440
0.866	22.0	0.931	0.042	0.075	0.920	0.046	0.027	5,260	1,650
0.875	22.2	0.943	0.042	0.085	0.931	0.046	0.028	5,310	1,730
0.901	22.9	0.972	0.042	0.085	0.959	0.046	0.029	5,470	1,850
0.938	23.8	1.013	0.042	0.085	1.000	0.046	0.031	5,690	2,060
1.000	25.4	1.080	0.042	0.085	1.066	0.046	0.033	6,070	2,330
1.023	26.0	1.105	0.042	0.085	1.091	0.046	0.034	6,210	2,460
1.062	27.0	1.138	0.050	0.103	1.130	0.056	0.034	7,010	2,550
1.125	28.6	1.205	0.050	0.103	1.197	0.056	0.036	7,420	2,860
1.188	30.2	1.271	0.050	0.103	1.262	0.056	0.037	7,840	3,110
1.250	31.7	1.339	0.050	0.103	1.330	0.056	0.040	8,250	3,530

(Continued)

Table 6. *(Continued)* **Internal Heavy Duty (Type B, Class 2) Spiral Retaining Ring Dimensions.**
See Text for Tolerances. *(Source, MIL27426 and Smalley Steel Ring Co.)*

Housing Diameter A		Ring Dimensions			Groove Dimensions			Thrust Capacity	
		Diameter G	Thickness F	Wall E	Diameter C	Width D	Depth D_1	Ring Shear[1]	Groove Yield[2]
inch	mm	Nom.	Nom.	Nom.	Nom.	Nom.	Nom.	lbs.	lbs.
1.312	33.3	1.406	0.050	0.118	1.396	0.056	0.042	8,660	3,900
1.375	34.9	1.471	0.050	0.118	1.461	0.056	0.043	9,070	4,180
1.439	36.5	1.539	0.050	0.118	1.528	0.056	0.045	9,490	4,580
1.456	37.0	1.559	0.050	0.118	1.548	0.056	0.046	9,610	4,730
1.500	38.1	1.605	0.050	0.118	1.594	0.056	0.047	9,900	4,980
1.562	39.6	1.675	0.062	0.128	1.658	0.068	0.048	12,780	5,300
1.625	41.2	1.742	0.062	0.128	1.725	0.068	0.050	13,290	5,740
1.653	42.0	1.772	0.062	0.128	1.755	0.068	0.051	13,520	5,960
1.688	42.8	1.810	0.062	0.128	1.792	0.068	0.052	13,810	6,210
1.750	44.4	1.876	0.062	0.128	1.858	0.068	0.054	14,320	6,680
1.812	46.0	1.940	0.062	0.128	1.922	0.068	0.055	14,820	7,050
1.850	47.0	1.981	0.062	0.158	1.962	0.068	0.056	15,130	7,320
1.875	47.6	2.008	0.062	0.158	1.989	0.068	0.057	15,340	7,560
1.938	49.2	2.075	0.062	0.158	2.056	0.068	0.059	15,850	8,080
2.000	50.8	2.142	0.062	0.158	2.122	0.068	0.061	16,360	8,620
2.062	52.3	2.201	0.078	0.168	2.186	0.086	0.062	21,220	9,040
2.125	53.9	2.267	0.078	0.168	2.251	0.086	0.063	21,870	9,460
2.188	55.5	2.334	0.078	0.168	2.318	0.086	0.065	22,520	10,050
2.250	57.1	2.399	0.078	0.168	2.382	0.086	0.066	23,160	10,500
2.312	58.7	2.467	0.078	0.200	2.450	0.086	0.069	23,800	11,280
2.375	60.3	2.535	0.078	0.200	2.517	0.086	0.071	24,440	11,920
2.440	62.0	2.602	0.078	0.200	2.584	0.086	0.072	25,110	12,420
2.500	63.5	2.667	0.078	0.200	2.648	0.086	0.074	25,730	13,080
2.531	64.2	2.700	0.078	0.200	2.681	0.086	0.075	26,050	13,420
2.562	65.1	2.733	0.093	0.225	2.714	0.103	0.076	29,940	13,760
2.625	66.6	2.801	0.093	0.225	2.781	0.103	0.078	30,680	14,470
2.688	68.2	2.868	0.093	0.225	2.848	0.103	0.080	31,410	15,200
2.750	69.8	2.934	0.093	0.225	2.914	0.103	0.082	32,140	15,940
2.813	71.5	3.001	0.093	0.225	2.980	0.103	0.084	32,880	16,700
2.834	71.9	3.027	0.093	0.225	3.006	0.103	0.086	33,120	17,230
2.875	73.0	3.072	0.093	0.225	3.051	0.103	0.088	33,600	17,880
3.000	76.1	3.204	0.093	0.225	3.182	0.103	0.091	35,060	18,300
3.062	77.7	3.271	0.111	0.281	3.248	0.120	0.093	42,710	20,130
3.125	79.3	3.338	0.111	0.281	3.315	0.120	0.095	43,590	20,990
3.157	80.1	3.371	0.111	0.281	3.348	0.120	0.096	44,040	21,420
3.250	82.5	3.470	0.111	0.281	3.446	0.120	0.098	45,330	22,510

(Continued)

Table 6. *(Continued)* Internal Heavy Duty (Type B, Class 2) Spiral Retaining Ring Dimensions.
See Text for Tolerances. *(Source, MIL27426 and Smalley Steel Ring Co.)*

Housing Diameter A		Ring Dimensions			Groove Dimensions			Thrust Capacity	
		Diameter G	Thickness F	Wall E	Diameter C	Width D	Depth D$_1$	Ring Shear[1]	Groove Yield[2]
inch	mm	Nom.	Nom.	Nom.	Nom.	Nom.	Nom.	lbs.	lbs.
3.346	84.9	3.571	0.111	0.281	3.546	0.120	0.100	46,670	23,650
3.469	88.0	3.701	0.111	0.281	3.675	0.120	0.105	48,320	25,710
3.500	88.8	3.736	0.111	0.281	3.710	0.120	0.105	48,820	25,980
3.543	89.9	3.781	0.111	0.281	3.755	0.120	0.106	49,420	26,550
3.562	90.5	3.802	0.111	0.281	3.776	0.120	0.107	49,690	26,940
3.625	92.0	3.868	0.111	0.281	3.841	0.120	0.108	50,560	27,670
3.750	95.2	4.002	0.111	0.312	3.974	0.120	0.112	52,310	29,690
3.875	98.3	4.136	0.111	0.312	4.107	0.120	0.116	54,050	31,770
3.938	99.9	4.203	0.111	0.312	4.174	0.120	0.118	54,930	32,850
4.000	101.5	4.270	0.111	0.312	4.240	0.120	0.120	55,800	33,930
4.125	104.8	4.369	0.111	0.312	4.339	0.120	0.120	57,540	34,990
4.250	107.9	4.501	0.111	0.312	4.470	0.120	0.120	59,280	36,050
4.330	109.9	4.588	0.111	0.312	4.556	0.120	0.120	60,400	36,730
4.500	114.2	4.768	0.111	0.312	4.735	0.120	0.120	62,770	38,170
4.625	117.4	4.899	0.111	0.312	4.865	0.120	0.120	64,510	39,230
4.750	120.6	5.030	0.111	0.312	4.995	0.120	0.123	66,260	41,300
5.000	126.9	5.297	0.111	0.312	5.260	0.120	0.130	69,740	45,950
5.250	133.2	5.559	0.127	0.350	5.520	0.139	0.135	83,790	50,100
5.375	136.5	5.690	0.127	0.350	5.650	0.139	0.135	85,780	51,290
5.500	139.7	5.810	0.127	0.350	5.770	0.139	0.135	87,780	52,480
5.750	145.9	6.062	0.127	0.350	6.020	0.139	0.135	91,770	54,870
6.000	152.3	6.314	0.127	0.350	6.270	0.139	0.135	95,760	57,260

[1] Based on a ring material shear strength of 120,000 PSI and a safety factor of 3.

[2] Based on a groove material yield of 45,000 PSI and a safety factor of 2.

Assembly of spiral retaining rings

Manual installation on an individual or low production basis within a housing bore or on a shaft is accomplished as follows. 1) Separate coils and insert end of ring into groove. 2) Wind ring into groove. 3) Inspect for proper seating in the groove. See *Figure 2.*

Production assembly is speeded up when special tooling or fixtures are designed. External installation on a shaft can be accomplished with a plunger and tapered plug. The plug, angled at *approximately 6 degrees,* is centered over the shaft end, and a loose fitting plunger pushes the ring into position over the tapered plug. See *Figure 3.*

Internal installation for production assembly is accomplished in a similar manner. A tapered bore sleeve acts as a ring contracting guide, and a plunger pushes the retaining ring into position (see *Figure 4*). Tooling for ring installation should have hardened working surfaces to minimize wear.

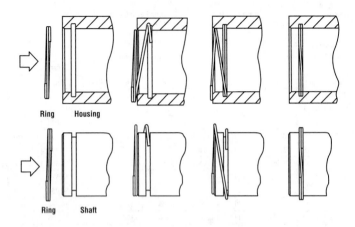

Figure 2. Manual installation of spiral retaining rings. (Source, Smalley Steel Ring Company.)

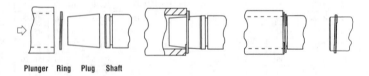

Plunger   Ring   Plug   Shaft

Figure 3. Production installation of external spiral retaining rings. (Source, Smalley Steel Ring Company.)

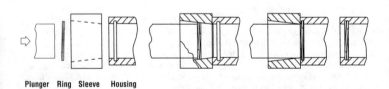

Plunger   Ring   Sleeve   Housing

Figure 4. Production installation of internal spiral retaining rings. (Source, Smalley Steel Ring Company.)

Tapered retaining rings

Tapered retaining rings are designed to provide shoulders for positioning and retaining machine components on shafts. Their tapered design, meaning that the retaining wall thickness is highest directly opposite the opening, and smallest just before the lugs, permits the ring to maintain practically constant circularity and pressure against the bottom of the groove.

Basic tapered retaining rings

Calculations for determining allowable thrust load and allowable load capacity of the groove wall are essentially the same as for spiral retaining rings (see formulas above). However, when calculating the allowable thrust load capacity of the ring, a safety factor (k) of 4 is recommended, because the loaded basic ring is subjected to both pure shear stress *and bending stress*. The values for ring shear strength (S_S), as taken from standard MS16624, are as follows.

Carbon steel or corrosion resistant steel:
  120,000 PSI ultimate shear strength for rings up to and including 0.672" shaft diameter.
  150,000 PSI ultimate shear strength for rings 0.688" and over shaft diameter.

Beryllium copper:
  110,000 PSI ultimate shear strength for all shaft diameters.

When calculating groove wall load capacity, a safety factor of 2 is recommended. The minimum distance between the outer groove wall and end of a shaft should be no less than three times the groove depth.

Care should be taken not to overexpand rings during installation. If an installed ring has play between the groove diameter and the inside ring diameter, the ring has most likely been overexpanded and will be prone to failure.

External tapered retaining rings

Appropriate safe rpm limits for basic external rings are as follows:

Shaft Diameter (inch)	0.125	0.500	1.00	2.00	3.00	4.00	6.00	8.00	10.00
Material	Maximum RPM								
Carbon Steel and Corrosion Resistant Steel	80,000	40,000	20,000	10,000	6,700	5,000	3,400	2,500	2,000
Beryllium Copper	50,000	25,000	13,000	6,400	–	–	–	–	–

Specifications for basic external tapered retaining rings are provided in **Table 7.** Groove depth is equal to shaft diameter minus groove diameter divided by 2. The minimum distance between the outer groove wall and end of the shaft should be no less than three times groove depth. Tolerances that may be applied to the Table are as follows (all dimensions in inches).

Free Inside Diameter (A). For shaft sizes 0.125 to 0.250 incl.: +0.002, –0.004. For sizes 0.276 to 0.500 incl.: +0.002, –0.005. For sizes 0.551 to 1.023 incl.: +0.005, –0.010. For sizes 1.062 to 1.500 incl.: +0.010, –0.015. For sizes 1.562 to 2.000 incl.: +0.013, –0.020. For sizes 2.062 to 2.500 incl.: +0.015, –0.025. For sizes 2.559 to 5.000 incl.: +0.020, –0.030. For sizes 5.250 to 6.000 incl.: +0.020, –0.040. For sizes 6.250 to 6.750 incl.: +0.020, –0.050. For sizes 7.000 to 7.500 incl.: +0.050, –0.130.

Thickness (T). For shaft sizes 0.125 to 0.156 incl.: ±0.001. For sizes 0.188 to 1.500

incl.: ±0.002. For sizes 1.562 to 5.000 incl.: ±0.003. For sizes 5.250 to 6.000 incl.: ±0.004. For sizes 6.250 to 7.500 incl.: ±0.008.

Groove Diameter (G). For shaft sizes 0.125 to 0.250 incl.: ±0.0015. For sizes 0.276 to 0.562 incl.: ±0.002. For sizes 0.594 to 1.023 incl.: ±0.003. For sizes 1.062 to 1.500 incl.: ±0.004. For sizes 1.562 to 2.000 incl.: ±0.005. For sizes 2.062 to 5.000 incl.: ±0.006. For sizes 5.250 to 6.000 incl.: ±0.007. For sizes 6.250 to 7.500 incl.: ±0.008.

Groove Width (W). For shaft sizes 0.125 to 0.236 incl.: +0.002, –0.000. For sizes 0.250 to 1.023 incl.: +0.003, –0.000. For sizes 1.062 to 2.000 incl.: +0.004, –0.000. For sizes 2.062 to 5.000 incl.: +0.005, –0.000. For sizes 5.250 to 6.000 incl.: +0.006. –0.000. For sizes 6.250 to 7.500 incl.: +0.008, –0.000.

Dimensions for metric external tapered retaining rings as specified in DIN 471 are given in **Table 15**.

Table 7. Basic Tapered External Retaining Ring Dimensions. See text for tolerances. *(Source, MS16624.)*

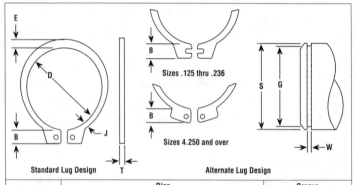

Shaft Diameter S		Ring					Groove	
inch	mm	Inside Diameter D	Lug Height B	Large Section E	Small Section J	Thickness T	Diameter G	Width W
0.125	3.2	0.112	0.046	0.018	0.011	0.010	0.117	0.012
0.156	4.0	0.142	0.054	0.026	0.016	0.010	0.146	0.012
0.188	4.8	0.168	0.050	0.025	0.016	0.015	0.175	0.018
0.197	5.0	0.179	0.056	0.026	0.016	0.015	0.185	0.018
0.219	5.6	0.196	0.056	0.028	0.017	0.015	0.205	0.018
0.236	6.0	0.215	0.056	0.030	0.019	0.015	0.222	0.018
0.250	6.4	0.225	0.080	0.035	0.025	0.025	0.230	0.030
0.276	7.0	0.250	0.081	0.035	0.024	0.025	0.255	0.030
0.281	7.1	0.256	0.080	0.038	0.025	0.025	0.261	0.030
0.312	7.9	0.281	0.087	0.040	0.026	0.025	0.290	0.030
0.344	8.7	0.309	0.087	0.042	0.026	0.025	0.321	0.030
0.354	9.0	0.320	0.087	0.046	0.029	0.025	0.330	0.030

(Continued)

Table 7. *(Continued)* **Basic Tapered External Retaining Ring Dimensions.**
See text for tolerances. *(Source, MS16624.)*

Shaft Diameter S		Ring					Groove	
		Inside Diameter D	Lug Height B	Large Section E	Small Section J	Thickness T	Diameter G	Width W
inch	mm							
0.375	9.5	0.338	0.088	0.050	0.030	0.025	0.352	0.030
0.394	10.0	0.354	0.087	0.052	0.031	0.025	0.369	0.030
0.406	10.3	0.366	0.087	0.054	0.033	0.025	0.382	0.030
0.438	11.1	0.395	0.088	0.055	0.033	0.025	0.412	0.030
0.469	11.9	0.428	0.088	0.060	0.035	0.025	0.443	0.030
0.500	12.7	0.461	0.108	0.065	0.040	0.035	0.468	0.040
0.551	14.0	0.509	0.108	0.053	0.036	0.035	0.519	0.040
0.562	14.3	0.521	0.108	0.072	0.041	0.035	0.530	0.040
0.594	15.1	0.550	0.109	0.076	0.043	0.035	0.559	0.040
0.625	15.9	0.579	0.110	0.080	0.045	0.035	0.588	0.040
0.669	17.0	0.621	0.110	0.082	0.043	0.035	0.629	0.040
0.672	17.1	0.621	0.110	0.082	0.043	0.035	0.631	0.040
0.688	17.5	0.635	0.136	0.084	0.048	0.042	0.646	0.047
0.750	19.0	0.693	0.136	0.092	0.051	0.042	0.704	0.047
0.781	19.8	0.722	0.136	0.094	0.052	0.042	0.733	0.047
0.812	20.6	0.751	0.136	0.096	0.054	0.042	0.762	0.047
0.875	22.2	0.810	0.137	0.104	0.057	0.042	0.821	0.047
0.938	23.8	0.867	0.166	0.110	0.063	0.042	0.882	0.047
0.984	25.0	0.910	0.167	0.114	0.064	0.042	0.926	0.047
1.000	25.4	0.925	0.167	0.116	0.065	0.042	0.940	0.047
1.023	26.0	0.946	0.168	0.118	0.066	0.042	0.961	0.047
1.062	27.0	0.982	0.181	0.122	0.069	0.050	0.998	0.056
1.125	28.6	1.041	0.182	0.128	0.071	0.050	1.059	0.056
1.188	30.2	1.098	0.182	0.132	0.072	0.050	1.118	0.056
1.250	31.7	1.156	0.183	0.140	0.076	0.050	1.176	0.056
1.312	33.3	1.214	0.183	0.146	0.076	0.050	1.232	0.056
1.375	34.9	1.272	0.184	0.152	0.082	0.050	1.291	0.056
1.438	36.5	1.333	0.184	0.160	0.086	0.050	1.350	0.056
1.500	38.1	1.387	0.214	0.168	0.091	0.050	1.406	0.056
1.562	39.7	1.446	0.235	0.172	0.093	0.062	1.468	0.068
1.625	41.3	1.503	0.235	0.180	0.097	0.062	1.529	0.068
1.688	42.9	1.560	0.235	0.184	0.099	0.062	1.589	0.068
1.750	44.4	1.618	0.237	0.188	0.101	0.062	1.650	0.068
1.772	45.0	1.637	0.237	0.190	0.102	0.062	1.669	0.068
1.812	46.0	1.675	0.238	0.192	0.102	0.062	1.708	0.068
1.875	47.6	1.735	0.239	0.196	0.104	0.062	1.769	0.068

(Continued)

Table 7. *(Continued)* **Basic Tapered External Retaining Ring Dimensions.**
See text for tolerances. *(Source, MS16624.)*

Shaft Diameter S		Ring					Groove	
inch	mm	Inside Diameter D	Lug Height B	Large Section E	Small Section J	Thickness T	Diameter G	Width W
1.969	50.0	1.819	0.245	0.200	0.106	0.062	1.857	0.068
2.000	50.3	1.850	0.239	0.204	0.108	0.062	1.886	0.068
2.062	52.4	1.906	0.266	0.208	0.111	0.078	1.946	0.086
2.125	54.0	1.964	0.266	0.212	0.113	0.078	2.003	0.086
2.156	54.8	1.993	0.266	0.212	0.113	0.078	2.032	0.086
2.250	57.1	2.081	0.267	0.220	0.116	0.078	2.120	0.086
2.312	58.7	2.139	0.267	0.222	0.118	0.078	2.178	0.086
2.375	60.3	2.197	0.267	0.224	0.119	0.078	2.239	0.086
2.438	61.9	2.255	0.268	0.228	0.120	0.078	2.299	0.086
2.500	63.5	2.313	0.268	0.232	0.122	0.078	2.360	0.086
2.559	65.0	2.377	0.268	0.238	0.125	0.078	2.419	0.086
2.625	66.7	2.428	0.268	0.242	0.127	0.078	2.481	0.086
2.688	68.3	2.485	0.268	0.246	0.129	0.078	2.541	0.086
2.750	69.8	2.543	0.310	0.248	0.131	0.093	2.602	0.103
2.875	73.0	2.659	0.308	0.256	0.133	0.093	2.721	0.103
2.938	74.6	2.717	0.308	0.260	0.136	0.093	2.779	0.103
3.000	76.2	2.775	0.308	0.264	0.138	0.093	2.838	0.103
3.062	77.8	2.832	0.298	0.252	0.131	0.093	2.898	0.103
3.125	79.4	2.892	0.308	0.272	0.141	0.093	2.957	0.103
3.156	80.2	2.920	0.308	0.274	0.143	0.093	2.986	0.103
3.250	82.0	3.006	0.308	0.280	0.145	0.093	3.076	0.103
3.346	85.0	3.092	0.308	0.286	0.147	0.093	3.166	0.103
3.438	87.3	3.179	0.308	0.292	0.148	0.093	3.257	0.103
3.500	88.9	3.237	0.328	0.285	0.148	0.109	3.316	0.120
3.543	90.0	3.277	0.328	0.288	0.149	0.109	3.357	0.120
3.625	92.0	3.352	0.328	0.296	0.153	0.109	3.435	0.120
3.688	93.7	3.410	0.330	0.302	0.156	0.109	3.493	0.120
3.750	95.2	3.468	0.332	0.310	0.160	0.109	3.552	0.120
3.875	98.4	3.584	0.330	0.318	0.163	0.109	3.673	0.120
3.938	100.0	3.642	0.342	0.318	0.163	0.109	3.734	0.120
4.000	101.6	3.700	0.352	0.318	0.163	0.109	3.792	0.120
4.250	108.0	3.989	0.395	0.318	.176	0.109	4.065	0.120
4.375	111.1	4.106	0.395	0.318	0.181	0.109	4.190	0.120
4.500	114.3	4.223	0.404	0.285	0.128	0.109	4.310	0.120
4.750	120.6	4.458	0.429	0.303	0.136	0.109	4.550	0.120
5.000	127.0	4.692	0.450	0.360	0.194	0.109	4.790	0.120

(Continued)

Table 7. *(Continued)* **Basic Tapered External Retaining Ring Dimensions.**
See text for tolerances. *(Source, MS16624.)*

Shaft Diameter S		Ring					Groove	
inch	mm	Inside Diameter D	Lug Height B	Large Section E	Small Section J	Thickness T	Diameter G	Width W
5.250	133.3	4.927	0.472	0.372	0.211	0.125	5.030	0.139
5.500	139.7	5.162	0.497	0.390	0.209	0.125	5.265	0.139
5.750	146.0	5.396	0.518	0.408	0.220	0.125	5.505	0.139
6.000	152.4	5.631	0.540	0.381	0.171	0.125	5.745	0.139
6.250	158.7	5.866	0.561	0.396	0.176	0.156	5.985	0.174
6.500	165.1	6.100	0.586	0.438	0.236	0.156	6.225	0.174
6.750	171.4	6.335	0.608	0.456	0.246	0.156	6.465	0.174
7.000	177.8	6.570	0.530	0.474	0.256	0.156	6.705	0.174
7.500	190.5	7.009	0.676	0.507	0.277	0.187	7.180	0.209
8.000	203.2	7.478	0.735	0.540	0.294	0.187	7.660	0.209
8.500	215.9	7.947	0.735	0.573	0.314	0.187	8.140	0.209
9.000	228.6	8.415	0.735	0.609	0.333	0.187	8.620	0.209
9.500	241.3	8.885	0.735	0.642	0.350	0.187	9.100	0.209
10.000	254.0	9.355	0.735	0.675	0.367	0.187	9.575	0.209

Except where noted, all dimensions in inches.

Internal tapered retaining rings

Specifications for basic internal tapered retaining rings are provided in **Table 8.** Groove depth is equal to groove diameter minus housing diameter divided by 2. The minimum distance between the outer groove wall and end of the shaft should be no less than three times groove depth. Tolerances that may be applied to the Table are as follows (all dimensions in inches).

Free Outside Diameter (A). For housing sizes 0.250 to 0.777 incl.: +0.010, –0.005. For sizes 0.812 to 1.023 incl.: +0.015, –0.010. For sizes 1.062 to 1.500 incl.: +0.025, –0.020. For sizes 1.562 to 2.000 incl.: +0.035, –0.025. For sizes 2.047 to 3.000 incl.: +0.040, –0.030. For sizes 3.062 to 3.625 incl.: ±0.055. For sizes 3.740 to 6.000 incl.: ±0.065. For sizes 6.250 to 7.000 incl.: ±0.080. For sizes 7.250 to 8.250 incl.: ±0.090.

Thickness (T). For housing sizes 0.250 to 1.500 incl.: ±0.002. For sizes 1.562 to 5.000 incl.: ±0.003. For sizes 5.250 to 6.000 incl.: ±0.004. For sizes 6.250 to 8.250 incl.: ±0.005.

Groove Diameter (G). For housing sizes 0.250 to 0.312 incl.: ±0.001. For sizes 0.375 to 0.750 incl.: ±0.002. For sizes 0.777 to 1.023 incl.: ±0.003. For sizes 1.062 to 1.500 incl.: ±0.004. For sizes 1.562 to 2.000 incl.: ±0.005. For sizes 2.047 to 5.000 incl.: ±0.006. For sizes 5.250 to 6.000 incl.: ±0.007. For sizes 6.250 to 8.250 incl.: ±0.008.

Groove Width (W). For housing sizes 0.250 to 0.312 incl.: +0.002, –0.000. For sizes 0.375 to 1.023 incl.: +0.003, –0.000. For sizes 1.062 to 2.000 incl.: +0.004, –0.000. For sizes 2.047 to 5.000 incl.: +0.005, –0.000. For sizes 5.250 to 6.000 incl.: +0.006, –0.000. For sizes 6.250 to 8.250 incl.: +0.008, –0.000.

Dimensions for metric internal tapered retaining rings as specified in DIN 472 are given in **Table 16.**

Table 8. Basic Tapered Internal Retaining Ring Dimensions. See text for tolerances. *(Source, MS16625.)*

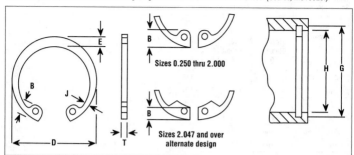

Housing Diameter H		Ring					Groove	
inch	mm	Diameter D	Lug Height B	Large Section E	Small Section J	Thickness T	Diameter G	Width W
0.250	6.4	0.280	0.045	0.025	0.015	0.015	0.268	0.018
0.312	7.9	0.346	0.055	0.033	0.018	0.015	0.330	0.018
0.375	9.5	0.415	0.062	0.040	0.020	0.025	0.397	0.019
0.438	11.1	0.482	0.080	0.049	0.020	0.025	0.461	0.029
0.453	11.5	0.498	0.086	0.050	0.030	0.025	0.477	0.029
0.500	12.7	0.548	0.114	0.052	0.035	0.035	0.530	0.039
0.512	13.0	0.560	0.114	0.053	0.035	0.035	0.645	0.039
0.562	14.3	0.620	0.132	0.055	0.035	0.035	0.656	0.039
0.625	15.9	0.694	0.132	0.060	0.035	0.035	0.685	0.039
0.688	17.5	0.763	0.132	0.063	0.035	0.035	0.732	0.039
0.750	19.0	0.831	0.142	0.070	0.040	0.035	0.796	0.039
0.777	19.7	0.859	0.146	0.074	0.044	0.042	0.825	0.046
0.812	20.0	0.901	0.155	0.077	0.044	0.042	0.862	0.046
0.860	22.0	0.961	0.155	0.081	0.045	0.042	0.920	0.046
0.875	22.2	0.971	0.155	0.084	0.045	0.042	0.931	0.046
0.901	22.9	1.000	0.155	0.087	0.047	0.042	0.958	0.046
0.938	23.8	1.041	0.155	0.091	0.050	0.042	1.000	0.046
1.000	25.4	1.111	0.155	0.104	0.062	0.042	1.066	0.046
1.023	26.0	1.136	0.155	0.108	0.054	0.042	1.091	0.046
1.062	27.0	1.180	0.180	0.110	0.065	0.050	1.130	0.058
1.125	28.6	1.249	0.180	0.116	0.067	0.050	1.197	0.058
1.181	30.0	1.319	0.180	0.120	0.058	0.050	1.255	0.058
1.188	30.2	1.319	0.180	0.120	0.058	0.050	1.262	0.058
1.250	31.7	1.388	0.180	0.124	0.062	0.050	1.330	0.058
1.259	32.0	1.388	0.180	0.124	0.062	0.050	1.330	0.058

(Continued)

Table 8. *(Continued)* **Basic Tapered Internal Retaining Ring Dimensions.**
See text for tolerances. *(Source, MS16625.)*

Housing Diameter H		Ring					Groove	
inch	mm	Diameter D	Lug Height B	Large Section E	Small Section J	Thickness T	Diameter G	Width W
1.312	33.3	1.456	0.180	0.130	0.062	0.050	1.390	0.058
1.375	34.8	1.526	0.180	0.130	0.063	0.050	1.481	0.058
1.378	35.0	1.526	0.180	0.130	0.063	0.050	1.464	0.058
1.438	36.5	1.596	0.180	0.133	0.066	0.050	1.528	0.058
1.450	37.0	1.616	0.180	0.133	0.065	0.050	1.548	0.058
1.500	38.1	1.600	0.180	0.133	0.066	0.050	1.594	0.058
1.562	39.7	1.734	0.202	0.157	0.078	0.062	1.658	0.068
1.575	40.0	1.734	0.202	0.157	0.078	0.082	1.671	0.068
1.625	41.3	1.804	0.227	0.164	0.082	0.052	1.725	0.068
1.653	42.0	1.835	0.227	0.167	0.083	0.052	1.755	0.068
1.688	42.9	1.874	0.227	0.170	0.085	0.062	1.792	0.068
1.750	44.4	1.942	0.234	0.171	0.083	0.062	1.858	0.068
1.812	46.0	2.012	0.234	0.170	0.084	0.062	1.922	0.068
1.850	47.0	2.054	0.234	0.170	0.085	0.062	1.962	0.068
1.875	47.6	2.072	0.234	0.170	0.085	0.062	1.989	0.068
1.938	49.2	2.141	0.234	0.170	0.085	0.062	2.056	0.068
2.000	50.0	2.210	0.240	0.170	0.085	0.078	2.122	0.068
2.047	52.0	2.280	0.250	0.185	0.091	0.078	2.171	0.086
2.062	52.4	2.280	0.250	0.188	0.091	0.078	2.186	0.086
2.125	54.0	2.350	0.260	0.195	0.096	0.078	2.251	0.086
2.165	65.0	2.415	0.264	0.199	0.098	0.078	2.295	0.086
2.188	55.6	2.415	0.264	0.199	0.098	0.078	2.318	0.086
2.250	57.1	2.490	0.270	0.203	0.099	0.078	2.382	0.086
2.312	58.7	2.560	0.270	0.206	0.100	0.078	2.450	0.086
2.375	60.3	2.630	0.270	0.207	0.102	0.078	2.517	0.086
2.440	62.0	2.702	0.280	0.209	0.103	0.078	2.584	0.086
2.500	63.5	2.775	0.280	0.210	0.103	0.078	2.648	0.086
2.531	64.3	2.775	0.280	0.210	0.103	0.078	2.681	0.086
2.562	65.1	2.844	0.280	0.222	0.109	0.093	2.714	0.103
2.625	66.7	2.910	0.280	0.226	0.111	0.093	2.781	0.103
2.677	68.0	2.980	0.300	0.230	0.113	0.093	2.837	0.103
2.688	68.3	2.980	0.300	0.230	0.113	0.093	2.848	0.103
2.750	69.8	3.050	0.300	0.234	0.115	0.093	2.914	0.103
2.812	71.4	3.121	0.300	0.230	0.115	0.093	2.980	0.103
2.835	72.0	3.121	0.300	0.230	0.115	0.093	3.006	0.103
2.875	73.0	3.191	0.310	0.240	0.120	0.093	3.051	0.103

(Continued)

Table 8. *(Continued)* **Basic Tapered Internal Retaining Ring Dimensions.**
See text for tolerances. *(Source, MS16625.)*

Housing Diameter H		Ring					Groove	
inch	mm	Diameter D	Lug Height B	Large Section E	Small Section J	Thickness T	Diameter G	Width W
2.953	75.0	3.325	0.310	0.250	0.122	0.093	3.135	0.103
3.000	76.2	3.325	0.310	0.250	0.122	0.093	3.182	0.103
3.062	77.8	3.418	0.310	0.254	0.125	0.109	3.248	0.120
3.125	79.4	3.488	0.310	0.259	0.129	0.109	3.315	0.120
3.149	80.0	3.523	0.310	0.262	0.129	0.109	3.341	0.120
3.156	80.2	3.523	0.310	0.262	0.129	0.109	3.348	0.120
3.250	82.5	3.623	0.342	0.269	0.135	0.109	3.446	0.120
3.346	85.0	3.734	0.342	0.276	0.140	0.109	3.546	0.120
3.469	88.1	3.857	0.342	0.286	0.144	0.109	3.675	0.120
3.500	88.9	3.890	0.342	0.289	0.142	0.109	3.710	0.120
3.543	90.0	3.936	0.342	0.292	0.142	0.109	3.755	0.120
3.562	90.5	3.936	0.342	0.292	0.142	0.109	3.776	0.120
3.625	92.1	4.024	0.342	0.299	0.150	0.109	3.841	0.120
3.740	95.0	4.157	0.342	0.309	0.155	0.109	3.964	0.120
3.750	95.2	4.157	0.342	0.309	0.155	0.109	3.974	0.120
3.875	98.4	4.291	0.370	0.319	0.160	0.109	4.107	0.120
3.938	100.0	4.358	0.370	0.324	0.161	0.109	4.174	0.120
4.000	101.8	4.424	0.370	0.330	0.166	0.109	4.240	0.120
4.125	104.8	4.558	0.370	0.330	0.171	0.109	4.365	0.120
4.250	108.0	4.691	0.370	0.336	0.180	0.109	4.490	0.120
4.331	110.0	4.756	0.405	0.343	0.180	0.109	4.571	0.120
4.500	114.3	4.940	0.405	0.351	0.181	0.109	4.740	0.120
4.625	117.5	5.076	0.405	0.360	0.183	0.109	4.865	0.120
4.724	120.0	5.213	0.405	0.370	0.183	0.109	4.969	0.120
4.750	120.6	5.213	0.405	0.370	0.183	0.109	4.995	0.120
5.000	127.0	5.485	0.435	0.390	0.186	0.109	5.260	0.120
5.250	133.3	5.770	0.455	0.408	0.196	0.125	5.520	0.039
5.375	136.5	5.910	0.455	0.408	0.198	0.125	5.650	0.139
5.500	139.7	6.066	0.455	0.408	0.198	0.125	5.770	0.139
5.750	146.0	6.336	0.455	0.408	0.198	0.125	6.020	0.139
6.000	152.4	6.620	0.455	0.408	0.196	0.125	6.270	0.139
6.250	158.7	6.895	0.485	0.429	0.211	0.156	6.530	0.174
6.500	165.1	7.170	0.485	0.438	0.219	0.156	6.790	0.174
6.625	168.3	7.308	0.485	0.447	0.221	0.156	6.925	0.174
6.750	171.4	7.445	0.530	0.456	0.224	0.156	7.055	0.174
7.000	177.8	7.720	0.530	0.474	0.232	0.156	7.315	0.174

(Continued)

Table 8. *(Continued)* **Basic Tapered Internal Retaining Ring Dimensions.**
See text for tolerances. *(Source, MS16625.)*

Housing Diameter H		Ring					Groove	
inch	mm	Diameter D	Lug Height B	Large Section E	Small Section J	Thickness T	Diameter G	Width W
7.250	184.1	7.995	0.560	0.480	0.236	0.187	7.575	0.209
7.500	190.5	8.270	0.560	0.507	0.247	0.187	7.840	0.209
7.750	196.8	8.545	0.560	0.523	0.255	0.187	8.100	0.209
8.000	209.2	8.820	0.600	0.540	0.262	0.187	8.360	0.209
8.250	209.5	9.095	0.600	0.558	0.270	0.187	8.620	0.209
8.500	215.9	9.285	0.600	0.579	0.277	0.187	8.680	0.209
8.750	222.2	9.558	0.600	0.591	0.286	0.187	9.145	0.209
9.000	228.6	9.830	0.600	0.609	0.294	0.187	9.405	0.209
9.250	235.0	10.102	0.600	0.625	0.299	0.187	9.668	0.209
9.500	241.3	10.375	0.735	0.642	0.304	0.187	9.950	0.209
9.750	247.7	10.648	0.735	0.658	0.309	0.187	10.190	0.209
10.000	254.0	10.920	0.735	0.675	0.315	0.187	10.450	0.209

Except where noted, all dimensions in inches.

Heavy duty tapered external retaining rings

A heavy duty version of the basic internal ring is thicker for more robust applications. Due to their increased thickness, the allowable thrust capacity formula is modified as follows.

$$P_S = (C_F \times D_O \times t \times S_S \times \pi) \div k$$

where P_S = Allowable thrust load capacity (lbs), based on shear strength of material
C_F = Conversion factor for thicker ring = 1.3
D_O = Shaft or housing diameter, in inches
t = Ring thickness, in inches
S_S = Shear strength of ring material, in PSI
k = Factor of safety (4 recommended).

The formula for allowable load capacity of the groove wall is similarly affected.

$$P_T = (C_F \times D_O \times D_1 \times S_Y \times \pi) \div k$$

where P_T = Allowable load capacity (lbs), based on groove deformation
C_F = Conversion factor for thicker ring = 2
D_1 = Groove depth, in inches
S_Y = Yield strength of the groove material, in PSI
k = Factor of safety (2 recommended).

Appropriate safe rpm limits for heavy duty tapered external rings are as follows:

Shaft Diameter (inch)	0.394	0.437	0.500	0.750	1.00	1.25	1.50	1.75	2.00
Material	Maximum RPM								
Carbon Steel and Corrosion Resistant Steel	80,000	69,000	65,000	40,500	30,000	23,000	18,500	15,500	14,000
Beryllium Copper	51,000	44,000	41,500	26,000	19,000	14,500	12,000	10,000	9,000

Dimensions for crescent type external heavy duty retaining rings are provided in **Table 9**. Groove depth is equal to shaft diameter minus groove diameter divided by 2. The minimum distance between the outer groove wall and end of the shaft should be no less than three times groove depth. Tolerances that may be applied to the Table are as follows (all dimensions in inches).

Free Inside Diameter (A). For shaft size 0.394: +0.003, –0.000. For sizes 0.473 to 1.000 incl.: +0.005, –0.010. For sizes 1.062 to 1.500 incl.: +0.010, –0.015. For sizes 1.562 to 2.000 incl.: +0.013, –0.020.

Thickness (T). For shaft sizes 0.035 to 0.669 incl.: ±0.002. For sizes 0.750 to 1.772 incl.: ±0.003. For sizes 1.938 to 2.000 incl.: ±0.004.

Groove Diameter (G). For shaft sizes 0.325 to 0.500 incl.: +0.001, –0.002. For sizes 0.591 to 1.000 incl.: +0.001, –0.003. For sizes 1.062 to 1.500 incl.: +0.002, –0.004. For sizes 1.562 to 2.000 incl.: +0.003, –0.004.

Groove Width (W). For shaft sizes 0.394 to 0.473 incl.: +0.003, –0.000. For sizes 0.500 to 0.669 incl.: +0.004, –0.000. For sizes 0.750 to 1.772 incl.: +0.005, –0.000. For sizes 1.938 to 2.000 incl.: +0.006, –0.000.

Groove Maximum Bottom Radii. For shaft sizes 0.394 to 1.000 incl.: 0.005 inch. For sizes 1.062 to 2.000 incl.: 0.010 inch.

Table 9. Heavy Duty Tapered External Retaining Ring Dimensions. See text for tolerances. *(Source, MS3217.)*

Shaft Diameter S		Ring					Groove	
inch	mm	Inside Diameter D	Lug Height B	Large Section E	Small Section J	Thickness T	Diameter G	Width W
0.394	10.0	0.362	0.101	0.068	0.039	0.035	0.368	0.040
0.473	12.0	0.435	0.101	0.088	0.053	0.042	0.444	0.047
0.500	12.7	0.460	0.120	0.090	0.050	0.050	0.468	0.056
0.591	15.0	0.543	0.130	0.102	0.057	0.050	0.555	0.056
0.625	15.9	0.575	0.130	0.106	0.059	0.050	0.588	0.056
0.669	17.0	0.616	0.130	0.112	0.062	0.050	0.629	0.056
0.750	19.0	0.689	0.180	0.127	0.077	0.078	0.704	0.086
0.787	20.0	0.689	0.180	0.127	0.077	0.078	0.740	0.086

(Continued)

Table 9. *(Continued)* **Heavy Duty Tapered External Retaining Ring Dimensions.**
See text for tolerances. *(Source, MS3217.)*

Shaft Diameter S		Ring					Groove	
inch	mm	Inside Diameter D	Lug Height B	Large Section E	Small Section J	Thickness T	Diameter G	Width W
0.875	22.2	0.804	0.180	0.148	0.083	0.078	0.821	0.086
0.984	25.0	0.906	0.180	0.151	0.084	0.078	0.925	0.086
1.000	25.4	0.906	0.180	0.151	0.084	0.078	0.938	0.086
1.062	27.0	0.978	0.220	0.161	0.090	0.093	0.998	0.103
1.125	28.6	1.036	0.220	0.169	0.095	0.093	1.059	0.103
1.181	30.0	1.087	0.220	0.176	0.098	0.093	1.111	0.103
1.188	30.2	1.087	0.220	0.176	0.098	0.093	1.111	0.103
1.250	31.7	1.150	0.220	0.185	0.103	0.093	1.174	0.103
1.312	33.3	1.208	0.220	0.192	0.106	0.093	1.214	0.103
1.375	34.9	1.268	0.220	0.200	0.110	0.093	1.291	0.103
1.378	35.0	1.268	0.220	0.200	0.110	0.093	1.291	0.103
1.500	38.1	1.380	0.280	0.218	0.123	0.109	1.406	0.120
1.562	39.7	1.437	0.280	0.228	0.127	0.109	1.468	0.120
1.575	40.0	1.437	0.280	0.228	0.127	0.109	1.480	0.120
1.750	44.4	1.608	0.290	0.254	0.140	0.109	1.650	0.120
1.772	45.0	1.608	0.290	0.254	0.140	0.109	1.669	0.120
1.938	49.2	1.782	0.314	0.280	0.154	0.125	1.826	0.139
1.969	50.0	1.782	0.314	0.280	0.154	0.125	1.850	0.139
2.000	50.8	1.840	0.314	0.290	0.160	0.125	1.880	0.139

Except where noted, all dimensions in inches.

Reduced section crescent type rings

These rings are applied radially and can be used advantageously when fast assembly for production is a primary consideration, and where comparatively small clearance diameters are desirable.

The formulas for determining allowable thrust load capacity (P_S) of the ring, and load capacity of the groove wall, are the same as for basic tapered retaining rings, with the following exception: the safety factor (k) must be doubled. Therefore, the safety factor for thrust load capacity would be 8 (2×4), and the safety factor for load capacity of the wall is 4 (2×2). Minimum distance between the outer groove wall and the end of the shaft should be at least two times groove depth.

Appropriate safe rpm limits for reduced section crescent type external rings are as follows:

Shaft Diameter (inch)	0.250	0.500	1.00	2.00
Material	Maximum RPM			
Carbon Steel and Corrosion Resistant Steel	60,000	25,000	12,500	6,000

Dimensions for reduced section crescent type rings are provided in **Table 10**. Groove depth is equal to shaft diameter minus groove diameter divided by 2. The minimum distance between the outer groove wall and end of the shaft should be no less than two times groove depth. Tolerances that may be applied to the Table are as follows (all dimensions in inches).

Free Inside Diameter (A). For shaft sizes 0.125 to 0.188 incl.: +0.002, –0.004. For sizes 0.219 to 0.437 incl.: +0.003, –0.005. For sizes 0.500 to 0.625 incl.: ±0.006. For sizes 0.687 to 1.000 incl.: ±0.007. For sizes 1.125 to 1.500 incl.: ±0.008. For sizes 1.750 to 2.000 incl.: ±0.010.

Thickness (T). For shaft sizes 0.125 to 1.500 incl.: ±0.002. For sizes 1.7500 to 2.000 incl.: ±0.003.

Groove Diameter (G). For shaft sizes 0.125 to 0.188 incl.: ±0.0015. For sizes 0.219 to 0.437 incl.: ±0.002. For sizes 0.500 to 1.000 incl.: ±0.003. For sizes 1.125 to 1.500 incl.: ±0.004. For sizes 1.750 to 2.000 incl.: ±0.005.

Groove Width (W). For shaft sizes 0.125 to 0.188 incl.: +0.002, –0.000. For sizes 0.219 to 1.000 incl.: +0.003, –0.000. For sizes 1.125 to 2.000 incl.: +0.004, –0.000.

Groove Maximum Bottom Radii. For shaft sizes under 0.500: 0.005 inch. For sizes 0.500 to 1.000 incl.: 0.010 inch. For larger sizes: 0.015 inch.

Table 10. Reduced Section Crescent Type Retaining Ring Dimensions. See text for tolerances.
(Source, MS16632.)

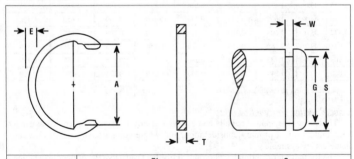

Shaft Diameter S		Ring			Groove	
inch	mm	Inside Diameter A	Large Section E	Thickness T	Diameter G	Width W
0.125	3.2	0.102	0.031	0.015	0.106	0.020
0.156	4.0	0.131	0.037	0.015	0.135	0.020
0.188	4.8	0.161	0.42	0.015	0.165	0.020
0.219	5.6	0.187	0.044	0.025	0.193	0.030
0.236	6.0	0.203	0.046	0.025	0.208	0.030
0.250	6.4	0.211	0.050	0.025	0.220	0.030
0.281	7.1	0.242	0.052	0.025	0.247	0.030

(Continued)

Table 10. *(Continued)* **Reduced Section Crescent Type Retaining Ring Dimensions.** See text for tolerances. *(Source, MS16632.)*

Shaft Diameter S		Ring			Groove	
inch	mm	Inside Diameter A	Large Section E	Thickness T	Diameter G	Width W
0.312	7.9	0.270	0.053	0.025	0.276	0.030
0.375	9.5	0.328	0.060	0.025	0.335	0.030
0.406	10.3	0.359	0.063	0.025	0.364	0.030
0.438	11.1	0.386	0.065	0.025	0.393	0.030
0.500	12.7	0.441	0.070	0.035	0.450	0.040
0.562	14.3	0.497	0.078	0.035	0.507	0.040
0.625	15.9	0.553	0.081	0.035	0.563	0.040
0.688	17.5	0.608	0.086	0.042	0.619	0.047
0.750	19.0	0.665	0.090	0.042	0.676	0.047
0.812	20.6	0.721	0.097	0.042	0.732	0.047
0.875	22.2	0.777	0.105	0.042	0.789	0.047
0.938	23.8	0.830	0.112	0.042	0.843	0.047
1.000	25.4	0.887	0.120	0.042	0.900	0.047
1.125	28.6	0.997	0.135	0.050	1.013	0.056
1.250	31.7	1.110	0.150	0.050	1.126	0.056
1.375	34.9	1.220	0.165	0.050	1.237	0.056
1.500	38.1	1.331	0.180	0.050	1.350	0.056
1.750	44.4	1.555	0.210	0.062	1.576	0.068
2.000	50.8	1.777	0.240	0.062	1.800	0.068

Except where noted, all dimensions in inches.

Grip lock external retaining rings

These smaller diameter rings are designed to use the friction force of high spring pressure to keep them securely against axial displacement from either direction under moderate thrust or vibration. They can be repositioned on the shaft or reused after disassembly.

Appropriate safe rpm limits for grip lock external rings are as follows:

Shaft Diameter (inch)	0.125	0.250	0.312	0.375	0.437	0.500	0.625	0.750
Material	Maximum RPM							
Carbon Steel and Corrosion Resistant Steel	80,000	77,000	58,000	51,000	44,000	40,000	32,000	25,000

Dimensions for grip lock rings, and allowable thrust for specific shaft sizes, are provided in **Table 11**. Groove depth is equal to shaft diameter minus groove diameter divided by 2. The minimum distance between the outer groove wall and end of the shaft should be as follows: for shaft sizes 0.250 to 0.437 incl.: 0.030 minimum; for size 0.500: 0.040 minimum; for size 0.625: 0.045 minimum; for size 0.750: 0.050 minimum. Tolerances that may be applied to the Table are as follows (all dimensions in inches).

Free Inside Diameter (A). For shaft sizes 0.078 to 0.138 incl.: +0.002, –0.003. For sizes 0.154 to 0.252 incl.: +0.002, –0.004. For sizes 0.310 to 0.440 incl.: +0.003, –0.005. For sizes 0.497 to 0.755 incl.: +0.004, –0.0.006.

Thickness (T). For shaft sizes 0.078 to 0.158 incl.: ±0.002. For sizes 0.185 to 0.503 incl.: ±0.003. For sizes 0.622 to 0.755 incl.: ±0.004.

Groove Diameter (G). Grooves are optional for all sizes 0.248 and larger. For shaft sizes 0.248 to 0.316 incl.: +0.005, –0.0015. For sizes 0.373 to 0.628 incl.: +0.001, –0.002. For sizes 0.745 to 0.755 incl.: +0.002, –0.003.

Groove Width (W). For shaft sizes 0.248 to 0.379 incl.: +0.003, –0.000. For sizes 0.434 to 0.755 incl.: +0.004, –0.000.

Table 11. Grip Lock External Retaining Ring Dimensions. See text for tolerances. *(Source, MS90707.)*

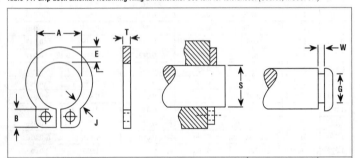

Shaft Diameter S		Ring					Groove		Allowable Thrust Load (lb)
inch	mm	Inside Diameter D	Lug Height B	Large Section E	Small Section J	Thickness T	Diameter G	Width W	
0.094	2.4	0.089	0.078	0.045	0.028	0.025	Sizes 0.094 through 0.187 are not recommended for use with grooves		8
0.125	3.2	0.120	0.078	0.070	0.048	0.025			10
0.156	4.0	0.150	0.078	0.079	0.051	0.025			12
0.187	4.8	0.181	0.097	0.086	0.052	0.035			20
0.250	6.3	0.230	0.097	0.101	0.057	0.035	0.240	0.041	23
0.313	7.9	0.290	0.141	0.114	0.073	0.042	0.303	0.048	25
0.376	9.5	0.354	0.141	0.125	0.075	0.042	0.361	0.048	30
0.437	11.1	0.412	0.151	0.138	0.083	0.050	0.419	0.056	40
0.500	12.7	0.470	0.158	0.140	0.082	0.050	0.478	0.056	45
0.625	15.9	0.593	0.180	0.175	0.100	0.062	0.599	0.069	60
0.750	19.0	0.706	0.233	0.176	0.104	0.062	0.718	0.069	75

Except where noted, all dimensions in inches.

Reduced section clip-on style retaining rings (circlips)

Also called "E" style retaining rings because of their similarity to that capital letter, these rings are popular in applications where light axial loads are encountered. Assembly time is minimal and these rings can be removed quickly without special tools.

Basic E style reduced section external clip-on ring

The formulas for determining allowable thrust load capacity (P_S) of the ring, and load capacity of the groove wall, are the same as for basic tapered retaining rings, but the safety factors (k) are different and must be increased by a factor of three. Thus, the safety factor for thrust load capacity would be 9 (3 × 3), and the safety factor for load capacity of the wall is 6 (3 × 2). Minimum distance between the outer groove wall and the end of the shaft should be at least two times groove depth.

Appropriate safe rpm limits for basic clip-on external rings are as follows:

Shaft Diameter (inch)	0.250	0.500	1.00
Material	Maximum RPM		
Carbon Steel and Corrosion Resistant Steel	25,000	14,000	6,500

Dimensions for basic E style rings are provided in **Table 12**. Groove depth is equal to shaft diameter minus groove diameter divided by 2. The minimum distance between the outer groove wall and end of the shaft should be no less than two times groove depth. Tolerances that may be applied to the Table are as follows (all dimensions in inches).

Free Inside Diameter (A). For shaft sizes 0.040 to 0.250 incl.: +0.001, –0.003. For sizes 0.312 to 0.500 incl.: +0.002, –0.004. For sizes 0.625 to 0.984 incl.: +0.003, –0.005. For sizes 1.188 to 1.375 incl.: +0.006, –0.010.

Thickness (T). For shaft sizes 0.040 to 0.062 incl.: ±0.001. For sizes 0.094 to 0.984 incl.: ±0.002. For sizes 1.188 to 1.375 incl.: ±0.003.

Groove Diameter (G). For shaft sizes 0.040 to 0.219 incl.: +0.002, –0.000. For sizes 0.250 to 0.984 incl.: +0.003, –0.000. For sizes 1.188 to 1.375 incl.: +0.005, –0.000.

Groove Width (W). For shaft sizes 0.040 to 0.125 incl.: +0.002, –0.000. For sizes 0.140 to 0.984 incl.: +0.003, –0.000. For sizes 1.188 to 1.375 incl.: +0.004, –0.000.

Groove Maximum Bottom Radii. For shaft sizes 0.040 to 0.062 incl.: sharp corners. For sizes 0.094 to 0.250 incl.: 0.005 inch. For sizes 0.312 to 0.438 incl.: 0.010 inch. For sizes 0.500 to 1.375 incl.: 0.015 inch.

Dimensions for metric external reduced section rings as specified in DIN 6799 are given in **Table 17**.

Bowed E style reduced section external clip-on ring

The bow in these rings allows them to effectively take up a degree of endplay. Formulas for allowable thrust and groove wall deformation are the same as for basic E style rings. Allowable rpm is also the same.

Dimensions for bowed E style rings are provided in **Table 13**. Groove depth is equal to shaft diameter minus groove diameter divided by 2. The minimum distance between the outer groove wall and end of the shaft should be no less than two times groove depth. Tolerances that may be applied to the Table are as follows (all dimensions in inches).

Free Inside Diameter (A). For shaft sizes 0.110 to 0.250 incl.: +0.002, –0.003. For sizes 0.312 to 0.500 incl.: +0.003, –0.004. For sizes 0.625 to 0.984 incl.: +0.003, –0.005. For sizes 1.188 to 1.375 incl.: +0.006, –0.010.

Thickness (T). For shaft sizes 0.110 to 0.125 incl.: ±0.001. For sizes 0.140 to 0.984 incl.: ±0.002. For sizes 1.188 to 1.375 incl.: ±0.003.

Table 12. Basic E Style Reduced Section External Retaining Ring Dimensions. See text for tolerances. *(Source, MS16633.)*

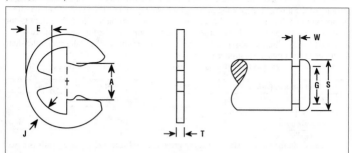

Shaft Diameter S		Ring				Groove	
inch	mm	Inside Diameter A	Large Section E	Small Section E	Thickness T	Diameter G	Width W
0.040	1.0	0.025	0.027	0.012	0.010	0.026	0.012
0.062	1.5	0.051	0.052	0.022	0.010	0.052	0.014
0.094	2.4	0.073	0.057	0.022	0.015	0.074	0.020
0.110	2.8	0.076	0.060	0.030	0.015	0.079	0.020
0.125	3.2	0.094	0.068	0.030	0.015	0.096	0.020
0.140	3.6	0.102	0.084	0.038	0.025	0.105	0.030
0.156	4.0	0.114	0.084	0.038	0.025	0.116	0.030
0.172	4.3	0.125	0.094	0.040	0.025	0.127	0.030
0.188	4.8	0.145	0.095	0.035	0.025	0.147	0.030
0.218	5.6	0.185	0.128	0.045	0.025	0.188	0.030
0.250	6.4	0.207	0.128	0.055	0.025	0.210	0.030
0.312	7.9	0.243	0.138	0.047	0.025	0.250	0.030
0.375	9.5	0.300	0.180	0.060	0.035	0.303	0.040
0.438	11.1	0.337	0.175	0.062	0.035	0.343	0.040
0.500	12.7	0.392	0.204	0.080	0.042	0.396	0.040
0.625	15.9	0.480	0.230	0.090	0.042	0.485	0.047
0.744	18.9	0.516	0.192	0.098	0.050	0.525	0.056
0.750	19.0	0.574	0.273	0.110	0.050	0.580	0.056
0.875	22.2	0.668	0.316	0.120	0.050	0.675	0.056
0.984	25.0	0.822	0.339	0.130	0.050	0.835	0.056
1.188	30.2	1.066	0.280	0.130	0.062	1.079	0.068
1.375	34.9	1.213	0.331	0.140	0.062	1.230	0.068

Except where noted, all dimensions in inches.

Table 13. Bowed E Style Reduced Section External Retaining Ring Dimensions. See text for tolerances. (Source, MS16633.)

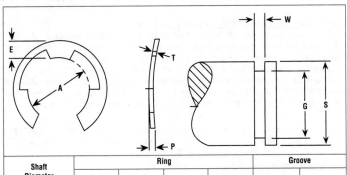

Shaft Diameter S		Ring				Groove	
inch	mm	Inside Diameter A	Large Section E	Thickness T	Bow Height P	Diameter G	Width W
0.110	2.8	0.076	0.155	0.010	0.030	0.079	0.035
0.125	3.2	0.094	0.073	0.010	0.030	0.096	0.022
0.140	3.6	0.102	0.090	0.015	0.033	0.105	0.025
0.156	4.0	0.114	0.090	0.015	0.035	0.116	0.027
0.172	4.4	0.125	0.099	0.015	0.037	0.127	0.029
0.188	4.8	0.145	0.101	0.015	0.038	0.147	0.030
0.219	5.6	0.185	0.132	0.015	0.051	0.188	0.040
0.250	6.4	0.207	0.167	0.025	0.058	0.210	0.049
0.312	7.9	0.243	0.136	0.025	0.058	0.250	0.047
0.375	9.5	0.300	0.188	0.035	0.068	0.303	0.060
0.438	11.1	0.337	0.183	0.035	0.068	0.343	0.060
0.500	12.7	0.392	0.213	0.042	0.084	0.396	0.073
0.625	15.9	0.480	0.240	0.042	0.089	0.485	0.077
0.744	18.9	0.516	0.202	0.050	0.100	0.525	0.085
0.750	19.0	0.574	0.283	0.050	0.100	0.580	0.085
0.875	22.2	0.668	0.326	0.050	0.100	0.675	0.085
0.984	25.0	0.822	0.349	0.050	0.100	0.835	0.085
1.188	30.2	1.066	0.297	0.062	0.124	1.079	0.107
1.375	34.9	1.213	0.348	0.062	0.124	1.230	0.107

Except where noted, all dimensions in inches.

Groove Diameter (G). For shaft sizes 0.110 to 0.125 incl.: +0.002, –0.000. For sizes 0.140 to 0.250 incl.: +0.003, –0.000. For sizes 0.312 to 0.984 incl.: +0.005, –0.000.

Groove Width (W). For all shaft sizes: +0.005, –0.000.

Groove Maximum Bottom Radii. For shaft sizes 0.110 to 0.250 incl.: 0.005 inch. For sizes 0.312 to 0.438 incl.: 0.010 inch. For sizes 0.500 and over: 0.015 inch.

Reinforced E style reduced section external clip-on ring

These rings have unusually large shoulders that allow them to withstand vibrations and shock loads better than basic rings. Formulas for allowable thrust and groove wall deformation are the same as for basic E style rings. Allowable rpm is also the same.

Dimensions for reinforced E style rings are provided in **Table 14**. Groove depth is equal to shaft diameter minus groove diameter divided by 2. The minimum distance between the outer groove wall and end of the shaft should be no less than two times groove depth. Tolerances that may be applied to the Table are as follows (all dimensions in inches).

Free Inside Diameter (A). For shaft sizes 0.094 to 0.125 incl.: +0.001, –0.003. For size 0.156: +0.002, –0.003. For sizes 0.188 to 0.312 incl.: ±0.003. For sizes 0.375 to 0.562 incl.: ±0.004.

Thickness (T). For all shaft sizes: ± 0.002.

Groove Diameter (G). For shaft sizes 0.094 to 0.188 incl.: +0.002, –0.000. For sizes 0.219 to 0.250 incl.: ±0.002. For sizes 0.312 to 0.562 incl.: ±0.003.

Groove Width (W). For shaft sizes 0.094 to 0.125 incl.: +0.002, –0.000. For sizes 0.156 to 0.562 incl.: +0.003, –0.000.

Groove Maximum Bottom Radii. For shaft sizes 0.094 to 0.250 incl.: 0.005 inch. For sizes 0.312 to 0.438 incl.: 0.010 inch. For sizes 0.500 and over: 0.015 inch.

Table 14. Reinforced E Style Reduced Section External Retaining Ring Dimensions. See text for tolerances. *(Source, MS16633.)*

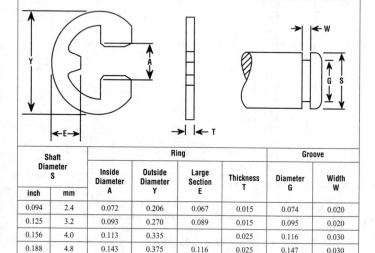

Shaft Diameter S		Ring				Groove	
inch	mm	Inside Diameter A	Outside Diameter Y	Large Section E	Thickness T	Diameter G	Width W
0.094	2.4	0.072	0.206	0.067	0.015	0.074	0.020
0.125	3.2	0.093	0.270	0.089	0.015	0.095	0.020
0.156	4.0	0.113	0.335		0.025	0.116	0.030
0.188	4.8	0.143	0.375	0.116	0.025	0.147	0.030

(Continued)

Table 14. *(Continued)* **Reinforced E Style Reduced Section External Retaining Ring Dimensions.**
See text for tolerances. *(Source: MS16633.)*

Shaft Diameter S		Ring				Groove	
inch	mm	Inside Diameter A	Outside Diameter Y	Large Section E	Thickness T	Diameter G	Width W
0.219	5.6	0.182	0.446	0.132	0.025	0.168	0.030
0.250	6.3	0.204	0.516	0.156	0.025	0.210	0.030
0.312	7.9	0.242	0.588	0.173	0.025	0.250	0.030
0.375	9.5	0.292	0.660	0.184	0.035	0.303	0.040
0.436	11.1	0.332	0.746	0.207	0.035	0.343	0.040
0.500	12.7	0.385	0.810	0.212	0.042	0.396	0.047
0.562	14.3	0.430	0.870	0.220	0.042	0.437	0.047

Except where noted, all dimensions in inches.

Table 15. Dimensions of Metric Normal Type Circlips for Shafts (DIN 471).

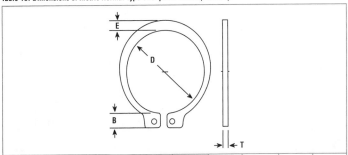

Shaft Diameter S	Inside Diameter D	Thickness T	Lug Height B Max.	Large Section E Approx.	Shaft Diameter S	Inside Diameter D	Thickness T	Lug Height B Max.	Large Section E Approx.
3.00	2.70	0.40	1.90	0.80	14.00	12.90	1.00	3.50	2.10
4.00	3.70	0.40	2.20	0.90	15.00	13.80	1.00	3.60	2.20
5.00	4.70	0.60	2.50	1.10	16.00	14.70	1.00	3.70	2.20
6.00	5.60	0.70	2.70	1.30	17.00	15.70	1.00	3.80	2.30
7.00	6.50	0.80	3.10	1.40	18.00	16.50	1.20	3.90	2.50
8.00	7.40	0.80	3.20	1.50	19.00	17.50	1.20	3.90	2.50
9.00	8.40	1.00	3.30	1.70	20.00	18.50	1.20	4.00	2.60
10.00	9.30	1.00	3.30	1.80	21.00	19.50	1.20	4.10	2.70
11.00	10.20	1.00	3.30	1.80	22.00	20.50	1.20	4.20	2.80
12.00	11.00	1.00	3.30	1.80	23.00	21.40	1.20	4.30	2.90
13.00	11.90	1.00	3.40	1.00	24.00	22.20	1.20	4.40	3.00

(Continued)

Table 15. *(Continued)* Dimensions of Metric Normal Type Circlips for Shafts (DIN 471).

Shaft Diameter S	Inside Diameter D	Thickness T	Lug Height B Max.	Large Section E Approx.	Shaft Diameter S	Inside Diameter D	Thickness T	Lug Height B Max.	Large Section E Approx.
25.00	23.20	1.20	4.40	3.00	72.00	67.50	2.50	8.20	6.80
26.00	24.20	1.20	4.40	3.10	75.00	70.50	2.50	8.40	7.00
27.00	25.10	1.20	4.60	3.10	77.00	72.50	2.50	8.50	7.20
28.00	25.90	1.50	4.70	3.20	78.00	73.50	2.50	8.60	7.30
29.00	26.90	1.50	4.80	3.40	80.00	74.50	2.50	8.60	7.40
30.00	27.90	1.50	5.00	3.50	82.00	76.50	2.50	8.70	7.60
31.00	28.70	1.50	5.10	3.50	85.00	79.50	3.00	8.70	7.80
32.00	29.60	1.50	5.20	3.60	87.00	81.50	3.00	8.80	7.90
33.00	30.50	1.50	5.30	3.70	88.00	82.50	3.00	8.80	8.00
34.00	31.50	1.50	5.40	3.80	90.00	84.50	3.00	8.80	8.20
35.00	32.20	1.50	5.60	3.90	92.00	84.50	3.00	8.80	8.20
36.00	33.20	1.75	5.60	4.00	95.00	89.50	3.00	9.40	8.60
37.00	34.20	1.75	5.70	4.10	97.00	89.50	3.00	9.40	8.60
38.00	35.20	1.75	5.80	4.20	98.00	89.50	3.00	9.40	8.60
39.00	35.80	1.75	5.90	4.30	100.00	94.50	3.00	9.60	9.00
40.00	36.50	1.75	6.00	4.40	102.00	94.50	3.00	9.60	9.00
41.00	37.50	1.75	6.20	4.40	105.00	98.00	4.00	9.90	9.30
42.00	38.50	1.75	6.50	4.50	107.00	98.00	4.00	9.90	9.30
44.00	40.50	1.75	6.60	4.60	108.00	98.00	4.00	9.90	9.30
45.00	41.50	1.75	6.70	4.70	110.00	103.00	4.00	10.10	9.60
46.00	42.50	1.75	6.80	4.80	112.00	103.00	4.00	10.10	9.60
47.00	43.50	1.75	6.80	4.90	115.00	108.00	4.00	10.60	9.80
48.00	44.50	1.75	6.90	5.00	117.00	108.00	4.00	10.60	9.80
50.00	45.80	2.00	6.90	5.10	118.00	108.00	4.00	10.60	9.80
52.00	47.80	2.00	7.00	5.20	120.00	113.00	4.00	11.00	10.20
54.00	49.80	2.00	7.10	5.30	122.00	113.00	4.00	11.00	10.20
55.00	50.80	2.00	7.20	5.40	125.00	118.00	4.00	11.40	10.40
56.00	51.80	2.00	7.30	5.50	127.00	118.00	4.00	11.40	10.40
57.00	52.80	2.00	7.30	5.60	128.00	118.00	4.00	11.40	10.40
58.00	53.80	2.00	7.30	5.65	130.00	123.00	4.00	11.60	10.70
60.00	55.80	2.00	7.40	5.80	132.00	123.00	4.00	11.60	10.70
62.00	57.80	2.00	7.50	6.00	135.00	128.00	4.00	11.80	11.00
63.00	58.80	2.00	7.60	6.20	137.00	128.00	4.00	11.80	11.00
65.00	60.80	2.50	7.80	6.30	138.00	128.00	4.00	11.80	11.00
67.00	62.60	2.50	7.90	6.40	140.00	133.00	4.00	12.00	11.20
68.00	63.50	2.50	8.00	6.50	142.00	133.00	4.00	12.00	11.20
70.00	65.50	2.50	8.10	6.60	145.00	138.00	4.00	12.20	11.50

(Continued)

Table 15. *(Continued)* Dimensions of Metric Normal Type Circlips for Shafts (DIN 471).

Shaft Diameter S	Inside Diameter D	Thickness T	Lug Height B Max.	Large Section E Approx.	Shaft Diameter S	Inside Diameter D	Thickness T	Lug Height B Max.	Large Section E Approx.
147.00	138.00	4.00	12.20	11.50	152.00	142.00	4.00	13.00	11.80
148.00	138.00	4.00	12.20	11.50	155.00	146.00	4.00	13.00	12.00
150.00	142.00	4.00	13.00	11.80	157.00	146.00	4.00	13.00	12.00

All dimensions in millimeters.

Table 16. Dimensions of Metric Normal Type Circlips for Bores (DIN 472).

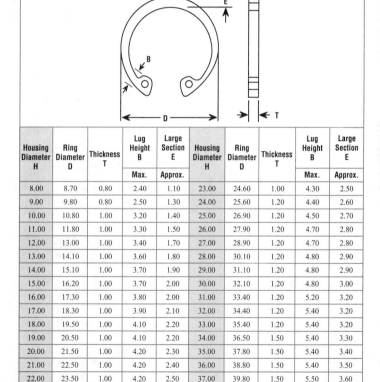

Housing Diameter H	Ring Diameter D	Thickness T	Lug Height B Max.	Large Section E Approx.	Housing Diameter H	Ring Diameter D	Thickness T	Lug Height B Max.	Large Section E Approx.
8.00	8.70	0.80	2.40	1.10	23.00	24.60	1.00	4.30	2.50
9.00	9.80	0.80	2.50	1.30	24.00	25.60	1.20	4.40	2.60
10.00	10.80	1.00	3.20	1.40	25.00	26.90	1.20	4.50	2.70
11.00	11.80	1.00	3.30	1.50	26.00	27.90	1.20	4.70	2.80
12.00	13.00	1.00	3.40	1.70	27.00	28.90	1.20	4.70	2.80
13.00	14.10	1.00	3.60	1.80	28.00	30.10	1.20	4.80	2.90
14.00	15.10	1.00	3.70	1.90	29.00	31.10	1.20	4.80	2.90
15.00	16.20	1.00	3.70	2.00	30.00	32.10	1.20	4.80	3.00
16.00	17.30	1.00	3.80	2.00	31.00	33.40	1.20	5.20	3.20
17.00	18.30	1.00	3.90	2.10	32.00	34.40	1.20	5.40	3.20
18.00	19.50	1.00	4.10	2.20	33.00	35.40	1.20	5.40	3.20
19.00	20.50	1.00	4.10	2.20	34.00	36.50	1.50	5.40	3.30
20.00	21.50	1.00	4.20	2.30	35.00	37.80	1.50	5.40	3.40
21.00	22.50	1.00	4.20	2.40	36.00	38.80	1.50	5.40	3.50
22.00	23.50	1.00	4.20	2.50	37.00	39.80	1.50	5.50	3.60

(Continued)

Table 16. *(Continued)* **Dimensions of Metric Normal Type Circlips for Bores (DIN 472).**

Housing Diameter H	Ring Diameter D	Thickness T	Lug Height B Max.	Large Section E Approx.	Housing Diameter H	Ring Diameter D	Thickness T	Lug Height B Max.	Large Section E Approx.
38.00	40.80	1.50	5.50	3.70	90.00	95.50	3.00	8.60	7.60
39.00	42.10	1.50	5.60	3.80	92.00	97.50	3.00	8.70	7.80
40.00	43.50	1.75	5.80	3.90	95.00	100.50	3.00	8.80	8.10
41.00	44.50	1.75	5.80	4.00	97.00	102.50	3.00	8.90	8.20
42.00	45.50	1.75	5.90	4.10	98.00	103.50	3.00	9.00	8.30
43.00	46.50	1.75	6.00	4.10	100.00	105.50	3.00	9.20	8.40
44.00	47.50	1.75	6.10	4.20	102.00	108.00	4.00	9.50	8.50
45.00	48.50	1.75	6.20	4.30	105.00	112.00	4.00	9.50	8.70
46.00	49.50	1.75	6.30	4.30	107.00	114.00	4.00	9.50	8.80
47.00	50.50	1.75	6.40	4.40	108.00	115.00	4.00	9.50	8.90
48.00	51.10	1.75	6.40	4.50	110.00	117.00	4.00	10.40	9.00
50.00	54.20	2.00	6.50	4.60	112.00	119.00	4.00	10.50	9.10
51.00	55.20	2.00	6.60	4.60	115.00	122.00	4.00	10.50	9.30
52.00	56.20	2.00	6.70	4.70	117.00	122.00	4.00	10.50	9.30
53.00	57.20	2.00	6.70	4.80	118.00	122.00	4.00	10.50	9.30
54.00	58.20	2.00	6.70	4.90	120.00	127.00	4.00	11.00	9.70
55.00	59.20	2.00	6.80	5.00	122.00	127.00	4.00	11.00	9.70
56.00	60.20	2.00	6.80	5.10	125.00	132.00	4.00	11.00	10.00
57.00	61.20	2.00	6.80	5.10	127.00	132.00	4.00	11.00	10.00
58.00	62.20	2.00	6.90	5.20	128.00	132.00	4.00	11.00	10.00
60.00	64.20	2.00	7.30	5.40	130.00	137.00	4.00	11.00	10.20
62.00	66.20	2.00	7.30	5.50	132.00	137.00	4.00	11.00	10.20
63.00	67.20	2.00	7.20	5.60	135.00	142.00	4.00	11.20	10.50
65.00	69.20	2.50	7.60	5.80	137.00	142.00	4.00	11.20	10.50
67.00	70.20	2.50	7.60	5.90	138.00	142.00	4.00	11.20	10.50
68.00	72.50	2.50	7.80	6.10	140.00	147.00	4.00	11.20	10.70
70.00	74.50	2.50	7.80	6.20	142.00	147.00	4.00	11.20	10.70
72.00	76.50	2.50	7.80	6.40	145.00	152.00	4.00	11.40	10.90
75.00	79.50	2.50	7.80	6.60	147.00	152.00	4.00	11.40	10.90
77.00	80.50	2.50	7.80	6.70	148.00	152.00	4.00	11.40	10.90
78.00	82.50	2.50	8.50	6.80	150.00	158.00	4.00	12.00	11.20
80.00	85.50	2.50	8.50	7.00	152.00	158.00	4.00	12.00	11.20
82.00	87.50	2.50	8.50	7.00	155.00	164.00	4.00	12.00	11.40
85.00	90.50	3.00	8.60	7.20	157.00	164.00	4.00	12.00	11.40
87.00	92.50	3.00	8.60	7.30	158.00	164.00	4.00	12.00	11.40
88.00	93.50	3.00	8.60	7.40	160.00	169.00	4.00	13.00	11.60

All dimensions in millimeters.

Table 17. Dimensions of Metric Retaining Washers for Shafts (DIN 6799).

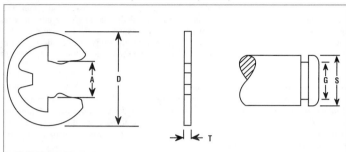

External Dia. D Max.	Inside Diameter A	Thickness T	Use for Shaft Sizes S	Groove Diameter G
4.25	1.28	0.40	2 - 2.5	1.50
4.80	1.61	0.50	2.5 - 3	1.90
6.30	1.94	0.60	3 - 4	2.30
7.30	2.70	0.60	4 - 5	3.20
9.30	3.34	0.70	5 - 7	4.00
11.30	4.11	0.70	6 - 8	5.00
12.30	5.26	0.70	7 - 9	6.00
14.30	5.84	0.90	8 - 11	7.00
16.30	6.52	1.00	9 - 12	8.00
18.80	7.63	1.10	10 - 14	9.00
20.40	8.32	1.20	11 - 15	10.00
23.40	10.45	1.30	13 - 18	12.00
29.40	12.61	1.50	16 - 24	15.00
37.60	15.92	1.75	20 - 31	19.00
44.60	21.98	2.00	25 - 38	24.00

All dimensions in millimeters.

Keys and Keyseats

Keys are used to transmit turning forces (torque) between a shaft and a hub. The key is held in place by a keyseat, which is a rectangularly shaped groove in a shaft or, less frequently, on a hub. The use of keys allows machinery components to be locked into place yet provides for disassembly without great difficulty. While offering many of the same attributes as a fixed splined shaft, they require much less machining and less adherence to accuracy.

Parallel and taper keys and keyseats

Standard sizes for parallel and taper (both plain and gib head) keys are provided in **Table 1**. A gib head key includes a wedge shaped shoulder that, once inserted, provides a surface against which a pry may be used for its removal. Head dimensions for gib head tapered keys are given in **Table 2**. Recommended key sizes for shafts of different diameters are shown in **Table 3**.

Keys are made from two materials—keystock or barstock. Keystock has a close positive tolerance and barstock has a broad, negative tolerance. Consequently, there are two standardized classes of fit. Class 1 provides a comparatively free fit and is for parallel keys produced from barstock. Class 1 fit tolerances are given in **Table 4**. Class 2 fit tolerances are shown in **Table 5**. Class 2 provides either clearance or interference on side fit for parallel rectangular and square keys, and both clearance and interference tolerances for taper keys. Class 2 is designed to be a tight fit and requires keystock. A third class of fit is recognized, but has not been standardized. Class 3 is intended to be a side interference fit, and it is generally accepted that the interference side fit tolerances for parallel rectangular and square keys given in Class 2 should be used for this application.

Precise alignment of keyways is critical for proper fit and reliability. Offset of centerline, parallelism, and lead angle (see *Figure 1*) are all covered. The tolerance for offset is 0.010 maximum. Parallelism tolerances are based on key and keyway tolerances. The tolerances for lead are related to keyseat length, and are as follows:

Keyseat Length	Lead (Maximum)
To 4 inches (incl.)	0.002"
Over 4 in. to 10 in. (incl.)	0.0005" per linear in. of keyseat
Over 10 in.	0.005".

Critical machining dimensions for keyways for square and rectangular parallel and taper keys are given in **Table 6**. The depth of the keyway in the shaft is indicated by dimension S on the Table, and dimension T is the distance from the top of the keyway hub to the bottom of the bore. An additional critical dimension, chordal height (Y), applicable to all keyways, is provided in **Table 7**.

Woodruff keys and keyseats

Cutters for Woodruff keyseats are covered in the Milling section of this book (see Tables 44 – 46 in the Milling section).

Critical dimensions for Woodruff keys are given in **Table 7**. The dimension *Y*, or chordal height, is the distance from the top of the sides of the keyseat and the top of the shaft. It is calculated with the following equation.

$$Y = (D - \sqrt{D^2 - W^2}) \div 2$$

where D = shaft diameter

   W = keyway width.

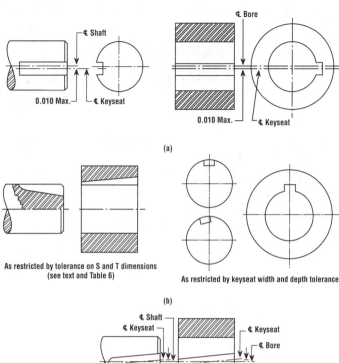

Figure 1. Alignment tolerances: (a) offset; (b) parallelism; (c) lead.

Dimension *T*, on the same Table, provides the distance from the bottom of the hub keyseat to the opposite side of the hub bore. It is calculated with the following equation:

$$T = ([D + H + \sqrt{D^2 - W^2}] \div 2) + C$$

where   *H* = hub seat depth measured from top of keyseat side to bottom of keyseat
        C = tolerance on the dimension (= ± 0.005). The values on the Table include a +0.005 tolerance.

Finally, dimension A on the Table is the height of the shaft and keyseat combined, calculated with the following formula:

$$A = D - Y + (P \div 2)$$

where P = key height from top of side of keyseat to top of key.

An additional dimension, the distance from the bottom of the shaft keyseat to the opposite side of the shaft (usually expressed as S) is not provided on the Table, primarily because one of the critical dimensions used in the equation changes not only with shaft diameter, but also with keyseat width. This dimension can be found in **Table 10** as dimension B. The value of S can then be calculated with the following equation:

$$S = D - Y + (B \div 2)$$

where B = keyseat depth from top of side of key to bottom of keyseat (from **Table 10**).

Basic dimensions of full-width Woodruff keys are given in **Table 8**, and **Table 9** provides dimensions for abbreviated width keys. Keyseat dimensions for Woodruff keys can be found in **Table 10**.

Chamfered keys and filleted keyseats

Although neither chamfered keys nor filleted keyseats are recommended, in some instances they may be preferred to reduce stress concentrations on the corners. Filleted keyseats, when used, should have the largest possible radius, but not so large as to induce excessive bearing stresses due to reduced area of contact. Keys, of course, must be chamfered a sufficient amount to clear the fillet. **Table 11** provides suggested fillet radius and chamfers for general conditions.

Table 1. Dimensions and Tolerances of Plain and Gib Head Keys.

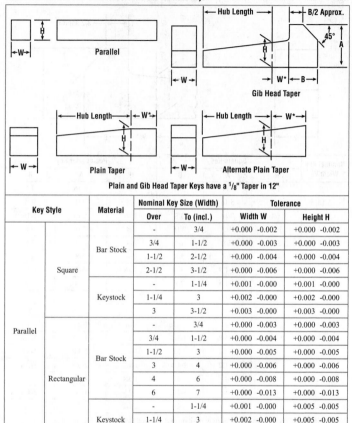

Plain and Gib Head Taper Keys have a $^1/_8$" Taper in 12"

Key Style		Material	Nominal Key Size (Width)		Tolerance	
			Over	To (incl.)	Width W	Height H
Parallel	Square	Bar Stock	-	3/4	+0.000 -0.002	+0.000 -0.002
			3/4	1-1/2	+0.000 -0.003	+0.000 -0.003
			1-1/2	2-1/2	+0.000 -0.004	+0.000 -0.004
			2-1/2	3-1/2	+0.000 -0.006	+0.000 -0.006
		Keystock	-	1-1/4	+0.001 -0.000	+0.001 -0.000
			1-1/4	3	+0.002 -0.000	+0.002 -0.000
			3	3-1/2	+0.003 -0.000	+0.003 -0.000
	Rectangular	Bar Stock	-	3/4	+0.000 -0.003	+0.000 -0.003
			3/4	1-1/2	+0.000 -0.004	+0.000 -0.004
			1-1/2	3	+0.000 -0.005	+0.000 -0.005
			3	4	+0.000 -0.006	+0.000 -0.006
			4	6	+0.000 -0.008	+0.000 -0.008
			6	7	+0.000 -0.013	+0.000 -0.013
		Keystock	-	1-1/4	+0.001 -0.000	+0.005 -0.005
			1-1/4	3	+0.002 -0.000	+0.005 -0.005
			3	7	+0.003 -0.000	+0.005 -0.005
Taper	Plain or Gib Head Square or Rectangular		-	1-1/4	+0.001 -0.000	+0.005 -0.000
			1-1/4	3	+0.002 -0.000	+0.005 -0.000
			3	7	+0.003 -0.000	+0.005 -0.000

* For locating position of dimensions H. Tolerance does not apply.
All dimensions in inches. Source: Extracted from B17.1-1967 (R1973), published by the American Society of Mechanical Engineers.

Table 2. Nominal Dimensions of Gib-Heads.

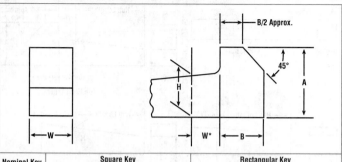

Nominal Key Width W	Square Key			Rectangular Key		
	H	A	B	H	A	B
1/8	1/8	1/4	1/4	3/32	3/16	1/8
3/16	3/16	5/16	5/16	1/8	1/4	1/4
1/4	1/4	7/16	3/8	3/16	5/16	5/16
5/16	5/16	1/2	7/16	1/4	7/16	3/8
3/8	3/8	5/8	1/2	1/4	7/16	3/8
1/2	1/2	7/8	5/8	3/8	5/8	1/2
5/8	5/8	1	3/4	7/16	3/4	9/16
3/4	3/4	1-1/4	7/8	1/2	7/8	5/8
7/8	7/8	1-3/8	1	5/8	1	3/4
1	1	1-5/8	1-1/8	3/4	1-1/4	7/8
1-1/4	1-1/4	2	1-7/16	7/8	1-3/8	1
1-1/2	1-1/2	2-3/8	1-3/4	1	1-5/8	1-1/8
1-3/4	1-3/4	2-3/4	2	1-1/2	2-3/8	1-3/4
2	2	3-1/2	2-1/4	1-1/2	2-3/8	1-3/4
2-1/2	2-1/2	4	3	1-3/4	2-3/4	2
3	3	5	3-1/2	2	3-1/2	2-1/4
3-1/2	3-1/2	6	4	2-1/2	4	3

Notes: Dimension *W of figure is for locating position of dimension H. Tolerance does not apply.
All dimensions in inches. Source: ANSI B17.1-1967 (R1973), published by the American Society of Mechanical Engineers.

Table 3. Recommended Square and Rectangular Key Sizes for Given Shaft Diameters.

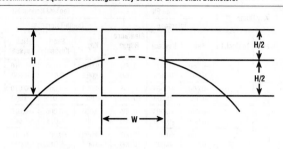

Nominal Shaft Diameter		Nominal Key Size		Nominal Keyseat Depth	
		Square Key W × H	Rectangular Key W × H	Square H/2	Rectangular H/2
Over	To (incl.)				
5/16	7/16	3/32 × 3/32	-	3/64	-
7/16	9/16	1/8 × 1/8	1/8 × 3/32	1/16	3/64
9/16	7/8	3/16 × 3/16	3/16 × 1/8	3/32	1/16
7/8	1-1/4	1/4 × 1/4	1/4 × 3/16	1/8	3/32
1-1/4	1-3/8	5/16 × 5/16	5/16 × 1/4	5/32	1/8
1-3/8	1-3/4	3/8 × 3/8	3/8 × 1/4	3/16	1/8
1-3/4	2-1/4	1/2 × 1/2	1/2 × 3/8	1/4	3/16
2-1/4	2-3/4	5/8 × 5/8	5/8 × 7/16	5/16	7/32
2-3/4	3-1/4	3/4 × 3/4	3/4 × 1/2	3/8	1/4
3-1/4	3-3/4	7/8 × 7/8	7/8 × 5/8	7/16	5/16
3-3/4	4-1/2	1 × 1	1 × 3/4	1/2	3/8
4-1/2	5-1/2	1-1/4 × 1-1/4	1-1/4 × 7/8	5/8	7/16
5-1/2	6-1/2	1-1/2 × 1-1/2	1-1/2 × 1	3/4	1/2
6-1/2	7-1/2	1-3/4 × 1-3/4	1-3/4 × 1-1/2	7/8	3/4
7-1/2	9	2 × 2	2 × 1-1/2	1	3/4
9	11	2-1/2 × 2-1/2	2-1/2 × 1-3/4	1-1/4	7/8
11	13	3 × 3	3 × 2	1-1/2	1
13	15	3-1/2 × 3-1/2	3-1/2 × 2-1/2	1-3/4	1-1/4
15	18	-	4 × 3	-	1-1/2
18	22	-	5 × 3-1/2	-	1-3/4
22	26	-	6 × 4	-	2
26	30	-	7 × 5	-	2-1/2

Notes: Values in nonshaded areas are preferred. All dimensions in inches. Source: ANSI B17.1-1967 (R1973), published by the American Society of Mechanical Engineers.

Table 4. Class 1 Fits for Square and Rectangular Parallel Keys.

Type of Key	Key Width		Side Fit			Top and Bottom Fit			
			Width Tolerance		Clearance Range	Depth Tolerance			Clearance Range
	Over	To (incl.)	Key	Keyseat		Key	Shaft Keyseat	Hub Keyseat	
Square	-	1/2	+0.000 -0.002	+0.002 -0.000	0.004 0.000	+0.000 -0.002	+0.000 -0.015	+0.010 -0.000	0.032 0.005
	1/2	3/4	+0.000 -0.002	+0.003 -0.000	0.005 0.000	+0.000 -0.002	+0.000 -0.015	+0.010 -0.000	0.032 0.005
	3/4	1	+0.000 -0.003	+0.003 -0.000	0.006 0.000	+0.000 -0.003	+0.000 -0.015	+0.010 -0.000	0.033 0.005
	1	1-1/2	+0.000 -0.003	+0.004 -0.000	0.007 0.000	+0.000 -0.003	+0.000 -0.015	+0.010 -0.000	0.033 0.005
	1-1/2	2-1/2	+0.000 -0.004	+0.004 -0.000	0.008 0.000	+0.000 -0.004	+0.000 -0.015	+0.010 -0.000	0.034 0.005
	2-1/2	3-1/2	+0.000 -0.006	+0.004 -0.000	0.010 0.000	+0.000 -0.006	+0.000 -0.015	+0.010 -0.000	0.036 0.005
Rectangular	-	1/2	+0.000 -0.003	+0.002 -0.000	0.005 0.000	+0.000 -0.003	+0.000 -0.015	+0.010 -0.000	0.033 0.005
	1/2	3/4	+0.000 -0.003	+0.003 -0.000	0.006 0.000	+0.000 -0.003	+0.000 -0.015	+0.010 -0.000	0.033 0.005
	3/4	1	+0.000 -0.004	+0.003 -0.000	0.007 0.000	+0.000 -0.004	+0.000 -0.015	+0.010 -0.000	0.034 0.005
	1	1-1/2	+0.000 -0.004	+0.004 -0.000	0.008 0.000	+0.000 -0.004	+0.000 -0.015	+0.010 -0.000	0.034 0.005
	1-1/2	3	+0.000 -0.005	+0.004 -0.000	0.009 0.000	+0.000 -0.005	+0.000 -0.015	+0.010 -0.000	0.035 0.005
	3	4	+0.000 -0.006	+0.004 -0.000	0.010 0.000	+0.000 -0.006	+0.000 -0.015	+0.010 -0.000	0.036 0.005
	4	6	+0.000 -0.008	+0.004 -0.000	0.012 0.000	+0.000 -0.008	+0.000 -0.015	+0.010 -0.000	0.038 0.005
	6	7	+0.000 -0.013	+0.004 -0.000	0.017 0.000	+0.000 -0.013	+0.000 -0.015	+0.010 -0.000	0.043 0.005

All dimensions in inches. Source: ANSI B17.1-1967 (R1973), published by the American Society of Mechanical Engineers.

Table 5. Class 2 Fits for Square and Rectangular Parallel Keys and Taper Keys.

Type of Key	Key Width		Side Fit			Top and Bottom Fit			
			Width Tolerance			Depth Tolerance			
	Over	To (incl.)	Key	Keyseat	CLearance/ INTerference	Key	Shaft Keyseat	Hub Keyseat	CLearance/ INTerference
Parallel Square	-	1 1/4	+0.001 -0.000	+0.002 -0.000	0.002 CL 0.001 INT	+0.001 -0.000	+0.000 -0.015	+0.010 -0.000	0.030 CL 0.004 CL
	1-1/4	3	+0.002 -0.000	+0.002 -0.000	0.002 CL 0.002 INT	+0.002 -0.000	+0.000 -0.015	+0.010 -0.000	0.030 CL 0.003 CL
	3	3-1/2	+0.003 -0.000	+0.002 -0.000	0.002 CL 0.003 INT	+0.003 -0.000	+0.000 -0.015	+0.010 -0.000	0.030 CL 0.002 CL
Parallel Rectangular	-	1-1/4	+0.001 -0.000	+0.002 -0.000	0.002 CL 0.001 INT	+0.005 -0.005	+0.000 -0.015	+0.010 -0.000	0.035 CL 0.000 CL
	1-1/4	3	+0.002 -0.000	+0.002 -0.000	0.002 CL 0.002 INT	+0.005 -0.005	+0.000 -0.015	+0.010 -0.000	0.035 CL 0.000 CL
	3	7	+0.003 -0.000	+0.002 -0.000	0.002 CL 0.003 INT	+0.005 -0.005	+0.000 -0.015	+0.010 -0.000	0.035 CL 0.000 CL
Taper	-	1-1/4	+0.001 -0.000	+0.002 -0.000	0.002 CL 0.001 INT	+0.005 -0.000	+0.000 -0.015	+0.010 -0.000	0.005 CL 0.025 INT
	1-1/4	3	+0.002 -0.000	+0.002 -0.000	0.002 CL 0.002 INT	+0.005 -0.000	+0.000 -0.015	+0.010 -0.000	0.005 CL 0.025 INT
	3	See Note	+0.003 -0.000	+0.002 -0.000	0.002 CL 0.003 INT	+0.005 -0.000	+0.000 -0.015	+0.010 -0.000	0.005 CL 0.025 INT

Note: Values for square keys to 3 $1/2$ inches (incl.) in width, and rectangular keys to 7 inches (incl.) in width.
All dimensions in inches. Source: ANSI B17.1-1967 (R1973), published by the American Society of Mechanical Engineers.

Table 6. Depth Control Values for Shafts and Hubs for Square and Parallel Keys.

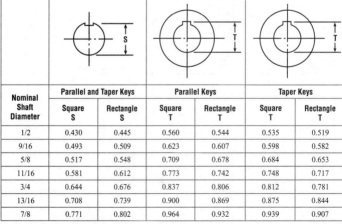

Nominal Shaft Diameter	Parallel and Taper Keys		Parallel Keys		Taper Keys	
	Square S	Rectangle S	Square T	Rectangle T	Square T	Rectangle T
1/2	0.430	0.445	0.560	0.544	0.535	0.519
9/16	0.493	0.509	0.623	0.607	0.598	0.582
5/8	0.517	0.548	0.709	0.678	0.684	0.653
11/16	0.581	0.612	0.773	0.742	0.748	0.717
3/4	0.644	0.676	0.837	0.806	0.812	0.781
13/16	0.708	0.739	0.900	0.869	0.875	0.844
7/8	0.771	0.802	0.964	0.932	0.939	0.907

(Continued)

Table 6. *(Continued)* Depth Control Values for Shafts and Hubs for Square and Parallel Keys.

Nominal Shaft Diameter	Parallel and Taper Keys		Parallel Keys		Taper Keys	
	Square S	Rectangle S	Square T	Rectangle T	Square T	Rectangle T
15/16	0.796	0.827	1.051	1.019	1.026	0.994
1	0.859	0.890	1.114	1.083	1.089	1.058
1-1/16	0.923	0.954	1.178	1.146	1.153	1.121
1-1/8	0.986	1.017	1.241	1.210	1.216	1.185
1-3/16	1.049	1.080	1.304	1.273	1.279	1.248
1-1/4	1.112	1.144	1.367	1.336	1.342	1.311
1-5/16	1.137	1.169	1.455	1.424	1.430	1.399
1-3/8	1.201	1.232	1.518	1.487	1.493	1.462
1-7/16	1.225	1.288	1.605	1.543	1.580	1.518
1-1/2	1.289	1.351	1.669	1.606	1.644	1.581
1-9/16	1.352	1.415	1.732	1.670	1.707	1.645
1-5/8	1.416	1.478	1.796	1.733	1.771	1.708
1-11/16	1.479	1.541	1.859	1.796	1.834	1.771
1-3/4	1.542	1.605	1.922	1.860	1.897	1.835
1-13/16	1.527	1.590	2.032	1.970	2.007	1.945
1-7/8	1.591	1.654	2.096	2.034	2.071	2.009
1-15/16	1.655	1.717	2.160	2.097	2.135	2.072
2	1.718	1.781	2.223	2.161	2.198	2.136
2-1/16	1.782	1.844	2.287	2.224	2.262	2.199
2-1/8	1.845	1.908	2.350	2.288	2.325	2.263
2-3/16	1.909	1.971	2.414	2.351	2.389	2.326
2-1/4	1.972	2.034	2.477	2.414	2.452	2.389
2-5/16	1.957	2.051	2.587	2.493	2.562	2.468
2-3/8	2.021	2.114	2.651	2.557	2.626	2.532
2-7/16	2.084	2.178	2.714	2.621	2.689	2.596
2-1/2	2.148	2.242	2.778	2.684	2.753	2.659
2-9/16	2.211	2.305	2.841	2.748	2.816	2.723
2-5/8	2.275	2.369	2.905	2.811	2.880	2.786
2-11/16	2.338	2.432	2.968	2.874	2.943	2.849
2-3/4	2.402	2.495	3.032	2.938	3.007	2.913
2-13/16	2.387	2.512	3.142	3.017	3.117	2.992
2-7/8	2.450	2.575	3.205	3.080	3.180	3.055
2-15/16	2.514	2.639	3.269	3.144	3.244	3.119
3	2.577	2.702	3.332	3.207	3.307	3.182
3-1/16	2.641	2.766	3.396	3.271	3.371	3.246
3-1/8	2.704	2.829	3.459	3.334	3.434	3.309
3-3/16	2.768	2.893	3.523	3.398	3.498	3.373
3-1/4	2.831	2.956	3.586	3.461	3.561	3.436

(Continued)

Table 6. *(Continued)* Depth Control Values for Shafts and Hubs for Square and Parallel Keys.

Nominal Shaft Diameter	Parallel and Taper Keys		Parallel Keys		Taper Keys	
	Square S	Rectangle S	Square T	Rectangle T	Square T	Rectangle T
3-5/16	2.816	2.941	3.696	3.571	3.671	3.546
3-3/8	2.880	3.005	3.760	3.635	3.735	3.610
3-7/16	2.943	3.068	3.823	3.698	3.798	3.673
3-1/2	3.007	3.132	3.887	3.762	3.862	3.737
3-9/16	3.070	3.195	3.950	3.825	3.925	3.800
3-5/8	3.134	3.259	4.014	3.889	3.989	3.864
3-11/16	3.197	3.322	4.077	3.952	4.052	3.927
3-3/4	3.261	3.386	4.141	4.016	4.116	3.991
3-13/16	3.246	3.371	4.251	4.126	4.226	4.101
3-7/8	3.309	3.434	4.314	4.189	4.289	4.164
3-15/16	3.373	3.498	4.378	4.253	4.353	4.228
4	3.436	3.561	4.441	4.316	4.416	4.291
4-3/16	3.627	3.752	4.632	4.507	4.607	4.482
4-1/4	3.690	3.815	4.695	4.570	4.670	4.545
4-3/8	3.817	3.942	4.822	4.697	4.797	4.672
4-7/16	3.880	4.005	4.885	4.760	4.860	4.735
4-1/2	3.944	4.069	4.949	4.824	4.924	4.799
4-3/4	4.041	4.229	5.296	5.109	5.271	5.084
4-7/8	4.169	4.356	5.424	5.236	5.399	5.211
4-15/16	4.232	4.422	5.487	5.300	5.462	5.275
5	4.296	4.483	5.551	5.363	5.526	5.338
5-3/16	4.486	4.674	5.741	5.554	5.716	5.529
5-1/4	4.550	4.737	5.805	5.617	5.780	5.592
5-7/16	4.740	4.927	5.995	5.807	5.970	5.782
5-1/2	4.803	4.991	6.058	5.871	6.033	5.846
5-3/4	4.900	5.150	6.405	6.155	6.380	6.130
5-15/16	5.091	5.341	6.596	6.346	6.571	6.321
6	5.155	5.405	6.660	6.410	6.635	6.385
6-1/4	5.409	5.659	6.914	6.664	6.889	6.639
6-1/2	5.662	5.912	7.167	6.917	7.142	6.892
6-3/4	5.760	5.885	7.515	7.390	7.490	7.365
7	6.014	6.139	7.769	7.644	7.744	7.619
7-1/4	6.268	6.393	8.023	7.898	7.998	7.873
7-1/2	6.521	6.646	8.276	8.151	8.251	8.126
7-3/4	6.619	6.869	8.624	8.374	8.599	8.349
8	6.873	7.123	8.878	8.628	8.853	8.603
9	7.887	8.137	9.892	9.642	9.867	9.617
10	8.591	8.966	11.096	10.721	11.071	10.696

(Continued)

Table 6. *(Continued)* **Depth Control Values for Shafts and Hubs for Square and Parallel Keys.**

Nominal Shaft Diameter	Parallel and Taper Keys		Parallel Keys		Taper Keys	
	Square S	Rectangle S	Square T	Rectangle T	Square T	Rectangle T
11	9.606	9.981	12.111	11.736	12.086	11.711
12	10.309	10.809	13.314	12.814	13.289	12.789
13	11.325	11.825	14.330	13.830	14.305	13.805
14	12.028	12.528	15.533	15.033	15.508	15.008
15	13.043	13.543	16.548	16.048	16.523	16.023

All dimensions in inches. Source: ANSI B17.1-1967 (R1973), published by the American Society of Mechanical Engineers.

Table 7. Critical Measurements for Woodruff Key and Keyways.

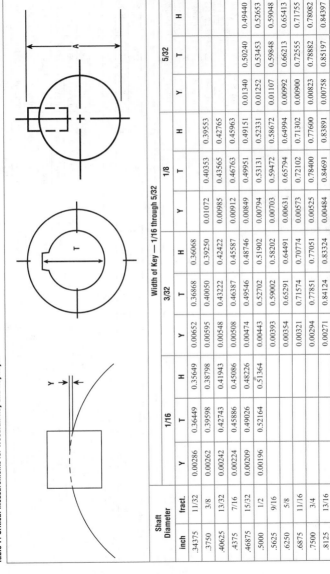

Shaft Diameter		1/16			3/32			1/8			5/32		
inch	fract.	Y	T	H	Y	T	H	Y	T	H	Y	T	H
.34375	11/32	0.00286	0.36449	0.35649	0.00652	0.36868	0.36068						
.3750	3/8	0.00262	0.39598	0.38798	0.00595	0.40050	0.39250	0.01072	0.40353	0.39553			
.40625	13/32	0.00242	0.42743	0.41943	0.00548	0.43222	0.42422	0.00985	0.43565	0.42765			
.4375	7/16	0.00224	0.45886	0.45086	0.00508	0.46387	0.45587	0.00912	0.46763	0.45963			
.46875	15/32	0.00209	0.49026	0.48226	0.00474	0.49546	0.48746	0.00849	0.49951	0.49151	0.01340	0.50240	0.49440
.5000	1/2	0.00196	0.52164	0.51364	0.00443	0.52702	0.51902	0.00794	0.53131	0.52331	0.01252	0.53453	0.52653
.5625	9/16				0.00393	0.59002	0.58202	0.00703	0.59472	0.58672	0.01107	0.59848	0.59048
.6250	5/8				0.00354	0.65291	0.64491	0.00631	0.65794	0.64994	0.00992	0.66213	0.65413
.6875	11/16				0.00321	0.71574	0.70774	0.00573	0.72102	0.71302	0.00900	0.72555	0.71755
.7500	3/4				0.00294	0.77851	0.77051	0.00525	0.78400	0.77600	0.00823	0.78882	0.78082
.8125	13/16				0.00271	0.84124	0.83324	0.00484	0.84691	0.83891	0.00758	0.85197	0.84397

Width of Key — 1/16 through 5/32

(Continued)

Table 7. (Continued) Critical Measurements for Woodruff Key and Keyways.

| Shaft Diameter | | Width of Key — 1/16 through 5/32 | | | | | | | | | | | | |
| inch | fract. | 1/16 | | | 3/32 | | | 1/8 | | | 5/32 | | |
		Y	T	H	Y	T	H	Y	T	H	Y	T	H
.8750	7/8				0.00252	0.90393	0.89593	0.00449	0.90976	0.90176	0.00703	0.91502	0.90702
.9375	15/16							0.00419	0.97256	0.96456	0.00656	0.97799	0.96999
1.0000	1							0.00392	1.03533	1.02733	0.00614	1.04091	1.03291
1.0625	1 1/16							0.00369	1.09806	1.09006	0.00578	1.10377	1.09577
1.1250	1 1/8							0.00348	1.16077	1.15277	0.00545	1.16660	1.15860
1.1875	1 3/16							0.00330	1.22345	1.21545	0.00516	1.22939	1.22139
1.2500	1 1/4							0.00313	1.28612	1.27812	0.00490	1.29215	1.28415
1.3750	1 3/8										0.00445	1.41760	1.40960
1.5000	1 1/2										0.00408	1.54297	1.53497
1.6250	1 5/8										0.00376	1.66829	1.66029

| Shaft Diameter | | Width of Key — 3/16 through 5/16 | | | | | | | | | | | | |
| inch | fract. | 3/16 | | | 7/32 | | | 1/4 | | | 5/16 | | |
		Y	T	H	Y	T	H	Y	T	H	Y	T	H
.5625	9/16	0.01608	0.60127	0.59327									
.6250	5/8	0.01439	0.66546	0.65746	0.01977	0.66788	0.65988						
.6875	11/16	0.01303	0.72932	0.72132	0.01786	0.73229	0.72429	0.02353	0.73447	0.72647			
.7500	3/4	0.01191	0.79294	0.78494	0.01630	0.79635	0.78835	0.02145	0.79905	0.79105	0.003410	0.80200	0.79400
.8125	13/16	0.01097	0.85638	0.84838	0.01500	0.86015	0.85215	0.01971	0.86329	0.85529	0.03125	0.86735	0.85935
.8750	7/8	0.01016	0.91969	0.91169	0.01389	0.92376	0.91576	0.01824	0.92726	0.91926	0.02885	0.93225	0.92425
.9375	15/16	0.00947	0.98288	0.97488	0.01294	0.98721	0.97921	0.01697	0.99103	0.98303	0.02681	0.99679	0.98879
1.0000	1	0.00887	1.04598	1.03798	0.01211	1.05054	1.04254	0.01588	1.05462	1.04662	0.02504	1.06106	1.05306
1.0625	1 1/16	0.00834	1.10901	1.10101	0.01138	1.11377	1.10577	0.01492	1.11808	1.11008	0.02350	1.12510	1.11710

(Continued)

Table 7. *(Continued)* Critical Measurements for Woodruff Key and Keyways.

Width of Key — 3/16 through 5/16

Shaft Diameter		3/16			7/32			1/4			5/16		
inch	fract.	Y	T	H	Y	T	H	Y	T	H	Y	T	H
1.1250	1 1/8	0.00787	1.17198	1.16398	0.01074	1.17691	1.16891	0.01406	1.18144	1.17344	0.02214	1.18896	1.18096
1.1875	1 3/16	0.00745	1.23490	1.22690	0.01016	1.23999	1.23199	0.01331	1.24469	1.23669	0.02093	1.25267	1.24467
1.2500	1 1/4	0.00707	1.29778	1.28978	0.00964	1.30301	1.29501	0.01263	1.30787	1.29987	0.01985	1.31625	1.30825
1.3750	1 3/8	0.00642	1.42343	1.41543	0.00876	1.42889	1.42089	0.01146	1.43404	1.42604	0.01799	1.44311	1.43511
1.5000	1 1/2	0.00588	1.54897	1.54097	0.00802	1.55463	1.54663	0.01049	1.56001	1.55201	0.01646	1.56964	1.56164
1.6250	1 5/8	0.00543	1.67442	1.66642	0.00740	1.68025	1.67225	0.00967	1.68583	1.67783	0.01517	1.69593	1.68793
1.7500	1 3/4	0.00504	1.79981	1.79181	0.00686	1.80579	1.79779	0.00897	1.81153	1.80353	0.01406	1.82204	1.81404
1.8750	1 7/8	0.00470	1.92515	1.91715	0.00640	1.93153	1.92353	0.00837	1.93713	1.92913	0.01311	1.94799	1.93999
2.0000	2	0.00440	2.05045	2.04245	0.00600	2.05665	2.04865	0.00784	2.06266	2.05466	0.01228	2.07382	2.06582
2.0625	2 1/16				0.00582	2.11933	2.11133	0.00760	2.12540	2.11740	0.01191	2.13669	2.12869
2.1250	2 1/8				0.00564	2.18201	2.17401	0.00738	2.18812	2.18012	0.01155	2.19955	2.19155
2.1875	2 3/16							0.00717	2.25083	2.24283	0.01122	2.26238	2.25438
2.2500	2 1/4							0.00699	2.31353	2.30553	0.01090	2.32520	2.31720
2.3750	2 3/8										0.01032	2.45078	2.44278

Width of Key — 3/8 through 9/16

Shaft Diameter		3/8			7/16			1/2			9/16		
inch	fract.	Y	T	H	Y	T	H	Y	T	H	Y	T	H
.9375	15/16	0.03913	1.00012	0.99212									
1.0000	1	0.03649	1.06526	1.05726									
1.0625	1 1/16	0.03419	1.13006	1.12206									
1.1250	1 1/8	0.03217	1.19458	1.18658	0.04428	1.19807	1.19007						

(Continued)

Table 7. (Continued) Critical Measurements for Woodruff Key and Keyways.

| Shaft Diameter | | Width of Key — 3/8 through 9/16 | | | | | | | | | | | | |
| inch | fract. | 3/8 | | | 7/16 | | | 1/2 | | | 9/16 | | |
		Y	T	H	Y	T	H	Y	T	H	Y	T	H
1.1875	1 3/16	0.03038	1.25887	1.25087	0.04176	1.26309	1.25509						
1.2500	1 1/4	0.02879	1.32296	1.31496	0.03953	1.32782	1.31982						
1.3750	1 3/8	0.02606	1.45069	1.44269	0.03573	1.45662	1.44862	0.04707	1.46093	1.45293			
1.5000	1 1/2	0.02382	1.57793	1.56993	0.03261	1.58474	1.57674	0.04289	1.59011	1.58211			
1.6250	1 5/8	0.02193	1.70482	1.69682	0.03000	1.71235	1.70435	0.03942	1.71858	1.71058	0.05023	1.72337	1.71537
1.7500	1 3/4	0.02033	1.83142	1.82342	0.02778	1.83957	1.83157	0.03647	1.84653	1.83853	0.04643	1.85217	1.84417
1.8750	1 7/8	0.01894	1.95781	1.94981	0.02588	1.96647	1.95847	0.03395	1.97405	1.96605	0.04318	1.98042	1.97242
2.0000	2	0.01774	2.08401	2.07601	0.02422	2.09313	2.08513	0.03175	2.10125	2.09325	0.04037	2.10823	2.10023
2.0625	2 1/16	0.01719	2.14706	2.13906	0.02347	2.15638	2.14838	0.03076	2.16474	2.15674	0.03909	2.17201	2.16401
2.1250	2 1/8	0.01667	2.21008	2.20208	0.02276	2.21959	2.21159	0.02983	2.22817	2.22017	0.03790	2.23570	2.22770
2.1875	2 3/16	0.01619	2.27306	2.26506	0.02210	2.28275	2.27475	0.02895	2.29155	2.28355	0.03678	2.29932	2.29132
2.2500	2 1/4	0.01574	2.33601	2.32801	0.02147	2.34588	2.33788	0.02813	2.35487	2.34687	0.03572	2.36288	2.35488
2.3750	2 3/8	0.01490	2.46185	2.45385	0.02032	2.47203	2.46403	0.02661	2.48139	2.47339	0.03379	2.48981	2.48181
2.5000	2 1/2	0.01414	2.5876	2.57961	0.01929	2.59806	2.59006	0.02526	2.60774	2.59974	0.03205	2.61655	2.60855
2.6250	2 5/8	0.01346	2.71329	2.70529	0.01836	2.72399	2.71599	0.02403	2.73397	2.72597	0.03049	2.74311	2.73511
2.7500	2 3/4				0.01751	2.84984	2.84184	0.02292	2.86008	2.85208	0.02907	2.86953	2.86153
2.8750	2 7/8				0.01674	2.97561	2.96761	0.02191	2.98609	2.97809	0.02778	2.99582	2.98782
3.0000	3							0.02098	3.11202	3.10402	0.02660	3.12200	3.11400

(Continued)

Table 7. *(Continued)* Critical Measurements for Woodruff Key and Keyways.

| Shaft Diameter | | 5/8 | | | Width of Key — 5/8 through 3/4 11/16 | | | 3/4 | | |
inch	fract.	Y	T	H	Y	T	H	Y	T	H
1.8750	1 7/8	0.05362	1.98563	1.97763						
2.0000	2	0.05008	2.11417	2.10617						
2.0625	2 1/16	0.04849	2.17826	2.17026						
2.1250	2 1/8	0.04700	2.24225	2.23425	0.05714	2.244771	2.23971	0.06838	2.25212	2.24412
2.1875	2 3/16	0.04559	2.30616	2.29816	0.05542	2.31193	2.30393	0.06629	2.31671	2.30871
2.2500	2 1/4	0.04427	2.36998	2.36198	0.05380	2.37605	2.36805	0.06434	2.38116	2.37316
2.3750	2 3/8	0.04186	2.49739	2.48939	0.05084	2.50401	2.49601	0.06077	2.50973	2.50173
2.5000	2 1/2	0.03969	2.62456	2.61656	0.04819	2.63166	2.62366	0.05758	2.63792	2.62992
2.6250	2 5/8	0.03775	2.75150	2.74350	0.04581	2.75904	2.75104	0.05471	2.76579	2.75779
2.7500	2 3/4	0.03598	2.87827	2.87027	0.04366	2.88619	2.87819	0.05212	2.89338	2.88538
2.8750	2 7/8	0.03438	3.00487	2.99687	0.04171	3.01314	3.00514	0.04977	3.02073	3.01273
3.0000	3	0.03291	3.13134	3.12334	0.03992	3.1399	3.13193	0.04763	3.14787	3.13987

All dimensions in inches.

Table 8. Woodruff Key Dimensions for Full-Length Keys.

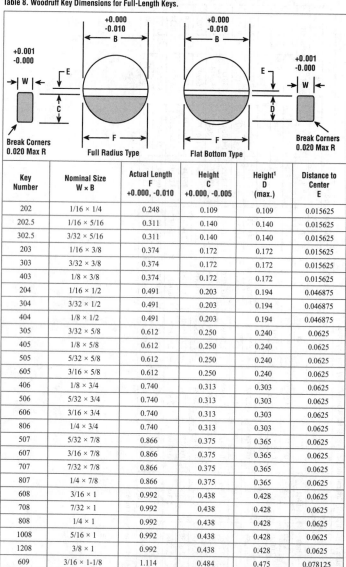

Key Number	Nominal Size W × B	Actual Length F +0.000, -0.010	Height C +0.000, -0.005	Height[1] D (max.)	Distance to Center E
202	1/16 × 1/4	0.248	0.109	0.109	0.015625
202.5	1/16 × 5/16	0.311	0.140	0.140	0.015625
302.5	3/32 × 5/16	0.311	0.140	0.140	0.015625
203	1/16 × 3/8	0.374	0.172	0.172	0.015625
303	3/32 × 3/8	0.374	0.172	0.172	0.015625
403	1/8 × 3/8	0.374	0.172	0.172	0.015625
204	1/16 × 1/2	0.491	0.203	0.194	0.046875
304	3/32 × 1/2	0.491	0.203	0.194	0.046875
404	1/8 × 1/2	0.491	0.203	0.194	0.046875
305	3/32 × 5/8	0.612	0.250	0.240	0.0625
405	1/8 × 5/8	0.612	0.250	0.240	0.0625
505	5/32 × 5/8	0.612	0.250	0.240	0.0625
605	3/16 × 5/8	0.612	0.250	0.240	0.0625
406	1/8 × 3/4	0.740	0.313	0.303	0.0625
506	5/32 × 3/4	0.740	0.313	0.303	0.0625
606	3/16 × 3/4	0.740	0.313	0.303	0.0625
806	1/4 × 3/4	0.740	0.313	0.303	0.0625
507	5/32 × 7/8	0.866	0.375	0.365	0.0625
607	3/16 × 7/8	0.866	0.375	0.365	0.0625
707	7/32 × 7/8	0.866	0.375	0.365	0.0625
807	1/4 × 7/8	0.866	0.375	0.365	0.0625
608	3/16 × 1	0.992	0.438	0.428	0.0625
708	7/32 × 1	0.992	0.438	0.428	0.0625
808	1/4 × 1	0.992	0.438	0.428	0.0625
1008	5/16 × 1	0.992	0.438	0.428	0.0625
1208	3/8 × 1	0.992	0.438	0.428	0.0625
609	3/16 × 1-1/8	1.114	0.484	0.475	0.078125

(Continued)

Table 8. *(Continued)* Woodruff Key Dimensions for Full-Length Keys.

Key Number	Nominal Size W × B	Actual Length F +0.000, -0.010	Height C +0.000, -0.005	Height[1] D (max.)	Distance to Center E
709	7/32 × 1-1/8	1.114	0.484	0.475	0.078125
809	1/4 × 1-1/8	1.114	0.484	0.475	0.078125
1009	5/16 × 1-1/8	1.114	0.484	0.475	0.078125
610	3/16 × 1-1/4	1.240	0.547	0.537	0.078125
710	7/32 × 1-1/4	1.240	0.547	0.537	0.078125
810	1/4 × 1-1/4	1.240	0.547	0.537	0.078125
1010	5/16 × 1-1/4	1.240	0.547	0.537	0.078125
1210	3/8 × 1-1/4	1.240	0.547	0.537	0.078125
811	1/4 × 1-3/8	1.362	0.594	0.584	0.09375
1011	5/16 × 1-3/8	1.362	0.594	0.584	0.09375
1211	3/8 × 1-3/8	1.362	0.594	0.584	0.09375
812	1/4 × 1-1/2	1.484	0.641	0.631	0.109375
1012	5/16 × 1-1/2	1.484	0.641	0.631	0.109375
1212	3/8 × 1-1/2	1.484	0.641	0.631	0.109375

[1] Key height tolerances for Flat Bottom Type are:

Keys with nominal lengths up to $1/2$ inch: +0.000, -0.005

Keys with nominal lengths $5/8$ inch to $1 1/2$ inch: +0.000, - 0.006.

All dimensions in inches. Source: Extracted from ANSI B17.2-1967 (R1972), published by the American Society of Mechanical Engineers.

Table 9. Woodruff Key Dimensions for Abbreviated-Length Keys.

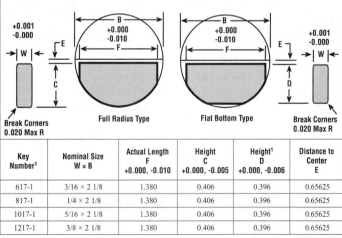

Key Number[1]	Nominal Size W × B	Actual Length F +0.000, -0.010	Height C +0.000, -0.005	Height[1] D +0.000, -0.006	Distance to Center E
617-1	3/16 × 2 1/8	1.380	0.406	0.396	0.65625
817-1	1/4 × 2 1/8	1.380	0.406	0.396	0.65625
1017-1	5/16 × 2 1/8	1.380	0.406	0.396	0.65625
1217-1	3/8 × 2 1/8	1.380	0.406	0.396	0.65625

(Continued)

Table 9. *(Continued)* **Woodruff Key Dimensions for Abbreviated-Length Keys.**

Key Number[1]	Nominal Size W × B	Actual Length F +0.000, -0.010	Height C +0.000, -0.005	Height[1] D +0.000, -0.006	Distance to Center E
617	3/16 × 2 1/8	1.723	0.531	0.521	0.53125
817	1/4 × 2 1/8	1.723	0.531	0.521	0.53125
1017	5/16 × 2 1/8	1.723	0.531	0.521	0.53125
1217	3/8 × 2 1/8	1.723	0.531	0.521	0.53125
822-1	1/4 × 2 3/4	2.000	0.594	0.584	0.78125
1022-1	5/16 × 2 3/4	2.000	0.594	0.584	0.78125
1222-1	3/8 × 2 3/4	2.000	0.594	0.584	0.78125
1422-1	7/16 × 2 3/4	2.000	0.594	0.584	0.78125
1622-1	1/2 × 2 3/4	2.000	0.594	0.584	0.78125
822	1/4 × 2 3/4	2.317	0.750	0.740	0.6250
1022	5/16 × 2 3/4	2.317	0.750	0.740	0.6250
1222	3/8 × 2 3/4	2.317	0.750	0.740	0.6250
1422	7/16 × 2 3/4	2.317	0.750	0.740	0.6250
1622	1/2 × 2 3/4	2.317	0.750	0.740	0.6250
1228	3/8 × 3 1/2	2.880	0.938	0.928	0.81250
1428	7/16 × 3 1/2	2.880	0.938	0.928	0.81250
1628	1/2 × 3 1/2	2.880	0.938	0.928	0.81250
1828	9/16 × 3 1/2	2.880	0.938	0.928	0.81250
2028	5/8 × 3 1/2	2.880	0.938	0.928	0.81250
2228	11/16 × 3 1/2	2.880	0.938	0.928	0.81250
2428	3/4 × 3 1/2	2.880	0.938	0.928	0.81250

[1] Keys with "-1" suffix have a shorter length (F) and a greater distance to center (E). Therefore, their height is less than exactly named keys without the "-1" suffix.
All dimensions in inches. Source: Extracted from ANSI B17.2-1967 (R1972), published by the American Society of Mechanical Engineers.

Table 10. Keyseat Dimensions for Woodruff Keys. (See Footnotes for Tolerances.)

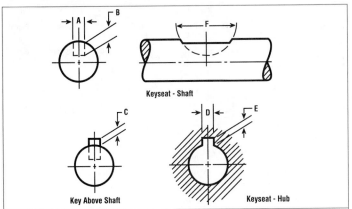

Keyseat - Shaft

Key Above Shaft

Keyseat - Hub

Key Number	Nominal Key Size	Keyseat - Shaft			Key Above Shaft[4]	Keyseat - Hub	
		Width[1] A	Depth[2] B	Diameter[3] F	C	Width[5] D	Depth[6] E
202	1/16 × 1/4	0.0630	0.0728	0.268	0.0312	0.0635	0.0372
202.5	1/16 × 5/16	0.0630	0.1038	0.330	0.0312	0.0635	0.0372
302.5	3/32 × 5/16	0.0943	0.0882	0.330	0.0469	0.0948	0.0529
203	1/16 × 3/8	0.0630	0.1358	0.393	0.0312	0.0635	0.0372
303	3/32 × 3/8	0.0943	0.1202	0.393	0.0469	0.0948	0.0529
403	1/8 × 3/8	0.1255	0.1045	0.393	0.0625	0.1260	0.0685
204	1/16 × 1/2	0.0630	0.1668	0.518	0.0312	0.0635	0.0372
304	3/32 × 1/2	0.0943	0.1511	0.518	0.0469	0.0948	0.0529
404	1/8 × 1/2	0.1255	0.1355	0.518	0.0625	0.1260	0.0685
305	3/32 × 5/8	0.0943	0.1981	0.643	0.0469	0.0948	0.0529
405	1/8 × 5/8	0.1255	0.1825	0.643	0.0625	0.1260	0.0685
505	5/32 × 5/8	0.1568	0.1669	0.643	0.0781	0.1573	0.0841
605	3/16 × 5/8	0.1880	0.1513	0.643	0.0937	0.1885	0.0997
406	1/8 × 3/4	0.1255	0.2455	0.768	0.0625	0.1260	0.0685
506	5/32 × 3/4	0.1568	0.2299	0.768	0.0781	0.1573	0.0841
606	3/16 × 3/4	0.1880	0.2143	0.768	0.0937	0.1885	0.0997
806	1/4 × 3/4	0.2505	0.1830	0.768	0.1250	0.2510	0.1310
507	5/32 × 7/8	0.1568	0.2919	0.895	0.0781	0.1573	0.0841
607	3/16 × 7/8	0.1880	0.2763	0.895	0.0937	0.1885	0.0997
707	7/32 × 7/8	0.2193	0.2607	0.895	0.1093	0.2198	0.1153
807	1/4 × 7/8	0.2505	0.2450	0.895	0.1250	0.2510	0.1310
608	3/16 × 1	0.1880	0.3393	0.102	0.0937	0.1885	0.0997

(Continued)

Table 10. *(Continued)* **Keyseat Dimensions for Woodruff Keys.** (See Footnotes for Tolerances.)

Key Number	Nominal Key Size	Keyseat - Shaft			Key Above Shaft[4] C	Keyseat - Hub	
		Width[1] A	Depth[2] B	Diameter[3] F		Width[5] D	Depth[6] E
708	7/32 × 1	0.2193	0.3237	0.102	0.1093	0.2198	0.1153
808	1/4 × 1	0.2505	0.3080	1.020	0.1250	0.2510	0.1310
1008	5/16 × 1	0.3130	0.2768	0.102	0.1562	0.3135	0.1622
1208	3/8 × 1	0.3755	0.2455	0.102	0.1875	0.3760	0.1935
609	3/16 × 1 1/8	0.1880	0.3853	1.145	0.0937	0.1885	0.0997
709	7/32 × 1 1/8	0.2193	0.3697	1.145	0.1093	0.2198	0.1153
809	1/4 × 1 1/8	0.2505	0.3540	1.145	0.1250	0.2510	0.1310
1009	5/16 × 1 1/8	0.3130	0.3228	1.145	0.1562	0.3135	0.1622
610	3/16 × 1 1/4	0.1880	0.4483	1.273	0.0937	0.1885	0.0997
710	7/32 × 1 1/4	0.2193	0.4327	1.273	0.1093	0.2198	0.1153
810	1/4 × 1 1/4	0.2505	0.4170	1.273	0.1250	0.2510	0.1310
1010	5/16 × 1 1/4	0.3130	0.3858	1.273	0.1562	0.3135	0.1622
1210	3/8 × 1 1/4	0.3755	0.3545	1.273	0.1875	0.3760	0.1935
811	1/4 × 1 3/8	0.2505	0.4640	1.398	0.1250	0.2510	0.1310
1011	5/16 × 1 3/8	0.3130	0.4328	1.398	0.1562	0.3135	0.1622
1211	3/8 × 1 3/8	0.3755	0.4015	1.398	0.1875	0.3760	0.1935
812	1/4 × 1 1/2	0.2505	0.5110	1.523	0.1250	0.2510	0.1310
1012	5/16 × 1 1/2	0.3130	0.4798	1.523	0.1562	0.3135	0.1622
1212	3/8 × 1 1/2	0.3755	0.4485	1.523	0.1875	0.3760	0.1935
617-1	3/16 × 2 1/8	0.1880	0.3073	2.160	0.0937	0.1885	0.0997
817-1	1/4 × 2 1/8	0.2505	0.2760	2.160	0.1250	0.2510	0.1310
1017-1	5/16 × 2 1/8	0.3130	0.2448	2.160	0.1562	0.3135	0.1622
1217-1	3/8 × 2 1/8	0.3755	0.2135	2.160	0.1875	0.3760	0.1935
617	3/16 × 2 1/8	0.1880	0.4323	2.160	0.0937	0.1885	0.0997
817	1/4 × 2 1/8	0.2505	0.4010	2.160	0.1250	0.2510	0.1310
1017	5/16 × 2 1/8	0.3130	0.3698	2.160	0.1562	0.3135	0.1622
1217	3/8 × 2 1/8	0.3755	0.3385	2.160	0.1875	0.3760	0.1935
822-1	1/4 × 2 3/4	0.2505	0.4640	2.785	0.1250	0.2510	0.1310
1022-1	5/16 × 2 3/4	0.3130	0.4328	2.785	0.1562	0.3135	0.1622
1222-1	3/8 × 2 3/4	0.3755	0.4015	2.785	0.1875	0.3760	0.1935
1422-1	7/16 × 2 3/4	0.4380	0.3703	2.785	0.2187	0.4385	0.2247
1622-1	1/2 × 2 3/4	0.5005	0.3390	2.785	0.2500	0.5010	0.2560
822	1/4 × 2 3/4	0.2505	0.6200	2.785	0.1250	0.2510	0.1310
1022	5/16 × 2 3/4	0.3130	0.5888	2.785	0.1562	0.3135	0.1622
1222	3/8 × 2 3/4	0.3755	0.5575	2.785	0.1875	0.3760	0.1935
1422	7/16 × 2 3/4	0.4380	0.5263	2.785	0.2187	0.4385	0.2247
1622	1/2 × 2 3/4	0.5005	0.4950	2.785	0.2500	0.5010	0.2560
1228	3/8 × 3 1/2	0.3755	0.7455	3.535	0.1875	0.3760	0.1935

(Continued)

Table 10. *(Continued)* **Keyseat Dimensions for Woodruff Keys.** (See Footnotes for Tolerances.)

| Key Number | Nominal Key Size | Keyseat - Shaft | | | Key Above Shaft[4] C | Keyseat - Hub | |
		Width[1] A	Depth[2] B	Diameter[3] F		Width[5] D	Depth[6] E
1428	7/16 × 3 1/2	0.4380	0.7143	3.535	0.2187	0.4385	0.2247
1628	1/2 × 3 1/2	0.5005	0.6830	3.535	0.2500	0.5010	0.2560
1828	9/16 × 3 1/2	0.5630	0.6518	3.535	0.2812	0.5635	0.2872
2028	5/8 × 3 1/2	0.6255	0.6205	3.535	0.3125	0.6260	0.3185
2228	11/16 × 3 1/2	0.6880	0.5893	3.535	0.3437	0.6885	0.3497
2428	3/4 × 3 1/2	0.7505	0.5580	3.535	0.3750	0.7510	0.3810

[1] Keyseat (Shaft) Width (A) tolerances are:
  Keys with nominal widths up to and including $5/32$ inch: +0.000, -0.0015
  Keys with nominal widths of $3/16$ inch: +0.000, -0.0017
  Keys with nominal widths $7/32$ to $1/4$ inch incl.: +0.000, -0.0018
  Keys with nominal widths of $5/16$ inch: +0.000, -0.0019
  Keys with nominal widths of $7/16$ to $3/4$ inch incl.: +0.000, -0.0020.
Width values are the maximum keyseat (shaft) width that will receive a key with the greatest amount of looseness consistent with the key's sticking in the keyseat. Minimum keyseat (shaft) width permits the largest shaft distortion acceptable when assembling maximum key in minimum keyseat.
[2] Keyseat (Shaft) Depth (B) tolerances are: +0.005, -0.000.
[3] Keyseat (Shaft) Diameter (F) tolerances are:
  Keys with nominal lengths up to and including $3/4$ inch: +0.000, -0.018
  Keys with nominal lengths $7/8$ to $1 1/8$ inch incl.: +0.000, -0.020
  Keys with nominal lengths $1 1/4$ to $1 1/2$ inch incl.: +0.000, -0.023
  Keys with nominal lengths $2 1/8$ to $3 1/2$ inch incl.: +0.000, -0.035.
[4] Key Height Above Shaft (C) tolerance: +0.005, -0.005.
[5] Keyseat (Hub) Width (D) tolerance: +0.002, -0.000.
[6] Keyseat (Hub) Depth (E) tolerance: +0.005, -0.000.
All dimensions in inches. Source: Extracted from ANSI B17.2-1967 (R1972), published by the American Society of Mechanical Engineers.

Table 11. Suggested Dimensions of Chamfered Keys and Filleted Keyseats.

| Keyseat Depth[1] | | Fillet Radius | 45° Chamfer |
Over	To (incl.)		
1/8	1/4	1/32	3/64
1/4	1/2	1/16	5/64
1/2	7/8	1/8	5/32
7/8	1 1/4	3/16	7/32
1 1/4	1 3/4	1/4	9/32
1 3/4	2 1/2	3/8	13/32

[1] Keyseat depth is measured from the bottom of the keyseat to the top of the side of the keyseat, and from the bottom of the keyseat hub to the top of the side of the hub keyseat. All dimensions in inches. Source: ANSI B17.1-1967 (R1973), published by the American Society of Mechanical Engineers.

Dowel pins

Dowel pins are popular devices for locating parts in a fixed position, and for precision aligning component parts. Standard hardened and ground dowel pins have long been favorites for these applications, and dimensions of standard dowel pins are provided in **Table 1**. Dimensions of metric dowel pins manufactured to ISO 8734 or DIN 6325 specifications are given in **Table 2**. Holes for dowel pins should be drilled and reamed to precise diameters, as the pins are held in place by interference created around their entire circumference.

In addition to traditional pins, variations of the basic pin have become widely available that lend versatility to the common dowel pin.

Vent dowels are designed to be used in blind holes. They employ a narrow helical groove running the length of the pin that prevents air locks under the pin. They are interchangeable with most standard dowels.

Pull-out dowels allow the pin to be reused. Spiral relief grooves run the length of the pin to provide uniform relief during insertion and removal. One end of the pin has a tapped hole into which a socket head screw is inserted, and a die hook is then positioned under the head of the screw to pry the dowel out of its hole. These pins are especially useful as they allow sections of a die to be removed without removing the entire die assembly from the press. They are available in diameters ranging from $1/4$" to 1".

Plain (or straight) pins are similar to dowel pins but are manufactured to their nominal diameter with a +0.000, −0.002 tolerance. Plain pins are produced from commercial wire or rod.

Grooved pins

The grooves in a grooved pin are formed by a swaging operation that enters the pin at 120° intervals. The introduction of the grooves deforms the pin and raises its diameter of the pin along their length. The expanded section, which varies in diameter with the diameter of the pin, the pin material, and the style of the groove, is a few thousandths of an inch larger than the nominal diameter. Holes for grooved pins should be drilled a few thousandths larger than the nominal diameter of the pin. Reaming is not required, but a small chamfer on the hole is recommended, especially in hardened steel and cast iron, or when using hardened groove pins. The pin is then pressed into the hole, and the expanded diameter of the pin compresses to lock the pin in the hole. Several varieties of grooved pins are available.

Type A, Full-length taper groove. These pins have three full-length tapered grooves and provide excellent locking properties and ease of assembly. It is often used in place of taper pins, set screws, or keys in sprockets, gears, collars, and knobs.

Type B, Half-length taper groove. The three tapered grooves in this pin run one-half the pin's length. Retention strength is not as good as with Type A pins, but assembly is easier.

Type C, Full-length parallel groove. The three parallel grooves run almost the full length of the pin. Parallel grooves are straight, and are evenly expanded through their length, while tapered grooves are large at the end and taper down to a small slit. The abbreviated length pilot, which is nominal diameter, aids installation.

Type D, Half-length groove reverse taper. The taper begins at its center and extends to one end of the pin.

Type E, Center groove. The parallel grooves on this pin are centered and run one-half the pin's length. Often used on "T" handles on valves, and as a cotter pin where center locking is required.

Type F (or Type U), Full-length parallel groove. Similar to Type C, but with a pilot on both ends that allows for hopper or automatic feeding.

Type G, Half-length parallel groove anchor. For use in blind or through holes. Annular groove on anchoring end allows for loop of tensioning spring, or spring clip.

Type H, Half-length parallel groove hinge. Used as a pivot stop or hinge pin.

Dimensions for grooved pins are found in **Table 3**, and expanded diameters are given in **Table 4**. Expanded diameters of grooved pins must be checked with a ring gage. Because the grooves are located at 120° intervals, neither a micrometer nor Vernier will provide an accurate reading.

Grooved studs

Grooved studs function in the same manner as grooved pins, but additional end loading resistance is provided by their head. They are used for a variety of applications including fastening brackets, control arms, knobs, and handles. Grooved studs produced by Driv-Lok, Inc., as pictured in *Figure 1*, are manufactured from low carbon steel and are zinc plated for corrosion resistance. Dimensions of standard round head grooved studs are provided in **Table 5**.

Spring pins

Slotted spring pins are rolled from strips of high-carbon or stainless steel. They are self-locking, nonprecision fasteners often used to replace dowels or stop pins. There are two different styles of spring pins: Style 1 (or Type A) pins have a 40° angle at each end of their break edge, while the slot edges on Style 2 (or Conventional) pins are parallel throughout their length. Dimensions of spring pins are given in **Table 6**. Dimensions for metric spring pins (DIN 1481) are provided in **Table 7**.

Spirally coiled spring pins

These pins are produced by coiling (or wrapping) the pin material tightly until a selected diameter is achieved. The coil is typically wrapped approximately $2\frac{1}{4}$ times, resulting in a pin that can be inserted into marginally out-of-round holes. Because their structure remains flexible, even after insertion, they are excellent for reducing shock and vibration loads. Coiled spring pins are used in a wide variety of applications.

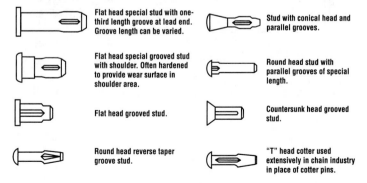

Figure 1. Special grooved studs manufactured by Driv-Lok, Inc.

Light duty pins can be used in most materials, including some plastics and ceramics. Standard duty pins are recommended for most nonhardened materials. Heavy duty pins are designed for severe, high strength applications. Unless otherwise specified, most manufacturers will supply standard duty pins. Dimensions for spirally coiled spring pins are given in **Table 8**.

Knurled pins

Knurled pins are popular for assembling parts with thin cross sections, and for use in plastics and die cast components. They are similar in concept to grooved pins because the serrated section is restricted to a nominal length on the pin. Knurled pins are usually roll formed, and the knurled section may have a straight knurl, a helical knurl, or a diamond knurl configuration. Examples of knurled pins are shown in *Figure 2*.

Clevis pins

Clevis pins offer extreme ease of assembly and, when properly secured with a cotter pin, they cannot loosen. They are recommended for applications that require frequent disassembly. Dimensions of standard clevis pins are provided in **Table 9**. Clevis pins are also manufactured with ball detent retaining systems, and with a groove that can accept a retaining clip or ring for retention.

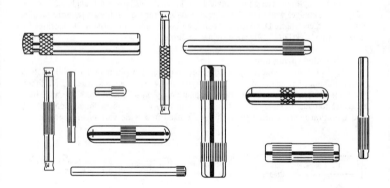

Figure 2. Examples of knurled pins. (Source, Driv-Lok, Inc.)

Table 1. Hardened and Ground Dowel Pin Dimensions. (See Notes.)

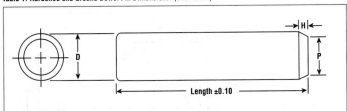

Nom. Size	Standard Pin Dia. D	Oversize[1] Pin Dia. D	Point Dia. P		Crown Height H		Recommended Hole Diameter[2]		Calculated Shear Strength[3]
	±.001	±.001	Max.	Min.	Max.	Min.	Max.	Min.	lb.
1/16	0.0627	0.0635	0.058	0.048	0.020	0.008	0.0625	0.0620	465
3/32	0.0940	0.0948	0.089	0.079	0.031	0.012	0.0937	0.0932	1,035
1/8	0.1252	0.1260	0.120	0.110	0.041	0.016	0.1250	0.1245	1,845
3/16	0.1877	0.1885	0.180	0.170	0.062	0.023	0.1875	0.1870	4,140
1/4	0.2502	0.2510	0.240	0.230	0.083	0.031	0.2500	0.2495	7,370
5/16	0.3127	0.3135	0.302	0.290	0.104	0.039	0.3125	0.3120	11,500
3/8	0.3752	0.3760	0.365	0.350	0.125	0.047	0.3750	0.3745	16,580
7/16	0.4377	0.4385	0.424	0.409	0.146	0.055	0.4375	0.4370	22,540
1/2	0.5002	0.5010	0.486	0.471	0.167	0.063	0.500	0.4995	29,460
5/8	0.6252	0.6260	0.611	0.595	0.208	0.078	0.6250	0.6245	46,020
3/4	0.7502	0.7510	0.735	0.715	0.250	0.094	0.7500	0.7495	66,270
7/8	0.8752	0.8760	0.860	0.840	0.293	0.109	0.8750	0.8745	90,190
1	1.0002	1.0010	0.980	0.960	0.333	0.125	1.0000	0.9995	17,810

[1] Oversize pins are intended for reassembly or replacement use.

[2] Hole diameters provided on Table are for hard materials such as steel and soft iron. For soft materials, reduce hole size by 0.0005.

[3] Single shear strength, in pounds, material and heat treatment per ANSI B18.8.2. Formula for calculation: 150,000 PSI × π × (nom. dia.)^2 + 4.

All dimensions in inches. Source: ASA B5.20 and manufacturer's specifications.

Table 2. Dimensions of Metric Dowel Pins.

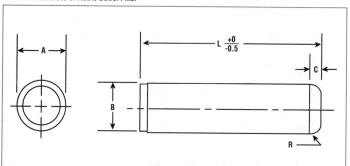

Nom. Size	Pin Dia. A		Point Dia. B		Crown Height C		Crown Radius R		Recommended Hole Size		Calculated Single Shear Strength	
	Max.	Min.	Max.	Min.	Max.	Min.	Max.	Min.	Max.	Min.	kN	lb.
3	3.008	3.003	2.9	2.6	0.8	0.3			3.000	2.987	7.4	1,670
4	4.009	4.004	3.9	3.6	0.9	0.4			4.000	3.987	13.2	2,965
5	5.009	5.004	4.9	4.6	1.0	0.4			5.000	4.987	20.6	4,635
6	6.010	6.004	5.8	5.4	1.1	0.4			6.000	5.987	29.7	6,650
8	8.012	8.006	7.8	7.4	1.3	0.5			8.000	7.987	52.5	11,850
10	10.012	10.006	9.8	9.4	1.4	0.6			10.000	9.987	82.5	18,550
12	12.013	12.007	11.8	11.4	1.6	0.6			12.000	11.985	119.0	26,700
16	16.013	16.007	15.8	15.3	1.8	0.8			16.000	15.985	211.0	47,450
20	20.014	20.008	19.8	19.3	2.0	0.8			20.000	19.983	330.0	74,000
25	25.014	25.008	24.8	24.3	2.3	1.0			25.000	24.983	515.0	116,000

All dimensions in millimeters. Material, ANSI B.18.85 alloy steel. Source: Unbrako/SPS Technologies, Inc.

Table 3. Grooved Pin Dimensions. (See Notes.)

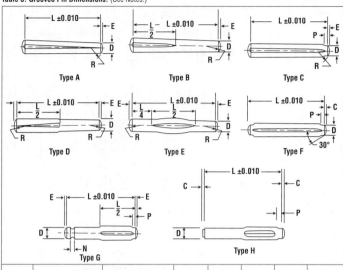

Type A   Type B   Type C

Type D   Type E   Type F

Type G   Type H

Nom. Size	Pin Diameter P		Recommended Hole Diameter		Crown Height H	Crown Radius R	Chamfer Length C	Pilot Length P	Neck Width N
	Max.	Min.	Max.	Min.	±0.005	±0.01	Min.	Ref.	±0.005
3/64	0.0469	0.1459	0.0478	0.0469	0.0000	-	-	-	-
1/16	0.0625	0.0615	0.0640	0.0625	0.0065	5/64	1/64	1/32	-
5/64	0.0781	0.0771	0.0798	0.0781	0.0087	3/32	1/64	1/32	-
3/32	0.0938	0.0928	0.0956	0.0938	0.0091	1/8	1/64	1/32	0.033
7/64	0.1094	0.1084	0.1113	0.1094	0.0110	9/64	1/64	1/32	0.033
1/8	0.1250	0.1230	0.1271	0.1250	0.1030	5/32	1/64	1/32	0.064
5/32	0.1563	0.1543	0.1587	0.1563	0.0170	3/16	1/32	1/16	0.064
3/16	0.1875	0.1855	0.1903	0.1875	0.0180	1/4	1/32	1/16	0.064
7/32	0.2188	0.2168	0.2219	0.2188	0.0220	9/32	1/32	1/16	0.096
1/4	0.2500	0.2480	0.2534	0.2500	0.0260	5/16	1/32	1/16	0.096
5/16	0.3125	0.3105	0.3166	0.3125	0.0340	3/8	3/64	3/32	0.127
3/8	0.3750	0.3730	0.3797	0.3750	0.0390	15/32	3/64	3/32	0.127
7/16	0.4375	0.4355	0.4428	0.4375	0.0470	17/32	3/64	3/32	0.190
1/2	0.5000	0.4980	0.5060	0.5000	0.0520	5/8	3/64	3/32	0.190

Notes: Length increments are by $1/8$ inch up to 1 inch, and by $1/4$ inch over 1 inch. Length of chamfered pins is measured overall, including chamfer. Length of crowned pins is $L + 2E$. Manufacturers supply each nominal diameter in several lengths. Typical lengths are as follows:

$3/64$ dia.: Lengths of $1/8$" to $5/8$". $1/16$ dia.: Lengths of $1/8$" to 1". $5/64$ dia.: Lengths of $1/4$" to 1". $3/32$ and $7/64$ dia.: Lengths of $1/4$" to 1 $1/4$". $1/8$ dia.: Lengths of $1/4$" to 1 $1/2$". $5/32$ and $3/16$ dia.: Lengths of $3/8$" to 2". $7/32$ dia.: Lengths of $1/2$" to 2 $1/2$". $1/4$ dia.: Lengths of $1/2$" to 2 $3/4$". $5/16$ dia.: Lengths of $5/8$" to 3 $1/4$". $3/8$ dia.: Lengths of $3/4$" to 3 $1/2$". $7/16$ dia.: Lengths of $7/8$" to 3 $3/4$". $1/2$ dia.: Lengths of 1" to 4".

All dimensions in inches. Source, ASA B5.20 and manufacturer's specifications.

Table 4. Expanded Diameters of Grooved Pins.

Groove Length	Nominal Diameter										
	Expanded Diameter										
	1/16	3/32	1/8	5/32	3/16	7/32	1/4	5/16	3/8	7/16	1/2
	±0.0015	±0.002	±0.0025						±0.003		
1/8	0.068	0.101	0.134	0.166	0.198	0.230	0.263	0.329	–	–	–
3/16	0.068	0.101	0.134	0.166	0.198	0.230	0.263	0.329	–	–	–
1/4	0.068	0.101	0.134	0.166	0.198	0.230	0.263	0.329	–	–	–
5/16	0.068	0.101	0.134	0.166	0.198	0.230	0.263	0.329	0.394	0.459	0.525
3/8	0.068	0.101	0.134	0.166	0.198	0.230	0.263	0.329	0.394	0.459	0.525
7/16	0.068	0.101	0.134	0.166	0.198	0.230	0.263	0.329	0.394	0.459	0.525
1/2	0.068	0.101	0.134	0.166	0.198	0.230	0.263	0.329	0.394	0.459	0.525
9/16	0.068	0.101	0.134	0.166	0.198	0.230	0.263	0.329	0.394	0.459	0.525
5/8	0.068	0.101	0.134	0.166	0.198	0.230	0.263	0.329	0.394	0.459	0.525
3/4	0.067	0.100	0.134	0.166	0.198	0.230	0.263	0.329	0.394	0.459	0.525
7/8	0.067	0.100	0.133	0.165	0.198	0.230	0.263	0.329	0.394	0.459	0.525
1		0.100	0.132	0.165	0.198	0.230	0.263	0.329	0.394	0.459	0.525
1 1/4		0.100	0.132	0.164	0.197	0.230	0.263	0.329	0.394	0.459	0.525
1 1/2			0.132	0.164	0.197	0.230	0.263	0.329	0.394	0.459	0.525
1 3/4				0.163	0.197	0.229	0.262	0.328	0.393	0.459	0.525
2				0.163	0.196	0.229	0.262	0.328	0.393	0.458	0.525
2 1/4						0.229	0.262	0.328	0.393	0.458	0.524
2 1/2						0.229	0.262	0.327	0.393	0.458	0.524
2 3/4							0.261	0.327	0.393	0.458	0.524
3									0.392	0.457	0.523
3 1/4							0.261		0.392	0.457	0.523
3 1/2									0.391	0.457	0.522

All dimensions in inches.

Table 5. Round Head Grooved Stud Dimensions.

Stud Number	Shank Dia.[1]	Recommended Drill Size	Head Diameter		Head Height	
			Max.	Min.	Max.	Min.
0	0.067	51	0.130	0.120	0.050	0.040
2	0.086	44	0.162	0.146	0.070	0.059
4	0.104	37	0.211	0.193	0.086	0.075
6	0.120	31	0.260	0.240	0.103	0.091
7	0.136	29	0.309	0.287	0.119	0.107
8	0.144	27	0.309	0.287	0.119	0.107
10	0.161	20	0.359	0.334	0.136	0.124
12	0.196	9	0.408	0.382	0.152	0.140
14	0.221	2	0.457	0.429	0.169	0.156
16	0.250	1/4	0.472	0.443	0.174	0.161

[1] Shank diameter tolerance is +0.000, -0.002.

Typical lengths are as follows.

Sizes 0 and 2: Lengths of $1/8$", $3/16$", and $1/4$". Size 4: Lengths of $3/16$ through $5/16$". Size 6: Lengths of $1/4$" through $3/8$". Size 7: Lengths of $5/16$" through $1/2$". Sizes 8 and 10: Lengths of $3/8$" and $1/2$". Sizes 12, 14, and 16: $1/2$".

All dimensions in inches. Source, Driv-Lok, Inc.

Table 6. Spring Pin Dimensions. (See Notes.)

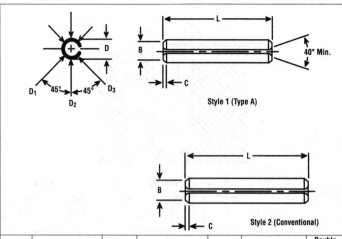

Style 1 (Type A)

Style 2 (Conventional)

Nom. Size	Pin Diameter[1] A		Chamfer B	Chamfer Length C		Thickness T	Recommended Hole Size		Double Shear Load[2]
	Min.	Max.	Max.	Min.	Max.	Basic	Min.	Max.	Min.
1/16	0.066	0.069	0.059	0.007	0.028	0.012	0.063	0.066	430
5/64	0.083	0.086	0.075	0.008	0.032	0.018	0.079	0.081	600
3/32	0.099	0.103	0.091	0.008	0.038	0.022	0.095	0.097	1,150
1/8	0.131	0.135	0.122	0.008	0.044	0.028	0.125	0.129	1,875
5/32	0.162	0.187	0.151	0.010	0.048	0.032	0.157	0.160	2,750
3/16	0.194	0.199	0.182	0.011	0.055	0.040	0.188	0.192	4,140
7/32	0.226	0.232	0.214	0.011	0.065	0.048	0.219	0.224	5,850
1/4	0.258	0.264	0.245	0.012	0.065	0.048	0.250	0.258	7,050
5/16	0.321	0.330	0.306	0.014	0.080	0.062	0.313	0.318	10,800
3/8	0.385	0.395	0.368	0.016	0.095	0.077	0.375	0.382	16,300
7/16	0.448	0.459	0.430	0.017	0.095	0.077	0.438	0.445	19,800
1/2	0.513	0.524	0.485	0.025	0.110	0.094	0.500	0.510	27,100

[1] Minimum pin diameter is calculated with the formula: $\frac{1}{3}(D_1 - D_2 + D_3)$. Maximum diameter to be measured with a Go Ring Gage.

[2] Material shear strength given for AISI 1070-1095 and AISI 420.

Manufacturers supply each diameter in several lengths. Typical lengths are as follows.

$\frac{1}{16}$ dia.: Lengths of $\frac{3}{16}$" to 1". $\frac{5}{64}$ and $\frac{3}{32}$ dia.: Lengths of $\frac{3}{16}$" to 1 $\frac{1}{2}$". $\frac{1}{8}$ dia.: Lengths of $\frac{5}{16}$" to 2". $\frac{5}{32}$ dia.: Lengths of $\frac{7}{16}$" to 2 $\frac{1}{2}$". $\frac{3}{16}$ dia.: Lengths of $\frac{1}{2}$" to 2 $\frac{1}{2}$". $\frac{7}{32}$ dia.: Lengths of $\frac{1}{2}$" to 3". $\frac{1}{4}$ dia.: Lengths of $\frac{1}{2}$" to 3 $\frac{1}{2}$". $\frac{5}{16}$ and $\frac{3}{8}$ dia.: Lengths of $\frac{3}{4}$" to 4". $\frac{7}{16}$ dia.: Lengths of 1" to 4". $\frac{1}{2}$ dia.: Lengths of 1 $\frac{1}{4}$" to 4".

All dimensions in inches. Source, Driv-Lok, Inc.

Table 7. Dimensions of Metric Heavy Spring Straight Pins (DIN 1481).

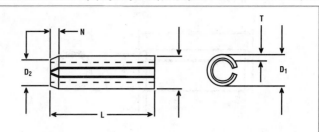

Hole Size	D_1 Open Dia.		D_2 Nose Dia.	N Nose Length	T Thickness
	Max.	Min.	Approx.		
M1	1.20	1.1	0.80	0.15	0.20
M1.5	1.70	1.6	1.10	0.25	0.30
M2	2.30	2.15	1.50	0.35	0.40
M2.5	2.80	2.65	1.80	0.40	0.50
M3	3.30	3.15	2.10	0.50	0.60
M4	4.40	4.2	2.80	0.65	0.80
M5	5.40	5.2	3.40	0.90	1.00
M6	6.40	6.2	3.90	1.20	1.25
M8	8.50	8.3	5.50	2.00	1.50
M10	10.50	10.3	6.50	2.00	2.00
M12	12.50	12.3	7.50	2.00	2.50

Recommended hole size is nominal diameter. All dimensions in millimeters.

Table 8. Spirally Coiled Spring Pin Dimensions. (See Notes.)

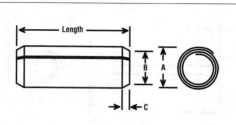

Hole Size	Standard Duty Pin Diameter[1] A		Chamfer B	Chamfer Length B	Recommended Hole Size		Double Shear Load[2]
	Min.	Max.	Ref.	Ref.	Min.	Max.	Min. lb.
1/32	0.033	0.035	0.029	0.024	0.031	0.032	75
3/64	0.049	0.052	0.045	0.024	0.046	0.048	170
1/16	0.067	0.072	0.059	0.028	0.061	0.065	300
5/64	0.083	0.088	0.075	0.032	0.077	0.081	475
3/32	0.099	0.105	0.091	0.038	0.093	0.097	700
7/64	0.114	0.120	0.106	0.038	0.108	0.112	950
1/8	0.131	0.138	0.121	0.044	0.124	0.129	1,250
5/32	0.163	0.171	0.152	0.048	0.155	0.160	1,925
3/16	0.196	0.205	0.182	0.055	0.185	0.192	2,800
7/32	0.228	0.238	0.214	0.065	0.217	0.224	3,800
1/4	0.260	0.271	0.243	0.065	0.247	0.256	5,000
5/16	0.324	0.337	0.304	0.080	0.308	0.319	7,700
3/8	0.388	0.403	0.366	0.095	0.370	0.383	11,200
7/16	0.452	0.469	0.427	0.095	0.431	0.446	15,200
1/2	0.516	0.535	0.488	0.110	0.493	0.510	20,000
5/8	0.642	0.661	0.613	0.125	0.618	0.635	31,000
3/4	0.768	0.787	0.738	0.150	0.743	0.760	45,000

[1] Pin diameters for Heavy Duty coiled spring pins are as follows:

Sizes $1/16$ through $1/8$: Min. Standard Duty pin diameter, -0.000. Max. Standard Duty pin diameter, -0.002".
Sizes $5/32$ through $3/4$: Min. Standard Duty pin diameter, -0.002". Max. Standard Duty pin diameter, -0.003".

Pin diameters for Light Duty coiled spring pins are as follows:

Sizes $1/16$ through $5/32$: Min. Standard Duty pin diameter, -0.000. Max. Standard Duty pin diameter, +0.001".
Sizes $3/16$ through $1/2$: Min. Standard Duty pin diameter, -0.000. Max. Standard Duty pin diameter, +0.002".

[2] Double shear load minimums are for Standard Duty Pins. For approximate double shear load minimums for Heavy Duty pins, increase value given by 33%. Light Duty double shear load minimums are approximately one-half the values for Standard Duty pins. Material is AISI 1070-1095 and AISI 420.

Manufacturers supply each diameter in several lengths. Typical lengths are as follows.

$1/16$ dia.: Lengths of $1/4$" to 1". $5/64$ dia.: Lengths of $5/16$" to 1 $1/4$". $3/32$ dia.: Lengths of $1/4$" to 1 $3/4$". $1/8$ dia.: Lengths of $5/16$" to 2". $5/32$ dia.: Lengths of $3/8$" to 2". $3/16$ dia.: Lengths of $1/2$" to 2 $1/4$". $7/32$ dia.: Lengths of $5/8$" to 2 $1/2$". $1/4$ dia.: Lengths of $5/8$" to 3 $1/2$". $5/16$ dia.: Lengths of $3/4$" to 4".

All dimensions in inches. Source, specifications published by Pivot Point, Inc., and other manufacturers.

Table 9. Clevis Pin Dimensions. (See Notes.)

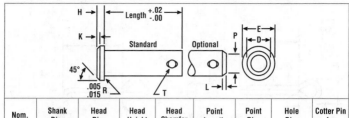

Nom. Dia.	Shank Dia. D	Head Dia. E	Head Height H	Head Chamfer K	Point Length L	Point Dia. P	Hole Dia. T	Cotter Pin for Hole Size
3/16	.181 - .186	.30 - .32	.05 - .07	.01 - .03	.035 - .055	.14 - .15	.073 - .088	1/16
1/4	.243 - .248	.36 - .38	.08 - .10	.02 - .04	.035 - .055	.20 - .01	.073 - .088	1/16
5/16	.306 - .311	.42 - .44	.08 - .10	.02 - .04	.049 - .071	.25 - .26	.104 - .119	3/32
3/8	.368 - .373	.49 - .51	.11 - .13	.02 - .04	.049 - .071	.32 - .33	.104 - .119	3/32
7/16	.431 - .436	.55 - .57	.14 - .16	.03 - .05	.049 - .071	.38 - .39	.104 - .119	3/32
1/2	.491 - .496	.61 - .63	.14 - .16	.03 - .05	.063 - .089	.43 - .44	.136 - .151	1/8
5/8	.616 - .621	.80 - .82	.19 - .21	.05 - .06	.063 - .089	.55 - .56	.136 - .151	1/8
3/4	.741 - .746	.92 - .94	.24 - .26	.06 - .08	.076 - .110	.67 - .68	.167 - .182	5/32
7/8	.866 - .871	1.02 - 1.04	.30 - .32	.08 - .10	.076 - .110	.79 - .80	.167 - .182	5/32
1	.991 - .996	1.17 - 1.19	.33 - .35	.09 - .11	.076 - .110	.92 - .93	.167 - .182	5/32

Notes: Manufacturers supply each diameter in several lengths. Typical lengths are as follows:
$3/16$ dia.: Lengths of $1/2$" through 3". $1/4$ dia.: Lengths of $1/2$" through 4". $5/16$ dia.: Lengths of $1/2$" through 4". $3/8$ dia.: Lengths of $3/4$" through 4". $7/16$ dia.: Lengths of 1" through 4". $1/2$ dia.: Lengths of 1" through 5". $5/8$ dia.: Lengths of 1 $1/4$" through 4". $3/4$ dia.: Lengths of 2" through 4".
All dimensions in inches. Source: Extracted from B18.8.1-1972, as published by the American Society of Mechanical Engineers, and manufacturer's specifications.

Table 10. Taper Pin Dimensions. (See Notes.)

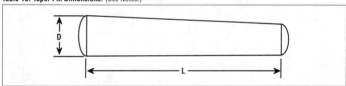

Pin Number	Basic Dia.[1]	Pin Number	Basic Dia.[1]	Pin Number	Basic Dia.[1]	Pin Number	Basic Dia.[1]
7/0	0.0625	2/0	0.1410	4	0.2500	9	0.5910
6/0	0.0780	0	0.1560	5	0.2890	10	0.7060
5/0	0.0940	1	0.1720	6	0.3410	11	0.8600
4/0	0.1090	2	0.1930	7	0.4090	12	1.0320
3/0	0.1250	3	0.2190	8	0.4920	13	1.2410

[1] The basic diameter of a taper pin is the large end diameter. Tolerances for taper pin diameters are based on the class of pin. The two classes, and their assigned tolerances, are:
Commercial Class: + 0.0013, -0.0007.
Precision Class: + 0.0010, -0.0000. Precision Class pins are available in sizes 7/0 through 10.
Taper is 1/4 inch per foot. The small end diameter of a taper pin can be calculated with the following formula:
Basic Diameter - (Pin Length × 0.0208). All dimensions in inches.

Table 11. Cotter Pin Dimensions.

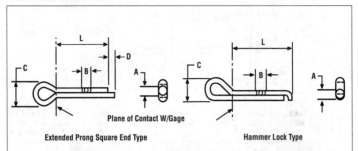

Plane of Contact W/Gage

Extended Prong Square End Type　　　**Hammer Lock Type**

Nom. Size	Total Shank Dia. A		Wire Width B		Head Dia. C	Extension D	Recommended Hole Size
	Max.	Min.	Max.	Min.	Min.	Min.	
1/21	0.032	0.028	0.032	0.022	0.06	0.01	0.047
3/64	0.048	0.044	0.048	0.035	0.09	0.02	0.062
1/16	0.060	0.056	0.060	0.044	0.12	0.03	0.078
5/64	0.076	0.072	0.076	0.057	0.16	0.04	0.094
3/32	0.090	0.086	0.090	0.069	0.19	0.04	0.109
7/64	0.104	0.100	0.104	0.080	0.22	0.05	0.125
1/8	0.120	0.116	0.120	0.093	0.25	0.06	0.141
9/64	0.134	0.130	0.134	0.104	0.28	0.06	0.156
5/32	0.150	0.146	0.150	0.116	0.31	0.07	0.172
3/16	0.176	0.172	0.176	0.137	0.38	0.09	0.203
7/32	0.207	0.202	0.207	0.161	0.44	0.10	0.234
1/4	0.225	0.220	0.225	0.176	0.50	0.11	0.266
5/16	0.280	0.275	0.280	0.220	0.62	0.14	0.312
3/8	0.335	0.329	0.335	0.263	0.75	0.16	0.375
7/16	0.406	0.400	0.406	0.320	0.88	0.20	0.438
1/2	0.473	0.467	0.473	0.373	1.00	0.23	0.500
5/8	0.598	0.590	0.598	0.472	1.25	0.30	0.625
3/4	0.723	0.725	0.723	0.572	1.50	0.36	0.750

All dimensions in inches. Source, ANSI B18.8.1-1972, as published by the American Society of Manufacturing Engineers.

Table 12. Dimensions of Metric Cotter Pins (DIN 94).

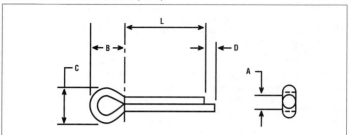

D Pin Shaft Dia.	C Head Height		B Head Width	O Overhang
	Max.	Min.	Approx.	Max.
1.00	1.50	1.60	3.00	1.60
1.60	2.80	2.40	3.20	2.50
2.00	3.60	3.20	4.00	2.50
2.50	4.60	4.00	5.00	2.50
3.20	5.80	5.10	6.40	3.20
4.00	7.40	6.50	8.00	4.00
5.00	9.20	8.00	10.00	4.00
6.30	11.80	10.30	12.60	4.00
8.00	15.00	13.10	16.00	4.00
10.00	19.00	16.60	20.00	6.30
12.00	24.80	21.70	26.00	6.30

All dimensions in millimeters.

Springs

Springs are used in a vast variety of mechanical devices, and choice of the proper type of spring, plus the best spring material for an application, is critical for success. The information in this section is courtesy of Century Spring Corporation, a leading supplier of springs for industrial use.

Materials

Mechanical properties of the most common spring wires are given in **Table 1**. Additional information about each of these materials follows.

High-carbon spring steels are the most popular of all spring materials. Generally speaking, these springs are the least expensive. However, they are not recommended for extreme temperatures, or for shock or impact loading. Minimum tensile strengths of ferrous spring materials are provided in **Table 2**.

Music wire, ASTM A 228. Cold drawn, possessing high and uniform tensile strength. Used for high quality springs and wire forms.

Hard drawn, ASTM A 277. Cold drawn. Used for average stress applications. Common choice for lower cost springs and wire forms.

High tensile, hard drawn, ASTM A 679. Cold drawn. Used for higher quality springs and wire forms.

Oil tempered, ASTM A 229. This material is cold drawn and heat treated before fabrication. It is a popular general purpose spring wire.

Carbon valve, ASTM A 230. Cold drawn and heat treated before fabrication, this material is suitable for cyclic operations.

Alloy spring steels are especially desirable for applications where high stress, shock, and impact loading are encountered. These materials can withstand higher and lower temperatures than the high-carbon spring steels.

Chrome vanadium, ASTM A 231. Cold drawn and heat treated before fabrication—this material is often selected where shock loads and moderately elevated temperatures are encountered.

Chrome silicon, ASTM A 401. Cold drawn and heat treated before fabrication. Often selected where shock loads and moderately elevated temperatures are encountered.

Stainless steel wire has steadily gained in popularity over the past few years, partially due to its ability to resist corrosion. The following materials can be safely used at elevated temperatures. Stainless steel spring materials are approximately 15% less stiff than carbon steel for a given wire size. Minimum tensile strengths of stainless steel wire spring materials are provided in **Table 3**.

AISI 302/304, ASTM A 313. Cold drawn, general purpose, corrosion and heat resistant. Magnetic in spring temper.

AISI 316, ASTM A 313. Cold drawn, heat resistant, and better corrosion resistance than AISI 302/304. Magnetic in spring temper.

17-7 PH, ASTM A 313 (631). Cold drawn and precipitation hardened after fabrication. High in strength, and good general purpose corrosion resistance. Slightly magnetic in spring temper.

Nonferrous alloy wire springs are usually made from copper base alloys that have good electrical properties and excellent resistance to corrosion. Although more expensive than high-carbon and alloy steel spring materials, they are frequently used in electrical

Table 1. Properties of Common Spring Materials. *(Source, Century Spring Corp.)*

	Material[1]	Nominal Analysis %	Tensile Properties[2]		Torsional Properties[3]		Temp. (Max.) °F/°C	Rockwell Hardness
			Minimum Tensile Strength	Modulus Elasticity	Design Stress %	Modulus in Torsion		
High-Carbon Spring Wire	Music Wire ASTM A 228	C .70-1.00 Mn .20-.60	230-399	30	45	11.5	250 / 121	C 41-60
	Hard Drawn ASTM A 277	C .45-.85 Mn .60-1.30	CLI 147-283 CLII 171-324	30	40	11.5	250 / 121	C 31-52
	High Tensile Hard Drawn ASTM A 679	C .65-1.00 Mn .20-1.30	238-350	30	45	11.5	250 / 121	C 41-60
	Oil Tempered ASTM A 229	C .55-.85 Mn .60-1.20	CLI 165-293 CLII 191-324	30	45	11.5	250 / 121	C 42-55
	Carbon Valve ASTM A 230	C .60-.75 Mn .60-.90	215-240	30	45	11.5	250 / 121	C 45-49
Alloy Steel Wire	Chrome Vanadium ASTM A 231	C .48-.53 Cr .80-1.10 V .15 min.	190-300	30	45	11.5	425/218.5	C 41-55
	Chrome Silicon ASTM A 401	C .51-.59 Cr .60-.80 Si 1.20-1.60	235-300	30	45	11.5	475/246	C 48-55
Stainless Steel Wire	AISI 302/304 ASTM A 313	Cr 17.0-19.0 Ni 8.0-10.0	125-325	28	30-40	10	550/288	C 35-45
	AISI 316 ASTM A 313	Cr 16.0-18.0 Ni 10.0-14.0 Mo 2.0-3.0	110-245	28	40	10	550/288	C 35-45
	17-7 PH ASTM A 313 (631)	Cr 16.0-18.0 Ni 6.5-7.5 Al .75-1.5	Cond. CH 235-335	29.5	45	11	650/343	C 38-57
Nonferrous Alloy Wire	Phosphor Bronze Grade A ASTM B 159	Cu 94.0-96.0 Sn 4.0-6.0	105-145	15	40	6.5	200/93.8	B98-104
	Beryllium Copper ASTM B 197	Cu 98.0 Be 2.0	150-230	18.5	45	7.0	400/204	C35-42
	Monel 400 AMS 7233	Ni 66.0 Cu 31.5 C/Fe	145-180	26	40	9.5	450/232	C23-32
	Monel K 500 QQ-N-286	Ni 65.0 Cu 29.5 C/Fe/A/Ti	160-200	26	40	9.5	550/288	C23-35

(Continued)

Table 1. *(Continued)* **Properties of Common Spring Materials.** *(Source, Century Spring Corp.)*

	Material[1]	Nominal Analysis %	Tensile Properties[2]		Torsional Properties[3]		Temp. (Max.)	Rockwell Hardness
			Minimum Tensile Strength	Modulus Elasticity	Design Stress %	Modulus in Torsion	°F/°C	
High-Temperature Alloy Wire	A 286 Alloy	Ni 26.0 Cr 15.0 Fe 53.0	160-200	29	35	10.4	950/510	C35-42
	Inconel 600 QQ-W-390	Ni 76.0 Cr 15.8 Fe 7.2	170-230	31	40	11.0	700/371	C35-45
	Inconel 718	Ni 52.5 Cr 18.6 Fe 18.5	210-250	29	40	11.2	1100/593	C45-50
	Inconel X-750 AMS 5698, 5699	N 73 Cr 15 Fe 6.65	No. IT 155 min. Spg. T 190-230	31	40	12	750-1100/ 399-593	C34-39 C42-48

[1] See text for information on method of manufacture, chief uses, and special properties of each material.
[2] Minimum Tensile Strength value is in ksi. Modulus of Elasticity value is PSI $\times 10^6$.
[3] Design Stress is expressed as a percentage of Minimum Tensile Strength. Modulus in Torsion value is PSI $\times 10^6$.

components and in applications subject to sub-zero temperatures. Copper base alloys are drawn to American Wire Gage sizes and are nonmagnetic.

Phosphor bronze, Grade A, ASTM B 159. Cold drawn. Has good corrosion resistance and electrical conductivity. About $1/3$ to $1/2$ as strong as steel.

Beryllium Copper, ASTM B 197. Cold drawn, and may be mill hardened prior to fabrication. Good corrosion resistance and electrical conductivity. About $1/2$ as strong as steel.

Monel 400, AMS 7233. Cold drawn. Good corrosion resistance at moderately elevated temperatures.

Monel K 500, QQ-N-286. Excellent corrosion resistance at moderately elevated temperatures.

High-temperature alloy wire is desirable for its ability to resist corrosion and withstand both high and sub-zero temperatures. It is nonmagnetic, and therefore well suited for such devices as gyroscopes, chronoscopes, and indicating instruments. Because of their high electrical resistance, high-temperature alloy wire springs should not be used for conducting electrical current.

A 286 alloy. Cold drawn, and precipitation hardened after fabrication. Good corrosion resistance at elevated temperatures.

Inconel 600, QQ-W-390. Cold drawn. Good corrosion resistance at elevated temperatures.

Inconel 718. Cold drawn, and precipitation hardened after fabrication. Good corrosion resistance at elevated temperatures.

Inconel X-750, AMS 5698, 5699. Cold drawn, and precipitation hardened after fabrication. Good corrosion resistance at elevated temperatures.

Table 2. Minimum Tensile Strength of Ferrous Wire Spring Materials. *(Source, Century Spring Corp.)*

Wire Diameter	Music Diameter	Hard Drawn	Oil Temper	Wire Diameter	Music Diameter	Hard Drawn	Oil Temper
inch	ksi	ksi	ksi	inch	ksi	ksi	ksi
0.008	399	307	315	0.046	309	249	-
0.009	393	305	313	0.047	309	309	248
0.010	387	303	311	0.048	306	306	247
0.011	382	301	309	0.049	306	306	246
0.012	377	299	307	0.050	306	306	245
0.013	373	297	305	0.051	303	303	244
0.014	369	295	303	0.052	303	303	244
0.015	365	293	301	0.053	303	303	243
0.016	362	291	300	0.054	303	243	253
0.017	362	289	298	0.055	300	242	-
0.018	356	287	297	0.056	300	241	-
0.019	356	285	295	0.057	300	240	-
0.020	350	283	293	0.058	300	240	-
0.021	350	281	-	0.059	296	239	-
0.022	345	280	-	0.060	296	238	-
0.023	345	278	289	0.061	296	237	-
0.024	341	277	-	0.062	296	237	247
0.025	341	275	286	0.063	293	236	-
0.026	337	274	-	0.064	293	235	-
0.027	337	272	-	0.065	293	235	-
0.028	333	271	283	0.066	290	-	-
0.029	333	267	-	0.067	290	234	-
0.030	330	266	-	0.069	290	233	-
0.031	330	266	280	0.070	289	-	-
0.032	327	265	-	0.071	288	-	-
0.033	327	264	-	0.072	287	232	241
0.034	324	262	-	0.074	287	231	-
0.035	324	261	274	0.075	287	-	-
0.036	321	260		0.076	284	230	-
0.037	321	258	-	0.078	284	229	-
0.038	318	257	-	0.079	284	-	-
0.039	318	256	-	0.080	282	227	235
0.040	315	256	-	0.083	282	-	-
0.041	315	255	266	0.084	279	-	-
0.042	313	254	-	0.085	279	225	-
0.043	313	252	-	0.089	279	-	-
0.044	313	251	-	0.090	276	222	-
0.045	309	250	-	0.091	276	-	230

(Continued)

Table 2. *(Continued)* **Minimum Tensile Strength of Ferrous Wire Spring Materials.** *(Source, Century Spring Corp.)*

Wire Diameter	Music Diameter	Hard Drawn	Oil Temper	Wire Diameter	Music Diameter	Hard Drawn	Oil Temper
inch	ksi	ksi	ksi	inch	ksi	ksi	ksi
0.092	276	220	-	0.130	258	-	-
0.093	276	-	-	0.135	258	206	215
0.094	274	-	-	0.139	258	-	-
0.095	274	219	-	0.140	256	-	-
0.099	274	-	-	0.144	256	-	-
0.100	271	-	-	0.145	254	-	-
0.101	271	-	-	0.148	254	203	210
0.102	270	-	-	0.149	253	-	-
0.105	270	216	225	0.150	253	-	-
0.106	268	-	-	0.151	251	-	-
0.109	268	-	-	0.160	251	-	-
0.110	267	-	-	0.161	249	-	-
0.111	267	-	-	0.162	249	200	205
0.112	266	-	-	0.177	245	195	200
0.119	266	-	-	0.192	241	192	195
0.120	263	210	220	0.207	238	190	190
0.123	263	-	-	0.225	235	186	188
0.124	261	-	-	0.250	230	182	185
0.129	261	-	-	0.3125	-	174	183

Table 3. Minimum Tensile Strength of Stainless Steel Wire Spring Materials. *(Source, Century Spring Corp.)*

Wire Diameter	Music Diameter	Hard Drawn	Wire Diameter	Music Diameter	Hard Drawn
inch	ksi	ksi	inch	ksi	ksi
0.008	325	345	0.022	296	-
0.009	325	-	0.023	294	-
0.010	320	345	0.024	292	-
0.011	318	340	0.025	290	330
0.012	316	-	0.026	289	325
0.013	314	-	0.027	287	-
0.014	312	-	0.028	286	-
0.015	310	340	0.029	284	-
0.016	308	335	0.030	282	325
0.017	306	-	0.031	280	320
0.018	304	-	0.032	277	-
0.019	302	-	0.033	276	-
0.020	300	335	0.034	275	-
0.021	298	330	0.035	274	-

(Continued)

Table 3. *(Continued)* **Minimum Tensile Strength of Stainless Steel Wire Spring Materials.**
(Source, Century Spring Corp.)

Wire Diameter	Music Diameter	Hard Drawn	Wire Diameter	Music Diameter	Hard Drawn
inch	ksi	ksi	inch	ksi	ksi
0.036	273	-	0.062	255	297
0.037	272	-	0.063	254	-
0.038	271	-	0.065	254	-
0.039	270	-	0.066	250	-
0.040	270	-	0.071	250	297
0.041	269	320	0.072	250	292
0.042	268	310	0.075	250	-
0.043	267	-	0.076	245	-
0.044	266	-	0.080	245	292
0.045	264	-	0.092	240	279
0.046	263	-	0.105	232	274
0.047	262	-	0.120	225	272
0.048	262	-	0.125	-	272
0.049	261	-	0.131	-	260
0.051	261	310	0.148	210	256
0.052	260	305	0.162	205	256
0.055	260	-	0.177	195	-
0.056	259	-	0.192	-	-
0.057	258	-	0.207	185	-
0.058	258	-	0.225	180	-
0.059	257	-	0.250	175	-
0.060	256	-	0.375	140	-
0.061	255	305			

Spring life

If a spring is not subjected to shock loads, rapid cycling, temperature extremes, corrosion, etc., and its total wire stress does not exceed the suggested maximum percentage of its parent material minimum tensile strength (see "Design Stress" in **Table 1**), then it should have an estimated life of between 100,000 and 1,000,000 cycles (deflections). A service life of over 1,000,000 cycles with infrequent failures may be expected if the wire maximum stress is reduced by approximately another 10% of the Table's recommended values.

Other than simply reducing the applied load, or deflection, there are many factors that contribute to extending a spring's service life. Specifications for critical applications that require maximum life with no failures in fatigue are of limited practical use, and should be replaced with specifications that detail the user's test conditions in terms of expected life in terms of empirically derived statistical failure rates. In all applications, it should be assumed that a certain percentage of all springs will fail early. Therefore, an attempt should be made to minimize any significant negative effect that an occasional early failure may have on the predicted performance of the spring.

One important method that can be used to extend the life of a spring is shot peening, which can extend the fatigue life of a spring by as much as 30%. Extension springs, however, are not a good candidate for shot peening, nor are compression springs with wire diameters under 0.040", due to buckling. The spring must be large enough to allow peening of the inside surfaces.

Although springs are sometimes made from square or rectangular wire, usually in an effort to increase the space efficiency of the design by shortening overall length while maintaining strength values, the formulas provided in this section, unless otherwise identified, are for round spring materials.

Compression and extension springs

When selecting a compression (*Figure 1*) or extension (*Figure 2*) spring, it is important to determine the length and spring rate (strength) required for the application. As a starting point, the length chosen should be approximately 30% longer (for compression springs) or 30% shorter (for extension springs) than the installed length of the spring, known as the *working length*. *Spring rate* is the load it takes to deflect (compress a compression spring, or stretch an extension spring) a spring a predetermined distance. Normally, the load is expressed in pounds and the distance is one inch. For example, if a spring is rated 40 lb/in., it will take 10 pounds to deflect the spring $1/4$", 40 pounds to deflect it 1", 80 pounds to deflect it 2", etc. Compression and extension springs share formulas for determining spring rate and wire stress. Stress, as discussed above, has a direct effect on spring life.

Spring rate: $\qquad\qquad\qquad R = G\,d^4 \div 8\,n\,D^3$,   or $R = P \div \Delta$

Wire stress–compression spring: $\quad S = 8\,P\,D\,K \div \pi\,d^3$,   or $S = 8\,R\,D\,K\,\Delta \div \pi\,d^3$

Wire stress–extension spring: $\qquad S = 8\,P\,D\,K \div \pi\,d^3 + S_1$, or $S = 8\,R\,D\,K\,\Delta \div \pi\,d^3$

where D = Mean outside diameter of spring (inches)
      d = wire diameter (inches)
      G = torsional modulus of elasticity (PSI)
          Spring steel = 11.5×10^6
          Stainless = 10×10^6
      K = Wahl curvature stress correction factor (*Figure 3*)
      n = number of effective coils
          Compression spring (see *Figure 4*)
          Extension spring: all coils in the spring ($n = N$)
      N = total number of coils in spring
      P = applied load (pounds)
      R = spring rate (lb/in.)
      S = wire stress (PSI)
      S_1 = stress due to initial tension (see *Figure 5*)
      Δ = deflection (inches)
      π = 3.1416.

Suggested maximum allowable spring-wire stress values can be found in **Tables 2** and **3**. Minimum tensile stress for spring wires varies with wire diameter, and a maximum of 30 to 45% of minimum allowable spring-wire stress—depending on material type—should be used as a corrected stress target for springs requiring a long fatigue life, and suggested design stress percentages may be taken from **Table 1**.

Stress in hooks on extension springs

Stresses in spring hooks are normally higher than in the spring body, because a bending stress occurs in addition to the wire tension stress. This takes place in the transition region

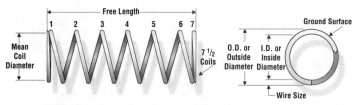

Figure 1. Compression spring. (Source, Century Spring Corp.)

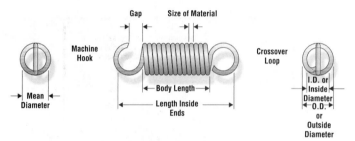

Figure 2. Extension spring. (Source, Century Spring Corp.)

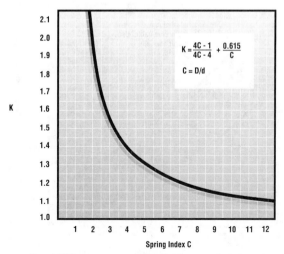

$$K = \frac{4C - 1}{4C - 4} + \frac{0.615}{C}$$

$$C = D/d$$

Figure 3. Wahl curvature stress correction chart for compression and extension springs. (Source, Century Spring Corp.)

Spring Characteristics	Type of Ends			
	Open	Open and Ground	Closed	Closed and Ground
Solid Length	$d(N+1)$	$d \times N$	$d(N+1)$	$d \times N$
Active Coils (n)	N	$N-1$	$N-2$	$N-2$
Total Coils (N)	n	N	$n+2$	$n+2$
Free Length	$(\pi N) + d$	$\pi \times N$	$(p \times n) + 3d$	$(p \times n) + 2d$

Figure 4. Equivalent number of coils for compression springs. N = total number of coils in spring, n = number of effective coils in spring. In open type springs, the spring and coils remain open, maintaining the helical wind shape of the spring. In closed type springs, the last coil at each end is bent back to touch the previous coil to create a flat base. The endings of a ground spring provide a more accurate flat base, but grinding reduces the solid length of the spring. (Source, Century Spring Corp.)

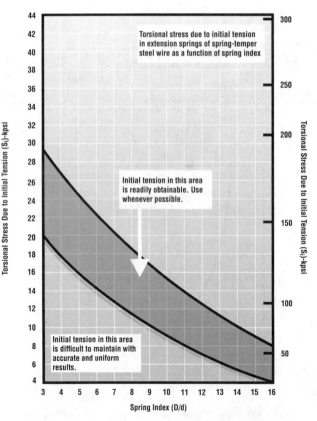

Figure 5. Initial tension chart for extension springs. (Source, Century Spring Corp.)

between the last body coil and the hook, and overstressed extension springs will normally fail in this region. The suggested allowable hook stress in torsion is 30 to 45% of the minimum tensile strength, and the suggested allowable for bending is 75% of the minimum tensile strength of the spring material. Maximum bending stress of a common extension spring hook or loop can be determined with the following equation:

$$S_A = ([32\,P\,R_1 \div \pi\,d^3] \times K_1) + (4\,P \div \pi\,d^3),$$

and the maximum torsion stress is found with the following equation.

$$S_B = (16\,P\,R_2 \div \pi\,d^3) \times ([4\,C - 1] \div [4\,C - 4])$$

where S_A = maximum bending stress at point A (see *Figure 6*)
S_B = maximum torsion stress at point B (see *Figure 6*)
R_1 = coil mean radius (inches)
R_2 = hook bend radius (inches)
$K_1 = (4\,C^2 - C - 1) \div (4\,C \times [C - 1])$
$C = 2\,R_1 \div d.$

Commonly used hook ends for extension springs are shown in *Figure 7*.

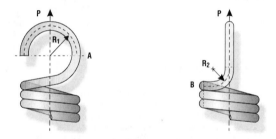

Figure 6. Stress areas in extension spring hooks. (Source, Century Spring Corp.)

Machine	Across Center	Double Full Loop	Side Hook	Extended

Figure 7. Common forms of extension spring hooks. (Source, Century Spring Corp.)

Torsion springs

Torsion springs are used to store and release angular energy, or to statically hold a mechanism in place by deflecting the legs about a centerline axis. A spring of this type will decrease in body diameter and slightly increase in body length when deflected in the preferred direction of the fabricated wind. The direction of the fabricated wind can be important for torsion spring applications due to the leg bearing/attachment locations having to be on the left or right side upon assembly. Torsion springs are normally supported by a rod (mandrel) that is coincident with the theoretical hingeline of the final product. *Figure 8* provides torsion spring terminology.

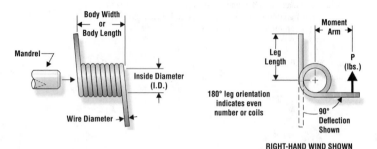

Figure 8. Torsion spring. (Source, Century Spring Corp.)

When selecting a torsion spring, it is important to calculate the torque that the spring is expected to develop. Torque is the applied force pushing perpendicularly on a leg or extension of the spring, multiplied by the distance from the centerline of the body of the spring. Torque, therefore, is equal to load times length. Next determine the angular deflection through which the spring leg is required to operate. Dividing the torque by the deflection angle will provide the spring constant rate.

$$C_R = T \div \theta$$

where C_R = spring constant rate
T = torque (inch pounds)
θ = deflection angle (degrees).

Torsion springs are normally classified by spring rate, and determining the rate is critical in selecting a spring from a manufacturer's catalog. Once springs with the desired rate are located, a selection can be made from the available springs to acquire a spring with the required inside diameter and body width. Torsion springs are also classified by "suggested maximum deflection," which provides limits that should not be exceeded if maximum spring life is to be achieved.

The initial unloaded relative-leg-angle orientation is determined by the number of coils in the spring body. As can be seen in *Figure 9*, an even number of coils provides flat legs, but an additional $1/8$ of a coil increases the unloaded relative-leg-angle orientation by 45°.

The hand orientation of a torsion spring can be determined by simulating the spring's end view with an individual's hand, as can be seen in *Figure 10*, which depicts a right hand wind spring.

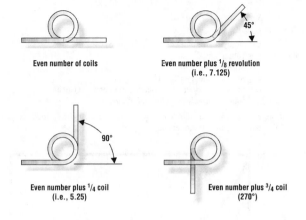

Figure 9. Leg-angle orientation in torsion springs is determined by the number of coils in the spring body. (Source, Century Spring Corp.)

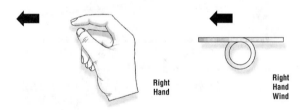

Figure 10. Torsion spring hand orientation. (Source, Century Spring Corp.)

Body lengthening due to deflection increases approximately one wire diameter for each complete leg revolution (360°). The body inside diameter decreases according to the following formula.

ID (loaded) = $(N \div [N + \text{rev}]) \times$ ID (unloaded)

where   ID = inside diameter of spring (inches)
        N = number of coils in spring
        rev = number of leg revolutions.

Once the inside diameter of the load-contracted spring body is determined, the maximum diameter of the supporting mandrel can be determined; usually it is set as 90% of load-contracted inside diameter.

Spring rate and stress for helical, round wire torsion springs can be determined with the following formulas (see earlier rate and stress formulas for symbols not identified below).

$$R_T = P\,L \div \theta \qquad \text{or} \qquad R = E\,d^4 \div 3888\,D\,N$$

$$S = 32\,P\,L\,K_F \div \pi\,d^3$$

where R_T = rate (in.lb/degree)
     L = moment arm in inches (see *Figure 8*)
     E = Young's modulus
        Spring steel = 30×10^6
        Stainless steel = 28×10^6
     D = mean body diameter = outside diameter – d (inches)
     K_F = stress correction factor (see *Figure 11*).

The suggested maximum allowable stress value for a torsion spring is considered to be 75% of the material's maximum tensile strength.

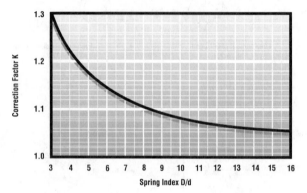

Figure 11. Stress correction factors for torsion springs. (Source, Century Spring Corp.)

Tapered springs

Conically shaped compression springs, commonly known as tapered springs as shown in *Figure 12*, are characterized not only by their shape, but also by the fact that their rate is normally nonlinear. There is a necessary increase in the applied force to compress the spring due to the flexibility of the larger diameter coils causing progressive contact with each other. This characteristic can be an advantage for spring-supported vibrating objects by reducing the resonant (bouncing) amplitudes commonly found in constant diameter spring support systems. Conical springs have more lateral stability and less tendency to buckle than regular compression springs, and their most beneficial characteristic is that they can be designed so that each active coil fits within the next coil, resulting in a solid height equal to one or two coils.

To calculate the highest stress at a given load or deflection for a tapered spring, the mean diameter of the largest active coil is used to replace the mean diameter (D) in the equation above for compression springs. The rate for a tapered spring becomes nonlinear once the larger diameter adjacent coils come into contact with one another during compression. This loss of active coils increases the stiffness of the spring, and spring rates given for taper springs are typically the approximate starting rate prior to coil contact.

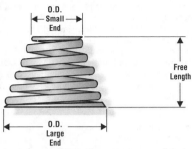

Figure 12. A conically shaped (tapered) compression spring. (Source, Century Spring Corp.)

Aside from any stress limits, the solid (minimum) length of a uniformly tapered, not fully collapsible spring with closed and ground ends can be estimated from

$$H = (N_{-2} \sqrt{d^2 - u^2}) + 2\, d$$

where H = final height, or length (inches)

N_{-2} = total number of coils, minus 2

u = largest outside diameter minimum smallest outside diameter divided by $2 \times N_{-2}$.

Die springs

Although die springs are primarily used in die machinery, they are well suited for many other applications where high static or shock-load stresses are required, or when maximum cycle life is important. A typical die spring is shown in *Figure 13.*

When used in die applications, it is a good rule to use as many springs as the die will accommodate to produce the desired load and cause the least amount of deflection. This will increase the service life and reduce chances of early failure. When selecting springs for a die operation, first decide if the springs will be used for short or long runs, average- or high-frequency cycling, or if they will be under high stress. The higher the rate of cycling, the higher the rate of fatigue caused failures. In slow oscillating dies or fixtures, it is possible to get good performance with springs operating at near maximum deflection. However, as the working speed increases, the life expectancy of the spring at that deflection decreases.

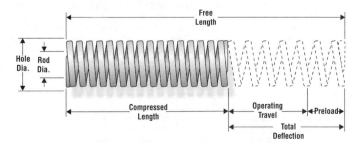

Figure 13. Die spring nomenclature. (Source, Century Spring Corp.)

Thermal effects are important to the performance and reliability of die springs. High heat has a pronounced negative effect on die spring operation, and **Table 4** provides load loss, as a percentage, caused by elevated temperature. The maximum service temperature for chrome-alloy steel is approximately 440° F (227° C).

Unlike other springs discussed to this point, die springs are usually fabricated from rectangular spring material, with rounded corners. Oil-tempered die springs usually have a very long service life if maximum deflection is held to about 25% of total spring length. Stresses for rectangular wire die springs can be calculated with the following equation.

$$S = (P\,D \div [b\,t\,\sqrt{b\,t}\,]) \times \beta$$

where b = wire width, radially (inches)
t = wire thickness (inches)
β = stress correction factor (see *Figure 14*).

Table 4. Temperature Effect on Die Springs. *(Source, Century Spring Corp.)*

Stress (PSI)	Approximate Percent Load Loss Due to Temperature								
	Carbon Steel SAE-1070				Chrome Alloy Steel				
	250° F 120° C	300° F 150° C	350° F 175° C	400° F 205° C	250° F 120° C	300° F 150° C	350° F 175° C	400° F 205° C	450° F 230° C
120,000	10	12	13	18	5	7	9	10	13
110,000	7	9	12	14	4	6	7	8	12
100,000	5	7	10	11	3	4	5	5	11
90,000	4	6	8	9	2	3	4	5	10
80,000	3	5	6	8	2	2	3	4	9
70,000	3	4	6	7	2	2	3	4	8
60,000	3	4	5	5	1	1	3	4	7
50,000	2	3	4	5	1	1	3	3	6
40,000	2	3	4	5	1	1	2	3	5

Marsh mellow springs

Marsh mellow, or cylindrical elastomer, springs are used in special applications where steel springs are inadequate. For example, they are useful in applications that require high shock and energy absorption, high vibration absorption, or where very large static load carrying demands are encountered. They excel where long life and quiet operation are desired. When installed and operated properly, more than one million cycles may be realized. March mellow springs have enjoyed considerable success in die related production operations. They are sensitive to elevated temperatures, however, and their use should be limited to a temperature range of –35° to 145° F (–37 to 63° C) to maintain a long service life.

The success of the design is due to a solid, primarily rubber core that has a hollow core and is contained in a very strong bias-ply fabric encasement. They have been very successful in replacing many standard size coil springs and nitrogen cylinders, and the larger diameter marsh mellow springs are capable of loads up to 17 tons.

Marsh mellow springs can be successfully deflected up to 40% of their free length, but a maximum compression of 30% is recommended for maximum life. The following guidelines provide suggested rate limits in cycles-per-minute (C.P.M.) for varying percentages of compression.

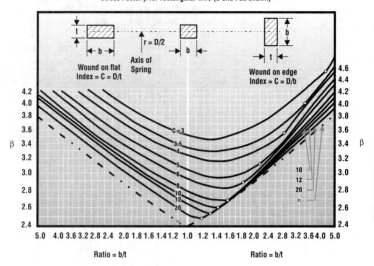

Figure 14. Stress factor β for rectangular wire. (Source, Century Spring Corp.)

	Maximum Percentage Compression						
	10	15	20	25	30	35	40
C.P.M.– 1 $^1/_4$ thru 2 $^1/_2$ O.D.	200	150	125	100	75	65	50
C.P.M.– 3 $^1/_4$ thru 6 O.D.	100	75	55	45	35	25	20

Suggested maximum free lengths for springs with outside diameters of up to 2 inches is 2 $^1/_2$ times the diameter. Maximum free lengths for larger diameters should not exceed twice the diameter. Marsh mellow springs may be stacked with a guide rod and washer shaped locators, preferably with Oilite sleeve bearings. Stacking two springs doubles the deflection for a given load, three springs triples the deflection, etc. The ends of the springs may be retained in place with short cylindrical protrusions. It should be noted that a compression set of 9% may occur over the life of the spring when operated at high deflections and/or cycle rates. Recommended mountings for marsh mellow springs are shown in *Figure 15*.

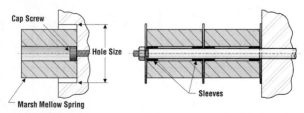

Figure 15. Mounting techniques for marsh mellow springs. (Source, Century Spring Corp.)

Table 5. Load/Deflection Relationship for Marsh Mellow Springs. *(Source, Century Spring Corp.)*

Hole Size	Spring O.D. × I.D.	Loads, in Pounds, as a Percentage of Deflection						
		10%	15%	20%	25%	30%	35%	40%
1 1/2	1 1/8 × 5/16	100	160	220	280	355	450	585
2	1 5/8 × 5/8	220	330	465	625	860	1180	1640
2 5/8	2 × 3/4	290	470	650	910	1250	1750	2450
3 1/4	2 1/2 × 3/4	570	920	1300	1780	2450	3450	4760
4 1/4	3 1/4 × 1	925	1550	2050	2770	3710	5150	7390
5 1/4	4 × 1	1550	2450	3540	4810	6570	9540	13410
6 1/2	5 × 1	2950	4600	6500	8850	11950	16670	23370
7 5/8	6 × 1	4500	6400	9580	12950	17620	24000	33460

Hole and spring size dimensions are in inches.

Wire Rope

Wire rope can be used to connect a power source and a load, when they are separated by a large distance. It is the first choice for many applications, and special ropes are provided for such diverse industries as logging and trawling. Typically, wire ropes are found employed in lifting operations such as elevators and cranes, and as control devices in instruments such as automotive accelerator and heater mechanisms. Due to the wide range of applications, they can range in diameter from very small to several inches. Much of the information in this section was contributed by Wire Rope Industries Ltd., who supply a wide range of products to industry.

Classes of wire rope

The class of a wire rope reflects the number of strands in the rope, plus the number of wires that comprise each strand. Thus, a 7 × 7 commercial quality wire rope will contain 7 strands (six exterior strands surrounding a single central strand), and each of the strands will contain 7 individual wires, arranged in the same symmetrical pattern. Unfortunately, the classification scheme is not strictly adhered to, and within several common classes several subclasses exist. For example, a 6 × 19 class rope might identify a rope with six strands and 19 wires in each strand, but it may also refer to any other six strand rope that has the same strength as a standard 6 × 19. Therefore, an order for a 6 × 19 rope, without further specifications, might result in the receipt of anything from a 6 × 15 wire rope, to a 6 × 26 wire rope. For this reason, when ordering wire rope, it is extremely important to exactly specify the rope that is desired.

Among the most popular classes of wire rope are 6 × 7, 6 × 19, 6 × 37, 8 × 19, 1 × 7 galvanized cable, and 7 × 7 and 7 × 19 galvanized and stainless steel aircraft cable. Strength and weight per foot values for these wires are provided in **Tables 1** through **6** in this section.

Table 1. Weights and Breaking Strengths for 6 × 7 Class IPS Wire Rope.

Diameter	Approximate Weight		Nominal Breaking Load[1]	
inch	lb./ft.	kg/m	Tons	kN
13/16	0.056	0.08	1.50	13
1/4	0.094	0.14	2.64	23
5/16	0.15	0.22	4.10	36
3/8	0.21	0.31	5.86	52
7/16	0.29	0.43	7.93	70
1/2	0.38	0.57	10.30	92
9/16	0.48	0.71	13.00	116
5/8	0.59	0.88	15.90	141
3/4	0.84	1.25	22.70	202
7/8	1.15	1.68	30.70	274
1	1.50	2.23	39.70	354
1 1/8	1.90	2.71	49.80	444
1 1/4	2.34	3.48	61.00	544
1 3/8	3.84	5.71	73.10	652
1 1/2	3.38	5.03	88.20	786

[1] Breaking load values are subject to a minus tolerance of 2.5%.

Table 2. Weights and Breaking Strengths for 6 × 19 Class Wire Rope.

Diameter	Approximate Weight				Nominal Breaking Load[1]			
	IPS Fiber Core		EIPS IWRC		IPS Fiber Core		EIPS IWRC	
inch	lb./ft.	kg/m	lb./ft.	kg/m	Tons	kN	Tons	kN
1/4	0.105	0.154	0.12	0.17	2.74	24.3	3.4	30
5/16	0.164	0.244	0.18	0.27	4.26	37.9	5.3	47
3/8	0.236	0.351	0.26	0.39	6.10	54.2	7.6	67
7/16	0.32	0.48	0.35	0.52	8.27	73.5	10.2	91
1/2	0.42	0.62	0.46	0.68	10.7	95.1	13.3	118
9/16	0.53	0.79	0.59	0.88	13.5	120	16.8	149
5/8	0.66	0.98	0.72	1.07	16.7	148	20.6	183
3/4	0.95	1.41	1.04	1.55	23.8	211	29.4	262
7/8	1.29	1.92	1.42	2.11	32.2	286	39.8	354
1	1.68	2.50	1.85	2.75	41.8	371	51.7	460
1 1/8	2.13	3.17	2.34	3.48	52.6	467	65.0	578
1 1/4	2.68	3.99	2.89	4.30	64.6	574	79.9	711
1 3/8	3.18	4.73	3.50	5.21	77.7	691	96.0	854
1 1/2	3.78	5.62	4.16	6.19	92.0	818	114	1014
1 3/4	5.15	7.66	5.67	8.44	124	1103	153	1361
2	6.72	10.00	7.39	11.00	160	1423	198	1762
2 1/4	8.51	12.66	9.36	13.93	200	1779	247	2199

[1] Breaking load values are subject to a minus tolerance of 2.5%. See text for explanation of abbreviations.
6 × 19 Class Wire Rope may have between 15 and 26 wires per strand.

Table 3. Weights and Breaking Strengths for 6 × 37 Class Wire Rope.

Diameter	Approximate Weight				Nominal Breaking Load[1]			
	IPS Fiber Core		EIPS IWRC		IPS Fiber Core		EIPS IWRC	
inch	lb./ft.	kg/m	lb./ft.	kg/m	Tons	kN	Tons	kN
1/4	0.105	0.15	0.12	0.17	2.74	24	3.4	30
5/16	0.164	0.24	0.18	0.27	4.26	37	5.3	47
3/8	0.236	0.35	0.26	0.39	6.10	54	7.6	67
7/16	0.32	0.47	0.35	0.52	8.27	73	10.2	91
1/2	0.42	0.62	0.46	0.68	10.7	95	13.3	118
9/16	0.53	0.79	0.59	0.88	13.5	120	16.8	149
5/8	0.66	0.98	0.72	1.07	16.7	148	20.6	183
3/4	0.95	1.41	1.04	1.55	23.8	211	29.4	262
7/8	1.29	1.92	1.42	2.11	32.2	286	39.8	354
1	1.68	1.29	1.85	2.75	41.8	371	51.7	460
1 1/8	2.13	3.16	2.34	3.48	52.6	467	65.0	578
1 1/4	2.63	3.91	2.89	4.30	64.6	574	79.9	711
1 3/8	3.18	4.73	3.50	5.21	77.7	691	96.0	854
1 1/2	3.78	5.62	4.16	6.19	92.0	818	114	1014

(Continued)

Table 3. *(Continued)* **Weights and Breaking Strengths for 6 × 37 Class Wire Rope.**

Diameter	Approximate Weight				Nominal Breaking Load[1]			
	IPS Fiber Core		EIPS IWRC		IPS Fiber Core		EIPS IWRC	
inch	lb./ft.	kg/m	lb./ft.	kg/m	Tons	kN	Tons	kN
1 3/4	5.15	7.66	5.67	8.44	124	1103	153	1361
2	6.72	10.0	7.39	11.0	160	1423	198	1762
2 1/4	8.51	12.66	9.36	13.9	200	1779	247	2199
2 1/2	10.50	15.62	11.6	17.2	244	2170	302	2668
2 3/4	-	-	14.0	20.8	-	-	361	3214
3	-	-	16.6	24.7	-	-	425	3783

[1] Breaking load values are subject to a minus tolerance of 2.5%. See text for explanation of abbreviations.
6 × 37 Class Wire Rope may have between 27 and 49 wires per strand.

Table 4. Weights and Breaking Strengths for 8 × 19 Class Wire Rope.

Diameter	Approximate Weight				Nominal Breaking Load[1]			
	TS Fiber Core[1]		EIPS IWRC		TS Fiber Core[2]		EIPS IWRC	
inch	lb./ft.	kg/m	lb./ft.	kg/m	Pounds	kN	Tons	kN
1/4	0.09	0.13	0.12	0.18	3,600	16	3.0	26
5/16	0.14	0.21	0.18	0.27	5,600	25	4.6	41
3/8	0.20	0.30	0.26	0.39	8,200	36	6.6	59
7/16	0.28	0.42	0.36	0.54	11,000	49	9.0	80
1/2	0.36	0.54	0.47	0.69	14,500	65	11.6	103
9/16	0.46	0.68	0.60	0.89	18,500	82	14.7	131
5/8	0.57	0.85	0.73	1.00	23,000	102	18.1	161
11/16	0.69	1.03	-	-	27,000	120	-	-
3/4	0.82	1.22	1.06	1.58	32,000	142	25.9	231
13/16	0.96	1.43	-	-	37,000	165	-	-
7/8	1.11	1.65	1.44	2.14	42,000	187	35.0	312
15/16	1.27	1.89	-	-	48,000	214	-	-
1	1.45	2.16	1.88	2.80	54,000	240	45.5	405
1 1/16	1.64	2.44	-	-	61,000	271	-	-
1 1/8	1.84	2.74	2.39	3.56	68,000	302	57.3	510
1 1/4	-	-	2.94	4.38	-	-	70.5	628

[1] Breaking load values are subject to a minus tolerance of 2.5%. [2] Elevator rope. See text for explanation of abbreviations. (Source, Wire Rope Industries Ltd.)

Table 5. Weights and Breaking Strengths for 1 × 7 Galvanized Wire Cable (ASTM A475 Class A Coating).

Diameter	Approximate Weight		Diameter of Coated Wires in Strand		Minimum Breaking Load			
					High Strength Grade		Extra High Strength	
inch	lb./1000 ft.	kg/304.8 m	inch	mm	Pounds	kN	Pounds	kN
3/16	73	33	0.062	1.57	2,850	12.68	3,990	17.75
7/32	98	44	0.072	1.83	3,850	17.13	5,400	24.02
1/4	121	55	0.080	2.03	4,750	21.13	6,650	29.58
9/32	164	74	0.093	2.36	6,400	28.47	8,950	39.81
5/16	205	93	0.104	2.64	8,000	35.59	11,200	49.82
3/8	273	124	0.120	3.05	10,800	48.04	15,400	68.50
7/16	399	181	0.145	3.68	14,500	64.50	20,800	92.52
1/2	517	234	0.165	4.19	18,800	83.63	26,900	119.66
9/16	671	304	0.188	4.78	24,500	108.98	35,000	155.69
5/8	813	369	0.207	5.26	29,600	131.67	42,400	188.61

(Source, Wire Rope Industries Ltd.)

Table 6. Weights and Breaking Strength for Galvanized and Stainless Steel 7 × 7 and 7 × 19 Aircraft Cable.

Diameter	Rope Construction	Approximate Weight		Minimum Breaking Load			
				Galvanized Cable		Stainless Steel Cable[1]	
inch		lb./100 ft.	kg/100 m	Pounds	kN	Pounds	kN
1/16	7 × 7	0.75	1.12	480	2	480	2
3/32	7 × 7	1.60	2.38	920	4	920	4
1/8	7 × 19	2.90	4.32	2,000	9	1,760	8
5/32	7 × 19	4.50	6.70	2,800	12	2,400	11
3/16	7 × 19	6.50	9.67	4,200	19	3,700	16
1/4	7 × 19	11.00	16.37	7,000	31	6,400	28
5/16	7 × 19	17.30	25.74	9,800	44	9,000	40
3/8	7 × 19	24.30	36.16	14,400	64	12,000	53

[1] Stainless steel types 302 and 304. (Source, Wire Rope Industries Ltd.)

Types of wire rope

Round strand ropes are the simplest of the true rope types, and are almost universal in use. They consist of 3 to 36 strands of wire laid in various arrangements around a core. The use of round wires laid in geometric patterns results in a round strand. Two typical arrangements are shown in *Figure 1*.

Strands are made up of two or more layers of wire around a center wire, and the wires in the layers may be cross-laid, meaning that the pitch or length of lay (see below for a explanation of lay and its effect on wire characteristics) will be longer for the outer layer. This construction is limited to small ropes and cords, to some nautical ropes, and to standing ropes in larger sizes. For all other purposes, ropes of equal-laid construction (meaning that each wire of a given diameter in every strand of the rope will be the same ultimate length as other wires the same diameter) are preferable. By this method, all layers of wires have the same pitch or length of lay. Therefore, each wire in every layer lies either in a bed formed by the interstices or valleys between the wires of an underlayer, or

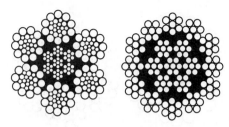

Figure 1. Round strand wire rope construction. (Source, Wire Rope Industries, Ltd.)

alternatively along the crown of an underlying wire. Because no layer of wire ever crosses over another, an equal-laid rope will maintain its diameter in service and have a more solid cross section and improved fatigue life when compared to ropes of the cross-laid variety. When wires are laid up in a strand in equal-laid construction, the arrangements of wire are referred to as *Warrington*, *Seale*, or *Filler*, depending on the way in which the wires are arranged. If a combination of two of these arrangements is used, the wire is described as having a *Combination* arrangement.

Warrington arrangement is made up of two different size wires laid over a center layer of wires of uniform diameter. The outer layer contains twice the number of wires contained in the inner layer. The use of two different size wires on the external layer provides an unusually smooth, round outer surface.

Seale arrangement contains an equal number of wires of two diameters, with the smaller diameter inner wires fitting snugly into the valley between the larger diameter outer wires. The irregular outer surface creates a very abrasive surface.

Filler arrangement is made up of two layers of wires of the same diameter, with the outer layer containing twice the number of wires of the inner layer, and a small wire filling the valleys between the layers. Filler wires exhibit good shock absorption characteristics.

Flattened strand ropes are built up of triangular or oval shaped strands. Certain types, specifically rotation resistant ropes, are built of a combination of oval-around-triangular, or sometimes oval-over-round strands. In the case of triangular shaped strands, the wires forming the strand are laid on a triangular shaped core formed of three or more round wires. The wires forming an oval strand are laid on an oval shaped center ribbon or around a group of usually 4 round wires laid parallel.

Because of the shape of the strands, flattened strand ropes have smaller synthetic cores than do round strand ropes. Hence, size for size, flattened strand ropes have approximately 10% greater metallic area with comparably increased break-load capabilities. As can be seen in *Figure 2*, flattened strand ropes spread frictional wear over a greater number of outer wires than do round strand ropes. In the flattened strand construction, wear is therefore more even, and loss of sectional area of the vital outside wires is much reduced. Because of the relatively smooth surface and circular cross section obtained with flattened strand ropes, the wear on sheaves and pulleys is materially reduced. When used in endless haulage systems, the increased external bearing surface enables grip and clips to be more securely employed, and clips are less likely to force flattened strand ropes out of shape.

Locked coil rope is completely different from both round strand and flattened strand

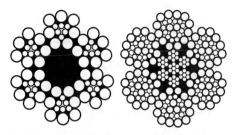

Figure 2. Flattened strand wire rope construction. (Source, Wire Rope Industries, Ltd.)

rope. Instead of a group of individual strands closed around a central core member, it consists of a single strand. However, the strand is built up of layer upon layer of wires. The center, or core, of a locked coil rope will consist of a concentric laid strand of round wires, and around this core will be laid one or more layers of shaped wires, with the outer layers always being interlocking. The shapes of all shaped wires in a locked coil rope will depend on the rope diameter and its intended use. Typical locked coil wire arrangements are shown in *Figure 3*.

Locked coil ropes have a higher breaking load capacity than stranded ropes of equal diameter ropes of the same grade. Because of their smooth external surface, depreciation in strength caused by frictional wear on drums or pulleys is greatly reduced. Because of their design, locked coil ropes are subject to less rotation and stretch than stranded ropes, but they require a larger bending ratio.

Concentric strand ropes are made up of round wires laid layer upon layer around a center wire. This simple strand construction is used exclusively for static purposes such

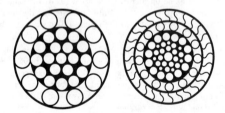

Figure 3. Locked coil wire rope construction. (Source, Wire Rope Industries, Ltd.)

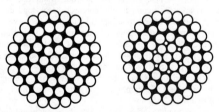

Figure 4. Concentric strand wire rope construction. (Source, Wire Rope Industries, Ltd.)

as suspension bridges, communication towers, and structural roof supports. Two general arrangements, shown in *Figure 4*, are used. One consists of concentric layers of wires of one size, arranged around a somewhat larger center wire. The other is made up of one or more concentric layers of wires, of one size, around a core group made up of an equal laid strand of Warrington, Seale, or Filler wire construction.

Standard abbreviations used for wire rope

Several abbreviations are used to describe wire rope construction and materials. Some of the most common are provided below.

EIP. Wire made from extra-improved plow steel.
FC. Wire with fiber core construction.
IPS. Wire made from improved plow steel.
IWRC. Wire with an independent wire-rope core.
MPS. Wire made from mild plow steel.
PS. Wire made from plow steel.
TS. Wire made from traction steel.
WSC. Wire with a wire-strand core.

Wire rope core material

The core forms the heart of the rope and is the component about which the main rope strands are laid. The core acts as a support for the strands, and prevents them from jamming against or coming into contact with each other under normal loading and flexings. The three core constructions described below are shown in *Figure 5*.

Fiber core is adequate for many types of service, and provides maximum flexibility and elasticity for the wire rope. Although generally made of hard fibers such as sisal or manila, man-made fibers such as polypropylene, polyester, and nylon are also used. Man-made cores are useful where conditions of rope use, such as attack by chemical agents, could cause failure of a natural fiber. Sisal and manila, when used, are impregnated with preservative lubricants during the cordage process. Generally speaking, fiber core ropes are best suited to less severe operations. Sometimes abbreviated FC.

Independent wire rope core is desirable where ropes are subjected to severe pressure while running over sheaves or winding onto drums. It should be used in conditions where temperatures may be damaging to fiber cores. In addition to offering more strength, elastic stretch is also as much as 30% less than may be experienced with a fiber core rope. Abbreviated IWRC.

Fiber Core   **Independent Wire Rope Core**   **Steel Strand Core**

Figure 5. Common wire rope core materials. (Source, Wire Rope Industries, Ltd.)

Strand cores consist of a strand of steel wires, nominally 7, 19, or 37 in number. It is sometimes used in running ropes of smaller diameters, and in guys, suspender ropes, etc., where extra strength, reduced stretch, and maximum resistance to weathering are required. Sometimes abbreviated WSC, or SC.

Wire rope lay

Lay describes three different characteristics of wire rope. It can indicate the direction of the strands in a rope, the relationship of the direction of the strands in the rope and the wires in the strands, and the distance along a length of rope required for one strand to make a complete spiral around the rope.

Regular (or ordinary) lay ropes have the wires in the strands laid in one direction, while the strands of the rope are laid in the opposite direction. The result is that the wire crowns run approximately parallel with the longitudinal axis of the rope. Such ropes have good resistance to kinking and twisting, and are therefore easy to handle. They are also able to withstand considerable crushing and distortion due to the short length of the exposed wires. As shown in *Figure 6*, regular lay ropes may have either a right lay or left lay, with right hand lay being the most common.

Right Lay Regular Lay   **Left Lay Regular Lay**

Figure 6. Regular lay construction. (Source, Wire Rope Industries, Ltd.)

Lang lay ropes have the wires in the strands and the strands in the ropes laid in the same direction. Consequently, the outer wires run diagonally across the top of the rope and are exposed for longer length than in a regular lay rope. Since the outer wires present greater wearing surface, lang lay ropes have greater resistance to abrasion. Due to the lay arrangement, they are also more flexible than regular lay ropes and possess greater resistance to fatigue. However, they are also more prone to kinking and untwisting than regular laid ropes, and they will not take the same abuse from distortion and crushing. To protect against untwisting, lang lay ropes should have both ends permanently fastened. Hence, they cannot be recommended for single part hoisting, nor should they be used with swivel end terminals. Most lang lay ropes have right lay. Lang lay construction is shown in *Figure 7*.

Right Lay Lang Lay   **Left Lay Lang Lay**

Figure 7. Lang lay construction. (Source, Wire Rope Industries, Ltd.)

Right lay and *left lay* ropes. A right lay rope is one in which the strands of the rope rotate to the right, while receding from the observer. A left lay rope is the opposite. Any rope can be manufactured with either lay, but only a few ropes are actually made left hand lay.

Zebra (or alternate) lay ropes have three strands with right hand lay and three with left hand lay. The six strands are positioned in the rope so that the right hand laid strands alternate with the left hand laid strands (see *Figure 8*). Use of alternate lay is quite limited.

Right Lay Zebra

Figure 8. Zebra lay construction. (Source, Wire Rope Industries, Ltd.)

Herringbone (or twin-strand) lay is a combination of right hand laid and left hand laid strands. It usually consists of two pairs of the right hand laid strands (lang lay), and two single left laid strands (regular lay). The two single left laid strands are separated by the pairs of right hand strands. *Figure 9* shows herringbone construction.

Left Lay Herringbone

Figure 9. Herringbone lay construction. (Source, Wire Rope Industries, Ltd.)

Measuring rope lay. One rope lay is the length along the rope that a single strand requires to make one complete spiral, or turn, around the core. It is an engineering factor in the design of a rope, and is carefully controlled during manufacture. Since there is often adjustment in rope lay during the initial break-in stages of a rope's usage, it is recommended that rope lay measurements be made after the initial loading for comparison purposes at succeeding periodic inspections.

One method of measuring rope lay is with ordinary blank white paper and a pencil. Firmly hold the paper on the rope and stroke the paper with the side of the pencil point, so that rope prints appear on the paper. By drawing a line through one strand of the print, counting off the number of strands in the rope, and then drawing another line on the print at the place where the same strand appears again, a measurement is established. Many inspectors have found that a crayon or marking stick and a roll of calculator tape are ideal for making a print at least three rope lays long. An average lay length can then be determined.

Changes in length of lay are usually gradual throughout the working life of a rope. It is important to compare current lay measurements with previous inspection results to note any sudden changes. An abrupt change in pattern can be a signal of an impending

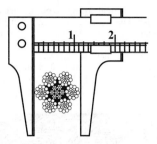

Figure 10. Correct method of rope measurement. (Source, Wire Rope Industries, Ltd.)

problem. As a rule, if lengthening of lay is noted, with loss of rope diameter, internal break-up or core destruction should be suspected. (See the next paragraph for information on properly measuring a rope's diameter.) When lengthening of lay is noted without loss of rope diameter, the rope is probably unlaying for some other reason, and further examination should be made to determine the cause. Unlaying sometimes results from operating a rope without having both ends secured to prevent rotation. An end swivel, for example, permits such rotation and unlaying. Another common problem of unlaying is worn sheaves. When the bottom of a sheave groove wears, it can restrict normal movement as the rope enters and leaves the groove. The result can be a build-up of twist that can change the length of lay.

The size of a wire rope is the diameter of the circle that will just enclose all the strands. The correct method to measure is to place a suitable dial caliper with increments of 0.001" over any pair of opposing strands, as shown in *Figure 10*. It is advisable to make at least three such measurements on a six strand rope, and four on an eight strand rope. Rope diameter should be measured over the extreme outer wires at no load, or at a load not exceeding 10% of the nominal rope-breaking load. Diameter tolerances are: for ropes less than $^5/_{16}$", –0, + 7% of diameter; for ropes $^5/_{16}$" and larger, –0, + 5%.

Working load limits for wire rope

The working load limit of a rope is the ratio of the nominal breaking load to the total working load. It is not possible to state an arbitrary working load factor for every wire rope application because the factor depends on individual working conditions. The working load factor for a new rope should be sufficiently high to allow the rope to render maximum service under normal fatigue stresses and abrasion, and provide dependability of operation throughout the rope's life. Higher working load factors should be chosen where human life or high material values are involved, where maximum rope life is demanded, where corrosion is severe, and where proper and frequent rope inspection is difficult to perform. The working load factors provided in **Table 7** are a general guide for acceptable safety, but it should be remembered that all individual applications should be studied to determine acceptable ratings.

Critical inspection points, which are points where failure would be most likely to occur, should be inspected frequently. These include the following.

Pick-up points. These are sections of rope that are repeatedly placed under stress when the initial load of each lift is applied, such as those sections in contact with sheaves.

Table 7. Suggested Working Load Factors for Wire Rope Applications. (See Text.)

Wire Rope Application	Suggested Working Load Factors[1]
Rope and Strand Guys (Static)	3 to 4
Main Ropes for Small Suspension Bridges	3 to 3 1/2
Suspender Ropes for Small Suspension Bridges	3 1/2 to 4
Aerial Tramway Track Cables	3 to 4
Aerial Tramway Haulage or Traction Ropes	5 to 6
Ski-lift Ropes	5 to 5 1/2
Clamshell Bucket Ropes	4 to 5
Excavator Ropes	5
Horizontal Haulage and Continuous Car Puller Ropes	4 to 5
Vertical and Incline Mine Hoist Ropes	5 (min.)
Industrial Hoist Ropes	6 (min.)
Crane Ropes—Bridge, Gantry, Jib, Pillar, Derrick, etc.	6 (min.)
Elevator Hoisting Ropes—Passenger	8 to 12
Elevator Hoisting Ropes—Freight	7 to 10
Hot Metal Crane Ropes	8 (min.)
Rotation Resistant Ropes (18 × 7 Class)	10 (min.)

[1] Load factors are for general information only. Every application should be studied for unusual stresses as well as equipment condition. (Source, Wire Rope Industries Ltd.)

End attachments. At each end of the rope, two items must be inspected. These are the fitting that is attached to the rope, or to which the rope is attached, and the condition of rope where it enters the attachment.

Equalizing sheaves. The section of the rope that is in contact with equalizer, or deflector, sheaves—as on bottom hoist lines—should receive careful inspection.

Drums. The general condition of the drum, and condition of the grooves in a grooved drum, should receive careful inspection, as should the manner in which the rope spools into the drum.

Sheaves. Every sheave in the rope system must be inspected and checked with a groove gage.

Heat exposure. Inspect for signs that a rope has been subjected to extreme heat, or to repetitive heat exposure.

Abuse points. Ropes are frequently subjected to abnormal scuffing and scraping, such as contact with a cross member of a boom. Look for bright spots on the rope, indicating contact.

Reel and drum capacity

The amount of rope that can fit onto an existing reel or drum can be anticipated in advance. The following formula predicts the rope capacity (*L*) in feet for any size drum or reel. The formula is based on uniform rope winding—it will not provide accurate results if the rope is wound nonuniformly on the reel. The dimensions in *Figure 11* should be expressed in inches.

$$L = (A + B) \times A \times C \times K$$

where L = rope length, in feet

        A = Depth of rope layer in inches = (H –B –M)/2 (see *Figure 10*)

        B = diameter of drum in inches

        C = width between reel flanges in inches

        $K = 0.2618 \div (D \times 2)$

        D = nominal rope diameter in inches

        H = height of reel flange in inches

        M = desired clearance (minimum clearance is one diameter of rope).

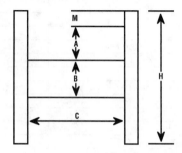

Figure 11. Reel and drum capacity variables. (Source, Wire Rope Industries, Ltd.)

Contents for Conversions Section

Table 1. SI Properties and Units

Physical and technical laws are represented by mathematical combinations of properties. The value of each physical property is the product of a numerical value and a unit of measurement.

value = numerical value × unit of measurement

The International System (SI), built upon a set of seven base units, enjoys widespread international use. Additional units used in special fields or in older literature are marked with an asterisk.

Property/Symbol	Unit	Relationship
Length x, y, z, s, l, r	**Meter (m)**	**SI Base Unit**
	astronomical unit (AU)*	$1 \text{ AU} = 149.600 \cdot 10^9 \text{ m}$
	light-year (lt-yr)*	$1 \text{ lt-yr} = 9.4605 \cdot 10^{15} \text{ m}$
	angstrom (Å)*	$1 \text{ Å} = 10^{-10} \text{ m}$
		(nuclear physics)
	(international) nautical mile (nm)*	$1 \text{ nm} = 1{,}852 \text{ m}$
plane angle α, φ	radian (rad)	$1 \text{ rad} = 1 \text{ m/m}$
	degree (°)	$1° = (\pi/180) \text{ rad}$
	perigon	$2\pi \text{ rad} = 360°$
solid angle Ω	steradian (sr)	$1 \text{ sr} = 1 \text{ m}^2 / \text{m}^2$
area A	square meter (m²)	
	are (a)	$1 \text{ a} = 100 \text{ m}^2$
	hectare (ha)	$1 \text{ ha} = 100 \text{ a} = 10^4 \text{ m}^2$
	barn (b)	$1 \text{ b} = 10^{-28} \text{ m}^2$
		(nuclear physics)
volume V	cubic meter (m³)	
	1 liter	$1 \text{ l} = 10^{-3} \text{ m}^3 = 10^3 \text{ cm}^3$
	solid cubic meter*	$1 \text{ solid cubic meter} = 1 \text{ m}^3$
		(timber industry)
	barrel*	$1 \text{ barrel} = 158.987 \text{ l}$
		(crude oil only)

Property/Symbol	Unit	Relationship
Mass m	**Kilogram (kg)**	**SI Base Unit**
	gram (g)	$1\ g = 10^{-3}\ kg$
	ton (t)	$1\ t = 10^3\ kg$
	hundredweight (cwt)*	$1\ cwt = 50\ kg$
	carat (ct)	$1\ ct = 0.2\ g$ (gemstones)
density ϱ	kg/m³	$1\ kg/m^3 = 10^3\ g/cm^3$
Time t	**Second (s)**	**SI Base Unit**
	minute (min)	$1\ min = 60\ s$
	hour (h)	$1\ h = 60\ min = 3{,}600\ s$
	day (d)	$1\ d = 24\ h$
frequency f, v	hertz (Hz)	$1\ Hz = 1$ cycle/s
rotational speed n	revolutions per minute (rpm)	$1\ rpm = 1$ revolution/60 s
angular frequency ω	1 cycle/s	
velocity	m/s knot (kn)*	$1\ m/s = 3.6\ km/h$
		$1\ kn = 1\ sm/h$
		$\quad = 0.5144\ m/s$
acceleration a	m/s²	
	standard gravity (g_n)	$g = 9.80665\ m/s^2$
angular velocity ω	rad/s	$1\ rad/s = 1/s$
angular acceleration α	rad/s²	$1\ rad/s^2 = 1/s^2$
force F,	newton (N)	$1\ N = 1\ kg\ m/s^2$
gravitational force G, F_G	dyne (dyn)*	$1\ dyn = 10^{-5}\ N$
	kilogram-force (kgf)*	$1\ kgf = 9.80665\ N$
torque M	newton-meter (N m)	$1\ N\ m = 0.102\ kgf\ m$
linear momentum p	N s	$1\ N\ s = 1\ kg\ m/s$
angular momentum L	N m s	$1\ N\ m\ s = 1\ kg\ m^2/s$
moment of inertia J	kg m²	
work W,	joule (J)	$1\ J = 1\ N\ m = 1\ W\ s$
energy E	kilowatt-hour (kWh)	$1\ kWh = 3.6 \cdot 10^6\ J$
heat quantity Q	electronvolt (eV)	$1\ eV = 1.602 \cdot 10^{-19}\ J$
	erg*	$1\ erg = 10^{-7}\ J$
	calorie (cal)*	$1\ cal = 4.1868\ J$
power P	watt (W)	$1\ W = 1\ J/s = 1\ V{\cdot}A$
	horsepower (hp)*	$1\ hp = 735.5\ W$

Property/Symbol	Unit	Relationship
pressure p	pascal (Pa)	$1\ Pa = 1\ N/m^2$
		$= 1\ kg/(s^2 \cdot m)$
	bar (bar)	$1\ bar = 10^5\ Pa = 10^3\ mbar$
	mmHg	$1\ mmHg = 133.322\ Pa$
	mm of mercury	(medicine)
	torr*	$1\ torr = 1.333224\ mbar$
	standard atmosphere	$1\ atm = 1.01325 \cdot 10^5\ Pa$
	(atm)*	$1\ atm = 760\ torr$
	technical atmosphere (at)*	$1\ at = 1\ kgf/cm^2$
		$= 0.981\ bar$
	meters of water column	$1\ mWS = 9806.65\ Pa$
	(mWS)*	
stress σ,	pascal (Pa)	$1\ Pa = 1\ N/m^2$
	N/m^2	$1\ MPa = 1\ N/mm^2$
modulus of elasticity E,		$1\ MPa = 0.102\ kgf/mm^2$
modulus of elasticity in shear G,		
modulus of compression K		
surface tension σ	N/m	$1\ N/m = 1\ kg/s^2$
dynamic viscosity η	pascal second (Pa·s)	$1\ Pa \cdot s = 1\ kg/(s \cdot m)$
	poise (P)*	$1\ P = 0.1\ Pa \cdot s$
kinematic viscosity ν	m^2/s	
	stoke (St)*	$1\ St = 10^{-4}\ m^2/s$

Electric Current Intensity I	**Ampere (A)**	**SI Base Unit**
current density j, J	A/m^2	
voltage U,	volt (V)	$1\ V = 1\ W/A$
electric potential φ		$1\ V = 1\ kg \cdot m^2/(A \cdot s^3)$
resistance R	ohm (Ω)	$1\ \Omega = 1\ V/A$
conductance G	siemens (S)	$1\ S = 1\ A/V = 1/\Omega$
electrical resistivity ϱ	ohm-meter (Ω-m)	$1\ \Omega\text{-}m = 1\ V \cdot m/A$
		$1\ \Omega\text{-}mm^2/m = 10^{-6}\ \Omega\text{-}m$

Property/Symbol	Unit	Relationship
conductivity κ	S/m	$1 \text{ S/m} = 1/(\Omega \cdot \text{m})$
charge Q	coulomb (C)	$1 \text{ C} = 1 \text{ A} \cdot \text{s}$
quantity of electricity Q	ampere-hour (Ah)	$1 \text{ Ah} = 3{,}600 \text{ C}$
surface charge density σ, electric flux density, electric displacement D	C/m²	
capacitance C	farad (F)	$1 \text{ F} = 1 \text{ C/V} = 1 \text{ A} \cdot \text{s/V}$
permittivity ε, dielectric constant ε	F/m	$1 \text{ F/m} = \text{A} \cdot \text{s}/(\text{V} \cdot \text{m})$
electric field strength E	V/m	$1 \text{ V/m} = 1 \text{ kg} \cdot \text{m}/(\text{s}^3 A)$
magnetic field strength H	A/m oersted (Oe)*	$1 \text{ Oe} = [10^3/(4\pi)] \text{ A/m}$
magnetic flux Φ	weber (Wb) maxwell (Mx)*	$1 \text{ Wb} = 1 \text{ V} \cdot \text{s} = 1 \text{ T} \cdot \text{m}^2$ $1 \text{ Mx} = 10^{-8} \text{ Wb}$
magnetic flux density B,	tesla (T) gauss (G)*	$1 \text{ T} = 1 \text{ Wb/m}^2 = 1 \text{ V} \cdot \text{s/m}^2$ $1 \text{ G} = 10^{-4} \text{ T}$
magnetic induction inductance L	henry (H)	$1 \text{ H} = 1 \text{ Wb/A} = 1 \text{ V} \cdot \text{s/A}$
permeability μ	H/m	$1 \text{ H/m} = 1 \text{ V} \cdot \text{s}/(\text{A} \cdot \text{m})$

Temperature T	**Kelvin (K)**	**SI Base Unit**
Celsius temperature t, ϑ	degrees Celsius (°C) degrees Fahrenheit (°F)*	$t/°\text{C} = T/\text{K} - 273.15$ $t/°\text{F} = 1.8 t/°\text{C} + 32$
heat quantity Q (also see "work")	joule (J)	$1 \text{ J} = 1 \text{ kg} \cdot \text{m}^2/\text{s}^2 = 1 \text{ W} \cdot \text{s}$
specific heat capacity c	J/(kg·K)	
thermal conductivity λ heat-transfer coefficient α	W/(m·K) W/(m²·K)	

Luminous Intensity I	**Candela (cd)**	**SI Base Unit**
luminance L	cd/m² stilb (sb)*	$1 \text{ sb} = 10^4 \text{ cd/m}^2$

Property/Symbol	Unit	Relationship
luminous flux Φ	lumen (lm)	1 lm = 1 cd·sr
illumination E	lux (lx)	1 lx = 1 lm/m^2
radiant flux density E_e	W/m^2	
radiant intensity I_e	W/sr	
radiance L_e	W/(m^2·sr)	
refracting power D	diopter (D)	1 D = 1/m (optical systems)
activity A	becquerel (Bq) curie (Ci)*	1 Bq = 1/s 1 Ci = 3.7 · 10^{10} Bq
absorbed dose D	gray (Gy) rad (rd)*	1 Gy = 1 J/kg 1 rd = 0.01 Gy
dose equivalent H	sievert (Sv) rem*	1 Sv = 1 J/kg 1 rem = 0.01 Sv
absorbed-dose rate $\dot{D}$	Gy/s	1 Gy/s = 1 W/kg
exposure dose J	C/kg roentgen (R)*	1 C/kg = 1 A·s/kg 1 R = 258 · 10^{-6} C/kg

Amount of Substance n	**Mole (mol)**	**SI Base Unit**
molar mass M	g/mol	1 g/mol = 10^{-3} kg/mol
molar volume V_m	l/mol	1 l/mol = 10^{-3} m^3/mol
molar heat capacity C_m	J/(mol K)	

SI Prefixes

Power	Name	Symbol	Power	Name	Symbol
10^{24}	yotta	Y	10^{-1}	deci	d
10^{21}	zetta	Z	10^{-2}	centi	c
10^{18}	exa	E	10^{-3}	milli	m
10^{15}	peta	P	10^{-6}	micro	μ
10^{12}	tera	T	10^{-9}	nano	n
10^{9}	giga	G	10^{-12}	pico	p
10^{6}	mega	M	10^{-15}	femto	f
10^{3}	kilo	k	10^{-18}	atto	a
10^{2}	hecto	h	10^{-21}	zepto	z
10^{1}	deka	da	10^{-24}	yocto	y

Table 2. Selected International Metric Terms, Units, and Symbols for Machine Tool Use.

Quantity	Selected SI Unit	SI Unit Symbol	Customary Unit	Customary Symbol	Conversion Factor[1]
Space - Distance					
Angular allowance for accuracy	Degree Minute Second	° ' "	Degree Minute Second	° ' "	1
Feed per tooth	Millimeter	mm	Inch	in.	25.4
Geometric tolerances					
Length-depth (general)					
Length-cable	Meter	m	Foot	ft	.30480
Surface finish-roughness height	Micrometer	μm	Millionth of an inch	μin.	0.2540
Threads-lead or pitch	Millimeter	mm	Threads per inch	th/in, tpi	.039370
Space - Area/Volume					
Area (general)	Square meter	m^2	Square feet	sq ft, ft^2	.092903
Capacity-reservoir	Liter	L	Gallon	gal	3.7854
Cubic content	Cubic meter	m^3	Cubic feet	cu ft, ft^3	.028317
Space - Time					
Acceleration	Meters per second per second	m/s^2	Inches per second per second	in./sec^2	.02540
Capacity-pump	Liters per minute	L/min	Gallons per minute	gal/min	3.7854
Feed per revolution	Millimeters per revolution	mm/r	Inches per revolution	in./rev	25.4
Feed rate	Millimeters per minute	mm/min	Inches per minute	ipm	25.4
Feed rate-rapid traverse	Meters per minute	m/min	Inches per minute	ipm	.02540
Flow, air-small	Liters per minute	L/min usually ≤1000	Cubic feet per minute	cu ft/min	28.3168
Flow, air-normal	Cubic meters per minute	m^3/min usually >1	Cubic feet per minute	cu ft/min	0.283
Flow-coolant	Liters per minute	L/min	Gallons per minute	gal/min	3.7854
Speed-cutting	Meters per minute	m/min	Feet per minute	ft/min	.30480
Speed-surface grinding wheel	Meters per second	m/s	Feet per minute	ft/min	.00508
Speed-spindle	Revolutions per minute	r/min	Revolutions per minute	rpm	1
Stroke rate	Stroke per minute	stroke/min	Stroke per minute	stroke/min	1
Periodic and Related Phenomena					
Frequency	Hertz	Hz	Cycles per second	-	1

(Continued)

Table 2. *(Continued)* **Selected International Metric Terms, Units, and Symbols for Machine Tool Use.**

Quantity	Selected SI Unit	SI Unit Symbol	Customary Unit	Customary Symbol	Conversion Factor[1]
Mechanics					
Energy	Joule	J	British thermal unit Calorie Foot-pound force	Btu cal ft lbf	1055.056 4.1868 1.3558
Moment of force (torque)	Newton meter	Nm	Pound-force-inch Pound-force-foot	lbf-in. lbf-ft	.1129 1.355
Force-clamping	Kilonewton	kN	Ton	ton	8.8964
Power-machine	Kilowatt	kW	Horsepower	HP	.7460
Load-table capacity Weight (mass)	Kilogram	kg	Pound	lb	.45359
Thrust in a given axis or Feed force-small	Newton	N usually <1000	Pound	lb	4.4482
Thrust in a given axis or Feed force-normal	Kilonewton	kN usually >1	Pound	lb	.0044
Power per revolution per minute	Watts per revolution per minute	W(r/min)	Horsepower per revolution per minute	HP/rpm	746
Material removal-machining	Cubic centimeters per minute	cm³/min	Cubic inches per minute	cu in./min	16.387
Pressure between surfaces	Kilonewtons per square meter	kN/m²	Pound per square inch	lb/in.²	6.8948
Pressure-liquid	Bar Kilopascal	bar 1 bar = 10⁵ Pa kPa 1 bar = 100 kPa	Pound per square inch	lb/in.²	.068948
Sawing rate	Square millimeters per minute	mm²/min	Square inches per minute	sq in./min in.²/min	645.16
Tensile strength	Kilopascal	kPa	Pounds per square inch	PSI	6.8948
Electricity					
Electric current	Ampere	A	Ampere	A	1
Electric resistance	Ohm	Ω	Ohm	ohm	1
Voltage	Volt	v	Volt	v	1

[1] To convert customary unit to metric, multiply the customary unit by the conversion factor. To convert metric unit to customary, divide the metric unit by conversion factor. This table was provided courtesy of Cincinnati Milacron.

Table 3. Inches and Fractions of Inches Expressed as Decimals of a Foot.

Fraction	-	1 In.	2 In.	3 In.	4 In.	5 In.	6 In.	7 In.	8 In.	9 In.	10 In.	11 In.
-	-	.0833	.1667	.2500	.3333	.4167	.5000	.5833	.6667	.7500	.8333	.9167
1/64	.0013	.0846	.1680	.2513	.3346	.4180	.5013	.5846	.6680	.7513	.8346	.9180
1/32	.0026	.0859	.1693	.2526	.3359	.4193	.5026	.5859	.6693	.7526	.8359	.9193
3/64	.0039	.0872	.1706	.2539	.3372	.4206	.5039	.5872	.6706	.7539	.8372	.9206
1/16	.0052	.0885	.1719	.2552	.3385	.4219	.5052	.5885	.6719	.7552	.8385	.9219
5/64	.0065	.0898	.1732	.2565	.3398	.4232	.5065	.5898	.6732	.7565	.8398	.9232
15/16	.0078	.0911	.1745	.2578	.3411	.4245	.5078	.5911	.6745	.7578	.8411	.9245
7/64	.0091	.0924	.1758	.2591	.3424	.4258	.5091	.5924	.6758	.7591	.8424	.9258
1/8	.0104	.0938	.1771	.2604	.3438	.4271	.5104	.5938	.6771	.7604	.8438	.9271
9/64	.0117	.0951	.1784	.2617	.3451	.4284	.5117	.5951	.6784	.7617	.8451	.9284
5/32	.0130	.0964	.1797	.2630	.3464	.4297	.5130	.5964	.6797	.7630	.8464	.9297
11/64	.0143	.0977	.1810	.2643	.3477	.4310	.5143	.5977	.6810	.7643	.8477	.9310
3/16	.0156	.0990	.1823	.2656	.3490	.4323	.5156	.5990	.6823	.7656	.8490	.9323
13/64	.0169	.1003	.1836	.2669	.3503	.4336	.5169	.6003	.6836	.7669	.8503	.9336
7/32	.0182	.1016	.1849	.2682	.3516	.4349	.5182	.6016	.6849	.7682	.8516	.9349
15/64	.0195	.1029	.1862	.2695	.3529	.4362	.5195	.6029	.6862	.7695	.8529	.9362
1/4	.0208	.1042	.1875	.2708	.3542	.4375	.5208	.6042	.6875	.7708	.8542	.9375
17/64	.0221	.1055	.1888	.2721	.3555	.4388	.5221	.6055	.6888	.7721	.8555	.9388
9/32	.0234	.1068	.1901	.2734	.3568	.4401	.5234	.6068	.6901	.7734	.8568	.9401
19/64	.0247	.1081	.1914	.2747	.3581	.4414	.5247	.6081	.6914	.7747	.8581	.9414
5/16	.0260	.1094	.1927	.2760	.3594	.4427	.5260	.6094	.6927	.7760	.8594	.9427
21/64	.0273	.1107	.1940	.2773	.3607	.4440	.5273	.6107	.6940	.7773	.8607	.9440
11/32	.0286	.1120	.1953	.2786	.3620	.4453	.5286	.6120	.6953	.7786	.8620	.9453
23/64	.0299	.1133	.1966	.2799	.3633	.4466	.5299	.6133	.6966	.7799	.8633	.9466

(Continued)

Table 3. *(Continued)* Inches and Fractions of Inches Expressed as Decimals of a Foot.

Fraction	•	1 In.	2 In.	3 In.	4 In.	5 In.	6 In.	7 In.	8 In.	9 In.	10 In.	11 In.
3/8	.0313	.1146	.1979	.2813	.3646	.4479	.5313	.6146	.6979	.7813	.8646	.9479
25/64	.0326	.1159	.1992	.2826	.3659	.4492	.5326	.6159	.6992	.7826	.8659	.9492
13/32	.0339	.1172	.2005	.2839	.3672	.4505	.5339	.6172	.7005	.7839	.8672	.9505
27/64	.0352	.1185	.2018	.2852	.3685	.4518	.5352	.6185	.7018	.7852	.8685	.9518
7/16	.0365	.1198	.2031	.2865	.3698	.4531	.5365	.6198	.7031	.7865	.8698	.9531
29/64	.0378	.1211	.2044	.2878	.3711	.4544	.5378	.6211	.7044	.7878	.8711	.9544
15/32	.0391	.1224	.2057	.2891	.3724	.4557	.5391	.6224	.7057	.7891	.8724	.9557
31/64	.0404	.1237	.2070	.2904	.3737	.4570	.5404	.6237	.7070	.7904	.8737	.9570
1/2	.0417	.1250	.2083	.2917	.3750	.4583	.5417	.6250	.7083	.7917	.8750	.9583
33/64	.0430	.1263	.2096	.2930	.3763	.4596	.5430	.6263	.7096	.7930	.8763	.9596
17/32	.0443	.1276	.2109	.2943	.3776	.4609	.5443	.6276	.7109	.7943	.8776	.9609
35/64	.0456	.1289	.2122	.2956	.3789	.4622	.5456	.6289	.7122	.7956	.8789	.9622
9/16	.0469	.1302	.2135	.2969	.3802	.4635	.5469	.6302	.7135	.7969	.8802	.9635
37/64	.0482	.1315	.2148	.2982	.3815	.4648	.5482	.6315	.7148	.7982	.8815	.9648
19/32	.0495	.1328	.2161	.2995	.3828	.4661	.5495	.6328	.7161	.7995	.8828	.9661
39/64	.0508	.1341	.2174	.3008	.3841	.4674	.5508	.6341	.7174	.8008	.8841	.9674
5/8	.0521	.1354	.2188	.3021	.3854	.4688	.5521	.6354	.7188	.8021	.8854	.9688
41/64	.0534	.1367	.2201	.3034	.3867	.4701	.5534	.6367	.7201	.8034	.8867	.9701
21/32	.0547	.1380	.2214	.3047	.3880	.4714	.5547	.6380	.7214	.8047	.8880	.9714
43/64	.0560	.1393	.2227	.3060	.3893	.4727	.5560	.6393	.7227	.8060	.8893	.9727
11/16	.0573	.1406	.2240	.3073	.3906	.4740	.5573	.6406	.7240	.8073	.8906	.9740
45/64	.0586	.1419	.2253	.3086	.3919	.4753	.5586	.6419	.7253	.8086	.8919	.9753
23/32	.0599	.1432	.2266	.3099	.3932	.4766	.5599	.6432	.7266	.8099	.8932	.9766
47/64	.0612	.1445	.2279	.3112	.3945	.4779	.5612	.6445	.7279	.8112	.8945	.9779

(Continued)

Table 3. *(Continued)* **Inches and Fractions of Inches Expressed as Decimals of a Foot.**

Fraction	-	1 In.	2 In.	3 In.	4 In.	5 In.	6 In.	7 In.	8 In.	9 In.	10 In.	11 In.
3/4	.0625	.1458	.2292	.3125	.3958	.4792	.5625	.6458	.7292	.8125	.8958	.9792
49/64	.0638	.1471	.2305	.3138	.3971	.4805	.5638	.6471	.7305	.8138	.8971	.9805
25/32	.0651	.1484	.2318	.3151	.3984	.4818	.5651	.6484	.7318	.8151	.8984	.9818
51/64	.0664	.1497	.2331	.3164	.3997	.4831	.5664	.6497	.7331	.8164	.8997	.9831
13/16	.0677	.1510	.2344	.3177	.4010	.4844	.5677	.6510	.7344	.8177	.9010	.9844
53/64	.0690	.1523	.2357	.3190	.4023	.4857	.5690	.6523	.7357	.8190	.9023	.9857
27/32	.0703	.1536	.2370	.3203	.4036	.4870	.5703	.6536	.7370	.8203	.9036	.9870
57/64	.0716	.1549	.2383	.3216	.4049	.4883	.5716	.6549	.7383	.8216	.9049	.9883
7/8	.0729	.1563	.2396	.3229	.4063	.4896	.5729	.6563	.7396	.8229	.9063	.9896
57/64	.0742	.1576	.2409	.3242	.4076	.4909	.5742	.6576	.7409	.8242	.9076	.9909
29/32	.0755	.1589	.2422	.3255	.4089	.4922	.5755	.6589	.7422	.8255	.9089	.9922
59/64	.0768	.1602	.2435	.3268	.4102	.4935	.5768	.6602	.7435	.8268	.9102	.9935
15/16	.0781	.1615	.2448	.3281	.4115	.4948	.5781	.6615	.7448	.8281	.9115	.9948
61/64	.0794	.1628	.2461	.3294	.4128	.4961	.5794	.6628	.7461	.8294	.9128	.9961
31/32	.0807	.1641	.2474	.3307	.4141	.4974	.5807	.6641	.7474	.8307	.9141	.9974
63/64	.0820	.1654	.2487	.3320	.4154	.4987	.5820	.6654	.7487	.8320	.9154	.9987

Notes: The first column provides fractions in $1/64$ inch increments. The second column gives the decimal equivalent of a foot for the fraction. For example, $17/64$ ths of an inch is equal to .0221 feet. For other columns, intersect the columns for the number of inches plus the fraction of an inch to find the decimal equivalent of a foot. For example, to find the decimal equivalent for 5 $1/8$" follow the 5 In. column down from the top until it intersects with the $1/8$" column on the left. The answer is .4271.

Table 4. Conversion of Inch Fractions to Millimeters and Decimal Inches. *(Source, Werkö, GmbH.)*

Inch Decimal	Inch Fraction	0 mm	+1 mm	+2 mm	+3 mm	+4 mm
0	0	0,0000	25,4000	50,8000	76,2000	101,6000
0.015625	1/64	0,3969	25,7969	51,1969	76,5969	101,9969
0.031250	1/32	0,7938	26,1938	51,5938	76,9938	102,3938
0.046875	3/64	1,1906	26,5906	51,9906	77,3906	102,7906
0.06250	1/16	1,5875	26,9875	52,3875	77,7875	103,1875
0.078125	5/64	1,9844	27,3844	52,7844	78,1844	103,5844
0.093750	3/32	2,3812	27,7812	53,1812	78,5812	103,9812
0.109375	7/64	2,7781	28,1781	53,5781	78,9781	104,3781
0.1250	1/8	3,1750	28,5750	53,9750	79,3750	104,7750
0.140625	9/64	3,5719	28,9719	54,3719	79,7719	105,1719
0.156250	5/32	3,9688	29,3688	54,7688	80,1688	105,5688
0.171875	11/64	4,3656	29,7656	55,1656	80,5656	105,9656
0.18750	3/16	4,7625	30,1625	55,5625	80,9625	106,3625
0.203125	13/64	5,1594	30,5594	55,9594	81,3594	106,7594
0.218750	7/32	5,5562	30,9562	56,3562	81,7562	107,1562
0.234375	15/64	5,9531	31,3531	56,7531	82,1531	107,5531
0.250	1/4	6,3500	31,7500	57,1500	82,5500	107,9500
0.265625	17/64	6,7469	32,1469	57,5469	82,9469	108,3469
0.281250	9/32	7,1438	32,5438	57,9438	83,3438	108,7438
0.296875	19/64	7,5406	32,9406	58,3406	83,7406	109,1406
0.31250	5/16	7,9375	33,3375	58,7375	84,1375	109,5375
0.328125	21/64	8,3344	33,7344	59,1344	84,5344	109,9344
0.343750	11/32	8,7313	34,1313	59,5313	84,9313	110,3313
0.359375	23/64	9,1281	34,5281	59,9281	85,3281	110,7281
0.3750	3/8	9,5250	34,9250	60,3250	85,7250	111,1250
0.390625	25/64	9,9219	35,3219	60,7219	86,1219	111,5219
0.406250	13/32	10,3188	35,7188	61,1188	86,5188	111,9188
0.421875	27/64	10,7156	36,1156	61,5156	86,9156	112,3156
0.43750	7/16	11,1125	36,5125	61,9125	87,3125	112,7125
0.453125	29/64	11,5094	36,9094	62,3094	87,7094	113,1094
0.468750	15/32	11,9063	37,3063	62,7063	88,1063	113,5063
0.484375	31/64	12,3031	37,7031	63,1031	88,5031	113,9031
0.500	1/2	12,7000	38,1000	63,5000	88,9000	114,3000
0.515625	33/64	13,0969	38,4969	63,8969	89,2969	114,6969
0.531250	17/32	13,4938	38,8938	64,2938	89,6938	115,0938
0.546875	35/64	13,8906	39,2906	64,6906	90,0906	115,4906
0.56250	9/16	14,2875	39,6875	65,0875	90,4875	115,8875
0.578125	37/64	14,6844	40,0844	65,4844	90,8844	116,2844
0.59375	19/32	15,0813	40,4813	65,8813	91,2813	116,6813
0.609375	39/64	15,4781	40,8781	66,2781	91,6781	117,0781
0.6250	5/8	15,8750	41,2750	66,6750	92,0750	117,4750
0.640625	41/64	16,2719	41,6719	67,0719	92,4719	117,8719
0.656250	21/32	16,6688	42,0688	67,4688	92,8688	118,2688
0.671895	43/64	17,0656	42,4656	67,8656	93,2656	118,6656
0.68750	11/16	17,4625	42,8625	68,2625	93,6625	119,0625
0.703125	45/64	17,8594	43,2594	68,6594	94,0594	119,4594
0.718750	23/32	18,2562	43,6562	69,0562	94,4562	119,8562

(Continued)

Table 4. *(Continued)* **Conversion of Inch Fractions to Millimeters and Decimal Inches.** *(Source, Werkö, GmbH.)*

Inch Decimal	Inch Fraction	0 mm	+1 mm	+2 mm	+3 mm	+4 mm
0.734375	47/64	18,6531	44,0531	69,4531	94,8531	120,2531
0.750	3/4	19,0500	44,4500	69,8500	95,2500	120,6500
0.765625	49/64	19,4469	44,8469	70,2469	95,6469	121,0469
0.781250	25/32	19,8438	45,2438	70,6438	96,0438	121,4438
0.796875	51/64	20,2406	45,6406	71,0406	96,4406	121,8406
0.81250	13/16	20,6375	46,0375	71,4375	96,8375	122,2375
0.828125	53/64	21,0344	46,4344	71,8344	97,2344	122,6344
0.843750	27/32	21,4313	46,8313	72,2313	97,6313	123,0313
0.859375	55/64	21,8281	47,2281	72,6281	98,0281	123,4281
0.8750	7/8	22,2250	47,6250	73,0250	98,4250	123,8250
0.890625	57/64	22,6219	48,0219	73,4219	98,8219	124,2219
0.906250	29/32	23,0188	48,4188	73,8188	99,2188	124,6188
0.921875	59/64	23,4156	48,8156	74,2156	99,6156	125,0156
0.93750	15/16	23,8125	49,2125	74,6125	100,0125	125,4125
0.953125	61/64	24,2094	49,6094	75,0094	100,4094	125,8094
0.968875	31/32	24,6063	50,0063	75,4063	100,8063	126,2063
0.984375	63/64	25,0031	50,4031	75,8031	101,2031	126,6031

This Table can be used for direct conversion of measurements up to 4 63/64 inches into millimeters. For example, to find the millimeter equivalent of 3 19/32 inches, first go to the column headed "+3 mm" and go down that column to the row that intersects 19/32 from the "Inch Fraction" column. It will be seen that the millimeter equivalent is 91,2813, with the comma used in metric expression serving the same purpose as a decimal (91.2813 mm).

Table 5. Microinches Converted to Microns (Micrometers). (1 microinch = 0.0254 microns.)

Microinches	Microns (micrometers)									
	0	1	2	3	4	5	6	7	8	9
0	-	0.025	0.051	0.076	0.102	0.127	0.152	0.178	0.203	0.229
10	0.254	0.279	0.305	0.330	0.356	0.381	0.406	0.432	0.457	0.483
20	0.508	0.533	0.559	0.584	0.610	0.635	0.660	0.686	0.711	0.737
30	0.762	0.787	0.813	0.838	0.864	0.889	0.914	0.940	0.965	0.991
40	1.016	1.041	1.067	1.092	1.118	1.143	1.168	1.194	1.219	1.245
50	1.270	1.295	1.321	1.346	1.372	1.397	1.422	1.448	1.473	1.499
60	1.524	1.549	1.575	1.600	1.626	1.651	1.676	1.702	1.727	1.753
70	1.778	1.803	1.829	1.854	1.880	1.905	1.930	1.956	1.981	2.007
80	2.032	2.057	2.083	2.108	2.134	2.159	2.184	2.210	2.235	2.261
90	2.286	2.311	2.337	2.362	2.388	2.413	2.438	2.464	2.489	2.515
100	2.540	2.565	2.591	2.616	2.642	2.667	2.692	2.718	2.743	2.769
110	2.794	2.819	2.845	2.870	2.896	2.921	2.946	2.972	2.997	3.023
120	3.048	3.073	3.099	3.124	3.150	3.175	3.200	3.226	3.251	3.277
130	3.302	3.327	3.353	3.378	3.404	3.429	3.454	3.480	3.505	3.531
140	3.556	3.581	3.607	3.632	3.658	3.683	3.708	3.734	3.759	3.785
150	3.810	3.835	3.861	3.886	3.912	3.937	3.962	3.988	4.013	4.039
160	4.064	4.089	4.115	4.140	4.166	4.191	4.216	4.242	4.267	4.293
170	4.318	4.343	4.369	4.394	4.420	4.445	4.470	4.496	4.521	4.547
180	4.572	4.597	4.623	4.648	4.674	4.699	4.724	4.750	4.775	4.801
190	4.826	4.851	4.877	4.902	4.928	4.953	4.978	5.004	5.029	5.055
200	5.080	5.105	5.131	5.156	5.182	5.207	5.232	5.258	5.283	5.309
210	5.334	5.359	5.385	5.410	5.436	5.461	5.486	5.512	5.537	5.563
220	5.588	5.613	5.639	5.664	5.690	5.715	5.740	5.766	5.791	5.817

(Continued)

Table 5. *(Continued)* Microinches Converted to Microns (Micrometers). (1 microinch = 0.0254 microns.)

Microinches	Microns (micrometers)									
	0	1	2	3	4	5	6	7	8	9
230	5.842	5.867	5.893	5.918	5.944	5.969	5.994	6.020	6.045	6.071
240	6.096	6.121	6.147	6.172	6.198	6.223	6.248	6.274	6.299	6.325
250	6.350	6.375	6.401	6.426	6.452	6.477	6.502	6.528	6.553	6.579
260	6.604	6.629	6.655	6.680	6.706	6.731	6.756	6.782	6.807	6.833
270	6.858	6.883	6.909	6.934	6.960	6.985	7.010	7.036	7.061	7.087
280	7.112	7.137	7.163	7.188	7.214	7.239	7.264	7.290	7.315	7.341
290	7.366	7.391	7.417	7.442	7.468	7.493	7.518	7.544	7.569	7.595
300	7.620	7.645	7.671	7.696	7.722	7.747	7.772	7.798	7.823	7.849
310	7.874	7.899	7.925	7.950	7.976	8.001	8.026	8.052	8.077	8.103
320	8.128	8.153	8.179	8.204	8.230	8.255	8.280	8.306	8.331	8.357
330	8.382	8.407	8.433	8.458	8.484	8.509	8.534	8.560	8.585	8.611
340	8.636	8.661	8.687	8.712	8.738	8.763	8.788	8.814	8.839	8.865
350	8.890	8.915	8.941	8.966	8.992	9.017	9.042	9.068	9.093	9.119
360	9.144	9.169	9.195	9.220	9.246	9.271	9.296	9.322	9.347	9.373
370	9.398	9.423	9.449	9.474	9.500	9.525	9.550	9.576	9.601	9.627
380	9.652	9.677	9.703	9.728	9.754	9.779	9.804	9.830	9.855	9.881
390	9.906	9.931	9.957	9.982	10.008	10.033	10.058	10.84	10.109	10.135
400	10.160	10.185	10.211	10.236	10.262	10.287	10.312	10.338	10.363	10.389
410	10.414	10.439	10.465	10.490	10.516	10.541	10.566	10.592	10.617	10.643
420	10.668	10.693	10.719	10.744	10.770	10.795	10.820	10.846	10.871	10.897
430	10.922	10.947	10.973	10.998	11.024	11.049	11.074	11.100	11.125	11.151
440	11.176	11.201	11.227	11.252	11.278	11.303	11.328	11.354	11.379	11.405
450	11.430	11.455	11.481	11.506	11.532	11.557	11.582	11.608	11.633	11.659

(Continued)

Table 5. *(Continued)* **Microinches Converted to Microns (Micrometers).** (1 microinch = 0.0254 microns.)

Microinches	Microns (micrometers)									
	0	1	2	3	4	5	6	7	8	9
460	11.684	11.709	11.735	11.760	11.786	11.811	11.836	11.862	11.887	11.913
470	11.938	11.963	11.989	12.014	12.040	12.065	12.090	12.116	12.141	12.167
480	12.192	12.217	12.243	12.268	12.294	12.319	12.344	12.370	12.395	12.421
490	12.446	12.471	12.497	12.522	12.548	12.573	12.598	12.624	12.649	12.675
500	12.700	12.725	12.751	12.776	12.802	12.827	12.852	12.878	12.903	12.929

Table 6. Microns (Micrometers) Converted to Microinches. (1 microinch = 0.0254 microns.)

Micron	Microinches									
	0	1	2	3	4	5	6	7	8	9
0.00	-	0.39	0.79	1.18	1.57	1.97	2.36	2.76	3.15	3.54
0.10	3.94	4.33	4.72	5.12	5.51	5.91	6.30	6.69	7.09	7.48
0.20	7.87	8.27	8.66	9.06	9.45	9.84	10.24	10.63	11.02	11.42
0.30	11.81	12.20	12.60	12.99	13.39	13.78	14.17	14.57	14.96	15.35
0.40	15.75	16.14	16.54	16.93	17.32	17.72	18.11	18.50	18.90	19.29
0.50	19.69	20.08	20.47	20.87	21.26	21.65	22.05	22.44	22.83	23.23
0.60	23.62	24.02	24.41	24.80	25.20	25.59	25.98	26.38	26.77	27.17
0.70	27.56	27.95	28.35	28.74	29.13	29.53	29.92	30.31	30.71	31.10
0.80	31.50	31.89	32.28	32.68	33.07	33.46	33.86	34.25	34.65	35.04
0.90	35.43	35.83	36.22	36.61	37.01	37.40	37.80	38.19	38.58	38.98
1.00	39.37	39.76	40.16	40.55	40.94	41.34	41.73	42.13	42.52	42.91
1.10	43.31	43.70	44.09	44.49	44.88	45.28	45.67	46.06	46.46	46.85
1.20	47.24	47.64	48.03	48.43	48.82	49.21	49.61	50.00	50.39	50.79
1.30	51.18	51.18	51.97	52.36	52.76	53.15	53.54	53.94	54.33	54.72
1.40	55.12	55.12	55.91	56.30	56.69	57.09	57.48	57.87	58.27	58.66
1.50	59.06	59.45	59.84	60.24	60.63	61.02	61.42	61.81	62.20	62.60
1.60	62.99	63.39	63.78	64.17	64.57	64.96	65.35	65.75	66.14	66.54
1.70	66.93	67.32	67.72	68.11	68.50	68.90	69.29	69.69	70.08	70.47
1.80	70.87	71.26	71.65	72.05	72.44	72.83	73.23	73.62	74.02	74.41
1.90	74.80	75.20	75.59	75.98	76.38	76.77	77.17	77.56	77.95	78.35
2.00	78.74	79.13	79.53	79.92	80.31	80.71	81.10	81.50	81.89	82.28
2.10	82.68	83.07	83.46	83.86	84.25	84.65	85.04	85.43	85.83	86.22
2.20	86.61	87.01	87.40	87.80	88.19	88.58	88.98	89.37	89.76	90.16

(Continued)

Table 6. *(Continued)* Microns (Micrometers) Converted to Microinches. (1 microinch = 0.0254 microns.)

Micron	Microinches									
	0	1	2	3	4	5	6	7	8	9
2.30	90.55	90.94	91.34	91.73	92.13	92.52	92.91	93.31	93.70	94.09
2.40	94.49	84.25	95.28	95.67	96.06	96.46	96.85	97.24	97.64	98.03
2.50	98.43	98.82	99.21	99.61	100.00	100.39	100.79	101.18	101.57	101.97
2.60	102.36	102.76	103.15	103.54	103.94	104.33	104.72	105.12	105.51	105.91
2.70	106.30	106.69	107.09	107.48	107.87	108.27	108.66	109.06	109.45	109.84
2.80	110.24	110.63	111.02	111.42	111.81	112.20	112.60	112.99	113.39	113.78
2.90	114.17	114.57	114.96	115.35	115.75	116.14	116.54	116.93	117.32	117.72
3.00	118.11	118.50	118.90	119.29	119.69	120.08	120.47	120.87	121.26	121.65
3.10	122.05	122.44	122.83	123.23	123.62	124.02	124.41	124.80	125.20	125.59
3.20	125.98	126.38	126.77	127.17	127.56	127.95	128.35	128.74	129.13	129.53
3.30	129.92	130.31	130.71	131.10	131.50	131.89	132.28	132.68	133.07	133.46
3.40	133.86	134.25	134.65	135.04	135.43	135.83	136.22	136.61	137.01	137.40
3.50	137.80	138.19	138.58	138.98	139.37	139.76	140.16	140.55	140.94	141.34
3.60	141.73	142.13	142.52	142.91	143.31	143.70	144.09	144.49	144.88	145.28
3.70	145.67	146.06	146.46	146.85	147.24	147.64	148.03	148.43	148.82	149.21
3.80	149.61	150.00	150.39	150.79	151.18	151.57	151.97	152.36	152.76	153.15
3.90	153.54	153.94	154.33	154.72	155.12	155.51	155.91	156.30	156.69	157.09
4.00	157.48	157.87	158.27	158.66	159.06	159.45	159.84	160.24	160.63	161.02
4.10	161.42	161.81	162.20	162.60	162.99	163.39	163.78	164.17	164.57	164.96
4.20	165.35	165.75	166.14	166.54	166.93	167.32	167.72	168.11	168.50	168.90
4.30	169.29	169.69	170.08	170.47	170.87	171.26	171.65	172.05	172.44	172.83
4.40	173.23	173.62	174.02	174.41	174.80	175.20	175.59	175.98	176.38	176.77
4.50	177.17	177.56	177.95	178.35	178.74	179.13	179.53	179.92	180.31	180.71

(Continued)

Table 6. *(Continued)* **Microns (Micrometers) Converted to Microinches.** (1 microinch = 0.0254 microns.)

Micron	Microinches									
	0	1	2	3	4	5	6	7	8	9
4.60	181.10	181.50	181.89	182.28	182.68	183.07	183.46	183.86	184.25	184.65
4.70	185.04	185.43	185.83	186.22	186.61	187.01	187.40	187.80	188.19	188.58
4.80	188.98	189.37	189.76	190.16	190.55	190.94	191.34	191.73	192.13	192.52
4.90	192.91	193.31	193.70	194.09	194.49	194.88	195.28	195.67	196.06	196.46
5.00	196.85	197.24	197.64	198.03	198.43	198.82	199.21	199.61	200.00	200.39

Table 7. Inches Converted to Millimeters (0.001" to 0.156")

0.001 to 0.039		0.040 to 0.078		0.079 to 0.117		0.118 to 0.156	
inch	mm	inch	mm	inch	mm	inch	mm
0.001	0.0254	0.040	1.0160	0.079	2.0066	0.118	2.9972
0.002	0.0508	0.041	1.0414	0.080	2.0320	0.119	3.0226
0.003	0.0762	0.042	1.0668	0.081	2.0574	0.120	3.0480
0.004	0.1016	0.043	1.0922	0.082	2.0828	0.121	3.0734
0.005	0.1270	0.044	1.1176	0.083	2.1082	0.122	3.0988
0.006	0.1524	0.045	1.1430	0.084	2.1336	0.123	3.1242
0.007	0.1778	0.046	1.1684	0.085	2.1590	0.124	3.1496
0.008	0.2032	0.047	1.1938	0.086	2.1844	0.125	3.1750
0.009	0.2286	0.048	1.2192	0.087	2.2098	0.126	3.2004
0.010	0.2540	0.049	1.2446	0.088	2.2352	0.127	3.2258
0.011	0.2794	0.050	1.2700	0.089	2.2606	0.128	3.2512
0.012	0.3048	0.051	1.2954	0.090	2.2860	0.129	3.2766
0.013	0.3302	0.052	1.3208	0.091	2.3114	0.130	3.3020
0.014	0.3556	0.053	1.3462	0.092	2.3368	0.131	3.3274
0.015	0.3810	0.054	1.3716	0.093	2.3622	0.132	3.3528
0.016	0.4064	0.055	1.3970	0.094	2.3876	0.133	3.3782
0.017	0.4318	0.056	1.4224	0.095	2.4130	0.134	3.4036
0.018	0.4572	0.057	1.4478	0.096	2.4384	0.135	3.4290
0.019	0.4826	0.058	1.4732	0.097	2.4638	0.136	3.4544
0.020	0.5080	0.059	1.4986	0.098	2.4892	0.137	3.4798
0.021	0.5334	0.060	1.5240	0.099	2.5146	0.138	3.5052
0.022	0.5588	0.061	1.5494	0.100	2.5400	0.139	3.5306
0.023	0.5842	0.062	1.5748	0.101	2.5654	0.140	3.5560
0.024	0.6096	0.063	1.6002	0.102	2.5908	0.141	3.5814
0.025	0.6350	0.064	1.6256	0.103	2.6162	0.142	3.6068
0.026	0.6604	0.065	1.6510	0.104	2.6416	0.143	3.6322
0.027	0.6858	0.066	1.6764	0.105	2.6670	0.144	3.6576
0.028	0.7112	0.067	1.7018	0.106	2.6924	0.145	3.6830
0.029	0.7366	0.068	1.7272	0.107	2.7178	0.146	3.7084
0.030	0.7620	0.069	1.7526	0.108	2.7432	0.147	3.7338
0.031	0.7874	0.070	1.7780	0.109	2.7686	0.148	3.7592
0.032	0.8128	0.071	1.8034	0.110	2.7940	0.149	3.7846
0.033	0.8382	0.072	1.8288	0.111	2.8194	0.150	3.8100
0.034	0.8636	0.073	1.8542	0.112	2.8448	0.151	3.8354
0.035	0.8890	0.074	1.8796	0.113	2.8702	0.152	3.8608
0.036	0.9144	0.075	1.9050	0.114	2.8956	0.153	3.8862
0.037	0.9398	0.076	1.9304	0.115	2.9210	0.154	3.9116
0.038	0.9652	0.077	1.9558	0.116	2.9464	0.155	3.9370
0.039	0.9906	0.078	1.9812	0.117	2.9718	0.156	3.9624

(Continued)

Table 7. *(Continued)* **Inches Converted to Millimeters (0.157" to 0.312")**

0.157 to 0.195		0.196 to 0.234		0.235 to 0.273		0.274 to 0.312	
inch	mm	inch	mm	inch	mm	inch	mm
0.157	3.9878	0.196	4.9784	0.235	5.9690	0.274	6.9596
0.158	4.0132	0.197	5.0038	0.236	5.9944	0.275	6.9850
0.159	4.0386	0.198	5.0292	0.237	6.0198	0.276	7.0104
0.160	4.0640	0.199	5.0546	0.238	6.0452	0.277	7.0358
0.161	4.0894	0.200	5.0800	0.239	6.0706	0.278	7.0612
0.162	4.1148	0.201	5.1054	0.240	6.0960	0.279	7.0866
0.163	4.1402	0.202	5.1308	0.241	6.1214	0.280	7.1120
0.164	4.1656	0.203	5.1562	0.242	6.1468	0.281	7.1374
0.165	4.1910	0.204	5.1816	0.243	6.1722	0.282	7.1628
0.166	4.2164	0.205	5.2070	0.244	6.1976	0.283	7.1882
0.167	4.2418	0.206	5.2324	0.245	6.2230	0.284	7.2136
0.168	4.2672	0.207	5.2578	0.246	6.2484	0.285	7.2390
0.169	4.2926	0.208	5.2832	0.247	6.2738	0.286	7.2644
0.170	4.3180	0.209	5.3086	0.248	6.2992	0.287	7.2898
0.171	4.3434	0.210	5.3340	0.249	6.32466	0.288	7.3152
0.172	4.3688	0.211	5.3594	0.250	6.3500	0.289	7.3406
0.173	4.3942	0.212	5.3848	0.251	6.3754	0.290	7.3660
0.174	4.4196	0.213	5.4102	0.252	6.4008	0.291	7.3914
0.175	4.4450	0.214	5.4356	0.253	6.4262	0.292	7.4168
0.176	4.4704	0.215	5.4610	0.254	6.4516	0.293	7.4422
0.177	4.4958	0.216	5.4864	0.255	6.4770	0.294	7.4676
0.178	4.5212	0.217	5.5118	0.256	6.5024	0.295	7.4930
0.179	4.5466	0.218	5.5372	0.257	6.5278	0.296	7.5184
0.180	4.5720	0.219	5.5626	0.258	6.5532	0.297	7.5438
0.181	4.5974	0.220	5.5880	0.259	6.5786	0.298	7.5692
0.182	4.6228	0.221	5.6134	0.260	6.6040	0.299	7.5946
0.183	4.6482	0.222	5.6388	0.261	6.6294	0.300	7.6200
0.184	4.6736	0.223	5.6642	0.262	6.6548	0.301	7.6454
0.185	4.6990	0.224	5.6896	0.263	6.6802	0.302	7.6708
0.186	4.7244	0.225	5.7150	0.264	6.7056	0.303	7.6962
0.187	4.7498	0.226	5.7404	0.265	6.7310	0.304	7.7216
0.188	4.7752	0.227	5.7658	0.266	607564	0.305	7.7470
0.189	4.8006	0.228	5.7912	0.267	6.7818	0.306	7.7724
0.190	4.8260	0.229	5.8166	0.268	6.8072	0.307	7.7978
0.191	4.8514	0.230	5.8420	0.269	6.8326	0.308	7.8232
0.192	4.8768	0.231	5.8674	0.270	6.8580	0.309	7.8486
0.193	4.9022	0.232	5.8928	0.271	6.8834	0.310	7.8740
0.194	4.9276	0.233	5.9182	0.272	6.9088	0.311	7.8994
0.195	4.9530	0.234	5.9436	0.273	6.9342	0.312	7.9248

(Continued)

Table 7. *(Continued)* **Inches Converted to Millimeters (0.313" to 0.468")**

0.313 to 0.351		0.352 to 0.390		0.391 to 0.429		0.430 to 0.468	
inch	mm	inch	mm	inch	mm	inch	mm
0.313	7.9502	0.352	8.9408	0.391	9.9314	0.430	10.9220
0.314	7.9756	0.353	8.9662	0.392	9.9568	0.431	10.9474
0.315	8.0010	0.354	8.9916	0.393	9.9822	0.432	10.9728
0.316	8.0264	0.355	9.0170	0.394	10.0076	0.433	10.9982
0.317	8.0518	0.356	9.0424	0.395	10.0330	0.434	11.0236
0.318	8.0772	0.357	9.0678	0.396	10.584	0.435	11.0490
0.319	8.1026	0.358	9.0932	0.397	10.0838	0.436	11.0744
0.320	8.1280	0.359	9.1186	0.398	10.1092	0.437	11.0998
0.321	8.1534	0.360	9.1440	0.399	10.1346	0.438	11.1252
0.322	8.1788	0.361	9.1694	0.400	10.1600	0.439	11.1506
0.323	8.2042	0.362	9.1948	0.401	10.1854	0.440	11.1760
0.324	8.2296	0.363	9.2202	0.402	10.2108	0.441	11.2014
0.325	8.2550	0.364	9.2456	0.403	10.2362	0.442	11.2268
0.326	8.2804	0.365	9.2710	0.404	10.2616	0.443	11.2522
0.327	8.3058	0.366	9.2964	0.405	10.2870	0.444	11.2776
0.328	8.3312	0.367	9.3218	0.406	10.3124	0.445	11.3030
0.329	8.3566	0.368	9.3472	0.407	10.3378	0.446	11.3284
0.330	8.3820	0.369	9.3726	0.408	10.3632	0.447	11.3538
0.331	8.4074	0.370	9.3980	0.409	10.3886	0.448	11.3792
0.332	8.4328	0.371	9.4234	0.410	10.4140	0.449	11.4046
0.333	8.4582	0.372	9.4488	0.411	10.4394	0.450	11.4300
0.334	8.4836	0.373	9.4742	0.412	10.4648	0.451	11.4554
0.335	8.5090	0.374	9.4996	0.413	10.4902	0.452	11.4808
0.336	8.5344	0.375	9.5250	0.414	10.5156	0.453	11.5062
0.337	8.5598	0.376	9.5504	0.415	10.5410	0.454	11.5316
0.338	8.5852	0.377	9.5758	0.416	10.5664	0.455	11.5570
0.339	8.6106	0.378	9.6012	0.417	10.5918	0.456	11.5824
0.340	8.6360	0.379	9.6266	0.418	10.6172	0.457	11.6078
0.341	8.6614	0.380	9.6520	0.419	10.6426	0.458	11.6332
0.342	8.6868	0.381	9.6774	0.420	10.6680	0.459	11.6586
0.343	8.7122	0.382	9.7028	0.421	10.6934	0.460	11.6840
0.344	8.7376	0.383	9.7282	0.422	10.7188	0.461	11.7094
0.345	8.7630	0.384	9.7536	0.423	10.7442	0.462	11.7348
0.346	8.7884	0.385	9.7790	0.424	10.7696	0.463	11.7602
0.347	8.8138	0.386	9.8044	0.425	10.7950	0.464	11.7856
0.348	8.8392	0.387	9.8298	0.426	10.8204	0.465	11.8110
0.349	8.8646	0.388	9.8552	0.427	10.8458	0.466	11.8364
0.350	8.8900	0.389	9.8806	0.428	10.8712	0.467	11.8618
0.351	8.9154	0.390	9.9060	0.429	10.8966	0.468	11.8872

(Continued)

Table 7. *(Continued)* Inches Converted to Millimeters (0.469" to 0.624")

\multicolumn{2}{c}{0.469 to 0.507}		\multicolumn{2}{c}{0.508 to 0.546}		\multicolumn{2}{c}{0.547 to 0.585}		\multicolumn{2}{c}{0.586 to 0.624}	
inch	mm	inch	mm	inch	mm	inch	mm
0.469	11.9126	0.508	12.9032	0.547	13.8938	0.586	14.8844
0.470	11.9380	0.509	12.9286	0.548	13.9192	0.587	14.9098
0.471	11.9634	0.510	12.9540	0.549	13.9446	0.588	14.9352
0.472	11.9888	0.511	12.9794	0.550	13.9700	0.589	14.9606
0.473	12.0142	0.512	13.0048	0.551	13.9954	0.590	14.9860
0.474	12.0396	0.513	13.0302	0.552	14.0208	0.591	15.0114
0.475	12.0650	0.514	13.0556	0.553	14.0462	0.592	15.0368
0.476	12.0904	0.515	13.0810	0.554	14.0716	0.593	15.0622
0.477	12.1158	0.516	13.1064	0.555	14.0970	0.594	15.0876
0.478	12.1412	0.517	13.1318	0.556	14.1224	0.595	15.1130
0.479	12.1666	0.518	13.1572	0.557	14.1478	0.596	15.1384
0.480	12.1920	0.519	13.1826	0.558	14.1732	0.597	15.1638
0.481	12.2174	0.520	13.2080	0.559	14.1986	0.598	15.1892
0.482	12.2428	0.521	13.2334	0.560	14.2240	0.599	15.2146
0.483	12.2682	0.522	13.2588	0.561	14.2494	0.600	15.2400
0.484	12.2936	0.523	13.2842	0.562	14.2748	0.601	15.2654
0.485	12.3190	0.524	13.3096	0.563	14.3002	0.602	15.2908
0.486	12.3444	0.525	13.3350	0.564	14.3256	0.603	15.3162
0.487	12.3698	0.526	13.3604	0.565	14.3510	0.604	15.3416
0.488	12.3952	0.527	13.3858	0.566	14.3764	0.605	15.3670
0.489	12.4206	0.528	13.4112	0.567	14.4018	0.606	15.3924
0.490	12.4460	0.529	13.4366	0.568	14.4272	0.607	15.4178
0.491	12.4714	0.530	13.4620	0.569	14.4526	0.608	15.4432
0.492	12.4968	0.531	13.4874	0.570	14.4780	0.609	15.4686
0.493	12.5222	0.532	13.5128	0.571	14. 5034	0.610	15.4940
0.494	12.5476	0.533	13.5382	0.572	14.5288	0.611	15.5194
0.495	12.5730	0.534	13.5636	0.573	14.5542	0.612	15.5448
0.496	12.5984	0.535	13.5890	0.574	14.5796	0.613	15.5702
0.497	12.6238	0.536	13.6144	0.575	14.6050	0.614	15.5956
0.498	12.6492	0.537	13.6398	0.576	14.6304	0.615	15.6210
0.499	12.6746	0.538	13.6652	0.577	14.6558	0.616	15.6464
0.500	12.7000	0.539	13.6906	0.578	14.6812	0.617	15.6718
0.501	12.7254	0.540	13.7160	0.579	14.7066	0.618	15.6972
0.502	12.7508	0.541	13.7414	0.580	14.7320	0.619	15.7226
0.503	12.7762	0.542	13.7668	0.581	14.7574	0.620	15.7480
0.504	12.8016	0.543	13.7922	0.582	14.7828	0.621	15.7734
0.505	12.8270	0.544	13.8176	0.583	14.8082	0.622	15.7988
0.506	12.8524	0.545	13.8430	0.584	14.8336	0.623	15.8242
0.507	12.8778	0.546	13.8684	0.585	14.8590	0.624	15.8496

(Continued)

Table 7. *(Continued)* **Inches Converted to Millimeters (0.625" to 0.780")**

0.625 to 0.663		0.664 to 0.702		0.703 to 0.741		0.742 to 0.780	
inch	mm	inch	mm	inch	mm	inch	mm
0.625	15.8750	0.664	16.8656	0.703	17.8562	0.742	18.8468
0.626	15.9004	0.665	16.8910	0.704	17.8816	0.743	18.8722
0.627	15.9258	0.666	16.9164	0.705	17.9070	0.744	18.8976
0.628	15.9512	0.667	16.9418	0.706	17.9324	0.745	18.9230
0.629	15.9766	0.668	16.9672	0.707	17.9578	0.746	18.9484
0.630	16.0020	0.669	16.9926	0.708	17.9832	0.747	18.9738
0.631	16.0274	0.670	17.0180	0.709	18.0086	0.748	18.9992
0.632	16.0528	0.671	17.0434	0.710	18.0340	0.749	19.0246
0.633	16.0782	0.672	17.0688	0.711	18.0594	0.750	19.0500
0.634	16.1036	0.673	17.0942	0.712	18.0848	0.751	19.0754
0.635	16.1290	0.674	17.1196	0.713	18.1102	0.752	19.1008
0.636	16.1544	0.675	17.1450	0.714	18.1356	0.753	19.1262
0.637	16.1798	0.676	17.1704	0.715	18.1610	0.754	19.1516
0.638	16.2052	0.677	17.1958	0.716	18.1864	0.755	19.1770
0.639	16.2306	0.678	17.2212	0.717	18.2118	0.756	19.2024
0.640	16.2560	0.679	17.2466	0.718	18.2372	0.757	19.2278
0.641	16.2814	0.680	17.2720	0.719	18.2626	0.758	19.2532
0.642	16.3068	0.681	17.2974	0.720	18.2880	0.759	19.2786
0.643	16.3322	0.682	17.3228	0.721	18.3134	0.760	19.3040
0.644	16.3576	0.683	17.3482	0.722	18.3388	0.761	19.3294
0.645	16.3830	0.684	17.3736	0.723	18.3642	0.762	19.3548
0.646	16.4084	0.685	17.3990	0.724	18.3896	0.763	19.3802
0.647	16.4338	0.686	17.4244	0.725	18.4150	0.764	19.4056
0.648	16.4592	0.687	17.4498	0.726	18.4404	0.765	19.4310
0.649	16.4846	0.688	17.4752	0.727	18.4658	0.766	19.4564
0.650	16.5100	0.689	17.5006	0.728	18.4912	0.767	19.4818
0.651	16.5354	0.690	17.5260	0.729	18.5166	0.768	19.5072
0.652	16.5608	0.691	17.5514	0.730	18.5420	0.769	19.5326
0.653	16.5862	0.692	17.5768	0.731	18.5674	0.770	19.5580
0.654	16.6116	0.693	17.6022	0.732	18.5928	0.771	19.5834
0.655	16.6370	0.694	17.6276	0.733	18.6182	0.772	19.6088
0.656	16.6624	0.695	17.6530	0.734	18.6436	0.773	19.6342
0.657	16.6878	0.696	176784	0.735	18.6690	0.774	19.6596
0.658	16.7132	0.697	17.7038	0.736	18.6944	0.775	19.6850
0.659	16.7386	0.698	17.7292	0.737	18.7198	0.776	19.7104
0.660	16.7640	0.699	17.7546	0.738	18.7452	0.777	19.7358
0.661	16.7894	0.700	17.7800	0.739	18.7706	0.778	19.7612
0.662	16.8148	0.701	17.8054	0.740	18.7960	0.779	19.7866
0.663	16.8402	0.702	17.8308	0.741	18.8214	0.780	19.8120

(Continued)

Table 7. *(Continued)* Inches Converted to Millimeters (0.781" to 0.936")

0.781 to 0.819		0.820 to 0.858		0.859 to 0.897		0.898 to 0.936	
inch	mm	inch	mm	inch	mm	inch	mm
0.781	19.8374	0.820	20.8280	0.859	21.8186	0.898	22.8092
0.782	19.8628	0.821	20.8534	0.860	21.8440	0.899	22.8346
0.783	19.8882	0.822	20.8788	0.861	21.8694	0.900	22.8600
0.784	19.9136	0.823	20.9042	0.862	21.8948	0.901	22.8854
0.785	19.9390	0.824	20.9296	0.863	21.9202	0.902	22.9108
0.786	19.9644	0.825	20.9550	0.864	21.9456	0.903	22.9362
0.787	19.9898	0.826	20.9804	0.865	21.9710	0.904	22.9616
0.788	20.0152	0.827	21.0058	0.866	21.9964	0.905	22.9870
0.789	20.0406	0.828	21.0312	0.867	22.0218	0.906	23.0124
0.790	20.0660	0.829	21.0566	0.868	22.0472	0.907	23.0378
0.791	20.0914	0.830	21.0820	0.869	22.0726	0.908	23.0632
0.792	20.1168	0.831	21.1074	0.870	22.0980	0.909	23.0886
0.793	20.1422	0.832	21.1328	0.871	22.1234	0.910	23.1140
0.794	20.1676	0.833	21.1582	0.872	22.1488	0.911	23.1394
0.795	20.1930	0.834	21.1836	0.873	22.1742	0.912	23.1648
0.796	20.2184	0.835	21.2090	0.874	22.1996	0.913	23.1902
0.797	20.2438	0.836	21.2344	0.875	22.2250	0.914	23.2156
0.798	20.2692	0.837	21.2598	0.876	22.2504	0.915	23.2410
0.799	20.2946	0.838	21.2852	0.877	22.2758	0.916	23.2664
0.800	20.3200	0.839	21.3106	0.878	22.3012	0.917	23.2918
0.801	20.3454	0.840	21.3360	0.879	22.3266	0.918	23.3172
0.802	20.3708	0.841	21.3614	0.880	22.3520	0.919	23.3426
0.803	20.3962	0.842	21.3868	0.881	22.3774	0.920	23.3680
0.804	20.4216	0.843	21.4122	0.882	22.4028	0.921	23.3934
0.805	20.4470	0.844	21.4376	0.883	22.4282	0.922	23.4188
0.806	20.4724	0.845	21.4630	0.884	22.4536	0.923	23.4442
0.807	20.4978	0.846	21.4884	0.885	22.4790	0.924	23.4696
0.808	20.5232	0.847	21.5138	0.886	22.5044	0.925	23.4950
0.809	20.5486	0.848	21.5392	0.887	22.5298	0.926	23.5204
0.810	20.5740	0.849	21.5646	0.888	22.5552	0.927	23.5458
0.811	20.5994	0.850	21.5900	0.889	22.5806	0.928	23.5712
0.812	20.6248	0.851	21.6154	0.890	22.6060	0.929	23.5966
0.813	20.6502	0.852	21.6408	0.891	22.6314	0.930	23.6220
0.814	20.6756	0.853	21.6662	0.892	22.6568	0.931	23.6474
0.815	20.7010	0.854	21.6916	0.893	22.6822	0.932	23.6728
0.816	20.7264	0.855	21.7170	0.894	22.7076	0.933	23.6982
0.817	20.7518	0.856	21.7424	0.895	22.7330	0.934	23.7236
0.818	20.7772	0.857	21.7678	0.896	22.7584	0.935	23.7490
0.819	20.8026	0.858	21.7932	0.897	22.7838	0.936	23.7744

(Continued)

Table 7. *(Continued)* **Inches Converted to Millimeters (0.937" to 1.00")**

0.937 to 0.952		0.953 to 0.968		0.969 to 0.984		0.985 to 1.000	
inch	mm	inch	mm	inch	mm	inch	mm
0.937	23.7998	0.953	24.2062	0.969	24.6126	0.985	25.0190
0.938	23.8252	0.954	24.2316	0.970	24.6380	0.986	25.0444
0.939	23.8506	0.955	24.2570	0.971	24.6634	0.987	25.0698
0.940	23.8760	0.956	24.2824	0.972	24.6888	0.988	25.0952
0.941	23.9014	0.957	24.3078	0.973	24.7142	0.989	25.1206
0.942	23.9268	0.958	24.3332	0.974	24.7396	0.990	25.1460
0.943	23.9522	0.959	24.3586	0.975	24.7650	0.991	25.1714
0.944	23.9776	0.960	24.3840	0.976	24.7904	0.992	25.1968
0.945	24.0030	0.961	24.4094	0.977	24.8158	0.993	25.2222
0.946	24.0284	0.962	24.4348	0.978	24.8412	0.994	25.2476
0.947	24.0538	0.963	24.4602	0.979	24.8666	0.995	25.2730
0.948	24.0792	0.964	24.4856	0.980	24.8920	0.996	25.2984
0.949	24.1046	0.965	24.5110	0.981	24.9174	0.997	25.3238
0.950	24.1300	0.966	24.5364	0.982	24.9428	0.998	25.3492
0.951	24.1554	0.967	24.5618	0.983	24.9682	0.999	25.3746
0.952	24.1808	0.968	24.5872	0.984	24.9936	1.000	25.4000

Table 8. Millimeters Converted to Inches (0.01 mm to 1.95 mm)

0.01 to 0.39		0.40 to 0.78		0.79 to 1.17		1.18 to 1.56		1.57 to 1.95	
mm	inch	mm	inch	mm	inch	mm	inch	mm	inch
.01	.00039	.40	.01575	.79	.03110	1.18	.04646	1.57	.06181
.02	.00079	.41	.01614	.80	.03150	1.19	.04685	1.58	.06220
.03	.00118	.42	.01654	.81	.03189	1.20	.04724	1.59	.06260
.04	.00157	.43	.01693	.82	.03228	1.21	.04764	1.60	.06299
.05	.00197	.44	.01732	.83	.03268	1.22	.04803	1.61	.06339
.06	.00236	.45	.01772	.84	.03307	1.23	.04843	1.62	.06378
.07	.00276	.46	.01811	.85	.03346	1.24	.04882	1.63	.06417
.08	.00315	.47	.01850	.86	.03386	1.25	.04921	1.64	.06457
.09	.00354	.48	.01890	.87	.03425	1.26	.04961	1.65	.06496
.10	.00394	.49	.01929	.88	.03465	1.27	.05000	1.66	.06535
.11	.00433	.50	.01969	.89	.03504	1.28	.05039	1.67	.06575
.12	.00433	.51	.02008	.90	.03543	1.29	.05079	1.68	.06614
.13	.00512	.52	.02047	.91	.03583	1.30	.05118	1.69	.06654
.14	.00551	.53	.02087	.92	.03622	1.31	.05157	1.70	.06693
.15	.00591	.54	.02126	.93	.03661	1.32	.05197	1.71	.06732
.16	.00630	.55	.02165	.94	.03701	1.33	.05236	1.72	.06772
.17	.00669	.56	.02205	.95	.03740	1.34	.05276	1.73	.06811
.18	.00709	.57	.02205	.96	.03780	1.35	.05315	1.74	.06850
.19	.00748	.58	.02283	.97	.03819	1.36	.05354	1.75	.06890
.20	.00787	.59	.02323	.98	.03858	1.37	.05394	1.76	.06929
.21	.00827	.60	.02362	.99	.03898	1.38	.05433	1.77	.06969
.22	.00866	.61	.02402	1.00	.03937	1.39	.05472	1.78	.07008
.23	.00906	.62	.02441	1.01	.03976	1.40	.05512	1.79	.07047
.24	.00945	.63	.02480	1.02	.04016	1.41	.05551	1.80	.07087
.25	.00984	.64	.02520	1.03	.04055	1.42	.05591	1.81	.07126
.26	.01024	.65	.02559	1.04	.04094	1.43	.05630	1.82	.07165
.27	.01063	.66	.02598	1.05	.04134	1.44	.05669	1.83	.07205
.28	.01102	.67	.02638	1.06	.04173	1.45	.05709	1.84	.07244
.29	.01142	.68	.02677	1.07	.04213	1.46	.05748	1.85	.07283
.30	.01181	.69	.02717	1.08	.04252	1.47	.05787	1.86	.07323
.31	.01220	.70	.02756	1.09	.04291	1.48	.05827	1.87	.07362
.32	.01260	.71	.02795	1.10	.04331	1.49	.05866	1.88	.07402
.33	.01299	.72	.02835	1.11	.04370	1.50	.05906	1.89	.07441
.34	.01339	.73	.02874	1.12	.04409	1.51	.05945	1.90	.07480
.35	.01378	.74	.02913	1.13	.04449	1.52	.05984	1.91	.07520
.36	.01417	.75	.02953	1.14	.04488	1.53	.06024	1.92	.07559
.37	.01457	.76	.02992	1.15	.04528	1.54	.06063	1.93	.07598
.38	.01496	.77	.03032	1.16	.04567	1.55	.06102	1.94	.07638
.39	.01535	.78	.03071	1.17	.04606	1.56	.06142	1.95	.07677

(Continued)

Table 8. *(Continued)* **Millimeters Converted to Inches (1.96 mm to 3.90 mm)**

1.96 to 2.34		2.35 to 2.73		2.74 to 3.12		3.13 to 3.51		3.52 to 3.90	
mm	inch	mm	inch	mm	inch	mm	inch	mm	inch
1.96	.07717	2.35	.09252	2.74	.10787	3.13	.12323	3.52	.13858
1.97	.07756	2.36	.09291	2.75	.10827	3.14	.12362	3.53	.13898
1.98	.07795	2.37	.09331	2.76	.10866	3.15	.12402	3.54	.13937
1.99	.07835	2.38	.09370	2.77	.10906	3.16	.12441	3.55	.13976
2.00	.07874	2.39	.09409	2.78	.10945	3.17	.12480	3.56	.14016
2.01	.07913	2.40	.09449	2.79	.10984	3.18	.12520	3.57	.14055
2.02	.07953	2.41	.09488	2.80	.11024	3.19	.12559	3.58	.14095
2.03	.07992	2.42	.09528	2.81	.11063	3.20	.12598	3.59	.14134
2.04	.08032	2.43	.09567	2.82	.11102	3.21	.12638	3.60	.14173
2.05	.08071	2.44	.09606	2.83	.11142	3.22	.12677	3.61	.14213
2.06	.08110	2.45	.09646	2.84	.11181	3.23	.12717	3.62	.14252
2.07	.08150	2.46	.09685	2.85	.11221	3.24	.12756	3.63	.14291
2.08	.08189	2.47	.09724	2.86	.11260	3.25	.12795	3.64	.14331
2.09	.08228	2.48	.09764	2.87	.11299	3.26	.12835	3.65	.14370
2.10	.08268	2.49	.09803	2.88	.11339	3.27	.12874	3.66	.14409
2.11	.08307	2.50	.09843	2.89	.11378	3.28	.12913	3.67	.14449
2.12	.08346	2.51	.09882	2.90	.11417	3.29	.12953	3.68	.14488
2.13	.08386	2.52	.09921	2.91	.11457	3.30	.12992	3.69	.14528
2.14	.08425	2.53	.09961	2.92	.11496	3.31	.13032	3.70	.14567
2.15	.08465	2.54	.10000	2.93	.11535	3.32	.13071	3.71	.14606
2.16	.08504	2.55	.10039	2.94	.11575	3.33	.13110	3.72	.14646
2.17	.08543	2.56	.10079	2.95	.11614	3.34	.13150	3.73	.14685
2.18	.08583	2.57	.10118	2.96	.11654	3.35	.13189	3.74	.14724
2.19	.08622	2.58	.10158	2.97	.11693	3.36	.13228	3.75	.14764
2.20	.08661	2.59	.10197	2.98	.11732	3.37	.13268	3.76	.14803
2.21	.08701	2.60	.10236	2.99	.11772	3.38	.13307	3.77	.14843
2.22	.08740	2.61	.10276	3.00	.11811	3.39	.13347	3.78	.14882
2.23	.08780	2.62	.10315	3.01	.11850	3.40	.13386	3.79	.14921
2.24	.08819	2.63	.10354	3.02	.11890	3.41	.13425	3.80	.14961
2.25	.08858	2.64	.10394	3.03	.11929	3.42	.13465	3.81	.15000
2.26	.08898	2.65	.10433	3.04	.11969	3.43	.13504	3.82	.15039
2.27	.08937	2.66	.10472	3.05	.12008	3.44	.13546	3.83	.15079
2.28	.08976	2.67	.10512	3.06	.12047	3.45	.13583	3.84	.15118
2.29	.09016	2.68	.10551	3.07	.12087	3.46	.13622	3.85	.15158
2.30	.09055	2.69	.10591	3.08	.12126	3.47	.13661	3.86	.15197
2.31	.09094	2.70	.10630	3.09	.12165	3.48	.13701	3.87	.15236
2.32	.09134	2.71	.10669	3.10	.12205	3.49	.13740	3.88	.15276
2.33	.09173	2.72	.10709	3.11	.12244	3.50	.13780	3.89	.15315
2.34	.09213	2.73	.10748	3.12	.12284	3.51	.13819	3.90	.15354

(Continued)

Table 8. *(Continued)* **Millimeters Converted to Inches (3.91 mm to 5.85 mm)**

| 3.91 to 4.29 | | 4.30 to 4.68 | | 4.69 to 5.07 | | 5.08 to 5.46 | | 5.47 to 5.85 | |
mm	inch	mm	inch	mm	inch	mm	inch	mm	inch
3.91	.15394	4.30	.16929	4.69	.18465	5.08	.20000	5.47	.21535
3.92	.15433	4.31	.16969	4.70	.18504	5.09	.20039	5.48	.21575
3.93	.15472	4.32	.17008	4.71	.18543	5.10	.20079	5.49	.21614
3.94	.15512	4.33	.17047	4.72	.18583	5.11	.20118	5.50	.21654
3.95	.15551	4.34	.17087	4.73	.18622	5.12	.20158	5.51	.21693
3.96	.15591	4.35	.17126	4.74	.18661	5.13	.20197	5.52	.21732
3.97	.15630	4.36	.17165	4.75	.18701	5.14	.20236	5.53	.21772
3.98	.15669	4.37	.17205	4.76	.18740	5.15	.20276	5.54	.21811
3.99	.15709	4.38	.17244	4.77	.18780	5.16	.20315	5.55	.21850
4.00	.15748	4.39	.17284	4.78	.18819	5.17	.20354	5.56	.21890
4.01	.15787	4.40	.17323	4.79	.18858	5.18	.20394	5.57	.21929
4.02	.15827	4.41	.17362	4.80	.18898	5.19	.20433	5.58	.21969
4.03	.15866	4.42	.17402	4.81	.18937	5.20	.20472	5.59	.22008
4.04	.15906	4.43	.17441	4.82	.18976	5.21	.20512	5.60	.22047
4.05	.15945	4.44	.17480	4.83	.19016	5.22	.20551	5.61	.22087
4.06	.15984	4.45	.17520	4.84	.19055	5.23	.20591	5.62	.22126
4.07	.16024	4.46	.17559	4.85	.19095	5.24	.20630	5.63	.22165
4.08	.16063	4.47	.17598	4.86	.19134	5.25	.20669	5.64	.22205
4.09	.16102	4.48	.17638	4.87	.19173	5.26	.20709	5.65	.22244
4.10	.16142	4.49	.17677	4.88	.19213	5.27	.20748	5.66	.22284
4.11	.16181	4.50	.17717	4.89	.19252	5.28	.20787	5.67	.22323
4.12	.16221	4.51	.17756	4.90	.19291	5.29	.20827	5.68	.22362
4.13	.16260	4.52	.17795	4.91	.19331	5.30	.20866	5.69	.22402
4.14	.16299	4.53	.17835	4.92	.19370	5.31	.20906	5.70	.22441
4.15	.16339	4.54	.17874	4.93	.19409	5.32	.20945	5.71	.22480
4.16	.16378	4.55	.17913	4.94	.19449	5.33	.20984	5.72	.22520
4.17	.16417	4.56	.17953	4.95	.19488	5.34	.21024	5.73	.22559
4.18	.16457	4.57	.17992	4.96	.19528	5.35	.21063	5.74	.22598
4.19	.16496	4.58	.18032	4.97	.19567	5.36	.21102	5.75	.22638
4.20	.16535	4.59	.18071	4.98	.19606	5.37	.21142	5.76	.22677
4.21	.16575	4.60	.18110	4.99	.19646	5.38	.21181	5.77	.22717
4.22	.16614	4.61	.18150	5.00	.19685	5.39	.21221	5.78	.22756
4.23	.16654	4.62	.18189	5.01	.19724	5.40	.21260	5.79	.22795
4.24	.16693	4.63	.18228	5.02	.19764	5.41	.21299	5.80	.22835
4.25	.16732	4.64	.18268	5.03	.19803	5.42	.21339	5.81	.22874
4.26	.16772	4.65	.18307	5.04	.19843	5.43	.21378	5.82	.22913
4.27	.16811	4.66	.18347	5.05	.19882	5.44	.21417	5.83	.22953
4.28	.16850	4.67	.18386	5.06	.19921	5.45	.21457	5.84	.22992
4.29	.16890	4.68	.18425	5.07	.19961	5.46	.21496	5.85	.23032

(Continued)

Table 8. *(Continued)* Millimeters Converted to Inches (5.86 mm to 7.80 mm)

5.86 to 6.24		6.25 to 6.63		6.64 to 7.02		7.03 to 7.41		7.42 to 7.80	
mm	inch	mm	inch	mm	inch	mm	inch	mm	inch
5.86	.23071	6.25	.24606	6.64	.26142	7.03	.27677	7.42	.29213
5.87	.23110	6.26	.24646	6.65	.26181	7.04	.27717	7.43	.29252
5.88	.23150	6.27	.24685	6.66	.26221	7.05	.27756	7.44	.29291
5.89	.23189	6.28	.24724	6.67	.26260	7.06	.27795	7.45	.29331
5.90	.23228	6.29	.24764	6.68	.26299	7.07	.27835	7.46	.29370
5.91	.23268	6.30	.24803	6.69	.26339	7.08	.27874	7.47	.29409
5.92	.23307	6.31	.24843	6.70	.26378	7.09	.27913	7.48	.29449
5.93	.23347	6.32	.24882	6.71	.26417	7.10	.27953	7.49	.29488
5.94	.23386	6.33	.24921	6.72	.26457	7.11	.27992	7.50	.29528
5.95	.23425	6.34	.24961	6.73	.26496	7.12	.28032	7.51	.29567
5.96	.23465	6.35	.25000	6.74	.26535	7.13	.28071	7.52	.29606
5.97	.23504	6.36	.25039	6.75	.26575	7.14	.28110	7.53	.29646
5.98	.23543	6.37	.25079	6.76	.26614	7.15	.28150	7.54	.29685
5.99	.23583	6.38	.25118	6.77	.26654	7.16	.28189	7.55	.29724
6.00	.23622	6.39	.25158	6.78	.26693	7.17	.28228	7.56	.29764
6.01	.23661	6.40	.25197	6.79	.26732	7.18	.28268	7.57	.29803
6.02	.23701	6.41	.25236	6.80	.26772	7.19	.28307	7.58	.29843
6.03	.23740	6.42	.25276	6.81	.26811	7.20	.28347	7.59	.29882
6.04	.23780	6.43	.25315	6.82	.26850	7.21	.28386	7.60	.29921
6.05	.23819	6.44	.25354	6.83	.26890	7.22	.28425	7.61	.29961
6.06	.23858	6.45	.25394	6.84	.26929	7.23	.28465	7.62	.30000
6.07	.23898	6.46	.25433	6.85	.26969	7.24	.28504	7.63	.30039
6.08	.23937	6.47	.25472	6.86	.27008	7.25	.28543	7.64	.30079
6.09	.23976	6.48	.25512	6.87	.27047	7.26	.28583	7.65	.30118
6.10	.24016	6.49	.25551	6.88	.27087	7.27	.28622	7.66	.30158
6.11	.24055	6.50	.25591	6.89	.27126	7.28	.28661	7.67	.30197
6.12	.24095	6.51	.25630	6.90	.27165	7.29	.28701	7.68	.30236
6.13	.24134	6.52	.25669	6.91	.27205	7.30	.28740	7.69	.30276
6.14	.24173	6.53	.25709	6.92	.27244	7.31	.28780	7.70	.30315
6.15	.24213	6.54	.25748	6.93	.27284	7.32	.28819	7.71	.30354
6.16	.24252	6.55	.25787	6.94	.27323	7.33	.28858	7.72	.30394
6.17	.24291	6.56	.25827	6.95	.27362	7.34	.28898	7.73	.30433
6.18	.24331	6.57	.25866	6.96	.27402	7.35	.28937	7.74	.30472
6.19	.24370	6.58	.25906	6.97	.27441	7.36	.28976	7.75	.30512
6.20	.24409	6.59	.25945	6.98	.27480	7.37	.29016	7.76	.30551
6.21	.24449	6.60	.25984	6.99	.27520	7.38	.29055	7.77	.30591
6.22	.24488	6.61	.26024	7.00	.27559	7.39	.29095	7.78	.30630
6.23	.24528	6.62	.26063	7.01	.27598	7.40	.29134	7.79	.30669
6.24	.24567	6.63	.26102	7.02	.27638	7.41	.29173	7.80	.30709

(Continued)

Table 8. *(Continued)* **Millimeters Converted to Inches (7.81 mm to 9.75 mm)**

| 7.81 to 8.19 ||	8.20 to 8.58 ||	8.59 to 8.97 ||	8.98 to 9.36 ||	9.37 to 9.75	
mm	inch	mm	inch	mm	inch	mm	inch	mm	inch
7.81	.30748	8.20	.32284	8.59	.33819	8.98	.35354	9.37	.36890
7.82	.30787	8.21	.32323	8.60	.33858	8.99	.35394	9.38	.36929
7.83	.30827	8.22	.32362	8.61	.33898	9.00	.35433	9.39	.36969
7.84	.30866	8.23	.32402	8.62	.33937	9.01	.35472	9.40	.37008
7.85	.30906	8.24	.32441	8.63	.33976	9.02	.35512	9.41	.37047
7.86	.30945	8.25	.32480	8.64	.34016	9.03	.35551	9.42	.37087
7.87	.30984	8.26	.32520	8.65	.34055	9.04	.35591	9.43	.37126
7.88	.31024	8.27	.32559	8.66	.34095	9.05	.35630	9.44	.37165
7.89	.31063	8.28	.32598	8.67	.34134	9.06	.35669	9.45	.37205
7.90	.31102	8.29	.32638	8.68	.34173	9.07	.35709	9.46	.37244
7.91	.31142	8.30	.32677	8.69	.34213	9.08	.35748	9.47	.37284
7.92	.31181	8.31	.32717	8.70	.34252	9.09	.35787	9.48	.37323
7.93	.31221	8.32	.32756	8.71	.34291	9.10	.35827	9.49	.37362
7.94	.31260	8.33	.32795	8.72	.34331	9.11	.35866	9.50	.37402
7.95	.31299	8.34	.32835	8.73	.34370	9.12	.35906	9.51	.37441
7.96	.31339	8.35	.32874	8.74	.34409	9.13	.35945	9.52	.37480
7.97	.31378	8.36	.32913	8.75	.34449	9.14	.35984	9.53	.37520
7.98	.31417	8.37	.32953	8.76	.34488	9.15	.36024	9.54	.37559
7.99	.31457	8.38	.32992	8.77	.34528	9.16	.36063	9.55	.37598
8.00	.31496	8.39	.33032	8.78	.34567	9.17	.36102	9.56	.37638
8.01	.31535	8.40	.33071	8.79	.34606	9.18	.36142	9.57	.37677
8.02	.31575	8.41	.33110	8.80	.34646	9.19	.36181	9.58	.37717
8.03	.31614	8.42	.33150	8.81	.34685	9.20	.36221	9.59	.37756
8.04	.31654	8.43	.33189	8.82	.34724	9.21	.36260	9.60	.37795
8.05	.31693	8.44	.33228	8.83	.34764	9.22	.36299	9.61	.37835
8.06	.31732	8.45	.33268	8.84	.34803	9.23	.36339	9.62	.37874
8.07	.31772	8.46	.33307	8.85	.34843	9.24	.36378	9.63	.37913
8.08	.31811	8.47	.33347	8.86	.34882	9.25	.36417	9.64	.37953
8.09	.31850	8.48	.33386	8.87	.34921	9.26	.36457	9.65	.37992
8.10	.31890	8.49	.33425	8.88	.34961	9.27	.36496	9.66	.38032
8.11	.31929	8.50	.33456	8.89	.35000	9.28	.36535	9.67	.38071
8.12	.31969	8.51	.33504	8.90	.35039	9.29	.36575	9.68	.38110
8.13	.32008	8.52	.33543	8.91	.35079	9.30	.36614	9.69	.38150
8.14	.32047	8.53	.33583	8.92	.35118	9.31	.36654	9.70	.38189
8.15	.32087	8.54	.33622	8.93	.35158	9.32	.36693	9.71	.38228
8.16	.32126	8.55	.33661	8.94	.35197	9.33	.36732	9.72	.38268
8.17	.32165	8.56	.33701	8.95	.35236	9.34	.36772	9.73	.38307
8.18	.32205	8.57	.33740	8.96	.35276	9.35	.36811	9.74	.38347
8.19	.32244	8.58	.33780	8.97	.35315	9.36	.36850	9.75	.38386

(Continued)

Table 8. *(Continued)* **Millimeters Converted to Inches (9.76 mm to 10 mm)**

9.76 to 9.80		9.81 to 9.85		9.86 to 9.90		9.91 to 9.95		9.96 to 10.00	
mm	inch	mm	inch	mm	inch	mm	inch	mm	inch
9.76	.38425	9.81	.38622	9.86	.38819	9.91	.39016	9.96	.39213
9.77	.38465	9.82	.38661	9.87	.38858	9.92	.39055	9.97	.39252
9.78	.38504	9.83	.38701	9.88	.38898	9.93	.39095	9.98	.39291
9.79	.38543	9.84	.38740	9.89	.38937	9.94	.39134	9.99	.39331
9.80	.38583	9.85	.38780	9.90	.38976	9.95	.39173	10.00	.39370

Table 9. Millimeters Converted to Inches (1 mm to 150 mm)

1 to 30		31 to 60		61 to 90		91 to 120		121 to 150	
mm	inch	mm	inch	mm	inch	mm	inch	mm	inch
1	0.03937	31	1.22047	61	2.40157	91	3.58268	121	4.76378
2	0.07874	32	1.25984	62	2.44094	92	3.62205	122	4.80315
3	0.11811	33	1.29921	63	2.48031	93	3.66142	123	4.84252
4	0.15748	34	1.33858	64	2.51969	94	3.70079	124	4.88189
5	0.19685	35	1.37795	65	2.55906	95	3.74016	125	4.92126
6	0.23622	36	1.41732	66	2.59843	96	3.77953	126	4.96063
7	0.27559	37	1.45669	67	2.63780	97	3.81890	127	5.00000
8	0.31496	38	1.49606	68	2.67717	98	3.85827	128	5.03937
9	0.35433	39	1.53543	69	2.71654	99	3.89764	129	5.07874
10	0.39370	40	1.57480	70	2.75591	100	3.93701	130	5.11811
11	0.43307	41	1.61417	71	2.79528	101	3.97638	131	5.15748
12	0.47244	42	1.65354	72	2.83465	102	4.01575	132	5.19685
13	0.51181	43	1.69291	73	2.87402	103	4.05512	133	5.23622
14	0.55118	44	1.73228	74	2.91339	104	4.09449	134	5.27559
15	0.59055	45	1.77165	75	2.95276	105	4.13386	135	5.31496
16	0.62992	46	1.81102	76	2.99213	106	4.17323	136	5.35433
17	0.66929	47	1.85039	77	3.03150	107	4.21260	137	5.39370
18	0.70866	48	1.88976	78	3.07087	108	4.25197	138	5.43307
19	0.74803	49	1.92913	79	3.11024	109	4.29134	139	5.47244
20	0.78740	50	1.96850	80	3.14961	110	4.33071	140	5.51181
21	0.82677	51	2.00787	81	3.18898	111	4.37008	141	5.55118
22	0.86614	52	2.04724	82	3.22835	112	4.40945	142	5.59055
23	0.90551	53	2.08661	83	3.26772	113	4.44882	143	5.62992
24	0.94488	54	2.12598	84	3.30709	114	4.48819	144	5.66929
25	0.98425	55	2.16535	85	3.34646	115	4.52756	145	5.70866
26	1.02362	56	2.20472	86	3.38583	116	4.56693	146	5.74803
27	1.06299	57	2.24409	87	3.42520	117	4.60630	147	5.78740
28	1.10236	58	2.28346	88	3.46457	118	4.64567	148	5.82677
29	1.14173	59	2.32283	89	3.50394	119	4.68504	149	5.86614
30	1.18110	60	2.36220	90	3.54331	120	4.72441	150	5.90551

(Continued)

Table 9. *(Continued)* **Millimeters Converted to Inches (151 mm to 345 mm)**

151 to 189		190 to 228		229 to 267		268 to 306		307 to 345	
mm	inch	mm	inch	mm	inch	mm	inch	mm	inch
151	5.94488	190	7.48031	229	9.01575	268	10.5512	307	12.0866
152	5.98425	191	7.51969	230	9.05512	269	10.5906	308	12.1260
153	6.02362	192	7.55906	231	9.09449	270	10.6299	309	12.1654
154	6.06299	193	7.59843	232	9.13386	271	10.6693	310	12.2047
155	6.10236	194	7.63780	233	9.17323	272	10.7087	311	12.2441
156	6.14173	195	7.67717	234	9.21260	273	10.7480	312	12.2835
157	6.18110	196	7.71654	235	9.25197	274	10.7874	313	12.3228
158	6.22047	197	7.75591	236	9.29134	275	10.8268	314	12.3622
159	6.25984	198	7.79528	237	9.33071	276	10.8661	315	12.4016
160	6.29921	199	7.83465	238	9.37008	277	10.9055	316	12.4409
161	6.33858	200	7.87402	239	9.40945	278	10.9449	317	12.4803
162	6.37795	201	7.91339	240	9.44882	279	10.9843	318	12.5197
163	6.41732	202	7.95276	241	9.48819	280	11.0236	319	12.5591
164	6.45669	203	7.99213	242	9.52756	281	11.0630	320	12.5984
165	6.49606	204	8.03150	243	9.56693	282	11.1024	321	12.6378
166	6.53543	205	8.07087	244	9.60630	283	11.1417	322	12.6772
167	6.57480	206	8.11024	245	9.64567	284	11.1811	323	12.7165
168	6.61417	207	8.14961	246	9.68504	285	11.2205	324	12.7559
169	6.65354	208	8.18898	247	9.72441	286	11.2598	325	12.7953
170	6.69291	209	8.22835	248	9.76378	287	11.2992	326	12.8346
171	6.73228	210	8.26772	249	9.80315	288	11.3386	327	12.8740
172	6.77165	211	8.30709	250	9.84252	289	11.3780	328	12.9134
173	6.81102	212	8.34646	251	9.88189	290	11.4173	329	12.9528
174	6.85039	213	8.38583	252	9.92126	291	11.4567	330	12.9921
175	6.88976	214	8.42520	253	9.96063	292	11.4961	331	13.0315
176	6.92913	215	8.46457	254	10.0000	293	11.5354	332	13.0709
177	6.96850	216	8.50394	255	10.0394	294	11.5748	333	13.1102
178	7.00787	217	8.54331	256	10.0787	295	11.6142	334	13.1496
179	7.04724	218	8.58268	257	10.1181	296	11.6535	335	13.1890
180	7.08661	219	8.62205	258	10.1575	297	11.6929	336	13.2283
181	7.12598	220	8.66142	259	10.1969	298	11.7323	337	13.2677
182	7.16535	221	8.70079	260	10.2362	299	11.7717	338	13.3071
183	7.20472	222	8.74016	261	10.2756	300	11.8110	339	13.3465
184	7.24409	223	8.77953	262	10.3150	301	11.8504	340	13.3858
185	7.28346	224	8.81890	263	10.3543	302	11.8898	341	13.4252
186	7.32283	225	8.85827	264	10.3937	303	11.9291	342	13.4646
187	7.36220	226	8.89764	265	10.4331	304	11.9685	343	13.5039
188	7.40157	227	8.93701	266	10.4724	305	12.0079	344	13.5433
189	7.44094	228	8.97638	267	10.5118	306	12.0472	345	13.5827

(Continued)

Table 9. *(Continued)* **Millimeters Converted to Inches (346 mm to 541 mm)**

346 to 385		386 to 424		425 to 463		464 to 502		503 to 541	
mm	inch	mm	inch	mm	inch	mm	inch	mm	inch
346	13.6220	386	15.1969	425	16.7323	464	18.2677	503	19.8031
347	13.6614	387	15.2362	426	16.7717	465	18.3071	504	19.8425
348	13.7008	388	15.2756	427	16.8110	466	18.3465	505	19.8819
349	13.7402	389	15.3150	428	16.8504	467	18.3858	506	19.9213
350	13.7795	390	15.3543	429	16.8898	468	18.4252	507	19.9606
351	13.8189	391	15.3937	430	16.9291	469	18.4646	508	20.0000
352	13.8583	392	15.4331	431	16.9685	470	18.5039	509	20.0394
353	13.8976	393	15.4724	432	17.0079	471	18.5433	510	20.0787
354	13.9370	394	15.5118	433	17.0472	472	18.5827	511	20.1181
355	13.9764	395	15.5512	434	17.0866	473	18.6220	512	20.1575
356	14.0157	396	15.5906	435	17.1260	474	18.6614	513	20.1969
358	14.0945	397	15.6299	436	17.1654	475	18.7008	514	20.2362
359	14.1339	398	15.6693	437	17.2047	476	18.7402	515	20.2756
360	14.1732	399	15.7087	438	17.2441	477	18.7795	516	20.3150
361	14.2126	400	15.7480	439	17.2835	478	18.8189	517	20.3543
362	14.2520	401	15.7874	440	17.3228	479	18.8583	518	20.3937
363	14.2913	402	15.8268	441	17.3622	480	18.8976	519	20.4331
364	14.3307	403	15.8661	442	17.4016	481	18.9370	520	20.4724
365	14.3701	404	15.9055	443	17.4409	482	18.9764	521	20.5118
366	14.4094	405	15.9449	444	17.4803	483	19.0157	522	20.5512
367	14.4488	406	15.9843	445	17.5197	484	19.0551	523	20.5906
368	14.4882	407	16.0236	446	17.5591	485	19.0945	524	20.6299
369	14.5276	408	16.0630	447	17.5984	486	19.1339	525	20.6693
370	14.5669	409	16.1024	448	17.6378	487	19.1732	526	20.7087
371	14.6063	410	16.1417	449	17.6772	488	19.2126	527	20.7480
372	14.6457	411	16.1811	450	17.7165	489	19.2520	528	20.7874
373	14.6850	412	16.2205	451	17.7559	490	19.2913	529	20.8268
374	14.7244	413	16.2598	452	17.7953	491	19.3307	530	20.8661
375	14.7638	414	16.2992	453	17.8346	492	19.3701	531	20.9055
376	14.8031	415	16.3386	454	17.8740	493	19.4094	532	20.9449
377	14.8425	416	16.3780	455	17.9134	494	19.4488	533	20.9843
378	14.8819	417	16.4173	456	17.9528	495	19.4882	534	21.0236
379	14.9213	418	16.4567	457	17.9921	496	19.5276	535	21.0630
380	14.9606	419	16.4961	458	18.0315	497	19.5669	536	21.1024
381	15.0000	420	16.5354	459	18.0709	498	19.6063	537	21.1417
382	15.0394	421	16.5748	460	18.1102	499	19.6457	538	21.1811
383	15.0787	422	16.6142	461	18.1496	500	19.6850	539	21.2205
384	15.1181	423	16.6535	462	18.1890	501	19.7244	540	21.2598
385	15.1575	424	16.6929	463	18.2283	502	19.7638	541	21.2992

(Continued)

Table 9. *(Continued)* Millimeters Converted to Inches (542 mm to 736 mm)

542 to 580		581 to 619		620 to 658		659 to 697		698 to 736	
mm	inch	mm	inch	mm	inch	mm	inch	mm	inch
542	21.3386	581	22.8740	620	24.4094	659	25.9449	698	27.4803
543	21.3780	582	22.9134	621	24.4488	660	25.9843	699	27.5197
544	21.4173	583	22.9528	622	24.4882	661	26.0236	700	27.5591
545	21.4567	584	22.9921	623	24.5276	662	26.0630	701	27.5984
546	21.4961	585	23.0315	624	24.5669	663	26.1024	702	27.6378
547	21.5354	586	23.0709	625	24.6063	664	26.1417	703	27.6772
548	21.5748	587	23.1102	626	24.6457	665	26.1811	704	27.7165
549	21.6142	588	23.1496	627	24.6850	666	26.2205	705	27.7559
550	21.6535	589	23.1890	628	24.7244	667	26.2598	706	27.7953
551	21.6929	590	23.2283	629	24.7638	668	26.2992	707	27.8346
552	21.7323	591	23.2677	630	24.8031	669	26.3386	708	27.8740
553	21.7717	592	23.3071	631	24.8425	670	26.3780	709	27.9134
554	21.8110	593	23.3465	632	24.8819	671	26.4173	710	27.9528
555	21.8504	594	23.3858	633	24.9213	672	26.4567	711	27.9921
556	21.8898	595	23.4252	634	24.9606	673	26.4961	712	28.0315
557	21.9291	596	23.4646	635	25.0000	674	26.5354	713	28.0709
558	21.9685	597	23.5039	636	25.0394	675	26.5748	714	28.1102
559	22.0079	598	23.5433	637	25.0787	676	26.6142	715	28.1496
560	22.0472	599	23.5827	638	25.1181	677	26.6535	716	28.1890
561	22.0866	600	23.6220	639	25.1575	678	26.6929	717	28.2283
562	22.1260	601	23.6614	640	25.1969	679	26.7323	718	28.2677
563	22.1654	602	23.7008	641	25.2362	680	26.7717	719	28.3071
564	22.2047	603	23.7402	642	25.2756	681	26.8110	720	28.3465
565	22.2441	604	23.7795	643	25.3150	682	26.8504	721	28.3858
566	22.2835	605	23.8189	644	25.3543	683	26.8898	722	28.4252
567	22.3228	606	23.8583	645	25.3937	684	26.9291	723	28.4646
568	22.3622	607	23.8976	646	25.4331	685	26.9685	724	28.5039
569	22.4016	608	23.9370	647	25.4724	686	27.0079	725	28.5433
570	22.4409	609	23.9764	648	25.5118	687	27.0472	726	28.5827
571	22.4803	610	24.0157	649	25.5512	688	27.0866	727	28.6220
572	22.5197	611	24.0551	650	25.5906	689	27.1260	728	28.6614
573	22.5591	612	24.0945	651	25.6299	690	27.1654	729	28.7008
574	22.5984	613	24.1339	652	25.6693	691	27.2047	730	28.7402
575	22.6378	614	24.1732	653	25.7087	692	27.2441	731	28.7795
576	22.6772	615	24.2126	654	25.7480	693	27.2835	732	28.8189
577	22.7165	616	24.2520	655	25.7874	694	27.3228	733	28.8583
578	22.7559	617	24.2913	656	25.8268	695	27.3622	734	28.8976
579	22.7953	618	24.3307	657	25.8661	696	27.4016	735	28.9370
580	22.8346	619	24.3701	658	25.9055	697	27.4409	736	28.9764

(Continued)

Table 9. *(Continued)* Millimeters Converted to Inches (737 mm to 930 mm)

| 737 to 775 | | 776 to 813 | | 814 to 852 | | 853 to 891 | | 892 to 930 | |
mm	inch	mm	inch	mm	inch	mm	inch	mm	inch
737	29.0157	775	30.5118	814	32.0472	853	33.5827	892	35.1181
738	29.0551	776	30.5512	815	32.0866	854	33.6220	893	35.1575
739	29.0945	777	30.5906	816	32.1260	855	33.6614	894	35.1969
740	29.1339	778	30.6299	817	32.1654	856	33.7008	895	35.2362
741	29.1732	779	30.6693	818	32.2047	857	33.7402	896	35.2756
742	29.2126	780	30.7087	819	32.2441	858	33.7795	897	35.3150
743	29.2520	781	30.7480	820	32.2835	859	33.8189	898	35.3543
744	29.2913	782	30.7874	821	32.3228	860	33.8583	899	35.3937
745	29.3307	783	30.8268	822	32.3622	861	33.8976	900	35.4331
746	29.3701	784	30.8661	823	32.4016	862	33.9370	901	35.4724
747	29.4094	785	30.9055	824	32.4409	863	33.9764	902	35.5118
748	29.4488	786	30.9449	825	32.4803	864	34.0157	903	35.5512
749	29.4882	787	30.9843	826	32.5197	865	34.0551	904	35.5906
750	29.5276	788	31.0236	827	32.5591	866	34.0945	905	35.6299
751	29.5669	789	31.0630	328	32.5984	867	34.1339	906	35.6693
752	29.6063	790	31.1024	829	32.6378	868	34.1732	907	35.7087
753	29.6457	791	31.1417	830	32.6772	869	34.2126	908	35.7480
754	29.6850	792	31.1811	331	32.7165	870	34.2520	909	35.7874
755	29.7244	793	31.2205	832	32.7559	871	34.2913	910	35.8268
756	29.7638	794	31.2598	833	32.7953	872	34.3307	911	35.8661
757	29.8031	795	31.2992	834	32.8346	873	34.3701	912	35.9055
758	29.8425	796	31.3386	835	32.8740	874	34.4094	913	35.9449
759	29.8819	797	31.3780	836	32.9134	875	34.4488	914	35.9843
760	29.9213	798	31.4173	837	32.9528	876	34.4882	915	36.0236
261	29.9606	799	31.4567	338	32.9921	877	34.5276	916	36.0630
762	30.0000	800	31.4961	839	33.0315	878	34.5669	917	36.1024
763	30.0394	801	31.5354	840	33.0709	879	34.6063	918	36.1417
764	30.0787	802	31.5748	841	33.1102	880	34.6457	919	36.1811
765	30.1181	803	31.6142	842	33.1496	881	34.6850	920	36.2205
766	30.1575	804	31.6535	843	33.1890	882	34.7244	921	36.2598
767	30.1969	805	31.6929	844	33.2283	883	34.7638	922	36.2992
768	30.2362	806	31.7323	845	33.2677	884	34.8031	923	36.3386
769	30.2756	807	31.7717	846	33.3071	885	34.8425	924	36.3780
770	30.3150	808	31.8110	847	33.3465	886	34.8819	925	36.4173
771	30.3543	809	31.8504	848	33.3858	887	34.9213	926	36.4567
772	30.3937	810	31.8898	849	33.4252	888	34.9606	927	36.4961
773	30.4331	811	31.9291	850	33.4646	889	35.0000	928	36.5354
774	30.4724	812	31.9685	851	33.5039	890	35.0394	929	36.5748
775	30.5118	813	32.0079	852	33.5433	891	35.0787	930	36.6142

(Continued)

Table 9. *(Continued)* Millimeters Converted to Inches (931 mm to 1000 mm)

931 to 944		945 to 958		959 to 972		973 to 986		987 to 1000	
mm	inch	mm	inch	mm	inch	mm	inch	mm	inch
931	36.6535	945	37.2047	959	37.7559	973	38.3071	987	38.8583
932	36.6929	946	37.2441	960	37.7953	974	38.3465	988	38.8976
933	36.7323	947	37.2835	961	37.8346	975	38.3858	989	38.9370
934	36.7717	948	37.3228	962	37.8740	976	38.4252	990	38.9764
935	36.8110	949	37.3622	963	37.9134	977	38.4646	991	39.0157
936	36.8504	950	37.4016	964	37.9528	978	38.5039	992	39.0551
937	36.8898	951	37.4409	965	37.9921	979	38.5433	993	39.0945
938	36.9291	952	37.4803	966	38.0315	980	38.5827	994	39.1339
939	36.9685	953	37.5197	967	38.0709	981	38.6220	995	39.1732
940	37.0079	954	37.5591	468	38.1102	982	38.6614	996	39.2126
941	37.0472	955	37.5984	469	38.1496	983	38.7008	997	39.2520
942	37.0866	956	37.6378	970	38.1890	484	38.7402	998	39.2913
943	37.1260	957	37.6772	971	38.2283	985	38.7795	999	39.3307
944	37.1654	958	37.7165	972	38.2677	986	38.8189	1000	39.3701

Table 10. Square Inch - Square Centimeter Conversion Table.

Sq. in.	in.² or cm²	Sq. cm
0.0002	0.001	0.0065
0.0003	0.002	0.0129
0.0005	0.003	0.0194
0.0006	0.004	0.0258
0.0008	0.005	0.0323
0.0009	0.006	0.0387
0.0011	0.007	0.0452
0.0012	0.008	0.0516
0.0014	0.009	0.0581
0.0016	0.01	0.0645
0.0031	0.02	0.1290
0.0047	0.03	0.1935
0.0062	0.04	0.2581
0.0078	0.05	0.3226
0.0093	0.06	0.3871
0.0109	0.07	0.4516
0.0124	0.08	0.5161
0.0140	0.09	0.5806
0.0155	0.1	0.6452
0.0310	0.2	1.2903
0.0465	0.3	1.9355
0.0620	0.4	2.5806
0.0775	0.5	3.2258
0.0930	0.6	3.8710
0.1085	0.7	4.5161
0.1240	0.8	5.1613
0.1395	0.9	5.8064
0.155	1	6.452
0.310	2	12.903
0.465	3	19.355
0.620	4	25.806
0.775	5	32.258
0.930	6	38.710
1.085	7	45.161
1.240	8	51.613
1.395	9	58.064
1.550	10	64.516
1.705	11	70.968
1.860	12	77.419
2.015	13	83.871
2.170	14	90.322
2.325	15	96.774
2.480	16	103.226
2.635	17	109.677
2.790	18	116.129
2.945	19	122.580
3.100	20	129.032
3.255	21	135.484
3.410	22	141.935
3.565	23	148.387
3.720	24	154.838
3.875	25	161.290
4.030	26	167.742
4.185	27	174.193
4.340	28	180.645
4.495	29	187.096
4.650	30	193.548
4.805	31	200.000
4.960	32	206.451
5.115	33	212.903
5.270	34	219.354
5.425	35	225.806
5.580	36	232.258
5.735	37	238.709
5.890	38	245.161
6.045	39	251.612
6.200	40	258.064
6.355	41	264.516
6.510	42	270.967
6.665	43	277.419
6.820	44	283.870
6.975	45	290.322
7.130	46	296.774
7.285	47	303.225
7.440	48	309.677
7.595	49	316.128
7.750	50	322.580
7.905	51	329.032
8.060	52	335.483
8.215	53	341.935
8.370	54	348.386

The center conversion value in bold faced type can be read as either square inches (in²) or square centimeters (cm²).

To convert 145.642 in² into cm², first find 145 in the bold face column and directly to its right the value in cm² is given (935.482). Next, convert 0.6 in² into cm² using the decimal column above (3.8710), then 0.04 (0.2581), and finally 0.002 (0.0129). The sum of the four is the conversion.

145.642 in² = 939.624 cm².

The values for converting 145.642 cm² into in² are:

145 cm² = 22.475 in²
0.6 cm² = 0.0930 in²
0.04 cm² = 0.0062 in²
0.002 cm² = 0.0003 in²
145.642 cm² = 22.5745 in²

(Continued)

Table 10. *(Continued)* Square Inch - Square Centimeter Conversion Table.

Sq. in.	in.² or cm²	Sq. cm	Sq. in.	in.² or cm²	Sq. cm	Sq. in.	in.² or cm²	Sq. cm	Sq. in.	in.² or cm²	Sq. cm
8.525	55	354.838	12.245	79	509.676	15.965	103	664.515	19.685	127	819.353
8.680	56	361.290	12.400	80	516.128	16.120	104	670.966	19.840	128	825.805
8.835	57	367.741	12.555	81	522.580	16.275	105	677.418	19.995	129	832.256
8.990	58	374.193	12.710	82	529.031	16.430	106	683.870	20.150	130	838.708
9.145	59	380.644	12.865	83	535.483	16.585	107	690.321	20.305	131	845.160
9.300	60	387.096	13.020	84	541.934	16.740	108	696.773	20.460	132	851.611
9.455	61	393.548	13.175	85	548.386	16.895	109	703.224	20.615	133	858.063
9.610	62	399.999	13.330	86	554.838	17.050	110	709.676	20.770	134	864.514
9.765	63	406.451	13.485	87	561.289	17.205	111	716.128	20.925	135	870.966
9.920	64	412.902	13.640	88	567.741	17.360	112	722.579	21.080	136	877.418
10.075	65	419.354	13.795	89	574.192	17.515	113	729.031	21.235	137	883.869
10.230	66	425.806	13.950	90	580.644	17.670	114	735.482	21.390	138	890.321
10.385	67	432.257	14.105	91	587.096	17.825	115	741.934	21.545	139	896.772
10.540	68	438.709	14.260	92	593.547	17.980	116	748.386	21.700	140	903.224
10.695	69	445.160	14.415	93	599.999	18.135	117	754.837	21.855	141	909.676
10.850	70	451.612	14.570	94	606.450	18.290	118	761.289	22.010	142	916.127
11.005	71	458.064	14.725	95	612.902	18.445	119	767.740	22.165	143	922.579
11.160	72	464.515	14.880	96	619.354	18.600	120	774.192	22.320	144	929.030
11.315	73	470.967	15.035	97	625.805	18.755	121	780.644	22.475	145	935.482
11.470	74	477.418	15.190	98	632.257	18.910	122	787.095	22.630	146	941.934
11.625	75	483.870	15.345	99	638.708	19.065	123	793.547	22.785	147	948.385
11.780	76	490.322	15.500	100	645.160	19.220	124	799.998	22.940	148	954.837
11.935	77	496.773	15.655	101	651.612	19.375	125	806.450	23.095	149	961.288
12.090	78	503.225	15.810	102	658.063	19.530	126	812.902	23.250	150	967.740

(Continued)

Table 10. *(Continued)* Square Inch - Square Centimeter Conversion Table.

Sq. in.	in.² or cm²	Sq. cm	Sq. in.	in.² or cm²	Sq. cm	Sq. in.	in.² or cm²	Sq. cm	Sq. in.	in.² or cm²	Sq. cm
23.405	151	974.192	27.125	175	1129.030	30.845	199	1283.868	34.565	223	1438.707
23.560	152	980.643	27.280	176	1135.482	31.000	200	1290.320	34.720	224	1445.158
23.715	153	987.095	27.435	177	1141.933	31.155	201	1296.772	24.875	225	1451.610
23.870	154	993.546	27.590	178	1148.385	31.310	202	1303.223	35.030	226	1458.062
24.025	155	999.998	27.745	179	1154.836	31.465	203	1309.675	35.185	227	1464.513
24.180	156	1006.450	27.900	180	1161.288	31.620	204	1316.126	35.340	228	1470.965
24.335	157	1012.901	28.055	181	1167.740	31.775	205	1322.578	35.495	229	1477.416
24.490	158	1019.353	28.210	182	1174.191	31.930	206	1329.030	35.650	230	1483.868
24.645	159	1025.804	28.365	183	1180.643	32.085	207	1335.481	35.805	231	1490.320
24.800	160	1032.256	28.520	184	1187.094	32.240	208	1341.933	35.960	232	1496.771
24.955	161	1038.708	28.675	185	1193.546	32.395	209	1348.384	36.115	233	1503.223
25.110	162	1045.159	28.830	186	119.998	32.550	210	1354.836	36.270	234	1509.674
25.265	163	1051.611	28.985	187	1206.449	32.705	211	1361.288	36.425	235	1516.126
25.420	164	1058.062	29.140	188	1212.901	32.860	212	1367.739	36.580	236	1522.578
25.575	165	1064.514	29.295	189	1219.352	33.015	213	1374.191	36.735	237	1529.029
25.730	166	1070.966	29.450	190	1225.804	33.170	214	1380.642	36.890	238	1535.481
25.885	167	1077.417	29.605	191	1232.256	33.325	215	1387.094	37.045	239	1541.932
26.040	168	1083.869	29.760	192	1238.707	33.480	216	1393.546	37.200	240	1548.384
26.195	169	1090.320	29.915	193	1245.159	33.635	217	1399.997	37.355	241	1554.836
26.350	170	1096.772	30.070	194	1251.610	33.790	218	1406.449	37.510	242	1561.287
26.505	171	1103.224	30.225	195	1258.062	33.945	219	1412.900	37.665	243	1567.739
26.660	172	1109.675	30.380	196	1264.514	34.100	220	1419.352	37.820	244	1574.190
26.815	173	1116.127	30.535	197	1270.965	34.255	221	1425.804	37.975	245	1580.642
26.970	174	1122.578	30.690	198	127.417	34.410	222	1432.255	38.130	246	1587.094

(Continued)

Table 10. *(Continued)* Square Inch - Square Centimeter Conversion Table.

Sq. in.	in.² or cm²	Sq. cm	Sq. in.	in.² or cm²	Sq. cm	Sq. in.	in.² or cm²	Sq. cm	Sq. in.	in.² or cm²	Sq. cm
38.285	247	1593.545	42.005	271	1748.384	45.725	295	1903.222	49.445	319	2058.060
38.440	248	1599.997	42.160	272	1754.835	45.880	296	1909.674	49.600	320	2064.512
38.595	249	1606.448	42.315	273	1761.287	46.035	297	1916.125	49.755	321	2070.964
38.750	250	1612.900	42.470	274	1767.738	46.190	298	1922.577	49.910	322	2077.415
38.905	251	1619.352	42.625	275	1774.190	46.345	299	1929.028	50.065	323	2083.867
39.060	252	1625.803	42.780	276	1780.642	46.500	300	1935.480	50.220	324	2090.318
39.215	253	1632.255	42.935	277	1787.093	46.655	301	1941.932	50.375	325	2096.770
39.370	254	1638.706	43.090	278	1793.545	46.810	302	1948.383	50.530	326	2103.222
39.525	255	1645.158	43.245	279	1799.996	46.965	303	1954.835	50.685	327	2109.673
39.680	256	1651.610	43.400	280	1806.448	47.120	304	1961.286	50.840	328	2116.125
39.835	257	1658.061	43.555	281	1812.900	47.275	305	1967.738	50.995	329	2122.576
39.990	258	1664.513	43.710	282	1819.351	47.430	306	1974.190	51.150	330	2129.028
40.145	259	1670.964	43.865	283	1825.803	47.585	307	1980.641	51.305	331	2135.480
40.300	260	1677.416	44.020	284	1832.254	47.740	308	1987.093	51.460	332	2141.931
40.455	261	1683.868	44.175	285	1838.706	47.895	309	1993.544	51.615	333	2148.383
40.610	262	1690.319	44.330	286	1845.158	48.050	310	1999.996	51.770	334	2154.834
40.765	263	1696.771	44.485	287	1851.609	48.205	311	2006.448	51.925	335	2161.286
40.920	264	1703.222	44.640	288	1858.061	48.360	312	2012.899	52.080	336	2167.738
41.075	265	1709.674	44.795	289	1864.512	48.515	313	2019.351	52.235	337	2174.189
41.203	266	1716.126	44.950	290	1870.964	48.670	314	2025.802	52.390	338	2180.641
41.385	267	1722.577	45.105	291	1877.416	48.825	315	2032.254	52.545	339	2187.092
41.540	268	1729.029	45.260	292	1883.867	48.980	316	2038.706	52.700	340	2193.544
41.695	269	1735.480	45.415	293	1890.319	49.135	317	2045.157	52.855	341	2199.996
41.850	270	1741.932	45.570	294	1896.770	49.290	318	2051.609	53.010	342	2206.447

Table 11. Cubic Inch - Cubic Centimeter Conversion Table.

Cu. in.	in.³ or cm³	Cu. cm	Cu. in.	in.³ or cm³	Cu. cm	Cu. in.	in.³ or cm³	Cu. cm	Cu. in.	in.³ or cm³	Cu. cm
0.0001	**0.001**	0.0164	0.0006	**0.01**	0.1639	0.427	**7**	114.709	1.892	**31**	507.999
0.0001	**0.002**	0.0328	0.0012	**0.02**	0.3277	0.488	**8**	131.096	1.953	**32**	524.386
0.0002	**0.003**	0.0492	0.0018	**0.03**	0.4916	0.549	**9**	147.484	2.014	**33**	540.773
0.0002	**0.004**	0.0655	0.0024	**0.04**	0.6555	0.610	**10**	163.871	2.075	**34**	557.160
0.0003	**0.005**	0.0819	0.0031	**0.05**	0.8194	0.671	**11**	180.258	2.136	**35**	573.547
0.0004	**0.006**	0.0983	0.0037	**0.06**	0.9832	0.732	**12**	196.645	2.197	**36**	589.934
0.0004	**0.007**	0.1147	0.0043	**0.07**	1.1471	0.793	**13**	213.032	2.258	**37**	606.321
0.0005	**0.008**	0.1311	0.0049	**0.08**	1.3110	0.854	**14**	229.419	2.319	**38**	622.708
0.0005	**0.009**	0.1475	0.0055	**0.09**	1.4748	0.915	**15**	245.806	2.380	**39**	639.095
			0.0061	**0.1**	1.6387	0.976	**16**	262.193	2.441	**40**	655.482
			0.0122	**0.2**	3.2774	1.037	**17**	278.580	2.502	**41**	671.869
			0.0183	**0.3**	4.9161	1.098	**18**	294.967	2.563	**42**	688.257
			0.0244	**0.4**	6.5548	1.159	**19**	311.354	2.624	**43**	704.644
			0.0305	**0.5**	8.1935	1.220	**20**	327.741	2.685	**44**	721.031
			0.0366	**0.6**	9.8322	1.281	**21**	344.128	2.746	**45**	737.418
			0.0427	**0.7**	11.4709	1.343	**22**	360.515	2.807	**46**	753.805
			0.0488	**0.8**	13.1096	1.404	**23**	376.902	2.868	**47**	770.192
			0.0549	**0.9**	14.7484	1.465	**24**	393.289	2.929	**48**	786.579
			0.061	**1**	16.387	1.526	**25**	409.677	2.990	**49**	802.966
			0.122	**2**	32.774	1.587	**26**	426.064	3.051	**50**	819.353
			0.183	**3**	49.161	1.648	**27**	442.451	3.112	**51**	835.740
			0.244	**4**	65.548	1.709	**28**	458.838	3.173	**52**	852.127
			0.305	**5**	81.935	1.770	**29**	475.225	3.234	**53**	868.514
			0.366	**6**	98.322	1.831	**30**	491.612	3.295	**54**	884.901

The center conversion value in bold faced type can be read as either cubic inches (in³) or cubic centimeters (cm³).
To convert 145.642 in³ into cm³, first find 145 in the bold face column and directly to its right the value in cm³ is given (2376.124). Next, convert 0.6 in³ into cm³ using the decimal column above (9.8322), then 0.04 (0.6555), and finally 0.002 (0.0328). The sum of the four is the conversion.
145.642 in³ = 2376.6445 cm³.

The values for converting 145.642 cm³ into in³ are:
145 cm³ = 8.848 in³
0.6 cm³ = 0.0366 in³
0.04 cm³ = 0.0024 in³
0.002 cm³ = 0.0001 in³
145.642 cm³ = 8.8871 in³.

(Continued)

Table 11. (Continued) **Cubic Inch - Cubic Centimeter Conversion Table.**

Cu. in.	in.³ or cm³	Cu. cm	Cu. in.	in.³ or cm³	Cu. cm	Cu. in.	in.³ or cm³	Cu. cm	Cu. in.	in.³ or cm³	Cu. cm
3.356	55	901.288	4.821	79	1294.578	6.285	103	1687.867	7.750	127	2081.157
3.417	56	917.675	4.882	80	1310.965	6.346	104	1704.254	7.811	128	2097.544
3.478	57	934.062	4.943	81	1327.352	6.407	105	1720.641	7.872	129	2113.931
3.539	58	950.449	5.004	82	1343.739	6.469	106	1737.028	7.933	130	2130.318
3.600	59	966.837	5.065	83	1360.126	6.530	107	1753.415	7.994	131	2146.705
3.661	60	983.224	5.126	84	1376.513	6.591	108	1769.802	8.055	132	2163.092
3.722	61	999.611	5.187	85	1392.900	6.652	109	1786.190	8.116	133	2179.479
3.783	62	1015.998	5.248	86	1409.287	6.713	110	1802.577	8.177	134	2195.866
3.844	63	1032.385	5.309	87	1425.674	6.774	111	1818.964	8.238	135	2212.253
3.906	64	1048.772	5.370	88	1442.061	6.835	112	1835.351	8.299	136	2228.640
3.967	65	1065.159	5.431	89	1458.448	6.896	113	1851.738	8.360	137	2245.027
4.028	66	1081.546	5.492	90	1474.835	6.957	114	1868.125	8.421	138	2261.414
4.089	67	1097.933	5.553	91	1491.222	7.018	115	1884.512	8.482	139	2277.801
4.150	68	1114.320	5.614	92	1507.610	7.079	116	1900.899	8.543	140	2294.188
4.211	69	1130.707	5.675	93	1523.997	7.140	117	1917.286	8.604	141	2310.575
4.272	70	1147.094	5.736	94	1540.384	7.201	118	1933.673	8.665	142	2326.963
4.333	71	1163.481	5.797	95	1556.771	7.262	119	1950.060	8.726	143	2343.350
4.394	72	1179.868	5.858	96	1573.158	7.323	120	1966.447	8.787	144	2359.737
4.455	73	1196.255	5.919	97	1589.545	7.384	121	1982.834	8.848	145	2376.124
4.516	74	1212.642	5.980	98	1605.932	7.445	122	1999.221	8.909	146	2392.511
4.577	75	1229.030	6.041	99	1622.319	7.506	123	2015.608	8.970	147	2408.898
4.638	76	1245.417	6.102	100	1638.706	7.567	124	2031.995	9.032	148	2425.285
4.699	77	1261.804	6.163	101	1655.093	7.628	125	2048.383	9.093	149	2441.672
4.760	78	1278.191	6.224	102	1671.480	7.689	126	2064.770	9.154	150	2458.059

(Continued)

Table 11. *(Continued)* Cubic Inch - Cubic Centimeter Conversion Table.

Cu. in.	in.³ or cm³	Cu. cm	Cu. in.	in.³ or cm³	Cu. cm	Cu. in.	in.³ or cm³	Cu. cm	Cu. in.	in.³ or cm³	Cu. cm
9.215	151	2474.446	10.679	175	2867.736	12.144	199	3261.025	13.608	223	3654.314
9.276	152	2490.833	10.740	176	2884.123	12.205	200	3277.412	13.669	224	3670.701
9.337	153	2507.220	10.801	177	2900.510	12.266	201	3293.799	13.730	225	3687.089
9.398	154	2523.607	10.862	178	2916.897	12.327	202	3310.186	13.791	226	3703.476
9.459	155	2539.994	10.923	179	2933.284	12.388	203	3326.573	13.852	227	3719.863
9.520	156	2556.381	10.984	180	2949.671	12.449	204	3342.960	13.913	228	3736.250
9.581	157	2572.768	11.045	181	2966.058	12.510	205	3359.347	13.974	229	3752.637
9.642	158	2589.155	11.106	182	2982.445	12.571	206	3375.734	14.035	230	3769.024
9.703	159	2605.543	11.167	183	2998.832	12.632	207	3392.121	14.096	231	3785.411
9.764	160	2621.930	11.228	184	3015.219	12.693	208	3408.508	14.158	232	3801.798
9.825	161	2638.317	11.289	185	3031.606	12.754	209	3424.896	14.219	233	3818.185
9.886	162	2654.704	11.350	186	3047.993	12.815	210	3441.283	14.280	234	3834.572
9.947	163	2671.091	11.411	187	3064.380	12.876	211	3457.670	14.341	235	3850.959
10.008	164	2687.478	11.472	188	3080.767	12.937	212	3474.057	14.402	236	3867.346
10.069	165	2703.865	11.533	189	3097.154	12.998	213	3490.444	14.463	237	3883.733
10.130	166	2720.252	11.595	190	3113.541	13.059	214	3506.831	14.524	238	3900.120
10.191	167	2736.639	11.656	191	3129.928	13.120	215	3523.218	14.585	239	3916.507
10.252	168	2753.026	11.717	192	3146.316	13.181	216	3539.605	14.646	240	3932.894
10.313	169	2769.413	11.778	193	3162.703	13.242	217	3555.992	14.707	241	3949.281
10.374	170	2785.800	11.839	194	3179.090	13.303	218	3572.379	14.768	242	3965.669
10.435	171	2802.187	11.900	195	3195.477	13.364	219	3588.766	14.829	243	3982.056
10.496	172	2818.574	11.961	196	3211.864	13.425	220	3605.153	14.890	244	3998.443
10.557	173	2834.961	12.022	197	3228.251	13.486	221	3621.540	14.951	245	4014.830
10.618	174	2851.348	12.083	198	3244.638	13.547	222	3637.927	15.012	246	4031.217

(Continued)

Table 11. *(Continued)* Cubic Inch - Cubic Centimeter Conversion Table.

Cu. in.	in.³ or cm³	Cu. cm	Cu. in.	in.³ or cm³	Cu. cm	Cu. in.	in.³ or cm³	Cu. cm	Cu. in.	in.³ or cm³	Cu. cm
15.073	247	4047.604	16.537	271	4440.893	18.002	295	4834.183	19.467	319	5227.472
15.134	248	4063.991	16.598	272	4457.280	18.063	296	4850.570	19.528	320	5243.859
15.195	249	4080.378	16.659	273	4473.667	18.124	297	4866.957	19.589	321	5260.246
15.256	250	4096.765	16.721	274	4490.054	18.185	298	4883.344	19.650	322	5276.633
15.317	251	4113.152	16.782	275	4506.442	18.246	299	4899.731	19.711	323	5293.020
15.378	252	4129.539	16.843	276	4522.829	118.307	300	4916.118	19.772	324	5309.407
15.439	253	4145.926	16.904	277	4539.216	18.368	301	4932.505	19.833	325	5325.795
15.500	254	4162.313	16.965	278	4555.603	18.429	302	4948.892	19.894	326	5342.182
15.561	255	4178.700	17.026	279	4571.990	18.490	303	4965.279	19.955	327	5358.569
15.622	256	4195.087	17.087	280	4588.377	18.551	304	4981.666	20.016	328	5374.956
15.683	257	4211.474	17.148	281	4604.764	18.612	305	4998.053	20.077	329	5391.343
15.744	258	4227.861	17.209	282	4621.151	18.673	306	5014.440	20.138	330	5407.730
15.805	259	4244.249	17.270	283	4637.538	18.734	307	5030.827	20.199	331	5424.117
15.866	260	4260.636	17.331	284	4653.925	18.795	308	5047.214	20.260	332	5440.504
15.927	261	4277.023	17.392	285	4670.312	18.856	309	5063.602	20.321	333	5456.891
15.988	262	4293.410	17.453	286	4686.699	18.917	310	5079.989	20.382	334	5473.278
16.049	263	4309.797	17.514	287	4703.086	18.978	311	5096.376	20.443	335	5489.665
16.110	264	4326.184	17.575	288	4719.473	19.039	312	5112.763	20.504	336	5506.052
16.171	265	4342.571	17.636	289	4735.860	19.100	313	5129.150	20.565	337	5522.439
16.232	266	4358.958	17.697	290	4752.247	19.161	314	5145.537	20.626	338	5538.826
16.293	267	4375.345	17.758	291	4768.634	19.222	315	5161.924	20.687	339	5555.213
16.354	268	4391.732	17.819	292	4785.022	19.284	316	5178.311	20.748	340	5571.600
16.415	269	4408.119	17.880	293	4801.409	19.345	317	5194.698	20.809	341	5587.987
16.476	270	4424.506	17.941	294	4817.796	19.406	318	5211.085	20.870	342	5604.375

Table 12. Foot - Meter Conversion Table.

Feet	ft or m	Meters	Feet	ft or m	Meters	Feet	ft or m	Meters	Feet	ft or m	Meters
0.0033	0.001	0.0003	0.0328	0.01	0.0030	22.966	7	2.134	101.706	31	9.449
0.0066	0.002	0.0006	0.0656	0.02	0.0061	26.247	8	2.438	104.987	32	9.754
0.0098	0.003	0.0009	0.0984	0.03	0.0091	29.528	9	2.743	108.268	33	10.058
0.0131	0.004	0.0012	0.1312	0.04	0.0122	32.808	10	3.048	111.549	34	10.363
0.0164	0.005	0.0015	0.1640	0.05	0.0152	36.089	11	3.353	114.829	35	10.668
0.0197	0.006	0.0018	0.1969	0.06	0.0183	39.370	12	3.658	118.110	36	10.973
0.0230	0.007	0.0021	0.2297	0.07	0.0213	42.651	13	3.962	121.391	37	11.278
0.0262	0.008	0.0024	0.2625	0.08	0.0244	45.932	14	4.267	124.672	38	11.582
0.0295	0.009	0.0027	0.2953	0.09	0.0274	49.213	15	4.572	127.953	39	11.887
			0.3281	0.1	0.0305	52.493	16	4.877	131.234	40	12.192
			0.6562	0.2	0.0610	55.774	17	5.182	134.514	41	12.497
			0.9843	0.3	0.0914	59.055	18	5.486	137.795	42	12.802
			1.3123	0.4	0.1219	62.336	19	5.791	141.076	43	13.106
			1.6404	0.5	0.1524	65.617	20	6.096	144.357	44	13.411
			1.9685	0.6	0.1829	68.898	21	6.401	147.638	45	13.716
			2.2966	0.7	0.2134	72.178	22	6.706	150.919	46	14.021
			2.6247	0.8	0.2438	75.459	23	7.010	154.199	47	14.326
			2.9528	0.9	0.2743	78.740	24	7.315	157.480	48	14.630
			3.281	1	0.305	82.021	25	7.620	160.761	49	14.935
			6.652	2	0.610	85.302	26	7.925	164.042	50	15.240
			9.843	3	0.914	88.583	27	8.230	167.323	51	15.545
			13.123	4	1.219	91.864	28	8.534	170.604	52	15.850
			16.404	5	1.524	95.144	29	8.839	173.885	53	16.154
			19.685	6	1.829	98.425	30	9.144	177.165	54	16.459

The center conversion value in bold faced type can be read as either feet (ft) or meters (m).

To convert 145.642 feet into meters, first find 145 in the bold face column and directly to its right the value in meters is given (44.196). Next, convert 0.6 feet into meters using the decimal column above (0.1829), then 0.04 (0.0122), and finally 0.002 (0.0006). The sum of the four is the conversion.

145.642 ft = 44.3917 m.

The values for converting 145.642 meters into feet are:

145 m = 475.722 ft
0.6 m = 1.9685 ft
0.04 m = 0.1312 ft
0.002 m = 0.0066 ft
145.642 m = 477.8463 ft.

(Continued)

Table 12. *(Continued)* Foot - Meter Conversion Table.

Feet	ft or m	Meters	Feet	ft or m	Meters	Feet	ft or m	Meters	Feet	ft or m	Meters
180.446	55	16.764	259.186	79	24.079	337.927	103	31.394	416.667	127	38.710
183.727	56	17.069	262.467	80	24.384	341.207	104	31.699	419.948	128	39.014
187.008	57	17.374	265.748	81	24.689	344.488	105	32.004	423.228	129	39.319
190.289	58	17.678	269.029	82	24.994	347.769	106	32.309	426.509	130	39.624
193.570	59	17.983	272.310	83	25.298	351.050	107	32.614	429.790	131	39.929
196.850	60	18.288	275.591	84	25.603	354.331	108	32.918	433.071	132	40.234
200.131	61	18.593	278.871	85	25.908	357.612	109	33.223	436.352	133	40.538
203.412	62	18.898	282.152	86	26.213	360.892	110	33.528	439.633	134	40.843
206.693	63	19.202	285.433	87	26.518	364.173	111	33.833	442.913	135	41.148
209.974	64	19.507	288.714	88	26.822	367.454	112	34.138	446.194	136	41.453
213.255	65	19.812	291.995	89	27.127	370.735	113	34.442	449.475	137	41.758
216.535	66	20.117	295.276	90	27.432	374.016	114	34.747	452.756	138	42.062
219.816	67	20.422	298.556	91	27.737	377.297	115	35.052	456.037	139	42.367
223.097	68	20.726	301.837	92	28.042	380.577	116	35.357	459.318	140	42.672
226.378	69	21.031	305.118	93	28.346	383.858	117	35.662	462.598	141	42.977
229.659	70	21.336	308.399	94	28.651	387.139	118	35.966	465.879	142	43.282
232.940	71	21.641	311.680	95	28.956	390.420	119	36.271	469.160	143	43.586
236.220	72	21.946	314.961	96	29.261	393.701	120	36.576	472.441	144	43.891
239.501	73	22.250	318.241	97	29.566	396.982	121	36.881	475.722	145	44.196
242.782	74	22.555	321.522	98	29.870	400.262	122	37.186	479.003	146	44.501
246.063	75	22.860	324.803	99	30.175	403.543	123	37.490	482.283	147	44.806
249.344	76	23.165	328.084	100	30.480	406.824	124	37.795	485.564	148	45.110
252.625	77	23.470	331.365	101	30.785	410.105	125	38.100	488.845	149	45.415
255.906	78	23.774	334.646	102	31.090	413.386	126	38.405	492.126	150	45.720

(Continued)

Table 12. *(Continued)* **Foot - Meter Conversion Table.**

Feet	ft or m	Meters	Feet	ft or m	Meters	Feet	ft or m	Meters	Feet	ft or m	Meters
495.407	151	46.025	574.147	175	53.340	652.887	199	60.655	731.627	223	67.970
498.688	152	46.330	577.428	176	53.645	656.168	200	60.960	734.908	224	68.275
501.969	153	46.634	580.709	177	53.950	659.449	201	61.265	738.189	225	68.580
505.249	154	46.939	583.990	178	54.254	662.730	202	61.570	741.470	226	68.885
508.530	155	47.244	587.270	179	54.559	666.011	203	61.874	744.751	227	69.190
511.811	156	47.549	590.551	180	54.864	669.291	204	62.179	748.032	228	69.494
515.092	157	47.854	593.832	181	55.169	672.572	205	62.484	751.312	229	69.799
518.373	158	48.158	597.113	182	55.474	675.853	206	62.789	754.593	230	70.104
521.654	159	48.463	600.394	183	55.778	679.134	207	63.094	757.874	231	70.409
524.934	160	48.768	603.675	184	56.083	682.415	208	63.398	761.155	232	70.714
528.215	161	49.073	606.955	185	56.388	685.696	209	63.703	764.436	233	71.018
531.496	162	49.378	610.236	186	56.693	688.976	210	64.008	767.717	234	71.323
534.777	163	49.682	613.517	187	56.998	692.257	211	64.313	770.997	235	71.628
538.058	164	49.987	616.798	188	57.302	695.538	212	64.618	774.278	236	71.933
541.339	165	50.292	620.079	189	57.607	698.819	213	64.922	777.559	237	72.238
544.619	166	50.597	623.360	190	57.912	702.100	214	65.227	780.840	238	72.542
547.900	167	50.902	626.640	191	58.217	705.381	215	65.532	784.121	239	72.847
551.181	168	51.206	629.921	192	58.522	708.661	216	65.837	787.402	240	73.152
554.462	169	51.511	633.202	193	58.826	711.942	217	66.142	790.682	241	73.457
557.743	170	51.816	636.483	194	59.131	715.223	218	66.446	793.963	242	73.762
561.024	171	52.121	639.764	195	59.436	718.504	219	66.751	797.244	243	74.066
564.304	172	52.426	643.045	196	59.741	721.785	220	67.056	800.525	244	74.371
567.585	173	52.730	646.325	197	60.046	725.066	221	67.361	803.806	245	74.676
570.866	174	53.035	649.606	198	60.350	728.346	222	67.666	807.087	246	74.981

(Continued)

Table 12. *(Continued)* **Foot - Meter Conversion Table.**

Feet	ft or m	Meters	Feet	ft or m	Meters	Feet	ft or m	Meters	Feet	ft or m	Meters
810.367	247	75.286	889.108	271	82.601	967.848	295	89.916	1046.588	319	97.231
813.648	248	75.590	892.388	272	82.906	971.129	296	90.221	1049.869	320	97.536
816.929	249	75.895	895.669	273	83.210	974.409	297	90.526	1053.150	321	97.841
820.210	250	76.200	898.950	274	83.515	977.690	298	90.830	1056.430	322	98.146
823.491	251	76.505	902.231	275	83.820	980.971	299	91.135	1059.711	323	98.450
826.772	252	76.810	905.512	276	84.125	984.252	300	91.440	1062.992	324	98.755
830.053	253	77.114	908.793	277	84.430	987.533	301	91.745	1066.273	325	99.060
833.333	254	77.419	912.074	278	84.734	990.814	302	92.050	1069.554	326	99.365
836.614	255	77.724	915.354	279	85.039	994.095	303	92.354	1072.835	327	99.670
839.895	256	78.029	918.635	280	85.344	997.375	304	92.659	1076.116	328	99.974
843.176	257	78.334	921.916	281	85.649	1000.656	305	92.964	1079.396	329	100.279
846.457	258	78.638	925.197	282	85.954	1003.937	306	93.269	1082.677	330	100.584
849.738	259	78.943	928.478	283	86.258	1007.218	307	93.574	1085.958	331	100.889
853.018	260	79.248	931.759	284	86.563	1010.499	308	93.878	1089.239	332	101.194
856.299	261	79.553	935.039	285	86.868	1013.780	309	94.183	1092.520	333	101.498
859.580	262	79.858	938.320	286	87.173	1017.060	310	94.488	1095.801	334	101.803
862.861	263	80.162	941.601	287	87.478	1020.341	311	94.793	1099.081	335	102.108
866.142	264	80.467	944.882	288	87.782	1023.622	312	95.098	1102.362	336	102.413
869.423	265	80.772	948.163	289	88.087	1026.903	313	95.402	1105.643	337	102.718
872.703	266	81.077	951.444	290	88.392	1030.184	314	95.707	1108.924	338	103.022
875.984	267	81.382	954.724	291	88.697	1033.465	315	96.012	1112.205	339	103.327
879.265	268	81.686	958.005	292	89.002	1036.745	316	96.317	1115.486	340	103.632
882.546	269	81.991	961.286	293	89.306	1040.026	317	96.622	1118.766	341	103.937
885.827	270	82.296	964.567	294	89.611	1043.307	318	96.926	1122.047	342	104.242

Table 13. Gallon-Liter Conversion Table.

Gallon	gal or L	Liter	Gallon	gal or L	Liter	Gallon	gal or L	Liter
0.0003	**0.001**	0.0038	0.0026	**0.01**	0.0379	1.849	**7**	26.498
0.0005	**0.002**	0.0076	0.0053	**0.02**	0.0757	2.113	**8**	30.283
0.0008	**0.003**	0.0144	0.0079	**0.03**	0.1136	2.378	**9**	34.069
0.0011	**0.004**	0.0151	0.0106	**0.04**	0.1514	2.642	**10**	37.854
0.0013	**0.005**	0.0189	0.0132	**0.05**	0.1893	2.906	**11**	41.640
0.0016	**0.006**	0.0227	0.0159	**0.06**	0.2271	3.170	**12**	45.425
0.0018	**0.007**	0.0265	0.0185	**0.07**	0.2650	3.434	**13**	49.210
0.0021	**0.008**	0.0303	0.0211	**0.08**	0.3028	3.698	**14**	52.996
0.0024	**0.009**	0.0341	0.0238	**0.09**	0.3407	3.963	**15**	56.781
			0.0264	**0.1**	0.3785	4.227	**16**	60.567
			0.0528	**0.2**	0.7571	4.491	**17**	64.352
			0.0793	**0.3**	1.1356	4.755	**18**	68.137
			0.1057	**0.4**	1.5142	5.019	**19**	71.923
			0.1321	**0.5**	1.8927	5.283	**20**	75.708
			0.1585	**0.6**	2.2712	5.548	**21**	79.494
			0.1849	**0.7**	2.6498	5.812	**22**	83.279
			0.2113	**0.8**	3.0283	6.076	**23**	87.064
			0.2378	**0.9**	3.4069	6.340	**24**	90.850
			0.264	**1**	3.785	6.604	**25**	94.635
			0.528	**2**	7.571	6.868	**26**	98.421
			0.793	**3**	11.356	7.133	**27**	102.206
			1.057	**4**	15.142	7.397	**28**	105.992
			1.321	**5**	18.927	7.661	**29**	109.777
			1.585	**6**	22.712	7.925	**30**	113.562

Gallon	gal or L	Liter
8.189	**31**	117.348
8.454	**32**	121.133
8.718	**33**	124.919
8.982	**34**	128.704
9.246	**35**	132.489
9.510	**36**	136.275
9.774	**37**	140.060
10.039	**38**	143.846
10.303	**39**	147.631
10.567	**40**	151.416
10.831	**41**	155.202
11.095	**42**	158.987
11.359	**43**	162.773
11.624	**44**	166.558
11.888	**45**	170.344
12.152	**46**	174.129
12.416	**47**	177.914
12.680	**48**	181.700
12.944	**49**	185.485
13.209	**50**	189.271
13.473	**51**	193.056
13.737	**52**	196.841
14.001	**53**	200.627
14.265	**54**	204.412

The center conversion value in bold faced type can be read as either gallons (gal) or liters (l).

To convert 145.642 gallons into liters, first find 145 in the bold face column and directly to its right the value in liters is given (548.885). Next, convert 0.6 gallons into liters using the decimal column above (2.2712), then 0.04 (0.1514), and finally 0.002 (0.0076). The sum of the four is the conversion.

145.642 gal = 551.3152 liters.

The values for converting 145.642 liters into gallons are:

145 l = 38.305 gal
0.6 l = 0.1585 gal
0.04 l = 0.0106 gal
0.002 l = 0.0005 gal
145.642 l = 37.4746 gal.

(Continued)

Table 13. *(Continued)* **Gallon-Liter Conversion Table.**

Gallon	gal or L	Liter	Gallon	gal or L	Liter	Gallon	gal or L	Liter	Gallon	gal or L	Liter
14.529	55	208.198	20.870	79	299.048	27.210	103	389.897	33.550	127	480.747
14.794	56	211.983	21.134	80	302.833	27.474	104	393.683	33.814	128	484.533
15.058	57	215.768	21.398	81	306.618	27.738	105	397.468	34.078	129	488.318
15.322	58	219.554	21.662	82	310.404	28.002	106	401.254	34.342	130	492.104
15.586	59	223.339	21.926	83	314.189	28.266	107	405.039	34.607	131	495.889
15.850	60	227.125	22.190	84	317.975	28.531	108	408.824	34.871	132	499.674
16.114	61	230.910	22.455	85	321.760	28.795	109	412.610	35.135	133	503.460
16.379	62	234.696	22.719	86	325.545	29.059	110	416.395	35.399	134	507.245
16.643	63	238.481	22.983	87	329.331	29.323	111	420.181	35.663	135	511.031
16.907	64	242.266	23.247	88	333.116	29.587	112	423.966	35.927	136	514.816
17.171	65	246.052	23.511	89	336.902	29.851	113	427.752	36.192	137	518.601
17.435	66	249.837	23.775	90	340.687	30.116	114	431.537	36.456	138	522.387
17.700	67	253.623	24.040	91	344.472	30.380	115	435.322	36.720	139	526.172
17.964	68	257.408	24.304	92	348.258	30.644	116	439.108	36.984	140	529.958
18.228	69	261.193	24.568	93	352.043	30.908	117	442.893	37.248	141	533.743
18.492	70	264.979	24.832	94	355.829	31.172	118	446.679	37.512	142	537.529
18.756	71	268.764	25.096	95	359.614	31.436	119	450.464	37.777	143	541.314
19.020	72	272.550	25.361	96	363.400	31.701	120	454.249	38.041	144	545.099
19.285	73	276.335	25.625	97	367.185	31.965	121	458.035	38.305	145	548.885
19.549	74	280.120	25.889	98	370.970	32.229	122	461.820	38.569	146	552.670
19.813	75	283.906	26.153	99	374.756	32.493	123	465.060	38.833	147	556.456
20.077	76	287.691	26.417	100	378.541	32.757	124	469.391	39.097	148	560.241
20.341	77	291.477	26.681	101	382.327	33.022	125	473.177	39.362	149	564.026
20.605	78	295.262	26.946	102	386.112	33.286	126	476.962	39.626	150	567.812

(Continued)

Table 13. *(Continued)* Gallon–Liter Conversion Table.

Gallon	gal or L	Liter	Gallon	gal or L	Liter	Gallon	gal or L	Liter	Gallon	gal or L	Liter
39.890	151	571.597	46.230	175	662.447	52.570	199	753.297	58.910	223	844.147
40.154	152	575.383	46.494	176	666.233	52.834	200	757.082	59.175	224	847.932
40.418	153	579.168	46.758	177	670.018	53.099	201	760.868	59.439	225	851.718
40.682	154	582.953	47.023	178	673.803	53.363	202	764.653	59.703	226	855.503
40.947	155	586.739	47.287	179	677.589	53.627	203	768.439	59.967	227	859.289
41.211	156	590.524	47.551	180	681.374	53.891	204	772.224	60.231	228	863.074
41.475	157	594.310	47.815	181	685.160	54.155	205	776.009	60.495	229	866.859
41.739	158	598.095	48.079	182	688.945	54.419	206	779.795	60.760	230	870.645
42.003	159	601.881	48.343	183	692.730	54.684	207	783.580	61.024	231	874.430
42.268	160	605.666	48.608	184	696.516	54.948	208	787.366	61.288	232	878.216
42.532	161	609.451	48.872	185	700.301	55.212	209	791.151	61.552	233	882.001
42.796	162	613.237	49.136	186	704.087	55.476	210	794.937	61.816	234	885.786
43.060	163	617.022	49.400	187	707.872	55.740	211	798.722	62.080	235	889.572
43.324	164	620.808	49.664	188	711.657	56.004	212	802.507	62.345	236	893.357
43.588	165	624.593	49.929	189	715.443	56.269	213	806.293	62.609	237	897.143
43.853	166	628.378	50.193	190	719.228	56.533	214	810.078	62.873	238	900.928
44.117	167	632.164	50.457	191	723.014	56.797	215	813.864	63.137	239	904.713
44.381	168	635.949	50.721	192	726.799	57.061	216	817.649	63.401	240	908.499
44.645	169	639.735	50.985	193	730.585	57.325	217	821.434	63.665	241	912.284
44.909	170	643.520	51.249	194	734.370	57.589	218	825.220	63.930	242	916.070
45.173	171	647.305	51.514	195	738.155	57.854	219	829.005	64.194	243	919.855
45.438	172	651.091	51.778	196	741.941	58.118	220	832.791	64.458	244	923.641
45.702	173	654.876	52.042	197	745.726	58.382	221	836.576	64.722	245	927.426
45.966	174	658.662	52.306	198	749.512	58.646	222	840.361	64.986	246	931.211

(Continued)

Table 13. *(Continued)* Gallon-Liter Conversion Table.

Gallon	gal or L	Liter	Gallon	gal or L	Liter	Gallon	gal or L	Liter	Gallon	gal or L	Liter
65.250	247	934.997	71.591	271	1025.847	77.931	295	1116.697	84.271	319	1207.546
65.515	248	938.782	71.855	272	1029.632	78.195	296	1120.482	84.535	320	1211.332
65.779	249	942.568	72.119	273	1033.417	78.459	297	1124.267	84.799	321	1215.117
66.043	250	946.353	72.383	274	1037.203	78.723	298	1128.053	85.063	322	1218.903
66.307	251	950.138	72.647	275	1040.988	78.987	299	1131.838	85.328	323	1222.688
66.571	252	953.924	72.911	276	1044.774	79.252	300	1135.624	85.592	324	1226.473
66.836	253	957.709	73.176	277	1048.559	79.516	301	1139.409	85.856	325	1230.259
67.100	254	961.495	73.440	278	1052.345	79.780	302	1143.194	86.120	326	1234.044
67.364	255	965.280	73.704	279	1056.130	80.044	303	1146.980	86.384	327	1237.830
67.628	256	969.065	73.968	280	1059.915	80.308	304	1150.765	86.648	328	1241.615
67.892	257	972.851	74.232	281	1063.701	80.572	305	1154.551	86.913	329	1245.401
68.156	258	976.636	74.497	282	1067.486	80.837	306	1158.336	87.177	330	1249.186
68.421	259	980.422	74.761	283	1071.272	81.101	307	1162.121	87.441	331	1252.971
68.685	260	984.207	75.025	284	1075.057	81.365	308	1165.907	87.705	332	1256.757
68.949	261	987.993	75.289	285	1078.842	81.629	309	1169.692	87.969	333	1260.542
69.213	262	991.778	75.553	286	1082.628	81.893	310	1173.478	88.233	334	1264.328
69.477	263	995.563	75.817	287	1086.413	82.157	311	1177.263	88.498	335	1268.113
69.741	264	999.349	76.082	288	1090.199	82.422	312	1181.049	88.762	336	1271.898
70.006	265	1003.134	76.346	289	1093.984	82.686	313	1184.834	89.026	337	1275.684
70.270	266	1006.920	76.610	290	1097.769	82.950	314	1188.619	89.290	338	1279.469
70.534	267	1010.705	76.874	291	1101.555	83.214	315	1192.405	89.554	339	1283.255
70.798	268	1014.490	77.138	292	1105.340	83.478	316	1196.190	89.818	340	1287.040
71.062	269	1018.276	77.402	293	1109.126	83.743	317	1199.976	90.083	341	1290.825
71.326	270	1022.061	77.667	294	1112.911	84.007	318	1203.761	90.347	342	1294.611

Table 14. Ounce - Gram Conversion Table.

Ounces	oz or gr	Grams
0.000035	**0.001**	0.0283
0.000071	**0.002**	0.0567
0.000106	**0.003**	0.0850
0.000141	**0.004**	0.1134
0.000176	**0.005**	0.1417
0.000212	**0.006**	0.1701
0.000247	**0.007**	0.1984
0.000282	**0.008**	0.2268
0.000318	**0.009**	0.2551

Ounces	oz or gr	Grams
0.000353	**0.01**	0.2835
0.000705	**0.02**	0.5670
0.001058	**0.03**	0.8505
0.001411	**0.04**	1.1340
0.001763	**0.05**	1.4175
0.002116	**0.06**	1.7010
0.002469	**0.07**	1.9845
0.002822	**0.08**	2.2680
0.003175	**0.09**	2.5515
0.003527	**0.1**	2.8350
0.007055	**0.2**	5.6699
0.010582	**0.3**	8.5049
0.014109	**0.4**	11.3398
0.017637	**0.5**	14.1748
0.021164	**0.6**	17.0097
0.246918	**0.7**	19.8447
0.028219	**0.8**	22.6796
0.031747	**0.9**	25.5146
0.035	**1**	28.350
0.071	**2**	56.699
0.106	**3**	85.049
0.141	**4**	113.398
0.176	**5**	141.748
0.212	**6**	170.097

Ounces	oz or gr	Grams
0.247	**7**	198.447
0.282	**8**	226.796
0.317	**9**	255.146
0.353	**10**	283.495
0.388	**11**	311.845
0.423	**12**	340.194
0.459	**13**	368.544
0.494	**14**	396.893
0.529	**15**	425.243
0.564	**16**	453.592
0.600	**17**	481.942
0.635	**18**	510.291
0.670	**19**	538.641
0.705	**20**	566.990
0.741	**21**	595.340
0.776	**22**	623.689
0.811	**23**	652.039
0.847	**24**	680.388
0.882	**25**	708.738
0.917	**26**	737.088
0.952	**27**	765.437
0.988	**28**	793.787
1.023	**29**	822.136
1.058	**30**	850.486

Ounces	oz or gr	Grams
1.093	**31**	878.835
1.129	**32**	907.185
1.164	**33**	935.534
1.199	**34**	963.884
1.235	**35**	992.233
1.270	**36**	1020.583
1.305	**37**	1048.932
1.340	**38**	1077.282
1.376	**39**	1105.631
1.411	**40**	1133.981
1.446	**41**	1162.330
1.482	**42**	1190.680
1.517	**43**	1219.029
1.552	**44**	1247.379
1.587	**45**	1275.728
1.623	**46**	1304.078
1.658	**47**	1332.427
1.693	**48**	1360.777
1.728	**49**	1389.126
1.764	**50**	1417.476
1.799	**51**	1445.826
1.834	**52**	1474.175
1.870	**53**	1502.525
1.905	**54**	1530.874

The center conversion value in bold faced type can be read as either ounces (oz) or grains (gr).

To convert 145.642 ounces into grams, first find 145 in the bold face column and directly to its right the value in grams is given (4110.680). Next, convert 0.6 ounce into grams using the decimal column above (17.0097), then 0.04 (1.1340), and finally 0.002 (0.0567). The sum of the four is the conversion. 145.642 oz = 4128.8804 gr.

The values for converting 145.642 grams into ounces are:
145 gr = 5.115 oz
0.6 gr = 0.21164 oz
0.04 gr = 0.001411 oz
0.002 gr = 0.00071 oz
145.642 gr = 5.328122 oz.

(Continued)

Table 14. *(Continued)* **Ounce - Gram Conversion Table.**

Ounces	oz or gr	Grams	Ounces	oz or gr	Grams	Ounces	oz or gr	Grams	Ounces	oz or gr	Grams
1.940	55	1559.224	2.787	79	2239.612	3.633	103	2920.001	4.480	127	3600.389
1.975	56	1587.573	2.822	80	2267.962	3.668	104	2948.350	4.515	128	3628.739
2.011	57	1615.923	2.857	81	2296.311	3.704	105	2976.700	4.550	129	3657.088
2.046	58	1644.272	2.892	82	2324.661	3.739	106	3005.049	4.586	130	3685.438
2.081	59	1672.622	2.928	83	2353.010	3.774	107	3033.399	4.621	131	3713.787
2.116	60	1700.971	2.963	84	2381.360	3.810	108	3061.748	4.656	132	3742.137
2.152	61	1729.321	2.998	85	2409.709	3.845	109	3090.098	4.691	133	3770.486
2.187	62	1757.670	3.034	86	2438.059	3.880	110	3118.447	4.727	134	3798.836
2.222	63	1786.020	3.069	87	2466.408	3.915	111	3146.797	4.762	135	3827.185
2.258	64	1814.369	3.104	88	2494.758	3.951	112	3175.146	4.797	136	3855.535
2.293	65	1842.719	3.139	89	2523.107	3.986	113	3203.496	4.833	137	3883.884
2.328	66	1871.068	3.175	90	2551.457	4.021	114	3231.845	4.868	138	3912.234
2.363	67	1899.418	3.210	91	2579.806	4.057	115	3260.195	4.903	139	3940.583
2.399	68	1927.767	3.245	92	2608.156	4.092	116	3288.544	4.938	140	3968.933
2.434	69	1956.117	3.280	93	2636.505	4.127	117	3316.894	4.974	141	3997.282
2.469	70	1984.466	3.316	94	2664.855	4.162	118	3345.243	5.009	142	4025.632
2.504	71	2012.816	3.351	95	2693.204	4.198	119	3373.593	5.044	143	4053.981
2.540	72	2041.165	3.386	96	2721.554	4.233	120	3401.942	5.079	144	4082.331
2.575	73	2069.515	3.422	97	2749.903	4.268	121	3430.292	5.115	145	4110.680
2.610	74	2097.864	3.457	98	2778.253	4.303	122	3458.641	5.150	146	4139.030
2.646	75	2126.214	3.492	99	2806.602	4.339	123	3486.991	5.185	147	4167.379
2.681	76	2154.564	3.527	100	2834.952	4.374	124	3515.340	5.221	148	4195.729
2.716	77	2182.913	3.563	101	2863.302	4.409	125	3543.690	5.256	149	4224.078
2.751	78	2211.263	3.598	102	2891.651	4.445	126	3572.040	5.291	150	4252.428

(Continued)

Table 14. *(Continued)* Ounce - Gram Conversion Table.

Ounces	oz or gr	Grams	Ounces	oz or gr	Grams	Ounces	oz or gr	Grams	Ounces	oz or gr	Grams
5.326	151	4280.778	6.173	175	4961.166	7.020	199	5641.554	7.866	223	6321.943
5.362	152	4309.127	6.208	176	4989.516	7.055	200	5669.904	7.901	224	6350.292
5.397	153	4337.477	6.243	177	5017.865	7.090	201	5698.254	7.937	225	6378.642
5.432	154	4365.826	6.279	178	5046.215	7.125	202	5726.603	7.972	226	6406.992
5.467	155	4394.176	6.314	179	5074.564	7.161	203	5754.953	8.007	227	6435.341
5.503	156	4422.525	6.349	180	5102.914	7.196	204	5783.302	8.042	228	6463.691
5.538	157	4450.875	6.385	181	5131.263	7.231	205	5811.652	8.078	229	6492.040
5.573	158	4479.224	6.420	182	5159.613	7.266	206	5840.001	8.113	230	6520.390
5.609	159	4507.574	6.455	183	5187.962	7.302	207	5868.351	8.148	231	6548.739
5.644	160	4535.923	6.490	184	5216.312	7.337	208	5896.700	8.184	232	6577.089
5.679	161	4564.273	6.526	185	5244.661	7.372	209	5925.050	8.219	233	6605.438
5.714	162	4592.622	6.561	186	5273.011	7.408	210	5953.399	8.254	234	6633.788
5.750	163	4620.972	6.596	187	5301.360	7.443	211	5981.749	8.289	235	6662.137
5.785	164	4649.321	6.632	188	5329.710	7.478	212	6010.098	8.325	236	6690.487
5.820	165	4677.671	6.667	189	5358.059	7.513	213	6038.448	8.360	237	6718.836
5.855	166	4706.020	6.702	190	5386.409	7.549	214	6066.797	8.395	238	6747.186
5.891	167	4734.370	6.737	191	5414.758	7.584	215	6095.147	8.430	239	6775.535
5.926	168	4762.719	6.773	192	5443.108	7.619	216	6123.496	8.466	240	6803.885
5.961	169	4791.069	6.808	193	5471.457	7.654	217	6151.846	8.501	241	6832.234
5.997	170	4819.418	6.843	194	5499.807	7.690	218	6180.195	8.536	242	6860.584
6.032	171	4847.768	6.878	195	5528.156	7.725	219	6208.545	8.572	243	6888.933
6.067	172	4876.117	6.914	196	5556.506	7.760	220	6236.894	8.607	244	6917.283
6.102	173	4904.467	6.949	197	5584.855	7.796	221	6265.244	8.642	245	6945.632
6.138	174	4932.816	6.984	198	5613.205	7.831	222	6293.593	8.677	246	6973.982

(Continued)

Table 14. *(Continued)* **Ounce - Gram Conversion Table.**

Ounces	oz or gr	Grams	Ounces	oz or gr	Grams	Ounces	oz or gr	Grams	Ounces	oz or gr	Grams
8.713	247	7002.331	9.559	271	7682.720	10.406	295	8363.108	11.252	319	9043.497
8.748	248	7030.681	9.595	272	7711.069	10.441	296	8391.458	11.288	320	9071.846
8.783	249	7059.030	9.630	273	7739.419	10.476	297	8419.807	11.323	321	9100.196
8.818	250	7087.380	9.665	274	7767.768	10.512	298	8448.157	11.358	322	9128.545
8.854	251	7115.730	9.700	275	7796.118	10.547	299	8476.506	11.393	323	9156.895
8.889	252	7144.079	9.736	276	7824.468	10.582	300	8504.856	11.429	324	9185.244
8.924	253	7172.429	9.771	277	7852.817	10.617	301	8533.206	11.464	325	9213.594
8.960	254	7200.778	9.806	278	7881.167	10.653	302	8561.555	11.499	326	9241.944
8.995	255	7229.128	9.841	279	7909.516	10.688	303	8589.905	11.535	327	9270.293
9.030	256	7257.477	9.877	280	7937.866	10.723	304	8618.254	11.570	328	9298.643
9.065	257	7285.827	9.912	281	7966.215	10.759	305	8646.604	11.605	329	9326.992
9.101	258	7314.176	9.947	282	7994.565	10.794	306	8674.953	11.640	330	9355.342
9.136	259	7342.526	9.983	283	8022.914	10.829	307	8703.303	11.676	331	9383.691
9.171	260	7370.875	10.018	284	8051.264	10.864	308	8731.652	11.711	332	9412.041
9.207	261	7399.225	10.053	285	8079.613	10.900	309	8760.002	11.746	333	9440.390
9.242	262	7427.574	10.088	286	8107.963	10.935	310	8788.351	11.782	334	9468.740
9.277	263	7455.924	10.124	287	8136.312	10.970	311	8816.701	11.817	335	9497.089
9.312	264	7484.273	10.159	288	8164.662	11.005	312	8845.050	11.852	336	9525.439
9.348	265	7512.623	10.194	289	8193.011	11.041	313	8873.400	11.887	337	9553.788
9.383	266	7540.972	10.229	290	8221.361	11.076	314	8901.749	11.923	338	9582.138
9.418	267	7569.322	10.265	291	8249.710	11.111	315	8930.099	11.958	339	9610.487
9.453	268	7597.671	10.300	292	8278.060	11.147	316	8958.448	11.993	340	9638.837
9.489	269	7626.021	10.335	293	8306.409	11.182	317	8986.798	12.028	341	9667.186
9.524	270	7654.370	10.371	294	8334.759	11.217	318	9015.147	12.064	342	9695.536

Table 15. KSI - MPa Conversion Table.

ksi	Conversion Value	MPa	ksi	Conversion Value	MPa	ksi	Conversion Value	MPa
0.14504	1	6.895	5.8015	40	275.79	11.458	79	544.69
0.29008	2	13.790	5.9465	41	282.69	11.603	80	551.58
0.43511	3	20.684	6.0916	42	289.58	11.748	81	558.48
0.58015	4	27.579	6.2366	43	296.47	11.893	82	565.37
0.72519	5	34.474	6.3817	44	303.37	12.038	83	572.26
0.87023	6	41.369	6.5267	45	310.26	12.183	84	579.16
1.0153	7	48.263	6.6717	46	317.16	12.328	85	586.05
1.1603	8	55.158	6.8168	47	324.05	12.473	86	592.95
1.3053	9	62.053	6.9618	48	330.95	12.618	87	599.84
1.4504	10	68.948	7.1068	49	337.84	12.763	88	606.74
1.5954	11	75.842	7.2519	50	344.74	12.909	89	613.63
1.7405	12	82.737	7.3969	51	351.63	13.053	90	620.53
1.8855	13	89.632	7.5420	52	358.53	13.198	91	627.42
2.0305	14	96.527	7.6870	53	365.42	13.343	92	634.32
2.1756	15	103.42	7.8320	54	372.32	13.489	93	641.21
2.3206	16	110.32	7.9771	55	379.21	13.634	94	648.11
2.4656	17	117.21	8.1221	56	386.11	13.779	95	655.00
2.6107	18	124.11	8.2672	57	393.00	13.924	96	661.90
2.7557	19	131.00	8.4122	58	399.90	14.069	97	668.79
2.9008	20	137.90	8.5572	59	406.79	14.214	98	675.69
3.0458	21	144.79	8.7023	60	413.69	14.359	99	682.58
3.1908	22	151.68	8.8473	61	420.58	14.504	100	689.48
3.3359	23	158.58	8.9923	62	427.47	15.954	110	758.42
3.4809	24	165.47	9.1374	63	434.37	17.405	120	827.37
3.6259	25	172.37	9.2824	64	441.26	18.855	130	896.32
3.7710	26	179.26	9.4275	65	448.16	20.305	140	965.27
3.9160	27	186.16	9.5725	66	455.05	21.756	150	1034.2
4.0611	28	193.05	9.7175	67	461.95	23.206	160	1103.2
4.2061	29	199.95	9.8626	68	468.84	24.656	170	1172.1
4.3511	30	206.84	10.008	69	475.74	26.107	180	1241.1
4.4962	31	213.74	10.153	70	482.63	27.557	190	1310.0
4.6412	32	220.63	10.298	71	489.53	29.008	200	1379.0
4.7862	33	227.53	10.443	72	496.42	30.458	210	1447.9
4.9313	34	234.42	10.588	73	503.32	31.908	220	1516.8
5.0763	35	241.32	10.733	74	510.21	33.359	230	1585.8
5.2214	36	248.21	10.878	75	517.11	34.809	240	1654.7
5.3664	37	255.11	11.023	76	524.00	36.259	250	1723.7
5.5114	38	262.00	11.168	77	530.90	37.710	260	1792.6
5.6565	39	268.90	11.313	78	537.79	39.160	270	1861.6

(Continued)

Table 15. *(Continued)* **KSI - MPa Conversion Table.**

ksi	Conversion Value	MPa	ksi	Conversion Value	MPa	ksi	Conversion Value	MPa
40.611	280	1930.5	97.175	670	-	162.44	1120	-
42.061	290	1999.5	98.626	680	-	165.34	1140	-
43.511	300	2068.4	100.08	690	-	168.24	1160	-
44.962	310	2137.4	101.53	700	-	171.14	1180	-
46.412	320	2206.3	102.98	710	-	174.05	1200	-
47.862	330	2275.3	104.43	720	-	176.95	1220	-
49.313	340	2344.2	105.88	730	-	179.85	1240	-
50.763	350	2413.2	107.33	740	-	182.75	1260	-
52.214	360	2482.1	108.78	750	-	185.65	1280	-
53.664	370	2551.1	110.23	760	-	188.55	1300	-
55.114	380	2620.0	111.68	770	-	191.45	1320	-
56.565	390	2689.0	113.13	780	-	194.35	1340	-
58.015	400	2757.9	114.58	790	-	197.25	1360	-
59.465	410	2826.9	116.03	800	-	200.15	1380	-
60.916	420	2895.8	117.48	810	-	203.05	1400	-
62.366	430	2964.7	118.93	820	-	205.95	1420	-
63.817	440	3033.7	120.38	830	-	208.85	1440	-
65.267	450	3102.6	121.83	840	-	211.76	1460	-
66.717	460	3171.6	123.28	850	-	214.66	1480	-
66.168	470	3240.5	124.73	860	-	217.56	1500	-
69.618	480	3309.5	126.18	870	-	220.46	1520	-
71.068	490	3378.4	127.63	880	-	223.36	1540	-
72.519	500	3447.4	129.08	890	-	226.26	1560	-
73.969	510	-	130.53	900	-	229.16	1580	-
75.420	520	-	131.98	910	-	232.06	1600	-
76.870	530	-	133.43	920	-	234.96	1620	-
78.320	540	-	134.89	930	-	237.86	1640	-
79.771	550	-	136.34	940	-	240.76	1660	-
81.221	560	-	137.79	950	-	243.66	1680	-
82.672	570	-	139.24	960	-	246.56	1700	-
84.122	580	-	140.69	970	-	249.46	1720	-
85.572	590	-	142.14	980	-	252.37	1740	-
87.023	600	-	143.59	990	-	255.27	1760	-
88.473	610	-	145.04	1000	-	258.17	1780	-
89.923	620	-	147.94	1020	-	261.07	1800	-
91.374	630	-	150.84	1040	-	263.97	1820	-
92.824	640	-	153.74	1060	-	266.87	1840	-
94.275	650	-	156.64	1080	-	269.77	1860	-
95.725	660	-	159.54	1100	-	272.67	1880	-

(Continued)

Table 15. *(Continued)* **KSI - MPa Conversion Table.**

ksi	Conversion Value	MPa	ksi	Conversion Value	MPa	ksi	Conversion Value	MPa
275.57	1900	-	310.38	2140	-	345.19	2380	-
278.47	1920	-	313.28	2160	-	348.09	2400	-
281.37	1940	-	316.18	2180	-	350.99	2420	-
284.27	1960	-	319.08	2200	-	353.89	2440	-
287.17	1980	-	321.98	2220	-	356.79	2460	-
290.08	2000	-	324.88	2240	-	359.69	2480	-
292.98	2020	-	327.79	2260	-	362.59	2500	-
295.88	2040	-	330.69	2280	-	365.49	2520	-
298.78	2060	-	333.59	2300	-	368.39	2540	-
301.68	2080	-	336.49	2320	-	371.30	2560	-
304.58	2100	-	339.39	2340	-	374.20	2580	-
307.48	2120	-	342.29	2360	-	377.10	2600	-

The center "conversion value" in bold faced type can be read as ksi or MPa. To convert 75 ksi to MPa, find 75 in the center column, then look to its right in the MPa column to determine that 75 ksi = 517.11 MPa. To convert 200 MPa to ksi, find 200 in the center column then look to its left in the ksi column to determine that 200 MPa = 29.008 ksi. 1 ksi = 6.894757 MPa. 1 MPa = 0.1450377 ksi. 1 psi = 6.894757 kPa. 1 psi = 6894.757 Pa.

Table 16. Horsepower - Kilowatt Conversion Table.

HP	HP or kW	kW	HP	HP or kW	kW	HP	HP or kW	kW	HP	HP or kW	kW
0.0013	0.001	0.0007	0.0134	0.01	0.0075	9.387	7	5.220	41.572	31	23.117
0.0027	0.002	0.0015	0.0268	0.02	0.0149	10.728	8	5.966	42.913	32	23.862
0.0040	0.003	0.0022	0.0402	0.03	0.0224	12.069	9	6.711	44.254	33	24.608
0.0054	0.004	0.0030	0.0536	0.04	0.0298	13.410	10	7.457	45.595	34	25.354
0.0067	0.005	0.0037	0.0671	0.05	0.0373	14.751	11	8.203	46.936	35	26.099
0.0080	0.006	0.0045	0.0805	0.06	0.0447	16.092	12	8.948	48.277	36	26.845
0.0094	0.007	0.0052	0.0939	0.07	0.0522	17.433	13	9.694	49.618	37	27.591
0.0107	0.008	0.0060	0.1073	0.08	0.0597	18.774	14	10.440	50.959	38	28.337
0.0121	0.009	0.0067	0.1207	0.09	0.0671	20.115	15	11.185	52.300	39	29.082
			0.1341	0.1	0.0746	21.456	16	11.931	53.641	40	29.828
			0.2682	0.2	0.1491	22.797	17	12.677	54.982	41	30.574
			0.4023	0.3	0.2237	24.138	18	13.423	56.323	42	31.319
			0.5364	0.4	0.2983	25.479	19	14.168	57.664	43	32.065
			0.6705	0.5	0.3728	26.820	20	14.914	59.005	44	32.811
			0.8046	0.6	0.4474	28.161	21	15.660	60.346	45	33.556
			0.9387	0.7	0.5220	29.502	22	16.405	61.687	46	34.302
			1.0728	0.8	0.5966	30.844	23	17.151	63.028	47	35.048
			1.2069	0.9	0.6711	32.185	24	17.897	64.369	48	35.794
			1.341	1	0.746	33.526	25	18.642	65.710	49	36.539
			2.682	2	1.491	34.867	26	19.388	67.051	50	37.285
			4.023	3	2.237	36.208	27	20.134	68.392	51	38.031
			5.364	4	2.983	37.549	28	20.880	69.733	52	38.776
			6.705	5	3.728	38.890	29	21.625	71.074	53	39.522
			8.046	6	4.474	40.231	30	22.371	72.415	54	40.268

The center conversion value in bold faced type can be read as either horsepower (HP) or kilowatts (kW). To convert 145.642 HP into kW, first find 145 in the bold face column and directly to its right the value in kW is given (108.126). Next, convert 0.6 HP into kW using the decimal column above (0.4474), then 0.04 (0.0298), and finally 0.002 (0.0015). The sum of the four is the conversion.

145.642 HP = 108.6048 kW.

The values for converting 145.642 kW into HP are:
145 kW = 194.448 HP
0.6 kW = 0.8046 HP
0.04 kW = 0.0536 HP
0.002 kW = 0.00271 HP
145.642 kW = 195.3333 HP

(Continued)

Table 16. *(Continued)* Horsepower - Kilowatt Conversion Table.

HP	HP or kW	kW	HP	HP or kW	kW	HP	HP or kW	kW	HP	HP or kW	kW
73.756	55	41.013	105.941	79	58.910	138.125	103	76.807	170.310	127	94.704
75.097	56	41.759	107.282	80	59.656	139.466	104	77.553	171.651	128	95.450
76.438	57	42.505	108.623	81	60.402	140.807	105	78.298	172.992	129	96.195
77.779	58	43.251	109.964	82	61.147	142.148	106	79.044	174.333	130	96.941
79.120	59	43.996	111.305	83	61.893	143.489	107	79.790	175.674	131	97.687
80.461	60	44.742	112.646	84	62.639	144.830	108	80.536	177.015	132	98.432
81.802	61	45.488	113.987	85	63.384	146.171	109	81.281	178.356	133	99.178
83.143	62	46.233	115.328	86	64.130	147.512	110	82.027	179.697	134	99.924
84.484	63	46.979	116.669	87	64.876	148.853	111	82.773	181.038	135	100.669
85.825	64	47.725	118.010	88	65.622	150.194	112	83.518	182.379	136	101.415
87.166	65	48.470	119.351	89	66.367	151.535	113	84.264	183.720	137	102.161
88.507	66	49.216	120.692	90	67.113	152.877	114	85.010	185.061	138	102.907
89.848	67	49.962	122.033	91	67.859	154.218	115	85.755	186.402	139	103.652
91.189	68	50.708	123.374	92	68.604	155.559	116	86.501	187.743	140	104.398
92.531	69	51.453	124.715	93	69.350	156.900	117	87.247	189.084	141	105.144
93.872	70	52.199	126.056	94	70.096	158.241	118	87.993	190.425	142	105.889
95.213	71	52.945	127.397	95	70.841	159.582	119	88.738	191.766	143	106.635
96.554	72	53.690	128.738	96	71.587	160.923	120	89.484	193.107	144	107.381
97.895	73	54.436	130.079	97	72.333	162.264	121	90.230	194.448	145	108.126
99.236	74	55.182	131.420	98	73.079	163.605	122	90.975	195.789	146	108.872
100.577	75	55.927	132.761	99	73.824	164.946	123	91.721	197.130	147	109.618
101.918	76	56.673	134.102	100	74.570	166.287	124	92.467	198.471	148	110.364
103.259	77	57.419	135.443	101	75.316	167.628	125	93.212	199.812	149	111.109
104.600	78	58.165	136.784	102	76.061	168.969	126	93.958	201.153	150	111.855

(Continued)

Table 16. *(Continued)* Horsepower - Kilowatt Conversion Table.

HP	HP or kW	kW	HP	kW	HP or kW	HP	kW	HP or kW	HP	HP or kW	kW
202.494	151	112.601	234.679	130.497	175	266.863	148.394	199	299.048	223	166.291
203.835	152	113.346	236.020	131.243	176	268.204	149.140	200	300.389	224	167.037
205.176	153	114.092	237.361	131.989	177	269.545	149.886	201	301.730	225	167.782
206.517	154	114.838	238.702	132.735	178	270.886	150.631	202	303.071	226	168.528
207.858	155	115.583	240.043	133.480	179	272.227	151.377	203	304.412	227	169.274
209.199	156	116.329	241.384	134.226	180	273.568	152.123	204	305.753	228	170.020
210.540	157	117.075	242.725	134.972	181	274.910	152.868	205	307.094	229	170.765
211.881	158	117.821	244.066	135.717	182	276.251	153.614	206	308.435	230	171.511
213.222	159	118.566	245.407	136.463	183	277.592	154.360	207	309.776	231	172.257
214.564	160	119.312	246.748	137.209	184	278.933	155.106	208	311.117	232	173.002
215.905	161	120.058	248.089	137.954	185	280.274	155.851	209	312.458	233	173.748
217.246	162	120.803	249.430	138.700	186	281.615	156.597	210	313.799	234	174.494
218.587	163	121.549	250.771	139.446	187	282.956	157.343	211	315.140	235	175.239
219.928	164	122.295	252.112	140.192	188	284.297	158.088	212	316.481	236	175.985
221.269	165	123.040	253.453	140.937	189	285.638	158.834	213	317.822	237	176.731
222.610	166	123.786	254.794	141.683	190	286.979	159.580	214	319.163	238	177.477
223.951	167	124.532	256.135	142.429	191	288.320	160.325	215	320.504	239	178.222
225.292	168	125.278	257.476	143.174	192	289.661	161.071	216	321.845	240	178.968
226.633	169	126.023	258.817	143.920	193	291.002	161.817	217	323.186	241	179.714
227.974	170	126.769	260.158	144.666	194	292.343	162.563	218	324.527	242	180.459
229.315	171	127.515	261.499	145.411	195	293.684	163.308	219	325.868	243	181.205
230.656	172	128.260	262.840	146.157	196	295.025	164.054	220	327.209	244	181.951
231.997	173	129.006	264.181	146.903	197	296.366	164.800	221	328.550	245	182.696
233.338	174	129.752	265.522	147.649	198	297.707	165.545	222	329.891	246	183.442

(Continued)

Table 16. (Continued) Horsepower - Kilowatt Conversion Table.

HP	HP or kW	kW	kW	HP or kW	HP	HP	HP or kW	kW	HP	HP or kW	kW
331.232	247	184.188	202.085	271	363.417	395.601	295	219.981	427.786	319	237.878
332.573	248	184.934	202.830	272	364.758	396.943	296	220.727	429.127	320	238.624
333.914	249	185.679	203.576	273	366.099	398.284	297	221.473	430.468	321	239.370
335.256	250	186.425	204.322	274	367.440	399.625	298	222.219	431.809	322	240.115
336.597	251	187.171	205.067	275	368.781	400.966	299	222.964	433.150	323	240.861
337.938	252	187.916	205.813	276	370.122	402.307	300	223.710	434.491	324	241.607
339.279	253	188.662	206.559	277	371.463	403.648	301	224.456	435.832	325	242.352
340.620	254	189.408	207.305	278	372.804	404.989	302	225.201	437.173	326	243.098
341.961	255	190.153	208.050	279	374.145	406.330	303	225.947	438.514	327	243.844
343.302	256	190.899	208.796	280	375.486	407.671	304	226.693	439.855	328	244.590
344.643	257	191.645	209.542	281	376.827	409.012	305	227.438	441.196	329	245.335
345.984	258	192.391	210.287	282	378.168	410.353	306	228.184	442.537	330	246.081
347.325	259	193.136	211.033	283	379.509	411.694	307	228.930	443.878	331	246.827
348.666	260	193.882	211.779	284	380.850	413.035	308	229.676	445.219	332	247.572
350.007	261	194.628	212.524	285	382.191	414.376	309	230.421	446.560	333	248.318
351.348	262	195.373	213.270	286	383.532	415.717	310	231.167	447.901	334	249.064
352.689	263	196.119	214.016	287	384.873	417.058	311	231.913	449.242	335	249.809
354.030	264	196.865	214.762	288	386.214	418.399	312	232.658	450.583	336	250.555
355.371	265	197.610	215.507	289	387.555	419.740	313	233.404	451.924	337	251.301
356.712	266	198.356	216.253	290	388.896	421.081	314	234.150	453.265	338	252.047
358.053	267	199.102	216.999	291	390.237	422.422	315	234.895	454.606	339	252.792
359.394	268	199.848	217.744	292	391.578	423.763	316	235.641	455.947	340	253.538
360.735	269	200.593	218.490	293	392.919	425.104	317	236.387	457.289	341	254.284
362.076	270	201.339	219.236	294	394.260	426.445	318	237.133	458.630	342	255.029

Table 17. Centigrade (Celsius) - Fahrenheit Conversion Table.

-459.4 to -250			-240 to -20			-15 to 19			20 to 42		
°C	°F/°C	°F	°C	°F/°C	°F	°C	°F/°C	°F	°C	°F/°C	°F
-273	-459.4	-	-151	-240	-400	-26	-15	9	-6.7	20	68.0
-268	-450	-	-146	-230	-382	-23	-10	14	-6.1	21	69.8
-262	-440	-	-140	-220	-364	-20	-5	23	-5.6	22	71.6
-257	-430	-	-134	-210	-346	-17.8	0	32	-5.0	23	73.4
-251	-420	-	-129	-200	-328	-17.2	1	33.8	-4.4	24	75.2
-246	-410	-	-123	-190	-310	-16.7	2	35.6	-3.9	25	77.0
-240	-400	-	-118	-180	-292	-16.1	3	37.4	-3.3	26	78.8
-234	-390	-	-112	-170	-274	-15.6	4	39.2	-2.8	27	80.6
-229	-380	-	-107	-160	-256	-15.0	5	41.0	-2.2	28	82.4
-223	-370	-	-101	-150	-238	-14.4	6	42.8	-1.7	29	84.2
-218	-360	-	-96	-140	-220	-13.9	7	44.6	-1.1	30	86.0
-212	-350	-	-90	-130	-202	-13.3	8	46.4	-.6	31	87.8
-207	-340	-	-84	-120	-184	-12.8	9	48.2	0	32	89.6
-201	-330	-	-79	-110	-166	-12.2	10	50.0	.6	33	91.4
-196	-320	-	-73	-100	-148	-11.7	11	51.8	1.1	34	93.2
-190	-310	-	-68	-90	-130	-11.1	12	53.6	1.7	35	95.0
-184	-300	-	-62	-80	-112	-10.6	13	55.4	2.2	36	96.8
-179	-290	-	-57	-70	-94	-10.0	14	57.2	2.8	37	98.6
-173	-280	-	-51	-60	-76	-9.4	15	59.0	3.3	38	100.4
-169	-273	-459.4	-46	-50	-58	-8.9	16	60.8	3.9	39	102.2
-168	-270	-454	-40	-40	-40	-8.3	17	62.6	4.4	40	104.0
-162	-260	-436	-34	-30	-22	-7.8	18	64.4	5.0	41	105.8
-157	-250	-418	-29	-20	-4	-7.2	19	66.2	5.6	42	107.6

(Continued)

Table 17. (Continued) Centigrade (Celsius) - Fahrenheit Conversion Table.

°C	43 to 65		66 to 88			89 to 210			212 to 430		
°C	°F/°C	°F	°C	°F/°C	°F	°C	°F/°C	°F	°C	°F/°C	°F
6.1	43	109.4	18.9	66	150.8	31.7	89	192.2	100	212	413.6
6.7	44	111.2	19.4	67	152.6	32.2	90	194.0	104	220	428
7.2	45	113.0	20.0	68	154.4	32.8	91	195.8	110	230	446
7.8	46	114.8	20.6	69	156.2	33.3	92	197.6	116	240	464
8.3	47	116.6	21.1	70	158.0	33.9	93	199.4	121	250	482
8.9	48	118.4	21.7	71	159.8	34.4	94	201.2	127	260	500
9.4	49	120.2	22.2	72	161.6	35.0	95	203.0	132	270	518
10.0	50	122.0	22.8	73	163.4	35.6	96	204.8	138	280	536
10.6	51	123.8	23.3	74	165.2	36.1	97	206.6	143	290	554
11.1	52	125.6	23.9	75	167.0	36.7	98	208.4	149	300	572
11.7	53	127.4	24.4	76	168.8	37.2	99	210.2	154	310	590
12.2	54	129.2	25.0	77	170.6	37.8	100	212.0	160	320	608
12.8	55	131.0	25.6	78	172.4	43	110	230	166	330	626
13.3	56	132.8	26.1	79	174.2	49	120	248	171	340	644
13.9	57	134.6	26.7	80	176.0	54	130	266	177	350	662
14.4	58	136.4	27.2	81	177.8	60	140	284	182	360	680
15.0	59	138.2	27.8	82	179.6	66	150	302	188	370	698
15.6	60	140.0	28.3	83	181.4	71	160	320	193	380	716
16.1	61	141.8	28.9	84	183.2	77	170	338	199	390	734
16.7	62	143.6	29.4	85	185.0	82	180	356	204	400	752
17.2	63	145.4	30.0	86	186.8	88	190	374	210	410	770
17.8	64	147.2	30.6	87	188.6	93	200	392	216	420	788
18.3	65	149.0	31.1	88	190.4	99	210	410	221	430	806

(Continued)

Table 17. *(Continued)* **Centigrade (Celsius) - Fahrenheit Conversion Table.**

°C	440 to 660 °F/°C	°F	°C	670 to 890 °F/°C	°F	°C	900 to 1120 °F/°C	°F	°C	1130 to 1350 °F/°C	°F
227	440	824	354	670	1238	482	900	1652	610	1130	2066
232	450	842	360	680	1256	488	910	1670	616	1140	2084
238	460	860	366	690	1274	493	920	1688	621	1150	2102
243	470	878	371	700	1292	499	930	1706	627	1160	2120
249	480	896	377	710	1310	504	940	1724	632	1170	2138
254	490	914	382	720	1328	510	950	1742	638	1180	2156
260	500	932	388	730	1346	516	960	1760	643	1190	2174
266	510	950	393	740	1364	521	970	1778	649	1200	2192
271	520	968	399	750	1382	527	980	1796	654	1210	2210
277	530	986	404	760	1400	532	990	1814	660	1220	2228
282	540	1004	410	770	1418	538	1000	1832	666	1230	2246
288	550	1022	416	780	1436	543	1010	1850	671	1240	2264
293	560	1040	421	790	1454	549	1020	1868	677	1250	2282
299	570	1058	427	800	1472	554	1030	1886	682	1260	2300
304	580	1076	432	810	1490	560	1040	1904	688	1270	2318
310	590	1094	438	820	1508	566	1050	1922	693	1280	2336
316	600	1112	443	830	1526	571	1060	1940	699	1290	2354
321	610	1130	449	840	1544	577	1070	1958	704	1300	2372
327	620	1148	454	850	1562	582	1080	1976	710	1310	2390
332	630	1166	460	860	1580	588	1090	1994	716	1320	2408
338	640	1184	466	870	1598	593	1100	2012	721	1330	2426
343	650	1202	471	880	1616	599	1110	2030	727	1340	2444
349	660	1220	477	890	1634	604	1120	2048	732	1350	2462

(Continued)

Table 17. *(Continued)* Centigrade (Celsius) - Fahrenheit Conversion Table.

°C	1360 to 1580		1590 to 1810			1820 to 2040			2050 to 2440		
	°F/°C	°F	°C	°F/°C	°F	°C	°F/°C	°F	°C	°F/°C	°F
738	1360	2480	866	1590	2894	993	1820	3308	1121	2050	3722
743	1370	2498	871	1600	2912	999	1830	3326	1127	2060	3740
749	1380	2516	877	1610	2930	1004	1840	3344	1132	2070	3758
754	1390	2534	882	1620	2948	1010	1850	3362	1138	2080	3776
760	1400	2552	888	1630	2966	1016	1860	3380	1143	2090	3794
766	1410	2570	893	1640	2984	1021	1870	3398	1149	2100	3812
771	1420	2588	899	1650	3002	1027	1880	3416	1160	2120	3848
777	1430	2606	904	1660	3020	1032	1890	3434	1171	2140	3884
782	1440	2624	910	1670	3038	1038	1900	3452	1182	2160	3920
788	1450	2642	916	1680	3056	1043	1910	3470	1193	2180	3956
793	1460	2660	921	1690	3074	1049	1920	3488	1204	2200	3992
799	1470	2678	927	1700	3092	1054	1930	3506	1216	2220	4028
804	1480	2696	932	1710	3110	1060	1940	3524	1227	2240	4064
810	1490	2714	938	1720	3128	1066	1950	3542	1238	2260	4100
816	1500	2732	943	1730	3146	1071	1960	3560	1249	2280	4136
821	1510	2750	949	1740	3164	1077	1970	3578	1260	2300	4172
827	1520	2768	954	1750	3182	1082	1980	3596	1271	2320	4208
832	1530	2786	960	1760	3200	1088	1990	3614	1282	2340	4244
838	1540	2804	966	1770	3218	1093	2000	3632	1293	2360	4280
843	1550	2822	971	1780	3236	1099	2010	3650	1304	2380	4316
849	1560	2840	977	1790	3254	1104	2020	3668	1316	2400	4352
854	1570	2858	982	1800	3272	1110	2030	3686	1327	2420	4388
860	1580	2876	988	1810	3290	1116	2040	3704	1338	2440	4424

(Continued)

Table 17. *(Continued)* Centigrade (Celsius) - Fahrenheit Conversion Table.

2460 to 2600			2620 to 2760			2780 to 2920			2940 to 4500		
°C	°F/°C	°F	°C	°F/°C	°F	°C	°F/°C	°F	°C	°F/°C	°F
1349	2460	4460	1438	2620	4748	1527	2780	5036	1616	2940	5324
1360	2480	4496	1449	2640	4784	1538	2800	5072	1627	2960	5360
1371	2500	4532	1460	2660	4820	1549	2820	5108	1638	2980	5396
1382	2520	4568	1471	2680	4856	1560	2840	5144	1643	2990	5414
1393	2540	4604	1482	2700	4892	1571	2860	5180	1649	3000	5432
1404	2560	4640	1493	2720	4928	1582	2880	5216	1926	3500	6332
1416	2580	4676	1504	2740	4964	1593	2900	5252	2204	4000	7232
1427	2600	4712	1516	2760	5000	1604	2920	5288	2482	4500	8132

Note: The center "conversion value" in bold faced type can be read as either Centigrade (Celsius) or Fahrenheit. To convert 75 °F to °C, find 75 in the bold faced column, then look to its left in the °C column to determine that 75 °F = 23.9 °C. To convert 200 °C to °F, find 200 in the bold faced column, then look to its right to determine that 200 °C = 392 °F.

Table 18. Units Conversion Table

To convert from	to	Multiply by	
abampere	ampere (A)	1.0	*E+01
abcoulomb	coulomb (C)	1.0	*E+01
abfarad	farad (F)	1.0	*E+09
abhenry	henry (H)	1.0	*E−09
abmho	siemens (S)	1.0	*E+09
abohm	ohm (Ω)	1.0	*E−09
abvolt	volt (V)	1.0	*E−08
acre-foot (U.S. survey)[a]	meter^3 (m^3)	1.233 489	E+03
acre-foot (U.S. survey)[a]	foot^3 (ft^3)	4.356 025	E+04
acre-foot (U.S. survey)[a]	gallon (U.S.)	3.258 533	E+05
acre (U.S. survey)[a]	meter^2 (m^2)	4.046 873	E+03
acre (U.S. survey)[a]	foot^2 (ft^2)	4.356 000	E+04
acre (U.S. survey)[a]	mile^2	1.562 000	E−03
acre (U.S. survey)[a]	yard^2	4.840 000	E+03
ampere-hour	coulomb (C)	3.60	*E+03
angstrom	meter (m)	1.000 000	*E−10
are	meter^2 (m^2)	1.000 000	*E+02
astronomical unit	meter (m)	1.495 980	E+11
atmosphere (normal)	pascal (Pa)	1.013 250	*E+05
atmosphere (technical = l kg$_f$/cm^2)	pascal (Pa)	9.806 65	*E+04
atmosphere	cm. of mercury	7.599 999	E+01
atmosphere	inch of mercury	2.992 126	E+01
atmosphere	foot of water	3.389 854	E+01
atmosphere	kilograms per meter^2 (kg/m^2)	1.033 300	E+04
atmosphere	pound/in.^{2} (PSI)	1.470 000	E+01
atmosphere	tons per ft^2	1.058 108	E+00
bar	pascal (Pa)	1.0	*E+05
bar	newton/meter^2 (N/m^2)	1.0	*E+05
barn	meter^2 (m^2)	1.0	*E−28
barrel (for crude petroleum, 42 gal)	meter^3 (m^3)	1.589 873	*E−01
barrel (cement)	pound cement	3.760 000	E+02
bags (cement)	pound cement	9.400 000	E+01
board foot	meter^3 (m^3)	2.359 737	E−03
board foot	inch^3 (in.^{3})	1.440 000	E+02
British thermal unit (International Table)	joule (J)	1.055 056	E+03
British thermal unit (mean)	joule (J)	1.055 870	E+03
British thermal unit (thermochemical)	joule (J)	1.054 350	E+03
British thermal unit (39°F)	joule (J)	1.059 670	E+03
British thermal unit (59°F)	joule (J)	1.054 800	E+03
British thermal unit (60°F)	joule (J)	1.054 680	E+03
British thermal unit	kilogram-calorie	2.520 000	E−02
British thermal unit	foot-pound (ft lbs)	7.775 000	E+02
British thermal unit	horsepower-hours	3.927 000	E−04
British thermal unit	kilogram-meters (kg-m)	1.075 000	E+02
British thermal unit	kilowatt-hour	2.928 000	E−04
Btu (thermochemical)/foot^2-second	watt/meter^2 (W/m^2)	1.134 893	E+04
Btu (thermochemical)/foot^2-minute	watt/meter^2 (W/m^2)	1.891 489	E+02
Btu (thermochemical)/foot^2-hour	watt/meter^2 (W/m^2)	3.152 481	E+00
Btu (thermochemical)/inch^2-second	watt/meter^2 (W/m^2)	1.634 246	E+06
Btu (thermochemical) in/s ft^2 °F (k, thermal conductivity)	watt/meter-kelvin (W/m K)	5.188 732	E+02

(Continued)

Table 18. *(Continued)* **Units Conversion Table**

To convert from	to	Multiply by	
Btu (International Table) in/s ft² °F (*k*, thermal conductivity)	watt/meter-kelvin (W/m K)	5.192 204	E+02
Btu (thermochemical) in/h ft² °F (*k*, thermal conductivity)	watt/meter-kelvin (W/m K)	1.441.314	E−01
Btu (International Table) in/h ft² °F (*k*, thermal conductivity)	watt/meter-kelvin (W/m K)	1.442.279	E−01
Btu (International Table)/ft²	joule/meter² (J/m²)	1.135.653	E+04
Btu (thermochemical)/ft²	joule/meter² (J/m²)	1.134.893	E+04
Btu (International Table)/h ft² °F (*C*, thermal conductance)	watt/meter²-kelvin (W/m² K)	5.678.263	E+00
Btu (thermochemical)/h ft² °F (*C*, thermal conductance)	watt/meter²-kelvin (W/m² K)	5.674.466	E+00
Btu (International Table)/pound-mass	joule/kilogram (J/kg)	2.326	*E+03
Btu (thermochemical)/pound-mass	joule/kilogram (J/kg)	2.324 444	*E+03
Btu (International Table)/lbm °F (*c*, heat capacity)	joule/kilogram-kelvin (J/kg K)	4.186.8	*E+03
Btu (thermochemical)/lbm °F (*c*, heat capacity)	joule/kilogram-kelvin (J/kg K)	4.184.000	E+03
Btu (International Table)/s ft² °F	watt/meter²-kelvin (W/m² K)	2.044.175	E+04
Btu (thermochemical)/s ft² °F	watt/meter²-kelvin (W/m² K)	2.042 808	E+04
Btu (International Table)/hour	watt (W)	2.930 711	E−01
Btu (thermochemical)/second	watt (W)	1.054 350	E+03
Btu (thermochemical)/minute	watt (W)	1.757 250	E+01
Btu (thermochemical)/minute	foot-pound/second (ft lb/s)	1.296 000	E+01
Btu (thermochemical)/minute	horsepower	2.356 000	E−02
Btu (thermochemical)/minute	kilowatt	1.757 000	E−02
Btu (thermochemical)/hour	watt (W)	2.928 751	E−01
Bushel (U. S.)	meter³ (m³)	3.523 907	E−02
calorie (International Table)	joule (J)	4.186 8	*E+00
calorie (mean)	joule (J)	4.190 02	E+00
calorie (thermochemical)	joule (J)	4.184	*E+00
calorie (15°C)	joule (J)	4.185 800	E+00
calorie (20°C)	joule (J)	4.181 900	E+00
calorie (kilogram, International Table)	joule (J)	4.186 800	*E+03
calorie (kilogram, mean)	joule (J)	4.190 020	E+03
calorie (kilogram, thermochemical)	joule (J)	4.184	*E+03
calorie (thermochemical)/cm² - minute	watt/meter² (W/m²)	6.973 333	E+02
cal (thermochemical)/cm²	joule/meter² (J/m²)	4.184	*E+04
cal (thermochemical)/cm² s	watt/meter² (W/m²)	4.184	*E+04
cal (thermochemical)/cm s °C	watt/meter-kelvin (W/m K)	4.184	*E+02
cal (International Table)/g	joule/kilogram (J/kg)	4.186 8	*E+03
cal (International Table)/g °C	joule/kilogram-kelvin (J/kg · K)	4.186 8	*E+03
cal (thermochemical)/g	joule/kilogram (J/kg)	4.184	*E+03
cal (thermochemical)/g °C	joule/kilogram-kelvin (J/kg · K)	4.184	*E+03
calorie (thermochemical)/second	watt (W)	4.184	*E+00
calorie (thermochemical)/minute	watt (W)	6.973 333	E−02
carat (metric)	kilogram (kg)	2.0	*E−04
centares (centiares)	meter² (m²)	1.0	*E+00
centigrams	gram	1.0	*E−02
centiliters	liter	1.0	*E−02

(Continued)

Table 18. *(Continued)* **Units Conversion Table**

To convert from	to	Multiply by	
centimeter	millimeter	1.0	*E+01
centimeter	inch	3.393 707	E−01
centimeter^3	foot^3 (ft^3)	3.531 466	E−05
centimeter^3	inch^3 (in.^{3})	6.102 374	E−02
centimeter^3	meter^3 (m^3)	1.000 000	E−06
centimeter^3	yard^3	1.307 950	E−06
centimeter^3	gallon (U.S.)	2.641 720	E−04
centimeter^3	liter	1.0	*E−03
centimeter^3	pint (liquid)	2.113 376	E−03
centimeter^3	quart (liquid)	1.057 000	E−03
centimeter^4 (moment of section)^b	inch^4	2.402 510	E−02
centimeter of mercury (0°C)	pascal (Pa)	1.333 22	E+03
centimeter of mercury	atmosphere	1.316 000	E−02
centimeter of mercury	foot of water	4.461 000	E−01
centimeter of mercury	kilogram/meter^2 (kg/m^3)	1.360 000	E+02
centimeter of mercury	pound/ft^2 (lb/ft^2)	2.785 000	E+01
centimeter of mercury	pound/inch^2 (lb/in^2)	1.934 000	E−01
centimeter of water (4°C)	pascal (Pa)	9.806 380	E+01
centimeters/second	foot/minute	1.969 000	E+00
centimeters/second	foot/second	3.281 000	E−02
centimeters/second	kilometer/hour	3.600 000	E−02
centimeters/second	meter/minute	6.000 000	E−01
centimeters/second	mile/hour	2.237 000	E−02
centipoise	pascal-second	1.0	*E−03
centistoke	meter^2/second (m^2/s)	1.000 000	*E−06
chain (engineer or ramden)	meter (m)	3.048	*E+01
chain (surveyor or gunter)	meter (m)	2.011 684	E+01
circular mil	meter^2 (m^2)	5.067 075	E−10
cord	meter^3 (m^3)	3.624 556	E+00
cup	meter^3 (m^3)	2.365 882	E−04
curie	bequerel (Bq)	3.70	*E+10
day (mean solar)	second (s)	8.640 000	E+04
day (sidereal)	second (s)	8.616 409	E+04
decagrams	gram	1.0	*E+01
decaliters	liter	1.0	*E+01
decameters	meter	1.0	*E+01
decigrams	gram	1.0	*E−01
deciliters	liter	1.0	*E−01
degree (angle)	radian (rad)	1.745 329	E−02
degree Celsius	kelvin (K)	$t_K = t_{°C} + 273.15$	
degree centigrade	kelvin (K)	$t_K = t_{°C} + 273.15$	
degree Fahrenheit	degree Celsius	$t_{°C} = (t_{°F} − 32)/1.8$	
degree Fahrenheit	kelvin (K)	$t_K = (t_{°F} + 459.67)/1.8$	
deg F h ft^2/Btu (thermochemical) (R, thermal resistance)	kelvin-meter^2/watt (K · m^2/W)	1.762 280	E−01
deg F h ft^2/Btu (International Table) (R, thermal resistance)	kelvin-meter^2/watt (K · m^2/W)	1.761 102	E−01
degree Rankine	kelvin (K)	$t_K = t_{°R}/1.8$	
dram (avoirdupois)	kilogram (kg)	1.771 845	E−03
dram (troy or apothecary)	kilogram (kg)	3.887 934	E−03
dram (avoirdupois)	grain	2.734 375	E+01

(Continued)

Table 18. *(Continued)* **Units Conversion Table**

To convert from	to	Multiply by	
dram (avoirdupois)	ounce	6.250 000	E−02
dram (U.S. liquid)	kilogram (kg)	3.696 691	E−06
dram (U.S. liquid)	ounce (liquid)	1.250 000	E−01
dyne	newton (N)	1.0	*E−05
dyne-centimeter	Newton-meter (Nm)	1.0	*E−07
dyne-centimeter^2	pascal (Pa)	1.0	*E−01
electron volt	joule (J)	1.602 190	E−19
EMU of capacitance	farad (F)	1.0	*E+09
EMU of current	ampere (A)	1.0	*E+01
EMU of electrical potential	volt (V)	1.0	*E−08
EMU of inductance	henry (H)	1.0	*E−09
EMU of resistance	ohm (Ω)	1.0	*E−09
ESU of capacitance	farad (F)	1.112 650	E−12
ESU of current	ampere (A)	3.335 600	E−10
ESU of electrical potential	volt (V)	2.997 900	E+02
ESU of inductance	henry (H)	8.987 554	E+11
ESU of resistance	ohm (Ω)	8.987 554	E+11
erg	joule (J)	1.0	*E−07
erg/centimeter^2-second	watt/meter^2 (W/m^2)	1.0	*E−03
erg/second	watt (W)	1.0	*E−07
farad, international U.S. (F$_{INT-US}$)	farad (F)	9.995 050	E−01
faraday (based on carbon 12)	coulomb (C)	9.648 700	E+04
faraday (chemical)	coulomb (C)	9.649 570	E+04
faraday (physical)	coulomb (C)	9.652 190	E+04
fathom (U.S survey)[a]	meter (m)	1.828 804	E+00
fathom	foot (ft)	6.0	*E+00
fermi (femtometer)	meter (m)	1.0	*E−15
fluid ounce (U.S.)	meter^3 (m^3)	2.957 353	E−05
foot	meter (m)	3.048	*E−01
foot	centimeter (cm)	3.048	*E+01
foot	inch (in.)	1.20	*E+01
foot	yard	3.333 333	E−01
foot (U.S survey)[a]	meter (m)	3.048 006	*E−01
foot^2	meter^2 (m^2)	9.290 304	*E−02
foot^3	meter^3 (m^3)	2.831 685	E−02
foot^3	centimeter^3 (cm^3)	2.832 000	E+04
foot^3	inch^3 (in^3)	1.728 000	E+03
foot^3	meter^3 (m^3)	2.832 000	E−02
foot^3	yard^3	3.703 703	E−02
foot^3	gallon (U.S.)	7.480 520	E+00
foot^3	liter	2.831 684	E+01
foot^3	pint (liquid)	5.984 415	E+01
foot^3	quart (liquid)	2.982 000	E+01
foot^4 (moment of section)[b]	meter^4 (m^4)	8.630 975	E−03
foot/hour	meter/second (m/s)	8.466 667	E−05
foot/minute	meter/second (m/s)	5.080	*E−03
foot/minute	foot/second (ft/s)	1.667 000	E−02
foot/minute	kilometer/hour	1.829 000	E−02
foot/minute	meter/minute	3.048 000	E−01
foot/minute	mile/hour	1.136 000	E−02
foot^3/minute	meter^3/second (m^{3}s)	4.719 474	E−04

(Continued)

Table 18. *(Continued)* **Units Conversion Table**

To convert from	to	Multiply by	
foot^3/minute	centimeter^3/second (cm^{3}s)	4.720 000	E+02
foot^3/minute	gallon (U.S.)/second	1.247 000	E−01
foot^3/minute	liter/second	4.720 000	E−01
foot/second	meter/second (m/s)	3.048 000	*E−01
foot/second	kilometers/hour	1.097 000	E+00
foot/second	mile/hour	6.818 000	E−01
foot^2/second	meter^2/second (m^{2}s)	9.290 304	*E−02
foot^3/second	meter^3/second (m^{3}s)	2.831 685	E−02
foot^3/second	gallon (U.S.)/minute	4.488 310	E+02
foot of water (39.2°F/4°C)	pascal (Pa)	2.989 067	E+03
foot of water (39.2°F/4°C)	atmosphere	2.949 980	E−02
foot of water (32°F/0°C)	inch mercury	8.826 710	E−01
foot of water	kilogram-force/meter^2 (kg-f/m^2)	3.048 000	E+02
foot of water	pound-force/fb^2 (lf-f/ft^2)	6.242 796	E+01
footcandle	lumen/meter^2 (lm/m^2)	1.076 391	E+01
footcandle	lux (lx)	1.076 391	E+01
footlambert	candela/meter^2 (cd/m^2)	3.426 259	E+00
foot-pound-force	joule (J) or Newton meter (Nm)	1.355 818	E+00
foot-pound-force	Btu	1.285 068	E−03
foot-pound-force	horsepower hour	5.050 505	E−07
foot-pound-force	kilowatt-hour	3.766 000	E−07
foot-pound-force/hour	watt (W)	3.766 161	E−04
foot-pound-force/minute	watt (W)	2.259 697	E−02
foot-pound-force/minute	horsepower	3.030 303	E−05
foot-pound-force/second	watt (W)	1.355 818	E+00
foot-pound-force/second	horsepower	1.818 182	E−03
foot-poundal	joule (J)	4.214 011	E−02
ft^2/h (thermal diffusivity)	meter^2/second (m^2/s)	2.580 64	*E−05
foot/second^2	meter/second^2 (m/s^2)	3.048	*E−01
free fall, standard	meter/second^2 (m/s^2)	9.806 65	*E+00
furlong	meter (m)	2.011 68	*E+02
gal	meter/second^2 (m/s^2)	1.0	*E−02
gallon (Canadian liquid)	meter^3 (m^3)	4.546 090	E−03
gallon (U.K.)	meter^3 (m^3)	4.546 092	E−03
gallon (U.K.)	gallon (U.S.)	1.200 950	E+00
gallon (U.S. dry)	meter^3 (m^3)	4.404 884	E−03
gallon (U.S. liquid)	meter^3 (m^3)	3.785 412	E−03
gallon (U.S. liquid)	centimeter^3 (cm^3)	3.785 412	E+03
gallon (U.S. liquid)	foot^3 (ft^3)	1.336 806	E−01
gallon (U.S. liquid)	inch^3 (in.^{3})	2.310 000	E+02
gallon (U.S. liquid)	yard^3	4.951 132	E−03
gallon (U.S. liquid)	liter	3.785 412	E+00
gallon (U.S. liquid)	pint	8.0	*E+00
gallon (U.S. liquid)	quart	4.0	*E+00
gallon (U.S. liquid)/day	meter^3/second (m^3/s)	4.381 264	E−08
gallon (U.S. liquid)	gallon (U.K.)	8.326 742	E−01
gallon (U.S. liquid)	pound (lb)	8.345 170	E+00
gallon (U.S. liquid)/minute	meter^3/second (m^3/s)	6.309 020	E−05
gallon (U.S. liquid)/minute	foot^3/second (ft^3/sec)	2.228 000	E−03
gallon (U.S. liquid)/minute	liter/second	3.785 412	E+00
gallon (U.S. liquid)/minute	foot^3/hour (ft^3/hr)	8.020 833	E+00

(Continued)

Table 18. *(Continued)* **Units Conversion Table**

To convert from	to	Multiply by	
gallon (U.S. liquid)/minute	kiloliter/hour	1.362 748	E+01
gallon (U.S. liquid)/minute	ton (U.S.)/24 hours	6.008 600	E+00
gamma	tesla (T)	1.0	*E−09
gauss	tesla (T)	1.0	*E−04
gilbert	ampere-turn	7.957 747	E−01
gill (U.K.)	meter^3 (m^3)	1.420 654	E−04
gill (U.S.)	meter^3 (m^3)	1.182 941	E−04
grade	degree (angular)	9.0	*E−01
grade	radian (rad)	1.570 796	E−02
grain (avoirdupois/apothecary/troy)	kilogram (kg)	6.479 891	*E−05
grain (avoirdupois/apothecary/troy)	ounce	2.285 714	E−03
grain-force	ounce-force	2.285 714	E−03
gram	kilogram (kg)	1.0	*E−03
gram	grain	1.543 236	E+01
gram	ounce	3.527 396	E−02
gram	pound (lb)	2.204 929	E−03
gram	troy ounce	3.215 075	E−02
gram/centimeter^3	kilogram/meter^3 (kg/m^3)	1.0	*E+03
gram/centimeter^3	ounce/in.^{3}	5.780 367	E−01
gram-force/centimeter^2	pascal (Pa)	9.806 65	*E+01
gram/liter	ounce/gallon (U.S.)	1.335 265	E−01
hectare	meter^2 (m^2)	1.0	*E+04
hectare	acre	2.471 044	E+00
hectare	foot^2 (ft^2)	1.076 000	E+05
hogshead (U.S.)	meter^3 (m^3)	2.384 809	E−01
horsepower (U.S. 550 ft lb-f/s)	watt (W)	7.456 999	E+02
horsepower (U.S.)	Btu/minute	4.243 561	E+01
horsepower (U.S.)	metric horsepower	1.013 870	E+00
horsepower (U.S.)	Newton meter/second (Nm/s)	7.456 999	E+02
horsepower (boiler)	watt (W)	9.809 500	E+03
horsepower (electric)	watt (W)	7.46	*E+02
horsepower (metric)	watt (W)	7.354 990	E+02
horsepower (water)	watt (W)	7.460 430	E+02
horsepower (U.K.)	watt (W)	7.457 000	E+02
hour (mean solar)	second (s)	3.600 000	E+03
hour (sidereal)	second (s)	3.590 170	E+03
hundredweight (long)	kilogram (kg)	5.080 235	E+01
hundredweight (short)	kilogram (kg)	4.535 924	E+01
inch	meter (m)	2.54	*E−02
inch	centimeter (cm)	2.54	*E+00
inch^2	meter^2 (m^2)	6.451 6	*E−04
inch^3	meter^3 (m^3)	1.638 706	E−05
inch^3	centimeter^3 (cm^3)	1.638 706	E+01
inch^3	foot^3 (ft^3)	5.787 037	E−04
inch^3	yard^3	2.143 347	E−05
inch^3	gallon (US)	4.329 004	E−03
inch^3	liter	1.638 707	E−02
inch^3	pint (liquid)	3.463 204	E−02
inch^3	quart (liquid)	1.731 602	E−02
inch^3/minute	meter^3/second (m^3/s)	2.731 177	E−07
inch^4 (moment of section)[b]	meter^4 (m^4)	4.162 314	E−07

(Continued)

Table 18. *(Continued)* **Units Conversion Table**

To convert from	to	Multiply by	
inch/second	meter/second (m/s)	2.54	*E−02
inch of mercury (0°C/32°F)	pascal (Pa)	3.386 389	E+03
inch of mercury (0°C/32°F)	atmosphere	3.342 106	E−02
inch of mercury (4°C/39.2°F)	foot of water	1.132 925	E+00
inch of mercury (60°F)	pascal (Pa)	3.376 850	E+03
inch of water (4°C/39.2°F)	pascal (Pa)	2.490 820	E+02
inch of water (4°C/39.2°F)	atmosphere	2.458 316	E−03
inch of water (60°F)	pascal (Pa)	2.488 400	E+02
inch pound-force	Newton meter (Nm)	1.129 848	E−01
inch/second^2	meter/second^2 (m/s^2)	2.54	*E−02
inch/second^2	foot/minute	5.0	*E+00
joule	watt-hour	2.777 778	E−04
joule	calorie	2.368 623	E−01
joule	Btu	9.479 896	E−04
joule	inch pound-force (in. lb-f)	8.852 357	E+00
joule	foot pound-force (ft lb-f)	7.375 621	E−01
kayser	1/meter (1/m)	1.0	*E+02
kelvin	degree Celsius	$t_C = t_K - 273.15$	
kelvin	degree Fahrenheit	(1.8 % [kelvin−273.15]) + 32	
kilocalorie (thermochemical)/minute	watt (W)	6.973 333	E+01
kilocalorie (thermochemical)/minute	Btu/minute	3.971 373	E+00
kilocalorie (thermochemical)/minute	horsepower	9.351 394	E−02
kilocalorie (thermochemical)/second	watt (W)	4.184	*E+03
kilocalorie (thermochemical)/second	Btu/second	3.971 373	E+00
kilogram	ounce	3.527 396	E+01
kilogram	pound	2.204 623	E+00
kilogram calorie	Btu	3.967 373	E+00
kilogram/centimeter^2	pound/in.^{2} (PSI)	1.422 334	E+01
kilogram meter^2 (moment of inertia)	pound foot^2 (lb ft^2)	2.373 036	E+01
kilogram meter^2 (moment of inertia)	pound inch^2 (lb in.^{2})	3.417 171	E+03
kilogram-force (kgf)	newton (N)	9.806 65	*E+00
kilogram-force meter	newton meter(Nm)	9.806 65	*E+00
kilogram-force-second^2/meter (mass)	kilogram (kg)	9.806 65	*E+00
kilogram-force/millimeter^2	pascal (Pa)	9.806 65	*E+06
kilogram-force/centimeter^2	pascal (Pa)	9.806 65	*E+04
kilogram-force/meter^2	pound foot^2 (lb/ft^2)	2.048 161	E−01
kilogram-force/meter^2	pound inch^2 (PSI))	1.422 334	E−03
kilogram-force/meter^2	pascal (Pa)	9.806 65	*E+00
kilogram-force/meter^2	newton/meter^2 (N/m^2)	9.806 65	*E+00
kiloliter	foot^3 (ft^3)	3.531 467	E+01
kiloliter	gallon (U.S.)	2.641 720	E+02
kilometer	foot (ft)	3.280 833	E+03
kilometer	mile	6.213 670	E−01
kilometer/hour	meter/second (m/s)	2.777 778	E−01
kilometer/hour	foot/minute	5.468 067	E+01
kilopond	newton (N)	9.806 65	*E+00
kilowatt	Btu/minute	5.690 708	E+01
kilowatt	horsepower	1.341 022	E+00
kilowatt hour	joule (J)	3.60	*E+06
kip (1,000 lbf)	newton (N)	4.448 222	E+03
kip/inch^2 (ksi)	pascal (Pa)	6.894 757	E+06

(Continued)

Table 18. *(Continued)* **Units Conversion Table**

To convert from	to	Multiply by	
knot (international)	meter/second (m/s)	5.144 444	E−01
lambert	candela/meter² (cd/m²)	3.183 099	E+03
langley	joule/meter² (J/m²)	4.184	*E+04
league, nautical (international and U.S.)	meter (m)	5.556	*E+03
league (U.S. survey)[a]	meter (m)	4.828 042	*E+03
league, nautical (U.K.)	meter (m)	5.559 552	*E+03
light year	meter (m)	9.460 55	E+15
link (engineer or ramden)	meter (m)	3.048	*E−01
link (surveyor or gunter)	meter (m)	2.011 68	*E−01
liter	meter³ (m³)	1.0	*E−03
liter	foot³ (ft³)	3.531 467	E−02
liter	inch³ (in³)	6.102 374	E+01
liter	gallon (U.S.)	2.641 720	E−01
liter	quart	1.056 688	E+00
liter/minute	foot³/hour (ft³/hr)	2.118 880	E+00
liter/minute	gallon/hour (U.S.)	1.585 031	E+01
lux	lumen/meter² (lm/m²)	1.0	*E+00
maxwell	weber (Wb)	1.0	*E−08
megapascal	1000 pounds/inch² (ksi)	1.450 377	E−02
meter	foot	3.280.833	E+00
meter	inch	3.937 000	E+01
meter	mile	6.213 670	E−04
meter	yard	1. 093 611	E+00
meter/minute	foot/minute	3.280 840	E+00
meter/minute	inch/minute	3.937 008	E+01
meter/minute	meter/second	1.666 667	E−02
meter/second	foot/second	3.280 840	E+00
meter/hour	foot/minute	5.468 067	E−02
meter³	centimeter³ (cm³)	1.0	*E−06
meter³	foot³ (ft³)	3.531 467	E+01
meter³	inch³ (in.³)	6.102 374	E+04
meter³	yard³	1.307 951	E+00
meter³	gallon (U.S.)	2.641 720	E+02
meter³	liter	1.0	*E+03
meter³	pint (liquid)	2.113 000	E+03
meter³	quart (liquid)	1.056 800	E+03
meter⁴ (moment of section)[b]	foot⁴ (ft⁴)	1.158 618	E+02
mho	siemens (S)	1.0	*E+00
microinch	meter (m)	2.54	*E−08
microinch	micron	2.54	*E−02
micron (micrometer)	meter (m)	1.0	*E−06
micron (micrometer)	inch (in.)	3.937 008	E−08
micron (micrometer)	microinch	3.937 008	E+01
mil	meter (m)	2.54	*E−05
mile, nautical (international and U.S.)	meter (m)	1.852	*E+03
mile, nautical (U.K.)	meter (m)	1.853 184	*E+03
mile (international)	meter (m)	1.609 344	*E+03
mile (U.S. survey)[a]	meter (m)	1.609 347	*E+03
mile (U.S. survey)	kilometer	1.609 347	E+00
mile² (international)	meter² (m²)	2.589 988	*E+06
mile² (U.S. survey)[a]	meter² (m²)	2.589 988	*E+06

(Continued)

Table 18. *(Continued)* **Units Conversion Table**

To convert from	to	Multiply by	
mile^2 (U.S. survey)^a	kilometer^2	2.589 998	E+00
mile/hour (international)	meter/second (m/s)	4.470 4	*E−01
mile/hour (international)	kilometer/hour	1.609 344	*E+00
milligram	grain	1.543 236	E−02
milliliter	ounce	3.381 402	E−02
millimeter	inch (in.)	3.937 000	E−02
millimeter^2	inch^2 (in.^{2})	1.549 997	E−03
millimeter^2/second	inch^2/hour (in.^{2}/hr)	5.580 011	E+00
millimeter^3	inch^3 (in.^{3})	6.102 374	E−05
millimeter of mercury (0°C)	pascal (Pa)	1.333 224	E+02
millipascal	newton/square meter (n/m^2)	1.0	*E−03
minute (angle)	radian (rad)	2.908 882	E−04
minute (mean solar)	second (s)	6.0	*E+01
minute (sidereal)	second (s)	5.983 617	E+01
month (mean calendar)	second (s)	2.268 000	E+06
newton	ounce-force (oz-f)	3.596 942	E+00
newton	pound-force (lb-f)	2.248 089	E−01
newton	kilogram-force	1.019 716	E−01
newton millimeter	ounce inch	1.416 119	E−01
newton/millimeter^2	pound/inch^2 (lb/in^2)	1.450 377	E+02
newton/centimeter^2	pound/inch^2 (lb/in^2)	1.450 377	E+00
newton/meter	pound/inch (lb/in.)	5.710 148	E−03
newton/meter	pound/foot (lb/ft)	6.852 178	E−02
newton meter	ounce inch-force (oz in.-f)	1.416 119	E+00
newton meter	foot pound-force (ft lb-f)	7.375 622	E−01
newton meter	kilogram meter (kg m)	1.019 716	E-01
newton/meter^2	pound/inch^2 (lb/in^2)	1.450 377	E−04
newton/meter^2	kilogram/meter^2	1.019 716	E−01
oersted	ampere/meter (A/m)	7.957 747	E+01
ohm, international U.S. (Ω_{INT-US})	ohm (Ω)	1.000 495	E+00
ohm-centimeter	ohm-meter (Ωm)	1.0	*E−02
ounce (U.K. fluid)	meter^3 (m^3)	2.841 307	E−05
ounce (U.S. fluid)	meter^3 (m^3)	2.957 353	E−05
ounce (U.S. fluid)	inch^3 (in.^{3})	1.804 688	E+00
ounce (U.S. fluid)	liter	2.957 353	E−02
ounce/second (U.S. fluid)	centimeter^3/second (cm^3/s)	2.957 353	E+01
ounce-force (avoirdupois)	newton (N)	2.780 139	E−01
ounce-force inch	gram-centimeter (gr-cm)	7.200 790	E+01
ounce-force-inch	Newton-meter (Nm)	7.061 552	E−03
ounce-mass (avoirdupois)	kilogram (kg)	2.834 952	E−02
ounce (avoirdupois)	troy ounce	9.114 583	E−01
ounce-mass (troy or apothecary)	kilogram (kg)	3.110 348	E−02
ounce- mass/yard^2	kilogram/meter^2 (kg/m^2)	3.390 575	E−02
ounce (avoirdupois)(mass)/inch^3	kilogram/meter^3 (kg/m^3)	1.729 994	E+03
parsec	meter (m)	3.083 740	E+16
pascal	kilogram/meter^2 (kg/m^2)	1.019 716	E−01
pascal	newton/meter^2 (N/m^2)	1.0	*E+00
pascal	pound/foot^2 (lb/ft^2)	2.088 543	E−02
peck (U.S.)	meter^3 (m^3)	8.809 768	E−03
pennyweight	kilogram (kg)	1.555 174	E−03
pennyweight	ounce (avoirdupois)	5.485 714	E−02

(Continued)

Table 18. *(Continued)* **Units Conversion Table**

To convert from	to	Multiply by	
perm (0°C)	kilogram/pascal-second-meter^2 (kg/Pa m^2)	5.721 350	E–11
perm (23°C)	kilogram/pascal-second-meter^2 (kg/Pa m^2)	5.745 250	E–11
perm (0°C)	kilogram/pascal-second-meter (kg/Pa s·m)	1.453 220	E–12
perm (23°C)	kilogram/pascal-second-meter (kg/Pa s·m)	1.459 290	E–12
phot	lumen/meter^2 (lm/m^2)	1.0	*E+04
pica (printer's)	meter (m)	4.217 518	E–03
pint (U.S. dry)	meter^3 (m^3)	5.506 105	E–04
pint (U.S. liquid)	meter^3 (m^3)	4.731 765	E–04
pint (U.S. liquid)	liter	4.731 765	E–01
point (printer's)	meter	3.514 598	*E–04
poise (absolute viscosity)	pascal-second (Pa s)	1.0	*E–01
pound	kilogram	4.535 924	E–01
pound	gram	4.535 924	E+02
pound-force (lbf avoirdupois)	newton (N)	4.448 222	E+00
pound-force inch	gram-force inch (gr-f in.)	4.535 924	E+02
pound-force inch	gram-force centimeter (gr-f cm)	1.152 125	E+03
pound-force inch	newton-meter (Nm)	1.129 848	E–01
pound-force foot	newton-meter (Nm)	1.355 818	E+00
pound-force foot/inch	newton-meter/meter (Nm/m)	5.337 866	E+01
pound-force inch/inch	newton-meter/meter (Nm/m)	4.448 222	E+00
pound-force/inch	newton/meter (N/m)	1.751 268	E+02
pound/inch^2	bar	6.894 757	E–02
pound/inch^2	newton/meter^2 (N/m^2)	6.894 757	E+03
pound/inch^2	pascal (Pa)	6.894 757	E+03
pound-force/foot	newton/meter (N/m)	1.459 390	E+01
pound-force/foot^2	pascal (Pa)	4.788 026	E+01
pound-force/inch^2 (PSI)	pascal (Pa)	6.894 757	E+03
pound-force/inch^2 (PSI)	pound-force/foot^2	1.44	*E+02
pound-force-second/foot^2	pascal-second (Pa s)	4.788 026	E+01
pound-mass (lbm avoirdupois)	kilogram (kg)	4.535 924	E–01
pound-mass (troy or apothecary)	kilogram (kg)	3.732 417	E–01
pound-mass-foot^2 (moment of inertia)	kilogram-meter^2 (kg m^2)	4.214 011	E–02
pound-mass-inch^2 (moment of inertia)	kilogram-meter^2 (kg m^2)	2.926 397	E–04
pound-mass/second	kilogram/second (kg/s)	4.535 924	E–01
pound-mass/minute	kilogram/second (kg/s)	7.559 873	E–03
pound-mass/foot^3	kilogram/meter^3 (kg/m^3)	1.601 846	E+01
pound-mass/inch^3	kilogram/meter^3 (kg/m^3)	2.767 990	E+04
pound-mass/gallon (U.K. liquid)	kilogram/meter^3 (kg/m^3)	9.977 633	E+01
pound-mass/gallon (U.S. liquid)	kilogram/meter^3 (kg/m^3)	1.198 264	E+02
pound-mass/foot-second	pascal-second (Pa/s)	1.488 164	E+00
poundal	Newton (N)	1.382 550	E–01
poundal/foot^2	pascal (Pa)	1.488 164	E+00
poundal-second/foot^2	pascal-second (Pa s)	1.488 164	E+00
quart (U.S. dry)	meter^3 (m^3)	1.101 221	E–03
quart (U.S. liquid)	meter^3 (m^3)	9.463 530	E–04
quart (U.S. liquid)	liter	9.463 530	E–01
rad (radiation dose absorbed)	gray (Gy)	1.0	*E–02

(Continued)

Table 18. *(Continued)* **Units Conversion Table**

To convert from	to	Multiply by	
rem (dose equivalent)	sievert (Sv)	1.0	*E−02
rhe	meter^2/newton-second (m^2/N s)	1.0	*E+01
rod (U.S. survey)^a	meter (m)	5.029 210	E+00
roentgen	coulomb/kilogram (C/kg)	2.579 76	*E−04
second (angle)	radian (rad)	4.848 137	E−06
second (sidereal)	second (s)	9.972 696	E−01
section (U.S. survey)^a	meter^2 (m^2)	2.589 998	E+06
shake	second (s)	1.0	*E−08
slug	kilogram (kg)	1.459 390	E+01
slug/foot^3	kilogram/meter^3 (kg/m^3)	5.153 788	E+02
slug/foot-second	pascal-second (Pa s)	4.788 026	E+01
statampere	ampere (A)	3.335 640	E−10
statcoulomb	coulomb (C)	3.335 640	E−10
statfarad	farad (F)	1.112 650	E−12
stathenry	henry (H)	8.987 554	E+11
statmho	siemens (S)	1.112 650	E−12
statohm	ohm (Ω)	8.987 554	E+11
statvolt	volt (V)	2.997 925	E+02
stere	meter^3 (m^3)	1.0	*E+00
stilb	candela/meter^2 (cd/m^2)	1.0	*E+04
stokes (kinematic viscosity)	meter^2/second (m^2/s)	1.0	*E−04
tablespoon	meter^3 (m^3)	1.478 676	E−05
teaspoon	meter^3 (m^3)	4.928 922	E−06
ton (assay)	kilogram (kg)	2.916 667	E−02
ton (long, 2,240 lbm)	kilogram (kg)	1.016 047	E+03
ton (metric)	kilogram (kg)	1.0	*E+03
ton (nuclear equivalent of TNT)	joule (J)	4.20	E+09
ton (register)	meter^3 (m^3)	2.831 685	E+00
ton (short, 2,000 lbm)	kilogram (kg)	9.071 847	E+02
ton (short, mass)	kilogram/second (kg/s)	2.519 958	E−01
ton (long, mass)/yard^3	kilogram/meter^3 (kg/m^3)	1.328 939	E+03
tonne	kilogram (kg)	1.0	*E+03
torr (mm H$_g$, 0°C)	pascal (Pa)	1.333 22	*E+02
township (U.S. survey)^a	meter^2 (m^2)	9.323 994	E+07
unit pole	weber (Wb)	1.256 637	E−07
watt/centimeter^2	watt/meter^2 (W/m^2)	1.0	*E+04
watt-hour	joule (J)	3.600 000	*E+03
watt-second	joule (J)	1.0	*E+00
yard	meter (m)	9.144	*E−01
yard^2	meter^2 (m^2)	8.361 274	E−01
yard^3	meter^3 (m^3)	7.645 548	E−01
yard^3	centimeter^3 (cm^3)	7.645 548	E+05
yard^3	foot^3 (ft^3)	2.700 000	E+01
yard^3	inch^3 (in.^{3})	4.665 600	E+04
yard^3	gallon (U.S.)	2.019 740	E+02
yard^3	liter	7.645 549	E+02
yard^3	pint (liquid)	1.616 000	E+03
yard^3	quart (liquid)	8.079 000	E+02
yard^3/minute	meter^3/second (m^3/s)	1.274 258	E−02
yard^3/minute	feet^3/minute (ft^3/min)	2.700 000	E+01
yard^3/minute	gallon (U.S.)/second	3.366 233	E+01

(Continued)

Table 18. *(Continued)* **Units Conversion Table**

To convert from	to	Multiply by	
yard^3/minute	liter/second	1.274 258	E+01
year (calendar)	second (s)	3.153 600	E+07
year (sidereal)	second (s)	3.155 815	E+07
year (tropical)	second (s)	3.155 693	E+07

Using the conversion table. To convert the unit in the left column to the unit in the center column, multiply the left column by the conversion factor in the right column. The far right hand exponential expression (E+ or E–) indicates how many places the decimal is to be moved before multiplying.

Example: Find the number of grams in three pounds. First, find "pound" in the left column, and then find "gram" in the center column. Next, using the E value, interpret the value in the right hand column. An E+ value means that the decimal point moves to the right the number of places indicated by the E value. For example, E+04 means that the decimal moves four places to the right. An E– value means that the decimal point moves to the left the number of places indicated by the E value. The conversion number 1.609344 E+03 is therefore equal to 1609.344, and the conversion number 4.731765 E–4 is equal to .0004731765.

The value in the right hand column for converting pounds to grams is 4.535924 E+02, so three pounds is equal to 3 × 453.5924 = 1,360.7772 grams.

* Asterisk indicates that the conversion factor is exact and all subsequent digits are zero.
a Based on the U.S. survey foot (1 ft = 1,200/3,937 m).
b Moment of inertia of a plane section about a specified axis.

This Conversion Factor Table has been compiled from several sources including ASTM E380-91a Standard Practice and Use of the International System of Units and additional sources.

(Entries in bold type refer to Tables)

(Entries in bold type refer to Tables)

(Entries in bold type refer to Tables)

(Entries in bold type refer to Tables)

(Entries in bold type refer to Tables)

(Entries in bold type refer to Tables)

(Entries in bold type refer to Tables)

(Entries in bold type refer to Tables)

(Entries in bold type refer to Tables)

(Entries in bold type refer to Tables)

(Entries in bold type refer to Tables)

(Entries in bold type refer to Tables)

(Entries in bold type refer to Tables)

(Entries in bold type refer to Tables)

(Entries in bold type refer to Tables)

(Entries in bold type refer to Tables)